TEMPERATURE

**Its Measurement and Control
in Science and Industry**

TEMPERATURE

Its Measurement and Control in Science and Industry

VOLUME FIVE

James F. Schooley
Editor-in-Chief

PART 1

Thermodynamic Methods   Scales

Fixed Points   Radiation

Pages 1–710

Cosponsored by

American Institute of Physics
Instrument Society of America
National Bureau of Standards

Published by **American Institute of Physics** *New York*

Copyright © 1982 American Institute of Physics 335 East 45th Street, NY, NY 10017

Individual readers of this volume and nonprofit libraries, acting for them, are permitted to make fair use of the material in it, such as copying an article for use in teaching or in research. Permission is granted to quote from this volume in scientific work with the customary acknowledgment of the source. Republication or systematic or multiple reproduction of any material in this volume is permitted only under license from AIP. Address inquiries to Office of Rights and Permissions, AIP.

Printed in the United States of America

Library of Congress Catalog Card No. 62-19138

ISBN 0-88318-403-6   ISSN 0091-9322

General Preface

The Sixth International Temperature Symposium was held in Washington, D.C. from March 15 to 18, 1982. Three sponsoring organizations pooled their considerable talents to ensure the success of this once-in-a-decade happening. The Instrument Society of America assumed the responsibility for the Symposium itself, which included, for the first time, a manufacturer's exhibit. The U.S. National Bureau of Standards organized the technical program under the impeccable direction of the Chief of its Temperature and Pressure Measurement and Standards Division, Dr. James F. Schooley. The contributors to that program produced camera-ready copies of their papers for publication by the American Institute of Physics in this, Volume 5 of *Temperature: Its Measurement and Control in Science and Industry,* Parts 1 and 2.

The preceding volumes in this series, dating back to 1939, have each represented major accomplishments in the documentation of thermometry state-of-the-art. We who have been involved with Volume 5 are confident that its 182 papers, authored by more than 400 individuals from 18 countries, will be regarded by you, the reader, as an equal among peers. This feeling persists despite the well-known difficulty of satisfying all of the people even some of the time—once every ten years?

Lawrence G. Rubin
Symposium General Chairman

Introduction

Progress in the science of thermometry since the publication of Volume Four of the symposium series, *Temperature: Its Measurement and Control in Science and Industry*, has been notable in several respects. Perhaps most significant has been an improvement in the accuracy of thermodynamic temperature measurement. Gas thermometry has been used to extend the lower limit of the International Practical Temperature Scale of 1968 (IPTS-68) and to demonstrate surprisingly large differences between the IPTS-68 and thermodynamic temperatures from 0 to 457 °C. An apparent deviation from thermodynamic temperatures as large as 0.5 °C has been discovered in the IPTS-68 in the range 600–1000 °C by means of radiation thermometry; the results of Johnson noise measurements have furnished corroborative evidence. The Johnson noise thermometer has become of especial metrological importance in the range from 4 K down to 0.01 K. Highly accurate techniques for thermodynamic thermometry now also include acoustic, nuclear orientation, gas refractive index, gas dielectric constant, and total radiation methods.

New levels of reproducibility have been achieved in many temperature fixed points, permitting corresponding reductions in the uncertainties of calibration temperatures. Accompanying this progress has been the development of sealed cells for improved practical realization of cryogenic triple-point temperatures. The concept of transportable, self-contained temperature reference devices has become a reality over a wide range of temperatures.

There has been noteworthy progress as well in the precision and versatility of temperature sensors. Particularly relevant to improvement in the precision of the IPTS-68 is an extension of the range of usefulness of the platinum resistance thermometer to temperatures as high as 1000 °C.

The use of computers and microcircuits in thermometry has become more prevalent since the 5th Symposium on Temperature. Not only have improved bridges, potentiometers, controllers, and calibration equipment reflected this trend, but a completely new thermometry method based on laser Raman scattering and employing on-line computers also has been developed. This method, known as Coherent Anti-Stokes Raman Spectroscopy (CARS) thermometry, is one of several laser-based techniques that permit sub-microsecond temperature measurements in gaseous media.

All of the papers published in this fifth volume of the *Temperature* series were submitted for presentation in the 6th Temperature Symposium. Not every advance in thermometry over the past decade was reflected in these submissions; however, the relatively large Program Committee successfully solicited contributions on all of the recent major advances in this broad and varied field.

Because of their relevance to a proposed replacement of the IPTS-68, Symposium papers on Thermodynamic Temperature Determinations and on Temperature Scales have been included preferentially in these respective categories. Highlighting such scale-related research has resulted, for example, in separating Johnson noise thermometry papers into one group concerned with accurate thermodynamic temperature measurement and a second group discussing industrial noise thermometry.

As has long been the custom, Symposium papers have been grouped under Temperature Fixed Points, Resistance Thermometry, Radiation Thermometry, or Thermocouple Thermometry according to their primary focus.

The papers on noise thermometry in industrial environments have been grouped in this volume with papers on nuclear quadrupole resonance, velocity-of-sound, semiconductor junction, and eddy-current thermometry to form the category Electronic Thermometry. The volume is completed by sections on Temperature Control, Calibration Techniques, and Thermometry for Special Applications.

Part 1 of this volume includes the following categories:
  Thermodynamic Temperature Determinations
  Temperature Scales
  Temperature Fixed Points
  Radiation Thermometry.
Part 2 contains the remaining sections, viz:
  Resistance Thermometry
  Thermocouple Thermometry
  Electronic Thermometry
  Temperature Control
  Calibration Techniques
  Thermometry for Special Applications.

Each paper in this volume was reviewed for technical content by one or more experts in the appropriate sub-specialty of thermometry. This review was arranged, and in many cases performed, by a member of the Symposium Program Committee. All papers were further reviewed by the Editor-in-Chief to promote consistency in style and notation. Camera-ready copy was prepared so that the American Institute of Physics could proceed promptly with photographic production of this volume. The copy was checked for accuracy during or immediately following the Symposium itself.

The cooperation of many people—the authors, the General and Program Committees of the Symposium, the technical reviewers, and the staff of the American Institute of Physics—was necessary to produce this fifth volume of *Temperature*. The Editor-in-Chief wishes to express his appreciation to all these people for making his duties less demanding by far than they might have been. To Sylvia C. Ramboz, who shepherded each manuscript through several stages to its successful publication, a special acknowledgment must be given; she provided levels of organization and care which were essential to the quality and completeness of this volume. Similarly, Dr. Jerry Jacobson of the American Institute of Physics must be recognized for his considerable and painstaking efforts in producing a book which is both accurate and attractive.

James F. Schooley
*Program Chairman and
Editor-in-Chief*

Acknowledgments

Symposium General Committee

Mr. Lawrence G. Rubin, *Chairman*
Francis Bitter National Magnet Laboratory
Massachusetts Institute of Technology
Cambridge, Massachusetts

Dr. Chun H. Cho
Fisher Controls Company
Marshalltown, Iowa

Mr. Charles T. Glazer
Instrument Society of America
Research Triangle Park, North Carolina

Mr. Robert H. Marks
American Institute of Physics
New York, New York

Dr. H. Preston-Thomas
Division of Physics
National Research Council of Canada
Ottawa, Ontario, Canada

Dr. James F. Schooley, *Chairman* and *Editor*
Temperature and Pressure Measurements
  and Standards Division
National Bureau of Standards
Washington, D. C.

Dr. John T. Scott
Publishing Services Branch I
American Institute of Physics
New York, New York

Mr. Edward D. Zysk
Engelhard Industries
Carteret, New Jersey

Symposium Program Committee

Dr. James F. Schooley, *Chairman* and *Editor*
Temperature and Pressure Measurements
  and Standards Division
National Bureau of Standards
Washington, D. C.

Dr. John Ancsin
Division of Physics
National Research Council of Canada
Ottawa, Ontario, Canada

Dr. A. C. Anderson
Physics Department
University of Illinois
Urbana, Illinois

Dr. Richard L. Anderson
Instrumentation and Controls Division
Oak Ridge National Laboratory
Oak Ridge, Tennessee

Dr. Ronald E. Bedford
Division of Physics
National Research Council of Canada
Ottawa, Ontario, Canada

Mr. Robert P. Benedict
Westinghouse Electric Corporation
Philadelphia, Pennsylvania

Mr. G. W. Burns
Temperature and Pressure Measurements
  and Standards Division
National Bureau of Standards
Washington, D.C.

Dr. Ared Cezairliyan
Thermophysics Division
National Bureau of Standards
Washington, D.C.

Dr. Chun H. Cho
Fisher Controls Company
Marshalltown, Iowa

Mr. N. Ralph Corallo
Becton Dickinson Company
East Rutherford, New Jersey

Mr. Robert D. Cutkosky
Temperature and Pressure Measurements
 and Standards Division
National Bureau of Standards
Washington, D.C.

Mr. H. L. Daneman
HLD Associates
Santa Fe, New Mexico

Dr. T. M. Dauphinee
Division of Physics
National Research Council of Canada
Ottawa, Ontario, Canada

Dr. J. Edrich
University of Colorado at Denver
Denver, Colorado

Mr. J. P. Evans
Temperature and Pressure Measurements
 and Standards Division
National Bureau of Standards
Washington, D.C.

Dr. G. T. Furukawa
Temperature and Pressure Measurements
 and Standards Division
National Bureau of Standards
Washington, D.C.

Mr. Randal Gauthier
RDF Corporation
Hudson, New Hampshire

Dr. L. A. Guildner
Temperature and Pressure Measurements
 and Standards Division
National Bureau of Standards
Washington, D.C.

Mr. Lawrence Howard
Boeing Aerospace
Seattle, Washington

Dr. Wilbur S. Hurst
Temperature and Pressure Measurements
 and Standards Division
National Bureau of Standards
Washington, D.C.

Dr. T. Kashiwagi
Center for Fire Research
National Bureau of Standards
Washington, D.C.

Dr. Edward Lange
Scripps Institute of Oceanography
La Jolla, California

Dr. B. W. Mangum
Temperature and Pressure Measurements
 and Standards Division
National Bureau of Standards
Washington, D.C.

Dr. Gene D. Nutter
Instrumentation Systems Center
University of Wisconsin
Madison, Wisconsin

Mr. E. R. Pfeiffer
Temperature and Pressure Measurements
 and Standards Division
National Bureau of Standards
Washington, D.C.

Dr. H. Preston-Thomas
Division of Physics
National Research Council of Canada
Ottawa, Ontario, Canada

Dr. Ray Radebaugh
Thermophysical Properties Division
National Bureau of Standards
Boulder, Colorado

Mr. M. L. Reilly
Temperature and Pressure Measurements
 and Standards Division
National Bureau of Standards
Washington, D.C.

Dr. J. R. Roberts
Radiation Division
National Bureau of Standards
Washington, D.C.

Dr. G. J. Rosasco
Temperature and Pressure Measurements
 and Standards Division
National Bureau of Standards
Washington, D.C.

Mr. Meyer Sapoff
Thermometrics, Inc.
Edison, New Jersey

Mr. Robert L. Shepard
Instrumentation and Controls Division
Oak Ridge National Laboratory
Oak Ridge, Tennessee

Dr. R. J. Soulen, Jr.
Temperature and Pressure Measurements
 and Standards Division
National Bureau of Standards
Washington, D.C.

Dr. Clayton A. Swenson
Department of Physics
Iowa State University
Ames, Iowa

Dr. James E. Zimmerman
Electromagnetic Technology Division
National Bureau of Standards
Boulder, Colorado

Other Contributors

Dr. Jerry L. Jacobson
Production Division I
American Institute of Physics
New York, New York

Mrs. Sylvia C. Ramboz
Temperature and Pressure Measurements
 and Standards Division
National Bureau of Standards
Washington, D. C.

Mr. Douglas Hetrick
 formerly with
Instrument Society of America
Research Triangle Park, North Carolina

Dr. Ralph Hudson
 formerly with
National Bureau of Standards
Washington, D. C.

Dr. A. W. K. Metzner
 formerly with
American Institute of Physics
New York, New York

Table of Contents

The Table of Contents is a cumulative listing of Parts 1 and 2.
Part 1 consists of pages 1–710 and Part 2, pages 711–1395.

SYMPOSIUM KEYNOTE ADDRESS

1	Temperature scales, the IPTS, and its future development	R. P. Hudson

I. THERMODYNAMIC TEMPERATURE DETERMINATIONS

9	The measurement of thermodynamic temperature	Leslie A. Guildner, Wilhelm Thomas
21	Measurements of thermodynamic temperature from 2.6 to 27.1 K	K. H. Berry
25	Measurements with a gas thermometer between 4 and 100 K	P. P. M. Steur, J. E. van Dijk, J. P. Mars, H. ter Harmsel, M. Durieux
33	Constant volume gas thermometry from 13.8 to 83.8 K	R. C. Kemp, L. M. Besley, W. R. G. Kemp
39	Constant volume gas thermometer for thermodynamic temperature measurements of the triple point of oxygen	H. Sakurai
43	Progress in NBS gas thermometry above 500 °C	Leslie A. Guildner, Robert E. Edsinger
49	Dielectric Constant Gas Thermometry (DCGT): A new method of accurate thermodynamic thermometry	D. Gugan
55	Surface-fitting of helium isotherms: Application to the temperature scale 2.6–27.1 K	D. Gugan
65	Primary acoustic thermometry: Principles and current trends	A. R. Colclough
77	^{4}He second and third virial coefficients from acoustical isotherms: The Helmholtz-Kirchhoff correction at temperatures below 35 K	Harmon H. Plumb
89	A refractive index thermometer for use at low temperatures	A. R. Colclough
95	Nuclear orientation thermometry from $\sim$0.001 to $\sim$1.2 K	H. Marshak
103	Radiometric measurement of thermodynamic temperature between 327 and 365 K	T. J. Quinn, J. E. Martin
109	Measurement of thermodynamic temperature with the NPL photon-counting pyrometer	P. B. Coates, J. W. Andrews
115	Noise thermometry at NBS using a Josephson junction	R. J. Soulen, Jr., Deborah Van Vechten
125	Errors in Johnson noise thermometry	G. Klempt
129	A high-accuracy noise thermometer for the range 100–150 °C	C. P. Pickup
133	Noise thermometry and related experiments at IMGC	L. Crovini, A. Actis
139	A new method of noise thermometry	M. Imamura, A. Ohto

II. TEMPERATURE SCALES

143	Vapor pressure of $D_2 + xHD$ and ^{20}Ne	G. T. McConville, D. A. Menke
145	Helium vapor pressure equations on the EPT-76	M. Durieux, J. E. van Dijk, H. ter Harmsel, P. C. Rem, R. L. Rusby
155	Fixed point combination and termination points for platinum resistance thermometer interpolation below 273.15 K	R. C. Kemp
159	Realization of the 1976 Provisional 0.5 K to 30 K Temperature Scale at the National Bureau of Standards	E. R. Pfeiffer, R. S. Kaeser

(Continued)

Page	Title	Authors
169	A photoelectric pyrometer temperature scale below 1064.43 °C and its use to measure the silver point	T. P. Jones, J. Tapping
175	Measurement of the thermodynamic temperature interval between the freezing points of silver and copper	M. Ohtsuka, R. E. Bedford
183	The NIM's photoelectric comparator and the realization of the IPTS-68 above the gold point	Zhao Qi, Den Sixiang, Sun Dinwen, Qiu Nairong, Li Zhenguo, Li Erming
191	The development of temperature standards at NIM of China	Ling Shankang, Zhang Guoquan, Li Ruisheng, Wang Zilin, Li Zhiran, Zhao Qi, Li Xumo
197	An international intercomparison of temperature standards of Asia/Pacific countries	T. P. Jones
201	Soviet standards of the unit of temperature for radiation pyrometry	I. I. Kirenkov, B. N. Oleinik, G. S. Ambrok, G. A. Krakhmalnikova
205	Realization of the triple point of water and the freezing points of tin and zinc at the National Institute of Standards (Egypt)	H. El-Shammaa, M. R. Moussa, M. H. Omar

III. TEMPERATURE FIXED POINTS

Page	Title	Authors
209	On the use of first-generation sealed cells in an international intercomparison of triple-point temperatures of gases	F. Pavese
217	Ten years of research on sealed cells for phase transition studies of gases at IMGC	F. Pavese, D. Ferri
229	The triple point of natural xenon	R. C. Kemp, W. R. G. Kemp, P. W. Smart
231	Thermal behaviour of thermometric sealed cells and of a multi-compartment cell	G. Bonnier, Y. Hermier
239	Reproducibility of the triple point of argon in sealed transportable cells	G. T. Furukawa
249	The triple points of equilibrium and normal deuterium	R. C. Kemp
251	Superconductive thermometric fixed points	J. F. Schooley, R. J. Soulen, Jr.
261	Realizations of the superconductive transition points of lead, indium, aluminium, zinc, and cadmium with SRM 767 devices	A. E. El Samahy, M. Durieux, R. L. Rusby, R. C. Kemp, W. R. G. Kemp
267	Temperature fixed points: Evaluation of four types of triple-point cell	J. D. Cox, M. F. Vaughan
281	Melting curves of H_2O	J. Ancsin
285	The effect of pressure on the water triple-point temperature	J. V. McAllan
291	Reproducibility of some triple point of water cells	George T. Furukawa, William R. Bigge
299	Triple point of gallium as a temperature fixed point	B. W. Mangum
311	An intercomparison of gallium fixed point cells	M. V. Chattle, R. L. Rusby, G. Bonnier, A. Moser, E. Renaot, P. Marcarino, G. Bongiovanni, G. Frassineti
317	Realization of the melting point of gallium	B. N. Oleinik, A. G. Ivanova, V. A. Zamkovets, N. N. Ergardt
321	The triple-point equilibria of succinonitrile: Its assessment as a temperature standard	M. E. Glicksman, P. W. Voorhees, R. Setzko
327	The triple point of rubidium: A temperature fixed point for biomedical applications	J. M. Figueroa, B. W. Mangum
339	Temperature references based on first order phase transition: Development and application	D. Rappaport, N. Karasikov, M. B. Roitberg
343	Realization of the triple point of indium in a sealed glass cell	S. Sawada

(Continued)

Part 1: pages 1–710; Part 2: pages 711–1395

347	A small transportable indium cell for use as a temperature reference	Magda Hanafy, M. R. Moussa, M. H. Omar
351	The use of the cadmium point to check calibrations on the IPTS	J. V. McAllan, J. J. Connolly
355	Investigation of the freezing temperature of cadmium	George T. Furukawa, Earl R. Pfeiffer
361	Measurement of the melting temperature of the copper 71.9% silver eutectic alloy with a monochromatic optical pyrometer	R. E. Bedford, C. K. Ma
371	Reference temperatures near 800 °C	J. V. McAllan
377	Radiance temperature of metals at their melting points as possible high temperature secondary reference points	A. Cezairliyan, A. P. Miiller, F. Righini, A. Rosso
383	Miniature thermometric fixed points for thermocouple calibrations	M. Tischler, M. J. Koremblit
391	On sealed freezing point cells	Zhu Ci-Zhun

IV. RADIATION THERMOMETRY

395	Temperature distribution measurement with a silicon photodiode array	T. Yamada, N. Harada, M. Koyanagi
401	A new method for temperature distribution measurement using multi-spectral radiance	Jiro Ohno
409	A broadband ratio pyrometer	J. L. Gardner, T. P. Jones, M. R. Davies
413	Single-band radiation thermometers: Harmonization of their calibration characteristics	Jiang Shichang, Wu Shuyuan, Ye Rongchang, Xu Liang
421	Establishing a practical temperature standard by using a narrow-band radiation thermometer with a silicon detector	F. Sakuma, S. Hattori
429	A photoelectric direct current spectral pyrometer with linear characteristics	B. Woerner
433	Ten years of high speed pyrometry at IMGC	F. Righini, A. Rosso
439	Microsecond and sub-microsecond multi-wavelength pyrometry for pulsed heating technique diagnostics	J.-F. Babelot, J. Magill, R. W. Ohse, M. Hoch
447	Two-color microsecond pyrometer for 2000 to 6000 K	G. M. Foley, M. S. Morse, A. Cezairliyan
453	Infrared temperature measurements of gas and dust explosions	K. L. Cashdollar, M. Hertzberg
465	A packaged, fiber-optic spectroradiometer for high temperature gases, with automatic readout	S. A. Self, P. H. Paul, P. Young
471	Radiation thermometry applied to the development and control of gas turbine engines	T. G. R. Beynon
479	Improvement of traceability for radiation pyrometers in the steel industry	K. Tamura, T. Iwamura, K. Kurita
485	Steel surface temperature measurement in industrial furnaces by compensation for reflected radiation errors	J. E. Roney
491	Two methods for simultaneous measurement of temperature and emittance by using multiple reflection and specular reflection, and their applications to industrial processes	T. Iuchi, R. Kusaka
505	Temperature measurement of steel in the furnace	Y. Tamura, M. Tatsuwaki, T. Sugimura, T. Yokoi, M. Sano, M. Koriki
513	Apparent emissivities of cylindrical cavities with partially specular conical bottoms	A. Ono
517	Numerical calculation of the effective emissivity by using a series technique	Yoshiko Ohwada
521	The effective temperature to express radiant characteristics of nonisothermal cavities	S. Hattori, A. Ono
529	A method for calibration of a precision thermal radiation density meter using a radiation source	V. A. Chistyakov, V. I. Gavrishchuk

(Continued)

Part 1: pages 1–710; Part 2: pages 711–1395

535	A practical-type fixed point blackbody furnace	F. Sakuma, S. Hattori
541	Experimental and theoretical study on the quality of reference blackbodies formed by lateral holes on a metallic tube	A. Ono, R. M. Trusty, D. P. DeWitt
551	Precision practical blackbody furnaces by a 3-zone temperature control method	I. Hishikari, T. Ide
559	−50 to +150 °C heat pipe blackbody sources for radiation thermometer calibration	Zhu Yingsong, Ma Hongqi, Wang Ronghua, Hua Chengsheng
567	Feedback stabilized tungsten strip lamp as a radiometric standard for photoelectric pyrometry	Zhu Ci-Zhun, Ju Hao-Rien
569	Estimation of the true temperature of targets by their thermal radiation based on Planck's law	G. S. Ambrok
575	Practical CARS temperature measurements	J. P. Taran, M. Péalat
583	A hardened CARS system for temperature and species-concentration measurements in practical combustion environments	G. L. Switzer, L. P. Goss
589	Pure rotational CARS thermometry	J. W. Fleming, A. B. Harvey, W. T. Barnes
595	CARS thermometry in reacting systems	J. F. Verdieck, J. A. Shirley, R. J. Hall, A. C. Eckbreth
609	CARS thermometry in an internal combustion engine	L. A. Rahn, S. C. Johnston, R. L. Farrow, P. L. Mattern
615	Temperature measurements for combustion diagnostics from high-resolution single-pulse CARS N_2 spectra	David Klick, K. A. Marko, L. Rimai
621	The use of rotational Raman scattering for measurement of gas temperature	M. C. Drake, C. Asawaroengchai, D. L. Drapcho, K. D. Veirs, G. M. Rosenblatt
631	Use of the vibrational Raman effect for gas temperature measurements	M. C. Drake, M. Lapp, C. M. Penney
639	Dynamic temperature measurements of flames using spontaneous Raman scattering	P. P. Yaney, R. J. Becker, P. D. Magill, P. Danset
649	Laser tomography for temperature measurements in flames	H. G. Semerjian, R. J. Santoro, P. J. Emmerman, R. Goulard
661	Recent advances in absorption spectroscopy of OH and their implications in rotational temperature measurements	Charles C. Wang, Dafan Zhou
665	A high speed non-intrusive temperature diagnostic for combustion processes	R. W. McCullough, G. B. Northam
677	Determination of the time evolution of the electron temperature profile of reactor-like plasmas from the measurement of blackbody electron cyclotron emission	P. C. Efthimion, V. Arunasalam, R. A. Bitzer, J. C. Hosea
687	Measurement of the electron temperature profile in a tokamak by observation of electron cyclotron emission using a Fourier transform spectrometer	F. J. Stauffer
693	Measurement of the central ion and electron temperature of tokamak plasmas from the x-ray line radiation of high-Z impurity ions	M. Bitter, S. von Goeler, M. Goldman, K. W. Hill, R. Horton, W. Roney, N. Sauthoff, W. Stodiek
705	Temperature measurement in a wall-stabilized arc used as a radiation standard in the vacuum ultraviolet	D. H. Nettleton

V. RESISTANCE THERMOMETRY

711	Automatic resistance thermometer bridges for new and special applications	R. D. Cutkosky
715	An automatic resistance thermometer bridge	C. G. M. Kirby
719	An automatic resistance thermometer bridge	N. L. Brown, A. J. Fougere, J. W. McLeod, R. J. Robbins

(Continued)

Part 1: pages 1–710; Part 2: pages 711–1395

729	A new range of high precision resistance bridges for resistance thermometry	P. C. F. Wolfendale, J. D. Yewen, C. I. Daykin
733	Designing accurate platinum RTD measuring systems for industry	J. R. Saffell
739	An automatic system for measuring Bowen ratio gradients using platinum resistance elements	L. J. Fritschen, J. R. Simpson
743	Evaluation and control of platinum oxidation errors in standard platinum resistance thermometers	Robert J. Berry
753	Oxidation, stability, and insulation characteristics of Rosemount standard platinum resistance thermometers	Robert J. Berry
763	The stability of commercially available high temperature platinum resistance thermometers of a 5 Ω silica cross type up to 961.93 °C	H. J. Jung, H. Nubbemeyer
771	Experiences with high-temperature platinum resistance thermometers	J. P. Evans
783	Stability of precision high temperature platinum resistance thermometers	Long Guang, Tao Hongtu
789	Practical high temperature resistance thermometry	J. V. McAllan
795	Investigation of the stability of small platinum resistance thermometers	B. W. Mangum, G. A. Evans, Jr.
803	Thermal hysteresis and stress effects in platinum resistance thermometers	D. J. Curtis
813	Construction of a laboratory working thermometer using industrial platinum resistance sensors	N. M. Bass
815	The calibration characteristics of industrial platinum resistance thermometers	J. J. Connolly
819	Interpolating equations for industrial platinum resistance thermometers in the temperature range from -200 to $+420$ °C	A. Actis, L. Crovini
829	The rhodium-iron resistance thermometer: Ten years on	R. L. Rusby
835	The state of development of planar germanium cryogenic thermometers	P. R. Swinehart
839	Platinum-cobalt alloy resistance thermometer for wide range cryogenic thermometry	T. Shiratori, K. Mitsui, K. Yanagisawa, S. Kobayashi
845	Carbon-glass sensors: Reproducibility and polynomial fitting of temperature vs resistance	B. W. Ricketson, R. Grinter
853	Carbon-glass thermometry in China	Yao Quanfa, Deng Daren, Ma Hongqi, Jiang Dehua, Ji Yunsong, Huang Xihuai
859	A thin platinum film for transient heat transfer studies	P. J. Giarratano, F. L. Lloyd, L. O. Mullen, G. B. Chen
865	Enhanced stability in precision interchangeable thermistors	T. H. LaMers, J. M. Zurbuchen, H. Trolander
875	The exactness of fit of resistance-temperature data of thermistors with third-degree polynomials	M. Sapoff, W. R. Siwek, H. C. Johnson, J. Slepian, S. Weber
889	Aging phenomena in nickel-manganese oxide thermistors	J. M. Zurbuchen, D. A. Case
897	Fast thermistor sensors for rapid reaction studies	R. L. Berger, B. Balko, T. R. Clem, W. S. Friauf
911	Tailoring PTC thermistor characteristics	R. E. Wendt, Jr.

VI. THERMOCOUPLE THERMOMETRY

915	Thermoelectric thermometry: A functional model	R. P. Reed
923	Testing of thermocouples for inhomogeneities: A review of theory, with examples	C. A. Mossman, J. L. Horton, R. L. Anderson
931	Validation diagnostics for defective thermocouple circuits	R. P. Reed
939	Failure of sheathed thermocouples due to thermal cycling	R. L. Anderson, R. L. Ludwig

(Continued)

953	Properties of some noble and base metal thermocouples at fixed points in the range 0–1100 °C	E. H. McLaren, E. G. Murdock
977	Decalibration of sheathed thermocouples	R. L. Anderson, J. D. Lyons, T. G. Kollie, W. H. Christie, R. Eby
1009	Thermocouple measurement uncertainty in compressor efficiency measurement: The effects of two uncertainty models	Ronald H. Dieck, Barbara G. Ringhiser
1019	Studies of sheathed thermocouple construction and installation in thermowells to obtain faster response	R. M. Carroll, K. R. Carr, R. L. Shepard
1025	Very low temperature thermocouple devices: Development and application techniques for temperature measurements	H. Armbrüster, W. P. Kirk, D. P. Chesire
1037	Differential type thermometer for measuring hot gas temperature	Jiro Ohno, Masakazu Nakamura, Yutaka Miyabe, Atsushi Kawasaki, Yukio Kanoshima
1043	Lining erosion measurements by sheathed multiple thermocouples through temperature transients	Y. Kawate, N. Nagai, M. Konishi, K. Yokoe, T. Horiuchi
1051	Miniature zircaloy-sheathed thermocouples for nuclear fuel-rod cladding temperature measurements	S. C. Wilkins
1061	2200 °C thermocouples for nuclear reactor fuel centerline temperature measurements	C. P. Cannon
1069	Temperature measurements with chromel/alumel thermocouples in a pressurized water reactor	P. Siltanen, T. Laaksonen, W. Joslin
1081	Temperature measurement in the WAGR	A. Thurlbeck
1097	Lifetime improvement of small-diameter sheathed thermocouples for use in high-temperature and thermal transient operations	R. W. McCulloch, J. H. Clift
1109	Thermocouples for measurements under conditions of high temperature and nuclear radiation	R. Schley, G. Metauer
1115	Proposed mechanism for the thermoelectric properties of nickel and some of its alloys near the Curie temperature	D. D. Pollock
1121	The nicrosil versus nisil thermocouple: Recent developments and present status	G. W. Burns
1129	The nicrosil versus nisil thermocouple: The influence of magnesium on the thermoelectric stability and oxidation resistance of the alloys	N. A. Burley, J. L. Cocking, G. W. Burns, M. G. Scroger
1147	Oxidation resistance and stability of nicrosil-nisil in air and in reducing atmospheres	T. P. Wang, C. D. Starr
1159	The nicrosil versus nisil thermocouple: A critical comparison with the ANSI standard letter-designated base-metal thermocouples	N. A. Burley, R. M. Hess, C. F. Howie, J. A. Coleman

VII. ELECTRONIC THERMOMETRY

1167	Signal processing techniques for temperature measurement	K. P. Shambrook
1173	A new nuclear quadrupole resonance standard thermometer	A. Ohte, H. Iwaoka
1181	Temperature profiling using multizone ultrasonic waveguides	L. C. Lynnworth
1191	Ultrasonic thin-wire thermometry for nuclear applications	H. A. Tasman, M. Campana, D. Pel, J. Richter
1197	Precision silicon transistor thermometer	A. Ohte, M. Yamagata, K. Akiyama
1205	Semiconductor junctions as cryogenic temperature sensors	M. Ganapati Rao
1213	Temperature measuring method by using the eddy current technique	K. Sano, T. Yamada, S. Ando, K. Watanabe
1219	A decade of progress in high temperature Johnson noise thermometry	T. V. Blalock, R. L. Shepard
1225	Application of noise thermometry in industry under plant conditions	H. Brixy, R. Hecker, K. F. Rittinghaus, H. Höwener
1239	High temperature noise thermometry for industrial applications	M. C. Decréton

(Continued)

Part 1: pages 1–710; Part 2: pages 711–1395

1245	Dual high temperature measurements using Johnson noise thermometry	T. R. Billeter, C. P. Cannon
1249	Johnson noise power thermometer and its application in process temperature measurement	T. V. Blalock, J. L. Horton, R. L. Shepard

VIII. TEMPERATURE CONTROL

1261	Modeling a closed loop control system	R. L. Fillmore
1265	A new generation of precision furnaces	C. A. Busse, C. Bassani
1275	A precision 4.2–300 K temperature controller using a genuine full-range sensor and inductive divider set-point coupled with a simple ac sensing bridge	Wang Zhensen, Deng Daren
1279	Temperature control at a high interference level: A case description	I. Karaila, J. Horelli
1283	A proposed pressure amplifier for a temperature control system	T. M. Kegel, D. E. Limbert

IX. CALIBRATION METHODS

1287	Automated temperature measurements from −183 to 2300 °C	M. H. Cooper, Jr., R. L. Anderson, C. A. Mossman
1293	Automation of a thermometer calibration facility	C. G. M. Kirby
1299	Automation of measurements in a low temperature laboratory	Craig T. Van Degrift, Robert S. Kaeser
1307	A highly stable calibration furnace for platinum thermometers up to 700 °C	Zhang Jipei, Shao Kaidi
1311	Calibration with confidence: The assurance of temperature accuracy	R. D. Collier

X. THERMOMETRY FOR SPECIAL APPLICATIONS

1317	Deep-ocean temperature measurement	T. M. Dauphinee
1327	Down-to-earth air temperature measurements during space shuttle earth atmosphere re-entry	T. M. Stickney, M. T. Stiles
1333	Cryogenic thermometry: A review of recent progress. II	L. G. Rubin, B. L. Brandt, H. H. Sample
1345	Design of a fluidic capillary pyrometer for contact duty at temperatures to 2750 °C	R. Michael Phillippi, Tadeusz M. Drzewiecki, Taki Negas, Harry S. Parker
1353	Diffusion thermometry, an engineering concept	M. Lamvik
1357	Response of installed temperature sensors	T. W. Kerlin, R. L. Shepard, H. M. Hashemian, K. M. Petersen
1367	Thermal response times of some cryogenic thermometers	D. Linenberger, E. Spellicy, R. Radebaugh
1373	Spectroscopic techniques for measuring the temperature of liquids in analytical instrumentation	Lemuel J. Bowie
1379	Imaging microwave thermography	J. Edrich, W. E. Jobe
1381	Thermal environments and thermal comfort: New instruments and methods	E. Mayer
1389	Temperature and thermodynamics of living matter	T. H. Benzinger
xxi	AUTHOR INDEX	
xxxi	SUBJECT INDEX	

Temperature scales, the IPTS, and its future development

R. P. Hudson

Bureau International des Poids et Mesures, Pavillon de Breteuil, F-92310 Sèvres, France

A brief review is given of the concept of an internationally-agreed temperature scale and the practical considerations which influence its construction and utilization. This sets the scene for a description of the current International Practical Temperature Scale (IPTS-68) and its presently-perceived deficiencies, also of new primary thermometry data and technical developments (sensors and fixed points) which will very probably form the basis for constructing a revised scale in the near future. Such a revision is also likely to include a substantial downwards extension of the temperature range covered by the scale, probably to 0.5 K and perhaps (though somewhat less probably) into the millikelvin region, and the possibilities and probabilities for this endeavor are set forth. Brief concluding remarks cover the relationship between state-of-the-art metrology and the thermometric requirements of science and industry; a possible conflict of viewpoints of temperature-scale producers and certain important users; and the concept of "secondary realizations" of an IPTS.

I. BASIC CONSIDERATIONS

Of all the basic physical quantities, *temperature* is, perhaps, the most difficult to measure accurately–so much so that it is rarely attempted. From time to time, however, epic experiments are undertaken to perfect a "thermodynamic" instrument, such as a speed-of-sound, noise, or classical gas thermometer. The results of ensuing measurements are then transferred to sensitive and reproducible instruments known as "secondary" thermometers which are employed within the framework of an agreed *temperature scale*. Such a temperature scale is a "recipe" for accurately and conveniently identifying (in good approximation) the thermodynamic temperature of a system. In developing such a scale, four basic features must be kept in mind; these are

(a) Definition
(b) Realization (and maintenance)
(c) Transfer
(d) Utilization

Definition and realization require reference temperatures (fixed points) plus instruments and procedures for interpolating between, and extrapolating beyond, such reference points.

For interpolation one requires a sensor which minimizes the inherent tendency to "nonuniqueness," that is to say, the tendency of different thermometers, calibrated in exactly the same way, to give different readings at temperatures between the fixed points. This arises because individual thermometers of the same type behave in slightly different ways which are not sufficiently well taken into account by the chosen interpolation equation. To minimize nonuniqueness the realization process must be well-tailored to the actual temperature dependence of the thermometric physical property X and the number of available fixed points. If the latter are rather few, the X versus T relation must have a rather simple physical form, otherwise the calibration "fit" will be a relatively crude one in between the reference points. A perfect fit is only possible (and that only in principle) *at* the fixed points and divergences between thermometers for the intervening regions constitute the nonuniqueness. If errors occur in realizing the fixed points, the nonuniqueness from laboratory to laboratory is even worse.

For extrapolation—which has to be done on faith (and physics), obviously—one looks for a device which utilizes a thermometric property varying with T according to a very-well-defined physical law. For example, the International Practical Temperature Scale is defined above the melting point of gold in terms of Planck's law and the radiant energy emerging from a blackbody. Nonuniqueness will arise to the extent that Planck's law does not actually apply to the cavities which we employ, i.e., departures from blackness, temperature dependence of the same, presence of temperature gradients, etc., will all introduce errors in the value derived for the temperature. A somewhat parallel situation exists at very low temperatures (where, however, an IPTS is not defined): in the microkelvin region one can do nothing, at present, but utilize the paramagnetism of certain nuclear systems and rely upon an extrapolation of Curie's law. The situation is different here from that of the optical region in that nonuniqueness will not arise so long as all use the same sensor material. But in both cases, limits exist to extending the procedure to ever further extremes of temperature: at high temperatures, the practical impossibility of making the necessary artifact for embodying the pertinent physical law; at low temperatures, first an increasing importance of heretofore neglected higher-order terms in the X-T relation and, beyond that, the onset of essentially temperature-independent behavior.

Such an approach as that just described for extrapolation is also useful, in principle, and perhaps even preferable, for *interpolation* application, but historically one has accepted empirical X versus T relations in the interests of practical convenience. Several physical laws that have been used with advantage in thermometry (e.g., Nyquist noise, ideal gas law, Curie law) are analytically very simple—certainly at the level concerned in practice—and such a feature would have obvious advantages in an interpolative application from the standpoints of simplicity of scale definition and realization, and uniqueness. Historically, it has only proven possible to simultaneously achieve high sensitivity, excellent reproducibility, and simplicity-of-law in terms of the liquid-in-glass thermometer. But in the low-temperature region, presently under active examination, a second example now seems likely to emerge (see below).

Although there would be some advantages if the scale-defining interpolation instrument were "practical" in a very broad sense of the term, that is not absolutely necessary: witness the operational deficiencies of the standard platinum resistance thermometer (PRT), such as fragility, rather large size, and rather long response time; and of the optical pyrometer, such as the need for unimpeded transmission of light. One strives for maximum reproducibility (commensurate with the resources of a national standardizing laboratory), within the framework of a measurement and transfer system which is substantially simpler, as well as more reproducible, than would be achievable through the routine employment of primary thermometry.

The "practical measurement of temperature" can (in principle) be effected in two distinct ways: in one, the scale is realized *in situ* and step (d) above follows immediately after step (b); step (c) is eliminated. In the other—which is, by far, the more common—the concept of a service (even if self-service) is introduced, where the utilization of the thermometer is physically quite distinct from the realization of the scale. This second procedure is inherently less accurate than the former, especially if the transfer instrument and the "utilization instrument" are separate entities. If, moreover, the instrument used in step (d) is of a different *type* than that concerned in step (c), it will usually be so because of "practical consideration"; it is frequently the case that some sacrifice of thermometric integrity has been consciously accepted in the interest of achieving operational convenience. The true extent of the degradation incurred in moving down the calibration chain can only be determined, however, by careful and sophisticated checking procedures, made in as direct a way as possible such as is done in "measurement assurance" procedures.

II. INTERNATIONALLY-AGREED SCALES

The original international agreement on temperature measurement was necessary in order to put thermodynamic-property measurements on a common basis internationally.[1,2] This required accomplishing steps (a) and (b) with, as essential features, a coverage of as wide a range of temperature as was then feasible and the selection of instruments and methods offering a high degree of sensitivity and reproducibility. Since the first formal agreement in 1927,[3] improvements and/or extensions have been undertaken from time to time, as is well described in the literature.[4] The responsible body is the International Committee of Weights and Measures (CIPM) and the "nuts and bolts" work of sponsoring and evaluating research and comparisons, and of formulating recommendations, is carried out by its Advisory Committee for Thermometry (CCT).

The most recent revision of IPTS was completed in 1967/68 and led to the promulgation of IPTS-68.[5] This was modified in 1975, giving rise to IPTS-68: Amended Edition of 1975.[6] The highly technical cerebrations, spiced with all-too-human subjectiveness, which led up to the production of IPTS-68 have been amusingly described[7] by the present President of the CCT before he reached that exalted position. I now turn to a discussion of the current status of the IPTS and the directions for its improvement and extension which are presently under active discussion.

III. IPTS-68

IPTS-68 is based upon the fixed points as listed in Table I, together with their assigned values of temperature. These

TABLE I. Defining fixed points of the IPTS-68.[a]

Equilibrium state	Assigned value of International Practical Temperature	
	T_{68} (K)	t_{68} (°C)
Equilibrium between the solid, liquid and vapor phases of equilibrium hydrogen (triple point of equilibrium hydrogen)[b]	13.81	−259.34
Equilibrium between the liquid and vapor phases of equilibrium hydrogen at a pressure of 33 330.6 Pa (25/76 standard atmosphere)[b,c]	17.042	−256.108
Equilibrium between the liquid and vapor phases of equilibrium hydrogen (boiling point of equilibrium hydrogen)[b,c]	20.28	−252.87
Equilibrium between the liquid and vapor phases of neon (boiling point of neon)[c]	27.102	−246.048
Equilibrium between the solid, liquid and vapor phases of oxygen (triple point of oxygen)	54.361	−218.789
Equilibrium between the solid, liquid and vapor phases of argon (triple point of argon)[d]	83.798	−189.352
Equilibrium between the liquid and vapor phases of oxygen (condensation point of oxygen)[c,d]	90.188	−182.962
Equilibrium between the solid, liquid and vapor phases of water (triple point of water)	273.16	0.01
Equilibrium between the liquid and vapor phases of water (boiling point of water)[e]	373.15	100
Equilibrium between the solid and liquid phases of tin (freezing point of tin)[e]	505.1181	231.9681
Equilibrium between the solid and liquid of zinc (freezing point of zinc)	692.73	419.58
Equilibrium between the solid and liquid phases of silver (freezing point of silver)	1235.08	961.93
Equilibrium between the solid and liquid phases of gold (freezing point of gold)	1337.58	1064.43

[a] Except for the triple points and one equilibrium hydrogen point (17.042 K) the assigned values of temperature are for equilibrium states at pressure $p_0 = 101\ 325$ Pa (1 standard atmosphere). In those cases where differing isotopic abundances could significantly affect the fixed point temperature, the abundances specified in Sec. III (Ref. 6) must be used.
[b] The term equilibrium hydrogen is defined in Sec. III (Ref. 6).
[c] Fractionation of isotopes or impurities dictates the use of boiling points (vanishingly small vapor fraction) for hydrogen and neon, and condensation point (vanishingly small liquid fraction) for oxygen (see Sec. III of Ref. 6).
[d] The triple point of argon may be used as an alternative to the condensation point of oxygen.
[e] The freezing point of tin [$t' = 231.9292$ °C, see Eq. (10) of Ref. 6] may be used as an alternative to the boiling point of water.

numbers are all based upon gas thermometry (plus, for the hydrogen region, some NBS acoustic thermometer data); the size of the degree (kelvin) is set by assigning the temperature 273.16 K to the triple point of water.[8]

The scale is defined from 13.81 K upwards, with three separate ranges involving two interpolation instruments and one extrapolation algorithm:

A. Platinum thermometer range

The PRT range runs from 13.81 to 273.15 K in one subregion and from 273.15 K to 630.74 °C in the other. The two subregions are distinguished by the procedures devised for the thermometer calibration. For the upper subregion the procedure is quite simple; for the lower region, the procedure is based upon four further subdivisions and is more complicated.[5]

The standard PRT is capable of achieving a reproducibility of the order of ± 0.02 mK. The procedures in IPTS-68 were devised to capitalize upon this quality. Unfortunately, the ultimate is seldom achieved in practice and reproducibility is frequently no better than ± 0.5 mK. Even more sobering is the recent establishing of the fact (see Fig. 1) that *realizations of the fixed points* can be so imperfect as to introduce relatively gross uncertainties (± 5 mK) into the process.[9] Thus there is ample room, logically, to tolerate some loss in *theoretical* accuracy if substantial improvements in simplicity are to hand, even though one may reasonably expect that the graphic demonstration of hitherto-unsuspected shortcomings will lead to improvements of practice in the delinquent laboratories.

Recent experiments in the lower regions of a definitive high-temperature gas-thermometry study at the U.S. National Bureau of Standards have shown that the fixed points between 0 and 420 °C (at least) probably need some downward adjustment in their assigned values[10] (see Fig. 2). These results have lately received very good support from NPL radiation thermometry measurements (which employ calorimetric detection at 2 K) carried out in a—so far—restricted region around 0 °C.[11] Evidence is to hand, too, that $T(68)$ deviates significantly from T below 100 K; the oxygen triple point, presently assigned the value 54.361 K, may be too high by as much as 0.01 K.[12] The smoothness and the thermodynamic accuracy below 30 K can also be notably improved (changes up to 9 mK), according to recent low-temperature gas thermometry measurements.[13] Throughout the world, several efforts in primary thermometry (gas and noise) below 0 °C offer the prospect of further improvement in our knowledge concerning thermodynamic temperature accruing in the very near feature.[14-16]

B. Thermocouple region

Between 630.74 °C and the gold point [$t(68) = 1064.43$ °C] IPTS-68 is defined by means of a second-order dependence upon temperature of the electromotive force (emf) of a Pt versus Pt/10%Rh thermocouple.[5] Studies in recent years have suggested that at the several-mK level of precision the emf/temperature relation is not adequately rendered by a polynomial of less than the *fifth* de-

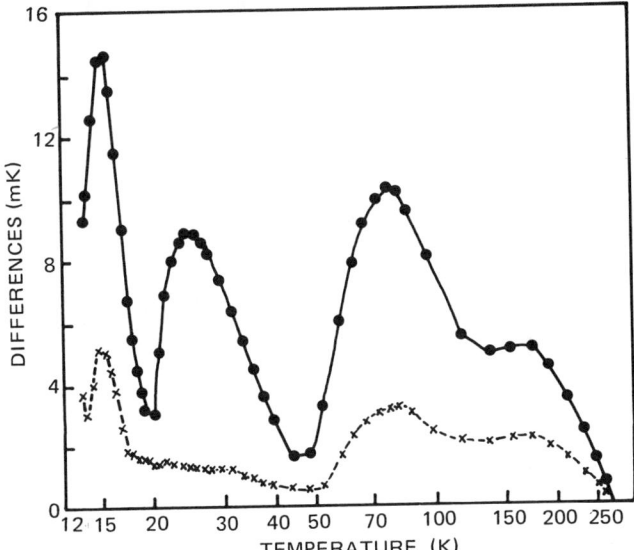

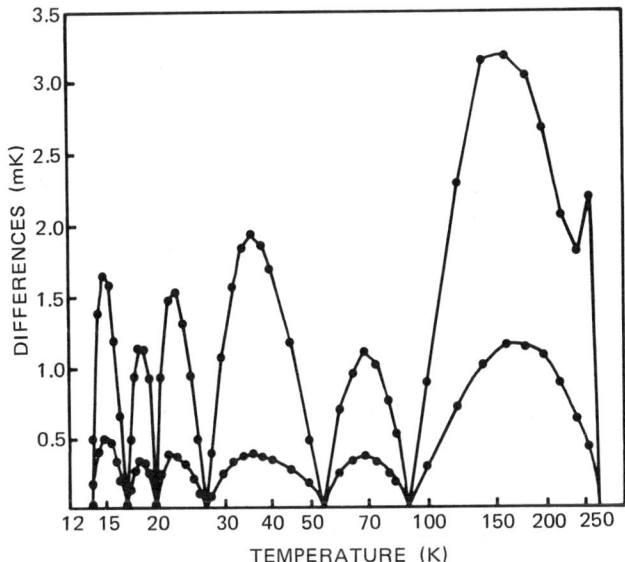

FIG. 1. Comparison of laboratory-reference-standard platinum resistance thermometers. Upper diagram: using parent-laboratory calibration. Lower diagram: after recalibration, in order to agree at the fixed points. In each diagram, upper curve gives the maximum divergence for any pair of thermometers and lower curve indicates the limit below which 68% of the differences lie [from S. D. Ward and J. P. Compton, Metrologia 15, 31 (1979)].

gree, however.[17] In comparisons with optical pyrometers ranging downwards from the gold point, too, several researchers are unanimous in concluding that IPTS-68 is substantially defective between 630 and 950 °C, with a maximum deviation of 0.5 K for $T - T(68)$ occurring in the region of 800 °C; see Fig. 3.[18] These findings have been supported by the results of noise thermometry.[19] The present calibration procedure is based upon three reference points only, and thus nonuniqueness is a problem.

Deficiencies stemming from metal imperfections are also significant: physical and chemical inhomogeneities in the wires in a region of temperature gradient give rise to a spurious emf which can result in a systematic error in the calibration. Such imperfections change continuously, too,

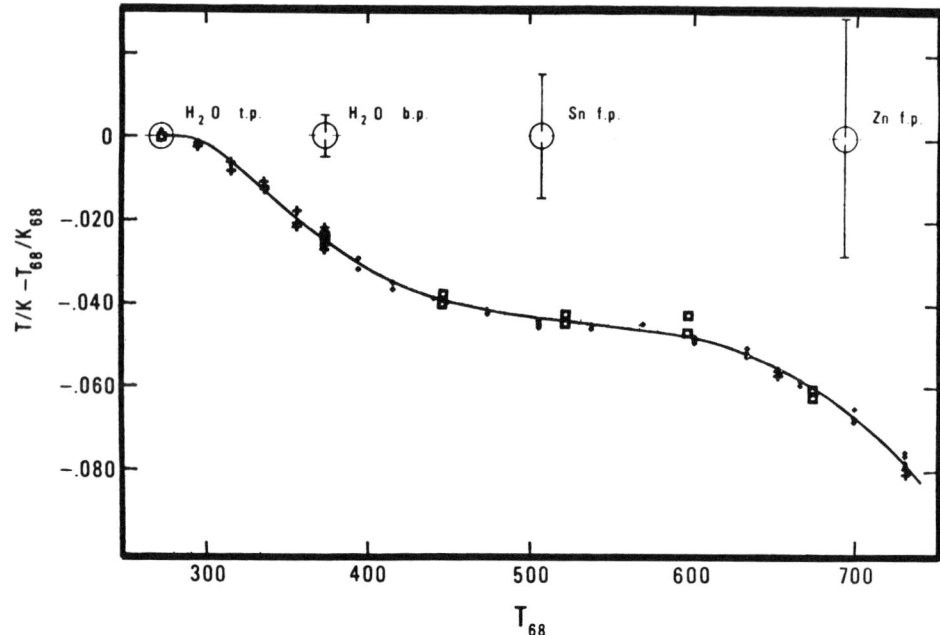

FIG. 2. The differences $T - T_{68}$ found via recent gas thermometry [from L. A. Guildner and R. E. Edsinger, J. Res. Natl. Bur. Stand. Sect. A **80**, 703 (1976)].

with use at elevated temperatures and the calibration will drift accordingly.

Jones[20] has concluded that, following calibration, the minimum uncertainty in a single reading of an unknown temperature, when the thermocouple is free of systematic errors, is ± 0.2 °C. In general use, the thermal history of the thermocouple affects its performance in a very complicated way and an indicated temperature may be uncertain by many tenths of a kelvin.[21] Improvement in this situation only seems to be feasible by extending the high-temperature, optical region (see below) downwards, or extending the PRT region upwards, or by a combination of both.

C. Optical pyrometer region

Above the gold point, T_{Au}, IPTS-68 is based upon measuring ratios of T_x / T_{Au}, via the spectral concentration of the radiance of a blackbody and invoking the Planck radiation law.[5] Realization is achieved through the agency of an optical pyrometer, although that is not specified in the text of the scale. Maintenance and transfer are accomplished through the use of tungsten strip lamps, evacuated for use below

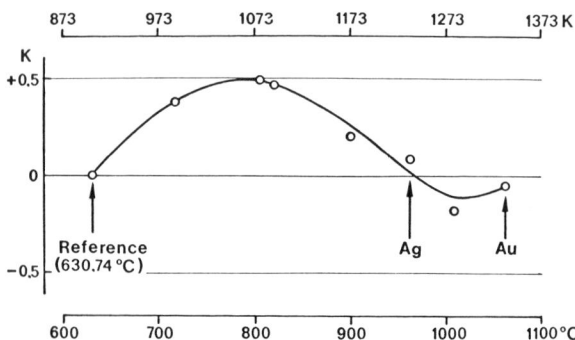

FIG. 3. The differences $T - T_{68}$ found via pyrometry [from J. Bonhoure, Metrologia **11**, 141 (1975)].

about 1600 °C and argon-filled for the higher-temperature region.[22]

In today's technology the visual pyrometer has been supplanted by a photoelectric successor. (This change, incidentally, led in quite short order to a reassessment of the temperature of the freezing point of platinum by some 4 K!)[23] Realization of the scale is beset by numerous subtleties of practice, including greyness of the "black" body; properties of filters, sectored disks, and of other components of an optical system such as detector linearity, lamp aging, polarization effects, etc. Current practice seems to be entirely adequate, however, for all practical needs.[24]

IV. EXTENSION DOWNWARDS OF THE IPTS

Between 0.5 and 5 K, tables of ^{3}He and ^{4}He vapor pressure exist (based upon gas thermometry) which are in wide use as semiofficial temperature standards.[25,26] These are presently undergoing revision.[27] Below 0.5 K, a host of thermometric methods are to hand (Fig. 4), including primary thermometers based upon anisotropy of γ-radiation from oriented nuclei or Johnson noise; nuclear paramagnetism susceptibility thermometers, which entail invoking the inverse-T Curie law and calibrating at one point; electron paramagnetic or resistance thermometers calibrated against He vapor presure or superconductive fixed points, the former being sometimes used in the "extrapolation mode"; and others.[28]

Thus, while below 0.5 K no official word is needed, perhaps, to guide the temperature-measuring community, and while from 0.5 to 5 K a measure of official guidance had been available in the past, for the technologically-important region of 5 to 13.81 K there is no practice-unifying international standard despite the unarguable need. A new IPTS will have to span this gap; to date, the burning question has been, "How?"

Thoughts are now crystallizing along the following lines[29]: an excellent transfer standard (from the points of

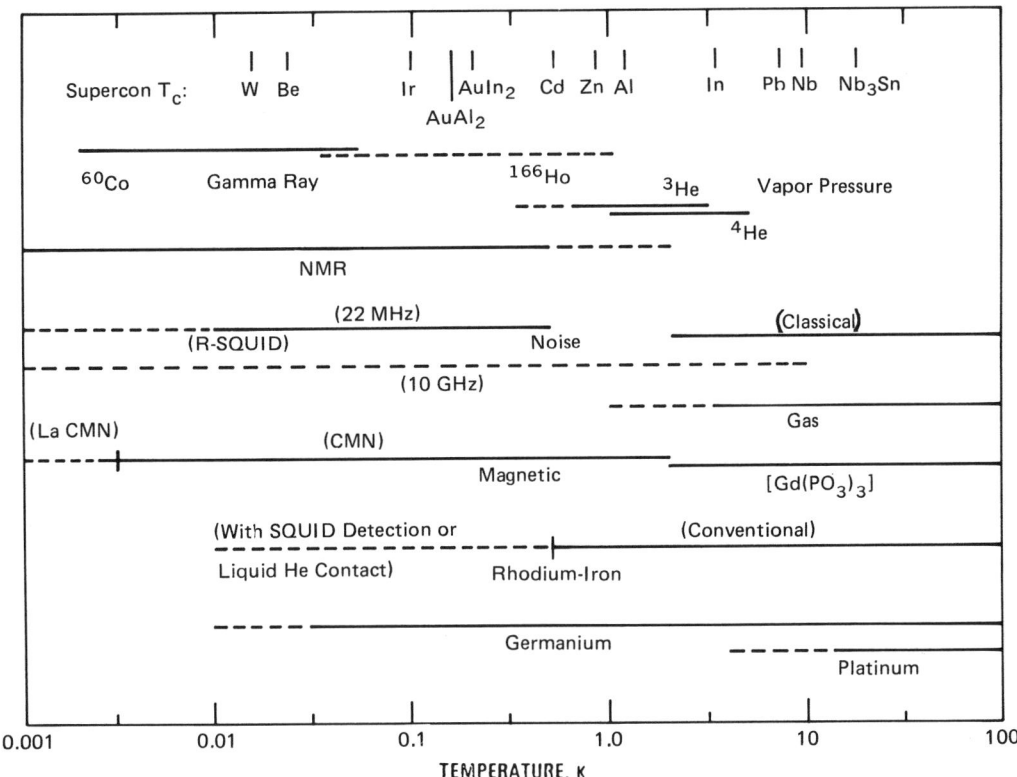

FIG. 4. Types of thermometer and their ranges of use for the region below 30 K.

view of quality, cost, and convenience) exists in the form of the Rh/0.5%Fe resistance thermometer.[30] (It *could* be used as a scale-defining interpolator but it requires many calibration points as its R versus T characteristic is not simple at the, say, ±0.2 mK level. There are problems of supply too which, undesirable in any case, are condemnatory for scale-defining applications.) The constant-volume gas thermometer, used simply as an interpolating instrument, can perform at the desired level, with only mildly restrictive design specifications plus a designated value for the second virial coefficient of helium. The operational problem is that of measuring pressure ratios accurately, e.g., an uncertainty of ±1 mK requires $\Delta p/p = 10^{-4}$ at 10 K, and pro rata. The only other contestant, the magnetic thermometer, is rather less clumsy and offers the advantage of an electrical—rather than mechanical—measurement, but it is less simple in application and less reliable in general operation. (The future may well bring substantial improvements in both: sensor properties and circuitry improvements for the magnetic thermometer, miniaturization and read-out sophistication for the gas thermometer.[31])

V. THE PLATINUM RESISTANCE THERMOMETER IN A NEW IPTS

A. Lower limit

Below 25 K, the sensitivity of a PRT falls off markedly. Measurement sensitivity can be restored, however, by increasing the measuring current above that permissible at higher temperatures without unduly increasing the self-heating. A more serious problem is nonuniqueness, although this may be due, partly, to the present inadequacies built into IPTS, the reference function being poorly representative of platinum in the low-temperature region and the fixed point values being in need of adjustment.[32]

At present, there are separate calibration procedures for the regions 13.81 to 20.28 K and 20.28 to 54.361 K, with 4-parameter, cubic functions being used for each.[5] Some simplification is evidently desirable, apart from the wish to reduce nonuniqueness. There would seem to be no difficulty in running an interpolating-gas-thermometer section up to the triple point of Ne at 24.56 K, say, should future analysis indicate that this would be advantageous. It is almost certain, however, that retention of the PRT down to 13.81 K will continue to entail calibration at two hydrogen boiling points, which runs counter to the current trend concerning this type of temperature reference (see below).

B. Upper limit

Reference has already been made to the desirability of removing the standard thermocouple from a future scale definition. The possibility of using the PRT right up the gold point was broached, in fact, by Callendar many decades ago, albeit before significant studies had been made on the effects of initial impurities, contamination, and strain. A great deal of research has been devoted, over the last 20 years to developing a suitable high-temperature platinum resistance thermometer.[33] Important design features include the size of wire, method of mounting, enclosure and mount material, and details of cleaning and filling. The nominal 2.5 Ω model offers the advantage of sensitivity, while the nominal 0.25 Ω sensor is more robust and is less afflicted with problems of leakage resistance and surface contamination effects. A high-temperature PRT will always be very expensive compared to the platinum-alloy thermocouple. It will always be, too, a higher-quality sensor, so the question comes down to one of benefit versus cost.

Another consideration is that the deterioration which occurs at elevated temperatures is a steep function of temperature, so that resistance and resistance-ratio changes occur much more slowly if usage is restricted to the region below the silver point T_{Ag} rather than to the region below T_{Au}. Thus in most recent times, debate upon the merits of the high-temperature PRT and its possible role in a new IPTS has tended to concentrate on consideration of its relative durability at these two fixed points and, as corollary, the consequences of lowering the Planck law "starting point" to T_{Ag}.[20]

Finally, the standards-setting body, CCT, has to become convinced that successful manufacture of a high-temperature PRT is not the exclusive reserve of one or two laboratories but is universally available at a "reasonable" level of expenditure. The debate continues.

VI. IMPROVEMENTS IN FIXED POINTS

Improvements are anticipated in both assigned values and the physical form of fixed points. The gas thermometry at both high and low temperatures already mentioned will undoubtedly lead to extensive revisions of "the numbers." In addition, all parties seem to be convinced that boiling points should be dropped, wherever possible, and solid-based phase transitions substituted, thus avoiding the added complication of accurate measurement of pressure. Going even further in this direction, numerous laboratories are studying sealed-cell *triple point* devices which offer the possibility of fixed points which are simple to realize and transfer, relatively inexpensive, and very convenient for checks and comparisons.[34]

One may expect to see the triple point of neon helping to define the next IPTS, the t.p. of argon (83.8 K) firmly established as a replacement for the triple point of oxygen, and perhaps even the t.p. of gallium[35] replacing the b.p. of water. As with boiling points, isotopically pure gases offer reproducibility advantages for triple points although elevated cost may be a negative feature. Impurities must be guarded against in t.p. cells; the argon t.p. (liquidus point), for example, can be lowered by between 2 mK and 3 mK per 100 ppm of impurities such as O_2, N_2, CO, and CH_4[36]; the oxygen t.p. is a little less sensitive to the presence of impurities, viz., -2 mK and $+1$ mK for 100 ppm of N_2 and argon, respectively.[36] The t.p. of deuterium, at 18.13 K, might be particularly useful, that is, if it led to the elimination of hydrogen boiling points, otherwise needed for scale definition. It is, of course, easily contaminated by hydrogen: its practicality is presently under investigation.[37] Similarly, the t.p. of krypton[38] at 115.764 K could play a very useful role in devising improved procedures for the low-temperature PRT region. Unfortunately, Kr has several abundant isotopes and is expensive even as the naturally-occurring mixture.

Since 1967 we have had no new values for the gold point except that which may be derived from a 1970 redetermination of the Stefan-Boltzmann constant; this suggested that the present value of 1064.43 °C is 0.31 K too high.[39] The National Bureau of Standards' high-temperature gas-thermometry program now looks as if it is to be halted at the silver point, but even that remains to be measured. In any case, as already discussed, the preservation of T_{Au} as a defining fixed point rests upon decisions yet to be made concerning high-temperature PRT development. In this connection, one may mention that in view of the presently elevated cost of gold, copper (M.Pt. = 1084.88 °C) is being put forward as a practical replacement, proponents insisting that this metal *can* be handled satisfactorily at no great added complexity of apparatus.[40] Finally, it should be noted that a possibly-tenfold improvement in pyrometry achieved in the last decade[41] is leading enthusiasts to the conclusion that a good thermodynamic determination of the antimony (or aluminum) point will suffice, to constitute the basis for *optical* redeterminations of any higher reference temperatures and thus to retire the gas thermometer from further service in that region.

VII. WHO NEEDS IT?

It is true, in general, that only in small areas of the vast expanse of research and applications (commercial, industrial, military, and domestic) does one require a measurement capability close to the state of the art for any physical quantity. (An exception to this generalization is, of course, the realm of time and frequency in military and space-exploring applications.) As far as temperature is concerned, needs are widespread for rather exacting temperature control. While this does not call for a knowledge of true temperature to the control accuracy, faithful reproduction of conditions and information transfer to other sites, even to other countries, requires some sort of worldwide system. (Even so, here we are more often talking of relative reproducibilities of a part in 10^3 to a few parts in 10^4 of whatever "practical temperature" constitutes the international standard, not 10^{-6} or 10^{-5}.) Note that reproducibility requirements alone do not entail that the international standard should bear any particular relation to thermodynamic temperature. It would be foolish to ignore the thermodynamic realities, however; otherwise all thermodynamic relationships would be compromised—and entropy would increase even more rapidly than the Supreme Being has ordained.

The accurate measurement of the physical properties of substances, especially those such as heat capacity which involve temperature derivatives, is one of the most demanding thermometric applications and commercial/industrial needs for such data at the 0.1% level or better—which are quite common—demand that the data-producer practice thermometry in which the slope of the measured parameter versus T curve possesses that same accuracy. In this case, moreover, T must be the true, thermodynamic temperature; thus we must know the relation between our "practical temperature" and T. These requirements cannot be met unless there is available to the researcher (or his thermometer supplier) a calibration service based upon thermometry which is superior to the application need, and it is this need for calibration services in our complex modern society which determines the ultimate accuracy requirement. Such a service is more difficult to establish for temperature than for any other physical quantity (although pressure might be said to give it a close run while mass cannot be much further behind) because of the quite large and seemingly-unavoidable "down-the-chain" deterioration, i.e., from national standards laboratory to major testing laboratory to supplier to final user. The scientist, engineer, or manufacturer will tap into the service chain wherever he needs it, but such a chain must exist and the head linch-pin—the national standards laboratory—must concern itself with millikelvins!

If falls to the international temperature standards bodies to ensure that these "millikelvins" are the same the world over. It was for reasons such as those discussed above that industrial nations decided more than a hundred years ago to ensure worldwide uniformity of the units of physical measurement by drawing up the Convention of the Metre and establishing the International Bureau of Weights and Measures (BIPM).

VIII. RAMIFICATIONS OF A SCALE REVISION

In addition to the intellectual momentum of statutory committees and the developing needs of the technical community, there exists a third imperative, that is, the desire within the instrumentation industry and within the preparation-of-critical-data community for a decent stability in the units of measurement. Changes such as those discussed above (for example, modifying the B.Pt. of water to 99.975 °C) would, if accepted and promulgated, lead to extensive revisions in a multitude of national and international reference tables, instrument scales, etc. Therefore, the question should be posed, are changes in numerical values going to be necessary in a revised IPTS? The suggested improvements and simplifications could, after all, be achieved by (i) revising the text of the scale to make it more easily realizable, and (ii) publishing from time to time a table of differences between the IPTS and thermodynamic temperatures. This Symposium would seem to furnish an ideal occasion for sounding opinions within the temperature-measurement community and for striving to achieve an optimum balance among the concerns of physics, aesthetics and pragmatics.

IX. SECONDARY REALIZATIONS OF THE IPTS

The international body which produces the International Practical Temperature Scale—the Comité Consultatif de Thermométrie (CCT), advisory to the International Committee for Weights and Measures (CIPM)—does not operate in a vacuum. Some reference has already been made to this fact in Sec. VIII. Many manufacturers feel bound to certify their products as being in conformity with the IPTS, especially in those countries which have laws and regulations to encourage that feeling! In many instances, however, the requirement is excessive to the application need. It is possible that both legal and general marketing requirements could be met at a lower but still documented level, if secondary realizations of the IPTS could be devised and given official (i.e., CIPM) standing.

The CCT has already engaged in rather lengthy discussion of this concept, focusing initially upon capsule-type PRT's for temperatures below 0 °C and various thermocouples for elevated temperatures. Of course, the IPTS is a rigorously defined entity so that purists are moved to argue that a "secondary realization" is a contradiction in terms. Pragmatists, on the other hand, feel that that point of view sets semantics above problem-solving. In any case, a lively and generally constructive discussion continues and this will, I believe, lead to some definitive proposals in the very near future.

[1] R. E. Bedford, High Temperatures—High Pressures, **11**, 135 (1979); this paper covers, from a 1978 perspective, much the same ground as the present review.
[2] For an historical review, see J. A. Hall, *Temperature: Its Measurement and Control in Science and Industry* (Reinhold, New York, 1955), Vol. II, p. 139.
[3] Septième Conférence Générale des Poids et Mesures, Compt. rend., Trav. et Mém. Bur. Inter. Poids et Mesures **18**, 94 (1930).
[4] See Appendix I of Ref. 5.
[5] Comité International des Poids et Mesures, Metrologia **5**, 35 (1969).
[6] Comité International des Poids et Mesures, Metrologia **12**, 7 (1976).
[7] H. Preston-Thomas, in Temperature: Its Measurement and Control in Science and Industry (Instrument Society of America, Pittsburg, 1972), Vol. IV, Pt. 1, p. 3.
[8] International Practical Temperature Scale of 1948 (amended edition of 1960); Onzième Conférence Générale des Poids et Mesures. Compt. rend., octobre 1960, Annexe 5, p. 124; English Text, J. Res. Natl. Bur. Stand. Sec. A **55**, 139 (1961).
[9] S. D. Ward and J. P. Compton, Metrologia **15**, 31 (1979).
[10] L. A. Guildner and R. E. Edsinger, J. Res. Natl. Bur. Stand. Sec. A **80**, 703 (1976).
[11] T. J. Quinn and J. E. Martin, in *Temperature: Its Measurement and Control in Science and Industry* (American Institute of Physics, New York, 1982), Vol. V, p. 103.
[12] R. C. Kemp, L. M. Besley, and W. R. G. Kemp, Metrologia **14**, 137 (1978).
[13] K. H. Berry, Metrologia **15**, 89 (1979).
[14] R. C. Kemp, L. M. Besley, and W. R. G. Kemp, in *Temperature: Its Measurement and Control in Science and Industry*, Vol. V, p. 33.
[15] P. P. M. Steur, J. E. van Dijk, J. P. Mars, H. ter Harmsel, and M. Durieux, in *Temperature: Its Measurement and Control in Science and Industry*, Vol. V, p. 25.
[16] C. P. Pickup, in *Temperature: Its Measurement and Control in Science and Industry*, Vol. V, p. 129.
[17] J. P. Evans, Metrologia **13**, 171 (1977).
[18] For a summary, see T. J. Quinn, L. A. Guildner, and W. Thomas, Metrologia **13**, 175 (1977).
[19] L. Crovini and A. Actis, Metrologia **14**, 69 (1978).
[20] T. P. Jones, Metrologia **4**, 80 (1968).
[21] E. H. McLaren and E. G. Murdock, Nat. Res. Counc. (Can.) Monograph NRCC 17407 (1979).
[22] See T. J. Quinn and J. H. Compton, Rep. Prog. Phys. **38**, 151 (1975) and references therein.
[23] T. J. Quinn and T. R. D. Chandler, *Temperature: Its Measurement and Control in Science and Industry*, (Instrument Society of America, Pittsburgh, PA, 1972), Vol. IV, Pt. 1, p. 295.
[24] T. P. Jones and J. Tapping, Metrologia **18**, 23 (1982); *Temperature: Its Measurement and Control in Science and Industry*, Vol. V, p. 173.
[25] F. G. Brickwedde, H. van Dijk, M. Durieux, J. R. Clement, and J. K. Logan, J. Res. Natl. Bur. Stand. Sec. A **64**, 1 (1960).
[26] R. H. Sherman, S. G. Sydoriak, and T. R. Roberts, J. Res. Natl. Bur. Stand. Sec. A **68**, 579 (1964).
[27] M. Durieux, J. E. van Dijk, H. ter Harmsel, P. C. Rem, and R. L. Rusby in *Temperature: Its Measurement and Control in Science and Industry*, Vol. V, p. 145.
[28] For a general review, see R. P. Hudson, H. Marshak, R. J. Soulen, and D. B. Utton, J. Low Tem. Phys. **20**, 1 (1975).
[29] Report of Working Group 4 to the C.C.T., 14th Session 1982 (to be published).
[30] R. L. Rusby, *Temperature: Its Measurement and Control in Science and Industry*, Vol. V, p. 517.
[31] C. T. van Degrift and W. J. Bowers, Jr. (to be published).
[32] R. C. Kemp, W. R. G. Kemp, and L. M. Besley, Metrologia **17**, 43 (1981).
[33] J. P. Evans, *Temperature: Its Measurement and Control in Science and Industry*, Vol. V, p. 459.

[34] F. Pavese and D. Ferri, *Temperature: Its Measurement and Control in Science and Industry*, Vol. V, p. 221.
[35] B. W. Mangum, *Temperature: Its Measurement and Control in Science and Industry*, Vol. V, p. 303.
[36] F. Pavese, Metrologia **14**, 93 (1978).
[37] R. C. Kemp and W. R. G. Kemp, *Temperature: Its Measurement and Control in Science and Industry*, Vol. V, p. 253.
[38] R. C. Kemp and W. R. G. Kemp, Metrologia **14**, 83 (1978).
[39] W. R. Blevin and W. J. Brown, Metrologia **7**, 15 (1971).
[40] R. E. Bedford, Document CCT/80-10 (B.I.P.M., 92310 Sèvres, France).
[41] See, for example, P. B. Coates and J. W. Andrews in *Temperature: Its Measurement and Control in Science and Industry*, Vol. V, p. 109.

The measurement of thermodynamic temperature

Leslie A. Guildner

National Bureau of Standards, Washington, D.C. 20234

Wilhelm Thomas

Physikalisch-Technische Bundesanstalt, 3300 Braunschweig, Federal Republic of Germany

The thermometers with which thermodynamic temperatures can be accurately realized are discussed and results are cited over the range from 2 mK to 1336 K. The deviations of scales from the most accurate realizations of thermodynamic temperatures are presented. Improvements have increased the accuracy of gas thermometry, which is still the most accurate means of determining thermodynamic temperatures from ca 2.5 to 1337 K, but improvements in other techniques have increased their accuracy to nearly the same level. At the extremes of the temperature range, both high and low, techniques other than gas thermometry provide more accurate thermodynamic temperatures.

1. Introduction

Thermodynamic temperature is a "fundamental physical quantity", that is, it is one of the quantities needed in describing and measuring physical phenomena. By its nature it is independent of the properties of any substance. It is measured and expressed in terms of a coefficient times a unit. The unit of thermodynamic temperature, the kelvin, is defined as 1/273.16 of the thermodynamic temperature of the triple point of water. Because temperature is an intensive quantity, the units are not additive. Thermodynamic temperatures must be determined by their ratios to the assigned value of the temperature of the triple point of water.

Other systems of temperature measurement that are constructed sequentially, such as the International Practical Temperature Scale (IPTS), are scales of temperature. The IPTS-68 utilizes the properties of materials for interpolating between temperatures of fixed points to which values have been assigned. Both the choice of values for the temperatures of the fixed points and the means of interpolation are intended to reproduce as closely as practicable the thermodynamic values as understood at a given time. Although thermodynamic temperature and the IPTS-68 have the triple point of water with an assigned value of 273.16 K in common, it must be clear that their units are not everywhere the same. This must be true since IPTS-68 temperatures have been found not to agree with thermodynamic temperatures at several points. Reflecting the success of the Comité Consultatif de Thermométrie in their construction of the IPTS, however, the kelvin of the IPTS varies with respect to the thermodynamic kelvin by no more than ± 0.4%.

The requirement that values of temperature be thermodynamic for calculations involving the Second Law of Thermodynamics, or involving statistical mechanics, is well understood. It is an objective that can only be approached as methods and techniques of measuring thermodynamic temperatures improve, and one may well ask, how accurately do the values need to be realized, and how accurately are they being realized?

No simple laboratory thermometers ("practical thermometers") give inherently thermodynamic values, but if they are precise, stable, and consistent with one another, their indications can be precisely related to thermodynamic temperatures that have been determined in metrological experiments of considerable complexity and duration. Clearly, there is little achieved by improving the accuracy of the thermodynamic measurements if the imprecision of the practical thermometers, or the variation between them, is already the dominant uncertainty. When we reflect that all advances in metrology find immediate application, we would have to consider that the ability to preserve and transfer the results by the practical thermometers is the appropriate limit for choosing our goals of accuracy in thermodynamic temperature measurements.

In principle, thermodynamic temperatures can be determined by measurements involving any Second Law process explicitly dependent upon thermodynamic temperature, or equivalently, upon statistical mechanical temperature. In fact, all of the techniques capable of accurate execution involve processes elucidated by statistics: (1) gas thermometry, (2) noise thermometry, (3) acoustic thermometry, (4) spectral radiation thermometry, (5) total radiation thermometry, (6) dielectric constant gas thermometry and (7) γ-ray anistropy thermometry. The expression governing the phenomenon on which each of these thermometers is based involves either RT or kT, where R is the molar gas constant, and k is Boltzmann's constant.

The determination of thermodynamic temperature fundamentally involves a ratio measurement. To be sure, thermodynamic temperatures have been measured at a single isotherm, with a larger uncertainty in the result because of the uncertainty of the gas constant (or the Boltzmann constant), which is invariably a part of the equation. The identical isotherm techniques used to measure thermodynamic temperature, can be used to determine R when the procedure is conducted at the temperature of the triple point of water. Thus entering a value of R into an equation for calculating a thermodynamic temperature amounts to employing a surrogate measurement at the triple point of water, whereby the required ratio is established. Any of the techniques carried out at state-of-the-art level use an explicit ratio measurement, which enhances the accuracy by cancellation of R (or k).

It is our purpose to select the outstanding experimental realizations of the processes cited above, to compare the levels of accuracy being attained, and to show how these measurements improve our knowledge of thermodynamic temperatures.

2. Gas Thermometry

We shall use "gas thermometry" in this paper to denote thermodynamic temperature measurements by those instruments that make direct use of the pV properties of a gas. Gas thermometers antedated the formulation of the laws of thermodynamics by more than a century, and the consistency of their results formed the basis of temperature standards even before the principles of thermodynamics were understood.

Thermodynamic temperatures are identical to the gas temperature obtainable in the limit of zero pressure from the equation of state of a real gas

$$pV = nRT[1 + B_V p/(RT) + ...] \quad (1)$$

where p is the pressure, V the volume, n the number of moles of gas, R the molar gas constant, T the thermodynamic or ideal gas temperature, and B_V the second volume virial coefficient. The expansion involves higher virial coefficients, but the range of pressure over which a gas thermometer is operated is generally restricted enough to make their effect insignificant.

Each type of gas thermometer is designed to be operated with certain experimental variables as constant as possible, so that in the expression for the ratio of T_2/T_1, they will be eliminated by cancellation. Many gas thermometers are designated by those experimental parameters maintained constant in their operation, such as constant volume, constant pressure,

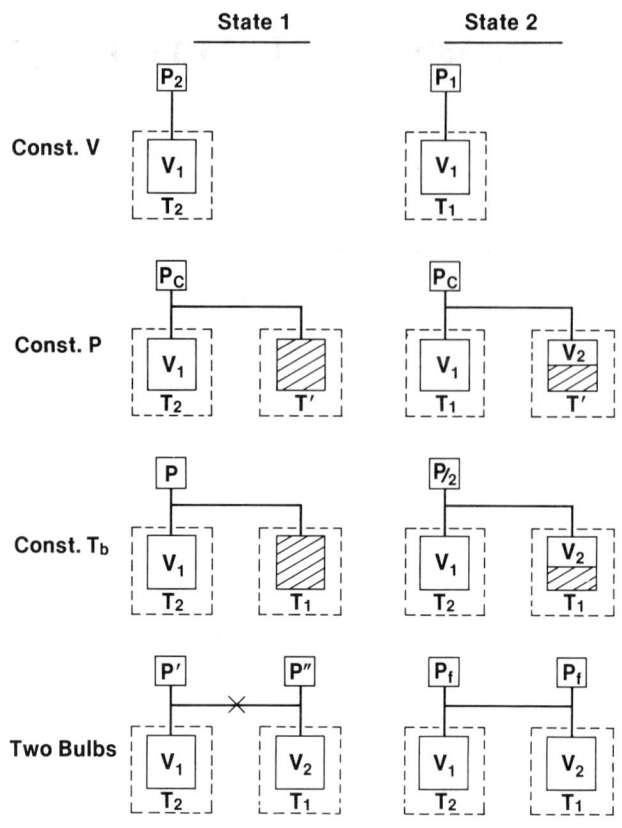

FIG. 1. Schematic illustration of different types of gas thermometers.

constant bulb temperature or isotherm gas thermometers. There are also two somewhat similar methods: the "method of two bulbs" and "absolute isotherm thermometry". As shown in Figure 1, the constant volume gas thermometer has a single bulb V_1 that is successively thermostated at temperatures T_2 and T_1. The constant pressure thermometer is comprised of a bulb of fixed volume V_1 and a second bulb of variable volume V_2. The bulb V_1 is successively thermostated at temperatures T_2 and T_1, and it is required that the order of the experiment and the volume of V_2 at the temperature T' be chosen in such a way that the pressure is kept constant. When V_1 is at the lower temperature of T_2 or T_1, V_2 is zero. The highest accuracy is achieved by choosing the temperature of the triple point of water for both T' and for the reference temperature T_1. (In the figure, $T_2 < T_1$.)

The constant bulb temperature thermometer uses two bulbs, V_1 and V_2, and two thermostats at temperatures T_2 and T_1, with V_2 of variable volume but equal to zero in state 1. In state 2 gas is drawn into V_2 in such a quantity that the final pressure (for maximum sensitivity) is half the initial pressure. The temperature of the second thermostat is at a reference temperature, T_1, with the highest accuracy of the thermodynamic temperature resulting when T_1 is the temperature of the triple point of water.

The method of two bulbs and the absolute isotherm thermometer both use the total volumes of V_1 and V_2 thermostated at temperatures T_2 and T_1, respectively. For this method in state 1, the bulbs are isolated from one another with pressures p' and p'', while in state 2 they are interconnected with a common pressure p_f. The absolute isotherm thermometer involves a series of additions of gas from the reference bulb V_2, thermostated at a reference temperature T_0. For the first step, V_1 is evacuated and V_2 is filled to a pressure p_i^1, after which the bulbs are interconnected with a final pressure p_f^1 in V_2. Gas is added in steps until the desired final pressure in V_1 is reached. The quantity of the added increments of gas is established from the measurements of the pressure and knowledge of the volume of the gas in the bulb V_2. The method provides for the separate measurement of the pressure, p_B^n, in V_1, hence the thermodynamic temperature to be measured need not be the temperature at which gas was added to V_1, but any difference causes a change of p_B^n from p_f^n.

An ideal gas thermometer has no thermal expansion and no mechanical deformation of the bulb, no volume of the connecting tubes, and is filled with an ideal gas. The equations of gas thermometers are established by equating the number of molecules of gas in each state, and are for the

Constant volume thermometer $\quad p_2/p_1 = T_2/T_1$

Constant pressure thermometer $\quad T_1 V_1 = T_2(V_1 - V_2)$
when $T' = T_1$ and $T_2 > T_1$
$T_1 V_1 = T_2 (V_1 + V_2)$
when $T' = T_1$ and $T_2 < T_1$

Constant bulb temperature thermometer $\quad T_1 V_1 = T_2 V_2$, for
$p_f = p_i/2$

Method of two bulbs thermometer $\quad T_1 V_1 = T_2 V_2 (p''-p_f)/(p_f-p')$

It is inherent in the concept of isotherm thermometry that the gases are not ideal. For the absolute isotherm thermometer, the number of moles of gas transferred from the reference bulb of volume V_2 is calculated from the equation of state. It is assumed that, at the reference temperature T_0, the second virial coefficient of the gas, B_r, is sufficiently well-known to avoid significant error. The number of molecules, N_B, transferred to the thermometer bulb V_1 is, in step 1,

$$N_B = \Delta N = \frac{p_i^1 V_2}{RT_0 [1 + B_r p_i^1/(RT_0)]} - \frac{p_F^1 V_2}{RT_0 [1 + B_r p_f^1/(RT_0)]} \quad (2)$$

Now the thermometer bulb at temperature T_B, containing N_B moles of gas at pressure p_B^1, has a "gas thermometer temperature" defined by $P_B^1 V_1/(RN_B)$, which by substituting N_B from the above is

$$p_B^1 V_1/(RN_B) = \frac{p_B^1 V_1 T_0}{V_2} \left[\frac{p_i^1}{1 + B_r p_i^1/(RT_0)} - \frac{p_f^1}{1 + B_r p_f^1/(RT_0)} \right]^{-1} \quad (3)$$

The pressure p_B^1 is not necessarily the same as p_f^1. The gas thermometer temperature can also be expressed in terms of the equation of state at the thermodynamic temperature T_B as

$$p_B^1 V_1/(RN_B) = T_B[1 + B(T_B)p_B^1/(RT_B) + \ldots]. \quad (4)$$

(If T_B is sufficiently low, terms with higher order virial coefficients may need to be included.) The amount of gas added to V_1 in the nth step is similarly calculated from the initial and final pressures in V_2 of that step, p_i^n and p_f^n, respectively, and the resulting pressure in V_1 at the temperature T_B, p_B^n, is also determined. The gas thermometer temperature depends primarily upon the pressure ratios p_i^n/p_B^n and p_f^n/p_B^n and the volume ratio V_1/V_2, and the thermodynamic temperature is derived as the intercept found by the extrapolation of the gas thermometer temperature vs. p_B^n.

Relative isotherm thermometry is equivalent to constant volume gas thermometry performed at a series of gas densities. After one finds the gas thermometer temperatures for a chosen temperature at each of a number of densities, they are extrapolated vs. the density of the gas or vs. the reference pressure. The value of the intercept is the thermodynamic temperature, and the slope is either directly, or proportional to, the difference of the second virial coefficients $B(T) - B(T_0)$ and possibly higher order terms.

Any real gas thermometer fails to operate ideally because it has a real, i.e. imperfect, gas for a thermometric fluid, together with numerous other practical problems:

(1) The thermometer bulbs are subject to thermal and pressure dilation and possibly significant mechanical creep.

(2) Connecting lines have finite volumes, the choice of their diameter being a compromise between the need for a small value to minimize the amount of gas contained in the lines and the need for a large value to maximize high vacuum conductance and to reduce thermomolecular pressures.

(3) There must be communication of the gas with a pressure measuring device. All modern gas thermometers make use of a pressure transducer. While the thermometer proper can be considered to end at a constant volume valve, care must be taken that the gas is not transferred into or from the transducer volume.

The basis for all gas thermometry is an accounting for the amount of gas present. In most cases, the procedures are designed to maintain the amount of gas constant, so that "n", the number of moles, cancels in the ratio. The number of moles is changed, however, and large errors may occur unless the effects of sorption and diffusion are adequately controlled, first by thorough cleanup of the system, second by use of highly purified helium for the thermometric fluid (the point being to use the gas least susceptible to sorption, and to purify it so that no sorbable impurities are present), and third, at high temperatures, by adoption of procedures to avoid diffusion of hydrogen through the walls of the bulb, either into or out of the system.

While in principle the values of thermodynamic temperatures should be in agreement for all types of gas thermometers, in fact more consistent and more precise results have usually been attained with constant volume thermometers. Uncertainties in evaluating bulb volumes inevitably contribute an added source of error for any other type of thermometer. This error should, however, be considerably smaller than the large variations reported. A much larger error can result from sorption effects; these may be smaller for constant volume thermometers because of their relatively simple construction.

The gas thermometry of highest accuracy published during this last decade comes from two sources: a combination of isotherm thermometry and constant volume gas thermometry of Berry [1], ranging from 2.6 K to 27 K, and the constant volume gas thermometry of Guildner and Edsinger [2] proceeding from 0 °C to 457 °C.

Berry's comprehensive discussion of his work at the National Physical Laboratory gives details and evaluation of the results. The scale, NPL-75, is recorded on three rhodium-iron thermometers and a

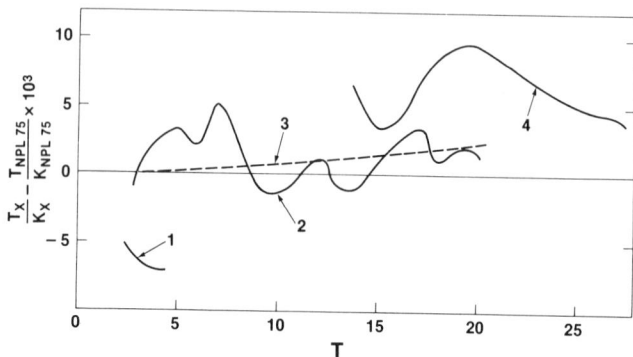

FIG. 2. Relation of other cryogenic scales to NPL-75, shown as $(T_x/K_x)-(T_{NPL-75}/K_{NPL-75})$; 1=$T_{58}$, 2=NBS 2-20, 3=$T_{76}$, 4=IPTS-68.

platinum resistance thermometer, and in terms of fixed points. The relationships to other scales (for example, T_{58}, NBS_{2-20}, EPT-76, and IPTS-68) are well established. The differences between these scales and NPL-75 are shown in Figure 2.

Results with the constant volume gas thermometer of the National Bureau of Standards have been recorded in terms of the difference from the IPTS-68, realized as an average from the indications of three, and sometimes four, calibrated standard platinum resistance thermometers. The results from 0 °C to 457 °C have been published as given in Figure 3. The total uncertainty (99% confidence) is about 8 ppm over the range of measurement. It may be of interest that it is planned to extend these measurements to the gold point.

The improvement in accuracy that has been achieved in gas thermometry with more sensitive measuring techniques and greater understanding of the problems has sometimes been cited as evidence indicating an untrustworthy character of the results. Yet gas thermometry has always offered the most accurate means of realizing thermodynamic temperatures over most of the temperature range. What is perhaps remarkable is how sensitive and potentially accurate other methods

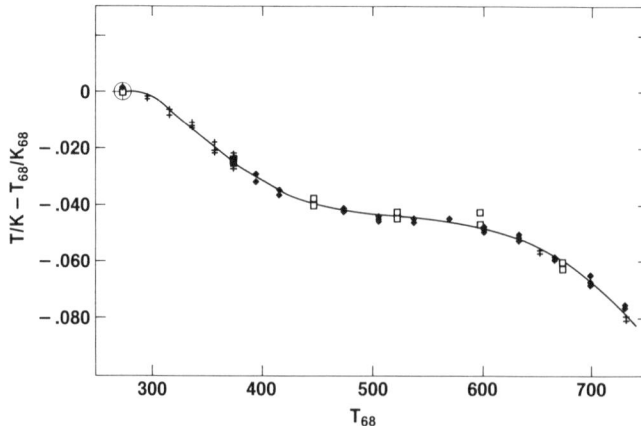

FIG. 3. The differences $(T/K)-(T_{68}/K_{68})$ found by NBS gas thermometer measurements between 273 and 730 K. Different symbols indicate different fiducial pressures.

have become. In spite of the short period of time over which some of the other methods have been used seriously as a means of realizing accurate thermodynamic temperatures, substantial systematic errors have been eliminated from those methods, and in parallel with the experience of gas thermometry, probably others will follow.

3. Noise Thermometry

The measurement of thermodynamic temperatures by noise thermometry depends upon the effect of Brownian motion of the electrons of a conductor. The rms value of the thermal agitation of the electrons, or Johnson noise, of a resistance R at a thermodynamic temperature T is given by Nyquist's equation, where the time average of the mean square voltage of the fluctuations, $<V^2>$, between the frequencies ν and $\nu + d\nu$ is

$$<V^2> = 4Rh\nu d\nu[\exp(h\nu/kT)-1]^{-1}, \quad (5)$$

where h is Planck's constant and k is Boltzmann's constant. When $h\nu \ll kT$, i.e., in the absence of significant quantum effects, $<V^2> = 4RkTd\nu$, which is the familiar expression. (This condition is readily fulfilled. The frequencies used in noise thermometry lie in the range 10^4 Hz $<\nu< 10^5$ Hz, so that the next term in the expansion of the exponent, $\frac{1}{2!}(\frac{h\nu}{kT})^2$, is <1 ppm for 0.33 mK <T< 33 mK. The practical lower limit of the system is imposed by the constraints of the measuring system.)

The fundamental imprecision of this thermometer is $\Delta T/T = 1/\sqrt{t\Delta\nu}$ where t is the time of actual averaging and $\Delta\nu$ is the frequency bandwidth.

The practical problems in noise thermometry arise from the facts that the fluctuation voltages are very small and require amplification over a frequency bandwidth; that there are other sources of thermal noise that must be excluded from the evaluation of $<V^2>$; that there are other kinds of noise to be excluded as well; and that the effective bandwidth must be accurately controlled and evaluated. Thus the measured quantity is represented by an integral

$$<V^2>_{meas} = \int_0^\infty (v_{noise} + \Sigma v_i)^2 \alpha_A F_A \, d\nu \quad (6)$$

where the v_i represent extraneous sources of noise voltage, α_A the amplification and F_A a filter by which the band width is controlled. For highest accuracy, the values of temperature must be evaluated by ratio measurements of the mean square voltages at the unknown temperature and at the temperature of the triple point of water, thus cancelling k and its attendant uncertainty.

The first measurements of thermodynamic temperatures by noise thermometry, reported by Garrison and Lawson [3], had large uncertainties (>0.1%) because of the noise instability of the amplifiers. A similar technique is presently used by Pickup at the National Measurement Laboratory of the CSIRO [4], where the noise voltage balance between two resistors at an unknown temperature and a reference temperature is evaluated by use of synchronous switching. A block diagram of his experimental apparatus is shown in Figure 4. In order to compare identical noise voltages, the resistances are adjusted so that $R_1T_1 = R_2T_2$. With constant channel bandwidth (a requirement), the shunt capacitances C_1 and C_2 must be adjusted so that $R_1C_1 = R_2C_2$. The typical bandwidth was 100 kHz. A second, higher range, bandwidth was introduced to allow a more exact balance of capacitance to be achieved.

With improved technology, the noise level and instability of amplifiers has been much reduced. Relative to a reference temperature measured on IPTS-68 near 25°C, Pickup measured a thermodynamic value of the oxygen boiling point 3 mK above IPTS-68 (90 K), and a thermodynamic temperature 8 mK larger than IPTS-68 at 97 K. If the value reported from NBS gas thermometry for 25 °C is used, the differences are reduced by 2 mK.

Crovini and Actis at the Istituto di Metrologia "G. Colonnetti" have reported thermodynamic values for temperatures between 629 °C and 962 °C [5] measured by a noise thermometer of a configuration similar to Garrison and Larrson. Their values of $t-t_{68}$ with estimated uncertainty at 99% confidence are shown later in this paper in Fig. 7. The authors propose to make use of noise thermometry for the evaluation of thermodynamic temperatures above the gold point.

The measurements of Klein, Klempt, and Storm [6] at the University of Munster take maximum advantage of the highest level of electronic technology to achieve low extraneous noise levels and long term stability of the amplifiers. They measured the noise voltage first

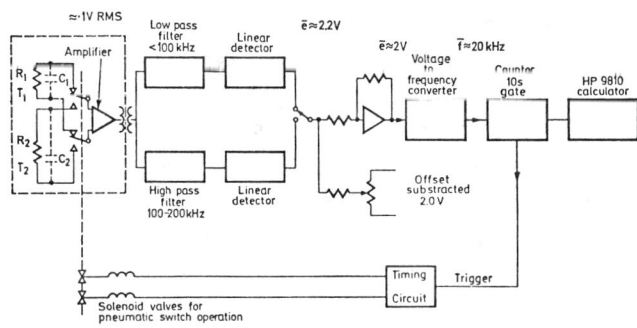

FIG. 4. Block diagram of the cross-correlated noise thermometer.

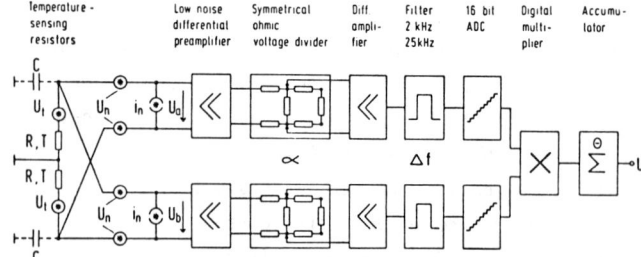

FIG. 5. Block diagram of the cross-correlated noise thermometer.

at one temperature and then at another. It is not required that $R_1T_1 = R_2T_2$, but the capacitance must be in the relation $R_1C_1 = R_2C_2$. A significant further reduction in the inherent uncertainty of the system is obtained by using a cross-correlating technique. The block diagram of their system is shown in Figure 5. The temperature sensing resistor 2R has its center point at ground and is connected in parallel to two amplifier channels a and b, made as nearly identical as possible. The noise of the amplifiers is equivalent to voltage sources u_n and current sources i_n (including the leads, which are shown to make essentially no contribution to the value of the noise voltage). The correlation technique reduces the effect of amplifier noise by a factor perhaps exceeding a thousand, which allows accurate measurements at temperatures as low as 2K. For a standard deviation of 10^{-5} in T, the measurements must be made for periods of 10^6s, first at the unknown temperature and then at the temperature of the triple point of water. For this technique of noise thermometry, non-ideal effects in the system are proportional to the resistance and can be eliminated by calculating F(R) from the equation

$$F(R) = \frac{\langle u \rangle}{\langle u_{tr} \rangle} \frac{R_{tr}}{R} \alpha_a \alpha_b T_{tr}[1 - \gamma\Delta(RC)] = T(1 + PR) \quad (7)$$

with α_a and α_b the attenuation factors of the a and b channels, and the unsubscripted terms the values of resistance and voltage at the unknown temperature T. If one measures F at the same temperature T for different values of R (and the related values of R_{tr}), the unknown temperature can be obtained from the intercept of F(R) extrapolated vs. R. The procedures can be executed so that in changing from T to T_{tr} the term $\gamma\Delta(RC)$ is of the order of 10^{-6}.

Klein, Klempt and Storm have reported measurements at seven cryogenic temperatures established by helium vapor pressures. From 2.1451 K to 4.2221 K, two of their values exceeded those of the EPT-76 by +0.0004 mK; the average difference of their values was +0.0002 K higher than the EPT-76. Their estimates of uncertainty are consistent with this result. They project a reduction in total temperature uncertainty from the presently attained 1×10^{-4} to around 1×10^{-5} at higher temperatures, and expect next to measure the temperature of the argon triple point, followed later by higher temperature measurements.

For very low temperatures, where even the cross-correlation technique fails to reduce the amplifier noise sufficiently, SQUID magnetometer noise thermometers or resistance SQUID (R-SQUID) noise thermometers can be employed where the amplifiers are based on the Josephson effect. By using such techniques the amplifier noise is reduced to a level equivalent to a noise temperature of 0.1 mK. Noise thermometers based on SQUID amplifiers have been used to measure temperatures as low as 0.01 K, and can be used for measurements up to 10 K. The upper temperature is necessarily limited because, except for the resistor, the components of the first stage amplifier must be superconducting.

In the apparatus of Soulen [7], at the National Bureau of Standards, the Josephson junction in a resistive SQUID noise thermometer is biased by a dc voltage, $\sim 10^{-11}$ V, which causes it to produce an ac signal in the audio-frequency range according to the Josephson equation

$$2eV = h\nu \quad \text{or} \quad V = 2.07 \times 10^{-15}\nu \quad (8)$$

where the voltage consists of the constant bias voltage plus the noise voltage generated across R in the circuit of Figure 6. This combined signal amplitude-modulates a radio frequency signal, which is then processed to give voltage noise information in terms of frequency. The frequency information is computer processed to give the variance

$$\sigma_\alpha^2 = \sum_1^N \frac{(\nu_i - \nu_{i+\alpha})^2}{2N} \quad \alpha = 1,2,3... \quad (9)$$

where ν_i is an individual frequency count and $\nu_{i+\alpha}$ is the frequency count α gate times later. When the measurements contain only the noise voltage of the resistor, the power spectrum is "white", as established by the fact that all the variances are equal. When that criterion is fulfilled, they can be related to temperature by

$$\sigma^2 = 2kRT/\phi_0^2\tau), \quad (10)$$

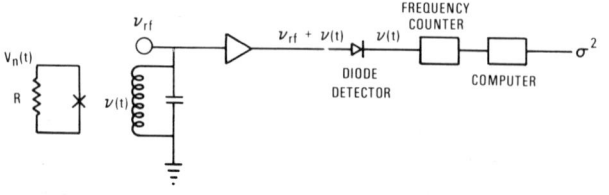

FIG. 6. Block diagram of the R-SQUID noise thermometer.

where τ is the gate time and ϕ_o is the Josephson constant equal to 2.07 fVHz^{-1}. This expression is free of the properties of any part of the system except the value of the resistance R.

The magnetometer SQUID functions as a null detector, so that the equations for the noise voltage measured by it contain parameters of other parts of the circuit. The seeming advantage of the R-SQUID is tempered by the fact that the resistance is an active element in the circuit, whereby the nature of the SQUID itself is involved at a second order level. In particular, the dc impedance of the SQUID is modified, thus affecting the apparent value of R. Likewise, the Josephson junction adds noise to the circuit. Both effects are increased by an increase in the ratio of the superconducting current to the biasing current. The effects can be eliminated by using the intercept found by extrapolation to zero of the noise frequency spectrum vs the square of the bias current.

It is the nature of the R-SQUID that its averaging time is ten times longer than the magnetometer noise thermometer. Both thermometers have a possibility of about a 10-fold reduction in measuring time (to achieve the same accuracy), which will allow a 1% statistical imprecision with an R-SQUID in 2000 s (τ = 1s). Both types of SQUID noise thermometer have a potential for realizing thermodynamic temperatures at very low temperatures with increased accuracy. The attained uncertainty of the SQUID magnetometer remains at a relatively high 3%. The present level of uncertainty for the R-SQUID does not exceed 0.5%.

4. Acoustic Thermometry

Sound is propagated adiabatically in an unbounded medium with a velocity given by $c^2 = B_s/\rho$ where the adiabatic bulk modulus is $B_s = -V(\frac{\partial p}{\partial V})_s$ and ρ is the density. For an ideal gas, $B_s = \gamma p$, where γ is the ratio of c_p/c_v for an ideal gas, and $\rho = pM/(RT)$ so that $c^2 = c_o^2 = \gamma RT/M$. For a real gas, used at pressures low enough that the equation of state is sufficiently accurate when written

$$pV = RT(1 + B_V\rho), \quad (11)$$

the speed of sound is

$$c^2 = c_o^2 (1 + A_1(T)p + ...) \quad (12)$$

where

$$A_1(T) = \frac{\gamma}{M} [2BT + \frac{4}{3} T \frac{dB(T)}{dT} + 4/15\ T^2 \frac{d^2 B}{dT^2}] \quad (13)$$

Thus acoustic thermometry, as noise thermometry, can be used to realize thermodynamic temperatures by isotherm measurements when the uncertainty in the value of R is acceptable. Acoustic thermometry, as gas thermometry, is affected by gas imperfection. This effect can be eliminated by evaluating c^2 as a function of p and extrapolating to zero pressure.

In reducing these principles to practice, one must take great care to balance competing choices in order to optimize the accuracy of the results. The speed of sound is typically determined in a cylindrical chamber, where the distances of resonance for a fixed frequency are determined by varying the path length. If low frequencies are used, the mode of propagation can be established, but very large boundary effects are encountered. It appears that these large effects can be determined accurately, however. If high frequencies are used, the modes of propagation are mixed. The boundary layer effects are much smaller, but the overall uncertainty is probably larger because of the difficulty of interpreting the results which rely on understanding the modes of propagation.

The first acoustic measurements of sufficient quality to be considered for the IPTS were performed by Plumb and Cataland [8]. These were high frequency measurements with thermodynamic values over the range from 2 to 34 K, without direct estimates of the uncertainties. The values of the provisional NBS 2-20 Scale derived from the earlier phase of this work has since been shown to exhibit a scatter of ±2.3 mK when evaluated by paramagnetic salt thermometry (Cetas) [9]. The scale deviates from NPL-75 by an average of 1.8 mK and tends to be slightly higher. The differences of later work, covering 7 to 34 K, are larger.

At the National Physical Laboratory, A.R. Colclough has measured the speed of sound of helium gas in a cylindrical container at low frequencies [10]. The resulting measurement system has well defined modes, such that "an unambiguous plane wave velocity" can be measured, with very large but accurately calculable boundary layer effects. The measured values of the velocity were adjusted by an amount depending on the Kirchhoff-Helmholtz absorption coefficient, α_{KH} according to the expression

$$\Delta c = (\alpha_{KH} c^2/\omega)(1 + 2 \alpha_{KH} c/\omega) \quad (14)$$

where $\alpha_{KH} = (\frac{1}{rc_o})[\nu^{1/2} + (\gamma-1)(K/(\rho c_p)^{1/2}][\omega/2]^{1/2} \quad (15)$

with r the cavity radius, ν the kinematic viscosity of the gas, K its thermal conductivity, c_p its specific heat at constant pressure, ρ its density, γ the ratio c_p/c_v, and ω the angular frequency.

The value of α_{KH} could be obtained as the difference of the measured absorption coefficient α and the "classical absorption coefficient α_o, where

$$\alpha_o = \frac{1}{2c^3} [\frac{4}{3} \nu + (\gamma+1) \frac{K}{\rho c_p}] \omega^2. \quad (16)$$

On the basis that η ($\equiv \rho \nu$), K and c_p are pressure independent, and that $\rho \propto p$ the values of α_{KH} were smoothed by plotting α_{KH}^{-1} vs $p^{1/2}$. They agreed with the values calculated from the critically evaluated values of ν, K, c_p, etc. within the uncertainties of the fit to a straight line.

Thus even though the value of $\Delta c/c$ was as high as 4×10^{-3} at the lowest pressures used, the error in Δc did not contribute significantly to the total uncertainty. Colclough measured the thermodynamic temperature of five cryogenic fixed points, with uncertainties evaluated at 1 to 4 mK (3σ), and additional systematic uncertainties from 0.5 to 1.7 mK. The values agree with Berry's gas thermometry within 2.2 mK at 17 K, but otherwise within about 1 mK at temperatures from 4.2 K to 20.27 K.

The NPL acoustic thermometer represents the most careful execution and analysis of a customary acoustic device up to the present time. The accuracy can be improved, however, as would be necessary to make a useful evaluation of the gas constant, or of higher thermodynamic temperatures.

The National Bureau of Standards' spherical resonator of Mehl and Moldover [11], is an acoustic device made in a shape that greatly reduces the boundary effects (to about 10% of those values that Colclough used). The spherical device permits the measurement of the resonant frequencies of the radial modes. The frequencies of 4 to 13 kHz are in the same range that Colclough used (3.3 to 7.25 kHz). Colclough reports standard deviations of c_0^2 in the range of 100 ppm. The observations of Mehl and Moldover have similar imprecision. It is expected that this instrument is also capable of greater accuracy, so that one may be able to measure useful values of thermodynamic temperatures with it at higher temperatures.

5. Spectral Radiation Thermometry

The measurement of thermodynamic temperature by a spectral photoelectric pyrometer depends upon an application of Planck's law. The ratio of the spectral radiances at two temperatures is given by

$$R = \frac{\varepsilon \, [\exp(\frac{c_2}{\lambda T_r}) - 1]}{\varepsilon_r \, [\exp(\frac{c_2}{\lambda T}) - 1]} \quad (17)$$

where subscript r refers to the reference state, ε's are emissivities, c_2 is the second radiation constant and λ is the mean effective wavelength.

Above the temperature of 457 °C, which is the upper temperature reported at present for the NBS gas thermometry, the only accurate modern thermodynamic temperature measurements have been made by spectral photoelectric pyrometers. This approach has become practicable as more sensitive detectors have become available, a fact that, in turn, allows evaluation of the non-linearity of the detectors and more precise and accurate values for the mean effective wavelength. It is by no means clear that the accuracy of such measurements has (except for a few special cases) exceeded the accuracy of the gas thermometry of Moser, Otto, and Thomas [12], performed in the 1950's at the Physikalisch-Technische Bundesanstalt. That work, however, was confined to fixed points over the range from 0 °C to 1064.43 °C, and particularly above 630 °C it is at temperatures between the fixed points that a study of the difference between IPTS and thermodynamic values has proved to be most interesting. Furthermore, use of spectral radiation thermometry has already led to the improved accuracy with which the difference of the thermodynamic temperatures of the freezing points of silver and gold is known, and there is a real prospect of much-increased accuracy overall in the near future.

Coates, Andrews and Chattle have recently reported results of work on a spectral photoelectric pyrometer at NPL [13]. They determined the differences between $t - t_{68}$ between 457 °C and 630 °C, based on the NBS gas thermometer value at 457 °C, and report an "overall uncertainty" increasing from 0.010 °C at 460 °C to 0.030 °C at 630 °C.

J. Bonhoure [14] at the Bureau International des Poids et Mesures used a spectral photoelectric pyrometer referred to the antimony point, and type S thermocouples, to measure the differences between thermodynamic temperature and the IPTS-68 at six temperatures ranging from 720 °C to 1064.43 °C. As references, the assigned values of the antimony point (903.89 K) and of the second radiation constant, c_2, (0.014388 m K) were used as specified by the IPTS-68. The reported standard deviations of the silver point and of the gold point were 0.13 K and 0.16 K, respectively. Further work is underway at BIPM to compare, by similar techniques, values of thermodynamic temperatures and the IPTS between 420 °C and 630 °C.

Numerous measurements have been made of the difference in the thermodynamic temperatures of the silver point and the gold point. The thermodynamic interval has been found to be appreciably less than the IPTS-68 interval. A value was reported by Quinn, Chandler and Chattle [15] as early as 1973, with that value and all the subsequent measurements in better agreement than would be expected from the estimated uncertainties. The determinations reported in more recent papers are considered by their authors to be the more accurate. On the basis of the freezing point of gold as 1064.43 °C, the thermodynamic temperature of the freezing point of silver was reported by Jung [16] 962.06 °C ± 0.1 °C and Ricolfi and Lanza [17] 962.05 °C ± 0.04 °C, compared to the IPTS-68 value for the silver point of 961.93 °C.

One may expect further significant contributions in this field from at least two laboratories, that of Jung at the Berlin Institut of the PTB and that of Jones and Tapping, at the CSIRO Division of Applied Physics in Sydney. We shall hear a report from Jones and Tapping at this Symposium, and regret that Jung considers the time premature for discussing his experiments.

In combination, the work cited above provides complete coverage of the difference between thermodynamic temperature and the IPTS-68 over the temperature interval from 457 °C to the gold point, although not as accurately as could be desired. There is additional work in this temperature region, viz. by Quinn, et al. [15], covering the range down from the gold point to 725 °C by spectral photoelectric pyrometry, and by Crovini, et al. [5], covering the range between 630 °C and 962 °C by noise thermometry. All these results with error bars for the total uncertainty estimated by the authors are presented in Figure 7.

6. Other Methods

There are other methods of measuring thermodynamic temperature, either less extensively developed or less widely used, that should be considered.

Gamma-ray Anisotropy Thermometry

At very low temperatures, nuclear γ-ray anistropy thermometry has been used by Marshak [18] at NBS to determine thermodynamic values between 12 and 35 mK. The measurement depends on the fact that at sufficiently low temperatures, the nuclear spin system of cobalt-60 is ordered when the nuclei are part of a single crystal of ^{59}Co. As a result, the γ-rays emitted in the process of nuclear decay are oriented with respect to the crystal axis. This anisotropy diminishes as the temperature increases, according to well-defined equations. The degree of ordering with respect to a cylindrical axis is defined by parameters $B_\lambda(I)$ given by

$$B_\lambda(I) = \sum_m (-1)^{I-m} [(2\lambda+1)(2I+1)]^{1/2} \begin{pmatrix} I & -I & \lambda \\ m & -m & 0 \end{pmatrix} a_m \quad (18)$$

where m = 2I+1 and a_m is the population of each of the m states. The index λ goes from 0 to 2I, with $B_0(I)$ = 1. The population for thermal equilibrium follows the

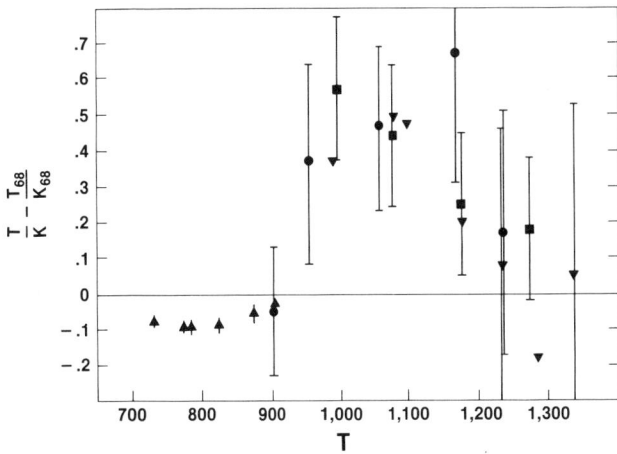

FIG. 7. $(T/K)-(T_{68}/K_{68})$ between 730 and 1337 K.
▲ Coates et al., ● Crovini et al., ■ Quinn et al.,
▼ Bonhoure. Uncertainties were not given for all of Bonhoure's values, but they are all presumably of a magnitude comparable to the ones shown.

Boltzmann distribution

$$a_m = \frac{e^{-E_m/kT}}{\Sigma_m e^{-E_m/kT}} \quad (19)$$

where E_m is the energy of a nuclear m-state, k the Boltzmann constant and T the thermodynamic temperature. Experimentally, the ratio of intensity at angle θ with respect to the c-axis of ^{59}Co of the 1.17 and 1.33 MeV γ-rays for the unknown low temperature and at a high temperature is determined, along with background counts according to

$$W(\theta) = \frac{C_c - B_c}{C_w - B_w} \quad (20)$$

where C's are total counts either cold or warm (subscript c and w) and B's are the background counts. This ratio can also be expressed as

$$W(\theta) = \Sigma_\lambda B_\lambda(I,T) U_\lambda A_\lambda Q_\lambda P_\lambda(\cos \theta) \quad (21)$$

where θ is the angle between the axis of orientation and the gamma ray. U_λ and A_λ are angular coefficients depending on the decay path. The angular dependence is expressed by $Q_\lambda P_\lambda(\cos \theta)$, where the $P_\lambda(\cos \theta)$ are Legendre polynominals and Q_λ accounts for the geometry of the detector. The sensitivity depends on the value of E_m/kT. The useful range with cobalt is restricted to an upper limit of about 40 mK, with a total uncertainty somewhat less than 2%, but it can be extended to very low temperatures with an estimated uncertainty of 7.5% at 2 mK.

Total Radiation Thermometry

Total radiation thermometers are calorimeters capable of accurately measuring the radiant heat flow from a blackbody. They have been built at NBS by Ginnings, Nuttal, and Reilly, at the NPL by Quinn and Martin [19], and at the NRLM (Japanese Standards

Institute) by A. Ono [20]. The thermodynamic temperature is established from the Stefan-Boltzmann Law, where the radiant power transferred between two surfaces at temperatures T_1 and T_2 with an effective emissivity ε_{12} is $\dot{q} = A\varepsilon_{12}\sigma(T_2^4 - T_1^4)$, with A the area and σ the Stefan-Boltzmann constant. (It should be of interest that our "surrogate ratio" to the water triple-point temperature still exists, inasmuch as $\sigma = 2\pi^5 k^4/(15c^2h^3)$.) There are formidable experimental problems to overcome in this experiment. The radiant power is measured with a calorimeter at liquid helium temperatures so that T_1^4 is insignificant compared with T_2^4. If the problems of achieving sensitivity, aperture alignment, ultra high vacuum and high emissivity are solved, there still remains the nearly intractable problem caused by diffraction of the radiation. The work at the NBS was dropped because of it. At the NPL, a special design was developed that, it is anticipated, will perform satisfactorily. The NPL instrument is operational, with initial results in hand for the Stefan-Boltzmann constant and some thermodynamic temperatures above 0°C. It is estimated that measurements of the steam point will be accurate within a few mK. The instrument at the NRLM is still under construction.

Dielectric Constant Gas Thermometry

The temperature dependence of the dielectric constant of a gas on its density has been successfully used by Gugan and Michel [21] for measurements of thermodynamic temperatures at low temperatures. The relative change in capacitance produced by the introduction of a gas at pressure p between the plates of a capacitor is $(C(p) - C(0))/C(0)$, where $C(p)$ and $C(0)$ are the capacitances at pressure p and under vacuum, respectively. Given that the dielectric constant of gas is ε, $\Delta C/C(0) = \varepsilon-1$ so that the observed value can be related to a modification of the Clausius-Mossotti equation that accounts for the effect of gas imperfection as

$$(\varepsilon-1)/(\varepsilon+2) = (A_\varepsilon/V)(1 + b/v + c/V^2 + \ldots) \quad (22)$$

where A_ε is the polarizability of the gas and V is the molar volume. This equation is combined with the equation for a real gas,

$$pV = RT(1+B/V + C/V^2 + \ldots), \quad (23)$$

To eliminate V, with an added term for the compressibility K of the capacitor to get

$$p = A_1\mu(1+A_2\mu+A_3\mu^2 + \ldots). \quad (24)$$

The expansion parameter, μ, is $(\varepsilon-1)/(\varepsilon+2)$ with a small increment involving K and

$$A_1 = (A_\varepsilon/RT+K/3)^{-1} \quad (25)$$

$$A_2 = B'/A_\varepsilon, \text{ where } B' = B-b \quad (26)$$

and within the limits of significance

$$A_3 = C/A_\varepsilon^2 \quad (27)$$

$$A_4 = D/A_\varepsilon^3 \quad (28)$$

etc.

Thus values of T can be determined from analyses of the values of the intercepts of isotherms, A_1, when values of A_ε, R and K are known. The polarizability of helium, A_ε, can be calculated from theory, with an estimated uncertainty of 1 ppm. In their laboratory at the University of Bristol, Gugan and Michel have determined values accurate within 0.3 mK in good agreement with Berry's NPL thermometry from 4.2 to 27.1 K. They estimate that the method should be useful over the range of 2.6 to 400 K with an uncertainty of 10 ppm.

Refractive Index Thermometry

A thermometer based on the temperature dependence of the index of refraction of a gas has been proposed and investigated in a preliminary way by A.R. Colclough [22] of the NPL. The principles are closely allied to the dielectric constant thermometer.

Gas Calorimetry Thermometry

The change in molar enthalpy of all ideal monatomic gases between any two temperatures T_1 and T_2 is the same. It is given by

$$H-H_0 = \int_{T_1}^{T_2} C_p dT \quad (29)$$

where C_p is the molar heat capacity at constant pressure. Since C_p is a constant for an ideal gas, equal to 2.5 R,

$$T_2-T_1 = (H-H_0)/C_p = (H-H_0)/(2.5R). \quad (30)$$

Thus by determining the differences in molar enthalpy of a real gas between 273.16 K and a fixed temperature T at a succession of densities, the enthalpy change between the temperature of the triple point of water and the temperature T for an ideal gas can be found by extrapolation of the enthalpy differences to zero density. The thermodynamic value of T can then be calculated from the change in temperature added to 273.16 K. An investigation based on this principle is being conducted at the University of Karlsruhe under the direction of Professor Ernst. There are as yet substantial experimental uncertainties, as well as the smaller but significant uncertainty in R itself.

Finally, a true Second Law thermometer can be based on the Clapeyron equation,

$$\frac{dp}{dT} = \frac{\Delta S}{\Delta V} = \frac{L_v}{T(V_g - V_\ell)} . \tag{31}$$

This can be integrated when L_v, the heat of vaporization, V_g, the volume of the gas and V_ℓ, the volume of the liquid are known as functions of temperature. As a practical matter, however, the uncertainties in the terms are too large, so that values of thermodynamic temperatures have never been determined accurately enough by this method to contribute significantly to an improved realization.

7. Summary

Over the temperature range from 5 K to the freezing point of gold (1337 K), gas thermometers presently offer the most accurate means available for realizing thermodynamic temperatures. In the ten years since the Fifth Temperature Symposium, previously existing thermodynamic thermometers have gained remarkably in accuracy, and some others have appeared "full blown" as accurate and reliable instruments, to the extent that the accuracy advantage of gas thermometry is substantially diminished. Because of their totally different systematic errors, the results obtained with these various kinds of thermometers are useful for comparison to assure essential absence of systematic errors for the different measurements.

Certain of these instruments - the γ-ray anisotropy thermometer and the Josephson junction noise thermometer - can operate at lower temperatures than is possible with a gas thermometer, with good accuracy. Other thermometers - the dielectric constant gas thermometer, the acoustic thermometer, and the cross-correlated noise thermometer - can be used to measure thermodynamic temperatures between 2 K to 5 K with greater ease and with accuracy comparable to a gas thermometer. Above 5 K, these instruments function at about the same relative accuracy as below, but the attainable accuracy of a gas thermometer becomes greater. At higher temperatures the more usual noise thermometer and the total radiation thermometer can approach the accuracy of good, but probably not the best, gas thermometry. Over the hotter part of the range, ca 700 K to 1337 K, spectral radiation thermometry can be used to make accurate measurements, but at a level of accuracy inferior to the best gas thermometry potentially available. Above the temperature of the gold point, thermodynamic temperatures can be measured by gas thermometers, but the advantage in accuracy lies with the spectral radiation thermometers.

References
[1] K.H. Berry, Metrologia 15, 89-115 (1979).
[2] L.A. Guildner and R.E. Edsinger, Jour. Res. Nat. Bur. Stds. 80A, 703-738 (1976).
[3] J.B. Garrison and A.W. Larson, Rev. Sci. Instr. 20, 785-794 (1949).
[4] C.P. Pickup, Metrologia 11, 151-159 (1975).
[5] L. Crovini and A. Actis, Metrologia 14, 69-78 (1978).
[6] H.-H. Klein, G. Klempt, and L. Storm, Metrologia 15, 143-154 (1979).
[7] R.J. Soulen, Jr., Proc. Fifth Int. Conf. on Noise, 249-259. (Ed.: Dietrich Wolf, Springer-Verlag, Berlin, 1978.)
[8] H.H. Plumb and G. Cataland, Metrologia 2, 127-139 (1966).
[9] T.C. Cetas, Metrologia 12, 27-40 (1976).
[10] A.R. Colclough, Proc. Roy. Soc. (London) A365, 349-370 (1979).
[11] J.B. Mehl and M.R. Moldover, J. Chem. Phys. 74, 4062-4077(1981).
[12] H. Moser, S. Otto, and W. Thomas, Z. Physik 147, 59-76 (1957).
[13] P.B. Coates, J.W. Andrews and M.V. Chattle, Document CC-80/27 du Comité Consultatif de Thermomètrie, Paris, 1980.
[14] J. Bonhoure, Metrologia 11, 141-150 (1975).
[15] Quinn, Chandler and Chattle, Metrologia 9, 44-46 (1973).
[16] H.J. Jung, Ins. Phys. Conf. Ser. No. 26, 278-286 (1975).
[17] T. Ricolfi and F. Lanza, High Temperature-High Pressure 9, 483-489 (1979).
[18] H. Marshak and R.J. Soulen, Low Temperature Physics LT-13, Vol. 4, 498-502 (Plenum Publishing Co., N.Y.).
[19] T.J. Quinn and J.E. Martin, this Symposium.
[20] A. Ono, Document CCT-80/15 du Comité Consultatif de Thermomètrie, Paris, 1980. See Ref. 13.
[21] D. Gugan and G.W. Michel, Metrologia 16, 149-167 (1980).
[22] A.R. Colclough, Metrologia 10, 73-74 (1974).

Measurements of thermodynamic temperature from 2.6 to 27.1 K

K. H. Berry

National Physical Laboratory, Teddington, Middlesex, England

A constant volume gas thermometry scale, NPL-75, has been produced which covers the range from 2.6 to 27.1 K. The measurements were made with reference to the normal boiling point of equilibrium hydrogen, its value of 20.271_2 K having been derived from P-V isotherm measurements. The total uncertainty in this figure, expressed as a 99% confidence limit, is ± 1.0 mK. Corrections for the non-ideality of the thermometric gas, ^{4}He, were made with virial coefficients derived from P-V isotherm measurements at eight temperatures in the range 2.6 to 27.1 K. NPL-75 is compared with IPTS-68, T_{58} and T_{XISU} (1975).

1. INTRODUCTION

About fifteen years ago when the Comité Consultatif de Thermométrie was busy preparing the final form of the then new International Temperature Scale, IPTS-68, in which it was proposed to extend the former scale from the normal boiling point of oxygen to the triple point of equilibrium hydrogen, there were several groups occupied making measurements of thermodynamic temperature below this range with the view that ultimately the International Scale would be extended to the helium vapour pressure regions and below. In the field of gas thermometry, however, there were problems. Values assigned to fixed points in the liquid hydrogen range were uncertain at the level of $\pm$ 10 mK and so there were no accurately known reference points on which to base constant volume gas thermometry measurements. There was also the problem pointed out by Rogers et al[1] concerning the available virial coefficient data for ^{4}He, the thermometric gas most commonly used in gas thermometry. Significant differences between the two most important sets of data[2,3] led to gas scales which were a function of the choice of data adopted. Some workers also pointed to the possibility of loss of thermometric gas from a gas thermometer by physical adsorption of gas on the walls of the gas bulb, this phenomenon becoming particularly significant below the critical temperature of the gas. These three problems in addition to the usual problems of gas thermometry associated with pressure measurement, deadspace corrections etc., compounded to make any gas thermometry measurements at low temperatures much less certain than the 1 mK accuracy in thermodynamic temperature which was thought to be desirable. It was against this background that NPL embarked on a programme of gas thermometry measurements, initially from the normal boiling point of hydrogen down to the helium vapour pressure region but later extended up to the normal boiling point of neon. Some preliminary measurements from this work were reported at the last Temperature Symposium in 1971 but the major text on this work may be found in reference 4.

The first priority was to determine absolutely the thermodynamic temperature of a fixed point in the liquid hydrogen range and for this purpose the normal boiling point of equilibrium hydrogen was chosen. The technique adopted was one of P-V Isotherm Thermometry, a method where the ratio of the gas pressure to the number of moles of gas contained in a bulb of constant volume is measured as a function of gas density at a constant temperature. The temperature of the isotherm is then determined by extrapolation of this ratio to zero density, the value of the intercept being given by (RT/V_B) where V_B is the volume of the gas bulb. With this method, the extrapolation to zero density overcomes the problem of the non-ideality of the gas. The value of the gas constant, R, is uncertain by about 30 ppm and so it might seem that any isotherm temperature would also be uncertain by this amount. We shall see in section 2, however, that it is possible in the case of P-V Isotherm Thermometry to design an experiment where the need for an accurate value of the gas constant is avoided. It would also appear that the volume of the gas bulb should be known to the same fractional accuracy as the temperatures themselves. This problem of absolute volume measurement can also be avoided as we shall see in section 2.

A temperature scale could have been derived from many P-V isotherms covering the range 2.6 to 27.1 K but such measurements are time consuming and so Constant Volume Gas Thermometry (CVGT) measurements were made to cover the whole range. With this method the temperature of a fixed mass of gas contained in a bulb of constant volume is calculated from a measurement of the ratio of the gas pressure at the unknown temperature to that at a reference temperature whose thermodynamic temperature is known, which in these experiments was the normal boiling point of equilibrium hydrogen. Clearly any uncertainty in the thermodynamic temperature of this fixed point is propagated to a CVGT scale based on it and so it was particularly important to have an accurate measurement of the reference point from the isotherm measurements. It is necessary in CVGT measurements to allow for the imperfections of the thermometric gas by including the virial coefficient contributions in the equation of state of the gas. New measurements of ^{4}He virial coefficients were made with a technique of relative isotherm thermometry at eight temperatures in the range 2.6 to 27.1 K.

2. METHODS OF P-V ISOTHERMS AND CVGT

2.1 Absolute P-V isotherm thermometry

In this experiment the absolute P-V isotherm at the normal boiling point of equilibrium hydrogen was measured by filling an evacuated bulb at the unknown temperature with ^{4}He gas from a reference volume immersed in a bath of melting ice. The product of R and the number of moles of gas N_B admitted to the bulb was calculated from the decrease in pressure in the reference volume. The resulting pressure P_B in the gas bulb was also measured. If we assume that the gas bulb was initially evacuated, after the first expansion the product RN_B may be calculated from

$$RN_B = [V_r/T_o] [P_i/\{1+(B_oP_i/RT_o)\}-P_f/\{1+(B_oP_f/RT_o)\}]$$
$$+ P_iD_i - P_fD_f \qquad (1)$$

where V_r is the volume of the reference volume whose temperature is T_o and B_o is the second virial coefficient of ^{4}He at this temperature. P_i and P_f are the initial and final values of the pressure in the reference bulb. If these pressures are less than 10^5 Pa, the virial equation may be terminated without any significant loss of accuracy after the term in B_o. D_i and D_f are the deadspace corrections defined here by

$$D \equiv \Sigma v/t \qquad (2)$$

where t is the temperature of an elemental volume v of the gas in the deadspace volume. (P_BV_B/RN_B) is then calculated from

$$(P_BV_B/RN_B) = [T_oP_BV_B/P_iV_r][\{1+(B_oP_i/RT_o)\}^{-1} -$$
$$- \{P_f/P_i\}\{1+(B_oP_f/RT_o)\}^{-1} +$$
$$+ (T_o/V_r)(D_i - D_fP_f/P_i)]^{-1} \qquad (3)$$

A set of values of (P_BV_B/RN_B) as a function of (N_B/V_B) was obtained by repeatedly adding more gas to the gas bulb from the reference volume. These data are related by

$$(P_BV_B/RN_B) = T_B[1+B(T_B)(N_B/V_B)+C(T_B)(N_B/V_B)^2+..] \qquad (4)$$

from which values for T_B and the virial coefficients may be calculated.

There are two important advantages in determining the number of moles of gas with a reference volume in preference to a method where the quantities of gas are measured by weighing. Firstly, when calculating (P_BV_B/RN_B), only measurements of ratios of pressure and volume are needed, which can be determined with a higher accuracy than measurements of their absolute values. Secondly, an accurate value of R is not required. A value with an accuracy of 0.1% is sufficient to allow the independent variable (N_B/V_B) to be calculated.

2.2 Relative isotherm thermometry

As we have already discussed, values of the virial coefficients were needed to cover the temperature range of the CVGT work. The second virial coefficient is not strongly dependent on temperature, and so it was only necessary to measure its value at a small number of temperatures. Absolute P-V isotherm thermometry could have been used for these measurements but once an accurate value of a reference temperature had been established, a much simpler method was possible. The product RN_B for a gas in the gas bulb at a temperature T_B may be calculated from a measurement of the gas pressure P_r when the gas bulb temperature is T_r, the temperature of the absolute isotherm already completed.

$$RN_B = [P_rV_B/T_r][1+B_r(P_r/RT_r)+(C_r-B_r^2)(P_r/RT_r)^2]^{-1}$$
$$+ (P_rD_i - P_BD_f),$$

and

$$(P_BV_B/RN_B) = [T_rP_B/P_r][\{1+B_r(P_r/RT_r) +$$
$$+ (C_r-B_r^2)(P_r/RT_r)^2\}^{-1} +$$
$$+ (T_r/V_B)(D_i-D_fP_B/P_r)]^{-1} \qquad (5)$$

An isotherm of (P_BV_B/RN_B) may then be measured as a function of (N_B/V_B), the virial coefficients and the temperature being calculated in the same way as for an absolute isotherm. The practical advantages of measuring a relative isotherm are considerable. The reference bulb at the ice-point is no longer needed thus avoiding both the lengthy filling procedure associated with absolute P-V isotherm thermometry and also the need to measure the ratio of the volumes of the two bulbs. Another advantage concerns the gas purity. As the bulk of the gas is now at a temperature of less than 27 K, impurities of oxygen and nitrogen which can be troublesome in the reference bulb when measuring absolute isotherms are now no longer so.

2.3 Constant volume gas thermometry

Once accurate measurements had been made of the reference temperature and the virial coefficients, the most direct approach to producing a temperature scale is with CVGT. The experimental method is identical to that of relative isotherm thermometry measured at only a single gas density but at more temperatures. When calculating the results, however, the deviations from ideal behaviour of the gas need to be accounted for by including virial coefficients in the calculations, whereas in isotherm thermometry the effect of imperfections in the gas is overcome by extrapolation to zero density. A temperature derived from CVGT is then calculated from an equation of the form

$$T = \frac{T_rP[1+B_r(P_r/RT_r)+(C_r-B_r^2)(P_r/RT_r)^2]}{P_r[1+B(P/RT)+(C-B^2)(P/RT)^2][1+(T_r/V_B)(D_r-DP/P_r)]} \qquad (6)$$

where the subscript r refers to measurements at the reference temperature.

3. GAS THERMOMETER DESIGN AND MEASURING EQUIPMENT

The gas bulb, which had a volume of about one litre, was substantially larger than those which had been used previously in low temperature gas thermometers in order that the uncertainties due both to the deadspace correction and gas adsorption would be reduced. The base of the gas bulb could be removed to allow tests to be made on the amount of gas adsorbed on its walls. This was achieved by introducing into the bulb several discs which served to increase the surface area by a factor of about 4.9 while causing little change to the volume. Investigations of this kind showed that any systematic effects due to gas adsorption amounted to less than 0.1 mK.

The reference volume for the absolute isotherm measurements had a volume of about six litres and was connected to the bulb as shown in Figure 1. The volumes of the regions α, β, γ and δ were approximately 0.7 cm^3, 3.3 cm^3, 13 cm^3 and 1.5 cm^3 respectively.

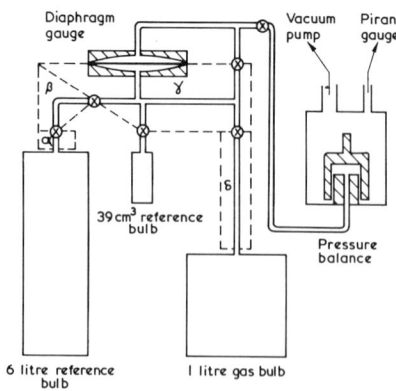

Fig. 1   Gas thermometer system

Gas pressure in the system was measured with a gas operated pressure balance in conjunction with an isolating diaphragm gauge. By making small adjustments to the temperature of the gas bulb, it was possible to arrange that the pressure difference across the diaphragm was small thereby producing little deflection. The diaphragm, therefore, also served to define the constant volume of the gas thermometer.

In actual gas thermometer runs the deadspace correction was calculated from the known volume and temperature distribution along each of the interconnecting tubes, gold-iron / chromel thermocouples being used to determine the temperature gradient along the region. It is difficult with this method, however, to assess the accuracy of the correction and so direct measurements of the correction were made with a gas thermometry technique applied directly to the pressure sensing tube which for this purpose was disconnected from the bulb and sealed at its lower end.

Associated with the deadspace correction in the δ region is the aerostatic head correction needed to be applied to all pressure measurements in the gas bulb, both corrections being inversely proportional to the temperature of the gas in the tube. The pressure measurements also needed to be corrected for the thermomolecular pressure difference in the pressure sensing tube resulting from the temperature gradient along the tube. These corrections were calculated from the Weber-Schmidt equation and compared with some measurements made directly on the tube.

4. RESULTS

4.1 Normal boiling point of equilibrium hydrogen

In order to determine the ratio of the volume of the reference volume to that of the gas bulb, both volumes were immersed in ice-baths and gas was expanded from the gas bulb to the reference volume. This ratio was measured six times and the measurements showed that there was no significant change in the ratio after thermally cycling the gas bulb between room temperature and 20 K. Two corrections needed to be made to this ratio. Firstly, there was the change in volume of the gas bulb on being cooled to 20 K which was calculated from the expansivity data of Kroeger [5]. Secondly, there was a correction for the pressure dilatation of the gas bulb. It was found experimentally that the volume changed by 20 ppm for a pressure change across the walls of one atmosphere.

The ratio of the volumes of the bulbs at their working temperatures was found to be $5.7227_6 (1 - 2 \times 10^{-10} P)$ where P is the pressure difference across the walls. An overall 99% confidence limit of 40 ppm is placed on this ratio.

Twelve isotherm points were recorded at densities in the range 0.8×10^{-4} to 6.2×10^{-4} mol/cm^3 and the usual method of least squares analysis was used to fit power series to these experimental data. The standard deviation of the residuals was 0.2 mK. Once the power series fits had been made, it was necessary to consider how closely the coefficients obtained from these series, whose order was necessarily low, compared with the actual coefficients of the infinite expansion of the virial equation. This is a difficult problem and for further discussion reference should be made to [4].

When all the random and systematic errors, estimated at a 99% confidence level, were combined in quadrature, the total uncertainty in the isotherm temperature was calculated to be $\pm$ 0.9 mK. The normal boiling point of equilibrium hydrogen realized at the NPL has a 99% confidence limit of $\pm$ 0.4 mK [6], and so the 99% confidence limit on the value of 20.271_2 K for the normal boiling point of equilibrium hydrogen is $\pm$ 1.0 mK.

4.2 Virial coefficient measurements

Relative isotherm thermometry was used to determine new values for the second virial coefficient of ^{4}He at temperatures of 2.60 K, 2.75 K, 3.33 K, 4.22 K, 7.20 K, 13.80 K and 27.10 K, the high density of readings at the lower temperatures being required as dB/dT is large in this region. The range of density covered by the isotherms is shown in Table I where it can be seen that the density ranges of the isotherms at 2.60 K and 2.75 K were necessarily limited; at low densities by the lower limit of operation of the pressure balance and at high densities by the saturation of the vapour.

The standard deviations of the residuals of the isotherm points were in the range 0.1 to 0.2 mK, the scatter in the data resulting mainly from thermal instability in the cryostat during the measurements. The measured values of the second and third virial coefficients are shown in Table I together with their associated random and systematic errors. The systematic uncertainty in B results from the uncertainty in the value of B at the reference temperature of the relative isotherm measurements. The second virial coefficient data may be represented by

$$B = 17.19 - 396.2/T - 48/T^2 \quad \text{in cm}^3/\text{mol}. \quad (7)$$

It was only at temperatures of 3.33 K, 4.22 K and 7.20 K that the effect of the third virial coefficient could be observed unambiguously. At the other temperatures, where the isotherm measurements were made over a smaller density range, it was only possible to estimate an approximate upper limit for the third virial coefficient. Such information is, however, needed when deciding on the appropriate density at which to operate a constant volume gas thermometer.

Table I : P-V isotherm results
(uncertainties are 99% confidence limits)

density range (**)	T (K)	B (cm^3/mol)	ΔB ran	ΔB sys	C (cm^3/mol)^2	ΔC ran
1.0 - 3.8	2.60	-142.5	0.8	0.2	4000 *	-
1.0 - 5.2	2.75	-133.2	0.5	0.2	1500 *	-
0.8 - 10.3	3.33	-105.8	0.5	0.2	1000	500
0.8 - 10.3	4.22	- 79.5	0.3	0.2	1200	300
0.7 - 8.8	7.20	- 39.0	0.5	0.2	900	300
0.7 - 5.0	13.80	- 11.7	0.2	0.2	200 *	-
0.8 - 6.2	20.27	- 2.4	0.2	0.0	300 *	-
0.9 - 4.4	27.10	+ 2.5	0.1	0.2	-100 *	-

* : these are only estimates of the upper limit on the magnitude of C
ran : random ; sys : systematic
** : (mol/cm$^3 \times 10^4$)

4.3 Constant volume gas thermometry results

The gas thermometer was operated at about 0.83 kPa/K which is the lowest density compatible with the minimum operating pressure of the pressure balance at the lowest temperature of the gas scale. The gas thermometry results were recorded on a set of three rhodium-iron resistance thermometers [7] and, to one of these, 78 gas thermometry measurements were transferred over the range 2.60 to 27.10 K. A tenth order polynomial was needed to give an adequate fit to the data and the standard deviation of the residuals was 0.2_0 mK. The scale recorded on these rhodium-iron thermometers is referred to as NPL-75. The best measure of its precision comes from the calculated value of the standard error in the curve fitted to the resistance data and for most of the range this was about 0.1 mK.

There were three main sources of systematic error and these are shown in Figure 2. The largest source comes from the uncertainty of 0.9 mK in the value of the reference temperature which introduces a systematic error in the gas scale of 44 ppm. There are also uncertainties introduced because of the virial coefficient data and at an almost insignificant level is the effect of adsorption. These three errors have been combined in quadrature and plotted in Figure 2.

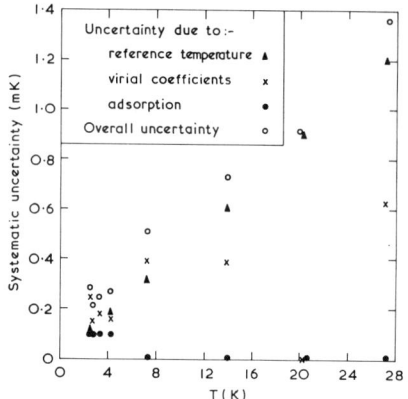

Fig. 2   Systematic uncertainties in NPL-75

5. COMPARISON OF NPL-75 WITH OTHER SCALES

The differences between T_{58}, IPTS-68 and NPL-75 are shown in Tables II and III respectively. Comparisons between NPL-75 and $T_{XISU}(1975)$ have been made at both ISU and NPL and these are shown in Figure 3. $T_{XISU}(1975)$ is a version of the magnetic scale of Cetas and Swenson[8] which has already been compared extensively with other temperature scales and so it provides a useful link between NPL-75 and these other scales.

6. CONCLUSION

A constant volume gas thermometry scale NPL-75 has been produced, based on the results of both P-V isotherm and constant volume gas thermometry experiments covering the range 2.60 K to 27.10 K. The random uncertainty in the scale, expressed as a 99% confidence limit, is ± 0.4 mK and it is estimated that the scale does not deviate from thermodynamic temperature by more than 0.3 mK, 0.9 mK and 1.4 mK at 4 K, 20 K and 27 K respectively. The three rhodium-iron thermometers used to record the scale have been stable within a few tenths of a millikelvin at 20 K and a tenth of a millikelvin at 4 K over a period of seven years.

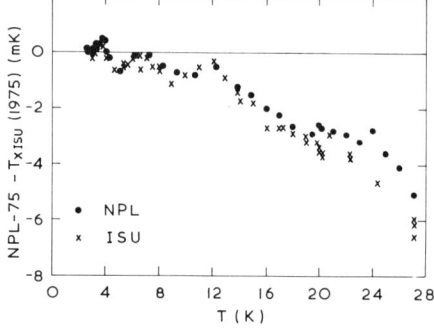

Fig. 3   Differences between NPL-75 and $T_{XISU}(1975)$

Table II : Comparison of NPL-75 with T_{58}

NPL-75** (K)	T_{58} (K)	ΔT (mK)
2.6054	2.6000	5.4
2.8059	2.8000	5.9
3.0063	3.0000	6.3
3.2066	3.2000	6.6
3.4069	3.4000	6.9
3.6070	3.6000	7.0
3.8071	3.8000	7.1
4.0071 *	4.0000	7.1
4.2221	4.2150	7.1

* : normal boiling point of ^{4}He
** : see reference 9

Table III : Comparison of NPL-75 with IPTS-68

IPTS-68 (K)	NPL-75 (K)	ΔT (mK)
13.8100 a	13.8035	6.5
14.1991	14.1937	5.4
14.6994	14.6955	3.9
15.4991	15.4955	3.6
14.9996	14.9961	3.5
16.2993	16.2947	4.6
17.0420 b	17.0356	6.4
17.6987	17.6909	7.8
18.3998	18.3909	8.9
18.9998	18.9904	9.4
19.6003	19.5908	9.5
20.2800 c	20.2712	8.8
21.6467	21.6389	7.8
23.0798	23.0733	6.5
24.0893	24.0835	5.8
25.2955	25.2907	4.8
26.4538	26.4493	4.5
27.1020 d	27.0979	4.1

a : triple point of e-hydrogen
b : '17 K' point
c : normal boiling point of e-hydrogen
d : normal boiling point of neon

7.   REFERENCES

1. Rogers J S, Tainsh R J, Anderson M S & Swenson C A : Metrologia, 4,47,(1968).
2. Keller W E :   Phys. Rev. 97,1,(1955).
3. Keesom W H and Walstra W K :
   Physica,7,985,(1940).
   Physica,13,225,(1947).
4. Berry K H : Metrologia,15,89,(1979).
5. Kroeger F R and Swenson C A : J. Appl. Phy. 48,853,(1977).
6. Compton J P   :   Temperature, Vol.4, part 1,195-209, Pittsburgh : Instrument Society of America, 1972.
7. Rusby R L : Temperature Measurement, 125-130, London, Institute of Physics, 1975.
8. Cetas T C and Swenson C A : Metrologia,8,46,(1972).
9. There are typographical errors in the 1st column in Table 11 of reference 4 which are now given correctly. (The comparison data given in Table II in this paper and Table 11 of reference 4 differ slightly from those given in Table 13 of reference 10. This change was made after more intercomparison data became available and after a small systematic error had been detected in the realization of T_{58} at NPL.)
10. Berry K H : NPL Quantum Metrology Report No.47, 1978.

Measurements with a gas thermometer between 4 and 100 K

P. P. M. Steur, J. E. van Dijk, J. P. Mars, H. ter Harmsel, and M. Durieux

Kamerlingh Onnes Laboratorium der Rijksuniversiteit Leiden, Nieuwsteeg 18, 2311 SB Leiden, The Netherlands

A gas thermometer for measuring thermodynamic temperatures between 4 and 100 K is described. Gas thermometer temperatures are compared with platinum resistance thermometers calibrated on the International Practical Temperature Scale of 1968 (IPTS-68) and with rhodium-iron resistance thermometers calibrated on the National Physical Laboratory (Teddington) Scale of 1975 (NPL-75). First results of the measurements, with NPL-75 at 27 K as a reference temperature for the gas thermometer, agree between 4 and 14 K with NPL-75 within 0.3 mK and deviate from it by + 1 mK between 16 and 27 K. Measurements above 27 K confirm that the IPTS-68 around 40 K is too high by about 9 mK. The measurements are being continued in order to obtain isochores at several densities between 4 and 100 K.

INTRODUCTION

In the temperature range between approximately 3 K and 1000 K gas thermometer is still the most common method for measuring thermodynamic temperatures. In the range below 273 K the two most complete and accurate gas thermometer experiments were those of Preston-Thomas and Kirby (1) between 273 K and 90 K and of Berry (2) between 27 K and 2.6 K. References to earlier experiments are given by thse authors. Preston-Thomas and Kirby determined the differences between the gas thermometer scale and IPTS-48. In Berry's by now already classical experiments, the gas thermometer results were given in terms of calibrated rhodium-iron resistance thermometers which now define the NPL-75 scale. Berry also determined the second virial coefficient of helium gas so that he did not have to rely on literature data for calculating the non-ideality correction of the gas.

In the last two decades other gas thermometer experiments below 100 K were undertaken in at least three laboratories: NML (3), PTB (4) and KOL (5). The first goal of these experiments was the establishment of laboratory temperature scales between about 4 K and 27 K in which range international scales were not yet defined or were expected to deviate appreciably from the thermodynamic temperature. With the introduction of the EPT-76 (6) the interpolation gas thermometer became one of the allowed instruments for realizing a practical scale between 4 K and 27 K. Later on, the report of a deviation of the IPTS-68 from the thermodynamic temperature of about 0.01 K at 54 K (7) and the proposed introduction of a new IPTS (8) stimulated the extension of gas thermometer measurements to temperatures above 27 K.

In this paper a gas thermometer experiment at KOL between 4 K and 100 K is described and results obtained so far are given. Gas thermometer temperatures between 4 K and 27 K are compared with the NPL-75 scale (2) and results between 14 K and 100 K with the IPTS-68. The results are also compared with noise thermometer data at 4.2 K (9), dielectric constant gas thermometer data between 4 K and 27 K (10) and preliminary gas thermometer data of NML between 14 K and 27 K (3).

EXPERIMENTAL SET-UP

The experimental set-up is very similar to that used by Berry (2). The low temperature part is shown in Fig. 1. The gas thermometer bulb A, with an inner volume of about 1003 cm^3, is made of OFHC copper. It consists of the can B, made out of a solid bar copper, the bottom plate C, and the top D to which the pressure sensing capillary E is soft soldered. The bottom plate C and the top D are connected to the can B by means of gold O ring seals. The wall of the can is 13.5 mm thick, the top plate of the can 25 mm thick and the bottom 35 mm. The various parts are electrolytically gold plated; the thickness of the gold layer is 5 μm. Two rhodium-iron resistance thermometers T_1 and T_2 and two standard platinum resistance thermometers T_3 and T_4 are placed in holes in the top and bottom of the gas thermometer.

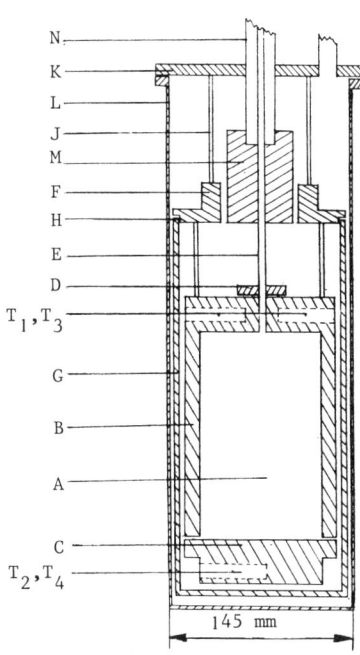

Fig. 1. The gas thermometer assembly.

The gas thermometer bulb is suspended by three thin-walled stainless steel capillaries from the solid piece of copper F, to which the copper shield G is connected by the screw fitting H. A germanium thermometer and a small ceramic platinum thermometer are placed in holes in F for temperature regulation below and above 30 K respectively. F is suspended by four stainless steel capillaries J from the top K of the vacuum can L.

The pressure sensing capillary E, with an inner diameter of 1 mm, is thermally anchored to the small copper plate M which forms together with the stainless steel tube N a separate vacuum jacket around the pressure sensing capillary. The plate M is separated from the top F of the shield, and its temperature can be measured and controlled by means of a germanium thermometer.

The separation between F and M originated from an earlier version of the apparatus in which there was a vapour pressure thermometer built in the gas thermometer; the vapour pressure sensing tube was thermally anchored at M and could thus be kept at a higher temperature than the gas thermometer, independently of the shield temperature. The supporting capillaries J are of adjustable length in order to be able to release tension on the capillary E, where it is soft soldered to D, as much as possible. The temperature distribution along the capillary between M and the top of the cryostat is measured with thirteen Au 0.03% Fe/chromel thermocouples.

The room temperature part of the apparatus is sketched in Fig. 2. Pressures are measured with a pressure balance; two piston-cylinder combinations are used, one for pressures between 2.1 kPa and 10.7 kPa and another for the higher pressures. The gas thermometer capillary E is connected to the pressure balance via valves 1 and 2 and a MKS-Baratron diaphragm pressure gauge with a total range of 1 Torr. The

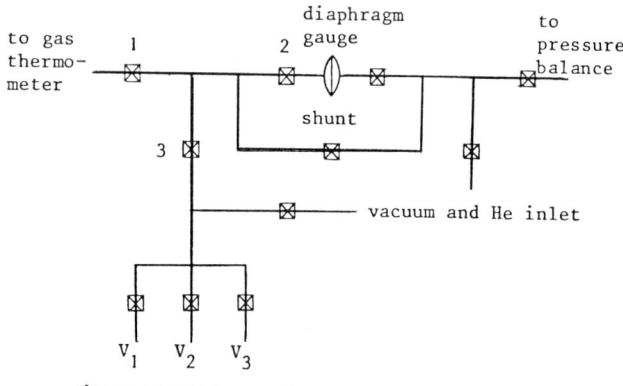

Fig. 2. The connections between gas thermometer, pressure balance and gas inlet.

type 315-1 diaphragm has been chosen because of its small (2 cm³) internal volume. The room temperature dead volume of the gas thermometer, including the volume of the diaphragm gauge, is approximately 7 cm³. Helium gas is admitted into the gas thermometer system through a copper capillary at 4.2 K. The vessels V_1, V_2 and V_3, with volumes 1 liter, 3 liter and 8 liter, can be used for measuring the amount of gas admitted into the gas thermometer. The ratios of the volumes of these vessels and of the gas thermometer bulb are determined by gas pressure measurements. The vessels are thermostated in a stirred water bath at room temperature; the temperature of the bath is determined with a standard platinum resistance thermometer.

The resistance of the platinum and rhodium-iron thermometers are measured by means of a classical DC potentiometric method.

EXPERIMENTAL PROCEDURE

In most cases the amount of gas in the gas thermometer was determined by means of the room temperature vessels V_1, V_2 and V_3 before and after a series of measurements. However, up to now most of the results were obtained with the apparatus used as a constant volume gas thermometer and only these measurements will be discussed.

During a measurement the shield temperature and the temperature of the capillary heat sink M are regulated. Pressure and resistance data were taken, after temperature equilibrium was established, for about two hours. The temperature drift during this period was in general not more than a few tenths of a millikelvin at the lower temperatures and 1 mK at temperatures above about 40 K.

Gas thermometer temperatures T_g were derived from the equation

$$T_g = T_r \frac{p(T_g)}{p_r} \frac{1+B(T_r)/V_m}{1+B(T_g)/V_m} \quad (1)$$

where T_r is a reference temperature, p the measured pressure, B the second virial coefficient of ^{4}He gas and V_m the molar volume of the gas.

For the second virial coefficient the equation given by Berry, ref. 2, p. 106, was used below 30 K and McCarty's equation, ref. 11, Eq. (8), above 30 K. However, a term

$$\delta B = (0.065 \times 10^{-6} - 19.12 \cdot 10^{-6} \text{ K}/T) \text{ m}^3/\text{mol} \quad (2)$$

was added to McCarty's expression for B in order to obtain at 30 K a smooth junction of B and dB/dT, with Berry's equation.

Changes in the volume of the gas thermometer bulb with temperature were calculated by using an expression for the expansion coefficient of copper obtained from a least squares fit to literature data for the expansion coefficient. (Use of, e.g., data for the expansion coefficient of copper given by White (12) would change the results for T_g in Tables I, II and III by less than 0.03 mK below 27 K, by +0.2 mK at 54 K and by +0.5 mK at 90 K).

Corrections were applied for the amounts of gas in the capillary E and in the room temperature dead volume, for the aerostatic head of the gas and for the thermomolecular pressure difference. The two standard platinum resistance thermometers in the gas thermometer were calibrated at KOL on the IPTS-68. Of the two rhodium iron thermometers one, number 226246, $R(0 °C) \simeq 100\ \Omega$, was calibrated at NPL on the NPL-scale, the other, number 223687, with $R(0 °C) \simeq 50\ \Omega$, was calibrated at KOL against NPL calibrated thermometers.

EXPERIMENTAL RESULTS

Before presenting the experimental results obtained so far with the apparatus described above, some results obtained with an earlier version of the gas thermometer will be given. In this version, the gas thermometer bulb had a volume of 207 cm³, it was also made of copper but the bottom plate was soft soldered and the bulb was not gold plated at the inside. Two platinum thermometers and four germanium thermometers were used for measuring the bulb temperature on IPTS-68 and EPT-76, respectively. The diaphragm pressure gauge was used with a Baratron type 315 GH symmetrical head which had an internal volume of 5 cm³ and a range of 1 Torr. Other parts of the apparatus were the same as described in the preceding sections.

The experimental results for the 207 cm³ bulb are given in Table I. The amount of gas in the gas thermometer was about 0.031 mol, corresponding to a sensitivity of the gas thermometer of about 1.26 kPa/K for series 207 to 211. After series 211 it was decreased to about 0.028 mol. The temperature T_{68} (KOL) in the table denotes the KOL realization of IPTS-68 (see ref. 13). The two platinum thermometers in the gas thermometer gave the same temperature T_{68} to within 0.2 mK, except in one case where it was 0.4 mK. The average of the two thermometers is used for T_{68} (KOL). For calculating the gas thermometer temperature T_g, the NPL-75 scale at 27.1 K was used as a reference. The difference between T_{68} (KOL) and T_{NPL-75} at 27.1 K was obtained from the dif-

Table I
Results obtained with the 207 cm³ gas thermometer bulb. $\Delta T = T_g - T_{68}$ (KOL). B is the second virial coefficient.

Series	T_{68} (KOL) K	ΔT mK	B cm³/mol
207	27.12	-5.3	2.51
208	77.85	1.1	11.23
209	27.12	-5.7	2.51
210	77.86	0.0	11.23
211	27.12	-5.2	2.51
213	27.13	-5.3	2.52
214	54.38	-5.2	9.56
215	63.18	1.4	10.38
216	27.13	-5.5	2.52
217	90.22	1.9	11.65
218	27.14	-5.5	2.52
219	77.88	3.7	11.23
220	83.80	3.9	11.45
221	27.13	-5.3	2.52

ferences T_{68}(NPL)$-T_{NPL-75}$ = 4.1 mK (2) and T_{68}(KOL)$-T_{68}$(NPL) = 1.3 mK (13). For the first filling of the gas thermometer (series 207 to 211) the average of series 207, 209 and 211 is used as the reference and for the second filling (series 213 to 221) the average of series 213, 216, 218 and 221. Values for B calculated as described above are given in the table. (The results for $T_g - T_{68}$(KOL) are also shown in Fig. 10.)

Three series of measurements were made with the 1003 cm³ gas thermometer bulb described before. Results of the first series, made in the summer of 1980, are given in Table II. For the temperature T_{68}(KOL) the average of the two platinum thermometers is used; the difference between the thermometers was never more than 0.3 mK. T_{NPL-75} is obtained from the NPL calibrated rhodium-iron thermometer number 226246. The amount of

Table II
Results obtained in the first series of measurements with the 1003 cm³ gas thermometer bulb.

Series	T K	ΔT_1 mK uncorrected T_r = 393 K	ΔT_2 mK uncorrected T_r = 393 K	ΔT_1 mK corrected T_r = 20.28 K	ΔT_2 mK corrected T_r = 20.28 K	ΔT_1 mK corrected T_r = 27.3 K	ΔT_2 mK corrected T_r = 27.3 K
100	20.27	-0.37	-9.77	0.01	-9.39	0.73	-8.67
101	27.28	-1.55	-6.95	-0.89	-6.29	0.08	-5.32
102	20.28	-0.57	-9.97	0.04	-9.36	0.76	-8.64
103	34.99		-11.84		-10.59		-9.35
104	20.28	-1.22	-10.62	-0.03	-9.43	0.69	-8.71
105	27.26	-2.80	-8.20	-1.05	-6.45	-0.08	-5.48
107	20.28	-1.41	-10.81	0.01	-9.39	0.73	-8.67
109	44.00		-12.47		-9.17		-7.61
110	44.00		-12.83		-9.47		-7.91
111	20.28	-1.68	-11.08	-0.03	-9.43	0.69	-8.71
114	20.28	-2.05	-11.45	-0.05	-9.45	0.67	-8.73
115	54.28		-11.07		-5.93		-4.00
116	20.28	-2.18	-11.58	0.05	-9.35	0.77	-8.63

$\Delta T_1 = T_g - T_{NPL-75}$ $\Delta T_2 = T_g - T_{68}$(KOL)

gas in the gas thermometer, roughly corresponding to a sensitivity of 1.33 kPa/K, was at first calculated from the amount of gas admitted into the gas thermometer from the calibrated vessels at room temperature. The corresponding values of $T_g - T_{68}$(KOL) and $T_g - T_{NPL-75}$ are given in Table II as "uncorrected results". It can be seen from the table that there is a gradual change with time in $T_g - T_{NPL-75}$ (and in $T_g - T_{68}$(KOL)) at 20.3 K. This is also shown in Fig. 3 where $T_g - T_{NPL-75}$ at 20.28 K

is plotted versus the date of the measurements. Apparently, there is a continuous loss of gas from the gas thermometer. Later this appeared to be caused by a leak in the connection between the pressure sensing capilary and the gas thermometer bulb. Taking a leak rate corresponding to -0.116 mK per day at 20.3 K (see Fig. 3), the temperatures T_g were corrected for the loss of gas. In columns 5 and 6 of Table II results.

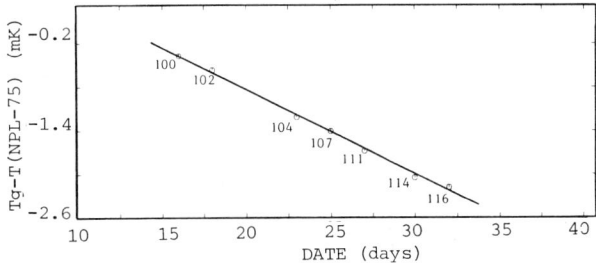

Fig. 3. The change of the gas thermometer temperature at 20.3 K with time (first series of measurements, see Table II). Numbers are series numbers. The drawn line corresponds to a rate of change of -0.116 mK per day.

corrected for the loss of gas are given; also, T_{NPL-75} at 20.28 K is chosen as a reference temperature, i.e. the average of $T_g - T_{NPL-75}$ for the seven series at 20.28 K is adjusted to zero. Because in the second series of measurements (see below) T_{NPL-75} at 27 K is used as a reference, results recalculated to $(T_g - T_{NPL-75})$ average = 0 at 27 K are also given in Table II (columns 7 and 8).

The second series of measurements with the 1003 cm³ bulb was made in February-March 1981. The leak in the pressure sensing capillary was repaired before this series. Results $T_g - T_{68}$(KOL) and $T_g - T_{NPL-75}$ with T_{NPL-75} at 27.1 K as a reference are given in columns 4 and 5 of Table III; somewhat arbitrarily, T_g is adjusted in such a way that the average of $T_g - T_{NPL-75}$ for the first three measurements at 27 K (series 2, 8 and 20) is equal to zero. In Fig. 4 $T_g - T_{NPL-75}$ at 27 K is plotted versus the date of the measurements. Again, there is an apparent loss of gas from the gas thermometer. For series 1 to 30 the change in T_g with time can be approximated by a constant rate of change of -0.0438 mK per day at 27.1 K (see Fig. 4). For series 31 to 40 a change of -0.23 mK per day at 27.1 K was assumed. Results $T_g - T_{68}$(KOL) and $T_g - T_{NPL-75}$ where T_g is corrected for these rates of change are given in columns 6 and 7 of Table III.

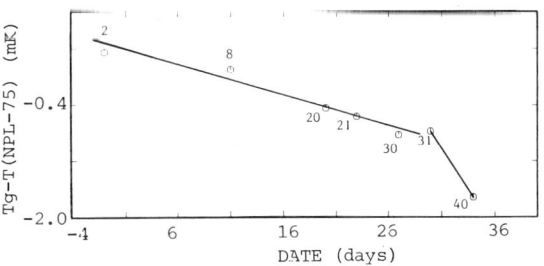

Fig. 4. The change of the gas thermometer temperature at 27 K with time (second series of measurements, Table III). The drawn lines correspond to rates of change of -0.0438 mK per day and -0.23 mK per day.

Table III
Results obtained in the second series of measurements with the 1003 cm³ gas thermometer bulb. T_{NPL-75} at 27 K is reference temperature.

Date	Series	T	ΔT_1	ΔT_2	ΔT_1	ΔT_2
		K	mK	mK	mK	mK
			uncorrected		corrected	
26.2.81	1	52.00		-6.7		-7.7
27.2.81	2	26.97	0.35	-5.1	-0.14	-5.6
4.3.81	3	52.00		-6.7		-7.2
5.3.81	4	44.01		-8.0		-8.4
6.3.81	5	54.37		-5.0		-5.4
9.3.81	6	54.39		-5.1		-5.2
10.3.81	7	35.29		-8.1		-8.1
11.3.81	8	27.10	0.10	-5.2	0.14	-5.2
12.3.81	9	20.27	0.99	-8.6	1.05	-8.5
12.3.81	10	23.53	0.64	-6.9	0.72	-6.8
13.3.81	11	54.40		-5.5		-5.3
16.3.81	12	54.39		-5.5		-5.0
17.3.81	13	4.24	0.14		0.19	
18.3.81	14	5.04	0.22		0.28	
18.3.81	15	6.05	-0.12		-0.04	
18.3.81	16	6.05	-0.09		-0.01	
19.3.81	17	7.07	-0.23		-0.13	
19.3.81	18	8.02	-0.08		0.04	
20.3.81	19	11.64	-0.35		-0.17	
20.3.81	20	27.09	-0.45		-0.02	
23.3.81	21	27.01	-0.57		0.00	
23.3.81	22	9.02	0.02		0.21	
24.3.81	23	10.03	-0.15		0.07	
24.3.81	24	13.00	-0.33		-0.04	
25.3.81	25	13.83	-0.31	-6.77	0.02	-6.44
25.3.81	26	15.00	0.15	-3.39	0.51	-3.33
26.3.81	27	16.03	0.32	-3.90	0.73	-3.49
26.3.81	28	17.00	0.47	-6.07	0.91	-5.63
27.3.81	29	18.01	0.41	-8.48	0.90	-7.99
27.3.81	30	27.01	-0.83		-0.09	
30.3.81	31	27.01	-0.78		0.09	
30.3.81	32	18.99	0.41	-9.66	1.06	-9.01
31.3.81	33	20.99	0.14	-8.94	0.96	-8.12
31.3.81	34	18.99	0.27	-9.77	1.05	-8.99
1.4.81	35	22.01	-0.04	-8.43	0.99	-7.40
1.4.81	36	22.99	-0.35	-8.33	0.77	-7.21
2.4.81	37	24.01	-0.66	-7.70	0.65	-6.39
2.4.81	38	24.99	-0.94	-7.15	0.47	-5.74
3.4.81	39	26.00	-1.31	-6.76	0.29	-5.16
3.4.81	40	27.01	-1.72	-6.97	0.00	-5.25

$\Delta T_1 = T_g - T_{NPL-75}$ $\Delta T_2 = T_g - T_{68(KOL)}$

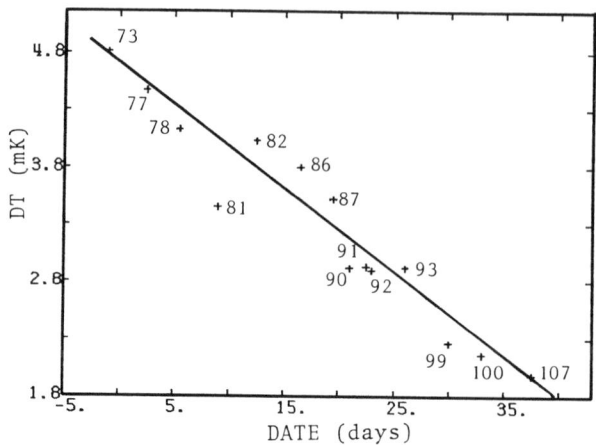

Fig. 5. The change of the gas thermometer temperature at 70 K with time (third series of measurements, Table IV). The drawn line corresponds to a rate of change of -0.0735 mK per day.

Table IV
Results obtained in the third series of measurements (27 K - 100 K) with the 1003 cm³ gas thermometer bulb. T_{NPL} at 27 K is reference temperature.

Date	Series	T	ΔT	ΔT	ΔT
		K	mK	mK	mK
			uncorrected	corrected	
					average of series 74 and 83 is reference
				series 74 is reference	
27.10.81	73	70.05	4.74	4.92	5.10
28.10.81	74	27.01	-5.53	-5.44	-5.37
29.10.81	75	34.01	-9.09	-8.93	-8.84
29.10.81	76	41.00	-8.13	-7.93	-7.82
30.10.81	77	70.06	4.40	4.83	5.01
2.11.81	78	70.07	4.08	4.73	4.91
4.11.81	79	67.00	3.37	4.14	4.31
5.11.81	80	62.01	0.87	1.64	1.80
6.11.81	81	70.09	3.41	4.32	4.50
9.11.81	82	70.09	3.99	5.16	5.34
11.11.81	83	27.00	-6.01	-5.51	-5.44
12.11.81	84	48.05	-9.94	-9.01	-8.89
12.11.81	85	54.99	-5.96	-4.87	-4.73
13.11.81	86	70.06	3.74	5.20	5.38
16.11.81	87	70.12	3.48	5.16	5.34
17.11.81	88	58.99	-2.43	-0.98	-0.83
17.11.81	89	64.95	1.18	2.81	2.98
18.11.81	90	70.06	2.91	4.70	4.88
19.11.81	91	70.12	2.90	4.80	4.98
20.11.81	92	70.12	2.88	4.82	5.00
23.11.81	93	70.12	2.90	5.06	5.24
23.11.81	94	65.28	1.37	3.41	3.58
24.11.81	95	67.02	1.25	3.42	3.59
25.11.81	96	68.63	2.22	4.51	4.69
26.11.81	97	77.18	2.79	5.41	5.61
26.11.81	98	85.02	2.19	5.12	5.34
27.11.81	99	70.13	2.23	4.69	4.87
30.12.81	100	70.13	2.14	4.82	5.00
1.12.81	101	90.21	2.94	6.48	6.71
1.12.81	102	92.01	2.08	5.73	5.97
2.12.81	103	94.02	2.05	5.83	6.07
2.12.81	104	96.01	1.95	5.87	6.12
3.12.81	105	98.14	2.25	6.31	6.59
3.12.81	106	100.04	2.06	6.24	6.50
4.12.81	107	70.13	1.96	4.97	5.15

$\Delta T = T_g - T_{68(KOL)}$

The third series of measurements with the 1003 cm³ bulb was made in October-December 1981 in the range 27 K to 100 K. Results $T_g - T_{68}$ (KOL), with again T_{NPL-75} at 27 K as a reference, are given in Table IV, column 4. During this series the 70 K point was repeated several times as a check on the amount of gas in the gas thermometer. Values of $T_g - T_{68}$ (KOL) for these points are plotted versus time in Fig. 5. There is an apparent loss of gas corresponding to 0.0735 mK (at 70 K) per day. Values of $T_g - T_{68}$ (KOL) corrected for this loss of gas are given in column 5 of Table IV. In column 6 the same data are given, but now referenced in the average of NPL-75 at the two points at 27 K.

Most likely, the gradual decrease in T_g of -0.044 mK per day at 27 K (Fig. 4) and -0.074 mK per day at 70 K (Fig. 5) was caused by a small leak in the gas thermometer. However, due to the rather high background of helium gas in the vacuum can, it was not possible to detect a leak of the required order of magnitude (5×10^{-8} atm cm³ per second) during operation of the gas thermometer. As these changes in T_g are quite regular, and only of the order of 1 ppm in T_g per day, the uncertainty in the corrected results due to this effect is very small.

COMPARISON OF RESISTANCE THERMOMETERS

After the first two series of measurements with the 1003 cm³ bulb reported above, a third standard platinum thermometer, LN 43, which was calibrated on T_{68}(KOL) and took part in the Ward and Compton scale comparison (13), and a second NPL calibrated rhodium-iron thermometer, number 229080, were added to the gas thermometer. Comparisons of the three platinum and three rhodium-iron thermometers were then made between 14 K and 30 K. Differences found between the two NPL calibrated rhodium-iron thermometers are plotted in Fig. 6, together with the result of comparisons of these thermometers made in 1977 and 1978, just after they were received from NPL. The cause of the systematic difference between the two thermometers above 18 K, reaching a maximum of 0.4 mK at 27 K, is not known; the stability of the thermometers between 1977 and 1981 apparently is excellent.

For the presentation of results $T_g - T_{NPL-75}$ in the preceding section it has been assumed that the average of the two rhodium-iron thermometers represents the NPL-75 scale and all data for T_{NPL-75}, which were obtained with thermometer 226246, have therefore been corrected according to the drawn line in Fig. 6. It is estimated that the error in T_{NPL-75} obtained in this way is not greater than 0.2 mK below 20 K and 0.4 mK at 27 K.

The measured differences T_{68}(KOL)$-T_{NPL-75}$, with T_{68}(KOL) the average of the three platinum thermometers and T_{NPL-75} the average of the two rhodium iron thermometers, are compared in Fig. 7 with values of T_{68}(KOL)$-T_{NPL-75}$ derived from data of Berry (2) and Ward and Compton (13). Platinum thermometers 1 and 2 differ between 16 K and 27 K by not more than 0.2 mK. The thermometer LN 43 shows in this range small systematic differences up to 0.3 mK from 1 and 2. The consistency of T_{68}(KOL)$-T_{NPL-75}$ measured in this experiment with the Berry and Ward-Compton data is quite satisfactory. It is estimated that the error in T_{68}(KOL) values given in the preceding section is not more than 0.5 mK above 16 K.

DISCUSSION OF THE GAS THERMOMETER DATA

In Fig. 8 the uncorrected and the corrected differences $T_g - T_{NPL-75}$, taken from Table III, are plotted versus T. For series 9 to 29 the correction for the change of T_g at the reference temperature is rather straightforward because the rate of change is constant. For series 31 to 40 the ad hoc assumption that there is a constant rate of change of the amount of gas in the gas thermometer is supported by the smoothness of the uncorrected and corrected data $T_g - T_{NPL-75}$ for series 32 to 39. The consistency of the corrected results of all series can be seen from Fig. 8.

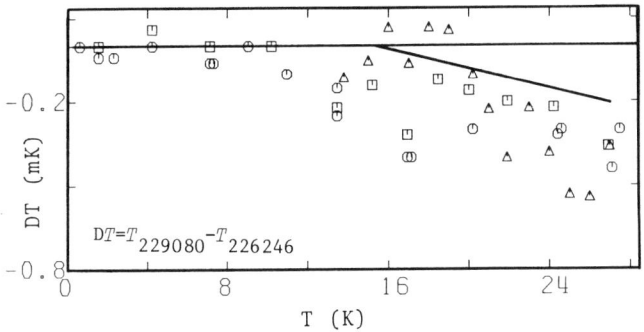

Fig. 6. Comparison of two rhodium-iron resistance thermometers. Temperatures are calculated from measured resistances using the NPL calibrations of 1977.
○ comparison of 1977   □ comparison of 1978
△ present comparison (1981).

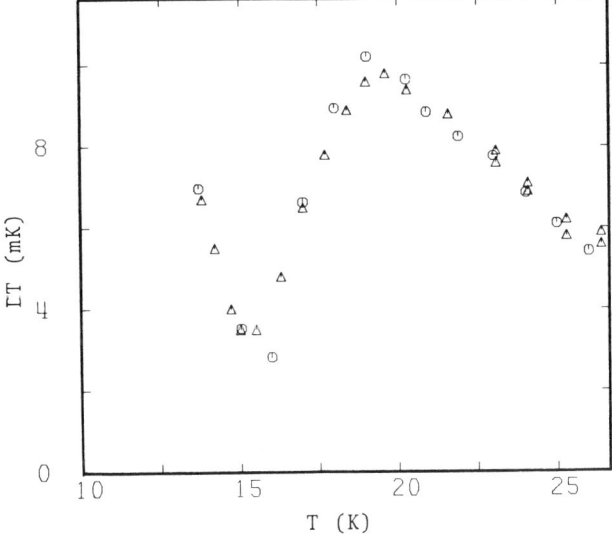

Fig. 7. Differences DT = T_{68}(KOL) $- T_{NPL-75}$. T_{68}(KOL) represents the average of the three platinum thermometers and T_{NPL-75} the average of the rhodium-iron thermometers 226246 and 229080. The circles represent the present comparison. The triangles are obtained from Berry's data T_{68}(NPL)$-T_{NPL-75}$ together with Ward and Compton's data T_{68}(KOL)$-T_{68}$(NPL).

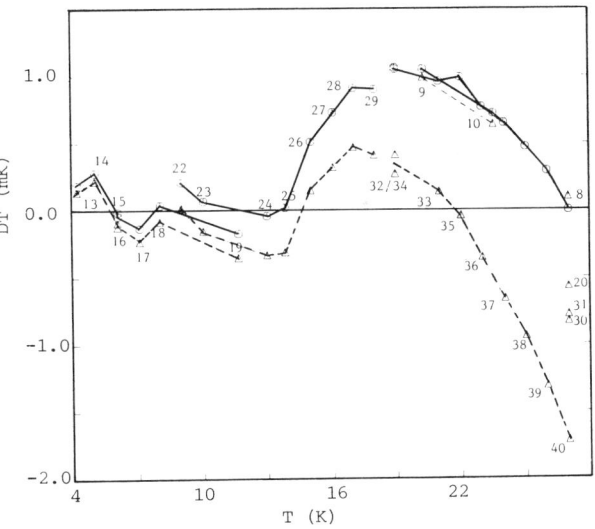

Fig. 8. Differences $T_g - T_{NPL-75}$, where T_g is the present gas thermometer temperature. (Second series; data taken from Table III.) The dashed line connects uncorrected points, the drawn line those corrected for the change in the reference temperature. Series are numbered as in Table III.

In Fig. 9 corrected differences $T_g - T_{NPL-75}$ are shown for all series. Between 4 K and 14 K the maximum difference is 0.28 mK at 5 K. Between 14 K and 27 K there is a maximum difference of 1.06 mK at 19 K. It is estimated that the inaccuracy in the data is not greater than 0.2 mK below 14 K and 0.3 mK above 14 K and, thus, that the differences between the present measurements and NPL-75 between 14 K and 27 K are real. The differences below 14 K do not exceed the combined uncertainty in the gas thermometer measurements and rhodium-iron thermometer calibrations.

The present data can, except with NPL-75 be compared with the noise thermometer data of Klein et al. (9) and with the results of the dielectric constant gas thermometer measurements of Gugan and Michel (10). Points $T_{noise}-T_{NPL-75}$ in Fig. 9 have been calculated from values of T_{noise} and corresponding ^{4}He vapour pressures given in Ref. 9 and $T_{58}-T_{NPL-75}$ given by Rusby and Swenson (Ref. 14 Table II). Gugan and Michel's data $T_{DCGT}-T_{NPL-75}$ have been taken from their Fig. 10, recalculated to 27.1 K as a reference temperature. Also included in Fig. 9 are provisional results of the gas thermometer experiment at NML (3), recalculated to 27.1 K as reference temperature.

Results T_g-T_{68} (KOL) taken from Tables I, II, III and IV are shown in Fig. 10. The points below 27 K have already been discussed in terms of the differences with NPL-75. As can be seen, differences between the various series of measurements with the 1003 cm^3 bulb above 27 K are about 1.2 mK at most (at 35 K and 54 K). Here the rather large differences are, in retrospect, caused by insufficient temperature equilibrium along the pressure sensing tube in series 103 and 115, Table II. For the third series of measurements the spread in the data points at the same or adjacent temperatures is less than 1 mK (see also Fig. 5).

A first, preliminary attempt was made to relate the results to room temperature (one point at 100 K in Fig. 10). The very good agreement with the other data is probably fortuitous.

For comparison, also $T_{Pt,NML}-T_{68}$ according to Kemp et al., where $T_{Pt,NML}$ denotes a corrected version of IPTS-68 fitted to NPL-75 at 27.1 K and to IPTS-68 at 83 K (7) is shown in Fig. 10.

CONCLUSION

A gas thermometer is in operation for measuring thermodynamic temperatures between 4 K and 100 K. Measured temperatures are compared with IPTS-68 and NPL-75.

An isochore has been measured between 4 K and 27 K, with temperature intervals of one kelvin. Gas thermometer temperatures calculated with NPL-75 at 27.1 K as a reference, agree with NPL-75 below 14 K within 0.3 mK but deviate from it by 1 mK between 16 K and 23 K. If for the measurments , between 27 K and 100 K NPL-75 at 27.1 K is used as a reference, the gas-thermometer temperatures are about 9 mK lower than IPTS-68 between 35 K and 45 K and 6 mK higher than IPTS-68 between 70 K and 100 K.

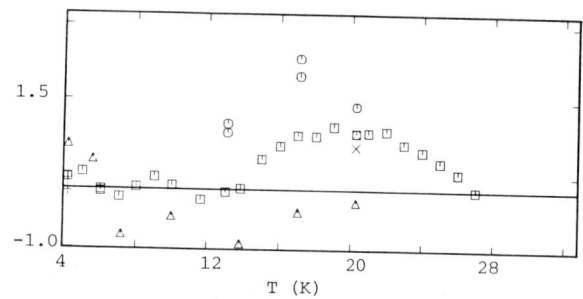

Fig. 9. Present gas thermometer temperatures (second series) compared with NPL-75 ; the point marked X gives the average result at 20.28 K in the first series of measurements (Table II).

+: $T_{noise} - T_{NPL-75}$ (9); Δ; $T_{DCGT}-T_{NPL-75}$ (10);

: $T_{gas}(NML) - T_{NPL-75}$ provisional results (3).

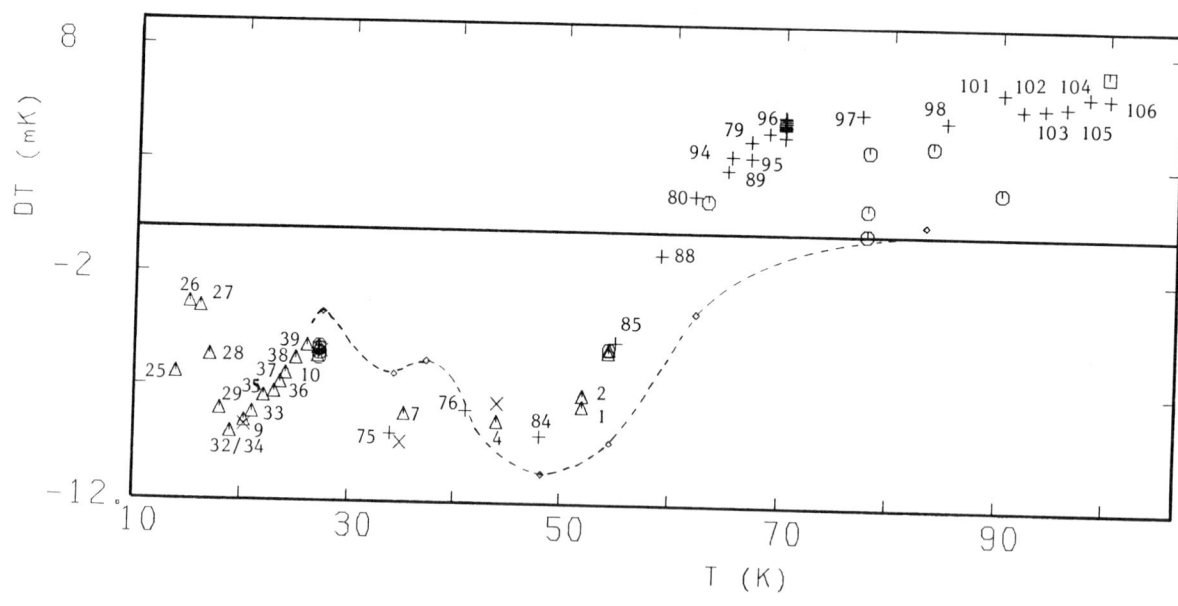

Fig. 10. Gas thermometer temperatures compared with IPTS-68, KOL realization. The reference temperature for the gas thermometer is T_{NPL-75} at 27.1 K ($T_{NPL-75}-T_{68}$(KOL) = -5.4 mK at 27.1 K). Point indicated , at 100 K, denotes a measurement with room temperature as a reference (3 Torr/K).

: 207 cm^3 bulb, Table I; X: 1003 cm^3 bulb, fierst series, Table II: Δ: second series, Table III.
+ : third series, Table IV; the dashed line denotes $T_{Pt,NML}-T_{68}$(NML).

ACKNOWLEDGMENTS

We wish to thank Mr. J. Mooibroek for his assistance in assembling the cryostat and Mr. E.J. Kruithof for his contribution in building the manometer and gas handling system and making preliminary measurements.

REFERENCES

1. H. Preston-Thomas and C.G.M. Kirby, Metrologia 4, 30 (1968).
2. K.H. Berry, Metrologia 15, 89 (1979).
3. R.C. Kemp, L.M. Besley, W.R.G. Kemp, and P.W. Smart, 13e Session du Comité Consultatif de Thermométrie, Sèvres, 1980, Document 69.
4. R.L. Anderson and W. Neubert, Temperature measurements 1975, Inst. of Phys., London and Bristol, Conference Series, number 26, p. 38.
5. J.E. van Dijk, E.J. Kruithof, P.P.M. Steur, H. ter Harmsel, and M. Durieux, 13e Session du Comité Consultatif de Thermométrie, Sèvres, 1980, Document 67.
6. The 1976 Provisional 0.5 K to 30 K Temperature Scale, Metrologia 15, 65 (1979).
7. R.C. Kemp, L.M. Besley, and W.R.G. Kemp, Metrologia 14, 137 (1978).
8. P. Giacomo, Metrologia 17, 69 (1981).
9. H.H. Klein, G. Kempt, and L. Storm, Metrologia 15, 143 (1979).
10. D. Gugan and G.W. Michel, Metrologia 16, 149 (1980).
11. R.D. McCarty, J. Phys. and Chem. Ref. Data 2, 923 (1973).
12. G.K. White, Proc. Thermal Expansion Symp., Lake of the Ozarks, 1973.
13. S.D. Ward and J.P. Compton, Metrologia 15, 31 (1979).
14. R.L. Rusby and C.A. Swenson, Metrologia 16, 73 (1980).

Constant volume gas thermometry from 13.8 to 83.8 K

R. C. Kemp, L. M. Besley, and W. R. G. Kemp

CSIRO Division of Applied Physics, Sydney, Australia 2070

A constant volume gas thermometer for use between 10 and 273 K has been constructed. Features include gas-flow temperature control and uninterrupted operation capability. Results are presented for isotherms measured at 27.1, 43.7, 54.3, and 83.8 K. Gas thermometry data from values agree with NPL-75 below 27.1 K, but deviate from IPTS-68 by up to 9 mK between 27 and 83.8 K. Values for the second virial coefficient of He^4 are given for the range 27 to 83.8 K. A comparison of gas thermometry, isotherm thermometry and earlier magnetic thermometry yields consistent results.

INTRODUCTION

It has been well known for some time that the International Practical Temperature Scale (IPTS)[1] between 13.81 K and 27.102 K differs from, and is not smooth with respect to, thermodynamic temperatures by up to 9 mK[2]. This was recognised by the International Committee of Weights and Measures (CIPM) who, pending the introduction of a new IPTS, introduced a provisional temperature scale EPT-76 from 0.5 K to 30 K which joined the IPTS at 27 K. Kemp et al.[3] further developed this work and using available thermodynamic data in the range 1-83K and IPTS-68 in the range 83-273.15 K constructed a practical temperature scale T_{PtNML} from 13.81 K to 273.15 K. In essence T_{PtNML} was constructed by joining the well established thermodynamic scale NPL-75[2], which has an upper limit at 27 K, to the IPTS-68 at 83 K using the magnetic susceptibility data of Cetas[4]. The differences between T_{PtNML} and IPTS-68 from 13.81 K to 83 K are as large as 12 mK. The validity of T_{PtNML} depends on the accuracies of the magnetic susceptibility data and of IPTS-68 above 83 K, both of which may be questionable. Whatever the merits of T_{PtNML} it does indicate that the data used, namely NPL-75, the magnetic susceptibility data and IPTS-68, are not consistent. In view of this we decided to obtain further thermodynamic data between 27 K and 273.15 K in an attempt to clarify the situation. The method chosen was constant volume gas thermometry and we constructed a suitable gas thermometer capable of operating over the range 10 K to 273 K. In this paper we present preliminary results obtained between 13.8 K and 83 K.

EXPERIMENTAL METHOD

The pressure P, volume V, temperature T and number of moles N of a gas are related by the equation

$$PV = NRT\left(1 + B\left(\frac{N}{V}\right) + C\left(\frac{N}{V}\right)^2 + \ldots\right) \quad (1)$$

where R is the gas constant and B,C are the 2nd and 3rd virial coefficients which depend on T. This equation may be rewritten as

$$PV = NRT\left(1 + B\left(\frac{P}{RT}\right) + (C - B^2)\left(\frac{P}{RT}\right)^2 + \ldots\right) \quad (2)$$

A constant volume gas thermometer (CVGT) may be used either as an interpolating instrument or to measure P-V isotherms. When used as an interpolating instrument equation (2) is used to calculate N at a known reference temperature. If the number of moles is kept constant and the temperature is changed to a new value, a measurement of the pressure enables the new temperature to be calculated. This method relies on a knowledge of the virial coefficients, in practice the 2nd virial coefficient B alone as the effects of the third virial coefficient C are small. There are, as far as we could discover, at least seven published sets of values of B for helium[2,5-10], some measured, some calculated and all of them different. The most recently measured values are those of Berry[2] for the range 2.6 K to 27.1 K and these we regard as the most reliable set. Unfortunately above 27 K the situation is less clear and none of the published values appear to be consistent with those of Berry[2]. The only solution appeared to be to measure values of B ourselves above 27 K using the method of P-V isotherm thermometry which utilizes equation 1. Values of PV/NR are plotted against N/V giving an isotherm which when extrapolated to zero gas density gives the temperature. The slope of the isotherm gives information concerning the virial coefficients. Isotherm thermometry requires the measurement of N, the number of moles of gas. A direct measurement of this is difficult and so we measured relative PV isotherms as described by Berry[2]. This technique relies on the temperature and virial coefficient being known for a particular reference value of T. This enables the number of moles N to be calculated. The temperature is then changed to the desired value, the pressure measured and values of PV/NR and N/V calculated. The procedure is then repeated for different gas densities. For measurements up to 83 K we have used as the reference point the boiling point of hydrogen which Berry[2] determined to be 20.2712 K by measuring an absolute isotherm. For measurements up to 273 K we intend using the triple point of water at 273.16 K as our reference point. Our strategy then is to measure relative isotherms, based on these two reference temperatures, at as many fixed points as possible and to join the results at some intermediate temperature, perhaps 90.188 K, the boiling point of oxygen. This will provide information on the virial coefficients up to 273.15 K so that a CVGT scale can be constructed.

A weakness of such a scale is that at the lower end it is based solely on the absolute isotherm measured by Berry[2] at the boiling point of hydrogen. The error of 0.9 mK given by Berry for this point is multiplied to 3.6 mK for a relative isotherm measured at 80 K. To reduce this uncertainty we intend to correlate the results with noise thermometry measurements. A suitable noise thermometer, designed by C.P. Pickup, is at present under construction in the Division.

APPARATUS

A complete description of the apparatus will appear elsewhere, nevertheless a brief description is useful here. In recent years a number[2,11,12] of constant volume gas thermometers have been constructed for operation at cryogenic temperatures. These have all been of similar design in that a cell, of one litre volume or less, enclosed in a radiation shield and/or vacuum can is suspended in a dewar either above or immersed in a suitable refrigerant. This design is suitable for temperatures close to that of the refrigerant but becomes less so for temperatures further removed. At higher temperatures for example it may take a considerable time for the gas thermometer to reach equilibrium and for measurements to be made. During this time the refrigerant may be exhausted or the conditions of temperature control may change as the level of refrigerant falls. In view of these considerations we decided to design a CVGT capable of operating between 10 K and 273 K and having the capacity for prolonged operation at any temperature in the range.

The construction of the gas thermometer is shown in Fig. 1. The cell [1] has a volume of one litre and is constructed of oxygen-free high conductivity copper. It consists of a copper cylinder about 135 mm long, having an internal diameter of 101.5 mm and a wall thickness of 10 mm. The cylinder is bolted to a copper base and its top opening is closed with a copper lid 10 mm thick which is bolted to the cylinder. Both joints are sealed with indium. The interior surfaces of the cell are gold plated, the gold layer being 0.05 mm thick. The exterior surface of the cell and all other copper surfaces in the apparatus are also plated with a thin layer of gold. A number of copper sleeves [2] for accommodating resistance thermometers are soft soldered to the outside of the cell. The leads from the thermometers are soldered to heat sinks [3] which are thermally bonded to, but electrically insulated from, the cell. The connections from the heat sinks to the outside of the cryostat are of 0.08 mm diameter copper wire insulated with quadruple formel varnish and protected against heat leaks by bonding them to the base of the cell. Approximately 0.5 m of wire are bonded to the cell with about 1 m of wire between this and the next bonding point. The cell is enclosed in a cylindrical radiation shield [4] which is attached to the base of the cell. The temperature of the cell may be electronically controlled to within ± 0.1 mK via either a platinum resistance thermometer (PRT) [9] or a carbon resistance thermometer [7] and a carbon heater [8]. The heat sink for the control system is provided by a gas cooled refrigerator [5] which is loosely coupled to the base of the cell via a stack of stainless steel shims [6]. The temperature of the refrigerator is controlled to within ± 1 mK by regulating the flow of cold gas through it. The cell, shield and refrigerator assembly are enclosed in a second radiation shield [13] whose temperature is electronically controlled via a PRT [14] and a manganin heater wound on the shield. The heat sink for this control system is provided by a second gas cooled refrigerator [11] which is loosely coupled to the base of the shield by a stack of stainless steel shims. The whole assembly is enclosed in a further two heat shields [17] [19] to which are attached gas-cooled refrigerators [16] [18]. The temperature of these refrigerators is controlled as before, using PRTs [12] as sensors.

The cell and the various heat shields are supported on stacks of stainless steel washers [22] so as to form long thermal path lengths between the various parts of the apparatus. The entire assembly is wrapped in 80 layers of aluminized mylar reflective insulating material [20] and enclosed in an aluminium vacuum can [21] which is continuously pumped.

The gas-cooled refrigerators on the cell and the second and third radiation shields are cooled with helium gas supplied through tubes [23] which pass through the bottom of the vacuum can in a vacuum insulated siphon and into a 60 litre liquid helium

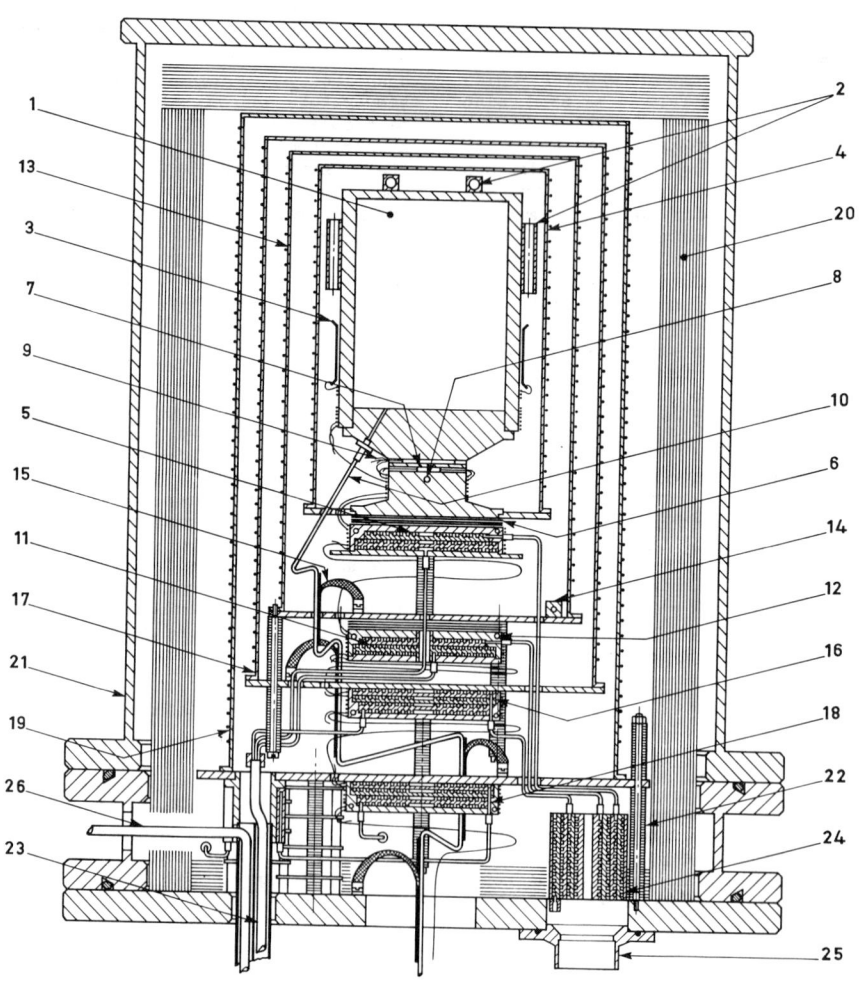

FIG. 1. Schematic diagram of gas thermometer.
Refer to text for explanation of number code.

storage dewar positioned below the CVGT. The helium usage is about 8 litres per day when the cell is maintained at 20 K. The helium dewar may be refilled in situ by means of a vacuum-insulated siphon [26] which is an integral part of the CVGT. The helium siphons are also shielded by radiation shields attached to the fourth gas cooled radiation shield surrounding the gas thermometer. The refrigerator on the fourth shield is cooled with cold nitrogen gas obtained from a 25 litre liquid nitrogen dewar adjacent to the CVGT.

The helium gas, after passing through the three refrigerators, passes through a fourth refrigerator assembly [24] which acts as a final heat sink to which all electrical leads are bonded before passing out of the CVGT through a common exit [25]. The electrical leads are also bonded to each of the radiation shield bases in turn with about 1 m of wire left between bonding points.

The pressure of the gas within the cell is measured via a 2 mm diameter thin-wall cupro-nickel capillary tube [10] which is connected through a demountable joint to the base of the cell. The capillary follows a fairly complicated path out of the CVGT assembly. A vertical section of the capillary is thermally anchored, via copper braid [15] soldered along the length of the vertical section, to the base of the second shield. The capillary is connected to the cell from the top of the vertical section. In operation the second shield is maintained at the same temperature as the cell so that the length of the capillary between the cell and the second shield is at a uniform temperature and no convection currents can occur. Thereafter vertical sections of capillary are thermally anchored to the base of each of the shields, which are maintained at successively higher temperatures, and the base of the vacuum can. Between the vertical sections the capillary rises from the bottom of one vertical section to the top of the next vertical section, which is at a higher temperature, to avoid the possibility of convection occurring in the gas contained in the capillary. This system also has the advantage that the temperature of the capillary over much of its length is controlled.

As already stated the thermometric gas used is pure helium. Initially, commercially available "pure" helium gas was used but was found by mass spectroscopy to contain up to 200 ppm of nitrogen. Eventually, pure helium was obtained by passing the commercially "pure" helium gas through a sintered filter maintained at 4.2 K.

A block diagram of the CVGT and the associated pressure measuring system is shown in Fig. 2. The thermometric gas was separated from the pressure measuring system by a capacitance manometer (Datametrics type 531 barocel) which was used only as a nulling device. The gas pressure was measured using a Ruska DDR 6000 pressure gauge which was continually calibrated with a Bell and Howell type 6-201-001 piston gauge which had been in turn calibrated by this Division's primary standard manometer.

RESULTS AND DISCUSSION

The analysis of results was made on a Hewlett-Packard 9845B desktop computer largely dedicated for use with the CVGT. Several important advantages arose from this facility. Firstly, the analysis of data was available within minutes of a measurement being made, which was of particular use in the proving stages of the experiment. Secondly, a very large program was written which incorporated all available values for virial coefficients and reference temperature scales and permitted the complete calculation of all necessary corrections, avoiding approximations wherever possible.

To check the operation of the CVGT we first used it as an interpolation instrument between 13.81 K and 27.102 K so that our results could be compared with the established CVGT scale NPL-75[2]. The CVGT was operated at a sensitivity of 1.4 kPa K^{-1}. The boiling point of hydrogen at 20.2712 K was used as a reference temperature and Berry's values for B were used[2]. The results are shown in Table 1. The reproducibility of these results was about ± 0.3 mK and the agreement with NPL-75 is within the combined error limits of the two sets of data.

Table 1

Comparison of CVGT Results with NPL-75

Fixed Point	Temperature on NPL-75 K	Temperature for CVGT K	$T_{CVGT}-T_{NPL-75}$ mK
Triple point of equilibrium hydrogen	13.8035	13.8042	0.7
17 K point of equilibrium hydrogen	17.0356	17.0354	− 0.2
boiling point of equilibrium hydrogen	20.2712	reference	−
boiling point of neon	27.0979	27.0968	− 1.1

We next measured relative isotherms at 27 K, 43 K, 54 K, and 83 K. The values of PV/NR and N/V obtained are listed in Tables 2 to 5 together with the isotherm temperatures and values of the second virial coefficient. We emphasise that these results are of a preliminary nature only and some of the isotherms are not complete. Nevertheless we feel that further measurements are unlikely to change the results significantly.

The analysis of the isotherms was carried out in a number of ways. We tried initially to deduce values of the 3rd virial coefficient but eventually concluded that the contribution from this coefficient at the lower gas densities was probably of the same order as the precision of the measurements. We concluded that meaningful values of C could not be obtained from these data and analysed the isotherms in terms of B only. In order to calculate the quantities PV/NR and N/V, values of B are required. We used a trial set of values for B based on the values of Berry[2] up to 27 K and White[9] up to 83 K. New values of B were then derived from the slope of the isotherms and then used to recalculate the isotherm in an iterative process. In fact any changes in B following the first iteration were very small.

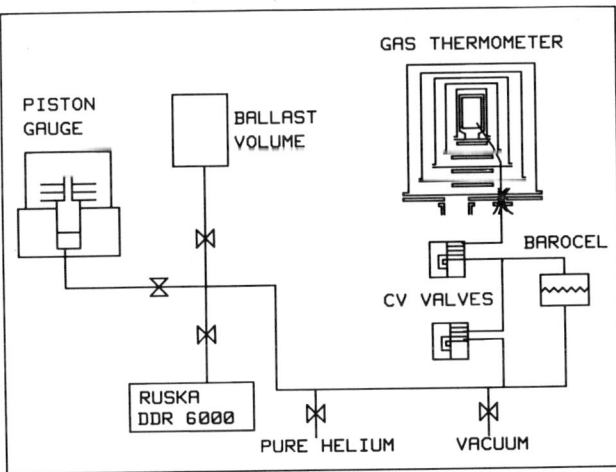

FIG. 2. Schematic diagram of measurement assembly.

Table 2

Results for 27 K Isotherm

P kPa	$\frac{PV}{RN}$	$\frac{PV}{RN}$ Nom	$\frac{N}{V}$	T_{68}
10.7158	27.1262	27.0902	4.7505	27.1282
14.5119	27.1324	27.0911	6.4322	27.1335
21.5766	27.0867	27.0932	9.5803	27.0854
33.9640	27.1102	27.0976	15.0673	27.1048
46.0541	27.1066	27.1011	20.4324	27.0977
56.1213	27.1099	27.1044	24.8958	27.0977
65.4096	27.1071	27.1070	29.0207	27.0923
75.3837	27.1097	27.1097	33.4427	27.0922
85.3566	27.1063	27.1125	37.8717	27.0860
99.3135	27.1106	27.1172	44.0572	27.0856

$B = 2.43 \times 10^{-6}$ m^3/mole

Isotherm Temperature T_{gas} = 27.0870 K
IPTS-68 T_{68} = 27.0922 K
$T_{68} - T_{gas}$ = 5.2 mK

Table 3

Results for 43.8 K Isotherm

P kPa	$\frac{PV}{NR}$	$\frac{PV}{NR}$ Nom	$\frac{N}{V}$	T_{68}
48.3453	43.7740	43.7731	132.827	43.7357
40.8099	43.7664	43.7664	112.143	43.7348
32.9783	43.7684	43.7590	90.6183	43.7443
25.0595	43.7589	43.7520	68.8739	43.7417

$B = 7.59 \times 10^{-6}$ m^3/mole

Isotherm Temperature T_{gas} = 43.7290 K
IPTS-68 T_{68} = 43.7348 K
$T_{68} - T_{gas}$ = 5.8 mK

Table 4

Results for 54 K Isotherm

P kPa	$\frac{PV}{NR}$	$\frac{PV}{NR}$ Nom	$\frac{N}{V}$	T_{68}
68.1281	54.4105	54.4105	15.0588	54.3371
56.0596	54.3978	54.3978	12.3937	54.3371
45.7862	54.4012	54.3856	10.1218	54.3527
36.0634	54.3913	54.3757	7.9742	54.3527
29.0453	54.3827	54.3669	6.4234	54.3529
21.4466	54.3251	54.3586	4.7480	54.3036

$B = 9.29 \times 10^{-6}$ m^3/mole

Isotherm Temperature T_{gas} = 54.3348 K
IPTS-68 T_{68} = 54.3371 K
$T_{68} - T_{gas}$ = 2.3 mK

Table 5

Results for 83.8 K Isotherm

P kPa	$\frac{PV}{NR}$	$\frac{PV}{NR}$ Nom	$\frac{N}{V}$	T_{68}
107.6033	83.9909	83.9908	154.078	83.8373
92.6685	83.9944	83.9688	132.687	83.8628
78.2192	83.9761	83.9505	112.023	83.8628
63.2086	83.9801	83.9301	90.5207	83.8871
48.0300	83.9602	83.9103	68.7999	83.8871
33.0937	83.8911	83.8911	47.4425	83.8372

$B = 11.08 \times 10^{-6}$ m^3/mole

Isotherm Temperature T_{gas} = 83.8463 K
IPTS-68 T_{68} = 83.8372 K
$T_{68} - T_{gas}$ = -9.1 mK

This procedure gave values of B at 27 K, 43 K, 54 K and 83 K. The value at 27 K of 2.43×10^{-6} m^3/mole agreed well with the value of Berry of 2.5×10^{-6} m^3/mole. We combined our values of B with those of Berry[2] below 27 K and derived an expression for B

$$B = T^{-0.25}(b_1 + b_2 T^{-0.5} + b_3 T^{-1}) \qquad (3)$$

where $b_1 = 6.648 \times 10^{-5}$
$b_2 = 2.823 \times 10^{-4}$
and $b_3 = -1.809 \times 10^{-4}$

which is valid from 13.81 K to 83 K. This expression allowed us to use the CVGT as an interpolation instrument and again using the boiling point of hydrogen as a reference point we obtained gas thermometer temperatures at 35 K, 41 K, 58 K, 71 K and 81 K. These results are shown in Table 6. To test the internal consistency of these results we used the data obtained by Cetas[4] up to 83 K on the magnetic susceptibility of manganous ammonium sulphate. The relationship between the temperature T and the magnetic susceptibility χ of certain paramagnetic salts can be described[13] by the relation

$$\chi = A + \frac{B}{T + D + \frac{E}{T}} \qquad (4)$$

where A, B, D and E are constants and may be determined from four pairs of (χ, T) values. To test the CVGT results we fitted the susceptibility data of Cetas to equation (4) at 13.8 K obtained from NPL-75 and 27.1 K,

Table 6

Comparison of Gas Thermometry Temperatures with IPTS-68
(Reference temperature 20.2714 K on NPL-75)

T_{gas}/K	$T_{gas} - T_{68}$/mK
13.8192	− 5.8
17.0824	− 6.6
24.5656	− 5.9
27.0878	− 4.7
34.8556	− 6.2
40.7669	− 6.6
58.5802	+ 1.3
81.5524	+ 8.6

54.5 K and 83 K from the CVGT results. The results of this exercise are shown in Fig. 3 where both the CVGT and magnetic temperatures are plotted. It will be seen that the two sets of data are reasonably consistent. We have as yet made no exhaustive investigation of the uncertainty of the CVGT results. Perhaps the major source of uncertainty is that of the reference temperature at 20 K. This has been given by Berry[2] as 0.9 mK. This translates to proportionately higher errors at the higher temperatures as shown by the error bars in Fig. 3.

CONCLUSIONS

Although the results reported here are of a preliminary nature we can draw tentative conclusions. Firstly, the results between 13.81 K and 27 K are further support for NPL-75. Secondly, the results from 27 K to 83 K give at the very least a strong indication of the direction followed by thermodynamic temperatures in this region. The agreement between the magnetic susceptibility data and the CVGT results is reassuring and indicates that the CVGT is functioning properly.

REFERENCES

1. The International Practical Temperature Scale of 1968 (Amended Edition 1975) Metrologia 12, 7 (1976).
2. K. H. Berry, Metrologia 15, 89 (1979).
3. R. C. Kemp, L. M. Besley and W. R. G. Kemp, Metrologia 14, 137 (1978).
4. T. C. Cetas, Metrologia 12, 27 (1976).
5. W. H. Keesom, Helium (Elsevier 1942) p.38.
6. F. G. Keyes, in "Temperature: Its Measurement and Control in Science and Industry", (Reinhold, New York, 1941) Vol. 1, p. 45
7. J. E. Kilpatrick, W. E. Keller, E. F. Hammel and N. Metropolis, Phys. Rev. 94, 1103 (1954).
8. J. E. Kilpatrick, W. E. Keller and E. F. Hammel, Phys. Rev. 97, 9 (1955).
9. D. White, T. Rubin, P. Camky and H. L. Johnston, J. Phys. Chem. 64, 1607-12, (1960).
10. R. D. McCarty, J. Phys. Chem. Ref. Data 2, 923 (1973).
11. R. L. Anderson and W. Neubert, in "Temperature Measurement 1975". Institute of Physics Conference Series No. 26 (Institute of Physics, London, 1975), p.38.
12. J. E. van Dijk, E. J. Kruithof, P. P. M. Steur, H. ter Harmsel and M. Durieux, CCT 1980 Doc. 67.
13. T. C. Cetas and C. A. Swenson, Metrologia 8, 46 (1972).

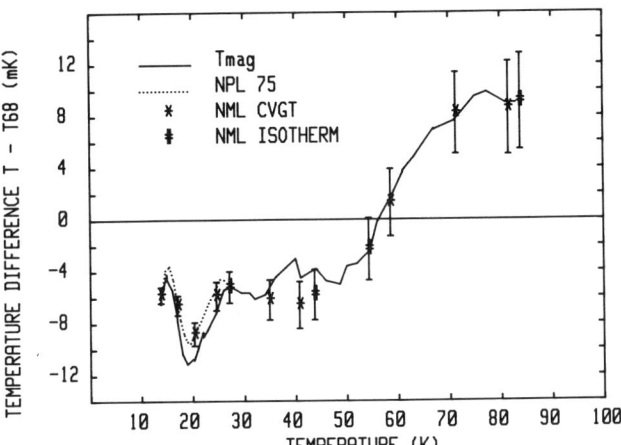

FIG. 3. Comparison of results from this work with those of magnetic susceptibility data of Cetas and gas thermometry data of Berry.

Constant volume gas thermometer for thermodynamic temperature measurements of the triple point of oxygen

H. Sakurai

National Research Laboratory of Metrology, 1-4, 1-Umezono, Sakura-Mura, Ibaraki, Japan

A constant volume gas thermometer is constructed to measure the thermodynamic temperature of the triple point of oxygen. The gas thermometer bulb (about 953 cm^3 in volume) is made of Pyrex glass, which is convenient to keep the surface smooth and clean and to change the volume in the same surface condition. To reduce the ambiguity of the molar distribution of gas in the pressure sensing capillary, a coaxial pipe is adopted, the outer pipe of which is used to obtain the average inverse temperature of the capillary from the pressure measurements. It is more accurate than the conventional thermocouple method. The results of a preliminary experiment are shown to estimate the accuracy.

INTRODUCTION

The International Practical Temperature Scale of 1968 was defined so as to be the closest scale to thermodynamic temperatures using the data mainly obtained by gas thermometry [1,2]. The recent development of thermodynamic temperature measurements, however, indicated discrepancies, especially in the high temperature region [3,4]. In the low temperature region, the technique of magnetic thermometry extended the measuring temperature up to 90 K [5] and it made more accurate comparisons between the IPTS-68 and thermodynamic temperature possible [6]. Those results indicate that the IPTS-68 does not accurately represent thermodynamic temperatures below 90 K. The maximum difference is thought to be nearly 6 mK [6]. On the other hand, thermodynamic temperature measurements of the condensation point of oxygen were done by several laboratories in the past [7] and the assigned value was also supported by the recent result [8]. From those points of view, thermodynamic temperature measurements below 90 K are necessary for the construction of a new accurate temperature standard.

The isotherm gas thermometer constructed by the National Physical Laboratory [9], the measurements of which were finished below 30 K, was one of the most accurate measurements in this temperature region. Berry pointed out that an isotherm gas thermometer is better than a constant volume one because the error of the reference point does not propagate to the measuring temperatures. The numbers of moles of the working gas at the triple point of water and at the relevant temperature or the ratio must, however, be determined within the accuracy needed. To determine these values becomes difficult as temperature rises, so that in the high temperature region a constant volume gas thermometer is superior to an isotherm one.

The dead space corrections are one of the most important problems in gas thermometry. The early gas thermometer of the National Research Laboratory of Metrology [10] was designed so as to put a diaphragm differential manometer in the low temperature region to reduce dead space corrections. In that case, the large temperature dependence of the balance point in the differential manometer was observed and the heat flow by helium gas to balance it made the temperature control difficult.

From those points of view, the constant volume gas thermometer was constructed to measure the thermodynamic temperature below 90 K. The main purposes of the experiments are to reduce the molar corrections in the dead space and the errors from the deformation of a gas bulb.

As a preliminary experiment the difference between the IPTS-68 and the thermodynamic temperature near 54 K was measured by this gas thermometer using the reference point near 90 K.

PRINCIPLE AND METHOD

From the view point of a gas thermometer, the state equation of non-ideal gas is written by

$$n = n_0 + \int \frac{P(x) A(x)}{R T(x) (1 + \Delta)} dx \quad \ldots \ldots (1)$$

where n_0 is the number of moles in the gas bulb. $P(x)$, $A(x)$, and $T(x)$ are the pressure, the cross sectional area and the temperature of a pressure sensing capillary at the position of x, respectively, R is the gas constant and Δ is the non-ideal gas correction, which is a function of pressure and temperature, n_0 can be expressed by $n_0 = P_0 V_0 / R T_0 (1 + \Delta)$ where P_0, V_0, and T_0 are the pressure, the volume, and temperature of the gas bulb, respectively. The integration of Eq. 1 is done from the gas bulb to the differential manometer. Thermodynamic temperature is determined from Eq. 1 by various methods using a reference temperature. In this experiment the conventional method of constant volume gas thermometry is applied except the estimation of the second term in Eq. 1.

In the case of helium as working gas and of a large gas bulb, the non-ideal correction in the integrand of Eq. 1 can be neglected. The position dependency of $P(x)$ consists of two parts, namely, the thermomolecular pressure effect and the hydrostatic pressure effect. The former depends on the temperature distribution, the diameter of the pressure sensing capillary and the pressure of the working gas. The large diameter capillary and relatively high pressure of the working gas are usually adopted to reduce these thermomolecular corrections.

The hydrostatic pressure effect gives a correction of 10 ppm to the integration for about 1 m vertical height at 100 kPa of helium gas in the temperature range above 50 K so that the first order approximation gives enough corrections to a gas thermometer.

Considering a system as shown in Fig. 1, the integration term of Eq. 1 can be rewritten as follows

$$\delta n \equiv \int \frac{P(x) A(x)}{R T(x)} dx = \int \frac{P_0 + P(x)}{R T(x)} A(x) dx$$

$$= \frac{P_0 v}{R T^{av}} \left(1 + \frac{g M h}{2 R T^{av}} \right) + \frac{P_0 v'}{R T^r} \quad \ldots (2)$$

where g, M, h and v are the gravitational acceleration, the molecular weight of the working gas, the capillary height and the capillary volume, respectively. T^{av} is the average inverse temperature defined by

$$1/T^{av} = \frac{1}{h} \int \frac{1}{T(x)} dx \quad \ldots (3)$$

The cross sectional area of the capillary is assumed to be constant in this calculation. The last term of Eq. 2 is the correction of the dead space, v', at room temperature, T^r, which includes the volume of the valves and the differential manometer. The first term of Eq. 2 depends only on the temperature distribution of the pressure sensing capilllary, so that the main part of Eq. 2 can be calculated from the measurement of the average inverse temperature. It is possible to determine this value by 'a gas thermometer' which has a bulb with the temperature distribution which is the same as that of the capillary. It was realized by a coaxial pipe in this experiment. This method is superior to the conventional thermocouple method. It is not always easy to determine the temperature distribution of the capillary precisely by thermocouples because of the poor thermal contact between thermocouples and the capillary and of the relatively high heat leakage through thermocouples.

EXPERIMENTAL APPARATUS

A Pyrex glass gas bulb of about 953 cm³ in volume, which was first used in this experiment, was well annealed to reduce the strain and to obtain a clean smooth surface in order to reduce uncertainties due to gas adsorption and the ambiguity of volume deformation. To measure the adsorption effect the ratio of the inner surface area to the volume must be changed while keeping the same surface condition. The glass bulb is convenient for this purpose compared with a copper bulb, the surface of which is usually mechanically treated. (Two other Pyrex glass bulbs of 700 cm³ and 500 cm³ in volume are being prepared for this experiment.)

The volume of the bulb was obtained from measurements of weight of water filled in the bulb as 953.645 ± 0.005 cm³. This was the standard volume used to estimate the dead space through this experiment. The leakage of the working gas (helium) from the gas bulb to the outside through the Pyrex glass wall was observed at room temperature to be less than $3 \cdot 10^{-8}$ Torr·l/sec which is negligibly small for this experiment. The thickness of the bulb wall is nearly 1 mm from the calculation of the diameter and the volume.

The bulb was surrounded by a gold-plated copper can, to which a standard platinum resistance thermometer calibrated by the fixed points of the IPTS-68 was attached. Helium gas was filled between the bulb and the can to obtain thermal equilibrium and to keep the same pressure as in the bulb.

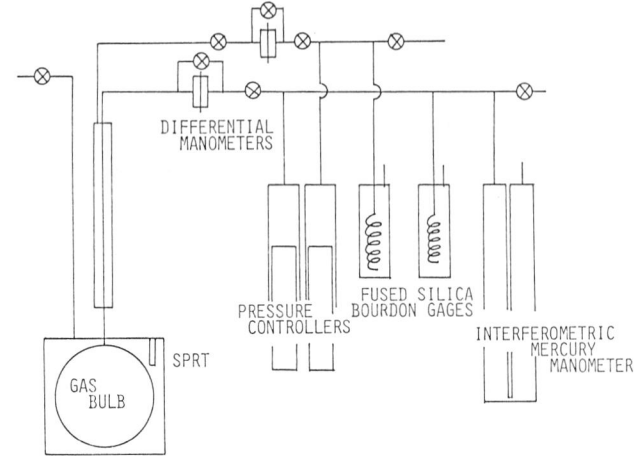

Fig. 1. Gas handling system

The thermal equilibrium can was surrounded by adiabatic temperature controlled copper shields. The conventional bath-type cryostat was used in this experiment.

A coaxial pipe was adopted for the pressure sensing capillary (10 mm in outer diameter) between the gas bulb and a diaphragm differential manometer at the room temperature which was used to separate the working gas from the pressure measuring instruments. About 100 kPa helium gas at room temperature was filled in an external pipe and the average inverse temperature along the inner capillary (0.72 mm in inner diameter, which was experimentally determined so as to get the thermolecularlr effect correction to be less than 0.2 mK) was measured from this gas pressure to correct the molar distribution of the working gas and the hydrostatic pressure effect by Eq. 2. The differential manometer (MKS Inst.) at room temperature was used to reduce the leakage of the helium from this pressure gage, i.e. a fused silica Bourdon gage (Texas Inst.). The gas handling system used for this experiment is shown in Fig. 1.

A differential manometer made of phosphor bronze was designed so as to reduce the dead space (Fig. 2) and the total volume including the valves and their connectors is 0.447 cm³ from the calculation of their dimensions. The displacement of the diaphragm by the pressure difference was detected by a capacitance bridge. The sensitivity is shown in Fig. 3. The stability of the differential manometer at the equilibrium point after and before 50 kPa was better than 0.2 Pa, but it has pressure dependency due to the deformation of the guard rings and the asymmetry of the electrodes, which was designed to reduce the dead space.

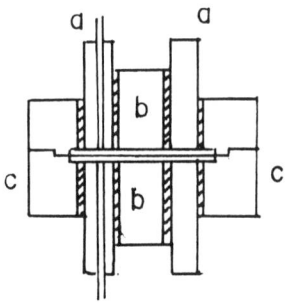

Fig. 2. Differential manometer
a: guard ring, b: electrodes, c: sensing diaphragm

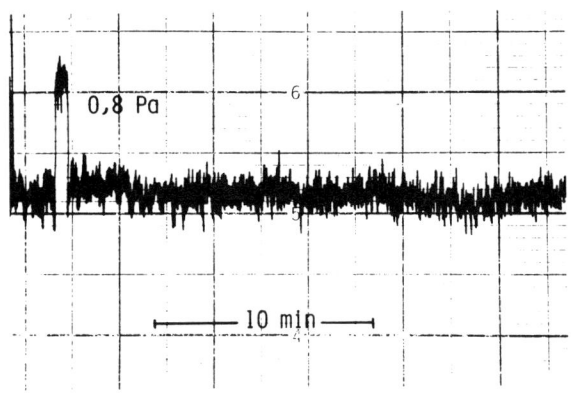

Fig. 3. Sensitivity of differential manometer

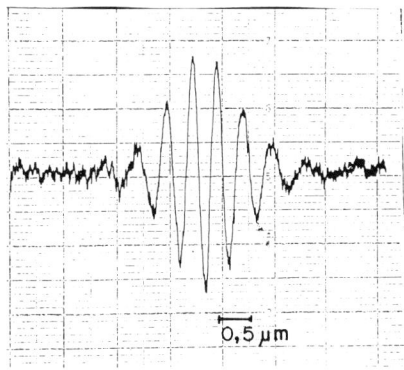

Fig. 4. Output signal of the white light interfreometer

Pressures of two different parts must be measured at the same time in this gas thermometer, the main gas bulb pressure and the pressure to measure the average inverse temperature of the capillary. The latter, mentioned above, was measured using commercially available pressure gages. The former was done by the interferometric mercury manometer, which was constructed for the early gas thermometer [10] below 14 K. It was modified to be measurable up to 100 kPa and to be possible to cover the interferometer with helium gas in order to reduce the optical path corrections. The corrections in the case of helium are 10 times smaller than that of air. The output signal of the white light interferometer, by which the mercury surface position was detected, is shown in Fig. 4. The signal was fluctuating according to the vibration of the mercury surface but the position can be determined within an accuracy of less than 1 μm. For the optical path corrections was used the linear approximation of the refraction index to the gas density. Atmospheric pressure was measured by another rough manometer to calculate the gas density. The maximum corrections, which occur at the maximum measuring pressure, are of order of 4 Pa for helium, so that the error seems to be less than 0,15 Pa at 100 kPa.

ELECTRIC INSTRUMENTS

The electric instruments in this experiment consisted of three parts, the platinum resistance thermometer, the temperature controller and the temperature measurements of the mercury and the dead space. These are conventional instruments.

The capsule type standard platinum resistance thermometer, which was situated at the thermal equilibrium can, was measured by a direct current comparator (Guildline Co. Model 9975). The standard resistor (Yokogawa Electric Works) was put in an oil bath (Guildline Co. Model 9732). The accuracy of the SPRT measurement is better than 0,1 mK using 1 mA measuring currents.

The temperature of the radiation shield was controlled by the adiabatic method, i.e. two pairs of thermocouples (Chromel-Alumel) were set between thermal equilibrium bulb can and the radiation shields to compare the temperature difference. PID controllers were used to maintain the temperature difference between them within 1 mK. The temperature drift measured by the SPRT of the thermal equilibrium can was less than 0,1 mK/h.

The temperature of the mercury in the interferometric manometer was measured by three thermistors (Yellow Spring Inst.), which were calibrated against the IPTS-68. The stabilities of these thermistors were checked by the gallium melting point instrument (YSI) and were well annealed before the calibration. The calibration errors including the stabilities of the thermistors are less than 0,5 mK for two months. The temperature of the mercury may be measured within 2 mK. The temperature distribution of the capillaries near the diaphragm manometers, where the coaxial pipe was not used, was measured by thermocouples. Those were automatically measured by the digital voltmeter (YEW Model 2501) with the scanner (Hewlet Packard Model 9345) and the corrections were calculated by the simple computer (HP Model 9815).

ESTIMATION OF ACCURACY

The accuracy of gas thermometry mainly depends on the pressure measurements, the dead space correction, the adsorption of the working gas to the bulb wall, and the ambiguity of the bulb volume. The estimated errors of this system are listed in Table I. The adsorption of the working gas depends on the surface condition of the gas bulb, the temperature, and the pressure. Above 50 K, in the region of which the van der Waals' force of helium is weak compared with the thermal vibration, the adsorption error seems to be small in the case of clean smooth wall and it was neglected in this table. Other errors, such as the ambiguity of the thermal expansion and the virial coefficients, are estimated from the references [11,12]. Those are to be experimentally determined in the near future.

A PRELIMINARY EXPERIMENT

1. Experimental Procedure

As preliminary experiment to estimate the errors, the thermodynamic temperature near the triple point of oxygen was measured using 90 K as a reference point (near the condensation point of oxygen) and the results were compared with the IPTS-68 measured by the SPRT. The procedure is as follows.

Before the experiment, the gas bulb was being evacuated for a few days. After the residual gas in the thermometer bulb became less than 0,05 Pa, helium gas was filled in the outer tube of the pressure sensing capillary at room temperature measuring the temperature of the capillary. The accuracy of the measurements is about 0,05 K and balancing the diaphragm manometer (MKS) at the zero point, the pressure (about 100 kPa) was measured by the fused silica Bourdon gage. The accuracy of these measurements was better than 40 Pa.

Then, a few Pascals of helium gas were filled in the thermal equilibrium copper can of the gas thermometer bulb, and the cryostat was cooled down to

Table I. SUMMARY OF UNCERTAINTIES

Source	Magnitude (σ)
1. Calibration of SPRT	
at c.p.O_2	0,2 mK
at t.p.O_2	0,1 mK
2. Pressure measurement (100 kPa)	0,5 Pa (0,45 mK)
3. Reproducibility of differential manometer	0,2 Pa (0,18 mK)
4. Dead space corrections	
pressure sensing capillary	0,3 mK
differential manometer (including the hydrostatic effect)	0,15 mK
5. Thermal expansion	7 % (0,14 mK)*
6. Thermomolecular effect	0,2 mK
7. Virial coefficients	4 % (0,2 mK)*
Total (3σ)	2,1 mK

*) The thermal expansion and the virial coefficients are not measured yet. Both are estimated from refs. 11 and 12.

Table II. PRELIMINARY RESULTS

Measuring quantity	Value	error (σ)
1. Reference temperature	90,279 56 K	0,2 mK
Initial pressure	71,118 22 kPa*	0,9 Pa
Average inverse temperature	211,304 K	50 mK
2. Final pressure	41,644 81 kPa*	0,6 Pa
Average inverse temperature	111,878 K	50 mK
SPRT scale	58,915 00 K	0,2 mK
Result (gas therm. scale)	58,905 2 K	1,5 mK**

*) In this experiment the interferometer of the mercury manometer was filled with air.
**) The errors of the thermal expansion and the virial coefficients are added to this value.

77 K by liquid nitrogen. Pure helium gas (commercially available 7 N) was then admitted through nitrogen trap into the gas bulb. The temperature was measured by the SPRT and the bulb temerature was first controlled near 90 K. The pressure of the bulb was measured by the interferometric mercury manometer and the fused silica Bourdon gage controlling the pressure so as to equilibriate the differential manometer. At the same time the pressure of the outer tube in the pressure sensing capillary was measured using another fused silica Bourdon gage. Then the bulb was cooled down to near 50 K. The same as 90 K was done. In this preliminary experiment, the interferometer of the mercury manometer was filled with air instead of helium.

2. Results and Discussion

As a preliminary experiment, the thermodynamic temperature was measured as mentioned above. The results and the measuring conditions are shown in Table II. The difference between IPTS-68 and the result is 10,2 mK at 58 K, which is larger than the value obtained by Kemp et al. [4.6]. The estimated errors are ± 4,5 mK (3 σ). These errors occurred in the pressure measurements because the interferometer of the mercury manometer was filled with air in this preliminary experiment. They will be reduced using helium instead of air as mentioned in the apparatus section.

CONCLUSION

The constant volume gas thermometer reported here was used to measure the triple point of oxygen. From the preliminary experiments, we believe it will be possible to determine the fixed point with the accuracy of order 1 mK after improving the optical path errors of the mercury mamometer. The adsorption effect of the working gas will also be estimated by changing the volume of the gas thermometer bulb in the same surface condition. The thermal expansion coefficient of the bulb is of order 10^{-6} K^{-1} near 90 K so this quantity needs only be known to 10 % to determine the triple point of oxygen within 1 mK. Those are left as future problems.

It is a pleasure to acknowledge the encouragement received from Dr. M. Morimura of the National Research Laboratory of Metrology.

REFERENCES

1. The International Practical Temperature Scale of 1968 Metrologia 5 (1969) 35
2. The International Practical Temperature Scale of 1968 Amended Edition of 1975: Metrologia 12 (1979) 7
3. Guildner L.A., Edsinger R.E.: NBS J. Res. 80A (1976 703)
4. Cetas T.C.: Metrologia 12 (1976) 27
5. Crovini L., Actics A.: Metrologia 14 (1978) 69
6. Kemp R.C., Besley L.M., Kemp W.R.C.: Metrologia 14 (1978) 137
7. Preston-Thomas H., Kirby C.G.M.: Metrologia 4 (1968) 30
8. Pickup C.P.: Metrologia 11 (1975) 151
9. Berry K.H.: Metrologia 15 (1979) 89
10. Mitsui K, Sakurai H, Mochizuki T: Temperature Vol.4 (Instrument Society of America 1972) 333
11. Keesom W.H.: Helium (Elsevier Pub. Co. 1942)
12. Touloukian Y.S.: Thermophysical Properties of Matter Vol.13 (Plenum Pub. Co. 1978)

Progress in NBS gas thermometry above 500 °C

Leslie A. Guildner and Robert E. Edsinger

National Bureau of Standards, Washington, D.C. 20234

Measurement of thermodynamic temperatures by gas thermometry above 500 °C requires the use of different apparatus than for lower temperatures. Extensive tests have been made of the thermostats and high temperature platinum resistance thermometers. Techniques to reduce the presence of hydrogen, caused by diffusion through the metals, have been established. After more than one thousand hours of operation at 962 °C, the thermostat was redesigned and rebuilt to improve its performance. Thermal expansion measurements of the bulb material by an interferometric technique are being carried out in a second thermostat designed to operate up to the gold point. Gas thermometer measurements have been made at low pressures over a considerable range of temperature, but especially at 660 °C to evaluate the attainable precision. Finally, a new gas thermometer has been built with a modified suspension system to improve temperature uniformity of the bulb. The measurement at an upper temperature limit of 1337 K (the freezing point of gold) remains the final goal of this project.

INTRODUCTION

The papers presented by us on Gas Thermometry at the Fourth, Fifth and now Sixth Temperature Symposia give a historical perspective of the efforts at the National Bureau of Standards to measure thermodynamic temperatures between 0 and 1064 °C by constant volume gas thermometry. The results are presented in terms of the differences from corresponding values on the IPTS-68. The first paper dealing with this program was presented at the Fourth Symposium [1]. While many of the ideas expressed there have survived, many have been changed, either basically or in detail. The first results at the steam point were presented in a paper delivered at the Fifth Symposium [2]. By the time of that Symposium most of the equipment and procedural changes had been made that were necessary to make the measurements up to 457 °C. Those results have been published [3], along with other papers that have given details of parts of the apparatus [4-11].

A later paper presented a discussion about optimizing the accuracy of the results by the choice of the pressure range of the measurements [12]. Based on the nearly constant relative uncertainty of the manometer over the range from 3.4 kPa to 130 kPa, one can expect a minimum uncertainty in the sum of the contributions from thermomolecular pressure, which is large at low pressures but decreases as p^{-2}, and gas imperfection, which is small at low pressures but is proportional to p. Inasmuch as the thermomolecular pressure can be measured with a total uncertainty of about 0.1%, the effect on thermodynamic temperatures up to the gold point (1337 K), can be expected not to exceed 0.5 ppm for a fiducial pressure of 3.37 kPa. The effect of gas imperfection is calculated from the difference $(B_2-B_1)/R$, where B_2 and B_1 are the density virial coefficients of the thermometric fluid at temperatures T_2 and T_1, and R is the molar gas constant. The pressures for 900 K and 1300 K corresponding to a fiducial pressure of 3.37 kPa are 11 kPa and 16 kPa, with differences of $(B_2-B_1)/R$ of 2.8×10^{-7} kPa^{-1} and 4.0×10^{-7} kPa^{-1} respectively. Thus the calculated effect of gas imperfection is 3.0 mK at 900 K and 6.4 mK at 1300 K for pressures consistent with a 3.37 kPa fiducial pressure, with an uncertainty at 99% confidence estimated at 10% at the lower temperature, and approaching 15% at the higher temperature. On balance, as low a pressure range as is consistent with high accuracy for the pressure ratio should be used.

In addition, given our ultimate purpose of measuring the thermodynamic temperature of the gold point, a new thermostat had to be used. To achieve a total uncertainty that is < 10 ppm, a greatly improved practical scale above 630 °C is also mandatory.

Our efforts have been directed toward the realization of these objectives. This paper discusses our findings and the changes made in the equipment and procedures to improve its operation.

HIGH TEMPERATURE THERMOSTAT

The thermostat is basically a furnace with more than the usual numbers of zone heaters and interior shields. It has a low composite thermal diffusivity, with a relaxation time of several hours. Attempts to probe the furnace for temperature uniformity, and to adjust the heater settings are very time consuming. The temperatures were measured by four high temperature platinum resistance thermometers. Moving a thermometer vertically had the effect of upsetting the temperature significantly, so that when combined with slow drifts in the overall temperature of the thermostat, an uncertainty in the temperature distribution amounting to 10 or 20 mK remained. The heat flows that occurred in the metal parts supporting the gas thermometer and the suspension structure were large enough to introduce angular assymetry in the temperature distribution. This same conduction made the temperature of the thermometer bulb responsive to the external cooling water used at the top, so that even the water had to be thermostated independently.

On an even more basic level, it is difficult to center the furnace with respect to the axis of its container sufficiently well to avoid significant temperature non-uniformity. The insulation in the annulus between the heater and the outer wall of the thermostat is 20 cm thick, so that the temperature drop is about 30 °C/cm. If the furnace were mechanically perfect, and remained so, there would still be a large effect for 1 mm misalignment.

On the other hand, the thermostat provided a high level of temperature stability that could be regulated within 1 mK over a period of time. This feature allowed evaluation of the experimental imprecision of the gas thermometer results when it was operated at low pressures.

Because we had become aware of the problem of hydrogen diffusion [3,11], the furnace was constructed to be gas tight except at the top. Two streams of purified argon were introduced: a distributed stream at the bottom of the insulated annulus, and a second, separately controlled stream into the bottom of the interior of the central portion. Rather prolonged periods of heating and purging were required before the hydrogen concentration that could be detected inside the gas thermometer by a residual gas analyzer reached a constant, minimum value. It proved to be possible to reduce the observed hydrogen concentration still further by elimination of all organic material, including the rubber parts of the pressure regulator, in the system conveying the argon to the furnace.

All the heated parts of the furnace were made of Inconel alloys, mostly Inconel 600. Despite the maintenance of a protective atmosphere, considerable oxidation occurred on the surfaces of all the exposed hot parts. Released by differential contraction upon cooldown, the oxide layer spalled off at room temperature and flew loose with considerable vigor. The original windings were not protected from the oxide, which is evidently conducting when hot. When the furnace was rebuilt, it was an important consideration to protect the windings from contamination. The arrangement used, shown in Fig. 1, combined several desirable features. The windings, which originally progressed around the furnace in a nearly horizontal helix, were made vertical, with single long insulators (A) interleaved by short vertical windings on each end (B) as band heaters.

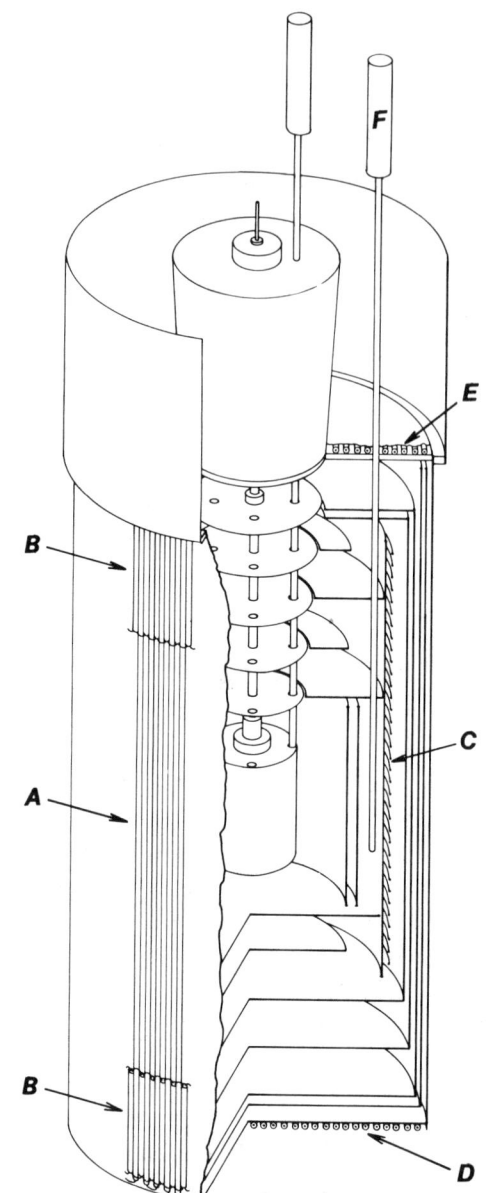

FIG. 1. Gas Thermometer Thermostat for temperatures in the range 450 °C to 1064 °C.

This arrangement exposed much less of the wire for oxide to penetrate than the fish spine insulators originally used, and the windings were also oriented so that the oxide particles would tend to fall past them to the bottom. As a further precaution, the heaters were protected from direct contact with the furnace by a thin Fiberfrax blanket. Of great value in the arrangement is the fact that the main heater can be controlled by quadrants (or any other useful subdivision). This allows compensation for off-centering of the furnace.

The furnace has a regulating heater (C) that distributes a small energy flux evenly over a large cylindrical region of the interior. The arrangement allows one to make small changes of the overall temperature as sensed by the regulating thermometer (F) or to compensate for other changes, but only under extreme conditions would it appreciably affect the relative temperature distribution. Overall, the temperature distribution depends on the adjustment of the various heaters, viz. a bottom heater (D), the main heater subdivided into vertical quadrants, and two circular band heaters, one at the top and one at the bottom, two heaters at the top, one on the furnace proper (E) and the other on the gas thermometer suspension, and two "tempering" heaters, one for the thermometers and one on the suspending tube of the gas thermometer. The power inputs at 660 °C and 960 °C were 730 w and 1500 w respectively with 1/2 to 1% of that total contributed by the regulating heater. It is evident that control within a mK involves exceedingly precise adjustment and subsequent stability of the various voltages, with the regulating heater accounting for the imperfections of the arrangement. We had to use a line conditioner of as high quality as possible, and the control circuits had to be carefully designed to reduce the imprecision typical of variable adjustments.

THERMAL EXPANSION MEASUREMENTS

The thermal expansion of the bulb material is determined by the change of interference fringes with temperature when a sample is used to separate the plates of a Fizeau interferometer. The thermal expansion thermostat was made as a furnace in order to avoid vibrations that could affect the interferometer. It has to cover the same temperature range as the gas thermometer thermostat. The two furnaces differ in that the entire expansion thermostat is vacuum tight and kept sealed; it is smaller and more massive; and the only hindrance to heat flow consists of gold and silver shields. The understanding gained by operating either of these furnaces contributes to techniques useful for operating the other one. A new interferometer furnace was assembled for the high temperature region with Inconel parts in place of the copper pieces originally used. The new furnace has a heating collar extending from the furnace shell into the cool region in order to assure adequate immersion of the platinum resistance thermometer. It was determined that the variations of temperature measured by the platinum resistance thermometer were real, because the resistance of the thermometer remained the same whether the heating collar was energized or not. Gradients in the sample chamber can be almost entirely eliminated by careful adjustment of the various heaters. Even though the main parts of the thermal expansion furnace are made of Inconel 600, they are not subject to appreciable oxidation because a carefully controlled inert environment of either helium or argon gas at a pressure of about 130 Pa is maintained in the system.

Measurements with a sample from the sheet used to make Bulb IV, of 80% platinum-20% rhodium alloy, have been made to 710 °C. The temperature is deduced from a quadratic scale, $W = 1 + At' + Bt'^2$, with a special thermometer built by us especially for the thermal expansion furnace. It has an alpha of 3.922×10^{-3} and is adequately stable for exact temperature measurements.

THERMOMOLECULAR PRESSURE

The thermomolecular pressure calculated from Eq. (10) of Ref. 3 at pressures consistent with a fiducial pressure of 3.37 kPa are as follows:

Table I

THERMOMOLECULAR PRESSURE ADJUSTMENTS

t (°C)	p (kPa)	p (Pa)	$\Delta p/p$ (ppm)
0	3.37	- .19	55
660.	11.3	4.27	378
780	13.0	5.22	402
962	14.8	6.41	433

The equipment for measuring thermomolecular pressure incorporates the Pt-10% Rh connecting tube of Bulb III. Thus, its diameter, has been accurately measured. The device can be lowered into the thermostat to allow it to be cleaned by vacuum bakeout. Preliminary tests have given adequate precision to allow a total uncertainty not exceeding 0.1 or 0.2%. A very simple equation can be written to describe the thermomolecular pressure, Δp, in the form

$$p\Delta p = C(T,T_1)$$

and the value of $C(T,T_1)$ can be calculated from

$$C(T,T_1) = C_0\left[\left(\frac{T}{T_0}\right)^{2n+2} - \left(\frac{T_1}{T_0}\right)^{2n+2}\right]$$

where T is the measuring temperature, T_1 is the ambient temperature of the apparatus, and $T_0 = 273.15$ K is the reference temperature used for determining η such that the viscosity of the gas can be expressed as

$$\frac{\eta}{\eta_0} = \left(\frac{T}{T_0}\right)^{n+\frac{1}{2}}$$

These relatively simple equations may need some refinement for 0.1% error, but the essentials of the phenomenon are expressed by them. It is clear that interpolation and even some extrapolation is reliable when the behavior is so well characterized.

HIGH TEMPERATURE PLATINUM RESISTANCE THERMOMETERS

Because of the need for reproducible "practical" thermometers on which to record the results of our gas thermometry and to transfer them to other laboratories, we have cooperated in the investigations of high temperature platinum resistance thermometers by Evans [13]. Numerous thermometers produced over a four year period have been evaluated in use as working instruments for gas thermometry measurements, and also by testing them for durability and stability following 1100 °C anneals for periods totaling as much as 1000 h. We are also producing new thermometers with a bifilar helix, of 2.5 Ω resistance at 0 °C, similar to the later design made by Evans.

High temperature platinum resistance thermometers typically are susceptible to failure from mechanical effects, progressive contamination of the sensor, or both. The design of the thermometer, shown in Fig. 2, minimizes the possibilities of mechanical disruption by the use of long capillaries that serve two functions: (1) They confine the leads so they cannot make contact with one another, and (2) They support the sensor in a position fixed with respect to the case. The capillaries are fixed in position with respect to the case by the arrangement in the header. As shown in Fig. 2a, the capillaries are attached to a fused silica ring (A), which is held in position because the guard lead (B) is inserted into a capillary (C) that provides a rigid brace across the intervening space. The sensor, shown in Fig. 2b, is suspended from the bottom of the capillaries by the leads which are immobilized by beads on either side of the constriction (D). The differential expansion of the leads is accommodated in the header by individual springs (E). The long capillaries also permit the use of an effective guard to reduce the effect of low electrical resistance of the silica [14]. The header allows the assembly of the thermometer with little exposed organic material, confined to the coolest part of the thermometer, while at the same time providing for comparatively easy reprocessing because it can be disassembled at the seal (F).

Our efforts to reduce contamination include the use of reference grade platinum for all the metal parts, further development of equipment and pickling procedures for cleaning the parts, and construction of equipment for vacuum bakeout that allows deep immersion of the thermometer with monitoring of the process by a residual gas analyzer.

Evans makes the point that within the present level of accuracy the differences between platinum sensors can be accounted for by a quadratic interpolating equation, viz. $W = 1 + At' + Bt'^2$, where W is $R(t)/R(0°C)$ [13]. It is important to recognize that this equation has been demonstrated to be valid at the aluminum point

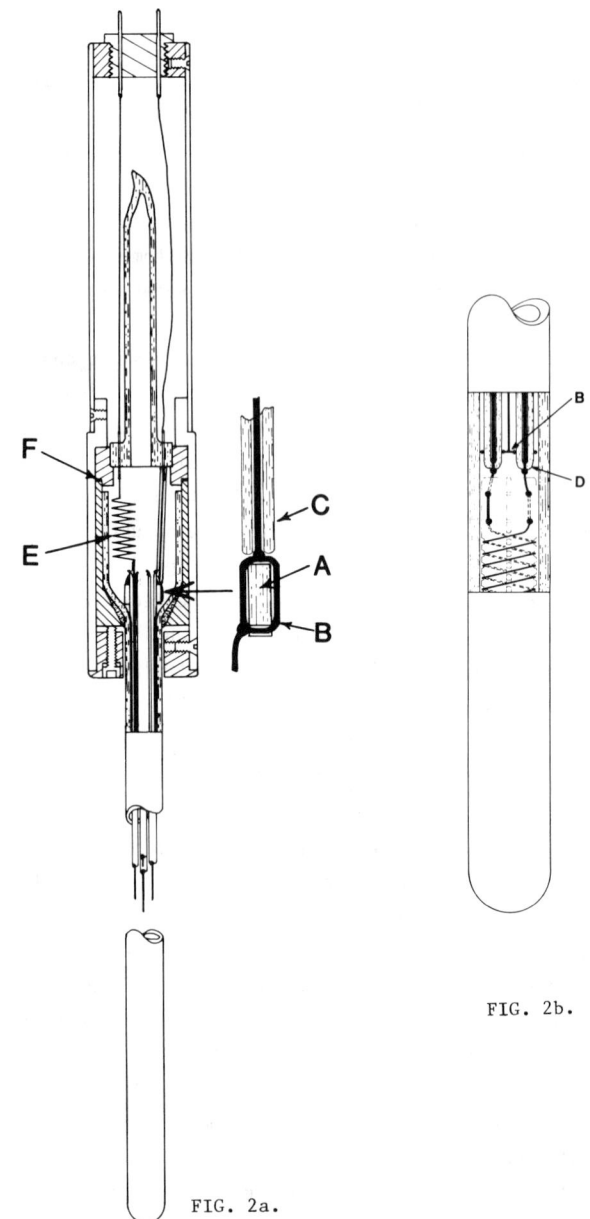

FIG. 2a.

FIG. 2b.

FIG. 2. High Temperature Platinum Resistance Thermometer.

for high temperature platinum resistance thermometers having A's ranging from 3.974×10^{-3} °C^{-1} to 3.985×10^{-3} °C^{-1}, a fact that requires B to be a function of A. The constants found by least squares for a linear equation, with data from 64 calibrations of 31 thermometers, gave

$$B = -1.5985499 \times 10^{-7} - 1.073269 \times 10^{-4} A$$

The standard deviation of B was 3.8×10^{-11} °C^{-2}, which indicates a high level of reliability in t' ($\sigma = 0.1$ mK at 100 °C and 2.5 mK at 500 °C), even with a considerably contaminated sensor.

The value of A is depressed by contamination which, if it occurs, tends to be progressive, particularly at higher temperatures. Therefore, the calibration of the thermometer tends to be unstable. On the other hand, some contaminated thermometers have become stable. The values of t' for these thermometers are not susceptible to added error because of the contamination.

GAS THERMOMETER PRECISION

The operation of the gas thermometer is more difficult above 500 °C than below because of the features discussed above. As a practical matter, it is more effective to fill the gas thermometer bulb at the upper temperature of a measurement sequence, and cool down slowly to the intermediate measuring points while maintaining the counterpressure outside the bulb nearly in balance with the pressure inside. In order to maintain cleanliness, the gas thermometer was thermostated in an ice bath for the fiducial measurement. With a fixed fiducial temperature, it became necessary to make a very careful adjustment of the initial temperature and loading pressure, so that a fiducial pressure in very close correspondence with 25.4 mm of Hg would result. (Small differences in pressure can be accounted for accurately, but large differences from the exact 1" gage block setting of the manometer involve excessive error, because they must be accounted for in our instrument either by changing the gage blocks to a composite stack, or by measuring the difference from the 1" pressure setting by the diaphragm pressure transducer.)

It was only after considerable experience that we were able to determine a combination of factors to give reliable measurements. When the gas thermometer was held for about a week at a temperature close to that of the aluminum point, and for the following week in an ice bath at a temperature of about -0.005 °C, the precision of the measurements and their susceptibility to drift from the effects of sorption could be evaluated. The five values at 660.44 °C had a relative standard deviation of the mean of 0.3 ppm at a pressure of 11.5 kPa, and five values at -0.005 °C had a relative standard deviation of the mean of 0.6 ppm at a pressure of 3.37 kPa. It is clear, therefore, that the results are not imperiled by a loss of precision in this lower pressure range, although they are subject to greater relative effects from absolute systematic errors. Measurements over pressure ranges for twice and four times the lowest fiducial pressure would be worthwhile.

Even with these data, asymmetry of the temperature distribution was present, with a maximum difference in thermometer readings of about 0.05 °C. At 961.9 °C, this difference increased to about 0.07 °C, although the results were less reliable because of increased thermometer instability. We could not eliminate the

FIG. 3. Gas Thermometer Assembly (Bulb V).

asymmetry, and ultimately concluded that a part of the problem arose from asymmetric heat loss through the suspension system of the gas thermometer.

CONSTRUCTION OF NEW GAS THERMOMETER BULB

A heavy structure was required to support the gas bulb and bulb case as originally designed, because of its large mass. We calculated that the bulb case could be cut down enough that it could be suspended by a centro-symmetric tube of reasonable thickness with no undue creep (< 0.01%) occurring over a period of a thousand hours at the gold point. Accordingly, a new bulb assembly was fabricated, as shown in Fig. 3. This assembly (Bulb V), which has not been installed as yet, incorporates the better features of prior assemblies, viz:

(1) A header which allows for the differential motion of the Pt-10% Rh connecting tube with respect to the Inconel suspending tube.

(2) Platinum leads welded every 25.4 mm along the tube to create a set of 20 Type S thermocouples for measuring the temperature distribution in the "dead space". The thermocouples were annealed by baking them at 1450 °C for 1 hour in air.

(3) A reference couple (Type S), running from the header to an ice bath, that may be combined electrically with each of the 20 couples to give emf's referenced to 0 °C.

(4) A counterpressure system, with the pressure applied from the header, thus maintaining a protective environment around the thermocouples, as well as a balanced pressure around the gas thermometer bulb.

The assembly also includes five shields that are a part of the tempering arrangement for the platinum resistance thermometers.

SUMMARY

The NBS gas thermometer and its peripheral equipment have been operated at various temperatures ranging up to 962 °C. Tests of the behavior of the thermostat, for temperature stability and uniformity, have led to modification of its design and also its control circuitry. Tests of the gas thermometer for measurements of thermodynamic temperature at 660 °C demonstrated that the results are precise at levels within 1 ppm at both 660 °C and 0 °C, for a fiducial pressure of 3.37 kPa.

The construction and testing of high temperature platinum resistance thermometers is being actively pursued to achieve a stable and precise practical scale on which to record the gas thermometer results. Measurements for other parameters, such as the thermal expansion and thermomolecular pressure, are under way. A new gas thermometer assembly has been constructed to measure thermodynamic temperatures in the range $450 < t < 1064$ °C at about 50 °C intervals.

REFERENCES

1. L.A. Guildner, TMCSI 3, Part I, Ch. 18 (Reinhold Publishing Corp., N.Y. 1962).
2. L.A. Guildner, R.L. Anderson, and R.E. Edsinger, TMCSI 4, Part I, Ch. 29 (Inst. Soc. Am. Pittsburgh 1972).
3. L.A. Guildner and R.E. Edsinger, Jour. Res. Nat. Bur. Stds. 80A, 703 (1976).
4. L.A. Guildner and H.F. Stimson, Rev. Sci. Instr. 34, 658 (1963).
5. L.A. Guildner and R.E. Edsinger, Jour. Res. Nat. Bur. Stds. 69C, 13 (1965).
6. R.L. Anderson and L.A. Guildner, Rev. Sci. Instr. 36, 615 (1965).
7. L.A. Guildner, H.F. Stimson, R.E. Edsinger, and R.L. Anderson, Metrologia 6, 1 (1970).
8. R.L. Anderson, L.A. Guildner, and R.E. Edsinger, Rev. Sci. Instr. 41, 1076 (1970).
9. R.E. Edsinger, L.A. Guildner, and R.L. Anderson, Rev. Sci. Instr. 42, 7 (1971).
10. L.A. Guildner, R.L. Anderson, and R.E. Edsinger, Jour. Res. Nat. Bur. Stds. 77A, 667 (1973).
11. L.A. Guildner and R.E. Edsinger, Jour. Res. Nat. Bur. Stds. 77A, 383 (1973).
12. L.A. Guildner, PTB-Mitteilungen 90, 41 (1980).
13. J.P. Evans, this Symposium.
14. R.D. Cutkosky, IEEE Trans. Inst. and Meas. IM-30 217 (1981).

Dielectric Constant Gas Thermometry (DCGT): A new method of accurate thermodynamic thermometry

D. Gugan

H. H. Wills Physics Laboratory, Royal Fort, Bristol, England

Thermodynamic temperature can be established by the well-proven procedure of isotherm measurements of the virial equation of state of a gas. Whilst simple in principle there is a difficulty in practice since it depends on a ratio of *extensive* parameters, i.e., the volume per mole, and thus on the distribution of gas throughout the *whole* of the apparatus. Recent advances in capacitance measurement allow this problem to be avoided since the local gas-density in the test environment can be obtained from a measurement of the dielectric constant of the gas, i.e., by a knowledge of an *intensive* parameter. This paper gives a resumé of the technique, some results of a detailed trial of the method, and some discussion of possible future developments including its potential for absolute thermometry.

1. INTRODUCTION

Dielectric Constant Gas Thermometry (DCGT) was originally developed in order to shed more light on IPTS-68 between 4.2 and 27.1 K, and in particular to allow an independent assessment of the RhFe resistance thermometer scale NPL-75 which Berry established from his conventional isotherm and Constant Volume Gas Thermometry (CVGT) measurements on ^{4}He[1]. A detailed discussion of the practice of DCGT has been given by Gugan and Michel[2], and the results confirm that NPL-75 is indeed an extremely accurate scale. The further examination of both the DCGT data and Berry's data by the method of surface-fitting given elsewhere in these proceedings[3] shows that a proper treatment of the data removes the one or two small questions that were left by the initial analysis, and one is led to the conclusion that DCGT is not only a simple, convenient and precise technique of gas thermometry, but also that it has potential as a true primary thermometer. This paper contains a general discussion of these points: it does not include a detailed description of DCGT, but it contains a brief summary of the principles for those to whom this new technique may not be familiar.

2. THEORY

2.1. The basic idea of DCGT is to combine the ideal-gas equation of state with the Clausius-Mossotti equation so as to eliminate the molar volume, V. In the simplest approximation we have

$$pV = RT, \text{ and } (\varepsilon-1)/(\varepsilon+2) = A_\varepsilon/V, \quad (1)$$

where ε is the dielectric constant of the gas at pressure p and temperature T, and where A_ε is the molar polarizability. Elimination of V gives

$$p = T(R/A_\varepsilon)\left[(\varepsilon-1)/(\varepsilon+2)\right]. \quad (2)$$

Since $(\varepsilon-1)/(\varepsilon+2)$ is an intensive parameter which can be measured under precisely controlled and uniform conditions, (2) shows that the DCGT technique avoids the great problems in conventional gas thermometry of gas contained in dead spaces not at temperature T, and of gas adsorbed on the walls of the apparatus.

An apparently important distinction between CVGT and DCGT is that while many gases are reasonably close to ideal, with small non-idealities which can be allowed for by writing the equation of state in the virial form (known to be exact in the limit of low density)

$$pV = (1 + B/V + C/V^2 + \ldots), \quad (3)$$

the polarizability A_ε, which is the essence of DCGT, arises from the finite size of the molecules (see e.g.[4]), i.e. from something which is a mere correction term in CVGT; moreover, different gases do not tend to the same limit of dielectric behaviour (cf. the ideal gas law) since they have different molecular structure. However, as we shall see, from a practical point of view the essential criterion for a useful system of thermometry is that the ratio $(\varepsilon-1)/(\varepsilon+2)$ of (2) can be measured accurately, which is true, and which certainly places DCGT on an equivalent footing with CVGT. It also proves to be true that for helium at least, the absolute value of A_ε is known to about 1 p.p.m. (see below) from very well established quantum-mechanical calculations, so that DCGT has effectively the same accuracy for establishing temperature as does conventional isotherm thermometry.

The effect of molecular interactions on the Clausius-Mossotti equation has been considered in detail by Hill, who applied the methods of statistical mechanics to an assembly of molecules in an electric field[5]. The effect of interactions can be expressed by a dielectric virial expansion

$$(\varepsilon-1)/(\varepsilon+2) = (A_\varepsilon/V)(1+b/V+c/V^2+\ldots) \quad (4)$$

which has the same range of validity as the ordinary virial expansion. The dielectric virial coefficients, b, c, etc., arise from the same pair, triplet interactions as give rise to the ordinary virial coefficients, B,C, etc., but they are about two orders of magnitude smaller: there is also a second-order effect of the electric field on B and C etc., but this proves to be several orders of magnitude smaller than could be detected in any practicable experiment on DCGT. The dielectric virial coefficients have a temperature dependence which theory suggests should be relatively about the same as for the ordinary virial coefficients, but there is no direct evidence on this point apart from what is known from DCGT about b(T) for ^{4}He[2]. A final point on (4) is that A_ε must itself ultimately have some temperature dependence because of thermal population of those excited states which also determine the polarizability: however, since these energy levels are typically about 20 eV above the ground state this effect is completely negligible at temperatures of interest here, say below 400 K.

Equations (3) and (4) could be combined to eliminate V, giving a series expansion analogous to (2). However, the dielectric constant is not directly available from experiment: what one measures in practice is the capacitance ratio for the capacitor with and without gas, the geometry of the electric field lines being the same in both cases. Since the capacitor is exposed to the pressure p in the first case, while p = 0 in the second, there is also an inevitable change of dimensions of the capacitor. The effective compressibility of the capacitor, K, depends on the detailed design: since it is necessarily of composite construction (e.g. conductor and insulator, at the very least) the compressibility is not the property of a single material, but, as we see later, by suitable design K can be made closely equal to the linear compressibility (i.e. one third of the bulk compressibility) of the major structural material of the capacitor. Taking this compressibility effect into account, the expanded form of (2) allowing for gas imperfections proves to be[2]

$$p = A_1\mu(1 + A_2\mu + A_3\mu^2 + \ldots), \quad (5)$$

where the expansion parameter, μ, which differs from $(\varepsilon-1)/(\varepsilon+2)$ only by small terms involving K, is directly available from experiment, and where the coefficients of (5) are given by

$$A_1 = (A_\varepsilon/RT + K/3)^{-1}, \quad (6)$$

$$A_2 = B'/A_\varepsilon, \text{ where } B' = B - b, \text{ and} \quad (7)$$

$$A_3 = (C - 2Bb + 2b^2 - c)/A_\varepsilon^2. \quad (8)$$

The contribution of b to (7) is small compared with that from B, but not negligible: however, the contributions of the dielectric virial coefficients to the terms in A_3 and above are negligible[2] so that we have

$$A_3 = C/A_\varepsilon^2, \quad (A_4 = D/A_\varepsilon^3 \text{ etc.}). \quad (9)$$

2.2. Analysis of DCGT isotherms at separate temperatures using (5) leads to values for the virial coefficients B', C, D etc., and to estimates of thermodynamic temperature via (6), in which the terms A_ε and K necessarily also occur. This causes some (soluble) problems of principle. In the initial analysis of DCGT data the isotherm temperature was recorded on the practical temperature scale NPL-75 which proves to be a very close approximation indeed to thermodynamic temperature[2]: writing the practical temperature as θ, we have

$$\Delta\theta = \theta - T \quad (10)$$

when on forming the quotient θ/A_1 we obtain

$$\theta/A_1 = (A_\varepsilon/R) + (K/3)T + (A_\varepsilon/R)(\Delta\theta/T). \quad (11)$$

Distinguishing between the separate terms in (11) causes some problems but proves to be practicable[2], as has been confirmed by a much more powerful data analysis which uses the method of surface-fitting. This is described in detail elsewhere[3], but the essential point is that when some reliable expression for the temperature dependence of the virial coefficients is known, the data for the separate isotherms can be combined into a single, large population of data points which can be weighted and least-squares fitted to a set of newly defined experimental parameters so as to yield a set of coefficients which are properly averaged over the whole sample population. For instance, an adequate representation of $B'(T)$ proves to be

$$B'(T) = B_1' + B_2'/T + B_3'/T^2, \text{ etc.} \quad (12)$$

so that the fit corresponding to (5), i.e. up to the third virial coefficient C(T), ($=C_1+C_2/T$), can be written[3]

$$p = (R/A_\varepsilon)[\theta\mu] - (KR^2/3A_\varepsilon^2)[\theta^2\mu] - \Delta\theta(R/A_\varepsilon)\mu$$
$$+ (B_1'R/A_\varepsilon^2)[\theta\mu^2] + (B_2'R/A_\varepsilon^2)[\mu^2] + (B_3'R/A_\varepsilon^2)[\mu^2/\theta]$$
$$+ (C_1R/A_\varepsilon^3)[\theta\mu^3] + (C_2R/A_\varepsilon^3)[\mu^3] \quad (13)$$

where the terms within square brackets in (13) are the newly defined fit variables, which are directly calculable from the experimental data. Surface fitting with (13) has many advantages over the single isotherm analysis based on (5)[3], mostly because the size of the sample population increases considerably faster than the number of coefficients of fit. The thermometric error term $\Delta\theta$ is indirectly available since it is the mean deviation of the points for each single isotherm from the optimal hypersurface which is fitted to all the isotherms.

2.3. The dielectric constant, ε, is a true bulk property, but the question arises whether surface effects at the capacitor plates can affect the results. Two cases have been considered[2], 'permanent' surface films independent of the pressure of the test gas, and adsorbed films of the test gas itself whose thickness varies with pressure. Both effects are diminished by using a capacitor with a fairly large gap, and the 'permanent' film can be shown not to cause problems of principle; the case of the variable adsorbed film is more difficult to analyse, but it appears to be true that one can always work under experimental conditions where the adsorbed film remains thin, i.e. not much more than a monolayer for pressures which remain below about a half of the saturated vapour pressure.

2.4. Equation (6) shows that measurement of a DCGT isotherm allows the thermodynamic temperature T to be calculated provided that the two quantities (A_ε/R) and K are known. In the experiments of Gugan and Michel[2], it was argued that K was negligibly different from the expected value for pure copper, so that thermodynamic temperature was established by a single calibration point which was chosen to be 20.2750 K: this calibration was equivalent to an experimental determination of (A_ε/R) and thus of A_ε, and it also allowed the error function $\Delta\theta(T)$ for NPL-75 to be estimated. The subsequent surface-fit analysis of the same data[3] justified the earlier assumption about the value of K, and also showed that no assumption is needed (cf.(13)); it also became clear that an apparent discrepancy between the experimental and theoretical values of A_ε for helium had arisen as a trivial misunderstanding of the nuclear-mass corrections which should be applied to obtain practical values of polarizability from values expressed in atomic units[2], and the later analysis clearly shows that one can rely on the theoretical value of A_ε for helium to be correct to about 1 p.p.m.[6] Since the value of R is believed to have an uncertainty of about 20 p.p.m.[7], it is clear that DCGT isotherms have essentially as firm a theoretical foundation for defining thermodynamic temperature as do conventional CVGT isotherms. Some possibilities for the use of DCGT in improving the value of R are discussed in section 4.

3. EXPERIMENTAL

The technical aspects of DCGT are considered in this section, with a commentary on a few points of particular interest: a detailed study of the important temperature range 4.2 - 27.1 K using helium gas has been presented elsewhere[2], and many points mentioned only in passing here are discussed in detail there.

3.1. The measurement of each isotherm point requires the accurate measurement (and control) of test-gas pressure, p, of capacitance under this gas pressure, C(p), and of empirical temperature, θ, on whatever practical thermometers are in use. The first and third of these present no problems out of the ordinary, and need not be considered further here, apart from the comments that since dead-spaces need not be avoided, the pressure sensing tubes can be of any convenient size to reduce thermomolecular effects, and that cold-spots (colder than the controlled temperature) are not only tolerable, but may even be advantageous in order to freeze-out unwanted condensible vapours: these represent small but real advantages of DCGT.

3.2. The accurate measurement of capacitance, $C(p)$, is the new technical feature of DCGT; and a typical requirement is to measure a capacitance of 10 pF to .01 p.p.m., or since in practice[2] it is convenient to measure the change in capacitance, $\Delta C = C(p) - C(p=0)$, with $\Delta C/C(0) \lesssim 0.001$ to measure ΔC to about 10 p.p.m. At first sight this may appear formidably difficult, but in fact the development of transformer bridges has made these requirements fairly easy to achieve using standard equipment: the measurement of the extremely small length changes needed to determine thermal expansion at low temperature is regularly made using apparatus similar to that necessary for DCGT, and there is plenty of evidence that these measurements are simple, sensitive and accurate, see e.g.[8,9]. The uncertainty of a well-designed transformer-bridge is usually about 1 p.p.m., and can be 2 or 3 orders of magnitude better than this in carefully designed systems (see e.g.[10]), so that measurement of the capacitance changes in DCGT is not a major problem: the errors arising from measurements of pressure are likely always to be the more serious limitation.

In the trial of DCGT made by Gugan and Michel[2], the major difficulty in the capacitance measurements arose from slow drifts in the internal standards of the commercial bridge which was used. There was a straightforward correction procedure for these drifts, but the size of the correction approached about twenty times the resolution of balance of the bridge and the net effect was to reduce the overall balance accuracy by a factor of two; moreover, these correction measurements added considerably to the time required to establish each isotherm point. Use of an inductive voltage divider with 1 in 10^8 resolution and a single, stable reference capacitor (probably kept in liquid helium, see 3.3. below) would avoid these difficulties.

3.3. The biggest unknown in the practice of DCGT concerns the mechanical stability of the test capacitor. It should be kinematically designed so that its dimensions change smoothly and reversibly with both temperature and pressure and independently of the insulating material, which must be kept away from any region of significant electric field. The symmetrical cross-capacitor is probably the design which could most easily meet these conditions[10], but the available capacitance is inconveniently small, $\varepsilon_0 \ln 2/\pi = 1.953..$ pF m^{-1}: it may be suitable for use at higher temperatures, say 80 - 400 K, where size is no particular disadvantage, especially since moderate gas pressure, $\sim$ 100 bar, would probably also be involved. However, cylindrical capacitors can be designed which meet the conditions reasonably well, and Fig.1 shows the design used by Gugan and Michel[2]. In this design the mica insulation, F, is kept out of the electric field between the high and low electrodes, B and A, by the interposed guard-ring, C. Other points are that the electrodes are both located on a reference plane on the guard ring by a spring-compression fitting which is partly, if not ideally, kinematic, that the capacitor is assembled so that the axes of the electrodes are coaxial (and thus insensitive to small lateral displacements), that all parts of the capacitor are made from the same piece of high conductivity copper, that the electrode surfaces are gold plated, and finally, that electrical connections are made with screened leads so that of the total of about 10 pF capacitance between the electrodes A and B, not more than 1 aF (i.e. .1 p.p.m.) was due to any cross-capacitance between the leads outside the cell.

The performance of the cylindrical capacitors was tested by comparing two similar capacitors against each other[2]. Their intrinsic stability at low temperature was within a few parts in 10^9, at least, over periods of many hours: the only exception to this appeared to be the effect of the sudden mechanical shock which resulted from opening a low-temperature needle valve used when pumping the capacitance cell to very low pressures. This sudden shock developed with time, perhaps from progressive damage of the valve seat, and ultimately proved to cause capacitance shifts of about 1 aF, i.e. about 30 or more times larger than the intrinsic stability of the capacitors. The effect caused some initial problems in the data analysis[2,3], since it led to a systematic error in the lowest pressure point in each set of isotherm measurements, and led, in effect, to the inclusion of a constant term, A_0, in (5). Once recognized, the effect can be accurately allowed for[2], but if discovered in time, it is easy to use procedures which avoid the problem. Apart from this very particular kind of shock, the capacitor seemed immune to the normal hazards of cryogenic operations, including even liquid helium transfer. The design shown in Fig.1 results in the expected thermal expansion and compressibility of the capacitor being equal to that for the electrode material, copper, within a small fraction of a percent. A direct measurement of the thermal expansion of the capacitors over the range 4 - 30 K showed them to lie within ten percent of the expected values for copper, while a direct comparison of the compressibility of the two capacitors showed them to differ from each other (and presumably from the ideal result) by not more than three percent of the compressibility of copper. Subsequent analysis using surface fitting[3] shows that the absolute value of compressibility was indeed extremely close to that for copper. As Gugan and Michel have discussed[2], there have been a number of experiments involving dielectric behaviour under pressure where the effective compressibility has proved to be vastly greater than the ideal value (100 or more times larger), and it is important to have established that simple designs can have close to ideal behaviour.

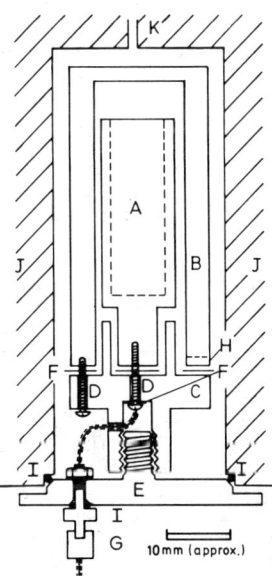

Fig.1. Schematic diagram of DCGT capacitor. A,B, low and high electrodes; C, earthed guard-ring; D, insulated screws; E, capacitor support plate and gas-cell cover-plate; F, mica sheet; G, screened coaxial lead (one of two); H, channels for gas flow (one of three); I, indium gasket; J, temperature controlled thermometer block; K, helium gas line. Two similar 10 pF capacitors were mounted in the thermometer block, one being the test capacitor, the other a vacuum reference capacitor[2].

Copper proved to be a suitable material for the capacitors. It combines a small compressibility ($\sim 7.1 \times 10^{-7}$ bar^{-1}) with high thermal conductivity, both of which are desirable. Annealed copper is mechanically soft, however, and the thermal expansion results[2], particularly when extended over larger temperature ranges (to $\sim$80 K, and near to room temperature) showed irregular and irreversible changes. In this region it might well be preferable to use a harder, metallurgically stable alloy instead of copper: beryllium-copper is one possibility, and Invar could be attractive for high temperature use where small thermal expansivity might be advantageous. There are also materials with much lower compressibility than copper, e.g. synthetic sapphire with K $\sim 2.5 \times 10^{-7}$ bar^{-1}, tungsten carbide ($\sim 1.3 \times 10^{-7}$ bar^{-1}), however, these are non-conductors, and thermal expansion experiments have shown that anomalous capacitance results can easily be obtained after putting a conducting film on to an insulating substrate[9], presumably due to occasional poor adhesion. There appears to be a potential advantage with the use of an insulator in that the capacitor can be made of a monoblock construction, with the active areas delimited by the conducting surface films: however, experience with the coaxial copper design suggests that this is not a very important benefit.

3.4. Gas purity is a slightly more serious problem in DCGT than in CVGT, since the polarizability of the different molecules likely to be present as contaminants varies by a factor of ten, or more. This was no problem in the previous experimental work because the helium gas used was kept clean by a 4.2 K cold-trap which condensed the impurities released by the warmer parts of the apparatus[2]. This procedure could also be adopted in experiments using helium at higher temperatures, but since the sensitivity of DCGT decreases at higher temperature (cf.(2)), it might sometimes be preferred to use a more polarizable gas (e.g. argon, with A_ε about eight times that of helium) when a large range of contaminants might need to be removed.

3.5. The experimental tests on DCGT show the procedures to be eminently practicable[2]. It proved possible to measure isotherm points at the rate of about one per hour (and at least one third of this time was occupied with the bridge calibration procedures which should in future be avoidable), and analysis of about ten points selected from a few minutes record of the chart recorder outputs from the capacitance bridge and the pressure balance showed that under a wide range of conditions the temperature resolution, $\delta T/T$, available from a balance point was about 3 p.p.m., cf. Fig.2. With present improvements in data-acquisition instrumentation one would expect significant improvements in precision and reduction in time to be easily available. The overall accuracy of an isotherm point was almost a factor of ten worse than the balance precision[3], presumably because of the random incidence of small systematic errors when calculating values of p, μ, (and θ) for use in (5) or (13). This presumption seems justified because the overall accuracy for a single point was only slightly worse than that calculated from the specification of the instruments used. The surface-fit analysis strongly suggests that the pressure balance was responsible for the excess random variation over what was expected[3], but there may also have been small, random irregularities in capacitor geometry on changing the gas pressure: in any case, it is clear that much of the accuracy lost for an individual point can be regained on averaging over the twenty or so points which define a complete isotherm, and there is a realistic prospect that one can expect to achieve a temperature resolution and ultimately an accuracy within a few p.p.m. with DCGT. In some cases this would be better than the transfer reliability of resistance thermometers.

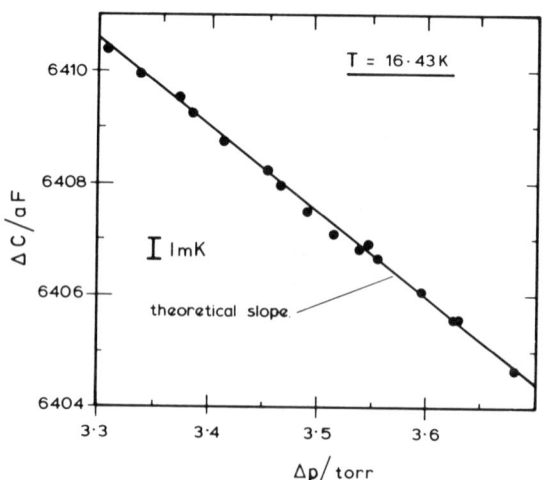

Fig.2. Precision of a DCGT balance point. Correlation between capacitance and pressure for one point on a DCGT isotherm at 16.43 K, corrected for very small temperature changes. These points were obtained from about five minutes of chart record under untypically noisy conditions. The theoretical slope depends only on the (known) behaviour of B(T). The required thermometric information is the position of the line, and in this case the uncertainty in position corresponds to about .05 mK in 16.5 K i.e. about 3 p.p.m. Both capacitance and pressure are measured in terms of differences from stable reference states[2], e.g. $\Delta C = C_1(p) - C_2(0)$, where $C_1(p)$ refers to the pressurised test capacitor and $C_2(0)$ to a nominally equal evacuated reference capacitor, cf. Fig.1.

4. PROSPECTS

4.1. DCGT has proved to be a technique which is relatively simple and convenient, and it is also robust in the sense that the conclusions do not depend significantly on the details of the data analysis[2,3]. In its first serious test it has produced results which are closely comparable with the best existing conventional CVGT/isotherm thermometry[1]; moreover, since it involves measurement of the dielectric constant, an intensive variable, it is likely that improvements in DCGT will be a great deal easier to make and more productive than those in CVGT. This is the familiar pattern of technological development.

DCGT shares with CVGT the technical problems arising from accurate pressure measurement, but this is because it shares with CVGT the advantage of being based on an exactly known ideal equation of state, with the small deviations from ideal behaviour being rather well known and capable of being extrapolated to zero. The evidence suggests that DCGT could be used up to about 400 K. Using helium (at up to about fifty atmospheres pressure) the known value of A_ε should enable absolute thermodynamic temperatures to be established with an uncertainty of only a few p.p.m. over the range $\sim$2.6 - 400 K. With argon the pressures would be lower by about a factor of eight, but the range would be only $\sim$ 80 - 400 K, and, of course, this could only be a relative thermometer, since there is no sufficiently reliable theoretical value of A_ε for argon. The range 2.6 - 400 K is an exceedingly large range for a single thermometer to cover; in practice the optimal requirements for different temperature ranges might make it preferable to work with 2 or 3 different test capacitors, but much of the ancillary equipment would be common. Apart from the (crucial) problem of capacitor behaviour at these higher temperatures, which still needs to be studied experimentally, the only addition required to the analysis given here would be to take account of the (weak) temperature dependence of the compressibility, K, and this can evidently be included in the surface-fitting procedure of (13).

4.2. Used as implied here, to establish thermodynamic temperature from isotherm measurements, DCGT has value as an accurate method of providing a wide range of reference temperatures. However, the technique can obviously also be used in a manner directly analogous to CVGT, as a single point measurement of temperature but using the known values of virial coefficients (and compressibility) to correct for non-ideal behaviour. Unlike CVGT, a DCGT single point measurement does not require the quantity of the test gas to be constant: gas can be admitted or removed as convenient, and problems of contamination over long periods do not really arise, since a suspect measurement can easily be repeated with a fresh charge of gas. In conventional CVGT it is usual to keep the pressure moderately low so as to reduce errors from uncertainties in the virial coefficients, however, this may be less serious in the DCGT analogue since it has proved relatively easy to extend and improve our knowledge of the virial coefficients up to $D(T)$[2,3]. One important point about comparing temperatures via the DCGT analogue of CVGT is that the change of vacuum capacitance with temperature of the test capacitor must be allowed for: this is given directly by the linear expansivity of the material the capacitor is made from, at least for a good design, but it is a big enough correction to be troublesome at the sub 1 mK level. Gugan and Michel made some preliminary trials of this single point DCGT technique, with the hope of studying the smoothness of NPL-75 at closer temperature intervals ($\sim$ 1K) than was possible from isotherm measurements: in principle their technique avoided the thermal expansion effect since they balanced the test capacitor against a nominally similar vacuum capacitor at exactly the same temperature[2]; however, in practice the two capacitors had a sufficiently large net difference that careful correction was necessary and although the preliminary data appeared good at the mK level, this was so much worse than the 0.1 mK level of the isotherms that the comparisons were not pursued.

4.3. The proven success of DCGT in establishing thermodynamic temperature is very encouraging for the prospects of Refractive Index Gas Thermometry (RIGT). Colclough has given some thought to this technique[11], and preliminary experimental work is in hand. The theoretical basis is evidently virtually identical with that of DCGT, with n^2 replacing ε in (1) etc., though the optical frequencies used mean that the frequency dependent polarizability must be considered; fortunately for helium, at least, this appears to cause no extra uncertainty relevant to thermometry[12].

The technical aspects, however, are extremely different from DCGT and involve laser interferometry, probably with a Fabry-Perot etalon[13]: for a five cm spacing a resolution of about 10^{-9} should be possible, which is much the same as for DCGT. The optical path length defined by the spacers can in principle be made the property of a convenient single material which, in contrast with DCGT, can be given a detailed, independent study outside the etalon in order to determine the compressibility and expansivity. This is certainly an advantage, but probably not a crucial one. A disadvantage of RIGT is that the wave length used is much smaller than the size of the optical cavity, so that problems of alignment and of signal access to the etalon become important: these appear non-existent for DCGT. The experience from thermal expansion work appears to be that capacitance techniques are more reliable as well as easier to use than optical interferometry, particularly at the lowest temperatures[8,9], however, the picture is not so clear for temperatures above about 30 K since it is probable that the optical work has not so far been stretched to its limit[13].

4.4. Finally, helium polarizability thermometry at the level of a few p.p.m. has an obvious potential for improving our knowledge of the gas constant, R, by combining a value of (A_ε/R) measured at the ice point with the theoretical value of A_ε[6,12] (cf.(5), (6)). It remains to be seen whether DCGT or RIGT would give the better accuracy, but it is already clear that DCGT offers a realistic prospect that one could improve upon the present 20 p.p.m. uncertainty in R[7] by almost an order of magnitude.

REFERENCES

1. K.H.Berry, Metrologia 15, 89 (1979).
2. D.Gugan and G.W.Michel, Metrologia 16, 149 (1980).
3. D. Gugan, these proceedings (1982).
4. C.J.F. Böttcher,'Theory of Electric Polarization', Elsevier, Amsterdam (1952).
5. T.L.Hill, J.Chem.Phys. 28, 61 (1958).
6. F.Weinhold, J.Phys.Chem. (to appear).
7. A.R.Colclough, Proc.R.Soc.London Ser.A365, 349 (1979).
8. G.K.White and J.G.Collins, J.Low Temp.Phys. 7, 43 (1971).
9. K.G.Lyon, G.L.Salinger and C.A.Swenson, J.App.Phys. 48, 865 (1977).
10. W.K.Clothier, Metrologia 1, 36 (1965).
11. A.R.Colclough, Metrologia 10, 73 (1974).
12. R.M.Glover and F.Weinhold, J.Chem.Phys. 65, 4913 (1976).
13. M.A.Norton, J.W. Berthold III and S.F.Jacobs, J.Appl.Phys. 47, 1683 (1976).

Surface-fitting of helium isotherms: Application to the temperature scale 2.6–27.1 K

D. Gugan

H. H. Wills Physics Laboratory, Royal Fort, Bristol, England

Dielectric Constant Gas Thermometry (DCGT)[1] has recently been shown to have an accuracy which is comparable with that of high quality, conventional isotherm thermometry.[2] Data from the two experiments were each originally analysed at a series of (eight) separate isotherm temperatures in order to obtain values of the virial coefficients and of thermodynamic temperature; however, a much improved analysis is possible by surface fitting, i.e., by fitting all the isotherms simultaneously, with some constraint on the form of the virial coefficients. The precision of analysis is improved because the ratio of the number of data points to the number of constraints is greatly increased, and this has a number of important benefits which are discussed. The results calculated from the DCGT data prove to be rather insensitive to the details of the analysis, and they confirm the published experimental value of the polarizability of ^{4}He (Ref. 1); the previous slight discrepancy with the theoretical value has been resolved by an explicit calculation of the nuclear mass effect, resulting in excellent agreement between theory and experiment. The implications of this for the accuracy of NPL-75 (2.6–27.1 K), and for the future of polarizability thermometer are discussed.

1. INTRODUCTION

The measurement of p-V isotherms is a well-known method for establishing thermodynamic temperature and for obtaining the temperature dependence of the virial coefficients $B(T)$, $C(T)$ etc., and Berry[2] has recently used what will be called the CVGT isotherm technique on ^{4}He in order to establish the temperature scale NPL-75 for the range 2.6 - 27.1 K. Gugan and Michell[1] have shown that a new technique of isotherm thermometry based on measurements of the dielectric constant of the gas in the test chamber has an accuracy about the same as the conventional technique, but with a number of significant technical advantages which arise from the measurement of an intensive parameter, the dielectric constant, ϵ, instead of an extensive ratio, the molar volume V. The dielectric constant gas thermometry (DCGT) used a RhFe thermometer carrying the NPL-75 scale, so that an independent test of the accuracy of this scale was possible: this was found to be very good, but there were a number of minor discrepancies which suggested that a more detailed analysis of the data of both experiments would be desirable. In particular, in the case of the new technique of DCGT, it seemed important to establish the firmness of the conclusions with respect to variations in the details of the analysis, and thus to confirm the validity of the method.

This paper is concerned with the results of computer-experiments which allow one to improve on the conclusions reached from the existing analysis, and which will be of value for similar work in future since there proves to be precision in the experimental data which the previous conventional analysis was unable to extract. The CVGT results[2] consist of 102 data points on 8 isotherms covering the range 2.60 - 27.10 K, while the DCGT results[1] comprise 186 data points on 8 isotherms between 4.23 - 27.17 K, 5 of them lying close to the CVGT isotherms. All these results were originally analysed by a conventional least-squares fitting procedure to obtain values of T and B (and in the case of DCGT, also C) for the separate isotherms, and the values of B (and C) at each temperature were then analysed in order to find appropriate formulae for the temperature dependence of the virial coefficients $B(T)$, $C(T)$. The conventional procedure applied in this manner neglects the constraints between the isotherms which must arise because $B(T)$, etc., are definite physical functions, not just arbitrary parameters, and while this was corrected to some extent in the DCGT analysis by constraining the isotherm fits with optimal forms of $B(T)$, etc., a more satisfactory procedure is to combine all the data so that the isotherm temperatures and the functional forms of the virial coefficients are determined simultaneously from a single, weighted, least-squares fit. This procedure, 'surface-fitting', is well known to mathematicians, see e.g. Clenshaw and Hayes[3], but it appears not to be much used by physicists: in fact it proves to be particularly appropriate to the analysis of isotherm thermometry, as will appear below, and in addition it is capable of revealing experimental information (on such things as the relative precision of the data under different experimental conditions) which is not usually obtainable from the simpler analysis.

2. THEORY

2.1. A resumé of the conventional isotherm analysis.

The analysis of CVGT isotherms is based on the familiar virial expansion, known to be exact in the limit of low densities,

$$pV = RT(1 + B/V + C/V^2 + \ldots), \quad (1)$$

while DCGT is based on the fact that there is a similar expansion of the Clausius-Mossotti function, the dielectric virial expansion,

$$(\epsilon-1)/(\epsilon+2) = (A_\epsilon/V)(1 + b/V + c/V^2 + \ldots) \quad (2)$$

where ϵ is the dielectric constant of gas at pressure p and temperature T, and A_ϵ is the molar polarizability. Equations (1) and (2) may be combined to yield an expression for a DCGT isotherm[1],

$$p = A_1 \mu(1 + A_2 \mu + A_3 \mu^2 + \ldots) \qquad (3)$$

where μ, the 'polarizability parameter' is a quantity available from experimental measurements of capacitance, and where the coefficients are given by

$$A_1 = (A_\epsilon/RT + K/3)^{-1} \qquad (4)$$

$$A_2 = B'/A_\epsilon, \text{ where } B' = B-b \qquad (5)$$

$$A_3 = C/A_\epsilon^2, \quad A_4 = D/A_\epsilon^3, \text{ etc.} \qquad (6)$$

The term in K in (4), the linear compressibility of the capacitor, arises inevitably because the test gas is under pressure: K is not the property of a material since the capacitor is a composite structure, but in a good design it has a value close to that calculated for an 'ideal' capacitor, when it is a very small though not negligible correction[1]. In CVGT, measurements of p as a function of V (or more strictly of N_B/V_B, where N_B is the number of moles of gas occupying the isothermal gas-bulb of volume V_B at the temperature T) lead via (1) to estimates of RT, B, C, D etc. for each isotherm, (the effect of R can be eliminated[2], by referring the measurements to the fundamental fixed point), while in DCGT measurements of $p(\mu)$ lead via (3) to estimates of $(A_\epsilon/RT+K/3)^{-1}$; B', C, D etc. DCGT contains three terms not found in CVGT: A_ϵ, K and b. b is very small (~ -0.1 cm^3 mol^{-1}) and causes no problems of principle; K is small, but together with A_ϵ it is intimately related to the estimate of thermodynamic temperature via (4). Experimentally, the isotherm temperature is measured on some practical temperature scale, θ, which differs by $\Delta\theta$ from the thermodynamic scale,

$$\Delta\theta(T) = \theta - T \qquad (7)$$

In the DCGT experiment, θ was a realization of NPL-75, and the error $\Delta\theta$ was studied by forming the quotient θ/A_1, which gives to excellent approximation

$$\theta/A_1 = (A_\epsilon/R) + (K/3)T + (A_\epsilon/R)(\Delta\theta(T)/T). \qquad (8)$$

Gugan and Michel[1] show that K in their work has the known value for pure copper, so that a single constraint on (8) is sufficient to determine (A_ϵ/R) and $\Delta\theta(T)$ uniquely: the constraint applied was to assume that $\Delta\theta(20.27 \text{ K}) = 0$, since this was the reference temperature for NPL-75, as well as being the most fully studied temperature in both isotherm experiments. The conclusions for $\Delta\theta(T)$ showed NPL-75 to be smooth and internally consistent at the level of about 1 mK (cf. Fig. 2 below) while the value obtained for A_ϵ was $517.257 \pm .025 \times 10^{-3}$ cm^3 mol^{-1}, compared with the then accepted theoretical value of $517.03_2 (\pm \sim .00_1) \times 10^{-3}$ cm^3 mol^{-1}. Since the major contribution to the experimental uncertainty was the value of the reference temperature (99% confidence limits of ± 1 mK), the difference between experiment and theory, while small, was nevertheless highly significant, and the chief motivation for this reanalysis was to see if the experimental value can be changed much: it cannot, and as we see later, the resolution of the discrepancy lies elsewhere (section 3.2.3.)

Values of B, B' and C from the single isotherm analysis were fitted to a variety of functional forms, the best representation for B (and B') being the expressions with constant coefficients,

$$B(T) = B_1 + B_2/T + B_3/T^2, \text{ etc,} \qquad (9)$$

and for C (from the DCGT work)

$$C(T) = C_2/T^2, \qquad (10)$$

with no evidence for a value of D different from zero.

2.2. <u>Surface analysis of CVGT isotherms.</u>
We discuss the surface-fitting of Berry's isotherm data first, to illustrate the general method in a simple form.

The experimental results[2] are values of p and (N_B/V_B) for each isotherm, together with θ_c, where θ_c is the estimate of thermodynamic temperature for the isotherm. Surface-fitting gives a different (and better) estimate of the true thermodynamic temperature, which we shall here label T. Writing $\Delta\theta = \theta_c - T$ in (1) we obtain to sufficient accuracy

$$(pV/R\theta - 1) \equiv Y = B/V + C/V^2 \ldots - \Delta\theta/\theta. \qquad (11)$$

Now, if we know B(T), C(T), etc., or at least know an adequate approximate analytic form for them, then the RHS of (11) can be expressed in a form common to all the isotherms. For instance, using B(T) and C(T) from (9) and (10), (11) becomes

$$[Y] = B_1[1/V] + B_2[1/V\theta] + B_3[1/V\theta^2] + C_2[1/V^2\theta]$$

$$-\Delta\theta/\theta,$$

$$\equiv B_1[X_1] + B_2[X_2] + B_3[X_3] + C_2[X_4] - \Delta\theta/\theta; \qquad (12)$$

this is a linear equation in the newly defined variables $[X_1]-[X_4]$ which can be least-squares fitted to the data from all the isotherms simultaneously. The least squares fit to $Y[X_1 \ldots X_4]$ clearly applies the constraint that $\overline{\Delta\theta/\theta} = 0$; this need not be true of course, but it represents no loss of information since the value of $\Delta\theta$ for each isotherm is retained as the mean deviation for the points of each isotherm from the hyper-surface of (12). The original estimates of thermodynamic temperature, θ_c can be corrected to $(\theta_c - \Delta\theta)$, and they can be renormalized to any convenient temperature. For some purposes it is convenient to express the deviation of the points from the hyper-surface as an equivalent deviation in temperature, and the root mean square of the weighted deviations is written below as 'RMSWDT'.

This is the basic outline of the surface-fitting procedure, but it is convenient to make a number of supplementary points here concerning the quality of such fits. The great advantage of surface-fitting arises because there is a large increase in the number of points fitted for only a modest increase in the number of variables: for the CVGT isotherm analysis we have an average of about 12 points on each isotherm to fit with three parameters using (1), but 102 points to fit with four parameters using (12). Naturally, this increases the internal consistency of the fit and also the precision of the fit parameters, as will be shown later, but the real advantage of this extra precision is that it becomes possible to assess the quality of the fit at a much higher level than before. If we introduce the variance function for the dependent variable of (12), S(Y), then we can parameterize the quality of a fit of order M to N data points as

$$\overline{\chi} (\equiv RMSCHI) = (\Sigma_i \chi_i^2 /(N-M-1))^{\frac{1}{2}}, \text{ where}$$

$$\chi_i^2 = \{(Y_i - Y_{calc})/S(Y_i)\}^2 \qquad (13)$$

For a good fit we know that $\overline{\chi}$ should be close to unity, and we also know that the relative standard deviation of the distribution of $\overline{\chi}$ is about $1/\sqrt{2N}$. Usually, one has only partial information on S(Y); the limitations arising from instrumental resolution can be fairly accurately calculated, but there are often additional, systematic contributions to S(Y) as one moves away from optimal experimental conditions. For instance in this work we find that the relative accuracy of the measurement of pressure falls off at higher pressures in a way which is not evident from the specification of the pressure balance used. This sort of behaviour can be detected in a large enough population by systematically removing some of the fitted points, e.g. by cutting the data to exclude the high (or the low) pressure (or temperature) points in the fit to (12), and by testing whether RMSCHI varies only by amounts which are statistically compatible with the size of the population. In most cases, and certainly here, the preferred surface fit is a weighted least-squares fit, with a weighting proportional to $(S(Y))^{-2}$, so that finding the appropriate semi-empirical form of $S(Y)$ requires iterative fitting procedures. This cutting procedure is also possible for single-isotherm fits [2], but statistical fluctuations make the interpretation doubtful when the sample is small. For instance, there is clear evidence from the surface analysis that $S(Y)$ contains a temperature dependent term which cannot be revealed by the pressure-cuts of the single-isotherm analysis: this should be included as a weighting factor in the further analysis to find the optimal form of $B(T)$ - if it were known. In surface fitting, however, no problem arises, since provided that an appropriate $S(Y)$ is used, the weighting is made automatically.

The problem of finding the appropriate form of $S(Y)$ is not independent of finding the appropriate functional form of Y itself, i.e. of deciding the number and nature of the variables $[X_i]$ in (12), however, there proves to be a range of functional forms which are acceptable (i.e. giving almost equally low, near-minimum values of RMSCHI) so far as the fit to a given number of data points is concerned, so that in practice it is not difficult to separate these two problems. On the other hand, the selection of the 'best' set of variables $[X_i]$, and the 'best' set of coefficients corresponding to them, cannot be decided merely on the basis of the least-squares minimisation procedure. The criteria here become more subjective and include, for instance, the requirement that the set of variables used lead to a set of coefficients whose values do not change significantly as the population of data points is cut. To achieve this may mean choosing sets of $[X_i]$ which do not achieve the very best fit to the data, though of course the value of RMSCHI must not be increased beyond the limit of its statistical variability. A functional form which has satisfactory stability of coefficients with respect to cutting can (and should) be made definite by selecting the best (weighted) average set of coefficients over the various cuts in the population, by applying these as a defined constraint, and by refitting with a reduced number of variables subject to this constraint. Thus, in (12) if this functional form is found to be satisfactory, and if the coefficient of the variable $[X_4]$ has an average value, say C_2^*, then the form of the second virial coefficient can be improved by using the constrained fit,

$$[Y - C_2^* (X_4)] = B_1[X_1] + B_2[X_2] + B_3[X_3] - \Delta\theta/\theta, \quad (14)$$

and so on. The uncertainty in the constraint propagates through the subsequent fits in a systematic way, which can be estimated by varying the magnitude of the constraint over a few values around the optimum. This procedure generates a branching set of linked coefficients, and these ultimately represent the real uncertainties in the data analysis: the final results of the various branches, e.g. for $\Delta\theta(T)$, form a population which while essentially reflecting the systematic uncertainties of the data analysis, nevertheless allow one to form a sample population corresponding to the random incidence of these systematic uncertainties, and thus to estimate the likely overall uncertainty of analysis. Much of the motivation for this work stem-

med from this, in particular connection with the determination of the polarizability, A_ϵ, from the analysis of DCGT isotherms, as mentioned in section 2.1.

A second stability criterion is the stability of the mean values of the coefficients with respect to changes in the choice of the variables $[X_i]$: for instance, in the detailed isotherm fits which are discussed later, it turns out that while $C(T)$ is well represented by $C(T) = C_2/T$, it can be significantly improved by the addition of one extra term, either C_1, or C_3/T^2 with nothing to choose between the values of RMSCHI. However, while the value of C_2 is little changed by the addition of the term involving C_1, it is greatly changed by that involving C_3: there proves to be no benefit from using both C_1 and C_3 terms, so C_1 is the preferred choice.

A further benefit from using the data population to define $S(Y)$ is that a criterion for rejection of data points can be arrived at. The contributions to RMSCHI (13) from the individual χ_i can be examined and tested for statistical self-consistency with far greater confidence than can data forming only a small population. In fact, no data need to be rejected from Berry's CVGT isotherm data, while there are 5 'outliers' in the DCGT isotherm data (see e.g. sections 3.1.2. and 3.2.1).

The procedures outlined so far amount to a parametrisation of the data: some further physical constraints can be applied to the values of the virial coefficients (as will be discussed elsewhere) but they are only weak in this temperature range. Fortunately, one of the results of greatest interest, $\Delta\theta(T)$, is not much affected by any error of parametrisation since the mean deviations of the isotherms are much the same for any smoothly-fitting hypersurface (cf. section 3.2.4). However, the quality of the fitted hypersurface is affected by these systematic deviations, and so are the values of the virial coefficients, B_1, etc., so to improve the fit, all the data points are corrected by the appropriate value of $\Delta\theta$, and then refitted: one such iteration is usually sufficient (cf. section 3.1.2).

Evidently, there is a great deal of personal judgment required in these surface-fitting procedures, and they can be justifiably described as computer experiments. However, it is important to recall that this judgment is exercised over matters which are usually either ignored or taken for granted in the standard analyses, and that the extra information is real information. As we shall see later, for instance, the analysis of CVGT isotherms by surface-fitting shows a several-fold improvement in precision over the original analysis; a ratio which is roughly comparable with that produced by a new generation of instrumentation.

2.3. <u>Surface analysis of DCGT isotherms</u>. The basic ideas are exactly the same as for the analysis of CVGT isotherms (2.2. above) though there are some extra complications which must be mentioned.

Firstly, for reasons discussed by Gugan and Michel[1], a term A_0 must in practice be added to the right hand side of (3), arising from an unexpected (and in general, avoidable) systematic error in the DCGT measurements. A_0 is constant for the group of points on an isotherm which correspond to one filling of the apparatus (between 4 and 11), but varies for the 2-4 such 'semi-isotherms' which together define the complete isotherm. If the difference between each point of a semi-isotherm and, say, the lowest point, $p'(\mu')$, is formed then (3) can be rewritten as

$$[\pi] \equiv [p - p'] = A_1[\Delta\mu] + A_1 A_2 [\Delta\mu^2] + A_1 A_3 [\Delta\mu^3] + \cdots \quad (15)$$

(where $\Delta\mu^n = \mu^n - \mu'^n$), which avoids most of the error from the A_0 term. This is of the same form as (3), and it is convenient here to use the notation of (3), though in fact both (3) and (15) have been used in the computer experiments on the DCGT data. Expanding (3)-(7) as in section 2.2. above gives

$$[p] = (R/A_\epsilon)[\theta\mu] + (B'R/A_\epsilon^2)[\theta\mu^2] + (CR/A_\epsilon^3)[\theta\mu^3] + ..$$
$$-(KR^2/3A_\epsilon^2)[\theta^2\mu] - \Delta\theta(R/A_\epsilon)\mu. \qquad (16)$$

Writing $B'(T)$ in the same form as $B(T)$, equation (9), and writing $C(T)$ and $D(T)$ in what prove to be the optimal forms $C(T) = C_1 + C_2/T$, $D(T) = D_2/T$ leads to

$$[p] = (R/A_\epsilon)[\theta\mu] - (KR^2/3A_\epsilon^2)[\theta^2\mu] - \Delta\theta(R/A_\epsilon)\mu$$
$$+ (B'_1 R/A_\epsilon^2)[\theta\mu^2] + (B'_2 R/A_\epsilon^2)[\mu^2] + (B'_3 R/A_\epsilon^2)[\mu^2/\theta] + (C_1 R/A_\epsilon^3)[\theta\mu^3] + (C_2 R/A_\epsilon^3)[\mu^3]$$
$$+ (D_2 R/A_\epsilon^4)[\mu^4]. \qquad (17)$$

This is similar to the expression for CVGT isotherms (12) though larger, 8 variables rather than 4, because there are two extra terms which did not enter (12) and because more virial terms are needed since the data extend to higher densities. As (17) shows, the extra terms (the first and second on the RHS of (17)) can be unambiguously separated by the surface fitting procedure, as contrasted with the single isotherm analysis which links them via (8), while the third of the terms linked via (8), the value of $\Delta\theta$ for each isotherm, is given by the mean deviation of the points for each isotherm from the hypersurface of (17), using the third term on the RHS of (17). It is an important point that the value of K can be deduced from the experimental data themselves, since there appears to be no way in which one can make an independent measurement of what is after all the effective compressibility of a composite structure [1].

A considerable problem in the analysis of the DCGT data arises from the systematic error term A_0 which enters each semi-isotherm: this is largely removed by taking the differences, as in (15), but one has introduced residual systematic errors, δA_0, into the population from what were random errors in the reference points $p'(\mu')$. These residual errors have dimensions of pressure and may conveniently be rewritten as δp_0 (3) or $\delta\pi_0$ (15): they can be corrected in the surface fits by studying the mean deviation of each semi-isotherm from the hypersurface for the whole population. The procedure is somewhat similar to the correction applied for the error term $\Delta\theta$ for each isotherm (see section 2.2. above) and will be discussed in more detail elsewhere: the two corrections, for δp_0 and for $\Delta\theta$ can be separated because of the difference in the form of their contribution to (17), and the procedure can be iterated if necessary.

3. RESULTS & DISCUSSION

The surface-fit procedures discussed above were carried out on a 32K PET microcomputer. Weighted, least-squares fits in up to eight variables were possible, using the method of orthonormal polynomials as discussed by Hall and Canfield [4]. More than 40 separate runs were made on the CVGT isotherm data (these runs are distinguished with an asterisk when it is convenient to refer to them individually) and more than 80 on the DCGT data, including a considerable amount of data cutting, of iteration, and of variation of constraints in each run. This leads to an enormous amount of numerical information, and, even of the useful information, only a small amount of perhaps most general interest will be summarised here. The analysis of the two sets of data was carried out in parallel, since one of the objectives was to examine the higher virial coefficients, $C(T)$, $D(T)$ for each set separately, and then to find whether a single set of coefficients could adequately represent both sets of data. However, in what follows, the results will be separated so far as possible into two parts: the re-examination of the CVGT isotherm data [2] (section 3.1.), and of the DCGT data [1] (section 3.2.).

3.1. Results of surface-fits to CVGT isotherm data.

3.1.1. After a considerable number of trials, a self-consistent form of $S(Y)$ was obtained which includes dependences on both the nominal isotherm temperature, θ_c, and the density (N_B/V_B)

$$S(Y) = (6E-6)(1 + 400/\theta_c^2)^{\frac{1}{2}}(1 + 2000(N_B/V_B))^{\frac{1}{2}} \qquad (18)$$

This optimal, semi-empirical expression was used to determine the weighting in nearly all the fits, except a few in which equal weighting was used in order to test the stability of the final results to this factor: the differences were not great, but the evidence definitely favours the weighted fits (cf. Table I).

3.1.2. Using (18), the distribution of deviations χ_i (equation (13)) was examined for a variety of good fits and found to be closely normal: e.g. for run 35*, the first cycle of fitting resulted in values of 1.72 and 0.358 mK for RMSCHI and RMSWDT respectively: after altering the isotherm temperatures by the calculated values of $\Delta\theta$ and refitting (cf. section 2.2.), the two fit parameters became 1.16 and 0.256 mK respectively, and the distribution of χ_i for the 102 data points was then 0-1.15, 69; 1.16-2.31, 29; 2.32-2.90 (max), 4. There are no unexpectedly large deviators, and this analysis confirms that the CVGT data form an excellently self-consistent data set, though from the point of view of further experiment the density dependence of $S(Y)$ in (18) perhaps deserves further examination.

3.1.3. In contrast to the original analysis of Berry (see e.g. his Table 4), surface-fits show that virial coefficients higher than the second are absolutely necessary; e.g. RMSCHI decreased from a value of 2.22 for fits involving $B(T)$ only, (9), to values of 0.96 (run 7*) and 1.03 (run 8*) when extra terms D_2/T or C_2/T respectively were added. The data do not allow much discrimination to be made between the possible forms of higher coefficient, but the DCGT data unequivocally require a term in C_2, when CVGT data show a residual dependence of C_2 on the level of cut of the density in equation (1) which implies a significant contribution from a term in $D(T)$. The final values of $C(T)$ and $D(T)$ are expressed by the values of D_2, C_2 and C_1, listed for runs 36* and 38* in Table I, where D_2 in particular is, of course, an 'effective' virial coefficient because of the truncation of (1) at this term.

3.1.4. Results for the second virial coefficient are collected in Table I. Values for B_1, B_2 and B_3 of (9) are given for the two preferred sets of higher virial coefficients in the runs 36* and 38*, both for the complete set of isotherms between 2.60-27.1 K ((i) and (ii)) and for the 6 highest temperatures, 3.33-27.1 K ((iii) and (iv)). Table I includes a set of data for a fit with equally weighted points, run 40*, and also the earlier results of the single isotherm analyses. The differences between the values of the coefficients for the different fits are not large, but

Table I. Value of B(T) from surface-fits to CVGT isotherm data

Run number (a)	RMS CHI (b)	RMS WDT (mK) (c)	D_2/K (d)	$C(T)=(C_1+C_2/T)$ /(cm^{6}mol^{-2}) C_1	C_2/K	$B(T)=(B_1+B_2/T+B_3/T^2)$/cm^{3}mol^{-1} B_1	B_2/K	B_3/K^2
36*:						(f)		
N=102 (i)	1.12	.24	.7E6	0	3450	17.250 ± .005	-397.69 ± .11 (f)	-48.1 ± .3 (f)
N=102 (ii)	1.12	.25	"	50	"	17.236 ± .004	-397.80 ± .10	-47.9 ± .2
N=76 (iii)	1.15	.25	"	0	"	17.276 ± .008	-398.35 ± .18	-45.4 ± .6
N=76 (iv)	1.15	.26	"	50	"	17.262 ± .007	-398.43 ± .19	-45.3 ± .6
38*:								
N=102 (i)	1.12	.26	.5E6	24	3590	17.244 ± .006	-397.75 ± .07	-48.1 ± .1
N=102 (ii)	1.13	.24	"	-43	3800	17.266 ± .006	"	"
N=76 (iii)	1.12	.26	"	24	3590	17.277 ± .007	-398.54 ± .09	-44.8 ± .3
N=76 (iv)	1.14	.25	"	-43	3800	17.299 ± .008	"	"
40*:								
N=102 (i)	1.14	.20	.5E6	24	3590	17.130 ± .022	-395.73 ± .19	-52.9 ± .9
N=102 (ii)	1.15	.20	"	-43	3800	17.157 ± .023	-395.77 ± .19	-53.0 ± .9
N=76 (iii)	1.23	.24	"	24	3590	-	-	-
N=76 (iv)	1.25	.24	"	-43	3800	-	-	-
'Original'*:								
N=102 (i)	-	-	0	-	-	17.19	-396.2	-48
N=102 (ii)	-	-	0	0	5420	17.20 ± .10	-397.8 ± 1.3	-52 ± 3
N=102 (iii)	-	-	0	0	"	17.29 ± .11	-398.7 ± 2.2	-49 ± 7
N=76 (iv)	-	-	0	0	"	17.22 ± .18	-398.3 ± 2.9	-50 ± 9
N=76 (v)	-	-	0	0	"	17.31 ± .18	-399.4 ± 3.9	-47 ± 13

(a) The computer runs are third-order constrained fits for the coefficients B_1, B_2, B_3, the constraints on C(T) and D(T) being as shown. Runs 36* and 38* are both a near-optimal form, while 40* is the same as 38* except that the optimal, semi-empirical weighting of the data (see text) has been replaced by an equal weighting. The runs labelled 'Original' derive from the earlier analysis of separate isotherms[2]; (i), C(T) is assumed finite but of no particular analytic form and the values of B are equally weighted in deriving the form of B(T). (ii) and (iv), data as (i), but assuming the form of C(T) obtained by Gugan and Michel[1]. (iii) and (v), as (ii) and (iv), but weighting the individual values of B according to their uncertainties as computed from the individual isotherm fits.

(b) RMSCHI is calculated for the deviations from the surface fit using in all cases the semi-empirical estimate of the variance of the individual points.

(c) RMSWDT is the weighted deviation from the surface fit expressed as a temperature deviation for each point (cf. the minimised quantity, which is a relative temperature).

(d) The fourth virial coefficient is assumed to have the form $D(T) = (D_2/T)$/cm^{9}mol^{-3}.

(e) The total number of data points is 102, divided between eight isotherms. The lowest two isotherms, at 2.60 and 2.75 K, are significantly below the range of the DCGT work, and they are omitted in the fits with N = 76 points, i.e. between 3.33 and 27.1 K. In both cases, the results quoted are averaged over four successive cuts which remove the points at high densities, the number of points fitted being 102, 88, 80 and 64 and 76, 62, 56 and 42 points respectively.

(f) The uncertainties quoted are standard errors in the mean estimated from the range of the values of the successive cuts. These appeared to have a random variation consistent with the precision calculated for the individual fits, except for run 40* ((iii) and (iv)), where B_i varied so strongly with the level of cut that no average value is meaningful.

they are clearly outside the apparent precision of the fits in some cases: however, the component parts of B(T) are strongly correlated, and when B(T) is calculated the differences between these various forms prove to be rather small. These differences are shown in Fig. 1, with respect to the results of run 38*(iii): curve (a), from run 36*(iii), which is expected to be the closest to the preferred fit, shows that the different values for D(T) have very small effect on the values subsequently calculated for B(T), while curves (b) and (c) show that the effect of using the coefficients from the wider temperature range causes larger, though still very small differences. Curves (d) and (e) from the equally-

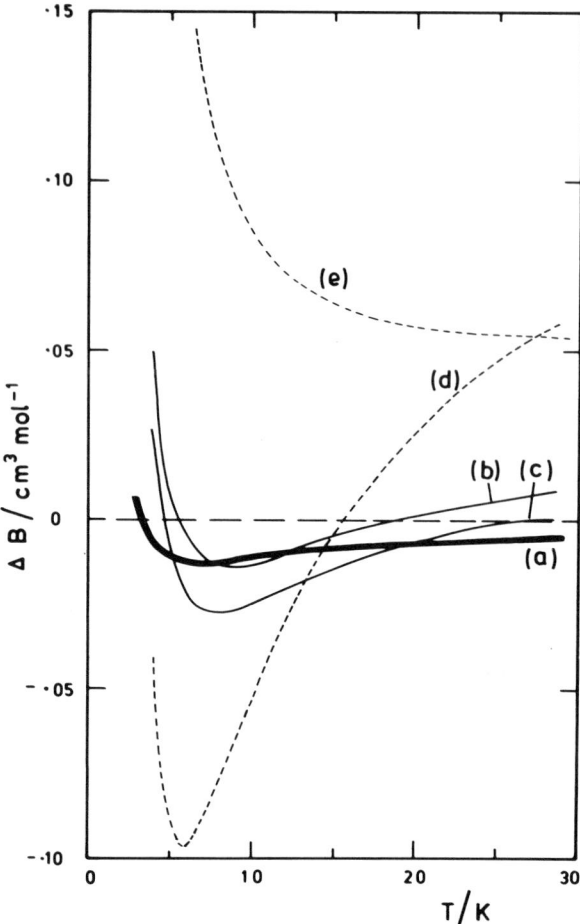

Fig. 1. Differences between functional forms of B(T) for range 4-28K. $\Delta B(T) = B_{ref}(T) - B(T)$, where $B_{ref}(T) = (17.277 - 398.54/T - 44.8/T^2) \text{cm}^3 \text{mol}^{-1}$, the preferred fit for the range 3.33-27.1 K, run 38*(iii) from Table I. Curves (a)-(e) show the difference in the form of B(T) for other, less satisfactory, fits as given in Table I: (a) run 36*(i); (b) run 38*(i); (c) run 36*(i); (d) run 40*(i) and (e), 'Original' (iv).

weighted fit 40*(i), and from the 'original', single-isotherm analysis respectively, show deviations which are much larger, though still within the limits implied by the uncertainties attached to the values of their coefficients. Not shown in Fig.1 is the comparison with run 38*(iv), which is evidently a horizontal line with $\Delta B = -.022 \text{ cm}^3\text{mol}^{-1}$. Overall it seems not unrealistic to suppose that an optimal set of coefficients can represent the values of B(T) for the range 4-27 K to something like $\pm .02 \text{ cm}^3 \text{mol}^{-1}$, which is a factor of about ten better than would be inferred from the single-isotherm analysis of the same data. The trends of curves (b) and (c) suggest that the deviations become much larger below 4 K: this is very probably because the three term expression (9) becomes inadequate at the lowest temperatures, but this possibility has not been explored beyond a few trials of three constant expressions with half-integral powers of T, none of which was an improvement on (9).

3.1.5. As explained in section 2.1., the mean deviations of the separate isotherms from the fitted hypersurface give the errors in the nominal isotherm temperatures, $\Delta \theta = \theta_c - T$, (7). From the values of RMSWDT for the surface fits quoted

in Table I, ~0.25 mK, we should expect to be able to define the mean values of $\Delta \theta$ to ~.05 mK, and we should not expect much variation as we vary the type of surface fit. That this is so is shown in Table II. The three columns for the surface fits are averages over different cut-levels of the data, while the columns labelled 'Single-isotherm fits' are the results of tests to confirm that the procedure is self-consistent by constraining single-isotherms (equation 1) with the virial constraints used in the surface fits. There is a large measure of consistency over all these results for $\Delta \theta$, and the best overall average value is probably just the average of the various fits, as listed in the final column. The meaning of these figures is that if Berry had used surface fitting in his original analysis[2], he would have assigned the values $\theta_c - \Delta \theta$ to his isotherm temperatures. The differences $\Delta \theta$ are extremely small, and this reanalysis should be regarded more as a confirmation of his work than as any criticism of it: in particular, the 99% confidence limits of ± 1.0 mK which Berry placed upon the accuracy of the hydrogen boiling point are fully confirmed, at least so far as the analysis of the data is concerned. Further conclusions one can draw about the accuracy of NPL-75 are discussed in section 3.2.4. below.

3.2. Results of surface-fits to DCGT isotherm data. The analysis of the DCGT isotherm data was more difficult than that of the CVGT data because of the need to correct for the effect of the systematic errors introduced by the term A_0 (see section 2.3.), and also because of the larger number of variables needed in the fit. It is not practicable to avoid the problem of choosing an appropriate, limited set of variables in (17) by merely increasing their number to cover all likely possibilities since there is a saturation point beyond which the value of RMSCHI does not decrease with increase in the order of fit[4]. Moreover, the uncertainties attached to the calculated fit parameters rise fairly rapidly with the order of fit, so that the other criteria for choosing the optimal form become useless. In many experiments, therefore, instead of minimising (17) with respect to the maximum number of variables, 8, it was decided to use a lower number, and to study the effect of variations in the least important terms by a series of constrained fits, with the coefficient of the constrained variable made to run through a range of selected values. This is different from the use of constrained fits discussed earlier (section 2.2.) because in that case the value of the constraint was chosen as a suitable average value to represent experiments over different selections of the data: in these present preliminary fits to the high order terms the procedure is more empirical. The resulting fit is really a self-consistent parametrization, but it is doubtful if it leads to serious error in the larger terms in (17). In any event, the range of calculated coefficients and resulting physical variables can be examined, much as for the CVGT data in Table I and Fig.1., and while the resulting uncertainty in B'(T) is greater than for B(T) it proves to be small enough that useful conclusions can be drawn about the difference $b(T) (= B(T) - B'(T))$, as will be discussed elsewhere.

Some general information about the DCGT fits is given in section 3.2.1., and in the following sections the three quantities of equation (8) which surface-fitting allows to be separated, K, (A_ϵ/R), and $\Delta \theta(T)$ are discussed separately.

3.2.1. The expression used for the variance of the data points depended on whether one was fitting (3) directly, or the difference expression (15). In both cases the expression for S(Y) contained a part arising from the known instrumental resolution[1], but the surface-fits revealed a further contribution causing S(Y) to increase with gas

Table II. Effect of surface fitting on CVGT isotherm temperatures

Isotherm temp. θ_c/K	$\Delta\theta = (\theta_c - T)/mK$					
	Surface-fit runs[a]			Single-isotherm fits[b]		Average Values
	36*(i-iv)	38*(i-iv)	40*(i-iv)	41*	42*	
27.0979	.21	.23	.26	.18 ± .04	.21 ± .07	.22 ± .02
20.2712	-.28	-.26	-.35	-.26 ± .04	-.30 ± .06	-.29 ± .02
13.8036	-.10	-.09	-.12	-.08 ± .01	-.09 ± .02	-.10 ± .01
7.1992	.32	.33	.36	.34 ± .03	.37 ± .04	.34 ± .01
4.2201	.02	-.01	.02	-.03 ± .03	-.07 ± .06	-.02 ± .02
3.3303	-.23	-.23	-.13	-.26 ± .01	-.28 ± .01	-.23 ± .04
2.7479	-.11	-.11	-.11	-.12 ± .02	-.11 ± .03	-.11 ± .01
2.6014	.15	.15	.14	.15 ± .04	.16 ± .04	.15 ± .01

(a) These give the systematic (weighted) deviations of the points for the separate isotherms from the surface fits of Table I. The values quoted are mean values over the separate fits and the several cuts of Table I. The <u>maximum</u> deviation for any fit was .10 mK from the mean, thus an overall estimate of the standard error in the means is about .01 mK. The effect of a large change in the weighting (run 40*) does not greatly change the general conclusion that a self-consistent analysis of the CVGT isotherm data leads to the small changes in the calculated isotherm temperatures which are listed in the final column.

(b) These give the fits for single isotherms, each fully constrained by virial coefficients as listed for runs 38* ((i)-(iv)) in Table I: run 41* uses the optimal, semi-empirical weights (see text), while in run 42* the points are weighted equally. The uncertainties quoted are for the weighted, least-squares fit to <u>all</u> the points on each isotherm.

density in a way rather similar to that discussed for the CVGT isotherm analysis in section 3.1.1. From the distribution of χ_i (13) using the semi-empirical forms of S(Y) the great majority of the data points were confirmed to form a statistically normal distribution, and the few outliers could be easily isolated: for instance from the series of runs (71-74) one finds that rejection of 2 points (χ_i = 5.57, 4.47) from the surface fits to the population of 183 reduces RMSCHI from 1.12 to 1.00, and that the distribution of χ_i among the fit to the remaining 181 points is 0-.99, 128; 1.00-1.99, 44; 2.00-2.97 (max), 9; which is certainly very close to a normal distribution. In fact, leaving the big deviators in the population does not greatly change the fit coefficients since the population is large, but on the whole it seems more logical to remove them. In all, five 'outliers' were removed from the DCGT data, the same in fact, as were omitted from the original single-isotherm analysis, but here with perhaps a stronger justification.

Some experiments were made to test whether the fit coefficients of (17) were dependent on the weighting: some dependence was found, but it is a small effect and does not call into question the use of the weighting based on the semi-empirical values of S(Y).

The corrections for the residual error from the A_0 term for the semi-isotherms, and for the error $\Delta\theta$ in the assigned isotherm temperature, θ, were applied regularly in the later fits. A typical improvement in the quality of fit for each correction applied once (run 27) was to reduce the value of RMSCHI from 1.45 initially (160 points) to 1.11 (Δp shift) and then to 1.03 ($\Delta\theta$ shift): naturally, there were also considerable reductions in the uncertainties of the fit coefficients.

The best fits were characterized by values of RMSWDT over the whole range 4.2-27.1 K of about 0.55 mK, compared with the value of 0.25 mK for the CVGT fits. Evidently the overall precision of a single temperature measurement is only about half as good in the DCGT work as in Berry's work, but some of this difference is certainly due to the capacitor instability which gives rise to systematic error term A_0, some is due to the fact that the DCGT isotherm at 13.8 K included the first measurements made with the technique which on subsequent analysis show more than twice the scatter which is statistically expected, and some is due to the fact that the DCGT measurements were extended to relatively high densities. Taking all these factors into account, one may reasonably expect the precision of the two techniques to be about the same: in fact, for the DCGT isotherms one can see that the effective thermometric precision, $\delta\theta/\theta$, is at present still about a factor of ten worse than the 3 p.p.m. which the analysis of the DCGT records shows to be available from the balance conditions for an individual datum (Gugan and Michel[1], especially Table 2).

3.2.2. The determination of the effective linear compressibility of the capacitor, K, was the first question examined using surface-fitting. The results of the run 1 using a sixth order fit of (17) (i.e. with C_1 and D_2 effectively assumed zero) and with no adjustments of any sort to the data gave a value for K of $-23.5 \pm 4.1 \times 10^{-13} Pa^{-1}$, as compared with the expected value (for copper) of $-24 \times 10^{-13} Pa^{-1}$ which indirect tests had suggested should be correct to about $\pm 0.8 \times 10^{-13} Pa^{-1}$ for the capacitor used in the DCGT experiments. The uncertainty in the calculated value of K is rather large here since the fit was not fully optimal, and since there were known systematic errors in the data which were not at this stage corrected: however, this fit clearly shows that there was no gross error in the single-isotherm analysis[1], where K was assumed to be equal to the value for copper. The 15 percent uncertainty would be considerably reduced in a better data-set, and this point was tested from time to time in subsequent surface-fits; however, since the evidence was that the best value for K was negligibly different from that for copper, K was constrained to the value for copper in the great majority of runs to find the optimal forms for the virial coefficients. Subsequent fits for K with constrained virial coefficients are obviously not independent tests, but at least they are necessary tests for self-consistency: for instance, in run 74(i) the calculated percentage differences of K from the value for copper as the data were cut were -0.9 ± 2.6, 0.4 ± 2.7, 1.6 ± 3.0, 3.6 ± 3.4, 5.6 ± 4.5 for populations of 181, 160, 146, 123 and 88 points respectively, while tests with non-optimal coefficients showed much less self-consistency.

3.2.3. The largest term in (17) is the one with the coefficient (A_ϵ/R). Table III gives the values calculated for this coefficient from five surface-fits (each with two sets of fully constrained virial coefficients chosen to represent the uncertainty in fitting the higher order terms in (17)) and also from five similarly constrained fits of the single isotherm at the nominal temperature of 20.275 K, while the values of $\Delta\theta(T)$ for the individual isotherms can be recovered by calculating their mean deviations from the hypersurface (17): the mean values of $\overline{(A_\epsilon/R)}$ from the surface-fits can therefore be normalised to the same temperature as the calculation of (A_ϵ/R) from the reference isotherm at 20.275 K, and this is done in Table III. The five sets of constraints (which will be discussed in more detail elsewhere) represent a range of constraints from what are believed to be the best surface-fits (runs 59, 68, 74, 75), as well as results which are believed to represent a realistic sample of barely acceptable fits. The twenty values of (A_ϵ/R) normalised to 20.275 K show a range of 0.6×10^{-5} cm^3 K J^{-1}, which is ten or more times the statistical uncertainty of an individual fit; however, this distribution of values is certainly a very much better indication of the reliability of the determination of the polarizability than is the result of the original analysis (also quoted in Table III), and by assuming that the best value of (A_ϵ/R) is the mean over this distribution, we now obtain a value of

$$\overline{(A_\epsilon/R)} = 6221.1_3 \pm .0_4 \times 10^{-5} \text{ cm}^3 \text{ K J}^{-1}$$

based on a reference temperature of 20.275 K. This value is in very close agreement with the result of the original single-isotherm analysis, but whereas in that work the best theoretical value was thought to be $6218.50_5 \times 10^{-5}$ cm^3 K J^{-1}, a very significant discrepancy (even allowing for the 99% confidence limits of ± 1 mK on the value of the reference temperature which gives to (A_ϵ/R) an extra experimental uncertainty of $\pm 0.3 \times 10^{-5}$ cm^3 K J^{-1}), recent work by Weinhold[5] which has taken explicit account of the effect of the finite nuclear mass gives a result in exact agreement with experiment. This has two important consequences for thermometry: firstly, a potentially serious discrepancy has been removed from the practice of polarizability thermometry - both DCGT and also the closely similar technique of Refractive Index Gas Thermometry proposed by Colclough[6], and secondly, it seems most reasonable to accept that this exact agreement between experiment and theory is not so much a test of quantum-mechanical calculations, as of thermodynamic accuracy of the reference temperature, as is discussed in section 3.2.4. below.

Table III. Values of $\overline{(A_\epsilon/R)}$ from DCGT surface-fits

$\overline{(A_\epsilon/R)} \times 10^5$/cm^3 K J^{-1} (a)			
Surface-fits - normalized to T_{ref} = 20.275 K		Single-isotherm fits at T_{ref} = 20.275 K	
Run 59 (i)	6221.160 $\pm$ .015	Run 68 (i)	6221.234 $\pm$ .036
(ii)	1.435 $\pm$ .012	(ii)	1.480 $\pm$ .047
74 (i)	1.247 $\pm$ .002	75 (i)	1.286 $\pm$ .020
(ii)	1.395 $\pm$ .006	(ii)	1.476 $\pm$ .020
78 (i)	0.986 $\pm$ .030	80 (i)	0.843 $\pm$ .060
(ii)	1.044 $\pm$ .048	(ii)	0.887 $\pm$ .090
83(a) (i)	0.922 $\pm$ .005	84(a) (i)	1.069 $\pm$ .050
(ii)	1.066 $\pm$ .008	(ii)	1.259 $\pm$ .063
83(b) (i)	0.884 $\pm$ .005	84(b) (i)	0.987 $\pm$ .059
(ii)	0.955 $\pm$ .017	(ii)	1.035 $\pm$ .088
Overall unweighted mean:	6221.13 $\pm$ .06	Original DCGT analysis:	6221.21 $\pm$ .04
Overall weighted mean:	6221.04 $\pm$ .06	Theoretical value[5]:	6221.10$_4$

(a) Each of the fits tabulated is a first-order fit with the virial coefficients fully constrained using two sets of values ((i) and (ii)) which represent their likely range of uncertainty. The values of $\overline{(A_\epsilon/R)}$ are averages over four cut-levels (181, 160, 146 and 124 points), and the standard errors are obtained from the range of these fits. The value of $\overline{(A_\epsilon/R)}$ for the surface-fits is optimised over all the isotherms, but the average value has been normalised to the temperature of the most completely studied isotherm, 20.275 K: the single-isotherm results refer to this temperature directly, and are averages over fits to 34, 25, 22 and 18 points.

3.2.4.
The smoothness of the temperature scale NPL-75 $\Delta\theta(T)$ is given by

$$\Delta\theta(T) = \theta(NPL-75) - T \qquad (7 \text{ bis.})$$

where T is the estimate of thermodynamic temperature calculated from the DCGT isotherm analysis based on the present surface-fits, and where $\theta(NPL-75)$ is the isotherm temperature on the NPL-75 scale as realised in our laboratory. The results are shown in Fig. 2 and in Table IV, where all the results have been adjusted to a reference temperature of 20.275 K, i.e. one is assuming for convenience that $\Delta\theta(20.275 \text{ K}) = 0$, as in section 3.2.3.

The dashed lines in Fig. 2 enclose the results of all the 20 fits listed in Table III; there is no significant dependence of $\Delta\theta$ on the choice of virial coefficients (i) or (ii) shown for the fits of Table III, but for reasons which will be discussed elsewhere, many of these fits constrain the value of B'(T) outside what the data suggest is possible, and the most reliable values for $\Delta\theta$ are shown by the cross-hatched area in Fig. 2 based on the results of runs 68, 74 and 75. The detailed results given in Table IV include results from two of the doubtful fits (78 and 80), while the results from run 59 are still not fully satisfactory since the analysis leads to a dependence of $\Delta\theta(27.17 \text{ K})$ on the cut-level, unlike the results from run 74. Runs 68 and 75 are fits to single isotherms using the same constraints as in the surface-fits 59 and 74: these also show a residual dependence of $\Delta\theta(27.17 \text{ K})$ on the cut-level, but since the isotherms are fitted separately there is not the same interaction with the other isotherms as in the surface-fit, 59.

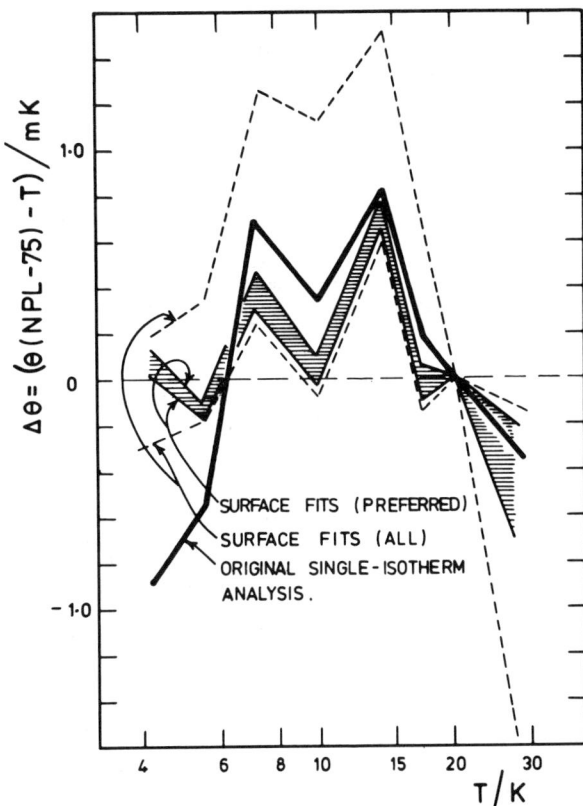

Fig. 2. The smoothness of the temperature scale NPL-75. The solid line shows the result of the original analysis of single DCGT isotherms[1]; the hatched area gives the results for the mean, plus and minus one standard error, for the three most satisfactory fits obtained from the surface-fit analysis (runs 68, 74, 75): the dashed lines give the range for all fits obtained from the surface analysis, including some which were not fully satisfactory (see text).

Table IV. The smoothness of the temperature scale NPL-75

Isotherm temp. T/K	$\Delta\theta^{(a)}$ = $(\theta(NPL-75)-T)/mK$						Original DCGT analysis[1]
	Surface-fit runs			Single-isotherm fits			
	59[b]	74	78	68[b]	75[b]	80[b]	
27.170	$-.78 \pm .17$	$-.11 \pm .05$	$-.48 \pm .03$	$-.6 \pm .6$	$-.6 \pm .6$	$-1.4 \pm .3$	$-.31 \pm .30$
20.275	0	0	0	0	0	0	0
17.023	$.66 \pm .05$	$.07 \pm .03$	$.29 \pm .03$	$-.02 \pm .02$	$-.12 \pm .06$	$.47 \pm .17$	$.19 \pm .19$
13.819	$.89 \pm .03$	$.63 \pm .01$	$.98 \pm .03$	$.82 \pm .07$	$.71 \pm .09$	$1.48 \pm .17$	$.82 \pm .29$
10.010	$.60 \pm .02$	$.05 \pm .02$	$.47 \pm .03$	$.06 \pm .02$	$-.08 \pm .08$	$1.05 \pm .20$	$.35 \pm .14$
7.203	$.62 \pm .02$	$.41 \pm .01$	$.76 \pm .03$	$.47 \pm .06$	$.26 \pm .08$	$1.26 \pm .23$	$.69 \pm .06$
5.552	$.14 \pm .01$	$-.16 \pm .01$	$.04 \pm .02$	$-.10 \pm .05$	$-.17 \pm .07$	$.36 \pm .06$	$-.55 \pm .06$
4.227	$.18 \pm .06$	$.13 \pm .01$	$.06 \pm .01$	$.05 \pm .10$	$-.01 \pm .13$	$-.20 \pm .11$	$-.88 \pm .05$
RMS CHI[c]	1.03_6	$.97_5$	1.10_9	—	—	—	—

(a) The values of $\Delta\theta$ are averages over cut levels of $\mu_{max} = 6, 5, 4$ all times 10^{-4}, and the standard errors are obtained from the range of these fits. These results (and others) are shown in Fig. 2: the results of runs 68, 74 and 75 are believed to be the most accurate estimate of $\Delta\theta$, as discussed in the text.

(b) The values at 27.17 K have a systematic dependence on the range of the fit.

(c) Averages over the three cut levels of (a) above.

The results for the original DCGT analysis are also shown on Fig. 2, where they can be seen to compare reasonably well with the present calculations: the general shape is rather similar, but the preferred fits show generally smaller changes than the original fits, especially at 4.22 and 5.55 K, where all the fits agree that the original single-isotherm analysis was wrong: this is a satisfactory conclusion, since the noise thermometry of Klein et al.[7] had suggested that NPL-75 was only about 0.2 mK below thermodynamic temperature at 4.22 K, a residual discrepancy which is certainly within the combined uncertainty of the two experiments. The present fits show most variability at 27.17 K, and there is also a strong suggestion from the surface-fits that much of the width between the extremes of the dashed lines in Fig. 2 is due to a 'pulling' effect from the value of $\Delta\theta$ at 27.17 K. The largest definite deviation (0.7 mK) is at 13.81 K: this is certainly implied by the data, but they were the first points measured and the values of RMSCHI for single-isotherm fits were always exceptionally high (~ 1.5 or more) clearly indicating a larger than usual random error. The only other apparently significant deviation is at 7.20 K, $\Delta\theta \sim 0.35$ mK. Both of these deviations are really extremely small, and they may in any case reflect only the realisation of NPL-75 in the DCGT experiment which, although it used the best, standard d.c. potentiometric techniques, relied upon a single Rh Fe thermometer from the canonical set which define the scale.

The accuracy of the reference temperature itself cannot, of course, be assessed independently of the determination of (A_ϵ/R), but as discussed in section 3.2.3., it is probably most realistic to accept that the present theoretical value of A_ϵ is correct, when the value of the gas constant, $R = 8.31441 \pm .00017$ J K^{-1} mol^{-1} (see e.g. Colclough[8]), leads to a value of the reference temperature of 20.2750_9 K, i.e. to an absolute value, $\Delta\theta_{ABS}(20.275) = -0.09$ mK, with uncertainties of $\pm 0.4_0$ mK arising from the 20 p.p.m uncertainty in R, and of $\pm 0.5_0$ mK from 99% confidence limits applied to the calculated values of section 3.2.3. These values are in excellent agreement with Berry's original estimate of the accuracy of NPL-75[2], and they also easily include the slight changes suggested by the recalculation of his data in section 3.1.5. above.

4. CONCLUSIONS

The object of this paper has been to present the relatively unfamiliar technique of surface-fitting as an effective method of improving the analysis of experimental data, and to illustrate its power by a reanalysis of data from extremely accurate experiments on isotherm thermometry.

The reanalysis of the CVGT isotherm results obtained by Berry[2] confirms the excellent internal consistency of his work, although it also suggests that the values he assigned to his isotherm temperatures should be corrected by amounts of a few tenths of a millikelvin (Table II). The surface-analysis suggests that the values of $B(T)$ have an uncertainty of only about ± 0.02 cm^3 mol^{-1}, at least for the range 4.2-27.1 K, which is about ten times more precise than Berry's original analysis suggested, but it also suggests that the three constant parametrization of $B(T)$, (9), is not adequate to represent the data at the lowest temperatures. The surface-fits also make it clear that the third virial coefficient is essential for the proper analysis of the CVGT data, and although the optimum values are obtained from the reanalysis of the DCGT data, the CVGT analysis gives definite evidence on the correct form of $C(T)$, information which is not uncovered without the use of surface fitting.

The reanalysis of the DCGT isotherm data[1] shows that the value of the effective compressibility of the capacitor used in the test cell can be deduced from the experimental measurements themselves, so that this term does not represent a difficulty of principle in the use of the technique. A wide range of possible fits to the data leads to little variation in the value of the absolute polarizability, A_ϵ (Table III), or of the form of the error in the temperature scale NPL-75, $\Delta\theta(T)$, Fig. 2. The mean experimental value of A_ϵ agrees exactly with the theoretical calculation, implying that the values of A_ϵ, R and thermodynamic temperature are self-consistent, within 99% confidence limits of about ± 0.7 mK at 20.275 K. The smoothness of NPL-75 is evidently extremely good, and Berry's original estimates of its accuracy are fully substantiated. Finally, this reanalysis shows that DCGT is a robust technique of thermometry, in the sense that it is rather insensitive to details of analysis, and in the sense that it is relatively easy to examine its internal consistency. Since it is a reliable technique as well as in many respects a convenient one, one may expect that DCGT and its close analogue Refractive-Index Gas Thermometry[6] will have an important part to play in developments of accurate low-temperature thermometry which are based on the theoretical value of the polarizability of the helium atom.

ACKNOWLEDGMENTS

I should like to thank the Department of Industry, which supported the original experimental work on DCGT, and also Dr. G.W. Michel for the pleasure of his enthusiastic participation in the project. I am grateful to Mr. R. Jewsbury for help with problems of computation, to Professor F. Weinhold for helpful correspondence on the theoretical value of the polarizability of helium, and to Professor R.G. Chambers for his continued interest. Finally, it is a pleasure to acknowledge the underlying influence of Professor C.A. Swenson in whose laboratory I absorbed some of the powerful fascination of thermometry, and where I discovered how easily capacitance may be measured.

REFERENCES

1. D. Gugan and G.W. Michel, Metrologia 16, 149 (1980). See also D. Gugan, these proceedings.
2. K.H. Berry, Metrologia 15, 89 (1979).
3. C.W. Clenshaw and J.G. Hayes, J. Inst. Maths. Applics. 1, 164 (1965).
4. K.R. Hall and F.B. Canfield, Physica 33, 481 (1967).
5. F. Weinhold, J. Phys. Chem. (to appear).
6. A.R. Colclough, Metrologia 10, 73 (1974).
7. H.H. Klein, H. Klempt and L. Storm, Metrologia 15, 143 (1979).
8. A.R. Colclough, Proc. R. Soc. London Ser. A365, 349 (1979)

Primary acoustic thermometry: Principles and current trends

A. R. Colclough

National Physical Laboratory, Teddington, Middlesex TW11 0LW, United Kingdom

In the past primary acoustic thermometry has been restricted for the most part to the range from 2.5 to 30 K owing to the uncertainty in the gas constant R and to the superiority of gas thermometry at high temperatures. Since R is now much better known and is likely to be better known still, and since sound velocity measurements can now be made with uncertainties approaching 2 parts per million (ppm), it is likely that acoustic thermometry will compete with gas thermometry over a much wider range of temperatures in future. The errors of conventional acoustic thermometry are analysed and past work reviewed. While conventional techniques cannot, on the evidence of previous measurements, be pushed beyond accuracies of ca. 25 ppm, unconventional techniques might be expected to improve on this by an order of magnitude. Three possible new approaches to acoustic thermometry are discussed: the spherical interferometers under development by Moldover and Mehl, unguided ultrasonic interferometers, and "hat-box" interferometers.

1 INTRODUCTION

The technique of acoustic thermometry is an old one. It was invented in 1873 by A M Mayer[1] who considered the possibility of measuring furnace temperatures with a long air-filled acoustic resonator. In more recent times the method has found application for primary temperature measurement in the very low temperature range from ca. 2.5 K to ca. 30 K. Its particular attraction for this purpose is that the directly measured thermometric quantity, the velocity of sound, is an intensive one and so independent of the amount of the thermometric gas employed. This contrasts favourably with the situation in conventional primary gas thermometry of the kind where pressure-volume isotherms are plotted as a function of molar density (pV/nR vs. n/V); here it is necessary to determine the number of moles n (or nR) and the volume of the thermometric gas in the bulb of the thermometer. It is these demanding measurements with their well-known systematic effects associated with dead spaces and adsorption which are responsible for gas thermometry being such a difficult technique[2]. Since dead space and adsorption effects are troublesome to control at very low temperatures where pressure sensing tubes descend through liquid helium baths, any alternative technique which avoids them becomes especially attractive.

However, many would now agree that the systematic effects characteristic of the various methods of acoustic thermometry employed in the past have been perhaps just as problematic as those of gas thermometry. Nevertheless, these effects are very different and so acoustic thermometry provides a useful cross check on the results of gas thermometry in a difficult range of temperature. It is also true to say that acoustic thermometry is now becoming better understood and that future work will proceed with fewer difficulties than in the past.

The method relies on the relation between the velocity of sound, c, in a gas at pressure p and the thermodynamic temperature, T:

$$c^2 = A_0(T) + A_1(T)p + A_2(T)p^2 + \ldots \quad (1)$$

where

$$A_0(T) = \gamma RT/M \quad (2)$$

$$A_1(T) = (2\gamma/M)(B + \frac{2TB'}{3} + \frac{2T^2 B''}{15}) \quad (3)$$

$$A_2(T) = (\gamma/RTM)(\frac{13C}{5} + \frac{16TC'}{15} + \frac{2T^2 C''}{15} - \frac{8B^2}{5}$$
$$+ \frac{98(TB')^2}{45} + \frac{8(T^2 B'')^2}{45} + \frac{8TBB'}{15}$$
$$+ \frac{4}{15}T^2 BB'' + \frac{56}{45}T^3 B'B'') \quad (4)$$

These are the "acoustic virial coefficients" expressed in terms of the virial coefficients proper, 1, B(T), C(T) etc. and their respective first and second temperature derivatives, B', C' and B'', C''[3,4]. M is the molar mass and R the gas constant per mole. Thus by plotting isotherms of measured values of $c^2(p)$ and extrapolating to zero pressure to obtain $c^2(0) = A_0(T)$, one can derive the thermodynamic temperature $Mc^2(0)/(\gamma R)$, γ taking the exact value of 5/3 for monatomic gases. The basic relation (1) follows from the virial equation of state for a real gas and the fluid dynamical equation $c^2 = B_s/\rho$ giving the sound velocity c in terms of the adiabatic bulk modulus of an ideal fluid B_s and its density ρ. Acoustic thermometers may in principle be calibrated at the triple point of water where T = 273.16 K by definition. This effectively evaluates the constant of proportionality $\gamma R/M$ between c^2 and T. In past work this has not been customary since it is difficult to design an instrument to work both at 273 K and at low temperatures. Instead, it has relied on a calculated value for $\gamma R/M$ using independently determined values for R and M. It can also be seen that a sound velocity measurement at the triple point of water effectively determines the gas constant when M is well defined.

Because the gradients of acoustic isotherms are typically gentle, pressure measurement is not usually a significant source of error in acoustic thermometry. Nor is the uncertainty in molar mass M when it needs to be used to calculate T. Generally ^{4}He is employed in the range 2.5 to 30 K with an uncertainty in M of less than 3 parts per million (ppm)[5], but at higher temperatures other inert gases of well defined molar mass might be chosen. In the case of Ar, for example, the molar mass can be determined to 5 ppm or better if measurements are made of the concentrations of the various isotopes present in the experimental sample.

Such an analysis was carried out for Quinn et al.[6] in a determination of the gas constant by acoustic means. Alternatively a pure monoisotopic sample of Ar could be employed.

Until recently the uncertainty in the gas constant has been of some importance in that its use to calculate T has set the upper limit of primary acoustic thermometry between 20 and 30 K. The hitherto generally accepted uncertainty of 93 ppm in R, representing three standard errors, derives from the 1973 adjustment of the fundamental constants by Cohen and Taylor[7]. This would give rise to a corresponding error of 93 ppm in acoustically measured temperatures which may be compared above about 20 K with a crude rule-of-thumb uncertainty of 50 ppm in temperatures determined by primary gas thermometry. Below about 20 K the uncertainties in gas thermometry increase so that a 93 ppm uncertainty in R would present a smaller relative disadvantage for acoustic thermometry. But the current uncertainty in R is arguably no greater than 25 ppm[8], a figure based on the recent acoustic determination by Colclough, Quinn and Chandler[9]. It represents an overall estimate of uncertainty obtained by combining in quadrature systematic uncertainties with the statistical uncertainty of three standard errors. Thus acoustic thermometry need no longer be limited to below 30 K for this reason, although an uncertainty of 25 ppm due to R remains appreciable. In future the uncertainty of the gas constant will probably fall to less than 10 ppm[8] in which case this source of uncertainty will become of secondary importance. However, if future acoustic thermometry can employ much more accurate methods of sound velocity measurement, workers will probably calibrate their thermometers at the water triple point to take advantage of this and uncertainties in R and M will become irrelevant. Then the most significant source of uncertainty at all temperatures will be the uncertainty in the measured value of sound velocity as it has been in previous work below about 30 K.

2 THE MEASUREMENT OF SOUND VELOCITY BY CONVENTIONAL INTERFEROMETRY

If one needs to measure low temperatures in a cryostat, it is not possible to employ time-of-flight methods using pulses since long paths would be required and these are difficult to accommodate. Problems can also arise with identifying the defining edge of the pulse since the sharper this is, the worse will be the problems of dispersion. Continuous wave methods are not used either, due to problems of path accommodation and the elimination of reflections. It is for these reasons that virtually all past work has resorted to interferometric techniques [10, 11, 12, 13, 14, 15].

The main problem for anyone embarking on acoustic thermometry is to correct for the systematic errors of interferometry. The particular systematic errors which arise depend upon the method of acoustic interferometry which is chosen. For example, with ultrasonic variable-path fixed-frequency interferometry such as that employed by Plumb et al.[12, 13] at the National Bureau of Standards (see Table I), the main difficulty

Table I

Some Past Work in Acoustic Thermometry

Work	Frequency (kHz)	Interferometer Type	Temperature Range (K)
Van Itterbeek[10] 1957	500	VP,US	4.2, 20
De Laet[11] 1960	3-15	VF,AF	2.1-4.2, 20,...
Plumb et al[12] 1966, 1977	1000	VP,US	2.3-20
Grimsrud & Werntz[14] 1967	1.9-2.6	VP,AF	1.2-3.8
Colclough[15] 1979	3.3-7.3	VP,AF	4.2-20

VP = Variable path, VF = variable frequency.
US = ultrasonic, AF = audiofrequency.

is to ensure that higher mode propagation does not lead to the measurement of a group velocity significantly different from the plane wave phase velocity appearing in the fundamental thermometric relation. It has been argued that it is very difficult to be absolutely sure that this effect is properly corrected for although it is unlikely to be large in practice[16]. But if, on the other hand, one operates at audiofrequencies, so avoiding this problem, one has to cope with significant boundary layer errors. In what follows, the various advantages and disadvantages of each method will be compared and a list of systematic errors drawn up.

Many different kinds of interferometer can be conceived of. Some of the less conventional ones will be discussed in Section 4. But among the conventional types there is still a variety of geometries and methods of operation and it can be difficult to assess which design is best for acoustic thermometry. In order to clarify the options available they are here classified according to the four following design choices:

(a) Variable path and fixed frequency
     OR
     Variable frequency and fixed path?

(b) Audiofrequency OR ultrasonic?

(c) One OR two transducers?

(d) What geometry?

We consider these options in turn. Tables II and III summarise the various advantages and disadvantages of the options presented in (a) and (b).

Choice (a): Variable path or variable frequency?

In the variable-path, fixed-frequency interferometer (as used by Plumb et al.[12,13], Grimsrud and Werntz[14] and Colclough[15]) a linear cavity dimension is changed, eg by withdrawing an acoustic reflector, to bring the cavity from one order of resonance to another at a constant frequency. Thus one only needs to measure changes in length to obtain wavelength and hence sound velocity. This is easier than accurately measuring absolute length as one has to in a fixed-path cavity

Table II

Relative Merits of Variable Path and Variable Frequency Interferometry

Variable Path & Fixed Frequency		Variable Frequency & Fixed Path	
Advantage	Disadvantage	Advantage	Disadvantage
No absolute length measurement is required	Complexity of moving parts	Simple cavity interfmtr.	Absolute length measurement required
All cavity parameters are constant for all orders of resonance			All cavity parameters change from one resonance to the next
End face losses can be distinguished from absorption losses			End face losses inseparable from transmission losses
Quartz transducers can be used at ultrasonic frequencies			Cannot use quartz transducer at US frequencies

Table III

Relative Merits of High and Low Frequency Interferometers

Audiofrequency		Ultrasonic	
Advantage	Disadvantage	Advantage	Disadvantage
Boundary layer errors measurable and calculable.....	but large and time consuming	Boundary layer errors small, though not negligible..........	but not well behaved?
			Serious problem of high mode analysis
	Quartz crystal transducer not available	Quartz crystal transducer available	Cavity alignment must be very accurate

(such as that used by de Laet[11]). But such variable-path cavities have moving parts and so are more complicated which is not a trivial consideration in a cryostat.

There are also two very important advantages of variable-path interferometry of an acoustic nature. Firstly, all the acoustic parameters which characterise the cavity (sound velocity with its small boundary layer dependence, absorption, reflection losses, reflection phase changes and hence effective cavity length, and the real and imaginary components of transducer impedance) are frequency dependent. Thus with the variable frequency method these change from one order of resonance to the next. Indeed, to second order, they change over a single resonance. With the variable path cavity, on the other hand, they remain constant at all orders of resonance since it is operated at constant frequency. This greatly simplifies the analysis of the acoustic resonances. The fact that with the variable-path instrument, absorption and reflection losses are constant confers the second very important advantage on the method. It means that they can be separated experimentally without recourse to theory by measuring the losses at two or more orders of resonance. Then one can check that loss mechanisms, especially those arising from the acoustic boundary layer, are as predicted by theory. This provides an invaluable check on the operation of the instrument and, if the instrument appears not to function normally, aids in diagnosis. Absorption coefficients can also be used directly to correct the measured sound velocity for the effect of the acoustic boundary layer.

Finally, if it is decided to operate at ultrasonic frequencies (see below), commitment to a fixed-frequency variable-path instrument allows one to employ a quartz crystal transducer operated at resonance where its intrinsic impedance is effectively zero. Thus a measurement of its impedance under load is effectively a direct measurement of the impedance of the loading resonant cavity and this greatly simplifies data processing.

Choice (b): Audio or Ultrasonic Frequencies?

Whether it is best to use audiofrequency or ultrasonic interferometry is a controversial matter. The author has argued in the past (1973) for acoustic frequencies[16]. He is still of this view, if less strongly, but the reasons have changed somewhat over the years. There used to be two main reasons why ultrasonic thermometry was thought to be problematic.

Firstly, there was no practical test one could make

to ensure that only the plane wave (00-order) mode propagated in the cavity. In fact it seemed quite probable that one would be measuring not the required plane wave phase velocity, but a group velocity corresponding to a group of unresolved higher modes. (Calculations of the likely magnitude of the resulting error in velocity and temperature measurement are given in Tables 4 and 5 of reference 16, but are based only on guesses about transducer behaviour). In the audiofrequency device the frequency is held below the first cut-off frequency of the cavity so that only the plane wave propagates. This requires that operating wavelengths are greater than 3.42 cavity radii for the cylindrical interferometer. It was at one time assumed that the presence of higher modes might be detected from changes in the separation of the resonances with order or from measurements of acoustic absorption coefficients. But the former assumption is incorrect as detailed analysis shows. Similarly, while increased absorption coefficients might be due to higher modes, their absence does not guarantee the absence of higher modes. All that can be done is to assume that the transducer executes "plausible" vibrations and to calculate the resulting error. Then an uncertainty could be quoted covering this error several times. However, the error can be made quite small (about 1 mk ca. 4.2 K perhaps) if very high frequencies are employed so that the various phase velocities of the higher modes are close to each other. But if frequencies are increased too much acoustic absorption, proportional to the square of the frequency at high frequencies, may increase rapidly and bring about a loss of sensitivity.

The second reason why ultrasonic interferometry was thought to be unreliable was that no theory existed giving the tolerances to which the cavity end faces needed to be in alignment. At high frequencies with very short ultrasonic wavelengths it is possible for resonance widths to approach optical dimensions given ideal geometry. But if the cavity has geometrical errors of this order could one be sure that one understood its behaviour as is required of a primary instrument? No answer was available in 1973. But the question was later treated by the author[17] who derived the following expression:

$$X << \lambda (\alpha 1 + \beta)/(5.8 \pi b) \quad (4)$$

where X is the angle of end face tilt, α the absorption coefficient, β the end face reflection loss, λ the wavelength and b the cavity radius. This provides a realisable criterion for adequate cavity alignment in the ultrasonic interferometer and shows that optical cavity alignment is generally satisfactory.

But another problem has arisen for ultrasonic interferometry since then. No absolutely clear explanation has been found for Plumb's discovery[18] that the boundary layer effect disappeared at low pressures (e.g. between 10 and 70 kPa at 1 MHz and ca. 30 K), but it may be due to the molecular mean free path approaching the acoustic boundary layer thickness. This effect can easily be detected by measuring the acoustic absorption (bearing out the earlier point on the importance of measuring one's losses). This raises the question as to how one should correct for the boundary layer effect on the sound velocity. It was originally thought that the small boundary layer effect at ultrasonic frequencies constituted an important advantage of the method. But an ultrasonic boundary layer effect which is not corrected for or which is wrongly corrected for in its absence can introduce a velocity error even larger than that which remains in an audio measurement after making an approximate correction for the greater audio boundary layer effect.

However, it may be the case that in practice the extra work required to control the larger boundary layer effect at low frequencies, while unproblematic in principle, imposes a considerable burden on the low frequency method which results in fewer data and so poorer statistical control. And certainly, the availability of quartz crystal transducers at ultrasonic frequencies is something that any audiofrequency worker will envy.

To summarise: ultrasonic thermometry is experimentally more convenient and productive, but can raise problems of interpretation while the opposite is the case with audiofrequency thermometry. Thus which is the best to pursue is a difficult matter of judgement.

Finally, there is another serious systematic error which the author has encountered in his audiofrequency work and which is not included in the above table because it may well occur in ultrasonic interferometry as well. This is due to the nonlinearity of the transducer used to excite the interferometer. This error has not been appreciated until recently although it has[15,19] been common knowledge for very many years that transducers, when driven sufficiently hard, are nonlinear. Now that its existence is known, it need not pose a serious threat either to ultrasonic or audiofrequency methods, but care is necessary to ensure that it is properly controlled.

Choice (c): One or Two Transducers?

Most workers use one transducer both to excite the acoustic resonances and to monitor them. Others (de Laet[11] and Grimsrud and Werntz[14]) use one to excite the resonances and another to monitor them. The former is simpler which is doubly important in a cryostat, but the latter has the advantage that data processing may be simplified since a pressure sensitive (as opposed to a velocity sensitive) transducer at the opposite end of the cavity to the exciting transducer produces symmetrical resonance peaks like the quartz crystal single transducer. (Single transducer interferometry was assumed in the preceding discussion.)

Choice (d): What Geometry?

In principle any geometry can be employed where a linear dimension of the resonator can be related to the acoustic wavelength. Needless to say the number of possible resonators where no change of dimension is called for, ie variable-frequency resonators, is larger than the number of variable-path resonators. For example a spherical resonator (a Helmholtz resonator) might be used such as that considered recently by Moldover and Mehl[20,21]. This has the advantage that there are no transmission boundary layer losses, but only reflection losses (see Section 4). However, when operating with a fixed frequency and a variable path as advised above, then one needs a geometry with a linear axis of symmetry. Resonators of rectangular or circular cross section have been employed at ambient temperatures, but the conventional cylindrical interferometer has always been favoured in cryostats. This is because it is difficult to manufacture a rectangular cavity accurately without assembling four separate walls and this introduces problems of sealing at low temperatures. Manufacture of the cylindrical cavity is, of course, simpler. But, even in this case, great care should be taken to ensure excellent and stable end face alignment at ultrasonic frequencies and at all frequencies that the wall roughness is small compared to the expected acoustic boundary layer thickness.

A Summary of Systematic Effects in the Conventional Method

Table IV lists the systematic effects of acoustic thermometry with some explanatory notes and references. Not all of the effects are problematic because in some cases their influence can be made negligible. (These are marked with an asterisk.) But all need to be

Table IV

The Systematic Effects of Conventional Acoustic Thermometry

	EFFECT	ERROR/ppm AF	ERROR/ppm US	NOTE
1	Higher modes propagate $dT/T \sim (X_{33})bk_{oo}^2$	0*	500?	1
2	Acoustic boundary layer $dc/c = \alpha_{BL}/k$	3000 (2)	200?	2,3
3	2nd order BL correction from Kirchhoff-Helmholtz 1st order BL correction	30 (0)	0 *	2,4
4	Molecular slip at wall	2 (0)	0 *	2,5
5	Temperature discontinuity at wall	4 (0)	0 *	2,6
6	Finite thermal conductivity	1 (0)	0 *	2,6,7
7	Wall smoothness	*	0	0	8
8	Wall elasticity		3	0	6,7
9	Wall adsorption	*	0	0	9
10	Finite sound amplitudes	*	0	0	10
11	Non-uniform cavity bore	*	0	0	11
12	End face misalignment	*	0	0	12
13	2nd order correction for using 1st order theory $\sim (\alpha/k)^2$	*	0	0	13
14	Blatt effect	*	0	0	14
15	Cavity length measurement (variable path assumed)		2	2	15
16	Resonance processing (geometrical processing of resonance circles or peaks assumed)		10-100	10-100	
17	Frequency measurement		0	0	
18	Pressure measurement (for isotherm plotting)		0	0	
19	Resistance thermometry (Rh-Fe or Pt)		0.3 mk	0.3 mk	
20	Nonlinearity of transducer	*	150 (20)	unknown	16,2
21	Isotherm fitting - effect of ignoring $A_2(T)p^2$		200 (0)	200 (0)	17,2

* Signifies that the error is negligible or eliminable

Notes to Table IV
1. Error calculated using Colclough, reference 16, paragraph (iii) page 96. It represents only a plausible guess. $X_{33} = 14.54$ is a constant characterising the 33-mode.
2. Figures in brackets indicate error remaining after making the correction.
3. See eg Colclough, reference 15 on p 353 eq. (6). It is assumed that α_{BL} is measured to 0.05 % at audiofrequences and used to correct c.
4. See reference to Shields, Lee and Wiley[22] in Quinn, Colclough and Chandler[6] on p 375.
5. See note 4 reference and p 402 of same paper.
6. See Henry[23] and Weston[28].
7. This extrapolates to zero with pressure and so does not affect isotherm measurements.
8. Protrusions and interstices on walls must be negligible compared to the boundary layer thickness. See p 376 in Quinn, Colclough, Chandler[6].
9. See reference of note 3, p 354.
10. This is easily checked by measuring velocities at various amplitudes.
11. Second order effects due to changes in the boundary layer effect could in principle arise.
12. See Colclough[17] especially eq (46).
13. See Appendix B of reference in Note 8. But see also Colclough, Quinn and Chandler[9] on p 128. One needs to think clearly whether to apply this correction.
14. See reference of note 2 on p 353. Also reference of note 8 p 410.
15. This assumes interferometric measurement with an unstabilized He-Ne laser.
16. This is typical of the author's experience, but with foresight the error might be eliminated.
17. It is easy to misinterpret an isotherm and to assume that apparent linearity implies that the $A_2(T)p^2$ term is negligible. This is not so and errors of this order can arise from linear extrapolation to zero pressure.

considered in a primary measurement. The errors are calculated for a typical audio frequency (AF) and a typical ultrasonic (US) thermometer. A question mark implies that there may be problems making corrections for the effect.
22623

3 PAST WORK IN ACOUSTIC THERMOMETRY

There exist two relatively early sets of measurements of the low frequency type due to De Laet[11] (following other preliminary work at Leiden) and to Grimsrud and Werntz[14]. De Laet used a simple cavity of fixed length and determined sound velocities from measurements of its resonant frequencies, a technique fraught with difficulties as described in the previous section. It was necessary to investigate end effects empirically by changing the cavity ends faces. Grimsrud and Werntz avoided these problems by using a cavity of variable path excited at some constant frequency. With the exception of a value of the boiling point of hydrogen from De Laet, neither of these investigations extended beyond the region between 2.0 and 4.3 K. Temperatures determined by De Laet were higher than temperatures measured on the helium-4 vapour pressure scale T_{58} by as much as 22 mK at 3.2 K and 12 mK at 4.2 K. This last figure compared with a 32 mK discrepancy found in the earlier Leiden work by using ultrasonic techniques, but the interferometry used there was too rudimentary to yield useful information at the accuracies currently of interest[10]. The acoustically determined temperatures of Grimsrud and Werntz obtained using both ^{3}He and ^{4}He as the thermometric gas exceeded temperatures measured on the T_{58} scale by between 1 and 7 mK over the range 1.2 to 3.8 K.

These results of Grimsrud & Werntz are in general agreement with the ultrasonic thermometry of Plumb and Cataland[12, 13] in the region in which they may be compared (see below). The temperatures determined by the latter using a variable path interferometer at a frequency of 1 MHz were higher than the T_{58} temperatures by between 5 and 12 mK in the region 2 to 5 K with a 10 mK difference at the normal boiling point of helium 4. These measurements of Plumb and Cataland extend over the region of 2 to 20 K at intervals of approximately 1 K forming a fully fledged ultrasonic temperature scale. Their scale was classified as provisional by the authors pending a clarification of its systematic uncertainties but has nevertheless achieved widespread use. Another set of results from the same workers[13,15] were later produced using a refined version of their original instrument.

More recently Colclough[15] has produced a set of six low frequency isotherms which can be accurately compared with the gas thermometry scale NPL-75 of Berry[24] (see Figure 1).

The instruments of Colclough and Plumb et al. are shown in Figures 2 and 3 respectively. The former's can be seen to consist of an optical interferometer and an acoustic interferometer both accommodated in a conventional liquid helium cryostat. The optical interferometer is used to measure changes in acoustic path and employs a 1 mW helium-neon laser with room temperature photocells triggering a bidirectional counter. The length measurement itself takes place isothermally in the acoustic interferometer region to

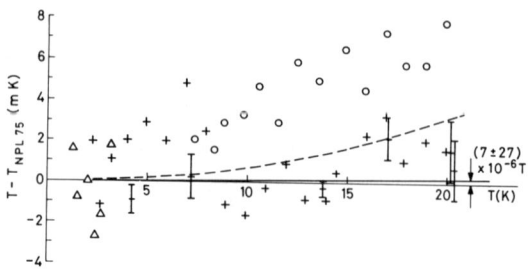

Figure 1. Comparison of the results of past work in acoustic thermometry with Berry's gas thermometer scale T - NPL75. Error bars represent 1σ. +, earlier work of Plumb et al., o, later work of Plumb et al., Δ, Grimsrud and Werntz, -----, T-XAC'.

avoid problems with the changing thermal contractions found in the conventional pushrod and external micrometer arrangement as used in the earlier measurements of Plumb et al. The moving mirror, the cube corner reflector J, travels on the rear face of the acoustic reflector, the piston H. The fiducial mirror, fixed relative to the transducer face, is the one-half half-silvered mirror L. Though not strictly necessary when extrapolating measured velocities to zero pressure, all optical length measurements were corrected for the refractive index of the thermometric gas. Above the acoustic and optical interferometer is an upper chamber P in which a bearing O slides. This houses a low temperature vacuum tight optical window sealing the end of the evacuated light entry and exit tube Q which also transmits the external drive to the bearing and thence, via gimbals M and pushrods K, to the piston H. The thermometric gas communicates with the instrument through the annulus between Q and the supporting tube W. Difficulties were experienced with gas instabilities in this space leading to loss of temperature control at sufficiently high pressures at any given temperature. This was much improved by winding the inner tube with nylon cord following the practice of Plumb et al. in response to similar difficulties.

The first cut-off wavelength Λ_{10} of a cylindrical guide or resonator is given by

$$\Lambda_{10} = 2\pi b/1.84 = 3.42b, \quad (5)$$

where b is the cavity radius. Thus, having decided to work below the first cut-off frequency to ensure that an unambiguous plane wave phase velocity is measured, one is constrained to use wavelengths greater than 3.42 cavity radii. It was decided, somewhat arbitrarily, that a little over five half waves should be a sufficient length for the cavity giving three more than the absolute minimum of two. Should any systematic deviation of measured velocity with path occur it would be likely to change as l changed from l = λ/2 to l>>λ/2, thus betraying its presence. To keep the cavity as short as possible it was desirable to use a bore of small radius. But then boundary layer errors would increase and the loading of the gas column on the transducer would decrease, leading to a drop in sensitivity. The radius b = 1 cm was finally chosen which just enabled the smallest cube corner available to be mounted on the rear face of the moving acoustic reflector. This resulted in a usable cavity length of 9.4 cm which allowed five resonances to be accommodated at a frequency of 0.9 of the cut-off frequency, f_{10}.

The acoustic cavity was made to exacting tolerances. The piston was chrome plated to prevent it wringing to the bore of the cavity and lapped to a fit of a few micrometres. The face of the piston was mirrored with an aluminium film so that variations in its alignment with position and orientation could be checked by an autocollimator. It was found to be square to the axis by between 7 and 15 microradians. Such a misalignment gives rise to no error in measured velocity and a reflexion loss of 10^{-6} at $\lambda = \lambda_{10}$ which is entirely

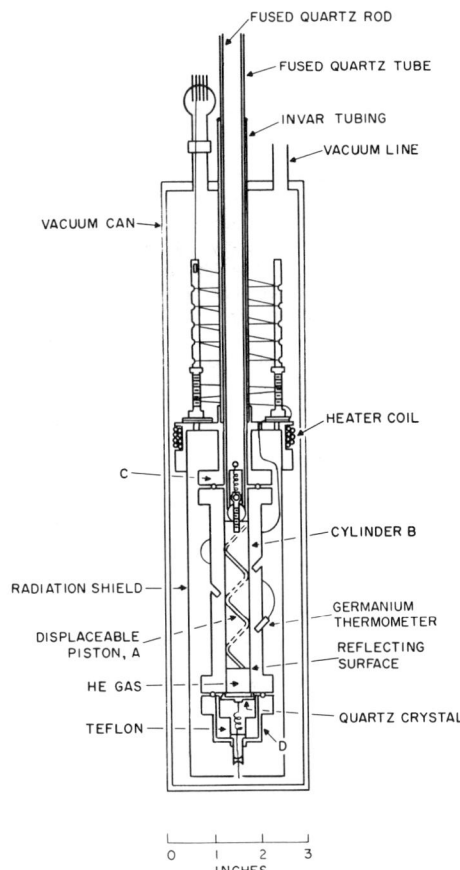

Figure 3. The ultrasonic thermometer of Plumb et al.

negligible compared to that of other mechanisms[17]. The moving coil-driven 'Duralumin' diaphragm F, 0.25 mm thick, was clamped between lapped steel blocks, subjected to a load of some 50 tons in a hydraulic press and annealed between the same blocks at 450 °C under a load of 20 kg for several hours. Its alignment was defined by clamping against a carefully machined annular surface. This surface was checked by placing a mirrored glass plate representing the diaphragm against it with a window in the mirroring looking into the mouth of the cavity. Thus two reflected autocollimator graticules, one from the substitute transducer and one from the piston face, could be compared. Their maximum relative misalignment was 28 microradians. Attributing all this to one end face implies another entirely negligible end face reflexion loss of approximately 4×10^{-6}.

The tolerances to which this instrument was built are perhaps excessively stringent, but since the effects of end face misalignments were not fully understood at the time, the author erred on the side of safety. Future workers need not go to so much trouble.

The temperature was stabilized by use of a four lead germanium resistance thermometer S as a two lead sensor in an a.c. Wheatstone bridge with compensating leads. The phase-sensitively detected output of the bridge

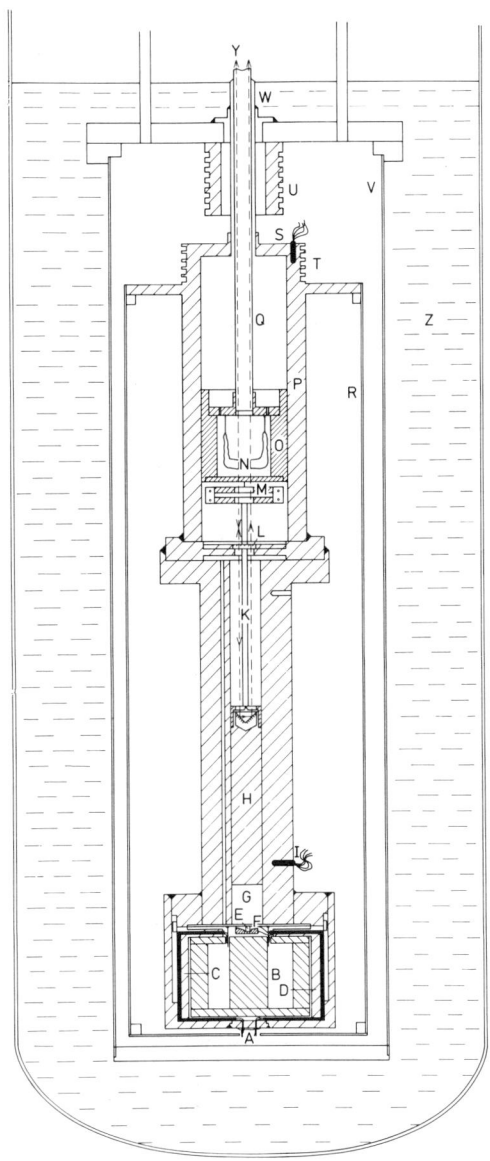

Figure 2. The audiofrequency acoustic thermometer of Colclough.

A Stycast seals, B permanent magnet assembly, C and D electrical lead screens, E PZT accelerometer, F transducer diaphragm, G acoustic cavity, H piston reflector, I Ge resistance thermometers, J cube corner reflector, K pushrods, L beam splitter, M gimbals, N optical window, O bearing, P upper chamber, Q moving tube, R radiation shield, S temperature controlling sensor, T thermal anchoring grooves (with heater), U 4.2 K thermal anchoring grooves, V vacuum can, W central supporting tube, Y laser beams, Z liquid He bath.

controlled the current in an electrical heater. Bridge earthing took place at one sensor lead to ensure good thermal contact with the interferometer. It is not necessary to control the temperature exactly for the duration of a whole isotherm point measurement (approximately an hour), but only during the plotting of the first and last resonances from which the velocity of sound is calculated. This required stabilities of ca. 0.1 mK for just a few minutes which were easily achieved. Any drift in temperature between the first and last resonances was corrected for.

Resonances were analysed by plotting admittance circles. This was done with the aid of a small velocity transducer mounted on the rear face of the acoustic transducer, in-phase and quadrature phase sensitive detectors and an x-y plotter. The circles enabled absorption coefficients to be measured as well as the exact point of resonance. The coefficients were subsequently used to make measured corrections for the effect of the acoustic boundary layer[15].

The instrument of Plumb employed a quartz crystal transducer driven on resonance at 1 MHz and used a rod and micrometer to measure reflector displacements. But in most other respects it was similar to the author's interferometer. The second instrument of Plumb et al. also employed optical interferometry to measure reflector displacements.

From Figure 1 it can be seen that the acoustically determined temperatures of Colclough deviate from those determined by Berry's gas thermometry by a statistically insignificant amount (7 ppm with a standard error of 27 ppm). Both the earlier and later results of Plumb et al. appear to have some systematic structure which is different in the two pieces of work. The earlier work is in good overall agreement with Berry's gas thermometry, but the later shows a significant departure of ca. 200 ppm. This might be due to the problem of higher modes discussed in the previous section, but there is no way of checking this. The broken line shows the magnetic thermometry scale T-XAC' fitted to Plumb's points[13]. The low frequency work of Grimsrud and Werntz agrees with all other work in its region of overlap, but the scatter approaches 1000 ppm and so the agreement is not significant at levels currently of interest. The three-standard-error statistical uncertainty of 81 ppm in the work of Colclough is about three times larger than the overall uncertainty of 25 ppm in the determination of the gas constant by Colclough, Quinn and Chandler using a similar instrument at 273 K. The latter instrument was developed after the acoustic thermometer and incorporated many lessons learned with the former. Its performance probably represents more realistically what can now be expected from low-frequency acoustic thermometry of a conventional kind.

In the next section a comparison will be made with less conventional approaches to acoustic thermometry and the ultimate accuracy which is likely to be possible in the future will be estimated. This should apply over a range of temperatures greatly exceeding that in which the technique has been used in the past.

4 UNCONVENTIONAL ACOUSTIC THERMOMETRY

The conventional low frequency variable path cylindrical resonator operated at audiofrequencies is capable of accuracies of the order of 25 ppm. Given that the exact condition of the Nth resonance can be identified to within 10^{-2} to 10^{-3} of the Nth resonance half width $\Delta_N = (\alpha l_N + \beta)/k$, one can see that transmission losses, ($\propto \alpha l_N$) and end face losses ($\propto \beta$) are going to limit its ultimate performance. Assuming that the maximum value of l_N is fixed by apparatus size and that c is fixed by gas choice, the ultimate achievable resolution will vary as root frequency given that both α and β are attributable almost entirely to boundary layer effects. For ^{4}He at 4.5 K and a pressure of 10^5 Pa in a cavity of radius 1 cm, α/cm and β are very roughly equal for typically used frequencies in the range 10^2 to 10^4 Hz (see Table V). Assuming for the last resonance that $l_N = 10$ cm, it can be seen that transmission losses are about ten times reflection losses. Since it is the separation of first and last resonances from which velocity is determined, it is clearly the case that the attainable velocity resolution is effectively limited by transmission losses. The same conclusion holds for ^{4}He and Ar at 273 K and a pressure of 10^5 Pa (Table V).

It is this thought that has moved some workers to look for unconventional interferometers where there are no significant axial transmission boundary layer losses. Three such devices will be briefly considered: the spherical interferometers of Moldover and Mehl, free propagation ultrasonic interferometers such as that employed by Sauder to determine the gas constant and "wide-cavity" or "hat box" interferometers.

4.1 Spherical interferometry

Radial modes in a Helmholtz spherical resonator are obviously unguided and so suffer no axial transmission losses due to guiding walls. Thus the only part which viscosity plays in attenuation is through the familiar internal absorption coefficient which is generally several orders of magnitude below typical boundary layer absorption coefficients at audiofrequencies (Table V). There is a reflection boundary layer loss due to thermal conduction at the spherical walls which is exactly the same as that seen in cylindrical interferometers since the walls of a resonator of a few cms diameter will have negligible curvature on the scale of the boundary layer thickness. One would expect, therefore, resolutions of about an order of magnitude better than attainable with conventional interferometers. Moldover and Mehl[20,21] have investigated this possibility and they conclude from their preliminary results that resolutions of the order of 3 ppm are attainable in a practical instrument. This development is of considerable importance for thermometry since it holds out the prospect of ppm levels of accuracy in our knowledge of thermodynamic temperature over a wide range of temperatures.

However, spheres have two possible disadvantages compared to cylindrical resonators. They cannot be manufactured in a variable-path form and, just as with cylindrical fixed-path resonators, the change of frequency from one order of resonance to the next causes all acoustical quantities defining the system to change. This means that at least two spheres of different diameters are required if such quantities as the boundary layer reflection phase changes are to be measured rather than calculated on the basis of a theoretical model. Moldover and Mehl solve the problem in just this way[21].

The second possible disadvantage is that any fixed-path resonator requires an absolute measurement of its critical dimensions in terms of which the wavelength is measured (cf the discussion of Section 2 on the interferometer of De Laet[11]). In the case of spherical interferometers it would be the volume which needed to be determined. If a spherical interferometer is designed to cover a range encompassing the triple-point of water this can be achieved acoustically simply by calibration at T = 273.16 K, though if the gas constant is to be determined in addition to thermodynamic temperatures an independent volume

Table V

Acoustic Losses for ^{4}He at 4.2 K and 273 K and for Ar at 273 K

^{4}He at 4.5 K and 10^5 Pa

Frequency (Hz)	β	$^b\alpha_{BL}$ (/cm)	α_{int}
10^2	1.4×10^{-4}	1.2×10^{-4}	3.9×10^{-10}
10^3	4.5×10^{-4}	3.8×10^{-4}	3.9×10^{-8}
10^4	1.4×10^{-3}	1.2×10^{-3}	3.9×10^{-6}
10^5	4.5×10^{-3}	3.8×10^{-3}	3.9×10^{-4}
10^6	1.4×10^{-2}	1.2×10^{-2}	3.9×10^{-2}
10^7	4.5×10^{-2}	3.8×10^{-2}	3.9×10^{0}

Values used in calculating the above losses:

$c_p = 8002$ J kg^{-1}K^{-1}, $\gamma = 2.453$,
$c = 108.2$ ms^{-1}, $\eta = 1.30 \times 10^{-6}$ Pa.s,
$K = 0.0106$ Wm^{-1}K^{-1}, $\rho = 14.50$ kgm^{-3}.

Table V - continued

^{4}He at 273 K and 10^5 Pa

Frequency (Hz)	β	$^b\alpha_{BL}$ (/cm)	α_{int}
10^2	3.0×10^{-4}	3.4×10^{-4}	5.2×10^{-10}
10^3	9.6×10^{-4}	1.1×10^{-3}	5.2×10^{-8}
10^4	3.0×10^{-3}	3.4×10^{-3}	5.2×10^{-6}
10^5	9.6×10^{-3}	1.1×10^{-2}	5.2×10^{-4}
10^6	3.0×10^{-2}	3.4×10^{-2}	5.2×10^{-2}
10^7	9.6×10^{-2}	1.1×10^{-1}	5.2×10^{0}

Values used in calculating the above losses:

$c_p = 5193$ Jkg^{-1}K^{-1}, $\gamma = 1.667$,
$c = 972.6$ ms^{-1}, $\eta = 1.870 \times 10^{-5}$ Pa.s,
$K = 0.145$ Wm^{-1}K^{-1}, $\rho = 0.1787$ kgm^{-3}.

Table V - continued

Ar at 273 K and 10 Pa

Frequency (Hz)	β	$^b\alpha_{BL}$ (/cm)	α_{int}
10^2	3.2×10^{-4}	3.6×10^{-4}	1.9×10^{-9}
10^3	1.0×10^{-3}	1.1×10^{-3}	1.9×10^{-7}
10^4	3.2×10^{-3}	3.6×10^{-3}	1.9×10^{-5}
10^5	1.0×10^{-2}	1.1×10^{-2}	1.9×10^{-3}
10^6	3.2×10^{-2}	3.6×10^{-2}	1.9×10^{-1}
10^7	1.0×10^{-1}	1.1×10^{-1}	1.9×10^{0}

Values used in calculating the above losses:

$c_p = 521.8$ Jkg^{-1}K^{-1}, $\gamma = 1.6706$,
$c = 307.8$ Ms^{-1}, $\eta = 2.1005 \times 10^{-5}$ Pa s,
$K = 1.6454 \times 10^{-2}$ wm^{-1}K^{-1},
$\rho = 1.78382$ kgm^{-3}.

determination will be required. But not all acoustic thermometers need to operate at such high temperatures where gas purity problems are much more likely than in the traditional acoustic thermometry range of 2.5 to 30 K. In such cases it might be necessary to determine resonator volume by weighing when filled with water or by pressure change measurements during volume-to-volume gas transfers at 273.16 K. These are the sort of demanding measurements characteristic of gas thermometry[24] or of the limiting density method for determining the gas constant[8] which, it was hoped, acoustic interferometry would afford a method of circumventing. Moreover, for use at low temperatures, it is necessary to correct the volume for thermal contraction. Moldover and Mehl[21] suggest the use of microwave resonance techniques to monitor the change in volume with temperature.

For a fuller discussion of the development of spherical resonators, the reader is referred to the work of Moldover and Mehl[20,21].

4.2 Unguided ultrasonic interferometry

Sauder has recently described an ultrasonic interferometer employing two arms which is being developed to measure the gas constant[25]. Its main feature of interest in the light of the preceding considerations is that it has no guiding walls and so no boundary layer effect for transmission. It is worth considering whether unguided ultrasonic beams, which under suitable conditions are self-collimating, might be employed for acoustic thermometry.

One possible approach to deriving the wavefield arising from a finite circular ultrasonic transducer is to treat the transducer as being located in the centre of an end face of a conventional interferometer or semi-infinite guide and to allow the cavity diameter to approach infinity while keeping the transducer diameter constant. In the limit of infinite radius one finds that all modes propagate and have the same phase velocity. The beam is well collimated in the near field (Fresnel) zone which extends, to a distance approximately given by (transducer diameter)^2/wavelength. This is about 1 m for a frequency of 1 MHz when $c = 3 \times 10^2$ m/s. Thus it is possible to avoid wall effects in an apparatus of practical dimensions.

But as frequency rises so does the loss in the bulk of the gas given by the familiar internal absorption coefficient α_{int} which increases as frequency squared. Assuming that maximum acoustic paths remain unchanged from values employed at low frequencies (l = 10 cm, say) and that it is still possible to resolve resonances to 10^{-2} to 10^{-3} of their half-widths, the appropriate figures for comparison of high and low frequency resolutions would be α/f and β/f. For ^{4}He at 4.5 K and a pressure of 10^5 Pa one finds that α/f at 1 MHz is about eight times less than at 1 kHz for a cavity of radius 1 cm while β/f is about thirty times less (Table V). At s.t.p. the situation is comparable both for ^{4}He and Ar. Although the assumptions on which these calculations are based are somewhat optimistic, they provide an adequate rough guide to what can be expected. Thus unguided ultrasonic interferometry is a plausible approach to the problem of wall absorption if a metrologically satisfactory wavefield can be produced.

4.3 "Hat Box" Interferometry

In Section 3 the author's low frequency interferometer was described having a radius b = 1 cm. However, it is possible in principle to employ a cavity of much larger radius, e.g. with b = 10 cm. This would reduce the boundary layer transmission losses by a factor of ten. If only two resonances are to be scanned, the cavity must be extensible to at least about 35 cm in length if the first higher mode is not to propagate. Large cavities like this are rather difficult to accommodate in off-the-shelf commercial cryostats, especially if allowance has to be made for a reflector actuation mechanism, transducer(s), wall thickness and radiation shields. The appetite for liquid helium of such a cryostat could be a matter of concern too in some laboratories. Special consideration would also have to be given to safety at operating pressures above atmospheric. Nevertheless, given the other costs and difficulties of primary measurements such an approach might well be considered practicable.

It is a simple matter to calculate the likely resolution of such an instrument. For ^{4}He at 4.5 K at a pressure of 10^5 Pa, $\Delta_1 = 3.5 \times 10^{-5}$ m and $\Delta_2 = 5.6 \times 10^{-5}$ m for a wavelength of 35 cm. Expressing the expected velocity resolution as $\delta \propto 2(\Delta_1^2 + \Delta_2^2)^{1/2}/(\lambda/2)$ where δ is the resolution to which one can locate the point of resonance on the resonance peak, one finds a velocity resolution of $3.7 \times 10^{-4} \delta$. If the exact point of resonance on a plotted impedance or admittance circle[26] of diameter 250 mm, say, is located to ± 0.5 mm, δ will be 2×10^{-3} corresponding to an uncertainty in cavity length of order 10^{-5} cm. Thus for ^{4}He at 4.5 K we would expect a velocity resolution better than 2 ppm on the basis of a single measurement. For ^{4}He and Ar at s.t.p. the corresponding figures are respectively 12 and 7 ppm for single measurements. By taking many measurements, of course, these uncertainties could be reduced further. Thus we see that this approach offers resolutions comparable to those obtainable with spheres. Similar performances would be achieved with two fixed path resonators of lengths $\lambda/2$ and λ, but if one preferred to use variable-path interferometry this technique, unlike spherical interferometry, would enable it to be done.

Thus it seems that several possibilities exist for reducing the uncertainties of acoustic thermometry by an order of magnitude. The technique is thus potentially as accurate as conventional gas thermometry.

5 CONCLUSIONS

The usefulness of acoustic thermometry above ca.30 K has been limited in the past due to the uncertainty in the value of the gas constant. Recently this has been reduced and there are good prospects that it will reduced further. This means that acoustic thermometry is likely to be used over the same wide temperature range as gas thermometry in future.

The systematic errors of conventional acoustic and ultrasonic techniques were assessed and compared and it was concluded that while alignment problems were no longer insoluble for the latter technique, higher mode errors were still difficult to control in principle although unlikely to be serious in practice. On the other hand, the lower frequency method was considerably more laborious.

The result of past acoustic and ultrasonic thermometry in the range 2.5 to 30 K have been reviewed and compared with what has been achieved by conventional gas thermometry. It appeared that past accurate work broadly confirmed the results of Berry's gas thermometry in this range to within the former's accuracy, but the accuracy was typically two to four times worse than that of gas thermometry.

Three new methods of acoustic thermometry, spherical interferometry, unguided ultrasonic interferometry and wide cylinder interferometry, were assessed. Simple calculations showed that their expected performance was impressive enough to reduce uncertainties in temperature measurements to the ppm level of accuracy. The practicability of this has already been demonstrated by Moldover and Mehl in the case of spherical interferometry.

At ambient temperatures this may well mean that primary thermometry will achieve resolutions and reproducibilities comparable with existing practical thermometry. At the lowest temperatures it will almost certainly exceed it.

Conventional wisdom has it that temperature is to be accorded the status of a basic physical quantity in spite of its conceptual dispensibility because absolute measurements of mechanical quantities such as pV or c cannot be made as accurately as pV or c ratios. (See e.g. De Boer[27]). For this reason temperatures are accorded a special dimension and measured relative to a standard reference state (the triple point of water). The link to mechanical quantities is made through the gas constant or Boltzmann's constant which are necessary to maintain dimensional homogeneity in physical equations by cancelling the dimension of temperature. These constants bear the uncertainty in the absolute energetic value of expressions such as RT and kT while T is relatively well defined. But as RT becomes defined to the ppm level of accuracy through acoustic (or other) measurements, the conventional wisdom will be challenged. When there is no longer good metrological reason to retain temperature as a physical quantity <u>sui generis</u>, will we still retain it for cultural reasons or, more plainly, for old time's sake? If, as one suspects, we do, it will show that the importance of the metrological account of the indispensibility of the concept of temperature was overestimated.

6 REFERENCES

1 A.M. Mayer, Phil. Mag. 45, 18, (1873).
2 K.H. Berry, see his contribution to this Symposium.
3 D.T. Grimsrud, Ph.D. Thesis (University of Minnesota, 1965).
4 N.K. Walstra, Physica, 13, 643 (1947).
5 International Union of Pure and Applied Chemistry, Commission on Atomic Weights and Isotopic Abundances, Pure and Appl. Chem. 52, 2349 (1980).
6 T.J. Quinn, A.R. Colclough and T.R.D. Chandler, Phil. Trans. Roy. Soc. Lond., 283, 367 (1976).
7 E.R. Cohen and B.N. Taylor, J. Phys. Chem. Ref. Data, 2, 663 (1973).
8 A.R. Colclough, "Methods for the Determination of the Gas Constant" in Proceedings of the Second International Conference on Precision Measurements and Fundamental Constants, N.B.S., Gaithersburg, 8-12th June, 1981, to be published.
9 A.R. Colclough, T.J. Quinn and T.R.D. Chandler, Proc. Roy. Soc. Lond. A368, 125 (1979),
10 A. Van Itterbeek, J. Acoust. Soc. Amer. 29, 584 (1957).
11 J. De Laet, Verh. K. Vlaam. Acad. Wet. 66, 22 (1960).
12 H. Plumb and G. Cataland, Metrologia, 2, 127 (1966).
13 L.M. Besley and W.R.G. Kemp, Metrologia, 13, 35 (1977).
14 D.T. Grimsrud and J.M. Werntz, Phys. Rev., 157, 181 (1967).
15 A.R. Colclough, Proc. Roy. Soc. Lond., A365, 349 (1979).
16 A.R. Colclough, Metrologia, 9 75 (1973).
17 A.R. Colclough, Acustica, 36, 257 (1976/1977).
18 H. Plumb, personal communication.
19 A.R. Colclough, Acustica, 142, 28 (1979).
20 J.B. Mehl and M.R. Moldover, J. Chem. Phys., 74, 4062 (1981).
21 M.R. Moldover and J.B. Mehl, "Spherical Acoustic Resonators: Promising Tools for Thermpometry and Measurement of the Gas Constant.", in Proceedings of the Second International Conference on Precision Measrements and Fundamental Constants, N.B.S., Gaithersburg, 8-12th June 1981, to be published.
22 F.D. Shields, K.P. Lee and W.J. Wiley, J. Acoust. Soc. Amer., 37, 4 (1965).
23 P.S. Henry, Proc. Phys. Soc. 43, 340 (1931).
28 D.E. Weston, Proc. Phys. Soc. B66, 695 (1953).
24 K.H. Berry, Metrologia, 15, 89 (1979).
25 W.C. Sauder, "An Ultrasonic Determination of the Gas Constant" in Proceedings of the Second International Conference on Precision Measurements and Fundamental Constants, N.B.S., Gaithersburg, 8-12th June, 1981. To be published.
26 A.R. Colclough, Acustica, 42, 18 (1979).
27 J. De Boer, Metrologia, 1, 158 (1965).

^{4}He second and third virial coefficients from acoustical isotherms: The Helmholtz-Kirchhoff correction at temperatures below 35 K

Harmon H. Plumb

National Bureau of Standards, Washington, D.C. 20234

Measurements of acoustical isotherms from 9 to 34 K have been extended up to 200 000 Pa and hence yield more accurate isotherm analyses because a determination of the quadratic pressure term permits a more accurate determination of the linear pressure term. The isotherm analysis has produced values for the ^{4}He second and third virial coefficients: They are, respectively, $B = 16.8925 - 383.095/T - 150.665/T^2$ (cm^3/mol) and $C = 5788/T$ cm^{6}mol^{-2}. The isotherm slopes (linear term in pressure) that have been measured experimentally are compared with those that have been calculated from values of the second ^{4}He virial coefficient, $B(T)$, which were determined in other, non-acoustical experiments. The close equality of these slope values indicates the inadequacy of the generally accepted "Helmholtz-Kirchhoff correction" (its theoretical derivation and/or its application to experimental measurements). The correction is usually involved to correct speed of sound measurement data in a confined tube to values that would have been measured in a free or open gas.

INTRODUCTION

The author has previously reported [1-7] the results of measuring acoustical isotherms at low temperatures, 2.3 K to 20.3 K. The emphasis was upon determining temperature values which resulted in a widely accepted provisional temperature scale[1]. The National Bureau of Standards Provisional Temperature Scale 2-20 K (NBS P2-20) (1965) was maintained at the National Bureau of Standards and was the basis for comparing, both nationally and internationally, values of temperature that were derived from other experiments. The scale was promulgated via the calibration of germanium resistance thermometers. More recently the above provisional scale has been replaced by an internationally approved temperature scale that is entitled The 1976 Provisional 0.5 K to 30 K Temperature Scale (EPT-76)[8]; this scale has resulted from the consideration and evaluation of many low temperature thermometry endeavors that have been conducted in the last three decades. Many of these endeavors continue to have merit in that they provide a basis for assessing the thermodynamic accuracy and smoothness of EPT-76. Since the NBS P2-20 (1965) was reported, additional acoustical isotherms have been measured at the National Bureau of Standards and the resultant information provides i. greater measurement accuracy, ii. a greater number of isotherms, iii. isotherms at higher temperatures than the author has previously reported, iv. means of more accurately analyzing isotherms, v. values for the second, $B(T)$, and third, $C(T)$, ^{4}He virial coefficients, and vi. a means for assessing a particular application of the "Helmholtz-Kirchhoff correction". Presumably the "Helmholtz-Kirchhoff correction"[7,9-12] should be of predictable significance to the NBS acoustical isotherm determinations. Our experimental results do not confirm the need for the correction. This paper is mainly concerned with the aforementioned items iv, v, and vi.

It is not the author's intention to present a comprehensive study of ^{4}He virial coefficients that have been measured experimentally or deduced theoretically. Rather, some will be presented that are of relatively recent vintage, that span a common temperature interval (2.3 K to 34 K) and that are derived from significantly different experiments.

THEORY: SPEED OF SOUND IN A GAS, VIRIAL COEFFICIENTS, AND TEMPERATURE

When sound is propagated through a gas, if the frequency of the sound is not too high relative to the mean collision frequency of the gas molecules, the propagation is nearly adiabatic and the speed of sound [13,14] is

$$W = \sqrt{E/\rho} = \sqrt{\frac{-v^2}{M}\left(\frac{\partial p}{\partial v}\right)_S} = \sqrt{-\frac{C_p v^2}{C_v M}\left(\frac{\partial p}{\partial v}\right)_T} \quad (1)$$

where ρ is the density of the medium and E is the adiabatic-bulk modulus defined by the relation

$$E = -v\left(\frac{\partial p}{\partial v}\right)_S = -\frac{C_p}{C_v} v \left(\frac{\partial p}{\partial v}\right)_T \quad (2)$$

The symbols v and p refer to the volume and pressure of the medium, and the partial derivative is at constant entropy or temperature, as indicated; C_p and C_v are specific heats at constant pressure and volume, respectively. If the conducting medium were an ideal gas, its molar equation of state would be expressed by

$$pv = RT \tag{3}$$

and there would result

$$\left(\frac{\partial p}{\partial v}\right)_T = -\frac{RT}{v^2} \tag{4}$$

Additionally the ratio of specific heats would be that of an ideal gas,

$$\left(\frac{C_p}{C_v}\right)_{\text{ideal gas}} = \frac{5}{3}$$

(if the gas is assumed to be monotomic), and Eq. (1) becomes

$$W_o^2 = \frac{5\,RT}{3M} \tag{5}$$

Equation 5 indicates the simple relationship between the temperature T and the speed of sound in an ideal gas W_o. (The gas constant R and the molecular weight M are considered to be accurately known constants.)

In actuality an ideal gas does not exist, so Eq. (5) must be modified for application to real gases. Of the two gases, ^{3}He and ^{4}He, which can be used for sound measurements at temperatures between 2 K and 20 K, ^{4}He has been selected for experimental measurements. Its equation of state can be assumed to be represented by the empirical relationship [15]

$$pv = RT\left(1 + \frac{B}{v} + \frac{C}{v^2} + \ldots\right) \tag{6}$$

where B and C are functions of the temperature called the second and third virial coefficients, respectively. If the quantity

$$\left(\frac{\partial p}{\partial v}\right)_T$$

is evaluated from Eq. (6) and the ratio C_p/C_v is evaluated for the real gas, Eq. (1) can be written [12,16,17, 19]

$$W^2 = \left(\frac{C_p}{C_v}\right)_{\text{ideal}} \frac{RT}{M}(1 + \alpha p + \beta p^2 + \ldots) \tag{7}$$

where

$$\alpha = \frac{1}{RT}\left[2B + \frac{4T}{3}\frac{dB}{dT} + (4/15)\,T^2\,\frac{d^2B}{dT^2}\right] \tag{8}$$

and

$$\beta = \frac{1}{R^2T^2}\left[\frac{13}{5}C + \frac{16}{15}T\frac{dC}{dT} + \frac{2}{15}T^2\frac{d^2C}{dT^2} - \frac{8}{5}B^2 + \frac{98}{45}T^2\left(\frac{dB}{dT}\right)^2 + \frac{8}{45}T^4\left(\frac{d^2B}{dT^2}\right)^2 + \frac{8}{15}TB\frac{dB}{dT} + \frac{4}{15}T^2B\left(\frac{d^2B}{dT^2}\right) + \frac{56}{45}T^3\left(\frac{dB}{dT}\right)\left(\frac{d^2B}{dT^2}\right)\right] \tag{9}$$

From Eqs. (5) and (7) we obtain

$$W^2 = W_o^2(1 + \alpha p + \beta p^2 + \ldots) \tag{10}$$

It is now apparent that, if one determines an isotherm of W^2 as a function of pressure, the intercept W_o^2 (P=o) of the extrapolated isotherm readily affords a means of calculating the isotherm temperature. Also, if the second virial coefficient (B) can be represented by a relatively simple function (e.g., $B = a + b/T + c/T^2$), then the second term of Eq. (10), $W_o^2\alpha$, can be evaluated from a number of isotherm slopes and yields values of B as a function of temperature. Furthermore, once B has been determined, the quadratic term, $W_o^2\beta$, permits the evaluation of the third virial coefficient (C).

In order to obtain values of the speed of sound as a function of pressure, one must determine the wavelength and the frequency. Several methods are possible but our work is based upon the creation of standing waves in a gas by means of an apparatus with a fixed frequency and a variable path. A quartz crystal, operating at its resonant frequency, radiates through ^{4}He gas to a reflector surface which is parallel to the quartz crystal; radiation is reflected back through the gas to the quartz crystal. If the spacing between the crystal and the reflecting surface is an integral number of half-wavelengths, the helium gas will be in resonance and a peak voltage will be measured across the quartz crystal. The reflecting surface is displaced through detected integral numbers of half-wavelengths and its displacement measured. Thus λ is determined, the frequency of the quartz crystal is measured and, consequently, W can be calculated. This value of W is, of course, the phase speed corresponding to whatever mode exists in the tube of gas. It is supposed that this mode is sufficiently like the plane wave mode to allow identification of W, for our purpose, as the free-space value. The mode may not be pure but, if this is the case, its components are spaced so closely that the ultrasonic interferometer does not resolve them. The cavity that contains the ^{4}He gas is a right circular cylinder with an X-cut quartz crystal covering one end of the cavity and the reflecting surface of a piston, which can slide freely within the cylinder, nearly fills the other end of the cylinder.

In the preceding the supposition was made that the energy transmission mode might be identified as the free-space value of W. Actually, a rigorous theory exists and the theory clearly suggests that the velocity of sound in a confining tube is less than a free-space value of velocity. Formulae are available so that measured confined values can be corrected to approximate free-space values. The corrective expression is associated with the effects of viscosity and thermal conduction that presumably affect the sound energy transmission at the interfaces of the helium gas and the containing solid surfaces. The theory upon which the correction is based commonly is called the "Helmholtz-Kirchhoff Correction". There have been some reported instances in which the applicability of the correction has been questioned. At the very least, it was suggested that the correction might be dependent upon surface finishes - degree of smoothness or polishing. The corrective equation is clearly dependent upon the frequency of the gas excitation, the diameter of the containing tube, values of viscosity and heat conduction; and the gas density or pressure.

The equations derived by Helmholtz [9] and Kirchhoff [10] are based upon the effects of viscosity and heat conduction at the boundary walls for a gas that travels along the length of a cylinder. The equations assume that the amplitude of the sound waves are uniform over the whole cross-section of the cylinder and predict that both the speed and the absorption differ from free space values. For the speed

$$W = W_f [1 - \frac{1}{2b} \sqrt{\frac{\eta_e}{\rho \pi f}}]$$

where W is the speed within a cylinder;
W_f is the free space speed
b is the radius of the tube
ρ_o is the density of the gas
f is the frequency of the sound
and η_e is an effective viscosity that is related to η, the real viscosity of a gas by

$$\eta_e = \eta [1 + \{(\gamma - 1) \sqrt{\frac{K}{C_p \eta \gamma}}\}]^2$$

γ is the ratio of specific heats
K is the thermal conductivity of the gas
and
C_p is the specific heat at constant pressure.

In principle the "Helmholtz-Kirchhoff Correction" predicts that raw isothermal data of the square of the velocity of sound in a gas as a function of the pressure will exhibit marked curvature with decreasing pressure. Presumably, appropriate application of the correction will result in linearizing the isotherm at low pressures and thus affect both the extrapolated isotherm intercept (from which values of absolute temperature can be calculated) and the isotherm slope (from which the second virial coefficient of the excited gas can be determined).

The acoustical data which has been measured and reported from our laboratory has never been corrected in accordance with the "Helmholtz-Kirchhoff Correction". Because of this there has been some discussion of both the data and its treatment in the literature and personal communications. The correction has not been applied because we were unable to detect measurable curvature in our raw data in the low pressure region; and also, because experiments involving different tube diameters produced identical isotherm data points[7] within the accuracy of the measurements. Consequently we believed that the correction was inappropriate for the measured isotherm data and that the theory involved in obtaining the "Helmholtz-Kirchhoff Correction" was inadequate, incorrect, or incomplete.

The data and data analysis that we have reported have been seriously limited, and we were aware of it, seriously limited because of the lack of isotherm data at higher pressures. Subsequently, the measurements were extended to higher pressures. This has been extremely helpful as will be obvious in what follows.

Acoustical thermometry's chief aim has been the determination of absolute temperature based upon the values of the gas constant, R, and the molecular weight of ^{4}He. (The assumption of a numerical value for R can be considered as equivalent to a calibration of the instrument at the defined temperature, 0° Celsius.) Indeed the aim of temperature-value determinations was a worthy goal and seemed to press the limitations of experimental measurements. However the extension of measurements to include higher pressures has permitted the calculation of both the second and third ^{4}He virial coefficients, which are second and third order quantities in equations 7 and 10. This accomplishment allows comparisons of acoustically determined temperature values and of derived virial coefficients with the same quantities as determined from gas thermometry, p-v isotherm measurements, and other comparable experiments.

The extension of acoustical thermometry to virial coefficient determinations permits additional assessments of the possible accuracy of methods and instrumentation. Specifically, acoustical isotherm slopes, and the quadratic pressure term, can be calculated from the values of second and third virial coefficients of various experimental and theoretical origins. Conversely, acoustically derived values of the second and third virials can be compared with values that originate from completely different kinds of experiments. Such comparisons enable an evaluation of the "Helmholtz-Kirchhoff Correction". If the correction were to be applied to the isotherm-slope data which are used in this paper, the slopes' magnitude would be substantially decreased above 13.9 K, and increased below 13.9 K.

DATA AND ANALYSIS

The desirability of extending acoustical isotherm measurements to higher pressures has been clearly recognized and stated. A function $[W^2,$ Eq. (10)$]$ which includes quadratic and higher order pressure terms cannot be well approximated by a linear term for low pressure data points only. This problem is exacerbated when the low pressure data are limited by low signal-to-noise ratios encountered as the acoustical mismatch between the vibrating quartz crystal and the helium gas increases with decreasing gas pressure. For the aforementioned reasons, the 1967-1968 isotherm data were measured in our laboratory up to two atmospheres.

It was not possible to conduct later (1972) experiments (2.3 K to 9.0 K) at substantially greater pressures because of experimental temperature limitations. The working gas, within the acoustical thermometer, was thermally exposed through the acoustical thermometer walls or boundaries to the surrounding refrigerant bath. The refrigerant bath temperature varied from about 4.2 K during the 9 K isotherm measurements to well below 2 K for the lowest isotherm temperatures. Because of the limitations of the experimental design, the working gas was subject to condensation if its pressure exceeded the vapor pressure of the surrounding liquid refrigerant.

The data that result from mathematical analyses of the isotherms are presented in Table I, in six columns. The values of temperature in Col. 1 are significant but do not constitute final values; they are, however, adequate for the present purposes. Column 6 indicates the range of pressures measured for the particular data points of each isotherm. In this context the isotherms at 9.9986 K and 21.0114 K are noteworthy because of their limited pressure range. Hence one would expect that the quadratic and linear terms for these two isotherms would be poorly defined and indeed such is the case. Column 2 contains the linear component of the isotherms, A_1 or $W_0^2\alpha$, while δA_1 of Col. 3 indicates the statistical uncertainty of A_1; Col. 4 contains the quadratic component of the isotherm, A_2 or $W_0^2\beta$, and δA_2 of Col. 5 is its uncertainty.

It should be clear at this point that the isotherm analyses do not extend beyond the quadratic term of Eq. 10. This increases the possibility of error, as would the elimination of the βp^2 term, but the author hopes that the omission of the terms γp^3 and above does not seriously affect the results of the analyses. Actually, analyses that permitted the evaluation of the γp^3 term did not yield a coefficient γ that was statistically defined.

Values of W_0^2 and δW_0^2 are included to afford a general comparison of successive terms. In all of the analysis the value $R = 8.314$ J·mol^{-1}·K^{-1} has been used for the gas constant.[18] It is possible to use another value for R but the indicated uncertainties of A_1 and A_2 far exceed any implied uncertainties in R. The value 4.0026 gmol^{-1} continues to be used for the molecular weight of ^{4}He.

A few remarks are appropriate for Table I. The algebraic signs for A_1 and A_2 do change in the temperature range of the isotherms, but they do not change at the same temperature value. (This could well be true for the possible A_3, A_4,.... terms that are not being considered at this time.) The uncertainties δA_1 and δA_2 for individual isotherms are clearly associated with the pressure range of measurements. We believe that the isotherm data below 4.2 K is currently less reliable than that above 4.2 K, and this is substantiated by the relative values of A_1, δA_1, A_2 and δA_2. The isotherms at 3.411 K and 7.199 K are of particular interest for they were determined in 1974[4] with a modified apparatus; displacement measurements were made through the use of a laser-interferometer instead of through the fused-quartz extension rod that was employed prior to 1973. However the 3.411 K and 7.199 K isotherms have been determined over a more restricted pressure range than those isotherms at adjacent temperatures. Except for the 3.411 K and 7.199 K isotherms, all of the isotherms including 8.9909 K and below were determined in 1972. Hence all of the data in Table 1 differs from that reported in 1966[1].

TABLE I
Acoustical Isotherm Terms

$$W^2 = W_0^2(1 + \alpha p + \beta p^2 + \ldots)$$

Temp K	$A_1(w_0^2\alpha)$ $m^2s^{-2}atm^{-1}$	δA_1 $m^2s^{-2}atm^{-1}$	$A_2(w_0^2\beta)$ $m^2s^{-2}atm^{-2}$	δA_2 $m^2s^{-2}atm^{-2}$	Pressure Range Atm	w_0^2 m^2s^{-2}	$\delta(w_0^2)$ m^2s^{-2}
2.3159	-6712.1	549.2	-29072.0	9219.0	.015 - .042	8017.50	7.60
2.8110	-6670.8	109.5	2916.0	1316.0	.015 - .065	9731.61	2.01
3.2103	-5180.1	64.1	-1453.6	352.8	.024 - .15	11113.77	2.52
3.4112	-4558.0	194.0	-2037.0	1262.0	.027 - .11	11809.37	6.96
4.2101	-3500.0	26.0	- 558.0	58.0			
5.0199	-2651.18	10.6	- 266.26	9.95	.035 - .97	17378.48	1.43
6.0527	-1935.22	10.6	- 38.43	14.3	.06 - .68	20953.86	1.58
6.9756	-1485.45	6.4	22.99	7.2	.06 - .79	24149.01	1.16
7.1985	-1344.8	41.22	- 3.65	58.1	.11 - .60	24920.58	6.54
8.0622	-1072.5	6.75	38.15	6.7	.08 - .89	27910.71	1.45
8.9909	- 816.57	6.3	51.40	5.7	.08 - .98	31125.59	1.42
8.9940	- 813.12	6.2	52.65	3.2	.12 - 1.8	31136.31	2.61
9.9986	- 552.97	45.76	33.11	35.1	.29 - .99	34614.15	13.82
10.9042	- 412.78	6.46	51.67	3.3	.30 - 1.6	37749.38	2.76
12.0180	- 235.56	6.63	46.12	2.9	.30 - 1.99	41605.17	3.23
12.9626	- 91.72	5.13	34.18	2.3	.32 - 1.95	44875.46	2.42
13.8102	- 9.50	6.80	35.60	2.9	.37 - 1.98	47809.59	3.47
14.0520	21.30	5.46	33.93	2.4	.32 - 1.96	48646.76	2.78
14.9913	120.43	7.44	27.39	3.1	.33 - 1.99	51898.48	3.88
16.0502	183.31	8.14	33.92	3.4	.35 - 1.97	55564.37	4.26
16.9681	273.50	6.39	24.96	2.8	.35 - 1.95	58742.20	3.28
18.0113	327.31	7.79	23.95	3.3	.34 - 1.95	62353.79	4.00
18.9433	389.09	6.89	18.93	3.0	.31 - 1.93	65579.98	3.47
20.0441	454.36	7.98	15.58	3.5	.31 - 1.92	69391.16	4.02
20.2772	465.37	7.42	15.31	3.2	.30 - 1.98	70198.11	3.73
21.0114	469.39	21.80	27.26	16.7	.20 - 1.10	72739.74	6.46
22.8268	530.17	7.48	28.26	3.3	.30 - 1.94	79024.55	3.76
27.1068	707.04	10.99	- 6.92	4.70	.32 - 2.00	93841.55	5.50
29.4118	758.66	4.54	5.90	1.58	.30 - 2.50	101821.0	2.67
33.9778	819.26	14.82	0.77	6.56	.30 - 1.97	117628.3	7.25

The present paper is not an exercise in correlating the data of 1966 and that of 1968, 1972, 1973, and 1974; that will be reported at a later time. This report presents the basis for the analysis that will be applied to all of the isotherms.

^{4}He SECOND VIRIAL COEFFICIENT B(T)

In Eq. 10, αW_0^2, or A_1, is the isotherm slope that can be expressed analytically through the application of Eq. 8. Previous analyses by the author in years past have shown that the second virial coefficient, as determined from acoustical thermometry, is adequately represented by the functional form

$$B_i = a_i + b_i/T + c_i/T^2 \qquad (11)$$

where the subscript i refers to constants that may be determined in different experiments.

When B and its derivatives are combined in Eq. 8 and thereafter in Eq. (10), there results

$$A_1 = \alpha W_0^2 = 2W_0^2(RT)^{-1}[a_i + 0.6\, b_i/T + (7/15) c_i/T^2] \qquad (12)$$

Equation (12) can be simplified in that $2W_0^2(RT)^{-1}$ is really a constant for all of the isotherms. Thus

$$A_1 = \alpha W_0^2 = 84.379[a_i + 0.6\, b_i/T + (7/15)\, c_i/T^2] \quad (13)$$

The coefficients were evaluated by fitting Eq. 13 with the A_1 and T data of Table I. The A_1 entries were weighted by $(\delta A_1)^{-2}$.

The result is

$$B = [16.8925(.1353) - 383.095(4.724)T^{-1} - 150.665(27.825)T^{-2}]$$

where the numbers in parentheses are the uncertainties of the respective coefficients.

In recent years two distinctly different experiments have been performed[22,23] that yielded values for the ^{4}He second virial coefficient, which covered nearly the same temperature range that is indicated in Table I, Col. 1, and furthermore the values were expressed in the functional form of Eq. 11. It is instructive to compare the coefficients a_i, b_i and c_i, as well as the values of B that are represented, to these reported by Berry of NPL (England)[22]. His experimentation may be broadly categorized as conventional P-V isotherm and gas thermometry investigations. Berry's reported second virial coefficient will be subscripted BERRY. Additionally, Gugan and Michel have reported[23] experimentation entitled, "Dielectric Constant Gas Thermometry". This work permitted an evaluation of the "smoothness" of a temperature scale T(NPL-75)[22] that resulted from Berry's work, and yielded values and functional forms for both the second and third ^{4}He virial coefficients. These virials will be subscripted GUGAN.

Thus there exist the following expressions:

$$B_{PLUMB} = 16.8925 - 383.095\ T^{-1} - 150.665\ T^{-2} \quad (14)$$

$$B_{BERRY} = 17.19 - 396.2\ T^{-1} - 48\ T^{-2} \quad (15)$$

$$B_{GUGAN} = 17.57(0.08) - 402.1(1.7)\ T^{-1} - 41.(6)\ T^{-2} \quad (16)$$

Values of the second virial coefficient B have been generated from Eqs. 14, 15, 16 and are listed in Table II. In addition, for facilitating comparisons, B_{KEESOM}, B_{KKHM}, and B_{KKH} are also listed. B_{KEESOM} refers to the second virial coefficient that is titled "Adopted"[24] in Keesom's book. B_{KKHM} refers to values reported by Kilpatrick, Keller, Hammel and Metropolis[25]; and B_{KKH} refers to those by Kilpatrick, Keller and Hammel[26]. Other values of the ^{4}He second virial could be added but those listed serve the author's purpose. B_{KEESOM}, B_{KKHM} and B_{KKH} have been used for many years as a basis for establishing gas thermometry scales, as well as for comparing values of temperature, e.g., fixed

TABLE II

Values of the ^{4}He Second Virial Coefficient (B)

All B (cm^3 mol^{-1})

T(K)	B_{PLUMB} cm^3/mol	B_{BERRY} cm^3/mol	B_{GUGAN} cm^3/mol	B_{KEESOM} cm^3/mol	B_{KKHM} cm^3/mol	B_{KKH} cm^3/mol
3.0	-127.55	-120.21	-121.02			
4.0	- 88.30	- 84.86	- 85.52	-80.96	-78.11	-81.73
5.0	- 65.75	- 63.97	- 64.49	-62.12	-59.14	-61.19
6.0	- 51.14	- 50.18	- 50.58	-49.12	-46.53	-47.53
7.0	- 40.91	- 40.39	- 40.71		-37.53	-37.78
8.0	- 33.35	- 33.08	- 33.33	-32.97	-30.78	-30.49
9.0	- 27.53	- 27.42	- 27.61		-25.53	-24.82
10.0	- 22.92	- 22.91	- 23.05	-23.32	-21.34	-20.29
11.0	- 19.18	- 19.22	- 19.32		-17.90	-16.60
12.0	- 16.08	- 16.16	- 16.22	-16.93	-15.04	-13.53
13.0	- 13.47	- 13.57	- 13.60		-12.63	-10.49
14.0	- 11.24	- 11.35	- 11.36	-12.31	-10.56	- 8.73
15.0	- 9.32	- 9.44	- 9.42		- 8.77	- 6.82
16.0	- 7.64	- 7.76	- 7.72	- 8.86	- 7.20	- 5.16
17.0	- 6.16	- 6.28	- 6.22		- 5.82	- 3.70
18.0	- 4.85	- 4.97	- 4.89	- 6.19	- 4.60	- 2.41
19.0	- 3.69	- 3.79	- 3.71		- 3.51	- 1.26
20.0	- 2.64	- 2.74	- 2.64	- 4.04	- 2.53	+ 0.25
21.0	- 1.69	- 1.78	- 1.67		- 1.65	0.70
22.0	- 0.83	- 0.92	- 0.79	- 2.27	- 0.85	1.53
23.0	- 0.05	- 0.13	+ 0.01		- 0.12	2.29
24.0	+ 0.67	+ 0.60	0.74		+ 0.54	2.98
25.0	1.33	1.26	1.42		1.15	3.62
26.0	1.93	1.88	2.04		1.71	4.20
27.0	2.50	2.45	2.62		2.23	4.73
28.0	3.02	2.98	3.16		2.71	5.22
29.0	3.50	3.47	3.65		3.15	5.67
30.0	3.95	3.93	4.12	+ 2.42	3.57	6.09
31.0	4.38	4.36	4.56			
32.0	4.77	4.76	4.96			
33.0	5.14	5.14	5.35			
34.0	5.49	5.49	5.71			
35.0	5.82	5.83	6.05			

points, that have been ascertained by different experimenters.

The primary purpose of Table II is to demonstrate the agreement that exists between the ^{4}He second virial coefficients that have been derived from experimental measurements of acoustical isotherms, p-v isotherms and gas thermometry, and dielectric constant gas thermometry. The experimental techniques, sources of error and uncertainty, corrective terms, etc., are greatly different.

Gugan[23] has reanalyzed Berry's isotherms[22] with the constraint of Gugan's ^{4}He third virial coefficient, C. The constants for B_{BERRY} are then

$$B_{BERRY} = 17.31(0.18) - 399.4(3.9)T^{-1} - 47.(13.)T^{-2}$$
(15a)

Table III is appropriate for it lists the temperature values at which Berry and Gugan measured isotherms from which ^{4}He virial coefficients could be determined. In comparison with the acoustical data of Table I, and the isotherm temperatures of Gugan in Table III, the emphasis of Berry's isotherms is 4.22 K and below. In Table I it is apparent that the slopes A_1 and the quadratics A_2 exhibit their greatest uncertainty at temperatures below 4.2 K; also in Table III, Gugan's isotherms do not extend below 4.2275 K. Because of the aforementioned circumstances, one might conclude that B_{BERRY} (Eq. 15) would probably yield the most accurate values for B below 4.2 K.

Table III was not included to afford a comparison of fixed point temperatures nor is there any reason to

TABLE III. Isotherm Temperatures

Berry	Gugan (NPL-75)[a]	Colcough (Batuecas's R)[b]
2.6014 K		
2.7479 K		
3.3303 K		
4.2201 K	4.2275 K	4.2212
	5.5519 K	
7.1992 K	7.2029 K	7.2003
	10.0108 K	
13.8036 K	13.8194 K	13.8032
	17.0232 K	17.0378
20.2712 K	20.2750 K	20.2728
27.0979 K	27.1688 K	

[a] Gugan assumed temperature values from calibrations on (NPL-75)

[b] Colcough has listed temperatures that would result from Batuecas's R value being accurate

TABLE IV

Acoustical Isotherm Slopes A_1 (Calculated from $B_i(T)$)

TEMP K	$A_{1\,PLUMB}$	$A_{1\,GUGAN}$	$A_{1\,BERRY}$
		100 $A_1(T)$ m^{2}s^{-2}Pa^{-1}	
3.00	-5.624	-5.410	-5.374
4.00	-3.744	-3.659	-3.634
5.00	-2.655	-2.618	-2.602
6.00	-1.946	-1.929	-1.919
7.00	-1.447	-1.439	-1.434
8.00	-1.077	-1.073	-1.072
9.00	-0.792	-0.788	-0.791
10.00	-0.565	-0.561	-0.566
11.00	-0.381	-0.376	-0.383
12.00	-0.229	-0.222	-0.231
13.00	-0.100	-0.091	-0.102
14.00	0.009	0.019	0.007
15.00	0.104	0.116	0.103
16.00	0.187	0.201	0.186
17.00	0.260	0.275	0.260
18.00	0.325	0.342	0.325
19.00	0.383	0.401	0.384
20.00	0.435	0.454	0.437
21.00	0.481	0.502	0.484
22.00	0.524	0.546	0.527
23.00	0.563	0.586	0.567
24.00	0.599	0.623	0.603
25.00	0.631	0.656	0.636
26.00	0.661	0.688	0.667
27.00	0.689	0.716	0.695
28.00	0.715	0.743	0.722
29.00	0.739	0.768	0.746
30.00	0.762	0.791	0.769
31.00	0.783	0.813	0.790
32.00	0.802	0.833	0.811
33.00	0.821	0.852	0.829
34.00	0.838	0.870	0.847
35.00	0.855	0.887	0.864

believe that nearly identical values of temperature (listed in Table III) constituted efforts to attain a thermal state (same hotness or coldness). The Table III is presented merely to indicate the number and approximate temperatures of isotherms as determined by recent investigators.

While the author has reservations about extrapolating empirical formulations of virial coefficients beyond the range of actual measurements, it is interesting to note the relative values of B_{BERRY} and B_{GUGAN} in comparison with B_{PLUMB} at values of temperature from 27 K to 35 K.

ACOUSTICAL ISOTHERM SLOPES

Table II, while it is important, is only one possible method of comparing second virial coefficients. Another method is the comparison of acoustical isotherm slopes, both experimental and functionally smoothed, with slopes that can be calculated using Eq. 13 from the constants of B_i, Eq. 11. The comparison is presented in Table IV. Except at the lowest temperatures, one is gratified by the agreement of A_i. However a further step is afforded in Table V.

Table V is patterned in accord with the author's interpretation of Colcough's Table IV[19]. It is assumed that the approximate temperatures of Colcough's Col. 1, Table V, are indeed very close to Colcough's temperature values as listed in Colcough's Table IV, Col. 2. $A_{1_{PLUMB}}$, $A_{1_{GUGAN}}$, and $A_{1_{PLUMB}}$ all were calculated using Eq. 13 from the constants of B_i, Eq. 11. $A_{1_{PLUMB}}$ Exp are the slopes (A_1) of Table I adjusted from the temperature values of Table I, Col. 1 to the temperature values of Colcough's Table V, Col. 1. Table V is not an attempt to reconcile or evaluate fixed point realizations; it is for the purpose of comparing both calculated and experimentally measured acoustical isotherm slopes as precisely as possible.

The parenthetic numbers of Table V are uncertainty estimates of adjacent A_i values. For values derived from Table I, the uncertainties throughout this report are statistical uncertainties representing one sigma. The uncertainties that are listed for $A_{1_{BERRY}}$ are taken from Colcough's Table IV.

It is apparent in comparing $A_{1_{PLUMB}}$exp with calculated A_1's that a serious discrepancy exists at 4.221 K. Otherwise the comparisons are remarkably close, especially between $A_{1_{PLUMB}}$ and $A_{1_{BERRY}}$. However comparisons of $A_{1_{COLCOUGH}}$ with all other A_1 indicate the probability of problems in the values of $A_{1_{COLCOUGH}}$ at temperatures higher than 7.2 K.

^{4}He THIRD VIRIAL COEFFICIENT C(T)

Additional information can be calculated from the data that are listed in Table I, namely, the ^{4}He third virial coefficient.

In the quadratic pressure term, Eq. 9, because B_i is known, e.g., Eq. 14, 15, 16, substitutions can easily be made for B, dB/dT, and d^2B/dT^2.

There results

$$A_2 = W_o^2 R^{-2} T^{-2}[(13/5)C + (16/15)T(dC/dT) + (2/15)T^2(d^2C/dT^2) - 8/5 \, a_i^2 - (16/5)a_i b_i T^{-1} - (6/5)b_i^2 T^{-2} - (8/3)a_i c_i T^{-2} - (32/15)b_i c_i T^{-3} - (8/9)c_i^2 T^{-4}] \quad (17)$$

which can be summarized in the form

$$A_2 T/.5141 = \sum C^+ + \sum B^+ \quad (18)$$

Because $A_2(T)$ and T are available from Table I and $\sum B^+$ is known from a_i, b_i, and c_i it is possible to fit

$$A_2(T) \, T/.5141 - \sum B^+ = f(1/T) \quad (19)$$

to a series expansion of 1/T: values of the left side of Eq. 19 are weighted by $(\delta A_2)^{-2}$. When the above is performed, there results

$$f(1/T) = 220.6(124.2) + 10418.8(1808.)T^{-1}. \quad (20)$$

In series fittings that included higher powers of T^{-1} the coefficients were uncertain. Because $f(1/T)$ is represented by $P + QT^{-1}$ it follows that $\sum C^+$ can be appropriately expressed in the same functional form. This form is readily obtained if the third virial coefficient is expressed as

$$C = m + n/T \quad cm^6 mol^{-2} \quad (21)$$

Hence

$$\sum C^+ = 2.6m + 1.8 \, n/T \quad (22)$$

TABLE V

Acoustical Isotherm Slopes (A_i) at Fixed Points

100 $A_1(T)$ $m^2 s^{-2} Pa^{-1}$

T	$A_{1_{COLCOUGH}}$ Exp.	$A_{1_{PLUMB}}$ Calc.	$A_{1_{GUGAN}}$ Calc.	$A_{1_{BERRY}}$ Calc.	$A_{1_{PLUMB}}$ Exp.
4.2212 K	-3.356(.031)	-3.457	-3.386	-3.363(.005)	-3.441(.026)
7.2003 K	-1.376(.012)	-1.365	-1.358	-1.354(.006)	-1.365(.041)
13.8032 K	-0.017(.003)	-0.011	-0.001	-0.012(.004)	-0.009(.007)
17.0378 K	0.244(.009)	0.263	0.278	0.263(.005)	
20.2728 K	0.417(.011) 0.426(.010)	0.448	0.468	0.450(.005)	0.459(.007)
27.098 K		0.692	0.720	0.698	0.698(.011)

By evaluating the coefficients of Eq. 20 and 22 one obtains for the ^{4}He third virial coefficient

$$C = 84.8(48.) + 5788(1004.)T^{-1} \text{ cm}^6\text{mol}^{-2} \quad (23)$$

where the coefficients' (m and n) uncertainties are in parenthesis.

The significance of the coefficient 84.8 is doubtful and its greatest relative contribution to C will occur at the highest temperatures; up to 12 K its value is less than the uncertainty of the coefficient n. Thus, for discussion in this paper at least, I shall omit the coefficient m of Eq. 21 from further consideration.

Since Gugan and Michel[23] determined values for the third ^{4}He virial coefficient from Dielectric Constant Gas Thermometry, their values afford a comparison. Gugan and Michel's preferred values for C are

$$C = \frac{5420. \pm 225.}{T} \text{ cm}^6\text{mol}^2. : \quad (21)$$

Other isotherm fitting procedures yielded

$$C = \frac{7200.}{T} - \frac{6300.}{T^2} \text{ cm}^6\text{mol}^{-2} \quad (22)$$

Comparisons of third virial coefficients are indications of the accuracy and quality of isotherm experimentation. For this reason third ^{4}He virial coefficients will be listed for Eq. 21, Eq. 22 and the $(C-2Bb_E)$'s for the eight isotherms that have been fitted by Gugan's "preferred method" [three other fitting procedures, with various values of $(C-2Bb_E)$ for each isotherm, are presented in the Gugan-Michel report]. These values are listed in Table VI, as well as values that were determined at particular isotherm temperatures by Berry[22], White[27], and Keesom[28].

ACOUSTICAL ISOTHERM QUADRATIC TERM

Similar to the treatment of $A_1(T)$ (whereby a calculated $A_1(T)$ may represent a smoothing of the $A_1(T)$ experimental values), $A_2(T)$ can be calculated based upon the B_i and C_i of different experiments or analyses. Furthermore, in the same way that non-acoustically derived B_i can be evaluated by employing them to calculate $A_1(T)$, which can then be compared with experimental acoustical $A_1(T)$ values, so also non-acoustically derived B_i and C_i can be employed to calculate $A_2(T)$. These $A_2(T)$ are then compared with $A_2(T)$ acoustical experimental. The $A_2(T)$ have been calculated, using Eq. 17, and are presented in Table VII.

TABLE VI

Values of the ^{4}He Third Virial Coefficient (C)

All C (cm^6 mol^{-2})

T(K)	C_{PLUMB} (Eq. 23a)	C_{GUGAN} (Eq. 24)	C_{BERRY}^a	C_{WHITE}	C_{KEESOM}	C_{GUGAN} $(C-2Bb_e)$	C_{GUGAN} (Eq. 25)
2.601	2225.		4000.				
2.748	2106.		1500.				
3.330	1738.		1000.(550.)				
4.22	1378.	1284.	1200.(300.)		1000.(130.)	1265.(69.)	1352.
5.55	1042.					869.(143.)	1092.
6.00	965.	903.					1025.
7.199	804.	753.	900.(300.)			811.(54.)	878.
8.00	724.	678.					802.
10.00	579.	542.				666.(372.)	657.
12.00	482.	452.					556.
13.804	419.	393.	200.			392.(258.)	489.
14.00	413.	387.					482.
16.00	362.	339.					425.
17.00	340.					721.(340.)	402.
18.00	322.	301.					381.
20.00	289.	271.					344.
20.271	286.	267.	300.			318.(45.)	340.
20.5	282.	264.		400.(100.)			336.
22.0	263.	246.					314.
24.0	241.	226.					289.
24.5	236.	221.		350.(100.)			284.
26.0	223.	208.					268.
27.098	214.	200.	100.			107.(79.)	257.
28.00	207.						
30.00	193.						

^{a}All Values of C_{BERRY}, except those with indicated uncertainties, are order of magnitude estimates.

TABLE VII

TEMP	A_{2PLUMB}	A_{2BERRY}	A_{2GUGAN}	A_{2EXPER}	δA_{2EXPER}
	100 $A_2(T)$ m^{-2}s^{-2}Pa^{-2}				
2.0	-11.131	-9.2442	-9.3426		
3.0	- 2.318	-2.0582	-2.0924		
4.0	- .6673	- .6168	- .6297		
5.0	- .2051	- .1966	- .2018	-.263	(.010)
6.0	- .0472	- .0486	- .0507	-.0379	(.014)
7.0	.0121	.0084	.0077	.0227	(.007)
8.0	.0344	.0306	.0305	.0377	(.007)
9.0	.0417	.0382	.0384	.0507	(.006)
10.0	.0425	.0396	.0399	.0327	(.035)
11.0	.0406	.0381	.0385	.0510	(.003)
12.0	.0376	.0355	.0359	.0455	(.003)
13.0	.0343	.0326	.0329	.0337	(.002)
14.0	.0311	.0296	.0299	.0335	(.002)
15.0	.0280	.0268	.0270	.0270	(.003)
16.0	.0253	.0242	.0244	.0335	(.003)
17.0	.0228	.0218	.0220	.0246	(.003)
18.0	.0205	.0197	.0199	.0236	(.003)
19.0	.0185	.0178	.0179	.0187	(.003)
20.0	.0168	.0161	.0162	.0154	(.003)
21.0	.0152	.0146	.0147	.0269	(.016)
22.0	.0138	.0132	.0133		
23.0	.0125	.0120	.0121	.0279	(.003)
24.0	.0114	.0109	.0110		
25.0	.0103	.0099	.0100		
26.0	.0094	.0090	.0091		
27.0	.0086	.0082	.0083	-.0068	(.005)
28.0	.0079	.0075	.0075		
29.0	.0072	.0069	.0069		
30.0	.0066	.0063	.0063		
31.0	.0060	.0057	.0057		
32.0	.0055	.0052	.0052		
33.0	.0050	.0048	.0048		
34.0	.0046	.0044	.0043	.0008	(.006)
35.0	.0042	.0040	.0040		

Naturally $A_2(T)$ can be calculated from a number of combinations of B_i and C_i. For the columns of Table VII, the second column is derived from B_{PLUMB} and C_{PLUMB}; the third, from B_{BERRY} and C_{GUGAN}; and the fourth, from B_{GUGAN} and C_{GUGAN}. Obviously both temperature extremes are not applicable for all of the B_i and C_i but represent extrapolations; since $A_2(T)$ is changing rapidly at the lowest temperatures, extrapolation errors in this temperature region can be serious.

Over most of the temperature range $A_2(T)$ is sufficiently smooth that $A_2(T)$, which have been determined experimentally from the acoustical isotherms, can be compared with nearest-integer-temperature values of A_2T of Cols. 2, 3, and 4. The experimental values and their uncertainties are listed in Col. 5 and 6.

Since $A_2(T)$ is a third order quantity with respect to W_o^2 or T and A_1, it is not surprising that A_2 calculated and A_2 experimental exhibit considerable variation in values. Even though from Table I, at 20 K, the ratios of W_o^2, A_1 and A_2 are 70,200., 465., and 15., it is still imperative that the quadratic term be determined as well as possible. If this is not done the error of not having determined A_2 will cascade into A_1 and thence W_o^2 or T.

Practically, the values of $A_2(T)$ in Col. 2, 3 and 4 are rather close and to a first approximation of a third order quantity, it probably does not matter which calculated representation of $A_2(T)$ one might choose to employ. The most interesting feature of Table VII is, without doubt, the rather close values of A_2 that have originated from three very different experiments.

DISCUSSION

It has been shown that the analysis of acoustical isotherms produced second and third virial coefficients for the working gas ^{4}He, that are precise in the temperature region 4 K to 35 K. The values for $B_{PLUMB}(T)$ that are reported here are not in disagreement with those reported earlier by Boyd, Larsen, and Plumb[6]. The 1968 values of B(T) were based upon acoustical isotherm data that were taken in our laboratory in 1965. The remarkable agreement of B_{BERRY} and B_{GUGAN} with B_{PLUMB} in Table II presents definitive second virial coefficient data that provide a clear target for theoretical investigations and calculations.

The third virial coefficients of ^{4}He given by

$$C_P = 5788(1004)T^{-1} \text{ cm}^6 \text{ mol}^{-2} \quad (23)$$

and

$$C_G = 5420(225)T^{-1} \text{ cm}^6 \text{ mol}^{-2} \quad (24)$$

are in close enough agreement that some might consider it fortuitous. However enough data were acquired in both experiments so that the 7% difference could be realistic.

ACOUSTICAL ISOTHERM SLOPES AND THE "HELMHOLTZ-KIRCHHOFF CORRECTION"

The fact that the $C_i(T)$, $B_i(T)$ and $A_1(T)$ (both the NBS experimental $A_1(T)$ and those calculated from $B_i(T)$) are in such excellent agreement is a most clear indication that the "H-K correction", in its predicted entirety, is not applicable to the acoustical isotherm data that have been measured in the NBS acoustical thermometry. The full "H-K correction" greatly exceeds the differences between the A_{1PLUMB} and A_{1BERRY}, and A_{1GUGAN}. In iteration, none of the NBS Acoustical Isotherm Data reported here or previously has been corrected by the "H-K Correction"; in Table 5 the $A_{1COLCOUGH}$ have been corrected and Colcough has reported the appropriateness of, and necessity for, correcting the NPL acoustical isotherm, as well as the confirmed accuracy of the "H-K Correction" for his experimentation. If Colcough's assertions are correct, it is most difficult to explain the discrepancies between $A_{1COLCOUGH}$ values and all of the other A_1 values at temperatures above 7.2 K. The discrepancies of $A_{1_{PLUMB}}$ values at 4.2 K and 7.2 K are clearly caused by the inadequacy of achieving a good representation of the linear term of the isotherms (the pressure ranges were not sufficiently high to characterize the quadratic term). At low temperatures, e.g., 4.2 K, the "HK Correction" is much smaller, in both W^2 units and equivalent T units, than it is at 13 K, 20 K, or 27 K. This possibly accounts for the more comparable values of $A_{1_{COLCOUGH}}$ at 4.2 K and 7.2 K.

If the "K-H correction" were to be applied to the NBS data, the consequence, as affecting the values of isotherm slopes $A_{1_{PLUMB}}$, would be substantial at the higher temperatures. At the particular temperatures,

10 K	.0158	$100A_1$	$m^2 s^{-2} Pa^{-1}$
18 K	.0276	$100A_1$	$m^2 s^{-2} Pa^{-1}$
27 K	.0434	$100A_1$	$m^2 s^{-2} Pa^{-1}$
34 K	.0592	$100A_1$	$m^2 s^{-2} Pa^{-1}$

the slopes $A_{1_{PLUMB}}$ would be decreased by the indicated amounts.

The author concludes that serious inadequacies exist in either the assumptions that were involved in deriving the "H-K Correction", or assumptions concerning its application. The apparent inadequacy of such a long-standing theory should present a challenge and stimulation for theoretical physics endeavor.

FUTURE INTENTIONS

Naturally, the determination of temperature values is still the prime purpose of acoustical thermometry. What has been discussed in this paper is not diversionary but rather most substantive for the analyses of acoustical isotherms. Equation 10 will be applied to the data analysis with the possible constraints of Equations 12 and 17 being employed singly or together. In the case of A_1, values of isotherm temperatures will be calculated based upon $A_i(T)$'s resulting from the second virial coefficients of Plumb, Berry and Gugan respectively. $A_2(T)$'s will be similarly formulated. One immediate anticipated benefit is the evaluation of the low temperature isotherms, where the range of pressures was restricted, employing the $A_{1_{BERRY}}$. The p-v isotherms that Berry measured from 4.2 K to 2.6 K undoubtedly enhance the accuracy of B_{BERRY} and consequently $A_{1_{BERRY}}$ below 4.2 K.

Actually more is to be benefited than just the analyses of the data that are represented in Table I. There exist 1965 acoustical isotherm data which were published previously, and also, isotherm measurements were conducted in 1974 and 1975. The basis for the latter measurements was a laser interferometer that indicated displacements, from which wavelengths are determined, instead of the customary fused quartz rod displacement scheme. Both the 1965 and 1974-5 data were determined in limited pressure ranges for different but, valid reasons. Thus the application of a quadratic term (constraint of Eq. 10 through the use of Eq. 17 [$A_2(T)$]) will be most valuable in analyzing and evaluating these "restricted isotherms".

After the work is completed, Deo volente, it is expected that all of the data (1964 through 1975) and its analysis will be published in the National Bureau of Standards Journal of Research.

REFERENCES

[1] H. Plumb and G. Cataland, Metrologia 2, 18 (1966).
[2] H. H. Plumb and G. Cataland, Science 150, 155 (1965).
[3] H. H. Plumb and G. Cataland, J. Res. NBS 69A, 375 (1965); G. Cataland and H. Plumb, J. Res. NBS 69A, 531 (1965).
[4] G. Cataland and H. H. Plumb, Metrologia 11, 161 (1975).
[5] G. Cataland and H. H. Plumb, J. Acoust. Soc. Am. 34, 1145 (1962).
[6] M. E. Boyd, S. Y. Larsen, and H. Plumb, J. Res. NBS 72A, 155 (1968).
[7] George Cataland and Harmon Plumb, NBS Technical Note 765 (1973).
[8] The 1976 Provisional 0.5 K to 30 K Temperature Scale, Metrologia 15, 65 (1979). This document is the Translation of the official French text of the "Echelle Provisoire de Température de 1976 entre 0,5 K et 30 K." The French version can be obtained from the Bureau International des Poids et Mesures F-92310 Sevres, France.
[9] H. Helmholtz, Verhandlungen des Naturhistorisch-medecinischen Variens zu Heidelberg 3,16 (1863).
[10] G. Kirchhoff, Ann. Phys. Lpz. 134, 177 (1868).
[11] A. Wood, Acoustics, p. 251, London and Glasgow (1940).
[12] T. J. Quinn, National Physical Laboratory Report QU5 (1969).
[13] A. Van Itterbeek: Progress Low Temperature Physics I, 355, North Holland Publ. Co. (1955).
[14] L. Bergman: Der Ultraschall, S.501. Stuttgart: Hirzel (1954).
[15] W. H. Keesom: Helium, p. 30 Amsterdam:(1942).

[16] N. K. Wallstra, Physica 13, 643 (1947).
[17] D. T. Grimsrud, Ph.D. Thesis, Univ. of Minnesota (1965).
[18] The value used for the gas constant, R, is an approximation to the value published in NBS Special Publication 398, Aug. 1974.
[19] A. R. Colcough, Proc. R. Soc. London. A. 365, 349 (1979).
[20] A. R. Colcough, Metrologia 15, 183 (1979).
[21] T. J. Quinn, A. R. Colcough and T. R. D. Chandler, Phil. Trans. R. Soc. London A, 243, 367 (1976).
[22] K. H. Berry, Metrologia 15, 89 (1979).
[23] D. Gugan and G. W. Michel, Final Report 1.10.78, H. H. Wills, Physics Lab., Univ. of Bristol, UK (1978); Metrologia 16, 149 (1980).
[24] W. H. Keesom, Helium, p. 49, Amsterdam (1942).
[25] J. E. Kilpatrick, W. E. Keller, E. F. Hammel, and Nicholas Metropolis, Phys. Rev. 94, 1103 (1954).
[26] J. E. Kilpatrick, W. E. Keller, and E. F. Hammel, Phys. Rev. 97, 9 (1955).
[27] D. White, T. Rubin, P. Camky, and H. L. Johnson, J. Phys. Chem. 64, 1607 (1960).
[28] W.H. Keesom and W.K. Walstra, Physica 7, 985 (1940); Physica 13, 225 (1947); see [22] and [23] above.

A refractive index thermometer for use at low temperatures

A. R. Colclough

National Physical Laboratory, Teddington, Middlesex, TW11 0LW, United Kingdom

The principles of refractive index thermometry are outlined and it is shown that straightforward interferometric measurements of the refractive index of a gas and its pressure should yield information about its thermodynamic temperature at the part per million level of resolution. The main systematic errors of the technique are those due to thermal contraction and hydrostatic compression of the thermometric gas cell in the interferometer. An analysis of these errors indicates that the accuracy to be expected from this technique is, though less than the expected resolution, comparable to that obtainable from other methods over a wide range of temperature. A practical refractive index thermometer built at the NPL is described and preliminary tests made of its performance are reported. The results of these tests are encouraging.

1 INTRODUCTION

The systematic errors of conventional isotherm gas thermometry and acoustic thermometry are now well known. The former suffers from problems of dead spaces and adsorption brought about by the need to control the amount of thermometric gas present in the thermometer's bulb[1]. The latter method was felt to be attractive in that it avoided these problems by measuring an intensive quantity, the velocity of sound, from which thermodynamic temperature could be inferred. However, this method too has many characteristic systematic errors which are difficult to correct for[2]. Recently two other intensive property methods have been considered: dielectric constant gas thermometry[3] and refractive index gas thermometry[4]. In this paper a description is given of a refractive index thermometer developed at the NPL and of preliminary tests of its performance.

The fundamental thermometric relation of refractive index thermometry is easily derived from the Lorenz-Lorentz law:

$$(n^2 - 1)/(n^2 + 2) = \alpha N_0 \rho/(3\varepsilon_0 M) \quad (1)$$

where n is the refractive index of the thermometric gas, ρ its density, α its atomic polarizability, M its molar mass, N_0 the Avogadro constant and ε_0 the permittivity of free space. This law is derived directly from the more familiar law of Clausius and Mossotti by substituting n^2 for the permittivity of the gas in the Clausius-Mossotti function. Since $(n-1) \ll 0$ for a gas one may write for sufficiently low pressure

$$n - 1 \propto \rho \quad (2)$$

which is the law of Gladstone and Dale. The quantity (n-1) is conveniently measured by an optical interferometer with a gas cell in the path of one of the interfering beams. This then gives a measure of density which, with a measurement of pressure, allows thermodynamic temperature to be deduced from Boyle's law.

But, as with all methods of thermometry depending on gas properties, allowance must be made for gas non-idealities at finite pressures. In practice the atomic polarizability will be a function of density since N interacting atoms will not generally have N times the polarizability of N non-interacting atoms.

One therefore writes

$$\alpha = \alpha(\rho) = \alpha_0(1 + x_1\rho + x_2\rho^2 + \ldots) \quad (3)$$

This yields directly a density expansion for the Lorenz-Lorentz function $L = (n^2-1)/(n^2+2)$. But for thermometric purposes a pressure expansion is more convenient. From the familiar virial equation of state for n' moles of a real gas

$$\rho = n'M/V =$$
$$(Mp/RT)(1 - (B/RT)p + ((2B^2-C)/(RT)^2)p^2 + \ldots) \quad (4)$$

whence

$$(n^2 - 1)/(n^2 + 2) =$$
$$(N_0\alpha_0/(3\varepsilon_0 RT))p(1 + L_1(T)p + L_2(T)p^2 + \ldots) \quad (5)$$

where

$$L_1(T) = (x_1 M - B)/RT \quad (6)$$

and

$$L_2(T) = (2B^2 - C - 2x_1 BM + x_2 M^2)/(RT)^2 \quad (7)$$

are the "refractive index virial coefficients" expressed in terms of the virial coefficients proper, $B = B(T)$, $C = C(T)$ etc. and the gas constant R. This is the fundamental thermometric relation of refractive index thermometry.

Generally the directly measured quantity is (n-1) rather than L, but the latter is easily calculated to sufficient accuracy from

$$L = (2/3)(n-1)(1 - (n-1)/6 - 2(n-1)^2/9 \ldots) \quad (8)$$

In order to calculate the resolution to be expected from this method, consider a gas cell of length 10 cm traversed twice by an optical beam of wavelength 633 nm from a He-Ne laser. ^{4}He gas has a refractive index of $1 + 3.49 \times 10^{-5}$ at s.t.p. and 633 nm whence to first order in pressure $(n-1) = 9.40 \times 10^{-8} p/T$ (p in Pa, T in K). This corresponds to a change in the 20 cm path of the optical cell of $1.88 \times 10^{-6} p/T$ cm or $2.97 \times 10^{-2} p/T$ optical fringes. Thus at 10 K one would expect to observe approximately 1000 fringes on filling the cell to a pressure of 3.3 atmospheres (1 standard atmosphere = 101325 Pa). Resolving a fringe to 0.1%, would yield a temperature resolution of about one part per million (ppm) or 10 microkelvins at 10 K and 0.3 mK at 273 K. When the uncertainty in pressure measurement is also taken into account this would rise to ca. 2 ppm (see section 2).

The above resolution is impressive by the standards of other methods. In practice the need to allow for the evaluation of refractive index virial coefficients will tend to lower the resolution, but this effect will easily be offset by taking a sufficient number of data points. Lorenz-Lorentz function or "refractive index" isotherms will be plotted as a function of pressure, L vs. p or L/p vs. p, and the first term of equation (5) extracted by least-squares fitting (see Figure 1).

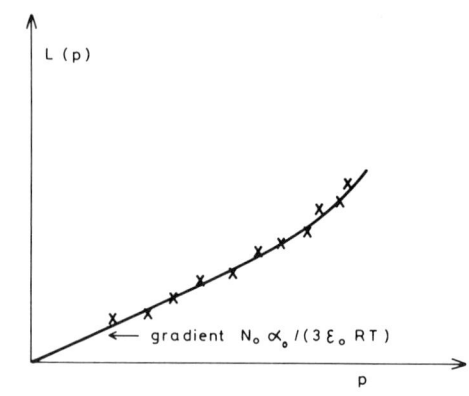

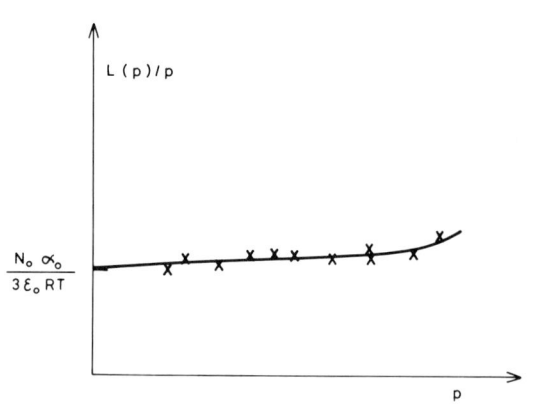

Figure 1. Lorenz-Lorentz function or "refractive index" isotherms.

In order to apply this method in the primary thermometric mode, it is necessary to know the thermometric constant of proportionality $N_o \alpha_o/(3\varepsilon_o R)$ between L and T^{-1} and the relationship between (n-1) and the fringe count. The constant of proportionality requires a knowledge of N_o, ε_o, R and α_o or, equivalently, the refractive index at s.t.p. The free-space permittivity ε_o is for present purposes exactly known and the uncertainty in N_o is ca. 15 ppm (three standard errors)[5]. That in R is arguably 25 ppm (three "standard errors" including an allowance for systematic uncertainty) but may be expected to fall to below 10 ppm in the future[2,6,7]. The uncertainty in α_o derived theoretically is probably several ppm.[9]

But for a primary measurement the best way to evaluate the constant of proportionality is by calibration at the triple point of water. This has the additional advantage of allowing for the effective length of the gas cell at s.t.p. and hence the ratio between (n-1) and fringe count. Some other reference temperature can be used for non-primary thermodynamic measurements. In both cases it is necessary to correct for the thermal contraction of the gas cell at lower temperatures, for its hydrostatic compression or expansion due to the pressure of the thermometric gas and for any impurities in the thermometric gas. It is these effects which in practice are likely to limit the accuracy of the method below the resolution of ca. 2 ppm.

The effect of thermal contraction

The thermal contraction of copper is shown in Figure 2. It can be seen that it amounts to about 3000 ppm between 273 K and liquid helium temperatures, to about 320 ppm between 90 K and liquid helium temperatures and almost ceases below 25 K. Since the directly observed thermometric quantity, the fringe count, is proportional to the cell length, these figures represent the fractional errors in temperature measurements made using a copper gas cell relative to calibration isotherms at 273 K, 90 K and 25 K respectively and uncorrected for thermal thermal contraction. For corrections made to 0.1% the errors become respectively 3, 0.3 and zero ppm.

At present literature values due to Kroeger and Swenson exist accurate to about 0.1%.[9] They argue that variations in expansion coefficient between different samples of high purity copper may be larger than this at the lowest temperatures, but there thermal contraction is negligible anyway. Above ca. 20 K sample variation is less than 0.1% but could be measured if it were thought to be necessary in a subsidiary experiment.

The effect of hydrostatic compression

Suppose that a refractive index thermometer has been calibrated at some temperature T_c and that its low pressure calibration isotherm slope is $S(T_c)$. Then to first order in pressure we have at some other temperature T:

$$T = T_c S(T_c)/S(T) \qquad (9)$$

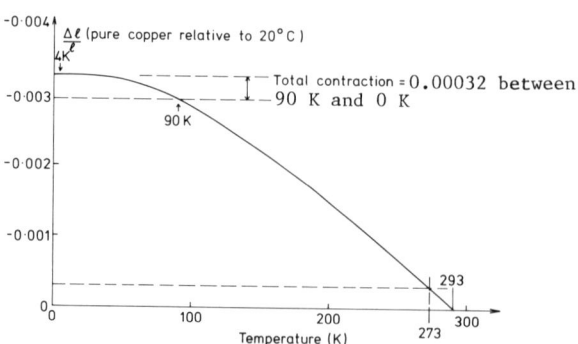

Figure 2. The thermal contraction of pure copper.

If it is now supposed that a spacer defining the length of the cell compresses linearly with a bulk modulus $K = K(T)$ one would expect to observe an apparent temperature

$$T' = T_c(S(T_c) - 1/3K(T_c))/(S(T) - 1/3K(T))$$
$$= T(1 - d(T_c/K(T_c) - T/K(T))) \quad (10)$$

where $d = 1/(3 \times 9.4 \times 10^{-8})$. Assuming a temperature-independent value of $K = 1.4 \times 10^{11}$ Pa one finds a fractional temperature error due to compression given by

$$(T' - T)/T = -2.5 \times 10^{-5}(T_c - T) \quad (11)$$

implying a maximum absolute error at $T_c/2$ if $T<T_c$.

Values of the resulting uncorrected temperature errors are given in Table I for various calibration temperatures. It can be seen that correcting for this effect to 1 % will leave maximum absolute errors of 4.7 mK at 136 K for T_c = 273 K, of 0.5 mK at 45 K when T_c = 90 K and of 0.04 mK at 12.5 K when T_c = 25 K.

These figures represent creditably small errors below 90 K when compared with past work in gas and acoustic thermometry.

Another way of dealing with hydrostatic compression errors is possible if one is prepared to interpolate in $1/T$ between two calibration temperatures T_{c1} and T_{c2} ($T_{c1}>T_{c2}$). If $K(T)$ is assumed to be constant then all observed slopes $S'(T) = S(T) - 1/3 K$ are in error by the same constant term so that

$$(S'(T) - S'(T_{c1}))/(S'(T_{c2}) - S'(T_{c1})) =$$
$$(1/T - 1/T_{c1})/(1/T_{c2} - 1/T_{c1}) \quad (12)$$

Thus compressional errors cancel to first order. If a small temperature dependence, $(1/K)dK/dT = -3 \times 10^{-4}$ in the case of copper, is allowed for it may be shown that

$$(T' - T)/T =$$
$$+ 7.6 \times 10^{-9}(T_{c1} - T)(T - T_{c2}) \quad (13)$$

resulting in a maximum absolute error at $(2T_{c1} + T_{c2})/3$ if $T_{c1}>>T_{c2}$. This would be 1.2 mK at ca.60 K for T_{c1} = 90 K and T_{c2} = 4.2 K. When corrected for to 10 %

Table I

The magnitude of the effect due to hydrostatic spacer compression

The effect is expressed in mK

T(K) =	4	12.5	25	45	90	136	273
T_c(K)							
273	27.3	82.5	157.0	259.9	417.2	471.9*	0
90	8.7	24.5	41.2	51.3*	0		
25	2.1	4.0*	0				

* marks the maximum compression error for a given T_c.

A constant bulk modulus $K = 1.4 \times 10^{11}$ Pa has been assumed for copper.

a negligible error would result.

The effect of impurities

Impurities represent a potentially more serious source of error in refractive index thermometry than in gas or acoustic thermometry. In gas thermometry adsorption of impurity species leads to an error in the number of moles in the thermometer bulb which is simply proportional to their original number concentration[1]. In acoustic thermometry the error in measured temperature is proportional to the error in molar mass attributable to impurities and hence is proportional to their concentrations and the differences in their molar masses from that of the pure thermometric gas. Since the molar masses of the impurity species are generally of the same order of magnitude as that of the thermometric gas, this source of error is manageable. But in the case of refractive index thermometry it is the number of impurity moles and their respective polarizabilities which matters. Since the polarizabilities of all impurity species are likely to be much greater than that of ^{4}He, it is correspondingly more important to control their concentrations. This will be no problem at liquid helium temperatures, but could be at higher temperatures. Perhaps the best way of ensuring that this problem is avoided is to pass all thermometric gas through a tube at 4.2 K before entering the gas cell and to provide a diffusive path back to this surface to enable slowly desorbing impurity species to be removed from the cell. This should not cause other problems of thermal instability if it is carefully designed.

2 THE INSTRUMENT

The NPL's refractive index thermometer is shown in Figure 3. It consists of a Michelson interferometer with a gas cell N enclosing most of one arm. The reference arm S is "bent" through 90° with the aid of a mirror V2 to enable it to be accommodated in the liquid helium cryostat and is situated in the insulating vacuum of the cryostat. The reference arm, the beam splitter V1 and the mirror carrier Z with its three-point mount U are sprung against the window T of the optical cell with four springs (not shown). The spacer P which defines the length of the gas cell is also sprung against the cell window so that the whole inteferometer is effectively located against the window. Thus any movement of the cell brought about by changes in temperature or gas pressure do not affect the relative positions of the interferometer components, though the light beams would take a different path through the instrument. Beam displacements parallel to the axes of the arms are not important, but tipping of the entry beam will cause a cosine error in the fringe count. For this reason all isotherms are compared with a two-point reference isotherm on the occasion of their measurement. Between these two measurements no disturbance of the cryostat is permissible.

The cell spacer is of OFHC copper as is the rest of the interferometer with the exception of the glass components and the "Kovar" tube on which the cell window and its graded glass seal are mounted. The spacer is nominally 10 cm long and consists of a cylinder with a conical end through which two "nostrils" have been drilled either side of the point which locates against the window. These allow the light beams to enter and leave the cell. Cube-corner reflectors are sprung against the ends of the spacers in the gas cell and reference arms.

The light source is a 1 mW He-Ne laser operating at 633 nm. The beams enter and leave the interferometer through a room temperature window W and holes Y in the radiation shields. It was found to be necessary to mount a thin glass window X on the outer 77 K shield J

All photocells were well shaded and the room was blacked out and internally illuminated with bulbs driven by a stabilized d.c. source. This prevented drift due to changing background light.

With a Michelson interferometer having equal arms it is not in principle necessary to stabilize the wavelength of the laser since changes in the number of wavelengths in the gas cell arm will be compensated by an equal change in the fiducial arm. However it is in practice inconvenient to equalise the arms to better than about 1 mm in length, or 2 mm in optical path. This is equivalent to approximately 3000 fringes which, for a wavelength change of only 1 ppm, will cause a shift of 0.003 fringes in the output. This could easily be seen on the x-y plotter as the laser mode-skipped on warming up. It was therefore stabilized using a tube heater controlled by the outputs of a polarizing beam splitter. The beam splitter separated the orthogonally polarized components of the light emerging from the rear of the laser which varied inversely one with the other and directly with the respective laser mode amplitudes. The separated components were intercepted by photocells whose outputs were passed to a servo system controlling the heater. This reduced mode jumping to a negligible level.

The interferometer completely enclosed in its isothermal shield had its temperature controlled by a heater wound on a bobbin M into which a controlling rhodium-iron resistance thermometer was inserted. The resistance of this thermometer was monitored by a four lead a.c. bridge (an NPL prototype of the Tinsley 5545 bridge) whose out-of-balance signal was passed to a power amplifier driving the heater. The required interferometer temperature was obtained by adjusting the bridge to the appropriate setting and control could generally be maintained indefinitely to better than 0.1 mK below 27 K with occasional manual trimming. The gas cell was furnished with four resistance thermometer wells so designed that 5 cm capsule rhodium-iron or platinum resistance thermometers could be inserted with the mid-points of their sensing elements level with the mid-point of the cell spacer. These resistance

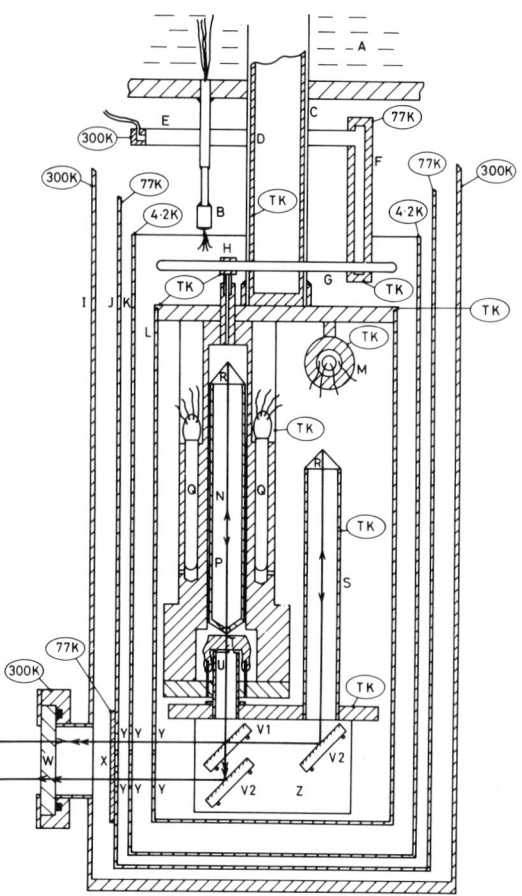

Figure 3. The refractive index thermometer

A: Liquid ^{4}He bath, B: Electrical lead-through seal, C: Stainless steel support tube, D: Copper heat exchange tube, E: Gas inlet pipe, 300 K to 77 K, F: Gas inlet pipe, 77 K, G: Horizontally coiled gas inlet pipe joining G and H, 77 K to T K, H: Gas cell inlet, I: 300 K cryostat wall, J: 77 K radiation shield, K: 4.2 K radiation shield, L: Isothermal radiation shield at T K, M: Block with heating coil and controlling sensor, N: Gas cell, P: Copper cell spacer, Q: Resistance thermometers, R: Cube-corner reflectors, S: Vacuum arm spacer, T: Gas cell window, U: Locating tube for mirror carrier with three-point mount against T, V1: Beam splitter, V2: Mirrors, W: 300 K window, X: 77 K window, Y: Apertures in radiation shields.

thermometers carried the gas thermometer scale NPL-75[1] and NPL realizations of the low-temperature fixed points of the IPTS-68. They were read with a four-lead a.c. bridge (Tinsley type 5545) but could also be switched to a d.c. potentiometer system (Tinsley stabaumatic type 58408). All leads to these thermometers were passed through the liquid helium bath A via a Stycast lead-through seal B and thermally anchored to the cell before connection to the thermometers.

The interferometer was suspended on a single thin-walled stainless steel tube C up which passed a copper heat exchange pipe D into the region of the helium bath. Exchange gas or condensing helium in this tube allowed rapid precooling of the interferometer which could then be quickly isolated by evacuating the tube. This enabled the simplicity of a modular tail dewar, with all its advantages for light beam access, to be combined with the speed of operation of a conventional cryostat where the experiment is housed in a can immersed completely in the helium bath.

The modular tail design also enabled the gas inlet and pressure sensing tube, E, F, G and H, to be brought out almost at the level of the gas cell without having to pass upwards through the helium bath. This had considerable advantages for pressure measurement. As in gas thermometry, measured pressures must be corrected for the hydrostatic head effect and this is difficult to do where inlet pipes are long and pass vertically through refrigerant baths. In the present apparatus it has been arranged that all temperature gradients occur in horizontal tubes. In the case of the horizontal tube E the gradient falls from 300 to 77 K and in the horizontal spiral G from 77 K at one end to T K at the other where T is the interferometer temperature. G also has a short portion anchored at 4.2 K by copper braid to ensure purity of the thermometric gas as described

to prevent radiative heating of the interferometer by 300 K radiation from the outer skin of the cryostat, I. The incoming beams enter the interferometer arms "behind" Figure 3, cross the cube corners "through" Figure 3 and depart "in front of" Figure 3. On emerging from the cryostat at different heights in front of the figure they are intercepted by two photocells. Since the emerging interfering beams will have different fringe phases, the two cell signals will be out of phase and may be passed to the two channels of an x-y plotter and used to plot interference ellipses. These enable one to interpolate between fringes which are also counted by a bidirectional counter.

The intensity of the laser was monitored by a third photocell which was illuminated by a stray reflection from one of the cryostat windows. Before being passed to the x-y plotter, the interference signals were electronically divided by the laser intensity signal to ensure that the plotted interference ellipses did not drift with any small changes in the laser output level.

in section 1. The vertical heavy copper tube F is held at ca. 77 K so that a very simple hydrostatic correction can be made. This tube carries a copper winding to act as a simple resistance thermometer to ensure that the copper braid anchoring to the liquid nitrogen bath is fully effective. Another larger, but equally simple hydrostatic correction is made for the gas head between H and the mid point of the gas cell. The head at ambient temperatures can also be corrected for, but is generally minute. Table II gives examples of hydrostatic head corrections.

Effects due to thermomolecular pressure gradients in the tubes E and G are virtually negligible since, in contrast to the case of gas thermometry, these tubes do not have to be narrow in order to reduce the size of dead space corrections. This is one of the important practial advantages of measuring intensive properties.

The pressure measurements are made with a Bell and Howell pressure balance consisting of a finely lapped spinning piston floating on the gas in a cylinder. Pressure is deduced from the cross section at the piston, its weight and the weights with which is loaded. The piston-cylinder assembly is covered with an evacuated glass dome equipped with a vacuum gauge so that absolute pressures are measured (or, rather, defined). The temperature of the assembly is measured to allow small corrections to be made for the thermal expansion of the piston cross section. Such a system is capable of absolute accuracies of the order of 10 ppm and linearities of the order of 1 ppm. Since measurements are made relative to a reference temperature isotherm, it is the latter figure which is relevant to calculating the uncertainties in temperature measurements arising from pressure measurement. A fuller account of the use of this instrument may be found in Berry's report of its application to primary gas thermometry[1].

The main difference between its use in this project and in Berry's is that in the latter the pressure balance was isolated from the thermometric gas by a capacitatively monitored diaphragm pressure gauge. This had the advantages that no thermometric gas was lost past the piston and that it enabled one to interpolate between the discrete pressures defined by the weights. On the other hand the diaphragm gauges called for regular painstaking recalibration and at pressures above one or two atmospheres tended to be unstable. For these reasons use of the diaphragm gauges was dropped and the piston gauges were connected directly to the gas cell.

Table II

The magnitude of hydrostatic head corrections

T (K)	ΔT_T (h = 10 cm) (mK)	ΔT_{77} (h = 5 cm) (mK)	ΔT_{293} (h = 10 cm) (mK)
273	0.4_7	0.8_4	0.4_4
90	0.4_7	0.2_8	0.1_5
25	0.4_7	0.0_8	0.0_4
4	0.4_7	0.0_1	0

Column 1 shows the temperature of the thermometric gas, column 2 shows the error due to a 10 cm head in the optical cell, column 3 shows that due to a 5 cm head in a tube at 77 K and column 4 shows that due to a 10 cm head at 20 °C. All effects are small and easily corrected for to a negligible uncertainty.

This raised another problem, however. In order to avoid having to rely upon the linearity of the x and y fringe channels, it was desirable always to make measurements at the same phase on each fringe ellipse e.g. at zero crossing, phase $\emptyset = 2N\pi$, on the x-channel (N = 0,1,2,...). For a given discrete pressure defined by the pressure balance, this would entail making small temperature changes about the nominal isotherm temperature. But, especially at higher temperatures, this would cause a temporary loss of stability which would take an inconveniently long time to die away. Thus it was decided to maintain a constant temperature and to allow the system to come to equilibrium generally at some incorrect phase. Then specially made small interpolation weights would be added to the pressure balance to bring it to the correct phase. The values required could be conveniently read off a correction chart as shown in Figure IV. Various methods were tried for defining the ellipse phase, but no significantly more effective method was found than simply calculating $\emptyset = \arcsin((a + x)/(2a))$ where 2a is the width of the ellipse in the x-direction, i.e. its projection onto the x-axis, and it has been assumed to lie symmetrically about the y-axis (Figure IV).

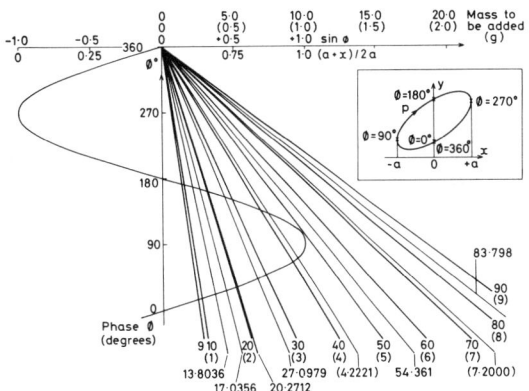

Figure 4. Calculation of the fringe phase from the ellipse ($\emptyset = \arcsin(a + x)/2a$ - inset) and of the masses to be added to the piston gauge to bring the phase to 360 degrees. The initial phase is read on the vertical axis and the required mass is read from the horizontal axis via the appropriate temperature line. Bracketed masses correspond to bracketed temperatures (0 - 10 K).

3 TESTS ON THE SYSTEM

The first requirement of the instrument is that it should be able to resolve the change in optical path brought about by the entry of the thermometric gas to approximately 0.001 fringe. Expressed to first order in p, the phase of the fringe pattern will be given by $\emptyset = 0.185$ p/T radians (p in Pa). At 27 K and 1 atmosphere, for example, a change of 0.001 fringe corresponds to a pressure change of 0.9 Pa and a temperature change of 0.24 mK. While these stabilities can easily be achieved for several minutes according to the instruments monitoring the cell, long enough to record all necessary readings, it needs to be established that the gas in the cell is stable to this degree and is not undergoing some convective process which causes optical problems. Similarly one needs to be sure that mechanical vibrations are not going to cause a significant degradation of resolution. Figure V shows a set of three fringe ellipses taken at 27 K and 1 atmosphere. Each has been described two or three times by the plotter pen as the cell pressure was reduced, though this is difficult to see. They have been mutually displaced in the y-direction in order to enable them to be distinguished. When the temperature and pressure were stable the red plotter pen was exchanged for a blue one which was placed on the paper for several minutes. The vertical lines show the range of the jitter of the blue pen at approximately zero

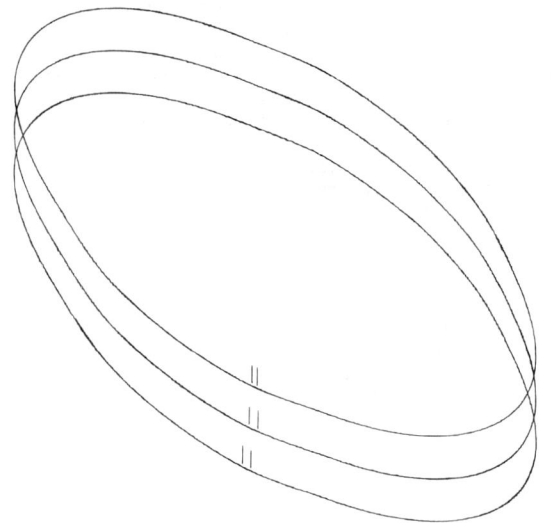

Figure 5. Three sets of experimental fringe ellipses taken ca. 27 K and 100 kPa. The vertical lines indicate the magnitude of the phase noise.

phase. The range is typically less than ± 2.5 mm in an ellipse of approximate width $2a = 25$ cm. This corresponds to a noise of 0.003 fringe, but since the centre of the noise can reliably be judged to ± 0.5 mm the resolution is probably as good as 0.0006 fringe. This demonstrates that the hoped-for resolution can be achieved in a practical instrument.

Several other tests have been made on the system. It is necessary to demonstrate that pressure and temperature cycling can be accomplished without changing the fringe count on returning to the initial conditions. A preliminary test at 20 K has shown that on pressure cycling at constant temperature from 0.1 atmospheres to 2 atmospheres and back the original phase is reproduced to $0.0006 \pm 0.002_2$ fringe. A temperature cycling test at a constant pressure of 0.1 atmosphere from 17 K to 27 K, then 4.2 K and back to 17 K again reproduced the initial phase to $0.002_3 \pm 0.002_5$ fringe. In both tests the uncertainties quoted are one standard error calculated from three measurements. Thus it seems that reproducibilities are comparable with the resolution of 0.001 fringe.

4 CONCLUSIONS

The principles and main systematic errors of refractive index thermometry have been discussed. Resolutions of 0.001 fringe are easily attainable in a practical optical interferometer without undue difficulty. For a 10 cm gas cell in the path of an interfering beam such a resolution implies a resolution in measured temperature of a few ppm. A Michelson interferometer housed in a liquid helium cryostat has been built at NPL in order to investigate this new technique. Resolutions of less than 0.001 fringe have been obtained and reproducibilities of this order have been demonstrated. These figures which relate only to preliminary tests encourage one to believe that this method of thermometry is capable of making a useful contribution to our knowledge of thermodynamic temperature.

5 REFERENCES

1. K.H. Berry, Metrologia, 15, 89 (1979).
2. A.R. Colclough, Contribution to this Symposium.
3. D. Gugan, contribution to this Symposium.
4. A.R. Colclough, Metrologia, 10, 73 (1974).
5. E.R. Cohen and B.M. Taylor, J. Phys. Chem. Ref. Data, 2, 663 (1973).
6. A.R. Colclough, "Methods for the Determination of the Gas Constant" in Proceedings of the Second International Conference in Precision Measurements and Fundamental Constants, NBS, Gaithersburg, 8-12th June, 1981, to be published.
7. A.R. Colclough, T.J. Quinn and T.R.D. Chandler, Proc. Roy. Soc. Lond., A368, 125 (1979).
8. R.M. Glover and F. Weinhold, J. Chem. Phys., 65, 4913 (1976).
9. F.R. Kroeger and C.A. Swenson, J. Appl. Phys. 48, 853 (1977).

Nuclear orientation thermometry from ~0.001 to ~1.2 K

H. Marshak

National Bureau of Standards, Washington, D.C. 20234

We have investigated γ-ray anisotropy thermometry using both ^{60}Co in cobalt single crystals (^{60}Co$\underline{Co}$) and ^{166m}Ho in Ho single crystals (^{166m}Ho$\underline{Ho}$) for their potential use in defining a low temperature scale covering the range from ~0.001 to ~1.2 K. The values of temperature derived from nuclear orientation thermometers are thermodynamic since they are deduced from the Boltzmann factor; viz. $\exp(-E_m/kT)$. The accuracy of the thermodynamic temperatures obtained depends upon how well one knows E_m and the uncertainties (both statistical and systematic) of the measurement. In the case of ^{60}Co$\underline{Co}$, E_m is known from NMR/ON (Nuclear Magnetic Resonance/Oriented Nuclei) measurements with an uncertainty of less than 1/1000. In the temperature range of 0.01 to 0.05 K we have compared the ^{60}Co$\underline{Co}$ thermometer to a Josephson junction noise thermometer, which also measures thermodynamic temperatures, and they agree within 0.5%. In the case of the ^{166m}Ho$\underline{Ho}$ thermometer, which covers the temperature region of ~32 mK to ~1.2 K, E_m has not been measured by NMR/ON and thus for the present it must be considered to be a secondary thermometer (e.g., as the susceptibility of the paramagnetic salt cerium magnesium nitrate).

INTRODUCTION

One of the most difficult and fundamental problems in research below 1 K is relating the observed effect to the thermodynamic temperature. Often the "temperature of" the sample can be measured using some secondary thermometer; however the relationship between these measured temperatures and thermodynamic temperatures can be difficult to establish. For the region above 0.5 K, temperatures close to thermodynamic can be obtained using the EPT-76 temperature scale[1]. To use this scale one has to rely on either: (a) secondary thermometers (resistance thermometers such as RhFe or Ge) obtained with a calibration certificate, or (b) one's own secondary thermometer calibrated by means of superconducting fixed points from NBS (SRM-767)[2] in conjunction with a suitable interpolating instrument (such as the paramagnetic salt CMN). There is no readily available thermodynamic thermometer for this region which could be incorporated into one's experiment. For the region below 0.5 K no internationally accepted temperature scale exists. Work is presently underway at several laboratories towards the development of a new temperature scale which would overlap EPT-76 and extend down to ~1 mK. At the National Bureau of Standards a preliminary temperature scale has been developed which covers the region from 10.2 to 519 mK. This scale is designated as NBS-CTS-1 (National Bureau of Standards-Cryogenic Temperature Scale-Version 1).[3] The temperatures for this scale are obtained using two thermodynamic thermometers, one a Josephson junction noise thermometer (JJNT)[4,5] and the other a nuclear orientation thermometer; specifically, a ^{60}Co$\underline{Co}$ single crystal γ-ray anisotropy thermometer[6-8]. In the region from 10.2 to 50.3 mK both thermometers are used to define the temperature scale. For the region from 50.3 to 519 mK the JJNT alone is used to define the temperature scale since the ^{60}Co$\underline{Co}$ thermometer loses sensitivity in that range. In order to use this scale, one only has option (b) available, i.e., to obtain superconducting fixed points from NBS (SRM-768)[9] and calibrate one's own secondary thermometers. However, in contrast to the situation with EPT-76 either of these primary thermometers can be incorporated into the experiment and thermodynamic temperatures measured directly. For temperatures less than 50 mK, the ^{60}Co$\underline{Co}$ thermometer has many advantages, e.g., it operates in a magnetic field as well as in zero field; it is physically small (1x1x10 mm) and can be easily attached (soldered) to the experimental package insuring good thermal contact; no wires are attached to it; the readout (counting) equipment is relatively cheap with digital output and finally it is simple to use. For the temperature region above 50 mK the JJNT similarly can be used directly in one's apparatus.

A second nuclear orientation thermometer is being investigated, namely the ^{166m}Ho$\underline{Ho}$ single crystal γ-ray anisotropy thermometer, which has the potential of yielding thermodynamic temperatures from ~32 mK to ~1.2 K. Thus, with only two γ-ray anisotropy thermometers, ^{60}Co$\underline{Co}$ and ^{166m}Ho$\underline{Ho}$, the potential exists to define a thermodynamic temperature scale from ~0.001 to ~1.2 K based solely on nuclear orientation thermometry. The results of this work are the subject of this paper.

THEORY

The degree of ordering of a nuclear spin system I with cylindrical symmetry about some axis can be specified by the nuclear orientation parameters $B_\lambda(I)$ defined as follows:

$$B_\lambda(I) = \sum_m (-1)^{I-m} [(2\lambda+1)(2I+1)]^{1/2} \begin{pmatrix} I & -I & \lambda \\ m & -m & 0 \end{pmatrix} a_m \quad (1)$$

where the m are the 2I+1 nuclear (spin) substates, with a_m being the population of each of these states. The index λ goes from 0 to 2I with $B_0(I) \equiv 1$.

There are many ways to change the populations of the nuclear substates, e.g., thermal, NMR and nuclear reactions. Nuclear orientation thermometry is concerned <u>only</u> with thermal methods, i.e. only with those methods which leave the nuclear spin system in <u>thermal equilibrium</u>. For these the populations are governed by the Boltzmann distribution and are given by

$$a_m = \frac{e^{-E_m/kT}}{\sum_m e^{-E_m/kT}} \quad (2)$$

where E_m are the energies of the nuclear m-states, k the Boltzmann constant and T the <u>thermodynamic temperature</u>. We designate the nuclear orientation parameter by $B_\lambda(I,T)$ for such systems. The most rapid changes of population caused by thermal effects occurs when $E_m \simeq kT$.

Thus, the entire basis for nuclear orientation thermometry lies in the Boltzmann factor, $\exp(-E_m/kT)$. Once the energies of the nuclear m-states are known, a measurement of any non-zero $B_\lambda(I,T)$ results in a unique value of the thermodynamic temperature. How well one can determine this value depends upon how well one knows E_m and how accurately one can measure any non-zero $B_\lambda(I,T)$.

There are many different ways to measure $B_\lambda(I,T)$ for a nuclear spin system. The method most often used is to measure the change in the γ-ray radiation pattern of a system of oriented radioactive nuclear spins. This subgroup of nuclear orientation thermometers are called γ-ray anisotropy thermometers.[10] We will deal solely with these in the rest of the paper.

The normalized directional distribution of γ-radiation emitted from a system of oriented radioactive nuclear spins with cylindrical symmetry is given by

$$W(\theta) = \sum_{\lambda=0}^{\lambda_{max}} B_\lambda(I,T) U_\lambda A_\lambda Q_\lambda P_\lambda(\cos\theta) \quad (3)$$

where θ is the angle between the orientation axis and the emitted γ-ray. The quantities U_λ and A_λ are the angular momentum deorientation and angular distribution coefficients respectively, and depend only upon the decay scheme (spins and multipole amplitudes) of the radioactive nucleus. The angular dependence of the radiation pattern is contained in the Legendre polynomials $P_\lambda(\cos\theta)$. Since the detector subtends a finite solid angle, the radiation pattern is "smeared out" and the $P_\lambda(\cos\theta)$ must be corrected. This is done by including the solid angle correction factors Q_λ. Only even values of λ enter into the summation since we are only concerned with the intensity distribution of the γ-ray and not its state of polarization. The maximum value of λ is equal to the lesser of the two values 2I' or 2L, where I' is the lowest nuclear spin in the decay sequence preceeding the observed γ-ray of multipolarity L.

How well we can deduce the thermodynamic temperature from Eq. 3 by measuring γ-ray intensities depends not only upon how well we know E_m, U_λ, A_λ, Q_λ and θ, but also on how accurately we can measure γ-ray intensities. In the case where the decay scheme is known completely, both U_λ and A_λ can be calculated exactly. The γ-ray intensity measurements are usually made using NaI or Ge detectors. Solid angle correction factors have been calculated for both types of detectors for various geometries, using Monte Carlo techniques: tabular results of these calculations can be found in the literature.[11] In addition, computer programs are available for odd-size detectors not listed in these tables. When necessary one can also experimentally determine Q_λ values. The overall uncertainty in Q_λ values can be reduced to $\leq 0.1\%$. The uncertainty in the $P_\lambda(\cos\theta)$ values, which come about from uncertainties in θ, can usually be minimized by careful experimental techniques. If E_m can be measured by NMR/ON (nuclear magnetic resonance on oriented nuclei) techniques, uncertainties of <0.1% are not unreasonable. Thus, in favorable cases (e.g. the ^{60}Co<u>Co</u> thermometer) values of $W(\theta)$ versus T can be calculated with an error in the range of a few tenths of a percent.

Measured values of $W(\theta)$ are determined using

$$W(\theta)_M = \frac{C_C - B_C}{C_W - B_W} \quad (4)$$

where C_C and C_W are the cold and warm γ-ray counts, and B_C and B_W are the cold and warm background γ-ray counts, all with the detector set at some angle θ. The cold counting is done when the γ-ray thermometer is at some unknown temperature T and the warm counting at a T sufficiently high so that the anisotropy is effectively zero. Measurements of $W(\theta)$ need only be done at one angle to determine a temperature, usually 0° or 90°, whichever gives the greatest change in the radiation pattern as a function of temperature. The uncertainty in $W(\theta)_M$, namely $\Delta W(\theta)_M$, will depend upon counting statistics, gain shifts in the counting system, background subtraction, background fluctuations, and small temperature changes in the thermometer during the measurement period. With care, it is possible to determine $W(\theta)_M$ to an uncertainty of <0.1%. The overall uncertainty in the deduced absolute temperature from such measurements, $W(0)_M \pm \Delta W(0)_M$, will not only depend upon $\Delta W(0)_M$, but also on all the other uncertainties in the parameters in Eq. 3 that have been discussed previously. In general, for those γ-ray anisotropy thermometers where the parameters are well known (e.g., the ^{60}Co$\underline{Co}$ thermometer), measurements of the thermodynamic temperature with an uncertainty of a few tenths of a percent are possible.

Before describing the two γ-ray thermometers of interest there is one other general aspect to discuss; namely, the useful temperature range and sensitivity (or response) function of a γ-ray anisotropy thermometer. The sensitivity function is defined to be

$$\partial W(\theta,T)/(\partial T/T) \qquad (5)$$

i.e. the change in $W(\theta,T)$ per fractional change in the temperature. In Fig. 1 we show the sensitivity functions for both the ^{60}Co$\underline{Co}$ and the ^{166m}Ho$\underline{Ho}$ γ-ray

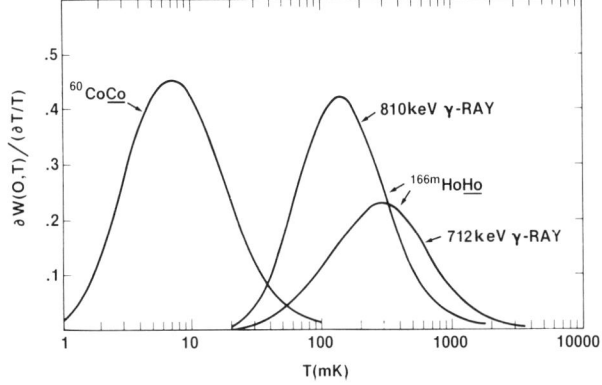

FIG. 1. The sensitivity function, $\partial W(0)/(\partial T/T)$, as a function of temperature for the ^{60}Co$\underline{Co}$ and ^{166m}Ho$\underline{Ho}$ single crystal γ-ray anisotropy thermometers.

anisotropy thermometers. The values used for the coefficients U_λ, A_λ and E_m for each thermometer are given later on when they are discussed in more detail. For simplicity we set the Q_λ equal to unity. In Figs. 2 and 3 we show the relevant parts of the decay scheme for ^{60}Co and for ^{166m}Ho. For the latter we only show the 9

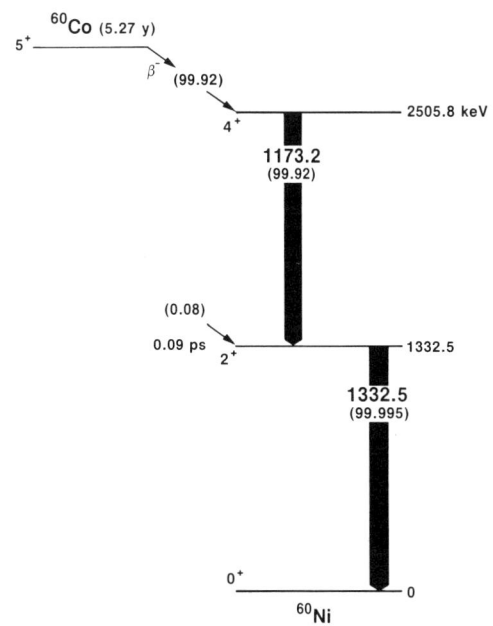

FIG. 2. Decay scheme of ^{60}Co. Some very weak transitions have been left out.

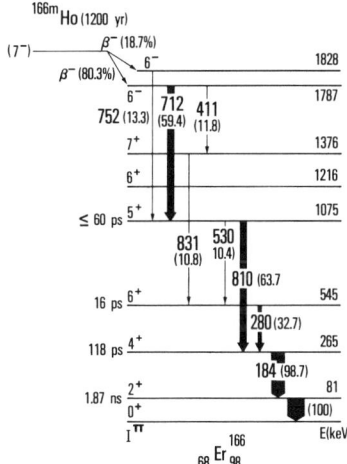

FIG. 3. Simplified decay scheme of ^{166m}Ho showing the nine most intense transitions.

most intense γ-rays. Theoretically for each γ-ray in the decay there is a sensitivity curve. Although ^{60}Co emits two γ-rays (1173 and 1332 KeV) in its decay, we only have one curve for it (Fig. 1) since they both have essentially the same $U_\lambda A_\lambda$ values. This is not the case for the γ-rays for ^{166m}Ho, which has a very complicated decay scheme. In Fig. 1 we show sensitivity curves for two of the more intense γ-rays (712 and 810 KeV). The useful range of a γ-ray anisotropy thermometer is given by $\partial W(\theta,T)/(\partial T/T) \geq 0.05$; for the ^{60}Co$\underline{Co}$ thermometer its useful range is from 1.3 to 50 mK, with its most sensitive region being around 7 mK. In the case of the

^{166m}Ho<u>Ho</u> thermometer, using both the 712 and 810 keV γ-rays, its useful range is from 32 to 1250 mK, with its greatest sensitivity around 150 mK.

Thus, by using only two γ-ray anisotropy thermometers, ^{60}Co<u>Co</u> and ^{166m}Ho<u>Ho</u>, we can cover a temperature range of 1.3 to 1250 mK.

^{60}Co<u>Co</u> γ-RAY ANISOTROPY THERMOMETER

Since the details of the ^{60}Co in hcp ^{59}Co single crystal γ-ray anisotropy thermometer have been reported in our previous publications[3,8] we will only touch on some of its more salient features here. Both it and the ^{166m}Ho<u>Ho</u> thermometer are prepared in similar fashions, with the activity introduced in-situ in the crystal by neutron capture in a reactor. This is possible because both elements occur monoisotopically in nature, viz., ^{59}Co and ^{165}Ho, and both have favorable thermal neutron capture cross sections. The useable life of a γ-ray anisotropy thermometer is about 2-3 times the half-life of the radioactive nucleus. Thus for the ^{60}Co<u>Co</u> thermometer its useful life is 10-15 years since its half-life is 5.3 years. The values used for the coefficients U_λ and A_λ for both γ-rays are, U_2 = 0.9393744, U_4 = 0.7977240, A_2 = -0.4477023 and A_4 = -0.3043790. These are obtained by assuming that both γ-rays are pure E2 (electric quadrupole) transitions. Although the 1332 keV ($2^+ \to 0^+$) transition is pure E2, the 1173 keV ($4^+ \to 2^+$) transition could have a small M3 (magnetic octupole) admixture. We have measured this admixture very carefully and have reduced the resulting uncertainty in the deduced temperatures from (±0.2 to -1.10%) to (±0.04 to -0.2%). Details are given in reference 3.

The values for E_m are calculated using

$$E_m/k = -a'm + P[m^2 - \frac{1}{3} I(I+1)] \qquad (6)$$

where a' and P, the magnetic hyperfine and quadrupole coupling constants respectively, have been measured by NMR/ON. The value for a' is -6.0668(34) mK and that for P is -2.9(7) μK.[12] Using these values to calculate E_m along with the $U_\lambda A_\lambda$ values given above, W(0) versus T can be calculated for a particular experimental set-up (i.e. for a set of Q_λ values). Thus a measurement of W(0) ± ΔW(0) results in a T ± ΔT value.

In order to test the ^{60}Co<u>Co</u> thermometer we have compared it to a Josephson junction noise thermometer — also a thermodynamic thermometer. The results of this comparison[3], which covered a temperature range of 10.2 to 50.3 mK, are shown in Fig. 4. Both thermometers yielded the same values within 0.5%, with the ^{60}Co<u>Co</u> thermometer systematically reading higher. Some, if not all, of this discrepancy could be explained by such things as Compton scattering, closure domains, etc.

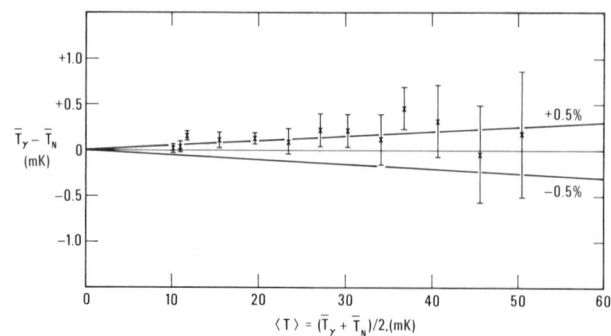

FIG. 4. The measured temperature difference, $\bar{T}_\gamma - \bar{T}_N$, between the ^{60}Co<u>Co</u> γ-ray anisotropy thermometer, $\bar{T}_\gamma$, and the Josephson junction noise thermometer, $\bar{T}_N$, versus the average temperature <T>. The errors shown are the combined standard deviations of the two thermometers.

However, better data is needed before any definite conclusions can be reached.

Plans are underway to recompare the thermometers at a reduced imprecision of ∼0.1% and also to extend the measurements to 5 mK. In this way we can measure the temperature dependence of any difference and perhaps relate it to a specific cause. Extending the measurements to 5 mK covers the most sensitive region of the ^{60}Co<u>Co</u> thermometer, thus high precision measurements are obtained in a much shorter counting time than that needed at higher temperatures.

^{166m}Ho<u>Ho</u> γ-RAY ANISOTROPY THERMOMETER

In contrast to the situation for the ^{60}Co<u>Co</u> thermometer where everything essential for the calculation of W(θ) is known (^{60}Co decay scheme, properties of the ^{60}Co<u>Co</u> system, etc.), much less is known about the ^{166m}Ho<u>Ho</u> thermometer, i.e., it is still in the developmental stage. At the start of this work (∼1977) no information existed on the system comprised of ^{166m}Ho in hcp Ho single crystals. In addition, although much was known about the decay scheme of the ^{166m}Ho nucleus at that time, i.e. spins and parities of most of the states, very few precise multipolarity (mixing ratio) measurements of the γ-ray transitions had been made and thus accurate values of U_λ and A_λ were not available. We now know the magnetic structure of the ^{166m}Ho<u>Ho</u> system[13] and also have some relatively accurate mixing ratio measurements for some of the more intense transitions[14]. However, although we do now have some measurements on the hyperfine coupling parameters for the ^{166m}Ho<u>Ho</u> single crystal system[14], and can thereby calculate E_m, no precise direct measurement (NMR/ON) has yet been done.

Monoisotopic holmium metal has the largest hyperfine interaction of all the rare earth metals. The hyperfine field, which is due to the unpaired 4f electrons, is 750 tesla! In comparison, the hyperfine field in cobalt metal is 21.8 T. This large difference, which gives rise to a large difference in the hyperfine splitting, is the reason that the holmium thermometer (when we incorporate radioactive ^{166m}Ho into it) operates at much higher temperatures than ^{60}CoCo thermometer. As mentioned previously, ^{166m}Ho can be incorporated substitutionally (in-situ) in holmium metal by neutron capture in a reactor. Its half-life is 1200 years, thus once we make a ^{166m}HoHo thermometer its useful life is 2400-3600 years!

Whereas cobalt is a fairly simple ferromagnet with a linear atomic magnetic spin structure, holmium is a very complicated ferromagnet with a helical atomic magnetic spin structure. For both materials the hyperfine field directions follow the atomic moment directions; thus for cobalt the hyperfine field lies equally in both directions along a single axis (c-axis), and for holmium the hyperfine fields form a helical spin structure. Since the γ-ray anisotropy thermometer does not depend upon nuclear polarization (odd B_λ-values), but only upon nuclear alignment (even B_λ-values), the uniaxial nature of the hyperfine fields in cobalt means that the ^{60}CoCo thermometer can operate in zero magnetic field. It can also, of course, operate in a magnetic field, the effective field at the nucleus being the sum of the applied field and the hyperfine field. Although the ^{166m}HoHo thermometer can operate in zero field, the multiaxial nature of the distribution of hyperfine fields substantially reduces the γ-ray anisotropy over that for a single axis system like cobalt. For maximum W(0) one is thus forced to use a magnetic field of sufficient strength to destroy the zero-field helical spin structure and reduce it to a single axis system. As expected the atomic magnetic properties of holmium metal are very anisotropic. The c-axis is the hard-axis of magnetization requiring fields of the order of 25 to 30 T to achieve magnetic saturation. The easy-axis of magnetization is the b-axis and only ∼0.5 T is needed to achieve saturation. Thus, one should position a single crystal of holmium metal with the applied field along a b-axis to most effectively use the ^{166m}HoHo thermometer.

As mentioned earlier, no NMR/ON measurement of E_m has yet been done on the ^{166m}HoHo system. This is mainly due to the difficulties in preparing a useable NMR/ON sample. Such a sample would have to have all the radioactivity (^{166m}Ho) located in a surface thickness of ∼0.1 μm. The reason for this is that the expected resonance frequency is very high ($\gamma_{hf} = \mu_I H_{hf}/Ih \simeq$ 3 GHz), resulting in a very shallow skin depth (∼0.1 μm).

One way to prepare such a sample would be by ion implantation using an isotope separator. However, the long half-life of ^{166m}Ho rules out this method because of the residue activity left in the isotope separator. There are other possible methods to prepare such a sample, but they are only in the "idea" stage.

We have, however, carried out an indirect measurement[14] of the hyperfine coupling parameters using the thermal technique. Namely, one measures the γ-ray anisotropy of a known γ-ray (the $A_\lambda U_\lambda$ are known) as a function of temperature (the temperature being determined from some other thermometer) and from this determines $B_\lambda(I,T)$, and thus E_m. In effect, we have calibrated the ^{166m}HoHo thermometer against another thermometer and thus it is not a primary thermometer, but a secondary thermometer. How accurate a secondary thermometer it is depends on how accurate the thermometer was that we used to calibrate it, and also on how well we know A_λ, U_λ, Q_λ, θ, and can measure W(θ). Although it is possible to calibrate the ^{166m}HoHo thermometer directly against the JJNT, such a calibration is not easy because of the latter's sensitivity to small magnetic fields. Instead we used two calibrated thin carbon thermometers. These thermometers were first calibrated from ∼10 to 500 mK in zero magnetic field against NBS-CTS-1 in another apparatus (the ^{3}He/^{4}He dilution refrigerator used for the comparison measurements[3]). They were then recalibrated in the second apparatus (the ^{3}He/^{4}He dilution refrigerator used for the holmium measurements) from ∼15 to 40 mK using the ^{60}CoCo thermometer. This was done both in zero magnetic field and as a function of the field to evaluate the magnetoresistive effect. As expected this effect was rather small, being the order of a few percent. These two thermometers along with a ^{166m}HoHo crystal were then mounted on the cold finger attached to the mixing chamber of the dilution refrigerator as shown in Fig. 5. Each carbon thermometer was mounted on a separate gold-plated OFHC copper strip. One strip was attached to the cold finger just above the holmium crystal and the other to the bottom face of the crystal. The holmium crystal itself was attached to the cold finger using indium solder. By using the two carbon thermometers in this way, they were removed from the most intense region of the magnetic field and could verify that there was no temperature gradient across the indium solder joint due to self-heating in the crystal. Although the uncertainty of the original calibration of the two thin carbon thermometers was estimated to be about ±1%, other sources of uncertainty (small temperature cycling changes, magnetoresistive effect, etc.) roughly double this estimate to about ±2%.

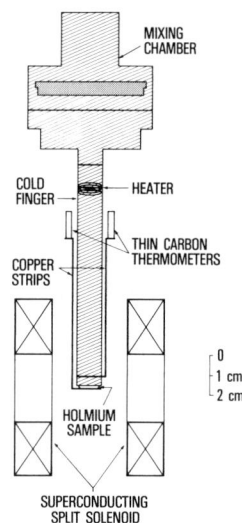

FIG. 5. Schematic drawing of the lower part of the ^{3}He/^{4}He dilution refrigerator used for the ^{166m}Ho<u>Ho</u> γ-ray anisotropy thermometry measurements.

Experimental values for the nuclear orientation parameters, $B_\lambda(I,T)_{exp}$ are obtained by measuring $W(0)$ and $W(\pi/2)$ as a function of temperature for a particular γ-ray. For most radioactive decays, usually only the $\lambda=2$ and $\lambda=4$ terms are necessary. The explicit expressions are

$$B_2(I,T)_{exp} = \frac{3W(0) - 8W(\pi/2) + 5}{7U_2 A_2 Q_2}$$

and (7)

$$B_4(I,T)_{exp} = \frac{4W(0) + 8W(\pi/2) - 12}{7U_4 A_4 Q_4}$$

Theoretical values for $B_\lambda(I,T)$ are obtained using Eqs. 1 and 2. The energy levels for the ^{166m}Ho<u>Ho</u> system are derived using Eq. 6, i.e., we assume that hyperfine interaction Hamiltonian has the same form for the ^{166m}Ho<u>Ho</u> system that it does for the ^{60}Co<u>Co</u> system. Thus, theoretical values for $B_\lambda(I,T)$ can be fitted to the experimental values (Eq. 7) by varying the hyperfine coupling parameter a' and P in Eq. 6.

Measurements of $W(0)$ and $W(\pi/2)$ were made at 17 temperatures ranging from 24 to 670 mK. The applied field used was 2T in order to insure that our disk shaped sample was completely magnetized. Since the γ-ray spectrum of ^{166m}Ho is so complex, a high resolution (but less efficient) Ge detector had to be used, rather than the more efficient (but poorer resolution) NaI detector used for ^{60}Co. Further details of these measurements can be found in reference 14. Although $W(0)$ and $W(\pi/2)$ data was obtained for 27 γ-rays, only 8 of the more intense (all those indicated in Fig. 3 except the 81 keV transition) γ-rays were used to determine a' and P. This was done by using a weighted least-square fit of $B_2(I,T)_{theo}$ to all 17 temperature points for all of the 8 γ-rays simultaneously. The coefficients derived from the fitting were a' = 137.1(22) mK and P = -0.58(25) mK. The fitting errors quoted only reflect the counting statistics. When the uncertainties in the temperature (±2%) and in the other parameters are included, we obtain a' = 137(5) mK and P = -0.6(6) mK. In Fig. 6 we show $W(0)$ results for the 712 and 810 keV γ-rays that were used in the sensitivity plots (Fig. 1).

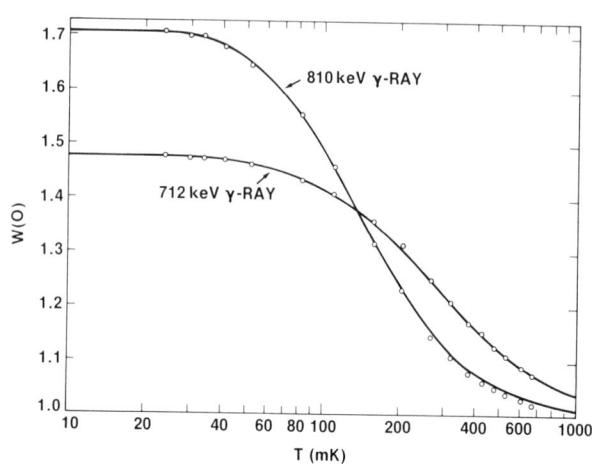

FIG. 6. Measured values (open circles) of $W(0)$ versus temperature for the 712 and 810 keV γ-rays. The standard deviation is less than or equal to the diameter of the circle in all cases. The solid curve is the theoretical fit (see text).

The circles are the measured values of $W(0)$ at the 17 temperatures, with $\Delta W(0)_M$ being less than or equal to the diameter of the circles. The overall agreement is within expected limits except for the higher temperature region of the 810 keV γ-ray where the experimental values are systematically low.

Further measurements are underway which will improve both the temperature calibration and the $U_\lambda A_\lambda$ values. The two thin carbon thermometers along with a germanium thermometer have been put into magnetic shields and calibrated against EPT-76 as well as NBS-CTS-1. The magnetic shield should eliminate the small magnetoresistive effect, and the germanium thermometer should serve as a check on the two carbon thermometers, since it is less susceptible to temperature cycling changes. These improvements should enable us to determine the temperature of the holmium sample within ±0.5%. More accurate values of $U_\lambda A_\lambda$ are being obtained both by better measurements and accounting for some small correction factors (pile-up, summing effects, etc.) in the treatment of the γ-ray data that were ignored in the initial work.

With these improvements and with careful measurement techniques it is estimated that temperatures can be measured by the ^{166m}Ho$\underline{\text{Ho}}$ thermometer with an uncertainty of ∼1% in the range from 32 to 1250 mK. These values will, of course, not be thermodynamic since the holmium thermometer was calibrated against secondary thermometers. Only when the energy splittings are measured directly, e.g. by NMR/ON, can the holmium thermometer be used to realize thermodynamic temperatures directly.

CONCLUSIONS

We have shown that, by using two nuclear orientation thermometers, the ^{60}Co$\underline{\text{Co}}$ and ^{166m}Ho$\underline{\text{Ho}}$ γ-ray anisotropy thermometers, we can measure temperatures from ∼0.001 to ∼1.2 K. Both of these thermometers follow a thermodynamic relation in which the temperature is derived from the Boltzmann factor, $\exp(-E_m/kT)$. How accurately these thermodynamic temperatures can be determined depends upon how well E_m is known and on how accurately one can measure the Boltzmann factor. In the case of the ^{60}Co$\underline{\text{Co}}$ thermometer, which covers the range from ∼1 to ∼50 mK, E_m is known from NMR/ON measurements with an uncertainty of less than 1/1000. Measurements of the Boltzmann factor within this same level of uncertainty are possible for this thermometer, particularly around 7 mK where it has its greatest temperature sensitivity (see Fig. 1). This thermometer, along with a JJNT (also a thermodynamic thermometer), has been used to define the lower part (≲50 mK) of NBS-CTS-1, with both thermometers yielding the same values within 0.5%. Improvements in the data acquisition system for the JJNT will enable us to recompare these two thermometers with an uncertainty of ∼0.1%. In the case of the ^{166m}Ho$\underline{\text{Ho}}$ thermometer, which covers the region ≳50 mK, no NMR/ON measurement of E_m has yet been made. However, E_m has been deduced from a calibration against a secondary thermometer. Thus, temperatures derived from ^{166m}Ho$\underline{\text{Ho}}$ measurements presently comprise a thermodynamically smooth but secondary scale accurate within about ±2%. Improvement in the calibration technique can be expected to reduce the level of inaccuracy to about ±1%. Because of the complexity of the nuclear decay scheme, measurements of the Boltzmann factor cannot be made to the same level of accuracy as those made using the ^{60}Co$\underline{\text{Co}}$ thermometer. Therefore one should examine other γ-ray anisotropy thermometers with simpler decay schemes than ^{166m}Ho and which cover the same temperature region.

The two thermometers discussed can be incorporated directly into low-temperature experiments, providing temperature data without the use of fixed points and without the possibility of calibration drift.

Use of yet a third γ-ray anisotropy thermometer, ^{54}Mn$\underline{\text{Al}}$ [15], would permit the measurement of temperature with a single technique over five orders of magnitude — from ∼10 μK to ∼1.2 K.

REFERENCES

[1] Comite International des Poids et Mesures, Metrologia 15, 65 (1979) and M. Durieux, D. N. Astrov, W. R. G. Kemp, and C. A. Swenson, Metrologia 15, 57 (1979).

[2] J. F. Schooley, G. A. Evans, Jr. and R. J. Soulen, Jr., Cryogenics 4, 193 (1980).

[3] R. J. Soulen, Jr. and H. Marshak, Cryogenics 7, 408 (1980).

[4] R. A. Kamper and J. E. Zimmerman, J. Appl. Phys. 42, 132 (1971).

[5] R. J. Soulen and H. Marshak, Proc. 1972 Applied Superconductivity Conference (1972) 588 IEEE, New York IEEE Publ. 72CH0682-5-TABSC.

[6] M. F. Cracknell and G. V. H. Wilson, Proc. Roy. Soc. A296, 71 (1966).

[7] T. L. Thorp, B. B. Triplett, W. D. Brewer, N. E. Phillips, D. A. Shirley, J. E. Templeton, R. W. Stark, and P. H. Schmidt, J. Low Temp. Phys. 3, 589 (1970).

[8] H. Marshak and R. J. Soulen, Proc. Low Temp Phys Plenum, New York, LT13 4, 498 (1974).

[9] R. J. Soulen and R. B. Dove, NBS Spec. Pub. 260-62 (1979), USGPO, Washington, D. C., Stock No. 002-002-02047-8.

[10] P. M. Berglund, H. K. Collan, G. J. Ehnholm, R. G. Gylling, and O. V. Lounasmaa, J. Low Temp. Phys. 6, 357 (1972); W. Weyhmann, in Methods of Experimental Physics, R. V. Coleman, ed. (Academic Press, New York, 1974), Vol. 11, p. 485; R. P. Hudson, H. Marshak, R. J. Soulen, Jr., and D. B. Utton, J. Low Temp. Phys. 20, 1 (1975).

[11] M. J. L. Yates, Nucl. Instr. Meth. 23, 152 (1963); see also 1964, α, β, and γ Ray Spectroscopy, ed. K. Siegbahn (North-Holland, Amsterdam) Appendix; and D. C. Camp and A. L. van Lehn, Nucl. Instr. Meth. 76, 192 and 87, 147 (1969).

[12] E. Zech, E. Hagn, H. Ernst, and G. Eska, Proc. IV International Conference on Hyperfine Interactions, North Holland, Amsterdam (1978) 342; and G. Eska, private communication.

[13] H. Marshak and B. G. Turrell, Solid State Comm. 30, 677 (1979).

[14] H. Marshak, Hyperfine Interactions 9, 1183 (1981).

[15] K. Ono, S. Kobayasi, M. Shinohara, K. Asahi, H. Ishimoto, N. Nishida, M. Imaizumi, A. Nakaizumi, J. Ray, Y. Iseki, S. Takayanagi, K. Terui, and T. Sugawara, J. Low Temp. Phys. 38, 737 (1980).

Radiometric measurement of thermodynamic temperature between 327 and 365 K[a]

T. J. Quinn

Bureau International des Poids et Mesures, Sèvres, France

J. E. Martin

National Physical Laboratory, Teddington, Middlesex, United Kingdom

Measurements have been made of the ratios of the total radiation emitted by a blackbody at a series of temperatures between 327 and 365 K to that emitted by a blackbody at the temperature of the triple point of water, 273.16 K. From these ratios the thermodynamic temperatures T were calculated, and, from measurements of the temperature of the blackbody made with platinum resistance thermometers calibrated on IPTS-68, values of $T - T_{68}$ were deduced. Preliminary results at $T = 327$, 344, and 365 K lead to values of $T - T_{68}$ of -13.6 mK, -20 mK, and -27.2 mK respectively. These results are in good agreement with earlier NBS gas thermometer measurements which first showed that IPTS-68 departs significantly from thermodynamic temperature in this range.

I. INTRODUCTION

This paper describes preliminary results of measurements of thermodynamic temperature made at three temperatures between 327 K and 365 K using a total radiation calorimeter. This work forms part of a larger project which includes a determination of the Stefan-Boltzmann constant and the measurement of thermodynamic temperature in the range 230 K to 692 K. In a paper given at PMFC-2 in 1981 (1) the present authors briefly described the apparatus and gave a preliminary result for the Stefan-Boltzmann constant.

The interest in measurements of thermodynamic temperature, T, and hence in the differences $T - T_{68}$, stem largely from the results of gas thermometry carried out at NBS by Guildner and Edsinger (2). These results showed that IPTS-68 departs from thermodynamic temperature by about 25 mK at 373 K and 80 mK at 733 K. Differences as large as this between T and T_{68} were unexpected and have, of course, important consequences in the formulation of the next version of the International Practical Temperature Scale. The aim of the present work is to make independent measurements of $T - T_{68}$ in this range and to achieve an accuracy of a few millikelvins — similar to that achieved by Guildner and Edsinger.

The total radiation, E(T), emitted by a blackbody at a temperature T is given by

$$E(T) = \sigma T^4 \text{ W m}^{-2} \quad (1)$$

where σ is the Stefan-Boltzmann constant. In a real experiment it is too difficult to measure accurately the whole of the radiation emitted by a blackbody over a hemisphere. It is, therefore, necessary to introduce into equation (1) a geometrical factor, 'g', which is a function of the size and disposition of the apertures used to define the beam of radiation being measured. We can thus write

$$E'(T) = g\sigma T^4 \text{ W} \quad (2)$$

In the paper referred to above[1], we described how a direct measurement of σ was made by measuring E'(273.16 K) using the same total radiation calorimeter as is now being used to measure T. The determination of σ, however, called for absolute measurements of both E'(273.16 K) and g. The accuracy aimed at was 1 part in 10^4 of σ, and the preliminary results described achieved a standard deviation of 1.4 parts in 10^4. The major uncertainties in the result were those related directly to the need to make absolute measurements. These uncertainties were: the emissivity of the blackbody radiator, 4 parts in 10^5; the absorptivity of the calorimeter, 7 parts in 10^5; absorption at the aperture edges, 8 parts in 10^5; diffraction, 7 parts in 10^5; scattering, 3 parts in 10^5 and mechanical measurements of the geometry of the apertures, 2.5 parts in 10^5.

The measurement of thermodynamic temperature does not require the absolute measurement of either radiant power or any geometrical factor for:

$$R(T/T_0) = \frac{E'(T)}{E'(T_0)} = \frac{E(T)}{E(T_0)} = \frac{T^4}{T_0^4} \quad (3)$$

where $T_0 = 273.16$ K, T is the unknown thermodynamic temperature and $R(T/T_0)$ is the measured ratio of radiant powers. The necessary and sufficient condition for equation 3 to hold is that the geometrical and other factors mentioned above are independent of radiator temperature, see 3.5 below. Our aim of achieving an accuracy of about 1 mK in measured values of T requires us to make measurements of $R(T/T_0)$ to an accuracy of about 1 part in 10^5, i.e. an accuracy about ten times better than was attempted in the absolute measurement of E'(273.16 K) for the determination of σ.

2. Principles of operation of the total radiation calorimeter

The equipment we are using to make measurements of T is illustrated in outline in Fig. 1. This differs in only a few minor respects from that shown in our paper at PMFC-2 and these result mainly from the change in use from an absolute measurement to a rather-more-precise ratio measurement. The principle of operation of the system is as follows: a blackbody radiator at an unknown thermodynamic temperature, T, irradiates an aperture system at a temperature close to 4.2 K, allowing a beam of thermal radiation to enter a second blackbody at a temperature near 2 K.

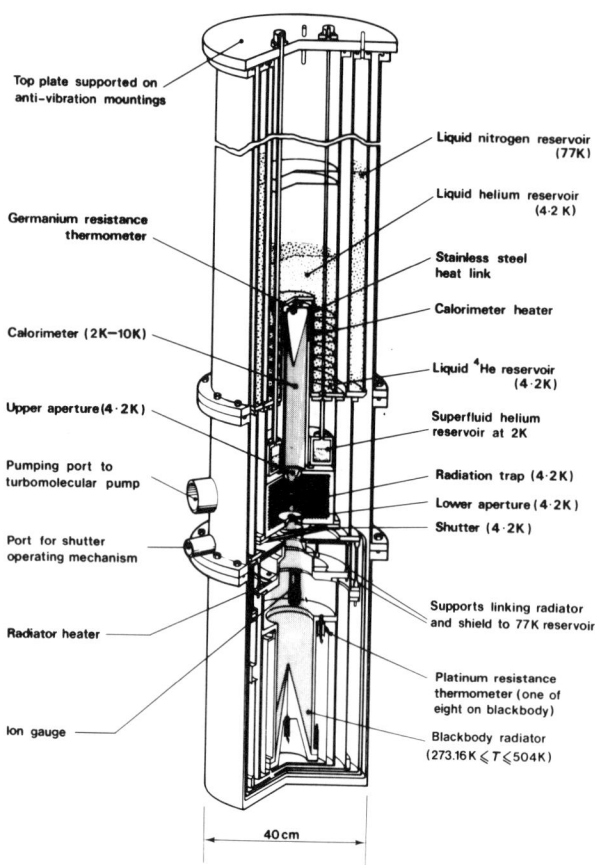

Fig. 1. An overall view of the radiation calorimeter assembly.

This second blackbody is attached, by a poorly-conducting heat link, to a reservoir containing superfluid helium which is maintained at a constant temperature close to 2 K, thus forming a heat-flow calorimeter. The thermal radiation absorbed by the calorimeter leads to an increase in its temperature until the heat arriving just balances the heat flow down the heat link to the constant-temperature reservoir at 2 K. The whole system is maintained at a pressure close to 1 µPa ($\approx 10^{-8}$ torr).

The temperature rise of the blackbody calorimeter - which is monitored by a germanium resistance thermometer - is equal to about 6.5 K for a radiator temperature of 365 K and about 3 K for a radiator temperature of 273 K. The calorimeter has a time constant of 4 minutes. When the calorimeter has reached its equilibrium temperature a cold shutter is closed, interrupting the thermal radiation from the radiator. At the same time, electrical power is applied to the calorimeter to maintain it at the same temperature as before. This electrical power is a precise measure of the thermal radiation power previously present. The balancing of the radiant power by electrical power is repeated as many times as is necessary to achieve the desired statistical uncertainty. The temperature of the radiator is then reduced to T_0 (which takes about a week) and a series of measurements of radiant power is made as before. The whole process is repeated to ensure that no drifts or other unwanted changes are taking place in the system.

During all of these measurement the International Practical Temperature of the radiator is measured using eight capsule-type platinum resistance thermometers previously calibrated on IPTS-68, Fig. 2. In this way the differences $(T - T_{68})$ can be established at each of the values of T for which $E(T)/E(T_0)$ has been measured. There are, of course, many other parameters and elements of the system that must be monitored during the measurements, such as : the pressure ; the temperatures of the shutter, radiation trap and apertures, various radiation shields, the superfluid helium reservoir and the calorimeter itself ; the various electrical control and power supplies which include the standard resistors and standard cells ; the helium boil-off rate, and the automatic liquid helium and liquid nitrogen filling systems. During measurements, most of the critical parameters are monitored and recorded, following a pre-determined sequence, by a desk computer.

3. Sources of uncertainty in the measured values of $T - T_{68}$

3.1 Diffraction and scattering

In the determination of T from measurements of the ratio $E(T)/E(T_0)$, diffraction and scattering need to be taken into account to the extent that, being wavelength dependent quantities, they are functions of T. Blevin (3) showed that for the purposes of calculating diffraction losses from a beam of thermal radiation, there exists an effective wavelength, λ_e, a function of T and geometrical parameters, such that the diffraction losses at this wavelength closely represent the overall diffraction losses for the total radiation. He showed that $\lambda_e(T) = Ac_2/T$, where c_2 is the second radiation constant of the Planck equation and A is a constant which has a value between 0.333 and 0.37, depending upon the details of the geometry of the diffracting aperture. For the present purposes we must take a value of 0.37 ; this leads to effective wavelengths which lie in the range 16.3 µm, for T = 327 K, to 14.6 µm, for T = 365 K.

In order to calculate the diffraction corrections which must be applied to a ratio measurement we must first evaluate the absolute diffraction losses for radiation of one temperature. We have already done this for 273 K radiation for the purposes of the Stefan-Boltzmann measurement and have arrived at a correction of 6.6 parts in 10^5 for 273 K radiation. The diffraction loss for radiation of any other temperature, $\Delta E(T)$, is given by $\Delta E(T) = \Delta E(T_0) \lambda_e(T)/\lambda_e(T_0)$.

In our earlier paper (1) we described how, following the principles outlined by Blevin, the geometry of the blackbody/apertures/calorimeter system was designed to minimize diffraction losses. This requires that the first aperture be irradiated hemispherically by the radiator and that the calorimeter be placed so that hemispherical collection of all radiation passing through the second aperture is ensured. Diffraction losses are calculated on the basis of evaluating the small departures from this ideal system.

Scattering, that is to say, radiation reaching the calorimeter that does not come directly in the geometrical beam or by diffraction, comes mainly from the radiation trap between the two apertures. Scattering was estimated (1) to contribute 13 parts in 10^5 to the measured 273 K radiation. Once again, for the purposes of a ratio measurement this is only important to the extent that it is a wavelength-dependent quantity.

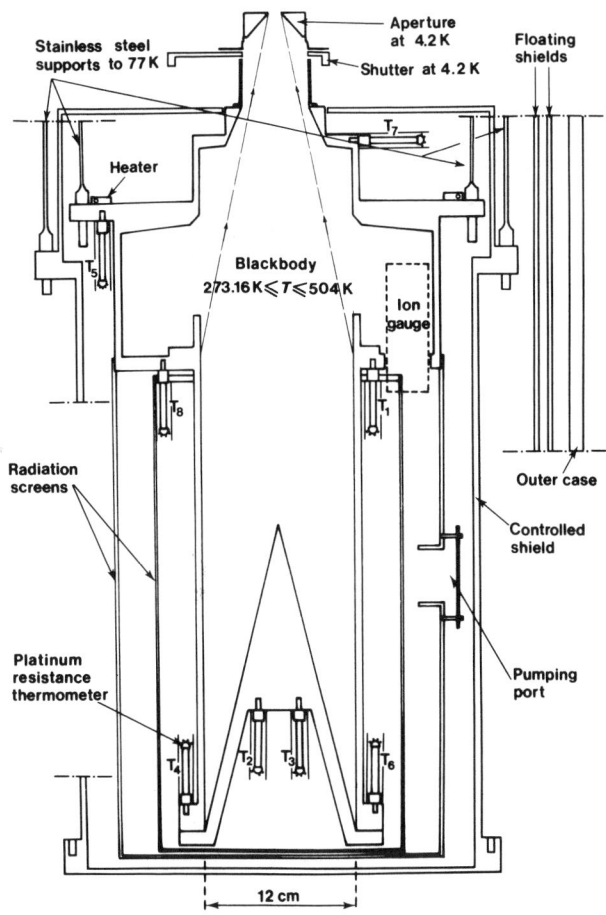

Fig. 2. The blackbody radiator showing the position of each of the eight platinum resistance thermometers used to establish T_{68}(radiator). The positions of the marginal rays of the beam of radiation viewed by the calorimeter are indicated.

Table I
Temperatures at eight different positions in the radiator measured by platinum resistance thermometers for two temperatures, (a) near T_0 and (b) near 327 K
See Fig. 2 for positions of thermometers.

Thermometer Position	Number	Measured temperature (a) T_{68} (K)	(b) T_{68} (K)	
Lower flange	T_4	272.0321	326.9044	
	T_6	272.0303	326.9056	Regions directly visible from calorimeter
Cone	T_2	272.0299	326.9022	
	T_3	272.0309	326.9040	
Upper cylinder	T_1	272.0302	326.9039	
	T_8	272.0310	326.9052	
Upper flange	T_5	272.0335	326.9165	
Top	T_7	272.0146	326.9125	

For the large set of apertures, the effective radiator temperature is given by

$$T_{68}(\text{radiator}) = 0.56\, T_{68}(\text{cylinder}) + 0.44\, T_{68}(\text{cone})$$

where $T_{68}(\text{cylinder}) = \frac{1}{4}(T_4 + T_6 + T_1 + T_8)$

and $T_{68}(\text{cone}) = \frac{1}{2}(T_2 + T_3)$

From which

T_{68} (radiator) : (a) = 272.030 7 K
(b) = 326.904 0 K

Measurements (4)(5) of the optical properties of the 3M-C-401 black used here show that it behaves as a flat, nearly-diffuse reflector, of reflectance 0.06, out to wavelengths of about 50 μm with a very small peak in reflectance near 10 μm. The reflectance is also a function of temperature, being rather lower at 77 K than at 365 K. At wavelengths beyond 50 μm the reflectance becomes increasingly specular and rises to about 8 % at 100 μm, increasing to 40 % beyond 350 μm. Integration of the product of the Planck function and the reflectance from 2 μm to 400 μm shows that the total reflectance for 365 K radiation is less than that for 273 K radiation by 0.7 % for paint at low temperatures and by 4 % for paint at 365 K. Taking into account multiple reflections, this could lead to a net change in the scattering of the radiation trap of a little more than 1 %. We have already mentioned that the scattering of 273 K radiation contributes 15 parts in 10^5 to the measured value of $E(T_0)$. A change in this scattering of 1 % leads, therefore, to a scattering contribution for 365 K radiation differing by 2 parts in 10^6 from that for 273 K radiation.

3.2 Emissivity of radiator and absorptivity of calorimeter

In the measurement of the Stefan-Boltzmann constant, we applied a correction of 8 parts in 10^5 and 30 parts in 10^5 for the emissivity of the radiator and the absorptivity of the calorimeter, respectively.

The changes in the optical properties of the radiator and absorber are directly proportional to changes in the reflectance of the black with which they are coated. A change of 0.7 % in surface reflectance leads to a change in absorptivity of the calorimeter of only 2 parts in 10^6 and can therefore be ignored, as can the change in emissivity of the radiator.

A correction that may be more difficult to quantify is that due to retroreflection. During experimental measurements of the absorptance of a replica of the calorimeter (1) some evidence came to light of the presence of a small retroreflectance at visible wavelengths. If this is confirmed it would affect directly the measurements of the Stefan-Boltzmann constant and indirectly the present measurements of T. But, as before, we are here interested only in the difference between the effects for radiation of slightly different spectral distribution and it seems likely that any retroreflection component would be proportional to the diffuse reflectance and would, therefore, introduce only a very small error in the measurement of T. It should be noted, however, that earlier measurements by Hsia and Richmond (6) showed no evidence of retroreflection for 3M's paint.

3.3 The measurement of T_{68} for the radiator

One of the most demanding requirements of the present work is for the correct value of T_{68} to be ascribed to the thermal radiation leaving the radiator. The thermometer emplacements T1 to T8 are indicated in the drawing of the radiator shown in Fig. 2. The assignment of an effective temperature to the radiator is made on the basis of a weighted average of the elements of the radiation leaving the different elements of the surface and reaching the calorimeter. In carrying out this calculation it is necessary to take into account the variation in effective emissivity and absorptivity of the elements of the internal surface of the radiator and calorimeter, respectively, as a function of position. In addition we must take into account the vignetting effect of the pair of apertures. Numerical integration shows that, for the large set of apertures being used for the present series of measurements, about 44 % of the radiation reaching the calorimeter comes from the cone and 56 % from the cylindrical walls of the radiator. Heat flows in the radiator have been reduced to very small levels by ensuring that the surrounding radiation screens are very close in temperature to that of the radiator itself. Heat flow down the cone is sufficiently small for the consequent temperature gradient to be insignificant. Typical temperature distributions at 273 K and 327 K are given in Table I. Some difficulty was encountered in obtaining a good temperature distribution at 365 K and this is thought to account for the increased scatter of the results at this temperature (see Fig 3 and item 2.2 of Table III).

3.4 The calorimeter

In addition to the error due to departure from blackness, the calorimeter is, in principle, susceptible to errors resulting from a difference in response between heating by radiant power and by electrical power. This was discussed in our earlier paper (1) where we argued that the fact of operating at very low temperatures reduced to negligible proportions most of the sources of uncertainty in this respect. In addition experiments had been carried out which showed that the response of the calorimeter did not depend upon which of the six heaters was used.

3.5 Apertures

For the measurement of the ratio $E(T)/E(T_0)$, we require that the physical dimensions of the apertures, their distance apart and the reflecting properties of their edges and entrance cones remain unchanged during the measurements of $E(T)$ and $E(T_0)$. In the present series of measurements the lower aperture has a diameter of 18 mm and the upper 26 mm. They are 100 mm apart. It is planned to repeat all of these measurements using a pair of smaller apertures, 10 mm and 18 mm in diameter. The apertures are maintained at liquid helium temperatures throughout the series of measurements and we have no evidence that any changes in physical dimensions take place.

3.6 Electrical measurements

We estimate that an uncertainty of less than 1 part in 10^5 arises from in our measurements of the ratio of electrical powers applied to the calorimeter. In addition to the tests of the external measurement system, we are also able to reverse the current in the heater and thus eliminate most of the thermal emf's in the low-temperature part of the heater circuit. The change in response of the calorimeter on reversal of the current is less than 2 parts in 10^5. Such a current reversal forms part of the regular measurement sequence and the mean values of current and voltage are taken for the calculation of the heater power. The effect on the resistance of the germanium thermometer of the magnetic field due to the heater current is estimated to be less than 1 part in 10^7 and can be ignored.

3.7 Gas conduction

The rate at which energy reaches the calorimeter by gas conduction from the radiator, when the shutter is open, is by no means negligible. The correction is calculated on the basis of pressure measurements made using the ion gauge placed inside the radiator (see Figs. 1 and 2) and assuming that most of the gas desorbed from the walls of the radiator is water vapour, and further assuming that the energy of adsorption of water molecules on the inner black surface of the calorimeter is 4 kcal per mole. This figure is estimated to be uncertain by some 50 %. A pressure of 1.3×10^{-6} Pa (1×10^{-8} torr), for example, in the radiator leads to a beam of molecules passing through the apertures and condensing on the surface of the calorimeter, liberating energy at a rate of 6.8×10^{-6} $E(T_0)$.

The pressures observed in the radiator increase from about 2.7×10^{-7} Pa ($\simeq 2 \times 10^{-9}$ torr) at 273 K to 6.5×10^{-6} Pa at 365 K. When the shutter is closed the pressure in the calorimeter is assumed to be less than 2×10^{-7} Pa since the pressure in the radiator does not change on opening the shutter. In calculating $E(T)$ the measured power is decreased by an amount equivalent to the rate of energy flow by gas conduction.

4. RESULTS

Results have so far been obtained at three temperatures, 327 K (54 °C) 344 K (71 °C) and 365 K (91 °C) ; measurements are proceeding at temperatures below 327 K. The present results are shown in Table II and Fig. 3 which also includes those of Guildner and Edsinger. The estimated uncertainties are shown in Table III. It is immediately clear that there is good agreement between the two sets of results.

The work is continuing.

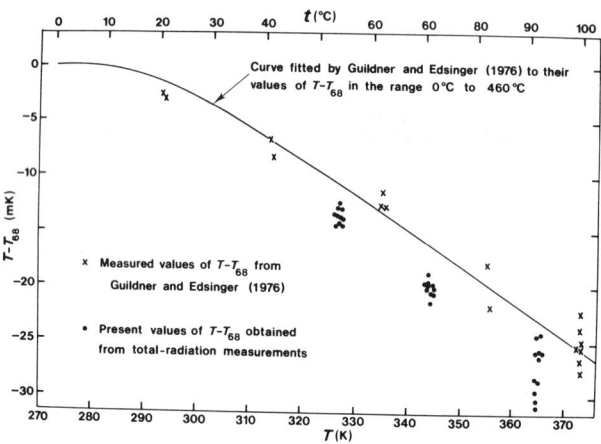

Fig. 3. $T-T_{68}$ in the range 273 K to 373 K according to Guildner and Edsinger (Ref 2) and the present preliminary results.

Table II Results		
T (K)	T - T_{68} (mK)	Uncertainty* (mK)
327	- 13.6	± 1.7
344	- 20.0	± 1.8
365	- 27.2	± 3.0
*evaluated in Table III		

Table III

Sources of uncertainty in $T - T_{68}$ (mK)

(evaluated at the 1 σ level)

1. Uncertainties not dependent upon T

   1.1 residual gas .... ± 0.5
   1.2 electrical measurements ...... ± 0.5
   1.3 T_{68} .... ± 0.5
   1.4 E (273.16 K) .... ± 0.7

2. Uncertainties dependent upon T

	327 K	344 K	365 K
2.1 Diffraction	± 1	± 1.2	± 1.4
2.2 Random uncertainties in ten measurements of E(T)	± 0.7	± 0.7	± 2.4

3. Combined uncertainties
   1 and 2 .... ± 1.7 ± 1.8 ± 3.0

ACKNOWLEDGMENTS

The authors are very pleased to acknowledge the many important contributions made in the early stages of this work by J.P. Compton. In addition we would like to acknowledge the able construction work carried out by E. Pinn, M. Rogers and E. Charles of the NPL workshop, the reflectivity measurements made on a model of the calorimeter by J. Geist and E. Zalewski of NBS, the development of the HP 85 computer programme by L. Forest, the many helpful discussions which have taken place with members of the NPL temperature section and finally the Directors of both BIPM and NPL who encouraged one of the authors (TJQ) to continue this work after his move to BIPM.

REFERENCES

(a) This work was carried out at the National Physical Laboratory.

1. T. J. Quinn and J. E. Martin, Proceedings of the 2nd International Conference on Precision Measurement and Fundamental Constants, NBS-1982.
2. L. A. Guildner and R. E. Edsinger, J. Res. NBS 80A, 703-738 (1976).
3. W. R. Blevin, Metrologia, 6, 39-44 (1970)
4. J. P. Compton, J. E. Martin and T. J. Quinn, J. Phys. D., 7, 2501-2510 (1974).
5. D. L. Stierwalt, Appl. Optics, 5, 1911-1915 (1966).
6. J.J. Hsia and J.C. Richmond, J. Res. NBS 80A, 189-220 (1976).

Measurement of thermodynamic temperature with the NPL photon-counting pyrometer

P. B. Coates and J. W. Andrews

Division of Quantum Metrology, National Physical Laboratory, Teddington, Middlesex, United Kingdom

Progress at NPL in the measurement of thermodynamic temperature above the zinc point by photon counting pyrometry is reviewed. A brief description of the instrument is given, with some of the design problems found. The magnitudes of the major sources of uncertainty in extending thermodynamic temperature measurement from the zinc to the gold point are estimated. Provisional results obtained in several parts of this programme are discussed.

INTRODUCTION

Since the last Symposium in this series, the programme of work undertaken within the Temperature section at the National Physical Laboratory, Teddington, has included an extensive investigation into the measurement of thermodynamic temperature using the technique of photon-counting pyrometry. The main aims of this project are :-

- the determination of the thermodynamic temperatures of the freezing points of pure metals of potential value as defining or secondary fixed points within the range 630 to 1700 °C
- the measurement of the deviation $(t - t_{68})$ of the International Practical Temperature Scale (IPTS-68) from thermodynamic temperature between the freezing points of zinc and gold
- the determination of the characteristics of high temperature platinum resistance thermometers, in terms of thermodynamic temperature, above 630 °C

This paper summarizes the status of this project, and presents developments since the last report, given in 1975 at the European Conference on Temperature Measurement [1].

Photoelectric pyrometry is a comparative technique, that is, the unknown temperature of a thermally radiating source is determined from the ratio of its spectral radiance at a particular wavelength to that from a blackbody source at a known reference temperature. As the relationship is given by a fundamental law of physics - Planck's law - the temperature obtained is thermodynamic in nature. However, no existing photoelectric pyrometer of standards quality is sufficiently sensitive at the defining fixed point of thermodynamic temperature, the triple point of water at 273.16 K, and higher reference temperatures with measured or assigned thermodynamic temperatures must be employed. Photoelectric pyrometry therefore has the status of a secondary thermodynamic technique.

The programme of work at NPL also includes the development of total radiation calorimetry, a technique based on Stefan's law. Progress in this project is described elsewhere in this Symposium by T.J.Quinn and J.E.Martin. It is hoped that it will yield accurate values for thermodynamic temperatures from about -30 °C up to the freezing point of tin (232 °C) at least, and possibly up to the zinc point, 419 °C. In conjunction with photon counting pyrometry, this would enable thermodynamic temperatures to be measured at NPL with radiative techniques from below 0 °C up to very high temperatures. At present, however, the only accurate thermodynamic temperature measurements below 630 °C are those obtained by Guildner and colleagues at the National Bureau of Standards in Washington, using classical gas thermometry. Values for the deviation $(t - t_{68})$ have been published [2] for temperatures from 0 °C up to 457 °C, and these results have been employed to provide reference temperatures for the programme of photon counting pyrometry.

It may also be noted that the major problem of pyrometry, the errors introduced by the unknown emissivity of radiating surfaces, is avoided in this work, as the sources of radiation used are nearly ideal blackbodies. The NPL photon counting pyrometer is of course also used to realize and disseminate the IPTS-68 at high temperatures, but these aspects of its application will not be covered here.

THE PHOTON-COUNTING PYROMETER

The original design of the photon-counting pyrometer was intended to achieve high sensitivity and accuracy at low radiance temperatures. However, because the instrument had also to measure high temperatures and to undertake the more routine tasks of lamp and thermocouple calibration, a number of compromises had to be accepted. As it has been described previously [1], its design and performance will be covered only briefly here, concentrating on the modifications which have been made since that time. The photon-counting pyrometer, shown in Figure 1, consists of three main sections :-

A) The optical system
Radiation from a source is collected by an off-axis ellipsoidal mirror of 60 cm focal length, limited by a circular stop of 11 cm diameter to give an aperture ratio of f/11. The focussed beam is reflected from a small plane mirror 8 cm in diameter, to produce an image of unit magnification on a field stop with a circular 0.75 mm diameter aperture. The pyrometer is aligned by moving a second plane mirror into the beam in front of this aperture, and observing the image produced on a graticule with a microscope. The

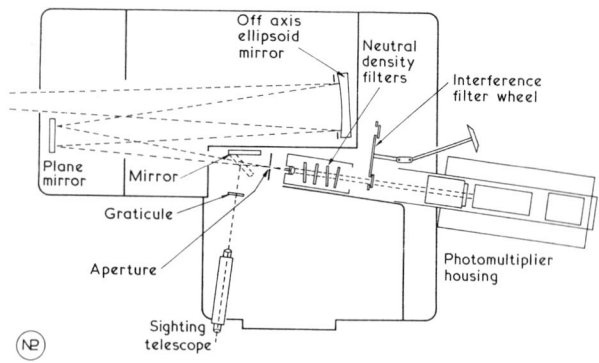

Figure 1 : The NPL photon-counting pyrometer

background signal from the detector is also measured with the mirror in this position. It would be preferable to work at the lowest radiance temperatures with larger target areas, but at high temperatures, and especially with tungsten ribbon lamps which may have filaments only 1.5 mm across, larger apertures raise practical difficulties. The use of reflecting rather than refracting optics produces a system of very high quality which will work over a wide wavelength range, and which has few optical surfaces, decreasing the proportion of scattered radiation.

B) The spectral radiance selection system

The beam passing through the 0.75 mm aperture is collimated by a small plano-convex lens, and the intensity adjusted into the normal operating range of the detector by inserting an appropriate combination of the four neutral density filters. This is not changed during the measurement of a radiance ratio, as it is difficult to eliminate completely the interactions within this section of the pyrometer. A narrow wavelength band is selected by one of five interference filters mounted on a rotatable wheel. These have peak transmissions at wavelengths between 660 and 820 nm, and bandwidths between 0.3 and 10 nm. Although a particular filter may be selected to suit a given application - the longer wavelength filters with broader bandwidths are obviously more suited to the measurement of the lowest temperatures - a fundamental guideline of this work is that all measurements should be made at more than one wavelength. Experience has shown that this provides a stringent test for the presence of systematic errors, especially of the more subtle types.

C) The photon detection system

An EMI photomultiplier type 9658 was selected as the photon detector, as it has useable sensitivity out to about 850 nm, and because venetian blind photomultipliers have stable gain characteristics. It is operated in the photon-counting mode, as this enables optimum sensitivity to be achieved, and the results may be accumulated very easily over the long measurement periods required at low source temperatures. The photomultiplier is contained in a commercial Peltier-cooled housing. At the normal operating temperature of -10 to -15 $^\circ$C, the background count rate is about 100 s^{-1}, so that the statistical error from this source is negligible in comparison to that from the signal count rate, which is usually greater than 10^4 s^{-1}. As the overall stability of the pyrometer is determined by the photocathode temperature, the cooling water for the Peltier unit is also temperature controlled.

The critical choice in the design of the pyrometer for this programme was that of the photon detector, as it was required to measure very low source temperatures while achieving a resolution of better than 0.01 $^\circ$C. The signal bandwidth is not important in many metrological applications, and the measurement time may be allowed to extend to 1-2 hours for critical observations.

In general, the performance of the detector may be limited by one of three mechanisms : the noise associated with background thermal radiation, the random statistics of the photons in the signal itself, and the current and voltage noise generated in the detector or its associated amplifier. The importance of each source of noise depends not only upon the detector, but also upon the optical throughput of the pyrometer, and its effective wavelength. Although it would appear sensible to adopt an infra-red detector in order to obtain long effective wavelengths, the investigation of the new set of optical components required would have constituted a major effort. In addition, the pyrometer was required to calibrate tungsten strip filament lamps up to very high temperatures at a wavelength around 655 nm. Only photomultipliers and silicon photodiodes were able to match these conflicting requirements. At the time, photomultipliers with sensitivity above 1000 nm appeared to be rather unstable as well as very expensive. Previous comparisons [3,4] of S-20 type photomultipliers and silicon photodiodes for pyrometric applications have come out strongly in favour of the latter, as these devices possess high quantum efficiencies at wavelengths beyond the cutoff thresholds of the S-20 and related photocathodes.

Calculations with the known optical throughput of the NPL pyrometer, assuming a 10 nm bandwidth for the interference filter, for different detectors indicated that a conventional venetian blind photomultiplier (type 9658) could be used to measure down to about 450 $^\circ$C, at an optimum wavelength of around 820 nm. The signal, as is usual with cooled photomultipliers, would be photon noise limited. The calculations were more complex with silicon photodiodes, as the minimum source temperature was limited by the noise generated in the detector amplifier, although this is usually determined by the shunt resistance of the photodiode [5,6,7]. With the performance figures available at the time for photodiode-amplifier combinations, the minimum measureable source temperature was found to be about 550 $^\circ$C, even though the optimum wavelength was around 1000 nm, where the quantum efficiency is typically 0.7-0.8. Improvements in components and operation at low temperatures [8] have reduced this to about 400 $^\circ$C.

The choice of a photomultiplier detector, although it enabled the lowest temperatures to be reached, brought with it a number of disadvantages. Photomultiplier systems are not only cumbersome and expensive, but the photon counting method severely limits the dynamic range of the pyrometer. Also, the first 9658 photomultiplier, in use in 1975, was found to be very linear in response, with no traces of any fatigue effect. After about 8 years of almost continuous operation, it developed an intermittent fault and has been replaced. The new tube, of the same type, is about six times more sensitive, but shows some fatigue. As the gain changes are quite small, about 0.2-0.3% at the most, they may be accurately corrected with a new technique [9,10] which determines the variations in the gain from simultaneous measurements of the photomultiplier count rate and anode current. These measurements are made under the control of a desktop computer which runs the experiment, so that the correction to the count rate may be made automatically.

Thin film neutral density filters were originally employed to set the count rate from the photomultiplier into the required operating range, as the transmission of this type is independent of wavelength over a wide range, and unaffected by variations in the ambient temperature. However, it has been found that the strong reflections from these and from the interference filters gave rise to two small systematic errors, dependent upon the neutral density filter combination in use. The first involved a reflection from the front of the interference filter onto the nearest neutral density filter, around the interference filter and its screen and onto the detector. As the spectral distribution of this radiation was different from that transmitted through the interference filter, it gave rise to an error in the apparent temperature of the source. It was eliminated with an additional blackened screen. The second error resulted from reflections from the neutral density filters which re-emerged from the pyrometer. With blackbody sources these were absorbed, but with standard lamps part was reflected from the filament, augmenting the normal signal. These reflections were eliminated by placing the filters at varying angles to the collimated beam, and introducing stops with 20 mm apertures between them. On further study, however, it was found that the angled filters had made the pyrometer sensitive to polarization in the radiation from the source. While this effect was not significant with blackbody sources, the radiation from tungsten filament lamps is typically 2-3% polarized, and detectable errors were produced. The thin film neutral density filters are being replaced with absorbing glasses, the reflections from which are much weaker.

MEASUREMENT UNCERTAINTIES

Unlike gas thermometry, many of the difficulties of photoelectric pyrometry tend to decrease at high temperatures. Although this project involves the extension of thermodynamic temperature measurement from 457 °C up to the gold point and above, the experiments would be simpler and the results more accurate if a higher reference temperature, for example 630 °C, were available. However, the uncertainties will be estimated in this section for the current projects with a starting temperature of 457 °C. From the Wien approximation to Planck's law, the error ΔT in the measured temperature T may be expressed in the form

$$\Delta T = \left(\frac{T}{T_o}\right)^2 \Delta T_o + T\left[\frac{T}{T_o} - 1\right]\frac{\Delta\lambda}{\lambda} + \frac{\lambda}{c_2}\cdot\frac{\Delta R}{R}$$

where T_o is the reference temperature and R the measured spectral radiance ratio. All experimental errors may be classified under one of the three terms appearing in this equation.

Uncertainty in the reference temperature
A fundamental limit to the measurement of thermodynamic temperatures by pyrometry is the uncertainty in the knowledge of the thermodynamic temperature of the reference temperature. In the present case, the

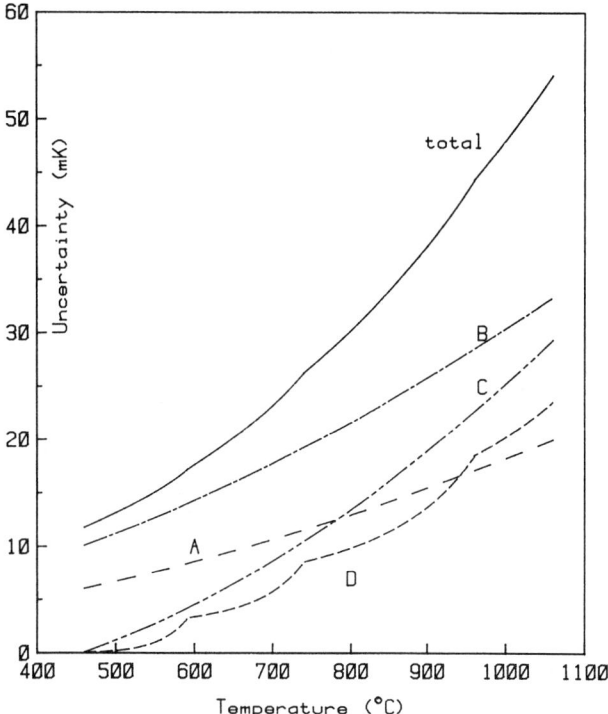

Figure 2 : Uncertainties in pyrometric measurements (A) from gas thermometry, (B) from reference temperature realization, (C) from wavelength errors, and (D) from non-linearities.

uncertainty given by Guildner and Edsinger [2] at 457 °C is 6 mK, with a confidence level of 99% (all uncertainties in this paper are given at this level). This will propagate at higher temperatures as $(T/T_o)^2$, and amounts to 15 mK at the gold point (Figure 2). In the present programme, however, the reference temperature is realized with a blackbody comparison system containing a calibrated platinum resistance thermometer. The uncertainty in this realization must of course also be included. The main source of error arises from the effects of temperature gradients in the furnace, and these are not easily measured or estimated. It is assumed, conservatively, that it contributes a further 10 mK to ΔT_o.

The uncertainty in the wavelength
In order to calculate radiance temperatures accurately from measured radiance ratios, it is necessary to allow for the finite bandwidth of real pyrometers by evaluating integrals of the form

$$I(T) = \int_{-\infty}^{+\infty} s(\lambda).t(\lambda).P(\lambda,T)d\lambda \qquad ...1$$

where $s(\lambda)$ is the spectral sensitivity of the pyrometer, principally that of the detector, $t(\lambda)$ is the transmission curve of the interference filter, and $P(\lambda,T)$ the Planck function. There has been considerable discussion in the literature [11,12,13,14,15] of convenient analytical forms to represent I(T). The traditional parameter, the effective wavelength, is defined by the relationship

$$R = I(T)/I(T_o) = P(\lambda_m,T)/P(\lambda_m,T_o) \qquad ...2$$

where λ_m is the mean effective wavelength between the temperatures T and T_o. For convenience in use, this is often converted to the limiting effective wavelength λ_e given by

$$1/\lambda_e = \lim_{T \to T_o} (1/\lambda_m) \quad \ldots 3$$

The concept of an effective wavelength, although apparently simple, is often confusing in precise applications. This is partly due to the fact that it is a function of the source temperature, and not determined solely by the characteristics of the pyrometer, as might be expected. In addition, the Planck function is often replaced by the Wien approximation in (2), either within the integrals I(T) and $I(T_o)$ or, for mathematical convenience, on the right hand side. Four slightly different expressions for λ_m therefore exist, but the distinction is not often made. For these reasons, effective wavelengths are now rarely employed for the purposes of photoelectric pyrometry of high accuracy.

There is little to choose between the other equations for I(T) that have been suggested, unless the transmission curve $t(\lambda)$ shows subsidiary peaks or wings with high transmission. In these cases, it may be necessary to introduce further terms into the approximation for I(T) [14,15]. However, it is much more important that the measured curves for $s(\lambda)$ and $t(\lambda)$ are shown to be sufficiently accurate, and this problem has received little attention. It is essential, of course, to ensure that these measurements are free of systematic errors. They should be made on the pyrometer itself, under normal operating conditions; if the measurement process is prolonged, temperature changes in the interference filter or in the detector may give rise to significant errors. In addition, it is necessary to ensure that a sufficient number of points have been taken, distributed sensibly over a wide wavelength range. The importance of the tails in the transmission curve $t(\lambda)$ is often underestimated; it has also been found that many filters have residual transmission peaks well away from the main peak, and these must be sought with some care. If $t(\lambda)$ is determined with a grating monochromator, it is necessary to distinguish between these and 'ghosts' produced by the grating, which are often of similar intensity. It is possible to check that the wavelength range is adequate by removing (or extending) part of one tail, and observing the effect upon the mean wavelength λ_o or the calculated temperature. For a given measurement accuracy, it is possible to check that a sufficient number of points have been measured by removing half and repeating the integration. This is an inefficient procedure, as it requires the measurement of many more points than strictly required, and it is rather sensitive to the order in which points were taken. A new computer-based procedure under development at NPL provides an optimal fit to the measured points with cubic splines, whose knots are automatically selected to give uniform uncertainties from each segment. A statistical analysis then enables an estimate of the error in λ_o or in T to be obtained.

This method rapidly checks that sufficient measurements have been taken, and avoids the need for redundant data. I(T) may be derived from an analytical integration of the splines, so that no further errors are introduced. Temperatures may be calculated from radiance ratios with a fast Newton-Raphson iterative procedure in the program, again with negligible errors.

The aim of the filter and detector calibration procedures at NPL is to achieve an overall uncertainty $\Delta \lambda_o$ less than 0.02 nm. The effect of such an error is shown in Figure 2, and amounts to an uncertainty at the gold point of 20 mK. It will be seen that the error from this source increases rapidly with the temperature interval covered, and is negligible for temperatures close to the reference temperature.

Errors in the radiance ratio

The fundamental limitation to the accuracy of radiance measurements, produced by the random statistics of photon detection, is of particular importance in photon counting pyrometry as it determines the minimum acceptable count rate. In practice, the counting time is adjusted for each measurement to give a total count of about 10^7 to 10^8, so that a statistical uncertainty of about 0.01 $^{\circ}$C in the temperature results. The long counting times required at low rates require that sources of thermal radiation, as well as the pyrometer, have very good temperature and mechanical stability.

The other major source of error in this category arises from non-linearities in the detection system. The dead time of the counting system, which determines the maximum acceptable count rate, is usually that of the pulse discriminator. It may be measured [1] with an uncertainty of less than 1 ns. This figure corresponds to an uncertainty of 0.01% at a count rate of 10^5 s^{-1}. The error is systematic, and, unlike the other sources of error, may be increased if several steps are required to cover a given temperature interval. This is often necessary as a result of the limited dynamic range of photon counting systems. The discontinuities in the slope of curve D, shown as an example in Figure 2, correspond to the reduction in the count rate required at each step. The other source of non-linearity error in photomultipliers arises from the fatigue effect. While procedures have been established for its correction [16,17,18], there is evidence [10,19] that the effect includes a very rapid initial change in gain, the effects of which may not be eliminated.

To summarize, the total uncertainty in the extension of thermodynamic temperature measurement from the zinc point to the gold point by photoelectric pyrometry would amount to about 50 mK. Of the contributory sources, only that from the realization of the reference temperature may be readily reduced in magnitude. The simplest way to achieve this would be to employ a fixed point as the reference temperature, for example the freezing point of zinc. Even if it were completely removed, however, the total uncertainty would only be reduced to about 35 mK. While this is considerably larger than the likely overall errors in gas thermometry, it is less by a factor of at least four than the existing uncertainties in this region. In addition, a set of pyrometric values would provide valuable support for the results of gas thermometry in the derivation of a new International Practical Temperature Scale. The uncertainty could of course be further reduced if an accurate value for a higher reference temperature were adopted.

THE PROGRESS OF WORK AT NPL

Measurement of metal freezing points

The techniques available for the measurement of the temperature interval between the freezing points of a pure metal and of gold are well established and documented in the literature [20,21]. Measurements have been carried out recently at NPL with ingots of aluminium, silver, copper [22] and palladium. Only the work on the latter will be described here. The freezing point of palladium is of particular interest as the value presently recommended by the CCT is based on measurements made nearly fifty years ago [23,24], although it is in regular use as part of the wire bridge technique for calibrating noble metal thermocouples. Although the overall uncertainty claimed for this technique is often as low as $\pm$ 2 $^{\circ}$C, it is by no means certain that the error in the value assigned to the palladium point itself is less than this. Investigations carried out recently have used the wire bridge technique to transfer this simple realization of the palladium point to a blackbody source via the calibrated thermocouple. The value for the freezing point of palladium obtained at NML [25] was 1555.0 $\pm$ 0.4 $^{\circ}$C; similar work at NPL supported by the EEC Community Reference Bureau (BCR) gave 1555.3 $\pm$ 0.4 $^{\circ}$C. However, apart from the uncertainties introduced by the thermocouples, it is not known how well the wire bridge technique realizes the palladium point, and this can only be resolved with a blackbody source immersed in a palladium ingot.

The blackbody system is shown in Figure 3. It is similar to that previously used for the measurements at NPL of the freezing point of platinum [26], and was contained in the same furnace. At NPL, it was found that it was more difficult to obtain good melting and freezing plateaux with palladium than with platinum, although the temperature was considerably lower. The sensitivity of the freezing point of palladium to the presence of oxygen appeared to be a minor problem. The main difficulty encountered was that of filling the crucible adequately, and the top of the blackbody tube was found to be exposed with one of the ingots used. The main effect was to make the shape of the melting curves sensitive to the heating rate and temperature gradient in the furnace; the freezing plateaux were much more reproducible and defined (Figure 4). They generally possessed an undercool of about 2-3 $^{\circ}$C. The preliminary result of this work, taken mainly from the freezes from one ingot, is 1555.3 $\pm$ 0.2 $^{\circ}$C, based on the IPTS-68 value for the gold point.

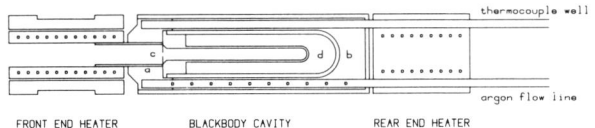

Figure 3 : Blackbody system for the measurement of the freezing point of palladium. (a) alumina cement, (b) alumina granules, (c) platinum disc with 1.5 mm diameter aperture, and (d) palladium ingot.

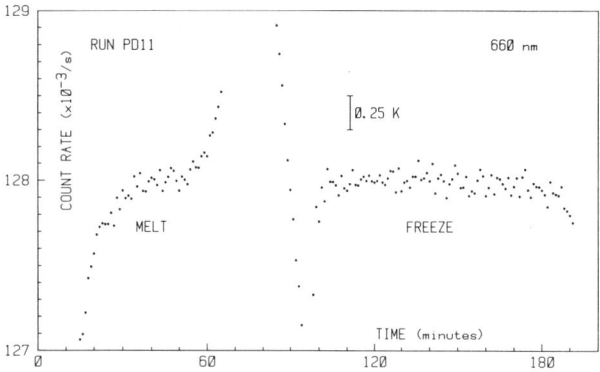

Figure 4 : Palladium point melting and freezing plateaux

Measurement of $(t - t_{68})$

To convert the values obtained in 4.1 into thermodynamic temperatures, that of the gold point is required. The first stage in the determination of this involved the measurement of $(t - t_{68})$ from 440 to 630 $^{\circ}$C, where no published values were available. The upper limit to this range is a natural break point, as up to this temperature t_{68} may be realized conveniently with a calibrated platinum resistance thermometer. In addition, the slope of t_{68} shows a serious discontinuity at this point [27]. To extend the gas thermometry results, the blackbody comparator assembly shown in Figure 5 was used. The radiance from the blackbody was measured with an interference filter with a 10 nm pass band centered at a wavelength of 812 nm, as a function of the temperature indicated by the platinum resistance thermometer inserted in the central well. A standard lamp, operated at a fixed current and base temperature, enabled variations in the sensitivity of the pyrometer to be measured and a correction made. To increase the stability of the lamp radiance, this was run at a temperature of about 1100 $^{\circ}$C, and the radiance reduced with neutral density filters. As the same blackbody source was used for all of the important spectral radiance measurements, several sources of systematic errors largely disappeared, for example the emissivity of the blackbody cavity and the size of source correction. The main difficulty was the estimation of the magnitude of the effects of temperature gradients on the temperature difference between the blackbody and the platinum resistance thermometer. It was assumed that this temperature difference would be negligible if the gradient were zero. This assumption remains unproven experimentally. The gradient was monitored with a small 100 Ω industrial platinum resistance thermometer which could be moved along one of the stainless steel tubes shown in Figure 5. The low thermal mass of this detector not only ensured that the reading quickly reached an equilibrium value, but that its movement along the tube produced little change in the gradient. Adjustments to the end heater settings allowed measured temperature gradients of about 1 mK/cm to be established. These remained stable even though the furnace temperature as a whole varied by about 0.1-0.2 $^{\circ}$C during a set of measurements.

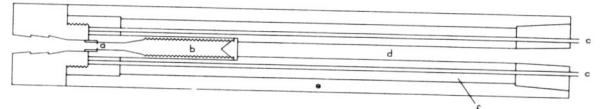

Figure 5 : Comparison blackbody assembly (a) platinum disc with 3 mm diameter aperture, (b) graphite blackbody, (c) stainless steel tubes, (d) platinum resistance thermometer well, (e) graphite case, and (f) silver ingot.

The results, which were presented [28] to the 13e Session of the CCT in 1980, are given in Figure 6. They remain provisional, as a detailed examination of the systematic errors involved in the experiment is still in progress. Recent tests have shown that errors produced by the presence of oxygen in the platinum resistance thermometer [29], and by the operation of this thermometer in the horizontal position, are negligible. It may be noted that the curve in Figure 6 is in reasonable agreement with that predicted from Guildner's results, and the requirement that it should

join the curve for $(t - t_{68})$ above 630 °C with a known

discontinuity in slope.

The thermodynamic temperature of the gold point
To extend the measurements described in the previous section to the gold point, a second blackbody comparator was constructed. This was similar to that shown in Figure 5, but the silver filling was of very high purity, and, to prevent contamination of this, the stainless steel tubes were omitted. To establish a

reference source at 630 °C a calibrated platinum resistance thermometer was placed in the well; the end heaters were adjusted so that the reading from this did not change when it was withdrawn by a distance of 5 cm. The spectral radiance of this source was then compared with that from a tungsten filament lamp, employed as a transfer standard, at a radiance temperature of about

770 °C. The platinum resistance thermometer was then removed, and the furnace temperature increased to the melting point of the silver ingot. Comparison with the filament lamp enabled the radiance ratio of the same blackbody at a known thermodynamic temperature about

630 °C and at the silver point to be established. Again, systematic errors are greatly reduced by the use of a single source. The silver point was finally compared directly to the NPL reference source at the gold point, during each set of measurements, in order to eliminate possible errors in the realization of the

silver point. At present, a complete set of readings has been completed only at a wavelength of 749 nm. The provisional value for the thermodynamic temperature of

the gold point is 1064.37 °C; this supports one of the conclusions reached in an analysis of earlier results [27], namely that the IPTS-68 values for the antimony and gold points are not seriously in error. The measurements have also given a value for the freezing point of silver, based on the IPTS-68 value for the

gold point, of 962.07 ±0.03 °C. This is in good agreement with other published values [20,21,30].

REFERENCES

1. P.B.Coates, 'Temperature Measurement', Inst. Phys. Conf. Ser. No.26, 238-243 (1975).
2. L.A.Guildner and R.E.Edsinger, J. Res. Natl. Bur. Stand., A 80 703-753 (1976).
3. D.R.Loveday, J. Opt. Soc. Amer., 52 1387-1398 (1962).
4. G.Ruffino, Appl. Optics, 10 1241-1245 (1971)
5. P.G.Witherell and M.E.Faulhaber, Appl. Optics, 9 73-76 (1970).
6. R.H.Hamstra and P.Wendland, Appl. Optics, 11 1539-1547 (1972).
7. R.C.Schaeffer, Appl. Optics, 15 2902-2905 (1976)
8. R.S.Neiswander and G.S.Plews, Appl. Optics, 14 1539-1547 (1975).
9. P.B.Coates, J. Phys. E, 8 189-191 (1975)
10. P.B.Coates and J.W.Andrews, J. Phys. E, 14 1164-1166 (1981).
11. H.J.Kostkowski and R.D.Lee, 'Theory and methods of of optical pyrometry', NBS Monograph 41 (1962).
12. H.J.Jung and J.Verch, Optik 38 95-109 (1973).
13. P.B.Coates, Metrologia, 13 1-5 (1977).
14. P.B.Coates, High Temperatures - High Pressures, 11 289-300 (1980).
15. G.Ruffino, High Temp.- High Press., 11 289-300 (1980).
16. I.Cantarell and I.Almodovar, Nucl. Instrum. Methods, 24 353-357 (1963).
17. J.L.Black and E.Valentine, Nucl. Instrum. Methods, 31 325-328 (1964).
18. A.M.Communetti, Nucl. Instrum. Methods, 37 125-134 (1965).
19. M.Yamashita, Rev. Sci. Instrum., 51 768-775 (1980).
20. H.J.Jung, 'Temperature Measurement', Inst. Phys. Conf. Ser. No.26, 278-286 (1975).
21. T.Ricolfi and F.Lanza, High Temperatures - High Pressures, 9 483-487 (1977).
22. P.B.Coates and J.W.Andrews, J. Phys. F, 8 277-285 (1978).
23. C.O.Fairchild, W.H.Hoover and M.F.Peters, J. Res. Natl. Bur. Stand., 2 931-944 (1929).
24. F.H.Schofield, Proc. Roy. Soc. A, 155 301-308 (1936).
25. T.P.Jones and K.G.Hall, Metrologia 15 161-163 (1979).
26. T.J.Quinn and T.R.D.C.Chandler, 'Temperature - Its Measurement and Control in Science and Industry' Vol.4 Part 1, 295-309 (1972).
27. P.B.Coates, 'Differences between IPTS-68 and thermodynamic temperature in the range 0 to 1064 °C', NPL Divisional Report QU56 (1980).
28. P.B.Coates, J.W.Andrews and M.V.Chattle, 'Measurement of $(t - t_{68})$ between 440 and 630 °C', Document CCT/80-27 (1980).
29. Berry, R.J., 1975, 'Temperature Measurement', Inst. Phys. Conf. Ser. No.26, 99-106.
30. F.Righini, A.Rosso and G.Ruffino, High Temp.- High Press., 4 471-475 (1972).

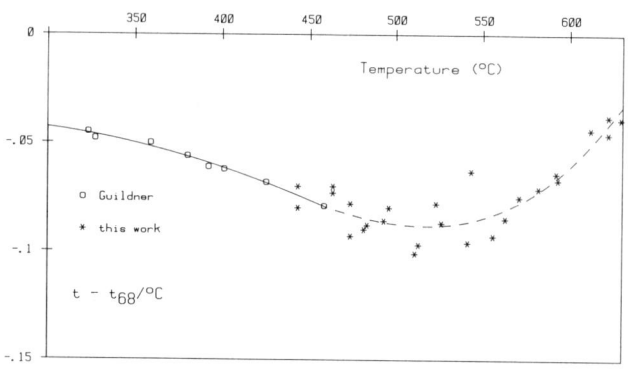

Figure 6 : Differences $(t - t_{68})$ up to 630 °C

Noise thermometry at NBS using a Josephson junction

R. J. Soulen, Jr. and Deborah Van Vechten*

Temperature and Pressure Measurements and Standards Division, National Bureau of Standards, Washington, D.C. 20234

We have been measuring the Johnson noise generated by a resistor using a Josephson junction. From these measurements we have developed a temperature scale extending from 0.01 to 0.52 K. The estimated inaccuracy at the lower end is ±0.5%; at the higher end it is reduced to ±0.2%. These estimates are based in part on comparisons with a γ-ray anisotropy thermometer from 0.01 to 0.05 K and with a paramagnetic salt from 0.03 to 0.52 K. We also describe how extraneous noise sources may be detected and suppressed.

A. INTRODUCTION

If the noise voltage $V_n(t)$ generated by a resistor R maintained at temperature T is squared and averaged over time, it can be shown that

$$<V_n^2> = \int_0^\infty S_n(\omega,T)d\omega \quad (1)$$

Here S_n is the spectral power density which, in general, depends on frequency and temperature[1]. At low frequencies the expression given in Ref. 1 for $S_n(\omega,T)$ approaches the familiar result

$$S_n(\omega,T) = 4kRT \quad (2)$$

This law is based on very fundamental considerations and therefore provides in principle a firm basis for establishing a temperature scale. Unfortunately, the noise voltage is so small that in practice electronic amplification must precede the averaging process. The amplifier contributes unwanted noise which in turn complicates the interpretation of the measured results. Many ingenious methods have been successfully employed[2] to reduce this problem to satisfactory levels. The spectral power density becomes exceedingly small at cryogenic temperatures, however. Thus at some temperature — about 2 K for present room temperature amplifiers — even these methods fail and a new solution must be found. Kamper and Zimmerman[3] showed in pioneering experiments that an amplifier based on the Josephson effect could provide the low-noise amplification. Moreover, this new method provides some unique advantages over conventional noise thermometry. Kamper and Zimmerman's original research has been extended and refined in a program established at the NBS; since then, a noise temperature scale extending from 0.01 K to 0.5 K, accurate to ≈0.5%, has been developed[4].

We will review herein the developments in this field in the last 10 years and systematically describe the effects which influence the Josephson junction noise thermometer at the level $\Delta T/T \sim 10^{-3}$. We will point out the anticipated experiments and procedures based on these results which may yield a temperature scale accurate to $\sim 0.1\% (10^{-3})$ from a few mK to 500 mK.

B. RESISTIVE SQUID NOISE THERMOMETER: DEFINITION AND OUTLINE OF MEASUREMENT TECHNIQUE

Consider the experiment shown schematically in Fig. 1. At the extreme lower left is a circuit, maintained at a cryogenic temperature T, which consists of a resistor R connected in parallel to a Josephson junction J. This combination is called an R-SQUID. The circuit is biased by a dc current I_0 which originates from a battery and resistor R_B at room temperature. The total voltage drop across R is given by

$$V(t) = V_0 + V_n(t) + V_T + V_J + V_p \quad (3)$$
$$= V_0 + V'(t) \quad (4)$$

where $V_0 = I_0 R$, the dc voltage drop across R due to the dc bias current. $V_n(t)$ is the Johnson thermal noise which provides the basis for the temperature measurement. V_T represents the thermal emf drop across R. V_J describes the noise voltage generated by the Josephson junction (n.b., the Josephson junction is <u>not</u> a noiseless device). Finally V_p takes into account any remaining noise voltage sources such as those caused by vibration, ground loops, etc. We assume that each of these voltages has an accompanying and well-defined spectral

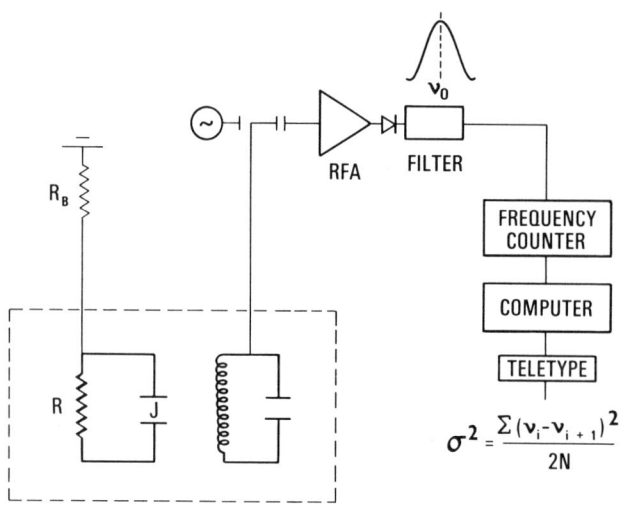

FIG. 1. Noise Thermometer Circuit. The circuit shown in the dashed box is maintained at a cryogenic temperature T, whereas the other components are at room temperature. The R-SQUID is formed by a resistor R (typically 10^{-5} Ω) connected in parallel with a point-contact Nb Josephson junction J. The circuit is biased by a dc current I_0 generated by a dry cell and a series resistor R_B. The voltage drop induced in R causes the Josephson junction to oscillate at an audiofrequency (typically 1-100 kHz). A radio frequency oscillator injects an rf current into a tank circuit coupled to the R-SQUID via a coaxial cable. The reflected power, phase or amplitude modulated by the af Josephson signal, is amplified by an rf amplifier RFA. Subsequent demodulation yields the af signal which is passed through a bandpass filter. At this point the af signal consists of the sum of the amplified signal from the Josephson junction and noise generated by the RFA. This noisy signal is presented to a frequency counter which repeatedly measures the frequency. An on-line computer calculates several variances (see text) which are used to detect the presence of extraneous noise signals and to define the noise temperature.

power density. Consequently, assuming that all terms of Eqn 3 are uncorrelated, the <u>total</u> spectral density is given by

$$S = S_0 + S_n + S_T + S_J + S_p \qquad (5)$$

where the subscripts again identify the type of voltage source generating the noise.

The total voltage given by Eqn. 4 causes the Josephson junction to generate a signal ($A \sin 2\pi\nu t$), where the frequency ν is determined by the Josephson equation

$$\nu = V\phi_0^{-1} = V_0\phi_0^{-1} + V'(t)\phi_0^{-1} = \nu_0 + \nu'(t) \qquad (6)$$

where $\phi_0 = h/2e$. Typically, I_0 is chosen so that ν_0 falls in the audio frequency range (1-100 kHz).

The amplitude A of this signal is too small to detect directly, but another property of the Josephson junction (parametric upconversion [5]) provides the necessary gain. To accomplish this, a radio frequency signal at a frequency ν_{rf} (typically, 24 MHz) is injected into a tank circuit coupled to the R-SQUID. The junction mixes the weak af signal with the strong rf signal. The rf amplifier (RFA in Fig. 1) amplifies this signal and, after demodulation, yields an af signal which is a stronger replica of that generated by the Josephson junction. The gain of the parametric upconversion process is simply $A' = \nu_{rf}/\nu$.

This amplification is not carried out without the addition of amplifier noise, however, so that final af signal available for processing is given by

$$V_{out} = V_A + (\nu_{rf}/\nu)A \sin 2\pi\nu t \qquad (7)$$

Here V_A is the af noise contributed by the overall amplifier system and is often referred to as "additive" noise. In this particular case this noise source depends on several parameters including the properties of the R-SQUID, the coupling between it and the tank circuit, and the particular room temperature amplifier used. We assume that, over the bandwidth of interest, the noise power spectral density of V_A is "white" (i.e., independent of frequency).

The presence of V_A obliges us to limit the af bandwidth before final processing of the af signal. Therefore, a bandpass filter with a center frequency ν_0 is interposed between the output of the RFA and the remaining circuitry. The bandpass filter is followed by a frequency counter and computer. The system is programmed to measure repeatedly the frequency of the af signal and to calculate several types of variances and to print out the results after N frequency counts.

The general definition of the variance and its relationship to the spectral power density which generates the frequency fluctuations is given by [6]

$$\sigma^2(N,r,\tau) = \langle \sum_{i=1}^{N} \frac{(\nu_i - \bar{\nu})^2}{N} \rangle \qquad (8)$$

$$= \left(\frac{N}{N-1}\right) \frac{1}{\pi\tau} \int_0^\infty S_\omega \frac{\sin^2 u}{u^2} \times$$

$$\{1 - \frac{\sin^2 Nru}{N^2 \sin^2 ru}\} \, du \qquad (9)$$

where $u = \omega\tau/2$, $\bar{\nu} = \sum_{i=1}^{N} \nu_i/N$, ν_i is an individual frequency measurement, τ is the gate time, and the brackets $\langle \rangle$ refer to an infinite time average. The uncertainty in σ^2 after N counts is given as $\sqrt{2/N}\,\sigma^2$. When the gate time is small, a round off error

(Shephard's correction) leads to an additional contribution to the variance[3].

The total time to make an individual frequency measurement, T, is the sum of the gate time τ of the counter plus the time needed for the system (frequency counter and computer) to rearm (often referred to as the "dead time"). We define the parameter $r \equiv T/\tau$, which has a minimum value of 1 (i.e., $T = \tau$; no dead time) and no maximum value. The infinite time average for the variance is approximated by the result for N frequency counts and thus an averaging time NT is required.

If the filter has a shape factor $|H(j\omega)|^2$, then

$$S_\omega(\omega,T) = \frac{S}{\phi_o^2}|H(j\omega)|^2 = \frac{|H(j\omega)|^2}{\phi_o^2} \times$$

$$\{S_o + S_n + S_T + S_J + S_p + S_A\} \quad (10)$$

where the factor $1/\phi_o^2$ comes about as a result of the voltage-to-frequency conversion of the Josephson junction. The term S_A represents the contribution of the additive noise. This noise source, in distinction to the others described in Eqn 10, is not converted into frequency fluctuations by the Josephson junction. The frequency fluctuations, and thus the contribution to the variance, arise from the general effect of amplitude noise on the triggering of a frequency counter. The inclusion of S_A into the formalism given here is quite legitimate, however[7]. Thus Eqn. 8 describes the mathematical algorithm used by the on-line computer to calculate the variance from the measured frequencies ν_i; while Eqns. 9 and 10 may be used to generate the theoretical expressions that ultimately convert the measured variances to a temperature value.

We have found it especially useful to program the computer to calculate five variances defined as special cases of the general expression given by Eqns. 8 and 9. Consider the quantities

$$\sigma^2(2,nr,\tau) \quad n = 1, 2, 3, 4 \quad (11)$$

and

$$\sigma^2(N,r,\tau) \quad (12)$$

The Allan variance, $\sigma^2(2,r,\tau)$ samples adjacent pairs of frequency counts and therefore is the least sensitive to long-term drifts. The quantity $\sigma^2(N,r,\tau)$ is more sensitive since it depends on a long-time average of the frequency. When the power spectral density is independent of frequency and when $|H(j\omega)|^2 = 1$, it is easy to show that all five of these variances are equal. However, other spectral power densities, e.g., 1/f noise, etc., have quite different dependences on N, τ, r[6]. Thus by measuring several variances, one may easily detect the presence of non-white noise sources. This greatly facilitates their elimination.

Now the evaluation of these variances for all of the spectral power densities given in Eqn. 10 and for an arbitrary filter function $|H(j\omega)|^2$ would be extremely complicated. We will therefore proceed by first evaluating the relative importance of the spectral power densities, removing the insignificant ones from consideration, and then evaluating the survivors for two specific filters used in the experiments.

C. SPECTRAL POWER DENSITIES

We now consider in turn each term appearing in Eqn. 10.

C-1. Noise in the dc Power Supply: S_o

There are two contributions to the noise in the dc circuit: white noise generated by the bias resistor at room temperature and noise generated in the battery. The bias resistor and the cryogenic resistor are two white noise sources in parallel, and thus the effective noise temperature, T_{eff}, is

$$T_{eff} = T + T_B(R/R_B) \quad (13)$$

Typical experimental parameters are: $R_B = 10^5 \Omega$, $R = 10^{-5}\Omega$, $T_B = 300$ K, in which case the correction to T amounts to 3×10^{-8}K. Clearly then the white noise contributed by the bias resistor is negligible.

The noise contributed by the battery has been investigated by Knott[8]. He demonstrated that, for several dry cells, the noise in the frequency range of interest here was equivalent to the thermal noise generated by a resistor of $10^3\Omega$ at 300 K. The correction to T due to this cause is thus 3 µK. We therefore are justified in setting $S_o = 0$ in the discussions which follow.

C-2. Thermal Noise Generated by R: S_n

This quantity is, of course, the desideratum in the experiment. The complete expression for S_n, which includes quantum-mechanical effects, was first given by Callen and Welton[1]

$$S_n(\omega,T) = h\nu[1/2 + (e^{h\nu/kT} - 1)^{-1}] \quad (14)$$

We may expand this quantity in powers of $h\nu/kT$, insert it into Eqn. (1) and integrate the frequency from zero to a cutoff frequency ν_c. The noise voltage is shown to be [9]

$$<V_n^2> = 4kRT\, \nu_c[1 + \frac{(h\nu_c/kT)}{36} + \ldots] \qquad (15)$$

Setting ν_c equal to the highest frequency used in the experiment reported herein (10^5 Hz) and using the lowest temperature (0.01 K), the second term in Eqn. 15 is clearly negligible. We may therefore set $S_n = 4kRT$ without any further reservations.

C-3. Thermal Effects: $\mathscr{S}_T$

The thermal emf, V_T, across R is given by

$$V_T = \mathscr{S}\Delta T \qquad (16)$$

where $\mathscr{S}$ is the thermopower and ΔT is the temperature gradient across R. Clearly, since we calculate frequency (and thus voltage) differences for the variance, a constant V_T generated by a constant temperature gradient is of no consequence. We have found that a well-designed R-SQUID incorporating proper thermal grounding techniques does not produce thermal emfs greater than 10^{-15} V (equivalent via Eqn 6 to a frequency of ~ 1 Hz). Moreover, this voltage does not change over time scales of several τ, so the four Allan variances (Eqn 11) are unaffected, although the long-term variance $\sigma^2(N,r,\tau)$ does occasionally sense a drift in the thermal emf. In these rare situations, the noise temperature has been determined only from the Allan variances.

Whereas constant or slowly-varying temperature gradients are generally not a serious problem, temperature oscillations generated either by an improperly-operating refrigerator or by a temperature control circuit with a very high feedback gain can generate two distinct and noticeable noise effects. To model these effects, we suppose that the thermal bath to which the R-SQUID is attached oscillates in temperature about an equilibrium value. The temperature gradient across the resistor then also oscillates; i.e.,

$$\Delta T = T_o \sin\omega_o t \qquad (17)$$

Using Eqn 16, and Eqn 6, we conclude that the Josephson audio frequency will be sinusoidal.

$$\nu(t) = \frac{\mathscr{S}T_o}{\phi_o} \sin\omega_o t \qquad (18)$$

The temperature oscillation of the bath also can produce a second effect. The voltage generated by the dc current I_o is $V_o = I_o R$, so that $\nu = V_o \phi_o^{-1}$. If R has a temperature dependence, dR/dT, then we can show that this also leads to a sinusoidal Josephson frequency

$$\nu(t) = \frac{I_o T_o}{\phi_o}(\frac{dR}{dT})\sin\omega_o t \qquad (19)$$

Thus the total effect may be written as

$$\nu(t) = [\frac{\mathscr{S}}{\phi_o} + \frac{I_o}{\phi_o}\frac{dR}{dT}] T_o \sin\omega_o t$$

$$= (A_T + A_R) T_o \sin\omega_o t \qquad (20)$$

We readily calculate the four Allan variances due to this sinusoidal time dependence

$$\sigma^2(2,r,\tau) = \sum_{i=1}^{N}\frac{(\nu_i - \nu_{i+1})^2}{2N}$$

$$= \frac{T_o^2}{2}\sin^2\omega_o\tau[2\cos\omega_o\tau - 1]^2 \qquad (21)$$

$$\sigma^2(2,2r,\tau) = \sum_{i=1}^{N}\frac{(\nu_i - \nu_{i+2})^2}{2N}$$

$$= \frac{T_o^2}{2}\sin^2\omega_o\tau[1 - 4\sin^2\omega_o\tau]^2 \qquad (22)$$

$$\sigma^2(2,3r,\tau) = \sum_{i=1}^{N}\frac{(\nu_i - \nu_{i+3})^2}{2N}$$

$$\qquad (23)$$

$$= \frac{T_o^2}{2}\sin^2\omega_o\tau[8\cos^2\omega_o\tau - 4\cos\omega_o\tau - 1]^2$$

$$\sigma^2(2,4r,\tau) = \sum_{i=1}^{N}\frac{(\nu_i - \nu_{i+4})^2}{2N}$$

$$\qquad (24)$$

$$= \frac{T_o^2}{2}\sin^2\omega_o\tau[16\sin^4\omega_o\tau - 20\sin^2\omega_o\tau + 4]^2$$

For the experimental conditions typical in the experiments to be described, $\omega_o \sim 10^{-2}$; 10^{-3} s $< \tau <$ 1s. We may therefore approximate $\sin\omega_o\tau \simeq \omega_o\tau$, and $\cos\omega_o\tau \sim 1$. In this case we find the simple result that

$$\sigma^2(2,nr,\tau) = \{A_R^2 + A_T^2\}(\omega_o T_o)^2 \frac{n^2\tau^2}{2} \qquad (25)$$

where $n = 1, 2, 3, 4$.

TABLE I

Characteristics of Two Noise Thermometers

LABEL	RESISTANCE MATERIAL	R ($\mu\Omega$)	dR/dT* (Ω/K)	$\mathscr{S}$ (V/K)	A_R	A_T(T=0.1 K)
MARK IV	$Cu_{99}Si_1$	12.9	2×10^{-7}	$\simeq 5 \times 10^{-8}T$**	10^3	2.5×10^6
MARK V	$Rh_{99.5}Fe_{0.5}$	17.4	11×10^{-7}	$-3 \times 10^{-6}T$***	5×10^3	1.5×10^8

* dR/dT was measured at NBS.

** Estimated from free-electron model.

*** J. E. Graebner, J. J. Rubin, R. J. Schutz, F. S. L. Hsu, and W. A. Reed, 2nd Conference on Magnetism and Magnetic Materials, AIP Conf. Series No. 24, 1975, p. 445.

We have used two R-SQUID noise thermometers which are almost identical in every respect (dR/dT, geometry, etc.) except for the fact that the resistors are composed of different alloys having rather different thermopowers. Table I gives the salient characteristics of the two thermometers. It is clear that for all temperatures and for both resistors, $A_T \gg A_R$. We also note that the effect on the variance of the thermopower is four orders of magnitude larger for the Rh-Fe than for the CuSi. We have occasionally observed a very small (< 1%) quadratic variation of the variances with n using the CuSi R-SQUID, but it was comparable with the statistical uncertainty and thus difficult to measure reliably. The effect was readily observable in the Rh-Fe R-SQUID, however, at higher temperatures where the thermopower increases rapidly. An illustration of the data taken when this effect was present is shown in Fig. 2. The quadratic dependence on n is clearly seen. In other data taken with this noise thermometer, the quadratic dependence on τ was also verified. The dc current was varied from 5 μA to 10 μA without effect, confirming that A_R is much smaller than A_T. Neither ω_0 nor T_0 were measured in these experiments, so the product was taken as a fitted parameter. It was found to vary from 4.8×10^{-8} to 2×10^{-7} as the temperature was increased from 100 mK to 500 mK. The data were taken without benefit of the insight provided by Eqn. 25, so a complete and accurate quantitative test of this interesting effect is yet to be performed. Suffice it to say, all the accurate noise thermometry at NBS has been based on the use of the CuSi alloy where this correction was insignificant.

C-4. Noise Generated by the Josephson Junction: S_J

Josephson junctions in general exhibit very interesting and complicated responses to applied dc and rf

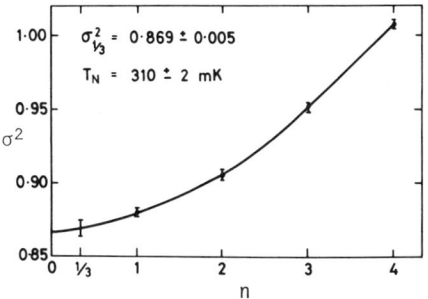

FIG. 2. Dependence of the Allan variance $\sigma^2(2,r,\tau)$ on the delay time, $n\tau$. The R-SQUID in these measurements consisted of a resistor (a $Rh_{99.5}Fe_{0.5}$ alloy) with a large thermopower. It is sensitive therefore to periodic temperature oscillations which produce a quadratic dependence of σ^2 on $n\tau$. (See text, C-2). That dependence is clearly borne out in these data. The R-SQUID used for the most accurate noise thermometry measurements contained a CuSi alloy instead, reducing this effect roughly by a factor of 10^4.

currents. Owing to the importance of these two effects on accurate noise thermometer temperatures, we have carefully measured the noise and impedance of the R-SQUID as functions of these parameters and have also developed a reasonably successful model for the observations. The results of this extensive investigation are reported elsewhere [10], so we only summarize them here. The impedance ρ of an R-SQUID i.e., the nonlinear combination of the resistor R and the impedance of the Josephson junction, may be expressed in the following way:

$$\rho^2 = R^2 + F \qquad (26)$$

where F is a complicated function of I_0, the critical current I_c of the Josephson junction, the amplitude and frequency of the impressed rf signal, and properties of the tank circuit. The noise of the circuit is proportional to $(d\rho/dI_0)^2$. When F is not small, this effect can cause S_J to be large (as much as 4 or 5 times S_n). We have measured the impedance and the noise of the R-SQUID and have a model which fits the data. By appropriate choice of the system parameters, however, we can reduce the magnitude of F to a few ppm, in which case $\rho = R$, and $S_J/S_n \sim 10^{-4}$. We therefore drop S_J from further consideration.

C-5. Noise Generated by Extraneous Noise Sources: S_p

One is never sure whether further extraneous noise remains to contaminate a noise thermometer measurement despite the best efforts at careful shielding, elimination of ground loops, etc. Two checks are routinely performed to increase our confidence in the experimental results: (1) comparison of the noise thermometer with other thermometers based on different fundamental physical principles, and (2) time correlation studies conducted to detect the presence of non-white noise.

The first check will be discussed later, and we will treat the latter now. As was mentioned earlier, we measure five variances defined by Eqns. 11 and 12. These effectively measure time correlations over time scales of τ, 2τ, 3τ, 4τ, and $N\tau$. To demonstrate the sensitivity of these quantities to various common types of noise spectral densities, consider the case $|H(j\omega)|^2 = 1$. Table II shows how $\sigma^2(2,r,\tau)$ and $\sigma^2(N,r,\tau)$ depend on N and τ. We see from Table II that after a few thousand counts, the results would be quite different. We routinely study the measured variance as a function of N and τ, and establish that the noise is white and originates solely from the presence of S_n and S_A (to be discussed). We therefore set $S_p = 0$ with the appreciation that these variances have proven to be an invaluable tool in the diagnosis and affirmation of the quality of the noise thermometer data.

TABLE II

Effect of Various Noise Spectra on the Variance

NOISE SPECTRUM	$\sigma^2(N,r,\tau)$	$\sigma^2(2,r,\tau)$
White: $S = h_0$	$h_0/2\tau$	$h_0/2\tau$
Flicker: $S = h_1\omega^{-1}$	$h_1 N \ln N/N-1$	$h_1 2\ln 2$
$S = h_2\omega^{-2}$	$h_2(2\pi)^2 \tau N/12$	$h_2(2\pi)^2 \tau/6$

C-6. Additive Noise: S_A

In the discussion following Eqn 10, we indicated that amplitude modulation noise superimposed on a carrier wave will cause "false" triggering of the frequency counter and thus generate an additional variance. This additive noise is dependent on the signal to noise ratio (S/N) of the combined signal. In the case of the R-SQUID, the S/N is dependent on several parameters. For example, the coupling between the tank circuit and the R-SQUID and the I_c of the R-SQUID strongly influence the S/N as do the frequency and amplitude of the rf bias. Many of these parameters are difficult to measure or to calculate accurately, and therefore we must leave the magnitude of this effect undetermined. As will be shown in a later section, the effect of additive noise is accounted for by its influence on experimental data.

C-7. Summary of Power Spectral Densities

Reviewing the results of the topics discussed in sections C-1 through C-6, we conclude that Eqn 10 may be reduced to

$$S_\omega(\omega,T) = \frac{|H(j\omega)|^2}{\phi_0^2} \{4kRT + S_A\} \qquad (27)$$

where the expression in the brackets is white (i.e., independent of frequency). Thus the entire frequency dependence for S_ω is determined by the filter. Under these simplifying, but quite realistic conditions, we now proceed to evaluate the influence of specific filters on the variance.

D. INFLUENCE OF THE FILTER ON THE VARIANCE

We have examined the effect of a one-pole Butterworth and a square-wave filter on the variance. For these two filters, the transfer functions are

$$|H(j\omega)|^2 = \frac{1}{1 + (\frac{\omega-\omega_0}{\omega_1})^2} \qquad (28)$$

$$|H(j\omega)|^2 = 1 \qquad |\omega| < \omega_1 \qquad (29)$$
$$= 0 \qquad |\omega| > \omega_1$$

Using these two cases, and Eqn 27, we have calculated the variances. The results for the one-pole Butterworth may be expressed analytically [7]:

$$\sigma^2(2,r,\tau)/\sigma_T^2 = 1 - \frac{1}{\tau}(\frac{2}{\omega_1} - c\omega_1) \times (3/4 - e^{-\omega_1\tau}) \qquad (30)$$

$$\sigma^2(N,1,\tau)/\sigma_T^2 = 1 - \frac{1}{\tau}(\frac{1}{\omega_1} - c\omega_1) \times (1 - e^{-\omega_1\tau}) \qquad (31)$$

$$\sigma^2(2,r,\tau)/\sigma_T^2 = 1 - \frac{1}{\tau}\left(\frac{1}{\omega_1} - c\omega_1\right) \times$$

$$(1 - e^{-\omega_1 \tau} + f(r)) \qquad (32)$$

where $c = S_A \phi_0^2/P_s S_n$, r is a measure of the dead time, P_s is the power of the signal (not the spectral power density), ω_0 is the center frequency of the filter, and ω_1 is the half-bandwidth of the filter. The correction due to finite dead time, $f(r)$, in the case of the tests to be reported here, may be neglected. Finally,

$$\sigma_T^2 = \frac{S_n}{2\phi_0^2 \tau} = \frac{2kRT}{\phi_0^2 \tau} \qquad (33)$$

We recognize σ_T^2 as the quantity which we seek to measure when using an R-SQUID as a noise thermometer. We therefore require that σ^2 be as close as possible to this value to minimize uncertainties due to corrections applied by using Eqns. 30, 31, 32.

To evaluate the variance for the case of the square-wave filter, numerical integration of Eqn. 9 using Eqns 27 and 29 is necessary. This was done for one set of parameters appropriate to the square-wave filter used in the experimental tests.

We used four one-pole Butterworth filters with progressively larger ω_1 to verify the predictions given in Eqns. 30, 31, 32, as well as the square-wave filter referred to above. The filters were used to process the output of the Mark IV (the CuSi resistor) R-SQUID while it was maintained successively at two temperatures (0.0156 K and 0.0994 K). One set of data was used to obtain a fitted value for c. With no other adjustable parameters (ω_1 and ω_0 were measured for each filter), the theoretical curves generated by Eqn. 31 were then drawn through the data. The results for the four one-pole Butterworth filters are shown in Fig. 3; those for the square-wave filter are shown in Fig. 4. Clearly the theory fits very well, and indicates that the correction to the variance due to the filter can be quantitatively evaluated.

In the experiments in which the noise thermometer data was taken, the gate time was set either to 1s or 0.1s. In this case the correction due to the filter was considerably less than the statistical imprecision of the measurement ($\sim 0.5\%$). Future experiments in which shorter gate times are anticipated will require a correction of 2-3% using these equations. Since this correction may be determined to $\sim 1\%$, however, we feel that any inaccuracy introduced by the filter will be considerably less than 0.1%.

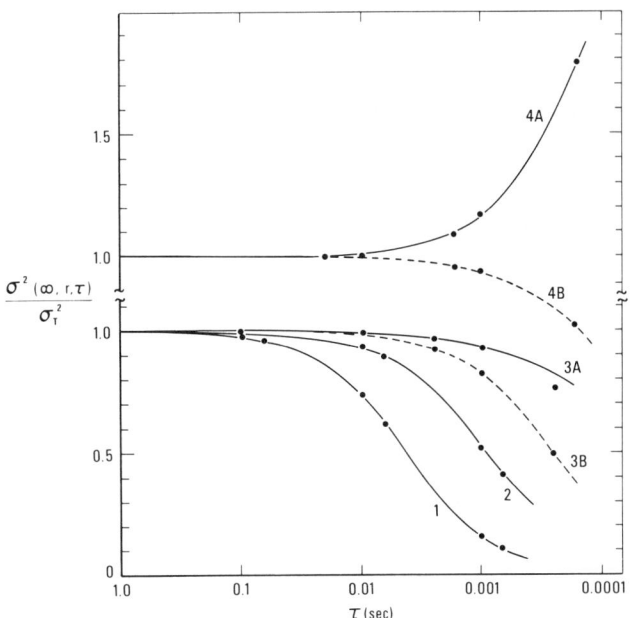

FIG. 3. Theoretical and experimental dependence of the variance σ^2 on the gate time τ for four one-pole Butterworth filters. We plot the variance normalized by $2\phi_0^2\tau/S_n$ versus τ for filters 1, 2, 3, and 4. Filters 3 and 4 were used at two temperatures. The parameter ω_0 for each filter was measured and ω_1 was obtained by fitting the measured transfer function to Eqn 28. The uncertainties in ω_1 given in the following table represent the uncertainty in the fit. The theoretical curves passing through the data points were generated by Eqn 31, using only one free parameter (see text).

Bandpass Characteristics

Curve #	$\omega_0(s^{-1})$	$\omega_1(s^{-1})$	T(mK)	Theoretical Curve
1	96,100	377 ± 2	15.6	——
2	98,600	1600 ± 10	15.6	——
3A	251,000	5700 ± 50	15.6	——
3B	251,000	5700 ± 50	99.4	---
4A	391,000	9860 ± 100	15.6	——
4B	391,000	9860 ± 100	99.4	---

Let us summarize the results of Sections C and D. We have considered the effect of several extraneous noise power spectra in addition to the thermal Johnson spectrum which is the one desired for noise thermometry. We have also considered the effect of the filter on the variance. We have discussed auxiliary measurements made with the same R-SQUID which indicate that reliable models for each effect may be made. We further indicate that we can choose a set of operating conditions which reduce <u>all</u> these unwanted effects to such a level that they influence the measured noise temperature by less than 0.1%. To this level of accuracy then, the conclusion of these two sections is that a very simple equation, Eqn. 33, may be used to relate the measured quantity, the variance, to T.

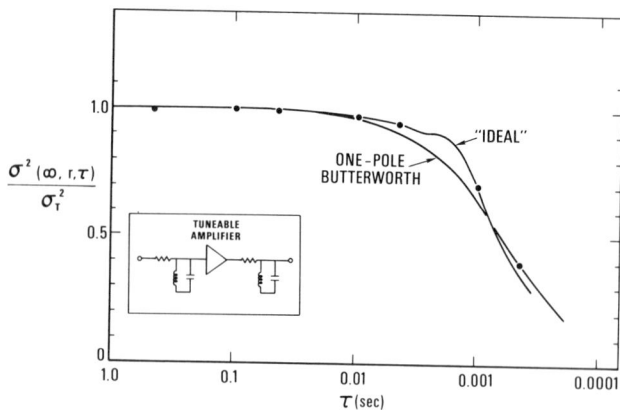

FIG. 4. Theoretical and experimental dependence of σ^2 on τ for a square-wave bandpass filter. Insert in lower left hand corner shows the circuit used to make an approximation to a square-wave filter. ($R = 2M\Omega$, $L = 100$ mH, $C = 120$ pF.) The amplifier A is a model UBM Rhode and Schwartz tuneable amplifier. These data were taken at $T = 99.4$ mK. The theoretical curve for the filter was generated by Eqn 29 with $\omega_1 = 2573$ s^{-1}. For comparison, the theoretical curve for a one-pole Butterworth filter with same value for ω_1 is shown. Note that the flatter curve corresponds to the square-wave ("ideal") filter.

E. NOISE THERMOMETER EXPERIMENTS

Encouraged by the fact that a temperature scale (albeit provisional) has been defined down to temperatures as low as 0.5 K (EPT-76), and noting that none exists below that value, H. Marshak and R.J. Soulen, Jr. have developed a scale below 0.5 K which is based on an intercomparison of a Co^{60} γ-ray anisotropy and R-SQUID noise thermometer. The most recent results have been published [4]. This comparison ranged from 0.01 K to 0.05 K, and the agreement was found to be 0.5% or better. This level was comparable with the combined statistical uncertainty of the two thermometers, so the agreement is considered to be good. Above 0.05 K this particular γ-ray thermometer loses sensitivity, but the noise thermometer was used to define several temperatures (recorded on two doped germanium resistance thermometers) from 0.05 K to 0.52 K (the superconductive transition temperature of Cd, the lowest defined fixed point on EPT-76). The results were shown to be consistent to $\simeq 0.2\%$ with a paramagnetic salt thermometer. The new scale, extending from 0.52 K to 0.01 K and maintained at NBS, is labelled NBS-CTS-1. It is distributed by a fixed point device, SRM 768 [11], which is described in another article in this symposium.

Efforts at NBS are presently underway to reduce the statistical and systematic errors of the R-SQUID thermometer to 0.1% and with this improved accuracy to compare the Co^{60} and R-SQUID thermometers from 0.005 K to 0.05 K. Finally, using another γ-ray thermometer which is described in another article in this Symposium, we expect to continue the comparison upwards, from 0.05 K to 0.5 K, or possibly higher, with similar imprecision.

CONCLUSION

The R-SQUID, taking advantage of the unique, low-noise amplification properties of the Josephson junction, has been used to extend noise thermometry to temperatures far below values reached through the use of conventional room-temperature measuring electronics. The R-SQUID noise thermometer, based on a fundamental physical law, has been compared with another thermometer (γ-ray anisotropy) which itself is based on an equally fundamental physical law. The comparison has thus far yielded a temperature scale accurate to $\pm 0.5\%$ from 0.01 K to 0.05 K. We see no fundamental impediment to the application of either principle to measure temperatures as low as a few μK, nor do we see any reason for the inaccuracy to be limited to $\pm 0.5\%$. Finally, we may remark that it is fortunate in temperature measurement when two quite different physical laws can be exploited in a single thermometric apparatus, and gratifying that in this case, they yield the same values of temperature within 0.5%. It is thus with considerable confidence that we anticipate the successful achievement of a temperature scale based on this comparison which extends from 0.005 K to 1 K with an inaccuracy of $\pm 0.1\%$.

REFERENCES

* Present address, Physics Department, Howard University, Washington, D.C. 20059.

[1] H.B. Callen and T.A. Welton, Phys. Rev. <u>83</u>, 34 (1951).

[2] See for example the review by T. Vaughn Blalock and Robert L. Shepard, Proc. Sixth Int'l. Conf. on Noise in Physical Systems, eds. P.H.E. Meijer, R.D. Mountain, and R.J. Soulen, Jr., NBS Spec. Publ. 614, 1981, p. 260.

[3] R.A. Kamper and J.E. Zimmerman, J. Appl. Phys. <u>42</u>, 132 (1971).

[4] R.J. Soulen, Jr. and H. Marshak, Cryogenics <u>7</u>, 408 (1980).

[5] J.E. Zimmerman and A.H. Silver, Phys. Rev. <u>167</u>, 167 (1978).

[6] J.A. Barnes, A.R. Chi, L.S. Cutler, D.J. Healey, L.B. Leeson, T.E. McGunigal, J.A. Mullen, W.L. Smith, R. Snyder, R.F.C. Vessot, G.M. Winkler, NBS Tech. Note 394 (1970), U.S. Govt. Printing Office, Washington, DC.

[7] R.J. Soulen, Deborah Van Vechten and H. Seppä, to be published. See also, L.S. Cutler and C.L. Searle, Proc. IEEE 54, 136 (1966).

[8] K.F. Knott, Electronic Letters 1, 132 (1965).

[9] R.P. Hudson, H. Marshak, R.J. Soulen, Jr., and D.B. Utton, Jour. of Low Temp. Phys. 20, 1 (1975); see pp. 39-40.

[10] See for example, D. Van Vechten, R.J. Soulen, Jr., and R.L. Peterson, in Proc. IC-SQUID 1980, Berlin, Germany (1980, p. 569), Editors, H.D. Hahlbohm and H. Lubbig, Publishers, Walter Gruyter and Co., Berlin, and references therein.

[11] R.J. Soulen, Jr. and R.B. Dove, NBS Special Publication 260-62 (1979).

Errors in Johnson noise thermometry

G. Klempt

Institut für Angewandte Physik der Universität Münster, D-4400 Münster, Federal Republic of Germany

A resistance in thermal equilibrium develops a noise voltage power spectrum 4 kTR which is directly proportional to the thermodynamic temperature with the other quantities either well known or measurable. The principal difficulties experienced in the past in measuring the small noise voltages and in constructing adequate electronic components have been overcome in the field of noise thermometry for temperature scale metrology. The remaining error sources—caused by the inherent noise of the preamplifiers—are discussed in detail in this paper. But they too can be eliminated by a correlation technique while still retaining simple operation.

There are two equations appropriate for basic thermometry in the temperature region from 1 K to 1500 K, i.e.

Gas equation $\quad p = kT \cdot n/V \quad$ (1)

Nyquist formula $\quad \overline{u_t^2} = kT \cdot 4R \cdot \Delta f \quad$ (2)

In both cases the measured values - the pressure p of an ideal gas and the mean square value of a resistor's thermal noise voltage u_t - are directly proportional to the thermodynamic temperature T. This gives the possibility of temperature measurements with completely different techniques. Both thermometers have error sources caused by the pressure transmitting tube and the noise voltage transmitting leads, respectively, to the measuring equipment. But more critical is the temperature dependence of system parameters as the number n of the gas particles and the bulb volume V in the case of gas thermometry or - to a minor degree - the resistance R and bandwidth Δf in the case of noise thermometry. An advantage of the noise thermometer therefore is that the thermal noise of a resistor is independent of its composition, the mass and nature of the charge carriers and their transport mechanism. In gas thermometry with real gases, however, difficult measurements of their thermodynamic data have to be done.

The noise thermometer like every other thermometer has its specific error sources. They can be illustrated at the simplified device shown in Fig. 1. One circuit model for Eq. 2 is a noise voltage generator u_t in series with a noiseless resistance of value R ohms. This generator is supplying the circuit with multifrequency noise. It is specified by its mean square value

$$\overline{u_t^2} = \int W_{ut}(f)\,df \quad (3)$$

with the spectral density or voltage noise power spectrum W_{ut}= 4kTR being independent of frequency into the microwave band. Here R is the real part of the sensor's impedance including the dielectric losses in the sensor leads capacitance C. Reactive components like C do not generate thermal noise.

u_t is amplified over a noise bandwidth Δf by a high-input-impedance, voltage-sensitive amplifier of gain V(f). The output voltage is squared and integrated over a

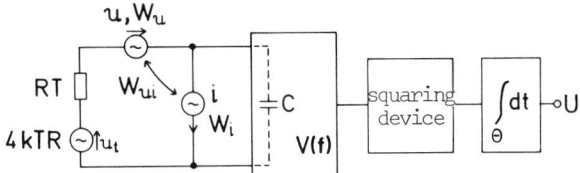

Fig. 1 Simple noise thermometer

time θ to obtain the mean value $\overline{U}$ of the squarer's fluctuating output voltage. All the amplifier noise, the thermal noise of the leads series resistance and isolator conductance, other noise of nonthermal origin and extraneous noise pickup can be taken into account by two equivalent noise sources, a voltage source u and a current i with corresponding power spectra W_u and W_i. Since the equivalent noise sources of an amplifier are not statistically independent the complex cross power spectrum W_{ui} must also be considered. The mean value of the squarer output voltage is given by

$$\overline{U} = K \int df \left| V(f)/(1 + j\omega RC) \right|^2 \cdot \left[4kTR + W_u + R^2 W_i + 2R \cdot \mathrm{Re}W_{ui} \right] \quad (4)$$

In this equation we have to consider five groups of error sources:

a.) The integrator output voltage U - averaged over a finite time θ - represents a stochastic process. U fluctuates around the expectation value $\overline{U}$ given by Eq. 4. A measure for the error of a single reading is the relative variance $\sigma/\overline{U}$. It decreases with measuring time θ and noise bandwidth Δf according to the Rice-formula

$$\sigma/\overline{U} = (\Delta f \cdot \theta)^{-1/2} \quad (5)$$

b.) Errors due to long-term transfer function drifts of the constant K, which depends on the squarer and integrator properties, and the amplifier gain V(f).

c.) Owing to the capacitive loading by the cable and amplifier input capacitance C, the effective gain $V(f)/(1 + j\omega RC)$ depends on R and C. Due to their temperature coefficients the gain becomes temperature dependent.

d.) Most noise thermometers - based on the device of Fig. 1 - are calibrated by the thermal noise voltage of a second resistor at the water triple point or another reference temperature. This involves an auxiliary measurement of the sensor resistance R, which can only be done with the desired accuracy in temperature scale metrology using a dc-voltage. So we must prove - in the frequency range of interest - that there is no frequency dependence due to the skin effect and dielectric losses of the resistor-substrate and the measuring cable insulation.

e.) In addition to the desired thermal noise 4kTR there are three additional, unwanted terms. Two of them depend on R in different ways. W_u can be measured with the amplifier in short-circuit condition (R = 0) taking into account that the noise bandwidth depends on R. In the same way the term $R^2 \cdot W_i$ can be separated from 4kTR by measuring the same temperature T at various R-values. The term $2R \cdot ReW_{ui}$, however, showing the same R-dependence as the thermal noise 4kTR, provides a systematic error which cannot be eliminated. The noise thermometer can only be calibrated by a single standard temperature (water triple point), if ReW_{ui} can be made sufficiently small.

W_u is by far the dominant error term in Eq. 4 given that some disturbances, which are not inherent in a noise thermometer system, can be avoided. The device must be shielded against external electromagnetic interference and protected against electrical background noise on the ac-power lines and ground lines by carefully filtered mains and using a differential amplifier technique to suppress common mode voltages.

CONTRIBUTIONS TO W_u

The voltage power spectrum W_u consists of the amplifier's inherent voltage noise and the thermal noise of the sensor leads series resistance. In correctly designed amplifiers with a junction field-effect transistor (JFET) in the input stage only its channel noise and the thermal noise of the feedback network contribute to the inherent noise of the amplifier. The thermal fluctuations of the JFET-channel resistance can be represented by an equivalent noise voltage source at the input of the device with a voltage power spectrum

$$W_o = 4kT_b R_{eq} \qquad (6)$$

Here R_{eq} is the equivalent noise resistance of the transistor, T_b = 290 K. In well designed modern units R_{eq} approaches the theoretical limit of $0.7/g_m$ (g_m = transconductance of the transistor) in the frequency range above 1 kHz. At lower frequencies there is excess 1/f noise. At high static drain currents R_{eq}-values of 30 Ω can be found and further lowered by connecting JFETs in parallel. The temperature dependent thermal noise of the leads resistance can be of the same magnitude as W_o if long and thin sensor leads are required to reduce heat leak in high precision thermometry.

Altogether, FET amplifier systems with a noise voltage power spectrum W_u down to $(0,4 \text{ nV})^2$/Hz are available nowadays. Compared with the thermal noise power spectrum of a 1 kΩ resistor at 300 K of $(4 \text{ nV})^2$/K this results in a temperature equivalent of 3 K.

CONTRIBUTIONS TO W_i

The main contributions to the system's noise current power spectrum are the gate noise current of the input FET, the thermal noise of the cable insulation and microphonic noise. For the FET's contributions to W_i the generally adopted results of Van der Ziel and others are:[1-3]

$$W_i = 2eI_g + 0.5\omega^2 \cdot C_{gs}^2 \cdot W_o + 4kT_c g_{11} \qquad (7)$$

The gate current noise consists of the shot noise of the gate dc-leakage current I_g, induced noise, which represents capacitive coupling to the gate circuit of thermal noise W_o in the channel via the gate-source capacitance C_{gs}, and thermal noise of the real part g_{11} of the FET's input admittance on device temperature T_c. In low noise JFETs with a leakage current in the order of some pA and a voltage noise spectrum W_o < $(1 \text{ nV})^2$/Hz the thermal noise $4kT_c g_{11}$ is by far the dominant one. With R = 1 kΩ the shot noise of I_g causes temperature errors of about 10 μK/pA, whereas an input conductance g_{11} of $10^{-9}\Omega^{-1}$, which is typical for ultra low noise multi-channel JFETs in the audio frequency range, contribute about 1 mK. Furthermore g_{11} increases as a square of the frequency starting between 10 kHz and 100 kHz, depending on the transistor chosen. This results from the capacitive coupling of the thermal noise of the parasitic series resistances on the drain and source side of the channel to the gate circuit. Up to date no systematic study has been carried out to understand the origin of the high-frequency independent gate current noise in the audio frequency range.

Another contribution to W_i is the thermal noise of the real part g_I of the input circuit insulation admittance with a noise current spectrum $W_I = 4kT_I g_I$. Here T_I is the mean noise temperature of the insulator conductances - i.e. the substrate of the sensor, the connecting cable, vacuum feedthroughs and switches including their dielectric losses and those of all material like varnish on the sensor leads or not completely removed soldering flux. The dielectric losses may strongly increase with frequency. The insulation conductance in the whole noise bandwidth must not exceed $10^{-9}\Omega^{-1}$ in order to keep the temperature error - caused by $R^2 \cdot W_I$ - below 1 mK with R = 1 kΩ and T_I = 300 K. This is equivalent to the insulating conductance and dielectric loss of a high quality mica or polystyrene capacitor of 50 pF which is the order of the cable capacity C. Using teflon as the only insulating material in the whole input circuit we found no measurable temperature error even with R = 10 kΩ. Insufficient insulation can cause temperature errors of some 100 mK.

A serious problem is the microphonic noise with power spectrum W_{im} which is produced by variations of the cable capacitance C when a thermovoltage U_{ts} in the input circuit is present. The capacity fluctuations of power spectrum W_{ic} can contribute considerably to W_i:

$$W_{im} = \omega^2 U_{ts}^2 W_{ic}$$

They are generated by mechanical vibrations, density fluctuations and convection of the heat conducting gas or acoustical oscillations in the case of low temperature thermometry. The microphonic term $R^2 \cdot W_{im}$ only remains negligible (<1 mK) at normal sound levels in the laboratory, if the thermo-voltages of the sensor leads are kept small - some 10 μV with pure metal wiring. They can reach some mV with alloy wiring in the region of the great temperature gradient. Furthermore the lower cut-off frequency of the measuring bandwidth must be set to some kHz, which is necessary anyway because of the electronic flicker noise and interference of technical frequencies.

CONTRIBUTIONS TO W_{ui}

The complex cross power spectrum of a FET's equivalent noise voltage and current is given by:

$$-W_{ui} = eI_g/g_m + j\omega C_{gs} \; 0.3 \; W_o \qquad (8)$$

The real part is caused by the shot noise of the leakage current I_g showing a strongly correlated noise component in the drain current. The imaginary part is determined by induced noise being proportional to the channel noise W_o. In modern JFETs with small I_g- and W_o-values the terms of W_{ui} can only be detected at high device temperatures - where I_g increases drastically - and at high frequencies, respectively. Therefore the systematic error term $2R \cdot eI_g/g_m$ in Eq. 4 is negligible with a theoretical temperature equivalent of some μK.

With no extraneous noise pickup, with very low-loss insulating dielectrics in the whole measuring input circuit and operating with ultra low noise JFETs in the input stage of a correctly designed preamplifier operating in a frequency range of some kHz to about 100 kHz, the term W_u in Eq. 4 produces a temperature equivalent of some K, $R^2 \cdot W_i$ of some mK and $2R \cdot ReW_{ui}$ of some μK with R in the $k\Omega$-region. The temperature error due to W_u can be determined experimentally with an uncertainty of about 10 mK only, because of the R-dependence of the noise bandwidth, the statistical character of the noise voltage u and its rather bad long-term stability. The latter is affected by amplifier temperature fluctuations and the influence of the thermal leads noise depending on the measuring and ambient temperature.

HIGH PRECISION NOISE THERMOMETRY

In the field of temperature scale metrology at medium and low temperatures an accuracy of some mK or even more is required in the development of an alternative to the gas thermometer. In such precise temperature measurements the elimination of all disturbing amplifier noise is essential. This can be best achieved by a correlation technique. Here the outputs from two amplifier channels, driven by the noise voltage from a single sensor are multiplied and averaged to reduce the uncorrelated noise from the two channels while emphasizing the correlated noise generated by the sensing resistor.

Fig. 2 shows the block diagram of a high precision noise thermometer, used in Münster.[4] The temperature-sensing resistor 2R, with centre point at earth potential, is connected in parallel to two identical highly linear and stable amplifier- and filter channels, using a 4-wire measurement technique. To achieve accurate measurements the apparatus is balanced to ground to avoid earthing problems and to reduce external interference voltage. For the same reason the noise thermometer is operated in a screened laboratory with filtered main leads. The digital part of the correlator is operated outside the screening connected to the analog-to-digital converters (ADC) by optic digital data transmission links.

For precise measurements a thorough theoretical analysis of the cross spectrometer is essential. If W_{uaub} indicates the complex cross spectrum of the preamplifier input voltages u_a and u_b, the mean output voltage $\bar{U}$ is given by:[5]

$$\bar{U} = L\alpha_a\alpha_b \int Re\left[V_a^* V_b \cdot W_{uaub}\right] df \qquad (9)$$

The constant L is determined by the ADCs properties. To avoid nonlinearity errors of the ADCs, the resistive voltage dividers are adjusted in such a way that the ADCs are equally modulated at the measurement of the unknown temperature T and at the calibration of the whole device at the water triple point temperature. The attenuation factor α must be independent of frequency to obtain the same noise bandwidth in all switching positions. V_a and V_b are the gains of the channels a and b. V_a^* means the conjugate complex of V_a. If both channels have an identical phase response $V_a^* V_b$ becomes a real number and only the real part of W_{uaub} contribute to $\bar{U}$. ReW_{uaub} was calculated carefully with equivalent circuits of the sensor and preamplifiers considering all inherent noise sources in this system.

Fig. 3 shows the symmetric preamplifier in some detail. In order to combine high input impedance with low noise and high gain, a special cascode stage is used with JFETs. The gain is stabilized and linearized by means of a very strong feedback via the resistors R_1 and Z. As the input signal voltage u_a is almost completely transferred to the source of the input Fet T_1 by the feedback circuit and via the source follower T_2 to its drain, no signal current can flow into the amplifier circuit via the gate-source and gate-drain admittance of T_1. This results in an extremely

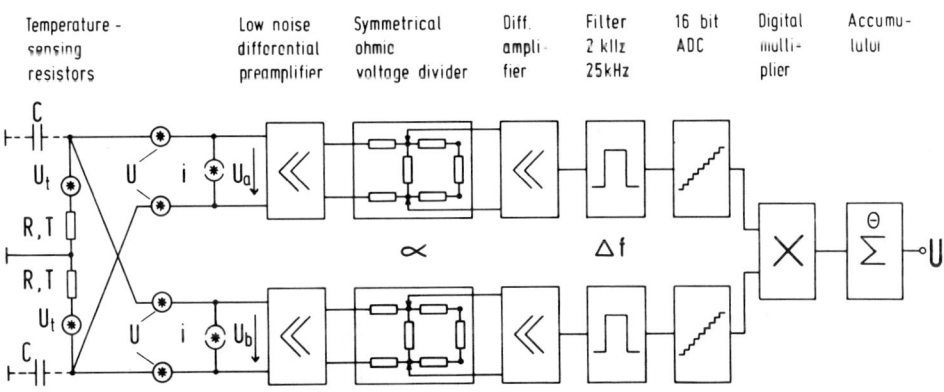

Fig. 2   Noise thermometer using correlation technique

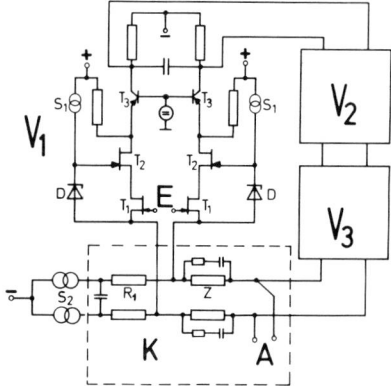

Fig. 3   Principle of the preamplifier

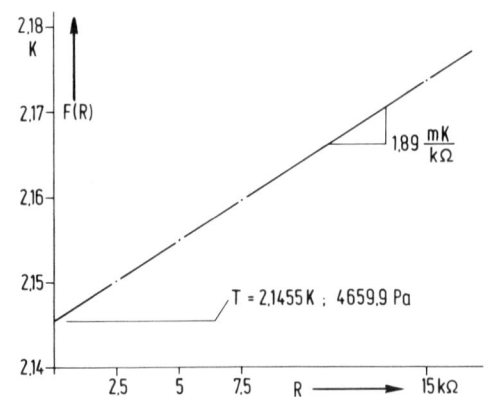

Fig. 4   F(R) at 2.1455 K

high input impedance over the whole noise bandwidth. With equal transfer and noise behaviour of the preamplifiers we obtain:

$$\text{Re}W_{uaub} = \frac{4kR}{1 + (\omega RC)^2}(T + B + AR) \quad (10)$$

In the case of a ultra high input impedance the coefficients are

$$A = (W_{ig} + \omega C\, \text{Im}W_{ui})/2k \quad (11)$$

$$B = (R_1 \cdot W_{ig} - \text{Re}W_{ui})/2k \quad (12)$$

in which the systems inherent noise voltage spectrum W_u no longer appears. Using a correlation technique and well designed amplifiers the uncorrelated noise sources u can be completely eliminated. The temperature dependent and unstable noise of the leads series resistance is therefore eliminated too, as long as the leads are brought out of the cryostat twice. This is necessary for the determination of 2R anyway. The noise of the ground line produces only common mode voltage at the inputs of the differential amplifiers and is therefore eliminated.

The noise sources u do not contribute to the mean value $\overline{U}$ but they enlarge the fluctuation of the reading. Its relative standard deviation now is given by a modified Rice-formula :

$$(6/\overline{U})^2 = \left[1 + (1 + W_u/4kTR)^2\right] \cdot (2\theta \cdot \Delta f)^{-1} \quad (13)$$

In the case of $W_u \ll 4kTR$ the Eq. 5 follows. This equation determines a lower limit for 4kTR. It should not be much smaller than W_u as the measuring time would become unnecessarily high for a required accuracy.

The noise sources i, however, cannot be eliminated since any current from either input will flow through the common resistance R to generate a voltage which appears at the input of each amplifier and hence is correlated. The sources i and their correlated components with u cause the contributions to the coefficients A and B in Eq. 10. Whereas B can be lowered to a few μK with low noise JFETs in correctly designed amplifiers, A has a value of about one mK/kΩ and each JFET connected in parallel.

As the theoretical estimation of A is not accurate enough, AR must be eliminated by experimental means in low temperature thermometry. For the measurement of an unknown temperature T the same sensing resistors with a very low temperature dependence are brought alternately to T and the water triple point temperature $T_{tr} = 273.16$ K. With Eqs. 9 and 10 for both temperatures we compute:

$$F(R) = \frac{U}{U_{tr}} \frac{R_{tr}}{R} \frac{\alpha_{tr}^2}{\alpha^2} T_{tr} \cdot \left[1 - \gamma \cdot \Delta(RC)\right] \quad (14)$$
$$= T(1 + PR)$$

which is a linear function of R. F is composed of the correlator output voltage ratio U/U_{tr}, the resistance ratio R/R_{tr}, the attenuation factor ratio α_{tr}/α and the water triple point temperature T_{tr}. The variations ΔR in R and ΔC in the lead capacity C parallel to R can be kept so small at the changeover from T to T_{tr}, that the correction term $\gamma \cdot \Delta(RC)$ is of the order of 10^{-6} with R below 5 kΩ, in a frequency range below 20 kHz and for measuring temperatues up to 400 K. If one measures F at the same temperature T for different values of R the temperature T can be obtained by extrapolating the graph F(R) to R = 0.

Fig. 4 shows an example. F has been measured with four different resistors at a temperature somewhat below the ^{4}He-lambda point. The linear law is well confirmed and, using the least squares method, the indicated temperature has been obtained with a total error of 0.3 mK, agreeing within 0.2 mK with the low temperature scale EPT-76. This measuring proceedure gives the thermodynamic temperature without any additonal corrections.

The error sources of Eq. 4 as well as those inherent in Eq. 9 - noted in b.) to e.)- can be made negligible compared with the statistical error a) by advanced circuit design and sophisticated measuring methods. According to Eq. 5 an accuracy of 10^{-3} for practical noise thermometers can be achieved with measuring times θ up to 10 sec and a noise bandwidth Δf of some 100 kHz. In noise thermometers for temperature scale metrology the accuracy can be raised to 10^{-5} (i.e. 3 mK at 300 K) with θ up to 10^6 sec and Δf of some 10 kHz using high accuracy electronic components only available in the frequency range below 50 kHz and stable temperature baths. Thus high precision noise thermometry is as time consuming as high precision gas thermometry, but offers an alternative means for the determination of a temperature scale.

REFERENCES

1. A. van der Ziel, Proc. IEEE 51, 461 (1963)
2. A. van der Ziel, J. W. Ero, IEEE Trans. Electr. Devices 11, 128 (1964)
3. W. C. Brunecke, A. van der Ziel, Trans. IEEE 13, 323 (1966)
4. H. H. Klein, G. Klempt, L. Storm, Metrologia 15, 143 (1979)
5. H. Bittel, L. Storm, Rauschen, Springer Verlag Berlin, Heidelberg, New York (1971)

A high-accuracy noise thermometer for the range 100–150 °C

C. P. Pickup

CSIRO Division of Applied Physics, Sydney, Australia 2070

Noise thermometry, based on the measurement of thermal, or Johnson, noise is a possible alternative to gas thermometry for the determination of thermodynamic temperatures. The arrangement adopted involves establishing equality between the noise voltages developed by two unequal resistances at different temperatures by alternately switching a measuring amplifier between them. The ratio of thermodynamic temperatures is then given by the inverse of the resistance ratio. The reference temperature is established by an ice bath and the other is adjustable in the range 50–150 °C by a controlled oil bath. Metal film resistors are used with mercury-wetted reed relays for switching. Errors due to stray capacitance shunting the noise resistances are eliminated by balancing simultaneously in two different frequency ranges. Lengthy averaging is required to reduce the statistical error. A series of 62 runs, each of 24 h, has been completed at approximately 135 °C. The mean noise temperature was 12 mK lower than the International Practical Temperature Scale of 1968 (IPTS-68) with standard deviation 2.4 mK which indicates considerably more random error than would be expected for the pre-detection bandwidth employed.

INTRODUCTION

Except for very low temperatures or extremely high frequencies, the electrical or Johnson noise due to thermal agitation in a conductor is universally accepted to be accurately described by the simple Nyquist formula[1]. Accordingly, noise thermometry represents one of the few practicable alternatives to gas thermometry for the precise determination of thermodynamic temperatures, although measurement of the very small noise signals presents formidable experimental problems and inevitably involves long averaging times.

This paper describes a system which should be able to establish thermodynamic temperatures to within a few mK in the range 100–150°C where the most recent gas thermometry[2] shows serious errors in IPTS-68.

OPERATING PRINCIPLE

Currently the most accurate system of noise thermometry uses the correlation amplifier technique[3] to make absolute measurements of the noise from a single resistor at the reference and unknown temperatures. With this system the electronic design must meet a number of very severe requirements, not the least of which is a long term gain stability comparable with the relative precision of the final result.

We have used a simpler system based fairly closely on a successful earlier design[1], which operated at the oxygen boiling point, using the switching principle pioneered by Garrison and Lawson[4]. In this arrangement a noise measuring channel is switched alternately between two unequal resistances at different temperatures. If the properties of the measuring channel are constant it is possible to establish a balanced condition in which the noise voltages from the resistors are equal and thus to determine the ratio of thermodynamic temperatures from the ratio of the resistances. Actually, slow variations of the measuring system need not cause significant errors assuming that they are not correlated with the switching operation. In practice there are several sources of correlated variation including finite recovery time from the transients caused by switching at the input of a high gain amplifier and systematic changes of frequency bandwidth and internal noise of the input stage due to changes of effective signal source resistance[1].

DETAILED DESCRIPTION

A block diagram of the noise thermometer is shown in Fig. 1. The resistance R_1 is maintained at the temperature to be measured (T_1) whilst R_2 is in an ice bath (T_2 = 273.15 K). Either resistance may be connected via mercury-wetted reed switches S_1 and S_2 to a fairly low noise pre-amplifier which uses a JFET type BFW11 in a cascade connection. This arrangement was selected mainly for high input resistance, and its equivalent noise resistance is approximately 450 ohms.

The amplified noise signals are fed through the shield surrounding the noise thermometer by a screened transformer and split by filters into two frequency ranges, 10–100 kHz and 100–200 kHz. The average of the absolute value of the signal in each range is determined

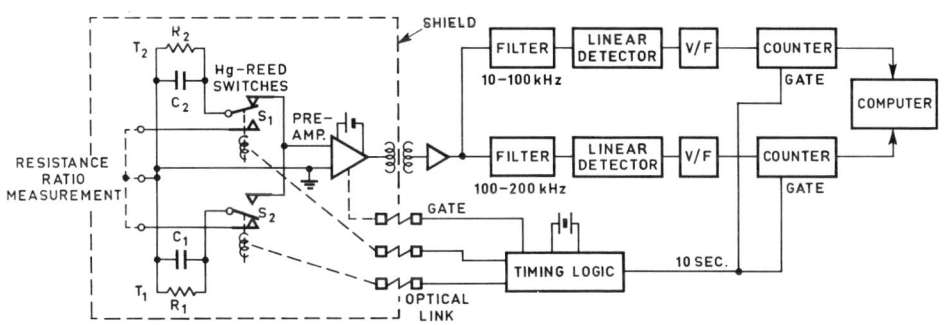

FIG. 1. Block diagram of switching noise thermometer

by counting the output of a voltage-to-frequency converter after linear detection. The data are collected and processed by a micro-computer system.

A crystal-controlled clock and counter chain provide timing signals for the system. In normal operation the switches change over each 15 s with the first 5 s being allowed after each switching operation for settling before integration starts. In addition, an early stage of the pre-amplifier is gated off during the switching period. Timing signals for this gating and the control of the reed switches are transferred through the surrounding shield by optical links with light guides to obviate the presence of paths which might introduce external interference.

Reference to Fig. 1 shows that the unselected noise resistance is normally short-circuited; however, the switching is arranged so that it is possible to measure the ratio R_1/R_2 (at 1 kHz) using a 7 decade inductive divider bridge. This is normally done at the start and finish of each run.

The temperature of the resistance R_1 is maintained by heated oil circulated from an external bath with the returning oil providing thermal shielding for an isothermal column extending upward into the electromagnetically shielded area. An IPTS-calibrated platinum resistance thermometer is provided for both measurement and control of the oil temperature. This temperature exhibits rapid fluctuations of the order of 10 mK, due presumably to the turbulent processes in the circulating pump. The mean value, however, is stable to within 1 mK indefinitely.

Resistance R_2 is surrounded by high purity crushed ice with the melt water drained from the bottom of the container. Tests conducted with a standard platinum resistance thermometer (inverted) in this ice bath showed variations generally less than 1 mK.

THERMOMETRIC RESISTANCES

It is desirable that the thermometric resistances should produce only thermal noise and have a conductance which is stable at temperatures up to 150°C and independent of frequency to 200 kHz.

Metal film resistances (Philips MR25) were used. The excess current noise of these would be well below the 10 kHz low frequency cut-off of the system and should be negligibly small since the dc gate current of the input FET is less than 2 pA.

Present technology does not permit the convenient measurement of the conductance ratio with suffient precision at frequencies as high as 200 kHz, and so extrapolation of audio frequency measurement is necessary. It should be noted that for the frequency range employed here the resistive film is much too thin to exhibit significant conductance changes due to skin effect. Changes, however, still might possibly arise from dielectric losses in the supporting substrate or protective lacquer, or from extreme spiral grooving of the film.

In order to provide even greater confidence in the ratio determination it was decided to use initially groups of nominally identical resistances (6.8 kΩ) in parallel, three for R_1 and two for R_2. This will result in a first order cancellation of any frequency effects which may be present; however, it does mean that balance is obtained for a 3/2 ratio of thermodynamic temperature, i.e. $T \approx 135$°C.

This is a reasonably convenient temperature to obtain a check on the scale and the particular resistances used are still adequately stable at this temperature.

MEASUREMENT PROCEDURE

Before the series of measuring runs was commenced (and also after), calibrating tests were performed with short circuits substituted for the noise resistors. This process provides a direct measurement of the differences in noise of the reed switches and the leads to the resistors at their normal operating temperatures which is subsequently used as a correction. In principle this process also provides reasonable corrections for certain possible extraneous noise signals arising from mechanical vibration or inductive coupling. When the resistances are replaced there will be some attenuation of the effective differences due to the shunt capacitances but this is less than 1% at 200 kHz and causes negligible effect on the corrections.

Also at this time the actual resistances of the leads at their operating temperatures can be measured. These resistances, of the order of 1 ohm, are needed to correct the measured resistance ratio which includes them.

In normal operation the bath temperature T_1 and the capacitance C_1 are varied until balance is obtained in both frequency ranges, thus ensuring that there will be no errors caused by the attenuation of the R_1C_1 and R_2C_2 time constants. The results of each 10 s integration are squared (because the detector is linear) and the differences between those from R_1 and R_2 (called U and L respectively) are stored by a dedicated micro-computer over the period of each run which normally lasts 24 hours.

At the end of each run, after the correction for the noise of the leads has been subtracted, there is normally a small residual unbalance $\overline{U-L}$ in both channels and the temperature T_1 is calculated from the formula[1]

$$T_1 = \frac{R_2}{R_1} T_2 \left[1 - p\left(\frac{7}{6} \left.\frac{\overline{U-L}}{\overline{U}}\right|_{LF} - \frac{1}{6} \left.\frac{\overline{U-L}}{\overline{U}}\right|_{HF} \right) \right] \quad (1)$$

Here $\overline{U}$ represents the net mean square noise from the system when switched to the upper (R_1) resistance and the factor p, which is taken as 1.15, reflects the reduction in the relative unbalance resulting from the noise of the measuring amplifier.

ERRORS DUE TO GATE NOISE CURRENTS

Since the resistances R_1 and R_2 have different values the switching noise thermometer is subject to errors due to noise currents originating in the input of the amplifier and also, to any systematic dependence of gain on source impedance resulting from feedback. The latter is effectively eliminated by the cascode arrangement.

Figure 2 shows the equivalent input circuit with the important noise current generators. The shunt resistance R_s arising from the input conductance of the FET and the insulation losses of the circuit was measured at 150 kHz, using a custom-build digital Q-meter, to be approximately 1000 MΩ. This may be considered to produce ordinary thermal noise corresponding to room temperature and, in the present system, results in an error of only 0.2 mK. It may be noted that the error arising from this source is more serious in a noise thermometer measuring cryogenic temperatures.

The gate leakage current I_g is approximately 2 pA and the shot noise arising from this may also be shown to give a completely insignificant error in this case.

The induced gate noise current is capacitively coupled from the channel of the FET and may be represented by

$$d\overline{i_g^2} = Bf^2 df \quad (2)$$

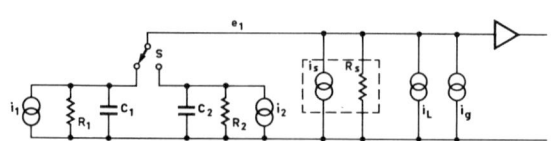

FIG. 2. Equivalent circuit in terms of noise current generators. Johnson noise in the thermometric resistances is represented by i_1 and i_2, R_s is the shunt resistance of the input circuit with noise i_s and i_L, i_g are respectively the shot noise due to the gate leakage current and the induced gate noise.

where B is a constant. For the frequencies and impedances involved here the correlation with the channel noise is pure imaginary and the voltages at the gate may be simply added in quadrature. The effect of this additional noise current is largest for the higher value resistance R_2 in the higher frequency range where it represents an apparent temperature increase of approximately 5 mK. This would be partly off-set by a proportionately smaller increase in the apparent temperature of R_1; however, it has been shown[1] that errors due to capacitively coupled noise currents at the gate, for the noise thermometer balanced simultaneously in two frequency ranges, cancel out completely in a second order analysis.

RESULTS

The most critical elements of the system are the switches; fortunately the performance of these can be tested by connecting a single resistor to both inputs and checking for bias. After a large number of experiments with different arrangements it was established that the mercury-wetted reeds were satisfactory.

Some 62 runs of 24 hours duration were completed and in all cases the standard deviations at the end of each run were in good agreement with theory for the pre-detection bandwidths employed, and, when converted to temperature, corresponded to approximately 12 mK.

For the set of 62 runs the mean difference

$$T_{IPTS} - T_{NOISE} = 12 \text{ mK at } 135°C$$

with the standard deviation of the overall mean being 2.4 mK.

The latter figure should be $\approx$ 1.5 mK for normal distribution, indicating the presence of additional sources of random errors.

This result is in the same direction but otherwise not in particularly good agreement with the results of recent gas thermometry; however, a number of break-downs towards the end of the series of runs raises the possibility of earlier undetected intermittent electronic problems.

REFERENCES

1. C. P. Pickup, Metrologia 11, 151 (1975).
2. L. A. Guildner and R. E. Edsinger, J. Res. NBS 80A, 703 (1976).
3. H. -H. Klein, G. Klempt and L. Storm, Metrologia 15, 143 (1979).
4. J. B. Garrison and A. W. Lawson, Rev. Sci. Instr. 20, 785 (1949).

Noise thermometry and related experiments at IMGC

L. Crovini and A. Actis

Istituto di Metrologia G. Colonnetti, Torino, Italy

High temperature noise thermometry is being developed at IMGC in a program for the determination of thermodynamic temperatures. The program has gone through several stages of development starting with measurements at the tin point, then followed by a thorough investigation in the range from 630 to 962 °C. The measurement methods and the results of these experiments are summarized in the first part of the paper. They formed the basis of design of the new high temperature noise thermometer, to be used in the range from 1000 to 1600 °C. In this new instrument, we still use a direct noise voltage comparison method where the balance is achieved through the adjustment of the resistance ratio at the input. The major sources of difficulties for high temperature applications are the insulation leakages and external electromagnetic interferences. In designing the instrument and the measurement method, we paid particular attention to these problems, and some details are given in the paper. Lastly, we describe a method for the determination of Boltzmann's constant which can be realized with our apparatus. It basically consists in the comparison of Johnson noise with a controlled binary random noise source.

INTRODUCTION

The noise thermometry program at IMGC is aimed chiefly at the determination of thermodynamic temperatures above 630 °C. Within this program, however, attention is paid to other applications, and, particularly, to a possible determination of Boltzmann's constant.

Three methods of noise thermometry have been considered: the first of them is the attenuation method, that was thoroughly discussed in a previous paper (1). It was introduced for the first time at NPL, Teddington (2) to circumvent the difficulty caused by an input impedance dependent positive feedback in the measuring amplifier. Therefore, the amplifier input is alternatively connected to two resistances of the same value, but at different temperatures. The noise at the output (amplified Johnson noise plus amplifier noise) is either directly recorded, or recorded after an attenuation by means of a calibrated voltage divider, depending on which of the two resistances is connected. The ratio of the thermodynamic temperatures of the resistances can be inferred from the particular attenuation ratio that equalizes the recorded signals. It is necessary, however, to apply a correction to account for the noise of the amplifier, which must be accurately measured. Two methods have been used to perform this operation, one entailing two, or more, resistances of different values, but at the same temperature (i.e., the ice point), the other with two equal resistances at two different, but known, temperatures. The latter is not suitable for absolute temperature determination (1), (3).

Whatever method is used, the measurement of the amplifier noise is a lengthy and delicate operation, and requires a careful control of all parameters that may influence the amplifier. In our experiment (3), a satisfactory stability of the amplifier noise was achieved over a 24-h period only when the temperature of the amplifier case was stabilized to within ±0.2 °C and the supply voltage to within ±0.1 %.

With the introduction of FET cascode amplifiers, particularly those using in the second stage a differential operational amplifier of sufficiently high gain-bandwidth product, the positive feedback can be made totally negligible particularly if the input resistances do not exceed 1000 Ω. Also the contribution of the noise generated by the amplifier input current is negligible. Therefore, it is again possible to make use of the method originally proposed by Garrison and Lawson (4), where the recorded output signals are equalized by adjusting the ratio of the two input resistances. At balance the following relationships must hold :

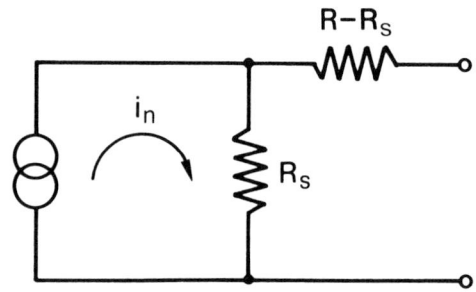

Figure 1: Injection of a noise current i_n in a resistance network, which presents an output resistance R.

$$R_1 T_1 = R_2 T_2, \quad (1)$$

and

$$R_1 C_1 = R_2 C_2 \quad (2)$$

where R_1, C_1, T_1 and R_2, C_2, T_2 are respectively the resistance, the total shunting capacitance and the thermodynamic temperature of the two resistors which are connected at the input. The second method of noise thermometry consists of this technique with a slight modification: the resistance ratio (R_1/R_2) is preadjusted in order to match as closely as possible the inverse ratio of the Kelvin temperatures on the IPTS-68, $(T_{68})_2/(T_{68})_1$. The fine balance is then obtained by means of the voltage divider. The resulting balance equation is as follows:

$$T_1 = a^2 \frac{R_1}{R_2} T_2 + (a^2 - 1) \frac{R_e}{R_1} T_2 \quad (3)$$

where a is the voltage attenuation ratio and R_e is the amplifier noise-equivalent resistance, referred to T_2. This method still requires the measurement of R_e, whereas the original method of Garrison and Lawson does not.

Actually a^2 is very close to unity and in practice R_e needs only to be determined to within a few percent. Since no resistance is changed during the experiment, trimming input capacitances can also be preadjusted in order to fulfill the requirements of Eq.(2), thus avoiding cumbersome double-balance procedures (4).

The most relevant results achieved at IMGC with these two methods are reported in a previous publication (3).

In the third method that has been considered at IMGC, the balance is achieved by injecting a known noise current, i_n, into the resistance at the lower temperature. With reference to the schematic diagram of Figure 1, and to Nyquist's relationship for Johnson noise, the resulting noise voltage across the resistance R, as seen by the amplifier, may be expressed as follows:

$$\overline{v^2} = (4kTR + \overline{i_n^2} R_S^2) \Delta f, \quad (4)$$

where k is Boltzmann's constant, Δf is the measurement bandwidth and $\overline{i_n^2}$ is the power spectral density of the injected noise current.

Provided a linear and reproducible noise current generator, with the same spectrum as Johnson noise, is made available, the current i_n can be calibrated in terms of thermodynamic temperature. Again two possibilities exist:
a) to use two reference temperatures, T_o and T_R ($T_o > T_R$) and always the same input resistance R. The outputs of the two resistors are balanced when i_n is set to a particular value. Then, this current is determined in terms of the temperature difference $(T_o - T_R)$. Any other value of i_n is obtained by adjusting a D.C. voltage at the input of a binary noise generator, in a way that causes the actual value, i_n, to be related to the current at the calibration through a linear parameter, b. Expressing

TABLE I: Results obtained with the IMGC noise thermometer in the range from 629 °C to 962 °C (3)

t_{68} (°C)	Number of measurements	$t_{noise} - t_{68}$ (°C)	Experimental uncertainty 99 % confidence limits	
			Statistical uncertainty (°C)	Total uncertainty (°C)
629	30	-0.05	0.15	0.18
629	5	-0.09	0.15	0.18
680	5	0.37	0.26	0.29
780	15	0.48	0.23	0.24
780	5	0.44	0.23	0.24
850	10	0.67	0.34	0.36
962	13	0.07	0.32	0.34

the current at the calibration in terms of the temperatures and of the resistances, we obtain:

$$\overline{i_n^2} = \frac{4kR}{R_S^2}(T_o - T_R) b . \qquad (5)$$

In the case of the example, b is equal to the square of the ratio between the actual D.C. voltage at the input of the binary noise generator and that used for the calibration;

b) to use only one reference temperature T_o (i.e., the triple point of water), but two different resistances R and R' (R' > R). The balance is achieved by feeding i_n to R, so that we have:

$$\overline{i_n^2} = \frac{4kT_o}{R_S^2}(R' - R) b . \qquad (6)$$

Considering now the balance condition at the input of the amplifier, we obtain two relationships. The first of them is

$$T = T_o + b(T_o - T_R) , \qquad (7)$$

resulting from the combination of Eqs.(4) and (5), and the second one is

$$T = T_o + b T_o (R' - R)/R , \qquad (8)$$

resulting from Eqs.(4) and (6).

RESULTS IN THE RANGE FROM 630 °C TO 962 °C

A full account of the results obtained in this range at IMGC is given in a previous publication (3). They are summarized in Table I.

Apart from any consideration on the temperature scale, which is not the aim of this paper, the results of Table I appear to be mostly affected by the statistical uncertainty. This was caused by the relatively short total integration time chosen, as each measurement required approximately 480 s. Looking at the original sequences of experimental data (not presented here), they exhibit a large range of values, giving the unpleasant feeling of poor accuracy. For instance, in a sequence of 15 measurements at 629 °C the results are scattered in a range of 1.20 K, whereas the standard deviation is 0.26 K. This last, however, compares extremely well with theoretical standard deviation (5), that for the IMGC noise thermometer can be expressed as follows (3):

$$\sigma_T = 2.7 \, T \, (\Delta f \, \tau_o)^{-1/2} \qquad (9)$$

τ_o being the integration time in seconds. For the case considered here $\sigma_T = 0.34$ K . Also the previous series of 15 measurements at 629 °C had given an experimental standard deviation just below theoretical value.

Furthermore, the results of both sequences, or the averages of group of them, pass the test of statistical significance within well acceptable limits. Such analysis would allow the rejection of a particular result without losing the information of the rest of the sequence. This is, indeed, the operational criterion of our noise thermometry: repeated short obser-

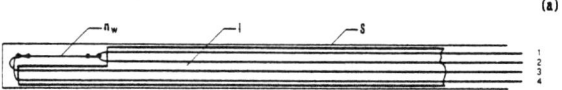

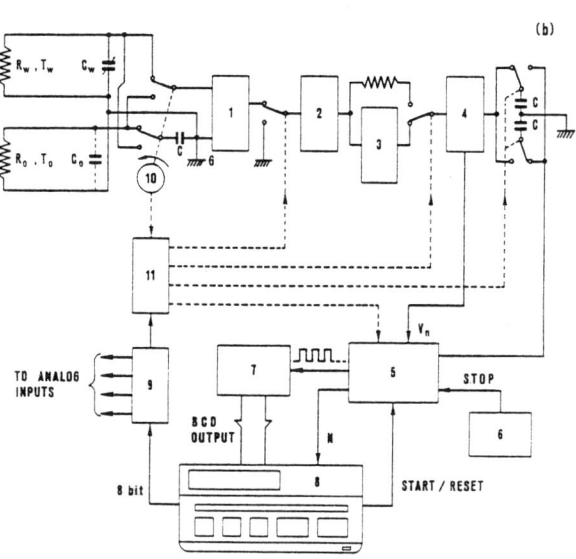

Figure 2: Schematic diagram of the sensor and the measuring apparatus. (a) S, platinum-rhodium shield; I, 99.7 % alumina four-bore insulator; 1,3,leads of the same materials as n_W (sensing wire) and the shield; 2,4,leads serving as the second leg of the thermocouple. (b) 1, low-noise amplifier; 2, 10-100 kHz filter; 3, voltage divider of output impedance R_D; 4 A.C. to D.C. converter; C, memory capacitors; 5, demodulator, D.C. amplifier, voltage-to-frequency converter and timer; 6, programmer; 7, UP/DOWN counter with display and BCD output; 8, programmable calculator; 9, multiplexer for connecting resistances, thermocouples, etc, to the analog data-acquisition section, and to program the switching sequence; 10, motor-driven input chopper. It provides the synchronizing input for the photocoupled switching unit, 11, that allows the noise signals from either R_W, or R_O, to be processed with the same apparatus. Excessive bursts of extra-noise cause V_n to exceed a given threshold, thus interrupting the observation for a given period of time, without changing the effective integration time. The number of interruptions N is recorded in the calculator to assess the validity of the determination.

vations in carefully reproduced conditions, which exhibit the expected standard deviation.

HIGH TEMPERATURE NOISE THERMOMETRY

Noise thermometry above 1000 °C is performed using the modified Garrison and Lawson technique (6). Figure 2 presents a schematical representation of both the sensor and the measuring apparatus. Johnson noise is generated in a thin Pt 10 % Rh wire, with four leads, of either pure platinum or Pt 10 % Rh alloy. The sensor resistance R_w ranges from 50 to 100 Ω. The leads, besides connecting the sensor to the amplifier (sensor leads), can be used as two thermocouples to measure the temperature of the two extreme of the wire. The sensing wire and the leads are insulated with high purity alumina and protected with a Pt 10 % Rh shield. In this way a comparison is obtained between the noise temperature and the IPTS-68, as realized with the thermocouples. The reference resistor R_o is kept in a thermostatic enclosure at about room temperature.

The input amplifier consists of a cascode circuit with four 2N3245 field effect transistors in parallel. It provides an equivalent noise resistance of about 47 Ω in the frequency bandwidth from 10 to 100 kHz. Input current noise is estimated to be less than 0.01 pA/$\sqrt{Hz}$, thus producing no appreciable effect on the rated input resistance.

A motor driven low-thermal chopper performs the input switching. All other switches are optically coupled analog gates. They are all synchronized with the input switch.

The amplified noise of either input resistances, is first filtered and, when necessary, balanced with the attenuator, then is converted to D.C. The residual difference of the D.C.voltages that appears on the two memory capacitors is amplified and converted into a frequency and then counted. The counter and the timer can be put in a stand-by condition, whenever an external interference is detected. Stand-by can last from 2 to 3 s to allow the interference to decay.

The measuring apparatus is equipped with a desk-top computer to perform the following operational functions:
a) to control the phase of the various sections of the synchronous switch;
b) to scan different analog inputs before starting the measurement sequence;
c) to start and reset the sequence (stop is provided by a suitable timer);
d) to collect and process output data, including the number of interferences that may have been occurred during the measurement;
e) to analyze the results and, particularly, to reject those measurements which appear to be affected by an excessive number of interferences.

The noise from the high temperature resistance must be corrected for the resistance of the leads. The four-lead arrangement provides two dummy leads, shorted together, having the same resistance and submitted to the same temperature distribution as the sensor leads. Connecting the dumming leads to the amplifier, we can measure a noise-equivalent temperature T_L. With reference to Figure 1, Eq.(3) is then modified as follows:

$$T_w = a^2 \frac{R_o}{R_w} T_o + (a^2-1)\frac{R_e}{R_o} T_o - T_L / (1+ \frac{\rho_L}{\rho_w} \frac{d_w^2}{d_L^2}), \quad (10)$$

where ρ_L and d_L are, respectively, the electrical resistivity and the diameter of that sensor lead which is not grounded and ρ_w and d_w the electrical resistivity and the diameter of the sensor wire.

Insulation leakage introduces a systematic error when the insulator temperature is different from that of the sensor. Therefore, the transition zone from the high temperature to below 1000 °C is made as short as possible (i.e. not exceeding 25 cm). Experiments with suitable dummy leads and insulators can provide a realistic estimate of this error, or a possible correction term.

PROPOSED METHOD FOR BOLTZMANN'S CONSTANT DETERMINATION

The redetermination of Boltzmann's constant have been considered at IMGC as a possible side-activity of noise thermometry, although no such program is underway at present. The measurement technique is based on the noise injection method. Operating in the same conditions as in Eq.(6), and assuming that $\overline{i_n^2}$ is measured independently, Boltzmann's constant k is determined through the following relationship:

$$k = \frac{R_s^2}{4T_o(R'-R)}\overline{i_n^2} - \frac{R'+R}{4T_o}\overline{i_{an}^2}, \quad (11)$$

where $\overline{i_{an}^2}$ is the spectral density of the amplifier input noise current.

The noise current i_n is generated by a variable level binary noise generator (7), (8). Its spectral density can be expressed as follows:

$$S(f) = I_o^2 \tau_c \frac{\sin^2(\pi f \tau_c)}{(\pi f \tau_c)^2}, \quad (12)$$

where τ_c is the period of the clock which controls the transition between two current levels differing by $2I_o$. At each clock pulse the transition occurs randomly, with a probability following Poisson's distribution. For $(\pi f \tau_c) \ll 1$, S(f) closely approaches the white noise spectrum. If this current is fed to the lower value resistance, so as to balance the noise of the other resistor, we obtain:

$$k = \frac{R_s^2}{4T_o(R'-R)} I_o^2 \tau_c - \overline{i_{an}^2} \frac{R'+R}{4T_o} \quad (13)$$

To investigate the feasibility of a measurement method based on Eq.(13), a circuit was realized (8) in which a D.C. voltage was modulated with FET switches, driven by a commercial binary noise generator. The clock period was set to 0.66 s, whereas the measurement bandwidth was 20 - 125 kHz. It was possible to compensate various sources of non-linearity in the modular (as, for instance, the channel resistance of the FET's) in such a way that a linearity of $\pm 0.01\,\%$ was achieved for a constant temperature operation to within $\pm 1\,°C$.

To obtain a determination of k, the resistances R, (see Figure 1) and R', the noise current generator and the amplifier should be placed in a well-shielded enclosure, where it is possible to keep the resistance at the temperature of the triple point of water for a very long period of time. A suitable noise current is injected in R_s, so that the output noise is the same for the two resistances. The peak current I_0 is determined from the D.C. voltage applied to the modulator and from the circuit resistance. The resistances R, R' and R_s can accurately measured at a suitable frequency in the measurement band. Lastly, the clock period can be measured with a counter. It would be desirable to reduce the measurement uncertainty of I_0, τ_c and the resistances to within $\pm 0.001\,\%$.

At present it is much more difficult to provide a waveform as close as necessary to that of the ideal random telegraph wave i.e., the binary noise source. Chiefly, the rise time of the wave, switching delays in the modulator and spikes generated by gate-to-channel capacitive couplings in the FET switches, may cause deviations from Eq.(12), which cannot easily be predicted and accounted for. Since these problems are less severe for lower clock frequency, and the upper cutoff frequency of the measurement band should be one twentieth of the clock frequency to ensure a flat spectrum and good gaussian distribution of amplitudes, we can forecast that the measurement bandwidth should not exceed 20 kHz. For the IMGC noise thermometer (3), the statistical uncertainty (standard deviation) in comparing the noise from the two resistors is expressed by the following relationship:

$$(\overline{\Delta v^2}/\overline{v^2}) = 2.4\,(\Delta f\,T_i)^{-1/2}\,, \qquad (14)$$

where T_i is the integration time. A $0.001\,\%$ imprecision (standard deviation) requires a continuous integration of 2.88×10^6 s, or 185 days. The only way to cover such a long period of time is with a fully automated measurement system, which is feasible with modern techniques. However, the temperature of the triple point of water can be maintained almost indefinitely within $\pm 0.00005\,\%$ in terms of the thermodynamic temperature.

REFERENCES

1. A. Actis, A. Cibrario, L. Crovini, in: Temperature its measurement and control in science and industry Volume 4, Editor in chief: H. H. Plumb, Part 1, 355, Instrument Society of America (Pittsburg) (1972).
2. H. Pursey, E. C. Pyatt, J. Sci. Instrum. 36, 260 (1959).
3. L. Crovini, A. Actis, Metrologia 14, 69-78 (1978).
4. J. B. Garrison, A. W. Lawson, Rev. Sci. Instrum. 20, 785 (1949).
5. S. O. Rice, Bell System Tech. J. 23, 282 (1944).
6. L. Crovini, High Temperatures - High Pressure 12, 253-259 (1980).
7. G. A. Korn, Random process simulation and measurements, Sect. 1-13 and Chapter 4, Mc Graw-Hill (N.Y.) (1966).
8. L. Crovini, A. Actis, Alta Frequenza 10, Volume XLIV, 617-327E (1975).

A new method of noise thermometry

M. Imamura and A. Ohte

Yokogawa Electric Works, Limited, Musashino-shi, Tokyo 180, Japan

A new method which determines thermodynamic temperatures without any temperature calibration is proposed. A shot noise diode is used in this method as reference noise source, instead of using a resistor at a reference temperature. The measured absolute temperature is directly proportional to the dc voltage developed across a sensor resistor by the shot noise diode bias current. A fully automatic prototype thermometer, incorporating a vacuum tube noise diode and a low-noise correlation amplifier, was developed and tested. The accuracy over the temperature range 77–1235 K was better than 0.3%. Several factors which affect the accuracy are discussed.

1. INTRODUCTION

The history of accurate noise thermometry dates back to 1949[1]. Noise thermometry offers the promise of constructing standard thermometers which measure temperature directly on the absolute thermodynamic scale. However, conventional noise thermometry methods require a temperature reference or temperature calibration, and are difficult to adapt for automatic measurements.

One conventional method [1] is as shown in Fig. 1. The thermal noise voltage of a sensor resistor is compared with that of a resistor at reference temperature, and the resistance of the reference resistor is mechanically varied to balance the noise voltages. The resistances of both sensor and reference resistors must be measured. Noise signal bandwidths of sensor and reference may differ due to stray capacitance; this must be compensated for.

Another method [2] measures the noise power in the sensor resistor. This does not require reference temperature and resistance measurements, but temperature calibration is always required.

A new method of noise thermometry that requires neither temperature reference nor temperature calibration has been proposed [3] and tested [4]. The aim was to produce a standard thermometer for ordinary laboratories, and to eliminate the need for periodical recalibration against a primary standard.

The accuracy of a prototype thermometer was better than 0.3% over the temperature range 77 to 1235 K. Temperature measurements by this method — using a shot noise diode as a reference noise source — are independant of sensor resistance or noise signal bandwidth. Automatic temperature measurement is possible. Automatic measurement is preferred in noise thermometry, since it is a statistical method, and data must be averaged.

2. PRINCIPLES OF NEW METHOD

Mean-square thermal emf $\overline{e_n^2}$ across a sensing resistor R due to Johnson noise is expressed by the well-known equation:

$$\overline{e_n^2} = 4kTR\Delta f \quad (1)$$

where k is Boltzmann's constant, Δf is signal bandwidth and T is absolute temperature (to be determined).

Shot noise is the fluctuation in DC current that flows over some energy barrier. Shot noise can be obtained from a vacuum tube noise diode. Mean-square shot noise current in a noise diode is given as

$$\overline{i_n^2} = 2qI_0\Delta f \quad (2)$$

where q is electron charge and I_0 is DC bias current in the diode. The noise current can be changed by controlling I_0, and can be switched on or off by forward- or reverse-biasing the diode. To control I_0, it is sufficient to vary filament temperature or filament current.

Figure 2 shows the principles of the new method using a shot noise diode as the reference noise source. Suppose that the DC current in the diode, and the superimposed shot-noise current-signal, is applied to the sensor resistor.

With the diode biased off, the mean square voltage $\overline{e_B^2}$ across a sensing resistor is due to Johnson noise alone;

$$\overline{e_B^2} = 4kTR\Delta f \quad (3)$$

With the diode biased on, the mean square voltage $\overline{e_A^2}$ becomes:

$$\overline{e_A^2} = 4kTR\Delta f + 2qI_0\Delta f R^2 \quad (4)$$

becase there is no correlation between the Johnson noise and the shot noise. If the DC current I_0 is adjusted so that the mean square voltage with the diode biased on is twice the mean square voltage (due to Johnson noise alone) with the diode biased off, i.e.:

$$\overline{e_A^2} = 2R\Delta f(2kT + qI_0R)$$
$$= 2\overline{e_B^2} = 2R\Delta f(4kT) \quad (5)$$

then the absolute temperature T is given by

$$T = (q/2k)I_0R = (q/2k)E_0 \quad (6)$$

where E_0 is the DC voltage developed across the sensor resistor by noise diode bias current I_0, and the coefficient q/2k is 5802.25 K/V.

In principle, this method has several advantages in consequence of (6): the DC voltage E_0 across sensor resistor R is a linear function of absolute temperature, and the coefficient comprises physical constants k and q. A reference resistor, kept at reference temperature, is not required. Electrical calibration is sufficient, temperature calibration is not required. Measurement is not affected by sensor resistance or noise signal bandwidth.

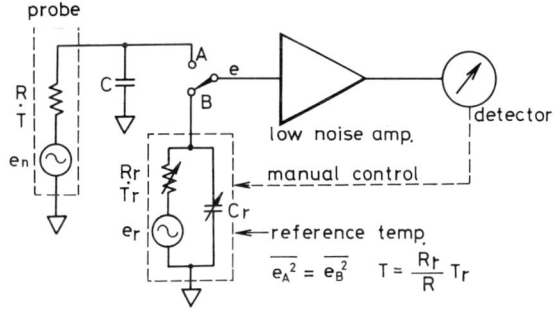

Fig. 1. Principles of Garrison's noise thermometer.

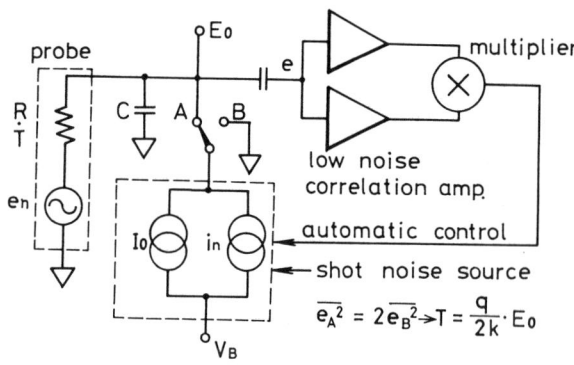

Fig. 2. Principles of new noise thermometer.

3. PROTOTYPE NOISE THERMOMETER

A fully automatic prototype noise thermometer was developed and tested (Fig. 3). The block diagram of the thermometer is shown in Fig. 4.

3.1 Noise diode.

The prototype noise thermometer uses a vacuum tube noise diode (LD342 NEC) as reference. Its characteristics are shown in Fig. 5. The solid line denotes anode current I_p vs. bias voltage V_p with filament voltages up to 2.7 V. The dotted line indicates that its conductance is proportional to I_p, and that — for sufficiently low conductance — plate voltage must be at least 10 V.

Another noise diode — Sylvania 5722 — was tested, but was not used because its conductance was not low enough to give sufficient temperature measurement accuracy in this application.

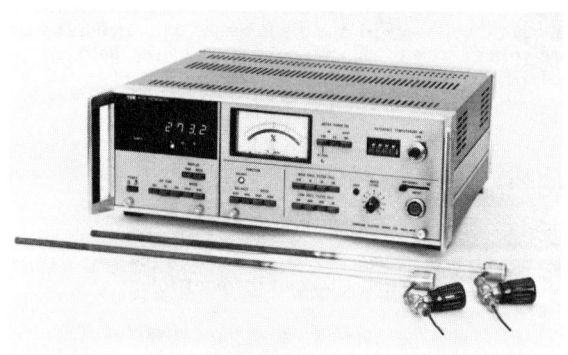

Fig. 3. Prototype noise thermometer and probes.

3.2 Preamplifier.

The low noise preamplifier circuit is shown in Fig. 6. The input FETs Q1 and Q3 (Sony Inc. type 2SK152) have 20 mS forward-transconductance at I_d = 8.2 mA. The equivalent input noise resistance of the FET is theoretically expressed as:

$$R_n = 2 * 0.7 / gm + (R_i // 2 R_f) = 104 \text{ (ohm)} \quad (7)$$

However, the value of R_n actually observed was about 160 ohms. The disparity is due to the noise contribution of the next stage.

L1 and L2 reduce the gain of the input stage at very high frequencies.

Common mode gain G_c of the preamplifier is given by

$$G_c = 2 R_f / R_i + 1 \quad (8)$$

G_c is adjusted to 29.011, (E_0 is amplified 29.011 times) so that the preamplifier output is exactly 5 mV/K. Figure 7 shows preamplifier unit and noise diode.

3.3 Amplifier.

With Garrison's method (Fig. 1), amplifier noise — the same for reference and probe measurements — cancels out. With our method, amplifier noise must be minimized.

The inside of the dotted line in Fig. 4 forms an ultra-low-noise correlation amplifier, and includes two low-noise differential preamplifiers, 3 dB attenuators, bandpass amplifiers, and wideband multiplier. When shot noise is superimposed on the Johnson noise, outputs of the preamplifiers are attenuated by the factor $1/\sqrt{2}$ so that multiplier output — proportional to the mean square noise signal voltage — remains the same. The attenuators are bypassed in the Johnson-noise-only case.

Bandpass amplifier gain is variable between 80 and 100 dB, and frequency-select switches determine HPF, LPF cutoff frequencies: 30 kHz - 100 kHz - 300 kHz - 1 MHz and 300 Hz - 1 kHz - 3 kHz - 10 kHz - respectively. The outputs of the bandpass amplifiers are multiplied together by a high speed multiplier (Analog Devices Inc. type 422J) for correlation. (Noise in CH1 does not correlate with that in CH2, so this noise is eliminated by multiplying the outputs of CH1 and CH2 together. This reduces the equivalent noise resistance of the preamplifiers from 160 ohms to less than one ohm). The four-wire input and multiplier also eliminate Johnson noise in the probe lead wires.

3.4 Control circuit.

The control circuit (see Fig. 4) adjusts the noise diode filament voltage (to adjust its bias current) so that equation (5) holds: The amplified mean square noise voltage due to Johnson noise is integrated over 50 ms, then half of the mean square noise voltage due to the combination of Johnson and shot noise is integrated in the opposite direction over the same interval (50 ms). The remainder in the integrator — a value proportional to the error in filament voltage — is fed to a PI controller connected (via an opto-coupler) to the filament driver.

Solid-state switches are used to change the polarity of the diode bias voltage and to bypass the $1/\sqrt{2}$ attenuators.

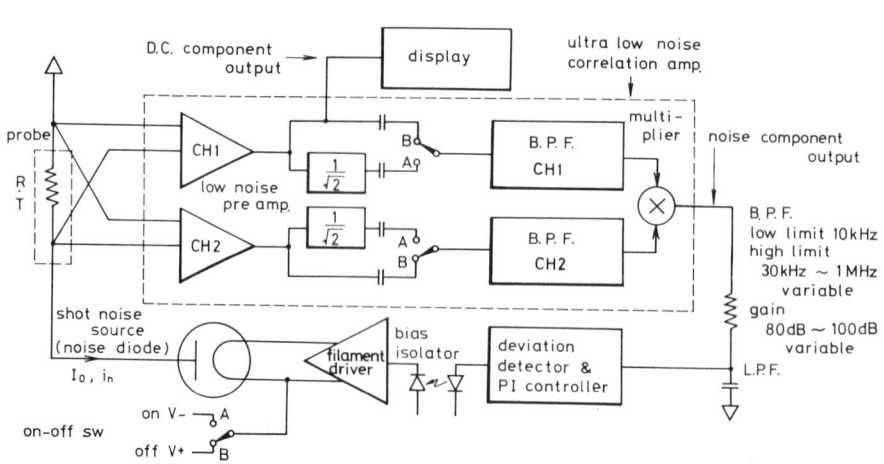

Fig. 4. Block diagram of a new noise thermometer.

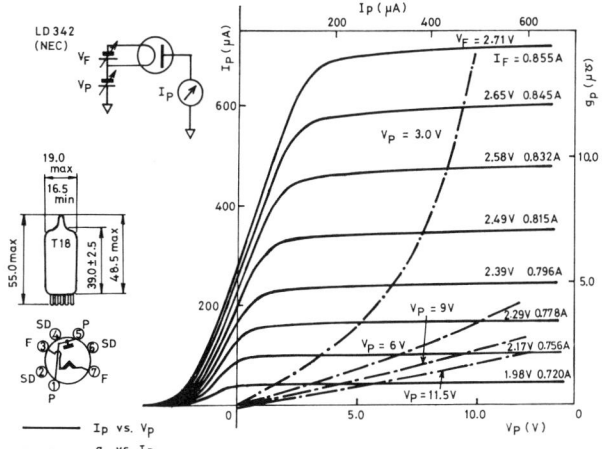

Fig. 5. Characteristics of noise diode.

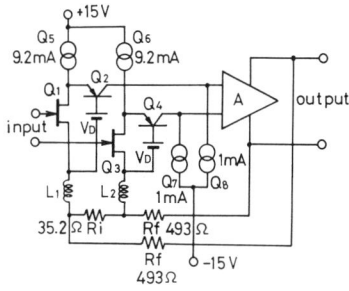

Fig. 6. Preamplifier circuit.

Fig. 7. Preamplifier unit and noise diode.

3.5 Display.

E_o — a DC voltage proportional to absolute temperature — is switched on and off every 50 ms. Synchronous demodulation is used to derive the amplified E_o signal from the "DC component output" applied to the display. The result is converted to a pulse width and then to a numerical value by A/D converter. Results are time-averaged — averaging time can be selected as one of 0.8, 8 or 80 sec. The results are indicated on a LED display, and can be sent to a computer via a GP-IB bus interface.

4. EXPERIMENTAL RESULTS

Two sensor probes with resistance values of 110 and 240 ohms respectively were made and tested (refer to Fig. 3.) Probe construction is shown in Fig. 8. The sensor resistor is a 0.06 mm diameter nickel-chromium wire bifilar-wound on a X-section alumina bobbin. To prevent oxidation of the sensor resistor, air was evacuated from the protective tube, and the tube was then filled with helium at atmospheric pressure for thermal conduction. The probe is a 6-wire configuration, two wires for channel 1, two for channel 2, and two for the noise diode, to eliminate the effects of lead wire impedance.

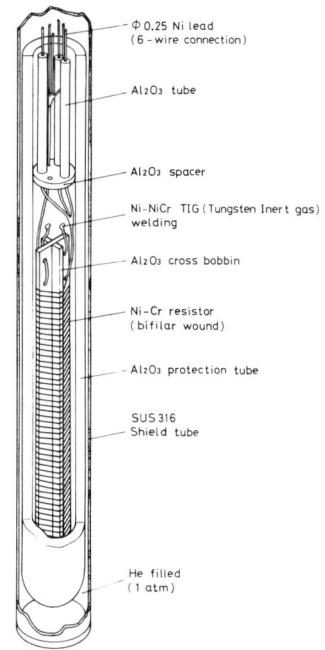

Fig. 8. Probe construction.

After electrical calibration, the thermometer was tested with the probes at several fixed-point temperatures — the temperature of liquid nitrogen (77.4 K), ice point (273.15 K), the freezing points of tin (505.12 K), zinc (692.73 K), antimony (903.89 K), and silver (1235.08 K). At 77.4 K and 273.15 K, bath temperatures were monitored with a standard thermometer. The freezing points of the metals were measured using SEMCO Instrument's crucibles.

Sensor resistances were almost constant at all temperatures (90 to 110% of their resistances at 0°C), but isolation resistance decreased at high temperatures (200 kilohms at 1200 K). More heat-resistant materials such as quartz must be used with future probes.

Measurements were repeated using low pass filter cutoff frequencies of 30 kHz, 100 kHz, 300 kHz, and 1 MHz. (The high pass filter was set to eliminate frequencies below 10 kHz — 1/f noise, noise induced from power bus, etc). Measurement data was averaged over 240 s — long enough to reduce statistical error. All test data were distributed within ±0.3% of measured absolute temperatures. The results are shown in Fig. 9.

Measurements at 0°C (with various bandwidths) were made using other probes which had resistances of between 100 ohms and 3 kilohms — this again verified that temperature readings were independant of sensor resistance as well as noise signal bandwidth (Fig. 10).

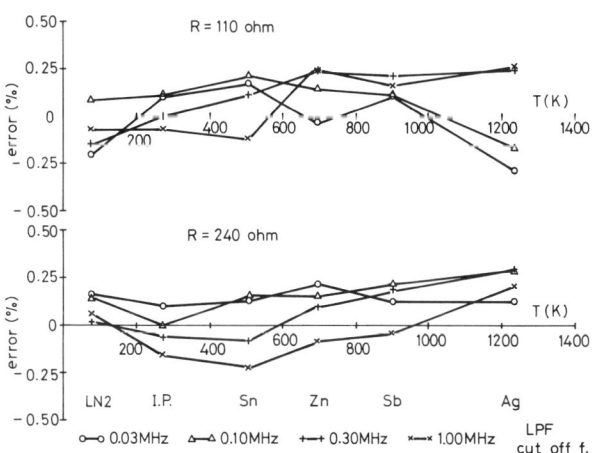

Fig. 9. Test results of noise thermometer.

M. Imamura and A. Ohte

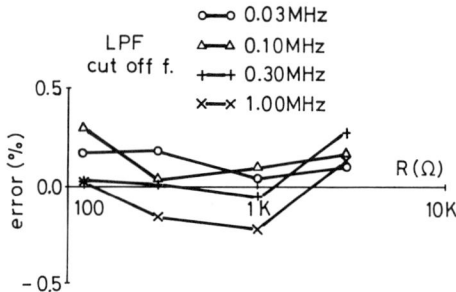

Fig. 10. Error at 0°C.

5. DISCUSSION OF ERRORS

The results include some error caused by several factors.

5.1 Statistical error.

Fluctuation in mean square thermal emf is expressed by the following ratio (Van der Ziel 1954).

$$\Delta \overline{e_n^2} / \overline{e_n^2} = 1/\sqrt{\Delta f \tau} \quad (9)$$

where τ is integration (averaging) time. With conventional methods, statistical error observed in T is twice this value. With the new method it is estimated as four times this value, as explained below. Equations (5) and (6) may be rewritten, including statistical error, as

$$2R\Delta f(2kT + qI_oR)(1 \pm \delta_1)$$
$$= 2R\Delta f * 4kT(1 \pm \delta_2) \quad (10)$$

$$T = qI_oR(1 \pm \delta_1)/(2k(1 \pm 2\delta_2 \mp \delta_1))$$
$$\simeq (qI_oR/2k)(1 \mp 2\delta_2 \pm 2\delta_1) \quad (11)$$

where δ_1 and δ_2 are errors caused in each half clock period (i.e. 50 ms);

$$\delta_1, \delta_2 = 1/\sqrt{\Delta f \tau/2} \quad (12)$$

As these have no correlation, the sum of δ_1 and δ_2 is $\sqrt{2}$ times (11), not twice. So statistical error is estimated as

$$\Delta T/T = 4/\sqrt{\Delta f \tau} \quad (13)$$

For example, when $\Delta f = 20$ KHz and averaging time = 240 s, statistical error is 0.18%. If bandwidth of 1 MHz is selected, the error becomes as small as 0.026%.

5.2 Power dissipated in probe.

Power P dissipated in probe (which causes self-heating) is:

$$P = E_o^2/R = (2kT/q)^2/R \quad (14)$$

A 100 ohm probe at 1000 K dissipates 297 μW. This value is comparable to that for a platinum resistance thermometer.

5.3 Noise diode.

Error caused by conductance of noise diode is expressed as

$$\varepsilon = \Delta T/T = Rg = RC_dI_o = C_dE_o$$
$$= 2C_dkT/q \quad (15)$$

where g is conductance of noise diode, and C_d is a coefficient that varies with bias voltage of the noise diode. From Fig. 5, with $V_p = 11.5$ V, coefficient C_d of the LD342 is about 1/300 (S/A), so ε is estimated to be less than 0.06% at 1000 K.

5.4 Probe impedance.

Probes have parasitic impedances (C or L shown as in Fig. 11). In this equivalent circuit, the amplifier input voltage due to Johnson noise e_n alone, is

$$e_B = e_n/(s^2LC + sCR + 1) \quad (16)$$

then with shot noise on, it becomes

$$e_A = (e_n + (R + sL)i_n)/(s^2LC + sCR + 1) \quad (17)$$

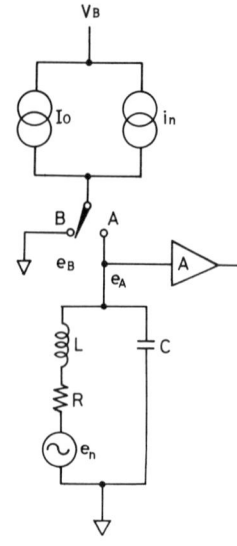

Fig. 11. Sensor parasitic impedance.

When the system is balanced, the equation $\overline{e_A^2} = 2\overline{e_B^2}$ holds, so

$$\overline{e_n^2} = 4kT\Delta f = \overline{i_n^2}|R + sL|^2$$
$$= 2qI_o \int_0^{\Delta f}\{R^2 + (2\pi fL)^2\}df \quad (18)$$

is obtained. Thus, C has no influence except to limit signal bandwidth. L reduces the observed temperature, so probe inductance must be minimised. Measured L for a 110-ohm bifilar-wound experimental probe is less than 0.3 μH, which causes 0.01% error for 1 MHz bandwidth.

Skin effect raises the resistance at very high frequencies, but its influence is negligibly small in practice.

5.5 Induced gate noise in FET's.

Johnson noise, caused by the FET channels, is applied to the sensing resistor due to gate-channel stray capacitance. This noise cannot be cancelled by the correlating circuit. So a compensation circuit was developed to double this FET noise when the noise diode is on. But the circuit does not work ideally, so the major error seems to be due to FET noise.

6. CONCLUSION

The experimental results demonstrate several advantages of a new method of noise thermometry. In particular, no temperature calibration or temperature reference is required, and automatic measurement is possible — these results have not been achieved before.

But the accuracy of the prototype thermometer is not yet sufficient for it to replace a standard PR thermocouple. To improve the accuracy of the thermometer, better noise diodes and FETs must be developed.

This method seems to offer the potential of wide application to noise thermometry.

REFERENCES

1 J. B. Garrison and A. W. Lawson, "An Absolute Noise Thermometer for High Temperatures and High Pressures", RSI, Vol. 20, No. 11, pp. 785-794, 1949.

2 C. J. Borkowski and T. V. Blalock, "A New Method of Johnson Noise Thermometry", RSI. Vol. 45, No. 2, pp. 151-162, 1974.

3 A. Ohte and M. Imamura, "Thermal Noise Thermometer", U. S. Patent 4,099,413; Jul. 11, 1978.

4 A. Ohte and M. Imamura, "Investigation of a New Method of Noise Thermometry", Digest of the 17th Annual Conference of the Society of Instrument and Control Engineers (SICE, Japan) pp. 153-154, 1978.

Vapor pressure of D_2 + xHD and ^{20}Ne

G. T. McConville and D. A. Menke

Monsanto Research Corporation, Mound*, Miamisburg, Ohio 45342

New vapor pressure measurements have been made using D_2 + 0.84% HD and ^{20}Ne. Temperatures were determined using the vapor pressure equations of Grilly and Furukawa and the 4.21 K He point to determine a resistance vs temperature function for the platinum thermometer. This function was compared to the original calibration. Another measurement of D_2 + 0.18% HD displays a smaller difference from the D_2 + 0.84% HD data than expected from the compilation of Woolley, Scott, and Brickwedde.

INTRODUCTION

This work is a product of establishing a temperature scale at our laboratory in the region between 12 and 30 K to measure vapor pressures of mixtures of ortho and para D_2.[1] We found that it was necessary to know the amount of HD in the samples to 0.01%. It appears that in all previous measurements of the vapor pressure of D_2, even in "chemically pure" samples, there was an unknown amount of HD. Thus, we have made measurements on a sample of normal D_2 + 0.84% HD, to be compared to the D_2 measurements of Grilly[2], and a new measurement of ^{20}Ne. This was done to improve on the thermometry used in the vapor pressure measurements of the neon isotopes[3].

RESULTS

In making vapor pressure measurements on D_2, one cannot use the H_2 fixed points for reference temperatures because the H_2 adsorbed in the system produces excess HD in the D_2 measurements. Thus, in our cryostat there has been an uncertainty at the low end of our temperature scale. A second problem showed up in the fitting of the neon data using the vapor-pressure equation of Furukawa[4] and the calibration of a Rosemont Engineering type 146L platinum resistance thermometer. As shown in Figure 1, the difference plot displays an oscillation which corresponds to about ±0.02 K. An oscillation of this sort usually means the fitting function does not describe the data. There are actually two fitting functions; the vapor pressure equation and the resistance vs. temperature function for the platinum thermometer.

A way to test the second function is to use the measured resistance at a given vapor pressure and calculate the temperature from the vapor pressure measurements of other workers. We compare with three sets of data which are all traceable to the NBS-1955 temperature scale. They are those of liquid ^{20}Ne, ^{22}Ne, and natural neon by Furukawa[4], of solid natural neon by Grilly[5], and the liquid and solid D_2 by Grilly[2]. The chemically pure research grade D_2 he used is now thought to have contained 0.8% HD. Therefore, we have made a new set of ^{20}Ne measurements and a set using research grade D_2 with 0.84% HD as measured with a gas chromatograph. The resistance of the type 146L thermometer was measured with the Leeds and Northrup 8079 type ER bridge.

The ^{20}Ne measurements using the Rosemont thermometer calibration are compared to the equation of Grilly with constants for ^{20}Ne using pressures corrected by adding 0.0936 times the vapor pressure difference between ^{20}Ne and ^{22}Ne to Grilly's smoothed values. The results are shown in Figure 2. The oscillation appears in the relative temperature of the Rosemont calibration. The extrapolation to temperatures below 20 K, the limit of Grilly's data, is uncertain. We have other evidence that below 20 K there is a temperature gradient between the vapor pressure bulbs and the platinum thermometer amounting to about 0.02 K at the triple point of normal D_2.

In the absence of an additional calibrated thermometer, a temperature scale of the temperature in the vapor pressure bulb against the resistance of the platinum thermometer was constructed in the following way. The vapor pressure was maintained constant to better than 1 part in 10^4 through a feed back loop from the

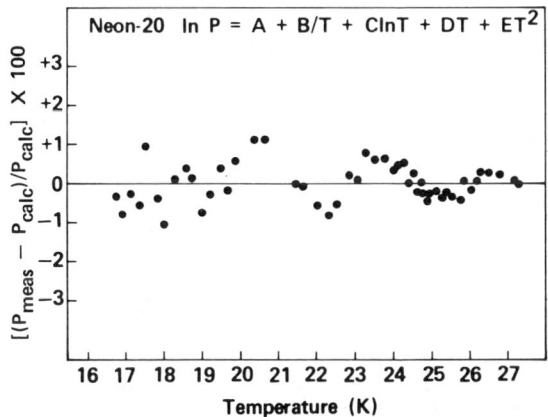

Figure 1 - Oscillation in neon vapor pressure when fitted to a standard vapor pressure equation.

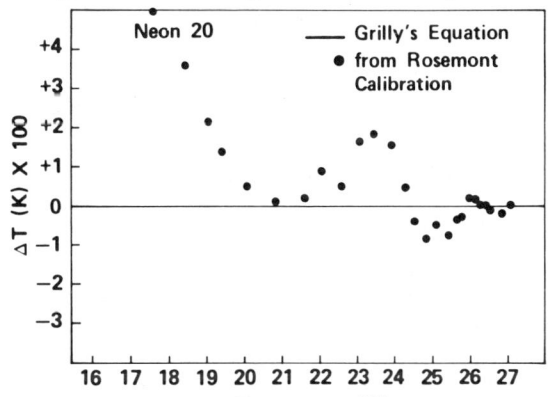

Figure 2 - Comparison of temperatures determined from Grilly's equations and from the calibration for thermometer #2788.

output of the capacitance manometer to a heater on the vapor pressure bulb. The resistance of the thermometer was measured at that pressure and the corresponding temperature was calculated from the respective vapor pressure equations of Furukawa or Grilly depending on the substance and phase. These pairs of resistances and temperatures and the resistance R_0 at 4.21 K were used to determine the parameters in Table I from

$$R = A + \sum_{n=1}^{5} B_n T^n \qquad (1)$$

for the type 146L (Serial #2788) thermometer.

Table I. Parameters for temperature vs. resistance equation

$R_0 = A = 0.17371$
$B_1 = -0.723691 \times 10^{-2}$
$B_2 = 0.259720 \times 10^{-2}$
$B_3 = -0.281698 \times 10^{-3}$
$B_4 = 0.184811 \times 10^{-4}$
$B_5 = -0.250266 \times 10^{-6}$

The scatter of the points from the different equations is shown in Figure 3 to be less than 5 mK. If only the points above 18.73 K are used to determine the B_n, the squares in Figure 3 represent temperature differences from Grilly's solid D_2 equation.

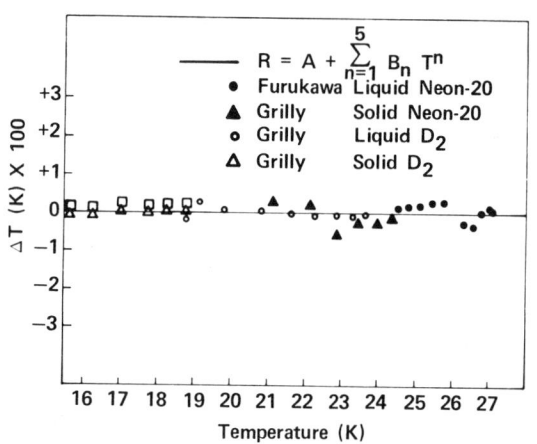

Figure 3 - A fit of temperature from several vapor pressure equations to measured resistance of the platinum thermometer using equation 1.

The resistance and temperature values in the Rosemont calculation table represents a data set that can be used to determine the B_n in equation (1) which are quite different than those determined from the vapor pressure data. An independent way of testing these points is to determine the effective 0°C resistance for each calibration temperature. One would expect $R_e(0°C)$ vs. T to be a smooth curve when determined by $R_e(0°C) = R_{cal}(T)(R_0/R_T)_{std.table}$.

The Rosemont calibration points for thermometer #2788 using IPTS-48 are shown in Figure 4. This curve is independent of equation (1) or any of the vapor pressure measurements. The variation from a monotonically decreasing function produces a variation in temperature within the stated accuracy of the Rosemont calibration of ±0.02 K. The oscillation in Figure 1 most likely resulted from trying to fit the vapor pressure to temperatures determined from a large graph of the points represented in Figure 4. The function in equation (1) with the B_n determined as in Figure 3 gives a much better representation of the temperature in the vapor pressure bulb.

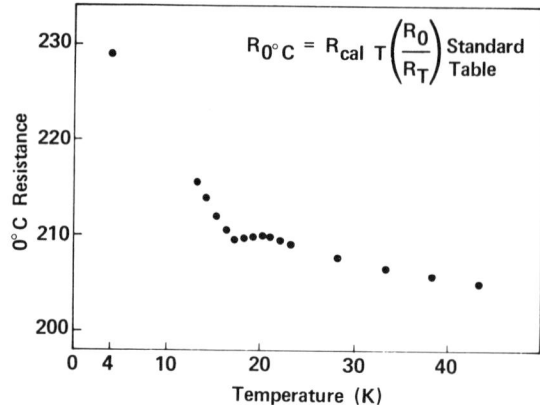

Figure 4 - Effective 0°C resistance from the calibration points of thermometer #2788.

The partial pressure of 0.84% HD in D_2 goes from 0.55 Torr at 16 K to 9.52 Torr at 23.67 K.[6] Thus, if one were to try and determine temperature using D_2 free of HD (not technically possible) and Grilly's equations, his determination would range from 17 mK too high at 16 K to 14 mK too high at 23.67 K. We have made differential measurements of nD_2 against eD_2 with HD concentrations of 0.18% and 0.14%, respectively, where considerable care was taken with getting the same amount of sample in each bulb. The $nD_2 + 0.84\%$ HD measurement was done with sample only in one bulb as was the case with the ^{20}Ne. Thus when the $nD_2 + 0.18\%$ HD was measured at the same pressures, the purer D_2 was 0.007 ± 0.003 mK higher in the liquid region but in the solid region the difference turned negative to -0.01 K at 16.2 K.

CONCLUSIONS

The measurements of ^{20}Ne and $D_2 + 0.84\%$ HD show that a consistent representation of the temperatures down to 16 K can be obtained from equation (1) in the region where there is large curvature in the resistance of the platinum thermometer. Below 16.2 K (D_2 vapor pressure of 50 Torr) the uncertainty in the vapor pressure measurement becomes significant. The use of equation (1) indicates that there is an oscillation in the calibration curve supplied by Rosemont Engineering. In trying to compare the vapor pressure of purer D_2 with the commercially available Research Grade D_2, temperature gradients in the present cryostat preclude more than a qualitative comparison to the partial pressure of HD calculated from Woolley, Scott and Brickwedde.

REFERENCES

*Mound is operated by Monsanto Research Corporation for the U.S. Department of Energy under Contract No. DE-AC04-76-DP00053.
1. G. T. McConville and David White, Physica B+C 107, 271 (1981).
2. E. R. Grilly, J. Am. Chem. Soc. 73, 843 (1951).
3. G. T. McConville, J. Low Temp. Phys. 15, 647 (1974).
4. G. T. Furukawa, Metrologia 8, 11 (1972).
5. E. R. Grilly, Cryogenics 2, 226 (1962).
6. H. R. Woolley, R. B. Scott, F. G. Brickwedde, J. Res. Natl. Bur. Stand. 41, 379 (1948).

Helium vapor pressure equations on the EPT-76

M. Durieux, J. E. van Dijk, H. ter Harmsel, and P. C. Rem

Kamerlingh Onnes Laboratory der Rijksuniversiteit Leiden, Nieuwsteeg 18, 2311 SB Leiden, The Netherlands

R. L. Rusby

Division of Quantum Metrology, National Physical Laboratory, Teddington, Middlesex, United Kingdom

New vapor pressure equations for ^{4}He and ^{3}He in which temperatures are expressed on the EPT-76 are presented. The derivation of the equations is described, after a short discussion of older helium vapor pressure equations. The new equations are proposed to the Comité Consultatif de Thermométrie for adoption and as a replacement of the presently used helium vapor pressure temperature scales.

INTRODUCTION

Helium vapour pressures are widely used as thermometric parameters at temperatures between about 0.5 K and 5 K. Before 1978 the recommended vapour pressure vs temperature table for ^{4}He was the 1958 ^{4}He scale (0.5 K – 5.2 K)[1] and the recommended vapour pressure equation for ^{3}He was the 1962 ^{3}He scale (0.2 K – 3.3 K)[2,3,4]. In 1978, the "1976 Provisional 0.5 K to 30 K Temperature Scale (EPT-76)" was adopted by the Comité Consultatif de Thermométrie (CCT)[5]. The definition of the scale has been published[6], together with a description of its derivation[7]. In Ref. 6 a table has been given of differences between T_{76} and temperatures derived from helium vapour pressures by using the 1958 ^{4}He scale and the 1962 ^{3}He scale.

At the time that the EPT-76 was adopted, the CCT agreed that new vapour pressure equations for ^{4}He and ^{3}He should be derived which relate the vapour pressures directly to T_{76}[8]. The use of such equations would evidently be more convenient than the use of the old scales together with the difference tables. For ^{4}He there would be the additional advantage that the vapour pressure temperature relation could be defined by equations instead of by a table as in the 1958 Scale.

In this paper such new vapour pressure equations, and their derivation, are given, after some remarks on the use and history of He vapour pressure equations.

The vapour pressure equations proposed in this paper have been submitted to the CCT as "vapour pressures on the EPT-76", with the designations "T_{76} (^{4}He)" and "T_{76} (^{3}He)" for the temperatures derived from the vapour pressures of ^{4}He and ^{3}He, respectively.

USE OF HELIUM VAPOUR PRESSURES IN THERMOMETRY

Vapour pressures of ^{4}He and ^{3}He are used for routine temperature measurements in many low temperature experiments as well as in thermometric measurements of the highest precision (see e.g. Refs. 9 and 10). Approximate values of vapour pressures P and derivatives dP/dT and $d \ln P/dT$ are given in Table I.

With simple mercury and oil manometers and cathetometers the obtainable accuracy in the pressure measurements is about 0.02 Torr and 0.002 Torr respectively. With the best commercial available capacitive diaphragm gauges the relative accuracy in the pressure measurements is about 0.1% to 0.01% depending on the possibility of repeated calibrations (e.g. against a pressure balance). For the pressure balance itself the relative accuracy in the pressure measurements is about 0.001% (above a pressure of 16 Torr). The corresponding accuracy in temperatures is 10^{-2} mK at 4.2 K for ^{4}He and at 0.5 K for ^{3}He (for relative accuracies in the vapour pressure measurements of 0.001% and 0.01%, respectively).

In practice, one has to apply corrections for the aerostatic head in the vapour pressure sensing tube and, at the lowest temperatures, for the thermomolecular pressure effect[2,11]. The first correction is in general less than 1 mK and can be calculated within a few percent, provided that the temperature distribution along the pressure sensing tube is measured. Alternatively, the aerostatic head correction can be determined from gas measurements in an auxiliary tube[9]. Some values for the thermomolecular pressure effect are given in Table II. For the best accuracy below the lambda

Table I
Vapour pressures of ^{4}He and ^{3}He and temperature derivatives.

^{4}He

T	P	dP/dT	$d \ln P/dT$
K	kPa	kPa/K	K^{-1}
1	0.016	0.15	9.70
2	3.13	9.19	2.94
3	24.0	36.0	1.50
4	81.6	82.3	1.01
5	188	137	0.73

^{3}He

T	P	dP/dT	$d \ln P/dT$
K	kPa	kPa/K	K^{-1}
0.5	0.21	0.30	14.4
1	1.17	5.57	4.74
2	20.1	36.6	1.82
3	82.4	92.4	1.12

Table II
Thermomolecular pressure corrections in mK for ^{3}He vapour pressure thermometry at various diameters of the pressure sensing tube (T_{hot} = 293 K).

T	2 mm	4 mm	10 mm	20 mm	40 mm
K					
0.5	17	9	2.8	0.9	0.3
0.6	6	2.3	0.5	0.14	0.04
0.7	2.0	0.6	0.11	0.03	0.01
1.0	0.1	0.03	< 0.01	< 0.01	< 0.01

point the use of ^{3}He would be preferred to ^{4}He, in order to avoid problems due to the superfluid ^{4}He film [2] and because of the higher value of dP/dT for ^{3}He.

EARLY ^{4}He VAPOUR PRESSURE SCALES

Early ^{4}He vapour pressure scales are compared with the proposed scale, T_{76} (^{4}He), in Fig. 1. In 1924, the

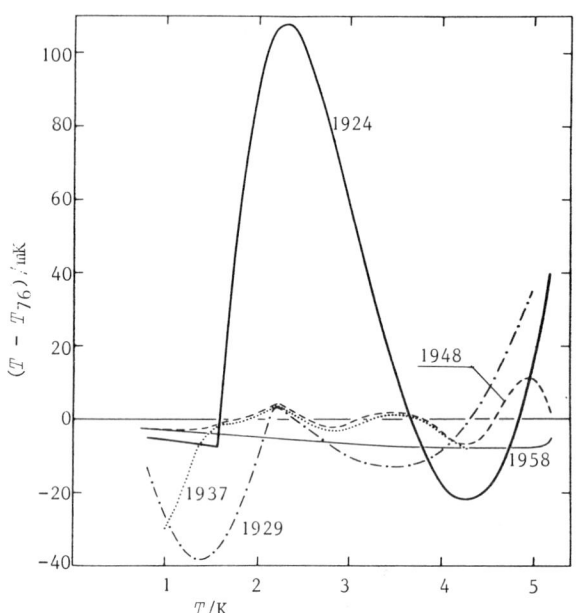

Fig. 1. Older liquid ^{4}He vapour pressure-temperature relations (T_n) compared with the EPT-76 (T_{76}). Data are taken from Table VI, Ref. 1 and Table III, Ref. 6.

λ-transition was not yet discovered and a simple analytic equation was used for the range from 1.6 K to 5.2 K. Note also the relative accuracy of the 1924 scale below 1.6 K, which was based on a thermodynamic calculation of the vapour pressure, compared to the later, empirical, relations of 1929, 1932 and 1937. The 1948 scale was the first of the helium scales that was internationally accepted as a temperature scale [12]. For more details on the early scales see, e.g. Ref. 13.

DERIVATION OF ^{4}He and ^{3}He VAPOUR PRESSURE EQUATIONS. GENERAL REMARKS

Before discussing the further developments of ^{4}He and ^{3}He vapour pressure relations, it is convenient to make some general comments on their derivation.

The relations between the helium vapour pressures and the temperature can be determined either from measurements with a primary thermometer (gas thermometers, acoustic thermometers and a noise thermometer have been used for this purpose, see below) or from a thermodynamic vapour pressure equation. The thermodynamic vapour pressure-temperature relation for a liquid with a mono-atomic vapour may be written as [14]

$$\ln P = -\frac{L_o}{RT} + \frac{5}{2} \ln T + i_o + \varepsilon(T)$$

$$- \frac{1}{RT} \int_0^T S_\ell \, dT + \frac{1}{RT} \int_0^P V_\ell \, dP \qquad (1)$$

where L_o is the heat of vaporization at $T = 0$, R is the molar gas constant, $i_o = \ln((2\pi m)^{3/2} k^{5/2}/h^3)$ the chemical constant, $\varepsilon(T) = \ln(PV_g/RT) - 2B/V_g - 3C/2V_g^2$ is a correction term representing the non-ideality of the vapour and the last two terms contain integrals of the molar entropy and the molar volume of the liquid along the saturation curve. As there is no direct experimental data available from which L_o can be obtained, this quantity has to be determined from a fit of Eq. (1) to experimental P, T data.

The usefulness of the thermodynamic equation then depends on the accuracy with which the last three "correction" terms in Eq.(1) can be obtained from experimental data on the virial coefficients of the vapour and the heat capacity and density of the liquid. The magnitudes of the correction terms, and of their uncertainties, increase rapidly with increasing temperature (see, e.g., Ref. 10, Table 1).

In the case of ^{4}He, Eq. (1) may be used for the calculation of T up to 1.5 K with an accuracy of 0.2 mK and up to T_λ with an accuracy of 1 mK, once L_o has been chosen. It may be noted that, since possible errors in the virial coefficients and the liquid heat capacity and volume are expected to vary smoothly with the temperature, a vapour pressure equation for ^{4}He calculated from Eq. (1) will, at least up to about 2.2 K, be to a high degree free from irregularities. Also, the thermodynamic equation will give the required shape of the vapour pressure curve near T_λ, with $d^2\ln P/dT^2$ going to minus infinity, better than an experimental P,T relation (see below). Because of the increasing uncertainties in the correction terms, the experimental P,T relation is more reliable than the calculated one in the range above T_λ.

In the case of ^{3}He, the contributions of the correction terms $\varepsilon(T)$ and $\int V_\ell \, dP$ are comparable to those for ^{4}He, but the contribution of the liquid entropy term is as large as 0.1 K even at 1 K (see Ref. 10, Table 1). Thus the use of experimental P,T data is to be preferred for deriving the vapour equation of ^{3}He down to as low a temperature as possible. At the time that the 1962 ^{3}He Scale was derived, however, accurate experimental P,T data for ^{3}He below 1 K did not exist and the authors of this scale were forced to use the thermodynamic equation for extrapolation below this temperature. For this purpose Eq. (1) was written in the form [3]

$$\ln P = i_o - \frac{a}{RT} - \frac{b}{R} + \frac{5}{2} \ln T$$

$$- \frac{1}{RT} \int_{T_m}^{T} dT' \int_{T_m}^{T'} \frac{C_\ell}{T''} d T'' + \frac{1}{RT} \int_0^P V_\ell \, dP + \varepsilon(T) \qquad (2)$$

where

$$a = L_o - \int_0^{T_m} C_\ell \, dT \qquad (3)$$

and

$$b = S_\ell (T_m) \qquad (4)$$

C_ℓ is the molar heat capacity of the liquid at saturation. T_m can be chosen arbitrarily (for $T_m = 0$, Eq. (2) is identical to Eq. (1)). The advantage of using Eq. (2) instead of Eq. (1) is, that experimental data for the heat capacity of the liquid are required only in the temperature range T_m to T where the equation is used; the disadvantage, of course, is the occurrence of two adjustable parameters (a and b) instead of one (L_o).

THE 1958 ^{4}He SCALE

The 1958 ^{4}He Scale was established after rather extensive discussions in which two scales, which differed by a few millikelvin, played a role. Only the basic principle of the scale will be explained here.

In Fig. 2 gas thermometer data available in 1958 are shown, together with calculated vapour pressure equations using Eq. (1) for various values of L_o. For

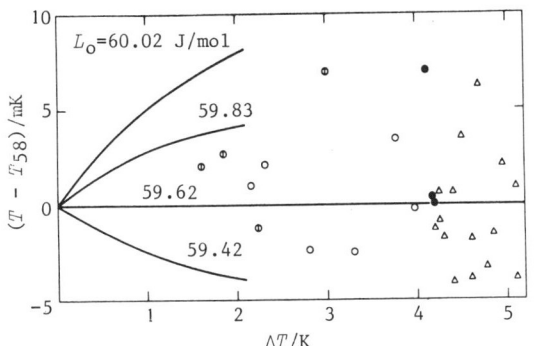

Fig. 2. Primary thermometer data available in 1958 (O Schmidt and Keesom [15], Δ Berman and Swenson [16] O Keller (^{4}He)[17] O Keller (^{3}He)[17]), and thermodynamic calculations of ^{4}He vapour pressure relations for various values of L_o, compared with T_{58}.

simplicity, and in order to avoid a discussion of the earlier helium scales, T_{58} itself is used as a reference in the figure. The problem of deriving the best scale was solved in 1958 in three steps: Firstly, a best value for the normal boiling point was chosen; as is evident from Fig. 2 much weight was given to the two nearly coinciding data points of Schmidt and Keesom [15]. Secondly, a thermodynamic calculation was made using Eq. (1), L_o being chosen as 59.62 J/mole. Thirdly, magnetic thermometer data between 1.5 K and 5.2 K were used for obtaining a smooth scale. The choice of L_o was made on the basis of best agreement with the gas thermometer data and consistency of the calculated P,T relation with the experimental P,T relation derived from magnetic thermometry in the range where they overlap (1.5 K to 2.2 K)[18]. (As will be clear from comparison with T_{76}, a much better scale would have been obtained in 1958 had the upper single data point of Schmidt and Keesom been chosen as a basis for the normal boiling point!).

THE 1962 ^{3}He SCALE

A complete account of the derivation of the 1962 ^{3}He Scale is given in Refs. 2-4. Only the principle of the derivation will be given here.

In 1962, no accurate P,T data based on primary thermometers were available for ^{3}He and therefore the ^{3}He scale was based, in the first place, on the 1958 ^{4}He Scale through accurate comparisons of ^{3}He and ^{4}He vapour pressures between the critical point of ^{3}He (3.3 K) and the lower limit of accurate ^{4}He vapour pressure measurements (0.9 K). The thermodynamic equation, Eq. (2), was then used for obtaining the vapour pressure-temperature relation below 0.9 K, the constants a and b being determined from a fit to the P,T data between 0.9 K and 2 K.

DEVELOPMENTS BETWEEN 1962 AND 1970

In 1966 Plumb and Cataland [19] published results of their measurements with an acoustical thermometer, which showed that T_{58} was too low between 2 K and 4 K by about 0.2%. This results was confirmed by acoustical thermometer data of Grimsrud and Werntz [20] and gas thermometry by Rogers, Tainsh, Anderson and Swenson [21] (see for a comparison of the various data with T_{58}, e.g., Ref. 21, Fig. 7).

MEASUREMENTS AFTER 1970. THE EPT-76

Several magnetic thermometer experiments were made with the purpose of establishing a temperature scale between 1 K and 30 K. In most of these experiments only "strong" magnetic salts, like manganese ammonium sulphate and gadolinium sulphate, were used, in which case three or four adjustable constants occur in the relation between the measured quantity and T [7,22]. In one of the magnetic thermometer experiments, however, cerium magnesium nitrate was used as a magnetic salt between 0.9 K and 2.6 K and chromium methylammonium sulphate and manganese ammonium sulphate between 0.9 K and about 30 K[23]. This resulted in a magnetic temperature scale with only two adjustable constants. When these constants were determined by fitting the magnetic data to the average of the NBS acoustic scale, a magnetic scale was obtained which deviated approximately linearly from T_{58}, the magnetic temperature being 7 mK higher at 4.2 K than T_{58} [24].

The scale $T_{XAc'}$, which was finally derived from these magnetic measurements, played an important role in the establishment of the 1976 Provisional 0.5 K - 30 K Temperature Scale (EPT-76) above 4 K [7]. The present discussion of the EPT-76 between 0.5 K and 5 K, and of helium vapour pressure equations which are consistent with this scale, will, however, be based on the following experiments:

1) Measurements of thermodynamic temperatures with a gas thermometer by Berry [25], from which the NPL-75 scale between 2.6 K and 27 K was derived, and with a noise thermometer by Klein, Klempt and Storm [19] at six temperatures between 2.15 K and 4.2 K. The noise thermometer temperatures were directly compared with ^{4}He vapour pressures; results $T_{noise} - T_{58}$, taken from Ref. 26, are shown in Fig. 3. The NPL-75 scale was defined in terms of a set of calibrated rhodium-iron resistance thermometers, which were compared with ^{4}He vapour pressures between 2.6 K and 4.2 K by Berry [25] (see Fig. 3).

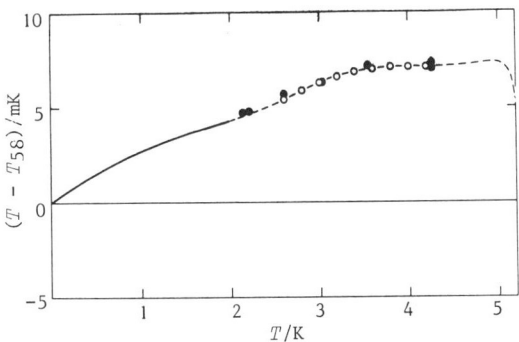

Fig. 3. Primary thermometer data available in 1976 (O Berry [25] gas thermometer, Klein et al. [9] noise thermometer) compared with T_{58}. Also shown are $T_{NPL-75} - T_{58}$ (2.6 K - 5.2 K) and $T_{X1} - T_{58}$ (2.0 K - 2.6 K) according to Rusby and Swenson, smoothed values from Table 2, Ref. 10 (dashed line), and the thermodynamic calculation with L_o = 59.83 J/mol (drawn curve).

2) A more extensive comparison of the NPL-75 scale with ^{4}He vapour pressures between 2.6 K and 5.2 K and with ^{3}He vapour pressures between 2.6 K and 3.3 K was made by Rusby and Swenson [10]. Their results for ^{4}He between 2.6 K and 4.2 K agreed within 0.1 mK with those of Berry. Smoothed differences $T_{NPL-75} - T_{58}$ taken from Table 2 of Ref. 10 are shown in Fig. 3. For the differences $T_{NPL-75} - T_{62}$ between 2.6 K and 3.3 K, see Table 2 and Fig. 9 of Ref. 10.

3) Rusby and Swenson also measured vapour pressures of ^{4}He and ^{3}He on a magnetic temperature scale, T_{X1}, based on cerium magnesium nitrate. The magnetic scale was defined between 3.1 K and 0.4 K. The two adjustable parameters in the scale were primarily obtained from a fit of the magnetic data to NPL-75 between 2.6 K and 3.1 K. This range for fitting was rather limited, but the consistency of the magnetic scale with the calculated ^{4}He vapour pressure equation (see below) between 1.4 K and 2.1 K (see Fig. 3, and, in more detail, Ref. 10, Fig. 8) and with the noise thermometry data of Klein et al. down to 2.15 K (see Fig. 3) gave confidence in the choice of the constants. Another important consideration was the agreement with the magnetic scale T_{X1SU} down to 0.95 K (see Ref. 10). For data $T_{X1} - T_{62}$ and for differences between T_{62} and T_{58} found by Rusby and Swenson see Ref. 10, Figs. 9 and 10, respectively.

Later on, a determination of the superconductive transition point of cadmium (0.52 K) with a noise thermometer by Soulen [7] agreed with the value of this transition point on T_{X1} within 1 mK and ^{3}He vapour pressures measured by Fisher and Brodale between 0.8 K and 1.5 K using a magnetothermodynamic method for determing temperatures, agreed with the Rusby Swenson data for $T_{X1} - T_{62}$ within about 0.5 mK [27].

4) New calculations of the vapour pressure-temperature relation of ^{4}He below 2.15 K, using Eq. (1) with best available data for the second virial coefficient of the vapour and the heat capacity and density of the liquid, were made by Van Dijk et al. [26] With $L_0 = 59.83$ J/mol the calculated relation was consistent with the experimental data for $T_{X1} - T_{58}$ between about 1.5 K and 2.1 K see Ref. 10 and Fig. 3.

5) Measurements of the vapour pressure of ^{3}He as a function of T_{X1} between 0.5 K and 1.8 K were made by El Samahy [28]. Temperatures T_{X1} were measured with calibrated rhodium-iron thermometers from NPL. These measurements were made in order to solve discrepancies between vapour pressure data and the T_{X1} scale below 0.8 K encountered by Rusby and Swenson [10]. At the lower temperatures the data of El Samahy for the 13 mm diameter vapour pressure sensing tube are consistent with an extrapolation of the vapour pressure-temperature relation above 1 K with Eq. (2), (see Fig. 4). At higher temperatures there is excellent agreement between the data of Rusby and Swenson and of El Samahy.

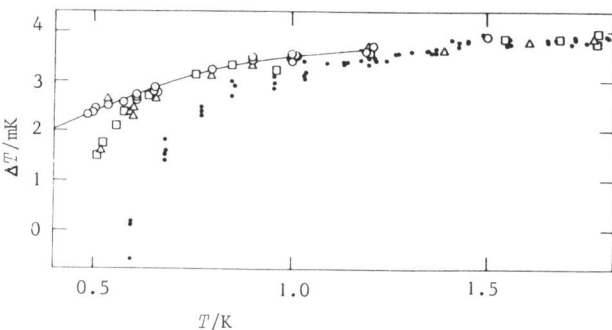

Fig. 4. Differences $T_{X1} - T_{62}$. T_{X1} is derived from rhodium iron thermometers calibrated on the T_{X1}-scale. T_{62} is derived from measurements of ^{3}He vapour pressures. ● Rusby and Swenson, 5 mm diameter vapour pressure sensing tube; △,□ El Samahy, 9 mm dimater tube; O El Samahy, 13 mm diameter tube. The drawn curve represents a thermodynamic extrapolation of the vapour pressure curve above 1 K. The figure is taken from Ref. 28.

INPUT DATA USED FOR DERIVING NEW ^{4}He and ^{3}He VAPOUR PRESSURE EQUATIONS ON THE EPT-76

For ^{4}He experimental P vs T_{X1} data of Rusby and Swenson [10] above the λ-point and the calculated vapour pressure relation of Van Dijk et al. between 0.45 K and 2.15 K with $L_0 = 59.83$ J/mol were used as input data for deriving the new equations. The value 59.83 J/mol was determined from a fit of the calculated vapour pressure relation to the Rusby and Swenson data between about 1.4 K and 2.1 K. The selected smooth data, which agree closely with $T - T_{58}$ values in Table 2 of Ref. 10, are given in Tables III and IV. Values $P_\lambda = 5041.8$ Pa, $T_\lambda = 2.1768$ K are taken from Ref. 10.

For ^{3}He experimental P vs T_{X1} data of Rusby and Swenson at the higher temperatures, and of El Samahy, at the lower temperatures, were used. The latter data were smoothed and extrapolated to 0.2 K by using Eq. (2). The selected smooth data, which correspond above 1.5 K to the $T - T_{62}$ values in Table 2 of Ref. 10, are given in Table V.

Table III
Input data used for fitting of ^{4}He vapour pressure equation below T_λ

T/K	$\ln(P/Pa)$
2.17680000E+00	8.52551840E+00
2.15000000E+00	8.45857200E+00
2.10000000E+00	8.32848500E+00
2.05000000E+00	8.19181400E+00
2.00000000E+00	8.04841600E+00
1.95000000E+00	7.89804300E+00
1.90000000E+00	7.74038000E+00
1.85000000E+00	7.57504700E+00
1.80000000E+00	7.40159600E+00
1.75000000E+00	7.21951500E+00
1.70000000E+00	7.02821500E+00
1.65000000E+00	6.82702500E+00
1.60000000E+00	6.61517800E+00
1.55000000E+00	6.39180200E+00
1.50000000E+00	6.15590300E+00
1.45000000E+00	5.90634600E+00
1.40000000E+00	5.64182900E+00
1.35000000E+00	5.36085600E+00
1.30000000E+00	5.06170000E+00
1.25000000E+00	4.74236100E+00
1.20000000E+00	4.40050700E+00
1.15000000E+00	4.03341100E+00
1.10000000E+00	3.63785600E+00
1.05000000E+00	3.21002800E+00
1.00000000E+00	2.74537000E+00
9.50000000E-01	2.23828870E+00
9.00000000E-01	1.68239950E+00
8.50000000E-01	1.06918571E+00
8.00000000E-01	3.88527070E-01
7.50000000E-01	-3.72455460E-01
7.00000000E-01	-1.23023340E+00
6.50000000E-01	-2.20623070E+00
6.00000000E-01	-3.32885840E+00
5.50000000E-01	-4.63664740E+00
5.00000000E-01	-6.18324630E+00
4.50000000E-01	-8.04572000E+00

CHOICE OF VAPOUR PRESSURE EQUATIONS FOR ^{4}He and ^{3}He

It seemed useful that the equations to be derived should cover the range from 0.5 K to the critical temperature, T_c, for ^{4}He and from 0.2 K to T_c for ^{3}He, the same ranges as used for the 1958 and 1962 scales. Further, it was necessary to use separate equations for ^{4}He below and above T_λ.

The first equations tried were of the form

$$T = \Sigma_{i=1}^{n} c_i (\ln P)^i \qquad (5)$$

Table IV
Input data used for fitting of ^{4}He vapour pressure equation above T_λ

T/K	$\ln(P/\text{Pa})$
2.17680000E+00	8.52551840E+00
2.20463000E+00	8.59322000E+00
2.30473000E+00	8.82484000E+00
2.40495000E+00	9.04072000E+00
2.50518000E+00	9.24296000E+00
2.60542000E+00	9.43322000E+00
2.70567000E+00	9.61283000E+00
2.80591000E+00	9.78276000E+00
2.90613000E+00	9.94393000E+00
3.00633000E+00	1.00971800E+01
3.10650000E+00	1.02432600E+01
3.20664000E+00	1.03828100E+01
3.30676000E+00	1.05163500E+01
3.40686000E+00	1.06443800E+01
3.50694000E+00	1.07672600E+01
3.60700000E+00	1.08854100E+01
3.70704000E+00	1.09991320E+01
3.80707000E+00	1.11087500E+01
3.90709000E+00	1.12145700E+01
4.00710000E+00	1.13168900E+01
4.10710000E+00	1.14159500E+01
4.20710000E+00	1.15119500E+01
4.22210000E+00	1.15260900E+01
4.30710000E+00	1.16050100E+01
4.40710000E+00	1.16954500E+01
4.50712000E+00	1.17834500E+01
4.60717000E+00	1.18691500E+01
4.70723000E+00	1.19526800E+01
4.80729000E+00	1.20341800E+01
4.90732000E+00	1.21137800E+01
5.00732000E+00	1.21915700E+01
5.00732000E+00	1.21915700E+01
5.10705000E+00	1.22676700E+01
5.13180000E+00	1.22864400E+01
5.17580000E+00	1.23199800E+01
5.19440000E+00	1.23340400E+01
5.19530000E+00	1.23347440E+01

Table V
Input data used for fitting of ^{3}He vapour pressure equation

T/K	$\ln(P/\text{Pa})$
3.30625000E+00	1.16396600E+01
3.25638000E+00	1.15882600E+01
3.19687000E+00	1.15260900E+01
3.10627000E+00	1.14293900E+01
3.00606000E+00	1.13191100E+01
2.90584000E+00	1.12048700E+01
2.80562000E+00	1.10862500E+01
2.70540000E+00	1.09627800E+01
2.60517000E+00	1.08339500E+01
2.50495000E+00	1.06991800E+01
2.40473000E+00	1.05578600E+01
2.30452000E+00	1.04092500E+01
2.20434000E+00	1.02525300E+01
2.10418000E+00	1.00867500E+01
2.00405000E+00	9.91079000E+00
1.90395000E+00	9.72334000E+00
1.80388000E+00	9.52279000E+00
1.70383000E+00	9.30724000E+00
1.60377000E+00	9.07432000E+00
1.50371000E+00	8.82109000E+00
1.40365000E+00	8.54384000E+00
1.30358000E+00	8.23776000E+00
1.20352000E+00	7.89661000E+00
1.10346000E+00	7.51198000E+00
1.00338000E+00	7.07233000E+00
9.03260000E-01	6.56131000E+00
8.03100000E-01	5.95489000E+00
7.02910000E-01	5.21614000E+00
6.02660000E-01	4.28486000E+00
5.02380000E-01	3.05533000E+00
4.52225000E-01	2.27116000E+00
4.02070000E-01	1.32133000E+00
3.51890000E-01	1.39040000E-01
3.21760000E-01	-7.24930000E-01
3.01680000E-01	-1.38550000E+00
2.81600000E-01	-2.13023000E+00
2.61520000E-01	-2.97762000E+00
2.41420000E-01	-3.95223000E+00
2.21320000E-01	-5.08775000E+00
2.11280000E-01	-5.72961000E+00
2.01230000E-01	-6.43050000E+00

These would allow easy calculations of temperatures from measured vapour pressures, but it soon appeared that they would require several more coefficients than equations of the more thermodynamic form

$$\ln P = c_1 \ln T + \Sigma_{i=2}^{n} c_i T^{i-3} \quad (6)$$

A next major question was, whether the equations should represent the logarithmic singularities at T_λ and T_c, $d^2 \ln P/d T^2$ being minus infinity at T_λ and plus infinity at T_c, which would require, of course, extra terms to be added to Eq. (6), or whether one should keep the vapour pressure equations as simple as possible for practical use, taking a slight loss of consistency with thermodynamics near T_λ and T_c for granted.

At first, least squares fits of the data for ^{4}He below and above T_λ and for ^{3}He in Tables III, IV and V were made with simple equations of the type of Eq. (6). For ^{4}He, the equations below T_λ and above T_λ were constrained to the values $P_\lambda = 5041.8$ Pa at $T_\lambda = 2.1768$ K, in order to avoid a discontinuity in P at T_λ. In these fits, and in all fits to be discussed later, a weight factor $dT/d \ln P$ was used. For ^{4}He above T_λ the EPT-76 value for the normal boiling point (4.2221 K) was used as a constraint. Residuals of the fits are shown in Figs. 5, 6 and 7. Maximum residuals are, e.g., 0.06 mK for a 7 parameter fit for ^{4}He from 0.5 K to T_λ, 0.24 mK for a 9 parameter fit above T_λ and 0.08 mK for a 7 parameter fit for ^{3}He between 0.2 K and 3.3 K. For ^{3}He this result is not surprising since the 1962 Scale was defined with a 7 parameter equation of the type of Eq. (6) and the present input data deviate from T_{62} in a smooth way.

For the ^{4}He equations the derivatives $d \ln P/dT$ at T_λ were, in general, different for the equations below and above T_λ. For example for the 7 parameter fit below T_λ $(d \ln P/dT)_\lambda = 2.468$ K^{-1} and for the 9 parameter fit above T_λ $(d \ln P/dT)_\lambda = 2.446$ K^{-1}. This was considered less desirable and, subsequently, fits were made with the value $(d \ln P/dT)_\lambda = 2.461$ K^{-1} as an additional constraint, the latter value being chosen on the basis of a thermodynamic calculation. Below T_λ an equation of the type

$$\ln P = \Sigma_{i=1}^{n} c_i T^{i-2} \quad (7)$$

was used, because it appeared that it yielded, for the same number of parameters, slightly better fits than Eq. (6). Above T_λ an equation earlier used by McCarthy[29] was used

$$\ln P = c_1 (1-\frac{T}{T_c})^{1.9} + \Sigma_{i=2}^{n} c_i (\frac{T}{T_c})^{i-3} \quad (8)$$

which had the advantage of representing correctly the thermodynamic behaviour of the vapour pressure near T_c. For the critical pressure of ^{4}He Kierstead gives the value $P_c = 1706.12 \pm 0.10$ Torr (227.463 kPa) which corresponds to $T_{58} = 5.18992$ K. With the difference $T_{76} - T_{58} = 5.4$ mK[10]) this yields $T_{76} = 5.1953$ K. These

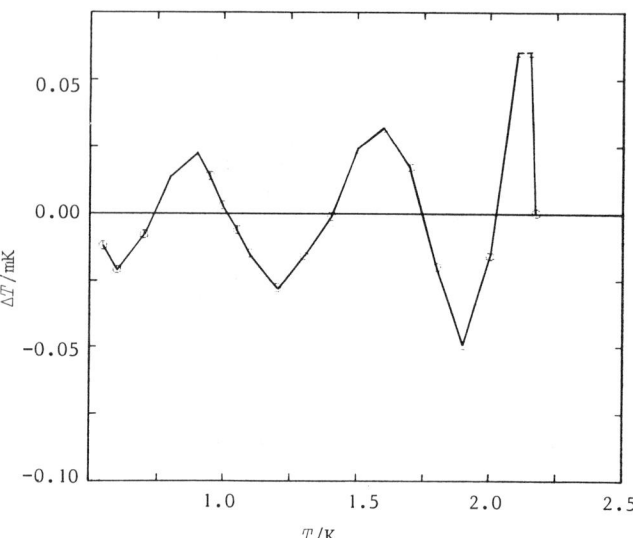

Fig. 5. Residuals $\Delta T = T_{calc} - T_{data}$ for a fit of the ^{4}He data below T_λ with Eq. (6) with 7 parameters. No constraint in $(d \ln P/dT)_\lambda$.

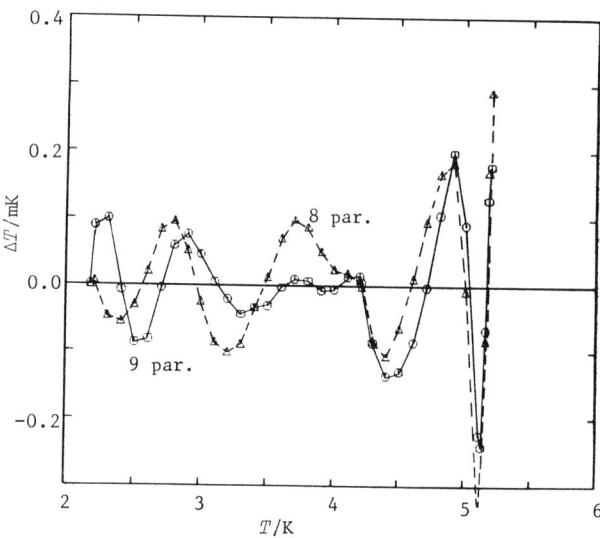

Fig. 6. Residuals $\Delta T = T_{calc} - T_{data}$ for fits of the ^{4}He data above T_λ with Eq. (6) with 8 and 9 parameters. No constraint in $(d \ln P/dT)_\lambda$.

values for P_c and T_c were used as a constraint in the fit. Residuals of the fits are shown in Figs. 8 and 9. It can be seen that the residuals are for the same number of parameters, about a factor of 2 larger near T_λ than for the fits without the constraint in the derivative at T_λ.

In the case of ^{3}He, fits without the $\ln T$ term gave slightly larger residuals than with this term included. Also, inclusion of a term $(1-(T/T_c))^{1.9}$ caused several more coefficients in the expansion to be required compared with Eq. (6). Both effects are probably due to the fact that the ^{3}He equation has to cover a range of a factor of about 16 in temperature.

In Fig. 10 differences T_{76} (^{4}He)$-T_{58}$ in the range 1.6 K to 2.8 K are plotted for representations of T_{76} (^{4}He) with Eq. (7), 8 parameter fit, below T_λ and with Eqs. (8), 10 and 11 parameter fits, above T_λ. It can be seen that the differences show a small kink at T_λ, which is the more pronounced above T_λ.

This is not due to a thermodynamic inconsistency in T_{58} but to the inability of the simple equations to represent the true shape of the vapour pressure curve near T_λ; this was checked by new thermodynamic calculations.

Hereafter, an attempt was made to use equations, which were still relatively simple, but which would correctly represent the behaviour of the vapour pressure near the λ-point and the critical points. Such equations were indeed found but, unfortunately, they were rather complex making them less suitable for the

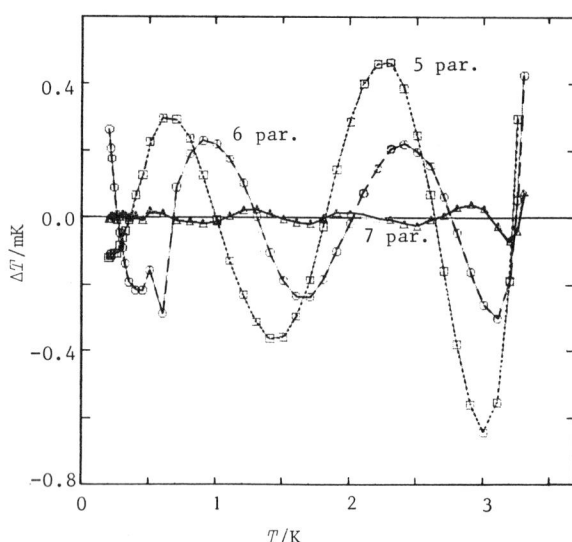

Fig. 7. Residuals $\Delta T = T_{calc} - T_{data}$ for fits of the ^{3}He data with Eq. (6) with 5, 6 and 7 parameters.

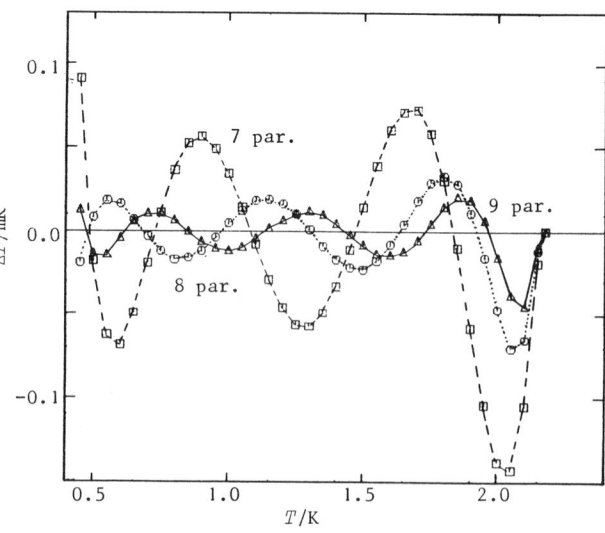

Fig. 8. Residuals $\Delta T = T_{calc} - T_{data}$ for fits of the ^{4}He data below T_λ with Eq. (7) with 7, 8 and 9 parameters $(d \ln P/dT)_\lambda = 2.461$ K^{-1}.

definition of temperature scales. Therefore, these equations will not be discussed in detail here, but only their principle and the results of the fits will be given.

For ^{4}He below the λ-point a term $F_\lambda(T)$ representing the effect of the anomaly in the liquid heat capacity at T_λ could easily be found and a fit to the input P, T data was made using the equation

$$\ln P - F_\lambda(T) = \Sigma_{i=1}^{n} c_i T^{i-2} \qquad (9)$$

The corresponding differences $T_{76}(^4\text{He})-T_{58}$ between 1.6 K and T_λ for the 7 parameter fit are given in Fig. 10. As could be expected, the slight irregularity in the $T_{76}(^4\text{He})-T_{58}$ curve found before has disappeared.

For the range above T_λ, several approaches were tried, but a term of the type of $F_\lambda(T)$ which represents the singularity at T_λ and at the same time does not disturb the fit at higher temperatures was not readily found. In connection with this difficulty, a different approach was used for taking the singularities at T_λ, as well as at T_c, into account. The vapour pressure was represented by the following expression

$$\ln P = (\Delta \ln P)_\lambda + (\Delta \ln P)_c + \Sigma_{i=2}^{n} c_i \left(\frac{T}{T_c}\right)^{i-3} \qquad (10)$$

where $(\Delta \ln P)_\lambda$ and $(\Delta \ln P)_c$ are small terms to be added to the simple expansion only in small temperature intervals near T_λ and T_c respectively. For the term $(\Delta \ln P)_\lambda$ an expression could be derived from the liquid heat capacity. The term $(\Delta \ln P)_c$ was written as $c_1 F_c(T)$ in which $F_c(T)$ is a specific function which contains a term $(1-(T/T_c))^{1.9}$. The coefficient c_1 was determined, together with the other coefficients in Eq. (10), in the least squares fit. Results of the fit between T_λ and 2.8 K are shown in Fig. 10. Again, as expected, the slight anomaly at T_λ has disappeared.

A similar fit with a small $(\Delta \ln P)_c$ term was used for ^{3}He.

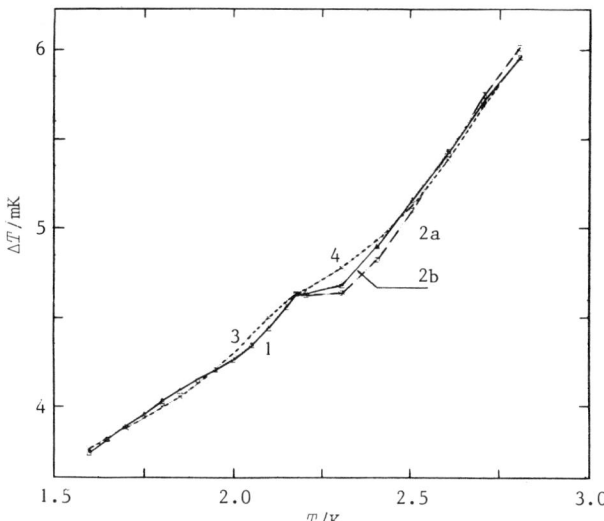

Fig. 10. $T_{76}(^4\text{He})-T_{58}$ for various representations of the ^{4}He vapour pressure equation on EPT-76 (1.6 K - 2.8 K).
1 below T_λ, Eq. (7), 8 parameters
2 above T_λ, Eq. (8), 10 parameters (2a) and 11 parameters (2b)
3 below T_λ, Eq. (9), 7 parameters
4 above T_λ, Eq. (10), 10 parameters
In all cases (d ln $P/dT)_\lambda$ is constrained at 2.461 K^{-1}. Eq. (7) with 8 parameters (curve 1) and Eq. (8) with 11 parameters (curve 2b) are proposed as the new ^{4}He scale on EPT-76.

The equation used in this case was

$$\ln P = (\Delta \ln P)_c + c_2 \ln T + \Sigma_{i=3}^{n} c_i T^{i-4} \qquad (11)$$

with $(\Delta \ln P)_c = c_1 F_c(T)$.

In Fig. 11 the magnitudes of the terms $(\Delta \ln P)_\lambda$ and $(\Delta \ln P)_c$ are shown. From the point of view of physics, Eqs. (10) and (11) are attractive for representing vapour pressure curves because the specific T_λ and T_c-terms are used only in regions close to these temperatures, where they have a physical meaning (in Eq. (8) the term $c_1(1-T/T_c)^{1.9}$ becomes large at temperatures far below T_c). Nevertheless, from the point of view of defining practical temperature scales, the slight improvement of the fits near T_λ and $T_{c\lambda}$ with Eqs. (9), (10) and (11) does not outweigh the increased complexity of the equations. They may, however, be useful in cases where a knowledge of the precise shapes of vapour pressure curves near T_λ or T_c are required. In the present context they show the magnitude of the errors made when fitting the data with the simple equations.

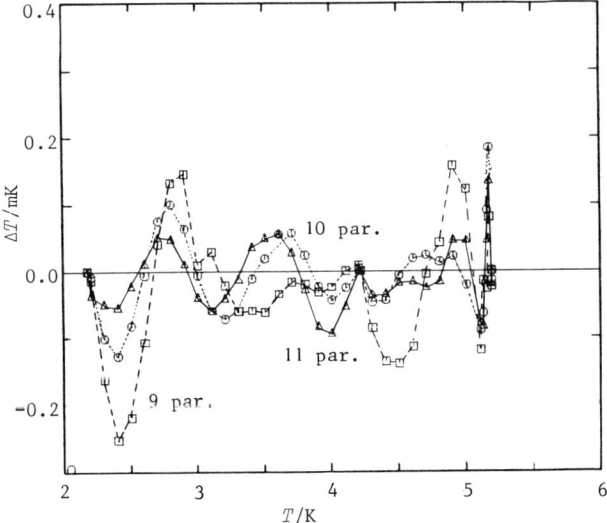

Fig. 9. Residuals $\Delta T = T_{calc} - T_{data}$ for a fit of the ^{4}He data above T_λ with Eq. (8) with 9, 10 and 11 parameters. (d ln $P/dT)_\lambda = 2.461$ K^{-1}.

CONCLUDING REMARKS ON THE VAPOUR PRESSURE EQUATIONS

In our opinion Eq. (7) with 8 parameters for ^{4}He between 0.5 K and 2.1768 K, Eq. (8) with 11 parameters for ^{4}He between 2.1768 K and 5.1953 K and Eq. (6) with 7 parameters for ^{3}He between 0.2 K and 3.3158 K are good compromises between accuracy and simplicity. Coefficients are given in Table VI; pressures and temperatures at the λ-point, boiling points and critical points are collected in Table VII. A proposal that these equations be adopted as the definitions of $T_{76}(^4\text{He})$ and $T_{76}(^3\text{He})$ has been sent to the CCT. In Fig. 12 corre-

Table VI
Coefficients of the vapour pressure equations

^{4}He 0.5 K to 2.1768 K, coefficients in Eq. (7)			^{4}He 2.1768 K to T_c, coefficients in Eq. (8)	
c_1	= −7.418 16	K	c_1 =	14.533 33
c_2	= 5.421 28		c_2 =	−30.932 85
c_3	= 9.903 203	K^{-1}	c_3 =	392.473 61
c_4	= −9.617 095	K^{-2}	c_4 =	−2 328.045 87
c_5	= 6.804 602	K^{-3}	c_5 =	8 111.303 47
c_6	= −3.015 460 6	K^{-4}	c_6 =	−17 809.809 01
c_7	= 0.746 135 7	K^{-5}	c_7 =	25 766.527 47
c_8	= −0.079 179 1	K^{-6}	c_8 =	−24 601.4
			c_9 =	14 944.651 42
			c_{10} =	−5 240.365 18
			c_{11} =	807.931 68
			T_c =	5.1953 K

^{3}He 0.2 K to T_c, coefficients in Eq. (6)		
c_1 =	2.254 84	
c_2 =	−2.509 43	K
c_3 =	9.708 76	
c_4 =	−0.304 433	K^{-1}
c_5 =	0.210 429	K^{-2}
c_6 =	−0.054 514 5	K^{-3}
c_7 =	0.005 606 7	K^{-4}

(For $\ln P$ in Eqs. (7), (8) and (9) read $\ln(P/\text{Pa})$)

Table VII
λ-point of ^{4}He and boiling points and critical points of ^{4}He and ^{3}He

λ-point ^{4}He	P = 5041.8 Pa	T_{76} (^{4}He) = 2.1768 K
b.p. ^{4}He	P = 101.325 kPa	T_{76} (^{4}He) = 4.2221 K
c.p. ^{4}He	P = 227.463 kPa	T_{76} (^{4}He) = 5.1953 K
b.p. ^{3}He	P = 101.325 kPa	T_{76} (^{3}He) = 3.1968 K
c.p. ^{3}He	P = 114.66 kPa	T_{76} (^{3}He) = 3.3162 K

The data for the λ-point, boiling point and critical point of ^{4}He were used as constraints in the fits (see text). P_{crit} (^{3}He) is taken from Behringer et al. (P_{crit} = 860 Torr)[31].

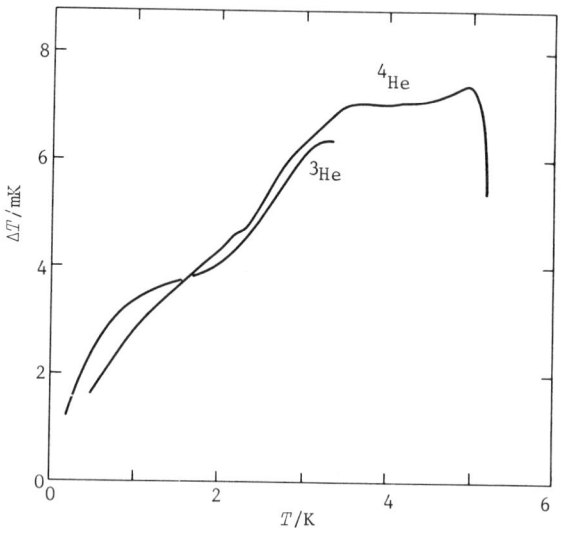

Fig. 12. T_{76} (^{4}He) − T_{58} and T_{76} (^{3}He) − T_{62} with T_{76} (^{4}He) and T_{76} (^{3}He) as proposed in this paper.
T_{76} (^{4}He) : Eq. (7), 8 parameters, Eq. (8) 11 parameters
T_{76} (^{3}He) : Eq. (6), 7 parameters

sponding differences T_{76} (^{4}He) − T_{58} and T_{76} (^{3}He) − T_{62} are shown. The differences between the two curves represent the differences between T_{62} and T_{58} found by Rusby and Swenson [10]. On the basis of these experiments it is expected that T_{76} (^{4}He) and T_{76} (^{3}He) as proposed here agree within 0.2 mK in the range where they overlap.

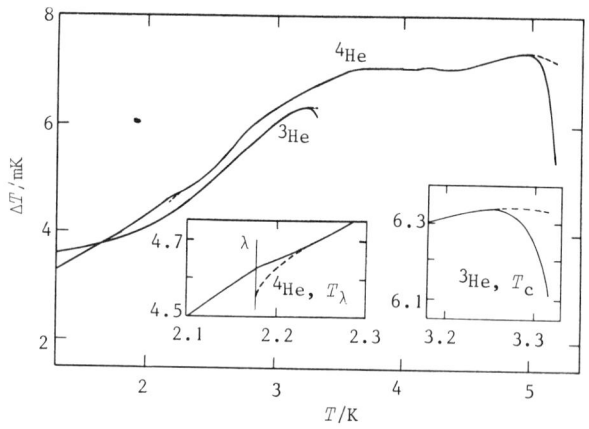

Fig. 11. Representation of T_{76} (^{4}He) and T_{76} (^{3}He) with the complex equations.
^{4}He below T_λ : Eq. (9), 7 parameters
^{4}He above T_λ : Eq. (10), 10 parameters
^{3}He : Eq. (11), 8 parameters
The dashed curves in the figure, and inserts, indicate temperatures according to $\ln P - (\Delta \ln P)_\lambda$ or $\ln P - (\Delta \ln P)_c$.

REFERENCES
1. The 1958 ^{4}He Scale of Temperatures, F.G. Brickwedde, H. van Dijk, M. Durieux, J.R. Clement and J.K. Logan, J. Res. Natl. Bur. Stand. 64A (1960) 1.
2. S.G. Sydoriak and R.H. Sherman, J. Res. Natl. Bur. Stand. 68A (1964) 547.
3. S.G. Sydoriak, T.R. Roberts and R.H. Sherman, J. Res. Natl. Bur. Stand. 68A (1964) 559.
4. T.R. Roberts, R.H. Sherman and S.G. Sydoriak, J. Res. Natl. Bur. Stand. 68A (1964) 567.
5. Comité Consultatif de Thermométrie, 12e session, Sèvres, 1978, p. T41.
6. The 1976 Provisional 0.5 K to 30 K Temperature Scale, Metrologia 15 (1979) 65.
7. M. Durieux, D.N. Astrov, W.R.G. Kemp and C.A. Swenson, Metrologia 15 (1979) 57.
8. Comité Consultatif de Thermométrie, 13e session, Sèvres, 1980.
9. H.H. Klein, G. Klempt and L. Storm, Metrologia 15 (1979) 143.
10. R.L. Rusby and C.A. Swenson, Metrologia 16 (1980) 73.
11. T.R. Roberts and S.G. Sydoriak, Phys. Rev. 102 (1956) 304.
12. H. van Dijk and D. Shoenberg, Nature 164 (1949) 151.
13. H. van Dijk and M. Durieux, Progr. in Low Temp. Phys. II (Ed. C.J. Gorter, North Holland Publ. Comp., Amsterdam, 1957) Ch. XIV.
14. H. van Dijk and M. Durieux, Physica 24 (1958) 1.
15. G. Schmidt and W.H. Keesom, Physica 4 (1937) 963.

16. R. Berman and C.A. Swenson, Phys. Rev. $\underline{95}$ (1954) 311.
17. W.E. Keller, Phys. Rev. $\underline{97}$ (1955) 1, $\underline{98}$ (1955) 1571 and $\underline{100}$ (1955) 1790.
18. H. van Dijk and M. Durieux, Physica $\underline{24}$ (1958) 920.
19. H.H. Plumb and G. Cataland, Metrologia $\underline{2}$ (1966) 127.
20. D.T. Grimsrud and J.H. Werntz, Phys. Rev. $\underline{157}$ (1967) 181.
21. J.S. Rogers, R.J. Tainsh, M.S. Anderson and C.A. Swenson, Metrologia $\underline{4}$ (1968) 47.
22. L.M. Besley and W.R.G. Kemp, Metrologia $\underline{13}$ (1977) 35.
23. T.C. Cetas and C.A. Swenson, Metrologia $\underline{8}$ (1972) 46.
24. C.A. Swenson, Metrologia $\underline{9}$ (1973) 99.
25. K. H. Berry, Metrologia $\underline{15}$ (1979) 89.
26. J.E. van Dijk, H. ter Harmsel and M. Durieux, Comité Consultatif de Thermométrie, 11e session, 1976, Doc. 9, and private comm. from J.E. van Dijk.
27. R.A. Fisher and G.E. Brodale, Proc. 16th Internat. Conf. on Low Temp. Phys., Los Angeles, 1981 (Physica B, to be published).
28. A.E. El Samahy, Thesis, University of Leiden (1979).
29. R.D. McCarty, J. Phys. and Chem. Ref. Data $\underline{2}$ (1973) 923.
30. H.A. Kierstead, Phys. Rev. $\underline{A3}$ (1971) 329.
31. R.P. Behringer, T. Doiron and H. Meyer, J. Low Temp. Phys. $\underline{24}$ (1976) 315.

Fixed point combination and termination points for platinum resistance thermometer interpolation below 273.15 K

R. C. Kemp

CSIRO Division of Applied Physics, Sydney, Australia 2070

A number of alternative fixed point combinations and lower temperature limits for platinum resistance thermometer interpolation below 273.15 K have been investigated. The results show that while it is possible to replace the boiling points of neon and hydrogen with the triple point of neon, the β-γ transition in oxygen and the deuterium triple point, there is some sacrifice of either the quality or the range of interpolation unless the helium boiling point is included. It is concluded that, pending further investigation of the deuterium triple point and the β-γ transition in oxygen, the hydrogen boiling points are best retained.

INTRODUCTION

In considering proposals for a replacement for the International Practical Temperature Scale (IPTS) in the region below 273.15 K it is worth speculating whether other sets of defining fixed points might provide equally good or better interpolation and easier realization. It would certainly be advantageous if boiling points could be avoided or relegated to the role of secondary points and replaced with triple points or solid phase transitions thus eliminating pressure measurements.

In addition, an alternative termination point might be more appropriate for a new set of defining fixed points than the existing one at 13.81 K. Any extension of the IPTS below 13.81 K may require a new interpolation instrument, for example a rhodium-iron thermometer, and it is profitable to consider alternative termination points for platinum resistance thermometer (PRT) interpolation. Various interpolation schemes have been discussed in the past by Kirby et al.[1], Ward[2] and Kemp et al.[3]. Kirby et al. and Ward showed that the differences ΔW between PRT resistance ratios, W, could be fitted using least squares techniques using expressions in W.

Kirby et al. used the expression

$$\Delta W = a(W - 1) + \sum_{i=1}^{5} b(\log_e W)^i \quad (1)$$

and determined the coefficients using the methods of least squares to fit the deviation at the seven fixed points of IPTS-68 plus a xenon point at 164 K.

Ward used expression (1) and also

$$\Delta W = a(W - 1) + b(W - 1)\log_e W + \sum_{i=1}^{4} C_i (\log_e W)^i \quad (2)$$

and

$$\Delta W = a(W - 1) + b(W - 1)^2 + \sum_{i=1}^{4} C_i (\log_e W)^i \quad (3)$$

The coefficients a, b, C_i were determined using least squares fitting weighted more heavily at the lower temperatures and not including the point at 164 K. Kemp et al. used these and other similar expressions but determined the coefficients by exact methods rather than by least squares.

Ward and Rusby[4] used expressions (1)-(3) and the method of least squares to investigate the use of three different combinations of fixed points as follows: (a) the usual fixed points with the neon boiling point replaced with the neon triple point, (b) further replacement of the two hydrogen boiling points by the deuterium triple point and the β-γ transition in solid oxygen and (c) terminating the interpolation procedure at the neon triple point.

RESULTS AND DISCUSSION

There are obviously many possible ways of combining the various interpolation procedures with different combinations of fixed points. We have investigated over fifty of these using both the method of least squares fitting favoured by Ward and the exact methods favoured by Kemp et al.[3]. It is impractical and perhaps unnecessary to discuss all the alternatives that we have investigated. For the present purposes we will restrict the discussion to the use of the expression

$$\Delta W = a(W_{ref} - 1) + b(W_{ref} - 1)^2 + \sum_{i=1}^{n} C_i (\log_e W_{ref})^i$$

with different fixed point combination and using both least squares fitting and exact methods to determine the coefficients a, b, C_i. For convenience we have set W_{ref} equal to $W_{Pt,NML}$, the reference function given by Kemp et al.[5] which is based on a real PRT.

This expression will be applied to the intercomparison data obtained by Ward and Compton[6] for 35 capsule PRTs. As a bench-mark or reference for our discussion on the quality of the interpolation methods, we show in Figs 1 and 2 the results of applying equation (3) to this data using both the exact and least squares fitting methods and the usual fixed points. In Figs 1 and 2 we plot the difference be-

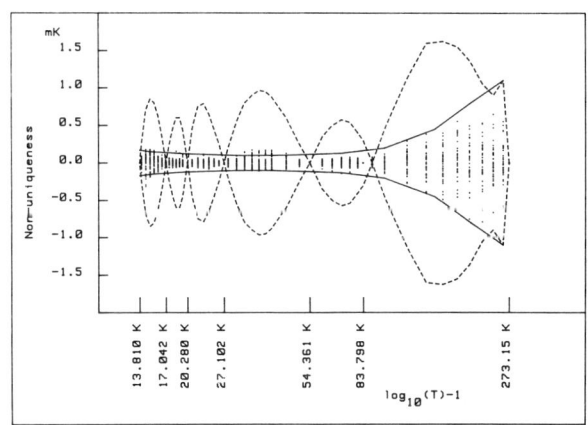

FIG. 1. Non-uniqueness of 35 PRTs for fixed points at 13.81 K, 17.042 K, 20.28 K, 27.102 K, 54.361 K, 83.798 K and with the coefficients of the deviation function determined by exact methods · non-uniqueness, --- envelope of IPTS-68 non-uniqueness, — experimental uncertainty.

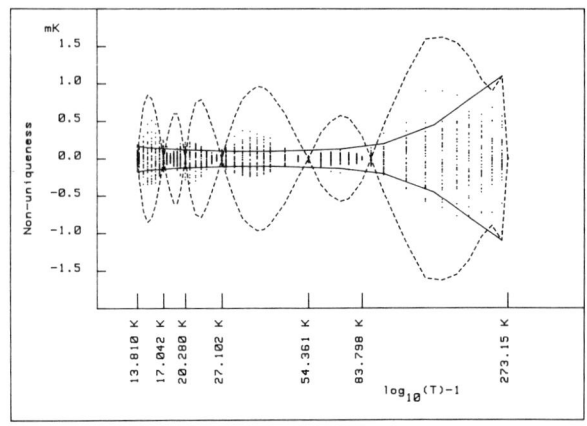

FIG. 2. Non-uniqueness of 35 PRTs for fixed points at 13.81 K, 17.042 K, 20.28 K, 27.102 K, 54.361 K, 83.798 K and with the coefficients of the deviation function determined by least squares fitting · non-uniqueness, --- envelope of IPTS-68 non-uniqueness, — experimental uncertainty.

tween individual thermometers, commonly called non-uniqueness, against temperature. The continuous lines in these and later plots represent the uncertainties of the intercomparison data as given by Ward and Compton[6].

The best interpolation is in our opinion shown in Fig. 1. Of the many alternatives studied we found only one which gave better interpolation. To achieve this we followed Kirby et al.[1] in adding the xenon triple point to the fitting procedure but used exact methods rather than least squares fitting. We also added the helium boiling point to the fitting procedure. The results of this exercise are shown in Fig. 3. The interpolation is better over the whole range than in Fig. 1. Excluding the helium boiling point from the fitting procedure makes the interpolation below 17 K marginally worse than in Fig. 1. This is because inserting a point at 160 K attempts, to some extent at least, to correct for errors of measurement as well as interpolation errors. For example, adding a fixed point at 240 K, where the errors of measurement are much greater than at 160 K, makes the non-uniqueness worse over the whole range by a factor of two.

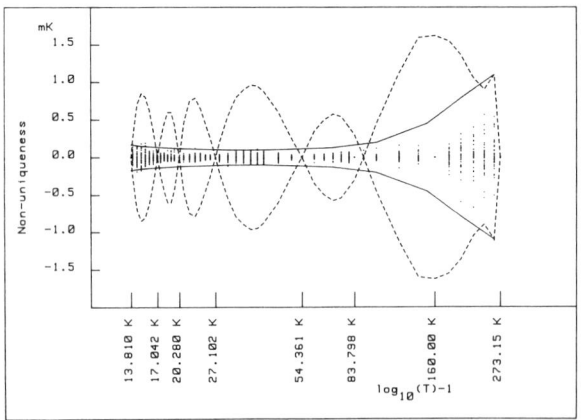

FIG. 3. Non-uniqueness of 35 PRTs for fixed points at 4.2 K, 13.81 K, 17.042 K, 20.28 K, 27.102 K, 54.361 K, 83.798 K, 160 K and with the coefficients of the deviation function determined by exact methods · non-uniqueness, --- envelope of IPTS-68 non-uniqueness, — experimental uncertainty.

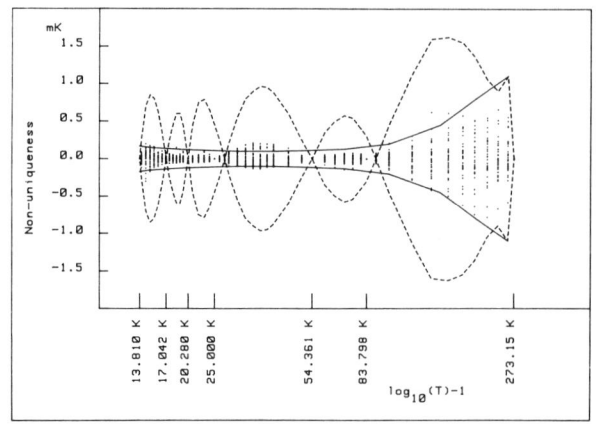

FIG. 4. Non-uniqueness of 35 PRTs for fixed points at 13.81 K, 17.042 K, 20.28 K, 25 K, 54.361 K, 83.798 K and with the coefficients of the deviation function determined by exact methods · non-uniqueness, --- envelope of IPTS-68 non-uniqueness, — experimental uncertainty.

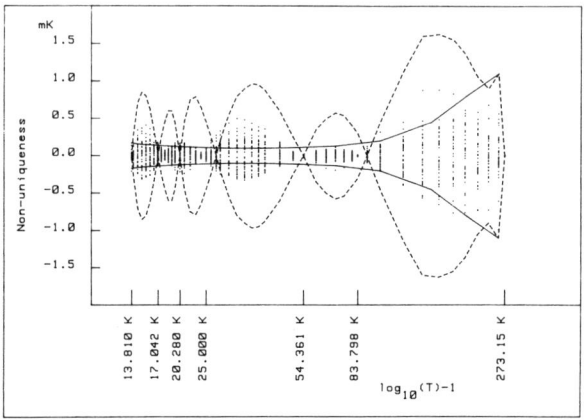

FIG. 5. Non-uniqueness of 35 PRTs for fixed points at 13.81 K, 17.042 K, 20.28 K, 25 K, 54.361 K, 83.798 K and with the coefficients of the deviation function determined by least squares fitting · non-uniqueness, --- envelope of IPTS-68 non-uniqueness, — experimental uncertainty.

Unfortunately the triple point of natural xenon is not really suitable as a fixed point as its melting range is very large because of the large numbers of xenon isotopes. However it is possible that the triple point of one of the less expensive xenon isotopes may be more suitable and work on this is underway in this Division.

We next attempted to replace some of the existing boiling points used as fixed points with triple points or solid phase transitions. The most obvious substitution is that of the neon triple point for the neon boiling point and the results of this substitution are shown in Figs 4 and 5 for exact and least squares fitting methods. A comparison of Figs 4 and 5 with 1 and 2 shows that there is little difference in interpolation when this substitution is made. The non-uniqueness is slightly, but not significantly, worse than when using the method of least squares. Evidently this substitution is perfectly feasible and must be seriously considered.

The next step involved replacing the two hydrogen boiling points at 17.042 K and 20.28 K with a single point at 18.5 K representing the deuterium triple point. The results of this step are shown in Figs 6 and 7 for exact and least squares fitting methods respectively.

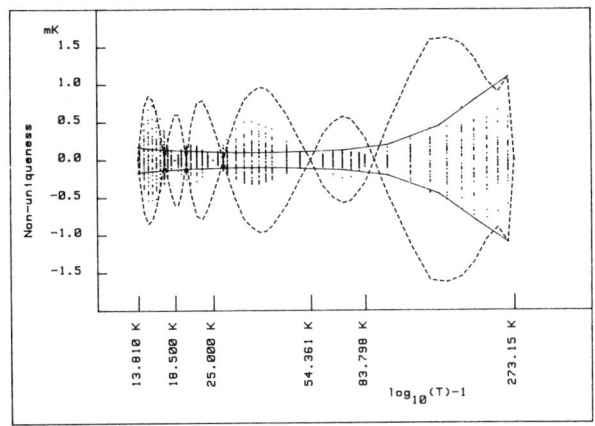

FIG. 6. Non-uniqueness of 35 PRTs for fixed points at 13.81 K, 18.5 K, 25 K, 54.361 K, 83.798 K and with the coefficients of the deviation function determined by exact methods
• non-uniqueness, --- envelope of IPTS-68 non-uniqueness, —— experimental uncertainty.

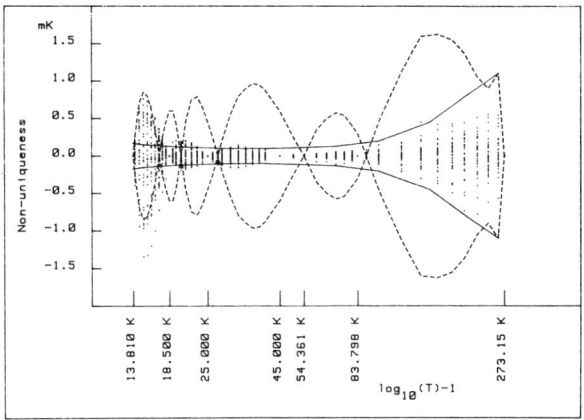

FIG. 8. Non-uniqueness of 35 PRTs for fixed points at 13.81 K, 18.5 K, 25 K, 45 K, 54.361 K, 83.798 K and with the coefficients of the deviation function determined by exact methods
• non-uniqueness, --- envelope of IPTS-68 non-uniqueness, —— experimental uncertainty.

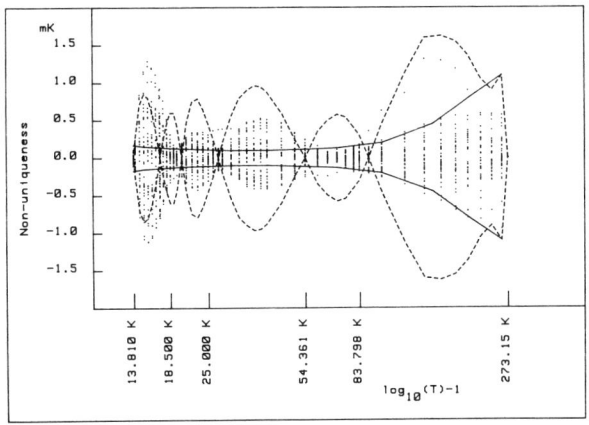

FIG. 7. Non-uniqueness of 35 PRTs for fixed points at 13.81 K, 18.5 K, 25 K, 54.361 K, 83.798 K and with the coefficients of the deviation function determined by least squares fitting
• non-uniqueness, --- envelope of IPTS-68 non-uniqueness, —— experimental uncertainty.

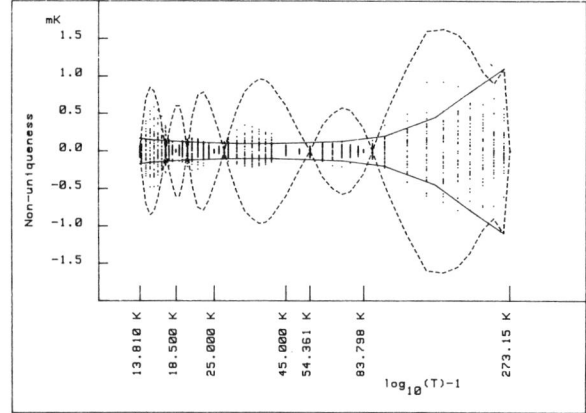

FIG. 9. Non-uniqueness of 35 PRTs for fixed points at 13.81 K, 18.5 K, 25 K, 45 K, 54.361 K, 83.798 K and with the coefficients of the deviation function determined by least squares fitting
• non-uniqueness, --- envelope of IPTS-68 non-uniqueness, —— experimental uncertainty.

In both cases the non-uniqueness is much worse at low temperatures particularly between 13.81 K and 18.5 K. The situation can be restored somewhat by introducing a point at 45 K representing the β-γ transition in solid oxygen and the results of this are shown in Figs 8 and 9. This improves the interpolation above 18.5 K to almost the level shown in Figs 1 and 2. However the interpolation below 18.5 K is still not very good.

We attempted to remove the problem below 18.5 K by terminating the interpolation procedure at 18.5 K and also at the triple point of neon. However this only moved the interpolating problem to higher temperatures. One possible solution is to terminate the interpolation at 18.5 K but to retain the hydrogen triple point in the fitting procedure. This provides interpolation from 18.5 K to 273.15 K as good as in Fig. 1 and does remove all boiling points from the scale.

Another alternative is to continue interpolation down to 13.81 K but to include the helium boiling point in the fitting procedure. It is customary to measure the so-called residual resistance of PRTs from time to time and this is usually done by comparison with a standard thermometer at 4.2 K, the helium boiling point. For our purposes this is sufficient, a precise realization of the helium boiling point not being necessary, as the sensitivity of a PRT at 4.2 K is ten times less than at 13.81 K and one hundred and sixty times less than at 54 K. Adding the point at 4.2 K to the fitting procedure produced a considerable improvement in interpolation between 13.81 K and 18.5 K for the exact case as shown in Fig. 10. There was no improvement when the least squares fitting method was used.

CONCLUSION

It is apparent that there are not many acceptable alternative defining fixed points for an IPTS below 273.15 K. The present neon boiling point can be replaced with the neon triple point. The neon[20] triple point[7] is even more suitable than that of natural neon as it has a melting range at least five times smaller and its use avoids any uncertainty concerning isotope composition.

The only remaining boiling points are those of hydrogen at 17.042 K and 20.28 K. These may be replaced by the deuterium triple point at 18.7 K and the β-γ transition in nitrogen at 43.8 K but only if the

R. C. Kemp

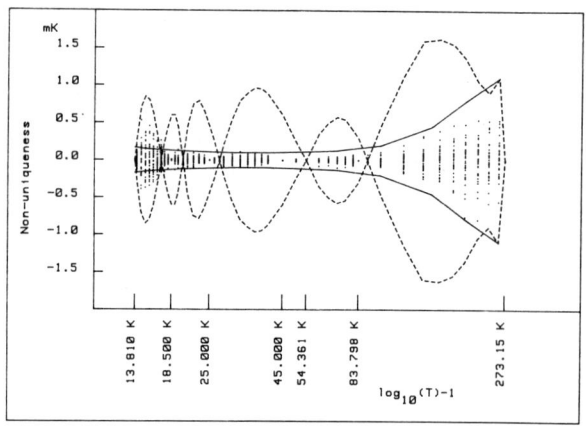

FIG. 10. Non-uniqueness of 35 PRT's for fixed points at 4.2 K, 13.81 K, 18.5 K, 25 K, 45 K, 54.361 K, 83.798 K, and with the coefficients of the deviation function determined by exact methods. ● non-uniqueness; --- envelope of IPTS-68 non-uniqueness; — experimental uncertainty.

helium boiling point at 4.2 K is added to the fitting procedure. Alternatively, the scale can be terminated at the deuterium triple point at 18.7 K, it is however necessary to include the hydrogen triple point in the fitting procedure to provide acceptable interpolation.

The use of the deuterium triple point and the β-γ transition in oxygen as defining fixed points is questionable as little work has been done on these points. However it is possible that, as a fixed point, the deuterium triple point may be of similar quality to the hydrogen triple point, provided that deuterium of suitable purity is obtainable. The β-γ transition is not particularly easy to realize[8] but it is possible to obtain good results. It is preferable at this stage that more work should be done on these points before they are incorporated in the IPTS. The hydrogen boiling points are, at this stage at least, better retained.

REFERENCES

1. C. G. Kirby, R. E. Bedford, and J. Kathnelson, Metrologia 11, 117-124 (1975).
2. S. D. Ward, CCT/78-36.
3. R. C. Kemp, W. R. G. Kemp, and L. M. Besley, Metrologia (to be published).
4. S. D. Ward and R. L. Rusby, CCT/80-52.
5. R. C. Kemp, L. M. Besley, and W. R. G. Kemp, Metrologia 14, 137-142 (1978).
6. S. D. Ward and J. P. Compton, Metrologia 15, 31-46 (1979).
7. R. C. Kemp and W. R. G. Kemp, Metrologia 17, 43 (1981).
8. J. A. Cowan, R. C. Kemp, and W. R. G. Kemp, Metrologia 12, 87-91 (1976).

Realization of the 1976 Provisional 0.5 K to 30 K Temperature Scale at the National Bureau of Standards

E. R. Pfeiffer and R. S. Kaeser

Center for Absolute Physical Quantities, National Bureau of Standards, Washington, D.C. 20234

The National Bureau of Standards presently disseminates a version of the "1976 Provisional 0.5 K to 30 K Temperature Scale" (EPT-76) which is maintained on two rhodium-iron resistance thermometers. Calibrations on the EPT-76 are made using a minicomputer-controlled measurement system. Maintenance of the scale is periodically checked against realization of the superconducting transition points of NBS SRM 767 (which comprise 5 of the 11 defining reference points of the EPT-76). In addition, versions of the EPT-76 derived from the NBS 2–20 K Scale (maintained on germanium resistance thermometers) and from the NBS version of the IPTS-68 (maintained on platinum resistance thermometers) have been realized and compared with the rhodium-iron based version over their respective overlapping regions. From those checks and comparisons, and from similar data from other sources, it is concluded that the EPT-76 is non-unique by as much as 1 mK at several places over its 0.5 to 30 K range. Thus, an uncertainty of ±1 mK has been assigned to the EPT-76 calibrations disseminated by NBS. These are tolerable limits considering the provisional status of the scale.

INTRODUCTION

The "1976 Provisional 0.5 K to 30 K Temperature Scale" (EPT-76)[1] is a provisional practical temperature scale recently defined by the International Committee of Weights and Measures (CIPM). It is recommended by the CIPM for international use between 0.5 K and 30 K until a new International Practical Temperature Scale can be adopted. The objectives in deriving the new scale were that it should be thermodynamically smooth, that it should be continuous with the International Practical Temperature Scale of 1968 (IPTS-68) at 27.1 K, and that it should agree with thermodynamic temperature as closely as the first two conditions allow. The derivation and development of the EPT-76 are described in a supplementary paper[2].

The EPT-76 (temperatures designated by T_{76}) is defined by temperatures assigned to 11 thermometric reference points, from which a realization of the scale may be obtained by various methods of thermodynamic interpolation. It is also permissible to realize the EPT-76 from certain previously existing scales (the IPTS-68, the ^{3}He and ^{4}He vapor pressure scales, and several laboratory scales) for which tables of differences from the EPT-76 are given. These differences are based on the results of comparisons of several laboratory temperature scales by Besley and Kemp[3] from 1 K to 30 K using germanium resistance thermometers (GRTs) and by Ward and Compton[4] above 13.8 K using platinum resistance thermometers (PRTs); these comparisons were performed under the auspices of the Consultative Committee for Thermometry (CCT). The CCT expected the different approved methods of realizing the EPT-76 to introduce slight ambiguities between realizations as a result of internal inconsistencies. Such ambiguities were considered acceptable, due to the provisional status of the scale and the fact that the EPT-76 does not replace the IPTS-68 in the region of overlap from 13.81 K to 30 K.

The National Bureau of Standards (NBS) provided calibrated thermometers for participation in each of the CCT-promoted comparisons. As a result, differences from the EPT-76 are defined for the NBS-maintained version of the IPTS-68 (NBS IPTS-68)[5] from 13.81 K to 30 K, carried on reference standard PRTs, and for the NBS Provisional Temperature Scale 2-20 K of 1965 (NBS 2-20)[6] from 2.3 K to 20 K, carried on reference GRTs. Thus, an NBS version of the EPT-76 between 2.3 K and 30 K can be constructed from previously existing NBS scales without realizing any of the EPT-76 defining reference points in that region. The portion of the EPT-76 below 2.3 K can be obtained from the easily realizable superconducting transition points of Cd, Zn, and Al from NBS Superconducting Fixed Point Device SRM 767[7,8,9] (defining EPT-76 reference points at 0.519 K, 0.851 K, 1.1796 K, respectively) in conjunction with a suitable thermodynamic interpolating instrument, such as a magnetic thermometer containing the paramagnetic salt cerous magnesium nitrate (CMN)[10].

In addition to realizing the EPT-76 in the above manner, another version of the EPT-76 was realized from rhodium-iron resistance thermometers (RIRTs)[11] calibrated on the National Physical Laboratory (England) gas thermometry (NPL-75) scale[12] and on the NPL CMN magnetic thermometry (T_{X1}) scale[13] -- additional scales for which differences from the EPT-76 are defined.

This paper briefly describes an NBS cryostat used for realizing the EPT-76 over its entire 0.5 K to 30 K range. Also described is a minicomputer-operated measurement system which provides automatic temperature control and data acquisition for resistance thermometer calibrations and for a variety of other measurements at cryogenic temperatures. Some results of an experiment comparing versions of the EPT-76 derived from different origins are presented. Also reported are the results of measurements performed on two NBS SRM 767 devices -- one device containing regular polycrystalline specimens, and the other a specially prepared device containing very carefully annealed specimens.

THE NBS EPT-76 CRYOSTAT

The NBS EPT-76 calibration cryostat is a ^{3}He cryostat[14] that has been modified to accommodate a gold-plated OFHC-copper comparator block mounted below the ^{3}He evaporator. The comparator block may be unscrewed from its mount to allow for easier, bench-top installation of temperature sensors. The block contains various size cylindrical wells into which capsule-type metallic (e.g., Pt, RhFe, PtCo) thermometers and GRTs of various manufacture may be inserted. During insertion, the thermometers are liberally coated with grease in order to enhance their thermal contact with the block. Numerous threaded holes in the block facilitate the attachment of GRTs installed in special mounts and/or other thermometric sensors, such as an NBS SRM 767 superconducting fixed-point device or a magnetic thermometer. In a typical run about 20 four-lead sensors are attached to the comparator block.

One hundred electrical leads run from a distribution box located in the laboratory environment into the cryostat and ultimately to the comparator block. The leads are thermally anchored to the cryostat at several important temperature zones before final thermal anchoring to the block. The leads terminate in gold-plated crimp sockets mounted in a nylon ring around the periphery of the block. Gold-plated connector pins, crimped to the thermometer leads, are simply plugged into the sockets to complete the electrical circuit. A manganin wire heater, bifilarly wound around the support mount of the block, is used for temperature control. The comparator block is surrounded by a polished copper radiation shield which, while not regulated in temperature, remains nearly at the temperature of the block. The block, block shield, and ^{3}He evaporator region are enclosed by a vacuum-tight can attached to the bottom of the pumped liquid ^{4}He reservoir. This vacuum-tight space may be filled with exchange gas, or evacuated, as needed. The vacuum-tight can, itself a radiation shield, is surrounded by two concentric polished copper radiation shields -- one attached to the (normally) unpumped liquid ^{4}He reservoir and the second attached to the liquid nitrogen reservoir. Besides the leads running to the comparator block, additional leads from the distribution box enter the cryostat and run to the top of the pumped ^{4}He reservoir. From there, they are distributed to various cryostat-monitoring thermometers located on that reservoir and on the various radiation shields.

Block temperatures approaching 1 K are obtained by pumping the liquid ^{4}He reservoir with a large booster pump backed by a mechanical vacuum pump. Temperatures to slightly below 0.5 K are achieved by operating the ^{3}He refrigerator in a recirculating mode; by operating in a "one-shot" cooling mode, a minimum temperature of about 0.31 K is reached. A supplemental heater located on the vacuum-tight shield surrounding the block may be activated, also, to obtain controlled temperatures as high as 100 K for calibrations on the IPTS-68.

AUTOMATIC TEMPERATURE CONTROL AND DATA ACQUISITION

A dedicated minicomputer-operated measurement system[15], coupled to the cryostat, provides automatic control of the comparator block temperature and of the acquisition and recording of measurement data. In addition to the minicomputer with 64 kbytes of memory, the hardware system consists of drive units for both rigid and floppy magnetic discs used for program and data storage, programmable power supplies, a precision nanovolt dc amplifier, a six-digit digital voltmeter, a digital clock, and several instrument scanners which connect the various sensors to the appropriate equipment. The minicomputer and peripheral instruments are connected via the IEEE-488 interface bus. A CRT terminal provides for input control of the computer and for temporary display of the measurement results. A printer provides a hard-copy output of the results. A FORTRAN-language program named "MOOSE", developed at NBS specifically for use in low-temperature laboratories, controls the scheduling of measurements and data storage.

Four-lead dc resistance measurements comprise the core of a calibration run. In the present experiment the resistance thermometer to be measured is connected in series with a reference resistor. The selected reference resistor, one of ten different reference resistors maintained in a temperature-controlled oil bath, has a resistance 10^3 to 10^4 times greater than the resistance to be measured. In this way most of the voltage applied to the thermometer circuit is distributed across the reference resistor. The circuit voltage is applied by a digital-to-analog converter power supply. During measurements the potential leads of the resistance thermometer are connected to the dc amplifier which is automatically set to a gain of either 10^3 or 10^4 depending on the calibration requirements for the particular thermometer. The amplifier output voltage is measured using the digital voltmeter. The current in the thermometer circuit is determined by measuring the voltage across the series reference resistor with the digital voltmeter directly connected. The amplified thermometer voltage and the reference resistor voltage are each measured a designated number of times, during which the circuit current is reversed in order to cancel the effect of thermal voltages in the circuit. The resistance of the thermometer is then calculated from the various average voltages, taking into account the reference resistor value and the amplifier gain. This resistance measurement cycle can be repeated up to a total of nine cycles, in which case the resistance average and its associated statistical uncertainty are reported.

The ten reference resistors (of magnitude 330 ohms, 1000 ohms, and thereafter decade multiples of each up to 10 megohms) are calibrated periodically by measuring a decaded bank of calibrated "NBS-type" dc resistance standards, all maintained in the same temperature-controlled oil bath.

Other computer-controlled tasks include low-frequency ac resistance measurements using a new automatic resistance bridge[16] developed at NBS, and mutual inductance measurements where the output of a mutual inductance bridge is connected to a lock-in amplifier and its output is read by the digital voltmeter. A heater-control task regulates the temperature of the comparator block to an assigned target temperature at a selected heater-control thermometer (or, alternatively, to a target resistance for an uncalibrated heater-control sensor). A helium level task monitors the unpumped liquid helium level and calculates the boil-off rate.

Prior to an experiment, a "TASKS" file containing up to 50 individual measurement tasks is constructed and stored in the computer memory. The file entries for a particular task include a task name identifying the sensor, information on the type of measurement (e.g., dc or ac resistance, heater-control, etc.), the sensor lead locations in the lead-connection matrix of the distribution box, and any other pertinent information needed for the task measurement. In the case of calibrated reference thermometers, such information would include a calibration table listing T- and R-values and the voltage, V, to be applied at the measured resistance; for uncalibrated resistance thermometers, a table listing R- and V-values must be placed in the task file if a variable applied voltage is to be automatically set. For measured values of R between the table values, the computer performs a four-point Lagrange polynomial interpolation of the table and assigns appropriate values to V and T. For heater-control tasks, the file contains a table of temperature vs. heat capacity for the region of the cryostat to be controlled. Also included in the TASKS file are the latest calibration values of the ten reference resistors and the bank of dc resistance standards.

The typical mode of operation in thermometer comparison experiments is to regulate the comparator block at a given temperature, to perform a calibration cycle with that temperature under fine regulation, and then to ramp to the next higher temperature -- all under computer control.

When fine temperature control has been achieved by meeting four control criteria residing in the measurement control program of the computer, a data cycle is begun automatically. At that time, one of the "NBS-type" dc resistance standards is measured one or more times in order to calibrate the system dc gains for the dc resistance comparison cycles to follow. In this way, the dc system can be corrected to compensate for any long-term drift which may have occurred since the last calibration of the reference resistors.

A "comparison cycle" usually consists of five sensor measurements in the following order: 1) a reference thermometer resistance measurement; 2), 3), 4) measurements on various sensors which are to be compared to the reference thermometer; 5) a heater-control task measurement. The heater-control task measures the reference thermometer resistance, which provides a second temperature for interpolating reference temperatures to the times of the intermediate comparisons, and also provides a measure of the temperature and temperature drift rate (relative to the latest previous reading of the same task) which are used to reset the heater power to maintain temperature control during the next comparison cycle. The drift rates during comparisons typically are less than 0.1 mK/min, except below 2 K where rates twice that amount are not unusual.

In a normal comparison-mode calibration of resistance thermometers, steps 2) - 4), above, would involve dc resistance measurements on three different sensors and "comparison cycles" would continue, while maintaining a fixed temperature, until all desired thermometers have been compared to the reference thermometer. Generally, this calibration data cycle, including the system dc gain recalibration described above, is performed twice at each calibration temperature in order to check the reproducibility of the data and to assure that at least one good calibration point is obtained for each compared thermometer.

Steps 2) - 4), above, could also be ac resistance measurements on one or more thermometers to provide either dc vs. ac comparisons or ac vs. ac comparisons. In the case of superconducting transition point measurements, steps 2) - 4) would represent repeated measurements of the mutual inductance bridge output from a single NBS SRM 767 device, and this "comparison cycle" would be done repeatedly while the temperature is swept slowly through the transition region.

As each task measurement is completed, a three-line data set comprised of the task number and title, date and time of measurement, and up to 12 measurement parameters is stored in the computer memory. This memory storage is temporary and is updated with new data from a subsequent measurement involving the same task. Permanent storage is obtained by recording the task measurement data on a floppy disc. Also, as each task measurement is completed, a single line of condensed data is written to the CRT terminal or to the printer; this provides an immediate record of the progress of the experiment and allows monitoring of the quality of the data.

After all measurements have been completed at a given temperature, the computer initiates a heating or cooling procedure in order to obtain the next designated calibration temperature. This procedure is automatic and involves periodic use of the heater-control task. Up to 10 designated temperatures may be stored for automatic operation until the next operator intervention, or the temperatures may be repeatedly incremented by a designated fixed amount.

This measurement system generally provides unassisted, around-the-clock operation during a calibration run except for occasional attending to liquid cryogen needs. However, operation in different temperature regions requires some manual assistance, such as starting or stopping the ^{3}He refrigerator, refilling the pumped liquid ^{4}He reservoir, etc. The entire process can be monitored and directed remotely, via standard telephone lines, using an acoustically coupled modem and a terminal.

ANALYSIS OF MEASUREMENT DATA

The data from any given experimental run might fill 10 or more eight-inch floppy discs. Prior to final analysis, the data are combined into a single data file on a rigid magnetic disc. The rigid disc has a capacity of about 19,000 three-line data sets and is more rapidly accessible than floppy discs.

The sensor comparison data are analyzed using a program called "DIGEST". This program recomputes all dc resistance values in double-precision arithmetic, taking into account the system gain drift factor obtained from measurements on the resistance standard at the beginning of each data cycle. Corrections are made for the effects of the finite input impedances of the amplifier and the digital voltmeter on the dc resistance circuit. The program also allows the original values for the reference resistors to be replaced with revised values and allows use of a revised calibration constant for the ac resistance bridge. For a calibrated thermometer, the temperature is recalculated from a calibration table or, alternatively, from coefficients of a fitted polynomial.

The main function of the DIGEST program is to screen the measurement data and perform linear interpolations of appropriate reference thermometer temperatures to obtain a temperature to be assigned to the measurement time of a compared sensor. The statistical uncertainties of the reference temperatures are similarly interpolated to obtain an uncertainty to be associated with the interpolated temperature.

A printed output of the comparison results is provided for each designated combination of reference thermometer and compared sensor. For comparisons involving resistance thermometers, the printout lists either the calibration table or the polynomial coefficients used for calibrated thermometers, and the values of the reference resistors and instrument parameters used during the analysis. The comparison results are then listed in tabular form: 1) The date, time, and duration of each comparison. 2) For the reference thermometer -- the interpolated temperature, temperature uncertainty, average rate of temperature drift between measurements, and the measuring current used. 3) For the compared thermometer -- the resistance, resistance uncertainty, and the measuring current used. 4) If the compared thermometer is a calibrated thermometer -- its temperature, temperature uncertainty, and the temperature difference between the two thermometers. 5) And finally, some information concerning the measurement complexity and the specific reference resistors used during the comparisons. For compared sensors other than resistance thermometers (e.g., mutual inductance devices), the DIGEST output is modified appropriately.

Additional output of DIGEST in the form of resistance (or mutual inductance) vs. temperature data may be stored in magnetic storage for later use with data plotting and/or data fitting routines.

SMOOTHNESS OF THE EPT-76 DERIVED FROM THE NBS 2-20 SCALE

The NBS provided three calibrated germanium resistance thermometers for participation in the CCT-promoted temperature scale intercomparison experiment carried out by Besley and Kemp[3]. The three thermometers (GRTs 76, 78, and 2817) were calibrated on the NBS 2-20 K temperature scale at 19 temperatures from 2.32 K to 18.94 K; on the NBS IPTS-68 at 13.81 K, 20.28 K and 27.102 K; and at the superconductive transitions of indium and lead. Besley and Kemp's results show an agreement among the NBS thermometers of about ± 0.2 mK at 10 K and below, and about ± 0.4 mK from 11 K to 20 K. Overall, Besley and Kemp claim "an uncertainty of 0.5 mK at the most, though a more likely value is 0.3 mK".

As a first step toward realizing the EPT-76 from the NBS 2-20 scale, we applied the EPT-76 defined differences to thermometers most closely linked to those used in Besley and Kemp's intercomparison experiment -- two germanium thermometers (GRTs 540 and 565) which served as secondary reference standards for maintaining the NBS 2-20 scale at NBS during the calibration of GRTs 76, 78, and 2817 (and during similar calibrations of numerous customer thermometers since 1965). The NBS 2-20 reference temperatures on GRTs 540 and 565 are defined with millikelvin precision; hence the resulting EPT-76 temperatures on these reference thermometers are not expected to be accurate to better than ± 0.5 mK in general. In applying the EPT-76 differences to the NBS 2-20 scale, some ambiguity exists concerning the value applicable to the 20 K calibration point -- the EPT-76 documents[1,2] list two values, -0.8 mK and +7 mK. We disregarded the -0.8 mK value traceable to Besley and Kemp's experiment because, apparently, no 20 K calibration point on the NBS 2-20 scale was supplied to them for GRTs 76, 78, and 2817. Instead we chose the +7 mK value applicable to "early NBS calibrations", a value which apparently emanates from a temperature scale comparison experiment by Anderson and Swenson[17].

The resultant EPT-76 data sets for GRTs 540 and 565 were separately fitted by a weighted least-squares method to a polynomial function of the popular form,

$$\log R = \sum_{n=0}^{m} a_n (\log T)^n, \qquad (1)$$

where R is the resistance at temperature T. Other approximation equations for GRT calibrations, together with some selection criteria, have been collected by Rindelhardt and Hegenbarth[18]. Our fitting program performs repeated fits to Eq. 1 for chosen m=m(min) to m=m(max) in any consistent integral increment. For each fit a printout lists the polynomial coefficients, their standard deviations, and the ratio of each coefficient to its standard deviation. Also listed are the differences in terms of R and T between the calibration data and the fitted polynomial, and the first and second derivatives of the polynomial in terms of the transformed variables (i.e., dlogR/dlogT and its derivative). The mean errors[18] in terms of logR, R, and T are computed as measures of the goodness of each polynomial fit. Finally, for a chosen polynomial fit, a table is generated which lists T, R, and the first and second derivatives of R with respect to T.

After a preliminary fitting with equal weights to obtain the general functional form, the data were refitted using weights = $(dT/dlogR)^2$. This weighting corresponds to minimizing $\Sigma(\Delta T)^2$, which seems appropriate for EPT-76 temperatures which are based on differences derived from Besley and Kemp's results. The results of these fittings for m=10 and m=11 are shown in Fig. 1, where the difference between the assigned EPT-76 temperature and the temperature obtained from the fitted polynomial is plotted against temperature. In some respects the polynomials with m=10 appear to provide optimal fits. The differences generally are within ±0.5 mK, the expected accuracy discussed earlier, and for each GRT the ratios of the coefficients to their standard deviations are about 3.0 or more, except for one coefficient where the ratio is about 1.7. However, the polynomials with m=10 give indications of being not well-behaved, in terms of their first and second derivatives, near the lowest calibration temperature. The polynomials with m=11 have derivatives which appear well-behaved over the entire 2.3 K to 20 K range, and slightly beyond, but the associated statistical parameters indicate that some overfitting has probably occurred -- for each GRT, the ratios of the coefficients to their standard deviations are less than 2.0 for about one-half of the coefficients. Also, the fitting within ±0.3 mK at so many temperatures seems unwarranted considering the expected accuracy of the data. We believe that the polynomials with m=10 very nearly smooth out the inaccuracies contained in this version of EPT-76, while the polynomials with m=11 to some extent accommodate those inaccuracies. Keeping in mind the shortcomings of each, polynomials with either m=10 or m=11 could provide acceptably smooth representations of EPT-76 data for GRTs 540 and 565.

The outstanding features of Fig. 1 are the prominent deviations which occur in the region 16 K to 18 K. Deviations of 1.0 - 1.3 mK exceed what would be expected on the basis of Besley and Kemp's uncertainty assessment and the precision of the original NBS 2-20 reference data. Such large deviations are not unique to GRTs 540 and 564, for we have observed similar deviations with m=10 polynomial fittings of the sets of 19 EPT-76 data points for GRTs 76, 78, and 2817 -- the actual thermometers from which the EPT-76 defined differences applicable to the NBS 2-20 scale were obtained, and on which the NBS 2-20 temperatures are defined with 0.1 mK precision.

Van Rijn, et al.[19] observed that polynomials of the kind given by Eq. 1, with m=12, are able to represent GRT calibrations against a smoothed magnetic temperature scale between 1.5 K and 30 K within ±0.2 mK. On that basis we conclude from our polynomial fittings that the EPT-76 obtained from the NBS 2-20 scale is smooth within ±0.5 mK, except in the region 16 K to 18 K where it is non-smooth (and therefore possibly in error) by as much as ±1.5 mK.

This mathematical method of investigating the smoothness of the EPT-76 derived from the NBS 2-20 scale cannot reveal exactly which of the suspect temperatures are seriously in error. That can be revealed only by further experimental comparisons of appropriate thermometers against a truly smooth EPT-76 or, possibly, by a re-examination of Besley and Kemp's data which served as the basis for the defined differences between the NBS 2-20 scale and the EPT-76.

AN EPT-76 COMPARISON EXPERIMENT

During a single experimental run we obtained EPT-76 comparison data from three sets of thermometers -- each set calibrated on a different laboratory scale: 1) Two germanium resistance thermometers (GRTs 540 and 565), the reference thermometers for NBS 2-20 calibrations, carried 20 calibration points which were converted to the EPT-76 and smoothed as described above. 2) Two platinum resistance thermometers (PRTs 85 and 87), calibrated on the NBS IPTS-68 and smoothed according to the IPTS-68 interpolation prescription, were converted to the EPT-76 by applying the defined differences from Ref. 1. 3) Two rhodium-iron resistance thermometers (RIRTs 78 and 79), obtained from the NPL, were calibrated on the NPL-75 gas thermometer scale[12] at 35 points from 2.6 K to 27 K and on the NPL CMN magnetic thermometer ($T_{\chi 1}$) scale[13] at 19 points from 0.45 K to 3.0 K. The calibrations on each scale were smoothed by means of polynomial fits supplied with the data. The NPL-75 data were converted to the EPT-76 by applying the defined differences from Ref. 1; by definition there is no difference between the NPL $T_{\chi 1}$ scale and the EPT-76. The EPT-76 temperatures obtained from these three sets of thermometers will be designated T_{76}(NBS 2-20), T_{76}(NBS IPTS-68), and T_{76}(NPL), respectively, to indicate their various origins.

The actual calibration temperatures on the different types of thermometers generally did not coincide; hence, the smooth representations of the various calibrations were used for making the comparisons. (Such smooth representations were desirable for another reason -- the four-point interpolation routine of our computer program interpolates more reliably between smooth data points). The polynomial fits with m=11, (with deviations as shown in Fig. 1, lower) were used for GRTs 540 and 565. Because the rhodium-iron thermometers carried the widest range of EPT-76 temperatures (0.5 K to 27.1 K), we selected one, RIRT 78, to serve as a master reference thermometer during this experiment and compared the other thermometers to it. The comparisons were made by means of four-lead dc resistance measurements.

The difference between T_{76}(NBS 2-20) realized from GRTs 540 and 565 and T_{76}(NPL) realized from RIRT 78 is shown in Fig. 2. Although both GRTs carrying T_{76}(NBS 2-20)

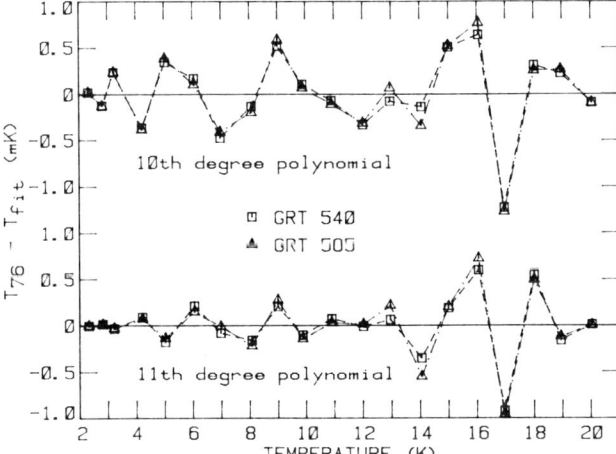

Fig. 1 Difference between EPT-76 temperatures and temperatures obtained from fitted polynomial functions for two germanium resistance thermometers which served as secondary standards for maintaining the NBS provisional 2-20 K temperature scale.

show a consistent pattern of deviation from T_{76}(NPL), the overall agreement lies within ±1 mK, which is probably acceptable considering the provisional nature of the EPT-76. The exceptional agreement at 3.4 K and 7.2 K precludes use of the superconducting fixed points of In and Pb from SRM 767 to determine which version of the EPT-76 should be preferred. The 1 mK disagreement in the region around 5 K might be resolved by a check of the calibrations against the ^{4}He vapor pressure scale, but that was beyond the scope of the present experiment.

The difference between T_{76}(NBS IPTS-68) realized from PRTs 85 and 87 and T_{76}(NPL) realized from RIRT 78 is shown in Fig. 3. Again the overall agreement is within about ±1 mK except below 15 K, which corresponds to the lower extreme of the IPTS-68 where the resistance of the PRTs is very small (0.04 ohm or less) and the measurement problems become quite severe. However, to produce a PRT temperature error of 1 mK at 13.8 K requires a resistance error of 200 ppm. We presently estimate that our dc resistance measurements are accurate to about ±10 ppm over the resistance range 0.1 - 10,000 ohms; outside that range a deterioration of the accuracy is expected, but we cannot quantify it at this time. (Of course, there may be errors other than just resistance measurement errors in a comparison experiment such at this; some such errors are discussed below.)

The temperatures realized from RIRTs 78 and 79 are compared in Fig. 4. Except for the region 1 K to 3 K, any disagreement appears to be due to random scatter in the measurements. The scatter increases with increasing temperature, as would be expected on the basis of the decreasing sensitivity of the thermometers. Their very low sensitivity above 15 K is a major disadvantage of using rhodium-iron thermometers to carry the full range of the EPT-76. For example, a 1 ppm error in resistance produces a temperature error of 0.08 mK at 25 K; hence, a total random error in resistance of about 6 ppm, distributed between the two RIRTs, could account for observed differences as large as 0.5 mK at temperatures above 24 K.

The apparently systematic disagreement occurring in the region 1 K to 3 K (Fig. 4) was first thought to be due to a large self-heating effect (which occurs in RIRTs in that temperature region) combined with improper thermal anchoring of one of the sensors. However, closer inspection of the data in the region 1.0 K to 1.2 K revealed that the two RIRTs sometimes agreed within ±0.1 mK but at other times disagreed by amounts approaching 1 mK, depending on the precise mode of comparison. In the region 1.2 K to 12 K the RIRTs were included in a "comparison cycle" which consisted of five thermometer measurements as described in the previous section on automatic data acquisition. Specifically, the cycle was of the form ABCDA, where A represents the reference RIRT, B and C represent two compared GRTs, and D represents the compared RIRT. In the region 0.50 K to 0.95 K, the RIRTs were compared in a simple ADA cycle in which the duration (or "complexity") of each resistance measurement was approximately one-half that of each measurement in the ABCDA cycle. In the region 1.0 K to 1.18 K and at 3.4145 K both modes of comparison were used. We now suspect that the large self-heating effect in RIRTs below 3 K, the relatively smaller heat capacity of

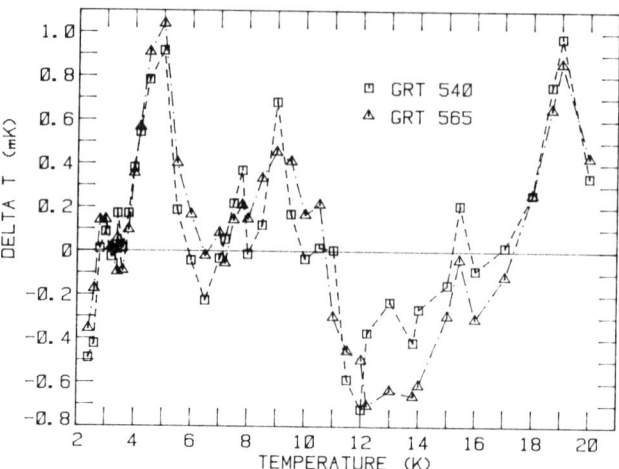

Fig. 2 Difference between T_{76}(NBS 2-20) realized from germanium resistance thermometers and T_{76}(NPL) realized from rhodium-iron resistance thermometer 78.

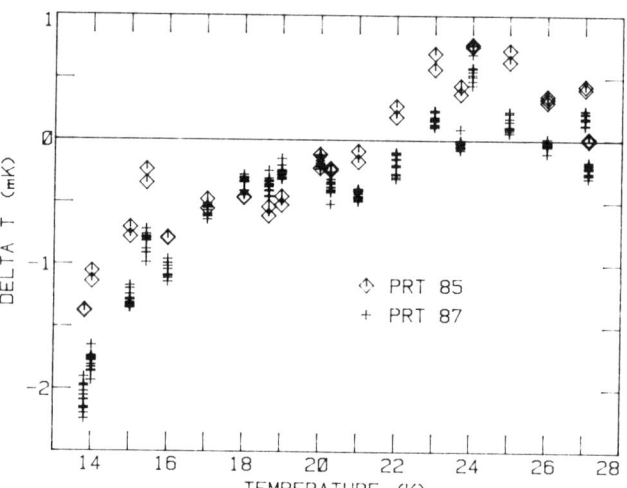

Fig. 3 Difference between T_{76}(NBS IPTS-68) realized from platinum resistance thermometers and T_{76}(NPL) realized from rhodium-iron resistance thermometer 78.

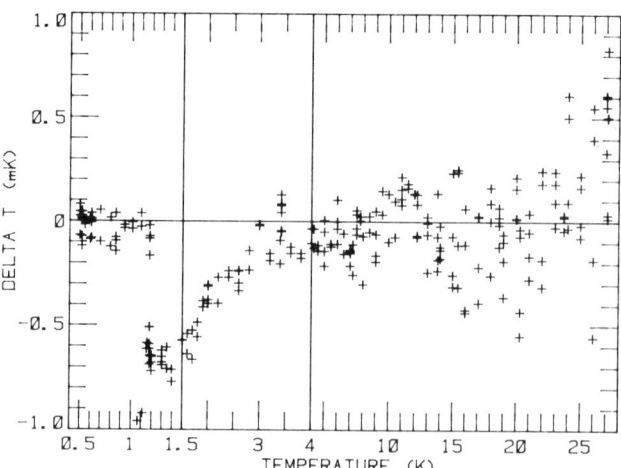

Fig. 4. Comparison of temperatures obtained from rhodium-iron resistance thermometers 78 and 79.

the comparator block, the non-symmetrical position of the compared RIRT (i.e., position D when grouped with GRT in the ABCDA cycle), and the longer duration of each resistance measurement, all combined to produce significant temperature fluctuations in the block during the longer "comparison cycle". The data analysis requirement of a constant rate of temperature drift between reference thermometer measurements probably was not achieved below 3 K with the longer comparison cycle. Above 3 K, the decreased self-heating of the RIRTs and the larger heat capacity of the block reduced such temperature fluctuations to a tolerable level. Above 12.2 K (where PRTs were included in the comparison routine), the compared RIRT was shifted to the symmetrical position C in the ABCDA measurement cycle.

The overall agreement of our two RIRTs is well within ±1 mK, the general limit of uncertainty suggested by the comparison results shown in Figs 1 and 2, and by the reproducibility limits of the superconducting fixed points which are discussed in the next Section.

SUPERCONDUCTING TRANSITION POINT REALIZATIONS

The superconducting transition points of Cd, Zn, Al, In, and Pb realized from NBS Superconducting Fixed Point Device SRM 767[7,8,9] constitute five of the eleven defining reference points of the EPT-76. The EPT-76 assigned temperatures for the five reference points are 0.519 K, 0.851 K, 1.1796 K, 3.4145 K, and 7.1999 K, respectively, although the SRM 767 devices were designed to provide temperature reproducibility within only ±1 mK. The use of these easily realizable reference points, in conjunction with a suitable magnetic thermometer as a thermodynamic interpolating device, constitutes an approved method of realizing the EPT-76 in the range 0.5 K to 7.2 K. Furthermore, a realization of these superconducting transitions alone can provide check points for calibrated reference thermometers which carry the EPT-76. In the present experiment the superconducting transition points from two SRM 767 devices were realized. One device, designated SRM 767-4, was a regular device containing high-purity, polycrystalline samples. The second device, designated SMR 767-Delta, was specially prepared by using very long annealing times in an attempt to obtain single-crystal samples which would exhibit extremely sharp superconducting transitions[20].

The superconducting transitions were measured by connecting the primary and secondary coils of the SRM 767 device to a mutual inductance bridge. The imbalance voltage of the bridge was fed into a lock-in amplifier, the output of which could be monitored visually on a meter or measured automatically using the digital voltmeter. The ambient magnetic field at the site of the superconductors was cancelled by a compensating field produced by a single large set of Helmholtz coils suspended in a gimbel-like mount outside the cryostat. This particular mounting arrangement allowed the axis of the coil pair to be tilted and rotated about the cryostat axis to obtain a proper orientation for nulling the ambient magnetic field. While regulating the temperature at the midpoint of the transition, the precise nulling field was established by visually monitoring the meter of the lock-in amplifier and varying the current and orientation of the compensating coils until the transition was moved to its maximum temperature.

Prior to collecting data on a superconducting transition, a bridge setting and an amplifier gain were selected to provide on-scale readings of the bridge imbalance signal through the entire transition. The transition measurements usually consisted of a pair of mutual inductance readings alternated with a reading of the reference thermometer, and this measurement sequence was performed repeatedly as the temperature was swept very slowly through the transition in both directions. The temperatures of the reference thermometer were interpolated to the times of the mutual inductance readings, and the resultant data were plotted to yield a transition curve for each sample.

The superconducting transition mid-point temperature, T_C, of each sample is listed in Table I. Temperatures are T_{76}(NPL) realized from reference thermometer RIRT 78. The width of the central 80 percent of each transition, W, is listed also. The values of W indicate that the transitions in four of the carefully annealed specimens of SRM 767-Delta were considerably sharper than the corresponding transitions in the polycrystalline samples. To some extent, the transition in the Pb specimen from SRM 767-Delta was sharper than its value of W indicates -- 70 percent of the transition occurred within an interval of 0.25 mK, but W was increased to 0.50 mK by a substantial contribution from a high temperature "tail" occurring in the remaining 30 percent. In contrast, the central 80 percent of the polycrystalline Pb transition from SRM 767-4 occurred uniformly across an interval of 0.35 mK.

The measured values of T_C for Pb, In, and Al fall within ±0.3 mK of the EPT-76 assigned values. For Zn the difference from the assigned values is as great as 1.0 mK, the maximum that might be expected on the basis of the ±1 mK irreproducibility of the SRM 767 devices. The Zn and Cd transitions from SRM 767-4 were determined during a similar earlier experiment and must be considered less accurate than the more recent measurements; the earlier results are included here mainly to compare the W-values.

In the case of Cd, our T_C results are 1.1 mK and 1.4 mK above the EPT-76 assigned value of 0.519 K. Such large deviations from the assigned value exceed the limit that would be expected from the SRM 767 irreproducibility alone. In fact, these large deviations seem partly due to the method used to assign an EPT-76 value to the Cd transi-

Table I. Transition temperatures, T_C, and transition widths, W, of single-crystal and polycrystalline versions of Superconductive Fixed Point Device SRM 767. Temperatures are T_{76}(NPL) from RIRT 78.

	EPT-76 T_C(K)	SRM 767-Delta T_C(K)	W(mK)	SRM 767-4 T_C(K)	W(mK)
Cd	0.519	0.52036	0.05	0.5201#	3.9#
Zn	0.851	0.85070	0.13	0.8500#	3.6#
Al	1.1796	1.17946	0.30	1.17940	1.6
In	3.4145	3.41438	0.25	3.41470	1.0
Pb	7.1999	7.20003	0.50	7.19979	0.35

Results from a preliminary experiment.

tion. The CCT assigned the value 0.519 K, attributed to some preliminary unpublished noise thermometry results at NBS, even though measurements at NPL on three SRM 767 devices had yielded an average T_c of 0.5197±0.0005 K on the NPL T_{X1} scale[2,13]. Although the NPL T_{X1} scale and the EPT-76 were defined to be equal, the NPL value was not considered to be inconsistent with the assigned value. Measurements on three SRM 767 devices at the Kamerlingh Onnes Laboratory (KOL) yielded T_c values on the NPL T_{X1} scale (hence on the EPT-76) ranging from 0.51965 K to 0.52035 K for Cd[21]. Our results on the Cd superconducting fixed point are consistent with the NPL and KOL results. Moreover, recent NBS noise thermometry measurements at the Cd transition yield a new T_c value close to 0.520 K[22].

DISCUSSION

In connection with our earlier discussion of the observed difference between T_{76}(NBS 2-20) and T_{76}(NPL), shown in Fig. 2, it was stated that we have no direct experimental evidence to prefer one EPT-76 version over the other. However, the polynomial smoothings of R vs. T data on reference GRTs 540 and 565 (Fig. 1) provide strong circumstantial evidence that T_{76}(NBS 2-20) may be significantly non-smooth, and therefore possibly in error, by as much as 1.0 - 1.5 mK. On the other hand, the polynomial fits supplied with the NPL-75 calibration data for RIRTs 78 and 79 indicate that those calibrations are smooth to better than 0.5 mK. Thus, from smoothness considerations alone, T_{76}(NPL) based on RIRTs 76 and 78 would be preferred over T_{76}(NBS 2-20) based on GRTs 540 and 565.

From the results shown in Fig. 2, after taking into account the differences between the smooth polynomial representation and the actual T_{76} calibration data (Fig.1, lower), we can construct an alternative set of differences for converting the NBS 2-20 scale to the EPT-76 (specifically to T_{76}(NPL)). These alternative differences are plotted in Fig. 5 together with the EPT-76 defined differences from Ref. 1. Starting from the results of an experiment comparing the NBS 2-20 scale to an ISU magnetic scale (XISU-75) by Anderson and Swenson[17], and applying two sets of published differences [1,3] to convert from the XISU-75 scale to the EPT-76, one can obtain another set of alternative differences between the NBS 2-20 scale and the EPT-76; these values, also, are plotted in Fig. 5 for comparison. No comparison data were taken at 2.3 K during the present experiment due to an oversight, so the EPT-76 defined difference at that temperature must be assumed to be correct.

It is probably correct to say that all three sets of differences plotted in Fig. 5 are in agreement within the combined uncertainties of the various comparison experiments. However, despite that general assessment, there are several temperatures at which the results of this research are in good agreement with the results derived from Anderson and Swenson's work but are substantially discrepant from the official EPT-76 differences (e.g., 1 mK discrepancy at 17 K). Furthermore, most of the substantial discrepancies occur at the appropriate temperatures in the appropriate directions to account for the non-smoothness shown in Fig. 1. Our

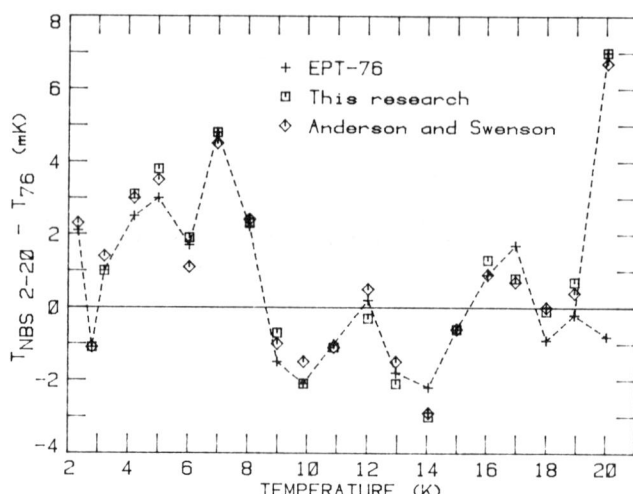

Fig. 5 Difference between the NBS provisional temperature scale and EPT-76, and alternative differences obtained from this research and from the temperature scale comparison results of Anderson and Swenson[17].

result at 20 K agrees with that derived from Anderson and Swenson's work and confirms our decision to use the +7 mK difference value (not the -0.8 mK alternative value) listed in the EPT-76 documents.

When our set of alternative EPT-76 difference values (Fig. 5) is applied to the 20 original NBS 2-20 calibration temperatures for GRTs 540 and 565, the resultant R vs. T data sets can be separately fitted to Eq. 1 to within ±0.3 mK with m=10 and to within ± 0.2 mK with m=11. Similar modification of the 19 original NBS 2-20 calibration temperatures for GRTs 76, 78, and 2817 (the thermometers which participated in the international temperature scale comparison) produces sets of R vs. T data which can be fitted to Eq. 1 to within ±0.2 mK with m=10. Such results provide strong evidence to sustain our earlier suspicions concerning the non-smoothness and possible error in T_{76}(NBS 2-20).

A similar EPT-76 comparison experiment was carried out at KOL by El Samahy[21], who compared T_{76}(NPL) carried on calibrated rhodium-iron thermometers with T_{76}(KOL) derived from a KOL magnetic scale carried on calibrated germanium thermometers. The latter version of the EPT-76 was obtained by means of defined differences from Ref. 1, also. The KOL comparison experiment showed differences between T_{76}(NPL) and T_{76}(KOL) as large as 2 mK for temperatures above 22 K, although the differences fell within ±1 mK at lower temperatures. El Samahy concluded that a slight revision of the EPT-76 defined differences is required to produce agreement of those two versions to ±0.6 mK between 2.6 K and 27 K.

CONCLUSIONS

The results of the present comparison experiment and of a similar comparison experiment at KOL indicate that the EPT-76, as derived from various existing laboratory scales by application of the defined differences, may be non-unique by as much as ±1 mK at several places over its 0.5 K to

30 K range. The non-uniqueness may be as large as ±2 mK near the extremes of certain laboratory scales. Versions of the EPT-76 constructed solely from the defining superconducting fixed points would be expected to be non-unique by as much as ±1 mK, also, due to an inherent irreproducibility of that amount in the NBS SRM 767 devices. It seems prudent, therefore, to assign an uncertainty of at least ±1 mK to the EPT-76 that is disseminated by NBS.

There is strong evidence to suggest that the EPT-76 version designated T_{76}(NPL) and carried on NPL-calibrated rhodium-iron thermometers is substantially smoother than T_{76}(NBS 2-20) carried on NBS reference germanium thermometers. Thus, T_{76}(NPL) over its full range, 0.5 K to 27 K, is expected to be smoother than any comparable version constructed from a combination of T_{76}(NBS 2-20), T_{76}(NBS IPTS-68), and the superconducting fixed points from SRM 767 (this combination method was discussed in the Introduction). Other practical considerations for favoring T_{76}(NPL), at present, are that NPL-calibrated rhodium-iron thermometers have been distributed to a number of national metrology laboratories and measurement results referenced to those thermometers have been reported already[13,21].

In view of the present results and above considerations, the National Bureau of Standards disseminates a version of the EPT-76 which consists of T_{76}(NPL) from 0.5 K to 27.1 K plus an extension to smoothly join the NBS IPTS-68 at 33.9 K (the next NBS IPTS-68 calibration temperature above 27.1 K). The scale is presently maintained on two rhodium-iron resistance thermometers. Maintenance of the scale is periodically checked against realizations of the defining reference points from NBS SRM 767 and against segments of the EPT-76 derived from the NBS 2-20 scale and from the NBS IPTS-68, maintained on reference germanium resistance thermometers and platinum resistance thermometers, respectively.

ACKNOWLEDGEMENTS

It is a pleasure to acknowledge many useful discussions with C. T. Van Degrift concerning the automatic data acquisition and analysis system, with H. H. Plumb about the NBS 2-20 K temperature scale and germanium thermometry, and with G.T. Furukawa about resistance thermometry measurements in general. G. A. Evans Jr. provided useful information about the preparation of the special SRM 767 device, and M. W. Reilly provided the least-squares polynomial fitting program which we adapted for use with germanium thermometer data. Conversations with C. A. Swenson concerning our comparison experiment, the development of the EPT-76 in general, and the development of alternative EPT-76 differences are gratefully acknowledged, also.

REFERENCES

1　The 1976 Provisional 0.5 K to 30 K Temperature Scale, Metrologia **15**, 65 (1979).

2　M. Durieux, D.N. Astrov, W.R.G. Kemp and C.A. Swenson, Metrologia **15**, 57 (1979).

3　L.M. Besley and W.R.G. Kemp, Metrologia **13**, 35 (1977).

4　S.D. Ward and J.P. Compton, Metrologia **15**, 31 (1979).

5　G.T. Furukawa, J.L. Riddle and W.R. Bigge, J. Res. Nat. Bur. Stand. (U.S.) **77A**, 309 (1973).

6　H.H. Plumb and G. Cataland, Metrologia **2**, 127 (1966).

7　J.F. Schooley, R.J. Soulen, Jr. and G.A. Evans, Jr., Nat. Bur. Stand. Spec. Pub. 260-44, SRM 767 (U.S. Dept. Commerce, Washington, D.C.) (1972).

8　R.J. Soulen, Jr., J.F. Schooley and G.A. Evans, Jr., Rev. Sci. Instrum. **44**, 1537 (1973).

9　J.F. Schooley, G.A. Evans, Jr. and R.J. Soulen, Jr., Cryogenics **20**, 193 (1980).

10　Cerous magnesium nitrate (CMN) may be considered the CCT-recommended paramagnetic salt for use below 1 K.

11　R.L. Rusby, Temperature Measurement, Inst. Phys. Conf. Ser. No. 26 (London 1975) p. 125.

12　K.H. Berry, Metrologia **15**, 89 (1977).

13　R.L. Rusby and C.A. Swenson, Metrologia **16**, 73 (1980).

14　H. Marshak and R.B. Dove, National Bureau of Standards Tech. Note 562 (U.S. Dept. Commerce, Washington, D.C.) (1970). The cryostat was designed such that the "working end" (^{3}He cooled region) could easily be modified for other types of experiments in that temperature region.

15　C.T. Van Degrift and R.S. Kaeser. A paper further describing this system will be presented at this Symposium.

16　R.D. Cutkowsky, IEEE Trans. **IM-29**, 330 (1980). See, also, this Symposium.

17　M.S. Anderson and C.A. Swenson, Rev. Sci. Instrum. **49**, 1027 (1978).

18　U. Rindelhardt and E. Hegenbarth, Cryogenics **15**, 355 (1975).

19　C. Van Rijn, M.C. Nieuwenhuys-Smit, J.E. Van Dijk, J.L. Tiggelman and M. Durieux, Temperature, Its Measurement and Control in Science and Industry, edited by H.H. Plumb (Instrument Society of America, Pittsburgh, 1972), Vol. **4**, p. 815.

20　We are indebted to J.F. Schooley for providing this specially prepared SRM 767 device. Further measurements on this device are reported by J.F. Schooley and R.J. Soulen, Jr. at this Symposium.

21　A.E. El Samahy, "Thermometry between 0.5 K and 30 K", Thesis, U. of Leiden (1979). The results of SRM 767 fixed point measurements are presented at this Symposium.

22　R.J. Soulen, Jr., private communication (1981).

A photoelectric pyrometer temperature scale below 1064.43 °C and its use to measure the silver point

T. P. Jones and J. Tapping

CSIRO Division of Applied Physics, Sydney, Australia 2070

A temperature scale below 1064 °C has been established with a photoelectric pyrometer relative to the freezing point of gold taken as 1064.43 °C. The uncertainties applying to this scale established on tungsten strip lamps are 0.025 °C at 1064 °C, 0.035 °C at 913 °C and 0.050 °C at 792 °C. The freezing point of silver measured on this scale, 961.980 °C ± 0.015 °C, differs significantly from other recent measurements. Investigations of likely systematic errors in this measurement did not locate any causes for the difference. One of the investigations led to a measurement of the freezing point of copper as 1084.890 °C ± 0.015 °C, which is not significantly different from recent determinations. An inconel heat-pipe blackbody using sodium as the working fluid has been shown to be satisfactory for the comparison of platinum resistance thermometers and photoelectric pyrometers at temperatures near the silver point.

INTRODUCTION

The inherent instabilities of Pt-Pt,10%Rh thermocouples has led to a recommendation by the Comité Consultatif de Thermométrie, Comité International des Poids et Mesures[1], that they be replaced as the means of interpolation in the International Practical Temperature Scale (IPTS) in the range 630.74 °C to 1064.43 °C. The uncertainty of the current IPTS 68[2] in this range is approximately 0.2 °C[3,4] and to warrant a change an order of magnitude improvement in this uncertainty is desirable.

The most likely means of replacing the thermocouple appears to be the extension above 630.74 °C of the region covered by resistance thermometry, and possibly also the extension below 1064.43 °C of the range of optical pyrometry. An important factor in the decision as to the junction of the regions to be covered by these instruments is the uncertainties of the scales established with each. Although reproducibilities of less than 0.01 °C have been claimed for selected resistance thermometers at the gold point[5,6], Marcarino and Crovini[7] have reported that high quality thermometers which they have tested had uncertainties of 0.003 °C at the silver point but were an order of magnitude worse at the gold point. Consequently the possibility of making the silver point the junction temperature should be investigated.

To provide information for discussion on this subject a temperature scale in the range 790 °C to 1064 °C has been established using a photoelectric pyrometer, and the uncertainties of the scale estimated. Also a measurement has been made of the silver point temperature which is important in itself, but also serves as a comparison between our scale and similar ones in other laboratories. Finally, to assist in the evaluation of the silver point as a suitable junction temperature, a comparison has been made between a resistance thermometer calibrated at the silver point and the pyrometer scale, using a heat-pipe blackbody.

REALIZATION AND UNCERTAINTY OF THE PHOTOELECTRIC PYROMETRY TEMPERATURE SCALE

Temperature scales below 1064.43 °C, the freezing point of gold, can be established using photoelectric pyrometers in a similar manner to that used to establish the IPTS 68 above the gold point[8]. The equation describing the practical establishment of temperature scales above and below the gold point is

$$\int_0^\infty L_\lambda(T) \cdot \tau_\lambda \cdot S_\lambda \cdot d\lambda = R \int_0^\infty L_\lambda(T_{Au}) \cdot \tau_\lambda \cdot S_\lambda \cdot d\lambda \quad (1)$$

where R is the spectral radiance ratio (greater than unity above 1064 °C and less below),

$L_\lambda(T)$ and $L_\lambda(T_{Au})$ are the spectral concentrations at wavelength λ (in vacuo) of blackbodies at temperature T(K) and at the freezing point of gold, T_{Au} (K),

τ_λ is the spectral transmission of the bandpass filter of the pyrometer,

and S_λ is the spectral responsivity of the remainder of the pyrometer.

The pyrometer used for the scale reported here is the Australian Photoelectric Pyrometer Mark 2 (APEP2), the design and operation of which have been described in detail elsewhere[8]. Briefly, APEP2 is a radiance comparator which measures the ratio of spectral radiances of two sources by viewing each in turn via a rotatable glass prism. A computer automatically operates the

Table I Equivalent temperature uncertainties in the realization of a photoelectric pyrometry temperature scale on tungsten strip lamps below 1064 °C

Temperature °C	1064	913	792
Ratio below gold point	1	1/8	1/64
Source of uncertainty	Uncertainty °C		
Gold point uncertainty	0.012*	0.009	0.008
High stability tungsten strip lamp (transferor)	–	0.014	0.020
High stability tungsten strip lamp (transferee)	0.018	0.025	0.042
Nonlinearity	–	0.004	0.007
Spectral responsivity	0.008	0.005	0.014
Accumulated random measurement	0.002	0.005	0.005
Overall uncertainty	0.024	0.032	0.050
(Rounded up to 0.005 °C)	0.025	0.035	0.050

* Fundamental realization of the freezing point of gold[8]

Table II Recent determinations of the freezing point of silver relative to the freezing point of gold taken as 1064.43°C

Reference		Determination
Quinn et al.[10]	(1973)	962.06 ± 0.2°C
Bonhoure[11]	(1975)	962.12 ± 0.39°C
Jung[12]	(1975)	962.06 ± 0.1°C
Coslovi et al.[13]	(1975)	962.10 ± 0.18°C
Ricolfi and Lanza[14]	(1977)	962.05 ± 0.04°C
Present determination (Rounded to 0.005°C)		961.980 ± 0.015°C

pyrometer, collects and processes the data, applies any necessary corrections and calculates the spectral radiance ratio of the two sources and also the temperature of one source in terms of the assigned temperature of the other. For the best accuracy, spectral radiance ratios are confined to the range one to eight. Photomultiplier anode currents are restricted by using absorption filters where necessary so that convenient corrections for the non-linearity of the photomultiplier can be applied.

The method of realization of the scale below 1064.43°C is similar to that described for the realization of IPTS 68 using APEP2 but absorption filters are not necessary in this range. Vacuum high stability tungsten strip lamps are calibrated at temperatures near 1064°C by comparing them with a gold point blackbody to give a first reference temperature. Lamps are then calibrated at a second reference temperature of about 913°C, which is equivalent to a spectral radiance of 1/8 of that of the gold point at 660 nm. These two references are then used to calibrate secondary lamps or measure other sources to spectral radiances of 1/64 of the gold point radiance, that is at temperatures from 1064°C to 792°C at 660 nm.

The uncertainties of the scale arise from contributions by the gold point blackbody, the tungsten strip lamp and APEP2. The details of the various components and their methods of evaluation have been discussed elsewhere[8,9] and the uncertainties for this scale derived in the same manner are given in Table I. All uncertainties are expressed as estimated 99% confidence limits and the overall uncertainties are taken as the square roots of the sums of the squares of the components. The overall uncertainties are rounded up to 0.005°C. It can be seen in Table I that the estimated overall uncertainties for the scale maintained on high stability tungsten strip lamps are 0.025°C at 1064°C, 0.035°C at 913°C and 0.050°C at 792°C.

FREEZING POINT OF SILVER

Recent determinations of the silver point are listed in Table II with uncertainties given as 99% confidence limits[15]. The three most precise of these[12,13,14] have been measured using spectral radiance ratios relative to blackbodies at the freezing point of gold. It was therefore decided to make an independent measurement of the silver point to provide a comparison between low temperature scales based on (1), as well as adding to the information on this important reference temperature.

Two silver point blackbody crucibles were constructed, the design being identical to that of the gold point crucible as described elsewhere[8]. Each crucible contained approximately 435 g of silver of nominal 99.999% purity. Sample 1 was supplied by Johnson Matthey (UK) and sample 2 by Cominco (Canada). The results of the manufacturers' spectrographic analyses are given in Table III. Because small amounts of dissolved oxygen cause large changes in the freezing point of silver, it is essential that this gas be removed, and that the molten metal be protected from it[16]. The crucibles were therefore filled in a vacuum by melting the metal ingots in a graphite filling unit which screwed into the crucibles in place of the lids, and then transferring the molten metal using gravity. The crucibles were cooled while still in the vacuum.

One furnace was used for both silver crucibles. Its design was similar to that of the five zone tube gold furnace previously described[17]. The blackbody crucible occupied the central zone, the remainder of the furnace tube being occupied by components made from graphite and sintered alumina. The ends of the furnace tube were normally covered with plates sealed by O-rings so that the tube was gas-tight. High purity argon gas, which was further purified by passing it over hot titanium was passed through the furnace tube from the back to the front to protect both the silver and the graphite from oxygen. When the front plate was removed for measurements graphite components in front of the crucible prevented atmospheric oxygen from reaching the silver.

A channel in the wall of the crucible was continued to the rear of the furnace to allow temperature profile measurements to be made along the crucible using a thermocouple. The heater zones of the furnace were adjusted to produce a minimum variation in measured temperature in the vicinity of the crucible. The silver was melted by raising and lowering the power to the central zone. The points were approached from about 5 to 10°C above and below the melting temperature at approximately 1°C per minute. The normal duration of a point was 25 min, but times up to 40 min were obtained. The supercool depressions prior to freezing were of the order of 1.5°C.

Measurements of the silver points were made in two ways; first by comparing them with calibrated tungsten strip lamps, and second by direct comparison with a gold point blackbody. The direct method eliminates errors which can be introduced by the strip lamps and should give a more accurate determination, but requires that the temperature plateaux for the two metals be coordinated. Strip lamps provide an unvarying reference which is more convenient to use, and is essential for studying the detailed nature of melting and freezing cycles for thermal analysis or for establishing the techniques necessary for direct comparison. The transfer via strip lamps also is equivalent to a comparison by substitution which will check for any errors which could be present in the direct comparison due to differences between the two viewing positions of the pyrometer.

Over a period of one year, seven independent sets of comparisons, making a total of 36 melts and 33 freezes, were made via five different lamps from the gold point to silver sample 1. The mean silver point temperature obtained, relative to a gold point of 1064.43°C, was 961.975°C. One set of comparisons using three lamps, eight melts and seven freezes was made on sample 2, and the mean silver point temperature obtained was 961.978°C. For the direct comparison silver sample 2 was compared to the gold point blackbody during six melts and seven freezes, and a mean silver point value of 961.983°C was obtained.

The two methods should in principle give exactly the same result, the main difference being the possible error introduced by changes in the tungsten strip lamps between the time of their comparisons to the silver and gold points. Since the strip lamps are used only as a transfer medium, systematic errors inherent in their use

Table III Spectrographic analyses of silver samples, impurities in parts per million

	Sample 1 Johnson Matthey (UK)	Sample 2 Cominco (Canada)
Bi	< 1	
Ca		0.1
Cd	< 1	
Cu	< 1	1
Fe	< 1	1
Mg	< 1	0.1
Na	< 1	
Pb		0.1
Si		0.2

Table IV Equivalent temperature uncertainties in the measurement of the freezing point of silver

Source of uncertainty	Uncertainty °C
Impurities in Ag	0.002
Impurities in Au	0.002
Temperature variations along cavity walls in Ag	0.005
Temperature variations along cavity walls in Au	0.005
Difference in emissivities of cavities	0.002
Difference in size of source effects	0.005
Spectral responsivity	0.007
Nonlinearity	0.003
Random measurement	0.003
Total uncertainty for direct comparison (Rounded up to 0.005°C)	0.013 / 0.015
Tungsten strip lamp uncertainties	0.010
Random measurement	0.003
Total uncertainty using tungsten strip lamp (Rounded up to 0.005°C)	0.017 / 0.020

should cancel. In both cases the systematic errors due to the blackbody cavities and the furnaces will tend to cancel, and the uncertainties for emissivity and size of source in Table IV account for possible incomplete cancellation.

For the physical dimensions of the blackbody cavity and aperture systems used, the calculated drop across the cavity walls in the area viewed is less than 0.001°C and no uncertainty is therefore included for this. The emissivity of the cavities was calculated to be 0.999 95 ± 0.000 03[17] which results in a decrease in the spectral radiance temperature of both points of 0.004°C ± 0.003°C. An uncertainty of 0.002°C was introduced into Table IV to account for slight changes in the physical dimensions between cavities and the uncertainty in the intrinsic emissivities of graphite at the two temperatures.

The size of source effects in the silver and gold furnaces should almost completely cancel, but to check this the effect was measured for both furnaces using the techniques already described[8]. The measurements indicated that the effects for both furnaces were the same within 0.005°C. Consequently the uncertainty relating to the difference in size of source effects between both furnaces was taken as 0.005°C in Table IV.

The most significant component uncertainty in the evaluation of the silver point temperature results from the uncertainty in the spectral responsivity of APEP2. This uncertainty is taken to be equivalent to a translation of the spectral responsivity of 0.05 nm, which results in an equivalent temperature uncertainty in the Ag point relative to the Au point of 0.007°C. The uncertainty in the linearity factor[8] for a ratio of approximately 1/4 has been measured as 0.005% which is equivalent to an uncertainty of 0.003°C in the silver point temperature. The random uncertainty associated with the measurement of a ratio of 1/4 is 0.003°C. The uncertainty due to the purity of the samples is difficult to estimate. The agreement between freezes and melts of the gold and both silver samples is approximately 0.002°C with insignificant differences between samples. This value, 0.002°C, was taken as the uncertainty due to impurity for both silver and gold in Table IV. Both impurity effects and the temperature difference along the cavity contribute to the melting and freezing ranges observed in the realization of fixed points. The maximum melting and freezing ranges observed for Ag and Au under optimum furnace power adjustment was 0.005°C. Although possibly an overestimate this value, 0.005°C, was taken as the uncertainty due to temperature variations along the cavity for both Au and Ag.

The long term uncertainty which relates to the use of tungsten strip lamps at the Au point has been detailed previously[8] and is stated as 0.018°C in Table I. In the measurement of the freezing point of silver relative to gold the tungsten strip lamp is used as a transfer device and the calibration at the gold point is compared with the silver point within days or at the most two weeks. This results in the cancelling or reduction of some of the component uncertainties. The uncertainties due to positioning, ambient temperature, and pin temperature correction remain the same as previously quoted whilst the uncertainties resulting from cleaning, resistance correction and current measurement are reduced. The overall uncertainty which relates to the use of a tungsten strip lamp in the measurement of the Ag point, 0.010°C, is stated in Table IV. The overall uncertainties are taken as the square roots of the sum of the squares of the components. The overall uncertainties applicable to the direct comparison of the silver and gold points are 0.015°C and, when tungsten strip lamps are used, 0.020°C.

The means obtained using tungsten strip lamps and by direct comparison are not significantly different. Consequently our best estimate of the temperature of the freezing point of silver is taken as the average of the three means, which, when rounded to the nearest 0.005°C becomes 961.980°C ± 0.015°C.

POSSIBLE SOURCES OF SYSTEMATIC ERROR

It can be seen from Table II that a difference of 0.09°C exists between the value of the silver point obtained here and the mean of other high precision determinations[12,13,14]. Because this difference is significant with respect to stated uncertainties, all aspects of our measurements were reviewed and all important parameters were remeasured. Some of the possible sources of systematic errors and the steps taken to check them will now be discussed.

Calculation of spectral radiance temperature

The value of the silver point, T(Ag), is obtained by solving equation (1) with T = T(Ag). In APEP2 the integrals are approximated using the method described by Coates[18], and then the equation is solved by an iterative process. The calculations for T(Ag) were checked using another computer program which derives the integrals either using Simpson's rule or a spline curve fitting process. All methods agreed within 0.001°C for T(Ag). The adequacy of the intervals between the measured values of τ_λ was confirmed by calculating the integrals using only alternate points, the result for T(Ag) differing by less than 0.001°C.

Coates[18] has shown that using the peak wavelength of τ_λ as the mean effective wavelength, and hence solving equation (1) as a single wavelength equation gives only a small error for symmetric narrow bandpass filters. This would provide a check on the calculations by eliminating the use of integrals. A measurement was therefore made using a filter with a peak wavelength of 656.2 nm and a half height bandwidth of 0.83 nm. The value for T(Ag) obtained using the peak wavelength was 961.974°C, with a calculated error[18] of less than 0.004°C.

Relative spectral response

The measurement of τ_λ and S_λ has been described[8]. If τ_λ is incorrect, this is likely to be due either to a systematic error in our wavelength scale or to undetected secondary transmission bands in the bandpass filter. To raise the value of T(Ag) by 0.09°C, the wavelengths used would need to be 0.6 nm too high, an error ten times our estimated uncertainty. Since the wavelengths were checked *in situ* using spectral lines such an error is unlikely. The shape of the transmission curve of the 660 nm filter was checked to better than one part in 10^4 around the peak and to a few parts in 10^6 in the wings, an accuracy adequate for our measurement[19]. As an additional check for the presence of secondary transmission bands and stray light in the pyrometer, measurements of spectral radiance ratios were made with high pass and low pass glass filters in the optical path. No significant effects were detected.

Computer simulation showed that because of the comparatively smooth and flat form of S_λ errors in its measurement generally have small effects on T(Ag). The exception to this was the long wavelength cut-off of the photomultiplier, where the bandpass filter still had a significant transmission, and so this region was measured carefully.

The silver point was also measured using two other bandpass filters as a check on τ_λ. One was the narrow band filter already mentioned, the other was similar to the 660 nm one, but its peak wavelength was 700 nm. Although the characteristics of the 700 nm filter were not as carefully measured as those of the 660 nm filter, they were sufficiently well known to establish that it gave the same result to within 0.02°C.

Ratio measurement

The accuracy of the spectral radiance ratio measurement depends upon the accuracy of the so-called linearity factor used to correct for the lack of linearity of the pyrometer. This factor, applied to the measured ratio, effectively reduces T(Ag) by 0.017°C ± 0.003°C. This uncertainty is small compared to the increase of 0.070°C which would be necessary to give T(Ag) = 962.07°C.

The linearity factors were checked by the flux addition method with the beamsplitter assembly previously described[8]. A variation of the method was then applied in which the images of the two lamps in the assembly were superimposed at the pyrometer field stop using a full reflector covering half the acceptance cone of the pyrometer instead of the usual 50% reflecting mirror covering the whole cone. Two types of full reflector were used, one a front surface aluminium type, and the other a glass prism. This experiment was to test for polarization effects in the 50% reflecting mirror. The results in all cases agreed with the original measurements.

It is interesting to note in passing that since the last test produces a split field at the photocathode of the photomultiplier, with each lamp illuminating half the field with constant irradiance, it indicates that the nonlinearity does not occur in the photocathode but in the dynodes.

FREEZING POINT OF COPPER

Some of the possible systematic errors in the measurement of the silver point could be detected by measuring another well known fixed point temperature. Since the freezing point of copper is the only other point which has been measured with sufficient accuracy by photoelectric pyrometry, a copper point blackbody was constructed for its measurement. The crucible was identical to those used for gold and silver and contained 340 g of high purity copper supplied by Asarco, USA. This metal is guaranteed to contain less than 10 ppm impurities. Since copper is subject to contamination by oxygen, precautions were taken in filling similar to those used for silver. A separate furnace was constructed, again similar to those used for the other metals.

It was found that, as reported by other workers[14,20] the melting temperature range of the copper was greater than the freezing range (typically 0.01°C for the melt and 0.002°C for the freeze) and so only freezing points were considered. Measurements were made by comparison to the gold point, with 11 comparisons via a lamp giving a mean of 1084.889°C and 15 direct comparisons giving a mean of 1084.893°C. Our best estimate of the temperature of the freezing point of copper is taken as the average of these two means, which when rounded to the nearest 0.005°C becomes 1084.890°C. The estimated uncertainty components applicable to the measurement of this point were similar to those for the silver point measurement; except that the uncertainty due to the purity of copper and the temperature variations along the cavity wall were each increased to 0.008°C, and the spectral responsivity component was reduced to 0.002°C. The overall uncertainty rounded up to the next 0.005°C was 0.015°C. The mean value obtained for the freezing point of copper, rounded to the nearest 0.005°C, was 1084.890°C ± 0.015°C.

The other recently measured values[14,20,21,22] of the freezing point of copper (relative to a value of 1064.43°C for the gold point) are shown in Table V with uncertainties given as 99% confidence limits. The temperature reported here agrees with these values within their uncertainties. The main implication of this is that the difference in the silver point value is not due to an error in the freezing point of gold, but because the radiance ratio between the copper and gold blackbodies is close to unity, and particularly because the measured copper point is higher than the previously reported values, it does not exclude errors in the measurement of radiance ratios or in wavelength measurement.

Table V Recent determinations of the freezing point of copper relative to the freezing point of gold taken as 1064.43°C

Reference	Determination
Righini et al.[21,22] (1972)	1084.88 ± 0.12°C
Coates and Andrews[20] (1978)	1084.87 ± 0.04°C
Ricolfi and Lanza[14] (1977)	1084.87 ± 0.02°C
Present determination (Rounded to 0.005°C)	1084.890 ± 0.015°C

COMPARISON OF PHOTOELECTRIC PYROMETRY WITH PLATINUM RESISTANCE THERMOMETRY

If the freezing point of silver is to be considered as the transfer temperature between platinum resistance thermometry and photoelectric pyrometry for a future IPTS it is essential that a comparison of the two instruments be practicable over a range of temperatures in the vicinity of the junction. Such a comparison would enable the continuity of both the scales and their first and second derivatives to be checked. This comparison was first attempted using an inconel clad copper block containing a thermometer well and a blackbody cavity in close proximity. The temperature variations in the block over the 100 mm length of the cavity could not be reduced to less than 0.2°C, and such a variation was too large to allow a comparison with sufficient accuracy. The comparison was then performed with a heat-pipe designed by Busse and constructed at the Joint Research Centre of the European Communities (JRC) Ispra, Italy. The blackbody heat-pipe, Fig. 1, is made of inconel with sodium as the working fluid. Its outside diameter is 62 mm and its overall length is 170 mm. The blackbody has an inside diameter of 14 mm and is 110 mm long including a 60° cone at its closed end; the aperture in the front of the blackbody is 3 mm in diameter. The platinum resistance thermometer is inserted from the rear of the heat-pipe

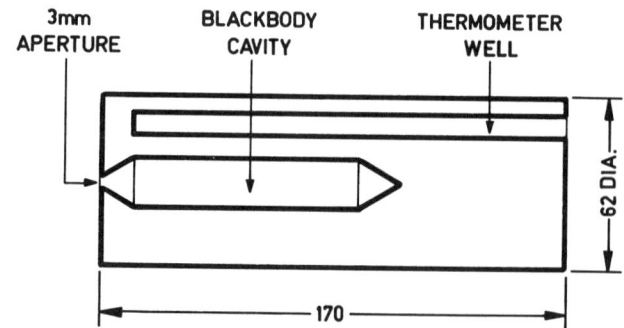

Fig. 1 Heat-pipe blackbody

into a well 160 mm long and 9 mm I.D. In use the well is situated centrally above the blackbody cavity.

The heat-pipe was used in a horizontal furnace 760 mm long with one main heater and with two end heaters behind and one in front of the heat-pipe. All heaters were bifilar wound to reduce electrical interference in the resistance thermometer. The aperture is situated 250 mm from the front of the furnace resulting in a maximum immersion of the platinum resistance thermometer of 500 mm from the rear of the furnace. Annular firebrick spacers were placed in front of the heat-pipe to form a cone with a ratio of length to diameter of 8/1. The rear of the furnace was filled with firebrick and kaowool and a silica guide tube for the thermometer. The guide tube was surrounded with a thin-walled inconel tube and this and the heat-pipe were earthed to reduce electrical interference in the thermometer.

After approximately 500 hours use at 960°C during preliminary experiments, the inconel heat-pipe distorted resulting in the well deflecting from its original axis and bending around one of its internal supports. This distortion reduced the maximum diameter of the thermometer sheath which could be inserted into the well to 5.3 mm. A special thermometer was constructed for this well using the techniques described by McAllan[23], with a Heraeus sensor type 1Pt50:K25.15, having a nominal ice point resistance of 50 Ω, and designed to have very low conduction and radiation losses. Resistance measurements were made with an 80 Hz bridge (Leeds and Northrup, Type 8078) operated from a battery supply to reduce electrical interference.

The power to each of the furnace zones was adjusted to maintain the blackbody at a temperature near the freezing point of silver. The adjustment of powers were not as critical as for the copper block blackbody but the temperature distribution within the heat-pipe was influenced by the relative powers in the front and rear heaters. At optimum adjustment, the temperature variations measured in the length of the well corresponding to the position of the blackbody were less than 0.02°C. Beyond the heat pipe thermal conduction errors appeared in the resistance thermometer measurement but did not influence the temperature measurements over the length of the blackbody.

The experiment was conducted by comparing the radiance of the blackbody cavity of the heat-pipe with that of a tungsten strip lamp calibrated at the gold point. Simultaneously the temperature of the heat-pipe was measured with the resistance thermometer inserted into the well. Twelve separate comparisons were made, with the resistance thermometer calibrated four times in a vertical silver melting point crucible[24] during the course of the measurements. The results were interpreted as an indirect measurement by the pyrometer of the silver point in the vertical furnace. Thus the pyrometer reading was converted to the temperature which would have been measured if the heat-pipe temperature was such that the resistance thermometer was at its silver point calibration resistance. The mean value of the silver point so obtained was 961.97°C, with 99% confidence limits of 0.035°C. This result is not significantly different from the direct measurement of the silver point by the pyrometer, and the experimental uncertainty is consistent with the estimated uncertainties of 0.03°C for the resistance thermometer measurements[23] and 0.025°C for the pyrometer readings.

This agreement shows that the intercomparison of pyrometer and resistance thermometer scales in the region of the silver point is feasible. It also indicates that the melting temperatures of the silver samples used in the blackbody crucibles and the vertical furnace are similar. Other sources of systematic errors in the pyrometer measurement of the silver point would be present also in the measurement of the heat-pipe blackbody temperature and therefore would not be detected.

The distortion of the heat-pipe prevented the use of a better quality resistance element, so we are unable to comment on comparisons with such a thermometer at the silver point. However the Heraeus element changed by the equivalent of 0.3°C at the silver point during the measurements which made it difficult to use compared to the pyrometer.

Whereas the design of the heat-pipe seems satisfactory in terms of its temperature uniformity and thermal coupling between the blackbody cavity and the thermometer well, it was reported to be difficult to construct and its structural integrity needs to be improved, particularly if it is to be used at higher temperatures. Because of these problems this work is to continue in a cylindrical gas-controlled sodium heat-pipe, also constructed at JRC[25].

CONCLUSION

It has been shown that photoelectric pyrometry can realize a temperature scale below 1064°C on high stability tungsten strip lamps with uncertainties of 0.025°C at 1064°C, 0.035°C at 913°C and 0.050°C at 792°C. The freezing point of silver has been measured to compare this scale with those realized in a similar way in other laboratories and the value obtained was 961.980 ± 0.015°C. Although this is the lowest uncertainty so far claimed for such a determination, the value obtained is significantly different from other pyrometry measurements. Extensive investigations have failed to locate any systematic errors in our determination. As one of the checks of technique, the freezing point of copper has been determined to be 1084.890°C ± 0.015°C, which is not significantly different from other recent measurements. It has been shown that an inconel heat-pipe blackbody using sodium as a working fluid is satisfactory for the comparison of platinum resistance thermometers and photoelectric pyrometers at temperatures near the silver point.

The significant difference between our value of the temperature for the freezing point of silver and others using photoelectric pyrometry is serious in that it shows that systematic differences of the order of 0.09°C can occur in very careful establishments of photoelectric pyrometry temperature scales below 1064°C. However we believe that the systematic difference will be resolved. We hope to conduct an intercomparison of tungsten strip lamps in the range 960°C to 1064°C with other laboratories which should identify specifically the cause of the difference. When the systematic error is removed we expect that the uncertainties which we have stated will still apply and it is these which should be considered in deciding future changes to the IPTS.

ACKNOWLEDGEMENT

It is a pleasure to acknowledge the assistance of S.R. Meszaros who constructed all blackbody crucibles and furnaces used in this investigation. We also wish to thank C.A. Busse and his colleagues at JRC, Ispra, for providing the heat-pipe blackbody.

REFERENCES

1. Comité International des Poids et Mesures: Comité Consultatif de Thermométrie 11e Session - p T8 (1976).
2. H. Preston-Thomas, Metrologia 12, 7 (1976).
3. E. H. McLaren and E. G. Murdock, Temperature Its Measurement and Control in Science and Industry, H. H. Plumb (Ed) Pittsburg: Instrument Society of America, 4, Part 3, 1543-1560 (1971).
4. T. P. Jones, Metrologia 4, 80 (1968).
5. J. P. Evans and S. D. Wood, Metrologia 7, 108 (1971).
6. N. Takiya, Bul. of the NRLM Series 33, 23 (1976).
7. P. Marcarino and L. Crovini, Temperature Measurement 1975, B. F. Billing and T. J. Quinn (Eds) Institute of Physics Conference Series 26, London and Bristol: Institute of Physics, 107-116 (1975).
8. T. P. Jones and J. Tapping, Metrologia (accepted).
9. T. P. Jones and J. Tapping, Metrologia 15, 135 (1979).
10. T. J. Quinn, T. R. Chandler and M. V. Chattle, Metrologia 9, 44 (1973).
11. J. Bonhoure, Metrologia 11, 141 (1975).

12. H. J. Jung, Temperature Measurement 1975, B. F. Billing and T. J. Quinn (Eds) Institute of Physics Conference Series 26, London and Bristol: Institute of Physics, 278-286 (1975).
13. L. Coslovi, A. Rosso and G. Ruffino, Metrologia 11, 85 (1975).
14. T. Ricolfi and F. Lanza, High Temp. - High Pressures 9, 483 (1977).
15. T. J. Quinn, L. A. Guildner and W. Thomas, Metrologia 13, 175 (1977).
16. G. Bongiovanni, L. Crovini and P. Marcarino, Metrologia 11, 125 (1975).
17. T. P. Jones and J. Tapping, Metrologia 8, 4 (1972).
18. P. B. Coates, Metrologia 13, 1 (1977).
19. P. B. Coates, High Temp. - High Pressures 11, 289 (1979).
20. P. B. Coates and J.W. Andrews, J. Phys. F. 8, 277 (1978).
21. F. Righini, A. Rosso and G. Ruffino, High Temp. - High Pressures 4, 471 (1972).
22. Comité International des Poids et Mesures: Comité Consultatif de Thermométrie 10e Session - p T47 (1974).
23. J. V. McAllan, "Practical high temperature resistance thermometry" (This Symposium).
24. J. V. McAllan, "Reference temperatures near 800°C" (This Symposium).
25. C. Bassani, C. A. Busse and F. Geiger, High Temp. - High Pressures 12, 351 (1980).

Measurement of the thermodynamic temperature interval between the freezing points of silver and copper

M. Ohtsuka[a] and R. E. Bedford

Division of Physics, National Research Council of Canada, Ottawa, Ontario, Canada, K1A 0R6

The thermodynamic temperature interval between the freezing points of silver and copper has been measured with a monochromatic optical pyrometer. Graphite blackbody cavities were immersed in ingots of silver and copper contained in graphite crucibles. The crucibles were placed in Inconel heat pipe liners (using sodium as the working substance) and heated in conventional wire-wound resistance furnaces. Freezing and melting plateaus lasting several hours and stable to within the photomultiplier drift were easily obtained. No significant difference was found between freezing and melting temperatures for either metal. The ratios of the spectral radiances of the freezing points of copper and silver were measured at five wavelengths defined by interference filters, three with 10 nm and two with 1 nm half widths. The copper-point radiance was reduced to near that of silver with rotating sectored disks. Relative to a copper point of 1358.03 K, the freezing point of silver was measured to be 1235.20±0.06 K, i.e., the copper-silver interval was found to be 0.12±0.06 K smaller than given in the IPTS-68. This is in agreement with previous measurements of the copper and silver freezing points relative to that of gold.

INTRODUCTION

In connection with a proposed revision of the International Practical Temperature Scale of 1968 (IPTS-68), and especially with the probable replacement of the platinum 10% rhodium/platinum (10/0) thermocouple by the platinum resistance thermometer (PRT) as a standard instrument in such a revision, it is imperative to know as accurately as possible the thermodynamic temperature intervals between the silver (Ag), gold (Au), and copper (Cu) freezing points. Each of these has been considered as a possible junction temperature between the PRT and optical pyrometer ranges. The two points not chosen as the junction will undoubtedly be named as primary or secondary reference temperatures in the new Scale, so the degree of thermodynamic compatibility* of practical temperatures within each range and between the two ranges will be influenced by the accuracy of these temperature intervals.

The most accurate method presently available for measuring these intervals is with a monochromatic optical pyrometer (MOP). Several authors have published the results of such measurements in this region. Quinn et al[1] measured the temperature of a blackbody cavity between 1060 K and 1337 K with both a calibrated MOP and calibrated 10/0 thermocouples. From these they estimated that the thermodynamic Au-Ag interval (ΔT_{Au}^{Ag}) is 0.12 K[2] smaller than defined by the IPTS-68 ($\Delta T_{Au}^{Ag}(68)$). Similar measurements by Bonhoure[3] using the freezing point of antimony rather than gold as reference gave $\Delta T_{Au}^{Ag}(68)-\Delta T_{Au}^{Ag}=0.13$ K or 0.19 K. The first value was from the mean of all of his measurements; the second value, which Bonhoure preferred, considered only the direct comparisons between the two temperatures. Note that in Quinn's experiment a silver point blackbody was not used, and in Bonhoure's neither silver nor gold was used. Coslovi et al[4] used blackbodies immersed in freezing silver and gold and using silver as reference obtained $\Delta T_{Au}^{Ag}(68)-\Delta T_{Au}^{Ag}=0.17$ K. A tungsten strip-filament lamp was used as an intermediate transfer between the silver and gold points. Their pyrometer differed from all others mentioned here in using a silicon photodiode rather than a photomultiplier tube as detector. In a similar experiment (involving also a strip lamp transfer) Ricolfi and Lanza[5] measured $\Delta T_{Au}^{Ag}(68)-\Delta T_{Au}^{Ag}=0.12$ K. They also measured the gold-copper interval (ΔT_{Au}^{Cu}) to be 20.44 K. Jung[6] measured the silver-gold interval relative to silver without any intermediate transfer with the result $\Delta T_{Au}^{Ag}(68)-\Delta T_{Au}^{Ag}=0.13$ K. Righini et al[7] used the same pyrometer as Coslovi et al[4] to measure $\Delta T_{Au}^{Cu}=20.45$ K[8]. Finally, Coates and Andrews[9] made a direct comparison between gold and copper blackbodies to obtain $\Delta T_{Au}^{Cu}=20.44$ K.

We report here a direct measurement of the thermodynamic temperature interval between the freezing points of copper and silver (ΔT_{Cu}^{Ag}) with a monochromatic photoelectric optical pyrometer. Although values of ΔT_{Cu}^{Ag} can be obtained by combining the results summarized above, this is to our knowledge the first direct measurement of ΔT_{Cu}^{Ag}.

APPARATUS

A schematic diagram of the optical pyrometer and associated readout is shown in Fig. 1. The matched objective achromatic lenses of 52 mm diameter and 508 mm focal length project an image of the target on to a circular field stop that defines the area of the target. An iris diaphragm between these lenses acts as

---

* "Thermodynamic compatibility" is used to mean that second and higher derivatives of T_p with respect to T are small, where T_p is the practical temperature on the revised IPTS.

an aperture stop that can be varied from 51 to 20 mm, corresponding roughly to f/10 to f/26. A set of interchangeable laser-drilled field stops of diameters ranging from 0.1 to 1.0 mm is available; most of these measurements used either 0.4 or 0.6 mm. The radiant flux that passes through the field stop is directed to the cathode of a photomultiplier tube (PMT) by a suitable optical system through a narrow-band interference filter that is mounted in a location where the beam is collimated. Five different filters were used: three of these have half-widths* of about 10 nm centred at 600.0, 657.9, and 781.4 nm respectively, and two have half-widths of about 1 nm centred at 655.7 and 666.9 nm respectively. The peak transmittances of these filters are 0.54, 0.62, 0.45, 0.34, and 0.39 respectively. The three 10 nm filters are 3-cavity with roughly rectangular pass bands; the two 1 nm filters are 2-cavity with somewhat triangular-shaped pass bands fairly sharply peaked. All filters are blocked outside the pass band to at least 10^{-4} and usually to 10^{-6}. A flash-opal diffuser is placed in front of the photocathode to reduce the possibility of the radiance measurements being affected by areal changes of sensitivity of the photocathode. We used both EMI 9558 (S-20 response) and EMI 9658 (extended S-20 response) PMT's, but mostly the former. The high voltage supply to the PMT provided regulation of 1 in 10^5. The cathode-first dynode voltage was held roughly constant at 157 V with a Zener diode, with inter-dynode resistors of 100 kΩ thereafter. The temperature of the PMT and of the interference filter was controlled to better than ± 0.1 K near 298 K by circulating temperature-regulated water through an enclosing jacket. The anode current of the PMT was measured from the voltage drop across a 1MΩ resistor, both with and without an associated operational amplifier working at unit amplification. We could detect no difference between the results in these two modes, but mostly the op-amp was used because it ensured that the anode-to-ground potential was independent of anode current. The anode current was read from a digital voltmeter, either directly, from a printer, or from a strip chart recorder that displayed the last three current digits after digital-to-analogue conversion. A neutral filter in the form of a rotating sectored disk can be inserted immediately in front (source side) of the field stop to reduce the radiant flux from the copper-point blackbody by a measured amount. Three disks of transmittances 0.25, 0.20, and 0.18 respectively were used.

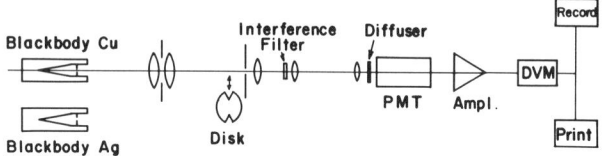

Fig. 1: Schematic diagram of optical pyrometer.

The freezing points of the metals were realized in identical blackbody crucibles of the sort shown in Fig. 2. The graphite blackbody cavity has the shape of a cylindro-cone with 60° apex angle, 13 mm internal diameter, and tapered lid with an aperture of 1.5 mm diameter. This cavity is surrounded by the freezing metal contained in a cylindrical graphite crucible. The central brace perforated with twelve 3 mm diameter holes supports the cavity and prevents its floating in the liquid metal. The metal extends well forward of the 1.5 mm radiating aperture to help ensure that the entire cavity including the region near the aperture is

---

* We use the term half-width to mean the full spectral width of the transmitted flux at half the peak transmittance.

maintained at the temperature of the melting or the freezing metal. The effective emissivity of the cavity was computed by the method of Bedford and Ma[10] to be 0.99997 ± 0.00003.

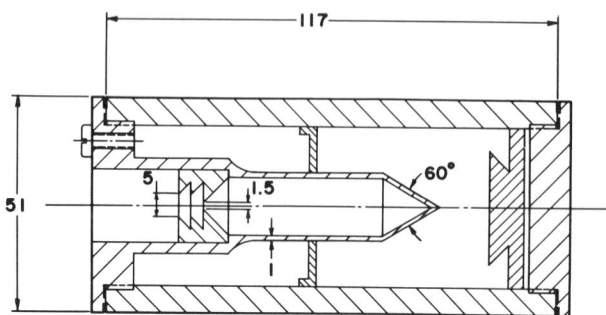

Fig. 2: Geometry of blackbody crucible (dimensions in millimetres).

The copper and silver blackbody crucibles were contained in resistance-heated furnaces of the sort shown in Fig. 3. Each furnace is 61 cm long by 30 cm diameter with a 9 cm I.D. aluminum oxide core on which about 45 turns of 1 cm by 0.04 cm Kanthal ribbon is uniformly wound along a 56 cm length. A sodium-filled Inconel heat pipe liner 9 cm O.D. by 5.2 cm I.D. by 30 cm long is placed centrally in the core. Inside the heat pipe is placed the crucible, behind which is a scavenger block of graphite and in front of which are scavenger rings of graphite. The temperature of the furnace is monitored with a 10/0 thermocouple enclosed in a fused silica sheath extending from the rear of the furnace to the back of the crucible. The rear of the core is insulated with Inconel heat shields and packed quartz wool, as is the front except for a 4 cm channel through which the pyrometer views the blackbody. The furnace is insulated with quartz wool and is water-cooled. Argon is fed from the rear along the core at about 300 cm^3/min to inhibit graphite oxidation. The furnaces are operated from voltage-regulated a.c. supplies: at equilibrium about 900 W is required for silver and 1.2 kW for copper. Both furnaces operated continually above 1100 K (occasional excursions below 1100 K on weekends) for many weeks with little deterioration of the graphite of the blackbody and with only occasional need for replacement of the scavenger graphite rings.

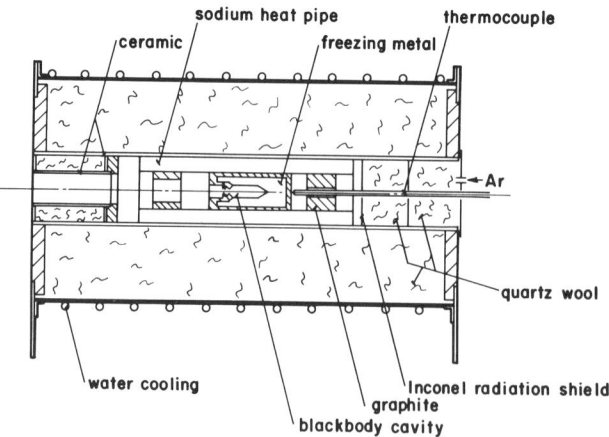

Fig. 3: Furnace used for realization of copper and silver freezing points.

The furnaces are mounted on rails in such a way that, once correctly aligned, the blackbody apertures can be repeatedly located on the optical axis of the pyrometer to within ± 0.05 mm. No problem was experienced with motion of the crucibles within the furnaces. Provision is made for precise adjustments in the lateral, axial, and vertical directions.

All of this system is mounted on a vibration-isolated optical table to ensure no relative motion between components and to remove any possible effects from vibration.

EXPERIMENTAL PROCEDURE

We fill each crucible by mounting it vertically in an auxiliary furnace with the screw in the crucible (Fig. 2) removed. The metal in the form of 8 to 12 mm diameter rods is placed in a graphite funnel, a spout in the bottom of which fits snugly into the screw hole. The funnel is capped with a tight-fitting graphite lid above which are placed several graphite disks. After melting, the metal flows into the crucible, filling it completely (about 800 g in the case of silver and about 650 g for copper). An excess of metal must be used in the funnel because the liquid metal has so high a surface tension that it will not flow through the spout without a head of several millimetres. We originally experienced some breakage of copper-filled crucibles but this difficulty was completely eliminated with the design in Fig. 2. The plug at the bottom is free to slide. Before filling the crucible we place a wafer of graphite about 1 mm thick between the plug and the removable bottom. After the crucible is filled, the wafer is removed. Thereafter the plug, the tapered portion of which is embedded in the metal, is free to move back and forth as the metal expands and contracts.

To obtain a freeze with either metal, the furnace temperature was allowed to drop at about 0.08 K/min until recovery from the supercool began, at which time the furnace power was increased to some pre-determined value that would maintain the furnace temperature a few degrees below the freezing temperature. A melt was obtained with an inverse procedure, the power reduction being made when the 10/0 monitoring thermocouple indicated a given interval before the onset of melting.

There was no difficulty in obtaining melting or freezing plateaus that lasted many hours. Typically, we used about a 1 h melt followed by a 3 to 4 h freeze. The amount of supercooling was typically about 1 K with both silver and copper.

The ratio of the spectral radiances of the freezing points of copper and silver was measured by moving each furnace alternately on to the optical axis of the pyrometer for equal periods of time (usually 3 minutes) after freezing or melting plateaus had been established for both metals. The sectored disk was moved into the beam when copper was being viewed to reduce the copper point radiance to near that of silver, thereby avoiding a correction for nonlinearity of the PMT. Occasionally the background was measured by blocking the radiation from the source with a stop on the pyrometer axis. A typical melting point comparison is shown in Fig. 4 and a freezing point comparison in Fig. 5. Most comparisons were made with freezing plateaus.

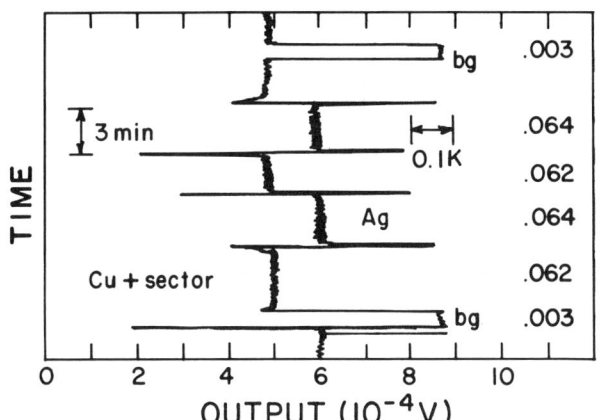

Fig. 5: Typical comparison of copper and silver freezing points using a sector disk with copper. Background (bg) measurements are also shown. Chart record is of last three digits of DVM output. The first three digits (V) are written along the right hand side.

When a PMT is first mounted in its housing its output initially drifts rapidly, the drift decreasing to a tolerable level after about a week. During the first few days of drift accurate measurements are impossible. We were able to maintain PMT stability by keeping the PMT continuously exposed to approximately the silver point radiance with a tungsten lamp when measurements were not in progress. Under these conditions PMT drift was almost constant at 0.007%/h.

RESULTS

If we denote by R the ratio of the spectral radiances of the copper and silver points, then

$$R = \frac{\int_{\lambda_1}^{\lambda_2} L_\lambda(T(Cu)) \tau_\lambda s_\lambda d\lambda}{\int_{\lambda_1}^{\lambda_2} L_\lambda(T(Ag)) \tau_\lambda s_\lambda d\lambda} \quad (1)$$

where $L_\lambda(T)$ is the Planck function, λ is the wavelength in air, τ_λ is the spectral transmittance of the interference filter, s_λ is the spectral sensitivity of the PMT, and λ_1 and λ_2 are the lower and upper wavelengths between which τ_λ is non-zero. It is assumed in Eq. (1) that the emissivities of the two blackbody crucibles are identical and wavelength independent, and that the transmittances of all other optical components of the pyrometer are wavelength

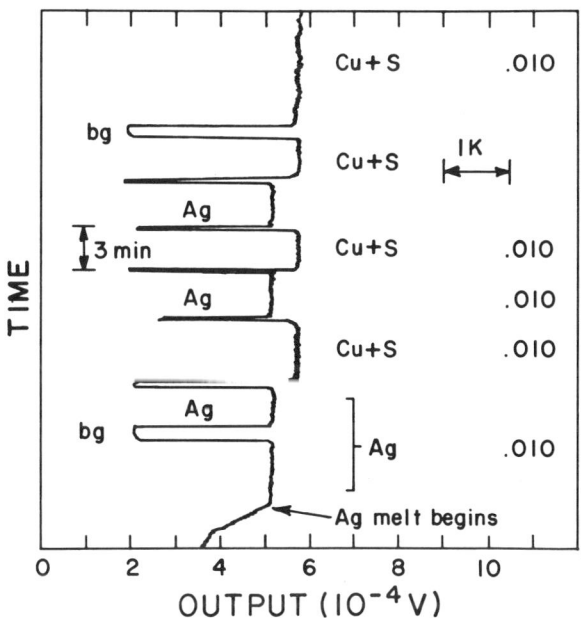

Fig. 4: Typical comparison of copper and silver melting points using a sector disk (S) with copper. Background (bg) measurements are also shown. Chart record is of last three digits of DVM output. The first three digits (V) are written along the right hand side.

independent. When the sectored disk of transmittance τ is used, the measured quantity is τR. Apart from a multiplicative constant $L_\lambda(T)$ is given by

$$L_\lambda(T) = \lambda^{-5} / [\exp(c_2/n\lambda T) - 1] \quad (2)$$

where n is the refractive index of air (taken to be 1.00028 at 22°C). If one or other of T(Cu) or T(Ag) is assigned, the other may be calculated from the measured ratio τR using Eq. (1).

The spectral transmittances of the interference filters were supplied by the manufacturer and were also measured by the Optics Section of the Division of Physics of NRC. There was close agreement between the two sets of measurements.

We have taken for T(Cu) the value 1358.03 K given in the list of secondary reference points associated with the IPTS-68[11] and calculated T(Ag) by iteration. For a first approximation to T(Ag) we used the value obtained from

$$\tau R = \frac{\tau L_{\lambda_0} T(Cu)}{L_{\lambda_0} T(Ag)} \quad (3)$$

where λ_0 is the centre wavelength of the filter pass band.

We have made many measurements of τR using the five filters and three disks. For the 10 nm filter with λ_0=657.9 nm and disk with τ∼0.25 the results are shown in Fig. 6. The dominant feature here is the dependence of τR on PMT anode current. The measured values decrease about 0.5% as the anode current decreases from 1μA to 30nA, but approach a constant value at currents below this. The same effect occurred with the other two 10 nm filters and with different disks. In an attempt to unravel this phenomenon we made the following tests: changed the cathode irradiance by using different sectored disks, by using no sectored disk, by varying the sizes of aperture and field stops, by attenuating both sources equally with auxiliary glass filters; changed the anode current for a constant cathode irradiance by varying the PMT voltage. We also changed PMT's. All of these data points (for disk with τ∼0.25) are included in Fig. 6, indicating that the effect is independent of the level of cathode irradiance. There was some indication that when a glass filter is used for τ instead of a disk the value of R was less dependent on current. Only a few measurements were made in this case because we could not measure the transmittance of the glass filter accurately enough to be of use in obtaining T(Ag). On the other hand, the nature of Fig. 6 does not result simply from the disk because when we measure the ratio $\tau L_\lambda(T(Cu))/L_\lambda(T(Cu))$ (i.e. measure τ) we find τ is independent of anode current, as shown in Fig. 7. This figure also indicates that any PMT nonlinearity is less than 0.03%, i.e. one-half the total spread of the measured points.

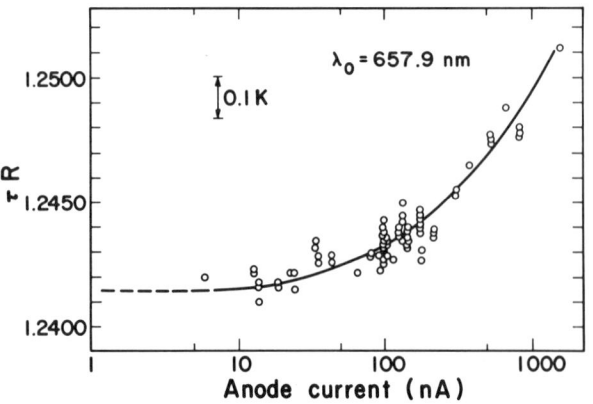

Fig. 6: Measured values of τR with 10 nm filter with λ_0=657.9 nm at different anode currents.

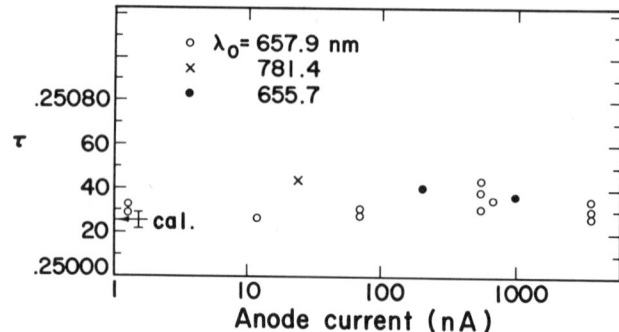

Fig. 7: Measured values of τ of sector disk, i.e. measured values of $\tau L_\lambda(T(Cu))/L_\lambda(T(Cu))$, at different anode currents. The calibrated transmittance of this disk (cal.) is also shown.

With the 1 nm bandwidth filters we obtain a constant value of τR unless the anode current is excessive (>300 nA). This is shown in Fig. 8 where normalized values of τR are plotted so that the results for both filters can be superposed.

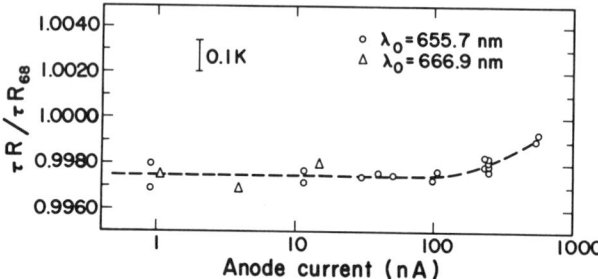

Fig. 8: Plot of normalized measured values of τR with two 1 nm filters at different anode currents. The normalization factor is $\tau R_{68} = \tau \exp(c_2/\lambda_0 T_{68}(Cu))/\exp(c_2/\lambda_0 T_{68}(Ag))$.

Figure 9 shows normalized curves representing the means of all of our measurements. There is a significant difference between the asymptotic values from the 10 nm filters and the 1 nm filters, equivalent to about 0.08 K.

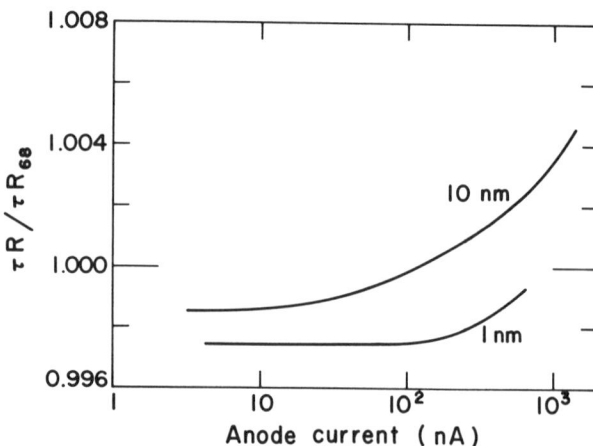

Fig. 9: Mean normalized values of all measured ratios τR as functions of anode current.

From these measured values of τR we calculated T(Ag) from Eqs. (3) and (1) in the manner described. The average results for each filter are given in Table I where also we list the corresponding values of

TABLE I: MEASURED VALUES OF SILVER FREEZING TEMPERATURE RELATIVE TO T(Cu)=1358.03 K AND DIFFERENCES OF THE MEASURED Cu-Ag INTERVAL FROM THAT GIVEN BY THE IPTS-68.

λ_0(nm)	half-width (nm)	T(Ag)(K)	$\Delta T_{Cu}^{Ag}(68) - \Delta T_{Cu}^{Ag}$ (K)
600.0	10	1235.20	0.12
655.7	1	1235.25	0.17
657.9	11	1235.15	0.07
666.9	1.3	1235.24	0.16
781.4	11	1235.15	0.07

$\Delta T_{Cu}^{Ag}(68) - \Delta T_{Cu}^{Ag}$. A comparison of our result with those of other authors is given in Table II. We see that there is good overall agreement.

UNCERTAINTIES

Many factors contribute to the overall uncertainty in our result:

a) Spectral transmittance of interference filter:
uncertainty in τ_λ is the major source of uncertainty in T(Ag). The uncertainty in the wavelength drive of the spectrophotometer used to measure τ_λ was ±0.3 nm, corresponding to ±0.05 K.

b) Radiance ratio:
uncertainty in the measurement of the spectral radiance ratio itself was about ±0.02%, corresponding to ±0.02 K. We have estimated this from the signal-to-noise ratio of the PMT output, which was rather less than ±0.02%, and from the consistency of several measures of τR on a given pair of melts or freezes, which was usually somewhat better than ±0.02%. Errors due to stray radiation were insignificant - the background reading was marginally larger than the dark current, was less than 0.1% of the signal, and did not vary significantly from day to day.

c) Sectored disk:
uncertainties in the transmittances of the sectored disks were at most ±.01$_4$%, corresponding to ±0.01 K. Consistency between results with different disks supported this. We also confirmed that the speed of rotation of the disk was unimportant.

d) PMT:
as indicated earlier, drift of the PMT output was controlled in such a way that it contributed little error to the measurements. Similarly, nonlinearity was not important. Fig. 7 shows that for a radiance ratio of 5, nonlinearity was less than 0.03%, so for the measured ratios near unity nonlinearity should be <0.01%. We have not measured s_λ of the PMT but have used the manufacturer's values which give s_λ decreasing approximately linearly as λ increases. Tests have shown[12] that varying the slope of s_λ vs λ from zero to twice the manufacturer's value affects T(Ag) by only ±0.01 K for the 10 nm filters and not at all for the 1 nm filters. This agrees with the analysis of Coates[13] who also points out that for this type of PMT the slope differences amongst various tubes are <20%. Thus the uncertainty caused by uncertainties in s_λ is negligible.

e) Blackbody cavity:
no significant error is associated with the blackbody cavity. The emissivity is known to very high accuracy, but in any case no error is incurred so long as the emissivity is the same for both cavities. Temperature gradients along the cavity were also not significant. We measured the temperature along the 30 cm heat pipe near 1200 K (when the crucible was not present) with a 10/0 thermocouple. Over 22 cm from the rear the temperature was constant to better than 0.1 K, thereafter dropping slightly by about 1 K at the mouth. The crucible was placed in this constant temperature zone. Normally the pyrometer was focused at a point slightly inside the cavity aperture. By moving the furnace axially about 20 mm we determined that the PMT output was independent of focal position so long as the aperture edge did not restrict the pyrometer aperture. Furthermore, Chen et al[14] show that in the rear of such a cavity the temperature of the internal graphite surface is << 0.01 K different from the metal freezing temperature. The pyrometer views only the rear part of the conical portion of the cavity, and Bedford and Ma[10] showed that even extreme temperature differences outside the viewed region will not materially affect the effective emissivity of the cone. We occasionally checked that the cavity aperture was properly located by moving the furnace vertically and laterally. The effective emissivity of the graphite face around the aperture is ~0.998, and the equivalent apparent change in temperature was easily detectable.

TABLE II: MEASUREMENTS OF COPPER AND SILVER FREEZING POINTS

Author	Cu (re Au) (K)	+Ag (re Au) (K)	Ag (re Cu) (K)	$\Delta T_{Au}^{Ag}(68) - \Delta T_{Au}^{Ag}$ (K)
*Quinn et al[1-2]		1235.20 ± 0.10		0.12 ± 0.10
*Bonhoure[3]		1235.20 ; 1235.25		0.12 ± 0.20 ; 0.17 ± 0.13
**Coslovi et al[4]		1235.22 ± 0.06		0.14 ± 0.06
**Ricolfi et al[5]	1358.02 ± 0.02	1235.20 ± 0.02		0.12 ± 0.02
Jung[6]		1235.19 ± 0.05		0.11 ± 0.05
**Righini et al[7-8]	1358.03 ± 0.10			
Coates and Andrews[9]	1358.02 ± 0.02			
Present work			1235.20 ± 0.06	0.12 ± 0.06++

* No silver fixed point used (or gold in the case of Bonhoure)
** Tungsten strip filament lamp transfer between the two fixed points.
+ All of the published values have been adjusted to a reference temperature of T(Au)=1337.58 K.
++ Difference is Cu-Ag which should be equal to Au-Ag.

f) Size-of-source:
in some pyrometric measurements a systematic error can arise if the two sources are of different size. Here the sources are of identical size, so no error occurs.

g) c_2:
the uncertainty in c_2 is too small to cause a significant error in T(Ag).

h) Impurities:
both the copper and silver were of nominal purity 0.999 999. We used two ingots of copper from a single lot and one ingot of silver (although a second from the same lot was used extensively for some preliminary associated measurements). Within the experimental uncertainty we detected no significant difference between the freezing and melting temperature of silver, nor between any of the freezing and melting temperatures of the two copper samples. We had samples of the as-received materials, and samples from one copper ingot and the preliminary silver ingot analyzed for impurities by quantitative dc-arc emission spectrography. The results, shown in Table III, confirm the manufacturers' purity claims, and show no contamination of the silver after ~ 600 h use above 1100 K. On the other hand, the iron and nickel content of the copper increased significantly, but after ~ 2200 h use above 1200 K the copper purity was still 0.999 99. We expect this to cause no significant change in the freezing temperature. We estimate that any uncertainty in the results due to solution of oxygen in the silver or copper is <0.01 K. The two copper ingots prepared at different times gave the same ratio, indicating that the freshly-prepared ingot had the same freezing temperature as the one used for several hundred hours. Direct comparison[12] between a freshly-prepared silver ingot and a used one also showed identical freezing temperatures. In addition, Bongiovanni et al[15] showed that a graphite crucible is sufficient to protect silver from contamination by oxygen to better than 0.01 K. They found, in fact, that bubbling argon containing 1% oxygen directly into the liquid silver depressed the freezing point by <4 mK.

i) Copper freezing point:
Uncertainties a)-h) determine the accuracy of the silver freezing point measured with respect to the assigned copper freezing point. To relate T(Ag) to T(Au) we must include any uncertainty in the value of T(Cu). From the pertinent references in Table II we conclude that the uncertainty in T(Cu) is <±0.02 K. This leads to an uncertainty in T(Ag) of <±0.01$_6$ K.

TABLE III: MEASURED IMPURITY ANALYSES (µg/g) OF AS-RECEIVED AND USED Cu AND Ag.

Impurity	Cu (as-received)	Cu (used)	Ag* (as-received)	Ag (used)
Fe	0.58	3.3	1.2	0.9
Ni	---	9.4	---	0.28
Si	0.028	0.39	0.2	0.017
Al	0.15	0.043	0.1	---
Pb	---	---	---	---
Mg	0.21	0.16	0.24	0.08
Cr	0.037	4.5	0.034	0.034

* Average of 3 samples. In an independent batch analysis Al, Si, V, Mn, Fe, Zn, As, Mg were detected totalling about 1 µg/g; P, Na, S, Cl, K, Ca, Ti were detected totalling about 0.4 µg/g.

These estimates of uncertainty are summarized in Table IV. The root mean square uncertainty is ±0.06 K, giving for our measurement of the silver freezing point the average of the values in Table I, 1235.20 ± 0.06 K. Thus we measure the Cu-Ag (and so the Au-Ag) interval to be 0.12 ± 0.06 K smaller than given by the IPTS-68.

TABLE IV: ESTIMATED UNCERTAINTIES IN MEASUREMENT OF SILVER FREEZING TEMPERATURE

Parameter	Uncertainty (K)
spectral transmittance of filter (τ_λ)	± 0.05
experimental radiance ratio (τR)	± 0.02
transmittance of sector (τ)	± 0.01
PMT spectral sensitivity (s_λ) and nonlinearity	<± 0.01
blackbody cavity (ε)	---
size-of-source	---
second radiation constant (c_2)	---
impurities	<± 0.01
uncertainty in T(Cu)	± 0.01$_6$
root mean square uncertainty	± 0.06

ACKNOWLEDGMENT

We thank Mr. C.K. Ma for his considerable assistance with all phases of this experiment.

REFERENCES

a Guest worker with the Division of Physics, N.R.C. Permanent address: Tokyo Institute of Technology, Tokyo, Japan.

1 T.J. Quinn, T.R.D. Chandler, and M.V. Chattle, Metrologia 9, 44-46 (1973).

2 The value 0.15 K published in (1) for $\Delta T_{Au}^{Ag}(68) - \Delta T_{Au}^{Ag}$ was corrected to 0.12 K to allow for the refractive index of air (T.J. Quinn, Metrologia 10, 115 (1974)).

3 J. Bonhoure, Metrologia 11, 141-150 (1975).

4 L. Coslovi, A. Rosso, and G. Ruffino, Metrologia 11, 85-87 (1975).

5 T. Ricolfi and F. Lanza, High Temperatures - High Pressures 9, 483-497 (1977).

6 H.J. Jung, Temperature Measurement 1975 (B.F. Billing and T.J. Quinn, eds., Conference Series Number 26, Institute of Physics, London, 1975) pp. 278-286.

7 F. Righini, A. Rosso, and G. Ruffino, High Temperatures - High Pressures 4, 471-475 (1972).

8 The value $\Delta T_{Au}^{Cu}=20.40$ K quoted in (7) was subsequently corrected (5) by 0.05 K to allow for impurities in the samples and for emissivity differences between the two blackbodies.

9 P.B. Coates and J.W. Andrews, J.Phys.F.: Metal Physics 8, 277-285 (1978).

10 R.E. Bedford and C.K. Ma, J. Opt. Soc. Am. 65, 565-572 (1975).

11 L. Crovini, R.E. Bedford, and A. Moser, Metrologia 13, 197-206 (1977).

12 R.E. Bedford and C.K. Ma, this Symposium.

13 P.B. Coates, Metrologia 13, 1-5 (1977).

14 Chen Hongpan, Chen Shouren, and Chu Zaixiang, Metrologia 17, 59-63 (1981).

15 G. Bongiovanni, L. Crovini, and P. Marcarino, Metrologia 11, 125-132 (1975)

The NIM's photoelectric comparator and the realization of the IPTS-68 above the gold point

Zhao Qi, Den Sixiang, Sun Dinwen, Qiu Nairong, Li Zhenguo, and Li Erming

National Institute of Metrology, Beijing, China

As a national primary standard, the photoelectric comparator at the National Institute of Metrology (NIM) is used to realize the IPTS-68 above the gold point. It makes use of a null detection system of sensitivity 0.03 °C in order to equalize the radiances of two sources. In this mode of operation it can calibrate a pyrometric lamp against a blackbody source at the gold point or against another lamp. By combining the radiances from two sources and comparing with the radiance of a third, it is possible to extend the scale upwards from the gold point. It operates at an effective wavelength of ca. 661 nm, which is determined to ±0.2 nm. An uncertainty of ±0.07 °C is achieved at the gold point and of ±1.2 °C at 2000 °C.

INTRODUCTION

According to the definition of the International Practical Temperature Scale of 1968 (IPTS-68), the temperature scale above 1064.43 °C is obtained from the following formula:

$$\frac{L_\lambda(T)}{L_\lambda(T_{Au})} = \frac{\exp(C_2/\lambda T_{Au})-1}{\exp(C_2/\lambda T)-1} \quad (1)$$

where $L_\lambda(T)$ and $L_\lambda(T_{Au})$ are the spectral radiances of blackbodies at wavelength λ and thermodynamic temperatures T and T_{Au} respectively, C_2 is the second radiation constant (0.014388 mK) and T_{Au} is the temperature at the freezing point of gold, 1337.58 K (1064.43 °C).

The formula (1) shows that the temperature scale above the gold point is not realized directly from Planck's law, but by the means of measuring ratios of spectral radiances. The reason for this is that the accuracy of absolute measurements of spectral radiance is unsatisfactory, but ratios of spectral radiances can be determined very accurately.

A primary photoelectric comparator has been developed by the NIM for realizing the IPTS-68 from the gold point up to 2000 °C. It can be used as a comparator to equalize sources or as summing pyrometer to combine them. A high stability tungsten strip lamp (made by G.E.C. England) and a cylindrical tungsten strip lamp (made in China) have been used as stable radiation sources for maintaining the IPTS-68 from 1064.43 °C to 2000 °C. A special blackbody furnace has been designed for realizing the freezing point of gold. The sensitivity of the photoelectric comparator is 0.03 °C at the gold point. The uncertainty in the realization of IPTS-68 is ±0.07 °C at the gold point and ±1.2 °C at 2000 °C.

PRINCIPLE AND CONSTRUCTION OF THE PRIMARY PHOTOELECTRIC COMPARATOR

Principle:

The principle of the comparator relies on the radiance equilibrium of two light sources, which is indicated by a null detector. There is no internal light source in the comparator, external sources only being used. Light emitted by the sources falls on the cathode of a photomultiplier after modulation by an optical modulator. If the radiances of the two sources are equal, then no alternating component is contained in the anode current of the photomultiplier.

Construction:

The optical system of the comparator is shown in Figure 1.

A comparator can compare the radiances of two sources employing two optical paths and be used as a summing pyrometer for extending the IPTS-68 up to 2000°C with three optical paths. Objective lenses 1, 2, 3 image the radiation sources B_1, B_2, B_3, on the windows A, B, C, respectively of the optical modulator 13, which converts the direct radiance signals into alternating ones. The optical guide 14 of 4 mm diameter transmits them and mixes them. The collecting lens 15 focuses light from the optical guide into a parallel light beam in which an interference filter 16 is positioned. The neutral filter 17 serves to keep the anode current of the photomultiplier within an optimal range for the temperature region 1400 °C-2000 °C.

After passing the collecting lens 18, spectral radiation arrives uniformly at the cathode surface of the photomultiplier 20. There is a mirror 5 in the middle path to compensate for the optical lengths of prisms 4, 6.

It is very important to ensure that the images of the light sources fall accurately on the appropriate slit windows of the optical modulator. There are two microscopes on the pyrometer. The microscope 11 is for sighting the light sources and 12 is for adjusting the modulator's slit windows and the corresponding graticule on the concave mirror 8 in order to align them. Using the 45° mirror 7 and the concave mirror 8, the images of the three light sources in the plane of the concave mirror 8 are simultaneously observed by the microscope 11. If we do not have the concave mirror, only the image of the light source in the middle path can be observed. A compensation glass 19 is used for aligning the modulator's slit windows and the graticule in the plane of the mirror B. In practical operation, it is often necessary to sight the sources onto the graticule of the concave mirror when the slit windows and the graticule are aligned. There are three squares engraved in the graticule (Figure 2); the distances between them and their dimensions are the same as those of the slit windows of the optical modulator.

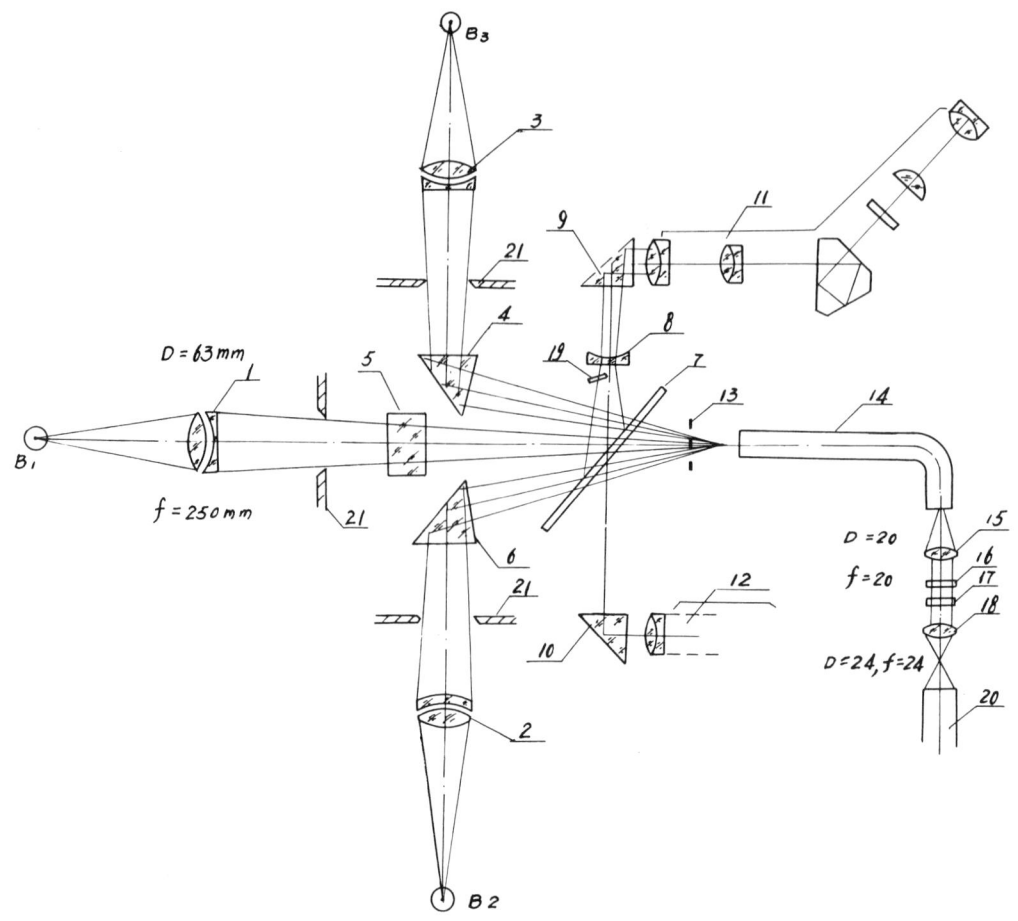

FIG. 1. Optical system of the primary photoelectric comparator.

B_1, B_2, B_3--sources; 1, 2, 3--objectives; 4, 6--78° prisms for changing the direction of radiation beam; 5--compensation mirror; 7--45° mirror; 8--graticule mirror; 9, 10--90° prisms; 11--microscope for sighting radiation source; 12--microscope for alignment; 13--optical modulator; 14--optical guide; 15, 18--lenses; 16--interference filter; 17--neutral attenuator; 19--compensation glass; 20--photomultiplier; 21--field aperture.

The optical modulator consists of three parts: the limiting plates, a slit and oscillatory plates cemented in a tunning form arm (Figure 3).

The three windows A, B, C, whose maximum areas are 0.5 mm x 0.5 mm, are framed by the knife edges of the slit, oscillatory plate and a limiting plate. The amplitude of the tuning fork is 0.24 mm and its frequency is 365.8 Hz. It has been demonstrated that the frequency is sufficiently stable. The slit and limiting plates are cemented together, but the oscillatory plates are cemented to an arm of the tuning fork. The distance between them should be as small as possible, e.g., <0.1 mm, as in Figure 3. The signals passing

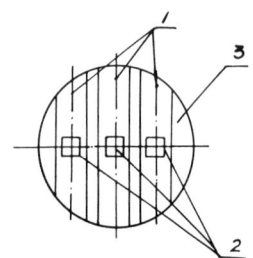

FIG. 2. Correct position of tungsten strip lamp images on the graticule mirror.
1. Images of tungsten strip lamp;
2. Squares engraved in the graticule;
3. The plane of graticule mirror.

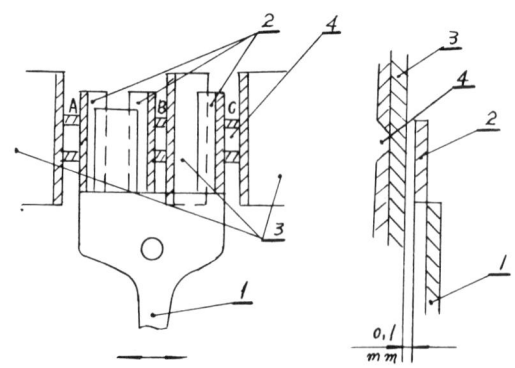

FIG. 3. Construction of optical modulator.
A, B, C--the windows of the slit; 1--one arm of the tuning fork; 2--oscillatory plates; 3--the limiting plates; 4--slit.

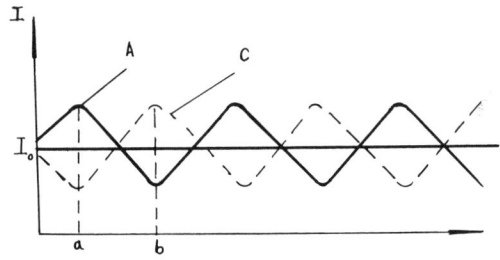

FIG. 4. Principle of optical modulator.

through the windows A, B, C depend upon instantaneous areas. When the oscillatory plates operate, the emerging signal is modulated. The beams passing the windows B and C are in phase, but those passing through A and C (A and B) are in antiphase. The two windows A and C (A and B) are only used for realizing the gold point and calibrating temperature lamps. When the light from the two sources is equal, the alternating signal is equal to zero as indicated by a null detector. This is because, the absolute values of their corresponding phototcurrents are just the same, but their phases differ by π.

Figure 4 shows the operating principle of the optical modulator. In a state of rest, the photocurrent of the photomultiplier is I_o and area of every window is 0.25 x 0.5 mm.

The optical guide consists of a bundle of thin glass fibers, the diameter of a single fiber being about 20-30 μm. It is very flexible and its spectral transmittance is nearly constant in the visible region. We can bend it according to our requirements. A light beam with an incidence angle less than 17° satifies the condition of total internal reflection in the optical guide. Thus signal losses are small. An optical guide rod was used in 1977, but the losses on the two side paths were rather large. The sensitivity did not meet the needs of the primary comparator; it only achieved about 0.1 °C at the gold point.

A photomultiplier tube of type EMI9658R has been employed as a detector in the instrument and was compared with type GDB-23 (made in China). Both of them have typical spectral response curves S-20. An interference filter (G-522-6600 made in U.S.) placed in the parallel beams serves to monochromatize the light. It has a peak wavelength of 661 nm, a half bandwidth 11 nm and a maximum transmittance of 40%. The current in the high voltage divider network of the photomultiplier should be more than ten times that of the anode, which is often less than or equal to 1 μA, to ensure a stable operating condition and reasonable linearity.

Electronic Measuring Circuit:

The electronic circuit (Figure 5) provides continuous indication of the spectral radiance difference between two compared sources, e.g., a standard light source and a tungsten strip lamp under calibration.

According to the magnitude and sign of the difference one may adjust the current in the lamp under calibration to obtain spectral radiance equilibrium. By this means calibration is effected against the standard light source.

After filtering out the DC component, the current of the photomultiplier is amplified by the preamplifier. The preamplifier is a low-noise one with high input impedance (about 5.1 MΩ) and high negative feedback. Its gain varies from 3 to 6. This amplifier also plays a role in impedance conversion for the following stage of amplification.

The frequency selection amplifier (FSA) selects the required frequency from various reference signals. The amplifier has a twin "T" bridge feedback network. By adjusting a potentiometer the resonance frequency of the twin "T" bridge is made equal to the frequency of the fork, which has an intrinsic frequency of 365.8 Hz. The gain of the FSA

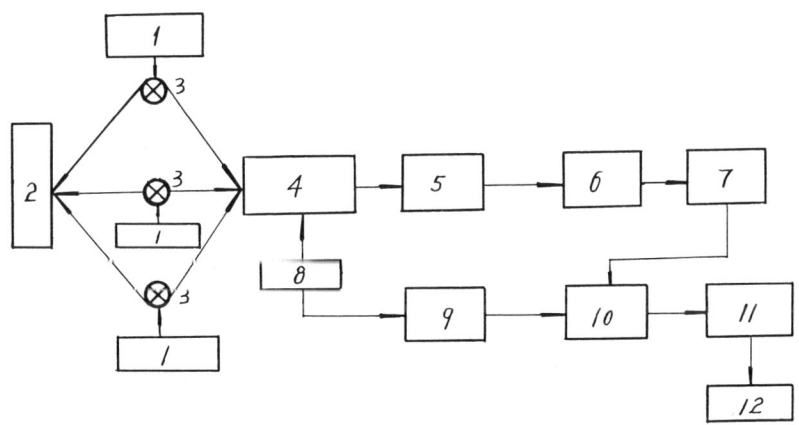

FIG. 5. Block diagram of the high temperature photoelectric comparator.
1. Power supply of stable current; 2. Digital voltmeter; 3. Tungsten strip lamp; 4. Optical modulator with tuning fork; 5. Photomultiplier; 6. Preamplifier; 7. Selected frequency amplifier; 8. Tuning fork oscillator; 9. Demodulation amplifier; 10. Phase sensitive amplifier; 11. Low-frequency filter; 12. Null detector.

is about 320 with bandwidth $\Delta f \cong 1$ Hz.

In order to make it oscillate, the fork is fed with sufficient energy to keep its amplitude constant. Under normal working conditions the amplitude should be 0.24 mm. The signal of the demodulation amplifier must be synchronous with that of the modulation. Therefore it is taken from the side coil of the fork. After its amplification this modulating signal becomes the reference signal for the phase sensitive amplifier (PSA).

The modulated AC signal is converted back to an analogue of the original DC signal by the phase sensitive amplifier. The magnitude and polarity of its output will depend on the magnitude of the input voltage and its phase, i.e., depend on the difference of the radiation of the two compared light sources. The null detector indicates its value and sign.

The output signal of the PSA passes through the low-frequency filter before entering the null detector. By adjusting a variable capacitance we can obtain different time constants. Usually the time constant used was 5 sec. A noise voltage of 8 mV at spectral radiance equilibrium at the gold point could be achieved. When the temperature changed by 0.03 °C, the output voltage rose to 18 mV, so that a resolution of 0.03 °C could be achieved.

The currents for the lamps were supplied by highly stabilized DC current sources ($dI/I < 2 \times 10^{-5}$). Values of current were obtained from a DVM and standard resistors.

REALIZATION OF THE FREEZING POINT OF GOLD

1. The blackbody and the freezing curve of the gold point:

A special blackbody furnace which moves on rails and can be accurately adjusted in any direction has been designed for realizing the gold point. It consists of two ceramic tubes. One of them is wound with a heating wire coil of nickel-chromium. The other, having a graphite crucible at its center and many graphite apertures in it to protect the crucible from oxidation, is positioned inside the former. There are four taps in the heating coil. According to requirements, we can connect some shunt resistors between the taps to adjust the furnace temperature to be uniform. The furnace with a length of 700 mm is surrounded by a water cooling tube. It takes 3.5 hours to heat the furnace from ambient temperature to 1064 °C and the power used is about 1.5 kw. The adjustment of the position of the furnace is quite convenient and a precision of position can be achieved of about 50 μm.

The design of the crucible depends on the following:

A. The cavity of the crucible must approximate a perfect blackbody as closely as possible. In order to achieve an emissivity close to unity, the ratio of length to diameter needs to be quite large.

B. The volume of the crucible must be large enough for realizing a stable freezing curve with a plateau of duration more than 10-15 minutes, although this is necessarily expensive.

C. The material chosen must posess high emissivity, high thermal conductivity and durability at high temperature. High purity is also important in order to keep the gold clean, and it must be possible to machine a long cavity from it with a thin inner wall.

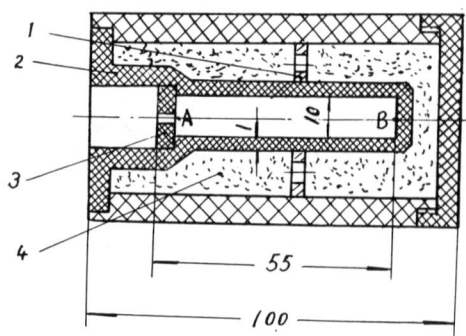

FIG. 6. Construction of the graphite cavity.
1. Holder; 2. Cavity; 3. Aperture; 4. Gold.

According to the above requirements, the crucible (Figure 6) has been made of graphite with an emissivity of 0.8 and a purity 0.9999. The cavity approximates to a perfect blackbody mainly by the geometrical design. As shown in Figure 6, its length is 55 mm. The diameter of the small aperture and the cavity are 2.5 mm and 10 mm, respectively. In order to ensure uniform temperature of the cavity the small aperature is enveloped in the freezing gold. There is a holder to protect the cavity from becoming distorted. The temperature difference between A near the aperture (Figure 6) and B near the bottom of the cavity was measured by the photoelectric comparator. It was found that the difference between A and B was less than the sensitivity of the instrument.

Using an equation of Quinn (1) the emissivity of the cavity is calculated in order to assess its performance. Taking the case of a uniform temperature along its length, the emissivity in the axial direction of the back wall through the aperture is given by following equation:

$$\varepsilon = 1 - \rho\frac{a^2}{D^2} - \rho^2 a^2 \int_{x=0}^{x=D} \frac{x(D-x)dx}{(1+x^2)^2\{1+(D-x)^2\}^2}$$

where

$$D = \frac{\text{length of cavity}}{\text{radius of cavity}} = 11$$

For obtaining the freezing curve of the gold point the experimental procedure could be divided into three steps:

(a) After cleaning the crucible, it was mounted in a vertical furnace and gold powder was melted several times.

(b) Then the crucible containing the gold was placed in the center of the horizontal furnace where there is a uniform temperature along the whole length of the crucible. The furnace was positioned accurately to align the axis of the crucible with the optical axis of the comparator with the aid of a special optical system.

(c) The temperature of the furnace was raised and lowered to melt and freeze the gold slowly.

The furnace and the reference lamp, the spectral radiance of which is matched with that of the gold point, were put symmetrically in two optical paths (Figure 1). Thus

Table I. Currents of the reference lamp during the freeze and melt of the gold.

Time (minutes)	Melt (A)	Freeze (A)
1	5.1509	5.1511
2	5.1512	5.1514
3	5.1512	5.1510
4	5.1514	5.1512
5	5.1514	5.1510
6	5.1511	5.1512
7	5.1515	5.1510
8	5.1512	5.1512
9	5.1511	5.1510
10		5.1510

Mean of melt: 5.1512 A
Mean of freeze: 5.1511 A
Difference of temperature between freeze and melt: 0.01 °C.

the phase difference of the corresponding modulated signals was π. A platinum 10% rhodium-platinum thermocouple placed in the back wall of the crucible was used to control the temperature of the furnace. The current of the reference lamp was adjusted so that it followed the spectral radiance of the blackbody when the difference of temperature between the furnace and the gold point was about 5 °C. During the plateau, the current of the reference lamp was recorded at intervals of one minute until readings were obtained, which are shown in Table I.

2. Calibrating a lamp at the gold point by the method of substitution:

The small changes of the currents in Table I show the good quality of the freezing curve, but the current does not reproduce the spectral radiance of the gold point in the reference lamp. This is because there are unavoidable small asymmetries in the optical paths to the reference lamp and the blackbody. To overcome this difficulty the reference lamp is then used to calibrate the gold point lamp which is substituted for the blackbody. In this way the asymmetries cancel. Because of the optical asymmetries, the calibration for null balance between blackbody and reference lamp will be:

$$\tau_1 \Omega_1 A_1 \int_{\lambda_1}^{\lambda_2} L_\lambda(T_{Au}) \tau_\lambda S_\lambda d\lambda = \tau_2 \Omega_2 A_2 \int_{\lambda_1}^{\lambda_2} \varepsilon_\lambda L_\lambda(T) \tau_\lambda S_\lambda d\lambda \quad (2)$$

where τ is the transmittance of the optical system; Ω is the solid angle received by the receiver; A is the surface area of the receiver; τ_λ is the spectral transmittance of the interference filter; S_λ is the spectral response of the photomultiplier; and ε_λ is the emissivity of the tungsten strip lamp.

Eq. (2) states that the radiant flux in the two optical paths is the same, the left being the radiant flux of the gold point and the right being the radiant flux of the reference lamp.

Defining geometrical factors K by

$$\tau \cdot \Omega \cdot A = K$$

Eq. (2) becomes:

$$K_1 \int_{\lambda_1}^{\lambda_2} L_\lambda(T_{Au}) \tau_\lambda S_\lambda d\lambda = K_2 \int_{\lambda_1}^{\lambda_2} \varepsilon_\lambda L_\lambda(T) \tau_\lambda S_\lambda d\lambda \quad (3)$$

According to the definition of brightness temperature, Eq. (3) may be written:

$$K_1 \int_{\lambda_1}^{\lambda_2} L_\lambda(T_{Au}) \tau_\lambda S_\lambda d\lambda = K_2 \int_{\lambda_1}^{\lambda_2} L_\lambda(T_s) \tau_\lambda S_\lambda d\lambda \quad (4)$$

$$\frac{\int_{\lambda_1}^{\lambda_2} L_\lambda(T_{Au}) \tau_\lambda S_\lambda d\lambda}{\int_{\lambda_1}^{\lambda_2} L_\lambda(T_s) \tau_\lambda S_\lambda d\lambda} = \frac{K_2}{K_1} = \frac{L_{\lambda e}(T_{Au})}{L_{\lambda e}(T_s)}$$

where λe is the effective wavelength.

Using Wien's law:

$$\frac{1}{T_{Au}} - \frac{1}{T_s} = \frac{\lambda e}{C_2} \ln \frac{K_2}{K_1} \quad (5)$$

In practice, it is very difficult of obtain $\tau_1 = \tau_2$, $\Omega_1 = \Omega_2$, $A_1 = A_2$. Therefore, usually $K_1 \neq K_2$.

Adjustment of the comparator has been carried out many times in our laboratory. A small difference between K_1 and K_2 still remains and the corresponding difference of temperature is about 1 °C. at the gold point which represents the error in the calibration of the reference lamp mentioned above. Using the method of substitution, the radiance of the gold point lamp instead of the blackbody cavity has been matched to the spectral radiance of the reference lamp keeping the same situation as in Eq. (2).

Thus Eq. (4) becomes:

$$K_1 \int_{\lambda_1}^{\lambda_2} L_\lambda(T_{sg}) \tau_\lambda S_\lambda d\lambda = K_2 \int_{\lambda_1}^{\lambda_2} L_\lambda(T_s) \tau_\lambda S_\lambda d\lambda \quad (6)$$

where T_{sg} is the brightness temperature of the gold point lamp.

From Eq. (4) and (6) one obtains:

$$K_1 \int_{\lambda_1}^{\lambda_2} L_\lambda(T_{sg}) \tau_\lambda S_\lambda d\lambda = K_1 \int_{\lambda_1}^{\lambda_2} L_\lambda(T_{Au}) \tau_\lambda S_\lambda d\lambda \quad (7)$$

We can see from Eq. (7) that T_{sg} equals T_{Au}. The current in the gold point lamp corresponds to the brightness temperature at the gold point at the effective wavelength λ_e. Thus the influence of asymmetry ($K_1 \neq K_2$) has been eliminated.

The effect of the size of source and stray light should also be considered. It can cause a large systematic error of a few degrees at the gold point. An auxiliary optical system (2) has successfully eliminated the systematic error.

3. Uncertainty of the gold freezing point:

The realization of the gold point and its transfer to the stable tungsten strip lamp were carried out from 1979 to 1981. The statistical uncertainty of the averaged current in the tungsten strip lamp is ±0.06 °C at the gold point (taking three times the standard deviation of arithmetic mean). The calculation, given in Table II, shows that the overall uncertainty is ±0.07 °C at the gold point.

EXTENSION OF THE TEMPERATURE SCALE AND ITS UNCERTAINTY

The method of "radiance summation" has been used for extending the temperature scale above the gold point. Here radiances from sources of equal known radiance temperatures are summed in the pyrometer to provide calibrations corresponding to a higher calculable temperature. This is the only time when all three optical paths of the pyrometer are used. The radiance equilibrium of the three radiation sources is obtained when the sum of the radiances in the windows B and C is equal to that in the window A. Because the two windows B and C are in phase, and the phase difference between A and B (A and C) is π, we have:

$$L_A = L_B + L_C$$

where L_A, L_B, L_C are the radiances corresponding to the windows A, B, C, respectively.

The process of radiance summation is divided into three steps:

(a) Windows A and B are open while C is closed. We arrange that $L_A = L_{Au}$ where L_{Au} is the blackbody radiance at the gold point. Then source B is adjusted to obtain $L_B = L_A = L_{Au}$.

(b) Windows A and C are open while B is closed. Exactly as in (a) we arrange $L_C = L_A = L_{Au}$.

(c) All three windows A, B and C are open. When $L_B = L_C = L_{Au}$ as before and $L_A = L_B + L_C$, then $L_A = 2L_{Au}$ so that

$$L_A/L_{Au} = 2$$

Thus the radiance ratio equals 2. Extending the scale once from T_{Au} to T_1, one obtains from Wien's law:

$$\frac{1}{T_{Au}} - \frac{1}{T_1} = \frac{\lambda_{e1}}{C_2} \ln 2$$

Then summing two equal sources of radiance temperature T_1, one can make a second extension to T_2:

$$\frac{1}{T_1} - \frac{1}{T_2} = \frac{\lambda_{e2}}{C_2} \ln 2$$

For the n-th extension, we have:

$$\frac{1}{T_{n-1}} - \frac{1}{T_n} = \frac{\lambda_{en}}{C_2} \ln 2$$

In this way the temperature scale was extended to 2000 °C.

Extending the temperature scale downwards, the radiance ratio is 1/2. That is:

for the first time:

$$\frac{1}{T_{Au}} - \frac{1}{T_1} = \frac{\lambda_{e1}}{C_2} \ln(1/2)$$

for the second time:

$$\frac{1}{T_1} - \frac{1}{T_2} = \frac{\lambda_{e2}}{C_2} \ln(1/2)$$

for the n-th time:

$$\frac{1}{T_{n-1}} - \frac{1}{T_n} = \frac{\lambda_{en}}{C_2} \ln(1/2)$$

An average effective wavelength λ_{en} has been used for realizing the temperature scale. The following expression for λ_{en} can be derived:

$$\lambda_{en} = \frac{C_2 \left(\frac{1}{T_{n-1}} - \frac{1}{T_n} \right)}{\ln \left\{ \frac{\int_{\lambda_1}^{\lambda_2} L_{T_n}(\lambda)\tau(\lambda)S(\lambda)\mu(\lambda)F(\lambda)d\lambda}{\int_{\lambda_1}^{\lambda_2} L_{T_{n-1}}(\lambda)\tau(\lambda)S(\lambda)\mu(\lambda)F(\lambda)d\lambda} \right\}}$$

A limiting effective wavelength λ_T is defined by:

Table II. Uncertainty of the gold point.

Error sources	°C
1. Deviation from absolute blackbody	0.012
2. Deviation from the gold point temperature of the bottom of the cavity through heat conduction	0.0034
3. Difference between melt and freeze	0.01
4. Statistical uncertainty of sighting temperature lamp and blackbody radiator, and random error	0.06
5. Electrical measurement	0.028
Overall uncertainty	±0.07

$$\lambda_T = \frac{\int_{\lambda_1}^{\lambda_2} L_T(\lambda)\tau(\lambda)S(\lambda)\mu(\lambda)F(\lambda)d\lambda}{\int_{\lambda_1}^{\lambda_2} \frac{1}{\lambda}L_T(\lambda)\tau(\lambda)S(\lambda)\mu(\lambda)F(\lambda)d\lambda}$$

here $L_T(\lambda)$ is the spectral radiance of a blackbody at temperature T, and at a wavelength λ; $\tau(\lambda)$ is the spectral transmittance of the interference filter; $S(\lambda)$ is the spectral response of the photomultiplier; $\mu(\lambda)$ is the transmittance of the optical guide; $F(\lambda)$ is the transmittance of the absorbing glasses.

$\tau(\lambda)$ and $S(\lambda)$ were measured many times by an accurate spectrometer (D-340) and an accurate monochromator respectively. The average value has been used. An absorbing glass $F_1(\lambda)$ was inserted in the front of the photomultiplier when the temperature of source reached 1797 K and 1907 K, and another glass $F_2(\lambda)$ for 2030 K, 2170 K and 2332 K.

The average values of the effective wavelengths are determined respectively by the following equations:

$$\overline{\lambda}_e = f_1(\overline{\tau}_\lambda, \overline{S}_\lambda, \overline{\mu}_\lambda, \overline{F}_\lambda)$$

$$\overline{\lambda}_T = f_2(\overline{\tau}_\lambda, \overline{S}_\lambda, \overline{\mu}_\lambda, \overline{D}_\lambda)$$

$\overline{\tau}_\lambda, \overline{S}_\lambda, \overline{\mu}_\lambda$ and $\overline{F}_\lambda$ are average values for $\tau(\lambda), S(\lambda), \mu(\lambda)$ and $F(\lambda)$ respectively. The error of the effective wavelengths could be calculated from the following equation:

$$d\lambda = \frac{\partial\lambda}{\partial\tau_\lambda}d\tau_\lambda + \frac{\partial\lambda}{\partial S_\lambda}dS_\lambda + \frac{\partial\lambda}{\partial\mu_\lambda}d\mu_\lambda + \frac{\partial\lambda}{\partial F_\lambda}dF$$

After calculating each term in the above equation independently, the uncertainty of the effective wavelength could be expressed by the root-mean-square:

$$\sigma_\lambda = \sqrt{(\sigma_{\tau_\lambda})^2 + (\sigma_{S_\lambda})^2 + (\sigma_{\mu_\lambda})^2 + (\sigma_{F_\lambda})^2}$$

From the result of the calculation it was found that: an uncertainty in the effective wavelength was obtained of about 0.2 nm at 661 nm.

$$\sigma_\lambda \simeq \pm 0.2 \text{ nm}$$

The variations of λe and λ_T over the whole temperature range from 1065 K to 2332 K were less than 0.3 nm. The standard deviation of the measurements was estimated to be ± 0.05 nm.
For example:

$\lambda e(1337.58 \text{ K}, 1397.10 \text{ K}) = 661.100$ nm
$\lambda e(1612.30 \text{ K}, 1699.56 \text{ K}) = 661.010$ nm
$\lambda e(2169.22 \text{ K}, 2330.23 \text{ K}) = 661.206$ nm
$\lambda_T(1337.58 \text{ K}) = 661.111$ nm
$\lambda_T(1699.56 \text{ K}) = 660.999$ nm
$\lambda_T(2330.23 \text{ K}) = 661.195$ nm

The temperature scale from 800 °C-1100 °C is realized using vacuum strip lamps and that from 1400 °C-2000 °C using gas-filled tungsten strip lamps.

The error of each successive temperature extension consists of:

(a) The error due to the temperature point T:

$$\sigma_{T_0} = (\frac{T}{T_0})^2 T_0$$

(b) The error due to the uncertainty of the effective wavelength:

$$\sigma_{\lambda e} = \frac{\Delta\lambda}{C_2}\ln 2 \, T^2$$

(c) The error due to error in sighting the temperature lamp, lamp stability, the reproducibility of the multiplier, and random errors. Three standard deviations of the arithmetic mean were accepted as a measure of the error σ_r.

(d) The error due to electrical measurements σ_v.

The total uncertainty of extending a temperature point may be expressed as:

Table III. Z233 Extension Result.

Extension No.:	1	2	3	4	5
Radiance ratio to L_{Au}:	?	4	8	16	32
Temperature °C:	1124	1189	1260	1339	1462

Experiment number:	current (A)				
1.	5.9598	6.5532	7.2483	8.0598	9.0063
2.	5.9603	6.5531	7.2496	7.2496	8.0592
3.	5.9604	6.5528	7.2472	8.0580	9.0042
4.	5.9603	6.5531	7.2475	8.0586	9.0040
5.	5.9606	6.5535	7.2484	8.0596	9.0040
Mean value:	5.9603	6.5531	7.2482	8.0590	9.0048
$\sigma_{n-1}/\sqrt{n}$ °C:	±0.015	±0.012	±0.042	±0.031	±0.047

Table IV. Q222 Extension Result

Extension No.:	6	7	8	9	10
Radiance Ratio to L_{Au}:	64	128	256	512	1024
Temperature °C:	1523	1633	1756	1896	2057
Experiment Number:	current (A)				
1.	12.531	13.805	15.350	17.231	19.548
2.	12.532	13.802	15.355	17.239	19.557
3.	12.351	13.805	15.354	17.241	19.559
4.	12.530	13.805	15.353	17.239	19.577
Mean Value:	12.531	13.804	15.353	17.238	19.555
$\sigma_{n-1}/\sqrt{n}$ °C:	±0.04	±0.06	±0.09	±0.18	±0.19

Table V. Uncertainty of Each Point for Extending the Scale.

Temperature °C:	1064	1124	1189	1260	1339	1426
Uncertainty °C:	±0.1	±0.12	±0.14	±0.19	±0.23	±0.28
Temperature °C:	1426	1523	1633	1756	1896	2057
Uncertainty °C:	±0.33	±0.41	±0.55	±0.68	±0.92	±1.2

$$\sigma = \sqrt{\sigma_{T_o}^2 + \sigma_{\lambda e}^2 + \bar{\sigma}_r^2 + \sigma_v^2}$$

$\bar{\sigma}_r$ is the largest term in the total uncertainty.

The uncertainty for each extended point was calculated as described in (a)-(d) above. The results are given in Table V.

CONCLUSIONS

The realization of the gold point and extension of the IPTS-68 above 1064 °C on stable tungsten strip lamps was carried out in 1979, 1980 and 1981. Extensive experimental work has proved that the present comparator is precise and reliable. By means of the substitution method and measurements with additional external apertures the respective influences of geometric factors and stray light were eliminated successfully. The calibrations of the vacuum lamps C654 and Z248 obtained in 1979 from PTB and in 1980 at NIM (China) agree to within 0.1-0.2 °C at the gold point. The calibration of the gas-filled tungsten strip lamp C682 obtained from NPL (1981) agreed with a preliminary at NIM within 2 °C at 2000 °C. But if the influence of the difference in lamp filament target areas used in NPL and in NIM was allowed for, the calibration difference would be expected to be smaller. The difference of the target area between NPL (d=0.75 nm) and NIM (1.4 mm x 1.4 mm) is quite large.

REFERENCES

1. T. J. Quinn, Temperature, Its Measurement and Control in Science and Industry, 4, Part 1, p.295.

2. Zhao Qi, Den Sixiang, Sun Dinwen, ACTA Metrologica Sinica, 2, No. 2, 1981.

The development of temperature standards at NIM of China

Ling Shankang, Zhang Guoquan, Li Ruisheng, Wang Zilin, Li Zhiran, Zhao Qi, and Li Xumo

National Institute of Metrology, Beijing, China

This paper describes the recent development of temperature standards at the National Institute Metrology (NIM). The emphasis is on the establishment of temperature scales and the making of standard instruments. Then the differences between the scales of NIM and other national standard laboratories are discussed. Some of the results are presented in this paper.

1. Introduction

As you may know, metrology in China has a long established history. When the First Emperor of Qin unified China in 221 B.C. a series of reforms and unification were adopted in order to consolidate the feudal economy and centralize the state power. One of these important measures was the unification of the system of Weights and Measures.

The bronze cultural relics of Shang Dynasty (1600 B.C.) found in recent excavations show they were finely made and symmetrically built. It shows the bright achievements of ancient metrology and metallurgical technology in China. It also shows that people at that time already knew how to determine temperature in accordance with its color.

During the long years afterwards, science in China as well as the economy was rather backward for various reasons. In the early nineteen fifties, after the founding of new China, more attention was paid to modern metrology, and so subsequent developments were achieved in this area.

The various primary standards, such as length, mass, time-frequency, electro-magnetism, temperature, etc., have been established one after another at NIM and a nation-wide calibration service net, including 29 provinces and cities and more than 2000 counties, has been formed for unifying physical quantites and their dissemination.

Our research work on the temperature standards was centered on the International Practical Temperature Scale of 1968 (IPTS-68), which is the basis for unifying temperature quantities. We have realized all the defining fixed points and developed some standard interpolating instruments specified in the IPTS-68. For a country with such a vast territory as China we must make great efforts to develop our temperature standards.

2. Radiation Pyrometry

Radiation pyrometry takes an important place in developing industry and scientific technology. According to the definition of the IPTS-68, the temperature scale above 1064.43°C is obtained from the Planck formula:

$$\frac{L_\lambda(T)}{L_\lambda(T_{Au})} = \frac{\exp(C_2/\lambda T_{Au}) - 1}{\exp(C_2/\lambda T) - 1}$$

where $L_\lambda(T)$ and $L_\lambda(T_{Au})$ are the spectral radiances of a blackbody at the wavelength λ and thermodynamic temperatures T and T_{Au} respectively. C_2 is the second radiation constant, 0.014388 m·K.

A standard photoelectric comparator has been developed recently at NIM for realizing the IPTS-68 in the region 1064.43-2000°C.

The principle of the comparator relies on the radiance equilibrium of two external light sources, which is indicated by a null detector. A tuning fork is used as an optical modulator to convert the direct radiance signals into alternating ones. After passing through the filter (λ_0 = 661 nm) the radiance comes to the cathode surface of a photomultiplier (EMI 9658R).

The ac current passes through a preamplifier, frequency selection amplifier and phase sensitive amplifier. After that the null detector indicates its value and sign. If the radiances of the two light sources are equal, then the null detector points to zero. The sensitivity of the instrument is about 0.03°C at the gold point. By combining the light from two sources and comparing with the radiance of a third, the scale may be extended to 2330 K from 1064.43°C or downwards to 1065 K. The extended scale has been kept on high-stability tungsten strip lamps (G.E.C. Gr. Britain) and cylindrical strip lamps (made in China). The overall uncertainty of realizing the freezing point of gold is ± 0.07°C. The uncertainty of realizing the scale (1064.43-1400°C) is ± (0.1-0.28°C) for the vacuum lamps and ± (0.33-1.2°C) for the gas-filled lamps in the region 1400-2000°C.

The uncertainty in determination of the effective wavelength is ± 0.2 nm.

The comparison between PTB and NIM on two vacuum lamps C 654 and Z248 showed agreement to within 0.10-0.20°C at the gold point. The calibration of the gas-filled tungsten strip lamp C682 obtained from NPL (1981) agreed with a preliminary at NIM within 2°C at 2000°C.

In cooperation with Yunnan Instrument Factory we are developing a precise photoelectric pyrometer as a secondary standard for calibrating the tungsten strip lamps. There is an internal light source in the pyrometer and we have adopted a fork modulator, filter (λ_0=660) and photomultiplier (GDB-23) made in China. The sensitivity of the instrument is better than 0.15°C at 1064°C.

Total radiation pyrometers are widely used in Chinese industry. The established calibration equipment consists of a blackbody furnace and a standard optical pyrometer providing overall uncertainty of ± 9°C at 2000°C.

Recently, more attention has been paid to developing the high stability vacuum lamp and blackbody lamp.

3. Thermoelectric Thermometry

Three kinds of standard thermocouple have been established at NIM: platinum-10% rhodium/platinum, platinum-13% rhodium/platinum and platinum-30% rhodium/platinum-6% rhodium. The α-coefficient for the platinum leg of the Pt-10%Rh/Pt thermocouple is 3.925×10^3, the rhodium content of the platinum-rhodium alloy leg is about $(9.99 \pm 0.01)\%$.

The temperature on the IPTS-68 can be realized to within about ± 0.2 K in the platinum-10% rhodium/platinum thermocouple range.

Two thermocouples (No. 62-6 and 74-03) were sent to BIPM for calibration in 1979, and agreed with results at NIM within 0.1-0.2°C, as in Table 1.

At present, with the help of NIM, Chinese manufacturers are producing the new platinum-10% rhodium thermocouples which are available with thermoelectric properties that fully agree with the reference tables of temperature versus emf (within specified tolerances by I.E.C. International Electrotechnical Commission). Therefore, the new thermocouples will now be formally mass produced and used, according to the reference tables recommended by I.E.C.

The standard platinum-30% rhodium/platinum-6% rhodium thermocouples are calibrated in a platinum blackbody cavity by means of the pyrometer. The results were checked by the fused wire method (using Pt and Pd wires). The results show that the difference between two methods would be 0.3°C at palladium and 1°C at platinum point. The overall uncertainty of the calibration is about $(0.2-0.3)\%$. This means 4.8°C at 1600°C.

The Chinese-manufactured Pt-13%Rh/Pt thermocouple has fine stable characteristics. The calibrations were carried out at the copper, antimony and zinc freezing points as usual. The calibration results of such thermocouples are in good agreement with those of Johnson Matthey thermocouples in the region 300-1100°C, which are given in Table II.

4. High Temperature Platinum Resistance Thermometer (HPRT)

The ITS-27 was initially defined by a Pt-10%Rh/Pt thermocouple in the region 630-1064°C and this definition was been maintained to the present time. Since the reproducibility of a thermocouple is not better than ±0.2°C, there is thus unsmoothness in the transfer at 630.74°C, which is the junction of the resistance thermometer and thermocouple range.

It was confirmed that as the standard interpolating instrument of the IPTS-68 the thermocouple should be replaced by a stable high-temperature platinum resistance thermometer.

TABLE II: THE DIFFERENCES OF emf BETWEEN CHINA MADE Pt-13Rh/Pt AND I.E.C. TABLE DATE

Temperature (°C)	emf of China -made thermo-couple (mV)	I.E.C. table date(mV)	Differences (μV)
300	2.396	2.400	-4
400	3.407	3.407	0
500	4.469	4.471	-2
600	5.581	5.582	1
700	6.742	6.741	1
800	7.950	7.949	1
900	9.204	9.203	1
1000	10.503	10.503	0
1100	11.846	11.846	0

A lot of researchers have been working on developing HPRT's for many years. Of course, they have accumulated a wealth of experience. We have started this work since 1977. The research work was quickened after participating with the activities of the CCT.

At present more than sixty of a new type of HPRT have been made. The sensor of the thermometer was designed to be a strip-shaped coil supported with notches.

We thought such design might decrease the strain on the wire to a minimum. The support was made of silica-glass or synthetic sapphire and the platinum wire of 0.4 mm in diameter was wound in a bifilar helix around the coil support.

The majority of the HPRT's have an ice-point resistance of about 0.25 ohm and their testing results are given in Table III. It is obvious that the average change in R_0 every 100 hours was about 1.3 mK. Before and after quenching at high temperature the changes in R_0 generally did not exceed ±1 mK.

The insulation resistance of these thermometers for the silica-glass support approached 160 M ohm at 961°C, 50 M ohm at 1000°C and 27 M ohm at 1064°C; for synthetic saphire support 270 M ohm at 961°C, 110 M ohm at 1000°C and 40 M ohm at 1064°C.

TABLE I: THE DIFFERENCES BETWEEN BIPM AND NIM OF Pt-10%Rh/Pt

Thermocouple No	62-6			70-03		
	NIM (mV)	BIPM (mV)	Difference (μV)	NIM (mV)	BIPM (mV)	Difference (μV)
Freezing points Gold (1064.43°C)	10.3284	10.3279	0.5	10.3333	10.3337	-0.4
Silver(961.93°C)	9.1440	9.1437	0.3	9.1479	9.1479	0
Antimony (630.755°C)	5.5488	5.5487	0.1	5.5514	5.5573	- -2.3

TABLE III. THE MAIN CHARACTERISTICS OF HPRT AND ITS TESTING

Thermometer No	R (ohm)	α ($\times 10^{-3}$)	Self-heating effect (mK/10mA)	Annealing time (h) Above 630°C	At 1070°C	Annealing at upper limit temperature, aver. change in R_0 (mK/100h)
80136	0.247116	3.9274	0.9	580	500	0.6
80176	0.249876	3.9274	0.9	406	350	0.8
80179	0.243495	3.9272	0.9	406	350	0.9
79501	0.243574	3.9261	0.9	563	457	1.3
79502	0.230301	3.9259	1.1	587	500	1.8
79486	0.255683	3.9269	1.5	529	450	1.1
79442	0.238272	3.9262	1.2	564	500	1.6
77226	0.253848	3.9274	1.1	409	350	1.5
78300	0.256738	3.9274	1.0	289	250	1.1
79536	0.250953	3.9266	1.3	532	425	2.1

The testing of stability for 32 HPRTs was aimed at judging the manufacturing procedure to provide good performance of these HPRTs. The results given in Table IV show that the average change in R_0 every 100 hours is within 2.0 mK for 72% (23 HPRTs) and within 2.5 mK for 88% (28 HPRTs).

We have tested high temperature platinum resistance thermometers for 2 years. This kind of HPRT performs satisfactorily. The NIM have accepted the suggestion from Dr. Quinn of BIPM this summer to deliver 20 HPRTs to some national laboratories for testing to confirm the quality of the thermometers.

5. Cryogenic Thermometry

Resistance Thermometers

For realizing and maintaining the IPTS-68, a Chinese capsule platinum resistance thermometer has been developed at NIM in cooperation with Yunnan Instrument Factory. The stability in R_0 of the thermometer is better than ±2 mK/year. Now, this work is still proceeding, and the stability in R_0 of thermometer is hoped to be improved.

TABLE IV: THE TESTING ON STABILITY OF 32 HPRTs

Change (mK)	Number of thermometer	Change (mK)	Number of thermometer
0-0.5	1	3.0-3.5	1
0.5-1.0	6	3.5-4.0	2
1.0-1.5	10	5.5-6.0	1
1.5-2.0	6		
2.0-2.5	5	Total thermometer:32	

In addition, the rhodium-iron and germanium resistance thermometers have also been investigated in cooperation with Yunnan Instrument Factory and Xinjiang Physics Institute respectively.

The standard rhodium-iron thermometer supplied by Yunnan Instrument Factory can be used for high-precision temperature measurement below 30 K. The R_0 is 50 ohm ± 2 ohm, the fractional sensitivity at 4.2 K and at 20 K are greater than 6×10^{-2}/K and 1.3×10^{-2}/K respectively. After many times thermal cycling the stability at 4.2 K was better than ± 0.3 mK. It will be considered as the secondary thermometer for maintaining the thermodynamic temperature of the acoustic thermometer of NIM.

The performance of the germanium thermometer provided by Xinjiang Physics Institute has been examined at NIM. The results of tests have shown that the curve of resistance against temperature is rather smooth. It can be employed for temperature measurement within the range of 1 K to 100 K. Six thermometers were used for tests of thermal cycling from 4.2K to 273.16 K and the changes of resistance of the thermometers were measured. At 4.2 K, better than ± 0.5 mK stability could be obtained by Blakemore's method.

Low Temperature Fixed Points

The lower part of the IPTS-68 (below 0°C) is based on six reproducible equilibrium states (defined fixed points). As an interpolating instrument of the IPTS-68 in the range of 273.15-13.81 K, the capsule platinum resistance thermometer is calibrated at these fixed points and one obtains the temperature values on the IPTS-68 in terms of specified equations. Therefore, realizing the fixed points is a key link for the establishment of the scale. These fixed points had been realized at NIM before 1974 and the reproducibility of them was in the region 0.4-4 mK. They could preliminarily meet the needs in science and industry at that time. A more accurate realization of the IPTS-68 is being carried on from this time forward. Better than 1 mK reproducibility will be aimed at.

It is well known that realizing boiling points requires high-precision pressure measurement and makes

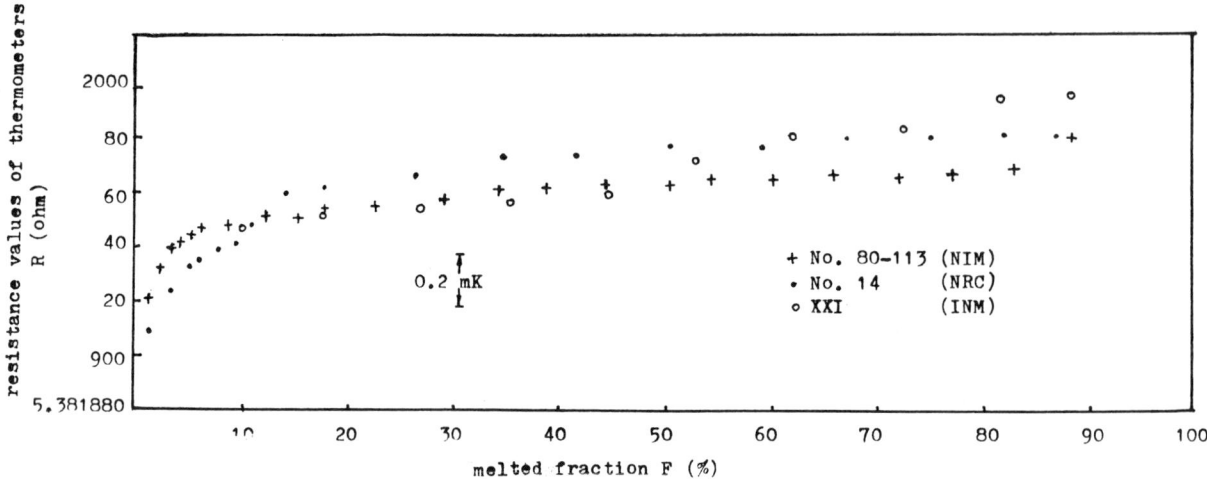

Fig. 1. The melting curve of argon

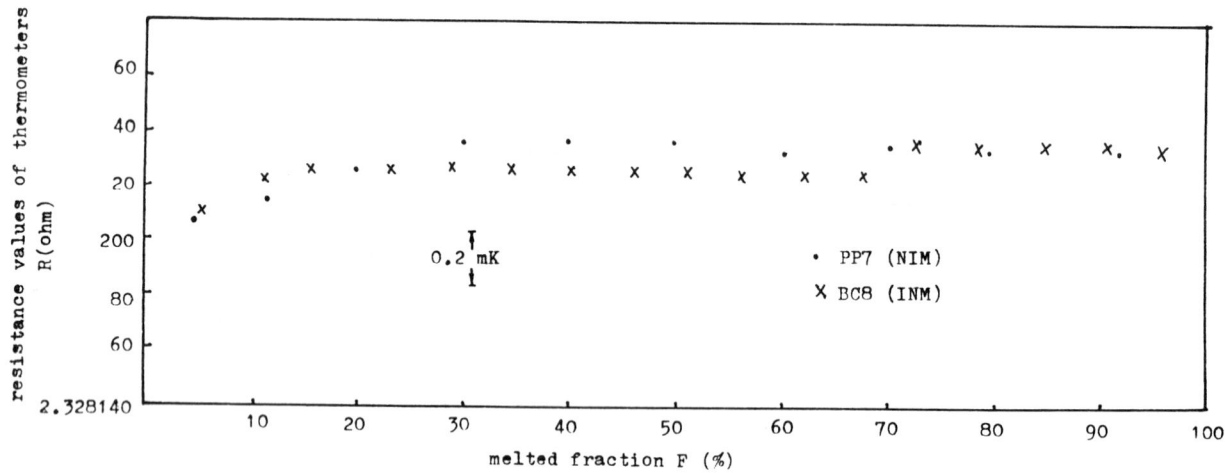

Fig. 2. The melting curve of oxygen

experiments complicated. The reproducibility of the temperature of the triple point is usually better than that of the boiling point. Especially, the sealed cell triple point has advantages: simple to use, easy to transport from one place to another, easy to keep high purity of gas in the cell and easy to compare with other laboratories. Therefore in recent years a lot of investigations have been carried out in various national standard laboratories. We have also started this work since 1979. Two sealed cells have been developed- argon and oxygen. A reproducibility of several tenths of mK was easily achieved by means of the calorimetric method.

With the triple point of argon, in thermal analysis, the results of experiments could be plotted as temperature against the inverse of the melted fraction F, and the temperature value of the triple point for the pure substance would be obtained for $1/F = 0$. In this way the reproducibility of the triple point of argon could be better than 0.1 mK. Our sealed cell of argon has been compared with those from NRC and INM. The results are shown in Fig. 1, and 0.1 mK and 0.3 mK agreement with that of NRC and INM respectively was found.

The oxygen sealed cells were filled with commercially available oxygen containing impurity of nitrogen of 3.1 ppm. The results of tests are shown in Fig. 2. A flat temperature plateau was found. The melting temperature width is less than 0.1 mK and a difference from that of INM is less than 0.2 mK.

Ultrasonic Manometer

The ultrasonic manometer used for high-precision pressure measurement has been developed at NIM. The uncertainty of the pressure measurement is not more than 20 μm Hg, and the measurement range is up to 800 mm Hg. It can be used either for acoustic thermometry or for vapor pressure thermometry.

REFERENCES

1. Ling Shankang, and Dai Leshan: Acta Metrologica Sinica, Vol. 2, No. 2, 1981.

2. Zhao Qi, Deng Xixiang, and Sun Dingwen: Acta

Metrologica Sinica, Vol. 2, No. 2, 1981.

3. Zhao Qi: IMEKO, TC-12, 1981.

4. T. Q. Quinn: <u>Temperature: Its Measurement and Control in Science and Industry</u> (ISA, Pittsburgh, 1972) Vol. 4, Part 1.

5. Evans J.P., Wood S.D.: Metrologia, 13, 171-172, 1977.

6. Crovini L, Actis A.: Metrologia, 14, 69-78, 1978.

7. F. Pavese, in Billing and Quinn (1975).

8. P.B. Coates, High Temperature-High pressures, Vol. 11, pp 119-134, 1979.

9. R.E. Bedford, High Temperature-High pressures, Vol. 11, pp 135-150, 1979.

10. J. Ancsin, Metrologia, 6, 53-56, 1970.

An international intercomparison of temperature standards of Asia/Pacific countries

T. P. Jones

CSIRO Division of Applied Physics, Sydney, Australia 2070

An intercomparison of the temperature standards in the fields of thermocouple and optical pyrometry of 10 Asia/Pacific countries is described. The travelling standards used were type S thermocouples and a vacuum tungsten strip lamp. Only one nation was shown to have unsatisfactory standards; the remainder agreed within their stated uncertainties. Results for the majority of the national calibrations fell within a range of 5 μV (0.5 °C) for thermocouples and 5 °C for the tungsten strip lamp.

INTRODUCTION

At the Seventh Conference of the Association for Science Cooperation in Asia (ASCA) held in Seoul in May 1979 it was decided that an intercomparison of temperature standards be undertaken, with CSIRO Division of Applied Physics providing the standards and coordinating the project. The Asia/Pacific Metrology Programme[1] (APMP) joined the intercomparison to ensure complete regional participation. APMP is a collaborative project aimed at establishing or improving standards of measurement in Asia/Pacific nations. Temperature was the first field in which an intercomparison was completed in this Programme but others are now underway in the fields of ac voltage and current, acoustics, dc resistance and voltage, length, mass, photometry, and time and frequency.

The intercomparison of temperature standards was initiated because many national standardizing laboratories in the Asia/Pacific region had recently established calibration facilities in the temperature field and the intercomparison was looked upon as a means of verifying that their standards were satisfactory and that their techniques of calibrating temperature measuring equipment were correct.

DESCRIPTION OF TRAVELLING STANDARDS

Two types of temperature standards were circulated for intercomparison, platinum-platinum, 10% rhodium thermocouples and a vacuum tungsten strip lamp for the calibration of optical pyrometers.

(i) Platinum-platinum, 10% rhodium thermocouples

Two type S (Pt-Pt, 10% Rh) thermocouples were used as travelling temperature standards for the checking of national thermocouple calibration facilities. To prevent damage during transport the thermocouples were placed in a slit in sheet foam which completely filled a rigid case. Before calibrations began, the wires of both thermocouples were cleaned in a series of acids and distilled water rinses and annealed at 1450°C for 1 hour by passing an electric current through the bare wires. They were then assembled in new 600 mm lengths of recrystallized alumina twin bore insulation and then isothermally annealed for 1 hour at 1100°C. This annealing sequence conforms with IPTS recommendations[2]. No cleaning or annealing of the thermocouples was permitted during the course of the intercomparison. The calibration was to be reported in a table indicating the thermocouple emf produced at 100°C intervals in the range 0° to 1000°C. The uncertainty of each stated emf was also requested.

(ii) Vacuum tungsten strip lamp

The lamp used as the temperature standard for the checking of national optical pyrometry calibration facilities was a Philips vacuum lamp, type W2GVV22i. The lamp was contained in a small box packed in foam within a larger box. The lamp was held in the smaller box in plastic foam, the centre of which was hollowed out in the form of the lamp. Detailed instructions were given on unpacking, mounting, and cleaning the glass window of the lamp. The voltage requirements of the dc lamp power supply were stated together with the change in spectral radiance temperature with change in current at each specified current. The polarity of connections to be made to the lamp was indicated. It was suggested that the lamp should be annealed on receipt by passing 8.5 amps through the lamp for one hour (T = 1500°C), to remove strain which may have been introduced by vibration during transport. The measurements to be performed consisted of measuring the spectral radiance temperature at the three specified currents stated in Table I. The time for the lamp to stabilize before measurements at each current was stated (20 minutes for lowest current to 1 minute for highest current). The results of the measurements were to be expressed as the spectral radiance temperature measured for each current together with the uncertainties pertaining to each result.

PARTICIPANTS

The scientists and national laboratories taking part in the intercomparison were:

Mr T P Jones
CSIRO Division of Applied Physics (CSIRO)
Sydney
AUSTRALIA   Thermocouple and tungsten strip lamp

Mr V P Wason/Mr K D Baveja
National Physical Laboratory (NPL)
New Delhi
INDIA   Thermocouple

Mr Suhud Hatmosuprobo
National Institute for Instrumentation (LIN)
Bandung
INDONESIA   Thermocouple and tungsten strip lamp

Dr Seiji Takata
National Research Laboratory of Metrology (NRLM)
Tsukuba
Ibaraki
JAPAN   Thermocouple and tungsten strip lamp

Dr Jong-Chul Park
Korean Standards Research Institute (KSRI)
Dae Jeon
KOREA   Thermocouple and tungsten strip lamp

The Controller
Standards & Industrial Research Institute of Malaysia
Shah Alam,   (SIRIM)
MALAYSIA   Thermocouple

Dr John Nicholas
Department of Scientific and Industrial Research
Physics and Engineering Laboratory   (PEL)
Lower Hutt
NEW ZEALAND   Thermocouple

Mr Teo Nam Kuan
Singapore Institute of Standards & Industrial Research
SINGAPORE   (SISIR)
               Thermocouple and tungsten strip lamp

Mr Pipat Panpaprai
Physics and Engineering Division   (PED)
Department of Science Service
Bangkok
THAILAND   Thermocouple and tungsten strip lamp

Mr Preecha Disathien
Thailand Institute of Scientific & Technological
Research
Bangkok   (TISTR)
THAILAND   Thermocouple

RESULTS

(i) Thermocouples

The order of calibration of Thermocouple No 1 was CSIRO, PEL, TISTR, PED, KSRI, CSIRO, SIRIM and CSIRO. The order of calibration of Thermocouple No 2 was CSIRO, SISIR, SIRIM, CSIRO, NPL, CSIRO, LIN and CSIRO. The measurements performed in the PED, Thailand, were completed by calibrating the thermocouples at the melting and freezing points of pure metals. The calibrations performed at PEL were by comparison with a standard resistance thermometer and were therefore restricted to temperatures up to 500°C. All other calibrations were performed by comparison with standard thermocouples. The temperature standards of CSIRO, NPL, NRLM and PEL were fundamentally established according to the IPTS-68[2] and those of other participants were traceable to the standards of USA and Australia. The results of the calibrations are shown in Figs. 1 and 2.

PEL and NPL quoted the repeatability of readings during their calibrations as their uncertainties, with magnitudes of approximately 1 μV. Other laboratories quoted uncertainties which applied to the accuracy of the thermocouples in subsequent use and contained a component to account for changes in the physical state of the thermocouple during short term use. For the initial calibration of the fully annealed thermocouples, CSIRO quoted an uncertainty, for 100 hours use in the temperature range 700° to 1000°C, of $(2 + 0.0004 E)$ μV, where E is the measured emf in μV[3]. The remaining laboratories stated uncertainties similar to or larger than CSIRO. The largest uncertainties were quoted by LIN and these varied from 12 μV at 200°C to 18 μV at 1000°C. The final CSIRO calibrations for both thermocouples at all temperatures differed from the initial calibrations by less than the stated uncertainty; the maximum difference being less than 4 μV. Since the CSIRO's emf measurements are accurate to a few tenths of a microvolt, these differences were due to systematic changes in the thermocouples during use and were probably due to the formation of rhodium oxide in the alloy wires[4]. Because of different temperature gradients in each calibrating furnace, it was not possible to estimate corrections to apply to each laboratory calibration to account for the changes in the physical states of the thermocouples. Consequently the CSIRO

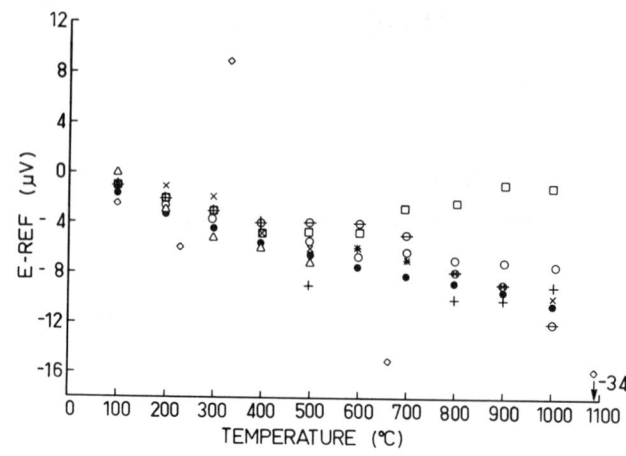

Fig. 1. Intercomparison of calibrations of thermocouple no. 1.
E = measured emf; REF = reference emf (NBS MONOGRAPH 125)
● CSIRO, AUSTRALIA 8/8/79   ○ CSIRO, AUSTRALIA 9/4/81
   (INITIAL)   (FINAL)
× NRLM, JAPAN   □ SIRIM, MALAYSIA
⊖ KSRI, KOREA   ◇ PED, THAILAND
△ PEL, NEW ZEALAND   + TISTR, THAILAND

initial calibrations with their associated uncertainties were taken as defining the expected emfs of the thermocouples under all conditions during the intercomparison.

(ii) Tungsten strip lamp

The intercomparisons were performed by comparison with national standards using photoelectric pyrometers at CSIRO, NRLM and KSRI and visual optical pyrometers at all other national laboratories. The effective wavelengths for the photoelectric pyrometers were 660 nm at CSIRO, 653 nm at KSRI and 650 nm at NRLM. The quoted manufacturer's effective wavelengths of the visual optical pyrometer were approximately 653 nm. All results were corrected to the median effective wavelength of 655 nm. The optical pyrometry temperature scales of CSIRO and NRLM were fundamentally established according to the IPTS-68[2] and those of other participants were traceable to the standards of USA and Australia. The results of the calibrations are shown in Table I, in the chronological order in which they were completed.

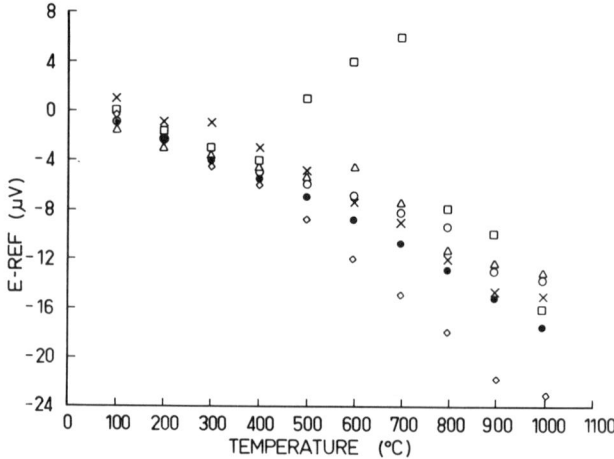

Fig. 2. Intercomparison of calibrations of thermocouple no. 2.
E = measured emf; REF = reference emf (NBS MONOGRAPH 125)
● CSIRO, AUSTRALIA 8/8/79   ○ CSIRO, AUSTRALIA 5/5/80
   (INITIAL)   (FINAL)
△ NPL, INDIA   □ SIRIM, MALAYSIA
◇ LIN, INDONESIA   × SISIR, SINGAPORE

TABLE I. Intercomparison of vacuum tungsten strip lamp 1579

Lamp current Laboratory	4.500 A	6.000 A	8.500 A
CSIRO, AUSTRALIA (27.7.79, INITIAL)	999° ± 1°C	1223° ± 1°C	1504° ± 2°C
NRLM, JAPAN	998° ± 5°C	1221° ± 5°C	1504° ± 5°C
KSRI, KOREA	1000° ± 3.5°C	1220° ± 4.5°C	1500° ± 5°C
PED, THAILAND	967° ± 16°C	1164° ± 59°C	1440° ± 55°C
SISIR, SINGAPORE	988° ± 10°C	1222° ± 8°C	1500° ± 8°C
LIN, INDONESIA	1001° ± 4°C	1225° ± 4°C	1505° ± 8°C
CSIRO, AUSTRALIA (15.8.80, FINAL)	999° ± 1°C	1223° ± 1°C	1504° ± 2°C

DISCUSSION

Figures 1 and 2 and Table I show that, for most laboratories, there are no significant errors. After the initial circulation of travelling standards, the only significant systematic errors present were in the thermocouple calibrations performed by SIRIM and PED, and the tungsten strip lamp calibration performed by PED. A review of the calibrations performed by PED indicated that their temperature standards in both thermocouple and optical pyrometry were not satisfactory. They are re-establishing their temperature standards but are not yet ready to repeat their measurements. The systematic errors in the SIRIM results occurred in the calibration of Thermocouple No 2 in the narrow range between 500° and 700°C, as shown in Fig. 2. They realized the existence of this error and noted it in their report, but were unable to repeat their calibration as the thermocouple had moved to the next laboratory. After the initial circulation, the other standard, Thermocouple No 1, was calibrated by SIRIM giving the improved results shown in Fig. 1.

Apart from the three calibrations which contained obvious systematic errors the remainder agree within their stated uncertainties. However the uncertainties quoted by LIN and SIRIM for thermocouple calibrations indicate that the techniques used are not satisfactory. These are being reviewed with the hope of reducing the magnitude of the uncertainties. With the exception of LIN, PED and SIRIM all thermocouple calibrations fall within the envelopes formed by the CSIRO initial calibrations and associated uncertainties.

If the PED results are removed from the optical pyrometry intercomparison, no significant systematic differences exist between the remaining laboratories. The result of SISIR for the low current calibration is being reviewed. With this one exception all calibrations fall within a temperature range of 5°C at each temperature.

The travelling standards are available for further laboratory intercomparisons. They will be sent to laboratories whose previous measurements are being reviewed, then they will be available on request.

CONCLUSION

With the exception of PED, all laboratories taking part had satisfactory standards in thermocouple and optical pyrometry. Further improvement in the calibration techniques for thermocouples are also necessary in LIN and SIRIM. The success of the intercomparisons indicated that the travelling standards used were satisfactory and these are now available for further intercomparisons. The systematic errors existing in the PED calibrations have been discovered in a very early stage of their establishment of temperature standards, before these errors were propagated from the laboratory, and thus a potential problem was prevented. The intercomparisons have increased the confidence of newly established national standardizing laboratories in their ability to calibrate temperature measurement equipment and has improved their international standing for both trade and scientific measurements.

ACKNOWLEDGEMENT

I have acted in these intercomparisons as coordinator and reporter, and it is my pleasure to acknowledge the fundamental contributions of scientists and national standardizing laboratories which I have previously listed.

REFERENCES

1. J. H. Buckingham, J. Phys. E., 14, 133 (1981).
2. H. Preston-Thomas, Metrologia, 12, 7 (1976).
3. R. E. Bentley and T. P. Jones, High Temp.- High Pressure, 12, 33 (1980).
4. E. H. McLaren and E. G. Murdock, Temperature Its Measurement and Control in Science and Industry, H. H. Plumb (Ed) Pittsburg: Instrument Society of America, 4, part 3, 1543-1560 (1972).

Soviet standards of the unit of temperature for radiation pyrometry

I. I. Kirenkov, B. N. Oleinik, G. S. Ambrok, and G. A. Krakhmalnikova

D. I. Mendeleyev Research Institute of Metrology (VNIIM), Leningrad, Union of Soviet Socialist Republics

A system of Soviet standards for radiation pyrometry made up of the primary standard of the unit of temperature and its related standards for the UV and IR regions is considered. The system of standards covers the temperature and wavelength ranges 600 to 3000 K and 0.3 to 4.5 μm, respectively.

The Soviet system of standards for the unit of temperature in the field of radiation pyrometry based on the recommendations accompanying the International Practical Temperature Scale of 1968 is in full accord with its latest version of 1975 [1]. This system comprises the primary standard for the visible region and standards for the ultraviolet and infrared regions related to the former.

For optical pyrometry above 1338 K the primary standard is used to realize the temperature scale by the photoelectric method over the visible region from 0.47 to 0.75 µm. The scale extrapolated above the gold point is fixed and maintained using Soviet-made tungsten-strip lamps Type "CN-10-300"[2]. Special standards are employed for the measurement assurance of UV and IR pyrometry [3]. Reference points of the temperature scale for these standards are obtained through comparison using a blackbody, with the scale interpolated from the reference points over the entire spectral range covered by them. For comparison between the IR special standard and the primary standards use is made of blackbodies at the antimony and gold freezing points and also of the nickel blackbody at intermediate temperatures measured by a standard thermocouple. A blackbody at an adjustable temperature measured by an UV pyrometer using tungsten strip lamps of the primary standard is utilized for the UV standard.

The standard of the unit of temperature for the UV radiation spectrum.

The primary standard developed at VNIIM in 1955-1972[4] comprises a set of measuring devices intended to realize, maintain and transfer the kelvin:
1. A blackbody source at 1337.58 K complete with its power supply circuit.
2. A photoelectric spectral comparator Type "CII-4K" and a mirror brightness doubler for extrapolation of the temperature scale.
3. Two groups of tungsten strip lamps calibrated over the wavelength range 0.47 to 0.75 µm, one of which is calibrated at the gold freezing point and the other over the range 1337.58 to 2800 K.

The operating principle of the standard involves the realization of the kelvin using a blackbody at the gold point and the transfer of the size of kelvin to a strip lamp employed as a reference standard using the spectral comparator Type "CII-4K". The transfer is made over the spectral range 0.47 - 0.75 µm and over the temperature range 1337.58 - 2800 K using a brightness doubler to extend the temperature range to 1900 K and absorption filters for further extension of the temperature scale to 2800 K.

For the temperature calculation use is made of the scale equation (Planck's formula):

$$\frac{B(\lambda,T)}{B(\lambda,T_{Au})} = \frac{\exp(C_2/\lambda T_{Au}) - 1}{\exp(C_2/\lambda T) - 1}$$

where T_{Au} = 1337.58 K is the thermodynamic temperature of the gold point, C_2 is the second radiation constant, T is the temperature to be measured, $B(\lambda,T)$ and $B(\lambda,T_{Au})$ are the blackbody spectral radiances at their temperatures T and T_{Au}, respectively.

Blackbody source at the gold point

For realizing the gold point use is made of a horizontal two-winding furnace whose temperature is uniform within 1 K per 100 mm. The crucible containing gold (Fig. 1 a) is placed in the furnace zone where the lowest thermal gradient occurs. Impurities in the gold total 0.002%. The ash content of graphite used for the blackbody source is 0.001%. The gold content of the source is about 180 grams. The correction for incomplete radiation and heat transfer over the cavity is 0.2 K. The estimated error of the correction is 0.1 K.

Spectral comparator Type "CII-4K"

The operating principle of the spectral comparator is based on the modulation method of brightness balancing. Radiation from the sources under comparison (a blackbody and a lamp or two lamps) passes through the monochromator and strikes the photomultiplier. In so doing, the modulator (an oscillating mirror) alternately directs light fluxes from the sources to be compared into the monchromator slit at a preset frequency. When substituting one flux for another of equal radiance, the total flux remains constant in time and space and the photomultiplier photoelectric current contains no variable component. When the radiances are different, the photocurrent incorporates a variable component whose frequency equals the modulation frequency. Upon amplifying and detecting, this component causes the meter pointer to deflect, which is indicative of unequal radiances.

The optical train of the Type "CII-4K" comparator [5] is shown in Fig. 2. Radiation from strip lamps is focussed on to a diaphragm which makes it possible to eliminate radiation from nonoperative parts of the sources under comparison in the initial stages of the optical system and thus to get rid of scattered light. A system of objectives directs the image of the source via a beam splitter on to the modulator mirror oscillating around its axis on the mirror galvanometer oscillograph principle. A holder accommodating absorption filters is placed in the path of parallel rays between the objectives. The beam splitter in the Type "CII-4K" device comprises two halves, with a blackened steel plate cemented between the prisms, where it is 0.1 mm or 0.3 mm thick, and extending outward. The protruding part of the plate impedes scattered light from the leading edge of the splitter and prevents the light from one lamp from striking the other one.

The monochromator aperture diaphragm is made coincident with the first extra-axial parabolic mirror and, with the modulator operating, is alternately overlapped by radiation from the two sources. At the monochromator output a photocell is located.

A double monochromator working over the wavelength range 0.47 to 1.0 µm is used in the Type "CII-4K" device. The linear dispersion is from 0.36 to 270 nm/mm. The monochromator is calibrated with hydrogen, mercury,

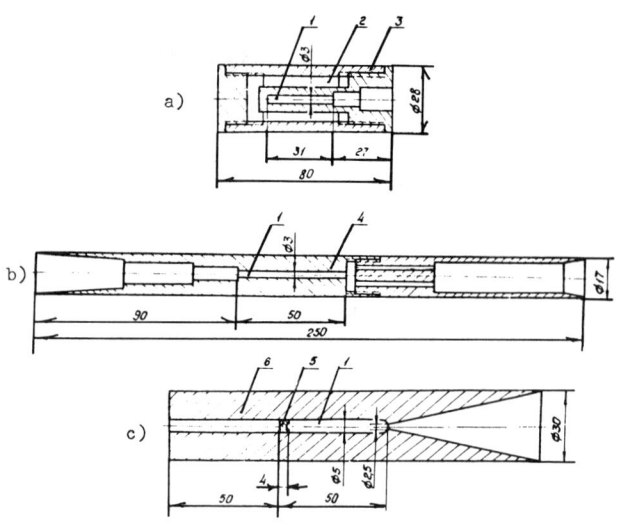

Fig. 1. Blackbody models for different spectral regions a) visible, b) ultraviolet, c) infrared 1 - radiating cavity; 2 - gold; 3 - graphite crucible; 4 - molybdenum-made model; 5 - nickel baffle; 6 - model made from oxidized nickel.

helium and krypton spectra, with a slit width of 0,1 to 0,2 mm. The use of a double monochromator eliminated the effect of scattered light and enabled an accurate determination of the effective wavelength. The threshold sensitivity of the Type "CII-4K" device is 0,02 - 0,05 K.

For extrapolating the temperature scale above the gold point use is made of a doubling system which is intended to preset the radiance ratio. It comprises two semi-transparent mirrors, one mirror dividing the auxiliary strip lamp radiation into two fluxes which

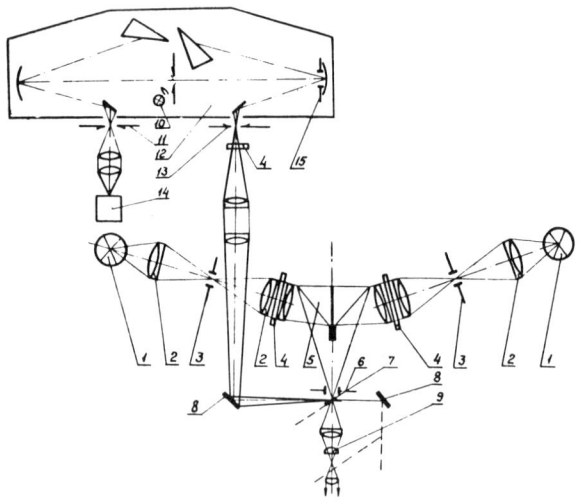

Fig. 2. Optical train of Type "CII-4K" comparator
1 - tungsten strip lamps; 2 - objective;
3 - intermediate diaphragm; 4 - filter holder;
5 - prism used as a beam splitter;
6 - diaphragm; 7 - modulator mirror;
8 - mirror; 9 - microscope; 10 - reversed ray lamp; 11 - exit slit; 12 - double monochromator; 13 - entrance slit; 14 - phototube; 15 - aperture diaphragm.

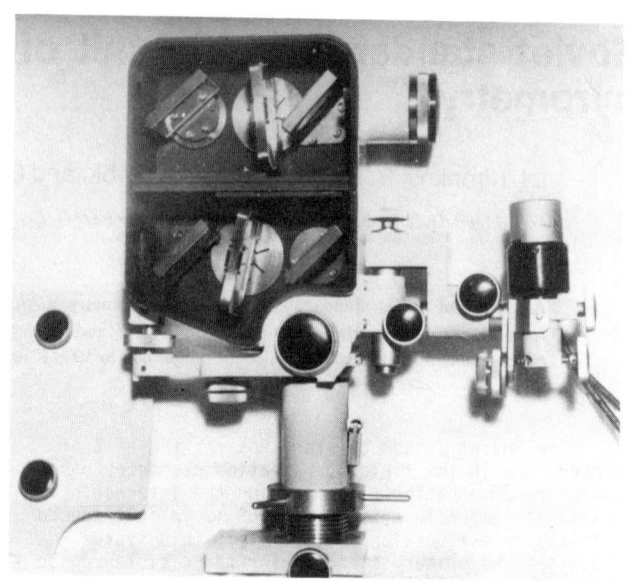

Fig. 3. Doubling system for extrapolation.

follow different paths and then are brought together using the other mirror. Shutters built into a baffle can completely block either one of the two fluxes or pass through both the fluxes to the spectral comparator, with radiance doubled in the latter case. Optical compensators are provided to vary the fluxes within small limits. (See Fig. 3.)

Beginning from 1900 K (or with about 1700 K at 0,75 μm), absorbing glasses are introduced at low wavelengths. These glasses make it possible to realize the scale up to 2700 - 2900 K. Grade "TC-2" glass is used for attenuation at 0,47, 0,55 and 0,67 μm, with "HC-13" and "IIC-2" glass employed at 0,656, 0,70 and 0,75 μm.

The size of the unit is transferred to reference standards, and strip lamps are calibrated with respect to radiance temperatures from 1337,58 to 2800 K over the wavelength range 0,47 to 0,75 μm. The error in the realization of the unit of temperature is 0,3 K (T = 1337,58 K) and those in the transfer of the size of this unit are 0,3 K, 0,8 K and 2,5 K at 1337,58, 1900 and 2800 K, respectively.

Standard of the unit of temperature for the UV region

In 1975 the national standard of kelvin, the unit of temperature, over the range 1800 to 3000 K for UV radiation was approved by the State Committee for Standards, Measures and Measuring Devices of the USSR [3]. The standard was developed at VNIIM in 1970-1975. The unit of temperature is realized by the optical method using a blackbody source at 1800 - 2500 K.

The standard comprises a set of measuring devices intended to realize, maintain and transfer the size of the unit of temperature.
- a blackbody source complete with its power supply circuit,
- a spectral comparator Type "CIIYΦ-1" working in the visible and near UV regions,
- tungsten strip lamps used for measuring the blackbody temperature.

The operating principle of this standard involves the realization of the kelvin using a blackbody at 1800 - 2500 K and transfer of the size of kelvin to secondary standard strip lamps through comparison with the device Type "CIIYΦ-1" over the range 1800 to 2500 K at 0,3 - 0,5 μm and by means of absorption filters over the range 2500 - 3000 K.

The size of the unit of temperature is realized by

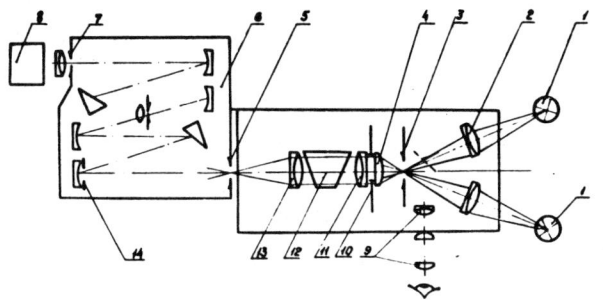

Fig. 4. Optical train of Type "СПУФ -1" comparator
1 - tungsten strip lamp; 2 - objective;
3 - mirror; diaphragm; 4 - modulator prism;
5 - entrance slit; 6 - monochromator;
7 - exit slit; 8 - photocell; 9 - microscope;
10 - diaphragm; 11 - objective; 12 - Dove
prism; 13 - objective; 14 - aperture diaphragm.

the special standard through measurements of the blackbody temperature using a strip lamp calibrated against the primary standard at 0.60 and 0.656 μm. The radiance corresponding to that of the blackbody at the temperatures specified above is attained by adjusting the current flowing through the strip lamp.

Blackbody Model

The blackbody under consideration (Fig. 1b) is a 50 mm long, 3 mm dia. cylinder-shaped cavity made from molybdenum and filled with pre-purified and heated argon to inhibit metal oxidation and evaporation. The blackbody is furnace-heated up to 2500 K with a 1200 mm long graphite coil (used as heater) that is thermally insulated with lamp black. For temperature adjustment, a thermostatic regulator Type "BPT-3" and a radiation pyrometer are provided. The effective emissivity of the blackbody is 0.999. The estimated temperature correction due to an imperfect blackbody is 0.2 - 0.5 K.

Spectral Comparator Type "СПУФ-1"

The optical train of the comparator is presented in Fig. 4. The spectral range of the device is 0.22 - 2.5 μm and its operating principle is based on the zero modulation method for the radiance balancing of radiation sources. Optical components of the comparator are achromatized for the spectral range 0.25 - 0.5 μm.

Radiation from strip lamps is directed by objectives to the modulator field diaphragm which is followed by a quartz biprism splitting the light beams from each lamp into two parts. In so doing, the extreme beams from each lamp are diaphragmed and the other two beams pass into the system intended to image the modulator field diaphragm in the plane of the monochromator entrance slit. Between the objectives, placed in the parallel beam, there is a Dove prism that turns the images through 90° so as to make the direction of the light beam out of parallel with that of the monochromator dispersion. Mounted after the monochromator is an objective that images the aperture diaphragm on the plane of the photocathode of the photomultiplier. The light beams are modulated using a quartz biprism vibrating at 760 Hz. The spectral comparator incorporates a double monochromator to reduce scattered light and to allow determination of the effective wavelength to a higher accuracy. Linear dispersion of the monochromator is 15 and 300 nm/mm at 0.25 and 0.65 μm, respectively. Photomultipliers are used as radiation detectors. Threshold sensitivity of the instrumentation is 0.02 - 0.05 K and 0.07 - 0.1 K for spectral regions down to 0.35 μm and 0.35 to 0.30 μm, respectively. Studies with the spectral comparator have shown that the strip lamp calibration error is 1.0 - 1.5 K.

Tungsten Strip Lamps

The filament of each lamp Type "СИ-10-300 у" is fabricated from tungsten and the envelope windows are made from uviol and sapphire. The envelopes are filled with krypton at 50 mm Hg. The lamps calibrated against the primary standard over the range 1800 to 2800 K are intended to determine the blackbody temperature over the range 1800 to 3000 K. For realizing the temperature scale within 2500 - 3000 K use is made of a set of filters composed of metallized quartz plates. The set comprises six filters which satisfy Foot's criterion and have transmittance covering the range 0.0020 to 0.0450.

The error due to the inaccuracy in determining the pyrometric attenuation is 3 K at 3000 K.

The size of the unit of temperature is transferred to strip lamps used as secondary standards, with the lamps calibrated with respect to radiance temperatures at 0.3 - 0.5 μm.

The errors that arise when realizing the unit of temperature are 1 and 1.5 K over the ranges 1800 to 2500 and 2500 - 3000 K, respectively, with residual systematic errors of 1 and 1.5 K at 1800 - 2500 and 2500 - 3000 K, respectively. The errors arising in the transfer of the size of the unit of temperature (calibration of secondary standards) are 1.5 and 2.5 K over the ranges 1800 - 2500 and 2500 to 3000 K, respectively.

Standard of the Unit of Temperature for the IR Region

In 1975 the State Committee for Standards, Measures and Measuring Devices of the USSR approved the national standard of kelvin, the unit of temperature, for IR radiation over the range 600 to 2300 K developed at VNIIM throughout 1961 - 1975[3].

The unit of temperature for this region is realized by the optical method using a blackbody at 600 - 1300 K.

The standard comprises a set of measuring devices intended to realize, maintain and transfer the unit of temperature:
1) A blackbody source complete with its power supply circuit,
2) a spectral comparator Type "СПИФ -1" working over the range 0.2 to 6.0 μm,
3) a group of three platinum-rhodium/platinum thermoelectric thermometers.

The operating principle of the standard consists in realizing the kelvin using a blackbody at 600 to 1300 K and transferring the size of the kelvin to a secondary standard strip lamp through comparison of the device Type "СПИФ -1" at 0.8 - 4.5 μm over the range 600 to 1300 K and using filters from 1300 - 2300 K.

As in the case of the standard for the UV region, the IR standard is traceable to the primary standard. For this purpose, the special standard incorporates platinum-rhodium/platinum thermoelectric thermometers calibrated against the primary standard. The lower radiation limit of the special standard is coincident with the upper limit of the primary standard (0.8 - 1.0 μm) which assures agreement over the spectral range covered.

Blackbody Model

The blackbody comprises a heating furnace built from two coaxial ceramic tubes with heater windings made from Ni-Cr alloy and with the space between the tubes filled with insulation. The outer tube winding is made up of individual sections so as to maintain the inner tube temperature to within 0.5 or 1 K over lengths of 4 to 5 cm and 6 to 7 cm, respectively. The inner tube accommodates a cylinder-shaped nickel blackbody (Fig. 1c). The blackbody surface is coated with

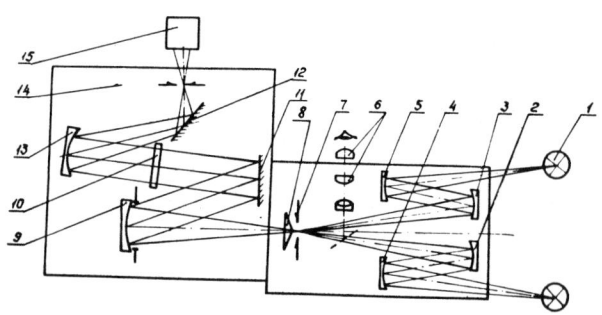

Fig. 5. Optical train of Type "СИИФ -1" comparator.
1 - tungsten strip lamps; 2,3,4,5 - aluminized spherical mirrors; 6 - microscope;
7 - field diaphragm; 8 - modulator;
9 - spherical mirror with aperture diaphragm;
10 - interference filter; 11, 12, 13 - mirrors;
14 - iris diaphragm; 15 - detector.

nickel oxide formed by heating for 15 hours at 1300 K. The effective emissivity of the blackbody is taken to be 0.997.

The residual systematic error due to an imperfect blackbody is not over 0.5 and 2 K for the ranges 600 to 1000 and 1000 to 2300 K, respectively.

Spectral Comparator Type "СИИФ -1"

As with the standards for the visible and UV regions, the comparator incorporates two optical channels (Fig. 5) and operates on the principle of comparison between two fluxes.

The outer optical system, composed of four aluminized spherical mirrors, focuses the images of two sources on a field diaphragm. The sources are tested for proper alignment using a microscope. Placed after the diaphragm is a small-size sapphire biprism with a refraction angle of 10° mounted on a phosphor fluoride bronze wire that moves the biprism perpendicular to the optical axis at a frequency close to 800 Hz. The biprism brings radiation fluxes from the sources under comparison into coincidence in space. Interference filters are used to separate the spectral ranges under study. Visual alignment in the IR region is possible due to the freedom from aberration in the external optical system. The spectral range of the standard covers filters and detectors as follows: photomultipliers up to 1.0 μm and PbTe photoresistor for the wavelength range 1 to 6 μm. The minimum and maximum threshold sensitivities of the detectors are 0.2 to 0.3 K and 0.03 - 0.05 K, respectively.

Thermoelectric Thermometers

The thermometers were calibrated at the antimony, silver, gold, tin and zinc freezing points. Values of the emf were calculated for the temperatures making up their proper ranges using a quadratic interpolation formula. The blackbody temperature measuring error is 0.1 K at 600 - 1300 K.

The size of the unit of temperature is transferred to secondary standard strip lamps calibrated with respect to radiance temperatures over the range 0.8 to 4.5 μm. The error that arises when realizing the unit of temperature is 1.0 K at 600 - 1300 K, with residual systematic errors of 1.5 K and 2.5 K over the ranges 600 - 1000 and 1000 - 2300 K, respectively. The transfer error (the secondary standard calibration) is 3 - 4 K at 600 to 1300 K.

The standards considered above assure calibration of industrial and laboratory measuring devices used in instrument-making, metallurgy, astrophysics, infrared engineering, etc.

REFERENCES

[1] Echelle Internationale Pratique de Température de 1968 (Edition amendée de 1975).
[2] G. A. Krakhmalnikova, Izmeritelnaya Tekhnika, 1970, 1, 44 (in Russian).
[3] G. S. Ambrok et al., Izmeritelnaya Tekhnika, 1976, 3, 35 (in Russian).
[4] I. I. Kirenkov et al., Izmeritelnaya Tekhnika, 1973, 4, 8 (in Russian).
[5] I. I. Kirenkov, G. A. Krakhmalnikova, Trudy Metrologicheskikh Institutov SSSR, 1977, 207 (267), 27 (in Russian).

Realization of the triple point of water and the freezing points of tin and zinc at the National Institute of Standards (Egypt)*

H. El-Shammaa, M. R. Moussa, and M. H. Omar**

National Institute of Standards (NIS), Cairo, Egypt

The triple point of water has been measured using four cells from two suppliers over a period of one year at the National Institute of Standards (NIS). Eight platinum resistance thermometers were used, five of which were calibrated at the National Bureau of Standards (NBS) in 1980. The intercomparison of these cells shows that their triple points agree within 0.2 m°C. Agreement between NBS and NIS measured values is within 0.2 m°C. The tin point was studied using three different cells, two loaned by NBS and one fabricated in this laboratory. The reproducibilities of the NBS cells as measured at NIS are of the order of 0.1 m°C for the high purity cell and 0.7 m°C for the other slightly lower purity cell. The NIS cell indicated a reproducibility of the order of 1 m°C. The zinc point was measured using one cell loaned by NBS. Our calibration data on SPRT's are on the average about 5 m°C lower than the calibrations obtained at the NBS using a high-purity zinc point cell.

1. INTRODUCTION

The National Institute of Standards in Egypt (NIS) has the responsibility for establishing, maintaining, and developing temperature standards of the nation. In order to meet this responsibility, it was necessary to develop measurement techniques in standard platinum resistance thermometry and to realize the triple point of water, freezing points of tin and of zinc on the International Practical Temperature Scale of 1968 (IPTS-68).

Through a joint project with the U.S. National Bureau of Standards (NBS), we were able to calibrate five of our standard platinum resistance thermometers (SPRT's) (2 capsule and 3 long stem) above 0°C in 1980. The thermometers were calibrated at the oxygen boiling point, the triple point of water and freezing points of tin and of zinc at the NBS.

We also received on loan from NBS one Jarrett Instrument Company triple point of water cell, two tin cells and one zinc cell. All of these cells and other cells of NIS are the subject of the study reported in this work.

This paper presents results on the realization of the triple point of water (TP) and the freezing points of tin and zinc. The differences between NBS and NIS measurements at these points are presented.

2. THE TRIPLE POINT OF WATER

2.1 Method and Procedure

In order to carry out measurements on the triple point of water, we modified a commercial top-loading freezer to accommodate four cells simultaneously. This modification was done in a way similar to that currently used at the NBS[1]. Tables I and II summarize the specifications of the cells and the SPRT's used in this study, respectively. The ice mantles were prepared using an immersion cooler as described by Evans and Sweger[2]. The cells were then kept in the modified freezer around 0 °C for a period of at least 24 hours before carrying out the measurements[3].

For each thermometer, a series of measurements was obtained on the different cells using two Tinsley AC bridges Type 5840. Both bridges were calibrated using a standard AC/DC resistor (ST.R) of 25 ohms supplied by the Tinsley Company. Measurements were taken at two different thermometer currents, namely, 1.0 and $\sqrt{2}$ mA, for the purpose of determining the resistance value at

Table I

Identification and Description of Triple Point of Water Cells

NIS assigned code	Purchase date	Manufacturer	Remarks
J-NIS-TP 1	1965	Jarrett	Large size
J-NIS-TP 2	Loaned by NBS	Jarrett	Large size
SP-NIS-TP 3	1965	Spembly	Small size
SP-NIS-TP 4	1965	Spembly	Small size

zero current. Closely fitting aluminum sleeves that were slightly longer than the SPRT coil were used to enhance the thermal contact.

2.2 Results and Discussion

The data obtained with the different SPRT's in all cells at 1 mA current are given in Table III.

It is clear from Table III that the maximum equivalent temperature difference between the four cells does not exceed 0.2 m°C. The reproducibility of the different SPRT's at the triple point of water for a period of one year is shown in Fig. 1. The obtained data are within 3 m°C for the long stem thermometers, except for NIS-5 SPRT. For the two capsule type SPRT's (NIS 6 and NIS 8), however, the reproducibility is better than 1 m°C.

Some of the wide scatter of the data for long stem thermometers is most probably due to its pre-exposure to high temperatures.

The difference between NIS and NBS values of R(0°C), at zero current and at "zero immersion depth," calculated from observed values of R(TP) is summarized in Table IV. It is to be noted that the measurements at the NIS are obtained using the 435 Hz AC bridge method while the measurements at the NBS are obtained by the DC bridge method. Some of the differences in the results are expected to be attributable, in addition to the difference in the resistance units, to the difference in the methods, the magnitude of the difference being dependent on the thermometer design and measurement conditions. For this paper, the difference in the methods has not been considered in the analysis of data. The ratios of the R(0°C) values obtained

Table II

Identification and Description of Standard Platinum Resistance Thermometers

NIS SPRT Code	Type	Serial Number	SPRT Class	Supplier Value for R_o	Date of original certificate	Stem Style
NIS.1	H. Tinsley	232771	1	25.5063	NPL/11.4.79	Long
NIS.2	Rosemount	232247	1	25.5598	NPL/13.8.64	Long
NIS.3	Rosemount	232234	1	25.564	NRC/Canada/ 25.1.65	Long
NIS.4	Rosemount	232259	1	N.A.*	NPL/13.8.64	Long
NIS.5	H. Tinsley	228855	1	24.3444	NPL/12.9.75	Long
NIS.6	Rosemount	231	1	N.A.	N.A.	Caps.
NIS.7	Leeds & Northrup	1645632	1	N.A.	N.A.	Long
NIS.8	H. Tinsley	1656981	1	N.A.	N.A.	Caps.

*N.A.: Not available.

Table III

Observed values of resistances of the eight standard platinum resistance thermometers in the four triple point of water cells*

NIS SPRT	J-NIS-2	J-NIS-1	SP-NIS-3	SP-NIS-4
1	25.50704	25.50705	25.50705	25.50703
2	25.56680	25.56679	25.56679	25.56680
3	25.56476	25.56476	25.56476	25.56475
4	25.54871	25.54871	25.54871	25.54871
5	24.34510	24.34509	24.34509	24.34510
6	25.53794	25.53794	25.53793	25.53791
7	25.54428	25.54435	25.54432	25.54428
8	25.53932	25.53935	25.53933	25.53933

*Values given represent the mean value of resistance measurements from February to June 1981 shown in Fig. 1. Although these readings are at 1 mA, the variations due to differences in self heating are expected to be not greater than 0.1 m°C because aluminum sleeves were made.

Table IV

Comparison of R(0°C) values of five SPRT's obtained at NIS and NBS*

SPRT Code	NIS Data R_o/Ω	NBS Data R_o/Ω	NIS/NBS $R_o(NIS)/R_o(NBS)$
NIS 1	25.505608	25.506616	0.999960481
NIS 3	25.563507	25.564180	0.999974674
NIS 5	24.343675	24.344601	0.999961963
NIS 6	25.536768	25.537631	0.999966207
NIS 8	25.538080	25.539032	0.999962724
		Mean	0.999965010
		Std. Dev.	± 0.000001143

*R_o is the zero current resistance at 0 °C.

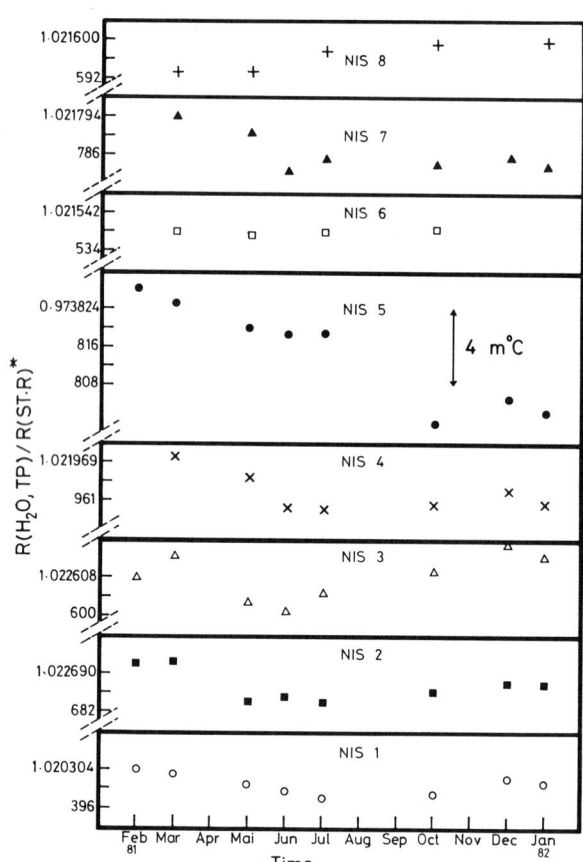

* TP = triple point
ST,R = standard resistor ($\approx$25 ohm)

FIG. 1. The average of the resistance ratio data [$R(H_2O,TP)/R(ST.R)$] during one year, 1981–1982.

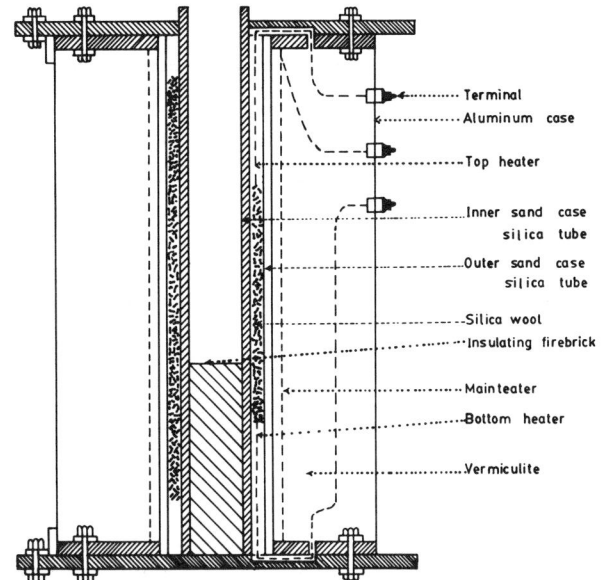

FIG. 2. Furnace used in the present work.

between NIS and NBS show the measurements to be consistent on the average to above 1 ppm or ±0.2 m°C.

3. THE TIN FREEZING POINT

3.1 Furnace and Tin-Point Cells

The furnace used for the study of both tin and zinc points is shown schematically in Fig. 2. This furnace was designed and constructed at NIS in 1974. It has 3 heaters, one is a main heater and two are guard heaters, placed at the top and bottom of the furnace. The main heater is energized by AC current through a temperature controller. It was possible to adjust the furnace in such a way to obtain a freezing plateau for at least four hours.

The specifications of the tin-point cells studied in this work are given in Table V. For more details on these cells, reference [4] is to be consulted.

Table V

Identification and description of tin point cells

NIS assigned code	Cell source	Tin purity	Sample source
NIS-Sn-1	NIS	6-9	Cominco
NIS-Sn-2	NBS (Sn70H)	6-9	NBS
NIS-Sn-3	NBS (Sn70B)	5-9	NBS

3.2 Results and Discussion

The cells were subjected to various melting and freezing cycles under a pressure of one atmosphere of dry argon gas. Usually, to melt the sample, the cell was kept 3 °C above the melting point for about 10 h. The run begins with reducing the furnace temperature to about 1 °C below the freezing point of tin, until the cell temperature approaches the freezing point. The cell is then removed from the furnace and the temperature is monitored by the SPRT on the recorder. This continues until the inversion in temperature is observed at the end of the supercool. Then, the cell is replaced back in the furnace[5]. The observed magnitude of the supercool ranged between 11 and 16 °C. The plateau is then followed on the chart recorder. Typical runs usually lasted for about four hours. Figure 3 shows some typical results using the three different cells.

During the freezing of tin, a test for reproducibility of measurements was carried out according to the following procedure. After an equilibrium reading, the SPRT was completely withdrawn from the cell, exposed to the ambient temperature and reinserted again into the cell. The process was repeated. Figure 4 shows that the SPRT comes to equilibrium in a short time and reproduces the freezing value within 0.1 m°C.

A study on the immersion characteristics of a long stem SPRT was carried out, and the results are shown in Fig. 5. It shows that the SPRT is adequately immersed when the tip is as high as 8 cm from the bottom of the thermometer well.

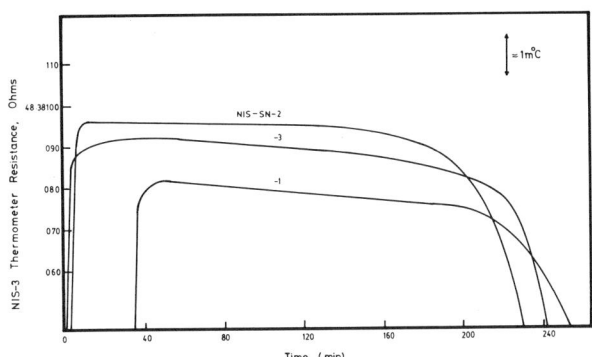

FIG. 3. Freezing curves of the different tin cells.

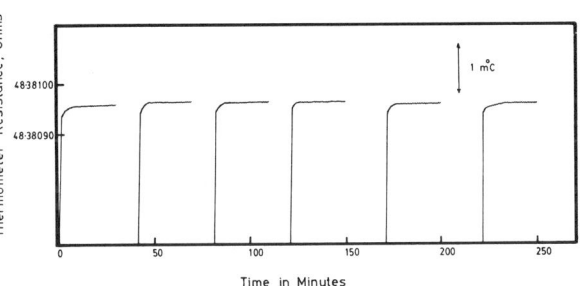

FIG. 4. Consecutive measurements of the resistance of a SPRT in the same tin freeze.

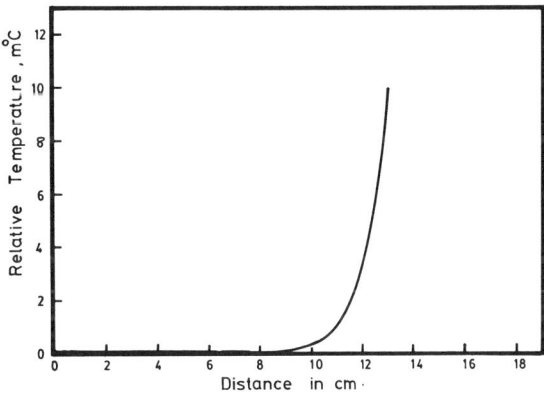

FIG. 5. Immersion characteristics of SPRT in a tin point cell.

Table VI

Comparison of calibrations of SPRT's at the tin point

SPRT Code	NIS Data			NBS Data			NIS-NBS
	$R(Sn)/\Omega$	R_0/Ω	$W = \frac{R(Sn)}{R_0}$	$R(Sn)/\Omega$	R_0/Ω	$W = \frac{R(Sn)}{R_0}$	ΔW
NIS 1	48.27583	25.505608	1.8927535	48.27773	25.506616	1.8927532	$+0.3 \times 10^{-6}$
NIS 3	48.38091	25.563507	1.8925772	48.38233	25.564180	1.8925829	-5.7×10^{-6}
NIS 5	46.07330	24.343675	1.8926190	46.07500	24.344602	1.8926167	$+2.3 \times 10^{-6}$

Table VII

Comparison of Calibrations of SPRT's at the Zinc Point

SPRT Code	NIS Data			NBS Data			NIS-NBS
	$R(Zn)/\Omega$	R_0/Ω	$W = \frac{R(Zn)}{R_0}$	$R(Zn)/\Omega$	R_0/Ω	$W = \frac{R(Zn)}{R_0}$	ΔW
NIS 1	65.51839	25.505608	2.5687837	65.52125	25.506616	2.5687943	-10.6×10^{-6}
NIS 3	65.65937	25.563507	2.5684805	65.66157	25.564180	2.5684989	-18.4×10^{-6}
NIS 5	62.52754	24.343675	2.5685333	62.53049	24.344602	2.5685567	-23.4×10^{-6}

The results show that the NIS-Sn-2 cell is of high quality, with a reproducibility of 0.1 m°C, while the other NIS-Sn-3 is reproducible only within 0.7 m°C. The NIS-Sn-1 indicated a reproducibility within 1 m°C. Accordingly, the first cell NIS-Sn-2 was chosen for carrying out a comparative test using the three long stem SPRT's which had been calibrated at NBS. The results are shown in Table VI. It is seen that the differences in the calibrations between NBS and NIS for the three long stem thermometers lies within about 1 m°C.

4. THE ZINC FREEZING POINT

The same procedure used with tin was also used for zinc, except that the freeze was initiated in the furnace without having to remove the cell momentarily out of the furnace. The study of the zinc point was restricted to one cell only, which was loaned from NBS and is labelled Zn74E. The observed reproducibility was found to be of the order 3 m°C. The data on the freezing point of zinc are given in Table VII. The above NIS data indicate that the zinc sample is not of the highest quality. The calibration results are on the average about 5 m°C lower than those obtained at the NBS using a zinc cell containing a sample that is at least 99.9999 percent pure. The results of the reproducibility test at the freezing point of zinc is shown in Fig. 6. The reproducibility of calibration is shown to be well within 1 m°C.

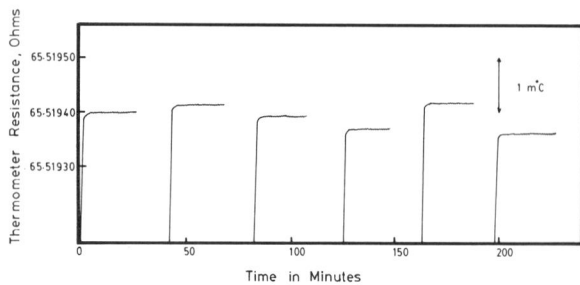

FIG. 6. Consecutive measurements of the resistance of a SPRT in the same zinc freeze.

5. CONCLUSIONS

The results show that at the NIS the measurement methods with SPRT's are reproducible within a few tenths of a m°C. The calibration of SPRT's can be performed in terms of the IPTS-68 within 1 mK at the tin point and within 5 mK at the zinc using the existing equipment. The use of a higher purity zinc point cell should improve the accuracy to within 1 mK.

ACKNOWLEDGMENTS

It is great pleasure to acknowledge the continuous help and support from NBS staff in the Temperature and Pressure Measurements and Standards Division, particularly Dr. J. F. Schooley, Dr. B. W. Mangum, and Dr. G. T. Furukawa. We are also indebted to Prof. Dr. Mahmoud Mokhtar, consultant to NBS-NIS project, for his contribution and useful suggestions during the progress of this work.

REFERENCES

* This work constitutes one part of project No. G-198 which is a joint project between National Bureau of Standards (NBS), Washington, D.C. and National Institute of Standards (NIS), Cairo, Egypt.
** Permanent address: American University in Cairo, Egypt.
[1] Riddle, J.L., Furukawa, G.T., and Plumb, H.H.: Platinum resistance thermometry, NBS Monograph 126, 126 pages, 1973.
[2] Evans, J.P., and Sweger, D.M., Immersion cooler for freezing ice mantles on triple-point of water cells, Rev. Sci. Instr. 40, 376 (1969).
[3] Berry, R.J., The temperature-time dependence of the triple point of water, Can. J. Phys. 37, 1230 (1959).
[4] Furukawa, G.T., Riddle, J.L., and Bigge, W.R., Investigation of freezing temperatures of National Bureau of Standards tin standards, in TMCSI Vol. 4, Part 1, 247-263 (Instrument Society of America, Pittsburgh, Pa. 1972).
[5] McLaren, E.H., Intercomparison of 11 resistance thermometers, at the ice, steam, tin, cadmium, and zinc points, Can. J. Phys. 37, 422 (1959).

On the use of first-generation sealed cells in an international intercomparison of triple-point temperatures of gases

F. Pavese

Istituto di Metrologia "G. Colonnetti," 10135 Torino, Italy

In 1978 the Comité Consultatif de Thermométrie decided to hold an International Intercomparison of fixed points, realized in small, transportable sealed-cells used as travelling standards, a technique developed in the last ten years; 25 of these cells, coming from six Laboratories and containing a total of seven gases (Ar, O_2, H_2, CH_4, N_2, Ne, and D_2) were circulated at BIPM of Sèvres and among 10 National Laboratories: ASMW (East Germany), IMGC (Italy), INM (France), NBS (USA), NIM (China), NML (Australia), NPL (United Kingdom), NRC (Canada), NRLM (Japan) and PRMI (Soviet Union). The organization and data analysis was the task of IMGC, which had been the promoter of the intercomparison. The relevant results collected so far on the differences between different cell models are reported in the paper.

INTRODUCTION

The calibration of platinum resistance thermometers of specified quality (SPRT) on the International Practical Temperature Scale (IPTS) in the cryogenic region consists of the measurement of their resistance values at a certain number of fixed temperatures, corresponding to the boiling and triple points of substances that are gaseous at room temperature. Although the Scale is based on thermodynamic states uniquely defined and on a specified quality of platinum, interlaboratory comparisons are necessary to check the degree of uniformity internationally achieved in the realization of the Temperature Scale.

In fact, the practical implementation of both the temperature fixed points and the interpolating instrument may perceivably affect the measured physical property; with platinum resistance thermometers it brings about a dispersion of the W versus T characteristic, which affects the uniqueness of the Scale between fixed point temperatures /1,2/, with fixed points it produces different temperature values for the same physical thermodynamic state in different measurements.

Several international intercomparisons have been promoted to control the uniformity of IPTS realizations in different laboratories /3,4,5,6/. An intercomparison requires travelling standards: only SPRT, calibrated on laboratory realizations of the IPTS (LAB-IPTS), were available for this purpose in the past, although they are very delicate instruments liable to instability when transported.

In the period following the 5th Symposium on Temperature extensive studies were made on the realization of fixed points in transportable sealed cells, down to the solid hydrogen temperature (see other papers at this Symposium for description and use of these sealed cells). As they proved to be accurate devices, the Comité Consultatif de Thermométrie decided in its 12th meeting in 1978 to sponsor an International Intercomparison of fixed points. The comparison had been proposed by the Istituto di Metrologia "G. Colonnetti" (IMGC) /7/, with the aim of comparing fixed point realizations in National Laboratories by using fixed points in small sealed cells, instead of capsule resistance thermometers, as travelling standards. Since these devices are both strong and stable in time, one set of them (one cell for each substance) would have been sufficient for the comparison (only triple points and solid-to-solid transitions could be studied in a sealed device at that time). However, as it was a new device, it was preferred to circulate more than one cell for each substance, when available, in order to also check the quality of the standards used.

This paper will not report results on temperature differences between LAB-IPTS realizations nor are all the data available yet. A comprehensive Final Report on the whole set of results of the Intercomparison will be issued by the Bureau International des Poids et Mesures after completion in 1982. However, it seemed important, at the half-way meeting of the participating Laboratories held in Paris in 1980, to take the opportunity of this Symposium to present the main results on the comparison of different cell models; few considerations on the behaviour of this device, and the possibility of using cells for primary realizations of fixed points in a new issue of IPTS will be added by the author.

The intercomparison was possible by a large cooperative effort of all the laboratories taking part in the measurements. The following staff has been involved in the intercomparison: H. Maas (ASMW), J. Bonhoure (BIPM), F. Pavese (IMGC), G. Bonnier (INM), G.T. Furukawa (NBS), Ling Shan-Kang (NIM), R. C. Kemp (NML), S.D. Ward and R.L. Rusby (NPL), J. Ancsin (NRC), K. Mitsui and Y. Koga (NRLM), D.N. Astrov and E.N. Ivanov (PRMI, Gosstandart) (legend of acronyms at the end of the paper).

INTERCOMPARISON CONFIGURATION

IMGC was entrusted with the organization of the intercomparison and the collection and analysis of results.

Six laboratories agreed to contribute sealed cells for circulation in the intercomparison; two others (NIM, PRMI) used sealed cells of their own construction for the comparison but did not circulate them. Table I shows the complete set of cells that were used in the intercomparison: the date of sealing of each cell is indicated when available. Only the first three of the seven substances listed in the Table, realize IPTS-68 fixed points. The other four were added to help in the studies of a future IPTS: nitrogen and deuterium have been circulated only since 1980 and only one cell for each of these gases has been available. Fig. 1 shows the schematic drawings of cell models.

The cells were sent to the laboratories according to their requests; in fact, not all laboratories were interested in measuring all cells. As the number of cells and laboratories involved was large and the cells are stable devices, a regular circulation pattern was not mandatory; however, an effort was made to send at least two models of cells for each of the substances (when available) and at the same time. The circulation started at the beginning of 1979 and will be completed in the first months of 1982. Although a large variety of unexpected events has randomly disturbed the regularity of the circulation pattern (involving 25 cells and 11 countries in the whole world), loss of substantial information could almost entirely be avoided.

Table II represents the distribution of the measurements, with indication of the date when they were taken. Few intercomparisons made before 1979, relevant for this intercomparison, are included.

A total of 25 cells and 7 substances were circulated; 6 additional cells were directly involved but not circulated. The 11 laboratories produced some 120 sets of independent measurements, totaling some 200 meltings. An average of four laboratories studied each of the circulating cells.

Table 1: Sealed cells involved in the intercomparison.

SUBSTANCE

Laboratory supplying cells	Argon 83.798 K (1)	Oxygen 54.361 K (1)	e-Hydrogen 13.81 K (1)	Methane 90.7 K (2)	Nitrogen 63.1 K (2)	Neon 24.6 K (2)	e-Deuterium 18.7 K (2)
BIPM	3AR (Feb 1977) (not circul.)			7CH4(Sep1977)			
IMGC	1AR(Jul1975) 2AR(May1978)	1O2(Sep1976) 8O2(Nov1978)	1HY(Aug 1980)	2CH4(Aug1976) 12CH4(Apr1979)	2N2(Feb 1980)	1NE(Jun 1977) 3NE(Feb1979)	1eD2(Nov1980)
INM	1AR(Sep1975)	8O2(Feb1976)					
NBS	M1(Feb1978)						
NIM		PP5(Jul1981) (not circul.)					
NRC	10AR(May1979)	15O2(Jun1979)	23HY(Aug1979)	18CH4(Aug1979)		12NE(Jun1979)	
NRLM	7803(Jun1978) 7801(Jun1978)(3)	7802(Jun1978)	**7801(Jun1978)**			NE01(Jul1978)(4) NE02(Jul1978)	
PRMI	(5) (not circul.)	(5) (not circul.)	(5) (not circul.)			(5) (not circul.)	

(1) IPTS-68 definition point; (2) approximate value (no official value); (3) circulated only at NBS; (4) circulated only at NPL; (5) PRMI cell is a unique multicomponent cell; **bold face**: reference cell.

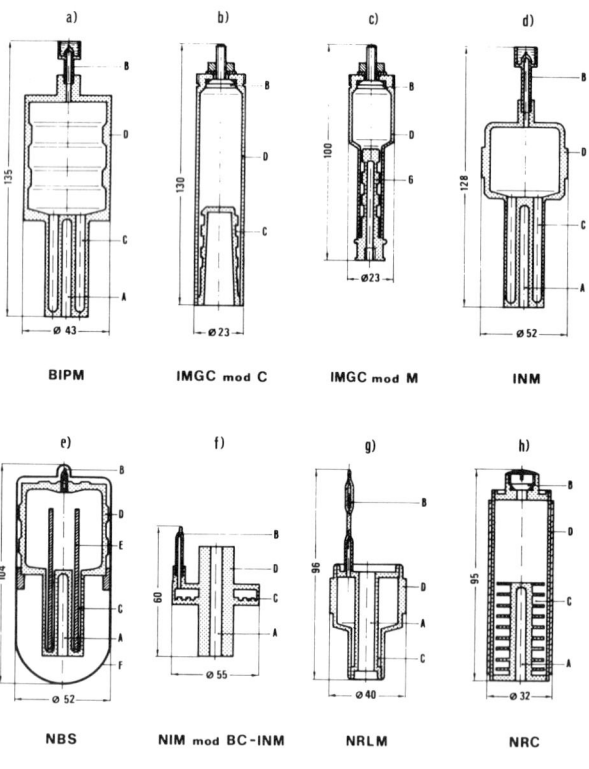

FIG. 1. Physical appearance of the sealed cells taking part in the intercomparison (not to scale). PRMI model is to be considered in addition: it is a multicomponent cell accomodating four gases: Ar, O_2, Ne and e-H_2.
Some of the IMGC cells were of a sligthly different model (mod. A and B) /8/.
Caption: A) thermometers(s) well: IMGC mod. C, NBS and NRLM cells require an adapter; IMGC mod. S has no thermometer well.
B) hermetic seal: b,c,h) indium seal; a,d,e,f,g) pinchoff tube;
C) thermometer block: b,g,h) copper; a,d,e,f) stainless steel; c) has no built-in thermometer block.
D) cell body: a,b,c,d,e,f) stainless stel; g) copper; h) copper with external stainless stel shell (all stainless steel with hydrogen filling).
E) copper equalizer: e) only.
F) copper outer shell: e) only.
G) copper body to transfer interface temperature to an external thermometer block/ c) only.

Each laboratory was asked to use the same thermometer for measurements on all cells - at least on those containing the same substance - in order to avoid the uncertainty due to thermometer calibration (only in few cases was this procedure not followed).

Consequently, the differences in the temperature value obtained with different cell models with respect to a reference cell can be directly obtained from the measurements: IMGC cells are taken here as references with argon, oxygen and neon, NRLM cell with hydrogen. Differences with respect to the reference cells are reported in Figs. 2 to 5, in a way that also represents the uncertainty associated with the comparisons. This kind of comparison was not possible with nitrogen and deuterium since only one cell of each gas was circulated.

Uncertainty derives from different sources:
a) reproducibility of measurements in each laboratory was obtained from assessment of the laboratory or from reproducibility of data from several melting curves made with the same cell;

b) fitting of the melting curves. The curves were all obtained with the calorimetric method, by intermittent

RESULTS OF THE INTERCOMPARISON

From the variety of information supplied by the intercomparison only data concerning cell performance will be reported here.

Laboratories made measurements with calorimetric apparatus used for studies and realization of IPTS fixed points; a description of the experimental equipment will be found in the relevant literature. The stability of each cell was verified at the originating laboratory.

Table II: Measurements for the International Intercomparison of fixed points by means of sealed cells (Jan 1982)

Laboratories		BIPM	IMGC	INM	NBS	NIM	NRC	NRLM	PRMI
ASMW	Ar O2 H2 Ne CH4		Nov 79 Feb 80 Mar 81 Mar 81 Dec 79				Feb 81 Jan 81	Nov 80 Dec 80	
BIPM	Ar CH4	Feb 80 Feb-Apr 80	Dec 79 Mar 80	Nov 79	Feb 80		Feb-Mar 80 Jan-Feb 80	Nov-Dec 79	
IMGC	Ar O2 H2 CH4 N2 Ne D2		Sep 75-May 80 Sep 76-Nov 81 Oct 80 Aug 76-Oct 81 Aug 80 Jun 77-Dec 81 Dec 80				Apr 80 Sep 80 Oct 80 Jul 80 Oct 80	Sep 80 Nov 80 Mar 79 Sep 80	
INM	Ar O2		Apr 80 Oct 80	Apr 80 May 80	Apr 80		Apr 80 May 80	Apr 80 May 80	
NBS	Ar		Aug 79		Feb-Mar 79		Feb 81	Jul 81	
NIM	Ar O2 CH4		Jul 81 Jul 81 Jul 81	Aug 81		Jul 81			
NML	Ar O2 H2 Ne D2		May 79 Nov 78	May 79 Sep 79			Nov 79 Nov 79 Dec 79	Jul 79 Oct 79	
NPL	Ar O2 H2 CH4 N2 Ne D2		Jan 78 Feb 78 Apr 78 Jun 81 Jun 80 Jan 82				Nov 81 Nov 81 Feb 80 Feb 80	May 79 May 79-Feb 80	
NRC	Ar O2 H2 CH4 H2 Ne D2	Aug 79	Dec 76 Dec 76-Oct 79 Dec 76-Aug 79 Nov 81 Jun 79 Nov 81	Dec 76 Dec 76-Nov 79	Aug 79		Jun 79 Jun 79 Sep 79 Aug 79 Sep 79	May 79 Oct 79 Jun 80 Oct 79	
NRLM	Ar O2 H2 CH4 Ne		Nov 81 Nov 81 Nov 81 Nov 81					Oct 77-Jun 81 Oct 78-Nov 81 Jul 78-Nov 81 Aug 79-Nov 81	
PRMI	Ar O2 H2 CH4 Ne		Oct 81 Oct 81 Oct 81 Oct 81 Oct 81						Oct 81 Oct 81 Oct 81

melting: therefore each curve consists of a number of experimental points obtained at different liquid-to-solid ratios F. By plotting these temperatures versus 1/F, one obtains the usual melting plot that shows a negative slope. The temperature value at 1/F=1 was used throughout the intercomparison to define the triple point (tp) temperature. More often, these temperature values must be obtained by extrapolation of the experimental points at lower F values and the difficulty in defining a smooth curve through them may limit the precision of the tp temperature value assigned to that plateau.

c) thermometer calibration. In the few cases where cells were measured with different thermometers (or with the same thermometer in a large time interval during which the R(0) value had changed), additional uncertainty derives from thermometer calibration.

It must be pointed out that, as a rule this uncertainty cannot be defined in a statistical way, since few measurements are generally involved and since a curve through the experimental points on a plateau cannot be fitted mathematically. Therefore the reported uncertainties associated with each difference are entirely related to the scatter range of the values available for this analysis, or to source a). However, the resulting figure was set at a value not less than ± 0.1 mK. Considering that the reported differences come from a couple of independent measurements of comparable precision, the uncertainty associated with the difference is 1.41 times that of each measurement.

The criteria used to set the uncertainty value for each laboratory are summarized here:

ASMW: precision assigned by the Laboratory. It corresponds to a 99.7% confidence level - which seems to be quite conservative.

BIPM: precision essentially limited by the "definability" of the melting plateau.

IMGC: precision given by the Laboratory on the basis of the internal consistency of a large number of measurements on meltings /8/. With hydrogen the value corresponds to a precision of ± 1 $\mu\Omega$

INM: set to the minimum value (0.15 mK). No specific information supplied by the Laboratory.

NBS: set to the minimum value. The measurements of the Laboratory show a reproducibility well within this limit.

NIM: precision limited by the "definability" of the melting plateau.

NML: set to the minimum value. No specific information supplied by the laboratory.

NPL: set to the minimum value, considering the accuracy given in Ref. 5.

NRC: set to the minimum value, considering the accuracy stated in most of the published studies on the fixed points of gases.

NRLM: set to the minimum value information supplied by the Laboratory.

PRMI: assigned by the Laboratory.

Analysis of Figs. 2 to 5 gives information about the accuracy which can be obtained from fixed points realized in sealed cells.

Furthermore, the intercomparison also led to some information concerning the uniformity of data that can be obtained with the same cell, when it is measured in different apparatuses and Laboratories. Plots of all the typical melting curves obtained with each cell in the various Laboratories have therefore been made. Fig. 6 gives such plots for some of the 25 cells involved in the intercomparison.

DISCUSSION OF RESULTS

Systematic differences between cells

Let us examine each gas separately.

ARGON: This gas has been the pivot of the intercomparison; therefore results on 40 sets of measurements, collected in Fig. 2, lead to 27 difference values. Results are excellent, except those from NML, where large and scattered differences were found - with no apparent reasons. Discarding these three data, the maximum observed differences are + 0.19 mK and - 0.26 mK. On the other hand, the ranges of the differences for individual cells are (+0.18, -0.05) mK, (+0.03, -0.12) mK, (+0.19, -0.05) mK, (+0.07, -0.09) mK. Only 6 out of 24 values are higher than 0.1 mK (11 are lower than 0.05 mK). In conclusion, no systematic differences within ± 0.1 mK were observed in the 11 Laboratories between the 9 cells.

OXYGEN: This gas, exhibits one of the flattest and most reproducible melting plateaux (on each sample), but systematic differences were found among some of the cells, confirming similar observations made at IMGC /8/ and at NRC /9/.

Cell 8-IMGC used as a reference for measurements at INM, NML, NRC and PRMI, was found at IMGC to be higher by 0.48 mK than reference cell 1-IMGC. Using this value, the measurements in the cited Laboratories have been corrected accordingly: this may add some uncertainty to the corrisponding differences reported in Fig. 3. The agreement between 1-IMGC, NRC and INM cells appears to be good, within (+0.29, -0.15) mK. However, the scatter of

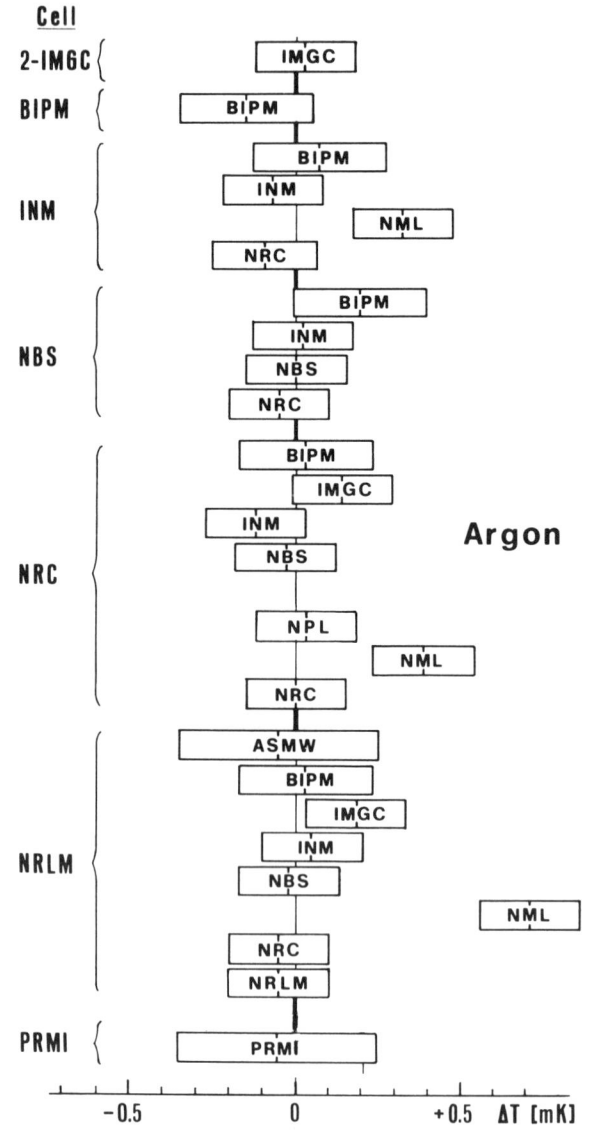

FIG. 2. ARGON: Differences between the cells made by different laboratories (left), by respect to the reference cell, as measured in different laboratories (inside the box). The box indicates the uncertainty limit. **Reference cell: 1-IMGC.**

measurements seems to be higher than for argon: for instance, with NRC cell it is (+0.29, -0.12) mK and with INM cell (+0.24, -0.15) mK.

PRMI cell appears to be higher than cell 1-IMGC by +0.59 mK and NRLM cell by about +2 mK.

This gas will require a careful analysis for use in the IPTS; results suggest that the correct value for the tp temperature is the lowest one: four cells agree on this value within ± 0.2 mK.

The only known reason for a rise of the oxygen tp temperature is contamination with argon; this binary mixture shows a peritectic at +0.7 K with argon concentrations up to 20%. This causes a 10 μK rise in tp temperature per vppm of argon, without separation of the solidus and liquidus lines, i.e. without increase of the melting range.

However, the content of argon in all the oxygen samples used in the cells, assessed by manufacturers or determined by specific analyses /8/ , was never higher than 10 vppm. Therefore it is doubtful that argon is the reason for the tp temperature increase, though no other explanation is available: it is difficult, to admit a systematic underestimation of the argon content in so many independent analyses of research-grade gases.

outside the combined uncertainties. The tp values appear more scattered than with the preceding gases, but it must be taken into account that submicroohm measurements are required for a resolution better than 0.25 mK.

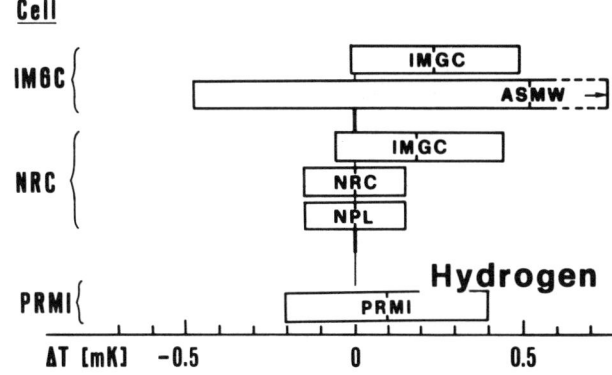

FIG. 4. HYDROGEN: Differences between the cells made by different laboratories (left), by respect to the reference cell as measured in different laboratories (inside the box). The box indicates the uncertainty limit. **Reference cell: 1-IMGC.**

METHANE: This substance was included in the intercomparison since its temperature value is very close to that of the normal boiling point of oxygen.

Results are not reported in this section since analysis of the data is not yet completed.

NEON: Neon was included in the intercomparison since its triple point could be used to replace its boiling point in a Temperature Scale. Fig. 5 collects the differences between cells, measured in six laboratories, with limiting values of (+0.18, -0.39) mK. However, since neon is a mixture of isotopes, one of the questions on the use of natural neon as a fixed point was the reproducibility of the isotopic composition, which has a strong influence on the tp temperature value. Results suggest some systematic differences among cells, IMGC cell differing from NRC cell by (+0.15 ± 0.15) mK, from

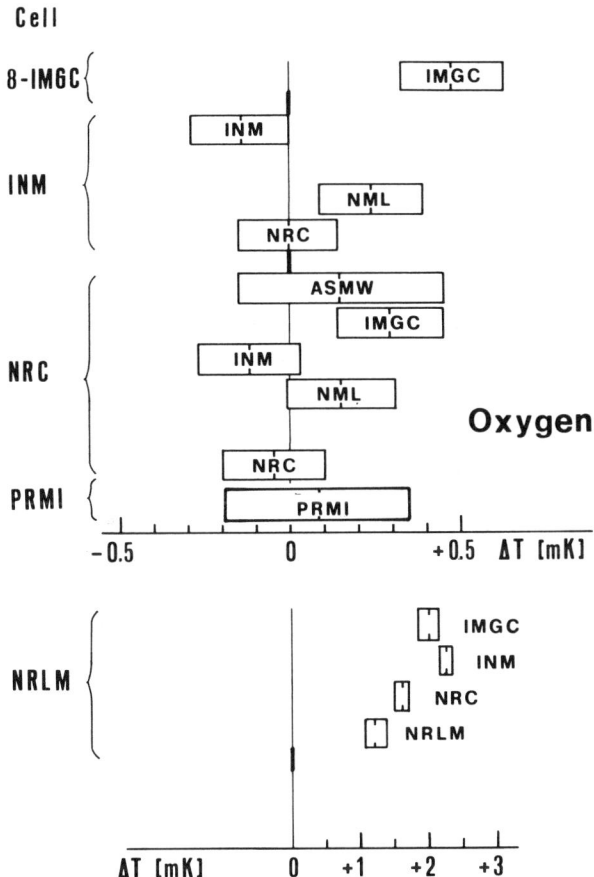

FIG. 3. OXYGEN Differences between the cells made by different laboratories (left), by respecto to the reference cell, as measured in different laboratories (inside the box). The box indicates the uncertainty limit. **Reference cell: 1-IMGC.**

HYDROGEN: A limited number of difference values are available at present (Fig. 4). Measurements at IMGC showed agreement between the three cells within their uncertainty. At ASMW agreement was again within the (large) uncertainty: two different thermometers were used, both calibrated at NPL, though a long time before (1975). Measurements at NPL show no difference between NRC and NRLM cells confirming measurements at NRC. PRMI multicomponent cell showed a tp temperature 0.44 mK lower than the reference cell, which is

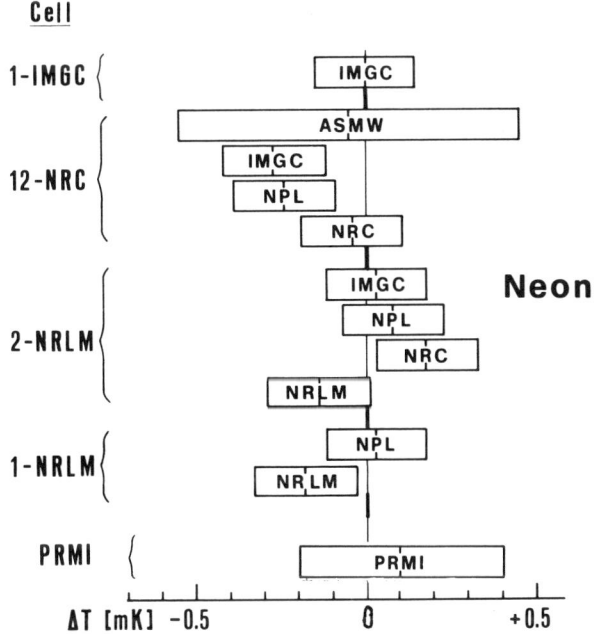

FIG. 5. NEON: Differences between the cells made by different laboratories (left), by respect to the reference cell, as measured in different laboratories (inside the box). The box indicates the uncertainty limit. **Reference cell: 3-IMGC.**

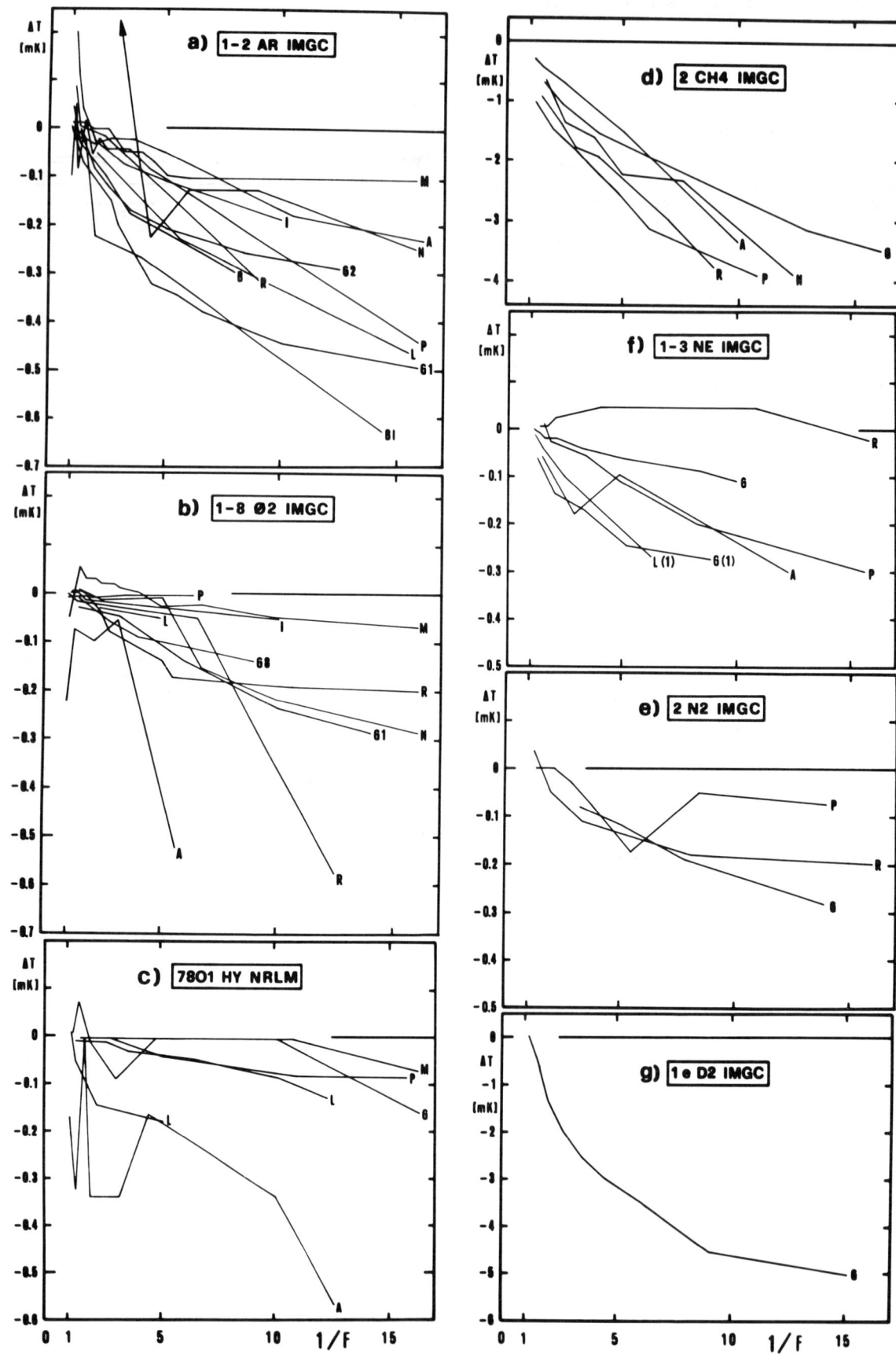

FIG. 6. Melting plateaux for the reference cells, as measured in different laboratories. F = melted fraction. ASMW = A; BIPM = BI; IMGC = G; INM = I; NBS = B; NIM = N; NML = M; NPL = P; NRC = R; NRLM = L.

NRLM cell by (-0.04 ± 0.15) mK and from PRMI cell by (+0.39 ± 0.3) mK.

NITROGEN, DEUTERIUM: These substances were included in the intercomparison because of their possible use in a future issue of IPTS: the studies on nitrogen in different laboratories have already shown that its triple point is very simple to realize and exhibits a very low melting range; the tp temperature of deuterium holds a key position as a substitute for the boiling points of hydrogen.

However, no data appear in this section, as only one cell of each of these substances was circulated.

Measurability of sealed cells

The intercomparison values presented in the former section permit the accuracy of the realization of fixed points in sealed cells to be assessed. Conclusions will be drawn in the next section.

The intercomparison represented a comprehensive test of the degree of dependance of results on the type and quality of the experimental apparatuses used in different laboratories, and on the design of the cells. This test has been particularly important, since these apparatuses were used for the studies of (conventional) fixed points of the past. Therefore, not only have the sealed devices been tested, but also the degree of uniformity achieved in cryogenic temperature measurements has been evaluated in National Metrology Laboratories.

The cells themselves represent a large variety of designs and materials (Fig. 1), with a large span in total mass (80 to 350 g), sealed sample quantity (0.02 to 0.3 mol), thermal response time (less than 1 to 60 min to recover within 0.1 mK) and ageing (few months to six years since sealing). The results, shown in Figs. 2 to 5, indicate that the influence of all these parameters on the accuracy of the fixed point realization, i.e. on the tp temperature value at $1/F=1$, is almost non existent.

A closer examination shows that the design and use of sealed cells influences the thermal behaviour of the sample during melting. The shape of the melting curve of each cell looks quite different in the measurements of different laboratories. It is impossible to report here the whole set of results for the 25 cells: Fig. 6 shows the melting curve that were obtained in different laboratories for each of the seven reference cells. The plots show that the same cell can yield different shapes. A similar situation occurred with other cell models, so that the variety of melting shapes appears to be almost independent of such parameters as the materials used in the cell body (all stainless steel, all copper, mixed materials) and the cell geometry; only a slight reduction in the spread of melting shapes seems to have been obtained when internal baffles (NRC design) were used.

The variety of melting shapes therefore must originate from the thermal hystory of the sample. Since the uncertainty associated with the extrapolation to $1/F=1$ can be seen to be mostly due to the shape of melting curve, achievement of an accuracy better than 0.1 mK depends on closer thermal control of the sample inside the cell -if sample purity can be kept to few parts per million.

CONCLUSION

The results reported here on the comparison of different sealed-cell models have shown that the fixed point of the substances investigated can be reproduced very accurately in national laboratories, much better than it was gathered from the data of previous international intercomparisons. The agreement of the realizations has been ± 0.1 mK with argon, ± 0.2 mK with neon on the other hand, ± 0.3 mK with hydrogen (owing probably to difficulty in submicrohm resistance measurements).

In the case of oxygen, high tp temperature values, tp to several millikelvin, were observed. Explanation of this fact by argon contamination cannot at present be considered as proved, although no other possible explanation is available.

Results on the individual nitrogen and deuterium cells available proved that the quality of their melting curve is high enough to include both substances among fixed-point candidates for a new IPTS (if purity problems will be solved for deuterium).

Another remarkable result was that no cell showed any shift of its temperature value during the time of the intercomparison (except perhaps CH4-BIPM), at the end of which most of the cells had been aged from three to more than six years.

Intercomparison of these cells also proved that their design has no sizable effects on results (10 different geometries and designs were tested), within the stated limits of accuracy and that these devices can reproduce the same temperature value when transported (they can be shipped with no special precautions) and used in different laboratories over a long period of time with any kind of experimental apparatus of the calorimetric type.

It has therefore been proved that international uniformity (within ± 0.2 mK between 14 and 90 K) can be achieved in the realization of temperature fixed points, a necessary condition of building a unique Temperature Scale.

However, the assessed limit of accuracy in the present intercomparison may not be entirely dependent on the quality of the cells used, but, more generally, might also be dependent upon the quality of the thermal and electrical measurements.

LEGEND OF ACRONYMS OF PARTICIPATING LABORATORIES

ASMW = Amt fur Standardisierung Messwesen und Warenprufung, DDR.
BIPM = Bureau International des Poids et Mesures, Sèvres
IMGC = Istituto di Metrologia "G. Colonnetti", Italy
INM = Institut National de Métrologie, France
NBS = National Bureau of Standards, U.S.A.
NIM = National Institute of Metrology, China
NML = National Measurements Laboratory, Australia
NPL = National Physical Laboratory, U.K.
NRC = National Research Council, Canada
NRLM = National Research Laboratory of Metrology, Japan
PRMI = Physicotechnical and Radiotechnical Measurements Institute, U.S.S.R.

REFERENCES

1 R.E. Bedford and C.K.Ma, Metrologia 6, 89 (1970).

2 F. Pavese and G. Demonti, Comptes Rendus Comitè Consultatif de Thermometrie, 12th Session (BIPM, 1978), Annexe T19.

3 M.P. Orlova, D.I. Sharevskaya, D.N. Astrov, I.G. Krutikova, C.R. Barber and J.C. Hayes, Metrologia 2, 6 (1966).

4 C.R. Barber and J.C. Hayes, Metrologia 2, 11 (1966).

5 S.D. Ward and J.P. Compton, Metrologia 15, 31 (1979).

6 L.M. Besley and W.R.G. Kemp, Metrologia 13, 35 (1977).

7 Comptes Rendu of 12th Session and Report to CIPM, Comptes Rendus Comité Consultatif de Thermométrie (BIPM, 1978).

8 F. Pavese and D. Ferri, "Ten years of research on phase transition of gases in sealed cells at IMGC", paper at this Symposium.

9 J. Ancsin, TMCSI 4, 211 (1972).

Ten years of research on sealed cells for phase transition studies of gases at IMGC

F. Pavese and D. Ferri

Istituto di Metrologia "G. Colonnetti," Torino, Italy

In 1969 a research program started at IMGC for the realization of the IPTS-68 in a permanent way. Since then, six types of sealed cells have been designed. Some 50 cells have been filled with a dozen substances, including all gases (and some light hydrocarbons) with triple point (tp) temperature below 100 K. The paper deals with the results obtained and the experience accumulated on the use of these devices as calibration fixed points for thermometry as well as for manometry. The temperature values of the tp obtained with these devices were reproducible, accurate, and stable with time: no drift was observed in cells almost seven years old.

INTRODUCTION

The state-of-the-art of the techniques for realizing low-temperature fixed points at the time of the 5th Temperature Symposium (1971) was still essentially the same that had led to the formulation in 1967 of the IPTS-68, where the defining fixed points are mostly boiling points (see papers in chapter B and C, section I, volume Four of the Proceedings /1/). Most of the relevant apparatuses (papers n.12 to 14, 17 to 19, 22 in /1/) were essentially vapour pressure cryostats, in which the vapour pressure bulb and the thermometers were both located in a massive copper block. Triple points were obtained using continuous melting or freezing technique and an accuracy around 1 mK was generally achieved.

In 1969 Ancsin /2,3/ began pioneering the realization of more accurate thermometric fixed points, by determining triple points adiabatically. He showed that while heating a partially melted sample its liquid portions always overheat, hence for reliable triple point determination "...a heating cycle must be followed by a non-heating period lasting as long as necessary for the sample to attain thermal equilibrium." /2/. The geometry of the sample holder in these experiments is quite different with respect to that of vapour-pressure thermometry, since: "...the triple point cell itself is designed so that the thermometer being calibrated is shielded from the heater by layers of liquid and solid." /4/.

When at the Istituto di Metrologia "G. Colonnetti" (IMGC) a research program was established in 1969 for the realization of IPTS-68 fixed points, triple points appeared to be the simplest way to approach the Scale realization since pressure measurements are not required and the quality of measurements can be generally assessed through the shape of the melting plateau, which depends on two parameters: a) thermal exchange of the substance with the surroundings; b) purity of the sample.

The latter appeared to be the main source of errors, as thermally-induced errors are easily kept to a minimum with the technique of adiabatic calorimetry /5/, which Ancsin had shown to be suitable also for accurate thermometry.

The effect of impurities in gases has been widely studied by chemical analysts and found quite large, so that the total amount of impurities must be kept below about 10 vppm, if the triple point is to be within $\pm$ 0.1 mK of the ideally pure substance. This involves not only the selection of reliable research-grade gas sources, but also handling problems. Simple calculations show that desorption of a monolayer of impurity gases from each $0.2\ m^2$ portion of geometrical surface (i.e. about 1 m^2 of effective, bright surface) causes a contamination of about 40 vppm in 5 L of gases, and that a 10^{-8} W (10^{-7} torr L/s) leak introduces about 2 vppm/day contamination in the same quantity of gas.

Therefore at IMGC we focused the attention on the proper handling of gases. Accordingly, the whole gas system inside the cryostat was made independent of it, so that it could be built separately with special, clean techniques and one gas system (the cell) was always used with each gas, in order to prevent cross-contamination. The only cryostat therefore necessary was built in 1969 and was designed so that the measuring cells in which each fixed points is realized can be easily interchanged (Fig. 1a) /6/.

The next step in order to reduce the possibility of gas contamination was the reduction of its handling: this could be achieved by leaving the sample of gas permanently sealed inside the cell. However, for long-term stability of these cells, impurity stability should be achieved, and consequently cleanliness of the cell was believed to be the key requirement.

The first measurements on a sealable cell began at IMGC in 1973 /8/, as three years were spent to develop expertise in clean techniques for cell fabrication and filling, with "the ultimate goal... to build high-pressure, self-contained, transportable fixed point cells, suitable for primary measurements." /7/.

The reason for developing a sealed device was not only to improve reliability, by avoiding gas manipulation, but also to realize permanent, ready-to-use fixed points.

This would avoid the need to rely only on stable thermometers. Furthermore, sealed cells permit direct comparison of fixed points between laboratories, since they are easily transported without risk of degradation.

In principle, not only triple points (and solid-to-solid transitions) can be measured with a sealed cell, but also boiling points and, more generally, vapour pressure curves, provided that they are used with a high stability and resolution differential pressure transducer, which can be connected to an external manometer. Recently at IMGC this requirement has also been given a preliminary solution.

THE CONCEPT OF IMGC SEALED CELLS

The concept of sealing a substance in a can for thermal studies is widely applied with liquid and solid materials. On the other hand, with gases only two studies, to our knowledge, based on this concept /9,10/ were made in the past, namely in 1956 and 1971. Both cells used then were made for long-stem thermometers (about 3 cm o.d. and 30 cm length), showed a very large heat of fusion (15 kJ and 4 kJ respectively) and were studied only with continuous melting (or freezing) technique. No assessment was made of the purity stability with time of the gas sample sealed therein.

The IMGC cells were made suitable for capsule thermometers and for measurements in an adiabatic environment. They were especially designed to obtain sample impurity stability (within a few part per million over a number of years) /8/. This is the main requirement to make this device truly permanent and is, in principle, the most difficult result to attain with gases.

At the time when these experiments began at IMGC, no available experiment could tell whether long-term purity could be mantained in sealed cells.

Consequently, IMGC cells were designed in such a way as to ensure clean machining and assembling of its parts (Fig. 1b); ultra-high vacuum procedures were then used to outgas the cell and manipulate the pure gas. Precautions were taken to prevent possible trapping of impurities during machining, introduction of impurities or trapping of gases during brazing or soldering, inadequate cleaning, increase of effective surface during machining or cleaning, and inadequate outgassing because of throttled pipings.

For cell fillings a sophisticated system was constructed, to manipulate research-grade gases (Fig. 2). The system is completely made up of ultra-high vacuum components. It includes a residual-gas analyzer, the routine use of which appeared to be fundamental /8/. The necessary clean-vacuum conditions which must be obtained prior to each cell filling can be obtained and monitored with this system. Additional impurity effects are thus excluded from contributiong to the uncertainty of results.

Account must be taken, however, that gases are purchased with a finite impurity content, that must be known accurately. The specifications of the gases bought in small bottles (0.5 to 1.5 L, containing 20 to 150 L of gas) from many different manufacturers were found to be far from reliable. Additional analyses by other laboratories (that consequently became routine practice, especially for possible air contamination) revealed that about 30% of the bottles were outside specifications.

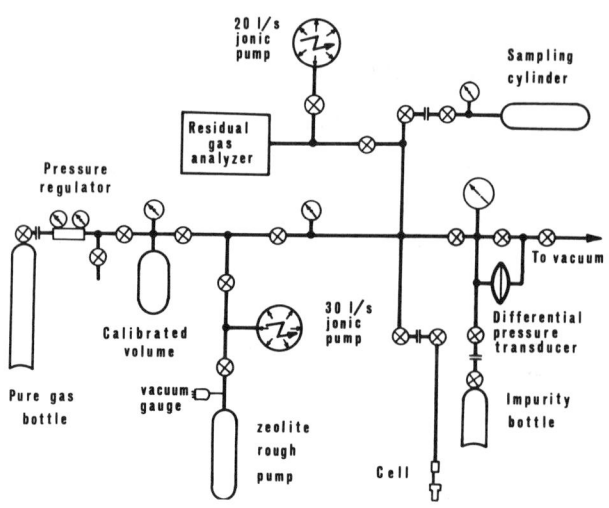

FIG. 2. System for ultra-pure gas handling.

MODELS OF IMGC SEALED CELLS

Long-stem model (mod. L)- The first model was a long-stem cell (Fig. 1b): it is a low-pressure model (0.5 MPa at ambient temperature with 1 L gas ballast). It has a very light sample container (14 g); the sealing device is at room temperature and it is made of an ultra-high vacuum valve, which can be reopened when necessary to change the gas sample. This arrangement makes the design of the cell quite complicated: since it involves a gas column between room temperature and the condensed sample, it is necessary to minimize the heat leak reaching the condensed sample mainly via its own vapour column. In addition, a vacuum jacket is necessary for the pressure tube to prevent cold spots during cooldown. The sample holder was made with a passthrough thermometer well; the diameter of its lower portion is smaller in order to limit the internal volume.

The whole cell was vacuum brazed by an external laboratory at 900°C after clean machining of parts /11/.

Three cells were sealed at IMGC with oxygen, argon /7,11,12,13/ and methane /14/ between 1973 and 1975. The results obtained since the first melting were so good that we immediately considered a simplified design, at the expense, if necessary, of some controlled performance degradation and of constructing the cells entirely at IMGC.

Miniature cells (mod. A,B,C,U)- The connection of the sample holder to the valve at room temperature is the substantial reason for the elaborate design of the L-cell. Therefore, sealing of the sample holder below E in Fig. 1b) would greatly simplify the cell.

This requires however a low-temperature seal and leads to a very large pressure increase when the cell is at room temperature. A satisfactory compromise was found between pressure, weight and size of the cell (Fig.3).

The following design principles were applied:
a) use of copper for the functional parts (i.e. thermometer(s) holder) and of stainless steel for the structural parts (with the additional purpose of decoupling the cell heater from the thermometer block);
b) high vacuum conductance in the connecting line during preparation for sealing;
c) minimum internal surface;
d) low-pressure filling method.

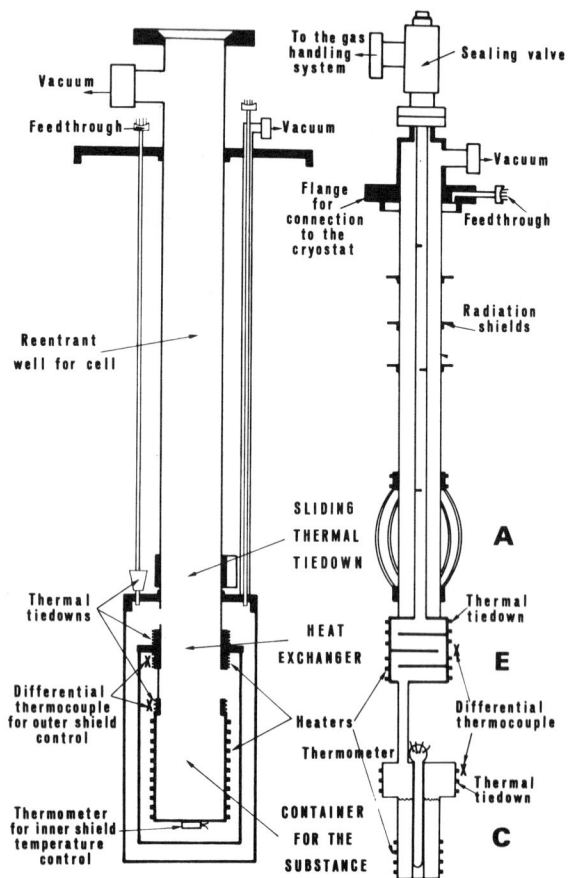

FIG. 1. a) Cryostat with reentrant well for accomodating the sealed cell.

b) Long-stem IMGC sealed cell. (not to scale; 1L ballast not shown)

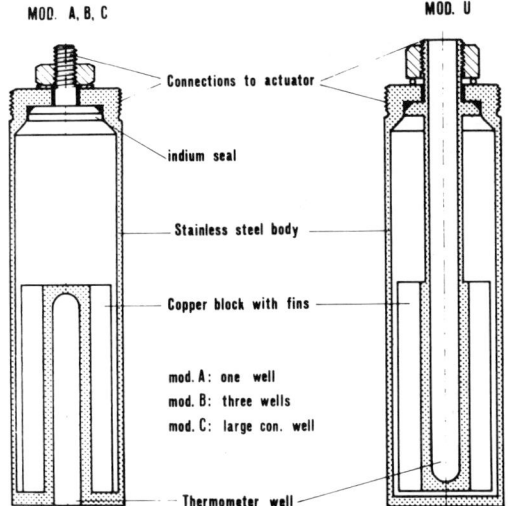
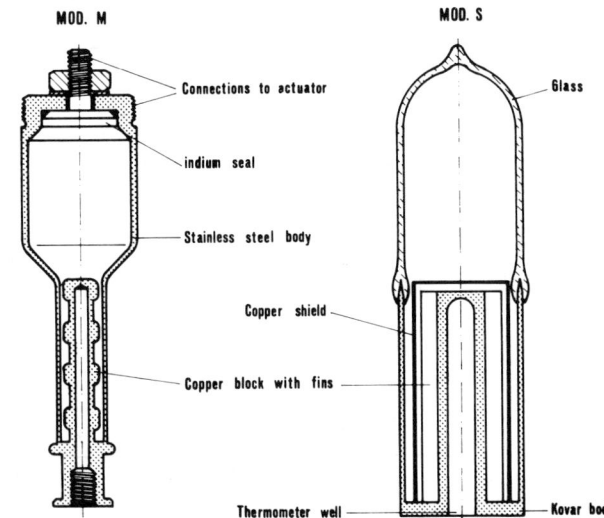

FIG. 3. Schematic diagrams of miniature IMGC sealed cells (not to scale).

Dimensions: diameter 23 mm; length: A,B,C,S 130 mm; U 150 mm; M 100 mm.

The first involved welding stainless steel to copper, a technological problem which had already been solved at IMGC. The second led to discarding pinch-off tubes, because they generally have a very large lenght-to-diameter ratio, in favour of an indium seal. The third avoided the use of baffles and of complicated inner shapes, and limited to three elements the functional parts of the cell /15/. Finally, the method of cell filling by cryogenic condensation was chosen.

In July 1975 the first miniature cell (holding one thermometer: mod. A) was made at IMGC and successfully sealed; to our surprise, results showed no expected degradation, and the cell behaved substantially as the L-cell.

Since then, 15 A-cells have been sealed; the total length of the cell was subsequently reduced and the geometry of the block changed, to allow a more versatile use of the thermometer holder, first by arranging three wells for standard capsule thermometers (mod. B)[2], then by separating the thermometer holder from the copper block, which shows only a large conical well for mating the holder (mod. C)[2]. Eventually, mod. U[2], having a 8 mm i.d. well passing through the indium seal was designed (Fig. 3), in order to make possible the use of cells also with long-stem thermometers, which must be inserted from the top. A summary of their characteristics is given in Table I.

Microcell (mod. M; temperature generator)-Recently a further effort was made, in order to simplify the cell configuration.

A functional analysis of the cell, which applies to any IMGC model, shows that it may be represented as a temperature generator (the solid-liquid interface) connected to a thermal load (the cell body and thermometer holder).

The temperature generator is a of a particular kind: in fact it can deliver only a limited amount of energy (the heat of melting of the enclosed sample) and shows quite a low static internal impedance (the melting range), but a generally much higher dynamic one (due to the speed of interface displacement being limited by the thermal conductivity of the sample).

The last characteristic limits the size of the thermal mass of the load, that can be attached to the temperature generator with respect to the amount of energy stored in the latter: in fact, the compensation for any thermal perturbation of the load derives from partial melting or freezing of the sample. Consequently, we considered the ratio of the heat of fusion of the gas to the specific heat of copper (figures for stainless steel are not much different), $X = H_f/c_p$, as a parameter to indicate how good is each gas in "driving" the temperature of its container: Table II collects such values for all gases of interest.

Table I: Models of miniature IMGC sealed cells.

Cell model	Type	Weight /g/	Quantity of substance /mol/	N. sealed	Total in.vol. /cm³/	Filling pressure /MPa/	Working volume c) /cm³/
L	long-stem	14 a)	0.2	3	1100 b)	20	5
A	miniature one-well	190	0.2	15	32	20	5
B	miniature three-well	180	0.15	17	22	20	4
C	miniature conical-well	165	0.1	10	20	10	2.5
U	miniature upper-well	200	0.15	3	27	10	5
M	microcell (temp.gen.)	80	0.05	1	8.5	4	1
S	miniature for liquids	100	0.5	1	50	-	10

a) sample can only; b) with 1 L ballast; c) around the block.

Table II: Thermal data for gases in cells as temperature generators.

Gas	T_{tp} /K/	H_f /kJ/mol/	c_p /J/K.g/	$X=H_f/C_p$ /kg.K/mol/	Minimum gas quantity M /mmol/
Carb.diox	216.581	8.4	0.3	28.0	20
Xenon	161.391	3.1	0.3	10.3	50
Krypton	115.764	1.5	0.25	6.0	100
Ethane	90.352	2.8	0.2	14.0	30
Propane	85.520	3.5	0.2	17.5	30
Methane	90.686	0.95	0.2	4.7	100
Argon	83.798	1.2	0.2	6.0	100
Nitrogen	63.146	0.72	0.15	4.8	100
Oxygen	54.361	0.44	0.1	4.4	100
Neon	24.562	0.33	0.01	33	15
Deuterium	18.678	0.22	0.002	110	5
Hydrogen	13.81	0.12	0.0013	90	6

H_f = heat of fusion of the gas; c_p = specific of copper.

The ratio X indicates the temperature increase of 1 kg of copper when heated by an amount of energy corresponding to the melting of 1 mol of gas, or the mass of the container (in kilograms) that causes 0.1% of the sample to melt (or freeze) when changing temperature by 1 mK.

The last column in Table II finally, shows the minimum quantity of sample that should be sealed in a cell in order to recover a 0.1 mK temperature change of 0.5 kg of thermal load with a liquid-to-solid ratio variation not larger than 0.1%.

From the above analysis and parameters, we concluded that there is advantage in physically separating the thermal load from the temperature generator. In fact the size of the latter can be considerably reduced, eliminating the thermometer block in our miniature models, the size of wich (chosen to accomodate at least three capsule SPRTs) almost completely determined the cell size. It was reduced (Fig. 3) to a small copper rod with thick baffles to suddivide the sample in thin layers, a technique that is thought to ensure the best thermal response speed from the cell over most of the melting range. The upper part of this new cell was left oversized (with respect to internal pressure requirements) to match the IMGC cell filling system.

The temperature generator is designed to be in thermal contact with the load only at the bottom surface, where it can be screwed tightly, e.g., to an external copper block containing the thermometers (Fig. 4).

One cell of this model was sealed in 1980 with nitrogen /29/.

Cell for liquids (mod. S) - In order to make some experiments on the triple point of water with miniature sealed cells, a special cell suitable for filling with substances that are liquid at ambient temperature has recently been built; it will be described when discussing results.

SUMMARY OF MEASUREMENTS AT IMGC

Some 46 small cells were built after the three long-stem ones, and about 400 melting plateaux have been recorded during the studies of a dozen of substances, carried out since 1975 (on oxygen and argon since 1973).

The relevant data obtained from these studies are collected in Table III, which is self-explanatory.

Results were very satisfactory, except for carbon dioxide, which was studied at different times between 1977 and 1979. With our cells this substance showed a very large melting range and a behaviour typical of impure samples, despite the fact that a number of analyses made with different gas-chromatographic and mass-spectrometric techniques had always certified a purity better than 99.999%; fractional distillation which was successful in purifying ethane and propane, did not improve results. Moreover, the samples very often showed a still unexplained, two-step behaviour during thermal re-equilibration following a heating cycle.

Some studies on the triple point of water were recently undertaken, in order to get acquainted with the melting behaviour of this substance over its melting range. In fact, surprisingly, no data on the complete melting behaviour of pure water could be found in the literature. The fact that triple-point values can be reproduced within 0.1 mK /27/, does not imply that the melting range of the melting plateau of water is 0.1 mK. Conventional water cells, with a heat of fusion of 200-250 kJ, are normally used in a range of 10-20% change of the liquid-to-solid ratio around about 50% of liquid fraction; here the slope of the melting curve, for reasonably pure substances, normally approaches a minimum.

On the other hand, it is generally recognized that the thermal characteristics of water are critically dependent on impurities and gases dissolved in it. To learn more about its melting behaviour a special miniature cell was built. It is made of a glass-to-metal joint in order to match the IMGC glass water-cell filling system /28/ (mod. S). A copper block (type B) was

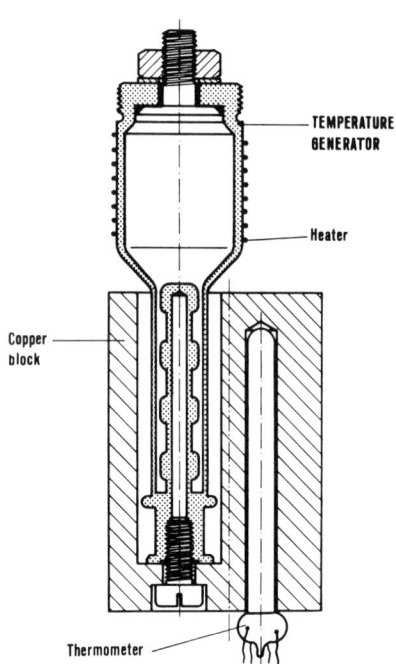

FIG. 4. Temperature generator mounted with an external thermometer holder.

Table III: Results on fixed points with IMGC sealed cells.

Ref.	Substance	Nominal purity	Fixed point	Temperature T /K/	Scale (3)	Reprod. T /K/	Pressure p /Pa/	Reprod. p /Pa/	Equilibration time /min/	Melting range[2] /mK/	mod.	Cells sealed Date	N°
13,14,16 17,18	Methane	99,995%	n.b.p.	111.656	IMGC	± 0.0003	101325	--			L	APR 76	∅
											A	AGO 76	1,2,3
											B	APR 77	4
			t.p.	90.6856	IMGC	± 0.0002	11696	± 0.7	90	5	B	APR 78	5
			s.s.t.	20.48	NBS	± 0.005			90	50	B(2),C,U	APR 79	10,11,12,13
19	Ethane	purif.to 99,996%	t.p.	90.352	IMGC	± 0.001			20	12	A	OCT 76	1
			s.s.t.	89.726	IMGC	± 0.001					A	DEC 76	2
			s.s.t.	89.834	IMGC	± 0.001					A	MAY 77	3
20	Propane	purif.to 99,95%	t.p.	85.520	IMGC	± 0.001			30	25	B	SEP 80	1
			t.p. (metastable)	81.226	IMGC	± 0.001							
7,11,12 13,15,21 22,23	Argon	99.9595%	n.b.p.	87.2952	IMGC	± 0.0002	101325	--			L	JUL 75	∅
											A	JUL 75	1
			t.p.	83.798	IPTS	± 0.0001	68890	± 1	20	0.3	C	MAY 78	2
23	Nitrogen	99.9595%	t.p.	63.146	IMGC	± 0.0001			20	0.5	B	DEC 79	1
											M	FEB 80	2
			s.s.t.	not studied									
7,11,12 13,18,22	Oxygen	99.998%	t.p.	54.361	IPTS	± 0.0001	146.25	± 0.1	10	0.2	L	JUL 75	∅
											A	SEP 76	1,2,3
			s.s.t.	43.803	NBS	± 0.001			60	2-5	B	DEC 77	4
			s.s.t.	23.888	NBS	± 0.005			30	20-40	C	JUL 78	5
											B(2),C,N	NOV 78	6,7,8
23,24	Neon	99.995%	t.p.	24.562	Mean	± 0.0002			10	0.7max	B	JUN 77	1
											C	JUN 78	2
											C	FEB 79	3
25	n-Deuterium	99.85%	t.p.	18.729	Mean	± 0.0002[4]			10	3	B	JUL 77	1
											B	FEB 78	2
26	e-Deuterium	99.85%	t.p.	18.678	Mean	± 0.0002			10	4	C	NOV 80	1
26	e-Hydrogen	99.9998%	t.p.	13.81	IPTS	± 0.0001			10	0.2	C	AGO 80	1
18	Carbon diox.	99.999%	t.p.	216.581	IMGC	± 0.002			120	50	A	JAN 77	1
											A	MAR 77	2
											B	APR 77	3
											U	JUL 77	4
											C	FEB 79	5
	Water[5]	distilled under vacuum	t.p.	273.16	IPTS	± 0.00015			20	2-4	B(2)	OCT 79	6,7
											S	JAN 81	1

[1] defined for temperature recovery at F=50% within reproducibility figure; [2] defined as temperature difference between 10% and 100% melted fraction; [3] IMGC = IMGC-IPTS-68 (above 54 K) NBS = NBS-IPTS-68; IPTS = definition point of IPTS-68; Mean = mean value from different laboratory realizations of IPTS-68; [4] when corrected for para-ortho conversion drift; [5] preliminary data.

mounted inside the kovar part of the joint. It is surrounded by a copper shield, which can act as an internal container around the block, to hold part of the filling water, when the cell is turned upside down (Fig. 3). All the copper parts were gold-plated, but this electrolytical plating was damaged by water during the initial vapour-rinsing procedure that precedes filling. It was carefully filled distilling water under vacuum in it.

Measurements taken in a period of six months were quite satisfactory and are reported hereinafter.

Argon, oxygen and methane where also studied as pressure fixed points, by using long-stem cells /13/.

QUALITY OF THE RESULTS OBTAINED AT IMGC

Although the quality of results could be described by simply stating the accuracy of the measurements, a more analytical way is preferred and this figure is therefore subdivided into three sections:
1) reproducibility (short term); 2) stability with time (long term); 3) systematic errors.

The first of these is listed in Table III. Appropriate references should be consulted for more information. Here we shall concentrate on the other two: stability with time is, of course, most essential, since lack of stability would greatly reduce the usefulness of sealed cells.

The magnitude of systematic errors is also important in assessing the reliability of results obtained with cells, and ultimately deciding, whether some of these cells could be used as primary standards.

Temperature fixed points

Triple points

It is desirable to intercompare realizations of fixed points amongst different laboratories to better assess the accuracy of the laboratory realizations. Intercomparison is particularly simple with sealed cells, as fixed points can quite safely be transported. These intercomparisons for IMGC fixed points have been done in 1975 and 1976 through travelling thermometers /3,21/ and in 1976 by shipping IMGC cells /22/; cells of IMGC and several other laboratories have also been intercompared between 1979 and 1982 under the auspices of the Comité Consultatif de Thermométrie: Ref. 29 should be consulted for data on systematic differences between fixed-point realizations among laboratories.

Let us examine now the results obtained at IMGC with different substances:

OXYGEN - Fig. 5 collects key measurements made with oxygen cells belonging to different batches. It can be observed that cells of the same group reproduce the same temperature but systematic differences exist between groups.

Fig. 6 represents these systematic differences: square boxes indicate the mean value for each set of intercomparisons and the bold-face figure of each group

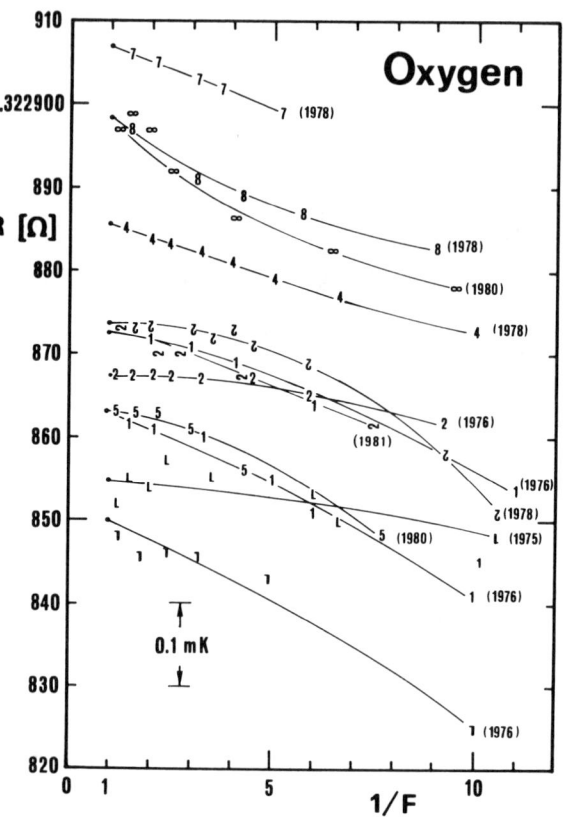

FIG. 5. Melting plateaux for different IMGC oxygen cells, made at different times. (cell number is indicated: L = long-stem) F = melted fraction.

indicates the mean value of the systematic difference relative to the L-cell, taken as reference. The accuracy of measured differences is ± 0.15 mK. The L-cell was taken as a reference since its value agrees with that of the cells of most of the other laboratories /29/.

It is difficult to find a reason for systematic differences observed between cells. IMGC used four different bottles of gas from three different sources. As indicated in Table IV, the analyses available for all gases show a very good purity and an argon content too low to explain the rise of the triple-point (tp) temperature (Table V is a compilation of the effect of impurities on triple-point temperatures). If argon is assumed to be the cause of the rise, then one has to assume also that a reliable analysis for its content is impossible, or that our cryogenic condensation method for filling the cells tends to concentrate argon impurities in the cells.

Apart from systematic differences, stability of the

Table IV: Impurity analyses on oxygen samples /vppm/.

Cell lot:	402,502			602,702,802	102,202,302	long-stem			
Source:	Matheson 99.999% nom.			Air Liquide 99.998% nom.	Air Liquide 99.998% nom.	Air Products 99.999% nom.			
Analysis:	Batch	LAB1	LAB2	LAB2	LAB2	Batch	NPL	LAB1	IMGC
Date:	1976	1977	1979	1978	1976	1974	1975	1975	1975
Nitrogen	4.4	3.2	7.6	1.8	8	8	< 5	5.3	-
ARGON	-	-	1.2	3.0	<10	<10	< 5	-	-
Methane	<0.05	-	-	0.7	-	< 0.5	-	-	-
Moisture	4.5	-	-	-	-	0.2	-	-	-
Carb.diox.	-	-	-	-	-	< 0.5	-	<1	-
Krypton	-	-	-	-	-	<10	< 5	-	< 2

Table V: Impurity effect on the triple point temperature /μK/vppm/

Impurity	CH_4	C_2H_6	C_3H_8	Ar	N_2	O_2	Ne	D_2	H_2
Methane	XX	-25	-45	-30	- 3	-30	--	--	--
Ethane	-60	XX	-45	--	--	--	--	--	--
Propane	-60	-25	XX	--	--	-50	--	--	--
Argon	-35	--	--	XX	- 4	+10	none	none	none
Nitrogen	-35	-10	--	-25★	XX	-20	- 8	none	none
Oxygen	-50	-10	-45	-20	-20	XX	--	none	none
Neon	--	--	--	none	--	- 1	XX	- 8	- 8
Deuterium	none	--	--	none	--	--	-10	XX	+ 6
Hydrogen	none★	--	--	none	--	--	-10	-4	XX
Krypton	+20	--	--	+ 5★	+25	- 5★	--	--	--
Carb.diox.	-60★	--	--	--	--	--	--	--	--
Hydr.deut.	--	--	--	--	--	--	--	- 1.5	+ 3

★ Value measured at IMGC; the remaining values are taken from literature.

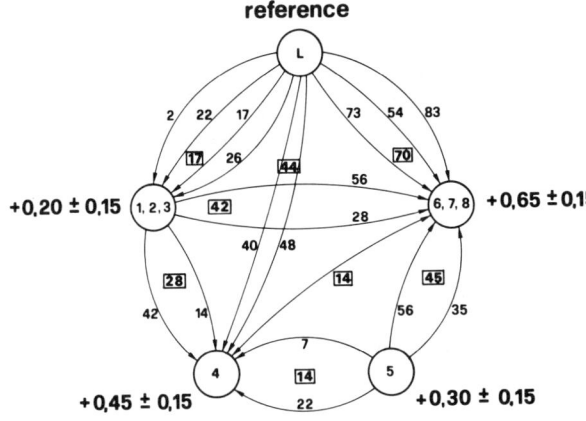

FIG. 6. Systematic differences among lots of oxygen cells at IMGC.
Numbers are in $\mu\Omega$; differences are positive when arrow points to the higher cell. Square boxes indicate the mean value of each path.

temperature value reproduced by each cell appears to be within the reproducibility of measurements in Figs. 5 and 6, i.e. ± 0.15 mK.

ARGON - Fig. 7 collects key measurements made with the three argon cells. No differences between cells (made with different bottles of gases) are measured and the stability of each cell can be seen to be within the reproducibility of the whole set of measurements, i.e. ± 0.1 mK.

METHANE - Fig. 8 collects again relevant data on the five group of cells which were made at different times with different bottles of gas (see Table III).

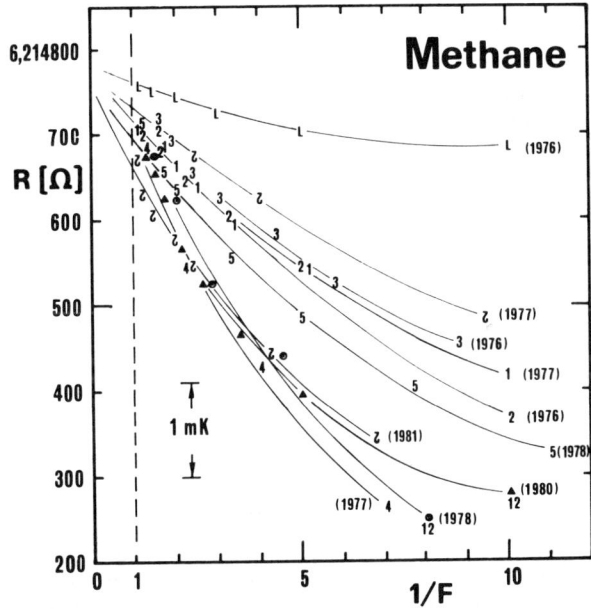

FIG. 8. Melting plateaux for different IMGC <u>methane</u> cells, made at different times. (cell number is indicated: L = long-stem) F = melted fraction.

With this gas, the melting range obtained with IMGC miniature cells was always larger than that obtained with the L-cell or that reported by other laboratories /29/. The reason for this is not clearly understood, since analyses of the gas samples always showed less than 10 vppm of total relevant impurities. On the other hand, we never observed spin-conversion instability, as was found in Japan /30/: thermal history of the sample only produced a small irreproducibility in the shape of the melting plateaux.

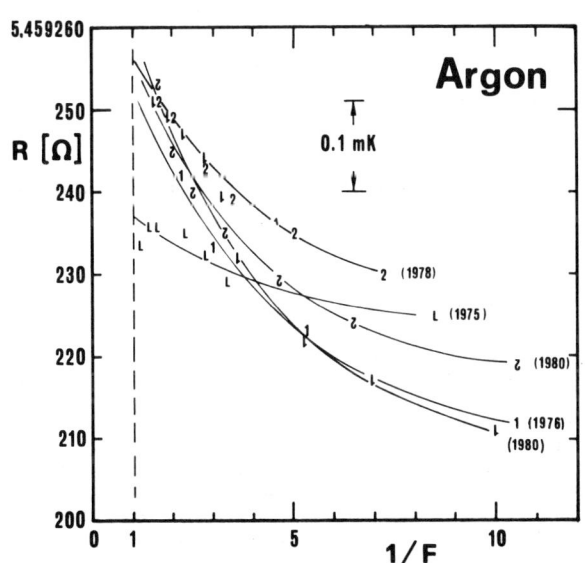

FIG. 7. Melting plateaux for different IMGC <u>argon</u> cells, made at different times. (cell number is indicated: L = long-stem) F = melted fraction.

All the cells were found to reproduce the triple point temperature at $1/F = 0$ within $\pm$ 0.2 mK and to be stable within the same interval (Fig. 8).

OTHER GASES- The other gases listed in Table II have been studied for shorter periods of time; drift of their tp temperature values has never been observed, since the date of their sealing.

This is particularly important with neon, which is sensitive to hydrogen and nitrogen impurities (Table V). In principle, both gases could diffuse with time in substantial amounts from the bulk material of the cell envelope. However, the three cells sealed so far at different times have always shown the same tp temperature value after more than three years since the first cell was sealed /23,29/.

With equilibrium deuterium and hydrogen cells, that also contain some activated hydrous ferric oxide as a catalyst[3], a longer time may be required to assess the stability of their tp temperature values with aging. In the case of deuterium the stability of the hydrogen content must be monitored over a longer time period, owing to possible desorption of hydrogen from the cell walls and decomposition of hydrogen deuteride, which is assumed to be the main initial impurity.

As regards ethane and propane, the cells contain samples specially purified by cryogenic distillation.

WATER - Although the results obtained so far with water must be regarded as preliminarry, they are nevertheless reported here, as they may be useful and as the small cell appeared to be suitable for measurements on liquids in this temperature range.

Melting ranges (defined for liquid fractions higher than 10%) between 2 and 4 mK were observed, but no correlation has so far been found between the slope of the melting plateau and the thermal history of the sample, the amount of the sample, or the geometry of the cell (which can be modified turning the cell upside down).

Measurements were made in a simple vacuum can, immersed in melting ice without additional adiabatic shields. Consequently, as thermal conditions were quite far from adiabatic and copper parts inside the cell lost part of the gold protection (which went floating into the water sample), results can be considered very good. The liquidus point agrees with the triple point obtained in a conventional glass triple point cell /28/ within $\pm$ 0.1 mK and no drift in this value was found in six months. The reaching of thermal equilibrium after each heating period was found quite short, compared with other substances (a large amount of power must be supplied to limit the total time required by each melting, owing the very high heat of melting of water).

Solid-to-solid-transitions

The very low vapour pressure and the lack of heat distributing baffles inside IMGC cells makes them unsuitable for easy realization of solid-to-solid phase transformations.

However the values measured for some of them are listed in Table III.

It must be recalled that the transition of methane and the lower one of oxygen are not intrinsically isothermal and that diffusion of the transition in the lattice is, for gases, an intrinsically slow process /31/.

Boiling points

Only long-stem cells are suitable for pressure measurements and only occasionally before 1980 were we interested in extensive measurements on vapour pressure.

Therefore only two normal boiling points were measured: methane in 1976 /14/ and argon in 1980 /18/ (Table III).

It is not possible to obtain the dew point or the boiling point with our cell by doing a complete vaporization experiment, as the liquid to vapour ratio cannot be altered during the course of a particular experiment.

However, reproducibilities of few tenths of a millikelvin were obtained for a temperature at 101325 Pa pressure.

Pressure fixed points

In using sealed cells as pressure standards the following possible difficulties should be considered:

a) volatile impurities may leave the triple point temperature unchanged yet may greatly influence the triple point pressure.

b) the derivative dp/dT values are generally high, except below the triple point. Temperature uncertainty will tend to lead to larger value of pressure uncertainty.

c) pressure within the cell must be measured while leaving it sealed.

d) for absolute pressure measurements, hydrostatic and thermomolecular pressure corrections must be appropriately applied.

Point c) is the most difficult to achieve in principle. No absolute high-pressure transducer, able to work at low temperatures, is commercially available; therefore the cell must be connected to a manometer at room temperature. A seal must then be provided with a differential pressure transducer, which must have very good zero stability and high resolution (no high accuracy in needed).

But since a low-temperature transducer of this kind is not commercially available either[4], it must be placed at room temperature too. Therefore only a long-stem cell can be used at present.

All three L-cells were used at different times at IMGC for measurements of the tp pressure (and for measurements at 101325 Pa) using oxygen, methane and argon /17/. Results are reported in Table III.

The following results can be considered with respect to the problems listed above:

a) An argon sample sealed in 1975 was measured in 1980 and found to give a triple point pressure within $\pm$ 1 Pa of that found by Ancsin /23/ and Bonhoure /33/.

b) The reproducibility of pressure values obtained at the triple point was essentially determined by the quality of the manometers used. However, thermal-drift rates caused by our thermal apparatus determined the accuracy of temperature and, hence, of pressure measurements at the boiling points.

c) It turned out that some commercially available pressure transducers can readily be adopted. Their null position can conveniently be checked by lowering the pressure on both sides of the diaphragm (by sufficiently cooling the cell and providing vacuum on the other side).

d) With the present IMGC cells, the hydrostatic head correction can be made with sufficient accuracy by assuming a linear temperature distribution of the gas column. The thermomolecular correction is very low, because the connection tube is not a capillary (8 mm).

DISCUSSION OF RESULTS

Realization of temperature fixed points

All gaseous substances with a tp temperature lower than 100 K, including light hydrocarbons, were measured at IMGC and were found to be equally suitable to make fixed points accurate within few tenths of a millikelvin, when their purity approaches 99.999%. The achievement of this purity depends on the substance: for some of them there are critical impurities, which could lead to substantial errors when in sizeable amount.

Many of the gases listed in Table III show some problem on this respect (see also Table V), which will be analyzed hereinafter:

METHANE - The effect of some relevant impurities on the tp temperature of methane is shown in Table V. We discovered some discrepancy in the published data for carbon dioxide. No experimental data close to pure methane are available for this mixture, but Donnelly and Katz /34/ suggested that "it is advantageous for extrapolating the existing data down to 90 K to assume that methane forms an eutectic mixture with carbon dioxide". On the other hand, Davis et al. /35/ disagreed with this assumption, because the liquidus line for only 0.2% CO_2 in CH_4 is still at 130 K. Our measurements /14/ showed that 10^4 vppm of CO_2 decrease the tp temperature of CH_4 by 0.6 mK. Further addition of CO_2 to CH_4 (up to 100 vppm) leaves the melting curve unaltered. Hence as the melting range is insensitive to these impurities, yet the curve is shifted, great care must be taken to avoid it.

ETHANE, PROPANE - Although these substances are commercially unavailable in high purity, it is relatively easy to purify them to better than 99.995% by cryogenic distillation. Purified samples can then be saved sealed in cells. In realizing the triple points care must be taken not to mistake the melting curves for some of the near by solid-to-solid phase transitions. See Table III and Ref. 19,20.

ARGON, NITROGEN - Both argon and nitrogen can be commercially obtained in very pure form. Thus when carefully transferred into the cells the question of impurities becomes academic. It may nevertheless be useful to make some comments.

For both substances most impurities lower the melting curves and widen the melting range. In the case of nitrogen, Kr and CO raise the melting curve /46/. The effect of Kr impurities in Ar is contraversial as conflicting results are published concerning the magnitude of the effect /12/.

Some troubles could potentially arise in detecting argon and methane in nitrogen, since both solid-liquid phase diagrams show an eutectic at a temperature relatively close to that of the triple point (-0.6 K at 20% Ar /36/ and -0.55 K at 24% CH_4 /37/). Both diagrams show in this region a very small separation between the liquidus and solidus lines, so that these impurities affect very little the melting range, though not tp temperature (Table V).

OXYGEN: Contrarily to other gases, with which we had never actually encountered systematic differences in the tp temperatures of different samples, differences up to + 0.7 mK were found with oxygen at IMGC (fig. 6). The only impurity known to rise the tp temperature of oxygen is argon. It forms mixtures with oxygen up to 20% Ar that show higher triple points up to about + 0.7 K (peritectic) /38/, with no increase of the melting range /39/. Incidentally, it also lowers - and broadens - the beta-gamma transition (15 μK/vppm) and the alpha-beta transition (20 μK/vppm) /40/: this fact could be used in checking for argon content. Therefore it is not generally possible to detect the presence of argon in oxygen from the melting range. Furthermore, an analysis of argon impurity in oxygen with other physico-chemical methods is not easy: mass-spectrometers require special techniques to eliminate oxygen, which would interact with the instrument /41, 42/; with current gas-chromatography the two components cannot be separated. Therefore special methods must be used for this kind of analysis, while it must be remarked that it is difficult to obtain a reliable analysis of commercial oxygen in small bottles. To avoid residual Ar impurities in commercial oxygen (produced by distillation of air) some researchers successfully used oxygen made by thermal decomposition of some compounds like $KClO_4$.

Nevertheless, several analyses were obtained at IMGC for all the samples used (Table IV), and we trust their reliability. They are showing very low argon amounts, so that the systematic differences which were measured - much larger than the reproducibility of the sample - cannot be explained.

NEON: Some problems may potentially arise also with this gas. As it is an isotopic mixture (9.2% ^{22}Ne, 90.5% ^{20}Ne and 0.27% ^{21}Ne), isotopic distillation may occur. This may take place during the production of the research-grade product, with consequent commercial samples having an isotopic composition different from the "natural" one, or it may happen during melting of the sample /42/, thus affecting the reproducibility of the tp temperature.

As to impurity effects, the neon triple point appears to be sensitive only to nitrogen /44/ and hydrogen isotopes /45/.

The effect of nitrogen in neon has been only recently observed by Ancsin /44/. It is similar to the effect of carbon dioxide on the tp temperature of methane - as indicated hereinbefore: after an initial lowering of the tp temperature down to -0.0023 K, saturation occurs for nitrogen concentrations higher that 150 vppm, with no further depression of temperature; this effect is observed for concentrations as high as 0.25% of nitrogen and the melting range is not affected at all (an eutectic-like behaviour). Nitrogen impurity in the sample must consequently be carefully checked.

With hydrogen and deuterium impurities, some data in the literature show an eutectic (at -0.3 K for 3% Ne in H_2 and at -0.2% for 2.3% Ne in D_2: see Table V), in contrast with Ancsin /43/ who claims no effect on the neon tp temperature for hydrogen concentrations up to 0.1%.

There is evidence however /29/ that commercially available natural neon has a reproducible liquidus point throughout the globe. Thus it seems to us unnecessary the use of ^{20}Ne as suggested by /46/.

DEUTERIUM - The tp temperature of deuterium is at a convenient temperature range for use in thermometry. Purity better than 99.9% is however not available commercially and therefore suitable impurity correction has to be applied to its measured triple point value /25/. A good analysis of the gas samples is not easily obtainable and the figure generally given are suspected of underestimating the impurity content. The main impurity is reported in the literature to be HD. This substance very slowly may convert into H_2+D_2 by metals - especially stainless steel - acting as a catalyst. Deuterium forms regular solutions with both hydrogen and hydrogen deuteride, but the correction for the tp temperature depression of the former is about double that required for HD (table V): when this correction amounts to several millikelvin, a good stability of these impurities is required.

Deuterium in spin equilibrium at room temperature tends to convert to the ortho- form when cooled; at the triple point, the spin-conversion speed is such that the tp temperature decreases by about 2 mK after 100 h /25/; the heat of conversion is too low to keep the sample overheated, but the temperature drift is too large for highest accuracy in measurements, since the drift effects are cumulative. A catalyst was therefore used to keep the sample in spin equilibrium.

HYDROGEN: The triple point of this gas was easily realized, since its tp temperature is insensitive to impurities except to its own isotopes and neon. The presence of these substances can easily be detected, as each of them produce a definite increase of the melting range. The use of a catalyst for spin conversion is necessary.

Pressure fixed points and vapour pressure scales

Not much experience has been gained on this subject, since studies were limited to a few substances.

However, measurements have shown that, when ever the triple point can be realized with a reproducibility near ±0.1 mK, the pressure value during melting is stable and uniquely defined to a degree that can be useful to pressure metrology /17/. Triple point pressure values ranging from 150 Pa to 518 KPa are available, in principle.

Measurements have also shown that there is advantage in realizing a vapour-pressure temperature scale with sealed devices. Vapour pressure scales are widely used in laboratories, and two of them (^{3}He and ^{4}He) were recently incorporated in the definition of the Echelle Provisoire de Température (EPT-76). A problem inherent in their realization is that impurities in the gas can alter the shape of the vapour pressure curve /44,47/ and hence produce a non-uniqueness in the Scale, as is realized in different Laboratories. The use of sealed devices improves reproducibility as the same sample of gas in always used; also, provided that very pure samples are enclosed in suitable cells, it can improve traceability among laboratories.

The sealed-cell device

After the experience gained with miniature sealed cells, the initial long-stem model shows only limited advantages.

One of the advantages is that this latter kind of cell can be easily reopened with its valve kept connected to the gas handling system during its use, so that different samples or gas mixtures can easily be studied. But different models of miniature cells, such as that used in Canada /29/ can also do that.

Another advantage is that only with long-stem cells the vapour pressure of the enclosed sample can be measured, until an accurate low-temperature pressure transducer is developed. Work in this direction is being made at NBS /48/.

The measurements carried out in the past years at IMGC seem to indicate that the techniques used in this Institute for the construction of the cells are adequate for keeping the impurity content of the gaseous samples stable over many years. Therefore the tp temperature (and pressure) can be reproduced at any time for calibration of thermometers.

With IMGC cells it was found possible to approach the triple point temperature value of the ideally-pure substances within few tenths of a millikelvin with commercial research-grade gases, with the exception of oxygen and, possibly, deuterium (see previous section). In addition, it was found that small amounts of impurities remained stable in the cells so that, also cells containing some impurities can still be used for calibrating thermometers, provided that their difference from a well established value is known.

The design of IMGC miniature cells proved to be suitable for calorimetric experiments, as discussed at the beginning of this paper. The actual behaviour of the condensed substance in a cell (including its exact distribution in the cell after condensation) is still a matter of pure speculation, since no visual observations has ever been made. Few attempts were made to calculate the thermal behaviour of a melting sample in a metal container: Tiggelman /43/ calculated the overheating trend of melting in a copper "bulb" of a spherical neon (and oxygen) sample; Bonnier /49/ recently made another calculation for thin cylindrical samples in stainless steel. However, in our opinion, reliable information about the cell design cannot be obtained with the oversimplifications necessary for solving equations.

We believe that the overall thermal behaviour of a cell during melting is dominated by the liquid phase, because the solid phase provides isothermal conditions for the thermometer holder; in the liquid phase, on the other hand, temperature gradients are easily introduced by heat fluxes through the cell, once sufficient liquid is present. These heat fluxes may be due to the heater or to heat leaks. Consequently, it is always advantageous to keep the sample in thin layers.

At IMGC we preferred to use fins instead of baffles in our cells, even though the latter are more effective in reaching thermal equilibrium rapidly after a heating period, because they substantially raise the internal surface area. As a consequence, and because of the much larger mass of the cell, IMGC miniature cells often gave rise to a melting range of the samples larger than those obtained with our L-cells or with cell models of other laboratories /29/ - for a comparable gas purity. Nevertheless, this distortion of the melting behaviour of the samples was always so small as to avoid any additional uncertainty to be added to the definition of the liquidus point. Therefore we gave priority to minimal internal surface, simple geometry and maximum room for thermometers. With the cell working only as a temperature generator, the thermometer holder is independent of cell geometry and this gives more freedom for optimization of the thermal behaviour and reponse time of the cell. As to the latter, we think that the criterion of keeping the sample in thin layers is ultimately the best one, since thermal re-equilibration time of the cell is related to the thermal contact of the sample with the container.

However, the geometry of the IMGC cells appeared to have a very small influence on the behaviour of melting. Six different geometries gave only minor changes in both the melting range and the thermal reponse time, with no sizeable effect on the temperature value of the liquidus point.

Also the quality of the apparatus used for measuring cells does not appear to have a critical effect, provided that the cell is placed in high vacuum and surrounded by a shield isothermal to such a degree as to reduce thermal exchange with the cell to less than a few tenths of a milliwatt. Cell design becomes much more important when the aim is to use the cell in the simplest apparatus, where thermal conditions are increasingly far from adiabatic. In this case, a suitable design can minimize the influence of external heat fluxes on the sample-thermometer interface. We found that with the IMGC design the cells can be used in an extremely simple apparatus.

A test has been made with methane and argon cells by mounting the cell in an aluminum spray-can acting as an adiabatic shield placed in an outer metal can filled with foam insulation (no vacuum insulation). The whole assembly has been immersed in liquid nitrogen and the shield regulated to the same temperature of the cell within few tenths of a kelvin through a differential thermocouple: temperature values well within 10 mK of the correct value were still obtained. Routinely a simpler cryostat than that shown in Fig. 1a is used. It has only one isothermal shield, regulated relative to the temperature of the cell within ±0.1 K; the cell itself is mounted on a stem similar to that of the cell of Fig. 1b (above the heat exchanger, which is replaced by a copper block, also regulated relative to the temperature of the cell through a differential thermocouple). In this way the cells can be interchanged also in this cryostat quite simply from the top flange.

CONCLUSIONS AND SUGGESTIONS FOR FUTURE DEVELOPMENT

The realization of fixed points represents the whole experimental work required to establish the Temperature Scale, once suitable commercial thermometers have been acquired and mathematical definition understood. The research work at IMGC has tried to show that permanent sealing of the gas in (metal) containers is possible and advantageous in two sectors: a) to improve traceability of fixed points, since the same sample of gas can be extensively measured in one laboratory and can next be sent to other laboratories for comparison; b) to simplify the realization of the Temperature Scale, since it is available indefinitively in time, once the fixed points are realized in sealed cells in a laboratory. This means that calibrations using sealed devices do not depend on the expertise of scientists well trained in the field of gas thermodynamic, but can rely on routine measurements with a device that can be used as a "black-box", following standardized - and simple - rules. In addition, each melting plateau permits a quite effective self-check of the overall quality of the measurements, except only with few gases for which a complete purity check is not possible through thermal analysis. This is important, moreover, from the point of view of those laboratories which are not likely to make the necessary investment for developing fixed points. They can obtain sealed cells from other laboratories or from commercial sources. In this way, an independent calibration of their thermometers is possible by direct application of the Scale definition.

This new opportunity is expected to have some impact also on legal metrology. At present, to be approved as a Secondary Laboratory, a laboratory must have its own standard thermometers certified by a Primary Institute, to establish "traceability". Sealed cells, which allow the Temperature Scale to be realized from its definition, make thermometer hierarchy unnecessary. This is a still unexplored area, which it would be desirable to investigate as soon as possible

Sealed devices are also expected to promote "home-made" calibrations. These were neither widespread since the realization of the IPTS is cumbersome, nor were they encouraged by official calibration laboratories. Sealed cells, have proved to be impervious to misuse at the accuracy level of about 0.01 K needed in most experiments. Their use is substantially similar to that of a thermometer comparison block, but the housing cryostat can be of lower qaulity, and the cells could allow for self-testing of the quality of the measurements.

It is clear from the preceding pages that several areas require further studies for better understanding of the thermal behaviour of the cells this could lead to improved cell design.

Producers of research-grade gases must be made to understand that the reliability of some of their products is not high enough. They should be encouraged to make available some specialty gases of higher purity than at present, although most of the present problems are actually connected with reliability of the gas purity and certification. This could allow future use of sealed cells in the field of reference materials /50/. Another promising area to be explored is the use of sealed cells as pressure standards or for the realization of vapour pressure scales. Much will depend on the development of an accurate - absolute if possible - pressure transducer capable of working over the whole cryogenic temperature range. This would allow fixed-point and vapour-pressure measurements to be made with a cell fully contained in a cryostat - and therefore would be simpler to use than long-stem cells.

REFERENCES

[1] Surfaces covered with only a monolayer of gas molecules can be obtained, under certain conditions, only in high vacuum.

[2] 17 B-cells, 10 C-cells and 3 U-cells have been sealed since then.

[3] It is dubious that ferric oxide is still hydrous after the activation process at 130 C under vacuum, when more than 10% in weigth of water is released by the catalyst.

[4] Few of them were especially developed for research in low-temperature gas thermometry in the past years.

1. Proceedings of 5th Symposium on Temperature, Its Measurement and Control in Science and Industry, H.H.Plumb Editor, Instr.Soc.of America, Pittsburg 1972 (herein referred as Temperature 4).
2. J. Ancsin, Metrologia 5, 77 (1969).
3. J. Ancsin, Metrologia 6, 124 (1970).
4. W.R.G. Kemp and C.P. Pickup, Temperature 4, pt. 1, 217 (1972).
5. see for example: E.F. Jr Westrum, G.T. Furukawa and J.P. McCullogh in: "Experimental Thermodynamics" (London, IUPAC, Butterworths) vol.1, 133.
6. IMGC - Annual Report A/23 (1970).
7. M.Durieux, W.R.G.Kemp, M.P.Orlova, C.A.Swenson, First Report (part 2) of Working Group 4 to the Comité Consultatif de Thermométrie, 1973.
8. F.Pavese and G.Cagna, Inst.Phys.Conf.Series (GB) 26, 70 (1976).
9. D.Ambrose, Brit.J. of Appl.Phys. 8, 32 (1957).
10. G.T.Furukawa, J.L.Riddle and W.R.Bigge, Temperature 4, pt 1, 231 (1972).
11. F.Pavese and G.Cagna, Alta Frequenza 44, 294E (1975).
12. F.Pavese, Metrologia 14, 93 (1978).
13. F.Pavese, Metrologia 17, 35 (1981).
14. F.Pavese, G.Cagna and D.Ferri, Proc. ICEC6, IPC Science and Technology Press, Guildford 1976, 281.
15. F.Pavese, G.Cagna and D.Ferri, ibidem, 205.
16. F.Pavese, D.Ferri and G.Cagna, Proc. INSYMET Conf. Bratislava 1976, 121.
17. F.Pavese, Metrologia 15, 47 (1979).
18. F.Pavese, Conf. on Temp. Measur. in Science and Industry, Karlovy Vary (CSSR), (1981).
19. F.Pavese, J.Chem.Thermodynamics 10, 369 (1978).
20. F.Pavese and L.M.Besley, J.Chem.Thermodynamics 13, 1095 (1981).
21. F.Pavese, G.Bonnier and J.Bonhoure, Doc. 76/32 to Comité Consultatif de Thermométrie, 1976.
22. J.Ancsin, Metrologia 14, 79 (1978).
23. F.Pavese, Doc. 80. 24 to Comité Consultatif de Thermométrie, 1980.
24. F.Pavese and C.Barbero, Proc. XVth Int.Congress on Refrigeration, Venezia 1979, vol. 1, 155.
25. F.Pavese and C.Barbero, Cryogenics 19, 255 (1979).
26. F.Pavese, Proc. XVIth Int. Conf. Low Temp. Phys., North-Holland ed. (Amsterdam 1981), 333.
27. J.A.Beattie,Tzu-Ching Huang and M.Benedict, Proc. Am.Acad.Arts Science 72, 137 (1937).
28. C.Bongiovanni, G.Frassineti and P.Marcarino, XXXIVth National Congress ATI 2, 235 (1979).
29. F.Pavese: "On the use of first-generation sealed cells in an international intercomparison of triple point temperatures of gases", see paper in this Conference.
30. A.Inaba and K.Mitsui, Jap.J.Appl.Phys. 18, 1183 (1979).
31. J.A.Cowan, R.C.Kemp and W.R.G.Kemp, Metrologia 12, 87, (1976).
32. J.Ancsin, Metrologia 9, 147 (1973).
33. Rapport du BIPM au CIPM, 25 (1981).
34. H.G.Donnelly and D.L.Katz, Industrial and Engineering Chem. 46, 511 (1954).
35. J.A.Davis, N.Rodewald and F.Kurata, A.I.Ch.E. Journal 8, 537 (1962).
36. H.M.Long and F.S.DiPaolo, Chem. Eng. Progr. Symp. Series 59, n.44, 30 (1964).
37. Z.Dokoupil, Physica 28, 309 (1962).
38. C.S.Barret, L.Meyer and J.Wasserman, J. Chem. Physics 44, 998 (1966).
39. J.Ancsin, Metrologia 9, 26 (1973).
40. E.L.Pace and R.L.Bivens, J.Chem.Physics 53, 748 (1970).
41. J.P.Compton and S.D.Ward, Analyst 99, 214 (1974).
42. DePaz M., F.Pavese and C.Cagna, Le Vide Suppl. n.169, 265 (1974).
43. J.L.Tiggelman, Thesis, Leiden 1973.
44. J.Ancsin, Metrologia 14, 1 (1978).
45. J.P.Brouwer, C.J.N.Van Den Meijdenberg,H.F.P.Knaap and J.J.M.Beenakker, Prod. 5th Symp. Thermophys. Properties, Newton 1970, 301.
46. R.C.Kemp and W.R.G.Kemp, Metrologia 17, 67 (1981).
47. J.Ancsin, Can.J.Phys. 52, 1521 (1974).
48. C.T.VanDegrift, W.J.Bowers, D.G.Wildes and P.B. Pipes, ISA Annual Conference 1978, 33.
49. G.Bonnier and Y.Hermier, Doc. 80/54 to Comité Consultatif de Thermométrie, 1980.
50. F.Pavese, Proc.Int.Symp. on Production and Use of Reference Materials, BAM ed. (Berlin 1980), 472.

The triple point of natural xenon

R. C. Kemp, W. R. G. Kemp, and P. W. Smart

CSIRO Division of Applied Physics, Sydney, Australia 2070

The realization of the triple point of natural xenon is described and the results of using different experimental techniques are discussed. It is concluded that the triple point should be further investigated to determine its suitability as a fixed point on the International Practical Temperature Scale.

INTRODUCTION

The measurement of temperatures below 273.15 K on the International Practical Temperature Scale of 1968 (amended edition 1975)[1] (IPTS-68) depends on interpolation between a number of reference temperatures or defining fixed points using a platinum resistance thermometer (PRT). The fixed points are boiling and triple points of pure substances and are not always located at convenient temperatures. In particular, there are at present no fixed points in the interval 90 K to 273 K. As a consequence of this, the non-uniqueness[2] or deviation between PRTs calibrated on IPTS-68 can be as much as 3 mK in this interval. The introduction of a fixed point in this range would reduce this rather large non-uniqueness. The xenon triple point occurs at the very convenient temperature of 161 K, lying as it does about midway between 90 K and 273 K. However, investigations[3-6] of the xenon triple point to date have indicated that it is not suitable for use as a fixed point. In particular, Ancsin[5] reported a melting range of 5 mK. Notwithstanding this, we decided to re-examine the suitability of the xenon triple as a fixed point because of its very convenient location.

RESULTS AND DISCUSSION

The apparatus used has been reported elsewhere[7], however a brief description of the apparatus might be useful here. The sample chamber is maintained in an evacuated and temperature-controlled can. It is also enclosed by a temperature-controlled radiation shield suspended in the vacuum. Heat leaks to the sample chamber are reduced to a very low level by adjusting the temperatures of the radiation shield and vacuum can. The temperature of the sample is measured using a Leeds and Northrup PRT No. 1731676. This PRT is contained in a copper sleeve along the axis of the cylindrical sample chamber. Attached to the sleeve are 15 copper discs which extend through the bulk of the sample ensuring that the PRT is in good thermal contact with the sample. To observe the triple point the sample is heated slowly towards the triple point and melted by applying heat in small pulses. After each pulse the temperature of the sample is allowed to reach equilibrium as indicated by the recorded trace of the test PRT.

The xenon used for this experiment was obtained from Matheson Gas Products, who supplied an analysis as follows: krypton < 50 ppm, nitrogen < 10 ppm, oxygen < 5 ppm, argon < 5 ppm, hydrogen < 5 ppm, hydrocarbons < 10 ppm; total impurity content < 50 ppm. Initial experiments indicated that the recovery time after a heat pulse was very long, being over one hour for large pulses. In view of this we decided to use small heat pulses of about 2.5% of the heat of transition followed by a recovery time of one hour. We also noticed that the shape of the melting curve was dependent on the method of heating used. That is, the apparent melting range depended both on the rate at which the sample was heated towards the triple point and also on the type of heat pulses used to melt the sample. This is illustrated by the two melting curves shown in Fig. 1.

Melting curve 1 was obtained by continuously heating the sample until its temperature was about 10 mK below the triple point and then using heat pulses of about 2.5% of the heat of transition and of 3 minutes

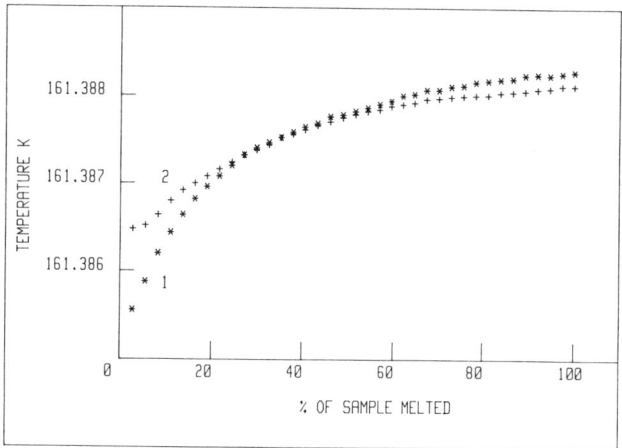

FIG. 1. Melting curve for natural xenon

duration, with a heater current of 10 mA and a recovery time of one hour, to melt the sample. The melting range for curve 1 is about 2.8 mK. Melting curve 2 was obtained by continuously heating the sample to about 100 mK below the triple point and then using heat pulses of 2.5% of the heat of transition and of 12 minutes duration, with a heater current of 5 mA and a recovery time of one hour, to melt the sample. The melting range for curve 2 was about 1.5 mK. An explanation for the difference in the melting curves is as follows: if heat is applied too quickly at the start of the melting curve, part of the sample melts while the bulk of the sample remains below the triple-point temperature. When the heating is switched off the liquid xenon re-freezes and the whole sample remains below the triple-point temperature. This effect results in an apparent melting curve which lies below the true curve at the start. If heat is applied slowly the whole sample is brought to the triple-point temperature before melting commences.

The melting curves obtained are certainly not as flat as for other fixed points, presumably due to natural xenon being made up of a mixture of isotopes. They are, however, rather better than have been reported previously.

In assigning a temperature to the xenon triple point we proceeded as follows. As no part of the melting curve was flat and as the shape of the curve, particularly at the start, depended on the experimental conditions, a mean melting temperature was determined from the final 80% of the melting curve. The temperature determined in this way from each curve was 161.3879 K on the version of IPTS-68 maintained at National Physical Laboratory (NPL) which in this range is identical to the version maintained at this Division. Assigning an uncertainty to this temperature is rather difficult. The reproducibility was better than ± 0.1 mK while the differences between PRTs in the region can be as much as 3 mK, due to the non-uniqueness noted above. Also the effects of isotopic composition are unknown. With these various uncertainties in mind we have assigned a temperature of (161.388 ± 0.001) K to the triple point of natural xenon.

This value may be compared with those obtained by other investigators which are given in Table I.

Table I

Values for the triple point of xenon

Michels & Prins[3] (1962)	161.375 K
Lovejoy[4] (1963)	161.392 K
Ancsin[5] (1978)	161.391 K
Inaba & Mitsui[6] (1978)	161.3918 K
This work	161.388 K

The agreement is reasonable bearing in mind the non-uniqueness of the IPTS-68 in this region is about 3 mK without allowing for any differences between the different scales used. In addition, other workers obtained melting ranges of up to 17 mK compared with the 1.5 mK reported here and the methods used to estimate the triple point temperature differed.

CONCLUSION

We conclude that while the triple point of xenon is not as good as the other triple points used in the IPTS-68, it does seem worthy of further investigation. It is likely that the relatively large melting range is due to xenon being a mixture of isotopes having different triple point temperatures. We intend to test this assumption by examining a single xenon isotope ^{136}Xe contained in a sealed cell.

REFERENCES

1. "The International Practical Temperature Scale of 1968" (Amended Edition 1975) Metrologia 12, 7 (1976).
2. S. D. Ward and J. P. Compton, Metrologia 15, 31 (1979).
3. A. Michels, Prins. C.: Physica 28, 101 (1962).
4. D. R. Lovejoy, Nature 197, 353 (1963).
5. J. Ancsin, Metrologia 14, 45 (1978).
6. A. Inaba and K. Mitsui, CCT 78 Doc. 26.
7. J. A. Cowan, R. C. Kemp and W. R. G. Kemp, Metrologia 12, 87 (1976).

Thermal behavior of thermometric sealed cells and of a multi-compartment cell

G. Bonnier and Y. Hermier

Institut National de Métrologie, CNAM, 75141 Paris Cédex 03, France

In order to improve the sealed cells realizing temperature fixed points, the thermal behavior of differently shaped devices has been studied at Institut National de Métrologie. A way of assembling several cells in the same housing has been investigated with success, both theoretically and experimentally. The device contains five different gases and is small enough (8 cm high and 6 cm in diameter) to be placed in an ordinary cryostat.

INTRODUCTION

In using sealed cells we noticed that the temperature rise during the heating process and the time to return to equilibrium are two important features of the cell design. Their study is important if one aims at ultimate accuracy.

Furthermore, our goal was to associate in the same housing several cells filled with different gases. We realized that in such a device the response time of each individual sealed cell should be short. Therefore, we tried to set up a theoretical model of their thermal behaviour according to the experimental results in order to point out the important design features. On this basis we built an improved type of individual cell and we assembled five such cells in the same housing.

Different shapes of sealed cells were studied at INM [1,2]. In each case (see Fig.1 to 4) the cell includes within a strong body (A), an amount of pure gas (B), a heat exchanger (C) between the thermometer and the sample, a room (D) to prevent damage from overpressure at room temperature, a pit (E) in which to place the thermometer and a copper tube (F) to be pinched for closing the cell.

THE FILLING SYSTEM (see Fig.5)

After chemical cleaning, the cell and the filling system was rinsed with pure gas. Between two rinses the system was pumped out using a mechanical (6) and a turbomolecular pump (7). Some pure gas was then admitted from the bottle (5), through the vessel (4). Pressure was controlled with a diaphragm gauge pressure transducer (2) and displayed (3).

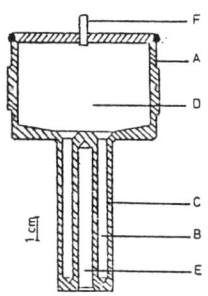

Fig. 1. Type A cell.
Mass of the cell: 270 g. The heat exchanger consists of vertical pits around the thermometer.

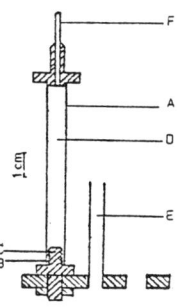

Fig. 2. Type MC cell.
Mass of the cell: 120 g. The heat exchanger is shaped as a screw thread.

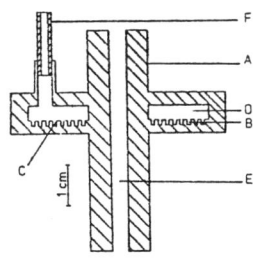

Fig. 3. Type BC cell.
Mass of the cell: 132 g. The heat exchanger has square cross section grooves.

After every rinse, the decrease of impurities was monitored using a mass spectrometer (8,9), connected to a chart recorder (10). Then, after pressurizing the entire system, the copper tube was closed and cut off by pinching. Thus the cell (1) is filled without being cooled. The cell was connected to the filling system through a demountable metallic connector. The mass difference of the cell plus connector, before and after filling, enabled us to calculate the amount of sample inside the cell.

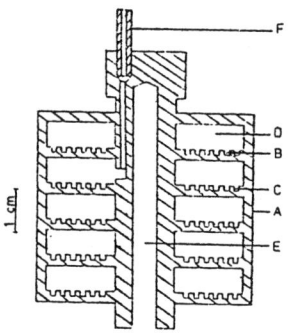

Fig. 4. Type BCM cell.
Mass of the cell: 340 g. The heat exchanger has square cross section grooves.

EXPERIMENTAL PROCEDURE

In order to measure temperatures under adiabatic conditions, the cell was suspended in an evacuated can (see Fig.6) with three nylon threads (diameter 0.1 mm) inside a radiation screen. Electric connections necessary for measurements are thermally anchored on the screen, by gluing. A differential thermocouple controlled the temperature difference between the screen and the cell to a few hundredths of a kelvin.

The Dewar was first filled with a coolant (liquid nitrogen or helium). To reach the triple points of oxygen and nitrogen we used solid nitrogen as coolant. Then helium gas was introduced under low pressure into the vacuum can in order to cool the cell below the triple point. After evacuation of helium, the temperature of the cell was raised, by electrical heating, to 500 mK below the triple point. The regulation of the screen temperature was adjusted in order to set up adiabatic conditions. Thermal drift of the cell was typically less than 1 mK per hour. At this stage heating of the sample was started by supplying successive amounts of energy and recording temperatures when temperature equilibrium is reached.

This fully automatic process was monitored by a micro-computer (see Fig.7). The heating time can be programmed and the time required to reach the temperature equilibrium deduced from the slope of the curve. Equilibrium is detected, using statistical criteria, by comparing successive sets of values displayed by the resistance thermometer. This routine can also calculate resistance values, overheatings, and response times. The statistical criteria used are more demanding than visual observation of a chart recorder and usually, in the presence of electrical noise, the time required to return to equilibrium is greater than a manually monitored one.

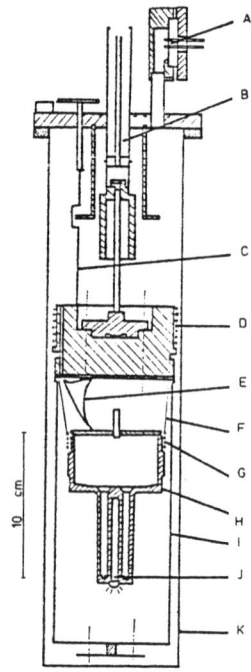

Fig. 6. Adiabatic calorimeter for sealed cells.
A Electrical feedthrough, B Pumping line, C Thermal link, D Heaters, E Differential thermocouple, F Nylon threads, G Heaters, H Sealed cell, I Radiation screen, J Thermometer, K Vacuum can.

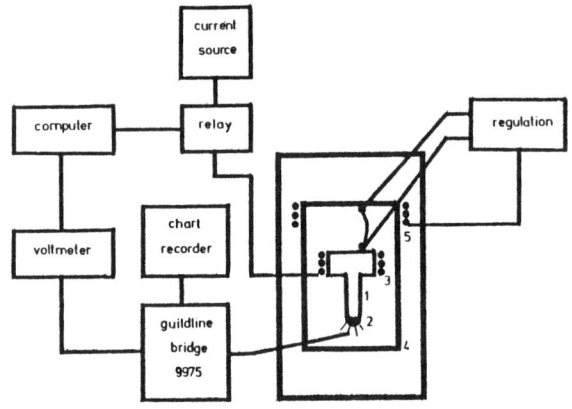

Fig. 7. Automatic monitoring of the experiment.

As described by [3] and subsequently confirmed by others (see for instance [4] and [5]), during melting the time required for the sample to approach equilibrium temperature within certain limits, after a heating period, increases while the temperature rise of the liquid portion of the sample during heating becomes larger with increasing liquid-to-solid ratio. As the return to equilibrium is an inverse exponential function [3] the achievement of perfect equilibrium will take an infinitely long time. It can, however, approach equilibrium, within say 0.1 or 0.01 mK, within reasonable time periods depending upon the sample purity and the sample container [6]. Because it is difficult to maintain adiabatic conditions indefinitely

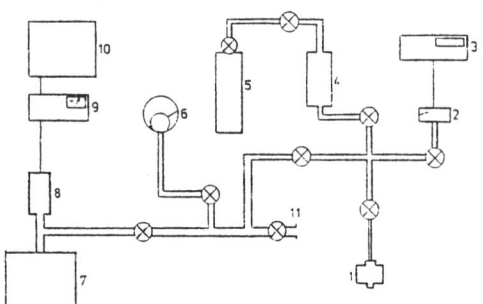

Fig. 5. Sealed cell filling system.

in an experimental apparatus, the accuracy of the experimental results are related to the time necessary for the sample to return to such "equilibrium" temperature.

We sought an explanation for such thermal behaviour and used the conclusions for improving the design of sealed cells.

PHYSICAL MODEL

Nomenclature

t	time, s
T	temperature displayed by the thermometer, K
T_f	temperature of fusion of the sample, K
L	latent heat of fusion, J g^{-1}
ρ	density, g cm^{-3}
C	heat capacity, J K^{-1}
λ	thermal conductivity, J s^{-1} cm^{-1} K^{-1}
α	thermal diffusivity, cm^2 s^{-1}
S	heat exchanger area, cm^2
X_o	mean thickness of the liquid formed layer, cm

As mentioned above the relationship between the temperature and the time, after a heat pulse, can be described by an inverse exponential function. The cell behaves, from a thermal point of view, like an R C circuit associating the thermal capacity C of the cell body to the thermal resistance $R_{th} = \frac{X_o}{\lambda S}$ of the liquid layer formed. This layer is located between the heat exchanger and the solid sample melting at the temperature T_f (see Fig.8).

The thermal balance at any time can be written:

$$CdT = -\frac{\lambda S}{X_o}(T-T_f)\,dt$$

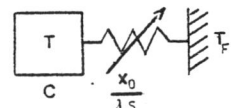

Fig. 8. Analog RC circuit.

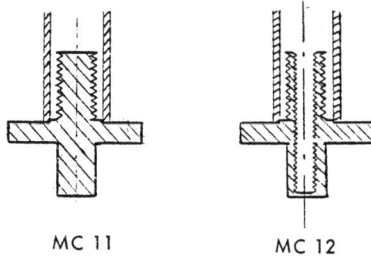

Fig. 9. Heat exchangers of type MC 11 and MC 12 cells.

The integrated equation yields a time constant that characterizes the cell:

$$\tau = \frac{X_o C}{\lambda S}$$

In a first approximation the heat capacity of the sample can be neglected (see Table I).

The solid sample temperature is uniform throughout and equal to the temperature of fusion T_f.

The thermal diffusivity of the stainless steel being much greater than any sample of liquid, the wall temperature can be considered uniform everywhere (see Table I).

STUDY OF THE RESPONSE TIME

In placing several cells in the same housing we tried reducing the size of type A cells. Two cells, MC 11 and MC 12, were built (see Fig.2). Each is made of a cylindrically shaped stainless steel tube (id 6 mm, od 7 mm). This tube is closed at both ends by threaded caps making a mechanical as well as a thermal contact with a copper adapter containing a capsule type platinum thermometer. Inside these cells the gas condenses in the volume between the heat exchanger and the cell wall. The heat exchanger of the MC 11 is made of a simple screw (see Fig.9).

The exchange surface area of 5.48 cm^2 is in contact with 3 × 10^{-6} mole of oxygen whereas the one of the MC 12 cell is 11.01 cm^2 in contact with 6.53 × 10^{-6} mole of oxygen.

Except for the heat exchangers, both cells are mechanically identical.

Table I. Comparison of the physical properties of stainless steel 18 Cr 8 Ni [7] and of the substances used for fixed points.

T_K	Substance	Fluid sample at its triple point			Stainless steel at the same temperature		
		λ (10^{-3} W cm^{-1} K^{-1})	$c\rho$ (J cm^{-3} s^{-1})	α (10^{-4} cm^2 s^{-1})	λ (10^{-3} W cm^{-1} K^{-1})	$c\rho$ (J cm^{-3} K^{-1})	α (10^{-4} cm^2 s^{-1})
273.15	H$_2$O	5.79	4.218	13.4	145	3.737	388
83.798	Ar	1.25	1.499	8.33	85.8	1.452	591
63.141	N$_2$	1.5	1.691	8.86	70.5	0.873	807
54.361	O$_2$	1.927	2.168	8.89	62.4	0.631	989
24.563	N$_e$	1.17	2.232	5.24	22.5	0.110	2045

Table II. Cell response time as a function of the amount of argon and oxygen.

Type of cell	Cell	Nature of the gas	Mass of the sample (g)	Latent heat of fusion (j)	τ_{exp} ** (s)
BC	BC 12	Ar	0.98	29	32
BC	BC 13	Ar	0.63	20	34
BC	BC 14	O_2	0.58	8	14
BC	NIMPP 5 *	O_2	0.66	9	15
BC	NIMPP 3 *	O_2	0.65	9	20
BC	NIMPP 15 *	O_2	0.74	10	24

* The cells NIMPP 5, NIMPP 3 and NIMPP 15 are of the INM's BC type. They were built, sealed and studied at the NIM's laboratory of Peking.
** Experimental response time.

RESULTS

The mean response time τ observed for the MC 11 cell was 328 seconds with a standard deviation of 32 seconds. For MC 12 type cell the mean of τ was 167 seconds with 35.2 seconds of standard deviation.

One MC 12 type cell was likewise filled with 2.75×10^{-3} mole of argon and the mean of τ was 112 seconds.

As pointed out by [3] it is possible to reduce the return time to equilibrium by subdividing the sample into small sections. The main effect of subdividing the sample is to improve the thermal coupling of the cell with the melting solid. This effect is due to the increase of the surface area and to the reduction of the thickness of liquid layer.

TYPE BC CELLS

As the response time is inversely proportional to the surface area, we built a cell designated BC (see Fig.3) which has an internal volume of 9.18 cm³ and a surface area of 24 cm². The gas condenses inside the grooves of 0.86 cm³ volume. The amount of sample is so adjusted that the volume of the liquid is less than 0.86 cm³. One can see in table II the response time versus the amount of sample for argon and oxygen.

TYPE BCM CELL

As the response time of the BC cell is sufficiently short we placed five such cells into one housing designated BCM (see Fig.4). The BCM cell, filled with five different substances, establishes five triple points. We began experiments by filling only one of the five compartments with argon and compared the response time of this cell with those of other types of cells previously studied (see Table III). The ratio $\tau S/C$ appears to be roughly constant confirming our assumptions about the thermal behaviour of the sealed cells. Then a second compartment was filled with oxygen and we compared the response time, versus the percentage of sample melted, of argon alone and of argon in the presence of a compartment filled with oxygen (see Table IV). The response time was essentially unchanged.

Then the compartments of the cell were filled with five different samples: water, argon, nitrogen, oxygen and neon. For details see Table V. Melting curves for each substance were obtained using a heating power of about 2.6 mW.

Table III. Response time study of differently shaped cells filled with argon.

Type of cell	Mass of the sample M (g)	Latent heat of fusion L (J)	Response time τ (s)	Heat exchanger area S (cm²)	Heat capacity C (J.K^{-1})	$\frac{C}{S}$ (J.K^{-1}.cm^{-2})	$\tau / \frac{C}{S}$
A	6.1	180	40	44	57	1.29	31.0
MC	0.115	3.4	112	6	28	4.67	24.0
BC	0.98	29	32	24	32	1.33	24.1
BCM	0.78	23	72	21	75	3.57	20.2

Table IV. Response time of a BCM cell as a function of the percentage of sample melted during the melting of argon.

Percent of sample melted	Response time Ar (s)*	Response time Ar+O_2 (s)**
25 %	64	61
50 %	57	59
75 %	82	63
100 %	86	85

* Cell filled with argon only.
** Cell filled with argon and oxygen.

THE TRIPLE POINT OF WATER

The purity of the distilled water is characterized by its electrical conductivity. The initial value of our bidistilled water was 3 µS cm^{-1}. After chemical cleaning, the cells are rinsed until the rinsing water conductivity reached a minimum of 8 µS cm^{-1}. To fill the type A cell, water is gently boiled in the cell to remove the volatile impurities and to adjust the amount of water sample. Then the copper tube was pinched. To fill type BCM cell the excess water is pumped at room temperature before sealing the cell. To realize the triple point, anti-freeze is used as a coolant. The mixture circulates in the dewar from a thermally regulated bath at around minus three degrees Celsius. Around the plateau the heat capacity of the cell is about 130 J K^{-1}. Figure 10 shows the percentage of sample melted as a function of thermometer resistance. Overheatings and the time to return to equilibrium after every heat pulse is plotted in Fig.11.

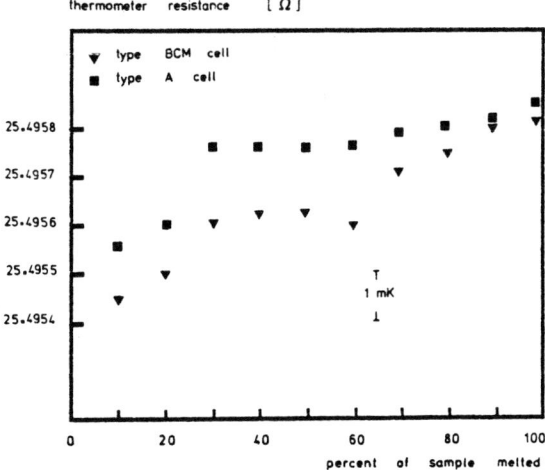

Fig. 10. Typical water melting curves.

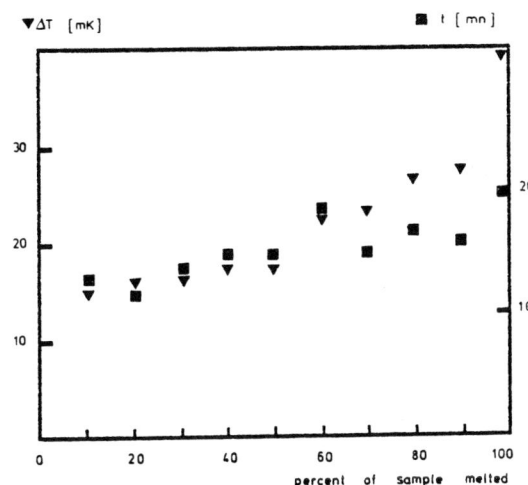

Fig. 11. Overheatings ΔT and times t to return to equilibrium for water.

Table V. BCM cell filled with five substances. Comparison between the theorical and the experimental response times.

	Sample mass	volume	Exchange surface area s (cm^2)	Experimental heat capacity c (JK^{-1})	Theorical response time τ (s) *	Experimental response time τ (s)
H_2O	1.3	1.3	36	125	30	170
Ar	1.11	0.784	25.7	75	116	75
N_2	0.54	0.622	22.44	55	82	80
O_2	0.78	0.597	21.02	46	57	54
N_e	0.45	0.373	17.46	8.3	21	60

* $\tau = \dfrac{X_o\, C}{\lambda\, S}$ is calculated with $X_o = 0.05$ cm.

THE OTHER TRIPLE POINTS

The four remaining compartments of the BCM cell were filled with gases supplied by the Air Liquide Company. The nominal impurity analyses of these samples are shown in Table VI.

Heat capacity of the filled cell at 273 K, 83 K, 63 K, 54 K and 23 K was 130 j/K, 75 j/K, 55 j/K, 45 j/K and 8 j/K respectively. Coolants used were anti-freeze at $-3°C$ for H_2O, liquid N_2 for Ar, solid N_2 for N_2 and O_2 and liquid He for Ne. Results are plotted in Figs 10 to 19.

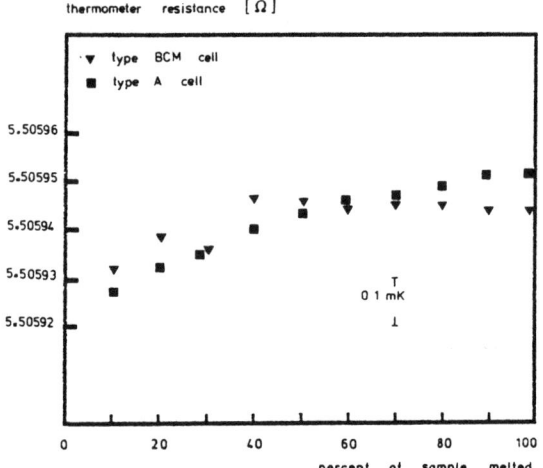

Fig. 12. Typical argon melting curves.

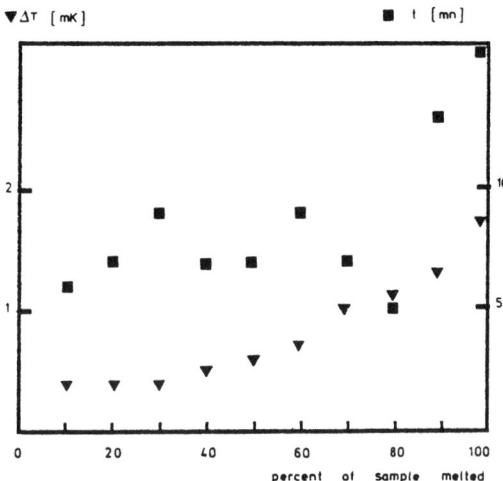

Fig. 13. Overheatings ΔT and times t to return to equilibrium for argon.

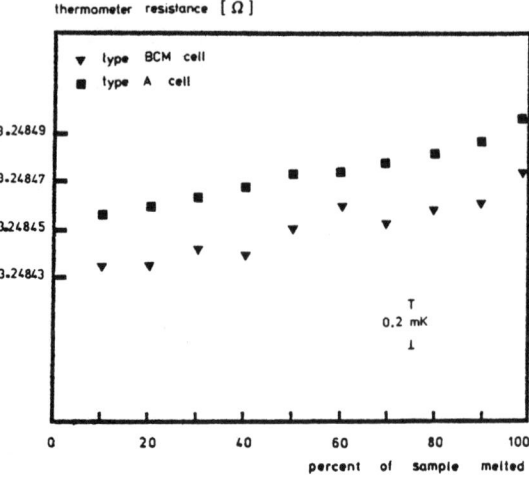

Fig. 14. Typical nitrogen melting curves.

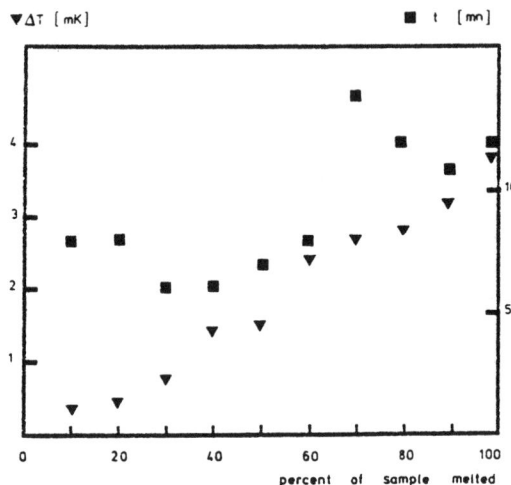

Fig. 15. Overheatings ΔT and times t to return to equilibrium for nitrogen.

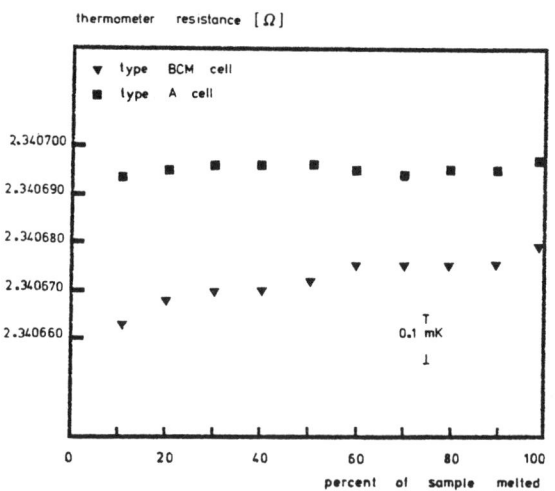

Fig. 16. Typical oxygen melting curves.

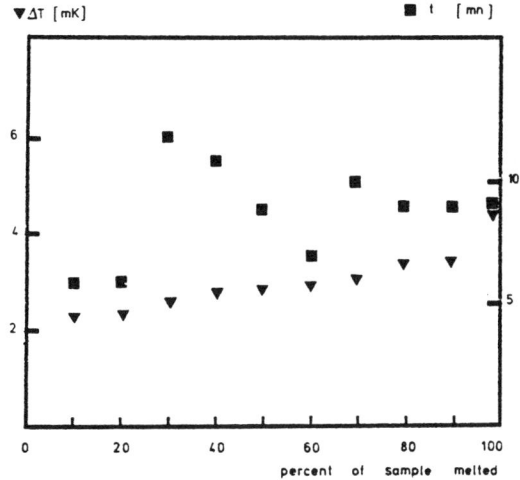

Fig. 17. Overheatings ΔT and times t to return to equilibrium for oxygen.

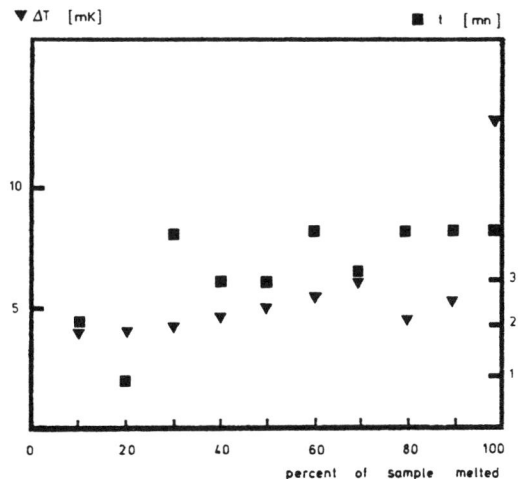

Fig. 19. Overheatings ΔT and times t to return to equilibrium for neon.

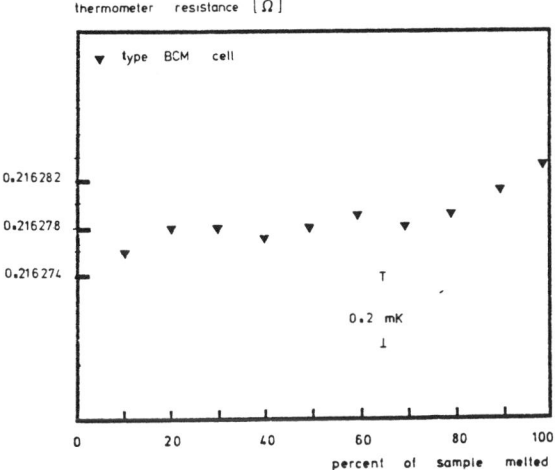

Fig. 18. Typical neon melting curve.

CONCLUSION

The platinum resistance thermometer used to study the melting curves indicates a good agreement between type A and the BCM cell. For oxygen, nitrogen and argon the differences in melting curves are less than 0.3 mK. For water the difference is as much as 1 mK. Except, may be, for argon, the melting curves seem to be lower for the BCM cell. These differences could be explained by the presence of impurities. Table V shows reasonable agreement between the experimental time and the theoretical response. The theoretical expression predicts the order of magnitude even for sophisticated shapes. The greater difference observed with the water can probably be explained by the relatively larger amounts of impurities.

We showed that it is possible to combine several cells within a single device. Each triple point can be realized although four different samples are in thermal contact with it.

We are now investigating the best choice of filling samples.

The authors wish thank Dr J. Ancsin for his help in editing this paper.

REFERENCES

1. G. Bonnier, and R. Malassis, Bull. Bureau National de métrologie 22, 19 (1975).
2. G. Bonnier, and Y. Hermier, Doc. CCT/80-54, in Comité Consultatif de Thermométrie, 13e session (1980), Annexe T1, BIPM, F-92310 Sèvres, France.
3. J. Ancsin, and M.J. Phillips, Metrologia [5], 77, (1969).
4. F. Pavese, Temperature Measurement, 1975 (Conf. Ser. No. 26, Inst. Phys., London, 1975), p. 70.
5. J.P. Compton, Temperature, Vol. IV, ed. Harmon H. Plumb (Instrument Society of America, Pittsburgh, 1972), Pt. 1, p. 195.
6. J. Ancsin, and M.J. Phillips, Metrologia [9], 147, (1973).
7. Russell B. Scott, Cryogenic Engineering, 1959, (D. Van Nostrand Company, 1959), p. 330.

Table VI. Nominal purities of gases used in BCM cell.

Ar	N_2	O_2	Ne
H_2O < 0.5	CO_2 < 10	Ar < 12	He < 80
N_2 < 0.5	CO < 5	N_2 < 5	N_2 < 15
O_2 < 0.2	O_2 < 5	H_2O < 3	O_2 < 3
CH_4 < 0.1	H_2O < 3	Kr+Xe < 3	H_2O < 3
		CO < 0.2	H_2 < 3
		CO_2 < 0.2	
		CH_4 < 0.2	
		H_2 < 0.1	

Reproducibility of the triple point of argon in sealed transportable cells

G. T. Furukawa

National Bureau of Standards, Washington, D. C. 20234

The reproducibility of the triple point of argon sealed in miniature pressure cells was investigated in calorimetric apparatus. The results obtained with samples of 99.9999 percent purity sealed in three cells of different designs, using two calorimetric cryostats, show that the triple point of argon can be reproduced well within ±0.1 mK. Measurements with six thermometers demonstrate that calibrations can be obtained consistent within the reproducibility of the fixed point.

1. INTRODUCTION

Boiling points[1] of substances have been employed as thermometric fixed points for many years[2,3,4]. One of the advantages of the boiling-point technique is that the boiling process usually purifies the thermometric substance further, the more volatile impurities being removed by distillation and the less volatile impurities being left in the boiler. The purified substance is allowed to reflux continuously where the thermometer is located, as in a hypsometer[5]. However, to evaluate the boiling point, the equilibrium pressure at the location of the thermometer must be known with appropriate accuracy for which enough time must be allowed for the molecular and hydrostatic pressures of all components of the measurement system to reach pressure equilibrium in a known thermal state so that their contributions to the pressure can be calculated accurately. Hence, the boiling-point technique is complex and is fraught with adjustments to be made to the readings of the manometer. On the other hand, once a suitable pressure measurement system is assembled and experience is obtained in its operation, measurement reproducibilities that correspond to about 0.1 mK can be achieved, particularly with substances such as hydrogen and neon which have high vapor pressure versus temperature sensitivities near their boiling points (both with dp/dT of over 29 kPa/K). A manometer system capable of measurement of mercury column to about 0.002 mm (about 0.3 Pa) has been described[6]; however, when an adequate pressure measurement system is not immediately available it would be expedient to develop wherever possible triple points and freezing points (or melting points) as temperature fixed points which do not require pressure measurements. To circumvent manometry the argon triple point was suggested as a substitute or as an alternative for the oxygen boiling point in the definition of the International Practical Temperature Scale of 1968 (IPTS-68)[7]. The suggestion as an alternative for the oxygen boiling point was accepted and included in the 1975 amended edition of the IPTS-68[4].

More recently, Pavese[8] and Bonnier[9] showed that the triple point of a high-purity gas can be observed with excellent reproducibility (better than 1 mK) in miniature pressure cells sealed with a small amount of the gas. The cells can be readily exchanged among different laboratories for comparison. At the NBS, the measurements of the triple point of argon in three pressure cells employing a calorimetric cryostat have been found to be reproducible to better than ±0.1 mK[10]. Because of this high reproducibility of the triple-point measurements of gases in sealed miniature pressure cells, the Comité Consultatif de Thermométrie sponsored a program for the International Comparison of Fixed Points by Means of Sealed Cells. The NBS has submitted for this international comparison a sealed argon cell. Thus far, as part of this program, the triple point of argon cells from the Institute of Metrology, "G. Colonnetti" (IMGC, Turin, Italy), the National Research Council (NRC, Ottawa, Canada), and the National Research Laboratory for Metrology (NRLM, Tsukuba, Japan) have been measured at the NBS. The results agree within ±0.1 mK with the results of three NBS cells. This paper describes the method used at the NBS for the realization of the triple point of argon in sealed cells using calorimetric cryostats. The results obtained using the calorimeter with the three NBS argon cells and with cells from the above three laboratories are compared.

2. METHOD AND APPARATUS

2.1 Calorimetric Cryostat

The measurement method used for the determination of the triple point of argon was very similar to that described in earlier publications for the calorimetric determination of purity and triple point of substances that can be melted in the calorimeter[11]. The essential part of the calorimetric apparatus is shown schematically in Fig. 1. (For greater details of the construction, see reference [12].) Two calorimeters (laboratory designation S and T) were used in the measurements. Their designs, except for wiring, were essentially the same; cryostat S had leads for only two platinum resistance thermometers, while cryostat T had leads for as many as seven thermometers. (Capsule-type platinum resistance thermometers that meet the specifications of the IPTS-68[4] were employed. Henceforth, for convenience these thermometers will be referred to by the abbreviation SPRT's.)

The calorimetric apparatus was designed to provide as nearly adiabatic conditions as possible for the system on which measurements are being made. All leads (B) entering the vacuum space, which is enclosed by the vacuum can (F), are first brought to the refrigerant temperature on the tempering ring (D) and then heated close to the adiabatic shield temperature on the floating copper cylinder (E). Electrically insulated copper strips are used for tempering the leads. The appropriate leads are tempered additionally on the adiabatic shield (G) before being attached to the copper tempering strips on the auxiliary isothermal shell (H). The adjacent surfaces of the adiabatic shield and auxiliary isothermal shell were gold plated to minimize heat transfer by radiation. The final tempering of the leads before they entered the SPRT is made on the sealed cell in the region where the argon sample would be condensed. The insulation resistance between all leads used in the measurements were regularly checked to be 1000 MΩ or greater. The temperatures of the adiabatic shield and the floating copper cylinder were controlled by automatic adjustment of power in the heaters wound on their surfaces, in conjunction with differential thermocouples of constantan and Chromel P wires. To integrate the temperatures, eight thermocouples were distributed between the adjacent surfaces of the auxiliary isothermal shell and the adiabatic shield. Three thermocouples were used between the adiabatic shield and the floating copper cylinder.

2.2 Sealed Argon Cell

To test the reproducibility of measurements of the triple point of argon in sealed cells, three cells of different designs were used. One of the sealed argon cells (designated Ar-NBS-1) is shown in Fig. 1, inside its auxiliary isothermal shell, suspended by means of a string inside the adiabatic shield. The cell was fabricated of stainless steel and has a capacity of about 50 cm³. The filling pressure used was about 9 MPa (0.2 moles) which corresponds to about 240 J of heat of fusion. The thickness of the copper tube (I) was designed so that the condensed argon would completely fill the cylindrical spaces up to about 2 mm above the top of the reentrant thermometer well. The thermometer well was designed approximately 1 mm larger than the typical diameter of capsule type SPRT's to accommodate a copper sleeve with a helical groove for venting air. By providing the vent for air, the SPRT could be easily inserted or removed from the well without causing possible "harm" to the thermometer. To enhance heat transfer between the parts, the dimensions were made to close tolerances and a thin film of stopcock grease was used on surfaces that were in contact.

The auxiliary isothermal shield (H), made of copper, served to protect the SPRT and the cell from stray thermal radiation and also to temper the measurement leads before they were finally tempered on the cell proper. Also, during the process of introducing heat into the cell, the isothermal shell provides convenient and better adiabatic control conditions which are helpful because during the heating period the stainless-steel cell is expected to have large

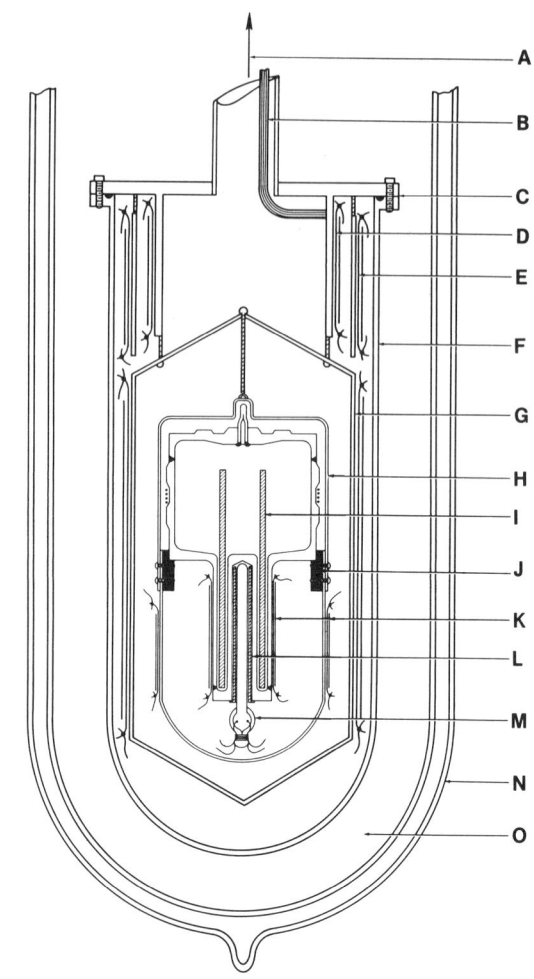

FIG. 1. Schematic of calorimetric apparatus with sealed argon cell.

- A. To vacuum pump, source of exchange gas, and vacuum feed-through for leads.
- B. Leads.
- C. Vacuum seal of gold gasket.
- D. Copper cylinder with copper strips for tempering of leads to refrigerant temperature.
- E. Copper cylinder and copper strips for tempering of leads nearly to the temperature of the adiabatic shield.
- F. Vacuum enclosure.
- G. Adiabatic shield with copper strips for tempering of leads to the temperature of the adiabatic shield.
- H. Auxiliary isothermal shell for the pressure cell.
- I. Copper tube.
- J. Copper girth ring soldered to the sample cell.
- K. Copper strips for tempering of leads to the temperature of the isothermal shell and of the pressure cell.
- L. Helically grooved copper sleeve for the re-entrant well of the sample cell. (The i.d. of the sleeve closely fits the SPRT.)
- M. Platinum resistance thermometer.
- N. Dewar.
- O. Liquid nitrogen space.

thermal gradients. The heater (illustrated as dots) is shown wound around the upper part of the cell. The auxiliary shell was attached by means of screws onto a copper girth ring (J) which was soldered on the cell; a thin film of stopcock grease was used on the surfaces that were in contact. (It is to be noted that, in the design, the heat capacity of the auxiliary isothermal shell and the degree of its thermal contact with the sample cell would depend on the quality of adiabatic control of the shield[13,14,15]. Where the adiabatic control is poorer the heat capacity of the auxiliary shell and the thermal resistance between the shell and the sample cell would be increased.) The heat capacity of the sealed cell containing the argon sample plus the auxiliary shell was approximately 100 J/K at 83.8 K; the measured heat of fusion for the sample was 250 J.

Except for the lower section where the condensed argon would collect, the design of the second argon cell (designated Ar-NBS-3) was very similar to that of the cell that is shown in Fig. 1. The second cell was designed to accommodate for calibration three capsule SPRT's. Hence, the lower section was somewhat larger in diameter and the copper tube (I) shown in Fig. 1 was also accordingly larger in diameter. Otherwise the general shape and dimensions of the second cell were the same as those of the cell shown in Fig. 1. The total internal volume was also about 50 cm^3. The same auxiliary isothermal shell was used for both cells. A second heater was wound around the lower section of cell Ar-NBS-3 to test whether the performance would be better with the heater near the bottom or near the top. Although tests showed that the results were not dependent on the location of the heaters, heaters were wound on all cells at the upper section so that any condensation would be driven to the bottom. The heat capacity of the sealed cell with the argon sample and the auxiliary shell was approximately 125 J/K at 83.8 K; the measured heat of fusion for the sample was 248 J.

The third cell (designated Ar-NBS-7) shown in Fig. 2 was designed somewhat simpler than the first two. The outer cylinder and the thermometer well were fabricated from commercially available stainless-steel tubes. The diameter of the outer cylinder is sufficiently large so that the lower section could be designed to accommodate as many as seven SPRT's. The copper cylinder (J) had six vertical holes of 3 mm diameter and a central hole that was slightly larger than the re-entrant tube for the thermometer. The internal volume was approximately 50 cm^3 similar to the other two cells. The cell was designed so that the condensed argon would fill the space between the thermometer reentrant tube and the copper cylinder and the six holes in the cylinder, as well as form a layer approximately 1 or 2 mm thick across the top. The reentrant well was designed to fit a helically grooved copper sleeve for the SPRT. The auxiliary isothermal shell was attached to the cell by means of a nut on a threaded part at the top. Surfaces in contact had a thin film of stopcock grease. The leads were tempered on insulated copper strips on the shell and on the cell. The heater was wound on the upper section of the cell. The heat capacity of the sealed cell with the argon sample and the auxiliary shell was close to 220 J/K at 83.8 K; the measured heat of fusion for the sample was 270 J.

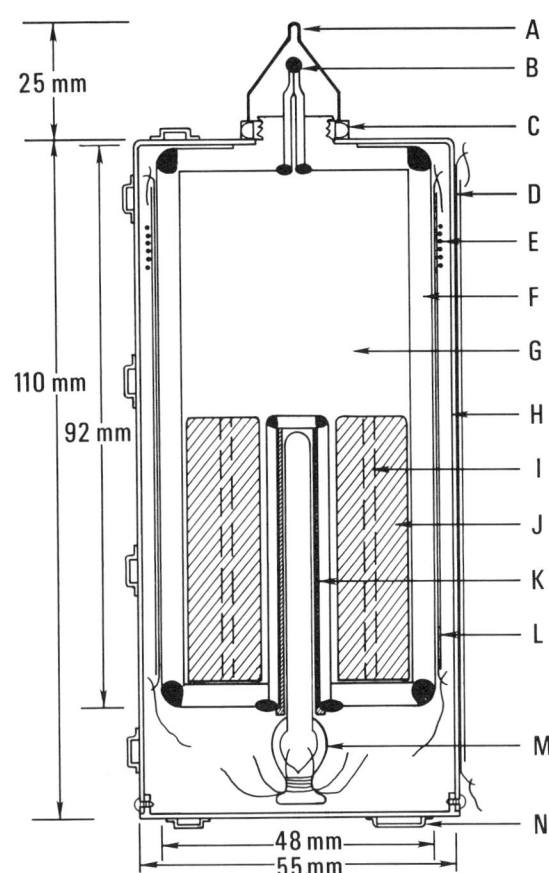

FIG. 2. Schematic of sealed argon cell of simple design, and its auxiliary isothermal shell.

A. Wire loop for hanging inside the calorimeter.
B. Pinched and electric arc sealed and severed capillary tube.
C. Nut for attaching auxiliary isothermal shell (H).
D. Electrically insulated copper strips for tempering leads on (H).
E. Heater.
F. Sample cell of stainless steel.
G. Sample space.
H. Isothermal auxiliary shell fabricated of copper and gold plated.
I. Vertical holes (3 mm diam).
J. Copper cylinder.
K. Helically grooved copper sleeve for the re-entrant well of the sample cell. (The i.d. of the sleeve closely fits the SPRT.)
L. Electrically insulated copper strips for tempering leads on the sample cell.
M. Platinum resistance thermometer.
N. Clips for inserting thermocouples.

As part of the international comparison of the triple points of pure gases sealed in cells similar to those described above, measurements were also made on argon cells received from IMGC, NRC, and NRLM. Only the results obtained will be reported in this paper for comparison with results obtained on NBS cells. Before measurements, auxiliary isothermal shells were specially constructed for each of the cells that were received and insulated copper tempering strips and heaters were attached to the cells and shells in the manner similar to the NBS cells.

2.3 Sample

According to the supplier, the argon sample used in the present work contained less than 1 ppm of impurities. On the basis of mass spectrometric analysis, the sample used in the earlier work was estimated to be not less than 99.998 percent pure[16]. The "slope" of the freezing curve obtained on the earlier sample and the relatively more constant melting points observed with the present sample indicate that the present sample is indeed close to 99.9999 percent pure or better. (See Section 4 for the estimate of the purity based on the present measurements and the ideal solution law).

The three cells described in Section 2.2 were filled directly from a high-pressure gas cylinder of the argon sample. Since it is necessary that both adsorbed and absorbed gases be removed as much as possible from the cells, before filling, the cells were first evacuated to 10^{-4} Pa for a number of days at about 500 °C, then small amounts of the argon sample were used to purge and evacuate the cells about ten times. Before cooling to room temperature, the cells were filled with the argon sample to about three atmosphere pressure so that, if any gas is absorbed, it would be argon. The cells were then evacuated and refilled finally to 9 MPa; the filling tube was pinched and was then severed by arc welding.

2.4 Thermometers

A total of six capsule SPRT's with NBS-IPTS-68 calibrations[17] were used. Three of the SPRT's (1812279, 1812282, 1812284), which were furnished recently to the National Physical Laboratory (NPL, U.K.) for comparison of various national temperature scales[18], had been calibrated three times before shipping to NPL and two times after they were returned. Also, these three SPRT's have calibrations on the NPL-IPTS-68. Two SPRT's (1774092 and 1774095) had been calibrated twice, the first at about the time of the earlier investigation of the triple point of argon[16] using a relatively large sealed system and the second just before the present series of investigations with miniature sealed cells. One SPRT (1842382) was calibrated only once. Table I summarizes the history of calibration of the SPRT's used in this work. The deviations are given relative to the latest calibration which was used to convert the readings to values of temperature. The calibrations on SPRT's 1774092 and 1774095 which have been used only at the NBS are extremely stable. The first calibrations on SPRT's 1812279, 1812282, and 1812284 are in closer agreement with the latest calibrations than the second and third calibrations, which themselves agree closely. Because of these fairly large differences between the times the SPRT's were shipped to NPL and returned, there arises an uncertainty in the comparison of the NBS and NPL scales of about 0.1 to 0.2 mK. However, on the whole, Table I shows that the calibrations have been made highly consistently and reproducibly over a number of years.

The capsule SPRT's were installed in the argon cells as shown in Figs. 1 and 2. To obtain the resistance ratio W (i.e., the resistance at the argon triple point to that at 0 °C) the capsule SPRT's were installed in special holders[15] for resistance measurements at the triple point of water (TP). All analyses of measurements are made in terms of the resistance ratio W.

2.5 Resistance Measurements

The resistances of the SPRT's were determined by means of a Guildline Model 9975 current comparator bridge[19] in conjunction with a 10-ohm reference resistor and a strip-chart recorder. The recorder tracings

TABLE I

Calibration History of Thermometers

Thermometer	Calibration Date and Deviation at 83.8 K in mK from Latest Calibration				
	2/2/77*	2/1/71			
1774092	0	+0.01			
1774095	0	−0.02			
	7/27/76*	6/18/76	8/29/74	8/9/74	7/18/74
1812279	0	−0.05	−0.15	−0.18	−0.09
1812282	0	−0.02	−0.15	−0.15	−0.04
1812284	0	−0.07	−0.16	−0.20	−0.05
	1/10/78*				
1842382	0				

*Latest calibration; hence, zero deviation is shown.

were estimated to about 0.3 mm. The amplifiers of both the bridge and the recorder were adjusted so that 1 mm on the chart corresponded to about 3×10^{-7} ohms. The reference resistor was thermostated at approximately 28 °C, near the maximum of its resistance versus temperature curve. The variation in the temperature of the resistor was estimated to be about ±2 mK on the basis of measurements on an industrial-type 100-ohm platinum resistance thermometer in the same thermostated enclosure. The principal problem with the reference resistor, when resistance measurements were being made at levels of 10^{-6} ohms or smaller, was the drift in its resistance with time. To circumvent the problem of drift in the reference resistor, the capsule SPRT's were calibrated at the TP just prior or very soon after the measurements on an argon cell were completed. The resistance ratios for the SPRT were calculated from the actually observed R(TP), if the date of its observation was close to that on which the triple point of argon was measured. Otherwise, the ratio was calculated from a value in a plot of R(TP) versus date that corresponded to the date of observation at the triple point of argon.

Since the temperature of the reference resistor was controlled near the maximum of its resistance versus temperature curve, the effect of small power differences on its resistance was considered to be negligible. To make negligible the error that would arise from variations in self heating in the SPRT's, measurements were made at two currents (1 and √2 mA) and the value at zero power (zero current) was obtained by linear extrapolation of the power dissipated in the SPRT. The validity of this linear extrapolation was checked, whenever a new installation of the argon cell was made in the calorimetric cryostat, by measurements at three currents (1, √2, and 3 mA).

2.6 Energy Measurements

A 12-volt lead storage battery was the source of electrical energy for the heater wound on the argon cell. During the heating period, the voltage across the heater and the current were determined using a digital voltmeter with seven-digit readout. The current was determined from a measurement of the voltage across a reference resistor of known value connected in series with the heater. The time interval of heating was determined with an analog clock operated using the NBS 60-Hz frequency standard. The clock was operated synchronously with the switch that turned the current on and off to the heater. (For details of circuitry see reference [20].) The energy increments that were introduced were relatively large (200 to 600 J) for the determination of heat capacity or heat of fusion of the sample but only about 10 to 30 J during measurements of the equilibrium temperatures at various small amounts of sample melted.

3. MEASUREMENT PROCEDURE AND METHOD OF ANALYSIS

The measurement procedure mentioned earlier was very similar to that used in the past in calorimetric determinations of purity and triple points of substances [11]. Briefly, with helium gas for heat exchange in the vacuum space and liquid nitrogen in the Dewar vessel (see Fig. 1), the argon sample cell was cooled to about 78 K. The helium gas was then pumped out to a high vacuum (about 0.001 Pa) and the automatic temperature control equipment was switched on to control the adiabatic shield at the temperature of the auxiliary isothermal shell of the sample cell. At about the same time, the automatic temperature control was switched on to control the temperature of the floating cylinder (E) at the temperature of the adiabatic shield. After temperature equilibrium was established, three heat capacity "points" were obtained at successively increasing temperatures below the triple point. Measurements were arranged so that the final temperature of the third heat capacity point would be at most 1 or 2 K below the triple point. The amount of electrical energy required to reach 1 or 2 K below the triple point was determined from the first two heat-capacity values. Then, to determine the heat of fusion of the argon sample, electrical energy was introduced continuously from the equilibrium temperature just below the triple point to just above it. After melting the sample, three heat capacity points were obtained at successively increasing temperatures in the liquid phase of argon. From the plot of the two sets of three heat capacity points the temperature dependence of the heat capacity of the argon cell plus sample and of the auxiliary isothermal shell in the temperature regions just below and just above the triple point was established. The heat of fusion L was then calculated from the total heat Q introduced from the temperature T_i just below the triple point to the temperature T_j just above it and the heat capacities in the two regions according to the relation:

$$L = Q - C_s(T_{tp} - T_i) - C_\ell(T_j - T_{tp}) \quad (1)$$

where C_s is the heat capacity in the region of solid argon at temperature $(T_{tp}+T_i)/2$, C_ℓ is the heat capacity in the region of liquid argon at temperature $(T_j+T_{tp})/2$, and T_{tp} is the triple-point value of argon.

To determine the triple point as a function of known amounts of argon melted, the sample was first cooled to a temperature below the triple point and adiabatic conditions established as previously described. After determining the equilibrium temperature T_1, the amount of heat Q_1 required to melt the desired initial fraction M_1 was calculated according to

$$Q_1 = M_1 \times L + C_s(T_{tp} - T_1), \quad (2)$$

where C_s is the heat capacity of the system in the region of solid argon at temperature $(T_{tp}+T_1)/2$. After the equilibrium temperature was determined at the fraction M_1 melted, energy Q_2 was added to obtain the equilibrium triple point at the fraction M_2 melted according to

$$Q_2 = (M_2 - M_1)L. \quad (3)$$

The energies to be added at other fractions melted were determined in the same manner. Usually, to simplify the procedure, the subsequent fractional increments melted were maintained the same by introducing energy Q_2 as in Equation (3). The electric power used was about 0.1 W during melting experiments and about 0.2 W during heat capacity determinations.

In experiments where the sample was initially partially frozen (i.e., when the sample was "cooled into the triple point"), the fractions melted corresponding to the observed equilibrium temperatures were determined after melting what was frozen. Hence, if T_2 is the equilibrium temperature just above the triple point after melting what was frozen, then the fraction M_1 melted that corresponds to the initial equilibrium temperature T_1 is

$$M_1 = [L - \Sigma Q_i - C_\ell(T_2 - T_{tp})]/L, \quad (4)$$

where ΣQ_i is the sum of energy increments added to completely melt the sample from temperature T_1 and C_ℓ is the heat capacity in the region of liquid argon at $(T_2 + T_{tp})/2$. The fractions melted at other energy increments that were introduced were calculated by reference to Equation (3), i.e.,

$$M_2 = M_1 + Q_2/L, \quad (5)$$

$$M_3 = M_2 + Q_3/L, \quad (6)$$

and similarly for other fractions melted. Again, to simplify the measurements and calculation, Q_2, Q_3, and other energy increments were about the same.

Temperatures were observed continuously starting approximately 15 minutes after the end of heating, since preliminary test measurements showed that temperature equilibrium was nearly reached in about 10 to 12 minutes after the end of heating. Usually the observations were made over 1 to 3 hours to make certain that temperature equilibrium was reached. Every observation included equilibrium measurements at 1 and $\sqrt{2}$ mA (occasionally at also 3 mA) and measurements of the bridge zero. For comparison, occasionally observations were made or repeated after overnight equilibration, and on a couple of occasions after equilibration over the weekend. The temperature differences were less than 0.1 mK when more than 10 percent was melted.

In this method, equilibrium temperatures are observed at various known fractions of argon melted. For an "ideal" system, the equilibrium between a liquid mixture and a pure solid phase is given by

$$\ln \frac{\lambda_\ell^o(T)}{\lambda_\ell(T)} = \frac{\Delta H}{R}\left(\frac{1}{T} - \frac{1}{T_o}\right), \quad (7)$$

where $\lambda_\ell^o(T)$ is the absolute activity of the pure liquid in equilibrium with the pure solid at temperature T and similarly $\lambda_\ell(T)$ the absolute activity of the liquid in the mixture[21]. T_o is the freezing point of the pure liquid, R is the molar gas constant, and ΔH is the average molar heat of fusion over the temperature range T to T_o. Since $\lambda_\ell^o(T)/\lambda_\ell(T)$ is always greater than unity, T is less than T_o. (Note: If there is solid solution formation, T is greater than T_o when the impurity concentration is higher in the solid than in the liquid.) From Gibbs-Duhem relation, for an ideal liquid mixture

$$\lambda_1 = (1 - x)\lambda_1^o, \quad (8)$$

where

$$x = N_2/(N_1 + N_2) \quad (9)$$

and N_1 is the number of moles of the major component (argon) and N_2 the number of moles of the minor component (impurities). Equation (7) may then be rewritten:

$$-\ln(1-x) = \frac{\Delta H}{RTT_o}(T_o - T). \quad (10)$$

When x is small (i.e., when the argon sample is highly pure), TT_o and T_o^2 are nearly equal and Equation (10) approximates to

$$x = \frac{\Delta H}{RT_o^2}(T_o - T). \quad (11)$$

For argon, $\Delta H/RT_o^2$ is approximately 0.0201 K^{-1}; hence,

$$x = 0.0201(T_o - T). \quad (12)$$

If it is assumed that the 1-ppm impurity claimed for the argon sample is in terms of volume, then $x=10^{-6}$; and (T_o-T) equals 0.05 mK. If it is assumed further that impurities remain segregated in the liquid, then at 10 percent melted, $x=10^{-5}$ and $(T_o-T)=0.5$ mK. Hence, by observing the equilibrium temperatures at different known amounts of argon melted, the impurity concentration in liquid argon, when completely melted, and the triple point T_o of pure argon may be evaluated.

In Equation (12) the impurity concentration given by mole fraction x varies inversely with fraction F melted; i.e., $x=x_o/F$, where x_o is the concentration when argon is completely melted. Substituting in Equation (12),

$$\frac{x_o}{F} = 0.0201(T_o - T), \quad (13)$$

or

$$(T_o - T) = \frac{49.8 x_o}{F} \quad (14)$$

The impurity concentration x_o is obtained from the slope of the plot of the observed equilibrium temperature versus 1/F. At $1/F=0$, $T=T_o$ the triple point of pure argon. The temperature at $1/F=1$ is that value when a vanishingly small amount of solid is present in equilibrium.

4. RESULTS AND DISCUSSION

The observed equilibrium temperatures that correspond to various percentages and reciprocals of fractions of argon melted are represented by the plots in Figs. 3, 4, and 5. To conserve space, the numerical values that correspond to the plotted values are not given here. Also, six other similar plots are not shown. The above figures represent results obtained on the three argon cells that are described in Section 2.2. Some of the observations at the small percentage melted were outside of the plot. Since they were not used in the analysis of the results presented in the present paper, no attempt was made to plot them. The straight lines

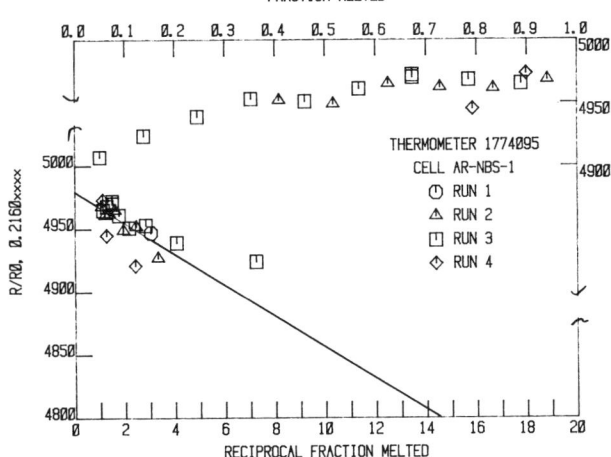

FIG. 3. Values of resistance ratios (W) at equilibrium temperatures of various amounts of sample melted; for thermometer 1774095, cell Ar-NBS-1, Cryostat S. The upper part of the figure shows the values of W (scale on the right) at various fractions of sample melted. The lower part shows the same values of W (scale on the left) plotted in terms of the reciprocals of the fractions melted.

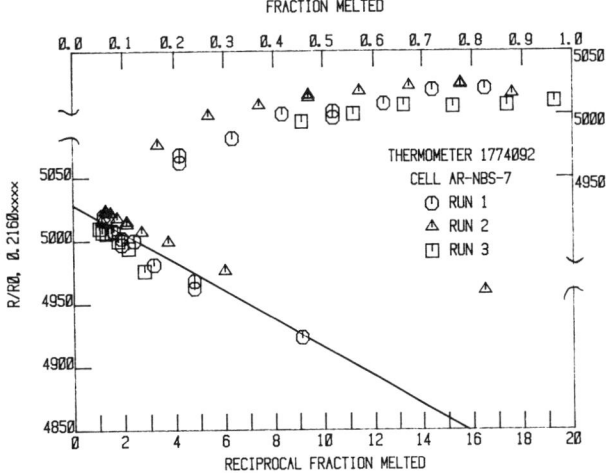

FIG. 5. Values of resistance ratios (W) at equilibrium temperatures of various amounts of sample melted; for thermometer 1774092, cell Ar-NBS-7, Cryostat T. The upper part of the figure shows the values of W (scale on the right) at various fractions of sample melted. The lower part shows the same values of W (scale on the left) plotted in terms of the reciprocals of the fractions melted.

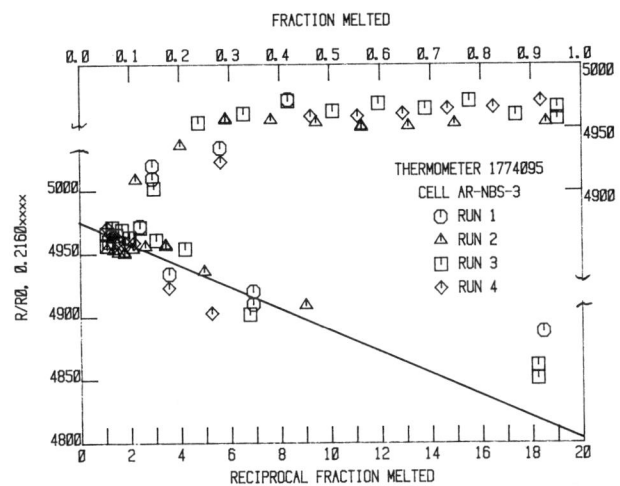

FIG. 4. Values of resistance ratios at equilibrium temperatures of various amounts of sample melted; for thermometer 1774095, cell Ar-NBS-3, Cryostat S. The upper part of the figure shows the values of W (scale on the right) at various fractions of sample melted. The lower part shows the same values of W (scale on the left) plotted in terms of the reciprocals of the fractions melted.

drawn for the T versus 1/F plots were obtained by fitting the data by means of the least squares method. The results of these analyses and those six other sets of similar measurements that are not shown are summarized in Table II. The values of W at 1/F=1 and 1/F=0 were converted to values of temperature employing the latest calibrations in terms of the NBS-IPTS-68[17] for the respective SPRT (see Section 2.4). Measurements (four more sets similar to those represented in Figs. 3 to 5) obtained on argon cells received from IMGC (Italy), NRC (Canada), and NRLM (Japan) are given in Table II for comparison.

The difference between the value 83.8003 K given in Table II and the value 83.798 K selected for the argon triple point in the 1975 amended edition of the IPTS-68 arises largely from temperature scale differences, because of the manner in which the NBS-IPTS-68 was devised to approximate the IPTS-68[22,17] and because the value 83.798 K is based on the best realization of the IPTS-68 and of the argon triple point at the time of the 1975 revision. [As an example (see reference [22]), in devising NBS-IPTS-68 the NBS realized oxygen boiling point was lowered by 1.9 mK.] Recently the NBS-IPTS-68 and the NPL-IPTS-68 were compared, the latter scale being based on the fixed points of the IPTS-68. The converted values of argon triple point temperatures in terms of the NPL-IPTS-68 (see Table II) were obtained from the published results of a comparison between the scale and the NBS-IPTS-68 [18]. Since the reproducibility of the argon triple point in different laboratories is expected to be about 0.1 to 0.2 mK, the difference between 83.797 K and 83.798 K is considered to arise also principally from differences in the realizations of the IPTS-68 based on the fixed points. As shown in Table I for 83.8 K, the NBS-IPTS-68 is considered to be more stable than the differences between the realizations of the IPTS-68 in different laboratories (see reference [18] for scale comparisons). It is expected that the NBS-IPTS-68[17] will be retained until a new IPTS is adopted possibly in 1988.

The values of the mole fraction of impurity given in the last column of Table II show generally close agreement for the three

TABLE II

Comparison of Argon Triple Point Measurements

Cell	Cryostat	Date[a]	Thermometer	Triple Point T,K(NBS-IPTS-68)		Deviation from Mean (1/F=0) mK	Impurity Mole Fraction 10^{-7}
				1/F=1	1/F=0		
AR-NBS-1	S	3/79	1774095	83.8002_{70}	83.8002_{97}	-0.0_{42}	5.7
Ar-NBS-3	S	8/78	1774092[b]	83.8002_{90}	83.8003_{06}	-0.0_{33}	3.4
Ar-NBS-3	S	8/78	1774095[b]	83.8002_{67}	83.8002_{88}	-0.0_{51}	4.0
Ar-NBS-3	S	11/78	1812279[c]	83.8002_{72}	83.8002_{90}	-0.0_{49}	3.6
Ar-NBS-3	S	11/78	1812282[c]	83.8002_{88}	83.8003_{06}	-0.0_{33}	4.1
Ar-NBS-3	T	1/79	1812279[d]	83.8003_{75}	83.8004_{03}	$+0.0_{64}$	5.3
Ar-NBS-3	T	1/79	1812282[d]	83.8003_{96}	83.8004_{19}	$+0.0_{80}$	4.9
Ar-NBS-3	T	1/79	1812284[d]	83.8003_{16}	83.8003_{41}	$+0.0_{02}$	5.0
Ar-NBS-7	T	5/79	1774092	83.8003_{75}	83.8004_{01}	$+0.0_{62}$	5.3
			Mean:	83.8003_{17}	83.8003_{39}		
				83.7970	83.7970 (NPL-IPTS-68)		
Ar-IMGC[e]	T	8/79	1774095[f]	83.8002_{37}	83.8002_{86}	-0.0_{53}	9.6
Ar-IMGC	T	8/79	1842382[f]	83.8003_{22}	83.8003_{71}	$+0.0_{32}$	9.7
Ar-NRC-10[e]	T	2/81	1774095	83.8002_{40}	83.8002_{56}	-0.0_{83}	3.2
Ar-NRLM-7801[e]	T	7/81	1774095	83.8002_{60}	83.8002_{70}	-0.0_{69}	1.8

[a] Month and year when measurements were completed.
[b] Thermometers 1774092 and 1774095 were together in cell Ar-NBS-3 in Cryostat S.
[c] Thermometers 1812279 and 1812282 were together in cell Ar-NBS-3 in Cryostat S.
[d] Thermometers 1812279, 1812282, and 1812284 were together in cell Ar-NBS-3 in Croystat T.
[e] IMGC = Institute of Metrology, "G. Colonetti" (Italy).
NRC = National Research Council (Canada)
NRLM = National Research Laboratory for Metrology (Japan).
[f] Thermometers 1774095 and 1842382 were together in cell Ar-IMGC in Cryostat T.

cells; i.e., the argon gas purity was closely retained in the filling process. The calculated impurity values based on the equilibrium temperature measurements indicate greater purity than the 99.9999 percent claimed by the supplier. The formation of a solid solution by the impurities with argon may be the cause of the apparently lower impurity concentration. The analysis of the data in terms of a solid solution will be the subject of a future paper.

The purities of argon in the NRC and NRLM cells are comparable to that used in the three NBS cells. The purity of argon in the IMGC cell is, however, somewhat lower (about 99.9999 percent according to equilibrium measurements).

The plots of equilibrium temperatures versus percentage melted show that above 20 or 30 percent melted the observed equilibrium temperatures change very little and that they are well within ±0.05 mK of their mean value. This arises from the high purity of the argon used. This would not be the case if the sample were less pure; e.g., referring to Equation (14), if the mole fraction impurity content x_0 is 10^{-5}, the equilibrium temperature at 50 percent melted would be approximately 0.4 mK lower than that at 90 percent melted. Also, because of the high purity of the sample the values of temperatures at 1/F=0 and 1/F=1 do not differ by more than 0.05 mK.

The results obtained in cryostat T seem on the average about 0.09 mK higher than those obtained with cryostat S. This deviation is greater than the differences obtained with different cells in either cryostat. The more recent measurements in cryostat T with argon cells supplied by IMGC, NRC, and NRLM show close agreement with values obtained in cryostat S. Hence, the relatively higher values with cryostat T are probably the result of small inconsistencies in the measurement system. One of the possibilities is that the reference resistor was at different temperatures when the measurements at the argon triple point and those at the TP of water were made. If there was a relatively long interruption in the power line, it is our normal practice to allow about one week for the resistor to come to equilibrium and make a new set of measurements at the TP of water. If there was a significant power line interruption, it occurred in our absence and was overlooked.

5. CONCLUSION

The results of the measurements show that the triple point of argon can be reproduced within about ±0.05 mK in a calorimetric cryostat. Several SPRT's can be effectively incorporated in the same sealed cell. Although in the measurements presented, the numerical value of the heat of fusion for the amount of argon sample used was about the same or slightly larger than the heat capacity of the sealed cell system, it is expected that equally accurate measurements can be obtained with the heat of fusion about one half or one third of the heat capacity. Also, the sample volume of the cell could be about 20 or 25 cm^3. For best results the cell should be large enough and designed so that the condensed sample would surround as much of the thermometer as possible. Cells of smaller volume are planned for future investigations of the triple points of other gases.

Where samples are highly pure (about 99.9999 percent or higher), the equilibrium temperatures in the range 30 to 90 percent melted agree within ±0.1 mK; these values of temperature agree also within the same temperature limit with calculated values at 1/F=1 or 1/F=0. Where samples are less pure, the calculated T at 1/F=0 would be expected to agree closely with temperature values obtained with very pure samples. However, because of possible solid solution formation and possible deviation from equilibrium of the measurements of samples with impurities, the results that are obtained by such procedures with samples of low purity should be considered cautiously.

Measurements of high precision can be obtained with sealed cells of pure gases. For comparison of results obtained over many months, continuous monitoring of reference devices should be part of the measurement process.

ACKNOWLEDGEMENT

The author is grateful to M. L. Reilly for help in the computer analysis of the data.

DISCLAIMER

Certain commercial equipment, instruments, or materials are identified in this paper in order to adequately specify the experimental procedure. Such identification does not imply recommendation or endorsement by the National Bureau of Standards, nor does it imply that the materials or equipment identified are necessarily the best available for the purpose.

REFERENCES

[1] Boiling point is defined as the equilibrium temperature of liquid and vapor phases of a substance at 1 standard atmosphere. Where isotopic composition influences the equilibrium temperature or where very small amounts of impurities are known to be present, the equilibrium temperature at 1 standard atmosphere for the substance can be better defined in terms of a vanishingly small vapor or liquid fraction.

[2] G. K. Burgess, The International Temperature Scale, J. Res. Nat. Bur. Stand. (U.S.) 1, 635 (1928).

[3] H. F. Stimson, International Practical Temperature Scale of 1948. Text revision of 1960, J. Res. Nat. Bur. Stand. (U.S.) 65a, 139 (1961).

[4] The International Practical Temperature Scale of 1968. Amended edition of 1975, Metrologia 12, 7 (1976).

[5] H. F. Stimson, Precision resistance thermometry and fixed points, Temperature Its Measurement and Control in Science and Industry, Vol. 2, (Reinhold Publishing Corp., New York, 1955), 141-168.

[6] J. Ancsin, Dew points, boiling points and triple points of "pure" and impure oxygen, Metrologia 9, 26 (1973).

[7] G. T. Furukawa, W. R. Bigge, and J. L. Riddle, Point triple de l'argon, Comité Consultatif de Thermométrie, T34, 106 (1971).

[8] F. Pavese, Realization of the IPTS-68 between 54.361 and 273.15 K and the triple points of oxygen and argon, Temperature Measurement 1975, Inst. Phys. Conf. Ser. 26, 70 (1975).

[9] G. Bonnier, Point triple de l'argon (83.798 K) référence de transfert, Bulletin BNM 6, 14 (1975).

[10] G. T. Furukawa, Réalization du point triple de l'argon dans une cellule scellée, Comité Consultatif de Thermométrie, T13, 100 (1978).

[11] A. R. Glasgow, Jr., G. S. Ross, A. T. Horton, D. Enagonio, H. D. Dixon, and C. P. Saylor and G. T. Furukawa, M. L. Reilly, and J. M. Henning, Comparison of cryoscopic determinations of purity of benzene by thermometric and calorimetric procedures, Anal. Chim. Acta 17, 54 (1957).

[12] G. T. Furukawa and M. L. Reilly, Heat capacity and thermodnamic properties of α-beryllium nitride, Be_3N_2, from 20 to 315 K, J. Res. Nat. Bur. Stand. (U.S.) 74A, 617 (1970).

[13] B. Persoz, Nouvelles méthodes de mesure de la chaleur spécifique vraie des métaux a haute température, Ann. Phys. 14, 237 (1940).

[14] E. D. West and D. C. Ginnings, An adiabatic calorimeter for the range 30 to 500 °C, J. Res. Nat. Bur. Stand. (U.S.) 60, 309 (1958).

[15] J. L. Riddle, G. T. Furukawa, and H. H. Plumb, Platinum resistance thermometry Nat. Bur. Stand. (U.S.), Monograph 126, 1973, 129 pages.

[16] G. T. Furukawa, W. R. Bigge, and J. L. Riddle, Triple point of argon, Temperature Its Measurement and Control in Science and Industry, Vol 4 (Instrument Society of America, Pittsburgh, 1972), 231-243.

[17] G. T. Furukawa, J. L. Riddle, and W. R. Bigge, The International Practical Temperature Scale of 1968 in the region 13.81 K to 90.188 K as maintained at the National Bureau of Standards, J. Res. Nat. Bur. Stand. (U.S.) 77A, 309 (1973).

[18] S. D. Ward and J. P. Compton, Intercomparison of platinum resistance thermometers and T_{68} calibrations, Metrologia 15, 31 (1979).

[19] N. L. Kusters, M. P. MacMartin, and R. J. Berry, Resistance thermometry with the direct current comparator, Temperature Its Measurements and Control in Science and Industry, Vol. 4 (Instrument Society of America, Pittsburgh, 1972), 1477-1485.

[20] E. F. Westrum, Jr., G. T. Furukawa, and J. P. McCullough, Adiabatic low temperature calorimetry, Experimental Thermodynamics, Vol. I, Calorimetry of Non-Reacting Systems (Plenum Press, New York, 1968), Chapter 5.

[21] E. A. Guggenheim, Thermodynamics, an advanced treatment for chemists and physicists (Interscience Publishers, Inc., New York, 1949), Chap. V.

[22] R. E. Bedford, M. Durieux, R. Muijlwijk, and C. R. Barber, Relationships between the International Practical Temperature Scale of 1968 and the NBS-55, NPL-61, PRMI-54, and PSU-54 temperature scales in the range from 13.81 to 90.188 K, Metrologia 5, 47 (1969).

The triple points of equilibrium and normal deuterium

R. C. Kemp

CSIRO Division of Applied Physics, Sydney, Australia 2070

The realization of the triple points of equilibrium and normal deuterium are described. It is concluded that the triple point of equilibrium deuterium would be suitable for use as a reference temperature on the International Practical Temperature Scale.

INTRODUCTION

The International Practical Temperature Scale of 1968 (amended edition 1975)[1] (IPTS-68) in the region below 273.15 K relies on interpolation between a number of reference temperatures or defining fixed points. The reference temperatures are boiling and triple points of pure substances. Triple points are easier to realise than boiling points and do not require an accurate pressure measurement. This was recognized by the introduction in 1975 of the argon triple point at 83.798 K as an alternative to the oxygen boiling point at 90.188 K. Similarly a possible alternative to the boiling point of neon at 27.102 K is the triple point of neon at 24.56 K. The remaining boiling points are those of equilibrium hydrogen at 20.28 K and 17.042 K measured at one atmosphere and 25/76 of one atmosphere respectively. It would be convenient if these points could be replaced with a suitable triple point. The only one in this region appears to be that of deuterium at 18.7 K about which little is known. This paper describes realizations of the triple points of both equilibrium and normal deuterium.

RESULTS AND DISCUSSION

In a similar manner to hydrogen, deuterium has the two modifications: orthodeuterium and paradeuterium. At 300 K equilibrium deuterium consists of 2/3 orthodeuterium and 1/3 paradeuterium and this is referred to as the normal composition whereas at the triple point the equilibrium composition is 98.86% orthodeuterium and 1.14% paradeuterium. Deuterium having this composition is thus in equilibrium at the triple point temperature. The use of equilibrium deuterium for a fixed point is to be preferred, as once having reached equilibrium the triple point temperature will not be affected by the further ortho-para conversion. It was decided therefore to initially investigate equilibrium deuterium.

The apparatus used has been described previously[2]; however, a brief description is useful. The chamber containing the sample is surrounded by a temperature-controlled radiation shield and this assembly is enclosed in an evacuated can whose temperature is also controlled. Heat leaks to the sample chamber are reduced to a very low level by adjusting the temperature of the radiation shield and vacuum can. To ensure that the deuterium in the sample chamber had the equilibrium composition the sample chamber was partially filled with a catalyst for the paradeuterium to orthodeuterium conversion. The catalyst used was hydrous ferric oxide which was activated by maintaining it at 120°C under vacuum for 48 hours. The design of the sample chamber had to be modified to allow the catalyst to be added and in fact a cell from an earlier apparatus[3] was used.

To realize the triple point the frozen sample was first heated continuously until its temperature was about 0.1 K below the triple point. The sample was then heated to the triple point and melted by applying heat in small pulses lasting for 3 minutes equal to about 6% of the heat of transition. After each heat pulse the sample chamber was allowed to reach equilibrium and the temperature was then recorded as representing a point on the melting curve. The melting curve is extremely flat, the melting range being about 0.2 mK. It is also reproducible to better than ± 0.1 mK. The triple point

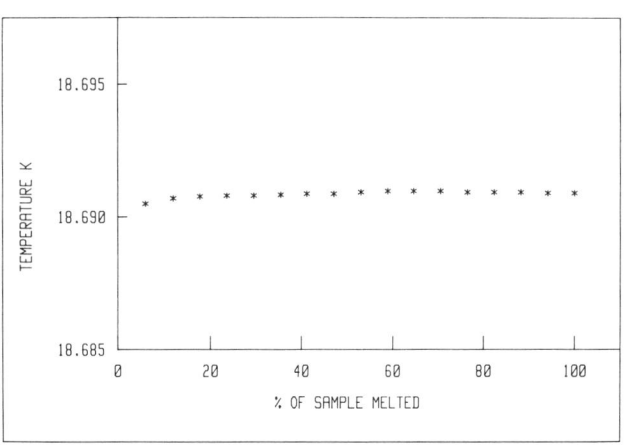

FIG. 1. Melting curve of equilibrium deuterium

temperature is 18.6909 K on the version of IPTS-68 maintained at the NPL at the time of the recent international intercomparison of platinum resistance thermometers[4]. This version agrees very closely with scale maintained at the Division of Applied Physics. A recent determination by Pavese[6] gives the temperature of this point as 18.678 K, some 13 mK lower than this work. This is a very large difference and may be due either to incomplete conversion of the sample used in this work or the presence of impurities. In assessing the uncertainty of realizing the triple point temperature there are two main sources of error to be considered. Firstly, whether the deuterium has in fact attained the equilibrium composition and secondly, the effects of any impurities present. The sample was maintained in intimate contact with activated catalyst throughout the experiment. It has been established[5] that the conversion of paradeuterium to orthodeuterium in the presence of a suitable catalyst is extremely rapid so that the sample is almost certainly in equilibrium, provided of course that the catalyst is effective. In any case small departures from equilibrium may not be too important. According to this work, the difference in temperature between the triple points of normal and equilibrium deuterium is about 18 mK which corresponds to a change in the orthodeuterium component from 66.7% to 98.9% or about a 1.7% change in composition per mK. The triple point is thus relatively insensitive to small changes in the ortho-para ratio.

The effect of impurities is difficult to establish. Firstly, in terms of the purities of the gases used in our other investigations on the IPTS, high purity deuterium was not obtainable. The sample used was supplied by Matheson Gas Products and was specified as having a deuterium content of not less than 99.5%; no impurity analysis was supplied. As the impurity content may have been quite large efforts were made to remove some of the possible impurities as follows. The volatile impurities helium and hydrogen may remain in the gaseous phase as the deuterium freezes. The vapour above a particular sample of frozen deuterium was pumped away which should remove any impurities in the gaseous phase.

Non-volatile impurities such as neon, oxygen and nitrogen are probably trapped on the cold surfaces of the filling tube before entering the sample chamber. This fact was utilized to remove or reduce the amount of such non-volatile impurities by withdrawing deuterium from the cold sample chamber into a 1 litre reservoir. The sample chamber was then heated and pumped out, in hopes of removing the trapped non-volatile impurities. The pumped sample was then recondensed into the cell.

The triple point temperature of this sample was, rather disappointingly, unchanged from the original value. Either the procedure was unsuccessful in removing impurities or those impurities removed did not effect the triple point temperature. This result is rather inconclusive and it is desirable that a detailed study of the effects of impurities on the triple point should be made.

The same apparatus, except that the sample chamber was unchanged and contained no catalyst, was used to realize the triple point of normal deuterium. The melting curve so obtained was not as flat as that of equilibrium deuterium and not very reproducible. This was attributed to some conversion to the ortho modification occurring in the cell. The melting curve obtained consisted of a region at the start where the melting temperature rose linearly until it reached a plateau. This is attributed to conversion to the ortho phase occurring initially where the deuterium is in contact with the surfaces of the sample chamber and lowering the melting temperature at the start of the plateau. The final part of the plateau is, on this reasoning, more characteristic of the unconverted or normal deuterium and from this part of the plateau we deduced a triple point temperature of (18.709 ± 0.001) K on the same version of IPTS-68. This may be compared with the value of 18.729 K reported by Pavese[7]. There is again a large difference of 20 mK which may have been due to partial conversion of the sample.

CONCLUSION

The melting curve of equilibrium reported here is flat and reproducible and appears promising for use as a reference temperature on the IPTS-68. However the large differences between this and other realizations must be investigated before this point can be recommended.

The melting curve of normal deuterium, in the apparatus used, cannot be considered as a reference temperature because of partial conversion to the equilibrium composition.

REFERENCES

1. "The International Practical Temperature Scale of 1968" (Amended Edition 1975) Metrologia **12**, 7 (1976).
2. J. A. Cowan, R. C. Kemp, and W. R. G. Kemp, Metrologia **14**, 83 (1978).
3. W. R. G. Kemp and C. P. Pickup, TIMCSI Vol 4, Part 1, 217.
4. S. D. Ward and J. P. Compton, Metrologia 15, 31 (1979).
5. A. Farkas, Orthohydrogen, Parahydrogen and Heavy Hydrogen (Cambridge University Press 1935) p.160.
6. F. Pavese, Private communication.
7. F. Pavese, CCT 80 Doc. 24.

Superconductive thermometric fixed points

J. F. Schooley and R. J. Soulen, Jr.

Center for Absolute Physical Quantities, National Bureau of Standards, Washington, D.C. 20234

We review the progress since the 5th Temperature Symposium in the development of temperature reference points based upon the transitions of various metal samples between the normal and superconductive states. Two superconductive fixed point devices, known as SRM (Standard Reference Material) 767 and 768, have become available from the National Bureau of Standards. One of these devices provides five of the temperature reference points for the 1976 Provisional 0.5 to 30 K Temperature Scale; the other provides the mechanism for transmitting an NBS cryogenic temperature scale covering the range 0.01 to 0.5 K. Current efforts in superconductive fixed-point research are devoted to evaluating superconductive transitions as possible reference temperatures for a replacement scale to succeed the IPTS-68.

INTRODUCTION

During the 5th Temperature Symposium, we presented the results of initial measurements on the superconductive transitions occurring in thirteen elements, compounds, and alloys[1]. Their superconductive transition temperatures in the absence of magnetic fields (T_c) ranged from 18.5 K to 0.1 K. We observed the superconductive transitions of the samples by monitoring the mutual inductance of coil pairs enclosing them; this technique was chosen for its relative ease of execution and to minimize sample stress and contamination[2]. The widths of the superconductive transitions of these samples varied widely, from less than one millikelvin for very pure, well-annealed, low-melting elements such as lead and indium to 2 kelvins for a sample of $Nb_{52}Ti_{48}$ alloy in an "as-received" condition. In the first paper[1], we also presented results obtained with several samples of Pb, In, Al, Zn, and Cd. These samples were prepared from materials of different purities; they were cast in vacuum and annealed for various times near their melting points. From our results, we concluded that these five elements could be employed as temperature reference points with irreproducibilities as small as 1 mK.

In this paper, we shall review the fabrication of the two multi-sample fixed point devices resulting from these initial experiments, labelled SRM 767 and 768, which are distributed via the NBS Office of Standard Reference Materials. We shall note the employment of these devices in thermometry at low temperatures, and we shall discuss the ultimate usefulness of superconductive transitions as defining temperatures in an IPTS.

SRM 767

A summary of the NBS experience in the development of the SRM 767 devices is contained in a recent publication[3]. Twenty-two prototype devices were constructed and cycled repeatedly to cryogenic temperatures over a period of more than one year prior to the fabrication of standard SRM 767 units. The sample T_c values, defined as the midpoint of the susceptibility change when the sample is warmed from the superconducting state to the normal state, and transition widths, W, defined as the temperature region over which the central 80% of the susceptibility change occurs, were recorded in terms of the resistance of one or more germanium resistance thermometers (GERT). Unsatisfactory samples of each of the five elements were found among the devices; these data have been omitted from the summary analysis given in Table I.

Table I. Analysis of Prototype SRM 767 Devices

Element	Pb	In	Al	Zn	Cd
Number of samples analyzed	19	18	19	20	19
Standard deviation of T_c (mK)	0.2	0.2	0.4	0.4	0.3
Range of transition widths (mK)	0.6 to 2.0	0.5 to 2.5	1.5 to 4.0	2.5 to 10.0	0.5 to 8.0
dT_c/dH (mK/µT)	0.04	0.06	0.06	0.09	0.10

In the measurements summarized above, we expected three major contributions to the overall standard deviation values listed in Table I: variations in our thermometric precision, arising from the use of a dc potentiometric system for measurement of the germanium resistance thermometers; errors in our cancellation of the earth's magnetic field; and sample-to-sample T_c variations. The magnitude of our thermometric imprecision was nearly as large as the listed standard deviation, leaving uncertain the size of the contributions from the other sources. However, our successful efforts to reduce ambient magnetic fields below 1 µT suggest that variations in the measured T_c values from this source are negligible. We note also that the variations observed in the widths of different sample transitions were not strongly reflected in T_c variations.

On the basis of the measurements described above, it was decided that all SRM 767 devices to be released for distribution should exhibit T_c values within ±1 mK of those of standard units retained at the NBS.

More than 100 cryogenic temperature fixed point devices bearing the designation SRM 767 have been distributed through the NBS Office of Standard Reference Materials since 1973. Each device incorporated samples of Pb, In, Al, Zn, and Cd enclosed within a mutual inductance coil pair. Several features should be noted from the publication describing initial work on the device[4].

High purity starting materials were used. The assays supplied with the SRM 767 materials are listed in Table II.

Thermal anchoring of the samples and of the electrical leads proved to be critical to the observation of non-hysteretic and reproducible transitions. Satisfactory results followed the use of stopcock grease in the sample-mounting holes and the application of conductive varnish to the mounted samples, in addition to the varnishing of about 15-cm lengths of each electrical leadwire to the mounting platform.

The use of low frequency (400 Hz or less), low amplitude (1 microtesla or less) measuring fields, and cancellation of the earth's magnetic field to within 1 microtesla during measurements were necessary for accurate fixed-point realizations.

Transition temperatures were defined as the midpoint between the normal and superconductive sample mutual inductance values. In case of a hysteretic transition, the midpoint of the transition during warming was to be used to define T_c.

Transition temperature values were given in terms of then-relevant temperature scales. No measurements of thermodynamic T_c values were implied by the quoted numbers.

Fig. 1 shows the sample arrangement for the SRM 767 device.

Table II. Assays of Bulk Starting Materials for SRM 767 Superconductive Thermometric Fixed Point Devices (ppm by weight)

ZINC SRM 682*		LEAD HPM 9284*		CADMIUM SRM 746*		ALUMINUM HPM 5831*		INDIUM JK 762*	
Cl	< 0.5	Cu	0.2	K	< 4	Si	1.0	Pb	3
O	< 0.5	Cd	0.1	Na	< 3	Cu	0.5	Tl	3
Si	< 0.5	Fe	0.1	O	< 2	Mg	0.5	Sn	1
Ca	< 0.2	Si	0.1	Pb	0.8	Ca	0.2	Cd	< 1
Na	< 0.2	Tl	0.1	Ca	< 0.6	Cr	0.2	Fe	< 1
Ti	< 0.2	Bi	< 0.1	Cr	0.4	Ag	0.1	Cu	nf
Cd	0.1	Ca	< 0.1	C	< 0.1			Ga	nf
Fe	0.1	Mg	< 0.1	Mg	0.1			Ni	nf
K	< 0.1	Ag	< 0.1	Cl	0.2			Ag	nf
Mg	< 0.1			Zn	0.1				
Ni	< 0.1			As	< 0.1				
				Rb	< 0.1				

*All of these materials are available from the NBS Office of Standard Reference Materials.

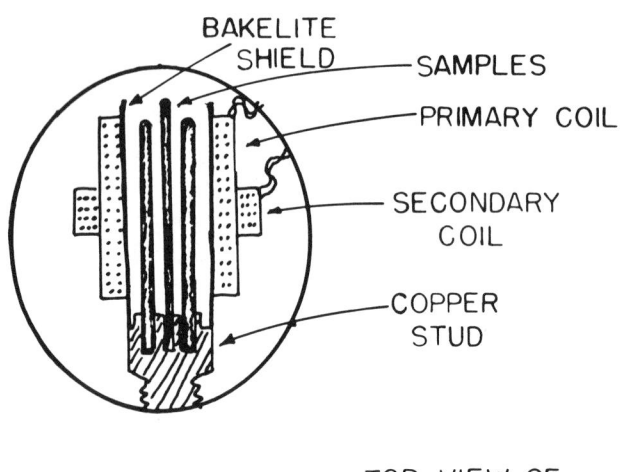
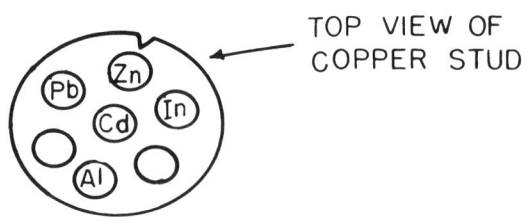

FIG. 1. Schematic of the sample mounting assembly for SRM 767. The mutual inductance coils are wound on bakelite formers. The primary coil contains 400 turns of #38 AWG copper wire and is 2.5 cm long, while the 1 cm long secondary coil contains 2000 turns of #40 AWG copper wire. The insert at the bottom shows the location of the individual samples relative to the indicator notch in the copper stud.

The mutual inductance bridge described in Ref. 3 is the Hartshorn bridge as used for studies of weakly-magnetic systems[5]. For the benefit of users of the SRM 767 device without access to such sensitive mutual inductance bridges, a simpler but equally useful bridge was developed and described [6], and still more recently an improved bridge was designed for this purpose[3]. This last version is shown in Fig. 2.

During the years since it was introduced, the SRM 767 has been studied by scientists working in other laboratories. The use of the SRM 767 for in situ calibrations of thermometers in heat capacity measurements has been discussed by Martin[7]. Cataland and Plumb[8] assigned temperature values on the NBS 2-20 K acoustic thermometer scale to the transition points for the SRM 767 Pb and In samples. Besley and Kemp utilized two of the devices in a comparison of some thirty low-temperature thermometers used in maintaining temperature scales throughout the world[9]. In their study, each of the test thermometers was measured at the Pb, In, and Al transition points. In work reported to the 1980 meeting of the Consultative Committee for Thermometry of the International Committee for Weights and Measures, Inaba and Mitsui observed the behavior of two SRM 767 devices using a rhodium-iron resistance thermometer[10]. The two SRM 767 Pb sample T_c values agreed within 0.2 mK; the In sample T_c values likewise agreed within 0.2 mK.

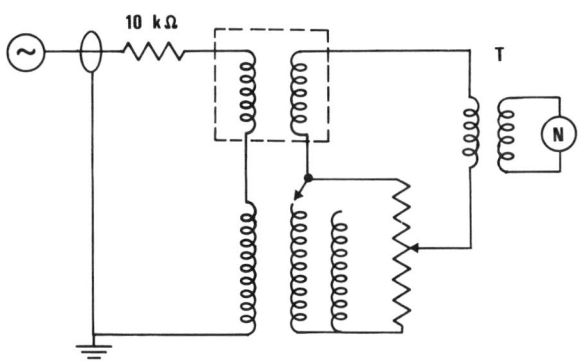

FIG. 2. Mutual inductance bridge for SRM 767 and 768. The circuit enclosed in dashes represents the device coils at cryogenic temperatures. A switch selects one of two secondary coils of the reference mutual inductance; the one with fewer turns provides a reference mutual inductance of 0.22 mH suited for use with SRM 768, the other one gives a mutual inductance of 5.1 mH for use with SRM 767. A ten-turn potentiometer (total resistance, 100 Ω) reduces the voltage developed by the standard mutual inductance until a bridge balance is achieved. A TRIAD G-4 transformer T with a turns ratio of ≅ 90, couples the circuit to a phase-sensitive detector, N. We have used a PAR Model 120 phase-sensitive detector with a PAR Model 112 pre-amplifier as well as preamplifiers and amplifiers of our own design with equal success. A series resistance, here shown as 10 kΩ, limits the magnetic field generated in the primary mutual inductance coil. Provision of a circuit to balance the small out-of-phase component of the voltage has not proved necessary.

The SRM 767 device was incorporated into the 1976 Provisional 0.5 K to 30 K Temperature Scale (EPT-76)[11]. This scale was derived to improve consistency and thermodynamic accuracy in temperature measurement below 30 K. Eleven reference points form the basis for the new scale. Among these are the transition temperatures for the SRM 767 device samples as measured on the Kelvin Thermodynamic Temperature Scale by several workers. The assigned T_c values are shown in Table III.

Table III. T_c values of SRM 767 device transitions as assigned in the 1976 Provisional 0.5 K to 30 K Temperature Scale

Reference Point	Assigned temperature T_{76} (K)
Superconducting transition point of Cd	0.519
Superconducting transition point of Zn	0.851
Superconducting transition point of Al	1.1796
Superconducting transition point of In	3.4145
Superconducting transition point of Pb	7.1999

(Superconducting transition point: the transition temperature between the superconducting and the normal state in zero magnetic field as given by the NBS-SRM 767.)

In a paper on the derivation and development of the EPT-76, the members of Working Group 4 of the Consultative Committee for Thermometry noted that various SRM 767 devices typically showed deviations as large as 0.5 mK with respect to the magnetic scale used in constructing the EPT-76[12]. This value lies within the ±1 mK tolerance set by NBS in producing these devices. However, in view of the importance of precise reference temperatures in any temperature scale the present authors have reconsidered the ±1 mK reproducibility criterion used in the production of SRM 767 devices and have decided to undertake experiments to determine the feasibility of preparing devices with T_c uniformities at the ±0.1 mK to ±0.2 mK level.

In an effort to improve the thermometry capability in the SRM 767 laboratory, the dc potentiometer system used to measure the germanium resistance thermometers has been replaced by an ac decade resistance bridge. In addition, a rhodium-iron resistance thermometer has been incorporated into the experiment; its resistance is measured by means of a new automatic resistance bridge designed by R. Cutkosky[13]. These steps have proved to be effective, as will be shown in a later section of this paper.

Table IV. Properties of Materials Used in SRM 768

Material	Residual Resistivity Ratio	Resistivity at 4.2 K ρ_0, Ω cm	Source	Sample Geometry	Purity
Tungsten	10^3	5×10^{-9}	Aremco Products, Inc. P. O. Box 429 Ossining, New York	Single crystal cylindrical rod, 0.13 cm diameter, 0.64 cm long. Cut from longer rods of same diam.	99.999%, nominal
Beryllium	79[+]	4×10^{-8}	Nuclear Materials, Inc. Cambridge, Mass	Irregular nuggets prepared by single vacuum distillation. Approximate dimensions 0.64 cm by 0.4 cm by 0.4 cm.	Detectable impurities (in ppm) are: Fe(3), Mn(5), Si(10), Al(15), N(5).
Iridium	2.5	4.4×10^{-6}	Material Research Corp. Route 303 Orangeburg, New York	Quarter-pie shapes obtained by cutting disks 0.23 cm thick from cylindrical rod of diameter 0.64 cm. The disk was then quartered. Boule was prepared by electron-beam zone refining.	Actual composition: 80% Ir; 20% Ru
$AuAl_2$	50	2×10^{-7}	Prepared at NBS[++] using 99.999% pure Al and Au	Rectangular parallelepiped, l = 0.66 cm, w = 0.22 cm, t = 0.22 cm. Spark cut from zone-refined boule.	
$AuIn_2$	50	2×10^{-7}	Prepared at NBS[++] using 99.999% pure In and Au	Rectangular parallelepiped, l = 0.66 cm, w = 0.22 cm, t = 0.22 cm. Spark cut from directionally solidified boule.	

[+]Dr. Ray Radebaugh, NBS Boulder, private communication.

[++]Prepared by Mr. Frank Biancaniello.

SRM 768

In view of the utility of the SRM 767 superconductive temperature reference points for cryogenic thermometry, assessing the feasibility of extending the range of superconductive fixed points to lower temperatures was a natural step. Accordingly, Soulen and Dove of the NBS initiated a study of superconductors with T_c values below 0.5 K. They found narrow, reproducible transitions among samples of tungsten, beryllium, iridium, and two compounds, $AuAl_2$ and $AuIn_2$. The corresponding T_c values range from 0.015 K to 0.2 K. Table IV shows the properties of the materials chosen for use in a new fixed-point device designated NBS SRM 768[14].

Owing to the mechanical hardness of these samples, it was not necessary to form them into long cylindrical shapes as was done with the SRM 767 samples. Instead, they were mounted inside a bundle of copper wires as shown in Fig. 3. The use of this technique provides good thermal equilibrium between the samples and the measuring coils even at the lowest transition temperature.

Superconductive transition temperatures of the SRM 768 samples were evaluated on an NBS cryogenic temperature scale, the NBS-CTS-1. This scale was derived from measurements of Johnson noise in a resistor and from nuclear orientation thermometry. Both methods produce temperatures soundly based in thermodynamics. The

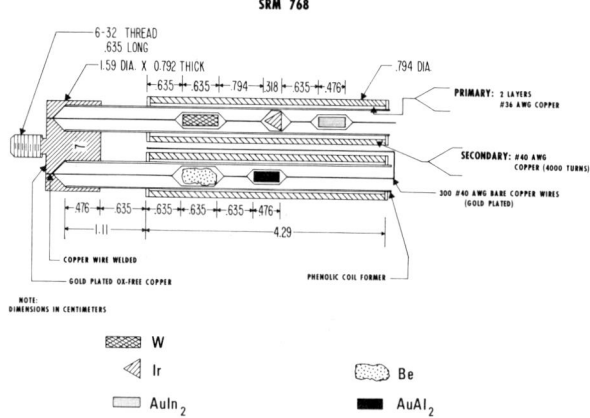

FIG. 3. SRM 768. The device is composed of two similar parts, each consisting of two or three samples bound in copper wires inside a pair of coils (primary and secondary). The four coils are connected in series opposition so as to minimize the total mutual inductance. The serial number of the unit (in this case, 7) is stamped on the end which is terminated with a 6-32 threaded stud designed for attachment to a cryostat. The threads are relieved near the body of the device so that it will bottom properly when screwed in, thereby establishing good thermal contact.

TABLE V

Reproducibility of later SRM 768 Units (Serial No. 7, 8, 9, 10, 11)

SAMPLE	T_c Run 1 4/21/78	T_c Run 2 4/24/78	T_c Run 3 4/28/78	T_c Run 4 5/3/78	T_c Run 5 6/14/78	T_c Run 6 6/20/78	T_c Run 7 6/27/78	T_c Run 8 6/30/78	T_c Run 9 7/6/78	T_c Run 10 7/11/78	T_c Run 11 7/12/78	Runs 1-4 s	Runs 1-4 $\overline{(T_c)}$	Runs 5-11 s	Runs 5-11 $\overline{(T_c)}$	$s/\sqrt{N}$	Width W	Hysteresis
	(mK)	(mK)	(mK)	(mK)	(mK)	(mK)	(mK)	(mK)	(mK)	(mK)	(mK)	(mK)	(mK)	(mK)	(mK)	(mK)	(mK)	(mK)
W-7	16.39	-	16.39	-	16.09	16.09	16.17	16.26	16.22	15.99	15.76	0.00	16.39	0.17	16.08	0.06	0.29	0.07
W-8	16.44	-	16.28	-	16.05	15.85	16.30	16.24	16.17	15.99	15.66	0.11	16.36	0.23	16.03	0.09	0.18	0.07
W-9	16.92	-	16.55	-	16.27	16.21	16.51	16.59	16.48	16.21	16.08	0.29	16.62	0.19	16.34	0.07	1.36	0.14
W-10	16.91	-	16.56	-	16.23	15.75	15.80	15.82	16.01	15.39	15.98	0.25	16.74	0.26	15.85	0.10	1.50	0.04
W-11	17.11	-	16.73	-	16.42	-	16.75	16.67	16.57	16.44	16.21	0.27	16.92	0.19	16.51	0.07	1.50	0.04
Be-7	23.31	23.35	23.19	-	23.06	23.06	23.14	23.02	23.04	23.01	22.97	0.06	23.25	0.03	23.02	0.01	0.13	<0.08
Be-8	23.25	23.29	23.04	-	23.12	23.04	23.08	22.97	23.04	23.04	23.01	0.13	23.19	0.05	23.04	0.02	0.25	<0.06
Be-9	23.12	23.27	22.95	-	22.61	22.41	22.43	22.20	22.39	22.38	22.72	0.16	23.11	0.17	22.44	0.06	0.27	0.27
Be-10	23.67	23.37	23.21	-	23.02	23.06	23.04	23.08	23.16	23.10	23.25	0.23	23.42	0.08	23.10	0.03	0.23	0.22
Be-11	23.33	23.67	23.35	-	23.27	23.23	23.18	22.99	23.04	22.97	23.01	0.19	23.45	0.12	23.10	0.05	0.30	0.18
Ir-7	99.33	99.33	99.36	-	99.41	99.37	99.37	99.21	99.37	99.47	99.40	0.017	99.34	0.08	99.37	0.03	0.83	<0.1
Ir-8	99.23	99.20	99.09	-	99.23	99.09	99.20	99.07	99.17	99.34	99.21	0.074	99.17	0.10	99.19	0.04	0.87	<0.08
Ir-9	99.00	98.96	98.97	-	99.05	98.95	98.98	99.20	99.13	99.11	99.07	0.021	98.98	0.09	99.07	0.03	0.90	0.13
Ir-10	99.07	98.97	98.97	-	99.16	99.03	98.96	98.84	99.15	99.10	99.01	0.058	99.00	0.11	99.03	0.04	1.20	0.21
Ir-11	98.93	98.84	98.90	-	99.01	98.91	98.91	98.72	99.13	99.02	98.88	0.046	98.89	0.13	98.94	0.05	1.10	0.22
AuAl$_2$-7	160.46	160.51	160.48	-	160.53	160.48	160.46	160.26	160.48	160.63	-	0.025	160.48	0.12	160.47	0.05	0.22	<0.05
AuAl$_2$-8	160.49	160.61	160.54	-	160.71	160.56	160.58	160.28	160.61	160.76	-	0.060	160.55	0.17	160.58	0.07	0.35	<0.05
AuAl$_2$-9	160.46	160.61	160.46	-	160.51	160.28	160.43	160.49	160.46	160.61	-	0.086	160.51	0.11	160.46	0.04	0.37	<0.03
AuAl$_2$-10	160.53	160.48	160.49	-	160.49	160.48	160.43	160.33	160.45	160.50	-	0.026	160.50	0.10	160.46	0.04	0.23	<0.03
AuAl$_2$-11	160.40	160.58	160.49	-	160.54	160.49	160.46	160.28	160.53	160.68	-	0.09	160.49	0.13	160.49	0.05	0.28	<0.03
AuIn$_2$-7	-	-	-	-	205.45	205.37	205.30	205.14	205.37	205.53	-			0.13	205.36	0.05	0.42	<0.04
AuIn$_2$-8	-	-	-	-	205.81	205.61	205.61	205.37	205.69	205.85	-			0.17	205.65	0.07	0.40	<0.06
AuIn$_2$-9	-	-	-	-	205.61	205.06	205.14	204.91	205.22	205.41	-			0.25	205.22	0.10	0.38	<0.04
AuIn$_2$-10	-	-	-	-	205.73	205.65	205.57	205.33	205.61	205.81	-			0.16	205.61	0.07	0.34	<0.04
AuIn$_2$-11	-	-	-	-	205.34	204.91	205.26	204.98	205.30	205.45	-			0.21	205.21	0.09	0.30	<0.04

methods were directly compared over the range 0.01 K to 0.05 K; the resulting temperatures agreed within ±0.5%[15]. Above 0.05 K, the noise thermometer was used alone to measure the device temperatures; the ^{60}Co nuclear orientation thermometer loses sensitivity above 0.05 K. In addition, the noise measurements were shown to be consistent with the Curie law as realized using a cerous magnesium nitrate thermometer. The agreement was found to be ±0.3% at 0.03 K and gradually decreased to a value of ±0.2% at the highest temperature (0.52 K). The results of this work were transferred to a calibrated germanium resistance thermometer which maintains NBS-CTS-1.

Five prototype SRM 768 devices (serial numbers 7, 8, 9, 10, 11) were mounted on a copper platform and the reproducibility of the T_c values examined during several thermal cycles. The results are summarized in Table V. The materials identified by a 7 (e.g., W-7, Be-7, etc.) are the ones incorporated in SRM 768 with serial number 7. The standard deviation, s, for each transition is given in the table. As was the case for preliminary checks of prototype SRM 767 units (see Table I and accompanying discussion), the major source of this uncertainty is probably due to imprecision of the measurement of the GERT. These preliminary experiments formed the basis for final design and calibration procedures for production of the SRM 768 units.

There are several interesting features of the SRM 768 superconductors. One of these is the relatively large supercooling in some samples when exposed to small magnetic fields, a result entirely consistent with the theory of superconductivity and the fact that the T_c is low. This sensitivity to magnetic fields limits the use of the SRM 768 to low- or null-field (less than 1 μT) environments. The supercooling effect does not reduce the precision of the T_c value of a particular sample, but makes the transition less convenient to use. In the cases of W and Be the supercooling effect was so large as to prompt the authors to spot-weld small buttons of high-purity Al to the samples as superconductive nucleation sites, thus reducing markedly the magnitude of this effect. It is worthwhile to re-emphasize, however, that supercooling in the SRM 768 samples is reduced to an insignificant level if the ambient magnetic field is reduced to 1 μT or less.

A summary of the properties of the SRM 768 devices is given in Table VI.

A disappointing feature of the SRM 768 device is the fact that different samples of the same material showed measurably different T_c values. This arises from the situation that, unlike the elements which make up the SRM 767 device, the SRM 768 materials commonly contain ferromagnetic impurities. Such impurities cause substantial changes in T_c, so that small variations in purity level from sample to sample are readily transformed into T_c variations. As a result of this problem, each of the SRM 768 devices is calibrated against the NBS-CTS-1 temperature scale and given an individual list of T_c values. Note that the reproducibility of a given sample T_c value is quite good, ranging from 0.1 mK to 0.2 mK. This is true even though the T_c values for a given material vary from sample to sample as shown in Table VI.

To date some 50 units of SRM 768 have been distributed through the auspices of the NBS Office of Standard Reference Materials. Although the units have been available only since 1979, they have already proved useful temperature reference sources in such diverse studies as spin polarized hydrogen[16], paramagnetic materials[17], and the properties of liquid ^{3}He[18].

Table VI
Summary of Properties of SRM 768

Material	Transition Temperature Will Lie Between	Typical Transition Width	Typical Reproducibility* Upon Thermal Cycling	dT_c/dH
	mK	mK	mK	mK/µT
W	15.0 - 17.0	0.7	0.20	.08
Be	21.0 - 24.0	0.2	0.10	.14
Ir	98.5 - 99.5	0.8	0.10	.08
$AuAl_2$	160.0 - 161.0	0.3	0.10	.10
$AuIn_2$	205.0 - 208.0	0.4	0.15	.08

*The values given are the averages of the standard deviations of several samples.

RECOMMENDED OPERATING CONDITIONS

1) Peak-to-peak magnetic field applied in primary coil: 1.0 µT (10 mG) for W transition, 0.46 µT (4.6 mG) for the others.
2) Heating generated with above conditions: 1.8×10^{-9} W and 7.5×10^{-11} W, respectively.
3) Ambient magnetic field kept below 1 µT.

In the last instance, controversies regarding the P-T phase diagram of superfluid and normal liquid ^{3}He were relieved by extrapolating NBS-CTS-1 (as embodied in the fixed points of W and Be) down to 0.5 mK.

Owing to its recent introduction, SRM 768 has received scant examination by other laboratories concerning its thermodynamic accuracy. Soulen[19] very recently summarized some preliminary information available from four laboratories. The results indicated that the T_c values assigned by these laboratories agreed within ±0.5% with those given on NBS-CTS-1 for all the materials in SRM 768 except tungsten. The aforementioned laboratories consistently assigned higher values (varying from 2% to 8%) to the W transitions. The basis for this discrepancy is under active investigation at the NBS. Recent research indicates that the germanium resistance thermometer used to transfer NBS-CTS-1 to the superconductive transitions has performed irregularly below 20 mK and thus has not maintained its (NBS-CTS-1) calibration.

Further independent testing and confirmation of these results are desirable. Fortunately, SRM 768 units are under study in several national standards laboratories.

SUPERCONDUCTIVE TRANSITIONS AS IPTS DEFINING FIXED POINTS

It is clear at the present time that the SRM 767 and 768 devices have an important place in the cryogenics laboratory. As discussed above, there is a growing literature on the use of these devices to provide reference temperatures in support of research laboratory temperature scales and to monitor the stability of low-temperature thermometers. For such purposes it is sufficient that the device samples provide reasonably precise and reproducible temperatures.

Despite the employment of the NBS SRM 767 transitions in the definition of the 1976 0.5 K to 30 K Provisional Temperature Scale, the prospective use of superconductive transitions in the definition of the International Practical Temperature Scale places more stringent requirements on the nature of the device transitions. All of the temperature reference points which define the IPTS-68 involve the liquid state of highly pure materials. What properties distinguish these points as uniquely suited for this purpose?

One can list several criteria by which a choice of IPTS defining fixed points can be made. We suggest the following:

1) the transition in question must occur where a temperature fixed point is particularly useful;
2) the realization of the transition temperature must be reasonably easy and economical;
3) the transition must occur over a temperature range which is narrow with respect to the limit of precision of the temperature scale;
4) a given device must yield the same temperature, within usefully precise limits, over a long period of time (in general, over a period of several years) and in spite of repeated use; and
5) there must exist a reliable recipe by which a skilled thermometrist can realize the transition at essentially the identical temperature exhibited by all other similarly-prepared devices.

Solid-state transitions never have been specified as defining fixed points for an IPTS. The absence of sharp, reproducible transition characteristics, in general thought to be connected with lattice imperfections and hysteresis, has thus far prevented such a step. In this section, we examine the class of superconductive transitions, and in particular those superconductors which have been most carefully studied, and we note which of the criteria listed above are met by this type of fixed point.

The first criterion listed above is easily met by superconductors as a class. Thermometry in cryogenics suffers from a lack of reliable fixed points in several temperature regions; notable among these are the range from 4 K to 13.8 K and the range below 0.5 K.

Criterion 2) is subjective, and thus bears some examination. The measurement of superconductive transitions involves a different technique from that employed in observing first-order phase transitions such as boiling points or melting points. Since no latent heat accompanies the second-order superconductive transition, use of a highly precise temperature controller is essential; on the other hand, a small sample suffices for the observation of either the magnetic susceptiblity or the resistivity change in the transition. In view of the low cost of superconductive samples, the ready availability of high-precision temperature control equipment in cryogenics, and the simplicity of detecting the superconductive transition magnetically, one must regard criterion 2) as satisfied by superconductors.

In our opinion, no superconductor has been shown to satisfy all of the three remaining criteria set out above for defining fixed points for the IPTS. In the remainder of this paper, we shall examine why this has been so in the past, we shall outline the steps we feel to be necessary to remedy the situation, and we shall note some progress which has already been made in our own laboratories.

Widths of Superconductive Transitions

In discussing criterion 3) above, one must explicitly acknowledge the solid-state nature of the superconductive transition. Melting-point, triple-point, and boiling-point transitions, if traversed slowly, show transition widths which are mainly functions of the sample purity. (Isotope effects can further complicate the situation in specific cases.) For superconductors in general, however, one must discuss the level of perfection of the crystalline lattice; therefore, sample purity is only one significant parameter. Another equally important characteristic is the state of lattice perfection of the particular sample under study.

The limit of precision with which temperature scales in cryogenics can be transmitted is somewhat less than 1 mK. Some superconductive transitions occur over a narrower range than 1 mK; examples include the 0.02 mK width of transitions in carefully-prepared gallium samples[20] and 0.10 mK width transitions in single-crystal, high-purity cadmium[21], as well as those narrow transitions noted in Tables I and V of this paper. In addition, individual samples of niobium have been found to possess transition widths as narrow as 1.0 mK to 1.4 mK[22,23]. Obviously all of these superconductors can be prepared in such a way as to exhibit usefully precise transitions for use in an IPTS, particularly when one notes that the midpoint of a superconductive transition can be located with a precision of about one-tenth of the transition width.

Thus criterion 3) is achievable for superconductors under circumstances of careful preparation and selection of samples, emphasizing especially the purity of the starting materials and the care in handling.

Reproducibility of T_c of Superconductive Samples

Criterion 4) listed above states that, to be considered as a high-quality fixed point, a given device must yield the same temperature within usefully precise limits over a long period of time and in spite of repeated use. As noted in Table I, the authors subjected some 20 devices containing Pb, In, Al, Zn, and Cd to repeated testing with the objective of examining their suitability with respect to just such a criterion. Despite the relative imprecision of the thermometry used in that study, each of the five materials exhibited a level of irreproducibility below 0.5 mK. Even better data for individual SRM 767 transitions have been reported by observers at three other laboratories[24].

Perhaps the most useful information about the long-term stability of the superconductive transitions in SRM 768 devices has been obtained at the NBS. A reference SRM 768 device, Serial No. 7, was studied for 24 thermal cycles over a three-year period. The average T_c values and standard deviations for the three higher T_c materials in this unit are: Ir(99.27 ± 0.12 mK), AuAl$_2$(160.45 ± 0.11 mK), and AuIn$_2$(205.03 ± 0.28 mK). The results for the Be and W transitions are shown in Fig. 4. The T_c for Be is found to be 23.08 mK with a standard deviation of 0.24 mK. Similarly, for W we obtain T_c = 15.37 ± 0.24 mK. In the former case the standard deviation is approximately three times larger than the temperature resolution of the GERT used to define temperatures in these experiments, whereas in the latter case the standard deviation is five times larger. The cause of this scatter and even the correct assignment

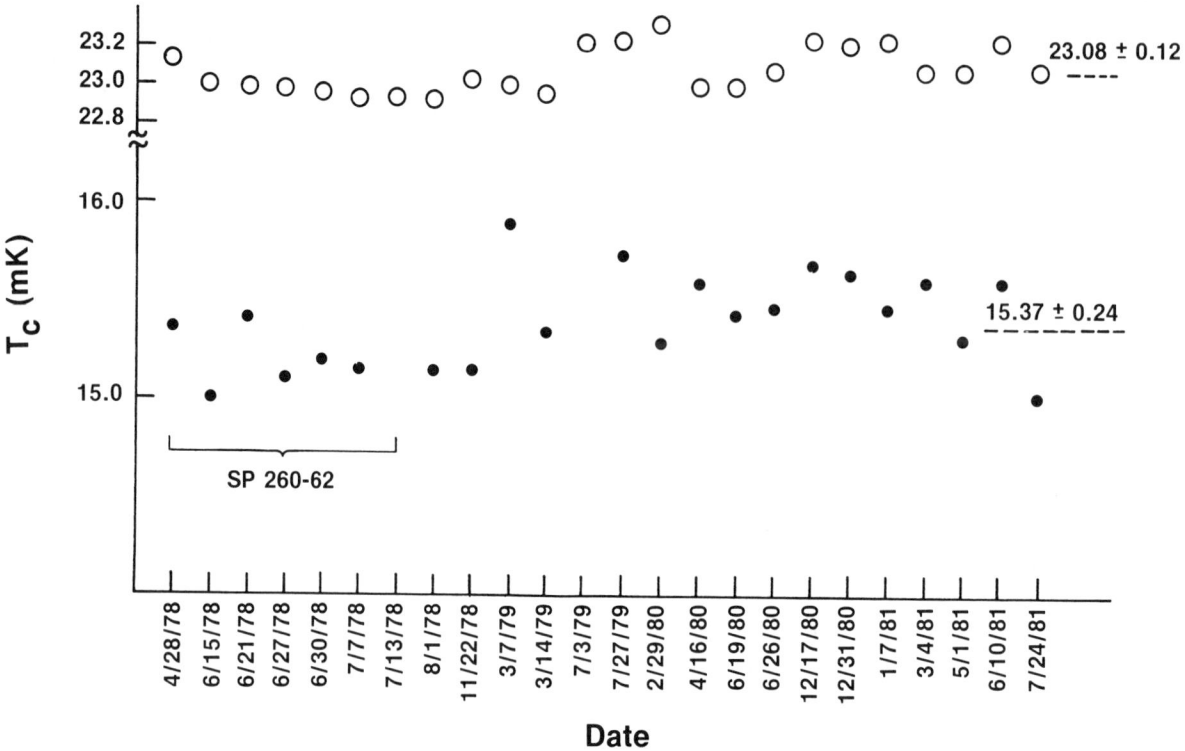

FIG. 4. Reproducibility of the Be and W superconductive transition temperatures in SRM 768 device Serial No. 7 as a function of time and thermal cycle.

for the T_c of W are not completely resolved, but recent evidence at NBS suggests an instability of the GERT as the cause. The results for tungsten notwithstanding, the reproducibility for the other fixed points is rather good.

It is the opinion of the authors that, at the narrow levels of reproducibility under discussion, any significant shift in T_c is likely to be accompanied by a measurable change in the width of the transition. Perhaps because of the relatively strain-free mounting procedure used in the SRM devices and the use of a non-contact method for detection of the transitions, we have not observed such changes in transition width in the NBS devices.

The evidence presented here indicates that superconductive devices are indeed useful in defining precise temperatures over long periods of time and after repeated use, thereby satisfying criterion 4).

Preparing Samples with Uniform T_c Values

It is a characteristic much to be admired that water triple-point cells, prepared according to well-known procedures, uniformly provide the same temperature within 1 mK. In fact, a recent study by Furukawa indicates that the level of irreproducibility among a group of some 15 such cells did not exceed 0.2 mK.[25]

There are studies of metal freezing-point cells, as well, in which the uniformity of the transition temperatures was examined for sizeable groups of such devices [26,27,28].

However, such studies have not been the rule for the low-temperature fixed-point cells employing gases such as hydrogen, neon, oxygen, nitrogen, and the like. Only with the advent of sealed, high-pressure cells for low-temperature triple points[29] and the international comparison of these cells[30] has the capability to demonstrate transition temperature uniformity at sub-millikelvin levels been realized in a coherent program.

Comparison at millikelvin or submillikelvin levels of the T_c values of superconductive fixed-point samples prepared from many different sources and by different techniques has never been attempted, so far as the authors are aware. Thus the fifth criterion of those proposed above stands untested.

It is the intention of the authors to undertake such a study as the next step in their program of development of superconductive fixed points.

It would appear to be a necessary precondition for the satisfaction of criterion 5) that samples of a superconductor prepared from a single source all exhibit suitably uniform T_c values. A modest attempt in this direction was made several years ago with respect to samples of Cd. Fig. 5, drawn from Ref. [21], indicates

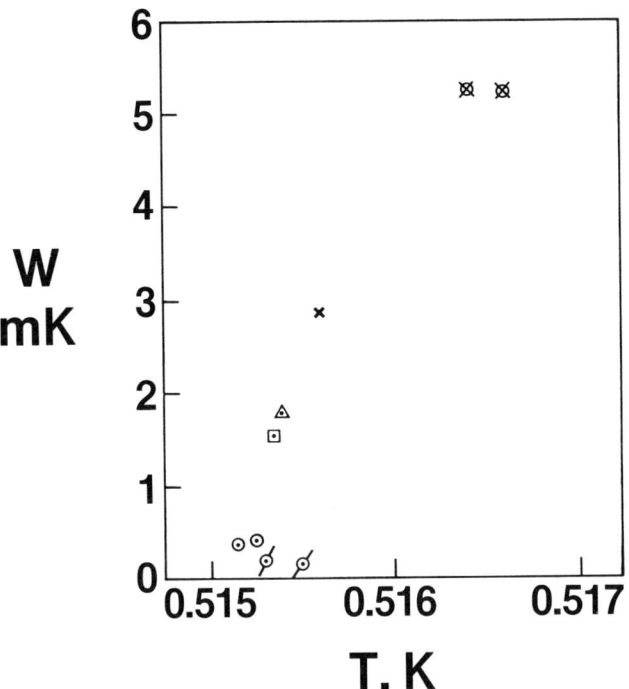

FIG. 5. Superconductive transition widths in zero magnetic field vs. midpoint temperatures for six cadmium samples: ⊙, center portion of single crystal; ⌀, fixed-point sample No. 2; ×, fixed-point sample No. 92; ▫, fixed-point sample No. 93; △, fixed-point sample No. 94; ⊠, fixed-point sample No. 95. (Taken from Ref 21. Note that the abscissa refers to the ^{3}He vapor pressure scale, T_{62}).

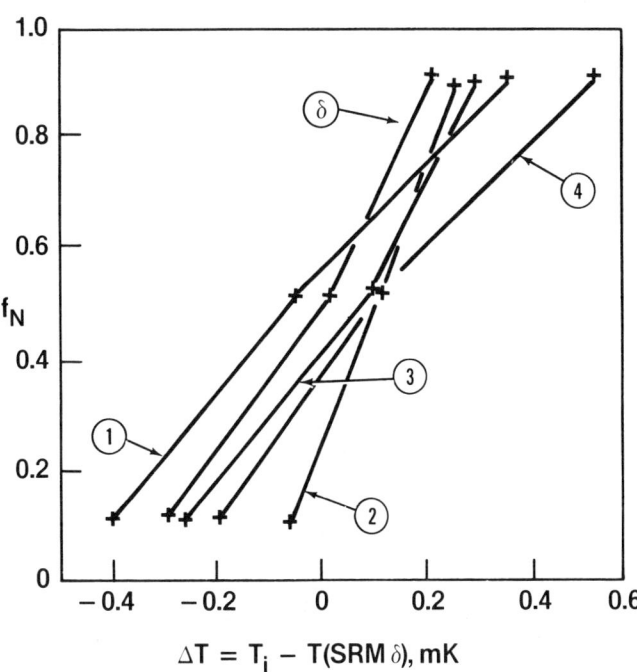

FIG. 6. Superconductive transitions in In. The ordinate shows the normal-state fraction of the sample, f_N, as derived from ac mutual inductance bridge measurements; only the central 80% of each transition is plotted. The abscissa shows the sample transition temperatures referred to the midpoint of sample δ (SRM 767-delta of Pfeiffer and Kaeser, This Symposium). Temperature data were derived from ac bridge measurements of a germanium resistance thermometer.

that samples of Cd prepared from NBS material (see Table II) exhibit T_c values identical within ±0.3 mK, so long as the transition width is less than about 2 mK. For larger widths, T_c appears to increase as $(W-2)/3$ mK.

Exploiting the improved thermometry now available in our SRM 767 laboratory, we have begun a systematic study of sample uniformity. A group of five In samples, all prepared from the same lot although at different times and all mounted on the same copper tempering block, were observed to possess T_c values agreeing within a range of 0.2 mK. The transitions are shown in Fig. 6. This experiment indicates that In samples might resemble those of Cd in showing rather uniform midpoint temperatures (i.e., T_c) despite varying widths.

In the same experiment, a group of five Pb samples was examined. The transitions are shown in Fig. 7. In this case, samples of different widths show substantial variation in T_c defined in the usual way. If a "recipe" for sample preparation and selection at the tenth-millikelvin reproducibility level is discovered, it may of necessity include a restriction on the maximum permissible width for Pb transitions.

In regard to T_c uniformity among SRM 768 samples, we present in Fig. 8 some data for tungsten transitions. Six SRM 768 units were included in a single experiment:

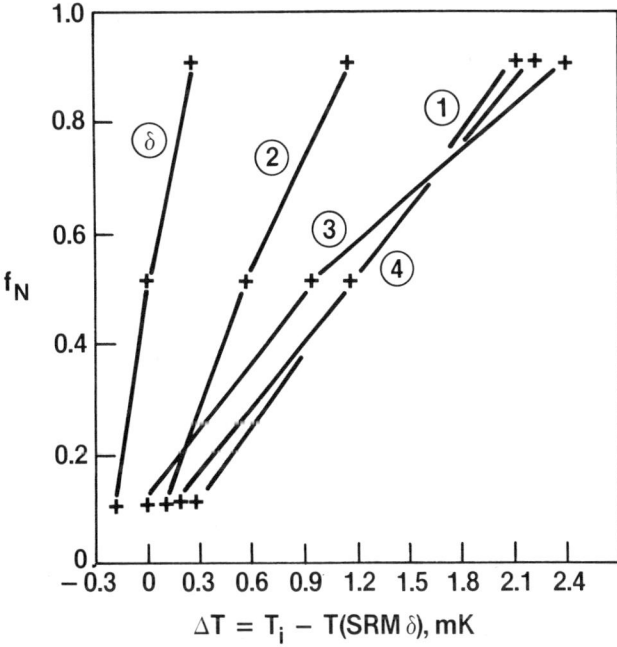

FIG. 7. Superconductive transitions in Pb. Abscissa and ordinate are defined as in Fig. 5. The reference sample is sample δ (SRM 767-Delta of Pfeiffer and Kaeser, This Symposium).

J. F. Schooley and R. J. Soulen, Jr.

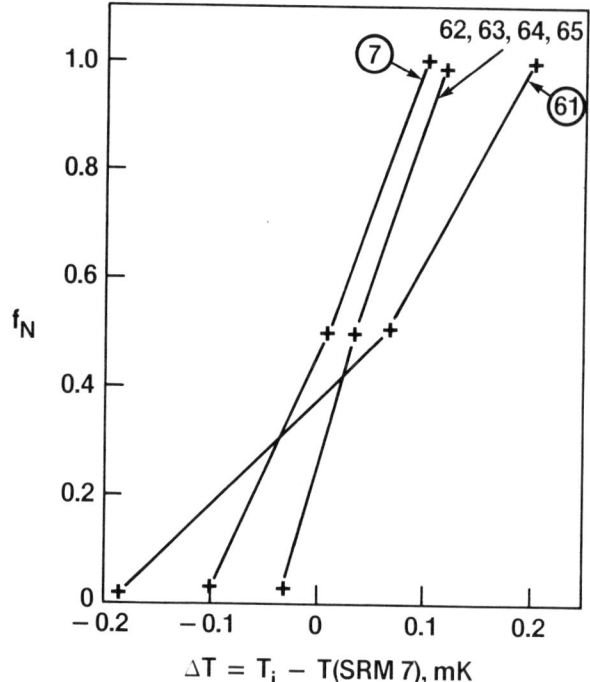

FIG. 8. Superconductive transitions of tungsten. The ordinate is defined as in Fig. 6. The abscissa is defined by the reference sample SRM 768, Serial Number 7.

serial nos. 7, 61, 62, 63, 64, and 65. The other T_c values were referred to serial no. 7, which consistently maintains a transition width of approximately 0.2 mK. Samples in serial nos. 62, 63, 64, and 65 were very sharp (width ~ 40 μK) and the T_c values were approximately 30 μK higher than that of serial no. 7. The broadest transition occurred in serial no. 61 where the width was 0.4 mK, and its T_c was observed to be about 60 μK higher than serial no. 7.

REFERENCES

1. J.F. Schooley and R.J. Soulen, Jr., Temperature, Its Measurement and Control in Science and Industry, 4, H.H. Plumb, Editor-in-Chief, Instrument Society of America, Pittsburgh, 169-175 (1972).
2. R.J. Soulen, Jr. and J.H. Colwell, J. Low Temp. Phys. 5, 325 (1971).
3. J.F. Schooley, G.A. Evans, Jr., and R.J. Soulen, Jr., Cryogenics, 193-199, April 1980.
4. J.F. Schooley, R.J. Soulen, Jr., and G.A. Evans, Jr., NBS Special Publication 260-44, 25, (1972).
5. R.A. Erickson, J.W.T. Dabbs, and L.D. Roberts, Rev. Sci. Instr. 25, 1178 (1954).
6. R.J. Soulen, Jr., J.F. Schooley, and G.A. Evans, Jr., Rev. Sci. Instr. 44, 1537 (1973).
7. D.L. Martin, Rev. Sci. Instr. 46, 1670 (1975).
8. G. Cataland and H.H. Plumb, Metrologia 11, 161 (1975).
9. L.M. Besley and W.R.G. Kemp, Metrologia 13, 35 (1977).
10. A. Inaba and K. Mitsui, paper CCT 80-19, available from the International Bureau of Weights and Measures, Pavillon de Breteuil, F-92310 Sevres, France.
11. The 1976 Provisional 0.5 K to 30 K Temperature Scale, Metrologia 15, 65 (1979).
12. M. Durieux, D.N. Astrov, W.R.G. Kemp, and C.A. Swenson, Metrologia 15, 57 (1979). See also Ref. 24.
13. R.D. Cutkosky, IEEE Trans. Instr. and Meas. IM-29, 330 (1980). See also his contribution to This Symposium.
14. R.J. Soulen, Jr. and R.B. Dove, NBS Special Publication 260-62, 37, April 1979.
15. R.J. Soulen, Jr. and H. Marshak, Cryogenics, 408-412, July 1980.
16. R. Jocheinsen, M. Morrow, A.J. Berlinsky, and W.N. Hardy, Phys. Rev. Lett. 47, 852-855 (1981).
17. W.E. Fogle, E.W. Hornung, M.C. Mayberry, and Norman E. Phillips, Proc. 16th Int'l Conf. on Low Temp. Physics, to be published.
18. E. Lhota, M.T. Manninen, J.P. Pekola, A.T. Soinne, and R.J. Soulen, Jr., Phys. Rev. Lett. 47, 590-592, 1981.
19. R.J. Soulen, Jr., Proc. 16th Int'l Conf. on Low Temp. Physics, to be published.
20. W.D. Gregory, Phys. Rev. 165, 556 (1968).
21. J.F. Schooley, J. Low Temp. Phys. 12, 421-437 (1973).
22. J.F. Schooley, Journal de Physique Colloque C6, supplément au No. 8, Tome 39, August 1978, p. 66-1169.
23. M. Durieux and G.P. van der Maij, communication No. 80-66 to the 13th Session of the Comite Consultatif de Thermometrie, 17-19 June 1980.
24. El Samahy, et al., This symposium.
25. G.T. Furukawa and W.R. Bigge, This symposium.
26. B.W. Mangum, This symposium.
27. G.T. Furukawa and E.R. Pfeiffer, This symposium.
28. J.M. Figueroa and B.W. Mangum, This symposium.
29. F. Pavese and D. Ferri, This symposium.
30. F. Pavese, This symposium.

Realizations of the superconductive transition points of lead, indium, aluminium, zinc, and cadmium with SRM 767 devices

A. E. El Samahy and M. Durieux

Kamerlingh Onnes Laboratory, Niewsteeg 18, Leiden, The Netherlands

R. L. Rusby

Division of Quantum Metrology, National Physical Laboratory, Teddington, Middlesex, United Kingdom

R. C. Kemp and W. R. G. Kemp

CSIRO Division of Applied Physics, National Measurement Laboratory, West Lindfield, Sydney, NSW 2070, Australia

The SRM 767 fixed point device provides a convenient means of realizing the superconductive transition temperatures of lead, indium, aluminium, zinc, and cadmium which lie in the range 7.2–0.5 K. Precise values have been assigned to these transition temperatures in the Provisional 0.5–30 K Temperature Scale of 1976, EPT-76, and so constitute five of the eleven reference points of that scale. It is therefore important to assess the variations in transition temperature which may occur between samples in different devices, and the experimental reproducibility which may be obtained with the devices. This paper brings together the results of measurements made at the Kamerlingh Onnes Laboratory, KOL, at the National Physical Laboratory, NPL, and at the National Measurement Laboratory, NML, which were obtained with a total of eight devices. Those measurements completed before May 1978 contributed to the choice of values assigned to the transition temperatures in the EPT-76.

INTRODUCTION

The supplementary paper [1] to the 1976 Provisional 0.5 K to 30 K Temperature Scale, EPT-76 [2], briefly describes the derivation of the values which are assigned in this scale to the superconductive transition temperatures, Tc, of Pb, In, Al, Zn and Cd as realized with NBS SRM 767 devices [3,4]. These values were based on data obtained at KOL, NML and NPL using a total of eight devices and on new NBS values for the Pb and In points [5] and the Cd point [6].

Little information has been published by users concerning the precision and reproducibility of these devices or the values of the transition temperatures (but see references 7 and 8), and the present paper is an attempt to remedy that situation. It describes our experiences in using the devices and gives our results which are now somewhat more extensive than they were in May 1978 when the EPT-76 was approved.

THE DEVICES AND THE REALIZATION OF THE FIXED POINTS

SRM 767 devices are obtainable from the Office of Standard Reference Materials, NBS, Washington DC, for the purpose of providing a set of stable reference points for thermometry in the range 0.5 K to 7.2 K. The manufacture of the device is described in the NBS special publication [3], and it suffices here to say that the samples of the five metals, which are small rods of about 30 mm long and 1.5 mm in diameter, are mounted on a copper stud which can be bolted to an experimental apparatus. The samples are physically protected by the former of the primary of the mutual inductance coil pair which is used for the detection of the transitions. One suitable circuit for this is described by Soulen et al [9].

As Tc is a function of magnetic field, it is necessary to compensate for the significant local static field or to shield the samples from it carefully with mu-metal or similar material [10] or, for less accuracy, to apply a correction using an assumed value for dTc/dH. It is also necessary to restrict the current in the primary coil and to check for its effect by observing Tc as a function of primary current.

The observation of a transition is a dynamical process, since it is necessary to probe several millikelvins above and below the midpoint to ensure complete coverage. Traces of mutual inductance versus temperature assist in establishing that this has been achieved. The average of the two extreme values of mutual inductance then gives the position of the midpoint, and the width, W, of the transition can be determined as the temperature change over which the central 80 per cent of the mutual inductance change occurred [3,4]. During a trace, temperature differences may exist between the sample and thermometer, giving rise to an apparent hysteresis, even if the transition is scanned over several minutes. We find, however, that in zero field the temperature of the midpoint, when measured in equilibrium, is stable and independent of the direction of approach. When this condition has been reached the compensating dc field can be trimmed (if shielding is not used) so that the maximum value of Tc is achieved, and the measuring field effect can be tested. A realization of the fixed point then results, and the measurement of Tc (or the calibration of a thermometer) can follow.

MEASUREMENTS AND RESULTS

Our experiences are with a total of eight devices: three each at KOL and NPL, and two at NML. The KOL results are well correlated with those of NPL through the use of common temperature scales, NPL-75 [11] and T_{X1} [12], as carried on NPL calibrated rhodium-iron resistance thermometers [13]. The NML results provide more detailed information on the reproducibility of the devices at the Pb, In and Al points, but as Tc values were obtained on the Iowa State University scale, T_{XISU}(1975) [14], there are uncertainties associated with comparing the NML values with those of KOL and NPL. Descriptions of the measurements made in the three laboratories now follow.

(i) Measurements at NPL

Three devices, serial nos. 111, 112 and 113, were first measured in 1976. They, and a home-made mutual inductance of about the same value as that of the devices, were mounted in the ^{3}He cryostat used in the CMN magnetic thermometry experiments of Rusby and Swenson [12]. Their mutual inductance bridge was adapted for the detection of the transitions, the empty coils being used as the opposing mutual inductance and the sensitivity being cut by a factor of 100 by connecting the secondary circuit across the whole of the autotransformer.

In the first experiment, SC1 of September 1976, Tc values for all the Pb, In and Al samples were obtained in terms of NPL-75 (for Pb and In) and T_{X1} (for Al) using a set of five calibrated rhodium-iron resistance thermometers. One Al sample gave identical Tc values on two consecutive days, and one In sample gave a value 0.1 mK lower after being warmed to room temperature, but no other reproducibility checks were made.

In the second experiment, SC2 of December 1976 and January 1977, the Zn and Cd transition temperatures were measured in terms of T_{X1} and the Pb points were repeated. The latter were found to be 0.0 mK, 0.3 mK and 0.7 mK lower than in SC1. The measurements were repeated twice, once with the helium reservoir pumped to 2.2 K to alter the thermal conditions, but the Tc values remained constant within 0.05 mK.

In the third experiment, SC3 of July 1977, all 15 Tc values were measured to test the reproducibility of the realizations, and for the Pb points, to check for possible drift in Tc. The SC3 results are given in Table I together with rough measurements of the transition widths and the observed differences in Tc from earlier values. The Pb points increased by a uniform 0.2 mK between SC2 and SC3, suggesting a systematic difference in realization rather than a lack of reproducibility in the Tc values. The differences from SC1 remain unexplained. The differences in the Cd points could have been due to difficulties in temperature control which were largely resolved in the later experiments. Apart from these observations the reproducibility of the devices, in so far as it was tested, appeared to be excellent. Similarly the maximum difference between Tc values for the three devices was 0.5 mK except that the Pb point in device no.111 (which was the most stable Pb point) lay 1.1 mK above the other two values. The last column in Table I gives an estimate of the uncertainty ΔT in the Tc values in terms of the temperature scales used. This quantity includes all statistical uncertainties at the level of 99 per cent confidence, errors in determining the midpoint positions and in the field compensation; ie all the errors involved in a realization, but not the thermodynamic uncertainty in the temperature scales. ΔT is smallest for In and Al, but is somewhat larger for Pb and Cd because of the greater difficulties of temperature measurement and control, and for Zn because of the width of the transitions.

(ii) Measurements at KOL

Measurements of the transition temperatures of the Pb, In and Al samples in the three devices, serial nos.102, 103 and 108, were first made in 1973 and 1974. However, although a reproducibility of 0.1 mK was reached when measurements were repeated in a period of one week, without thermal cycling, the data which were obtained could not easily be related to the data obtained at NPL or NML, or to later data at KOL, mainly because temperatures were determined with germanium resistance thermometers the calibration of which could not be related to NPL-75 or T_{X1} with sufficient accuracy.

In a second experiment, in 1979, all Tc values, transition widths and transition curves were determined. The devices were bolted to a copper block in which two rhodium-iron and several germanium thermometers were mounted. The block was surrounded by a shield in a ^{3}He cryostat, and was connected to it by a weak thermal link so that the block slowly followed temperature variations of the shield. By appropriate adjustment of the electronically regulated shield temperature, the block temperature could be kept constant, or varied, with a resolution of 0.01 mK.

The measurements at each transition point consisted in general, of:
1) A preliminary recording of the inductance change and the block temperature, as indicated by a germanium resistance thermometer during a passage of the transition. At the midpoint the currents in three perpendicular sets of coils outside the cryostat were adjusted such that the transition temperature had its maximum value and the magnetic field at the site of the sample was zero.
2) A measurement of the transition curve, i.e. the inductance change versus temperature, with increasing and decreasing temperature. The transition curves were derived from recordings of the inductance change and the block temperature versus time. The rate of temperature change was usually 1 mK in 30 to 90 min during the passage of the central 80 per cent of the transition, increasing to 1 mK per 5 min in the tails of the transition.
3) Measurements of the transition temperatures. The temperature of the block was adjusted to the transition midpoint and, after checking the currents in the field compensator for maximum Tc, the resistances of two rhodium-iron thermometers were measured. Two determinations of Tc were made for each sample, T_{c1} with the midpoint approached from the low temperature side, and T_{c2} approached from the high temperature side.

Experimental results are given in Table II and Figs. 1, 2 and 3. The Tc values given in Table II are those measured with one of the rhodium-iron thermometers, and are averages of the two measurements T_{c1} and T_{c2}.

Differences $T_{c1} - T_{c2}$ are between -0.12 mK and +0.08 mK

Table I T_c values obtained at NPL in terms of NPL-75 for Pb and In and T_{X1} for Al, Zn and Cd in the third, SC3, experiment. The widths, W, of the transitions are given, as are the differences between SC3 values and those obtained in previous runs, SC2 and SC1. ΔT is the uncertainty in the realization of each point, estimated at 99% confidence.

		111	112	113	ΔT
Pb	T_c	7.2010 K	7.1999 K	7.1999 K	0.0004 K
	W	0.0015	0.0005	0.0005	
	SC3-SC2	0.0002	0.0002	0.0002	
	SC3-SC1	0.0002	-0.0001	-0.0005	
In	T_c	3.4148	3.4147	3.4149	0.0003
	W	0.0008	0.0009	0.0015	
	SC3-SC1	0.0000	0.0001	-0.0002	
Al	T_c	1.1795	1.1796	1.1800	0.0003
	W	0.0027	0.0027	0.0026	
	SC3-SC1	-0.0001	-0.0001	0.0000	
Zn	T_c	0.8517	0.8513	0.8516	0.0004
	W	0.0107	0.0110	0.0107	
	SC3-SC2	0.0000	0.0002	0.0001	
Cd	T_c	0.5198	0.5196	0.5198	0.0005
	W	0.0022	0.0007	0.0022	
	SC3-SC2	0.0003	0.0000	0.0004	

and are believed to be due to experimental inaccuracies and not to hysteresis. Temperatures measured with the second rhodium-iron thermometer were within 0.1 mK equal to those obtained from the first, except in two cases where the differences were 0.2 mK. Temperatures are on the NPL-75 scale for Pb, In and Al and on the T_{X1}-scale for Zn and Cd.

The spread in Tc values between the three devices was smaller than 0.3 mK for Pb, In and Al but considerably larger for Zn and Cd (see Table II). The widths of the transitions, W, are also given in Table II. The transition curves for Pb, In and Al are quite similar for the three devices. Examples are shown in Fig. 1. For Zn and Cd large differences in widths between devices occur; the transition curves for all three devices are given in Figs. 2 and 3. Both the transition temperatures Tc and the transition curves were reproducible within the measurement inaccuracy (0.05 to 0.1 mK) when measurements were repeated on the same or, in the case of Zn for devices 102 and 103, on consecutive days.

For checking the effect of the primary current in the inductance, a transition curve of Cd (where the magnetic field effect is largest) was measured with a primary current of 100 μA (corresponding to a magnetic induction of 2 μT) instead of the usual current of 20 μA. The decrease in transition temperature of 0.14 mK agrees well with the calculated effect of 0.17 mK.

From the transition curves there was no evidence of hysteresis effects. Small differences between the inductance versus temperature curves measured with increasing and decreasing temperatures (see, e.g., the curves for In in Fig. 1) are believed to be due to changing thermal emf's in the potential leads of the germanium thermometer. (During the passage of the transition the germanium thermometer indication was recorded, without reversal of the measuring current).

A transition curve of Cd (for device 102) was also measured without field compensation. As appears from Fig. 3 there was, in this case, a hysteresis of 0.2 mK. The decrease in transition temperature was 5.4 mK, which corresponds to a magnetic induction of 51 μT.

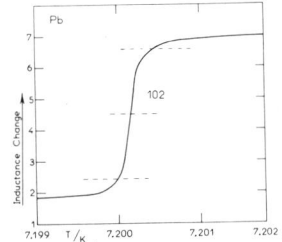
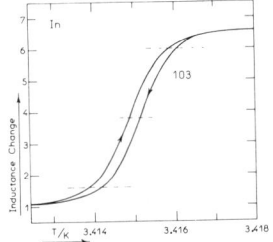

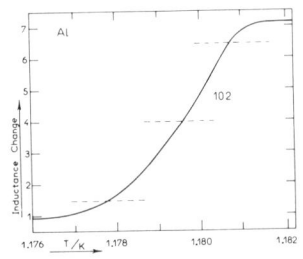

Fig. 1. Inductance change versus temperature during the passage of the superconductive transition for Pb, In and Al at zero field. The dashed lines indicate the midpoints and ± 40% of the transitions. (The small difference between the curves for increasing and decreasing temperature for In is not a real hysteresis, see text. The inductance change is in arbitrary units).

Table II. Transition temperatures T_c and widths of transition W, measured at KOL for three SRM 767 devices (nos.102, 103 and 108). Temperatures are on the NPL-75 scale for Pb, In and Al and on the T_{X1}-scale for Zn and Cd. T_c is the temperature at the midpoint of the transition and is the average of two determinations T_{c1} and T_{c2} (see text), δT is the spread in T_c values for the three devices, and W is the width of the transitions. All measurements were in zero field and for a primary current of 20 μA except where indicated.

		102	103	108	δT
Pb	T_c	7.2001$_5$ K	7.2001$_5$ K	7.2001 K	0.0$_5$ mK
	$T_{c1}-T_{c2}$	-0.0001	0.0000$_5$	0.0000	
	W	0.0005	0.0009	0.0003	
In	T_c	3.4145$_5$	3.4148$_5$	3.4147$_5$	0.2$_5$
	$T_{c1}-T_{c2}$	0.0000	-0.0001	0.0000	
	W	0.0014	0.0019	0.0009	
Al	T_c	1.1794$_5$	1.1794$_5$	1.1796$_5$	0.2
	$T_{c1}-T_{c2}$	0.0000	0.0000	0.0000$_5$	
	W	0.0029	0.0025	0.0025	
Zn (i)	T_c	0.8502$_5$	0.8511$_5$	0.8505$_5$	0.9
	$T_{c1}-T_{c2}$	-0.0001	0.0000$_5$	-0.0001	
	W	0.0033	0.0089	0.0061	
(ii)	T_c	0.8503	0.8510		
	$T_{c1}-T_{c2}$	0.0000$_5$	0.0000$_5$		
	W	0.0030	0.0086		
Cd (i)	T_c	0.5200	0.5203$_5$	0.5196$_5$	0.6$_5$
	$T_{c1}-T_{c2}$	0.0000	-0.0000$_5$	0.0000	
	W	0.0029	0.0062	0.0006	
(ii)	T_c	0.5198$_5$			100 μA
	$T_{c1}-T_{c2}$	0.0000$_5$			
	W	0.0026			
(iii)	T_c	0.5146			without field compensation
	$T_{c1}-T_{c2}$	0.0002			
	W	0.0029			

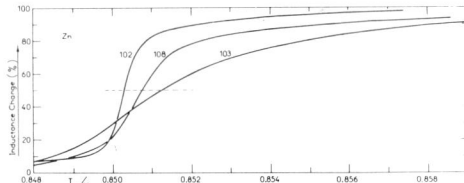

Fig. 2. Inductance change versus temperature for the superconductive transition of Zn measured with three devices in zero field.

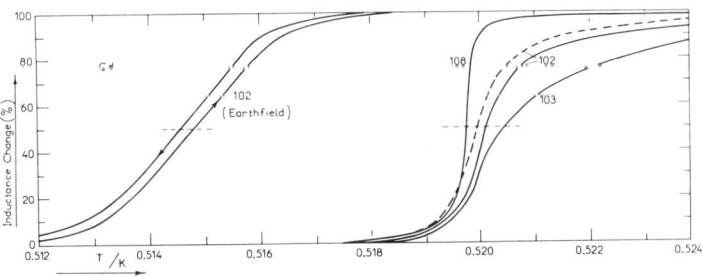

Fig. 3. Inductance change versus temperature for the superconductive transition of Cd measured with three devices in zero field. For device no.102 the transition was also measured with a primary current of 100 μA (ac induction about 2 μT) instead of 20 μA (dashed line) and without field compensation (marked 'Earth field').

(iii) Comparison of the KOL and NPL results

The Tc values for the devices measured at KOL and NPL are compared in Fig. 4 and the widths of the transitions are given. There are no obvious systematic differences between the KOL and NPL sets of data, but in some cases significant differences between devices occur.

For Pb the Tc values of five of the devices are within 0.3 mK but the sixth, no.111, lies 1.0 mK higher than the average of the rest, and its width is also significantly greater. For In the transition temperatures are all within 0.4 mK. For Al the widths are all 2.5 mK or more, but five of the Tc values are within 0.2 mK, the sixth being 0.4 mK higher.

For Zn and Cd the spreads in Tc values are 1.4 mK and 0.7 mK respectively, but Fig. 5 shows that, within the uncertainty of the measurements, the transition temperature is linearly related to the width. An approximately linear relation of about the same slope has been observed by Schooley [15]. The existence of this correlation implies that by defining Tc for Zn and Cd as the transition temperature at zero (or some specified) width, the spread between devices would be reduced considerably. More devices should, however, be investigated before such a procedure could be recommended.

(iv) Measurements at NML

Two devices, nos.110 and 115, were used in the international intercomparison of germanium resistance thermometers [8] which was conducted at NML in 1973-75. The devices and a home-made mutual inductance, equivalent in configuration and value to the devices, were mounted in a modified version of the cryostat used by Cetas [16] for magnetic thermometry above 1 K. The magnetic salt holder was replaced by a copper block suspended from the original block by a copper rod. The sensors and heaters used for temperature control were mounted on the upper block so as to reduce stray magnetic fields near the devices on the lower block. The devices were shielded from external fields by means of an annealed mu-metal shield placed outside the enclosing dewars. The superconductive transitions were detected by means of a simple mutual inductance bridge, the output from which was displayed on a chart recorder so that the centre of the transition could readily be located and monitored while resistance measurements were made.

This apparatus was used to obtain Tc values for the Pb, In and Al samples in terms of $T_{XISU}(1975)$ [14] and also to determine the resistance, at the transition temperatures, of all the germanium thermometers included in the international intercomparison. The Tc values given in Table III were obtained using germanium thermometer serial no. 2247, which was calibrated on $T_{XISU}(1975)$ and had a proven record of reliability [16]. The reproducibility of the complete measuring system was $\pm$ 0.3 mK at the Al point, and $\pm$ 0.2 mK at the Pb and In points. These limits include errors in resistance measurement, temperature control, thermometer instability, errors in setting at the mid-point of the transitions and the reproducibility of the devices themselves, but they do not include the effects of thermal cycling. In order to compare the Tc values of Table III with those of Tables I and II, the differences between the temperature scales must be taken into account. It is believed that these are negligible at the In point, that 0.4 mK should be added to $T_{XISU}(1975)$ at the Al point, and that 0.4 mK should be subtracted at the Pb point to convert the temperatures to the NPL scales [14]. If this is correct, the results for In and Al agree well with those for the NPL and KOL devices, but the Pb points

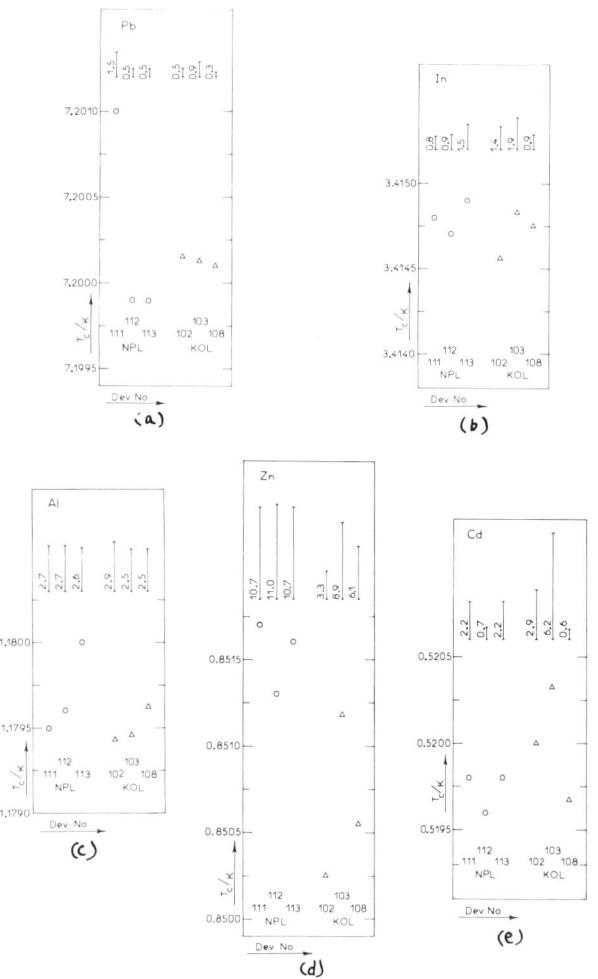

Fig. 4. NPL and KOL data for the superconductive transition temperatures obtained with a total of six devices. Figs. 4a, b, c, d and e give the results for Pb, In, Al, Zn and Cd, respectively. The transition widths, in millikelvins, are indicated in the upper part of the figures.

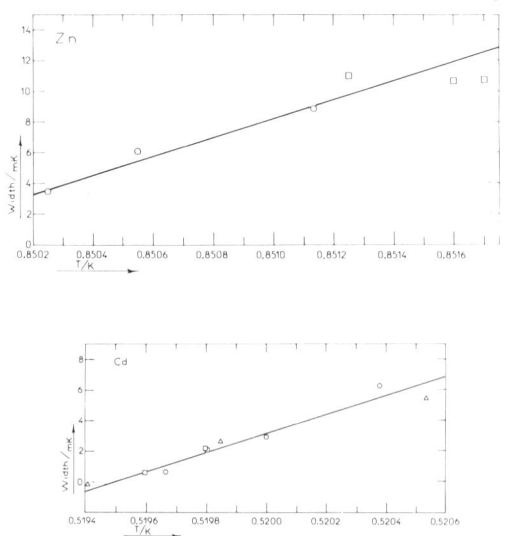

Fig. 5. Dependence of the superconductive transition temperatures of Zn and and Cd on the transition width W. o represents KOL data, □ represents NPL data, and Δ represents data taken from Schooley [15].

are both about 0.5 mK below the values plotted in Fig. 4. These comparisons are of doubtful validity, however, and it is more significant to note that the two Pb points differ by only 0.1 mK, while the differences between the In and Al points are both 0.5 mK. This compares with the maximum differences between the NPL and KOL devices, of 0.4 mK for In and 0.6 mK for Al.

The NML devices were also used in three further runs in 1974 each time incorporating a different set of germanium thermometers. While it is not possible to establish the reproducibility under thermal cycling for each device separately because of the change of thermometers, the relative reproducibility of the two devices can be deduced. Specifically, the quantity ΔT_c, the difference between $T_c(115)$ and $T_c(110)$, can be calculated using the germanium thermometers as intermediaries. The 17 measurements of ΔT_c obtained at the Pb point lay in the range -0.2 mK to +0.5 mK with an average of +0.1 mK. Thus there was no evidence of irreproducibility outside the combined experimental uncertainty of ±0.3 mK. For In, 18 measurements gave values of ΔT_c in the range -0.1 mK to -1.3 mK and on average of -0.9 mK. The spread in these values is greater than would be expected from the experimental uncertainty, again ±0.3 mK. Taking the average ΔT_c for each of the three runs gave -0.6 mK, -1.0 mK and -1.0 mK. The first value is consistent with the difference, -0.5 mK, calculated from Table III, but the second and third suggest that one of the In samples suffered a shift of about 0.4 mK between the first and second runs.

Table III T_c values obtained at NML for Pb, In and Al in terms of $T_{XISU}(1975)$ for devices nos. 110 and 115.

Element	110	115
Pb	7.1998 K	7.1997 K
In	3.4151	3.4146
Al	1.1793	1.1798

CONCLUSIONS

The data presented here have provided precise information on the realization of the superconductive transition points for eight SRM 767 devices. The results may be summarised as follows:

a) For the six devices of NPL and KOL the transition temperatures were measured on the same temperatures scales. The spreads in Tc values were 1.1 mK, 0.4 mK, 0.6 mK, 1.5 mK and 0.7 mK for Pb, In, Al, Zn and Cd, respectively. (For Pb the spread is reduced to 0.25 mK if one device is excluded). If the NML data on Pb, In and Al for two devices are included the spreads increase to 1.7 mK, 0.5 mK, and 0.9 mK for Pb, In and Al, respectively, but the increases may plausibly be ascribed to uncertainties in the differences between the temperature scales used at NML and in the other two laboratories. Although spreads as found for these devices will not be serious for most users, they are larger than is desirable for using the devices in thermometry of the highest accuracy. The data suggest, however, that for Zn and Cd the spread could be reduced by a factor of two or three by recalculating the measured Tc values to a certain standard transition width.

It may be mentioned here that superconductive transitions in pure single crystals have been studied at NBS and the transitions have been found to be narrower than for polycrystalline material [6,15]. Suggestions have been made for preparing SRM 767 devices with single crystal material of the five metals. These devices could then be circulated among various thermometry groups for testing and for comparisons of laboratory temperature scales.

b) Regarding the stability of the devices, it was found that changes in the transition temperatures were less than 1 mK for three devices after one year (NPL), for two devices on repeated cycling between room temperature and low temperatures (NML), and for three devices after four years (KOL). Most of the changes that have been found could be attributed to experimental uncertainties, but for one Pb sample a real change of -0.7 mK between two measurements, with a half year interval, was found.

c) When the EPT-76 was established, best values had to be assigned to the five superconductive transition points. This task was beset with difficulty because transition temperatures had been measured in various laboratories on different temperature scales. The present NPL and KOL data provide accurate values of Tc for six devices on the NPL-75 scale (Pb, In) and the T_{X1}-scale (Al, Zn, Cd). In Table IV the assigned values for the transition temperatures in the EPT-76 are given, together with the average T_{76} values derived from the NPL and KOL data using the scale differences published in Reference 2. It can be seen that the assigned values agree within 0.4 mK with the new average values, except for Cd where the difference is 0.9 mK. The assigned T_{76} value for the cadmium point, however, was not simply determined using the T_{X1}-scale, but was assigned after considering the average NPL value of T_{X1} for their three devices (0.5197 K ± 0.0005 K), and an independent noise thermometer determination by Soulen (0.5190 K ± 0.0006 K) [1].

Table IV Assigned EPT-76 values of the superconductive transition temperatures of Pb, In, Al, Zn and Cd, and average experimental values for the six SRM 767 devices at KOL and NPL. In the last column the average experimental values are recalculated to T_{76} in accordance with Reference 2.

	Assigned value T_{76}/K	Average value for six devices		
		T_{NPL-75}/K	T_{X1}/K	T_{76}/K
Pb	7.1999	7.2000*		7.2003*
In	3.4145	3.4148		3.4149
Al	1.1796		1.1794	1.1794
Zn	0.851		0.8511	0.8511
Cd	0.519		0.5199	0.5199

* Average with device 111 not included.

REFERENCES

1. M Durieux, D N Astrov, W R G Kemp and C A Swenson, Metrologia 15, 57 (1979).
2. The 1976 Provisional 0.5 K to 30 K Temperature Scale, Metrologia 15, 65 (1979).
3. J F Schooley, R J Soulen and G A Evans, NBS Special Publication 260-44, US Department of Commerce, Washington DC, 1972.
4. J F Schooley and R J Soulen, Advances in Cryogenic Engin., Vol 17, Ed. Timmerkaus, Plenum Press, New York and London, 1972, p 192.
   Also Cryogenics, 20, 193 (1980).
5. G Cataland and H H Plumb, Metrologia 11, 161 (1975)
6. R J Soulen, private communication.
7. D L Martin, Rev. Sci. Instrum. 46, 1670 (1975).
8. L M Besley and W R G Kemp, Metrologia 13, 35 (1977)
9. R J Soulen, J F Schooley and G A Evans, Rev. Sci. Instrum. 44, 1537 (1973).
10. D L Martin and R L Snowden, Rev. Sci. Instrum. 46, 523 (1975).
11. K H Berry, Metrologia 15, 89 (1979).
12. R L Rusby and C A Swenson, Metrologia 16 73 (1980).
13. R L Rusby, Temperature Measurement, Inst. of Phys. Conf. Series No.26, Inst. of Phys., London, 1975, p 125.
14. M S Anderson and C A Swenson, Rev. Sci. Instrum. 49, 1027 (1978).
15. J F Schooley, Journal of Low Temperature Physics, 12, 421 (1973).
16. T C Cetas, Metrologia, 12, 27 (1976).

Temperature fixed points: Evaluation of four types of triple-point cell

J. D. Cox and M. F. Vaughan

Division of Quantum Metrology, National Physical Laboratory, Teddington, Middlesex TW11 0LW, United Kingdom

Triple points are invariant temperature fixed points. Use of the triple points of water, phenoxybenzene, 1,3-dioxolan-2-one and n-icosane in providing temperature fixed points, accurate to 1 mK or better, is described: Triple-point cells containing 1,3-dioxolan-2-one have outstanding potential for calibrating or checking thermometers (near 37 °C) in biomedical laboratories. To secure good reproducibility of triple-point temperature (realization-to-realization, laboratory-to-laboratory and cell-to-cell), the working substances in triple-point cells need to be very pure. Criteria for purity and means for their implementation are discussed. Values for the respective triple-point temperatures and for other pertinent constants are presented. Experimental methods for realizing the triple-point condition and maintaining it for weeks are described; the use of simple equipment and non-specialist skills is emphasized, thus ensuring the applicability of NPL's triple-point cells to a wide variety of thermometric problems in the temperature range 0–37 °C.

1 INTRODUCTION

The temperature variation of the physical properties of substances is the basis of practical thermometry: the variation may be continuous, or it may be discontinuous, as when a substance undergoes a first-order change of state. The use of certain characteristic temperatures to define practical temperature scales dates back to the beginning of quantitative thermometry. In more modern times various characteristic temperatures, the so-called temperature fixed points (TFPs) have played an important role in the International Practical Temperature Scale. Thus the current version[1], IPTS-68, names thirteen "defining fixed points", whose collective role is to define the scale between 13.81 K and 1064.43 °C (with extrapolation beyond); it also mentions 33 "secondary reference points", whose exact role is not stated.

Of the defining fixed points[1], (a) four are triple points (for which the pressure, as well as the temperature, is uniquely fixed by the intrinsic properties of the working substance itself), (b) four are freezing points (for which the temperature is slightly dependent on the pressure and on the nature of any confining gas), and (c) five are the temperatures at which a chosen pressure is exerted by the saturated vapour of the working substance. Since temperatures falling in category (c) are very sensitive to pressure, they scarcely merit the description "fixed point", in our opinion! But be that as it may, defining fixed points of category (c) are the hardest to realize, and the use of alternates for two of them is permitted[1] under IPTS-68.

Concerning the function of the secondary reference points cited in the text of IPTS-68, it is evident that the listing of just 33 such points can have no fundamental significance, as Working Group II of the Consultative Committee for Thermometry (responsible for IPTS) has recently added 22 more to the list[2]; rather it seems that the listed substances represent a selection from the literature of those characteristic temperatures which have been rather extensively studied and for which the results are reasonably trustworthy. As to the unstated metrological role of these secondary reference points, we suppose it to include (i) the calibration of thermometers, wherever this cannot conveniently be done by use of the defining fixed points and interpolation procedures (e.g. because the sensor is built into apparatus or its usable temperature range is narrow), and (ii) the calibration-checking of thermometers that have already been calibrated (e.g. where the construction or mode of use of a thermometer makes it likely that calibration will not remain valid for long). In the remainder of this article we envisage this latter role as the most relevant one for the triple-point devices we shall describe: such devices provide an early warning to the user that his thermometer should be returned to a calibration laboratory for re-examination.

We deal in this paper with the triple-points of water, phenoxybenzene, ($C_{12}H_{12}O$), 1, 3-dioxolan-2-one ($C_3H_4O_3$) and n-icosane ($C_{20}H_{42}$). The first named is unique in having three aspects, (i) it is the basis for defining the kelvin, (ii) it is a defining fixed point on IPTS-68, (iii) it is widely used to check the R_0 value of a calibrated platinum resistance thermometer (PRT). The second substance named above is listed[1,2] as affording a secondary reference point: its triple-point temperature is important in oceanography, climatology and physical chemistry. The third and fourth named substances have been developed by us to provide reference temperatures near to 37 °C, to meet the pressing need of researchers in the bio-medical field, particularly enzymologists.

2 METHODS FOR CHECKING A THERMOMETER'S CALIBRATION

2.1 Use of (i) TFP's and (ii) isothermal comparison with reference thermometers

In this discussion we exclude the special methods used at very low and very high temperatures; for the intermediate temperature range, we consider the means by which the possessor of a calibrated thermometer (liquid-in-glass, thermocouple, resistance) may check from time to time that the calibration is still valid. There are two basic possibilities, which may be used separately or in combination, (i) comparing the reading of the thermometer when inserted into a certified TFP device with the known temperature value of the device, or (ii) comparing the reading of the thermometer with that of a reference thermometer in close proximity to it in an isothermal medium; with this method, the assumptions are that the user has access to a superior (in terms of its quality of calibration) thermometer, and that temperature fluctuations of the isothermal medium are negligible compared with the desired accuracy of intercomparison.

In terms of flexibility, method (ii) is better than method (i), since any temperature(s) can be selected for the calibration-checking. But in terms of speed and convenience the use of TFPs is to be preferred. Means for establishing an isothermal zone in stirred-liquid baths or furnaces, as required by method (ii), are described in many texts, e.g. reference 3. Means for establishing TFPs in general have also been outlined, for example in earlier Symposia in this series; detailed prescriptions for realizing four triple points are given later in this paper.

2.2 Use of freezing points and triple points, as TFPs.

It will be recalled that a triple point is the unique point in the temperature-pressure diagram for a pure substance at which solid, liquid and vapour phases co-exist, i.e. the junction point of the vapour-solid, vapour-liquid, and liquid-solid curves in the p, T equilibrium diagram; the p and T co-ordinates at the triple point are characteristic of the substance. The freezing point of a substance is more arbitrary, being that point on the p, T-curve for liquid-solid equilibrium which corresponds to some prescribed pressure, conventionally 101 325 Pa. Experience with a given pure metal indicates that the repeatability of temperature measurement is about the same at its freezing point as at its triple point; the absolute difference between the two temperatures is obtainable by either measurement or calculation from the Clapeyron equation:

$$\frac{dp}{dT} = \Delta_{fus}H/T\Delta_{fus}V \qquad (1)$$

where $\Delta_{fus}H$ is the molar enthalpy of fusion and $\Delta_{fus}V$ is the change in molar volume due to fusion.

On the other hand, experience with molecular liquids indicates superior repeatability of temperature measurement at the triple point as compared with the freezing point, when the required pressure of 101 325 Pa is exerted by means of a gas. The reason is the comparatively high solubility of most gases in most molecular liquids, leading to a lowering of the freezing point, ΔT, according to the equation:

$$\Delta T = Cx \qquad (2)$$

where x is the mole fraction of dissolved gas and C is a constant. This gas-dissolution effect can be circumvented by use of a piston or diaphragm to exert the pressure, or can be made reproducible by using some means for ensuring that the pressurizing gas fully saturates the liquid before freezing begins, thus creating a thermodynamically definable state. In general, though, these complications are unwanted by experimenters, so that triple points have a clear advantage over freezing points in terms of convenience, as well as repeatability. A related advantage of triple points, which accrues when the triple-point device is permanently sealed, is that adventitious contamination of the charge of working substance with an environmental impurity (e.g. water vapour, grease, solder, plastics) cannot occur, nor can any chemical interaction take place between the charge and air.

It remains to discuss the effect of impurities that are intrinsically present in a charge of working substance, i.e. species extra to a dissolving gas or adventitious impurities. Let us suppose that all impurities present conform to the same simple law given by, or related to, equation (2), so that their effect on equilibrium can be expressed in terms of a single composition variable, e.g. by the total mole fraction of impurities, x. We can now speak of the p, T, x equilibrium surface, of which the relevant parameter is $(\partial T/\partial x)_p$. In general, it seems unlikely that the value of $(\partial T/\partial x)_p$ when p = 101 325 Pa will differ greatly from the value at the triple-point pressure, p_{tp}, assuming the latter to lie below 101 325 Pa. (If $p_{tp} >$ 101 325 Pa, then the substance has no freezing point at the conventional pressure, in which event the present comparison of freezing points with triple points has no meaning, as for carbon dioxide[4]). So unless the p, T, x surface has an unexpected bulge below p ~ 10^5 Pa, there will be no clear-cut advantage of a freezing point over a triple point, or vice versa, arising from impurities in the working substance.

3 CHARACTERISTICS OF TRIPLE-POINT CELLS

Whilst triple-point devices can in principle take many forms, our own experience is limited to cylinders with re-entrant wells, as described in an earlier paper[5]; we shall therefore refer from now on to triple-point "cells". A consequence of this geometry is the existence of a hydrostatic head (typically 20 cm) acting on the working substance at the level of the temperature sensor. For water triple-point cells the head will cause the equilibrium temperature to be slightly lower than that for zero head (because dp/dT in equation (1) is negative, $\Delta_{fus}V$ being negative), whilst for the organic working substances studied by us the head will cause the equilibrium temperature to be slightly higher than that for zero head (because dp/dT is positive, $\Delta_{fus}V$ being positive). The effect of the head can be absorbed into the certified temperature for an individual cell, if it is prescribed that sensors should be positioned at the bottom of the re-entrant well.

3.1 Purity of the working substance

3.1.1 Theoretical considerations

The great majority of triple-point cells described in the literature have had highly purified working substances within them: a eutectic mixture could conceivably be used as working substance, but McAllan's work[6] on the freezing characteristics of eutectics indicates the importance of the kinetics of freezing and diffusion, so that temperature readings are more time-dependent than is the case with pure substances; the treatment that follows relates to the latter only.

For a batch of working substance having an impurity content of less than 10^{-3} mole fraction (i.e. a dilute solution), equation (2) is a convenient starting point for discussion; in full, it takes the form

$$T_e^* - T_e' = x_2/[\Delta_{fus}H/R(T_e^*)^2] \qquad (3)$$

where T_e' is the equilibrium temperature of the actual sample when an infinitesimal amount of solid has deposited, T_e^* is the corresponding temperature for a completely pure sample, x_2 is the mole fraction of all liquid-soluble impurities in the batch of working substance and R is the gas constant. For convenience we introduce the cryoscopic constant $A = \Delta_{fus}H/R(T_e^*)^2$, which has the dimension K^{-1}. (The quantity A should more formally be called the "first cryoscopic constant, on a mole fraction basis" to distinguish it from a quantity K_f, which is sometimes called the "cryoscopic constant"[7]). For a given substance the value of T_e^* is a constant of nature. Clearly T_e' for a given batch of a working substance should be close to T_e^* if good inter-cell reproducibility of temperature is sought. But T_e' should also be close to T_e^* to meet the equally important requirement that the equilibrium temperature of a triple-point cell should not be markedly dependent on the extent to which the charge is frozen. The reason for this can be readily seen from equation (4), developed from equation (3):

$$T_e^* - T_e = x_2/F[\Delta_{fus}H/R(T_e^*)^2] \quad (4)$$

Here T_e is the equilibrium temperature when the mass fraction of the charge that is liquid is F. The assumption behind equation (4) is that all impurities remain in the liquid phase as solidification progresses; more realistic assumptions are discussed later.

For a fixed value of F, the depression of triple-point temperature is seen from equation (4) to be proportional to the product of x_2 and $(T_e^*)^2$ and inversely proportional to $\Delta_{fus}H$. The last-named quantity is itself equal to the product of the specific enthalpy of fusion and the molar mass. It follows that the effect of impurities on triple-point temperatures is the less the higher the molar mass of the working substance, other factors being equal.

It seems to us that an important criterion of the quality of a triple-point cell is the change in equilibrium temperature when F varies between certain limits: we suggest that F = 0.75 and F = 0.25 are realistic limits within which thermodynamic equilibrium ought to be attainable for all working substances. Suppose it is decided that the equilibrium temperature of a triple-point cell should not fall by more than 1 mK when F changes from 0.75 and 0.25. By use of equation (4) and values of A it is then possible to estimate the maximum permissible value of x_2, for a working substance of interest, since by this quality criterion

$$x_2 = 3.75 \times 10^{-4} A \quad (5)$$

Examples of the resulting values of x_2 are shown in Table I, which summarizes data for some inorganic substances, metals and organic compounds. From the column headed "ex eqn (5)" (which assumes that impurities are insoluble in the solid) it is seen that the maximum permissible value of impurity content to meet the 1 mK-criterion stated above ranges from 0.69×10^{-6} mole fraction for zinc to 30×10^{-6} mole fraction for n-icosane. The effect of high molar mass (see previous paragraph) is well illustrated by the data for n-decane ($C_{10}H_{22}$) and n-icosane ($C_{20}H_{42}$). The effect of molecular symmetry is well illustrated by the data for cyclopentane (a rather symmetrical, "globular" molecule) and its methyl homologue; there is a 20-fold change in A, reflecting the much higher triple-point temperature and lower molar enthalpy of fusion of the more symmetrical molecule: were these substances to be used in triple-point cells, the purity requirements for cyclopentane would be much higher than for its homologue.

In general, it is apparent from Table I that the total of impurities in a batch of working substance needs to be in the parts-per-million range, to ensure that the equilibrium temperature will change by no more than 1 mK when the charge changes from one-quarter frozen to three-quarters frozen. The purity requirement

TABLE I

Purity requirements for working substances of triple-point cells

Working substance	Cryoscopic constant A/K^{-1}	10^6 x maximum permissible values of x_2	
		ex eqn (5)	ex eqn (6), with K = 5
Argon	0.0294	11	29
Carbon dioxide	0.0222	8.3	??
Water	0.00972	3.7	9.7
Mercury	0.00503	1.9	5.0
Gallium	0.0073	2.7	7.3
Indium	0.00212	0.79	2.1
Zinc	0.00184	0.69	1.8
Cyclopentane	0.0023	0.86	2.3
Methylcyclopentane	0.0488	18	49
Benzene	0.0152	5.7	15
Phenoxybenzene	0.0230	8.6	23
1,3-Dioxolan-2-one	0.0165	6.2	16
n-Butanol	0.0331	12	33
n-Decane	0.058	22	58
n-Icosane	0.081	30	81

is evidently much stricter for the metals than for n-icosane. However, all the data for impurity contents in Table I are estimates, based on the assumption that thermodynamic equilibrium is achievable and that impurities are insoluble in the solid phase; it is now necessary to reconsider the latter assumption. For this purpose, a modified form of equation (4) is introduced[8], namely

$$T_e^* - T_e = \frac{x_2}{A} \left[\frac{1}{F - \left(\frac{1}{1-K}\right)} \right] \quad (6)$$

Here K is the distribution coefficient of impurities between the liquid and solid phases of a partially frozen system. For non-infinite values of K, equation (6), unlike equation (4), admits the possibility that impurities may dissolve in the solid phase.

It is an obvious consequence of the intense purification needed to produce a batch of substance with impurities in the parts-per-million range that impurities will tend to have (physico)chemical properties close to those of their host. Hence residual impurities ("true" impurities, not "adventitious" ones) may very well form solid solutions with the host, especially if fractional freezing has been used as the purification technique. Three special cases are noteworthy: (i) when $K \rightarrow \infty$, impurities are insoluble in the solid, (ii) when $K \rightarrow 1$, impurities have very little effect on freezing temperature, as happens in a mixture of species with differing isotopic compositions, (iii) when $K < 1$, the solid phase actually becomes enriched in impurities as freezing proceeds. But for the situations more commonly encountered a K value between 2 and 100 might well apply. For $K = 5$, the 1 mK-criterion discussed above requires x_2 not to exceed $A \times 10^{-3}$ - see Table I. Hence the purity requirement embodied in equation (5), for the case that $K = \infty$, may be relaxed somewhat when solid solutions are formed between host and impurities, e.g. with $K = 5$. Unfortunately, the formation of solid solutions gives rise to a drawback, namely the more sluggish attainment of thermodynamic equilibrium, due to slower diffusion in the solid phase than in the liquid phase. The article by McLaren[9] is of interest in this regard. He quotes examples of the extent to which p.p.m. impurities in metals vitiate the attainment of equilibrium freezing temperatures; he also gives an example where impurities segregate preferentially in the solid phase at certain freezing rates. Evidently many subtle phenomena connected with nucleation and diffusion may affect the rate of attainment of equilibrium when the working substance has a significant impurity content. The same phenomena have a bearing on the purification of the charge to be used as filling for a triple-point cell. Ideally the purification method would not utilize liquid/solid equilibria, but our experience with organic substances shows that fractional freezing is the only feasible technique, as discussed in the next section.

3.1.2 Practical considerations

It has been shown above that to meet the criterion of triple points stable to 1 mK over a reasonable range of F values it is necessary to fill the cells with materials having purities between 99.99 and 99.999 mole %, depending on the substance. A general method for determining such purities is cryoscopy, especially for organic substances, where reliable alternative methods of determining impurities at the p.p.m. level are unavailable. However, the simplest cryoscopic method, that based on equation (3), will be inapplicable unless an accurate value of T_e^* is available, which is rarely the case. (The other datum needed is a value for A, but this can usually be estimated with sufficient accuracy.) Moreover, our experience with organic substances shows that measurements of freezing-point depression under a gas pressure of ca. 10^2 kPa are often vitiated by dissolved gas, when the mole fraction of genuine impurity is 10^{-4} or less. This problem can be avoided by cryoscopy in the absence of a confining gas (i.e. under triple-point conditions): it is then possible to make reproducible measurements of triple-point temperature for the state when a very small amount of solid has separated, as required by equation (3). Such experiments reveal whether a purification procedure is progressively yielding a purer product, but they will not afford a quantitative measure of x_2 until T_e^* has been established to ± 1 mK. This is best achieved by regarding T_e^*, like x_2, as an unknown to be solved for via equation (4). In principle this requires at least two measurements of T_e, for a corresponding number of known values of F.

Two experimental methods are now described for securing meaningful measurements of T_e for various known values of F, then for solving for x_2 and T_e^*. The need for an effective equation-solving technique arises from the hyperbolic form of equation (4), and the effect of experimental errors in T_e and F on the solution. In other contexts, this problem has been handled by regarding 1/F as the variable, but we have not pursued this path: in fact, our preferred variable is the quantity (1-F), which is the fraction of the charge that is solid. In method (i), known values of (1-F) are obtained by continuous slow cooling of a nucleated charge, whereas in method (ii), known values of (1-F) are obtained by controlled undercooling of the charge. A common approach is used to solve for x_2 and T_e^*. It consists in plotting a family of calculated curves of $(T_e^* - T_e)$ versus (1-F) for various assumed values of x_2, employing equation (4) with known values of $\Delta_{fus}H/R(T_e^*)^2$ for the substance in question; an approximate value of T_e^* suffices for the derivation of $\Delta_{fus}H/R(T_e^*)^2$. By way of illustration, Fig 1 shows two such calculated curves for phenoxybenzene. An experimental curve of T_e versus (1-F) is then plotted on the same scale. The experimental plot is superimposed over the family of calculated curves, in order to select the one that most nearly matches in slopes and curvatures, for a good part of the experimental range. The corresponding value of x_2 is noted, and T_e^* is evaluated from the ordinate of T_e relative to that of $(T_e^* - T_e)$ for the calculated curves. The experimental techniques were as follows.

(i) A batch of working substance is placed in a glass tube equipped with a removable (standard-taper joint) head bearing a high-vacuum tap and a re-entrant pocket for a PRT. The cell is outgassed by cycles of freezing and thawing, then the triple-point state is created with the minimum of solid present (Section 3.2). The tube is placed in an environment controlled ca. 4 K below the triple-point temperature, with a heat-transfer regime that permits the charge to freeze fully over ca. 24 hours. Temperature is automatically measured hourly by means of a PRT and an a.c. bridge. (More details of our thermometric equipment are given in reference 5.) The fall in temperature for a fairly pure charge will be sufficiently small over the whole freezing

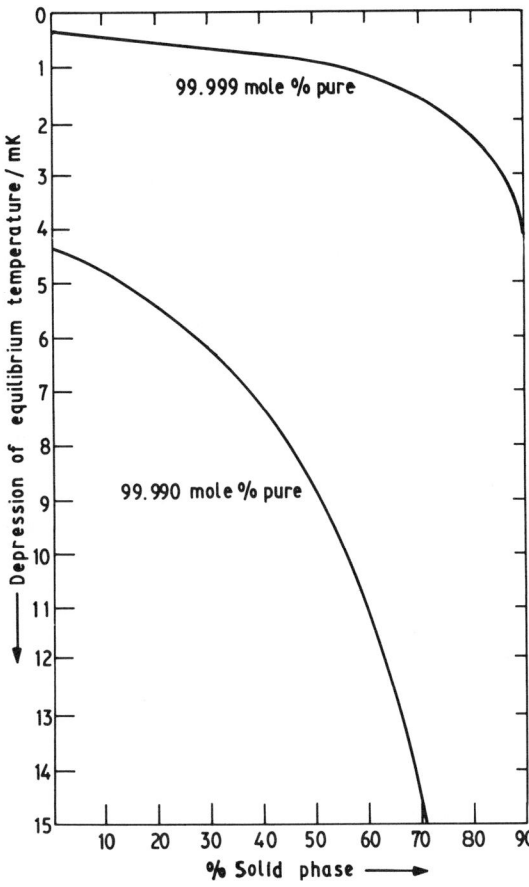

Fig. 1. Phenoxybenzene triple point. Calculated depression of equilibrium temperature as a function of percentage of charge that is solid, for two sample purities.

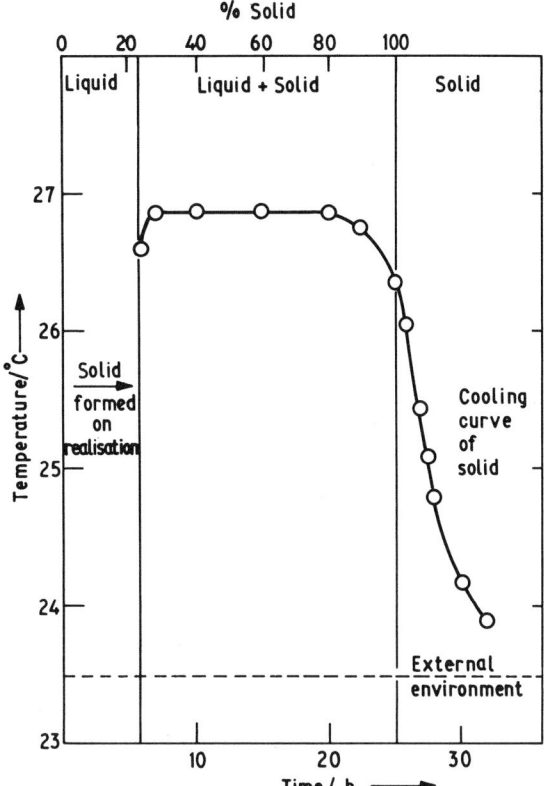

Fig. 2. Phenoxybenzene triple-point cell. Realization of triple-point, followed by cooling, with an environment at 23.5 °C.

range for the rate of loss of enthalpy to the fixed-temperature environment to be considered constant. Intermediate values of (1-F) can then be deduced by proportional parts, with allowance for the small amount of solid initially present. The experiment is therefore analogous to the well-known procedure of measuring time-temperature freezing curves: the main differences from traditional procedure are the use of triple-point conditions, the use of very slow cooling, and the non-use of a stirrer. The absence of agitation permits more accurate measurement of temperature (to 10^{-4} K) for a longer range of (1-F) values than would otherwise be the case. The temperature excursion for a typical experiment is illustrated in Fig 2.

(ii) For those substances which can be undercooled with ease (Section 3.2), a chosen value of (1-F) can be created by nucleating an undercooled system, by control of the extent of undercooling. Measurement of the equilibrium temperature after nucleation then yields the corresponding value of T_e needed for the curve-matching technique described above. The relationship between extent of undercooling and (1-F) for a given cell of the type described in (i) can be established by either calculation[10] or special experiments. For example, for a particular cell containing phenoxybenzene it was shown that 2.5% of the charge froze, after nucleation, for every kelvin by which the charge was undercooled. Fig 3 illustrates how T_e was observed to depend on (1-F) for a particular cell, when values of (1-F) were derived first by method (i) (experimental points shown as crosses), then by method (ii) (experimental points shown as circles). This batch of material is seen to be somewhat impure, comparable with the lower of the two calculated curves shown in Fig 1.

3.2 Realization of the triple-point condition

It is possible to approach the equilibrium triple-point state by either partial melting of the solid or partial freezing of the liquid, in the presence of vapour. When the latter approach is used, undercooling is commonly experienced before spontaneous freezing of the liquid begins; alternatively, an experimenter may decide not to wait for spontaneous crystallization but may induce crystallization by shock treatment (mechanical or thermal) of the metastable, undercooled liquid. The presence of nuclei and/or particular surface states of the container walls must play a role in the initiation of crystallization, but so also do the intrinsic physical properties of the working substance. Thus we have found that n-icosane undercools but little, whereas water, phenoxybenzene and 1,3-dioxolan-2-one form metastable states which may persist for weeks at temperatures many kelvins below the equilibrium temperatures. This tendency to undercool can be exploited in the determination of

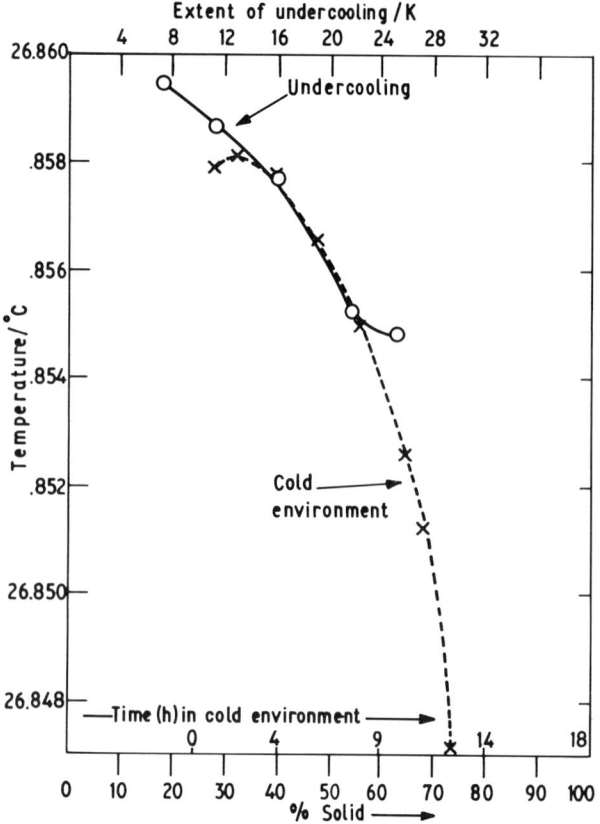

Fig. 3. Phenoxybenzene triple-point cells. Observed equilibrium temperature as a function of percentage of charge that is solid; (i) crosses refer to a cell in a cold environment, (ii) circles refer to a cell subjected to controlled undercooling.

sample purities by thermal measurements (section 3.1.2), and in the ready realization of the triple-point, as will be described in section 3.2.2.

Hitherto, two general methods have been developed for realizing the triple-point condition from the liquid state - the sheath method and the mush method. Both methods are used at the NPL and also a third method for certain organic compounds, which we call "the outer sheath method (film)".

3.2.1 The inner sheath method

In the sheath method, here called the "inner-sheath" method, the contents of the cell are liquefied and cooled to a temperature just above the triple-point temperature. A solid sheath is then created around the re-entrant well by filling it with a coolant at a much lower temperature (e.g. for a water triple-point cell either (a) powdered solid carbon dioxide[11-15] or (b) a metal rod cooled in liquid nitrogen[16]). When the sheath is of the required thickness, say 1/3 of the way to the outer wall of the cell, the coolant is replaced by a warm liquid at a temperature higher than that of the triple point. This releases the sheath, so that when the cell is given a twist, the sheath rotates around the re-entrant well.

When the coolant is first removed, the solid attached to the re-entrant well is at a very low temperature. The warm liquid first brings the sheath up to the triple-point temperature and then any excess heat serves to melt the solid. Thus, providing the contact liquid is warm enough to melt only some of the solid, this inner core of the cell (i.e. the molten film, glass re-entrant well, contact liquid and thermometer) automatically equilibrates at the triple-point temperature. One drawback to this method is that the triple-point temperature may be established rather slowly for substances with poor thermal conductivity (Table II), because very cold solid sheaths will take some time to equilibrate.

Another possible problem is inter-diffusion of the two liquid layers (within and without the solid layer), if the sheath has inadequate thickness or it becomes broken. The outer layer of liquid, which extends to the outer wall of the cell is too close to the environment to be exactly at the triple-point temperature, so inter-diffusion will cause temperature fluctuations. Further, if the melted inner liquid layer is purer than the outer liquid layer, temperature fluctuations may result from the compositional gradient.

3.2.2 The outer sheath method (mush)

As already mentioned, many molecular liquids can be undercooled by several kelvins. If such liquids are contained in a triple-point cell, when crystallization occurs either spontaneously or as a result of shock (e.g. by application of a cold spot to the wall of the cell, or by shaking the cell), the network of crystals spreads rapidly throughout the cell and the temperature inside quickly rises to the triple-point value. The phenomenon forms the basis of the so-called "mush" method of realizing the triple point[11,17].

One problem with this method is that the mush may not be "anchored" in the liquid and in a short time may sink to the bottom of the cell or, in the case of water, may rise to the surface. The mush can be stabilized (i) by leaving the cell at the undercooled temperature for a short period (effective for water triple-point cells) or (ii) by placing the cell in an enclosure controlled at a temperature just below that of the triple point. Under these conditions, the mush thickens on the outer wall of the cell forming an outer sheath. For this reason, we refer to it as the "outer sheath method (mush)." In this method, the triple point is self-regulating in that the solid formed either during the initial crystallization or during the subsequent thickening of the sheath releases just enough enthalpy of crystallization to bring the temperature inside the sheath (i.e. of the liquid working substance, re-entrant well, contact liquid and thermometer) to the triple-point value.

Another problem that sometimes arises with this method is the deposition of appreciable amounts of solid on the re-entrant well. If this has poor thermal conductivity, rapid approach to temperature equilibrium is hindered, which is particularly undesirable in the region near the thermometer. In such cases, before temperature measurements are made, a warm metal rod is inserted in the re-entrant well several times, to melt this solid material.

3.2.3 The outer sheath method (film)

Some liquid organic compounds, when cooled, deposit solid phase on the walls of the containing vessel the moment the temperature drops below the triple-point value. Despite the release of enthalpy of crystallization, the temperature continues to fall, arresting at ca. 0.5 K below the triple-point temperature. Some substances may stay for a long while at this pseudo-equilibrium temperature, where presumably the rate of heat liberation by crystallization just balances the rate of heat loss to the environment. Other substances (e.g. n-icosane) remain at this pseudo-equilibrium temperature for only a few minutes; then the solid particles on the cell wall suddenly coalesce to form a coherent film, and the temperature inside the cell rises rapidly to the equilibrium value. This film can be considered an outer sheath and thus we call the method the "outer sheath method (film)".

3.3 Maintenance of a constant temperature

It is self-evident that maintenance of a constant temperature in the well of a triple-point cell requires that the heat flux from or to the outer walls of the cell should not be excessive or fluctuating, also that any heat flux which bypasses the working substance of the cell (e.g. down the thermometer wall or through connecting leads) should be minimal. The function of the working substance is not only to provide a reproducible temperature but also to buffer the thermometer against local thermal disturbances while that temperature is being measured. There are two differing ways of achieving good buffering: (i) if the working substance is a good thermal conductor, there should be a good thermal linkage between the thermometer and the working substance, so that any thermal disturbance is rapidly dissipated throughout the system as a whole, but (ii) if the working substance is a poor thermal conductor, there is merit in having a relatively poor thermal linkage, so that the thermometer does not sense a thermal disturbance at the boundary of the cell - with this approach it is doubly important to minimize heat leaks down the thermometric well.

The type of thermometric inconstancy just discussed may be termed "fluctuation". On a longer time scale (hours) the term "drift" is appropriate and for longer still (days) the term "longevity" applies. The most likely causes of drift (after the 30-minute period needed to establish thermal equilibrium in the first place) are slow annealing of the crystals and progressive melting/freezing of the charge, with a concomitant change in the distribution of impurities between the solid and liquid phases (section 3.1); careful realization of the triple-point condition will control the first problem, whilst use of a high-quality cell filling and a good arrangement for minimizing heat flux to or from the cell will control the second. The current through the sensor of a PRT is one source of heat flux to a cell; the current should be maintained constant for the optimum performance of the combination of PRT + triple-point cell. For a given heat flux, longevity is principally a function of the enthalpy of fusion of the charge of working substance: a large mass and a working substance with a high specific enthalpy of fusion both favour longevity. The data in the third column of Table II show values of specific enthalpy of fusion varying by a factor of 4.

Finally, one other aspect of longevity should be mentioned, namely that the system should be chemically stable, i.e. that the working substance does not decompose and there is no interaction (chemical change, dissolution) between the working substance and the materials of construction with which it is in contact.

From the practical point of view the maintenance of the triple-point condition depends on placing the cell in an enclosure that minimizes the heat flux to or from the cell. Since, in general, the temperature of such an enclosure will be slightly higher or lower than the triple-point temperature, the direction of the net heat flux is important in maintaining the phases in equilibrium, as will be discussed next.

3.3.1 Direction of heat flux

Even if the heat flux is low, the triple point will not be maintained for very long if the flux is in the wrong direction. For example a thin sheath in an enclosure at a slightly higher temperature than that of the triple-point will soon collapse, whereas with a thick sheath subject to cooling, the disposition of liquid in the cell will become ineffective. In both cases drift will be high and longevity poor. Contrariwise, favourable directions of heat flux will improve any initial shortcomings in the dispositions of solid and liquid layers. For example, with organic substances it has been found convenient to start with a comparatively thin sheath and to mount the cell at a temperature slightly lower than that of the triple point, then to insert a warm rod in the re-entrant well from time to time to melt off the extra solid formed. This rejuvenation is indicated in Fig 5 by the words "warm rod".

3.3.2 Magnitude of heat flux

The low thermal conductivity of most molecular compounds causes the corrective response to an external thermal fluctuation to take place comparatively slowly. But at least the disposition of solid and liquid layers will not be radically altered if the corrective response takes place slowly. With triple-point cells, therefore, low thermal conductivity of the working substance is not a disadvantage, as has been suggested[18] but a positive advantage in securing low drift and good longevity.

In general, longevity is favoured by keeping the difference in temperature between the cell at the triple point and its immediate environment under close control. In the case of water, the triple-point condition is usually maintained by keeping the cell in an open bath of crushed ice[11-16]. With most organic compounds a comparable procedure is rarely feasible for reasons of cost and safety.

Other types of triple-point cell are normally mounted in a thermostatted enclosure. There is merit in having a semi-insulated barrier between the cell and the thermostatted enclosure. For triple-point temperatures near the ambient a Dewar flask lined with cotton wool is a convenient barrier. However, a problem with such an enclosure is that it has an appreciable enthalpy; to ensure that not too much heat is removed from or transmitted to the cell during the early equilibration period, it is important that such an enclosure is pre-conditioned with a dummy cell[19].

3.4 A user's requirements for an effective triple-point cell

The physical principles discussed in sections 3.1 - 3.3 may be summarized in the following desiderata for a triple-point cell and its associated equipment, from the user's standpoint:
(i) quick and easy realization of the triple-point condition
(ii) good temperature repeatability from day to day
(iii) good temperature constancy during a working day
(iv) long-term reliability.

How far these requirements are met by the four types of cell developed to date at NPL is discussed in Sections 4 to 7. Pertinent thermophysical data for the working substances of the four types of cell are summarized in Table II.

TABLE II

Selected thermophysical constants

Substance	T_{tp} K	$\Delta_{fus} h$ J.g^{-1}	$\lambda(l, T_{tp})$ W.m^{-1}.K^{-1}
Water	273.160	333.5	0.569
Phenoxybenzene	300.021	101.15	0.135
1,3-Dioxolan-2-one	309.474*	113.6	[0.15]
n-Icosane	309.641*	247.4	0.150
Gallium	302.923	80.3	28.1

* Provisional values from the present work, corrected to zero current through the measuring PRT, but uncorrected for hydrostatic head.

4 WATER TRIPLE-POINT CELLS

NPL's cells have been designed specifically to accommodate standard-pattern PRTs. Thus the dimensions of the NPL cells are:- height 28 cm; diameter 4 cm; well depth, 33.5 cm; well diameter 1.2 cm. They are of borosilicate glass.

These cells are filled by the method described by Ambrose et al.[20] and are certified by comparison against master water triple-point cells held at NPL. After the triple point of a candidate cell has been realized by the sheath method, it is placed in crushed ice and left 24 hours to reach temperature equilibrium. Usually four or more candidate cells are evaluated at one time. Their triple points are compared with that of a master cell, using a PRT, daily for 3 days. NPL currently has three master cells which are frequently inter-compared, and whose performances over many years have been monitored. The certificate states "The temperature of the cell did not differ from that given by the reference cell by more than 1×10^{-4} K". Very few cells fail to meet this standard and consequently fail to be certified.

4.1 Purity

Water having a purity suitable for triple-point cells can be prepared by distillation of deionized water. An important matter is to ensure that the empty glass cell is of the required cleanliness before filling. Our earlier procedure is described in reference 16 and the current procedure in reference 20.

4.2.1 Realization of triple point; sheath method

The generally accepted method for realizing the triple point of water is the sheath method[11-16]. The method recommended in the NPL certificate is similar to that described by Foster[14], whereby crushed solid carbon dioxide is used to create a solid sheath round the re-entrant tube.

4.2.2 Realization of triple point; mush method

Earlier attempts to use the mush method for the realization of the triple point of water were abandoned because the mush tended to float to the surface and the

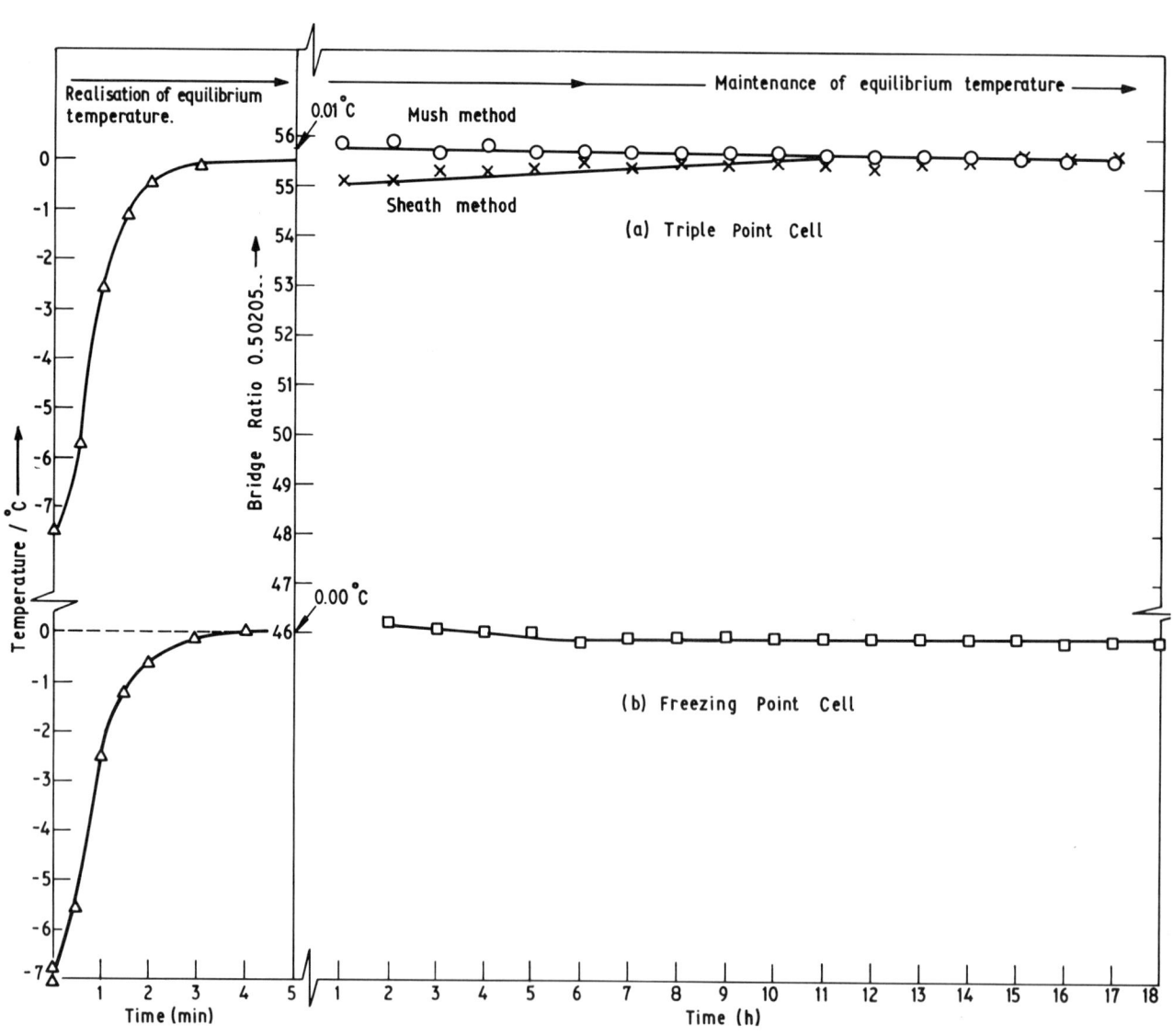

Fig. 4. Realization and Maintenance of equilibrium temperatures.
(a) Water triple-point cell. (b) Water freezing-point cell.

triple point could be maintained for only a few hours. In NPL's alternative method the mush is stabilized by conversion into an outer sheath (section 3.2.2) and the triple point can then be maintained for many days (see Fig 5).

Generally is it difficult to undercool water below -9°C but at temperatures between this and 0°C the liquid remains metastable unless subjected to shock. The cell containing liquid water is placed in a bath at -8°C (brine/ice mixture or methanol cooled with solid carbon dioxide) and temperature stability is reached after about 30 minutes. When the cell is given a slight shake, crystallization spreads rapidly throughout the cell and the temperature in the cell rises quickly to the triple-point (see Fig 4). The cell is returned to the -8°C bath for about 15 minutes (not less than 10 minutes or the mush may still be too soft; not greater than 20 minutes or too much ice may form). The cell is then briefly immersed in water (20°C) until the mushy sheath is released from the outer wall and rotates when the cell is given a twist. After being inverted three times to ensure isothermal conditions, it is placed in a Dewar flask packed with crushed ice and is ready for use.

4.2.3 Comparison of sheath and mush methods

One advantage claimed for the sheath method is that the melted layer of water next to the re-entrant tube is ultrapure and this gives a higher value for the triple point[12]. However more recently Berry[13] discounted this effect and concluded that the real value of the inner layer of water is in having a continuous ice-water interface close to the thermometer. Whatever speculative advantage the sheath method may have in this respect, the results of our own experiments show very little difference between the sheath and mush methods (see Fig 4; a difference of 1×10^{-7} in the bridge ratio[5] is roughly equivalent to a 0.1 mK difference in temperature).

On the practical side, most of the advantages are with the mush method. Unlike the sheath method, which requires a supply of solid carbon dioxide or liquid refrigerant, all that is required is a supply of ice and sodium chloride. Furthermore it requires less time and less supervision. Another practical advantage is that several cells can readily be treated at the same time. Table III shows that the values of the the bridge ratio for a PRT in each of 12 cells are all within the equivalent of 0.2 mK of the expected ratio 0.502 055 5, within 1 hour of the realization of the triple points.

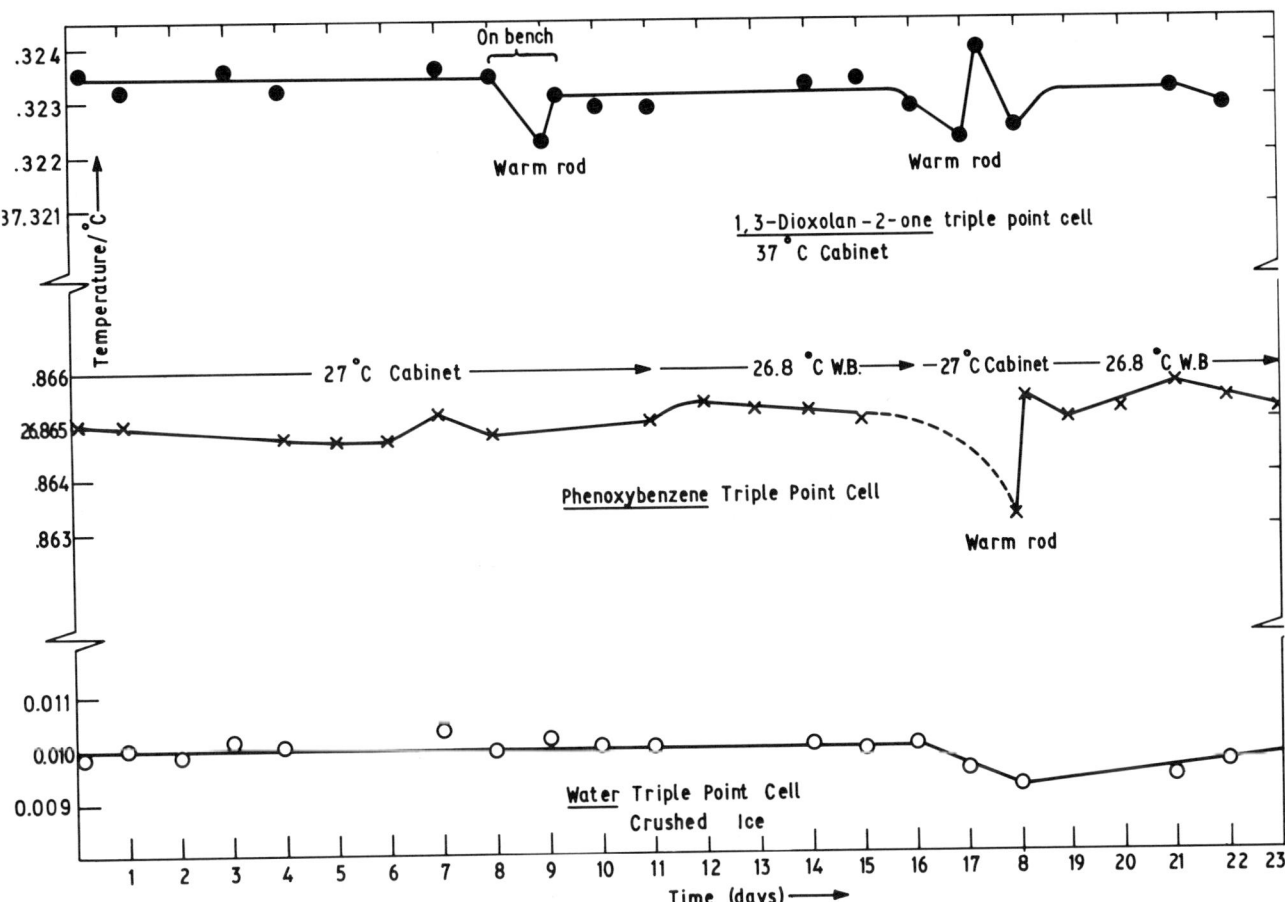

Fig. 5. Long-term maintenance of triple-point. (a) Water. (b) Phenoxybenzene. (c) 1,3-Dioxolan-2-one.
The letters W.B. refer to use of a water bath.

J. D. Cox and M. F. Vaughan

TABLE III

Reproducibility of water triple point
(in terms of bridge ratio) between
different cells

Cell Number	Bridge ratio after 1 h
84	0.502 055 5
85	0.502 055 6
86	0.502 055 5
87	0.502 055 5
88	0.502 055 4
89	0.502 055 3
90	0.502 055 2
91	0.502 055 4
92	0.502 055 7
93	0.502 055 4
94	0.502 055 7
96	0.502 055 6

TABLE IV

Repeatability of "ice point" (in terms
of bridge ratio) for a single cell

Run No.	Bridge ratio after 1 h
1	0.502 046 2
2	0.502 045 8
3	0.502 046 0
4	0.502 045 8
5	0.502 045 9
6	0.502 045 6
7	0.502 045 9
Mean	0.502 045 9
Std.Dev.	0.000 000 2
(equivalent to 0.2 mK)	

4.3 Maintenance of triple point

After realization by either method, the triple-point temperature takes a few hours to reach its equilibrium value (see Fig 4) and can then be preserved for weeks by placing the cell in crushed ice (Fig 5). These are good conditions for maintaining the triple point. By both methods the sheath is comparatively thin and the fact that ice melts at atmospheric pressure at a temperature lower than that of the triple point means that the heat flux is in the right direction to ensure the sheath grows. Freshly crushed ice should be added to the outer bath daily, to ensure longevity.

4.4 Water freezing-point cells

Sometimes the "ice point" is required. For instance with some liquid-in-glass thermometers it is recommended that the zero of the scale should be periodically checked against the ice point. A common method of realizing it is by mixing distilled water and distilled-water ice, and accurate realizations have been made in this way[21]. However this procedure is not as reliable as it sounds, since the mixture takes some time to reach equilibrium; the temperature may vary depending on the ratio of ice to water and the temperature may be lowered if impurities are dissolved from the atmosphere, containing vessel etc. Furthermore, not all laboratories have supplies of distilled-water ice. At the NPL, therefore, we have prepared some experimental water freezing-point cells, which since they are sealed do not become contaminated.

Cells similar to the water triple-point cells were thoroughly cleaned, filled with distilled water saturated with air, and sealed off under atmospheric pressure. The freezing temperature is realized by the mush method and is maintainable for several hours by placing in a Dewar flask packed with crushed ice (see Fig 4). With these cells the equilibrium freezing temperature declines as the amount of ice increases, and drifts as large as 4 mK have been observed. This is probably because (i) the air dissolved in the liquid water acts as an impurity and as ice forms may be left in the liquid phase in a state of supersaturation, and (ii) changes in the amount of substance in the gas phase and of the volume available to the gas cause pressure drifts. Nevertheless if a rigid realization procedure is followed, realization-to-realization variations can be kept below 1 mK (see Table IV).

5 PHENOXYBENZENE TRIPLE-POINT CELLS

The triple point of phenoxybenzene (diphenyl ether) at 26.87°C is one of the Secondary Reference Points listed in IPTS-68[1]. This particular TFP has not been widely used. Reasons for this have been suggested by Mangum[18], who stated that phenoxybenzene is difficult to purify and has a low thermal conductivity, and by Sostman[22] who opined that this triple-point temperature is known to only ±10 mK and that equilibrium points of organics are generally unsatisfactory (because of the difficulty of obtaining materials of sufficient purity), are brief in plateau duration, and require delicate equipment and considerable operator skill. These criticisms may have been true of some earlier attempts to make and use this type of triple-point cell, but our experience has been quite different. We have found that it is not too difficult to prepare phenoxybenzene of the required purity[23] and have developed methods for realizing and maintaining the triple point at 26.869°C, using simple materials and apparatus available in most laboratories.

5.1 Purity

In some earlier work at the NBS, Schwab and Wichers[24,25] reported that they deduced the purity of the phenoxybenzene contents of their triple-point cells from cryoscopic measurements, which were not detailed. In our work, phenoxybenzene was purified as already described[5,23], whilst sample purities were assessed by the cryoscopic techniques described in Section 3.1.2. Methods (i) and (ii) were both applicable. The value of the cryoscopic constant was taken from Furukawa et al.[26]. Typical results are shown in Fig 6 for a specimen of phenoxybenzene at two stages of purification.

5.2.1 Realization of triple point; inner sheath method

Takata et al.[27] used the sheath method with phenoxybenzene, and initially we used a similar method: the contents of the cell were melted and kept just above 27°C, then the sheath was created by repeated insertions of a metal rod, cooled in liquid nitrogen, into the re-entrant well. This method was found to have three disadvantages.

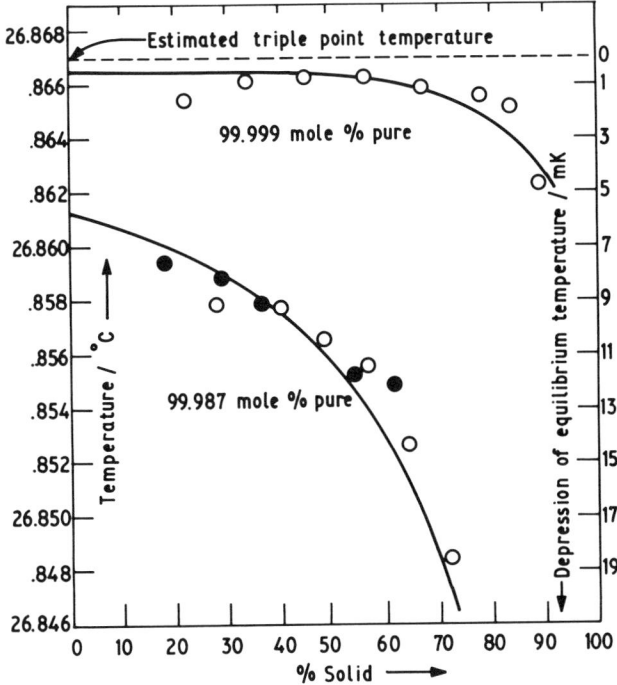

Fig. 6. Phenoxybenzene triple-point cells. Matching of calculated curves to experimental points for two specimens; (i) open circles refer to use of a cold environment, (ii) solid circles refer to use of controlled undercooling.

(a) The sheath was brittle and tended to break easily. (Crystal mats of organic compounds are mechanically weaker than those of ice.)
(b) Unlike water where the sheath tends to float to the top, isolating the two liquid layers either side of the sheath, with phenoxybenzene the sheath tended to sink, allowing mixing of the two layers.
(c) Once the sheath had been created, the temperature in the cell took several hours to reach equilibrium.

It was decided therefore to use the more convenient mush method.

5.2.2 Realization of triple point; outer sheath method (mush)

Liquefied phenoxybenzene is well mixed before being placed in a cell in an ice+water mixture, where it is allowed to cool. When the temperature reaches ca 2°C, crystallization occurs spontaneously throughout the cell. Under these conditions some while elapses before equilibrium is reached but if, after crystallization, a warm metal rod is inserted into the re-entrant tube three times and the cell is subsequently inverted, equilibrium is obtained in a matter of minutes (see Fig 7).

5.3 Maintenance of triple point

5.3.1 Environment of melting phenoxybenzene

By analogy with the convenient procedure used to maintain water triple-point cells, a phenoxybenzene cell in the triple-point condition was kept in a bath of melting phenoxybenzene. However, the triple-point temperature was maintained for only about 2 hours (Fig 7) and then declined rapidly. The likely reasons for the lack of success were (i) the external phenoxybenzene was less pure than the internal specimen, giving rise to a temperature gradient that increased as melting proceeded, (ii) there was a considerable amount of liquid in the melting phenoxybenzene, so that the interface with the cell was bathed in liquid, giving an (unwanted) effective method of heat transfer.

5.3.2 Dewar flask lined with cotton wool

Once the triple point had been realized, cells were placed in a Dewar flask lined with cotton wool. After treatment with the warm metal rod, the outer sheath was comparatively thin, so to ensure that it grew inwards an efflux of heat was essential. Therefore the Dewar flask was placed in an environment whose temperature was lower than that of the triple point.

Three different environments have been investigated (see Fig 7)
(a) on the bench at ca. 20°C, when the triple point was maintained for about 10 hours
(b) in a water bath thermostatted at 26.7°C, when the triple point was maintained for at least four days
(c) in a cabinet controlled at ca. 27°C, when the triple point was maintained for at least 11 days.

By a combination of (b) and (c), with occasional insertions of a warm metal rod into the re-entrant well to melt some of the solid phase formed, the triple point was maintained for 23 days (Fig 5). It is likely that the triple point could be maintained for much longer, with comparatively little effort.

6 1,3-DIOXOLAN-2-ONE TRIPLE-POINT CELLS

1,3-Dioxolan-2-one (ethylene carbonate) was found to have crystallizing properties similar to those of phenoxybenzene; consequently procedures for its purification and triple-point realization were developed along similar lines.

6.1 Purity

Since a value for the cryoscopic constant was available[28] and batches could be undercooled by at least 20 K, it has been possible to utilize both method (i) and method (ii) (section 3.1.2) for monitoring the progress of purification. Fig 8 illustrates results for two specimens. Details of purification methods will be published in due course.

6.2 Realization of triple point

The outer sheath (mush) method (section 3.2.2) was used to realize the triple point of dioxolanone. The triple point can be realized by cooling until crystallization occurs spontaneously, or by applying a cold spot to the cell. In addition the crystallization of dioxolanone can be initiated by shaking an undercooled cell. Many variations of procedure have been investigated but the following has been adopted as the standard one for realizing the triple point of dioxolanone.

The cell is immersed quickly and completely in water at 80°C and left for 1 h. The cell is next inverted twice and placed in a beaker of water at 20°C until it reaches temperature equilibrium (15-20 minutes). When shaken, crystallization spreads rapidly throughout the cell, which is then placed in a Dewar flask lined with cotton wool (cf. Section 5.3.2). Constant temperatures were achieved within 1 h. By this method, the triple point of a particular cell was realized 23 times; the mean temperature was 36.324°C, with a standard deviation of 0.5 mK.

Under these conditions the triple point is very nearly reached after 30 minutes. However, as with phenoxybenzene, the approach to equilibrium can be speeded up by the insertion of a warm metal rod in the re-entrant well (see Fig 9).

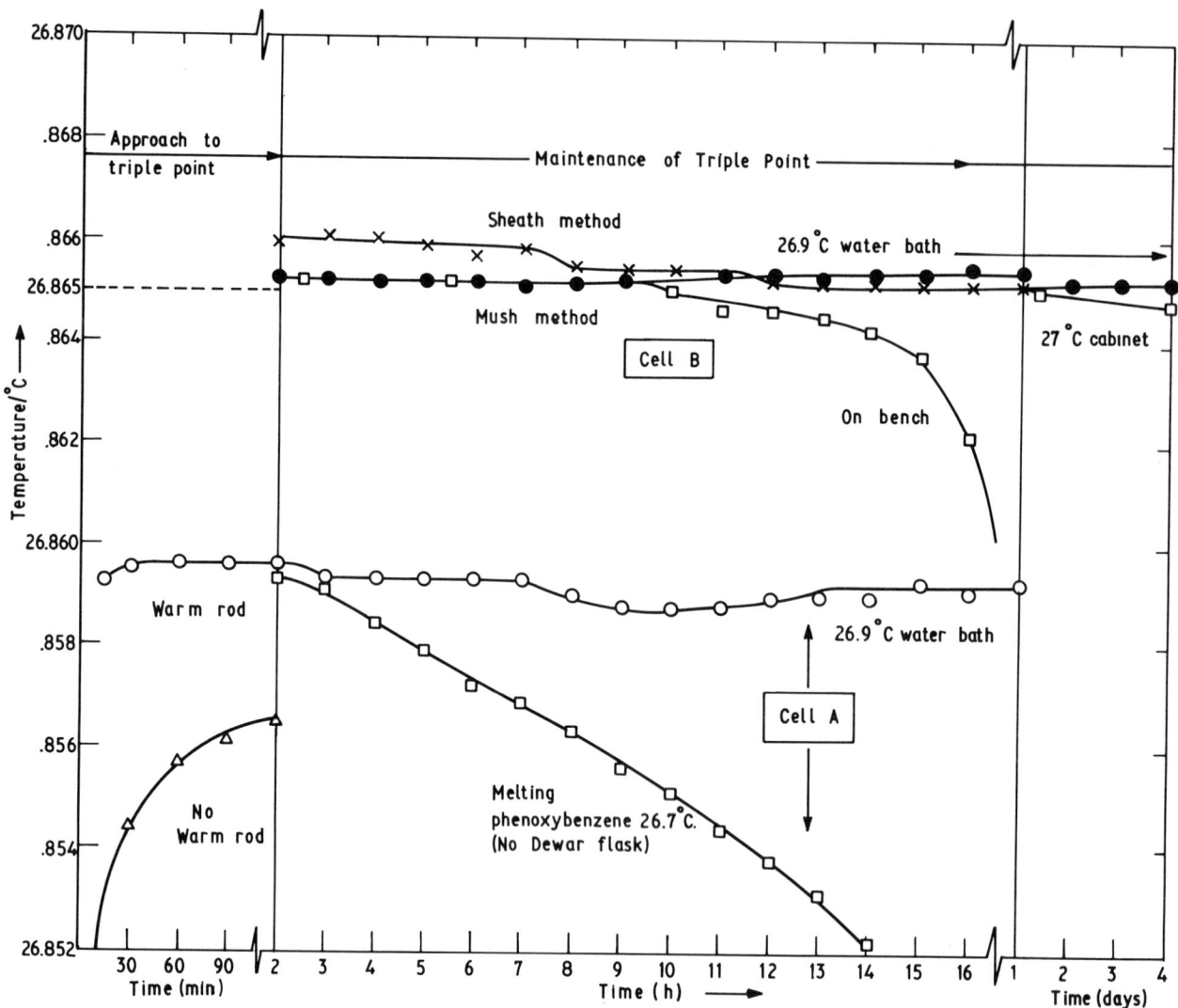

Fig. 7. Phenoxybenzene triple-point cells A and B. Approach to, and maintenance of, equilibrium temperature.

6.3 Maintenance of triple point

The temperature enclosure mostly used to contain the cell was a Dewar flask lined with cotton wool. Experiments were conducted with the Dewar flask (a) on the bench at ca. 20°C, (b) on the bench after pre-conditioning (c) in a water bath at (35.9 ± 0.3)°C and (d) in a cabinet at ca. 37°C. Procedure (a) gave a lifetime of one working day, (c) gave a lifetime of one working week. The results of (a), (b) and (c) are shown in Fig 9 whilst those of (d) are in Fig 5.

7 n-ICOSANE TRIPLE-POINT CELLS

Many n-alkanes show polymorphism, and for some of them crystal-to-crystal transitions occur too close to the melting temperature for the substance to be used as a TFP: fortunately, n-icosane does not show polymorphism[29].

The crystallizing behaviour of n-icosane is different from that of the three substances already discussed, in that spontaneous crystallization invariably occurs as the liquid is cooled 0.5 K or less below the triple-point temperature. This difference in behaviour has important consequences for purification, purity control by cryoscopy, and realization of the triple point.

7.1 Purity

Icosane was tedious to purify by fractional freezing, due to its crystallizing behaviour and the fact that the available starting material contained isomers and congeners. Nevertheless this method was developed to give adequately pure samples of n-icosane.

Although it was not possible to examine the purity of the contents of the cells by controlled undercooling, it was possible to deduce sample purities from cooling curves (section 3.1.2), using the published cryoscopic constant[30]. By way of example, Fig 10 shows the results for two specimens of n-icosane.

Details of purification will be given elsewhere.

7.2 Realization of triple point

The triple point was realized by the outer sheath method (film) (section 3.2.3).

The repeatability of measurement of the triple-point temperature and the time taken to reach equilibrium depend greatly on the exact cooling schedule used. After much investigation, the following method was adopted.

The cell is immersed quickly and completely in water at 80°C, until the contents are liquid, when it

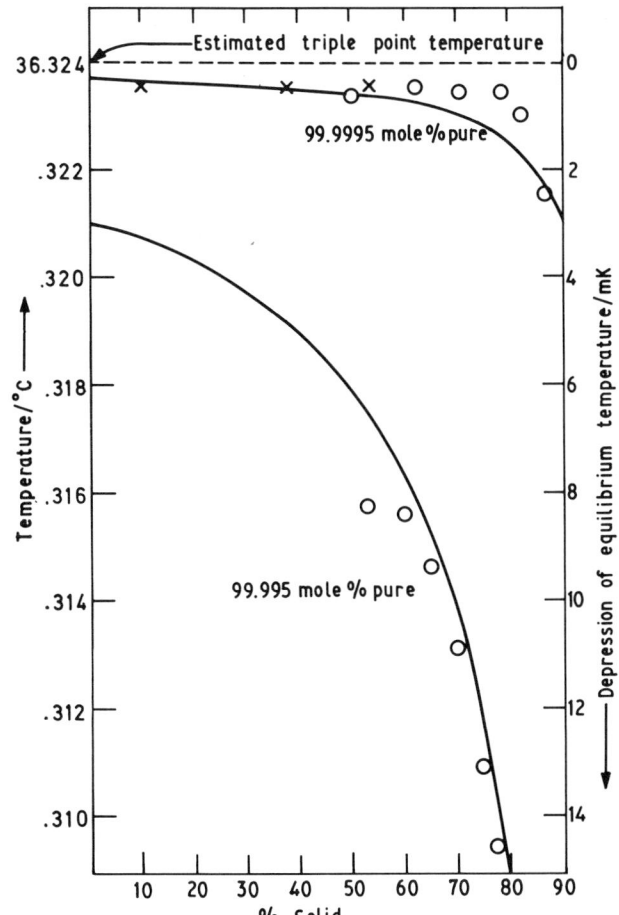

Fig. 8. 1,3-Dioxolan-2-one triple-point cells. Matching of calculated curves to experimental points for two specimens; (i) circles refer to use of a cold environment, (ii) crosses refer to use of controlled undercooling.

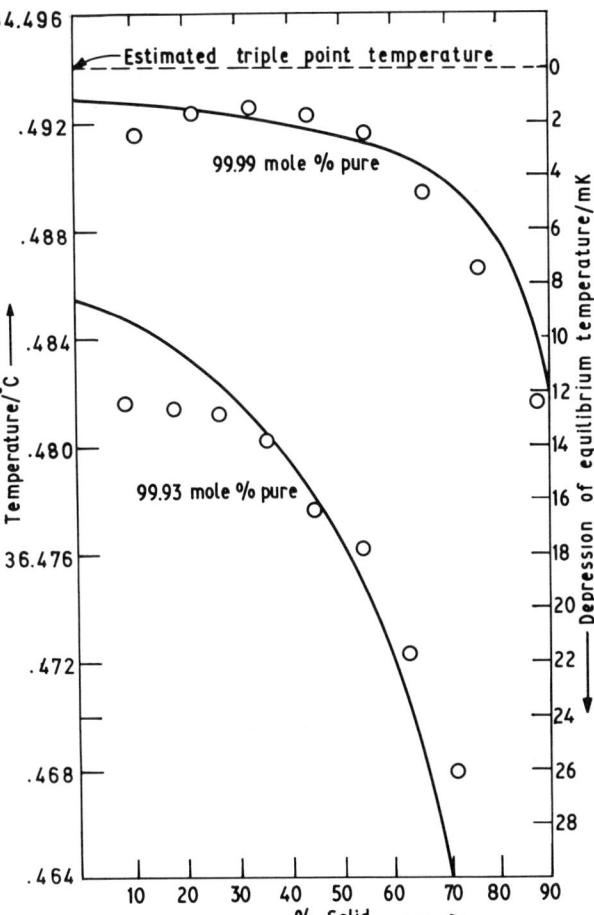

Fig. 10. n-Icosane triple-point cells. Matching of calculated curves to experimental points for two specimens.

NOTE: Figure 9 will be found on the following page.

is inverted twice. It is next placed in a liquid bath initially at ca 35°C, and the two are allowed to cool together. After 10-15 minutes the temperature in the cell arrests itself 0.2-0.5 K below the triple-point temperature, and a few minutes later crystallization spontaneously occurs. The cell is then transferred to the temperature enclosure (section 7.3). By this method the triple point of one cell was realized on 15 separate occasions; after 1 hour from realization the mean temperature was 36.491°C, with a standard deviation of 0.5 mK.

However, later experiments showed that full temperature stabilization may take up to 4 hours to achieve. The insertion of a warm rod does not appear to accelerate the approach to equilibrium in this case (Fig 9).

7.3 Maintenance of triple point

To maintain the triple-point condition, the standard enclosure of a Dewar flask lined with cotton wool was used. Even with the most favourable environment tried, when the Dewar flask was placed in a water bath thermostatted at 35.9°C, the triple-point temperature began to decline after 2 days (Fig 9). For longer service it may be necessary to match the temperature of the water bath closer to the triple-point temperature (36.49°C).

REFERENCES

1. H Preston-Thomas, Metrologia, 12, 7, (1976).
2. L Crovini, R E Bedford and A Moser, Metrologia, 13, 197, (1977).
3. J A Hall and C R Barber, Notes on applied science, No. 12. Calibration of temperature measuring instruments, 3rd edition, Her Majesty's Stationery Office, London (1964).
4. C H Meyers and M S van Dusen, J. Res. Nat. Bur. Stand., 10, 381, (1933).
5. J D Cox and M F Vaughan, Metrologia, 16, 105, (1980).
6. J V McAllan, Temperature. Its measurement and control in science and industry, Vol 4, Instrument Society of America, Pittsburgh, (1972), 265.
7. J A Riddick and W B Buner. Organic solvents. Physical properties and methods of purification. 3rd edition, Wiley-Interscience, New York (1970), 48.
8. Cahiers de thermodynamique chimique (edited by P Clechet, M Ducros and H Tachoire), Ecole Nationale Supérieure de Techniques Avancées, Paris (1979), 166.
9. E H McLaren. Temperature. Its measurement and control in science and industry. Vol 4, Instrument Society of America, Pittsburgh (1972), 185.

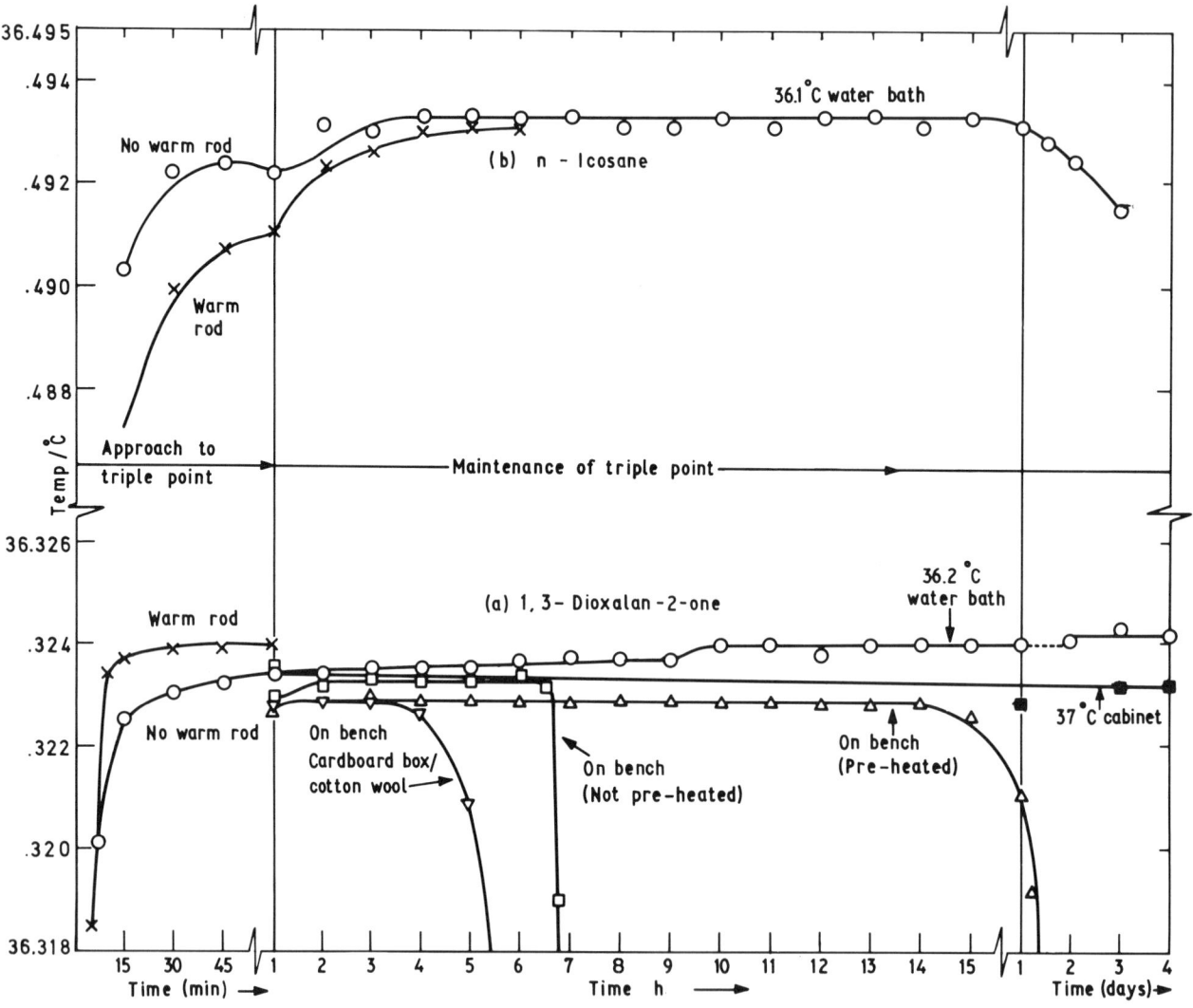

Fig. 9. Approach to, and maintenance of, equilibrium temperature. (a) 1,3-Dioxolan-2-one triple-point cell. (b) n-Icosane triple-point cell.

10 J D Cox and M F Vaughan, to be published.
11 W P White. J. Am. Chem. Soc. 56, 20 (1934).
12 H F Stimson. Temperature. Its measurement and control in science and industry, Vol 1, Reinhold Publishing Corp., New York (1955), 159.
13 R J Berry. Can. J. Phys. 37, 1230 (1959).
14 R B Foster. Temperature. Its measurement and control in science and industry, Vol 4, Instrument Society of America, Pittsburgh (1972), 1403.
15 L L Sparks and R L Powell. ibid. p. 1415.
16 C R Barber, R Handley and E F G Herington. Brit. J. Appl. Phys., 5, 41, (1954).
17 J L Thomas. J. Res. Nat. Bur. Stand., 12, 323, (1934).
18 B W Mangum. NBS Special Publication 481, 711, (1977).
19 D P Enagonio, E G Pearson and C P Saylor. Temperature. Its measurement and control in science and industry. Vol 3 Part 1. Reinhold Publishing Corp., New York (1962), 219.
20 D Ambrose, R R Collerson and J H Ellender. J. Phys. E., 6, 975, (1973).
21 J L Thomas. Temperature. Its measurement and control in science and industry. Vol 1. Reinhold Publishing Corp., New York, (1941), 159.
22 H E Sostman. Rev.Sci.Instrum., 48, 127, (1977).
23 M F Vaughan. NPL Report Chem 86 (1978) 36.
24 E Wichers. Comptes Rendus 16e Conférence de l'Union Internationale de Chimie Pure et Appliquée, (1951) 70.
25 F W Schwab and E Wichers. Comptes Rendus 15e Conférence de l'Union Internationale de Chimie Pure et Appliquée, (1949) 113.
26 G T Furukawa, D C Ginnings, R E McCoskey, and R A Nelson. J.Res.Nat.Bur.Stand., 46, 195, (1951).
27 S Takata, H Uchiyama, T Sugiyama, S Yamashita and I Shimizu. Bull.Natl.Res.Lab.Metrol., 18, 15, (1969).
28 S K Gross and C Schuerch, Anal. Chem., 28, 277, (1956).
29 M G Broadbent. J.Res.Nat.Bur.Stand.A, 70, 481, (1966).
30 M D Tilicheev, V P Peshkov and S G Yuganova, J.Gen.Chem.USSR, 2, 1341, (1951).

Melting curves of H₂O

J. Ancsin

National Research Council of Canada, Division of Physics, Ottawa, Ontario, Canada, K1A 0R6

Melting curves of H_2O have been obtained at the triple point pressure and at 1 atmosphere pressure of air, O_2, He, Ar, CH_4, Kr, Xe, and CO_2. A linear relationship has been obtained between the depression of the liquidus point of ice and the solubility of these substances in water. A relationship has also been found for the lowering of the liquidus point of ice as a function of pressure exerted by CH_4 up to 41 atm. A melting curve for standard sea water was also obtained.

INTRODUCTION

We have developed a system for calibrating capsule type Platinum Resistance Thermometers (PRT) in a comparison block that accommodates seven thermometers and one of a set of cryogenic triple point cells. The utility of the system would be greatly enhanced if it included a water triple point cell because one could conveniently obtain the resistance of the PRT's at the triple point of water without removing them from the comparison block. Normally they would be removed to measure their value at that temperature in a conventional triple point of water cell. As we will show these all-metal-cells of the sort made by several national standards laboratories can be used in this way and can also be used without any difficulty to determine the melting curves of ice (solid H_2O). We have performed the following experiments to elucidate the behaviour of melting ice.

EXPERIMENTAL DESCRIPTION

The apparatus used in these experiments has already been described[1] except for the following modifications: the sample chamber was replaced by a Cu comparison block that is suspended by three thin cotton threads. It weighs 434 g and contains seven capsule PRT's and a triple point cell (≈ 200 g). The adiabatic heat shield is surrounded by an additional heat shield that intercepts most heat leaks due to thermal radiation and residual exchange-gas conduction sufficiently well that the cryostat is capable of calibrating thermometers from 13 K to the steam point. Without this outer shield the sample temperature stability was not sufficiently good for reliable experimentation, above about 200 K. Thermometer resistances are measured with a Kuster's Current Comparator Resistance Bridge. For the melting curve of ice experiments we replaced the comparison block with a light weight copper cell holder consisting of a thin copper pipe held in place by cotton threads. Its upper section was partially split along its axis so that a triple point cell could be slid into it (indium-coated on the inside) and clamped with a screw to make good heat contact between the cell and its holder. The cell heater, all of the differential thermocouple junctions and the thermometer are permanently attached to this cell holder. Changing the cell is thus simply done by loosening the clamping screw, removing the cell, and replacing it with another without having to touch the thermometer or any of the electrical wiring.

CELLS AND THEIR FILLING

Our all-metal cells were originally developed for use as cryogenic fixed points. For that purpose they had to be strong enough to hold gases at high pressure (≈ 100 atmospheres), at room temperature, yet have high enough heat conductance to expedite the reaching of thermal equilibrium of a poorly heat conducting substance at low temperatures. After testing several cell designs we finally adopted a cell made of copper having a number of closely spaced copper baffles thermally tied to the thermometer well. The cell was then enclosed by a stainless steel jacket to strengthen it (see Fig. 1). It can be sealed with an In tipped Cu screw. This sealing arrangement makes it possible to open the cell (by simply removing this seal-screw), refill it with another sample of any suitable substance, and then simply reseal. The sealing screw is reconditioned each time by re-melting its deformed In tip.

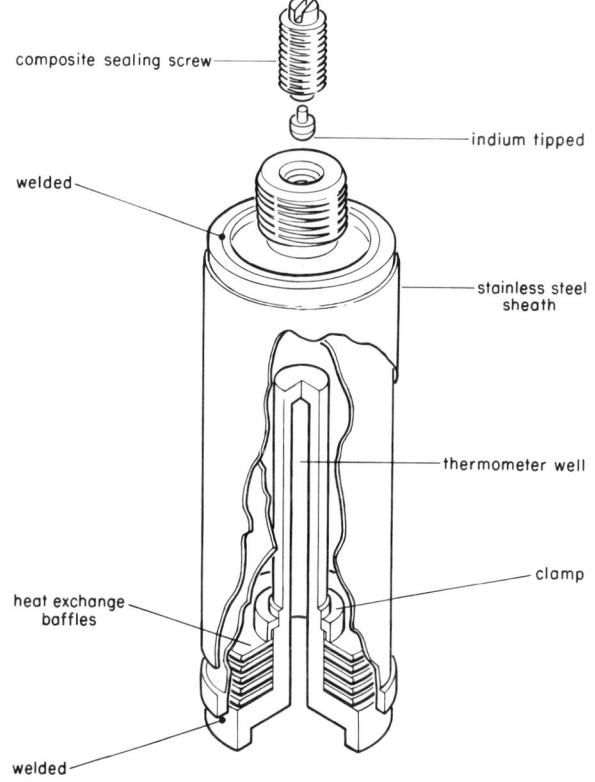

Fig. 1 Triple point cell.

To determine the effect of various gases on the melting curve of ice we first purged the cell with a given gas. This was done by blowing the selected gas gently into the opened cell using a hypodermic syringe that was connected through a rubber tube to the pressure regulator of the high pressure gas bottle. The mass of the purged cell containing only the purging gas at ambient pressure was then determined. Afterwards, 1 to 5 cm^3 of doubly distilled, de-ionized water was injected into the cell using a clean hypodermic syringe. As the injection of water may have resulted in the infusion of some air the cell was purged again, with the gas and then sealed and finally its mass redetermined. Masses were measured with a Mettler # 315 balance, sensitive and accurate to better than 1 mg. The difference of 15 mg between an air filled and nitrogen filled cell for instance could thus be easily determined. When a melting experiment was completed the mass of the cell was redetermined. Masses before and after the experiment usually agreed within one or two milligrams (this is one of the conditions necessary for a reliable heat of fusion determination). Afterwards, to remove the filling gas from the cell, the seal screw was removed, and the cell heated to the boiling point of water, on top of a hot plate. As the water inside the cell came to a boil steam could be clearly seen escaping from the cell. This escaping steam carried with it the filling gas out of the cell. It was kept steaming for a few minutes and then without removing it from the hot plate it was resealed and re-mounted in the cryostat. The subsequent melting curve is the topmost curve in Fig. 2.

THE MEASUREMENTS

The experimental procedure was that used in a conventional adiabatic heat capacity determination. After supplying a given amount of heat to the sample, its final temperature is determined from its temperature vs time curve. The time for this final temperature determination was kept sufficiently long, so that one could ascertain that the sample temperature remained constant within 0.1 mK for a prolonged time period. This cooling period could be as short as a few minutes or as long as 10-15 hr depending on the shape of the cooling curve. The heating-cooling period was then repeated.

The temperature change of a solid sample caused by a given amount of supplied heat is equivalent to a resistance change (ΔR) of PRT, and the heat supplied is equivalent to a heating time (Δt) because the heating current and the voltage drop across the sample heater were kept constant. The ratio $[\Delta t/\Delta R]$ was then calculated. Several such ratios were determined before the sample reached the melting temperature. These ratios increased proportionally to the heat capacity (C) of the sample system (i.e., sample + cell + holder + thermometer) as

$$C = [\Delta t/\Delta R] \, Vi \qquad (1)$$

Once the value of the temperature change became less than expected for an applied heating time (Δt), it was assumed that melting had begun. The <u>melting time</u>, $(\Delta t)_{melt}$, for a particular heating period was taken to be the <u>heating time</u>, $(\Delta t)_{heat}$, minus the time required to heat the sample system through the resulting temperature change (ΔR), i.e.,

$$(\Delta t)_{melt} = (\Delta t)_{heat} - [\Delta t/\Delta R] \, \Delta R , \qquad (2)$$

where $\Delta t/\Delta R$ is the mean value for the interval ΔR obtained from a plot of $\Delta t/\Delta R$ versus R. The total melting time (Tm) was then taken to be the sum of all such $(\Delta t)_{melt}$ values i.e.

$$Tm = \sum_{i=1}^{k} [(\Delta t)_{melt}]_i . \qquad (3)$$

The corresponding F value (where F is the fraction of sample melted) was then calculated from

$$F_j = \frac{\sum_{i=1}^{j} [(\Delta t)_{melt}]_i}{Tm} . \qquad (j \leq k) \qquad (4)$$

The latent heat of fusion (Q) is given by

$$Q = \frac{Vi}{M} Tm \qquad (5)$$

where M is the mass of the sample. Normally the first F value was of the order of 0.1%.

Power was supplied from a Hewlett-Packard DC power supply. The heating time was controlled by a darkroom interval timer. The required heating time was pre-set on the timer, which in turn energized a relay that closed the heater circuit for the chosen time interval. The voltage drop across the 4-terminal heater and across a 1 Ω standard resistor (in series with the heater) was measured with a Keithley 171 Digital Volt Meter. The amount of heat supplied was then numerically equal to

$$Q = V_H \, V_i (\Delta t)$$

where V_H, V_i are the voltages measured across the heater, and the 1 Ω standard respectively and Δt is the time set on the interval timer. Voltages were measured to four digits and the time to $\pm$ 0.01 sec using an

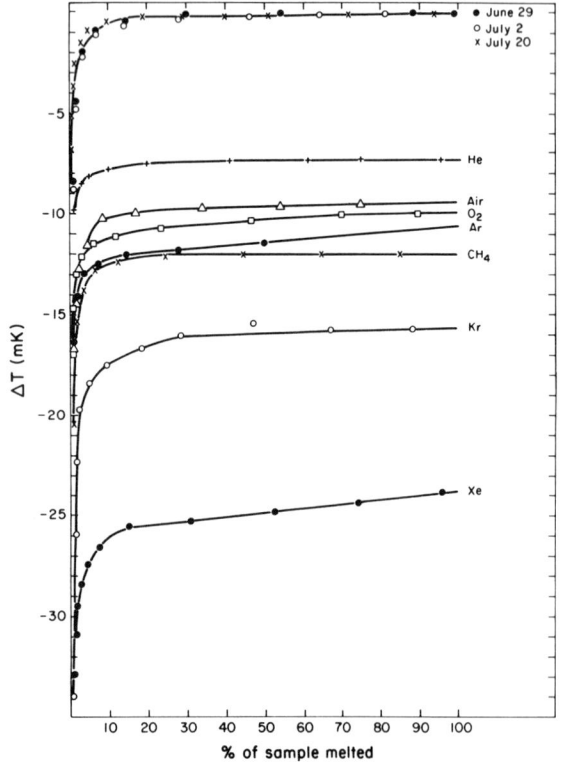

Fig. 2 Melting curves of ice.
The upper curve is due to melting under its triple point pressure, and the rest of the melting curves are under 1 atm. pressure exerted by the indicated substances. Temperatures are relative to the triple point of H_2O.

electronic stop-watch connected across the relay contacts that energized the heater circuit. Total melting times were typically between 600 and 3000 sec.

DATA

Figures 2 and 3 show typical melting curves of ice. The upper curve in Fig. 2 is the melting curve of pure ice at its triple point pressure. Its liquidus point agrees within 0.1 mk with the triple point of water as maintained by our conventional glass water triple point cell.

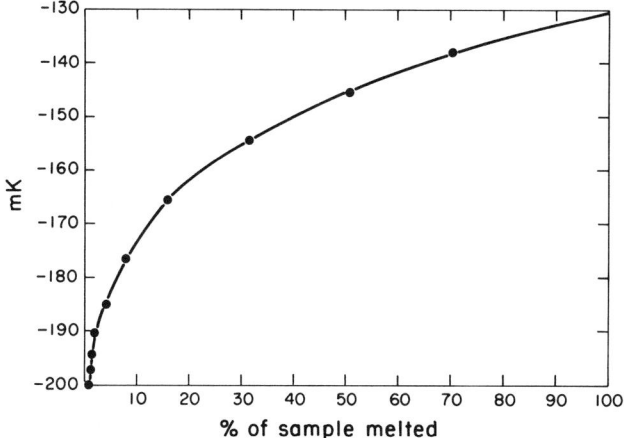

Fig. 3 Melting curve of ice at 1 atm. CO_2 pressure. Temperature is relative to the triple point of H_2O.

This curve is made up of data points obtained using a cell that originally contained different gases over water, (CO_2, air, Kr) yet after removing these gases, as described earlier, the samples yield a unique triple point.

The rest of the curves are melting curves obtained with the water sample sealed in the cell under 1 atm. of the indicated gas. Because the gas filling took place at the prevailing ambient pressure, these curves were produced by correcting each data point by multiplying it by the factor $T_1 P_o / T_o P_1$ where T and P are the temperature and pressure and the subscript "1" refers to the ambient and "o" to the 0°C condition respectively, in order to get values at 1 atm. pressure and 0°C.

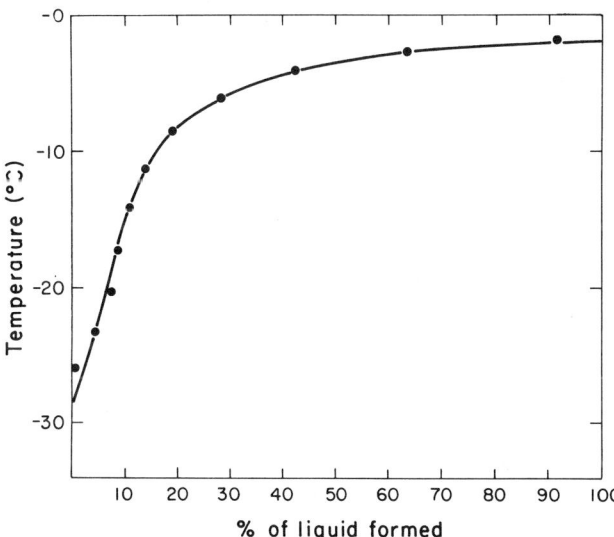

Fig. 4 Melting curve of standard sea water.

As standard sea water[6] was also available to us we measured its melting curve under 1 atm. air pressure, out of curiosity (see Fig. 4).

To examine the behaviour of compound cells (i.e. a single cell containing more than one substance) we attempted to obtain melting curves of both water and methane (CH_4) using cells that contained 1, 5 and 41 atm. of CH_4 and some water. The main purpose of these measurements was to see whether a single cell could yield more than one fixed point. Figure 5 shows the liquidus point temperature of ice as a function of CH_4 pressure in the cell. The curve results from the effect of pressure as well as of impurity, i.e., the absorbed CH_4 will act as impurity whereas the rest will apply pressure on the melting ice. These effects could be separated if desired by running similar experiments using substances of different solubilities.

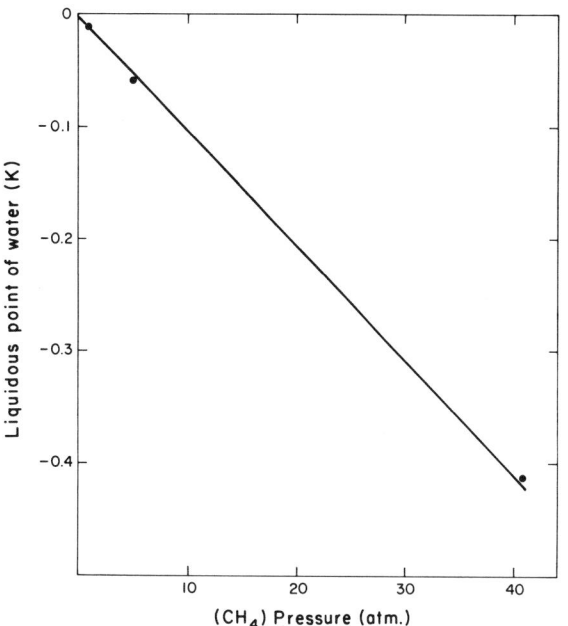

Fig. 5 Liquidus point of ice as a function of pressure exerted by CH_4.

In trying to obtain the melting curve of methane using a sample having 5 atm. of CH_4 over water, the result appeared to be a wide heat capacity maximum with peak about 4 times the heat capacity of the sample. The cell with 41 atm. of CH_4 filling pressure however, did yield a reasonable melting curve as seen in Fig. 6. Here the upper curve is the melting curve of CH_4 that had been sealed in a cell for 2 years after purifying it by fractional distillation. The lower curve is the melting curve of CH_4 in our compound cell, i.e. having 0.8 g of pure H_2O and 41 atm. of CH_4 straight from the high purity commercial methane bottle. This melting curve, then, does yield a liquidus point that agrees well within 1 mK with that of the purified substance. On the other hand the melting curve of ice in this compound cell was shifted by 0.413 K. It seems then that water influences the melting behaviour of CH_4 far less than vice versa. One could correct for this shift once it is known. More serious, however, the presence of this amount of CH_4 slows down the reaching of thermal equilibrium at any point along the melting curve of this ice sample to such an extent as to make the use of this cell for calibration purposes impractical. This melting curve consisting of 10 data points took 3 days to complete. Had the CH_4 been absent the same curve would have been obtained in no more than 3 hours.

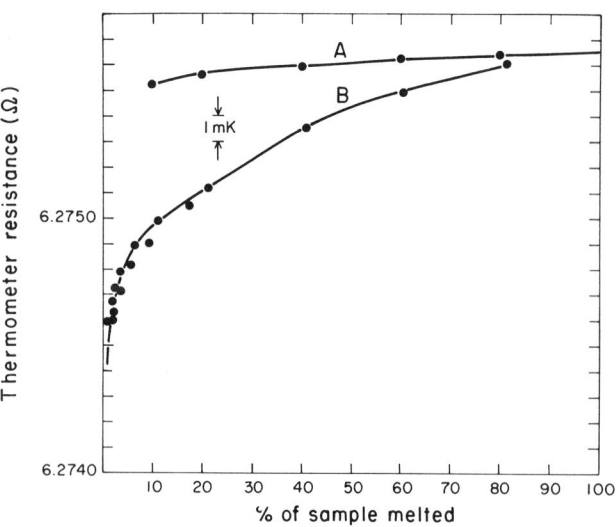

Fig. 6 Melting curves of CH_4.
A) purified CH_4
B) unpurified CH_4 + water

Figure 7 shows the lowering of the melting point of ice by 1 atm. of the indicated impurities as a function of their respective solubilities in water (solubility values were taken from reference 2). The solubility is in cm^3 of substance at 1 atm. and 0°C per 100 g of water. It is interesting to note that even CO_2 is well behaved in this figure. The vertical intercept indicates the pressure effect only. From the same data replotted on a larger scale in Fig. 8 we see that 1 atm. pressure lowers the melting point of ice by 7.3 mK. This compares well with the long-accepted value[3] of 7.4 mK.

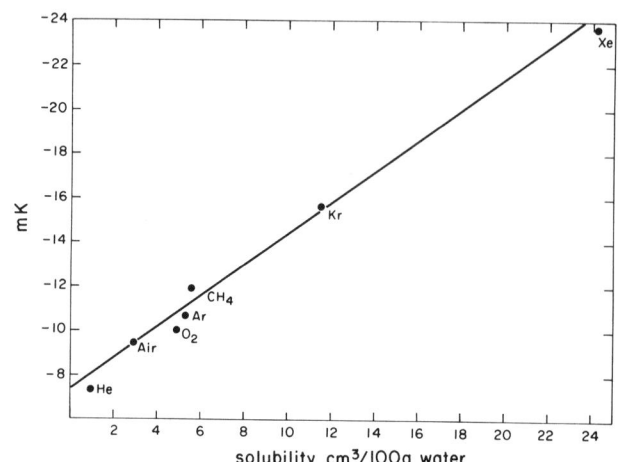

Fig. 8 Triple point lowering of ice by 1 atm. of the indicated substances as a function of their solubility.

For the three samples making up the upper curve of Fig. 2 the latent heats of fusion were calculated to be 334.9 j/g 334.3 j/g and 335.3 j/g giving an average value of 334.8 j/g. This may be compared with the values 332.5 j/g and 333.5 j/g obtained by Jacobs[4] and by Dickinson and Osbourne[5] respectively.

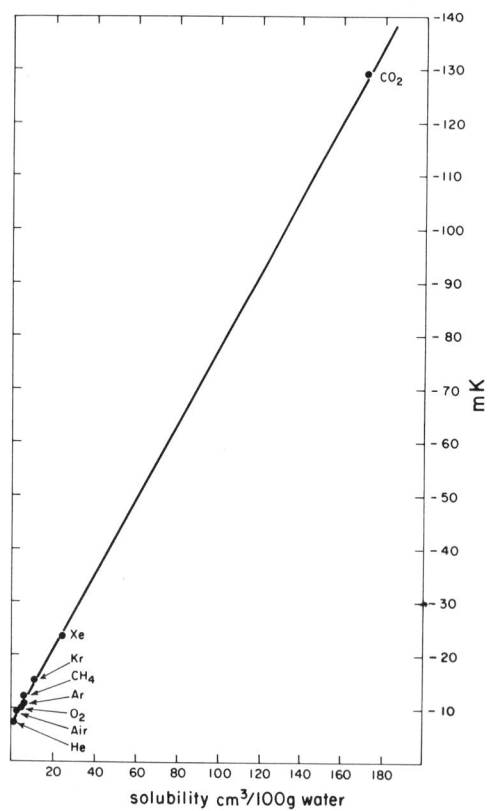

Fig. 7 Triple point lowering of ice by 1 atm. of the indicated substances as a function of their solubility.

REFERENCES

(1) Metrologia 14, 1 (1978).

(2) Lange's Handbook of Chemistry 12th ed. McGraw Hill, p. 10-3.

(3) Ann. der Physik 5, 1, 341 (1929).

(4) Trans. Faraday Soc. 31, 813 (1935).

(5) Bull. Bur. Standards, 12, 49, (1915).

(6) Standard sea water of 35.0000 °/oo practical salinity (19.3739 °/oo chlorinity) has by definition an electrical conductivity ratio of unity at 15°C with a KCl solution containing a mass of 32.436 g KCl in a mass of 1 Kg of solution. Ampules of such standard sea water samples were prepared at the Institute of Ocean Sciences, Wormley, England and distributed to various laboratories (amongst them the NRC) who participated in establishing the Practical Salinity Scale of 1978.

The effect of pressure on the water triple-point temperature

J. V. McAllan

CSIRO Division of Applied Physics, Sydney, Australia 2070

Pressure-related effects for the air-free ice-water system were measured using resistance thermometry with ac techniques having a sensitivity approaching the Johnson noise limit of about $2\,\mu$K. A value of 7.474 ± 0.01 mK was obtained on changing from ice at ambient pressure to the triple point, which gives dT/dP = 7.423 $(\pm 0.01)\times 10^{-8}$ K Pa^{-1}. A direct comparison gave the difference between the triple point and melting air-saturated ice as 9.85 mK. The variation of temperature with height in a triple point cell has been found to be 8.1×10^{-4} K m^{-1} and all measurements were above the expected value of 7.3×10^{-4} K m^{-1}. Experiments investigating this discrepancy are described. Surface tension varies with crystal size and causes a temperature variation, and the magnitude of this is discussed and a simple method given for estimating its size in stable triple point cells. The water triple point is the fundamental reference for all thermometry, and the limits placed by these effects on its realization are discussed, and their relevance to the determination of other reference freezing points.

INTRODUCTION

There are several effects capable, at least in principle, of producing variations of 100 µK or more in the freezing and melting temperatures of highly purified materials. Because 100 µK is the order of the reproducibility of many reference freezing points, a detailed investigation of the effects is of interest; there is special interest in the results of these effects on the water triple point, which provides the basic reference for all temperature measurements. A cell used for maintaining the triple point of water can be kept under almost constant conditions by immersion in crushed ice and such a cell is especially suited for the study of such small effects.

There have been several precise and detailed studies of the thermal behaviour of the water triple point using dc techniques involving thermocouples or platinum resistance thermometers with sensitivities of better than 100 µK. The most recent and probably the most thorough of these investigations was by Berry[1] who achieved a sensitivity of 20 µK under optimum conditions using a long time-constant galvanometer as the detector. All these investigators used glass cells of similar design, incorporating a re-entrant well for the thermometer. The arrangement used in this laboratory for long-term investigations of such cells is shown in Fig. 1.

An ac resistance thermometer bridge developed in this laboratory[2,3] when used with a chart recorder and filter to produce a long time constant for the system, can measure differences over a period of many hours with a sensitivity comparable to the Johnson noise in the resistance thermometer. The Johnson noise in a 25.5 ohm resistance thermometer at the water triple point is equivalent to 0.8 µK if a 1 mA measuring current and a time constant of 60 s are used. The equivalent noise of the bridge-thermometer combination is $\sqrt{2}$ times this, i.e. 1.1 µK.

While the sensitivity can be improved by longer time constants, such improvements make severe demands on the stability of the measuring system and the long time constant masks any systematic short-term variations. Since the instrument now approaches the optimum obtainable with resistance thermometry, and the reproducibility of the water triple point is an ultimate limit for thermometry, this laboratory has investigated various effects such as crystal size and migration of impurity which must affect the temperature. This paper deals with pressure-related effects.

THE PRESSURE-TEMPERATURE RELATION

The equilibrium temperature between a pure solid and its melt is affected by pressure according to the Clapeyron relation

$$\frac{dT}{dP} = \frac{\Delta V}{\Delta S} = \frac{T\Delta V}{L}$$

where ΔV is the volume change on melting associated with an entropy change $\Delta S = L/T$, L being the latent heat.

This relation can be checked experimentally by varying the gas pressure over the liquid-vapour surface and measuring the associated change in temperature. The effect is small (for most materials less than 10 mK for an atmosphere) but has been measured by McLaren[4,5] and confirmed with an accuracy of a few percent for the metals Bi, Cd, In, Pb, Sn, Sb and Zn. The accuracy was limited both by sensitivity of temperature measurement and by knowledge of ΔV. Some of these results have been confirmed with similar accuracy in various national standardising laboratories.

Similar measurements can be made by varying the pressure over the ice-water system, and indeed the prediction and measurement of this effect for ice by the Thomson brothers[6,7] was the first verification of a prediction in thermodynamics.

MEASUREMENT OF dT/dP FOR WATER-ICE

Procedures

A triple point cell was constructed with a teflon-plug vacuum valve on the usual filling tube. The cell was cleaned with acid and steam cleaned in the usual way, filled by a single distillation from degreased tap water and degassed by repeated shaking and pumping. The elaborate procedures we use for primary standard cells were not followed, and the cell temperature was about 125 µK below our best cells with about four times their electrical conductivity as measured by Misch's method[8]. Mantles of ice were frozen on the central thermometer well using solid CO_2, and after the ice was about 4 mm thick the well was washed several times with ice-cold distilled water. This washing melted the mantle against the well, so that the mantle rotated when the cell was twisted sharply. Such techniques were first described by White[9].

The cell was placed in a clear plastic container (Fig. 1) and held centrally by foam plastic spacers, providing an air gap of about 10 mm between cell and

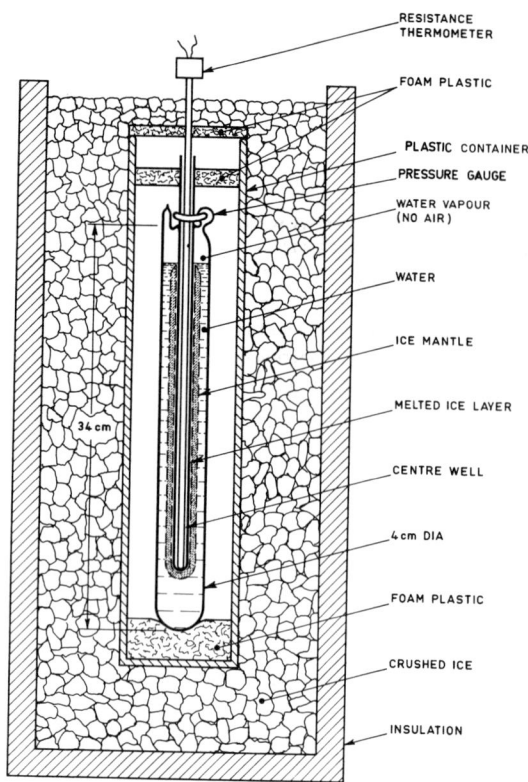

FIG. 1. Experimental arrangement in water triple point determinations.

container. This reduced the heat transfer between cell and surrounding ice bath by a factor of 10 (to less than 40 µW per cm of mantle height) so that the ice-water interface in the cell was very close to equilibrium conditions, and reduced markedly the freezing of the free water surface caused by heat transfer through the water vapour space, which occurs when cells are placed directly in the ice bath.

These are the normal procedures used by this laboratory with water triple cells, and they produce a mantle with small ice crystals whose temperature, depressed initially by about 100 µK, rises over one or two days as the crystals grow. The large, flat, strain-free crystals which are finally obtained maintain a very stable temperature for periods of months. The effect of imposed pressure should be the same for the small initial crystals as for the large ones.

Air was admitted to the cell through a liquid air cooled trap (to remove CO_2) but some erratic early results were attributed to movement of vapour to the cold trap, and in later measurements this was disconnected after the initial filling. The pressure change was the ambient pressure minus the vapour pressure of water at the triple point (611 Pa). When a thermometer was placed in a cell, even after precooling in ice, it melted a small amount of the mantle next to the well and this altered the impurity concentration in that water layer. The resulting small temperature changes died away over a period of some hours, and so a thermometer was placed in the cell for at least three hours before the pressure was altered, except when measuring a fresh ice mantle.

A second cell was constructed using normal vacuum taps, cleaning with a laboratory glassware cleaning solution and rinsing several times with distilled water. The cell was filled with the laboratory distilled water and after the taps were greased the water was degassed by repeated shaking and pumping. The water had an obvious grease film on the surface, but the temperature of a well aged ice mantle was only 370 µK below a very good cell. The cell was left for six hours after placing a thermometer in the well, to reduce impurity diffusion effects, before measuring the effect of a change of pressure.

Results

For the first and purer cell, a freshly prepared ice mantle with small crystals gave a value during the first three hours after freezing of

$$\frac{dT}{dP} = 7.424(\pm 0.02) \times 10^{-8} \text{ K Pa}^{-1}$$

the accuracy being limited by the drifting temperature. The same ice after ageing for three days under vacuum gave

$$\frac{dT}{dP} = 7.423 \text{ and } 7.422(\pm 0.01) \times 10^{-8} \text{ K Pa}^{-1}$$

in two separate determinations.

It was thought that some air might have dissolved during the measurements of atmospheric pressure and the chemical solution would give a large value for dT/dP. The normal exposure to air was for about one hour, and exposure for 22 hours produced a rise of less than 70 µK, due to the overall instrument instability, so that the solution effects over 22 hours seemed to be less than 0.1% of the pressure effect. After resetting the instrument zero, evacuation for 18 s and then for a further 10 s returned the temperature to within 2 µK of the triple point value obtained before admitting air 25 hours previously, and the final air value just before evacuation gave

$$\frac{dT}{dP} = 7.406 \times 10^{-8} \text{ K Pa}^{-1}.$$

This value, after 25 hours of air diffusion, was less than 0.02×10^{-8} K Pa^{-1} different from that obtained after the normal exposure of one hour, and consequently the diffusion effects in one hour should be negligible.

Three determinations of dT/dP were made with the impure cell, giving 7.416, 7.421 and 7.432×10^{-8} K Pa^{-1}, and a mean of 7.423×10^{-8} K Pa^{-1}. The noise on the chart recording was about five times that of the other cell, presumably partly at least because of the impurity of the greasy cell. There was no other obvious effect.

The best value determined in this laboratory is considered to be

$$\frac{dT}{dP} = 7.423(\pm 0.01) \times 10^{-8} \text{ K Pa}^{-1},$$

equivalent to a temperature change on going from the ice point (101 325 Pa) to the triple point (611 Pa) of 7.476 ± 0.01 mK. Present results, and all earlier values, are listed in Table I.

Comparison with earlier measurements

A Bunsen ice calorimeter is calibrated by measuring the change in volume on melting ice with a known amount of electrical energy. This is a direct measurement of $\Delta V/L$, and depends on the electrical units and the density of mercury at 0°C. The best value of the calibration constant of the calorimeter was obtained by Ginnings and Corruccini[10], and when this is corrected from international to absolute joules and to a modern value[11] for the density of mercury at 0°C (13 505.080 kg.m^{-3}) the value obtained is

$$\frac{\Delta V}{L} = 272.005(\pm 0.06) \times 10^{-12} \text{ m}^3 \text{ J}^{-1}.$$

This gives $\frac{dT}{dP} = \frac{T \Delta V}{L} = 7.4299(\pm 0.0016) \times 10^{-8}$ K Pa^{-1}

and gives a temperature change of $7.4829(\pm 0.0016)$ mK on changing from ice point to triple point pressure.

TABLE I

Temperature changes from the melting point of ice (101 325 Pa pressure) to the triple point of water (611 Pa vapour pressure)

Observer	Quantity	Value (mK)	Triple to ice (no CO_2)(mK)
Beattie et al. (14)	Chemical effect, air + CO_2 (calculated)	2.440 ± 0.012	-
"	Chemical, air, no CO_2 (calculated)	2.380 ± 0.012	-
"	Triple to ice + CO_2	9.81 ± 0.05	9.87 ± 0.05
Stimson (13)	Triple to ice + CO_2	9.97 ± 0.1(?)	9.91 ± 0.1(?)
Thomas (15)	Triple to ice + CO_2 (at 750 mm)	9.72 ± 0.1	9.91 ± 0.1
Moser (12)	Pressure on triple cell	7.43 ± 0.05	9.81 ± 0.05
Ginnings et al. (10)	Ice calorimeter	7.4829 ± 0.0016	9.863 ± 0.012
This work	Pressure on triple cell	7.476 ± 0.01	9.854 ± 0.02
"	Triple to ice (+ CO_2)	9.85 ± 0.1	9.91 ± 0.1

The Bunsen ice calorimeter gives a true dP/dT value at one atmosphere pressure whereas the direct measurements give an average $\Delta P/\Delta T$ from 0 to 1 atmosphere. But no non-linear effects are known, especially over such a narrow pressure range.

The temperature effect was measured directly by Moser[12] and Stimson[13], the value obtained by Moser being 7.43 ± 0.05 mK on increasing the pressure from the triple point to one atmosphere. This gave

$$\frac{dT}{dP} = 7.38(\pm 0.05) \times 10^{-8} \text{ K Pa}^{-1}.$$

Stimson constructed a special cell which he compared with a group of normal triple point cells, and then introduced CO_2 free air which bubbled through a sintered glass seal at the bottom to saturate the water with air in a reasonable time. He did not publish a value for the initial pressure effect but obtained 9.97 mK for the difference between the melting point of ice and the triple point, corrected to the normal CO_2 content, which implies 9.91 mK for air free of CO_2.

The difference between melting point of ice and triple point was measured as 9.81 ± 0.05 mK by Beattie et al.[14] and 9.85 ± 0.1 mK (corrected from 750 mm Hg) by Thomas[15] (both used thermocouples). Thomas froze the water from outside, forming fine ice needles throughout the triple cell, and waited only 24 hours before comparing with the ice cell. Modern experience would anticipate that his results would be slightly low.

The pressure effect was calculated by Moser (7.471 mK) and Beattie et al. (7.47 ± 0.05 mK) but their agreement is fortuitous since they used different values for every term in the Clapeyron relation. Uncertainties in the values of the density and latent heat of ice make this calculation unreliable.

Beattie et al.[14] have calculated a value of 2.440 ± 0.012 mK for the effect of saturation with air at 0°C, when the air had the normal content of CO_2. Removal of CO_2 would reduce the value by 2.4% giving 2.380 ± 0.012 mK, and these values can be used to compare measurements of the pressure effect with those of the difference between ice point and triple point.

The scatter in all the results is only 0.10 mK, and the most accurate values (the present ones and those of Ginnings and Corruccini) differ by only 0.007 mK, giving confidence in the reliability of the experimental techniques.

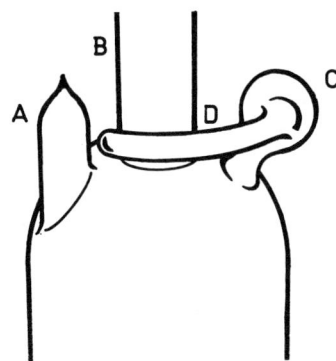

FIG. 2. Residual pressure gauge

A Final sealing tube
B Thermometer well
C McLeod gauge bulb (~ 1 cm³)
D Capillary (2 mm diameter).

Residual pressure in triple cells

If the Johnson noise limit is to be the major limitation on water triple cell measurements, pressure effects must be reduced below 1 μK, so that the residual air pressure must be less than 10 Pa (10^{-4} atm). An estimate of the residual pressure can be made by inverting the cell and trapping vapour in the sealing tube. The head of water compresses the gas and condenses water vapour, and volume changes give an estimate of residual pressure. This can be made quantitative by attaching a small McLeod gauge to the cell (Fig. 2). This gauge consists of a small sphere of about 1 cm³ with 30 mm of 2 mm diameter capillary tubing which is bent around the thermometer well to make it less susceptible to breakage. It is attached to the cell by a further short piece of capillary tubing, which provides a precise volume when closed off by water. The hydrostatic head can be varied to change sensitivity, and capillarity corrections must be made. Typically P = 0.026 × (length of air in capillary) × (hydrostatic head + 15) where P is the residual pressure in pascals and the lengths are measured in mm.

A typical cell has a residual pressure of 2 Pa, giving a depression of the triple point of 0.15 μK. This device can be used to monitor the pressure in a cell during evacuation, to ensure that all dissolved gas has been removed, and can also detect any gas released during the final sealing of the cell, or any long term release of gas from the glass. The main difficulty in its use is caused by the large rate of change of vapour pressure with temperature. If good temperature uniformity is not maintained during the measurements quite erroneous results can be obtained.

HYDROSTATIC HEAD EFFECTS

The hydrostatic head of liquid produces an additional pressure, and the temperature change can be calculated from the Clapeyron relation where the pressure is ρgh, ρ being liquid density and h the depth below the free liquid surface. The size of this effect is listed in the IPTS(68) for a series of pure metals and for water and has been confirmed experimentally, initially by McLaren[4,5] and later by many standardising laboratories, but the sensitivity and reproducibility of the results were only 5 to 10%, as far as can be judged from the published data.

The aim of measurements on fixed points is to obtain equilibrium values, but the presence of a gravitational field places a constraint on any system which has vertically separated crystals or a vertical crystal-liquid interface. In accordance with Le Chatelier's principle, the system will change to minimise this constraint, and true equilibrium will only be achieved when all liquid is separated from all solid by one horizontal interface. The measured vertical gradients are an indication of the response of the system to the constraint. They require a finite flow of heat through the material, which is not provided from outside the system, and should exist until all interfaces are horizontal. Most materials contract on freezing, and should be hot at the bottom. This is achieved by the bottom freezing faster than the top, releasing excess latent heat to provide the temperature gradient. During melting the upper layers will melt faster, producing the same effect. The differential growth rate for a zinc sample with a shell of solid 3 cm in diameter is about 0.37 nm (i.e. one atomic layer) s^{-1} per cm of height.

This slow growth cannot be observed directly in metals, but is inferred from the temperature gradient. An attempt was made to photograph the growth in water triple point cells under carefully controlled conditions. This was unsuccessful presumably because the growth is very slow, and outside perturbations were too large, even when photographing in a room kept at 0 ± 1°C.

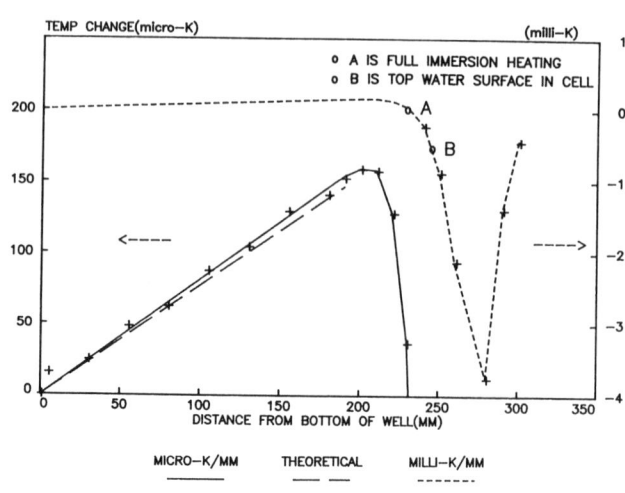

FIG. 3. Temperature gradients in water triple points.

Measurement of vertical gradients in triple cells

The Clapeyron relation predicts a gradient of 7.314×10^{-4} K m^{-1} for water. Measurements in this laboratory always give a larger gradient, for the same ice mantle when two weeks old and three months old, for different triple cells and using different resistance thermometers. Typical results are shown in Fig. 3.

The temperature of the Meyers-type sensor used in this work is raised about 300 µK by a 1 mA measuring current, and the value must be extrapolated to zero current by using two currents and assuming RI^2 heating and linear gradients. A smaller current reduces the heating but not the instrument noise, so the signal to noise ratio is worse. 1 mA seems a good compromise, but whenever the thermometer is moved in the cell the heating must be determined (it can vary by a few percent) and the noise level of the whole measurement is equivalent to 3 µK.

To test for conduction from the room, the water in the thermometer well was made level with that in the cell and the thermometer was lowered through the ice bath into the cell. The indicated temperature fell to 4 mK below the triple point value before rising to follow the cell gradient (Fig. 3). This implies that all heat conduction from the room had been eliminated. As a further check, solid CO_2 was packed around the exposed thermometer stem when the sensor was just under the surface of the water, but no conduction effect was seen. There was no effect from switching on room lights except for a light directly over the cell, but measurements were usually made with the room almost completely dark.

The triple point cell provides an extremely stable environment, partly because ice and water are opaque to thermal radiation from the room. With the thermometer fully immersed and water in the well level with the free water surface in the cell, there was no effect greater than 10 µK (the noise level in this experiment) when the cell was removed from the ice bath and held in a dark room at 20°C, or in a brine bath at -15°C.

Attempts were made to measure the gradient using three other resistance bridges; an 80 Hz bridge designed in this laboratory by C.P. Pickup, a Leeds and Northrup G.4 Mueller Bridge with a Guildline galvanometer amplifier, and a Guildline Kusters type current comparator resistance bridge. All were being used beyond their design specifications and while all detected the gradient, the signal to noise ratios were inadequate to decide if the discrepancy was real.

The water in the well provides a different dielectric to the air around the thermometer and the leads are unshielded inside the glass stem. Calculations indicated that any effects caused by this would be negligible with ac bridges, but a measurement was made with the thermometer at the bottom of the well, and then the water was siphoned out until it was only 20 mm above the element. The indicated temperature fell by < 5 µK, whereas the gradient anomaly required a rise of > 30 µK.

A further possible source of error was that the thermometer leads inside the glass sheath were acting as an r.f. aerial and that when the thermometer was lifted less leads were shielded by the silvered dewar which contained the ice bath, and there was more heating of the sensor by r.f. Varying the diameter of the dewar from 60 to 10 cm, and hence the efficiency of the shielding, had no effect. A small coil was placed around the thermometer stem and fed from an r.f. generator, but no effect was seen at low r.f. levels although at higher levels the indication of the resistance bridge showed resonance effects in its filtering system. No resonance effect was seen until the r.f. was at 10^4 times that of the background level, and such resonances should appear as scatter in the measurements if very large background variations occurred during a gradient search, but should not cause a systematic change in the gradient.

Gradients of 8.1×10^{-4} K m^{-1} were measured on several occasions, but some noisier sets of measurements gave values near 7.6×10^{-4} K m^{-1}. No reason for the discrepancy is known, and the most likely systematic error, conduction along the leads, is certainly not the cause with the experimental arrangement used.

CRYSTAL SIZE EFFECTS

Surface tension at a crystal surface produces a pressure-related temperature change. This is given by the Gibbs-Thomson relation

$$\Delta T = \frac{T}{\rho L} \gamma_{SL}\left(\frac{1}{R_1} + \frac{1}{R_2}\right)$$

where L is latent heat per unit mass, ρ is the density, γ_{SL} is the surface tension, and R_1 and R_2 are the principal radii of curvature, and ΔT is the depression of the melting point T. Berry[1] discussed this as a cause of the rise in temperature of triple cells during the first days after the freezing of an ice mantle. In this laboratory solid CO_2 is used to freeze the mantles but it is placed in a protecting glass tube and not directly in the well, so that the temperature gradients in the freezing ice are much less severe. Under these conditions a cell is only 100 to 200 µK low, 8 minutes after the freezing is complete. However the temperature falls slightly (by about 30 µK) in the next hour. This is attributed to the increase of impurity segregated at crystal boundaries during the initial rapid growth of the very small crystals. After this time, a slow rise in temperature occurs while the crystals grow and the impurity diffuses along crystal boundaries into the bulk liquid. If the solid CO_2 is put directly into the well, the initial depression is much larger and the small drop is masked.

A well aged ice mantle has many crystals with an area of about 0.5 cm² exposed to the liquid. There are a few very large crystals, perhaps 2 cm² in surface area, and in some regions a group of smaller crystals of perhaps 0.05 cm². The surface energy differences between these crystals should cause small temperature differences, and it was considered that this would be a cause of variation in hydrostatic gradient measurements.

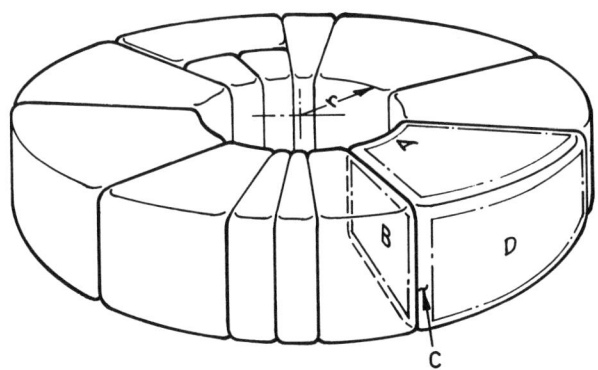

FIG. 5. Schematic view of crystal in aged ice mantle

A Negative temperature at well
B Flat sides
C Curved edges
D Positive curvature

For the top surface, 12 crystals give a well temperature

$$\Delta T = 2.5 \times 10^{-6}\left(1 - \frac{24}{\pi}\right) = -17 \text{ µK.}$$

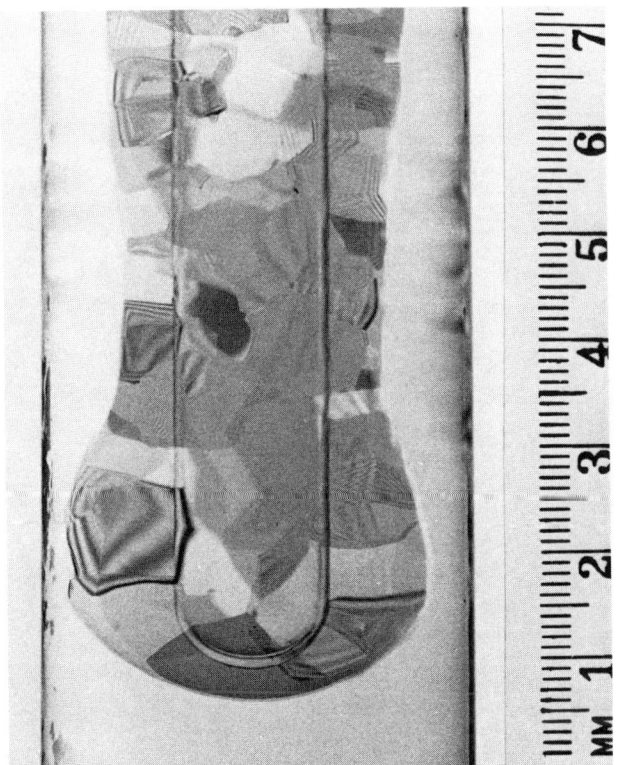

FIG. 4. Aged ice mantle (viewed in polarised light.

Many crystals in a well aged ice mantle (Fig. 4) extend from the thermometer well to the bulk liquid and the situation can then be represented as in Fig. 5. Calculation of the temperature depression was made assuming that the inner and outer surfaces were concentric with the well, and that the radial surfaces were flat, but that all were rounded at the edges. This would imply that the inner surface had a negative curvature, which is unstable. Such inner surfaces are produced only when the well is chilled, and the curved crystal edges very quickly reappear on the inside surfaces. This geometrical effect will tend to produce crystals flatter on the inside than outside, so that the inner surface is slightly hotter, and for crystals of 6 mm diameter this should be about 1 µK, whereas for 1 mm crystals it will be about 7 µK, and could contribute to the rise of temperature seen when the inner surface is melted.

Small crystals are more depressed below the equilibrium temperature, but they subtend a smaller angle at the centre, and control a smaller part of the well surface. If the depression is summed around the circumference of a well, it is found that

$$\Delta T \approx \frac{dT}{dP} \frac{\gamma_{SL}}{r}\left(1 - \frac{2n}{\pi}\right)$$

where r is the radius of the well and n is the number of crystals around the circumference. For the case of uniform crystals, this expression gives the same depression of the inner surface as the exact calculations within 1 µK, for crystals varying from 1 mm (n = 37) to 18 mm (n = 2). However the useful feature is the estimate in the practical case of a mixture of sizes, without the measurement of individual crystals.

When measuring vertical gradients in the present experiments certain positions gave consistent departures from the general gradient. These were seen to be regions of smaller crystals, but the thermometer sensor was 3 cm long and measured an average over too large a portion of the mantle to make a practicable quantitative comparison of temperature variation with crystal numbers. Such corrections would have to be made in any calibration where sensitivity approaches the theoretical noise limit.

Similar effects must occur in the metal fixed points. McLaren and Murdoch[5] advanced them as an explanation for small systematic differences ($\lesssim 600\ \mu K$) between different freezing techniques with Sb, where decanting experiments showed that sharp dendritic interfaces persisted for a large part of a freeze. For most metals of interest dT/dP is about half the value for water, and γ_{SL} is 2 to 4 times as large (Turnbull[16]).

So the overall effect should be slightly larger, but the times available for crystals to grow are typically hours instead of weeks. Freezing techniques are designed to place a layer of solid on the well and on the outside of the crucible. Most freezing occurs at the outer wall, where large crystals grow over the small ones, but the thermometer measures the temperature near the well where growth rates are very small and crystals and dendrites of about 1 mm diameter can exist for most of the freeze. If attempts are made to measure freezing points to much better than 100 μK, such departures from equilibrium must be considered.

CONCLUSION

The variation of the temperature of the ice-water interface with pressure was found to be

$$\frac{dT}{dP} = 7.423(\pm 0.01) \times 10^{-8}\ K\ Pa^{-1},$$

in good agreement with the value of $7.4299(\pm 0.0016) \times 10^{-8}\ K\ Pa^{-1}$ derived from the work of Ginnings and Corruccini. The variation of temperature with hydrostatic head was always found to be larger than the expected $7.314 \times 10^{-4}\ K\ m^{-1}$ and a value of $8 \times 10^{-4}\ K\ m^{-1}$ is consistent with the experiments. No reason for this discrepancy was found.

Simple methods for estimating residual pressure and crystal size effects are given. With careful experimental techniques it is now possible to use the very simple water triple point cell very close to the thermodynamic limits.

REFERENCES

(1) R.J. Berry, Can. J. Phys. 37, 1230 (1959)
(2) A.M. Thompson and G.W. Small, Proc. IEEE 118, 1622 (1971)
(3) J.J. Connolly, J.V. McAllan and G.W. Small, Temperature, Its Measurement and Control in Science and Industry 4 (2), 1487 (1972)
(4) E.H. McLaren, Temperature, Its Measurement and Control in Science and Industry 3 (1), 185 (1962)
(5) E.H. McLaren and E.G. Murdoch, Can. J. Phys. 46, 369 (1968)
(6) J. Thomson, Cambridge and Dublin Mathematical Journal 5, 248 (1850)
(7) W. Thomson, London Edinburgh Dublin Philosophical Magazine 37, 123 (1850)
(8) E.S. Misch, Rev. Sci. Instrum. 33, 569 (1962)
(9) W.P. White, J. Am. Chem. Soc. 56, 20 (1934)
(10) D.C. Ginnings and R.J. Corruccini, J. Res. Nat. Bur. Stand. 38, 583 (1947)
(11) A.H. Cook, Phil. Trans. R. Soc. London Ser. A 254, 125 (1961)
(12) H. Moser, Ann. Phys. 5 (1), 341 (1929)
(13) H.F. Stimson, J. Wash. Acad. Sci. 35, 201 (1945)
(14) J.A. Beattie, H. Tzu-Ching and M. Benedict, Proc. Am. Acad. Arts Sci. 72, 137 (1938)
(15) J.L. Thomas, J. Res. Nat. Bur. Stand. 12, 323 (1934)
(16) D. Turnbull, J. Appl. Phys. 21, 1022 (1950)

Reproducibility of some triple point of water cells

George T. Furukawa and William R. Bigge

National Bureau of Standards, Washington, D.C. 20234

The reproducibility of some triple point of water cells was investigated by platinum resistance thermometry. The standard deviation of measurements with a single cell was found to be better than ±0.01 mK. The range of temperatures observed with different cells was about 0.2 mK. The cells with more residual air tended to give lower temperatures. The cells of high quality gave temperatures within the range 0.05 mK.

1. INTRODUCTION

The triple point of water (temperature of equilibrium between solid, liquid, and vapor phases of water) is the single temperature common to two principal temperature scales of science and technology. First, the Tenth General Conference of Weights and Measures adopted in 1954 [1] the Kelvin Thermodynamic Temperature Scale (KTTS) based on the assignment of a value of temperature to one fixed point, for which the triple point of water was selected and assigned 273.16 K. Second, in the International Practical Temperature Scale of 1968 (Amended Edition of 1975) [2], which is based on the assigned values of the temperatures of thirteen equilibrium states between phases of pure substances (defining fixed points) and on standard interpolation instruments calibrated at these temperatures, the triple point of water is the most important defining fixed point to which the values 273.16 K and 0.01 °C are assigned. By definition, t = T - 273.15 K, where t is the Celsius temperature; hence, the unit of Celsius temperature, degrees Celsius (°C), is equal in magnitude to the unit of temperature, the kelvin, K. (In this paper, these temperature units and scales will be used interchangeably as defined.) The temperature 0 °C is no longer defined as the temperature of the ice point, although it can be prepared to approximate 0 °C closely; 0 °C is now defined as 0.01 °C below the triple point temperature of water.

The standard interpolating instrument of the IPTS-68 between 13.81 K (-259.34 °C) and 903.89 K (630.74 °C) is the platinum resistance thermometer of specified design and measured properties. (Henceforth, for convenience platinum resistance thermometers that meet the specifications of the IPTS-68 will be referred to by the abbreviation SPRT's.) The values of temperatures in the interval are expressed as the functions of measured resistance ratios: $W(T) = R(T)/R(0°C)$, where $R(T)$ is the observed resistance at temperature T and $R(0°C)$ is the resistance at 0 °C. At the National Bureau of Standards (NBS), where about 80 to 100 SPRT's are calibrated each year, the resistance of the SPRT at 0 °C is calculated from the resistance measurement at the triple point of water [3]. (Henceforth, the triple point of water will be abbreviated TP.) For calibration of the SPRT's the resistances are measured at the fixed points (normally, in the case of long-stem type SPRT's, after annealing for four hours at about 480 °C) in the sequence TP, zinc point, TP, tin point, TP, oxygen point, and TP for every SPRT. The resistance ratios W corresponding to the readings at the zinc, tin, and oxygen points are calculated from the readings at the TP obtained immediately after the readings for each of the points. In the application of such calibrated SPRT, as described above, the temperature evaluation of resistance measurements at "unknown temperatures" requires resistance measurements also at the TP. Hence, it is extremely important that there is information on the reproducibility of the triple point of water in practical routine calibration and in the application of SPRT's in temperature measurements, as well as in the definition of the KTTS. Moreover, although below 0 °C the effect of error in R(0°C) on W(T) becomes reduced, above 0 °C the effect of error in R(0°C) becomes amplified (e.g., W is about 3.5 at 631 °C). If the future IPTS involves the extension of the SPRT resistance ratio to the gold point [4], where W is about 4.6, the effect of error in R(0°C) that corresponds to 1 mK causes about 4.6 mK error at the gold point. (For discussion regarding the importance of using R(0°C) determined immediately after measuring R(T) to obtain W(T), see reference [5].)

This paper describes the results of comparison of some commercially available triple point of water cells as routinely prepared and used. Methods are described for testing residual air in the cells. The cells totaling 21 in number were part of those used in several laboratories at the NBS; some were not normally used because of excessive amount of air in the cell, and some were borrowed from a supplier of TP cells. Since the cells are breakable, the number of existing old cells is very limited and any old cells that do remain may have been recognized early to be unsatisfactory. Nevertheless, cells that were available were included in the comparison so that if a substandard cell is in the group the results of the measurements could show how to quickly recognize it.

2. TRIPLE POINT OF WATER CELLS

As long as the triple point of water cell can accommodate and have adequate depth of immersion for thermometers, its size and shape is not critical. However, cells that are suitable for SPRT's and those that are commercially available generally take the two forms shown in Fig. 1. There may be slight variations in the design depending on how the cell was cleaned and was sealed after filling with pure, air-free water. The thermometer well is usually constructed of uniform bore tubing of 8 to 22 mm i.d. or larger and 250 to 325 mm depth, excluding the well extension for type B cells which is usually standard bore glass tubing of slightly larger i.d. than that of the well and 80 to 100 mm long depending upon the depth of the thermometer well. For best results, the i.d. of the thermometer well should be as close to the o.d. of the SPRT as possible within practical limits, including the straightness of the SPRT and well. Where the i.d. of the well is somewhat larger than the o.d. of the SPRT, aluminum or glass sleeves are used in the thermometer well to improve the thermal contact of the SPRT with the TP cell [6]. The o.d. and the length of the outer glass shell are 51 to 64 mm or larger and 340 to 350 mm, respectively, depending upon the dimensions of the thermometer well. The extension arm of type A cell serves as a handle and as a means for resting the cell on the edge of an ice bath [6]; it

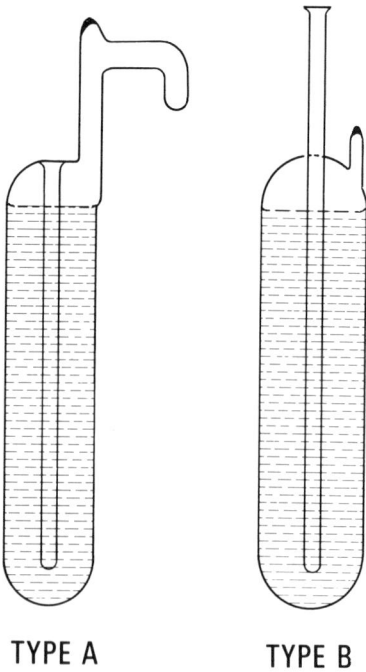

Fig. 1: Triple point of water cells with wells for platinum resistance thermometers. The cells contain pure air-free water.

can serve also as a MacLeod gauge for estimating the amount of air inside the TP cell. In the early design of the cell, an auxiliary vessel was attached to the extension arm as an integral part of the cell [7]. Periodically the system was inclined to transfer all of the water into the auxiliary vessel and the water distilled back into the cell, leaving behind any non-volatile material that was dissolved from the glass in the water. For type A cells, pure, air-free water is introduced through an extension where the seal as shown in Fig. 1 is made. Similarly for type B cells, water is introduced through an extension where the seal is made on the small-diameter tubular appendage shown in Fig. 1.

The IPTS-68 requires TP cells to be sealed and contain only water of high purity of the isotopic composition of ocean water. An increase of 0.001 moles of ^2H per 100 moles ^1H corresponds to an increase of 0.04 mK in the temperature of the TP. The continental surface water has normally about 0.001 moles less of ^2H per 100 moles ^1H than ocean water. Depending upon how the water is purified for the TP cell, the isotopic composition may be modified. Also, the isotopic composition at an ice-water interface in a TP cell is slightly dependent on the way the cell was frozen [2]. No attempt was made in this work to verify the isotopic compositions of the water in the TP cells that were investigated. It is most probable that the effect of the variations in the isotopic composition is smaller than the precision of the measurements that was obtained in the present investigation.

Although the cells may exhibit, as the water in the cell is gently moved from one end to the other, the clicking sound characteristic of the "water hammer" effect caused by the liquid water striking against the vessel, most cells have some residual air because in the manufacture of TP cells a small amount of air that remains in the water is difficult to remove [8]. The sound becomes "sharper" with smaller amounts of residual air. When the TP cell is cooled closer to 0 °C, the solubility of air in water increases and the vapor pressure of water decreases; consequently, the sound becomes sharper also at the lower temperatures. However, a small amount of air, e.g., 133 Pa partial pressure, causes only a small depression of the TP temperature. Assuming Henry's law and the air contains 0.03 percent carbon dioxide, the effect of solution of the gases is -0.00244 °C per atm or -3×10^{-6} °C for 133 Pa of residual gases. The effect of the pressure is -0.00747 °C per atm or -1×10^{-5} °C for 133 Pa. (The effect of solubility and pressure of air on the TP temperature was taken from reference [7]). The total effect then becomes 13×10^{-6} °C or 0.013 mK for 133 Pa of residual air.

For type A cells the amount of residual air in the cell can be estimated by using the extension arm as a MacLeod gauge. Figure 2 shows the steps to follow to entrap the air in the cell. From position 2a approximately 26 cm³ of vapor is compressed into a "bubble" in Fig. 2b by a hydrostatic head of water which is about 35 cm for the cells used in the study. If v cm³ is the volume of the entrapped bubble, then, assuming that the air follows the ideal-gas law, the pressure of the residual air can be estimated by:

$$p(\text{Pa}) = 3400v/26 \qquad (1)$$

If the volume of the entrapped bubble (Fig. 2b) is 1 cm³, then the residual pressure is approximately 131 Pa which would, as shown above, cause about 0.01 mK lowering of the temperature (see reference [9]).

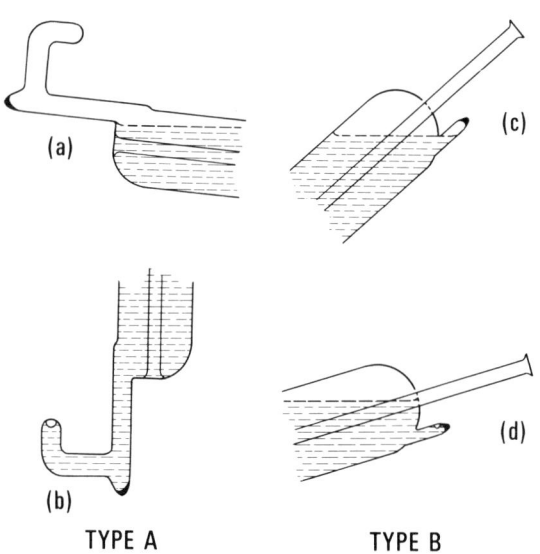

Fig. 2: Procedure for entrapping gas bubbles in triple point of water cells. Type A: position cell as in a; gently tip bottom of cell upward as in b entrapping gas. Type B: position cell as in c; gently tilt bottom of cell upward to position d; any further tilting will cause the bubble to escape.

For type B cells, it is stated that if the entrapped volume in the extension, where the cell is sealed, can be compressed to 1/3 by the cell water, the effect of air on the TP temperature is less than 0.0005 °C [10] (see Fig. 2c and 2d). It is to be noted that with type B cells, the entrapped volume may include excess volume of water vapor because the hydrostatic head of water that confines the entrapped volume is not greater than the vapor pressure of water. However, with type A cells, the entrapped volume can be confined in the extension arm by a hydrostatic head of water (about 35 cm) that is greater than the vapor pressure of water. (If type B cells are provided with an inverted U appendage similar in principle to type A cells, the residual air can be better estimated.)

TABLE I

Identification and Description of Triple Point of Water Cells

Cell	Type	Serial No.	Thermometer Well I.D. mm	Approximate Bubble Size Diameter, mm	Date Sealed with Water
A	A	JJ-176	11	6	5/58
B	A	JJ-303	11	15	7/59
C	A	380	9	1	9/63
D	A	382	9	<0.5	5/65
E	A	556	9	1	5/68
F	A	407	11	6	5/65
G	A	651	11	16	6/71
H	A	871	11	<0.5	3/81
I	A	878	11	1	3/81
J	A	839[a]	13	4	11/78
K	A	840	13	2	11/78
L	A	894	13	2	2/81
M	A	895	13	5	2/81
N	A	906	13	14	3/81
O	A	910	13	2	3/81
P	A	912	13	3[b]	3/81
Q	B	926	11	4[b]	2/81
R	B	933	11	3[b]	3/81
S	B	941	13	2[b]	3/81
T	B	942	13	2[b]	3/81
U	B	944	16	4[b]	3/81

[a] Broke near end of comparison work.

[b] These bubbles were under a few cm of hydrostatic head of water; hence, they include excess volume of water vapor.

As mentioned earlier, the clicking sound, caused by the water hammering against the cell wall, indicates absence of or very little air in the cell. Table I identifies the TP cells that were compared. The cells were tested for presence of the clicking sound and the size of bubble (diameter in mm) that can be entrapped at room temperature. All of the cells, except for three (cells B, G, and N), gave the clicking sound as the cells were gently tilted along their axes. Also, bubbles, indicative of presence of air, could be entrapped in the extensions of all of the cells. In two cells (D and H) a magnification of 4X was needed to decide whether a bubble was entrapped or the bubble was in the glass. In some type A cells, the size of the bubble that can be entrapped seemed to vary from less than 1 to 3 mm or so depending upon the previous temperature of the cell and whether just previous an entrapped bubble was forced to dissolve completely by the hydrostatic pressure of water. Three of the cells, that did not click on gentle tilting, clicked when the cells were quickly tilted. When type A cell was maintained in position b for overnight, bubbles of about 6 mm completely disappeared by dissolving in the water. Larger bubbles (cells B, G, and N) became smaller and they too would be expected eventually to "disappear" given enough time by "transferring" into the upper vapor space. As mentioned earlier, in type B cells enough hydrostatic head could not be exerted to overcome excess water vapor bubble; hence, a slight forward motion was given to the water to see whether the bubble could be collapsed or made smaller. Although the bubble appeared to become smaller, the water motion also caused the bubble to escape from confinement. Consequently, the test for residual air in type B cells was made by comparing the sharpness of the clicking sound with type A cells. It was estimated that none of the type B cells that were in the group had air in amounts greater than what corresponds to a 5 mm bubble in the type A cell.

3. PREPARATION OF THE CELLS

Since it was not the intention of this work to determine how long one must wait before the cell temperature settles to a steady value, ice mantles were frozen and the cells were stored in an ice pack at least one week before testing. One week of storage was considered adequate for any temperature changes to be reduced to negligible amounts [11].

Prior to freezing the mantles on the thermometer wells of the TP cells, the wells were dried and closed with rubber stoppers and the cells were precooled by packing in ice in a large container insulated with plastic foam. (Precooling yields more uniform mantle along the thermometer well.) The precooled cells were then removed from the ice pack, leaving chambers in the ice pack undisturbed for later re-insertion, and mantles were frozen by filling the thermometer wells with crushed Dry Ice and maintaining the well full for about 12 to 15 minutes [3]. After observing that the mantle was uniformly distributed along the length of the thermometer well, the cells were returned to the ice pack. The remaining Dry Ice continued to freeze more mantle until it all evaporated, which resulted in the mantle near the bottom of the thermometer well to be somewhat thicker than other parts of the mantle. The wells were filled with ice water and the cells were stored in the ice pack, as mentioned above, for at least one week prior to measurements. Glass rods, initially at room temperature, were inserted into the wells of all of the TP cells to form the "inner melt."

As the ice in the insulated storage container melted, more was added. The storage container was sufficiently insulated so that it was 6 or 7 days before enough water melted to cause freezing of water inside the TP cell. When ice began to form inside the TP cells, the cells were removed from the storage container and placed in a temporary ice pack. The storage container was then drained free of water and repacked with finely divided ice. Cylindrical recesses were then cut into the packed ice, using metal tubes of diameter slightly larger than the cells. The cells were slipped into the recesses and covered with more ice. Whenever the ice mantle in the TP cell grew excessively on storage in the ice pack, the excess was melted by inserting the cell in warm water or exposing the cell directly to hot running water from the laboratory faucet. By such procedures ice mantles on the TP cells were preserved for measurements over number of weeks. However, because of prolonged storage any impurities that were segregated into the outer liquid during the preparation of the ice mantle are expected to have diffused into the inner melt. For the present study no attempt was made to determine the effect of diffusion of impurities into the inner melt or to cause the outer liquid to replace the inner melt by inverting the TP cell. For measurements, a separate smaller ice bath was used that accommodated only one cell at a time (see reference [6] for details of arrangement).

4. RESISTANCE MEASUREMENTS AND PROCEDURE

Measurements were performed employing an SPRT (S/N 1805523) with the head modified to receive DNC connector and a 400 Hz AC bridge designed by Cutkosky [12]. The output of the bridge was recorded continuously with a strip chart recorder. With 1 mA thermometer current, the sensitivities of the recorder amplifier and the bridge amplifier were adjusted so that about 300 mm width on the recorder chart corresponded to 1 mK. The recorder tracings were estimated to about 0.2 to 0.3 mm; hence, at 1 mA thermometer current, the temperature readings that were estimated corresponded to about 1 µK (1 x 10^{-6} K).

Although the inner melt had already been formed earlier by inserting a glass rod into the thermometer well, just prior to making measurements another glass rod was inserted into the well to make certain that the ice mantle was "free". The mantle was considered free when it moved by slight tilting of the TP cell. The buoyant force of the ice mantle causes the temperature where the mantle is in contact with the bottom of the thermometer well to be colder; hence, it is believed that this colder temperature eventually in turn causes the mantle to stick by freezing some of the water in the inner melt. For a TP cell with thermometer well of 13 mm i.d. (15.4 mm o.d.) and 6-mm thick ice mantle (about 130 cm^3 of ice), the buoyant force exerted by the ice on the hemispherical surface at the bottom of the thermometer well is estimated to cause the temperature of contact to be about 21 μK colder. In testing the presence of inner melt, a sharp rotational impulse may cause the ice mantle to break loose and rotate about the well, indicating that an inner melt is present, but the mantle may soon stick when the rotation ceases. With a certain cell, temperatures obtained after breaking the mantle loose to rotate have been found to be 0.02 mK colder than when the mantle was tested to be free by just tilting the TP cell. Hence, before the SPRT was inserted into the well a glass rod initially at room temperature was always momentarily inserted first to free the ice mantle.

When an SPRT was left undisturbed to read the TP-cell temperature continuously for several hours (or overnight) with 1 mA or $\sqrt{2}$ mA current, a gradual decrease in temperature was observed as the ice mantle gradually began to stick to the thermometer well. Under certain conditions, the $\sqrt{2}$ mA current seemed to have nearly offset the sticking process, since even after 17 hours the SPRT readings were, although lower, within about 0.005 mK of the readings when the mantle was free. Aside from the current in the SPRT, the length of time before the effect of sticking became thermometrically detectable seemed to depend on the thickness of the inner melt, the size of the ice mantle, the diameter of the thermometer well, and the degree of heat transfer between the TP cell and the ice bath (e.g., whether the ice pack had water or no water or whether there was an air space provided between the cell and the bath).

When the mantle becomes stuck to the thermometer well, the thermometer indication is expected to be affected by the temperature of the outer ice-water interface of the mantle, i.e., without "ice bridge" between the mantle and the TP-cell wall. Also, in some cases, where there is significant amount of impurities in the TP cell, the lower temperature of the outer ice-water interface is expected to cause the mantle to stick earlier to the well. However, in practice, where many SPRT's are tested consecutively in the TP cell, the process of inserting the SPRT from the ice bath, where it is precooled, into the TP cell transfers enough energy to melt the ice around the thermometer well. Hence, where many SPRT's are measured successively at short intervals, there may not be enough time for the mantle to stick to the thermometer well.

The measurement procedure was as follows. The aluminum bushing [6], which is normally used to improve the thermal contact of the SPRT with ice-water interface in DC measurements, was not used to avoid any possible effect on the readings with the AC method. With every TP cell, a set of observations comprised equilibrium tracings obtained with 1 and $\sqrt{2}$ mA thermometer currents at two settings each of the bridge to obtain the sensitivity of the measurements at each current. One of the two tracings for each current was close to the bridge zero, which was recorded as a tracing with the bridge current off, while keeping all other operating conditions the same. From such set of readings the SPRT resistance at "zero current" was obtained [6]. Moreover, each of the resistance readings was adjusted to the same 26.5 cm immersion of the SPRT (the height from the surface of water in the TP cell to the mid-point of the SPRT coil). Since the equilibrium conditions of both the TP cell and SPRT were disturbed prior to every measurement, at least one hour was allowed for every set of readings to assure that a new equilibrium was attained. The chart speed was set to 5 cm hr^{-1} so that it was necessary to obtain over 1 cm or 12 min of tracing for each bridge or current setting to be certain that an equilibrium condition was reached.

Whether the measurements were to be made on another TP cell or to be repeated on the same TP cell the SPRT was always pre-cooled in the ice bath before being inserted into the TP cell. While the SPRT was in the ice bath, the ice mantle of the TP cell was checked to be free and, if not, a "warming rod" was inserted into the thermometer well to free the mantle. Also, following each set of readings, the ice mantle in the TP cell was tested to be free. In some cases, measurements were obtained continuously overnight without disturbing the SPRT. The first set of readings was obtained in the morning without testing to see whether the ice mantle was free. However, a second set of readings was made immediately after with the mantle free.

The TP cells that were compared had thermometer wells of 9, 11, 13, and 16 mm i.d. The cells with larger i.d. required longer times for equilibrium to be established. For example, when the thermometer current was changed from 1 to $\sqrt{2}$ mA, approximately 2, 3, 4, and 7 min were required for the SPRT to reach equilibrium for thermometer wells of 9, 11, 13, and 16 mm i.d., respectively. When borosilicate glass sleeves were used with wells of 13 mm i.d. to improve the thermal contact, the SPRT came to equilibrium in about 2 min when the current was changed from 1 to $\sqrt{2}$ mA.

5. RESULTS AND DISCUSSION

Figure 3 summarizes in measurement groups a to n the comparisons of TP cells listed in Table I. The cells are identified at the top and the dates of the measurements are given at the bottom of the plots of each group. The observations on the TP cells within each group are shown relative to each other. The measurements of groups a to f were obtained during April and May, of groups g to h during August and September, and of group i during November, and of groups j to n during December, all during the year 1981. The ice mantles of the TP cells that were investigated during each of the above four periods were prepared and stored at least one week before the start of each measurement period. Some of the TP cells that were investigated during the April to May period were not among the later measurement periods because they were returned to those people from whom they were borrowed.

The measurements were grouped into units of not longer than one week to avoid adjusting for the effect on the results by the drifts in the reference resistor and in the SPRT that were used. As mentioned earlier, efforts were made to estimate the recorder tracings to about 1 μK or about 4 parts in 10^9 for an SPRT with R(0°C) of about 25.5 ohms. Our standard resistor has been found to drift about 1 ppm per year or 3 parts in 10^9 per day. Considering the drifts in the reference resistor and SPRT and the actual results that were obtained (about 1 part in 10^8) in closely spaced repetitive measurements, the grouping of measurements into intervals not longer than about one week seemed reasonable. The measurements of each group included one or more reference TP cells.

On the basis of preliminary measurements, cells K, M, P, and R were selected to be reference cells; one or more of these cells were included among the measurements in each group. Since cell M was one of the borrowed cells that was returned after the April to May measurement period, most measurements were obtained on reference cells K, P, and R. In measurement groups g and h, the repeatability of measurements on TP cells P and R, respectively, was determined (see Fig. 3). The estimated standard deviation of 15 readings obtained over 5 days with cell P was found to be ±0.008 mK. Similarly, the estimated standard deviation of 12 readings obtained over 4 days with cell R was found to be ±0.007 mK. Cells P and R were compared directly in measurement group i, and, as shown in Table II, the temperature of cell R was found to be about 0.010 mK higher than cell P. On the other hand, in measurement group f, where

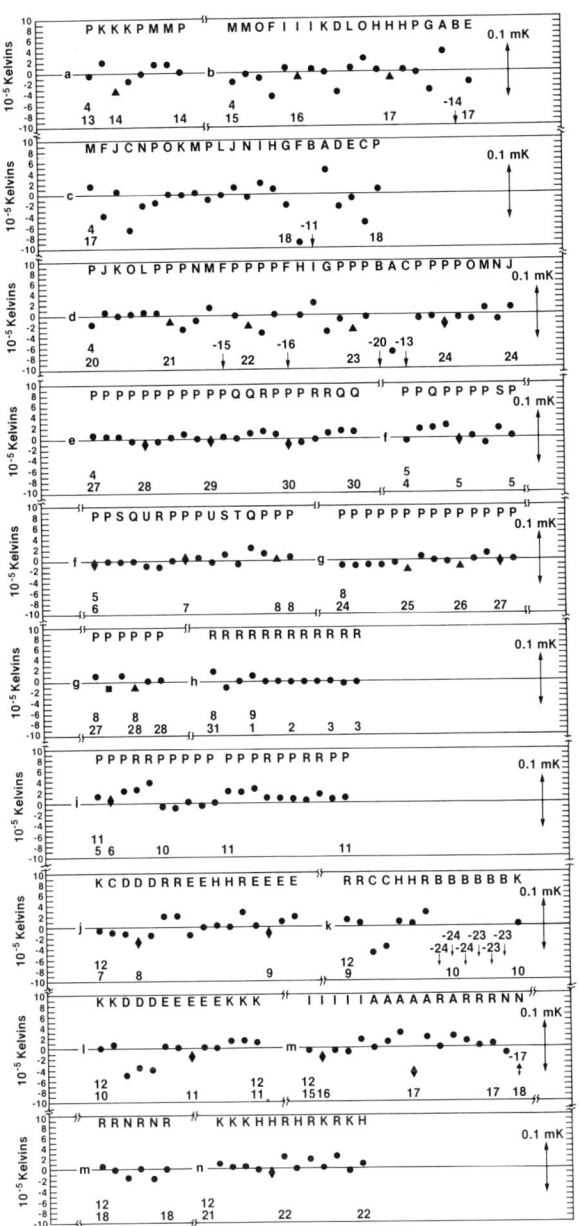

Fig. 3: Comparison of temperatures of triple point of water cells. In each plot, the cells are identified at the top and the measurement dates are given at the bottom, all in 1981. Results of measurement groups a to n are shown. The minus figures associated with the downward pointed arrows indicate the number of 10^{-5} K below the zero line for the TP cell.

TABLE II

The deviations of the temperatures of each reference triple point of water cell from their averages in each measurement group

Measurement Group[b]	Reference Cells[a]			
	K m°C	M m°C	P m°C	R m°C
a	−0.004	+0.010	−0.007	
b	+0.004	−0.007	+0.003	
c	−0.002	+0.008	−0.007	
d	−0.001	+0.012	−0.011	
e			−0.003	+0.003
f			+0.009	−0.009
i			−0.005	+0.005
j	−0.014			+0.014
k	−0.005			+0.005
n	−0.009			+0.009

[a] The figures represent the deviation of the average reading for the TP cell from the average of average readings of all reference TP cells in the measurement group (see Fig. 3). Readings of the reference TP cells are not equally distributed. Readings with symbols △, ◇, and □ are not included in the averages.

[b] Measurement groups g, h, l, and m had only one reference TP cell; hence, they are not included in this tabulation.

these two cells were measured along with other cells, the results on cell R in Table II are shown to be 0.018 mK lower. However, in this case only one measurement on cell R is compared with eleven measurements on cell P. In measurement group e, where three measurements on cell R are compared with eleven measurements on cell P, the results are in close agreement with those measurements of group i. The range of temperatures of reference cells K, M, and P was found in measurement group a to be 0.017 mK (see Table II). Hence, the measurements on the reference cells are consistent within the range of 0.01 or 0.02 mK.

Table III shows the deviations of each of the TP cells (except reference cells K, M, P, and R) from the reference cells. The figures were obtained by averaging the deviations found for the TP cell in each of the measurement group. The deviation for the cell in a measurement group was the average deviation, if more than one observation was obtained for the cell in the measurement group, from the average observation found for the reference cell or reference cells in the group. The deviation figures of Table III show that, except for cells B, C, and F, the temperatures of the TP cells that were tested are within a range of about 0.05 mK. On the basis of the statement in the IPTS-68 document [2]: "The differences in the temperatures of properly constructed cells, from various sources, will not exceed 0.2 mK", all cells that were compared in the present work would be acceptable. However, for high-precision measurements the TP cells should fall well within the band of 0.05 mK. McLaren [13] in the comparison of nine type B TP cells found them to have a range of 0.05 mK, after excluding two cells with temperatures that were lower than the remaining cells of the group by about 0.1 to 0.2 mK.

The possible effect of residual air in the temperature of the TP cell was described in Section 2. It was shown that 133 Pa of air (1 cm^3 of air bubble in type A cell) would cause the temperature to be lowered about 0.01 m°C. The results on cells G and N show that even with relatively large volume of residual air the temperature is lowered only by a small amount, confirming the above calculated figure. The very low value of temperature found with cell B with its relatively large amount of residual air may be attributable to the added effect of possible presence of impurities.

TABLE III

The deviation of the temperature of some triple point of water cells from the reference TP cells

Cell	Measurement Group[a]	Deviation from Reference Cell(s)[b] m°C
A	$bcdm$	+0.007
B	$bcdk$	−0.172
C	$cdjk$	−0.065
D	$bcjl$	−0.032
E	$bcjl$	−0.009
F	bcd	−0.094
G	bcd	−0.027
H	$bcjkn$	−0.000
I	$bcdm$	+0.012
J	cd	+0.007
L	bcd	−0.005
N	cdm	−0.014
O	bcd	+0.006
Q	ef	+0.011
S	f	+0.013
T	f	−0.003
U	f	−0.003

[a] The letters represent the measurement groups in which readings were obtained on the TP cell.

[b] The figures represent the average of the deviations found for the TP cell from the reference cell(s) in each of the measurement groups.

The temperature for cell F should fall within the range of 0.05 mK; however, the observations on the cell have been somewhat erratic and the results deviate more than expected. It may contain relatively large amounts of impurities. Additional test measurements will be needed to determine the source of the erratic results.

In some measurements, after making the last observation near the close of working day, the TP cell and SPRT were left undisturbed and a continuous record of temperature in the cell was obtained overnight at either 1 or $\sqrt{2}$ mA through the SPRT. In the morning a complete set of readings (i.e., readings at 1 and $\sqrt{2}$ mA plus determinations of sensitivities and bridge zero) was obtained without disturbing the system. This was followed by readings obtained using the usual procedure (i.e., transferring the SPRT into an ice bath, inserting a warming rod into the TP cell to free the mantle, checking the ice mantle to be free, and re-inserting the SPRT into the TP cell). The first reading obtained in the morning without disturbing the system is identified by a triangle △ or a diamond ◇ depending upon whether the SPRT readings were recorded overnight with 1 or $\sqrt{2}$ mA, respectively. When an observation during the working day was found to have been taken with the mantle not "free", the observation was plotted as a square □. Following every observation the ice mantle was checked whether it was free or stuck. The mantle was found in every case following overnight recording to be stuck or not as free as after a warm rod was inserted in the thermometer well. The ice mantle may have become less free because during the night the part of the mantle in contact with the bottom of the thermometer well was converted to conform more closely with the contact surface. As shown in Fig. 3 most of the readings obtained after recording overnight at $\sqrt{2}$ mA are fairly close to but always lower than the readings obtained after "freeing" the ice mantle. Hence, the former readings may be under "nearly free" conditions. On the other hand, where measurements were obtained after 1 mA was in the SPRT overnight, the first readings in the morning were lower than those obtained after $\sqrt{2}$ mA was in the SPRT overnight. In those two cases (see measurement group d), where the readings with the mantle free are lower than the first readings of the morning, it is very likely that a small fragment of ice had become lodged in the plastic foam used to cushion the SPRT at the bottom of the thermometer well. (It was found that small fragments of ice can readily become embedded in soft plastic foam. Under such circumstances the foam was replaced with another one.)

6. CONCLUSIONS

By making certain that the ice mantle surrounding the inner melt is free, the temperature measurements in a TP cell can be made with a precision of ±0.01 mK or better. Because of the buoyant force of the ice mantle against the bottom of the thermometer well, the ice mantle when left undisturbed can eventually become somewhat stuck to the well. The sharp rotational motion of the TP cell used to test for the presence of inner melt may cause the ice mantle to break loose from the "stuck state" and indicate only a momentarily free ice mantle. When the rotational motion of the ice mantle ceases, it may soon stick back onto the thermometer well. The presence of a free mantle can be tested best by slightly tilting the TP cell to change the center of gravity of the mantle. If the mantle is free, it will move.

If the water is pure, a small amount of residual air (about 133 Pa) reduces the temperature of a TP cell by only about 0.01 mK. In a type A cell, 133 Pa of residual air appears as a noticeably large amount - an air bubble of about 1 cm^3 can be entrapped in the extension. Also, when the cell is gently tilted back and forth with the thermometer well nearly horizontal, with 133 Pa of residual air, the clicking sound that would be caused by the water hammer action may be "cushioned" by the air or the sound would be relatively dull. The clicking sound becomes sharper with decreasing amounts of residual air or by reduction of cell temperature, which causes more air to be dissolved in the water and the water vapor pressure to become less. Therefore, when the TP cell is tested at room temperature for the clicking sound of the water hammer action, if the sound is sharp, the amount of residual air is most likely less than 133 Pa and the lowering of the TP cell temperature is less than 0.01 mK.

Although measurements were obtained on only one very old TP cell (A) that fell within the 0.05 mK band, borosilicate glass is a suitable material for making TP cells. An initially good cell should last as long as it is not broken.

ACKNOWLEDGMENTS

The authors are grateful to James L. Cross of the Jarrett Instrument Company for the loan of some TP cells.

DISCLAIMER

Certain commercial equipment, instruments, or materials are identified in this paper in order to adequately specify the experimental procedure. Such identification does not imply recommendation or endorsement by the National Bureau of Standards, nor does it imply that the materials or equipment identified are necessarily the best available for the purpose.

REFERENCES

[1] H. F. Stimson, International Practical Temperature Scale of 1948. Text revision of 1960, J. Res. Nat. Bur. Stand. (U.S.) 65A, 139 (1961).

[2] The International Practical Temperature Scale of 1968. Amended edition of 1975, Metrologia 12, 7 (1976).

[3] G. T. Furukawa, J. L. Riddle, and W. R. Bigge, The International Practical Temperature Scale of 1968 in the region 90.188 K to 903.89 K as maintained at the National Bureau of Standards, J. Res. Nat. Bur. Stand. (U.S.) 80A, 477 (1976).

[4] J. P. Evans, High temperature platinum resistance thermometry, in Temperature, Its Measurement and Control in Science and Industry, Vol. 4, ed. by H. H. Plumb, Instrument Society of America, Pittsburgh, 1972, pp. 899-906.

[5] R. J. Berry, Control of oxygen-activated cycling effects in platinum resistance thermometers, Temperature Measurement 1975, Institute of Physics Conference Series No. 26, pp. 99-106, London 1975.

[6] J. L. Riddle, G. T. Furukawa, and H. H. Plumb, Platinum resistance thermometry, Nat. Bur. Stand. (U.S.), Monogr. 126, 129 pages (April 1973).

[7] H. F. Stimson, The measurement of some thermal properties of water, J. Wash. Acad. Sci. 35, 201 (1945).

[8] J. A. Ferguson, Realization of the triple point of water, J. Phys. E, Sci. Instr. 3, 447 (1970).

[9] Instructions for the preparation and use of triple-point-of water cells, Jarrett Instrument Company, Inc., May 1972.

[10] Manual, "Equiphase" triple point of water cell, Foxboro Analytical, A123767, Rev. B.

[11] R. J. Berry, The temperature-time dependence of the triple point of water, Can. J. Phys. 37, 1230 (1959).

[12] R. D. Cutkosky, An a-c resistance thermometer bridge, J. Res. Nat. Bur. Stand. (U.S.) 47C (Engr. and Instr.), 15 (Jan-June 1970).

[13] E. H. McLaren, The freezing points of high purity metals as precision temperature standards. I. Precision measurements with standard resistance thermometers, Can. J. Phys. 35, 78 (1957).

Triple point of gallium as a temperature fixed point

B. W. Mangum

National Bureau of Standards, Washington, D.C. 20234

The triple-point temperature of high-purity gallium has been determined to be 29.77398 °C using five standard platinum resistance thermometers (SPRTs), recently dried and then calibrated on the IPTS-68, and using ten samples of gallium from three commercial sources. All data obtained on the highest-purity sample have a standard deviation of ±0.00014 °C and the systematic uncertainty is estimated to be ±0.0006 °C. Three of the samples investigated were in all-plastic cells and seven samples were in steel cells with Teflon containers for the gallium and with Teflon-coated stainless-steel thermometer wells. Intercomparisons of the triple-point temperatures of all ten samples, each of a different lot, were made for several different gallium mantles of each sample.

INTRODUCTION

We report here the results of the continuation of our investigation[1] of the suitability of the triple point of pure gallium as a defining temperature fixed point of the International Practical Temperature Scale[2]. Although the abundance of gallium in the earth's crust is only about 5 to 15 parts per million and is widely dispersed, gallium is readily available commercially at very high purity and at a relatively low price. Consequently, samples of gallium suitable for use in fixed-point cells are readily available to anyone.

Gallium was first isolated[3] in 1875 and several melting-point temperature determinations were subsequently made[4-8]. Since gallium became commercially available at high purity, it has been used in numerous studies of melting behavior[9-14] and there have been several determinations of its melting-point or freezing-point temperature[13,15-18] and of its triple-point temperature[1]. Its pressure dependence has also been measured[1,17].

The investigation reported here was undertaken (1) to determine the possible variation of the triple-point temperature with different gallium mantles of the same sample for 10 samples from 3 sources, (2) to investigate the variation of triple-point temperature from lot to lot, (3) to determine the triple-point temperature of the highest-purity sample to the highest possible accuracy using five standard platinum resistance thermometers (SPRTs)[19] recently dried and calibrated, and (4) to investigate the use of steel cells relative to all-plastic cells.

EXPERIMENTAL
Gallium Samples

The ten gallium samples studied in this investigation were obtained from three sources - Alcoa, Alusuisse and Eagle-Picher.* Three of the samples were the same as those on which we reported earlier[1]. Of the ten samples studied, seven were from Alcoa, two were from Alusuisse and one was from Eagle-Picher. The six new samples from Alcoa and the one new sample from Alusuisse were stated by the manufacturers to be 99.9999+% pure and 99.99999% pure, respectively. The sources, lot numbers and nominal purity of the samples are listed in Table I. One sample from each of the three sources of gallium, lot numbers 3809B, F17/220 and J-57-76, constituted the three samples of gallium on which we reported earlier[1]. The purities as specified by the suppliers are difficult, if not impossible, to verify.

Table I. Specifications of Gallium Samples

Source	Lot Number	Nominal Purity
Alcoa	3809B	99.9999+% (Semiconductor grade)
	3854	99.9999+% (Semiconductor grade)
	3855	99.9999+% (Semiconductor grade)
	3856	99.9999+% (Semiconductor grade)
	3860	99.9999+% (Semiconductor grade)
	8002	99.9999+% (Semiconductor grade)
	8005	99.9999+% (Semiconductor grade)
Alusuisse	F17/220	99.9999+% (Microwave grade)
	F17/252	99.99999+%
Eagle-Picher	J-57-76	99.99999%

Gallium and Water Triple-Point Cells
Gallium Triple-Point Cells

Although the all-plastic cells described previously[1] have the very desirable feature that the possibility of metallic contamination of the gallium from components of the cells is eliminated, they have the undesirable features that plastics are permeable to gases and moisture and are much less rugged than steel cells. The latter features are not serious problems in most standards laboratories, where facilities for pumping the all-plastic cells (during their use) and for filling the cells with dry argon (after their use) are usually available. Since such facilities might not be available for many possible users, however, we decided to test the feasibility of using steel cells, which could be evacuated and sealed. These would be easier to use than the all-plastic cells and would require fewer facilities for their proper operation and the realization of the gallium triple-point temperature.

The all-plastic cells containing samples 3809B, F17/220 and J-57-76 have been described in detail previously[1] and will not be described again here. The cells for the other seven samples were very similar in design to the all-plastic cells. The material of construction, as stated above, however, was different except the Teflon container in which the gallium was held. The outer cases of the cells were Teflon-coated steel. The cap assembly, which consisted of the cap, the re-entrant thermometer-well tube, the pumping tube and a valve, was constructed of stainless steel and, with the exception of the valve, was Teflon-coated. The cap was attached to the outer case of the cell by means of an O-ring seal and was held in place by 8 screws passing through the outer rim of the cap into the outer case. The inner container was made of Teflon. Its design was very similar to that used in the all-plastic cells, the difference being the addition of a splash cap. The latter was hollow with each side of the cap containing a small hole through which the space above the sample could be evacuated. The holes were located π radians apart.

The thermometer well (Teflon-coated stainless-steel tube, 3/8 inch in diameter with 0.005 inch thick walls) of each cell contained a welded-in-place and Teflon-coated stainless-steel bushing, 0.032 inch thick and 5 cm long, in the bottom of the well to provide contact between the SPRTs and the gallium.

Before the cells were assembled, the Teflon and Teflon-coated components were soaked overnight in aqua regia, thoroughly rinsed in distilled water and then air dried. The cap assembly was not attached to the base of a cell until the cell had been filled with the gallium sample.

The cells were filled with the gallium samples in a glove box containing a dry argon atmosphere. The liquefied gallium was poured into the Teflon containers of the cells directly from polyethylene squeeze bottles in which the gallium had been sealed under argon by the suppliers. Approximately 900 grams of gallium were put into each cell. After a gallium sample was poured into a cell, the cap assembly was attached to the base of the cell, the valve closed and the gallium solidified by decreasing the temperature of the glove box to 29 °C, or lower, and inserting a liquid-nitrogen-cooled copper rod into the thermowell several times. The solidification process was monitored by a thermometer which was periodically placed in the thermowell.

Water Triple-Point Cells

One triple point of water cell, B-11-803, obtained from Jarrett, Inc. (USA) was used in all of the determinations of the gallium triple-point temperature.

Equipment
Thermometers

Five SPRTs purchased from Leeds and Northrup were used in this investigation. Two of them (L&N 8167-25 Series) had platinum elements that consisted of a single layer helix of bifilarly-wound platinum wire on a mica cross. The other three SPRTs (L&N 8163 Series) had platinum elements consisting of a coiled helix of platinum wire mounted on a mica cross. Pertinent information regarding the five SPRTs is given in Table II, in which S/N refers to the serial number and NBS I.D.# refers to the NBS identification number.

Prior to this investigation, the heads of the SPRTs had been modified for ac measurements by the removal of the external copper leads, provided by the manufacturer for dc measurements, and the installation of BNC connectors mounted in bakelite caps. These were connected to the diagonal pairs of platinum wires which came up the stems of the SPRTs from the helical elements in a square array and passed through hermetic seals. By this arrangement, one coaxial cable served as the current leads and the other served as the potential leads. The SPRTs' resistances were insensitive to lead positions.

We had some evidence before this study began that some of the SPRTs probably contained moisture. Consequently, SPRTs 089, 090, 369, 374 and 375 were opened, connected to a high vacuum system and evacuated. Once a good vacuum was obtained, the SPRTs were heated to 225 °C and maintained at that temperature for about 2 to 3 days, but, in any case, until the pressure was down to $\leq 10^{-5}$ Pa. When the pressure reached the range of 10^{-5} to 10^{-6} Pa, the furnace was de-energized and the SPRT in the furnace permitted to cool overnight. Then, a mixture of 90% Ar and 10% O_2 was admitted into the SPRT to a pressure of 50 kPa and the SPRT sealed. The SPRTs

Table II. Description of SPRTs Used in the Investigation

SPRT S/N	NBS I.D.#	SPRT Type	Date Calibrated	Calibration Constants α	δ
1808364	089	8163	Jan. 1977	3.925683×10^{-3}	1.497137
1808366	090	8163	May 1977	3.926196×10^{-3}	1.496878
			Nov. 1979	3.926191×10^{-3}	1.496847
1808369	369	8163	May 1977	3.926181×10^{-3}	1.496823
			Nov. 1979	3.926166×10^{-3}	1.496850
1846677	374	8167-25	Jan. 1977	3.926390×10^{-3}	1.496519
			Nov. 1979	3.926354×10^{-3}	1.496427
1846729	375	8167-25	Jan. 1977	3.926079×10^{-3}	1.496538
			Nov. 1979	3.926073×10^{-3}	1.496586

were then calibrated, without any further annealing, by the NBS Platinum Resistance Thermometer Calibration Laboratory at measuring currents of 1 and 2 mA, with extrapolations to zero current. The zero-current calibrations were then used in all calculations of temperature, thereby eliminating self-heating effects in all of the measurements.

The thermometers had a slightly smaller amount of self-heating in the steel cells than they did in the all-plastic cells. For example, with a measuring current of 1 mA, SPRTs 374 and 375 had self-heatings of ∼0.85 m°C in the all-plastic cells and ∼0.75 m°C in the steel cells. Similarly, SPRT 369 had a self-heating of ∼2.15 m°C in the all-plastic cells and ∼2.00 m°C in the steel cells.

Constant Temperature Bath

The constant temperature bath and the control system used in this investigation is the same as that described previously[1]. Through the use of this system, temperatures were maintained uniform and constant to ±1 m°C, or better.

Resistance Bridges

An ac resistance-ratio bridge[20] and a dc resistance-ratio bridge[1] were used to measure the resistances of the SPRTs. The ac bridge, designed and built at NBS[1], operates at 400 Hz and is stated to be in error by no more than 3 µΩ when used with a 100 Ω standard resistor[20]. The resolution obtained with this bridge, operating at measuring currents of 1 and $\sqrt{2}$ mA and using a standard resistor of 100 Ω as the reference resistor, was ±0.15 µΩ when the output signal was displayed on a strip-chart recorder.

Since a dc Mueller bridge was used in calibrating the SPRTs, dc techniques were also used in making measurements in this investigation. A Guildline Current Comparator, Model 9975, was used for this purpose. This bridge has an automatic current reversal feature, with reversal periods of 4, 8 or 16 seconds. All measurements made with this bridge during this investigation were made with a 4 second reversal period. The dried thermometers had been previously checked in the triple-point of water cell at 4 and 16 second reversal periods and no differences were observed in the bridge readings. The inaccuracy of the current comparator was stated by the manufacturer to be ≤2 parts in 10^7 plus 1 digit in the last (8th) dial. Using a 3-second time constant and averaging the strip-chart recording of the bridge output for 10 minutes gave a bridge resolution of ± 1.5 $\mu\Omega$ (± 15 $\mu°C$) at measuring currents of 1 and $\sqrt{2}$ mA.

Only the ac bridge was used for the investigation of the immersion characteristics of the SPRTs in the different types of cells and for the study of the melting behavior of the samples. Both the ac and the dc bridges were used for direct comparisons of the different gallium samples and for the measurements of the triple-point temperature.

Standard Resistor

The standard resistor used as the reference resistor for both resistance bridges was a 100 Ω precision resistor, Model No. HA518, which we designated as H19, manufactured by Vishay. It was encased in an aluminum block which, in turn, was enclosed by and thermally shielded from a large copper container, the temperature of which was controlled at 27.75 $\pm$ 0.1 °C through the use of a mercury thermostat. Based on measurements of similarly enclosed resistors, a temperature variation of about ± 3 m°C over a period of several days was estimated for H19. Using this estimate of temperature variation and using an estimated temperature coefficient of resistance of about 0.1 to 0.3 ppm/°C for the resistor at its regulated temperature, values which are based on the manufacturer's specifications, the variation of resistance of H19 was calculated to be $\leq \pm 0.1$ $\mu\Omega$. Although the resistance value of the standard resistor is not used in the determination of temperatures, the resistor was calibrated at 400 Hz and at dc by the Absolute Electrical Measurements Section of the NBS on 9 February 1977 and on 21 March 1978. Over the period of time between calibrations, the 400 Hz and the dc resistances of H19 increased by 0.12 and by 0.10 ppm, respectively.

Vacuum System

The vacuum and gas handling system, by means of which the gallium cells were evacuated during the experiments and, for the nylon cells, filled with argon upon completion of an experiment, consisted of a mechanical rotary pump, a mercury manometer, an oil manometer, two alcohol-solid-carbon-dioxide mixture cold traps, manometer bypasses and some valves as described in detail previously[1]. Since $dT/dP = -2.011$ m°C/atm for gallium, it is not necessary to have a high vacuum in order to realize experimentally the triple-point temperature. With the simple system described above, the pressures obtained were sufficiently low that the uncertainty in the triple-point temperature from this source was <0.1 $\mu°C$, well below our resolution. During the investigation reported here, the manometers were closed off from the remainder of the system and bypassed. A cold trap was located between the gallium cell or cells connected to the system and the other parts of the system. This ensured that the gallium was protected from contamination, even if the manometers were open to the system. All argon gas entering the gallium cells passed through the cold traps.

Measurements and Procedures
Preparation and Maintenance of Water Triple Points

Only one water triple-point cell was used in this investigation. It was obtained in February 1979. Triple points were prepared through the use of a liquid-nitrogen-cooled copper tube followed by the use of a heat pipe cooled by a solid carbon dioxide and ethyl alcohol mixture, as described in a previous publication[1]. Cracking of the mantle was avoided (1) by using only modest amounts of dry ice in the operation of the heat pipe (in order to give a slow, steady growth of the mantle), and (2) by terminating the freeze after the mantle reached about 1 cm in thickness, as viewed through the water. Several hours after terminating the freeze, the thermometer well of the cell was appropriately filled with chilled water, a small piece of foam rubber was placed in the bottom of the well to cushion the SPRTs from shock when they were being inserted into the well, and a chilled aluminum bushing, 5 cm long, was inserted into the well to improve thermal contact between the inner liquid-solid interface and the SPRTs. The bushing fitted relatively snugly into the well and around the SPRTs. The cell so prepared was kept in a Styrofoam jacket which in turn was kept packed in ice in a dewar. During the measurement period, the cell was checked at least every morning to ensure that the ice mantle was free to rotate and not frozen to the thermometer well.

Preparation and Maintenance of Gallium Triple Points

In preparation for a comparison of the gallium triple-point temperatures of a given set of mantles of the samples, the following procedures were used. The all-plastic cells, with an atmosphere of argon above the gallium, were kept in ∿50 °C oil overnight in order to totally melt the gallium. The cells were then removed from the hot oil, placed in air, and allowed to cool to ∿29 °C. A mantle was then prepared by first initiating a freeze of the gallium by repeated insertions into the thermometer well of the cell a liquid-nitrogen-cooled copper tube, and then, after initiation of the freeze, putting the cell of solidifying gallium into a dewar containing a small amount of ice at the bottom. The cell was then covered at the top so that the mantle grew upward and outward from the bottom part of the thermometer well. The freezing process from initiation to completion required several hours. After the solidification was complete, the gallium triple point was prepared by placing the cell, containing the solid gallium under an atmosphere of argon, in an oil bath at ∿60 °C. The hot oil was more or less continuously pumped into the thermometer well, maintaining an average temperature of ∿40 °C, to ensure that there was an inner and an outer liquid-solid interface the full length of the column of gallium. After 20 minutes, during which time about 25 to 50% of the sample was melted, the cell was placed in a constant temperature oil bath at a temperature ∿10 m°C above the gallium triple-point temperature. Before measurements began, the cell was connected to the vacuum system and evacuated. Pumping of the cell continued throughout the measurements.

The gallium samples in the steel cells were similarly treated, except that they were always kept under a vacuum and except that in the preparation of the gallium triple point, the cells were kept in the 60 °C oil for only 6 minutes. They too were pumped during measurements.

After completion of comparison measurements on a set of mantles, the gallium samples were totally melted again and the process described above repeated, beginning with the preparation of a new set of mantles.

Following the comparison of the gallium samples, the triple-point temperature was determined from a set of measurements on the Alusuisse sample (F17/220) in the all-plastic cell. For these measurements, the triple-point was prepared as described above.

Thermal contact to the SPRTs was provided by the oil in the thermometer wells in the all-plastic cells and by the oil plus the bushings in the steel cells.

Temperature Measurements

The immersion study, the melting behavior and the comparison of the samples were made with only one SPRT, but the determination of the triple-point temperature involved the use of 5 SPRTs.

The triple-point temperature was determined from resistance ratio measurements of the 5 SPRTs in the Alusuisse sample (F17/220) at the triple point in the all-plastic cell, and in the water triple-point cell. Normally, 4 SPRTs were cycled through the gallium and water triple-point cells each day. The measurement sequence consisted of an SPRT being preheated in the oil bath containing the gallium cell for at least 30 minutes, then placed in the gallium cell. After the SPRT had been in the gallium cell for at least 30 minutes, its ac and dc resistance ratios were measured. Following these measurements, the SPRT was removed from the cell, and then precooled in an ice bath for at least 30 minutes before being placed in the water triple-point cell. After being in the cell for at least 30 minutes, the ac and dc resistance ratios of the SPRT in the water triple-point cell were then measured.

Each time the resistance ratio of an SPRT was measured, its power dependence was determined by making measurements at 1 and $\sqrt{2}$ mA of measuring current. A 10 minute integration time was used for both ac and dc measurements and, in addition, a 3 second time constant was used for the dc measurements. The zero-power values of the SPRTs were used in the calculation of the triple-point temperature.

SPRT Immersion and Hydrostatic Head Effects

The immersion characteristics of the SPRTs in the all-plastic cells have been reported previously[1]. The behavior in the steel cells was very similar, as shown in Fig. 1. Measurements were made with SPRT 374 using the Cutkosky bridge[20]. Measuring currents of 1 and $\sqrt{2}$ mA were used, with extrapolation to zero current. The triple point for the immersion study was prepared as described earlier. Measurements were made on extraction and on insertion of the SPRT, using increments of 0.5 to 1 cm. When the SPRT was fully inserted into the thermometer well, the center of the platinum sensing element was approximately 13.5 cm below the top surface of the gallium. The temperature of the bath was ~10 m°C above the gallium triple-point temperature. As seen from Fig. 1, the effects of the temperature outside the cell were no longer discernible when the tip of the thermometer was within about 7.5 cm of the bottom of the well.

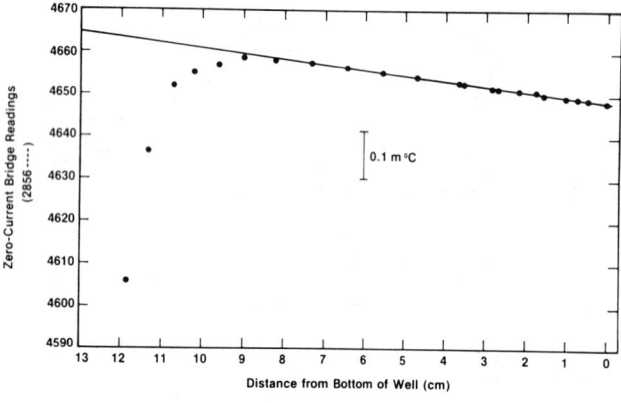

FIG. 1. Immersion characteristics of SPRT 374 in a steel triple-point cell containing gallium sample 3860 at the triple-point temperature. The data have not been corrected for hydrostatic head effects. The solid line, calculated from the measured pressure dependence of the melting point, has a slope corresponding to the effects of the hydrostatic pressure of the column of liquid gallium, i.e., -12 μ°C per cm of liquid gallium.

The hydrostatic head effect, i.e., the change of the melting temperature with the distance of the SPRT from the bottom of the well, is in excellent agreement with the results reported earlier[1], both the direct measurements of the effect and that calculated from the results of the melting temperature versus pressure experiments.

Gallium Melting Experiments

Melting behavior of the gallium in the all-plastic cells has been reported previously[1]. The melting behavior of the gallium in the steel cells was measured in a fashion similar to that used for the all-plastic cells. An oil bath with a temperature held constant at ~40 °C was used for these measurements. The thermometer used in this phase of the study was SPRT 374, at a measuring current of 1 mA. From consideration of the design of the cells, it was clear that the principal point of thermal contact between the bath and the gallium would be at the bottom of the cell. Consequently, in order to reduce heat flow at that point and to prevent premature melting of the gallium at the bottom, a Styrofoam jacket was fitted to the bottom half of the cell. With such an arrangement, at least 3 melting curves were obtained for each cell, with at least 2 of the 3 melts being obtained for the samples after the mantles had been prepared in the standard way, as described earlier. This was to check for reproducibility of the melting behavior. Some melting curves were also obtained for samples after the triple points had been prepared. The purpose of the latter was to check for stability of the triple-point temperature and to check for its independence of the fraction of gallium melted.

Direct Comparisons of the Different Gallium Samples

Intercomparisons of the triple-point temperatures of the 10 gallium samples were made for 6 different mantles of each sample. This was accomplished by making a direct comparison of each sample with the Alusuisse sample (F17/220) in the all-plastic cell. These measurements did not involve measurements at the water triple point. After completion of the comparison of the triple-point temperatures of a given set of mantles, the gallium in each cell was totally melted and then frozen again in the way described earlier. The triple point of each new mantle was then prepared in the standard manner and a comparison of the triple-point temperatures of that set of mantles was conducted. The thermometer involved in these measurements was SPRT 375. Both ac and dc techniques were used, with measuring currents of 1 and $\sqrt{2}$ mA being employed to permit extrapolation to zero current for comparison purposes. Intercomparisons of the samples were made two at a time, with 5 dc and 2 ac sets of comparisons being made on each pair of mantles. A full day of measurements was required for a comparison of a pair of gallium mantles.

Measurement of the Gallium Triple-Point Temperature

As described earlier, 5 SPRTs were used in these measurements. Seven determinations were made with SPRT 089, 19 with each of SPRT 090 and SPRT 369, and 20 with each of SPRT 374 and SPRT 375. Two different gallium mantles of the Alusuisse sample (F17/220) in the all-plastic cell were involved in the measurements. The cell was pumped continuously and kept in a bath maintained ~10 m°C above the triple-point temperature. The measurements were conducted over a period of about two months.

RESULTS AND ANALYSIS
Immersion and Hydrostatic Head Results

The results of SPRT 374 immersion and the hydrostatic head measurements on gallium sample 3860 in one of the steel cells are shown in Fig. 1. These results are in agreement with those reported previously[1] for gallium in the all-plastic cells. When the tip of the thermometer (SPRT 374) was within about 7.5 cm of the bottom of the well, the thermometer was not influenced by the external environment of the cell. Below that depth of immersion, only the hydrostatic head effects were being observed and these results yield a value of -12 μ°C per cm of gallium, at the triple-point temperature, in agreement with the value of dT/dP = -2.011 m°C/atm reported previously[1].

Melting Behavior of the Gallium Samples

As indicated earlier, at least two melting curves were obtained for each sample of gallium in the steel cells after the gallium had been frozen in the usual way. The general features of the curves, a typical one being shown in Fig. 2, were very similar to those obtained for the gallium samples in the all-plastic cells and which have been reported previously[1]. After the cells, with the SPRT, were placed in the 40 °C bath, the temperature, on a fairly coarse scale (uppermost curve of Fig. 2), very rapidly attained a nearly constant value at which it remained for about 8 hours (7 hours for the all-plastic cells). It then fairly slowly increased by about 0.5 °C over the next 1 1/2 to 2 hours before increasing very rapidly and approaching the bath temperature. On a scale having 250 times better resolution than that just described, the temperature (middle curve of Fig. 2) began increasing fairly slowly some 4 1/2 hours after melting began. The rate of change in temperature continuously increased until the increase became quite rapid some 7 hours after melting began. On a scale having yet 100 times greater resolution, the temperature (the bottommost curve) began increasing rather rapidly some 4 hours after melting began.

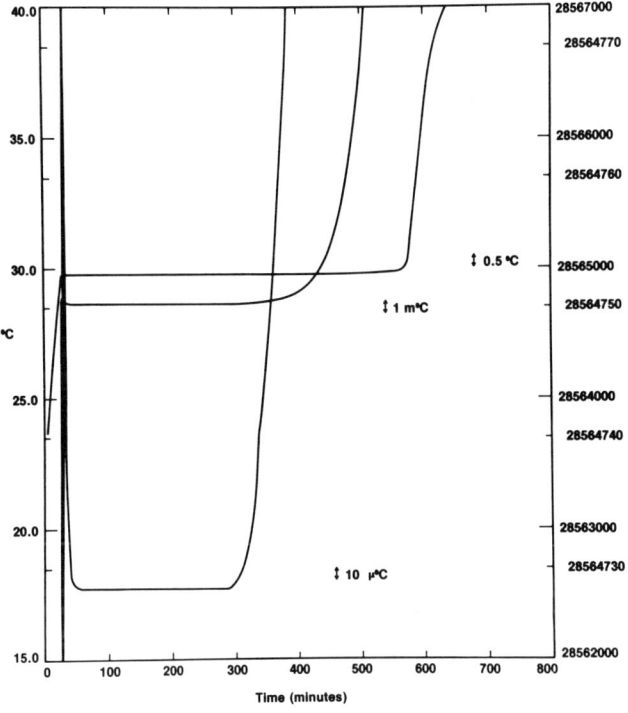

FIG. 2. Melting curve at three levels of resolution obtained with SPRT 374 for sample 3856 in a steel cell in a 40 °C bath following a standard freeze. The temperature scale on the left-hand side of the figure is associated with the upper curve; the scale running from 28562000 to 28567000 on the right-hand side of the figure is associated with the middle curve; and the other scale on the right-hand side is associated with the lowest curve.

We attribute this melting behavior to the melting of the gallium mantle from the top of the cell downward. Upon complete melting of the gallium, the sample temperature approached the bath temperature exponentially. The melting behavior of each sample was reproducible and the details of the melting curves at the highest resolution were strongly characteristic of each sample. There are at least two important factors which influence the shape of these curves. One is the impurities and their distribution, as determined by the sample's prior thermal treatment, and the second is the establishment of the hydrostatic head, which varies from cell to cell. The latter factor affects the initial portion of the melting curve and the first factor affects the melting range of the sample. Although most melting curves were very similar, both in shape and in melting range, to those reported earlier for the gallium samples in the all-plastic cells[1], some samples definitely had a larger melting range. We interpreted this as probably being due to impurities introduced into the sample through pin-holes in the Teflon covering the thermometer well since 6 of the 7 samples were from the same supplier and were presumably of the same purity. The lower triple-point temperatures observed for those same samples confirm the presence of more impurities in those samples. Although the results from the melting curves agree qualitatively with the results of of comparison of the triple-point temperatures, it is difficult to make a quantitative comparison of the samples based purely on the melting behavior.

Melts were obtained also for gallium mantles (of the less pure samples) which had been rapidly frozen. As expected, the curves were characterized by a flat plateau due to a uniform distribution of the impurities present, rather than by the non-flat "plateau" characteristic of impure samples.

A melting curve was obtained also for a gallium sample in a steel cell which had been prepared for measurement of the triple-point temperature. After preparation of the triple point, the cell was connected to the vacuum system and continuously pumped, an SPRT was placed in the thermometer well, the assembly placed in a 40 °C oil bath, and the gallium melted. As observed and reported previously[1] for gallium in the all-plastic cells, the temperature in the thermowell remained perfectly constant, to within our resolution of ±1.5 µ°C, until an abrupt rise several hours after the cell was placed in the 40 °C oil, which we attribute to penetration of the mantle by liquid gallium. This constancy of the triple-point temperature is important in that when the triple point of a sample, whether in an all-plastic cell or in a steel cell, is prepared as described earlier, the temperature of that triple point is independent of time and of the fraction of gallium melted.

Intercomparison of Gallium Samples

For a given set of gallium mantles, at least five comparison measurements using dc techniques were made for each sample, relative to the Alusuisse sample (F17/220) in the all-plastic cell, and two comparison measurements were made using ac techniques. Such comparisons were made on 6 different mantles of each sample, with the exception of the Eagle-Picher sample (J-57-76) for which only five different mantles were investigated by dc techniques. The results are given in Table III. A given number in the column headed dc under each sample is the mean value of the five dc comparison measurements for a given mantle; that number in the same row of the Table as just discussed and in the column headed ac under each sample is the mean value of the two ac comparison measurements for that same mantle. The other numbers in a given column refer to the mean values of the measurements for the different mantles of the sample.

It can be seen that the results obtained for the Alcoa (3809B) and the Eagle-Picher (J-57-76) samples (samples in all-plastic cells) are in good agreement with those obtained previously[1]. All of the samples in the steel cells have triple-point temperatures which are lower than those in the all-plastic cells. Since all of the Alcoa samples in the steel cells were supposedly of equal purity and the same as that of the Alcoa sample (3809B) in the all-plastic cell and since the Alusuisse samples were supposedly of the same purity, we conclude that the Teflon coating of the stainless steel thermowells must have had some pinholes, which then allowed the gallium to become contaminated. Samples 3854 and 8002 appear to have considerably more impurities than the other samples.

Table III. Results of direct comparisons of triple-point temperatures for different mantles of each of the 10 gallium samples using both ac and dc techniques. Samples F17/220, 3809B and J-57-76 were in all-plastic cells; the remaining samples were in steel cells. The temperature differences are expressed in µ°C.

ΔT(µ°C) F17/220-3809B		ΔT(µ°C) F17/220-(J-57-76)		ΔT(µ°C) F17/220-F17/252		ΔT(µ°C) F17/220-3854		ΔT(µ°C) F17/220-3855		ΔT(µ°C) F17/220-3856		ΔT(µ°C) F17/220-3860		ΔT(µ°C) F17/220-8002		ΔT(µ°C) F17/220-8005	
ac	dc	ac	dc	ac	dc	ac	dc	ac	dc	ac	dc	ac	dc	ac	dc	ac	dc
19	35	4															
				37	79	92	146	60	101	14	76	21	105	163	192	20	119
-35	-26	-14	9	-37	15	48	121	-26	17	23	79	5	65	150	217	14	88
30	14	26	36	54	116	106	141	44	86	27	92	62	107	247	293	84	134
10	17	70	109	23	67	137	190	52	88	62	101	64	136	110	184	40	89
-12	-18	9	-5	-24	42	89	132	46	90	20	58	50	87	168	224	32	79
33	31	22	58	7	82	118	180	58	88	48	108	92	130	230	291	49	96
Mean±S.D. (S_D)																	
22±36	9±25	44±56	41±45	10±35	67±35	98±30	152±27	39±32	78±31	32±19	86±18	49±32	105±27	178±51	234±48	40±25	101±21

The scatter among the data obtained by dc techniques for a given mantle of a given sample is comparable to the scatter among the data for the different mantles of the same sample. Only two measurements were made by ac techniques on each mantle, but they showed much better agreement than did the dc results. The scatter among the data for the different mantles of a given sample, however, was comparable to that observed among the dc results. This may indicate a slight variability of the triple-point temperature with mantle preparation. By looking at the ac and dc results for a given mantle of a sample, it is clear that the ac measurements were affected by the presence of the metal surrounding the thermometer in the steel cells, as expected.

Two conclusions can be drawn from the comparison study. The first is that the gallium samples in the steel cells are less pure than those in the all-plastic cells, as deduced by the fact that their triple-point temperatures are lower. This is thought to be due to contamination from the steel thermowells because of the presence of pin-holes in the Teflon covering of the thermometer wells. Some samples appear to be more impure than others. These results are in qualitative agreement with the melting behavior of the different samples.

The second conclusion is that the scatter among the dc results (probably the most reliable results since the thermometer readings are influenced by the presence of metal near the platinum sensing element when ac techniques are used) for different mantles of the same sample is about the same as the scatter of the measurements on a given mantle of a sample and is about ±50 μ°C. This indicates that if indeed different mantles, each prepared in the usual way (as described earlier), of a given sample give rise to different triple-point temperatures, the scatter in those values is no greater than the scatter in the measurements on a given mantle. Thus, one would conclude that the scatter observed in these experiments is just that due to the SPRT and the resolution of the bridge.

Gallium Triple-Point Results

The gallium triple-point temperature ascertained in this investigation was determined from measurements on the Alusuisse sample (F17/220) in the all-plastic cell using five SPRTs. The temperatures determined for each SPRT were calculated from the zero-power resistance ratios using the standard interpolating formula specified by the IPTS-68, the SPRT calibration constants (given in Table II) provided by the NBS Platinum Resistance Thermometer Calibration Laboratory, and the appropriate corrections described below. The calculated temperatures obtained by both ac and dc techniques for the Alusuisse sample in the all-plastic cell are given in Table IV. These data for SPRTs 090, 369, 374 and 375 are plotted in Fig. 3. As indicated earlier, a few measurements were also made of the triple-point temperature of gallium sample F17/252 in a steel cell. Those calculated temperatures are also shown in Table IV. Since the results of a calibration of an SPRT are expressed relative to the SPRT resistance at the ice-point of water, $R_{H_2O}(0)$, and since our reference is the triple point of water, $R_{H_2O}(t.p.)$, it was necessary in deriving the gallium triple-point temperature to calculate $R_{H_2O}(0)$ of an SPRT from the measured $R_{H_2O}(t.p.)$. The temperature experienced by an SPRT in the water triple-point cell was not 0.01 °C, however, because of hydrostatic head effects. Thus, the resistances of the SPRTs is the gallium and water triple-point cells were corrected for the depression of the true triple-point temperatures due to the hydrostatic pressures exerted by the columns of liquids. For the water triple-point cell, the depression is 7 μ°C per cm of water above the point of measurement[2]. Since the height of the water column in our triple point of water cell, cell B-11-803, was approximately 28 cm, the depression amounted to 196 μ°C. The variations in the amount of ice comprising

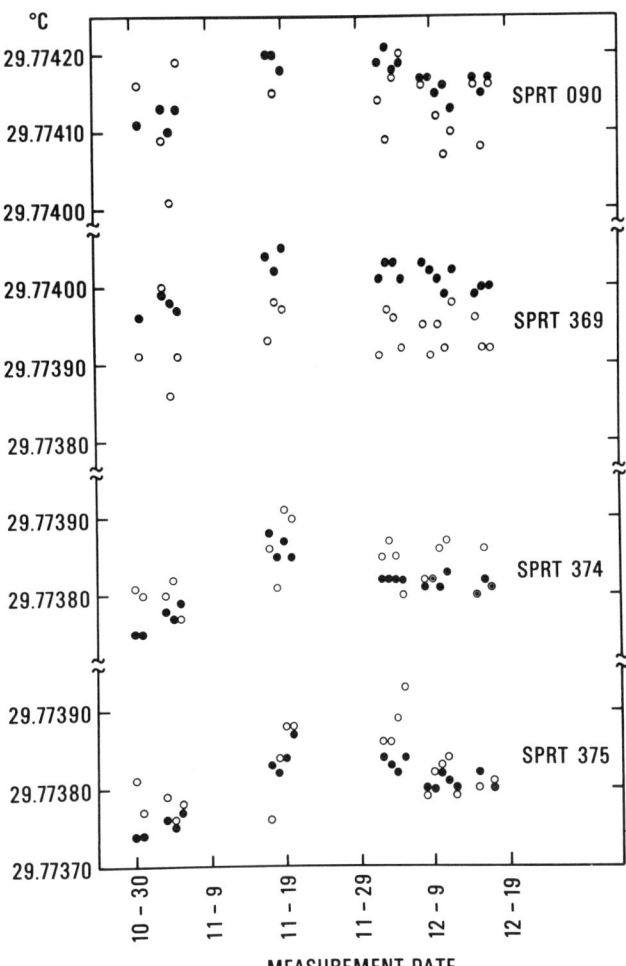

FIG. 3. Plots of the triple-point temperatures of gallium sample F17/220, obtained with SPRTs 090, 369, 374 and 375, as a function of the measurement date. Data plotted are from Table IV. ● represent the ac results, ○ represent the dc results, and ⊖ represents ac and dc results that occur at the same point.

the mantle could cause an uncertainty of ±0.5 cm in the height of the water column, leading to an uncertainty of ±3.5 μ°C in the calculated hydrostatic pressure correction. The zero-power resistance of an SPRT determined in the triple-point of water cell and experiencing a temperature of 0.009804 °C was thus corrected to 0 °C by dividing that resistance by the appropriate resistance ratio, the W(t).

The depression of the temperature in the gallium triple-point cell due to the column of liquid gallium at the triple-point temperature was calculated from the results reported previously[1] and confirmed in this investigation as 12 μ°C per cm of liquid gallium. The height of the liquid gallium column in the all-plastic cell was approximately 13 cm and that height results in a depression of 156 μ°C. The height of the liquid gallium column in the steel cell was approximately 13.5 cm, resulting in a temperature depression of 162 μ°C. These values of depression, either 156 μ°C or 162 μ°C, were added to the temperatures actually measured in the respective gallium triple-point cell (and calculated as indicated above for the appropriate resistance ratio).

We have included in Table IV the temperatures determined by using ac techniques. Note the fairly good agreement of these values with those determined by dc

Table IV. Tabulation of triple-point temperatures (°C) determined for samples F17/220 (in an all-plastic cell) and F17/252 (in a steel cell) by using ac and dc techniques. See text for details of calculation.

SPRT:	089		090		369		374		375	
Date	ac	dc	ac	dc	ac	dc	ac	dc	ac	dc
					Data for F17/220					
10/30/80			29.77411	29.77416	29.77396	29.77391	29.77375	29.77381	29.77374	29.77381
10/31/80			29.77413	29.77409	29.77399	29.77400	29.77375	29.77380	29.77374	29.77377
11/03/80			29.77410	29.77401	29.77398	29.77386	29.77378	29.77380	29.77376	29.77379
11/04/80			29.77413	29.77419	29.77397	29.77391	29.77377	29.77382	29.77375	29.77376
11/05/80			29.77420	29.77420	29.77404	29.77393	29.77379	29.77377	29.77377	29.77378
11/17/80			29.77420	29.77415	29.77402	29.77398	29.77388	29.77386	29.77383	29.77376
11/18/80			29.77418	29.77418	29.77405	29.77397	29.77385	29.77381	29.77382	29.77384
11/19/80							29.77387	29.77391	29.77384	29.77388
11/20/80			29.77419	29.77414	29.77401	29.77391	29.77385	29.77390	29.77387	29.77388
12/02/80			29.77421	29.77409	29.77403	29.77397	29.77382	29.77385	29.77384	29.77386
12/03/80			29.77418	29.77417	29.77403	29.77396	29.77382	29.77387	29.77383	29.77386
12/04/80			29.77418	29.77420	29.77401	29.77392	29.77382	29.77385	29.77382	29.77389
12/05/80			29.77417	29.77416	29.77403	29.77395	29.77382	29.77380	29.77384	29.77393
12/08/80			29.77417	29.77417	29.77402	29.77391	29.77381	29.77382	29.77380	29.77379
12/09/80			29.77415	29.77412	29.77401	29.77395	29.77381	29.77382	29.77382	29.77382
12/10/80			29.77416	29.77407	29.77399	29.77392	29.77381	29.77386	29.77380	29.77384
12/11/80			29.77413	29.77410	29.77402	29.77398	29.77383	29.77387	29.77382	29.77383
12/12/80			29.77417	29.77416	29.77399	29.77396			29.77381	29.77379
12/15/80			29.77415	29.77408	29.77402	29.77392	29.77380	29.77380	29.77380	29.77380
12/16/80			29.77417	29.77416	29.77400	29.77392	29.77382	29.77386	29.77382	29.77381
12/17/80							29.77381	29.77381	29.77380	
12/18/80	29.77422	29.77412								
12/18/80	29.77421	29.77424								
12/18/80	29.77421	29.77417								
12/18/80	29.77422	29.77422								
12/19/80	29.77420	29.77415								
12/19/80	29.77411	29.77408								
12/19/80	29.77421	29.77418								
					Data for F17/252					
01/14/81					29.77399	29.77386	29.77381	29.77388	29.77377	29.77375
01/16/81					29.77397	29.77388	29.77377	29.77376	29.77376	29.77372

Table V. Average triple-point temperature (°C), standard deviation (S_D) and standard deviation of the mean ($S_{\overline{D}} = S_D/\sqrt{n}$ where n = number of data points) for each SPRT in the Alusuisse gallium sample (F17/220) in the all-plastic cell determined by ac and dc techniques.

SPRT	ac			dc		
	Temp. (°C)	S_D	$S_{\overline{D}}$	Temp. (°C)	S_D	$S_{\overline{D}}$
089	29.774197	±0.000039	±0.000015	29.774166	±0.000055	±0.000021
090	29.774163	±0.000031	±0.000007	29.774137	±0.000051	±0.000012
369	29.774008	±0.000024	±0.000005	29.773938	±0.000040	±0.000009
374	29.773814	±0.000035	±0.000008	29.773835	±0.000037	±0.000008
375	29.773805	±0.000036	±0.000008	29.773825	±0.000047	±0.000011
Avg.	29.773997	±0.000162	±0.000018	29.773980	±0.000143	±0.000016

techniques. Although ac measurements of temperature are generally less reliable than dc measurements, we believe the agreement confirms that the SPRTs were dry. Their ac behavior at the triple point of water lends further evidence to this interpretation.

We have listed in Table V the average temperature, its standard deviation (S_D) and the standard deviation of the mean ($S_{\overline{D}}$) derived from ac and dc measurements for each of the 5 SPRTs in the Alusuisse sample (F17/220) in the all-plastic cell. The total spread of 0.34 m°C among the dc values from the 5 SPRTs would at first glance seem rather surprising. Differences may be expected among the thermometers, however, and they may arise from three sources. For measurements of a given SPRT, the uncertainty arising from calibration errors (calibrated against fixed points of the IPTS-68 as maintained at NBS) may be as large as ±0.21 m°C. Another source of error is the possible presence of moisture on the thermometer insulation. The error from this effect would vary from thermometer to thermometer. Although we think the error from this source is small in our case, it has been demonstrated[21] that it can cause an error at the triple point of water equivalent to +0.0 or -0.5 m°C without the thermometer having a detectable "wet kick." Then, of course, different SPRTs may indicate discrepant temperatures at fixed points intermediate to the calibration points due to the use of real, non-ideal, materials. Here again, we would expect the error from this source to be small compared to 0.1 m°C near 30 °C. Considering these sources of error, then, we might expect our SPRTs to indicate temperatures of a given intermediate fixed point near 30 °C that differ by approximately 0.4 m°C to 0.5 m°C. Thus, the differences in the temperatures indicated by the different SPRTs are not inconsistent with these error estimates.

As indicated previously[1], it is thought that the Alusuisse sample (F17/220) in the all-plastic cell is the purest sample that we have and the effects of impurities in that sample would be to depress the melting temperature by less than 0.01 m°C. The triple-point temperature of that sample, then, most closely represents the triple-point temperature of pure gallium. That value determined in this investigation through the use of 5 SPRTs is 29.77398 °C with a standard deviation of all data points of ±0.00014 °C and a standard deviation of the mean of ±0.00002 °C. The standard deviation of the data for individual SPRTs range from ±0.000037 to ±0.000055 °C.

The measured triple-point temperature of sample F17/252 in a steel cell is given in Table VI and is consistent with the direct comparison results. Table VII compares the dc results of the two Alusuisse samples.

A comparison of the gallium triple-point temperatures obtained with the same SPRTs used in this investigation and in that reported previously[1] is given in Table VIII. Note that the temperatures obtained in this investigation are lower than those reported earlier. We attribute this to the presence of moisture in the SPRTs when the previous results were obtained. The magnitude of the change is not inconsistent with this assumption.

Discussion of Errors

The triple-point temperature of gallium measured on the IPTS-68 in this investigation is subject to uncertainties from several potential sources of systematic errors. They arise from SPRT calibrations, bridge inaccuracy, impurities, moisture on the SPRT insulation, and variations in isotopic composition.

The uncertainties in realizing the fixed points of the IPTS-68 which were used in the calibration of the SPRTs are ±0.1 m°C for the triple point of water[2,19] and ±1 m°C for the freezing points of tin and zinc[19]. The combination of these results gives an uncertainty from calibration of ±0.4 m°C. The estimated inaccuracy of the dc bridge, discussed earlier, corresponds to an uncertainty in the temperature measurements of ±0.16 m°C. In consideration of the arguments given earlier, the uncertainty resulting from impurities in the Alusuisse sample in the all-plastic cell is estimated to be less than ±0.01 m°C[1]. Although there is the possibility of the presence of moisture on the SPRT insulation[21], we believe that in view of our treatment of the SPRTs prior to the triple-point temperature measurements, the uncertainty from this source is negligible. Variations in isotopic composition of the gallium samples would affect the triple-point temperature, but since there is no information available regarding the magnitude of such an effect, it is not possible to make a realistic estimate of the uncertainty arising from this source. We would expect this uncertainty to be small, however, since the isotopic ratios $^{69}Ga/^{71}Ga$ of the Alcoa and the Alusuisse samples used to fill the cells were generally in very close agreement[22] and the differences bore no correlation with the measured triple-point temperatures. The source of the uncertainty in the triple-point temperature of water, mentioned above under discussion of errors in calibration, is the variation of the triple-point temperature with isotopic composition. Although

Table VI. Average triple-point temperature (°C), standard deviation (S_D) and standard deviation of the mean ($S_{\overline{D}} = S_D/\sqrt{n}$ where n = number of data points) for 3 SPRTs in the Alusuisse gallium sample (F17/252) in the steel cell determined by ac and dc techniques.

SPRT	ac			dc		
	Temp. (°C)	S_D	$S_{\overline{D}}$	Temp. (°C)	S_D	$S_{\overline{D}}$
369	29.773980			29.773870		
374	29.773790			29.773820		
375	29.773765			29.773735		
Avg.	29.773845	±0.000106	±0.000043	29.773808	±0.000073	±0.000030

Table VII. Comparison of triple-point temperatures for Alusuisse gallium samples in the all-plastic cell (sample F17/220) and in the steel cell (sample F17/252) (dc results).

SPRT	All-plastic cell (°C)	Steel cell (°C)	ΔT (m°C)
369	29.773938	29.773870	0.068
374	29.773835	29.773820	0.015
375	29.773825	29.773735	0.090
Avg.	29.773866	29.773808	0.058

Table VIII. Comparison of present dc results on triple-point temperature with those reported previously for the Alusuisse sample in the all-plastic cell (sample F17/220).

SPRT	Values from present study (°C)	Previously reported values (°C)	ΔT (m°C)
089	29.77417	29.77421	0.04
374	29.77384	29.77395	0.11
375	29.77383	29.77402	0.19
Avg. (from listed SPRTs above)	29.77395	29.77406	0.11
Avg., S_D (5 SPRTs for present results and 3 SPRTs for previous study)	29.77398±0.00014	29.77406±0.00011	0.08

this uncertainty would normally be included again in temperature measurements with SPRTs, it is inappropriate to include this uncertainty a second time here since the water triple-point cell used in these measurements came from the same source as those used in the SPRT calibrations. The total uncertainty from all of these sources for which we can make estimates is ±0.6 m°C.

The systematic and random uncertainties comprise the total uncertainty and is about ±0.7 m°C for our measurements.

SUMMARY AND CONCLUSIONS

An intercomparison of 10 gallium cells using one SPRT, and a new determination of the triple-point temperature of high-purity gallium using five newly-dried and then calibrated SPRTs were made. The 10 cells consisted of three all-plastic cells studied previously[1] (one containing Alusuisse sample F17/220, one Alcoa sample 3809B, and one Eagle-Picher sample J-57-76) and seven steel cells (one containing an Alusuisse sample and six containing Alcoa samples). The Alusuisse sample (F17/220) in the all-plastic cell has the highest-purity and its triple-point temperature was determined to be 29.77398 °C. From the results of the intercomparison of the triple-point temperatures of the samples and of the melting curves, we confirmed our previous assessment of the relative purity of the three samples in the all-plastic cells and determined that the samples placed in the steel cells were already or became contaminated. We believe they became contaminated by contact with the stainless steel thermowells by way of pin-holes in the Teflon coating. Consequently, we conclude that all-plastic cells or cells with non-metallic thermowells should be used in preparing triple-point cells for gallium. All triple-point temperature data obtained with the five SPRTs on the highest-purity sample, the Alusuisse sample (F17/220) in the all-plastic cell, have a standard deviation of ±0.00014 °C (and a standard deviation of the mean of ±0.00002 °C), although the standard deviations of the data for individual SPRTs range from ±0.000037 °C to ±0.000055 °C (and standard deviations of the means of data for individual SPRTs range from only ±0.000008 °C to ±0.000021 °C). We estimate the systematic uncertainty to be ±0.6 m°C. The differences in triple-point temperatures indicated by the different SPRTs are due to calibration errors and/or to different behavior of the platinum sensing elements at points intermediate to the calibration fixed points of the IPTS-68. From these results, we conclude that more fixed points (perhaps Ga, In and Cd triple points) should be defined for the next IPTS for the use of those who desire high precision and accuracy.

The hydrostatic pressure effects were measured and were in agreement with the previously reported[1] pressure dependence of the melting point.

ACKNOWLEDGMENTS

We gratefully acknowledge the help in the preliminary phases of this work of Dr. D. D. Thornton.

REFERENCES

* Certain commercial equipment, instruments, or materials are identified in this paper in order to adequately specify the experimental procedure. Such identification does not imply recommendation or endorsement by the National Bureau of Standards, nor does it imply that the materials or equipment identified are necessarily the best available for the purpose.
1. Mangum, B.W. and Thornton, D.D., Metrologia 15, 201 (1979).
2a. Comptes rendus des séances de la Treizième Conférence Générale des Poids et Mesures, Annexe 2, p. A1 (1967-1968).
2b. Comité Consultatif de Thermométrie, 8e Session, 1967, Annexe 18.
2c. Metrologia 5, 35 (1969).
2d. Comptes rendus des séances de la Quinzième Conférence Générale des Poids et Mesures, Resolution 7, p. 105, Annexe 2, p. A1 (1975).
2e. Metrologia 12, 7 (1976).
3. Lecoq de Boisbaudran, P.E.: Comp. Rend. 81, 493 (1875).
4. Lecoq de Boisbaudran, P.E.: Comp. Rend. 82, 1036 (1876); Comp. Rend. 83, 611 (1876).
5. Richards, T.W., Boyer, S.: J. Am. Chem. Soc. 43, 274 (1921).
6. Craig, W.M., Drake, G.W.: J. Am. Chem. Soc. 56, 584 (1934).
7. Roeser, Wm. F., Hoffman, J.I.: J. Res. NBS 13, 673 (1934).
8. Hoffman, J.I.: J. Res. NBS 13, 665 (1934).
9. Jach, J., Sebba, F.: Trans. Faraday Soc. 50, 226 (1954).
10. Boedtker, O.A., La Force, R.C., Kendall, W.B., Ravitz, S.F.: Trans. Faraday Soc. 61, 665 (1965).
11. Pennington, P.R., Ravitz, S.F., Abbaschian, G.J.: Acta Met. 18, 943 (1970).
12. Pashaev, B.P.: Fiz. Tverd Tela 3, 416 (1961).
13. Wenzl, H., Mair, G.: Z. Phys. B21, 95 (1975).
14. Abbaschian, G.J., Ravitz, S.F.: J. Cryst. Growth 28, 16 (1975).
15. Thornton, D.D., Mangum, B.W.: Procès Verbaux du CCT, Document No. CCT/76-13, June, 1976.
16. Lau, S., Schlott, P.: Procès Verbaux du CCT, Document No. CCT/76-23 June, 1976.
17. Sostman, H.E.: Rev. Sci. Instrum. 48, 127 (1977).
18. Thornton, D.D.: Clin, Chem. 23, 719 (1977).
19. Riddle, J.L., Furukawa, G.T., Plumb, H.H.: Platinum resistance thermometry, NBS Monograph 126, 1973, Supt. of Documents, U.S. Government Printing Office, Washington, D.C. 20402.
20. Cutkosky, R.D.: J. Res. NBS 74C, 15 (1970).
21. Berry, R.J.: Metrologia 2, 80 (1966).
22. Gramlich, J.W., Machlan, L.A., Private communication.

An intercomparison of gallium fixed point cells*

M. V. Chattle and R. L. Rusby

Division of Quantum Metrology, National Physical Laboratory, Teddington, Middlesex, TW11 0LW, United Kingdom

G. Bonnier, A. Moser, and E. Renaot

Institut National de Metrologie, Conservatoire National des Arts et Metiers, 292 Rue Saint-Martin, 75141 Paris, France

P. Marcarino, G. Bongiovanni, and G. Frassineti

Istituto di Metrologia G. Colonnetti, Strada delle Cacce 73, 10135 Torino, Italy

Cells for the realization of the melting and triple points of gallium (near 29.77 °C) have been prepared at the NPL, INM, and IMGC. Three cells, one of each manufacture, have been measured at each of the three laboratories using both a locally-calibrated standard platinum resistance thermometer and one circulated with the cells after calibration at NPL. The results of the intercomparison are presented.

INTRODUCTION

It has been shown [1] that a cell for realising either the triple or melting point of gallium is a very reproducible fixed point device near 29.77 °C. Potentially, it has several important applications as a secondary reference point in, for example, the calibration of standard platinum resistance thermometers, and in the checking of medical thermometers. However, to establish its validity as a fixed point, its reproducibility when manufactured to different designs and with gallium from different sources, needs to be determined.

The work described in this paper concerns the manufacture and characterisation of gallium cells by three national standards laboratories, viz, National Physical Laboratory (NPL-United Kingdom), Institut National de Metrologie (INM-France) and Istituto di Metrologia G Colonnetti (IMGC-Italy). Each laboratory manufactured its own cell using a slightly different design and a different source or purity of gallium.

A cell from each laboratory, together with a long-stem standard platinum resistance thermometer supplied and calibrated by NPL, was circulated to all three laboratories. In each laboratory the cells were measured at the melting and triple points of gallium, (ie containing argon at 1 atmosphere, and under vacuum, respectively), using both the NPL thermometer and a locally-calibrated thermometer. In this way the reproducibilities of the gallium cells were measured and compared, as also were the NPL, INM and IMGC interpolations of IPTS-68 at around 29.77 °C.

DESCRIPTION OF CELLS

In all three cells the gallium was contained in an inner vessel of PTFE, which was accommodated in an outer case of nylon or glass suitable for suspension in a liquid bath. The re-entrant well into which the thermometer fitted (with an inner metal lining sleeve in the NPL cell and a metal tube in the IMGC cell) was also made of nylon or PTFE. In the NPL and INM cells the outer nylon cases were sealed to the re-entrant wells and a valve was fitted at the top of each cell so that the atmosphere above the gallium could be maintained as desired. The IMGC cell was constructed differently, the plastic container of gallium being fitted into a glass outer case, to the top of which was connected, through an O-ring seal, a metal cap. A metal lining tube for the thermometer was sealed through the cap, which was also provided with a pumping connection.

The INM cell was supplied suspended near the closed lower end of a long brass cylinder which contained sufficient oil to bring the level to a centimetre or so above the top of the cell; the cylinder was intended to be suspended in a stirred liquid bath.

The essential details of the cells are summarised in Table I; their constructions are indicated in figures 1(a), 1(b) and 1(c).

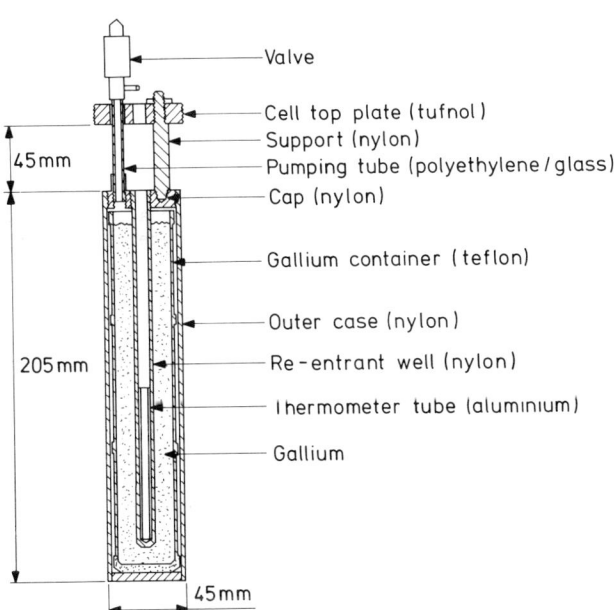

FIG 1(a) NPL GALLIUM POINT CELL

---

* This work was organised and partially funded by the Community Bureau of Reference, Commission of the European Communities, Brussels, Belgium.

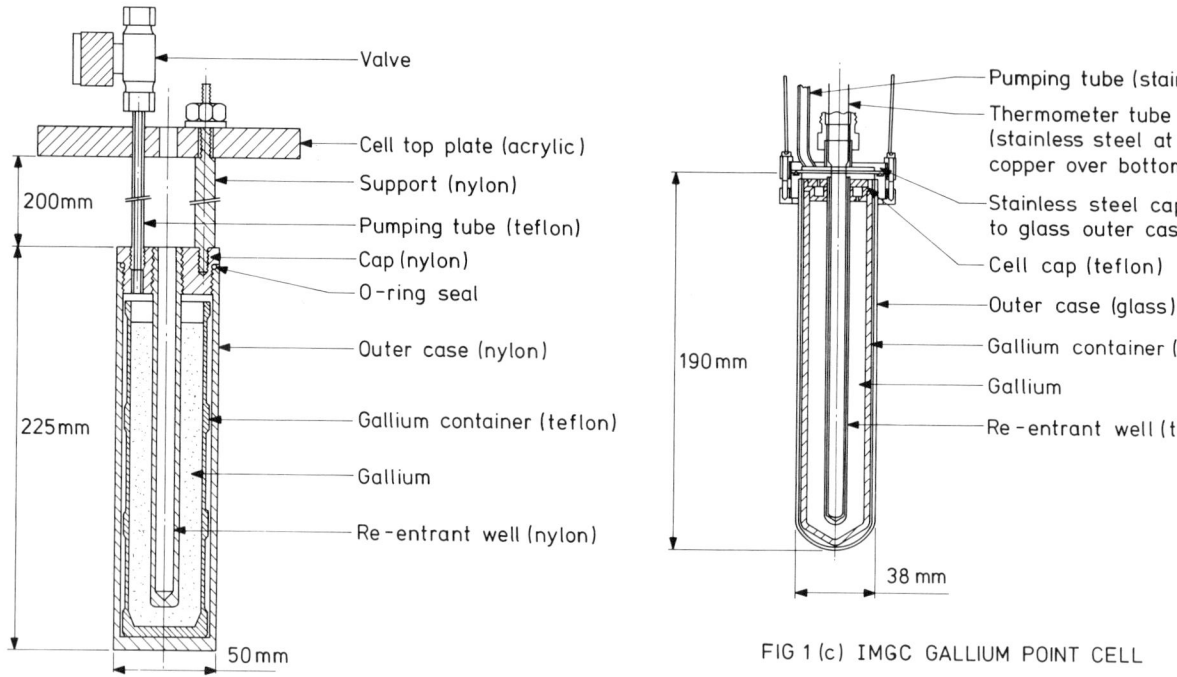

FIG 1 (b) INM GALLIUM POINT CELL

FIG 1 (c) IMGC GALLIUM POINT CELL

Table I. Details of cells

	NPL cell	INM cell	IMGC cell
Source of gallium	Cominco	Alusuisse	Alusuisse
Purity	99.9999%	99.9999%	99.99999%
Approximate quantity	780 g	660 g	500 g
Inner gallium container:			
material	PTFE	PTFE	PTFE
internal diameter	34 mm	31 mm	28 mm
length	180 mm	174 mm	170 mm
Re-entrant well:			
material	nylon	nylon	PTFE
internal diameter	12 mm	9 mm	9 mm
Thermometer sleeve:			
material	aluminium	none fitted	copper
internal diameter	7.5 mm	-	8 mm
Distance from gallium surface to bottom of re-entrant well	165 mm	170 mm	143 mm

REALIZATION OF THE GALLIUM POINTS

All three cells had been extensively studied in their manufacturing laboratories before being supplied for the comparison. Each laboratory, therefore, provided information upon the depth to which the cell should be immersed in the bath, the temperature of the bath and any special procedures recommended for use with its cell. In the case of the NPL and IMGC cells, small cylindrical electric heaters were supplied to fit into the thermometer wells, and it was recommended that as soon as a thermometer in the well indicated that the cell temperature was close to the melting point, the thermometer should be removed and replaced, for the specified period, by the heater. The details are shown in Table II.

Two standard thermometers were used for each set of measurements, one being that which was circulated with the cells (Tinsley type 5187 SA, serial number 238681, referred to as 'BCR thermometer'), the other being a 'local' thermometer which had been calibrated in the laboratory where the measurements were being made. Each cell in turn was set up in a temperature-controlled oil bath, in accordance with the procedure recommended. Three realizations of the gallium point were then made with each cell, each realization following preparation of the cell from a fully frozen ingot condition. The $R(0^{o}C)$ values of the thermometers were determined from triple point of water measurements made before, and after, each realization of the gallium point, and for the purpose of calculating the gallium temperature, the mean of the two values of $R(0°C)$ for each thermometer was used.

The measurements in each realization were made in the following sequence:

    melting point - BCR thermometer
    triple point   - BCR thermometer
    triple point   - 'local' thermometer
    melting point - 'local' thermometer
    melting point - BCR thermometer

The first measurement in the set was made when a chart recorder monitoring the resistance of the BCR thermometer in the cell indicated that the plateau temperature had been fully reached; this generally occurred within 1.5 to 4 hours after the completion of the preparation of the cell. The second melting point measurement with the BCR thermometer was made several hours after the first; agreement between the two was always within 0.1 mK. All measurements (including those at the triple point of water) were made using two different measuring currents (1 mA and 2 mA), and were then extrapolated to zero current to minimize the effects of thermometer self-heating. In all three laboratories, direct current resistance bridges were used for the measurements.

Except when triple points were being measured the cells were kept under high purity argon at atmospheric pressure, the latter being measured with a suitable gauge connected to the system. For triple point measurements the cells were evacuated, the system being pumped continuously while the triple point measurements were being made.

Corrections were applied to both melting and triple point temperatures to allow for the effect of the hydrostatic pressure of the liquid gallium at the level of the thermometer sensors (0.012 mK per cm of gallium). In the case of the melting points, additional corrections (1.98 mK per atmosphere) were applied to allow for departures of atmospheric pressure from one standard atmosphere.

RESULTS

The detailed results have been listed in Tables III, IV and V, while Table VI summarises the mean values. Tables VI and VII contain further summaries.

Table II. Recommendations for use

	NPL cell	INM cell	IMGC cell
Bath temperature	30 - 31 °C	31 - 32 °C	30 - 31 °C
Depth of bottom of thermometer well below bath liquid surface:	220 mm	375 mm	250 mm
Contact liquid in thermometer well	oil	oil	water
Special procedure to melt gallium around the re-entrant well:	electric heater: 10 W for 30 minutes	none	electric heater: 10 W for 20 minutes

Table III. NPL Measurements

		Melting Point/°C		Triple Point/°C	
		BCR Thermometer	NPL* 'Local' Thermometer	BCR Thermometer	NPL* 'Local' Thermometer
NPL Cell		29.77168	29.77153	29.77367	29.77367
		29.77160	29.77171	29.77362	29.77374
		29.77157	29.77162	29.77359	29.77365
	MEAN	29.77162	29.77162	29.77363	29.77369
INM Cell		29.77166	29.77187	29.77370	29.77366
		29.77158	29.77172	29.77368	29.77361
		29.77161	29.77170	29.77360	29.77367
	MEAN	29.77161	29.77176	29.77366	29.77364
IMGC Cell		29.77169	29.77180	29.77365	29.77371
		29.77174	29.77185	29.77381	29.77382
		29.77174	29.77191	29.77369	29.77382
	MEAN	29.77172	29.77185	29.77372	29.77379

* Tinsley type 5187 SA - serial number 227028; calibrated at NPL.

Table IV. INM Measurements

		Melting Point/°C		Triple Point/°C	
		BCR Thermometer	INM* 'Local' Thermometer	BCR Thermometer	INM* 'Local' Thermometer
NPL Cell		29.77173	29.77190	29.77365	29.77392
		29.77177	29.77195	29.77362	29.77398
		29.77167	29.77185	29.77367	29.77385
	MEAN	29.77172	29.77190	29.77365	29.77392
INM Cell		29.77147	29.77175	29.77359	29.77360
		29.77139	29.77173	29.77370	29.77362
		29.77151	29.77170	29.77352	29.77376
	MEAN	29.77146	29.77173	29.77360	29.77366
IMGC Cell		29.77175	29.77194	29.77379	29.77398
		29.77190	29.77192	29.77396	29.77398
		29.77187	29.77203	29.77414	29.77404
	MEAN	29.77184	29.77196	29.77396	29.77400

* Leeds & Northrup type 8167-25 - serial number 1792417; calibrated at INM

Table V. IMGC Measurements

		Melting Point/°C		Triple Point/°C	
		BCR Thermometer	IMGC* 'Local' Thermometer	BCR Thermometer	IMGC* 'Local' Thermometer
NPL Cell		29.77151	29.77186	29.77354	29.77379
		29.77153	29.77185	29.77356	29.77379
		29.77156	29.77190	29.77355	29.77380
	MEAN	29.77153	29.77187	29.77355	29.77379
INM Cell		29.77173	29.77203	29.77357	29.77386
		29.77167	29.77200	29.77349	29.77384
		29.77161	29.77201	29.77348	29.77388
	MEAN	29.77267	29.77201	29.77351	29.77386
IMGC Cell		29.77168	29.77200	29.77364	29.77393
		29.77170	29.77197	29.77366	29.77394
		29.77171	29.77196	29.77367	29.77394
	MEAN	29.77170	29.77198	29.77366	29.77394

* Leeds & Northrup type 8167-25 - serial number 1773054; calibrated at IMGC

UNCERTAINTY OF RESULTS

Following the method published [2] in the BIPM Draft Recommendation on the Statement of Uncertainties, the estimates in Table IX were made in the three laboratories. Category B uncertainties arose, principally, from uncertainties in the realization of the defining fixed points of IPTS-68, from differences in interpolation between standard platinum resistance thermometers and from systematic uncertainties in the measurement of resistance ratios.

ANALYSIS OF RESULTS

a) Considering the overall average temperature of each cell, as determined by the 3 laboratories, involving measurements made with 4 different thermometers:

	Melting Point	Triple Point
NPL Cell	29.77171 °C	29.77370 °C
INM Cell	29.77170 °C	29.77365 °C
IMGC Cell	29.77184 °C	29.77384 °C

This indicates agreement between the temperatures of the cells to within 0.19 mK, the triple points averaging 1.98 mK higher than the melting points.

b) Table VII shows that the differences between the measurements made in the 3 laboratories using the BCR thermometer (with its NPL-calibration) are all within 0.30 mK, the majority being well within 0.2 mK.

Table VI. Summary of all measurements

	Measuring Laboratory	Mean Melting Point/°C		Mean Triple Point/°C	
		BCR Thermometer	'Local' Thermometer	BCR Thermometer	'Local' Thermometer
NPL Cell	NPL	29.77162	29.77162	29.77363	29.77369
	INM	29.77172	29.77190	29.77365	29.77392
	IMGC	29.77153	29.77187	29.77355	29.77379
	Overall Mean	29.77162	29.77180	29.77361	29.77380
INM Cell	NPL	29.77161	29.77176	29.77366	29.77364
	INM	29.77146	29.77173	29.77360	29.77366
	IMGC	29.77167	29.77201	29.77351	29.77386
	Overall Mean	29.77158	29.77183	29.77359	29.77372
IMGC Cell	NPL	29.77172	29.77185	29.77372	29.77379
	INM	29.77184	29.77196	29.77396	29.77400
	IMGC	29.77170	29.77198	29.77366	29.77394
	Overall Mean	29.77175	29.77193	29.77378	29.77391

Table VII. Comparison between measurements in different laboratories using the BCR thermometer

Differences between temperatures measured with the BCR thermometer in different laboratories

		Melting Point/°C	Triple Point/°C
NPL cell	NPL - INM	- 0.00010	- 0.00002
	INM - IMGC	+ 0.00019	+ 0.00010
	IMGC - NPL	- 0.00009	- 0.00008
INM cell	NPL - INM	+ 0.00015	+ 0.00006
	INM - IMGC	- 0.00021	+ 0.00009
	IMGC - NPL	+ 0.00006	- 0.00015
IMGC cell	NPL - INM	- 0.00012	- 0.00024
	INM - IMGC	+ 0.00014	+ 0.00030
	IMGC - NPL	- 0.00002	- 0.00006

Table VIII. Comparison between measurements in different laboratories using 'locally-calibrated' thermometers

Differences between temperatures measured with 'locally-calibrated' standard thermometers

		Melting Point/°C	Triple Point/°C
NPL cell	NPL* - INM	- 0.00028	- 0.00026
	INM - IMGC	+ 0.00003	+ 0.00013
	IMGC - NPL*	+ 0.00025	+ 0.00013
INM cell	NPL* - INM	- 0.00005	- 0.00001
	INM - IMGC	- 0.00028	- 0.00020
	IMGC - NPL*	+ 0.00033	+ 0.00021
IMGC cell	NPL* - INM	- 0.00017	- 0.00024
	INM - IMGC	- 0.00002	- 0.00006
	IMGC - NPL*	+ 0.00019	+ 0.00018

* Average of BCR and NPL 'local' thermometers.

Table IX. Estimated uncertainties

	Category A (random) (3 standard deviations)	Category B (systematic) (99% confidence)	Overall Uncertainty (99% confidence)
NPL	+ 0.3 mK	+ 0.4$_5$ mK	+ 0.5$_5$ mK
INM	+ 0.4 mK	+ 0.4$_5$ mK	+ 0.6 mK
IMGC	+ 0.3 mK	+ 0.4$_5$ mK	+ 0.5$_5$ mK

Taking average differences for all 3 cells (using both melting, and triple, points):

	difference	standard deviation
NPL - INM	- 0.04 mK	± 0.14 mK
INM - IMGC	+ 0.10 mK	± 0.17 mK
IMGC - NPL	- 0.06 mK	± 0.07 mK

c) Table VIII compares 'locally-calibrated' thermometers and is, effectively, a comparison of national IPTS-68 temperature scales at 29.77 °C. The maximum difference is 0.33 mK, and the averages are:

	difference	standard deviation
NPL - INM	- 0.17 mK	± 0.11 mK
INM - IMGC	- 0.05 mK	± 0.15 mK
IMGC - NPL	+ 0.21 mK	± 0.07 mK

The results set out in Tables IV and V confirm that the BCR thermometer (with its NPL calibration) gives results which are systematically about 0.3 mK below those obtained using INM and IMGC 'local' thermometers.

These differences are probably due, largely, to differences in calibration at the 3 laboratories (i.e. due to variations in the temperatures realised by the water triple point cells and the zinc and tin freezing point cells used in those calibrations). However, they may also be due, in part, to differences between the interpolation of IPTS-68 by the BCR and NPL thermometers (both are of Type 5187SA, manufactured by H Tinsley & Co. Ltd.), and those used at INM and IMGC, (Leeds & Northrup type 8167-25). The design and construction of these thermometers is significantly different, as is indicated by their self-heating effects (Tinsley thermometers 2.8 mK for 1 mA at the triple point of water - Leeds & Northrup thermometers 0.5 mK for 1 mA).

Nevertheless, the differences between the results obtained using the 'locally-calibrated' thermometers are not inconsistent with the uncertainty estimates given by the calibrating laboratories and listed above.

CONCLUSIONS

It has been demonstrated that the three gallium cells, made to slightly different designs and containing gallium from two different sources, gave the same fixed point temperatures within 0.2 mK, the triple points being, on average, 1.98 mK higher than the melting points. These results confirm the suitability of the gallium fixed point cell as a highly reproducible fixed point device.

The results of measurements made in the 3 laboratories using the same calibrated standard thermometer agreed, on average, to within 0.1 mK, the largest difference being 0.2 mK.

Similar measurements made with 'locally-calibrated' thermometers agreed, on average, to within 0.2 mK, the largest difference being 0.3 mK.

REFERENCES

1. Mangum, B.W., Thornton, D.D.: Metrologia 15, 201 (1979).
2. Giacomo, P.: Metrologia 17, 73 (1981).

Realization of the melting point of gallium

B. N. Oleinik, A. G. Ivanova, V. A. Zamkovets, and N. N. Ergardt

D. I. Mendeleyev Research Institute of Metrology, Leningrad, Union of Soviet Socialist Republics

This report presents the measurement results of the gallium melting point and its reproducibility for a one and a half year period. It has been found that the melting point of 99.997% pure gallium is (29.7704 ± 0.0004) °C.

INTRODUCTION

The last five years are marked by a heightened interest for realization of the melting (freezing) point and triple point of gallium. This interest is due to the possibility of achieving an adequately reproducible fixed point within the range of room temperatures for improving the calibration accuracy of thermometers used in the interval from 0 to 150 °C, particularly thermistors and quartz thermometers.

The first results on realization of the gallium freezing point obtained at VNIIM were published in 1975. The study of the gallium fixed point was continued with the purpose of producing a hermetic cell for gallium, an automated thermostatic device and determining the optimal melting and solidification conditions to improve the reproducibility of the phase transition temperature.

This report presents the results of determination of the gallium melting temperature value in two hermetic cells and its reproducibility over one and a half years. The data of preliminary studies of the melting conditions and the temperature field of the cell are shown.

EXPERIMENTAL DETAILS AND RESULTS

Gallium of home production was used, its purity being 99.9997% with the following impurity content according to manufacturer's data: Ni - 2×10^{-6}%, Zn - 2×10^{-6}%, Cu - 8×10^{-7}%, Al - 1×10^{-5}%, Pb - 2×10^{-6}%, Mg - 1.2×10^{-5}%, Fe - 5×10^{-5}%, Si - 6×10^{-5}%. Since gallium expands considerably on freezing, fluoroplastic - 4 was chosen for the cell fabrication. This material is highly plastic, it does not interact with the metal and it ensures the cell tightness. Figure 1 represents two design versions of the fabricated hermetic cells.

The smaller cell has an internal diameter of 34 mm and a depth of the thermometer well immersion into the metal of 130 mm; the diameter of the larger cell is 59 mm and the depth of the well immersion is 180 mm.

The thermometer well with an internal diameter of 9.5 mm and an external diameter of 14 mm is made together with the cap as a single piece. The smaller cell is inserted in a metal case with a clearance along the lateral surface of about 1 mm for equalizing temperature along its height. The cell cap is fixed to the case so as to provide a hermetic seal. In the larger cell, a metal bushing was applied which was attached to the cap by screws to secure the cap and the cell. In the cap assembly, provision was made for an outlet to evacuate and fill the cell with gas at a preset pressure.

The cells were filled with gallium in a glove box under a dry argon atmosphere.

For gallium melting, a special automated thermostatic device was designed. The internal thin-wall metal shell of this appliance has a distributed wire

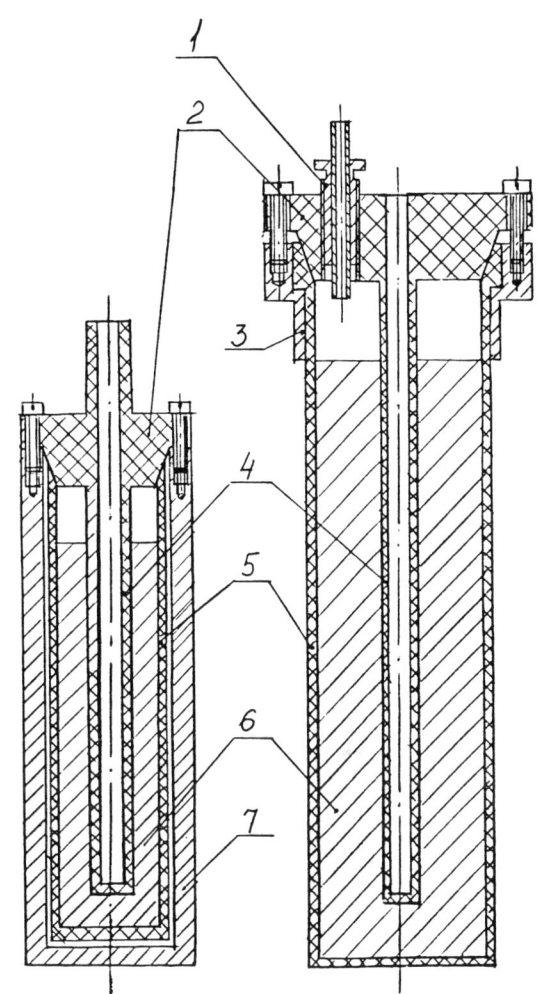

FIG. 1. 1 - outlet for gas evacuation and pumping; 2 - cap assembly with thermometer well; 3 - metal bushing; 4 - thermometer well; 5 - fluoroplastic cell; 6 - gallium; 7 - metal case.

heater. The cell is placed inside the shell. The air in the clearance between the cell and the shell is stirred by a mixer mounted under the bottom of the cell. Preliminary heating is conducted at a prescribed rate until the melting temperature is attained.

During the melt plateau, the desired temperature of the heater is held constant within ± 0.02 K. The thermostatic device together with the control unit is 170 x 200 x 470 mm in dimensions.

The effect of thermal conduction losses from the thermometer on its readings and the optimal depth of immersion were determined in the smaller cell by measuring the gallium "melting" temperature during the same plateau at different depths of the thermometer into the well. Table I shows the results of the measurements made with the use of a standard platinum thermometer with a 10 Ω resistance at a current of 2 mA.

Table I.

Distance from the bottom to the middle of the sensing element	Resistance Readings
mm	Ω
25	11.018405
30	11.018394
35	11.018386
40	11.018402
45	11.018410
50	11.018423
55	11.018448
60	11.018492
65	11.018492
25	11.018386

The measurements were performed by the compensation method using a potentiometer P - 348. The standard deviation of the resistance measurement was estimated from the measurements of the reference coil resistance and was found to be $\pm\ 0.4 \times 10^{-5}$ Ω which is equivalent to $\pm\ 1 \times 10^{-4}$ K for a platinum thermometer with a 10 Ω resistance. It may be believed that at distances of the middle of the sensing element from the bottom up to 40 mm there is no heat loss effect, thus the depth of the

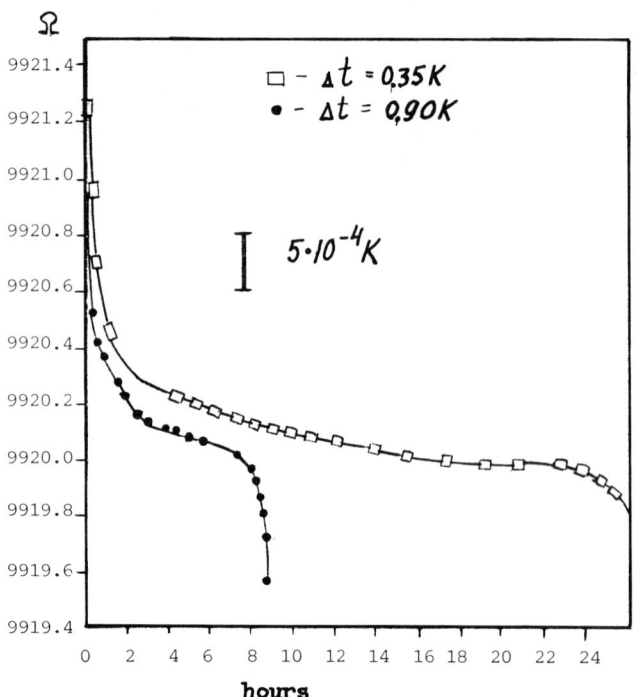

FIG. 2. Melting curve at Δt = 0.35 K and 0.90 K. The ordinate axis is the thermistor resistance.

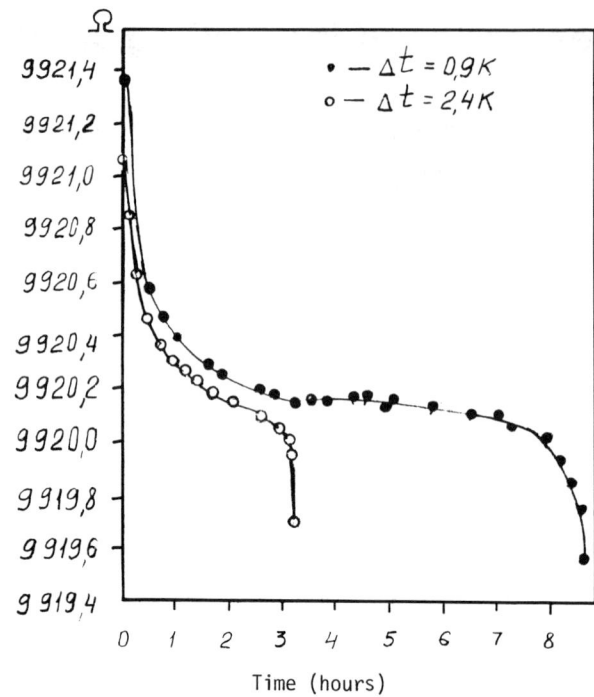

FIG. 3. Melting curve at Δt = 0.90 K and 2.40 K. The ordinate axis is the thermistor resistance.

thermometer immersion is sufficient. It seemed unnecessary to conduct such investigations in the larger cell.

Thereupon, experiments were conducted to determine the effect of filling the thermometer well with liquids on the thermometer readings and on the temperature gradient of the well at a plateau. Measurements were performed with air, water, or oil filling the well. It was established that the differences in results obtained by filling the well with water or oil was within the accuracy of measurements. Measurements in the air-filled thermometer well showed variations of the resistance thermometer readings equivalent to (1 to 1.5) $\times 10^{-3}$ K. This result may be accounted for by the increase in thermal resistance of the clearance between the metal and thermometer when filled with air and by heat from the upper part of the cell. To evaluate the results and to choose the best conditions for measurement of the gallium melting point, a study was carried out on the temperature distribution along the well during the plateau period when the well was filled with air and with water. The measurements of the well temperature distribution were made in the larger cell with a 46 Ω platinum resistance thermometer, 28 mm in length and 3 mm in diameter, with a spacing of 10 mm to a height of 60 mm from the bottom (the zone of the sensitive element of standard platinum resistance thermometers) and with a spacing of 30 mm to a height of 150 mm from the bottom (the water level). The measurements were conducted with a temperature of 32.5 °C in the thermostated device. It has been found that the temperature increases along the well height. In the well filled with water, the temperature was constant within the accuracy of measurements at heights up to 60 mm from the bottom; at a height of 120 to 150 mm from the well bottom, the temperature increased by (3 to 9) $\times 10^{-3}$ K. With air filling the well, the temperature increased by 1 $\times 10^{-3}$ K at a height of 60 mm; at a height of 150 mm the increase was 2 $\times 10^{-2}$ K.

Similar measurements of the temperature gradient in the smaller cell were performed with the aid of a stable thermistor, type CT 4 - 16, having a resistance of about 10 Ω; it had a diameter of 2.5 mm and a length of 13 mm. The thermistor resistance measurements were conducted by the potentiometer method using a potentiometer P - 345. The standard deviation of the resistance measurements was $\pm 2 \times 10^{-2}$ Ω, which is equivalent to $\pm 5 \times 10^{-5}$ K at the gallium melting point. As a result, it was found that in the thermometer well filled with air there is a temperature increase relative to the well bottom, but the temperature variation along the height up to 80 mm from the bottom does not exceed 3×10^{-4} K.

It was finally concluded that in measurement of the gallium melting point, the thermometer well should be filled with water or oil.

The next stage was devoted to investigation of the effect of the gallium melting rate upon the melting temperature value. In all the experiments a single procedure of rapid gallium solidification was applied. It was done by placing the cell with melted gallium in a vessel with ice and water mixture and introducing this mixture into the thermometer well. As shown in McLaren's study[1], in rapid solidification there occurs a uniform distribution of impurities which must have a favorable effect on the next melting plateau.

The gallium melting rate is determined by the intensity of heat supply to the cell, that is by the difference, Δt, between the thermostatic device and gallium melting temperatures. The experiments were performed at three melting rates corresponding to the temperature differences Δt = 0.35 K; 0.90 K; 2.4 K. The investigations of the smaller cell were performed in a controllable oil bath with the use of a thermistor CT 4 - 16. The thermistor resistance stability during the measurements was defined by measuring its stability at the triple point of water. The averaged measurement results for three gallium melting rates are shown in Table II and are represented graphically in Figs 2 and 3.

Table II.

Temperature difference $\Delta t(K)$	Thermistor resistance during melting plateau $R(\Omega)$	Time of measurement Time (hours)
0.35	9920.120	16 to 18
0.90	9920.126	5.5 to 5.8
2.40	9920.203	1.4 to 1.8

While the temperature coefficient of the thermistor resistance at 30 °C is -4%/K, it is at 0 °C -5%/K.

The results of several measurements at the lowest melting rates are given in Table III.

As seen from the melting curves in Figs. 2 and 3, the variation of the gallium melting rate is accompanied by changes not only in the duration of the melting plateau, but also in its nature, namely in the slope and scatter of readings. This fact complicates the comparison of the results. Nevertheless, the data obtained attest to the influence of the thermostat temperature during the melting plateau period on the plateau level, though the divergence of the mean values for each of the melting rates is only 2 - 3 times as much as the reproducibility of the plateau temperature values for one melting rate. The explanation of the observed variations by an additional heat supply from the thermostat is not confirmed by the experiments.

In [2] it is noted that for cells several times smaller than those here described the effect of the thermostat temperature on the melting plateau level is within the accuracy of measurements (being $\pm$ 0.0002 K). In [3] it is pointed out that for cells commensurable with ours the variation of the temperature difference in the thermostat from Δt = 0.1 K to Δt = 0.3 K pro-

Table III.

Date	Temperature difference $\Delta t(K)$	Thermistor resistance during $R(\Omega)$	Time of measurement Time(min)	Scatter during melting plateau $\times 10^{-4}$ (K)	Thermistor resistance at triple point of water $R_{tpw}(\Omega)$
03.08.80	0.37	9920.097	60	0.4	37965.667
06.08.80	0.31	9920.105	50	0.2	37965.495
07.08.80	0.30	9920.151	45	0.5	37965.721
10.08.80	0.40	9920.137	50	0.3	37965.681
24.08.80	0.34	9920.150	65	0.3	-
25.08.80	0.35	9920.145	60	1.3	-
15.09.80	0.35	9920.104	30	0.02	37965.745
16.09.80	0.37	9920.137	50	0.9	-
21.09.80	0.37	9920.122	50	1.2	-
24.09.80	0.37	9920.131	30	1.1	37965.443
25.09.80	0.39	9920.114	25	0.5	37965.443

Table IV.

Thermometer N°	Number of measurements	Standard deviation
1731	68	1.1×10^{-4} K
1748	52	1.4×10^{-4} K
1792	51	1.5×10^{-4} K

Table V.

Thermometer N°	Date	Gallium melting temperature value °C
9	6.12.79	29.7703
	7.12.79	29.7704
	20.11.80	29.7702
	21.11.80	29.7702
	18.05.81	29.7704
	average	29.7703
8	26.11.80	29.7702
13	26.11.80	29.7706
average		29.7704 ± 0.0004

duces a change in the plateau level of no more than 1×10^{-3} K. Our results are not at variance with these data.

SUMMARY

Our experimental studies in the most part were aimed at the determination of reproducibility of the gallium melting point temperatures in two cells. The measurements were performed with the use of three reference platinum resistance thermometers. In the experiments, the thermometer well was filled with water and the temperature in the thermostatic device was 30.12 °C, which corresponds to $\Delta t = 0.35$ K. The experiment was conducted over one and a half years. The results are given in Table IV.

The melting point of 99.9997% pure gallium was defined from measurements obtained by the use of three standard platinum resistance thermometers: Nos. 8, 9, and 13. After extrapolation of the data to zero current, the gallium melting point was found to be (29.7704 ± 0.0004) °C. The data are presented in Table V.

The value obtained for the melting point of 99.9997% pure gallium and its reproducibility are in agreement with the results reported in [3] and [4].

In conclusion it may be stated that in the range of room temperatures the reference melting point of pure gallium makes it possible to improve the calibration accuracy of precision thermometers: thermistors, quartz thermometers, and others used in this temperature range.

REFERENCES

[1] F. Weinberg and E.H. McLaren, The solidification of dilute binary alloys, 1962, Ottawa, Canada.
[2] D.D. Thornton, Gallium melting point standard: A determination of the liquid-solid equilibrium temperature of pure gallium on the International Practical Temperature Scale of 1968, Clin. Chem. 23, p. 719 (1977).
[3] H.E. Sostman, Melting point of gallium as a temperature calibration standard, Rev. Sci. Instrum. 48, 127 (1977).
[4] D.D. Thornton and B.W. Mangum, Comité Consultatif de Thermometrie, 11e session, p. 76 (1976).

The triple-point equilibria of succinonitrile: Its assessment as a temperature standard

M. E. Glicksman and P. W. Voorhees

Materials Engineering Department, Rensselaer Polytechnic Institute, Troy, New York 12181

R. Setzko

Optical Information Systems, Exxon Enterprises Incorporated, Elmsford, New York 10523

The organic material succinonitrile [$NC(CH_2)_2CN$] was tested for its suitability as a temperature standard. There is a need for a number of evenly spaced temperature references in the 0 to 100° range, and the development of the succinonitrile standard is an attempt to meet part of this need. Results show that succinonitrile can be purified to a high degree, and that the triple-point temperature of this pure material is 58.0805 ± 0.004 °C, which is reproducible to ± 0.0001 K. The material is inexpensive, relatively non-toxic, and displays good chemical stability when stored hermetically within a borosilicate glass apparatus over periods greater than two years. The material processing sequences, physical characteristics and response under various thermal environments of a triple-point cell containing 200 g succinonitrile are described in detail.

INTRODUCTION

Most temperature measurements are referenced to the International Practical Temperature scale (IPTS).[1] This scale was created to allow for a more precise and convenient means for measuring temperature than would be possible using the thermodynamic temperature scale. In addition, it was designed to correspond closely with the thermodynamic temperature scale. An agreed upon sequence of so-called temperature "fixed points", which provides for the correspondence between the IPTS and the thermodynamic temperature scale, forms the operational basis of practical thermometry. These fixed points range at present from the triple point equilibrium of hydrogen through the melting point of gold.[1]

The fixed points of the IPTS are established under specialized thermodynamic conditions at which two or more phases of a pure material coexist in equilibrium. It is essential that the proper material be chosen for use in a fixed-point apparatus so that a suitably stable temperature can be produced. Basically, the substance must possess material properties which allow a close approximation to be achieved in practice to a reversible equilibrium temperature, even under the inherently nonequilibrium conditions with which a temperature reference cell would be used. It has been found, for example, in numerous metals which crystallize in simple centrosymmetric crystal structures that these nonequilibrium effects can be reduced to inconsequential levels. This accounts for the widespread use of pure metals for temperature reference cells. Organic materials, in contrast, have generally proven to be less satisfactory as temperature standards due mainly to both the difficulty in obtaining and maintaining sufficiently pure material and the complex crystallization kinetics associated with most organic compounds. We, however, have succeeded in constructing a temperature reference cell using an organic material which provides better reproducibility than most of the organic materials in use as secondary reference standards of the IPTS.[2]

MATERIAL PROPERTIES

The material chosen for the investigation was succinonitrile [$NC(CH_2)_2CN$]. The physical properties of succinonitrile are well suited for use in a triple-point cell. A summary of the properties is shown in Table I. The manner in which these properties allow for a reproducible and stable triple point is discussed below.

A true equilibrium triple-point condition exists when each of the three conjugate phases is in equilibrium with one another. In practice, it is often the difficulty in achieving the solid-liquid phase equilibrium which prevents the establishment of a true

Table I. Properties of Succinonitrile [$NC(CH_2)_2CN$]

Property	Value	Ref.*
Molecular Weight	80.092	6
Density of Solid	1016 kg/m^3**	6
Density of Liquid	988 kg/m^3**	8
Thermal Expansivity of Solid	-5.6×10^{-4}/K	9
Thermal Expansivity of Liquid	-8.1×10^{-4}/K	8
Shear Viscosity	2.6×10^{-3} pascal-sec.	8
Kinematic Viscosity	2.6×10^{-6} m^2/sec.	***
Surface Tension (Liquid-Vapor)	46.78 mJ/m^2	8
Surface Tension (Solid-Liquid)	8.9 mJ/m^2	10
Refractive Index (Solid)	1.4340**	11
Refractive Index (Liquid)	1.4150**	11
Equilibrium Temperature (Triple Point)	331.24 K****	3
Latent Heat of Fusion	4.78×10^7 J/m^3	6
Entropy of Fusion	1.45×10^5 J/m^{3}K	6
Heat Capacity of Solid	1913 J/kgK	6
Heat Capacity of Liquid	2000 J/kgK	6
Thermal Conductivity of Solid	0.225 J/mKs	10
Thermal Conductivity of Liquid	0.223 J/mKs	10
Thermal Diffusivity of Solid	1.16×10^{-7} m^2/sec	***
Thermal Diffusivity of Liquid	1.12×10^{-7} m^2/sec	***

* indicates value derived from properties cited in reference.
** indicates value of property at melting point, 331.23 K.
*** indicates property derived from other properties in the table.
****approximate value published prior to this work.

triple-point temperature. Succinonitrile has certain characteristics which allow solid-liquid equilibrium to be achieved quite easily. For example, the solid-liquid interface never forms crystallographic facets, ensuring uniform interfacial temperatures at all points on the cell's mantle. Succinonitrile also has rapid interfacial molecular attachment kinetics ($\mu \doteq 20$ cm/sec °C)[3], which limits the undercooling accompanying interface motion normally experienced during triple-point cell operation to small levels ($\doteq 10^{-5}$°C). The capillarity (Gibbs-Thomson) coefficient is sufficiently small ($\Gamma = 1.31 \times 10^{-6}$ cm °C) that any change in the radii of curvature of the grains comprising the mantle produces a small temperature shift from the bulk equilibrium triple-point temperature. Finally, it has been demonstrated[4] that it is feasible to zone refine succinonitrile to purities greater than 99.9999%. This level of purity helps to limit the variation of interfacial temperature to minor departures from the true triple-point temperature. Of all the material characteristics mentioned above, it is the ability to achieve and maintain a highly purified specimen which is the most critical and difficult step in using organic materials as temperature reference standards.

The triple-point temperature of succinonitrile occurs in a practically useful range, because there is a paucity of reference temperatures in the important range 0-100°C. Only recently,[5] in fact, has a high-quality standard been established in that range, namely the gallium melting point cell, which melts at 29.77°C IPTS-68.[5] Succinonitrile, if accepted, would add another high-quality temperature reference in this temperature range.

Besides the desirable interfacial solidification characteristics possessed by succinonitrile, its mechanical and chemical properties make it suitable for use in a triple-point cell. Succinonitrile's β-phase exists from about -45°C to the triple point as a disordered plastic crystalline BCC structure.[6] The extreme plasticity manifested by this solid allows it to flow easily and transmit pressure under minimal stress gradients. This characteristic reduces the chances of accidentally breaking the triple-point cell when the solid is melted without allowing the liquid to expand freely into its vapor space. More fundamentally, the easy flow characteristics of solid β-phase succinonitrile prevents the formation of large pressures within the cell which would alter the solid-liquid equilibrium. In addition, succinonitrile contracts on freezing, which precludes cell breakage as the mantle freezes during operation. The chemical characteristics of succinonitrile prove that it is relatively inert, reasonably non-toxic, and neither decomposes, changes structure, nor recontaminates in vitro over a period of at least two years. Finally, one aspect of this material, which is a requirement for easy commercial application, and widespread use, is the low cost of succinonitrile.[7] The unpurified material is significantly less expensive than most materials used in triple-point temperature reference cells. Consequently, large (kilogram) cell sizes would be both practical and economical.

CELL CONSTRUCTION

The borosilicate apparatus that was used to prepare the material prior to its introduction into the triple-point cell is shown in Figure 1. This unit was thoroughly cleaned with a series of organic solvents, followed by 24-hour soakings in sulfuric acid and aqua regia (75% HCl, 25% nitric).

99.9% succinonitrile (SCN) was initially processed by vacuum distillation and zone refining in an apparatus similar to the one shown in Figure 1 to a purity of about 99.99%. This material was then transferred under vacuum through port A into distillation bulb B. Port A was then sealed off with a gas-oxygen flame.

The SCN in bulb B was then heated to circa 70°C in an external water bath and allowed to distill slowly (16 hours) into the zone refining tube under vacuum. The tube was then sealed off at point C.

The SCN was further purified by another stage of zone refining. The zone refining tube was slowly pulled by a motor through four heater coils spaced approximately 10 cm apart, in a manner to cause molten zones of SCN to move from point D toward point C. To move four zones over the length of the tube required approximately 24 hours. After every 10 complete passes, some of the SCN in the tube was melted and allowed to flow into the triple-point cell via connecting tube D. The liquid succinonitrile was made to contact the walls of the triple-point cell and shaken vigorously to pick up any remaining impurities. It was then melted back into the zone refining tube and allowed to solidify. Zone refining was then continued. A total of 42 complete passes of four zones were made.

The purified material was then partially melted by heating the zone refining tube and transferred to the triple-point cell, which was then sealed off at the constriction D, leaving most of the impurities sequestered in the unmelted part of the zone tube.

Figure 2 shows the SCN triple-point cell. Its overall design is similar to that of a conventional water triple-point cell, with some notable exceptions.

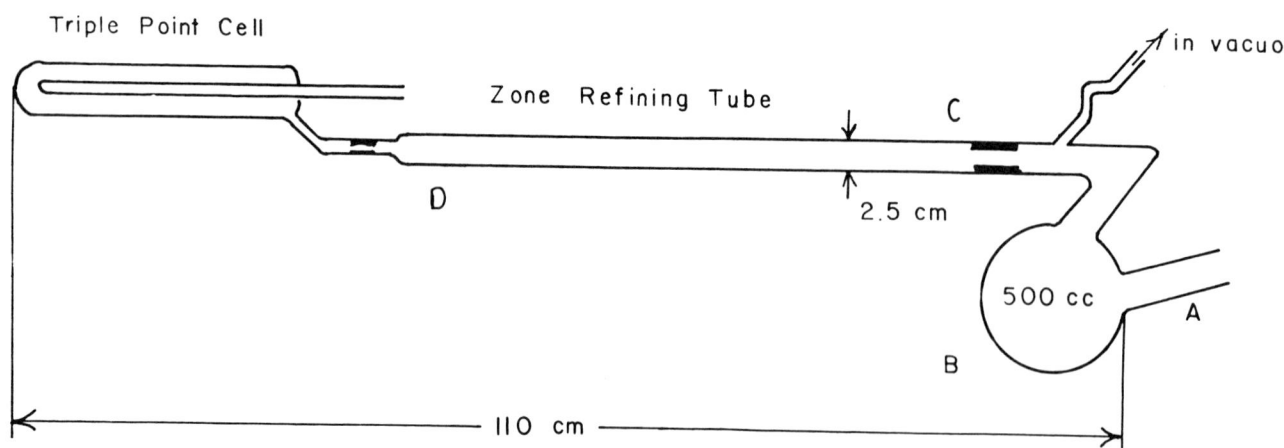

Figure 1. Integrated apparatus for purification and encapsulation of succinonitrile into a triple-point cell. A-filling port; B-distillation bulb; C-hermetic seal to zone refining tube; D-hermetic seal to triple-point cell.

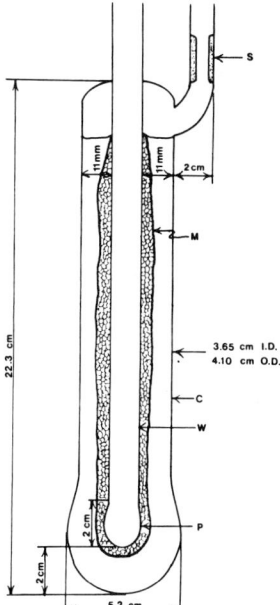

Figure 2. Schematic view of cross-section through an operating triple-point cell at bath temperatures greater than 58.080°C. At bath temperatures less than 58.080°C solid forms on the outer wall, as well as across the liquid-vapor interface. S-hermetic seal; M-crystalline mantle; C-outer wall; W-thermometer well; P-protuberance for mechanical support of mantle.

The protuberance P located at the end of the thermometer well W is designed specifically to counteract the tendency of the denser solid mantle to slide down the thermometer well to the bottom of the cell. The outer wall of the cell follows the profile of the thermometer well so that a relatively constant distance is maintained between the two walls.

DESCRIPTION OF THE EXPERIMENTAL SETUP

The environment for the triple-point cell was a temperature controlled bath. The fluid in the bath was stirred constantly and the temperature was controlled by a resistance heating element responding to a Tronac temperature controller. An 81% ethylene glycol and water mixture both matched the index of refraction of succinonitrile and served as the fluid in the thermostat. A glass window in the side of the thermostat chamber allowed the cell to be viewed clearly with the unaided eye or observed at magnifications up to 25X with a Wild M5A microscope. During the experimental runs, the window was covered to prevent stray radiation from affecting the resistance thermometer. Temperature of the water-glycol bath could be controlled routinely to ± 0.0003K over a period of several hours.

Temperature measurement: temperatures were sensed with a calibrated long-stem platinum resistance thermometer (Leeds & Northrup 8167-25B), and measured with a Smith resistance bridge (L&N 8079 type ER) and a nano-volt null detector (L&N 9828-2). The calibration of the thermometer was checked daily with a triple-point ice/water cell. The accuracy of this bridge and thermometer combination was ± 0.004K and the precision was 0.0001K.

EXPERIMENTAL PROCEDURES AND TESTS

To determine specimen purity achieved within the cell, the liquidus temperature, T_ℓ, and the solidus temperature, T_s, were measured. The cell was melted completely and then immersed fully into the thermostat. The temperature was set near the triple-point temperature. A small crystal of SCN was nucleated on the wall of the cell by touching its outer surface with a cold metal rod. The small crystal was then viewed through the microscope to observe whether it was growing or melting. The temperature of the surrounding bath was carefully adjusted until it was not possible to determine, over a 5-hour period, whether the crystal was melting or growing. The temperature at this point was chosen as the <u>liquidus</u> temperature.

The solidus temperature was measured by placing a completely solidified cell into the controlled temperature bath and observing the initiation of melting at the grain boundaries. "Tubes" of liquid generally first formed at the grain boundary edges and vertices and then as the bath temperature was raised about 0.6 mK the network of "tubes" began to extend gradually onto grain faces. The temperature at which the grain faces melted completely was chosen as the <u>solidus</u> temperature.

The procedures outlined above led to the following values:
a) T_ℓ = 58.0800 ± 0.0003°C,
b) T_s = 58.0801 ± 0.0003°C.

The limits of error reflect the temperature stability of the thermostat. Therefore, due to the error induced by the thermostat's temperature instability, we are forced to adopt the very conservative value of $T_\ell - T_s$ = 0.001K. As a result, the purity calculated on this basis represents the maximum possible solute content.

If we assume that Raoult's law holds, then the purity of the SCN can be calculated.[4] We thereby estimate that this cell contained SCN of at least 99.9996% purity.

To prepare the SCN triple-point cell for thermometric purposes, it was first immersed in a water bath at circa 70°C to melt the material completely. The cell was then transferred to the temperature-controlled bath and allowed to approach equilibrium. The triple-point cell was immersed in the thermostat so that the liquid-vapor interface was at least 10 cm below the liquid level of the thermostat. The triple point cell was further insulated from the cooler room temperature by a 2" vapor space above the liquid level in the thermostat and a 2" thick styrofoam thermostat cover. These precautions were sufficient to prevent significant influence of room temperature on the temperature sensed by the thermometer within the cell.

For the tests, the bath temperature was set in the range of -0.3 to +0.3K around the triple-point temperature of 58.0805°C. Henceforth, the bath temperature will be indicated relative to the triple point of SCN.

The solid mantle was produced by pouring water at about 22°C into the thermometer well. A thin solid shell forms within about 30 seconds, permitting the water to be quickly pumped or drained out of the thermometer well. Paraffin oil was used as the thermally conducting medium in the thermometer well. The oil was preheated to 58°C to preserve the already formed mantle. Paraffin oil was used instead of water to reduce heat loss due to evaporation. The upper level of the oil was usually chosen to be in between the top surface of the SCN and the outer wall of the cell. This amount of oil in the thermometer well makes full use of the mantle without directly contacting the outer wall of the triple-point cell.

RESULTS AND DISCUSSION

As a result of 7 runs over a range of bath temperatures from -0.3K to +0.3K we can now establish that the measured average triple-point temperature of

Table II. Triple Point Temperature Using Only Plateau Temperatures at a Number of Different Ambient Temperatures

Ambient Temp. 10^{-3}K	Plateau Temp. (°C)	Plateau Duration (Hrs.)
-10	58.0804	3.0
-50	58.0807	3.0
-100	No Plateau	-
-200	No Plateau	-
-300	No Plateau	-
+10	58.0806*	1.0
+50	58.0806	1.5
+100	58.0807	1.0
+200	58.0805	1.5
+300	58.0801	1.0

----

*Approximate Plateau Temperature because thermometer removed and replaced midrun.
Average Triple-Point Temperature = 58.0805°C
(not including head correction)(331.2305K)
Standard Deviation of the Triple Point Temperature
S.D. = 8.63 x 10^{-5}K

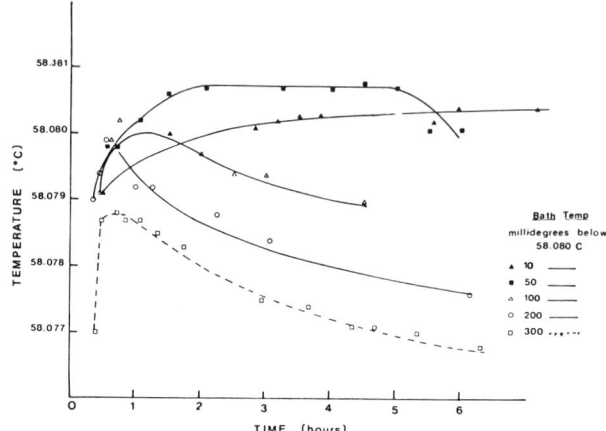

Figure 4. Sensed triple-point temperature versus time for various external bath temperatures below 58.080°C

SCN is 58.0805 ± .004°C, corresponding to a true triple point temperature of 58.0800 ± .004°C, accounting for a 0.k mK head effect, and a purity of at least 99.9996%. The standard deviation of 7 temperature plateaus is ±0.0001K. As noted before, this temperature lies conveniently near the middle of the 0-100° temperature range where good standards are needed. In order to establish the reproducibility, stability over time, and the dependence of the sensed triple-point temperature on the bath temperature, a number of additional experiments were performed (see Table II).

Before experiments could be conducted to determine any of the previously mentioned intrinsic properties of the cell, it was first necessary to determine the dependence of the sensed triple-point temperature on the immersion depth of the platinum resistance thermometer. This test was essential to ensure that thermometer stem heat losses were inconsequential. The results are shown in Figure 3. These measurements were taken with the bath temperature 15 mK below the triple-point temperature, about 3 hours after the production of a solid mantle. It will be shown later that these test conditions ensured a stable triple-point temperature to ± 0.2 mK for at least 24 hours. Figure 3 clearly shows that the thermometer does not begin to sense the outside environment to a tolerance of circa 0.2 mK until the thermometer tip is raised 3 cms from the bottom of the thermometer well. A Clausius-Clapeyron shift in the sensed triple-point temperature as a function of decreasing immersion depth cannot be detected with the instrumentation used in this experiment. In the case of SCN the sensed triple-point temperature will vary by 3.3 x 10^{-5}K per cm of immersion depth, far below the resolution of our instrumentation for a 1-cm change in immersion depth. Therefore any changes sensed in the triple-point temperature as a function of immersion depth must be ascribed to stem loss effects. The measured temperature is then a slowly decreasing function of immersion depth up to about 10 cm of thermometer pull-out, after which the sensed triple-point temperature drops rapidly. Stem losses can be neglected therefore with full insertion of the RTD when the bath temperature is at -15 mK. The independence of sensed triple-point temperature on immersion depth permits a precise determination of the characteristic of the cell.

Temperature standards should be relatively independent of the bath temperature in which it is used. This was tested by measuring the triple-point temperature as a function of time for various bath temperatures. These results are summarized in Figures 4 and 5.

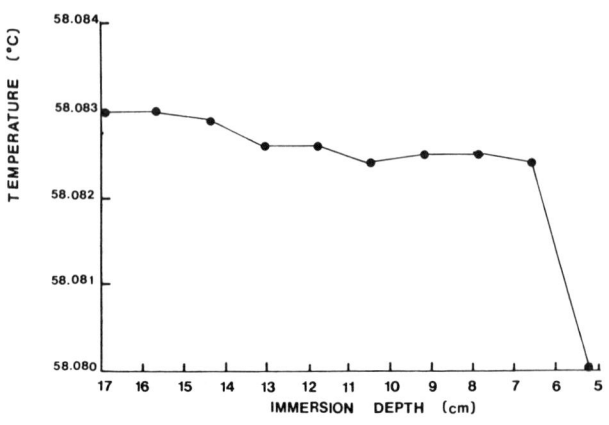

Figure 3. Sensed triple-point temperature versus immersion length of platinum RTD in thermometer well. Maximum insertion in well occurs at 17 cm.

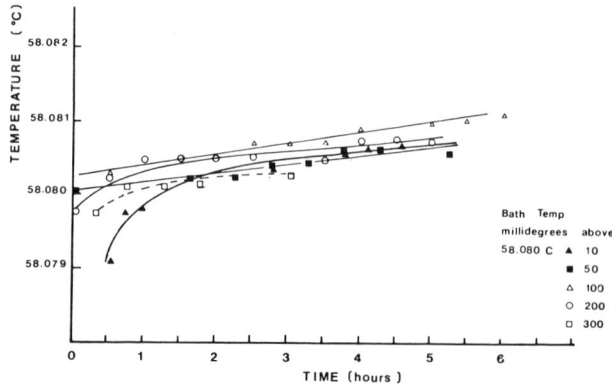

Figure 5. Sensed triple-point temperature versus time for various external bath temperatures above 58.080°C.

Figure 4 shows the sensed triple-point temperature as a function of time for various bath temperatures below the triple point temperature of 58.0800°C. Basically, each curve exhibits a transient rise in temperature, followed by a region of constant temperature, and then by a decrease in temperature. The causes of these general characteristics are discussed below.

The initial rise in sensed triple-point temperature is due primarily to the time required to raise the platinum resistance thermometer to the cell temperature. The short term rise in temperature could probably be reduced further by using a temperature probe with a smaller thermal mass or preheating the thermometer to a temperature closer to the triple-point temperature.

The cause of the region of constant temperature and the subsequent decrease in the temperature are closely related. The duration of the temperature plateau is clearly a function of the bath temperature. In the run at the -10 mK bath temperature, the plateau, although not shown on the plot, actually lasted at least 24 hours, and at each succeedingly cooler bath temperature the plateau decreased in duration. The explanation for this behavior is relatively straightforward. It was noticed, for bath temperatures below the triple-point temperature, that the liquid-vapor interface was evaporatively cooled, and the solid formed had the tendency to grow across the liquid-vapor interface and then down the sides of the cell, eventually coating the outer walls. As the mantle and solid on the outer wall grew towards each other, each solid-liquid interface rejected impurities into the liquid remaining between the advancing interfaces. As the solute content in the liquid increased, the temperature of each solid-liquid interface decreased, thus causing the sensed triple-point temperature to fall. At the lower bath temperatures the rate of growth of these interfaces towards each other increases, driven by the increased rate of conduction of heat from the cell to the environment. Therefore, the solute in the liquid accumulates at a faster rate at the lower ambient temperature, thus further contracting the plateau and causing the measured temperature to fall faster. In the experimental run at -10 mK, the growth of the mantle and the outer solid shell was extremely slow. In addition, the upper surface of the SCN was frozen almost throughout the entire run. Although this condition could destroy the triple-point conditions in the strict thermodynamic sense, it is the special properties of SCN which allow the cell to remain at a fixed temperature. Specifically, it is the plastic properties of the solid phase which allows the bridge to deform plastically thus keeping pressure at the solid-liquid interface reasonably close to the vapor pressure of SCN. Therefore, to a good approximation, the chemical potentials of SCN in each of its three phases remain equal, and the measured triple-point temperature is, as can be seen from the data, independent of the growth of a solid layer separating the liquid and vapor phases.

When the bath temperature was below the triple-point temperature the mantle tended to fracture at grain boundaries and separate into large pieces. These pieces promptly fell to the bottom of the cell and eventually formed thermal "shorts" between the cell's outer wall and the thermometer well. When the bath temperature was set to -10 mK, the mantle did not collapse for 24 hours. The duration of mantle integrity decreased as the bath temperature was decreased. This effect contributed to some extent to the short duration of the temperature plateaus at the cooler bath temperatures.

When the bath temperature is above the triple-point temperature, the cell performance is as shown in Figure 5. Clearly, there is less scattering of the data and the duration of the temperature plateaus is greater than in the cases where the bath temperature is below the triple-point temperature. This is caused, primarily, by the absence of the solid phase on the outer walls of the cell, and increased mechanical stability of the mantle.

When the bath temperatures are above the triple-point temperature evaporational cooling ceases at the liquid-vapor interface. As a result, the solid phase does not grow across upper surface and down the inside of the outer wall, as was the case when the bath temperature was below the triple-point temperature. Thus, the solute accumulation problem was not encountered and the sensed temperature did not change measurably after a given time of cell operation.

The mantle also remained intact throughout the duration of each run. Since the bath temperature was above the triple-point temperature, the mantle slowly melted towards the thermometer well. The time for the mantle to melt completely back to the thermometer well limited the operating time of the triple-point cell at the higher ambient temperatures. Since the mantle remained intact, the stem of the RTD was completely enclosed by the mantle during the entire run, and thus the triple-point temperature did not change appreciably over time.

CONCLUSION AND FUTURE INVESTIGATIONS

A number of conclusions can be drawn from this study:

- It is possible to construct a high quality, low-cost temperature reference standard using an organic material. The cell we constructed using succinonitrile displayed a triple-point temperature of 58.0800°C ± 0.004K, based on experimental determinations with the ambient temperatures displaced as much as ± 0.3K about the triple point.
- Other, perhaps previously overlooked, organic materials could be considered for use in temperature reference cells. The use of organic materials for triple-point temperature cells might be particularly desirable since they often can be obtained at low cost, and many melt at temperatures where additional reference points are needed.
- The current results indicate that it should be possible, with certain modifications in cell design and more extensive (staged) zone refining of SCN, to improve even further the performance of the cell described in this paper.
- A test of the long-term stability of SCN is needed. Little is known at present about the chemical stability and behavior of SCN sealed in vitro over long periods of time. A separate investigation in our laboratory indicates that significant changes do not occur even after two years of repeated melting and freezing[4] cycling in an hermetically sealed borosilicate glass cell.
- An independent assessment of the performance of the SCN triple-point temperature cell should be carried out at other laboratories to ensure that the measurements reported here are reproducible under rather general circumstances.

ACKNOWLEDGEMENT

The authors gratefully acknowledge the support provided by the Division of Materials Research, National Science Foundation, under contract DMR77 21628, which permitted detailed studies of the solid-liquid interface, including the effects of minor solute additions. In addition, our thanks go to the Materials Processing in Space Program Office of the National Aeronautics and Space Administration, which supported our studies of the ultra-purification of succinonitrile under contract NAS8-32425.

REFERENCES

1. The International Practical Temperature Scale of 1968 amended edition 1975: Metrologia, 12, 7 (1976).
2. B.W. Mangum, Clinical Chemistry, 23, 4, 711 (1977).
3. M.E. Glicksman, R.J. Schaefer, J.D. Ayers, Met. Trans., 7A, 1747 (1976).
4. S.C. Huang, M.E. Glicksman, Acta Met., 29, 701 (1981).
5. D.D. Thornton, Clinical Chemistry, 23, 4, 719 (1977).
6. C.A. Wulff, E.F. Westrum, J. Phys. Chem., 67, 2376 (1963).
7. Aldrich Chemical Catalog, 856 (1981-82).
8. M.J. Timmermans and Mme. Hennaut-Roland, J. de Chimie Physique, 34, 693 (1937).
9. H. Fontaine and M. Bee, Bull. Soc. Franc. Mineral. Crystallogr., 95, 441 (1972).
10. R.J. Schaefer, M.E. Glicksman and J.D. Ayers, Phil. Mag., 32, 725 (1975).
11. R.M. MacFarlane, E. Courtens, and T. Bischofberger, Mol. Cryst. Liq. Cryst., 35, 27 (1976).

The triple point of rubidium: A temperature fixed point for biomedical applications

J. M. Figueroa* and B. W. Mangum

Center for Absolute Physical Quantities, National Bureau of Standards, Washington, D.C. 20234

In order to test the feasibility of using the triple point of rubidium as a thermometric fixed point in biomedical applications, a study of the melting and freezing behavior of this metal was conducted. An investigation of the reproducibility of the plateau temperatures of a group of six rubidium cells, filled under vacuum, was made. The triple-point temperature of pure rubidium was estimated by fitting the experimental data to a hyperbolic equation under the hypothesis of the theory of dilute solutions. The triple-point temperature was established to be 39.265 ± 0.014 °C, at the 99.7% confidence interval.

INTRODUCTION

Melting and freezing temperatures of high purity metals are widely used as reference points in high precision thermometry [1,2,3]. The freezing temperatures of zinc, tin, gold and silver play an important role in the definition of the International Practical Temperature Scale of 1968 (IPTS-68)[4] and the freezing-point temperatures of a large number of metals are available as secondary reference points [4,5]. Recently, the melting point of gallium has been developed as an easy-to-use reference temperature standard[6,7] for use in the clinical laboratory and to provide a point needed for frequent checks of calibrations of electronic thermometers used in the biomedical temperature range. Those thermometers frequently have thermistors, which are very sensitive ceramic semiconductors, as temperature probes. Since it is possible that a thermistor's calibration may drift significantly in a relatively short period of time (depending on the type of thermistor and its environment), it is desirable to check their calibration often [8].

Biomedical laboratories need temperature fixed points as references for specific tests, for checking thermometers for calibration drift, and for calibration of thermometers. There is also a need for a temperature fixed point near body temperature. A possibility for one such point is the melting/freezing point or the triple point of rubidium. In order to test the feasibility of using rubidium to provide a thermometric fixed or reference point in biomedical and other applications, a study of the melting and freezing behavior of this metal was conducted. An investigation of the reproducibility of the plateau temperatures of melting curves of a group of six rubidium cells filled under vacuum was made.

As part of the study, some calculations were made on the temperature versus time data, using the theory of dilute solutions to predict the impurity content of the sample as well as to determine the triple-point temperature of pure rubidium.

Rubidium (from Latin rubidus-red, dark red) was discovered by Kirchhoff and Bunsen in 1861 when they were studying the spectra of a sample of Rosena lepidolite; they observed new lines in the red region of the spectrum [9]. Rubidium is not as rare an element as was thought several years ago. It is now considered to be the 16th most abundant element in the Earth's crust [10]. At room temperature, rubidium is a soft silvery-white, metallic element which ignites on contact with air and reacts violently with water, burning the liberated hydrogen. Extreme caution is recommended in its handling.

EXPERIMENTAL
The Rubidium Sample

The rubidium sample used in the course of the present work, with a nominal purity of 99.9%, was obtained from a commercial source at the highest purity level available. The one kilogram sample of the metal was supplied in a stainless steel container equipped with bellows-sealed valves and compression-type metal-to-metal fittings to minimize contamination during handling and transfer of the metal.

Among the different impurities that can be found in alkali metals, the alkaline-earth and other alkali metals are distinguished, because they constitute the major impurities[11]. Rubidium is always found associated with potassium and cesium in the mineral ores from which they are obtained. For instance, lepidolite, a mineral ore from which a Cs(3%)-K(77%)-Rb(20%) alloy is extracted, may have a total amount of Rb up to 3% by weight [12]. Some other ores and minerals such as amazonite, muscovite and leucite, depending on the area of origin, usually contain Rb_2O in amounts ranging from 0.3% to 1.1% by weight. Because of the similarity of the chemical and physical properties of alkali metals, particularly those of Rb and Cs, separation and purification of the metals are difficult. Consequently, Cs is the primary contaminant of Rb, even in the highest-purity samples of Rb.

Some non-metallic elements are also found as important contaminants of high-purity commercial samples of Rb. Among the most important are oxygen, nitrogen, silicon and carbon[11]. The first one has important effects on the thermal behavior of rubidium. A study of the depression of the rubidium melting point as a function of oxygen impurity has been determined by Weatherford et al.[13].

Although no attempt was made to verify the purity of the rubidium sample used in our experiments, a supplier's analytical report, based on a spectrographic analysis of the chloride salt, of the levels of impurities is summarized in Table I. Several techniques[14,15,16,17] are available for and have been applied to purity determinations of high purity substances, however, under the assumptions of the theory of dilute solution. Such techniques are based on the analysis of the temperature versus time of the melting and freezing curves and on the depression of melting and freezing

TABLE I
Analytical report for high purity rubidium
Callery Chemical Company

Element	Impurity Content (ppm)	Element	Impurity Content (ppm)
Ag	<1	Mn	<1
Al	10	Mo	<7
B	<7	Na	1
Ba	<4	Ni	<7
Be	<1	Pb	<7
Ca	7	Rb	Balance
Co	<7	Si	98
Cr	<7	Sn	<7
Cs	980	Sr	<1
Cu	<3	Ti	<7
Fe	<7	V	<1
K	7	Zr	<14
Mg	11		

points as criteria of purity. The results of our purity analysis based on these methods are discussed in a subsequent section.

Because of the difficulty in purification of alkali metals, no attempt was made of further purification of the commercial high-purity sample used in our study.

The Rubidium Triple-Point Cells

The one kilogram sample of rubidium was transferred by the supplier to six cells of equal volume and to a seventh cell of larger volume. The cells, supplied by us, were made of stainless steel with a bellows valve in the filling tube. Figure 1 shows a schematic drawing of the cells with their dimensions. The dimensions of the large cell are the same as those shown, except that the length is 11.1 cm longer. Stainless steel was selected as the material for the cells because of its good properties of being able to contain alkali metals up to temperatures of 1000 °C with no noticeable deterioration[18], and also because of its mechanical strength. The re-entrant thermometer well was made of stainless steel tubing, 6.35 mm in diameter and with 0.25 mm thick walls. The cells were specially designed to be used in a constant-temperature oil bath.

Unfortunately, the small volume cells were not filled with equal amounts of rubidium as desired. Table II shows the amounts of rubidium in each cell, determined by weighing the cells before and after they were filled. That table also shows the volume occupied by the corresponding amount of rubidium in the solid phase (at 20 °C) and in the liquid phase (at the melting-point temperature). Those volumes were calculated from the specific gravity of the metal at 20 °C (1.532 g/cm^3) and near the melting point (1.475 g/cm^3)[9]. Although some cells were almost full, there was enough space for the 2.54% free expansion[9,19] of the metal upon melting since the Rb was transferred into the cells in the liquid phase.

After the components of the cells were machined, they were washed well with soapy water, rinsed with tap water and then immersed in a 10% solution of hydrochloric acid in distilled water for 30 minutes. Then, they were rinsed with distilled water and finally air dried. Plastic gloves were used in handling the cell parts. The components were then welded together and a second washing of the interior of each cell was made by pouring 10 cm^3 of dilute hydrochloric acid into each cell, shaking the cells for five minutes, removing the acid from the cells, pouring 50 cm^3 of distilled water into the cells, shaking them and removing the water. After repeating this last step ten times for each cell, the cells were air dried and then vacuum pumped for six hours. The cells were sent under vacuum to the Rb supplier to be filled. The filling procedure was carried out under the supplier's own techniques for

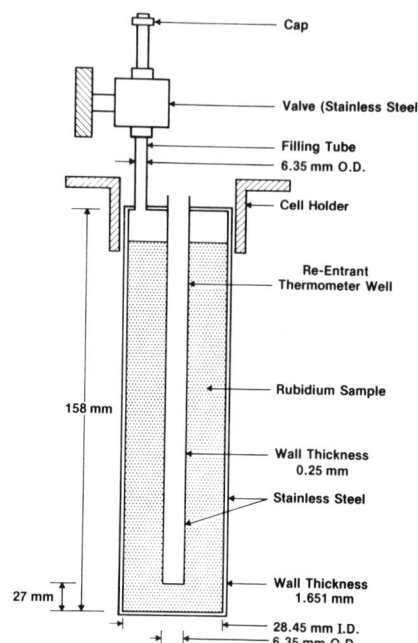

FIG. 1. Diagram of the rubidium cell. The diagram is not drawn to scale.

TABLE II
Quantity of Rb in each cell

Cell #	Amount of Rb in cell (g)	Volume occupied by Rb in each cell (cm^3)		Measured volume of cell (cm^3)
		Solid (20 °C)	Liquid (~39 °C)	
011	101	65.93	68.48	97±1
012	141	92.04	95.59	97±1
013	140	91.38	94.92	97±1
014	121	78.98	82.03	97±1
015	137	89.43	92.88	97±1
016	137	89.43	92.88	97±1

handling high-purity alkali metals. This involved filling the cells under vacuum and then closing the valves while the cells were still under vacuum and the Rb was in the liquid state.

Apparatus
Constant Temperature Bath

A constant temperature oil bath with a 10 liter capacity was used to provide a temperature gradient to drive the melting and freezing experiments of the rubidium samples. Clear mineral oil with a 17 centistokes viscosity at 40 °C was used as the fluid for the bath.

The oil bath's temperature was controlled by the combined effects of the cooling and heating systems of the bath. Cooling was obtained by means of pressurized air which had been previously chilled by passage through a copper coil immersed in an ice-water bath. The heater was controlled by a proportional controller which used a thermistor as the sensing element. Through the use of this system, it was possible to maintain uniform temperatures constant to ± 1 m°C or better during a one-day period.

Data Acquisition System

Temperature measurements during melting and freezing experiments were made by means of a data acquisition system which consisted of (1) a microcomputer, (2) a digital voltmeter (DVM), (3) a battery operated constant-current source, (4) a standard resistor (10 kΩ or 1 kΩ) and (5) a calibrated bead-in-glass-probe thermistor used as a thermometer.

A general purpose BASIC program designed to operate the system with up to six thermistors at the same time was used. The program provided enough flexibility so that the operator could choose the rate of thermistor resistance measurements, a standard resistor of the desired value and one of several outputs of the current source. In our case, 10 kΩ for the standard resistor and 10.8 μA for the current were the best choices in order to get a maximum resolution in our measurements. Only one thermistor was used in these measurements.

The resistance value for the thermistor at a given temperature was obtained from the known resistance of the standard resistor and the ratios of the measured voltage values across the potential leads of the thermistor and across the potential leads of the standard resistor when a constant current (10.8 μA) was flowing through them. The voltage values used in obtaining the ratios were the averages of the absolute magnitudes of the two voltage values measured with opposite current directions. This was done in order to correct for thermal emf's. Also, each one of those voltage values was the average of ten voltage measurements obtained during a 25 seconds period of time.

The thermistor's resistance values were used to calculate the corresponding temperature values, using the calibration constants previously obtained during a calibration experiment.

The DVM used for voltage measurements had a resolution of 1 μV in the 100 mV range. Such a resolution corresponded to a temperature resolution of 0.4 m°C at 39 °C when the thermistor's temperature sensitivity of -0.4 m°C/μV @ 10.8 μA (at 39 °C) was taken into account.

In the temperature range in which the experiments were performed (37-41 °C), the thermistor's temperature sensitivity, at a current of 10.8 μA, changed from -0.333 m°C/μV at 37.0 °C to -0.392 m°C/μV at 41 °C. Therefore, the temperature resolution of the system was between 0.33 and 0.39 m°C in that temperature range.

Thermistor Calibration

A bead-in-glass probe thermistor, 2.3 mm in diameter and 6.2 mm in length, was used as the temperature sensor, in conjunction with the data acquisition system described previously.

The thermistor has been calibrated several times during the past few years of use in this laboratory. The agreement among the different calibrations on this particular thermistor is very good as can be seen from Table III, in which the resistances at given temperatures for different calibrations are shown. Table IV shows the equivalent temperature differences among those calibrations.

The calibrations of the thermistor were made by comparison with a standard platinum resistance thermometer (SPRT) in a temperature controlled oil bath. The equation

$$\frac{1}{T} = A + B\,(\ln R) + C\,(\ln R)^3 \qquad (1)$$

where T is the absolute temperature in kelvins as measured with the SPRT and R is the thermistor resistance in ohms at that temperature, was fitted [20] to the data. Values for the coefficients A, B and C as obtained from each calibration are shown in Table V.

The uncertainty in realizing temperatures with the SPRT, in terms of the IPTS-68 as maintained at NBS, was no more than ± 0.001 °C. The uncertainty in thermistor measurements was about ± 0.0005 °C as a result of (1) calibration, ± 0.00038 °C (largest deviation when equation (1) was fitted to the data), (2) possible drift of thermistor between calibrations, ±0.00015 °C (@ 40 °C), (3) possible variations in the self-heating of the thermistor during measurements, ± 0.0001 °C, and (4) limiting precision of the measurement, ± 0.0002 °C. Consequently, the overall uncertainty in the temperatures is ± 0.0011 °C.

Experimental Procedure

By heating or cooling a rubidium sample in a constant temperature bath, the melting or freezing of

TABLE III

Thermistor calibration history (for measuring current of 10.8 μA)

TEMPERATURE (°C)	DR/DT (1978) (Ω/°C)	DR/DT (1980) (Ω/°C)	RESISTANCE VALUES (Ω)			
			June 1977	July 1978	December 1980	August 1981
10	-1118.77	-1118.83	23,066.55	23,065.011	23,066.0506	23,065.0661
20	-657.82	-657.37	14,406.89	14,406.644	14,407.2082	14,406.9356
30	-398.07	-398.09	9,247.43	9,247.354	9,247.5728	9,247.3170
40	-247.36	-247.37	6,086.57	6,086.412	6,086.4996	6,086.5353
50	-157.52	-157.52	4,099.78	4,099.579	4,099.6313	4,099.7228
60	-102.18	-102.18	2,821.04	2,820.844	2,820.9044	2,821.0339

TABLE IV

Resistance and equivalent temperature differences among thermistor calibrations at 10.8 μA

Temperature (°C)	ΔR R(81) - R(77) (Ω)	ΔT T(81) - T(77) (m°C)	ΔR R(81) - R(78) (Ω)	ΔT T(81) - T(78) (m°C)	ΔR R(81) - R(80) (Ω)	ΔT T(81) - T(80) (m°C)
10	-1.485	+1.33	+0.054	-0.05	-0.985	+0.88
20	+0.284	-0.43	+0.480	-0.73	-0.273	+0.42
30	-0.113	+0.28	-0.037	+0.09	-0.256	+0.64
40	-0.034	+0.13	+0.124	-0.51	+0.036	-0.15
50	-0.057	+0.37	+0.144	-0.91	+0.092	-0.58
60	-0.006	+0.05	+0.190	-1.86	+0.130	-1.27

TABLE V

Values of the calibration constants of the thermistor used in the study (coefficients A, B, C in equation (1)) for the different calibrations performed

Calibration Date	Coefficients		
	A ($\times 10^{-3}$)	B ($\times 10^{-4}$)	C ($\times 10^{-8}$)
June 1977	1.1170448	2.319623	8.3176230
July 1978	1.1193198	2.312071	8.0165468
December 1980	1.1169668	2.31975	8.3126765
August 1981	1.1166033	2.3203044	8.2946785

the metal was obtained. Continuous monitoring of the temperature in the cell's thermometer well allowed us to obtain temperature vs. time curves for those experiments. In this section, techniques and procedures are described which were used to determine the level of reproducibility for the melting and freezing behavior.

Melting Experiments

Prior to melting the samples, the Rb was solidified as described in the section on Freezing Experiments. Except for the first melting of each sample, a melting experiment was always preceded by a freezing experiment. The melting behavior was found to be the same for mantles prepared from induced and from normal freezes, whether they were slow or quenched freezes.

With the bath temperature set at 41.000 ± 0.005 °C and controlled with a variation of $\leq \pm 1$ m°C during a one-day period, a rubidium cell was immersed in the oil bath to initiate melting. In order to reduce the heat flow from the bath to the bottom part of the cell, a Styrofoam jacket was fitted to the cell, covering the bottom 85%. Care was taken to completely fill the thermometer well with oil and to prevent the formation of air bubbles in it. The thermistor probe was inserted to the bottom of the thermometer well. The immersion of the cell in the oil bath was such that a thin layer of oil was continuously running over the top of the cell. Seven melts were performed on each cell under these conditions. Another set of melting experiments was conducted under different conditions for two of the cells. In the latter experiments, smaller temperature gradients were employed to induce the melt; the bath temperature was $\sim$ 0.5 °C, $\sim$ 0.2 °C or $\sim$ 0.1 °C above the melting temperature of the sample. In these cases, the Styrofoam jacket was not used. Also, the tops of the cells were immersed approximately 7.5 cm below the top surface of the oil. Before the initiation of a melting experiment, the bath temperature was measured with the same thermistor used to monitor the melting experiment. A room temperature measurement was then made by inserting the thermistor in the thermometer well of the cell prior to the cell's immersion in the oil bath. Temperature measurements were made every two minutes during the experiment. The metal sample was considered melted when the temperature reading of the thermistor in the thermometer well reached the bath temperature value.

Freezing Experiments

A freezing experiment was always preceded by a melting experiment. After a melting experiment, the cell was placed in an oil bath at a temperature of 45 °C to 55 °C in order to keep the metal liquid until freezing experiment was performed, usually the next day.

Freezing of the sample was induced by immersing the cell in the oil bath controlled at a temperature lower than the rubidium freezing point. Six freezes were made on each sample with a bath temperature of 37.000 ± 0.005 °C. In those experiments, the same Styrofoam jacket used in the melting experiments was used to reduce the heat flow from the bottom part of the cell to the bath.

In order to induce nucleation and produce a layer of solid rubidium around the thermometer well, a copper rod which had been cooled in liquid nitrogen was inserted into the well. This was systematically done in the first six freezes of each of the cells. The seventh freeze of each cell was conducted without the use of the cold rod in order to see if that affected the observed melting behavior.

A group of experiments under different temperature gradients ($\sim$ 0.5 °C, $\sim$ 0.2 °C and $\sim$ 0.1 °C) was performed on a set of two cells after the seven rounds of experiments just described. In these latter experiments, the Styrofoam jacket was not used since the gradients were small and a cold rod was not used to initiate the freezes since no essential differences in freezing behavior had been observed if the cold rod were used or not. The only differences observed between the induced and the normal freezes were the shorter recalescent periods and the somewhat shortened freezing curves due to the additional heat loss to the cold copper rod in the induced freezes.

Immersion

The immersion characteristics of the thermistor in the rubidium cells were determined in experiments in which the cells were immersed such that a thin layer of oil continuously flowed over the tops of the cells. The temperature of the oil was about 1.7 °C above the melting point of the rubidium. Once melting began and the "plateau" was reached, the temperature of the oil bath was reduced to a value about 6 m°C above the temperature indicated by the thermistor in the bottom of the thermometer well. This essentially stopped the melt and permitted a determination of the immersion characteristics. (The temperature at the bottom of the well was unchanged during the course of the experiment.) The temperature of the thermistor was then measured as a function of its distance from the bottom of the well, with measurements being made upon extraction of the thermistor by increments of about 1 cm. The results of one such experiment are shown in Fig. 2.

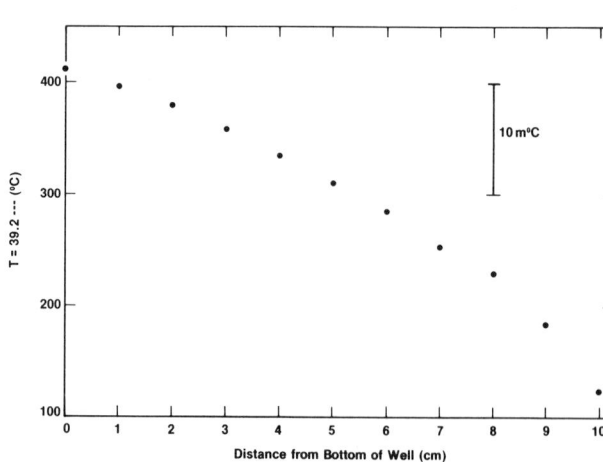

FIG. 2. Immersion results of the thermistor in rubidium cell #13.

RESULTS AND ANALYSIS

For purposes of analysis and discussion of the results, we are going to report on the six cells in two groups: A and B. Group A is composed of those cells with numbers 11 and 14 and Group B consists of cells with numbers 12, 13, 15 and 16. The only "a priori" difference between these two groups of cells is that they were filled with different amounts of rubidium, but from the same lot. Rubidium in cells numbered 11 and 14 occupy 70% and 80%, respectively, of the total cell volume. On the other hand, Rb in group B cells occupy between 92% and 94% of the cell volume.

Results of Immersion Study

The immersion results are somewhat baffling and not understood. We know from use of our thermistors in other types of cells and in fluids that only a few centimeters immersion (≤ 5 cm) are required for their indicated temperature to be within a few tenths of a millidegree of the "true" temperature of the medium in which they are situated. Thus, the behavior observed in these immersion experiments is not due to the thermistors themselves, but must be due to the rubidium and/or the steel cells with the stainless-steel thermometer wells, filled with oil to provide thermal contact between the metal and the thermometer. Since the pressure dependence of the melting point[19] is only 18 μ°C/cm of liquid Rb, we would not have been able to detect the hydrostatic head effects and that is not what was being measured. The thermistor did not fit snuggly into the thermometer well so that may have contributed to the behavior, although the temperature indicated by the thermistor was stable to within the instrumental resolution for 6 to 8 minutes of measurements.

Results from Melting Experiments

Figure 3 shows a typical melting (temperature vs. time) curve for those experiments conducted with a 1.7 °C temperature gradient and with a Styrofoam jacket on the cells.

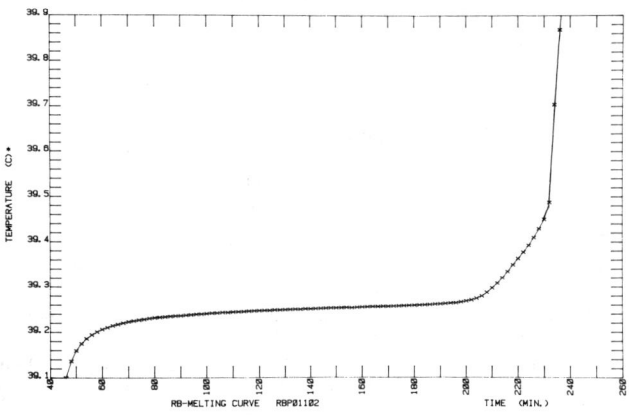

FIG. 3. Melting curve of rubidium sample #11 plotted as a function of time.

Both groups of cells show the same kind of melting features as those illustrated in Fig. 3. That is to say, a region of rapid temperature change followed by a region of decreasing rate of change, which then approached a constant value, was characteristic of the first 40 to 60 minutes of the melting experiments. That stage was followed by the flattest part of the curve, a "plateau", where the rate of change of temperature with time was almost constant and where it attained its smallest value along the melting curve (0.0004 ± 0.0002 °C/ minute for cells of group A and 0.0008 ± 0.0004 °C/minute for cells of group B). Prior to the complete liquefaction of all the metal, there was a period of about 30 minutes when the rate of change of temperature increased to a larger, constant value (0.010 °C/minute to 0.012 °C/minute). This latter behavior was characteristic of those experiments in which the Styrofoam jacket was used on cells and is a consequence of the last amount of solid rubidium melting around the thermometer well. The jacket provided a thermal shield around the bottom of the cell, causing the major heat flow to be from the top of the cell to the bottom in such a way that the last pieces of solid were localized in the bottom of the cell, presumably around the thermometer well. There is, however, a minimum amount of solid required to provide good thermal shielding of the thermometer against the bath temperature. When the amount of solid is less than or equal to this minimum, the thermometer starts sensing the bath temperature, accounting for the behavior of the measured temperature in the last stage of the melt. The completion of the melt was easily recognized because the rate of change of temperature suddenly increased, and the cell's temperature quickly reached the bath temperature. The time at which all of the metal was in the liquid phase was determined as that corresponding to the last measurement before a big jump in the temperature occurred. That time is associated with the value F=1 for the fraction (F) melted.

An intercomparison of the mean temperatures of the plateaus, i.e., the flattest part of the melting curves of the cells, is summarized in Table VI where the average temperatures, and their standard deviations, for thirty data points equally spaced along each curve are given. Those thirty experimental points, which represent the plateau in each melting experiment, were selected as follows. Different successive groups of thirty consecutive experimental points were fitted by least squares to the straight line

$$T = T_1 + a_1 t \qquad (2)$$

where T is the measured temperature at the time t, T_1 is the temperature of the first point of the group of thirty points and a_1 is a constant which corresponds to the slope of the line. The group of data which gave the best least squares fit, i.e., minimized the quadratic expression

$$Q = \sum_{i=1}^{30} \{T_i - T_1 - a_1 t_i\}^2 \qquad (3)$$

was selected as representative of the plateau of the melting curve. That selection assured that the data points comprising the plateau are closely clustered around a straight line that represents the flattest part of the curve.

The temperatures of the plateaus, calculated as described above and given in Table VI, show a spread of from 0.010 °C to 0.016 °C for each cell, a total spread of 0.060 °C for all of the measurements on all of the cells, a spread of 0.022 °C for cells of Group A, and a spread of 0.020 °C for cells of Group B.

Cells in group A had higher plateau temperatures and had plateaus which were flatter and lasted longer than those for cells of group B.

The principal cause of the difference of the plateau temperatures is different amounts of impurities in the cells. Although the cells were filled from the same lot of rubidium, some contamination could have occurred during the filling process. Also, the cells themselves could have contained some contaminants. The melting curves were analyzed in an attempt to establish the impurity content of each sample. As will be shown later, the results indicate a higher purity for cells of group A than for those of group B.

As suggested by the theory of dilute solutions[21, 22,23], either temperature vs. time or heat content vs. temperature[24] data from a melting or freezing experiment of a high-purity substance can be used to estimate

TABLE VI

Average plateau temperatures and their standard deviations obtained from melting curves.
Bath temperature = 41.0 °C (1.7 °C temperature gradient), Styrofoam jacket used.

Experiment No.	Cell #011 (°C)	Cell #012 (°C)	Cell #013 (°C)	Cell #014 (°C)	Cell #015 (°C)	Cell #016 (°C)
01	39.2423 ±0.00070	39.2118 ±0.00308	39.2028 ±0.00446	-	39.1971 ±0.00404	39.2128 ±0.00365
02	39.2536 ±0.00063	39.2038 ±0.00277	39.2122 ±0.00244	39.2452 ±0.00087	39.2123 ±0.00228	-
03	39.2544 ±0.00069	39.2064 ±0.00305	39.2148 ±0.00157	39.2348 ±0.00211	39.2049 ±0.00276	39.2113 ±0.00269
04	39.2571 ±0.00057	39.2056 ±0.00361	39.2031 ±0.00340	39.2385 ±0.00175	39.1984 ±0.00408	39.2171 ±0.00184
05	39.2530 ±0.00060	39.2147 ±0.00158	39.2166 ±0.00132	39.2422 ±0.00122	39.2042 ±0.00230	39.2062 ±0.00321
06	-	39.2034 ±0.00383	-	39.2428 ±0.00118	39.2059 ±0.00228	39.2143 ±0.00199
07	39.2557 ±0.00059	39.2125 ±0.00205	39.2003 ±0.00368	39.2370 ±0.00162	39.1983 ±0.00381	39.2092 ±0.00324
$\overline{T}$ (Avg. of Avg.) σ	39.2526 ±0.00527	39.2083 ±0.00458	39.2083 ±0.00704	39.2401 ±0.00395	39.2030 ±0.00545	39.2118 ±0.00384
$\Delta T = T_{MAX} - T_{MIN}$	0.0148	0.0113	0.0163	0.0104	0.0152	0.0109
$(T_{MAX})_{MAX}$	39.2571					
$(T_{MIN})_{MIN}$	39.1971					

the amount of impurities in the sample as well as to estimate the equilibrium temperature of the solid-liquid interface of the pure material. This estimation is based on a set of hypotheses[14] under which it is possible to obtain an analytical relationship between the observed temperatures T and the corresponding fraction melted, F, of the given substance. The relationship has the form of a hyperbolic equation[14,23,24],

$$T = T_0 - \frac{BX_2''}{F} \quad (4)$$

where T_0 is the solid-liquid equilibrium temperature for the pure substance, X_2'' is the total mole fraction of impurities in the sample (for a two component system) and B is the cryoscopic constant, i.e.,

$$B = \frac{RT_0^2}{\Delta H_F} \quad (5)$$

where R is the gas constant and ΔH_F is the latent heat of fusion of the major component. The value reported by Martin[25] for this quantity was used in our calculations.

In Table VII, we give the triple-point temperature of the rubidium samples obtained by a least squares fitting of equation (4) to the temperature vs. time data of the melting curves. To fit the equation to the data, we used the method reported by H. Kienitz[26], in which the quadratic expression

$$Q = \sum_{i=1}^{N} \{(T_i - T_0) F_i + A\}^2 \quad (6)$$

is minimized. In this equation, $A = BX_2''$. The mean value of the triple-point temperature obtained for each cell over the seven experiments, performed with a 1.7 °C temperature gradient (bath temperature = 41.000 °C) and while using the Styrofoam jacket to insulate the cells, is also shown in Table VII.

The parameters of equation (6), (and hence of Eq. (4)) were determined by using only the first 50% of the melting curve to fit the equation. This set of data corresponds to the values of F from 0 to 0.5. It is in this region where (a) the rate of change of temperature with fraction melted (dT/dF) is largest, so that the temperature dependence on F is more sensitive and, thus, defines a preferable region in which to make the calculation[26], and (b) no anomalous behavior in the melting curve was observed. The equation was fitted by first selecting the times at which F=0 and F=1 occurred, and then assuming that F was proportional to the time. The F=1 value was selected graphically from a temperature vs. time plot by picking out the time at which dT/dt started changing rapidly, with the temperature of the sample approaching the bath temperature. Several fittings with different data points associated with the value F=0 were tried and that which minimized the quadratic expression Q, Eq. (6), was selected to be the value for F=0.

If equation (4) describes the situation in our rubidium samples, then the equilibrium temperature at any given time during melting is proportional to 1/(fraction of sample melted), i.e., 1/F. That means that for constant heat flow to the sample, or for a constant temperature gradient, the equilibrium temperature during melting is directly proportional to 1/time. Thus, another way of displaying the melting curve and perhaps an easier way to graphically determine the liquidus temperature is to plot the temperature as a function of 1/time. Such a plot is shown in Fig. 4. The liquidus

TABLE VII

Rb triple-point temperatures, T_o, calculated by fitting $T = T_o - A/F$, no solid solution, to the experimental data. Bath temperature = 41.00 °C. Styrofoam jacket used.

Experiment #	Cell #011 (°C)	Cell #012 (°C)	Cell #013 (°C)	Cell #014 (°C)	Cell #015 (°C)	Cell #016 (°C)
01	39.2565	39.2367	39.2342	-	39.2322	39.2486
02	39.2631	39.2385	39.2393	39.2621	39.2362	-
03	39.2621	39.2435	39.2366	39.2620	39.2413	39.2346
04	39.2635	39.2372	39.2366	39.2570	39.2393	39.2330
05	39.2615	39.2372	39.2435	39.2603	39.2360	39.2329
06	-	39.2468	-	39.2579	39.2298	39.2423
07	39.2663	39.2389	39.2377	39.2573	39.2236	39.2386
T_o(Average)	39.2622	39.2398	39.2380	39.2594	39.2340	39.2386
n, σ_T	6, ±0.0032	7, ±0.0038	6, ±0.0032	6, ±0.0023	7, ±0.0060	6, ±0.0062
$T_{MAX} - T_{MIN}$	0.0098	0.0101	0.0093	0.0051	0.0177	0.0157

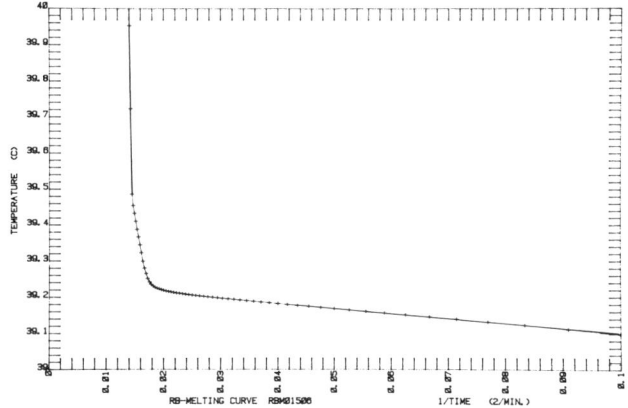

FIG. 4. Melting curve of rubidium sample #15 plotted as a function of 1/time.

temperature so determined graphically for rubidium is greater than that determined algebraically, as described in the preceding paragraph, by approximately 20 m°C. The reason for this difference is that the graphical determination emphasizes the last part of the melting curve whereas the analytical determination emphasizes the first part of the melting curve.

Temperatures obtained for the triple point of Rb in cells of group A were higher than those calculated for cells of group B. The average value over the seven cycles of experiments for cells in group A is 0.025 ± 0.003 °C greater than the average value for cells in group B.

We give in Table VIII the calculated values for the liquidus temperature, T_L, obtained from equation (4) by taking F=1, i.e.,

$$T_L = T_o - BX_2" \quad (7)$$

The quantities T_o and $BX_2"$ were determined from the least squares fit of Eq. (6) to the data. These liquidus temperatures for samples in group A are about 30 ± 5 m°C greater than the corresponding values for the samples in group B.

Table IX gives the calculated values of the purity of the samples investigated. These values were obtained in the analysis of the data using Eq. (4). As shown in Table IX, samples of group A appear to have a higher purity than samples of group B. The purity values reported in Table IX are very high in comparison with the values given by the supplier. Therefore, either our sample comes from a high-purity lot and purity levels are better than those reported by the supplier or our purity determination technique is inadequate and/or insensitive. The latter is more probable since it seems unlikely that the purity is as high as that calculated. The relatively large melting ranges observed for the samples tend to confirm this.

If Cs is present, thermodynamical equilibrium conditions require a solid-soluble solution of it in the rubidium sample and several studies support this situation[25,27,28,29]. Since Cs is the major contaminant of Rb, its effect on the melting behavior of Rb should be dominant. When a Mastrangelo[13] type equation is fitted to the temperature vs. time data in order to take account of a solid solubility of Cs in Rb, the results for the triple-point temperature, T_o, and for the purity of the sample are comparable to those obtained when no solid solution is assumed. Furthermore, the value obtained for the distribution coefficient K is very small. This may be a consequence of a rapid melt in which the period of observation is so short that the impurity may not remain in equilibrium amounts in the two phases[30].

A segregation of impurities may not occur when a sequence of fast freezes and melts is conducted. Then the observed temperature vs. time melting curves correspond to a pseudo-equilibrium situation in which the measured temperature values are lower than the melting temperature of pure rubidium but the indicated purity is higher than the actual purity.

TABLE VIII

Liquidus temperature, calculated from data by fitting $T(F) = T_0 - A/F$. $T_L = T(1) = T_0 - A$.
No solid solution. Bath temperature = 41.00 °C. Styrofoam jacket used.

Experiment #	Cell #011 (°C)	Cell #012 (°C)	Cell #013 (°C)	Cell #014 (°C)	Cell #015 (°C)	Cell #016 (°C)
01	39.2504	39.2216	39.2176	-	39.2126	39.2305
02	39.2560	39.2202	39.2220	39.2519	39.2226	-
03	39.2552	39.2235	39.2221	39.2506	39.2208	39.2202
04	39.2574	39.2200	39.2200	39.2472	39.2189	39.2214
05	39.2549	39.2246	39.2254	39.2505	39.2180	39.2186
06	-	39.2274	-	39.2497	39.2142	39.2248
07	39.2593	39.2230	39.2189	39.2476	39.2145	39.2241
T_L (Average)	39.2555	39.2229	39.2210	39.2496	39.2174	39.2233
n, σ_{T_L}	6,±0.00300	7,±0.00261	6,±0.00278	6,±0.00184	7,±0.00372	6,±0.00424

TABLE IX

Results of purity determinations (% mole fraction of Rb), by fitting $T = T_0 - A/F$ to data.
No solid solution. Bath temperature = 41.00 °C. Styrofoam jacket used.

Experiment #	Cell #011 (%)	Cell #012 (%)	Cell #013 (%)	Cell #014 (%)	Cell #015 (%)	Cell #016 (%)
01	99.9983	99.9959	99.9955	-	99.9947	99.9951
02	99.9982	99.9950	99.9953	99.9972	99.9963	-
03	99.9981	99.9946	99.9961	99.9969	99.9945	99.9961
04	99.9983	99.9953	99.9955	99.9974	99.9945	99.9968
05	99.9982	99.9960	99.9950	99.9974	99.9951	99.9961
06	-	99.9948	-	99.9978	99.9958	99.9959
07	99.9981	99.9957	99.9949	99.9974	99.9950	99.9951
Purity (Average)	99.9982	99.9953	99.9954	99.99735	99.9951	99.99585
n, σ_P	6,±0.000089	7,±0.000553	6,±0.000431	6,±0.000295	7,±0.000685	6,±0.000657
$10^5 X_2''$	1.8	4.7	4.6	2.7	4.9	4.2

We conducted another set of melting experiments on the six cells in which (1) the temperature gradient between the sample and the bath was about 0.1 °C (bath temperatures = 39.400 °C) instead of 1.7 °C (cf. Table VII), (2) the Styrofoam jacket was not used, and (3) the cells were immersed such that the tops of the cells were about 7.5 cm below the top surface of the oil in the oil bath. The results are given in Table X. As seen in this table, the average triple-point temperatures T_0 for three melts of the samples were higher by amounts ranging from 0.008 °C to 0.015 °C and the calculated mole fractions of impurities X_2'' range from 1.13×10^{-5} to 1.70×10^{-5} (i.e., a higher purity is indicated).

The liquidus temperatures calculated from the same experiments are listed in Table XI. As was the case with the triple-point temperatures, the liquidus temperatures were, for all samples, several m°C higher for this set of experiments. The reasons for these effects are uncertain. This could indicate a better equilibrium distribution of impurities between the two phases of Rb. It is unlikely to be due to immersion effects of the thermistors, however, since only about 5 cm of immersion of the thermistors in other types of cells are adequate for an uncertainty of ≤ 1 m°C. The presence of the Styrofoam jacket may also possibly contribute to the effect.

TABLE X

Rb triple-point temperature, T_0, and purities of samples calculated by fitting $T = T_0 - A/F$ to the data. Bath temperature = 39.40 °C. Styrofoam jacket not used. Cells deeply immersed.

Experiment #	Cell #011 (°C)	Cell #012 (°C)	Cell #013 (°C)	Cell #014 (°C)	Cell #015 (°C)	Cell #016 (°C)
01	39.2678	39.2480	39.2556	39.2715	39.2479	39.2524
02	39.2698	39.2511	39.2494	39.2667	39.2549	39.2543
03	39.2721	39.2431	39.2538	39.2667	39.2526	39.2454
T_0 (Average)	39.2699	39.2474	39.2529	39.2683	39.2518	39.2507
σ (ST. DEV.)	±0.0022	±0.0040	±0.0032	±0.0028	±0.0036	±0.0047
$10^5 X_2''$	1.19	1.30	1.57	1.15	1.70	1.13
Purity, %	99.9988	99.9987	99.9984	99.9990	99.9983	99.9988

TABLE XI

Liquidus temperature calculated from data by fitting the equation $T = T_0 - A/F$. $T_L = T_0 - A$. Bath temperature = 39.400 °C. No Styrofoam jacket used.

Experiment #	Cell #011 (°C)	Cell #012 (°C)	Cell #013 (°C)	Cell #014 (°C)	Cell #015 (°C)	Cell #016 (°C)
01	39.2631	39.2435	39.2507	39.2678	39.2424	39.2491
02	39.2662	39.2461	39.2438	39.2625	39.2483	39.2493
03	39.2672	39.2381	39.2468	39.2678	39.2457	39.2411
T_L (Average) σ	39.2655 ±0.0021	39.2426 ±0.0041	39.2471 ±0.0035	39.2660 ±0.0031	39.2455 ±0.0030	39.2438 ±0.0048

Pressure Effects on Melting

The reported[19] experimental value for the pressure dependence of the melting point, dT/dP, is 18.1 m°C/atmosphere, which is equivalent to 27 µ°C/cm of liquid Rb. This dT/dP value gives a hydrostatic head correction for the temperature at the bottom of the thermometer well of ~ 0.35 m°C which is of the order of magnitude of our thermometer resolution. The experimental value[19] for the pressure dependence of the melting temperature is compatible with that obtained from the Clausius-Clapeyron equation

$$\frac{dT}{dP} = \frac{T_0 \Delta V}{\Delta H_F} \qquad (8)$$

where ΔV is the change in molar volume upon melting. The value calculated from this is about 20 m°C/atm. (obtained by using T_0 = 312.415 K, the value reported in reference 19 for ΔV, and the value given in reference 25 for ΔH_F, the heat of fusion). Thus, the pressure effect on the observed temperatures will be negligible at our level of resolution.

Results from Freezing Experiments

Freezing curves may not be as suitable as melting curves for purity and triple-point temperature determinations. Besides the supercooling effect, the attainment of equilibrium is more difficult during crystallization than during melting[24]. This can be explained in terms of the crystallization rates in the cooling curves[31]. Nonetheless, some reproducible points of the freezing curve can be used as reference points for the calibration of thermometers.

In our experiments, we found that the maximum temperature of the freezing curve, which was reached 20 to 30 minutes after the freeze was initiated, was quite reproducible. In Table XII, we list those maximum temperatures for each experiment with each cell and their average values. Cells of group B show a lower maximum freezing temperature than those of group A, consistent with the melting results.

CONCLUSIONS

Due to its higher reproducibility and lower impurity content, cell #11 has been chosen as the most

TABLE XII

Maximum observed temperature in Rb freezing curves.
[* = cold rods were not used]
[** = Cells without Styrofoam jacket, 0.1 °C temperature gradient]

Experiment #	Cell #011 (°C)	Cell #012 (°C)	Cell #013 (°C)	Cell #014 (°C)	Cell #015 (°C)	Cell #016 (°C)
01	39.2564*	39.242	39.246	39.258	39.246	39.234
02	39.2692	39.233	39.210	39.261	39.243	39.242
03	39.2668	39.242	39.230	39.254	39.242	39.238
04	39.2638	39.244	39.234	39.267	39.246	39.233
05	39.2633	39.243	39.238	39.267	39.240	39.223
06	39.2630	39.233	39.231	39.265	39.250	39.235
07	39.2578*	39.233	39.232	39.255*	39.241	39.239
11	39.2597*,**			39.255*,**		
12	39.2567*,**			39.254*,**		
14	39.2559*,**			39.256*,**		
15	39.2562*,**			39.258*,**		
T(Average) (n) n=# of experiments	39.2652 (5) 39.2571 (6)*	39.2386 (7)	39.2316 (7)	39.2620 (6) 39.2556 (5)*	39.2440 (7)	39.2349 (7)
σ(ST. DEV.)	±0.0027 ±0.0014*	±0.0053	±0.010	±0.0053 ±0.0015*	±0.0035	±0.0061
$\Delta T = T_{MAX} - T_{MIN}$	0.0062 0.0038*	0.011	0.036	0.013 0.004*	0.010	0.019

representative of pure Rb. Consequently, the triple-point temperature of pure rubidium is calculated as the average of the nine T_0 values obtained from the melting experiments conducted on this cell. The value assigned to T_0 is 39.265 ± 0.014 °C (312.415 ± 0.014 K), at the 99.7% confidence interval. This value agrees to within ≤ 1 m°C with the value for the average maximum temperature obtained in the freezing experiments reported in Table XII.

The value reported in the literature for the melting point of rubidium is 39.32 ± 0.02 °C (312.47 ± 0.02 K)[25], a value which was obtained by means of a calorimetric technique, and by assuming a solid-soluble impurity of Cs in the sample. It is unclear whether this reported value is for the melting point at a pressure of one atmosphere or is for the triple point. There is uncertainty also as to which temperature scale was used — the IPTS-68 or the IPTS-48. Assuming that the reported value is that of the melting point (at one atmosphere pressure) on the IPTS-68, then including the 18 m°C correction for pressure effects, our result for the melting point of Rb (at 1 atm.) is 0.04 ± 0.02 °C lower than that reported by Martin[25]. If Martin used the IPTS-48, then our value is only 0.03 ± 0.02 °C lower than his. If we assume that the reported value is that of the triple point and that the Scale used was the IPTS-48, then our value is 0.05 ± 0.02 °C lower than his. In either case, there appears to be a small difference between our value and the value reported by Martin. At least two factors may have contributed to that difference. Martin used a calorimetric technique whereas we used a dynamic method. In addition, Martin used a different region of the melting curve than we did in making calculations of the melting point.

Although they were different for each cell, plateau temperatures for a given cell were reproducible to ± 0.020 °C, at the 99.7% confidence interval. Similarly, the maximum temperatures of freezes had irreproducibilities of ≤ ± 0.015 °C at the same confidence interval (when the result of experiment No. 2 in cell #13 is considered to be an outlier and is discarded). Consequently, temperatures of plateaus and/or freezing maxima of Rb can serve as temperature reference points, at the uncertainty levels mentioned above, when the techniques described above are employed. Although the irreproducibilities of the Rb reference temperatures are about ± 0.02 °C, at the 99.7% confidence interval, these reference points of Rb fill the need of many biomedical applications for a reference point near human body temperature.

In order to obtain a better reproducibility and accuracy in the triple-point temperature, further investigations would be useful. For instance: (a) long term reproducibilities of Rb melting and freezing behavior are needed in order to establish the possible ageing behavior of the cells and its effect on the reference temperature; (b) it might be useful to study the effects of different temperature gradients, under which melting or freezing experiments are conducted, on the reproducibility and stability of the freezing and melting behavior; and (c) it would be worthwhile to conduct experiments with higher purity rubidium in order to improve the reproducibility of the reference points, both in melting and freezing experiments, if the purer material were available.

ACKNOWLEDGMENTS

We want to thank Dr. D.D. Thornton, who participated in some of the initial design of some aspects of the apparatus used in these experiments, and Dr. George T. Furukawa, for helpful discussions.

REFERENCES

* Guest Worker. Permanent address is Escuela Superior De Fisica y Matematicas, Instituto Politecnico Nacional, Mexico, 14, D. F., Mexico. Also supported by Consejo Nacional De Ciencia y Tecnologia.

1. B.W. Mangum and D.D. Thornton, Metrologia 15, 201-215 (1979).
2. G.T. Furukawa, J.L. Riddle, W.R. Bigge, and E.R. Pfeiffer, "Application of Some Metal SRM's as Thermometric Fixed Points", to be published.
3. E.G. McLaren, "The Freezing Points of High Purity Metals as Precision Temperature Standards", in American Institute of Physics, Temperature, Its Measurement and Control in Science and Industry, Vol. 3, Part 1, Reinhold Publishing Corp., New York, N.Y., 1962, pp. 185-198.
4. The International Practical Temperature Scale of 1968, Amended Edition of 1975, Metrologia 12, 7 (1976).
5. L. Crovini, R.E. Bedford, and A. Moser, Metrologia 13, 197 (1977).
6. Henry E. Sostman, Rev. Sci. Instrum. 48, 127 (1977).
7. G.N. Bowers, Jr., B.W. Mangum, D.D. Thornton, H.E. Sostman, and S.R. Inman, in "Gallium Melting Point Standard", NBS Special Publication 481, June 1977.
8. B.W. Mangum in "Gallium Melting Point Standards", NBS Special Publication 481, June 1977.
9. F.M. Perel'man, "Rubidium and Caesium". MacMillan Co., New York, 1965.
10. Handbook of Chemistry and Physics. CRC. 51st Edition (1970-71), p. B-27.
11. J.W. Mausteller, F. Tepper, and S.J. Rodgers, "Alkali Metal Handling and System Operating Techniques", Gordon & Breach, N.Y. 1967.
12. F. Tepper, in "The Alkali Metals" (International Symposium on The Alkali Metals). Nottingham, England, July 1966, Special Publication No. 22, p. 370, The Chemical Society, Burlington House, London, 1967.
13. W.D. Weatherford, Jr., R.K. Johnston, M.L. Valtierra and J.W. Rhoades, "Handling of Lithium and Rubidium Metals", ASD-TDR-413, Wright-Patterson AFB, Ohio (1963).
14. W.J. Taylor and F.D. Rossini, J. Res. Nat. Bur. Stands. 32, 197 (1944).
15. A.R. Glasgow, Jr., A.J. Streiff and F.D. Rossini, J. Res. Nat. Bur. Stands. 45, 355 (1945).
16. C.P. Saylor, Anal. Chim. Acta 17, 36 (1957).
17. S.V.R. Mastrangelo and R.W. Dornte, J. Am. Chem. Soc. 77, 6200 (1955).
18. T.D. Brotherton, O.N. Cole, and R.E. Davis, "Properties and Handling Procedures for Rubidium and Caesium Metals", Trans. Met. Soc. AIME, 24, 287-292 (1962).
19. Dan McLachlan, Jr. and E.G. Ehlers, J. Geophysical Research 76, 2780-2789 (1971).
20. OMNITAB II User's Reference Manual, NBS Technical Note 552, October 1971.
21. E.W. Washburn, J. Am. Chem. Soc. 38, 653 (1910).
22. G.N. Lewis and M. Randall, "Thermodynamics and Free Energy of Chemical Substances", McGraw-Hill Book Co., New York, N.Y., 1923, p. 238.
23. W.P. White, J. Phys. Chem. 24, 393 (1920).
24. W.M. Smit, Recuil Trav. Chim. des Phys. Bas. 75, 1309 (1956).
25. Douglas L. Martin, Canadian J. Phys. 48, 1327 (1970).
26. H. Kienitz, Anal. Chim. Acta 17, 43 (1957).
27. M. Hansen, Constitution of Binary Alloys, McGraw-Hill Book Co., New York, NY, 1958, pp. 578-579.
28. E. Rinck, Compt. Rend. 205, 135-137 (1937).
29. B. Böhm and W. Kenn, Z. Anorg. Chem. 243, 69-85 (1939).
30. John P. McCullough and Guy Waddington, Anal. Chim. Acta 17, 80 (1957).
31. W. M. Smit, Anal. Chim. Acta 17, 23 (1957).

Temperature references based on first order phase transition: Development and application

D. Rappaport

Galai Laboratories Limited, Migdal Haemek, Israel 10500

N. Karasikov and M. B. Roitberg

Technion Israel Institute of Technology, Solid State Institute, Haifa, Israel 32000

During the last years, special attention has been given to the subject of first order phase transitions, especially those phase transitions between semiconductor and insulator in ferroelectric materials (i.e., rare earth doped $BaTiO_3$). During phase transitions of the first order there is an abrupt change in the lattice and in the physical parameters such as conductivity. This change takes place within a very narrow temperature range. This interesting feature gave impetus to using the phase transition point (Curie point) in doped ferroelectric material to build temperature references where the material itself will stabilize at the phase transition point. The described technology enables the changing of the phase transition point in the range 300–500 K. Thus, temperature reference points can be prepared for any temperature in this range. In order to construct a good temperature reference two basic types of construction are possible: (a) adiabatic construction in which the temperature is extremely stable and it can even function as a secondary temperature reference, or (b) isothermal construction in which there is a large radiating area and the temperature is in equilibrium with the surroundings. The article reviews the theoretical background of this type of phase transition, discusses the pertinent properties, and describes novel designs and technologies of adiabatic and isothermal temperature references based on the use of this phase transition point.

INTRODUCTION

During the last few years special attention has been given to the subject of first order phase transitions and especially those phase transitions between semiconductor and insulator states. In ferroelectric material (i.e., rare earth doped $BaTiO_3$), when a phase transition of the first order occurs there is an abrupt change in the lattice geometry and in physical parameters such as conductivity.

This interesting feature gave us the idea to use the phase transition point (Curie point) as a secondary temperature reference, the idea being that the material itself will stabilize at the phase transition point due to the high non-linearity of loss near the phase transition temperature.

At Galai Laboratories we have developed the technology for manufacturing doped ferroelectric material. Different stoichiometries yield phase transitions in the range of 300 - 500 °K. Thus a fixed point temperature reference can be made within that range.

In order to construct a good temperature reference point one should use an adiabatic construction in which there are no heat losses and the temperature is extremely stable and reproducible as required for a fixed point secondary reference temperature. However, for some applications, such as infra-red pyrometer calibration, a radiating area of fixed temperature is essential. For this application an isothermal construction is introduced having a large radiating area.

The concept described here has been used by us in various applications: as a fixed temperature black body source, as a calibration source for infra-red pyrometers and as a contact calibration source for thermocouples. Before going into application details, we will first review the theoretical background for the use of the phase transition point as a temperature reference point.

The IPTS-68 fixed point temperature standards include mostly phase transitions in liquid phase. The most common secondary fixed point based upon a liquid/solid phase transition is the ice-water phase transition in which the two phases are in equilibrium at 0 °C. However, as discussed by Schooley [1], there are potentially good temperature reference points in a phase transition of some property in the solid state. Use of clean and well defined material can yield excellent repeatability of the temperature of phase transition. We have studied this concept for several years and after achieving promising results we decided to utilize phase transitions in the solid state as temperature reference points in commercial, industrial-oriented instruments and to apply them to the measurement of high temperatures required in the industry. Special attention is to be given to references suitable for remote calibration of infra-red instrumentation.

Potentially Useful Phase Transitions

In order to enable easy operation of the temperature reference points, our main effort was given to phase transitions that can be generated electromagnetically. Thus three different phase transitions were considered:
  a) ε (dielectric constant) phase transition - The TANDEL phenomenon [2,3] (see note).
  b) R (resistance) phase transition - Positive temperature coefficient resistor.
  c) μ (permeability) phase transition.

All three phase transitions can yield good temperature reference points since the material will self-stabilize at the temperature of phase transition. However, in order to achieve extremely high accuracy, it is preferable to use a phase transition of the first order [4] in which the change in the property is abrupt rather than continuous. Examples of such a phase transition are:

The ferroelectric/paraelectric phase transition of $BaTiO_3$ [5] at 120 °C and application to the TANDEL device.
The semiconductor/insulator phase transition of Nd doped $BaTiO_3$ polycrystal and application in PTC (Positive Temperature Coefficient) resistors [6].
The ferromagnetic/paramagnetic phase transition [7] of Ni at 1051 °C.

Thermodynamic Background

Two types of phase transition are considered [4]; the first order, in which the first temperature derivative (entropy) of Landau's free energy is not

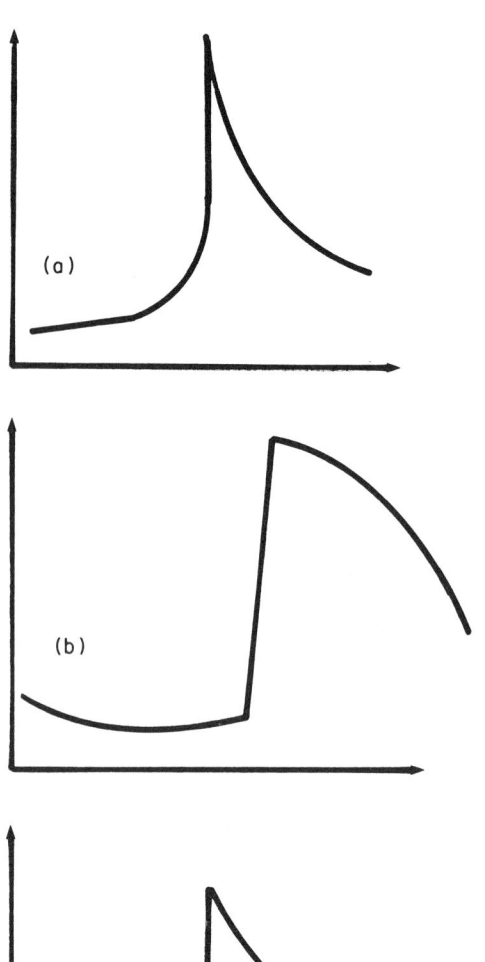

FIG. 1. Loss behavior of the different physical principles applied to temperature reference. (a) Static dielectric construction near a first order phase transition. (b) Resistance of a Positive Temperature Coefficient Semiconductor. (c) Permiability of metal near a first order magnetic phase transition.

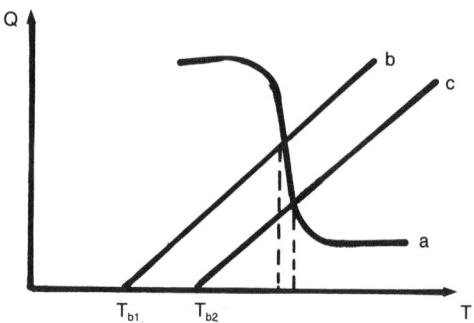

FIG. 2. Heat balance of phase transition material. An ambient temperature change of $T_{b2}-T_{b1}$ yielded a much smaller equilibrium change, and thus stabilization is achieved.

a -- Q_{in} -- internal dissipation near phase transition

b -- Q_{out} for T_{b1} -- heat loss of active material for T_{b1} ambient

c -- Q_{out} for T_{b2} -- heat loss of active material for T_{b2} ambient

relevant property enables a wide dynamic range of dissipation power within a narrow range of material temperature. Thermal equilibrium is such that heat balance is maintained, and the internal dissipation in the active material equals the heat loss from the active material to the surroundings. The heat loss changes with ambient temperature, but the change is much smaller than the range of internal dissipation. Thus a temperature very close to the phase transition will be maintained for any ambient, and a stabilization of temperature is achieved.

The stabilization of temperature is proportional to the abruptness of the property change causing the loss. Figure 1 shows qualitatively the behavior of the relevant property (a - TANDEL, b - PTC, c - Magnetic).

The internal dissipation, which is the heating process, is given by:

$Q_{in} \propto \varepsilon'' E^2$ in TANDEL where E is the electric field

$Q_{in} \propto \frac{1}{R} v^2$ for PTC where v is the voltage applied

$Q_{in} \propto \frac{1}{\mu''} B^2$ magnetic, where B is the magnetic field

The dependence of these losses on T is given in Fig. 2a. Near the phase transition, the loss is reduced considerably, the slope is dictated by the sharpness of the phase transition.

As explained, thermal equilibrium is achieved when transfer of heat to the surroundings, Q_{out}, compensates for the internal dissipation: $Q_{in} = Q_{out}$.

Assuming $Q_{out} = \alpha(T-T_b)$ where T_b is the ambient temperature, we get Q_{out} as in Fig. 2b,c for two ambient temperatures. Analyzing the graphs, we conclude that the abrupt change of Q_{in} attenuates the influence of ambient changes upon equilibrium and thus excellent stability of equilibrium is achieved at the phase transition temperature. Naturally, working in quasi-stationary surroundings without stimulated drifts of air is preferable.

continuous, and a second order phase transition in which only the second temperature derivative (heat capacity) starts the discontinuity.

It can be shown [4] that the change of relevant parameters is abrupt in first order phase transition. Therefore, this type of phase transition should be preferred due to higher non-linearity and thus better stabilization can be achieved.

As for choosing between adiabatic or isothermal mode [9], the adiabatic mode will maintain higher temperature, closer to the phase transition and thus better stabilization will be achieved.

Basic Principle of Auto-Stabilization

All phase transitions mentioned above will self-stabilize at the phase transition point when subjected to suitable electric (magnetic) excitation. The mechanism is as follows. The abrupt change of the

Note: TANDEL is an acronym for Temperature Autostabilization Nonlinear Dielectric Element.

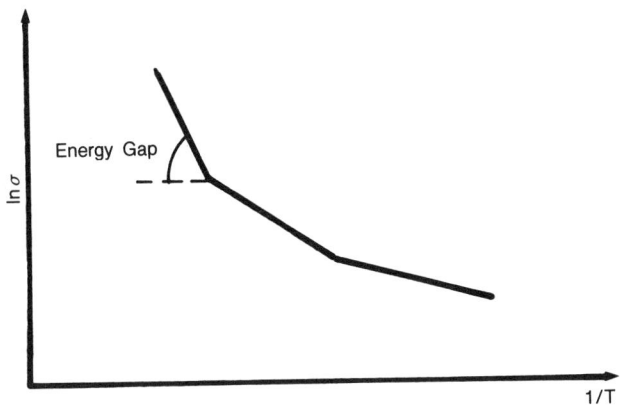

FIG. 3. Conductivity versus T. At high temperature the energy gap dominates.

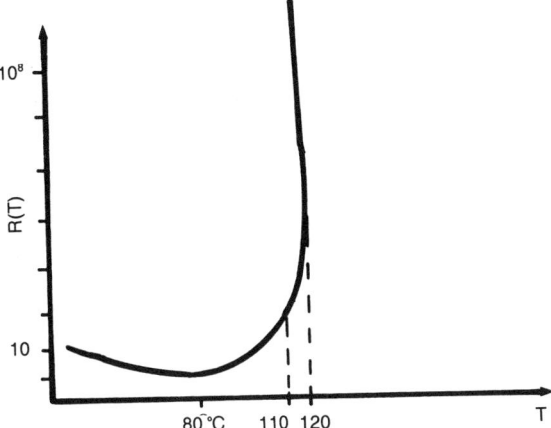

FIG. 4. Experimental results for PTC, showing resistance of active material (polycrystalline of Nd doped $BaTiO_3$) as a function of temperature. The abrupt change near the phase transition is apparent.

Temperature Ranges for the Three Types of Phase Transition

The effect described so far has an inherent parasitic loss mechanism - Joule heating due to the finite resistivity of the material. The general behavior of the conductivity with T is as shown in Fig. 3.

As is apparent from Fig. 3, the conductivity increases with temperature. For ferroelectrics (like $BaTiO_3$), the conductivity is quite large at 500 K and thus cannot be neglected. The effect of the conductivity is to deteriorate the abruptness of the ε and R phase transitions and thus to eliminate the possibility of using TANDEL or PTC as temperature references above 700 K.

This is not the case with the behavior of μ. Thus at higher temperatures one should use the magnetic loss which disregards the electrical conductivity and can maintain very sharp phase transitions at temperatures higher than 1000 K.

The Construction

So far we have been concerned only with phase transition characteristics of the material itself. However, overall performance is strongly dependent upon the type of construction chosen. The active material can be placed in a gold coated, optically reflecting cylindrical enclosure that is constructed adiabatically. In this case, we used an infra-red pyrometer to measure temperature, and find a stability of 0.01 - 0.05 °C at temperatures in the range 100-150 °C under ±5% variations of supply voltage and ambient temperature. However, with such a construction, remote measurement for infra-red calibration is quite difficult because of the low losses, i.e., only a small amount of energy is radiated out of the construction. For infra-red applications, a quasi isothermal construction is preferable since one can have a large radiating area and still maintain good temperature stability of 0.1 - 0.5 °C.

Choice of Mechanism

After analyzing the potential approaches, we decided that for infra-red calibration where temperatures up to 500 K are sufficient, the best temperature reference which will combine stability with compactness will be the one based upon PTC. This device can be operated from any AC/DC voltage source and the active material can be manufactured in various sizes and geometries.

The PTC Material

At our laboratories, we performed experiments with certain polycrystalline materials changing the stoichiometry and dopants.

The material was tested for the R(T) behavior as well as for durability under electrical shocks. We have succeeded in achieving an excellent material that is very robust, stable, and with a very sharp slope near the phase transition. Experimental results are given in Fig. 4, for a Nd doped $BaTiO_3$.

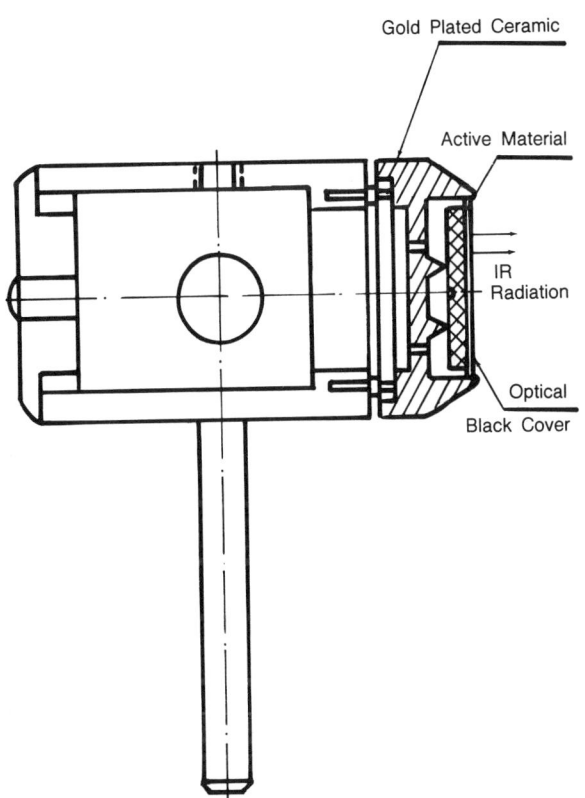

FIG. 5. Isothermal Black Body, BB100, designed for remote measurement of infra-red detectors and pyrometers.

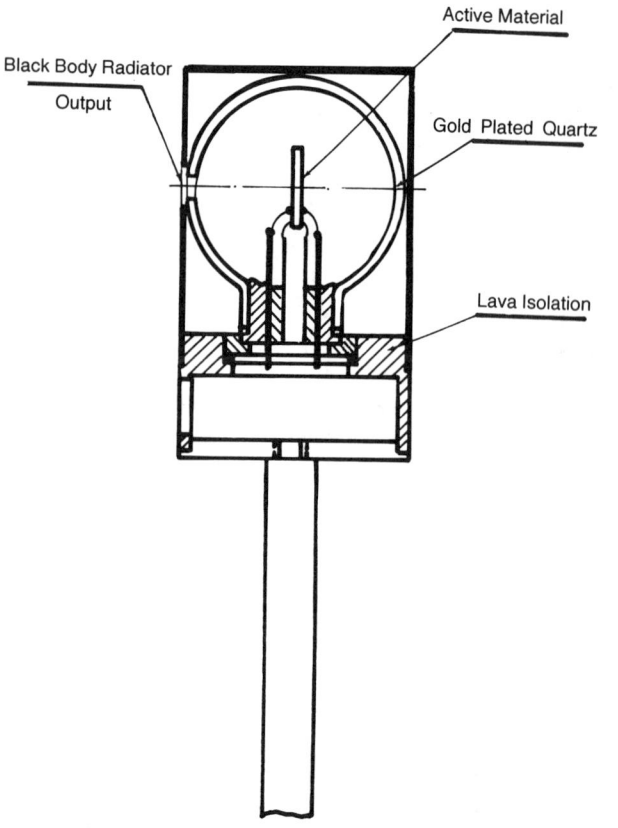

FIG. 6. Adiabatic Black Body, BB101.

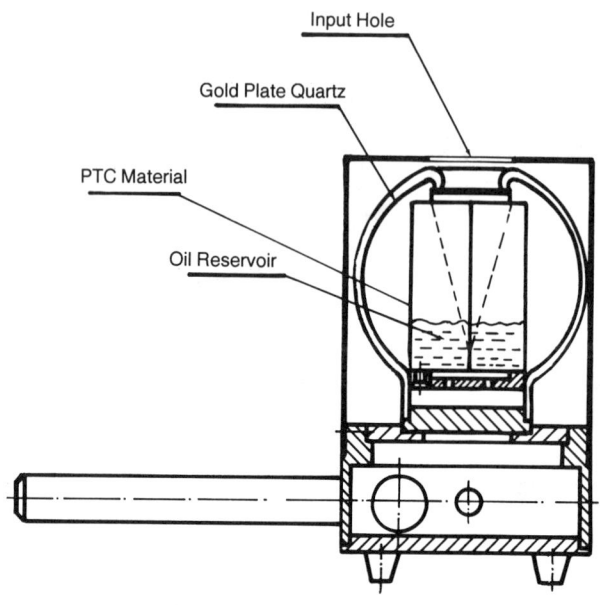

FIG. 7. Temperature reference for thermocouples, BB102. Designed for dual use as contact and remote reference.

By changing the stoichiometry, the phase transition can be "tuned" in the range of 60 °C - 200 °C.

The material is so robust that simply connecting it to the mains will transfer it within minutes to the phase transition point.

Illustration of Temperature References that are being Produced

Three basic constructions for infra-red temperature references are being produced: Isothermal - Fig. 5, Adiabatic - Fig. 6, as well as a special version for calibrating thermocouples - Fig. 7.

These temperature references were tested and proved to be stable to better than 0.01 °C. The stability was measured in a calorimeter and final tests were performed using a precision contact thermometer, an infra-red pyrometer and an AGA thermovision.

SUMMARY

We have introduced a series of temperature reference points, based upon the phase transition in PTC material. These references have been designed especially for infra-red application and a temperature range from ambient to 500 K. Future works and developments will extend the series to higher temperature ranges and to further applications.

REFERENCES

[1] J.F. Schooley, Inst. Phys. Conf. Ser. No. 26, pp. 49 (1975.
[2] A. Glanc, IEEE Trans. on Elec. Dev. ED-16, No. 6, p. 540, June 1969.
[3] A. Glanc, et al., J. Appl. Phys. 35, No. 6, p. 1870 June 1964.
[4] C.B. Kittle, "Int. to Solid State Phys.", John Wiley and Sons, New York, 1971.
[5] M.B. Roitberg, et al., $BaTiO_3$-Seminar of the Academy of Sciences USSR, Moscow, 1973.
[6] O. Saburi, J. Phys. Sci. Japan 14, No. 9, September 1959.
[7] M. Hansen, "Construction of Binary Alloys", McGraw Hill, New York, 1958.
[8] M.B. Roitberg, et al., Sov. Phys. Crystallogra. 14, No. 5, p. 814, 1970.
[9] N. Karasikov and M.B. Roitberg, Ferroelectrics 33, p. 217, 1981.

Realization of the triple point of indium in a sealed glass cell

S. Sawada

National Research Laboratory of Metrology, 1-4, 1-Umezono, Sakura-Mura, Ibaraki, Japan

The triple point of indium was realized in sealed glass cells for the purpose of standard platinum resistance thermometry. Three samples of pure indium (about 700 g) from different sources were placed in cylindrical glass cells (external diameter, 35 mm; height, 200 mm; inner diameter of the thermometer well, 11 mm), which were sealed after evacuating lower than 0.01 Pa. The cell surrounded by a hollow block of aluminum was set in an electric furnace. The melting and freezing curves were obtained using three long stem-type Pt resistance thermometers which had been calibrated at the zinc point, the tin point, and the triple point of water. The estimated value of the triple point of indium from the melting and freezing plateaus gave 156.6296 °C ± 0.0003 °C on the IPTS-68. The range of super-cooling in the freezing point experiments was as small as 1 °C.

INTRODUCTION

The boiling point of water (100 °C), which is being used as one of the important defining fixed points of the International Practical Temperature Scale of 1968 (IPTS-68),[1] is less suitable as a defining fixed point, since it requires a complicated measurement of the vapor pressure and its error affects the reproducibility of the temperature, and also since the equilibrium condition at a boiling point may not be realized easily. In the text of the IPTS-68 the freezing point of tin is recommended as the alternative to the boiling point of water, but its higher temperature causes some problems as follows.

For instance, the calibration of a capsule type platinum resistance thermometer (PRT) used for low temperature measurement involves the measurements at the temperatures of seven fixed points below 0 °C besides that at 100 °C. In this case, the adoption of the freezing point of tin instead of the boiling point of water to get a better accuracy is not practically recommendable, because the capsule type PRT has platinum lead wires with a lead-glass seal which is not suitable for heating at the freezing point of tin. So, the precise calibration of a capsule type PRT at 100 °C has been performed by comparison in a furnace with a long stem type standard PRT which has been calibrated at the freezing point of tin.

This paper reports the construction and the performances of the triple point cells of indium which were developed recently at the National Research Laboratory of Metrology. The temperature of the triple point of indium is within the working range of the capsule type PRT and its reproducibility is expected to be as good as or better than that of the boiling point of water.

TRIPLE POINT CELL OF INDIUM

Use of a graphite crucible for a sample vessel is believed to be the most reliable method to realize a metal freezing point. In the text of the IPTS-68, a graphite crucible is adopted in the realization of all of the metal freezing points. In this case, particular attention must be paid to avoid degradation of the purity and oxidation of a metal sample; use of the highest purity graphite for the crucible; covering the metal surface with a graphite powder; keeping the crucible in a glass or quartz cylinder with a lid, and filling with an inert gas. Nevertheless, during a long term use, slight oxidation and contamination are expected which would cause an error in the fixed point temperature. The new triple point cell reported here will solve these problems completely.

The cell is basically a Pyrex cylinder, 200 mm in height, 35 mm in outside diameter, and with an axial thermometer well, 11 mm in diameter. Its volume is about 100 cm^3. For the preparation of the cell, it was washed successively with an acid, a detergent and distilled water and was dried, then the cell was evacuated and heated in an electric furnace at 350 C for about 10 hours, as shown in Fig. 1. Dry argon was admitted into the cell to about 0.1 MPa to prevent air from entering, then a proper quantity (about 700 g) of the indium sample was filled carefully through the upper tube which was opened by cutting. The tube was closed again and the cell evacuated to a high vacuum. The temperature was raised to a few degrees below the indium point in order to remove gases adsorbed on the sample, then the heater power was increased to melt the metal. The equipment was left in the above condition for several hours until gases in the cell were removed completely. The cell was sealed at the stem after cooling to room temperature.

SAMPLE METAL OF INDIUM

Three kinds of indium sample were purchased from three manufacturers. The nominal purity, the shape and the impurities in the sample are shown in Table 1. Three triple point cells were constructed with each of the cells containing the different samples of indium.

EXPERIMENTS

For protection against mechanical shock and to equalize the temperature distribution, the glass cell is held in an aluminum container. Then, as shown in Fig. 2, it is slipped into a glass cylinder, the upper part is filled with glass wool to reduce heat loss.

The electric furnace employed in the experiment was one that is used for the realizations of the zinc point and the tin point in the calibration of the standard PRT's at the NRLM. The furnace has two separate heaters (double coiled) and its temperature is regulated to the desired temperature by adjusting the heater power. An aluminum block was placed inside of the furnace to equalize the temperature distribution of the furnace, particularly around the cell. A PRT was

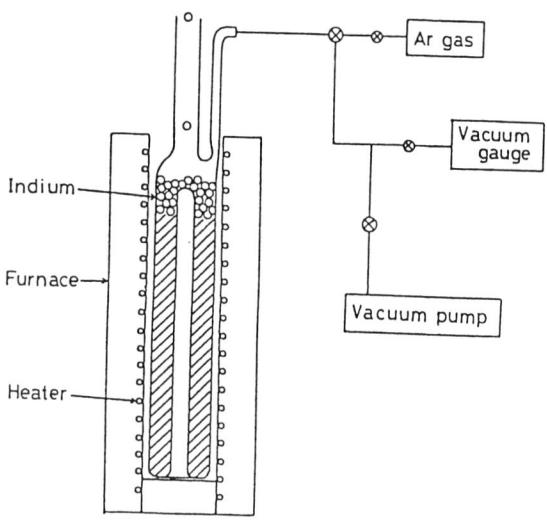

Fig. 1. Process for making the indium cell.

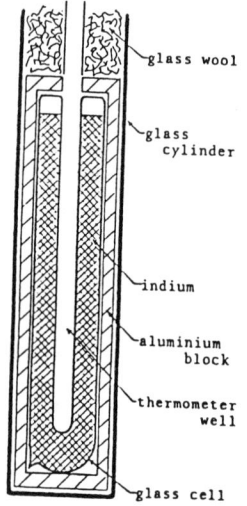

Fig. 2. Setup of the triple point of indium.

point cell of indium was placed in the furnace and its temperature was held a few degrees above the indium point until melting was completed. Then, the furnace was allowed to cool naturally by switching off the heater power, and when the temperature approached the

Table 1. Impurity contents in indium samples.

Cell	Makers	Purity (Nominal)	Shape	Impurity metals					
				Ca	Tl	Cu	Al	Mg	Si
1	A	99.9999 %	tear drop	<0.3	<0.3	<0.1	–	–	*
2	B	99.9999 %	ball (2~3mmø)	–~±	–	–	–	–~±	–~±	**
3	C	99.999 %	ball (2~3mmø)	–	–	±~+	–~±	–~±	–	**

*: chemical analysis   Unit: p.p.m
**: spectroscopic analysis
–: not detected
±: comparable with noise level
+: slightly detected

Table 2. Values of α, δ for platinum resistance thermometers used in this experiment.

PRT	Lead Wire mmø	α	δ
No 6707	Pt 0.4	3.9264538×10^{-3}	1.497 159 2
No 6708	Pt 0.4	3.9262264×10^{-3}	1.496 711 1
No 6401	Au 0.4	3.9230321×10^{-3}	1.497 600 5

$$t_{68} = t' + 0.045 \left(\frac{t'}{100}\right)\left(\frac{t'}{100}-1\right)\left(\frac{t'}{419.58}-1\right)\left(\frac{t'}{630.74}-1\right)$$

$$t' = \frac{1}{\alpha}\left(\frac{R_{t'}}{R_0}-1\right) + \delta\left(\frac{t'}{100}\right)\left(\frac{t'}{100}-1\right)$$

mounted in the block to monitor its temperature.

Three long stem type PRT's were employed in the precise measurement of the temperature of the cells. These thermometers were calibrated just before the experiment against the zinc point, the tin point, and the water triple point, and their constants, α, δ in the equation in the IPTS-68 were determined as shown in Table 2. One of them, No. 6401 which was found not to conform to the IPTS-68 specifications ($\alpha < 3.9 < 3.9250 \times 10^{-3}$) was used as well, but it did not cause any significant difference at the indium point as described later.

The Mueller bridge (Leeds & Northrup G-III) was used to measure the resistance of the standard PRT's, and the sensitivity of the measurement at the water triple point was 10 mm in deflection of a galvanometer versus one position change of the 10^{-5} dial of the bridge.

As for the adequacy of thermometer immersion in the cells, measurements were made by varying the depth of the thermometer for 2~3 cm in the thermometer well for each of the cells. No differences were obtained.

In the preparation for the measurement, the triple

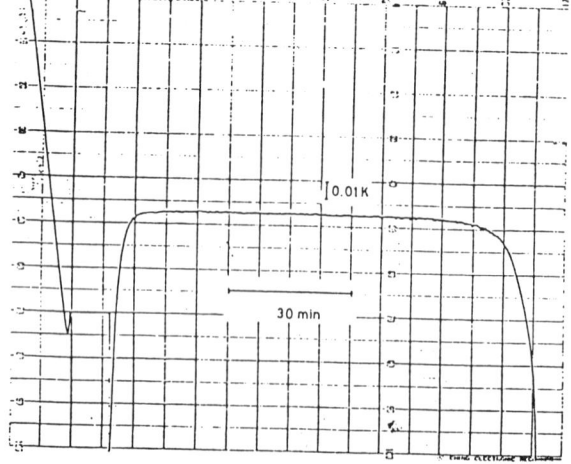

Fig. 3. Freezing curve

indium point the heater power was applied again. The furnace was so regulated that its temperature would be held a few degrees below the freezing point of indium. The temperature of the cell, after passing the supercool, increased very rapidly and reached the temperature plateau. Figure 3 shows the freezing curves of indium obtained in the experiment. The range of supercooling of indium in the freezing process was about 1 K, and its triple point was realized easily.

RESULTS AND DISCUSSION

1. Evaluation of the Temperature of Different Indium Samples

Figure 4 shows the melting curves and the freezing curves of the three cells that were obtained. As is evident, Cell 1 does not show any slope in its melting curve and the melting point is in good agreement with the freezing point. From this, it is estimated that the purity of Cell 1 is satisfactorily high. In comparison with Cell 1, the other two cells show some slope in their melting curves, and the temperature differences between the beginning of melting and its end are 3 mK for Cell 2, and 5 mK for Cell 3, respectively. These results indicate the presence of significant amounts of impurities. On the other hand, the freezing curves of all the cells show their flat plateaus. The temperatures of freezing points for Cell 2 and Cell 3 are lower than that of Cell 1 in conformity with amount of impurity present. Significant temperature differences were found to exist among the three cells.

2. Evaluation of the Temperature for Different Thermometers

Figure 5 shows the result of temperature measurement of the freezing point with three standard PRT's. R_{In} is the resistance of thermometer at the freezing point of indium, and R_0, at 0°C, was derived from the measurement at the triple point of water. Each point of R_0 in Fig. 5 indicates the averaged value of two measurements taken before and after the measurement of the indium point. In general, the plateau on the freezing curve can be maintained for 45 min to several hours by adjusting the temperature difference between the sample metal and the aluminum block in the furnace. No dependence is found among the measured values of the freezing point on the duration of the plateau. (45 min ~ a few hours). The short horizontal lines at the right end of Fig. 5 indicate the averaged values of R_{IN}/R_0 for each cell. The differences among them are consistent with those of Fig. 4, and they show that they do not depend on the thermometers used.

The temperature values of the triple point of indium were derived from the results of the measurements and are listed for each of the thermometers in Table 3. The temperature values on the IPTS-68 have a slight spread due to the different characteristics of the thermometers. This is because the thermometer No. 6401, which did not conform to the IPTS-68 specifications, was also included.

As is evident from Table 3, PRT No. 6607 shows slightly lower values than the others. It can be assumed that the constants and of No, 6607 could not have changed because of the high stability of R_0 values that were observed during the measurement. It is not clear from only this experiment whether the difference comes from systematic errors in the measurement or from the different characteristics of the thermometers. The results of No. 6401, which did not conform to the IPTS-68 specifications, do not show any significant differences from those of the others.

Among the estimated temperature of the triple point of indium in the three cells, there are significant differences such as -0.5 mK for Cell 2, -1.3 mK for Cell 3 compared with Cell 1.

In conclusion, the temperature of the triple point of indium is estimated to be the temperature obtained with Cell 1 in which the metal is supposed to have the

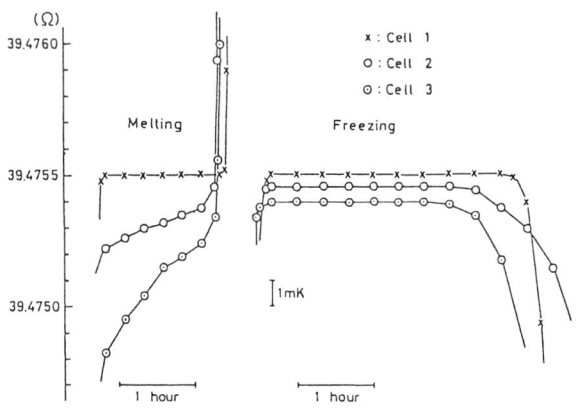

Fig. 4. Plateaus of melting curves and freezing curves.

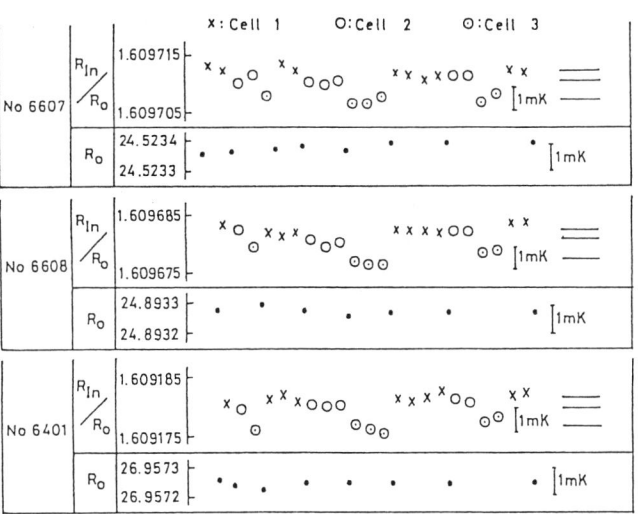

Fig. 5. R_0 and R_{In}/R_0 measured by three platinum resistance thermometers for three different indium samples.

Table 3. Triple point temperature values for three samples measured by three platinum resistance thermometers.

PRT	Cell 1		Cell 2		Cell 3	
	Temp (°C)	Std dev (mK)	Temp (°C)	Std dev (mK)	Temp (°C)	Std dev (mK)
No 6607	156.6293	0.16	156.6288	0.20	156.6280	0.17
No 6608	156.6297	0.20	156.6293	0.32	156.6284	0.27
No 6401	156.6298	0.17	156.6293	0.30	156.6285	0.21
Mean	156.6296	0.30	156.6291	0.38	156.6283	0.32

higest purity, and its value is as follows,

$$156.6296 \pm 0.0003 \text{ °C}.$$

This value also agrees with the temperature value of the freezing point of indium at 1 atm pressure, which is one of the secondary reference points of the IPTS-68, within the limit of error after applying a pressure correction to 1 atm. (0.0049 °C/atm).[2,3]

3. Effect of Thermal Contact

The degree of thermal contact of the thermometer with its surroundings has a great effect on the measurement accuracy. As the cells reported here were constructed for the calibration of a capsule type PRT, their thermometer wells are 11 mm in diameter which leave considerable space for a stem type PRT (8 mm in diameter). In order to investigate the effect of the thermal contact, the measurements were performed with and without silicon oil in the space. As shown in Fig. 6, the effect of the 3 mm space is negligible.

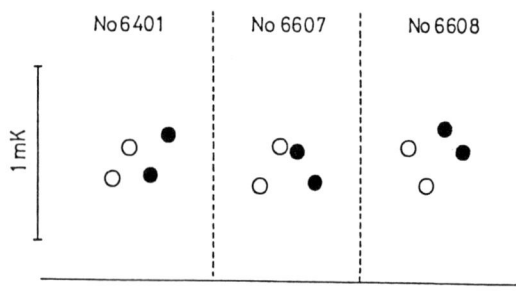

Fig. 6. Influence of thermal contact in the thermometer well. ○: with oil ●: without oil

The existence of good temperature uniformity in the furnace and good thermal contact are also supported by the fact that No. 6401 thermometer showed the same result as the others notwithstanding its gold lead wires which are presumed to give a larger heat flow than the platinum lead wires of the others and the fact that the temperatures of the thermometer with and without silicon oil in the thermometer well were the same.

CONCLUSIONS

Indium triple point cells were constructed entirely of borosilicate glass instead of graphite which is conventionally used in metal freezing point cells. The triple point of indium was investigated as a possible alternative for the boiling point of water.

Its advantages are as follows:
(1) As the cell is seald in vacuum, the sample is free from contamination and oxidation,
(2) As it is portable, an intercomparison by circulation is possible,
(3) Since the range of supercooling of indium is only about 1 K, which is much smaller than that of tin, the measurement procedure is also simpler.

The triple point of indium can be realized with high accuracy near 100°C, the triple point cell is expected to provide an important reference point for practical use.

REFERENCES

1. The International Practical Temperature Scale of 1968; Metrologia 5, 35 (1969).
2. E. H. McLaren and E. G. Murdock; Can. J. Phys. 38, 111 (1960).
3. E. H. McLaren; Can. J. Phys. 36, 1131 (1958).

A small transportable indium cell for use as a temperature reference*

Magda Hanafy, M. R. Moussa, and M. H. Omar**

National Institute of Standards (NIS) Cairo, Egypt

Preliminary melting point results are reported on two small-size indium cells which were constructed for use as transportable temperature reference cells. The cells are made of Teflon (external diameter 23 mm, height 65 mm, and inner diameter of the thermistor well 3.5 mm) and contain about 70 g of indium. The space above the metal is filled with dry argon gas under a pressure of about one atmosphere. For measurement, each cell was inserted in a thermally insulated copper block whose temperature was regulated by means of electric heaters and a precision proportional temperature controller. Melting curves were obtained using a calibrated precision thermistor. Initial tests using indium of 99.99% purity gave a melting point of 156.65 °C ± 0.02 °C on the International Practical Temperature Scale of 1968 (IPTS-68). The simplicity of cell design and the ease in its manipulation and handling make it an useful temperature reference device for use in industry and in research laboratories.

1. INTRODUCTION

The use of the indium melting (or freezing) point as an alternative to the water boiling point has been widely recommended since it lies midway between the freezing point of tin and the boiling point of water. The realization of such a point is of practical importance for calibration purposes. Also, there is a definite need for having small transportable temperature reference cells which can be used in industry and in research as calibration devices. A joint program was initiated between the U.S. National Bureau of Standards (NBS) and the National Institute of Standards (NIS) for developing transportable cells using different substances, namely, mercury, normal heptane, Freon-11 and indium, to cover a portion of the temperature scale. This paper reports the construction and performance of two indium cells prepared at NIS.

2. APPARATUS AND MEASURING PROCEDURE

a. The Cell

Figure 1 shows the main parts of the cell as well as the cell dimensions. The cells used in this work were made of Teflon while the cap and thermistor well were made of nylon. The design is similar in principle to the gallium cells used by Thornton[1]. Care was taken in machining the cell parts to insure that no dirt or contamination was introduced. The different parts of the cell were washed in an ultrasonic bath (using successively detergent solution, distilled water, and ethyl alcohol) before the filling process took place.

A special filling system was used which is shown schematically in Fig. 2. The filling system is housed in a metal chamber which has a provision for evacuation and a controlled flow of dry argon gas. The empty cell was inserted in an aluminum block heated electrically (heater 2) to a temperature of about 180 °C. A glass tube containing the pure metal was heated by another heater (heater 1) until the indium melted and flowed into the cell under the combined action of gravity and argon gas pressure.

The cell was allowed to cool in an atmosphere of dry argon gas. Then, the inlet and outlet holes were closed by a matching teflon rod glued by a fast drying epoxy resin. The upper nylon cap was then glued by the same epoxy resin. Tests on this cell design indicated that the nylon parts started to deteriorate during the heating of the cell to the indium temperature.

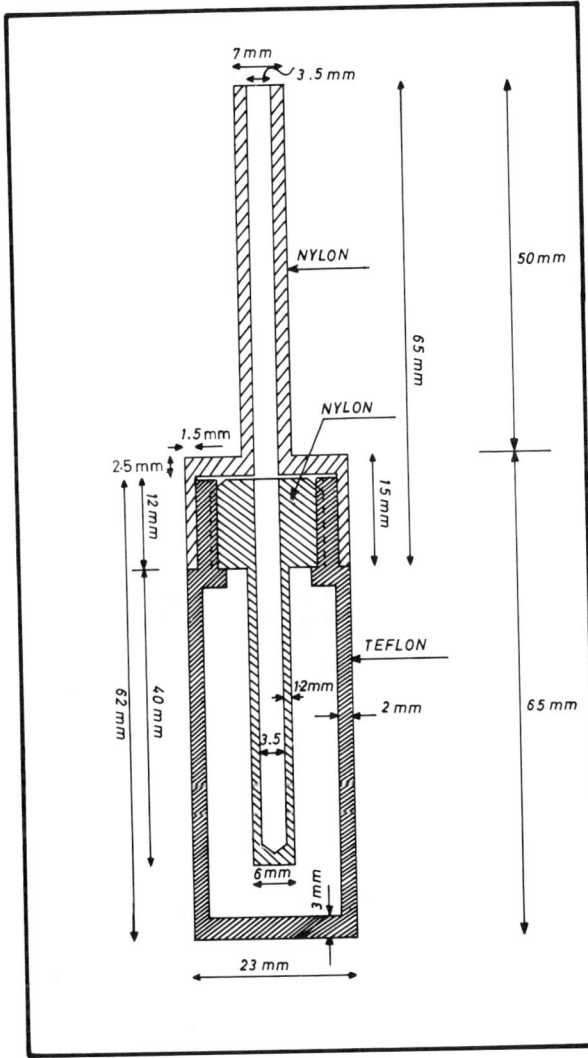

FIG. 1. Indium cell.

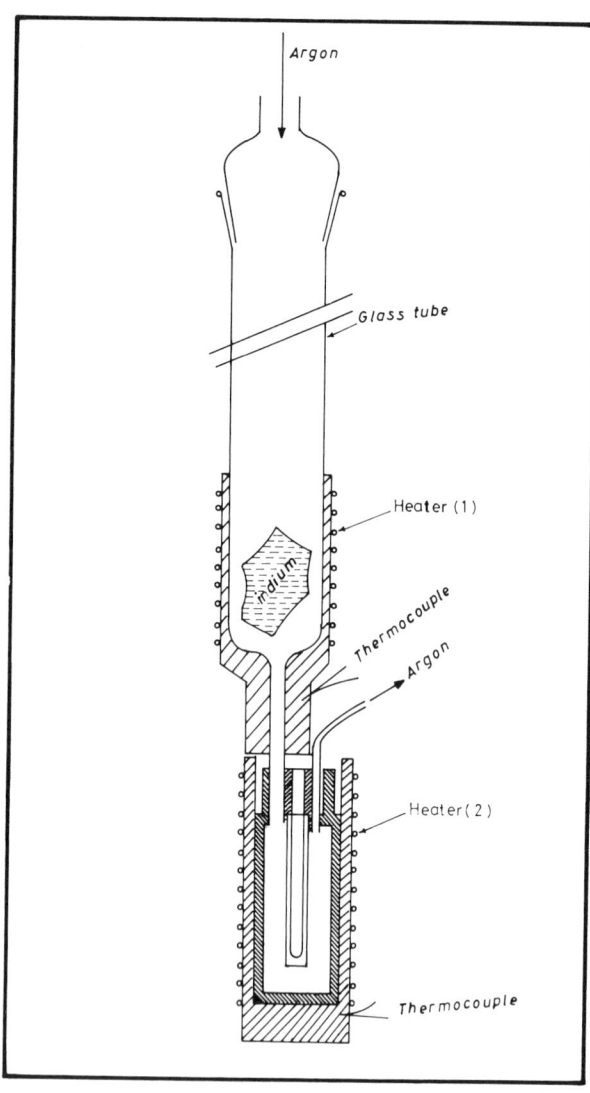

FIG. 2. Indium filling system.

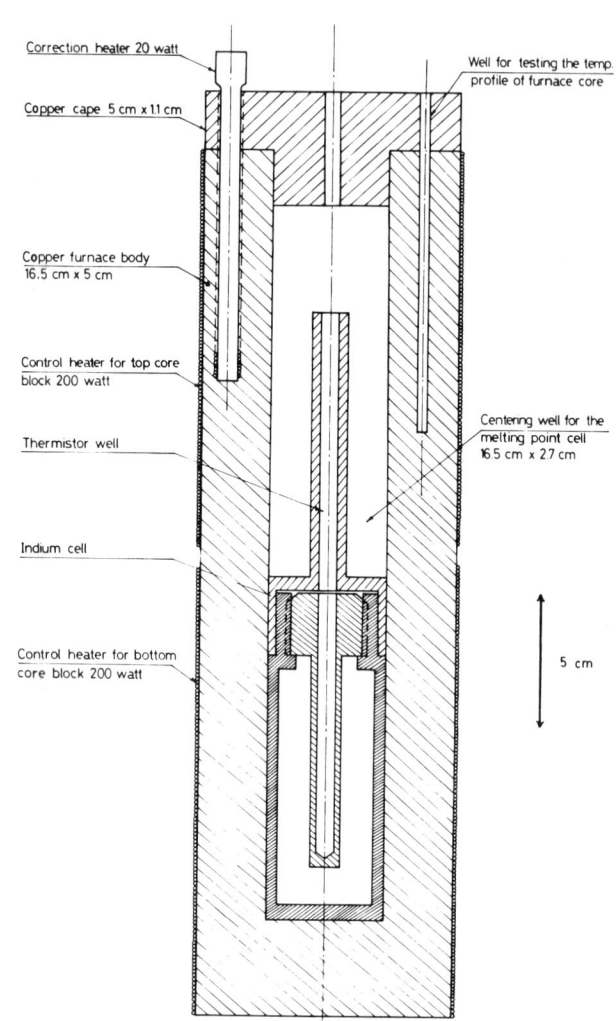

FIG. 3. Schematic of furnace body and core.

b. The Furnace

The cell was inserted in a thermally insulated copper block around which two heaters are wound non-inductively. The maximum power dissipation in each is 200 watts. A small correction heater (20 watt) was inserted at the top of the block to compensate for heat losses. Details of the furnace are shown in Fig. 3. A precision proportional temperature controller was used and the block temperature could be maintained within 0.02 °C of a preset value.

c. The Temperature Measuring Circuit

Two measuring circuits which operate in parallel are shown in Fig. 4. In the first, a Keithley (Model 225) constant current source is used to supply the required current to the thermistor circuit. The thermistor was connected in series with a 10 kΩ precision resistor***. The voltages across the thermistor and the resistor were measured by means of a Fluke digital voltmeter (Type 8502A). From these measurements the thermistor resistance was computed. In the second, a precision voltage source (Knick, Type S14) was used to counter balance the voltage generated across the thermistor. The potential difference was amplified by a Keithley null-detector (Type 150B). The amplified signal was fed to a chart recorder.

Thermistors**** used in this work are of the bead-in-glass probe type with an external diameter of about 2 mm. Their nominal resistance at room temperature is of the order of 120 kΩ which drops to about 1.3 kΩ at the indium point. Provision was made to reverse the current in the entire circuit and the average of voltage values was determined.

d. Measuring Procedure

Melting curves were obtained in the usual way. In the melting experiment, when the cell temperature very nearly reached the indium point, the block temperature was held slightly above the melting point ($\simeq 0.5$ °C). While observing the plateau on the recorder, several checks were made to insure that the current in the thermistor was stable. With the described measuring circuit, it was possible to monitor temperature variations of about 5 m°C.

3. RESULTS

In this work several runs were carried out on the melting of the two indium cells. Both cells were identical in form and contained indium of 99.99 percent purity.***** The first cell (No. I) contained 70 gm of indium while the second cell (No. II) contained 65 gm.

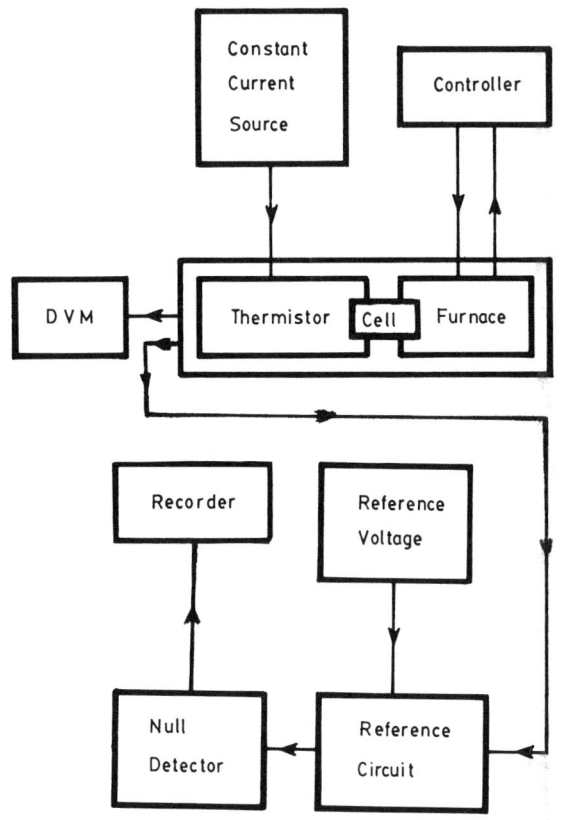

FIG. 4. Block diagram of the used circuit.

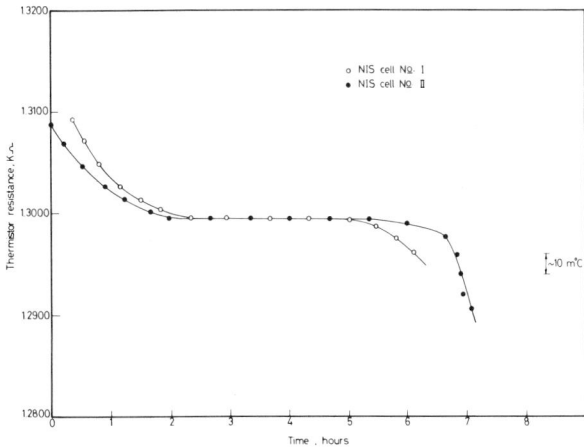

FIG. 5. Indium melting curves of 4N grade using DA103 thermistor type N-J-08817.

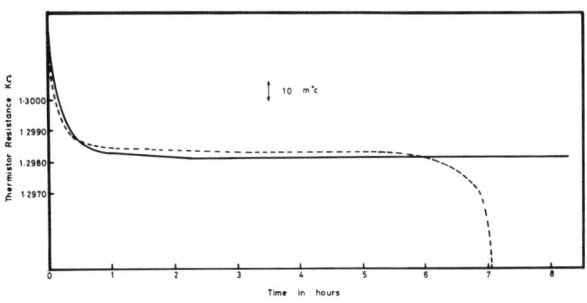

FIG. 6. Indium melting curves of cell no. II after recalibration of the thermistor.

Table I. Indium Melting Point using two different cells filled with the same sample of indium of nominal purity of 99.99 percent

Date	Melting Temperature of Indium
For Cell No. I	
16-3-81	156.66
22-3-81	156.65
28-3-81	156.66
21-4-81	156.66
22-4-81	156.66
7-6-81	156.64
For Cell No. II	
22-4-81	156.66
25-4-81	156.65
26-4-81	156.66
28-4-81	156.64
7-6-81	156.61
8-6-81	156.63
27-7-81	156.62
28-7-81	156.63
29-7-81	156.63
4-8-81	156.62

Several thermistors were used and a study on their behavior was conducted. It was found that the thermistors were rather stable during measurements and could reproduce well up to a power dissipation of 10 μW. One thermistor was found to give non-reproducible results, but it was discovered later that it had been accidentally subjected to an excessively high voltage.

Calibration of the thermistors was done against a standard platinum thermometer which was calibrated at the NBS in 1980. The uncertainty of temperature measurement during each individual melting is estimated to be less than 5 m°C. The maximum difference between melting points for different rates of heating (i.e., 1 to 7 hour total melting times) did not exceed 0.02 °C.

In total, 16 runs were carried out, 6 on cell No. I and 10 on cell No. II. The data obtained are shown in Table I. Melting curves for both cells are shown in Fig. 5. It was possible in some cases to maintain a plateau for more than seven hours as illustrated in Fig. 6.

4. CONCLUSIONS

Although relatively impure indium was used, the described cell proved to be useful as a temperature reference cell with temperature accuracy of the order of ±10 m°C. The cell design proved to be convenient and easy to fabricate in laboratories with modest facilities.

Higher accuracy could be obtained by using higher grade indium, and work is currently in progress in this direction. Also it is planned to make the cell entirely of Teflon as nylon was found to deteriorate. Teflon proved to be adequate for making such cells but care must be taken in designing the cells to reduce the effect of its relatively low thermal conductivity with respect to the small thermal capacity of such a cell. Tests will also be conducted using a small platinum thermometer instead of a thermistor.

When considering higher accuracies, one must correct for the effect of deviation from 1 atm of argon gas pressure on the melting point; McLaren found that this effect is about 0.0049 °C/atm[3]. Corrections must be applied also for the hydrostatic head of liquid indium, and for thermometer self-heating.

5. ACKNOWLEDGEMENTS

It is a great pleasure to acknowledge the continuous help and support from the NBS staff in the Temperature and Pressure Measurements and Standards Division, namely, Dr. J. F. Schooley, B. W. Mangum and G. T. Furukawa.

We are also indebted to Prof. Dr. Mahmoud Mokhtar, consultant to the NBS-NIS project, for his contribution and useful suggestions during the progress of this work

* This work constitutes one part of project No. G-198 which is a joint project between National Bureau of Standards (NBS), Washington, D.C. and National Institute of Standards (NIS), Cairo, Egypt.

** Permanent address American University in Cairo, Egypt.

REFERENCES

*** Manufactured by Julie Research Labs, Type 211W-6TST

**** Thermistor Type CSP 60 DB 124M/S, manufactured by Thermometrics, Inc., Edison, N.J. 08817.

***** Indium received from Indium Corporation of America.

[1] D. D. Thornton: The gallium melting point standard: A determination of the liquid-solid equilibrium temperature of pure gallium on the International Practical Temperature Scale of 1968, Clinical Chemistry 23, 719 (1977).

[2] E. H. McLaren: Can. J. Phys. 36, 1131 (1958).

[3] B. W. Mangum: The gallium melting point standard: Its role in our temperature measurement system, Clinical Chemistry 23, 711 (1977).

[4] S. Sawada: Realization of the triple point of indium in sealed glass cell, CCT/80-21.

The use of the cadmium point to check calibrations on the IPTS

J. V. McAllan and J. J. Connolly

CSIRO Division of Applied Physics, Sydney, Australia 2070

In this laboratory a resistance thermometer usually is calibrated on the IPTS using two separate freezes of zinc, cadmium, and tin. The cadmium measurements can reveal systematic errors which may affect the other points differently, e.g., through inferior electrical insulation. There have been design changes in the products of some manufacturers over the years, making a check on systematic errors highly desirable, and routine repetition of the zinc and tin points is not an effective method for detecting such errors. The determinations of the cadmium freezing temperature show the reproducibility of the IPTS as a working scale transmitted under routine calibration procedures. Details are given of the procedures used and of variations found among 43 IPTS thermometers of eight types which gave a mean cadmium freezing temperature of 321.108 14 °C (S.D. 0.71 mK). Thermometers made by Meyers were on average 0.3 mK higher than the mean but no other systematic effects were found. Industrial standard thermometers ($\alpha = 0.003\,85$) gave a mean temperature 1.0 mK below that of the IPTS thermometers and it is suggested that the IPTS restrictions on thermometer constants above 0 °C are unnecessarily severe. Systematic errors in cadmium point determinations on the IPTS are discussed.

ROUTINE CALIBRATION PROCEDURES

When fixed points used for resistance thermometer calibrations are investigated initially by a specialist temperature measurement laboratory, extreme care is taken over the measurements. However, in routine thermometer calibrations such attention (and time) is not lavished on measurements and the laboratory uses a simplified procedure which is assessed to give a calibration on the IPTS within the required accuracy. Usually, platinum resistance thermometers are measured at the appropriate IPTS fixed point temperatures, sometimes by direct measurement such as a zinc point determination, and sometimes by comparison with a calibrated thermometer. An assessment of calibration accuracy is based primarily on the reproducibility of the fixed points in the initial investigation by that laboratory, and little account is taken of subsequent variations in thermometer design which can produce changes such as inferior electrical insulation at high temperatures or increased thermal conduction errors.

This laboratory feels that it is desirable to know the accuracy of a temperature measurement, intermediate between fixed point temperatures, such as would be made by the user of the thermometers which we calibrate. For several years, we have calibrated 'long stem' thermometers for use above 0 °C at the freezing point of cadmium as well as at the freezing points of zinc and tin. As a result there has been an accumulation of a substantial number of determinations of the cadmium freezing temperature under routine conditions of calibration.

The procedures used to establish the freezing point of Cd are much the same as those recommended for the zinc point in the IPTS (1968). The sample crucible is made from pure graphite with a graphite or glass well and allows approximately 165 mm immersion of the midpoint of the thermometer sensor below the metal surface. The induced freezing technique[1] is used, where the thermometer is removed from the sample after the freeze nucleates, and is replaced after cooling in air for about one minute.

For most of the thermometers the resistance was monitored with a resistance bridge[2,3] operating at 320 Hz and a few dc measurements were made with a Leeds and Northrup G4 Mueller bridge. An ac bridge permits continuous recording and easy checking of current heating effects and immersion errors. At any freezing point the dc calibration was higher than that using ac by from 4 to 7 parts in 10^7, and at the cadmium point the overall dc and ac calibrations differed by less than 0.2 mK. Some calibrations before 1970 involved only a Mueller bridge, and after the ac characteristics were well known some later thermometers had no dc measurement but were calibrated with an 80 Hz bridge.

The melting behaviour of the cadmium samples is checked periodically and in all cases the 99% melting range has been less than 1 mK and no part of the melt has been more than 1 mK away from the freezing value.

THE CADMIUM POINT AS A MONITOR OF THERMOMETER VARIABILITY

The calibrations of 43 thermometers of IPTS-68 quality have been analysed to look for significant differences. All thermometers were annealed at 470 °C to 500 °C before calibration (our usual procedure) and thermometer details are given in the Table. Thermometers constructed by N.M. Bass of this Division (formerly the National Measurement Laboratory) were intended for use in other sections of this laboratory and were given routine calibrations without special treatment. The thermometers reported in the table are of eight different designs, and three separate manufactures of the Meyers single coil design are included.

The mean cadmium freezing temperature determined by all the thermometers was 321.108 14 °C with a standard deviation of 0.71 mK. The largest deviation from the mean was 1.7 mK. The eight thermometers constructed by C.H. Meyers gave a mean of 321.108 44 °C with a standard deviation of 0.55 mK, which is significantly different from the rest at the 90% confidence level, but most of the difference is due to two thermometers which gave values more than 1 mK high. No other significant variations due to manufacture or style of construction were found and in particular, while the results from Leeds and Northrup Meyers single coil thermometers were 0.33 mK high this was not significant because of the small sample, and those made by Bass were 0.23 mK low which again is not significant.

There was no significant variation of cadmium temperature with the IPTS calibration constants α or δ, but while the mean δ value was 1.4966 all thermometers constructed by Meyers had δ values below 1.4965 and

TABLE I

Details of IPTS Quality Thermometers

Manufacturer	Construction type	Number tested	Cadmium temperature
			321.xxxx
Leeds & Northrup Co	Meyers design - single coil on mica	4	1085
Leeds & Northrup Co	Meyers design - coiled coil on mica	3	1076
C.H. Meyers	-single coil on mica	8	1084
N.M. Bass	Meyers design - single coil on mica	8	1079
N.M. Bass	Strelkov design, double helix silica	2	1082
R.E.C.	Type 162C, coil in ceramic, (s.s. sheath)	4	1080
R.E.C.	E109 sensor, coil in ceramic mounted by N.M. Bass	4	1082
Chino	NRLM design, coiled coil on silica	2	1079
Minco	coil in ceramic, (s.s. sheath)	2	1086
H. Tinsley & Co	Barber design, coil in glass	6	1082

all the REC-E105 and Minco ceramic insulated sensors had δ values above 1.4969. Thermometers of other types had δ values above and below the mean.

There were no errors due to thermal conduction, for example from the use of stainless steel sheaths instead of glass or silica. However it was noticed that thermometers with a narrow stainless steel sheath gave low cadmium point temperatures, unless an aluminium sleeve about 50 mm long was inserted into the glass protective well to surround the sensor region. This reduced the zinc point value and hardly affected the other points, so that the calculated cadmium temperature increased to the usual value. It was considered that the sleeve prevented air in the protecting well from reaching the sensor region and that the anomaly (high by as much as 5 mK in zinc) was caused by surface oxidation of the steel. Thermometers with wide stainless steel sheaths did not show this effect, presumably since there was no room for air in the well, but they did show larger thermal conduction effects than the other thermometers. This error in the zinc point measurements was not detected through problems in the zinc point, but through the unusual cadmium point values, illustrating the utility of this calibration procedure.

The ceramic insulated sensors were no different from other sensors at the cadmium point, in contrast to the systematic variations of about 1 mK which are measured in the range 0 to 250°C using comparison bath methods, as reported elsewhere in this symposium[4].

THERMOMETERS WITH LOW α COEFFICIENTS

The IPTS-68 limit of 0.003 925 on α was set to ensure that the scale is as reproducible as possible. The above methods have been used to calibrate a group of six thermometers from Leeds and Northrup, used as industrial standards up to 500°C. These have stainless steel sheaths and ceramic insulated sensors, and have α values with mean 0.003 856 3 (standard deviation 0.000 000 5) and low δ values with mean 1.4954 (S.D. 0.000 4). These thermometers gave a mean cadmium temperature of 321.1072°C (S.D. 1.7 mK), but two of them showed some instability at the zinc point and had values 2 and 4 mK respectively below the mean cadmium value. The other four had a mean of 321.108 15°C (S.D. 0.50 mK).

This indicates that thermometer stability during calibration is very important in obtaining a good representation of the IPTS, and also that the purity of the platinum is of little significance in the range 0 to 420°C. The departures reported at the lower part of this range are almost certainly due to differential expansion effects and are not related to purity. Measurements of industrial thermometers at higher temperatures, reported elsewhere in this symposium[5], support this view, and indicate that at high temperatures contamination of highest purity platinum may produce more serious effects than with platinum having α of 0.003 85.

THE CADMIUM POINT VALUE ON IPTS-68

The recommended value for the cadmium point is based on the work of McLaren[5,6,7], and his mean value of 321.032 50 on IPTS-48 is equivalent to 321.107 75 on IPTS-68, and has a standard deviation of 0.75 mK. The agreement with the present result of 321.108 15 (S.D. 0.71 mK) is very satisfactory. Values at this laboratory over 20 years have always indicated a slightly higher value than McLaren's, although well within the experimental uncertainties.

There are two sources of error which might cause such differences. Firstly, the melting range of the cadmium samples appears slightly smaller than that used by McLaren, indicating that our metals are purer and presumably have a higher freezing temperature. The present results could be higher than McLaren's by at most 0.3 mK from this cause, and presumably would be closer to the correct cadmium point temperature.

The second error would be due to a low thermometer insulation resistance at the zinc point which is not present at the cadmium point. The faulty zinc value would produce a high δ and hence a high value for the cadmium point temperature. Most of McLaren's 11 thermometers were Meyers style single coil made by Leeds and Northrup and by Meyers; the present results for thermometers from these sources are higher than our average. If they are considered separately they increase the discrepancy to 0.68 mK and if they are excluded the average only falls by 0.08 mK. Consequently, while an error of this type may be present it certainly must occur in nearly all the thermometers we have tested.

CONCLUSION

The use of the cadmium point in routine IPTS calibrations as well as the zinc and tin points has provided a monitor of systematic errors in the calibration techniques, detecting small errors in the other fixed point calibrations. It has shown also that the IPTS in this temperature region is transferred from the primary calibration laboratory with high reproducibility, a standard deviation of 0.7 mK being obtained for 43 thermometers of eight different designs.

Thermometers constructed by C.H. Meyers probably give temperatures a few tenths of a mK above other thermometers in this region, but no other significant differences were detected. Thermometers of good quality with industrial standard α coefficients (0.003 85) gave temperatures within 1 mK of the IPTS standards at the cadmium point.

The mean IPTS value obtained for the cadmium point 321.108 15°C (S.D. 0.7 mK), is almost identical to the value originally obtained by McLaren on which the IPTS recommended value is based, but uses a wider variety of thermometers and slightly purer metal.

REFERENCES

1. E. H. McLaren, Temperature, Its Measurement and Control in Science and Industry $\underline{3}$ (1), 185 (1962).
2. A. M. Thompson and G. W. Small, Proc. IEEE $\underline{118}$, 1622 (1971).
3. J. J. Connolly, J. V. McAllan and G. W. Small, Temperature, Its Measurement and Control in Science and Industry $\underline{4}$ (2), 1487 (1972).
4. J. J. Connolly, "The calibration characteristics of industrial platinum resistance thermometers", this symposium.
5. J. V. McAllan, "Practical high temperature resistance thermometry", this symposium.
6. E. H. McLaren, Can. J. Phys. $\underline{35}$, 1086 (1957).
7. E. H. McLaren, Can. J. Phys. $\underline{36}$, 1131 (1958).
8. E. H. McLaren, Can. J. Phys. $\underline{37}$, 422 (1959).

Investigation of the freezing temperature of cadmium

George T. Furukawa and Earl R. Pfeiffer

National Bureau of Standards, Washington, D.C. 20234

The freezing points of five cadmium cells, which were prepared using samples from two different sources, were found to agree within ±0.1 m°C. The calibration, during a single freeze, of six standard platinum resistance thermometers (SPRT's) at temperatures all within ±0.1 m°C is demonstrated. Measurements with eight SPRT's gave an average freezing-point temperature of 321.1082 °C. The results show that the cadmium point is suitable for testing the consistency of calibration of SPRT's on the International Practical Temperature Scale of 1968 and that it is also a suitable alternative to the zinc point for calibrating SPRT's that are used below 321 °C.

INTRODUCTION

At the National Bureau of Standards (NBS) capsule-type platinum resistance thermometers are not calibrated at the zinc point as is routinely done with long-stem type platinum resistance thermometers, because of possible damage and the electrical leakage across the metal-glass electrical lead seal at such high temperatures [1]. (Henceforth, any reference to platinum resistance thermometer indicates a thermometer of standards quality that meets the International Practical Temperature Scale of 1968 (IPTS-68) specifications [2]. For convenience such thermometers will be referred to by the abbreviation SPRT.) Instead, for the solution of the quadratic relation,

$$W(t') = R(t')/R(0°C) = 1 + At' + Bt'^2 , \quad (1)$$

for the SPRT temperature scale above 0 °C, the coefficient B is assumed, based on the average B of about 200 long-stem type SPRT's. In Eq. (1), R is the electrical resistance of the SPRT at temperature t' on the SPRT scale. The coefficient A is then obtained from calibrations at the triple point of water and the tin point. However, the range of values of B of the 200 SPRT's suggests, in using the average value of B, an uncertainty at 100 °C of about ±1 m°C and an uncertainty in α of about ±3 x 10^{-8}(°C)$^{-1}$, where

$$\alpha = A + 100B . \quad (2)$$

Moreover, any uncertainty in α would also be propagated below 0 °C down to the lower temperature limit 13.81 K of the SPRT scale [3]. (t = T - 273.15K; the temperature units °C and K will be used interchangeably in this paper [2]).

To reduce the above uncertainty, the cadmium freezing point (321 °C) was selected as a possible alternative to the zinc point for the calibration of capsule-type SPRT's. Also, the cadmium point would be an excellent calibration point for checking the calibration of long-stem type SPRT's obtained from measurements at the triple point of water and the tin and zinc points. Error of calibration at either the tin or zinc point of about 0.5 m°C or larger is expected to be detected from measurements at the cadmium point. This paper describes the investigation on the reproducibility of cadmium freezing points and the results of calibration of a set of long-stem type SPRT's at the triple point of water and tin, zinc, and cadmium points.

1. Cadmium Samples

The freezing-point investigations were conducted on cadmium samples from two sources. One sample, designated SRM 746, was received from the Office of Standard Reference Materials (OSRM) of the NBS. The material is normally issued by the OSRM as a vapor-pressure standard in the form of rods 6.4-mm diameter and 64-mm length. Enough rods were obtained to assemble two freezing-point cells (see Section 3). The SRM 746 samples were prepared by a commercial supplier from a selected lot of cadmium, which was first vacuum distilled twice and then zone refined by at least twenty passes. The zone refined units were homogenized into one lot and cast into billets which were then extruded into 6.4-mm diameter rods using a tungsten carbide die. The rods were cut into 45.7-cm lengths, etched, washed, and dried. The rods were received by the OSRM packaged in plastic tubes. On the basis of chemical analyses (emission and mass spectrometric) and residual resistivity ratio (ratio of electrical resistance at 0 °C to that at 4.2 K) the SRM 746 samples are considered by the OSRM to have a purity greater than 99.999 percent (see reference 4 for more details on the preparation and chemical analyses of SRM 746). The second sample was obtained from COMINCO in the form of rods (1.3-cm diameter and 20-cm length) from which three freezing-point cells were prepared. The supplier claimed the sample to be 99.9999 percent pure. No chemical analysis was supplied with the sample.

2. Assembly of Cadmium Freezing-Point Cells

The assembled cadmium freezing-point cells with borosilicate-glass holders (Fig. 1) were very similar in dimensions and arrangements of components to those of tin and zinc [1]. The procedure for filling the graphite cells with cadmium samples by melting under protective argon atmosphere using induction heating was similar to that used previously with zinc samples [4]. Similar to zinc, the vapor pressure of cadmium (about 15 Pa [5]) at its melting point is high enough to require argon gas to reduce vaporization losses. (For details of the filling procedure and assembly of cadmium freezing point cells, see reference 4.)

The two freezing-point cells that were assembled using SRM 746 and the three cells using samples obtained from COMINCO are identified in Table I. In order to accommodate the special holders for capsule-type SPRT's which are 11.1 mm O.D. [6], the thermometer wells of two graphite cells (COM 78-1 and COM 78-2) were 11.2 mm I.D. The wells of other three cells had 7.5 mm I.D. However, standard long-stem type SPRT's which have about 7.4 mm O.D. sheaths were employed for comparison of the cells and for obtaining the freezing point on the IPTS-68. Hence, special graphite sleeves of 7.5 mm I.D. were fitted snugly inside the larger wells to provide better thermal contact for the SPRT's.

3. Measurement Apparatus

3.1 Furnace for Melting and Freezing

The tube furnace that was used with the cadmium freezing-point cells was the same as that used previously with aluminum-point cells [7]. The furnace had four temperature-control zones. The temperatures were controlled automatically employing thermocouples and heaters associated with each zone. The main zone where the graphite cell was located was about 20 percent longer than the cell. There were one guard zone at the bottom and two guard zones at the top where most of the heat loss occurred. The guard zones at the bottom and top were able to maintain the temperature of the main zone uniform to ±0.01 °C at 320 °C.

3.2 Thermometry

For temperature measurements, ten long-stem type SPRT's which meet IPTS-68 specifications were used. The "heads" of two of the SPRT's were modified to accommodate BNC connectors and were used with a 400 Hz AC bridge designed by Cutkosky at the NBS [8]. These two SPRT's were calibrated in terms of the IPTS-68 several years ago. A strip-chart recorder was connected to the output of the bridge and the tracings on the chart were estimated to better than 1 micro-ohm (about 10 μK). The AC measurement system was employed principally for comparing the freezing points of different cells and for recording the freezing curve. The AC measurement system was used also for monitoring the cadmium freeze. Where freezing points of the cells were compared, resistance values that correspond to "zero power" were used in the comparison to avoid the effect of variations in the self heating of the SPRT. The zero power values were obtained from measurements at 1 and $\sqrt{2}$ mA by linearly extrapolating to zero current the readings that correspond to the square of the two currents [6]. This procedure was followed for both the cadmium-point cell and the triple point of water cell readings.

The other eight SPRT's were newly calibrated in terms of the IPTS-68 and were used with one of the cadmium-point cells (OSRM 78-2) to obtain the freezing point on the IPTS-68. The SPRT resistances were measured with a DC current comparator bridge in conjunction with a stable reference resistor [9]. A strip-chart recorder was also used with the bridge and readings were obtained at 1 and $\sqrt{2}$ mA and the zero-power values were calculated. The tracings on the strip-chart recorder were estimated to 10 μK or better. To obtain the measured resistance ratios, R(Cd)/R(0°C), at the cadmium point for the SPRT's, measurements on the thermometers were also made at the triple point of water. For the final value of the freezing point of cadmium, adjustment was made for the hydrostatic head of the liquid which amounted to 17.5 cm (0.845 m°C). For comparison, the temperature value of the cadmium point was determined also with the AC measurement system and the two SPRT's with BNC connectors.

3.3 Control and Measurement of Pressure

The freezing-point pressure was maintained at 1-atm pressure within about ±10 Pa by controlling the helium-gas pressure in the cell by means of a manostat, similar in design to that described by Brombacher, et al. [10]. Helium gas was used in the freezing-point cell to enhance thermal conductivity and to provide an inert atmosphere for the cadmium sample. For the investigation of the effect of external pressure on the freezing point the manostat was operated at other pressures, also with ±10 Pa stability. Pressures were read using a calibrated dial gauge. The estimated uncertainty of the readings was about ±10 Pa.

4. Experimental Procedures

For comparison of the freezing points or freezing-plateau temperatures of the cadmium-point cells, a

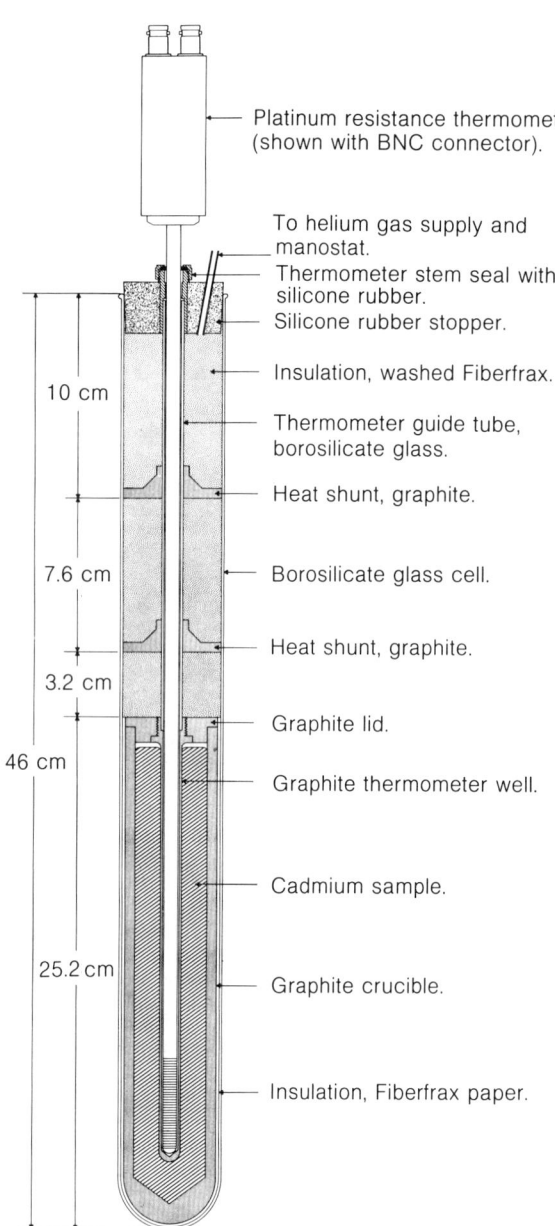

FIG. 1. Freezing-point cell of cadmium.

Table I

Identification of Cadmium Samples and Designation of Cadmium-Point Cells

Cell Designation	Sample	Approximate Mass g	I.D. of Thermometer Well, mm
OSRM 78-1	SRM 746	1400	7.5
OSRM 78-2	SRM 746	1400	7.5
COM 78-1	COMINCO	1670	11.3
COM 78-2	COMINCO	1670	11.3
COM 78-3	COMINCO	1400	7.5

complete or almost complete freezing curve was obtained for each of the cells. As mentioned earlier, for these comparison measurements the AC bridge was used with an SPRT. Readings at approximately the maximum in the freezing curves were compared. For the analysis, the readings were converted to a common depth of immersion of the SPRT of 17.5 cm (the height from the middle of the SPRT coil to the liquid metal surface in the cell). To minimize the effect of resistance drift in the SPRT and of the reference resistor on the comparison of the cells, the readings of resistances at the cadmium point were converted to W, the ratio of the resistance at the cadmium point, $R(Cd)$, to that at 0 °C, $R(0°C)$. For this purpose, readings $R(TP)$ at the triple point of water were obtained (and converted to $R(0°C)$) for every reading at the cadmium point. (See reference 1 for details on how $R(TP)$ is converted to $R(0°C)$.)

Liquid cadmium was found to behave very similarly to liquid zinc and to supercool not more than 0.5 °C. Hence, the freezing-point experiments were conducted as follows. The sample was melted overnight by controlling the furnace temperature 5 °C above the melting point. Next morning the SPRT was inserted into the cell to monitor the freezing process. The manostat was activated to control the helium-gas pressure in the freezing-point cell at 1 atm. To start the freeze, the furnace temperature control was set to control 5 °C below the freezing point. When the SPRT indicated that recalescence has occurred, the SPRT was removed and two borosilicate glass rods were inserted successively into the thermometer well for about 5 minutes each to induce an inner freeze immediately next to the thermometer well. During this time, the furnace temperature control was adjusted to control 1 °C below the freezing point. After withdrawing the second glass rod, the SPRT was reinserted into the thermometer well to monitor the freezing process. Figure 2 shows typical results of monitoring a complete freeze; the results for each of the five cells are shown.

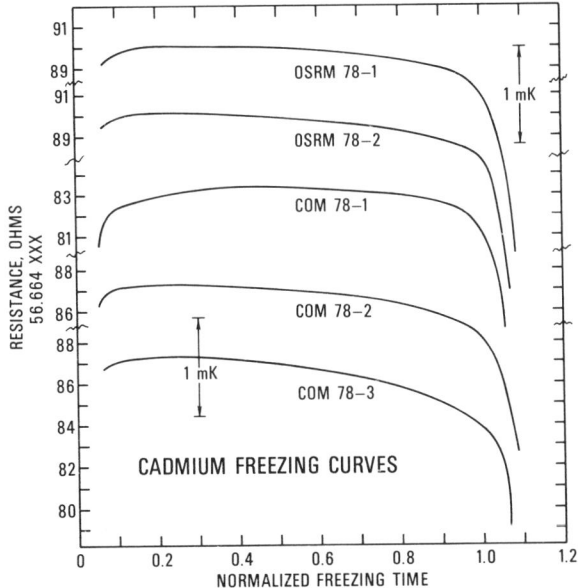

FIG. 2. Typical freezing curves that were obtained with the five cadmium-point cells.

5. Experimental Results

5.1 Effect of Pressure on the Freezing Temperature

The effect of helium-gas pressure on the freezing point of cadmium was investigated over the range 95 kPa to 105 kPa. A dial gauge that was calibrated in terms of a mercury manometer was used. The results obtained with each of the five cells are shown in Fig. 3, plotted relative to 1-atm pressure.

The pressure effect on the freezing temperature was also calculated employing the Clapeyron equation:

$$dT/dp = T_f (V_\ell - V_s)/L ; \qquad (3)$$

where V_ℓ and V_s are the specific volumes of the liquid and solid cadmium, respectively, at the freezing point T_f; L is the latent heat of fusion; and dT/dp is the pressure coefficient of T_f. Using 594.258 K for T_f [2], 0.125 cm^3/g for V_ℓ [11], 0.119 cm^3/g for V_s [11], and 54.3 J/g for L [12], the calculated pressure coefficient of the freezing temperature becomes 6.2 mK/atm or 61 µK/kPa. These values are consistent with those reported earlier by McLaren [13]. The calculated coefficient is shown as a straight line in Fig. 3 which agrees closely with the experimental observations. (It is to be noted that the Clapeyron relation yielded the derivative dT/dp at 1 atm. It is expected that the effect of pressure is very nearly constant over the pressure range of measurements.) For any adjustments of pressure effects, the value 6.2 mK/atm or 48.3 µK/cm of liquid cadmium (at 321 °C) was selected.

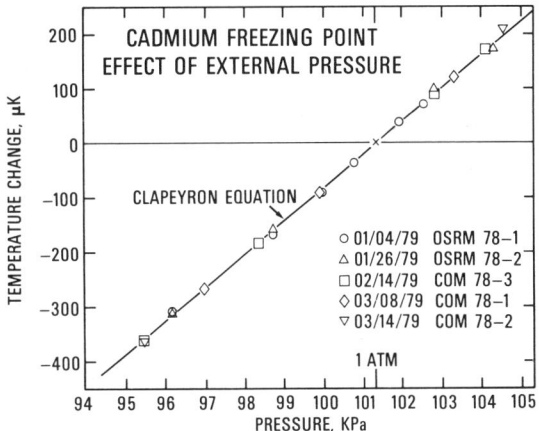

FIG. 3. Effects of external pressure (helium gas pressure) on the freezing point of cadmium. The line is based on the Clapeyron relation at 1 atm which was assumed to be the same over the range of measured pressure shown.

5.2 Comparison of Freezing Points

As described in Section 4, for the comparison of the freezing points of the five cadmium-point cells, freezing curves were obtained for each of the cells and the maximum readings of the curves were taken for comparison. Examples of typical curves are shown in Fig. 2 for each of the cells. The small change in temperature over the complete range of freeze that is shown for the five cells indicates that they contain samples of high purity [6]. Although two glass rods plus the SPRT were initially inserted into the thermometer well to form a thin mantle of solid cadmium around the well and therefore enhance the rate of temperature equilibration in the cell, there were long periods of temperature increase at the beginning of the freezing curves. Figure 4 shows chronologically the maximum readings for each of the freezes that were obtained with the cells. Along with the plots of $R(Cd)$ are given the plots of corresponding values of $R(0°C)$ that were observed and of the values of resistance ratio W that were calculated using the matching pairs of $R(Cd)$ and $R(0°C)$. The values of W are also summarized in Table II. Although values of $R(Cd)$ and $R(0°C)$ show significant change with time, which are attributed to drifts in the reference

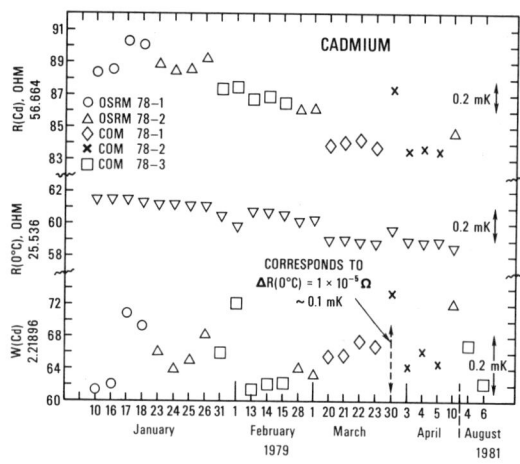

FIG. 4. Maximum freezing-point readings for the five cadmium-point cells. Values of R(Cd), R(0°C), and W(Cd) are plotted. The effect on W(Cd) of 0.1 mK error in R(0°C) is shown.

Table II

Comparison of Cadmium Freezing-Point Cells

Date	W	Date	W
OSRM 78-1		**OSRM 78-2**	
1-10-79	2.21896614	1-23-79	2.21896661
1-16-79	621	1-24-79	639
1-17-79	709	1-25-79	682
1-18-79	693	2-28-79	641
		3-01-79	633
Mean:	659	4-10-79	721
Std. Dev.	± 49		
	∿±0.136 mK	Mean:	661
		Std. Dev.	±34
Dev. from GM	-2		∿±0.094 mK
	∿-0.006 mK		
		Dev. from GM	0
			∿ 0.000 mK
COM 78-1			
3-20-79	2.21896655	**COM 78-3**	
3-21-79	657	1-31-79	2.21896660
3-22-79	675	2-01-79	721
3-23-79	668	2-13-79	613
		2-14-79	621
Mean:	664	2-15-79	622
Std. Dev.	± 9	8-04-81	670
	∿±0.025 mK	8-06-81	623
Dev. from GM	3	Mean:	647
	∿ 0.008 mK	Std. Dev.	±39
			∿±0.108 mK
COM 78-2		Dev. from GM	-14
3-30-79	2.21896734		∿-0.039 mK
4-03-79	644		
4-04-79	663	GRAND MEAN (GM)	2.21896661
4-05-79	647		
Mean:	672		
Std. Dev.	±42		
	∿±0.116 mK		
Dev. from GM	11		
	∿ 0.030 mK		

resistor and the SPRT, the values of W(Cd) are shown to vary less. The plot shows that a small error in R(0°C), e.g., 0.1 mK, contributes significantly to the error in W(Cd). The results show that the freezing points of the five cadmium point cells agree within about ±0.1 m°C.

5.3 Cadmium Point on the IPTS-68

To determine the cadmium point on the IPTS-68 the resistances of eight newly calibrated SPRT's were measured at the cadmium point and at the triple point of water. Before proceeding with these measurements, the performance of the cadmium cell when a number of SPRT's are inserted successively for measurements was tested.

At the NBS the process of calibration of SPRT's with fixed-point devices, such as the tin and zinc cells, involves successive insertion and reading of the thermometer resistances at the equilibrium temperature [4]. The first reading is obtained with a check SPRT which is then followed successively by usually six SPRT's to be calibrated. To avoid excessive freezing of the metal, the SPRT's, except for the initial check SPRT reading, are preheated in an auxiliary furnace so that the SPRT would be at a temperature slightly below the freezing point when inserted into the thermometer well. The SPRT is not inserted when it is at a temperature above the freezing point to avoid melting the inner freeze around the thermometer well which is essential to the measurements. Finally, the check SPRT is read again to assure that the equilibrium temperature had not changed significantly. For tin and zinc cells the change is usually not more than 0.1 or 0.2 m°C [1]. The same check SPRT is used with every freeze to make certain that the current calibration process is consistent with the previous ones. A different check SPRT is used with different types of calibration devices.

To test the performance of a cadmium-point cell under similar operating procedure as those used with tin and zinc cells, measurements were made repeatedly with the same SPRT using the AC bridge. After an equilibrium reading was obtained in the cadmium cell with the SPRT, the thermometer was removed and inserted into an auxiliary preheat furnace. After several minutes the SPRT was removed from the furnace and inserted back into the cadmium cell and equilibrium resistance readings obtained. This was repeated six times; the results are shown in Fig. 5. The total change in the observed resistances during seven hours since initiation of the freeze is shown to correspond to about 0.2 m°C.

The immersion characteristic of an SPRT was investigated in the cadmium cell with which the cadmium point was to be determined using eight calibrated SPRT's. The procedure involved observation of the SPRT resistance (at zero power, to avoid effects of variations in self heating) at various depths of immersion in the cell. The AC measurement system was used for the test. The results showed the readings to track the effect on the temperature by the hydrostatic head of liquid cadmium up to about 5 cm from the bottom of the thermometer well. Hence, it was assumed that SPRT's of similar construction which were used should be adequately immersed when inserted to the bottom of the thermometer well.

Measurements were made with the eight SPRT's having determined that satisfactory multiple measurements can be successively obtained and that the SPRT's would be adequately immersed in the cadmium-point cell. These results are summarized in Table III. The values of temperatures on the IPTS-68 with the eight SPRT's agree with ±0.1 mK (standard deviation of the mean). The value 321.1082 °C is in close agreement with the values reported by McLaren and Murdock [13,14] and the value 321.108 °C given in the IPTS-68 document [2]. The values obtained with the AC measurement system, although slightly higher, are in good agreement with the values obtained using the DC measurement system. The mean values obtained with the eight SPRT's show a range of about 0.9 m°C. This range is larger than expected, particularly since the calibration process with its "checks" should not cause such differences [1]. The most likely source of the variations is insulation

TABLE III
Cadmium Freezing Point Determination with Ten Thermometers
(Cd Cell: OSRM 78-2)

PLATINUM RESISTANCE THERMOMETER
Resistance Ratio, $W = R(Cd)/R(0°C)$

FREEZE NO. / DATE	1752305	1698755	1761950	1803100	1803104	1810371	1852743	1868896	1805523[a]	1852739[a]
1 11/07/81										2.2189182_{47}
										2.2189177_{43}
										2.2189176_{25}
										2.2189173_{91}
2 11/10/81	2.2189913_{10}	2.2186992_{07}	2.2190270_{61}				2.2188076_{32}	2.2189931_{76}		
	2.2189912_{76}									
	2.2189901_{81}									
3 11/11/81	2.2189926_{24}	2.2187009_{17}	2.2190271_{20}				2.2188089_{43}	2.2189944_{39}		2.2189182_{67}
	2.2189914_{00}									
	2.2189921_{86}									
4 11/12/81	2.2189937_{74}						2.2188094_{44}	2.2189966_{22}		
5 11/13/81	2.2189935_{80}	2.2187018_{18}	2.2190294_{53}							2.2189192_{26}
6	2.2189940_{34}	2.2187011_{07}	2.2190302_{60}				2.2188097_{73}	2.2189963_{81}	2.2189673_{29}	2.2189193_{42}[b]
7 11/21/81				2.2191140_{15}	2.2190354_{62}	2.2186621_{86}			2.2189664_{50}	
8 11/23/81				2.2191140_{28}	2.2190363_{86}	2.2186628_{23}			2.2189658_{97}	
9 11/24/81	2.2189939_{65}								2.2189672_{61}	
11/25/81	2.2189946_{57}									
Mean W:	2.2189926_{35}	2.2187007_{62}	2.2190284_{73}	2.2191140_{22}	2.2190359_{24}	2.2186625_{04}	2.2188089_{48}	2.2189951_{54}	2.2189667_{34}[d]	2.2189182_{78}[d]
t, °C	321.1082_{31}	321.1076_{01}	321.1084_{79}	321.1084_{65}	321.1081_{61}	321.1077_{73}	321.1083_{14}	321.1084_{16}	321.1085_{20}[d]	321.1086_{74}[d]
Std. Dev. of Mean (m°C)	$±0.1_{22}$	$±0.1_{53}$	$±0.2_{26}$	$±0.0_{02}$	$±0.1_{28}$	$±0.0_{88}$	$±0.1_{30}$	$±0.2_{27}$	$±0.0_{95}$	$±0.0_{80}$
Range (m°C)	1.2_{40}	0.7_{23}	0.8_{86}	0.0_{04}	0.2_{56}	0.1_{76}	0.5_{93}	0.9_{55}	0.3_{97}	0.5_{41}
Mean of Eight Thermometers: 321.1082$_{05}$ °C; Std. Dev. of Mean: ±0.1$_{05}$ m°C; Range: 0.8$_{78}$ m°C.										
Dev. from Mean (m°C)	$+0.0_{26}$	-0.6_{04}	$+0.2_{74}$	$+0.2_{60}$	-0.0_{44}	-0.2_{32}	$+0.1_{09}$	$+0.2_{11}$	$+0.3_{15}$	$+0.4_{69}$

[a] Thermometers used with Cutkosky AC Bridge.

[b] Measurements made 11/20/81 to test freeze after one-week hiatus.

[c] The values of W were converted to values of temperatures using thermometer calibrations and adjusted to standard conditions by correcting for immersion depth in the cadmium cell (17.5 cm = 0.845 m°C).

[d] Not included in the mean, nor in the range.

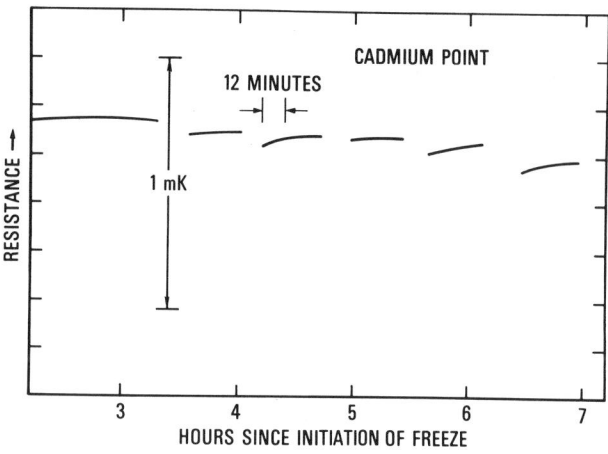

FIG. 5. Consecutive measurements of the resistance of an SPRT in a single cadmium freeze. AC bridge was employed. Following each equilibrium resistance reading, the SPRT was removed and inserted into an auxiliary preheat furnace and then quickly inserted back into the cell. The data show the somewhat cooled SPRT to come to equilibrium rapidly.

leakage caused by moisture; either the moisture was already present inside the thermometer or was generated by decomposition of the mica insulation [15]. Similar interpretation was given by McLaren and Murdock [14] for the range of 1.3 m°C they found for a group of SPRT's at the cadmium point. To check this interpretation we plan to "dry" the SPRT's and to use silica insulated SPRT's for future measurements. Variations at the cadmium point may also service to indicate the "wetness" of the SPRT.

CONCLUSION

The comparison measurements show that the five cadmium-point cells have freezing points that agree within ±0.1 m°C. The agreement in the cadmium point values of eight SPRT's (range: 0.9 m°C) demonstrates that consistency of calibrations at the triple point of water and the tin and zinc points for different SPRT's can be tested from measurements at the cadmium point. The AC bridge measurements yielded values only slightly higher than those obtained using the DC bridge method. Additional comparison measurements are needed before any relation between the results of the two methods can be developed.

At this point, the immersion characteristics of capsule type SPRT's in the special holder still require careful testing in cells COM 78-1 and COM 78-2. If the immersion proves to be inadequate, the holder must be redesigned to reduce axial heat conduction and to improve radial heat conduction. On the basis of the results with the tin-point cell with which the capsule type SPRT is routinely calibrated, the present holder is expected to show adequate immersion.

The relatively long temperature recovery period at the beginning of the freeze requires further investigation. McLaren [13,14,16] showed similar observations with various metal freezes, particularly with less pure samples. Perhaps our cadmium samples are less pure than expected or had become contaminated.

Platinum resistance thermometers with electrical insulation properties better than mica would help evaluate thermometric fixed points more precisely.

ACKNOWLEDGEMENT

The authors are grateful to M. L. Reilly for help in the computer analysis of literature data on the freezing point of cadmium.

DISCLAIMER

Certain commercial equipment, instruments, or materials are identified in this paper in order to adequately specify the experimental procedure. Such identification does not imply recommendation or endorsement by the National Bureau of Standards, nor does it imply that the materials or equipment identified are necessarily the best available for the purpose.

REFERENCES

1. Furukawa, G.T., Riddle, J.L., and Bigge, W.R.: The International Practical Temperature Scale of 1968 in the region 90.188 K to 903.89 K as maintained at the National Bureau of Standards, J. Res. Nat. Bur. Stand. (U.S.) 80A, 477 (1976).
2. The International Practical Temperature Scale of 1968. Amended Edition of 1975. Metrologia 12, 7 (1976).
3. Furukawa, G.T., Riddle, J.L., and Bigge, W.R.: The International Practical Temperature Scale of 1968 in the region 13.81 K to 90.188 K as maintained at the National Bureau of Standards, J. Res. Nat. Bur. Stand. (U.S.) 77A, 309 (1973).
4. Furukawa, G.T., Riddle, J.L., Bigge, W.R., and Pfeiffer, E.R.: Application of some metal SRM's as thermometric fixed points, NBS Special Publication 260-XX, pending.
5. Nesmeyanov, A.N.: Vapor Pressure of the Elements (Academic Press, Inc., New York, N.Y., 1963).
6. Riddle, J.L., Furukawa, G.T., and Plumb, H.H.: Platinum resistance thermometry, Nat. Bur. Stand. (U.S.), Monogr. 126, 129 pages (April 1973).
7. Furukawa, G.T.: Investigation of freezing temperatures of National Bureau of Standards aluminum standards, J. Res. Nat. Bur. Stand., (U.S.), 78A, 477 (1974).
8. Cutkosky, R.D.: An a-c resistance thermometer bridge, J. Res. Nat. Bur. Stand. (U.S.), 74C, 15 (1970).
9. Kusters, N.L., MacMartin, M.P., and Berry, R.J.: Resistance thermometry with the direct current comparator bridge, in Temperature, Its Measurement and Control in Science and Industry, Vol. 4, ed. by H.H. Plumb, Instrument Society of America, Pittsburgh, 1972, pp. 1477-1485.
10. Brombacher, W.G., Johnson, D.P., and Cross, J.L.: Mercury barometers and manometers, Monograph 8, 59 pages (May 1960).
11. Gmelins Handbuch der Anorganischen Chemie, Achte Auflage, Cadmium, System-Nummer 33, (Verlag Chemie, Berlin, 1959).
12. Wagman, D.D., Jobe, T.L., Domalski, E.S., and Schumm, R.H.: Temperatures, pressures, and heats of transition, fusion, and vaporization, American Institute of Physics Handbook, pp. 4-222 to 4-261, third edition (McGraw-Hill Book Company, New York, (1972).
13. McLaren, E.H.: The freezing points of high purity metals as precision temperature standards. II. An investigation of the reezing temperatures of zinc, cadmium, and tin, Can. J. Phys. 35, 1086 (1957).
14. McLaren, E.H. and Murdock, E.G.: The freezing points of high-purity metals as precision temperature standards. VIII b. Sb: Liquidus points and alloy melting ranges of seven samples of high-purity antimony; temperature-scale realization and reliability in the range 0-631 °C, Can. J. Phys. 46, 401 (1968).
15. Furukawa, G.T. and Mangum, B.W.: Insulation support for SPRTs, Comité Consultatif de Thermométrie, 13th Session (June 17-19, 1980), Document No. 80-35.
16. McLaren, E.H.: The freezing points of high-purity metals as precision temperature standards. IV. Indium: Thermal analysis of three grades of cadmium, Can. J. Phys. 36, 1131 (1958).

Measurement of the melting temperature of the copper 71.9% silver eutectic alloy with a monochromatic optical pyrometer

R. E. Bedford and C. K. Ma

National Research Council of Canada, Division of Physics, Ottawa, Ontario, Canada, K1A 0R6

The thermodynamic temperature of the melting point of the eutectic alloy copper 71.9 wt.% silver has been measured with a monochromatic optical pyrometer relative to the freezing point of silver. The fixed points were realized in identical graphite crucibles (in which a high quality blackbody cavity was immersed in the metal ingot) in identical furnaces containing sodium-filled heat pipe liners to provide isothermal environments around the crucibles. The ratio of the spectral radiances of the eutectic and silver at their transformation temperatures was measured at four wavelengths (599.4, 656.8, 657.4, and 781.6 nm) defined by three 10 nm and one 6 nm half-width interference filters. Three rotating sectored disks were used to reduce the silver radiance to near that of the eutectic. The melting temperature of the eutectic was taken to be that corresponding to the peak in a histogram of percentage of melting time taken to cross equal small temperature intervals. The mean value for the eutectic melting temperature from two eutectic ingots and three silver ingots was 1053.12±0.10 K relative to a silver point of 1235.20 K. The eutectic melting temperature was also measured on the IPTS-68 with a calibrated platinum 10% rhodium/platinum and a calibrated platinum 13% rhodium/platinum thermocouple. The mean value was 1052.72±0.10 K. These results give $T - T_{68} = 0.40 \pm 0.14$ K at 1053 K. The eutectic was found to freeze at temperatures from 0.05 to 0.3 K below the melting temperature.

1. INTRODUCTION

It is known that the International Practical Temperature Scale of 1968 (IPTS-68) departs widely from thermodynamic temperatures (T) in the range from 900 K to 1337 K where values of $(T-T_{68})$ reach a maximum of about 0.5 K near 1070 K. For any forthcoming revision of the IPTS-68, where removal of the platinum 10% rhodium/platinum (10/0) thermocouple as a standard instrument in favour of the platinum resistance thermometer (PRT) is contemplated, it is critically important that these values of $(T-T_{68})$ be well documented so that suitable interpolation for the PRT can be devised. Measurements of $(T-T_{68})$ in this range have been based upon both optical pyrometry[1-2] and noise thermometry[3]. The pyrometer experiments used blackbody cavities of which the temperature T_{68} was measured with calibrated 10/0 thermocouples and the temperature T with a pyrometer. They do not give an absolute measure of T: Quinn et al[1] measured T in the range 1000 K to 1337 K relative to the IPTS-68 assigned value for the freezing point of gold (1337.58 K); Bonhoure[2] measured T from 904 K to 1338 K relative to the freezing point of antimony taken as 903.89 K. For comparison, these results should be converted to a common reference temperature[4], although in this case the resulting changes are much smaller than the experimental uncertainties. Bonhoure obtained $(T-T_{68}) = 0.5_4 \pm 0.2$ K at 1050 K as compared with 0.6_5 K by Quinn et al, but at this temperature Quinn et al report an increasing uncertainty due to decreasing sensitivity and increasing radiance ratio. The noise thermometry experiment of Crovini and Actis[3] used a metal block comparator for which T was measured absolutely with the noise thermometer and T_{68} with a calibrated PRT. Their results agree well with those of Bonhoure and give $(T-T_{68}) = 0.4_7 \pm 0.2_4$ K at 1050 K.

The melting point of the copper 71.9 wt.% silver (Cu/Ag) eutectic alloy at 1053 K lies very near to the temperature where $(T-T_{68})$ is a maximum. By measuring the thermodynamic temperature of this fixed point relative to that of another fixed point with an optical pyrometer we can obtain a value of $(T-T_{68})$ that avoids direct reliance on transfer from a calibrated thermometer. Such a measurement of $(T-T_{68})$ at 1053 K provides a useful supplement to the above-mentioned experiments. In addition this method has the advantage that blackbody conditions are more nearly approached. A cavity that is almost completely surrounded by a metal ingot that is melting or freezing will be very nearly isothermal because its temperature is closely controlled by the latent heat of fusion of the ingot. In experiments without fixed points, however, the furnace itself must be relied upon to establish a uniform temperature over the cavity, which is more difficult. Furthermore, in the latter case, a thermometer must be inserted into the cavity to measure T_{68}, with the possibility that temperature uniformity will be disturbed through conduction of heat along the thermometer and its sheath. The Cu/Ag eutectic freezing point has occasionally been used as a secondary reference point for thermocouple calibration. The most detailed study of its behaviour is that of Bongiovanni et al[5] who examined with a PRT the effect of different rates of freezing and melting on the measured liquidus temperatures. They concluded that the freezing temperature is rate dependent but that the melting temperature is not and is reproducible to within 0.01 K. This poses no limitation at all for an optical pyrometer which has far greater uncertainty at 1050 K.

We have used a monochromatic optical pyrometer to measure the ratio of the spectral radiance of a blackbody at the freezing point of silver to that of a blackbody at the melting point of the Cu/Ag eutectic. By assigning a value to the silver point we can calculate the thermodynamic temperature of the Cu/Ag eutectic point. By combining this with a thermocouple measurement of the eutectic melting point, we obtain a value for $(T-T_{68})$ at 1053 K.

2. APPARATUS

2.1 Optical Pyrometry

The monochromatic optical pyrometer used has been described by Ohtsuka and Bedford[6]. For these measurements it was used mostly at f/10 with a 0.4 mm field stop and an EMI 9658 photomultiplier tube (PMT) run at 1000 V as detector. Monochromaticity was provided by three interference filters with half-widths (i.e. full spectral width between half peak-transmittance wavelengths) of about 10 nm centred at 599.4, 657.4, and 781.6 nm respectively and one filter with half-width of about 6 nm centred at 656.8 nm. The first three are 3-cavity filters with approximately rectangular pass bands; the fourth is a 2-cavity filter with somewhat gaussian-shaped pass band, but more sharply peaked. All are blocked to better than 10^{-4} outside the pass bands. Sufficient precision in the measurements cannot be obtained with very narrow (~ 1 nm) filters and a PMT at the low spectral radiance level of a blackbody at 1050 K. The temperature of the PMT and interference filter was controlled to better than ± 0.1 K at 289 K by enclosing these components in a housing through which temperature-regulated water was circulated. A noticeable reduction in dark current and a marginal improvement in signal-to-noise level is obtained with decreasing PMT temperature, but at too low a temperature condensation of water vapour on various glass surfaces is troublesome. The choice of 289 K at which to maintain the PMT was a useful compromise.

Since the previous description[6], the PMT output circuit has been modified to include a microcomputer. As well as being displayed on a strip chart recorder via a digital-to-analogue converter for a visual record of events, the PMT output is directed to the microcomputer which is programmed to print the time, the average (v) of n individual DVM readings, the standard deviation (σ) of the n readings $\left(\sigma^2 = (n-1)^{-1} \sum_{i=1}^{n} (v_i - v)^2\right)$, and the minimum and maximum readings in the group of n readings. Each individual DVM reading is a true integration of the PMT output over 8.3 s. After each group of n readings the microcomputer will ignore a chosen number (m) of readings (m=o up) to allow time for manual moving of furnaces and/or the sectored disk. Rotating sectored disks of approximate transmittances 0.050, 0.065, and 0.085 were used to reduce the silver point spectral radiance to near that of the eutectic point.

The fixed points were realized in the same furnaces using the same type of graphite blackbody crucibles as described and illustrated previously[6]. The novel feature of the furnaces is the Inconel heat pipe liner containing sodium as the working fluid that is used to provide an extended near-isothermal zone ($\Delta T < 0.1$ K over 22 cm) in which the blackbody crucible is located. The effective emissivity of each blackbody cavity (illustrated in Fig. 1) was computed by the method of Bedford and Ma[7] to be 0.999 97 ± 0.000 03. For an experiment of this type where spectral radiance ratios are measured it is, of course, not necessary to know the emissivity exactly so long as it is the same for both cavities.

The furnaces are mounted on rails in such a way that, once correctly aligned, the blackbody apertures can be repeatedly located on the optical axis of the pyrometer to within ± 0.02 mm. Provision is made for precise adjustments in the lateral, axial, and vertical directions.

All of this system is mounted on a vibration-isolated optical table to ensure no relative motion between components and to remove any possible effects from vibration.

2.2 Thermocouple Thermometry

To obtain a value of $(T-T_{68})$ at the Cu/Ag eutectic point we must also measure the melting point of the eutectic on the IPTS-68. The vertical furnace used for this measurement and for other thermocouple fixed point calibrations is 23 cm diameter and has a central alumina core 46 cm long by 6.4 cm internal diameter on which a single, uniformly-spaced Nikrothal ribbon heater is wound over 38 cm. A 30.5 cm long by 3.4 cm internal diameter Inconel heat pipe liner having sodium as the working fluid is located centrally in the alumina core. The bottom 7.6 cm of the core below the heat pipe, as well as the body of the furnace, is insulated with packed fused silica wool. From the top of the heat pipe to the top of the furnace extends a 7.6 cm long by 2.5 cm internal diameter alumina tube supported by three metal pegs on an Inconel annulus resting on the pipe — the peg support minimizes conduction of heat from the pipe. Between the outer core and this alumina tube are Inconel rings and silica wool for insulation and shielding as at the bottom. Each metal fixed-point ingot is contained in a high-purity graphite crucible of internal diameter 1.7 cm and length 8.4 cm. Each crucible is held in a fused silica closed-end tube 46 cm long by 2.2 cm internal diameter which can be inserted into the heat pipe so that the crucible is located with its base 7.6 cm from the bottom of the heat pipe. Above the crucible in the silica tube layers of silica wool alternate with graphite disks to act as insulation, radiation shields, and oxygen getters. Down the centre of this tube extends a closed-end, thin-walled, fused silica well of internal diameter 0.4 cm that is free to move through the crucible lid into (or out of) the ingot when it is molten. The thermocouple is positioned in this well with its tip 0.1 cm from the bottom. The volume of metal in the crucible is about 16 cm^3 at the melting point. With this system one fixed point is easily replaced by another in the same furnace.

For a fixed point calibration, the crucible assembly is mounted in the furnace, the ingot melted, the thermocouple well lowered into the ingot, and then the thermocouple emf recorded during a slow freeze followed by a slow melt. The reference junction is maintained in a Dewar of shaved ice and distilled water. Emfs were measured with a Guildline Type 9180 vernier potentiometer with least steps of 0.1 µV and with a recording potentiometer with sensitivity of 2 µV/cm as null indicator. Emfs were also simultaneously measured with a Keithley Model 181 Nanovoltmeter having 0.01 µV resolution and 0.1 µV accuracy (after calibration) on the 20 mV range. The Nanovoltmeter was connected to a microcomputer which printed the thermocouple emf in the same fashion as for the PMT output. The Nanovoltmeter is not an integrating voltmeter. On the 20 mV range it gives 4 readings per second, and we printed (usually) the average of 125 readings. These averages agreed with the potentiometer output to within ± 0.1 µV.

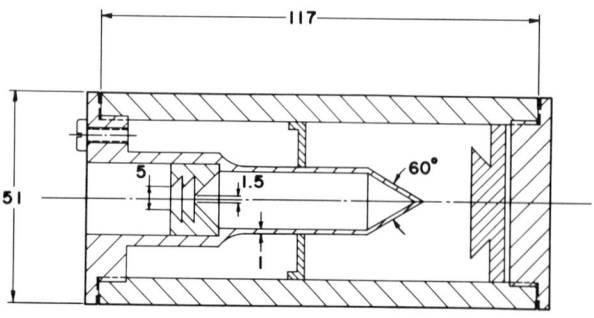

Fig. 1: Blackbody crucible (dimensions in millimeters). The graphite plug inside the rear screw cap is free to slide inside the crucible.

3. EXPERIMENTAL PROCEDURE

3.1 Optical Pyrometry

The silver crucibles were filled as described previously[6] using a graphite-enclosed funnel mounted above the crucible in a filling furnace. From the mass of silver required to completely fill the crucible at the melting point (~ 770 g) we could estimate very closely the masses of silver and copper needed to fill the eutectic crucible, taking as the eutectic composition 28.1 wt% Cu and 71.9 wt% Ag (total ~ 720 g). Appropriate masses of silver (nominal purity 0.999 999) and copper (nominal purity 0.999 99+) in the form of small diameter rods were placed in the funnel which was mounted above the crucible in the same manner as before[6]. A solid graphite plug, with volume the same as that expected for the residual metal left in the funnel after filling, was inserted into the crucible between the sliding plug and the bottom of the blackbody cavity. After the copper and silver had been melted into the crucible, the graphite plug was replaced with the residual metal remaining in the funnel. The crucible was then mounted in the blackbody furnace and heated to above the melting point of copper for about 30 min to ensure that any residual copper in the ingot was completely melted and to homogenize the ingot composition. To ascertain that this was so we submitted two samples, one cut from each end of an ingot, for measurement of the composition at the conclusion of the experiment. Within the measurement uncertainty (~ ± 1%) both samples had the same composition which, in turn, was the same as the eutectic composition.

We first obtained several preliminary eutectic melts and freezes to find the furnace power levels required for given rates of approach to the transformation temperature and to discover what unusual effects, if any, we might encounter with the eutectic. Bongiovanni et al[5] show that both the freezing and melting plateaus may slope by up to 0.2 K depending upon the rate of approach in the case of freezing and the history of the previous freeze in the case of melting. Further, the melting temperature may be higher than the immediately previous freezing temperature by up to 0.3 K. In our case where the detector is a PMT working near its limit of accuracy it is difficult to separate true eutectic effects, especially slopes of plateaus, from those associated with drift and general signal-to-noise level of the PMT. The signal-to-noise level, as given by the standard deviations of groups of n readings, was typically ± .07 K for λ 599.4, ± 0.05 K for λ 656.8, ± .04 K for λ 657.4, and ± .05 K for λ 781.5. We had no difficulty in maintaining a melt or a freeze for several hours and they were neither as flat nor as reproducible as those of silver, in agreement with Bongiovanni et al[5]. Typical depths of supercool were 0.1 to 0.2 K for the eutectic as compared with 0.6 to 0.9 K for silver. There was an indication that the freezing temperature was rate-dependent and that the melting temperature was not, also in agreement with Bongiovanni et al[5]. In the face of this, we followed their recommendations regarding preferred rates of furnace temperature change (generally < 0.05 K/min) in approaching the transformation temperature, in taking the melting temperature as the definitive point, and in defining the melting temperature as the barycentre of the peak in a melting histogram. The total melting time was taken to be from about 0.1 K below the plateau to about 0.1 K above it. It turns out that the choice of melting time, so long of course as it includes the whole plateau, has no significant effect on the resultant melting temperature.

Three typical melting histograms for the Cu/Ag eutectic (one for each of the three 10 nm filters used) are shown in Fig. 2. In Fig. 2 we have plotted the percentage of the total melting time in PMT anode current intervals of 0.02, 0.04, 0.04 nA (equivalent to 0.06, 0.03, 0.05 K respectively) for the three filters respectively. These intervals are just those equivalent to the signal-to-noise ratio with each filter as judged from the values of σ from groups of n readings. We found σ to be approximately independent of n for n between 5 and 20. We see from the histograms that more than 80% of the melting occurs within a temperature interval smaller than 0.2 K.

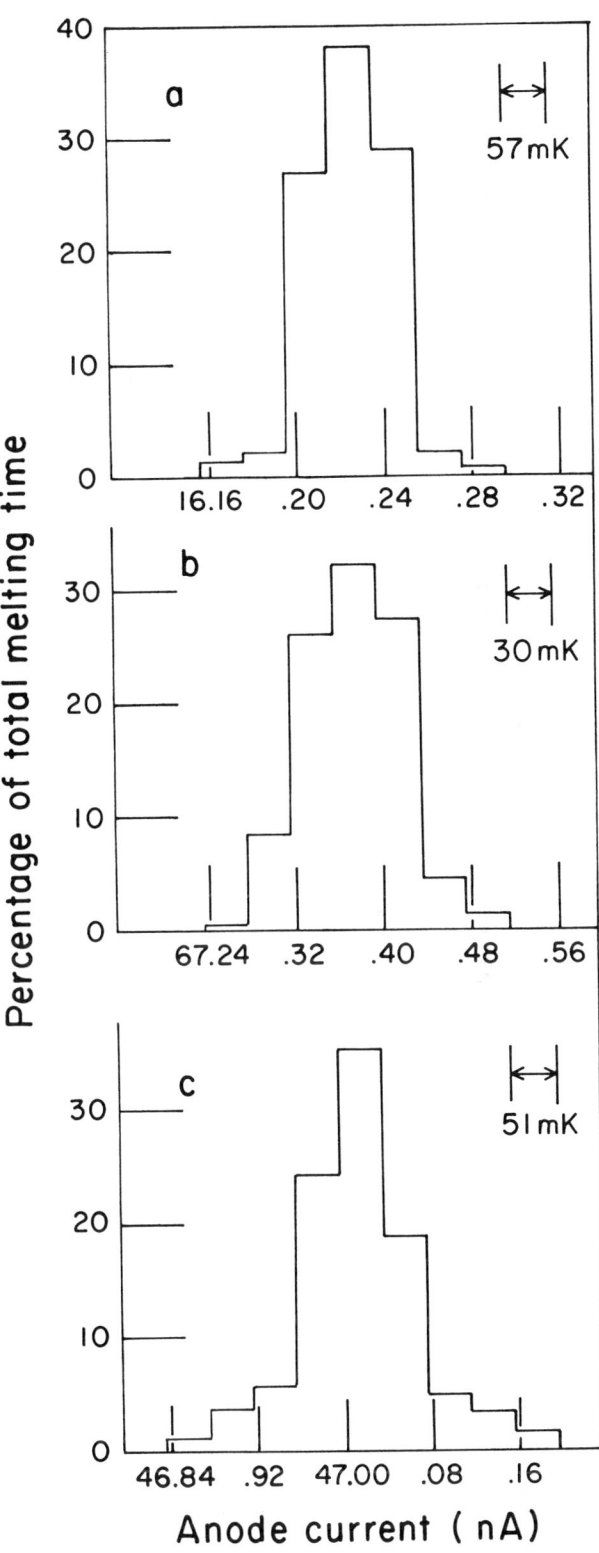

Fig. 2: Three typical melting histograms for the copper 71.9 wt% silver eutectic as measured with the optical pyrometer: a) λ 599.4 nm; b) λ 657.4 nm; c) λ 781.6 nm.

For comparison, two melting histograms for silver (as stepped down by the sectored disk) are shown in Fig. 3. We note that the apparent melting range is narrower than for Cu/Ag but not dramatically so. The true melting range is undoubtedly narrower (as we will see later for thermocouple measurements); it is obscured here by the limitations of the PMT. A plot of this kind cannot distinguish between alloy melting range and PMT drift and noise.

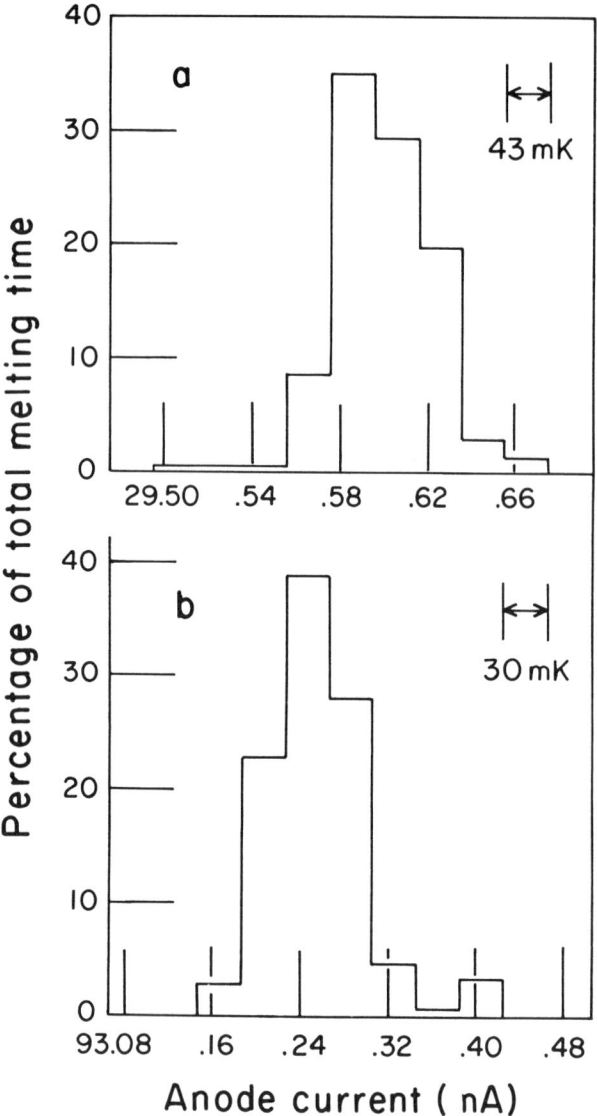

Fig. 3: Two typical melting histograms for silver as measured with the optical pyrometer: a) λ 599.4 nm; b) λ 657.4 nm.

In plotting these histograms, where the melting curves have not always been observed continuously because of alternate viewing of Cu/Ag and Ag by the pyrometer, we have apportioned the "non-viewing time" according to the nanovolt interval distribution before and after the break. Any uncertainty thereby introduced into the histogram is slight.

We measured the ratio of the spectral radiance of freezing or melting silver as attenuated by the sectored disk to that of the melting eutectic by moving each furnace alternately on to the optical axis of the pyrometer. Some of these measurements involved alternating the furnaces for equal short intervals (usually 3 min) but these could introduce systematic errors when the eutectic plateau has significant slope, depending upon where along the plateau they were made.

The preferred method was to alternate the furnaces at longer intervals (~ 15 min each) and to follow both the silver and eutectic transformations from beginning to end. In this way complete histograms for both could be obtained, allowing a comparison between the silver freezing point and the defined eutectic transformation temperature. This procedure is much slower, of course, because it does not permit change of components (filter, disk, etc) during a given pair of plateaus. For the most part we used n=5 or 10 and m=0 in the microcomputer program, and simply ignored the first average after interchange of furnaces. An example of a comparison of a eutectic melt with both a silver melt and freeze is shown in Fig. 4, where the eutectic melting time was almost six hours. Note that the apparent melting range of the eutectic is ~ 0.1 K whereas no significant silver melting or freezing range is detectable, and that the silver melting and freezing temperatures are the same to within the uncertainty of the measurements.

The PMT background reading was measured occasionally. The background was from 0.5% to 3% of the signal (depending upon which filter and disk were used), but was stable for long periods to within about 1%, so background fluctuations contributed no significant uncertainty to the results. The background reading was usually marginally larger than the PMT dark current.

The spectral radiance ratio of a first eutectic ingot (Cu/Ag I) to either of two silver ingots (Ag II, Ag III) was measured using each of the three 10 nm interference filters with each of the three sectored disks. Subsequently, the spectral radiance ratio of a second eutectic ingot (Cu/Ag II) to two silver ingots (Ag III, Ag IV) was measured with the 6 nm filter and one 10 nm filter (λ 657.4) using one sectored disk (τ ~ 0.050). Then the spectral radiance ratios of Cu/Ag I to Cu/Ag II, Ag II to Ag III, and Ag III to Ag IV (Ag IV of nominal purity 0.999 99) were measured. We also measured the freezing point of the eutectic relative to silver, and intercompared eutectic freezing points. As already pointed out, however, the freezing point was less reproducible than the melting point.

For all of these measurements, to ensure that the aperture of the blackbody cavity was on the optical axis of the pyrometer, we occasionally recorded the cavity radiance for small, discrete horizontal and vertical movements of the furnaces over a range of 2-4 mm. A drop from 0.2 to 0.6% in radiance occurred when the aperture was off the axis, so that it was easy to attain proper alignment. Figure 5 shows a typical change in pyrometer output when the 1.5 mm blackbody aperture is moved across the optical axis when the field stop is 0.4 mm diameter. Generally, we found that the alignment was not disturbed on interchanging crucibles in the furnaces.

The spectral transmittances of the interference filters were measured with a Cary Model 501 spectrophotometer.

3.2 Thermocouple Thermometry

The melting and freezing points of the copper/silver eutectic on the IPTS-68 were measured with a standard 10/0 thermocouple and with a 13/0 thermocouple. The thermocouples were calibrated at the freezing points of antimony, silver, and gold (using for T(Sb) the value 903.905 K). The 10/0 thermocouple calibration was calculated both by a quadratic interpolation between the fixed points as defined in the IPTS-68 and by differences from standard reference tables[8]; the 13/0 thermocouple calibration was obtained from differences from standard reference tables[8]. The freezing point calibrations and the measurement of the eutectic fixed points were all made in the same furnace, described in Sec. 2.2. For each thermocouple the fixed points were taken in the order eutectic, antimony, silver, gold. For antimony, which may undercool up to 20 K, the freezes were induced when the furnace temperature was within 1 K of the freezing

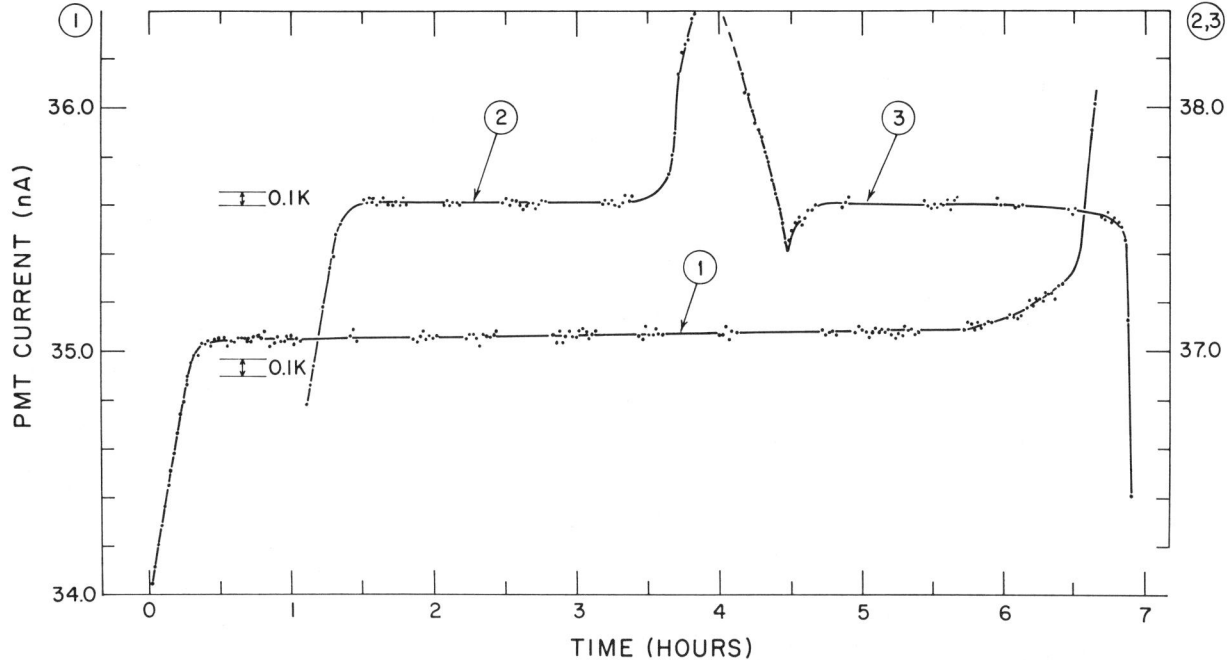

Fig. 4: A typical copper 71.9 wt% silver eutectic melt (1) (λ 656.8 nm) simultaneous with a silver melt (2) followed by a silver freeze (3). With silver a sectored disk with τ ~ 0.050 was used. Each point is the average of five 8-second integrations of the PMT anode current. The background reading during the measurements was constant at 0.38 nA.

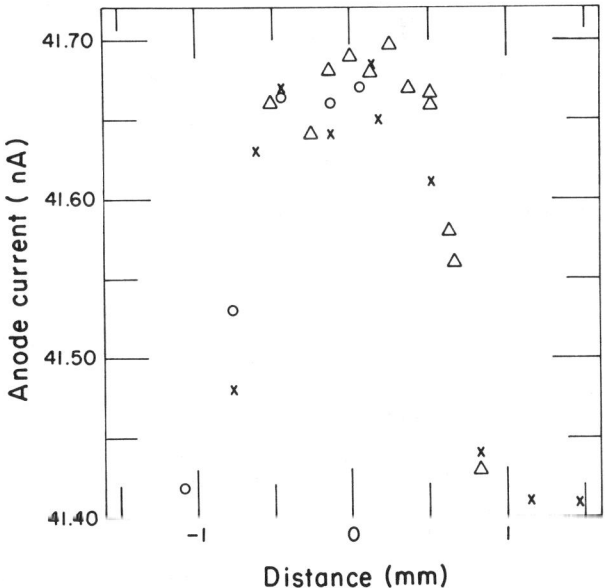

Fig. 5: Pyrometer output for different positions of blackbody aperture relative to the optical axis of the pyrometer. The three different symbols show three consecutive passes across aperture.

temperature by withdrawing the crucible assembly into the furnace throat until solidification started and then replacing it. Typical melting and freezing times for all four metals were 40 to 60 min.

The eutectic melting temperature was obtained from a melting histogram of percentage of total melting time in emf intervals of 0.1 µV (equivalent to 9.3 mK for 10/0 and 8.2 mK for 13/0) as read from the microcomputer output. The standard deviation of each group of 125 individual readings was 0.02 µV (~ 2 mK), but this was about the same as the noise level of the Nanovoltmeter, so is not a true indicator of the fixed point stability. Figure 6 shows these histograms for each thermocouple. We see that roughly 50% of the sample (about the central maximum) melted over a width of 0.04 K, and 87% over a width of 0.11 K. This latter is roughly the same as found with the optical pyrometer.

We also show in Fig. 7 a melting histogram for silver as measured with each thermocouple. In this case emf intervals of 0.05 µV (equivalent to 4.4 mK for 10/0 and 3.8 mK for 13/0) are used. We see that the peaks are narrower than for Cu/Ag, or for Ag as measured with the optical pyrometer. About 90% of the sample melts over a width of .03 to .04 K about the maximum. This supports our earlier statement that the PMT limits observation of the true Ag melting range in the pyrometry measurements. Even here, the true Ag melting range is likely smaller, the observation being limited by both the thermocouple and the Nanovoltmeter. Bongiovanni et al[5] take the melting range for Ag to be the temperature interval corresponding to 40% of the melting time about the peak. In our case the 40% melting range is about 8 mK from Fig. 7 (the true range is undoubtedly smaller) as compared with 3 and 2 mK for Bongiovanni's two samples.

These melting range data from Figs. 2, 3, 6, 7 are summarized in Table I.

Histograms similar to those for silver were obtained for both antimony and gold. Freezing and melting temperatures with antimony, silver, and gold agreed to within 0.1 µV, whereas with the Cu/Ag eutectic the freezing temperatures were 3 to 4 µV (~ 0.3 K) lower.

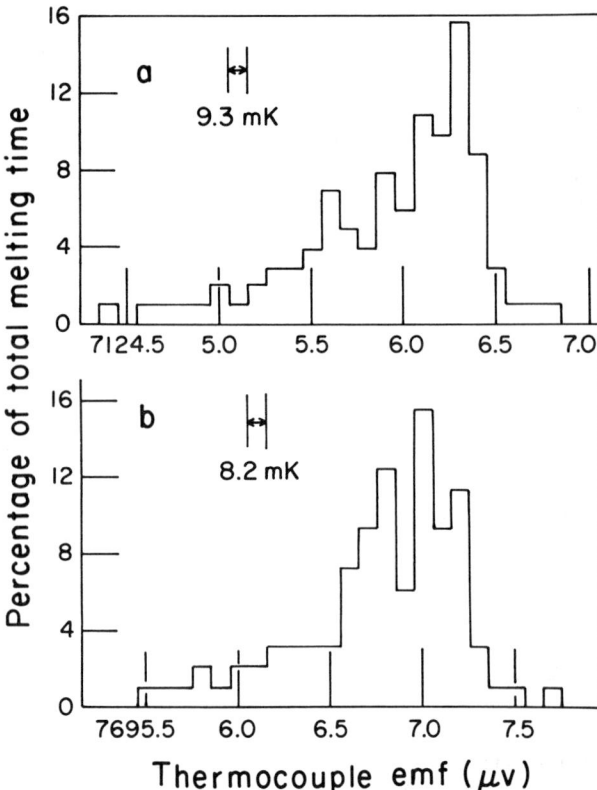

Fig. 6: Two melting histograms for the copper 71.9 wt% silver eutectic as measured with thermocouples: a) 10/0; b) 13/0.

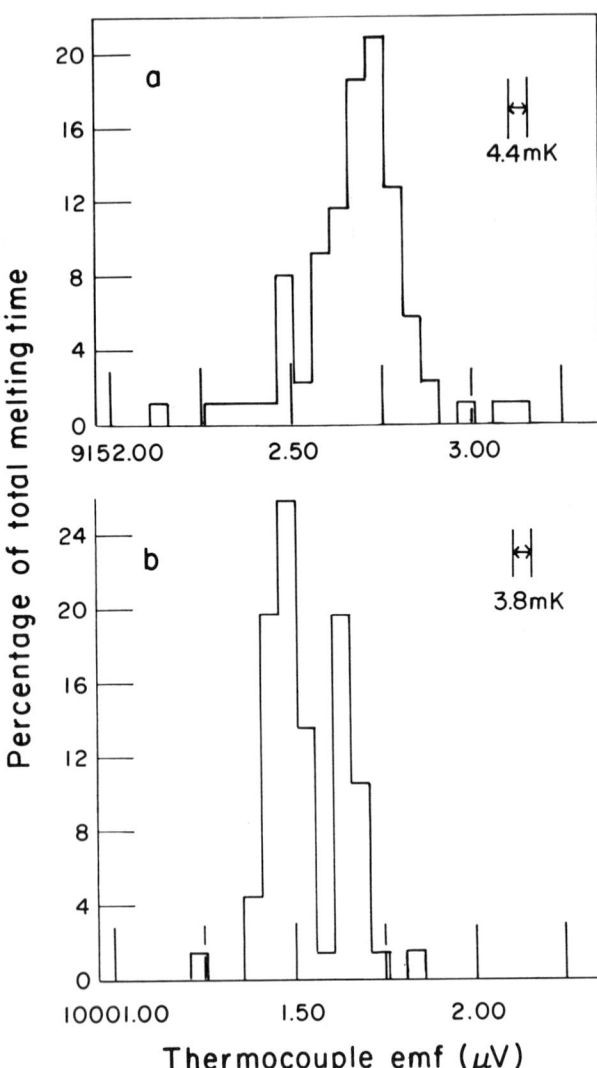

Fig. 7: Two melting histograms for silver as measured with thermocouples: a) 10/0; b) 13/0.

4. RESULTS FOR $(T-T_{68})$

The spectral radiance $L_\lambda(T)$ of a blackbody at temperature T and wavelength λ in a medium of refractive index n is given by Planck's equation

$$L_\lambda(T) = c_1\lambda^{-5}/[\exp(c_2/n\lambda T)-1] \quad (1)$$

where c_1 is the first radiation constant, c_2 is the second radiation constant (taken to be 14388 μm·K in the IPTS-68), and n = 1.000 28 for air at 295 K. For the comparison of the silver point radiance attenuated by a neutral disk of transmittance τ with the eutectic point radiance we may write

$$\int_{\lambda_1}^{\lambda_2} L_\lambda(T(eu))\tau_\lambda s_\lambda d\lambda = \tau \int_{\lambda_1}^{\lambda_2} L_\lambda(T(Ag))\tau_\lambda s_\lambda d\lambda \quad (2)$$

where T(eu) is the temperature of the melting eutectic, τ_λ is the spectral transmittance of the interference filter, s_λ is the spectral sensitivity of the PMT, λ_1 and λ_2 are the lower and upper wavelengths between which τ_λ is non-zero, and where we have assumed that the blackbody cavity emissivities are equal and wavelength independent and that the transmittances of all other optical components of the pyrometer are wavelength independent. By substituting for $L_\lambda(T)$ from Eq. (1) in Eq. (2) and assigning a value to T(Ag), we obtain an integral equation for the unknown temperature T(eu) in terms of the measured radiance ratio. We solved this integral equation by iteration using as a first approximation for T(eu) the value obtained from $L_{\lambda_0}(T(eu)) = \tau L_{\lambda_0}(T(Ag))$, where λ_0 is the centre wavelength of the filter pass band. Iteration was continued until the calculated and measured ratios agreed to within 1×10^{-6} (~ 0.1 mK).

We have not measured s_λ, but have solved Eq. (2) using the following approximations for s_λ, where s_λ is written in the form $s_\lambda/s_{\lambda_0} = 1 + s_1(\lambda-\lambda_0)$: (a) constant; (b) manufacturer's values (which are almost linear across the filter pass bands); (c) linear with slope $s_1 = \frac{1}{2}$ and 2 times the manufacturer's value. These calculations showed that T(eu) is insensitive to s_λ, as might be expected because s_λ is a slowly varying function of λ and appears under both integrals. For the approximations to s_λ listed, the maximum change in T(eu) was 0.12 K; for the slopes $s_1 = \frac{1}{2}$ and 2 times the manufacturer's value, the maximum change in T(eu) was 0.06 K. Coates[9] found that the maximum deviation from the manufacturer's values for s_1 for a group of EMI 9558 PMT's was 20%. The range of s_1 for EMI 9658 PMT's is likely to be similar, so the uncertainty in T(eu) when the manufacturer's values for s_λ are used is unlikely to be larger than ± 0.01 K.

The mean value of T(eu) calculated for each filter and disk relative to T(Ag)=1235.20 K is given in Table II. It is evident from the table that the melting temperature is higher and more reproducible than the freezing temperature, and that results for different filters differ significantly, but not systematically, from one another. Reasons for this may be: the uncertainties in τ_λ; the PMT anode currents for filter

TABLE I. Width of Histogram Peak for Different Percentages of Sample Melted

		optical pyrometry		
	% melted	width (K)		
		λ599.4	λ657.4	λ781.5
Cu/Ag	78			.15
	85		.09	
	89			.26
	94	.17		
Ag	86	.12		
	90		.09	

		thermocouple thermometry	
	% melted	width (K)	
		10/0	13/0
Cu/Ag	45	.037	
	54		.041
	70		.057
	74	.083	
	87	.12	.10
Ag	40	.009	
	46		.008
	60		.012
	65	.018	
	88	.039	
	93		.027

599.4 nm are lower by factors of from 2 to 4 (< 20 nA as compared with 40 to 70 nA) than for the other two 10 nm filters, so the uncertainty is somewhat larger for this filter; the effect observed by Ohtsuka and Bedford[6] may operate to produce higher values of T for lower anode currents. The relatively large melting range of the eutectic, however, precluded the possibility of measuring any dependence of T(eu) on anode current.

The intercomparison of the two Cu/Ag ingots (I and II) prepared from the same stock material showed, from the mean of four comparisons at three wavelengths, that the melting temperature of I was 0.06 ± 0.05 K higher than that of II. This is just barely significant, but the same result is evident in Table II for λ 657.4 where the temperature for Cu/Ag I is found 0.05 K higher than for Cu/Ag II. The comparison of Ag II with Ag III (both of nominal purity 0.999 999) gave the melting temperature equal to the freezing temperature to within the accuracy of measurement, and T(III)-T(II) = 0.01 ± .02 K on the average. This difference, which is insignificant, was measured only at 657.4 nm both with and without the sectored disk. Similarly, using the sectored disk and two wavelengths, we found for the Ag III versus Ag IV (nominal purity 0.999 99) comparison, T(III)-T(IV) = 0.00 ± .02 K. It is useful to know that silver with 10 ppm impurity content is just as satisfactory for an optical pyrometry fixed point as silver with 1 ppm impurity content as there is almost a factor of two difference in cost.

The mean value for the thermodynamic melting temperature of the Cu/Ag eutectic is 1053.12 K. This is obtained by giving equal weight to each of the four filters and each of the two eutectic ingots, and taking cognizance of the systematic difference of 0.06 K between Cu/Ag I and Cu/Ag II. The freezing temperature averages 0.05 K lower, in contrast with a difference of almost 0.3 K measured with the thermocouples. This we attribute to the relatively more rapid freezing in the latter case (see reference 5) where the ingot cooled several degrees below the freezing temperature. Similarly, the variation in freezing temperature in Table II is related to the rate at which the freezing occurred.

The agreement in T(Ag) between Ag II, III, IV indicates that no contamination of the Ag occurred because the different ingots experienced different periods of use. This is partially confirmed by the

TABLE II. Measured Values of the Thermodynamic Temperatures of the Melting and Freezing Points of the Copper 71.9 wt% Silver Eutectic.

		T(K)				
τ		Cu/Ag I			Cu/Ag II	
		599.4 nm	657.4 nm	701.6 nm	656.8 nm	657.4 nm
.08529	*					
	m	1053.22	1053.14	1052.96		
	f	1053.17	1053.07	1052.93		
.06495	m	1053.23	1053.12	1053.02		
	f	1053.26	1053.05	1053.00		
.05017	m	1053.23	1053.03	1053.05	1053.22	1052.98
	f	1053.17	1052.87	—		

* m: melting temperature
  f: freezing temperature

TABLE III. Measured Impurity Analyses (µg/g) of As-Received and Used Materials.

Impurity	Ag (I, II, III)* (as-received)	Ag (I) (used)	Ag (IV)** (as-received)	Ag (III)*** (used)	Cu/Ag (I)*** (used)
Fe	1	0.9	3	1	0.3
Ni	—	0.3	—	1	—
Si	0.2	0.02	3	2	0.2
Al	0.1	—	0.4	—	—
Pb	—	—	0.6	—	—
Mg	0.2	0.08	1	0.1	—
Cr	0.03	0.03	0.03	—	—

* Average of 3 samples of nominal purity 0.999 999. In an independent batch analysis Al, Si, V, Mn, Fe, Zn, As, Mg were detected totalling about 1 µg/g; P, Na, S, Cl, K, Ca, Ti were detected totalling about 0.4 µg/g.

** Average of 2 samples of nominal purity 0.999 99.

*** Samples from either end of the ingot produced identical analyses.

TABLE IV. Estimated Uncertainties in the Measurement of T(eu) and T_{68}(eu)

Parameter	Uncertainty (K)			
	599.4 nm	656.8 nm	657.4 nm	781.6 nm
spectral transmittance of interference filter (τ_λ)	±.08	±.07	±.07	±.06
measurement of PMT anode current (eutectic)	.035	.025	.02	.025
measurement of PMT anode current (silver)	.035	.025	.02	.025
transmittance of sector (τ)	.01	.01	.01	.01
spectral sensitivity of PMT (s_λ)	.01	.01	.01	.01
possible differences in Cu/Ag ingots	.03	.03	.03	.03
uncertainty in assigned value for T(Ag)(±0.02)	$.01_5$	$.01_5$	$.01_5$	$.01_5$
root mean square uncertainty in T(eu)	±.10	±.09	±.08	±.08
root mean square uncertainty in T_{68}(eu)	±.10			

impurity analyses in Table III obtained by quantitative dc-arc emission spectrography. The ingot Ag I was analyzed after ~ 600 h use above 1100 K and Ag III after ~ 2500 h use above 1175 K. Ag II was used under similar conditions.

From the 10/0 thermocouple measurements of the eutectic melting temperature we calculate T_{68}(eu) = 1052.67 K from quadratic interpolation and 1052.70 K by differences from the standard reference table. The difference between these is insignificant, but the former should be used according to the definition of the IPTS-68. The result with the 13/0 thermocouple was 1052.77 K, so that the mean value for T_{68}(eu) is 1052.72 K. This is in excellent agreement with Crovini and Marcarino's result[10] of 1052.75 K.

Combining these measurements, we find $T-T_{68}$ = 0.40 K at 1053 K.

The chief uncertainty in the measurement of T(eu) arises from the uncertainty in the interference filter transmittance. The uncertainty in the wavelength drive of the spectrophotometer used to measure τ_λ is ± 0.3 nm, leading to ± 0.08, 0.07, 0.07, 0.06 K at 599.4, 656.8, 657.4, 781.6 nm respectively. We note also that the pass bands of two of these filters have shifted slightly over about 2 years since they were used in a previous experiment[6]. This shift (0.6 and 0.5 nm) is double the wavelength uncertainty of the spectrophotometer, although the repeatability of measurement was 0.2 nm. Were this shift spurious, values of T(eu) for the filters would agree more closely.

The statistical uncertainty in the PMT output (i.e. the value of σ for groups of 5 or 10 readings) was about 0.1_3, 0.08, 0.06, 0.08% at 599.4, 656.8, 657.4, 781.6 nm respectively, corresponding to 0.07, 0.05, 0.04, 0.05 K respectively. However, when comparing the silver freezing curve with the eutectic melting curve with alternate observations of up to 15 minutes each, the uncertainty in the mean was less than half the above values, in rough agreement with $\sigma/\sqrt{n}$.

The transmittances τ of the sectored disks were known to within 0.01_4%, corresponding to about 0.01 K. The absence of systematic differences in values of $T(eu)$ with the three disks confirmed that no significant error arose from τ.

Uncertainties associated with s_λ and the effect of impurities have already been discussed; they are at most 0.01 K.

It is a little difficult to assess the uncertainty in the value for $T_{68}(eu)$. The text of the IPTS-68 asserts that measurements with a thermocouple may not be better than ± 0.2 K. In this special case of fixed point calibrations, however, the accuracy appears to be higher. Our two thermocouples gave values of $T_{68}(eu)$ differing by 0.1 K. Our mean value agrees with that of Crovini and Marcarino[12] to within 0.03 K. Thus we guess the uncertainty in $T_{68}(eu)$ is at most ± 0.1 K and is probably nearer ± 0.05 K.

All of these uncertainties are summarized in Table IV.

Our results then are

$$T(eu) = 1053.12 \pm 0.10 \text{ K}$$

$$T_{68}(eu) = 1052.72 \pm 0.10 \text{ K}$$

$$T - T_{68} = 0.40 \pm 0.14 \text{ K at } 1053 \text{ K}.$$

This value for $T-T_{68}$ is lower than that of Bonhoure[2] and of Crovini and Actis[3], but not significantly so. Within the respective limits of error, the three values are in agreement.

ACKNOWLEDGMENTS

We thank Miss M. Ohtsuka for measuring the transmittances of the interference filters, Dr. A. Robertson and Mr. A. Cameron for loan of the spectrophotometer facilities, and Mr. H.B. MacPherson and Mr. V. Clancy of the Division of Chemistry for the impurity analyses and measurement of the eutectic composition. We also thank Miss P. Desloges for the typing and laying out of the manuscript, not only of this but of all the papers to the Symposium from the Heat and Thermometry Section of the National Research Council.

REFERENCES

1. T.J. Quinn, T.R.D. Chandler, and M.V. Chattle, Metrologia 9, 44-46 (1973).
2. J. Bonhoure, Metrologia 11, 141-150 (1975).
3. L. Crovini and A. Actis, Metrologia 14, 69-78 (1978).
4. R.E. Bedford and C.K. Ma, 11e Session, Comité Consultatif de Thermométrie, Annexe T22, pp. T186-T190 (1976).
5. G. Bongiovanni, L. Crovini, and P. Marcarino, High Temperatures - High Pressures 4, 573-587 (1972).
6. M. Ohtsuka and R.E. Bedford, this Symposium, Paper y (1982).
7. R.E. Bedford and C.K. Ma, J. Opt. Soc. Am. 65, 565-572 (1975).
8. R.E. Bedford, C.K. Ma, C.R. Barber, T.R. Chandler, T.J. Quinn, G.W. Burns, and M. Scroger, Temperature, Its Measurement and Control in Science and Industry (Instrument Society of America, Pittsburg, 1972), Vol. 4 (H.H. Plumb, ed.), 1585-1602.
9. P.B. Coates, Metrologia 13, 1-5 (1977).
10. L. Crovini and P. Marcarino, 10e Session, Comité Consultatif de Thermométrie, Annexe T7, pp. T79-T82 (1974).

Reference temperatures near 800 °C

J. V. McAllan

CSIRO Division of Applied Physics, Sydney, Australia 2070

Procedures are described for calibrating resistance thermometers near 800 °C. The melting temperature of Ag-Cu eutectic alloy is $t_{68} = 779.898$ °C ($t' = 779.416$) with a reproducibility of ± 0.01 °C and an overall uncertainty of ± 0.03 °C when determined by the IPTS-68 resistance thermometer scale extrapolated from the zinc point. The alloy must be frozen very slowly before each melting point determination and this requires very precise temperature control. An alternative, the freezing point of NaCl, was investigated and found to be $t_{68} = 802.31$ °C on the same temperature scale with a reproducibility of ± 0.02 °C and an overall uncertainty of ± 0.05 °C. This point is very difficult to realize because of the infrared transparency and poor thermal conductivity of salt, and because water cannot be removed completely. A further method, useful only for comparison calibrations, is to provide a very stable temperature through a pressure-controlled heat pipe which has sodium as the working fluid.

INTRODUCTION

The temperatures at which metals freeze are the main reference points for thermometric calibration above 0°C but there are no metal elements which freeze between 660°C (Al) and 962°C (Ag). A good reference point is desirable for thermometers which will not be used above 800°C, to avoid instabilities involved in subjecting them to Ag point temperatures, and also so that laboratories can compare their scales in this region where there is a major difference between the thermocouple scale of IPTS-68 and the thermodynamic scale.

Platinum resistance thermometers appear capable of accuracies of about 0.01°C in this range and this paper describes three approaches for an 800°C calibration.

THE Ag-Cu EUTECTIC POINT

The properties of metal eutectics have been studied extensively using modern metallurgical theories and techniques, and a very clear analysis of eutectic freezing was given by Jackson and Hunt[1]. The limitations of eutectics as thermometric references were outlined by McAllan[2] who used an alloy of Al-CuAl$_2$ to provide a reference near 550°C. Bongiovanni et al.[3] studied the Ag-Cu eutectic and both studies confirmed the metallurgical predictions that the freezing temperature would depend strongly on the freezing conditions because the components in the liquid have to separate by diffusion to form the two solid eutectic phases. The IMGC freezing data[3] has been used with the metallurgical theory of Jackson and Hunt[1] to obtain an estimate of the equilibrium temperature.

Although the freezing temperature is affected by details of the freezing process, the liquid boundary layer always adjusts so that the solid which is formed is very close to the eutectic composition[1]. Such solid will melt close to the true eutectic temperature; thus the melting point temperature is not strongly dependent on the overall composition nor on details of the previous freeze. Measurements in this laboratory have confirmed that Ag-Cu behaves very similarly to Al-CuAl$_2$ and that the melt following a slow freeze gives a very reproducible temperature. The procedure recommended in Ref. 2 for Al-CuAl$_2$ has proved very satisfactory for Ag-Cu.

Apparatus and procedures

Measurements were made with a platinum resistance thermometer ($R_0 = 25$ Ω) from Chino Works Ltd, and an ac resistance bridge (Thomson and Small[4]) with a chart recorder as described by Connolly et al.[5]. This system was more than adequate for these measurements since long term changes in the freezing point of Ag-Cu are at least 50 times greater than the stability and sensitivity achieved when measuring the water triple point[6] with this equipment. No ac interference effects were detected when the furnace power was turned off for a short time.

Materials used were ASARCO high purity copper and COMINCO 69 grade silver. The eutectic composition is not known from metallurgical studies to better than 0.1%; an alloy of 28.12% by weight copper was used, and its behaviour indicated that it was very close to the eutectic value. The sample was contained in a standard 200 mm long crucible of United Carbon UT-6 grade graphite; such a crucible is normally used for primary fixed points. The general arrangement is similar to that used in many laboratories and has been described elsewhere[7], but some improvements have been effected. To permit evacuation and yet prevent silica from contacting the metal, a graphite well was used with a sealed silica well inside it. This ensured the control of the argon gas purity and reduced the risk of contamination during ingot mixing at high temperatures, but had the disadvantage of poor thermal conduction from metal to thermometer. All silica components were roughened to prevent heat loss by light piping and were cleaned in dilute HF, then in HCl, and finally in water, to lessen devitrification by residual alkali.

Graphite felt and discs were used for thermal insulation and were degassed under vacuum at 1000°C and saturated with argon before use. (Graphite felt absorbs large amounts of moisture during exposure to the atmosphere and requires prolonged initial degassing.) The felt was separated from the walls of the silica container by a thin layer of silica wool (Refrasil) to prevent devitrification of the container. To keep the oxygen content as low as possible the metals were degassed at 800°C and then dry argon was introduced before raising to 1050°C to produce the alloy. Oxygen is known to affect the freezing points of Ag and Cu, although its effect on the alloy is unknown.

Gravitational segregation of components of different densities can occur with repeated melting and freezing[2] and although the 15% density difference between Ag and Cu is much smaller than that in the Al-CuAl$_2$ system, it was felt desirable to stir the alloy at intervals. The assembly was placed with the bottom of the ingot in an r.f. heater coil, and the bottom was heated to near 1100°C, while the top reached about 850°C. Waves of red were seen rising on the sides of the crucible which indicated that severe convection of the alloy was being achieved; this was considered a more effective method of ensuring homogeneity than that used previously[2] when a furnace had excessive power in the bottom heater. Because the procedure was rather hazardous and the density differences were not great, it was not carried out after each melt but was done twice in a set of 13 melts.

Because the freezes must be very slow (to reduce departures from equilibrium) the furnace control must be very stable. The main heater was controlled using a resistance-thermometer detector system and drifts in set point appeared to be less than 0.02°C per week, although this was very difficult to determine. The end heaters were controlled using differential Pt-Pt13%Rh thermocouples and only Pt wire experienced the main furnace gradients so that Rh-oxide effects were eliminated. Large transients occurred at the ends of the furnace when the main controller was reset to change between melt and freeze; the end controllers were usually offset for a short time to compensate for this. The uniformity and stability requirements for this fixed point are much more severe than for a normal pure metal such as Ag where a stabilised power supply with manually adjusted heater voltage is usually sufficient.

Melting and freezing behaviour

Results of the measurements are given in Table I. Freezes were varied in duration from 10 to 140 hours, with one fast (about 10 min) freeze. The freezing temperatures did not follow a clear pattern, and appeared to depend on details of nucleation as well as on freezing rate, as was reported earlier for Al-CuAl$_2$[2]. The melts which followed the freeze were much more reliable and were higher in temperature. The temperature of the last 10 to 15% of each melt rose erratically from the stable value, indicating that the main crystal structure had melted and only off-eutectic high melting point solid remained. Such a rise occurs in pure metals when the last few crystals separate and hot liquid reaches the inner well, but it seems more pronounced in eutectic systems. Secondary arrests on the melting curve after the initial melt was complete were described by Bongiovanni et al.[3] and attributed to residual primary solid, but were not seen in the experiments reported here. Variation in melting time from 2 to 4 hours had no effect on any parameter measured.

The initial part of the melt was much more depressed by the rate of the preceding freeze than were the later stages, but all melts approached a similar temperature before the final rapid rise. The initial depression was attributed to curvature effects on the lamellar eutectic surfaces. Fast freezing produces finer lamellae which have a lower equilibrium temperature[1] and which will melt first. After the first r.f. stirring the ingot was frozen rapidly (in about 10 min), but was held at about 776°C for 24 hours before remelting to allow the furnace to regain equilibrium. The melt following this freeze was below the others by about 0.025°C in the initial stages but it approached the more usual values as the melt progressed so that some of the eutectic solid was near equilibrium composition even at this freezing rate.

It is considered that the best estimate of the equilibrium eutectic temperature is given by the last region of the melt before the commencement of the rapid rise, for two reasons. Firstly, the final crystals to melt should have the thickest lamellae (and highest equilibrium temperature) and be closest to the temperature of the ideal eutectic. Secondly, because most impurities depress the freezing and melting points and segregate to the boundaries of growing crystals, such impure boundary regions should melt before purer ones, so that later stages of the melt should be at the temperature of the purest material.

The melt was recorded from the time the rate of rise fell below 0.01°C/min until it again rose to this rate; outside this region the temperature changed very rapidly. The flattest region of the temperature-time melting curve was extrapolated to meet the 100% axis and the temperature obtained was considered the best estimate of the equilibrium eutectic value, and is recorded in Table I. The melting curve after the first freeze is shown in Fig. 1, and is typical. The extrapolated 'best value' is shown as point A. Other criteria, derived from a differential curve (inversion histogram) are discussed later.

The temperature is above the IPTS-68 range for the resistance thermometer, so the IPTS calibration at the ice, tin (232°C) and zinc (419°C) points was extrapolated to 780°C; the extrapolated values obtained are sensitive to small errors in the calibration. Results can be given in terms of t', a simple quadratic relation (equation 11 of IPTS-68) or t_{68}, which is t' plus a quartic term designed to compensate for departures of t' from thermodynamic temperature (equation 10 of IPTS-68). Values in Table I are in t_{68} temperatures. The present results yield t_{68} = 779.898°C (t' = 779.416) as the mean of 13 measurements, with a standard deviation of 0.009°C. The melt following the fast freeze gave a value of t_{68} = 779.884°C but this was also given by another melt, and these values could not be discarded on statistical grounds, although the melt following a fast freeze is expected to be depressed below the equilibrium value, as discussed above.

TABLE I

Details of Ag-Cu Eutectic Measurements

No.	Freeze Duration (hr)	Extrapolated Eutectic Temp. (779.xxx)	Histogram Peak Temp. (779.xxx)	% of Melt elapsed up to Peak	Diff. between Temps. (°C)	1% Range (°C)
1	13	884	873	61	0.011	0.155
2	20	893	875	62	0.018	0.230
3	20	894	870	62	0.024	0.200
4	10	893	869	58	0.024	0.155
5	78	903	890	69	0.013	0.195
6	140	906	895	61	0.011	0.075
7	< ¼ *	884	874	72	0.010	0.195
8	45	895	883	55	0.012	0.100
9	68	896	891	53	0.005	0.140
10	50	910	894	57	0.016	0.160
11	28	912	900	58	0.012	0.080
12	13	905	877	52	0.028	0.225
13	30	896	875	51	0.021	0.195
Mean	−	898	882	59	0.016	0.162
St. Dev.		0.009	0.011	6	0.007	0.052

* followed by 24 hour at 776°C.

An alternative method of obtaining a reproducible temperature is to plot a histogram showing the percentage of time spent in consecutive temperature intervals. Figure 2 is the histogram corresponding to Fig. 1. In the present experiments such a histogram provides a clear maximum temperature, ("B" in Figs. 1 and 2) and is given in Table I together with the time taken to reach this value. The histogram peak is below the extrapolated 'best' value by a mean value of 0.016°C, with a standard deviation of 0.007°C. The mean amount melted by the peak stage was 59%. This maximum value was not quite as reproducible as the extrapolated value, presumably because it referred to an earlier part of the melt and was more affected by the non-equilibrium features of the previous freeze.

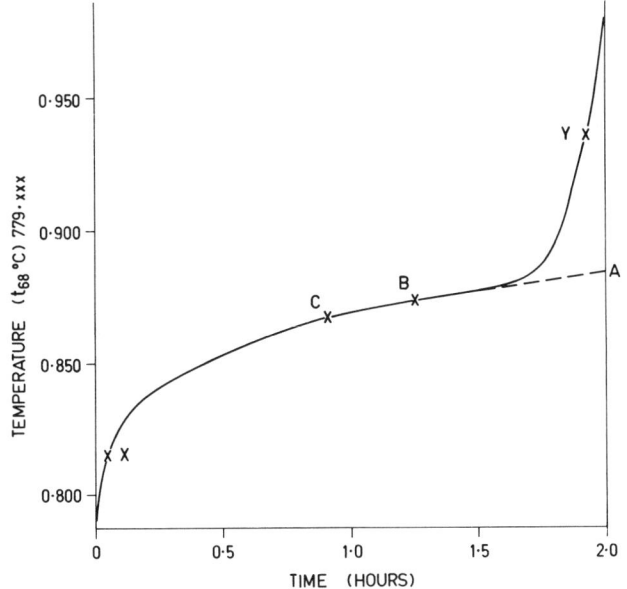

FIG. 1. Melting curve after 13 hour freeze (first freeze) - details as in Fig. 2.

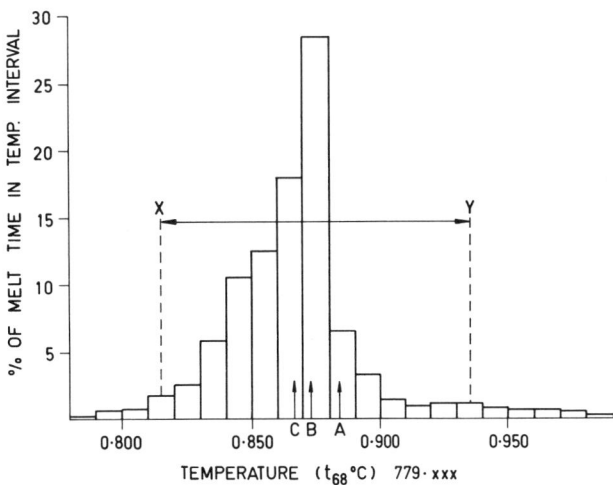

FIG. 2. Histogram of time spent in 10 mK intervals during freeze of Fig. 1
"A" - extrapolated value ('best' estimate of equilibrium temperature)
"B" - peak of histogram (flattest part of normal curve)
"C" - centroid of histogram
"X" and "Y" - limits of the 1% range.

The histogram displays the rate of change of the melt, and minor arrests on an otherwise smooth melt would appear as minor peaks on the histogram. No such arrests were apparent in these experiments, although some sharp rises which occurred in the last stages of some melts appeared as gaps in the histogram. A useful index is the '1% range', the range of temperatures over which at least 1% of the total melting time is spent in each increment of the histogram. Table I shows this 1% range, using increments of 10 mK on the histogram, and this is equivalent to the range of temperature over which the rate of change of the melt is less than 10 mK per % of melt time. The 1% range has a mean of 0.16°C, with a standard deviation of 0.05°C.

A third method of assessing a melting temperature is to use the centroid of the histogram, which is most useful when a sample has several peaks indicating segregation of impurities. This is indicated by "C" on Figs. 1 and 2. Because of the smooth curves, indicating high purity material and uniform freezing conditions, this value was always below the peak value in the present arrangements. The difference was 13 mK after the fast freeze (No. 7) but in all other cases was below 10 mK.

A, B, and C appear to have little connection in Fig. 1, but Fig. 2 shows their close relation. Again, the positions of X and Y show that Fig. 1 is of no value in assessing such impurity parameters.

Comparison with earlier measurements

The only measurements on the Ag-Cu eutectic which have comparable precision to the present ones and utilise modern metallurgical theories are those obtained at IMGC[3]. The 1% melting ranges are very similar, the two curves published in Ref. 3 having ranges of 0.18°C although the second sample showed a secondary arrest. The peak on the histogram always occurred before 50% of the sample melted (average 36%) whereas in the present work it was always after 50% (average 59%). In part this might be due to the use of different criteria for the beginning and end of the melt, especially since IMGC reported arrests after most of the primary eutectic was melted. Their values of (peak - 50% melted) temperature differences should be similar to the peak-centroid differences of the present work, but are of opposite sign and vary from 8 to 63 mK. This is also probably due to secondary arrests and the very long temperature range included in their histogram calculations.

The peaks in the published histograms are not nearly as sharp as those of the present work, despite their wider temperature unit of 15 mK, implying that their materials are not as pure, and their second sample showed histogram secondary peaks typical of impurity effects. However the 1% range for the first sample, although it is difficult to assess from the published histogram, appears to be very similar to the present values.

Bongiovanni et al. estimated the difference between their melting peak and the true equilibrium value by analysing one freeze very thoroughly, using the metallurgical theory of Jackson and Hunt[1]. They estimated that the equilibrium value was above their melting peak by 10 mK ± 10 mK. The theory is for rate dependent effects in the eutectic and does not consider impurities, which would be expected to increase the difference. The extrapolated value of the present work is higher than the peak by 16 mK ± 7 mK, implying that it is close to the difference expected by Bongiovanni et al.

Values for the IMGC determination of the Ag-Cu eutectic are given by Marcarino and Crovini[8]. Only two of their thermometers were used with the eutectic, the others being compared in a copper block with these two thermometers whose values for the histogram peak were t_{68} = 779.828°C (t' = 779.347) and t_{68} = 779.868°C (t' = 779.386). These should be compared with the histogram maximum in the present work of t_{68} = 779.882°C with a standard deviation of 0.011°C (Table I, column 4).

The 0.040°C difference between the two IMGC thermometers[0] is larger than the uncertainty of ± 0.007°C associated with the determination of the peak quoted in §5 of the paper by Bongiovanni et al.[3] or the standard deviation of 0.012°C of Table 5 of their published results. The difference arises not only from the difficulties in realising this fixed point but also from the problems in obtaining adequate stability in their thermometers and accuracy in their calibration. Even reported values of the Al freezing point (660°C) using resistance thermometry differ by 0.010°C[7,9]; the requirements for measurements at 780°C are much more severe. In the light of these difficulties the difference of 0.034°C between the mean histogram peak of present work and the mean of the IMGC thermometers must be taken as good agreement between the resistance thermometry scales of the two laboratories.

The Ag-Cu point requires very demanding experimental techniques and it was considered simpler by Marcarino and Crovini[8] to carry out conventional thermometer comparisons as much as possible. However it is only by measurements on a fixed point that it is possible to compare the scales of the two laboratories in this region, unless thermometers are transported between laboratories with risk of calibration changes due to vibrational strain. Since most high temperature thermometers are designed with a very open structure to minimise electrical insulation leakage they are very prone to such vibration induced instability.

THE FREEZING POINT OF NaCl

Salts are a possible alternative to metal eutectics as a source of reference temperatures. Many simple salt systems are available and, if salt eutectics are included, it is possible to select a possible reference temperature close to any desired temperature. Salts are of interest not only as temperature references but also as materials for heat storage and transfer and similar properties are of interest for both purposes. A simple salt was used, thereby avoiding the equilibration problems of eutectics; NaCl was chosen because it could be obtained commercially with a purity that was much higher than other salts. Major problems were expected from impurities and from heat transfer and the precautions taken and results obtained are described below.

Salt purity

The NaCl was Merck "Suprapur" grade. The manufacturer's estimate was checked by spark source mass spectrometry, and K and Ca content were also checked by flame emission spectroscopy. These measurements were carried out by the Lucas Heights Research Establishment of The Australian Atomic Energy Commission, who considered that the values obtained would be high rather than low; so impurity effects should not be greater than those calculated from these analyses. The major impurities are listed in Table II.

As expected K was the major impurity, but K, Ca and Sc were well above the manufacturer's estimate while Ba, Cs and Rb were low. Levels of Br and S (possibly as sulphate) were low compared to the metal ions, but the likely impurities OH^- and H^+ could not be measured.

For a pure metal such as Zn or Sn such impurity levels would change the freezing point temperature by at most 2 mK and give a melting range from 10 to 20 mK. The van t'Hoff calculation of freezing point depression for NaCl gives

$$\Delta T/X = (RT^2/L)(1 - k)$$
$$= 0.34 \text{ mK per part per million (molar)}$$

where ΔT is the temperature depression caused by X mole of impurity, T is the freezing temperature, R the gas constant, L the latent heat and k the distribution coefficient (ratio of solubility in solid to solubility in liquid).

The value 0.34 mK/ppm is a maximum obtained by assuming complete insolubility in the solid (k = 0) and it implies a depression of not more than 19 mK for 56 ppm impurity. The solubility of the major impurities in the solid should be very high, especially for K, and so a depression of only a few mK was anticipated. The phase diagrams (Levin et al.[10]) for K and Ca were used to estimate k and obtain depressions of 1.4 mK (0.034 mK/ppm) for K and 2.4 mK (0.27 mK/ppm) for Ca. Thus this sample of NaCl was expected to provide freezing points of good quality, comparable to the best melting points obtained with the Ag-Cu eutectic although not as good as very high purity materials such as Sn.

However it is known that water adsorbed on most salts will dissociate and that OH^- is especially difficult to remove. This would depress the freezing point as would an addition of NaOH, whose phase diagram indicates a depression of about 0.2 mK/ppm. Moreover, water

TABLE II

Impurity in NaCl (Parts Per Million)

Element	Manufacturer's Estimate	Measured
Br	-	1.5
S	-	1
Mg	0.05	0.2
Ca	0.1	9
Sr	0.1	0.3
Ba	5	< 0.2
Sc	-	4
K	5	40
Rb	1	< 0.1
Cs	5	< 0.1
Total	16	56

(or OH^-) is extremely corrosive in the presence of molten salts or salt vapour, and so several procedures were used to remove water from the system. Traces of a few parts per million would be serious both for temperature reproducibility and corrosion effects.

Sample preparation

The crucible assembly was similar to that used for the Ag-Cu experiments. The crucible was made from Union Carbide ATJ grade and was cleaned by boiling in HCl and baking in vacuo to remove volatile metallic chlorides (Rumbo and McAllan[11]). This procedure was known to be effective in removing Ca, Al and Fe from the graphite. The main concern in preparing the samples was the removal of water.

Three samples were used. NaCl sublimed appreciably at 550°C in vacuo, so argon gas was used as much as possible at temperatures above 450°C, with evacuation for short periods only. The first and second samples were dried by evacuation at room temperature and then at different elevated temperatures. The procedures were repeated twice to introduce enough NaCl (about 240 g) to fill the crucible.

Temperature measurements on the first sample took 21 days and on the second 14 days. In these times nearly 1 g of NaCl sublimed and deposited on cooler parts of the system. All silica components which had been hot were very badly devitrified, indicating that residual moisture had been in contact with the NaCl and silica, and that OH^- attack had occurred. Evacuation can drive off H^+ in the form of HCl as well as H_2O so that it is very difficult to remove the last traces of OH^- by this relatively simple method.

The third sample was saturated with dry HCl to reduce the dissociation of H_2O. The HCl was generated from the same quality NaCl as the freezing point samples using degassed H_2SO_4, and was transferred through an aerosol trap and two simple distillations using cold tubes in liquid nitrogen. The dry HCl could be moved around the system simply by varying the temperature of the cold tubes, and after each charge of HCl was removed the system was evacuated thoroughly through a liquid nitrogen trap. The sample was evacuated for 24 hours at room temperature, 100 and 200°C to remove as much water as possible and was then treated with separate charges of 20 kPa of HCl at 450, 500 and 700°C, and the salt was then melted with HCl present. Extra NaCl was added and the whole process repeated. The well was then pushed into the salt with most of the thermal insulation in place and using a further change of HCl. This final HCl was removed after cooling to 500°C and the sample was evacuated while cooling to room temperature before adding dry argon.

NaCl in the third sample sublimed as before during measurements over 21 days but there was no sign of devitrification of silica. The HCl saturation technique seems a satisfactory method of removing OH⁻ from chlorides, thereby stopping severe devitrification, and it is necessary if a sample is to be used for extended periods at 800°C. The technique is awkward and hazardous and is one serious disadvantage of the NaCl freezing point procedures.

Thermal properties

NaCl is a poor absorber of 800°C thermal radiation and has very low thermal conductivity and diffusivity compared with metals. Consequently the liquid-solid interface is not closely coupled by conduction or radiation to the graphite well, and the well receives radiation directly from the wall of the crucible. Any temperature change or gradient along the crucible disturbs the temperature of the well and is not effectively attenuated by the latent heat of the melting interface.

Most salts expand considerably on melting and NaCl expands by over 25% by volume. This will cause a considerable movement of liquid in the crucible and also will produce large voids in an ingot which is freezing uniformly from the outside. If the crucible is hot at the top the freezing salt will have a large concave depression at the top centre as the liquid falls, and this will reduce the effective immersion of the well in the salt. There is a tendency in the later stages of ingot melting for crystals to fall from the well. The well is absorbing radiation and presumably is transferring heat to the crystals which melt free. The lowest region in the crucible contains the loose crystals and is therefore the last to melt, and stays close to the melting temperature despite radiation transfer from the hot walls. It was not clear if there would be an advantage in the standard technique introduced by McLaren of chilling the well after a freeze had nucleated. This technique produces a thin layer of solid on the well while most of the freezing occurs at the outer wall and it was thought that a thin inner layer might have enough control by conduction to produce a suitably stable freezing temperature. Experiments were carried out to assess this technique.

Thermal behaviour

A commercial platinum resistance sensor (Rosemount type 1050) was mounted in a roughened silica tube with fine leads and insulators designed to minimise conduction errors. This thermometer was used to investigate the effects of furnace imbalance and of melting and freezing rates.

To simplify the procedures used for Ag-Cu, stabilised power supplies were used without automatic temperature controllers. The temperature gradients in the ingot proved very sensitive to end heater power, presumably due to the very poor thermal diffusivity of the salt. There was a temperature maximum after about 15% of the freeze had occurred, but the temperature fell steadily after this unless the furnace was held very close to the freezing temperature, and simple power stabilisation proved inadequate. Melts started at temperatures above but within 0.1°C of the freezes but were very broad, a range of 0.4°C being typical, with large vertical gradients occurring in the later stages of the melt. This was presumably because the remaining solid had fallen to the bottom of the crucible. At this stage there were erratic variations in temperature of about 0.02°C either due to convection or to stirring by the moving solid.

Part of this melting behaviour could have been due to segregation of OH⁻ impurity in the salt, but it was clear that very close temperature control would be required and therefore the furnace controls used for the Ag-Cu studies were used for sample 3. Measurements were made with the same thermometer used for the Ag-Cu work so that it is possible to compare the behaviour of the NaCl and Ag-Cu under very similar experimental conditions. It was desired to obtain uniformity in the furnace and small temperature differences between furnace and crucible; this made the melts and freezes very long, varying from 4 to 24 hours. Even in a 24 hour melt the range was 0.08 K, the 50 mm point rose rapidly in temperature 2 hours before the temperature finally rose at the bottom of the well, and the erratic variations were still present but were smaller. There was no obvious difference between a freeze where the thermometer was undisturbed after nucleation and another where it was removed and replaced after about 30 seconds, thus chilling a layer of solid on the well.

The first stable region on the melt and the hottest portion of the freeze were taken as the best values, undisturbed by radiation effects from the walls. The reproducibility of this series of measurements was much improved over those on sample 2, the standard deviation of freezes falling from 0.12 to 0.022°C and of melts from 0.11 to 0.013°C. The mean of 4 melt values was given by the IPTS-68 resistance thermometer equations as $t_{68} = 802.323$°C and the mean of 4 freezes as $t_{68} = 802.303$°C. The best value for the NaCl point is considered to be 802.31°C with a reproducibility of ± 0.02°C and overall uncertainty of ± 0.05°C.

However it is difficult to decide if the chosen portion of the melting curves is the correct one, and although the reproducibility is comparable to that of the Ag-Cu eutectic, the stable portion of both melts and freezes is very short. Since the furnace control problems are even more severe, and the HCl treatment for OH⁻ removal is much more difficult than the procedures required for Ag-Cu, NaCl could not be recommended as a fixed point in preference to the Ag-Cu eutectic.

POSSIBLE IMPROVEMENTS

The most useful improvement in the realisation of these fixed points would be to produce more uniform and stable furnaces. It is possible to use sealed heat pipes to smooth furnace gradients. Such a heat pipe with sodium as working fluid in an inconel container would be satisfactory at 780°C and suitable heat pipes are available commercially. A heat pipe provides very high thermal conductivity through the latent heat of vaporisation of the working fluid, as long as the container walls are not too thick, but it does not have a very high thermal inertia so that it will follow variations in the furnace temperature. Consequently a sealed heat pipe does not relax the requirements on the main control system although end heater control need not be as precise.

It may be possible to relax the furnace requirements by using a pressure regulated heat pipe of the type developed by Bassani et al.[12]. In such a system power variations alter the amount of vapour which reaches the water-cooled condenser, but have little effect on the temperature, which is controlled by vapour pressure. A thermally insulated ballast volume can provide a reference pressure very easily. Precision pressure measurement is not necessary for comparisons, and the system is then very much simpler than the complex Ag-Cu or NaCl systems.

Measurements have been made on such a heat pipe in this laboratory, over the range 500-700°C, and stability of ± 2 mK per day can be achieved, with a zone about 250 mm long uniform to < 10 mK. The uniformity and stability appear better as the temperature is increased, and measurements are continuing at higher temperatures.

However, such an arrangement can only be used to compare laboratory scales by transporting thermometers between laboratories, with considerable risk of introducing instabilities through strain from vibration. Separate determinations of a fixed point such as Ag-Cu provide a reliable and permanent comparison of such scales. An example of such a use of a fixed point is given elsewhere in this Symposium by McAllan and Connolly[13].

CONCLUSION

The melting temperature of Ag-Cu was measured by extrapolation of the resistance thermometer scale of IPTS-68 and was found to be t_{68} = 779.898°C (t' = 779.416°C) with a standard deviation of 0.009°C in 13 measurements, and an estimated uncertainty of 0.03°C including calibration errors. The criterion for assessing the best value is different from that used in the IMGC investigations, but when the IMGC criterion is used the present value is 0.034°C above the IMGC mean.

The best value of the freezing point of NaCl on the same scale is t_{68} = 802.31°C with an overall uncertainty of ± 0.05°C. Both these fixed points are very difficult to realise, but the Ag-Cu eutectic is a simpler and better system than NaCl.

It is possible that a pressure-controlled heat pipe using sodium as the working fluid will provide a much more convenient comparison device than the above fixed points. However it does not provide as straightforward a comparison between laboratories as does a fixed point, such as the Ag-Cu eutectic reported here.

Acknowledgement: The author wishes to express his appreciation of the advice and assistance given by Dr J.S. Cook on the problem of removing water contamination from the salt.

REFERENCES

1. K. A. Jackson and J. D. Hunt, Trans AIME 236, 1129 (1966).
2. J. V. McAllan, Temperature, its Measurement and Control in Science and Industry 4(1), 265 (1972).
3. G. Bongiovanni, L. Crovini and P. Marcarino, High Temp. - High Pressures 4, 573 (1972).
4. A. W. Thomson and G. W. Small, Proc. IEEE 118, 1622 (1971).
5. J. J. Connolly, J. V. McAllan and G. W. Small, Temperature, its Measurement and Control in Science and Industry 4(2), 1487 (1972).
6. J. V. McAllan "The effect of pressure on the water triple point temperature" This Symposium.
7. J. V. McAllan and M. M. Ammar, Temperature, its Measurement and Control in Science and Industry 4(1), 275 (1972).
8. P. Marcarino and L. Crovini, Inst. Phys. Conf. Ser. No. 26, 107 (1975).
9. G. T. Furukawa, W. R. Bigge, J. L. Riddle and M. L. Reilly, Inst. Phys. Conf. Ser. No. 26, 389 (1975).
10. E. M. Levin, C. R. Robbins and H. F. McMurdie, Phase Diagrams for Ceramists, Amer. Ceramic Soc. (1964) supplements (1969, 1975).
11. E. R. Rumbo and J. V. McAllan, Laboratory Practice 24, 73 (1975).
12. C. Bassani, C. A. Busse and F. Geiger, High Temp. - High Pressures 12, 351 (1980).
13. J. V. McAllan and J. J. Connolly, The use of the cadmium point to check calibrations on the IPTS" This Symposium.

Radiance temperature of metals at their melting points as possible high temperature secondary reference points[a]

A. Cezairliyan and A. P. Miiller

Thermophysics Division, National Bureau of Standards, Washington, D.C. 20234

F. Righini and A. Rosso

CNR Istituto di Metrologia "G. Colonnetti," Strada Cacce 73, Torino, Italy

A summary is given of the measurements performed at the National Bureau of Standards and at the Istituto di Metrologia for the determination of the radiance temperature at two wavelengths (near 0.65 and 1 μm) of selected metals (Fe, Pd, Ti, Zr, V, Nb, Mo, and Ta) at their melting points. The melting temperature of the metals studied ranged from about 1800 to 3300 K. All the experiments were performed with a pulse-heating technique, in which the specimen was heated from room temperature to its melting point in less than one second by the passage of an electrical current pulse through it. Millisecond-resolution optical pyrometry was used for temperature measurements. The measurements show constancy and reproducibility of the radiance temperature at the melting point for a given metal, irrespective of the initial surface conditions of the specimen, and of operational conditions. The results suggest the possibility of the use of the radiance temperature of selected metals at their melting point as secondary reference points.

1. INTRODUCTION

In optical pyrometry, realization of secondary reference points usually involves blackbody cavities, operating at the solid-liquid equilibria of selected pure metals with high melting temperatures. Successful operation of blackbody cavities at high temperatures becomes increasingly difficult with increasing temperature. The difficulties are the result of severe problems that are created at high temperatures because of increased heat transfer, chemical reactions, evaporation, loss of mechanical strength and electrical insulation, etc.

Blackbody cavities that operate reliably near the gold point (1337.58 K) can be constructed without any major problem. However, successful operation of blackbody cavities at temperatures above about 1800 K becomes considerably difficult, and above 2500 K, becomes extremely difficult if not almost impossible.

In an earlier study,[1] it was noted that the radiance temperature[b] of niobium at 0.65 μm during its initial melting period was constant and reproducible. Subsequent similar measurements on other selected metals,[2-8] both at 0.65 and 1 μm, have confirmed the above observation. This finding has suggested that the technique of measuring radiance temperature at the melting point of selected metals may be a simpler and possibly a more accurate alternative to the use of elaborate blackbody cavities for performing secondary calibration on instruments and for conducting in-situ checks on complicated measurement systems at high temperatures.

The objective of this paper is to summarize the results of the radiance temperature of selected metals (Fe, Pd, Ti, Zr, V, Nb, Mo, Ta) at their melting point measured with a millisecond-resolution dynamic technique and to discuss the possibility of the use of the radiance temperature as high temperature secondary reference points. The measurements at 0.65 μm were performed at the National Bureau of Standards (NBS), and those at 1 μm were made at the Istituto di Metrologia "G. Colonnetti" (IMGC).

2. METHOD

The method, used at both NBS and IMGC, is based on rapid resistive self-heating of the specimen from room temperature to its melting point in less than one second by the passage of an electrical current pulse through it; and on measuring specimen radiance temperature with a high-speed optical pyrometer. A schematic diagram showing the arrangement of the specimen and the pyrometer is presented in Fig. 1. Details regarding the construction and the operation of the two measurement systems are given in the literature: the NBS

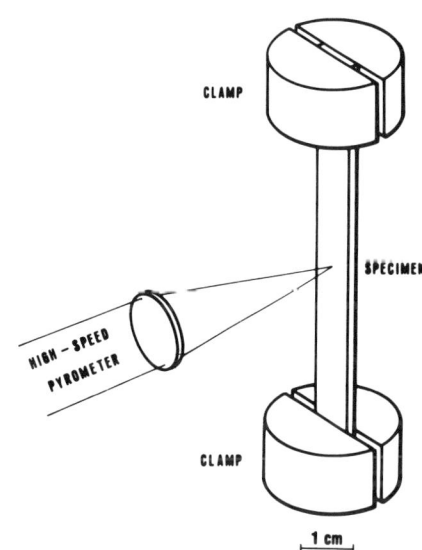

FIG. 1. Schematic diagram showing arrangement of the specimen, clamps, and pyrometer. Dimensions are not to scale.

system[9,10] and the IMGC system[11,12]. At NBS, temperature is measured at the rate of 1200 measurements per second with a photoelectric pyrometer[13] operating at an effective wavelength near 0.65 μm. The bandwidth of the interference filter is 10 nm, and the circular area viewed by the pyrometer is 0.2 mm in diameter. At IMGC, temperature is measured at the rate of up to 13,000 measurements per second with an infrared pyrometer[14] operating at an effective wavelength near 1 μm. The bandwidth of the interference filter is 50 nm, and the circular area viewed by the pyrometer is 0.2 mm in diameter.

3. MEASUREMENTS

In this section, a summary of procedures and parameters relevant to the measurements of radiance temperature at the melting point of selected pure metals which have been reported earlier[1-8], is presented.

The specimens were in the form of strips with the following nominal dimensions: length, 35-75 mm; width, 3-6.3 mm; and thickness 0.25 mm. For a given metal, specimens with different surface roughnesses were used in order to establish the insensitivity of the radiance temperature at the melting point to the initial surface conditions of the specimen. Before the experiments, the surface of most specimens was treated with different grades of abrasive, yielding roughnesses (RMS) in the range 0.2-0.95 μm. Specimens in "as received" condition had a surface roughness of about 0.1 μm.

The measurements were performed with the specimen either in vacuum or in an inert gas (argon) environment at about atmospheric pressure. In most cases, the specimens were annealed by the manufacturer; in some cases the specimens were heat treated in the laboratory.

Different heating rates, ranging from 400-20,000 K·s^{-1}, were used in the experiments. The heating period for the specimens (from room temperature to the melting point) ranged from 0.04 to 1.2 s. Duration of the melting plateau ranged from 0.008 - 0.17 s.

Figure 2 shows the variation of radiance temperature (near 0.65 μm) as a function of time near and at the melting point for five typical experiments on niobium. These experiments were selected to represent different specimen surface and operational conditions. All the curves with the exception of "A" show spikes of varying magnitude just before the onset of melting. The spike is due to the change in the normal spectral emittance of the specimen surface as it starts to melt. The specimen heating rate in experiments corresponding to curves "B", "C", "D", and "E" were four times slower than that of "A". Thus, the lack of a spike in curve "A" may be attributed to the inability of the system to

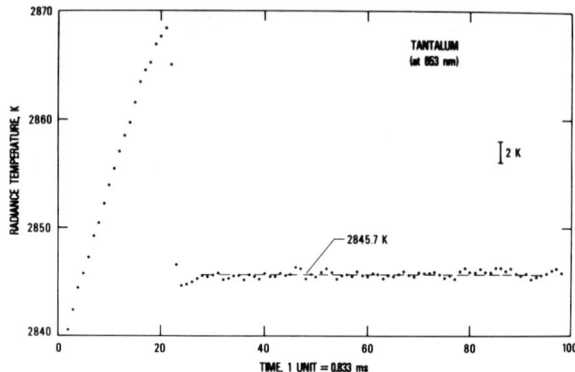

FIG. 3. Variation of radiance temperature (at 653 nm) as a function of time near and at the melting point of a tantalum specimen (surface roughness: 0.2 μm).

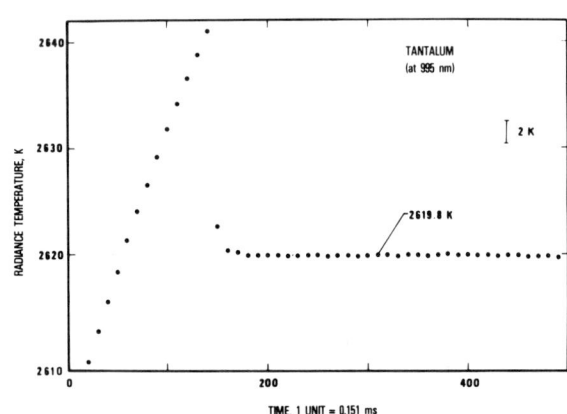

FIG. 4. Variation of radiance temperature (at 995 nm) as a function of time near and at the melting point of a tantalum specimen (surface roughness: 0.5 μm). One temperature out of ten is plotted.

detect it. It is interesting to observe that, regardless of the initial surface and operational conditions, radiance temperature at the melting plateau is approximately the same (within ±1 K) for all the specimens.

Figures 3 and 4 show the variation of radiance temperature as a function of time near and at the melting point for tantalum specimens near 0.65 μm and 1 μm, respectively. Constancy of the radiance temperature at the melting plateau for both wavelengths is evident.

4. RESULTS

The radiance temperature of selected metals at their melting points as well as other pertinent information corresponding to measurements near 0.65 μm and 1 μm are presented in Tables I and II, respectively. All temperatures reported in this paper are based on IPTS-68[15].

The radiance temperature at the melting point of a given specimen was obtained by averaging the measured radiance temperatures along the plateau. Typically, the standard deviation of an individual temperature from the average was in the range 0.2-0.5 K at 0.65 μm and was in the range 0.1-0.3 K at 1 μm. Ultimately, a final value for each metal (listed in Tables I and II)

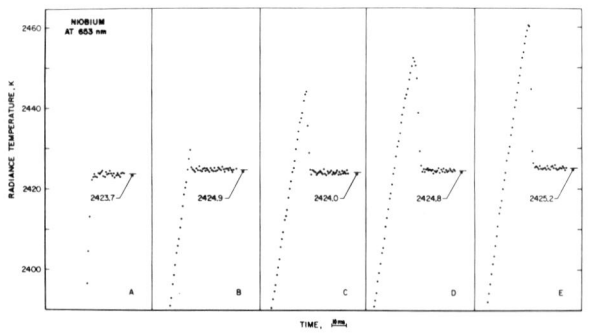

FIG. 2. Variation of radiance temperature (at 653 nm) of niobium as a function of time near and at its melting point for five typical experiments representing specimens with different surface roughnesses. Correspondence between curve codes and surface roughnesses (in μm) are as follows: A, 0.1; B, 0.1; C, 0.4; D, 0.5; and E, 0.9.

Table I. Summary of Measurements of Radiance Temperature (at 653 nm) at the Melting Point of Selected Metals.

Metal	Ref.	Purity (%)	Number of Specimens	Radiance Temp. (K)	Avg. Abs. Deviation (K)	Max. Abs. Deviation (K)	Est. Max. Uncertainty (K)
Iron	2	99.95	13	1670	0.8	1.7	± 6
Palladium	7	99.99+	12	1688	0.5	0.9	± 5
Titanium	6	99.9+	10	1800	0.4	0.6	± 6
Zirconium	3	99.98	13	1940	0.3	0.7	± 8
Vanadium	8	99.9+	14	1992	1.2	1.9	± 7
Niobium	1	99.9+	12	2425	0.6	1.2	± 10
Molybdenum	4	99.95	14	2531	0.6	1.2	± 8
Tantalum	5	99.99+	15	2846	1.0	1.7	± 8

Table II. Summary of Measurements of Radiance Temperature near 1 μm at the Melting Point of Selected Metals.

Metal	Ref.	Purity (%)	Effective Wavelength (nm)	Number of Specimens	Radiance Temp. (K)	Avg. Abs. Deviation (K)	Max. Abs. Deviation (K)	Est. Max. Uncertainty (K)
Titanium	6	99.9+	997	11	1711	0.8	1.3	± 6
Vanadium	8	99.9+	993	15	1875	0.3	0.5	± 7
Molybdenum	4	99.95	995	14	2331	0.7	1.2	± 8
Tantalum	5	99.99+	995	14	2620	0.4	0.9	± 8

was obtained by averaging the results of the individual specimens of that metal. Average absolute deviation (listed in Tables I and II) of the individual results from the final value for a metal is in the range 0.3–1.2 K at 0.65 μm and is in the range 0.3–0.8 K at 1 μm. The corresponding maximum deviations are in the range 0.6–1.9 K at 0.65 μm and are in the range 0.5–1.3 K at 1 μm. As examples, the deviations from the final value for molybdenum are presented in Figs. 5 and 6 corresponding to measurements near 0.65 μm and 1 μm, respectively.

In order to determine the trend of measured temperatures at the plateau, temperatures for each experiment were fitted by a linear function in time using the least squares method. The slopes of the linear functions did not show any significant bias with respect to direction during the initial melting period. Temperature difference between the beginning and end of the plateau was generally less than 1 K.

Intercomparison of the results of radiance temperature measured by different investigators at wavelengths slightly different from each other requires a

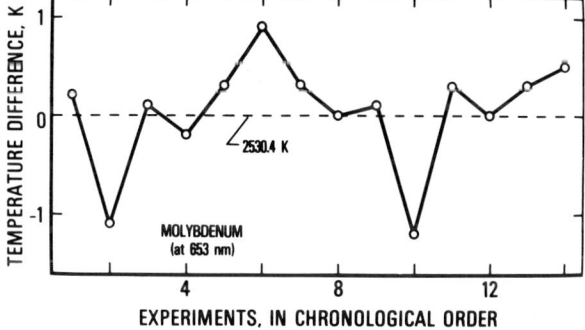

FIG. 5. Difference of radiance temperature (at the melting point of molybdenum, at 653 nm) for individual experiments from their average value of 2530.4 K.

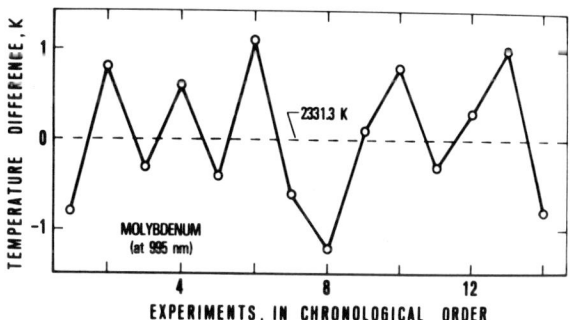

FIG. 6. Difference of radiance temperature (at the melting point of molybdenum, at 995 nm) for individual experiments from their average value of 2331.3 K.

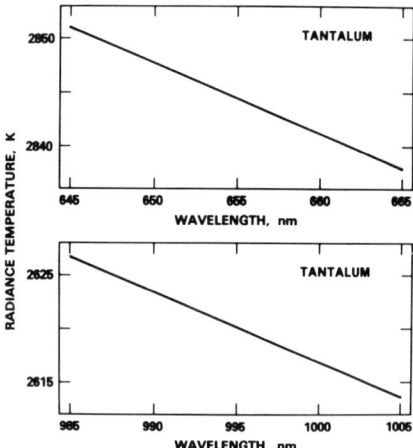

FIG. 7. Wavelength dependence of the radiance temperature of tantalum at its melting point near 0.65 and 1 μm [computed with the use of Eq. (1)].

Table III. Typical estimated uncertainty in radiance temperature measurements (at 2500 K and at 0.65 μm) with the high-speed photoelectric pyrometer at NBS.

Source	Uncertainty (K)
Standard lamp calibration	3
Drift in standard lamp calibration	1
Radiation source alignment	2
Neutral density filter calibration	1
Window calibration	1
Pyrometer calibration stability	2
Effective wavelength calibration	2
Total uncertainty (root sum square of above items)	5

Table IV. Typical estimated uncertainty in radiance temperature measurements (at 2400 K and at 1 μm) with the high-speed infrared pyrometer at IMGC.

Source	Uncertainty (K)
Gold-point calibration	1.4
Neutral density filter calibration	1.5
Window calibration	1
Radiation flux ratio calibration	4.5
Pyrometer calibration stability	2
Effective wavelength calibration	2
Total Uncertainty (root sum square of above items)	6

knowledge of the dependence of radiance temperature on wavelength. The following relation, derived from Planck's law, may be used to obtain the dependence of radiance temperature $T_r(\lambda)$, on wavelength λ, of a metal at its melting point:

$$T_r(\lambda) = c_2 \left[\lambda \ln\left(1 + \frac{\exp(c_2/\lambda T_m) - 1}{\varepsilon(\lambda)}\right) \right]^{-1} \quad (1)$$

where c_2 is the second radiation constant (0.014388 m·K), T_m is the melting point and $\varepsilon(\lambda)$ is the normal spectral emittance at the melting point. In the present work, the quantity $\varepsilon(\lambda)$ for a given metal is evaluated from a linear relation between the two values corresponding to the two wavelengths (near 0.65 and 1 μm). At each wavelength, $\varepsilon(\lambda)$ is obtained from the measured radiance temperature at that wavelength and the melting point. As an example, the wavelength dependence of the radiance temperature of tantalum is presented graphically in Fig. 7 for two wavelength regions (near those used in the experimental measurements). The results indicate that variation of radiance temperature with wavelength is nearly linear over the wavelength ranges considered, and that radiance temperature decreases with increasing wavelength. The slope of radiance temperature with respect to wavelength is: -0.26 K·nm^{-1} for titanium, -0.35 K·nm^{-1} for vanadium, -0.60 K·nm^{-1} for molybdenum, and -0.66 K·nm^{-1} for tantalum.

A limited amount of data exist in the literature on radiance temperature (or normal spectral emittance) of metals at their melting point. In general, the differences in radiance temperature (between results given in this paper and those reported in the literature) are within 10 K; in a few cases differences as large as of the order 20 K have been noted. The details are discussed in the individual papers cited in Tables I and II.

5. ESTIMATE OF ERRORS

The details of sources and estimates of errors in temperature measurements in dynamic experiments with the present system are given in earlier publications: the NBS system[9], and the IMGC system[14,16]. Summaries of typical estimated uncertainties (combined random and systematic errors) in radiance temperature measurements with the high-speed pyrometers pertinent to the work described in this paper are given in Tables III and IV for measurements at 0.65 μm and 1 μm, respectively.

The estimated uncertainty in radiance temperature measurements at both wavelengths and at about 2400-2500 K is not more than 6 K. At lower temperatures, near 1500 K, the estimated uncertainty is about 4 K.

Measured radiance temperature at the melting point depends on the purity of the specimen. The effect of impurities in the metals studied on the measurements is estimated to be in the range 0.5-2 K.

The estimated maximum uncertainty (due to both pyrometry and specimen impurities) in radiance temperature measurements at the melting point of selected metals is in the range 5-10 K and is given in Tables I and II for measurements at 0.65 μm and 1 μm, respectively.

6. DISCUSSION

Results of dynamic measurements on selected metals have demonstrated the constancy and reproducibility of the radiance temperature (generally within ±1 K) at the melting point of each metal both at about 0.65 and 1 μm. It has been also shown that radiance temperature at the melting point is independent of the initial surface conditions (roughness) of the specimen.

Radiance temperatures (at melting point) covered in this paper are in the range 1670 K (iron) to 2846 K (tantalum) at about 0.65 μm, and are in the range 1711 K (titanium) to 2620 K (tantalum) at about 1 μm. Unquestionably, tungsten will be an important addition to the list extending the upper temperature by several hundred degrees. Also, it will be desirable to fill the gap between 1992 K (vanadium) and 2425 K (niobium). Hafnium is a candidate. However, difficulties in obtaining pure hafnium may discourage its use. Radiance temperature at the melting point of hafnium (containing 3 wt.% zirconium) near 0.65 μm is 2236 K, and is found to be constant and reproducible[17].

Compared to the difficulties encountered in the operation of blackbody cavities at temperatures above 1800 K, the technique of measuring radiance temperature at the melting point of selected metals may be a more practical and a more accurate approach for performing secondary calibration on instruments. This technique may also be useful for conducting in-situ checks on complicated measurement systems at high temperatures.

The present work has demonstrated the advantages and the reliability of radiance temperature measurements in applications to high-speed pyrometry. Application to conventional pyrometry (time response about 1 s or slower) would require the development of techniques for the maintenance of stable metallic surfaces over tens of seconds during solid-liquid transitions.

Additional work is needed to accurately establish the wavelength dependence of radiance temperature at the melting point of selected metals. Measurements of radiance temperature at 0.1 μm intervals in the wavelength range of interest is likely to be adequate for this purpose.

For pyrometry and related work, accurate data are needed on radiance temperature at the melting point of at least five or six metals in the temperature range 1500–3200 K. The final selection of metals for high temperature secondary reference points will depend on the results of measurements of radiance temperature at several wavelengths. It will also depend on several factors such as, favorable material properties (vapor pressure, machinability), and availability and cost of pure specimens.

REFERENCES

a. The work performed at NBS was supported in part by the US Air Force Office of Scientific Research.
b. Radiance temperature (sometimes referred to as brightness temperature) of the specimen surface is the temperature at which a blackbody has the same radiance as the surface, corresponding to the effective wavelength of the measuring pyrometer.

1. A. Cezairliyan, J. Res. Nat. Bur. Stand. (US) 77A, 333 (1973).
2. A. Cezairliyan, and J.L. McClure, J. Res. Nat. Bur. Stand. (US) 79A, 541 (1975).
3. A. Cezairliyan, and F. Righini, Rev. Int. Hautes Tempér. et Réfract. 12, 201 (1975).
4. A. Cezairliyan, L. Coslovi, R. Righini and A. Rosso, in Temperature Measurement, B.F. Billing and T.J. Quinn, eds. (Conference Series No. 26, Institute of Physics, London, 1975), p. 287.
5. A. Cezairliyan, J.L. McClure, L. Coslovi, F. Righini, and A. Rosso, High Temperatures-High Pressures 8, 103 (1976).
6. F. Righini, L. Coslovi, A. Cezairliyan, and J.L. McClure, in Proceedings of the Seventh Symposium on Thermophysical Properties, A. Cezairliyan, ed. (Am. Soc. Mech. Eng., New York, 1977), p. 312.
7. A.P. Miiller, and A. Cezairliyan, High Temperature Science 11, 41 (1979).
8. A. Cezairliyan, A.P. Miiller, F. Righini, and A. Rosso, High Temperature Science 11, 223 (1979).
9. A. Cezairliyan, M.S. Morse, H.A. Berman, and C.W. Beckett, J. Res. Nat. Bur. Stand. (US) 74A, 65 (1970).
10. A. Cezairliyan, J. Res. Nat. Bur. Stand. (US) 75C, 7 (1971).
11. F. Righini, A. Rosso, and G. Ruffino, High Temperatures-High Pressures 4, 597 (1972).
12. F. Righini, A. Rosso, and L. Coslovi, in Proceedings of the Seventh Symposium on Thermophysical Properties, A. Cezairliyan, ed. (Am. Soc. Mech. Eng., New York, 1977), p. 358.
13. G.M. Foley, Rev. Sci. Instrum. 41, 827 (1970).
14. L. Coslovi, F. Righini, and A. Rosso, Alta Frequenza 44, 592 (1975).
15. International Committee for Weights and Measures, The International Practical Temperature Scale of 1968, Metrologia 5, 35 (1969).
16. F, Righini, and A. Rosso, in Temperature: Its Measurement and Control in Science and Industry, J.F. Schooley, Ed., Vol. 5 (Am. Inst. Phys. New York, 1982).
17. A. Cezairliyan, and J.L. McClure, J. Res. Nat. Bur. Stand. (US) 80A, 659 (1976).

Miniature thermometric fixed points for thermocouple calibrations

M. Tischler and M. J. Koremblit

Instituto Nacional de Tecnología Industrial, C.C. 157, 1650 San Martin, Provincia de Buenos Aires, Argentina

Low cost miniature graphite crucibles (0.12 cm^3) filled with corresponding metal ingots have been constructed for thermocouple primary calibrations. The two legs of the thermocouple are plugged into the graphite at appropriate perforations at the top and bottom of the crucible, closing thereby the thermoelectric circuit through the crucible itself without touching the metal ingot. The law of intermediate metals, coupled with a low crucible-mass to metal-mass ratio, ensure proper thermocouple performance and long lasting plateaus (typically 25 min). The thermocouple calibration is performed by running both melting and freezing plateaus. Increasing with furnace rate of heating or cooling, a drag effect may be observed, which tends to raise or lower the melting or freezing plateaus, respectively, as well as tending to increase the span of the plateaus. Nevertheless, the average values of EMF determined during a corresponding melt and freeze pair of runs, are reproducible to $\pm 1\,\mu$V and agree with the IPTS-68 at least within 0.3 K. An advantage of the miniature fixed points is the possibility of calibrating thermocouples *in situ* under the presence of strong temperature gradients (a situation often found in practice) which may render useless a conventional laboratory calibration. Results are presented for In, Sn, Cd, Pb, Zn, Sb, Al, Ag, Au, and Cu.

INTRODUCTION

Although, roughly speaking, the use of thermocouples for measurements above room temperature does not provide a better than 0.5 K accuracy, they are widely employed for temperature measurement and control in science and industry. In particular, even in the case of the standard thermocouple, it is pointed out[1] that, in general use, an accuracy better than 0.2 K cannot be expected because of the continually changing chemical and physical inhomogeneities in the wires in the region of temperature gradients. Moreover, for not specially selected thermocouples, mostly of industrial quality, these inhomogeneities may produce measurement errors amounting to several kelvins, in spite of, and because of, having been carefully calibrated under ideal laboratory conditions characterized by a large immersion depth in a gradientless temperature region.

Thermocouple calibration procedures based on comparisons against standard thermometers, themselves calibrated according to the IPTS-68, are well known. The accuracy of realization of the IPTS-68 over most of its range is much better than necessary for thermocouple calibrations, and since the implementation of the IPTS-68 is rather difficult and expensive for most thermometry laboratories, looking into ways of setting up a convenient temperature scale to a secondary level of accuracy seems quite to the point. This has long since been recognized and has prompted the CCT to recomend work in this direction[2].

The fixed-point method of thermocouple calibration, is based on realizing freezing-point cells of very pure materials, for which the phase transition temperatures are specified by the IPTS-68[1]. Typically, these cells consist of pure graphite crucibles filled with about one kilogram of high-purity metal in which a graphite tube (calibration well) is permanently immersed. Thermocouples can be calibrated by measuring their electromotive force (EMF) outputs with their measuring junctions inside the calibration well, while the metal ingot is freezing.

Although such fixed-point cells are rather expensive, because of the large amounts of pure metal and graphite required, they are so far indispensable for the calibration of very accurate thermometers, such as standard platinum resistance thermometers, since the temperature inside the calibration wells is reproducible and remains fixed within a few millikelvins for long periods of time. On the other hand, because of the above mentioned intrinsic accuracy limitations of thermocouples, such fixed-point cells are certainly overdimensioned, and can be dispensed with for most thermocouple calibrations, as will be shown below.

A third method of calibrating thermocouples is the wire-bridge method[3], in which a thin wire of pure metal, several millimeters long, connects the two legs of the thermocouple forming, at the same time, the measuring junction. The temperature of the furnace, in which the measuring junction is immersed, is increased while the thermocouple EMF output is recorded until the thermoelectric circuit opens as the metal bridge melts. A change in slope in the recorded EMF vs. time curve may be observed as the metal bridge begins to melt, getting to a point where the EMF remains more or less constant for a short time just before the thermoelectric circuit breaks. This indicates the value of EMF to be assigned to the melting temperature of the particular wire used. The accuracy of the procedure is mainly limited by the contamination of the metal bridge by the thermocouple wire, and the furnace's tendency to raise the temperature of the thermocouple legs, weakly opposed by the latent heat absorbed by the melting bridge.

The new primary procedure to calibrate thermocouples, described below, combines the advantages of both the fixed point method and the wire-bridge method, rendering calibrations that are at least as accurate as thermocouples themselves can be. Moreover, it provides the possibility of calibrating thermocouples "in situ", thereby subjecting their entire length to the actual temperature gradients.

MINIATURE THERMOMETRIC FIXED-POINT

Figure 1 is a photograph of a miniature thermometric fixed point (MTFP) constituting the measuring junction of the thermocouple to be calibrated. It consists basically of a small crucible machined out of a 6 mm diameter rod of spectrographic graphite, filled with a pure-metal ingot. The main details of the device are given by the cross sectional views in Figure 2. The perforations at the bottom and lid of the crucible are used to plug in the legs of the thermocouple, thereby completing the thermoelectric circuit through the electrically-conducting graphite, yet without touching the metal ingot. This prevents contaminating the metal ingot, and permits repeated use of the MTFP with the same thermocouple or exchanging thermocouples.

The thermocouple calibration is performed in a temperature-controlled furnace in order to slowly pass through the phase-transition point of the metal ingot, while recording the EMF as a function of time. The resulting melting and freezing plateaus, are much like those obtained when using conventional fixed-point cells. The tightly fitting crucible lid has a 0.5 mm diameter perforation to vent occluded gases as the temperature is raised. The mechanical characteristics of graphite are ideal to construct a thin walled receptacle, which ensures a low graphite-heat-capacity to metal-heat-capacity ratio. Chemical compatibility with most metals, with melting points below and including copper, is favorable, just as are the thermal and thermoelectrical properties[4]. The latter include a relatively high thermal conductivity value of 0.32 cal/cm sec °C, and a low thermoelectric power[5], 3 μV/K, both of which contribute significantly to the performance of the device as will be presently discussed. The useful life of the crucible is limited by oxidation, which in normal atmosphere begins at about 450 °C, and mechanical damage which results after numerous insertions of thermocouple wires into the corresponding bores. Yet, the replacement of the little metal ingot into a new crucible is quite simple, and the low cost of the entire device is self evident.

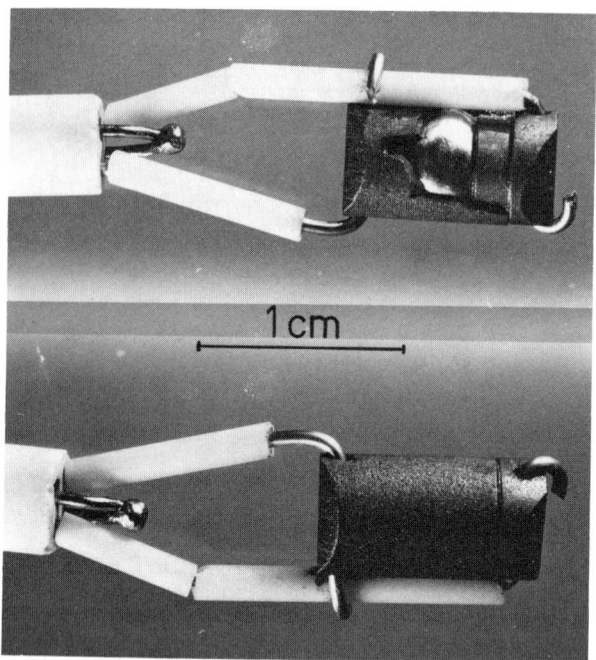

Fig. 1. Photographs of miniature thermometric fixed points (MTFP) attached to the two legs of the thermocouple, forming thereby the measuring junction. The upper photograph shows the MTFP partially broken in order to expose inner details. The measuring junction of the reference thermocouple assembled in the same ceramic rod, is also shown.

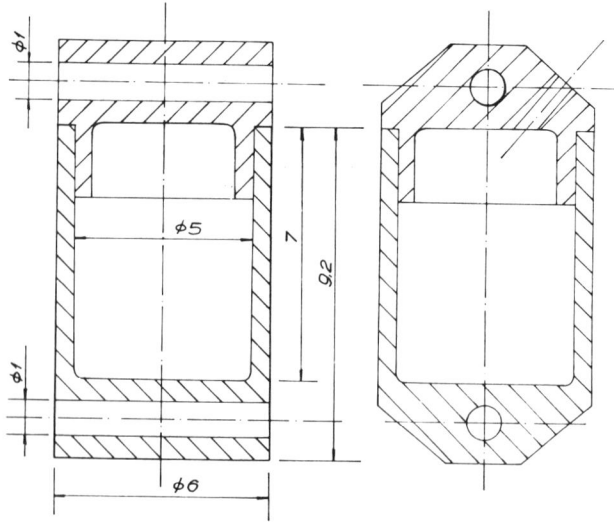

Fig. 2. Cross-sectional drawings of the graphite crucibles used for the MTFP. The dimensions are in millimeters.

Figure 3 represents a simplified equivalent thermoelectric circuit for the thermocouple, with a MTFP attached to it as its measuring junction. It is intended as a model for analysis of the device's performance. Point 2 represents the junction of one leg of the thermocouple, A, with a portion of graphite, G, for instance, at the top of the crucible. Point 3 stands for the middle point of the crucible, assumed to be at the temperature of the metal ingot, T_0, and point 4 is the junction of the graphite portion, G, at the bottom of the crucible with the other leg of the thermocouple, B. Points 2 and 4 are assumed to be at temperatures Δ_A and Δ_B above temperature T_0, respectively, representing temperature inhomogeneities that may be present on the crucible itself, due to heat exchange with both the metal ingot (while it is melting or freezing) and the furnace walls.

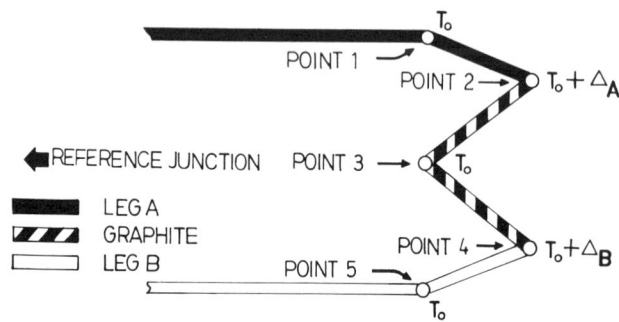

Fig. 3. Rough equivalent thermoelectric circuit of a graphite MTFP forming the measuring junction of a thermocouple with one leg of material A, and the other leg of material B.

During a melt, the furnace raises the crucible's temperature ($\Delta_A > 0$ and $\Delta_B > 0$) opposed by the steady absorption of latent heat. This tends to keep the crucible's temperature close to T_0. During a freeze the opposite happens, so that $\Delta_A < 0$ and $\Delta_B < 0$. The absolute values of Δ_A and Δ_B during a melt, are equal to the corresponding absolute values during a freeze, assuming conditions of symmetry (e.g. the furnace temperature, T_f, is stabilized symmetrically above or below the fixed-point temperature, T_0, when melting or freezing).

Assuming for convenience that the temperatures of points 1 and 5 are arbitraly kept equal to the temperature, T_0, of the metal ingot (which for ideally-homogeneous materials produces no additional EMF) it follows that the EMF output of the thermoelectric circuit, V_0, may be written as:

$$V_0 = V_{A/B}(T_0) + (dE/dT)_A \Delta_A - (dE/dT)_G (\Delta_A - \Delta_B)$$
$$- (dE/dT)_B \Delta_B \quad (1)$$

where $V_{A/B}(T_0)$ is the EMF output of thermocouple A/B when its measuring junction is at temperature T_0, and $(dE/dT)_i$ is the thermoelectric power of material i, with i standing for A, B, or G. Eq. (1) may be rewritten, by adding and subtracting $(dE/dT)_B \Delta_A$, as

$$V_0 = V_{A/B}(T_0) + [(dE/dT)_A - (dE/dT)_B] \Delta_A$$
$$- [(dE/dT)_G - (dE/dT)_B] (\Delta_A - \Delta_B) \quad (2)$$

Referring all thermoelectric powers to leg B of the thermocouple, just as it is usually done with pure platinum, it follows that the EMF output will be in error by the amount

$$F = (dE/dT)_A \Delta_A - (dE/dT)_G (\Delta_A - \Delta_B) \quad (3)$$

The quantity F is a measure of a drag effect due to the furnace and is related to the temperature inertia of the MTFP. The temperature inertia of an ideal fixed point is infinite, in the sense that its temperature is not affected by heat exchange with the furnace or temperature bath, which is exactly set off releasing or absorbing latent heat during the change of phase. The corresponding value of F is therefore exactly zero since Δ_A and Δ_B vanish. On the other hand, real fixed points have a finite temperature inertia, so that the absolute values of F will increase the more the furnace temperature, T_f, lags the fixed-point temperature, T_0. According to the symmetry arguments given above, the quantity F obtained during a melting plateau changes sign during a corresponding freezing plateau, so that the average of the corresponding values of EMF should be free of this source of error. Moreover, as it has been experimentally verified in the case of the ten fixed points described below (see Fig. 5), it is possible to write

$$F(T_f) = S \, (dE/dT)_A \, (T_f - T_0) \quad (4)$$

where S is independent of T_f. This is a first order approximation valid within the measuring uncertainty of $\pm 1\,\mu V$, for $|T_f - T_0| < 5$ K. By combining equations (3) and (4) it follows that

$$S = \frac{\Delta_A}{(T_f - T_0)} - (dE/dT)_G / (dE/dT)_A \frac{(\Delta_A - \Delta_B)}{(T_f - T_0)} \quad (5)$$

In the particular case of a Pt10%Rh/Pt thermocouple, the thermoelectric power, $(dE/dT)_A$, varies from 8 $\mu V/K$ at about 150 °C to about 12 $\mu V/K$ at the copper point, while $(dE/dT)_G$ for pure graphite remains practically constant at 3 $\mu V/K$ for the same temperature range. Consequently the second term in Eq. 5 may safely be neglected, as it also contains the difference $(\Delta_A - \Delta_B)$, which on symmetry grounds may be expected to be very small as compared to $(T_f - T_0)$. It follows therefore that

$$S = \Delta_A / (T_f - T_0), \quad (6)$$

S may therefore be interpreted as the temperature difference that exists between the thermocouple legs and the metal ingot, as the furnace temperature, T_f, remains stabilized one degree above the fixed-point temperature, T_0, during a melting plateau. Values of S for each of the MTFPs constructed, derived from the experimental data, are given in Table II.

In order to get some additional insight about the parameters governing the MTFP's performance, a simple heat-flow model may also be devised, relating the quantity S just defined, with the latent heat of fusion, L, the duration of the change of phase, τ, the rate of heat flow between two parts of the system, ϕ, and the resistance to heat flow between these two parts, R. Starting from the definition of R in terms of ϕ, given by

$$R = \Delta T / \phi \quad (7)$$

where ΔT is the temperature difference between the corresponding two parts of the system, it is possible to write

$$L/\tau = (T_f - T_0)/(R_f + R_m) \quad (8)$$

and

$$L/\tau = S(T_f - T_0)/R_m \quad (9)$$

where R_f is the resistance to heat flow between the furnace and the MTFP, and R_m is the corresponding resistance between the metal ingot and the thermocouple legs. The latter includes the resistance of the graphite portion, and the surface-contact resistance between the graphite and the metal ingot. Combining Eqs. (8) and (9) we get

$$R_m = S \, R_f / (1 - S). \quad (10)$$

Based on these relations it is possible to conclude that the duration of the plateau, τ, will be proportional to the reciprocal of the difference $(T_f - T_0)$, and will increase as thermal insulation between the furnace and the MTFP increases. This will in turn tend to decrease the value of S, (i.e. the drag effect) just as will the relatively high value of graphite's thermal conductivity, and good thermal contact between the crucible and the ingot. Different values of S may therefore be expected for different fixed points, as the thermal contact with graphite, thermal conductivity, latent heat, undercooling, and speed of nucleation vary from one metal to another.

Due to the small size of the metal ingots, the ratio of surface-tension energy to latent heat is relatively high compared to conventional fixed points. This could be a source of systematic error, which conceivably acts in the same direction as an increase of pressure, namely, by shifting the phase-transition temperature. This effect will probably become noticeable only with very small ingots.

CONSTRUCTION OF MTFPs

Miniature thermometric fixed points like the one described above, have been constructed for In, Sn, Cd, Pb, Zn, Sb, Al, Ag, Au, and Cu. The crucibles with a total weight of 0.34 g, and a useful inner volume of approximately 0,12 cm^3, were turned out of a 6 mm diameter rod of Ringsdorff RW0, spectrographic graphite. In order to prevent oxidation of the graphite, all the calibrations were run in furnaces under protective nitrogen atmospheres. The metal ingots were prepared by melting an excess of high-purity metal directly into an open crucible. The resulting ingots were then turned with regular steel tools in order to accomodate the lid, taking into account the expected effects of thermal expansion relative to the graphite, as well as the volumetric shrinkage present during a liquid-solid phase transition. This ensures the best possible thermal contact between the metal and the crucible's walls. Metal shavings produced while turning the ingots were spectrographically analysed. The main impurities detected are given in Table I.

TABLE I. MAIN IMPURITIES OF THE METALS USED FOR THE MTFPs, AS DETERMINED BY MEANS OF SPECTROGRAPHIC ANALYSIS, IN PARTS PER MILLION

	In	Sn	Cd	Pb	Zn	Sb	Al	Ag	Au	Cu
Al	300	30	30	30	30	30		30	30	30
Ag	5	3		10		10	3		30	50
Bi				30		30			30	
Ca	50	7		30		30	30	50		
Cu	5	30	50	10	1	3	50	1	1	
Fe	100	500	300	100	500	50	100	500		
Mg	10	30	70	30	10	50	50	30	10	10
Mn							10			
Ni		100					100			
Pb		500								
Sb		500			300		300	300		300
Si	300	100	70	100	70	100	500	70	70	70
Sn	30									

THERMOCOUPLES

The MTFPs were used to calibrate a Pt10%Rh/Pt thermocouple with which all the results given below were measured. The thermocouple, after the usual electrical annealing in air[7] at 1450 °C, was mounted in a 1 m long, silimanite, four-bore-ceramic rod, along with a second Pt10%Rh/Pt thermocouple, cut from the same batch of wire, and annealed in the same way. This second thermocouple (referred to below as the reference thermocouple), was used to monitor the furnace temperature. Its measuring junction was located 20 mm from the mid-point of the MTFP, (see Fig. 1). The two thermocouples were again annealed, after assembly, in a furnace at 1200 °C, and slowly cooled to room temperature over a period of several hours.

In order to protect the thermocouple from possible contamination with the metallic vapors given up by Sb, Cd, or Zn, the ceramic rod supporting the thermocouples was sealed into a closely fitting Pyrex-glass tube approximately 60 cm long. By heating the tip of the glass tube to its softening point, it was pressed onto the protruding thermocouple wires, leaving extensions just long enough to attach to them the MTFP, in order to make the measuring junction. All this was done after finishing the calibration at the Cu, Au, Ag and Al points. The thermocouple protected with the Pyrex tube was then calibrated at the Sn, In, Pb, Sb, Zn, and Cd points.

EMF MEASUREMENTS

Electrical measurements of the reference thermocouple were performed with a Fluke 8502A multimeter having a 1 μV resolution. A Hewlet Packard 704B calibration-standard/differential multimeter with a 0.1 μV resolution, was used to measure the EMF for the thermocouple under calibration. The latter piece of equipment provided an appropriately amplified signal which was used to drive a Chessell 301E recorder on which the freezing and melting plateaus were recorded. The two voltmeters were frequently recalibrated against a Fluke 335D DC standard. As a result a $\pm 1\,\mu V$ accuracy for all the EMF measurements has been estimated.

FURNACES

All the calibrations were performed in two furnaces. One of them, a Thermco "Ranger 3000", used for the Cu, Au, Ag, Al, Sb, and Zn points, has a cylindrical heating chamber approximately 1.5 m long by 12 cm inner diameter. It possesses three electronically controlled heating zones, with which it is possible to obtain a 70 cm long central region with temperature inhomogeneities less than 0.5 K. In order to

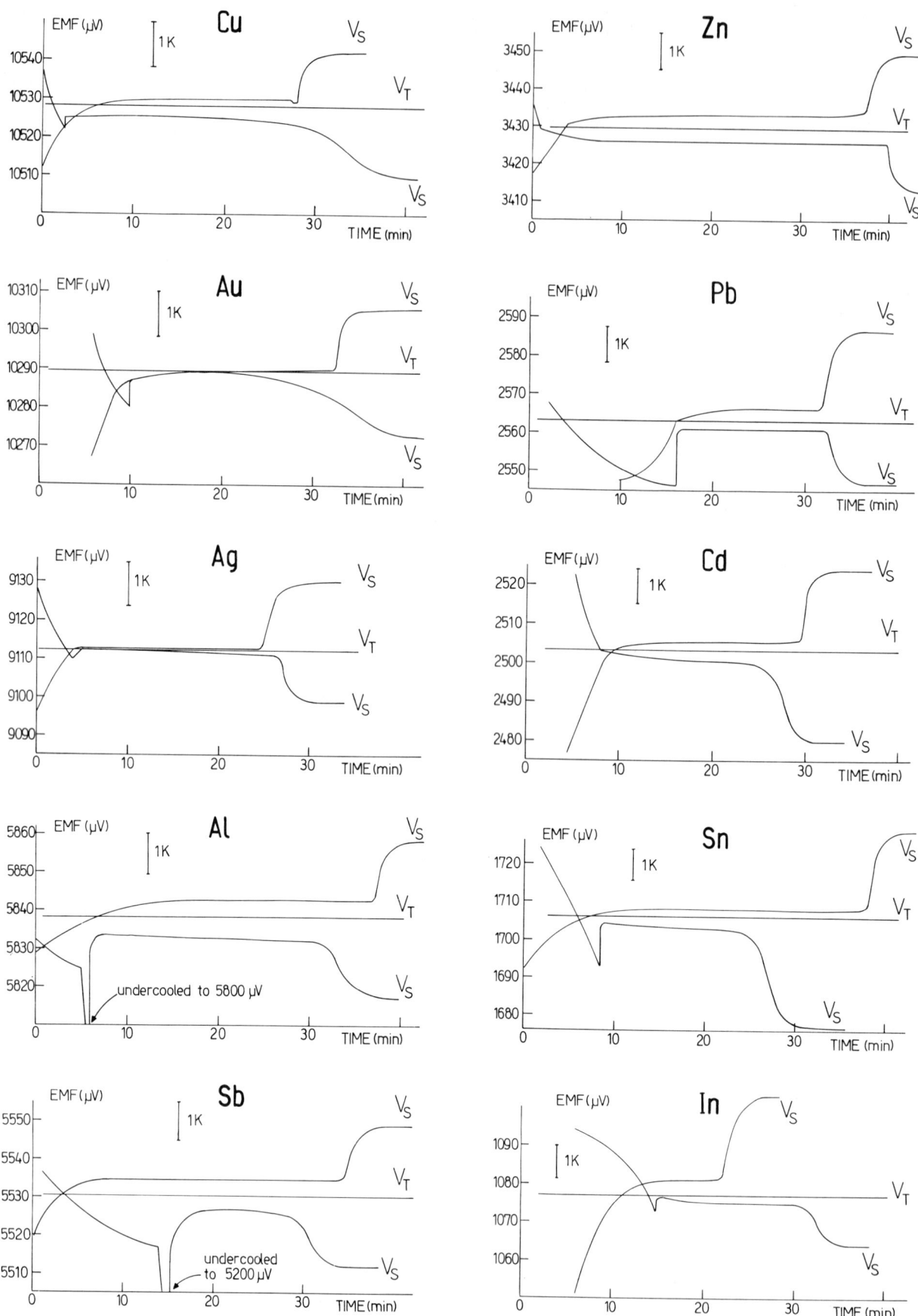

Fig. 4. Examples of melting and freezing plateaus for each of the fixed points constructed. V_T is the true EMF corresponding to the fixed-point temperature determined by eliminating the drag effect, as described in relation to Fig. 5. V_S is the EMF output of the thermocouple after the plateau is over and corresponds to the temperature at which the furnace is stabilized.

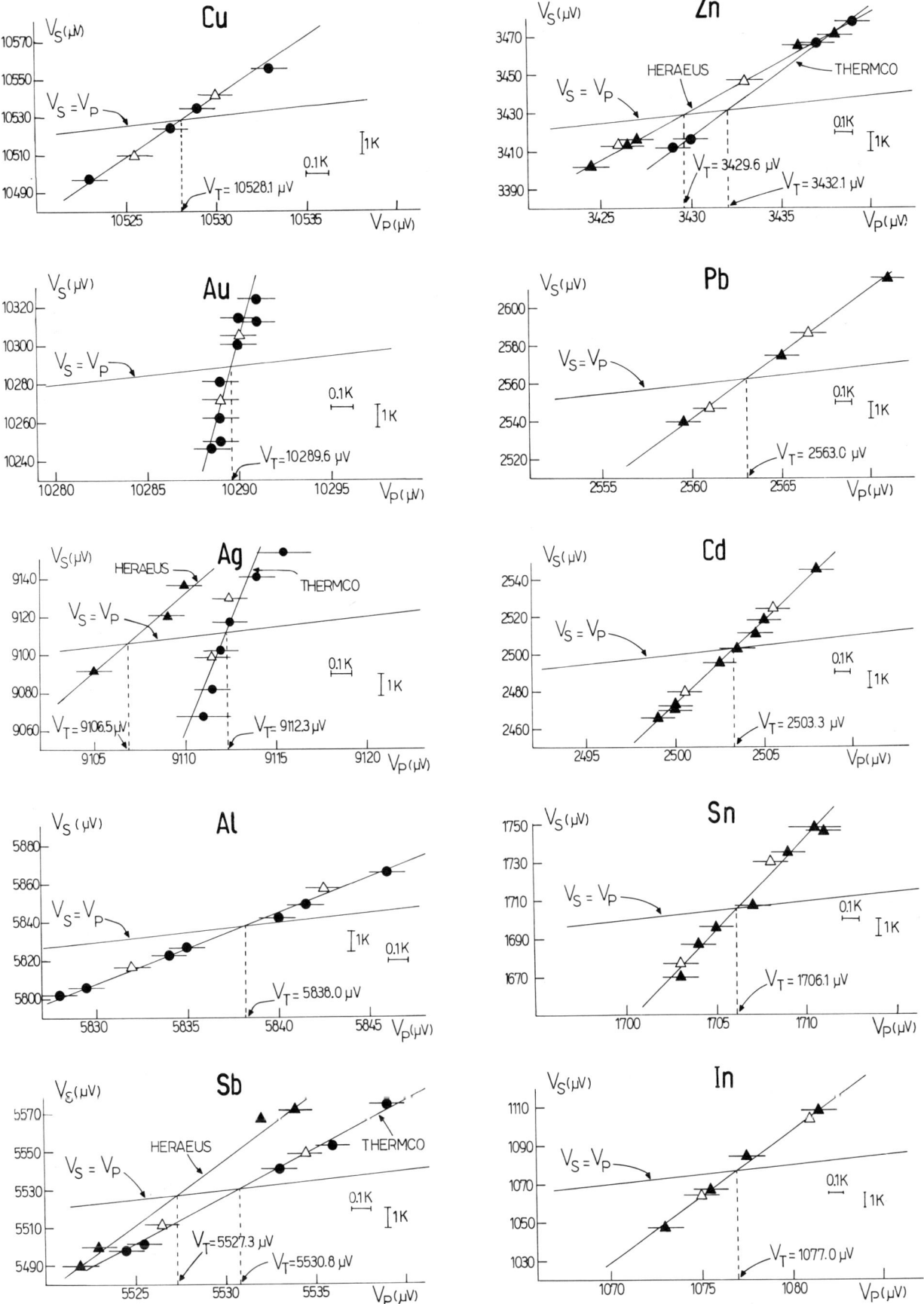

Fig. 5. Plots of V_P vs. V_S for each of the MTFPs. V_P is the EMF value of the flat portion of the plateau, and V_S is the corresponding EMF value after the plateau is over, when the MTFP reaches the temperature at which the furnace has been stabilized. The intersection of the resulting curve with the $V_P = V_S$ line, determines the EMF value, V_T, to be assigned to the fixed-point temperature. The open triangles represent points derived from the plateaus shown in Fig. 4.

provide such a uniform central region, the outer heating zones are kept at a higher temperature (typically 20 K higher), compensating thereby for axial heat-losses to the environment. The resulting temperature profile is then a very flat central region extending over more than 50 cm, with temperature differences less than 0.5 K, followed by a hump approaching to a maximum of about 20 K above the central region, prior to a steady decline to ambient. The second furnace used for the temperature calibrations with the Ag, Sb, Zn, Pb, Cd, Sn, and In MTFPs, is a Heraeus, ROK/F 4/60, with a cylindrical heating chamber 60 cm long by 4 cm diameter. A steel tube, 3 cm outer diameter by 3 mm wall thickness, was introduced to function as a temperature equalizer. Notwithstanding, the axial temperature distribution in this furnace had an approximately parabolic shape, exhibiting a 1.5 degrees drop at 5 cm from the center, and a 6 degrees drop at 10 cm from the center, after stabilization of the center temperature at 400 °C.

A protective atmosphere was mantained in both furnaces, by means of a low flow of nitrogen (approximately 0.2 l/min), through a fused quartz tube. This tube (1.2 m long by 10 mm inner diameter) was immersed in the furnace, with a 30 cm length protruding into room temperature. This portion was used for rapid cooling of the MTFP by withdrawing the thermocouple, still under protective gas, before its exchange by another MTFP, in order to continue calibration at another temperature. Thermocouple withdrawal along the quartz tube was also useful to induce nucleation of MTFPs presenting large undercooling effects. As soon as nucleation started (as evidenced by a sudden rise in EMF output) the thermocouple was quickly returned to its immersed position.

CALIBRATION PROCEDURE

The thermocouple with the MTFP attached to it, as shown in Fig. 1, was introduced into the quartz tube containing the protective atmosphere, and the measuring junction placed close to the center of the furnace (where the temperature distribution was most uniform). The furnace temperature, T_f, was stabilized with the help of the reference thermocouple, at several values, $T_f = T_i$, $i = 1,2,...$, above and below the corresponding fixed point temperature, T_0, differing from the latter by values approaching to 10 K. The corresponding melting and freezing plateaus were then recorded for each temperature, T_i, as functions of EMF vs. time. These plateaus showed the drag effect, as anticipated. As the furnace was still in the process of reaching the stable state for temperature T_i, and while the change of phase was already under way, the drag effect showed up by increasing the plateau's span, until a flat portion, within $\pm 0.5\,\mu V$, appeared as the furnace finally stabilized. The duration of the phase transition, typically 25 minutes, was roughly proportional to the reciprocal of the difference $(T_i - T_0)$, as expected from Eq. (8). For large enough differences (typically > 10 K) the flat portion did not show up at all, since the furnace did not have enough time to reach equilibrium before the change of phase was over. The EMF value at the flat portion of the plateau, V_P, was taken as a first approximation to the true value, V_T. Figure 4 shows examples of melting and freezing plateaus for each of the MTFPs constructed. The symbol V_S indicates the EMF output of the thermocouple after the plateau was over, when its measuring junction, i.e. the MTFP, reached the corresponding temperature, T_i, at which time the furnace was stabilized.

In order to eliminate the drag effect as a source of error, a plot was made of the V_P vs. V_S EMF values. The intersection of the resulting curve with the $V_P = V_S$ line, determined the EMF value, V_T, which would have been measured had the furnace been stabilized precisely at the fixed point temperature T_0 (in which case no drag effect error is to be expected). Fig. 5 represents the V_P vs. V_S curves determined for each of the MTFPs during the calibration. The open triangles represent the values derived from the corresponding plateaus shown in Fig. 4. As follows from Fig. 5, the V_P vs. V_S curve could always be approximated, within the uncertainty of $\pm 1\,\mu V$, by a straight line, for all of the ten fixed points tried. This fact indicates that just two points, for instance running one melting and one freezing plateau, are enough to eliminate the drag effect, rendering very accurate calibrations.

The reciprocal of the slope of the straight line determined for each MTFP from Fig. 5, is equal to the quantity S appearing in Eqs. (4), (5), and (6), and provides an estimate of the temperature differences between the thermocouple legs and the metal ingot during a plateau, when the furnace is stabilized 1 K above T_0. Measured values of S for each of the MTFPs constructed are given in Table II, along with the mass of metal used for each of them, the corresponding amounts of latent heat of fusion[6], and estimated values of the resistances to heat flow, R_f and R_m. The latter values were calculated employing Eqs. (8) and (10), with values of $(T_f - T_0)$ and τ, derived from the curves presented in Fig. 4. Average values were determined using both the melting and freezing plateaus (except for Sb, for which the duration of the freezing plateau was considered to be abnormally short as a result of the large undercooling). Due to the oversimplified model employed, the resistance values obtained are only suggestive. The decrease of the R_f resistance with temperature rise is due to an ever-increasing share of radiative heat-transfer. The influence of this effect on the R_m resistance is also noticeable, although the presence of metal-oxide layers in the metal-graphite interface, and the degree of surface contact caused by hydrostatic pressure and surface tension, are doubtlessly other determining factors for the differences observed between one MTFP and another. This explains, for instance, the very small value of R_m obtained for gold as compared with aluminum.

TABLE II

MTFP	Melting point IPTS-68 (°C)	S	Mass of metal (grams)	Latent heat (Joules)	R_f (°C/W)	R_m (°C/W)
Cu	1084.88	0.16	0.96	200	9.2	1.8
Au	1064.43	0.029	2.1	126	15.	0.44
Ag	961.93	0.043	1.1	128	13.	0.60
Al	660.46	0.27	0.29	115	24.	8.9
Sb	630.755	0.20	0.78	127	23.	4.2
Zn	419.58	0.19	0.79	86	36.	8.6
Pb	327.502	0.15	1.3	31	58.	10.
Cd	321.108	0.11	0.96	54	45.	5.6
Sn	231.968	0.10	0.84	50	81.	9.0
In	156.634	0.15	0.84	24	72.	13.

Besides the calibration procedure just described, which achieves relatively high accuracies, simpler calibration procedures employing MTFPs have also been tried. The latter are performed without the need of stabilizing the furnace's temperature close to the corresponding fixed-point temperature, by just letting it pass through it, even at the relatively high rates of about 5 K/minute. By recording the EMF value corresponding to the point where the change in slope appears while melting, V_m, and the maximum EMF value measured during a freeze, V_f, just after undercooling is over, it is possible to estimate the value V_T corresponding to the fixed-point temperature with uncertainties less than 0.5 K. Due to the fact that each MTFP has characteristics of its own (see Figs. 4 and 5), for some fixed points, it is convenient to take $V_T = V_m$, while for others more accurate values are obtained by taking $V_T = (V_m + V_f)/2$. The MTFPs for Cu, Sb, Pb, Sn and In correspond to the first case, because recovery after undercooling is not very good for them when the difference $(T_f - T_0)$ is very large, rendering low values of V_f. For the aluminum MTFP the average value has to be taken, due to the large drag effect that it presents. In the case of the Au, Ag, Zn, and Cd MTFPs, both procedures lead to the same values of V_T, with uncertainties of about 0.2 K. For gold and silver this is so because of the very low values of S (see Table II) that the corresponding MTFPs present, while Zn and Cd show no undercolling and very sharp slope changes as melting or freezing starts.

CALIBRATIONS "IN SITU"

In practice, one of the main drawbacks of using thermocouples for exact temperature measurements is that the chemical and physical inhomogeneities, that are generally present in the thermocouple wires, generate unpredictable electromotive forces as they are subject to temperature gradients. The measured EMF will not only be a function of the temperature of the measuring and reference junctions, but will also depend upon the particular temperature distribution along the thermocouple. It follows therefore that a conventional calibration, for instance by the comparison method or the fixed-point method, may be meaningless for applications where large temperature gradients are present, unless the thermocouple happens to be sufficiently homogeneous and stable. The MTFP provides the possibility of calibrating thermocouples "in situ", exposed over their entire length to the actual temperature gradients. Moreover, a thermocouple may permanently include a MTFP as its measuring junction, selected in such a way that its phase-transition temperature approaches the expected temperature range of the intended application. This arrangement permits frequent, in service checks of thermocouple performance, detecting thereby changes in the thermocouple's properties, such as drift, which may then be accounted for.

Disposable MTFPs, available as small capsules including a pure metal, with wire extensions or some other means to be attached to any size of thermocouple wire, to be used for "in situ" industrial calibrations, are also conceivable due to their low cost.

For the present work, the furnace temperature gradient had a noticeable effect on the thermocouple EMF output, presumably due to chemical inhomogeneities in the wires. This became apparent by calibrating the thermocouple with the Ag, Sb and Zn MTFPs in both furnaces. It was determined (see Fig. 5) that the measured EMF values were lower when using the Heraeus furnace as compared to using the Thermco furnace. The observed differences were approximately 5.5 μV at the Ag point, 3.5 μV at the Sb point, and 2.5 μV at the Zn point. These values are compatible with the expectation that corresponding differences for other fixed points should vary linearly with temperature, approaching zero at room temperature. The corresponding straight line, labeled as "corrections for gradient differences" is represented in Fig. 7. The measured values of EMF, referred to the NBS Thermocouple Calibration Tables[7], are plotted in the same figure, as full triangles in the case of measurements performed in the Heraeus furnace, and as full circles, in the case of measurements performed in the Thermco furnace. The open symbols plotted in the same figure were obtained from corresponding measured points, by applying the correction for gradient differences assuming linearity.

The obviously smooth curves that may be drawn through the plotted points, one corresponding to the Heraeus furnace and the other to the Thermco furnace, as well as the comparison performed with standard thermometers described below, seem to confirm the validity of the corrections applied, and stress the usefulness of the MTFPs for "in situ" calibrations.

COMPARISONS WITH IPTS-68

The thermocouple calibration employing MTFPs, has been compared between room temperature and the copper point with several standard thermometers. Between room temperature and the Sb point, the comparison was performed employing two platinum resistance termometers (PRT), one Tinsley type 5187 SA, calibration class I, calibrated at NPL, and the other a Heraeus scientific thermoresistor model PTB, Normal type, calibrated at PTB. Resistance measurements were performed using a Tinsley inductively coupled, double ratio bridge, type 5650.

The comparison was continued between the Sb point and the gold point, employing two standard Pt10%Rh/Pt thermocouples calibrated at the PTB. An additional comparison was performed between 800 °C and 1100 °C, employing a Schmidt and Haensch disappearing-filament optical pyrometer, calibrated in our laboratory between 800 °C and 1250 °C, with an estimated uncertainty of 0.5 K. The optical pyrometer was calibrated according to the IPTS-68, employing a gold-point black body and using calibrated sectored disks to extend the calibration above and below the gold point.

The same Thermco furnace (used for the thermocouple calibration with the high-temperature MTFPs) was also used to heat a 50 cm long by 6 cm diameter, ceramic black body, with a 1 cm diameter sighting hole, into which the thermocouple that had been calibrated with the MTFPs was immersed through an off-axis bore (Fig. 6). The very uniform temperature distribution over the whole black body, achieved with this three-zone furnace, coupled with a 0.9999 effective emittance as calculated according to Quinn[8], ensured meaningfull pyrometric measurements.

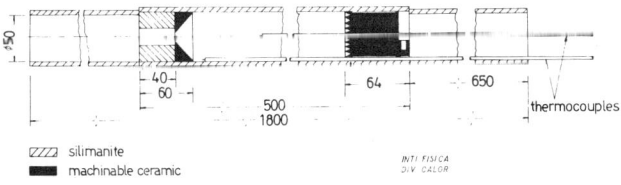

Fig. 6. Cross-sectional drawing of the ceramic black body used for the comparison of the MTFP-calibrated thermocouple with the optical pyrometer. Dimensions are in millimeters.

The main advantage of employing the same furnace for the thermocouple calibration with the MTFPs, and the pyrometric comparisons, is that the thermocouple was subject to the same temperature gradients in both cases. All differences caused by spurious EMFs, due to thermocouple inhomogeneities, were thereby eliminated. This is a clear case of "in situ" calibration made possible by the MTFP method.

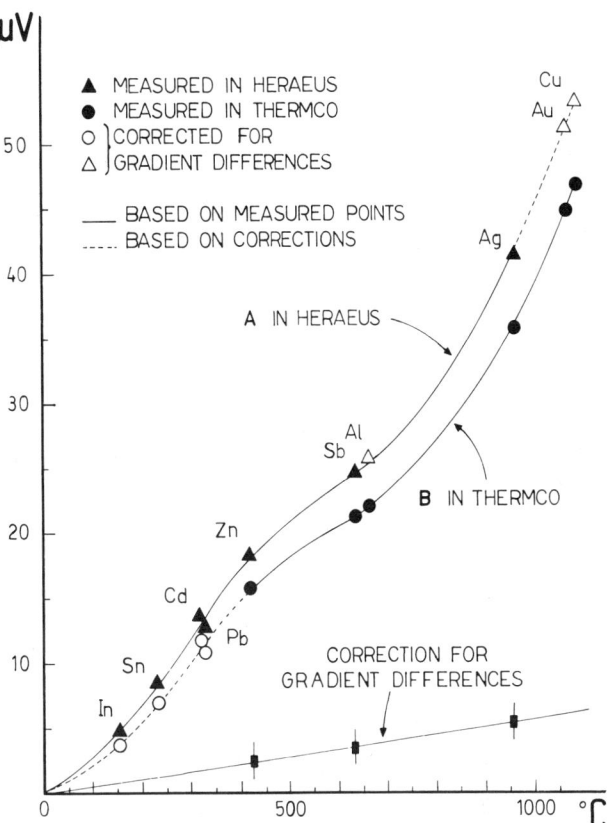

Fig. 7. Thermocouple's EMF output referred to the NBS Thermocouple Reference Tables[7]. The differences between the EMF outputs under the gradients present in both furnaces used, are given by the straight line plotted, and are equal to the differences between curve A and curve B. Uncertainty of the measured points: ± 1 μV.

An uncertainty of approximately 0.7 K has been assigned to the comparison with the optical pyrometer, based on compounding in quadrature the 0.5 K uncertainty of the pyrometer's calibration, with 0.5/2 = 0.25 K resulting from the maximum 0.5 K temperature difference that was observed over the black-body's length, with the observer's measuring standard deviation of 0.3 K, and with the 0.1 K uncertainty resulting from the EMF measurements.

The comparison between the standard thermocouples or the platinum resistance thermometers, and the thermocouple calibrated with the MTFPs, was performed in a Heraeus, ROK 6.5/60-TPK, cylindrical furnace, with a 60 cm long by 6.5 cm diameter heating chamber, containing a stainless-steel, equalizing block, approximately 25 cm long by 6 cm diameter. The resulting temperature profile was similar to the one prevailing in the Heraeus furnace used for calibrating the thermocouple with the MTFPs. Consequently, the calibration-curve, A, given in Fig. 7, is to be used for this comparison, due to the inhomogeneities of the thermocouple legs, just as calibration-curve, B, applies to the optical pyrometer comparison.

A total uncertainty of 0.2 K has been estimated for the comparison with the platinum resistance thermometers. This resulted from compounding the uncertainty of the EMF measurements with the uncertainties due to temperature gradients along the equalizing block. The latter arise from the dissimilar size of the PRT's sensing coil, which integrates over a length of about 5 cm, compared to the smaller thermocouple measuring junction, placed at approximately the mid-point of the thermometer coil. The uncertainty of the comparison using the standard thermocouples, was estimated to be approximately 0.4 K, obtained by compounding in quadrature the 3 μV uncertainty of the original calibration of the standard thermocouples, and the 1 μV uncertainty of our EMF measurements.

The differences detected in all the comparisons were always less than the estimated uncertainties, and were always positive, in the sense that the thermocouple calibrated with the MTFPs, measured temperatures systematically higher than those indicated by standard instruments. Moreover, the differences measured with the standard thermocouples and the optical pyrometer were always less than 0.3 K, indicating the likelihood that their corresponding uncertainties were overestimated.

CONCLUSIONS

The procedure for primary thermocouple calibrations based on miniature thermometric fixed points, described in the present work, is simple and inexpensive. Furthermore, it permits "in situ" calibrations, eliminating the source of systematic error caused by the presence of inhomogeneities along the thermocouple legs. The accuracy of the calibrations, as measured by its repeatability for In, Sn, Cd, Pb, Zn, Sb, Al, Ag, Au, and Cu, is assuredly less than 0.2 K. The accuracy of the calibrations, as measured by the departures from the IPTS-68, at the corresponding fixed-point temperatures, is better than 0.2 K up to the Sb point, and probably better than 0.3 K up to the Cu point.

ACKNOWLEDGMENTS

The authors are indebted to Professor R. Steinberg for his continuous support, and to Lic. R.T. Suguer for collaborating during the comparisons with the IPTS-68. Thanks are also due to Mr. R.E. Pastega for turning the crucibles and the metal ingots, to Vega y Camji S.A.I.C. for supplying the Pt10%Rh/Pt thermocouples used for this work, and to Ms. S. Mari for typing the manuscript.

REFERENCES

[1] Preston-Thomas, H., The International Practical Temperature Scale of 1968. Amended Edition of 1975. Metrología **12**, 7 (1976).

[2] Recommandation du Comité Consultatif de Thermometrie presentée au Comité International des Poids et Mesures, Comité Consultatif de Thermometrie, 12e Session - 1978 (9-10 mai) (Bureau International des Poids et Mesures, Sèvres, France) p. T-5.

[3] Henning, F., Moser, H., *Temperaturmessung*, 3rd Edition (Springer Verlag, Berlin, 1977) p. 117.

[4] Ringsdorff High Purity Graphite in the Laboratory, published by Ringsdorff-Werke GmbH, Bonn-Bad Godesberg, printed by Bonner Universitäts Buchdruckerei, Nr. SP428 e/17.

[5] Franks, E., High-Temperature Thermocouples Using Nonmetallic Members, *Temperature, its Measurement and Control in Science and Industry*, Vol. 3, Part 2 (C.M. Herzfeld ed., Reinhold Publishing Corporation, N.Y., 1962) p. 189.

[6] Smithells, C.J., Editor, *Metals Reference Book*, 5th edition, (Butterworths, London & Boston, 1976) p. 186.

[7] NBS Monograph 125, *Thermocouple Reference Tables Based on the IPTS-68* (U.S. Department of Commerce, NBS, 1974) pp. 14, 18-28.

[8] Quinn, T.J., The calculation of the emissivity of cylindrical cavities giving near black-body radiation, Brit. J. Appl. Phys., **16**, 973(1965).

On sealed freezing point cells

Zhu Ci-Zhun

National Institute of Measurements, Szechuan, People's Republic of China

The equilibrium temperatures of temperature scale defining fixed points and second reference points depend upon the pressure and the impurity level of the reference material. It is expensive to use noble metals, especially when the thermometer is calibrated against one point in each furnace. In this paper we describe the design and fabrication technology of sealed freezing point cells and, in addition to this, propose the possibility of calibrating high temperature PRT's by using one furnace for several fixed points and by using cheap reference material in place of expensive silver and gold.

INTRODUCTION

It is known that resistance or thermocouple thermometers can be calibrated correctly only when equilibrium state temperatures of the IPTS, such as those at Au, Ag, and Zn points as well as other recommended values such as those at the Cu and Sb points have been realized accurately. Usually the freezing points of high purity material at 1 standard atm are accepted as the equilibrium state temperatures, but in the real measurement situations the effects of pressure and purity must be considered. The recommendations made in the IPTS supplementary information could not eliminate adequately the effect of pressure and material contamination problems. To avoid the possible contamination by the vapors of another fixed-point substance that might be present, it would be desirable to have furnaces for each fixed point. However, this will necessarily involve large investments and much room. The effect of pressure can be corrected after measuring it, while the effect of decline in purity is difficult to account for. Whenever a sample of known purity is available, all this could be overcome by using a sealed freezing point cell. By such isolation of the fixed point material from the surroundings, the pressure would be constant (e.g., one standard atm. for the freezing point) and the fixed point can be precisely reproduced. It is both economical and easy to place in operation one single furnace for many temperature points.

DESIGN CONSIDERATIONS

The Choice of Fixed Point Material

As regards the three fixed points defined by the IPTS in the range of 630.74 - 1064.43 °C, gold and silver are noble metals, so that the material which is used in the cell, amounting to one or more kilograms, is very expensive. Furthermore, the difference in this range of the temperature scale from the thermodynamic temperature scale $\Delta T = T_I - T_T$ is far greater than the difference in other ranges, having a poor thermodynamic smoothness, resulting in part from the distribution of fixed points. A better choice of fixed points, on which the calibration of high temperature PRT's or Pt-Pt/Rh thermocouples is based, should improve the situation.

The junction point connecting the two ranges 0 - 630.74 and 630.74 - 1064.43 °C of the temperature scale is near the Sb point. In the former range the 630.74 °C is obtained by extrapolation. In order to obtain a better accuracy of the PRT temperature scale at this point, the Sb point should be chosen as one of calibration points.

The upper limit of the PRT temperature range joins the temperature range defined by the radiation thermometer. When we make a choice of the junction temperature, we should consider the sensitivity and uncertainty of photoelectric pyrometry at that point. Taking the silver point for illustration, the sensitivity and uncertainty of the measuring instruments currently used in radiation thermometry or photoelectric pyrometry, are much worse than those at the gold point. Moreover, if we anneal the thermometer at 1100 °C to improve its stability, the PRT has still much potential for use at the silver point. Additionally, considering gold and silver to be expensive noble metals, ordinarily we reduce the amount of material to a minimum, which means reducing the dimensions of the blackbody cavity. In turn, this limits the sensitivity of the radiation thermometry instruments. With the advances made in the technology of purifying and deoxidizing copper, the PRT has provided the possibility of replacing the Au point with the Cu point. The advantage of doing so is easily seen.

Fundamentally, it is the poor thermodynamic smoothness of the thermocouple thermometer interpolation formula that causes a large ΔT in the range of 630.74 - 1064.43 °C, which results from an unreasonable distribution of fixed points. If we can eventually find an appropriate point near 800 °C to replace the Ag point, the situation will improve. When we have to make a choice between Ag-Cu eutectic and NaCl salt, it seems better to choose the NaCl salt because of the better reproducibility of its chemical composition and phase behavior. But alternatively, the Ag-Cu eutectic has the advantage in thermal conductivity and supercooling. In order to forego the expensive material of silver, it is better to choose the NaCl salt to provide a point near 801 °C after some special problems associated with NaCl have been solved.

CONSTRUCTION

The construction of the sealed freezing point cell is shown in Fig. 1. A crucible made of high purity graphite (impurity less than 0.01%) is installed into a quartz tube. A cover over the crucible helps reduce the heat loss from axial radiation, consequently the thermal gradient in the axial direction has been reduced. The cell is filled with argon (99.99% purity). The thickness of the fixed point substance surrounding the thermometer well in the cell is greater than the well radius. The immersion depth of the thermal well in the fixed point substance is about 10 cm. The isothermal volume in the cell is considered to be ample.

The specially designed heating furnace is temperature regulated and controlled continuously and precisely in the range of 400 - 1100 °C. The region of uniform

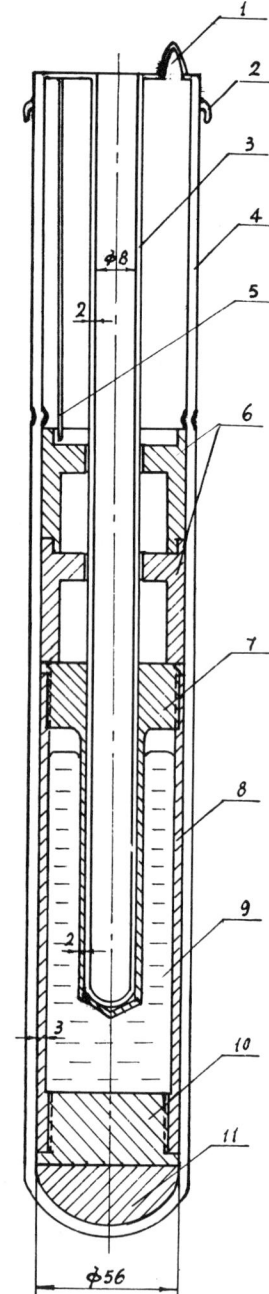

Fig. 1. Structure scheme of the cell.
1 - gas outlet; 2 - cell ear; 3 - thermal well; 4 - quartz tube; 5 - quartz tube; 6 - graphite jacket; 7 - crucible cavity; 8 - crucible cylinder; 9 - fixed point substance; 10 - crucible bottom cover; 11 - filling element.

temperature slightly shifts at different temperatures. A prior measurement was made to determine the best position for the cell to be placed so that the furnace can be used at various temperatures. The arrangement of the heating furnace and the cell is shown in Fig. 2.

FABRICATIONS

The fabrication process consists of two steps: providing a cleaned cell prior to sealing pumping, filling of gas and sealing of the cell.

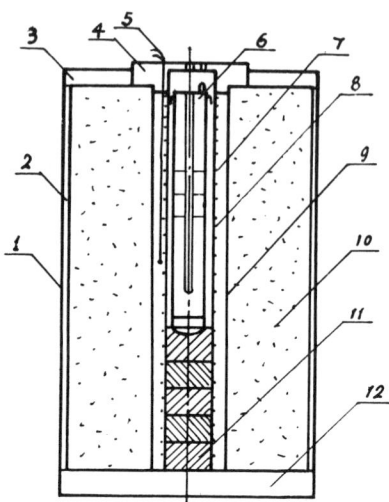

Fig. 2. Mounting diagram. 1 - outer jacket; 2 - heat isolation cylinder; 3 - top cover; 4 - upper cover; 5 - thermocouple; 6 - point cell; 7 - heating wire; 8 - internal ceramic pipe; 9 - external ceramic pipe; 10 - heat isolation material; 11 - graphite; 12 - heat isolation plate

The Preparation of the Cell

A careful cleaning was given to all parts of the cell. The next step was to remove any kind of ions such as that of H, O, and metals, and eliminate water vapor. Care must be taken while introducing the fixed point substance into the crucible to prevent oxidation. It was necessary to be sure that there was enough immersion of the well with the fixed point substance around it. When the cover of cell is sealed, one must make sure to obtain good thermal contact between the quartz tube and the graphite wall of the well. A tube for pumping-filling gas with appropriate length and aperture was provided.

The Pumping, Filling and Sealing of the Cell

We connected the cleaned cell to the system as shown in Fig. 3, and put the cell into the isothermal region of the furnace. We pumped the system to a vacuum of 1×10^{-2} Torr at room temperature, then while maintaining the vacuum, purged it with argon several times. Warming and pumping the system at a vacuum of 10^{-2} Torr, we ensured that the temperature was raised gradually and the cell was pumped continuously. The boiling point of antimony falls to 540 °C at 10^{-3} Torr and to 620 °C at 10^{-2} Torr; this limit should be considered in selecting the highest pumping temperature for the antimony cell.

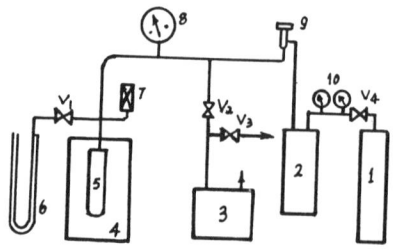

Fig. 3. Schematic drawing of cell pumping system.
1 - Ar; 2 - buffer; 3 - pump; 4 - heating furnace; 5 - cell; 6 - U-shape water manometer; 7 - vacuum gauge; 8 - precision pressure meter; 9 - needle valve; 10 - pressure meter.

It was better to raise the temperature slowly than to degas the absorbed gas more rapidly so that no melting or excessive evaporation would occur. When a vacuum of 10^{-2} Torr was reached, the argon was filled at the high temperature to a pressure a little higher than the ambient pressure. The system was allowed to equilibrate at this pressure for a moment, then pumped, refilled, and equilibrated again; finally the system was pumped to a vacuum of 10^{-2} Torr and filled with argon to a pressure a little higher than the ambient pressure. We raised the temperature of the furnace to the temperature of phase equilibrium point and it was maintained there. Based on calculation, the system pressure was adjusted to 1 standard atmosphere. The valve was closed, the pressure was maintained, then the temperature profile and the temperature plateau of the equilibrium state were measured. After this was finished, the furnace temperature was reduced while monitoring the pressure change of the system. Just as the system pressure became lower than the atmospheric pressure, the pumping outlet was sealed by using a H_2-O_2 or CH_4 flame.

MEASUREMENT

The Measurement of the Temperature Profile Inside the Well

The No. 78-03 Pt 10% Rh/Pt thermocouple working standard was used as the temperature-measuring sensor, the thermal EMF was measured by a UJ-35 precision potentiometer. As the plateau of equilibrium state was reached, the thermocouple was placed longitudinally in the cell to measure the temperature distribution. A temperature distribution in one of the Cu point cells is shown in Fig. 4. Each data point was an arithmetic average of two measured values. The uniform region of temperature with ±0.02 °C deviation in the Sb cell had a length of 45 mm, which is greater than that of the PRT sensor. In the measurement, the actual immersion depth of the resistance thermometer is determined by the position of uniform temperature region.

MANOMETRY

From Fig. 3 we can see that the precision pressure meter indicates the pressure in the system when the valve V_1 is closed. When the meter indicates 1 atm. usually a supplementary argon pressure has been added. (where the altitude is high above sea-level) to ensure that the equilibrium state temperature is precisely at 1 standard atm. The added argon pressure which has been known by calculation is indicated at the U-shaped water manometer and is adjusted by vernier needle valve 9. One branch of the water manometer is open to the atmosphere. We used a standard mercury barometer of ±13µ uncertainty to measure the local atmospheric pressure at that time; however, 1 mm measurement uncertainty of height is adequate for the water manometer. A possible error in the amount of gas supplied resulted from the uncertainties in determining the relative position of the liquid level of crucible and that of thermometer's sensor in the well and from the error in the density of the fixed point substance at the equilibrium state temperature. In other words, the error in the amount of gas supplied is related to the error of determination of the hydrostatic pressure of the fixed point substance.

The Measurement of the Temperature Plateau at the Equilibrium State

We measured the temperature plateau of Sb and Cu cells using two home-made high temperature PRT's (No. 80009 and No. 79-03, respectively). The electrical measuring instrument used is a Guildline 9975 current comparator for resistances. The 1Ω and 10Ω standard resistors which were used for the determination of resistance ratio were placed in the 9732 thermostat; the oil temperature was stable to ±0.001°C at 20 °C.

Based on the most suitable techniques for temperature increase, decrease and control and the data which were obtained previously, the heating furnace was program-controlled to make the cell reach a temperature plateau. The thermometer was preannealed and the immersion depth of the thermometer was determined in accordance with the position of the isothermal region in the well.

During the repetitive measurements for Sb and Cu cells, the thermometer had been inserted repeatedly. The temperature plateaus were constant to $\pm 3 \times 10^{-6}$ for the Sb point and to $\pm 1 \times 10^{-5}$ for the Cu point in terms of R_x/R_s. The values obtained are shown in Tables 1 and 2.

Table I. Temperature Plateau for the Sb Point

Sequence of measurements	Ser. No. of thermometers	Duration (min.)	R_x/R_s average value
1	80-009	38	0.884303_8
2	80-009	29	0.884307_0
		Mean	0.884305_4

Table II. Temperature Plateau for the Cu Point

Sequence of measurements	Ser. No. of thermometers	Duration (min.)	R_x/R_s average value
1	79-03	25	1.1766999_6
2	79-03	12	1.1767017_4
3	79-03	15	1.1766894_7
4	79-03	36	1.1766968_8
		Mean	1.1766979_4

The results in the tables show that the reproducibility of the temperature plateaus (average values) of the Sb cell is within 5 mK, while that of Cu cell is within 10-20 mK. In the most favorable cases, the latter had a plateau reproducible to ±5 mK.

CONCLUSIONS AND COMMENTS

As was stated before, for the Sb point and the Cu point (including the NaCl point between them) one common heating furnace can be used. The repetitive measurements

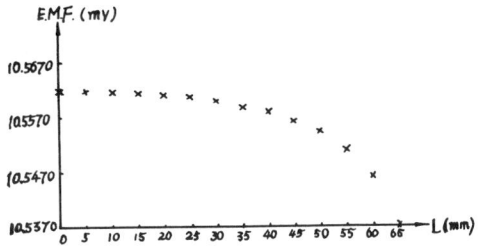

Fig. 4. Temperature distribution inside the well of the copper point cell.

made during a few months after completing the cell have shown that the cells can provide stable equilibrium temperatures. The good reproducibility of the Cu point cell has proved the possibility that the Cu point can be used as an upper limit point for calculating high temperature PRT interpolation equations, replacing the Au point or the Ag point.

With regards to the uncertainties of the equilibrium state temperatures of the cells, we have done some test and analysis, but there is much effort yet to be made. We have used another commercial thermometer made by the Chino Manufacturing Factory in Japan. The measurement results showed that there were small deviations, but they were within the uncertainty limit of the thermometer. Some other unsolved special problems, caused by excessive supercooling of NaCl as much as 168 °C, requires further investigation. The studying and establishing of high temperature PRT interpolation formula by making use of several sealed freezing point cells will be attempted in the near future.

Temperature distribution measurement with a silicon photodiode array

T. Yamada, N. Harada, and M. Koyanagi

Technical Research Center, Nippon Kokan K.K., Kawasaki, Japan

This paper describes an application of the self-scanning photodiode array to the temperature distribution measurement in the steel industry. Investigating the characteristics of a photodiode array and making various field tests in the steel works with the micro-computer based test system, the authors have demonstrated that the photodiode array sensor is of practical use in the temperature distribution measurement. Good results have been obtained in the application of the test system to steel making and hot rolling processes.

1. INTRODUCTION

In the rolling and heating processes of steel, non-contact temperature measurement has been established with the use of automatic brightness pyrometers based on the silicon diode.

In order to analyze the heat transfer process and to improve the quality and the productivity of rolled products, some non-contacting measurement procedure is required for determination of the temperature distribution of the products during processing. So far, such methods, as scanning a mirror mechanically [1] or arranging many detectors [2], have been used.

On the other hand, the self-scanning silicon photodiode array (linear array), in which multi-photosensing elements are arranged linearly and scanned electronically, was introduced into the industry about ten years ago and has been applied in a wide variety of non-contact measurements. In the steel industry it has been successfully used on the process line to continuously locate the edge of hot or cold product strip [3, 4, 5].

The silicon photodiode array could inherently be a temperature distribution sensor because the basic diode is a silicon photosensitive cell and its spectral sensitivity is in the red end of the visible spectrum and in the near infrared. Recently, a few attempts to apply the photodiode array to the temperature distribution measurement have been made [6, 7].

With a micro-computer based test system, the authors have investigated and evaluated such characteristics of a linear array as temperature calibration, non-uniformity, and crosstalk. We have obtained good results in the application of the test system to steel making and hot rolling processes.

2. GENERAL DESCRIPTION OF THE LINEAR ARRAY

The linear arrangement of the sensing elements on the single chip is shown in Fig. 1. The spatial resolution of the array is dependent on size, separation, and number of elements. Its signal output is dependent not only on the area of each element but also on the product of exposure time and the radiation intensity.

Fig. 2 is a schematic diagram of the temperature

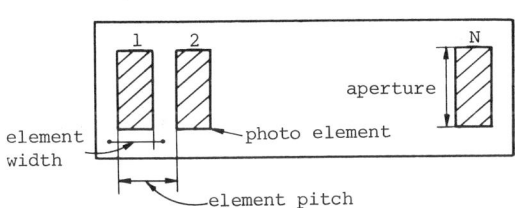

Fig. 1. Photodiode elements

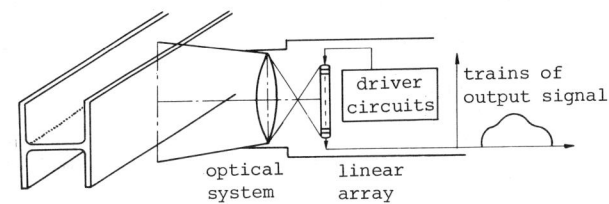

Fig. 2. Schematic diagram of the linear array camera

measurement process using the linear array. The radiation pattern of an object is focused by the optical system onto the linear array within a camera and trains of analog electrical pulses each having an amplitude proportional to the radiation intensity received by a given element are transmitted by the camera electronics.

3. CONSTRUCTION OF THE TEST SYSTEM

In order to verify the applicability of the linear array to the temperature distribution measurement, the test system with a micro-computer shown in Fig. 3 was developed.

Table I. Specifications of the linear array used in the test system

Number of elements	1024
Aperture	464μm
Element pitch	28μm
Element width	16μm
Non-uniformity	±10%
Output voltage	0∿4V
Clock frequency	1MHz

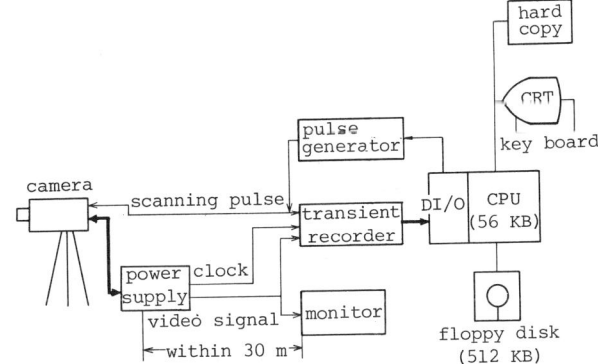

Fig. 3. Block diagram of the test system

Pulse generator provides scanning pulse and video signal is stored through transient recorder and DI/O into computer system

The following points are taken into consideration.

(1) Each output signal is stored into computer system.
(2) The temperature distribution data obtained can be effectively processed.
(3) The scanning frequency (1/exposure time) can be easily adjusted.

The specifications of the linear array used in this system are listed in Table I.

In Fig. 3, the pulse generator provides a suitable scanning frequency pulse and the trains of video output signal are produced by drive circuits within the camera and amplified. When the computer sends the input command to the camera circuits, the output signal is immediately digitized and stored in the transient recorder and the computer memory.

Previously programmed software of various kinds such as routines for data acquisition, temperature transformation, and effective signal processing are stored on the floppy disk.

As the entire system is designed to be portable, the field measurements are easily performed. If necessary, the acquired data can be stored on the floppy disk, and the data analysis can be performed in a laboratory.

4. PRINCIPLE OF TEMPERATURE DISTRIBUTION MEASUREMENT

Under the black body radiation condition at thermodynamic temperature T, the output voltage $V(T)$ of linear array is expressed by the following formula:

$$V(T) = G \cdot F(f) \cdot A \int \tau_\rho(\lambda) \cdot \tau_e(\lambda) \cdot S(\lambda) \cdot L(\lambda,T) d\lambda \quad (1)$$

where,
- G: voltage gain of the amplifier
- $F(f)$: a function of the scanning frequency f
- A: a function of the lens aperture
- $\tau_\rho(\lambda)$: spectral propagance of the light path
- $\tau_e(\lambda)$: spectral propagance of the lens
- $S(\lambda)$: relative spectral sensitivity of the linear array (as is shown in Fig. 4 [8])
- $L(\lambda,T)$: spectral radiance of a black body of thermodynamic temperature T at wavelength λ

If the medium of the light path is dry air and the wavelength interval is $0.3 \sim 1.2$ μm (sensitivity range of linear array), the spectral propagance of the light path may be considered as unity.

Then, the relative value of the video signal output voltage $V(T)$ of the linear array at fixed optical condition and constant scanning frequency is expressed as follows:

$$V'(T) \propto \int \tau_e(\lambda) \cdot S(\lambda) \cdot L(\lambda,T) d\lambda \quad (2)$$

According to Planck's law, spectral radiance, $L(\lambda,T)$, of a black body is expressed as follows:

$$L(\lambda,T) = C_1 \lambda^{-5}/(\exp(C_2/\lambda T)-1) \quad (3)$$

where,
- C_1 = the first radiation constant
- C_2 = the second radiation constant

When $\lambda T < 3 \times 10^{-3}$ (m·K), $\exp(C_2/\lambda \cdot T) \gg 1$ and Planck's formula of radiation as given in Eq.(3) may be approximated by Wien's formula:

$$L(\lambda,T) = C_1 \cdot \lambda^{-5} \cdot \exp(-C_2/\lambda T) \quad (4)$$

Introducing the relationship of Eq.(3) and the values of $S(\lambda)$ shown in Fig. 4 into Eq. (2), the relationship between temperature and signal output of linear array is approximately calculated as follows:

$$V'(T) = A_0 \cdot C_1 \cdot \lambda_e^{-5} \cdot \exp(-C_2/\lambda_e \cdot T) \quad (5)$$

or

$$T = \frac{C_2}{\lambda_e} / \log(A_0 \cdot C_1 / V'(T) \cdot \lambda_e^5) \quad (6)$$

where,
- A_0: a function of the lens, lens aperture, and scanning frequency
- λ_e: effective wavelength

Because the output signal amplitude must be in inverse proportion to the scanning frequency [9], the normalized signal output is expressed as follows:

$$V_n = \frac{fs}{fr} R(F_r, F_s) V_s \quad (7)$$

where,
- V_n: normalized signal of V_s
- V_s: output voltage at scanning frequency fs and lens aperture F_s
- F_r: reference lens aperture
- fr: reference scanning frequency
- R: signal amplitude ratio dependent on F_r and F_s

Substituting Eq.(7) into Eq.(6), the temperature, using any scanning frequency and any lens aperature, can be calculated as follows:

$$T = \frac{C_2}{\lambda_e} / \log(A_0' \cdot C_1 / V_n \cdot \lambda_e^5) \quad (8)$$

The temperature distribution can be calculated by substituting each signal output of the linear array into Eq.(8).

5. CHARACTERISTICS AND PERFORMANCE OF THE LINEAR ARRAY

Various characteristics and performances of the linear array as applied to the temperature distribution

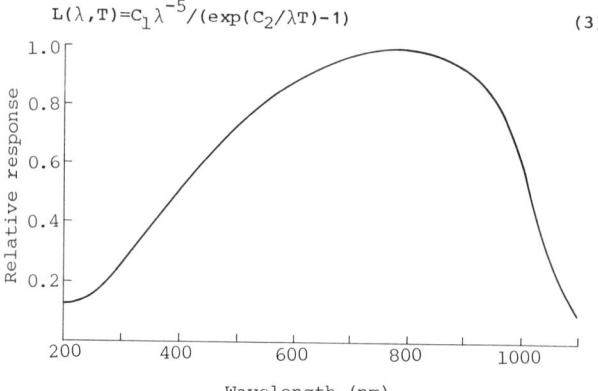

Fig. 4. Relative spectral response [8]

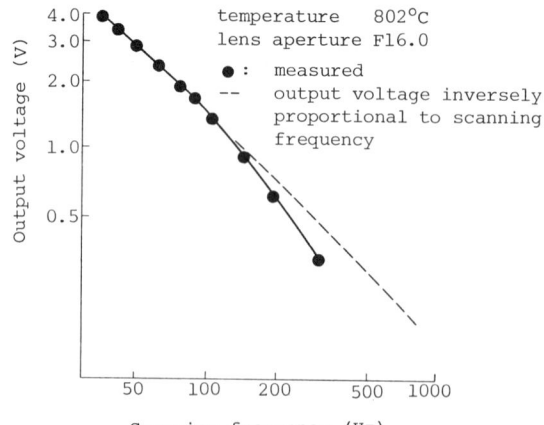

Fig. 5. The relation between output voltage of the linear array and the scanning frequency at constant temperature

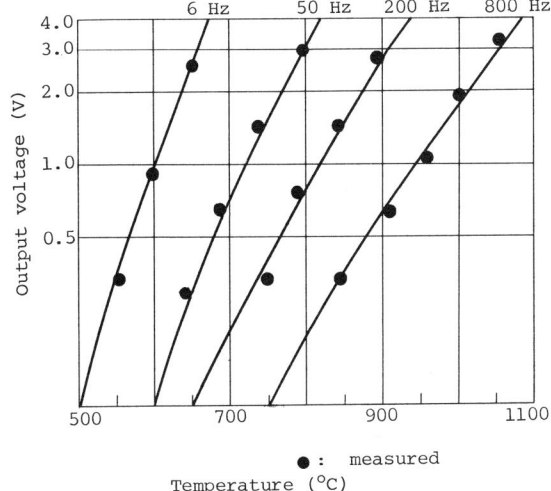

Fig. 6. Calibration curves of the linear array as a function of the scanning frequency

measurement were investigated and evaluated.

5-1 Relationship between temperature and signal output

The relation between source temperature and the output signal voltage of the linear array was measured as a function of the scanning fequency and the lens aperture. A black body furnace was used as radiation source. Fig. 5 shows the relation between the output signal voltage and the scanning frequency at constant temperature. The output voltage greater than 0.7 is approximately in inverse proportion to scanning frequency and Eq.(7) nearly holds true.

Fig. 6 shows the relation between output voltage and temperature from 500°C to 1100°C measured by adjusting the scanning frequency.

A wide range calibration curve shown in Fig. 7 was calculated by using Eq.(8) and data measured at three temperatures (653, 802, and 953°C). The effective wavelength, λ_e =0.936 μm, was selected to make the temperature error between source temperature and calculated temperature less than ±5°C.

In the steel industry, a wide range temperature measurement under 1700°C is desired. The linear array sensor can cover the range over 500°C. Temperature over 1100°C can be also measured with the use of an attenuator for the optical system.

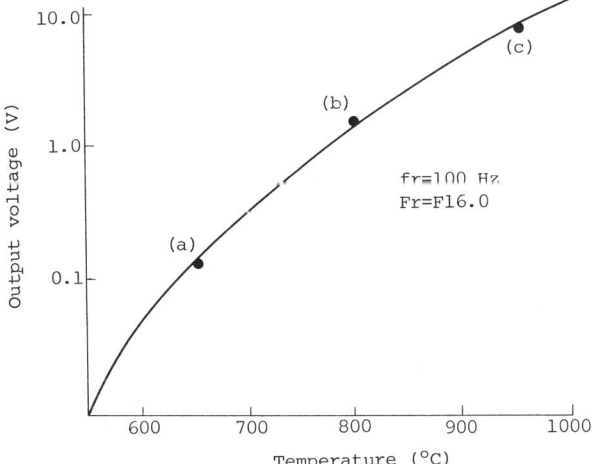

Fig. 7. A wide range calibration curve

5-2 Scanning frequency and temperature resolution

The scanning frequency plays an important role not only in the wide range temperature measurement but also in the high resolution temperature measurement.

The accuracy with which the output signal amplitude can be recorded digitally is limited by the analogue-to-digital conversion process. Fig. 8 shows the calculated temperature resolution corresponding to the least count of an 8-bit A/D converter for several temperatures and scanning frequencies.

In the steel making process, the temperature of the hot material varies, not only along the length but even across the width of the material. By varying the scanning frequency according to the object temperature, high resolution measurement is possible.

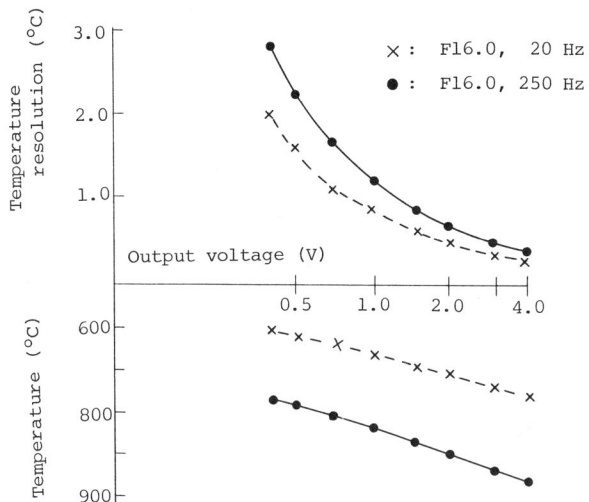

Fig. 8. Semi-logarithmic plot of the temperature resolution (above) and the temperature to be measured (below) as a function of the output voltage

5-3 Diode-to-diode sensitivity variation

The linear array has been considered not to be suitable for a temperature distribution sensor because the manufacturer's specifications typically state ±10% as an upper bound for the variation of the diode-to-diode response within a given array.

Fig. 9 (above) shows the effect of the non-uniformity on the temperature error at each temperature to be measured. These calculated results indicate only a ±1% temperature error is produced by a ±10% non-uniformity in diode-to-diode response. Fig. 9 (below) shows the experimental results of the non-uniformity about the middle 500 elements which were recorded with the linear array camera focused through the window of the black body furnace. These two results suggest that the temperature error due to the non-uniformity is less than ±5°C.

Fig. 10 shows an example of a waveform diagram of the non-uniformity at 795°C.

The effect of the non-uniformity is less than first imagined and considering the advantages of the linear array in the temperature measurement, this error is acceptable.

5-4 Crosstalk

The elements of linear array are not necessarily independent of each other due to so-called crosstalk [10]. The effect of the diode crosstalk was studied by placing a knife edge across a portion of the black body furnace in a direction perpendicular to that of the array. Fig. 11 is an example of the step response. In the range from 550°C to 900°C, the number of elements required to make transition from 10% to 90% is about 7. When a linear array is used to locate the edge of a hot strip, it would be difficult to acquire high accuracy unless this effect is taken into consideration.

Consequently, much attention must be paid to the temperature measurement in the edge portion of a hot material, and spatially high resolution is needed when measuring a small object.

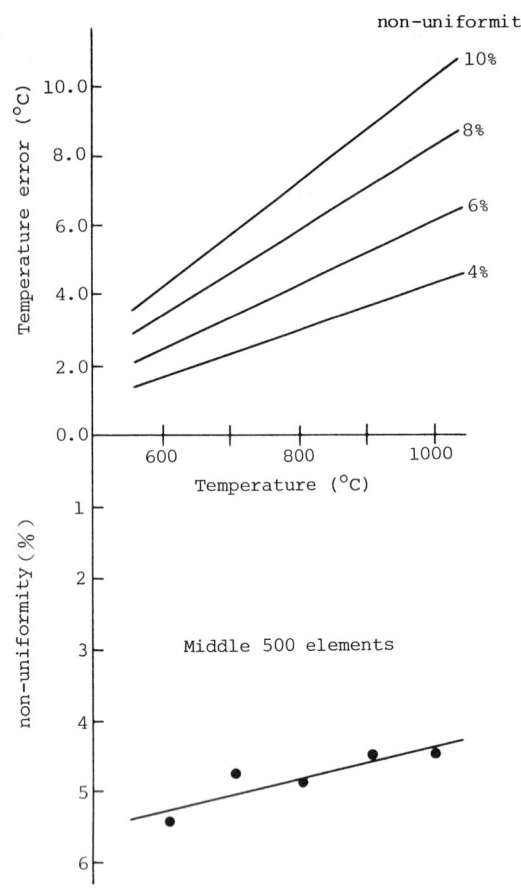

Fig. 9. The effect of the non-uniformity on the temperature error by calculation (above) and experiment (below)

Non-uniformity(●)=2 $(V_{max}-V_{min})/(V_{max}+V_{min})$

where, V_{max}: maximum voltage of the 500 elements
V_{min}: minimum voltage of the 500 elements

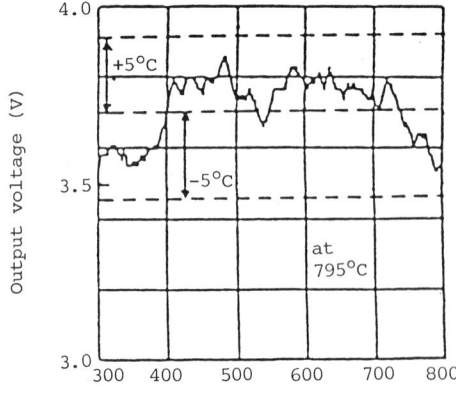

Fig. 10. An example of a wave form diagram of the non-uniformity

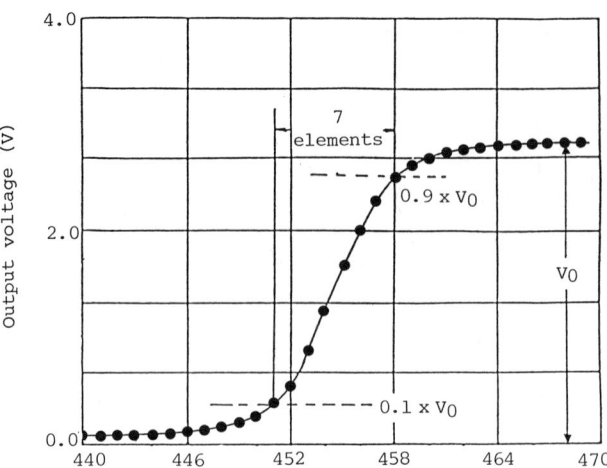

Fig. 11. An example of the step response of constant temperature

5-5 Background noise

The background noise associated with the dark current of the linear array was studied as a function of the variation of its ambient temperature. Fig. 12 shows the relation between the scanning frequency and the background noise level as a function of the ambient temperature. Output levels are the mean value of the output signals for the condition that the linear array is optically shielded. The results indicate that the operating temperature range between 5°C to 50°C and with the scanning frequency more than 10 Hz, the background noise is less than 1% FS and almost negligible in the temperature measurement.

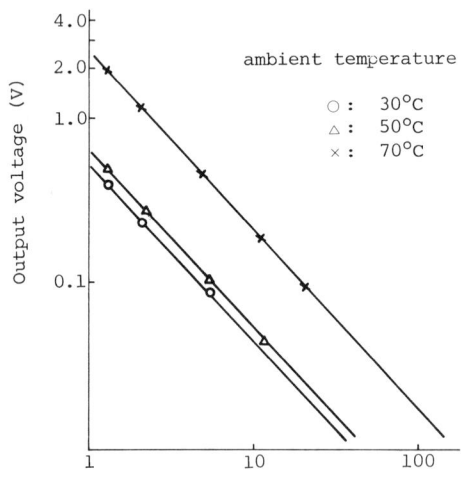

Fig. 12. The relation between background noise and scanning frequency as a function of the ambient temperature

6. EXAMPLES OF THE FIELD TESTS

The test system shown in Fig. 3 has been used for analytical and operational purpose at various locations in our steel mill. Much useful information about our process lines, previously unknown, has been obtained.

6-1 H-shaped steel

H-shaped steel is manufactured through hot rolling [11]. Undesirable residual stress in its flange section as tension and in its web section as compression adversely affects the functional performance of the product. It also creates problems in further processing. The residual stress is mainly caused by the large temperature gradients which are present on the web and flange during rolling and cooling [12, 13]. Hence, measurement of the temperature distribution in the flange and web section is significant.

Curve (a) of Fig. 13 shows the temperature distribution sampled just after the moving H-shaped bar was milled by the finisher. The temperature difference between corresponding locations across the flange portion, shown by curve (b) is obtained through signal processing. The temperature differences are caused by contact with rolls during rolling and overflow of the roll cooling water on the web face. From these data, the cooling condition in the process line could be diagnosed or controlled.

In Fig. 14, several temperature distributions of flange section, sampled from a long bar running and stopping repeatedly while hot sawing, are displayed. The temperature in each part of the flange could be calculated for analytical use.

6-2 Hot strip

In the hot rolling process, quality control automation has been achieved by a computer control system including the new rolling model [14]. Control based on temperature measurement not only along the length but also across the width of the strip has saved energy and at the same time improved product quality.

Temperature measurement toward the edge of the strip is important to insure that strips passing through the roughing and finishing mills are within the desired temperature range throughout the entire width. At the coiler, temperature distribution measurement

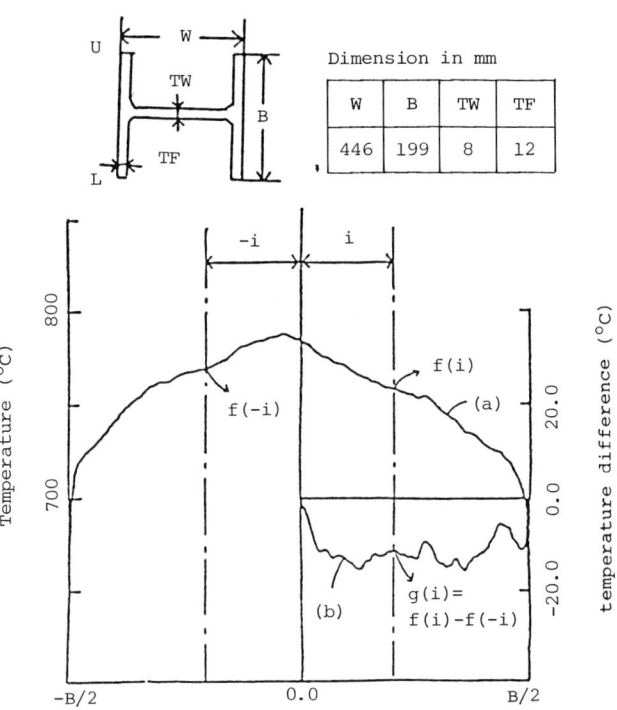

Fig. 13. Temperature distribution across the flange part of the H-shaped steel (a) and temperature difference between corresponding locations (i,-i) (b)

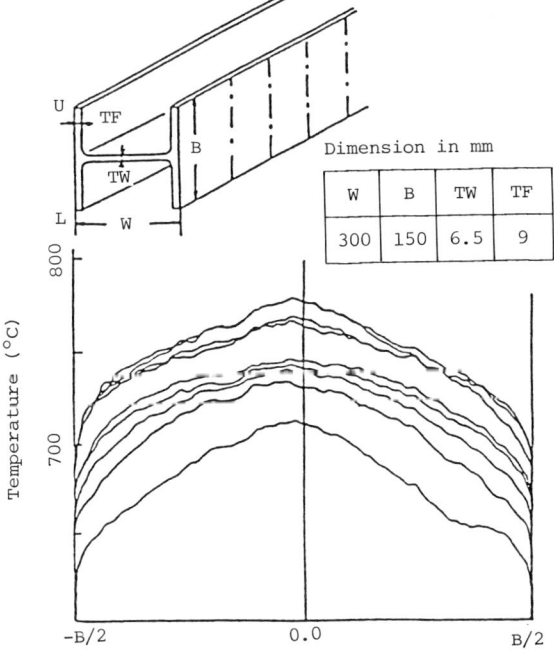

Flange part of the H-shaped steel

Fig. 14. Temperature distributions of the flange part of the H-shaped steel at hot sawing

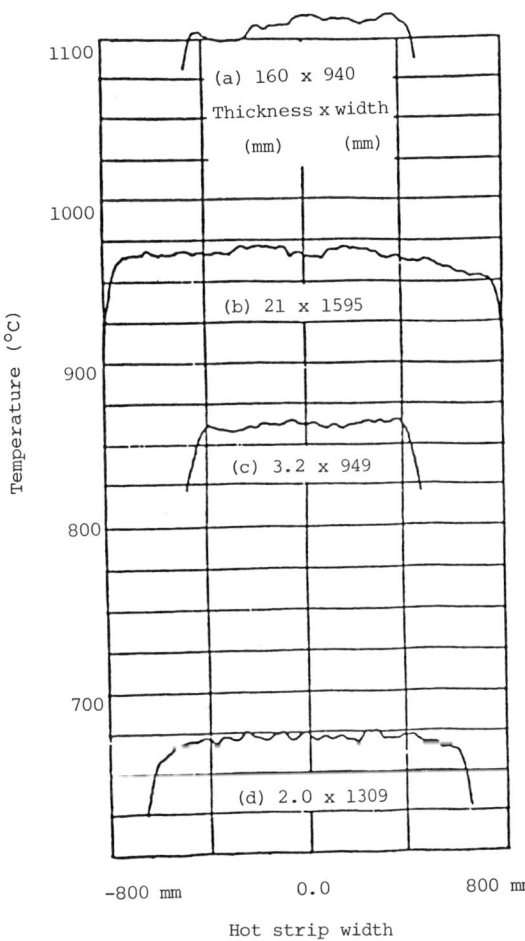

Fig. 15. Temperature distributions of the hot strip from above, at the rougher (a), at the finisher between No. 1 and No. 2 stand (b), at the exit end of the finisher (c), and at the coiler (d)

assures that strip has the desired metallurgical properties.

Using the linear array system, the temperature distribution measurements in the hot rolling process were carried out. Some results are shown in Fig. 15. At each location, the size of the strip is different.

At the rougher, the temperature distribution indicates that the free side of the strip is 10°C higher than the drive side.

At the location between the No. 1 and No. 2 stand of the finisher it was possible to obtain an accurate temperature distribution in spite of interference by steam and water.

At the coiler, the temperature variation in the center of the pattern is caused by spray cooling.

In Fig. 15, the steep temperature gradient near both edges can be observed.

Fig. 16 is the comparison of the temperature indicated by both the linear array and the silicon radiation pyrometer located at the exit end of the finisher. These data show the close agreement between temperatures obtained with the linear array system and those obtained with the silicon radiation pyrometer.

The results indicate that the linear array system is applicable throughout the hot rolling process line.

7. CONCLUSION

The temperature measurement system with a microcomputer based linear array is characterized as follows.

1) Wide range temperature measurement with one sensor is obtained by controlling the scanning frequency.
2) Simultaneous measurement of temperature and size. From the temperature profile, much information, such as edge position, center position, width average temperature and peak temperature position, can be easily obtained.
3) Reliability comparable to the conventional silicon radiation pyrometer.

Developing the micro-computer based test system, investigating and evaluating the characteristics and performances and making various field tests, the authors have assured that the linear array sensor is of practical use in the temperature distribution measurement in the steel industry.

REFERENCES

1. T. P. Murray, Iron and Steel Engineer. 54, 40 (1977).

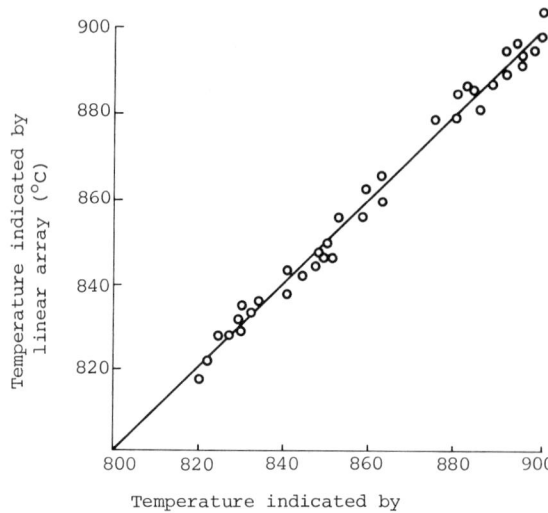

Fig. 16. Comparison of the temperature indicated by the silicon-radiation pyrometer and the linear array at the exit end of the finisher

2. J. Ihlefeldt, H. J. Kopineck, and W. Tappe, Stahl u. Eisen. 100, 474 (1980).
3. D. L. Burk, and N. F. Simic, Iron and Steel Engineer, 55, 43 (1978).
4. D. Larin, Association of Iron and Steel Engineers Annual Convention, (1978).
5. M. Guegan, Revue de Metallurgie-CIT, 309 (1981).
6. P. Bohlander, Stahl u. Eisen, 97, 927 (1977).
7. H. J. Kopineck, Stahl u. Eisen, 101, 267 (1981).
8. EG&G Reticon, Solid State Image Sensing Camera System catalogue.
9. G. Horlick and E. G. Codding, Analytical Chemistry, 45, 1490 (1973).
10. R. Hopwood, The society of photo-optical instrumentation engineers for SPIE's technical symposium east, April 1980.
11. I. Oyama, T. Irie, and co-workers, Nippon Kokan technical report, 59, 129 (1973).
12. T. Kusakabe and Y. Mihara, The Journal of the Iron and Steel Institute of Japan, 65, 1375 (1979).
13. T. Kusakabe and Y. Mihara, The Journal of the Iron and Steel Institute of Japan, 65, 1383 (1979).
14. S. Fujii, H. Sasao, and co-workers, Nippon Kokan technical report, 85, 19 (1980).

A new method for temperature distribution measurement using multi-spectral radiance

Jiro Ohno

Fundamental Research Laboratory, Nippon Steel Corporation, Kawasaki 211, Japan

A new method for measuring the temperature distribution of a small hot area is described. The temperature distribution function x(T), which is defined as the area of the source having temperature T, can be calculated from the spectral radiance emitted by the source. A comparison between the x(T) for a silicon carbide source and the x(T) calculated from the multispectral radiance measurements shows the peak temperature of the source was determine with an accuracy of 3%.

1. INTRODUCTION

In radiation thermometry, the temperature within the field of view is usually assumed to be uniform. However, when a heat-generating area that is small compared to the field of view is measured, it is impossible for the measured value to be representative of the temperature because of the temperature distribution. Although the temperature distribution can be measured with a scanning-type apparatus such as an infrared camera, such a method has the disadvantages that a large-scale apparatus is necessary and that measurement accuracy is affected by the uncertainty of the emissivity of the surfaces being measured.

For some purposes, it is sufficient to determine only the temperature distribution. The temperature measurement at a weld or at the reference point for the measurement of the thermal diffusion coefficient by the laser flash technique are two such examples.

In this paper, a new method of measuring temperatures is proposed, in which the temperature distribution function of the object to be measured is determined from the spectral radiance emitted by the source. The principle of this method and examples of measurement are described. An advantage of the method is that the temperature distribution function can be determined even if the object to be measured vibrates, provided that it remains within the field of view.

The relationship between the radiance and the temperature distribution function is expressed by a Fredholm integral equation. The author has devised a relaxation method using a Dynamic Programming approach and obtained satisfactory results.

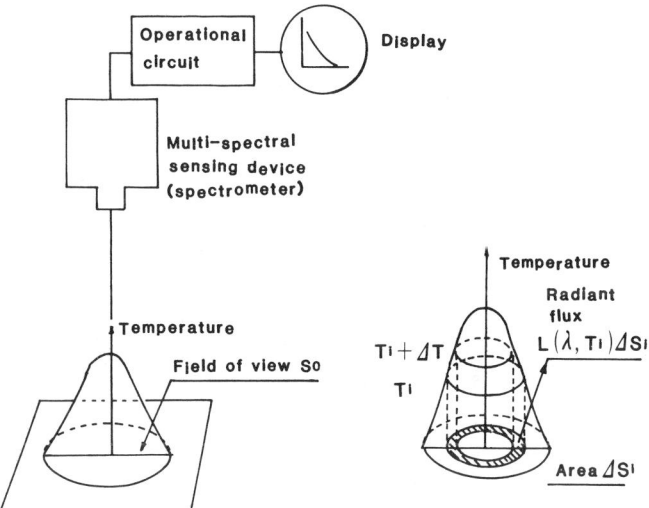

(a) sensing system

(b) radiance emitted from ΔS_i over which the temperature ranges from T_i to $T_i + \Delta T$

Fig. 1 Principle of the temperature distribution measurement using multispectral radiant power.

2. PRINCIPLE OF THE MEASUREMENT

The principle of the measurement is illustrated in Fig. 1. As shown in Fig. 1(a), a radiant source with a temperature distribution is placed within the field of view of a multi-spectral sensing device. For simplicity of explanation, the emissivity of the source is assumed to be 1.0. The spectral radiance detected by the sensor, $b(\lambda)$, can be expressed by

$$b(\lambda) = \iint_S L(\lambda,T)\,dx\,dy \qquad (1)$$
$$= \sum_i L(\lambda,T_i)\Delta S_i ,$$

where S is the area of the field of view, λ is the effective wavelength for the measurement, T is the thermodynamic temperature of an element of area of the source, dS, and $L(\lambda,T_i)$ is the Planck radiation function. As shown in Fig. 1(b), ΔS_i is an area over which the temperature ranges between T_i and $T_i + \Delta T$. The temperature distribution function#, $x(T)$, is defined as

$$x(T) = \frac{dS}{dT} . \qquad (2)$$

ΔS_i can then be expressed as a function of $x(T)$ as follows

$$\Delta S_i = \int_{T_i}^{T_i+\Delta T} x(T)\,dT . \qquad (3)$$

Upon substituting Eq.(2) into Eq.(1), one obtains the following Fredholm integral equation of the first kind

$$b(\lambda) = \int_0^{T_{max}} x(T)L(\lambda,T)\,dT , \qquad (4)$$

$$T_o \geq T \geq 0 \quad \text{for} \quad x(T) \geq 0$$
$$T_{max} > T_o > T \quad \text{for} \quad x(T) = 0$$

\# Strictly speaking this is the <u>inverse</u> of the spatial temperature distribution function. It is the distribution of area with respect to temperature and is proportional to the probability of obtaining the temperature value T in a radiometric measurement of a random point within the source area S. $x(T)$ can also be regarded as the radiance weighting function required to produce the observed spectral distribution from Plankian radiators of different temperatures (see Eq.(4)).

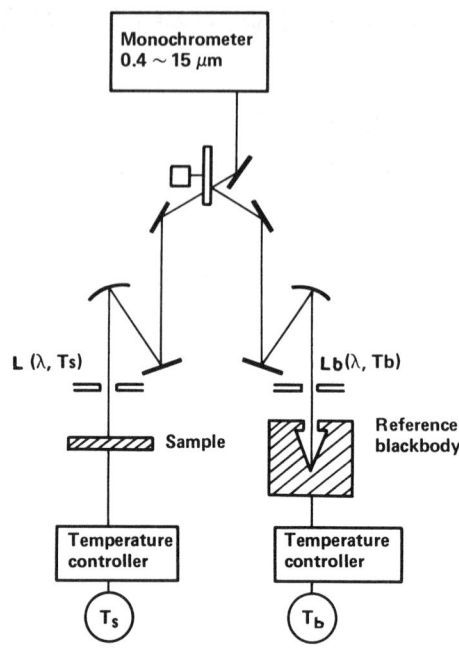

Fig. 2 Configuration of the measuring system.

where T_o is the maximum temperature in the field of view. If values of b are measured as a function of λ, then $x(T)$ can be determined from Eq.(4).

3. ANALYSIS OF THE MEASUREMENT SYSTEM

3.1 Measurement System

Fig. 2 illustrates the configuration of the measurement system. A double beam spectrometer is used to alternately measure the radiant flux from the sample and that from a reference blackbody cavity. The temperature of the blackbody is adjusted so that its spectral radiance distribution is almost the same as the sample. The spectral radiance of the sample can be calculated from the ratio of the measured radiant flux from the sample to that from the blackbody and the temperature of the blackbody. With this system, the spectral radiance of a sample can be determined over a broad wavelength range. The measurements are almost independent of the absorbtance of the atmosphere.

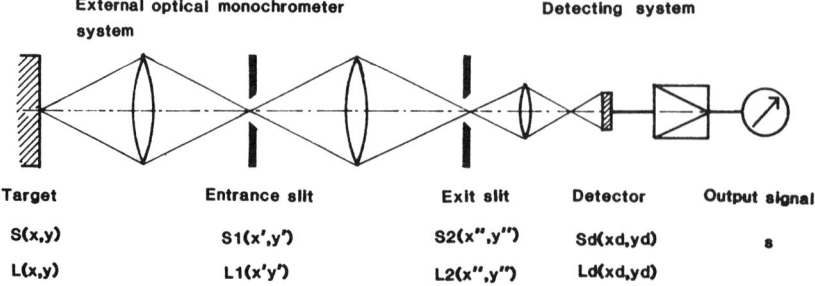

Fig. 3 Optical configuration of the measuring apparatus.

3.2 Spacial Distribution of Optical Gain

The optical configuration of the measurement apparatus is shown in Fig. 3.

The radiance $L(x,y)$ emitted by a small area $dS(x,y)$ of the source and the radiance $L_2(x'',y'')$ of an image of dS at the exit slit of the monochrometer are related by

$$L_2 = \tau L, \qquad (5)$$

where τ is the optical transmissivity between the target and the exit slit.

Applying thin lens theory, the image at the exit slit, dS_2, is given by

$$dS_2 = \left(\frac{f}{a}\right) dS, \qquad (6)$$

where a is the optical path length between the source and the lens, and f is the focal length of the lens.

The output signal, ds, corresponding to the radiance emitted from dS_2 is given by

$$\begin{aligned}ds &= \int_{S_d} d\Phi(x_d,y_d,\theta,\phi,\lambda) \cdot R_\phi(x_d,y_d,\theta,\phi,\lambda) dS \\ &= \int_{S_d} L_2(x_d,y_d,\theta,\phi,\lambda) dS_2 \cdot \\ &\quad \frac{\cos\theta}{\ell^2} \cdot R_\phi(x_d,y_d,\theta,\phi,\lambda) dS_d, \end{aligned} \quad (7)$$

where $d\Phi(x_d,y_d,\theta,\phi,\lambda)$ is the incident flux from dS_2 to the small area dS_d at the point x_d,y_d of the detector and $R_\phi(x_d,y_d,\theta,\phi,\lambda)$ is the responsivity at the point x_d,y_d, S_d is the area of the detector, ℓ is the distance between dS_2 and S_d, and θ is the incident angle to dS_d.

If $\ell \gg 1$, and $\theta \simeq 0$ then Eq.(7) can be simplified to

$$\begin{aligned}ds &= L_2(x'',y'',\lambda) dS_2 \cdot \frac{1}{\ell^2} \cdot \int_{S_d} R_\phi(x_d,y_d,\theta,\phi,\lambda) dS_d \\ &= L(x,y,\lambda) dS \cdot \tau(x,y,\lambda) \cdot \left(\frac{f}{a\ell_o^2}\right) \cdot \int_{S_d} R_\phi dS_d, \end{aligned} \quad (8)$$

where $L_2(x'',y'',\lambda)$ is the radiance emitted by dS_2, and ℓ_o is the distance between the exit slit and the detector.

Defining the optical gain as $p(x,y,\lambda)$, Eq.(8) can be written as

$$ds = L(x,y,\lambda) dS \cdot p(x,y,\lambda), \qquad (9)$$

where

$$p(x,y,\lambda) = \tau(x,y,\lambda) \cdot \left(\frac{f}{a\ell_o^2}\right) \cdot \int_{S_d} R_\phi dS_d,$$

Eq.(9) states the qualitative relationship between the radiance at the point x,y on the source and the output signal from the monochrometer.

Integrating Eq.(9) over the whole area of the source, the total signal is given by

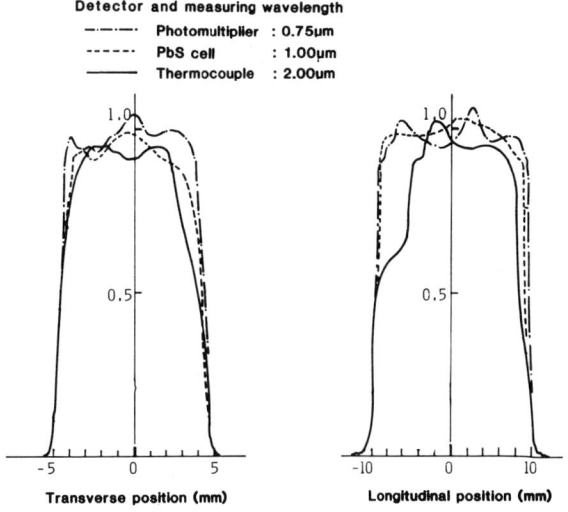

Fig. 4 Examples of the measurement of optical gain, $p(x,y,\lambda)$.

$$s(\lambda) = \int_S L(x,y,\lambda) \cdot p(x,y,\lambda) dS, \qquad (10)$$

which is a modified form of Eq.(1).

3.3 Measurement of Optical Gain

The optical gain of the double beam spectrometer was measured over the wavelength range 0.40 to 15 μm. Three different detectors were used: a photomultiplier, a PbS cell, and a thermocouple. The source was a 12 W incandescent lamp which was mounted on a XY stage so that it could be moved in both a horizontal and vertical direction. A collimated narrow beam (1 mm in diameter) was focused on the entrance slit of the spectrometer.

An example of the optical gain measured with the three detectors is shown in Fig. 4. The abscissas indicate the relative position of the source and the ordinates are the relative output signal corresponding to the optical gain, $p(x,y,\lambda)$. The field of view was chosen to be a rectangle 8 mm wide and 16 mm long. These dimensions correspond to an area such that the deviation of the output signal of each detector remained within ± 10% of its average value. Under this condition the optical gain was be assumed to be uniform and Eq.(10) was simplified to

$$b(\lambda) = \frac{s(\lambda)}{p} = \int_S L(x,y,\lambda) dS, \qquad (11)$$

which coincides with Eq.(1).

3.4 Numerical Calculation [1,2]

Eq.(4) can be expressed in matrix form as

$$b = L \cdot x. \qquad (12)$$

A least mean square approximation to the

difference between the measured and calculated values of b(λ) as a function of wavelength can be carried out after selection of a suitable initial value of c. The following equation is obtained.

$$x = (L'L+\alpha I)^{-1}(L'b+\alpha c), \quad (13)$$

where L' is the transpose matrix of L, I is the unit matrix, and α is a constant. Some kind of iterative relaxation method is required for the solution of Eq.(13).

3.5 Characteristics of Eq.(13)

The integral range and the number and selection of measurement wavelengths affect the accuracy of the solution of Eq.(13). These characteristics were examined using a computer. First b(λ) was calculated from a known temperature distribution. Then x(T) was calculated using the iterative Dynamic Programming method and compared with the original distribution.

(1) Integral range

The integral range, especially the upper limit, T_{max}, affects the accuracy of the calculated x(T). If T_{max} coincides with the maximum of the temperature distribution, T_o, the highest accuracy is obtained, but this cannot be known in advance. Therefore it is convenient to choose T_{max} to be larger than the expected T_o. Fig. 5 shows the results of two calculations. In Fig. 5(a), x(T) changes discontinuously at T_o and the error becomes large near T_o. On the other hand, Fig. 5(b) shows a continuous x(T) for which large errors do not occur.

(2) Wavelength range

It is necessary to select a wavelength range that is consistent with the temperature range of the source, especially T_o. When the relationship between wavelength range and calculating accuracy was studied, it was found that the accuracy near T_o increased when short wavelengths were selected and that the accuracy near T_o generally deteriorated when long wavelengths were used. It is desirable to select a range of short wavelengths since, in practical use, it is often most important to know the distribution near T_o. It was found to be advantageous to select a short wavelength limit that corresponds to a power that is several orders of magnitude below the maximum value of b(λ).

(3) Emissivity

If the emissivity, ε, of the source is constant, independent of temperature and wavelength, the effect of emissivity can be eliminated if x(T) is normalized as will be understood from Eq.(4). If ε is a function of temperature, x(T) generates an error in proportion to the deviation from the mean of ε at that temperature. If ε is a function of wavelength, the error Δx, is expressed as

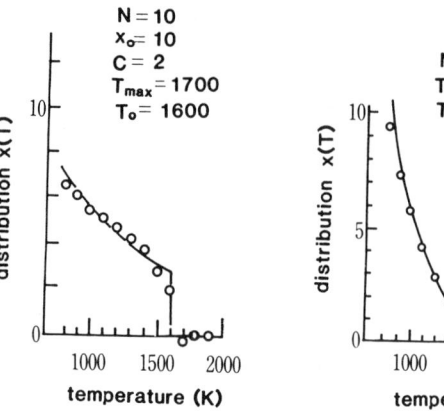

Fig. 5 Examples of results calculated by the Dynamic Programming iterative method.
——— original, ∘∘∘∘ calculated.
Spectral range is 0.5 to 2.5 μm.
Discontinuous $x(T) = x_o \exp(-CT/T_o)$,
continuous $x(T) = \cos(\pi T/T_o)+1$.

$$\Delta x = L^{-1}\left\{\frac{\Delta\epsilon_\lambda}{\epsilon_o} \cdot b(\lambda)\right\}, \quad (14)$$

where ϵ_o is the mean and $\Delta\epsilon_\lambda$ is the deviation. Because L^{-1} is constant, its magnitude can be estimated at approximately $(\Delta\epsilon_\lambda/\epsilon_o)$. However, it cannot be expressed in a general form.

4. MEASUREMENT

The following experiments were conducted to confirm the principle of the method. Both the spectral radiant power from a sample and its temperature distribution were determined. Then the temperature distribution function was calculated from the measured spectral radiant powers using Eq.(13). The calculated temperature distribution was then compared with the measured distribution to examine the accuracy of the method.

4.1 Isothermal Disc

In the first experiment, the sample was coated with an optical black paint (ε=0.97). A stop with a circular aperture 10 mm in diameter was placed above the sample. The temperature of the sample was measured with a thermocouple. The radiant flux from the heated sample was compared with that from the reference blackbody using the measuring system illustrated in Fig. 2.

Fig. 6 shows the measured radiant power ratio, $g(\lambda) = L_b(\lambda, T_s)/L_b(\lambda, T_b)$. The blackbody temperature, T_b, was held at 200 °C and the sample temperature, T_s, was set at 150, 200, and 250 °C.

Next, the radiance of the sample, $b(\lambda) = g(\lambda) \cdot L_b(\lambda, T_b)$ was calculated from g(λ) and the known blackbody temperature, T_b. Finally, x(T) was determined using Eq.(13). The results of the calculation for a sample temperature of 250 °C are shown in Fig. 7.

If both the measurements and the calculation are carried out accurately, x(T) should be a delta function having its peak at

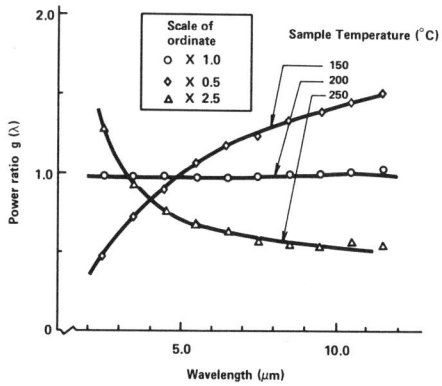

Fig. 6 Measured values of the power ratio g(λ). Sample is the isothermal disc having an emissivity of 0.97.

T_s. The calculated values, indicated in Fig. 7 by the dotted line, have a rather broad peak at 270 °C which is somewhat higher than the sample temperature of 250 °C. The fact that the distribution function does not become a delta function is an indication of the limitation of the accuracy of the calculation.

Table 1. shows the correlation between the measured sample temperature and the peak value of x(T) for the three sample temperatures. In each case the calculated peak temperature of x(T) was slightly higher than the measured value. In each case the distribution function showed a relatively broad peak.

4.2 Sample with Temperature Distribution

(1) measurement of spectral radiant power, b(λ)

In the second experiment, the sample was a silicon carbide rod having a symmetric concave central region as shown in Fig. 8. When an electric current flows through the sample, the central region with the reduced diameter becomes very hot which produces a temperature gradient along the longitudinal axis of the sample. The sample was heated by

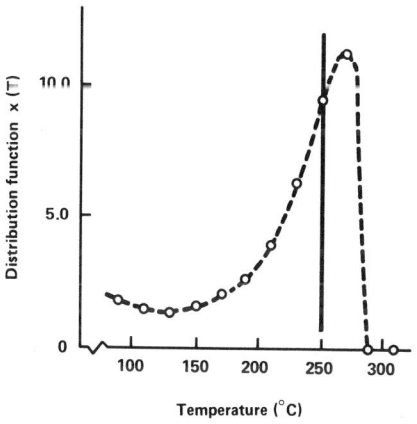

Fig. 7 An example of x(T) calculated for the isothermal disc at 250°C.

Table 1 Comparison between the calculated and measured peak temperature of the isothermal disc.

Thermocouple temperature (°C)	Peak Temperature of x(T) (°C)	Error (°C)
150	157	7
200	212	12
250	270	20

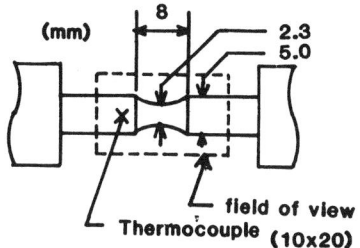

Fig. 8 Geometry of the silicon carbide sample with a symmetric concave central region.

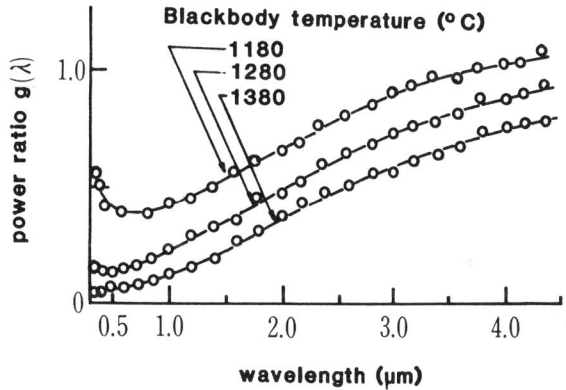

Fig. 9 Measured values of the power ratio g(λ). Sample is the silicon carbide rod.

a current from a high-precision DC stabilized power supply having an setting accuracy of 0.1% and a stability of 0.01%. No additional temperature control was required to provided stable conditions for the experiment.

Fig. 9 shows the results of the measurement of the radiant power ratio. The temperature distribution on the sample was kept constant and the temperature of the reference blackbody was set at 1180, 1280, and 1380 °C. The temperature of the blackbody was selected so that g(λ) did not vary greatly with wavelength. A mean value for the radiant power, $b_m(\lambda)$, was calculated from Eq.(15)

$$b_m(\lambda) = \frac{a}{3} \cdot \sum_{i=1}^{3} g(\lambda, T_{bi}) L(\lambda, T_{bi}) , \quad (15)$$

where T_{bi} is the temperature of the reference

Table 2 Data for calculation the power ratio g(λ).

$<b_c/b_m>=0.807\times10^{-4}$

wavelength	observed g(λ)			blackbody radiance (W/mm²·μm)			b_m	b_c
	blackbody temp. °C			blackbody temp. °C				
(μm)	1180	1280	1380	1180	1280	1380	$<g(\lambda)L(\lambda,T)>$	Eq. (12)
0.4	0.550	0.118	0.029	1.986×10^{-1}	9.834×10^{-1}	4.011×10	1.14×10^{-2}	0.965×10^{-6}
0.8	0.392	0.170	0.087	1.502×10^3	3.342×10^3	6.749×10^3	5.81×10^2	0.480×10^{-2}
1.0	0.412	0.223	0.121	5.873×10^3	1.114×10^4	1.954×10^4	2.42×10^3	0.198×10^{-1}
2.0	0.623	0.473	0.350	2.656×10^4	3.658×10^4	4.853×10^4	1.694×10^4	1.40×10^{-1}
3.0	0.850	0.684	0.557	1.872×10^4	2.336×10^4	2.845×10^4	1.595×10^4	1.28×10^{-1}
4.0	1.012	0.860	0.762	1.066×10^4	1.270×10^4	1.486×10^4	1.100×10^4	0.825×10^{-1}

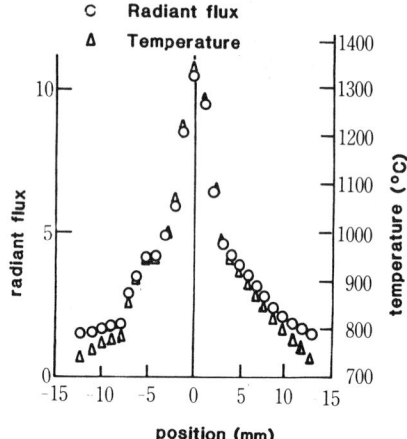

Fig. 10 Radiant power distribution and the temperature distribution of the silicon carbide rod at 3.2 μm.

blackbody furnace, $L(\lambda,T_{bi})$ is the spectral radiance of the reference blackbody, and a is a constant specific to the equipment.

(2) measurement of the temperature distribution function, x(T)

The sample was placed on an XY stage. A fixed stop with a 0.5 mm aperture was mounted above it. A Type S thermocouple, made from 0.3 mm diameter wire, was attached to the sample at the point indicated in Fig. 8. The stage was adjusted so that the sample remained centered beneath the aperture as it was moved in the direction along its axis. The axial distribution of the flux from the aperture was measured as the sample was moved in the longitudinal direction beneath the plate.

Fig. 10 shows the results of the measurement of the radiant power distribution at 3.2 μm. Temperatures were calculated from the radiant power distribution using the temperature indicated by the thermocouple as a reference and assuming the emissivity of the sample to be 0.7 in the near infrared [3,4].

The temperature distribution function over the field of view was determined by multiplying the axial temperature distribution by the projected area ratio. The results are also shown in Fig.10. x(T) is found to have a sharp peak near 1300 °C. Substituting this x(T) into Eq.(12) yields a calculated value for the radiant power distribution, $b_c(\lambda)$. Table 2 shows values of $g(\lambda)$, $b_m(\lambda)$ and $b_c(\lambda)$. Values for $L(\lambda,T)$ were taken from numerical tables [5].

Fig. 11 shows the ratio of $b_c(\lambda)/b_m(\lambda)$. This ratio should be independent of wavelength, however as shown in Fig. 11, the value of the ratio gradually decreases with increasing wavelength. Possible causes for this result are (i) that the emissivity of the sample changed with wavelength and temperature, (ii) that an error occurred in the measurement of x(T). The b(λ) used for the calculations was a value obtained by multiplying $b_m(\lambda)$ by the mean of the values of Fig. 11, i.e. 0.807×10^{-4}.

5. CALCULATION OF x(T)

As shown in Fig. 10, the temperature distribution of the sample has a sharp peak near 1350 °C. According to computer simulation studies, the accuracy of T_0 increases when a peak occurs at the maximum temperature

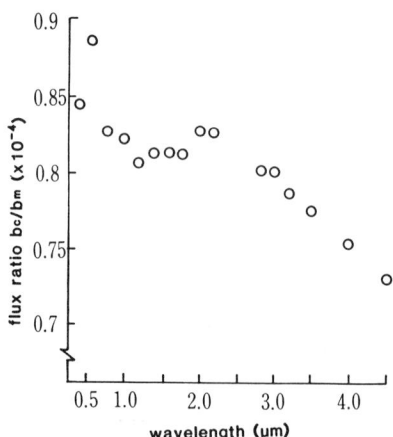

Fig. 11 Comparison of the measured radiant power, $b_m(\lambda)$, and calculated radiant power, $b_c(\lambda)$.

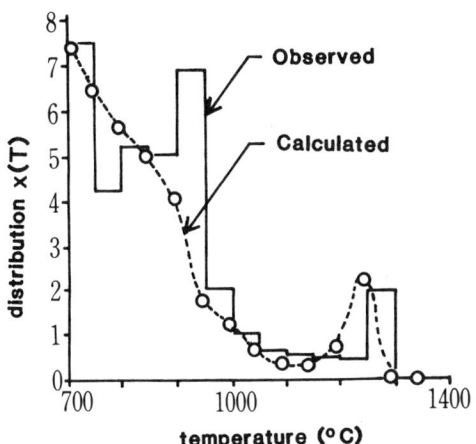

Fig. 12 Comparison of the observed x(T) and calculated x(T) after 100 iterations. Spectral range is 0.5 to 4.0 μm.

in the field of view. Fig. 12 shows a comparison of calculated and measured values of x(T). Calculating conditions are also indicated in the legend. The calculation did not converge well because of the complexity of the distributional form. However, the results of the calculation clearly show a peak at 1280 °C and are in fair agreement with the measured values except for the regions of lowest temperature. As already revealed by preliminary examination, the accuracy of the lower temperature region of x(T) deteriorates when the wavelength range is selected on the short wavelength side, and the overall accuracy of x(T) fluctuates when it is selected on the long wavelength side. Further, if the comparison between measured and calculated values is made only near T_o, the measurement accuracy is 3% of T_o, 5% of x(T), and approximately 1% of b(λ).

6. PROBLEMS IN PRACTICAL USE

So far, the principle of the method for temperature distribution measurement and examples to confirm it have been explained. Some simplification is conceivable in putting this method into practical use. In actual temperature distribution control, there is often an optimum distribution and the process is so controlled as to obtain it as far as possible; that is, temperature distribution control is done by simply treating deviations from the optimum distribution. In such a case, the initial value of Eq.(13) is regarded as the optimum distribution and the measuring wavelength and integral range can be selected accordingly. If a comparison is made between the optimum radiance distribution, $b^*(\lambda)$, and the measured value, $b(\lambda)$, then the direction of the deviation from the optimum distribution, $x^*(T)$, can be known without calculating x(T). For example, the following δ is considered.

$$\delta = \sum_i \alpha_i \{b_i(\lambda)/b_i(\lambda_o) - b_i^*(\lambda)/b_i^*(\lambda_o)\} . \quad (16)$$

If α_i is properly selected, it is possible to find a simple linear relation between T_o and δ over a limited range.

7. CONCLUSION

In radiation thermometry, a new method for temperature distribution measurement was examined. The principle of the method was confirmed and actual measurements were carried out with the following results:

(i) A new method was proposed, in which the temperature distribution function of an object of measurement is determined from the wavelength distribution of the radiant power. A constant emissivity, independent of temperature and wavelength, was assumed for simplicity.

(ii) If the temperature distribution function, x(T), is defined as x(T) = dS(T)/dT, the relationship between x(T) and the wavelength distribution of radiant power is given by a Fredholm integral equation of the first kind in which the Planck radiation function is the kernel. x(T)ΔT represents an area in the field of view for which the temperature lies between T and T + ΔT. S(T) denotes an area in which the temperature lies between 0 and T.

(iii) Some relationships between the integral range, the wavelength range, and the accuracy of the calculated x(T) were examined. Criteria for selecting these parameters were proposed.

(iv) The measuring accuracy of this new method is approximately ±5%. This accuracy can be improved further depending upon the form of x(T).

This measurement technique involves a new concept of the relationship between the spatial temperature distribution of a source and its radiant spectral distribution.

REFERENCES

[1]. S. Twomey, J. Frank. Inst., 279, 95 (1965).

[2]. W.L. Smith, Appl. Opt., 9, 1993 (1970).

[3]. S. Silberman, JOSA, 38, 989 (1948).

[4]. Y.S. Touloukian, Thermophysical Properties of High Temperature Solid Materials, (Macmillan, New York, 1967), Vol. 5, p. 131.

[5]. M. Pivovonsky, Tables of Blackbody Radiation Functions, (Macmillan, New York, 1961).

A broadband ratio pyrometer

J. L. Gardner, T. P. Jones, and M. R. Davies

CSIRO Division of Applied Physics, Sydney, Australia 2070

The design and construction of a broadband ratio pyrometer is described. The design characteristics necessary to allow a spectral radiance ratio measurement to be made with 0.25% imprecision are given. The ratio pyrometer uses silicon detectors with a spectrally selective beam-splitter to produce wavelength separations between channels. The response of each detector is digitized by a microprocessor which also controls instrument operation and temperature evaluation. The ratio pyrometer measured greybody temperatures from 600 to 900 °C to better than 0.5%. Unsatisfactory measurements were obtained when oxide surface layers were forming on stainless steel, inconel and aluminium, but for most surfaces reported, e.g., copper, cast iron, oxidized stainless steel, mild steel, and inconel, the temperature errors of the basic pyrometer were less than 2%. These errors were reduced to less than 1% by introducing, for each surface, a spectral emissivity ratio that was constant, applicable over the full temperature range covered.

INTRODUCTION

The need to measure surface temperatures in an industrial environment has led to the development of a six-wavelength pyrometer[1] which operates under computer control. That instrument showed the practicability of using broadband filters and of calculating the effective wavelength[2] of each channel of the pyrometer as a function of source temperature. The six-wavelength instrument was shown to measure satisfactorily temperatures of practical surfaces for which spectral emissivities were approximately a linear function of wavelength over the wavelength range of the instrument, 0.75 to 1.7 μm. However, for other practical surfaces whose spectral emissivities were not linear it was shown that the six-wavelength bands produced no more accurate surface temperatures than the two bands of shorter wavelength when used as a ratio pyrometer.

The photoelectric measurement of the spectral radiance ratio to evaluate temperature seems to have been first discussed by Campbell[3] in 1925. Since that time accounts of numerous designs of ratio pyrometers have been published and reviewed (eg by Reynolds[4]). Most ratio pyrometers have been constructed with relatively narrow spectral bands in both channels and with the assumption that monochromatic conditions apply. To ensure a satisfactory ratio of signal to noise for radiometric measurements at relatively low temperatures, broad spectral pass bands are desirable. The use of broad-band ratio pyrometers has been considered[5,6], but their accuracies, both on greybody surfaces and on surfaces of variable spectral emissivity, have been difficult to predict because the effective wavelengths of each pass band have not been used. Our experience with multiwavelength pyrometry has revealed design features which could be applied to such a pyrometer. The design principles for ratio pyrometers and a particular design, for use in the temperature range 600°C to 900°C, are discussed in this paper.

RATIO PYROMETER DESIGN PRINCIPLES

Accurate radiometry is required to yield precise temperature measurement with ratio pyrometry[7]. In a pyrometer with filters sufficiently narrow to be considered monochromatic at wavelengths λ_1 and λ_2 the relative error in measuring the ratio, R, of the source radiance at the two wavelengths is related to the temperature error at temperature T(K) by

$$\Delta R/R = (C_2/\Lambda T) \cdot (\Delta T/T) \quad (1)$$

where Λ is the two-colour effective wavelength given by

$$1/\Lambda = 1/\lambda_1 - 1/\lambda_2. \quad (2)$$

In deriving the relationship (1), Ruffino[7] and others have assumed that Wien's law applies and that the surface emissivity is equal at the two wavelengths. As two-colour effective wavelengths are typically an order of magnitude greater than those of monochromatic brightness pyrometers, a much greater radiometric precision is required in ratio pyrometers.

Such an increase in precision can be achieved by using broad bandpass filters. Because these filters can no longer be considered monochromatic, an effective wavelength, λ_e, is required for each filter (as distinct from the two-colour effective wavelength Λ) defined by the relationship[2]

$$L_b(\lambda_e, T) = (\int_0^\infty S(\lambda) \cdot L_b(\lambda, T) \cdot d\lambda / \int_0^\infty S(\lambda) \cdot d\lambda \quad (3)$$

where $L_b(\lambda, T)$ is the Planck blackbody radiance at temperature T and wavelength λ and $S(\lambda)$ the spectral responsivity of the entire system of optical elements, filter and detector. Note that this definition of effective wavelength and the value obtained differ from that of monochromatic pyrometry, where the term applies to the comparison of different source temperatures with one filter[8]. The two-colour effective wavelength, now temperature dependent, may be obtained by substituting effective wavelengths from (3) into (2).

A basic assumption of ratio pyrometry is that the spectral emissivity of the surface is the same for both wavelengths. It has often been stated (eg Hornbeck[9]) that the performance of a ratio pyrometer may be improved by reducing the wavelength separation of the two channels, so that the spectral emissivity is more nearly equal at the two wavelengths. However, the temperature error depends on the slope of the spectral emissivity curve, as can be readily demonstrated by considering a surface where the spectral emissivity is approximately linear with slope m, or, where a non-linear spectral emissivity is approximated by the slope taken as that between the values at wavelengths λ_{e1} and λ_{e2}. The ratio of emissivities at λ_{e2} and λ_{e1} is given by

$$\varepsilon_2/\varepsilon_1 = 1 + (m/\varepsilon_1) \cdot (\lambda_{e2} - \lambda_{e1}) \quad (4)$$

Application of the ratio pyrometer assumes that this ratio is unity, so that the relative error in the radiance ratio is given by

$$\Delta R/R = (m/\varepsilon_1) \cdot (\lambda_{e2} - \lambda_{e1}). \quad (5)$$

Substitution of (5) and (2) into (1) gives the relative temperature error as

$$\Delta T/T = (m \cdot \lambda_{e2} \cdot \lambda_{e1} \cdot T)/C_2 \cdot \varepsilon_1. \quad (6)$$

It is clear from (6) that, for a given slope m, the error obtained with ratio pyrometry due to non-constant emissivity is reduced by using a short wavelength in both channels. The separation of the wavelengths affects only the two-colour effective wavelength and hence from (1) the precision with which the measurement of spectral radiance must be performed for a required temperature precision. It may be advantageous to choose wavelengths where the slope of spectral emissivity between those wavelengths is minimal for particular surfaces, but general applications are best served by using two closely spaced channels at short wavelengths.

REALIZATION OF THE DESIGN PRINCIPLES

The most obvious broadband detector to use in a pyrometer is a silicon photodiode with no additional spectral filtering. Effective wavelengths calculated for a typical silicon photodiode response were near 0.85 μm for temperatures in the region of 1000°C. A relative error of 0.25% in the radiance ratio was considered achievable through both calibration and measurement stages and an error of 0.5% in temperature measurement was regarded as desirable. These values lead to a two-colour effective wavelength of 22 μm and that implies a minimum separation of 33 nm between the effective wavelengths of the two channels near 0.85 μm. This separation is almost achievable by using available silicon photodiodes enhanced in different spectral regions (eg Silicon Detector Corp. (USA) SD-444-11-11 and SD-444-12-12) with a spectrally non-selective beamsplitter, or by directing the beam reflected from one silicon photodiode onto a second photodiode. However, the desired separation was obtained by using a 3-layer spectrally selective beamsplitter (two 66 nm layers of TiO_2 separated by 108 nm of SiO_2), operated at an angle of incidence of 10°.

Figure 1 shows an optical diagram and electronic block diagram of the ratio pyrometer. Both photodiodes were used in the photovoltaic mode, with the amplifier outputs connected to voltage-to-frequency converters. A microprocessor-controlled counter was used to acquire the signal from each channel simultaneously, with the count time programmed from a crystal-controlled clock and capable of variation from 25 ms to 27 s. In this manner a system with good accuracy over a wide dynamic range was achieved. As shown in Fig. 1 a third count to indicate the ambient device temperature was also acquired to track changes in the response of the channels due mainly to drift in gain in the voltage-to-frequency converters. The microprocessor controlled a motor-driven shutter to obtain background readings and also operated a serial communications line. During calibration the microprocessor responded to the control of an external computer. Once calibrated, the pyrometer operated independently with the serial line used to display results and to accept commands setting the time of measurement, number of readings averaged and spectral emissivity ratio to be applied as a correction to the measured radiance ratio.

A Schott RG 580 red filter was placed in the collimated beam ahead of the detectors. A negligible contribution to the emitted signal was expected at shorter wavelengths but reflected visible light could be a significant source of error if not removed by such a filter.

CALIBRATION

The relative spectral responsivities of both channels were calibrated against a gold-black bolometer, using a Leiss double monochromator with flint glass prisms and a quartz halogen lamp. These measurements were performed with the ratio pyrometer maintained at 20°C and 40°C without significant variations occurring in the relative spectral responsivities of both channels. The data obtained are plotted in Fig. 2. Effective wavelengths calculated from these spectral responsivities are plotted in Fig. 3, together with the two-colour effective wavelength, as a function of source temperature. The reduction in wavelength with increasing temperature helps to offset the reduced accuracy at higher temperatures. The linearity of each channel was measured by a flux-addition method using a series of neutral filters to be better than 0.1% over the signal range equivalent to a temperature range of 600°C to 900°C.

The band responsivities of both channels were determined by relating the measured responses to the spectral radiances of a blackbody (Infrared Industries, USA, Model 464), calculated at the blackbody temperature, measured with a thermocouple, and for the effective wavelengths of both channels appropriate for this temperature. The ratio of band responsivities was then

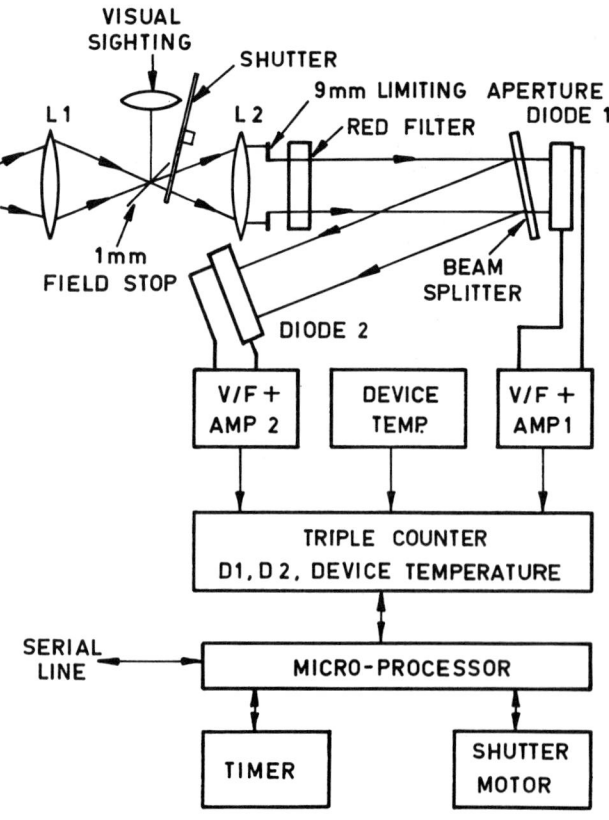

Fig. 1. Optical diagram and electronic block diagram of ratio pyrometer.

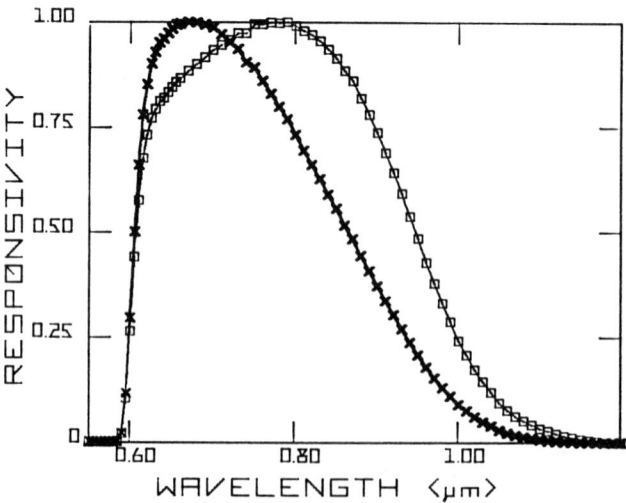

Fig. 2. Relative spectral responsivities of both channels

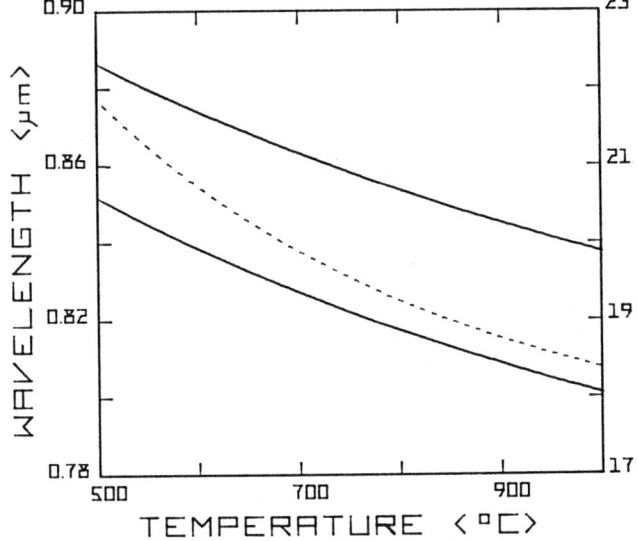

Fig. 3. Effective wavelengths of both channels (solid line, left ordinate) and two-colour effective wavelength (dashed line, right ordinate) for the pyrometer.

stored in the microprocessor program as a factor to convert response ratio to spectral radiance ratio. The variation of band responsivity ratio with ambient device temperature was measured to be a linear function which decreased by 0.6% over the range 20°C to 40°C. This relationship was stored to make corrections for device temperature during the measurement mode. Target temperatures were calculated from the measured spectral radiance ratio using a polynomial expression for inverse target temperature as a function of the spectral radiance ratio. This function was fitted to the spectral radiance ratio calculated using the effective wavelengths and target temperatures in the range 600°C to 900°C. A quadratic expression was found sufficient, having a maximum deviation of 0.1% in temperature in the range considered.

The measurement of spectral and band responsivities and the calculation of effective wavelength were only necessary to evaluate the design characteristics and instrument performance. It would be possible to calibrate a similar instrument without measuring the spectral characteristics by measuring the instrument response ratio for the channels at various blackbody temperatures throughout the range[6]. Calibration would be necessary at a number of points over the temperature range to ensure a satisfactory conversion of response ratio to temperature in the measurement mode.

RESULTS

The performance of the instrument was checked by measuring the temperature of a blackbody source over the temperature range 600°C to 900°C. The blackbody source was that used for calibration and thus these measurements essentially were a test of the program used in the measurement mode and the sufficiency of the calibrated spectral and relative band responsivities. The blackbody temperatures measured with the ratio pyrometer were compared with the associated thermocouple readings and errors of less than 3°C were obtained over the range 600°C to 900°C.

A furnace was constructed to test the measurement of temperatures of surfaces of practical materials under laboratory conditions up to 900°C. The assembly consisted of a stainless-steel cylinder, heated electrically. The sample was bolted to the front of the cylinder in such a position that no furnace radiation was reflected into the pyrometer. A spring-loaded thermocouple was placed in a hole along the axis of the stainless-steel block with the junction in contact with the rear face of the 1 mm thick sample. The temperature drop across the sample and oxidized front surface has been computed and measured with samples of different thicknesses to be less than 10°C[1].

The temperature of a number of samples were measured while heating and cooling over a 3 to 4 hour period. All samples were initially cleaned with alcohol and polished with emery cloth. Some samples were remeasured without cleaning the oxide layer formed during the initial run. The errors in the surface temperatures of various materials as measured by the ratio pyrometer (ie ratio pyrometer reading - thermocouple reading) are plotted in Figs. 4 to 7 as a function of surface temperatures.

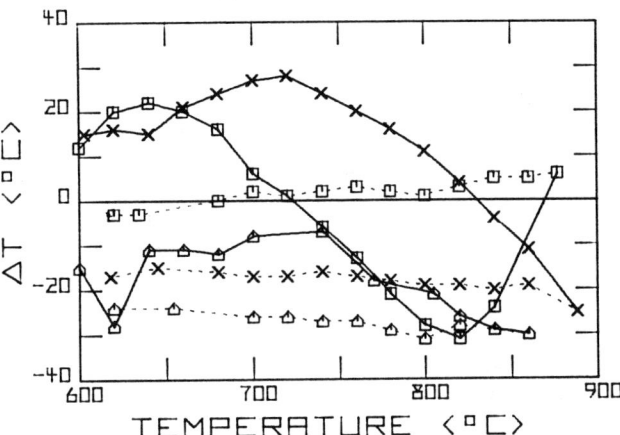

Fig. 4. Error, $\Delta T(°C)$, in surface temperature measured with ratio pyrometer for
× stainless steel, initially unoxidized
△ mild steel, initially unoxidized
□ inconel, initially unoxidized
(solid line heating, dashed line cooling).

Figure 4 shows that the error in the surface temperature measured by the ratio pyrometer varies as the surfaces of stainless steel, mild steel and inconel become oxidized. However once a stable oxide layer is formed the temperature error during a cooling cycle shows much less variation as the temperature is reduced. When these samples were reheated, the temperature error measured on the oxidized surface is shown in Fig. 5 to have less variation with changing temperature than when unoxidized. The errors for these three materials lie

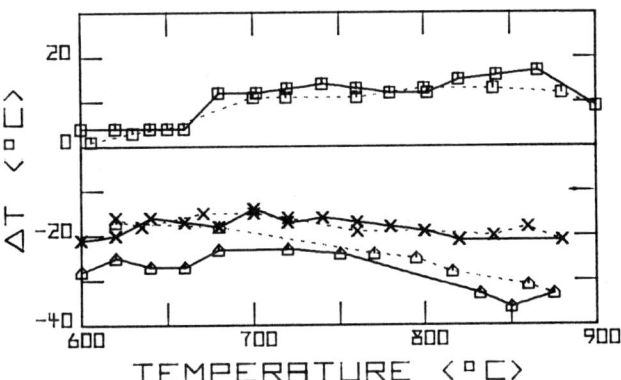

Fig. 5. Error, $\Delta T(°C)$, in surface temperature measured with ratio pyrometer for
× stainless steel, oxidized
△ mild steel, oxidized
□ inconel, oxidized
(solid line heating, dashed line cooling).

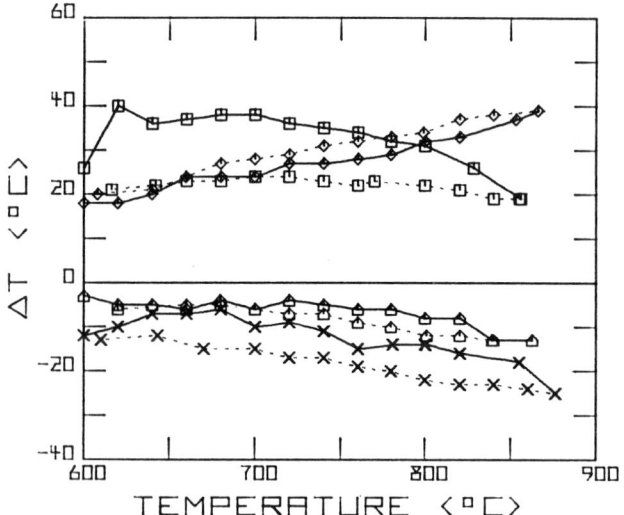

Fig. 6. Error, ΔT(°C), in surface temperature measured with ratio pyrometer for
  □ brass,   initially unoxidized
  △ copper,   initially unoxidized
  × cast iron, initially unoxidized
  ◇ platinum
(solid line heating, dashed line cooling).

within ± 2% approximately. As shown in Fig. 6 the errors which result when the surface temperature of copper is measured with the ratio pyrometer are stable at about 1%, for cast iron the error is fairly stable at 2% whilst for brass and platinum the errors are variable up to 4%.

The surface temperature of aluminium was measured in the range 500°C to 620°C and as shown in Fig. 7 the error was greater for the unoxidized surface (3%) than for the oxidized (less than 2%). The surface of the aluminium was scraped clean of oxide and remeasured at approximately 620°C and whereas the response in each channel decreased by approximately one half, the measured temperatures did not change significantly.

Figures 4 to 7 show the error that would result if the usual assumption is made that the spectral emissivities at the effective wavelength of both channels are the same and consequently cancel when a ratio is taken. In the series of measurements performed on various materials, the surface temperature of the sample was taken to be that measured by the thermocouple. It was then possible to compute the ratios of spectral emissivities for the channels which were necessary to yield the correct surface temperatures. This ratio could then be used for subsequent temperature measurements on a similar surface at a similar temperature. To be useful in practice, this emissivity ratio should be constant over the temperature range of interest. Of the materials tested only those which showed large variations in errors with variation of surface temperatures were considered unsuitable for the application of spectral emissivity ratios to correct the ratio pyrometer. Consequently the instrument described would not be suitable for the measurement of brass and unoxidized stainless steel, mild steel, inconel and aluminium. Systematic errors of less than 1% would occur if constant spectral emissivity ratios were applied over the range of temperatures 600°C to 850°C for all other materials tested. Typical spectral emissivity ratios applied were 1.005 for oxidized inconel, 0.988 for oxidized mild steel and 1.015 for platinum. If the temperature range was restricted to say 100°C the errors involved were reduced to less than 0.5%.

CONCLUSION

The aim of the design, to produce an optically simple ratio pyrometer using broad spectral pass bands and the proven stability of silicon detectors, has been successfully achieved for certain surfaces. The effective wavelengths of both channels of the ratio pyrometer were made as short as practicable, consistent with radiometric accuracy, to minimize effects of variation in spectral emissivity. In the range 600°C to 900°C the resultant errors were less than 0.5% for greybodies and less than 1% for copper and 2% for mild steel, stainless steel, cast iron and inconel. The pyrometer gave larger errors for brass, platinum and aluminium, and was considered unsuitable in its basic state for measurement on these three materials.

After application of a spectral emissivity ratio at the two wavelengths, entered via a keyboard on the serial communication line, the instrument has been shown to measure most of the practical surfaces tested with an error of less than 1% over the range 600°C to 900°C. A single spectral emissivity ratio cannot be applied if an oxide surface layer is forming as for stainless steel and inconel. However, for most materials tested, if the temperature range is restricted to say 100°C a spectral emissivity ratio can be applied which would result in measurement errors of less than 0.5%. The success of the instrument is shown in that despite a decrease of 50% in spectral radiance when an aluminium surface was scraped clean of oxide, the indicated temperature remained the same.

ACKNOWLEDGEMENT

It is a pleasure to acknowledge the assistance of S.R. Meszaros who constructed the instrument and W.G. Sainty who deposited the three-layer beamsplitter.

REFERENCES

1. J. L. Gardner, T. P. Jones and M. R. Davies, High Temp. - High Pressures (in press).
2. J. L. Gardner, Appl. Opt. 19, 3088 (1980).
3. N. R. Campbell, J. Sci. Instrum. 2, 177 (1925).
4. P. M. Reynolds, Br J. Appl. Phys. 15, 579 (1964).
5. H. Herne, Br J. Appl. Phys. 4, 374 (1953).
6. B. B. Brenden, Temperature, Its Measurement and Control in Science and Industry, C. M. Herzfeld (Ed) Pittsburgh: Instrument Society of America, 3, part 2, 429-433 (1962).
7. G. Ruffino, Temperature Measurement 1975, B. F. Billing and T. J. Quinn (Eds) Institute of Physics Conference Series 26, London and Bristol: Institute of Physics, 264-272 (1975).
8. H. J. Kostkowski and R. D. Lee, Temperature, Its Measurement and Control in Science and Industry, C. M. Herzfeld (Ed) Pittsburgh: Instrument Society of America 3, part 1, 449-481 (1962).
9. G. A. Hornbeck, Appl. Opt. 5, 179 (1966).

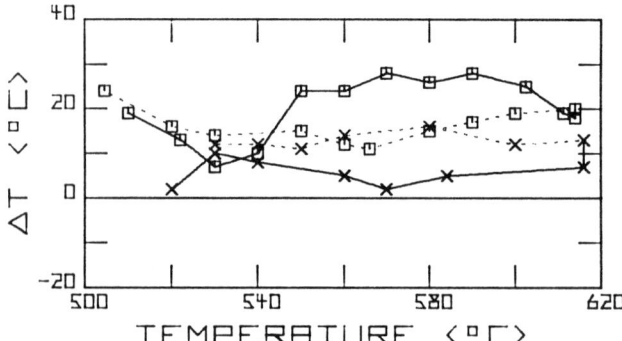

Fig. 7. Error, ΔT(°C), in surface temperature measured with ratio pyrometer for
  □ aluminium, unoxidized
  × aluminium, oxidized
(solid line heating, dashed line cooling).

Single-band radiation thermometers: Harmonization of their calibration characteristics

Jiang Shichang, Wu Shuyuan, Ye Rongchang, and Xu Liang

Shanghai Institute of Process Automation Instrumentation, Shanghai, China

Investigations have been made on the calibration characteristics of single-band radiation thermometers utilizing a silicon photocell as the detector. The representation of the thermometer errors is derived from the calibration equation. The relation between the mean effective wavelength and other factors associated with the detector and the filter is established. The optimum operating conditions for the detector and their fulfillment are discussed. A "harmonized" effective wavelength is introduced for a given measuring span, and the corresponding reference table is calculated for industrial use. Experiments show that 75% of China-made photocells, selected through a certain procedure, are acceptable, with the result that the output of a thermometer using such a photocell is in agreement with the reference table to ±0.3%.

I. INTRODUCTION

The electro-optical system of the radiation thermometer (see Fig. 1) is critical to its calibration characteristics, for it receives directly the radiant power and converts it to useful signals. However, there are a number of uncertain factors here, such as the non-agreement of the spectral characteristics of the individual detectors, filters, and other optical parts, which result in the non-agreement of the calibration characteristics of radiation thermometers of the same kind. It is desirable that radiation thermometers, like industrial thermocouples and resistance thermometers, have a common reference table. For this reason, we have derived the calibration equation, established the representation of various errors and the relations among the parameters of this type of instrument, analysed the factors that influence the conformity of the calibration characteristics, and presented some criteria for the sorting of photocells.

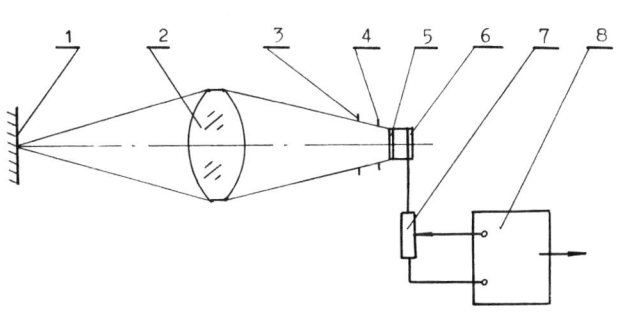

Fig. 1. Schematic diagram of the radiation thermometer.

1. Blackbody radiation source
2. Objective lens of K8 glass
3. Aperture diaphragm
4. Field diaphragm
5. HWB_3 filter
6. Silicon photocell detector
7. Equivalent load resistance Z
8. Signal processing circuit

II. CALIBRATION EQUATION FOR THE RADIATION THERMOMETER

The calibration equation for a radiation thermometer represents the relationship between its output signal and the temperature of the measured blackbody surface. Since, in our case, use is made of a short wavelength, and the temperature to be measured is not supposed to be very high, Wien's formula is employed in the following analyses.

$$L_{\lambda,b}(\lambda, T) = c_1 \cdot \lambda^{-5} \cdot \exp\left(-\frac{c_2}{\lambda \cdot T}\right) \quad (1)$$

where $L_{\lambda,b}(\lambda, T)$ is Wien's function at temperature T, in K; λ is the wavelength; and c_1 and c_2 are the first and the second radiation constants, with c_2 equal to 0.014388 m·K.

The radiant power emitted from the blackbody and entering the photodetector is

$$\Phi = A \cdot \omega \cdot \int_0^\infty L_{\lambda,b}(\lambda, T) \cdot \tau_0(\lambda) \cdot \tau_f(\lambda) \cdot d\lambda \quad (2)$$

where Φ is the radiant power input, in watts; A is the photodetector aperture area, projected normal to the axis; ω is the solid angle of the radiant power on the detector; the product $A \cdot \omega$, known as the geometric extent of the optical system, is a constant for a given radiation thermometer; $L_{\lambda,b}(\lambda, T)$ is the spectral radiance of the blackbody; $\tau_0(\lambda)$ is the spectral transmittance of the lens system; and $\tau_f(\lambda)$ is the spectral transmittance of the filter, which, in our case, is of type HWB_3, an infrared-transmitting glass (see Fig. 2).

The output signal i of the detector in response to the radiant power input is

$$i = A \cdot \omega \cdot \int_0^\infty R(\lambda) \cdot L_{\lambda,b}(\lambda, T) \cdot \tau_0(\lambda) \cdot \tau_f(\lambda) \cdot d\lambda \quad (3)$$

where $R(\lambda)$ is the spectral responsivity of the photodetector, which is defined as

$$R(\lambda) = \frac{di}{d\Phi(\lambda)}, \text{ in amperes/Watt.} \quad (4)$$

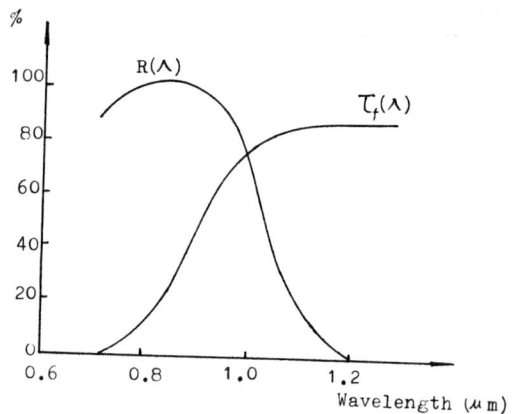

Fig. 2. Typical $R(\lambda)$ and $\tau_f(\lambda)$ curves.

When irradiance levels are below a certain value, we will have

$$R(\lambda) = \text{const.} \quad (5)$$

The current i produces a voltage signal V_T across the equivalent load resistance Z.

$$V_T = i \cdot Z. \quad (6)$$

When the radiation thermometer views a blackbody at temperatures T_1 and T_2, we have respectively

$$V_{T_1} = Z \cdot A \cdot \omega \int_0^\infty R(\lambda) \cdot L_{\lambda,b}(\lambda, T_1) \cdot \tau_0(\lambda) \cdot \tau_f(\lambda) \cdot d\lambda \quad (7)$$

and

$$V_{T_2} = Z \cdot A \cdot \omega \int_0^\infty R(\lambda) \cdot L_{\lambda,b}(\lambda, T_2) \cdot \tau_0(\lambda) \cdot \tau_f(\lambda) \cdot d\lambda. \quad (8)$$

Combining Eqs. (7) with (8) gives

$$\frac{V_{T_1}}{V_{T_2}} = \frac{\int_0^\infty R(\lambda) \cdot L_{\lambda,b}(\lambda, T_1) \cdot \tau_0(\lambda) \cdot \tau_f(\lambda) \cdot d\lambda}{\int_0^\infty R(\lambda) \cdot L_{\lambda,b}(\lambda, T_2) \cdot \tau_0(\lambda) \cdot \tau_f(\lambda) \cdot d\lambda}. \quad (9)$$

From Eq. (9) we obtain

$$V_{T_1} = V_{T_2} \cdot \exp\left[\frac{c_2}{\lambda_{1,2}} \cdot \left(\frac{1}{T_2} - \frac{1}{T_1}\right)\right] \quad (10)$$

where $\lambda_{1,2}$, the mean effective wavelength, is in the following form:[1,5]

$$\lambda_{1,2} = \frac{c_2 \left(\frac{1}{T_2} - \frac{1}{T_1}\right)}{\ln \dfrac{\int_0^\infty R(\lambda) \cdot L_{\lambda,b}(\lambda, T_1) \cdot \tau_0(\lambda) \cdot \tau_f(\lambda) \cdot d\lambda}{\int_0^\infty R(\lambda) \cdot L_{\lambda,b}(\lambda, T_2) \cdot \tau_0(\lambda) \cdot \tau_f(\lambda) \cdot d\lambda}} \quad (11)$$

When T_1 and T_2 both approach a temperature T, the mean effective wavelength of the radiation thermometer is called the limiting effective wavelength, expressed as

$$\lambda_e(T) = \frac{\int_0^\infty R(\lambda) \cdot L_{\lambda,b}(\lambda, T) \cdot \tau_0(\lambda) \cdot \tau_f(\lambda) \cdot d\lambda}{\int_0^\infty \frac{1}{\lambda} R(\lambda) \cdot L_{\lambda,b}(\lambda, T) \cdot \tau_0(\lambda) \cdot \tau_f(\lambda) \cdot d\lambda}. \quad (12)$$

Eq. (10) is the very calibration equation of the radiation thermometer, from which the output voltages at various temperature points can be obtained provided that V_{T_2} is known.

It has been assumed in the deduction of Eq. (10) that $R(\lambda)$, being a constant, is independent of Φ. This condition must therefore be satisfied when the silicon photocell is in operation. We have established the relation between the output current and the radiant power received by the silicon photocell, as shown in Fig. 3 [1,2], and $R(\lambda)$ is the first derivative of the curve in this figure. In order to satisfy Eq. (5), the silicon photocell must operate in the linear section of the curve.

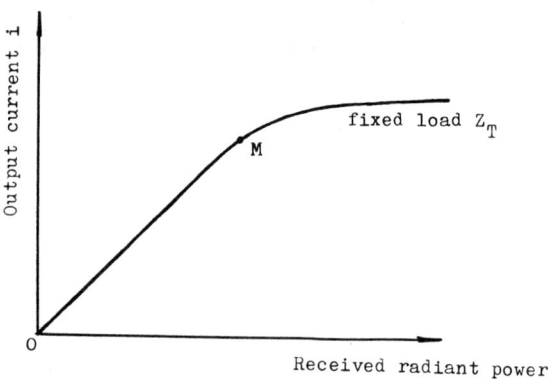

Fig. 3. The relation between the radiant power received by a silicon photocell and its output signal.

However, the output currents of different photocells used at present differ greatly from each other. To establish a common reference table, therefore, measures must be taken to confine the output signals of the thermometer to a definite range. To this end, V_{T_2} of Eq. (10) should be fixed. It is also necessary to ensure that Eq. (10) is still valid when the temperature varies. Even at the upper limit of the span, T_2, at which the radiation thermometer receives maximum radiant power, the silicon photocell should nevertheless operate in the linear region of the curve (the OM section in Fig. 3). Fig. 4 presents the load characteristics of the silicon photocell.

With the silicon photocell exposed to constant radiation, the voltage drop across the load resistance Z varies with the value of Z, and a rule should therefore be set up to determine the load resistance Z.

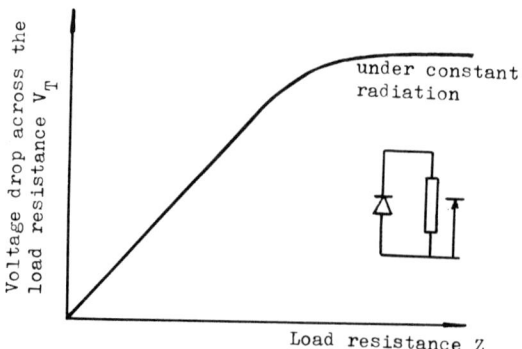

Fig. 4. Load characteristic curve of the silicon photocell.

Two factors should be taken into account in the determination of the value of the load resistance. Firstly, Z should be capable of providing a partial voltage drop at the upper limit of the span, T_2, equal to the corresponding value specified in the reference table. Secondly, it should be ensured that, with the chosen load resistance, the photocell will operate in the current mode.

This practice has two advantages: (1) The adjustment can be easily made by changing the output of the potentiometer which serves as Z; (2) The optical performance of the radiation thermometer is not affected, for, once the temperature measuring span is fixed, there is no need to adjust the aperture diaphragm, the field diaphragm or other optical and mechanical parts for different silicon photocells.

The long-term stability of the radiation thermometer and the compensation for the environmental temperature variations are dealt with in another paper by the authors [1].

III. REPRESENTATION OF ERRORS

Substitution from Eq. (3) into Eq. (6) gives the output signal of the radiation thermometer used in measuring the temperature of a blackbody:

$$V_T = Z \cdot A \cdot \omega \cdot \int_0^\infty R(\lambda) \cdot L_{\lambda,b}(\lambda,T) \cdot \tau_0(\lambda) \cdot \tau_f(\lambda) \, d\lambda. \quad (13)$$

It can be seen from Eq. (13) that the accuracy of the radiation thermometer depends on quite a few factors. Variations in any one of these may cause a change ΔV in the output signal V_T. In fact, the inaccuracy of the measuring instruments used in calibrating the radiation thermometer, the non-ideality of the calibration source as a "blackbody", and the drift with time of the load resistance Z and of the responsivity of the silicon photocell, and the slight slack in mechanical parts, will make contributions to the error of the measurement, ΔT, but most of these factors are functionally unrelated to each other. Therefore, we have to work out the general representation of the absolute conformity error ΔT caused by whatsoever factors.

When ΔV is not very large, ΔT may be expressed as

$$\Delta T = \frac{dT}{dV} \cdot \Delta V. \quad (14)$$

Since

$$\frac{dT}{dV} = \frac{1}{(dV/dT)},$$

we have

$$\frac{dV}{dT} = \frac{Z \cdot A \cdot \omega \cdot d\left[\int_0^\infty R(\lambda) \cdot L_{\lambda,b}(\lambda,T) \cdot \tau_0(\lambda) \cdot \tau_f(\lambda) \, d\lambda\right]}{dT}$$

$$= \frac{c_2}{T^2} \cdot Z \cdot A \cdot \omega \cdot \int_0^\infty \frac{1}{\lambda} \cdot R_0(\lambda) L_{\lambda,b}(\lambda,T) \cdot \tau_0(\lambda) \cdot \tau_f(\lambda) \, d\lambda. \quad (15)$$

Multiplying each side of Eq. (12) by the corresponding side of Eq. (15), we obtain

$$\lambda_e(T) \frac{dV}{dT} = \frac{c_2}{T^2} \cdot V. \quad (16)$$

In consequence, we have

$$\frac{\Delta T}{T} = \frac{\lambda_e(T)}{c_2} \cdot \frac{\Delta V}{V} \cdot T. \quad (17)$$

Eq. (17) represents the measuring error of the radiation thermometer caused by ΔV.

It can be seen from Eq. (17) that for a given relative change $\Delta V/V$ of the radiation thermometer, there is a relative error $\Delta T/T$, which is proportional to the measured temperature T. c_2 is a constant, and $\lambda_e(T)$ may also be regarded as a constant if the temperature range is not too wide.

Eq. (17) holds true for all ΔV, no matter what the cause is. It is therefore appropriate to use it as a representation of the conformity error of the radiation thermometer with respect to the reference table.

IV. ESTABLISHMENT OF THE REFERENCE TABLE

Eq. (10) represents the relation of the radiation thermometer outputs at temperatures T_1 and T_2 (T_2 being specified in practice). If V_{T_1} is to be determined numerically, the two quantities V_{T_2} and $\lambda_{1,2}$ must be evaluated.

A. Determination of V_{T_2}

The load resistance Z for different temperature measuring spans of the radiation thermometer is determined through tests on a large number of silicon photocells.

The values of the output voltage at the upper limit of the measuring span, V_{T_2}, are listed in Table I. With these values, the silicon photocell will operate in the linear section of its characteristic curve, and the resulting radiation thermometer will be conveniently used in conjunction with a measuring potentiometer or a recorder.

Table I. The values of V_{T_2}

Measuring span (°C)	Output voltage V_{T_2} at the upper limit of the span (mV)
500 − 800	10
700 − 1100	20
800 − 1200	20

B. Determination of the mean effective wavelength

Two methods are used here to determine the mean effective wavelength. The first one is to perform calculations, using Eq. (11), into which the specific $\tau_f(\lambda)$ and $R(\lambda)$ values of the optical parts are applied. The lens and the protective glass are of China-made K8 glass. As its spectral transmittance is flat within the responsive band of the silicon photocell, it has negligible influence on the mean effective wavelength. Typical $\tau_f(\lambda)$ and $R(\lambda)$ curves are shown in Fig. 2. The advantage of this method lies in the high accuracy that the calculation provides, because the integral region can be finely divided by a computer and the radiant power of the blackbody can be made in exact agreement with Wien's formula. It is, however, affected by the inaccuracy of the measurements of $\tau_f(\lambda)$ and $R(\lambda)$.

The second method of determining $\Lambda_{1,2}$ is to substitute the signal produced by the photodetector receiving the radiant power, into the following equation, which is equivalent to Eq. (11):

$$\Lambda_{1,2} = \frac{c_2(\frac{1}{T_2} - \frac{1}{T_1})}{\ln \frac{V_{T_1}}{V_{T_2}}} \quad (18)$$

Where V_{T_1} and V_{T_2} are the outputs of the radiation thermometer when it is viewing a blackbody at temperatures T_1 and T_2, respectively. This method has the advantage that the $\Lambda_{1,2}$ obtained is directly related to the blackbody radiation source, thus avoiding the use of different measuring instruments and minimizing the accumulative errors. Its drawback is that the accuracy of the $\Lambda_{1,2}$ thus obtained is affected by the non-ideality of the radiation source used and the uncertainties of the measuring instrument. It should be added that here the values of V_{T_1} should be obtained with the value of V_{T_2}, the output of the radiation thermometer at the upper temperature limit of the span, adjusted to be in conformity with that listed in Table I. Fig. 5 is the measuring system.

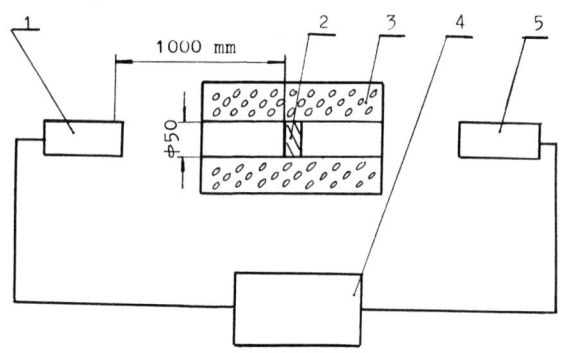

Fig. 5. Schematic diagram of the measuring system

1. Radiation thermometer
2. Furnace target
3. Blackbody furnace
4. Measuring potentiometer
5. Standard thermometer

It is obvious that these two methods would give different $\Lambda_{1,2}$ values, which in turn would lead to different thermometer outputs. Table II presents the maximum discrepancies in the thermometer outputs, using the different $\Lambda_{1,2}$ values obtained by the two methods.

It can be seen that the discrepancies are not greater than 4.03 °C in the 800 - 1200 °C span. However, in the 500 - 800 °C span, the maximum discrepancy reaches 7 °C (not shown in Table II), which is unacceptable for a reference table.

Table II. Maximum discrepancies in thermometer outputs in the 800-1200 °C span, using different $\Lambda_{1,2}$ values obtained by two methods.

Photocell No.	$\Delta V = V_m - V_c$ (mV)	$\Delta T = \frac{dT}{dV} \cdot \Delta V$ (°C)
80-10	-0.0014	-0.22
80-23	-0.0048	-0.79
80-25	-0.0246	-4.03
80-24	-0.0098	-1.61

C. Final determination of the mean effective wavelength

In our application of the first method, two instruments of different natures (the photospectrometer and the monochromater) were used to measure $\tau_f(\lambda)$ and $R(\lambda)$ respectively. As a result, there were bigger errors in the determination of $\Lambda_{1,2}$ by the first method than by the second method, because the blackbody source involved in the latter method was just what was to be used in the final verification tests. We decided therefore that the average voltage output value of the photocell at any particular temperature point, obtained by substitution into Eq. (18) of the $\Lambda_{1,2}$ value determined by the first method, is weighted 25%, and that the corresponding average voltage output value of the photocell resulting from the second method is weighted 75%. The mean of the weighted values is then supposed to be the radiation thermometer output at the temperature point concerned. Applying these output values to Eq. (18), we obtain what we call the weighted mean effective wavelength value at any given temperature point with respect to the upper limit of the span. (See Table III, in which, by way of illustration, only data for the 800 - 1200 °C span are given.) Owing to errors associated with actual measurements, the $\Lambda_{1,2}$-T relation thus obtained is not very smooth, but it is acceptable for industrial purposes.

Table III. The weighted mean effective wavelength values.

T (°C)	Weighted mean effective wavelength with respect to the upper limit of the span (μm)
800	0.982
850	0.979
900	0.976
950	0.973
1000	0.976
1050	0.972
1100	0.975
1150	0.972
1200	

In addition, the expression of the mean effective wavelength should not be made too complicated, considering that the thermometer, as an industrial instrument, should be characterised by ease of operation and of mass production by the manufacturers. For these reasons, we take the arithmetic average of the weighted mean effective wavelength values for each span as the __harmonized__ effective wavelength, λ_{eh} (see Table IV). Applying the

Table IV. The harmonized effective wavelength for each span.

Span (°C)	Harmonized effective wavelength λ_{eh} (μm)
500 - 800	0.990
700 - 1100	0.978
800 - 1200	0.976

corresponding harmonized effective wavelength value to Eq. (10), we produce a reference table for the span.

The introduction of the harmonized effective wavelength does give rise to an additional error, but it could be reduced if the harmonizing point (i.e. the temperature at which the thermometer output is made to agree with the corresponding value specified in the reference table) and the photocell were well chosen.

Table V. Errors caused by the deviation of the harmonized effective wavelength from the mean effective wavelength.

Span (°C)	$\Delta\Lambda_m = \lambda_{eh} - \Lambda_{1,2}$ (μm)	ΔT_1 (K)	$\frac{\Delta T}{T}$ (%)
700 - 1100	0.005	2.0	0.15
800 - 1200	0.006	1.75	0.17

D. Errors associated with the uncertainty of the mean effective wavelength

Before the calculation of the reference table, an analysis should be made of the errors associated with the uncertainty of the mean effective wavelength.
Uncertainty of $\Lambda_{1,2}$ leads to an error ΔT_1.

$$\Delta T_1 = \frac{\partial T_1}{\partial \Lambda_{1,2}} \cdot \Delta \Lambda_{1,2} = \frac{\frac{\partial V_{T_1}}{\partial \Lambda_{1,2}}}{\frac{\partial V_{T_1}}{\partial T_1}} \cdot \Delta \Lambda_{1,2}.$$

From Eq. (10) we obtain

$$\frac{\partial V_{T_1}}{\partial T_1} = \frac{\partial \left\{ V_{T_2} \cdot \exp\left[\frac{c_2}{\lambda_{1,2}}\left(\frac{1}{T_2} - \frac{1}{T_1}\right)\right]\right\}}{\partial T_1}$$

$$= V_{T_1} \cdot \frac{c_2}{\lambda_{1,2}} \cdot \frac{1}{T_1^2}$$

and

$$\frac{\partial V_{T_1}}{\partial \lambda_{1,2}} = \frac{\partial \left\{ V_{T_2} \cdot \exp\left[\frac{c_2}{\lambda_{1,2}}\left(\frac{1}{T_2} - \frac{1}{T_1}\right)\right]\right\}}{\partial \lambda_{1,2}}$$

$$= V_{T_1} \cdot \frac{c_2}{\lambda_{1,2}^2} \cdot \left(\frac{1}{T_1} - \frac{1}{T_2}\right).$$

Hence

$$\Delta T = \frac{T_1}{T_2} \cdot (T_2 - T_1) \cdot \frac{\Delta \Lambda_{1,2}}{\Lambda_{1,2}} \quad (19)$$

Eq. (19) gives the errors due to the introduction of the harmonized effective wavelength, and they are presented in Table V, where $\Delta\Lambda_m$ denotes the maximum deviation of the harmonized effective wavelength from the mean effective wavelength in the span.
With T_2 and $\Delta\Lambda_{1,2}(T)/\Lambda_{1,2}(T)$ fixed, and assuming that

$$\frac{d(\Delta T)}{dT_1} = 1 - 2\frac{T_1}{T_2} = 0, \quad (20)$$

we find that ΔT_1 is at its maximum when $T_1 = \frac{1}{2} T_2$. In other words, the maximum error appears at the midpoint of the span, where

$$\Delta T_1 = \frac{1}{4} T_2 \cdot \frac{\Delta \lambda_{1,2}}{\Lambda_{1,2}}. \quad (21)$$

Eq. (21) provides the foundation for the choice of the harmonizing point — if it is positioned at the center of the span, ΔT_1 will be minimized.

E. Formation of the reference table

With Eq. (10), we calculate the reference table for each span, using the V_{T_2} values in Table I and the λ_{eh} values in Table IV.

Taking the derivative of Eq. (10) with respect to T and substituting λ_{eh} for $\Lambda_{1,2}$, we obtain the differential voltage of the radiation thermometer

$$\frac{dV_T}{dT} = \frac{d\left[\exp\frac{c_2}{\lambda_{eh}}\left(\frac{1}{T_2} - \frac{1}{T}\right)\right] \cdot V_{T_2}}{dT}$$

$$= V_T \cdot \frac{c_2}{\lambda_{eh}} \cdot \frac{1}{T^2}. \quad (22)$$

For the sake of universality, T is used here instead of T_1, which is, _per se_, an arbitary temperature.

The reference tables (abridged) are given in Tables VIa, VIb, and VIc. In these tables are also given the relative sensitivity values of the radiation thermometer, $(\frac{dV_T}{dT})/V_T$, for the convenience of the user or manufacturer of the instrument.

$$\left(\frac{dV_T}{dT}\right)/V_T = \frac{c_2}{\lambda_{eh}} \cdot \frac{1}{T^2}. \quad (23)$$

Table VIa. The reference table for the 500-800 °C span (abridged)

t (°C)	V (mV)	$\dfrac{dV}{dT}\Big/V$ (1/K)	$\dfrac{dV}{dT}$ (mV/K)
500	0.0522	0.0243	0.0012
600	0.4496	0.0191	0.0085
700	2.4867	0.0153	0.0382
800	10.0000	0.0126	0.1262

Table VIb. The reference table for the 700-1100 °C span (abridged)

t (°C)	V (mV)	$\dfrac{dV}{dT}\Big/V$ (1/K)	$\dfrac{dV}{dT}$ (mV/K)
700	0.2446	0.0155	0.0038
800	1.0006	0.0128	0.0128
900	3.2196	0.0107	0.0344
1000	8.6211	0.0090	0.0782
1100	20.0000	0.0078	0.1560

Table VIc. The reference table for the 800-1200 °C span (abridged)

t (°C)	V (mV)	$\dfrac{dV}{dT}\Big/V$ (1/K)	$\dfrac{dV}{dT}$ (mV/K)
800	0.4800	0.0128	0.0061
900	1.5476	0.0107	0.0166
1000	4.1526	0.0090	0.0377
1100	9.6501	0.0078	0.0754
1200	20.0000	0.0068	0.1358

V. THE CONFORMITY ERRORS

The conformity error is defined as $\dfrac{\Delta T}{T}$, where ΔT is the deviation of the thermometer indication from the value specified in the relevant reference table, and T is the blackbody temperature under discussion.

The conformity errors for the 700 - 1100 °C and the 800 - 1200 °C spans are presented in Table VII. Since our primary concern was with these spans, the data for the 500 - 800 °C span are omitted.

Our experiments have shown that 75 percent of the China-made photocells are acceptable so long as they are sorted in such a way as is discussed below, and that the conformity errors of the thermometers using such photocells fall within ± 0.3%.

Table VII. Conformity errors of the radiation thermometer.

Span (°C)	Conformity error ($\Delta T/T$)
700 - 1100	± 0.3 %
800 - 1200	± 0.3 %

VI. THE SORTING OF SILICON PHOTOCELLS

As has been described, the differences in the mean effective wavelength result in the conformity errors. Such differences arise as they do from the non-identity of the spectral response of the photocells and of the spectral transmittance of the filters. Since glass filters are used in our case, and they always come from the same batch, the differences in their spectral transmittance are negligible. Therefore, we concentrate our attention on the spectral response of the silicon photocells. However, it was time-consuming to measure this parameter directly, so we measured the thermometer output instead, on condition that the spectral response curves of the silicon photocells are roughly the same.

It can be seen from Eq. (18) that λ_{12} is a single-value function of V_{T_1}/V_{T_2}, the ratio of the signals at the upper and lower limits. The silicon photocells can then be sorted in terms of a constraint on this value. When V_{T_2}, the voltage output at the upper limit of the span, is made to agree with the value specified in Table I, the maximum conformity error will appear at the lower limit of the span. The photocell sorting procedure may then be reduced to seeing whether its signal at the lower limit meets the requirement set forth in Table VIII, where ΔV is the allowable deviation of the photocell output at the lower limit from the corresponding value specified in the reference table.

In fact, from Eq. (17) we have

$$\frac{\Delta V}{V} = \frac{c_2}{\lambda_e T} \cdot \frac{\Delta T}{T}. \qquad (24)$$

It can be seen from Eq. (24) that, with $\Delta T/T$ fixed, the higher the temperature that the radiation thermometer is to measure, the stricter the demand on, or the smaller the value of, $\Delta V/V$.

Table VIII. Requirements for sorting the silicon photocells

Span (°C)	$\Delta V/V$	ΔV (mV)	Allowable range of voltage (mV)
700 - 1100	0.045	0.011	0.234 - 0.256
800 - 1200	0.041	0.019	0.461 - 0.50

VII. VERIFICATION OF THE CHARACTERISTICS CONFORMITY

A. The sorting of the photocells

The silicon photocells are sorted in accordance with the requirements in Table VIII. The measuring system is shown in Fig. 5. The results are presented in Tables IXa and IXb. A brief summary is given in Table IXc.

Table IXa. Data for sorting the silicon photocells, 800 - 1200 °C span.

Photocell No.	Output at 800 °C (mV)
80-1	0.4868
80-5	0.5065
80-16	0.4693
80-12	0.4601
80-23	0.4912
80-24	0.4820
80-25	0.4833
80-7	0.5318
80-6	0.4896
80-20	0.4818
80-9	0.5050
80-21	0.4852
80-14	0.4951
80-11	0.5054
80-10	0.5074
80-8	0.4935
75-3	0.4998
75-4	0.4724
77-F	0.4910
77-G	0.4796

Table IXb. Data for sorting the silicon photocells, 700 - 1100 °C span.

Photocell No.	Output at 700 °C (mV)
80-1	0.2406
80-5	0.2560
80-16	0.2374
80-12	0.2390
80-23	0.2464
80-24	0.2409
80-25	0.2472
80-6	0.2481
80-20	0.2471
80-9	0.2456
80-21	0.2475
80-14	0.2524
80-10	0.2521
80-8	0.2522

Table IXc. Summary of the sorting of the silicon photocells

Span	700 - 1100 °C	800 - 1200 °C
Number of test photocells	14	20
Number of acceptable photocells	13	15
Acceptability	92.9 %	75.0 %

B. Verification tests

A sorted silicon photocell is fixed into the radiation thermometer, and a blackbody furnace is used as the radiation source. The harmonizing point is chosen at the midspan. It will be observed that the farther the temperature is away from the harmonizing point, the larger the deviation of the output from the reference table. Therefore, a check at the lower limit as well as at the upper limit, is enough to verify the conformity of the thermometer characteristics.

C. Results of the verification tests

The results of the verification tests are shown in Tables Xa and Xb for the 800 - 1200 °C and Tables Xc and Xd for the 700 - 1100 °C span respectively.

It can be seen that all the silicon photocells used in the verification tests meet the requirement set forth in Table VII, with the exception of the photocell No. 80-12, which was later found to have been damaged.

Table Xa. Results of the verification test at 1200 °C (1473.15 K), the upper limit of the 800 - 1200 °C span.

(harmonizing point: 1000 °C)

Photocell No.	Values in reference table	Deviation from ref. table		
		ΔV (mV)	ΔT (K)	$\Delta T/T$ (%)
80-1		-0.2148	-1.58	-0.1
80-16		+0.2148	+2.2	+0.15
80-12		+0.7200	+5.3	+0.35
80-23		-0.3589	-2.6	-1.7
80-24		-0.0043	-0.03	-0.002
80-25	20 mV	-0.0407	-0.03	-0.002
80-6	dV	+0.2640	+1.94	+0.13
80-20	($\frac{dV}{dT}$=0.136 mV/K)	-0.0680	-0.1	-0.03
80-21		+0.2736	+2.0	+0.1
80-14		-0.1882	-1.38	-0.09
80-8		-0.3337	-2.5	-0.2
75-3		+0.2159	+1.6	+0.1
75-4		+0.3916	+2.9	+0.2
77-F		-0.0289	-0.21	-0.01
77-G		+0.3773	+2.78	+0.19

Table Xb. Results of the verification tests at 800 °C (1073.15 K), the lower limit of the 800 - 1200 °C span.

(harmonizing point: 1000 °C)

Photocell No.	Values in reference table	Deviation from ref. table		
		ΔV (mV)	ΔT (K)	$\Delta T/T$ (%)
80-1		-0.0062	-1.0	-0.1
80-16		-0.0064	-1.1	-0.1
80-12		-0.0106	-1.74	-0.16
80-23		+0.0008	+0.13	+0.01
80-24	0.480 mV	-0.005	-0.8	-0.1
80-25	dV	+0.0043	+0.7	+0.01
80-6	($\frac{dV}{dT}$=0.0061 mV/K)	-0.0074	-1.1	-0.1
80-20		-0.0027	-0.3	-0.03
80-21		+0.0102	+1.7	+0.16
80-14		+0.0037	+0.5	+0.01
80-8		+0.0096	+1.5	+0.13
75-3		+0.0028	+0.45	+0.04
75-4		+0.0014	+0.23	+0.02
77-F		+0.0007	+0.12	+0.01
77-G		+0.0005	+0.1	+0.01

Table Xc. Results of the verification test at 1100 °C (1373.15 K), the upper limit of the 700 - 1100 °C span.

(harmonizing point: 900 °C)

Photocell No.	values in reference table	Deviation from ref. table		
		ΔV (mV)	ΔT (K)	$\Delta T/T$ (%)
80-12		+0.8912	+5.7	+0.4
80-23	20 mV	+0.1193	+0.76	+0.05
80-25	dV	-0.0348	-0.22	-0.02
80-6	($\frac{dV}{dT}$=0.156 mV/K)	-0.0553	-0.35	-0.03
80-8		-0.0405	-0.26	-0.02
80-14		+0.0520	+0.34	+0.02

Table Xd. Results of the verification test at 700 °C (973.15 K), the lower limit of the 700 - 1100 °C span. (harmonizing point: 900 °C)

Photocell No.	Values in reference table	Deviation from ref. table		
		ΔV (mV)	ΔT (K)	$\Delta T/T$ (%)
80-12	0.245 mV	-0.005	-1.3	-0.1
80-23		-0.0051	-1.31	-0.1
80-25	$\frac{dV}{dT}$ = 0.0038 mV/K	+0.0062	+1.6	+0.2
80-6		+0.0023	+0.6	+0.06
80-8		+0.0012	+0.32	+0.03
80-14		+0.0061	+1.6	+0.16

VIII. CONCLUSION

Our experiments have proved that the deduction of the equations, the selection of the harmonizing point, and the silicon photocell sorting procedure are all appropriate. The reference tables are useful. With the acceptability of the photocells no less than 75%, the conformity errors of the radiation thermometers do not exceed ± 0.3 %.

REFERENCES

1. Jiang Shichang et al., "Model WFH Infrared Radiation Thermometer", SIPAI Report (1978).
2. Barber and T. Land, Temperature, Its Measurement and Control in Science and Industry, Vol. 3, Part 2, p. 391, Reinhold Publishing Corporation (1962).
3. Jiang Shichang et al., "Performance and Applications of Silicon Photocells", SIPAI Report (1979).
4. H. J. Kostowski and R. D. Lee, Temperature, Its Measurement and Control in Science and Industry, Vol. 3, Part 1, p. 449, Reinhold Publishing Corporation (1962).
5. D. R. Lovejoy, ibid., Part 1, p. 502.

Establishing a practical temperature standard by using a narrow-band radiation thermometer with a silicon detector

F. Sakuma and S. Hattori

National Research Laboratory of Metrology, Sakura-mura, Ibaraki 305, Japan

This paper proposes a 900 nm narrow-band radiation thermometer using a silicon photodiode, which is calibrated by a set of practical-type fixed point blackbody furnaces, for use as a practical temperature standard of improved accuracy. This radiation thermometer covers the temperature range from 420 to 2000 °C and its focal range extends from 40 cm, at which the target area is 3 mm in diameter, to infinity. The image of the measured plane is focused by an objective lens on an aperture of 0.75 mm diameter. The light travels through an interference filter having its maximum transmission at a wavelength of 900 nm and a half-width of 14 nm, and reaches a silicon photo-cell to be converted to an electric current. A resolution of 1 °C at 420 °C and better than 0.01 °C above 600 °C can be obtained. The relation between the output voltage V(T) of this radiation thermometer and the temperature T of the target can be expressed by $V(T) = C \cdot \exp\{-c_2/(AT+B)\}$, where $c_2 = 0.014\,388$ m·K. When the calibration to determine the coefficients A, B, and C is made by using three practical-type fixed point blackbody furnaces of Al, Ag, and Cu with accuracy of 0.3 °C, a temperature scale with ±0.5 °C accuracy can be realized with this thermometer in the temperature range from 600 °C to 1100 °C.

INTRODUCTION

Radiation thermometers commonly used are calibrated by comparison with thermocouples, where the accuracy is of the order of several degrees C. This paper proposes a practical temperature standard of improved accuracy realized by a narrow-band radiation thermometer using a silicon photodiode and a set of practical-type fixed point blackbody furnaces[1].

A narrow wavelength band around 900 nm is selected as the measuring wavelength of the radiation thermometer. The temperature range is from 420°C to 2000°C, and the resolution is about 1°C at 420°C and better than 0.01°C above 600°C. The relation between the output signal voltage V of the radiation thermometer and the temperature T of the target is expressed as follows: $V(T) = C \cdot \exp\{-c_2/(AT+B)\}$, where c_2 is the second constant of radiation. The coefficients A, B and C are determined by the calibration with equal to or more than three fixed-point blackbody furnaces, and the instrument is then used as a standard radiation thermometer. In the case of interpolation, a temperature scale of ± 0.5°C accuracy is established by this method.

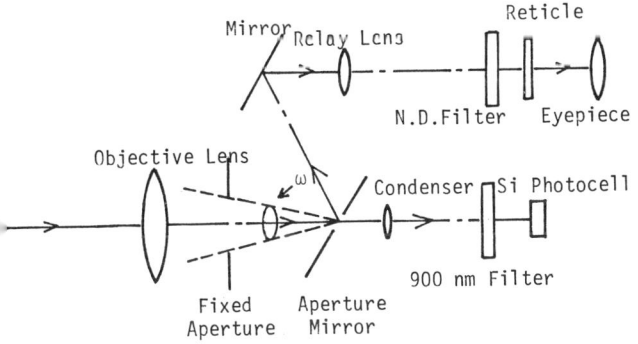

Fig. 1. Optical System of the 900 nm narrow-band radiation thermometer using a silicon photodiode.

DESCRIPTION OF THE RADIATION THERMOMETER

Optical system and photodetector

Figure 1 shows the optical system of the narrow-band radiation thermometer. The focal range extends from 40 cm, where the target area is within a circle of 3 mm diameter, to infinity. The image of the measured plane is focused by the objective on a mirror with an aperture of 0.75 mm diameter. The light passed by the aperture travels through a condenser, through an interference filter, of which the maximum transmission wavelength is 900 nm and the half-width (full bandwidth at half of peak transmittance) is 14 nm, and then reaches a silicon photodiode operated in the photovoltaic mode to be converted into an electric current. This photocurrent is converted by an FET amplifier into a voltage signal and measured by a digital voltmeter. The reflected light from the mirror is reflected by another mirror to a finder system, and the image is focused on a cross-hair reticle by a relay lens. Using the image observed through an eyepiece, the optical axis of the thermometer is set on the target. A ring lighting and a neutral density filter are provided to facilitate sighting on a low and high radiance target, respectively.

Generally the measurement of the radiance is regarded as independent of the working distance l. In detail this effect can be expressed approximately by[2]

$$\frac{L_2}{L_1} = \tau_{ob} \cdot (1 + \frac{3}{4} \cdot \tan^2\beta) \cdot (1 - \frac{3}{4} m^2 \tan^2\beta), \quad (1)$$

where L_1 and L_2 are the radiance of the target and its image, respectively, τ_{ob} is the transmittance of the objective, 2β is the apex angle of the radiant flux incident on the image ($\tan\beta = 0.154$), and m is the magnification which varies from 1/4 ($l=40$ cm) to 0 ($l=\infty$). So, L_2/L_1 varies only 0.1 % when l varies from 40 cm to ∞.

Figure 2 is a photograph of this radiation thermometer. The thermometer itself is 430 × 180 × 80 mm³ in size and 4 kg in mass. A level, a power switch, a

Fig. 2. Photograph of the narrow-band radiation thermometer using a silicon photodiode.

zero adjuster, and a ring lighting device can be seen in the photograph.

Table I lists the main specifications of the radiation thermomter

Selection of measuring temperature range

The output of the radiation thermometer V_0 is approximately expressed as follows

$$V_0 = R_f \cdot R(\lambda_m) \cdot L_{\lambda,b}(\lambda_m, T) \cdot \omega \cdot S \cdot \tau_o \cdot \tau_m \cdot \Delta\lambda , \quad (2)$$

where R_f is the feed-back resistance of the amplifier, $R(\lambda_m)$ is the responsivity of the radiation detector, ω is the solid angle of radiant flux incident on the field-defining aperture as shown in Fig. 1, S is the area of the aperture on the mirror, τ_m, λ_m and $\Delta\lambda$ are the maximum transmittance, its wavelength and the bandwidth of the interference filter, respectively, τ_o is the transmittance of the optical system other than the filter, and $L_{\lambda,b}(\lambda_m, T)$ is Planck's formua for the blackbody spectral radiance at the wavelength λ_m and temperature T.

The relation between R_f, T and V_0 can be calculated from Eq. (2).

$V_0 = 0.1$ mV when $R_f = 100$ MΩ and $T = 400°C$,
$V_0 = 2$ V when $R_f = 100$ kΩ and $T = 2000°C$

where the instrument configuration described in the previous section is considered and following values are assumed:

Table I Specifications of the 900 nm narrow-band radiation thermometer using a silicon photodiode.

Temperature range	420°C - 2000°C
Wavelength	
transmission maximum	900 nm
full width at half maximum	14 nm
Minimum target diameter	3 mm (at 400 mm distance)
Field of view	0.537°
Resolution	
above 600°C	better than 0.01°C
at 420°C	1°C
Detector	silicon photodiode Hamamatsu TV S874-5K

$R=0.4$ A/W, $\lambda_m=900$ nm, $\Delta\lambda=14$ nm, $\omega=0.08$ sr, $S=4.42\times10^{-7}$ m^2, $\tau_m=0.5$ and $\tau_o=0.9$.

The silicon photo-diode S874-5K (Hamamatsu TV) has the following characteristics: 1) parallel resistance of about 10 GΩ is expected; 2) the spectral responsivity abruptly decreases beyond 900 nm; and 3) the temperature coefficient of the responsivity increases beyond 900 nm. Considering these three conditions, together with the calculated result described above, and also that 4) the effective wavelength commonly used in a factory is around 900 nm and feed-back resistances for four ranges, that is, 100 kΩ, 1 MΩ, 10 MΩ, 100MΩ are selected and the temperature range from 400°C to 2000°C is covered. In the temperature range below 500°C, however, the level of the photo-current is so low that 1) the resolution is limited by the noise of the amplifier, and that 2) the effect of the temperature drift of the zero of the amplifier increases.

CHARACTERISTIC EQUATION

It is assumed that the output voltage V_r of this radiation thermometer using a silicon photodiode can be related to the target temperature T by the following empirical equation

$$V_r = C \exp\left(-\frac{c_2}{AT+B}\right) , \quad (3)$$

or

$$T = \frac{c_2}{A(\ln C - \ln V_r)} - \frac{B}{A} , \quad (4)$$

where $c_2 = 0.014\,388$ m·K, A, B and C are the characteristic constants of each radiation thermometer determined by the constitution condition and the spectral characteristics of the radiation thermometer. By using Eq. (4), the temperature T can directly be obtained from the output voltage V_r. Coefficients A, B and C can be calculated from the measured values of V_r at equal to or more than three points of temperature. The calculating procedure is this. Equation (4) can be transformed as follows;

$$a \cdot T \cdot \ln V_r + b \cdot \ln V_r + c = T , \quad (5)$$

where

$$A = \frac{a \cdot c_2}{\frac{b}{a} + c} , \quad (6)$$

$$B = \frac{b \cdot A}{a} , \quad (7)$$

$$C = \exp\left(\frac{1}{a}\right) . \quad (8)$$

Coefficients a, b and c can easily be calculated from the data of $V_r(T)$ by the least squares fit to Eq. (5).

Accuracy of approximation

The output voltage V of a radiation thermometer can be related to the blackbody of temperature T and the effective wavelength λ_e by the Planck equation

$$V_p = \frac{C_e}{\lambda_e^5 \left(\exp\left(\frac{c_2}{\lambda_e \cdot T}\right) - 1\right)} , \quad (9)$$

and

$$V_w = \frac{C_e}{\lambda_e^5} \exp\left(-\frac{c_2}{\lambda_e \cdot T}\right) , \quad (10)$$

for the Wien's approximation, where C_e is a constant. In the case that $\lambda_e = 900$ nm, the temperature difference between V_p and V_w is less than 0.01°C below 1100°C and 0.03°C when T=1500°C. The effective wavelength λ_e of a common radiation thermometer is a function of the temperature, and it is known that λ_e of a narrow-band radiation thermometer can be given by the following equation from the consideration of the effective wavelength of the optical pyrometer.[3]

$$\lambda_e = A_e + \frac{B_e}{T} , \qquad (11)$$

Using the values measured by the narrow-band radiation thermometer at three fixed point blackbodies of Cu (1084.88°C), Ag (961.93°C), and Al (660.46°C), coefficients A, B and C of Eq. (3) and coefficients A_e, B_e and C_e of Eqs. (9) and (11) are determined. The values V_r, V_p and the difference $V_r - V_p$ are calculated by substituting these coefficients into each equation. The solid line in Fig. 3 shows the difference and the broken line shows $V_r - V_w$ where V_w is obtained by substituting the same values of A_e and B_e into Eq. (10). This figure shows that the difference between Eq. (3) and Eq. (10) is much less than that between Eq. (3) and Eq. (9). This leads to the appreciation that Eq. (3) causes as large error as Eq. (10).

Influence of nonlinear characteristics of detector responsivity

In the case the output signal V is distorted by some cause into the form

$$V_{nl} = \alpha \cdot V^{1+\delta} , \qquad (12)$$

after substituting Eq. (3) into Eq. (12), V_{nl} is transformed into

$$V_{nl} = c_{nl} \cdot \exp(-\frac{c_2}{A_{nl}T + B_{nl}}) , \qquad (13)$$

where

$$A_{nl} = \frac{A}{1+\delta} , \qquad (14)$$

$$B_{nl} = \frac{B}{1+\delta} , \qquad (15)$$

$$C_{nl} = \alpha \cdot C^{1+\delta} . \qquad (16)$$

This implies that Eq. (3) allows for the distortion as in Eq. (12). When the responsivity of the detector and the amplification factor of the amplifier vary at constant rate for every n-fold of the radiation flux incident to the detector, this causes the equal effect on the output signal with the distortion expressed in Eq. (12). So, even when the nonlinearities of the detector and the amplifier cannot be neglected, Eq. (3) can be regarded as having the effect of excluding the influence of the nonlinearity they have in the form of Eq. (12).

In the manufacturer's catalogue, the responsivity of the silicon photodiode used here provides good linearity in the range from 1 pA to 1 mA of the photo-current (which corresponds to the range from R_f=100 MΩ and V=0.1 mV to R_f=10 kΩ and V=10 V). From these consideration, Eq. (3) is expected to give a good approximation to the output characteristics.

CHARACTERISTICS

Observation of the fixed points of blackbodies

Figure 4 shows examples of freezing curves of practical-type fixed point blackbodies measured by the narrow-band radiation thermometer using a silicon photodiode. In this figure, (a) is the freezing curve of an Al point blackbody where supercooling of about 1°C is observed and (b) is that of a Zn point blackbody (419.58°C). From these curves, it is found that around the Zn point, which is almost the lower limit of detection, noises equivalent to 1°C are involved in the output signal and that above 600°C the influence of the noises decreases to less than 0.01°C.

The incline of the freezing curves at the Zn point in the figure is caused by the temperature drift of the zero of the amplifier which is negligible at the Al point.

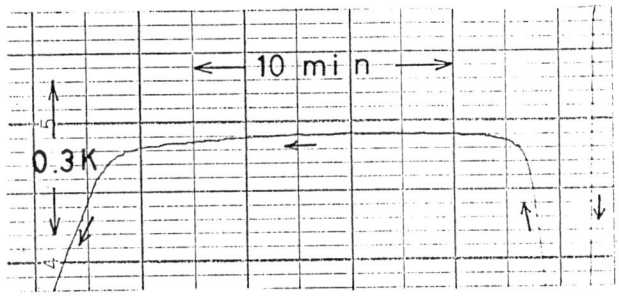

(a)

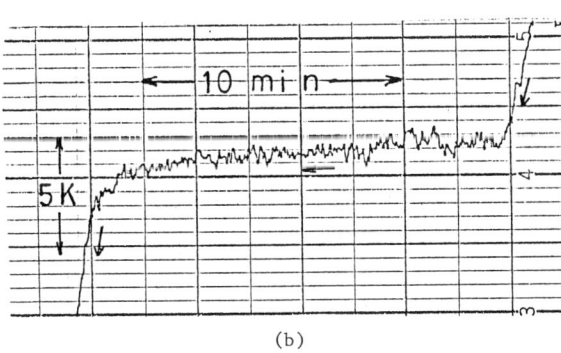

(b)

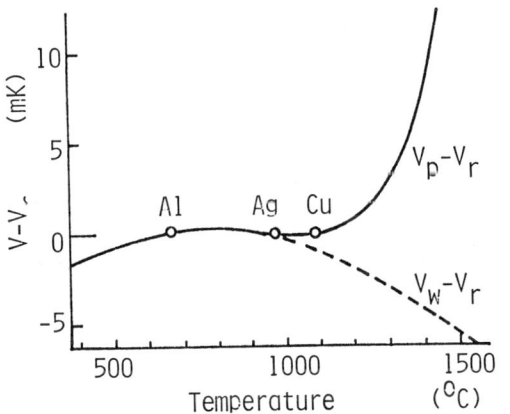

Fig. 3. Difference of the calculated output voltage between Eq.(3) and Eq.(7) with Eq.(9) (solid line) and that between Eq.(3) and Eq.(8) with Eq.(9) (broken line). Data of Cu, Ag and Al point blackbodies are used to determine the coefficients.

Fig. 4. Typical record of freezing curve of the practical-type fixed point blackbody measured by the narrow-band radiation thermometer. Metal for fixed point is (a) Al and (b) Zn.

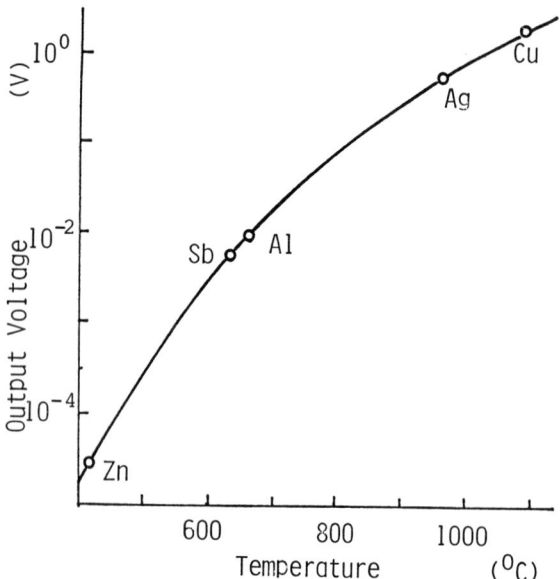

Fig. 5. Relation between the target temperature and output voltage of the narrow-band radiation thermometer. The regression curve of Eq. (3) is calculated from measured values at fixed points indicated by open circles.

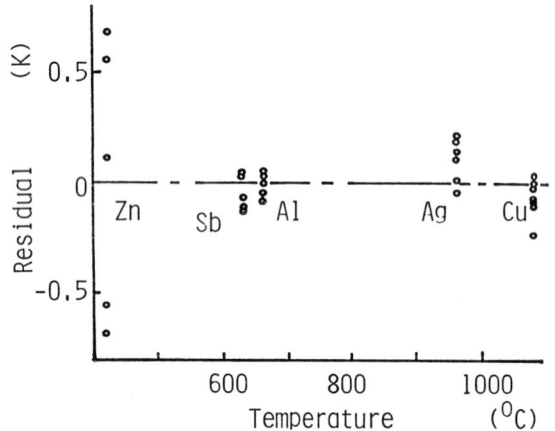

Fig. 6. Residual of the measured values from a least-squares fit to Eq. (3).

Calibration against the reference temperatures

Figure 5 presents the relation between the temperature and the output signal voltage of the radiation thermometer, calibrated by five practical-type fixed point blackbodies of Zn, Sb, Al, Ag and Cu, and calculated by the regression representation of Eq. (3). Each freezing point was measured seven times, and three data points out of thirty-five were discarded because of discrepancy from others. The regression curve is expressed by calculating the coefficients A, B and C by the least-squares method using the remaining thirty-two data points. Figure 6 shows the residuals from the regression curve. In this figure, the dispersion of the measured values of the Zn point is comparatively large. This is caused by the noise and the drift of the amplifier zero point. The dispersion of other points is small.

From this measured result, it is found that the output voltage at the Cu point is about 10^5 times as large as that at the Zn point and that the measured values at each fixed point deviate little from the regression curve. The measured output signal is about 50 % larger than that roughly estimated at the design in section "Selection of measuring temperature range". So the condition at the design is sufficiently satisfied.

Effect of the ambient temperature

Tables II and III show the result measured at the three fixed point blackbodies of Al, Ag and Cu at various ambient temperatures. Table II lists the mean signal voltage V of the measured values V_i at each fixed point, the ratio of the residual ΔV_i to V, $\Delta V_i/V$ and the coefficients of Eq. (3) calculated from the mean signal voltage and the measured values at each ambient temperature. Table III lists the result of the analysis of variance of the ratio $\Delta V_i/V$ of Table II. The variance of repetition is pooled in the variance of the error.[4]

The variation with the ambient temperature is due mainly to the temperature coefficient of the detector responsivity. From $\Delta V_i/V$ of Table II, the temperature coefficients of the radiation thermometer can be calculated: 0.148 %/°C (0.081°C/°C) at the Al point, 0.066 %/°C (0.063 °C/°C) at the Ag point and 0.068 %/°C (0.079°C/°C) at the Cu point. The temperature coefficient at the Al point is larger than those at the other points. The analysis of variance in Table III indicates that the factor R (room temperature) and the interaction F × R (fixed point × room temperature) are on 99 % level of significance. From the variance of the error E in Table III, the standard deviation can be evaluated to be 0.043 %, and the corresponding temperature values are 0.023°C at the Al point, 0.041°C at the Ag point and 0.050°C at the Cu point.

Table II Mean signal voltage, residual and regression coefficients A, B and C at various room temperatures.

Blackbody	Mean signal voltage [V]	residual/mean output voltage[%] room temperature		
		21.5°C	23.5°C	26.5°C
Al 1	0.009912	−0.337	−0.054	0.319
Al 2		−0.361	−0.024	0.460
Ag 1	0.62946	−0.197	−0.008	0.176
Ag 2		−0.135	0.013	0.153
Cu 1	2.0206	−0.218	0.000	0.185
Cu 2		−0.129	0.010	0.173
A (nm)		895.6	897.8	895.9	893.3
B (nm K)		5065	4255	5394	7082
C (V)		262650	258250	262140	267320

Table III Analysis of variance for the ratio of residual to mean signal voltage. The asterisks ** indicate 99 % level of significance.

Factor	Degree of freedom	Variance	Variance ratio
Total	17		
Fixed point (F)	2	0.0003	0.16
Room temperature (R)	2	0.3217	169**
F × R	4	0.0326	17**
Error (E)	9	0.0019	

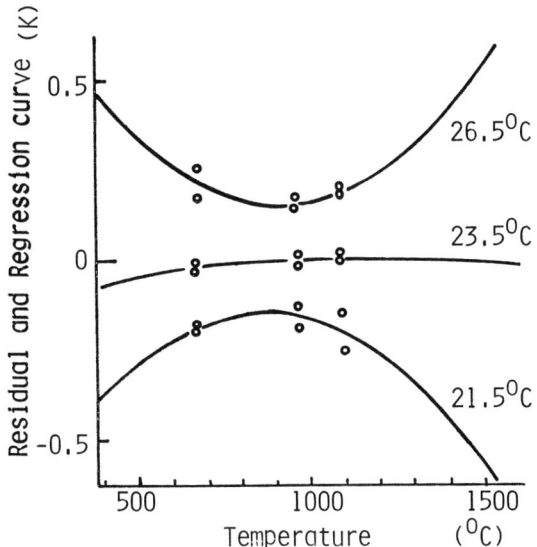

Fig. 7. The regression curve of the residuals target temperature for selected ambient temperatures. Zero is the regression curve calculated from all data.

Figure 7 shows the residuals and the corresponding regression curve calculated at each ambient temperature. The zero is referred to the regression curve calculated from all data. From this figure, it is seen that the influence of the ambient temperature on the extrapolation becomes large as the temperature departs from the three fixed points, and the temperature coefficient of responsivity at 1500°C is about 0.22°C/°C (0.11 %/°C). This value remains only a 20 % increase over the mean temperature coefficient (0.094 %/°C) of the responsivity at three fixed-points.

From these experimental results, it is expected that if the calibration is carried out at constant ambient temperature, the coefficients A, B and C can be determined with good accuracy and that in the measurement after the calibration the influence of the ambient temperature is determinde by the temperature coefficient of the responsivity (about 0.1 %/°C).

ACCURACY OF THE TEMPERATURE SCALE

Effective wavelength

The output voltage V_c of the radiation thermometer depends on the spectral radiance $L(\lambda,T)$ of the target, the spectral responsivity $R(\lambda)$ of the detector and the spectral transmittance $\tau(\lambda)$ of the optical system of the measuring instrument and is given by

$$V_c(T) = C_0 \int_0^\infty W(\lambda,T) \, d\lambda \quad , \quad (17)$$

$$W(T) = L(\lambda,T) \cdot R(\lambda) \cdot \tau(\lambda) \quad , \quad (18)$$

where C_0 is a constant determined by the solid angle the radiation thermometer sees, by the field-defining aperture and by the amplification factor of the amplifier. The spectral radiance $L(\lambda,T)$ is approximated by Wien's equation. The spectral transmittance $\tau(\lambda)$ can be expressed by the product of τ_o, $\tau_i(\lambda)$ and $\tau_s(\lambda)$, where τ_o is the transmittance of the optical system other than filters and is assumed to be independent of the wavelength. The spectral transmittance of the interference filter $\tau_i(\lambda)$ is approximated by using the characteristic equation of the interference filter

$$\tau_i(\lambda) = \frac{0.45}{1+F \sin^2(\frac{0.9 \times 10^{-6} \cdot \pi}{\lambda})} \quad , \quad (19)$$

where the maximum transmittance is 0.45, corresponding wavelength is 900 nm and F is the Finesse. The spectral transmittance of the sharp-cut filter $\tau_s(\lambda)$ is approximated by fitting some data points of the filter manufacturer's catalogue to a sort of hyperbolic function.

$$\tau_s(\lambda) = \frac{0.88}{1+\exp\{-0.68 \times 10^8 \times (\lambda - 0.8 \times 10^{-6})\}} \quad . \quad (20)$$

When the radiation on the longer wavelength side of the filter cannot be neglected, it is possible to install the infrared absorbing filter, whose characteristics $\tau_l(\lambda)$ are given by

$$\tau_l(\lambda) = \frac{0.82}{1+\exp\{-2.82 \times 10^7 \times (1 \times 10^{-6} - \lambda)\}} \quad . \quad (21)$$

Figure 8 shows the characteristics R_n and R_o of the spectral responsivity $R(\lambda)$ taken from the detector manufacturer's new and old catalogues, respectively. They differ by about 100 nm in the characteristics of the infrared region above 1 μm. The relative spectra S_{ij} of $W(\lambda,T)$ of Eq.(18) at the Al point when F=1500 (half-width 14.8 nm) are shown in the same figure.

$$S_{ij} = \frac{W(\lambda, T_{Al})}{W(900nm, T_{Al})} \quad , \quad (22)$$

where i = n and o denote using the new and old catalogue values, respectively, and j=1 and 0 denote the states with and without the infrared absorbing filter, respectively. From this figure, it is known that S_{oO} of the old catalogue responsivity without an infrared absorbing filter has a long tail to the longer wavelength side.

Equation (3) is an empirical expression relating target temperature to instrument output voltage, and was selected because of its simplicity and convenience of applicaiton. Equation (17) gives the correct relationship, but is more difficult to use. It will now be

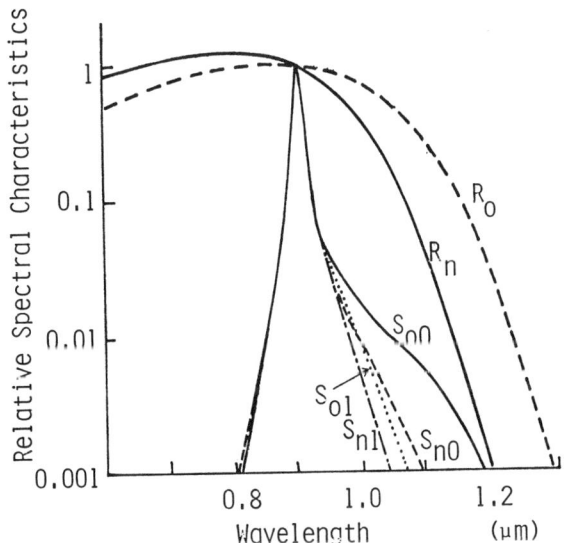

Fig. 8. Relative spectral responsivity $R_n(\lambda)$ and $R_o(\lambda)$ as given in the silicon photodiode manufacturer's new and old catalogue, and apparent relative spectral radiance on the output side of the silicon detector at the Al point.

shown that Eq. (3) is sufficiently accurate for our purpose. Figure 9 shows the temperature value corresponding to the difference $V_r - V_c$ between Eq. (3) and (17) where the coefficients A, B and C of Eq. (3) are determined from the calculated values of V_c at the three fixed points, Cu, Ag and Al. This figure indicates that Eq. (3) represents the output voltage with better accuracy than 0.02°C in the temperature range of the interpolation. The error in the range of the extrapolation up to 1500°C remains within 0.3°C.

When the detector with the responsivity as given by the new catalogue values and the infrared absorption filter are used to decrease the tail of the longer wavelength side, the extrapolation can be carried out with even better accuracy.

Calibration error

The influence of the calibration error depends on the temperature, which is serious in the extrapolated region. A systematic error and an accidental error are considered separately.

The practical-type fixed point blackbody furnaces are constructed alike. The error in the effective emissivity of the blackbody cavity of the furnace is evaluated as - 0.001 ± 0.0005 and the error in the temperature of the furnace as - 0.1 ± 0.1°C.[1] The systematic error is composed of the error in the effective emissivity (-0.001) and of the error in the temperature of the cavity (-0.1°C). The curve of the broken line in Fig. 10 represents the error at each temperature introduced by the systematic error when the temperature is represented by Eq. (3).

On the other hand, the accidental error can be classified in terms of 1) the difference of the effective emissivity between blackbody furnaces; 2) the difference of the error in the temperature between five fixed point blackbody furnaces; 3) the dispersion of the error in the temperature arising from the change of the condition of measurement; and 4) the dispersion of the measured output of the radiation thermometer arising from the change of the ambient temperature at calibration. These are the four main kinds of error. The influence of each error on the interpolation and extrapolation by Eq. (3) is summed up to an error of the variation of the output signal of the radiation thermometer for 1) and 4), and an error of the variation of the temperature of the blackbody furnace for 2) and 3).

The standard deviation of the output signal variation σ_V/V (ratio to the output signal) can be estimated to be 0.05 %, and that of the temperature variation σ_T, to be 0.05°C from the result of estimating the accuracy

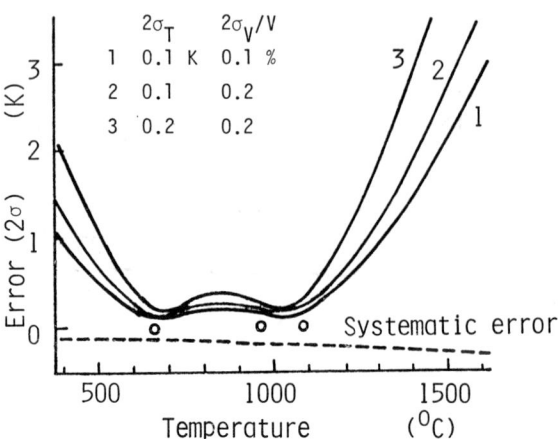

Fig. 10. Accidental error (solid line) with the standard deviation of temperature variance σ_T and that of the output signal variance σ_V/V, and systematic error (broken line).

of the blackbody furnace[1] and from that of the analysis of variance in Table III. The solid lines in Fig. 10 show the relation between the temperature and twice the standard deviation (2σ) when the temperature is represented by Eq. (3). Here the conditions are that σ_V/V is 0.05 % and 0.1 % and that σ_T is 0.05°C and 0.1°C.

From this figure it is found that 1) the systematic error of the fixed point blackbody furnaces has little dependence upon the temperature and does not increase much when extrapolated, 2) the errors at 600°C and at 1000°C reflect the errors of calibration at the fixed points; comparing the values, the error in the interpolation range is seen to remain within twice the error of calibration at the fixed points, and 3) the accidental error increases rapidly in the extrapolation range; to obtain the accuracy of 3°C at 1500°C, it is necessary to calibrate with the accuracy that σ_T=0.05°C and σ_V/V=0.1 % (or that σ_T=0.1°C and σ_V/V=0.5 %).

As the temperatures of the Cu and Ag points are close together, their errors have a larger influence on the calibration error in the extrapolated region. The larger the difference of their errors becomes, the more the instrument calibration error increases. Figure 11 shows an example of them. The solid line presents the case when the difference of their errors is large; e_T(Cu)=0.1°C, e_V(Cu)=-0.001·V, e_T(Ag)=-0.1°C, e_V(Ag)=0.001·V, e_T(Al)=-0.1°C and e_V(Al)=0.001·V: while the broken line presents the case when the difference is small; e_T(Cu)=-0.1°C, e_V(Cu)=0.001·V, e_T(Ag)=-0.1°C, e_V(Ag)=0.001·V, e_T(Al)=0.1°C and e_V(Al)=-0.001·V. Here e_T and e_V express the error in the temperature of the blackbody furnace and that in the output signal of the radiation thermometer, respectively.

From this consideration it is found that good accuracy can be expected also in the extrapolated regions when care is taken in measuring the Ag and Cu points to decrease the differnce of the errors. For that purpose, one method is to complete the measurement of both points in a short time when the room temperature change can be neglected.

Overall accuracy

Table IV shows in order the estimated components of error that determine the accuracy when the temperature scale is established using Eq. (3) with the narrow-

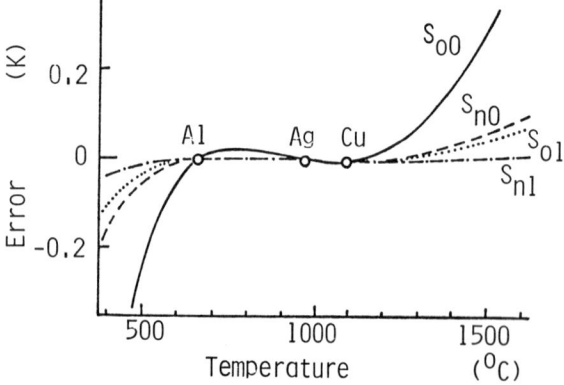

Fig. 9. Error ΔT_{rc} between the numerically calculated value by Wien's equation and the value regressed from the calculated values at the Al, Ag and Cu points.

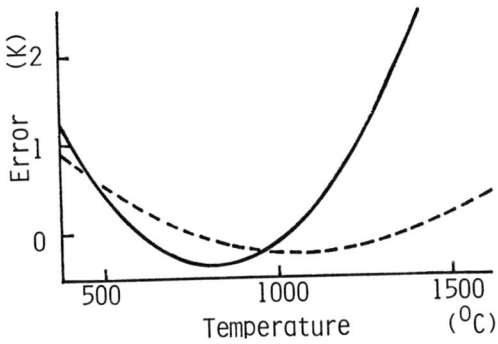

Fig. 11 Influence of the error at the fixed points on the extrapolation. The conditions are; $e_T(Cu)=0.1°C$, $e_V(Cu)=-0.001 \cdot V$, $e_T(Ag)=-0.1°C$, $e_V(Ag)=0.001 \cdot V$, $e_T(Al)=-0.1°C$ and $e_V(Al)=0.001 \cdot V$ for solid line: $e_T(Cu)=-0.1°C$, $e_V(Cu)=0.001 \cdot V$, $e_T(Ag)=-0.1°C$, $e_V(Ag)=0.001 \cdot V$, $e_T(Al)=0.1°C$ and $e_V(Al)=-0.001 \cdot V$ for broken line, where e_T and e_V express the error in the temperature of the furnace and that in the output signal of the radiation thermometer, respectively.

band radiation thermometer with a silicon detector calibrated by the practical-type fixed point blackbody furnaces. The error of measuring the fixed point given in this table is calculated assuming that the measurement is repeated three times and that the change of the ambient temperature at measurement is within ±0.1°C.

Table V lists the uncertainty at each temperature estimated on the basis of the consideration of the previous section supposing that the calibration at the fixed points is carried out with such uncertainty. This table also lists the uncertainty when the allowed change of ambient temperature at use is within ± 5°C of the temperature at calibration.

From Table V it is estimated that it is possible to realize the temperature scale with an uncertainty of ±0.5°C in the interpolation range (from 600°C to 1100°C) and of ±1°C in the range from 500°C to 1300°C.

It will be necessary to study further the long term stability of the radiation thermometer.

Table IV Error in the calibration of the narrow-band radiation thermometer using a silicon photodiode.

	ΔT [°C]	$\Delta V/V$ [%]	Condition
Systematic error			
realizing the fixed point	-0.1	-0.1	
approximating with Eq.(3)		negligible	400-1500°C
spectral response of the radiation thermometer		negligible	500-1500°C
Accidental error			
realizing the fixed point (2σ)	± 0.1	± 0.05	
measuring the fixed point (2σ)		± 0.06	*
Total	± 0.1	± 0.08	

* three-time measurement, where the change of room temperature is within ± 1 °C.

Table V. Uncertainty of the temperature scale by the narrow-band radiation thermometer using a silicon photodiode. (units in °C)

Temperature range	500	600-1100	1300	1500
Systematic error	-0.15	-0.20	-0.25	-0.30
Accidental error	±0.4	±0.2	±0.8	±1.8
Uncertainty of temperature scale	±0.55	±0.4	±1.05	±2.1
Error when room temperature change ±5 °C	±0.6	±0.5	±0.7	±1.1
Uncertainty at use	±1.2	±0.9	±1.8	±3.2

CONCLUSION

An investigation has been made of a method of establishing a temperature scale by using a narrow-band radiation thermometer with a silicon detector of measuring wavelength 900 nm referred to practical-type fixed point blackbody furnaces, and of its accuracy. This method for establishing a standard has the feature that special skillness is not needed for the realization. This method is effective in obtaining a temperature standard for calibrating radiation thermometers at a manufacturer of radiation thermometers and at a standards laboratory of a factory.

As the output signal of the radiation thermometer depends on the ambient temperature, an IC temperature senser is installed near the silicon photodiode. It is planned that the correction of the ambient temperature will be made in the course of the conversion of the output voltage to the temperature scale by using a microcomputer.

The range of the reliable temperature scale by this method is now limited from 500°C to 1300°C. Extending the temperature range is possible in principle.

With 1) detailed investigations and improvement of the quality of the narrow-band radiation thermometer, 2) development of a higher temperature practical-type fixed point blackbody furnace, 3) improvement in the accuracy of the practical-type fixed point blackbody furnaces, and 4) the study of the difference between this temperature scale and the international practical temperature scale (IPTS), the temperature range can be extended. These are left for the future work.

REFERENCES

1. "Practical-type Fixed Point Blackbody Furnace" F. Sakuma and S. Hattori, in Temperature, Its Measurement and Control in Science and Industry, Vol.5
2. T. Shimozuma, et al. "Ondo Keisoku" ("Temperature Measurement"), published by Soc. Instrum. Control Engineers, Tokyo, 1981, p.186. (in Japanese)
3. A more general expansion* is $1/\lambda_e = (a/T) + b$. In both expansions, λ_e has a close value to each other in the condition that $a/bT \ll 1$ or $B/AT \ll 1$. This condition is satisfied for the narrow-band radiation thermometer.
*. D. R. Lovejoy, Temperature, Its Measurement and Control in Science and Industry, Vol.3 Part 1. p.502. Reinhold Publishing Co., 1962.
4. In this experiment, two sets of Al, Ag and Cu fixed point blackbody furnaces were used. By the analysis of variance, the effect of repeated measurement, including the influence of the difference between the blackbody furnaces, is judged to have no significant effect. The variance is therefore pooled in the variance of the error.

A photoelectric direct current spectral pyrometer with linear characteristics

B. Woerner

Institut für Kernenergetik und Energiesysteme der Universität, 7000 Stuttgart 80, Federal Republic of Germany

A new photoelectric pyrometer is described, based on the principles of the standard pyrometer used at the Physikalisch Technische Bundesanstalt for the realization of the IPTS 68 above 1337 K. The instrument combines the advantages of the PTB measurement method with flexibility and convenience for general laboratory application. A detector system is employed which is characterized by a highly linear relationship between measured radiance and output signal. This is achieved by a specially designed S-20 vacuum photocell in combination with high quality dc electronics. The photocell output is linear over a range of 6 decades (from 10^{-14} to 10^{-8} A) corresponding to a temperature span from 900 K up to 2100 K at a wavelength of 650 nm. The advantages of this system, especially the easy calibration procedure, are discussed. The electronic and optical design of the instrument meet the requirements for a high quality pyrometer. The limitations of stability and repeatability were investigated in detail on a series of instruments. The resulting uncertainties are reported. A resolution of less than 0.02 K at 1337 K and an estimated total error of about 0.1 to 0.2% of temperature were obtained. A comparison measurement performed during extended tests at PTB showed agreement with the standard pyrometer to within a few 10^{-1} K on strip lamps up to 2500 K and to within less than 0.1 K on a black radiator up to 1700 K.

INTRODUCTION

The measurement of absolute temperatures requires a reference to the temperature fixed-points and interpolation procedures defined by the International Practical Temperature Scale (IPTS). Above the freezing point of gold (1337 K) the IPTS 68 is defined by Planck's law and is realized by radiance measurements on blackbodies.

At the Physikalisch Technische Bundesanstalt Braunschweig (PTB) extensive investigations have been carried out on the development of a photoelectric standard pyrometer which enables one to calculate the blackbody temperature above the gold point by means of the quasi-constant sensitivity of the instrument. A special type of vacuum photocell serving as an extremely linear detector was developed as a consequence of these investigations. The results obtained with this standard pyrometer and its special advantages are described by Kunz and Kaufmann[1].

The principle of the linear characteristic of this pyrometer was adapted to develop a pyrometer offering the high degree of instrument-accuracy needed today at metrological institutes and in industry laboratories. As a result of this development an instrument was designed in cooperation with PTB and is being manufactured, complying with these requirements but also having convenient size and handling characteristics.

PRINCIPLES OF OPERATION

A linear detector - such as a photocathode or a photodiode - produces a photocurrent proportional to the incident light. Having assured a sufficient constancy of the sensitivity versus intensity as well as versus time within a certain range, the detector system allows measurements of light intensity ratios in terms of current ratios I_1/I_2. Together with an optical system this is also valid for radiance ratios L_1/L_2.

$$\frac{L_1}{L_2} = \frac{I_1}{I_2} \qquad (1)$$

When the measurements are performed on a blackbody at temperatures T_1 and T_2 in a narrow wavelength interval, one can write, using Planck's equation,

$$\frac{I_1}{I_2} = \frac{[e^{c_2/\lambda_{1,2} T_1} - 1]^{-1}}{[e^{c_2/\lambda_{1,2} T_2} - 1]^{-1}} \qquad (2)$$

where $\lambda_{1,2}$ is the "mean effective wavelength" of the system relating T_1 and T_2. Hence it follows that

$$T_2 = \frac{c_2}{\lambda_{1,2}} / \ln \left[\frac{I_1}{I_2} (e^{c_2/\lambda_{1,2} T_1} - 1) + 1 \right]. \qquad (3)$$

The data I_1 and T_1 required for an application of Eq. (3) are determined by a measurement of a photocurrent $I_1 = I_{cal}$ at a source of a known blackbody temperature $T_1 = T_{cal}$ (Reference point). For any measured photocurrent $I_2 = I$ one can now calculate the respective temperature $T_2 = T$ if $\lambda_{1,2}$ is well known.

$\lambda_{1,2}$, the "mean effective wavelength" or "mathematical pyrometer wavelength" as it is denominated by Kunz[1] can be calculated for any pair of temperatures T_{cal} and T by numerical evaluation of the integral which describes the production of the photocurrent

$$I(T) \sim \sum_{\lambda=0}^{\infty} L_\lambda(\lambda,T) \tau_F(\lambda) s(\lambda) \Delta\lambda \qquad (4)$$

with the detector sensitivity distribution $s(\lambda)$ and the relative filter transmittance function $\tau_F(\lambda)$. $s(\lambda)$ and

$\tau_F(\lambda)$ can be determined by in situ-measurements. Using Eq. (4), the photocurrents $I(T_{cal})$ and $I(T)$ are calculated and inserted into Eq. (2). When the right-hand term of this equation is tabulated as a function of λ within the region near the center wavelength of the filter, $\lambda_{1,2}$ is simply found by interpolation. Or, using Wien's approximation in Eq. (2), $\lambda_{1,2}$ can be calculated directly from the relation

$$\lambda_{1,2} = c_2 \left(\frac{1}{T} - \frac{1}{T_{cal}} \right) / \ln \frac{I_{cal}}{I} \quad . \tag{5}$$

Since for T_{cal} = const $\lambda_{1,2}$ varies slightly with T a function $\lambda_{1,2}(T)$ must be established for each wavelength filter used by computation for several pairs of temperature according to the above scheme. The iteration needed now for solving Eq. (3) requires only a few steps.

For practical application of the linear pyrometer Eq. (3) can be written in the form

$$T = \frac{c_2}{\lambda_{1,2}} \bigg/ \ln \left[\frac{\varepsilon \cdot \rho \cdot \tau \cdot I_{cal}}{I} \left(e^{\frac{c_2}{\lambda_{1,2} T_{cal}}} - 1 \right) + 1 \right], \tag{6}$$

with corrections for the emittance ε of the target, reflectance ρ and transmittance τ of optical media in the light path. Additive signal corrections ΔI can be done by inserting $I' = I - \Delta I$ instead of I into Eq. (6), if compensation must be made for reflections from the target.

SPECIAL ADVANTAGES OF LINEAR OPERATION

As a consequence of the above described operation principles, the pyrometer provides certain special features such as:

- the output signal is the measured photocurrent which corresponds to the target radiance; radiance temperatures are easily computed using Eq. (6);
- calibration reduces to a measurement at only one reference temperature;
- no internal reference lamp is required; calibration check can be done directly with respect to the laboratory reference source; if a gold point blackbody is used, one can realize the temperature scale within the instrument's accuracy;
- additional wavelength filters can be used, since the resulting effective wavelengths can be calculated easily and with a high precision from in-situ measurements;
- the dependance of the effective wavelength on the target temperature can be taken into account by suitable algorithms and by iterational computing;
- the evaluation of the measured photocurrents at different wavelengths is only a question of software. Thus intensity and ratio pyrometry can be performed as well as radiometric measurements;
- the easy calibration procedure makes it easy to use different optics or field stops or even aperture stops; this results in a high adaptibility for laboratory applications.

OPTICAL AND ELECTRONIC DESIGN

An optical schematic of the measuring head is shown in Fig. 1. The objective lens forms an image that is focused on the field stop. The light going through the hole is collimated. The filters, aperture stop and detector are arranged in the collimated light beam. Interference filters and neutral density filters can be changed by means of two filter wheels and the stops and the objective lenses are easily exchangable. The target image in the field-stop plane can be adjusted visually through a telescope. Pilot light adjustment is also provided. When using an objective focal length of 140 mm, target diameters of 0.4 up to 1.3 mm are obtained at distances of 480 up to 1100 mm. By the use of different focal lengths (110 up to 200 mm) and target field-stop diameters (0.1 up to 1.5 mm) a wide range of optical adaption is provided.

Special care has been taken to minimize optical and mechanical sources of error that could affect the reproducibility of the instrument response. The effect of size of source which is produced by image defects and stray light effects was made less than 0.1 % of radiance when the diameter of the radiation source is reduced from 20 mm to 1.3 mm, at a target diameter of 1 mm. Distance effects are also less than 0.1 % over the whole normal distance range from 400 up to 1500 mm. Filter wheel movement, in connection with usual filter-inhomogeneity, causes errors not more than 0.02 % of radiance. If the incident radiation is linear polarized, a maximum error of 0.07 % can occur. Mechanical deformation effects are carefully avoided.

A vacuum photocell with an S-20 spectral response is used as detector. It was specially developed [a] and is manufactured exclusively for this application. It has

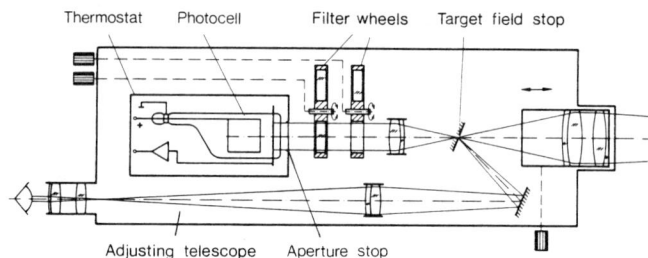

Fig. 1: Schematic of the optical arrangement of the Linearpyrometer

a 25 mm diameter semitransparent cathode and is equipped with a guard ring. The current-to-voltage converter consists of an electrometer dc-amplifier [b] with feedback resistors of 10^8 up to 10^{12} ohms, switched by reed relays.

Dark currents of some 10^{-15} to 10^{-14} A are observed with this arrangement. Dark current noise is about a few 10^{-16} A when the time constant is 10 s. Thus the lower end of current range is 10^{-13} A as determined by resolution and short-term stability. The upper end is reached at about 10^{-8} A when small time-dependent reversible changes of sensitivity are observed. This corresponds to a usable temperature range from 1000 K up to 2100 K at a wavelength of 650 nm, or from 970 K up to 2370 K at 750 nm at optical and filter-conditions as given in Table 1.

The photocell and the amplifier are mounted in a temperature-controlled box. This is due to the considerable temperature coefficients of these components. By a slight amount of heating, humidity effects at surfaces of the insulations are avoided.

The low impedance signal (voltage) is measured with a built-in digital voltmeter with a resolution of 2.5×10^{-4} up to 2.5×10^{-5}, depending on the signal level. Together with digital information about the selected range and filter wheel positions, the measurement data can be processed by a simple desk-top computer or by other means.

LINEARITY AND STABILITY TESTS

For the test of the linearity the well-known method of radiance-addition by means of a semitransparent mirror was applied. With this method, measurements on two sources can be performed separately one after another, leading to photocurrents I_1 and I_2, or together at the same time (when both paths of light are opened), leading to a current I_3. For a linear detector, $I_3 = I_1 + I_2$. If $I_1 \approx I_2$, the relative linearity error can be defined as $(I_3 - (I_1 + I_2))/0.5 I_3$. The test accuracy was restricted to an estimated value of 0.02 % because of resolution and stability limits of the lamps and the detector system and by what is presumed to be a slight interaction between the two optical paths by stray light. The measured nonlinearities for each doubling of the intensity at wavelenghts of 650 up to 750 nm were of the order of 0.01 % up to 0.04 % with arbitrary sign, depending on instrument and range being tested. This is due to the superposition of the nonlinearities of photocathode, amplifier and feed-back resistors. Higher nonlinearities may occur at photocurrents above about 10^{-8} A, thus indicating the limit of linear operation of the photocathode, or when a voltage above about 1 V appears at the highest-valued resistors ($3 \times 10^{12} \Omega$). The latter is a consequence of the resistors voltage-coefficient and must be avoided either by voltage restriction or by addition of several separate resistors. Thus over a range of photocurrents from 10^{-13} up to 10^{-9} A a nonlinearity of less than ±0.04 % was found, when doubling the radiance. Over a span of 2 decades this leads to a maximum error of 0.28 %, if only one sign occurs, or to a probable error of 0.1 %, if the deviations were thought to be independent. A mean value of about 0.2 % can be estimated to be the most probable and was used for the calculation of temperature errors in Table 1.

Stability is the other important property to be assured by tests. This is because the temperature measurements must be performed at different time and normally under more or less different conditions, compared to the reference measurement. Some data concerning stability versus varying optical conditions were given above. These tests were performed on a very high confidence level by using a blackbody[2,3] having a heat pipe as walls, thus obtaining an extremely homogenous temperature distribution. Additionally, the effect of stray light was investigated by viewing on a black hole surrounded by a large bright area. When referred to the bright-area radiance, an amount of 0.2 % was transmitted into the measurement path of light. This value is subject to a possible increase, depending on the state of objective lens and protection window cleanliness, by a factor of 5 or more.

Filter transmittance uncertainties may be caused by ambient temperature changes or by changes of the optical surfaces. Temperature coefficients of the types of neutral density filters used were measured and found to be less than a few 10^{-5}/K. Interference filters showed coefficients for peak transmittance of about 0.03 %/K but higher values were observed for different types (which seems to be dependent on the kind of blocking glasses used). Wavelength shift coefficients are usually about +0.02 nm/K. These values ought to be taken into account when measurements of high precision must be performed in an insufficiently temperature-controlled room. Any long-term changes of transmittance, e.g. due to depositions, are automatically controlled when a calibration check is performed.

Stability of the detector system versus ambient temperature is achieved by the use of a well-insulated detector box that is temperature-controlled to within ±0.05 K. When considering detector system temperature-coefficients of about -0.2 up to -1 %/K, depending on used wavelength and resistor, the resulting photocurrent errors are less than 0.1 %.

Extended tests were performed concerning the drifts of dark current and signal at various levels of photocurrent. The dark current changes were found to be within a few 10^{-16} A over a period of months. Slight deviations of the same order are observed for some minutes after high-level illumination of the cathode. Short term drift is within 0.05 % over 8 hours at currents less than 10^{-9} A. At the highest permitted photocurrent of 10^{-8} A, a reversible sensitivity decrease of 0.1 % was observed within the first hour at a wavelength of 650 nm. It is interesting to note that the recovery of the cathode after high illumination takes place even during a high illumination at another wavelength; this may be due to the layer structure of the multialkalicathode which partly can be considered as a system of independent detectors.

Long-term drift of the photocurrent (including possible slight changes of the interference filter and the tungsten strip lamp that was used as reference) was found to be within 0.2 % over several months.

Table 1 Characteristic data of the Linearpyrometer LP2 and temperature error components resulting from the reported uncertainties (values in brackets) under the following conditions: target diameter 0.8 mm, distance 500 mm, entrance aperture 45 mm, wavelength 650 nm (HBW 10 nm, τ_{max} = 60 %).

Temperature (K)	1000	1200	1337	1600	1919	2400*)
Photocurrent (A)	10^{-13}	4×10^{-12}	2.7×10^{-11}	4×10^{-10}	4×10^{-9}	4×10^{-10}*)
Time constant 90 % (s)	20	6	6	6	1	1
Resolution**) (K)	0.05	0.015	0.02	0.03	0.05	0.07
Noise equivalent (K)	0.15 / 0.03+)	0.025 / 0.01+)	0.01	<0.01	<0.01	<0.01
Individual Errors (K)						
calibration point [0.5 K]	0.20	0.28	0.35	0.5	0.78	1.22
effective wavelength [0.1 nm]	0.06	0.05	0.03	-	0.07	0.20
non-linearity [±0.2 %/100]	0.16	0.13	0.08	-	0.18	0.58
dark current drift [±0.1 %]	0.05	0.07	0.08	0.12	0.18	0.29
short term drift [0.1 %/h]	0.05	0.07	0.08	0.12	0.18	0.29
long term drift [±0.2 %/month]	0.1	0.14	0.16	0.24	0.36	0.58
thermal and optical effects most probable error [0.4 %]	0.18	0.26	0.32	0.46	0.73	1.15
Sum of error values (K)	0.80	1.0	1.1	1.44	2.48	4.3
probable error (K)	0.34	0.44	0.52	0.74	1.17	1.9

*) measured with neutral density filter ND2 (τ = 0.01)
**) most unfavourable values, obtainable with $4^3/_4$ digit scale-length instrument
+) with use of an interference filter with 700 nm (HBW 20 nm, τ_{max} = 50 %)

ACCURACY AND TEMPERATURE RESOLUTION

The measurement uncertainty is given by the combination of the individual errors described above with the uncertainties resulting from the "mean effective wavelength" determination and the absolute error of the reference point used for calibration.

For a 0.8 mm target diameter at 500 mm distance and an entrance aperture diameter of 45 mm at a filter wavelength of 650 nm (half-power bandwidth 10 nm, maximum transmitance 60 %), the resulting uncertainties of the temperature at a 95 % confidence level were calculated and are summarized in Table 1. A wavelength uncertainty of 0.1 nm and a reference-point error of 0.5 K at a calibration temperature of 1600 K were assumed, which may be observed by means of a well-calibrated vacuum tungsten strip lamp.

As Table 1 shows, the maximum errors do not exceed 0.1 % up to 0.2 % of temperature. The most probable error can be assumed to be less than 0.1 %.

The prototype of the Linearpyrometer described was tested in detail at PTB. The results of comparison measurements between this instrument and the PTB standard pyrometer were reported by Kunz[4] and are given in Table 2. For these measurements, the gold-point blackbody was used as reference. The deviations reported are remarkably smaller than the total errors given in Table 1. This is partly due to much higher accuracy of a primary calibration as it was performed. Another reason may be smaller non-linearity errors than the above assumed, according to a value of 0.5×10^{-5} at a doubling of intensity as is reported by Kunz[4]. The relatively increased errors at the measurements on a tungsten strip lamp when compared to those on a blackbody are a consequence of the non-ideal optical conditions associated with the use of strip lamps. The resolution reported in Table 2 is obtained by an external digital voltmeter (DVM), which allows the full signal/noise ratio of the pyrometer detector system to be used. The corresponding values given in Table 1 are obtained with the built-in DMV.

The signal/noise ratio at low temperatures can be further improved by the use of an interference filter with a broader bandwidth and a somewhat longer center wavelength; resolution values up to 0.01 K at 1000 K can be obtained.

CONCLUSIONS

A pyrometer was developed combining the advantages of a linear operating system with a set-up of a high optical and electronic performance. It was demonstrated by extensive tests on a first series of instruments that a high level of accuracy was reached. Temperature errors between 0.1 and 0.2 % and less have been verified and can be maintained by the simple one-point calibration check allowed by this system. When operated carefully under conditions as they usually are in standard-institutes, remarkably improved accuracy is achieved. As was shown by the PTB comparison[4], the instrument thus enables a realization of the IPTS above 1100 K within an uncertainty of only some tenths of a degree, when a gold-point blackbody or even when a well-calibrated lamp is available. Thus the pyrometer can be provided to serve as a reference or even as a standard instrument.

On the other hand, the instrument's high adaptability with respect to optics and wavelengths, together with its extremely well-defined relation between signal and radiance, makes it profitable for various special research applications such as polarization pyrometry, laser absorption pyrometry[5] and photometric or optical property measurements.

ACKNOWLEDGEMENTS

The author wishes to thank Dr. H. Kunz from Physikalisch Technische Bundesanstalt Braunschweig for his extensive cooperation in design, development and tests of this instrument. The development was supported and the instrument is manufactured by the Institut fuer Kerntechnik und Energiewandlung e.V., Stuttgart.

Table 2 Results of test measurements at PTB[4]. ΔT is the temperature error compared to the PTB standard pyrometer a) on a black body; b) on a tungsten strip lamp.

T (K)	I (A)	R (Ω)	τ (63 %) (s)	δT (mK)	ΔT (K) a	ΔT (K) b
1100	0.8×10^{-12}	2×10^{12}	10	20	-0.01	0.08
1338	2.7×10^{-11}	2×10^{10}	3	3	±0.00 (Ref.)	0.10
1551	2.6×10^{-10}	2×10^{10}	3	3	—	-0.10 / -0.21
1700	0.9×10^{-9}	3×10^{8}	0.3	2	0.07	—
1900	3.5×10^{-9}	3×10^{8}	0.3	2	—	-0.25 / -0.40
2500	6.3×10^{-10}*	3×10^{8}	0.3	5	—	1.08

*) measured with neutral density filter ND2

I   Photocurrent
R   Load resistance
τ = RC, time constant of the current to voltage converter
δT Temperature resolution (obtained with external DVM)

REFERENCES

a) Heimann GmbH, D 6200 Wiesbaden, Germany

b) Teledyne Philbrick Inc. USA

1. Kunz, H. and Kaufmann, H.J.: Photoelectric direct current standard pyrometers and their calibration at PTB. Inst. Phys. Conf. Ser. No. 26 (1975), Chapter 5, pp. 244-255

2. Wörner, B.: Betrieb eines Wärmerohr-Schwarzen Körpers im Temperaturbereich zwischen 500 °C und 1100 °C. Inst. fuer Kernenergetik und Energiesysteme, University Stuttgart, Bericht Nr. 5-184 (1974)

3. Neuer, G., and O. Brost: Heat Pipes for the realisation of isothermal conditions at temperature reference sources. Inst. Phys. Conf. Ser. No. 26 (1975), Chapter 7, pp. 446-452

4. Kunz, H.: Transportables Linearpyrometer zur primären Darstellung der Temperaturskala. Physikalisch Technische Bundesanstalt, Germany, PTB Jahresbericht 1979 pp. 165-166

5. Kunz, H.: On the state of ratio pyrometry with laser absorption measurements. Inst. Phys. Conf. Ser. No.26 (1975), pp. 273-277

Ten years of high speed pyrometry at IMGC

F. Righini and A. Rosso

CNR Istituto di Metrologia "G. Colonnetti" strada delle Cacce, 73-10135 Torino, Italy

A research program on high speed pyrometry was started at IMGC in the early seventies, as a necessary complementary effort to the measurement of thermophysical properties by the pulse technique. Two high speed pyrometers were developed with some unique characteristics: a) monochromatic operation either in the visible or in the near infrared using silicon photodetectors; b) time resolution in the microsecond range for one of the instruments; c) good temperature resolution and calibration stability; d) wide temperature range starting from 800 K. The characteristics, performance, calibration, and accuracy of the instruments are reviewed by presenting the experience accumulated during a decade on various technological aspects (selection of components, use of interference filters, internal lamp stability, etc.). New instruments currently under development (a scanning pyrometer and an autoranging pyrometer) are briefly described.

INTRODUCTION

Dynamic techniques for the measurement of thermophysical properties at high temperatures have gained wide acceptance in recent years on account of their potential advantages and instrinsic simplicity. High speed pyrometry is an integral part of these methods and the development of accurate instruments with millisecond or microsecond time resolution is a necessary step for further advances. In the last decade an accurate high speed pyrometer was developed by Foley [1] for the apparatus of the National Bureau of Standards; other instruments were built for the systems at the Lawrence Livermore Laboratory [2] and at the Karlsruhe Joint Research Centre of the European Economic Community [3].

A research program on the measurement of thermophysical properties at high temperatures by a subsecond pulse heating technique was also started at the Istituto di Metrologia "G. Colonnetti" (IMGC) in 1970 using a method successfully developed by Cezairliyan at the National Bureau of Standards. The experimental method required high temperature measurements with submillisecond time resolution and the construction of a high speed pyrometer began as a necessary part of the development of the experimental apparatus. A system for fast high temperature measurements, including the first version of a millisecond time resolution pyrometer was described in 1972 [4]. Experience with the instrument dictated several fundamental changes: a new version of the high speed pyrometer, its performance and calibration procedures were presented a few years later [5]. The design criteria and the performance of an accurate pyrometer with microsecond time resolution were published in 1979 [6]. The present paper intends to review the development of high speed pyrometry at IMGC during the decade, by presenting both the performance of the instruments and the experience accumulated in several years of use and calibration. A final section will focus on new high speed pyrometers currently under development.

DESCRIPTION AND PERFORMANCE

The two mentioned pyrometers have been routinely used by the Thermophysical Property Group of IMGC in its research activity. In the following paragraphs, the instruments will be distinguished according to their time resolution: millisecond [5] or microsecond [6], respectively. Considerable experience has been gained in using both of them in measurements of thermophysical properties with millisecond time resolution [7, 8, 9]; in these cases an RC network was connected at the output of the microsecond time resolution pyrometer to reject high frequency noise. This millisecond mode of operation will be reviewed here; details of the performance with microsecond time resolution may be found in an earlier publication [6].

In high speed pyrometry important information is gained by some knowledge of the signal to be measured (shape, time behavior, etc...). The transformation of the time domain signal into a frequency domain spectrum by Fourier analysis permits the assessment of the uncertainties caused by a particular choice of the high frequency limit of the measuring pyrometer [6]. The application of this concept to typical subsecond pulse experiments reveals that a frequency cutoff at 1 kHz causes a relative uncertainty of 10^{-4}. In the millisecond mode of operation both pyrometers exibit a frequency cutoff slightly above 1 kHz and the uncertainties due to this high frequency limit are of the same order of magnitude of the intrinsic instrument noise.

The high speed pyrometers developed at IMGC are monochromatic instruments that may be operated either in the visible or in the near infrared. The detector is a commercial silicon photodiode (EG&G model SGD-444) operated in the photoconductive mode. Both instruments are based on the optical scheme shown in fig. 1; some geometric characteristics are reported in table I. The pyrometers operate in the DC mode, with the assumption that the drift of the detector and of the associated amplifiers is negligible during the short measurement period. The detector output is fed to an operational amplifier: a detailed electronic circuit analysis including detector-amplifier matching criteria is reported in an earlier publication [6]. Two computer-operated rotary solenoids activate two shutters, controlling the opening of the target or reference optical channels or the measurement of the detector dark current. The reference lamp is always kept at a constant current during the pyrometer operation. The pyrometer output before and after the pulse experiment (zero and reference measurements) is measured with a high resolution digital voltmeter with long integration times to smooth the noise.

The high speed pyrometers may be used at any wavelength within the responsivity band (400-1100 nm) of the detector by mounting an appropriate interference filter. Operation of the pyrometers at a particular wavelength presents advantages and disadvantages depending on the specific use of the instrument. In more general terms the advantages of operation in the visible (650 nm) are:
a) good detector linearity;
b) good temperature resolution;
c) known emissivity values.
The advantages of use in the near infrared (900-1000 nm) are:
a) good detector responsivity;
b) high radiation input;
c) good signal-to-noise ratio;
d) wide temperature ranges;
e) reference lamp calibration at the gold point.

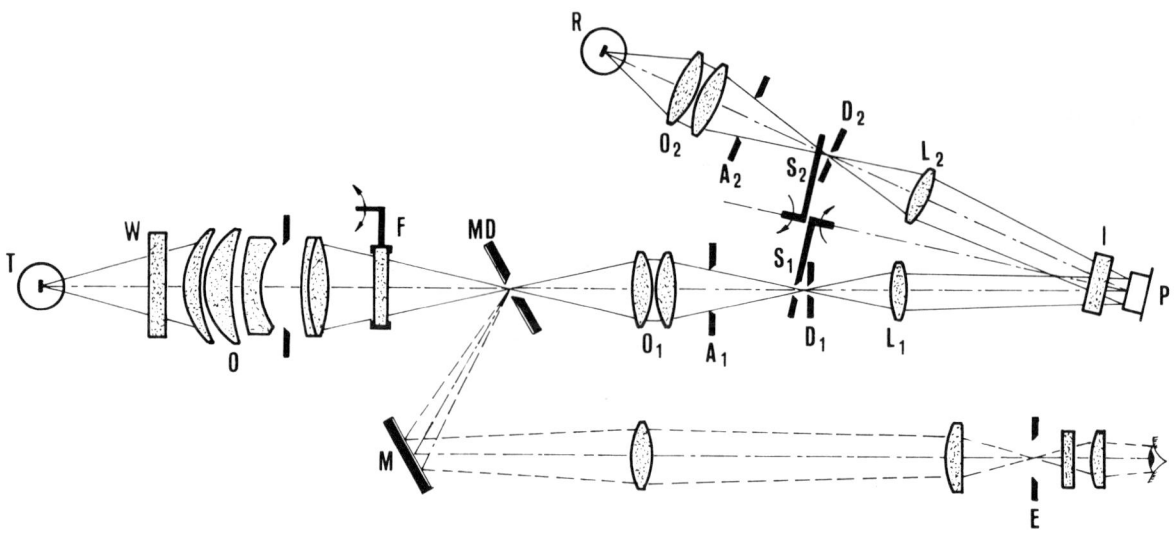

Fig. 1 Optical system of the high speed pyrometer. T, target; R, reference lamp; I, interference filter; P, photodetector; F, removable neutral-density filter; W, experimental chamber window inserted in the target optical path; O, O_1, O_2, objective lenses of the target and reference optical paths; A_1, A_2, aperture stops; D_1, D_2, field stops; S_1, S_2, shutters operated by rotary solenoids; L_1, L_2, field lenses, MD, mirror and diaphragm, M, mirror of the viewing channel; E, eyepiece of the viewing channel.

Table I. Geometric characteristics of the high speed pyrometers.

Characteristic	Millisecond pyrometer	Microsecond pyrometer
target area diameter	0.2 mm	0.8 mm
semiaperture angle	5.0°	5.3°
distance target-objective lens	230 mm	210 mm
detector area	1 cm^2	1 cm^2

Monochromatic operation is achieved with a wideband (50-80 nm) interference filter. Filters have been very carefully selected for their sharp cutoff and absence of side bands (relative transmission outside the pass-band less than 10^{-4}). Considering the difficulty of making accurate measurements of the low signal level outside the transmission band, an alternative method of filter selection has been implemented: the filters and their measured effective wavelength are used in accurate determinations of the gold-copper interval, using gold and copper point crucibles available in the laboratory /10/. Any side band (that changes the effective wavelength value) is immediately shown as a fairly large difference of this well known temperature interval. Incriminated filters can be examined more thoroughly to assess both the magnitude and the position of the side band. The millisecond pyrometer has been operated either at 900 nm or at 1000 nm; the microsecond pyrometer has been used at 550, 650, 750, 850, 900 and 1000 nm. The characteristics of the instruments at the most used wavelengths are reported in table II.

An important concept in high speed pyrometry is that the instrument cannot be considered by itself but always in connection with a data acquisition system. A typical combination is shown in fig. 2. Since the main use of these pyrometers is in automated systems with no manual intervention, the characteristics of both systems form a complementary package with consequences on the final result. For instance, the temperature resolution of the pyrometer depends on two main factors: instrument noise and resolution of the associated analog-to-digital converter. The converter to be used depends on the phenomen to be studied and on a trade-off between conversion speed and the value of the least significant bit. At IMGC all millisecond time resolution experiments are performed with a multiplexed data acquisition system (8 channels, 20 kHz) that uses a 14-bit analog-to-digital converter (least significant bit 0.61 mV). Microsecond time resolution measurements use another system (16 channels, 250 kHz) with a 12-bit converter (least significant bit 2.5 mV).

As indication of the capabilities of the overall system, typical radiance temperature plateaux of molybdenum and niobium at their melting point are reported in fig. 3. The molybdenum plateau has been measured with the millisecond pyrometer operating at 995 nm; the niobium plateau has been obtained with the microsecond pyrometer working at 658 nm.

The best temperature resolution is obtained in the upper part of the fundamental range which occurs for most wavelengths around 2000 K. There are no upper limits to the temperature range, except for the availability of the necessary neutral-density filter and the problem of its calibration. The low temperature limit of the instruments is defined by the acceptable signal-to-noise ratio. In most cases the fundamental temperature range starts from 1300 K, but measurements have also been performed from 800 K /9/, by inserting an external amplifier between the pyrometer and the data

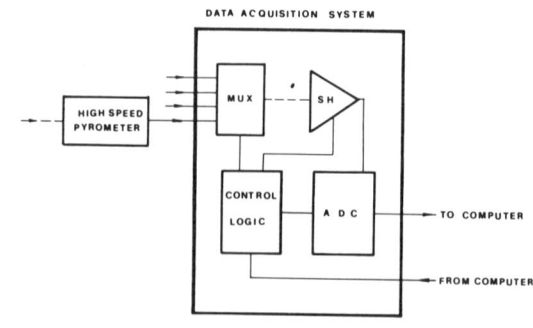

Fig. 2 Typical data acquisition system for measuring the high speed pyrometer output.

Table II. Wavelength dependent characteristics of the high speed pyrometers.

	Nominal center wavelength (nm)	Bandwidth (nm)	Reference lamp signal	Fundamental temperature range (K)
millisecond pyrometer	900	82.5	123 mV (at 1337.58 K)	1300-2100
	1000	50	62 mV (at 1337.58 K)	1300-2500
microsecond pyrometer	650	50	54 mV (at 1552.1 K)	1550-2450
	900	82.5	281 mV (at 1337.58 K)	1300-1900
	1000	50	242 mV (at 1337.58 K)	1300-2050

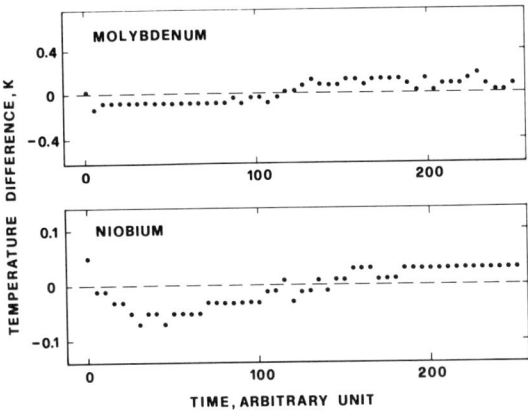

Fig. 3 Typical behavior of the high speed pyrometers during measurements of the radiance temperature of molybdenum (at 995 nm, millisecond pyrometer) and niobium (at 658 nm, microsecond pyrometer) at their melting points. The difference between the experimental point and the average plateau temperature, represented by the dotted zero line, is plotted. The radiance temperatures corresponding to the zero line are 2331.4 K for molybdenum and 2420.4 K for niobium. One time unit 0.15 ms.

acquisition system.

Further specific details on the technical characteristics of the pyrometers may be found in earlier publications /5, 6/.

CALIBRATIONS AND ACCURACY

A high speed pyrometer is considered a secondary instrument and the most straightforward method of calibration is a direct comparison against a calibrated lamp. This easy approach presents several hidden problems such as:
a) availability, cost and stability of the lamp calibrations;
b) effective wavelength of the lamp calibration (lamps are generally calibrated with a narrow-band pyrometer and their effective wavelength may change very little in the whole temperature range; high speed pyrometers generally use wideband interference filters);
c) differences in target areas, lamp orientation and alinement.

The main limitation is clearly the availability of calibrated lamps: in the comparison method the pyrometer use is restricted to the region around 650 nm on account of the practical unavailability of calibrated lamps at other wavelengths.

At IMGC the high speed pyrometers are calibrated as primary instruments by executing at each operating wavelength the following four steps:
1) effective wavelength calibration;
2) reference lamp calibration;
3) linearity calibration;
4) neutral-density filter calibration.

This approach is more time consuming than comparison against a lamp, but complete control of the calibration effort is mantained within the laboratory with the possibility of several crosschecks.

The pyrometers described here may be operated either at high speed (millisecond or microsecond time resolution) or at low speed (a few measurements per second). Accurate measurements have confirmed the equivalence of these two methods of operation. All calibrations are performed at low speed and the pyrometer output is measured with a high resolution digital voltmeter. Both the digital voltmeter and the high speed data acquisition system must be calibrated against a common source to assure compatibility between their readings.

Effective wavelength calibration

The effective wavelength and its temperature dependence are determined through an automatic procedure developed at IMGC /11/. A monochromator and the pyrometer are connected on-line with a computer. The spectral region of interest is automatically scanned at a fixed high temperature to obtain the system responsivity as a function of wavelength. The temperature dependence of the effective wavelength is computed according to the method described by Lee /12/. Such a temperature dependence is taken into account when computing temperatures from pyrometer readings. Experience accumulated through the years has indicated that the effective wavelength calibration is very stable. The largest cause of variation was the different angle of mounting when the interference filter was removed to use the pyrometer at another wavelength. This was corrected by placing each interference filter in a holder, which is precisely screwed in front of the detector. In this way filters can be exchanged between pyrometers and returned later to the instrument with an accurate position reproducibility. Repeated measurements of the effective wavelength with removal of the filter and realinement of the pyrometer indicated variations of the effective wavelength of less than 0.1 nm.

Reference lamp calibration

The reference lamp of the high speed pyrometer is calibrated at a fixed temperature and kept constant during the operation of the instrument. The current to the lamp (1-2 A) is furnished by a high stability power supply operated in constant current mode. The gold point (1337.58 K) is used as reference temperature whenever possible, but for some ranges and wavelengths the calibration point must be at other temperatures (see table II).

At 1337.58 K a direct calibration is performed by placing the high speed pyrometer in front of a graphite blackbody crucible surrounded by freezing gold. A small furnace (total length 400 mm) with a large aperture was especially designed for this purpose. At the same time of the goldpoint blackbody calibration (performed approximately once a year) this value is transferred to a high stability vacuum lamp (General Electric model 10/V). This lamp is used during experimental runs to check the pyrometer internal reference. This control is performed weekly during normal measurements and daily when very accurate data are needed.

The pyrometer internal reference is affected by a slow drift caused by a small mechanical movement of the filament, which brings slightly different areas in front of the detector. This displacement tends to sight the detector on filament areas away from the center and requires small increases of the lamp current to be corrected. Another factor of instability is the evaporation of the lamp tungsten filament that may condense on the output window. This effect is much smaller than the first one because the true temperature of the lamp filament is low.

A calibrated lamp is used when the reference point must be at a temperature different from 1337.58 K. This procedure requires the calibration of another pyrometer /13/ working at

Table III. Total uncertainty in temperature measurements under blackbody conditions.

Source of error	Total uncertainty (K)								
	Operation at 650 nm			Operation at 900 nm			Operation at 1000 nm		
	1500K	2000K	3000K	1000K	2000K	3000K	1000K	2000K	3000K
Reference lamp calibration	0.5	1.0	2.1	0.2	0.7	1.5	0.2	0.7	1.5
Effective wavelength calibration	0.1	0.4	2.1	0.1	0.6	2.1	0.1	0.5	1.8
Linearity and neutral-density filter calibration	0.1	0.2	1.2	0.1	0.3	1.7	0.1	0.3	1.9
Pyrometer calibration stability	1.0	1.5	2.5	1.0	1.0	2.0	1.0	1.0	2.0
Total uncertainty (root sum square of above items)	1.1	1.8	4.1	1.0	1.4	3.7	1.0	1.3	3.6

the same wavelength, the transfer of this calibration to a high stability strip lamp and the calibration of the reference lamp of the high speed pyrometer with the strip lamp. There are two main problems associated with this procedure: the first one is the different target areas of the two pyrometers. When the strip lamp filament is not uniform this requires a correction based on the thermal mapping of the filament and on the difference between the target areas /5, 6/. The second one is due to the inevitable difference in effective wavelength between the lamp calibration and the high speed pyrometer. Even if the same interference filter is used (which is advisable to minimize the effect) a small correction is often required.

Linearity calibration

Commercially available silicon detectors are rated for a nonlinearity of approximately 1-2 % for seven orders of magnitude of input radiation. High speed pyrometers are generally used on a 3-4 decade input range, but nevertheless need a linearity calibration. This is performed by relating the signal ratio to the radiation flux ratio, using the superposition method. Arbitrary radiation flux ratios are formed with two tungsten strip lamps and added through a summing cube. An automated computer procedure has been developed at IMGC /14/ to perform this calibration in the fastest possible way without a reduction in accuracy in comparison to manual procedures. Accurate linearity calibrations of photodetectors are routinely performed in a few hours and the final curve of radiation flux ratios versus signal ratios has an estimated relative uncertainty of 10^{-5} over the whole measurement region.

The results obtained at IMGC on the wavelength dependence of the linearity of photoconductive silicon detectors confirm the data of Jung /15/ on photovoltaic detectors: the linearity is very good in the visible (relative deviation less than 10^{-5}) and worsens toward the infrared. Around 900 nm the deviation from linearity is approximately 0.2 % and becomes 1.5-2 % around 1000 nm.

Neutral-density filter calibration

The amount of light reaching the detector above the maximum level allowed by the fundamental range must be reduced to avoid saturation of the amplifiers. This is done by inserting a neutral-density filter in the target optical path. The best results have been obtained by placing large neutral-density filters in front of the pyrometer between the objective lens and the target. This position has the additional advantage of reducing the stray radiation level. The neutral-density filter is rigidly fixed to a mechanical support with good position reproducibility and is mounted at a slight angle to avoid interreflections between optical surfaces.

Different neutral-density filters can be used to optimize the pyrometer operation or a filter can be selected to obtain the best pyrometer performance in a predetermined temperature range. The neutral-density filters are calibrated directly on the high speed pyrometer using a tungsten strip lamp at constant temperature in the pyrometer fundamental range. The ratio between measurements with and without the filter gives the transmission. The accuracy of filter calibrations depends on the accuracy of the linearity calibration which must be performed in advance.

Accuracy evaluation

Two main factors must be considered in the evaluation of the accuracy of high speed temperature measurements: instrument accuracy and experimental conditions. The former may be evaluated quantitatively by considering the uncertainties remaining after a careful instrument calibration. Sources and estimates of uncertainties in high speed pyrometry are given in detail in earlier publications /5, 6/. Specific items of this analysis have been recomputed in the present work

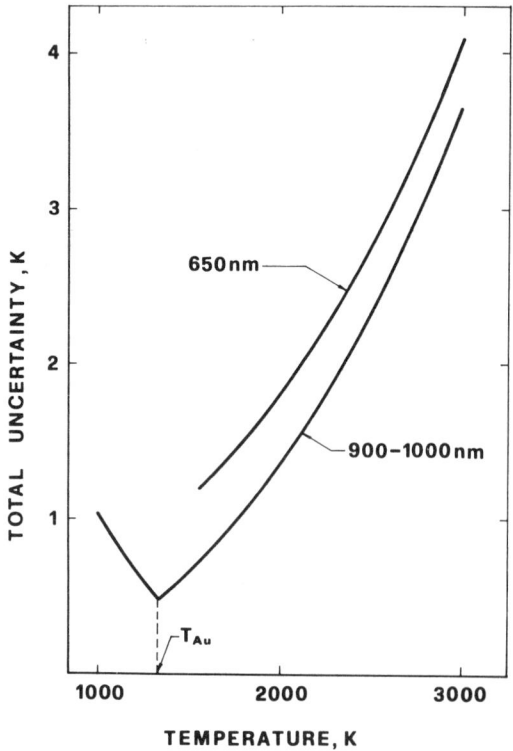

Fig. 4 Total uncertainty of high speed pyrometer measurements under blackbody conditions for three operating wavelengths.

because some of the earlier figures are obsolete on account of improved calibration methods /14/ or slight modifications of the instruments. The total uncertainty of pyrometer measurements under blackbody conditions is presented in table III and in fig. 4 for the most used operating wavelengths (650, 900 and 1000 nm).

The final accuracy of high speed temperature measurements will strongly depend on experimental conditions such as specimen temperature nonuniformity, surface or blackbody target, electromagnetic interference effects, coating of chamber window during the experiment, etc... Quantitative figures cannot be given here because their magnitude is strictly dependent on the operational characteristics of the system where the high speed pyrometer is used.

FUTURE DEVELOPMENTS

The present section refers to instruments presently under test to verify their final performance. It describes a scanning pyrometer for measurements of temperature profiles and an improved microsecond time resolution pyrometer with autorange capability. Both developments are connected to a research program for the measurement of transport properties by a dynamic method. In this new technique the specimen is self-heated from room temperature to the maximum temperature in times of the order of 30 s. The heat conduction losses toward the cooled clamps establish temperature profiles on the sample. The experimental technique requires the real time measurement of the evolving temperature profiles using the scanning capability of a high speed pyrometer with microsecond time resolution.

Scanning pyrometer

The microsecond time resolution instrument developed earlier /6/ has been transformed in a high speed scanning pyrometer according to the scheme of fig. 5. The present capabilities of the optical system permit measurements over a length of 60 mm with a total transmission loss (due to different path length and defocussing) of less than 1% at the ends of the specimen. The associated data acquisition system (12 bits, 250 kHz) must effectively freeze the temperature profile in time so that no significant temperature changes occur during the measurement. Assuming typical experimental conditions (temperature range 3000 K, total experiment duration 30 s), the average heating rate of the specimen is 100 K/s. During the time of the profile measurement (1 ms) the specimen temperature changes by 0.1 K, a negligible quantity under the experimental conditions. The profile is determined by 250 data points (1 measurement every 4 μs) with the pyrometer target area partially overlapping itself for redundancy.

Autorange pyrometer

A new high speed pyrometer has been designed and is presently being tested. Its main characteristics are:
a) microsecond time resolution and improved signal-to-noise ratio;
b) single channel pyrometer, with a thermostatted silicon photodiode replacing the reference channel;
c) autorange capability.

Significant technological improvements in electronic and electrooptic components are incorporated in this instrument. The high frequency cutoff is around 100 kHz with a noise reduction of one order of magnitude with respect to the previous version.

The concept of a single channel (comparator) pyrometer has been widely used in instruments built in standard laboratories. These slow instruments generally rely on an outside reference (a calibrated high stability lamp) and use the pyrometer optics and detector to compare the unknown with the reference. The new high speed pyrometer takes a further step in trying to substitute the calibrated reference lamp with a calibrated detector. This approach results in a major simplification of the instrument because the reference channel and its associated equipment may be eliminated. The validity of this solution depends on the minimization of thermal effects that change the responsivity of the photodetector and effect the drift of amplifiers. The detector-amplifier housing

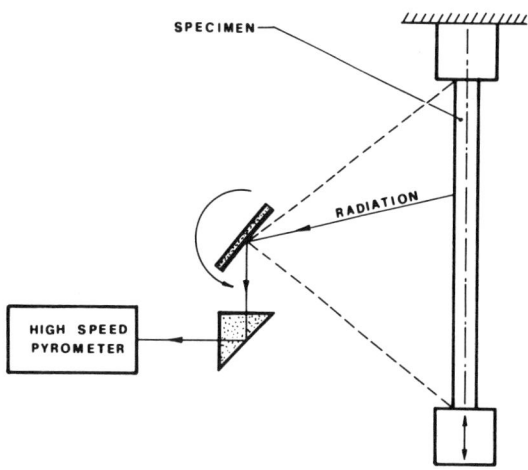

Fig. 5 Optical scheme of the high speed scanning pyrometer.

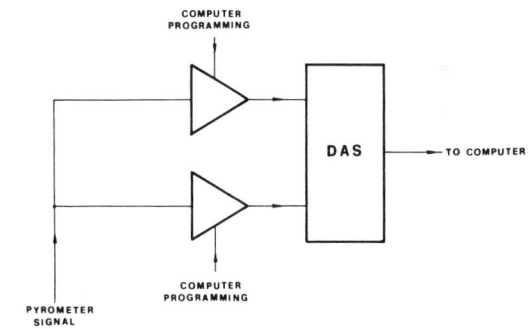

Fig. 6 Block diagram of the autoranging data acquisition system.

is thermostatted using Perltier elements. Tests are in progress to determine the stability and accuracy of this solution.

Dynamic experiments for the measurement of thermophysical properties cover wide temperature ranges (i.e. 1000-3000 K). These measurements have traditonally been performed by subdividing the temperature interval in several overlapping subranges, on account of limitations due to high speed pyrometers. The use of silicon detectors operating in the near infrared has partially improved this situation because the combination of the pyrometer fundamental range (1300-2000 K) and an extended range (1800-3500 K) with an appropriately chosen neutral-density filter practically covers any necessity. But this solution still requires the availability of a fairly large number of neutral-density filters to choose from, the calibration of the chosen filter and the performance of at least two experiments when only one might be sufficient. Since the radiation input is reduced, an additional problem is the degradation in pyrometer performance with increased inaccuracy (due to the neutral-density filter calibration) and poorer temperature resolution. The new instrument under development attempts to solve these problems by adopting an autoranging scheme in the associated data acquisition system, as shown in fig. 6. For millisecond time resolution measurements one programmable amplifier only might be sufficient; in the present case (microsecond time resolution) two programmable amplifiers are needed because their settling time during a range change (approximately 100 μs) would disturb the temperature-time sequence necessary for computing the heating rate. An additional advantage of the double amplifier scheme is that the analog-to-digital converter might always work in the high part of its range (i.e., from 5 V to 10 V) with improved resolution. The software programming of the amplifiers requires at the beginning a range selection and the decision whether increasing or decreasing signals are sampled; during continuous measurements the sampled data are compared to a predetermined level and the unused amplifier is programmed on the appropriate next range to be ready for data collection as soon as the other one bypasses the set level.

CONCLUSIONS

The instruments developed at IMGC during the past decade have demostrated the feasibility and the good performance of high speed pyrometers based on silicon detectors. High speed pyrometry has been developed originally as a support activity for another research effort. This means that not every aspect of this technique has been investigated, but mainly those areas of current need, where an instrument with specific characteristics was mandatory for further advances. The effort has been characteristic of a metrological laboratory concerned mainly with high quality instrumentation, with limited attention paid to performance at lower levels than those requested by a primary laboratory. The demands from potential application fields (diagnosis for laser-material interactions, studies of fast chemical reactions, measurement of thermophysical properties, etc...) are gradually changing this picture, with more attention being paid to technological advances that could be incorporated to simplify the instruments. The future research activity and the corresponding developments will probably proceed along two parallel areas:
a) high performance instruments for specialized applications (scanning pyrometry, extended autoranging starting from low temperature, etc...);
b) simplified instruments for technological applications where fast and accurate high temperature measurements are needed with noncontact methods.

ACKNOWLEDGEMENTS

Several people contributed their expertise to the high speed pyrometry efforts at different times during the decade. G. Ruffino introduced silicon detectors in radiation pyrometry, contributed to the first high speed pyrometer and provided expert advice and guidance during the early years. L. Coslovi developed important parts of the microsecond pyrometer and was responsible for its electronic design. A. Cibrario developed the data acquisition systems and the electronic part of the new instruments presently under development.

REFERENCES

/1/ G.M. Foley; Rev. Sci. Instrum. 41, 827 (1970).

/2/ G.R. Gathers, J.W. Shaner, and R.L. Brier; Rev. Sci. Instrum. 47, 471 (1976).

/3/ R.W. Oshe, J.F. Babelot, P.R. Kinsman, K.A. Long, and J. Magill; High Temp. High Press. 11, 225 (1979).

/4/ F. Righini, A. Rosso, and G. Ruffino; High Temp. High Press. 4, 597 (1972).

/5/ L. Coslovi, F. Righini, and A. Rosso; Alta Frequenza 44, 592 (1975).

/6/ L. Coslovi, F. Righini, and A. Rosso; J. Phys. E: Sci. Instrum. 12, 216 (1979).

/7/ F. Righini, A. Rosso, and L. Coslovi; in "Proceedings of the Seventh Symposium on Thermophysical Properties" A. Cezairliyan Ed. (A.S.M.E., New York, 1977) p. 358.

/8/ F. Righini, and A. Rosso; High Temp. High Press. 12, 335 (1980).

/9/ F. Righini, A. Rosso, and L. Coslovi; in "Proceedings of the Eighth Symposium on Thermophysical Properties" J.V. Sengers Ed. (A.S.M.E., New York, 1981).

/10/ F. Righini, A. Rosso, and G. Ruffino; High Temp. High Press. 4, 471 (1972).

/11/ F. Righini; La Termotecnica 25, 610 (1971).

/12/ R.D. Lee; Metrologia 2, 150 (1966).

/13/ G. Ruffino, F. Righini, and A. Rosso; in "Temperature. Its Measurement and Control in Science and Industry" H.H. Plumb Ed., Vol. 4, Part 1 (I.S.A., Pittsburg, 1972) p. 531.

/14/ L. Coslovi, and F. Righini; Appl. Opt. 19, 3200 (1980).

/15/ H. J. Jung; Metrologia 15, 173 (1979).

Microsecond and sub-microsecond multi-wavelength pyrometry for pulsed heating technique diagnostics

J. -F. Babelot, J. Magill, and R. W. Ohse

Commission of the European Communities Joint Research Center, Karlsruhe Establishment, European Institute for Transuranium Elements, Postfach 2266, D-7500 Karlsruhe, Federal Republic of Germany

M. Hoch

Department of Materials Sciences and Metallurgical Engineering, University of Cincinnati, Cincinnati, Ohio 45221

The extension of thermophysical property measurements of refractory materials far beyond their melting point up to temperatures of the order of 10 000 K required the development of new transient-type dynamic pulse heating techniques in the microsecond and submicrosecond time scale. Fast transition rates and large temperature gradients emphasized the construction of optical pyrometers of adequate temporal and spatial resolution. Unknown spectral emissivity data led to the development of high speed multi-wavelength pyrometers. A sub-microsecond multi-wavelength pyrometer, applying a diode matrix digitizer for data acquisition is described, allowing a time resolution of 10 ns and a spatial resolution of the order of 100 μm.

INTRODUCTION

The incentive for microsecond and sub-microsecond time resolution in photoelectric pyrometry, put forward within the last decade, is mainly given by the fast transition rates observed when extending thermophysical property measurements of refractory materials far beyond their melting point up to temperatures of 5000 K or even 10000 K. In safety risk assessment as an example, emphasized by nuclear reactor technology, equation of state studies of oxide fuel materials had to be extended to at least 5000 K leading to evaporation rates of the order of 1 m/s. These extreme conditions brought forward an entirely new range of experimental problems which required the development of new transient-type dynamic pulse heating techniques with adequate high speed diagnostics.

Each of the four main approaches of dynamic pulse heating taken by now, i.e. volume heating by rapid resistive heating (exploding wire technique), pulse electron and neutron beam heating and finally surface heating by flash and laser pulse heating, introduced their own specific problems and thus requirements on optical pyrometry in order to guarantee adequate reliability and accuracy in the property measurement[1,2].

Volume heating techniques are limited in case of resistive heating to conductive materials or to adequate beam penetration in electron and neutron heating. Measurements on electrically non-conductive materials are more difficult since the required surface heating by a radiant heat flux or laser beam produces large radial and axial temperature gradients. Because of the large temperature gradients over the heated area adequate spatial resolution is required in addition to the high temporal resolution due to the necessarily small time scale of the experiment at fast transition rates. The limit of applicability of optical pyrometry is determined by optical absorption of the emitted light in the gas jet which is partly ionized at these high temperatures. Finally, the lack of high temperature spectral emissivity data, and their possible change by surface composition changes in multi-component systems led to the development of high speed multi-wavelength pyrometers.

Apart from the requirements on response time of the detector special attention had to be paid to the recording electronics. Depending on the thermophysical property to be measured and the heat generating technique applied, temperature resolution is either required only for the final steady state temperature region or over the entire temperature range. In the first case, an advanced transient recorder, using a diode matrix or a scan converter tube, was found to present the most successful solution whereas in the second case fast logarithmic amplifiers or range multiplexed linear amplifier systems were applied.

The submicrosecond multiwavelength pyrometer described in this paper has been especially designed for high temperature resolution in the upper temperature region and high spatial resolution necessary because of the larger temperature gradients involved in transient laser heating.

PYROMETER REQUIREMENTS

The main requirements of optical pyrometry for equation of state studies of multicomponent systems such as nuclear fuel materials under transient laser pulse heating conditions up to 5000 K and higher are essentially:

- high temperature resolution of the upper steady state temperature and its fluctuations, of the order of 5 degrees at 5000 K,
- submicrosecond time resolution to study and resolve fast transient phenomena and measure transition rates,
- high spatial resolution to handle the extreme radial and axial temperature gradients occurring under transient laser pulse heating,
- simultaneous pyrometric measurements at different wavelengths to solve the problem of the variations of the spectral emissivity with wavelength,
- calibration on the basis of the International Practical Temperature Scale (IPTS) extended towards 5000 and 10000 K.

Temperature and time resolution

The various concepts of high-speed optical pyrometry for thermophysical properties studies can be subdivided into two main classes[3]:

- the first technique concentrates on high resolution of the final steady state temperature, and possible fluctuations of this temperature during the short pulse time
- the second technique has been designed to provide maximum resolution over the whole or distinct portions of the large temperature range.

For equation of state studies, mainly the first class of high-speed pyrometry is concerned. The pyrometer described in this paper has been especially designed to meet these requirements in both temperature and time resolution.

Spatial Resolution

Laser or in general beam heating necessarily involves large radial and axial temperature gradients given by the peak temperature, focal spot diameter and absorption length. A central temperature of 5000 K and focal spot diameter of 2 mm as an example leads to radial gradients of the order of 50 000 degrees per cm. Pyrometric temperature measurements require a finite surface area. In order to avoid large deviations of the measured average temperature from the required peak or central temperature, the measured area has to be kept small enough. If, in case of a Gaussian power profile, applying focal spot diameters of 2 mm, the radial power decrease should not exceed 2% of its peak power, the target diameter, selected for temperature measurement, is already restricted to 100 μm.

Simultaneous measurements at different wave-lengths

The problem arising now is the determination of the temperature from the light intensity signals recorded. Spectral emissivity data are usually required since the surface cannot be considered as a black or grey body. Such data, however, are usually known only in the lower temperature region.

a) Single wave-length pyrometry
The emissivity is assumed either to be constant with temperature, or has been measured previously, for example using an integrating sphere technique. For UO_2, the emissivity was determined at the melting point, using the plateau in the cooling curve, and then assumed to be constant in the liquid range. This value of 0.84 was confirmed by independent experiments[4]. Knowing the spectral emissivity, measurements at only one wave-length are sufficient to determine the temperature, and the two- or multi-wave-length signals are for verification.

b) Two-wave-length pyrometry
The self-consistent two-wave-length method of Lincoln and Pettit[5] was applied to $(U,Pu)O_2$. The method assumes a linear dependence of the emissivity with the temperature

c) Six-wave-length pyrometry
The newly constructed pyrometer described further in this paper permits the simultaneous measurement at six wave-lengths. The calculational procedure used to obtain the temperature is described in the next paragraph.

MULTI-WAVE-LENGTH OPTICAL PYROMETRY

Temperature measurements above 1000°C (the gold point) are carried out optically and are based on the Planck-Wien equation

$$L_b(\lambda,T) = C_1 \lambda^{-5} \exp(-C_2/\lambda T) \qquad (1)$$

where $L_b(\lambda,T)$ is the spectral radiance of a blackbody at temperature T and wave-length λ and C_1 and C_2 are the first and second radiation constants. If the radiator is not a blackbody then

$$E_\lambda L_b(\lambda,T) = L_b(\lambda,T_1) \qquad (2)$$

where E_λ is the spectral emissivity and T_1 is the spectral radiance temperature, i.e. the temperature of an equivalent blackbody having the same spectral radiance as the surface of emissivity E_λ. Combining equations (1) and (2) give

$$\frac{\lambda \ln E_\lambda}{C_2} = \frac{1}{T} - \frac{1}{T_1} \qquad (3)$$

Equation (3) is the basic working equation of optical pyrometry in that it connects the spectral emissivity with the spectral radiance temperature and the blackbody temperature. The variation of emissivity with wavelength can be expressed in the form

$$E_\lambda = a + b\lambda + c\lambda^2 + \ldots \qquad (4)$$

where a,b,c,... are functions of temperature.

Standard Pyrometers and Multi-Wave-length Pyrometry

Equation (3) is the operating equation of the single wave-length, mostly 0.65 μm, disappearing filament optical pyrometer. To obtain E_λ one measures T and T_1 or conversely to obtain T one requires T_1 and E_λ.

If one measures at n wave-lengths one obtains the n values $T_1, T_2, \ldots T_n$ together with n equations of type (3), which yield the n+1 unknowns T and $E_{\lambda 1}, E_{\lambda 2} \ldots E_{\lambda n}$. The measurements thus allow the determination of n-1 coefficients in equation (4). Thus in a two-colour pyrometer one constant in equation (4) can be determined (the other must be equal to zero) and in a three-colour pyrometer two constants can be determined, etc...

The commercial instruments take E_λ = a (indep. of wave-length) for a two-colour pyrometer and $E_\lambda = a+b\lambda$ for a three-colour instrument. This, however, is not necessary since in a two-colour pyrometer one could e.g. take $E_\lambda = c\lambda^2$ if the assumption satisfied the physical situation. From the above discussion on multi-wave-length pyrometry, from a determination of the values $T_1, T_2 \ldots T_n$ one can calculate T and the coefficients in equation (4). Commercial instruments do not do this.

Emissivity variations

At the temperatures concerned here there is not much known about the variation of E_λ with wave-length and temperature. In one set of experiments carried out at λ = 0.65 μm T_1 and T were measured and a relation of the form $T-T_1 = n+mT_1$ was indicated where n and m were constants. Inserting this into equation (3) gives

$$\frac{\lambda \ln E_\lambda}{C_2} = - \frac{n+mT_1}{\{T_1(1+m)+n\}T_1} \qquad (5)$$

Thus a fairly complex variation of E_λ with T is observed and indicates that a linear variation $E_\lambda = a+bT$ does not support the experimental results.

The variation of E_λ with T can also be observed through the total emissivity E_t. In grey bodies E_λ is assumed to be constant except where there is a cut-off band.

Experimental data[6-9] has shown that C_p/E_t is constant for metals and ceramics. In tungsten this was shown to be true over a temperature range of 2000 K (1000 to 3000 K) and this implies a complicated temperature variation of E_t since the expression for the specific heat generally takes the form

$$C_p = 3RD(\theta/T) + bT + dT^3 \qquad (6)$$

The method chosen here is to determine the wavelength dependency of E_λ at constant temperature. The first term a in eq.(4) can be eliminated by using the intensity ratio. The other coefficients however require a least square analysis depending on an adequate number of measurements. Empirically a six-wave-length measurement proved to be adequate.

Intensity ratio pyrometry

The application of equation (3) at six different wave-lengths allows us to construct 15 possible two-colour pyrometers. The two-colour arrangement allows the elimination of one constant in equation (4) and use of different wave-length combinations provides a check on the other coefficients since they should lead to the same value of T. First the measurements were carried out on a tungsten filament lamp (between 1500 and 2500 K) and then on liquid UO_2 up to 5000 K.

If the spectral emissivity $E\lambda_m$ at the various wave-lengths λ_m is not known, but only the spectral radiance $L(\lambda_m, T_n)$, as measured by the photodetector, an approach in two steps can be taken. The filtered radiation gives a photodetector signal proportional to the spectral radiance, which can be expressed in the form

$$J_\lambda = P_\lambda E_\lambda \lambda^{-5} e^{-C_2/\lambda T} \qquad (7)$$

where J_λ is the photocell current, P_λ the calibration factor, and E_λ the emissivity at wave-length λ. P_λ is determined by a calibration with a standard blackbody, and contains the first radiation constant C_1. E_λ can be expressed as a function of λ, according to (4). In any case, $b\lambda \ll a$, $c\lambda^2 \ll a$, ... (in two-colour pyrometry one even takes $b = c = ... = 0$), and (7) can be developed

$$\ln\left\{\frac{J_\lambda}{P_\lambda} \lambda^5\right\} = \ln a + \frac{b}{a}\lambda + \frac{c}{a}\lambda^2 + \ldots - \frac{C_2}{T\lambda} \qquad (8)$$

or

$$\lambda \cdot \ln\left\{\frac{J_\lambda}{P_\lambda} \cdot \lambda^5\right\} = -\frac{C_2}{T} + \lambda \ln a + \lambda^2 \frac{b}{a} + \lambda^3 \frac{c}{a} + \ldots \qquad (9)$$

The first step of the mathematical procedure is then to plot the term $\lambda \cdot \ln\left\{\frac{J_\lambda}{P_\lambda} \lambda^5\right\}$ versus λ.

An examination of the plot indicates the significant coefficients, and the wave-length range in which the emissivity equation is valid. The final temperature evaluation is then restricted to this range. For tungsten, as shown in Fig.1, a parabolic variation, i.e. linear variation for the emissivity (coefficients a and b) was found to be sufficient between 480 and 850 nm. A measurement at 400 nm was not considered for the temperature determination.

In fact one can always find a wave-length range where the emissivity variation is sufficiently linear. Then only two coefficients, a and b, are needed. Applying (8) at two wave-lengths λ_1 and λ_2, and substracting, leads to

$$\ln\left\{\frac{J_{\lambda_1}}{P_{\lambda_1}}\lambda_1^5\right\} - \ln\left\{\frac{J_{\lambda_2}}{P_{\lambda_2}}\lambda_2^5\right\} = \frac{b}{a}(\lambda_1 - \lambda_2) - \frac{C_2}{T}\left\{\frac{1}{\lambda_1} - \frac{1}{\lambda_2}\right\} \qquad (10)$$

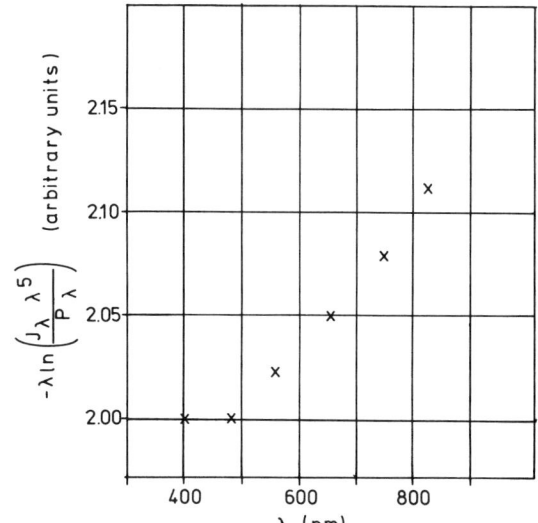

Fig.1: Photodetector signal J_λ measured on tungsten at six wavelengths, plotted according to Eq.(9).

or

$$\frac{1}{\lambda_1 - \lambda_2} \ln \frac{J_{\lambda_1} \lambda_1^5/P_{\lambda_1}}{J_{\lambda_2} \lambda_2^5/P_{\lambda_2}} = \frac{b}{a} + \frac{C_2}{T\lambda_1\lambda_2} \qquad (11)$$

The second step, according to (11), is to plot the left-hand side against $1/\lambda_1\lambda_2$, for all possible combinations of two wave-lengths. A least square analysis now yields b/a and C_2/T, i.e. permits the determination of the temperature T.

Number of wave-lengths

Ideally, measurements at as many wave-lengths as possible would be desirable, but for practical reasons the number of wave-lengths should be kept to a minimum. For the least square plot of Eq.(8), measurements at 3 wave-lengths yield 3 different pairs, which are not enough, whereas measurements at 4 wave-lengths lead to six pairs, giving 6 points, which is reasonable. Plotting Eq.(9), measurements at 4 wave-lengths give only 4 points. However 5 points at least are necessary to check the parabolic variation of the left-hand side of Eq. (9). Assuming an irregular point in addition, measurements at six wave-lengths are necessary.

EXPERIMENTAL REALIZATION

Optical concept

The role of the optical part of the pyrometer is to conduct the radiation emitted by the surface to be measured, after splitting it into different wave-lengths, onto the detectors. A schematic display of the experimental set-up is shown in Fig.2.

a) Imaging of the surface

In order to measure the temperature of a surface, an image of this surface is formed on a micro-aperture at the entrance of the pyrometer. The micro-aperture diameter and the magnification of the imaging system determine the spatial resolution of the device.

For measurements on UO_2 and $(U,Pu)O_2$, Ohse et al.[10-12] used an aperture of 400 µm with a magnification of 2 to 2.5, i.e. a spatial resolution of 160 to 200 µm. The imaging of the surface onto the micro-

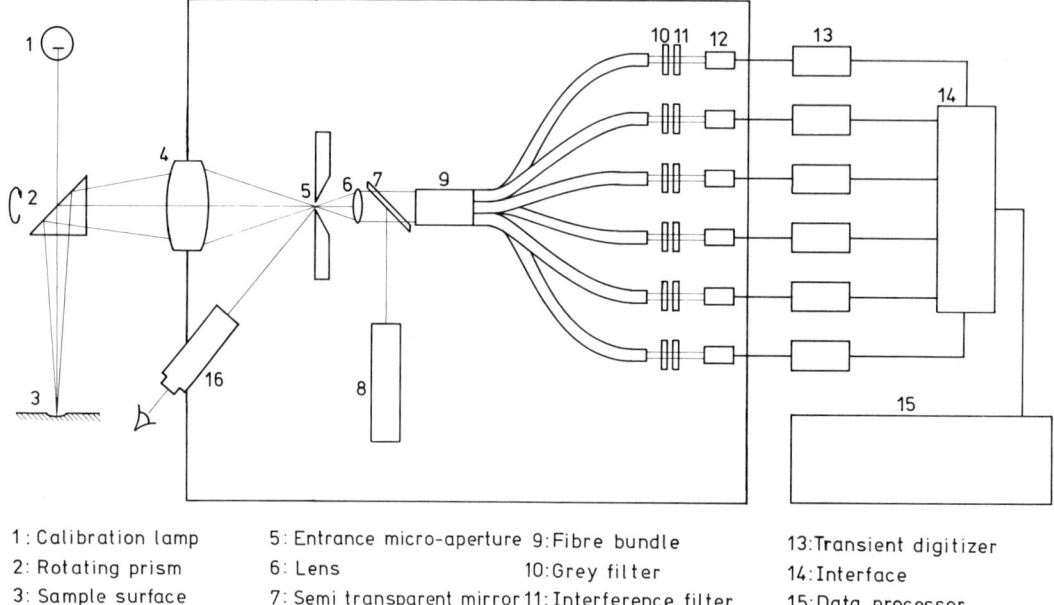

1: Calibration lamp   5: Entrance micro-aperture 9: Fibre bundle   13: Transient digitizer
2: Rotating prism   6: Lens   10: Grey filter   14: Interface
3: Sample surface   7: Semi transparent mirror 11: Interference filter 15: Data processor
4: Objective   8: HeNe laser   12: Photodiode   16: Telescope

Fig.2: Experimental set-up for a high temporal and spatial resolution six wavelength pyrometry.

aperture was done by an objective. The same arrangement is used for the new pyrometer described here (Fig.2). The calibration is done under the same experimental conditions as during the measurements, simply replacing the surface by the tungsten ribbon of a calibrated lamp.

b) Wave-length selection

In order to form a parallel beam, the distance of the microaperture to the lens is adjusted to its focal length. The focal length of the lens and the aperture of the optical imaging system determine the diameter of the beam, its divergence is given by the focal length and the micro-aperture diameter. The characteristics of all the components are chosen such that the beam diameter, when reaching the detectors, covers a significant part of the photosensitive surface. The beam is split before reaching the detectors, inserting an interference filter into each light path to select the wave-length. Ohse et al.[11,12] used two channels, separated by half-mirrors with a set of four interchangeable filters on each path having a half-width of 10 nm and a peak transmission wave-length determined to $\pm$ 2 nm. Grey filters were also foreseen for extending the range of temperature measurement, to ensure the photomultipliers operated on the linear part of their characteristic. The transmissivity of the grey filters is determined during the calibration procedure. For a six-wave-length pyrometer as discussed here, the fiber optics is of advantage, because the beam splitting into more than four channels with the help of half-mirrors becomes too laboursome. The parallel beam enters the fiber bundle, randomly split into six paths. One additional path is foreseen for the He-Ne laser, used for alignement.

c) Adjustment of the pyrometer on the target

In the case of pulse heating, the area of constant temperature is very small, and in addition to good spatial resolution, very good adjustment is required. The parallel beam of the He-Ne alignment laser is focussed through the lens of the pyrometer (Fig.2) onto the micro-aperture. By imaging through the objective, a luminous spot is obtained on the surface, indicating the measurement area. A second method for focussing the image of the surface on the micro-aperture uses a viewing telescope by which the centering of the image of the area to be measured on the micro-aperture is checked. When applying laser pulse heating, the viewing telescope permits direct checking of the centering of the pyrometer during the shot.

Detectors

For microsecond or sub-microsecond pyrometry, the most important characteristic of the detector is its rise time. For the visible range, there are photomultipliers and photodiodes with nanosecond rise time. But the rise time of the pyrometer depends of course in addition on the amplification circuit and on the data acquisition system.

a) Photomultipliers

Typically, the maximum of the spectral sensitivity curve of a photomultiplier is below 600nm; the sensitivity curve can however be extended in the infrared region. Ohse et al.[11,12] used EMI photomultipliers (type 9558 QB), maintained at 24 $\pm$ 0.1°C by a thermostat. The rise time was 10 ns. The linearity range was of two orders of magnitude, and thus grey filters of 1/5, 1/30, and 1/100 transmissivity were applied.

b) Photodiodes

In the new pyrometer, Si-photodiodes are used, with a rise time of 5 ns. Compared to the previously employed photomultipliers, their spectral sensitivity curve is shifted towards higher wave-lengths, i.e. in the direction of the maximum blackbody emission at the temperatures of interest. An other advantage is that the linearity range of a Si-photodiode is 6 orders of magnitude, and if any, one grey filter should be enough to measure temperatures in the range 2000-7000 K. At these temperatures, the loss in sensitivity compared to photomultipliers is not so important, because of the high blackbody emission. Table I summarizes some typical figures of different

Table I: Comparison of the properties of various types of detectors.

Detector	Spectral range (µm)	Rise time (ns)	Responsivity (A/W)	Dark current (nA)	Noise equival. power(W/√Hz)
vacuum photodiode photocathode $Cs_3Sb(S4)$	0.3 -0.6	0.2	0.035 (at 0.437µ)	0.5	
vacuum photodiode photocathode $Na_2KC_sSb(S20)$	0.3 -0.8	0.4	0.01 (at 0.698µ)	0.2	
photomultiplier (multialkali, side-on type)	0.16-0.95	1.2	$6.4 \cdot 10^5$ (at 0.45µ)	2	$1.4 \cdot 10^{-16}$
photomultiplier photocathode AgOCs(S1)	0.3 -1.1	1.2	360 (at 0.8µ)	50	$3.1 \cdot 10^{-13}$
photomultiplier photocathode $Na_2KC_sSb(S20)$ extended red	0.32-0.92	12	$1.2 \cdot 10^4$ (at 0.63µ)	10	
Germanium photodiode	0.5 -1.8	0.3	0.97 (at 1.5µ)		10^{-12}
Germanium Photodiode (ultra fast)	0.5 -1.8	< 0.08	0.15 (at 1.5µ)		10^{-10}
Silicon photodiode (ultra-fast)	0.3 -1.1	< 0.035	0.2 (at 0.72µ)		10^{-10}
Silicon photodiode	0.4 -1.1	1	0.5 (at 0.8µ)	2.5	$5.7 \cdot 10^{-14}$
Silicon photodiode (avalanche type)	0.4 -1.1	2	75 (at 0.9µ)	100	$1.5 \cdot 10^{-14}$

types of detectors. The choice of the detector depends strongly on the special application, and the user should refer to a "Buyers'Guide" for a complete review and description.

Data acquisition

After amplification, the output signals of the photodetectors are recorded on the diode matrix of transient digitizers, applying one digitizer per channel. The data are then stored in a digital memory for further processing in a computer. The diode matrix has 512x512 elements. The signal is stored with 512 values in a predefined array in the computer. For a 250 µs-pulse, this limits the time resolution to 0.5 µs.

The theoretical temperature resolution was calculated to be 5 K. The application of dual digitizers, using scan converter tubes instead of diode matrix, allows an exact comparison of two signals and larger flexibility with regard to the number of values per signal and their distribution over the pulse time.

THE INTERPRETATION OF PYROMETRIC SIGNALS

Laser Surface Heating Experiments

In order to produce steady state surface temperatures from 3000 K - 8000 K using laser radiation, power densities 10^4 to 10^9 Wcm^{-2} are required[13]. Under vacuum conditions, energy is consumed primarily in evaporation of the irradiated surface. A well-defined correlation of the thermophysical and thermodynamic properties as a function of temperature, for example the evaporation rate necessary for EOS studies of fuel materials, can only be obtained by well-defined spatial and temporal intensities profiles in the focal spot of the laser beam. This can be achieved by operating the laser in the single transverse mode and producing a laser intensity profile which is Gaussian in space and square wave in time. Although these conditions are necessary, they are not sufficient to guarantee a correct measurement of the surface temperature. The large spatial and temporal temperature gradients produced within the heated target area imply that the pyrometer receives light from a finite volume and over a finite time and that the measured temperature is not necessary equal to the "time" surface temperature but some average value over space and time.

a) Temporal temperature gradients

An important characteristic time-scale in laser surface heating experiments is given by $t_1 = k/v_{rec}^2$ where k is the thermal diffusivity of the condensed phase and v_{rec} is the surface recession speed. This is the time during which an element of volume undergoes heating before it is evaporated. This heating time is to be compared with the time required to establish equilibrium within the surface layer. The interaction of the laser radiation with the condensed phase and the energy transfer process can be written in the following form

```
                    ~10⁻¹³ s  PLASMONS
        <10⁻¹⁴ s              ⇗⇘                ~10⁻⁷ s  VAPOUR
LASER ─────────→ HOT ─────────────────→ PHONONS ⇌         ATOMS
PHOTONS           CARRIERS ⇌                              
                         ~10⁻¹¹ s
                         ~10⁻¹² s  IMPACT
                                   IONISATION
```

Thus the incident laser radiation is absorbed by electrons in the condensed phase. The energy is then passed onto the lattice via electron-phonon collisions. The bulk phonons then transmit their energy to the surface atoms giving rise to evaporation. The electron-electron and phonon-phonon collision times are usually smaller than the electron-phonon collision time such that one can consider the electrons and phonons as separate statistical subsystems. In equilibrium each of the subsystems electrons, phonons etc. will have the same temperature. In transient heating experiments each of the subsystems can still have a well-defined temperature provided study state has been reached. However the temperature of the different subsystems will in general not be equal and care must be exercised in the interpretation of the pyrometer signal.
A specific example of the above phenomena is to be found in the laser annealing literature where two theories of annealing have been put forward, one based on melting, the other on the function of a non-equilibrium electron-hole plasma[14,15]. Using this latter theory, experiments on Si are interpreted in the following way: Irradiation of silicon

surfaces using greater than band gap laser radiation gives rise to a high density of electron hole pairs. The electrons so produced have high energy and penetrate deep into the material before giving up their kinetic energy to thermal energy of the lattice (i.e. the electron-phonon-collision time is high). In addition because the penetration depth of the electrons is large, the increase in lattice temperature can be quite small. In such a situation the radiation recorded by the pyrometer is characteristic of the high density electron hole plasma rather than the lattice. Another example can be found in the laser evaporation experiments on UO_2 for vapour pressure determination. Calculations indicate that due to the high recession speed of the evaporating surface (~ 1 ms^{-1} for UO_2 at 5000 K) there may be no time for the solid-liquid transition, and that evaporation could be taking place directly from a superheated solid. The radiation recorder by the pyrometer would then be characteristic of a superheated solid rather than an equilibrium liquid surface.

b) Radial and axial temperature gradients

In laser surface heating experiments, the radial temperature gradient across the heated target area is caused by the spatial power profile in the focal spot of the laser beam. For vapour pressure measurements on nuclear fuel materials, as an example, the rate of evaporation is calculated from the central depth of the crater, the temperature and the exposure time. The central temperature T_c has to

Table II: Decrease in surface temperature over area viewed by pyrometer+

T_c(K)	3000	3500	4000	4500	5000
ΔT(K)	5	8	10	13	16

+Laser beam half-width = 1/2 mm, area viewed by pyrometer = 160 μm, material UO_2

be calculated from the measured average temperature over a small surface area. High spatial resolution is required to keep the temperature decrease T over this area sufficiently small. Table II shows the results for typical parameters used previously.[16] Thus it can be seen, that using the above parameters, the correction due to radial temperature gradients may be neglected even at 5000 K. It is obvious that the error increases with decreasing focal spot diameter.

Axial temperature gradients in the thin heated layer near the surface are a consequence of the high evaporation rates. In steady state, a temperature gradient is set up such that the heat conducted from the bulk to the surface provides the energy required for evaporation. Assuming Fourier's law of heat conduction one can solve the one dimensional heat diffusion equation with a distributed source term in a reference frame moving with the evaporating surface.

The steady state solution subject to the appropriate boundary conditions may be shown to be[17,18]

$$T(s) = \left\{T_o + \frac{I}{\rho v C_p(\beta-1)}\right\} \exp\left(-\frac{vs}{a}\right) - \frac{I}{\rho v C_p(\beta-1)} \exp(-bs) \quad (10)$$

where T(s) is the temperature at a distance s from the moving surface, a is the thermal diffusivity and $\beta = ab/v$. The resulting temperature profile is shown in Fig. 3 for various values of the absorption coefficient b.[18]

The surface cooling effect caused by front surface evaporation is seen clearly. The positive temperature gradient into the material implies heat is being diffused from the bulk to the surface to support evaporation. It is also clear that temperature gradients into the material introduce an error in the surface temperature measurement. The intensity of light emitted from the surface contains contributions from the higher temperature material behind the surface. Solution of the radiation transfer equation for this temperature error yields the results shown in Table III[13].

Thus for the parameters used in the calculation it follows that the effects of axial temperature gradients may be neglected up to 5000 K. As the surface temperature increases, however, the evaporation rate increases exponentially and the temperature gradients near the surface become very high. When the gradients occur over a few mean free paths of the electrons or phonons one can no longer use Fourier's law of heat conduction which contains a relationship between the heat flux and the local temperature gradient. Instead one must develop a non-local relationship between the heat flux and the temperature gradient by solving the Boltzman equation.

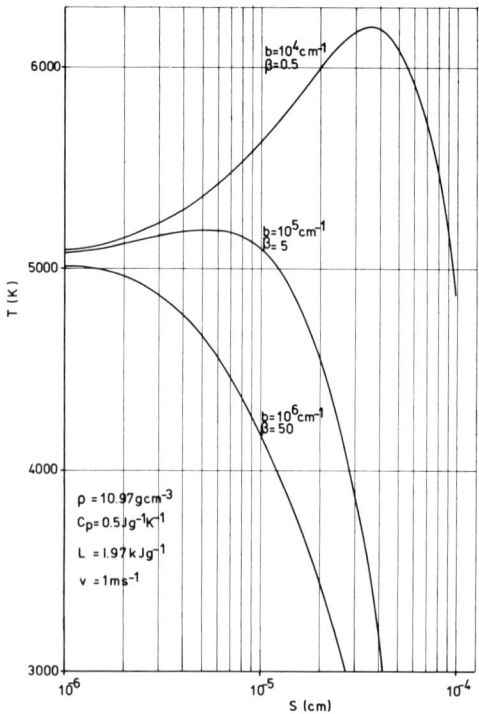

Fig. 3. Temperature profile dependence on the absorption coefficient during steady state laser evaporation of UO_2.

Table III: Temperature error at 5000 K for various absorption coefficients+

b (cm^{-1})	5x10^3	1x10^4	5x10^4	1x10^5	5x10^5	1x10^6
$\frac{\Delta T}{T}$ (%)	− 13	− 8	− 0.8	− 0.1	+ 0.2	+ 0.1

+wavelength of pyrometer radiation considered 602 nm

c) Absorption of light in the gas jet

The degree of thermal ionisation in the laser induced gas jet may be calculated using the Saha equation which provides a relationship between the electron density and the density and temperature of the gas. The absorption of the laser radiation or the radiation coming from the surface in this gas jet as a function of temperature and density is given by[13]

$$b(cm^{-1}) \simeq \frac{10^{-7} n_T Z^2 \exp{(L_v - I_{eff})/kT}}{T^2 (L_v/kT)^3} \quad (11)$$

Table IV: Degree of ionization, and photon mean free path for 1.06 μm photons as a function of temperatures for UO_2

T_s(K)	T_g(K)	n_s(cm^{-3})	n_g(cm^{-3})	ΔI(eV)	I_{EFF}(eV)	b^{-1}(cm)	α(%)
7500	5000	9.5×10^{21}	3.2×10^{21}	1.4	4.1	1.4×10^{-3}	0.6
7000	4667	5.3×10^{21}	1.4×10^{21}	1.1	4.4	1.6×10^{-2}	0.4
6000	4000	1.3×10^{21}	4.5×10^{20}	0.76	4.7	6.0×10^{-1}	0.15
5000	3333	1.5×10^{20}	0.5×10^{20}	0.36	5.1	1.4×10^{2}	0.05
4000	2667	8.8×10^{18}	2.9×10^{18}	0.14	5.36	2.7×10^{5}	0.001

where T_s = temperature at the surface
T_g = temperature of the gas
n_s = saturated vapour density
ΔI = reduction of the ionization potential
b = absorption coefficient for light with wavelength 1.06 μm
α = degree of ionization

where Z is the ionic change and I_{eff} the effective ionisation potential. The degree of ionisation and photon mean free path (=b^{-1}) for 1.06 μm photons for UO_2 vapor at different temperatures are shown in Table IV.

If we impose the condition that no more than 5% of the radiation should be absorbed in the gas jet and assume that the region over which absorption takes place is approximately equal to focal spot diameter = 1 mm, then a value for the absorption coefficient of b = 0.5 cm^{-1} is obtained. From Table IV this yields a limiting surface temperature for UO_2 of around 6000 K. These calculations have assumed photoionisation and inverse Bremsstrahlung are the main processes of absorption in the gas jet. Bound-bound transitions, however, can reduce this limiting temperature further.

Specific Problems in Resistive Heating Experiments

In the exploding wire technique a capacitor bank is discharged through a rod shaped sample and evaporation is prevented by a high pressure background gas such as argon. By measuring the energy input, volume expansion and the surface temperature as a function of time one can obtain the equation of state of the material concerned. For experiments on liquid metals above 3000 K sample heating must occur on a microsecond timescale to avoid collapse of the liquid metal column. The pyrometric temperature measurement is complicated by anomalous surface heating caused by the skin effect, unknown emissivities, temperature and opacity effects of the boundary layer between the background gas and the metal surface.

Temperature Gradient produced by the Skin Effect

At high frequencies currents are limited initially to a region close to the surface and surface temperatures much higher than the bulk value are obtained. Since the resistivity of liquid metals increases with temperature the surface becomes less conducting in comparison to the bulk and after a certain time the temperature profile throughout the wire cross section is almost uniform. Calculations on tantalum[19] show that after 12-14 μs into the current pulse the surface temperature has risen to 2500 K and the maximum temperature difference throughout the wire was about 80 K. After melting had occured (~17 μs) the temperature difference was down to less than 30 K and the skin effect could be neglected. For materials with lower resistivity or using higher heating rates this effect may be important.

Opacity Effects

Evaporation of the metal surface in the exploding wire technique is avoided by the high pressure background gas. Nevertheless in a very narrow region near the surface there will be a mixture of hot argon and metal atoms and radiation from this region may effect the surface temperature measurement.

CONCLUSION

The requirements of dynamic pulse heating techniques and recent developments in fast electronics, and fiber optic systems, has led to the construction of a new generation of pyrometers capable of measuring the quantity of light emitted by a surface, simultaneously at different wave lengths, with high spatial and temporal resolution. The multiwavelength measurements are in general used to overcome the problem of unknown emissivity and transmissivity. Two-colour pyrometry is applied to grey bodies, the temperature is determined from the ratio of the light intensity at two different wavelengths. Measurements at six wavelengths, as applied here, are necessary to determine the emissivity dependence on wavelength at constant temperature. The simultaneous measurement of these six wavelengths was found necessary because of the non-reproducibility in temperature, due to slight variations in the sample surface and laser pulse energy. An investigation of the relaxation times in the condensed phase (electron-electron, electron-phonon etc.) reveals that for experiments carried out on the microsecond timescale the different subsystems electrons, phonons are characterised by the same temperature and thus the temperature of the surface layer is well defined. Errors introduced into the temperature measurement due to the large temperature gradients produced in laser evaporation experiments have been shown to be negligible for UO_2 up to around 5000 K. Beyond this temperature the pyrometer no longer receives light from the surface but from the ionised gas or plasma jet.

REFERENCES

1. R.W. Ohse, J.-F. Babelot, C. Cercignani, P.R. Kinsman, K.A. Long, J. Magill and A. Scotti, J. Nucl. Mater. 80 (1979) 232.
2. R.W. Ohse, J.-F. Babelot, A. Frezzotti, K.A. Long, J. Magill, C. Cercignani and A. Scotti, High Temp. Sci. 13 (1980) 35.
3. A. Cezairliyan, J.-F. Babelot, J. Magill and R.W. Ohse, Dynamic temperature measurements by optical pyrometry, in Theory and Practice of Radiation Thermometry, Ed. D.P. DeWitt and G.D. Nutter, to be published.
4. M. Bober, and H.U. Karow, Measurements of Spectral Emissivity of UO_2 above the Melting Point, Proc. 7th Symp. on Thermophysical Properties (NBS Gaithersburg, 1977), New York (1977), pp. 344-350.
5. R.C. Lincoln, R.B. Pettit, High Temp.-High-Press., 5 (1973) p. 421.

6. H.V.L. Narasimhamurty, A.S. Iyer and M. Hoch, J. Phys. Chem., 69 (1965) 1420.
7. M. Hoch, A. Silberstein and H. Chapman, Thermal Conductivity of Aluminum Oxide, in Proceedings of Fourth Symposium on Thermophysical Properties, The American Society of Mechanical Engineers, New York, 1968, p. 150.
8. C.K. Jun and M. Hoch, Thermal Conductivity and Total Emittance of Tantalum, Tungsten, Rhenium, Ta-10%, T_{111}, T_{222} and W-25Re in the Temperature Range 1500-2800 K, in High Temperature Technology, Butterworths, London, 1969, p. 535.
9. M. Hoch, High-Temperature Thermophysical Properties of Tungsten Molybdenum, Niobium and Tantalum, in Thermodynamics of Nuclear Materials 1974 (Proc. Symp. Vienna, 1974) Vol. II, IAEA, Vienna (1975) 113.
10. R.W. Ohse, P.G. Berrie, H.G. Bogensberger, E.A. Fischer, in Thermodynamics of Nuclear Materials 1974 (Proc. Symp. Vienna, 1974) Vol. I, IAEA, Vienna (1975) 307.
11. R.W. Ohse, P.G. Berrie, G.D. Brumme and P.R. Kinsman, Advances in Vapour Pressure Studies over Liquid Uranium Plutonium Oxides up to 5000 K in Plutonium and Other Actinides, Eds. H. Blank and R. Lindner, p. 19, North-Holland Publishing Company, Amsterdam, 1976.
12. R.W. Ohse, J.-F. Babelot, G.D. Brumme and P.R. Kinsman, Rev. Int. Hautes Temp. Refract. Fr. 15 (1978) 319.
13. Ya. B. Zel'dovich and Yu. P. Raizer, Physics of Shock Waves and High Temperature Phenomenon (W.D. Hayes, R.F. Probestein, Eds.) Vol. 1, Academic Press, New York 1967.
14. J.A. Van Vechten, R. Tsu and F.W. Saris, Phys. Lett. 74A, (1979) 422-426
15. J.A. Van Vechten, R. Tsu, F.W. Saris and D. Hoonhout, Phys. Lett. 74A, (1979) 417-421
16. R.W. Ohse, J.-F. Babelot, P.R. Kinsman, K.A. Long, J. Magill, High Temp.-High. Press., 11 (1979) 225.
17. F.W. Dabby and U.C. Peak, IEEE J. Quantum Electron 8 (1972) 106-111.
18. J. Magill, C. Ronchi, J.-F. Babelot, K.A. Long and R.W. Ohse, High Temp.-High.Press., 12, (1980) 503-514.
19 G.R. Gathers, J.W. Shaner and R.L. Brier, Rev. Sci. Instrum. 47, (1976) 471.

Two-color microsecond pyrometer for 2000 to 6000 K[a]

G. M. Foley,[b] M. S. Morse, and A. Cezairliyan

Thermophysics Division, National Bureau of Standards, Washington, D.C. 20234

An accurate pyrometer has been developed for measurements on solid and liquid specimens in the range 2000 to 6000 K. The pyrometer measures spectral radiance temperature near 0.65 and 0.9 μm. Fiber optic cables transmit the radiation to silicon diode detectors. In each channel, a set of three linear amplifiers with high-speed automatic gain switching give high resolution and wide temperature range. Signals are digitally recorded with 0.1% resolution at 1.5 μs intervals. Linearity of measurements of radiance has been confirmed using a calibrated tungsten strip lamp and the sun. Characteristics and performance of the pyrometer are discussed and results on pulse heating of metallic specimens beyond the melting point are presented.

1. INTRODUCTION

The properties of liquid metals and other electrically conducting substances in the temperature range 2000 to 10,000 K have become of increasing importance in recent years. Because of the severe problems of steady-state methods at these extreme temperatures, dynamic techniques are required to make the measurements. A review of pyrometric methods, both photoelectric and photographic, used for dynamic measurements of high temperatures before 1970 was published in 1972[1]. Almost all of this work was of a preliminary nature. During the following decade, accurate millisecond-resolution photoelectric methods for pyrometry were developed[2,3] in connection with accurate measurements of thermophysical properties of solids at temperatures up to about 4000 K.

Recent interest in properties up to 10,000 K has required the development of microsecond-resolution pyrometers. A few techniques of a preliminary nature for the dynamic measurement of temperature with microsecond resolution have been developed for experimentation on rapidly heated specimens[4-9]. Each pyrometer was custom designed for the specific measurement system.

In the present paper, an accurate microsecond-resolution pyrometer is described which is developed specifically to measure the temperature of a specimen heated rapidly by the passage of a high current pulse produced by the discharge of a high voltage capacitor bank. The system is designed to measure selected thermophysical properties of both solid and liquid substances which are electrically conducting in the range 2000 to 6000 K.

The salient features of the pyrometer are:
1. It is sufficiently fast and precise to justify the recording of its signals by a digital data acquisition system at regular intervals as short as 1.5 μs with a resolution of 0.1 percent.
2. It measures radiance temperatures in two narrow spectral regions, 0.65 μm and 0.9 μm.
3. In order to minimize the effect of electromagnetic interference, the electronics of the pyrometer are entirely outside the shielding of the room in which the capacitors and all of the discharge circuits are placed. Radiation selected by the optics in each of the two spectral regions is transmitted by fiber optic cables through the shielding walls to the detectors and their associated electronic circuits.
4. The pyrometer uses high-speed automatic gain switching of linear amplifiers to achieve high stability, high resolution and wide temperature range.

2. DESIGN OF THE PYROMETER

2.1. Optics

Figure 1 shows the optical arrangement of the pyrometer. An objective lens O forms two images of the target T at the field stops S1 and S2 by means of the beam splitter B. Filter F1 passes radiation near 0.9 μm to S1, and F2 passes radiation near 0.65 μm to S2; each filter has a bandwidth of 0.03 μm. This design permits the two field stops to be focussed independently to minimize the effects of chromatic aberration. The field stops are 0.5 mm diameter and the magnification is unity, so the target is the same size. The radiation passing through each field stop falls on the end of one of two fiber optic cables, C1 or C2, for transmission to silicon diode photodetectors. The stops are so close to the ends of the cables that the stops, and not the cables, define the target areas. A mirror M can be swung into the path between objective lens and beam splitter to form the image of the target at a reticle R. The experimenter aims and focuses the instrument by viewing the reticle and target image through a microscope.

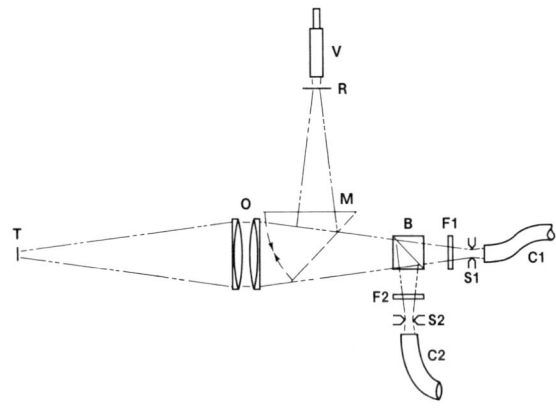

FIG. 1. Schematic diagram of the optical arrangement of the two-color microsecond pyrometer.

2.2. Electronics

The silicon diode detectors used were specially chosen for good stability and linearity. The photocurrents are converted to voltages and amplified by a series of operational amplifiers. The analog output of each amplifier channel is applied to a 10-bit high-speed analog-to-digital converter (ADC). Other ADCs digitize voltages proportional to the heating current and to the voltage drop in the specimen. Another signal which gives the ranges of the pyrometer channels is also digitized. Various digitizing intervals from 1.5 to 512 µs may be chosen. 1024 10-bit words of digital data may be stored from each of five data channels in a semiconductor memory.

The digital recording system limits the temperature resolution of the pyrometer. If the photocurrents were amplified at fixed gain before digitization, there would be adequate resolution, as compared with the precision and accuracy, at the upper end of any selected temperature range, but not enough at lower temperatures, even when the specimen emitted sufficient energy for accurate pyrometry. Logarithmic amplifiers theoretically would give wide range and acceptably uniform resolution. In tests, however, several high-speed logarithmic amplifiers drifted so rapidly that the results of steady state calibrations could not be used to interpret the data from experiments which were performed only a few minutes before or after the calibration.

The pyrometer uses linear amplifiers with high-speed automatic gain switching. The output of each channel is switched through as many as three overlapping ranges. This procedure uses fast, stable components, and results in a response which approximates the logarithm of the input. Changes in amplifier gain are completely negligible. The precision is limited by short term zero drift which on the most sensitive (lowest temperature) range is less then 0.05% over an hour. The useful range of the amplifier extends over more than 3 decades of photo-current. Figure 2 is a block diagram of the analog electronics of each channel.

The output voltages of amplifiers A2 and A3 are compared with an 8 volt reference. If an amplifier output voltage exceeds the reference, that stage is switched out of the amplifier chain to reduce the gain; if it falls below the reference, the stage is switched in again, increasing the gain. The transresistance of amplifier A1 in each channel is 30,000 V/A and the gain of each amplifier A2 and A3 is 11. Switching from low to midrange occurs when the photocurrent rises above 2.2 µA, and from midrange to high range at a photocurrent of 24 µA. The maximum output on the high temperature range is reached at a photocurrent of 0.37 mA. The corresponding radiance temperature ranges are:

Range	0.65 µm Channel	0.9 µm Channel
Low	1960 - 2520 K	1545 - 2030 K
Mid	2500 - 3460 K	2015 - 2920 K
High	3430 - 5700 K	2885 - 5500 K

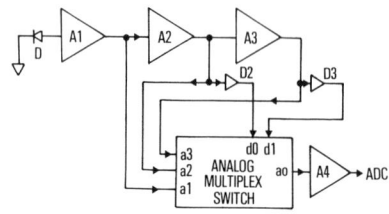

FIG. 2. Block diagram of the analog electronics of each channel of the two-color microsecond pyrometer.

3. DETERMINATION OF TEMPERATURE

3.1. Spectral Radiance Temperature

The current in the photodiode detector in each channel of this pyrometer is

$$I = 2\ kn A\omega c \int (\tau(\lambda)/(\lambda^4 (\exp(C_2/\lambda T) - 1)))\ d\lambda \qquad (1)$$

where

k = ratio of international coulomb to electronic charge, 1.5922×10^{-19}
n = quantum efficiency of detector, assumed to be 1
A = target area, 1.96×10^{-3} cm^2
ω = solid angle at target, 0.041 steradian
c = velocity of light, 3×10^{10} cm·s^{-1}
$\tau(\lambda)$ = transmission of pyrometer optics at specified wavelength
λ = wavelength, cm
C_2 = Planck's second radiation constant, 1.4388 cm·K
T = spectral radiance temperature, K

The product of the values given for the six constants preceding the integral sign will be called k_1 and has the value 7.677×10^{-13} A·cm^3.

In actual use of the pyrometer, the value recorded in the digital array is a multiple of the photodiode current plus some zero offset:

$$B = k_2 I + B_0 \qquad (2)$$

whence

$$B - B_0 = k_1 k_2\ S(\lambda,T) = k_3\ S(\lambda,T) \qquad (3)$$

where $S(\lambda,T)$ is the result of numerical integration of Planck's equation for the measured spectral transmission of the channel. Equation (1) is usually approximated for routine temperature measurement by a single term, containing a "mean effective wavelength", with the following result:

$$T_1 = C_2/(\lambda_e \log_e (1 + k_4/(B_x - B_0))) \qquad (4)$$

where

λ_e = mean effective wavelength for the channel
$k_4 = k_3 \lambda_e^{-4}\ \Delta\lambda$
$\Delta\lambda$ = effective bandwidth of the channel

The derivation and values of these constants are given in Sect. 4 below.

Numerical integration is now quite practical and convenient by digital computer. The following procedure for calculating temperatures was used to confirm that Eq. (4) gives results of satisfactory accuracy:

A. Measure the relative transmission of each channel.
B. Measure B and B_0 at several values of T of a calibrated strip lamp or other source of known spectral radiance temperature. Use these data in Eq. (3) to estimate several values for k_3. The weighted mean of these estimates, using $(B-B_0)$ as the weighting factor, is taken as the best possible estimate of k_3.
C. When a value of $(B_x - B_0)$ is obtained from a source, first approximate the temperature from Eq. (4) above.
D. Let $i = 1$.
E. Calculate
$C_i = k_3 S(\lambda, T_i)$
$B_i = k_3 S(\lambda, T_j)$ where $T_j = T_i + 10$

$$T_{i+1} = T_i + 10\ (B_x - B_0 - C_i)/(B_i - C_i)$$

F. Set $i = i + 1$ and go to E above, unless T_{i+1} is negligibly different from T_i. If so, T_i is the unknown temperature.

The result is then compared with T_1 from Eq. (4).

3.2. Color Temperature

"Color temperature" is the temperature of the blackbody whose spectral distribution approximates the relative spectral distribution of radiance observed in the real target. If λ_1 and λ_2 are the effective wavelengths of the two channels of the microsecond pyrometer, the ratio of spectral radiances at the two wavelengths of a blackbody, having temperature T_c, can be approximated from Planck's equation as

$$R_c = k_1 \lambda_2^4 (\exp(C_2/\lambda_2 T_c) - 1) / k_2 \lambda_1^4 (\exp(C_2/\lambda_1 T_c) - 1) \quad (5)$$

Since the spectral radiance temperatures T_1 at λ_1 and T_2 at λ_2 are the temperatures of blackbodies having, at those wavelengths, the same radiances as the target, then the ratio of the target radiances,

$$R_t = k_1 \lambda_2^4 (\exp(C_2/\lambda_2 T_2) - 1) / k_2 \lambda_1^4 (\exp(C_2/\lambda_1 T_1) - 1) \quad (6)$$

Making $R_c = R_t$

$$(\exp(C_2/\lambda_1 T_c) - 1)/(\exp(C_2/\lambda_2 T_c) - 1)$$
$$= (\exp(C_2/\lambda_1 T_1) - 1)/(\exp(C_2/\lambda_2 T_2) - 1) \quad (7)$$

For any particular values of spectral radiance temperatures T_1 and T_2 this transcendental expression can readily be solved for T_c by iteration. The corresponding equation resulting from Wien's approximation, which is generally acceptable for wavelengths below 1 μm and temperatures below 3000 K, is

$$T_c = (\lambda_1 - \lambda_2)/(\lambda_1/T_2 - \lambda_2/T_1) \quad (8)$$

This equation also gives a good starting value for the solution of Eq. (7).

4. CHARACTERISTICS AND PERFORMANCE

4.1. Spectral Transmission

The spectral transmission of each channel of the pyrometer, which is needed for the numerical integration described in Sect. 3.1 above and for the calculation of the mean effective wavelength as in Sect. 4.2 below, was measured using a tungsten strip lamp, grating monochromator and silicon diode detector assembly. The ratio of detector response through the pyrometer to that through the monochromator alone was calculated for each wavelength, and the value normalized to the maximum transmission for each channel. The precision was about 0.5% of the maximum transmission. Numerical integration of these data gives the bandwidths of ideal filters with rectangular passbands which would transmit the same amount of energy as the real channels do. The results are 0.030697 μm for the short wavelength channel, and 0.028387 μm for the long wavelength channel.

4.2. Mean Effective Wavelength

The evaluation of Eq. (4) above gives an estimate of spectral radiance temperature which is quite satisfactory. The equation, rearranged, is equivalent to

$$B_j = k_4 / (\exp(C_2/\lambda_e T_j) - 1) \quad (9)$$

The mean effective wavelength, λ_e, is defined here as that quantity which, when put in Eq. (9) for two values of temperature T_j, for instance T_1 and T_2, gives a ratio B_1/B_2 which is the same as the ratio of the pyrometer detector currents measured with sources whose true spectral radiance temperatures are T_1 and T_2. The mean effective wavelength can be calculated by determining k_3 as Sect. 3.1 (B) above and using Eq. (3) to find the latter ratio.

The mean effective wavelength will differ slightly for every pair of temperatures used to define it. When one temperature is 2500 K and the other is varied from 2000 K to 6000 K, the mean effective wavelength of the short wavelength channel varies from 0.6547 to 0.6541 μm, and that of the long wavelength channel varies from 0.9006 to 0.9003 μm. Taking median values from these ranges, the effective wavelength method gives results in the entire 1800-6000 K region which differs by a small fraction of a kelvin from those obtained by numerical integration.

4.3. Efficiency

From Eq. (3)

$$B - B_0 = k_3 S(\lambda, T) = 7.677 \times 10^{-13} S(\lambda, T) \quad (10)$$

in which all the factors except the maximum transmission of the pyrometer can be measured or calculated. The normalized transmission of the pyrometer is known from the measurements described in Sect. 4.1 above, and these data can be used in numerical integration to calculate values for $S(\lambda, T)/\tau$, where τ is the maximum transmission of the pyrometer channel. Measurements of known temperatures give values in the range 0.055-0.059 for the maximum transmission of the 0.65 μm channel, and in the range 0.093-0.094 for the 0.9 μm channel. Separate measurements of the transmissions of the fiber cables, and estimates of the transmissions of other components give 0.034 for the maximum transmission of the short wavelength channel and 0.075 for the long wavelength channel. It appears that the actual losses are smaller than the estimates used in the latter calculation, and that no significant unrecognized losses are present.

5. CALIBRATION

5.1. Source

Two gas-filled tungsten strip lamps, selected for good stability of spectral radiance as indicated by measurements at 0.65 μm, were calibrated by the Radiometric Physics Division of the National Bureau of Standards. The calibrations covered the range 1873-2573 K at 0.6546 μm and the range 1800-2425 K at 0.9099 μm. These wavelengths were the best estimates available, at the time the calibrations were requested, for the mean effective wavelengths of the pyrometer channels. The results are consistent with the spectral emissivity of tungsten reported in the literature.

5.2. Example of Calibration

The pyrometer was used to record the radiance from a calibrated lamp at eight values of filament current identical with those specified on the certificate of calibration. Each measure of response, together with an appropriate measure of dark response, was used in Eq. (9) above to estimate a value for k_4. These values were used in turn to calculate a weighted mean for each channel, using the measured response as the weighting factor. The weighted means for k_4 were then used in Eq. (4) to attribute spectral radiance temperatures to the lamp. The differences between the calibration values and the measurements lay between 0.1 to 2.0 K for the 0.65 μm channel and between 0.1 and 0.6 K for the 0.9 μm channel for each of eight spectral radiances.

These experiments were repeated five weeks apart. When the earlier mean values for k_4 were used to calculate the lamp temperatures from the later data, the differences from the calibration values ranged from 0.3 K to 2.1 K for the 0.65 μm channel and from 0.5 K to 2.2 K for the 0.9 μm channel.

6. ACCURACY

The individual sources of error in the use of the pyrometer are discussed in the following sections and summarized in Sect. 6.11 and Table I.

6.1. Calibration Error of Secondary Standard

The reports of calibration give estimates of the accuracy at two temperatures, 2073.15 K and 2573.15 K, for each wavelength. One of the lamps is used as a working standard in periodic recalibration of the pyrometer. The other is reserved for use at long intervals to find how much the working standard has drifted. The two lamps were compared after three and one half years. The spectral radiance of the working standard did not change more than 0.6% at 0.6546 μm nor more than 0.7% at 0.90045 μm as compared with that of the reserved lamp.

6.2. Time Response

The amplifier output reaches 95% of its full scale response to a step change of radiance at the pyrometer input within 1.75 μs. The low frequency response was checked with a high-speed rotating shutter and a steady light source. This assured that there was no rise or droop of the response more than 10 μs after the shutter was opened. The high frequency response was checked with a light emitting diode supplied by a square wave source of high impedance.

6.3 Geometry of Calibration Instrument and Pyrometer

The calibrating instrument measures the spectral radiance of the strip lamp over a solid angle having a vertex of 7.16 degrees in the vertical plane and of 3.58 degrees in the horizontal plane. The pyrometer accepts light from a cone having a vertex angle of 13.1 degrees. During calibration it was found that the lamps showed an increase of spectral radiance of as much as 0.61% when the measurement was made at an angle of 7 degrees off the normal to the strip. The measurement was only made with respect to rotation around the long axis of the strip, but there is no reason to expect that some increase of radiance would not occur at similar angles in other directions. The average increase of radiance at an angle of 7 degrees was closer to 0.25%. In no case was a lower radiance observed off the normal to the strip.

The excess of the solid angle accepted by the pyrometer above the angle accepted by the calibrating instrument forms an annulus which is centered near 7 degrees, and the fraction of the total light accepted by the pyrometer which is contained in this annulus is around 0.7, so that the spectral radiance of the lamp as viewed by the pyrometer may be as much as 0.4% greater than the radiance detected by the calibrating instrument.

6.4. Scattered Light

Scattered light gives rise to the "size of source effect"[2] because light from outside the nominal target area affects the output of the pyrometer. The estimation of the error due to scattered light seems best made by measuring the response of the pyrometer to the radiation from a hot surface containing a well defined hole a little larger than the nominal target, with care to avoid reflections from sources external to the pyrometer. This experiment was performed using a tantalum strip containing a round hole 0.75 mm in diameter. The response to the light scattered from the strip surrounding the hole was less than 2.7% at 0.65 μm and less than 3.4% at 0.9 μm of the response when the pyrometer was aimed at the surface adjacent to the hole. Because this error varies greatly depending upon the character of the target, it is not included in the summary of errors in Table I. The error from this source will be smaller for targets which have uniform brightness over areas larger than 0.75 mm diameter, and where the surrounding areas are not dark. If the target were an isolated disk 0.75 mm in diameter, surrounded by areas which did not radiate at all, the possible radiance temperature errors at 0.65 μm would increase from the value of 3.7 K at 2500 K shown in Table I to 8.5 K, and from the tabulated value of 32 K at 5500 K to 50 K. In the same case, the errors at 0.9 μm would increase from 3.6 K to 9.5 K at 2025 K and from 19 K to 80 K at 5500 K.

6.5. Effective Wavelength

The spectral response of the instrument used to calibrate the secondary standard lamps is different from that of the pyrometer. Because the emissivity of the tungsten strip in the lamps is fairly low, and varies with both temperature and wavelength[10], this difference can give rise to error in the transfer of the strip lamp calibration to the pyrometer.

Defining $C(\lambda, T_c)$ as the result of integration of Planck's equation at the spectral radiance temperature T_c reported as the result of calibration, and $C(\lambda, T)$ as the result of integrating Planck's equation using the emissivity of the tungsten strip at wavelength λ and true temperature T, and in each case using the spectral transmission of the calibrating instrument, the true temperature of the tungsten strip will be the temperature T which is a solution of the equation:

$$C(\lambda, T) = C(\lambda, T_c) \qquad (11)$$

In the same way, if $P(\lambda, T_p)$ is the result of the corresponding integration using the spectral radiance temperature T_p and the spectral transmission of the pyrometer, and $P(\lambda, T)$ the result of integration using the true temperature T, the spectral transmission of the pyrometer and the spectral emissivity of the tungsten lamp filament, the spectral radiance temperature T_p given by the pyrometer will be the solution of the equation:

$$P(\lambda, T_p) = P(\lambda, T) \qquad (12)$$

Data on the emissivity of tungsten over the temperature range 2000-2800 K and the wavelength range 0.65-0.8 μm[10] were used in a least squares procedure to obtain the following equation relating emissivity with wavelength and temperature:

$$\varepsilon(\lambda, T) = 0.597222 - 0.178284\,\lambda - 2.34105 \times 10^{-5}\,T \qquad (13)$$

expressing the wavelength λ in micrometers. This equation was then used in the integrals of Eqs. (11) and (12) with the spectral data for the calibrating instrument and the pyrometer to find a solution for those equations. The result is that the error due to the difference in effective wavelength of the 0.65 μm channel in the range 1875-2575 K is less than 0.1 K. The spectral transmission of the 0.9 μm channel was not so well known when the calibration was requested, with the result that the error, due to the difference in effective wavelength, for this channel is as large as 5.17 K at 2423.1 K.

The error can be calculated by Eqs. (11-13), so that the calibration data can be corrected for the difference in effective wavelength. Perturbation of the coefficients of Eq. (13) shows that the value of the correction should be accurate to within 0.1% at 2423.1 K, and more accurate at other temperatures in the calibrated range of the lamps.

6.6. Drift of Sensitivity and Zero

Both these sources of error are completely negligible by comparison with others. Repeated measurements of the gains of the amplifiers showed stabilities exceeding 0.01%. The drift of the amplifier zero on the most sensitive range is less than 0.05% per hour. During calibration or use of the pyrometer, the zero responses of each channel on every range which will be used are recorded less than five minutes before the experiment.

6.7. Linearity at High Radiation Intensity

Temperatures above the range for which tungsten strip lamps can be calibrated must be estimated by extrapolation. It is necessary that the detectors and amplifiers be linear up to the maximum temperatures for which the pyrometer is designed, in this case 6000 K. To confirm the linearity of the detectors and associated electronics the pyrometer was twice used to measure the attenuation of neutral filters, once with a source having an intensity in the range of calibration below 2500 K, and again using the sun, which has an intensity near the maximum. The two results differed by less than 2%.

6.8. Noise

The noise at the output of the amplifers was measured with a wide band oscilloscope. This measurement was confirmed by calculating the standard deviation of individual digitized samples from the mean of a hundred samples when the pyrometer was aimed at a steady source, a tungsten strip lamp. The noise is insignificant on the high range of both channels, and on the midrange of the 0.9 μm channel. It is also insignificant on the upper part of the midrange of the 0.65 μm channel, and on the upper part of the low range of the 0.9 μm channel. The contribution of the measured noise to the uncertainty of temperature in the low part of the midrange of the 0.65 μm channel and to the uncertainty in the low ranges of both channels is given in Table I.

6.9. Quantization

The analog-to-digital converters have 10-bit resolution, so that the amplifier output is recorded with a resolution of 0.1% of full scale on each range. This causes maximum uncertainty at the lower end of each range, and minimum uncertainty at the upper end of each.

6.10. Transfer Error

The errors caused by drift, noise and quantization affect each measurement twice, once when the calibration is made and once when the measurement is made. The square root of the sum of the squares of these errors is calculated for the temperatures in the range in which calibration is performed., The value for the highest temperature at which calibration is performed, 2500 K at 0.65 μm and near 2025 K at 0.9 μm, is used for all higher temperatures. The errors of extrapolation are separately accounted for under non-linearity.

6.11. Summary of Accuracy at Various Temperatures

Table I gives the estimate of error in spectral radiance as a percentage of the radiance for the temperature and channel involved. The square root of the sum of the squares of the various individual errors is taken to be the probable accuracy of spectral radiance for the pyrometric system as a whole, and this number is converted to an uncertainty in temperature. To summarize the accuracy of the pyrometer, data are given for the low and high end of each range in each channel. Since the temperature ranges of the two channels are different, twelve estimates are given for seven different temperatures, six of the estimates referring to each of the two channels.

7. EXAMPLE OF RAPID TEMPERATURE MEASUREMENT

To demonstrate the measurement of temperature under dynamic conditions, a niobium rod, 25 mm long and 1.6 mm diameter, was heated rapidly with the use of the capacitor discharge system described elsewhere[11]. The results of the radiance temperature measurement by the pyrometer are shown in Fig. 3. The plateaus in the temperature plot show the melting of the niobium. It can be seen that the pyrometer followed the entire melting, and continued to show reasonable values of spectral radiance temperature for several hundred degrees into the liquid phase.

The entire time, from room temperature to near 3500 K, was only 100 μs. Melting of the niobium took place in about 15 μs. The average radiance temperatures measured at the melting point of niobium are 2431 K at 0.65 μm and 2250 K at 0.9 μm. The difference at the two wavelengths is due to the difference in normal spectral emittance of the metal at the two wavelengths. The value for spectral radiance temperature at 0.65 μm at melting in this experiment compares favorably with the value of 2425 K measured earlier during slower heating by a system with millisecond resolution[12]. Taking the value of 2750 K for the melting temperature of niobium[13], the normal spectral emittance of niobium is calculated to be 0.348 at 0.65 μm and 0.275 at 0.9 μm.

8. CONCLUSION

We have shown that a pyrometer approaching microsecond resolution has good accuracy and stability over a very wide temperature range. This instrument allows the measurement of radiance temperature of materials up to about 6000 K. Appropriate design permits the use of fiber optic cables in an accurate, stable pyrometer.

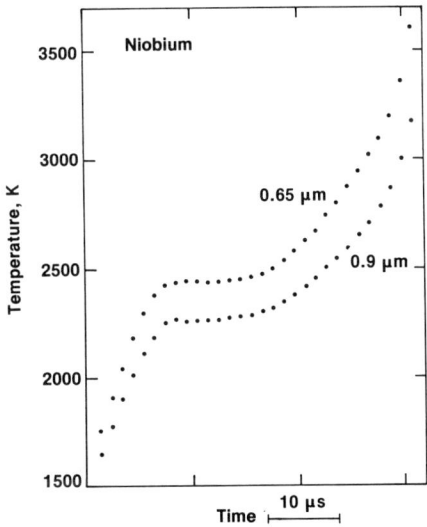

FIG. 3. Radiance temperatures of a rapidly heating niobium specimen measured with the two-color microsecond pyrometer. Plateaus in temperature indicate melting of niobium. Data on the right of the plateaus are for liquid niobium.

TABLE I. Estimated radiance errors[1] of two-color microsecond pyrometer

	0.65 μm Channel					
	Low Range		Mid Range		High Range	
Temperature, K	2000	2500	2500	3450	3450	5500
Second Standard, %	0.7	1.0	1.0	1.0	1.0	1.0
Acceptance Angle, %	0.4	0.4	0.4	0.4	0.4	0.4
Effective Wavelength, %	0	0	0	0	0	0
Amplifier Drift, %	0.01	0.02	0.01	0.02	0.01	0.02
Nonlinearity, %	0	0	0	0.2	0.2	2.0
Total Noise, %	2.5	0.5	0.2	0	0	0
Quantization, %	1.0	0.2	1.0	0.2	1.0	0.2
Transfer, %	2.7	0.5	0.5	0.5	0.5	0.5
Root Mean Square, %[2]	3.9	1.3	1.6	1.2	1.6	2.3
Error, K[3]	7.1	3.7	4.4	6.2	9.9	32

	0.9 μm Channel					
	Low Range		Mid Range		High Range	
Temperature, K	1550	2025	2025	2900	2900	5500
Secondary Standard, %	1.0	1.1	1.1	1.1	1.1	1.1
Acceptance Angle, %	0.4	0.4	0.4	0.4	0.4	0.4
Effective Wavelength, %	0.1	0.1	0.1	0.1	0.1	0.1
Amplifier Drift, %	0.01	0.02	0.01	0.02	0.01	0.02
Nonlinearity, %	0	0	0	0.2	0.2	2.0
Total Noise, %	2.5	0.5	0.2	0	0	0
Quantization, %	1.0	0.2	1.0	0.2	1.0	0.2
Transfer, %	2.7	0.5	0.5	0.5	0.5	0.5
Root Mean Square, %[2]	4.0	1.4	1.6	1.3	1.6	2.4
Error, K[3]	6.0	3.6	4.2	6.9	8.6	19

[1] Estimates of errors are expressed as a percentage of radiance for a given temperature and channel.

[2] Root mean square of the listed individual errors.

[3] Root mean square error expressed in temperature.

REFERENCES

a) This work was supported in part by the U.S. Department of Energy through the Argonne National Laboratory.
b) Consulting Physicist, Canal Winchester, Ohio 43110.
1. A. Cezairliyan, in Temperature, Vol. 4, Part 1, H.H. Plumb, ed. (Instrument Society of America, Pittsburgh, 1972) p. 657.
2. G.M. Foley, Rev. Sci. Instrum. 41, 827 (1970).
3. F. Righini, A. Rosso and G. Ruffino, High Temperatures-High Pressures 4, 597 (1972).
4. I. Ya. Dikhter and S.V. Lebedev, High Temperature (USSR) 8, 51 (1970).
5. T. Korneff, Rev. Sci. Instrum. 42, 1561 (1971).
6. G.R. Gathers, J.W. Shaner, and R.L. Brier, Rev. Sci. Instrum. 47, 471 (1976).
7. U. Seydel, H. Bauhof, W. Fucke, and H. Wadle, High Temperatures-High Pressures 11, 35 (1979).
8. R.W. Ohse, J.F. Babelot, P.R. Kinsman, K.A. Long, and J. Magill, High Temperatures-High Pressures 11, 225 (1979).
9. L. Coslovi, F. Righini, and A. Rosso, J. Phys. E: Sci. Instrum. 12, 216 (1979).
10. L.N. Latyev, V. Ya. Chekhovskoi, and E.N. Shestakov, High Temperatures-High Pressures 2, 175 (1970).
11. A. Cezairliyan, M.S. Morse, G.M. Foley, and N.E. Erickson, in Proceedings of the Eighth Symposium on Thermophysical Properties, J.V. Sengers, ed. (Am. Soc. Mech. Eng., New York, 1981).
12. A. Cezairliyan, J. Res. Nat. Bur. Stand. (U.S.) 77A, 333 (1973).
13. A. Cezairliyan, High Temperatures-High Pressures 4, 453 (1972).

Infrared temperature measurements of gas and dust explosions

K. L. Cashdollar and M. Hertzberg

Pittsburgh Research Center, Bureau of Mines, U.S. Department of the Interior, Pittsburgh, Pennsylvania 15236

This paper presents temperature measurements of gas and dust explosions. Instrumentation used includes a commercial rapid-scan spectrometer and two multi-wavelength infrared pyrometers developed at the Bureau of Mines. The rapid-scan spectrometer has a wavelength range of 1.7 to 4.8 μm and a maximum rate of 800 scans per second. One of the pyrometers is a three-wavelength near-infrared instrument with a time response of about 20 μsec, and it measures continuum radiation from particles in the flame at wavelengths of 0.8, 0.9, and 1.0 μm. The second infrared pyrometer measures continuum radiation from hot particles at four wavelengths (1.6, 2.3, 3.8, and 5.0 μm) and band emission from carbon dioxide gas at 4.4 and 4.6 μm. It has a time response of about 25 μsec. Temperatures for small and large scale methane-air explosions were measured with the rapid-scan spectrometer, based on the absolute radiance and on emission-absorption measurements using a blackbody source. Measurements of both gas and dust temperatures for small-scale dust explosions were made with the rapid-scan spectrometer and the two pyrometers. The two pyrometers have also been used to make temperature measurements within full-scale mine coal dust explosions.

INTRODUCTION

The most significant thermodynamic parameter that characterizes the final equilibrium state of an exothermic substance or combustible mixture is the flame or explosion temperature reached as a result of its chemical transformation to the lowest free energy state. The potential hazard of such substances or mixtures is directly related to that temperature in several ways. First, there are the questions of whether the compound is flammable or detonable, or whether the mixture composition is within or beyond its limits of flammability. Those conditions are often predictable in terms of whether the final flame temperature is above or below some critical value which characterizes a given class of fuels.[1,2] If the mixture or substance is flammable, the magnitude of the hazard also scales with that final temperature. For example, the final explosion pressure resulting from a substance's ignition and subsequent deflagration in an adiabatic system of constant volume is governed mainly by its final equilibrium temperature. For a constant pressure system, such as a commercial burner, where the combustion process is localized and stabilized by flow control, the flame temperature is still the most important intensive variable. It determines the effectiveness of the flame in its normal function of melting, calcining, welding, etc.; and the spatial distribution of temperature in the flame's burned gas region or its enclosure determines the performance of the combustor or furnace that is powered by the flame. Finally, the ability to control many of the pollutants generated by the combustion process is also a sensitive function of the flame temperature and its distribution within the combustor. Accordingly, there is a continuing interest in the accurate measurement of the high temperatures present in the burned gas regions of flames or explosions.

The most direct method of temperature measurement is the contact thermocouple, and thermocouples have naturally found widespread use in research and industry. However, for the higher temperatures that characterize most flames or explosions, the thermocouple is limited by several factors: its finite melting point, its relatively slow time response, its large radiation correction, the complications associated with its catalytic recombination activity,[3] and its tendency to accrete condensible materials in a multiphase system.[4]

For flames or explosions, a more convenient means of measuring temperatures involves radiation methods. For very high temperatures or for rapidly changing, nonsteady state systems, radiation methods are the only ones readily available. The radiation method most used in the past for normal gas flames or burners in steady-state is the "line reversal" method.[5,6,7,8] In its ideal form, the line reversal method uses a blackbody source whose radiance is viewed through the burned gases of the flame. The background blackbody radiance may be viewed in any region of the spectrum where there is measurable absorption and emission from the flame. If the energy level distribution associated with the absorbing and emitting flame specie is in thermal equilibrium with the flame temperature, and if the blackbody source is at that same temperature, then the spectral radiation field within the flame is in equilibrium with the radiation field traversing through it from the blackbody. In that equilibrium state, flame radiance and blackbody radiance become indistinguishable from one another. If the flame in the foreground is hotter than the blackbody source in the background, then the flame appears brighter than the source because more radiation is emitted from the specie's upper energy level than is absorbed by the specie's lower energy level. If the flame is cooler than the source, then the flame appears darker than the blackbody source because the lower energy level absorbs more radiation than is emitted from the upper level. It is only when flame and source temperatures are equal that the flame brightness in the foreground matches the source brightness in the background. This "line-reversal" temperature measurement is independent of the actual value of the flame's absorptivity-emissivity, as long as the flame is not completely transparent in the chosen wavelength range. The method usually used to assure a measurable interaction between the two radiation fields is to seed the flame with a sodium salt and to view the spectral radiance of the sodium D-lines at 589 nm. In the traditional method, the spectrum of the flame and a tungsten ribbon lamp is observed visually. The lamp intensity is varied until the Na D-lines disappear against the lamp continuum. The lines are then at the same intensity as the continuum. The flame temperature is calculated from the known temperature of the calibrated ribbon lamp, corrected for the emissivity of the tungsten. An additional complication is caused by the fact that burner flames are traditionally viewed from the side, through a cooler zone consisting of entrained air mixing with the burned gas species. A cautious interpretation of the data or a more careful selection of wavelengths becomes necessary in order to correct for such self-absorption effects. While this line reversal technique can be used for steady-state or slowly changing flames, it is usually not applicable to rapid explosions since the blackbody or lamp cannot be varied in intensity rapidly enough to follow the changing explosion radiance.

Another early method, which is more easily adapted to explosions or other flame systems in which the temperature changes rapidly, is the emission-absorption method[6] first used by Schmidt[9] and first applied with modern equipment by Silverman.[10] One directly measures (at any convenient wavelength) the absorption or transmission for a fixed path of flame and simultaneously measures the same flame's absolute spectral radiance at the same wavelength. If the absorbing specie is in thermal equilibrium at the flame temperature, its emissivity is equal to its absorptivity and is therefore also known. Now the spectral radiance emitted from an ideal blackbody is given by Planck's Law:

$$L_\lambda = \frac{2hc^2}{\lambda^5} [e^{hc/\lambda kT}-1]^{-1}, \quad (1)$$

where L_λ is the radiance in energy per unit area per unit time per steradian per unit wavelength interval, h is Planck's constant, c is the speed of light, λ is the wavelength of the radiation, k is Boltzmann's constant, and T is the absolute temperature. For non-blackbodies,

$$H_\lambda = \varepsilon L_\lambda = \frac{2\varepsilon hc^2}{\lambda^5} [e^{hc/\lambda kT}-1]^{-1}, \quad (2)$$

where H_λ is the radiation emitted and ε is the emissivity. In this emission-absorption method, H_λ and ε are measured at a known wavelength λ; and therefore T can be calculated. Since the combustion products from hydrocarbon flames contain sufficient quantities of water vapor and carbon dioxide, the emission bands of H_2O and CO_2 at 2.7 µm and of CO_2 at 4.4 µm are convenient wavelengths to use.[11] This method also requires a cautious interpretation of the data if it is used for a burner flame that is viewed through a zone containing cooler, self-absorbing product gases mixed with air.

In contrast to burner flames, the spherical combustion wave of an explosion maintains a sharp discontinuity between burned products and premixed reactants. The view of hot emissive products generated in the fireball of such explosions is thus unobscured by self-absorption from products cooled by mixing with air. There may still be some narrow absorption lines from atmospheric H_2O and CO_2 if the optical path is not inerted (by nitrogen), but this is less of a problem.

If the product gases are optically thick or opaque at the observed wavelength, the emissivity is close to unity and the transmission measurement is superfluous. In this case, the temperature can be calculated from the measured radiance using Eq. (2) with $\varepsilon = 1$. For hydrocarbon gas flames, the emissivity reaches unity for relatively short path lengths of CO_2 at the 4.4 µm band.[11]

The present paper will present some representative data for CH_4-air explosions in various sized chambers. Temperatures are calculated from the CO_2 absolute radiance at the 4.4 µm band and from absorption-emission measurements at the 2.7 µm H_2O band and the 4.4 µm CO_2 band.

This paper will also present data for the temperatures of phase-heterogeneous systems that contain both gaseous and condensed-phase species. Although the multiphase system is considerably more complex than a homogeneous gas mixture, it is nevertheless of considerable practical significance. All dust explosions are phase-heterogeneous, as are solid-phase detonations and rocket engines. Most of the coal produced throughout the world is eventually pulverized and consumed in the burner flames of furnaces. These flames are phase-heterogeneous and contain both gaseous and condensed phase reactants and products. This heterogeneous problem was briefly addressed, mostly theoretically, in an earlier symposium.[12] In the present paper the problem is mainly addressed experimentally. An extensive amount of data has been accumulated for the radiation temperatures of phase-heterogeneous systems, mostly air-dispersed carbonaceous dusts. Some of this data will be presented here. Emphasis will be placed on the technique used, which involves the measurement of "particle continuum temperatures", obtained by fitting the measured continuum radiance at several (or many) wavelengths to the best-fit Planck curve. The data also show that the radiance level of the CO_2 band peak at 4.4 µm is invariably higher than the continuum level even if the system has essentially unit emissivity for both particles and CO_2. This is interpreted to mean that the gas temperature is hotter than the particle temperature, and the CO_2 gas temperatures are then calculated from the differential radiance at the 4.4 µm band.

GAS EXPLOSION TEMPERATURES WITH A RAPID-SCAN SPECTROMETER

Homogeneous gas explosions were studied in two constant-volume, spherical combustion chambers of 0.61 and 3.66 m diameter and in a 9-liter cylindrical chamber (0.20 m diameter and 0.31 m high). Premixed methane-air reactants were centrally ignited, and the spherical combustion waves from the developing explosions were viewed through sapphire windows with a rapid-scan spectrometer[13],[14] whose optical system is shown in Fig. 1. The scanning wheel can be rotated at up to 33.3 rev sec^{-1}. It contains 24 equally spaced corner mirrors and thus provides up to 800 scans sec^{-1}. During the experiments described in this paper, the scanning wheel was rotated at a slower speed, providing a scan time of 12.5 msec per spectrum when observing explosions in the cylinder and the 0.61 m sphere and 62.5 msec in the 3.66 m sphere. This instrument is a double-pass spectrometer, and the radiation is recorded simultaneously by both a short-wavelength and long-wavelength detector. The 1.7 to 3.2 µm radiation is sensed by an indium arsenide (short λ) detector, and the 3.1 to 4.8 µm radiation is sensed by an indium antimonide (long λ) detector. Both are maintained at liquid nitrogen temperature (77 K).

The detector signals were amplified and then recorded on a high speed tape recorder. Details of the data processing and storage are found in references 14 and 15. To convert the data to absolute radiance, the primary calibration standard was a blackbody furnace with a 31 cm diameter spherical cavity, a 7 cm aperture, and a maximum temperature of 1500 K. A feedback-controlled silicon carbide infrared radiation source was used for routine checks of instrument sensitivity.

Typical absolute radiance data from an explosion of 9.1% CH_4 in air are shown in Fig. 2. This is for a relatively small fireball with a burned gas path length of 25 to 30 cm-atm in the 9-liter chamber.[14] Even for this relatively short optical path, the CO_2 gas at 4.4 µm is optically thick[16],[17] and thus has an emissivity of close to one. Therefore, the gas temperature can be calculated from Eq. (2). This temperature is shown in Fig. 2 as the 2,000 K Planck curve. The narrow self-absorption line at about 4.25 μm is due to atmospheric CO_2, and the other self-absorption dip at about 2.7 μm is due to atmospheric H_2O. These appear due to uncontrolled variations in the amounts of atmospheric H_2O and CO_2 in the optical path during flame measurements as compared to the amounts present during calibration. It is only at 4.4 μm that the gas has an emissivity of one. The H_2O bands at shorter wavelengths have a much lower emissivity and thus do not reach the Planck curve. (The pyrometer wavelength positions shown in the figure are for reference only and will be discussed in a later section of the paper.)

A more complete record of the entire radiance growth from a large-scale explosion[18,19] in the 3.66 m diameter (25,700 liter volume) sphere is shown in Fig. 3. This is a graphical summary presented in three-dimensional perspective. The absolute spectral radiance (vertical axis) is plotted as a function of wavelength (horizontal axis). Numerous curves are plotted for various times (perspective axis) from ignition at time zero to complete combustion at about 1.1 sec. The mixture contained a stoichiometric methane-air mixture diluted with 10% excess nitrogen at a total initial pressure of 1 atm. Complete combustion occurred as the fireball just filled the entire chamber, generating a peak pressure of 7.0 atm. For the 366 cm diameter sphere, the maximum burned gas optical path was approximately 2500 cm-atm, and each spectrum shown in Fig. 3 is labelled with its corresponding burned gas optical path. For this data, the amount of normal atmospheric CO_2 in the optical path was the same for both calibration and flame measurements, so the narrow CO_2 self-absorption line of Fig. 2 does not appear in Fig. 3. However, there is still some self-absorption at the H_2O band since the amount of water vapor in the atmosphere is more variable than the amount of CO_2.

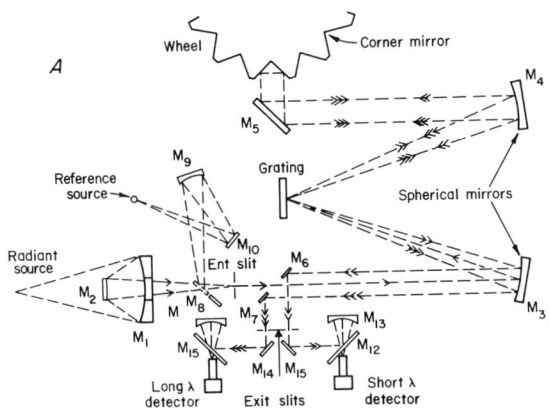
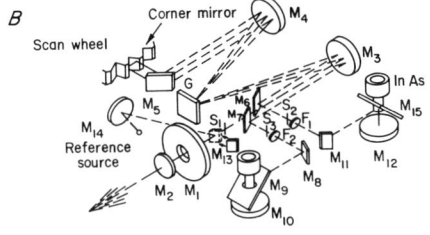

FIG. 1. Optical schematic of rapid-scan spectrometer, A. Top view, B. perspective view.

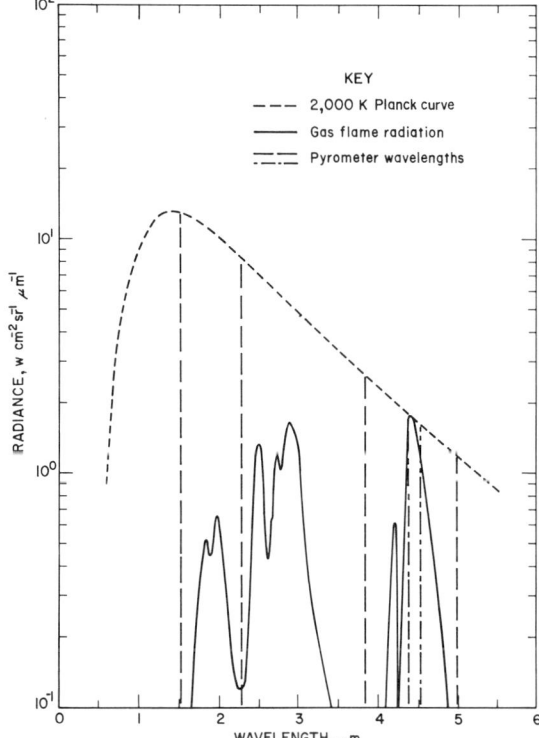

FIG. 2. IR spectral radiance of a CH_4-air explosion in the 9-liter chamber, measured with the rapid-scan spectrometer.

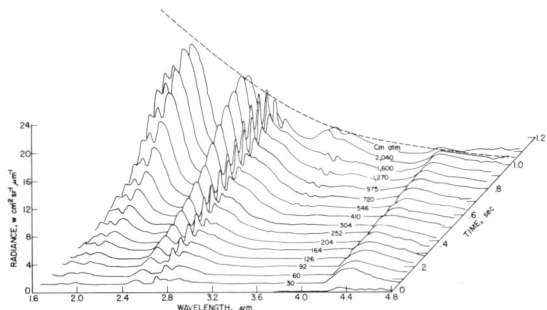

FIG. 3. Time dependent spectral radiance growth of a CH_4-air explosion in the 3.66 m diameter sphere.

Additional data[18] obtained in the 0.61 m diameter (120 liter volume) sphere with the rapid-scan spectrometer are shown in Fig. 4. The absolute radiance growth pattern in the 4.4 to 4.6 μm CO_2 band is shown in detail for a stoichiometric CH_4-air mixture diluted with 10% nitrogen. The growth pattern is resolvable into two components: an initial state of emissivity growth and a later stage of adiabatic compression. The initial radiance growth was rapid and approached unit emissivity for fireball depths of about 30 cm-atm at about 70 msec. while combustion was still at nearly constant pressure. Accordingly, the 4.4 to 4.6 μm band was used to obtain flame temperatures from the radiance data alone throughout the remainder of the combustion process. The slower, subsequent radiance growth after 70 msec reflected only the adiabatic compression process. The measured fireball temperature in Fig. 4 during the early stage of constant pressure combustion was quite close to the adiabatic constant pressure value, $(T_{ad})_p$. As the fireball grew and the pressure increased, the temperature increased continually until about 230 msec. Buoyancy and non-adiabatic losses prevented the measured final explosion temperature from reaching its constant volume adiabatic value $(T_{ad})_v$. Buoyancy caused the top of the fireball to contact the sphere before downward propagation was complete. Thus, convective cooling of

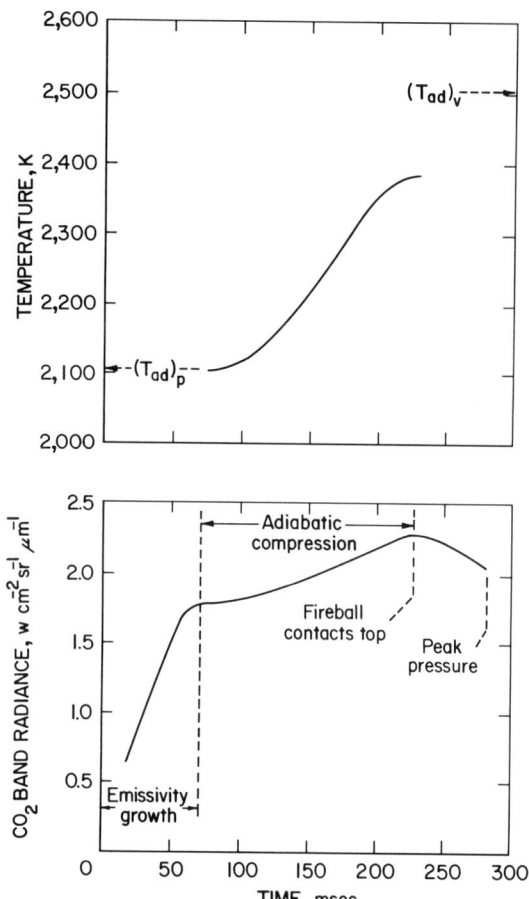

FIG. 4. Temporal growth of the CO_2 emission band radiance and the flame temperature for a CH_4-air explosion in the 0.61 m diameter sphere.

the top of the fireball started before the peak pressure was attained, causing the CO_2 band to display a characteristic self-absorption after 230 msec. (This is also seen in the last spectrum in Fig. 3.) Beyond that time of contact, it is no longer possible to use the CO_2 band to measure fireball temperatures; therefore, the temperature at peak pressure was not known.

Similar data obtained from the 4.4 μm CO_2 band absolute radiances for explosions in the 3.66 m diameter sphere are shown in Fig. 5. They are compared with adiabatic flame temperature values calculated for the varying stages of compression.[18] The measured values were nearly identical to the adiabatic values, especially for the near-stoichiometric mixture (9.8% CH_4) whose flame speed was fast enough to ensure near-spherical, near-adiabatic combustion. Measured values were at most some 75 K cooler even for the leanest mixture shown. For these data, only temperatures that were measured prior to the time of contact with the top of the sphere are shown. Beyond that time of contact, the peak radiance at 4.4 μm becomes severely diminished by self-absorption from the convectively cooled boundary layer in contact with the wall.

The most accurate data[15] were obtained with the 0.61 m diameter spherical vessel and the arrangement shown in Fig. 6. For these experiments, the optical path from the combustion chamber to the rapid-scan spectrometer was inerted with N_2 gas, thus eliminating the problem of self-absorption lines from atmospheric CO_2 and H_2O. Also, both the emission and absorption of the fireball were measured. When the optical chopper blocks the blackbody radiation

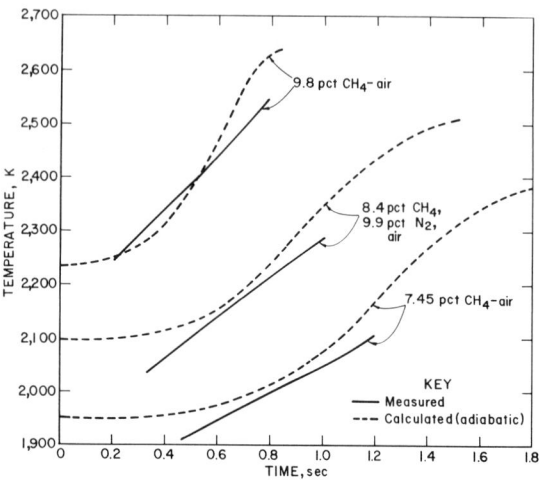

FIG. 5. Measured CH_4-air flame temperatures from the 3.66 m diameter sphere compared with calculated adiabatic flame temperatures.

source, the absolute spectral radiance of the explosion fireball is measured directly, as in the previous experiments. During alternate scans, however, the blackbody radiation transmitted through the fireball is added to the fireball radiance to give the total signal. Subtracting the fireball radiance from the total radiance and dividing by the intrinsic blackbody radiance gives the transmission. The emissivity is equal to the absorption, which is one minus the transmission. Knowing the emissivity ε and the radiance H_λ at any wavelength allows the temperature to be calculated using Eq. (2). To measure both the emission (flame radiance) and the absorption, the optical chopper is phase locked to the spectrometer scan period so that the blackbody source is blocked on alternate scans, with each spectral scan being 12.5 msec.

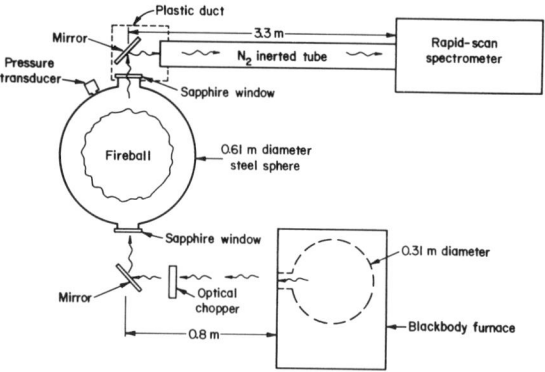

FIG. 6. Schematic of emission-absorption measurements of CH_4-air explosions with the rapid-scan spectrometer.

The time dependent growth of the absolute spectral radiance from a stoichiometric CH_4-air explosion in the 0.61 m sphere is shown in Fig. 7. The nitrogen inerting of the optical path has now eliminated even the atmospheric H_2O self-absorption around 2.7 μm as compared to the data in Fig. 3. Note that as emission depths thicken, the contributions from the higher vibrational-rotational states within all bands increase markedly as the band peaks begin to saturate near unit emissivity. There is then a marked broadening and overlapping of bands and a flattening of band peaks as their wings become congruent with the Planck function.

Temperatures calculated by the emission-absorption method and the corresponding pressure are shown as a function of time in Fig. 8. The data show good agreement between temperatures calculated at the relatively thin 2.7 μm H_2O band and those calculated at the 4.4 μm CO_2 band. In both cases, data were calculated over a band of wavelengths, including the wings of the emission bands. For the CO_2 band, the fireball becomes opaque beyond 50 msec and alternate spectral scans give identical signals.[15] For these later times, the emissivity at 4.4 μm is one, the transmission is zero, and its measurement is superfluous. The measured temperatures agree fairly well with the calculated, adiabatic values shown by the dashed curve.

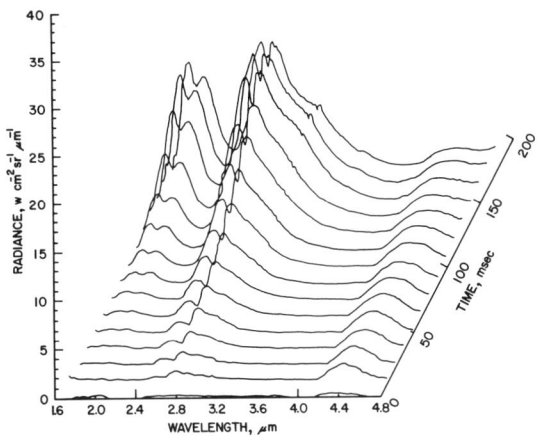

FIG. 7. Time dependent spectral radiance growth of a CH_4-air explosion in the 0.61 m diameter sphere, (adapted from Fig. 12 of reference 15).

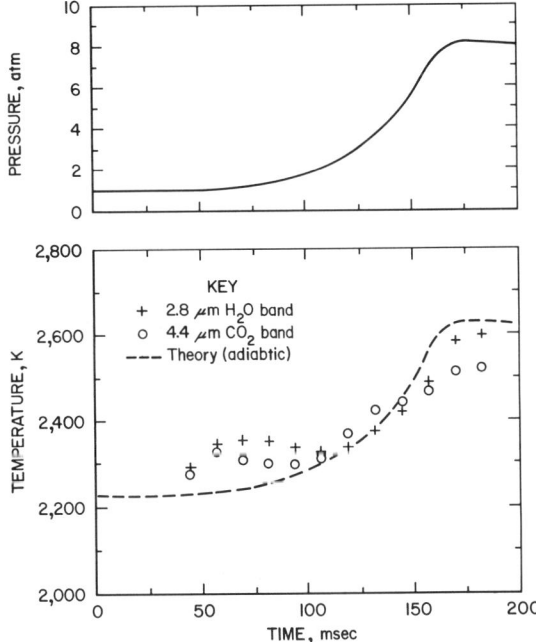

FIG. 8. CH_4-air explosion pressure and temperatures measured at the H_2O and CO_2 bands by the emission-absorption method (adapted from Figs. 14 and 15 of reference 15).

DUST EXPLOSION TEMPERATURES WITH A RAPID-SCAN SPECTROMETER AND TWO MULTI-WAVELENGTH PYROMETERS

The use of the rapid-scan spectrometer to observe explosions was extended to observations of phase heterogeneous systems containing both gaseous and particulate matter. Fig. 9 shows the spectral radiance from an explosion of Pittsburgh seam coal dust in air at a concentration of 500 g m^{-3} in the 9-liter chamber. Details of the measurements are in reference 14. The continuum radiation from the hot particles matched well to a 1,400 K Planck curve. The gas temperature as measured by the absolute radiance at the 4.4 μm CO_2 band is about 1700 K. These data did not have an inerted optical path and show self-absorption by atmospheric H_2O at 2.7 μm and by CO_2 at 4.25 μm. In these tests, the dust cloud was ignited as it was dispersed; and the absolute radiance in Fig. 9 was measured before the dust reached the sapphire window on the top of the chamber.[14] However, even if the absolute radiance is not known, the particle continuum temperature can be found from the best Planck curve fit to the relative radiance data.

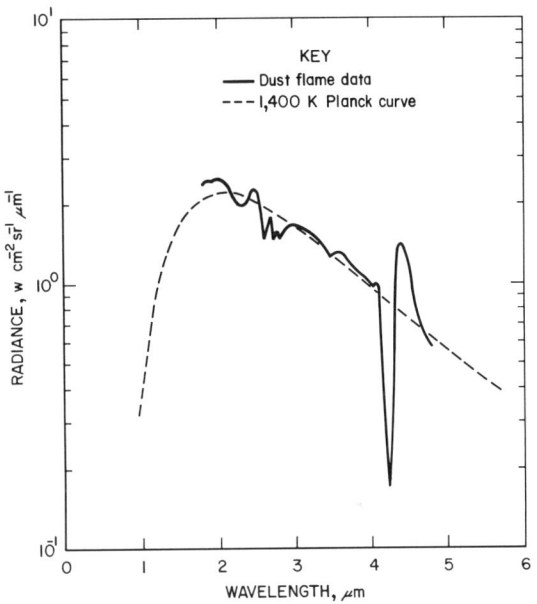

FIG. 9. IR spectral radiance for a coal dust explosion in a 9-liter chamber.

Based on the results of the previous studies with the rapid-scan spectrometer, two multi-wavelength infrared (IR) pyrometers were developed by the Bureau of Mines to observe dust explosions. The first pyrometer is a six-wavelength (6λ) instrument[20,21] that measures both particle and gas temperatures. It was developed both for laboratory use and for use within full-scale mine dust explosions. This 6λ pyrometer uses room temperature lead selenide (PbSe) photoconductive detectors and infrared interference filters to measure flame radiation simultaneously at all 6 wavelengths. An explosion-proof version of the 6λ pyrometer used in the Bureau of Mines experimental mine is shown in Fig. 10. Dust flame radiation enters through the sapphire window, and then an arsenic trisulfide lens focuses the radiation on the filters and PbSe detectors located on the front face of the detector-amplifier box. The 4.8 degree viewfield of the pyrometer is virtually common for all six detectors if the dust flame is optically thick close to the window. Far from the window, the detector fields of view diverge at an angle of 1.9 degrees from the center axis.

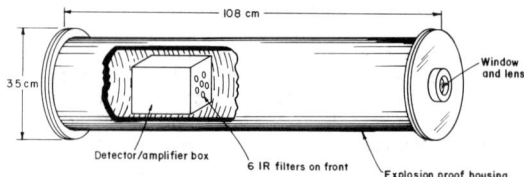

FIG. 10. Mine version of the six-wavelength (6λ) pyrometer.

A smaller version of the 6λ pyrometer is used to measure explosion temperatures in laboratory-scale flammability chambers. This model observes the flame through a sapphire window in the wall of the flammability chamber and therefore does not have to be explosion-proof. An aperture over the lens limits the viewfield to 4.0 degrees. As with the mine version, the viewfields of all 6 detectors are virtually identical for flames that are optically thick close to the window.

The electrical circuit for the 6λ pyrometer is described in previous publications.[20,21] Briefly, each of the 6 channels is a bridge circuit with one dark PbSe detector and a second detector exposed to the incoming radiation. The dark detector provides some temperature compensation for the circuit. This is needed because these photoconductive detectors' dark resistances vary with slight changes in ambient temperature. When the second PbSe detector views infrared radiation, its resistance decreases; and the resulting signal is amplified in two stages to provide an output voltage that is linearly proportional to the incident radiation. The intrinsic time response of the PbSe detectors is about 2 μsec, but the response of the pyrometer is limited to about 20 to 30 μsec due to the slower frequency response of the amplifiers. When viewing short time duration explosions, it is not necessary to mechanically chop the incoming radiation as would be necessary for longer time measurements.

Four of the wavelengths (1.57, 2.30, 3.84, and 5.00 μm) of the 6λ pyrometer were chosen to measure particle continuum radiation. These wavelengths correspond to minima in the radiation from a hydrocarbon gas flame as seen in Fig. 2. The other two wavelengths (4.42 and 4.57 μm) are at the CO_2 gas emission band. The narrow bandwidths of the six channels range from 0.10 to 0.26 μm. These bandwidths are narrow enough so that the effective wavelength of a channel does not change significantly when observing continua at different temperatures.

The second Bureau of Mines pyrometer is a three-wavelength (3λ) instrument that has been described in previous publications.[21,22] An explosion-proof version of the 3λ pyrometer is shown in Fig. 11. Flame radiation enters through a quartz window and is split by two cube beamsplitters into three parts, each of which passes through an interference filter to a corresponding silicon photodiode. The 3 wavelengths are 0.80, 0.90, and 1.00 μm and the bandwidths are 0.01 μm each. For this pyrometer also, the bandwidths of the 3 channels are narrow enough so that the effective wavelengths do not change significantly when observing continua of different temperatures. This permits the extrapolation of the blackbody calibration beyond the temperature range of the blackbody furnace. The output of each photodiode is fed directly into an operational amplifier, thereby providing a linear output over several orders of magnitude of input radiation. The measured time response of the detector-amplifier combination ranges from 10 to 30 μsec for different versions of the pyrometer containing different amplifiers. A laboratory version of the 3λ pyrometer views explosions through a window in the flammability chamber. The 3 detectors

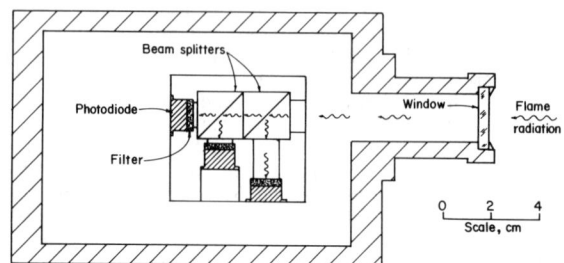

FIG. 11. Mine version of the three-wavelength (3λ) pyrometer.

in this version have identical fields of view of 4 degrees. The explosion-proof version (Fig. 11) of the 3λ pyrometer has been used to measure particle temperatures within full-scale mine explosions, and its detectors have identical viewfields of 8 degrees.

Other researchers[23] have used two-wavelength pyrometers in the visible or near-infrared region that operate on the same principle as the 3λ and 6λ pyrometers of this report. The advantage of having 3 or more wavelengths to calculate the particle temperature is that the standard deviation of the temperature fit to the Planck curve can be calculated. In addition, the 6λ pyrometer can measure gas temperatures in dust flames, which the two-wavelength pyrometers cannot do. The advantage of the pyrometers over the rapid-scan spectrometer is that there is much less data so that the temperature calculations can be done on a small, programmable calculator rather than on a computer. In addition, the pyrometers have a much faster response time and are more compact and more rugged. In some cases, however, the complete spectrum from the rapid-scan spectrometer is useful in better understanding the entire explosion phenomenon.

Since the voltage output from each channel of either the 3λ or 6λ pyrometer is linear with input radiation, Eq. (2) can be revised to:

$$b_\lambda V_\lambda = a\varepsilon L_\lambda = \frac{2a\varepsilon hc^2}{\lambda^5}\left[e^{hc/\lambda kT}-1\right]^{-1}, \quad (3)$$

where V_λ is the voltage at one of the wavelengths, a is a constant scale factor independent of wavelength, and b_λ is a calibration factor for each wavelength. The constant, a, is a function of the fraction of the field of view of the pyrometer that is filled by the flame and the attenuation of the flame radiation by intervening, cold, unburned dust. The calibration factor, b_λ, is measured for each of the channels by calibrating each pyrometer with the blackbody cavity radiation source over the temperature range of 1000 to 1500 K. The measured linearity of the pyrometers allows extrapolation to higher temperatures.

A cloud of burning dust particles emits continuum radiation according to Eq. (2). The emissivity of a single particle is less than one and may vary with wavelength, but the combined emitted and scattered radiation from an optically thick cloud of dust particles produces an effective emissivity of close to one, independent of wavelength. Even for an optically thin cloud, the radiation should be gray (i.e. emissivity independent of wavelength) if the burning dust particles are significantly larger than the pyrometer wavelengths.

In addition to the particle continuum radiation from a carbonaceous dust explosion, there are also discrete emission bands from gases, especially water vapor and carbon dioxide. For an optically thick flame that contains both hot particles and gases, the IR spectrum would be a continuum with superimposed emission or absorption bands depending on whether the gases are hotter or cooler than the particles (see Fig. 9).

For most of the observations with these two pyrometers, the dust cloud flame was optically thick and the emissivity was assumed to be gray for the temperature calculations. The signal attenuation due to a dust coating on the window was experimentally found to be independent of wavelength, as expected for particles larger than the pyrometer wavelengths. Therefore, in Eq. (3), the scale factor, a, could be combined with the emissivity, ε, into a single scale factor, $a\varepsilon$, that is independent of wavelength. The relative radiance at each wavelength is the measured voltage multiplied by the blackbody calibration factor. For the particle temperature calculations, the measured relative radiation at the dust continuum wavelengths is fitted by least squares to Eq. (3) to determine the two unknowns: the scale factor and the temperature. Note that the particle temperature calculations are made independently for the two pyrometers.

After the scale factor, $a\varepsilon$, has been determined for the 6λ pyrometer, the CO_2 gas temperatures can be calculated at the 4.42 and 4.57 μm wavelengths for optically thick flames by assuming the same scale factor (i.e., gray emissivity) and using Eq. (3) and the measured radiance. For dust explosions that are optically thin at the dust wavelengths but still optically thick at the CO_2 band, only the particle temperatures can be calculated since the scale factors, $a\varepsilon$, would be different for the dust and for the gas.

Even for optically thick flames, there are complications because the two pyrometer channels, 4.42 and 4.57 μm, observe radiation from both the particles and the CO_2 gas. Because of the high emissivity per molecule of gas at 4.42 μm, the gas usually reaches unit emissivity over a shorter path length than the dust; and therefore this channel would observe mainly gas radiation. At 4.57 μm the gas has a lower emissivity per molecule and this pyrometer channel may reach unit emissivity for the gas and for the dust over a similar path length. This means that this channel would measure radiation from both the gas and the dust and therefore measure a temperature that is some average of the gas and dust temperatures. An additional complication is that the gases in an explosion may cool as the flame front hits the window, and the resulting self-absorption would reduce the measured gas flame radiation, especially at 4.42 μm. As a result of the above complications, the true gas temperature in the explosion is assumed to be the higher of the two temperatures calculated at 4.42 and 4.57 μm if the temperatures differ significantly. For some confined dust explosions at high pressures, the 4.4 μm CO_2 gas band extends to longer wavelengths and even the 5.00 μm channel of the pyrometer may measure significant amounts of gas radiation in addition to particle radiation. In this case the particle temperature is calculated from the data of the first 3 channels of the pyrometer.

Examples of the temperature calculations are shown in Figs. 12 and 13, where the computed flame radiance, $b_\lambda V_\lambda/a\varepsilon$, is plotted versus the wavelength on a semi-logarithmic scale. Fig. 12 shows the 6λ pyrometer data from a full-scale mine coal dust explosion. (This data was from an earlier version[21] of the pyrometer in which the third wavelength was 3.46 μm rather than 3.84 μm.) The least squares fit of the 4 dust radiation data points to Eq. (3) is obtained by minimizing the squares of the deviations of the calculated scale factors at each wavelength; this is equivalent to minimizing the squares of the deviations of the measured radiance from the Planck curve on the semi-logarithmic plot. The resulting calculated temperature is 1220 K with a standard deviation of 20 K. The calculated average gas temperature is 1410 K, but the possible self-absorption at 4.42 μm may mean that the true gas temperature is closer to the value of 1460 K calculated at 4.57 μm.

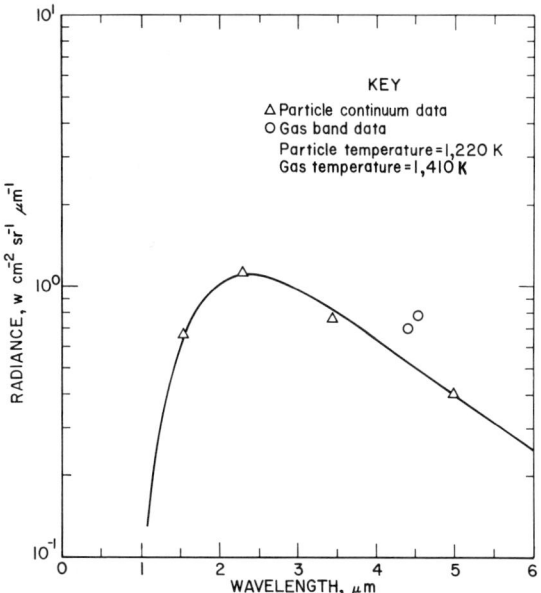

FIG. 12. IR radiance from a mine coal dust explosion, measured by the 6λ pyrometer.

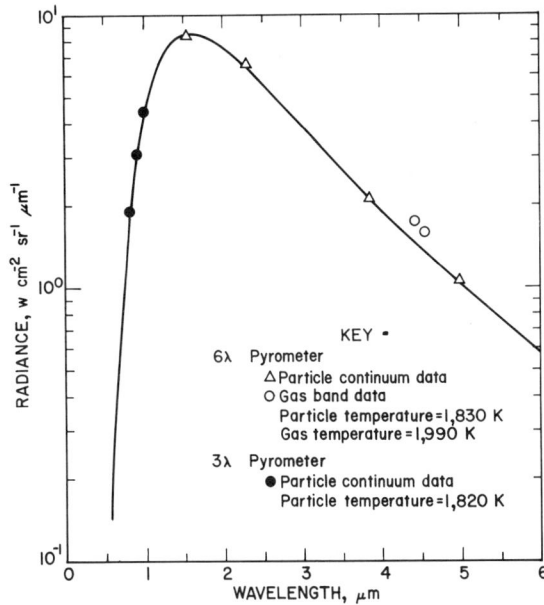

FIG. 13. IR radiance from a coal dust explosion in an 8-liter chamber, measured by the two pyrometers.

A second example of the temperature calculations is shown in Fig. 13. For this data, the 3λ pyrometer and 6λ pyrometer simultaneously observed a coal dust explosion through a sapphire window in the top of an 8-liter flammability chamber.[24] For each pyrometer, the measured relative radiation, $b_\lambda V_\lambda$, is divided by the calculated average scale factor, $a\varepsilon$, for that pyrometer and the combined radiance data are plotted in the figure. The calculated particle temperature is 1830 K for the 6λ pyrometer and 1820 K for the 3λ pyrometer. The plotted Planck continuum curve is for both temperatures. The calculated average gas temperature measured from the excess radiation at the CO_2 band is 1990 K.

Not all of the data obtained with the 6λ pyrometer is as good as the data in Figs. 12 and 13. For some confined dust explosions at high pressures, the 4.4 μm CO_2 gas band extends to longer wavelengths and even the 5.00 μm channel of the pyrometer may measure significant amounts of gas radiation in addition to particle radiation. In this case the particle temperature is calculated using the data from the first 3 channels of the pyrometer. For dust explosions that are optically thin at the dust wavelengths but still optically thick at the 4.4 μm CO_2 band, only the particle temperature can be calculated since the scale factors, aε, would be different for the dust and for the gas.

The 3λ and 6λ pyrometers have been used to simultaneously observe laboratory coal dust explosions in the closed 8-liter chamber. The details of the chamber and experimental procedure are found in reference 24. In brief, the dust was dispersed by a pulse of air; and the resulting, uniform dust cloud was ignited. The two pyrometers viewed the flame radiation through a sapphire window in the top of the vessel.

Fig. 14 shows the time history of such a dust explosion (425 g m^{-3} of Pittsburgh seam coal). The absolute pressure trace in Fig. 14A shows the initial rise in pressure due to the air pulse that dispersed the dust. Then the pressure remained constant at 1 atm absolute until the dust was ignited by electrically activated chemical matches at about 0.3 sec. The maximum explosion pressure was 6.9 atm, and then the pressure gradually dropped as the combustion products cooled. A representative trace from one of the pyrometer channels is shown in Fig. 14B. This radiance signal at 1.57 μm wavelength reached a peak when the flame front arrived at the window, shortly before peak pressure. Due to buoyancy, the flame front would reach the top window before reaching the bottom of the chamber. The additional, later burning in the bottom of the chamber caused the peak in the pressure trace to occur later than peak radiance. For comparison, a platinum-rhodium thermocouple (TC) trace is shown in Fig. 14C. The thermocouple wire was 1 mil, or 25 μm, and the bead size was about 2 mil, or 50 μm. This thermocouple responds fast enough to observe the explosion and measures a temperature similar to that measured by the pyrometers.

Flammability data for explosions of a narrow size distribution of Pittsburgh seam coal dust in air in the 8-liter chamber are shown in Fig. 15. The dust had an intrinsic surface mean diameter, $\bar{D}_s$, of 5 μm and mass mean diameter, $\bar{D}_w$, of 7 μm, although the agglomerated size (when the dust was dispersed in the chamber) was larger.[24] This Pittsburgh seam coal had a volatility of about 35 pct according to the ASTM D3175 standard test. The measured pressure ratio in Fig. 15C was the maximum explosion pressure (absolute) divided by the initial pressure (about 1 atm absolute). The data show no flame propagation, and therefore no pressure rise, below 100 g m^{-3}. Between 100 and 200 g m^{-3} there is a rapid increase in the explosion pressure. In a previous publication,[24] the authors chose a pressure ratio of 2.0 as a criterion of significant flame propagation; this gives a lean flammable limit of 130 g m^{-3} for this coal dust. At the higher concentrations, the pressure curve levels off as all the oxygen is consumed. This is shown in Fig. 15B, which is a graph of the decimal fraction of the oxygen in the air that is consumed during the explosion. This was calculated from the measured amount of oxygen left in the chamber after the explosion.

The pyrometer temperatures shown in Fig. 15A were calculated at the time of maximum radiance for each explosion. The particle temperatures were calculated separately for the 3λ and 6λ pyrometers, and these temperatures agreed to within experimental error at the higher concentrations. At the lower dust concentrations (below 200 g m^{-3}), the temperatures

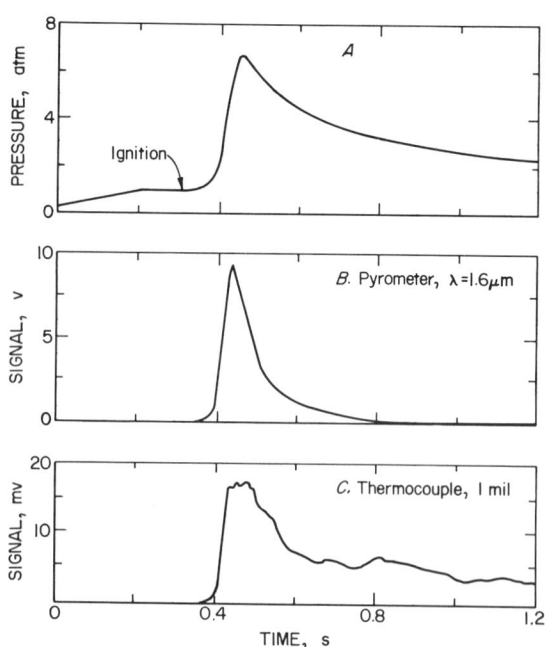

FIG. 14. Signal from one pyrometer channel compared with thermocouple signal and pressure trace for a dust explosion in an 8-liter chamber.

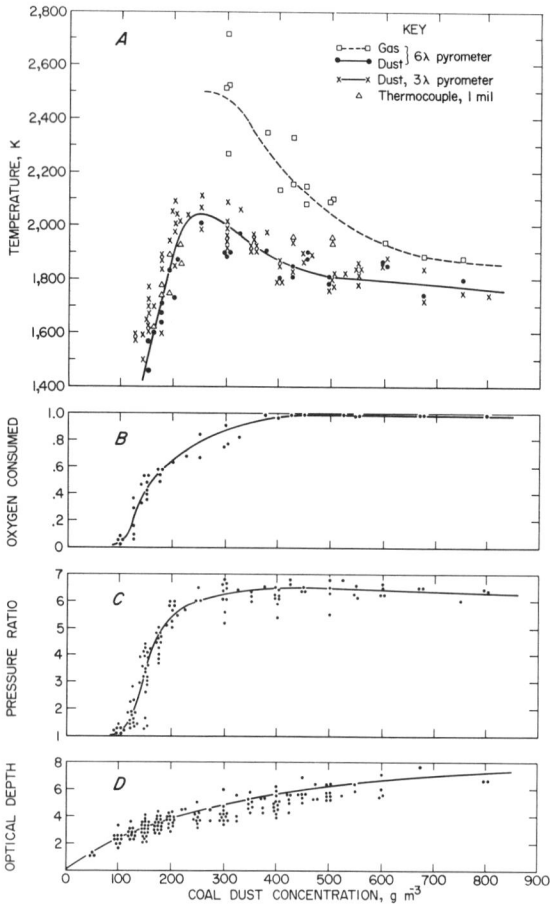

FIG. 15. Flammability data for 5 μm coal dust in an 8-liter chamber.

calculated from the 3λ pyrometer were significantly higher than those from the 6λ pyrometer. This was probably due to the fact that, as the lean flammable limit concentration was approached, the flame became more filamentary and did not fill the entire chamber. This would result in a broad range of temperatures in the chamber, and the 3λ pyrometer would measure only the hottest particles because radiation is a very steep function of temperature at wavelengths shorter than the Wien peak. At the longer wavelengths of the 6λ pyrometer, radiation is not as steep a function of the temperature, and the pyrometer would observe more of an average temperature.

The gas temperatures could not be calculated accurately at dust concentrations below 300 g m^{-3} because the dust becomes less optically thick and therefore has a lower emissivity than the 4.4 μm gas band. The variation in dust cloud emissivity with dust concentration can be seen from the data of Fig. 15D. The optical depth, τ, is defined as

$$\tau = -\ln t, \quad (4)$$

where t is the transmission through the dust cloud. An optical dust probe[25,26] measured the transmission (at λ = 0.94 μm) over a 3.8 cm path length through the unburned dust cloud just prior to ignition. Then the optical depth in Fig. 15D was calculated based on the computed transmission for a 20 cm path length. The real optical depth of the fireball would depend on its size at maximum radiance, when the temperatures were calculated. An approximate value of the emissivity of the fireball is

$$\varepsilon = 1 - e^{-\tau} . \quad (5)$$

This definition of emissivity neglects scattering or assumes that in-scattering and out-scattering cancel. The emissivity increases and approaches one as the optical depth increases. At the higher dust concentrations, where the flame propagation is rapid, the fireball is probably at least 20 cm in depth, and the emissivity calculated from Eq. (5) and the optical depth of Fig. 15D are probably accurate. However, at the lower concentrations, near the lean flammable limit, the flame is filamentary and the fireball depth is probably less than 20 cm. Therefore, the calculated flame emissivity would be an overestimate. Note that the emissivity calculated from the dust probe transmisson is for comparison only and is not used in the temperature calculations.

For the data of Fig. 15A, the maximum dust particle temperature and presumably the maximum gas temperature occur at a concentration of about 250 g m^{-3}; this value is close to the concentration at which the burning velocity is a maximum for dusts in this size range.[27] It is also approximately the concentration at which the amount of combustible volatiles from the coal is at a stoichiometric ratio with respect to the air. The temperatures measured in this constant volume chamber are higher than those of atmospheric flames due to adiabatic compression. The gas temperatures for these explosions are from one hundred to several hundred degrees higher than the dust particle temperatures. This tends to confirm a combustion mechanism in which the coal particles devolatilize, generating hydrocarbon gases which combust in the gas phase. The dust particles remain cooler because of their continuing endothermic pyrolysis and the fact that the seat of the exothermic reaction is in the gas phase at a significant distance from the particle.

The thermocouple measurements shown in Fig. 15A were made with the same 1 mil (or 25 μm) TC wire of Fig. 14C. These are uncorrected TC temperatures. At the high dust concentrations, there should be almost no emissivity correction for the TC because it is surrounded by hot particles. As expected, the TC measures a temperature that is intermediate between the gas and particle temperatures measured by the pyrometers. At the lower concentrations, as the dust cloud becomes optically thin, the emissivity correction for the TC would be larger. The uncorrected TC readings at the lower concentrations are comparable to the particle temperatures.

Flammability data from the 8-liter chamber for minus 120 mesh (<125 μm) Pittsburgh seam coal dust are shown in Fig. 16. This very broad distribution had $\bar{D}_S$ = 27 μm and $\bar{D}_W$ = 45 μm. For this dust, the 3λ pyrometer consistently measured a higher particle temperature than the 6λ pyrometer, even at the high concentrations. This probably resulted from a broad range of temperatures for the particles in the explosion. At wavelengths (0.8 to 1.0 μm) shorter than the Wien peak, the 3λ pyrometer would have observed a temperature strongly weighted by the hottest particles. At the longer wavelengths (1 to 5 μm) beyond the Wien peak, the 6λ pyrometer would have measured a less weighted average temperature.

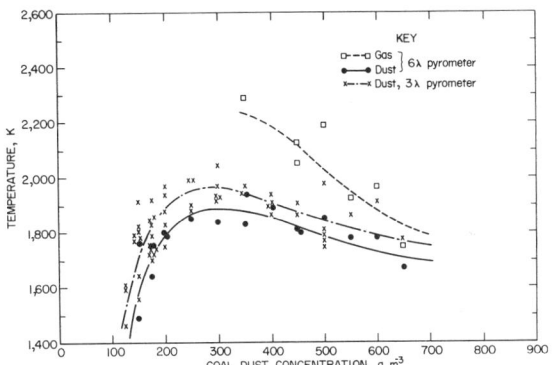

FIG. 16. Flammability data for a broad size distribution (minus 120 mesh) of coal dust in an 8-liter chamber.

Both the 3λ and the 6λ infrared pyrometers have also observed numerous full-scale dust explosions in the Bureau's experimental mine. Temperatures have been calculated as a function of time for several locations in the mine entry. The 6λ pyrometer data showed that the gas temperatures were usually several hundred degrees hotter than the dust particle temperatures, and that the maximum difference coincided with the passage of the active flame front. Later, as the flame cools, the gas and particle temperatures approach a common value.

TEMPERATURE ERRORS

The accuracy of the temperature measurements with the rapid-scan spectrometer and the two multi-wavelength, infrared pyrometers depends first on the calibration with the blackbody cavity furnace. It has a listed absolute accuracy of ±7 K or ±0.5% at 1500 K. For the gas explosion temperature measurements, the spectrometer was usually calibrated before each test. The two pyrometers were calibrated about every six months. For the 3λ pyrometer, calibration variations over a two year period corresponded to a temperature uncertainty of ±0.7% or ±10 K at a temperature of 1500 K. For the 6λ pyrometer, calibration variations over a three year period corresponded to a temperature uncertainty of ±1% or ±15 K at 1500 K.

Additional errors in the temperature measurements come from uncertainties in the emissivity of the gas or dust in an explosion. As discussed previously, measurements of absolute radiance from a gas explosion may be affected by absorption in cooler, intervening gas. Therefore, temperature calculations are not made after the flame front reaches the viewing window (see Figs. 4 and 5). A partial evaluation of the accuracy of the spectrometer gas temperature measurements was made by comparing them to the predicted adiabatic calculations, although the temperatures in any experimental study would be expected to be slightly lower than the adiabatic values due to various losses. An evaluation of the precision of the absorption-emission gas temperature measurements can be made from the scatter of the data in Fig. 8.

The pyrometer particle continuum temperature measurements were based on the assumption of gray emissivity and are thus subject to any variations in emissivity with wavelength. The best evidence for gray emissivity at high dust concentrations is the good agreement in temperature for the two pyrometers for explosions of narrow size distributions of dust (see Fig. 15) which could be expected to have an intrinsically narrow temperature range. For high concentrations (above 200 g m^{-3}) of two narrow sizes of Pittsburgh coal dust, the average temperature difference (and standard deviation) between the two pyrometers was $T_{3\lambda} - T_{6\lambda} = 35 \pm 57$ K. The precision of the pyrometer particle temperature measurements can be seen from the scatter in the data in Fig. 15A, although part of the scatter from test to test is due to real temperature variations correlated to the pressure variations in Fig. 15C. At the lower dust concentrations, the scatter becomes larger due to different methods of averaging by the two pyrometers and possible deviations from gray emissivity at the shorter optical depths. The 6λ pyrometer gas temperature measurements have larger uncertainties because they include the errors from the particle continuum temperature calculations plus additional errors due to the possible differences in emissivity between the gas and dust. Again, the scatter in data in Fig. 15A gives an estimate of the precision of the gas temperature measurements in dust explosions.

CONCLUSION

The data presented show that the IR radiation method is a convenient way of obtaining temperatures of gas explosions with fast time resolution. Rapid-scan spectrometer data were presented for methane-air explosions in both small-scale and large-scale spherical chambers. An analysis of the data showed that explosion temperature measurements based on the absolute spectral radiance at the 4.4 µm CO_2 band were in good agreement with thermodynamically calculated, adiabatic values. Also, essentially similar temperatures were measured by the absorption-emission method at the optically thinner 2.7 µm H_2O band.

Rapid-scan spectrometer IR radiation measurements were also used to determine temperatures in heterogeneous coal dust explosions. A three-wavelength and a six-wavelength pyrometer were then especially developed for temperature measurements within full-scale mine dust explosions. Dust particle temperatures were calculated based on the best Planck curve fit to the IR continuum radiation. For the spectrometer and 6λ pyrometer, gas temperatures were calculated by using the differential radiation at the 4.4 µm CO_2 band. In all of the dust explosions observed, gas temperatures were significantly higher than particle temperatures. For dust explosions with temperatures below the melting point of platinum-rhodium, rapid-response thermocouples also measured temperatures that were a realistic average between the gas and particle temperatures measured by the 6λ IR pyrometer. The advantages of the pyrometers over the spectrometer were that they were more compact, easier to use, and explosion-proof. The 3λ pyrometer was the simplest and easiest to use, but it measured only the particle continuum temperature. The 6λ pyrometer measured both the particle and gas temperatures for dust explosions.

ACKNOWLEDGMENTS

The authors wish to acknowledge the contributions of A. L. Furno, A. L. Johnson, M. J. Sapko, and W. F. Donaldson of the Pittsburgh Research Center (PRC) to the measurements of gas explosion temperatures. In particular, the authors wish to thank Mr. Johnson for permission to use data from his master's thesis. The measurements of dust explosion temperatures included the assistance of W. F. Donaldson, C. D. Litton, J. J. Opferman, and J. Leff, all of PRC.

REFERENCES

1. M. G. Zabetakis, S. Lambiris, and G. S. Scott, "Flame Temperatures of Limit Mixtures" Seventh Symposium (International) on Combustion (Butterworths, London, 1959), pp. 484-487.
2. M. Hertzberg, The Theory of Flammability Limits, Bureau of Mines Report of Investigations 8127, 15 pp. (1976).
3. R. M. Fristrom and A. A. Westenberg, Flame Structure (McGraw-Hill, New York, 1965), p. 151.
4. T. A. Milne and J. E. Beachey, Exploratory Studies of Flame and Explosion Quenching, Appendix H, (Bureau of Mines Research Contract No. H0122127), June 1976, OFR 112-77, available from NTIS, Springfield, Va., PB 268 695/AS.
5. C. Fery, "Sur la Temperature des Flammes" Compt. Rend. Acad. Sci. 137, 909 (1903).
6. R. H. Tourin, Spectroscopic Gas Temperature Measurement (Elsevier, Amsterdam, 1966), pp. 25-30, 34-38.
7. H. P. Broida, "Experimental Temperature Measurements in Flames and Hot Gases" Chap. 17 in Temperature, Its Measurement and Control in Science And Industry, Vol. 2, ed. H.C. Wolfe, (Reinhold, New York, 1955), p. 278.
8. A. G. Gaydon, and H. G. Wolfhard, Flames: Their Structure, Radiation, and Temperature, 4th ed., (Chapman and Hall, London, 1979), chap. 10.
9. H. Schmidt, "Prufung der Strahlungsgesetze der Bunsenflamme" Ann. Physik. 29, 971-1028 (1909).
10. S. Silverman, "The Determination of Flame Temperatures by Infrared Radiation" J. Opt. Soc. Am. 39, 275-277 (1949).
11. A. G. Gaydon, The Spectroscopy of Flames, 2nd ed., (Chapman and Hall, London, 1974), pp. 224-225.
12. J. M. Adams, "The Spectral Comparison Method for Temperature Measurement in Two-Phase Flames" Chap. 57 in Temperature, Its Measurement and Control in Science and Industry, Vol. 4, Part 1, ed. H. H. Plumb, (Instrument Society of America, Pittsburgh, 1972).
13. S. A. Dolin, H. A. Kruegle, and G. J. Penzias, "A Rapid-Scan Spectrometer That Sweeps Corner Mirrors Through the Spectrum" Appl. Opt. 6, 267-274 (1967).
14. M. Hertzberg, C. D. Litton, W. F. Donaldson, J. M. Kuchta, and A. L. Furno, The Spectral Growth of Expanding Flames, Bureau of Mines Report of Investigations 7779, 38 pp. (1973).
15. A. L. Johnson, Measurements of Infrared Brightness Temperatures with a Rapid-Scan Spectrometer System, M. A. thesis, Dept. of Physics and Astronomy, Brigham Young University, Provo, Utah, Dec. 1976.
16. C. C. Ferriso, "High Temperature Spectral Absorption of the 4.3-Micron CO_2 Band" J. Chem. Phys. 37, 1955-1961 (1962).
17. R. H. Tourin, "Measurements of Infrared Spectral Emissivities of Hot Carbon Dioxide in the 4.3-µm Region" J. Opt. Soc. Am. 51, 175-183 (1961).

18. M. Hertzberg, A. L. Johnson, J. M. Kuchta, and A. L. Furno, "The Spectral Radiance Growth, Flame Temperatures, and Flammability Behavior of Large-Scale, Spherical, Combustion Waves" Sixteenth Symposium (International) on Combustion (The Combustion Institute, Pittsburgh, 1977), pp. 767-776.
19. A. L. Johnson, A. L. Furno, and J. M. Kuchta, Infrared Spectral Radiances and Explosion Properties of Inhibited Methane-Air Flames, Bureau of Mines Report of Investigations 8246, 22 pp. (1977).
20. K. L. Cashdollar, M. Hertzberg, and C. D. Litton (assigned to U.S. Department of the Interior), Multichannel Infrared Pyrometer, U.S. Patent 4,142,417 (March 6, 1979).
21. K. L. Cashdollar and M. Hertzberg, "Infrared Pyrometers for Measuring Dust Explosion Temperatures" Opt. Eng. $\underline{21}$, 82-86 (1982).
22. K. L. Cashdollar, "Three-Wavelength Pyrometer for Measuring Flame Temperatures" Appl. Opt. $\underline{18}$, 2595-2597 (1979).
23. A. B. Ayling and I. W. Smith, "Measured Temperatures of Burning Pulverized-Fuel Particles..." Combust. Flame $\underline{18}$, 173-184 (1972).
24. M. Hertzberg, K. L. Cashdollar, and J. J. Opferman, The Flammability of Coal Dust-Air Mixtures, Bureau of Mines Report of Investigations 8360, 70 pp. (1979).
25. I. Liebman, R. S. Conti, and K. L. Cashdollar, "Dust Cloud Concentration Probe" Rev. Sci. Instr. $\underline{48}$, 1314-1316 (1977).
26. K. L. Cashdollar, I. Liebman, and R. S. Conti, Three Bureau of Mines Optical Dust Probes, Bureau of Mines Report of Investigations 8542, 26 pp. (1981).
27. M. D. Horton, F. P. Goodson, and L. D. Smoot, "Characteristics of Flat, Laminar Coal-Dust Flames" Combust. Flame $\underline{28}$, 187-195 (1977).

A packaged, fiber-optic spectroradiometer for high temperature gases, with automatic readout[a]

S. A. Self, P. H. Paul, and P. Young

Mechanical Engineering, Stanford University, Stanford, California 94305

A packaged, automated instrument for temperature measurements in high temperature gases is being developed, based on the emission-absorption (or line-reversal) technique. It employs optical fibers to transmit radiation from a standard lamp to the gas, and from the gas to a detector. Miniature choppers are used to generate a repetitive sequence of three signals, one from the lamp, a second from the gas, and a third transmitted from the lamp through the gas. Following electronic demodulation, the three signals are used to calculate the gas temperature from a simple algorithm in a minicomputer or microprocessor, which output the temperature with a time resolution of 1 msec. The instrument is intended for use with gases seeded with Na or K in the temperature range 2000–3000 K. Spectral selection of a resonance line is accomplished with a narrow band interference filter or Fabry-Perot etalon. Provision is made for simultaneously generating signals at a wavelength detuned from the line, which extends its use to the case of particle-laden gases.

INTRODUCTION

The spectroscopic emission-absorption or line-reversal technique is an old established, non-intrusive method for measuring gas and flame temperatures above the range accessible to thermocouples or other probe techniques. Numerous variations of the basic technique have been proposed and used, as reviewed by Tourin.[1]

In the basic method, the gas is transilluminated by radiation from a calibrated tungsten ribbon lamp, and this light, together with radiation from a suitable emission line from the gas is focused onto the entrance slit of a monochromator.

Usually a resonance line from an alkali metal, lying suitably in the visible region, is used, the gas being seeded to a suitable concentration with a convenient compound of the metal.

In the original line-reversal technique the spectral response is observed either by eye, with the exit-slit of the monochromator removed, or from the output of a detector as the monochromator is slowly scanned across the line. If the lamp's brightness temperature T_L is lower than the gas temperature T_G, the gas radiation is seen as an emission peak against the continuum background from the lamp, whereas if $T_L > T_G$ the line is seen as an absorption dip in the continuum. The basic procedure is then to adjust the lamp current until the line is invisible against the background (the reversal condition), in which case $T_G = T_L$.

This basic method has several drawbacks. First, the judgement of the reversal condition is subjective and is compounded by the fact that when the temperature of the gas is inhomogenous along the optical path, a true reversal is not obtained. Second the measurement is slow and yields only a long time average value of T_G. Third, the method is restricted to gas temperatures less than the maximum (~2300°C) at which tungsten lamps will maintain a stable calibration.

While several modified techniques have been employed to overcome some of these drawbacks, it is generally true that radiometric techniques have mostly been restricted to scientific investigations of laboratory scale combustion systems having an environment characteristic of a physics laboratory. The extension of such techniques for measurements on larger scale industrial systems poses a number of problems related to obtaining optical access and, particularly, of interfacing a rigid and bulky optical system requiring critical alignment with an inflexible combustion rig in an industrial environment. For instance, some of the problems of making measurements in combustion MHD generator systems have been discussed by Self.[2]

The aim of the present work is to develop a self-contained, packaged and versatile instrument that can readily be interfaced with larger scale combustion flow systems. The use of optical fibers eases the problem of coupling the optical system to the rig and allows the instrument package to be mounted remotely. Moreover, by using miniature choppers and modern electronics, a direct read out of temperature can be obtained, with a time resolution of 1 msec, that is not limited to temperatures below the lamp temperature.

The initial motivation of this work was directed to the measurement of temperature in combustion MHD flows, where the temperature may range up to 3000 K and the gas is heavily seeded with potassium, typically 1% K by mass. However it is also directed to more general combustion applications where the flow would be seeded to a much lower level of sodium (say) to facilitate the measurement.

THEORETICAL BASIS

The theoretical basis of the emission-absorption technique may be summarized as follows.

The power flux of blackbody radiation across an area dA, in a solid angle $d\Omega$ and in a wavelength interval $d\lambda$ is

$$dP = N_\lambda^b(T) \, dA \, d\Omega \, d\lambda \quad , \qquad (1)$$

where the spectral radiance is given by

$$N_\lambda^b(T) = C_1 \lambda^{-5} \exp(-C_2/\lambda T) \quad , \qquad (2)$$

and $C_1 = 1.1909 \times 10^{-16}$ watt m^2 steradian^{-1}
$C_2 = 1.4380 \times 10^{-2}$ m K

are the first and second radiation constants. Equation (2) is Wien's approximation to the Planck function, valid when $\exp(C_2/\lambda T) \gg 1$, which is well satisfied for visible wavelengths and $T < 3000$ K.

Considering radiation propagating in the x direction through the gas, the increase in spectral radiance dN_λ across the elementary slab dx at x is the sum of the increase due to emission and the decrease due to absorption (scattering being neglected),

$$dN_\lambda = e_\lambda \, dx - \alpha_\lambda N_\lambda \, dx \quad . \tag{3}$$

Here e_λ (watts m^{-4} steradian^{-1}) is the volumetric spectral radiance of the gas emission and α_λ (m^{-1}) is the monochromatic absorption coefficient.

By Kirchoff's law, for a gas in local thermodynamic equilibrium at temperature T_G

$$e_\lambda = \alpha_\lambda N_\lambda^b(T_G) \quad . \tag{4}$$

Then from Eqs. (3) and (4) we have the differential equation for radiation transfer in the form

$$\frac{dN_\lambda}{dx} = \alpha_\lambda N_\lambda^b(T_G) - \alpha_\lambda N_\lambda \quad . \tag{5}$$

Note that in an inhomogeneous gas, α_λ is a function of x through the variation of composition and temperature, while $N_\lambda^b(T_G)$ is a function of x through the variation of $T_G(x)$. Also α_λ and N_λ^b are functions of λ, with α_λ being a rapidly varying function in the vicinity of a spectral line and often taken as having a Voigt profile.

Equation (5) can be integrated across a slab of gas extending from $x = 0$ to $x = \ell$ to give

$$N_\lambda(\ell) = \exp-\tau_\lambda(\ell) \int_0^\ell \alpha_\lambda(x) N_\lambda^b(x) \exp \tau_\lambda(x) dx$$
$$+ N_\lambda(0) \exp-\tau_\lambda(\ell) \tag{6}$$

where the optical depth is defined by

$$\tau_\lambda(x) = \int_0^x \alpha_\lambda(x) \, dx \quad . \tag{7}$$

In Eq. (6) the first term is the radiation emitted by the gas that reaches $x = \ell$, taking account of absorption, while the second term is the radiation reaching $x = \ell$ from the radiation $N_\lambda(0)$ incident at $x = 0$, taking account of absorption. For a gas of uniform composition and temperature, Eq. (6) can be written explicitly:

$$N_\lambda(\ell) = [1 - \exp - \tau_\lambda(\ell)] N_\lambda^b(T_G)$$
$$+ [\exp - \tau_\lambda(\ell)] N_\lambda^b(0) \quad , \tag{8}$$

where $\tau_\lambda(\ell) \to \alpha_\lambda \ell$ in this case.

When the gas is optically thick, $\tau_\lambda(\ell) \gg 1$, e.g. with heavy seeding or near the center of a strong line, the radiation exiting the gas, $N_\lambda(\ell)$, is due primarily to gas emission (first term) and originates predominantly from the region near $x = \ell$; very little of the incident radiation (second term) penetrates to $x = \ell$. On the other hand when the gas is optically thin, $\tau_\lambda(\ell) \ll 1$, e.g. with light seeding or remote from line center, the gas contributes little to $N_\lambda(\ell)$, and the incident radiation is transmitted with little attenuation.

Now suppose the incident radiation is derived from a source of brightness temperature T_L. Then for a uniform gas the spectral radiance exiting the gas when transilluminated by the lamp is

$$N_\lambda(\ell) = [1 - \exp - \tau_\lambda(\ell)] N_\lambda^b(T_G)$$
$$+ [\exp - \tau_\lambda(\ell)] N_\lambda^b(T_L) \quad . \tag{9}$$

Clearly, when $T_L = T_G$,

$$N_\lambda(\ell) = N_\lambda^b(T_G) = N_\lambda^b(T_L) \quad , \tag{10}$$

which is the reversal condition. For a uniform gas the reversal condition is satisfied simultaneously at all wavelengths. However, the maximum sensitivity in setting $T_L = T_G$ increases with $\tau_\lambda(\ell)$ i.e. with heavy seeding and at line center.

The actual signal voltage at the detector output from the gas transilluminated by the lamp may be written

$$V_{G+L} = [1-\exp-\tau_\lambda(\ell)] N_\lambda^b(T_G) K_2 S \delta A \delta \Omega \delta \lambda$$
$$+ [\exp - \tau(\ell)] N_\lambda^b(T_L) K_1 K_2 S \delta A \delta \Omega \delta \lambda \tag{11}$$

Here δA is the effective beam area, $\delta \Omega$ is the effective solid angle, $\delta \lambda$ is the effective spectral passband, K_1 is the transmission of the transmitting optics (between lamp and gas), K_2 is the transmission of the receiving optics (between gas and detector) and S is the detector sensitivity in volts/watt (assumed linear). If $\delta \lambda$ is large enough that α_λ varies significantly, then V_{G+L} should be determined by an integral of $N_\lambda(\ell)$ over the passband.

Two points of practical importance should be made regarding Eq. 11. The effective beam area δA is determined by the limiting field stop of the optical system (usually the monochromator entrance slit) while $\delta \Omega$ is determined by the limiting aperture stop (usually set by the F-number of the monochromator). It is essential that the product $\delta A \delta \Omega$ be identical in each term in Eq. (11), otherwise gross errors in temperature measurements will result. Physically, this means that the transmitting optics be designed so that every element of gas contributing to the emission reaching the detector must be transilluminated by radiation from the lamp. It should also be noted that while K_2, the transmission of the receiving optics appears in both terms of Eq. (11) and is thus immaterial, the transmission K_1 of the transmitting optics appears only in the second term (the radiation received from the lamp). Thus in using Eq. (11) as a basis for measurement, K_1 must be known either from measurement or estimation. For instance, assuming the transmitting optics consists of a simple lens and window the four surfaces will give a total reflective loss of ~16% which corresponds to an effective reduction of lamp temperature of ~2% or 60 K at 2800 K. The correction can be significantly reduced by using AR coated optics. As an alternative, the lamp brightness temperature can be calibrated with a pyrometer sighted through the transmitting optics before or after the hot gas is present, in which case one takes $K_1 = 1$ in Eq. (11). It is also evident that any uncontrolled change in K_1 due to input window contamination during the course of a measurement can lead to significant errors.

The foregoing discussion has given the theoretical basis for spectroradiometric methods, with particular reference to the conventional line reversal technique for a uniform gas. When the gas temperature (or seed concentration) is nonuniform along the line of sight, exact reversal is not obtained simultaneously at all wavelengths, and the method yields some weighted average temperature. The particular weight function will depend on the temperature distribution and the optical depth i.e. the exact wavelength at which the reversal condition is taken.

AUTOMATED METHODS

Several modifications of the basic line-reversal technique have been employed[3,4,5] which lend themselves to an automated instrument with a shortened time resolution.

Because tungsten lamps take a relatively long time to equilibrate when their current is changed, it is clear that the lamp temperature should be held constant, and one should abandon the null (reversal) method in favor of one utilizing the out of balance signals obtained when the lamp radiation is alternately blocked and unblocked with a chopper disk. This allows one to measure temperatures above the maximum lamp temperature, though as a general rule maximum accuracy is obtained when $T_L \gtrsim T_G$.

Two methods[4,5] involve making measurements at two wavelengths, essentially on and off the line, by scanning a monochromator or filter. However such mechanical scanning is slow and limits the time resolution to greater than ~1 sec. Measurements at two wavelengths can be made simultaneously by using two detectors. However, because of the problem of drift in the relative sensitivities of detectors, especially photomultipliers, this is generally not a desirable practice. It is better to use a single detector to measure the separate signals in rapid sequence and to demodulate the signal electronically into separate channels.

A successful automatic instrument has been developed by Vasil'eva et al[3] and extensively used on combustion MHD flows of relatively large size. This instrument uses a fixed lamp temperature, a single photomultiplier detector and a single wavelength selected by a monochromator to be on or to the short wave side of the Na 589.0 nm resonance line, depending on the optical depth. The method relies on generating a repetitive sequence of three signals through the monochromator onto the detector. The first, S_L, is from the lamp alone, the second, S_G, is from the gas emission alone (the first term in Eq. (11)) and the third, S_{G+L}, is from the gas transilluminated by the lamp (both terms in Eq. (11)).

From Eq. (11) it is readily shown that

$$T_G = T_L \left\{ 1 - \left(\frac{\lambda T_L}{C_2}\right) \ln\left[\frac{K_1 S_G}{S_G + S_L - S_{G+L}}\right] \right\}^{-1} . \quad (12)$$

Thus a measurement of the three signals allows T_G to be calculated in terms of T_L. The sequence of signals is demodulated by electronic gating to yield the signals in three separate channels. These signals are then digitized and inputted to a minicomputer which calculates T_G from the algorithm (12) in real time.

An important question in this, and similar methods concerns the generation of the signal from the lamp alone, in the absence of the gas. Ideally this signal should be transmitted over the same optical path as the others, but it is obviously impossible to remove and replace the hot gas from the path repetitively. In principle it would be possible to determine a value for S_L initially, before the hot gas flow is established, and store this value for use throughout a measurement run. However, it is clearly desirable to continuously generate S_L signals throughout the measurement run so as to compensate for drifts in lamp temperature or detector sensitivity.

In the Soviet instrument[3] a lamp signal S_L is generated continuously from a secondary lamp, nominally identical to the primary one, which is situated on the receiving side of the system and coupled into the optical path transmitting the signals S_G and S_{G+L} by a beam splitter ahead of the monochromator. The sequence of three signals is then generated by two synchronized choppers, with appropriate blade design, one on the transmitting side to periodically block the primary lamp, the other on the receiving side to periodically block the secondary lamp. The signal from the secondary lamp is equalized to that from the primary lamp prior to establishment of the hot gas flow. This system is, of course, vulnerable to relative changes of brightness temperatures of the two lamps during a measurement run. The chopper disks rotate at 3,000 rpm, which gives an individual signal duration of 3 msec and a measurement resolving time for the three signal sequence of ~10 msec.

FEATURES OF THE INSTRUMENT

The instrument being developed is functionally similar to the Soviet instrument, just described, but incorporates a number of novel features with the principal aim of producing a versatile, packaged instrument which can be located remotely from the test rig.

A schematic of the optical configuration is shown in Fig. 1. By using optical fibers rather than fixed elements for the measurement path, the instrument can be packaged and located remotely from the test rig. The use of fibers also simplifies the optical access to the gas flow and increases versatility in that only the lens/port combination coupling the fibers to the flow is specific to a particular system.

In Fig. 1, radiation from the standard tungsten strip lamp L is coupled via an F.1 lens into a 1 mm diameter step index quartz fiber (Numerical Aperture 0.2) to transmit radiation to the test rig where it is coupled through the flow by transmitting and receiving lenses into a receiving fiber.

The design of the coupling optics will depend on the particular configuration and conditions of the test rig. Typically the fiber ends and lenses will be mounted rigidly in tubes which slide in port tubes attached to the channel walls and have screw adjustments for alignment. It is convenient to use the lenses themselves (with suitable gaskets) as windows sealing the flow, and normally a slight flow of dry nitrogen will be needed to purge the ports and prevent contamination of the lens surfaces.

A magnified image of the end of the transmitting fiber is produced at the center of the flow channel and a demagnified image of this is produced at the end of the receiving fiber. To satisfy the requirement that all elements of the gas contributing an emission signal into the receiving fiber are transilluminated by lamp radiation, two features are incorporated. First, the receiving fiber is made of smaller diameter (e.g. 0.6 mm) than the transmitting fiber; second, the receiving lens has an aperture stop (A.S.) so that the numerical aperture for collection is smaller than that for illumination.

A reference path, identical to the measurement path (apart from the fiber length), is provided inside the instrument package. This is generated from the back surface of the strip lamp, and conveniently avoids the use of a secondary lamp to establish the lamp signal S_L on a continuous basis. The provision of a dummy set of transmitting/receiving optics in the reference path ensures that the reflection losses in the two paths will be closely equal and also facilitates the use of a single chopper C_1 to alternately chop the two paths. Any slight difference in transmission between the two paths can be zeroed by

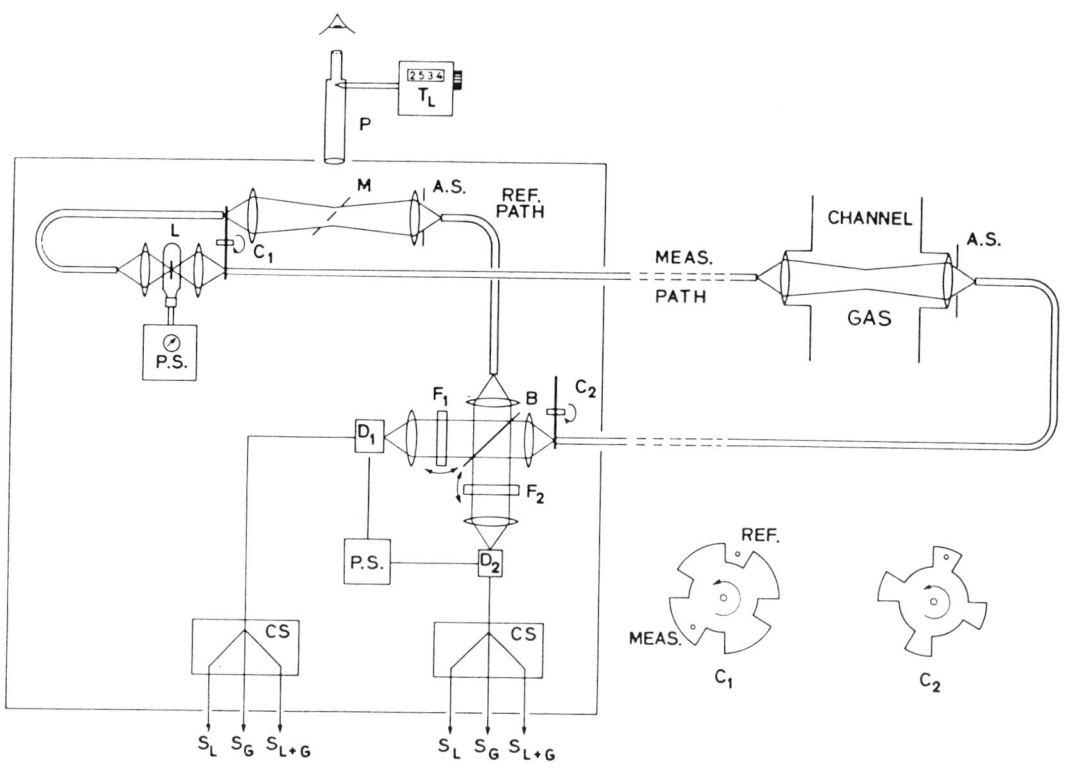

Fig. 1 Schematic of Optical System

adjusting the aperture stop in the reference path to give equal signals prior to the establishment of the hot gas flow.

Provision is made for checking the effective brightness temperature of the lamp, as measured at the image in the flow (i.e. taking account of the transmission loss K_1) by inserting, temporarily, the high reflectivity mirror M into the reference path. This allows a calibrated optical pyrometer to be sighted onto the image of the fiber end at the center of the dummy channel without disturbing the optical system.

The optical signals at the receiving ends of the measurement path and reference path fibers are each collimated by F.1 lenses and combined by a pellicle beam splitter. This provides two parallel optical channels each of which is spectrally filtered (F_1, F_2) and condensed by lenses onto separate, identical detectors (D_1, D_2) fed from a common power supply (P.S.). The detectors are silicon photoconductive diodes having integral amplifiers. These have a higher quantum efficiency than photomultipliers in the visible, are more stable and have much less sensitivity to ambient magnetic fields (for MHD applications).

In keeping with the aim of engineering a compact instrument, it is planned to avoid the use of a grating monochromator for spectral selection, as is common practice for such measurements. Instead we are experimenting with either narrow band multilayer dielectric filters or a Fabry-Perot etalon, depending on the application.

For combustion MHD applications, where the gas is heavily seeded with potassium, it is convenient to work on the short wavelength side of the shorter resonance line at $\lambda = 766.5$ nm. Because of the heavy seeding the resonance lines are very broad and usually self-reversed at line center due to the presence of cooler boundary layers. To obtain a reasonable weighted spatial average of the temperature with minimal correction for the boundary layer, and yet obtain good sensitivity of measurement, it is desirable to work at a wavelength, suitably detuned from line center, such that the optical depth $\tau_\lambda(\ell)$ has a value of 1/2 or thereabouts.[6,7] Typically this requires a detuning of 1-2 nm from line center. Currently we are using a 2-cavity filter of half width (FWHM) of 0.3 nm centered at 766.6 nm at normal incidence, and which can conveniently be tuned to shorter wavelengths by tilting through a small angle.

For measurements in more conventional combustion systems, where it is desirable to use only enough alkali metal seed to facilitate the measurement (e.g. to give an optical depth $\tau_\lambda(\ell) \lesssim 1$ at line center) we plan to use the shorter wavelength resonance line of sodium at 589.0 nm. In this case the line will be relatively narrow and higher spectral resolution is required. Since evaporated dielectric filters are not readily available with bandwidths less than ~0.3 nm, we are currently experimenting with an air-spaced, thermally stabilized, Fabry-Perot etalon for this application. This has a free spectral range of 1.8 nm, a finesse of 75 and hence a passband (FWHM) of 0.025 nm, with a peak transmission of 75%. It is piezo-electrically tunable and used in conjunction with a subsidiary 3-cavity dielectric filter of bandwidth 1.2 nm centered at 589.2 nm. This ensures single mode transmission of the Fabry-Perot which can be electronically set anywhere in the range 589.2 – 588.0 nm spanning the resonance line 589.0 nm and its short wavelength wing.

For either application, using either the resonance lines of potassium or sodium, the second detector channel will be filtered with a narrow band-dielectric filter detuned sufficiently to the short wavelength side of the line (e.g. by 10 nm or more) to a region where, in the absence of particulates the gas would be essentially transparent. For gas flows where the particle loading is sufficiently low that the extinction due to scattering and absorption by the particles is not too great, it appears possible to use the three signals from this second channel to correct the gas temperature measurement for the effects of particle extinction and emission and also to obtain information on the particle loading and temperature.

To produce the required multiplexed sequence of three signals in each channel, two synchronized choppers C_1, C_2 are used, whose blade slot configuration is shown inset in Fig. 1. Chopper C_1 serves to sequentially pass lamp radiation to the reference path (S_L), then to block lamp radiation to both paths (S_G) and finally to pass lamp radiation to the measurement path (S_{L+G}). Chopper C_2 serves to block radiation from the measurement path during the first interval, and to pass radiation from the measurement path during the second and third intervals. By locating the chopper blades close to the fiber ends, it is possible to utilize miniature choppers (~1" diam blades) and achieve a minimum sequence time or data cycle of 1 msec.

Each signal occupies $120° \pm \varepsilon$ of a data cycle, where ε is some small phase angle fluctuation inherent in the choppers. To accomplish sequential multiplexing it is necessary to lock the choppers to the same frequency with a specified relative phase angle.

An electronic system has been designed and tested to provide the necessary chopper control and monitoring, and to perform the data demultiplexing. A schematic circuit diagram is shown in Fig. 2.

be adjusted by the phase shifter. Further, by comparison with a pre-set memory, an error flag is generated should either chopper go out of phase during operation. By this means the choppers operate stably and reliably to provide sequential signal multiplexing.

By synchronously demultiplexing the signal via gated averaging and sample and hold circuits, all three signals are simultaneously available for asynchronous acquisition by a minicomputer to compute the temperature via the algorithm (12). To facilitate computer acquisition an eight input analog multiplexer under computer control has been added. The eight inputs consist of the six demultiplexed data signals (three from each detector channel) together with signals from the lamp current shunt and the pyrometer current shunt.

DISCUSSION

The present status of development of the instrument is that two optical systems have been breadboarded, one designed for potassium seeded combustion MHD flows, the other for combustion flows lightly seeded with sodium. Also, a complete electronics system has been constructed and tested satisfactorily.

Laboratory tests are in progress to establish the operating characteristics and the measurement uncertainties over a reasonably wide range of temperature and seed concentration using laboratory burners. To simulate the temperatures and potassium seed concentrations appropriate to combustion MHD conditions, a special bench top burner is being constructed. For sodium seeded combustion flows a

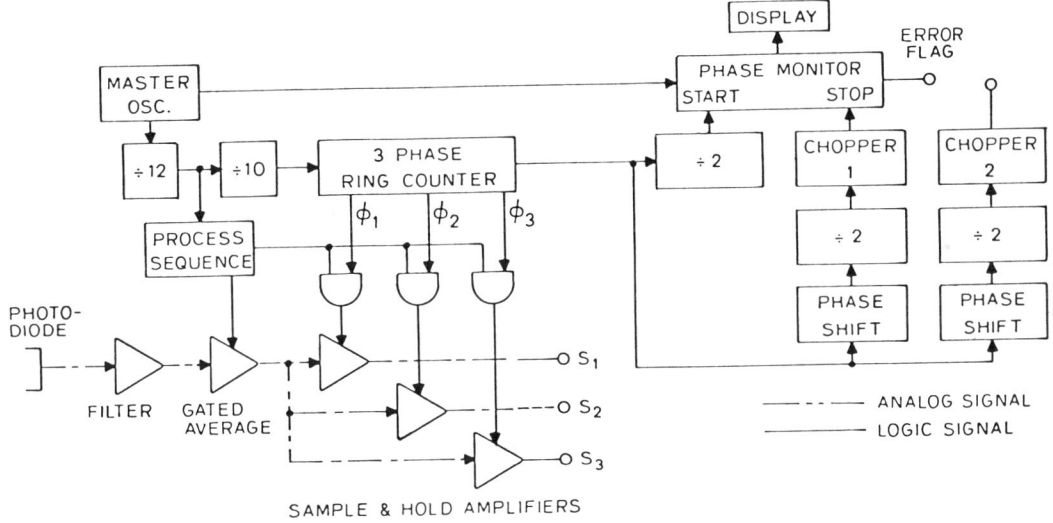

Fig. 2 Schematic of Electronic System

The multiplexed analog signals from each photodiode detector are fed to a gated averager and thence to three parallel sample and hold amplifiers. Intersignal transitions are gated out, as there is some overshoot due to bandwidth limitations. An average is imposed to remove frequencies greater than the sampling rate, to allow the use of sample and hold circuits.

System timing is derived from a high frequency crystal oscillator, divided down to provide stability. By using a frequency 720 times that of the chopper drives, and counting the number of cycles between the leading edge of the phase reference and that of a chopper sync output, it is possible to measure accurately the chopper phase angle, which can

Perkin-Elmer slot burner with an atomizer for seeding with salt solution is being employed.

In either case, temperature measurements from the instrument will be compared with simultaneous measurements from a conventional emission-absorption apparatus employing fixed optics and a monochromator for spectral selection. In addition, in the case of the burner simulating MHD conditions, where the potassium concentration is known, a sensitive cross check of the temperature can be made by measurements of the electron concentration using submillimeter interferometry[8],[9] through the use of the Saha equilibrium relation.

REFERENCES

a) This work is supported by the National Science Foundation and the U.S. Air Force Office of Scientific Research.

1. Tourin, R. H., *Spectroscopic Gas Temperature Measurement*, Elsevier, New York (1966).

2. Self, S. A., Diagnostic Techniques for Combustion MHD Systems, AIAA International Meeting, May 1980 (AIAA-80-0926).

3. Vasil'eva, I. A., V. V. Kirillov, I. A. Maksimov, G. P. Malyushonok, and V. B. Novosadov, "Measurement of a Plasma's Temperature by a Spectroscopic Method with Continuous Automatic Detection," High Temperature, Vol. 11, No. 4, July-August 1973.

4. James, R. K., and C. H. Kruger, "Boundary Layer Profile Meausrements in the Electrode Wall of a Combustion-Driven MHD Channel," 17th Symposium on Engineering Aspects of Magnetohydrodynamics, Stanford (1978).

5. Lyddon Thomas, D., "An Automatic Remotely Operated Sodium D-Line Reversal Temperature Measuring Technique," Combustion Flame, 12, 569-574 (1968).

6. Strong, H. M., and F. P. Bundy, "Measurement of Temperature in Flames of Complex Structure by Resonance Line Radiation. I. General Theory and Application to Sodium Line-Reversal Methods," and II. Sodium Line-Reversal by High Resolution Spectroscopy," J. Applied Phys., 25, 1521-1526 and 1527-1530 (1954).

7. Daily, J. W., and C. H. Kruger, "Effects of Cold Boundary Layers on Spectroscopic Temperature Measurements in Combustion Gas Flows," J. Quant. Spectros. Radiat. Transfer, 17, 327-338 (1977).

8. Self, S. A., Reigel, F. O., Clements, R. M., and James, R. K., "Electron Concentration Measurements in Combustion MHD Flows by Submillimeter Laser Interferometry," Journal of Energy, Vol. 1, 1977, pp. 206-211.

9. Annen, K. D., Kuzmenko, P. J., Keating, R., and Self, S. A., "Electron and Positive Ion Measurements in MHD Combustion Plasmas with Phosphorous Addition," Journal of Energy, Vol. 5, 1981, pp. 31-38.

Radiation thermometry applied to the development and control of gas turbine engines

T. G. R. Beynon

Land Pyrometers Limited, Wreakes Lane, Dronfield, Sheffield, United Kingdom

The realization of continuous non-contact temperature mapping of rotating components in gas turbines has necessitated a new generation of radiation thermometers. This paper describes some of the applications—the benefits obtained, and the problems overcome—in relation to both aircraft engines and ground-based (power generation) turbines. Radiation thermometers are described which will operate in ambient temperatures from $-60\,°C$ to $+550\,°C$ under 30 atmos. pressure and continuous vibration, giving response times of a few microseconds and with angular fields of view down to a few milliradians. (Typical measurement range is 600–1100 °C).

1. INTRODUCTION

The last decade has brought strenuous and continuing efforts to improve the thermal efficiency of gas turbines. These efforts have been motivated, in the civil sector, by rising fuel prices and, in the military sector, by a continuing desire for higher power/thrust to weight ratios.

Better thermal efficiency almost inevitably implies higher gas temperatures at the turbine inlet, and consequently a much harsher working environment for "hot-end" components. Particularly severe combinations of temperature/stress are encountered by rotating parts - the turbine blades and, sometimes, the disk into which the blades are rooted. Acceptable working life for these components can be ensured only by sophisticated design with exotic materials and/or elaborate cooling geometries.

A very pressing need has arisen for precise temperature measurement on rotating components during development of new machines, and also for temperature monitoring of these components during service.

Radiation thermometry (pyrometry) has, despite many difficulties, proved capable of meeting both requirements. As a development tool it has obvious advantages over embedded thermocouples with slip-rings or telemetry. Moreover, pyrometry is the _only_ technique which has been (or indeed offers any serious prospect of being) applied successfully in service.

At the present time, pyrometry is an accepted development tool on all kinds of turbines and an accepted in-flight control/monitor for military aircraft engines. In-service application on large power-generating turbines appears imminent, as do a variety of quality-control applications. There is a serious prospect of in-flight use in civil aero engines. The in-service possibilities in smaller power turbines and automotive/marine engines remain largely unexplored.

Gas turbines represent a uniquely difficult application of radiation thermometry. The thermometers developed for use on gas turbines represent a quantum leap in performance beyond that of conventional industrial instruments.

This paper describes the measurement requirements, the problems, the solutions, and the instruments themselves, illustrated with some of the more interesting (perforce mainly non-military) applications with which the author has been personally involved.

2. APPLICATION REQUIREMENTS

On turbine blades there are three basic measurement requirements: (a) Blade profiling, (b) Hot blade detection and (c) Blade averaging. These correspond (loosely) to the application categories development, monitoring, and control.

In blade profiling the objective is detailed thermal mapping of individual blades - and indeed of the whole blade array. This sort of complete information is invaluable in engine development - in particular in the development of blade cooling.

In hot-blade detection the requirement is to detect individual blades which are running hotter than their neighbours in the array. Often only _differential_ temperatures are required. Hot-blade detection applied at production pass-out, and subsequently in service, can give warning of defective blade cooling in advance of catastrophic failure.

In blade averaging one measures an average temperature for the blade array. The essential requirement is generally a signal which is uniquely related to engine condition - rather than a true spatial average and/or temperature on an absolute scale. Average blade temperature is of considerable interest as an in-flight control/top-temperature-limiting parameter in aero engines - where it can offer significant benefits over (conventional) jet-pipe thermocouple temperatures.

On turbine disks the requirement has usually been for an average (i.e. circumferential average) temperature as a function of radius. Sometimes detailed mapping is of interest in the vicinity of the blade roots. On certain large turbines the rate of disk heating on start-up is of critical interest. [Disk measurements, to date, have represented a small fraction of the total applications. The remainder of this article concentrates on blades].

The turbine blade application requires thermometers measuring from $\sim 550\,°C$ to $\sim 1000\,°C$ (exceptionally to 1300 °C); with narrow, well-defined, fields of view; with response times down to a few microseconds; in physical forms suitable for accessing machines from helicopter engines to giant turbogenerators; resistant to the extremes of temperature, vibration, and thermal shock encountered inside these engines; and often - aero applications being the obvious case - with exceptional and proven reliability.

3. THE THERMOMETRY PROBLEMS AND THEIR SOLUTION

For the vast majority of turbine blade applications the optimum thermometer has proved to be the simple silicon cell instrument. This is aimed at the blade array which effectively self-scans as the rotor spins. However, the realisation of high measurement accuracy and resolution, coupled with equipment reliability, has required solution of a multitude of problems.

3.1 Access

The first problem is how to obtain access to a target - the turbine blades - buried deep within the machine. Two basic accessing arrangements have developed - Figs. 1 and 2.

In Fig. 1 viewing is direct: a "sighting tube", fastened to the inner engine casing, supports a long,

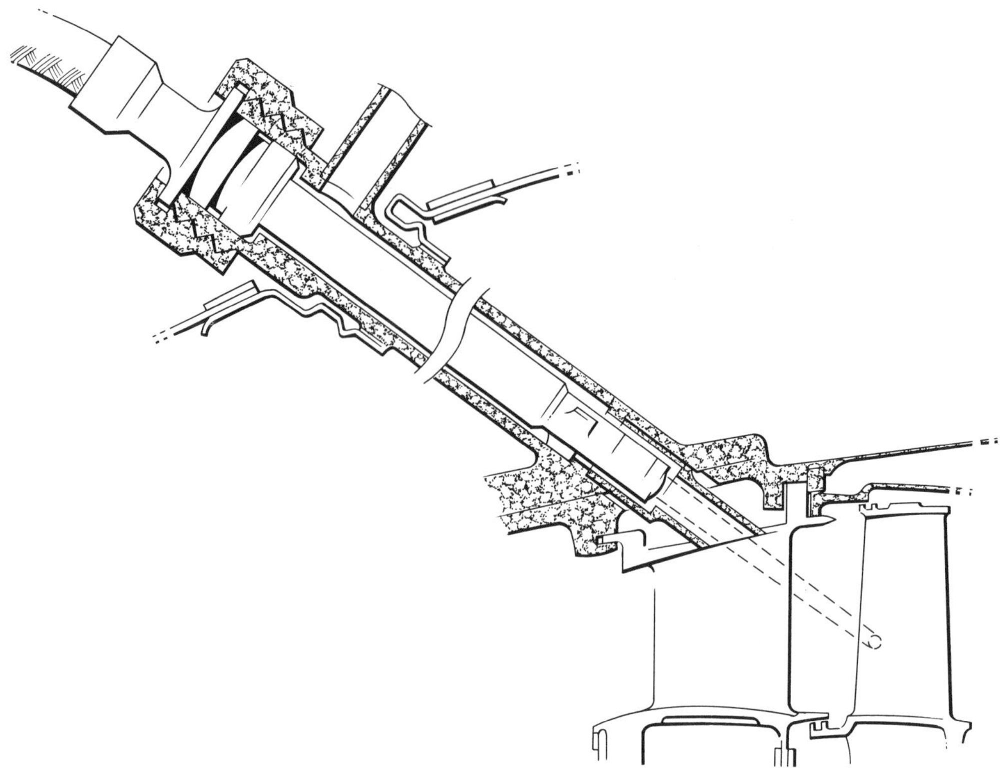

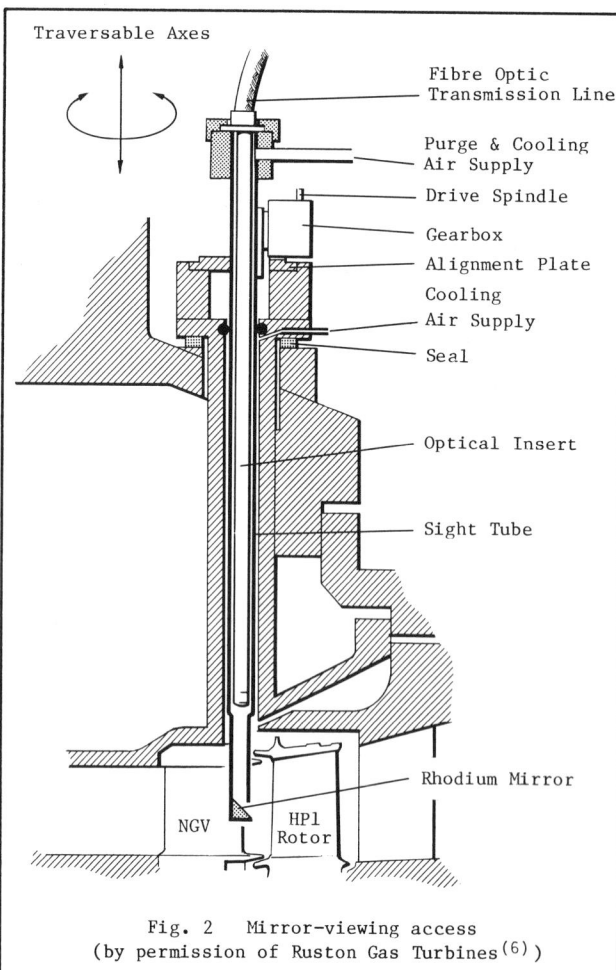

Fig. 1 Typical Direct Viewing Access

Fig. 2 Mirror-viewing access
(by permission of Ruston Gas Turbines[6])

thin pyrometer "optic head" – often on stand-off fins as shown. The pyrometer sights between adjacent stator vanes on to the blades – preferably looking downstream. Air, from the compressor or an external supply, is bled into the annulus between the optic head and the sighting tube, providing both cooling and purging. In aero applications the sighting tube may bridge the by-pass duct and be fastened, in a semi-flexible manner, to the outer engine casing.

In Fig. 2 viewing is indirect, by means of a mirror located in the end of the sighting tube. The sighting tube projects into the gas stream between adjacent stator vanes. Again the optic head is stood off from the sighting tube on fins and high pressure air is bled into the annulus. This air cools and purges both the lens and the mirror.

The mirror approach is extremely useful in development: it allows the engine to be accessed radially (which is usually simpler); facilitates measurement at different blade heights simply by racking the whole optic head/sighting tube assembly; and, by rotating this assembly, sometimes permits measurements on both upstream and downstream blades, as well as the adjacent stator vanes. For example in Fig. 3 two mirror-viewing instruments cover three stages of blades and vanes. [A third instrument views the first stage blade shrouds directly].

Note, finally, that in both Fig. 1 and Fig. 2, the pyrometer can be removed in seconds – by unscrewing a single nut. Installations incorporating a pressure valve, to permit pyrometer removal with the engine running, have been devised.

3.2   Environment

The inside of a gas turbine presents a singularly hostile environment to an opto-electronic instrument – notably in respect of ambient temperature, thermal shock, and vibration. The most fruitful approach to this problem has been to locate only the optical part (optic head) of the thermometer within the engine and to

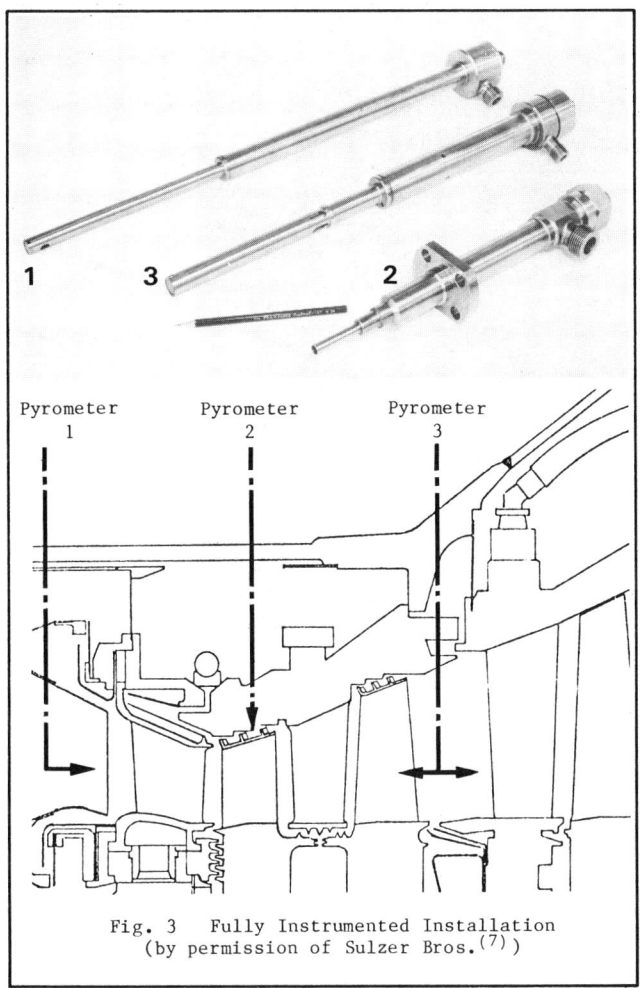

Fig. 3 Fully Instrumented Installation (by permission of Sulzer Bros.[7])

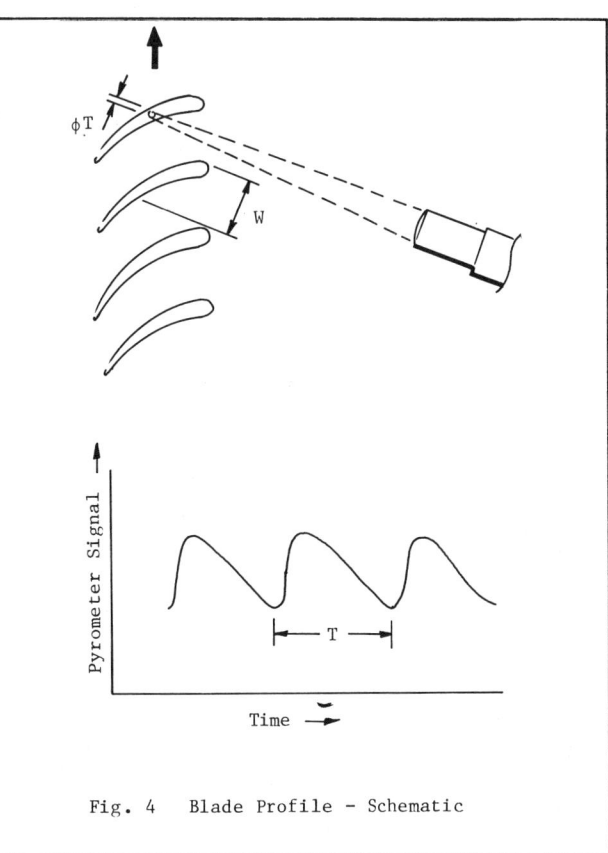

Fig. 4 Blade Profile - Schematic

extract the optical signal via a short, flexible, fibre-optic light guide.

Purpose-built instruments, in this configuration, with sapphire optics, welded-hermetic optic heads, and specially constructed light guides have proved reliable in even the most extreme application - i.e. in-flight use on military aircraft engines. Some of these instruments are described below.

Mirrors projecting into the gas-stream are liable to surface attack - particularly if facing upstream. In duration trials, rhodium faced mirrors have suffered significant surface erosion after $\sim$ 50-100 hours running. However they have proved quite suitable for development work.

3.3 Optics fouling

It is essential to keep external optical surfaces - lens and mirror (if fitted) - clean. This is usually accomplished by air purging. Theoretical and modelling studies - in particular of geometries using the layer attachment or "Coanda" principle - have provided reliable design guidelines, at least for direct viewing installations.

Arrangements of this type have proven effective and reliable in development work and in in-flight monitoring on military aircraft. For example, on the pyrometer fitted as standard equipment on one military aircraft engine, signal loss amounts to roughly a 3 °C error in 100 hours flying. It is thought that much of this loss results from contaminants in the purge air itself. Further, space restrictions on this engine compromise the purge geometry quite severely.

Lens fouling presents little obstacle to pyrometry on manned ground-based turbine installations given (as noted above) easy pyrometer removal (for cleaning) without stopping the engine. Whether lens fouling will preclude pyrometry on civil aero engines and/or remote ground turbines remains to be seen. The writer is not aware of any serious studies and mistrusts a priori judgements - particularly in respect of hot-blade detection.

Mirrors have generally been kept clean by getting as high a flow velocity as possible in the sighting tube side aperture. This approach has been satisfactory in development work. In long-term running, surface erosion may be a greater problem than fouling. At the moment, mirror systems are generally regarded as convenient development or diagnostic tools only. Recent developments in mirror construction (below) may change this view.

3.4 Bandwidth

As the rotor spins, the pyrometer scans the blade array. There will generally be a temperature variation from leading to trailing edge on each blade, so the pyrometer output signal is basically a sawtooth on a d.c. pedestal (Fig. 4): the discontinuities correspond to transition of the "target spot" from one blade to the next.

Fairly obviously, good <u>resolution</u> in the scan can only be obtained if:
(a) $\phi_T \ll W$ and
(b) $\tau \ll T$
where ϕ_T is the target spot diameter; W is the viewable blade width; τ is the electronic response (rise) time of the pyrometer; T is the time for one blade to traverse the pyrometer optical axis.

In profiling work ϕ_T has usually been taken as $\sim W/5$ with a "matching" electronic response $\tau \sim T/5$. This has typically required $\phi_T \sim 1$ to 10 mm and $\tau \sim 4$ to 40 µs depending on engine type. [For hot blade detection $\phi_T \sim W/2$; $\tau \sim T/2$ is sufficient].

These response times are quite beyond the capabilities of conventional industrial thermometers

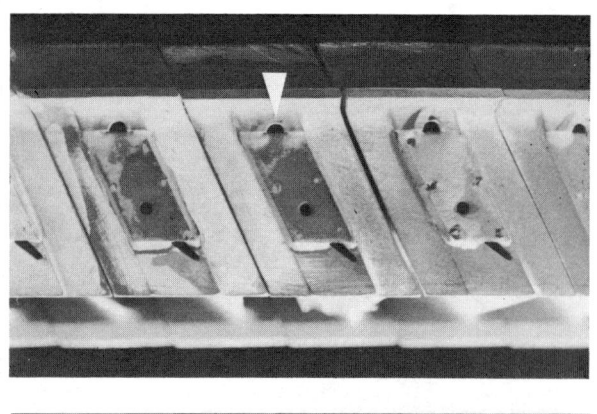

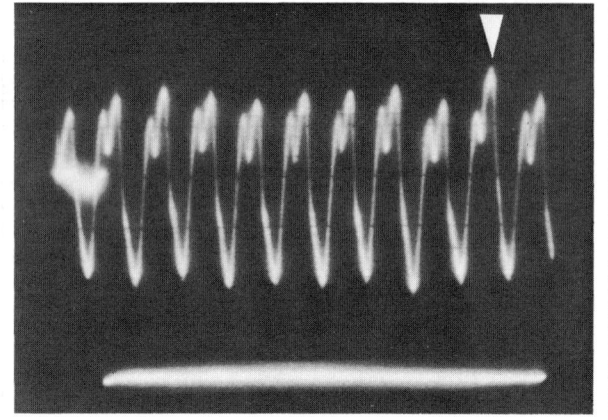

Fig. 5 Resolved Turbine Blades and Detection of a
Single Overheated Blade
(by permission of Ruston Gas Turbines[6])

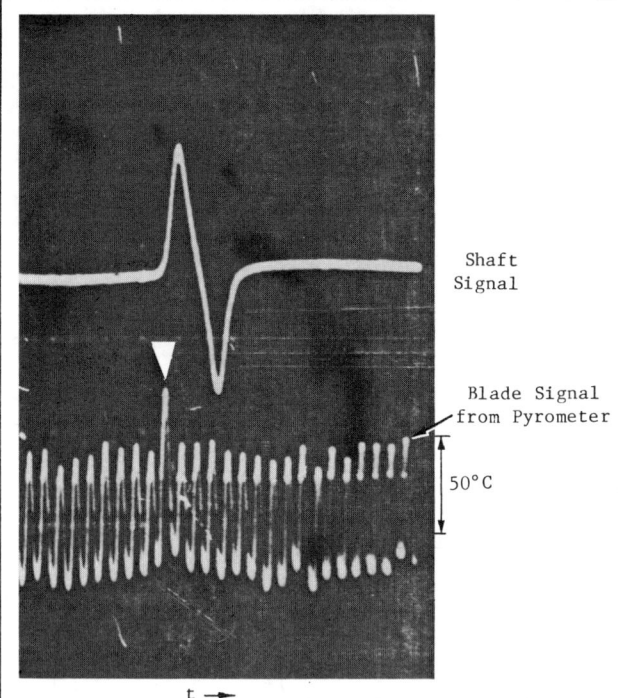

Fig. 6 Detection of a Single Overheated Blade
(by permission of Rolls Royce[9])

(where millisecond responses are "fast"), and have required the development of radically new detector/amplifier configurations.

Profiling, at reasonable resolution, is now possible on most engines under test-bed conditions. The thrust of new development is towards:

(a) Profiling/hot blade detection capability under in-flight conditions where the detector/amplifier may experience extreme ambient temperature excursions.

(b) Simple, rugged, unconditionally stable systems for hot-blade detection, on-site, on ground-based turbines.

For example, Fig. 5 shows well resolved temperature profiles on cooled first-stage blades on a prototype industrial turbine. The central cooling channel, in the blade, is quite apparent.

Fig. 5 shows detection of a single overheated blade - which was subsequently found to have a defective cooling channel. This discovery has lead to a revision of manufacturing method, and consideration of pyrometry as a quality control tool.

Fig. 6 shows detection of an overheated blade on a military development engine (details not released). The blade was subsequently found to have a blocked cooling channel.

3.5   Emissivity/spurious signal sources

Doubts concerning reflected radiation and sight-path transparency plagued early work in turbine pyrometry. Theoretical studes, experimental trials, and the accumulation of substantial experience have resolved these questions to the extent that:

(a) A majority of applications can be identified where these effects are not a serious problem.

(b) In the remaining applications, effective suppression/compensation techniques are becoming available.

The surface emissivity of a well oxidised turbine blade (to a silicon cell pyrometer) is typically about 0.85. Reflection, by the "target" blade, of radiation originating from its surroundings raises the "effective" value[2] of this emissivity. Provided the source of the reflecting radiation is emission from blades and vanes adjacent to the target blade, this increase in effective value is predictable to within tight limits and is not a problem.

For example, given the above proviso, one can generalise that the effective emissivity, in respect of blade averaging, will be 1.0 ± approximately 0.05 [1 & 3] This translates to a temperature uncertainty of less than ±5 °C. On the other hand, one expects a small variation in effective emissivity across each blade and profiling studies must account for this or incur profile errors ~ ±10 °C.

Real difficulties only arise when combustor radiation contributes significantly to the reflected component. This is primarily a problem on first stage blades in aero engines with annular combustors. Techniques, based on simultaneous measurement at two or more wavelengths, which can correct for reflected combustor radiation are becoming available[4].

On turbines with discrete combustors it has generally been possible to avoid serious combustor reflection problems - even on first stage - by suitable location of the pyrometer. Accuracy can sometimes be verified by moving the pyrometer off the blade onto an adjacent (upstream) vane, and referencing to a thermocouple. e.g. Fig. 7 shows a comparison of first stage vane temperatures as measured by a pyrometer (assuming unit emissivity) and an embedded thermocouple.

The first step, in resolving the question of sight

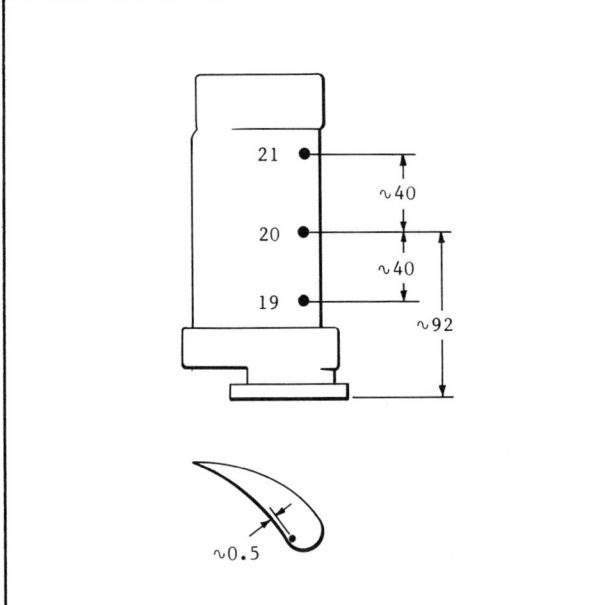

Fig. 7 Comparison of Pyrometer and Thermocouple Temperatures on First Stage Vane (by permission of Brown Boveri[8])

TIME	PYROMETER READING	THERMOCOUPLE 20
1725	810 °C	815 °C
1730	810 °C	818 °C
1735	815 °C	820 °C
1742	812 °C	815 °C
1746	810 °C	815 °C

path transparency, is to choose an operating wavelength band where the gas in the sight path is transparent. Reasonably accurate calculations of sight path gas opacities are now possible [5]. The silicon cell operating wavelength band is, in the vast majority of situations, free from significant gas opacity effects.

However, hot particles emit at all wavelengths and, on traversing the sight path, will inevitably cause spurious signal elevations. Because this "noise" is all positive-going, a slow response system (blade averaging) can give an incorrect measurement. However experience has shown hot particles to be a serious problem in only a limited number of situations - for example slam accelerations in military aircraft engines. Given a fast response pyrometer, particle spikes are, in general, easily distinguished from a "clean" signal. Several systems have been devised to electronically suppress particle spikes in real time; the most sophisticated of which can extract a clean blade signal from almost continuous particle interference [2].

4. THE NEW INSTRUMENTS

Present capabilities of commercially available equipment are indicated below, and some development areas noted.

4.1 Standard ground-based equipment

"Transducers" specified for use on ground-based turbines usually comprise: optic head; flexible light guide; detector/amplifier module.

Optic heads are of stainless steel construction, electron-beam or laser welded, with sapphire optics. The lens assemblies will withstand both continuous temperatures and rapid temperature changes in -60 °C to +550 °C. They are vacuum tight to 33 atmospheres. Their long-term durability is well proven.

Heads of broadly similar appearance to that in Fig. 1 are available in a wide variety of different lengths and diameters. Heads with a flange-type fixture at the lens end are also available. Sighting distances vary from about 0.1 to 1.0 metres with target spot diameters about 2 to 20 mm.

Heads are usually supplied with an appropriate air-purged sighting tube. Aside from the cooling/purging function, the sighting tube provides a useful "design interface" between the pyrometer supplier and the turbine engineer - in that its external dimensions are easily tailored to suit the particular application.

Flexible light guides are sealed, armoured, glass multifibre units developed from military aircraft designs, where durability has been well proven in more than 17,000 flying hours. Maximum operating temperature is usually 250 °C and lengths up to about 3 metres are possible without significant signal attenuation.

Detector/amplifier units are conventionally based on the silicon cell detector with a high speed, low noise/drift amplifier. Output is a high level voltage directly related to (but non-linear with) target temperature. Exponential rise times down to 1 µs (150 kHz) are available. Maximum operating temperatures are typically 85 °C. Designs using a chopper-stabilised front-end amplifier ensure unequivocal long-term stability.

Typical signal to noise performance is shown in Fig. 8. Here the ordinate "N.E.T." is noise + drift expressed as an equivalent uncertainty in target temperature. (The signal available increases rapidly with target temperature).

The figure is drawn for a transducer with optical throughput $F = 1.0$ mm^2. N.E.T. is inverse linear with F. N.E.T. values can be extrapolated to other optical throughputs using the approximate relation.

$$F \simeq \frac{\phi_T^2 \times \phi_L^2}{S^2}$$

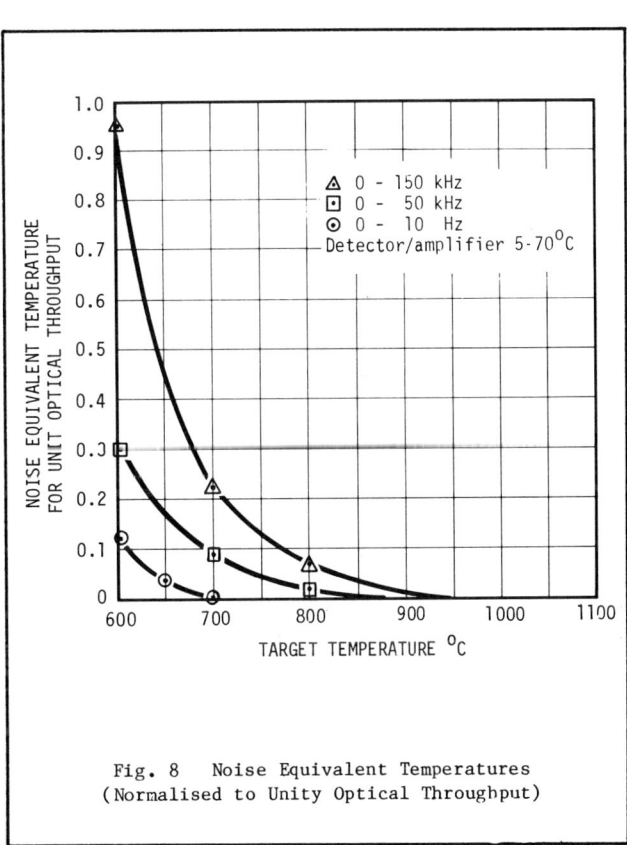

Fig. 8 Noise Equivalent Temperatures (Normalised to Unity Optical Throughput)

Table I
Design Parameters for a recently supplied system

Designation	GBX: 10-1500-15/550/MLA
Sighting Distance	1.5 metres
Lens Aperture	15 mm
Target Diameter	10 mm max. 5 mm preferred
Bandwidth	50 kHz
Linearization Range	600-1100 °C
Outputs	Fast linear/slow linear/raw

where
    ϕ_T is the target spot diameter
    ϕ_L is the lens aperture
    S is the sighting (strictly focus) distance

In profiling work F values from about 0.01 to 0.01 mm^2 are typically obtainable, permitting low noise profiles for target temperatures above 700 °C or so.

Table 1 summarises design parameters for a particular transducer for use on a large power-generating turbine. Figs. 9, 10 and 11 show target spot definition, N.E.T., and transient response actually measured for this instrument.

Linearization units are available to convert the transducer signal to a reading directly in temperature - e.g. 10 mV/°C. A typical specification might be range 550 °C to 1000 °C with maximum conformity error ±3 °C. Linearization is possible without significant loss in bandwidth.

A variety of units are available to process the transducer signal. These range from very simple averagers and/or slew-rate limiters (to give a circumferential mean rotor temperature, or perhaps a <u>vane</u> temperature) to very sophisticated digital units which give detailed blade profiles and are able to "reject" transient noise due to flame and/or carbon particles.

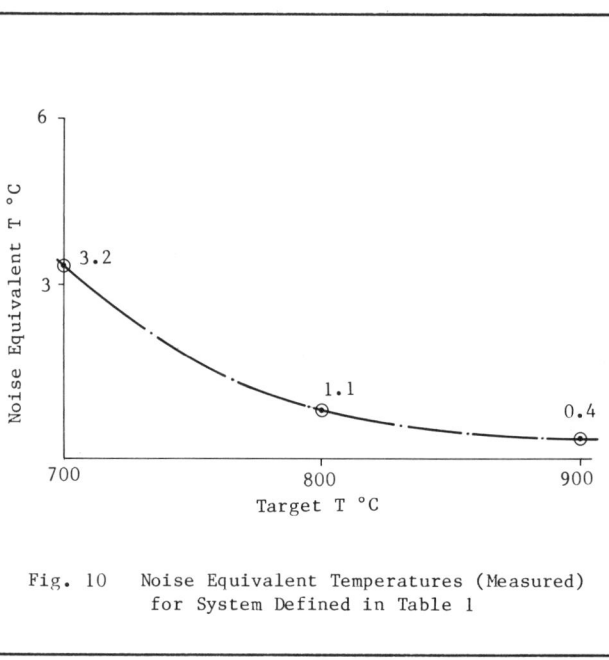

Fig. 10   Noise Equivalent Temperatures (Measured) for System Defined in Table 1

4.2   Standard in-flight equipment

Transducers fundamentally very similar to those described above are available in flight-rated form. Optic heads are identical to those described above. Light guides are usually similar, although flight-approved units with fused terminations - giving ambient temperature capability 350 °C to 400 °C have recently become available. Detector/Amplifer units are miniaturised and can be rated to around 125 °C ambient. The noise/bandwidth performance of these flying systems is quite comparable with that described above for ground-based equipment.

In some in-flight systems the detector and amplifier are, of necessity, separated - the detector being engine mounted and the amplifier cockpit mounted. [Detectors capable of operation at 200 °C are available while amplifiers are generally limited to about 125 °C]. This inevitably leads to "pick-up" at some level and N.E.T. performance suffers. It also restricts the attainable bandwidth. However the arrangement has proved quite adequate for in-flight blade averaging.

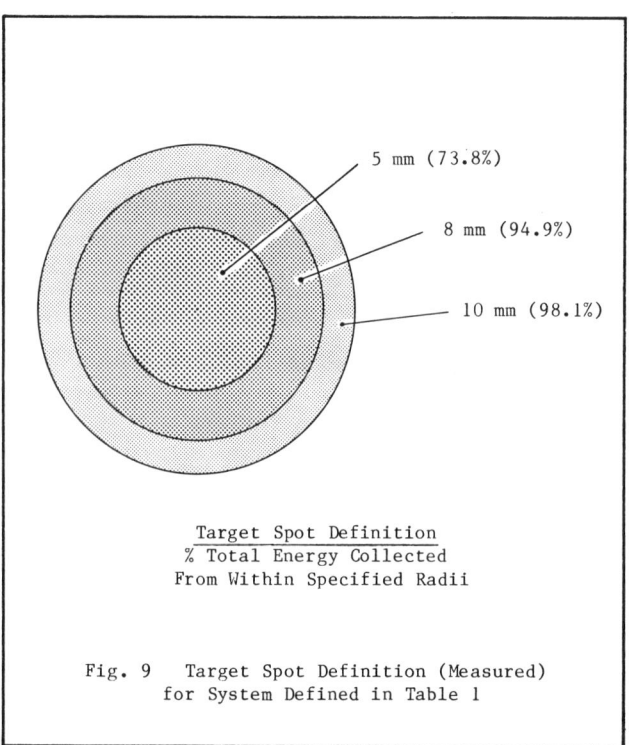

Fig. 9   Target Spot Definition (Measured) for System Defined in Table 1

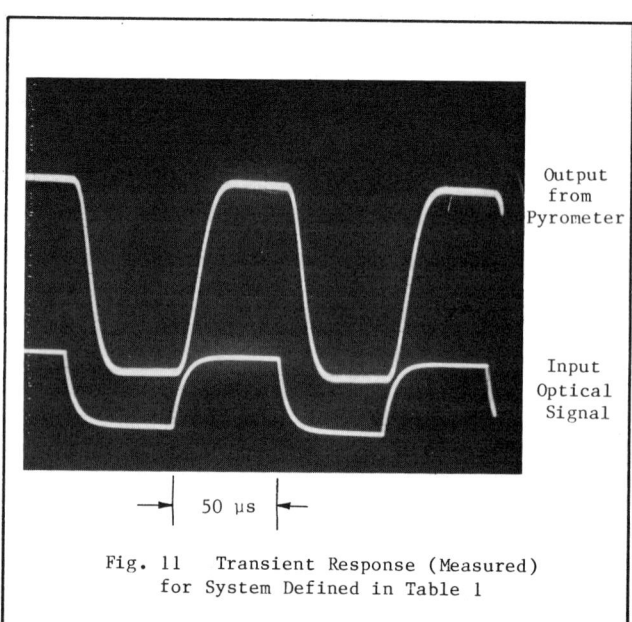

Fig. 11   Transient Response (Measured) for System Defined in Table 1

4.3 State-of-the-art equipment

Many equipment developments are taking place. Some examples are given below - all of which are commercially available on the understanding that they are "state-of-the-art" and have not (yet) received the sort of testing which has been applied to the standard items above.

(a) Very low attenuation light guides, capable of piping the optical signal right away from the turbine to a control room/instrument bay environment have been supplied for experimental use.
(b) Detector/amplifier units with bandwidths to about 450 kHz (~ 2 μs) have been shown possible given closer control of ambient temperature.
(c) Detector/amplifier units capable of operation at 200 °C are possible now, albeit with poor noise levels. Intensive development is taking place with a medium-term objective of a low-noise 250 °C unit.
(d) Sapphire-faced mirrors - which should be highly resistant to surface erosion - have been successfully manufactured. Engine trials are scheduled.
(e) Optic heads capable of withstanding 1000 °C are in development.
(f) Multi-detector systems capable of simultaneous two-dimensional temperature mapping have been used experimentally.

REFERENCES

(1) Barber R. "A radiation pyrometer designed for in-flight measurement of turbine blade temperatures". ASME. 690432. (1969)

(2) Beynon T.G.R. "Turbine pyrometry - an equipment manufacturer's view". ASME. 81-GT-136. (1980)

(3) Douglas J. "High speed turbine pyrometry in extreme environments". ASME, "Measurement methods in Rotating Turbomachinery". (1980)

(4) Atkinson W.H. and Strange R.R. "Pyrometer measurements in the presence of reflected radiation". ASME. 76-HT-74. (1976)

(5) N.A.S.A. "Handbook of Infra-red Radiation from Combustion Gases". (1973)

(6) Smith M.K.D., Ruston Gas Turbines. Private communication (1981). (Further results to be presented in "The Design and Test of Air-cooled blading for an Industrial Gas Turbine". J.M. Harris and M.K.D. Smith at ASME Gas Turbine Conference and Exhibit, London, 1982).

(7) Marriot C.A., Sulzer Bros. Private communication "Pyrometer results with prototype GT10 Gas Turbine" (1981)

(8) Eggmann T., Brown Boveri, 1981. Private communication "Prüfung eines Pyrometers in der Gasturbine 9C in Münchenstein" (Brown Boveri internal report 1975)

(9) Doughty J., Rolls Royce, 1980. Private communication (Military engine - details not released).

Improvement of traceability for radiation pyrometers in the steel industry

K. Tamura, T. Iwamura, and K. Kurita

Kawasaki Steel Corporation, Tokyo, Japan

A traceability system for radiation pyrometers has been established. In this system a radiation pyrometer is calibrated beforehand by fixed points, and it is used as the primary standard pyrometer. In the range of 800 °C–1400 °C the automatic optical pyrometer is used as the primary standard and the Si-cell pyrometer as the secondary standard. In the range between 200 °C–800 °C the PbS-cell pyrometer is used as the primary standard. Major error factors of this traceability have been investigated, and the amount of each error factor has been measured quantitatively. The inaccuracy of this traceability system is then estimated to be smaller than ± 2 °C.

1. INTRODUCTION

Kawasaki Steel, as well as other steel making companies, employs many radiation pyrometers for the temperature measurement of steel. In an effort to measure a true temperature as precisely as possible, the company has improved a traceability system for radiation pyrometers, as follows:
"fixed point blackbody furnace → primary standard pyrometer → blackbody furnace → secondary standard pyrometer"
A conventional traceability system, which was used before the improvement, consisted as follows:
"thermocouple → blackbody furnace → radiation pyrometer"
The error of the conventional traceability was about ± 1 % of temperature indication of a pyrometer, which was insufficient for steel production. This error primarily comes from the fact that the effective emissivity of the blackbody furnace considerably deviates from unity, and secondarily from the fact that a temperature difference tends to occur between the target for a pyrometer and the measuring-point for a thermocouple.
Since the improved traceability system uses a radiation pyrometer as a standard instead of a thermocouple (i.e., radiation pyrometer → blackbody furnace → radiation pyrometer), the deviation of emissivity from unity makes much less effect on the calibration accuracy. A study has shown, indeed, that the error resulting from the emissivity deviation can be reduced to an amount several times smaller than that of the conventional traceability system. In addition, other error factors of the improved traceability have been investigated. These mainly come from the features of the fixed point blackbody furnace, and the primary and secondary standard pyrometers. By a number of calculations and experiments these errors have been shown to be acceptably small.

2. ERROR FACTORS OF TRACEABILITY

2.1. Effective Emissivity and Transfer Error

The effective emissivity $\varepsilon(\zeta_1)$ in a cavity made of diffusely reflecting wall, is given by the equation (1), as a result of interreflection. The equation represents the superposed state of emission and multiple reflections.

$$\varepsilon(\zeta_1) = \varepsilon_0 f(\zeta_1) + (1-\varepsilon_0) \int_s K(\zeta_1,\zeta)\varepsilon(\zeta)ds(\zeta) \quad (1)$$

where, $f(\zeta_1) = \Psi_b(T(\zeta_1))/\Psi_b(T(\zeta_0))$

$\Psi_b(T(\zeta_1))$ = blackbody radiation at a point ζ_1
$\Psi_b(T(\zeta_0))$ = blackbody radiation at a standard point ζ_0
$T(\zeta_1)$ = temperature at a point ζ_1
ε_0 = emissivity of the wall
$K(\zeta_1,\zeta)$ = kernel function
s = total area of the inner wall of the cavity

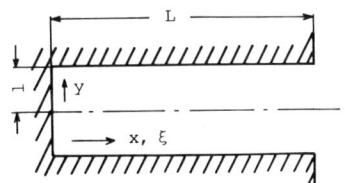

Fig.1. Configuration of the cylinder of the blackbody furnace.

For a cylindrical cavity which is shown in Fig.1, the area of integration is simplified and the following equations are obtained.[1]

$$\varepsilon(x) = \varepsilon_0 f(x) + (1-\varepsilon_0)\int_0^L K(x,\xi)\varepsilon(\xi)d\xi \\ + (1-\varepsilon_0)\int_0^1 K(x,y)\varepsilon(y)dy \quad (2)$$

$$\varepsilon(y) = \varepsilon_0 f(y) + (1-\varepsilon_0)\int_0^L K(y,x)\varepsilon(x)dx \quad (3)$$

where,

$$K(x,\xi) = \frac{1}{2}\left(1 - \frac{|x-\xi|\{(x-\xi)^2+6\}}{\{(x-\zeta)^2+4\}^{3/2}}\right) \quad (4)$$

$$K(x,y) = \frac{2xy(x^2-y^2+1)}{\{(x^2+y^2-1)^2+4x^2\}^{3/2}} \quad (5)$$

$$K(y,x) = \frac{2x(x^2-y^2+1)}{\{(x^2+y^2-1)^2+4x^2\}^{3/2}} \quad (6)$$

The calculation of the effective emissivity has been carried out by using an iteration method[1,2] with the following conditions.
(1) Shape of cavity: Length(L) = 500 mm
                     Radius   = 30 mm
(2) Temperature distribution: Six patterns along the cylinder of the furnace, which have been

TEMPERATURE © 1982 American Institute of Physics 479

actually measured, are shown in Fig.2. (Within any cross-section normal to the axis of the cylinder the temperature is assumed to be constant.)

(3) Emissivity of the wall: Low = 0.3
                               High = 0.8
(4) Wavelength: 0.65 μm, 1.0 μm and 2.0 μm

In Fig.2 ΔT_w denotes the maximum difference between the temperature of the cylindrical wall and that of the back wall.

(2) Within the experimental range, the transfer error is almost independent of the temperature patterns. That means, even if the effective emissivity does change in large amount depending on the temperature pattern, this change has almost no effect on the transfer error, ΔT_r.

(3) ΔT_r can be reduced, furthermore, if a higher emissivity material is used for the furnace; for example SiC, instead of Al_2O_3 ($\varepsilon_0 \approx 0.3$), which is presently used.

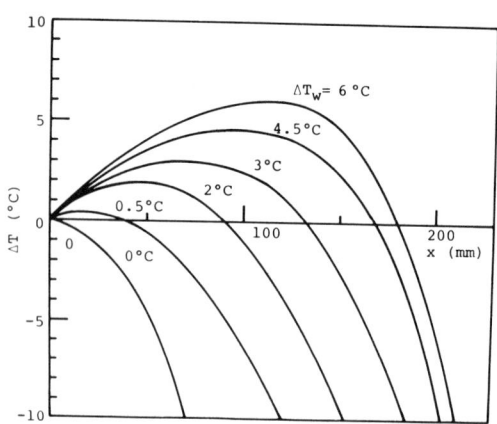

Fig.2. Temperature distribution along the cylinder in the blackbody furnace.
$\Delta T = T(x) - T(x_0)$
$T(x)$ = wall temperature at x
$T(x_0)$ = target temperature (x=0)
ΔT_w = maximum value of ΔT

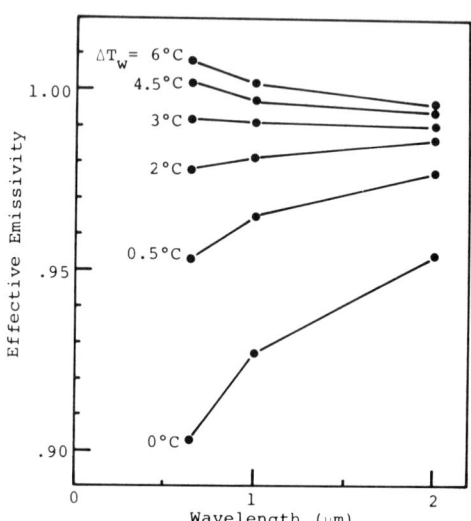

Fig.3. Effective emissivities of the blackbody furnace. Calculation in case of a low emissivity wall. $\varepsilon_0 = 0.3$, $T(x_0) = 1000$ °C

The results of the effective emissivity calculation are shown in Fig.3 for the low emissivity wall and also in Fig.5 for the high emissivity wall. Fig.4 and Fig.6 show the errors of the brightness temperature, which are calculated from the values of the effective emissivity in Fig.3 and Fig.5 using the equation (7).

$$\frac{1}{T} - \frac{1}{S_\lambda} = \frac{\lambda}{C_2} \ln\varepsilon \qquad (7)$$

where,   T = true temperature, K
           S_λ = brightness temperature, K
           λ = wavelength
           C_2 = second radiation constant
           ε = effective emissivity

The calculations of the effective emissivity have indicated the following results.

(1) The effective emissivity of the cavity is shown to be very dependent upon the temperature distribution along the wall. The dependence is greater at 0.65 μm than at 2 μm, and it is also greater at $\varepsilon_0 = 0.3$ (low emissivity) than at $\varepsilon_0 = 0.8$ (high emissivity).

(2) The most suitable temperature pattern of the blackbody furnace is in the neighborhood of $\Delta T_w = 3$ °C for which the effective emissivity approaches unity and it becomes almost independent of wavelength.

And then the "transfer error" from the primary to the secondary pyrometer, which should be a function of the emissivity deviation and of their effective wavelengths, has been investigated. The results are:

(1) In the case that the primary instrument is an optical pyrometer (whose effective wavelength is 0.65 μm) and the secondary instrument is a Si-cell pyrometer (1 μm), the value of the transfer error ΔT_r, is smaller than 0.5 °C (Fig.4).

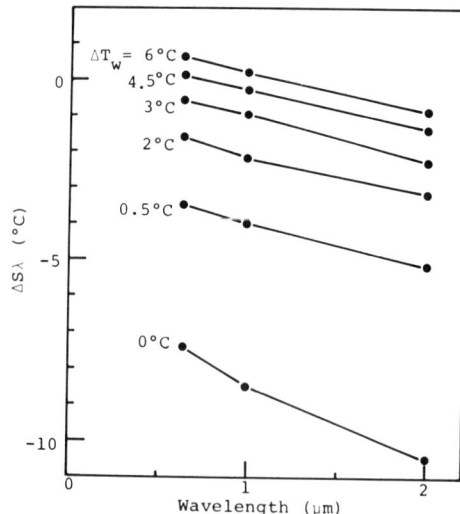

Fig.4. Errors of brightness temperature. (Calculated from the effective emissivity values in Fig.3)

$\Delta S_\lambda = S_\lambda - T(x_0)$

S_λ = brightness temperature
$T(x_0)$ = true temperature

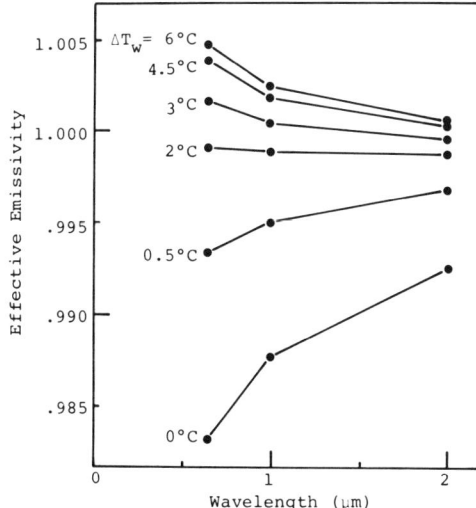

Fig.5. Effective emissivities of the blackbody furnace. Calculation in case of a high emissivity wall, $\varepsilon_0 = 0.8$, $T(x_0) = 1000$ °C

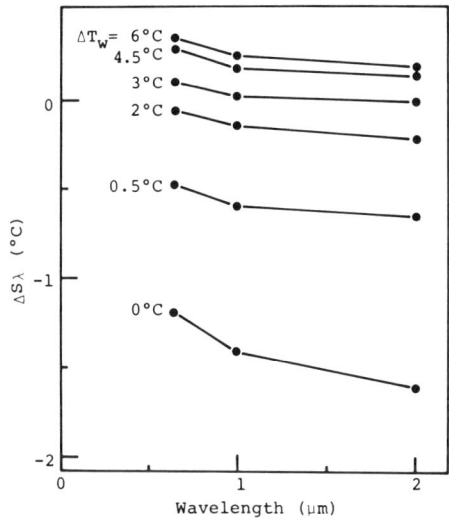

Fig.6. Errors of brightness temperature. (Calculated from the effective emissivity values in Fig.5)

2.2. Error of Radiation Pyrometer

In addition to the "transfer error", the accuracy of the traceability is largely determined by the error factors of the radiation pyrometer itself: stability of the pyrometer, size-of-source effect, and reliability of interpolated curves between the fixed points. The magnitude of each factor has been checked as follows:
(1) Stability: The stability of the pyrometer has been verified through highly reproducible measurement of the fixed point blackbody frunace.
(2) Size-of-source effect: Since the target size of the fixed point blackbody furnace is much smaller than that of the blackbody furnace, the variation of the target size causes error. The degree of the error depends upon the kind of a pyrometer. The PbS-cell pyrometer, whose size-of-source effect is relatively large, has been examined and a quantitative analysis on the effect has been obtained.

(3) Reliability of interpolation: The reliability of the interpolation and extrapolation procedures required to obtain calibration at temperatures other than the fixed points affects the accuracy of the traceability. For the Si-cell pyrometer, the interpolation and extrapolation can be performed with high accuracy because of the extremely linear response of the detector.

3. ACCURACY OF TRACEABILITY

3.1. Fixed point Blackbody Furnace

The effective emissivity of the fixed point blackbody furnace calculated with the above mentioned method, has shown a value higher than 0.999. It is estimated from this value that the inaccuracy of the furnace temperature, ΔT_1, is within ± 0.3 °C if pure metals (> 99.99 percent) are used for the fixed points.

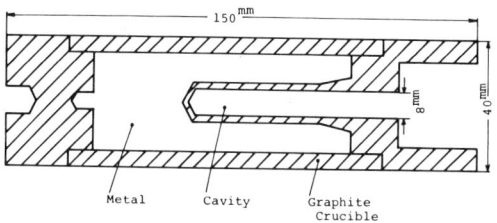

Fig.7. Configuration of the blackbody radiator in the fixed point furnace.

The experimental condition is as follows. As is shown in Fig.7 the graphite crucible is put into a quartz tube, and it is heated in argon gas, which prevents the graphite from oxidation. The fixed points being used are shown in Table I.[3,4,5,6] (note: Although Cu-Ag and Cu-Al are alloys -that means their eutectic points are used-, they have shown the same reproducibility as the other pure metals.)[3]

3.2. Primary and Secondary Standard

(A). 800 °C - 1400 °C
   primary standard   - automatic optical pyrometer
   secondary standard - Si-cell pyrometer

The primary (automatic optical) pyrometer has been calibrated at the Ag and Cu points of the fixed point furnace. The inaccuracy of the primary pyrometer, ΔT_2 (optical), is estimated to be smaller than ± 1 °C because $\Delta T_2 = \Delta T_1 +$ other factors < ± 0.3 °C ± 0.5 °C (ΔT_1 is the inaccuracy of the fixed point furnace). Fig.8 shows an example of the calibration, in which heating and cooling curves of the fixed points are measured with the automatic optical pyrometer.

The secondary (Si-cell) pyrometer has been calibrated with the primary pyrometer. The inaccuracy of the secondary pyrometer, $\Delta T_3(Si)$, is equal to the sum of the inaccuracy of the primary pyrometer and the transfer error: $\Delta T_3 = \Delta T_2(\text{optical}) + \Delta T_r = \pm 1.0$ °C ± 0.5 °C. That means $\Delta T_3(Si)$ does not exceed ± 2.0 °C. The Si-cell pyrometer is regarded to be an optimum secondary standard because it shows long term stability and good linearity.

(B). 200 °C - 800 °C
The PbS-cell pyrometer, the primary standard in this range, has been calibrated at six points of the fixed point furnace. These were fixed points 3 through 8 in Table I. The inaccuracy of the primary standard, $\Delta T_2(\text{PbS})$, has been estimated to be smaller than ± 2 °C ($\Delta T_2(\text{PbS}) = \Delta T_1 +$ other factors ≤ ± 0.3 °C ± 1.5 °C).

Table I. List of Fixed Points

	No.	Fixed Points	Temperature, °C
Above 800 °C	1	Cu	1084.88
	2	Ag	961.93
200 °C to 800 °C	3	Cu-Ag	779.6
	4	Al	660.46
	5	Cu-Al	548.26
	6	Zn	419.58
	7	Pb	327.502
	8	Sn	231.9681

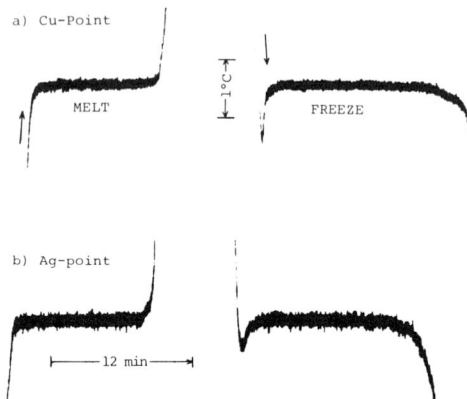

Fig.8. Heating and cooling process of the fixed point blackbody furnace measured with an automatic optical pyrometer.

It has been found that the larger value of the "other factors" for the PbS-cell (1.5 °C) than those in the automatic optical one (0.5 °C) mainly comes from two factors: its short term stability and large size-of-source effect. The procedures taken to insure that these two factors remain acceptably low are as follows.
Stability:
  To check the drift of output voltages of the PbS-cell pyrometer, the pyrometer is calibrated once a day at one of the fixed points.
Size-of-source effect:
  A quantitative analysis of the size-of-source effect has been made by the following experiment. A "plane" radiation source (i.e., A hot plate whose emissivity is about 0.9.) is adopted instead of a blackbody furnace, because this experiment deals with relatively large size objects. The lens of the PbS-cell pyrometer is focused on a diaphragm which is set in front of the hot plate. The size-of-source effect is measured by the change of temperature indication of the pyrometer when the diaphragm is removed. Similarly the target-distance effect is measured by the change of temperature indication when the target separation distance is changed while focusing the lens on the diaphragm. (The target-distance effect includes to some extent the effect of atmospheric absorption.)
  In this measurement it is necessary to avoid the interreflection between the hot plate and the diaphragm. Such cares as placing the diaphragm on the hot plate and constructing the diaphragm's surface with a low emissivity material (for example aluminum plate) are effective for that purpose.
  These effects have been measured with the following conditions.
  (1) Target size(D): 8, 12, 15, 18, 20, 25, 40, and 80 mm (for size-of-source effect)
      40 mm (for target-distance effect)
  (2) Target distance(H): 600, 800, 1000 and 1200 mm (for both effects)
(note: The target size is equal to the inner diameter of the diaphragms.)
  The results for the PbS-cell pyrometer are shown in Fig.9, from which following conclusions are obtained.
  (1) There exists a "critical size" for each target distance. When the target size is larger than this size, the size-of-source effect becomes significantly small.(Within the experimental range it is smaller than ± 0.5 °C.) The critical size for the target distance of 600 mm, for example, is about 20 mm.
  (2) There is an appropriate target distance at which the size-of-source effect can be cancelled out by the target distance effect. That is, even if the target size of a blackbody furnace is larger than that of the fixed point blackbody furnace, no calibration is required by selecting an appropriate target distance. The condition of D = 60 mm and H = 800 mm (blackbody), for example, has almost the same value of $\Delta U/U$ as that of D = 20 mm and H = 600 mm (fixed point blackbody).
Because the value of $\Delta T_2(PbS)$ between 200 °C and 800 °C is as large as that of $\Delta T_3(Si)$ above 800 °C, the PbS-cell pyrometer is also used as the secondary standard for that temperature range.

3.3. Resultant Accuracy

The inaccuracy of the improved traceability system ($\Delta T_3(Si)$ above 800 °C and $\Delta T_2(PbS)$ between 200 °C and 800 °C) is smaller than ± 2 °C which is sufficient for steel temperature measurements.

3.4. Further Improvement of Traceability

At present each of the secondary standards (above 800 °C) is calibrated against the primary standard once a year. In future it is planned that secondary standards will be compared repeatedly with

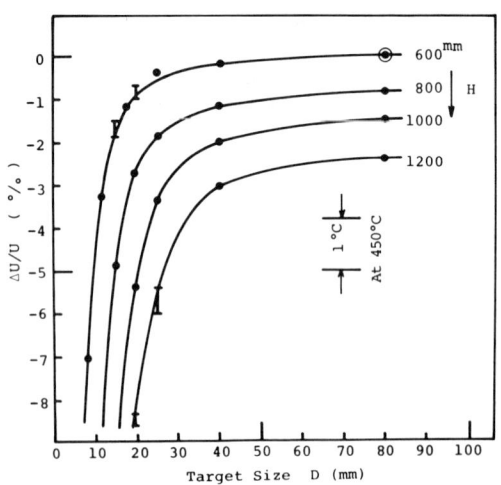

Fig.9. Dependence of the change of incidence power ($\Delta U/U$) on target size (D) and distance (H). (The standard point is chosen arbitrarily at L = 600 mm and D = 80 mm.)
$\Delta U = U - U_0$
U_0: incidence power at the standard point.
U : incidence power when the target size or the distance is changed.

each other. By that comparison, the accuracy of secondary standards will be checked more frequently and as a result calibrations to the primary standard may be done less frequently.

4. CONCLUSION

A traceability system has been established which uses a fixed point blackbody furnace as the standard for radiation pyrometers. At the range between 800 °C and 1400 °C, an automatic optical pyrometer has shown high accuracy as the primary standard. Because the secondary (Si-cell) pyrometer has shown long term stability and good linearity, the traceability system in this range is easy to establish. Below 800 °C, on the contrary, a PbS-cell pyrometer, the primary standard in this range, has shown a relatively large errors.

As a consequence, when the PbS-cell pyrometer is used as the primary standard, the magnitude of those errors must be measured carefully.

REFERENCES

1. H.Yamamoto, OYO BUTURI 38, 618 (1969).
2. E.M.Sparrow, L.U.Albers and E.R.G.Eckert, J.Heat Transfer 84, 73 (1962).
3. J.V.McAllan, Temperature, Its Measurement and Control in Science and Industry (Instrument Society of America, Pittsburgh, 1972) Vol.4, Part 1, p.265.
4. G.Ruffino, High Temperatures-High Pressures 9, 253 (1977).
5. H.Preston-Thomas, Metrologia 12, 7 (1976).
6. L.Crovini, R.E.Bedford and A.Moser, Metrologia 13, 197 (1977).

Steel surface temperature measurement in industrial furnaces by compensation for reflected radiation errors

J. E. Roney

Jones and Laughlin Steel Corporation, Pittsburgh, Pennsylvania 15227

Optimum heating control is achieved when steel temperatures rather than zone temperatures are measured. Traditionally, zone temperatures were used to control furnaces because reflected radiation from flames, heating tubes, and hot furnace walls caused large errors (> 100 °C) in the measurements made with radiation pyrometers aimed directly at the steel surface. Dual pyrometer methods are described which compensate for the error caused by reflected radiation. These pyrometers are positioned at a safe distance from the product passline to insure long term reliability. Accurate measurement can be made of slab and billet temperatures in reheat furnaces and strip temperature in galvanizing lines. Dual pyrometer installations on these process lines are described. Equations with sample calculations are shown to estimate the effect of various parameters on the measurement accuracy.

Until recently furnace zone temperatures were used to control the heating in our furnaces. Operating rates were based on experience for the product size and zone temperature. Zone temperature measurements rather than steel temperatures were used because of the difficulty in obtaining accurate surface temperature measurements in a furnace. For a pyrometer aimed directly at the steel surface measurement errors of over 100°C could be caused by radiation from flames, hot furnace walls and heating tubes.

With zone control, productivity and fuel rate are not at the optimum because the steel temperature is usually higher than required. Quality may also be inconsistent because of the variability in the actual steel temperature.

Direct measurement of steel surface temperature can be made by aiming the pyrometer through a cooled sight tube with a cooled flange near the steel surface. The intensity of the furnace radiation is reduced with each reflection between the flange and steel surface. Analysis of this method shows that two of the many parameters that affect the pyrometer's error are the ratio of flange height above the surface to the flange radius and an experimentally determined number that is a measure of the specularity of the surface [1]. Because the flange tube is near the steel passline it may be subject to damage and it interferes with maintenance.

In order to reduce the error caused by the reflected radiation we use the signal from a second pyrometer to compensate for the error of the pyrometer aimed at the steel. By this method the pyrometers and their accessories can be placed far away from the passline to achieve long term reliability. Several different versions of the dual pyrometer systems will be described. The version selected depends on the particular measurement conditions.

BACK-OFF METHOD

As shown in Fig. 1, one pyrometer is aimed at the furnace wall [2]. Then a portion of the signal of the pyrometer aimed at the wall is subtracted from the signal of the pyrometer aimed at the steel surface. The relative magnitude of each of these signals is shown on the rays representing the radiation received by the dual pyrometer. The three terms correspond to the emitted and reflected radiation from the steel surface and the radiation from the furnace wall.

Conceptually:

$$Wa = \frac{\varepsilon_s f(T_s)}{\varepsilon_p} + \frac{(1 - \varepsilon_s)\, \varepsilon_F f(T_F)}{\varepsilon_p} - \frac{\varepsilon_F f(T_F)}{\varepsilon_R} \qquad (1)$$

where:

Wa	= Net radiation received by the dual pyrometer
ε_s and ε_F	= Effective emissivity of the steel surface and furnace wall, respectively
$(1 - \varepsilon_s)$	= Effective reflectivity of the steel surface
ε_p and ε_R	= Emittance settings of the pyrometers aimed at the steel surface and at the wall, respectively
$f(T_s)$ and $f(T_F)$	= Radiation emitted by a black body at the steel temperature T_s and wall temperature T_F, respectively (Wien's approximation).

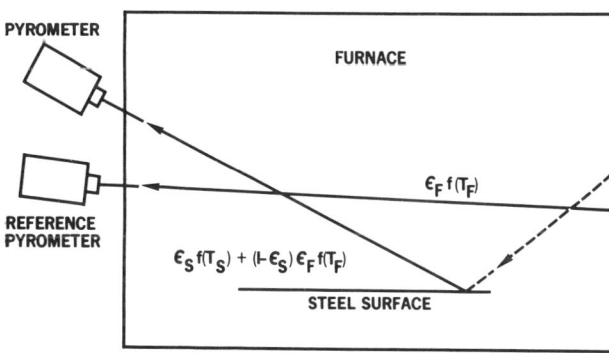

FIG. 1. DUAL PYROMETER - BACK-OFF METHOD

To simplify the equations used to describe the dual pyrometer concepts, the two pyrometers are assumed to have identical response characteristics. In addition, the solid angle subtended by the target area of the steel surface is assumed to be exactly equal to that of the reference area on the furnace wall. The furnace itself is assumed to be a black body cavity. Then to cancel the reflected error, the second and third terms of equation (1) must be equal. This requires that the reference pyrometer's emittance setting be at:

$$\varepsilon_R = \frac{\varepsilon_p}{1 - \varepsilon_s} \quad (2)$$

With the reference pyrometer's emittance set at the value given by equation (2), the radiation detected by the reference pyrometer is the same as the radiation reflected from the steel surface. They may not always be the same if the products of the view factor and area for the reflected radiation from the steel surface and from the wall are not equal. These products which can be considered to be contained in ε_F may not vary equally in slab reheat furnaces where over 95% of the heat transfer is by radiation from the flames. For best accuracy, the steel temperature measurement should be near its end close to the wall area. But at the slab end the heating and cooling may be quite different than near the center of the furnace.

COOLED REFERENCE TARGET

To get improved compensation for the reflected radiation, we have placed a cooled reference plate in our reheat furnace near the measured area of the slab [3]. This target plate reflects radiation from both the flames and the furnace wall. As shown in Fig. 2, one pyrometer is aimed at the slab surface and the reference pyrometer is aimed at the cooled reference target. The oxidized target surface has nearly the same reflectance as the oxidized slab surface in the furnace.

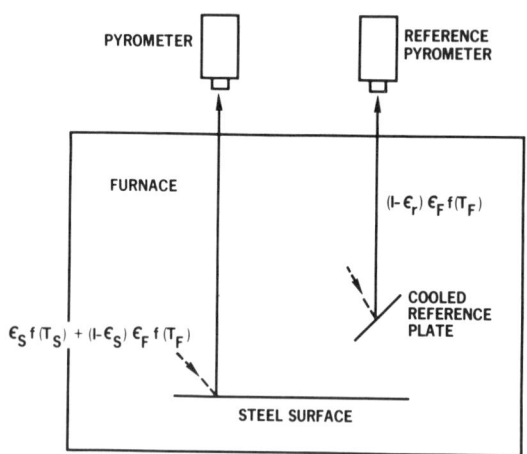

FIG. 2. DUAL PYROMETER - COOLED REFERENCE TARGET METHOD

The equation for this method is:

$$W_a = \frac{\varepsilon_s f(T_s)}{\varepsilon_p} + \frac{(1 - \varepsilon_s) \varepsilon_F f(T_F)}{\varepsilon_p} - \frac{(1 - \varepsilon_r) \varepsilon_F f(T_F)}{\varepsilon_R} \quad (3)$$

where:

ε_r = Effective emissivity of the reference target

To cancel the reflected radiation error:

$$\varepsilon_R = \varepsilon_p \frac{1 - \varepsilon_r}{1 - \varepsilon_s} \quad (4)$$

When the reference target emittance equals the steel surface emittance, $\varepsilon_R = \varepsilon_p$, both pyrometers are set at the same emittance. If the target surface temperature is too high it can cause an error in the measurement. This error will be less than 5°C if the target temperature is lower than 650°C at a steel temperature of 815°C. For convenience we use water cooling which keeps the surface temperature below 100°C. An experimental furnace with overhead firing similar to reheat furnaces was used to determine the dual pyrometer error caused by tilting the reference target from the plane of the slab surface. A slab with a thermocouple embedded in its surface was placed in the furnace with the pyrometer aimed at 45° to the slab surface. The cooled reference target could be tilted up to 45° from the plane of the slab surface with only a 5°C error in the dual pyrometer temperature indication with the error increasing to 55°C with the reference target at 90° to the slab. Thus, to keep the error from target position less than 5°C, we install the reference targets at an angle of 45° or less. At this angle dust and dirt slide off the target surface.

These dual pyrometer systems have been installed on three zones of each of the three reheat furnaces at both our Cleveland and Indiana Harbor Works. They can be mounted on the side walls to aim underneath the flames or they can be mounted on top the furnace to aim between the flames. Even if a pyrometer wavelength is selected that can look through clear flames excess fuel mixtures or incomplete mixing or burning of the fuel will produce luminous carbon particles which will interfere with the measurement.

Fig. 3 shows a cut-a-way view of a furnace zone with the pyrometer heads mounted on the roof. One pyrometer is aimed at the cooled reference target mounted on the burner wall, the other pyrometer aims at the slabs as they pass beneath it. Each pyrometer's signal is equally amplified, then the signal of the pyrometer aimed at the reference target is subtracted from the signal of the pyrometer aimed at the slab. Finally, the resulting signal goes to the meter, recorder, and control system. All our dual pyrometer systems use a similar subtraction circuit. Pyrometer emittance setting, ε_p and ε_R are fixed by the gain adjustment of the amplifier for each pyrometer.

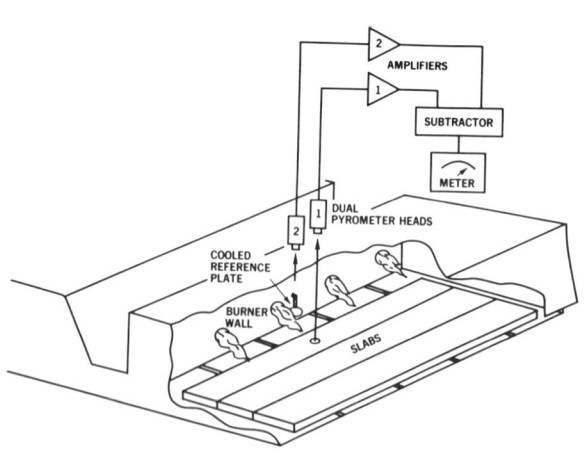

FIG. 3. DUAL PYROMETER ON TOP ZONE OF REHEAT FURNACE

To withstand the severe environment of the reheat furnace, the cooled reference target is ruggedly built and then mounted against the burner wall as shown in Fig. 4. Mounting the target against the burner wall provides additional support and an out of the way location. Yet it is still close to the measurement area of the slab.

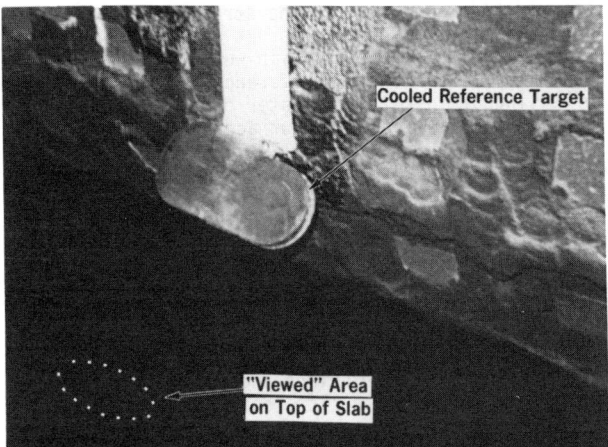

FIG. 4. COOLED REFERENCE TARGET MOUNTED ON FURNACE BURNER WALL

Bottom as well as top slab surface temperatures are measured with the dual pyrometers. Firing of the zones can thus be controlled independently to get the desired slab temperature. No specially designed cooled reference plate is needed in the bottom zone because a section of water cooled support pipe with the insulation removed is used as the reference target.

On-line performance is checked with an embedded thermocouple in the slab surface. As the slab passes through the furnace, the thermocouple is fed through the rear charging door. When the thermocouple junction passes near the pyrometer target, the two measurements are compared. Further checks were also made with a thermocouple embedded in a small test slab. This test slab was attached to a water cooled pipe and then pushed through a side door during the heat-up when the slabs were not moving. Gas flow, air flow, and air/fuel ratios were varied to confirm the dual pyrometer's performance for various firing conditions. The pyrometer's indication of the surface temperature varied less than ±11°C for gas flow variations of 90% of range, air flow changes of 65% of range, and air/fuel ratios from 0.6 to 1.6. Emittance differences of the oxidized surface for various grades of steel were measured. These differences will cause the pyrometer's indication to vary only ±6°C.

Estimated measurement inaccuracy of ±18°C is the root-sum-square (RSS) of the individual errors in the system. These errors are the individual pyrometer inaccuracy in the temperature range we use of ±9°C for each of the two pyrometers, the errors caused by furnace firing variations of ±11°C and the errors caused by emittance variations of ±6°C. Although the inaccuracy of each pyrometer is ±9°C, its reproducibility is ±4°C. Furnace firing variations, fuel and air flow and air/fuel ratios are normally within one-half of those described above. As a consequence, the reproducibility of the measurement of the slab temperature in the furnace is estimated to be ±10°C.

Surface scale temperature is measured by the dual pyrometer instead of the desired steel surface temperature underneath the scale. Slabs enter the furnace with a thin tight scale which grows to a thickness of about 3 mm at the exit end of the intermediate zone. In the first (primary) zone, the scale is thin and the rate of heat flow is high. Computer modeling analysis indicates that the temperature of the scale surface is an average of 55°C higher than the steel surface. In the intermediate zone, the scale is thicker, and the rate of heat flow into the slab is low, the temperature difference averages 28°C. When using the dual pyrometers to control the furnaces, these differences should be used to calculate the heat flow into the slab and to determine control strategy.

Dual pyrometer measurements have improved control and operation of the furnaces. Slabs of different thicknessess in the furnace at the same time cannot be heated to the same temperature, but the dual pyrometer will show when the thick ones are too cold and the thin ones are too hot. Dripping, melted scale from overheating the surfaces damages the insulation on the support pipes, causes build-up on the furnace bottom, and causes defects because melted scale is difficult to remove from the steel surface during rolling. The melting of the scale can be prevented by not allowing the surface temperature to go over 1370°C. During mill delays the furnace firing is reduced. Before the use of dual pyrometers for steel temperature monitoring and control, the firing cutback was not enough because the "Heater" was concerned that the slabs would be too cold to roll. Now with the dual pyrometer measurement, he knows and maintains the slabs at a predetermined temperature. This change alone has resulted in a 4% fuel savings.

Dual pyrometer signals cannot be used in a standard feedback control system because of the variability of slab thickness in the furnace. A thick cool slab would cause high firing while a thin hot slab would reduce the firing. To control this process a list of the dimensions of the slabs in the furnace and the desired dual pyrometer temperatures as a function of operating parameters are stored in a microprocessor. If the slabs are not at the desired temperature when measured by the dual pyrometer, the furnace firing is changed. This microprocessor control is now being installed in the furnace control system. When it is in operation, further improvements in furnace control and fuel rate should result.

CLOSED-END TUBE

Cooled reference targets cannot be placed in some furnaces because of their design, nor are cooled targets needed in all furnaces. If the zone temperature is uniform; no flames, no heating tubes, etc., only a measurement of the zone temperature is needed to correct the steel temperature measurement for the error caused by furnace radiation that is reflected from the surface. This zone temperature measurement can be made with a thermocouple and then the correction calculated. Instead of using a thermocouple to obtain a correction measurement, a correction signal can be obtained from a pyrometer aimed into a closed-end tube as shown in Fig. 5. This correction signal given by Wien's approximation can then be used to subtract the reflection error from the signal of the pyrometer aimed at the steel surface.

The equation for the closed-end tube method is:

$$Wa = \frac{\varepsilon_s f(T_s)}{\varepsilon_p} + \frac{(1-\varepsilon_s)\ \varepsilon_F f(T_F)}{\varepsilon_p} - \frac{f(T_F)}{\varepsilon_R} \quad (5)$$

To cancel the reflection error:

$$\varepsilon_R = \frac{\varepsilon_p}{(1-\varepsilon_s)\ \varepsilon_F} \quad (6)$$

We use the closed-end tube method at our Indiana Harbor Works to measure strip temperature on two galvanizing lines. Strip temperature is measured just beyond the heating zone exit. There are no heating tubes beyond the exit at the pyrometer position, but the zone temperature at this measurement location is high enough to cause a reflection error.

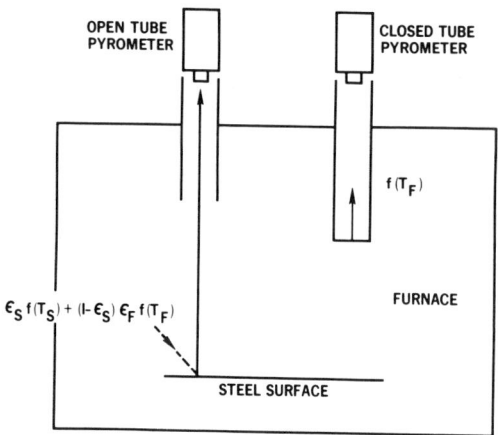

FIG. 5. DUAL PYROMETER - CLOSED TUBE METHOD

By accurately measuring and controlling the strip annealing temperature to the minimum needed, both the heating and cooling requirements are minimized. This lower strip temperature allows us to maximize productivity by operating at the fastest possible line speed.

SPECULAR REFLECTION

The last dual pyrometer method to be described can be used for specularly reflecting (mirror-like) surfaces. For this method one pyrometer is aimed at an angle to the smooth strip. Then because the reflected radiation's angle of reflection is equal to its angle of incidence, the area (reference area) of the furnace wall from which the reflected radiation is coming is known. Another pyrometer as shown in Fig. 6 measures the radiation from this reference area. It is assumed that the furnace wall radiation is equal both in the direction of incidence to the steel surface and in the direction of the reference pyrometer.

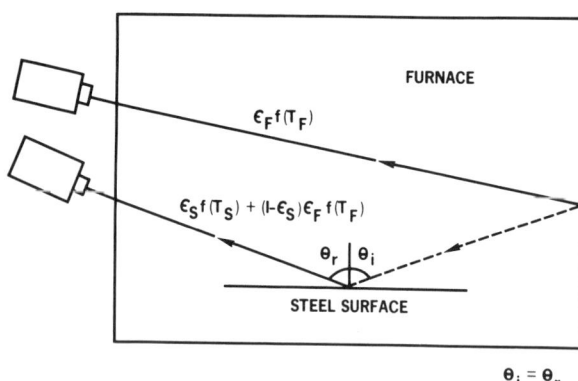

FIG. 6. DUAL PYROMETER - SPECULAR REFLECTION METHOD

The equation for this specular reflection method is:

$$W_a = \frac{\varepsilon_s f(T_s)}{\varepsilon_p} + \frac{(1 - \varepsilon_s) \varepsilon_F f(T_F)}{\varepsilon_p} - \frac{\varepsilon_F f(T_F)}{\varepsilon_R} \qquad (7)$$

To cancel the reflection error:

$$\varepsilon_R = \frac{\varepsilon_p}{1 - \varepsilon_s} \qquad (8)$$

By placing a black body furnace at the reference area both strip temperature and its emittance can be measured [4]. However, if the strip emittance is known, a reference furnace is not needed.

Another modification which we have used on a pilot line is to make the reference area cool and non-reflecting by installing a water cooled conical cavity in place of the black body furnace. This change makes the second term of equation (7) insignificant and eliminates the need for the reference pyrometer.

ERROR CALCULATION

When planning a dual pyrometer system it is essential to be able to estimate measurement accuracy before the installation. For the Back-Off method, let S be the differential signal obtained from the dual pyrometer. Using Wien's approximation for the spectral radiance of the steel surface and the furnace wall, S can be represented to a close approximation by:

$$S = G_p/\varepsilon_p \{\varepsilon_s \exp[-C_2/\lambda_p T_s] + (1-\varepsilon_s) \varepsilon_F \exp[-C_2/\lambda_p T_F]\}$$

$$- G_R/\varepsilon_R \{\varepsilon_F' \exp[-C_2/\lambda_R T_F']\} \qquad (9)$$

where:

G_p and G_R = The response coefficients for the pyrometer aimed at the steel surface and for the pyrometer aimed at the furnace wall, respectively. Calibration information as well as other measurement constants of the pyrometers are contained in these factors.

λ_p and λ_R = The effective wavelengths of the pyrometer aimed at the steel surface and the pyrometer aimed at the furnace wall, respectively

ε_p and ε_R = The emittance setting of the pyrometer aimed at the steel surface and the pyrometer aimed at the furnace wall, respectively

ε_s and ε_F = The effective emissivities of the steel surface and the furnace wall, respectively

T_s and T_F = The temperature of the steel surface and the furnace wall, respectively

ε_F' and T_F' = The effective emissivity and temperature of the furnace wall which may not be the same as ε_F and T_F, respectively

Equation (9) can be rearranged in a more useful form for error analysis.

$$S = G_p/\varepsilon_p \{\varepsilon_s \exp(-C_2/\lambda_p T_s)\} \cdot$$

$$\{1 + \frac{(1 - \varepsilon_s) \varepsilon_F}{\varepsilon_s} \exp[-\frac{C_2}{\lambda_p}(\frac{1}{T_F} - \frac{1}{T_s})]$$

$$- \frac{G_R}{G_p} \frac{\varepsilon_p}{\varepsilon_R} \frac{\varepsilon_F'}{\varepsilon_s} \exp[-C_2(\frac{1}{\lambda_R T_F'} - \frac{1}{\lambda_p T_s})]\} \qquad (10)$$

Similar equations can be derived for the other three methods. Reference pyrometer emittance setting for the method used as given by equations (2), (4), (6), and (8) should be substituted for ε_R. As can be seen in equation (10), to get accurate reflection error compensation, the magnitude of the last and next-to-last terms should be small compared to one. These two terms are smaller for a high effective emissivity, ε_s, than for a low effective emissivity. When $T_s = T_F$ the exponential term is one. When the steel temperature, T_s, is much higher than the furnace temperature, T_F, the magnitude of these terms is small compared to one. The following examples show the error in dual pyrometer compensation that can occur for some installations.

1. Slab temperature in a reheat furnace using the cooled reference method.
   where: $T_s = 1260°C$, $T_F = T_F' = 1620°C$

   $\varepsilon_s = 0.85$, $\varepsilon_p = \varepsilon_r = \varepsilon_R = 0.80$

   $\varepsilon_F = \varepsilon_F' = 0.2$, $G_p = G_R$

   $\lambda_p = \lambda_R = 0.9\ \mu m$

   The dual pyrometer temperature measurement is 1256°C, for an error of -4°C. T_F and ε_F represents the flame temperature and emittance, because most of the radiation is from the flame rather than the furnace walls. If the wavelength is changed to 3.9 μm the error is 27°C for the same conditions. For other flame and slab temperatures, the error may be less at the longer wavelength. We chose the short, 0.9 μm wavelength because the silicon detector used at this wavelength is more rugged and stable than the longer wavelength detectors.

2. Strip temperature in the heating zone of a galvanizing line where the heating tubes and furnace walls are much hotter than the strip. Specular Reflection method.
   where: $T_s = 540°C$, $T_F = 980°C$, $T_F' = 974°C$

   $\varepsilon_s = \varepsilon_p = 0.55$, $\varepsilon_R = \dfrac{\varepsilon_p}{1 - \varepsilon_s}$

   $\varepsilon_F = \varepsilon_F' = 1$, $G_p = G_R$

   $\lambda_p = \lambda_R = 0.9\ \mu m$

   The dual pyrometer measurement is 744°C, an error of 204°C. This large error occurs because over 99% of the radiation received by the pyrometer is reflected radiation. A slight error in the measurement of the reflected radiation (less than the 1% pyrometer temperature measurement accuracy) causes a large error in the dual pyrometer measurement. Obviously a dual pyrometer measurement is not feasible at this location.

3. Strip temperature beyond the exit of the heating zone of the galvanizing line using the Closed-End Tube method.
   where: $T_s = 700°C$, $T_F = 700°C$, $T_F' = 692°C$

   $\varepsilon_s = \varepsilon_p = 0.55$, $\varepsilon_R = \dfrac{\varepsilon_p}{(1 - \varepsilon_s)\ \varepsilon_F}$

   $\varepsilon_F = \varepsilon_F' = 1$, $G_p = G_R$, $\lambda_p = \lambda_R = 0.9\ \mu m$

   The dual pyrometer temperature measurement is 707°C an error of 7°C. The small compensation error of 7°C occurs in this example because only 45% of the radiation received by the pyrometer is reflected radiation.

CONCLUSION

Dual pyrometer systems have provided temperature measurement in many locations where accurate measurements could not be obtained by other means. These steel temperature measurements improve quality, productivity, and reduce the fuel rate. Nevertheless, as shown in the example error calculations, dual pyrometers should not be used where the reflected radiation is many times higher than the emitted radiation. To estimate the installed accuracy, error calculations should be made which determine the effects of pyrometer wavelength, emittance variations, and pyrometer accuracy.

REFERENCES

1. T. Iuchi, et al, Transactions ISIJ, Vol. 16, 1976, "Temperature Measurement System of Steel Strips in a Continuous Annealing Furnace".
2. R. Barber, Industrial and Process Heating, February, 1967, "Furnace Load Temperature Measurements During the Heating Process".
3. J. E. Roney, U.S. Patent No. 4,144,758, "Radiation Measurement of a Product Temperature in a Furnace", Assignee: Jones and Laughlin Steel Corporation.
4. T. Iuchi, U.S. Patent No. 4,172,383, "Method and Apparatus for Simultaneous Measurement of Both Temperature and Emissivity of a Heated Material, Assignee: Nippon Steel Corporation.

Two methods for simultaneous measurement of temperature and emittance using multiple reflection and specular reflection, and their applications to industrial processes

T. Iuchi and R. Kusaka[a]

Fundamental Research Laboratories, Central R & D Bureau, Nippon Steel Corporation, 1618 Ida, Nakahara-ku, Kawasaki 211, Japan

For radiation thermometry, the two problems of emittance and stray radiation from the surroundings always exist. The authors have developed two methods which can simultaneously measure both the temperature and emittance. One method, named "MA," utilizes the multiple reflection of radiation within a cavity. The second method, named "TERM," utilizes the specular reflection characteristics of the object being measured. The latter method has a special feature of eliminating stray radiation at the same time. These methods are closely related to each other from the viewpoint of interreflection. Both methods have resulted in new instruments which give not only the true temperature, but also surface information associated with the emittance of the object examined. In this paper, the principle of each method and experimental results are described.

INTRODUCTION

Radiation thermometry can be widely used where the measurement by a contact method such as with a thermocouple is impossible or undesirable, because the surface temperature of an object can be measured quickly and without contact. However, the reliability and accuracy of radiation thermometry is not yet high for low emittance materials or in the case of varying emittance. This is due to the fact that problems of emittance and stray radiation from the surroundings always exist. In order to reduce the emittance problem, several methods or techniques have been suggested. For example, Drury et al. [1] developed a pyrometer with a hemispherical reflector which makes the surface of the object to be measured into a pseudo-blackbody. Kelsall [2] suggested an emittance-compensated pyrometer utilizing specular reflection. Svet [3] investigated a new ratio pyrometer which eliminated the need for actually measuring the ratio, which in turn greatly improved the precision and dependability of the temperature measurements. These methods have been aimed at making radiation thermometry free from emittance variation, or reducing emittance effects.

On the other hand, the authors have focused their efforts on the development of simultaneous measurement of both temperature and emittance. Two methods of measurement have been developed. One method, named "MA" utilizes the multiple reflection of radiation within a cylincrical cavity. The second method, named "TERM", utilizes the specular reflection characteristics of the object being measured. The latter method has the special feature of eliminating stray radiation at the same time. The two methods are closely related to each other from the viewpoint of interreflection. Both methods have resulted in new instruments which give not only the true temperature, but also surface information associated with the emittance of the object examined.

In this paper, the principle of each method and experimental results are described and discussed. Online systems which apply these methods to several steel processes are introduced and several examples of the online measurement associated with the operation and control of processes are described.

PRINCIPLES OF MEASUREMENT

Multiple Reflection Method

The unknown quantities of temperature and emittance can be uniquely determined if two simultaneous equations, each including the two quantities, are available. The authors obtained these equations by utilizing the multiple reflection of radiation in a cylindrical cavity [4] which has a specularly-reflecting gold-plated interior surface. The fundamental phenomena are illustrated in Fig. 1. In Fig. 1(a), the open cylindrical cavity is placed above the object to be measured. There is a gap, H, between the object and the bottom opening of the cavity. In Fig. 1(b), the cylindrical cavity with a base surface

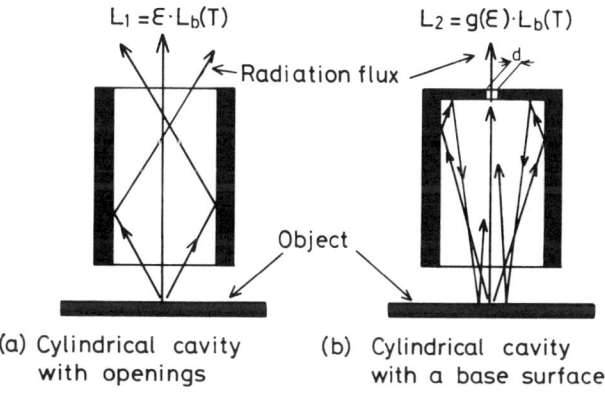

Fig. 1 Fundamental phenomena of interreflection of radiation between a cylindrical cavity and an object.

having a small aperture of diameter, d, is placed above the object. The aperture is much smaller than the diameter of the cavity opening. Identical radiometers located at the upper end of each cavity detect the radiance which comes through the cavity opening in Fig. 1(a) and the aperture in Fig. 1(b), respectively.

In Fig. 1(a), the radiance, L_1, is proportional to the emittance, ε, of the object as given in Eq.(1).

$$L_1 = \varepsilon \cdot L_b(T) \tag{1}$$

where T is the surface temperature of the object and $L_b(T)$ represents the blackbody radiance at temperature T.

On the other hand, in Fig. 1(b), the radiation from the object in multiply reflected between the object and the cavity so that the radiance increases. Accordingly, the detected radiance, L_2 is

$$L_2 = g(\varepsilon) \cdot L_b(T) \tag{2}$$

where $g(\varepsilon)$ represents the apparent emittance of the object, which is a function of ε.

The following assumptions are made in order to formulate $g(\varepsilon)$:

(i) The radiation by the cavity surface can be neglected if its temperature is kept much lower than that of the object. (This is possible by water-cooling or air-cooling.)

(ii) The effective reflectance, ρa, of the cavity with a base surface is constant if the reflectance of the interior surface of the cavity and the cavity dimensions (diameter 2R and length L) are constant.

(iii) The emittance of the object is independent of direction.

(iv) Reflected radiation from the cavity is reflected by the object. The fraction reflected back into the cavity is $(1-\varepsilon) \cdot q$ where the coefficient, $q (0<q<1)$, depends upon the diameter of the cavity, the distance H, and the reflection characteristics of the object surface.

(v) Only a negligible amount of radiation from the region surrounding the object and cavity reaches the detector.

From the above assumptions, L_2 is derived as

$$L_2 = \varepsilon \cdot L_b(T) \cdot \sum_{n=0}^{\infty} (\rho a \cdot (1-\varepsilon) \cdot q)^n \tag{3}$$

where the zeroth term in the series corresponds to the radiance of the object itself and successive terms correspond to the increased radiance due to nth interreflection between the cavity and the object. Because the product $\rho a \cdot (1-\varepsilon) \cdot q$ is less than one, Eq.(3) can be written as

$$L_2 = \frac{\varepsilon}{1-\rho a \cdot (1-\varepsilon) \cdot q} \cdot L_b(T) \tag{4}$$

$$= \frac{(\alpha+1) \cdot \varepsilon}{\varepsilon + \alpha} \cdot L_b(T) \tag{5}$$

where

$$\alpha = \frac{1 - \rho a \cdot q}{\rho a \cdot q} \quad (\geq 0) \tag{6}$$

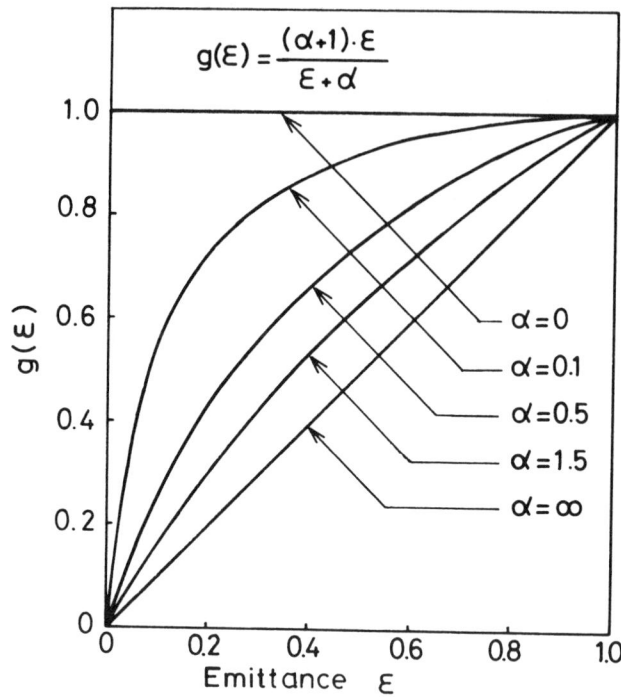

Fig. 2 Relation between $g(\varepsilon)$ and ε. (Cf. Eq. (7)).

Accordingly, $g(\varepsilon)$ becomes

$$g(\varepsilon) = \frac{(\alpha+1) \cdot \varepsilon}{\varepsilon + \alpha} \tag{7}$$

The parameter, α, is independent of ε, but depends upon the gap, H, the reflectance, ρa, the cavity dimensions, and the reflection characteristics of the object.

In Fig. 2, the characteristic curve of $g(\varepsilon)$ is shown for various values of α. It is obvious that, for a given ε, $g(\varepsilon)$ increases as α becomes smaller. That is, $g(\varepsilon)$ increases with increasing ρa and q.

The most important point of this method is that α remains constant even though ε changes. By solving Eqs.(1) and (5) simultaneously, both the temperature and the emittance of the object can be obtained.

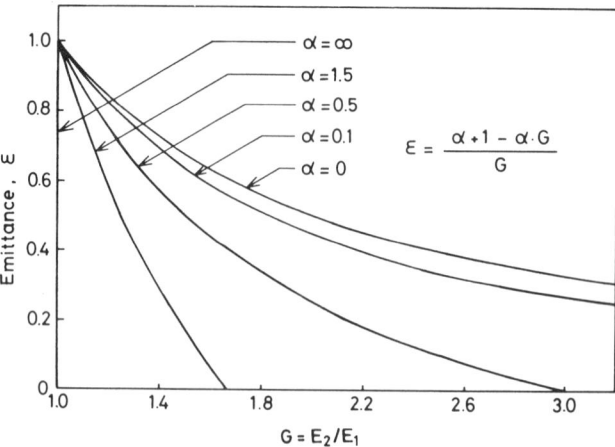

Fig. 3 Relation between ε and G. (Cf. Eq. (9)).

This can be done as follows: Let G equal the ratio L_2/L_1. Then the following equation results.

$$G = \frac{\alpha + 1}{\varepsilon + \alpha} \quad (8)$$

Eq.(8) can be rewritten as

$$\varepsilon = \frac{\alpha + 1 - \alpha \cdot G}{G} \quad (9)$$

The emittance is obtained from Eq.(9). The emittance is generally a nonlinear function of G as shown in Fig. 3.

The temperature can then be obtained from Eq.(1) upon substituting the value of ε obtained above.

The principle of the measurement is realized by the apparatus illustrated in Fig. 4. A cylindrical cavity with an open top is placed above the object to be measured. A fan-shaped rotating chopper is placed close to the open top of the cavity. There is a slit of width, d, in the chopper blades. A radiometer detects the radiance from the object through this slit. The specular surfaces of the cavity interior and the chopper blade are gold plated.

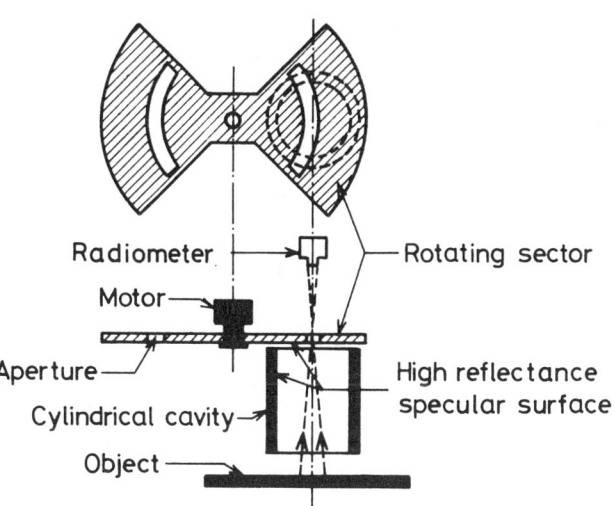

Fig. 4 Schematic diagram of the measurement apparatus. (MA).

When the upper opening of the cavity is not covered by the chopper, the situation is equivalent to Fig. 1(a). At that time the signal from the radiometer, E_1, is given by

$$E_1 = K \cdot \varepsilon \cdot F(T) \quad (10)$$

where K is the response coefficient for the radiometer. This factor contains the constant of calibration as well as other measurement constants including the solid angle of the target and the effective wavelength and bandwidth of the instrument, ε is the emittance of the object surface, and F(T) is the quantity $\exp[-C_2/\lambda_e T]$ from the Wien approximation of the spectral radiance of a blackbody. (C_2 is the second radiation constant and λ_e is the effective wavelength of the radiometer.)

When the opening is covered by the chopper, the signal from the radiometer, E_2, is given by

$$E_2 = K \cdot g(\varepsilon) \cdot F(T) \quad (11)$$

The two signals, E_1 and E_2, are alternately detected as the chopper rotates. Temperature and emittance can be simultaneously derived from these signals. A block diagram of the measuring system is shown in Fig. 5.

The method is named "MA" (Modulation Aided).

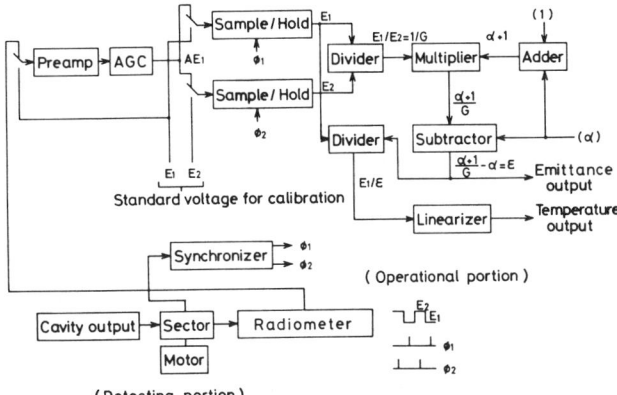

Fig. 5 Block diagram of a measuring system.

Specular Reflection Method

When the surface of an object is optically smooth and flat, namely a perfect specular reflecting surface, the following equation is satisfied

$$\varepsilon_\theta + \rho_\theta = 1 \quad (12)$$

where ε_θ is the emittance of the object in the direction of angle θ with respect to the normal to the object surface and ρ_θ is the reflectance of the object for the flux incident from the direction $-\theta$. The principle of the method will be described with reference to Fig. 6.

A radiometer and a blackbody are disposed symmetrically at an angle θ with respect to the normal to the surface of the object. If the temperatures of

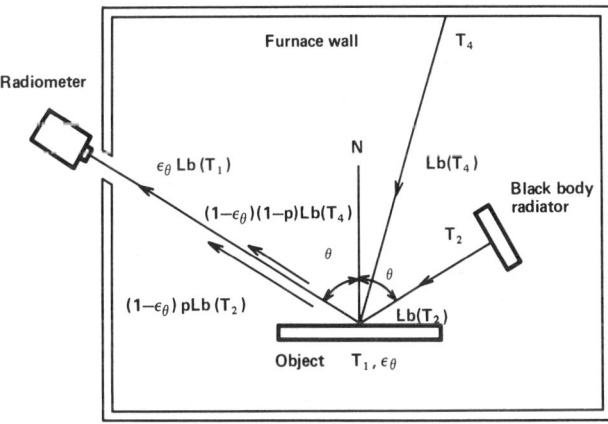

Fig. 6 Schematic diagram of the measuring method. (TERM).

the object and the blackbody radiator are T_1 and T_2, respectively, the output signal from the radiometer can be expressed as

$$E_2 = K \cdot \{\varepsilon_\theta \cdot F(T_1) + (1-\varepsilon_\theta) \cdot F(T_2)\} \quad (13)$$

where the quantities K, $F(T_1)$, and $F(T_2)$ are the same as for Eq.(10) above. The first term on the right side of Eq.(13) is associated with the radiance from the object itself. The second term is associated with the radiance of the blackbody, at temperature T_2, which is specularly reflected to the radiometer from the surface of the object.

When the temperature of the blackbody radiator is changed to T_3, the output signal from the radiometer, E_3, is given by

$$E_3 = K \cdot \{\varepsilon_\theta \cdot F(T_1) + (1-\varepsilon_\theta) \cdot F(T_3)\} \quad (14)$$

Equations (13) and (14) can be solved simultaneously for ε_θ and $F(T_1)$.

$$\varepsilon_\theta = 1 - \frac{E_2 - E_3}{K \cdot \{F(T_2) - F(T_3)\}} \quad (15)$$

$$F(T_1) = \frac{F(T_3) \cdot E_2 - F(T_2) \cdot E_3}{(E_2 - E_3) - K \cdot \{F(T_2) - F(T_3)\}} \quad (16)$$

From equations (15) and (16) it will be seen that, for the case of the perfect specularly reflecting plane surface, the emittance and the temperature of the object surface can be determined independently by this method.

In general, however, because the surface of a real object has a certain amount of roughness, the condition of the measurement is not ideal. This situation causes the specular components of the reflectance to be reduced and a diffuse component to appear. The fraction of reflectance that is specular can be expressed through the use of a "specular reflection factor", p ($0 \leq p \leq 1$), such that the real specular reflectance is

$$\rho_\theta \cdot p = (1 - \varepsilon_\theta) \cdot p \quad (17)$$

(The factor $(1-p)$, correspondingly, represents the fraction of the reflectance that is diffuse.)

For the real object, which is placed in a furnace having a uniform wall temperature of T_4, equations corresponding to Eqs.(13) and (14) above are

$$E_2 = K \cdot \{\varepsilon_\theta \cdot F(T_1) + (1-\varepsilon_\theta) \cdot p \cdot F(T_2) +$$
$$(1-\varepsilon_\theta) \cdot (1-p) \cdot F(T_4)\} \quad (18)$$

$$E_3 = K \cdot \{\varepsilon_\theta \cdot F(T_1) + (1-\varepsilon_\theta) \cdot p \cdot F(T_3) +$$
$$(1-\varepsilon_\theta) \cdot (1-p) \cdot F(T_4)\} \quad (19)$$

As before, the first term on the right side of Eq.(18) and (19) is associated with the radiance from the object itself. The second term is associated with that fraction of the radiance of the blackbody at temperature T_2 or T_3 which is specularly reflected to the radiometer from the surface of the object. The third term is associated with the stray, background radiance that is diffusely reflected to the radiometer from the surface of the object. The background radiance is assumed to be that of a blackbody at the furnace temperature T_4.

Solving equations (18) and (19) simultaneously for ε_θ and $F(T_1)$ will give the following expressions

$$\varepsilon_\theta = 1 - \frac{E_2 - E_3}{K \cdot p \cdot \{F(T_2) - F(T_3)\}} \quad (20)$$

$$F(T_1) = $$
$$\frac{p \cdot [F(T_3) \cdot E_2 - F(T_2) \cdot E_3] + (1-p) \cdot F(T_4) \cdot (E_2 - E_3)}{(E_2 - E_3) - p \cdot K \cdot \{F(T_2) - F(T_3)\}} \quad (21)$$

It can be seen from Eq.(20) that ε_θ can be determined independently of T_1 and T_4 provided that these temperatures and p remain constant over the time interval required to obtain E_2 and E_3.

Although the value of T_1 calculated by means of Eq.(21) is independent of the emittance of the object (provided that p remains constant over the time interval required to obtain E_2 and E_3), it is apparent that the temperature will be in error by an amount which depends upon p, T_4, and temperatures selected for the two blackbodies, T_2 and T_3.

Substituting the expression for ε_θ in Eq.(20) into Eq.(19) will result in

$$F(T_1) = \frac{E_3}{K \cdot \varepsilon_\theta} - \frac{1-\varepsilon_\theta}{\varepsilon_\theta} \cdot p \cdot F(T_3) -$$
$$\frac{1-\varepsilon_\theta}{\varepsilon_\theta} \cdot (1-p) \cdot F(T_4) \quad (22)$$

which is a useful form for computation. Fig. 7 is a block diagram of the computation for effectively implementing the principle of the measurement.

This method is named "TERM" (Temperature and Emittance measurement be Reflection Method).

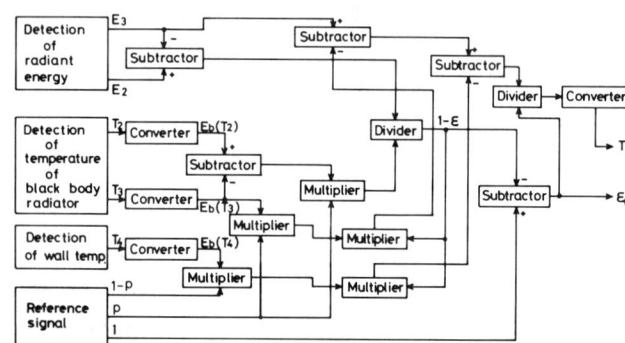

Fig. 7 Block diagram of a measuring system.

EXPERIMENTS FOR MA

Method of Measurement

An experimental apparatus shown in Fig. 4 was manufactured. The brass cavity (inside diameter of 50 mm and length of 150 mm) is a co-axial double cylinder through which cooling water flows. The diameter of the fan-shaped chopper is 220 mm and the width of the slit in the blade is 5 mm. The detector is PbS which is sensitive over the wavelength region of 1.8 to 3.0 μm.

A sample to be measured (diameter of 100 mm) was heated on a furnace. The temperature of each sample was measured using a Type K thermocouple spot-welded to its surface. The two signals, E_1 and E_2, were detected and their ratio, G, was computed. The emittance of the sample at temperature, T, was calculated from the following equation.

$$\varepsilon = \frac{E_1}{K \cdot F(T_1)} \qquad (23)$$

The parameter α was selected so that the ε's of Eqs. (9) and (23) were equal.

Confirmation of Principle of the Method

Samples of cold rolled steel sheet, stainless steel sheet, and aluminum plate were prepared with surfaces having various conditions of roughness. The relation between ε and G was obtained over the temperature range 200 to 900 °C.

It is clear from Table I that the parameter, α is nearly constant for samples of a given material in spite of large variation in ε. On the other hand, different materials exhibit different α's because the surface conditions are different. The principle of the method was experimentally confirmed.

Table I Experimental confirmation of the measurement principle.

Sample	Temperature range (°C)	Emittance range ε	$\alpha + \Delta\alpha$	RMS α (μm)	$\left|\frac{\Delta\varepsilon}{\varepsilon}\right|$	$\left|\frac{\Delta T}{T}\right|$
Cold rolled steel (dull)	200 - 600	0.19 - 0.81	0.53 ± 0.04	1.329	0.11	0.01
Stainless steel (BA1)	300 - 900	0.20 - 0.74	0.25 ± 0.02	0.078	0.06	0.01
Stainless Steel (BA2)	300 - 900	0.21 - 0.73	0.34 ± 0.02	0.190	0.06	0.01
Aluminum (specular)	300 - 500	0.06 - 0.08	0.38 ± 0.02	0.351	0.28	0.026
Aluminum (shot blast)	300 - 500	0.30 - 0.38	0.75 ± 0.04	8.541	0.05	0.005

Analysis of Error

The error due to fluctuation of the measured values is analyzed. When α and G fluctuate by amounts $\Delta\alpha$ and ΔG, respectively, the error, $\Delta\varepsilon$, of ε is computed from Eq.(9) as follows:

$$\Delta\varepsilon = -\frac{G-1}{G} \cdot \Delta\alpha - \frac{\alpha+1}{G^2} \cdot \Delta G \qquad (24)$$

The temperature error, ΔT, can be expressed as

$$\Delta T = \frac{-\lambda_e \cdot T^2}{C_2} \cdot \frac{\Delta\varepsilon}{\varepsilon} \qquad (25)$$

By substituting the expression for $\Delta\varepsilon$ in Eq.(24) into Eq.(25), the error, ΔT, can be expressed as

$$\Delta T = \frac{\lambda_e \cdot T^2}{C_2} \cdot [\frac{G-1}{\varepsilon \cdot G} \cdot \Delta\alpha + \frac{\alpha+1}{\varepsilon \cdot G^2} \cdot \Delta G] \qquad (26)$$

In Eqs.(24) and (26), the first term on the right side shows the error due to variation of α, and the second term shows the variation due to G, respectively. While the former increases with G, the latter decreases with G^2. Accordingly, larger G causes less error. This implies that the selection of the cavity dimensions is important.

It is obvious from Eq.(26) that the temperature error is reduced as λ_e becomes shorter.

In Table I, the relative errors of ε and T are computed as a function of $\Delta\alpha$. Errors $|\Delta\varepsilon/\varepsilon|$ and $|\Delta T/T|$ are within 10% and 1%, respectively, except for the specular aluminum plates which have very low emittances.

Characteristics of the Cavity

The reflection characteristics of the cavity are closely related to its dimensions, surface condition of its interior and of the object to be measured, and the gap, H. The cavity should have the following characteristics.

1) Large G
2) Small variation of G with H
3) A shape that is easily constructed and an inside surface that is easily maintained.

As described below, a long cylindrical cavity is highly suitable.

In order to investigate the effect of cavity shape, both cylindrical and hemispherical cavities were constructed. The disk type upper base of the cylindrical cavity could be moved vertically inside the cavity. A coated steel sample was selected. The emittance of the sample remained constant during the measurements.

In Fig. 8, measured values of G are plotted as a function of the normalized gap distance, H/R. It is clear that the cylindrical cavity is superior to the hemispherical cavity. It produces a larger G and a smaller variation of G with H.

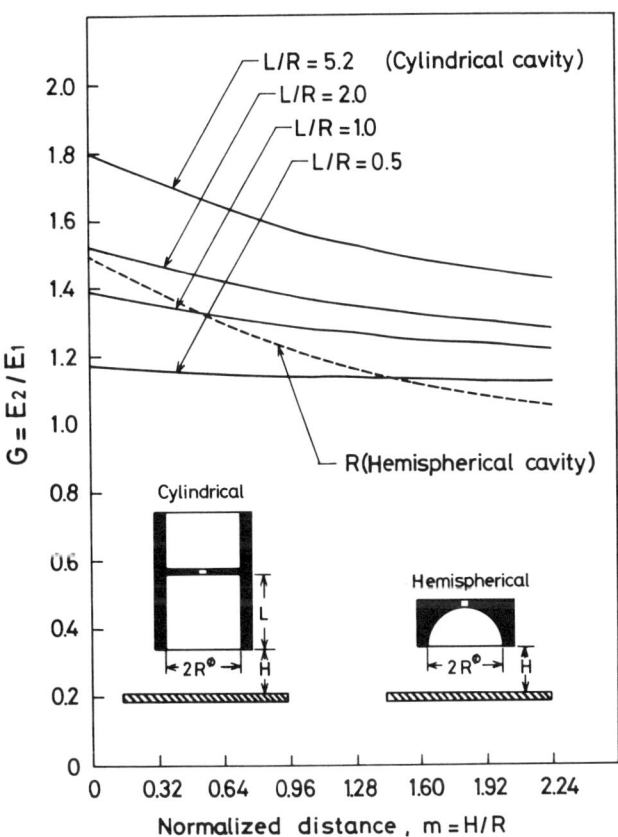

Fig. 8 Experimental relation between $G = E_2/E_1$ and $m = H/R$ for cylindrical and hemispherical cavities.

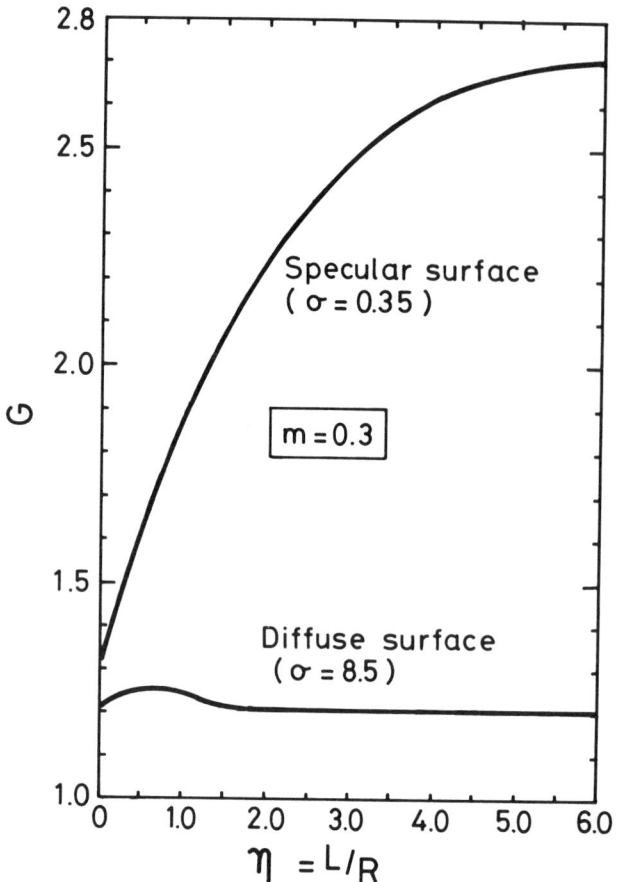

Fig. 9 Experimental relation between G and $\eta = L/R$ for specularly reflecting and diffusely reflecting samples.

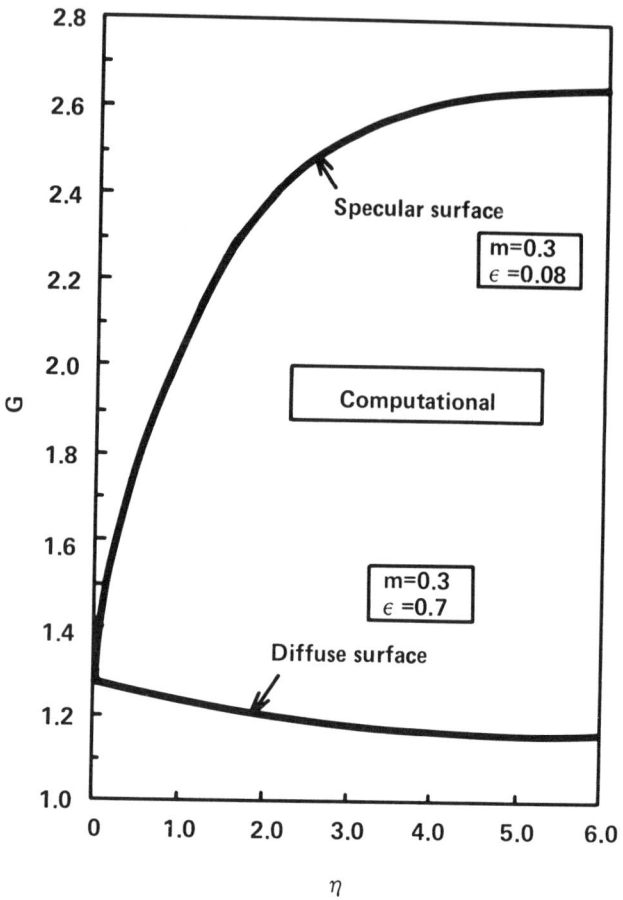

Fig. 10 Computational relation between G and η for specularly reflecting and diffusely reflecting surfaces.

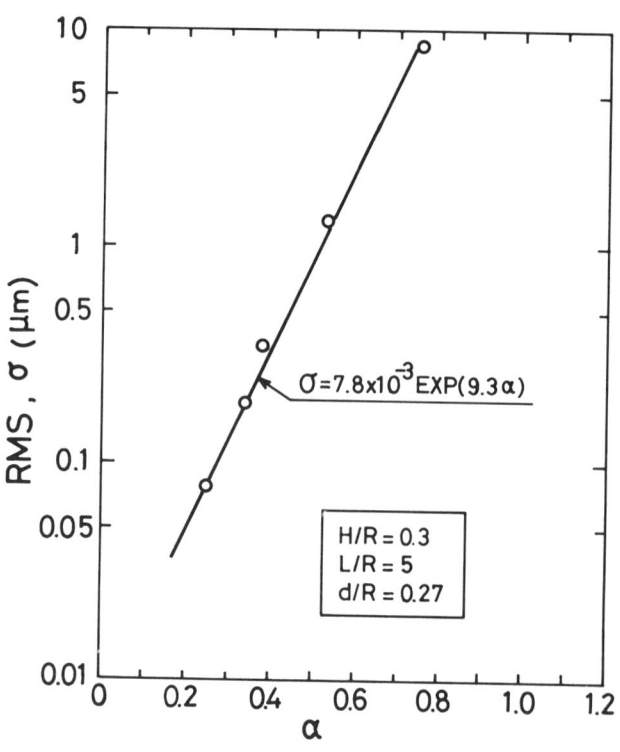

Fig. 11 Experimental relation between α and σ.

In Fig. 9, measured values of G for two aluminum samples are plotted as a function of the normalized cavity length, $\eta = L/R$. One sample is an aluminum plate having specular reflection characteristics (root-mean-square roughness, $\sigma = 0.35$ μm), and the other is a shot-blasted aluminum plate having diffuse reflection characteristics ($\sigma = 8.5$ μm).

For the specular sample, the value of G increases rapidly as η increases from 0. The rate at which G increases gradually diminishes in the region of η greater than 4. On the other hand, for the diffuse sample, the value of G gradually decreases with increasing η.

Fig. 10 shows the values of G as a function of η calculated for a specular object surface and a diffuse object surface. (See the appendix for details of the calculations.) The computed curves correspond well with the experimental results shown in Fig. 9.

It is clear that the reflection characteristics are strongly affected by the surface roughness. RMS roughness [5], σ, was measured for every specimen. The relation between α and σ is plotted in Fig. 11 together with the empirical relation

$$\sigma = 7.8 \times 10^{-3} \exp(9.3\alpha) \qquad (27)$$

EXPERIMENTS FOR TERM

Determination of p

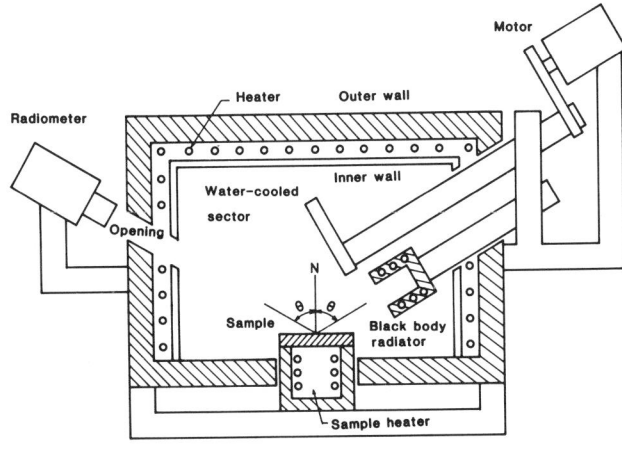

Fig. 12 Simulation furnace to measure p and f (TERM).

A simulation furnace was set up. Fig. 12 schematically shows the experimental arrangement. The inner wall of the furnace is made of thin steel plate and is of boxlike shape. The entire surface of the inner wall was coated with black paint to have an emittance of 0.95. The ceiling and side wall portion of the inner wall are heated by three separate heaters. The inner wall temperature, T_4, can be controlled independently so that it is substantially uniform over the entire surface. The temperature of the inner wall was monitored with Type K thermocouples placed at three locations on the ceiling and side wall portions.

An opposite pair of side walls of the furnace were each provided with an opening 50 mm wide and 100 mm long. A blackbody radiator was inserted through one opening and a radiometer through the other. The two were symmetrically disposed at an angle of 56° with respect to the normal to the sample surface.

The blackbody radiator was a hollow graphite cylinder having an inner diameter of 50 mm and a depth of 125 mm. It was maintained at 368 ± 1 °C using a three-mode controller. Its temperature was detected with a Type K thermocouple mounted in the bottom surface of the radiator. A blackened, water-cooled sector was placed between the sample and the radiator in order to cover the opening of the radiator as required. The solid angle subtended by the sector from the sample surface was 0.05π sr.

Three radiometers were used. The detecting elements were PbS (λ_e = 2.2 μm), InSb (λ_e = 5.0 μm), and a thermistor bolometer (λ_e = 8.0 μm).

To determine the factor p in the absence of significant background radiation, a series of experiments were performed with T_4 kept near 40 °C. The experimental procedure was as follows. Each sample was mounted on the sample heater and inserted into the unheated furnace. The surface temperature of the sample (100 mm in diameter) was measured with a Type K thermocouple which was spot-welded to the sample surface. Each sample was set at an arbitrary temperature, T_1, in the range from 250 to 450 °C. To determine the emittance, ε_θ at T_1, the water cooled sector was rotated in front of the opening of the blackbody radiator. The measured value of ε_θ under these conditions is

$$\varepsilon_\theta = \frac{E_1}{K \cdot F(T_1)} \quad (28)$$

When the sector is rotated away from the blackbody opening, the output signal of the radiometer is given by Eq. (18).

$$E_2 = K \cdot \{\varepsilon_\theta \cdot F(T_1) + (1 - \varepsilon_\theta) \cdot p \cdot F(T_2)\} \quad (18)$$

Combining Eqs. (28) and (18) by eliminating $E(T_1)$ and solving for ε_θ gives

$$\varepsilon_\theta = 1 - \frac{E_2 - E_1}{p \cdot K \cdot \{F(T_2)\}} \quad (29)$$

The specular reflection factor, p, is determined such that the value of ε_θ computed from Eqs. (28) and (29) are the same for each sample at every temperature.

Table II lists the results of measurements of various samples of cold rolled sheet, stainless steel sheet, and aluminum plate with a roughened surface. It will be seen that, for each sample, the calculated values of p are nearly constant for a given wavelength and tend to increase with increasing wavelength.

The variation in calculated values of emittance that is caused by small variations in p can be estimated from these data.

By rearranging Eq. (29), p can be expressed as

$$p = \frac{E_2 - E_1}{(1 - \varepsilon_\theta) \cdot K \cdot F(T_2)} \quad (30)$$

By differentiating Eq. (30) with respect to ε_θ one can express the variation $\Delta\varepsilon$ with Δp as

$$\frac{\Delta\varepsilon}{\varepsilon_\theta} = \left(\frac{1}{\varepsilon_\theta} - 1\right) \cdot \frac{\Delta p}{p} \quad (31)$$

Table II Experimental results of p for cold rolled steel sheet, stainless steel sheet and aluminum plate.

Sample	λ (μm)	T_1 (°C)	ε_θ	p	$\bar{p}$	Δp	$\Delta\varepsilon/\varepsilon_\theta$	$\Delta T/T_1$
Cold rolled steel sheet	2.2	326 327 369 390 410 410	0.316 0.443 0.477 0.495 0.690 0.820	0.90 0.91 0.94 0.94 0.94 0.89	0.92	±0.03	0.07	0.007
Stainless steel sheet (bright annealed)	2.2	306 327 390 410 390 369	0.261 0.300 0.305 0.356 0.338 0.339	0.94 0.99 1.00 1.00 0.98 0.98	0.98	±0.02	0.06	0.006
Shot-blast treated aluminum plate (#250)	2.2	306 368 410 368	0.383 0.424 0.440 0.434	0.091 0.098 0.109 0.097	0.098	±0.009	0.15	0.015
Cold rolled steel sheet	5.0	221 275 326 368 430	0.144 0.146 0.151 0.170 0.287	0.97 0.96 0.97 0.97 0.98	0.97	±0.01	0.07	0.007
Stainless steel sheet (bright annealed)	5.0	221 326 368 430	0.155 0.181 0.178 0.181	0.98 0.99 0.98 0.99	0.99	±0.01	0.06	0.013
Shot-blast treated aluminum plate (#250)	5.0	221 326 368 430 326	0.412 0.423 0.438 0.462 0.453	0.14 0.14 0.15 0.15 0.15	0.15	±0.006	0.25	0.057
Cold-rolled steel sheet	8.0	296 326 368 431 296	0.138 0.139 0.147 0.195 0.376	0.99 0.99 0.98 0.99 1.00	0.99	±0.01	0.06	0.024
Stainless steel sheet (bright annealed)	8.0	296 326 368 431	0.184 0.176 0.164 0.174	1.01 1.00 1.00 1.01	1.00	±0.005	0.03	0.009
Shot-blast treated aluminum plate (#250)	8.0	326 368 431 368 326	0.451 0.464 0.483 0.464 0.463	0.27 0.26 0.26 0.26 0.26	0.26	±0.005	0.02	0.009

The relative variation of the temperature with respect to the emittance can be expressed by the well known formula

$$\frac{\Delta T}{T_1} = \frac{-\lambda_e \cdot T_1}{C_2} \cdot \frac{\Delta \varepsilon}{\varepsilon_\theta} \quad (32)$$

Substitution of Eq. (31) into Eq. (32) will give

$$\frac{\Delta T}{T_1} = \frac{-\lambda_e \cdot T_1}{C_2} \cdot \left(\frac{1}{\varepsilon_\theta} - 1 \right) \cdot \frac{\Delta p}{p} \quad (33)$$

Taking $\bar{p}$ to be the mean of the calculated values of p for each sample at a given λ_e, and Δp to be the value given in Table II, the values of $\Delta\varepsilon/\varepsilon_\theta$ and $\Delta T/T_1$ appearing in columns 8 and 9 were calculated using Eqs. (31) and (33) and the values of ε_θ obtained from Eq. (29).

From Table II it will be seen that p approaches 1 with increasing wavelength, λ_e, and that the relative variation, $\Delta p/p$, decreases as λ_e increases.

Because the emittance and temperature determined by this method depend upon ε_θ and λ_e in addition to $\Delta p/p$ as shown in Eq. (31) and (33), the measurement accuracy at long wavelengths is not necessarily excellent. For example, for the data in Table II the relative error of the emittance is least at $\lambda_e = 8$ μm, while the relative error of the temperature is least at $\lambda_e = 2.2$ μm.

Fig. 13 shows the relation between the true emittance (Eq. (28)) and emittance obtained by this method (Eq. (29)) for measurements made at 2.2 μm on many samples of cold rolled steel sheet. In this experiment, $\bar{p}$ of 0.92 was used. It can be seen that the two values of emittance are in good agreement in spite of the large variation in emittance and the formation of an oxide film on the sample surface during heating.

Fig. 14 similarly shows the relation between the true temperature and that obtained by this method. The uncertainty is about ± 5 °C.

From Table II, and Figs. 13 and 14, we have shown that, for the same kind of steel plate, once the value

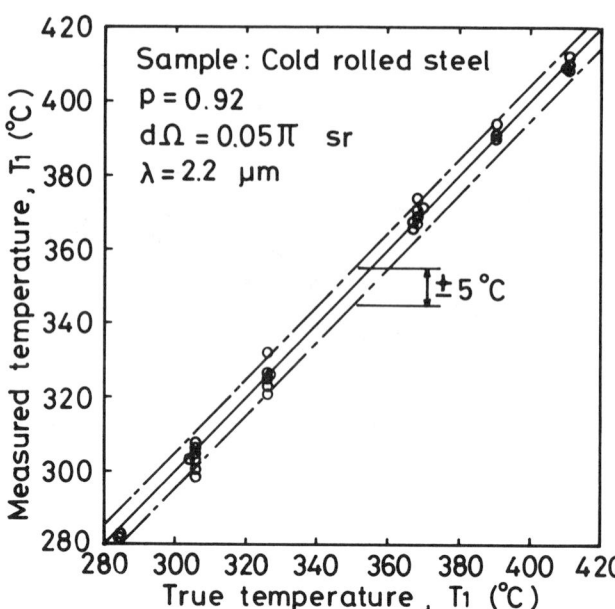

Fig. 14 Comparison of the true temperature and the temperature measured by TERM.

of the p is established, the temperature and emittance can be measured simultaneously and accurately, regardless of wide variations in emittance.

Next the temperature, T_4, was controlled at 390°C. The opening of the blackbody radiator was covered with the water-cooled sector. For this series of measurements, each sample was water-cooled so that its surface temperature remained near room temperature. Under these conditions the output signal from the radiometer, E_s, provides information on the fraction of the radiation from the furnace wall that is received by the radiometer after being diffusely reflected from the surface of the sample.

$$E_s = (1-\varepsilon_\theta) \cdot (1-p) \cdot \phi \cdot F(T_4) \quad (34)$$

where ϕ is a function of the solid angle, $d\Omega$ subtended by the sector from the measured area on the sample surface.

Because the emittance of each sample had been previously measured, the "background factor", f, defined as

$$f = (1-p) \cdot \phi \quad (35)$$

could be computed from the measurements.

The results of a series of measurements in which the solid angle subtended by the cooled sector was varied from 0.05π to 0.20π sr are given in Table III and plotted in Fig. 15. It will be seen that for each sample, f approaches zero as λ_e increases, or as the sample approaches a more specularly reflecting surface. In addition, an increase in $d\Omega$ makes f small. It is therefore possible to control the error in the temperature determined by this method by choosing the optimum position for the cooled sector.

Now let us describe the method for determining the placement of the sector. When the sector covers the blackbody radiator and the true temperature of the sample is T_1, the apparent temperature, T_a, can be determined from

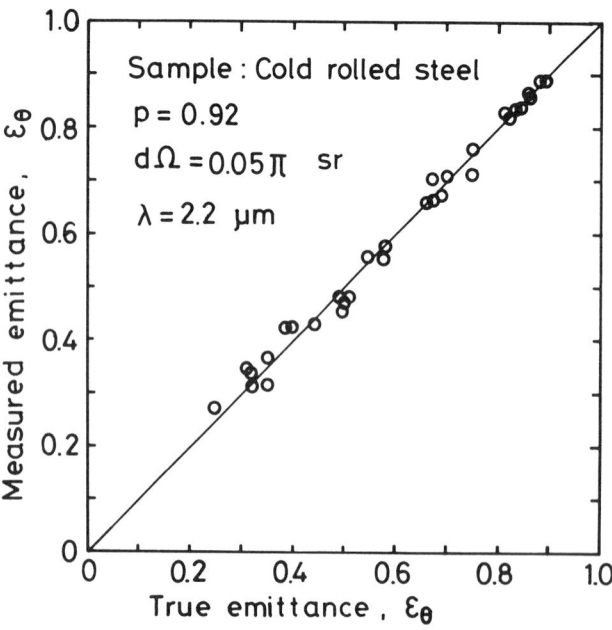

Fig. 13 Comparison of the true emittance and the emittance measured by TERM.

$$F(T_a) = F(T_1) + \frac{1-\varepsilon_\theta}{\varepsilon_\theta} \cdot f \cdot F(T_4) \quad (36)$$

If the maximum acceptable error in the temperature is ΔT, where $T_a = T_1 + \Delta T$, then

$$F(T_1 + \Delta T) = F(T_1) + \frac{1-\varepsilon_\theta}{\varepsilon_\theta} \cdot f \cdot F(T_4) \quad (37)$$

Solving Eq.(36) for f, and substituting the minimum value expected for the emittance, ε_{min}, will yield the maximum value, f_{max} of f, as follows

$$f_{max} = \frac{\varepsilon_{min}}{1-\varepsilon_{min}} \cdot \frac{F(T_1 + \Delta T) - F(T_1)}{F(T_4)} \quad (38)$$

Giving specific values of T_1 and T_4, and selecting λ_e will determine the f_{max} for the temperature measurement error within $\pm\Delta T/2$. Then the solid angle, $d\Omega_{min}$, corresponding to f_{max} can be determined from data of f vs $d\Omega$.

Practically, the maximum value of T_4 should be selected because the temperature of a furnace wall is not uniform.

Specific computations for cold rolled steel sheet and stainless steel sheet will be described. Table IV lists the f_{max} and $d\Omega_{min}$ necessary for restricting ΔT to within 10°C for $T_1 = 700°C$ and $T_4 = 600$, 700, and 800°C. From Fig. 15, it will be seen that f decreases as λ_e increases, thus reducing the effect of the background radiation. However, the $d\Omega_{min}$

Table III Experimental results of the relation between f and $d\Omega$ for cold rolled steel sheet, stainless steel sheet and aluminum plate.

Sample	Effective wavelength λ (μm)	Emittance ε_θ	Solid angle $d\Omega$ (sr)	Diffuse reflection factor $f = (1-p)\phi$	Specular reflection factor, p
Cold rolled steel	2.2	0.25 ~ 0.82	0.05 π 0.15 π 0.20 π	0.061 ± 0.01 0.04 ± 0.003 0.03 ± 0.005	$d\Omega = 0.05$ π sr 0.92
	5.0	0.15 ~ 0.40	0.05 π 0.15 π 0.20 π	0.048 ± 0.005 0.03 ± 0.004 0.015 ± 0.003	0.97
	8.0	0.08 ~ 0.38	0.05 π 0.15 π 0.20 π	0.021 ± 0.001 0.013 ± 0.001 0.009 ± 0.001	0.99
Stainless steel	2.2	0.30 ~ 0.36	0.05 π 0.15 π 0.20 π	0.023 ± 0.007 0.017 ± 0.004 0.005 ± 0.002	0.98
	5.0	0.15 ~ 0.19	0.05 π 0.15 π 0.20 π	0.008 ± 0.002 0.004 ± 0.002 0.002 ± 0.001	0.99
	8.0	0.15 ~ 0.19	0.05 π 0.15 π 0.20 π	0.003 ± 0.001 0.0012± 0.0005±	1.00
Aluminum (rough surface)	2.2	0.23 ~ 0.26	0.05 π 0.15 π 0.20 π	0.902 ± 0.009 0.550 ± 0.005 0.299 ± 0.003	
	5.0	0.25 ~ 0.28	0.05 π 0.15 π 0.20 π	0.850 ± 0.006 0.440 ± 0.004 0.212 ± 0.004	
	8.0	0.27 ~ 0.29	0.05 π 0.15 π 0.20 π	0.740 ± 0.00 0.290 ± 0.00 0.105 ± 0.00	

necessary for restricting ΔT to a specific value becomes smaller as λ_e decreases. The capability of reducing $d\Omega_{min}$ is very important from the technical view point. For stainless steel sheets, it is possible to make $d\Omega_{min}$ very small.

The experimental relation between f and θ_a is plotted in Fig. 16, where θ_a is an average slope angle which is one of the definitions of surface roughness as shown in Fig. 17. This shows that if θ_a of a sample is known, the value of f can be determined.

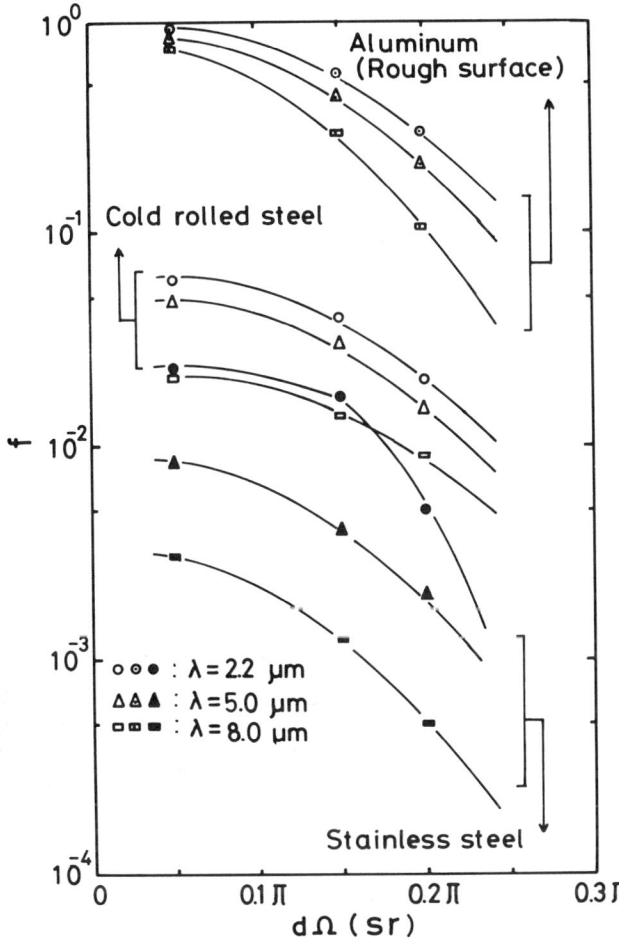

Fig. 15 Experimental relation between f and $d\Omega$.

Table IV Computational results of f_{max} and $d\Omega_{min}$ necessary for restricting ΔT within 10°C.

Specimen	Effective wavelength λ (μm)	ε_{min}	Temperature of specimen, T_1(°C)	Temperature of wall, T_4(°C)	f_{max}	$d\Omega_{min}$ (sr)
Cold rolled steel	2.2	0.25	700	600 700 800	0.051 0.024 0.0127	0.10 π 0.19 π 0.22 π
	5.0	0.15	700	600 700 800	0.012 0.005 0.004	0.21 π 0.25 π 0.26 π
	8.0	0.08	700	600 700 800	0.004 0.0017 0.0014	0.25 π 0.27 π 0.28 π
Stainless steel	2.2	0.30	700	600 700 800	0.066 0.031 0.016	0 0 0.15 π
	5.0	0.15	700	600 700 800	0.0076 0.0054 0.0046	0.07 π 0.12 π 0.14 π
	8.0	0.15	700	600 700 800	0.0043 0.0037 0.0031	0 0 0.05 π

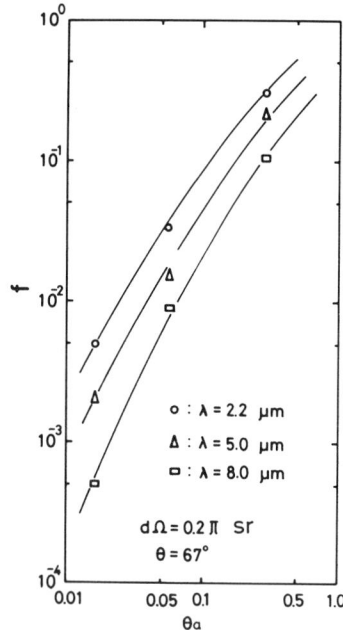

Fig. 16 Experimental relation between f and θ_a in case of $d\Omega = 0.2\pi$ sr, $\theta = 67°$.

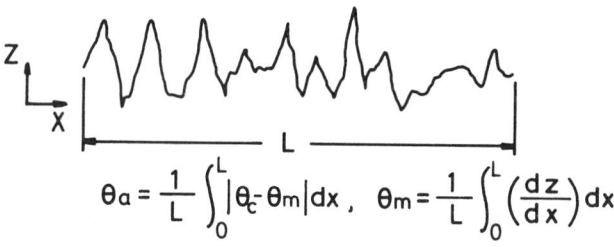

Fig. 17 Definition of average slope angle, θ_a.

ONLINE APPLICATIONS TO STEEL PROCESSES

Online use of MA

The application of this method to the temperature measurement of a coated steel strip is an appropriate example, though there are many processes to which this method is applicable. For this process, the parameter α is very constant because the coated strip has a uniform surface roughness.

The outline of the process for making the coated steel strip is shown in Fig. 18 and is briefly described as follows. After passing through a continuous annealing furnace, the steel strip enters the zone where the coating is applied and dried. The control of the coating process was based upon the temperature of the steel strip at the outlet of the dryer. Accurate measurement by a conventional radiation thermometer was impossible because of the striking changes in the emittance of the coating. Therefore, a contact-type thermometer was used intermittently and the dryer temperature was manually controlled. As a result, the condition of the coating was unstable and the operation of the process was difficult.

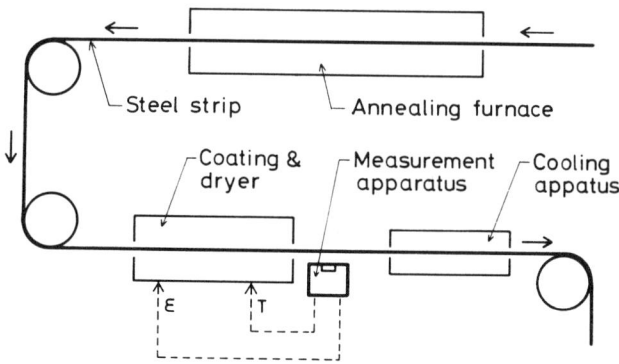

Fig. 18 Outline of the film coated steel making process.

To such a process, an online MA system has been applied, making possible continuous measurement of both temperature and emittance. The online system was installed at the outlet of the dryer.

A fused silica window was placed across the opening of the cavity to keep the interior surface clean. The outside of the window was purged with dry N_2 gas to prevent an accumulation of dust. The transmittance of the window was checked every two months, causing no trouble for practical use.

To prevent infiltration of dust, all of the detection apparatus, including the cavity, radiometer, and optical system were placed in a water-cooled box in which the pressure was kept several tens of torrs above atmospheric.

In Fig. 19, the results of typical online experiments are collected, where (a) indicate the readings of the strip temperature, Tr, by the contact-type thermometer, (b) are the emittance readings determined with the online MA system, and (c) are differences, ΔT, between T and Tr, where T is the temperature indicated by the MA system. At the lower portion of (c), the coil number of the strips is given, where S1-S5 and Th1-Th6 identify the coated steel strips and the threading strips, respectively.

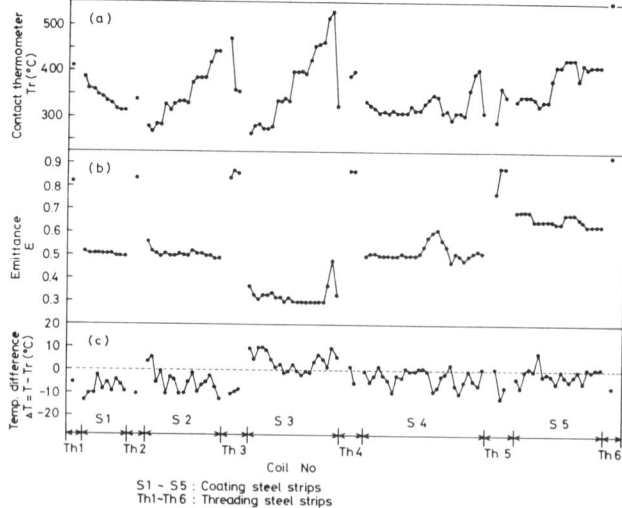

Fig. 19 Results of online experiments in the film coated steel making process.

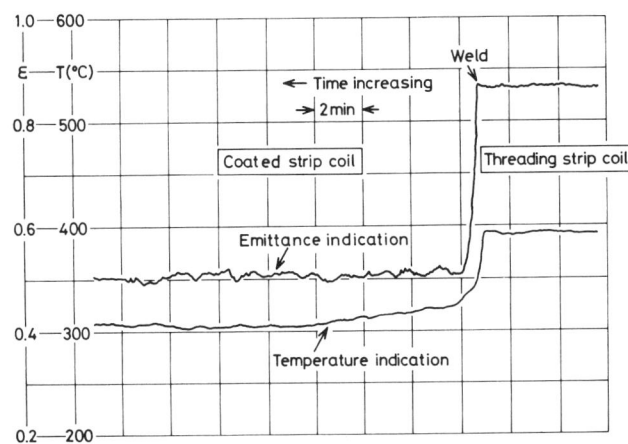

Fig. 20 An example of online measurement near a welded joint portion between threading and coated steel sheets.

The authors have previously shown [6], that the total estimated measurement error for the contact-type thermometer, at a line speed of about 100 m/min, was ±5°C when used by a skillful engineer.

Fig. 19 clearly shows that while the strip temperature ranged between 300°C and more than 500°C (see (a)), the emittances indicated large changes from 0.3 to 0.9 for all sorts of coils (see (b)). On the other hand, the temperature difference, ΔT, was almost within ±10°C. Each ΔT of the coil S3 in the temperature region below 300°C was slightly higher than the others. This was caused by the low sensitivity of the radiometer in the range of low emittance and low temperature. The emittances of the threading coils were about 0.8 because their surfaces were blackened from repeated use. From these results it was concluded that the online system succeeded in practical use.

Two additional examples of the online measurements, which represent typical features of the MA method, will be introduced. Fig. 20 shows the measured data near a welded joint between a threading coil and a coated steel coil. The values of ε and T changed abruptly in the vicinity of the joint. The ε of the threading steel was 0.87 in this case. Because of its high emittance, the threading steel was a good absorber of heat and its temperature is higher than the coated strip. On the other hand, the ε of the coated steel was about 0.51 and its temperature is lower by about 80°C than the threading steel. These measurements provided reliable information for the stable operation of the dryer.

Fig. 21 is an example of the measurement where the dryer temperature increases by several steps, each ranging from 50 to 80°C. The temperature of the coated steel increased with increasing dryer temperature. The emittance of the coated steel, however, showed no change until the strip temperature reached 470°C. Above that temperature, the ε rapidly incresed. The phenomenon corresponded with an abrupt change in the surface condition of the coating material. This information was a control variable for the quality control of the coating.

Online use of TERM

Recently, new furnaces for annealing of continuous steel sheet have been constructed. These furnaces were designed to lower energy costs through improved thermal efficiency. Unlike conventional furnaces in which the sheet is heated in an atmosphere of reducing gases, the new furnaces use rapid heating by direct heating burners in the first zone. This zone is called the Non-Oxidizing Furnace (NOF). However, the weak oxidizing atmosphere in the NOF causes wide variations of the emittance of the steel sheet. The control of the burners of the NOF is based upon the temperature of the sheet as it passes into the second zone, the Reducing Furnace (RF).

Fig. 22 shows schematically the four zones of the new furnace for annealing continuous steel sheet. In the last three zones the sheet is annealed in an atmosphere containing H_2, N_2 and CO so that the oxide film on the steel sheet is reduced and the emittance is stabilized when it leaves the annealing furnace.

A TERM system was installed at the throat between the NOF and the RF as shown in Fig. 23. The two blackbody radiators were controlled to temperatures T_2 and T_3, respectively. The solid angle subtended from the sheet by the opening of each radiator was 0.08π sr. The radiometer alternately produced signals E_2 and E_3 by optical scanning. The detecting element of the radiometer was PbSe ($\lambda_e = 3.45 \pm 0.5$ μm).

Online experiments were done for four kinds of cold rolled steel sheet. The desired temperature for the sheet as it entered the RF was from 250 to 650°C. The two blackbody radiators were set to temperatures

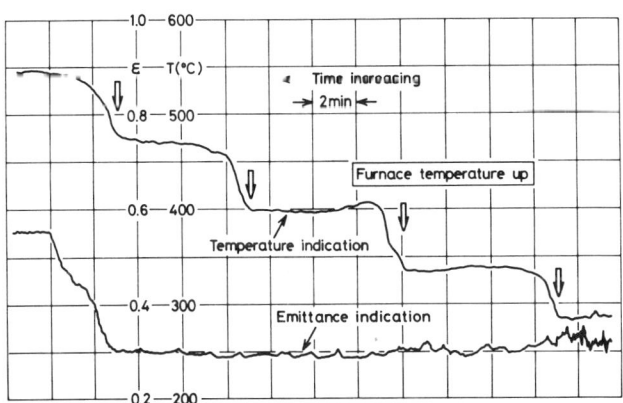

Fig. 21 An example of online measurement in case of the dryer temperature increasing.

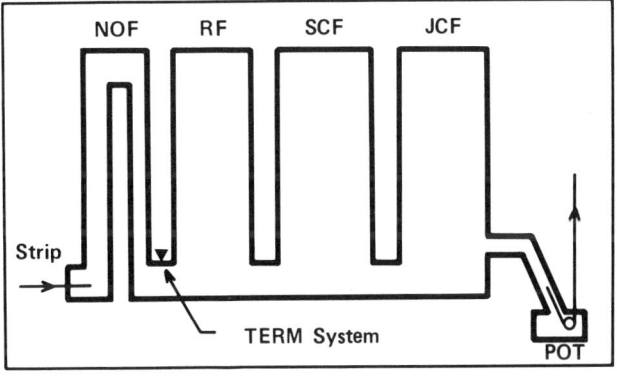

Fig. 22 A continuous annealing furnace composed of four zones, where TERM was introduced.
NOF : Non-Oxidizing Furnace,
RF : Reducing Furnace,
SCF : Slow Cooling Furnace, and
JCF : Jet Cooling Furnace.

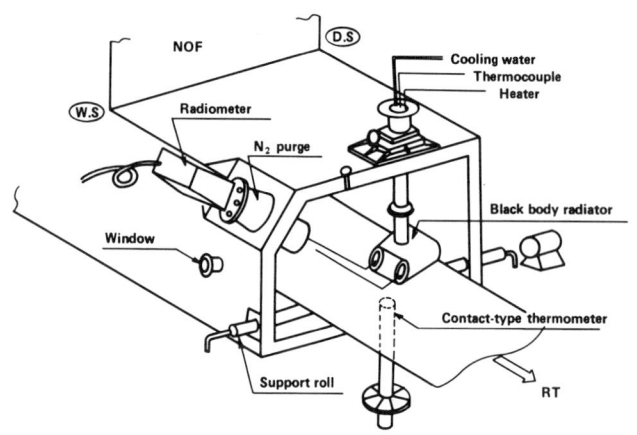

Fig. 23 TERM system installed at a throat between NOF and RF.

Table V Online measurement results of $\Delta p/p$ and ΔT for sheets of a different kind.

Sample	$\Delta p/p$	Emittance	Temperature error ΔT (°C)
A	± 0.051	0.2 – 0.4	±13.7
B	± 0.027	0.2 – 0.4	± 3.6
C	± 0.025	0.2 – 0.4	± 2.1
D	± 0.039	0.2 – 0.4	± 5.8

450°C and 40°C. The temperature of the surroundings, T_4, was assumed to be 400°C. Values for the emittance and the specular reflection factor, p, were obtained in such a manner that the temperatures determined by the TERM system coincided with the readings of the contact-type thermometer used to intermittently monitor the temperature of the bottom surface of the steel sheet.

Table V shows the relative variation in p, range of emittance, and the temperature error resulting from the use of a constant value of p for each kind of steel sheet. Except for the steel sheet specified as A, the temperature error remained within ±10°C. If a conventional radiometer having the same PbSe detecting element is used, the variation of emittance from 0.2 to 0.4 will cause a temperature error of about ±50°C.

The use of the TERM system for the control of the NOF temperature has significantly reduced the effects of such disturbances as changes in line speed or sheet thickness. Moreover, the measurement of the emittance provided useful information about the oxide film formation on the sheet as it passed through the NOF. By contributing to more stable control of the new furnace for annealing continuous steel sheet, it has helped to improve the quality of the steel product and to lower the energy costs required to produce it.

CONCLUSIONS

By utilizing interreflection of radiation, two methods of simultaneous measurement of temperature and emittance have been developed.

The one method, named "MA", is a passive method using multiple reflection between a cylindrical cavity and the surface of the object to be measured. The other method, named "TERM" is an active method utilizing the specular reflection characteristics of the surface to be measured and blackbodies as standard radiators. For the MA method it is necessary that a parameter, α, be kept constant within some tolerance regardless of emittance change. The same situation exists for the TERM method where a parameter, p, must remain constant.

It was confirmed that these two parameters were closely correlated with the surface roughness of an object. TERM has a special feature of eliminating stray radiation at the same time, so that this method especially meets the requirements for radiation thermometry in an environment such as the inside of an industrial furnace filled with stray radiation. Online systems of both methods were applied to the measurement of temperature and emittance in steel making processes where the temperature errors were within 1% and the emittance errors were within 10%.

Acknowledgements

The authors would like to thank Dr. J. Ohno, Senior Researcher, and Mr. F. Tanaka, Research Engineer of the Fundamental Research Laboratories of Nippon Steel Corporation for their useful suggestions and discussions during the basic study of this paper. They would like to express their hearty thanks to Messrs. H. Hirata, Manager, and K. Watanabe, Assistant Manager, of Plant Engineering & Technology of Nippon Steel Corporation, and T. Kawamura, Senior Assistant to General Manager, of Research & Development Office of Plant & Machinery Division of Nippon Steel Corporation, and T. Shibata, Assistant Manager, of Engineering & Design Office of Nagoya Works of Nippon Steel Corporation for their contribution to the development of online systems of these methods.

APPENDIX

The principle of the MA method is based upon condition that parameter, α, is constant regardless of the value of emittance of the surface of the object. As shown in Eq. (a), α is a function of both ρa and q.

$$\alpha = \frac{1-\rho_a \cdot q}{\rho_a \cdot q} \qquad (a)$$

where ρa is the effective reflectance of the cylindrical cavity and q is the fraction of radiation from the cavity that is reflected from the object surface back into the cavity.

The effective reflectance of a cylindrical cavity having specularly reflecting interior surface was derived from Lin's analysis [7].

$$\rho a = \rho - 2 \cdot \eta \cdot (1-\rho)^2 \cdot \sum_{\ell=1}^{\infty} \rho^\ell \cdot [\sqrt{1+(\tfrac{\eta}{\ell})^2} - (\tfrac{\eta}{\ell})] \qquad (b)$$

where

ρ = the reflectance of the interior surface of the cavity,

η = L/R, where L is the length of the cavity and R is the radius of the cylinder, and

ℓ = the number of reflections of radiation within the cavity.

When the object has a perfect diffusely reflecting surface, the coefficient, q, can be expressed as follows [8]

$$q(\text{diff}) = F_{1H} = \tfrac{1}{2}\{2 + m^2 - \sqrt{(2+m^2)^2 - 4}\,\} \quad (c)$$

where

F_{1H} = geometric factor which represents the fraction of radiation that leaves the surface of the object and arrives at the cavity opening, and

m = H/R, where H is the distance between the surface of the object and the opening of the cavity.

The curve, shown in Fig. 10, of G for a diffusely reflecting object surface was calculated from

$$G = \frac{\alpha + 1}{\varepsilon + \alpha} \quad (d)$$

together with Eqs. (a), (b) and (c) using $\rho = 0.95$, $m = 0.3$ and $\varepsilon = 0.7$.

When the object has a perfect specularly reflecting surface, it is assumed that the main portion of the detected radiance comes from only a few interreflections which are in the direction near the normal to the object surface. As a consequence, q can be expressed as

$$q(\text{spec}) = q(\text{diff}) \cdot \cos(\theta)$$
$$= F_{1H} \cdot \cos[\tan^{-1}(\tfrac{1}{m+\eta})] \quad (e)$$

where $\tan(\theta)$ is defined as R/(L+H).

The curve, shown in Fig. 10, of G for a specularly reflecting surface was calculated from Eq. (d) together with Eqs. (a), (b), and (e) using $\rho = 0.95$, $m = 0.3$, and $\varepsilon = 0.08$. In Eq. (b), the upper limit for interreflections, ℓ, was 4.

REFERENCES

a) Present Address: School of Energy Engineering, Toyohashi University of Technology, Tenpaku-cho, Toyohashi, 440 Japan.

1. M.D. Drury, K.P. Perry and T. Land, J.I.S.I. 169, 245 (1951).

2. D. Kelsall, J. Sci. Instrum. 40, 1 (1963).

3. D. Ya Svet, Temperature, Its Measurement and Control in Science and Industry (Instrument Society of America, Pittsburgh, 1972), Vol. 4, Part 1, p. 587

4. T. Iuchi, 8th ACTA IMEKO, 249 (1979).

5. P. Beckmann and A. Spizzichino, The Scattering of Electromagnetic Waves from Rough Surface (Pergamon Press, 1963), p. 187.

6. T. Iuchi, J. Ohno and R. Kusaka, Trans. I.S.I.J. 16, 195 (1976).

7. S.H. Lin, J. Heat Transfer C87, 299 (1965)

8. E.M. Sparrow and R.D. Cess, Radiation Heat Transfer (Books/cole Publishing Company, California, 1966) Chapter 4.

Temperature measurement of steel in the furnace

Y. Tamura and M. Tatsuwaki

Head Office, Sumitomo Metal Industries Limited, Osaka, Japan

T. Sugimura and T. Yokoi

Central Research Laboratories, Sumitomo Metal Industries Limited, Amagasaki, Japan

M. Sano

Wakayama Steel Works, Sumitomo Metal Industries Limited, Wakayama, Japan

M. Koriki

Kashima Steel Works, Sumitomo Metal Industries Limited, Kashima, Japan

Two methods for the measurement of the temperature of steel in the reheating furnace have been developed and put into operation in Sumitomo Metal Industries Limited. The first is radiation pyrometry using a cooled sight tube. We established a design method for the cooled sight tube for hot rolled steel. The suitability of this method is examined in an on-line test. The second is the measurement of the time-temperature pattern of a particular object in the furnace. We called this system the heat-resistant data logger which moves with the steel through the furnace. The features of this heat-resistant data logger are its simple construction, its large capacity for storing data, and its ability to measure the temperature of rotating objects.

INTRODUCTION

In recent years many computer control systems for reheating furnaces have been put into operation for controlling the quality of steel and for energy saving in the steel industry. For these systems it is necessary to measure the temperature of the steel in the furnace. One of our controlled rolling systems has a reheating furnace at the intermediate stage of rolling. In this case the temperature of the slabs as they are extracted from the furnace is very critical, and must be controlled within ±10°C of required temperature (about 900°C). Methods for measurement of the temperature of the steel in the furnace were proposed by Barber [1], and the temperature measurement for the reheating furnace was reviewed by Dixon [2]. Following Dixon, the proximity temperature sensor, which was developed by British Steel Corporation is now used within BSC. Using this method the temperature of the steel is estimated indirectly from the surface temperature of the proximity sensor and the heat flux through the sensor. For accurate temperature measurement it is desirable to measure the temperature of the steel directly. We established the method for designing the cooled sight tube for radiation pyrometry and developed a practical measuring system for the hot rolled slab in the furnace.

It is necessary to measure the time-temperature pattern of the steel in order to verify that the temperature model of the computer control agrees with the actual process. The time-temperature pattern is usually measured by long thermocouples (about 60 m for our reheating furnaces) attached to the slab to be measured. Because this method has several disadvantages, other methods have been studied. A data gathering system which moves with the steel in the furnace was developed by Emschermann et al[3]. This apparatus is protected from the hot surroundings by a heat insulator, a water jacket and a Dewar flask. In another field the same method which uses digital memories was reported by Yokota et al [4]. Because these devices were developed mainly for the purpose of monitoring the temperature, the total amount of the data which they can gather at once is too small for the analysis of the mathematical model of the computer control. We have developed a data gathering system which is constructed simply and can store a large amount of data. The maximum capacity of the prototype model, which can be operated in 1200°C for 4 hours, is 4096 numbers of total data. Its dimensions are 300 mm H x 400 mm W x 680 mm L. We call this data gathering system the "Heat-resistant Data Logger". This method was applied not only to the temperature measurement of the slabs in the reheating furnace but also to the rotating pipes in the heat treatment furnace.

CONTINUOUS TEMPERATURE MEASUREMENT OF SLABS IN REHEATING FURNACE

Measurement Method

Temperature measurement errors in the reheating furnace may be large if proper care is not taken because the radiation pyrometer measures not only the radiation from the hot slab but also the reflected component of the radiation from the wall which is hotter than the slabs. It is possible to shield the slab from the incident radiation at the point of measurement by a water cooled probe. When this method is applied in practice, the major problem is the provision of an adequate shield. In order to develop an adequate shield there are three important factors to be considered.

1. The construction of the cooled sight tube. How to determine the dimensions of the cooled sight tube

in order to decrease the reflected component to within the required accuracy.

2. Selection of the wavelength of the radiation pyrometer

3. A method which can distinguish the slabs from the hearth of the reheating furnace. In this type of reheating furnace, the slabs are continuously transferred from the inlet of the furnace to the outlet. As a consequence, the radiation pyrometer with the cooled sight tube sequentially measures the temperature of the slab and that of the hearth of the furnace.

The construction of the cooled sight tube is shown in Fig. 1. Three coaxial steel tubes are welded in order to construct the framework of the probe, and many studs are welded on the outer surface of the steel tube. Then the outside of the probe is formed using a high-alumina castable refractory which is commercially available. The main constituents of the castable refractory are about 70% alumina and about 25% silica. Using this design the temperature of the inside and base of the probe can be kept sufficiently lower than the temperature of the slab so that the radiation from the probe does not produce significant measurement error.

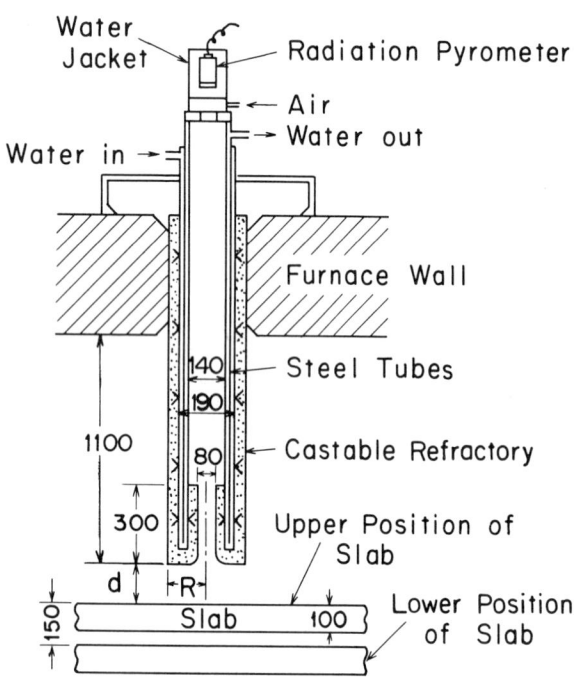

Fig. 1. Probe configuration of radiation Pyrometer with cooled sight tube

(R=150mm d=150mm)

The optical characteristics of the pyrometer, the degree of the tilt likely to be encountered and the reflecting properties of the slab, must all be considered in order to determine the maximum distance to radius ratio (d/R) for the probe (See Fig. 1). Because our reheating furnace uses the walking hearth mechanism, it is assumed that the effect of the tilt can be ignored. The reflecting properties of the hot rolled slab have not been completely investigated, but from our temperature measurement experience with the hot rolled process it is assumed that the hot rolled slab has diffuse reflectivity. The reflectivity of the slab is 1-ε, where ε is the emissivity of the slab. From the results of some experimental work carried out on a variety of hot rolled steel, we know the emissivity of the slab to be about 0.8. It is also assumed that the base of the cooled sight tube is the perfect absorber. From above assumptions, the decreasing degree of the reflecting component by the cooled sight tube can be calculated by the model shown in Fig. 2.

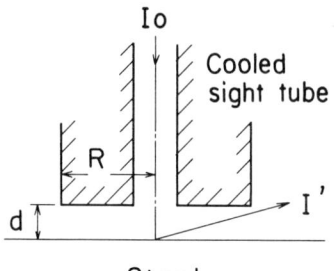

Fig. 2. The model for estimating the effect of the cooled sight tube

$$K = (1 - \varepsilon) \times (1 - \frac{1}{\sqrt{1+(d/R)^2}}) \quad (1)$$

K is really the "effective reflectivity" of the slab.

The characteristics of the radiation pyrometer can be expressed as follows.

$$E = \frac{\varepsilon \cdot A}{\lambda_e^5} \cdot \exp(-\frac{C_2}{\lambda_e T}) \quad (2)$$

where E is the output of the pyrometer, ε is the emissivity of the body to be measured, T is temperature [K], C_2 is the second radiation constant (0.014388 m·K), λ_e is the effective wavelength of the pyrometer and A is a proportionality constant which can be determined through calibration of the pyrometer.

Because the radiation from the gas in the furnace can be ignored, the radiation which is detected by the radiation pyrometer has two components. The first is the radiation from the hot slab.

$$E_s = \frac{\varepsilon \cdot A}{\lambda^5 e} \cdot \exp(-\frac{C_2}{\lambda_e T_s}) \quad (3)$$

where T_s is the temperature of the slab. The second is the radiation from the furnace wall that is reflected into the pyrometer by the slab.

$$E_r = \frac{K \cdot A}{\lambda^5 e} \cdot \exp(-\frac{C_2}{\lambda_e T_w}) \quad (4)$$

where K is defined in equation (1) and T_w is the temperature of the furnace wall which can be expressed as below.

$$T_w = T_s + \Delta T \quad (5)$$

Temperature (T_I), indicated by the radiation pyrometer, is expressed as below.

$$\varepsilon_I \cdot \exp(-\frac{C_2}{\lambda_e T_I}) = \varepsilon \cdot \exp(-\frac{C_2}{\lambda_e T_s}) + K \cdot \exp(-\frac{C_2}{\lambda_e T_w}) \quad (6)$$

where ε_I is the emissivity compensating value of the pyrometer. When the temperature conditions are fixed, the temperature measurement error can be calculated from equations (1) and (6) as a function of the distance to radius ratio (d/R) of the probe.

There will be an optimum wavelength for the pyrometer such that the measurement error becomes minimum, but we used the silicon cell detector for the pyrometer because of its stability. The effective wavelength of the silicon cell detector at 900°C is 0.95 µm. When the temperature of the slab is 1173 K and ε_I is equal to ε, the measurement errors of the pyrometer are shown in Fig. 3. If the permissible error is determined, the maximum value of the distance to radius ratio (d/R) can be obtained from Fig. 3. When 1% error is allowed, that is +9°C error at 900°C, the temperature conditions of the reheating furnace and the resultant maximum value of d/R are shown in Table I. The gap between the probe and the slab must be determined from the viewpoint of safe operation of the furnace. It is 150 mm for our furnace, so the diameter of the probe is determined to be 300 mm for the soaking zone.

indicated by radiation pyrometer changes with the movement of the slab as shown in Fig. 4. The reading differences between the upper and lower positions can be calculated theoretically from the distance between the cooled sight tube and the slab, the temperature of slab and the temperature of the wall. The slab temperature and the wall temperature were measured by the thermocouples. The relationship between the measured reading difference and the calculated value is shown in Fig. 5. Since the calculated values agree closely with the measured values, it is proven that the design of the cooled sight tube is useful in practice for hot rolled steel.

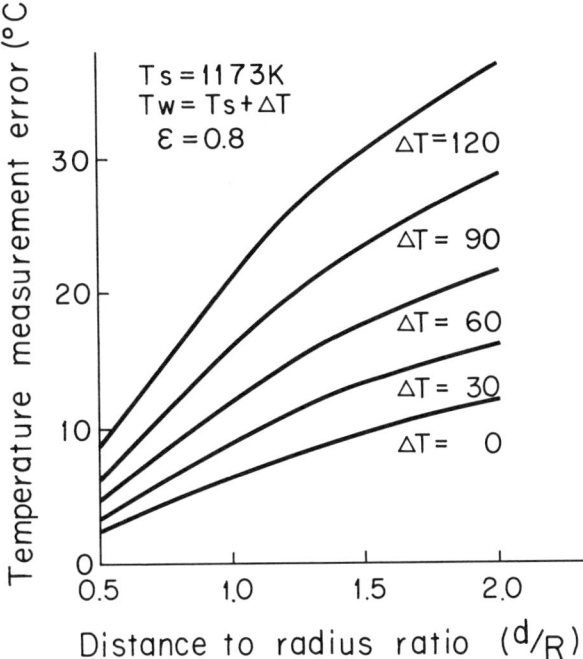

Fig. 3. The calculated relationship between temperature measurement error and the distance to radius ratio (d/R) for probe.

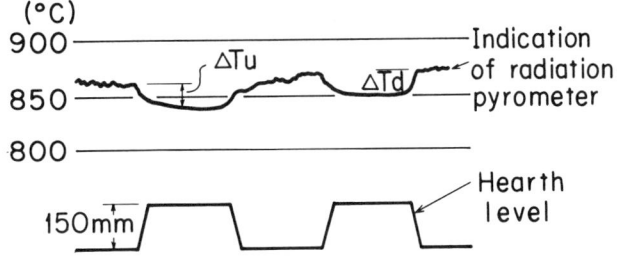

Fig. 4. Temperature variation of the radiation pyrometer when the slab is lifted up or down at same position

Table I. Temperature condition and maximum value of d/R

Position of pyrometer	Third heating zone	Soaking zone
Temperature of slab: Ts	900°C	900°C
Temperature of wall: Tw	975°C	930°C
Maximum value of d/R	0.75	1.0

1% measurement error can be allowed.

The slabs can be distinguished from the hearth of the furnace by the γ-ray object detector placed near the pyrometer and tracking signal of the process computer control system.

Results of Measurement in the Reheating Furnace

First the design of the cooled sight tube was examined. The slabs can be lifted or lowered by 150 mm at the same position in the reheating furnace by the moving mechanism of furnace. Of course the temperature

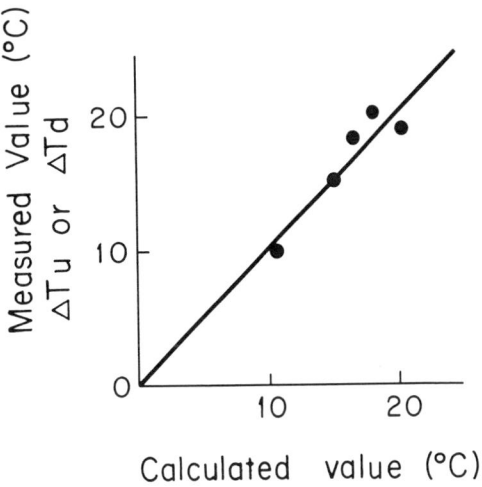

Fig. 5. Temperature indication difference when the slab is lifted up or down

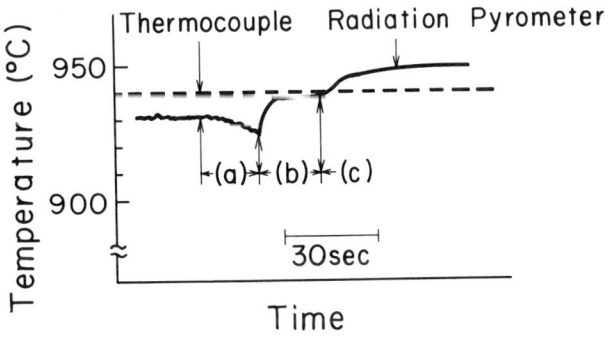

Fig. 6. Temperature indication of the radiation pyrometer. (a) slab is lifted up (b) slab is advanced (c) slab is lowered.

Fig. 6 shows the actual temperature reading obtained by the radiation pyrometer. When the slab is lifted slowly, the temperature reading is decreased because the surface of the slab is cooled by the convection heat transfer of the air. When the slab is moved forward, the temperature reading increases and agrees with the temperature of the thermocouple attached to the surface of the slab during the test. Because the thermocouple does not measure the temperature of the slab exactly beneath the cooled sight tube, the temperature indicated by the thermocouple remains constant. When the slab is lowered, the temperature reading naturally rises. When the slab is moved forward, the proper temperature signal is transferred to the computer system by sampling the temperature reading of the radiation pyrometer synchronized with the movement of the hearth and the signal of the γ-ray object detector. The temperature indicated by the radiation pyrometer must be higher than the thermocouple, but both are coincident. The reason for this is because it is presumed that the surface of the slab is cooled by the air and the emissivity of the slab may not coincide with the emissivity compensation value (0.8) of the radiation pyrometer.

Application to Control System

The slab temperature obtained by the radiation pyrometer is used for the input to the mathematical model in the computer control system. The temperature of the slabs in the furnace is calculated using this model at fixed intervals, and the furnace temperatures of all zones are set by the computer in order to obtain the required extracting temperature for all of the slabs. Also the flow rate of fuel and air for all zones are set by the computer in order to settle the furnace at the required temperature faster. The computer system compares the calculated temperature of the model with the temperature indicated by the radiation pyrometer, and adjusts the parameters of the mathematical model in order that both temperatures agree. The results of the computer control are satisfactory and the extracting temperatures are controlled within ±10°C of the required temperature.

HEAT-RESISTANT DATA LOGGER

Method of Measurement

The heat-resistant data logger contains a data capsule which collects and stores the temperature data, a heat-shielding unit and a data readout unit. The sheathed thermocouples are connected to the data capsule. Then it is inserted into the heat shielding unit which is placed on the steel. They are all inserted into the furnace. At preset intervals the output of the thermocouples are scanned and converted to digital form by an A/D converter. The digital data are all stored in integrated circuit memories in the data capsule. The temperature of the reference junction of the thermocouples are measured by the transistor temperature sensor and also stored in IC memories. After the heated steel is extracted from the furnace, the heat shielding unit is recovered. The stored data in the data capsule are read out by dedicated readout unit.

Heat Shielding Unit

The most important part of the heat-resistant data logger is the heat shielding unit. The electronic circuitry must be shielded from high temperature. A typical heat shielding unit is made from heat insulating material, a water jacket and the Dewar flask [3]. The temperature of the electronic circuitry in the Dewar flask can be easily kept below 70°C. Since the shape of the Dewar flask is usually cylindrical, the number of large scale integrated circuits which can be put into the cylindrical vessel are much smaller than in the square or rectangular vessel. Since a large capacity for storing data is one of the most important factors for our heat-resistant data logger, we adopted a very simple construction for the heat shielding unit shown in Fig. 7. The heat-shielding unit is made from ceramic fiber heat insulator and one or two water jackets. The features of this heat shielding unit are as follows.

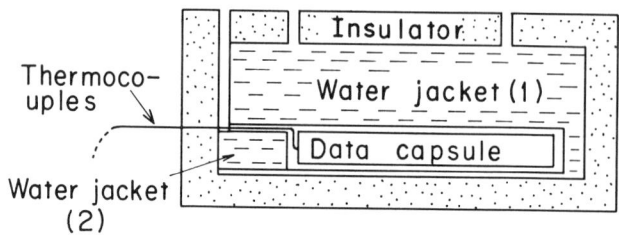

Fig. 7. The construction of the heat-resistant data logger for slabs

(1) The construction is very simple.

(2) The thermal design of the heat shielding unit is very simple.

(3) The electronic circuitry must be operated at 100°C because the cooling of the data capsule is mainly done through the evaporation of the water.

Since the heat shielding unit must be designed correctly for each application, it is very important to be able to build the heat shielding unit easily. There are two important factors to consider in designing the heat shielding unit.

(1) The optimum thickness of the thermal insulator

(2) The heat shielding unit must be easily designed for use in many applications so that the heat shielding unit can withstand various given conditions.

The height of the heat-resistant data logger is limited by the physical dimensions of the furnace. When the total height (H_t) is constant (300 mm for this study), there is a optimum insulator thickness for the heat shielding unit. If the thickness of the insulator is decreased, the heat mass of water in water jacket increases but the heat flow into the water becomes high. The amount of water evaporation is calculated when the height of the water jacket, (H_w) in Fig. 8(a), is changed. The thermal conductivity of the insulator is 0.14 Kcal/m·hr·°C at 650°C which is the average temperature of the insulator. The results are shown in Fig. 8(b) when the temperature surrounding the heat shielding unit is 1200°C. The optimum value of H_w/H_t is from about 0.6 to 0.7. From this result we decided that the thickness of the ceramic fiber insulator should be 50 mm. Since the dimensions of the prototype heat shielding unit are 300 mm H x 400 mm W x 680 mm L, the dimensions of the water jacket should be 200 mm H x 300 mm W x 580 mm L. The performance of the heat shielding unit is determined experimentally. The heat shielding unit is inserted into the furnace in the laboratory the temperature of which is controlled at 1200°C. After extracting the heat shielding unit, the amount of water in the water jacket is measured. The experimental results are shown in Fig. 9. Before heating the total amount of water in the water jacket was 27.2ℓ.

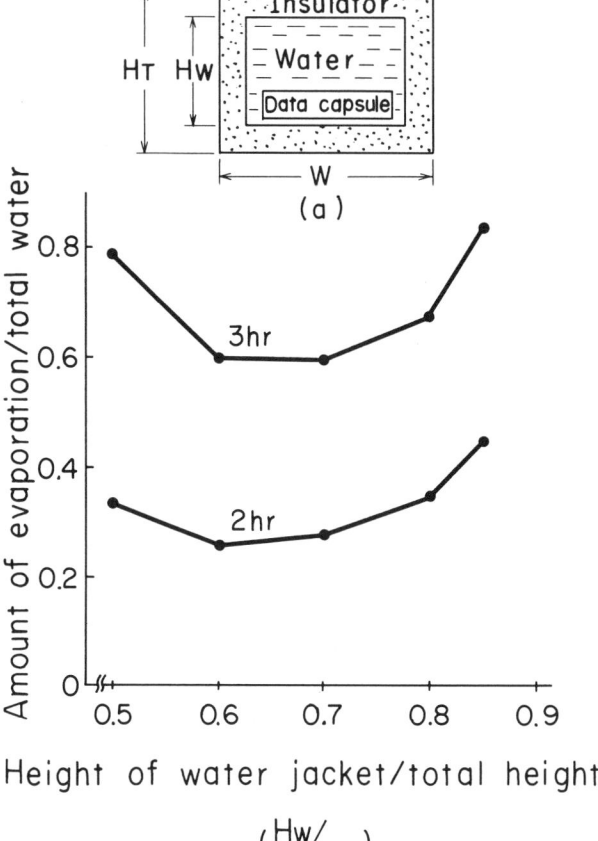

Fig. 8. Amount of evaporation to total water ratio is calculated. Furnace temperature is 1200°C.

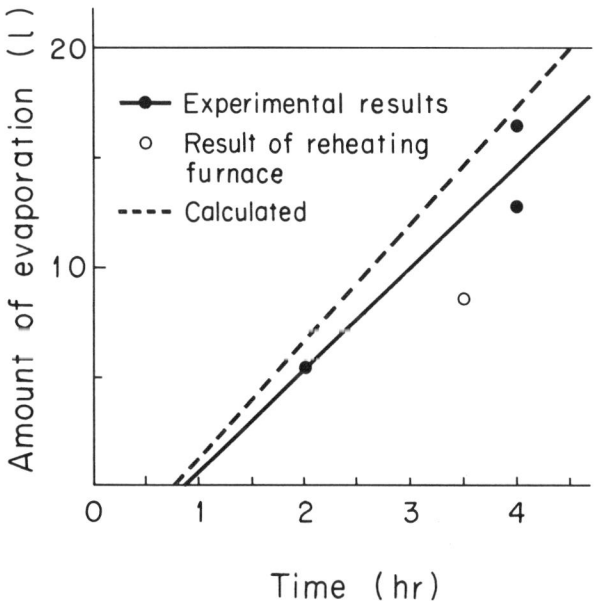

Fig. 9. Amount of evaporation versus time interval in the furnace. Furnace temperature is 1200°C

If the evaporated water was less than 20ℓ, all parts of the data capsule remained in the water, and the temperature of the data capsule was kept at 100°C. From these results it was decided that this heat shielding unit could be used in the reheating furnace. The data obtained in the reheating furnace for hot strip mill is also plotted in Fig. 9. Because the reheating furnace temperature is not constant, the time interval in the furnace is modified as if heat shielding unit was always in 1200°C. The evaporation curve calculated from the heat transfer analysis is also shown in Fig. 9 as a dotted line. From Fig. 9 it is proven that the calculated and the two experimental data agree with each other. It is believed that the heat shielding unit can be easily designed by the method described above for the different shapes and dimensions expected in other applications.

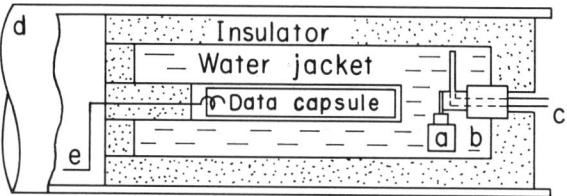

a : Counter weight   b : Rotary joint
c : Outlet of the steam   d : Steel pipe
e : Sheathed thermocouples

Fig. 10. The construction of heat-resistant data logger for rotating pipe

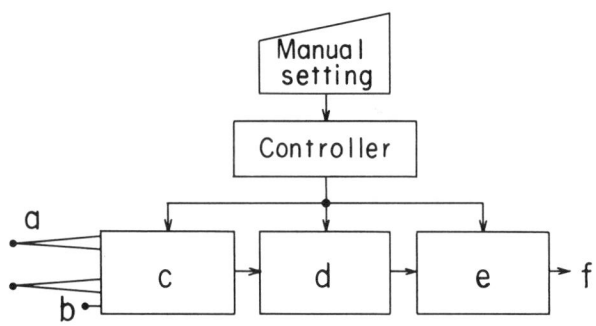

a : Sheathed thermocouples

b : Temperature sensor for reference junction compensation

c : Analog multiplexer

d : A/D converter (10-bit)

e : IC memories (12bit×4096)

f : To data readout unit

Fig. 11. Block diagram of data capsule

The construction of the heat shielding unit for the rotating pipe is shown in Fig. 10. Since the cooling water must not be spilled from the heat shielding unit when the pipe rotates, the outlet of water vapor is always maintained at the top by the rotary joint and counter weight as shown in Fig. 10. The thermal design of the heat shielding unit is also done by the method described above.

Data Capsule

The block diagram of the data capsule is shown in Fig. 11. The measuring interval can be manually preset by the switches in the data capsule. The outputs of the thermocouples and the temperature of the thermocouple reference junctions are scanned by the analog multiplexer. They are converted to digital form by a 10-bit A/D converter and stored in the digital memories. The digital data word has twelve bits, ten bits are used for data and two bits are used to check the conditions of the data capsule. The power for the data capsule is supplied from a battery.

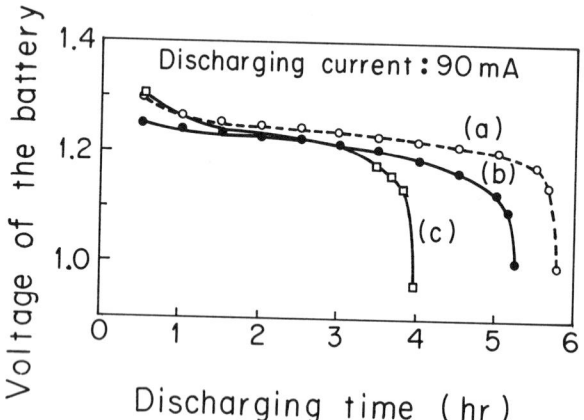

Fig. 12. High temperature performance of nickel-cadimum battery

(a) at room temperature
(b) 2nd use at 100°C
(c) 12nd use at 100°C

The data capsule must satisfy the following characteristics.

(1) All elements must be operated by a small amount of power from the battery.

(2) All elements must withstand 100°C during operation.

The active elements in the data capsule are special grade CMOS IC (Complimentary Metal Oxide Semiconductor Integrated Circuit) which can be operated at 100°C with a small amount of power. All other components are carefully selected to withstand the 100°C temperature. Because there is no battery which is guaranteed to operate at 100°C, we investigated the high temperature characteristics of a nickel-cadmium battery. A plot of output voltage vs discharge time is shown in Fig. 12. From this investigation it was found that the nickel-cadmium battery could be successfully operated at 100°C.

The specifications of the prototype data capsule are shown in Table II. The maximum amount of stored data can be increased to 8096 in the same size capcule.

Data Readout Unit

Data stored in the data capsule are read by the dedicated readout unit. During readout time, the readout unit and the data capsule are controlled by a microcomputer in the readout unit. The microcomputer has the following functions.

(1) Data transfer from the data capsule.

(2) Reference junction compensation of type K thermocouple.

(3) Linearization of type K thermocouple.

The combined inaccuracy of the data capsule and data readout unit is ±3°C.

Applications of Heat-resistant Data Logger

Examples of applications of the heat-resistant data logger are temperature measurement of slabs, blooms and beam-blanks in the reheating furnace, pipes in the heat treatment furnace, pipes in the plastic coating process and so on. The time-temperature data obtained with the logger have been used for development and improvement of the mathematical model for computer control and for analysis of the process to obtain optimum energy savings and product quality control. We

Table II. Specifications of the prototype data capsule

Numbers of channels	12
Sampling interval	1, 2, 4, 8 min or 1, 2, 4, 8, 16, 32 sec selectable
Data storage	Digital IC RAM (CMOS) 12 bit x 4096
A/D converter	10 bit
Thermocouple	Type K
Power supply	Nickel-cadmium battery 7.2 V 450 mA·H x 2 Supply current : 18 mA at measurement 0.1 mA at storage
Dimension	80 mm H x 80 mm W x 500 mm L

describe two examples below, one is the temperature measurement of the slab in the reheating furnace and the other is the temperature measurement of the rotating pipe in the heat treatment furnace.

The reheating furnace for slabs in the hot strip mill at the Kashima Steel Works is a walking beam type. The slab is lifted and moved forward by the walking beam. The heat-resistant data logger is placed on the slab as shown in Fig. 13 and charged into the reheating furnace. Temperatures of the slab are measured and stored in the logger when the slab is in the furnace.

The temperature sensors are the sheathed thermocouples in which thermo-elements are type K. The outer diameter of the sheath is 4.8 mm and the material of the sheath is 25%Cr-20%Ni heat-resistant steel. The measuring junction of the sheathed thermocouple is fixed to the slab in the drilled hole by using the filling pipe which is made from the same material as the slab. (See Fig. 13.) Because the maximum temperature of the reheating furnace is about 1200°C, the sheathed thermocouples are checked after the module is removed from the furnace. There is no abnormal appearance on the surface of the sheath and the thermal emfs at elevated temperatures are all within the allowance of the type K thermocouple. An example of the measurement is shown in Fig. 14. Temperatures obtained with three of the thermocouples are shown in figure, they are located at the center of the slab in the longitudinal direction. The size of slab is 240 mm thick, 950 mm wide and 9410 long. Various temperatures, calculated by the mathematical model in the computer which controls the reheating furnace are also shown in Fig. 14 as a dotted line. Above about 1,000°C calculated temperatures are slightly lower than the measured temperatures. The reason for this difference is that the mathematical model only calculates the radiation heat transfer between the slab and the furnace wall. To obtain better agreement between the calculated and observed temperatures it is also necessary to consider the convection heat transfer between the slab and gas or flame. The calculated temperatures of the surface of the slab agree with the measured temperatures, but the calculated

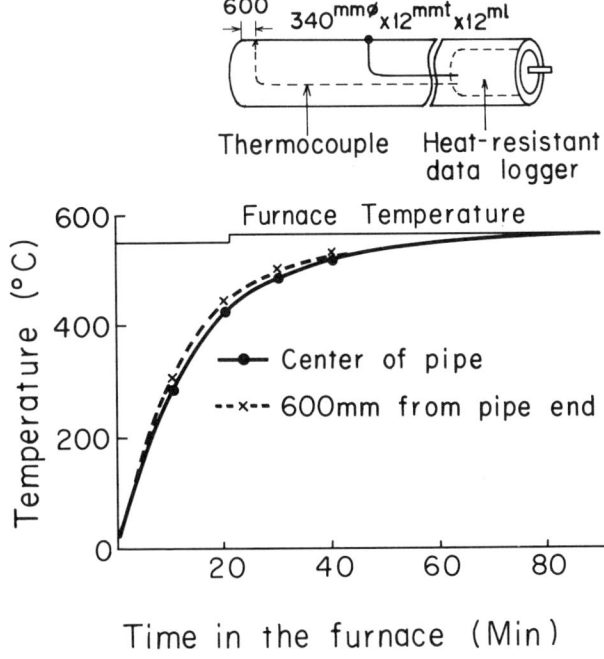

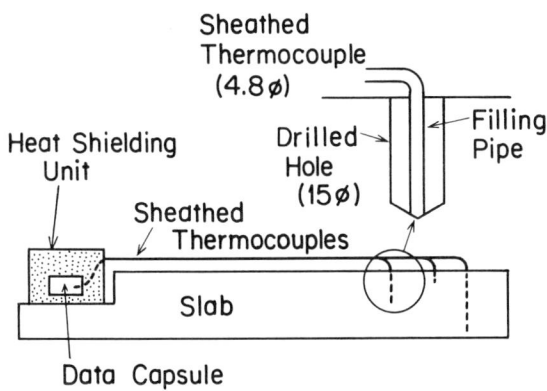

Fig. 13. Method of attachment of sheathed thermocouple to the slab

Fig. 15. The example of time-temperature pattern of rotating pipe measured by the heat-resistant data logger

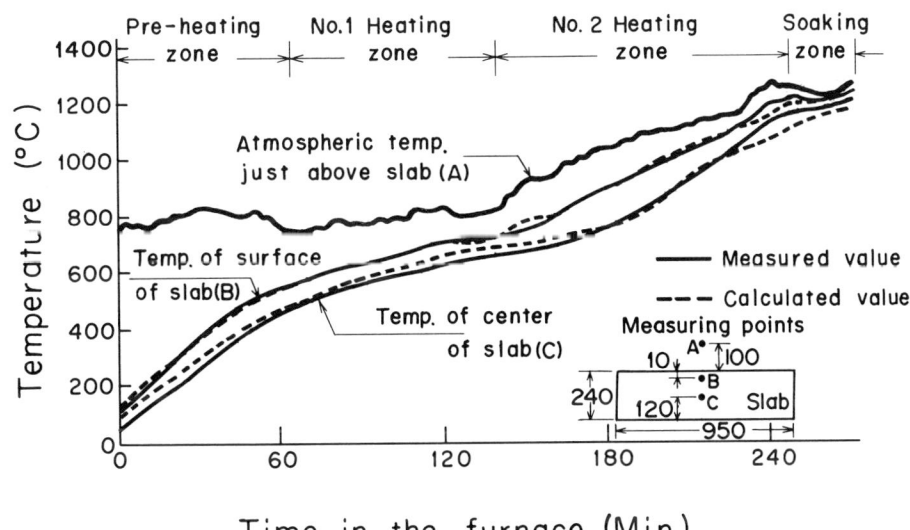

Fig. 14. The example of temperature measurement of slab by heat-resistant data logger. (Dimensions are in mm.)

Tamura *et al.*

temperatures at the inner measuring point are slightly higher than the measured temperatures. This reason is that the thermal conductivity of the slab used in the mathematical model may be slightly higher than the actual value. The rate of temperature rise in No.1 heating zone is lower than other heating zones, so the temperature of the preheating zone can be decreased without affecting the temperatures of the slab at the end of the No.1 heating zone. With this operation more efficient heating can be obtained.

The desired mechanical properties of the steel pipes can be attained by heat treatment. The pipes are usually rotated when they are heated and cooled in the heat treatment process because it is necessary to get a uniform temperature distribution throughout the circumference. The temperature distribution in the longitudinal direction must be also uniform. Since there is no proper continuous temperature measurement during the on-line heat treatment process due to the rotation of pipes, the heat-resistant data logger shown in Fig. 10 is applied. The heat-resistant data logger is inserted into the pipe and fixed to it. An example of the time-temperature pattern during the tempering process is shown in Fig. 15. The results obtained from these measurements are as follows. (1) The uniformity of the temperature distribution in the circumferential direction can be achieved by rotating the pipe. (2) The temperature distribution in the longitudinal direction of the pipe depends strongly upon the temperature distribution of the furnace. Because the heat transfer from the furnace to pipe in the soaking zone is small, the temperature uniformity of the furnace in the heating zone is very important as well as in the soaking zone. By adjusting the temperature distribution in the heating zone of the furnace, good temperature distribution as shown in Fig. 15 was obtained. (3) The tempering time can be shortened by increasing the temperature and keeping the temperature uniform in the heating zone.

CONCLUSION

Two methods for measurement of the temperature of the steel in the furnace have been developed and put into operations in Sumitomo Metal Industries, Ltd. One is for continuous temperature measurement of slabs in the reheating furnace and the other is for logging time-temperature data of an object during a particular process. We discussed mainly the design method of the cooled sight tube for the radiation pyrometry and the heat shielding equipment for the heat-resistant data logger. Also some applications were described. From the measurements discussed in this paper improved models for on-line computer control of the furnace operation have been successfully developed. The heat-resistant data logger has proven to be very useful for the analysis of the heating process in our facilities. This method may also be useful in other industrial applications.

REFERENCES

[1] R. Barber, Industrial and Process Heating, Feb. (1967)

[2] J. Dixon, Temperature Measurement and Control, 92 (1978)

[3] H. H. Emschermann, B. Fuhrmann and D. Huhnke, Stahl u. Eisen 96, 1290 (1976)

[4] T. Yokota, H. Kinoshita and S. Uyeda, Bulletin of the Earthquake Research Institute 55, 75 (1980)

Apparent emissivities of cylindrical cavities with partially specular conical bottoms

A. Ono

National Research Laboratory of Metrology, Umezono Sakura, Ibaraki 305, Japan

The apparent emissivities of diffuse cylinders having partially specular conical bottoms are calculated by the Monte Carlo method. The conical surfaces intermediate between diffuse and specular are characterized by the uniform specular-diffuse model. It is pointed out that apex angles of the cone near 60°, 90°, and 180° should be avoided because successive specular reflections on the cone can considerably reduce the apparent emissivity. Special attention is given to apex angles of 120°, 70°, and 50°. For an angle of 120°, the apparent emissivity increases linearly with the degress of specularity. For angles of 70° and 50°, however, the apparent emissivity can have minima at intermediate degrees of specularity. An advantage of choosing 120° as an apex angle is stressed and the physical explanation is provided. In order to evaluate accurately the apparent emissivity, it is important to specify the surface emissivity of the cone and its degree of specularity. Guidelines are given for constructing blackbody cavities with apparent emissivities greater than 0.99 and 0.999. It is suggested that an effort to make a conical surface more specular can be an approach to a perfect blackbody even if the conical surface is not black.

INTRODUCTION

Theoretical evaluation of blackbody cavities has been well studied for fundamental cavity shapes assuming that the cavity surfaces are diffuse.[1,2] For cavities composed of totally specular surfaces, absorbing characteristics of cylinders and cones have been investigated.[3,4] However, real surfaces are often neither totally diffuse nor specular, but rather, intermediate between them.

Radiative properties of matter depend on the surface condition as well as the bulk property. Surface roughness influences the angular distribution of reflected radiation.[5,6] As wavelengths of radiation become longer, reflection from a rough surface becomes more specular. Hence, the quality of blackbody cavities should be evaluated taking into account partially specular reflection when the cavities are used for the calibration of infrared and total radiation thermometers.

In order to extend the cavity quality evaluation to surfaces intermediate between diffuse and specular, we have recently applied a Monte Carlo method with the aid of random numbers.[7] The radiative properties of surfaces were characterized by the uniform specular-diffuse model in which the reflected radiation consists of two components, a specular and a diffuse one, and the directional-hemispherical surface reflectivity (and consequently, the directional surface emissivity) has the same value for any angle of incidence (and emission).

The present paper provides an application of the Monte Carlo method to cylindrical cavities with conical bottoms where the cylindrical surfaces are diffuse and the conical surfaces are intermediate as characterized by the uniform specular-diffuse model. Such modeling may be a good approximation for real cavities where the cylindrical surfaces are spirally grooved or baffled and the conical surfaces are as machined.

MONTE CARLO CALCULATION

When Kirchhoff's law is applied to an isothermal opaque cavity, it states that the apparent directional emissivity at a location on the cavity wall is equivalent to the apparent reflectivity of the cavity when radiation is incident at the location in the reverse direction. This is valid for surfaces having any type of radiative properties – not just diffuse.[7,8]

In the Monte Carlo calculation special attention was directed to the fractions of radiation entering a cavity that escape through the cavity aperture after undergoing a certain number of reflections within the cavity. Radiation entering a cavity was considered to consist of many ray bundles. When ray bundles collide with the cavity wall, some of them are absorbed and others are reflected according to the value of the absorptivity. If reflected, ray bundles leave the surface according to the angular distribution of surface reflectivity and collide again somewhere with the cavity wall. Therefore the history of each ray bundle was determined in such a way that the selections of either absorption or reflection processes on the surface were made by the aid of random numbers. The paths of the rays were pursued until the ray bundles were absorbed or escaped through the cavity aperture.

A cylindrical cavity with a conical bottom (cylindro-cone) is schematically illustrated in Fig. 1 where the total depth (from the opening to the conical apex) is L, the cylindrical diameter is d, and the total angle of the apex is θ. Ray bundles were incident parallel to the cylindrical axis onto the cone. The incident points were uniformly distributed over the cone in order to have an average emissivity of the cavity bottom. The uniform distribution was also made by the aid of random numbers.

Traces of a ray bundle that underwent reflections three times within the cavity before it escaped are shown in Fig. 1. First, the ray bundle was reflected from the cone, denoted by 1, and second, it collided with the cylinder, denoted by 0, and was reflected. Third, it was reflected again on the cylinder and finally, it escaped through the cavity aperture. Such a ray bundle was labeled by a binary number as (100). The binary numbers indicate detailed history on what part of the cavity wall ray bundles were reflected from

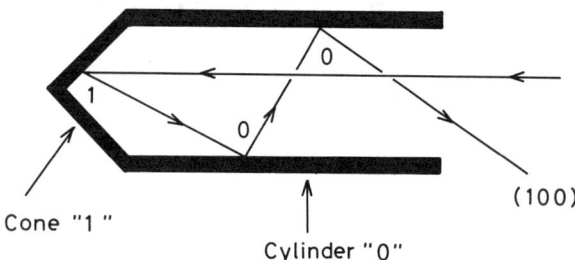

Fig.1 Schematic of cavity consisting of a diffuse cylinder "0" and a partially specular cone "1". The total depth is L, the cylinder diameter is d, and the apex angle of cone is θ. The path of a ray bundle is denoted as (100) by a binary number.

in sequence until their escape. Ray bundles were labeled, classified, and counted according to their own history.

In order to carry out efficiently the Monte Carlo calculation, the directional-hemispherical reflectivity of the surface (abbreviated as surface reflectivity) was normalized to 1 (i.e. no absorption) keeping the angular distribution of reflected radiation fixed (i.e. no change of the fraction of specular reflection component). Let $F(n)$ be the fraction of ray bundles entering the cavity that undergoes a history denoted by a binary number, n, and r_0 and r_1 be the surface reflectivities of the cylinder and cone, respectively. Then the apparent reflectivity of the cavity is expressed as

$$\rho = F(1)r_1 + F(10)r_0 r_1 + F(11)r_1^2$$
$$+ F(100)r_0^2 r_1 + F(101)r_0 r_1^2 + F(110)r_0 r_1^2 + F(111)r_1^3$$
$$+ \cdots , \qquad (1)$$

and the apparent emissivity of the cavity is given by

$$\epsilon = 1 - \rho . \qquad (2)$$

Figure 2 shows the apparent emissivity versus the apex angle for a cylindro-cone of the depth-to-diameter ratio 3. The surface emissivities, e, of the cylinder and cone were varied, but kept equal. The degree of specularity on the conical surface was taken to be 30%.

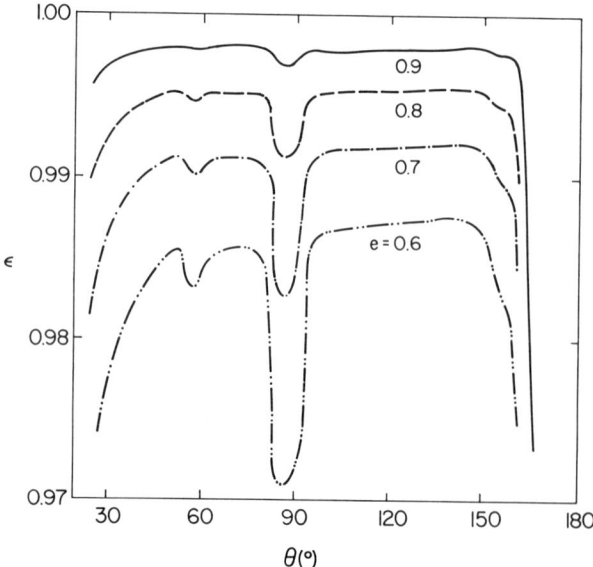

Fig.2 Apparent emissivities versus the apex angle for a cavity of L/d = 3 and $r_{1s}/r_1 = 0.3$.

The apparent emissivities have minima around the apex angles of 60 and 90°, and show sharp decreases beyond 150°. These minima and sharp decreases correspond to the escape of ray bundles undergoing successive specular reflections three times, twice, and once on the cone, respectively. It is seen that the minima of the apparent emissivities are deeper for lower surface emissivities. It was also shown that they were deeper for higher degrees of specularity. The apex angles near 60, 90, and beyond 150° should be avoided so as not to deteriorate the quality of the cavity.

It is reasonable to give attention to apex angles of 120, 70, and 50° at which successive specular reflections on the cone confine the ray bundles to the deepest part (i.e. closest to the bottom) of the cylinder. Figures 3, 4, and 5 illustrate the fractions, $F(n)$, of ray bundles for a cavity of L/d = 3 where the apex angles are 120 and 70°. The abcissas indicate the fraction of the specular reflection component, (r_{1s}/r_1), where r_{1s} is the specular component of the surface reflectivity.

First, in Figs. 3(a) and (b), the F(1) decrease linearly with increasing degree of specularity and reach zero at the totally specular condition. The reason is that the direct escape after a reflection is due to a diffuse reflection on the cone. As far as the first reflection is concerned, a higher degree of specularity confines more ray bundles within the cavity. It should be noted that the F(1) at $r_{1s}/r_1 = 0$ (totally diffuse condition) are equivalent to the view factors of the cavity aperture at the bottom (about 0.025).

Second, attention is given to the fractions associated with two reflections in Figs. 3(a) and (b). For the wider apex angle, the F(10) is much larger than the F(11) because many ray bundles experience a second collision with the cylinder than with the cone. For the narrower apex angle, in contrast, more ray bundles are apt to collide again with the cone. It is interesting to note that the F(11) for an apex angle of 70° has a maximum at an intermediate degree of specularity. It suggests that a certain combination of first specular and second diffuse reflections results in the most probable escape of ray bundles. It should be noted that the F(10) at $r_{1s}/r_1 = 1$ (totally specular condition) for the cavity of $\theta = 120°$ is related to the view factor of the aperture at the deepest part of the cylinder (about 0.008).

Third, the F(n) associated with three reflections are shown in Figs. 4(a) and (b). For $\theta = 120°$, the F(100) is dominant. The ray bundles of (100) are related to those which approach the cavity aperture gradually during two reflections on the cylinder. It is interesting to note that the F(101) becomes comparable with the F(100) at higher degrees of specularity. The ray bundles corresponding to (101) are those which are reflected back to the cone once again. The value of F(101) at the totally specular condition (about 0.010) is related to the view factor of the aperture at the deepest part of the cylinder when viewing through the mirror-like conical bottom. For $\theta = 70°$, the ray bundles undergoing successive reflections on the cone become more significant at higher degrees of specularity. It should be noted that the F(111) also has a maximum at an intermediate degree of specularity.

Finally, the F(n) associated with four reflections are illustrated in Figs. 5(a) and (b). They will be useful to calculate accurately the apparent emissivities of the cylindro-cone with arbitrary surface reflectivities.

The apparent emissivities of the cavities were calculated through Eqs. (1) and (2) taking into account up to five reflections within the cavities. The results for $\theta = 120$ and 70° are plotted against the degree of specularity in Figs. 6(a) and (b), respectively, where the surface reflectivities of the cylinders are 0.2 for both apex angles. As the degree of specularity increases, the apparent emissivites for $\theta = 120°$ increase almost linearly owing mainly to the

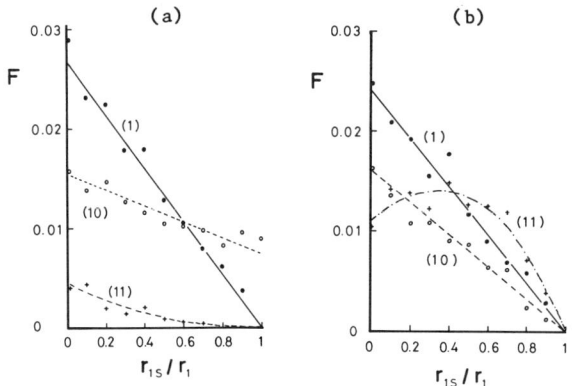

Fig.3 Fractions of ray bundles escaping after 1 and 2 reflections plotted against the degree of specularity for a cavity of L/d = 3. (a) is for $\theta = 120°$ and (b) is for $\theta = 70°$.

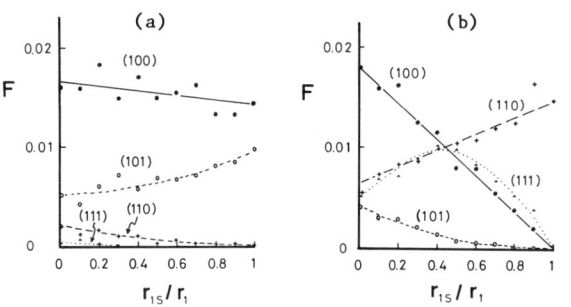

Fig.4 Fractions of ray bundles escaping after 3 reflections plotted against the degree of specularity for a cavity of L/d = 3. (a) is for $\theta = 120°$ and (b) is for $\theta = 70°$.

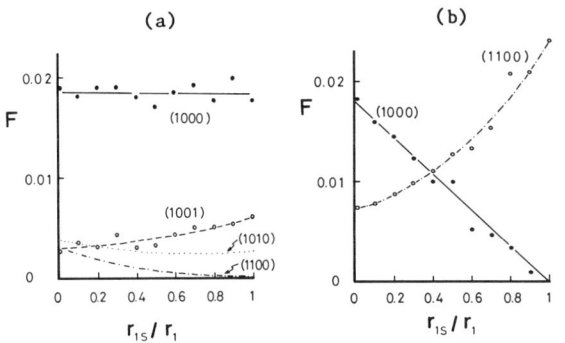

Fig.5 Fractions of ray bundles escaping after 4 reflections plotted against the degree of specularity for a cavity of L/d = 3. (a) is for $\theta = 120°$ and (b) is for $\theta = 70°$.

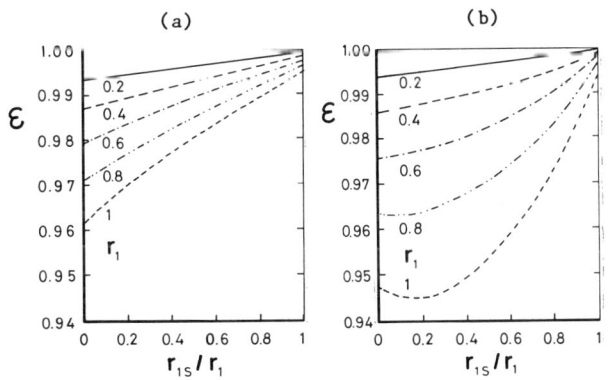

Fig.6 Apparent emissivities versus the degree of specularity for a cavity of L/d = 3 and $r_o = 0.2$. Figure (a) is for $\theta = 120°$ and (b) is for $\theta = 70°$.

linear decrease of the F(1). At the totally specular condition, the apparent emissivities are close to unity. It is interesting to note that a totally specular and completely reflecting (no absorbing) conical bottom gives a higher apparent emissivity (0.995) than a totally diffuse and almost black ($r_1 = 0.2$) conical bottom does (0.994).

For $\theta = 70°$, some of the apparent emissivities have minima because the F(11) and F(111) had maxima at intermediate degrees of specularity. It is interesting to note that the apparent emissivities are lower than those for $\theta = 120°$ at almost all degrees of specularity, especially for the higher surface reflectivities. For $\theta = 50°$, the apparent emissivities were also calculated taking into account up to five reflections. The apparent emissivities were lower than those for $\theta = 70°$ at almost all conditions, and the minima at intermediate degrees of specularity were deeper.

Figure 7 shows the apparent emissivities of a deeper cylindro-cone, L/d = 10 and $\theta = 120°$, where the cylinder surface emissivity is 0.8. The apparent emissivities increase almost linearly with the degree of specularity. It should be noted that a totally specular and completely reflecting conical bottom gives a higher apparent emissivity (0.9997) than a totally diffuse and almost black ($e_1 = 0.8$) conical bottom does (0.9995). Figures 8(a) to (d) show how the apparent emissivities depend on the cylinder surface emissivity, e_o. In spite of the large surface emissivity change, 0.5 to 1.0, the apparent emissivities vary little. The variations are less than 0.0002 in emissivity units when the cone surface emissivity is 0.6. If the cone surface emissivity, its degree of specularity, and the cylinder surface emissivity are estimated as $e_1 = 0.8 \pm 0.05$, $r_{1s}/r_1 = 0.5 \pm 0.1$, and $e_o = 0.8 \pm 0.2$, respectively, the apparent emissivity is determined as $\epsilon = 0.99975 \pm 0.0001$. Such an evaluation accuracy of the apparent emissivity is needed for a precise determination of the Stefan-Boltzmann constant by the absolute measurement of blackbody total radiation.[9,10]

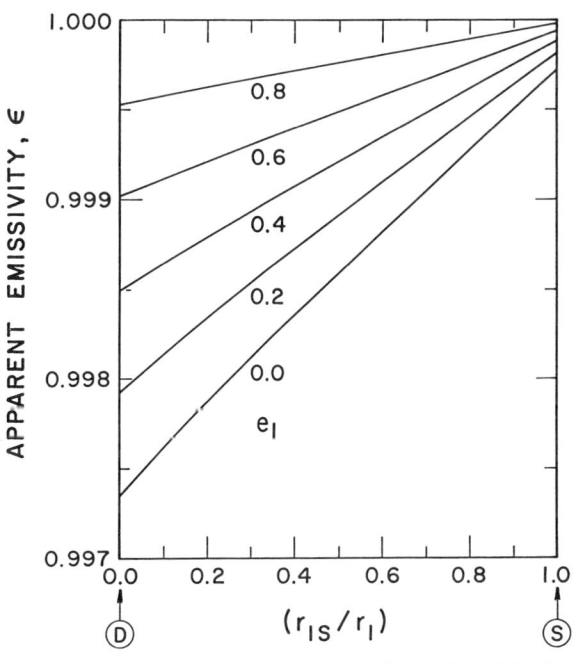

Fig.7 Apparent emissivities versus the degree of specularity for a cavity of L/d = 10, $e_o = 0.8$, and $\theta = 120°$.

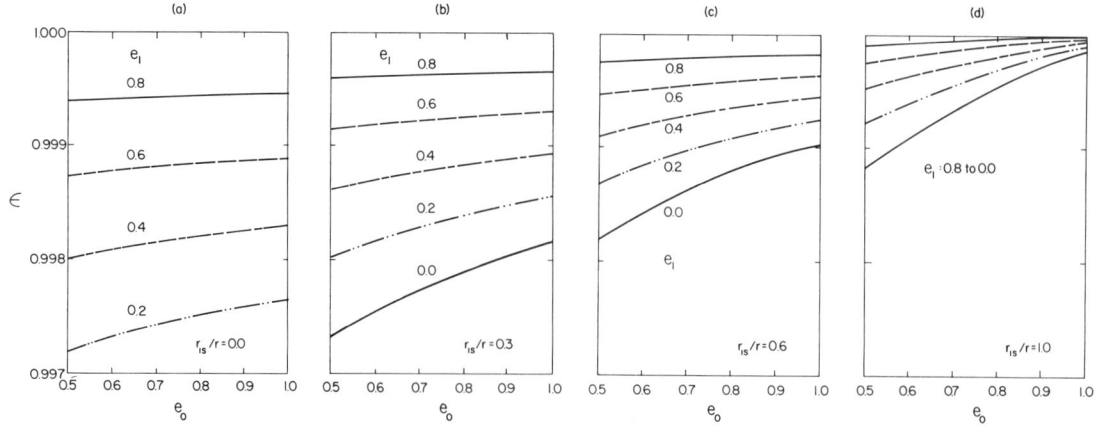

Fig.8 Apparent emissivities versus the surface emissivity of the cylinder for a cavity of L/d = 10 and $\theta = 120°$. Figures (a), (b), (c), and (d) are for the degree of specularity 0.0, 0.3, 0.6, and 1.0, respectively.

SUMMARY

The apparent emissivity of cylindro-cones has been calculated by the Monte Carlo method. Several guidelines are given for the construction of blackbody cavities for practical and standards use. An advantage of choosing 120° as an apex angle is clear because one can guarantee that the apparent emissivity is not less than that of a diffuse cavity of the same shape and the same surface emissivity whatever the degree of specularity. An apparent emissivity higher than 0.99 for practical use can be obtained when the depth-to-diameter ratio is 3 and the surface emissivities are 0.7. An apparent emissivity not less than 0.999 for standards use is obtained when the cavity depth-to-diameter ratio is 10 and the surface emissivities are 0.6. For an accurate evaluation of the apparent emissivity, the cone surface emissivity and its degree of specularity should be specified precisely, but the cylinder surface emissivity is not required so precisely.

For cavities of $\theta = 120°$, the apparent emissivities increased with the degree of specularity of the conical bottom. A cavity with a totally specular and completely reflecting (mirror-like) conical bottom gave a higher apparent emissivity than the corresponding totally diffuse cavity having the same surface emissivity on the cone as on the cylinder. This is clearly understood from a consideration of view factors. The lowest value of view factors of the cavity aperture among the surfaces of the cylindro-cones can be realized at the cylinder surface closest to the bottom, rather than at the apex of the bottom, except for sharp apex angles. That is, that part is darkest in illumination and, consequently, brightest in emission.[11] A mirror-like conical bottom of $\theta=120°$ guides incident radiation onto the darkest part of the cavity by a specular reflection. Therefore, a higher degree of specularity gives a higher apparent emissivity, and a totally specular condition gives a higher apparent emissivity than the totally diffuse condition does. An effort to make a conical surface more specular can be an approach to a perfect blackbody even if the conical surface is not black.

ACKNOWLEDGEMENT

The author would like to express his sincere thanks to Professor D. P. DeWitt of the School of Mechanical Engineering, Purdue University, for his helpful discussions in preparing the paper while on sabbatical leave at his laboratory.

REFERENCES

1. R. E. Bedford, *Temperature, Its Measurement and Control in Science and Industry* (Instrument Society of America, Pittsburgh, 1972), Vol. 4, Part 1, p. 425-434.

2. F. O. Bartell and W. L. Wolfe, Appl. Opt. 15, 84-88 (1976).

3. S. H. Lin and E. M. Sparrow, Appl. Opt. 4, 277-283 (1965).

4. T. J. Quinn, Infrared Physics 21, 123-126 (1981).

5. R. C. Birkebak and E.R.G. Eckert, J. Heat Transfer 87, 85-94 (1965).

6. K. E. Torrance and E. M. Sparrow, J. Heat. Transfer 87, 283-292 (1965).

7. A. Ono, J. Opt. Soc. Am. 70, 547-554 (1980).

8. S. Takata, J. Illum. Eng. Inst. Jpn. 51, 702-713 (1967) (in Japanese).

9. W. R. Blevin and W. J. Brown, Metrologia 7, 5-29 (1971).

10. A. Ono, Proceedings of the Second International Conference on Precision Measurement and Fundamental Constants, held at the National Bureau of Standards, Gaithersburg, Maryland, 1981 (To be published).

11. R. E. Bedford and C. K. Ma, Opt. Soc. Am. 65, 565-572 (1975).

Numerical calculation of the effective emissivity by using a series technique

Yoshiko Ohwada

National Research Laboratory of Metrology, 1-1-4, Umezono, Sakura-mura, Niihari-gun, Ibaraki 305, Japan

Effective emissivities of double-cones and cylindro-cones having diffuse surfaces are computed by an iterative procedure and also from effective reflectances. An upper bound for the truncation error in the calculation of the effective reflectance is $2\rho^{n+1}I_n/(1-\rho)$ where ρ is the surface reflectance and I_n is the nth order coefficient of the effective reflectance.

INTRODUCTION

Effective emissivities of double-cones and cylindro-cones have been investigated in detail by R.J. Chandos and R. E. Chandos,[1] and Bedford and Ma.[2,3] In this paper effective emissivities of cavities with diffuse surfaces, where the directional emissivity and reflectance are independent of angle or direction, are computed by an iterative procedure and also from effective reflectances.[4,5] The calculation of the effective reflectance is developed from the approach of DeVos,[6] Takata,[7] Bartel and Wolfe,[8] Quinn[9,10] and Ono[11,12] to compute up to a desired order approximation. Obtaining coefficients in a series for the effective reflectance allows the calculation of the effective emissivity for arbitrary surface emissivities. Coefficients in such series have been listed in tables up to the tenth order for cylindrical cavities, and conical cavities,[4,5,9~11] and here they are listed for double-cones and cylindro-cones. Numerical calculation is carried out by using the zonal approximation method described in Refs. 2, 3, 13.

CALCULATION

A schematic drawing of a double-cone(D_c) and cylindro-cone(C_y) is shown in Fig. 1. Distances perpendicular to the cavity axis from the apex are denoted by the coordinate X and distances along the cavity axis from the apex are denoted by the coordinate Y. Numerical calculation is performed for L=3, 5, 8, R=0.5, 1.0, K=0.1 to 1.0 in steps of 0.1, X=0.1, 0.25, 0.50, and Y<L_1 where K=L_1/L and K=1.0 signifies conical cavities with a lid of internal radius R. The zonal approximation method is used to replace integrals by summations and to carry out analytical calculation of angle factors.[2,3,13]

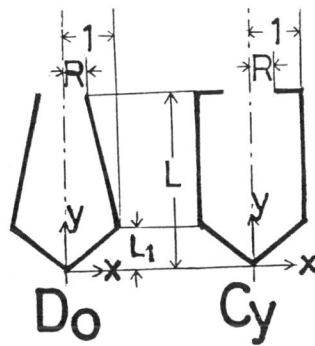

Fig. 1. Geometry of the double-cone(D_o) and cylindro-cone(C_y).

We first calculate the effective emissivity by an iterative procedure and evaluate its accuracy. Consideration is given to the radiant interchange problem within a cavity with isothermal surfaces. The radiation per unit area leaving an area element dQ at position Q at a temperature T is denoted by B(Q). The quantity B(Q) equals the emission plus the reflected incident radiation. The effective emissivity E(Q) is defined by E(Q)=B(Q)/$\sigma_0 T^4$ where σ_0 is the Stefan-Boltzmann constant. An nth order approximation E_n(Q) to E(Q) is given by adding the fraction of the radiation reflected in a hemisphere at the nth reflection, when the radiation is incident on dQ, to the (n-1)th order approximation E_{n-1}(Q). Hence there is the following relation for $n\geq 2$ and Q=-N',...,-1,0,1,...,N:

$$E_n(Q)=E_{n-1}(Q)+\rho\sum_{P=1}^{N}[E_{n-1}(P)-E_{n-2}(P)]\cdot\Delta f(Q,P) \quad (1)$$

with $E_0(Q)=0$, $E_1(Q)=\varepsilon$, where the surfaces of the cavity are subdivided into N bands, -N',...,-1,0, refer to evaluation points, ρ is the surface reflectance, $\Delta f(Q,P)$ is the angle factor between a band ΔP at position P and an infinitesimal area element dQ, ε is the surface emissivity, and $E_\infty(Q)=E(Q)$. The angle factor $\Delta f(Q,P)$ is given by

$$\Delta f(Q,P)=|f(Q,P)-f(Q,P-1)| \quad (2)$$

where $f(Q,P)$ is the angle factor between a circular disk at P and dQ, and may be written as

$$f(Q,P)=|\cos\alpha\cdot f_1(Q,P)\pm\sin\alpha\cdot f_2(Q,P)|, \begin{cases}+(X_q<X_p)\\-(X_q>X_p)\end{cases} \quad (3)$$

(see Ref. 14), where α is the angle between the cavity axis and dQ, and (X_p,Y_p) and (X_q,Y_q) are the coordinates of P and Q. For dQ on the conical portion α equal the half angle of the apex, for dQ on the cylindrical portion $\alpha=0$, for dQ on the lid $\alpha=\pi/2$. The angle factors $f_1(Q,P)$ and $f_2(Q,P)$ are given by

$$f_1(Q,P)=|Y_q-Y_p|\times([X_q^2+X_p^2+(Y_q-Y_p)^2]\times\{[X_q^2+X_p^2+(Y_q-Y_p)^2]^2-4X_q^2X_p^2\}^{1/2}-1)/(2X_q), \quad (4)$$

$$f_2(Q,P)=\{1-[X_q^2+(Y_q-Y_p)^2-X_p^2]\times\{[X_q^2+(Y_q-Y_p)^2+X_p^2]^2-4X_p^2X_q^2\}^{1/2}\}/2. \quad (5)$$

Uncertainties, σ, are assigned to the effective emissivities by examining the convergence as the band width is decreased.[4,5] The effective emissivities are

Table I. Effective emissivities $E(X) \pm \sigma$. The value of σ given in the table is the uncertainty in the last displayed digit of $E(X)$.

			D_o		C_y	
L	K	X	$\varepsilon=0.50$	$\varepsilon=0.75$	$\varepsilon=0.50$	$\varepsilon=0.75$
3	0.3	0.1	0.97333±5	0.99209	0.91027±3	0.97151
		0.5	0.96771±8	0.98975±1	0.89258±9	0.96368±2
	0.7	0.1	0.97928±3	0.99445	0.9351±2	0.98082±2
		0.5	0.95736±3	0.98697	0.87283±8	0.95823±2
	1.0	0.1	0.98524±9	0.99584	0.9522±2	0.98536±2
		0.5	0.94769±8	0.98269±2	0.8505±3	0.94878±5
5	0.3	0.1	0.99206±2	0.99779	0.96987	0.99141
		0.5	0.98960±4	0.99694	0.95998±6	0.98807
	0.7	0.1	0.99495±1	0.99869	0.98313±7	0.99522
		0.5	0.98529±3	0.99620	0.95208±8	0.98665±1
8	0.3	0.1	0.99797	0.99941	0.99186±1	0.99767
		0.5	0.99715±2	0.99915	0.98808	0.99659
	0.7	0.1	0.99886	0.99968	0.99598±2	0.99881
		0.5	0.99599±1	0.99902	0.98580±5	0.99632

listed in table I along with σ for $\varepsilon=0.50$, 0.75, X=0.1, 0.5, and $Y<L_1$ where σ smaller than 1×10^{-5} is omitted and a position Q whose Y value is smaller than L_1 is denoted only by X. It is pointed out in Ref. 3 that normal effective emissivities of cavities with a given aperture and overall length vary little with the shape. In Fig. 2 the effective emissivity E(X) of D_o with L=3 and R=0.5 is compared with E(X) of C_y with L=3 and a lid of internal radius R=0.5 for $\varepsilon=0.5$, X=0.10, 0.25, 0.50, and $Y<L_1$. Differences between the effective emissivities of the two cavities for the same value of K were smaller than 3×10^{-3}. For $K\geq0.3$, E(X) of C_y was higher than E(X) of D_o. This is presumably due to the contribution of the effective emissivity of the lid where the effective emissivity is rather high.[2] The cylindro-cone C_y has a lid and D_o does not. At K=0.2, E(X) of C_y was smaller than E(X) of D_o only by $3\sim4\times10^{-5}$. At K=0.1, E(X) of C_y was smaller than E(X) of D_o by 6×10^{-4}.

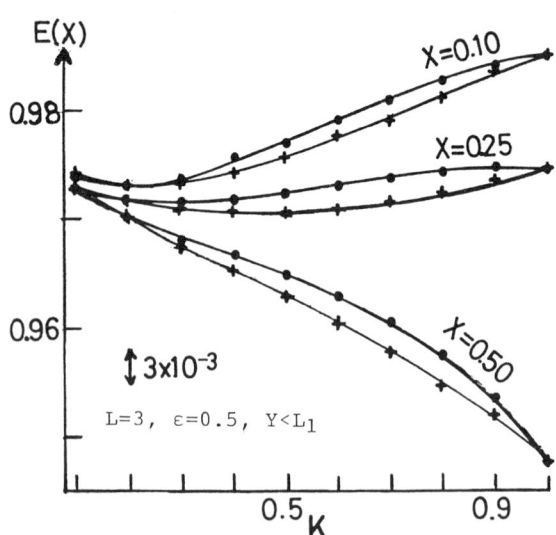

Fig. 2. Comparison between a double-cone(D_o) and Cylindro-cone(C_y), ($Y<L_1$), ●; D_o, +; C_y.

Next, the effective reflectance and absorptance are calculated. Consideration is given to the radiant interchange problem of a cavity at a temperature of absolute zero. The effective reflectance R(Q) is defined as the fraction of the radiation emerging from a cavity when the radiation is incident on dQ on the inside wall of the cavity. The calculation of R(Q) is summarized as follows: if the nth order approximation to R(Q) is denoted by $R_n(Q)$,

$$R_n(Q)=R_{n-1}(Q)+\rho \sum_{P_1=1}^{N} \sum_{P_2=1}^{N} \Delta F(Q,P_2^{n-1},P_1) \cdot \Delta f(P_1,A) \quad \text{for } n\geq 2 \quad (6)$$

(see Ref. 4) where $R_1(Q)=\rho \cdot \Delta f(Q,A)$, $\Delta F(Q,P_2^{n-1},P_1)$ is a function describing the fraction of the initially incident radiation on dQ that travels from ΔP_2 to ΔP_1 at the $(n-1)$th reflection, $\Delta f(P_1,A)$ is the angle factor between dP_1 and the aperture, and $R_\infty(Q)=R(Q)$. The last term of (6) expresses the fraction that leaves the cavity through the aperture at the nth reflection. The next relation (7) must be satisfied according to the physical meaning of $\Delta F(Q,P_2^n,P_1)$, since both sides of (7) express the fraction of the initially incident radiation on dQ that travels ΔP_2 to ΔP_1 at the nth reflection.

$$\Delta F(Q,P_2^n,P_1)=\rho \sum_{P_3=1}^{N} \Delta F(Q,P_3^{n-1},P_2) \cdot \Delta f(P_2,P_1) \quad \text{for } n\geq 2 \quad (7)$$

where $\Delta F(Q,P_2^1,P_1)=\rho \cdot \Delta f(P_2,P_1) \quad \text{for } P_2=Q$
$\qquad\qquad\qquad\qquad = 0 \qquad\qquad\quad \text{for } P_2 \neq Q$.

The effective absorptance A(Q) is defined as the fraction of the radiation absorbed by the wall when the radiation is incident on dQ. The nth order approximation $A_n(Q)$ to A(Q) is given by

$$A_n(Q)=A_{n-1}(Q)+(1-\rho)\sum_{P_1=1}^{N}\sum_{P_2=1}^{N}\Delta F(Q,P_2^{n-1},P_1) \quad \text{for } n\geq 2 \quad (8)$$

(see Ref. 4) where $A_1(Q)=1-\rho$, and $A_\infty(Q)=A(Q)$. The last term of (8) expresses the fraction of the radiation absorbed by the wall at the nth impingement on the wall. The quantity $A_n(Q)$ is equivalent to $E_n(Q)$,[4] and their numerical values agreed with one another within 1×10^{-14} for L=3, K=0.3, and $\varepsilon=0.5$.

Finally, E(Q) is obtained from R(Q) and we estimate a truncation error in the calculation of R(Q) when reflections after the nth are neglected. If a quantity $G_n(Q)$ which satisfies the relation

$$\rho^n G_n(Q) = \sum_{P_1=1}^{N} \sum_{P_2=1}^{N} \Delta F(Q,P_2^n,P_1), \quad (9)$$

is introduced for $n \geq 1$, then

$$R_n(Q)=\sum_{k=1}^{n} \rho^k [G_{k-1}(Q)-G_k(Q)] \quad \text{for } n\geq 1, \quad (10)$$
$$A_n(Q)=(1-\rho)\sum_{k=1}^{n}\rho^{k-1}G_{k-1}(Q) \quad \text{for } n\geq 1, \quad (11)$$

where $G_0(Q)\equiv 1$.[4] There is the relation among $A_n(Q)$, $R_n(Q)$, and $\rho^n G_n(Q)$,

$$A_n(Q)+R_n(Q)+\rho^n G_n(Q)=1. \quad (12)$$

For $n \gg 1$,

$$\rho^n G_n(Q) \simeq 0, \quad (13)$$
$$A_n(Q) (=E_n(Q)) \simeq 1-R_n(Q), \quad (14)$$

(see Ref. 4) which is a commonly used relation. An example of the correspondence between $E_n(Q)$ and $1-R_n(Q)$ is given in table II for L=3, K=0.3, R=0.5, X=0.1, $Y<L_1$, and $n\leq 10$ where Q is denoted only by X since $Y<L_1$. Upon setting $I_k(Q)=G_{k-1}(Q)-G_k(Q)$,

$$R_n(Q)=\sum_{k=1}^{n}\rho^k I_k(Q). \quad (15)$$

Table II. The values of $G_n(X)$, $E_n(X)(=A_n(X))$, and $1-R_n(X)$, (D_o, $L=3$, $R=0.5$, $K=0.3$, $X=0.1$, $Y<L_1$). The quantities $E_n(X)$ and $R_n(X)$ were computed for $\varepsilon=0.5$.

n	1	2	3	4	5	6	7	8	9	10
$G_n(X)$	0.97809	0.95074	0.91676	0.87987	0.84194	0.80435	0.76778	0.73254	0.69877	0.66649
$E_n(X)$	0.50000	0.74452	0.86337	0.92066	0.94816	0.96131	0.96760	0.97060	0.97203	0.97271
$1-R_n(X)$	0.98905	0.98221	0.97796	0.97566	0.97447	0.97388	0.97360	0.97346	0.97339	0.97336

Table III. Coefficients $I_k(X)$ of the effective reflectance, (C_y, $R=1.0$, $Y<L_1$).

L	K	X\k	1	2	3	4	5	6	7	8	9	10
3	0.3	0.1	0.08052	0.09874	0.10696	0.10203	0.09165	0.0797	0.0682	0.0579	0.049	0.041
		0.5	0.10793	0.11316	0.11291	0.10099	0.08730	0.0743	0.0630	0.0532	0.045	0.038
	0.7	0.1	0.05210	0.07019	0.08300	0.09000	0.09116	0.0875	0.0806	0.0721	0.063	0.054
		0.5	0.11800	0.15134	0.14231	0.12003	0.09719	0.0774	0.0613	0.0484	0.038	0.030
	1.0	0.1	0.04178	0.04810	0.05519	0.06168	0.06633	0.0685	0.0682	0.0657	0.062	0.057
		0.5	0.14930	0.17501	0.15233	0.11930	0.09052	0.0684	0.0519	0.0398	0.031	0.024
5	0.3	0.1	0.02327	0.03075	0.03870	0.04512	0.04952	0.0518	0.0523	0.0514	0.049	0.047
		0.5	0.03314	0.04121	0.04978	0.05520	0.05719	0.0566	0.0544	0.0514	0.048	0.045
	0.7	0.1	0.01301	0.01707	0.02084	0.02456	0.02824	0.0317	0.0349	0.0374	0.039	0.041
		0.5	0.03410	0.05429	0.06759	0.07306	0.07286	0.0694	0.0643	0.0587	0.053	0.048
8	0.3	0.1	0.00647	0.00797	0.00972	0.01163	0.01363	0.0156	0.0176	0.0194	0.021	0.022
		0.5	0.00945	0.01174	0.01433	0.01725	0.02016	0.0227	0.0248	0.0264	0.027	0.028
	0.7	0.1	0.00335	0.00404	0.00468	0.00531	0.00595	0.0066	0.0073	0.0081	0.009	0.010
		0.5	0.00936	0.01430	0.01963	0.02467	0.02890	0.0321	0.0342	0.0353	0.036	0.035

Table IV. Coefficients $I_k(X)$ of the effective reflectance, (D_o, $R=0.5$, $Y<L_1$).

L	K	X\k	1	2	3	4	5	6	7	8	9	10
3	0.3	0.1	0.02191	0.02735	0.03398	0.03689	0.03793	0.0376	0.0366	0.035	0.034	0.032
		0.5	0.03002	0.03135	0.03749	0.03873	0.03861	0.0375	0.0361	0.035	0.033	0.032
	0.7	0.1	0.01432	0.02146	0.02865	0.03424	0.03822	0.0406	0.0416	0.042	0.041	0.040
		0.5	0.03570	0.04838	0.05235	0.05161	0.04945	0.0468	0.0441	0.041	0.039	0.037
	1.0	0.1	0.01158	0.01366	0.01860	0.02263	0.02659	0.0297	0.0320	0.034	0.035	0.035
		0.5	0.05068	0.05632	0.05520	0.05097	0.04778	0.0446	0.0418	0.039	0.037	0.035
5	0.3	0.1	0.00601	0.00779	0.01002	0.01218	0.01418	0.0159	0.0173	0.018	0.019	0.019
		0.5	0.00866	0.00994	0.01239	0.01481	0.01682	0.0183	0.0192	0.020	0.020	0.020
	0.7	0.1	0.00337	0.00502	0.00672	0.00843	0.01019	0.0120	0.0137	0.015	0.017	0.018
		0.5	0.00927	0.01608	0.02170	0.02544	0.02756	0.0285	0.0287	0.029	0.028	0.027
8	0.3	0.1	0.00164	0.00199	0.00239	0.00283	0.00329	0.0038	0.0043	0.005	0.005	0.006
		0.5	0.00241	0.00278	0.00320	0.00375	0.00439	0.0051	0.0058	0.007	0.007	0.008
	0.7	0.1	0.00085	0.00114	0.00143	0.00172	0.00201	0.0023	0.0026	0.003	0.003	0.004
		0.5	0.00243	0.00394	0.00566	0.00742	0.00909	0.0106	0.0118	0.013	0.014	0.014

coefficients $I_k(Q)$ are computed iteratively by using (6) and (7). The numerical calculation was performed for $X=0.1$, 0.5, and $Y<L_1$. Computed values of $I_k(X)$ are listed in tables III and IV for C_y with $R=1.0$ and D_o with $R=0.5$, respectively. The effective emissivity is approximated by $1-R_n(X)$.

Error due to a truncation of terms after the nth in

$$R(Q) = \sum_{k=1}^{\infty} \rho^k I_k(Q) \quad (16)$$

is estimated as follows:

$$\sum_{k=n+1}^{\infty} \rho^k I_k(Q) \approx \rho^{n+1} I_{n+1}(Q)/(1-\rho) \quad (17)$$

$$\approx \rho^{n+1} I_n(Q)/(1-\rho). \quad (18)$$

From the comparison between values obtained by performing iterative processes (6) and (8) until a difference between $1-R_n(X)$ and $A_n(X)$ became smaller than 1×10^{-12} and values obtained by performing (6) up to the nth order,

$$\sum_{k=n+1}^{\infty} \rho^k I_k(X) \leq 1.2 \rho^{n+1} I_{n+1}(X)/(1-\rho) \quad \text{for } n\geq 5. \quad (19)$$

which is the same result as that in Ref. 5.

Furthermore,

$$\sum_{k=n+1}^{\infty} \rho^k I_k(X) \leq 2\rho^{n+1} I_n(X)/(1-\rho) \quad \text{for } n\geq 1. \quad (20)$$

REFERENCES

1. R. J. Chandos and R. E. Chandos, Appl. Opt. 13, 2142-2152(1974).
2. R. E. Bedford and C. K. Ma, J. Opt. Soc. Am. 65, 565-572(1975).
3. R. E. Bedford and C. K. Ma, J. Opt. Soc. Am. 66, 724-730(1976).
4. Y. Ohwada, J. Opt. Soc. Am. 71, 106-111(1981).
5. Y. Ohwada, Appl. Opt. 20, 3332-3335(1981).
6. J. C. DeVos, Physica(Utrecht), 20, 669-689(1954).
7. S. Takata, J. Illum. Eng. Inst. Jpn., 51, 702(1967).
8. F. O. Bartell and W. L. Wolfe, Appl. Opt. 15, 84-88(1976).
9. T. J. Quinn, Br. J. Appl. Phys. 18, 1105-1113(1967).
10. T. J. Quinn, Comite Consultatif De Thermometrie, CCT/80-45(1980).
11. A. Ono, J. Opt. Soc. Am. 70, 547(1980).
12. A. Ono, this volume.
13. R. E. Bedford and C. K. Ma, J. Opt. Soc. Am. 64, 339-348(1974).
14. E. M. Sparrow and R. D. Cess, *Radiation Heat Transfer*, (Books/Cole, Belmont, Calif., 1966).

The effective temperature to express radiant characteristics of nonisothermal cavities

S. Hattori and A. Ono

National Research Laboratory of Metrology, Umezono Sakura, Ibaraki 305, Japan

The spectral radiance of a nonisothermal cavity is expressed by a new quantity, effective temperature T_e and by the cavity emissivity, ϵ_c through the Planck relation $L(\lambda) = \epsilon_c L_b(\lambda, T_e)$. The effective temperature is charged with the effect of temperature distribution since the cavity emissivity is defined as the effective emissivity when the cavity approaches an isothermal condition. Effective temperatures are calculated by numerical integration for various diffuse, cylindrical cavities which have temperature distributions along the cylindrical axes that are depressed by up to 10% at the apertures. For small temperature gradients, the effective temperature is independent of the wavelength and temperature. For large temperature gradients, a theoretical approach considering the strong nonlinear behavior of blackbody spectral radiance is made and an empirical equation is presented to provide approximate effective temperatures for nonisothermal cavities including the wavelength and temperature dependence. The equation has three parameters, the values of which are identified for various diffuse cylindrical cavities. A procedure using the effective temperature that minimizes calibration errors is presented for calibrating with a nonisothermal cavity radiation thermometer operating at different wavelengths.

INTRODUCTION

Blackbody cavities with large apertures are often used to calibrate industrial radiation thermometers operating at different wavelengths. Such cavities can have large temperature distributions (nonisothermal) and can be poor approximations to a perfect blackbody. Temperature distribution effects on the cavity quality are significant for such nonisothermal cavities. In the field of radiation thermometry, the effective emissivity has been frequently used as a quantity to express radiant characteristics of isothermal and non-isothermal cavities.[1-8] For a given temperature distribution within a cavity, however, the effective (spectral) emissivity explicitly depends on the wavelength of radiation and the temperature through the Planck relation for blackbody spectral radiance. Therefore effective emissivities have to be evaluated (calculated) at every wavelength and every temperature associated with the radiation thermometers to be calibrated.

To avoid such complicated processes of cavity evaluation, one of the authors (S.H.) has introduced a new concept of the effective temperature of a cavity which expresses the radiant characteristics of non-isothermal cavities.[9,10] The effective temperature of the cavity depends neither on the wavelength of the radiation nor on the temperature of the cavity to the first-order approximation of temperature differences within a cavity; it depends only on the temperature distribution.

In this paper a new concept of the effective temperature is reviewed and a theoretical approach is made to obtain an empirical equation which gives wavelength dependence of the effective temperature beyond the first-order approximation. An application of the empirical equation is made to cylindrical diffuse cavities with temperature distributions that have up to 10% depressions at the apertures. Usefulness of the effective temperature of cavity is demonstrated for the calibration of radiation thermometers with a nonisothermal cavity.

THE EFFECTIVE TEMPERATURE

The Spectral Radiance of a Nonisothermal Cavity

Figure 1 shows a nonisothermal opaque cavity in which the temperature distribution is expressed by $T(x)$. Assuming that the temperature of the surroundings is low enough to neglect thermal radiation incident into the cavity, radiation exitent from a location x_o on the cavity wall in the direction D is considered to consist of two contributions. The first contribution is radiation intrinsically emitted by the surface, and the second is that which comes from other surface locations of the cavity wall and is reflected at the surface in the direction D. Then the spectral radiance $L_o^D(\lambda)$, at a location x_o in a direction D is expressed by the integral equation,

$$L_o^D(\lambda) = e_o^D L_b(\lambda, T(x_o)) + \frac{1}{\cos\theta_o^D} \int_S d\Omega_o^1 r_o^{1,D} L_1^o(\lambda) \cos\theta_o^1, \quad (1)$$

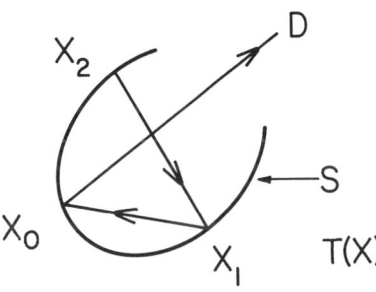

Fig.1 Schematic of a nonisothermal cavity.

where θ_o^D is the angle between the normal to the surface at x_o and the direction D, θ_o^1 similarly defined, $d\Omega_o^1$ the solid angle element at x_o in the direction x_1, and S indicates that the integration is carried out over the cavity wall. The radiative properties of the surface are characterized by the directional emissivity e_o^D and by the partial reflectivity $r_o^{1,D}$ which is defined as the fraction of radiation reflected at x_o in the direction D per unit solid angle when radiation is incident from the direction x_1. Although the directional emissivity and the partial reflectivity are functions of wavelength and temperature, notations for the dependence have been omitted for brevity. In Eq. (1), L_b is the Planck relation for blackbody spectral radiance given by

$$L_b(\lambda,T) = (c_1/\lambda^5)[\exp(c_2/\lambda T) - 1]^{-1}, \quad (2)$$

where $c_1 = 1.19096 \times 10^{-16}$ W·m^2 sr^{-1} and $c_2 = 0.014388$ m·K.

The iterative solution of the integral equation, Eq. (1), is given by a series expression,

$$L_o^D(\lambda) = e_o^D L_b(\lambda,T_o)$$

$$+ (1/\cos\theta_o^D)\int_S d\Omega_o^1 r_o^{1,D} \cos\theta_o^1 e_1^o L_b(\lambda,T(x_1))$$

$$+ (1/\cos\theta_o^D)\int_S d\Omega_o^1 r_o^{1,D} (\cos\theta_o^1/\cos\theta_o^1)$$

$$\int_S d\Omega_1^2 r_1^{2,o} \cos\theta_1^2 e_2^1 L_b(\lambda,T(x_2))$$

$$+ \ldots, \quad (3)$$

where the second term on the right-hand side is the contribution of the reflected radiation that is originally emitted at x_1 where the temperature is $T(x_1)$, and then incident on x_o. The third term is the contribution of reflected radiation that is originally emitted at x_2 where the temperature is $T(x_2)$, reflected at x_1, and then incident on x_o. Combining the nested integrals, Eq. (3) can be written in a form of a single surface integral,

$$L_o^D(\lambda) = \int_S W(x) L_b(\lambda,T_x) ds_x, \quad (4)$$

where ds_x is the surface element at x, $T(x)$ is abbreviated as T_x, and $W(x)$ is the weighting factor which expresses the radiant contribution of the surface element at x to the spectral radiance. The weighting factor depends on the cavity geometry and on the radiative properties of the cavity wall, but does not depend on the wavelength nor on the temperature.

The Effective Emissivity

The effective emissivity, ϵ_e, of a nonisothermal cavity is defined as the spectral radiance ratio between the cavity and a blackbody at a reference temperature T_o:

$$\epsilon_e = L_o^D(\lambda)/L_b(\lambda,T_o) \quad (5\text{-a})$$

$$= \int_S W(x) L_b(\lambda,T_x) ds_x / L_b(\lambda,T_o). \quad (5\text{-b})$$

Considering that the position x_o in the cavity will be viewed by a radiation thermometer, the temperature, T_o, is called the cavity temperature. It is seen in Eq. (5-b) that the effective emissivity of a nonisothermal cavity depends on the wavelength and the cavity temperature, as well as, on the temperature distribution through the Planck relation for blackbody spectral radiance.

In order to distinguish the effective emissivity of an isothermal cavity from that of a nonisothermal cavity, we call the former the cavity emissivity. Therefore the cavity emissivity, ϵ_c, is defined by

$$\epsilon_c = L_o^D(\lambda)/L_b(\lambda,T_o) \quad (6\text{-a})$$

$$= \int_S W(x) ds_x. \quad (6\text{-b})$$

The cavity emissivity, of course, does not depend on the wavelength nor the temperature.

It will be useful to note the definition of the radiance temperature, S, of the nonisothermal cavity,

$$L_o^D(\lambda) = L_b(\lambda,S). \quad (7)$$

The Effective Temperature

The effective temperature of a nonisothermal cavity, T_e, a new concept introduced here, is defined by the equation,

$$L_o^D(\lambda) = \epsilon_c L_b(\lambda,T_e), \quad (8)$$

where ϵ_c is the cavity emissivity the nonisothermal cavity would have if it were isothermal, that is, if $T_x = T_o$. In Eq. (8) the temperature distribution effect on the spectral radiance is concentrated in the effective temperature. The effective temperature should be clearly distinguished from the radiance temperature defined by Eq. (7). When the temperature distribution of a cavity diminishes to be isothermal, the effective temperature approaches the cavity temperature. However, the radiance temperature does not approach the cavity temperature because $\epsilon_c \neq 1$.

Combining Eqs. (4), (6-b), and (8), the effective temperature is expressed as a ratio of integrals,

$$L_b(\lambda,T_e) = \frac{\int_S W(x) L_b(\lambda,T_x) ds_x}{\int_S W(x) ds_x}. \quad (9)$$

The effective temperature is numerically calculated through Eq. (9). View factors associated with several basic shapes of cavity have been given in analytical forms by Bedford and Ma.[5-7] Using these view factors and taking into account the first five terms in Eq. (3), i.e. up to four interreflections within a cavity, numerical calculations of the effective temperature and the effective emissivity were made through Eqs. (9) and (5-b), respectively, for cylindrical diffuse cavities. A result is illustrated in Fig. 2 for a cylindrical cavity where the length-to-radius ratio, a/r, is 8 and the surface emissivity, e, is 0.75 (consequently, the cavity emissivity is 0.9956). The cavity temperature at the cylinder bottom is 1000K and the temperature distribution is quadratic with a depression at the aperture of 100K. That is,

$$T_x = 1000\left\{1 - 0.1(x/a)^2\right\}, \quad (10)$$

where x is the distance from the cavity bottom along the cylindrical axis. The effective temperature decreases with the wavelength and approaches a constant 998.5K at a long wavelength limit while the effective emissivity increases with the wavelength and approaches a constant 0.9941. It should be noted that the effective temperature varies by 0.3 K from 998.9 to 998.6

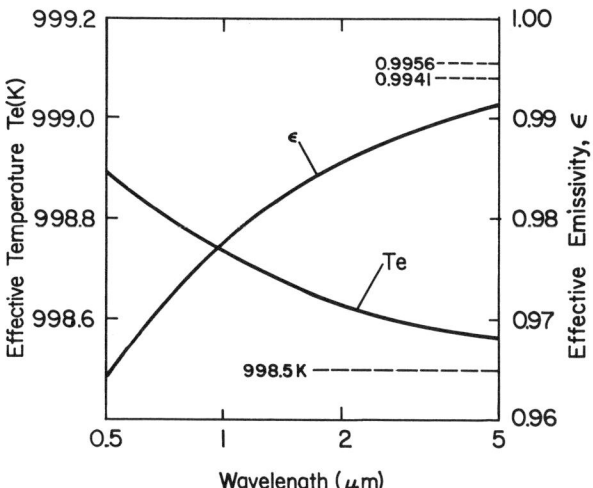

Fig.2 Effective temperature and effective emissivity of a cylindrical diffuse cavity where $e = 0.75$, $(a/r) = 8$, and $T_x = 1000[1-0.1(x/a)^2]$.

(only 0.03% of the temperature) in the spectral range 0.5 to 5μm while the effective emissivity varies by 0.03 from 0.96 to 0.99 (3% of the emissivity).

The effective temperature can be also calculated by a statistical method (Monte Carlo method). The method is useful for cavities where the surface is intermediate between diffuse and specular and/or the surface emissivity is low. An application to a lateral hole formed on a metallic tube will be shown elsewhere in this volume[11].

The First-Order Approximation

When the temperature differences within the cavity are small, the effective temperature is expected to be close to the cavity temperature. Since the Planck relation for blackbody spectral radiance can be developed to the first order of the temperature differences as

$$L_b(\lambda,T_e) \simeq L_b(\lambda,T_o) + \frac{\partial}{\partial T_o}L_b(\lambda,T_o)(T_e - T_o) , \qquad (11)$$

and

$$L_b(\lambda,T_x) \simeq L_b(\lambda,T_o) + \frac{\partial}{\partial T_o}L_b(\lambda,T_o)(T_x - T_o), \qquad (12)$$

Equation (9) can be reduced to give the effective temperature as

$$\Delta T_e \simeq \frac{1}{\epsilon_c}\int_S W(x) \Delta T_x ds_x , \qquad (13)$$

where $\Delta T_e = T_o - T_e$ and $\Delta T_x = T_o - T_x$ are the depressions of effective temperature and local temperature, respectively, from the cavity temperature.

Equation (13) has a significant feature for the effective temperature that the depression of effective temperature depends neither on the wavelength of radiation nor on the cavity temperature. It depends only on the temperature distribution. From Eqs. (5-a), (8), and (11) we can express the effective emissivity by

$$\epsilon_e \simeq \epsilon_c[1 - \frac{\partial}{\partial T_o}\ln L_b(\lambda,T_o)\Delta T_e] . \qquad (14\text{-}a)$$

For Wien's approximation, the expression is

$$\epsilon_e \simeq \epsilon_c[1 - (c_2/\lambda T_o)(\Delta T_e/T_o)], \qquad (14\text{-}b)$$

and for the Rayleigh-Jeans' approximation,

$$\epsilon_e \simeq \epsilon_c(1 - \Delta T_e/T_o). \qquad (14\text{-}c)$$

EMPIRICAL FORMULA FOR THE EFFECTIVE TEMPERATURE

When a cavity has a large temperature distribution such as given by Eq. (10) and wavelengths are shorter than that at the peak of the blackbody spectral radiance, the effective temperature of the cavity cannot be assumed to be independent of wavelength, as is seen in Fig. 2. The first-order approximation is no longer valid because blackbody spectral radiance $L_b(\lambda,T)$ has strong nonlinearity with respect to temperature, especially at short wavelengths. In order to obtain an empirical formula which provides the wavelength and temperature dependence of the effective temperature, a theoretical approach is made taking into account the nonlinear behavior of blackbody spectral radiance.

Basic Relations about the Effective Temperature

Wien's approximation is used for $L_b(\lambda,T)$ because the nonlinear behavior is significant at short wavelengths. The equation determining the effective temperature is reduced to

$$\epsilon_c \exp(-c_2/\lambda T_e) = \int_S W(x) \exp(-c_2/\lambda T_x)ds_x. \qquad (15)$$

Nondimensional quantities, $t_e, t_a, F(x)$, and m, are introduced which are defined as

$$T_e = T_o(1-t_e) , \qquad (16)$$

$$T_x = T_o[1-t_a F(x)] , \qquad (17)$$

where

$$t_a = (T_o - T_a)/T_o , \qquad (18)$$

and

$$c_2/\lambda T_o = \gamma/m , \qquad (19)$$

where γ is a constant, $\gamma = 4.965$. The quantity, t_e, is the relative depression of the effective temperature from the cavity temperature. The quantity, t_a, is the relative depression of the temperature at the aperture. Considering a cylindrical cavity with the length-to-radius ratio of (a/r) as an example, $F(x)$ is the depression of temperature distribution along the cylindrical axis which is normalized to $F(0) = 0$ and $F(a) = 1$. The quantity, m, is related to the product between the wavelength and the cavity temperature which is normalized as $m = 1$ at the peak of the blackbody spectral radiance.

It is assumed that the relative depression of effective temperature, t_e, and the relative depression of local temperature, $t_a F(x)$, are much smaller than 1;

$$(1 - t_e)^{-1} \simeq 1 + t_e . \qquad (20)$$

$$[1 - t_a F(x)]^{-1} \simeq 1 + t_a F(x) . \qquad (21)$$

Then Eq. (15) is reduced to

$$\epsilon_c \exp(-\gamma t_e/m) = \int_S W(x) \exp[-\gamma F(x)t_a/m]ds_x. \qquad (22)$$

It should be noted that, in order to keep Eq. (22) valid at short wavelengths, we do not assume that the arguments of the exponential, $(\gamma t_e/m)$ and $[\gamma F(x)t_a/m]$, are much smaller than 1. If such an assumption is made, Eq. (22) is reduced to be equivalent to Eq. (13) which is the first-order approximation.

Since t_e is determined by $t_a, m, F(x)$, and $W(x)$, it may be written as $t_e(t_a, m, F, W)$. As is seen in Eq. (22), however, there is a special relation between (t_e/m) and (t_a/m). A given ratio of (t_a/m) provides the same results for (t_e/m); for an arbitrary constant, p,

$$t_e(t_a,m,F,W)/m = t_e(pt_a,pm,F,W)/pm. \qquad (23)$$

Letting $p = \theta/t_a$ where θ is a selected value of t_a, the effective temperature is expressed for an arbitrary t_a as

$$t_e(t_a,m,F,W) = (t_a/\theta)t_e(\theta,m\theta/t_a,F,W). \qquad (24)$$

Now it is useful to define a normalized relative depression of effective temperature, f, so that $f = 1$ at $m = 1$;

$$f(t_a,m,F,W) = t_e(t_a,m,F,W)/t_e(t_a,1,F,W) . \qquad (25)$$

Combining Eqs. (24) and (25), an expression for t_e for arbitrary t_a and m is obtained as

$$t_e(t_a,m,F,W) = t_e(\theta,1,F,W)(t_a/\theta)f(\theta,m\theta/t_a,F,W). \qquad (26)$$

Equation (26) suggests that t_e can be analytically obtained for arbitrary t_a and m using the value of t_e at selected values of $t_a = \theta$ and $m = 1$ if a universal function is provided for f.

Empirical Equation for the Normalized Function f

The normalized function, $f(t_a,m,F,W)$, can be explicitly written using Eq. (22);

$$f(t_a,m,F,W) = \frac{m \ln[\varepsilon_c^{-1}\int W(x)\exp[-\gamma F(x)t_a/m]ds_x]}{\ln[\varepsilon_c^{-1}\int_S W(x)\exp[-\gamma F(x)t_a]ds_x]} . \qquad (27)$$

For an arbitrary temperature distribution which has its maximum at the cylinder bottom, i.e., $F(x) > 0$ and $t_a > 0$, the normalized function behaves with respect to m as follows:

$$f(t_a,0,F,W) = 0 \qquad \text{for } m=0 , \qquad (28)$$

$$f(t_a,1,F,W) = 1 \qquad \text{for } m=1, \qquad (29)$$

$$(\partial f/\partial m) > 0 \qquad \text{for any } m, \qquad (30)$$

$$\lim_{m\to\infty}(\partial f/\partial m) = 0 \qquad \text{for } m = \infty. \qquad (31)$$

The function $f(t_a,m,F,W)$, is shown against $\log m$ in Fig. 3 for a cylindrical diffuse cavity. It was obtained by numerical integration of Eq. (15) when the cavity length-to-radius ratio is 4, the surface emissivity is 0.75 (consequently, the cavity emissivity is 0.9815), and the temperature distribution is given as $F(x) = (x/a)^2$ and $t_a = 0.1$. (The curve shape of Fig. 3 and the general conditions for the function remind us of the Fermi-Dirac distribution function in statistical mechanics.) It is reasonable to introduce an empirical equation for the function, f, in the form

$$f(t_a,m,F,W) = c[\exp[-\alpha(\log m-\beta)] + 1]^{-1}, \qquad (32)$$

where α, β, and c are constants. Considering the condition of Eq. (29), it is written as

$$f(t_a,m,F,W) = (1 + B)/[(1/m)^A + B], \qquad (33)$$

where the constants, A and B, are dependent on $F(x)$ and $W(x)$. Thus the relative depression of effective temperature, t_e, is given by

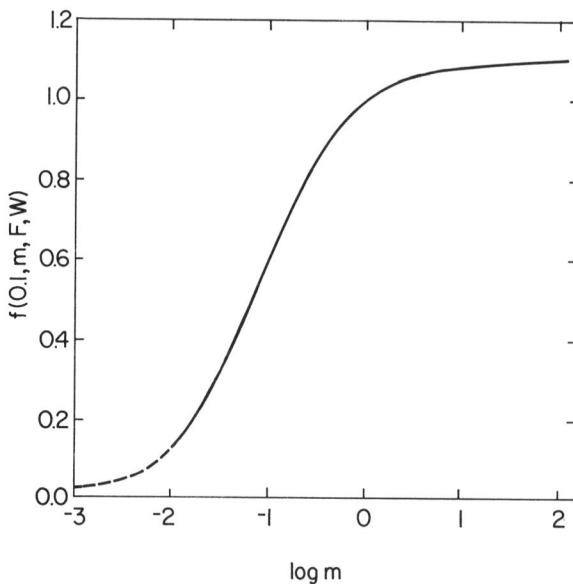

Fig.3 Relation of normalized function, f, to $\log m$ for a cylindrical diffuse cavity where $e = 0.75$, $(a/r) = 4$, $t_a = 0.1$, and $n = 2$.

$$t_e(t_a,m,F,W) = t_e(\theta,1,F,W)(t_a/\theta)(1 + B)/[(t_a/\theta m)^A + B]. \qquad (34)$$

This is an analytical expression for the wavelength and cavity temperature dependence of the effective temperature. It also analytically includes the magnitude of the temperature distribution. Although the accuracy of the empirical equation, Eq. (33), should be checked by the numerical integration of Eq. (9), the discussion in this section is applicable not only to cylindrical cavities but also to cavities in different geometrical forms.

APPLICATION TO CYLINDRICAL DIFFUSE CAVITIES

The empirical equation has been applied to cylindrical, diffuse cavities and the accuracies have been checked by numerical integrations. The length-to-radius ratio of the cavity and the surface emissivity were varied as $(a/r) = 2, 4,$ and 8 and $e = 0.5$, 0.75, and 0.9, respectively. The temperature distributions along the cylindrical axis were chosen in the form of an n-th degree polynomial,

$$\Delta T_x = \Delta T_a (x/a)^n . \qquad (35)$$

The degree n was varied as $n = 1, 2,$ and 3. As a selected value of the relative depression of temperature at the aperture, a 10% depression was chosen, i.e. $\theta = 0.1$.

Introducing a parameter $C = t_e(\theta,1,F,W)/\theta$ and considering $\theta = 0.1$, Eq. (34) can be written in the convenient form,

$$\Delta T_e = C\Delta T_a (1 + B)/[(\Delta T_a/34.5\lambda T_o^2)^A + B] , \qquad (36)$$

where numerical values associated with T and λ should be given in units of kelvin and meter, respectively. It should be noted that the first term of the denominator in Eq. (36) becomes 1 when the wavelength is at the peak of blackbody spectral radiance, $\gamma T_o = 2.90 \times 10^{-3}$ m·K, and the relative depression of the temperature is 10%, $\Delta T_a/T_o = 0.1$. The three parameters, A, B, and C in Eq. (36), were determined for each cavity by the effective temperatures at three wavelengths corresponding to m = 0.1, 1, and 10 which were obtained by numerical integrations of Eq. (15). The parameters are listed in Table I with the cavity emissivities obtained by Sparrow et al.[12]

Normalized relative depressions of effective temperature, $f(0.1,m,F,W)$, are shown in Fig. 4 for a cavity of $(a/r) = 4$, $e = 0.75$, and $n = 2$. The solid and dashed lines correspond to the empirical equation, Eq. (36), and the numerical integral, Eq. (9). In the spectral range $0.1 \leq m \leq 2$, the empirical equation provides $(T_o - T_e)$ with an accuracy better than 3% when the relative depression of temperature distribution at the aperture is less than 10%. The same accuracy was obtained for other cavities listed in Table I. In the spectral range $m \geq 2$, the empirical equation provides greater values of $(T_o - T_e)$ than the numerical integration does because Wien's approximation in Eq. (15) becomes poor. Considering the weak dependence on wavelength, effective temperatures beyond $m = 2$ should be replaced with the value at $m = 2$, $f(0.1,2,F,W)$, which is shown by the dashed-dotted line in Fig. 4. Then the accuracy of the empirical equation is better than 3%. The effective temperature corresponding to $m = 2$ can be a good approximation to Eq. (13) when the temperature gradient is small.

Extension of the empirical equation, Eq. (36), has been investigated to temperature distributions which are given as

$$\Delta T_x = \sum_{n=1}^{N} \Delta T_{a,n} (x/a)^n , \quad (37)$$

where $\Delta T_{a,n}$ is the component of the temperature depression at the aperture which corresponds to the n-th degree. For a small temperature gradient, the depression of effective temperature, ΔT_e, is linear with respect to the depression of temperature, ΔT_x, as is seen in Eq. (13). Therefore it is reasonable to extend Eq. (36) as follows.

$$\Delta T_e = \sum_{n=1}^{N} C_n \Delta T_{a,n} (1+B_n) / \left[(|\Delta T_{a,n}|/34.5\lambda T_o^2)^{A_n} + B_n \right] , \quad (38)$$

where A_n, B_n, and C_n are parameters provided in Table I

Table I Parameters of the empirical equation for cylindrical diffuse cavities

a/r	n	e = 0.5			0.75			0.9		
		A(n)	B(n)	C(n)	A(n)	B(n)	C(n)	A(n)	B(n)	C(n)
2	1	1.10	9.1	0.158	1.09	8.0	0.080	1.08	7.3	0.032
	2	1.03	9.6	0.093	1.03	8.9	0.047	1.02	8.4	0.019
	3	1.01	9.5	0.063	1.01	9.0	0.032	1.00	8.7	0.013
	$\epsilon_c =$	0.8394			0.9389			0.9785		
4	1	1.05	11.6	0.138	1.04	10.4	0.066	1.04	9.6	0.026
	2	0.93	10.4	0.067	0.95	10.4	0.030	0.96	10.6	0.012
	3	0.97	10.9	0.041	0.96	10.7	0.018	0.96	10.6	0.007
	$\epsilon_c =$	0.9460			0.9815			0.9936		
8	1	1.00	15.9	0.092	1.00	14.5	0.042	0.99	13.8	0.016
	2	0.93	15.1	0.033	0.93	14.8	0.014	0.93	14.6	0.005
	3	0.93	13.0	0.017	0.93	12.8	0.007	0.92	12.6	0.002
	$\epsilon_c =$	0.9880			0.9956			0.9984		

for the corresponding degree of the polynomial. It should be noted that the absolute value is taken for $\Delta T_{a,n}$ in the denominator of Eq. (38), considering the possibility of its having a negative value.

The accuracy of the empirical equation, Eq. (38), has been checked for selected quadratic and cubic temperature distributions expressed by

$$F_2(x,x_o) = [(x-x_o)^2 - x_o^2]/[(a-x_o)^2 - x_o^2], \quad (39)$$

and

$$F_3(x,x_o) = [(x-x_o)^3 + x_o^3]/[(a-x_o)^3 + x_o^3], \quad (40)$$

where x_o is a shift of temperature distribution on the x axis. The quadratic distribution, Eq. (39), can be a good approximation to a real cavity where the location of the maximum temperature is far from the cylinder bottom. Effective temperatures were calculated for a cavity where $(a/r) = 4$, $e = 0.75$, and the temperature distributions were given by Eqs. (39) or (40) with a 10% total depression at the aperture. Results of the effective temperature are illustrated against x_o in Fig. 5 when the wavelength is at the peak of the blackbody spectral radiance (m=1). The solid and dashed lines correspond to the empirical equation, Eq. (38), and the numerical integration, Eq. (9). Over the range $-1 \leq (x_o/a) \leq 0.3$ the empirical equation provides the effective temperature with an accuracy better than 5% of the $(T_o - T_e)$ values corresponding to $x_o = 0$. The same accuracy was obtained for other cavities listed in Table I when the relative depression of temperature at the aperture is less than 10%.

CALIBRATION OF RADIATION THERMOMETERS

When the details of the temperature distribution of a cavity such as given by Eq. (37) are known, the spectral radiance of the cavity is fully expressed by the effective temperature through Eqs. (8) and (38). Then radiation thermometers operating at different wavelengths are calibrated against the nonisothermal cavity knowing the radiance temperature at the corresponding wavelengths through Eq. (7). For a cavity where details of temperature distribution are not well known, an alternative calibration procedure has been considered which uses an experimentally determined effective temperature of the nonisothermal cavity.

The cavity temperature, T_o, is measured by a standard contact thermometer such as a thermocouple which is located close to the target point of the radiation thermometers, and the radiance temperature, $S(\lambda_o)$, of

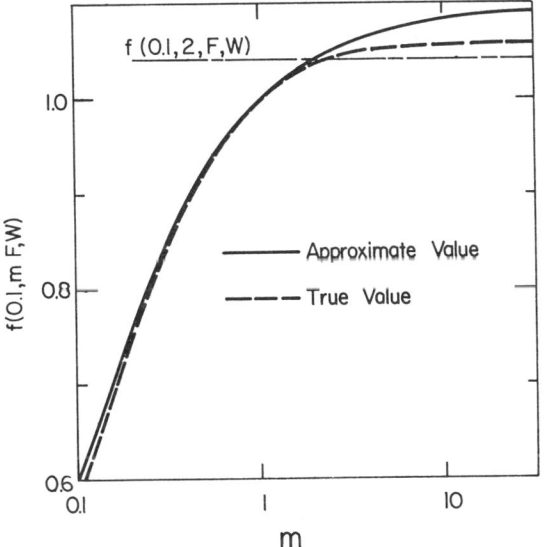

Fig.4 Comparison of normalized function, f, between an empirical equation (solid line) and numerical integration (dashed line) for the same cavity as in Fig.3.

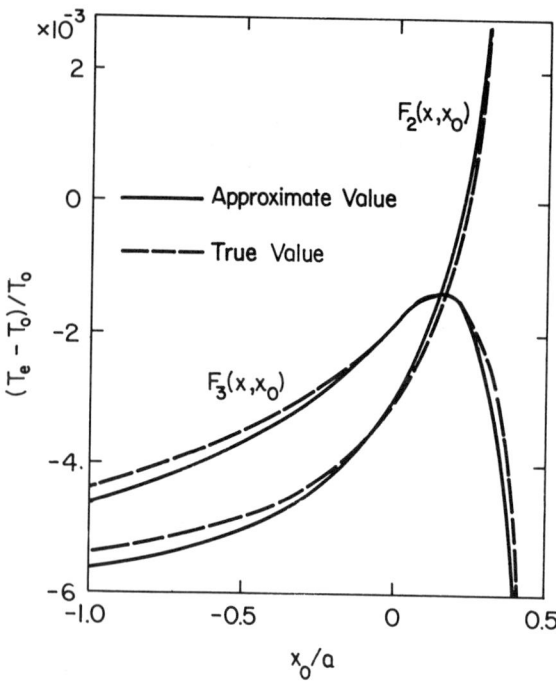

Fig.5 Effective temperatures obtained by an empirical equation (solid lines) and by numerical integration (dashed lines) for different forms of temperature distribution. The cavity condition is that e = 0.75, (a/r) = 4, and $T_a = 0.9 T_o$.

the target point is measured by a standard radiation thermometer whose wavelength is λ_o. Assume that the cavity emissivity, ϵ_c, is available from the literature such as References.[1,5-7,12-14] Then the effective temperature of the cavity, $T_e(\lambda_o)$, at the wavelength λ_o is determined by $T_o, S(\lambda_o)$, and ϵ_c through Eqs. (7) and (8). Recall that the spectral dependence of the effective temperature is weak for a small temperature gradient. Assuming that $T_e(\lambda)$ at a different wavelength, λ, is equal to $T_e(\lambda_o)$, the radiance temperature of the cavity $S(\lambda)$ at a wavelength λ is calculated through Eqs. (7) and (8). Thus radiation thermometers are calibrated against the nonisothermal cavity at different wavelengths.

Calibration errors caused by assuming $T_e(\lambda) = T_e(\lambda_o)$ were calculated for the same cylindrical diffuse cavity as used in Fig. 2 (a/r=8 and e=0.75) where the cavity temperature was 1000K and the temperature distribution was quadratic. The wavelength of a standard radiation thermometer that uses a silicon photocell[15] was taken to be $\lambda_o = 0.9 \mu m$. In Fig. 6 the calibration errors are illustrated versus the wavelength of a radiation thermometer to be calibrated. The upper and lower solid lines correspond to 10% and 1% depressions of temperature at the aperture, respectively. In the spectral range 1 to 10μm the calibration errors are less than 0.2K and 0.003K for 10% and 1% depressions, respectively.

In order to investigate the usefulness of the effective temperature, calibration errors have been compared among different calibration procedures using the same cylindrical diffuse cavity as before. Consider a calibration procedure where the effective emissivity is assumed to be independent of λ. The effective emissivity $\epsilon_e(\lambda_o)$ at the wavelength λ_o is experimentally determined by T_o and $S(\lambda_o)$ through Eqs. (5-a) and (7). Assuming that $\epsilon_e(\lambda)$ at a different wavelength, λ, is equal to $\epsilon_e(\lambda_o)$, the radiance temperature of the cavity $S(\lambda)$ is calculated through (5-a) and (7). Thus radiation thermometers are calibrated against the nonisothermal cavity at different wavelengths. Calibration procedures like this are common. Commercially available blackbody radiators are specified in such a way that the cavity temperature is measured by a contact thermometer embedded in the radiator and the emissivity value is provided by the manufacturer at a wavelength (or not specified).

Calibration errors of the procedure are shown by the dashed lines in Fig. 6. The upper and lower dashed lines correspond to 10% and 1% depressions of temperature at the aperture respectively. The calibration errors are greater than 1K and 0.1K at 2μm and approach 10K and 1K beyond 5μm for the respective temperature depressions.

The dotted lines in Fig. 6 show the calibration errors of a procedure where no correction is made for the cavity emissivity and temperature distribution, i.e. the cavity is regarded as a perfect blackbody of temperature T_o. The cavity temperature T_o is measured by a standard contact thermometer. (No need for a standard radiation thermometer.) Then the calibration errors are given by the differences between the cavity temperature, T_o, and the radiance temperatures, $S(\lambda)$, which are calculated through Eqs. (7), (8), and (9). The upper and lower dotted lines correspond to 10% and 1% temperature depression, respectively.

In Fig. 6 the upper dashed line crosses the upper dotted line near 3μm. This means that, for a large temperature gradient, no correction for a cavity can be better than an inadequate correction neglecting the spectral dependence of effective emissivity. The first calibration procedure which uses the new concept of the effective temperature is more accurate than the other two procedures by 1 to 2 orders of magnitude in temperature.

CONCLUSIONS

A real blackbody formed in a cavity is more or less far from a perfect blackbody from two points of view. The first is that a cavity is not a complete enclosure. The second is that a cavity is not completely isothermal. The two points of imperfection deteriorate the radiant characteristics of the cavity.

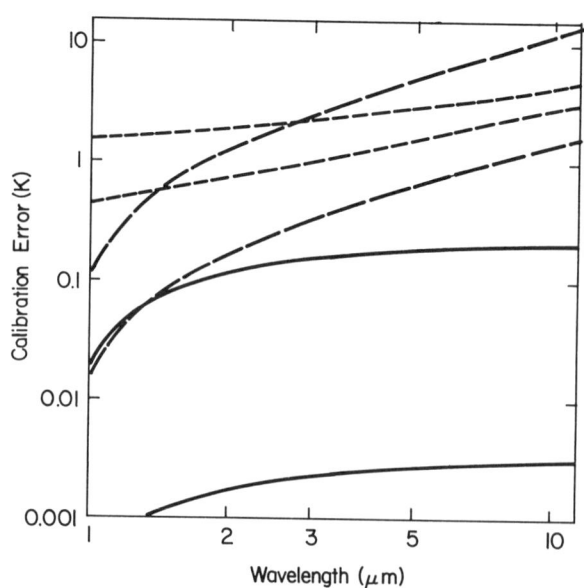

Fig.6 Calibration errors of spectral radiation thermometers for different calibration procedures. The corresponding upper and lower lines are for $t_a = 0.1$ and 0.01, respectively.

The radiant characteristics of nonisothermal cavities studied heretofore concentrated these points together in the effective emissivity of cavity. In the present work, however, the two points were separately focused on the cavity emissivity and the effective temperature, respectively. It is reasonable that the effect of temperature distribution can be represented not by the effective emissivity but by a quantity which has a dimension of temperature as was done here. Therefore a theoretical approach becomes feasible.

One of the conclusions is that the effective temperature is dependent neither on the wavelength nor the cavity temperature to the first-order approximation (Eq. (13)). Then the wavelength and temperature dependence of the spectral radiance of cavity is derived simply from the Planck relation for blackbody spectral radiance (Eq. (8)). It suggests that a nonisothermal cavity behaves like an isothermal cavity which has an effective emissivity of ϵ_c and a cavity temperature of T_e. The effective temperature can be regarded as the true temperature of the cavity.

Beyond the first-order approximation, an empirical equation which provides approximate effective temperatures was introduced taking into account the strong nonlinear behavior of blackbody spectral radiance. Although the physical reason why such a universal function as Eq. (33) fits for the effective temperature is not yet clear, it is useful to be able to have analytically radiant characteristics of nonisothermal cavities. As long as the temperature differences are confined within 10%, the accuracy of the empirical equation is better than 5% in a spectral range (m > 0.1) wide enough for practical interest.

ACKNOWLEDGMENT

The authors would like to express their sincere thanks to Professor D. P. DeWitt, School of Mechanical Engineering, Purdue University for his help in preparing the paper during the time that one of the authors (A.O.) was on sabbatical leave at his laboratory.

REFERENCES

1. R. E. Bedford, *Temperature, Its Measurement and Control in Science and Industry*, H. H. Plumb Editor-in-Chief (Instrument Society of America, Pittsburgh, 1972) Vol. 4, Part 1, pp. 425-434.

2. E. M. Sparrow, Appl. Opt. 4, 41 (1965).

3. C. L. Sanders, Metrologia 3, 119 (1967).

4. P. Campanaro and T. Ricolfi, Appl. Opt. 5, 1271 (1968).

5. R. E. Bedford and C. K. Ma, Opt. Soc. Am. 64, 339 (1974).

6. R. E. Bedford and C. K. Ma, ibid. 65, 565 (1975).

7. R. E. Bedford and C. K. Ma, ibid. 66, 724 (1976).

8. F. O. Bartell and W. L. Wolfe, Appl. Opt. 15, 84 (1976).

9. S. Hattori,, Transactions of the Society of Instrument and Control Engineers 15, 847 (1979), in Japanese.

10. S. Hattori, ibid. 16, 539 (1980), in Japanese.

11. A. Ono, R. M. Trusty, and D. P. DeWitt, to be published in this volume.

12. E. M. Sparrow, L. U. Albers, and E.R.G. Eckert, J. Heat Transfer, 84, 73 (1962).

13. A. Ono, J. Opt. Soc. Am. 70, 547 (1980).

14. Y. Ohwada, ibid. 71, 106 (1981).

15. F. Sakuma and S. Hattori, to be published in this volume.

A method for calibration of a precision thermal radiation density meter using a radiation source

V. A. Chistyakov

Mendeleyev Research Institute of Metrology, Leningrad, Union of Soviet Socialist Republics

V. I. Gavrishchuk

Research Institute for Labour Protection in Agriculture, Oryol, Union of Soviet Socialist Republics

Radiative heat transfer in a coaxial radiator-diaphragm-radiometer system with consideration for shadow zones is studied. Equations for shadow factors are derived. Analysis of procedural and instrumental errors of the radiometer calibration made using cylindrical or plane blackbody sources is given.

The problem of a higher measurement accuracy for the thermal flux density q over a wide range of change has become increasingly urgent recently. However, no measurement assurance, as a rule, for the thermal flux density meters used for this purpose is provided. Besides, the calibration error of q meters, in particular when calibrated using a radiation source giving near-blackbody radiation is too great. This arises from our inadequate knowledge of thermal parameters of a radiation source and improper choice of the instrument calibration procedure.

In this paper consideration is being given to the theoretical grounds underlying the calibration procedure for a precision thermal flux density meter (radiometer) with a hemispherical angle of view using a blackbody source. This procedure is suitable for any combination of a coaxial and diffuse radiating surface and a radiometer both with and without an intermediate diaphragm.

If the emitting area of the radiation source is A_s, its uniform temperature T_s and emissivity ε_s, the thermal flux density sensed by the receiving surface area A_R of the radiometer is

$$q = \sigma \varepsilon_s \varepsilon_R (T_s^4 - T_R^4) F_{s,R} \frac{A_s}{A_R} \qquad (1)$$

where σ is the Stefan-Boltzmann constant, ε_R and T_R are the emissivity and temperature of the radiometer receiving surface, respectively, $F_{s,R}$ is the angular coefficient between the source and the radiometer.

If individual sections of the emitting surface, as distinct from the entire source, have approximately equal T_s and ε_s (such as the bottom of a cylindrical cavity), the radiometer angle of view is limited to sighting the homogeneous surface using an intermediate diaphragm. Then the diaphragm opening whose parameters are T_s and ε_s is assumed to be a true source and $F_{d,R}$, the angular coefficient specific to the geometry of the diaphragm-radiometer system, and A_d, the diaphragm opening area, are substituted instead of $F_{s,R}$ and A_s, respectively, into Eq. (1).

A decrease in the radiometer angle of view results in a considerable drop of the radiation density under measurement which is impermissible. For this reason a radiometer practically always sights the emitting surface featuring non-uniform distribution of T_s and ε_s. In this case, precise determination of q by traditional methods becomes impossible. The thermal flux density sensed by the radiometer can be determined with reasonable accuracy on a computer using the zonal method in which the source and radiometer surfaces are divided into elements whose parameters are known. Then the heat transfer between the radiometer and the source elements is determined in succession and the total heat flux between these devices is found through the summation over all the elements of the radiometer. For finding the mean flux density q_m sensed by the radiometer, the total heat flux calculated is related to the complete area of the radiometer receiving surface. When using a cylindrical blackbody, the heat flux is determined singly for the cavity bottom and lateral surfaces. Thereupon the total power is related to the entire receiving surface of the radiometer so as to find q_m.

For the radiometer and the cavity bottom forming a system of two coaxial and diffuse radiating discs, the power of heat transfer between the circular surface elements is found from

$$d^2 Q_{i,j} = \sigma \varepsilon_{\text{eff}\,i}\, \varepsilon_j (T_o^4 - T_j^4) dF_{i,j} dA_i,$$

where subscripts i and j are specific to the cavity and the radiometer, respectively, ε_{eff} is the effective emissivity, T_o is the bottom center temperature.

From the algebra of angular coefficients[1] it is evident that

$$dA_i dF_{i,j} = A_{i+1}(F_{i+1,j+1} - F_{i+1,j}) - A_i(F_{i,j+1} - F_{i,j}), \quad (2)$$

where A_i, A_{i+1} are the disc areas, $F_{i+1,j+1}$, $F_{i+1,j}$, $F_{i,j+1}$, $F_{i,j}$ are the angular coefficients between the discs, subscripts i, i+1 and j, j+1 specific to the discs whose radii are r_i, $r_i + \Delta_i$ and r_j, $r_j + \Delta_j$, respectively (with subscripts Δ_i, Δ_j denoting the subintervals over the source and radiometer radii).

Substituting the areas and angular coefficients in Eq. (3) presented below, we obtain after calculations

$$dF_{i,j} dA_i = \frac{\pi}{2} \left(\sqrt{(r_{i+1}^2 + r_j^2 + h^2)^2 - 4 r_{i+1}^2 r_j^2} + \sqrt{(r_i^2 + r_{j+1}^2 + h^2)^2 - 4 r_i^2 r_{j+1}^2} - \sqrt{(r_{i+1}^2 + r_{j+1}^2 + h^2)^2 - 4 r_{i+1}^2 r_{j+1}^2} - \sqrt{(r_i^2 + r_j^2 + h^2)^2 - 4 r_i^2 r_j^2} \right),$$

where h is the distance between the radiometer and the cavity bottom.

When limiting the radiometer angle of view by an intermediate diaphragm, shadow zones to be taken into account in a specific instance of heat transfer between two circular surface elements are formed, with due consideration given for shadow factor $K_{i,j}$. For the purpose of determining $K_{i,j}$, consider Fig. 1 where 1 - cavity bottom, 2 - diaphgram, 3 - radiometer. The heat transfer between elements of elementary rings d^2A_i and d^2A_j less the diaphragam is described by

$$d^4Q_{i,j} = \sigma\varepsilon_{eff,i}\varepsilon_j(T_o^4-T_j^4)\cos^2\alpha_1 \, d^2A_i d^2A_j/\pi \, S_1^2 +$$
$$\sigma\varepsilon_{eff,i}\varepsilon_j(T_o^4-T_j^4)\cos^2\alpha_3 \, d^2A_i d^2A_j/\pi \, S_2^2 \quad (3)$$

where S_1 and S_2 are the distances between the lower element of a j-th elementary ring of the radiometer and the lower and upper elements of an i-th elementary ring of the cavity, α_1, α_3 are the angles between normal n to the radiometer surface and lines S_1 and S_2.

The diaphragm cancels the first term in Eq. (3) and hence the heat transfer power for shadowed elements $d^4Q^*_{i,j}$ is

$$d^4Q^*_{i,j} = \sigma\varepsilon_{eff,i}\varepsilon_j(T_o^4-T_j^4)\frac{\cos^2\alpha_3}{S_2^2}d^2A_i d^2A_j.$$

Then the shadow factor is

$$K_{i,j} = \frac{S_1^2\cos^2\alpha_3}{S_1^2\cos^2\alpha_3 + S_2^2\cos^2\alpha_1} \quad (4)$$

From Fig. 1 it is clear that $\cos\alpha_1 = h/S_1$, $\cos\alpha_3 = h/S_2$, $S_1 = \sqrt{h^2+(r_j-r_i)^2}$, $S_2 = \sqrt{h^2+(r_j+r_i)^2}$. Then Eq. (4) takes the form

$$K_{i,j} = \frac{S_1^4}{S_1^4 + S_2^4}$$

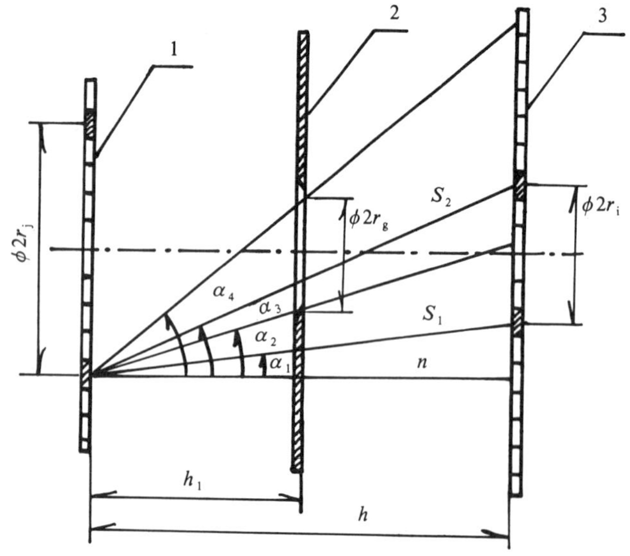

Fig. 1. Diagram of heat transfer in the cavity bottom-diaphragm-radiometer system. 1 - radiometer; 2 - diaphragm; 3 - cavity bottom.

or

$$K_{i,j} = \frac{[h^2+(r_j-r_i)^2]^2}{[h^2+(r_j-r_i)^2]^2+[h^2+(r_j+r_i)^2]^2}, \quad (5)$$

Considering the shadow factor, the heat transfer formula for two shadowed circular elements is described by

$$d^2Q_{i,j} = K_{i,j}\sigma\varepsilon_{eff,i}\varepsilon_j(T_o^4-T_j^4)dF_{i,j}dA_i \quad (6)$$

For finding the moment when to introduce the shadow factor into Eq. (6), let us direct our attention back to Fig. 1. According to Fig. 1,

$$\alpha_1 = \text{arctg}[(r_j-r_i)/h], \; \alpha_2 = \text{arctg}[(r_j-r_g)/h_1],$$
$$\alpha_3 = \text{arctg}[(r_j+r_i)/h], \; \alpha_4 = \text{arctg}[(r_j+r_g)/h_1],$$

where α_2 and α_4 are the angles between normal n and the lines connecting the lower point of a j-th circular element of the radiometer to the lower and upper points of the diaphragm opening.

By theory, angles α_1, α_2, α_3, and α_4 can vary from -90 to +90°, with the minus sign used for the angles counted off clockwise from the normal and the plus sign for those counted off counter-clockwise. It is more convenient to compare the angles varying from 0 to 180°. For this purpose, we add 90° to each one of the above angles. Then the angles will be counted off from a straight line in the plane of the drawing lying on the radiometer surface. Using this new notation, the above angles will be

$$\alpha'_1 = 90° \pm \alpha_1, \begin{array}{c}+(r_j>r_i)\\-(r_j<r_i)\end{array} ; \; \alpha'_2 = 90° \pm \alpha_2, \begin{array}{c}+(r_j>r_g)\\-(r_j<r_g)\end{array}$$

$$\alpha'_3 = 90° + \alpha_3 \qquad \alpha'_4 = 90° + \alpha_4.$$

If $\alpha'_2 < \alpha'_1 < \alpha'_4$ and $\alpha'_2 < \alpha'_3 < \alpha'_4$, the j-th circular element of the radiometer will fully "sense" the i-th circular element of the cavity. In this case, the shadow factor $K_{i,j}=1$. If $\alpha'_1 < \alpha'_2$ and $\alpha'_3 > \alpha'_4$, the j-th circular element of the radiometer does not "sense" the i-th circular element of the cavity and there is no heat transfer between them. If $\alpha'_2 \leq \alpha'_1 \leq \alpha'_4$ and $\alpha'_3 \geq \alpha'_4$ or $\alpha'_1 \leq \alpha'_3 \leq \alpha'_4$ and $\alpha'_1 \leq \alpha'_2$, the j-th circular element of the radiometer "senses" the i-th circular element of the cavity as showed; in this case, the shadow factor is determined from Eq. (5) and introduced into Eq. (6).

For the lateral surface of the cavity and the radiometer (SEe Fig. 2 where 1 - lateral surface of the cavity; 2 - diaphragam; 3 - radiometer) forming a coaxial system, the heat flux between the surface elements is determined from

$$d^2Q_{m,j} = K_{m,j}\sigma\varepsilon_{eff,m}\varepsilon_j(T_o^4-T_j^4)dF_{m,j}dA_m, \quad (7)$$

where index m is representative of an elementary circular strip on the lateral surface of the cavity. Using the algebra of angular coefficients, we find that according to Fig. 2

$$dF_{m,j}dA_m = \frac{\pi}{2}(\sqrt{(R_o^2+r_j^2+h_m^2)^2 - 4R_o^2 r_j^2} +$$
$$+\sqrt{(R_o^2+r_{j+1}^2+h_{m+1}^2)^2 - 4R_o^2 r_{j+1}^2} -$$

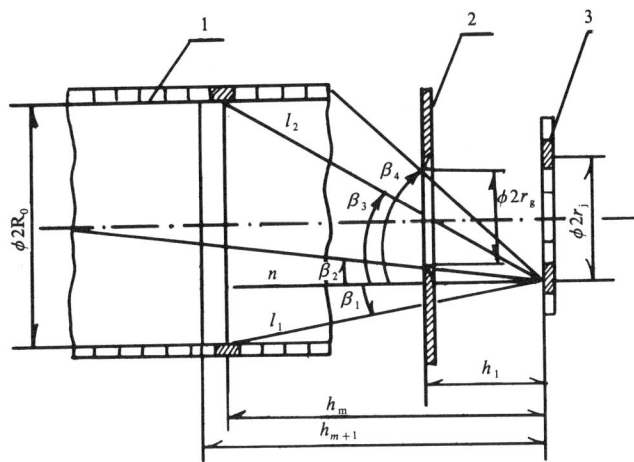

Fig. 2. Diagram of heat transfer in a system formed by the lateral cylinder surface of the cavity, diaphragm and radiometer. 1 - cavity; 2 - diaphragam; 3 - radiometer.

$$-\sqrt{(R_o^2+r_{j+1}^2+h_m^2)^2 - 4R_o^2 r_{j+1}^2}$$

$$-\sqrt{(R_o^2+r_j^2+h_{m+1}^2)^2 - 4R_o^2 r_j^2}$$

where R_o is the cavity radius, h_m is the distance from the radiometer to an elementary strip labelled m, $h_{m+1} = h_{m+1} + \Delta h$, Δh - is the subinterval over the cavity lateral surface.

The shadow factor $K_{m,j}$ is determined in the same way as $K_{i,j}$. According to Fig. 2,

$$K_{m,j} = \frac{[h_m^2+(R_o-r_j)^2]^2}{[h_m^2+(R_o-r_j)^2]^2 + [h_m^2+(R_o+r_j)^2]^2} .$$

The conditions for introducing $K_{m,j}$ into Eq. (7) are similar to those for introduction of $K_{i,j}$ into Eq. (6); angles β_1, β_2, β_3, and β_4 corresponding to angles α_1, α_2, α_3 and α_4, with $\beta_2 = \alpha_2$, $\beta_4 = \alpha_4$ and

$$\beta_1 = \text{arctg}\ [|R_o-r_j|/h_m] ,$$

$$\beta_3 = \text{arctg}\ [(R_o+r_j)/h_m] .$$

It should be noted here that if $r_j > R_o$, the cavity outlet plays a role similar to the opening of the intermediate diaphragm.

The thermal radiation flux Q received by the radiometer from the cavity is $Q = Q_o + Q_L$ where Q_L and Q_o are the thermal radiation fluxes from the lateral surface and cavity bottom, respectively.

These fluxes are calculated from

$$Q_o = \sum_1^{axB} K_{i,j}\sigma\varepsilon_{\text{eff},i}\ \varepsilon_j(T_o^4-T_j^4)dF_{i,j}dA_i ,$$

$$Q_c = \sum_1^{axc} K_{m,j}\sigma\varepsilon_{\text{eff},m}\ \varepsilon_j(T_o^4-T_j^4)dF_{m,j}dA_m ,$$

where a, b and c are the numbers of subdivisions into surface elements over the radiometer receiving surface, the bottom and wall of the cavity, respectively.

The mean flux density sensed by the radiometer is $q_m = Q/A_R$.

When using the procedure suggested, the radiometer calibration errors are combined from the procedural and instrumental errors.

The procedural errors stem from the fact that the equations for the shadow factors $K_{i,j}$ and $K_{m,j}$ are derived approximately when the space problem of heat transfer is substituted by a plane problem. The correctness of the uses of these factors is without question when $K_{i,j} = K_{m,j} = 0$; $K_{i,j} = K_{m,j} = 1$. In this first case the diaphragm is closed whereas in the second one it is open so much that all the zones of the radiometer "sense" all the zones both of the bottom and lateral surface of the cavity. To be checked is the case when $0 < K_{i,j} < 1$ and $0 < K_{m,j} < 1$.

In the specific case, for instance, the equation for $K_{i,j}$ is derived for elementary circular elements i and j that lie in the plane perpendicular to that of the diaphragm passing through the axes of the coaxial system built up from a cavity, a diaphragm and a radiometer and hence the above equation is valid for their associated circular elements.

It would be more correct to consider interaction between each circular element of the radiometer and their associated elements of the cavity. However, the analytical derivation of an equation describing such an interaction is fairly elaborate and, besides, the computation of the heat transfer becomes a serious problem.

Therefore, a careful check was made upon the equations for $K_{i,j}$ and $K_{m,j}$ and hence, on the account taken of the influence of the intermediate diaphragm upon the calibration accuracy. The check was carried out both for a specific system built up from a cavity, a diaphragm and a radiometer and for another system composed of two coaxial discs and an intermediate diaphragm, with parameters of the latter system substantially different from those of the former.

The first system employed a blackbody using a 80 mm diameter, 430 mm long cavity. The intermediate diaphragm opening and the radiometer receiving surface were 30 and 27 mm in diameter, respectively. The distance of the cavity outlet edge to the diaphragm was 60 mm and that of the diaphragm from the radiometer determined for three cases when the extreme points of the radiometer receiving surface sighted the bottom, center, and edge of the cavity outlet was 559, 308 and 68 mm, respectively.

The checking procedure is set forth below. The flux density q_{pr} was determined for the above cavity-diaphragm-radiometer system by a precision method in which the diaphragm opening works as an actual cavity whose bottom center temperature and emissivity are the same as those of real cavity.

The q_{pr} obtained was compared with the flux density q_m calculated by the zonal method[1] when a real cavity had equal T and ε suitable for its center. Subintervals over the lateral surface of the cavity, its bottom and the radiometer receiving surface were taken to be 10, 1 and 0.5 mm, respectively.

With these parameters of the system and the above intervals, comparison was made for three cases of the radiometer sighting the cavity bottom, center and outlet edge.

The deviation of q_m from q_{pr} was + 1.5, -1.6, and -2.7%, respectively.

The parameters of the other system tested were: the source disc, the diaphragm opening and the radiometer radii 6.2 and 4 cm, respectively, and the distance from the radiometer to the disc was 5 cm. With 50, 100 and 200 subintervals over the disc and the radiometer radii, the deviations of q_m from q_{pr} were +4.2, +2.2 and +1.1%, respectively.

The results obtained indicate that the accuracy of q_m is strongly dependent on that of the computational solution. Based on the Type M-222 computer time of not over 15 minutes and requirements on the calibration accuracy for the radiometer involved, sub-intervals exceeding 100 do not seem to be advisable.

The instrumental errors in q_m are combined from those arising in the determination of Q_0, Q_L and A_R.

Considering that the equations to determine Q_0 and Q_L are similar in structure and dimensions, their calculation errors should be determined using one equation.

Basic calculation errors for Q_0 or Q_L are those in the computation of:
 shadow factor δK,
 Stefan-Boltzmann constant $\delta\sigma$,
 cavity effective emissivity $\delta\varepsilon_{eff,i}$ or $\delta\varepsilon_{eff,m}$,
 emissivity of radiometer receiving surface $\delta\varepsilon_j$,
 cavity operating temperature δT_0,
 radiometer receiving surface temperature δT_R,
 product $F_{i,j}A_i$ specific to the mutual geometry of the cavity-radiometer system, δF.

Errors δK and δF include those in measurements of geometrical parameters of the cavity-diaphragm-radiometer system and calculation errors for $K_{i,j}$ and $F_{i,j}A_i$. The measuring error for the geometrical parameters is 0.1% and the computer calculation errors are less than 0.01% which results in the maximum value of the relative error in the determination of $K_{i,j}$ and $F_{i,j}A_i$ corresponding to the minimum i and j of not over 0.2%.

When calculating the heat flux, the computed value of the Stefan-Boltzmann constant $\sigma = (5.66961 \pm 0.00017) \times 10^{-8} W/M^2 K$ was used. According to [2], the maximum error in the computed value of σ is less than 0.02%. The error in the calculation of the cavity local effective emissivity was computed using the intrinsic emissivity error for the cavity wall material equal to 10% for a cylindrical blackbody whose parameters were given above. Figures 3 and 4 present the distribution of effective emissivity ε_{eff} and temperature T over the cavity walls.

Calculations performed on the Type M-222 computer by the method described in [2] for the cavity bottom yielded $\delta\varepsilon_{eff,i} \leq 0.2\%$. However, for the cavity walls, particularly for the areas close to its hole, this error is substantially greater. In this case it is also strongly dependent on the cavity wall temperature measuring error. The calculations show that for the third case, when the radiometer sights the cavity outlet edge, the error $\delta\varepsilon_{eff,m}$ mounts up to 1.5%.

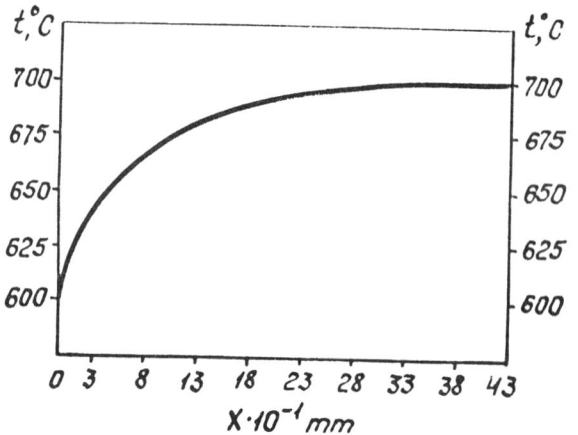

Fig. 3. Temperature distribution along the wall of the cylindrical blackbody from its opening to bottom.

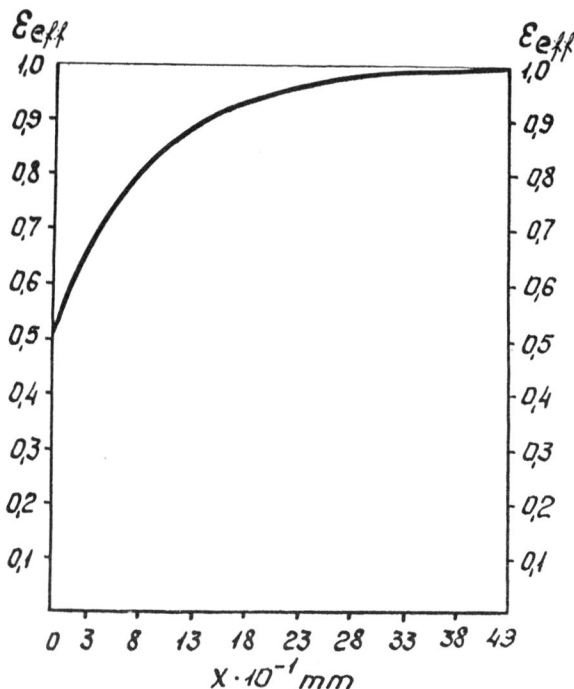

Fig. 4. Distribution of the effective emissivity along the wall of the cylindrical blackbody.

For radiometers having relatively small plane receiving surface (not over 30 mm in diameter), the distribution of ε_j and T_j is practically homogeneous. For instance, the emissivity of the Type "PON - 1" radiometer developed at the Institute for Engineering Thermal Physics of the Ukrainian SSR Academy of Sciences is $\varepsilon_R = 0.95 \pm 0.01$, i.e., $\delta\varepsilon_j = 1\%$. The radiometer receiving surface temperature can be determined using a Chromel-Copel thermocouple to an accuracy of 2 K which gives an error in the determination Q_0 at $T_0 = 973$ K and $T_R \leq 400$ K not over 0.06% while that for the cavity temperature, δT_0, at $T_0 = 975$ K is no more than 1 K which gives an error in Q no more than 0.4%.

Considering that the sign of the systematic errors discussed above is uncertain, the uneliminated systematic error in the calculation of Q_0 and hence of Q_L, is found from the law relating to the determination of the random quantity errors. With the uneliminated systematic error in the calculation of Q_0 and Q_L denoted by δQ_0, and δQ_L, respectively, we may write that

$$\delta Q_0 = \sqrt{(\delta K)^2 + (\delta\sigma)^2 + (\delta\varepsilon_{eff,i})^2 + (\delta\varepsilon_j)^2 + (\delta T_0)^2 + (\delta F)^2} = 1.1\%$$

$$\delta Q_L = 1.9\%.$$

Because of the radiometer sighting the cavity outlet edge in the case considered above $Q_0 \approx Q_L$, δQ will be 1.5%.

The total uneliminated systematic error of the calculation of q_m will also be 1.5% because the error in A_R is rather small.

For calculation against the incident flux, when ε_j is excluded from the calculation, the total uneliminated systematic error of the calculation of q_m is some $\approx 1.1\%$.

Based on the procedural errors considered above, the total calculation error for q_m when calibrating against the incident or absorbed fluxes is 3.8 and 4.2%, respectively.

REFERENCES

[1] R. Siegel and G. Howell, Thermal Radiation Heat Transfer (1972).
[2] R.E. Bedford and C.K. Ma, J. Opt. Soc. Am. 64, 339 (1974).
[3] R.E. Bedford, Advances in Geophysics 14 (1970). Precision radiometry.
[4] V.I. Gavrishchuk and V.A. Chistyakov, Izmeritelnaya Tekhnika 10, 41 (1980), in Russian.

A practical-type fixed point blackbody furnace

F. Sakuma and S. Hattori

National Research Laboratory of Metrology, Sakura-mura, Ibaraki 305, Japan

A set of fixed point blackbody furnaces of simple design and easy operation has been constructed to serve as a standard for calibrating a silicon narrow-band radiation thermometer. Metals used for fixed points are Cu, Ag, Al, Sb, and Zn, which cover the temperature range from 420 to 1085 °C. Each furnace of horizontal type, 10 cm in diameter and 25 cm in length, houses a graphite crucible which is mounted in a stainless steel capsule. The crucible contains 26 cm³ metal of 99.999% nominal purity and is provided with a cylindrical cavity of 1 cm diameter and 4.6 cm length, of which the emissivity is 0.999 ± 0.000 5. Electric power of 400 W can raise the temperature of the capsule to 1100 °C in about an hour and a half, and 240 W can keep the capsule at this temperature. The plateau duration is longer than 10 min. The error of the cavity temperature at each fixed point is estimated to be −0.1 ± 0.1 K, and the accuracy of the fixed point realized for calibrating the 900 nm silicon narrow-band radiation thermometer to be 0.3 K, which can give a practical standard of better accuracy than ±0.5 K in the temperature range from 600 to 1100 °C.

INTRODUCTION

A fixed point blackbody furnace is an effective means for calibrating a radiation thermometer. Hitherto, such a furnace has been made only for best accuracy, so it is not commonly used because of its complicated structure and troublesome operation. This paper deals with a set of fixed point blackbody furnaces of simple design and easy operation for calibrating a 900 nm silicon narrow-band radiation thermometer[1] made for establishing a practical temperature standard of not the best, but better accuracy than, for example, using a ribbon strip lamp.

These furnaces serve as the traceability for radiation thermometers: Common radiation thermometers of wide field of view and wide band-width are calibrated by a comparative blackbody furnace of wide aperture, which in turn is calibrated by the silicon narrow-band radiation thermometer. This thermometer is calibrated by the fixed point blackbody furnaces, so the fixed point blackbody furnaces are the reference for the system. As the required accuracy for the silicon radiation thermometer is 1 K, better than 0.5 K of accuracy is necessary for the fixed point blackbody furnaces.

Table I lists the sorts and main specifications of the furnaces.

CONSTRUCTION OF THE BLACKBODY FURNACE

Structure of the furnace

The fixed point blackbody furnace of the horizontal cylinder type, shown in Fig. 1, is composed of a graphite crucible which constitutes the blackbody cavity, a stainless steel (SUS 310S) capsule for heating the crucible, and a heat insulating part which prevents heat from flowing out of the furnace.

The details of the graphite crucible are shown in Fig. 2. The cavity is a cylinder of 1 cm diameter and about 4.6 cm length with a conical end whose apex angle is 120°. This angle has the advantage of minimizing the decrease of the effective emissivity due to the specular reflection.[2] The cavity is designed so that its emissivity is about 0.999. The inner volume is 26 cm³. The fixed point metals are of 99.999 % nominal purity. The empty crucible is first baked in a vacuum, then metal is cast into the crucible in a vacuum for Cu, Ag and Al, or in an argon atmosphere for Sb and Zn. The part around the screw is thickened to improve its durability.

Figure 3 shows radial and axial cross sections of the capsule containing the graphite crucible in it. In the figure, (a) is a first design, is used for experiment, and is improved in durability into (b) and (c). In (a), insulating pipes are buried in twelve equispaced longitudinal grooves and in them heater winding (0.5 mm diameter Kanthal wire, total length about 1 m, and of 15 Ω resistance) is embedded. On the other hand, in (c) the heater wire of 0.8 mm diameter is wound in spiral around the capsule made longer for temperature distribution. And (b) has both types of heater winding. The stainless steel aperture of 6 mm diameter is placed before the cavity and in (b) and (c) is screwed down to the capsule.

An alumina reflecting ring is placed before the aperture for optical alignment. One or two hundred

Table I
Specifications of the practical-type fixed point blackbody furnaces.

Sorts and temperatures	
copper (Cu)	1084.88 °C
silver (Ag)	961.93 °C
aluminium (Al)	660.46 °C
antimony (Sb)	630.755 °C
zinc (Zn)	419.58 °C
Amount of metal	26 cm³
Aperture diameter	6 mm
Cavity emissivity	0.999 + 0.000 5
Error at fixed point	− 0.1 ± 0.1 °C
Plateau duration	10 min
Furnace dimensions	
diameter	10 cm
length	25 cm
Furnace weight	3 kg
Power consumption	400 W (Maximum)

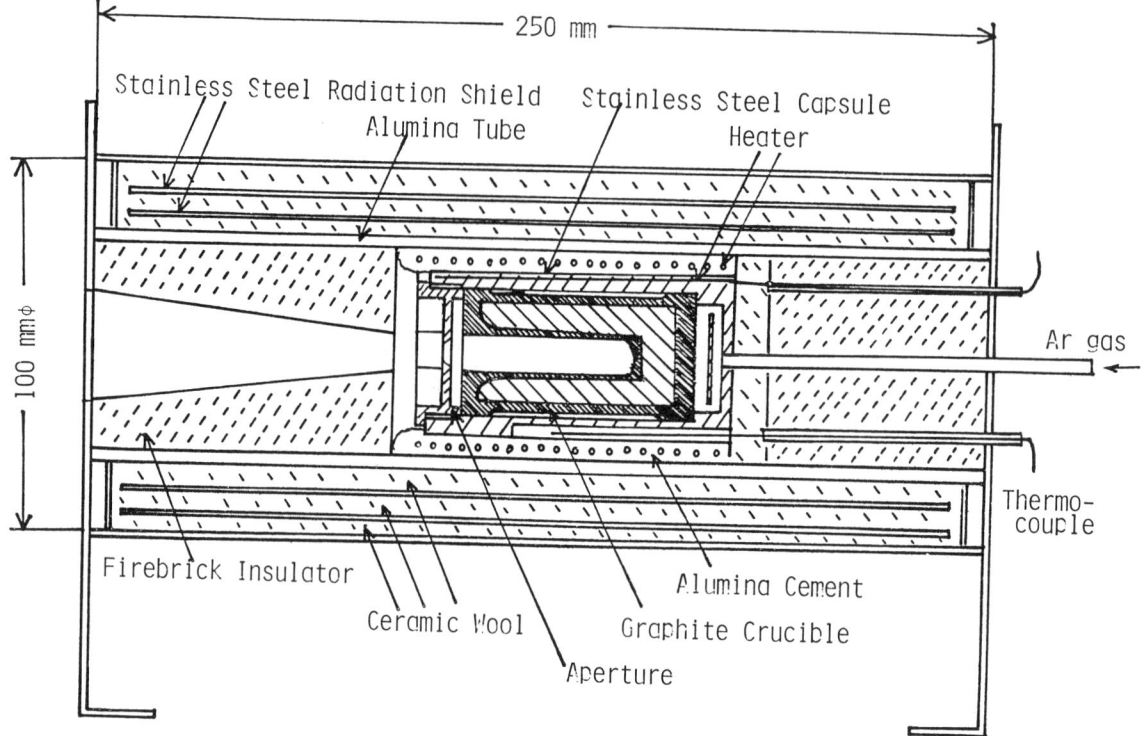

Fig. 1. Practical-type fixed point blackbody furnace

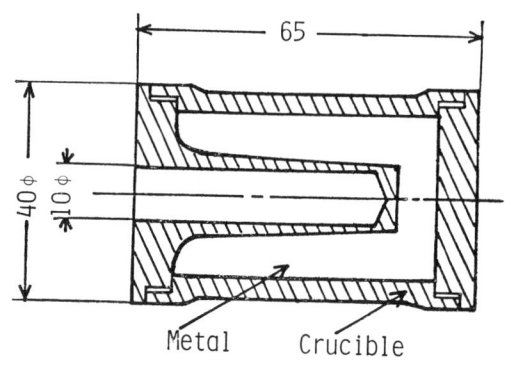

Fig. 2. Graphite Crucible (dimensions in mm)

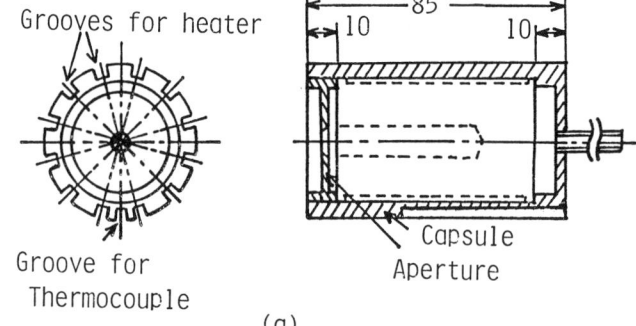

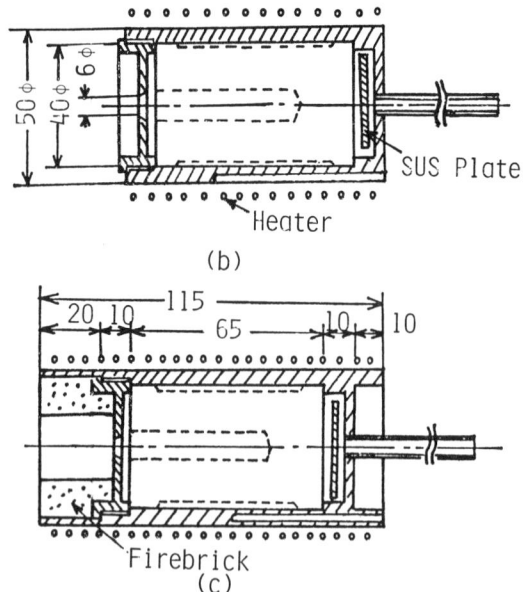

Fig. 3. Stainless Steel Capsule of different types. (dimensions in mm)

cm³/min argon gas is blown through a stainless steel pipe from the rear to prevent the graphite crucible from oxidation while heating. A stainless steel plate is placed in front of the exit of argon gas so that the oxidation of the crucible by remaining oxygen may be scattered.

Antimony is a poisonous metal, so care was taken in using the Sb point blackbody furnace.

Temperature control of the furnace

The freezing points were realized by the following method. First, the metal in the crucible was completely melted, the furnace was maintained for a while at a temperature higher than the melting temperature, and then it was cooled at a rate of a few °C/min.

The fixed point blackbody furnaces were operated by the aid of a controlling circuitry composed of a small-type SCR power regulator and an on-off temperature controller with temperature indication of the moving coil type as shown in Fig. 4. The temperature is monitored by a thermocouple set in the capsule. When operation switch S is connected to (a) side, the controller keeps the crucible temperature at a temperature T_0, 10 to 20°C higher than the freezing point of the metal in the crucible. The freezing point is observed after connecting the switch (S) to (b) side. A variable resistor (VRb) determines the cooling rate and the plateau duration t. Input electric power of 400 W can raise the capsule temperature to 1100°C in one hour and a half after power switch (S_0) is on, and the power is determined by a variable resistor (VRa). Once the variable resistors are adjusted for a set temperature (T_0), the fixed point can be observed by operations of the power switch S_0 and operation switch S.

A protective circuit is added to the controlling circuitry. The control system with the furnace forms one body as shown in Fig. 5.

FREEZING CHARACTERISTICS

Freezing curves

In Fig. 6, a recorder chart of a typical freezing curve on the Ag point blackbody is illustrated. The temperature was measured by a 900 nm silicon narrow-band radiation thermometer. Arrows in the figure indicate the corresponding temperature intervals of 0.2 K. It should be noted that supercooling of about 0.5 K was observed.

For five metals, plateau duration longer than ten minutes was easily realized. Supercooling of about 1 K was observed at the Cu and Al points, and larger than 10 K at the Sb point. At the Zn point, the resolution of the radiation thermometer was so poor (about 1 K) that supercooling could not be observed.

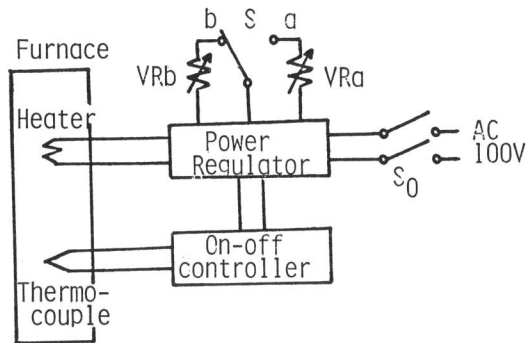

Fig. 4. Electric Circuitry for furnace temperature control.

Fig. 5. Photograph of the furnace

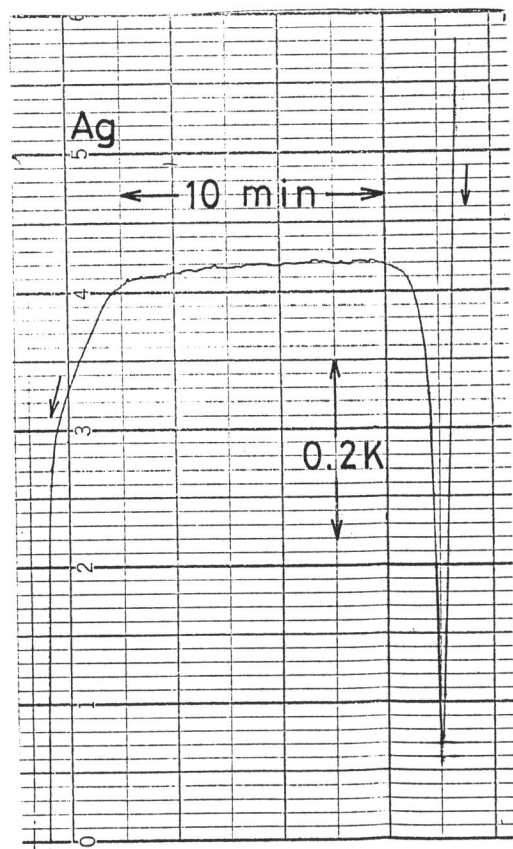

Fig. 6. Example of recorder chart of freezing Ag measured by the 900 nm silicon narrow-band radiation thermometer.

Error in realizing freezing points

Assuming that the bottom end of the cavity is flat and that radiation into the cavity from outside is small, the heat flux density at the base is expressed as follows;

$$\Phi \simeq \frac{\Omega_0}{\pi} \varepsilon_0 \sigma_S T^4 + \frac{\Omega_a - \Omega_0}{\pi} \varepsilon_a \varepsilon_0 \sigma_S (T^4 - T_a^4) \quad (1)$$

where Ω_0 and Ω_a are the solid angles when the aperture and the opening of the cavity are viewed from the base plane ($\Omega_0 = \pi 3^2/45^2$ and $\Omega_a = \pi 5^2/45^2$), respectively, σ_S is the Stefan-Boltzmann constant ($\sigma_S = 5.6696 \times 10^{-8}$ W/m^{2}K^4), ε_0 is the emissivity of the cavity wall, ε_a is the emissivity of the inner surface of the aperture, and T and T_a are the temperatures of the base and the aperture, respectively. The value of Φ in Table II is the heat flux density at the base of each fixed point blackbody for $\varepsilon_0 = 0.9$, $\varepsilon_a = 0.7 \pm 0.2$ and $T_a = 0.95$ T $\pm$ 0.09 T. The temperature gradient $d\theta/dx$ in the table is calculated by using Φ and the thermal conductivity κ of the fixed point metal. The temperature gradient across the metal causes the plateau of the freezing curve in Fig. 6 to slope. The change of the temperature distribution in the cavity, and the change of the impurity concentration in the course of freezing are other possibilities to give rise to the incline.

The first line in Table III shows the average value θ_1 of the freezing range of the plateau over five measurements and the second line shows its standard deviation σ_1. The value θ_1 is represented by the product of the incline $d\theta/dt$ at the center of the plateau of the freezing curve and the plateau duration t.

The error in realizing the freezing point, denoted by ΔT_f in the table, calculated from the next equation, is this error plus the temperature difference θ_2 between both sides of the graphite wall (2.5 mm thick and thermal conductivity κ as 0.6 W/cm·K) of the cavity base calculated from heat flux density ϕ in Table II:

$$\Delta T_f = (\frac{\theta_1}{2} + \theta_{2a}) \pm \sqrt{(\frac{\theta_1}{2} + \sigma_1)^2 + \theta_{2b}^2} . \quad (2)$$

where θ_{2a} and θ_{2b} are the first and the second terms respectively of θ_2 in Table III. Data for Zn are omitted in Table III because the measurement sensitivity was too poor to determine the inclination.

Any of the three causes of the inclination above mentioned is liable to increase the error in the course of freezing, so the temperature is considered to be nearer to the true value at the earlier stage of freezing. Considering these factors, the error at around the peak value of the freezing curve just after recovery from supercooling is expected to be less than ΔT_f in Table III.

ESTIMATION OF ACCURACY

Cavity emissivity

Assuming the cavity is isothermal, the effective emissivity ϵ of the cavity in Fig.3 is estimated to be 0.998 8 and 0.998 5 for wall emissivities of 0.9 and 0.85, respectively.[3] Here it is also assumed that the cavity is a diffuse cylinder with a flat end and the depth-to-radius ratio l is 9.

When the cavity has an aperture and two thirds of the opening area is covered with the aperture at the same temperature as the wall of the cavity, the emissivity of the cavity with the aperture ϵ_S is approximately expressed as follows,[4]

$$\epsilon_S = \epsilon_0 - \frac{2}{3}\frac{1-\epsilon_0}{l^2} - \frac{2}{3}(1-\epsilon_0)^2 I_2' , \quad (3)$$

where I_2'=0.043 for l=9. For ϵ_0=0.9 and 0.85, ϵ_S=0.999 5 and 0.999 3, respectively.

In this blackbody furnace, an aperture of 6 mm diameter is placed at the opening, and about two thirds of the opening area is covered with the aperture. However, the aperture is a little off from the cylinder, so the effect is somewhat less than that calculated from Eq.(3). Considering these effects, the emissivity of the cavity of this fixed point blackbody may be estimated to be 0.999 ± 0.000 5.

Long term stability and reproducibility

Fixed point furnaces of Ag, Sb and Zn have been so far used for calibrating resistance thermometers and thermocouples. The fixed point metals in these blackbody furnaces are completely covered with graphite crucibles, so it is expected that the degradation of the purity while using them may be small, and that long term instability may not be observed.

The same is the case with the Cu point. Copper might be oxidized which would shift the freezing point, but it is expected that at high temperature the oxidized copper may be reduced by graphite (or carbon mono-oxide) and that Cu may be free from accumulating oxides.

On the other hand, oxidation may be a problem in case of the Al point. At high temperature the oxidation would not proceed because oxygen is consumed by graphite, while it is possible that oxides produced at lower temperature may be accumulated in repeated melting and freezing.

Investigation on how the accumulation of oxides would influence the freezing temperature and on how oxides would be accumulated will be further work.

Furnaces of different types were compared. The prototype with the capsule (a) in Fig. 3 which had been used for two years and the improved new (b) were measured alternately at the Al point. Their difference was negligible in comparison with their standard deviation (0.04 K). The same was the case with the improved capsules of types (b) and (c) at the Ag point. At both points, the plateau duration of the new models became shorter, but it had no influence on the accuracy and other specifications.

Accuracy in calibrating radiation thermometers

There is a report that the temperature depression of the freezing point of Al is estimated to be 0.06 K from the phase diagram assuming that the impurities are 100 ppm of Cu, Fe and Si in the typical abundance proportions of 1:1:3.[5] However, the depression of the freezing point varies according to the kind and content of impurities, and is not yet comprehended precisely.

Here, assuming that the content of impurities in the fixed point metal in the crucible is less than 100 ppm, and that the depression of the freezing point by impurities is 0.05 ± 0.05 K, the error ΔT of the effective temperature (a representative temperature when temperature distribution exists in the cavity)[6] of this fixed point blackbody furnace compared to the fixed point is estimated. The result is shown in Table IV. The value ΔT in this table is the temperature depression mentioned above plus ΔT_f in Table III, and the uncertainty expresses the square root of the square sum. From this calculated result, the error of realizing the fixed point is estimated to be -0.1 ± 0.1 K.

For calibrating radiation thermometers, the influence of the error of the emissivity of the cavity is added. This error is shown in Table IV as ΔT_ϵ for the

Table II
Temperature gradient across the metal at the cavity base.

	Cu	Ag	Al	Sb	Zn
Heat flux density at the cavity base: ϕ (W/m²)	949 +164	650 +113	212 +37	187 +32	64 +11
Thermal conductivity: κ (W/cm K)	3.30	3.55	2.11	0.167	1.00
Temperature gradient of frozen metal at the base: $d\theta/dx$ (K/cm)	0.029 +0.005	0.018 +0.003	0.010 +0.002	0.112 +0.019	0.006 +0.001

Table III
Errors estimated from plateau gradient.

	Cu	Ag	Al	Sb
Plateau range: $\theta_1 = d\theta/dt \cdot t$ (K)	0.030	0.031	0.023	0.028
Standard deviation of θ_1: σ_1 (K)	0.019	0.019	0.004	0.015
Temperature drop at cavity base: θ_2 (K)	0.039 +0.007	0.027 +0.005	0.009 +0.002	0.008 +0.001
Errors estimated from plateau gradient: ΔT_f (K)	-0.054 +0.034	-0.043 +0.035	-0.021 +0.016	-0.022 +0.029

Table IV
Estimated errors in freezing points.

	Cu	Ag	Al	Sb
Errors at realizing the freezing point: ΔT (K)	−0.104 +0.060	−0.093 +0.061	−0.071 +0.052	−0.077 +0.058
Errors due to effective emissivity: ΔT_ε (K)	−0.113 +0.057	−0.095 +0.048	−0.055 +0.027	−0.051 +0.026
Twice standard deviation: 2σ (K)	+0.162 −	+0.154 −	+0.090 −	+0.110 −
Total errors of three times measurement: E (K)	−0.22 +0.13	−0.19 +0.12	−0.13 +0.08	−0.13 +0.09

effective wavelength of 900 nm. In this table also shown is twice the standard deviation of the data when measuring one set of blackbody furnaces is repeated a total of seven times in two days using a 900 nm silicon narrow-band radiation thermometer. And the error E is estimated for the three-time average in considering the standard deviation.

In the temperature range from 600°C to 1100°C, the indication change of the silicon narrow-band radiation thermometer is 0.05 to 0.1 K when ambient temperature varies 1 K. In this error E is included the influence of ambient temperature.

From these results of calculation, the accuracy of calibrating the silicon narrow-band radiation thermometer is estimated to be 0.3 K.

CONCLUSION

A set of practical-type fixed point blackbody furnaces for calibrating a 900 nm silicon narrow-band radiation thermometer was constructed. The accuracy in realizing the fixed points was estimated to be 0.3 K. When calibrated by these fixed point furnaces at three fixed points of temperature, the radiation thermometer is expected to have the accuracy of 0.5 K in the interpolation range as reported elsewhere[1].

The authors wish to express their acknowledgement to Dr. Akira Ono in this laboratory for valuable discussions and to persons of Hayashi Denko in charge of co-operating in a trial manufacture of practical-type fixed point blackbody furnaces.

REFERENCES

1. "Establishing a practical temperature standard by using a narrow-band radiation thermometer with a silicon detector", F. Sakuma and S. Hattori, in this volume.
2. A. Ono, J. Opt. Soc. America, 70-5, 547 (1980)
3. Y. Ohwada, J. Opt. Soc. America, 71-1, 106 (1981)
4. T. J. Quinn, Brit. J. Appl. Phys., 18, 1105 (1967)
5. J. V. McAllan and M. M. Ammar, in Temperature, Its Measurement and Control in Science and Industry, Vol.4, Part 1, 275 (Instrum. Soc. America, Pittsburgh, 1972)
6. "The Effective Temperature to Express Radiant Characteristics of Nonisothermal Cavities", S. Hattori and A. Ono, in this volume.

Experimental and theoretical study on the quality of reference blackbodies formed by lateral holes on a metallic tube

A. Ono,[a] R. M. Trusty,[b] and D. P. DeWitt[c]

Thermophysical Properties Research Laboratory, School of Mechanical Engineering, Purdue University, West West Lafayette, Indiana 47907

The apparent emissivities of lateral holes formed on a tantalum tube are evaluated experimentally by the integral blackbody method and theoretically by a Monte Carlo analysis. Viewing the lateral holes 10° to 20° horizontally off normal is recommended for highest quality. For such angles of view, the apparent emissivities are experimentally determined by extrapolating the radiances of different-sized lateral holes to a view factor of zero, the condition at which the hole would be a perfect blackbody. The apparent emissivities are obtained with uncertainties less than 0.01 emissivity units for surface emissivities of 0.15 to 0.5. A Monte Carlo method is applied to the apparent emissivity calculation of the lateral holes. The radiative properties of partially specular metallic surfaces are characterized by the uniform specular-diffuse model and multiple reflections (up to 20) of radiation within the tube are considered. The calculation results agree with the experimental observations within their uncertainties. The usefulness of the present Monte Carlo method is further demonstrated by comparison with the DeVos method for cavity quality evaluation. The effect of temperature distribution along the tube on the cavity quality is also investigated by the Monte Carlo method using a new concept, effective temperature.

INTRODUCTION

The integral blackbody method has been used to measure emissivities of metals at high temperatures from the visible to the infrared. The sample, in the form of a long-thin walled tube, is heated by the passage of electrical current. A lateral hole near the mid-length serves as the reference blackbody. Because of the close proximity between the sample target and the lateral hole, an isothermal condition can be achieved. Hence, the ratio of the radiance from the sample target to that from the lateral hole is a measure of the emissivity. The uncertainty in the quality of the blackbody can be a significant factor in the uncertainty of the measurement.

The integral blackbody method has been carefully studied beginning with the work of DeVos[1] and later Larrabee[2] for the classical measurements on tungsten. In 1961, Riethof[3] used this method for refractory metals measurements at temperatures up to 2800 K and over the spectral region 0.3 to 2.5 μm. Recently, one of the authors (D.P.D.)[4,5] has applied this method to emissivity measurements of nonconducting materials where the sample is placed in a metallic tube just behind another lateral hole. Special attention was given to the sample-blackbody isothermal requirement and the spectral region was extended to 15 μm. Preliminary measurements on semitransparent materials have been also made by this method.[6]

In the infrared, clean surfaces of metals have high reflectivity and, consequently, emissivity as low as 0.1 or 0.05 depending on the surface roughness and the material. The degree of specularity depends on the wavelength and the surface roughness; at longer wavelengths, reflection from a rough surface becomes more specular. A black and diffuse surface is no longer a good approximation to metals in the infrared. One of the authors (A.O.)[7,8] has applied a Monte Carlo method to radiant characteristics analysis of cavities. The method is useful for cavities which are composed of partially specular and/or low emissivity surfaces.

The present work provides experimental and theoretical evaluations of the quality (apparent emissivity) of lateral holes serving as reference blackbodies in the integral blackbody method. Different-sized lateral holes are prepared on a uniform temperature region of a metallic tube. A technique of extrapolating the hole radiances to a view factor of zero is presented to determine experimentally the apparent emissivity of lateral holes.

A theoretical approach to predicting lateral hole quality is made by the Monte Carlo method. Multiple reflections of radiation (up to 20 times) within a metallic tube are taken into account and a surface intermediate between diffuse and specular is characterized by the uniform specular-diffuse model.

EXPERIMENTAL EVALUATION

Knowledge of the apparent emissivity or quality of the blackbody reference hole is important for high accuracy. Assuming that isothermal conditions between the sample and blackbody targets are adequately approximated, we must evaluate systematic errors caused by

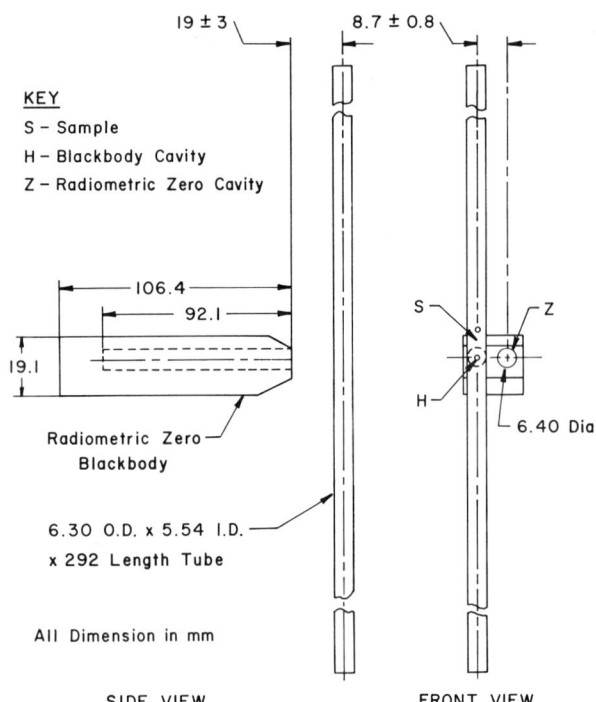

Fig.1 Configuration of sample target S, blackbody lateral hole H, and cold cavity for radiometric zero Z. The dimensions are in mm.

uncertainties in quality of the reference hole.

An experimental configuration of a tantalum tube with a lateral hole for the integral blackbody method is illustrated in Fig.1. The tantalum tube was heated by direct passage of electrical current in a high vacuum about (about 3×10^{-4} Pa). The temperature distribution along the tube was measured by an optical pyrometer the wavelength of which was 0.65 μm. The temperatue uniformity was good over the mid-portion of the tube, as shown later, except for the cross-section through the lateral hole where the local concentration of the electrical current due to the presence of the hole generates additional Joulian heat. Since the temperature increment was less than 1 K at 1273 K and about 5 K at 1873 K for a 1.3 mm diameter hole and the hotter area was limited near the hole, the effect of the temperature irregularity in analyzing the hole emissivities was neglected.

Consider the blackbody reference hole consisting of a lateral hole on a tube (H). Using the argument that a high quality hole will have low apparent reflectivity, it follows that a ray of radiant flux incident upon the hole in the normal direction represents the worst viewing condition. The apparent reflectivity will increase with higher reflectivity and will decrease with a higher degree of diffuseness of the tube inner surface. Obviously, the larger the lateral hole, the higher the apparent reflectivity. The extent to which these factors influence the quality of a lateral hole on tantalum tubes are experimentally investigated and an experimental method is presented to evaluate the apparent emissivities of lateral holes.

The Effect of Viewing Direction

The lateral hole-tube configuration used to determine the effect of viewing direction on the hole quality is shown in Fig. 2. Three holes of 1.7 mm dia. are spaced at 10 mm intervals near the mid-length of a long tantalum tube (6.30 mm OD x 5.54 mm ID and 292 mm long). The tube is mounted within the Multiproperty Apparatus[4,5] such that the emissometer optic axis (see Fig. 3) and the normal to the lower hole (L) are col-

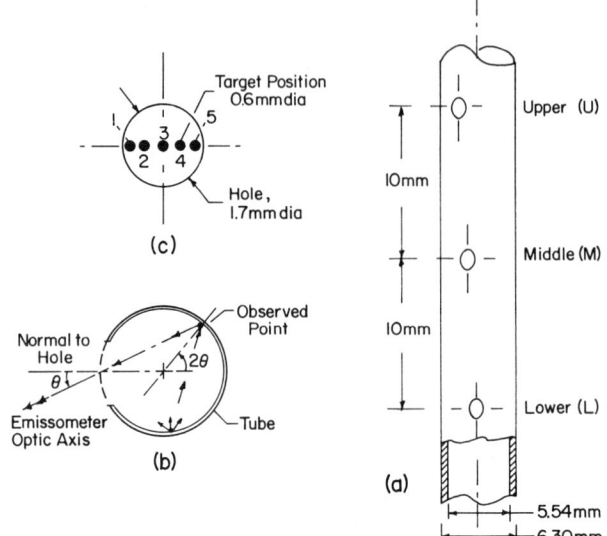

Fig.2 Lateral hole-tube configuration for viewing angle effect. (a): Front view of a tube with three lateral holes in 1.7 mm diameter. (b): Cross sectional view of the tube at a level of either Upper or Middle Hole. The angle of viewing θ is shown for target position 3 in the following figure, (c). (c): Front view of a lateral hole for identifying target positions 1 to 5 on the hole. The target area is drawn smaller than the actual size.

inear. The middle (M) and upper (U) lateral holes are drilled normal to the tube surface but at 10° and 20° increments from the lower (L) lateral hole. That is, the emissometer optic axis and the normal to each hole have included angles of 10° and 20°, respectively. The angle of viewing, θ, is illustrated in Fig. 2(b) for target position 3—see Fig. 2(c)—which corresponds to the condition of the target coinciding with the hole center. It should be noted that the angle of viewing ranges ±6° around θ for near normal viewing because of the finite area of target (0.6 mm in diameter). The angle of viewing, θ, can be conveniently changed in a controlled manner by changing the target positions (1-5). This change is accomplished by slight adjustment of mirror M1 (Fig. 3), and visual sighting of the tube hole at the plane of the aperture. Through simple geometric relations, the five target positions for each of the three lateral holes can be related to the angle of viewing.

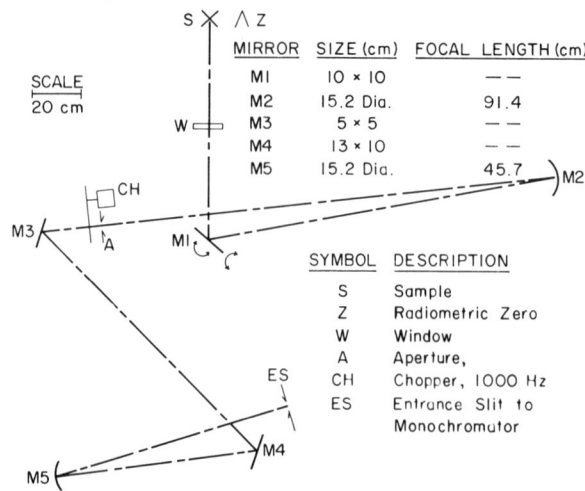

Fig.3 Schematic of the emissometer optical system.

The results of experiments conducted on the tantalum tube at 1273 K are presented in Fig. 4. The radiance distribution, in arbitrary units for each curve, as a function of angle of viewing, θ, is shown for selected wavelengths over the range 0.7 to 7 μm. Note also that the spectral emissivity for the tantalum surface determined from experiments on the tube as described later is identified with each spectral radiance distribution curve. The relationship between the target positions for each of the lateral hole locations (L,M,U) is also shown near the abscissa.

The effect of emissivity on the radiance distribution curves of Fig. 4 is immediately evident. The upper three curves for emissivities greater than 0.16 have similar behavior, namely, a slight depression about the normal direction ($\theta \simeq 0°$). For $\theta > 10°$, the distribution is flat indicating that, for this region, the quality of the lateral hole could be independent of the direction of viewing. For the two lower curves representing the longer wavelengths and the lower emissivity surfaces, the depression about the normal direction is quite substantial. The quality of the hole is maximized for angles of viewing beyond $10°$. The data points for the lower hole (position 5) show scatter and the curve minima are far from $0°$, probably as a consequence of a slight alignment shift at the initial tube mounting. The downward trend with increasing angle of viewing for the upper hole beyond $20°$ should be noted.

It can be concluded that the spectral radiance for different angles of viewing is very much dependent upon the nature of the tube inner surface. For a low emissivity surface which is likely to have a high degree of specularity, it is necessary to view a lateral hole at least $10°$ from the normal. This may be achieved by controlling the direction of viewing relative to the lateral hole normal and/or the position of the target relative to the hole center.

Determination of the Apparent Emissivity

The lateral hole-tube configuration used to determine experimentally the apparent emissivity of lateral holes is shown in Fig. 5. Three holes, spaced at 10 mm intervals near the mid-length of a long tantalum tube

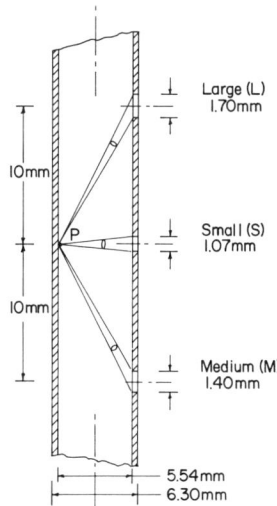

Fig.5 Lateral hole-tube configuration for determining the apparent emissivity of lateral hole. The tube has three different-sized holes on a vertical line.

(6.30 mm OD x 5.54 mm ID x 292 mm long), are of diameters 1.07 mm (S), 1.40 mm (M) and 1.70 mm (L). The tube is mounted within the Multiproperty Apparatus such that the angle of viewing is $15°$ thereby assuring that the quality of the hole is maximized from a directional consideration. A useful concept for estimating the influence of the lateral hole on the quality is the geometric view factor of the lateral hole with respect to the point directly behind it. Referring to Fig. 5, the view factor for point P is determined from the solid angles subtended with respect to P for the small, large, and medium holes. From geometric considerations, assuming negligible tube wall thickness, the view factors for points behind the small, medium and large holes are 0.0133, 0.0175 and 0.025, respectively. It should be recognized that these view factors represent the deviation from a quality of unity if the inner surface were black and diffuse. A real surface could, however, have deviations greater or less than this amount. It is interesting to note, however, that the presence of the holes neighboring the small hole does have a significant effect (33%) on the view factor with respect to the point behind the small hole.

Radiance measurements were made for each lateral hole and the apparent emissivities were experimentally determined by the following analysis. The radiance measurements (proportional to the emissometer signal output) are normalized by the radiance observed for the small hole and plotted as a function of the view factor for the corresponding hole. A straight line is fitted to the three points and extrapolated to a view factor of zero where the imaginary hole would be a perfect blackbody. The radiance ratio (ordinate) intercept is then used to renormalize each of the radiance ratios of the three holes to obtain the apparent emissivity of that hole. Fig. 6 shows the linear behavior of the apparent emissivity with the view factor for three selected wavelengths. The results for experiments conducted at 1445 K and 1875 K are presented in Table I for selected wavelengths in the range 0.7 to 5 μm. Note that the spectral emissivity for the tantalum surface also listed in the table was determined by measurements of radiance ratios between the tube outer surface and the small hole and corrected by the apparent emissivity of the hole also listed in the table.

The data of Table I are represented in Fig. 7 where apparent emissivity for the three different-sized lateral holes is plotted as a function of the emissivity of the tube outer surface. It is not unreasonable to assume that the inner surface emissivity is similar to that of the outer surface since the outer surface is in the as-received (not polished) condition. Although scatter of the individual points is notice-

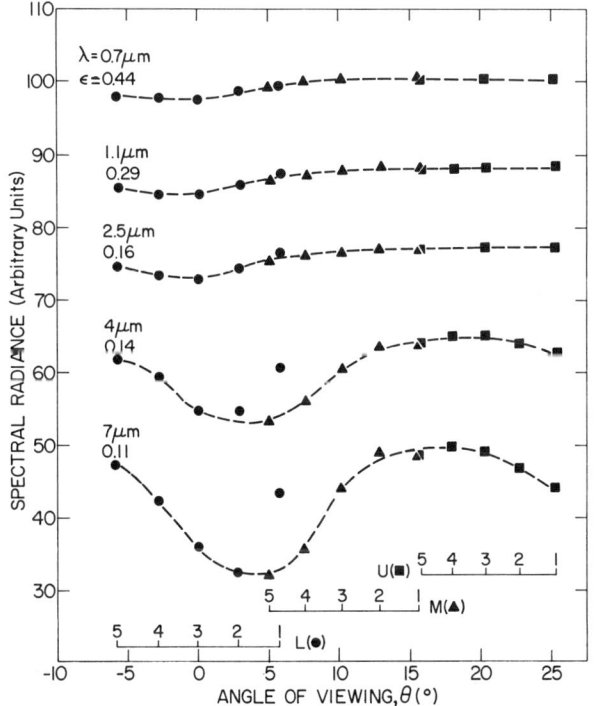

Fig.4 Viewing angle effect on the spectral radiance of lateral hole

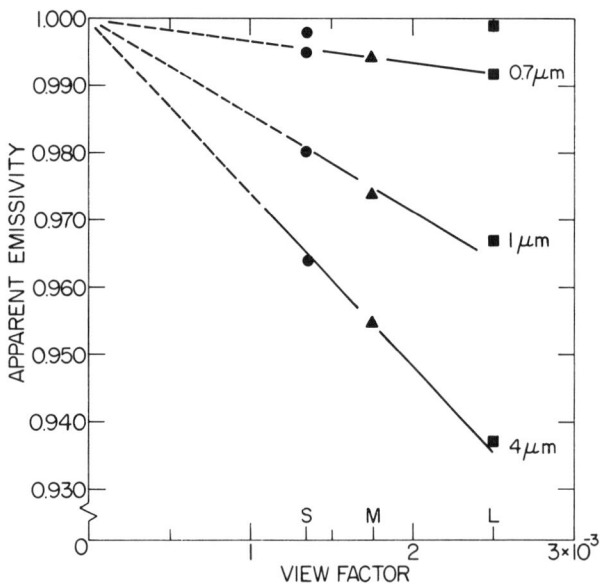

Fig.6 Extrapolation of the spectral radiances to a view factor of zero and the linear behavior.

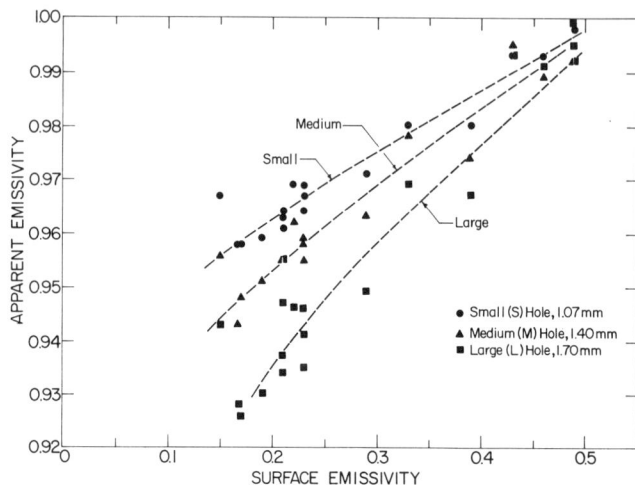

Fig.7 Relation of the apparent emissivity of lateral hole to the tube surface emissivity (experiment).

Table I Quality of the lateral holes (Fig. 5) at selected wavelengths for 1445 and 1875 K.

T(K)	λ(μm)	Apparent Emissivity Hole Designation			ε_λ
		S	M	L	
1445	0.7	0.995	0.992	0.992	0.49
	0.7	0.998	0.995	0.999	
	0.8	0.993	0.989	0.991	0.46
	1	0.980	0.974	0.967	0.39
	2	0.971	0.963	0.949	0.29
	3	0.964	0.955	0.935	0.22
	3	0.967	0.958	0.941	
	3	0.969	0.959	0.946	
	4	0.964	0.955	0.937	0.20
	4	0.963	0.955	0.934	
	4	0.961	0.947	0.929	
1875	0.7	0.993	0.995	0.993	0.43
	1	0.980	0.978	0.969	0.32
	2	0.970	0.963	0.948	0.21
	3	0.959	0.951	0.930	0.18
	4	0.958	0.948	0.926	0.16

able, the trends indicated by the dashed lines are evident and the deviations are less than 0.01 in emissivity units. The effect of hole size is more significant at lower emissivity values; at a surface emissivity of 0.2, the apparent emissivity of the large hole is nearly 0.93 and that of the small hole is 0.96. The deviations from unity are significant. For surface emissivities greater than 0.5, however, the hole size had almost no effect and the quality of the lateral holes are greater than 0.99. It is concluded that the quality of lateral holes cannot be regarded as perfect with low surface emissivities and adequate corrections should be made to the reference blackbody quality in emissivity measurements.

THEORETICAL EVALUATION

The Method of Calculation

The Monte Carlo Calculation

The factors influencing quality (apparent emissivity) of lateral holes on a metallic tube include hole diameter, tube diameter, tube length, direction of viewing, the radiative properties of the tube inner surface, and temperature distribution along the tube. The first four of these factors are geometric and can be readily specified. The fifth factor is difficult to specify since, in general, knowlege of the specular vs. diffuse nature of surfaces is lacking. Metals have high reflectivity (consequently, low emissivity) in the infrared when the surfaces are clean. Hence, reflection of radiation on metal surfaces is partially specular rather than totally diffuse. The fraction of the specular component depends on the surface roughness as well as on the wavelength of radiation. The uniform specular-diffuse model can be a good approximation to such surfaces in which the reflected radiation consists of two components--a specular and a diffuse one-- and the directional hemispherical reflectivity has the same value for all angles of incidence. The relation between reflectivities is

$$r = r_s + r_d , \qquad (1)$$

where r is the (directional-hemispherical) surface reflectivity, and r_s and r_d designate the magnitudes of the specular and diffuse components, respectively.

In order to estimate theoretically apparent emissivities of lateral holes on a metallic tube, multiple reflections of radiation within the tube must be considered taking into account partially specular reflection. Assuming the inner surface of the tube is characterized by the uniform specular-diffuse model, a Monte Carlo method has been applied to calculate the apparent reflectivity of lateral holes. When Kirchhoff's law is applied to an isothermal, opaque cavity, it can be shown that the cavity emissivity* at a location on a cavity wall in a direction is equivalent to the cavity reflectivity* where radiation is incident at the location from the opposite direction. This is valid for surfaces having any type of radiative properties, not just diffuse.[7]

Radiation entering a lateral hole was considered to consist of many ray bundles. When ray bundles collide with the tube wall, some of them are absorbed and others are reflected. Ray bundles which were reflected leave the surface to collide again somewhere with the tube wall. Therefore the history of each ray bundle was determined in such a way that the selections of either absorption or reflection processes on the surface--of either specular or diffuse reflection and of the direction of reflection, if diffusely reflected--were made by the aid of random numbers.

---

*The terms cavity emissivity and cavity reflectivity are used to stress that they are the apparent emissivity and the apparent reflectivity, respectively, for an isothermal cavity.

Thus ray bundles were pursued until they were absorbed or escaped out through the apertures of the tube.

In order to compute efficiently the cavity reflectivity for arbitrary surface reflectivities, the directional hemispherical reflectivity of the surface was normalized to 1 (i.e. no absorption) keeping the angular distribution of reflected radiation fixed (i.e. no change of the fraction of specular reflection component). Letting $N_i(k)$ be the number of ray bundles entering a lateral hole that escape through the i-th aperture (holes or tube ends) after undergoing reflections k times within the tube, the fraction is given by

$$F_i(k) = N_i(k)/N, \qquad (2)$$

where N is the total number of incident ray bundles. Then the contribution of the i-th aperture to the cavity reflectivity is calculated for an arbitrary surface reflectivity, r, as

$$\rho_i(r) = \sum_{k=1}^{\infty} F_i(k) r^k. \qquad (3)$$

The (total) cavity reflectivity, $\rho_c(r)$, and the cavity emissivity, $\epsilon_c(r)$, are given by

$$\rho_c(r) = \sum_{i=1}^{I} \rho_i(r), \qquad (4)$$

where I is the number of apertures associated with the tube and

$$\epsilon_c(r) = 1 - \rho_c(r). \qquad (5)$$

Results obtained by the Monte Carlo method statistically fluctuate since the number of escaping ray bundles is subject to the binomial distribution. The standard deviation of $N_i(k)$ is given by the square root of its value; that is

$$\Delta N_i(k) \simeq N_i(k)^{1/2}, \qquad (6)$$

for $N_i(k) \ll N$. Thus the statistical error of the cavity reflectivity is estimated as

$$\Delta_s \rho_c(r) \simeq \left[\rho_c(r^2)/N\right]^{1/2}. \qquad (7)$$

It should be noted that the statistical error for a surface reflectivity of r is associated with the cavity reflectivity for a surface reflectivity of r^2.

In the Monte Carlo computation, pursuing ray bundles was stopped at a finite number of reflections. When stopped at the K-th reflection, the effect of reflections beyond it may be estimated assuming that ray bundles continue to escape at the fraction of F(K). Thus the truncation error (Δ_t) of the cavity reflectivity is given by

$$\Delta_t \rho_c(r) \simeq F(K) r^{K+1}/(1-r). \qquad (8)$$

The hole-tube configuration used for the calculation is the same as that for the experiment as shown in Fig.5. Although the total length of tube was 29 cm in the experiment, a uniform temperature distribution was confined within a mid-portion of the tube 14 cm long, outside of which the temperature decreased rapidly. Hence, in the calculation, the tube was taken to be 14 cm long. Since the target area of the emissometer on the tube was 0.6 mm in diameter, ray bundles were incident uniformly on the corresponding circular area.

It is useful to investigate in advance, what influence successive specular reflections have on the cavity reflectivity of the lateral holes. Ray bundles undergoing successive specular reflections escape back through the lateral hole if they are incident with special angles given by

$$\Theta = 90°\left[1-2m/(n+1)\right], \qquad (9)$$

where n and m are positive integers which correspond to the number of reflections and to the number of times the ray bundle experiences a complete turn about the inner cylindrical surface, respectively. Recall that a parallel beam of 0.6 mm diameter includes angles of incidence of $15 \pm 6°$ when entering a lateral hole on a tube of 5.54 mm inner diameter. This suggests that there are some ray bundles which escape after undergoing successive specular reflections 4, 6, 8 and 11 times. These reflections correspond to incident angles of 18.0, 12.9, 10.0, and 15.0°, respectively.

The Effective Temperature

In order to analyze radiant characteristic of nonisothermal cavities, a new quantity of effective temperature has been introduced[9]. The spectral radiance of a nonisothermal cavity at a location Z_0 on the cavity wall in a direction D is fully expressed by the cavity emissivity, ϵ_c, and the effective temperature, T_e, as

$$L_0^D(\lambda) = \epsilon_c L_b(\lambda, T_e), \qquad (10)$$

where L_b is the Planck relation for the blackbody spectral radiance. It should be noted that the temperature distribution effect on the spectral radiance is concentrated in the effective temperature instead of the apparent emissivity; recall that the cavity emissivity is defined as the apparent emissivity when the cavity is assumed to be isothermal. Let the temperature at Z be T(Z) and the temperature at the target point be T_0 (cavity temperature). Then the difference between the effective temperature and the cavity temperature, $\Delta T_e = T_e - T_0$, is a measure of the temperature differences effect. When a cavity approaches the isothermal condition, ΔT_e diminishes.

Considering the multiple reflections of radiation within a cavity and the reciprocal law of the partial reflectivity, it can be shown that ΔT_e is expressed by a summation of nested integrals to the first order of temperature differences;

$$\Delta T_e = \frac{1}{\epsilon_c} \sum_{k=1}^{\infty} \Delta t_k, \qquad (11)$$

and

$$\Delta t_k = \int_W d\Omega_0^1 r_0^{D,1} \int_W d\Omega_1^2 r_1^{0,2} \int_W \cdots \int_W d\Omega_{k-1}^k r_{k-1}^{k-2,k} \epsilon_k^{k-1} \Delta T(Z_k), \qquad (12)$$

where $d\Omega_0^1$ is the solid angle element at a location Z_0

in the direction Z_1, $\Delta T(Z)$ is the temperature variation given by $T(Z) - T_0$, W indicates that the integration is carried out on the cavity wall, e_k^{k-1} is the directional emissivities of the wall surface at Z_k in the direction Z_{k-1}, and $r_0^{D,1}$ is the partial reflectivity of the wall surface which is defined as the fraction of radiation reflected at Z_0 in the direction Z_1 per unit solid angle when radiation is incident from the direction D.

When the radiative properties of the wall surface are characterized by the uniform specular-diffusive model, Δt_k is reduced to

$$\Delta t_k = \Delta T_k (1-r) r^k, \qquad (13)$$

where ΔT_k is defined by

$$\Delta T_k = \int_W d\Omega_1^0 r_0^{D,1} \int_W d\Omega_1^2 r_1^{0,2} \int_W \cdots \int_W d\Omega_{k-1}^k r_{k-1}^{k-2,k} \Delta T(Z_k), \qquad (14)$$

where the surface reflectivity r is normalized to 1. Equation (14) suggests that a Monte Carlo method can be applied to the calculation of ΔT_k. The nested integral shows a reflection process that radiation entering the cavity at Z_0 from the direction D is multiply reflected in the cavity according to the partial reflectivity and at the k-th reflection it reaches a location Z_k where the temperature variation is $\Delta T(Z_k)$. Therefore ΔT_k is computed by the aid of random numbers as

$$\Delta T_k = \frac{1}{N} \sum_{n=1}^{N} \Delta T(Z(n,k)), \qquad (15)$$

where $Z(n,k)$ is the Z-coordinate where the n-th ray bundle collides with the inner tube surface at the k-th reflection.

The effective temperature of a lateral hole is calculated through Eqs. (11), (13), and (14) by the Monte Carlo method and the apparent emissivity of a lateral hole is corrected for a temperature distribution along the tube as

$$\epsilon \simeq \epsilon_c \left[1 + \frac{\partial}{\partial T_0} \ln L_b(\lambda, T_0) \Delta T_e \right]. \qquad (16)$$

Calculation Results

Computational Results

Table II(a) shows computed data when 10000 ray bundles were incident on the medium hole horizontally $15°$ off normal and vertically normal; the fraction of the specular reflection component is 60% ($r_s/r = 0.6$). The three holes (Large, Small, and Medium) and the tube ends (Upper and Lower) are abbreviated as LH, SH, MH, UE, and LE, respectively. The numbers of ray bundles, N_1 to N_5, that escape through the holes and ends, and the total number, N_t, are identified for the corresponding number of reflections, k, which the rays undergo within the tube until escaping. In spite of the statistical fluctuations it is readily seen that the numbers for the Medium Hole, N_3, are the largest for k < 10 since the ray bundles remain still near the incident hole. Note that N_3 decreases with increasing k having several peaks at k = 4, 6, 8, and 11. As was discussed before, these peaks correspond to the escape of ray bundles that underwent successive specular reflections. The numbers of ray bundles other than N_3

Table II

(a) Numbers of ray bundles escaping through the lateral holes and ends for total incident ray number of 10000.
 $\theta = 15°$, $\psi = 0°$, and $(r_s/r) = 0.6$.

k	N_1	N_2	N_3	N_4	N_5	N_t	ΔT_k
	LH	SH	MH	UE	LE	TOTAL	(K)
1	0	1	73	0	0	74	0.04
2	1	1	62	1	2	67	0.15
3	1	7	54	2	2	66	0.30
4	6	4	383	4	10	407	0.47
5	5	8	32	4	12	61	0.47
6	7	8	107	2	10	134	0.87
7	5	8	35	6	18	72	1.07
8	14	8	43	8	32	105	1.22
9	16	7	22	6	21	72	1.39
10	8	11	24	6	28	77	1.39
11	9	7	31	5	24	76	1.72
12	20	6	27	14	19	86	1.89
13	7	2	14	12	30	65	2.01
14	9	8	14	13	24	68	2.15
15	9	7	18	15	34	83	2.26
16	16	6	13	14	32	81	2.38
17	17	4	8	9	28	66	2.52
18	7	6	13	18	34	78	2.63
19	15	1	9	13	36	74	2.73
20	9	6	10	8	50	83	2.79

(b) Cavity emissivities, reflectivities and contributions of the holes and ends.

r	ρ_1 x100	ρ_2 x100	ρ_3 x100	ρ_4 x100	ρ_5 x100	ρ_c x100	ϵ_c x100	ΔT_e (K)
	LH	SH	MH	UE	LE	TOTAL		
0.1	0.00	0.00	0.08	0.00	0.00	0.08	99.92	0.01
0.2	0.00	0.00	0.18	0.00	0.00	0.19	99.81	0.01
0.3	0.00	0.01	0.32	0.00	0.00	0.34	99.66	0.03
0.4	0.00	0.01	0.53	0.00	0.01	0.56	99.44	0.05
0.5	0.01	0.02	0.86	0.01	0.02	0.93	99.07	0.09
0.6	0.03	0.05	1.37	0.02	0.05	1.52	98.48	0.16
0.7	0.07	0.09	2.18	0.05	0.14	2.53	97.47	0.28
0.8	0.18	0.19	3.49	0.14	0.38	4.37	95.63	0.52
0.9	0.54	0.43	5.70	0.43	1.22	8.33	91.67	0.89
1.0	1.81	1.16	9.92	1.60	4.46	18.95	(81.05)	(0.00)

increase with k since the ray bundles will gradually approach the other holes and ends during successive reflections. Note that N_1 and N_2 reach maxima and then decrease; conversely N_4 and N_5 appear to be approaching maxima.

Beyond k = 10 the effect of the tube ends, rather than the Medium Hole, becomes dominant. Although ray bundles are concentrated near the incident hole at early reflections, they are distributed more uniformly during later reflections. Then the effect of the tube ends becomes more important because the areas of the tube ends are much larger than that of the holes. The numbers of ray bundles for the Lower End (N_5) are larger than that for the Upper End (N_4) because the incident hole (MH) is located closer to the Lower End than to the Upper End (see Fig.5). When k is reaching 20, the numbers, N_1, N_2, and N_3, aproach proportionality with their respective hole areas. This suggests that the distribution of ray bundles is reaching a uniform distribution over the mid-portion of the tube.

The cavity reflectivities and emissivities are listed in Table II (b) with the contribution of each hole and end for increments of 0.1 of the surface emissivity. The identified values were calculated from Table II (a) through Eqs. (2) to (5). For low surface reflectivities, the contribution of the Medium Hole

(ρ_3) is dominant; the contributions of other holes and ends share no greater than 10% of the (total) cavity reflectivity for r less than 0.6. For high surface reflectivities, the contributions of other than the Medium Hole become important; they reach a 30% share of the (total) cavity reflectivity for r = 0.9.

The cavity reflectivity increases rapidly with increasing surface reflectivity. When r approaches 1, ρ should reach 1. However, the identified value of ρ is much lower than 1 because the calculation was stopped at a finite number of reflections. Up to k = 20, 8105 ray bundles still remain unidentifed in the tube. For r less than 0.9, however, the truncation errors are not serious. They are estimated through Eq. (8) to be 0.009 (11% of ρ) and 0.0004 (0.9% of ρ) in emissivity units for r = 0.9 and 0.8, respectively. For r less than 0.7, the truncation errors are negligible. The statistical errors are evaluated through Eq. (7) to be 0.002 (2.5% of ρ) and 0.0015 (3% of ρ) in emissivity units for r = 0.9 and 0.8 respectively. For r less than 0.7 the errors are less than 0.001 in emissivity units and negligible.

It will be useful to note that the computer CPU time was about 30 seconds to obtain the results of Table II(a) and (b) using the CDC 6500/6600 system of Purdue University.

Incident Angle Dependence

The cavity emissivity of lateral holes were calculated varying the angle of incidence horizontally and vertically. The relations of the cavity emissivity versus the horizontal angle of incidence are illustrated in Fig. 8 for selected surface reflectivities and fractions of specular reflection component. Considering that the hole size was 1.70 mm in the experiment (see Fig. 2), ray bundles were incident on the Large Hole with horizontal angles of incidence ranging 0 to 30° and vertically normal. The cavity emissivities have deep minima around 0° where a specular reflection results in direct escape of ray bundles through the incident hole. With increasing angle of incidence up to 10°, direct escape diminishes. Beyond 10° the cavity emissivities reach maxima and remain almost constant until the beginning of the decrease near 20°. There are shallow minima around 30° where two successive specular reflections result in escape through the incident hole. The deterioration of cavity emissivity around 0° and 30° is more significant for a higher degree of specularity and a higher surface reflectivity. Those trends have all been found in the experimental results (see Fig. 4).

Comparing the experimentally obtained ratios between the minimum and maximum radiances with the calculation results, the fractions of specular reflection component on the inner surface of the tantalum tube were estimated as ~ 0.1 at 0.7 to 2.5μm, 0.2 to 0.3 at 4μm, and 0.5 to 0.6 at 7μm. The theory and experiment suggest that lateral holes used as a reference blackbody be viewed horizontally 10 to 20° off normal to have the highest cavity emissivities that are almost independent of the angle of viewing.

The relations of the cavity emissivity versus the vertical angle of incidence are shown in Fig. 9 for various surface reflectivities and fractions of specular reflection component. Ray bundles of 5000 total number were incident on the Medium Hole for each vertical angle of incidence ranging 0 to 2.5° with the horizontal angle at 15°. The cavity emissivities have minima around 0°, increase with the vertical angle of incidence, and approach constant values beyond 2°. The minima are deeper for greater fractions of the specular reflection component and higher surface reflectivities. Those minima result from the escape of ray bundles which undergo successive specular reflections of 4, 6, 8, 11 times and so on. Beyond 2° such escape completely disappears because of sufficient inclination. It should be noted that the vertical angle dependence is much weaker than the horizontal angle dependence. Vertically normal incidence is not serious for a surface reflectivity lower than 0.8 and a fraction of specular reflection component less than 0.6. It was experimentally observed that long-wavelength radiance variations less than 1% occurred along a vertical diameter of a lateral hole; this may correspond to this effect.

Extrapolation Tests

In the previous section, the apparent emissivities of different-sized lateral holes were experimentally determined by the extrapolation of radiances to a view factor of zero. Recall that, in identifying the view factor with respect to the point just behind a hole, the presence of the neighboring holes was taken into account. This approach assumes that the tube inner surface was black and diffuse. For metal surfaces such as tantalum, however, this assumption is not realistic.

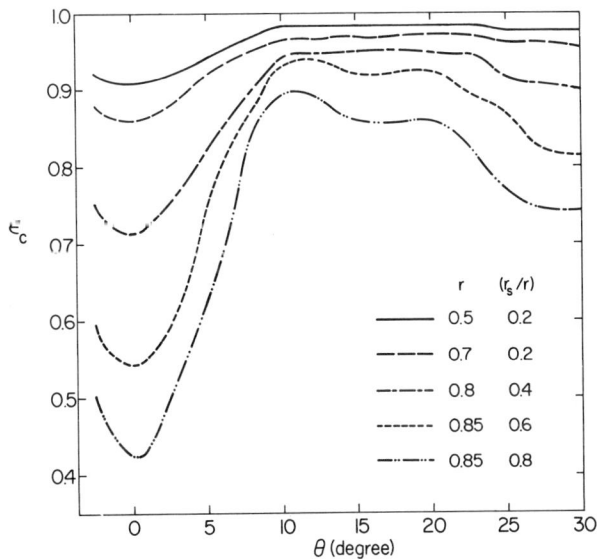

Fig.8 Horizontally incident angle dependence of the apparent emissivity of lateral hole.

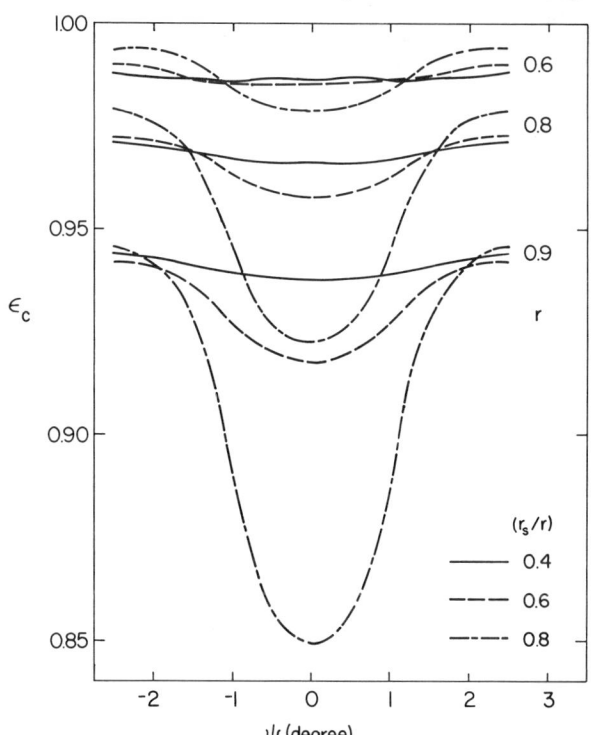

Fig.9 Vertically incident angle dependence of the apparent emissivity of lateral hole at $\theta = 15°$.

Therefore, the validity of the extrapolation technique has to be tested for a low surface emissivity and a high degree of specularity.

Extrapolation tests were made with ray bundles incident on the three different-sized lateral holes horizontally 15° off normal and vertically normal. Results of extrapolation tests are shown in Fig. 10 for fractions of the specular reflection component of 0.4 and 0.6. When the surface reflectivity is lower than 0.7, the extrapolations to a view factor of zero give cavity emissivity values ranging 1.00±0.005. Including a surface reflectivity of 0.8, the extrapolations fall within a range of 0.995±0.01. The same results were also obtained for other fractions of the specular reflection component of 0.0 (totally diffuse) and 0.2. The reason why the extrapolations for higher surface reflectivities are likely to give lower cavity emissivities is that the effect of tube ends depresses the cavity emissivities for different-sized lateral holes by the same amount.

It is interesting to extend the extrapolation test to a very high degree of specularity. When the fractions of the specular reflection component were 0.8 and 1.0 (totally specular), extrapolations to a view factor of zero provided cavity emissivities confined within 0.995±0.005 and 0.99±0.01, respectively, for surface reflectivities less than 0.7. When including a surface reflectivity of 0.8, they spread out to a range of 0.975±0.025. The reason why such good extrapolation accuracies are obtained for very high degrees of specularity is that the ray bundles of parallel incidence are reflected to diverge because of the cylindrical mirror effect of the metal tube and spread out during successive reflections wide enough compared with the lateral hole diameters.

Evaluation of Cavity Emissivities

The cavity emissivity of the Medium Hole is illustrated in Fig. 11 against the fraction of the specular reflection component. Ray bundles were incident with a horizontally incident angle of 15°. The upper and lower lines for each surface reflectivity correspond to vertically incident angles of 2.5 and 0.0°, respectively, which represent the maxima and minima of the curves shown in Fig. 9. Although the upper and lower lines coincide when the surface is nearly diffuse, they deviate from another with increasing fraction of specular reflection component. The upper lines eventually converge to unity at the totally specular reflection

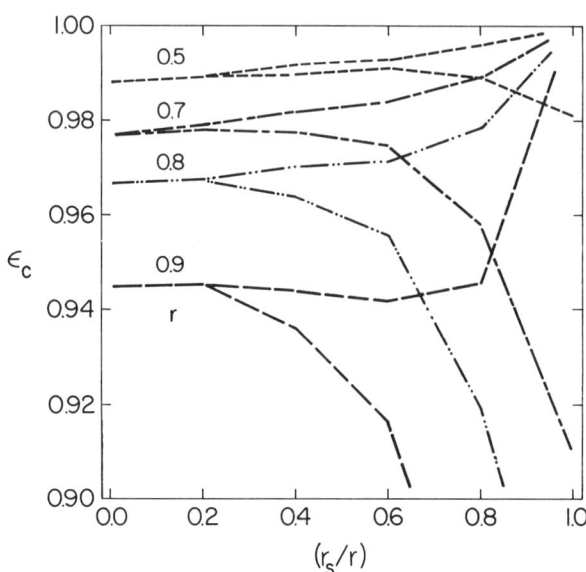

Fig.11 Relation of the apparent emissivity of lateral hole to the degree of specularity.

condition while the lower lines decrease. These two lines indicate the upper and lower limits of the cavity emissivity evaluation.

From an experimental point of view, it is not easy to determine the vertical angle of viewing with a high accuracy, e.g. better than 1°. Moreover, radiation is observed with a diverging angle (conical-shape rays) that depends on the optical system. In the present experiment the total angle of the radiation cone was 6°. Therefore the uncertainty of the cavity emissivity should be estimated by the upper and lower limits as shown in Fig. 11. The uncertainty is larger for a higher surface reflectivity and a higher degree of specularity.

The cavity emissivity of the Medium Hole viewed horizontally 15° off normal is illustrated against the surface emissivity in Fig.12. When the vertical angle of viewing ranges 0. to 2.5° and the degree of specularity ranges 0 to 60%, the cavity emissivity is confined within the shaded zone. For an 80% degree of specularity, the cavity emissivity is shown by the upper and lower dashed curves that correspond to vertically viewing angles of 2.5 and 0.0°, respectively. For totally specular reflection, the cavity emissivity is shown by the dashed-dotted curves. The shaded zone indicates that the cavity emissivity of the Medium Hole viewed horizontally 15° off normal can be evaluated as 0.90±0.02 (for a surface emissivity, e = 0.05), 0.93±0.012 (for e = 0.1) 0.95 ±0.01 (for e = 0.15), 0.965±0.007 (for e = 0.2), and 0.98±0.005 (for e = 0.3) when the surface specularity is estimated as lower than 60%. Beyond e = 0.35 and e = 0.6, the cavity emissivity becomes greater than 0.98 and 0.99, respectively. When a surface has a very high degree of specularity, an uncertainty of the fraction of specular reflection component results in a large error in evaluating the cavity emissivity of lateral holes, especially with low surface emissivities. With a high surface emissivity greater than 0.5, however, it can be seen in Fig.12 that the cavity emissivity of the Medium Hole is not less than 0.98, irrespective of the degree of specularity and the vertical angle of incidence.

The dotted curve in Fig. 12 shows the apparent emissivity of the Medium Hole that was experimentally obtained by the extrapolation technique (see Fig. 7). The deviation of the curve from the shaded zone is less than 0.005 in emissivity units. Taking into account the deviation and scatter in the extrapolation, it can be concluded that the theoretical results are in good agreement with the experimental results.

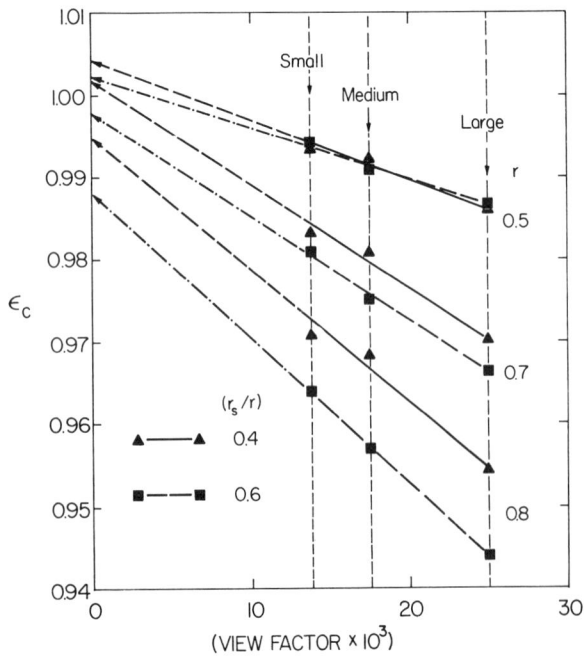

Fig.10 Tests of the extrapolation technique.

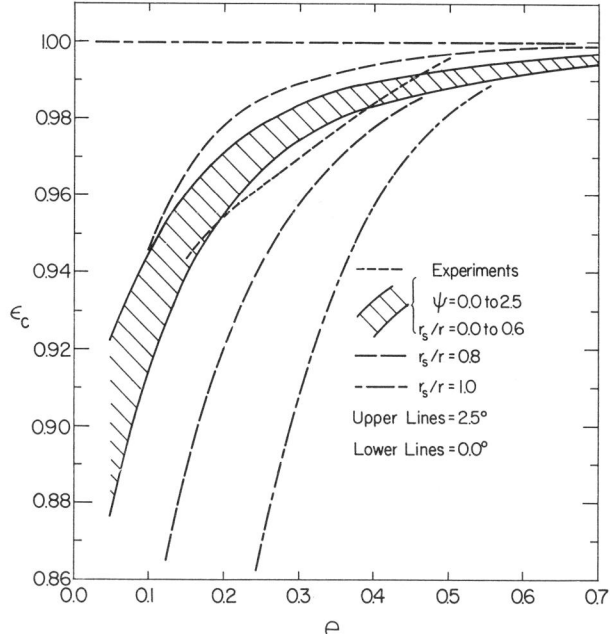

Fig.12 Relation of the apparent emissivity of lateral hole to the tube surface emissivity (calculation).

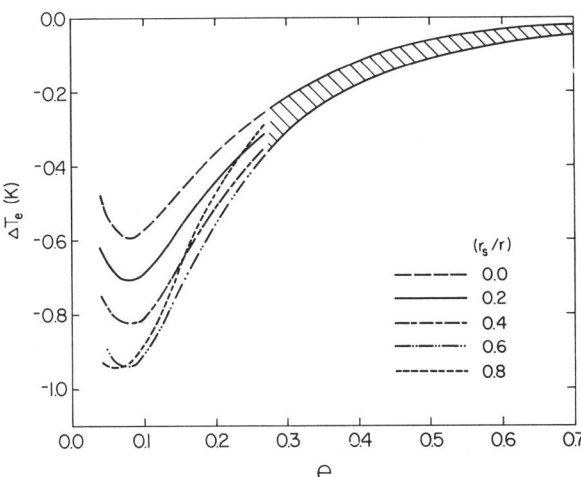

Fig.13 Effective temperature depression of lateral hole caused by a temperature distribution along the tube.

Effect of Temperature Distribution

In the experiments, the tantalum tubes were heated by direct electrical current in a high vacuum. The longitudinal temperature distributions were of quadratic form over the mid-portion of tube. When the temperature was 1100°C at the mid-point of a tube (29 cm long), the following equation was a good approximation to the temperature distribution within a region of 14 cm long:

$$T_z = 1100(^\circ C) - 0.52(^\circ C/cm^2)z^2 . \quad (17)$$

For this temperature distribution, the ΔT_k were computed through Eq. (14) by the Monte Carlo method. Computational results of ΔT_k are shown in Table II(a) when the ray bundles were incident on the Medium Hole horizontally 15° off normal and vertically normal for a 60% degree of specularity condition. With increasing times of reflection, ΔT_k increases because ray bundles are distributed to colder regions during successive reflections. The depressions of effective temperature, ΔT_e, identified in Table II(b) were computed through Eqs. (11) and (13).

Results on the Medium Hole for various fractions of specular reflection component are illustrated in Fig. 13 against the surface emissivity. The shaded zone shows that all results for different fractions of specular reflection component fall within the zone. For total specularity, the depression of effective temperature vanishes. Beyond the minima found near a surface emissivity of 0.1, the depressions of effective temperature diminish with increasing surface emissivity. The effect of temperature distribution is weaker for a higher surface emissivity (consequently a lower surface reflectivity) because the isothermal region close to the incident hole contributes more to the spectral radiance of the hole. In contrast, the nonisothermal (colder) regions far from the incident hole contribute more significantly to the spectral radiance when the surface reflectivity is higher. Although the depression of effective temperature depends on the degree of specularity, it is less than 1K even for very high surface reflectivity. It is interesting to note that the effect of temperature distribution is maximized at an intermediate degree (60%) of specularity. This suggests that such a combination of diffuse and specular reflections has ray bundles distributed furthest from the incident hole with the fewest reflections.

Corrections to the cavity emissivity of the Medium Hole were made through Eq. (16) at the corresponding wavelength using the surface emissivities and the degrees of specularity for the as-received tantalum tube. The maximum correction was at 2.5μm where the depression of apparent emissivity was only 0.002 in emissivity units. It is concluded that the temperature distribution in the present experiments was uniform enough to have good quality of lateral holes.

CONCLUSIONS

It was experimentally shown that the angle of viewing is important to the quality of lateral holes. Viewing 10 to 20° horizontally off normal is recommended for highest apparent emissivity that is independent of the viewing angle. For such angles of view, the apparent emissivities were experimentally determined by extrapolating the radiances of different-sized lateral holes to a view factor of zero. The extrapolation technique is based on the idea that the extrapolated radiance is blackbody radiance. The apparent emissivities were determined with uncertainties less than 0.01 in emissivity units for surface emissivities 0.15 to 0.5.

The direct heating method for metallic tubes can realize a uniform temperature region at the mid-portion of the tube, and the lateral holes formed there are assured to be precisely isothermal. As far as the authors know, this is the first report which experimentally determines the apparent emissivity of a cavity by emission measurements. This experimental determination was accomplished on the basis that well-specified, different-sized cavities could be prepared under a precise isothermal condition by virtue of electrical direct heating in a vacuum.

A Monte Carlo method was applied to the apparent emissivity calculation of lateral holes. The radiative properties of partially-specular metallic surfaces were characterized by the uniform specular-diffuse model and twenty multiple reflections within a tube were taken into account. The calculations gave the apparent emissivity of a lateral hole (medium size) with a truncation error less than 0.01 in emissivity units and a statistical error less than 0.002 for surface emissivities higher than 0.1 and for 10000 incident ray bundles. The calculation was also able to predict the apparent emissivity of the lateral hole with an uncertainty less than 0.01 in emissivity units for surface emissivities higher than 0.15 in spite of the large uncertainty on the degree of specularity ranging 0 to 60%. No significant deviations were evident between the theoretically and experimentally obtained results for the incident angle dependence of apparent emissivity and for the apparent emissivities themselves.

The quality of lateral holes on a metallic tube has been theoretically studied by DeVos[10] for the spectral emissivity measurement of tungten. It is interesting to compare the calculation results between the earlier (DeVos) and the present methods. The earlier method takes into account partially specular reflection (Fig.3 to 5 in Ref.10) and multiple reflections up to twice (Eq. (17) in Ref.10) for a hole-tube configuration similar to Fig.5 that has only one lateral hole. The first-order approximation in the earlier work is equivalent to $N_3(1), N_4(1)$, and $N_5(1)$ in the present work (see Table II (a)), and the second-order approximation corresponds to $N_4(2)$ and $N_5(2)$ but does not include $N_3(2)$. If the earlier method of evaluation is applied to the present hole-tube configuration, it will provide apparent reflectivity values that are much lower than those by the present method, since the former neglects higher-order multiple reflections. For the Medium Hole specified in Table II(a), the earlier method gives the apparent reflectivity 0.005, 0.006, and 0.007 for surface reflectivities of 0.7, 0.8, and 0.9, respectively, while the present method gives 0.024, 0.040, and 0.074 for the respective surface reflectivities. The earlier method underestimates the apparent reflectivity and consequently, overestimates the apparent emissivity. For the same hole (Medium Hole), the earlier method gives the apparent emissivity 0.995, 0.994, and 0.993 which correspond to 0.976, 0.960, and 0.926, respectively, given by the present method for the respective surface reflectivities. If an accuracy better than a few percent is desired in the emissivity measurement, adequate corrections should be made for the quality of lateral holes used as reference blackbodies.

It should also be noted that the successive specular reflections of ray bundles are important for a high degree of specularity. It is clearly seen in Table II that the escape of ray bundles undergoing four successive specular reflections, $N_3(4)$, gives a dominant effect on the apparent emissivity, especially with high surface reflectivities. If a metallic tube is formed not in a circular but in a regular triangular cross section as was used by DeVos, three successive specular reflections result in total escape when ray bundles are incident normal to a side of the regular triangular tube.

The Monte Carlo calculation also provided insight on the vertically incident angle dependence of the apparent emissivity and gave enough evidence to verify the extrapolation technique on which the experimental determination of apparent emissivity was based. The temperature distribution effect on the apparent emissivity of lateral holes was investigated by the Monte Carlo method using a new quantity, the effective temperature. It was found that the effect was very small and the tubes used in the present experiment were sufficiently long.

The theoretical and experimental results illustrated the importance of lateral hole features on the reference blackbody quality. Precise determinations of emissivity of metals using the integral blackbody method can only be assured if proper attention is given to the use of a lateral hole.

ACKNOWLEDGMENT

The authors would like to express sincere thanks to Dr. R. E. Taylor, Head, Thermophysical Properties Research Laboratory, School of Mechanical Engineering, for his help in using the Multiproperty Apparatus.

REFERENCES

a) Visiting Scholar; On sabbatical leave from the National Research Laboratory of Metrology, Ibaraki 305, Japan.

b) Formerly Engineering Assistant; Present address is Lawrence Livermore National Laboratory, University of California, CA 94550.

c) Professor, School of Mechanical Engineering.

1. J. C. DeVos, Physica 20, 690-714 (1954).

2. R.D. Larrabee, J. Opt. Soc. Am. 49, 619-625 (1959).

3. T. R. Riethof and V. J. DeSantis, *Measurement of Thermal Radiation Properties of Solids* (J. C. Richmond, Editor) NASA SP-31, pp. 565-584 (1963).

4. D. P. DeWitt, R. E. Taylor, and T. K. Riddle, *Proceedings of the 7th Symposium on Thermophysical Properties*, Amer. Soc. Mech. Eng., NY, pp. 256-264 (1977).

5. P. E. Johnson, D. P. DeWitt, and R. E. Taylor, AIAA Journal 19, 113-120 (1981).

6. D. P. DeWitt, A. Ono, R. E. Taylor, and H. M. James, "High Temperature Emissivity of Conducting and Nonconducting Materials" *Proceedings of the 8th Symposium on Thermophysical Properties*, Amer. Soc. Mech. Eng., NY, 1981. (To be published).

7. A. Ono, J. Opt. Soc. Am. 70, 547-554 (1980).

8. A. Ono, to be published in this volume.

9. S. Hattori and A. Ono, to be published in this volume.

10. J. C. DeVos, Physica 20, 669-689 (1954).

Precision practical blackbody furnaces by a 3-zone temperature control method

I. Hishikari and T. Ide

Chino Works Limited, 1-26-2 Nishi-shinjuku, Tokyo 160, Japan

In order to meet the progress of industrial radiation thermometers, we have developed compact and easy-to-operate precision practical blackbody furnaces, each having a large aperture (50 mm) and high cavity emissivity. These blackbody furnaces can calibrate radiation thermometers over a range from room temperature to 1500 °C with an accuracy of ±0.25% of readings. They are classified into cylindrical and spherical blackbody furnaces. As a result of producing a calculated temperature distribution by controlling the temperature of respective cavity walls by a three-zone method, the effective cavity emissivity was 0.998 in cylindrical blackbody furnaces and 0.995 in spherical blackbody furnaces. These effective cavity emissivity values were evaluated by comparing the cavity temperature measured with calibrated thermocouples with the brightness temperature measured at wavelengths of 0.65 and 0.90 μm with radiation pyrometers previously calibrated at the freezing points of aluminum, silver, and copper. The warm-up time from room temperature to 1500 °C is 1.5 h and the short-term stability and the long-term stability are 0.1 K/min and 0.5 K/h, respectively, after warm-up. These precision practical blackbody furnaces provide a temperature program function, and the calibrated values of radiation thermometers are automatically printed out every 10 K step over the working temperature range.

INTRODUCTION

Radiation pyrometry is now rapidly spreading in industrial fields in Japan. Its introduction is particularly remarkable in the iron and steel making industry. Fully automated steel making systems require a radiation pyrometer for no-contact temperature measurement as a temperature sensor in continuous heat treatment processes.

CHINO WORKS, LTD. has developed all types of industrial radiation pyrometers and supplies these instruments to the iron and steel making industry, electronic industry, mining industry, automobile, petroleum, chemical, and other industries to meet these demands. These industrial radiation pyrometers comprise optical pyrometers, thermistor-bolometer radiation pyrometers, thermopile radiation pyrometers, PbS radiation pyrometers, silicon radiation pyrometers, two-color pyrometers, scanning pyrometers, etc.

As the utility of radiation pyrometers increases, the demanded accuracy becomes as high as ±0.5 % in online measurements. In order to meet this demanded accuracy, radiation pyrometers are required to be reliable to better than ±0.25 % in their production process when the emissivity compensation error is taken into consideration. This demand is very severe.

Assuming that the repeatability of radiation pyrometers is ±0.15 %, only ±0.2 % is allowed for the calibration error of a temperature scale. The temperature scale calibration error comprises an error caused by the effective emissivity of a blackbody furnace, an error of the standard thermocouple for measuring the cavity temperature, and an operating error during calibration. If the operating error and the standard thermocouple error are each presumed to be ±0.1 %, the temperature error caused by the effective emissivity of the blackbody furnace is allowed to be only ±0.14 %.

An indicating temperature error of a radiation pyrometer caused by emissivity depends upon index "n" of pyrometer ($\doteqdot C_2/\lambda T$). The smallest index "n" in radiation pyrometers to be calibrated is 4 in total radiation pyrometers within normal working temperature ranges. Accordingly, the effective emissivity of the blackbody furnace must be larger than 0.994 because of $(1 - 0.0014)^4$.

The lens diameter of radiation pyrometers for industrial use is less than 40 mm in most cases. Assuming that the parallax is ±5 mm when focusing on the blackbody furnace, an aperture diameter of at least 50 mm is required for practical blackbody furnaces.

Many good papers have been published regarding the effective emissivities of blackbody furnaces.[1-19] However, the cavity emissivity is often calculated under the isothermal condition.

Since the depth-to-radius ratio (L/D) of practical blackbody furnaces is comparatively small, and a nonisothermal temperature distribution is unavoidably produced from the target area of the cavity wall to the aperture, it is impossible to realize an isothermal condition. Thus, the blackbody furnace construction often results in low quality. The effective cavity emissivity of the low quality blackbody furnace is not unity, and also it is not fixed since the thermal equilibrating condition of the cavity is affected by the set temperature, room temperature, time at temperature, etc.

In addition, since the emissivity depends upon the measuring temperature and measuring wavelength, it cannot be determined precisely. It can, of course, be determined by comparing the low quality blackbody furnace with a blackbody lamp or a fixed point blackbody furnace by using a precision radiation pyrometer.

In order to obtain a calibration accuracy of better than ±0.2 % when n=4 in a low quality blackbody furnace with the effective cavity emissivity of 0.96, the radiation pyrometer's measuring wavelength is limited within ±20 % of the measuring wavelength at which the effective cavity emissivity was determined.[20]

When this technique is adopted, many transfer precision radiation pyrometers with different wavelength must be prepared. A silicon radiation pyrometer is the only available transfer pyrometer with a stability of better than ±0.1 % annually.

On the other hand, the cavity temperature of a high quality blackbody furnace can be measured with a thermocouple, and various types of radiation pyrometers can be calibrated with high accuracy, irrespective of their measuring wavelengths. A desirable temperature distribution is obtainable by positively utilizing the unavoidable temperature distribution on the cavity wall, and thus, the ideal effective cavity emissivity can be achieved.[2] A temperature control system is indispensable for the realization of this temperature distribution.

In this paper a desirable temperature distribution on the cavity walls is theoretically calculated for a cylindrical cavity and a spherical cavity, and a construction of blackbody furnace and temperature controlling method of the cavity walls are described. Then, an experimental confirmation of the realized effective cavity emissivity of the blackbody furnace and also the operating characteristics and features of this furnace are described.

THEORY

According to Kirchhoff's law, in the interior of a cavity composed of an opaque material, mutual thermal radiation occurs between respective points on the internal walls of the cavity.

When the internal wall surface of the cavity is completely diffuse, the radiant exitance at arbitrary point "x" on the wall surface is obtained as a composite value of the radiant emittance at point x and reflected flux at point x irradiated from all other point "ξ" of the cavity which can see the point x.

Assume that the emissivity of the material of the cavity wall is ϵ and the temperature of the cavity wall is nonisothermal, and that T(x) is the temperature at point x. Then the radiant exitance $\Phi(x)$ at point x is expressed by the following equation:[2,6]

$$\Phi(x) = \epsilon E\{T(x)\} + (1-\epsilon) \int_S K(x,\xi)\Phi(\xi)dS(\xi) \quad (1)$$

where $E\{T(x)\}$: Radiant exitance of a blackbody at point x with temperature T(x);
$K(x,\xi)$: Angle factor determined by the geometric form of the cavity;
dS : Area element of cavity wall.

The optimum temperature distribution for the cylindical cavity and the spherical cavity can be obtained by using this fundamental expression.

Cylindrical cavity

For the cylindrical cavity, let y be the coordinate of the target from the center in the radial direction, $\Psi(y)$ be the radiant exitance at point y, x and ξ be the coordinates on the walls from the target to the aperture, and $\Phi(x)$ be the radiant exitance at point x, as shown in Fig. 1 (a). Then from equation (1) we obtain:

$$\Phi(x) = \epsilon E\{T(x)\} + (1-\epsilon) \int_0^1 K(x,\xi)\Phi(\xi)d\xi$$
$$+ (1-\epsilon) \int_0^1 K(x,y)\Psi(y)dy \quad (2)$$

$$\Psi(y) = \epsilon E\{T(y)\} + (1-\epsilon) \int_0^1 K(y,\xi)\Phi(\xi)d\xi \quad (3)$$

Each angle factor can be expressed as follows:[6,11]

$$K(x,\xi) = \frac{1}{2}\left[1 - \frac{|x-\xi|\{(x-\xi)^2+6\}}{\{(x-\xi)^2+4\}^{3/2}}\right]$$

$$K(x,y) = \frac{2xy(x^2-y^2+1)}{\{(x^2+y^2-1)^2+4x^2\}^{3/2}}$$

$$K(y,\xi) = \frac{2\xi(\xi^2-y^2+1)}{\{(\xi^2+y^2-1)^2+4\xi^2\}^{3/2}}$$

The effective cavity emissivity of the target is obtained from the simultaneous integral equations (2) and (3) where $\bar{\epsilon}$ is defined as

$$\bar{\epsilon} = \Psi(y_0)/E\{T(y_0)\} \quad (4)$$

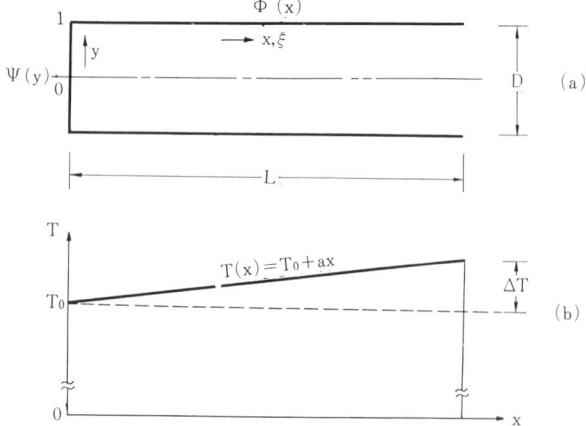

Fig. 1. Schematic of a cylindrical cavity; (a) Coordinates; (b) Ideal temperature distribution of the cylindrical cavity walls.

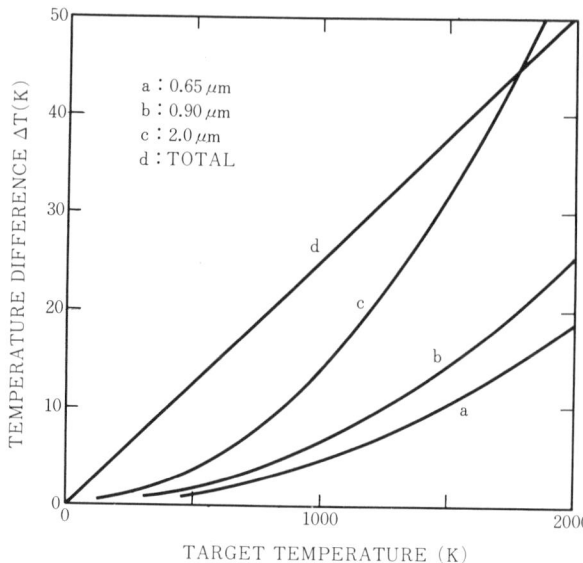

Fig. 2. Temperature difference ΔT of a cylindrical cavity for effective emissivity of unity. $\Delta T = T_{aperture} - T_{target}$

The kernel functions K(x,y), K(y,ξ) of these simultaneous integral equations show singularities at the corners of the target.[6,11] However, these singularities can be avoided and the convergence of the numerical integration improved, thereby enhancing the calculation accuracy, by using integration by parts.[11]

Thus, optimum conditions can be obtained by using this solution method when the temperature distribution is nonisothermal. As a precondition, the target is assumed to be isothermal, and the temperature difference ΔT between the target and the aperture that will cause the effective cavity emissivity to become unity obtained by applying a linear temperature distribution to the cavity walls as shown in Fig.1 (b).

The calculation procedure is as shown below:
(1) Obtain $\bar{\epsilon}$ with reference to temperature T_0 and the measuring wavelength under the following conditions:

$$L/D = 4, \quad \epsilon = 0.75,^{21} \quad T(x) = T_0 + ax$$

Temperature difference ΔT : arbitrary

(2) Determine the temperature difference ΔT to obtain $\bar{\epsilon} = 1$ by interpolation from the above calculation results.

Fig. 2 shows the results thus obtained. The illustrated temperature difference ΔT can be expressed by a very simple form with reference to the target temperature T_0.

$$\Delta T_\lambda = \alpha \lambda T_0^2 \quad \text{when measuring wavelength is } \lambda; \quad (5)$$

$$\Delta T = \beta T_0 \quad \text{when total radiation is employed}; \quad (6)$$

where, $\alpha = 7.05 \ (m^{-1} \cdot K^{-1})$.
$\beta = 0.0251$

The effective cavity emissivity becomes unity if the conditions in equations (5) and (6) are satisfied, however, a temperature drop is unavoidable at the aperture, in practice. Accordingly, an error of $\bar{\epsilon}$ was evaluated, assuming that an ideal temperature distribution given by equation (5) or (6) was set to the target temperature midway between the target and the aperture.

In other words, even if an optimum temperarure distribution can be realized up to L/D = 2 only, when L/D = 4, the error in $\bar{\epsilon}$ is limited to only -0.1 %.

The above results were obtained assuming that the intrinsic emissivity of the cavity wall is 0.75. Even if this emissivity changes over a range from 0.60 to 0.90, the resultant $\bar{\epsilon}$ error is within ±0.1 %.

Accordingly, if an ideal furnace wall temperature distribution can at least be expended to half of the cavity on the target side in a cylindrical cavity with L/D = 4, in which the intrinsic emissivity of the wall is between 0.6 and 0.9, the effective emissivity of the cavity is expected to be between 0.998 and 1.001.

Spherical cavity

If the cavity is spherical, the kernel function is $1/4\pi$.[2] Let $d\omega$ be the solid angle when an arbitrary surface element dS located at point x is viewed from the center of the cavity and let ω_1 be the solid angle when the aperture is viewed from the center of the cavity as shown in Fig. 3(a). Then the radiant exitance $\Phi(x)$ can be expressed from equation (1) as follows:[2]

$$\Phi(x) = \epsilon E\{T(x)\} + \frac{\epsilon(1-\epsilon)\int_0^{4\pi-\omega_1} E\{T(\xi)\}\,d\omega}{4\pi - (1-\epsilon)(4\pi-\omega_1)} \quad (7)$$

Accordingly, the effective emissivity $\bar{\epsilon}$ of the cavity is expressed by the following equation when the positional coordinate of the target is x_0:

$$\bar{\epsilon} = \frac{\Phi(x_0)}{E\{T(x_0)\}} = \epsilon + \frac{\epsilon(1-\epsilon)\int_0^{4\pi-\omega_1} \frac{E\{T(\xi)\}}{E\{T(x_0)\}}\,d\omega}{4\pi - (1-\epsilon)(4\pi-\omega_1)} \quad (8)$$

Since the integrand representing the temperature distribution does not include x, the generation point of the temperature distribution, which contributes to the effective emissivity of cavity, is equivalent at all positions on the internal spherical wall,[2] unlike in the cylindrical cavity.

If $\bar{\epsilon}$ is unity when the furnace wall temperature is constant at T_0 from the target to point x_1 on the aperture side and constant at $T_0 + \Delta T$ from point x_1 to the aperture as shown in Fig. 3 (b),[2] the following equation is obtained from equation (8):

$$\frac{E(T_0 + \Delta T)}{E(T_0)} = \frac{\omega_1(1-\epsilon) + 4\pi\epsilon(1-x_1)}{\epsilon\{4\pi(1-x_1) - \omega_1\}} \quad (9)$$

$L/D = 4$ and $\epsilon = 0.75$,

then $E(T_0 + \Delta T)/E(T_0) = 1.044$ when $x_1 = 0.5$

and $E(T_0 + \Delta T)/E(T_0) = 1.090$ when $x_1 = 0.75$ (10)

The following conditions are now obtainable from the measuring wavelength after ΔT is obtained from the above equations:

$$\Delta T = \frac{\alpha \lambda T_0^2}{1 - \alpha \lambda T_0} \quad \text{when the measuring wavelength is } \lambda; \quad (11)$$

$\Delta T = \beta T_0$ when total radiation is employed: (12)

where, $\alpha = 2.97$ (m$^{-1}\cdot$K^{-1}), $\beta = 0.0108$ when $x_1 = 0.5$
$\alpha = 6.02$ (m$^{-1}\cdot$K^{-1}), $\beta = 0.0219$ when $x_1 = 0.75$

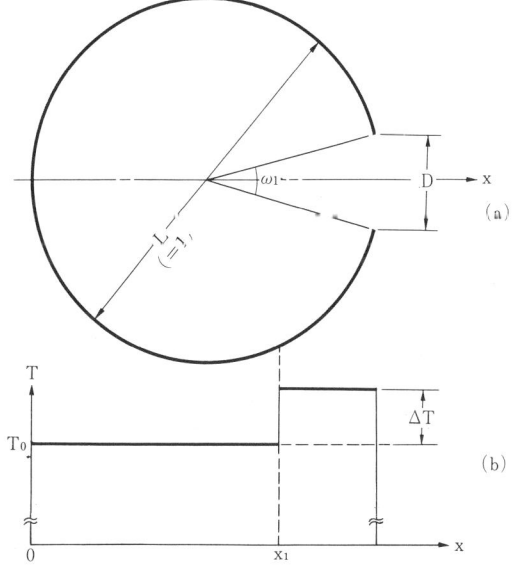

Fig. 3. Schematic of a spherical cavity; (a) Coordinates; (b) Ideal temperature distribution of the spherical cavity walls.

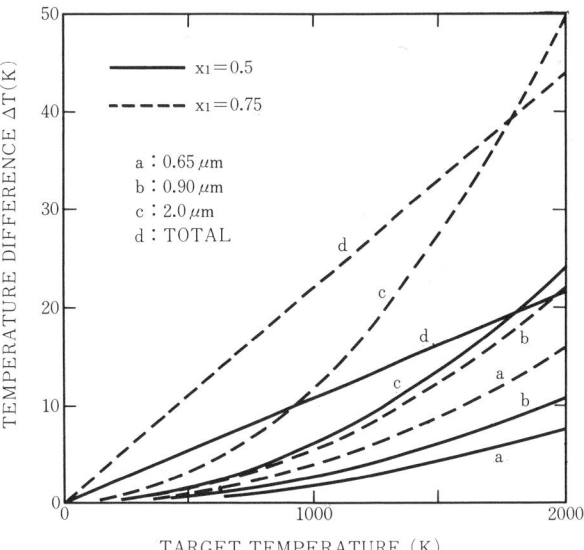

Fig. 4. Temperature difference ΔT of a spherical cavity for effective emissivity of unity. $\Delta T = T_{aperture} - T_{target}$

Fig. 4 shows ΔT obtained from equation (11) as a function of the target temperature. A stepwise change of the temperature distribution over an arbitrary portion of the furnace wall is not practical. However, if an expected temperature distribution is obtained when x_1 is between 0.5 and 0.75, Fig. 4 can be applied. Assuming that x_1 changes from 0.5 to 0.75, the effect on $\bar{\epsilon}$ is estimated to be from +0.6 % to –0.3 %.

The above conclusion was drawn by setting the intrinsic emissivity of the cavity material equal to 0.75. Even if it changes from 0.60 to 0.90, a resultant error is within –0.2 to 0.1%. Accordingly, if an ideal temperature distribution can be realized within a range from 1/2 to 1/4 of the sphere on the aperture side as viewed from the target when the spherical cavity is made of a material having the intrinsic emissivity of 0.6 to 0.9 and also L/D is 4, it is expected that its effective emissivity will be between 0.995 and 1.007.

CONSTRUCTION

As described above, a desirable effective cavity emissivity is obtainable by applying an ideal temperature distribution to the cavity wall in a practical blackbody furnace having a large aperture diameter. As a means of realizing such a temperature distribution, we adopted a three zone temperature control system. This system will be described for a cylindrical cavity and for a spherical cavity.

Construction of Cylindrical Cavity

As shown in Fig. 5, a silicon carbide cylindrical tube having an inner diameter of 50 mm and a length of 400 mm is employed as a cavity. Also, the target having 120° faces on both sides is welded to the center. This cavity is held by a equalizing tube made of silicon carbide of 80 mm ID and 750 mm length, and three diaphragms of 50 mm ID are fixed in this tube with equal spaces. As a result of the above construction, energy is transferred from both ends supported by the equalizing tube to the cavity by thermal conduction, and a cavity effect to be produced in a 3 mm space between the equalizing tube and the cavity tube. Thus, the temperature distribution on the cavity wall is made uniform. The target is designed as a wedge type because the thermal conduction from the cavity tube becomes effective, and also the cavity action with the cavity wall is accelerated.[15] The diaphragms prevent the convection of air introduced from the aperture.

Heaters are mounted via a space on the outer circumference of the equalizing tube. A Kanthal heater wire or a silicon carbide heating rod is employed depending upon the working temperature range. The blackbody furnace using the Kanthal wire heaters is normally used within a range from room temperature to 1100 °C, while the blackbody furnace using silicon carbide heating rods is operated from 500 °C to 1500 °C. These heaters are divided into three in the axial direction of the equalizing tube and arranged symmetrically about the target installed at the center.

I. Hishikari and T. Ide

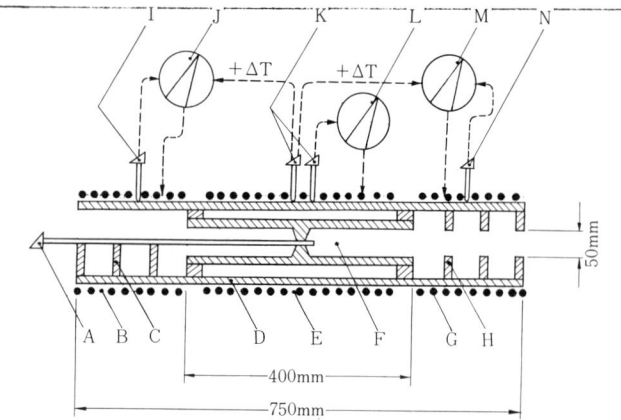

Fig. 5. Outline of a cylindrical blackbody furnace and a 3-zone temperature control method; A- Standard thermocouple; B, G-Sub-heaters; C-Holders of standard thermocouple; D-Equalizing tube; E- Main heater; F-Cavity; H- Diaphragms; J, M-Sub-controllers; I,K,N- Thermocouples for temperature control; L- Main controller.

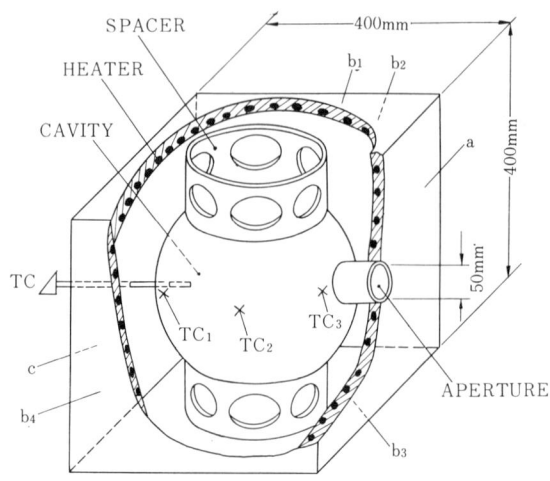

Fig. 6. Outline of a spherical brackbody furnace and method of zone division of a hexahedral heater (a-Front zone; $b_1 \sim b_4$-Center zone; c-Rear zone).

The tips of temperature control thermocouples encased in alumina sheaths along the outer circumference of the equalizing tube make contact with the center and both sides of the equalizing tube via heaters. The thermocouple installed at the center is connected to a main temperature controller to control the power of the heater mounted in the center zone. All of the other thermocouples are led to sub-temperature controllers to control the power of the heaters mounted in both side-zones, so that the difference of mutual electro-motive forces between the center and each side becomes the setting value. As a result, the equalizing tube shows linear temperature distribution from the center to the thermocouple installation points on both sides.

The temperature distribution value should strictly be changed according to the set temperature of the blackbody furnace. However, it is determined as a constant value from the practical reason of simplifying the configuration of the temperature control system and reducing costs.

Through an open hole at the center of the target, the standard thermocouple encased in a 6 mm OD high purity alumina sheath, projects 30 mm into the cavity to measure its temperature.

Originally this standard thermocouple made contact with the rear face of the target. But it is possible that a temperature difference can be produced between the faces of the target due to such a contact, so that the cavity temperature is not accurately represented. Therefore the above projection system of thermocouple is employed.

This installation method has a practical advantage. When the set temperature of the blackbody furnace changes, the temperature distribution in the cavity is unbalanced, and the effective emissivity deviates from the specified value, causing the silicon carbide and alumina sheath to be clearly discriminable visually at the target. As the temperature distribution in the cavity is balanced, the silicon carbide and alumina sheath cannot be identified from each other visually, even if they are observed through a red glass filter. Accordingly, a worker can confirm that the cavity radiation is being realized correctly when the alumina sheath tip cannot be observed visually.

Construction of Spherical Cavity

As shown in Fig. 6, a spherical cavity is made of 200 mm ID and 5 mm thick silicon carbide and a 50 mm dia. aperture is provided, while an open hole is also provided on the target part opposite to the above 50 mm dia. aperture so as to insert the standard thermocouple.

This cavity is supported on both upper and lower sides by a spacer having large holes on the side face and installed inside a hexahedral heater. This spacer has an effect of reducing the contact area with the cavity to decrease local temperature gradients. The holes on the side face of this spacer accelerate a cavity effect produced inside the heater and make uniform the temperature distribution on the cavity wall.

The hexahedral heater employs a Kanthal wire when the working temperature of the blackbody furnace ranges from room temperature to 1100 °C, while it employs a silicon carbide heating rod when the working temperature of the blackbody furnace ranges from 500 °C to 1500 °C.

The Kanthal wire is fixed into the groove on a square alumina plate to make a heating plate, and six heating plates compose the cubic hexahedral heater. A 60 mm dia. hole is provided at the center of the front heating plate to observe the cavity.

In addition, an open hole is provided at the center of the rear heating plate to insert the standard thermocouple. The front heating plate, rear heating plate and four center heating plates comprise three zones. Control thermocouples are installed at certain parts of each zone. The control thermocouples touch the surface of the outer wall of the spherical cavity. The center zone and rear zone are set to the same temperature, while the front zone temperature is set higher. This temperature difference is maintained constant for the same practical reason as in the cylindrical cavity.

The silicon carbide heating rods are vertically installed along the outer circumference of the cavity, and also other heating rods are horizontally installed at the upper and lower parts of the cavity through spaces between the above heating rods to make a hexahedral heater. Respective heating rods are electrically connected to each other on every face to make three zones.

The standard thermocouple projects into the furnace from the target by 30 mm for the same reason as in the cylindrical cavity, and it measures the typical temperature of the cavity.

EXPERIMENT

Two experimental methods were attempted to determine the effective cavity emissivity. The temperature distribution on cavity wall was measured by thermocouples, and the effective cavity emissivity was calculated based on the measured temperature distribution. Then the temperature values were checked by using a precision radiation pyrometer calibrated against fixed point blackbodies.

For this experiment, two blackbody furnaces were employed. The cylindrical cavity is a 500 °C to 1500 °C blackbody furnace using a silicon carbide heater, while the spherical cavity is a room temperature to 1100 °C blackbody furnace using a Kanthal heater.

Measurement of Temperature Distribution on Cavity Wall and Calculation of Effective Emissivity

After equilibrating the blackbody furnace at certain temperatures, the temperature on the cavity wall was measured by thermocouples.

For measuring the temperature distribution, type B thermocouples of 0.5 mm dia. wire encased in 6 mm OD × 1 m long high purity alumina sheaths with open tips were employed. The thermocouple tip project 15 mm from the sheath to measure properly the cavity wall temperature by good thermal contact of the hot junction to the cavity surface without any thermal conduction from the sheathed portion.

Two thermocouples were prepared for this measurement. One was straight, while the other was bent almost perpendicularly at 55 mm from the tip. The cylindrical cavity was measured with the straight thermocouple, while the spherical cavity was measured with both straight and bent thermocouples. An error of each thermocouple was corrected, in advance.

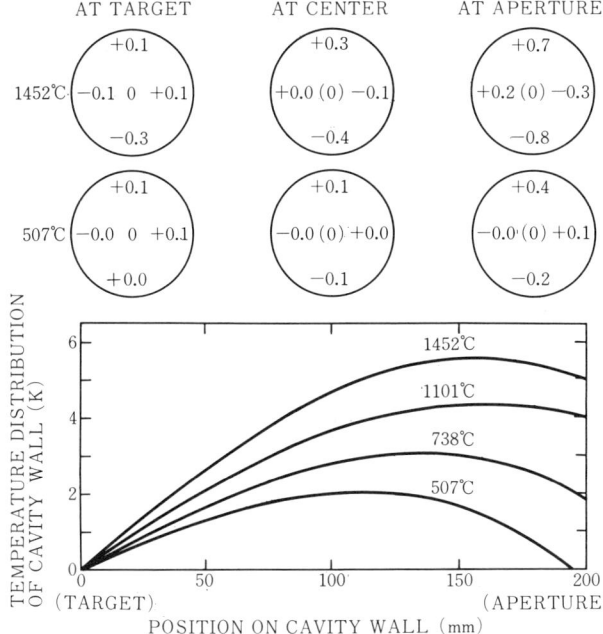

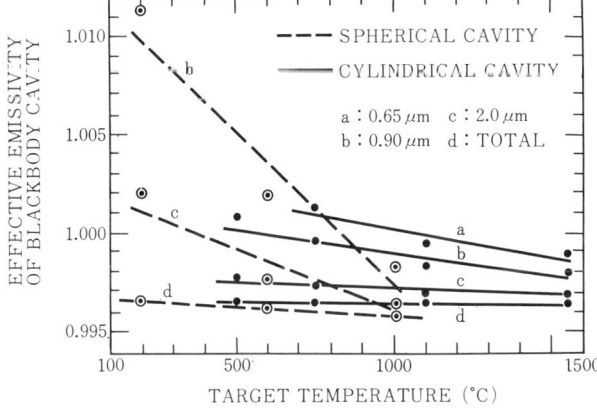

Fig. 7. Temperature distribution of a cylindrical cavity walls at several set temperatures (K).

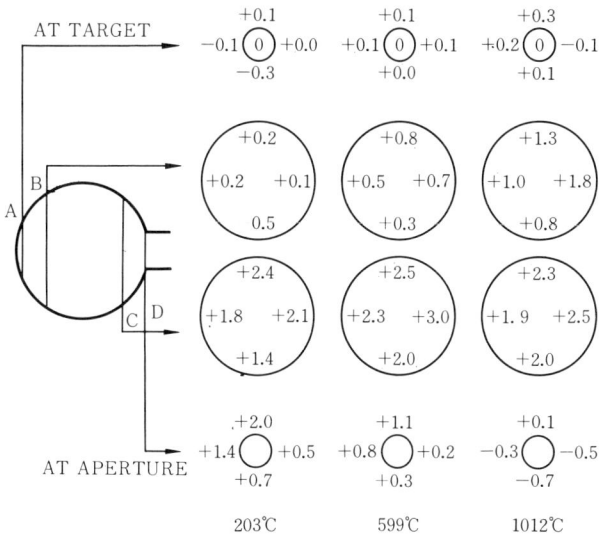

Fig. 8. Temperature distribution of a spherical cavity walls at several set temperatures (K).

In the case of the cylindrical cavity, the temperature distributions were measured on the target and the cavity wall in the axial direction. They were measured at 4 points; upper, lower, and both sides of the cavity center. The temperature differences are comparatively small on the target. A distinct temperature difference was detected on the upper face and lower face of the cavity wall due to the convection of the air introduced from the aperture. This temperature difference was larger nearer the aperture. It appeared more distinctly at higher set temperatures of course.

Fig. 7 shows the temperature distribution at the target, center and aperture as well as the axial temperature distribution. For the axial temperature distribution, a mean value on the upper and lower face was adopted, since the temperature distribution is the same on both sides of the cavity wall.

The temperature distribution of the spherical cavity was measured on the target viewed from the aperture, periphery of the aperture, and 2 peripheral points of the intermediate shell, 4 points in total. These points were specified as A, B, C and D faces as viewed from the target, and each face was measured at the top, bottom, and both sides, 4 points in total. Face A was measured by the straight thermocouple, while the other faces were measured by the bent thermocouple.

Fig. 8 shows the results at each temperature. All values are indicated by a temperature difference from the center of the target. According to these results, the temperature differences are least distinctive on face A of the target. They tend to increase as the measuring points approach the aperture as B → C → D. The temperature is clearly higher on face C than on faces A and B. The temperature differences over the target are larger than those of the target of the cylindrical cavity. This is presumably because the cylindrical cavity has a equalizing tube, while the spherical cavity is directly facing the heating plate.

The effective emissivities of the respective cavities were calculated based on the above results. The successive substitution method was applied to the cylindrical cavity by using equations (2) and (3). For the spherical cavity, equation (8) was applied, and measured values were used on four equally divided faces. Fig. 9 shows the effective emissivity of the cavities thus calculated.

According to these results, the effective emissivity of the cylindrical cavity is distributed over a range of 0.996 to 1.002, and it tends to decrease as the working temperature becomes higher. This tendency is more noticeable in the spherical cavity where the effective emissivity ranges from 0.995 to 1.012.

Comparative Measurements of Effective Emissivity with Fixed Point Blackbodies

The blackbody furnace and small fixed point blackbodies[22-25] were compared by using a precision radiation pyrometer.[26]

The radiation pyrometer and fixed point blackbody employed for this experiment were manufactured by CHINO based on research and development by the National Research Laboratory of Metrology, Japan, for the purpose of enhancing the accuracy of traceability of radiation pyrometers.

As shown in Fig. 10, the fixed point blackbody was manufactured as follows: A graphite cavity is provided inside a high quality

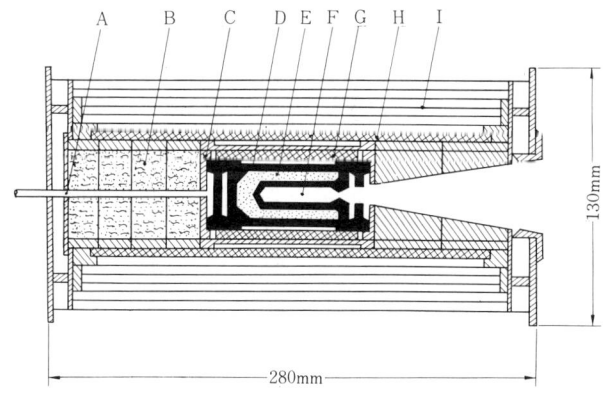

Fig. 9. Estimation of the effective emissivity of the blackbody cavity based on measuring results of the temperature distribution.

Fig. 10. Schematic cross section of the fixed point blackbody; A-Argon inlet; B-Ceramic wool; C-Stainless steel case; D-Graphite crucible; E-Fixed point substance; F-Cavity; G-Heater; H-Ceramic tube; I-Radiation shields.

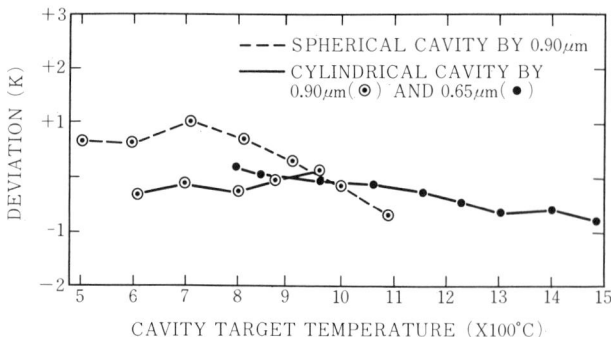

Fig. 11. Comparison of a radiation pyrometer with a standard thermocouple at measurement of the blackbody furnace.

graphite crucible, the interior of the crucible is filled with fixed point substances, and argon gas is injected to protect graphite and fixed point substances from being oxidized. This furnace is compactly assembled for practical reasons.

The cavity is 50 mm in length, and a 10 mm aperture is provided with a 5 mm lid with a hole. The effective cavity emissivity was calculated as 0.999 when the cavity was isothermal.

Zinc (Zn), antimony (Sb), aluminum (Al), silver (Ag), and copper (Cu), each having purity higher than 0.9999, were employed as fixed point substances and crucibles, each having an internal volume of 26 cm^3, were filled with these substances respectively. These crucibles are encased in stainless steel with a heater embedded on the outer circumference, and argon gas is fed from the rear face of the case.

Three fixed point blackbodies of Al, Ag, and Cu were employed for this experiment. The plateau at the fixed point can be maintained for 30 minutes maximum by adjusting the heater power. However, a duration of about 8 minutes was used for practical reasons. The operation was repeated more than 80 times in one year. The change in each fixed point did not exceed ±0.1 K.

The radiation pyrometer for comparison measurement has its measuring wavelength range limited by an interference filter and a silicon (Si) photo voltaic cell is employed as detecting element.

Two radiation pyrometers were prepared with respective measuring wavelengths of 0.65 μm and 0.90 μm. The halfwidth of the interference filter is 18 nm. The diameter of the objective lens is 40 mm. Its field of view is 5 mrad. A resolution of better than 0.01 K above 800 °C is obtained with the 0.65 μm pyrometer, and above 600 °C with the 0.90 μm pyrometer.

The pyrometer outputs change with the ambient temperature, but the output drift is only about 0.01 K when the ambient temperature changes 1 K. These instrument also show excellent long-term stability with ±0.4 K annually.

The 0.90 μm pyrometer was calibrated by using the Wien's law, assuming that the outputs of the pyrometer are proportional to the incident radiant flux from the fixed point blackbody furnaces of Al, Ag, and Cu. On the other hand, the 0.65 μm pyrometer was calibrated with a previously calibrated tungsten ribbon lamp, and this value was corrected by measurements at the Ag and Cu points.

The standard thermocouple installed in the blackbody furnace cavity was calibrated at the IPTS-68 fixed points. The expected accuracy is ±0.5 K over a range from 0 °C to 1100 °C and ±2.0 K from 1100 °C to 1500 °C.

The comparison experiment was executed by focusing the radiation pyrometer on the target of the blackbody furnace, and reading the electromotive force of the standard thermocouple installed in the cavity and the radiation pyrometer output concurrently, when the cavity temperature was stable.

Fig. 11 shows the results. This figure shows the indicated difference of the radiation pyrometer from the standard thermocouple at each target temperature. The spherical blackbody furnace was compared by the radiation pyrometer of 0.90 μm, while the cylindrical blackbody furnace was compared by both the 0.90 μm and 0.65 μm radiation pyrometers. Results agree within ±1 K. The radiation pyrometer indication tends to decrease as the blackbody furnace temperature becomes higher. This tendency is more noticeable in the spherical blackbody furnace than in the cylindrical blackbody furnace. This agree with calculated results from the temperature distribution in Fig. 9. The emissivity of the spherical blackbody furnace is measured to be 0.995 to 1.016, while that of the cylindrical blackbody furnace is 0.996 to 1.000. These values do not agree with Fig. 9 but the order of magnitude is correct.

Slight incoincidence of experimental values may be caused mainly by the calibration accuracy of the standard thermocouple and an error in the small fixed point blackbodies. In any case, these blackbody furnaces can be considered as practical ones having a sufficient effective emissivity.

OTHER FEATURES

A practical blackbody furnace should have a compact construction and be easy to operate and easy to maintain. The temperature difference over the cavity was not specified to be proportional to the set temperature of the furnace but was preset as a constant value. Practical features will be described below.

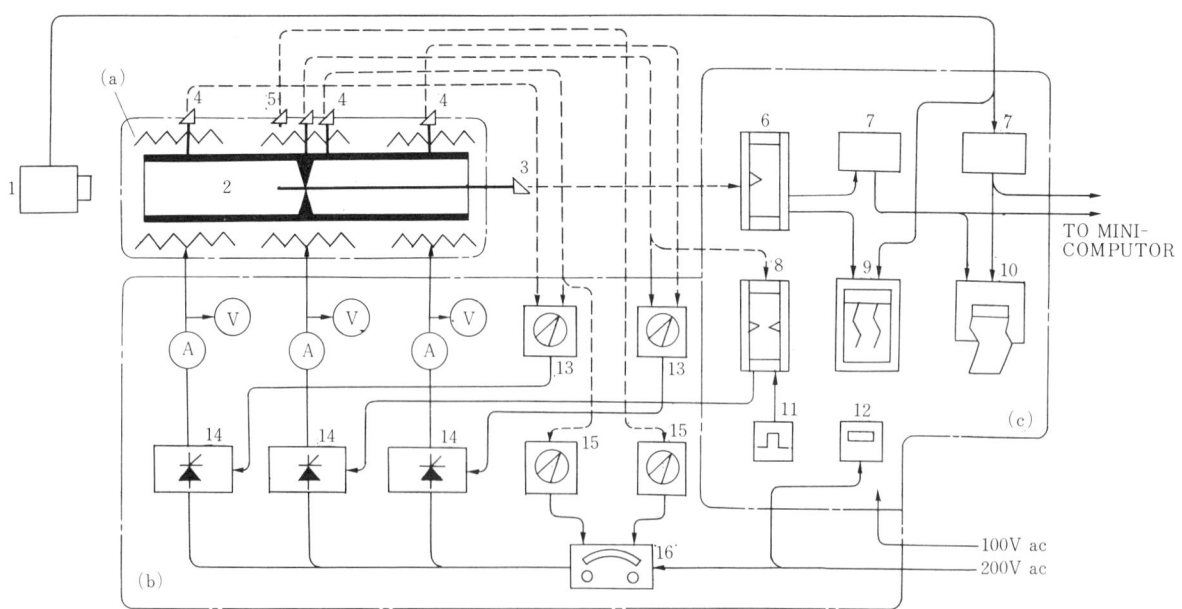

Fig. 12. Schematic diagram of a precision practical blackbody furnace with 3-zone temperature control system; (a) Furnace; (b) Instruments installed in rear panel; (c) Instruments installed in front panel; 1. Radiation pyrometer to be caribrated; 2. Cavity; 3. Standard thermocouple; 4. Thermocouples for temperature control; 5. Thermocouple for alarm; 6. Temperature converter; 7. Digital indicators; 8. Main controller; 9. Monitor recorder; 10. Printer; 11. Pulse generator; 12. Watthour meter; 13. Sub-controller; 14. Thyrister; 15. Alarm; 16. Braker.

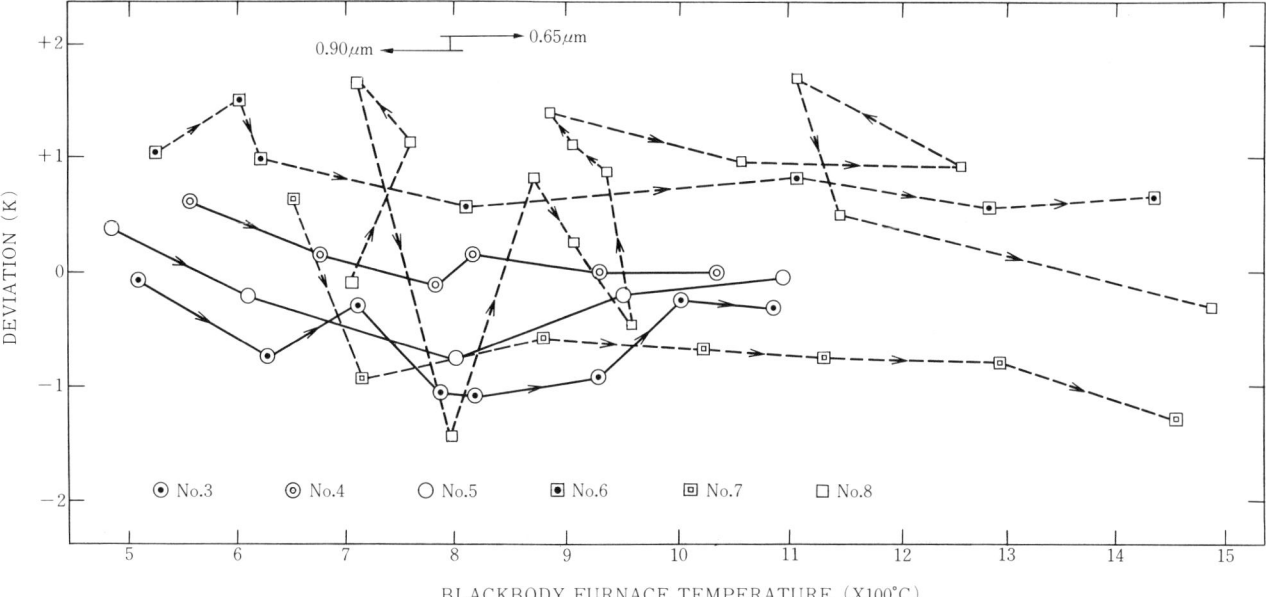

Fig. 14. Working accuracy of blackbody furnaces. No. 3, 4 and No. 5 are Kanthal wire heater type with a working range from room temperature to 1100°C and No. 6,.7 and No. 8 are silicon carbide heating rod type with a working range from 500°C to 1500°C. No. 4 and No. 8 are spherical blackbody furnaces, while the others are cylindrical blackbody furnaces.

Ancillary Equipment of Blackbody Furnace

As shown in Fig. 12, the cylindrical cavity and spherical cavity furnaces are assembled in cabinets having the same appearance. The cabinet consists of upper and lower portions. The furnace is mounted in the upper part, and instruments are mounted in the lower part. These instruments consist of three systems to control the cavity temperature, a system to measure, display, and print the cavity temperature, and an output display system for the radiation pyrometer. Various devices required for operating the blackbody furnace are arranged on the front operation panel, while the devices which are not frequently required are mounted on the rear panel.

The main controller is settable by pulse signals. It allows the operator to execute automatic program control by pulse drive as well as manual setting. The set value is changeable at a rate of 0.5 K to 1.0 K per minute so as not to unbalance the temperature distribution of cavity.

Two mV sub-controllers without a reference junction compensation circuit are provided to maintain the specified temperature difference between zones. They attain their purposes so long as the reference points of the control thermocouples are at the same temperature.

The control thermocouples are connected to each controller by extension wires. The same thermocouple wire is also connected from the standard thermocouple to the temperature converter, and the compensation accuracy of ±0.1 K on the reference junction was obtained by using a small oven.

The thermocouple signals and radiation pyrometer outputs are introduced to an analog recorder for monitor, digitally monitored on a meter, and then led to a printer. In the automatic program mode, the printer operates each time the blackbody furnace temperature changes 10 K.

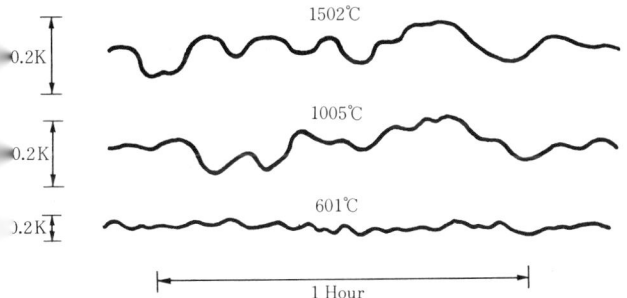

Fig. 13. Short-term stability of a blackbody furnace measured with a 0.90 μm radiation pyrometer.

Stability of Cavity Temperature

All of the blackbody furnaces reach the maximum working temperature from room temperature within 1.5 hours, and the temperature distribution in the cavity is stabilized within 30 minutes thereafter. The stabilization time is 10 to 20 minutes when the blackbody furnace temperature increases every 100 K. However, the stabilization time for cooling may be from 0.5 to 3 hours depending upon the set temperature. As shown in Fig. 13, temperature fluctuations after the stabilization are usually within 0.2 K per hour. This fluctuation is caused by mutual interference between the three controllers.

Life of Blackbody Furnace

Four blackbody furnaces of this type were installed 4 years ago, and three were installed in the next year. Thus, seven blackbody furnaces are now operating. The operation time of those installed first has exceeded 20,000 hours on average, and of the second group has reached 15,000 hours.

According to these experiences, the life of the blackbody furnace is determined by damage to the heater and radiation tube (cavity). This has been detected in a furnace operated above 1100 °C, and no damage has been detected in those operated below 1100 °C.

Regarding the silicon carbide heater of the furnace operated at 1500 °C maximum. The resistance value increases 40 % after 3,000 hours on an average. One or several heaters are damaged almost concurrently at 6,000 to 7,000 hours. Similarly, either the equalizing tube or the radiation tube is cracked or abruptly broken after 6,000 to 8,000 hours, and it must be changed, an operation requires one hour.

The standard thermocouple is recalibrated once every 6 months. If the radiation tube is broken, the corresponding thermocouple may be mechanically damaged, and so is replaced.

Working Accuracy

The blackbody furnace cannot be used until its temperature distribution equilibrates at the specified value. Accordingly, a timer is mounted on the furnace panel, although skilled operators often judge the working time empirically.

In order to confirm the operator's judgment, 6 blackbody furnaces were measured with both a radiation pyrometer and the standard thermocouple when the operator judged equilibrium had been reached after arbitrarily setting the furnace temperature. Fig. 14 shows a typical result. Differences between thermocouple and pyrometer were less than ±2 K at working temperature of 500 °C to 1500 °C, although this value differs somewhat for other furnaces.

Calibration System

Ten blackbody furnaces are being employed for calibrating every type of radiation pyrometer. Of these furnaces, seven are of a 3-zone temperature control type, while the others are of 1-zone temperature control type (oil bath type, air bath type, and Tanmann type electrically heated graphite radiator). Radiation pyrometers can be calibrated from -50 °C to 3000 °C by these units.

The calibration temperature of these blackbody furnaces and the outputs of the radiation pyrometers to be calibrated are introduced into a mini-computer via a random access multiplexer and automatically processed.

The random access multiplexer transmits the tag number of a blackbody furnace and the data to the mini-computer each time the furnace temperature changes 10 K. If it was simultaneously accessed by two or more furnaces, they are called in the order of lower number according to the program.

Data is stored in mini-computer and, on command, it generates the maximum probability calibration of the radiation pyrometer, and displays it on a CRT. In addition, it produces a hard copy and loads all data into a magnetic tape. When the output characteristic of the radiation pyrometer is represented by hard copy, it is possible that a smooth curve is not obtained due to poor resolution. In such a case, an X-Y plotter is employed. In addition to the tabulation of calibrated results, it is also utilized for automatic production of scale plates of receiving instruments.

This system can be automated so that the calibration work is automatically done without attendance by operators.

CONCLUSION

As a result of testing the practical blackbody furnaces described herein, the following conclusions were drawn:

(1) It was theoretically proved that an effective cavity emissivity of unity with sufficient accuracy is obtainable by increasing the cavity wall temperature from the target to the aperture.

(2) The 3-zone temperature control system was applied to the cylindrical and spherical cavity radiators so as to realize the theoretically calculated wall temperature distribution, and the desired temperature distribution was confirmed to be obtained by thermocouple measurement.

(3) In order to confirm the effective emissivities of these furnaces, comparison measurements were made with a fixed point blackbodies and a radiation pyrometers. The experimental values showed a tendency to decrease as the set temperature of the blackbody furnace increased. They agreed with the theoretical values within the accuracy of ±0.1 %.

(4) A correct emissivity value for the blackbody furnace is obtained after the cavity wall temperature distribution has converged to the specified value. This condition can be automatically obtained or entrusted to the operator's decision. By optionally changing the preset temperature values of several blackbody furnaces, the accuracy of operator's setting was found to be ±0.25 % of the temperature. This may be regarded as the overall accuracy of this equipment.

(5) All types of blackbody furnaces reach the maximum working temperature from room temperature within 1.5 hours, and they are measured thirty minutes after they have reached the maximum working temperature. Their lives were determined by the heating element and cavity radiator. When these furnaces are used below 1100 °C, no damage was detected after 20,000 hours. However, when they are employed at 1100 °C to 1500 °C, exchange of both heating element and radiation tube was needed every 5,000 hours.

(6) The blackbody furnaces are programmed for automatically calibrating radiation pyrometers with the aid of a mini-computer.

ACKNOWLEDGEMENT

The authors would like to thank Dr. S. Takata, Hokkaido University, Japan, Dr. H. Toyota, University of Tokyo and Dr. S. Hattori, National Research Laboratory of Metrology, Japan, for their suggestions and comments.

REFERENCES

1. J.C. De Vos, Physica 20, 669 (1954)
2. M. Jyotaki, Res. Electrotechnical Laboratory Japan 558, 36 (1957)
3. E.M. Sparrow, L.U. Albers, and E.R.G. Eckert, J. Heat Transfer C84, 73 (1962)
4. E.M. Sparrow, and S.L. Lin, Int. J. Heat Mass Transfer 8, 769 (1965)
5. F.E. Nicodemus, Appl. Opt. 4, 767 (1965)
6. B.A. Peavy, J. Res. Nat. Bur. Stand. Sec. C70, 139 (1966)
7. S. Takata, J. Illum. Eng. Inst. Japan 51, 702 (1967)
8. F.E. Nicodemus, Appl. Opt. 7, 1359 (1968)
9. M.L. Fecteau, Appl. Opt. 7, 1363 (1968)
10. J.S. Toor, and R. Viskanta, Int. J. Heat Mass Transfer 11, 883 (1968)
11. H. Yamamoto, Oyo Buturi Japan 38, 618 (1969)
12. E.M. Sparrow, and R.P. Heinsch, Appl. Opt. 9, 2569 (1970)
13. R.E. Bedford, Temperature, Its Measurement and Control in Science and Industry (Instrument Society of America, Pittsburgh, Pa. 1972), Vol 4, Part 1, 425
14. R.J. Chandos, and R.E. Chandos, Appl. Opt. 13, 2142 (1974)
15. A. Ono, J. Opt. Soc. Am. 70, 547 (1980)
16. Y. Ohwada, J. Opt. Soc. Am. 71, 106 (1981)
17. E.M. Sparrow, Appl. Opt. 4, 41 (1965)
18. F.J. Kelly, and D.G. Moore, Appl. Opt. 4, 31 (1965)
19. R.P. Heinisch, Temperature, Its measurement and Control in Science and Industry (Instrument Society of America, Pittsburgh, Pa. 1972), Vol 4, Part 1, 435
20. I. Hishikari, Paper presented at the 43rd Temperature Symposium of Soc. Inst. Cont. Eng. Japan (1979)
21. R.J. Thorn, and G.H. Winslow, Temperature, Its measurement and Control in Science and Industry (Reinhold Pub. Co. New York 1962) Vol. 3, Part 1, 421
22. H. Ito, Trans. Soc. Inst. Cont. Eng. Japan 12, 127 (1976)
23. S. Hattori, Report of the N.R.L.M. Japan 27, 86 (1978)
24. F. Sakuma, and S. Hattori, Trans. Soc. Inst. Cont. Eng. Japan 18, 52 (1981)
25. I. Hishikari, H. Nagarekawa, and T. Shimizu, Paper presented at the 20th conference of Soc. Inst. Cont. Eng. Japan (1981)
26. A. Kawamura, K. Hayashi, and S. Hattori, Trans. Soc. Inst. Cont. Eng. Japan 15, 71 (1979)

−50 to +150 °C heat pipe blackbody sources for radiation thermometer calibration

Zhu Yingsong, Ma Hongqi, and Wang Ronghua

Shanghai Institute of Process Automation Instrumentation, Shanghai, China

Hua Chengsheng

Chinese Space Technology Research Institute, Beijing, China

Two horizontally-laid cylindro-conical heat pipe blackbodies for thermometric purposes are described, both of which are of the same cavity size, 60 mm in diameter and 300 mm long. One of them is an ammonia/stainless steel heat pipe which operates between −50 and +50 °C; the other is a water/copper heat pipe used in the +40 to +150 °C region. Theoretical considerations of the design are given, and *in situ* measurement using differential copper/constantan thermocouples shows that the temperature uniformity over a 200 mm length from the cone apex is ±0.03 °C in the +110 to +150 °C range in the case of the water/copper heat pipe; ±0.01 °C in the −20 to +50 °C range in the case of the ammonia/stainless steel heat pipe; ±0.06 °C in the +110 to +40 °C range in the case of the water/copper heat pipe; and ±0.02 °C in the −20 to −40 °C range in the case of the ammonia/stainless steel heat pipe. The temperature controllability is ±0.01 °C for 20 min, excluding the transient period. The total effective emissivities of the conical part and that at about four tenths of the cylindrical part adjacent to the cone are calculated to be greater than 0.999.

INTRODUCTION

The development of heat pipe technology makes it possible to innovate some of the thermometer calibration sources, and a great deal of work has been published in this respect [1-4].

As the operating temperature range of quite a few types of radiation thermometer has been extended down to below 0°C, it has become most necessary to have sub-zero blackbody calibration sources in a thermometric laboratory. The conventional facility of this kind comprises blackbodies immersed in liquid baths. Although this can provide fairly good temperature uniformity (from a few hundredths of a kelvin to a few millikelvins, depending on the design of the apparatus) [5,6], complicated arrangements must be made for high-speed liquid circulation, and, in the case of a horizontal layout, for sealing and even for refilling/removal of the liquid according to the thermal contraction/expansion conditions. However, with horizontally-laid heat pipe sources, we have none of these problems to worry about, and the design will be simpler, with fairly good temperature uniformity, relatively small heat capacity and heat inertia, and pollution of the environment as well as vibration and noise will not be a "headache" at all.

In the following sections heat pipe facilities that have recently been developed at SIPAI for −50 - +150°C radiation thermometer calibrations will be described. Both theoretical considerations and experimental results will be presented.

DESIGN CONSIDERATIONS

1. Construction of the heat pipes

As far as we know a heat-carrying liquid that can cover the whole operating range −50 to +150°C can hardly be found. (Incidentally, this is also true of a bath.) We have therefore to use two heat pipes of the same size: One is an ammonia/stainless steel heat pipe operating between −50 and +50°C; the other is a water/copper heat pipe for the +40 - +150°C region.

Each of the two blackbody heat pipes (see Figs. 1 - 5) is made up of an outer cylinder and an inner cylindro-conical hollow body, the cylindrical part being 2 mm thick, the front ring flange 8 mm thick and the rear cover 3 mm thick. The cavity is 60 mm in diameter and 300 mm in total length with a 60° apex angle cone. The cone-bottom is designed to withstand the high vapor pressure of ammonia which will reach above 20 bar at +50°C. The enclosed surfaces are tightly covered with two layers of 300 mesh screen of copper and stainless steel as wicks respectively; and narrow sectored screens are fixed at regular intervals to bridge the inner and the outer cylinder wicks. Such an arrangement of wicks is capable of overcoming the 102 mm height difference and making the horizontally-laid heat pipe work isothermally in spite of its heater/cooler positions. Two long wells (6 mm i.d. and 300 mm long) are arranged in the vapor space. One of them accommodates three 46-ohm platinum resistance thermometers (3.5 mm in diameter and 25 mm

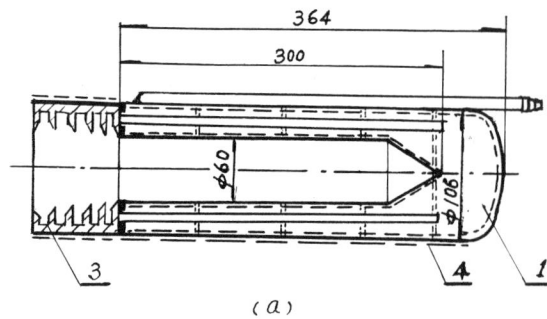

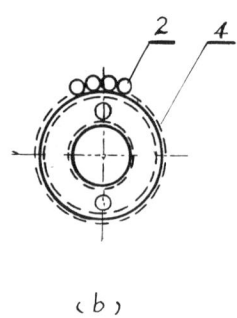

Fig. 1. Design of the water/copper heat pipe blackbody device for +40 – +150 °C radiation thermometer calibration

(a) Longitudinal section; (b) cross section.
1. Water/copper heat pipe blackbody
2. Fast cooler on top
3. Heat compensator with a set of diaphragms for reducing free air convection effect
4. Thin-film constantan heater

2. Temperature uniformity

As is well known, an ideal blackbody must have good temperature uniformity along its cavity walls. Therefore, factors that may lead to temperature non-uniformity should be considered in designing a heat pipe calibration source, and measures should be accordingly taken in its construction.

(i) Because of the non-uniform arrangement of the energy sources in our design wherein the heater and cooler are attached to the lower and upper parts of the outer cylinder, the circular vapor pressure gradient which causes the vapor to flow upward along the circular path might make more contribution to the temperature non-uniformity than the vapor pressure gradient in the axial direction where the energy source is relatively well-distributed. From the formula for vapor pressure drops in heat pipe theory [7] and the Clausius-Clapeyron equation we have

$$\Delta T_v = \frac{R_u T_v^2}{\lambda P_v M} \Delta P_v$$

$$= \frac{R_u T_v^2 \mu_v Q L}{\lambda^2 P_v M K A_v \rho_v} \quad (1)$$

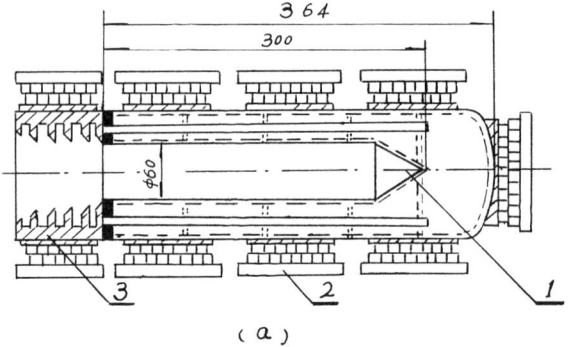

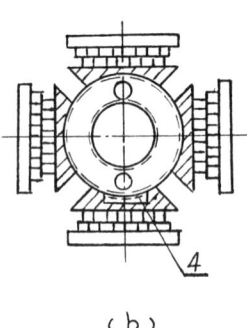

Fig. 2. Design of the ammonia/stainless steel heat pipe blackbody device
(a) Longitudinal section; (b) cross section.
1. Ammonia/stainless steel heat pipe blackbody
2. Semiconductor Peltier thermoelectric refrigerator
3. Heat compensator with a set of diaphragms for reducing free air convection effect
4. Thin-film constantan heater

in length) for temperature measurement, control and alarm. The reference junctions of 14 copper/constantan differential thermocouples that are to be used for determination of the temperature uniformity are located in the other well.

To reduce heat inertia a thin-film constantan heater is attached to the outer surface of the water/copper heat pipe with a fast cooler on top. The latter is composed of four 6 mm o.d. copper tubes connected in series. When the water/copper heat pipe blackbody needs fast switching from its upper temperature limit to its lower limit, cooling water is allowed to run through these tubes.

The ammonia/stainless steel heatpipe has a set of two-stage semiconductor Peltier thermoelectric refrigerator/heater around its outer surface. It serves as a heater when the heat pipe needs fast transition from its lower temperature limit to its upper limit, and vice versa. The controlling power is likewise supplied by a thin-film constantan heater, but it is attached to the lower part of the heat pipe (see Fig. 2).

A specially designed accessory, through which dry nitrogen or another inert gas can be uniformly distributed and led out to protect the cavity from dew and frost, is installed in the fore part of the ammonia/stainless steel heat pipe blackbody when it is used below room temperature.

Fig. 3. Photo of the assembled water/copper heat pipe blackbody block.

Fig. 4. Photo of the assembled ammonia/stainless steel heat pipe blackbody block.

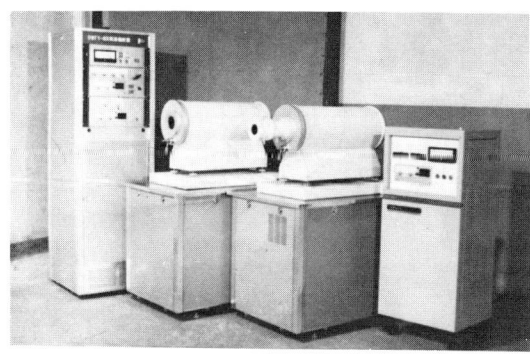

Fig. 5. Photo of the outward appearance of the two heat pipe blackbody devices. In the front is the ammonia/stainless steel heat pipe blackbody; in the rear is the water/copper device.

Thus, for the water/copper heat pipe the maximum temperature difference of the vapor along the circular path will be $\Delta T_v = 1.8 \times 10^{-6}$ K at +50°C; and for the ammonia/stainless steel heat pipe it will be $\Delta T_v = 1.7 \times 10^{-7}$ K at -50°C.

From this we can see that in our case the effect of ΔT_v is negligible.

(ii) Next, the heat transfer between the vapor and the outer surface of the cavity wall and the outward heat conduction along the cavity wall should be taken into account. For simplification, we neglect the free air convection effect on the inner surface of the cavity. (This will be further discussed below.) Using the heat transfer model of one-dimensional rod [8], (see Fig. 6) an equation can be obtained:

$$\Delta T_x = T_x - T_v = (T_0 - T_v) \frac{\cosh m (x - l)}{\cosh m\, l} \quad (2)$$

where

$$m = \sqrt{\frac{\alpha u}{k F}}$$

$$\alpha = \frac{k_w}{\delta_w}$$

$$k_w = \frac{\beta - \varepsilon}{\beta + \varepsilon} k_1 \quad \text{(See Ref. 9)}$$

$$\varepsilon = \frac{\pi f M' d'}{4} \quad \text{(See Ref. 10)}$$

From Eq. (2) we can calculate that in the case of the water/copper heat pipe the temperature difference is smaller than 0.1 °C along the cavity wall 5 cm from beyond its mouth, and that the ammonia/stainless steel heat pipe should have only a 0.01 °C difference 2 cm from beyond its open end.

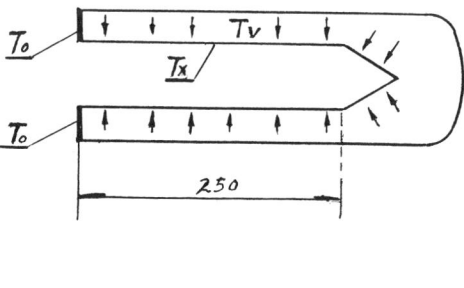

Fig. 6. Heat transfer model for the cavity walls with its vapor.

T_0 — temperature of the front ring flange

T_v — vapor temperature

T_x — temperature of the cavity walls in a distance x from open end of cavity

(iii) As free air convection creates temperature differences throughout the cavity space, the convection heat transfer coefficient along the longitudinal direction and the circumference of the cavity wall will change. Obviously, such a complicated heat transfer problem cannot be analytically solved. We have then to evaluate this effect through the following procedures:

(a) By measuring the differences between the power supplied with the cavity mouth open and that with the cavity mouth closed when the heat pipe is in both cases in equilibrium and the ambient temperature is constant, we can obtain the total heat flux loss ($Q_{rad} + Q_{conv}$) caused by radiation and convection. Subtracting Q_{rad}, which can be theoretically calculated, from this, we will have the heat flux loss through free convection, Q_{conv}, and the average free convection heat flux density q_{aver} over the inner surface of the cavity.

(b) It is assumed that the maximum free convection heat flux density q_{max} appears near the open mouth, which is supposed to be equal to $2q_{aver}$; and that the minimum heat flux density q_{min} in a certain depth, which is supposed to be equal to $\frac{1}{2}q_{aver}$.

Using the following heat conduction equation [8]

$$\Delta T_x = T_v - T_x = \left(\frac{\delta_w}{k_w} + \frac{\delta}{k}\right) q = f(q) \quad (3)$$

we have $\Delta T_{max} = f(q_{max}) - f(q_{min})$, which will be 0.06°C at +150°C in the case of the water/copper heat pipe and which will be 0.075°C at -50°C in the case of the ammonia/stainless steel heat pipe.

From the above analyses we come to the conclusion that the main causes of temperature non-uniformities over the cavity surface are the axial heat leakage along the cavity walls and the free air convection effect. That is why the cavity in our case should be long enough to ensure the temperature uniformity in a longer part of the cylinder near the conical bottom and to increase the effective emissivities. Furthermore, in the front of the cavity is installed a heat compensator with a series of diaphragms whose temperature is adjusted close to that of the heat pipe itself so as to reduce the free air convection effect.

Besides, in the front of the heat compensator there is a water-cooled rotatable gate with a maximum aperture 60 mm in diameter (not shown in Figs. 1 and 2) to provide a proper background radiation.

EXPERIMENTAL RESULTS

1. Temperature distributions

As is shown in Fig. 7, fourteen pairs of copper/constantan differential thermocouples 0.2 mm in diameter were soldered on the upper and the lower parts of the cavity inner surface. The constantan wires had been selected through homogeneity tests. All the reference junctions of the fourteen differential thermocouples are inserted into one of the long wells, 200 mm deep from the open end of the well. A sufficient length of each wire should be laid tightly along the cavity wall before it is extracted through the cavity mouth.

In Tables I and II only the data of 5 representative positions on each level are presented. In Tables III and IV the temperature uniformity data for the whole operating range are shown, which were measured at two levels, over 200 mm and 260 mm from the cone apex respectively.

Care must be taken to properly locate the reference junctions of the differential thermocouples. Since the long well into which they are placed is in the vapor region which provides the best isothermal condition of all parts of the heat pipe, the two junctions of each differential thermocouple are generally subject to the same temperature changes. We once tightly attached with epoxy all the reference junctions at the "n-7" measuring point of the cavity surface (see Fig. 7), only to fail to obtain stable and correct values of the temperature differences.

From the data in Tables I - IV we observe that the small temperature differences between the upper and lower levels confirm our theoretical prediction as elucidated above, and that the tight-contact-with-the-surface arrangement of the two different energy sources is feasible.

2. Transient behavior and stability

With commercial DWK-702 three-mode (PID) controllers (China-made) the two heat pipe blackbodies maintain in the whole temperature range a stability of ± 0.01°C/20 min after the transient time is over.

(i) For the water/copper heat pipe blackbody: The transient time from room temperature up to its upper limit (+150°C) is 50 minutes.

The opposite transition from its upper limit (+150°C) down to +40°C requires 1 hr and 50 min with the help of the fast cooler when the cooling water is at 30°C in summer. Should the fast cooler be not used, the transient time would be four times longer.

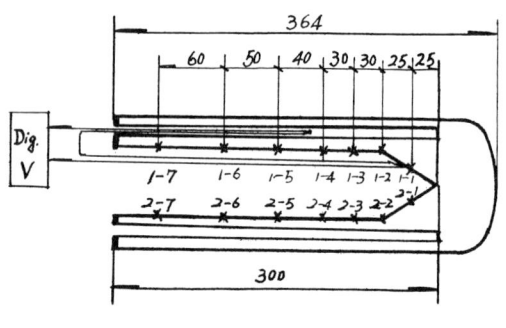

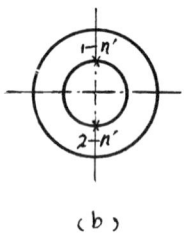

Fig. 7. Positions of 14 copper/constantan differential thermocouples for measuring the temperature uniformity of the cavity inner surfaces.
(a) Longitudinal view of the measured points;
(b) Cross sectional view of the positions.

The transient time for a 10°C step upward or downward transition is 15 - 20 minutes.

(ii) For the ammonia/stainless steel heat pipe blackbody:

Owing to the additional heat capacity of the semiconductor Peltier thermoelectric refrigerators it takes a longer transient time than the water/copper heat pipe blackbody before a $\pm 0.01°C/20$ min stability is maintained.

An hour is required to switch from -40°C up to a +30°C point.

The transient time from +50°C down to -50°C is 2.5 hr when the cooling water for the two-stage semiconductor Peltier thermoelectric refrigerator is at about 30°C in summer. If the cooling water is at such lower temperatures as are encountered in spring, autumn and winter, this transient time will be shorter, e.g. 1 hr.

For a 10°C step upward or downward transition 40 minutes is required.

Table I. The measured temperature differences between various positions on the surface of the ammonia/stainless steel heat pipe blackbody cavity. (For the position designations, see Fig. 7. The vapor temperature is used as reference.)

$$\Delta t = t_{x'} - t_v \quad °C$$

A \ B	n-1	n-4	n-5	n-6	n-7
at +50 °C (blackbody)					
1-n' upper	-0.01	-0.01	-0.01	-0.01	-0.02
2-n' lower	-0.01	-0.01	-0.02	-0.03	-0.05
at +40 °C (blackbody)					
1-n' upper	-0.00	-0.00	-0.00	-0.00	-0.00
2-n' lower	-0.00	-0.00	-0.00	-0.00	-0.01
at 0 °C (blackbody)					
1-n' upper	-0.00	-0.01	-0.00	-0.01	+0.01
2-n' lower	-0.01	-0.01	-0.01	-0.00	-0.00
at -40 °C (blackbody)					
1-n' upper	-0.01	-0.02	+0.00	-0.03	+0.07
2-n' lower	-0.03	-0.03	-0.02	-0.03	-0.02
at -48 °C (blackbody)					
1-n' upper	-0.01	-0.03	+0.01	-0.03	+0.14
2-n' lower	-0.06	-0.08	-0.06	-0.07	-0.04

Note:

A - Axial position;
B - Perpendicular position.

Table II. The measured temperature differences between various positions on the surface of the water/copper heat pipe blackbody cavity. (For the position designations see Fig. 7. The vapor temperature is used as reference.)

$$\Delta t = t_x - t_v \quad °C$$

A \ B	n-1	n-3	n-5	n-6	n-7
at +150 °C (blackbody)					
1-n' upper	-0.02	-0.01	-0.05	-0.08	-0.07
2-n' lower	-0.04	-0.02	+0.00	-0.02	-0.10
at +140 °C (blackbody)					
1-n' upper	-0.04	-0.03	-0.02	-0.03	-0.11
2-n' lower	-0.05	-0.03	-0.01	-0.05	-0.17
at +100 °C (blackbody)					
1-n' upper	-0.05	-0.01	-0.01	-0.01	-0.13
2-n' lower	-0.07	-0.01	+0.00	-0.04	-0.21
at +50 °C (blackbody)					
1-n' upper	-0.10	-0.04	-0.01	-0.04	-0.17
2-n' lower	-0.07	-0.01	+0.01	-0.02	-0.18
at +40 °C (blackbody)					
1-n' upper	-0.08	-0.04	-0.00	-0.05	-0.14
2-n' lower	-0.06	-0.01	+0.01	-0.02	-0.13

Note:

A - Axial position;
B - Perpendicular position.

EFFECTIVE EMISSIVITIES OF THE BLACKBODIES

The surfaces of the two blackbody cavities were coated with F-26 black paint (China-made) 20 - 30 μ in thickness, the total hemispherical emissivity of which is 0.88. Bedford's method for isothermal cavities[11] was employed in the calculation of the effective emissivity distribution along the cavity walls, and the results are depicted in Fig. 8. The data thus obtained show that the total effective emissivities of the cavity walls within 150 mm from the cone apex are greater than 0.999.

Table III. The temperature uniformities of the cavity's inner surface within 200 mm and 260 mm from the cone apex in the whole temperature range of the ammonia/stainless steel heat pipe blackbody.

°C

Distance / Temperature	200 mm	260 mm
+50	± 0.01	± 0.02
+40	± 0.01	± 0.01
+30	± 0.01	± 0.01
+20	± 0.01	± 0.01
+10	± 0.01	± 0.01
0	± 0.01	± 0.01$_5$
−10	± 0.01	± 0.02$_5$
−20	± 0.01	± 0.04
−30	± 0.01$_5$	± 0.05
−40	± 0.02	± 0.05
−50	± 0.05	± 0.10

Table IV. The temperature uniformities of the cavity's inner surface within 200 mm and 260 mm from the cone apex in the whole temperature range of the water/copper heat pipe blackbody.

°C

Distance / Temperature	200 mm	260 mm
+150	± 0.03	± 0.07
+140	± 0.03	± 0.07$_5$
+130	± 0.03	± 0.08
+120	± 0.03	± 0.08$_5$
+110	± 0.03	± 0.10
+100	± 0.04	± 0.11
+ 90	± 0.05	± 0.13
+ 80	± 0.05$_5$	± 0.16
+ 70	± 0.05$_5$	± 0.16
+ 60	± 0.06	± 0.13
+ 50	± 0.05$_5$	± 0.10
+ 40	± 0.05	± 0.07$_5$

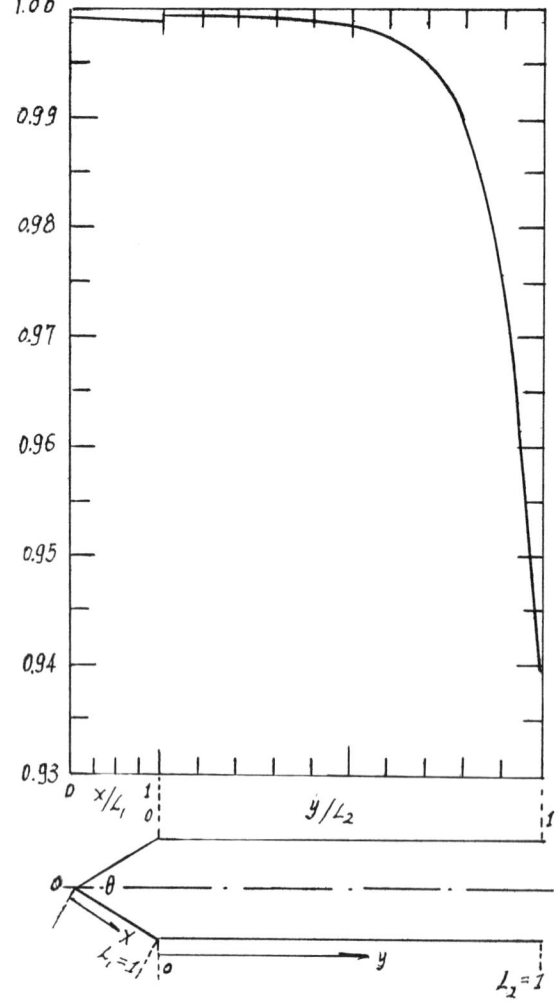

Fig. 8. Effective emissivities of surfaces of the two cavities (calculated as isothermal cavities) with $\varepsilon = 0.88$, $L_1 = 60$ mm, $L_2 = 248$ mm, and $\theta = 60°$.

CONCLUSIONS

The application of heat pipes to horizontally-laid thermometric blackbodies in the range −50 − +150 °C is satisfactory. Even under the conditions of free air convection in a cavity 60 mm in diameter and 300 mm in total length with no lid at its mouth, we still have a uniformity better than ± 0.06°C within a space region 200 mm from the cone apex in the whole temperature range (+40 − +150°C) of the water/copper heat pipe blackbody and better than ± 0.02°C in the temperature range +50 − −40°C of the ammonia/stainless steel heat pipe blackbody.

These two blackbodies are suitable for the calibration of precision radiation thermometers in thermometric laboratories.

ACKNOWLEDGEMENTS

The authors would like to express their sincere thanks to Mr. Hou Zengqi and Mr. Li Tinghan for their contributions to the construction of the heat pipes. Acknowledgement is also due to Mr. Chu Zaixiang for his help in calculating the effective emissivities.

REFERENCES

1. C. A. Busse et al, Temperature Measurement 1975, Institute of Physics Conference Series 26, London, 428-438 (1975).

2. G. Neuer and O. Brost, Temperature Measurement 1975, Institute of Physics Conference Series 26, London, 446-452.

3. C. Bassani and C. A. Busse, Proceedings of 2nd International Heat Pipe Conference, Report ESA SP-112, Vol. 1 (Noordwijk: ESTEC) 537-544 (1976).

4. C. Bassani et al, High Temperature-High Pressure, 12, 351-356 (1980).

5. Susumu Hammori, Transactions of the Soc. of Instr. and Contr. Engineers, (Japanese), 15, 774-779 (1979).

6. R. E. Bedford, Canad. J. Phys. 38, 1256-1278 (1960).
7. D. Chisholm, The Heat Pipe, Mills & Boon Ltd., (1971).
8. E. R. G. Eckert, Heat and Mass Transfer, McGraw-Hill Book Company (1959).
9. P. Dunn and D. A Reay, Heat Pipes, 2nd ed., Pergamon Press, Oxford, 127-128 (1978).
10. B. D. Marcus, Theory and Design of Variable Conductance Heat Pipe, NASA CR-2018, 49 (1972).
11. R. E. Bedford and C. K. Ma, J. Opt. Soc. Am., 65, 565-572 (1975).

NOMENCLATURE

(Many symbols are defined as they appear in the text.)

ΔT_v Temperature difference of vapor phase, in K.
T_v Temperature of vapor, in K.
R_u Universal gas constant = 8.3×10^3 J/K kg mol.
μ_v Dynamic viscosity of vapor, in N s/m^2.
Q Heat flux, in W.
L Length of vapor path, in m.
λ Latent heat of vaporisation, in W s/kg.
P_v Pressure of vapor, in Bar.
ΔP_v Pressure drop of vapor, in Bar.
M Molecular weight of working fluid.
K Wick permeability, in m^2.
A_v Cross-sectional area of vapor flow path, in m^2.

ρ_v Density of vapor, in kg/m^3.
T_0 Temperature of the front flange, in K.
x Axial coordinate from the open mouth, in m.
l Length of the cylindrical part of the cavity, in m.
ΔT_x Temperature difference between vapor and cavity surfaces in a distance x, in K.
α Coefficient of heat transfer between vapor and cavity wall, in W/m^2 K.
u Perimeter of the cavity's inner cylinder, in m.
F Cross-sectional area of the cavity wall, in m^2.
k Thermal conductivity of cavity cylinder, in W/ m K.
k_w Thermal conductivity of wicks, in W/ m K.
k_l Thermal conductivity of liquid phase of working fluid, in W/m K.
δ Thickness of cavity cylinder, in m.
δ_w Thickness of wicks, in m.
β $= (1 + \dfrac{k_s}{k_l})/(1 - \dfrac{k_s}{k_l})$

(See Ref. 9, p. 111).

k_s Thermal conductivity of wick wire, in W/m K.
ε Volume fraction of solid phase of wicks.
f Crimping factor = 1.05 (See Ref. 10, p.49).
M' Wires per inch.
d' Diameter of wick's wire, in m.
q Heat flux density, in W/m^2.

Feedback stabilized tungsten strip lamp as a radiometric standard for photoelectric pyrometry

Zhu Ci-Zhun and Ju Hao-Rien

National Institute of Measurement, Dayi Post-box 659, Szechuan, People's Republic of China

Accuracies of intercomparisons of temperature standards in the photoelectric pyrometer range are low because of the instability of tungsten strip lamps. The feedback stabilized tungsten strip lamp described in this paper can provide a highly stable radiance output, making it possible to establish a new transfer standard for pyrometry.

INTRODUCTION

Realizations of the IPTS-68(75) above the gold point with photoelectric pyrometers are preserved and transferred by tungsten strip lamps. In order to take advantage of the high precision and high sensitivity of photoelectric pyrometers, temporal stability of these lamps is needed. Many studies have been directed to improving the behavior of the lamps by improving their structure. The achievement of a high stability lamp with blackbody characteristics which has been suggested recently (T.J. Quinn, et al.) constitute a great advance. However, the yield of satisfactory lamps from a production run is low and requires careful selection. Even with such improvement, the lack of stability of tungsten strip lamps still limits the precision of photoelectric pyrometers.

Tungsten strip lamps are conventionally used to produce radiance temperatures that are assumed to be realized when calibrated values of current are passed through it. The current sources used for strip lamps are held stable with 1 to 10 ppm, but the achievable radiances of the strip lamps are stable only within 1 to 10 parts per thousand. This level is really not satisfactory for precise applications. As is well known, in fact, not only the lamp current affects the radiance output of the lamp, but also non-electric factors, such as thermal etching of filament, leakage of gas through the envelope, variations of thermal heat losses from the lamp, and effects related to the orientation of the lamp. Sufficient control of all of these factors by improving the construction and the manner of utilization of the lamp will be very difficult. An alternative approach to this problem is detection of the radiance of the tungsten strip lamp with a silicon photodiode (PIN type), and feedback stabilization of the output radiance of the lamp to a set value. This feedback stabilized tungsten strip lamp can replace the usual strip lamp in the application discussed here and in other applications as well.

PRINCIPLE

The total power dissipated by the strip lamp is given by

$$P = P_1 + P_2 + P_3$$

where P is the electric power applied to the lamp, $P = f(I,R)$, P_1 is the radiant energy emitted from the lamp strip, $P_1 = f(T,\varepsilon)$, P_2 is the thermal conduction of the lamp leads, $P_2 = f(K, dT/dx)$, and P_3 is the thermal convection loss in the gas around the strip, $P_3 = f(T,T_a)$.

In the above equation, I is the lamp current, R is the resistance of the lamp, ε is the spectral emissivity of the strip surface, K is the thermal conductivity, dT/dx is the temperature gradient at the joint of the strip and the lead, and T_a is the ambient temperature.

In order that the lamp operate according to "temperature determined by the current", P_2 and P_3 must be constant, as well as the characteristics of the lamp, R, ε, and K. Only then is the calibration function of the temperature lamp given by $I = f(T)$, where I is the lamp current, and T is the radiance temperature. In reality, the above assumption is not fulfilled precisely. If on the other hand, the current passing through the lamp strip is adjusted to achieve a constant radiance, then the radiance temperature should also be constant. The principle is shown schematically in Fig. 1.

The radiance B emitted from R of the lamp is detected with a silicon photodiode, converted to the voltage V_B, and added to the applied voltage V_E, which was pre-set, based on the radiance temperature calibration. Then the sum is directed to the feedback amplification circuit K. If $|V_B|>|V_E|$, the circuit will reduce the current applied to the lamp R. If $|V_B|<|V_E|$, the circuit will increase the current. The net effect is feedback stabilization of the output of the silicon photodiode about the set point value V_E.

The difference between V_B and V_E is dependent on adjustable parameters of the feedback circuit (such as the loop gain). The stability of the radiance depends upon the stability of those parameters which involve the quantum efficiency of the silicon photodiode, the set point value V_E, the operational amplifier drift, etc. The uncertainty of the constant radiance temperature depends mainly on the reliability of the value determined for V_E. One advantage of this approach is the transformation of the estimation of the effects of the non-electrical disturbances into techniques for treating electrical parameters. The accurate estimation of the former is usually difficult, while the latter are well enough developed to realize the high precision of the photoelectric pyrometer. With this approach, improvements in the construction of tungsten strip lamps may not be needed to obtain satisfactory behavior.

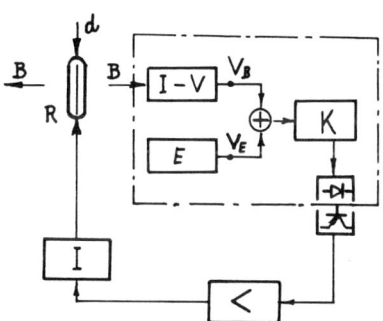

FIG. 1. Schematic diagram of the adjustment system.

REALIZATION

The results of the test by us have shown that a highly stable radiance can be achieved with the described feedback stabilized system. For instance, a stable radiance temperature of 1000 °C can be achieved within the range of ± 0.01 °C. The stability of the quantum efficiency of selected silicon photodiodes used with care also falls within the range of ± 0.01% (the temperature of the photodiode is controlled within ± 0.02 °C). The set point value V_F is determined by a primary calibration based on the gold point blackbody. In order to improve the quality of the stabilization circuit, a pre-feedback parameter has been adopted. The instabilities are negligible when the parameters of the adjustable system are reasonably chosen. In order to protect the system from disturbances depending on power fluctuations in the system, a separating couple between the weak signal link and the power driven link has been used.

With this approach, the relation between the photoelectric current and the radiance temperature in the radiator system replaces the relation between the lamp current and the radiance temperature as the calibration equation for the strip lamp.

FUTURE PROSPECTS

The set point values V_F corresponding to various radiance temperatures from the radiator system will be determined by reference to a gold point blackbody. The set point values will be scaled up and down from the gold point with the aid of two radiator systems as described above and the pyrometer. The calibrations of the set point voltage of the radiator system can then be used as a primary standard or secondary standard to calibrate the photoelectric pyrometer and the other radiator systems.

When the given value of the set point voltage and the quantum efficiency of the photodiode are very well determined after long-observations, it is possible to calculate the temperatures corresponding to the photoelectric current in the radiator system, based on predetermined value of the quantum efficiency of the photodiode. The IPTS above the gold point may be realized with this method, without need for added calibration tests.

Estimation of the true temperature of targets by their thermal radiation based on Planck's law

G. S. Ambrok

D. I. Mendeleyev Research Institute of Metrology (VNIIM), Leningrad, Union of Soviet Socialist Republics

A linear differential equation to determine the spectral distribution of emissivity, given a derivative with respect to the radiance temperature, is presented. The basic problem amounts to determining the constant of integration for the differential equation involved. For estimating the true temperature of the lamp strip, the results of dc experimental calibration of three tungsten strip lamps at the radiance temperatures are used.

Estimation of the true temperature or determination of the emissivity of targets in the course of measurements is a major problem of pyrometry. Attempts to solve this problem were made by De Witt and Kunz whose method is based on simultaneous measurements of temperature ratios and emissivities at two wavelengths [1]. Ruffino has presented an algorithm for calculating the true temperature based upon emissivities for two wavelengths known for a particular temperature [2]. Zhagullo has proposed to present the logarithm of emissivity as a polynomial and to determine the true temperature when the coefficients of this polynomial are known [3]. Tingwaldt and Hoffmann suggested in the thirties to use a parameter dependent on the difference of reciprocal radiance temperatures to estimate the true temperatures. A modified version of this parameter has frequent applications in the studies by Svet [4].

In this paper it is proposed to use a derivative with respect to the reciprocal radiance temperature for evaluation of the true temperatures rather than the parameter dependent on the difference of the reciprocal radiance temperatures. An equation relating true temperature T to radiance temperature S_λ has the form

$$\frac{\varepsilon_\lambda}{\exp(\frac{C_2}{\lambda T})-1} = \frac{1}{\exp(\frac{C_2}{\lambda S_\lambda})-1} \qquad (1)$$

where λ is the wavelength of monochromatic radiation,
ε_λ is the emissivity,
C_2 is the radiation constant.
Equation (1) can be rewritten as

$$\frac{1}{T} = \frac{1}{S_\lambda} + \frac{\lambda}{C_2} \ln y \qquad (2)$$

where $y = \varepsilon_\lambda + (1-\varepsilon_\lambda) \exp(-\frac{C_2}{\lambda S_\lambda}) \qquad (3)$

On differentiating Eqn. (2) with respect to λ so as to eliminate the unknown true temperature, we obtain

$$\frac{d\ln y}{d\lambda} + \frac{1}{\lambda} \ln y = \frac{C_2}{\lambda} \cdot \frac{dS_\lambda}{d\lambda} \cdot S_\lambda^{-2} \qquad (4)$$

The latter is a linear differential equation with respect to $\ln y$ and can serve to determine the spectral distribution of the emissivity ε_λ if the derivative of the radiance temperature with respect to wavelength is known. Solving Eqn. (4), we obtain

$$\ln y = \frac{DC_2}{\lambda} + \frac{C_2}{\lambda} \int \frac{dS_\lambda}{d\lambda} \cdot S_\lambda^{-2} \cdot d\lambda \qquad (5)$$

where D is the constant of integration having the same dimension as the reciprocal temperature.

Eqn. (5) is naturally compatible with (2), if the integral in (5) is calculated and the constant of intergration is taken to be $1/T$. Hence, for an experimental calculation of ε_λ from the spectral distribution of the radiance temperature, the constant of integration D should be determined.

If the reciprocal radiance temperature found experimentally for a particular spectral region is presented as a third-degree polynomial of the form

$$\frac{1}{S_\lambda} = \frac{A}{C_2} \lambda^3 + \frac{C}{C_2} \lambda + \frac{E}{C_2}, \qquad (6)$$

the derivative will be

$$\frac{dS_\lambda}{d\lambda} S_\lambda^{-2} = 3 \frac{A}{C_2} \lambda^2 + 2 \frac{B}{C_2} \lambda + \frac{C}{C_2} \qquad (7)$$

Substituting (7) into Eqn. (5) we obtain an expression for the spectral distribution of $\ln y$

$$\ln y = \frac{DC_2}{\lambda} + A\lambda^2 + B\lambda + C \qquad (8)$$

From this expression it will be obvious that the effect of the constant D on the emissivity determined experimentally is less prominent in the infrared spectral region. We rewrite Eqn. (3) as

$$\varepsilon_\lambda = \frac{y - \exp(-\frac{C_2}{\lambda S_\lambda})}{1 - \exp(-\frac{C_2}{\lambda S_\lambda})} \qquad (9)$$

or

$$\varepsilon_\lambda = \frac{\exp[A\lambda^2 + B\lambda + C + \frac{DC_2}{\lambda}] - \exp(\frac{C_2}{\lambda S_\lambda})}{1 - \exp(-\frac{C_2}{\lambda S})} \qquad (10)$$

when (8) is taken into account.

For small values of the product λS_λ, that is, for the region where Wien's law holds, it is evident from (10) that

$$\ln \varepsilon_\lambda = \frac{DC_2}{\lambda} + A\lambda^2 + B\lambda + C \qquad (11)$$

Hence, one can calculate the spectral distribution y or ε_λ from coefficients A,B,C and integration constant D found experimentally or determine the true temperature from formula (2).

For determining the true temperature of tungsten strip lamps, use is made of the results of experimental calibration of three tungsten strip lamps for radiance temperatures at 23 A D.C., that is, at a true temperature constant for each lamp. These results are presented in Table I.

Table I. Radiance temperatures of standard lamps

Lamp Nos	Current, A	Radiance temperature (K) at different λ (μm)							
		0.300	0.325	0.350	0.400	0.470	0.550	0.656	0.750
38	23	2326	2319	2311	2296	2271	2212	2194	2147
48	23	2332	2323	2316	2301	2273	2245	2195	2117
49	23	2364	2355	2347	2329	2303	2273	2222	2173

The strip lamps were calibrated at VNIIM on special comparators to better than 3 K. The derivatives related to the square of the radiance temperature were computed from radiance temperatures. These derivatives were smoothed on a computer by the least squares technique and presented as a polynominal Eqn. (7). Smoothing was carried out to reduce random errors which occurred in the calibration of strip lamps. In Table II are listed both the derivatives calculated at the wavelengths averaged for the values measured and the coefficients A, B, C of the polynomial for the three lamps. The points calculated for the derivatives with respect to wavelength of the radiance temperature, and the smoothed curve fitting polynomial (7) are given in Fig. 1 for lamp N° 38. Fig. 2 presents the approximating curves for the three lamps and a curve fitting the radiation of tungsten [5] at 2400 K attenuated due to the transmission of glass. The last-named curve for $\frac{dS_\lambda}{d\lambda} S_\lambda^{-2}$ designated in Fig. 2 by W was obtained from (7) and (11) using the tabulated emissivities of tungsten approximated by a polynomial as in (11), with D = 0.

Let us assume that for all the three strip lamps the constant D is zero. Then we can determine the conditional emissivity of the tungsten lamp strip and the strip conditional temperature designated as T* (with $y = \varepsilon_\lambda$) from Eqns. (11) and (2), respectively.

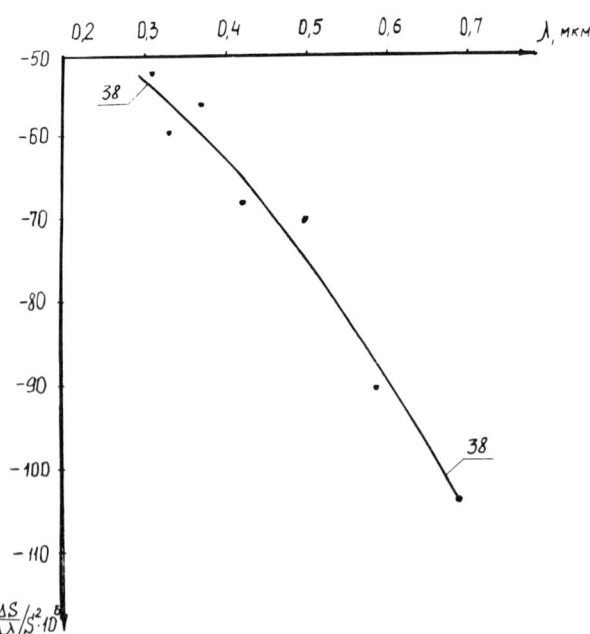

FIG. 1. Derivatives of radiance temperatures for tungsten strip lamp No. 38 and temperature curve smoothed over by the least squares technique.

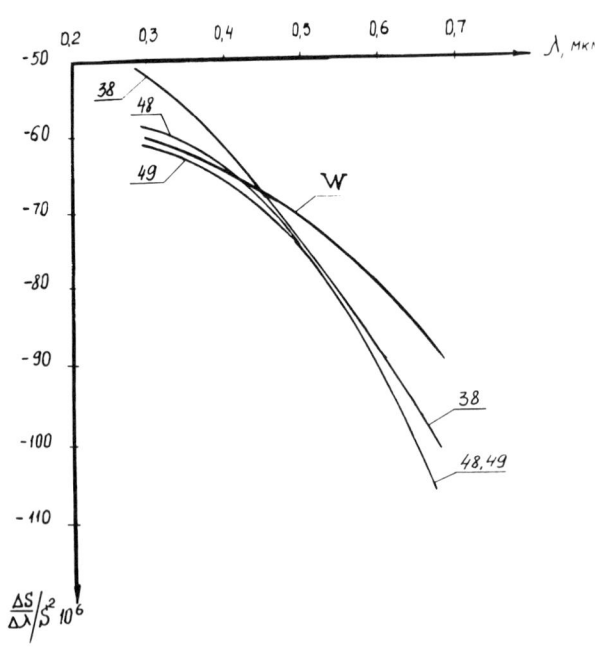

FIG. 2. Curves featuring radiance temperature derivatives for three lamps and radiance temperature derivatives fitting the tabulated emissivity of tungsten at 2400 K.

Table II. Eqn. (7) evaluated for standard lamps

λ, μm	0.3125	0.3375	0.375	0.435	0.510	0.603	0.703
38 $\frac{\Delta S}{\Delta \lambda} S^{-2} \cdot 10^6$	-51.9	-59.7	-56.5	-68.4	-71.2	-92.0	-106.1
	A = -0.6861		B = 0.0678		C = -0.615		
48 $\frac{\Delta S}{\Delta \lambda} S^{-2} \cdot 10^6$	-66.4	-52.0	-56.2	-76.4	-68.5	-97.5	-108.3
	A = -1.122		B = 0.7567		C = -0.9868		
49 $\frac{\Delta S}{\Delta \lambda} S^{-2} \cdot 10^6$	-64.6	-57.8	-65.8	-69.2	-71.6	-95.2	-107.9
	A = -1.220		B = 0.9750		C = -1.135		

In Table III are given values of T* obtained as specified above for different wavelengths of the three strip lamps; Fig. 3 shows values of the conditional emissivity for these lamps and the tabulated experimental emissivity of tungsten of two temperatures, with adjusted values of tungsten emissivity ε_λ corrected for transmittance of the lamp glass taken to be 0.92.

It is clear from Table III that values of T* have good repeatability for each lamp to the mean T_m^*. At the same time, Fig. 3 shows a marked scatter in absolute values of the conditional emissivity for each lamp compared to the tabulated values of the tungsten emissivity. The conditional emissivity of lamp N° 48 agrees with the tabulated values much better. Constant D for this lamp can be assumed to approach zero and the conditional temperature T_m^* for the lamp to be closest to its true temperature. Besides, there is a better repeatability of emissivities for all the lamps with increased wavelength which agrees with formula (11).

Let us estimate now constants D. Considering that effective values of the emissivities for different lamps are close to one another and slightly dependent upon temperature because the lamp strips are made from tungsten and denoting the constants of integration and conditional emissivities by D_1, D_2, D_3 and ε_1^*, ε_2^*, ε_3^*, respectively, for lamps Nos. 38, 48 and 49, we can write three equations (12) based on Eqn. (11). Thereupon substitute values of coefficients A,B,C from

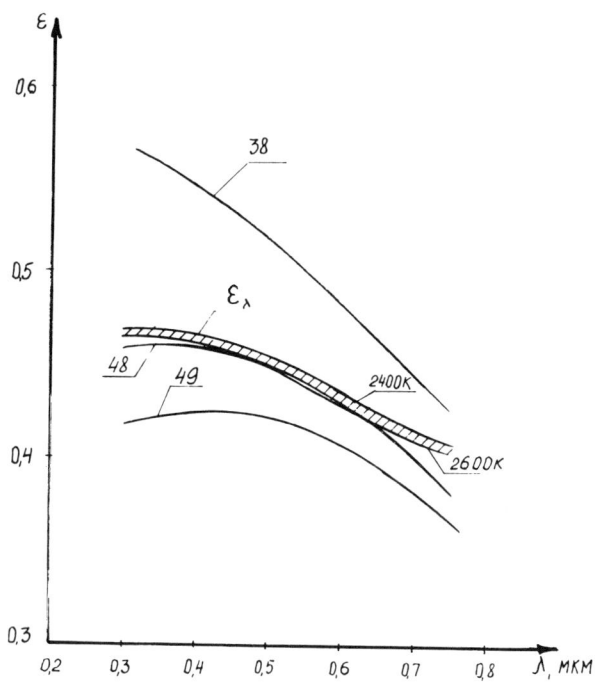

FIG. 3. Conditional emissivities for three lamps and emissivities of tungsten at 2400 and 2600 K.

Table III. T* vs λ determined for standard lamps

λ, μm		0.3	0.35	0.4	0.47	0.55	0.656	0.75
38	T*, K	2401.9	2402.0	2402.4	2401.3	2403.0	2401.0	2402.2
	T_m^* = 2402.0							
48	T*, K	2433.6	2434.2	2436.0	2431.9	2434.8	2431.9	2433.3
	T_m^* = 2433.7							
49	T*, K	2480.1	2480.8	2480.2	2479.5	2482.0	2478.9	2480.8
	T_m^* = 2480.3							

Table IV. Values of D_i vs λ

λ, μm	D_1	D_2	D_3
.465	$\underline{-4.431 \times 10^{-6}}$	0.326×10^{-6}	2.496×10^{-6}
.425	-4.491×10^{-6}	$\underline{0.266 \times 10^{-6}}$	2.409×10^{-6}
.35	-4.458×10^{-6}	0.299×10^{-6}	$\underline{2.448 \times 10^{-6}}$
Mean	-4.458×10^{-6}	0.299×10^{-6}	2.442×10^{-6}

Table V. Conditional and true temperatures for standard lamps

Lamp No.	38	48	49
T*, K	2402.0	2433.7	2480.3
T, K	2428.1	2432.5	2465.8
T − T*, K	26.1	−1.2	−14.5

Table II into these expressions and obtain a number of values for the differences $D_i - D_j$ (13).

$$\begin{aligned} \frac{D_1 C_2}{\lambda} + \ln \varepsilon_1^* &= \frac{D_2 C_2}{\lambda} + \ln \varepsilon_2^* \\ \frac{D_1 C_2}{\lambda} + \ln \varepsilon_1^* &= \frac{D_3 C_2}{\lambda} + \ln \varepsilon_3^* \\ \frac{D_2 C_2}{\lambda} + \ln \varepsilon_2^* &= \frac{D_3 C_2}{\lambda} + \ln \varepsilon_3^* \end{aligned} \quad (12)$$

$D_1 - D_2 = -4.757 \times 10^{-6}$; $D_1 - D_3 = -6.9 \times 10^{-6}$;
$D_2 - D_3 = -2.143 \times 10^{-6}$ (13)

These differences were obtained by averaging over seven wavelengths. Hence, if constant D for one lamp is known, we can find constants D for other lamps or for other temperatures.

Now go back to Fig. 2 from which it is clear that curves for lamps Nos. 38 and 48 intersect curve W at 0.465 and 0.425 μm, respectively, and the one for lamp No. 49 is practically coincident with curve W over the range 0.3 to 0.4 μm. The experimental curve for tungsten W [5] at 2400 K was approximated over the range 0.3 to 0.75 μm by a second-degree polynomial using the least squares technique; in so doing, the coefficients involved are:

A = −0.4445, B = 0.1468, C = 0.7576.

And now, based on Eqns. (7) and (11) we can write an expression for the three lamps mentioned above.

$$A_w \lambda^2 + B_w \lambda + C_w + \ln \tau = \frac{D C_2}{\lambda} + A \lambda^2 + B \lambda + C \quad (14)$$

where τ is the transmittance of the lamp glass.

Eqn. (14) makes it possible to calculate constants D_1, D_2, D_3 at 0.465, 0.425 and 0.35 μm, respectively. The results are given in Table IV where the values underlined are calculated from Eqn. (14) and the rest are found from the above differences (12). The mean values of the constants are presented in the last line of Table IV.

The maximum deviations from the mean values of D given in Table IV do not result in temperature departures more than 0.2 K when calculating corrections to conditional temperatures T*. Thereupon, we can calculate corrections to ε* and T* using Eqns. (11) and (12) so as to approximate the true temperatures of the lamp strips. In Table V are listed the conditional and true temperatures (the former with constants D neglected) calculated as specified above.

Although the conditional temperature correction is slightly over 1% at most, it should be borne in mind that only three examples are considered in this paper, as applied to a material such as tungsten in inert atmosphere (a strip lamp), with its radiating properties clearly understood.

Constant D is not eliminated either for two- or three-color spectral ratio pyrometers. This can easily be shown by substituting Eqn. (11) into the formulae for differences between the color and true temperatures. Hence, the estimation of the true temperature involves the determination of constant D. Besides, a device used for spectral measurements of the radiance temperature should have a low instrumental error so as to obtain smooth derivatives.

REFERENCES

[1] D. P. De Witt, H. Kunz, Theory and technique for surface temperature determinations by measuring the radiance temperatures and the absorptance ratio for two wavelengths. In: *Temperature, Its Measurement and Control in Science and Industry*, (Instrument Society of America, Pittsburgh, 1972), Vol. 4, pp. 599-610.

[2] D. Ruffino, Surface temperature measurement through radiance temperature and emissivity at two wavelengths, High Temperatures - High Pressures, 1979, $\underline{11}$, 2, 221-224.

[3] O. M. Zhagullo, The concept of conventional temperature in modern pyrometry, Teplofizika Vysokikh Temperatur, 1970, $\underline{8}$, 6, 1260-1264 (in Russian).

[4] Optical methods of temperature measurements in metallurgy, Ed. by D. Ya. Svet, "Nauka" Publishers, Moscow, 1979 (in Russian).

[5] Radiating properties of solids, Ed. by A. E. Sheindlin, "Energia" Publishers, Moscow, 1974 (in Russian).

Practical CARS temperature measurements

J. P. Taran and M. Péalat

Office National d'Etudes et de Recherches Aérospatiales, 92320 Châtillon, France

We give a review of nonintrusive temperature measurements taken in various reactive media using Coherent anti-Stokes Raman Scattering (CARS). Small flames, a large scale combustor, and a hydrogen plasma generator were studied. Experimental difficulties encountered and results are presented.

I. INTRODUCTION

CARS has become the established technique for remote, spatially resolved, non intrusive temperature and concentration measurements in flames[1-5] and discharges.[6,7] This technique is still cumbersome to use, however. Progress is being made in the areas of reliability, measurement accuracy, spatial resolution, sensitivity, etc... In this communication, we present some of the most recent results obtained in various experimental situations using our CARS spectrometer.

We recall that CARS is observed when two collinear laser beams of frequencies ω_1 and ω_2 (with $\omega_1 > \omega_2$) are focused in a sample having a Raman-active vibrational mode of frequency ω_v. If $\omega_1 - \omega_2 = \omega_v$, a signal is then generated in the forward direction at $\omega_3 = 2\omega_1 - \omega_2$. Since the theoretical foundations of CARS measurements have already been discussed extensively,[8-10] we shall not dwell on this subject here. Rather, instrumental difficulties and their solutions will be discussed.

II. CARS SPECTROMETER

The spectrometer is a complex and versatile instrument which serves numerous distinct purposes.[11] It is capable of operation at moderate resolution but also in the broadband mode for multiplex spectroscopy.[12] The spatial resolution is adjustable using a BOXCARS beam geometry.[13] If necessary, nonresonant background cancellation[14] can also be used.

Figure 1 is a schematic diagram of the instrument. This is a commercial instrument which was designed so as to be portable. All the components for the lasers and beam-combining optics are bolted onto a cast-aluminium table of 0.5 m x 1.5 m. This unit weighs about 50 kg. The passively Q-switched Yag oscillator, with two 75 mm-long amplifiers and one type II KD*P doubler delivers up to 200 mJ at 532 nm in 10 ns pulses at 1 and 10 Hz. The output is single frequency over 99 % of the shots and presents a spectral jitter under ± 0.01 cm^{-1}. These characteristics are possible only through the use of a stable oscillator cavity with temperature-controlled etalons for line-narrowing. A delay line of 5 m length allows proper coverage of the 6 mm diameter amplifier rods by the beam coming from the oscillator, thanks to diffraction alone, at any repetition rate. This avoids the use of a telescope of adjustable magnification to keep up with the change in beam diameter and divergence when the repetition rate is changed. In addition, only the central, bell-shaped portion of the diffraction pattern is amplified. A telescope with magnification close to 1 is introduced in front of the first frequency doubler at 1 Hz or below ; in addition, a low power negative lens is inserted between the two amplifiers to compensate for their thermal lensing at 10 Hz.

The dye laser chain, composed of one oscillator and one amplifier, is pumped by the Yag second harmonic. A second doubler is used to convert the remainder of the IR emerging from the first doubler. About 20 mJ of second harmonic are thus produced in a low quality beam and are used to pump the dye laser oscillator. Approximately 25 % of the green from the first doubler is also split off to pump the amplifier stage. The oscillator can be tuned with a fixed, high incidence grating and a rotating mirror. The linewidth is 0.7 cm^{-1} and it can be reduced to 0.07 cm^{-1} through insertion of a prism beam expander. This operation only causes a slight shift of the line centre and maintains the cavity alignment precisely. With 50 mJ pump energy in the amplifier, the dye chain delivers from 1 to 8 mJ of tunable radiation in a diffraction-limited beam over the useful CARS range of 560-700 nm, regardless of the linewidth. The tuning is driven by a stepping motor which allows both a continuous sweep from 500 to 800 nm in coarse steps of 0.25 cm^{-1} and limited sweeps of 6 nm about the coarse drive setting in fine steps of 0.012 cm^{-1}.

For multiplex CARS, the dye laser cavity is switched to a configuration where a 100 % back mirror replaces the grating/rotating mirror combination. A Fabry-Pérot etalon of 3 μm thickness is then used for the tuning ; this etalon is slightly wedged so that the tuning is easily achieved by translation of the etalon, and that a one to one correspondence is established between wavelength and position. The finesse is about 10, giving an effective laser linewidth of 40 cm^{-1}. The fraction of Yag second harmonic ω_1 unused for pumping the dye, i.e. typically 100 to 150 mJ, and the dye laser ω_2 undergo some manipulation prior to their focussing into the sample. Both are initially horizontally polarized. They are expanded to a diameter of 7 mm by means of telescopes (that for the dye is corrected for chromatism). The ω_1 beam is passed through a tilted parallel plate with a 50 % coating on one side and a 100 % coating on the other, both of which cover a limited area on the plate. A simple translation of the plate allows passage from the single beam configuration to the two beam (BOXCARS) configuration for improved spatial resolution[13] without loss of alignment. A quarter-wave plate and a half-wave plate can also be inserted for background cancellation.[14] The quarter-wave plate is used to have an exact linear polarization at the sample by a suitable compensation of the ellipticity introduced by the various beam handling optics and windows. The half-wave plate is rotated so as to have an angle of approximately 60° between the ω_1 and ω_2 beams at the sample. The ω_2 beam is passed through two parallel plates which allow one to translate it without changing its direction. A Glan-Thompson prism is used to make the polarization strictly linear and horizontal. The ω_1 and ω_2 beams are combined on a dichroic mirror which is held on a sturdy mount offering 10 µrad alignment sensitivity, which is just sufficient given the beam diffraction angles of about 100 µrad. A small fraction ($\simeq 5$ %) of the beams is subsequently split off to pump the reference channel.

The detection assembly, including the reference leg, is installed on a separate table of 50 cm x 150 cm which is also portable (Fig. 2). All focusing lenses are AR-coated air-spaced achromats. In scanning CARS, the anti-Stokes signals are filtered by means of compact double monochromators, preceded by dichroic filters to prevent breakdown on the entrance and intermediate diaphragms and detection is done with PM tubes. The latter are mounted in the same rack as the electronic signal processing unit to reduce noise pick-up problems. The light is thus piped from the monochromators to the detectors by means of 1 mm-diameter fibers. The photocurrent pulses are treated by the electronic unit, which gates them, calculates their ratios, square roots and their average for a fixed number n of shots (n = 1 to 10 in practice). The electronic unit also rejects shots which do not fall within ± 35 % of the mean in the reference leg, and tunes the dye laser after the n shots have been collected.

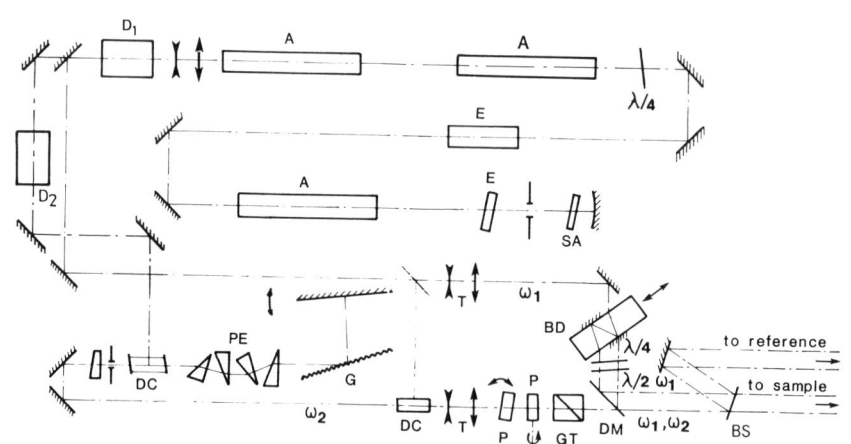

Fig. 1 - Laser source assembly. A : Nd Yag amplifier. BD : parallel plate for production of parallel beams for BOXCARS (BOXCARS arrangement shown, translation of the plate allows passage to collinear arrangement without loss of alignment). BS : beam splitter for reference channel. D1, D2 : KDP doublers. DC : dye cell. DM : dichroic mirror. E : Fabry-Pérot etalon. G : grating. GT : Glan Thompson prism. P : AR coated parallel plate for beam translation. PE : prism expander. SA SA : saturable absorber. T : telescope. $\lambda/4$, $\lambda/2$ are quarter-wave and half-wave plates respectively. ω_1 : "laser" beam. ω_2 : "Stokes" beam.

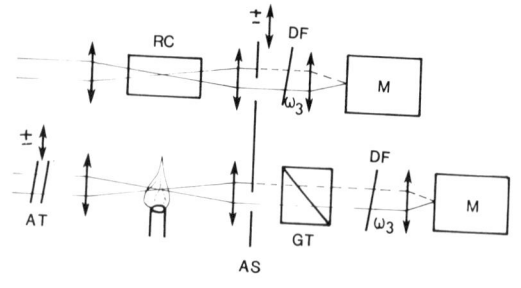

Fig. 2 - Schematic of sample and reference channels. AS : movable aperture stop for operation with parallel beams or crossed beam (crossbeam position shown here). AT : movable attenuators. DF : dichroic filters. M : monochromator and detector. RC : reference cell.

For multiplex CARS, a spectrograph and an optical multichannel analyzer (OMA2 from PAR) are used. The dispersive element in the spectrograph is a 2100 lines/mm, aberration-corrected concave holographic grating with f = 750 mm. The net spectral resolution is 1 cm^{-1}. Both signal and reference spectra are recorded simultaneously on the same vidicon and ratioed channel by channel ; square roots, and averages if necessary, are subsequently calculated. Recording the reference spectrum is a vital requirement since the dye laser spectrum is not reproducible and exhibits appreciable modulation.

In both forms of spectroscopy, the anti-Stokes signal levels in the sample and reference channels are adjusted at approximately 10^4 photoelectrons per shot, or per camera pixel near line maximum, which corresponds to a Poisson uncertainty of about 1 %. Higher fluxes may cause saturation, lower fluxes result in unacceptable uncertainty levels. The sample and reference channels are matched carefully, especially in BOXCARS experiments. The reference cell contains 20 bars of argon ; at higher pressures, difficulties are encountered with breakdown and phase matching. Note that a Glan-Thompson prism is used in the signal leg to isolate the proper anti-Stokes polarization component when background suppression is done. Of course, no polarization filtering is then performed in the reference channel.

III. MEASUREMENT OF ROVIBRATIONAL EXCITATION OF H_2 IN A DISCHARGE

The study of rovibrational populations in tenuous discharges is an impossible task using mechanical probes. This study is also exceedingly difficult using absorption/emission methods. CARS, however, offers an interesting measurement potential over an appreciable temperature and density range. This was demonstrated recently[7] in an H_2 discharge (Fig. 3) designed for H^- production.[15] The spectra are shown in Figures 3 and 4 for a total pressure of 0.13 mbar, without the discharge (Fig. 4a) and with a discharge voltage of 90 V and current of 3 A (Fig. 4b). Collinear CARS without background cancellation was used for maximal signal strength and sensitivity. Note that, since the spatial resolution of collinear CARS is not excellent, the amplitudes of the $v = 0$ lines may come out slightly stronger because of some signal contribution from the cold H_2 surrounding the generator. Only the central portions of the first four Q lines were plotted. Horizontal bars on the plots give the theoretical heights of these lines assuming Boltzmann equilibria at 290 K (Fig. 4a) and 475 K (Fig. 4b) for the rotation. Uncertainties in temperature measurements are 5 K and 15 K respectively. From the line intensities with the discharge turned on, we deduce that molecular H_2 constitutes approximately 90 % of the gaseous mixture, the rest being composed of ions, radicals and electronically excited H_2. The comparison of the relative amplitudes of Q(1) lines in $v = 0$, $v = 1$ and $v = 2$ (Fig. 5) also gives a measure of the vibrational excitation.

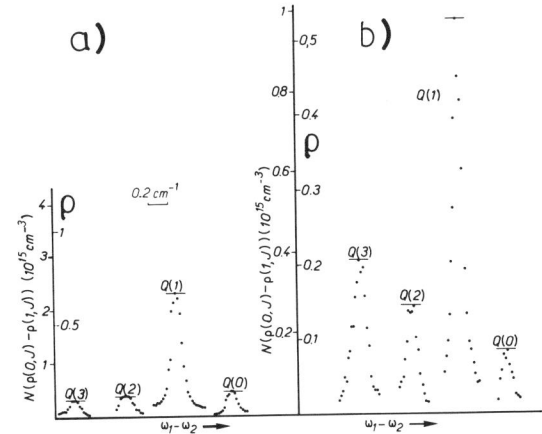

Fig. 4 - Profiles of lines Q(0) to Q(4) for the $v = 0 \rightarrow 1$ fundamental transition of neutral H_2 :
(a) without the discharge ;
(b) with a discharge of 90 V, 3 A.
The strongest portions of the lines only are shown, in steps of 0.02 cm^{-1}, and with 10 consecutive measurements averaged at each point. The spectral resolution of the dye laser was 0.07 cm^{-1}.

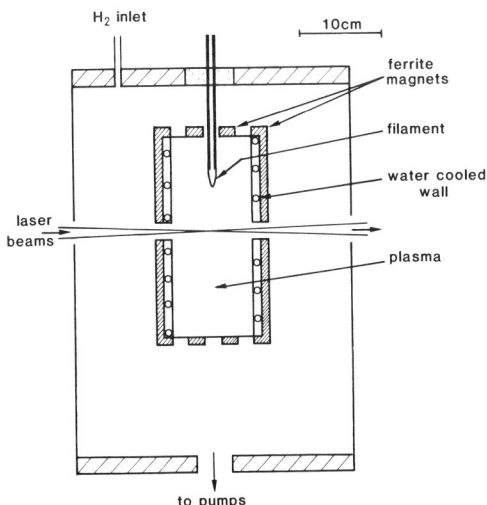

Fig. 3 - Schematic of the plasma generator, drawn to scale.

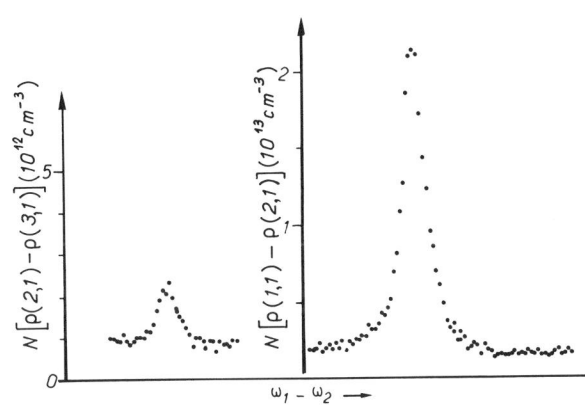

Fig. 5 - Profiles of excited states lines, with 30 measurements averaged at each point.

The line associated with $v = 3$ could not be detected. In this preliminary work, the detection sensitivity on $v = 2$ is about 10^{12} cm^{-3}. A sensitivity of 10^{11} cm^{-3} is technically feasible and should be demonstrated in the near future.

The present results give :
$N(v=0) = 1.9 \; 10^{15}$ cm^{-3},
$N(v=1) = 4.5 \; 10^{13}$ cm^{-3}
and
$N(v=2) = 4.4 \; 10^{12}$ cm^{-3},
assuming
$N(v=3) = 10^{12}$ cm^{-3}.

The assumption on $N(v=3)$ is necessary since it is the difference between the number densities of the Raman levels which is measured in CARS experiments.[7-10] These results clearly show that the vibrational equilibrium established is non-Boltzmann (Fig. 6).

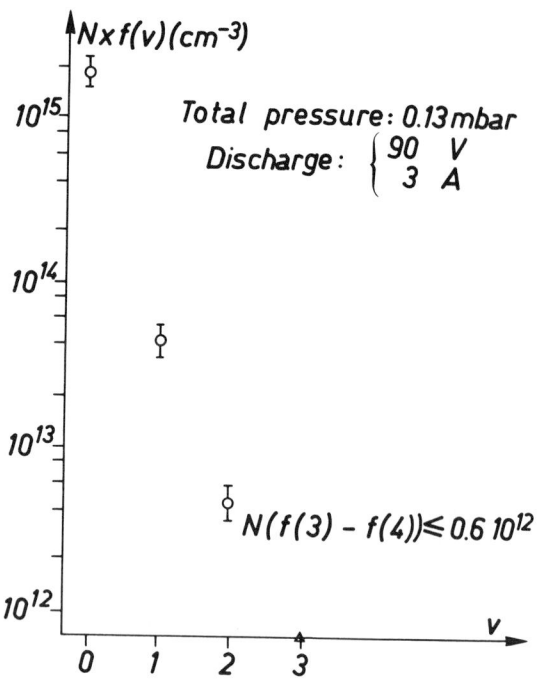

Fig. 6 - Distribution of vibrational populations in the discharge.

IV. TEMPERATURE AND H_2 CONCENTRATION MEASUREMENTS IN AN ETHYLENE-AIR BUNSEN FLAME.

H_2 gas plays an important role in the chemistry and energy balance of flames. This was demonstrated recently in a premixed Bunsen flame burning ethylene in air.[16] By recording CARS spectra of the Q-branch of the H_2 formed by the pyrolysis of ethylene, we were able to obtain the static temperature profile and the H_2 concentration distribution. A BOXCARS optical arrangement was used to give a spatial resolution of about 1 mm. The spectral analysis was similar to that done in the plasma, although we only monitored peak line intensities. In the data processing, special precautions had to be taken with regard to line broadening mechanisms and concentration calibration. The burner itself was designed so as to give good flame stability. It had a diameter of 10 mm. It was operated at a C/O ratio of 0.57 and mass flow rates of 8.9 mg/s for C_2H_4 and 76.2 mg/s for air ; this gave a flame cone height of 26 mm at 296 K and 0.98 bar, corresponding to a burning velocity close to 17 cm/s. The results are presented in Figs. 7 and 8.

Most interesting is a comparison of these results with thermodynamic equilibrium calculations of adiabatic flame temperature and stable product composition. Using the water-gas equilibrium and neglecting dissociation, we find for our flame an adiabatic flame temperature of 1940 K and a composition of 14.9 mol% CO, 4.0 mol% CO_2, 10.2 mol% H_2O and 8.6 mol% H_2. There is strikingly good agreement between the final measured temperature in the burnt gas and the calculated temperature, and in particular, between the hydrogen molar fraction measured just beyond the reaction zone and its thermodynamic equilibrium value of 8.6 %. This agreement leaves no doubt that the CARS technique is indeed suitable for spatially resolved concentration measurements in reactive media. However, more checks have to be carried out, e.g. CARS measurements of CO, CO_2 and H_2O, to find out if water-gas equilibrium is established or not. Furthermore, the hydrogen rotational temperatures should be compared to N_2 rotational temperatures. Hence, at the present moment, we regard our measured hydrogen concentration as semi-quantitative ; it will be improved upon once better knowledge of lineshapes and widths becomes available.

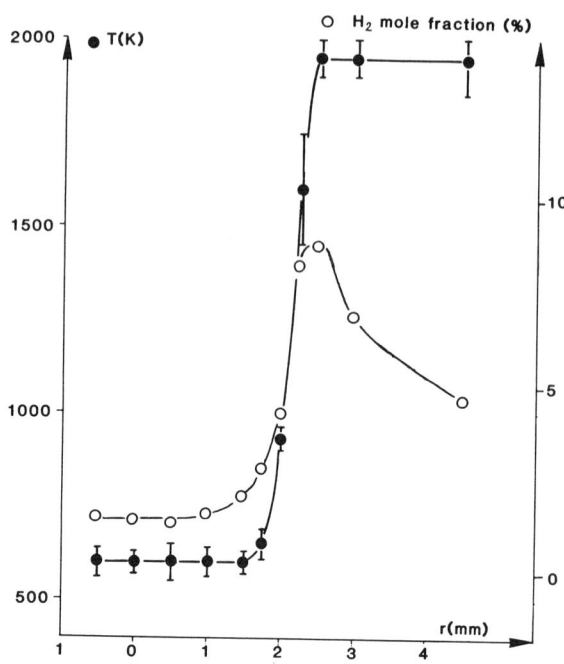

Fig. 7 - Rotational temperatures ● and hydrogen molar percentages O as a function of radial distance from burner axis. Height above burner exit plane : 11.5 mm.

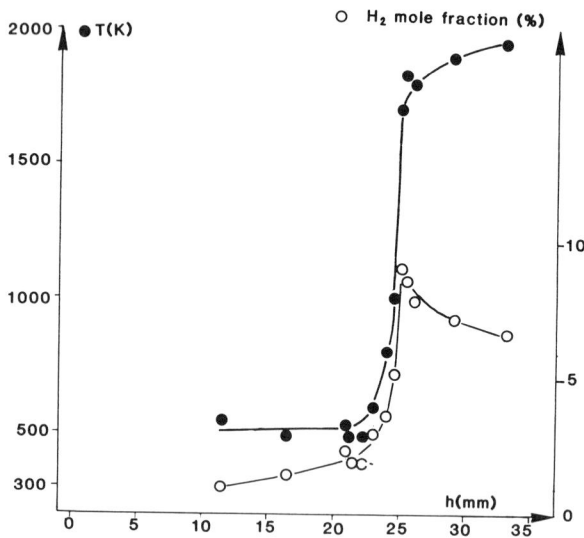

Fig. 8 - Rotational temperatures ● and hydrogen molar percentages O as a function of distance from burner exit plane along burner axis.

As to now, our results demonstrate remarkably well, besides oxidation of hydrogen in the burnt gas, the effect of hydrogen diffusion from the reaction zone into the fresh gas. It is quite surprising that the hydrogen molar fraction is about 1 % at h = 11.5 mm (i.e. 14.5 mm downwards from the flame top and at a radial distance of 2.5 mm from the flame cone) and as high as 3.7 % at a distance of 2 mm below the flame top. Hydrogen is thus enriched in the central flow line. A similar effect, the enrichment of heavier hydrocarbons in the central flow line of a Bunsen flame, was described by Flossdorf et al.[17] However, the enrichment of hydrogen in the fresh gas has some implications on the heat balance in the fresh gas. It is quite conceivable that the pre-reaction zone hydrogen originates from the recombination of hydrogen atoms. This would, as a quick calculation shows, for a hydrogen molar fraction of 1 % give a temperature rise of the fresh gas of about 80 K and hence partially account for our observed temperature rise of about 200 K, compared to ambient, at h = 11.5 mm. Even if the molecular hydrogen itself was the main diffusing species, it would still carry some enthalpy and thus still act as a "heat recirculating chemical carrier", though, of course, in a much less efficient way than the hydrogen atom. The rise in H_2 concentration near the reaction zone with increasing height above the burner exit plane could also be an explanation for the well established fact that the burning velocity is not a constant over the total flame area.[17]

V. INSTANTANEOUS MEASUREMENT OF TEMPERATURE IN A WELL-STIRRED REACTOR

The well-stirred reactor presents interesting challenges to the CARS diagnostician : can one measure the magnitude of the temperature fluctuations and the gas residence time ? None of these measurements can be undertaken properly using conventional probing techniques, yet they are of extreme interest to the theorist.

A well-stirred reactor was built for these studies.[18] The reactor is cylindrical in shape with 5.5 cm id. and 3 cm height. A mixture of methane and air is injected through an array of small holes placed on a central injection column. Exhaust is through five 1 cm-diameter holes drilled in the top flange.

The CARS beams are passed horizontally through holes of 6 mm diameter which are diametrically opposed in the cylinder walls. The temperature study was done in the multiplex mode, using a BOXCARS configuration which gave under 1 mm spatial resolution. Numerous instantaneous spectra were recorded. That presented in Fig. 9 is typical. This spectrum is shown with a theoretical spectrum calculated for a temperature of 1700 K. The single shot temperature measurement accuracy is better than $\pm$ 50 K for this series of experiments. The temperature measured is close to the adiabatic temperature (2100 K), the drop being easily accounted for by losses to the walls. The standard deviation in the measured peak intensity (16 %) receives contributions from the local temperature fluctuations as well as from our measurement uncertainty. The measurement uncertainty comes largely from beam wandering caused by the gradients in the flame. Originally, a large source of imprecision was the mixing of hot flame gases with cold air in the vicinity of the optical ports. This caused amplitude fluctuations in the range of 1 to 4, and distortions of the spectral contour which seriously affected the temperature measurement accuracy. The cure was to apply fused silica windows to the ports. Heating of the walls limits burner operation to 5 minutes, which enables one to collect about 500 spectra.

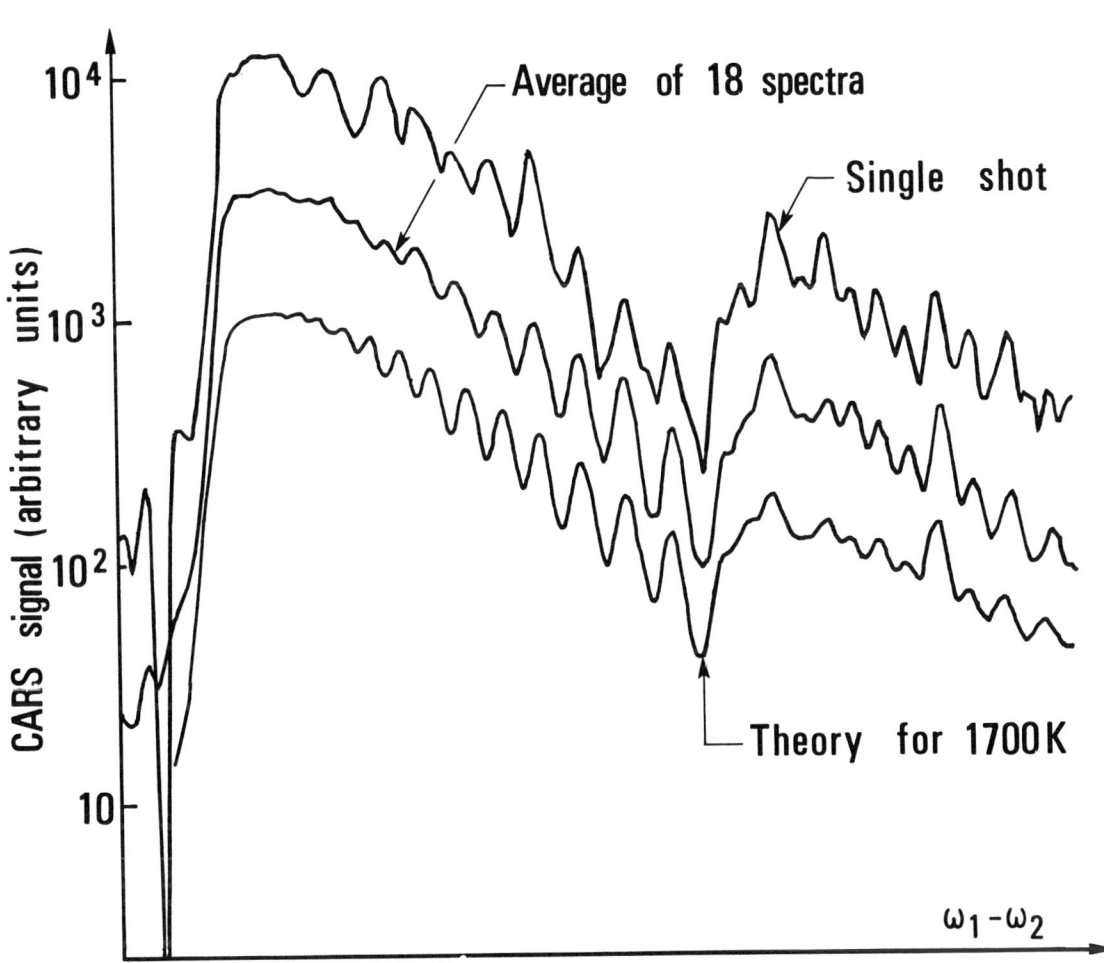

Fig. 9 - Single shot BOXCARS spectrum of N_2 in well-stirred reactor and calculated spectrum assuming T = 1700 K. The peak amplitude corresponds to about 5.10^3 photoelectrons per OMA channel. Spectral resolution is about 1 cm^{-1}. Gated intensifier stage.

The residence time was monitored using a tracer injected in the fresh gases immediately before the burner for periods of 30 ms separated by 30 ms. The tracer which we intented to use originally was CF_4, but this gas was rapidly abandoned because of its chemical reactivity in the burner above 1500 K, and also because of the rapid drop in CARS line intensity as a function of temperature due to broadening (Fig. 10). Instead, we used CO_2 which offers two sharp and intense Q branches which do not broaden as a function of temperature (Fig. 11). The dye laser was operated narrowband and was tuned approximately 0.2 cm^{-1} away from the ν_1 mode so that a small frequency jitter would not cause excessive signal change. For a sample of fixed concentration, the standard deviation on the CO_2 concentration measurement is thus 15 %. Figure 12 shows the exponential rise of the excess CO_2 concentration in the burner following beginning of its injection. In the well-stirred reactor, the rise and decay times are in principle equal. The residence time deduced from this curve (1.8 ms) is in good agreement with that (1.7 ms) obtained by dividing the burner volume (71 cm^3) by the volumetric flow rate used in that particular run (41 10^3 cm^3/s at 1700 K using 7.82 g/s of air and 0.404 g/s of CH_4) ; 0.627 g/s of CO_2 was introduced during the injection periods.

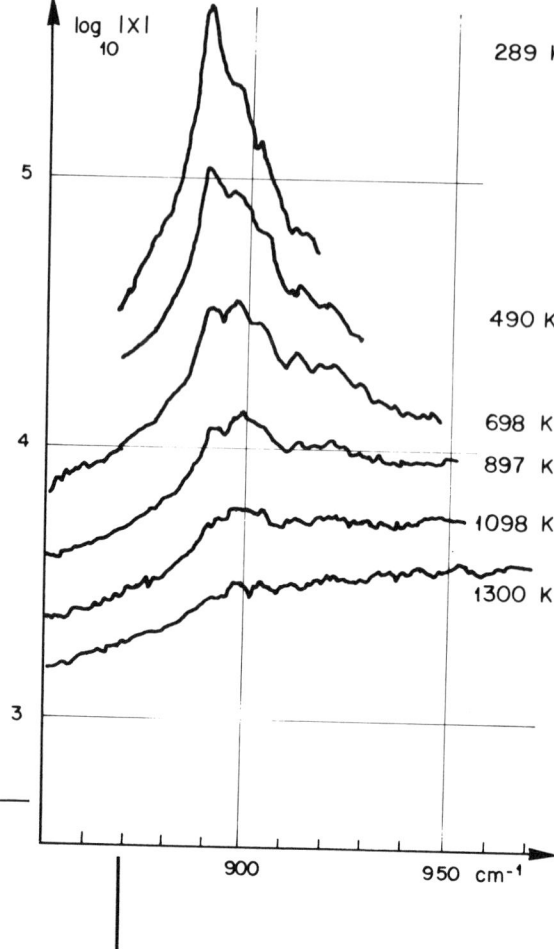

Fig. 10 - Spectrum of pure CF_4 in furnace at atmospheric pressure and for several temperatures, recorded at 0.1 cm^{-1} resolution. Scanning CARS was used, with 4 measurements averaged at each point.

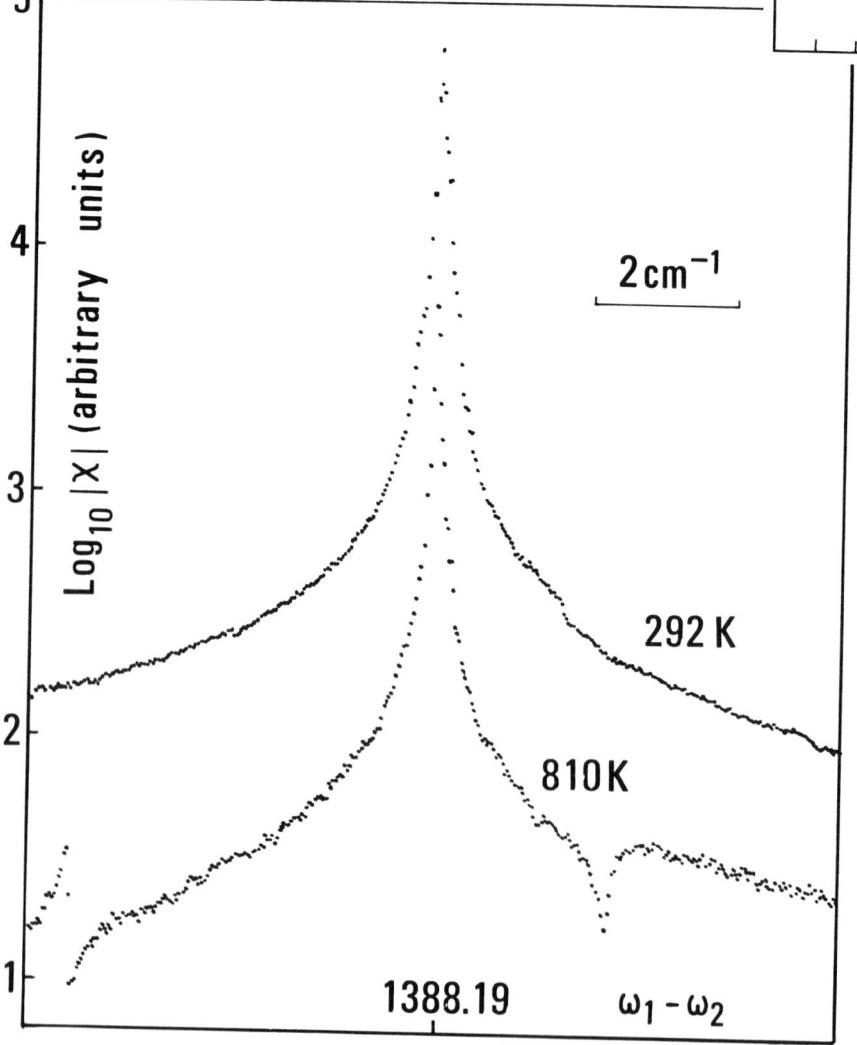

Fig. 11 - Spectrum of pure CO_2 at atmospheric pressure for two values of temperature. Same conditions as for Fig. 10.

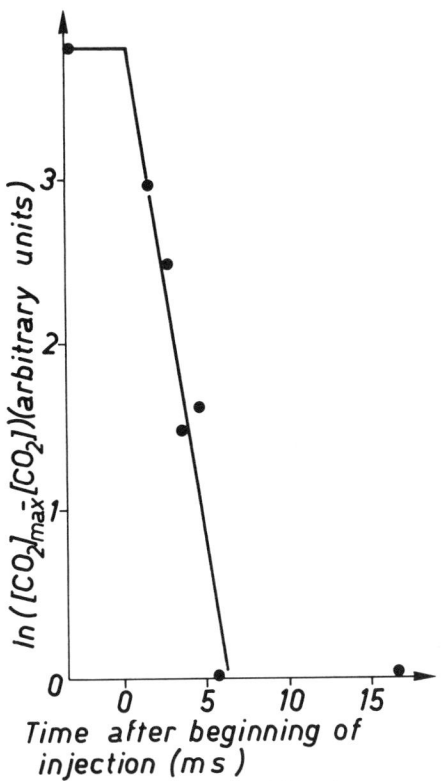

Fig. 12 - Rise in CO_2 concentration in reactor following beginning of injection ; 100 measurements are averaged for each data point ; repetition rate : 2 Hz ; the data are presented after subtraction of maximal CO_2 concentration during injection ; opening and closing times are about 0.3 ms.

VI. TEMPERATURE MEASUREMENTS IN A SIMULATED TURBOMACHINE COMBUSTOR

Large scale combustors are difficult to tackle by CARS, since environment is harsh and optical access is limited. Therefore, many precautions have to be taken to protect the optical set-up from noise, dust and oil vapors. The feasibility of making temperature and concentration measurements by CARS in a kerozene-fueled burner was tested at SNECMA.[19] The burner produces a flow which is approximately rectangular, with a size of 15 x 50 cm, and a mass flow rate approaching 2 kg/s for atmospheric pressure operation. The CARS optical tables were installed in a test room near to that housing the combustor (Fig. 13), along with the power supplies and electronics. Single shot BOXCARS spectra were taken at various locations in the burner, with a quality approaching, but not as good as, that of Fig. 9. A small percentage of spectra was lost in the primary zone (near the injectors) because of fuel droplets and turbulence, which cause spectral envelope distortion and attenuation.

Figure 14 presents a temperature profile recorded in the primary zone. This temperature is a time average over about 20 consecutive single shot measurements recorded in a 20-minute period (the data acquisition rate then was close to 2 per minute). The profile clearly shows the drop in temperature, which tends to a limiting value of 500 K close to that of the preheated air used in that particular run.

We have also studied O_2 and CO, trying to obtain their spatial distributions. The detectivity using BOXCARS without background cancellation was about 3 % in mole fraction for these species in the flame, at the time of these experiments. With background cancellation, the detectivity was further degraded by a factor of two. These results will be published elsewhere.[19]

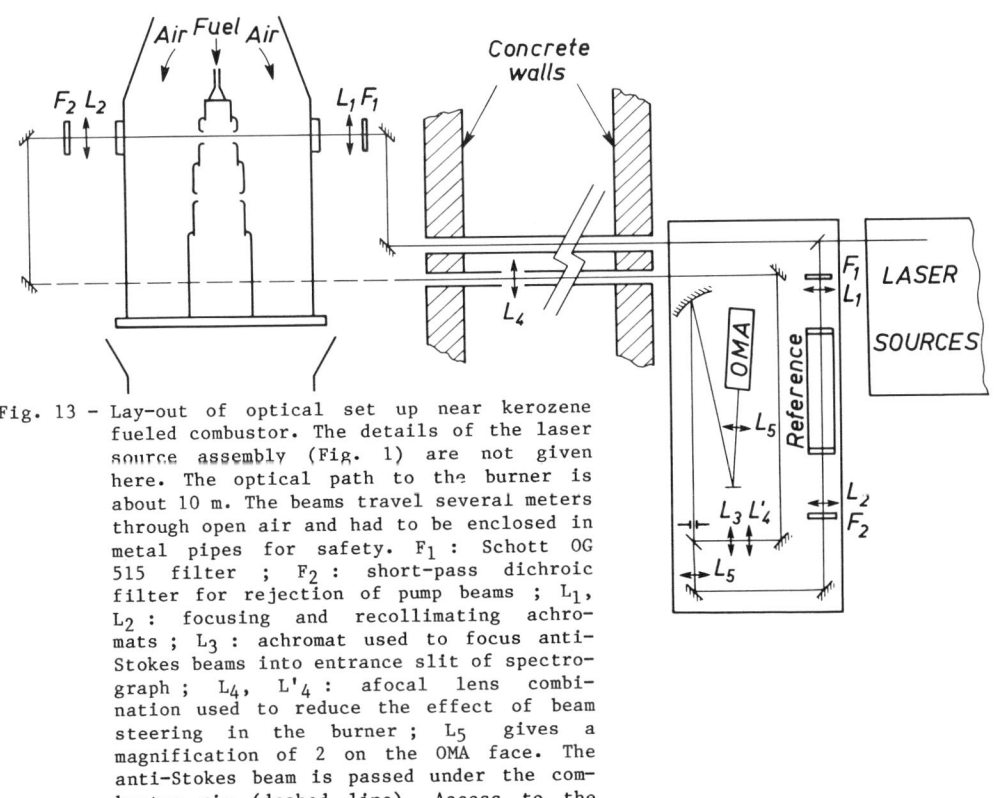

Fig. 13 - Lay-out of optical set up near kerozene fueled combustor. The details of the laser source assembly (Fig. 1) are not given here. The optical path to the burner is about 10 m. The beams travel several meters through open air and had to be enclosed in metal pipes for safety. F_1 : Schott OG 515 filter ; F_2 : short-pass dichroic filter for rejection of pump beams ; L_1, L_2 : focusing and recollimating achromats ; L_3 : achromat used to focus anti-Stokes beams into entrance slit of spectrograph ; L_4, L'_4 : afocal lens combination used to reduce the effect of beam steering in the burner ; L_5 gives a magnification of 2 on the OMA face. The anti-Stokes beam is passed under the combustor rig (dashed line). Access to the primary reaction zones is through air dilution holes in the can.

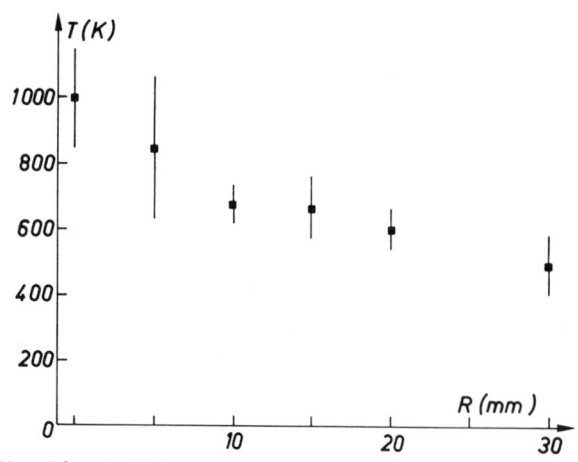

Fig. 14 - Radial temperature profile in primary zone. R is measured from burner midplane; note that the pump beams cross the plane of the air injection port at R = 50 mm. The error bars represent the standard deviation in the measurements and reflect contributions from real temperature fluctuations and from measurement uncertainty.

VII. CONCLUSION

A large number of experimental results was accumulated over a period of 18 months. This was made possible by the quality of components now available for CARS diagnostics, so that quick installation and reliable operation have become possible. A wide variety of experiments can be undertaken, and extremely useful information can be extracted from flames and plasmas. Only a few of these have been presented here; molecular lasers, piston engines,[5,20] chemical vapor deposition, etc... can also be studied successfully.

VIII. ACKNOWLEDGMENTS

The authors are especially grateful to J. Bédué and P. Gastebois for their interest in CARS, which led to the common research effort on the SNECMA combustor, and for their active collaboration during the various phases of this work. Special thanks are due to R. Bailly, P. Bouchardy, J. Gayrard, P. Gicquel, J.L. Babigeon for experimental assistance and useful conversations.

REFERENCES

1 - P.R. Régnier and J.P.E. Taran, Appl. Phys. Letters 23, 240 (1973).
2 - F. Moya, S.A.J. Druet, and J.P.E. Taran, Optics Commun. 13, 169 (1975).
3 - A.C. Eckbreth and R.J. Hall, Combustion and Flame 36, 87 (1979).
4 - G.L. Switzer, W.M. Roquemore, R.B. Bradley, P.N. Schreiber, and W.B. Roh, Applied Optics 18, 2243 (1979).
5 - I.A. Stenhouse, D.R. Williams, J.B. Cole, and M.D. Swords, Applied Optics 18, 3819 (1979).
6 - J.W. Nibler, J.R. Mc Donald, and A.B. Harvey, Opt. Commun. 18, 371 (1976).
7 - M. Péalat, J.P.E. Taran, J. Taillet, M. Bacal, and A.M. Bruneteau, J. Appl. Phys. 52, 2687 (1981).
8 - J.W. Nibler and G.V. Knighten, "Coherent Anti-Stokes Raman Spectroscopy", Topics in Current Physics, Chapter 7, Ed. A. Weber, Springer Verlag, Berlin, New York 1977.
9 - S.A.J. Druet and J.P.E. Taran, "Coherent AntiStokes Raman Spectroscopy", Chemical and Biochemical Applications of Lasers, Ed. C.B. Moore, Academic Press, New York, 1979; ibid, Progress in Quantum Electronics, Pergamon, New York, London, Vol. 7, N° 1 (1981).
10 - M.D. Levenson and J.J. Song, "Coherent Raman Spectroscopy", Topics in Current Physics, Chapter 7, Ed. M.S. Feld and V.S. Letokhov, Springer Verlag, Berlin, New York, 1981.
11 - M. Péalat, J.P.E. Taran, and F. Moya, Opt. Laser Technol. 21 (February 1980);
B. Attal, M. Péalat, and J.P.E. Taran, J. Energy, 4, 135 (1980).
12 - W.B. Roh, P.W. Schreiber, and J.P.E. Taran, Appl. Phys. Letters 29, 174 (1976).
13 - A.C. Eckbreth, Appl. Phys. Letters 32, 421 (1978).
14 - L.A. Rahn, L.J. Zych, and P.L. Mattern, Opt. Commun. 30, 249 (1979).
15 - M. Bacal and G.W. Hamilton, Phys. Rev. Letters 42, 1538 (1979).
16 - K. Muller-Dethlefs, M. Péalat, and J.P.E. Taran, Ber. Bunsenges. Phys. Chem. 85, 803 (1981).
17 - J. Flossdorf, W. Jost, and H. Cg. Wagner, Ber. Bunsenges. Phys. Chem. 78, 378 (1974).
18 - P. Bouchardy, P. Gicquel, M. Péalat and J.P.E. Taran, to be published.
19 - J. Bédué, P. Gastebois, R. Bailly, M. Péalat and J.P.E. Taran, to be published.
20 - D. Klick, K.A. Marko, and L. Rimai, Appl. Optics 20, 1178 (1981).

A hardened CARS system for temperature and species-concentration measurements in practical combustion environments[a]

G. L. Switzer and L. P. Goss

Systems Research Laboratories, Incorporated, 2800 Indian Ripple Road, Dayton, Ohio 45440-3696

A system employing the technique of Coherent Anti-Stokes Raman Spectroscopy (CARS) has been constructed to permit non-intrusive combustion diagnostics in the harsh environment of practical combustion systems. Features incorporated in the system design include the capability of obtaining simultaneous temperature and concentration information on multiple gas species, temporally resolved true time-averaged data at rates of up to 20 Hz, and high spatial resolution within the sampled volume. In order to handle the potentially large number of CARS optical signals economically, optical fibers are employed to accomplish a parallel-to-serial conversion of the light pulses between the hardened optical system and a remotely situated spectrometer/detector combination. A description of the system and its capabilities is presented along with a quantitative description of such parameters as measurement accuracy, precision, and spatial resolution.

INTRODUCTION

The potential of the coherent anti-Stokes Raman spectroscopy (CARS) technique for non-perturbing optical diagnostics in practical combustion media has continued to develop over the past few years.[1-4] At the Air Force Wright Aeronautical Laboratories/Aero Propulsion Laboratory, a program contributing to this development has been directed toward employing the CARS technique to obtain temperature and number-density information in the harsh environment produced by simulated practical combustors. Experiments conducted early in this program with a ruby-laser-based collinear optical system operating at 1 Hz[5] demonstrated the ability of a CARS system to perform reliably in such hostile environments. Data from these initial experiments indicated that improvements in the CARS-diagnostic-system capabilities were needed. For these improvements a system capable of high spatial resolution was required to provide more spatially precise single-shot simultaneous temperature and number-density data as well as higher repetition rates. With higher repetition rates, sufficient data could be obtained in a short period of time to allow true time-averaged measurements of these quantities to be developed from their probability distribution functions (pdf's).

This paper contains a description of an environmentally hardened CARS system constructed to incorporate those improvements which were suggested by earlier experimentation. Among other improvements to system performance, the system design has been extended to include the capability of performing temperature and number-density measurements simultaneously on more than one gas species. The results of measurements performed during the initial check-out of this system in the laboratory are presented to give an indication of the precision and accuracy of which the system is presently capable.

OPTICAL-SYSTEM CONFIGURATION

The hardened CARS system is based upon a Quanta-Ray DCR1A frequency-doubled neodymium-YAG laser (which has a maximum repetition rate of 20 Hz) and a folded BOXCARS[6-7] optical configuration which permits very high spatial resolution. The optical layout of the system is shown schematically in Fig. 1 and can be traced as follows. The Quanta-Ray laser is supported above a 1.2 m × 1.83 m Newport Research Corp. breadboard table top. Its 532-nm (green) output is folded down by periscope P1 to an optic plane 7.6 cm above the table. Approximately 50% of the nominally 185 mJ of green output is passed through beam splitter BS1 and focused by lens DL1. This focused beam is split in half by beam splitter BS2. The portion reflected from BS2 is used for optical pumping of dye cell DC1 which, combined with mirrors DM1 and DM2, forms a broad-band (150 cm^{-1}), concentration-tuned dye-laser oscillator. The remaining green pump energy transmitted through BS2 is used to pump dye cell DC2 for dye-laser amplification (optionally, the energy used to pump the amplifier can be diverted and used to drive a second-frequency dye oscillator for multiple-species operation). The Stokes-radiation beam thus produced passes through a beam-expander telescope formed by dye lenses DL2 and DL3 in order that its divergence may be adjusted to produce a coincident focus of the dye and green beams in the sample volume.

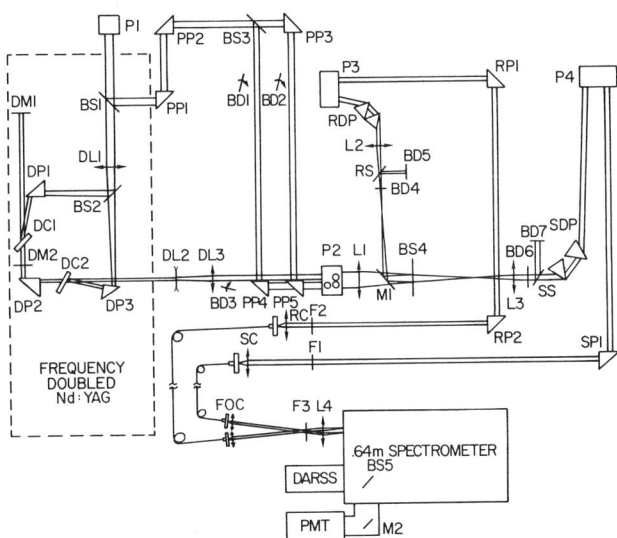

FIG. 1. CARS-Optical-System Schematic.

The remaining green beam, formed from the 50% of the input energy reflected from beam splitter BS1, is split in half by beam splitter BS3 after passing through a delay-equalization path formed by pump prisms PP1 and PP2 to compensate for a delay experienced in the dye-laser geometry. The beam reflected at BS3 is passed parallel to the table toward pump prism PP4. The transmitted portion reflected by PP3 is inclined such that as it leaves prism PP5, it is vertically aligned with--but displaced 2.5 cm below--the green beam leaving PP4. It is this separation of the two green "pump" beams which forms the basis for the

BOXCARS[8] configuration. The dye-laser beam is made parallel and coplanar with the pump beam from prism PP4 but separated horizontally by 1 cm, causing the formation of the folded BOXCARS geometry.

The three-beam configuration thus formed is elevated to a height of 38 cm by periscope P2 which consists of a 50-mm right-angle prism on the bottom and two 25-mm prisms on the top. One of the upper prisms controls the pointing of the pump and dye pair, while the other directs the single pump beam toward a 10-cm-diam., 50-cm-focal-length lens L1. Since the upper two prisms on periscope P2 are independently adjustable in their vertical separation, the CARS geometry is continuously adjustable from collinear (with properly reconfigured dye placement) to BOXCARS with a maximum crossing full angle of $\approx$ 10 deg. The three beams converge from L1 through beam splitter BS4 toward a sample volume located midway between BS4 and a 7.6-cm-diam., 40-cm-focal-length collimating lens L3. The spacing of the components BS4 and L3 is such that the sampled volume can be scanned $\pm$ 11 cm about the center of a 47-cm-diam. combustion tunnel. After collimation at L3, most of the green and dye energies are absorbed in beam dumps BD6 and BD7. The anti-Stokes signals generated in the sample volume are spatially separated from any remaining green light by passing them through sample-dispersing prisms SDP. After dispersion, separation of spectral components takes place over a 3-m path traversing perisocpe P4, sample prism SP1, and neutral-density filter F1; and, finally, the anti-Stokes energy is focused onto a 200-μm-diam. fused-silica fiber-optic waveguide by the numerical-aperature-matching sample-collection optic SC.

In order to compensate for fluctuation in the CARS signals caused by laser-power fluctuations or slight beam misalignments, a parallel referencing scheme is employed. The medium in which the reference signal is generated is controlled to produce either a resonant or a nonresonant CARS signal, depending upon whether integrated intensity or spectral normalization of the sample signal is desired. This reference leg begins with the 15% energy reflection from beam splitter BS4. A description similar to that given for the operations performed in the sample beam path can be given for the reference leg ending with the signal being focused into a reference fiber by the collection optic RC.

The lengths of the sample and reference fiber-optic waveguides are critical to overall system operation and are determined using two criteria. First, in order to minimize environmental effects found in practical combustion situations upon the sensitive optical detectors, the spectrometer and detectors are separated from the optical system and operated in a controlled environment. To accomplish this separation, a minimum of 20 m of fiber optics is presently required. Second, for reasons to be discussed in conjunction with the data-collection electronics, the reference fiber optic is made 20 m ($\approx$ 100-ns optical transit time) longer than the sample. The diameter of these fibers (200 μm) also has an impact upon system performance. Since the CARS signals transmitted through the fibers lose their spatial coherence, the degree to which these signals can be refocused upon the detectors is limited to a minimum which is the fiber diameter, resulting in lowered system spectral resolution.

Upon exiting the optical fibers, the sample and reference beams are collected by fiber-optic collimators FOC, pass through band-pass filter F3, and are focused in a horizontal plane by the f-number (F/5.2) matching lens L4. The parallel beams pass through the horizontal entrance slit of an Instruments SA HR-640 spectrometer, and their spectral contents are imaged at separate locations upon a Tracor Northern TN-1710 DARSS system detector head. Beam splitter BS5 directs 30% of the CARS signals into an EMI D551 photomultiplier tube which is employed for performing integrated intensity measurements.

The input-optics platform to the spectrometer which contains the fiber-optic collimators FOC, filter F3, and lens L4 also contains a system feature which greatly improves the available dynamic range of the multichannel detection system. The usable, linear dynamic range of the DARSS system for the short-pulsed optical phenomena being studied has been measured to be $\approx$ 20. In order to extend this factor to the order of at least 300, which can be anticipated in a turbulent combustion environment, the sample CARS signal may be split after being collimated into two, three, or more different proportions. These beams are then displaced across the DARSS detector as described previously. The largest portion of the sample beam is monitored for the highest temperature indications, and the weakest split is adjusted to permit observation of the coolest temperatures present. Thus, with the proper selection of splitting ratios, one of the multiple signals will be located within the usable linear range of the DARSS at all available temperatures.

The optical system--with the exception of the spectrometer and detectors--is enclosed by a light-weight protective cover which provides isolation from acoustic vibrations, thermal gradients, and atmospheric dust. The NRC honeycomb table top is supported on pneumatic castered wheels which allows the system to be easily transported.

DATA-COLLECTION ELECTRONICS

Once the sample and reference CARS optical signals have been converted by the detectors, the data are conditioned, collected, and stored for subsequent analysis by the data-collection electronics outlined in Fig. 2. The spectral information obtained by the Diode Array Rapid Scan Spectrometer (DARSS) is digitized by the Tracor Northern 1710 Multichannel Analyzer System. The 512 channels of data produced are then transferred via computer-interfacing circuitry to a Modular Computer Systems (MODCOMP) 7840 Classic Series Computer where permanent storage is provided on 10-MByte disks. Machine-language software has been developed for the PDP-11 based Tracor system which controls and synchronizes the firing of the laser, handshaking with the MODCOMP computer, and integrated as well as spectral-data transfer. The data transfer is accomplished within 15 msec, allowing ample time for DARSS signal integration during normal 10-Hz (100-msec) operation.

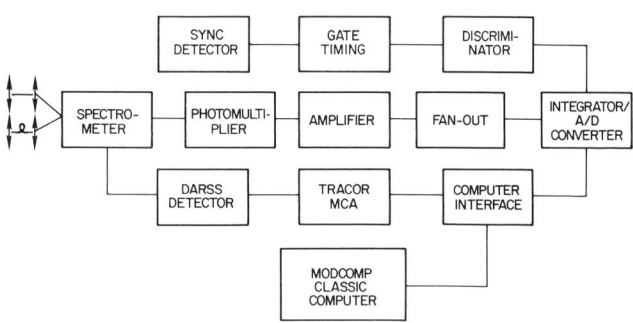

FIG. 2. Signal-Detection and Data-Collection Electronics Configuration.

All integrated intensity information is obtained by the common photomultiplier tube and is processed in a serial fashion. This feature is made possible by a parallel-to-serial conversion in the $\approx$ 10-nsec FWHM CARS optical pulses arriving at the detector. That is, since the reference fiber optic is made 20 m longer than the sample, the reference signal arrives at the PMT $\approx$ 100-nsec after the sample signal. The series of pulses thus formed can be amplified and is then duplicated by a LeCroy Research Systems Model 428F Linear FAN-OUT. The duplicated signal trains are applied to separate analog inputs of a LeCroy 2249SG 12-Channel Analog-to-Digital Converter (ADC). A PIN diode detector sensing the green output of the laser synchronizes the generation of a series of properly timed and spaced gate signals which are applied to gate inputs on the ADC. Since each channel of the ADC must be gated into operation, the time of arrival of a gate pulse determines which signal of the analog input pulse train will be digitized in a given ADC channel. Once the integrated data have been digitized, they are passed via the computer interface to MODCOMP disc storage. For the six channels of data presently employed, the digitization-plus-transfer cycle is accomplished within 100 μsec.

Although the system description to this point has been concerned with only two signals—one sample and one reference—the capability exists for up to six different optical inputs to be integrated and digitized in the manner described. Thus, optical inputs for up to three molecular species (three sample plus three reference signals) can be accommodated. The likelihood of performing simultaneous multiple-species measurements is increased due to the factor of ≈ 100 increase in sensitivity gained by use of photomultiplier rather than DARSS detection. Since it is only necessary to employ the DARSS detector in order to obtain spectral information on one species (e.g., nitrogen) for temperature determination, much weaker signals will suffice to furnish simultaneous multiple-species number-density information. Thus, the increased PMT sensitivity can be employed to offset the power requirements imposed on the additional Stokes lasers (and ultimately the Nd:YAG laser) which must be configured into the optical system.

SYSTEM-PERFORMANCE CHARACTERIZATION

A series of controlled experiments was conducted in order to characterize the CARS-system measurement capabilities. For this initial measurement program, the following system configuration and parameters were employed. Laser operation was at 10 Hz, with a 10-nsec FWHM pulse width. A single species (nitrogen) was observed to obtain simultaneous temperature and number-density information. A folded BOXCARS optical geometry defined by an 8.7-deg. full angle between the green beams and the dye beam which lies at 1.5 deg. to their plane was employed to obtain near-maximum spatial resolution. An ambient-air reference, producing resonant nitrogen CARS signals, was used for normalization of the integrated-intensity data only (i.e., no spectral normalization was employed). The spectral dispersion on the DARSS detector was measured to be 2.7 cm^{-1}/channel, contributing to an overall system spectral resolution of 7.2 cm^{-1}. Intensifier gating of the DARSS detector was employed for a period of 10 μsec.

The following experiment was performed in order to characterize the spatial resolution of the system which can be expected at operational beam-energy densities. A "slab" of nitrogen gas was formed from a 3-cm-long nozzel having a rectangular cross section of 0.27 mm × 10.9 mm. The tip of the nozzel was located 2.4 mm below the sample volume, and the slab was translated along the optic axis through the common foci of the CARS beams in such a way that the beams passed through the thinnest dimension of the slab. The effect of passing the pure nitrogen through the sample volume was enhanced by applying a continuous purge of helium to the focal area to remove the contribution of atmospheric nitrogen from the measurement. The result of this experiment is shown in Fig. 3. These data indicate that > 97% of the CARS signal is generated within a 2-mm path along the optic axis. A similar data set, obtained by rotating the nozzel 90 deg. and translating the thin dimension of the slab perpendicular to the optic axis through the common foci, indicated that > 98% of the signal generation is obtained within 0.8 mm. Assuming symmetry in the transverse dimensions, these numbers suggest a sample volume on the order of ≈ 1.3 mm^3. It is, however, important to note that these measured dimensions, which describe the spatial resolution of this geometry, were possibly affected by such factors as the alignment of the slab to the optic axis or a thickening of the nitrogen slab beyond 0.27 mm as the gas exited the nozzel. For these reasons the indicated dimensions should be considered as maximum values only, until a more precise measurement scheme can be designed.

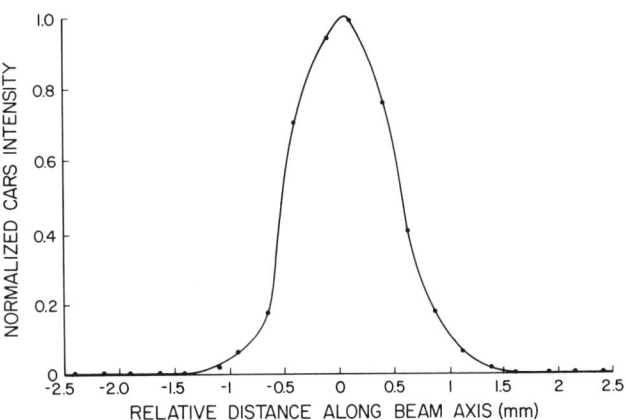

FIG. 3. Axial Distribution of CARS Generation through Sample Volume.

An indication of the CARS system capability to determine molecular concentration at room temperature is given in Table I. This table contains the results of nitrogen-concentration measurements (based upon integrated-intensity data and the Ideal Gas Law) performed on a sample cell filled to 1 atm. pressure at 294 K with the stated proportions of nitrogen in oxygen. The measured concentrations represent the average of up to 10 groups of 40 single-shot concentration determinations each. The precision of these measurements (defined by one standard deviation) reflects the ≈ 5% shot-to-shot intensity fluctuations experienced in the CARS signals after normalization. Although the measured concentration values in Table I agree with the mixed values well within the stated margins of error, it is believed that the noticeable decrease in accuracy (defined by 1 − |mixed−measured|/mixed) experienced in the 10% measurement is caused by the onset of background, non-resonant CARS contributions from oxygen. This effect could be reduced by employing a form of background reduction.[9]

TABLE I

AMBIENT-TEMPERATURE N_2-CONCENTRATION MEASUREMENT ACCURACY

SAMPLE MIXTURE (N_2 IN O_2 ± 1.5%) (%)	MEASUREMENT		
	CONCENTRATION (%)	PRECISION (±) (%)	ACCURACY (%)
80	77.8	3.7	97
60	59.3	3.0	98
40	39.0	2.1	98
20	20.6	1.1	97
10	11.2	0.6	89

Although the folded BOXCARS configuration provides excellent spatial resolution, it has the disadvantage of being a less efficient CARS generator than planar BOXCARS. The effect of this geometry upon signal strength is quite pronounced at the large beam-crossing angles employed in this system. However, as indicated by the spectrum in Fig. 4 (a 2014-K spectrum with a ± 40-K fitting error estimated from the degree to which a nonlinear least-squares fitting routine[10] could match calculated spectrum to the measured data obtained in a premixed propane flame), there is sufficient signal-generation capability to permit observation of the temperatures that can be expected from the combustion of most common fuels.

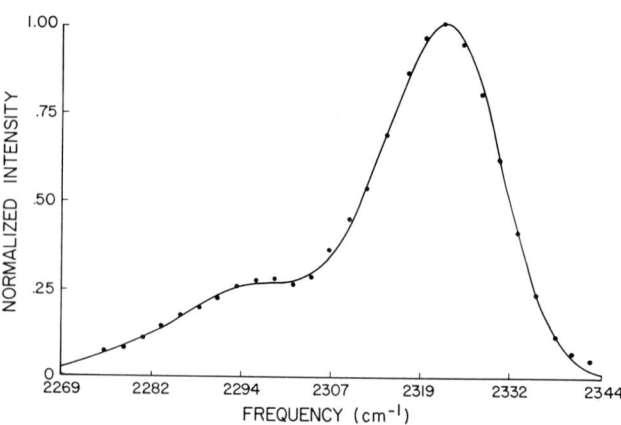

FIG. 4. Single-Shot N_2 Spectrum in Premixed Propane Flame, 2014 ± 40 K. Calculated (-), Experimental (·).

To obtain a measure of accuracy in simultaneous temperature and number-density determinations, a simple "hot-air" furnace was constructed. The furnace, consisting of a 2.5-cm-i.d. × 20-cm-long stainless-steel tube, was supported above a 10-cm-long premixed propane flame and centered on the CARS sample volume. The ends of the tube extended sufficiently beyond the combustion region that no combustion products (or effects other than temperature) would influence the CARS signals produced. A chromel-alumel thermocouple placed within 2 mm of the sample volume indicated a furnace temperature of 965 K with drifts of ± 10 K. Based upon this average temperature and assuming only a 1/T density dependence, the nitrogen number density sampled would be ≈ 5.94 × 10^{18} molecules/cm³. While the thermocouple was monitoring the furnace temperature, data for 1,000 single-shot CARS measurements were recorded. Figure 5 is an example of the single-shot spectra obtained during this experiment. The pdf's

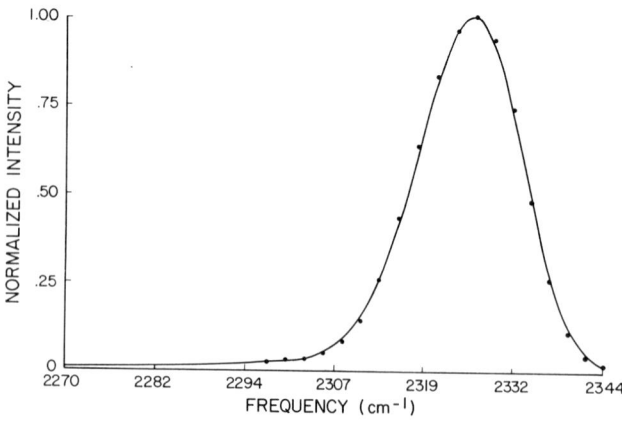

FIG. 5. N_2 Spectrum from Hot-Air Furnace, 980 ± 35 K. Calculated (-), Experimental (·).

developed from these 1,000 samples for furnace temperature and number density are shown in Figs. 6 and 7, respectively. They indicate a CARS-derived mean temperature of 894 K with a standard deviation of ± 66.9 K (± 7.5%) and a mean number density of 6.12 × 10^{18} molecules/cm³ with a standard deviation of ± 0.54 × 10^{18} molecules/cm³ (± 8.8%). The marginal agreement between the thermocouple and CARS temperatures suggests the possibility of systematic error in some measured parameter. It is believed that this error will be greatly reduced by refining the determination of such parameters as the system spectral resolution (which influences the shape of the calculated spectra); such refinements will come about through further testing and calibration. However, by comparing the averaged quantities of the thermocouple and CARS data, a measurement accuracy of ≈ 92.6% for temperature and ≈ 97.1% for number density was indicated in these preliminary measurements.

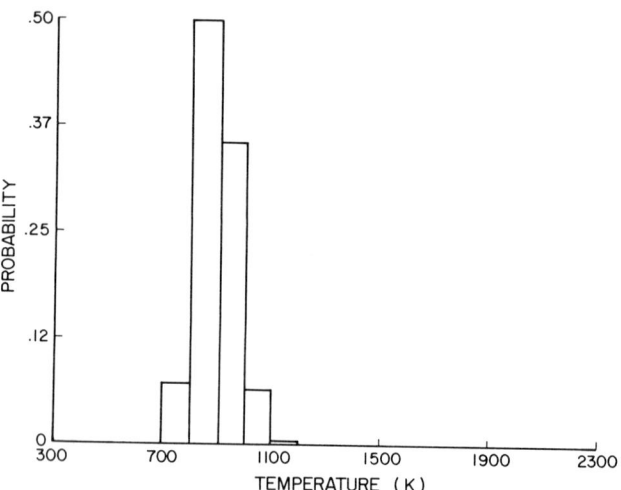

FIG. 6. Probability Distribution Function for Hot-Air-Furnace Temperature.

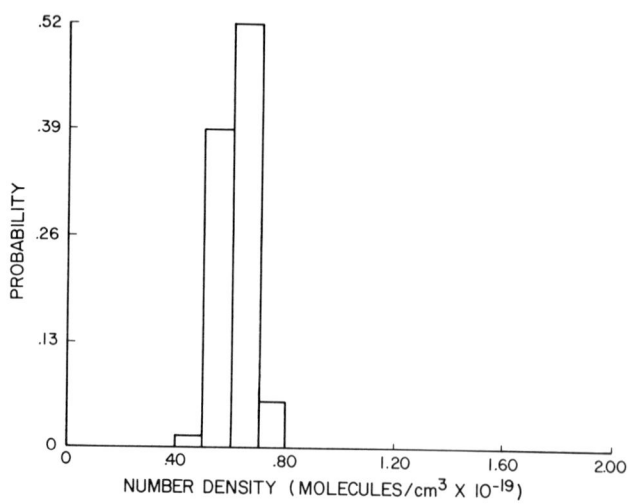

FIG. 7. Probability Distribution Function for Hot-Air-Furnace N_2 Density.

CONCLUSIONS

The ability of the hardened CARS optical diagnostic system described here to perform simultaneous temperature and number-density measurements on a single gas species with a high degree of spatial and temporal resolution has been demonstrated in the laboratory. System performance obtained during these measurements indicates the capability of the system in obtaining simultaneous multiple-species CARS information in the environments presented by practical combustion systems. Implementation and demonstration of this and the additional system capabilities that have been described are planned.

Although the determinations of measurement accuracy and precision presented here were not intended to be rigorous or exhaustive, it is felt that they do support a high level of confidence in the ability of the CARS technique to provide valuable diagnostic information on combustion processes.

REFERENCES

a) Work sponsored by Air Force Wright Aeronautical Laboratories Aero Propulsion Laboratory under Contract F33615-80-C-2054.

1. G. L. Switzer, W. M Roquemore, R. B. Bradley, P. W. Schreiber, and W. B. Roh, "CARS Measurements in a Bluff-Body Stabilized Diffusion Flame," Appl. Opt. 18, 2343 (1979).

2. A. C. Eckbreth, "CARS Thermometry in Practical Combustors," Comb. Flame 39, 133 (1980).

3. B. Attal, M. Paelat, and J-P Taran, "CARS Diagnostics of Combustion," J. Energy 4, 135 (May-June 1980).

4. I. A. Stenhous, D. R. Williams, J. B. Cole, and M. D. Swords, "CARS Measurements in an Internal Combustion Engine," Appl. Opt. 18, 3819 (1979).

5. G. L. Switzer, L. P. Goss, W. M. Roquemore, R. P. Bradley, P. W. Schreiber, and W. B. Roh, "Application of CARS to Simulated Practical Combustion Systems," J. Energy 4, 209 (1980).

6. Y. Prior, "Three-Dimensional Phase Matching in Four-Wave Mixing," Appl. Opt. 19, 1741 (1980).

7. J. A. Shirley, R. J. Hall, and A. C. Eckbreth, "Folded BOXCARS for Rotational Raman Studies," Opt. Lett. 5, 380 (1980).

8. A. C. Eckbreth, "BOXCARS: Crossed-Beam-Phase-Matched CARS Generation in Gases," Appl. Phys. Lett. 32, 421 (1978).

9. L. A. Rahn, L. J. Zych, and P. L. Mattern, "Background-Free CARS Studies of Carbon Monoxide in a Flame," Opt. Comm. 30, 249 (1979).

10. L. P. Goss, G. L. Switzer, and P. W. Schreiber, "Flame Studies with the Coherent Anti-Stokes Raman Spectroscopy Technique," Paper No. AIAA-80-1543 presented at the 15th AIAA Thermophysics Conference, Snowmass, CO, 14-16 July 1980.

Pure rotational CARS thermometry

J. W. Fleming and A. B. Harvey

Naval Research Laboratory, Washington, D.C. 20375

W. T. Barnes[a]

Dresser Industries, Houston, Texas 77042

The use of rotational CARS as a diagnostic probe is presented. A folded BOXCARS optical configuration permits the acquisition of spectral information down to a few wavenumbers Raman shift for both broadband single-pulse experiments and narrow-bandwidth scanning. Results are shown for the measurement of rotational temperature from ambient N_2 gas and practical combustion systems where both N_2 and O_2 prove to be useful probe species. Vibrational temperatures in the higher temperature range are measured via pure rotational transitions in excited vibrational levels. Limitations as well as merits of the technique are discussed.

The use of lasers as non-intrusive temperature probes has become an alternative technique in areas where physical probes are not acceptable. Of the Raman type scattering methods Coherent Anti-Stokes Raman Spectroscopy (CARS) has proven to be a very valuable diagnostic tool for temperature measurement (1). CARS provides excellent spatial resolution because of the small focal volume of crossed, focused probe beams. Recent efforts have centered on the analysis of low frequency Raman shifts with CARS. Although this region contains information on pure rotational Raman transitions and low frequency molecular vibrations, the pure rotational area has received the most attention to date (2, 5).

One of the primary considerations for temperature probes for applications such as combustion, chemical reactions or turbulent systems is the ability to monitor as many parameters as possible in a fixed amount of time. Although vibrational CARS data from a multiplex or broad-band system gives one adequate temporal behavior, the frequency information is limited to those vibrational and rovibrational transitions which can be detected via the broad-band source used, usually the gain curve of a dye laser. Of course the possibility of having several probe lasers is one alternative but the expense and complexity of such a system limits its practicality. Since nearly all species exhibit pure rotational Raman transitions in the region below a few hundred wavenumbers, one can, in principle, detect nearly all materials of interest with a single shot of one laser dye. Although this could potentially cause confusion of spectral information due to the large number of lines, the possibility of using very narrow laser linewidths could reduce the interference because of the inherently narrow rotational transitions.

There are several experimental configurations which may be used to obtain CARS spectra at very low Raman shifts. The common interference problem in this region is the frequency similarity of all beams, laser probes and signal. Furthermore, since a dye laser is used as one laser source the amplified spontaneous emission (ASE) from the dye cavity generally overlaps the frequency of the desired signal.

For this reason, a crossed beam arrangement must be employed. One method is to use two perpendicularly polarized beams, ω_1 and ω_2 ($\omega_1 > \omega_2$) crossed at a small angle (2). This arrangement relies on spatial rejection of ω_2 and polarization rejection of ω_1, since the signal at ω_3 ($\omega_3 = 2\omega_1 - \omega_2$) has the same polarization as ω_2. This non-phased matched arrangement is fairly simple even though some signal is sacrificed to achieve better spatial separation. However, it is very difficult to obtain data below 20 cm^{-1}.

Of the crossed beam phased matched configurations "folded BOXCARS" (4) is the most effective. This is a three beam phase matched scheme which has excellent spatial resolution and which allows complete separation of the laser beams ω_1, ω_2 and ω_3 ($\omega_1 = \omega_2 > \omega_3$) from the signal ω_4 by spatial filtration even at zero wavenumber shift.

The experimental arrangement employed in our studies is shown in Figure 1. The frequency doubled output of a Nd:YAG laser (532 nm) was utilized for the pump beams ω_1 and ω_2. The frequency tripled output (355 nm) was used to pump a Molectron DL-14 dye laser capable of producing a 50-200 kW probe beam at ω_3.

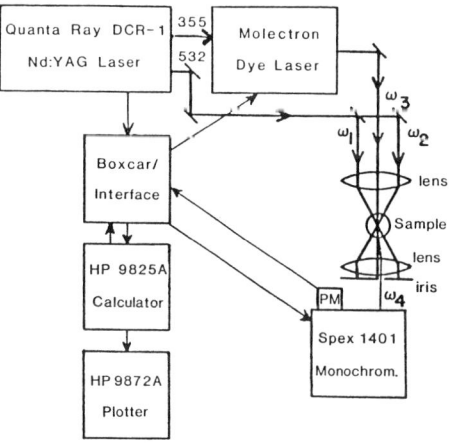

FIG. 1. Experimental Apparatus.

The fundamental, second and third harmonics of the Nd:YAG laser were separated using a Pellin-Broca prism. Delay lines were incorporated in the pump beams, ω_1 and ω_2, to compensate for dye rise time and difference in path lengths. The two pump beams were produced by splitting the second harmonic with a 50% transmitting dichroic reflector. The focusing and collecting lenses were both 10-cm diameter, 15-cm focal length simple planoconvex types. The three input beams were made parallel to and equidistant from the optical axis passing through the lens centers. The optical axis and pump beams, ω_1 and ω_2, were in a plane which was parallel to the optical table. Probe beam ω_3 and the optical axis defined a second plane perpendicular to the first. Following the collecting lens, the signal at ω_4 appeared in the ω_3-optical axis plane on the opposite side of the axis from ω_3 but equidistant from the axis.

Figure 2 diagrams the optical configuration. Under typical conditions, all beams made an angle of $\sim 7.5°$ (almost any angle can be arranged) with the optical axis. Following the collecting lens, an iris was used to isolate the signal, which was then passed into a Spex 1401 double monochromator and detected using a RCA 1P28 photomultiplier specially configured for linear response (6). The monochromator served only as a broad band filter and not as a resolution element. An HP 9825A calculator synchronously scanned the dye laser and monochromator, and also processed data from an interfaced laboratory constructed boxcar. Data was recorded on cassettes and hard copy could be obtained from an HP 9872A plotter peripheral.

To obtain rotational temperature information use is made of the temperature dependence of the total rotational CARS profile. The generated ω_4 beam may be described as (7)

$$P_4 \propto \left(\frac{\omega_4}{n_4}\right)^2 \left| 3 \chi^{(3)} \right|^2 (P_1 P_2) P_3 \quad (1)$$

where P_1, P_2, P_3 and P_4 are powers of the respective beams, ω_4 is the generated frequency, n_4 the refractive index at frequency ω_4 and $\chi^{(3)}$ the total third order nonlinear susceptibility.

The total susceptibility can be expressed by

$$\chi_T = \chi_N + \chi_R' + i\chi_R'' \quad (2)$$

where χ_N is the nonresonant component, χ_R' is the real part of the resonant component, and χ_R'' is the imaginary part of the resonant component. Thus for $|\chi_T|^2$

$$\chi^*\chi = \chi_N^2 + 2\chi_N\chi_R' + (\chi_R')^2 + (\chi_R'')^2 \quad (3)$$

The value for χ_N is essentially a constant, while χ_R' and χ_R'' are obtained as

$$\chi_R' = \Sigma C(R)(\omega_R^2 - \Delta^2) \quad (4)$$

and

$$\chi_R'' = \Sigma C(R) \, 2\Gamma_R \Delta \quad (5)$$

Summations are over all Raman resonances with corresponding frequencies ω_R; Δ is the Raman shift such that $\Delta = \omega_1 - \omega_3$; Γ_R is the Raman linewidth; and $C(R)$ is given by

$$C(R) = \frac{A\omega_R}{(\omega_R^2 - \Delta^2)^2 + (2\Gamma_R\Delta)^2} \quad (6)$$

in which the amplitude factor A is given by

$$A = \frac{64}{45} \frac{\Pi^4}{\hbar c} g_J \frac{\gamma^2 3(J+1)(J+2)}{2(2J+3)} \frac{N_T}{Q_R}$$
$$\cdot \exp(-F(v,J)hc/kT)\left[1 - \exp(-\omega(v,J)hc/kT)\right] \quad (7)$$

In the expression for the amplitude factor for a $J + 2 \rightarrow J$ selection rule, g_J is the nuclear spin weighting, γ is the anisotropy of the molecular polarizability, J is the rotational quantum number, v is the vibrational quantum number, N_T is the total number density, Q_R is the rotational partition function, $F(v,J)$ is the relative energy of the state (v,J) and $\omega(v,J)$ is the frequency of the transition from (v,J) to $(v,J+2)$.

Using these equations, $|\chi|^2$ may be calculated as a function of Raman shift, Δ. Since the frequency bandwidth of the laser source is much broader than the Raman linewidth Γ_R, the values for $|\chi|^2$ must be convolved with the laser lineshape.

Figure 3 presents the room temperature spectra of air of which the major two constituents are N_2 and O_2. Figures 4 and 5 show N_2 and O_2, respectively, both at room temperature and one atmosphere.

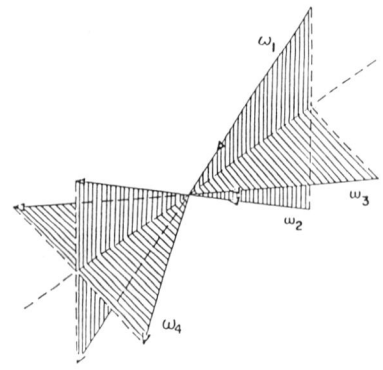

FIG. 2. Optical configuration of folded BOXCARS. In this arrangement $|\omega_1| = |\omega_2|$ but for their vectors $\underset{\sim}{k_1} \neq \underset{\sim}{k_2}$.

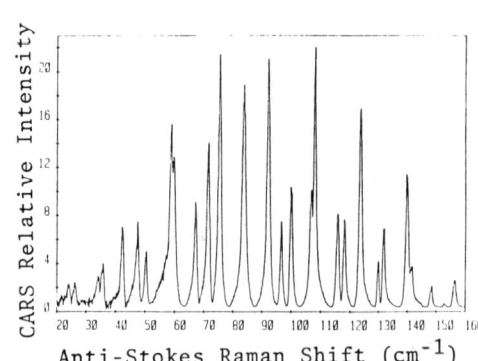

FIG. 3. Pure rotational CARS spectrum of laboratory air.

In comparing Figures 3, 4 and 5, it will be noticed that although there are several rotational lines in the region below 150 cm^{-1} Raman shift for each molecule, there are only about four pairs of lines which severely overlap. In our configuration an effective overall linewidth for the CARS system is 0.3 cm^{-1} (HWHM). A major concern if one chooses to use only modestly narrow sources then, is the problem of spectral interference in a multicomponent system. An excellent choice at this point, particularly from the point of air-fed combustion studies, is the possibility of examining a major component that does not directly participate in the combustion process, i.e., N_2. Vibrational temperature measurement via CARS using N_2 in flames is now a well accepted diagnostic tool (1).

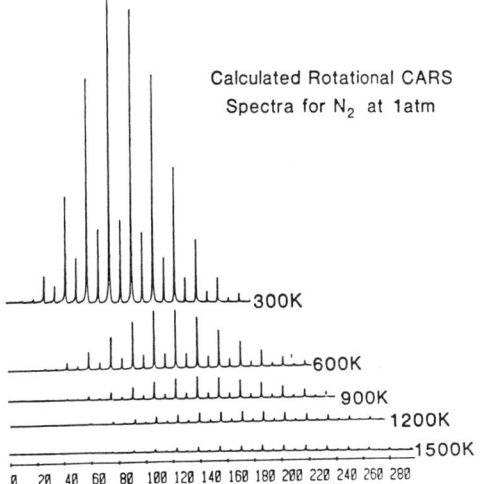

FIG. 6. Calculated rotational CARS profile as a function of temperature. Each spectrum has been corrected for intensity relative to 300 K.

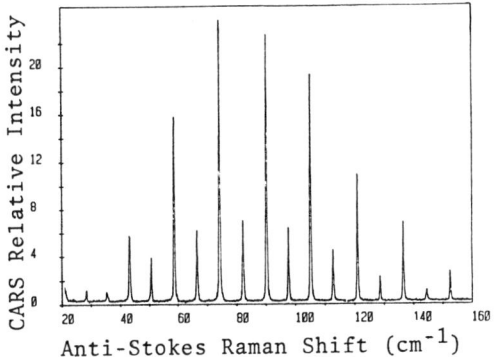

FIG. 4. Pure rotational CARS spectrum of N_2 at 700 torr and 297 K.

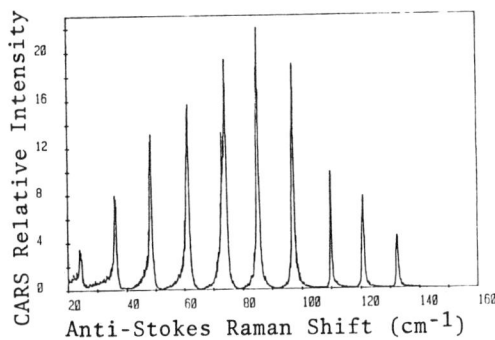

FIG. 5. Pure rotational CARS spectrum of O_2 at 700 torr and 297 K.

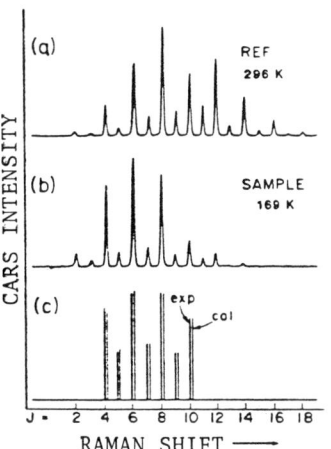

FIG. 7. Broadband rotational CARS spectra of (a) room-temperature reference N_2 gas (b) cold N_2 gas. Calculated and measured intensities for b are compared in (c) which gave a best-fit of 173 K for a thermocouple value of 169 K. (From Murphy and Chang, Ref. 3, reprinted with permission.)

The pure rotational temperature dependence of N_2 is shown in Figure 6. For purposes of comparison, each spectra has been corrected for number density to show the marked squared dependence.

The temperature region below 300 K has been examined by Murphy and Chang using pure rotational CARS (3). A broadband dye laser and optical multichannel analyzer were used to obtain very good agreement between computed CARS profiles and experimental spectra from samples of cold N_2 gas. Figure 7 shows spectra of room temperature N_2 at 296 K and cold N_2 at 169 K. The sample gas temperature was obtained by comparison of experimental intensities to computed intensities over a wide range of temperatures. Figure 8 presents the results of several measurements in the range 169 to 296 K. Both single-shot averages from 10 pulses and series of 10 averaged spectra are shown. Standard errors of ± 5 K are attributed mainly to signal recovery from the OMA.

The experimental and calculated pure rotational profile for CO_2 is presented in Figure 9. In the calculation only ground state vibrational levels are considered which give fairly good agreement with the experimental spectra. Because of nuclear spin statistics odd J lines are missing in the ground state but both even and odd J are present from v = 1. These pure rotational transitions from vibrationally excited states or hot-bands are very apparent in CO_2 as can be observed in Figure 10.

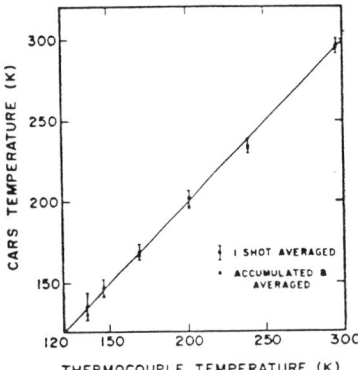

FIG. 8. Calculated best-fit temperature of N_2 via CARS versus thermocouple value for the range 135-296 K. The average and rms deviations for a series of 10 single-pulse spectra as well as the average for each series of 10 averaged spectra is shown. (From Murphy and Chang, Ref. 3, reprinted with permission.)

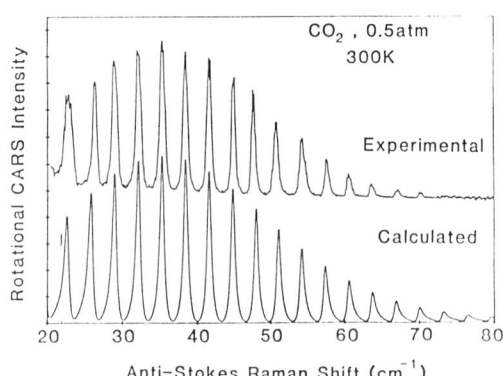

FIG. 9. Calculated and experimental rotational CARS spectra of CO_2 at 370 torr and 300 K.

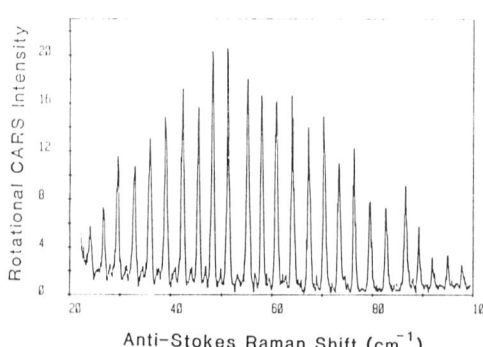

FIG. 10. Rotational CARS spectra of CO_2 at 700 torr and 603 K. The weaker but very apparent hot-bands can be observed alternating with the even ground state lines.

In the measurement of rotational temperature by CARS use is made of the temperature dependence of the total pure rotational profile. By virtue of the small energy spacing of rotational levels, this method has reasonably high sensitivity to temperature, especially at temperatures near or well below ambient as has been shown. However, a serious drawback to pure rotational CARS is the loss of sensitivity in the rotational envelope at high temperatures. As the temperature increases, higher rotational levels become more populated at the sacrifice of lower levels which effectively flattens the shape of the rotational envelope and thus causes a reduction in sensitivity to high temperature measurements and an increase in the species concentration detection limit. The decrease in high temperature sensitivity occurs because temperature changes have a less discernible effect on the broadened rotational profile. The detection limit increases because the maximum intensity is decreased with increasing temperature, thus diminishing the maximum observed signal-to-noise ratio.

Much of the diminished temperature sensitivity may be recovered by employing "hot bands" which begin to appear in the pure rotational spectrum at elevated temperatures. This concept was first introduced by Harvey (8) and Barrett and Harvey (9) for temperature measurements in electric discharges by spontaneous Raman scattering. Hot bands and the ground rotational transitions appear at separate frequencies because of vibration-rotational interaction which increases with rotational quantum, J. For a homonuclear diatomic, the rotational energy may be expressed as (10)

$$F(J) = B_v J(J+1) - D_v J^2(J+1)^2 + H_v J^3(J+1)^3 + \ldots \quad (8)$$

For selection rule $\Delta J = +2$ for pure rotational transitions, the frequency shift of the associated transition $J+2 \to J$ can be written as

$$\Delta \nu_{rot} = F(J+2) - F(J)$$
$$= (4B_v - 6D_v)(J+3/2) - 8D_v(J+3/2)^3 + \ldots \quad (9)$$

Since the centrifugal stretching term, D_v, and higher order coefficients are usually several orders of magnitude smaller than the rotational term, D_v, and higher order coefficients are usually several orders of magnitude smaller than the rotational term B_v, then $\Delta \nu_{rot}$ reduces to

$$\Delta \nu_{rot} = 4B_v(J+3/2). \quad (10)$$

B_v depends on the internuclear distance and is thus a function of the vibrational quantum number v and the anharmonicity α_e according to

$$B_v = B_e - \alpha_e(v+1/2) + \gamma_e(v+1/2)^2 + \ldots \quad (11)$$

Again higher order terms can be neglected. The separation of frequency shifts between vibrational levels becomes

$$\Delta_v \equiv \Delta(\Delta \nu_{rot})_{v,v+1}$$
$$= 4B_v(J+3/2) - 4B_{v+1}(J+3/2)$$
$$= 4(J+3/2)(B_v - B_{v+1}) = 4(J+3/2)\alpha_e \quad (12)$$

For the case of N_2, $\alpha_e = 0.0174$ cm^{-1} so that at room temperature $J_{max} = 8$ (the most intense line in the rotational profile) and $\Delta_v = 0.66$ cm^{-1} but at combustion temperatures around 1800 K $J_{max} = 24$ and $\Delta_v = 1.77$ cm^{-1}. Figure 11 shows the relative CARS instensity for J = 26 to 29 as a function of temperature. Spectra has been offset for clarity and each has a different relative scale. Calculations were performed on the HP9825A. Figure 12 shows an experimental spectrum obtained in a methane/air flame from a laboratory burner. The calculated temperature of 1500 ± 75 K was obtained from a visual best fit of temperatures in this range. The calculated value compares favorably to a radiation corrected thermocouple value of 1441 K at the probe site.

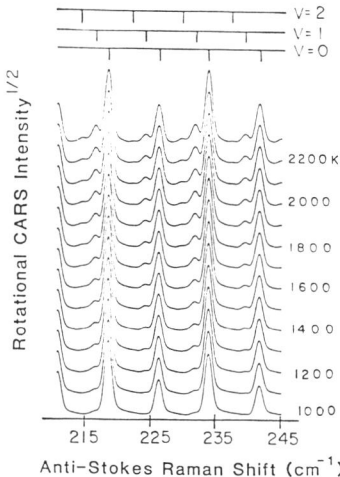

FIG. 11. Calculated rotational CARS spectrum of N_2 for J = 26 to 29 showing the dependence of the pure rotational transitions in vibrationally excited states on temperature.

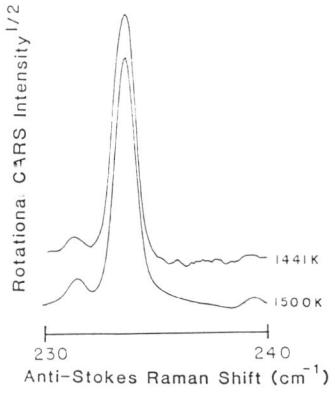

FIG. 12. Experimental CARS spectra for N_2 in a methane/air flame from a laboratory burner. The calculated curve is that obtained from a calculation using 1500 K compared to a radiation corrected thermocouple reading of 1441 K.

The method of hot-band analysis as a vibrational temperature probe for O_2 has recently been reported by Teets and Bechtel (11). A hydrogen-oxygen flame spectrum is shown in Figure 13 where an oxygen vibrational temperature of 2980 ± 110 K is calculated for a measured adiabatic temperature of 3074 K. Similar assumptions of vibrationally independent rotational cross section and Raman linewidth are made as in our study. However, in a methane/oxygen flame the vibrational temperature for O_2 was found to be 3300 ± 150 K as compared to an adiabatic temperature of 2916 K. Several sources of error were considered but the exact nature of the discrepancy is being investigated.

In considering the breakdown of this technique as in any laser technique, there are certain conditions which must be considered as to whether they are going to be a serious limitation to the experiment. A partial list of considerations include laser saturation, Stark effects on the observed spectra, turbulence effects, and beam overlap. The dependence of the cross section and linewidth on v and J and specifically the temperature, pressure and J dependence of the linewidth are molecular parameters which must be accounted for. Finally, for multicomponent systems the interference of spectrally close lines and/or the nonresonant susceptibility on a weak transition or for a minor species probe molecule must be taken into account.

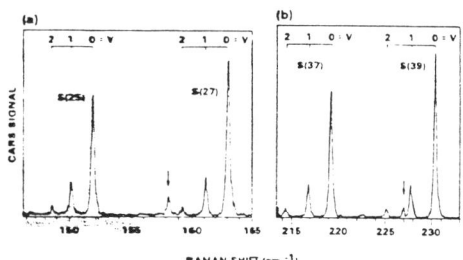

FIG. 13. CARS spectra of a hydrogen-oxygen flame. The arrows indicate electronic Raman scattering from 0 atoms. Rotational hot-bands from O_2 gave a temperature of 2980 ± 110 K compared to an adiabatic temperature of 3074 K. (From Teets and Bechtel, Ref. 11, reprinted with permission.)

In summary, the pure rotational CARS technique is an extremely useful nonintrusive temperature probe. Rotational temperatures from well below room temperature to those present in practical combustion systems can be measured. In addition, the presence of pure rotational transitions from excited vibrational levels can be used to measure vibrational temperatures in combustion environments. It should be mentioned that the latter technique does indeed measure vibrational temperatures whereas the method based on the rotational profiles measures rotational temperatures. The assumption is usually made that most combustion systems are rotationally and vibrationally equilibrated. Thus, these systems should yield the same temperature for both techniques. This method could be especially useful then for those systems in which the two temperatures are not the same, for example, in an electric discharge (12).

The authors would like to thank Dr. L. P. Goss for his early participation and continuing interest in our work and those other acknowledged contributors to this paper.

REFERENCES

a) NRC/NRL Postdoctoral Research Associate 1980-81. Current address: Dresser Industries, Houston, Texas
1. A. C. Eckbreth and P. Schreiber in *Chemical Applications of Nonlinear Raman Spectroscopy*, A. B. Harvey, ed., Chapter 2, Academic Press, New York (1981). See also W. M. Tolles, J. W. Nibler, J. R. McDonald and A. B. Harvey, *Appl. Spec.* **4**, 253 (1977).
2. L. P. Goss, J. W. Fleming and A. B. Harvey, *Opt. Lett.* **5**, 345 (1980).
3. D. V. Murphy and R. K. Chang, *Opt. Lett.* **6**, 233 (1981).
4. J. A. Shirley, R. J. Hall and A. C. Eckbreth, *Opt. Lett.* **5**, 380 (1980).
5. I. R. Beattie, T. R. Gilson and D. A. Greenhalgh, *Nature* **276**, 378 (1978).
6. F. E. Lytle, *Anal. Chem.* **46**, 545A (1974).
7. The general expression shows explicit four beam dependence. The subsequent derivation can be derived from general expressions, i.e., J. W. Fleming and C. S. Johnson, *J. Raman Spec.* **8**, 284 (1979).
8. A. B. Harvey, "Measurement of Vibrational and Rotational-Translational Temperatures Independently from Pure Rotational Raman Spectra" in *Laser Raman Gas Diagnostics*, edited by M. Lapp and C. M. Penney, Plenum Press, New York, p. 147 (1974).
9. J. J. Barrett and A. B. Harvey, *J. Opt. Soc. Am.* **65**, 392 (1975).
10. G. Herzberg, *Molecular Spectra and Molecular Structure I. Spectra of Diatomic Molecules*, D. Van Nostrand Co., Inc., Princeton, p. 553 (1950).
11. R. E. Teets and J. E. Bechtel, *Opt. Lett.* **6**, 458 (1981).
12. W. M. Shaub, J. E. Nibler and A. B. Harvey, *J. Chem. Phys.* **67**, 1883 (1977).

CARS thermometry in reacting systems*

J. F. Verdieck, J. A. Shirley, R. J. Hall, and A. C. Eckbreth

United Technologies Research Center, East Hartford, Connecticut 06108

CARS (Coherent anti-Stokes Raman Spectroscopy) diagnostic techniques provide accurate temperature measurements in environments ranging from controlled laboratory flames to hostile systems such as gas turbine combustors and internal combustion engines. CARS is an optical wave mixing process wherein incident laser beams at frequencies ω_1 and ω_2, with a frequency difference appropriate to the molecular species being probed, interact to generate the CARS signal at $\omega_3 = 2\omega_1 - \omega_2$. Through analysis of the frequency distribution of the CARS spectrum, temperature measurements can be obtained. Concentration measurements derive from the intensity of the CARS radiation or, in certain cases, from its spectral shape. CARS is a remote, nonperturbing technique which determines temperature in a matter of seconds with good, spatial resolution (of the order of millimeters). Under some circumstances, CARS signal levels are intense enough to permit "instantaneous," single pulse temperature measurements, i.e., $\sim 10^{-8}$ seconds. Because CARS is a coherent, nonlinear optical technique, it offers decided advantages over conventional incoherent spectroscopic methods. These advantages derive both from the strong CARS signal and its laser-like beam character which furnish excellent discrimination against interferences such as combustor luminosity and laser-induced interferences, which are often quite severe. In this paper, the fundamentals of CARS thermometry from both theoretical and experimental points of view, with emphasis on the nitrogen molecule, will be discussed in detail. Areas covered will include the accuracy of CARS thermometry, computer-generated CARS spectra, linewidth considerations and high pressure effects. Additionally, the use of other molecular thermometric probes for CARS diagnostics, H_2, O_2, H_2O, CO, and CO_2, will be shown. Two important experimental techniques, BOXCARS, a particular laser beam configuration which achieves good spatial resolution, and single-shot thermometry, wherein the CARS spectrum from a single pulse is recorded, will be treated. The utilization of CARS thermometry will be illustrated by its application in several different flame systems and in practical combustor devices.

INTRODUCTION

Temperature measurements utilizing spectroscopic techniques such as absorption and emission spectroscopy are well established methods which have served well over many decades. More recently, Raman scattering and fluorescence, using laser sources, have been applied with considerable success. These spectroscopic techniques offer several advantages over probe thermometry such as remote operation, nonperturbation of the sample, and extremely high temperature capability. The major disadvantage of spectroscopic thermometry is that the system under investigation must be "seen", i.e., be visible through some type of optical ports.

This paper describes a more recent thermometric method based upon a coherent, nonlinear optical technique named CARS (for coherent anti-Stokes Raman spectroscopy). In addition to the advantages listed above, the CARS method offers several others; good accuracy (better than 1% demonstrated with nitrogen CARS thermometry), rapid measurement and processing (typically less than ten seconds), and most importantly, the capability for a high degree of spatial resolution (better than one millimeter). Additionally, the exciting potential for performing virtually instantaneous (in 10^{-8} seconds) temperature measurements is available with single pulse CARS. In this paper the term reacting system shall be loosely interpreted to mean combustion systems. The extension of CARS measurement to other types of chemically reacting systems is easily made. It should be emphasized, that in combustion, the single most important parameter to measure is the temperature. Given the initial conditions, and the temperature at a point in a combusting system, the products, heat release, and other parameters can be estimated quite well. For this reason, highly accurate, spatially precise measurements of temperature are a necessity in understanding and controlling combustion processes.

CARS thermometry has been applied to many diverse systems ranging from carefully controlled laboratory flames to such difficult cases as burning propellant strands. Internal combustion engines, gas turbine combustor cans, furnaces, shock tubes, and plasmas are among the several devices which have been probed by the CARS method. Some of these applications will be covered in detail, following a discussion of the theoretical and experimental aspects of CARS measurements. A comparison of the CARS spectra of several molecular species, of particular interest in combustion, which have been used for thermometry is also presented. The effect of high pressure on the CARS nitrogen spectrum is discussed to emphasize the dramatic spectral changes which occur with pressure increase.

CARS FUNDAMENTALS: THEORETICAL CONSIDERATIONS

Temperature measurements derived from spectroscopic techniques ultimately depend upon the Boltzmann equation, which, of course, determines the energy level population distribution with temperature. This fundamental dependence is also true of CARS temperature measurements; however, because of the nonlinear nature of the CARS effect, the derivation of temperature is somewhat more complicated. This complication makes the fitting of computer-modelled CARS spectra with experimental CARS spectra mandatory. For this reason an analytical modelling capability is of equal importance to the experimental program in any CARS installation.

CARS is a coherent, nonlinear process which involves the mixing of four optical fields in a material medium (liquid, gas, or solid). In the usual CARS experiment, two of the fields are derived from the same laser beam, called the pump beam, of frequency ω_1. The other beam, the Stokes beam, has frequency ω_2, which is less than ω_1. The top portion of Fig. 1 sketches one experimental means of mixing

ω_1 and ω_2 to generate ω_3, the CARS frequency. The middle section of Fig. 1 demonstrates the photon energy conservation requirement for CARS; namely, that

$$\omega_3 = 2\omega_1 - \omega_2 \qquad (1)$$

Note also that $\omega_1 - \omega_2 = \omega_v$. When ω_v corresponds to a Raman-active vibrational-rotational transition in the molecule of interest, the CARS effect is strong, because of a quantum-mechanical resonance which will not be discussed here. Because of this resonance, the CARS effect is usually much stronger than the conventional spontaneous Raman effect. Equally important is the fact that the CARS radiation is generated in a laser-like beam nearly identical to the input beams. This means that, unlike Raman scattering, the entire CARS beam can be collected. Furthermore, this provides good discrimination against background, if aperturing is employed. Obviously, this provides a tremendous advantage when making measurements in highly luminous combustion devices. This point will be illustrated in more detail in the section on CARS applications.

The bottom portion of Fig. 1 depicts two different ways to generate an entire CARS spectrum. In the first of these, the pump frequency ω_1 is combined with a narrowband Stokes frequency, ω_2, which is tuned, generating the CARS spectrum stepwise. The second method shown employs a broadband Stokes laser, ω_2, which in combination with ω_1, generates the entire CARS spectrum, ω_3, from each laser pulse. This latter method offers the potential for making temperature measurements from a single, 10^{-8} second, laser pulse, as will be shown later.

- Approach

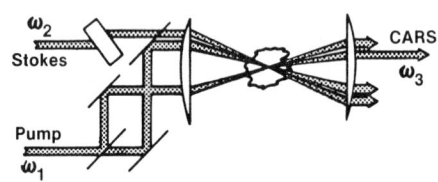

- Energy level diagram

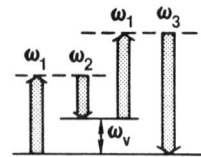

- Spectrum

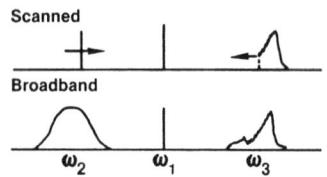

Fig. 1. Coherent anti-Stokes Raman spectroscopy (CARS).

The thereotical basis for CARS will be briefly sketched; for a more detailed, yet readable, account, the reader may refer to Ref. 1, and to the references contained therein. The physical origin of the CARS effect arises from the nonlinear response of molecules (both the electronic and nuclear portions) to the incident electric fields. This response is a macroscopic polarization (a charge distortion) that is proportional to the product of the three electric field amplitudes. This time dependent polarization then acts as an oscillating dipole source to generate the CARS radiation. The material parameter which relates the macroscopic polarization to the product of the electric fields is named the third order nonlinear susceptibility, $\chi^{(3)}$. It is a complex quantity and depends upon all molecules present in the sample. The electronic portion of $\chi^{(3)}$ is nearly frequency independent (unless an electronic transition is present) and is termed the nonresonant susceptibility, χ^{nr}. In contrast, the nuclear response of $\chi^{(3)}$ is strongly frequency dependent and complex, and is called the resonant susceptibility. The total susceptibility is written as

$$\chi^{(3)} = \sum_j (\chi' + i\chi'')_j + \chi^{nr}, \qquad (2)$$

where the bracketed term, the resonant portion, is summed over all transitions, j. In terms of $\chi^{(3)}$ and the incident laser intensities, the CARS intensity is given by (for the case of plane wave excitation, which is sufficient for our use here)

$$I_3 = K \left(\frac{\omega_3}{n_3}\right)^2 I_1^2 I_2 \left|\chi^{(3)}\right|^2 L^2, \qquad (3)$$

where n_3 is the refractive index at frequency ω_3; I_1 and I_2 are the laser intensities at frequencies ω_1 and ω_2 respectively. L is an interaction length over which the input laser beams are strongly focused, and phase matched. K is a constant. Phase matching is achieved when the vector equation

$$\vec{k}_3 = 2\vec{k}_1 - \vec{k}_2 \qquad (4)$$

is satisfied. The wave vector k is defined by $|k_i| = n_i \omega_i / c$, n_i is the refractive index at frequency ω_i. Equation (3) assumes the form shown when phase matching is satisfied. Phase matching is mainly a problem of directing the laser beams appropriately and will be discussed in detail in the experimental section.

The third-order susceptibility can be derived quantum-mechanically and expressed in spectroscopic terms as

$$\chi^{(3)} = \frac{2N}{\hbar} \sum_j \frac{(\alpha_j)^2 \Delta_j}{2(\omega_1 - \omega_2 - \omega_j) - i\Gamma_j} + \chi^{nr} \qquad (5)$$

where χ^{nr} has been separated from the resonant susceptibility, the summed term. The sum is often restricted to so-called Q-branch transitions, with $\Delta J = 0$ (no change in rotational quantum number) because the Q-branch transitions often dominate the CARS spectrum. In equation (5), N is the number density of the molecular species of interest; α_j, ω_j, and Γ_j are the polarizability matrix element (related to the Raman cross section), the transition frequency, and the isolated Raman linewidth respectively, of transition Q(j). The term Δ_j is the Boltzmann population difference and is responsible for the temperature dependence of $\chi^{(3)}$. Extracting the temperature from a CARS signal, which depends upon the squared modulus of $\chi^{(3)}$, is not a simple task and the assistance of computed model spectra is a necessity, as previously stated. Several examples of computer-model, predicted CARS spectra which demonstrate considerable sensitivity to temperature will be shown in a later section. In simple terms, the temperature is derived from the spectral distribution, i.e., the "shape" of the CARS signal. What factors determine the shape requires closer examination. Based upon consideration of equation (2), there are two limiting cases which determine the shape of a CARS spectrum as shown in Fig. 2. For the "strong signal" case, it is assumed that the resonant susceptibility is large compared to χ^{nr}. This leads to the resonance type of line shape indicated. An example of the strong case would be that of a majority species such as nitrogen in an air-fed combustion system. CARS spectra of the strong line case provide a basis for accurate temperature measurement. The other extreme, the "weak line" case, occurs when χ^{nr} is large compared to the resonant susceptibility. The desired resonant CARS signal then appears as a modulation or dispersion-type spectrum on the χ^{nr} background. This type of signal can be extremely useful for determining the concentration of a species by fitting the shape of the spectrum, without making an absolute signal intensity measurement. This topic will be treated briefly in a later section of this paper. A more complete, detailed description of the modelling of CARS spectra can be found in Refs. 2, 3. Those papers present the calculation of $\chi^{(3)}$ from spectroscopic constants and the Boltzmann distribution of energy levels. Instrumental factors such as laser linewidths and spectrometer resolution are taken into consideration.

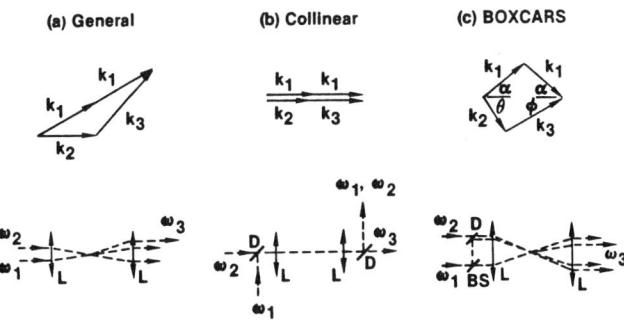

Fig. 3. Illustration of CARS phase-matching approaches. Subscripts denote beams: 1, pump; 2, Stokes; 3, anti-Stokes.

high intensities). Although CARS has been observed with continuous lasers (Ref. 4), high peak power, pulsed lasers are almost universally employed for the generation of CARS, particularly for gas phase diagnostics. This is the type of system which will be described in this section after a short description of phase matching.

In order to obtain the maximum CARS signal, the incident laser beams must be so aligned that the wave mixing process is properly phased. The CARS signal generated at a certain point will be in phase with CARS generated at other points in space leading to a constructive buildup of signal (rather than destructive interference, if out of phase). In Fig. 3a is shown the pictorial representation of equation (4), the general condition for phase matching. Underneath the vector diagrams are shown the experimental arrangement of the beams. In a gas, the refractive index changes little with wavelength, and the collinear arrangement of Fig. 3b satisfies the phase-matching condition. Although very easy to implement, the collinear configuration has very poor spatial resolution, as demonstrated by Eckbreth (Ref. 5). The length of sample interrogated by the CARS effect depends upon the focal length of field lens and can be quite large. For example, a 10 cm lens yields a sample length of only ~ 0.07 cm, but a 100 cm lens (which may be required in some instances) generates a sample length of ~ 7 cm. Clearly, using a long focal length lens, a collinear CARS measurement over a small scale burner would sample both hot and cold regions and, hence, would be considerably misleading. Recognizing this, Eckbreth (Ref. 5) devised a crossed-beam phase-matching scheme illustrated in Fig. 3c, called BOXCARS because of the shape of the diagram. In this case, CARS is generated only at the point where the three beams cross, which can have a dimension along the lens axis of ~ 1 mm, and less in the transverse direction. A further advantage of BOXCARS is that the CARS beam is almost completely spatially (hence spectrally) separated from the input laser beams, and is easier to isolate than the collinear case. Complete separation of the CARS beam is achieved if the BOXCARS diagram is folded (hence, folded BOXCARS) about the axis as shown in Fig. 4b. The arrangement of the incident and exiting beams, also shown, clearly demonstrates how the ω_3 beam is completely separated from the ω_1 and ω_2 beams. This spatial, and consequent spectral, separation is particularly advantageous for rotational CARS experiments (Ref. 6), where the spectral separation

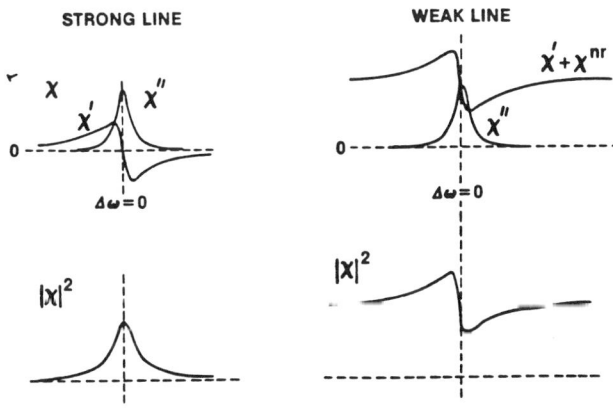

Fig. 2. Illustration of strong and weak line CARS intensity profiles.

CARS FUNDAMENTALS: EXPERIMENTAL CONSIDERATIONS

CARS is a coherent, nonlinear optical process, as stated previously. Laser sources are required in order to achieve signal coherence (obtain phase matching), and nonlinearity of response (requires

from input frequencies is small. Rotational CARS spectra obtained by use of folded BOXCARS are discussed in a later section. There are several other types of CARS phase-matching schemes which will not be discussed here. These methods are described and referenced in Ref. 1.

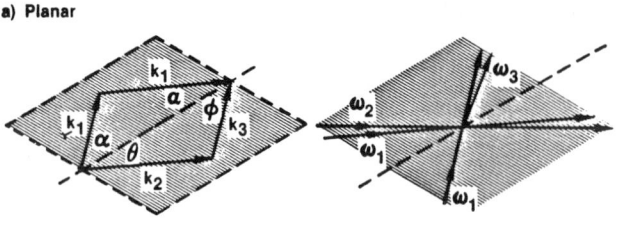

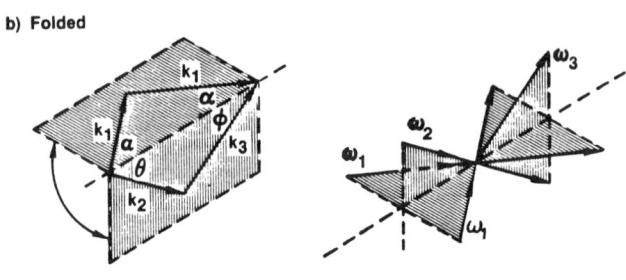

Fig. 4. CARS crossed-beam phase-matching approaches showing the phase-matching diagram and actual geometry of the optical beams for (a) planar and (b) folded BOXCARS. Subscripts denote beams; 1, pump, 2, Stokes, 3, anti-Stokes.

A typical broadband CARS arrangement for combustion diagnostics is schematically illustrated in Fig. 5 (Refs. 1, 7, 8). The neodymium:YAG laser provides two 2nd harmonic beams at 532 nm by frequency-doubling the fundamental 1.06 μm output from the laser, then frequency-doubling the residual 1.06 μ emerging from the first doubler. The primary 2nd harmonic, i.e., after the first doubler, is typically 200 mJ, while the secondary, i.e., from the second doubler, is about an order of magnitude less. The secondary passes through a slab of KG3 Schott glass placed at the Brewster angle to absorb any remaining 1.06 μm. The secondary is then directed by mirrors and focussed to pump, slightly off axis, the Stokes dye cell oscillator. The output from the oscillator is amplified in a second dye cell, pumped by splitting a portion of the primary (typically 33%) with a beamsplitter as shown. The Stokes laser and 532 nm pump component each pass through Glan laser polarizers to ensure polarization purity, which is important when performing polarization sensitive CARS experiments. Each beam passes through expansion or contraction telescopes whose function will be detailed shortly. The Stokes beam passes through a rotatable optical flat and then, in folded BOXCARS as shown, directly to the focussing lens. In planar BOXCARS (Ref. 5) the Stokes beam would pass through a dichroic mirror used to reflect one of the primary pump components as depicted in Fig. 5. One advantage of folded BOXCARS is the elimination of this dichroic element. Another advantage of folded over planar BOXCARS is the complete angular and spatial separation

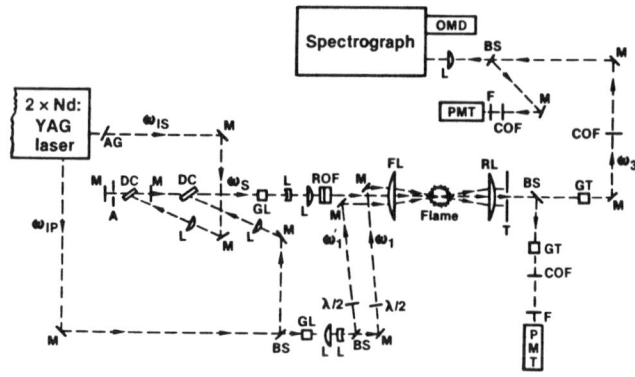

Fig. 5. Schematic of CARS experiment arrangement.

of the CARS beam which eliminates the need to disperse or filter the CARS beam from one of the pump components. The primary pump beam is split at a second beamsplitter (50%) to produce the two components for BOXCARS. These beams bass through low order half wave plates which control the polarization orientation of the pump components. The pump components are directed by mirrors to the focussing/crossing lens. If the three beams are aligned parallel to each other, they will, by definition, cross at the focal point of the lens. However, due to wavefront sphericity and chromatic aberation, they may not necessarily waist at the crossing point. The adjustable telescopes are used to position the beam waists at the crossing point. At high laser intensities, the beams are readily visualized near the focal region from the room air Rayleigh scattering. The primary component waists will coincide only if the distances travelled by each pump component from the second beamsplitter to the focussing lens are equal. The telescopes, depending on their magnification, also permit the focal diameters to be varied. Rotation of the optical flat in the Stokes beam permits its displacement on the focussing/crossing lens, permitting the phase-matching angle θ to be varied. After passing through the crossing point, the four beams i.e., CARS at ω_3, ω_1, ω_1', ω_2, are recollimated by a second lens of generally the same focal length as the focussing lens. The unwanted frequencies are trapped and the spatially separated CARS beam is split into two components, not necessarily equivalent, by a beamsplitter. The component split off at right angles passes through a Glan Thomson polarization analyzer set normal to the polarization of the resonant mode CARS signal. Thus, only the nonresonant CARS signal, acutally some fraction thereof, is transmitted and is monitored after spectral filtering by a photomultiplier. This signal is used for in-situ referencing which will not be treated here. The portion of the CARS radiation transmitted by the beamsplitter passes through a polarization analyzer and is directed to a spectrograph fitted with an optical multichannel detector (OMD), or a monochromator equipped with a photomultiplier for spectral scanning. The second polarization analyzer may allow any contributions to the CARS radiation, i.e., isotropic and anisotropic resonant modes, nonresonant signals, to pass. Or, it may be oriented to suppress detection of one or two of the above. A small fraction of the CARS beam is split off at a glass slide before the spectrograph and sent to a photomultiplier tube fitted with an appropriate narrowband interference filter. This signal is averaged on a BOXCAR averager and used to monitor and "tweak" the alignment of the CARS system. It is relatively immune to angular and translational motions of the CARS beam which occur during peaking up the alignment. The alignment of a CARS system should never be monitored through a narrow angular and spatial acceptance, such as a slit, where signal loss due to slight steering can be misinterpreted for alignment detuning. The spatial resolution of a given configuration is readily checked by generating CARS from within a translatable, thin microscope slide cover with the laser beams suitably attenuated to prevent optical damage to the glass. Microscope slide covers are generally less than 200 μm thick and allow the pointwise CARS signal contribution to be ascertained. Upon integration of the pointwise contributions, the spatial resolution can be found. In general, it is far better practice to measure the spatial resolution than to calculate it.

One advantage of broadband CARS is the simplicity of the Stokes dye laser which requires no elements for tuning and spectral condensation. A flowing dye cell, oriented at Brewster's angle, resides within a planar Fabry-Perot oscillator cavity. Because the dyes amenable to 532 nm pumping typically exhibit high gain, slightly off-axis pumping works quite well and leads to very good beam quality, i.e., low angular beam divergence, in contrast to that often obtained with transverse pumping. The dye spectrum is centered at the desired wavelength by selecting the dye appropriately and by adjusting its concentration. Solvent tuning can also be employed.

With the planar Fabry-Perot dye oscillator arrangement, bandwidths vary from 100 to 200 cm^{-1}, depending on the pump energy and dye employed. Binary dye mixtures are often used to improve dye conversion efficiency in the desired wavelength region. For single pulse CARS diagnostics or for spectral scanning of laminar flames, it is important that the dye spectrum be smooth and reproducible from pulse to pulse. The dye cavity is purposely designed to have a high Fresnel number and to accommodate as many modes as possible to "fill in" the spectrum. The pulse to pulse spectral stability of the dye laser has been examined on the OMD and found to be fairly good. Single pulse dye spectra display an irregular fine structure with an amplitude variation of ± 5%, presumably due to spatial "hole burning" in the dye laser. This fine structure averages out in time to produce spectrally-smooth profiles. High quality, averaged CARS spectra are readily achieved, but single pulse CARS spectra may exhibit some spurious structure, particularly at moderate spectral resolution, < 1 cm^{-1}. At lower resolutions, > 1 cm^{-1}, this fine structure is spectrally smeared and is not a problem. In broadband CARS, either the linewidth of the pump laser or the resolution of the monochromator/spectrograph determines the ultimate resolution of the spectrum. 1-m monochromators typically have a limiting resolution of about 0.5 cm^{-1} in the visible. 2xNd lasers with intracavity etalons have linewidths in the 0.1 to 0.4 cm^{-1} range, and 0.8 cm^{-1} without. CARS spectra so obtained thus have a limiting resolution between 0.5 and 1 cm^{-1}. This moderate resolution is generally more than sufficient for diagnostics. Furthermore, it is important to note that the large pump laser linewidth is not detrimental in regard to the strength of the CARS radiation. For broadband CARS generation, it is easy to show that the spectrally integrated intensity of a CARS transition is independent of the pump laser linewidth.

Narrowband CARS systems for combustion diagnostics (Ref. 9) are conceptually similar to that described above with the obvious exception of the tunable Stokes source. For narrowband Stokes work, a variety of dye laser configurations exist, are commercially available or readily assembled. The oscillator sections are generally transversely-pumped to obtain gains high enough to overcome the insertion losses engendered by the spectral condensation scheme employed. The Hansch design (Ref. 10) of a circular telescope, large two dimensional grating has generally given way to one dimensional expansion schemes employing a grazing incidence grating (Ref. 11) or multiple-prism beam expanders (Ref. 12). The one-dimensional expanders require a grating large in one dimension only and greatly reduce cost.

The output from the monochromater can be detected by a photomultiplier tube or with an optical multi-channel analyzer OMA (Ref. 13). Photomultiplier signals are usually averaged with a boxcar integrator. Detection with an OMA is a better method, because the entire spectrum from each CARS pulse, is captured and many spectra may be summed over a desired number of pulses. Given sufficient signal strength, a single CARS pulse (typically 10 nanoseconds) can provide enough information to extract the temperature, as stated previously. Moreover, if the OMD signal is processed with an on-line laboratory minicomputer which can store model predicted CARS spectra, then temperature measurements can be made in a matter if seconds (10 seconds corresponds to 100 laser pulses). This capability is presently available at UTRC.

CARS THERMOMETRY

Temperature can be derived from a CARS spectrum because the vibrational - rotational quantum state populations depend upon temperature. Statistical mechanics tells us that the probability, P (v,J) that a state with vibrational and rotational quantum numbers, v and J, is populated is given by (to first order of approximation)

$$P(v,J) = \frac{g_J (2J+1) e^{-E_v/RT} e^{-B_v J(J+1)/RT}}{Q_v Q_R(v)} \quad (6)$$

where g_J is the nuclear spin degeneracy; Q_v and $Q_R(v)$ are respectively the vibrational and rotational partition functions for level v; E_v is vibrational energy and B_v is the rotational constant in state v; k is the Boltzmann constant and T the equilibrium temperature. For most molecules of interest in combustion the vibrational CARS spectrum is dominated by Q-branch transitions, whose frequencies are given by (for a diatomic molecule)

$$\omega_{vJ} = \omega_e - 2\omega_e \chi_e (v+1) - \alpha_e J(J+1) \quad (7)$$

Thus, the two equations (6) and (7), through the basic equations for the CARS effect discussed in the theoretical section, determine the shape of the resonant CARS spectrum. Experience in this laboratory has demonstrated that temperature determination from experimental CARS spectra requires fitting to computer-generated spectra (Refs. 2, 3, 14). The possible exception might be for hydrogen CARS spectra, where, because of the large vibration-rotation interaction, $\alpha_e = 2.933$ cm^{-1}, the Q-branch transitions are well separated and temperature can be found from relative peak heights. However, hydrogen is present in abundance in combustion systems only for special cases. A more useful CARS probe, present in large quantity in all air-supported combustion, is the nitrogen molecule.

CARS NITROGEN THERMOMETRY

In addition to the nearly ubiquitious presence of nitrogen, the spectroscopic parameters, including the Raman linewidth, needed for accurate CARS spectral modelling, are quite well known. That the N_2 vibrational CARS spectrum provides an excellent basis for thermometry can be appreciated through examination of the computer calculated temperature dependence of the N_2 spectrum shown in Fig. 6. These calculations, carried out by Hall (Ref. 3), correspond to multiplex (broadband Stokes) CARS employing a 0.8 cm^{-1} bandwidth ω_1, a 150 cm^{-1} bandwidth Stokes ω_2, and a 1 cm^{-1} instrumental slit function. Clearly, these spectra exhibit pronounced temperature sensitivity, particularly at higher temperatures, and show much spectral detail even though the instrumental resolution is not particularly high. These spectra all correspond to the "strong line" case. It is assumed that nitrogen is present at 70%, hence, there is little contribution from the nonresonant susceptibility background although its inclusion in the calculations is important for accurate thermometry. Some notable features should be pointed out. At room temperature a single narrow peak, without structure at this resolution, is observed. This peak broadens with increasing temperature and shows resolved Q-branch transitions corresponding to Q(J = 20) to Q(J = 40) at about 1000 K. The peaks observed correspond to even-J Q-branch transitions; the odd-J transitions, are reduced in intensity by one-fourth (nuclear statistical factors are squared in CARS, because χ is squared). A second vibrational band, a "hot band," $\nu=1$ to $\nu=2$, becomes noticable above 2000 K, but is not shown in Fig. 6. For the moderate resolution conditions stated (and readily achieved) for Fig. 6, vibrational CARS thermometry would be most accurate for measurements above $\sim$ 1000 K. Given better instrumental resolution, and lasers of smaller bandwidth (both requiring greater cost and effort), more spectral detail would be observable at lower temperatures, which would increase the accuracy of CARS measurement at lower temperatures. In contrast, under conditions of poorer spectral resolution, when the fine structure is lost, temperature still can be found from the width of the fundamental and/or the relative strength of the hot band. From the preceding discussion, it is seen that vibrational CARS thermometry can be applied over a wide range of temperature with a varying degree of accuracy.

As an example of the type of accuracy that can be achieved with CARS N_2 thermometry, experiments were performed at UTRC in premixed flat flame burners (Refs. 14,15). Comparison of the CARS determined temperatures was made with radiation-corrected thermocouples (in turn calibrated by sodium line reversal). Figure 7 shows a scanned, broadband N_2 CARS spectrum

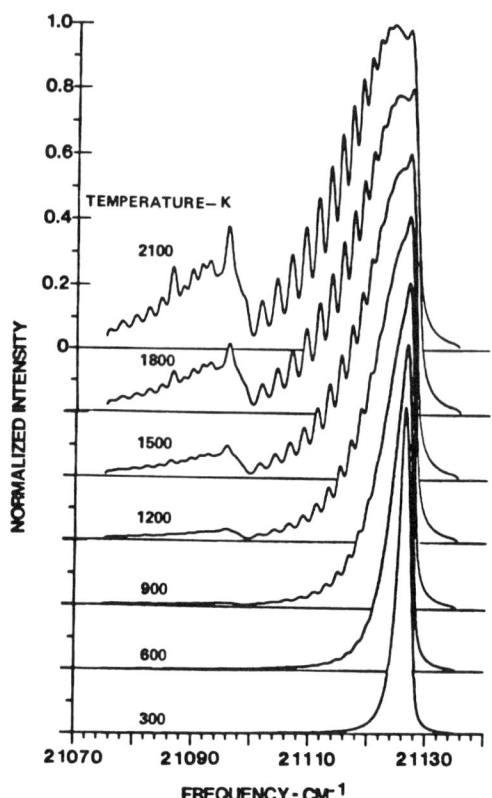

Fig. 6. Temperature variation of the CARS spectrum of N_2 for a pump linewidth of 0.8 cm^{-1}, a Stokes bandwidth of 150 cm^{-1} and a 1 cm^{-1} slit width.

taken in a clean (non-sooting) methane/air flame at a point where the temperature, measured by the thermocouple, was determined to be 2110 K. The dotted curve is the computer model, least mean squares fit to the experimental spectrum; the best fit temperature is 2104 K $\pm$ 9 K. This corresponds to an accuracy of better than 1.0%. In order to achieve this type of accuracy for the theoretical fit, very accurate spectral constants for the N_2 vibrational-rotational energy levels are required. In particular, the Raman linewidths, which enter into the expression for the resonant susceptibility equation (5), must be known as a function of temperature and J-state. Fortunately, accurate measurements of the Raman linewidths of N_2 have been obtained by Rahn and co-workers (Ref. 16) at 300 K and 1700 K using high resolution stimulated Raman gain spectroscopy. Equally important, a theoretical model for Raman linewidths, developed by Bonamy and colleagues (Ref. 17), is in good agreement with the experimental values. This model, anchored by the experimental points, permits extrapolation to higher temperatures for estimates of the Raman linewidths, with an error probably less than 20%. Moreover, the T and J dependence of the linewidth is weak at higher temperatures. The sensitivity of the CARS determined temperature to the Raman linewidth has been examined by Eckbreth and Hall (Ref. 1). At 2100 K, it is estimated that the temperature error is less than 25 K (hence, $\sim$ 1%) for a 20% error in the Raman linewidths. Thus, the sensitivity to linewidth is not large, and extrapolated values should be quite adequate.

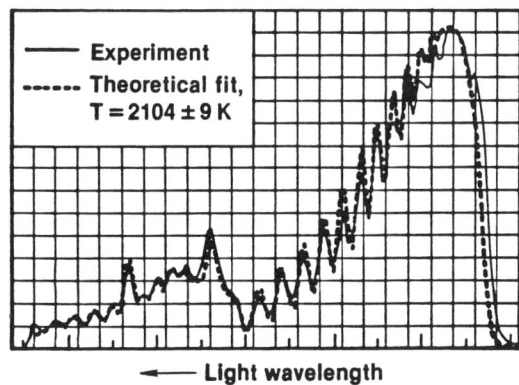

Fig. 7. Comparison of experimental (solid line) and theoretical least squares fit (dashed) N_2 CARS spectra at a measured temperature of 2110°K.

To illustrate the application of CARS thermometry under adverse conditions, measurements in a highly sooting flame will be described (Ref. 3). BOXCARS, with a spatial resolution of about 0.3 by 1.0 mm, was employed in a laminar propane/air diffusion flame. The axial and radial variation of temperature in this flame are shown in Figs. 8, 9, respectively. The quality of the sooting flame CARS spectra is equivalent to that obtained from clean flames. The significance of CARS temperature measurements in the highly luminous, sooting flame is that the CARS method succeeds where conventional Raman methods would fail because laser-induced particulate incandescence (Ref. 18) would swamp the weak Raman signal. In contrast, the CARS signal, which is completely collected, is sufficiently intense to determine temperature from a single, 10^{-8} second pulse. An example of single shot CARS spectra will be presented under the section dealing with applications.

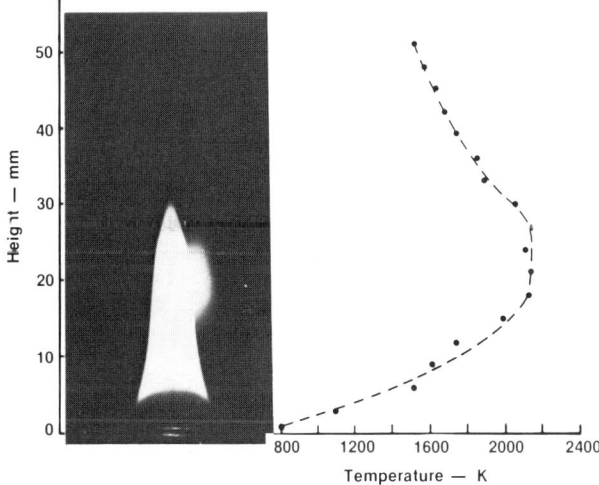

Fig. 8. Axial temperature profile determined in a laminar propane diffusion flame.

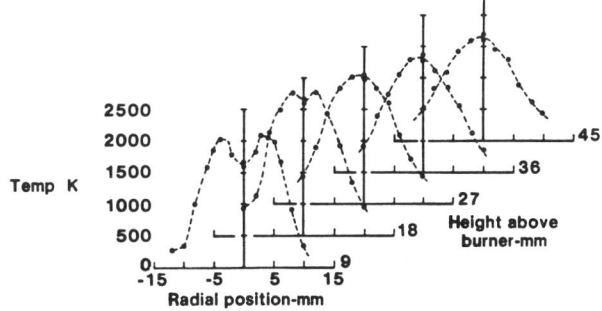

Fig. 9. Radial temperature profiles determined by CARS in a laminar propane diffusion flame.

PRESSURE EFFECTS IN N_2 THERMOMETRY

To achieve maximum usefulness as a diagnostic tool, CARS must be extended to pressures much higher than one atmosphere. Combustion at high pressure is important in such devices as gas turbines, internal combustion engines (especially diesels), and rockets. The pressure encountered in such systems ranges from a few tens of atmospheres to well over 100 atmospheres. If the theoretical CARS analysis used to explain one atmosphere CARS flame spectra were employed at high pressure, it would predict that the overall bandwidth of the specturm would increase linearly with pressure. The unmodified theory would also predict that the $1 \to 2$ "hot band" would merge with the fundamental band, and therefore would be lost as a sensitive indicator of temperature.

The effect of pressure on N_2 CARS spectra has been investigated experimentally and theoretically over the pressure range of one to 100 atmospheres (Ref. 19). A similar study has been performed by Roland and Steele (Ref. 20) for pressure up to 30 atmospheres. The room temperature studies of Hall, Eckbreth, and Verdieck show experimentally that, after an initial narrowing of the Q-branch band over the first five to ten atmospheres, the width of the band remains constant with increasing pressure to 100 atmospheres. Figure 10 shows collinear CARS Q-branch spectra of nitrogen at one and 100 atmospheres scanned with a monochromator spectral width of ~ 0.4 cm^{-1}. The dotted line, the theoretical fit, demonstrates the excellent agreement of the computer model modified to include "collision-induced band narrowing". In simple terms, collisional narrowing of the band occurs when the collision frequency (a function of pressure) becomes comparable to the spacing between individual Q-branch transitions. When this happens, the molecule moves rapidly through many J states and appears to spend, on the average, most of the time in a most probable J-state, defined by the Boltzmann distribution. The phenomenon has also been observed in NMR (Ref. 21), and Raman spectroscopy (Ref. 22). For greater detail on the quantum mechanical treatment of collisional narrowing in CARS spectra, the interested reader may consult (Refs. 1 and 19). Figure 11 illustrates more dramatically the importance of collisional narrowing by a comparison of a collision-narrowed line with a line calculated not including narrowing. Note that for the latter case a very broad line, ~ 10 cm^{-1}, results.

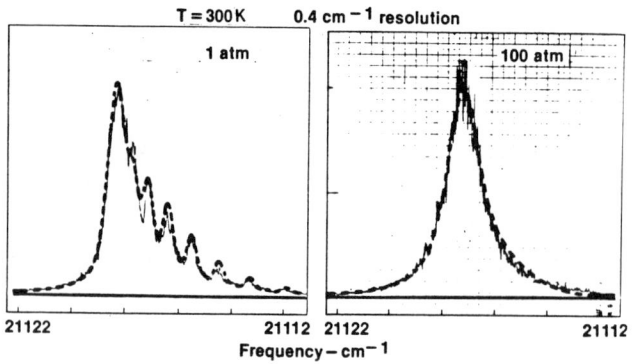

Fig. 10. CARS signatures of 300°K N_2 at one and one hundred atmospheres pressure. The solid line is the experimentally-scanned spectrum in each case, the dashed line the theoretical prediction.

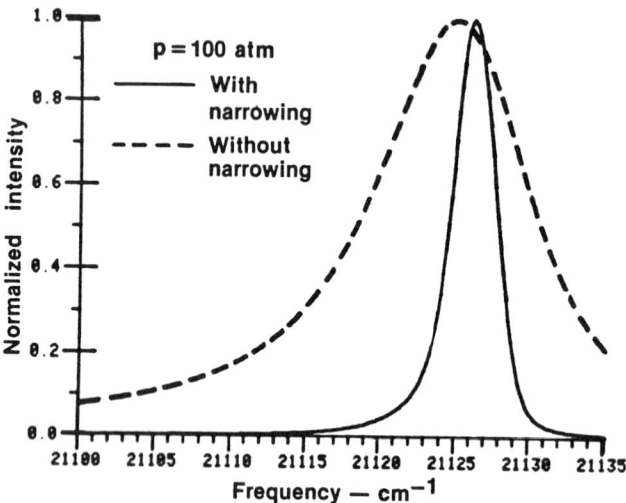

Fig. 11. Comparison of calculated 100 atm signature with and without collisional narrowing — with narrowing, --- without narrowing.

The effect of high pressure on high temperature CARS spectra has been initiated at UTRC; some results are shown in Fig. 12 (Ref. 23). Experimental results are shown at ∼ 1600 K for pressures of ∼ 8 and 28 atmospheres. The model predicted spectra are presented for comparison. It is noted that although the rotational structure collapses over this pressure change, the hot band persists; hence, the potential for temperature measurements at high pressure still exists. Obviously the area of high pressure, high temperature CARS spectra must be investigated more carefully and thoroughly in order to assess CARS diagnostics for the several important high pressure applications listed earlier.

PURE ROTATIONAL CARS THERMOMETRY

Pure rotational CARS may offer advantages over vibrational-rotational CARS at high pressures, and also at low temperatures (under 1000 K, at low pressures). The reason for this is that the pure rotational lines have better spectral separation, 4B; where B is the rotational constant. In N_2, this amounts to 8 cm^{-1}. Pure rotational CARS presents the

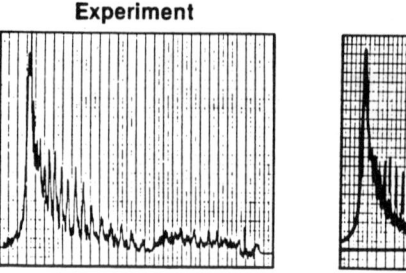
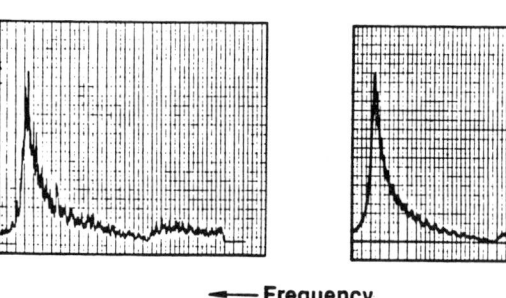

Fig. 12. High pressure CARS spectra for N_2 at 1630°K.

same classic problem that pure rotational Raman spectroscopy does; the frequency shifts from the incident radiation are small, and the CARS radiation is difficult to separate spectrally from the input beams. One method, employed in conventional Raman spectroscopy is to employ a double (or triple) monochromator. In this laboratory the necessary spectral separation of the CARS beam is achieved by the complete spatial separation of the beams, without a dispersing optic, through use of the three-dimensional phase-matching technique called "folded BOXCARS" (Ref. 6), discussed earlier in the experimental section. Referring back to Fig. 4, it can be seen that the CARS beam, ω_3, emerges well separated angularly from the input beams. In this manner, the pure rotational CARS spectrum of room air (one atmosphere, 300°K) was recorded by spectrally scanning the isolated ω_3 beam (Fig. 13). This experimental spectrum is in good agreement with the theoretical model (Ref. 6).

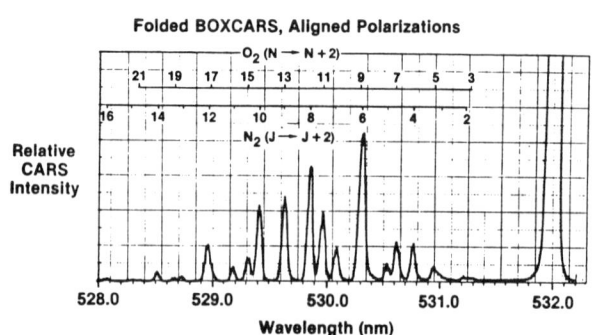

Fig. 13. Pure rotational CARS spectra of air.

CARS THERMOMETRY WITH OTHER MOLECULES

At UTRC, several other molecules of interest to combustion science have been examined for CARS thermometry and concentration measurements. CARS spectra have been obtained for hydrogen, oxygen, carbon monoxide, carbon dioxide, water, and methane. Space does not permit a discussion of all of these; hydrogen, water, and carbon dioxide CARS spectra are selected for discussion with application to temperature measurement.

Computed CARS spectra of hydrogen at four different temperatures are illustrated in Fig. 14. The vibrational-rotational interaction, α, is much larger in hydrogen (2.993) than in nitrogen (0.0187); therefore, the spectrum is simple (few lines), and quite spread out (over nearly 200 cm^{-1}). The alternation in intensity due to nuclear spin statistics is quite apparent. Recall that ortho/para forms of hydrogen bear statistical weights of 3:1. However, in CARS spectroscopy, population differences are squared (because χ is squared), therefore, the intensity contrast between adjacent lines is quite large, and hence useful for temperature measurements. For comparison, experimental CARS spectra for H_2 recorded on an OMA and taken in a H_2/air diffusion flame, are presented in Fig. 15. The temperatures listed were obtained from comparison of the peak height ratios of Q(1), Q(3), and Q(5) transitions. As a practical application, temperature profiles of a slot-shaped H_2/air diffusion flame were determined from H_2 CARS spectra, and also from O_2 CARS spectra (O_2 CARS spectral temperatures are obtained in manner analogous to that for N_2), and are displayed in Fig. 16. The CARS derived temperatures are in fairly good agreement with a radiation-corrected thermocouple. A detailed discussion of all aspects of the hydrogen CARS studies is found in Ref. 24.

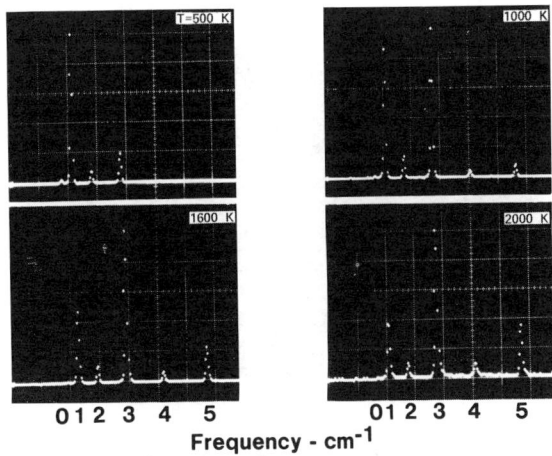

Fig. 15. CARS spectra of H_2 in H_2-air diffusion flame at temperatures determined from relative strengths of indicated Q-branch transitions. Frequency scale corresponds to 0.60 cm^{-1} per dot.

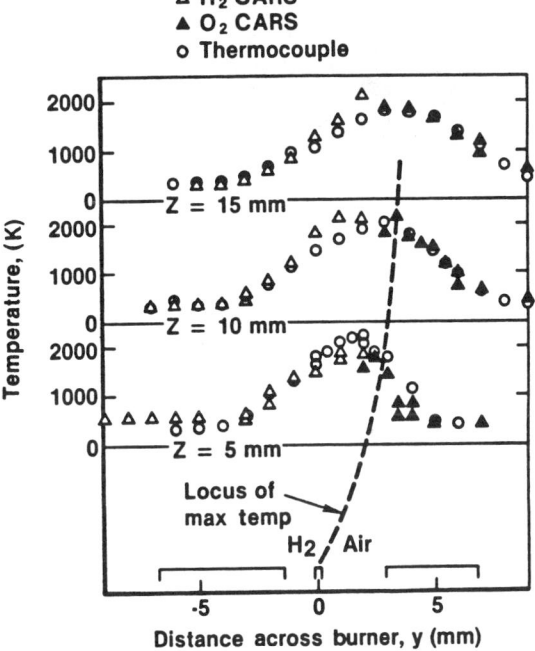

Fig. 16. Temperature measurements in a flat H_2-air diffusion flame. Symbols: circles, radiation-corrected thermocouple; open triangles, H_2 CARS; and solid triangles, O_2 CARS. Dotted curve is locus of maximum temperature.

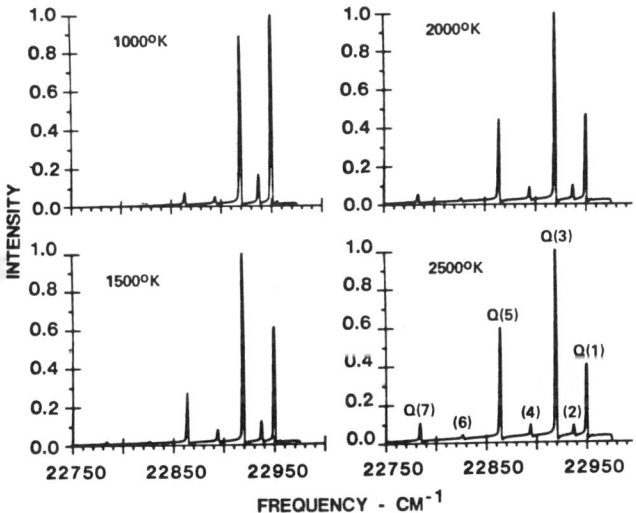

Fig. 14. Theoretical CARS spectra of H_2 over the temperature range $1000°K \leq T \leq 2000°K$.

Because water is a major product of air supported combustion of hydrogen-containing fuels, it can serve as a thermometric probe molecule, if present in sufficient abundance. Moreover, the measurement of water concentration as a function of distance in a combustion system could yield chemical kinetic information and a better understanding of the combustion process.

The water molecule is an asymmetric top with three vibrational modes, of which the ν_1 symmetric stretch is strongly Raman allowed, with a Raman shift of 3657 cm^{-1}. Because it is an asymmetric top, the rotational structure is complex and the rotational energy levels cannot be expressed in closed form. However, the structure of the Q-branch of the ν_1 mode is tractable, and Hall, et al. (Ref. 25) have modelled the CARS spectrum. Figure 17 displays the comparison

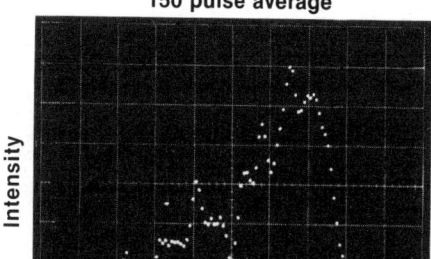

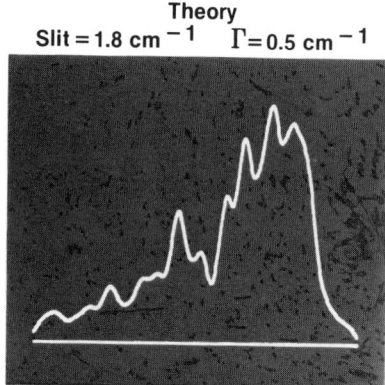

Fig. 17. Comparison of calculated CARS spectrum of pure H_2O with spectrum measured in a heated cell at 773°K.

between computed and experimental CARS spectra of water in a static, heated cell at 773 K. Because Raman linewidth data were not available for this calculation, a value of 0.5 cm^{-1}, independent of J was selected on the basis of the good agreement achieved between the calculated and experimental aspects. A similar comparison of theory and experiment is shown in Fig. 18 for H_2O in a 1700 K premixed methane/air flame. Because of the higher temperature, the assumed Raman linewidth was estimated to be 0.2 cm^{-1}. In these calculations, nearly one thousand vibrational-rotational Q-branch transitions were included in the computed spectrum. Because the shape of the water spectrum is quite sensitive to temperature, the water molecule may offer the potential for thermometry over a wide range of temperature.

The other major product of hydrocarbon combustion is carbon dioxide. CO_2, a linear triatomic molecule, has three normal vibrational modes, a symmetric stretch (ν_1), an antisymmetric stretch (ν_3), and doubly-degenerate bend, (ν_2). Moreover, an accidental degeneracy of ν_1 and $2\nu_2$, the prototype case of Fermi resonance, mixes these two states and makes both transitions from the ground state Raman active. Not only these two states are in Fermi resonance, but several other combinations, occurring up the vibrational state ladder, are also in Fermi resonance which makes the problem of calculating CO_2 vibrational energy levels difficult. In spite of this complexity, theoretical CARS spectra for CO_2 have been calculated by Hall (Ref. 26).

BOXCARS spectra of CO_2 have been obtained by Eckbreth (Ref. 1) in several propane-air and CO-air flames, as shown in Fig. 19 where a measured thermocouple temperature is given for each spectrum. The experimental signature displays a fundamental band at a Raman shift of about 1388 cm^{-1}, assigned to the transition between the ground vibrational state and the first symmetric stretch state. A number of hot bands, originating in vibrationally excited initial states, appear at larger shifts. The relative strength of the hot bands is seen to be moderately sensitive to temperature, making the molecule potentially attractive for thermometry.

Figure 20 compares the experimental CO_2 CARS spectrum at 1520 K with a theoretical spectrum based on the isolated line approximation (Ref. 26). As can be seen, the agreement is fairly good, except for the highest hot bands. In the theoretical calculation, vibrational energy levels, rotational constants, and polairzability matrix elements were computed following the treatment of Courtoy (Ref. 27). This basically involves diagonalizing the system Hamiltonian for each set of near resonance states. It was found, however,

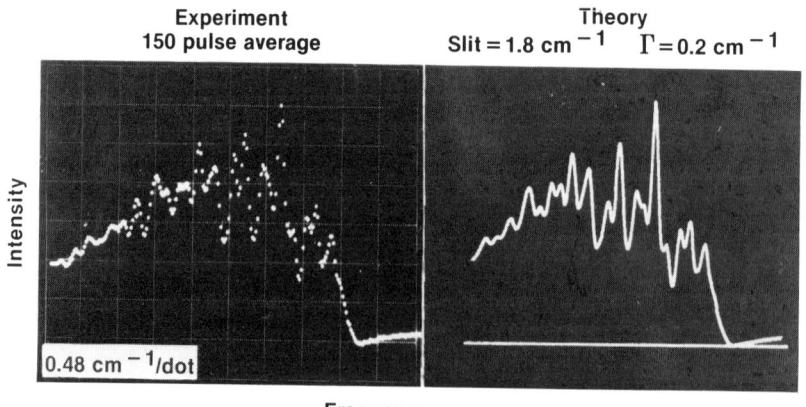

Fig. 18. Comparison of experimental and theoretical CARS spectra of water vapor in a flame at 1700°K. A best-fit pressure-broadened linewidth $\Gamma = 0.19$ cm^{-1} was inferred for all transitions in the calculation.

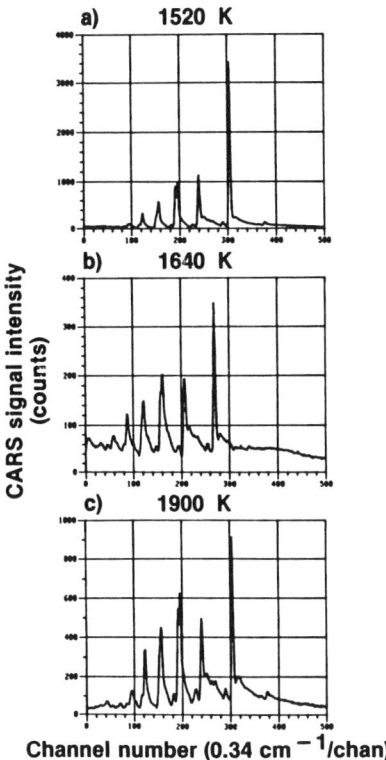

Fig. 19. Experimental carbon dioxide spectra at selected temperatures.

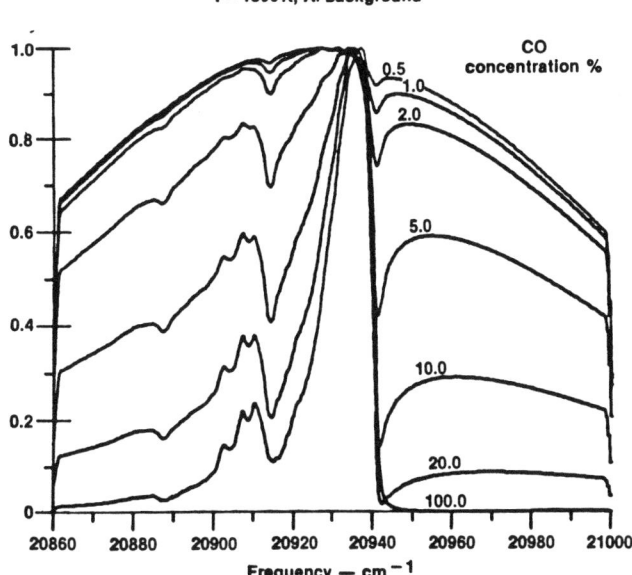

Fig. 21. Computed CARS spectral behavior of CO at various concentrations at 1800°K.

that better agreement with experiment resulted from the use of tabulated values for the rotational constants (Ref. 28). Similar agreement was obtained with the other experimental spectra of Fig. 19. Further investigations, mainly theoretical, will be needed before CO_2 CARS can be regarded as a reliable diagnostic tool.[2] For example, the calculations display some sensitivity to assumed Raman linewidth, and more information concerning the magnitude of these quantities is needed. Also, the high degree of line overlap within each band may make the isolated line approximation questionable even at one atmosphere pressure.

CARS CONCENTRATION MEASUREMENTS

Although this review is intended to deal primarily with CARS temperature measurements, species concentration measurements from CARS will be included for completeness. CARS measurements of species concentration in combustion are made in either of two quite different ways; from the absolute intensity of the CARS signal, or in certain concentration ranges, from the shape of the spectrum. The latter case, which will be presented here, results when the resonant and nonresonant contributions are comparable and the weak signal limit applies, shown in Fig. 2. This behavior can be better understood by reference to Fig. 21, which displays a set of model predicted spectra of carbon monoxide at 1800 K. The concentration values range from 0.5 percent to 100 percent; however, the useful range extends to ∿ 30 percent at the high end. Beyond this value some sort of reference is required to scale the intensity. Below ∿ 0.5 percent level, the resonant signal disappers into the baseline.

Spectral shape fitting for concentration measurements has been demonstrated experimentally in CO, as illustrated by the two spectra in Fig. 22. The upper portion is the experimental CARS spectrum (dotted line) of a 2.1 percent CO (in argon) calibrated gas mixture at 300 K. The thin solid line is the computer generated CARS spectrum calculated for the same concentration, and is in excellent agreement. The bottom part of Fig. 22 displays a similar comparison for CO in a $CH_4/O_2/Ar$

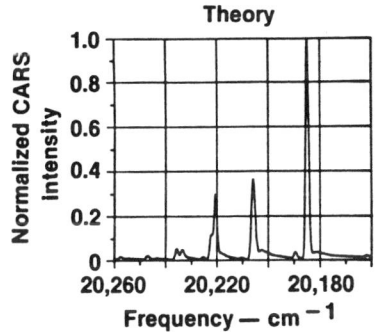
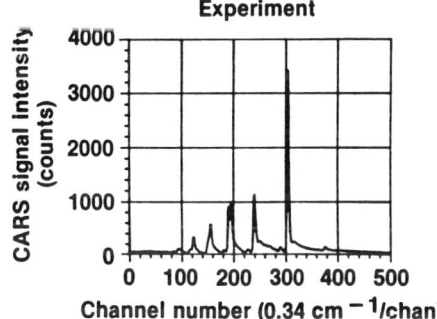

Fig. 20. Comparison of experimental and calculated CARS carbon dioxide spectra.

flame doped with CO. The flame temperature was 2100 K. The theoretical spectrum was calculated for 3.6 percent CO, which was determined with a quartz microprobe and NDIR CO detector. Again the agreement is good. For more detailed information on CO concentration measurements, including a discussion of other CARS methods, the reader may consult Ref. 29. Concentration measurements by the CARS spectral shape fitting technique have been applied to oxygen (Ref. 30) and nitrogen (Ref. 3). Spectral fitting could be used for measuring water vapor concentrations as well.

Before closing the discussion of concentration measurement by spectral shape, it must be emphasized that both the temperature, and the nonresonant susceptibility, must be known in order to determine the concentration of the desired species. Because the nonresonant susceptibility changes little (10-15%) through the air fed combustion reaction going from reactants to products, intelligent estimates can be made for χ^{nr}, if the location, and temperature at that point, are known. From the experiments shown above, it is clear that concentration measurement by spectral shape fitting is a viable technique, easily applied, where appropriate.

PRACTICAL APPLICATIONS OF CARS THERMOMETRY

In this section some illustrations of CARS measurements, mainly temperature measurements, in industrial scale combustion devices are listed. A detailed description of CARS measurements performed at UTRC on a gas turbine test combustor concludes this final section.

Apparently the first application of CARS measurements on a practical, large-scale combustion device was performed by Taran and co-workers at ONERA in 1978 (Ref. 31). Temperature and concentration measurements of N_2 and CO_2 were made in a simulated turbomachine combustor burning kerosene. More recently this group has made measurements in a commercial gas turbine combustion can. Similar measurements at Wright-Patterson AFB have been carried out in a large scale diffusion flame which was stabilized with a bluff body (Ref. 32).

CARS measurements in an internal combustion engine have been made by Stenhouse et al. (Ref. 33). The CARS input laser pulses were synchronized with the engine cycle to generate the CARS spectrum in time steps, and could be positioned at known points on the engine cycle. More recently, workers at Ford Motor Company Research have measured temperature and carbon monoxide concentration in a research scale, single cylinder engine (Ref. 34). A special type of non-collinear phase matching was employed which achieved a spatial resolution of ~ 2 mm along the beam and 100 μm transverse to the beam. Single pulse, 10^{-8} second, CARS spectra were obtained, from which temperature was determined. CARS measurements also have been made in a commercial diesel engine (Ref. 35) at Komatsu Corporation. Further measurements in internal combustion engines using CARS are scheduled at many laboratories throughout the world.

In experiments at UTRC, Eckbreth (Ref. 7) made CARS measurements in a 0.5 m diameter combustion tunnel fitted with a variety of burners, one of which was a gas turbine (JT-12) combustor. CARS temperature measurements (using BOXCARS) from nitrogen were made in the primary zones of flames and in the exhaust. Both gaseous (propane) and liquid (Jet A) fuels were used.

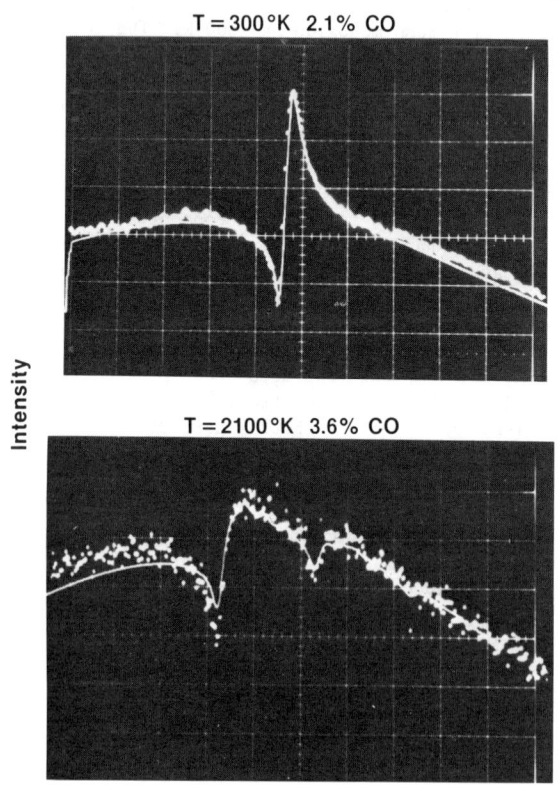

Fig. 22. Comparison of experimental and computed (solid line) CARS spectra at 300°K (2.1% CO) and 2100°K (3.6% CO).

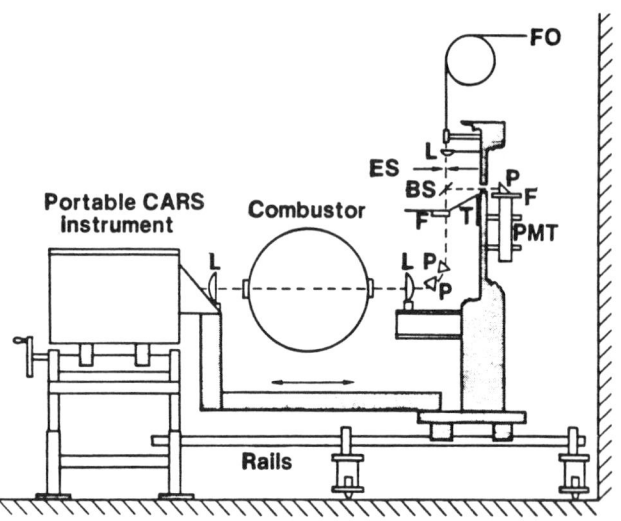

Fig. 23. Cross-section through combustor test tunnel indicating the CARS transmitter on the left and CARS receiving optics on the right. Abbreviations are defined in the text.

The experimental scheme is shown in cross-section in Fig. 23. The pump and Stokes lasers, along with the requisite optics, are contained on a one by two meter optical pallet placed near the test tunnel. Lenses focus and collect the BOXCARS beams through the tunnel windows. It should be noted that the flame in the test combustor is so luminous that most conventional optical methods would fail. The emergent laser beams are manipulated in a receiver which spectrally disperses the beams, and traps the unwanted ω_1 and ω_2 frequncies. A reference PMT detects a small fraction of the CARS signal and serves to monitor the optical alignment of the input optics. The CARS beam is focussed into a fiber optic (20 m long) and piped out to a control room. The control room furnishes a much quieter environment for delicate instrumentation, such as the spectrometer, and the optical multichannel analyzer which is subject to microphonics.

As examples of CARS measurements from this experiment, Fig. 24 displays averaged N_2 CARS system at two different locations downstream of the burner exit face. These CARS spectra correspond to 100-150 laser pulses, or to a 10-15 second average. The spectrum at x = 6 cm was made in the fuel spray, hence, the relatively cool temperature. The second location, at 39 cm, is in the flame, and is much hotter. Figure 25 demonstrates that temperature can be determined in a highly luminous, noisy combustion system, from a single, 10^{-8} second, pulse spectrum. Comparison is made with a 130 pulse average spectrum. The single pulse spectrum displays photon statistical noise (shot noise); however, it is of sufficiently good quality to allow instantaneous temperature measurement. The single shot measurement capability, together with the high spatial resolution of BOXCARS, fulfills an obvious need for studies in highly turbulent, inhomogeneous combustion systems.

In the UTRC CARS Laboratory, in East Hartford, a multipurpose portable CARS apparatus is being assembled for use at the Government Products Division (GPD) of Pratt & Whitney Aircraft. When complete, measurements will be made in the exhaust of a production run 1130 (a modified F-100) gas turbine engine. In addition to N_2 temperature measurements, the concentration of water, carbon dioxide, and oxygen will be determined, as well as some measure of total unburned hydrocarbons. These measurements are scheduled to take place before the end of 1981.

The application of CARS diagnostics to both research scale and production line combustion devices is increasing rapidly. Academic, government, and industrial research laboratories are, or soon will be, performing measurements in internal combustion engines, gas turbine combustors, furnaces, chemical process streams, and propellant burning strands. The equipment necessary for CARS is commercially available, and special methods for employment near noisy environments can be developed, as illustrated by the previous discussion. The capabilities of the CARS technique will be further improved by experimental modification, and more importantly, from advances in computer data processing (including better modelling of spectra). Considering the great amount of knowledge that conventional diagnostic methods have provided, CARS should increase our understanding of fundamental and applied combustion processes even further.

Acknowledgements

The authors wish to thank the Environmental Protection Agency, the Army Research Office, the Air Force Office of Scientific Research, NASA, and the Office of Naval Research for supporting portions of research carried out at UTRC.

References

1. A. C. Eckbreth and R. J. Hall, "Coherent Anti-Stoke Raman Spectroscopy (CARS): Application to Combustion Diagnostics" to appear in Laser Applications (Vol. V), ed. by R. K. Erf; Academic Press.

2. R. J. Hall, Combust. Flame, 35, 47, (1979).

3. A. C. Eckbreth and R. J. Hall, Combust. Flame, 36, 87 (1979).

4. J. J. Barrett and R. F. Begley, Appl. Phys. Lett., 27, 129 (1975).

5. A. C. Eckbreth, Appl. Phys. Letts., 32, 421 (1978).

6. J. A. Shirley, R. J. Hall and A. C. Eckbreth, Opt. Letts., 5, 380 (1980).

7. A. C. Eckbreth, Combust. Flame, 39, 133 (1980).

8. A. C. Eckbreth and R. J. Hall, Combust. Sci. and Tech., 25, 175 (1981).

9. R. L. Farrow, R. E. Mitchell, L. A. Rahn and P. L. Mattern, AIAA Paper No. 81-0182, AIAA, New York, New York.

10. T. W. Hansch, Appl. Opt. 11, 895 (1972).

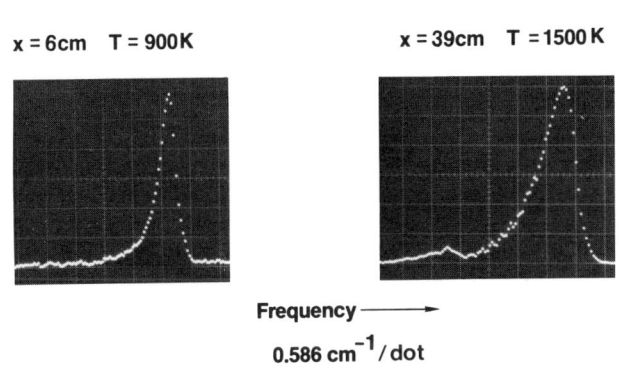

Fig. 24. Spatial variation of temperature from averaged CARS spectra of N_2 in swirl burner with Jet A fuel.

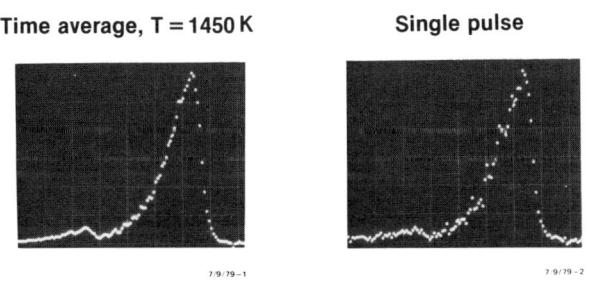

Fig. 25. Comparison of averaged and single pulse N_2 CARS spectra in swirl burner with refractory back wall fueled with Jet A at an overall equivalence ratio of 0.8.

11. M. G. Littman, Opt. Letts. 3, 138 (1978).

12. F. J. Duarte and J. A. Piper, Opt. Comm., 35, 100 (1980).

13. The Optical Multichannel Analyzer is Manufactured by the PAR Company; OMA is a registered trademark.

14. R. J. Hall, Appl. Spectrosc., 34, 700 (1980).

15. A. C. Eckbreth and R. J. Hall, Opt. Eng., 20, 494 (1981).

16. L. A. Rahn, A. Owyonng, M. E. Coltvin, and M. L. Koszykowski, "The J Dependence of Nitrogen "Q" Branch Linewidths," Proc. of VII International Conference on Raman Spectroscopy, Ottawa, Canada (1980).

17. J. Bonamy, L. Bonamy, D. Robert, J. Chem. Phys., 67, 4441 (1977).

18. A. C. Eckbreth, J. Appl. Phys., 48, 4473 (1977).

19. R. J. Hall, J. F. Verdieck and A. C. Eckbreth, Opt. Comm., 35, 69 (1980).

20. C. M. Roland and W. A. Steele, J. Chem. Phys., 73, 5924 (1980).

21. N. Bloembergen, E. M. Prucell and R. V. Pound, Phys. Rev., 73, 679 (1948).

22. A. D. May, J. C. Stryland and G. Varghese, Can. J. Phys., 48, 2331 (1970).

23. J. A. Shirley, R. J. Hall, J. F. Verdieck and A. C. Eckbreth, New Direction in CARS Diagnostics for Combustion, AIAA Paper No. 80-1542 (1980).

24. J. A. Shirley, R. J. Hall and A. C. Eckbreth, "Investigation of the Feasibility of CARS Measruements in Scramjet Combustion," Technical Report R79-954390-8 under Contract NASI-15491.

25. J. A. Shirley, R. J. Hall and A. C. Eckbreth "Investigation of the CARS Spectrum of Water Vapor," Proceedings of Laser 1980 International Conference, New Orleans (1980).

26. R. J. Hall and A. C. Eckbreth, Opt. Eng., 20, 494 (1981).

27. C. P. Courtoy, Can. J. Phys., 35, 608 (1957).

28. L. S. Rothmann and W. S. Benedict, Appl. Opt., 17, 2605 (1978).

29. A. C. Eckbreth and R. J. Hall, Comb. Sci. and Tech., 25, 176 (1981).

30. J. A. Shirley, R. J. Hall and A. C. Eckbreth, Proceedings of 16th JANNAF Meeting, CPIA Publication 309, 487 (1979).

31. B. Attal, M. Pealat and J-P. E. Taran, CARS Diagnostics at Combustion, AIAA Paper No. 80-0282.

32. G. L. Switzer, L. P. Goss, W. M. Roguemore, R. P. Bradley and P. W. Schreiber, AIAA Paper No. 80-0353, AIAA, New York.

33. I. A. Stenhouse, D. R. Willams, J. B. Cole and M. D. Swords, Appl. Opt., 18, 3819 (1979).

34. D. Klick, K. A. Marko and L. Rimai, Appl. Optics, 20, 1178 (1981).

35. K. Kajiyama, Komatsu Ltd., Hiratsaka, Japan, Private Communication.

CARS thermometry in an internal combustion engine

L. A. Rahn, S. C. Johnston, R. L. Farrow, and P. L. Mattern

Sandia National Laboratories, Livermore, California 94550

Coherent anti-Stokes Raman spectra of nitrogen have been measured in an operating internal combustion engine. Time-averaged data were acquired during the compression and power strokes of a propane-fueled research engine operating under homogeneous-charge conditions. Temperatures from 655 K to greater than 2600 K were determined from a nonlinear least squares fit of the CARS spectra. In addition, results using polarization background suppression and in situ referencing techniques are discussed.

Introduction

Applications of Raman spectroscopic techniques to thermometry have contributed significantly to combustion research[1-3]. Recent interest in these techniques has been motivated by the possibility of making measurements that are highly resolved both spatially and temporally, yet leave the delicate interaction between chemistry and fluid dynamics unperturbed. The topic of this paper concerns the application of a specific form of non-linear light scattering--coherent anti-Stokes Raman spectroscopy (CARS)--to measure gas temperature in an operating internal-combustion engine.

Spontaneous (linear) Raman spectroscopy has been applied to internal-combustion engines under motored conditions for fuel-air mixing measurements[4], and to firing engines under homogeneous-charge operating conditions for measurement of burned gas temperature[5]. However, this technique is of limited applicability in stratified-charge and diesel engine combustion, where incomplete mixing of fuel and air leads to diffusion flame burning. Under these conditions spontaneous Raman signals are easily masked by high background luminosity generated from sooting or droplet combustion[6]. This type of problem, encountered in practical combustion devices, has encouraged the development of a more powerful technique, coherent anti-Stokes Raman spectroscopy[7,8,9].

This paper presents time-averaged CARS data obtained from a propane-fueled research engine operating under highly reproducible homogeneous-charge conditions. Temperatures were extracted from these data using a least squares fit to theoretical spectra. In addition to these temperature measurements, preliminary results using polarization techniques for background suppression[10,11] and in situ referencing[12,13] are reported. These results include spectra which were obtained near the rich operating limit of the research engine, where soot can be formed.

Experiment

The CARS experiment is shown schematically in Fig. 1. Three linearly-polarized laser beams are incident on the focusing lens in a three-dimensional phase matching geometry[14]. Measured laser energies were used to normalize the CARS signal for each laser pulse. For in situ referencing, the CARS signal was split into two beams before being analysed with a polarizer. Scattered laser light was separated from the signal with a dielectric filter and monochromator. The outputs from GaAs-photocathode PMT detectors were integrated, digitized, and read by a computer after each laser pulse. Data were acquired by synchronizing the laser to the engine crank angle and averaging the CARS signal for up to ten laser pulses at each spectral increment of the dye laser and monochromator.

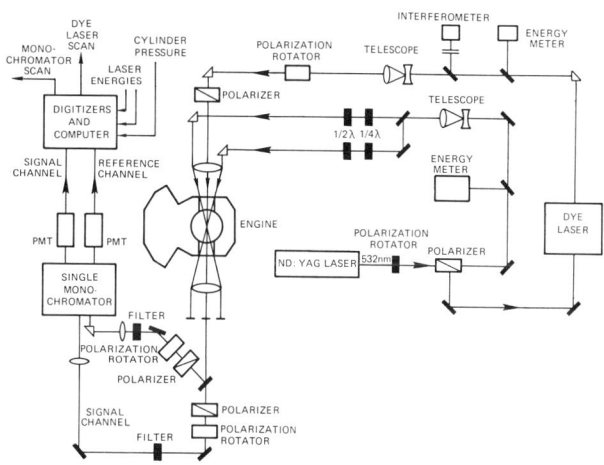

Fig. 1. A schematic of the experimental apparatus for a polarization CARS experiment in an internal-combustion engine. The experiment incorporates 3-dimensional phase matching and two separately analysed detection channels for in situ referencing.

An engine designed to provide a simplified combustion chamber geometry with full optical access was used in these investigations[4]. The combustion chamber was a right circular cylinder, the top of which contained a large sapphire window for shadowgraph photography. Two opposing sides of the cylinder wall were fitted with 12.5 mm aperture windows for the CARS lasers. The compression ratio of the engine was 5.4 to 1. Engine speed was maintained at 1200 rpm by an electric motor. A shaft encoder with a one crank-angle-degree resolution was used to keep track of crank shaft position and to provide timing signals to the laser pulse-control electronics. A single-discharge capacitive ignition system produced a spark from an electrode mounted in the cylinder wall.

Several steps were taken to minimize cycle-to-cycle variation in engine performance. First, fuel and air were mixed thoroughly prior to induction through a shrouded intake valve. Second, the valve geometry was adjusted to impart a high-velocity swirling motion to the inducted charge, thereby reducing burn duration and decreasing the influence of turbulence on flame propagation. Third, to ensure that the combustion products from the previous cycle had been adequately purged from the cylinder (and thus did not mix with the fresh charge), the engine was fired every second compression stroke. This combination of procedures resulted in peak cylinder pressures with better than 1% repeatability in both magnitude and crank angle.

Temperature Results

Rotational-vibrational Raman spectra of common diatomic molecules are typically used for gas-phase laser thermometry[15]. For these experiments, the ro-vibronic spectra of nitrogen were chosen for measurement and subsequent analysis, since the concentration of nitrogen is high and relatively constant during combustion in an air-breathing engine. Examples of CARS spectra of the nitrogen 'Q' branch obtained during the power stroke of the engine with an approximately stoichiometric fuel-to-air ratio are shown in Fig. 2. The data (solid lines) were measured in (a) at 10 degrees after ignition in the unburned charge prior to flame arrival, and in (b) in the burned gases at 40 degrees after ignition. In the latter, both the first and second vibrationally excited states of nitrogen are evidenced by the hot bands at 2300 cm^{-1} and 2270 cm^{-1}, respectively. Ignition was at 8 degrees before top dead center, with the peak pressure occurring 23 degrees later. For these spectra, 700 spectral positions were measured covering 90 cm^{-1}. Input laser polarizations were adjusted to be parallel (no background suppression) and only one detection channel was used. Spectral resolution of the experiment was determined by the laser linewidths, and was 0.5 cm^{-1}. Similiar spectra were obtained at eight other engine crank angles (up to 190 degrees after ignition) to provide temperature information during the compression and power strokes.

A potential difficulty in applying the CARS technique to engine combustion is the steering or defocusing of the input laser beams by the sharp density gradients generated by the flame, resulting in a loss of CARS signal. This effect was observed to occur for only six out of the 190 crank angle degree (CAD) interval that was studied. It is possible that meaningful CARS data may be acquired even under these conditions by using the polarization methods briefly described later in this paper.

In Fig. 2 the dashed lines are nonlinear least squares fits of calculated CARS spectra to the data. Agreement between data and fit for these spectra, as well as for others not shown, is very good. Best-fit temperatures are 655 K and 2622 K in Fig. 2(a) and (b) respectively. The statistical precision of these fits is high, 12 K and 19 K for one standard deviation respectively. The absolute accuracy of the temperatures, however, is expected to be dominated by systematic errors discussed below. Temperatures derived from all the data are plotted versus the crank angle relative to the spark crank angle in Fig. 3. The fact that the temperatures are not a smooth function of crank angle is indicative of the spacially nonuniform combustion process which resulted from convection-dominated flame travel through the swirling charge in the engine cylinder[16]. At the rela-

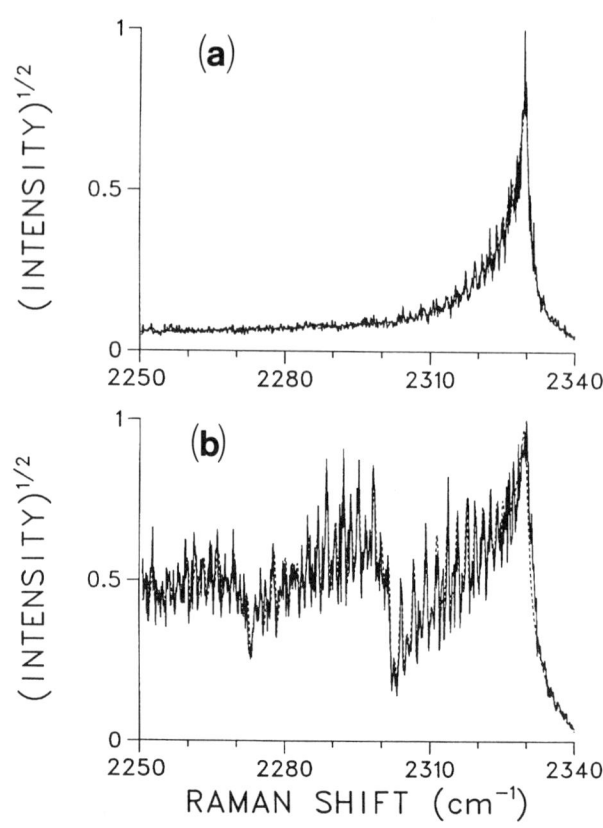

Fig. 2. CARS spectra (solid lines) of nitrogen in an internal-combustion engine taken (a) 10 degrees after ignition and (b) 40 degrees after ignition with cylinder pressures of 7.1 atm and 15.6 atm, respectively. The dashed line is a least squares fit to the data with a temperature of 655 K and 2622 K in (a) and (b) respectively.

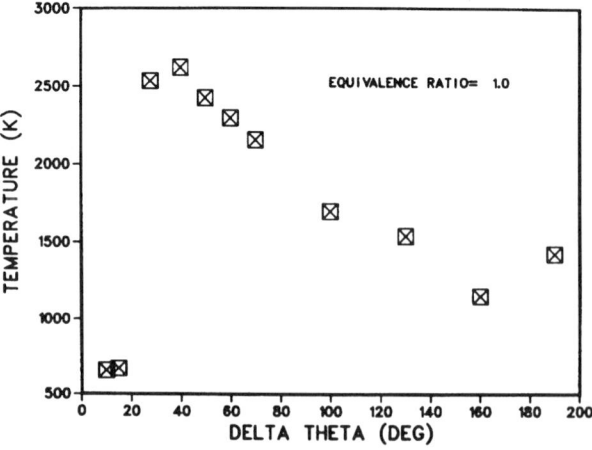

Fig. 3. Temperatures determined by fitting CARS data are plotted as a function of crank angle degrees after ignition. The measurements were made in the center of the combustion chamber while the engine was operated on a nearly stoichiometric mixture of propane and air.

tively high swirl number (about 8) which occurred in the research engine at top dead center, the flame spread much more rapidly along the in-cylinder streamline direction than normal to the flow direction. A typical shadow-graph of the flame contour is shown in Fig. 4. The CARS measurements were made at the center of the cylinder shown in the figure. During the power stroke, combustion gases having different thermal histories (and thus temperatures) flowed through the measurement volume, resulting in the increase in temperature at large values of delta theta shown in Fig. 3.

Results from this experiment also show that there is excellent consistency in the values of temperature obtained from repeated measurements (reproducibility of 2 K has been observed). This agreement is due primarily to the small cyclic variation in engine performance.

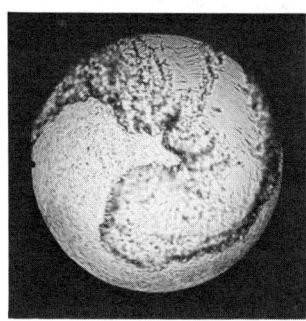

Fig. 4. Shadowgraph photo of the flame in the engine cylinder. The high swirl in the cylinder dominates the path of the flame through the unburned charge.

Error analysis

There are several sources of systematic error that must be considered in the estimation of the absolute accuracy of the derived temperatures. Wavelength-dependent variations in signal throughput and dye laser alignment can contribute systematic errors due to improper weighting of the hot bands. Also, cyclic variation in gas temperature at the in-cylinder measurement location during data acquisition would result in systematic error toward lower temperatures. This bias is a consequence of temperature-dependent density fluctuations which are reflected nonlinearly in the amplitude of the CARS signal.

Assumptions made in the calculation of the CARS spectra also can contribute systematic errors. The procedure used to fit the CARS data includes a calculation of transition linewidths as a function of J, the rotational quantum number. Previous applications of this code resulted in predicted temperatures which were within 7 to 33 K of corrected thermocouple measurements in atmospheric flames[13]. However, the assumed J-dependence[17] has not been confirmed for temperatures above 2000 K. Moreover, the calculation does not yet include the effects of collisional narrowing[18,19] at high pressures. Although the latter effect is small in the measurements reported here (as evidenced by the excellent fits), pressure-dependent lineshape corrections are anticipated to be significant in the analysis of similar experiments on diesel engine combustion. Work to quantify or eliminate these sources of error is in progress.

In summary, although it is difficult to be confident in the accuracy of the error analysis, it is clear that systematic errors are largest for the highest temperatures presented in Fig. 3, and are estimated to be ±50 K.

Polarization Techniques

Significant advances have been made recently in the development of polarization CARS techniques for combustion diagnostics[11,13]. For example, the capability of rejecting nonresonant electronic background[10,11] has been shown to result in an increase in the detection sensitivity for minor species.

Oudar and Shen[12] have demonstrated in liquids that when the background dominates the CARS signal, the light rejected by the polarizer can be used to normalize the background-free signal. However, when the background and Raman signals are of the same order, as in the case of nitrogen thermometry in the engine, the reference and Raman channels will be significantly heterodyned. Farrow et. al[13] have demonstrated an alternate polarization technique for in situ referencing. In this method, two separate CARS signals are derived from a beam splitter and individually analyzed for the rejection of either the Raman or nonresonant component of the susceptibility. This method provided an accurate internal reference, resulting in an improved signal-to-noise ratio for time-averaged concentration and thermometry measurements during combustion.

Recent applications[20] of this normalization method to time-resolved measurements have demonstrated a single shot signal-to-noise ratio of ∿100. These results suggest that it may be possible to design a four channel CARS apparatus, using two narrowband lasers, which will determine temperature and density simultaneously, with a time resolution of ∿10 ns. With such a system, statistical information about temperature and density, as well as correlations between them, could be determined.

Results of a preliminary application of these CARS polarization techniques to the study of engine combustion are shown in Figures 5 and 6. Background-free CARS spectra of carbon monoxide and hydrogen obtained 60 CAD after ignition with a fuel-to-air ratio near the rich operating limit are plotted in Figure 5. For comparison, data acquired (a) without, and (b) with background rejection are shown. The use of in situ referencing with the method of Farrow et al.[13] is demonstrated in Fig. 6. These data (solid line), acquired at 40 CAD after ignition, were averaged for ten laser shots, just as those shown in Fig. 2. The dashed line in Fig. 6 is a calculated spectrum for a best fit temperature of 2378 K with a statistical error of 9.6 K (This temperature cannot be compared with the data of Fig. 3 because of different engine operating conditions). The improvement in signal-to-noise ratio with the use of in situ referencing is apparent from the reduction of the statistical errors from 19 K to 9.6 K at high temperatures.

Internal referencing has considerable merit for time-averaged temperature measurements. Variations in signal due to changing laser power and alignment are corrected to first order, since they also occur in the reference channel. Wavelength-dependent changes in throughput are compensated by normalization to a spectrum obtained with the nonresonant signal selected in both channels. Also, since the nonresonant signal is proportional to the square of the gas density, the temperature er-

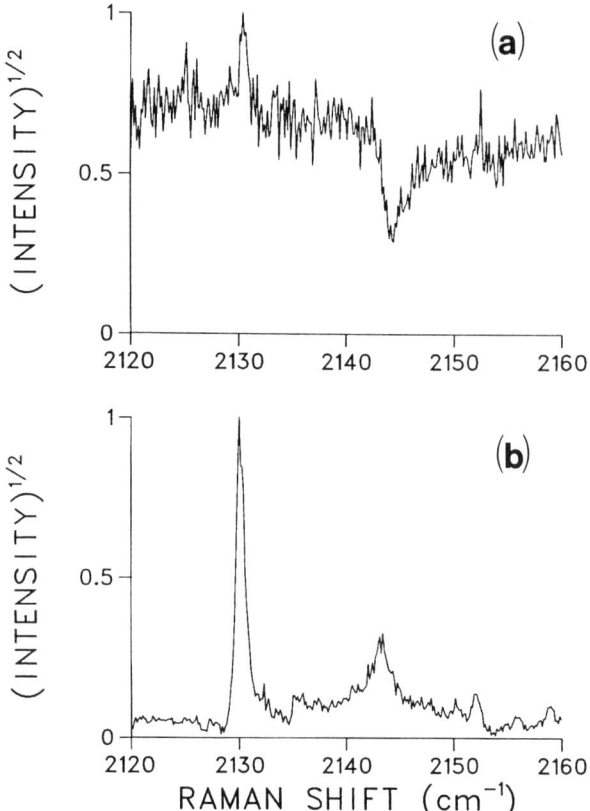

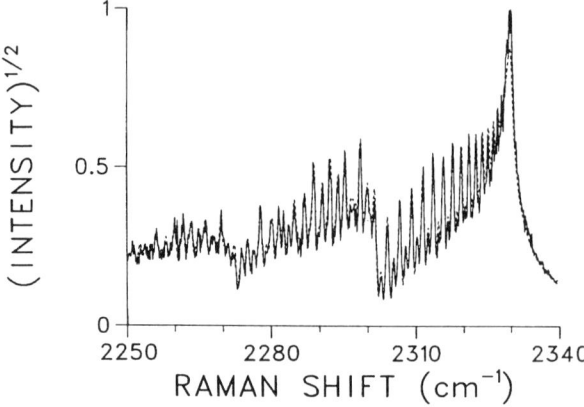

Fig. 5. CARS spectra of carbon monoxide (2143 cm^{-1}) and hydrogen (2130 cm^{-1}) (a) before and (b) after the incorporation of background-rejection techniques. These data were measured at 60 crank angle degrees after ignition while operating near the rich combustion limit for propane/air. The cylinder pressure at the time of the measurement was 13.7 atm and 20 laser pulses were averaged per spectral point.

Fig. 6. In situ referenced CARS spectrum (solid line) of nitrogen taken 40 degrees after ignition. The calculated fit (dashed line) is for a temperature of 2378 K.

ror due to the density weighting effect discussed above is significantly reduced. In addition, if single shot measurements prove practical, this technique would allow measurements in the presence of significant beam steering and defocusing. Finally, this method appears to offer the possibility of obtaining simultaneous temperature and density data at a higher rate than is now achievable with broadband CARS techniques[21].

Conclusions

Accurate, time-averaged temperature measurements have been made using polarization CARS techniques in an operating internal-combustion engine. Excellent agreement between computed and experimental nitrogen spectra has been demonstrated. The analysis of measurements taken at higher temperatures and pressures will require the inclusion of additional detailed linewidth data and pressure-narrowing effects in the computer code used to fit the experimental data.

In addition, the application of internally-referenced polarization CARS techniques have been demonstrated in a research engine. These methods have shown increased accuracy in time-averaged temperature and density measurements, and may allow real time, single-pulse thermometry under the adverse conditions that occur in stratified-charge or diesel engines.

ACKNOWLEDGEMENTS

The authors thank Dennis B. Sparger and Art Van Hook for their technical support in building and maintaining the experiment. Also, the helpful suggestions and careful review of this paper by Dr. J. E. M. Goldsmith and Dr. R. M. Green of Sandia are gratefully acknowledged.

This work was funded jointly by the U. S. Department of Energy and the Motor Vehicle Manufacturers Association of America.

REFERENCES

1 Marshall Lapp and C. Murray Penney, "Analysis of Raman and Rayleigh Scattered Radiation," in Methods of Experimental Physics, Vol. 18, Part B, edited by R. J. Emrich (Academic, New York, 1981), pp.408-433.

2 Sylvie A. J. Druet and Jean-Pierre E. Taran, "CARS Spectroscopy," Prog. in Quant. Elect. 7, 1-72 (1981).

3 Robert J. Hall and Alan C. Eckbreth, "Coherent Anti-Stokes Raman Spectroscopy (CARS): Application to Combustion Diagnostics," in Laser Applications, Vol.V, edited by R. K. Erf (Academic, New York, to be published).

4 S. C. Johnston, "Precombustion Fuel/Air Distribution in a Stratified Charge Engine Using Laser Raman Spectroscopy," Transactions of the SAE, Vol. 88, Paper No. 790092, pp.353-370 (1979).

5 J. R. Smith, "Instantaneous Temperature and Density by Spontaneous Raman Scattering in a Piston Engine," paper AIAA-80-1359, AIAA 13th Fluid and Plasma Dynamics Conference, July, 1980.

6 L. A. Rahn, P. L. Mattern, and R. L. Farrow, "A Comparison of Coherent and Spontaneous Raman Combustion Diagnostics," The Eighteenth Symposium (International) on Combustion (The Combustion Institute, 1981), p. 1533.

7 P. R. Regnier and J.-P. E. Taran, "On the Possibility of Measuring Gas Concentrations by Stimulated Anti-Stokes Scattering," Appl. Phys. Lett. 23, 240 (1973).

8 I. A. Stenhouse, D. R. Williams, J. B. Cole, and M. D. Swords, "CARS Measurements in an Internal Combustion Engine," Appl. Opt. 18, 3819 (1979).

9 David Klick, K. A. Marko, and L. Rimai, "Broadband Single-pulse CARS Spectra in a Fired Internal Combustion Engine," Appl. Opt. 20, 1178 (1981).

10 A. F. Bunkin, S. G. Ivanov, and N. I. Koroteev, "Gas Analysis by Coherent Active Raman Spectroscopy with Polarization Discrimination," Pis'ma Zh. Tekh. Fiz. 3, 450-455 (1977) Sov. Tech. Phys. Lett. 3, 182 (1977).

11 L. A. Rahn, L. J. Zych, and P. L. Mattern, "Background-free CARS Studies of Carbon Monoxide in a Flame," Opt. Comm. 39, 249 (1979).

12 J.-L. Oudar, R. W. Smith, and Y. R. Shen, "Polarization Sensitive Coherent Anti-Stokes Raman Spectroscopy," Appl. Phys. Lett. 43, 398 (1979).

13 R. L. Farrow, R. E. Mitchell, L. A. Rahn and P. L. Mattern, "Crossed-Beam Background-Free CARS Measurements in a Methane Diffusion Flame," AIAA Paper No. 81-0182, AIAA New York (1981).

14 Yehiam Prior, "Three Dimensional Phase Matching in Four Wave Mixing," Appl. Opt. 19, 1741 (1980).

15 M. Lapp and D. L. Hartley, "Raman Scattering Studies of Combustion," Combustion Science and Technology, 1976.

16 Peter O. Witze and Fernando R. Vilchis, "Stroboscopic Laser Shadowgraph Study of the Effect of Swirl on Homogeneous Combustion in a Spark-Ignition Engine," SAE International Congress and Exposition, Paper No. 810226, February, 1981.

17 L. A. Rahn, A. Owyoung, M. E. Coltrin, M. L. Koszykowski, "The J Dependence of Nitrogen "Q" Branch Linewidths," Proceedings of the VIIth International Conference on Raman Spectroscopy, edited by W. F. Murphy (North-Holland, New York, 1980), p. 694.

18 B. A. Alekseyev, A. Grasiuk, V. Ragulsky, I. I. Sobel'man, and F. Faizulov, "Stimulated Raman Scattering in Gases and Gain Pressure Dependence," IEEE J. Quant. Elect. QE4, 654 (1968).

19 R. J. Hall, J. F. Verdieck, and A. C. Eckbreth, "Pressure-Induced Narrowing of the CARS Spectrum of N_2," Opt. Comm. 35, 69 (1980).

20 R. L. Farrow and L. A. Rahn, to be published.

21 Won B. Roh, Paul Schreiber, J. P. E. Taran, "Single Pulse Coherent Anti-Stokes Raman Scattering," Appl. Phys. Lett. 29, 174 (1976).

Temperature measurements for combustion diagnostics from high-resolution single-pulse CARS N₂ spectra

David Klick, K. A. Marko, and L. Rimai

Scientific Research Laboratories, Ford Motor Company, Dearborn, Michigan 48121

Temperatures from 800 to 2800 K were measured in two combustion systems by applying an integral ratio method to coherent anti-Stokes Raman scattering (CARS) spectra of nitrogen. Full nitrogen spectra with spectral resolution of 0.6 cm^{-1} were acquired from single 10 nsec laser pulses with spatial resolution in the order of 1 mm^3. Temperature measurements were made within the cylinder of an internal combustion engine, where cyclic variation and combustion inhomogeneity require such high temporal and spatial resolution. Precision of the method was determined in the post-flame region of a stable burner: The mean and standard deviation of 20 single-pulse temperature readings were found to be 2000±79 K. Considerably more precision was possible in the stable burner using multiple-pulse-averaged CARS spectra. Temperature measurements from such spectra were compared for accuracy with temperatures derived from laser anemometry determinations of flow velocities, and agreement was within 100 K.

INTRODUCTION

The 1981 Nobel Prize in Physics was shared by N. Bloembergen for his work in laser spectroscopy. One of the nonlinear effects for which he provided the theoretical understanding (1), coherent anti-Stokes Raman scattering (CARS), now finds wide use as a combustion diagnostic technique 15 years after it was first discovered (2) in this laboratory. The first proposal for the application of CARS to gases (3) in 1973 was quickly followed by a number of early investigations (4-7) in laboratory devices. As a temperature measurement technique for hostile combustion environments, CARS is now nearing routine use in practical devices such as automotive engines (8, 9) and jet turbine combustors (10-12).

CARS has come into common use, despite the added complexity and expense associated with nonlinear optical techniques, due to a number of advantages (13, 14): Configurations are possible which give full spectral information with high signal strength, yielding concentration and temperature for majority species like nitrogen, while retaining excellent time and space resolution. In addition, CARS experiments can be made insensitive to interfering light from flames (such as soot luminosity or radical luminescence), from laser scattering, and from laser-induced fluorescence. CARS remains a substantially nonintrusive probe as long as electrical breakdown does not occur.

METHOD

The temperature measured within an engine cylinder is expected to show cycle-to-cycle variations similar to those commonly measured for pressure (on the order of 10% under normal operating conditions). These cyclic fluctuations are thought to be due to random flow and ignition disturbances and imperfect mixing of exhaust gas residuals. For a practical engine, therefore, cycle-resolved measurements of temperature are necessary. This is accomplished with the apparatus shown in Fig. 1. A Nd:YAG pump laser firing synchronously with the engine at 10 Hz (1200 rpm) pumps a dye laser to produce a broadband Stokes beam, allowing the entire CARS spectrum of the nitrogen in the sample to be acquired on each laser pulse and yielding a single-cycle time resolution of 10 nsec. Spatial resolution is also important in the turbulent, inhomogeneous engine environment. By using a 2-beam, noncollinear phase-matching technique (7,15) a cylindrical interaction volume (diameter ~ 100 μm, length along beams ~ 1 mm) is attained, with some degree of tolerance to beam steering by the nonuniform refractive-index field.

High spectral resolution is desirable for making accurate temperature estimates. Spex monochromators (.75 m double or 1.25 m single) with 600 grooves/mm ruled gratings used in 4th order disperse the CARS signal, providing spectral resolution of 0.6 cm^{-1} for burner studies and 1.2 cm^{-1} for engine studies. An optical multichannel analyzer (OMA) processes the single-pulse signals from the sample and reference cell, providing background-subtracted sample and reference spectra in 30 sec. A temperature

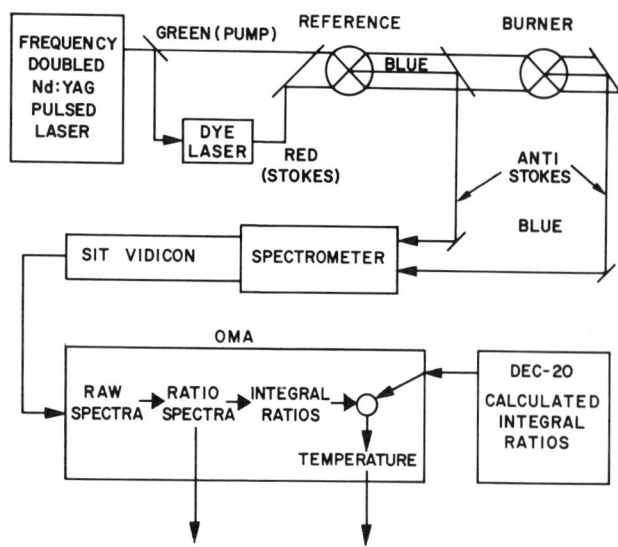

FIG. 1. Block diagram of the CARS temperature apparatus. The resonant anti-Stokes spectrum of N_2 from the burner (or engine) is normalized by the broadband dye profile, found by simultaneously acquiring the nonresonant CARS spectrum of CO_2 in a reference cell. The resulting ratio spectra can be rapidly compared to calculated spectra by the integral-ratio method to yield temperatures.

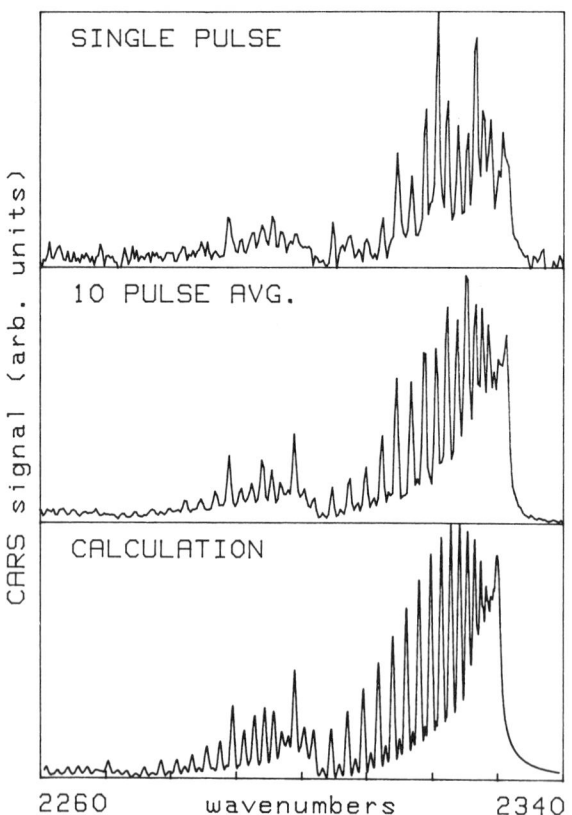

FIG. 2. CARS N_2 Q-band intensity plotted against wavenumber shift from the pump. The single-pulse and ten-pulse-averaged experimental spectra are from a point in the post-flame region of a stable burner. The calculation is for a temperature of 2000 K and slit width of 0.6 cm^{-1}.

determination from the divided spectrum is performed on or off line in another 30 sec. Typical spectra from the stable burner are shown in Fig. 2. Single-pulse spectra are somewhat noisy, but accumulating ten pulses on the vidicon face (as one can confidently do, given the stable temperature field of the burner) provides very good agreement with a calculated spectrum for nitrogen at 2000 K, also shown in Fig. 2.

One can visually compare the averaged data in Fig. 2 to calculations at different temperatures in order to determine the measured temperature. These high-resolution data are so similar to the calculations that there is no question that the experiment is working properly and that the estimated temperature is correct. However, visual inspection is not sufficiently quantitative and not an automated method for extracting temperatures. A least-squares-fitting procedure is excessively time consuming. And one would experience difficulty with either method in estimating a temperature for noisy single-pulse data as in Fig. 2. An integral-ratio method (16) is therefore used to rapidly extract temperatures from experimental spectra.

A measure of vibrational temperature, to be referred to as T(1), can be obtained from the ratio of the integrated first hot band (from 2272 to 2302 cm^{-1}) to the integrated cold band (from 2302 to 2332 cm^{-1}). A measure of rotational temperature T(2), useful for cooler nitrogen (T < 1500 K) when the hot bands are diminished, can be obtained from the ratio of the area under the higher rotational lines of the cold band (2302 to 2317 cm^{-1}) to the area under the lower lines (2317 to 2332 cm^{-1}). For 600 < T(K) < 1000, one can use a rotational integral ratio even closer to the band head at 2330 cm^{-1}, denoted T(3).

Calculations of CARS spectra (see the Appendix for details of the calculation) were performed over a range of temperatures for two triangular slit widths, 0.6 and 1.5 cm^{-1}. Integral ratios were taken at each temperature, resulting in Fig. 3 for the smaller slit width and Table I for the larger one. The influence of spectral resolution was minor over the useful range of T(1-3), changing the temperature found from a given integral ratio by less than 50 K. The effect was smallest for the vibrational temperature T(1). Changing the rotational line width from 0.05 cm^{-1} to 0.1 cm^{-1} had no effect.

For nitrogen molecules in thermal equilibrium, one expects the rotational and vibrational temperatures found from multiple-pulse spectra to agree. In Fig. 4 are plotted T(1-3) for spectra taken at points across the burner flame front. Integral ratios were calculated from the data, and temperatures were extracted from Fig. 3. Agreement at high temperatures is excellent. At low temperatures the agreement is slightly poorer due to the use of numerically small integral ratios that are affected by noise.

It is often found that experimental spectra fail to match the calculations near the cold-band head at 2330 cm^{-1}. In many cases, some collinear signal from room-temperature nitrogen contaminates the spectra.

TABLE I. Integral ratios for the nine sets of integration limits from Table II. The integrals were performed on calculated spectra (smoothed with a 1.5 cm^{-1} slit function) at the listed temperatures.

T(K)	T(1)	T(2)	T(3)	T(4)	T(5)	T(6)	T(7)	T(8)	T(9)
600	.008	.026	.079	.015	.034	.085	.028	.054	.124
800	.013	.044	.142	.020	.047	.135	.033	.062	.168
1100	.032	.085	.254	.045	.060	.228	.063	.093	.256
1400	.076	.141	.367	.101	.128	.323	.132	.139	.346
1700	.151	.206	.473	.191	.187	.413	.237	.197	.430
1900	.216	.253	.536	.267	.229	.467	.324	.239	.480
2100	.292	.300	.593	.353	.273	.517	.421	.283	.526
2300	.375	.347	.645	.446	.317	.563	.524	.326	.569
2800	.605	.462	.754	.692	.425	.662	.792	.433	.659

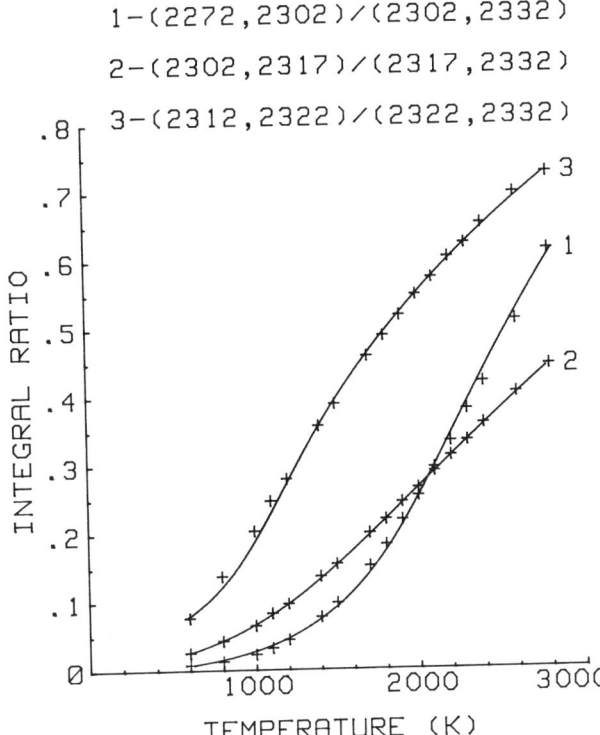

FIG. 3. Integral ratios of portions of calculated CARS N_2 spectra plotted against the temperature assumed in the calculations. Three sets of integration limits are shown at top. The vibrational temperature T(1) is used for post-flame temperatures above 1500 K. At lower temperatures, vibrational integral ratios are too small, and the rotational temperatures T(2) or T(3) are used. The curve that was fit to the points gave temperature as a third-order polynomial in the square root of the integral ratio.

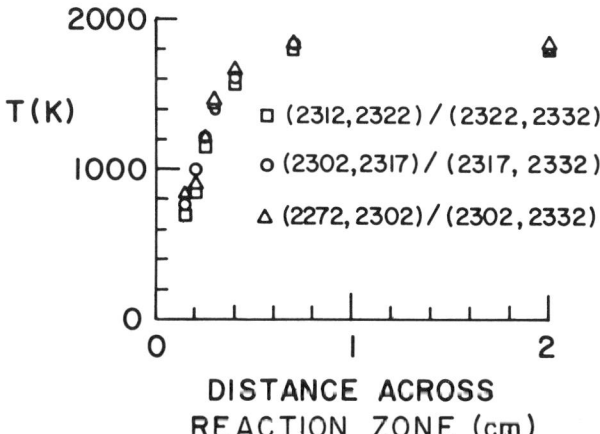

FIG. 4. Agreement of vibrational temperatures [triangles, T(1)] with rotational temperatures [circles, T(2) and squares, T(3)] extracted from the same spectra taken at various points across the reaction zone of a stable burner.

The cold collinear signal is quite strong due to the efficiency of its phase-matching and the fact that it derives from the denser cold nitrogen. Also at the band head, the collisionally-broadened rotational lineshapes overlap even at atmospheric pressure (for example, in the burner), allowing cross-relaxation among the first few rotational lines. Increasing the pressure (for example, in the engine) leads to pressure-induced narrowing effects near the band head (17). (The engine cycle may be approximated by a tenfold compression at constant temperature, followed by a tenfold increase in temperature at constant volume. While the pressure may reach 100 atmospheres, the frequency of collisions, which governs pressure-induced narrowing, reaches only about 30 times the STP frequency.) Fortunately, spectra are affected by the cold collinear contribution and pressure-induced narrowing only very near the band head, which can be removed by careful choice of integration limits in the integral-ratio method. The integration limits for T(4-6) of Table II are similar to T(1-3) except that 4 cm^{-1} at the cold-band head are removed. Similarly for T(7-9), 6 cm^{-1} are removed. For the rest of the data presented in this paper, the cold-band head was removed before estimating temperatures.

The temperatures provided by the integral ratio method are relatively insensitive to the problems one encounters in CARS studies. Nonetheless, these problems should be enumerated. There are uncertainties in the calculation: rotational and pump laser linewidths are not measured, and the triangular slit function is only an approximation. Variation of these parameters over a reasonable range does not change the extracted temperature. The nonresonant CARS background was assumed to be negligible in the calculations of nitrogen spectra, and it is not noticeable in our experimental spectra. It comprises an unknown fraction of the noisy baseline seen in experimental spectra (see Fig. 2).

Another contributor to a nonzero baseline is the vidicon background. The Stokes beam is blocked to record a background spectrum, which is taken within a minute of the signal spectrum. During that minute there can be a change in the vidicon dark current, which introduces a bias after subtraction. The presence of a bias affects integral ratios, but the bias was always small for the measurements described here. Photon noise can be appreciable for single-pulse spectra, since each element of the intensified vidicon array (spaced at about 3 elements/cm^{-1}) will typically contain less than 100 counts. Such noise least affects the vibrational temperature, for which the most elements are contained in the integral ratio.

Before one arrives at a measured temperature from an experimental spectrum, there are a number of choices to be made about handling the data, all of which affect the final result. Removal of the cold-band head has already been discussed, as has the choice of using vibrational or rotational temperatures. One can smooth the reference spectrum to various degrees. Such choices are made on the basis of experience. Integral limits for the data are found by calibrating the vidicon with the widely spaced lines of the nitrogen O-bands, after cross-polarizing the pump and Stokes beams. A more accurate procedure, however, if multiple-pulse spectra are available, is to use the rotational lines of the Q-band spectrum itself. This is one advantage to having high-resolution spectra, even if the final result comes from an integral over many wavenumbers. Other advantages are the sharp pass bands provided for the integrals and the ability to compare every

TABLE II. Integration limits in wavenumbers for calculating temperatures nine ways. T(1), T(4), and T(7) are vibrational temperatures and the others are rotational temperatures. T(1-3) include the whole spectrum, T(4-6) exclude some of the cold-band head (see text), and T(7-9) exclude more of the cold-band head.

	T(1)	T(2)	T(3)	T(4)	T(5)	T(6)	T(7)	T(8)	T(9)
low	2272	2302	2312	2276	2302	2312	2276	2302	2312
mid	2302	2317	2322	2302	2315	2320	2302	2314	2319
high	2332	2332	2332	2328	2328	2328	2326	2326	2326

rotational line in the experimental spectra to calculations.

RESULTS

The precision and accuracy of the method determine its applicability to various temperature measurement situations. Experiments to measure the precision or repeatability of the method were performed. Twenty single-pulse and five 10-pulse-averaged spectra were acquired at a point deep in the post-reaction zone of the stable burner. Here the temperature field is quite homogeneous and stable, with fluctuations expected to be less than the upper limit on velocity fluctuations, 2%, established by laser anemometry measurements. Any substantial variation in CARS temperature is therefore due to the experimental method.

Results are given in Table III. A vibrational temperature $T(4)$ and two rotational temperatures $T(5,6)$ were extracted from spectra with cold-band heads removed. In single-pulse spectra, the rotational temperatures are sensitive to noise and thus can be quite different from the vibrational temperature (e.g. spectrum 2, where $T(6)$ was over 4000 K, and was therefore eliminated). However, the single-pulse means and the 10-pulse-averaged spectra show good agreement among the vibrational and rotational temperatures, as in Fig. 4. Rotational temperatures are not used in this hot post-flame region, so the following discussions of precision will involve only the vibrational temperature.

The single-pulse precision was found to be about 4% (2000 ± 79 K). Temperatures from the 10-pulse-averaged spectra were within 30 K of the single-pulse mean. The distribution of single-pulse temperatures from the burner is plotted in Fig. 5, along with a distribution of single-pulse temperatures measured in the engine (listed in Table IV). The engine distribution is broadened by cycle-to-cycle variation in temperature, as well as by the imprecision of the method (which, from the burner data, should be about ± 80 K). From the comparison of distributions in Fig. 5, it appears that the precision of the single-pulse method is adequate for studying the cyclic variation of in-cylinder temperature that is typically found in engines.

Parenthetically, some description of the engine is in order. It is a Waukesha CFR single-cylinder engine that is provided with diametrically opposed quartz windows of 15 mm clear aperture. At our 10:1 compression ratio, the field of view remains clear at TDC (top dead center or $0°$). Readings were taken at the chamber center, but the focal volume can be translated along the 8 cm diameter. The stoichiometric ($\phi = 1.00 \pm .01$) propane/air mixture was ignited at the optimum moment ($15°$ before TDC), and the measurement took place well after combustion at $60°$ after TDC. The 10% variation in single-pulse localized temperature readings seen in our engine was observed under similar operation of another engine using the spontaneous Raman technique (18).

The stable burner (19) consists of a 16 x 100 mm rectangular channel bisected by a 3 mm cylindrical rod that acts as a flame holder. A lean ($\phi \sim 0.6$) methane/air mixture flows vertically upward at constant velocity (about 60 cm/sec) below the flame holder. The flame stabilizes in a U shape centered on the flame holder. Since measurements are taken where end effects are minimized, the flame is essentially 2-dimensional in a plane perpendicular to the holder. Multiple-pulse CARS temperature and 2-component laser anemometry (20) measurements were made on a fine grid, and an interpolation scheme allows the temperature and velocity to be given at any point (21).

Furthermore, one can find flow tubes (paths followed by elements of fluid) by finite-difference integration of the interpolated velocity. Fig. 6 follows two such flow tubes, A and B. On the left are plotted position coordinates, where the flame holder is at (0,0) and one side wall is at x = 8 mm. On the

TABLE III. Temperatures (K) extracted from 20 single-pulse spectra and 5 ten-pulse-averaged spectra from a spot in the post-flame region of the stable burner.

spectrum	T(4)	T(5)	T(6)
1	1892	2041	2411
2	1897	2064	----
3	2001	1937	2319
4	2098	2429	2356
5	2013	2021	1756
6	2001	2233	2219
7	1886	1672	1459
8	2042	2297	1949
9	1971	2247	2014
10	1983	2151	2077
11	2012	2196	2233
12	1969	1894	1698
13	2026	2140	2312
14	1969	2165	2622
15	2031	2194	1958
16	2165	2087	2150
17	2065	1904	2080
18	2140	2025	2460
19	1951	2159	2171
20	1883	2014	2044
mean	2000	2093	2120
st.dev.	79	167	282
range	282	757	1163
ten-pulse-averaged spectra			
1	2006	1907	1957
2	2010	1990	1972
3	1969	1889	1776
4	1988	1856	1829
5	1996	1920	1945
mean	1994	1912	1896
st.dev.	16	50	88
range	41	134	196

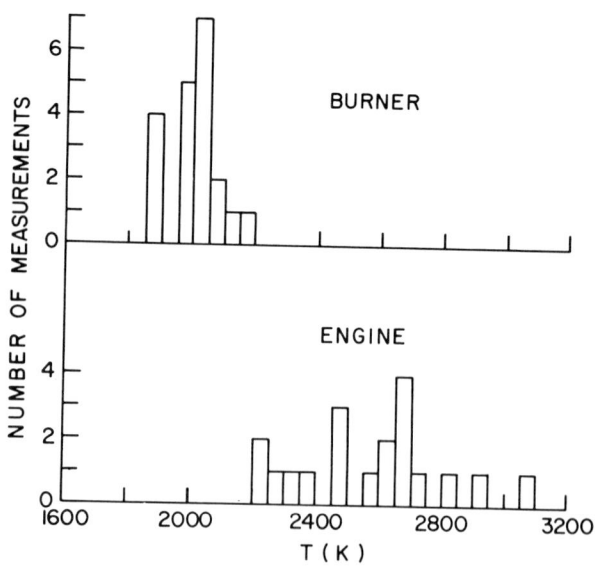

FIG. 5. Histograms of single-pulse temperatures from the stable burner and the engine. The standard deviation of the burner distribution (80 K) is indicative of the precision of the single-pulse CARS temperature measurement. This precision is adequate for engine studies, where the measured standard deviation for cyclic variation in temperature is 230 K.

TABLE IV. Temperatures (K) extracted from 19 single-pulse spectra from the center of the engine cylinder. Some temperatures are above 2800 K and are thus extrapolations rather than interpolations of the calculations in Table 1.

spectrum	T(7)	T(8)	T(9)
1	2815	2604	2937
2	2575	2462	2810
3	2618	2247	2129
4	2240	2020	2282
5	2676	2454	2656
6	2669	2418	2122
7	2680	2304	3331
8	2673	2770	2324
9	2730	2572	2246
10	2476	2626	2171
11	2225	2297	1874
12	2625	2914	2641
13	2271	2236	2041
14	2326	2570	1997
15	2451	2239	1976
16	2465	2444	2781
17	3070	2796	3007
18	2913	2537	2347
19	2380	2425	2536
mean	2573	2470	2432
st.dev.	230	222	405
range	845	894	1457

right are plotted CARS temperatures interpolated along the flow tubes, as well as a "flow temperature" derived from velocity measurements. The gas remains within a flow tube so one can calculate the density at any point, knowing the area of the tube and the gas velocity. Assuming constant pressure (a good approximation at velocities far below the speed of sound), the ideal gas law allows one to calculate the temperature at any point relative to a reference point along the tube. Flow temperatures are normalized to the CARS temperature at a reference height (y = 10 mm in Fig. 6) in the homogeneous post-flame region. At other points, agreement between flow and CARS temperatures is within 100 K, for temperatures as low as 800 K. Most of the disagreement must be ascribed to the inaccuracies in establishing flow tube areas, rather than imprecision in multiple-pulse CARS measurements.

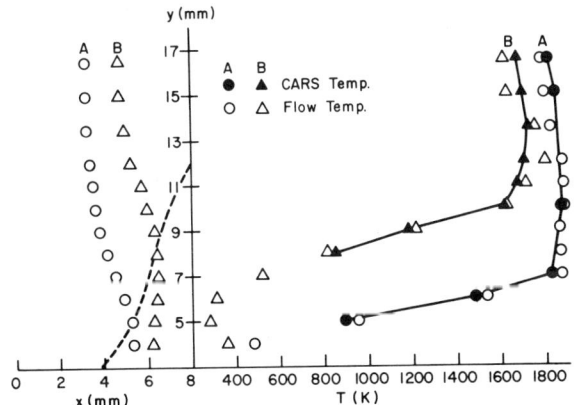

FIG. 6. On the left, a plot of two burner streamlines obtained from integrated laser velocity measurements. A is nearer the flame holder at (0,0) and B is nearer the wall at x = 8 mm. The dashed line is the 1000 K isotherm, which is parallel to and very close to the visual flame front. On the right, CARS and flow temperatures (see text for explanation) are plotted against height for A and B. Away from the normalization height at y = 10 mm, agreement is within 100 K.

CONCLUSIONS

In concert with earlier thermocouple comparisons in the post-flame zone, these velocity measurements establish the absolute accuracy of the multiple-pulse CARS temperature probe to within 100 K over the temperature range 800-2000 K. From Table III, the multiple-pulse precision or repeatability is within 30 K. The multiple-pulse CARS temperature measurement was judged suitably exact for combustion/flow studies in the stable burner (21). For in-cylinder engine temperature measurements, single-pulse readings are required because of cycle-to-cycle temperature fluctuations measured to be as much as 500 K away from the mean (σ = 230 K). The CARS method was found to have adequate precision for engine temperature studies: In the stable burner, the single-pulse measurements were all within 165 K of the mean (σ = 80 K).

Besides allowing single-pulse measurements, the high-spectral-resolution broadband system described here allows the rapid acquisition of scanned-quality multiple-pulse spectra. Simple visual comparison of such high-quality spectra to calculations is important for fostering faith in the data. The integral-ratio method is a rapid and apparently precise way of extracting the vibrational and rotational temperatures of a spectrum, which are in good agreement. With more powerful computer hardware, the time for a broadband CARS/integral-ratio temperature measurement could be reduced to seconds from the present minute, allowing large amounts of data to be collected for measurements of probability density functions of temperature.

APPENDIX

This appendix presents some details of the calculations. For comparison to the experimental spectra shown here, the nitrogen calculations assumed a homogeneous temperature, an etalon-narrowed pump laser linewidth of 0.1 cm^{-1}, molecular rotational linewidths of 0.1 cm^{-1}, and the absence of any nonresonant background. The ith component of the anti-Stokes field of frequency ω = ck at a point $\underline{r}$ beyond the focus can be written as an integral over the probed volume V for the angular distribution of the signal and a double sum over the small range of pump frequencies for the spectral distribution of the signal,

$$E^i(\omega) = \frac{k^2}{r} e^{i(kr-\omega t)} \int_V \Gamma_o^2(\underline{r}')\Gamma_s(\underline{r}')e^{-i\underline{k}\cdot\underline{r}'} d^3r' \quad \times \quad (1)$$

$$\times \sum_{\omega_o}\sum_{\omega_o'} \chi_{ijk\ell}(\omega_o,\omega_o',\omega_o+\omega_o'-\omega)E_o^j(\omega_o)E_o^k(\omega_o')E_s^\ell(\omega_o+\omega_o'-\omega)$$

where E_o is the pump field, E_s is the Stokes field, and χ_{ijkl}^o is the third-order susceptibility tensor, which is assumed to be invariant in space (constant temperature assumption). A Bessel-function solution of the volume integral containing Γ_o and Γ_s (angular distributions of the input beams) was presented earlier (15) which found that simple two-beam input geometries can provide longitudinal spatial resolution of about 1 mm. The phase-matching integral need not concern us further since we are interested here in spectral content.

The double sum over pump frequencies simplifies to a more tractable single sum under certain assumptions. For generality, we include a nonresonant background susceptibility χ^{NR} containing also resonances outside the spectral region of interest. Any susceptiblity component can be written

$$\chi = \chi^o(\omega_o,\omega) + \chi^o(\omega_o',\omega) + \chi^{NR} .$$

The sums on ω_o and ω_o' in Eq. (1) can then be simplified for linearly polarized pump and Stokes fields:

$$\sum_j [\sum_{\omega_o} E_o^j(\omega_o)] \times \sum_{\omega_o'} E_o^j(\omega_o') [2\chi_{ijj\ell}^o(\omega_o',\omega) + \chi_{ijj\ell}^{NR}] E_s^\ell(\omega_o + \omega_o' - \omega).$$

By assuming that the Stokes field varies slowly over a frequency range comparable to the pump linewidth, E_s can be factored out of the sum. Upon squaring the amplitude in Eq. (1), the cross terms vanish if we assume that the individual laser modes have random phases. The time-averaged anti-Stokes power is then

$$\langle |E^1(\omega)|^2 \rangle \propto \sum_j |E_s^\ell(2\langle\omega_o\rangle - \omega)|^2 \times \qquad (2)$$

$$\times \sum_{\omega_o} |E_o^j(\omega_o)|^4 \times |2\chi_{ijj\ell}^o(\omega_o,\omega) + \chi_{ijj\ell}^{NR}|^2 .$$

The sum over ω_o is just a convolution of the pump laser and rotational line shapes. Eq. (2) is equivalent to results used by Hall (22) with the assumptions we have stated.

REFERENCES

1. N. Bloembergen, *Nonlinear Optics* (Benjamin, New York, 1965).
2. P.D. Maker and R.W. Terhune, Phys. Rev. A: 137, 801 (1965).
3. P.R. Regnier and J.P.E. Taran, Appl. Phys. Lett. 23, 240 (1973).
4. F. Moya, S.A.J. Druet, and J.P.E. Taran, Opt. Commun. 13, 169 (1975); W.B. Roh, P.W. Schreiber, and J.P.E. Taran, Appl. Phys. Lett. 29, 174 (1976).
5. A. C. Eckbreth, *Proceedings of the 17th Intl. Combustion Symposium, Leeds, England,* 975 (1978).
6. R.F. Begley, A.B. Harvey, R.L. Byer, and B.S. Hudson, Am. Lab. 6, 11 (1974); J.W. Nibler, J.R. McDonald, and A.B. Harvey, Opt. Commun. 18, 371 (1976).
7. K.A. Marko and L. Rimai, Opt. Lett. 4, 211 (1979).
8. I.A. Stenhouse, D.R. Williams, J.B. Cole, and M.D. Swords, Appl. Opt. 18, 3819 (1979).
9. David Klick, K.A. Marko, and L. Rimai, Appl. Opt. 20, 1178 (1981).
10. G.L. Switzer, W.M. Roquemore, R.B. Bradley, P.W. Schreiber, and W.B. Roh, Appl. Opt. 18, 2343 (1979).
11. M. Pealat, J.P.E. Taran, and F. Moya, Opt. Laser Technol. 12, 21 (1980).
12. A.C. Eckbreth, Combust. Flame 39, 133 (1980).
13. David Klick, K.A. Marko, and L. Rimai, *1981 SAE Congress and Exposition, Detroit* (Society of Automotive Engineers, 400 Commonwealth Drive, Warrendale, PA 15096), paper 810227.
14. L.A. Rahn, P.L. Mattern, and R.L. Farrow, *Proceedings of the 18th Intl. Combustion Symposium, Waterloo, Ontario* (1980).
15. L.C. Davis, K.A. Marko, and L. Rimai, Appl. Opt. 20, 1685 (1981).
16. L. Rimai, K.A. Marko, and David Klick, IEEE J. Quant. Elec. QE-17, 154 (1981).
17. R.J. Hall, J.F. Verdieck, and A.C. Eckbreth, Opt. Commun. 35, 69 (1980).
18. J.R. Smith, *AIAA 13th Fluid and Plasma Dynamics Conference, Snowmass, Colorado,* paper AIAA-80-1359 (1980).
19. Adapted from a design by F. Clauser of Cal. Tech. (personal communication).
20. K.A. Marko and L. Rimai, To be published in Opt. Lett. (March, 1982).
21. L. Rimai and K.A. Marko, Submitted to the 19th Intl. Combustion Symposium, Haifa, Israel (1982).
22. R.J. Hall, Comb. and Flame 35, 47 (1979).

The use of rotational Raman scattering for measurement of gas temperature

M. C. Drake
General Electric Research and Development Center, Schenectady, New York, 12301

C. Asawaroengchai
Thai Metalcote Company Limited, Bangkok, Thailand

D. L. Drapcho
Argonne National Laboratory, Argonne, Illinois 60439

K. D. Veirs and G. M. Rosenblatt
Los Alamos National Laboratory, Los Alamos, New Mexico 87545

Rotational Raman scattering is a new laser diagnostic method capable of remote, non-perturbing, spatially and temporally resolved temperature measurements. It offers the advantages of extended temperature range and high accuracy. Theoretical and experimental advances in rotational Raman temperature measurements which have occurred in the last ten years are summarized. Equations are presented for the calculation of temperature from rotational Raman intensities which include theoretical corrections to the rotational Raman cross section due to centrifugal distortion and vibrational anharmonicity. Examples of the application of these equations to experimental rotational Raman and CARS spectra from diatomic molecules demonstrate the importance of these corrections for accurate temperature determinations. A review of recent rotational Raman (and CARS) temperature measurements compared with more established temperature measurement techniques show a temperature accuracy of $\sim$2–5 % is possible from 10–3000 K in a variety of adverse vapor environments.

I. Introduction

Vapors and gases often are under conditions (high temperature, chemically corrosive, easily perturbed flow) which discourage conventional probe temperature measurements.[1,2] Laser diagnostic techniques offer an alternative which, because they require no physical probe device, can extend the range of temperature measurement capabilities. However, before these laser techniques can be routinely and reliably applied, their theoretical foundations must be understood and their results experimentally verified by comparison with probe measurements under conditions where probes are applicable.

For gas phase temperature measurements the most widely applied laser technique has been vibrational scattering from spontaneous Raman[3-10] and coherent anti-Stokes Raman processes.[4-9] However, rotational scattering from Raman and CARS offers distinct advantages of low-temperature capabilities and higher temperature accuracy and thus is applicable to a wider range of practical applications, from cooled supersonic-nozzle beams to high temperature combusting flows.

Although the potential of determining temperatures from rotational Raman intensities was recognized in the 1930's,[11] it saw little application for quantitative temperature measurements until the early 1970's,[12] after the development of high powered lasers and photoelectric detectors. In addition, the recent discovery of new nonlinear optical techniques such as coherent anti-Stokes Raman scattering (CARS) and Raman gain spectroscopy, as well as technical advances in multichannel detectors, have further extended the capabilities of rotational Raman methods. Experimental capabilities have advanced to the point where the limit in the accuracy of temperatures measured by rotational Raman techniques can lie in uncertainties in theoretical Raman linestrength calculations where corrections for centrifugal distortion and vibrational anharmonicity are appreciable.

Thus, it is appropriate in this volume to summarize the application of rotational Raman temperature measurements which has occurred in the last decade. Emphasis will be on recent advances in the theory and correction factors necessary for accurate temperature measurements, and on experimental results which demonstrate the temperature accuracy and precision currently obtainable.

II. Theory

Only aspects of rotational Raman (and CARS) intensity theory necessary for temperature measurements from diatomic molecules are summarized here. More detailed descriptions of Raman[11,13,14] and CARS[15,18] theory are available elsewhere, but they generally neglect correction factors to the rotational Raman cross section due to centrifugal distortion and vibrational anharmonicity.

A. Spontaneous Rotational Raman Theory

1. *Scattering from ground vibrational levels.*

The rotational temperature determination can be based upon the thermal distribution of population amongst the various rotational levels of the ground vibrational state given by

$$N_{OJ} = \frac{(2J+1) N_o \, g(J) \, e^{-F_o(J) c_2/T}}{Q_r} \quad (1)$$

where N_{OJ} is the number density in a given rotational level J of the ground vibrational level v=0, N_O is the total number density in the ground vibrational state, g(J) is a nuclear spin degeneracy factor, Q_r is the conventional rotational partition function, $F_O(J) = B_O J(J+1) - D_O J^2(J+1)^2 + \ldots$ is the energy (in cm^{-1}) of level J above J=0, and c_2 = hc/k is the second radiation constant. Nuclear spin and molecular constants are known for many diatomic[13,19,20,21] and some polyatomics[13,21,22] molecules.

The rotational number density, N_{OJ}, is related to the (total integrated) intensity of a rotational Raman transition by[3,5,14,23]

$$I_{OJ} = N_{OJ} \, (d\sigma/d\Omega) \, I_o \, \Omega \, \ell \, \epsilon \quad (2)$$

where I_{OJ} is the energy flux of the Raman scattering from rotational level J in vibrational level v=0:

$d\sigma/d\Omega$ is the differential rotational Raman cross-section; and I_o is the energy flux density of the incident radiation. Instrumental factors in eqn. 2 include the solid angle Ω and the laser beam length ℓ over which the Raman scattering is collected and the spectrometer and detector efficiency ϵ.

The ϵ term is a numerical constant C_{exp} divided by a frequency dependent term $R(\nu)$. The constant C_{exp}, necessary for determining molecular concentration, but not temperature, can be determined by calibration Raman measurements at known temperature and concentration. The frequency dependent spectral response $R(\nu)$ can be calibrated by comparing measured intensities of a tungsten strip lamp with the true emitted lamp intensity as a function of frequency.[23]

The rotational Raman cross-section for a particular rotational transition in the ground vibrational level is[23]

$$d\sigma/d\Omega = C \left(\frac{2\pi\nu}{c}\right)^4 [S(J)/(2J+1)] f_{oo}(J) \beta_o^2 \quad (3)$$

where C is a numerical constant dependent only upon the laser polarization and collection direction, $S(J)$ is the rigid-rotor line strength of the transition, $f_{oo}(J)$ is a correction factor arising from the rotation-vibration interaction, and β_o^2 (also called γ_o^2) is the square of the molecular anisotropy in the vibrational ground state. Because of vibrational anharmonicity, β depends upon the vibrational quantum number discussed in the next section.

Absolute values of rotational Raman cross sections for N_2, O_2, H_2, CO, and CO_2[14,24] and relative cross sections for other diatomic and simple polyatomics[5,25] are available. The nuclear degeneracy $g(J)$ equals 1 for heteronuclear diatomics and equals $(I+1)/(2I+1)$ or $I/(2I+1)$ for homonuclear diatomics in either the odd or even J rotational levels, depending upon electronic symmetry. Values for the transition line strength, $S(J)$, can be found in Herzberg[19] and Weber.[13] For N_2, H_2 and other diatomic molecules in $^1\Sigma$ electronic states, there are two cases. For pure rotational Stokes scattering (as well as for Stokes vibrational-rotational S branches)

$$S(J) = (J+1)(J+2)/(2J+3) \quad (4)$$

For pure rotational anti-Stokes scattering (as well as for Stokes vibrational-rotational O branches)

$$S(J) = (J-1)J/(2J-1) \quad (5)$$

where J is the rotational quantum number of the initial state.

Technically, eqns. 4 and 5 do not apply to O_2 which has a $^3\Sigma_g^-$ electronic ground state.[21] Interactions between the unpaired electron spin with the rotational energy levels affects the energy levels and splits each rotational Raman transition in a triplet.[26] An exact treatment of the rotational linestrength S for O_2 has been worked out[27] and the calculated spectra compares well with room temperature experimental spectra.[28] For practical temperature measurements, however, eqns. 4 and 5 are reasonably accurate for O_2 rotational linestrengths for transitions with $J_i > 3$.[27,28]

The correction factor $f_{oo}(J)$ arises because the molecular internuclear distance r increases as the value of J increases due to centrifugal distortion. This change in r can result in a change in β and hence in the rigid rotor line strength.[25,29-31] Using first-order perturbation theory, James and Klemperer have shown for pure rotational Stokes scattering[29]

$$f_{oo}(J) = [1+(4/\chi)(B_e/\omega_e)^2(J^2+3J+3)]^2 \quad (6)$$

where

$$\chi = \beta_e/(r_e \beta_e')$$

where β_e is the polarizability anisotropy and β_e' is its first derivative with respect to internuclear distance evaluated at r_e. The constants B_e, ω_e, and J assume their usual spectroscopic definitions. For pure rotational anti-Stokes scattering, J must be replaced by J-2 in eqn. 6. Experimental values of χ for N_2, O_2, and CO are available [χ_{N_2} = +0.45 ± 0.09, χ_{O_2} = +0.23 ± 0.07, and χ_{CO} = +0.27 ± 0.13][31]. For these molecules the value of $(B_e/\omega_e)^2$ [7.2 x 10^{-7}, 8.4 x 10^{-7}, and 7.9 x 10^{-7} respectively] is so small that the $f_{oo}(J)$ correction does not become important until J>23 assuming that the measured intensities are accurate to 1%. For H_2 and D_2 the value of $(B_e/\omega_e)^2$ [1.96 x 10^{-4} and 9.4 x 10^{-5} respectively] is large enough that the correction for $f_{oo}(J)$ is significant even at J=0. Recent ab initio calculations[32] have shown that the first-order perturbation theory results for $f_{oo}(J)$ are not accurate for H_2 and D_2, both because of the large value of $(B_e/\omega_e)^2$ and because they neglect the second derivative of the polarizability anisotropy with respect to internuclear distance.[32] However, accurate ab initio calculated values of $f_{oo}(J)$ for H_2 and D_2 have been calculated and are reproduced in Table I.[32]

Table 1

Rotational Raman Intensity Correction Factors[a]
For 480.0 nm Excitation

J	f_{oo}^S		f_{oo}^{AS}	
	H_2	D_2	H_2	D_2
0	[1.012][b]	[1.006][b]		
1	1.027	1.014		
2	1.051	1.026	[1.012][b]	[1.006][b]
3	1.082	1.041	1.027	1.014
4	1.122	1.061	1.051	1.026
5	1.171	1.086	1.082	1.041
6	1.231	1.114	1.122	1.061
7	1.300	1.148	1.171	1.086
8	1.382	1.186	1.231	1.114
9	1.475	1.229	1.300	1.148
10	1.583	1.278	1.382	1.186

a. From ref. 32
b. Normalized to the equivalent form from first-order perturbation theory.

Combining equations (1), (2), (3) and converting the scattered intensity from energy flux I_{oJ} to photon flux I_{oJ}^* gives

$$\frac{-F_o(J)c_2}{T} = \ln\left(\frac{I_{oJ}^* R(\nu)}{\nu^3 g(J)S(J)f_{oo}(J)}\right) + \ln\left(\frac{hc^4 Q_r}{(2\pi)^4 C C_{exp}I_o\Omega\ell N_o\beta_o^2}\right) \quad (7)$$

(Relative peak intensities are essentially the same as relative integrated intensities for isolated lines of constant peak width which are dominated by the spectrometer resolution.) The second term on the right in eqn. 7 is a collection of constants for any experiment and will not concern us further. Defining $I_{corr}(J)$ equal to the first term in brackets in eqn. 7 results in a final simplified equation relating the temperature to the corrected intensity

$$\frac{-F_o(J)c_2}{T} = \ln\left(I_{corr,o}(J)\right) + \text{constant} \quad (8)$$

Thus, a plot of $\ln(I_{corr}(J))$ vs $F_o(J)$ yields a slope of $-c_2/T$.

2. Scattering from excited vibrational levels.

When vibrational levels other than v=0 are populated (such as in high temperature systems), the pure rotational Raman spectra consists of a series of lines from each populated vibrational level. The spacing, δ_v, between Stokes pure rotational Raman lines arising from different vibrational levels with the same initial value of J is given by[12,24]

$$\delta_v(J) = 4 (J_{lower} + 3/2) \alpha_e v + \ldots \quad (9)$$

where α_e is the rotation-vibration interaction constant[19,20] and should not be confused with the molecular polarizability. For H_2, this spacing is large compared to typical experimental spectral resolution and no problems arise, but for other molecules such as N_2 and O_2, the spacing is comparable to the typical spectrometer resolution and lines from different v levels overlap. Because the amount of overlap decreases as J increases, the presence of pure rotational transitions from vibrationally excited molecules at high temperatures can result in large systematic errors in rotational temperatures calculated from eqn. 8.[14,25,33]

Accurate temperatures can be derived by multiplying $I_{corr,o}(J)$ in eqn. 8 by an additional correction factor $\eta(J)$ given by[33]

$$\eta(J) = 1 / [1 + (\beta_1/\beta_o)^2 n_1 X_1(J) + (\beta_2/\beta_o)^2 n_2 X_2(J) + \ldots] \quad (10)$$

The n_v term is the fraction of molecules in vibrational level v and $X_v(J)$ is the fraction of the intensity from the v th vibrational level that contributes to the v=0 measured rotational intensity. For high J values ($\gtrsim 50$) there is an additional contribution which must be included in eqn. 10 for intensity of transitions originating in excited vibrational levels in the next higher J level (ie. v, J+1 → v, J+3).[33] Values of $X_v(J)$ are computed from calculated $\delta_v(J)$ values and the spectrometer slit function experimentally measured by scanning an isolated rotational Raman line. Since the correction factor $\eta(J)$ is strongly temperature dependent, temperatures are calculated by an interative procedure.[14,25,33]

The $(\beta_v/\beta_o)^2$ terms in eqn. 10 were not included in previous analyses[25,31,32] of rotational Raman intensities and correct for the vibrational level dependence of the square of the molecular polarizability anisotropy. Expanding β in a Taylor series about r_e gives

$$\beta_v = \beta_e + \beta_e' \langle r-r_e \rangle_v + \tfrac{1}{2} \beta_e'' \langle r-r_e \rangle_v^2 + \ldots \quad (11)$$

where $\beta_e' = (\partial \beta/\partial r)_e$ and $\beta_e'' = (\partial^2 \beta/\partial r^2)_e$

For an anharmonic potential well in diatomic molecules the average value of $\langle r-r_e \rangle_v$ differs from zero.[31,34] It can be evaluated in terms of conventional spectroscopic constants by first order perturbation theory.[31,34]

$$\langle r-r_e \rangle_v = r_e [(3 B_e/\omega_e) + (\alpha_e/2B_e)] (v + \tfrac{1}{2}) \quad (12)$$

The average value of $\langle r-r_e \rangle_v = C_R(v+1/2)$. Dropping the β_e'' and higher terms in the expansion for β_v results in a simple expression for β_v/β_o as a function of v.

$$\beta_v/\beta_o = 1 + v [\beta_e' C_R / (\beta_e + 1/2 \beta_e' C_R)] \quad (13)$$

Using values of β_e and β_e' taken from recent experimental measurements[31] and theoretical calculations[32] and spectroscopic constants from ref. 20, eqn. 13 yields values of $(\beta_1/\beta_o)^2$ and $(\beta_2/\beta_o)^2$ of 1.35 and 1.75 for H_2, 1.03 and 1.06 for N_2, 1.07 and 1.15 for O_2, and 1.06 and 1.11 for CO.[33]

In addition to calculations of η, the correction factor $(\beta_v/\beta_o)^2$ is also necessary for determining vibrational temperatures based upon ratios of pure rotational intensities from excited vibration state compared to ground vibrational level. From eqns. 2 and 3[34]

$$\frac{I_{corr,v}(J)}{I_{corr,o}(J)} = \frac{N_{v,J}}{N_{o,J}} \left(\frac{\beta_v}{\beta_o}\right)^2 \quad (14)$$

where it is assumed that the two lines are close enough so that $R(v)$ and v^3 terms cancel.

B. <u>Rotational Coherent Anti-Stokes Raman Scattering Theory</u>

Coherent anti-Stokes Raman scattering (CARS) is a multiphoton technique closely related to spontaneous Raman scattering but with significant experimental advantages.[8,15-18] In CARS, input photons at v_1 and v_2 from two laser sources (one of which is independently tunable) interact with the third order-nonlinear molecular susceptibility $\chi^{(3)}$ to generate a collimated beam of photons at the anti-Stokes frequency $v_3 = 2v_1 - v_2$. This mixing process occurs at all input frequencies and angles because of non-resonance processes, but the intensity at v_3 is greatly enhanced when (1) the energy difference $hv_1 - hv_2$ equals an energy level difference in the molecule which corresponds to a classical Raman transition and (2) linear photon momentum is conserved. CARS theory, summarized in several reviews,[15-18] is considerably more complex than that for spontaneous Raman scattering. In spontaneous Raman scattering, spectra are a simple incoherent addition of individual line intensities each linearly proportional to the incident laser energy and population of the initial level. Since CARS spectra result from the square of the absolute value of the local third-order nonlinear susceptibility, constructive and destructive interference effects occur with neighboring transitions and the background susceptibility. These interference effects complicate CARS data reduction when many transitions are closely spaced in frequency (i.e. in a vibrational Q branch) which requires that temperature measurements be made by fitting the experimental spectra to computer-generated ones.[18]

However, under conditions where the nonresonant contribution to $\chi^{(3)}$ is negligible and for an isolated line on resonance, the CARS signal in a pure rotational CARS signal, I_3, is proportional to[18,35]

$$I_3 \propto I_1^2 I_2 [\Delta N_{vJ} (d\sigma/d\Omega) \gamma^{-1}]^2 \quad (15)$$

where I_1 and I_2 are the intensities of the incident laser beams, ΔN_{vJ} is the population difference between the initial and final levels, $d\sigma/d\Omega$ is the conventional spontaneous Raman cross section defined in eqn. 3, and γ is the homogeneous Raman linewidth. In rotational CARS the population difference factor is given by[36]

$$\Delta N_{vJ} = N_{vJ} (1 - \exp^{-\Delta F(J)c_2/T}) \quad (16)$$

From eqn. 15 and 16, a temperature equation similar to eqn. 8 for spontaneous Raman scattering can be obtained but because of the more complex dependence of ΔN_{vJ} on temperature, the more common procedure for determining temperatures from rotational CARS transitions in v=0 is to compare the measured spectrum with spectra calculated as a function of temperature.[37]

At high temperatures rotational CARS spectra also include contributions from vibrationally excited molecules. These transitions are less troublesome for rotational CARS analysis of v=0 rotational intensities because the N_{vJ}^2 intensity dependence makes excited vibrational transitions much less important in CARS spectra than spontaneous Raman spectra and the much higher spectral resolution easily obtained in CARS reduces the overlap of excited vibrational transitions with ground state transitions. However, at sufficiently high temperatures pure rotational CARS transitions from vibrationally excited molecules are observable and can be used to measure vibrational temperatures using eqn. 17.

$$\frac{I_{corr,v}(J)}{I_{corr,o}(J)} = \left[\frac{N_{v,J}(1-e^{-\Delta F_1 c_2/T})}{N_{o,J}(1-e^{-\Delta F_o c_2/T})}\left(\frac{\beta_v}{\beta_o}\right)^2\left(\frac{\gamma_{o,J}}{\gamma_{v,J}}\right)\right]^2 \quad (17)$$

C. Comparison of Rotational Raman Theory to Vibrational Raman Theory

Essentially the same theoretical analysis described here for rotational Raman transitions applies to vibrational Raman transitions as well. The major difference is that vibrational matrix elements depend upon the change in polarizability with vibrational coordinate $\partial\alpha/\partial Q$ instead of the polarizability anisotropy β.[11,13,14,18] Vibrational Raman scattering for diatomic molecules is dominated by Q ($\Delta J=0$) branch transitions whose linestrength factors, S, and linestrength correction factors, f_{o1}, have been determined for N_2, O_2, and H_2.[33] The f_{o1} correction has been found to be less significant than f_{oo} for accurate temperature measurements.[33] In the harmonic oscillator approximation, the vibrational dependence of the vibrational Raman cross-section equals (v+1).[13,19] The effect of vibrational anharmonicity on these vibrational cross-sections (analogous to the β_v/β_o correction term for rotational Raman scattering) and its effect on derived temperatures has not been determined but is probably small based upon the excellent agreement between temperatures measured by vibrational Raman and vibrational CARS with other techniques.

Despite theoretical similarities, the methods of deriving temperatures from experimental vibrational spectra are considerably different from that for rotational spectra. Individual vibrational Q branch Raman transitions from different rotational levels lie close together and generally have not been resolved in experimental spectra from either spontaneous Raman scattering or CARS. (However, sufficient resolution is available with CARS or stimulated Raman gain techniques.) Usually temperature determinations from vibrational spectra are based upon comparisons of the shape of the experimental profile with theoretical profiles. The calculated profiles are convolutions of many spectral lines[10,18] and are sensitive functions of the spectrometer slit function. This is especially troublesome in vibrational CARS where the convolution must include the spectral energy density of both source lasers, the Raman linewidths, as well as the instrumental slit function.[18] Also, as mentioned earlier, CARS spectra can exhibit constructive and destructive interference effects for closely spaced transitions.[18] Thus, temperature analysis for rotational Raman spectra are simpler than for vibrational Raman spectra (particularly for CARS) which, in some cases, may result in more accurate temperature determinations from rotational spectra.

III. SPECTRAL EXAMPLES AND TEMPERATURE DETERMINATIONS

A few examples of experimental rotational Raman and rotational CARS spectra may help to illustrate the temperature measurement techniques described above. Particular attention will be paid to the effect of f_{oo}, η and (β_v/β_o) corrections on temperature determinations. Although the effects of f_{oo} and η corrections are discussed in the recent literature,[12,25,33] the effect of β_v/β_o on the η calculation has not been considered heretofore.

Figure 1 shows rotational Raman scattering from N_2 at room temperature and in a laminar H_2-air diffusion flame. Figure 2 shows the application of eqn. 8 to the calculation of rotational temperatures from the measured peak intensities in Fig. 1. The temperature is determined from the slope computed by a linear least-squares fit. The calculated N_2 rotational temperatures are 301 ± 5 K and 1583 ± 15 K, where the quoted uncertainties represent ±2σ. In the room temperature N_2 spectra, the linestrength correction for centrifugal distortion, $f_{oo}(J)$, has a negligible effect on the computed temperature because the correction factors are small (for N_2) for small values of J. The η correction is also negligible at low temperatures because of the negligible population of N_2 excited vibrational levels. Thus at reasonably low temperatures (< 1000 K) the accuracies in temperatures determined from spontaneous rotational Raman spectra is a function only of experimental inaccuracies in measuring peak intensities and in calibration of the relative spectral response of the optical system.

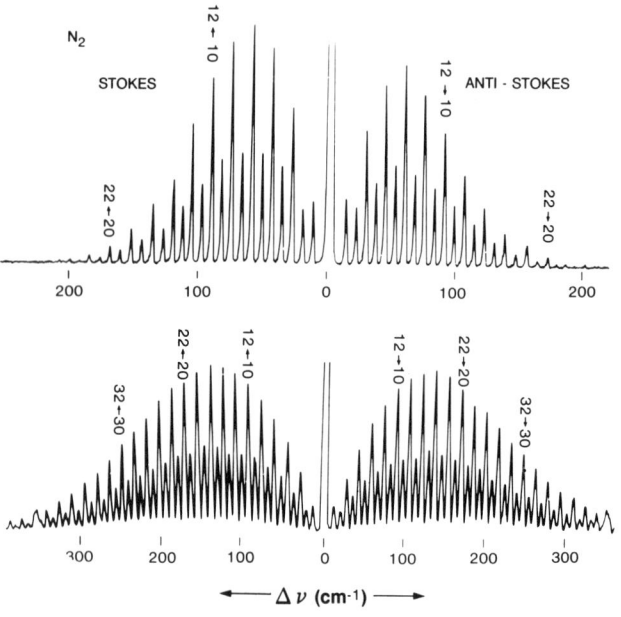

FIG. 1. Rotational Raman spectra of N_2(g). Top curve at room temperature, 1 atm, 15.8 W excitation at 488 nm, 0.6 cm^{-1}/sec scan rate, 3 sec time constant. Bottom curve in a cool region of a H_2 diffusion flame, 1583 K, 14 W excitation at 488 nm, 3 cm^{-1} resolution, 0.25 cm^{-1}/sec scan rate, 1 sec time constant. From Drake and Rosenblatt, ref. 12. Note the wavelength scale change.

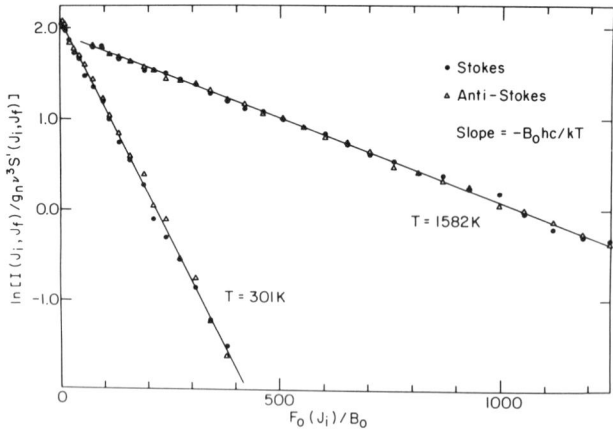

FIG. 2. N_2 rotational temperature determination plot for the spectra in Fig. 1. $T_{rot}(N_2)$ = 301 ± 5 K, $T_{rot}(N_2)$ = 1583 K ± 13 K. Uncertainty is two standard deviations calculated by a linear least-squares computer fit of the slopes. From Drake and Rosenblatt, ref. 12.

However, at elevated temperatures the f_{oo} and η corrections are important and can easily limit the accuracies of temperatures determined from rotational Raman spectra. For example, in the bottom spectrum in Figure 1 the temperature computed without these corrections is 1555 K. Including the f_{oo} correction factor lowers the computed temperature by $\sim 1\%$ while the η correction raises it by $\sim 3\%$ resulting in a final computed temperature of 1583 ± 15 K.

At still higher temperatures the magnitude of the η correction for O_2 and N_2 increases rapidly because the population of excited vibrational levels increases exponentially with temperature. The effects of the η and the β_v/β_o anharmonicity corrections are illustrated in Figure 3. The intensities were measured from an O_2 rotational Raman spectrum in a premixed H_2-O_2 flame. In Figure 3 the experimental data points were plotted without these corrections. The curved line is calculated from the data including the η and β_v/β_o corrections and contrasts with the straight line fit which neglects these corrections. The η correction calculated assuming $\beta_v/\beta_o=1$ raised the computed temperature from 1964 K to 2126 K. Including the linestrength correction factors for vibrational anharmonicity $(\beta_v/\beta_o)^2$ quoted earlier for O_2[30] increases the calculated intensities for vibrationally excited transitions, decreases the η factor, (see eqn. 10) and increases the computed temperature to 2138 K $\pm$ 75 K. At temperatures where vibrationally excited molecules are significantly populated the accuracy of rotational temperatures derived from eqn. 8 and 10 are limited by the accuracy in determining $\eta(J)$ because of its exponential dependence on temperature and its sensitivity to the width and shape of the spectrometer slit function.

For H_2, the situation is quite different. Previously pure rotational Raman intensities were measured from v=0 and v=1 for several different rotational transitions[38] in a laminar H_2 diffusion flame in air. The temperature determined from pure rotational Raman intensities in ground vibrational levels was 1890 ± 80 K and that derived from relative rotational Raman intensities of v=1 compared to v=0 was 1900 ± 150 K.[38] In neither case were the linestrength correction factors included in the calculation. From Figure 4 it is clear

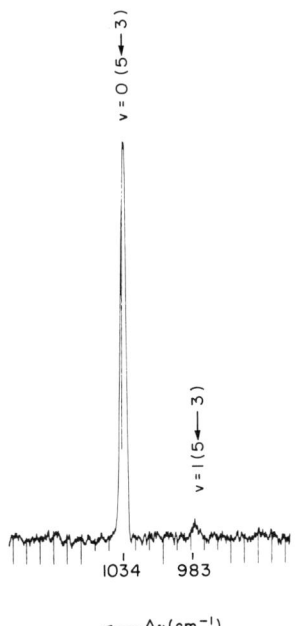

FIG. 4. Part of the rotational Raman spectrum of H_2 in a H_2 diffusion flame. Laser excitation 14 W at 488 nm, 4.5 cm^{-1} resolution, 0.25 cm^{-1}/sec scan rate, 3 sec time constant. From Drake and Rosenblatt, ref. 38.

that rotational transitions from vibrationally excited H_2 molecules are well separated from the v=0 transitions and so the η correction is negligible. However, as Table 1 indicates, the correction factors $f_{oo}(J)$ for H_2 are sizable even at J=0. Including the $f_{oo}(J)$ term in eqn. 8 lowers the calculated temperature from pure rotational transitions in the ground vibrational level to 1710 ± 80 K. Vibrational temperatures determined from relative intensities of v=1 and v=0 pure rotational H_2 transitions using eqn. 14 and the $(\beta_1/\beta_o)^2$ values listed earlier are lowered to 1740 K. Thus the rotational and vibrational temperatures for H_2 are in agreement as expected although two different linestrength correction factors had significant effects on the calculated temperature values.

There are fewer examples of temperatures determined from rotational CARS spectra than from spontaneous rotational Raman spectra. However, recently Teets and Bechtel[39] measured temperatures from O_2 rotational CARS spectra (eg. see Figure 5) from H_2-O_2 and CH_4-O_2 premixed flames. By using an equation similar to eqn. 17 and assuming $\beta_v=\beta_o$ and $\gamma_o=\gamma_v$, they calculated a temperature in the CH_4-O_2 flame of 3300 ± 150 K which was obviously in error since it far exceeded the adiabatic flame temperature of 2916 K.[39] This discrepancy is resolved by including the $(\beta_v/\beta_o)^2$ correction factor (calc. from eqn. 13) for the relative rotational Raman cross sections due to vibrational anharmonicity which lowers the computed temperature from 3300 K to 2990 ± 150 K which agrees within experimental error with the adiabatic flame value (2916 K).[34]

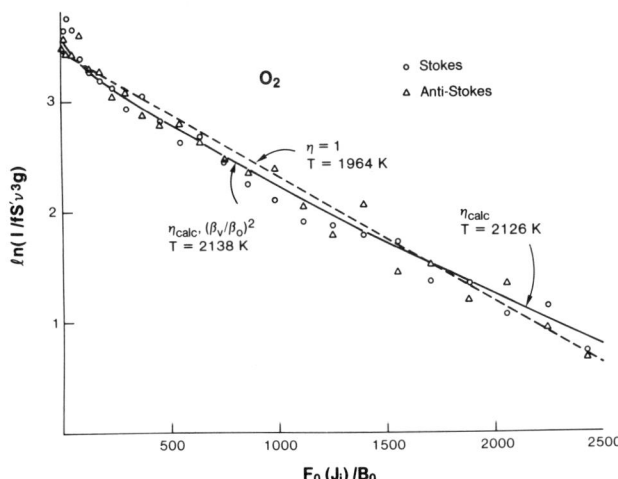

FIG. 3. Plot of temperature determination from an O_2 rotational Raman spectrum in an H_2-O_2 premixed flame. The dashed straight line is from temperature calculation without the $\eta(J)$ correction (T = 1964 K $\pm$ 80 K). The solid line is calculated to fit the experimental points by subtracting $\ln(\eta(J))$ from the straight line where $\eta(J)$ was calculated assuming $(\beta_v/\beta_o)^2 = 1.0$ (T = 2126 K $\pm$ 75 K). Including the β_v/β_o correction in the $\eta(J)$ term results in a line indistinguishable on this scale with the solid line but with a higher calculated temperature (2138 $\pm$ 74 K).

IV. CONCLUSIONS AND SUMMARY OF APPLICATIONS

A. Spontaneous Rotational Raman

1. Low temperature results (T < 1000 K)

Rotational Raman scattering is one of only a very few nonintrusive, nonperturbing laser diagnostic techniques capable of measuring temperatures at room temperature and below. As demonstrated earlier, accurate temperatures can be determined without correction factors from rotational Raman spectra of N_2, O_2 or other

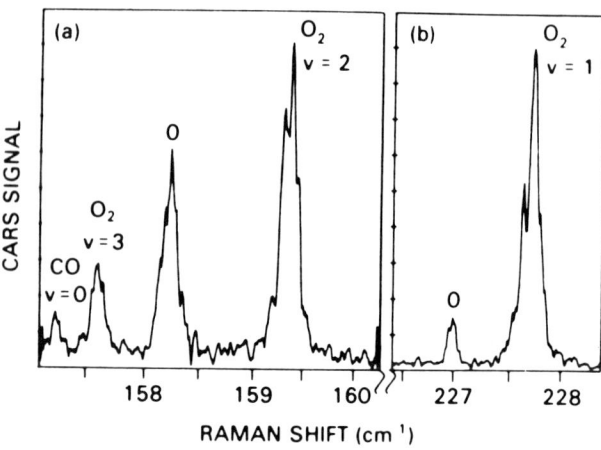

FIG. 5. Narrow-band CARS spectra in a methane-oxygen flame. The vertical scales in (a) and (b) are different. From Teets and Bechtel, ref. 39 with permission.

non-hydrogen containing diatomic molecules at low temperature. For the special cases of H_2 and D_2 the centrifugal distortion linestrength correction factors $f_{oo}(J)$ are large and must be included for accurate temperature derivations but reliable values of $f_{oo}(J)$ from ab initio calculations are listed in Table 1.

Not surprisingly, the early use of rotational Raman techniques were for nonperturbing temperature measurements in static gas[40] and in wind tunnels[41] and for remote atmospheric sensing.[42] More recent applications include temperature measurements from 10 K - 300 K in cooled supersonic beams of N_2[43,44] and CO_2.[45] Comparisons with temperatures calculated or measured by more established methods gave excellent agreement with the rotational Raman temperature measurements throughout the range. Typical Raman experimental uncertainities were ± 2 K.[43,44] Near room temperature, values measured by Raman scattering were equal to that measured by thermocouples within 1 K in laboratory tests. For remote atmospheric monitoring, field experiments demonstrate a temperature accuracy of 1% for a 10 sec. measurement at a range of 1 km with a spatial resolution of 100 m.[46] Above room temperature, rotational Raman measurements at 750 K in a CO_2 electric discharge[47] and up to 1200 K in chemical vapor deposition flow reactors[48] demonstrate its use for temperature and temperature gradient measurements in highly reactive systems.

2. High temperature results (T > 1000 K)

The preceding theoretical sections have demonstrated the importance of f_{oo}, η and β_v/β_o correction factors for accurate temperature determinations from rotational Raman intensities at high temperatures. When temperatures are determined from relative rotational Raman intensities from the vibrational ground state of O_2 or N_2, the correction for rotational transitions from vibrationally excited molecules (η correction) is particularly important. The analysis of the bottom spectrum of Figure 1 indicates that with proper care rotational temperatures can be measured with a precision (±2 σ) of ~ ± 1 %. (The precision decreases in absolute terms at high temperatures because of the increased number of rotational levels populated, decreased intensity from any one rotational transition, and increased background emission.) However the accuracy is no better than ± 2 % because of the uncertainty in the experimental slit function and the η correction factor. This conclusion is supported by the agreement (within 3 %) in temperatures measured by rotational Raman and vibrational Raman scattering from N_2 and O_2 in H_2-air, H_2-O_2, and CH_4-air flames[12,25,38,49] and in temperatures measured by rotational Raman and vibrational Raman scattering from N_2 with OH absorption and Na D line reversal in laminar H_2-N_2-O_2 flames where care was taken to assure a homogeneous temperature across the burner for meaningful line-of-sight results.[50]

Since the temperature accuracy from N_2 or O_2 rotational Raman intensities is determined primarily by the η correction when eqn. 10 is used, better accuracy may be obtained by increasing the experimental spectral resolution to resolve all lines from excited vibrational levels.[51] However, this procedure decreases signal intensities resulting in a temperature uncertainty (± 2 % at 2100 K)[51] similar to those reported here. Instead, a different analysis procedure grouping bands containing many rotational transitions[52] or contour fitting the entire rotational spectrum may lead to better accuracy. Temperatures measured from rotational Raman scattering from H_2 at high temperatures are not influenced by η corrections but are subject to large f_{oo} corrections. The accuracy of temperatures determined from H_2 rotational Raman intensities have not been checked against other techniques but agreement was obtained with vibrational temperatures measured from H_2 rotational spectra such as Figure 4.

Vibrational temperatures can be measured also from the relative intensities of pure rotational Raman transitions from vibrationally excited molecules. This technique was first experimentally demonstrated from spectra of vibrationally excited CO_2[47] and N_2[53] in an electric discharge. Others[54,50,25] have used the technique for flame temperature measurements. However, only recently[34] has the importance of $(\beta_v/\beta_o)^2$ correction factors been recognized (see Theory section). When the correction factor $(\beta_v/\beta_o)^2$ are included, accurate temperatures should be obtainable from pure rotational Raman scattering from different vibrational levels. This accuracy has not been conclusively demonstrated experimentally but is suggested by the agreement between rotational and vibrational temperatures determined for H_2 in a laminar H_2-air diffusion flame (see section on temperature determinations.)

3. Future directions for Raman

Improved sensitivity and temperature accuracy may be feasible by increasing signal intensity and decreasing the effects of background radiation by taking advantage of the nearly periodic nature of rotational Raman spectra. Simultaneous observation of a number of rotational lines using an interferometer whose free spectral range equals the spectral spacing between rotational Raman lines has been used for temperature measurements in room air[55] and in an electrical discharge.[56] A similar multiplex advantage occurs with Fourier transform data analysis[57] of conventionally obtained spectra.

Time-resolved temperatures from rotational Raman scattering measurements have been demonstrated in room temperature transient gas flows,[58] in turbulent H_2-air flames[59] and in a rocket exhaust plume[52] using a pulsed laser and a gated multichannel detector (or a series of photomultipliers). For time-resolved measurements temperature accuracy is usually experimentally limited by the relatively small number of detected photons.[10] Similarly, the simultaneous two-dimensional image of temperature in slightly heated gas jets or a one-dimensional image of temperature along a line in a flame or high temperature flow should be currently feasible based upon similar vibrational Raman experiments.[60,61]

B. Rotational CARS

1. Low temperature results

For temperature measurements, the theoretical analysis for rotational CARS spectra has been shown here to be considerably simpler than for vibrational CARS. In spite of this, the use of rotational CARS has received much less attention than vibrational CARS due to the experimental difficulty of separating the rotational CARS signals from the incident lasers. Although double monochromators,[62] four color,[63] and crossed polarization[62,64] techniques have been used in the past to record rotational CARS spectra in room temperature gases, each suffers from severe disadvantages.[36] However, the use of a folded BOXCARS configuration[36,65,66] spatially separates the CARS signal beam permitting excellent rejection down to Raman shifts of only a few (<10) cm^{-1} and appears to be the current method of choice for rotational CARS measurements. With folded BOXCARS, high quality rotational CARS spectra from N_2 (and from air) at room temperature have been published showing good agreement with a theoretically calculated spectrum for N_2[66] but no analysis of temperature errors quoted.[66]

Only one published paper[65] quantitatively assesses rotational CARS for room temperature measurements. Instantaneous temperature measurements (10 ns temporal resolution) of cold N_2 were made from rotational CARS spectra generated with a broadband Stokes laser and a multichannel detector. Raman linewidths were derived from Jammu[67] assuming a $T^{-\frac{1}{2}}$ temperature dependence. Temperature measurements from 135-296 K showed excellent agreement with thermocouple results. For example, at a thermocouple temperature of 169 K, a series of 10 consecutive single shot rotational CARS spectra gave an average temperature of 170.2 K with an rms standard deviation of ± 5 K. The standard deviation was attributed primarily to noise introduced in recovering the multichannel detector signals[65] so even more accurate time averaged measurements may be feasible with rotational CARS generation using a narrow band scanning Stokes laser and photomultiplier detection.

A major limitation to wide application of rotational CARS for temperature measurements, particularly at low temperature, is the lack of information on spontaneous Raman linewidths as a function of rotational quantum number and their dependence on temperature, pressure, and surrounding gas composition. Experimental[68] and theoretical[69] results indicate that pressure broadening is much larger for rotational Raman transitions than for vibrational Raman Q branch transitions.[13] The limited data that are available on rotational Raman linewidths and their pressure broadening coefficients are summarized in ref. 13. The use of stimulated Raman gain (or loss) techniques[70,8] may make linewidth data more accurate and much more extensive.

2. High temperature results

To our knowledge, only three papers[36,39,81] have appeared which present rotational CARS spectra from high temperature gases. A rotational CARS spectrum containing 22 rotational transitions from the ground vibrational level of N_2 was measured in a premixed methane-air flame at 1600 K (measured by a thermocouple).[36] This demonstrates that high quality spectra can be obtained, but no quantitative temperature analysis was performed.[36]

The other rotational CARS temperature measurement (example shown in Figure 5 and discussed in detail previously) involved rotational transitions from ground and vibrationally excited O_2 molecules in CH_4-O_2 and H_2-O_2 premixed flames.[39] This technique is capable of high temperature accuracy when the proper linestrength correction for vibrational anharmonicity is included, although it is limited to temperatures where excited vibrational levels are significantly populated. (This can occur at reasonably low temperatures for molecules with low lying vibrational states, i.e. CO_2). Lack of knowledge of Raman linewidths and the uncertainties in the β_v/β_o correction are the likely factors currently limiting temperature accuracy.

C. Comparison of Rotational and Vibrational Scattering From Spontaneous Raman and CARS

1. Spontaneous Raman

Theory and experiments[12,25,50] have shown that spontaneous rotational Raman scattering has advantages over vibrational Raman scattering which include: (1) greater sensitivity due to about one order of magnitude larger rotational Raman cross section, (2) statistically more precise temperature determinations because of the large number of separate transitions measurable, (3) capability of independently measuring rotational and vibrational temperatures and (4) a wider range in which temperature measurements can be measured.

In direct experimental comparisons[12,25,50] in H_2 and CH_4 flames, temperatures based upon rotational Raman scattering had lower imprecision (1-4 %) than those based upon vibrational Raman scattering (3-9 %). The rotational Raman temperature measurement inaccuracies quoted are about a factor of 2 larger than the calculated temperature imprecision because of uncertainties in the η correction. Ways of reducing the experimental uncertainty in the η correction described earlier should further enhance the advantages of rotational Raman over vibrational Raman as a temperature probe in the 1000 - 2000 K range. At still higher temperatures, rotational Raman scattering from vibrationally excited molecules may provide comparable temperature accuracy to that measured by vibrational Raman techniques.

2. CARS

Theoretically, rotational CARS should have significant advantages over vibrational CARS for temperature measurements which include all four of the advantages mentioned earlier for spontaneous rotational Raman as well as (1) well separated lines which simplifies spectral computations and comparison of experiment and theory, (2) somewhat more extensive data on rotational Raman linewidths and (3) the smaller scanning range needed for temperature measurements made from rotational transitions in ground and excited vibrational levels.

These factors should make rotational CARS techniques more accurate than vibrational CARS at least for low temperature applications. Rotational CARS intensities for comparable laser powers and geometries are predicted to be an order of magnitude larger than vibrational CARS intensities from N_2 at room temperature,[36] but this has not been obtained in practice. (Note that the order of magnitude difference in Raman cross section which would give rotational CARS a 100 times greater intensity is partially offset by the small population difference for rotational CARS transitions which lowers rotational CARS intensities relative to vibrational CARS intensities.)

Rotational temperatures can be measured at reasonably low temperatures from the N_2 Q branch vibrational CARS spectra either by comparing partially resolved spectra with spectra taken at known temperature[71] or theoretically calculated[72] or by line intensities of the more completely resolved Q branch transitions at high J.[73] Although these approaches have been used to estimate a temperature of 30-40 K in a molecular beam,[72] 600 K in a flame,[71] and several other flame temperatures,[74] reliable low temperature measurements require sufficient resolution to resolve individual transitions in the vibrational Q branch. An excellent example is the vibrational CARS spectra of O_2 N_2, H_2, D_2, and ethylene in seeded supersonic nozzle beam expansion at temperatures as low as 11 K.[75]

For the special case of H_2 where the rotational Q branch lines are easily separated and do not interfere, rotational temperatures can be measured over a wide temperature range. Indeed, the first gas temperature measurement using CARS was made with this technique in a flame.[76] At one flame position the CARS measured temperature (1350 ± 30 K) was in excellent agreement with a radiation corrected thermocouple reading of 1340 ± 20 K. This agreement may have been somewhat fortuitous because of the large sample volume of the colinear CARS and the presence of flame temperature gradients as well as the neglect of centrifugal distortion Q branch correction factors f_{oJ}. For spontaneous Raman analyses of H_2 Q branch intensities, including the f_{oJ} corrections was found to raise H_2 temperatures by 2 - 3 %[33] and the correction to CARS H_2 intensities should result in essentially the same temperature correction.

At elevated temperatures it is not clear, either theoretically or experimentally, whether rotational or vibrational CARS methods will yield the most accurate temperatures. However, it has been clearly demonstrated by comparison of experimental with theoretical vibrational CARS spectra of N_2 that vibrational CARS can result in temperature determinations with high precision (i.e. 2104 ± 9 K from analysis of Figure 3 in ref. 77). Typically quoted temperature accuracies[8,77] for vibrational CARS are larger (25-50 K) because of a variety of theoretical and experimental uncertainties possibly related to the proper Raman linewidth.[78,79,35] This accuracy of 25-50 K was determined by comparing temperatures derived from CARS vibrational Raman spectra from N_2 in premixed CH_4/air flames with temperatures measured by radiation corrected thermocouples. There was excellent agreement although the CARS measurements were typically ~ 40 K higher at 2100 K.[78] It is possible that this small discrepancy could result from the neglect of anharmonicity in the vibrational Raman cross section as a function of v for vibrational Q branch lines - a factor which is presently unknown.

Time resolved vibrational CARS temperature measurements have been demonstrated for a wide range of high temperature applications (atmospheric flames internal combustion engines, and high pressure, liquid-fueled combustors).[8] Even the internal energy distributions of transient species generated by laser photolysis were recently measured by vibrational CARS.[80] Whether rotational CARS will occupy a significant place in time-averaged or time-resolved temperature measurements at elevated temperatures is currently unknown, but it is clearly an area warranting investigation.

Final Note

Perhaps it is appropriate to end with this note. The last decade of theoretical and experimental advances have demonstrated that either rotational Raman or rotational CARS techniques can provide accurate (~ 2 - 5 %) measurements of temperature from 10 K - 3000 K in a variety of adverse vapor environments. We are confident the next decade will bring a wide range of applications for these "new" temperature measurement tools.

REFERENCES

1. J.W. Hastie, High Temperature Vapors: Science and Technology, Academic Press, New York 1975.
2. J.W. Hastie, ed., Characterization of High Temperature Vapors and Gases, NBS SP-561, U.S. Government Printing Office Washington, 1979.
3. M. Lapp and C.M. Penney, eds., Laser Raman Gas Diagnostics, Plenum Press, New York, 1974.
4. M. Lapp and C.M. Penney in Adv. in Infrared and Raman Spectrosc, R.J.H. Clark and R.E. Hester, eds., Heyden and Sons Ltd., London, 1977, vol. 3, chap. 6.
5. A.C. Eckbreth, P.A. Bonczyk, and J.F. Verdieck, Appl. Spectros. Reviews 13, 15 (1977). Also, Prog. Energy Combust. Sci., 5, 253 (1979).
6. R. Goulard, Combustion Measurements, Academic Press, New York, 1976.
7. D.R. Crosley, ed., Laser Probes for Combustion Chemistry, American Chemical Soc., Washington 1980.
8. A.C. Eckbreth, Eighteenth Symposium (International) on Combustion, The Combustion Institute, Pittsburgh, 1981, p. 1471.
9. S. Lederman, J. Prog. Energy Combustion Sci., 3, 1 (1977).
10. M.C. Drake, M. Lapp, and C.M. Penney, this volume.
11. G. Placzek in Handbuch der Radiologie, E. Marx, ed., Akad. Verlagsgesellschaft, Leipzig, 1934, p. 209. [English transl. by A. Werbin, U.C.R.L. Transl. No. 526(L)].
12. M.C. Drake and G.M. Rosenblatt in ref. 2, p. 609.
13. A. Weber, in A. Anderson, ed., The Raman Effect vol. 2 Applications, Marcel Dekker, New York, 1973, p. 543.
14. H.W. Schrötter and H.W. Klöckner in Raman Spectroscopy of Gases and Liquids, A. Weber, ed., Springer Verlag, New York, 1979, p. 123.
15. S. Druet and J.P. Taran in Chemical and Biological Applications of Lasers, C.B. Moore, ed., Academic Press, New York, 1979, p. 187.
16. J.W. Nibler and G.V. Knighten, in Raman Spectroscopy of Gases and Liquids, A. Weber, ed., Springer-Verlag, New York, 1979, p. 253.
17. J.W. Nibler, W.M. Shaub, J.R. McDonald, and A.B. Harvey, in Vibrational Spectra and Structure, J.R. Durig, ed., Elsevier, New York, 1977, vol. 6, p. 173.
18. R.J. Hall, Combustion Flame 35, 47 (1979).
19. G. Herzberg, Molecular Spectra and Molecular Structure. I. Spectra of Diatomic Molecules, Van Nostrand, Princeton, 1950.
20. K.P. Huber and G. Herzberg, Constants of Diatomic Molecules, Van Nostrand Reinhold, New York, 1979.
21. A. Weber, in Raman Spectroscopy of Gases and Liquids, A. Weber, ed., Springer-Verlag, New York 1979, p. 71.
22. A. Weber, J. Chem. Phys. 73, 3952 (1980).
23. M.C. Drake and G.M. Rosenblatt in ref. 2, p. 609.
24. C.M. Penney, R.L. St. Peters, and M. Lapp, J. Opt. Soc. Amer. 64, 712 (1974).
25. M.C. Drake and G.M. Rosenblatt, Combustion and Flame 33, 179 (1978).
26. D.L. Renschler, Jr., J.L. Hunt, T.K. McCubbin, Jr., and S.R. Polo, J. Mol. Spectrosc. 31, 173 (1969).
27. D. Lepard, Canad. J. Phys. 48, 1664 (1970).
28. K. Altman, G. Strey, J.G. Hochenbleicher, and J. Brandmüller, Z. Naturforsch 27A, 56 (1972).
29. T.C. James and W. Klemperer, J. Chem. Phys. 31, 130 (1959).
30. A.D. Buckingham and A. Szabo, J. Raman Spectrosc. 7, 46 (1978).
31. C. Asawaroengchai and G.M. Rosenblatt, J. Chem. Phys. 72, 2664 (1980).
32. L.M. Cheung, D.M. Bishop, D.L. Drapcho, and G.M. Rosenblatt, Chem. Phys. Lett. 80, 445 (1981).
33. M.C. Drake, C. Asawaroengchai, and G.M. Rosenblatt in ref. 7, p. 231.
34. M.C. Drake, "Rotational Raman Intensity Correction Factors Due to Vibrational Anharmonicity", submitted for publication, 1982.
35. M.A. Yuratich and D.C. Hanna, Mol. Phys. 33, 671 (1977) and M.A. Yuratich, Mol. Phys. 38, 625 (1979).
36. J.A. Shirley, R.J. Hall, J.F. Verdieck, and A.C. Eckbreth, AIAA Paper 80-1542 (1980). Presented at the AIAA 15th Thermophysics Conf., July 14-16, 1980, Snowmass, Colo.
37. D.V. Murphy and R.K. Chang, Opt. Lett. 6, 233 (1981).
38. M.C. Drake and G.M. Rosenblatt, Chem. Phys. Lett. 44, 313 (1976).
39. R.E. Teets and J.H. Bechtel, Optics Lett. 6, 458 (1981).
40. R.S. Hickman and L.H. Liang, Rev. Sci. Instr. 43, 796 (1972).
41. A.R. Bandy, M.E. Hillard, and L.E. Emory, Appl. Spect. 27, 421 (1973).
42. S. Saltzman in ref. 3, p. 177.
43. W.D. Williams and J.W.L. Lewis, AIAA J 13, 1269 (1975).
44. R.A. Hill, C.W. Peterson, A.J. Mulac, and D.R. Smith, J. Quant. Spectros. Radiat. Transfer 16, 953 (1976).
45. I.F. Silvera and F. Tommasini, Phys. Rev. Lett. 37, 136 (1976).
46. H. Inaba in Laser Monitoring of the Atmosphere, E.D. Hinkley, ed., Springer-Verlag, New York, 1976, p. 153.
47. J.J. Barrett and A. Weber, J. Opt. Soc. Amer. 60, 70 (1970).
48. G.H. Miller, A.J. Mulac, and P.J. Hargis, Jr. in ref. 2, p. 1135.
49. M.C. Drake and J.W. Hastie, Combustion Flame, 40, 201 (1981).
50. M.C. Drake, L. Grabner, and J.W. Hastie, in ref. 2, p. 1105.

51. W. Stricker and T. Just, VDI-Berichte Nr. 346, 111 (1979).
52. W.D. Williams, H.M. Powell, R.L. McGuire, L.L. Price, J.H. Jones, D.P. Weaver, and J.W.L. Lewis, in Turbulent Combustion, L.A. Kennedy, ed., American Inst. of Aeronautics and Astronautics, New York, 1978, p. 273.
53. A.B. Harvey in ref. 3, p. 147.
54. C.J. Dasch and J.H. Bechtel, Opt. Lett. 6, 36 (1981).
55. J.J. Barrett, Optical Engineering 16, 85 (1977).
56. J.J. Barrett and A.B. Harvey, J. Opt. Soc. Amer. 65, 392 (1975).
57. C.M. Penney, unpublished.
58. J.R. Smith and W.H. Giedt, Int. J. Heat Mass Transfer 20, 899 (1977).
59. W.D. Williams, H.M. Powell, L.I. Price, and G.D. Smith, "Laser-Raman Measurements in a Ducted, Two-Stream Subsonic H_2/Air Combustion Flow," AEDC-TR-79-74, Dec. 1979.
60. B.F. Webber, M.B. Long, and R.K. Chang, Appl. Phys. Lett. 35, 119 (1979).
61. L. Socket, M. Lucquin, M. Bridoux, M. Crunelle-Cras, F. Grase, and M. Delhaye, Combustion Flame 36, 109 (1979).
62. C.M. Roland and W.A. Steele, J. Chem. Phys. 73, 5919 (1980).
63. I.R. Beattie, T.R. Gilson, and D.A. Greenhalgh, Nature 275, 378 (1978).
64. L.P. Goss, J.W. Fleming, and A.B. Harvey, Opt. Lett. 5, 345 (1980).
65. D.V. Murphy and R.K. Chang, Opt. Lett. 6, 233 (1981).
66. J.A. Shirley, R.J. Hall, and A.C. Eckbreth, Opt. Lett. 5, 380 (1981).
67. K.S. Jammu, G.E. St. John, and H.L. Welsh, Can. J. Phys. 44, 797 (1966).
68. A.D. May, J.C. Stryland and G. Varghese, Canad. J. Phys. 48, 2331 (1970).
69. R.G. Gordon, J. Chem. Phys. 45, 1649 (1966).
70. A. Owyoung and L.A. Rahn, IEEE J. Quant. Elect. QE-15, 25D (1979).
71. I.R. Beattie, J.D. Black, and T.R. Gilson, Combustion and Flame 33, 101 (1978).
72. P. Huber-Walchli, D.M. Guthals, and J.W. Nibler, Chem. Phys. Lett. 67, 233 (1979).
73. F. Moya, S. Druet, M. Pealat, and J.P.E. Taran, AIAA Paper 76-29, 1976.
74. W.B. Roh, P.W. Schreiber and J.P.E. Taran, Appl. Phys. Lett. 29, 174 (1976).
75. P. Huber-Walchli and J.W. Nibler, J. Chem. Phys. 76, 273 (1982).
76. F. Moya, S.A.J. Druet, and J.P.E. Taran, Opt. Commun. 13, 169 (1975).
77. R.J. Hall and A.C. Eckbreth, Opt. Engineering 20 494 (1981).
78. R.J. Hall, Appl. Spectrosc. 34, 700 (1980).
79. R.L. St. Peters, Opt. Lett. 4, 401 (1979).
80. J.J. Valentini, D.S. Moore, and D.S. Bomse, Chem. Phys. Lett. 83, 217 (1981).
81. See also paper by J.W. Fleming, W.T. Barnes, and A.B. Harvey in this Volume.

Use of the vibrational Raman effect for gas temperature measurements

M. C. Drake, M. Lapp, and C. M. Penney

General Electric Research and Development Center, Schenectady, New York 12301

The basic properties of the Raman effect that make it desirable for temperature measurements are described. Because relative atomic or molecular energy level populations are determined directly, this optical light scattering method permits straightforward analysis for temperature in systems at thermal equilibrium and for "population" temperatures in non-thermal equilibrium systems. Pulsed Raman scattering provides the capability for measuring gas phase temperatures with high temporal (≤ 1 μs) and spatial (<0.1 mm³) resolutions for a wide range of conditions, from cooled fluid expansions to flames and gas discharges. Experimental aspects of pulsed vibrational Raman temperature measurements are discussed, including instrumentation, system calibration, reduction and interpretation of data, and determination of accuracy and precision. Examples of the use of vibrational Raman scattering for difficult measurement environments, including turbulent combustion and shock tube flows, illustrate the broad applicability as well as the current capabilities and limitations of the technique.

I. Introduction

Although the Raman effect was discovered in the 1920's,[1] its practical application for temperature measurements in gases has been limited until recently by the weakness of the effect. However, technical advances within the last twenty years in high power laser sources, multichannel spectroscopy, and sensitive photon counting detectors,[2,3] have greatly expanded capabilities for determining temperatures from Raman spectra. In the 1970's numerous demonstrations of Raman temperature measurements were reported ranging from cooled supersonic molecular beams[4] to high temperature flames[3,5-7] combustion systems,[8] shock tubes,[9] and gaseous discharges.[10] Based upon these very encouraging results, Raman scattering may well have a major impact on temperature measurements in these and many other gas phase systems in the future.

In this paper, aspects of Raman theory necessary to understand the basics of Raman temperature measurements are presented. Raman diagnostic methods for steady-state systems are reviewed briefly, and the use of pulsed vibrational Raman scattering for time-resolved temperature measurements in turbulent flames is discussed in some detail. In this discussion, we will emphasize the experimental and analytical methods necessary for accurate temperature measurements, the precision and accuracy currently obtainable, and the advantages of pulsed vibrational Raman scattering for gas-phase temperature measurements.

II. Theory of the Spontaneous Raman Effect

Spontaneous Raman scattering is one of many light scattering processes [particle (often called Mie) scattering, Rayleigh scattering, fluorescence, coherent anti-Stokes Raman scattering (CARS),...] which can occur when gas molecules interact with an incident laser beam. (See Table 2 of Ref. 7.) The oscillating electric field of the laser beam perturbs the electron cloud of the molecules causing an induced oscillating polarization in the molecule. This induced polarization is proportional to the electric field of the incident light and the molecular polarizability, which in turn is modulated by rotational, vibrational, or electronic motions in the molecule. Thus, scattered radiation from these molecules occurs at the incident laser frequency (termed Rayleigh scattering) and at frequencies shifted by energy level differences in the molecule (Raman scattering). (See Fig. 1 for a schematic illustrating these effects.) The intensity of the Raman scattering provides a measure of molecular concentrations while Raman bandshape analysis provides a measure of temperature. Note the strong temperature dependence of the rotational Raman scattering from N_2, with shifts less than ± 400 cm^{-1}, and of the vibrational Raman scattering from N_2 centered at ± 2331 cm^{-1}.

Most spontaneous Raman temperature measurements of high temperature vapor systems have used vibrational Raman scattering, often from N_2 molecules. (Since temperature measurements from rotational Raman scattering and from CARS - a four-wave optical mixing process -

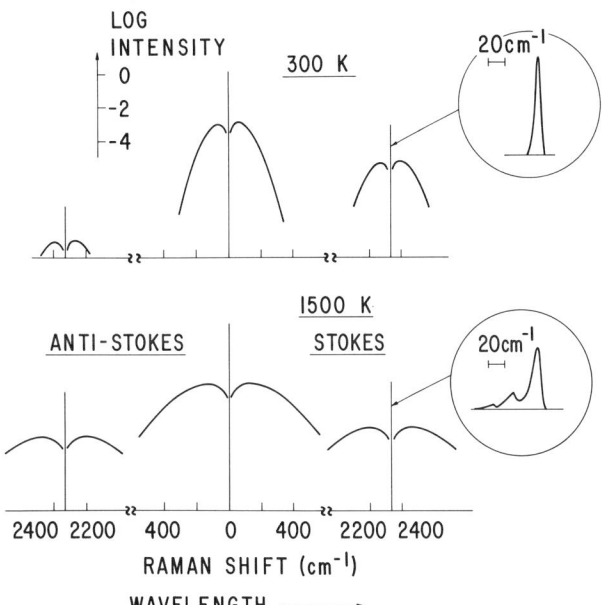

FIG. 1. Raman and Rayleigh scattering from N_2 at ambient (300 K) and elevated (1500 K) temperatures for an exciting laser line in the midvisible. The central unshifted peak corresponds to Rayleigh scattering, which is flanked by rotational Raman scattering represented here by wing envelopes of the rotational line peak intensities. The vibrational Q-branches on the Stokes and anti-Stokes sides are shown at the characteristic Raman shifts for N_2 of 2331 cm^{-1}. These Q-branches are surrounded by weaker vibrational bands called the O-and S-branches, shown also by wing envelopes. Note that relative intensities are drawn on a logarithmic scale and that large breaks occur along the wave number and wavelength axes. The spectral contours of the Q-branches shown in the two inset diagrams are presented on a linear scale and have been calculated using a triangular spectrometer slit function with 6 cm^{-1} (~ 0.18 nm) FWHM.

are discussed elsewhere in this volume,[11] only vibrational Raman scattering will be considered here.) Although the theory of Raman scattering has been well-established,[12] its implementation for wide-spread diagnostic use requires careful attention to experimental issues concerning the shapes and intensities of the spectral lines and bands characteristic of the probed molecules.

A. <u>Vibrational Band Contours Used for Temperature Analysis</u>. In Fig. 2, four common methods of determining temperature from N_2 vibrational Raman scattering are summarized. All of the techniques rely on the increased population of N_2 molecules in excited vibrational levels as the temperature is increased. The transitions from these excited levels are shifted in frequency as a result of the molecular anharmonicity, and thus give rise to the characteristic "sawtooth" shape for most diatomic vibrational Raman spectra used in high temperature diagnostics [the fundamental series $(v + 1 \leftarrow v)$ Stokes Q-branch $(\Delta J = 0)$; see inset diagrams of Fig. 1 also]. This shape can be obtained from the convolution of the radiant flux S for a rotational line with the spectral slit function that corresponds to the acquisition of experimental Raman scattering band contour data. Thus[13]

$$S(v,J) \sim \frac{g(2J+1)(v+1)\omega_R^4 C_o}{Q_{rot} Q_{vib}} \exp\left[-\frac{c_2}{T} G(v,J)\right] \quad (1)$$

Here, the factor g accounts for the effect of nuclear spin, J is the rotational and v the vibrational quantum number, ω_R is the wave number of the fundamental Raman-scattered line $(v+1, J \leftarrow v, J)$, C_o contains that part of the scattering cross section not already written out in Eqn. 1, Q_{rot} and Q_{vib} are the rotational and vibrational partition functions, $c_2 = hc/k = 1.439$ cm K is the second radiation constant, and $G(v,J)$ is the molecular term value.

Equation 1 is important for the analysis of Raman temperature probe techniques because it tells us, for steady-state analyses, the contour which must be fit (either fully, as in part (b) of Fig. 2, or over selected bandpasses, as in part (c) or (d) of that figure) in order to determine parametrically the value of gas temperature - while, for transient analyses, it tells us what spectral slit width or bandpass to employ for the monochromator or filters utilized in order to pass the desired amount of each spectral signature through that slit or bandpass. Depending upon specific experimental circumstances, either total contours or specified (temperature-sensitive) fractions of these contours can be utilized for pulsed temperature analyses, and the resultant experimental accuracy for temperature determination can therefore depend strongly upon accurate knowledge of these shapes.

The vibrational amplitude factor $(v+1)$ in Eq. (1), corresponding to Stokes scattering, is obtained from the harmonic oscillator solution to the wave equation; for the anti-Stokes signal $(v+1 \rightarrow v)$, this $(v+1)$ factor is replaced by the factor v.[14] Although generally applicable to diatomic molecules, certain limitations exist for this equation. Thus, it includes both anharmonic and vibration-rotation interaction terms in the description of the molecular energy levels, but does not include depolarization effects nor overlapping lines from S- or O-branches $(\Delta J = \pm 2)$, although these (usually small) effects can be added in at the sacrifice of some additional complexity. Furthermore, possible changes in molecular polarizability and polarizability derivatives in excited vibrational energy levels, and therefore on the value of the cross section, have not been evaluated to date.

In temperature diagnostic experiments for which carefully-determined spectral slit functions were used to convolute values of the radiant flux S found from Eq. (1), use of curve-fitting procedures to fit the experimental data to the computed spectral contours produced good agreement, often to within several percent, with independently-determined values.[15] For polyatomics, contour analysis can become more complicated, but re-

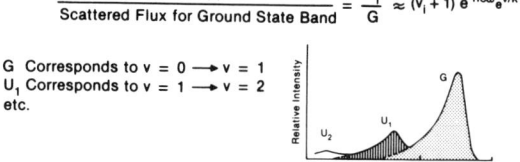

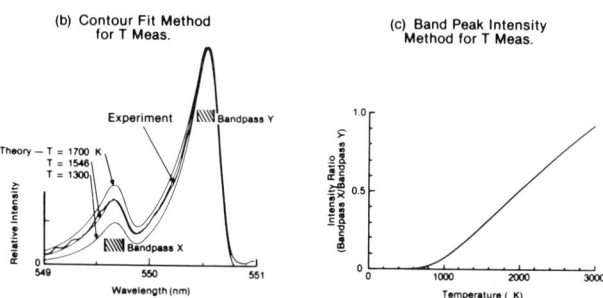

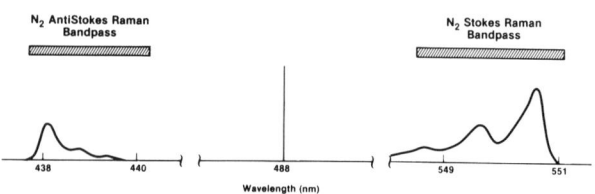

FIG. 2. Illustrations of Four Methods for Temperature Determination From N_2 Vibrational Raman Scattering. (a) Band Area Method. The expression shown determines temperature from the relative integrated intensities of the upper state (or "hot") bands and that of the ground state band. Here, v is the vibrational quantum number, h is Planck's constant, c is the speed of light, ω_e is the vibrational constant, k is Boltzmann's constant, and T is temperature. (b) Contour Fit Method. This figure shows a least-squares fit of a Raman contour obtained from N_2 in a laminar premixed flame whose temperature was found by detailed thermocouple measurements to be within several percent of that determined by the Raman fit. (c) Band Peak Intensity Method. In order to illustrate the potential use of spectral interference filters for temperature measurement with vibrational Q-branches, this curve gives the temperature sensitivity of the intensity ratio for two band passes (as shown in part (b)) at the peaks of the contour. (d) Stokes/Anti-Stokes Intensity Method. The ratio of Stokes to anti-Stokes vibrational Raman scattering has a temperature sensitivity similar to method (c) with the additional advantage of wide spectral bandpasses for maximum collected signal intensity.

sults applicable to temperature diagnostics have been accomplished. Using a simplified analysis of the CO_2 spectrum, Lapp et al[16] found good agreement with thermocouple-measured data near ambient temperatures. Using a more refined treatment of this molecule, Stevenson[17] has extended this technique to flame temperatures. Similarly, H_2O vapor Raman signatures have been studied as a function of temperature.[17,18]

B. <u>Comparison of Temperature Measurement Methods Based on Detector Type</u>. For photomultiplier detection, the band area or band contour analysis techniques (methods a and b in Fig. 2) are most suitable for time-averaged temperature measurements (i.e. in laminar flames) using continuous wave lasers and slow spectrometer scans. Methods c and d are most suitable for time-resolved temperature measurements (i.e. in turbulent flames) using pulsed lasers and fixed spectrometer detector bandpasses with pairs (or more) of gated detectors. Most time-resolved Raman temperature measurements (and the one used in this work) are based, in particular, upon the Stokes/anti-Stokes method (d in Fig. 2) pioneered first for temperature diagnostics by Lederman.[19] For this method, the very broad detector bandpass maximizes the number of Raman photons collected, and hence minimizes the statistical error in the temperature measurement. The temperature is found from the ratio of the numbers of detected photons in the Stokes and anti-Stokes channels, N_S and N_{AS}, respectively. Thus, from Eq. (1),

$$\frac{N_S}{N_{AS}} = K \left(\frac{\omega_S}{\omega_{AS}}\right)^3 \exp\left[\frac{c_2}{T} \Delta G\right],$$

or

$$T = c_2 \Delta G / \left[\ln \frac{N_S}{N_{AS}} + 3 \ln \frac{\omega_{AS}}{\omega_S} + \ln(1/K)\right] \quad (2)$$

where K is the ratio of the spectrometer-detector system response at the Stokes wave number ω_S to that at the anti-Stokes wave number ω_{AS} and ΔG is the Raman shift for a fundamental band (i.e., the energy level difference between the ground and first excited vibrational level, given by $\omega_e - 2\omega_e x_e + (13/4) \omega_e y_e$ for a diatomic molecule in terms of standard spectroscopic constants[20]) The sensitivity of the temperature measurement predicted by Eq. (2) depends upon the temperature range over which it is utilized as well as an estimate of the photon economy for the overall measurement (to be considered in the next section). Thus, although the exponential dependence of temperature upon the Stokes/anti-Stokes scattered light ratio predicts a high sensitivity for relatively low temperatures, insufficient photons appear in the anti-Stokes signal under these conditions for precise measurements.[21]

For detection by means of <u>multiple-spectral-channel detectors</u>[2] (as for example, with an Optical Multichannel Analyzer OMA® detector*), this situation is altered. These devices in their most advanced form combine optical image intensifiers with two-dimensional detectors, such as a vidicon or a diode array; they can thereby accept spectral information from an extended bandpass (such as an entire vibrational Raman band) over an array of spatial positions simultaneously. They can be gated to provide short (< 1 μs) time resolution for which the noise produced can be very much smaller than one count per channel. Thus, multiple-spectral-channel detectors mate well with pulsed laser sources. When these detectors are used for temperature sensing, methods a, b, and c (band area method, contour analysis, or ratio of two bandpass intensities) are all suitable. Method d, the vibrational Stokes/anti-Stokes intensity ratio is not very practical here because of the very large spectral span between the two bands. (Multiple-spectral-channel detectors have not yet achieved widespread use for spontaneous Raman scattering experiments, because of their cost and complexity relative to phototube detection - problems which should abate with time. They are already used extensively for time-resolved CARS measurements, since temperatures are often determined in this technique from analysis of entire vibrational contours generated by a pulsed narrowband laser in conjunction with a pulsed broadband laser.[22])

---

*OMA® is a registered trademark of Princeton Applied Research Corp., Princeton, NJ.

C. <u>Photon Economy and Experimental Accuracy</u>. Here, we address the question: Are enough photons scattered to make a detectable measurement? For a transition (which, for the processes we are considering, can be a vibrational band, or a rotational line within a vibrational band, or a pure rotational line) corresponding to a differential scattering cross section[23] σ (more precisely denoted as dσ/dΩ, and usually expressed in units of cm^2/sr^{-1}) we find the number of photons N actually detected during the measurement time to be

$$N = Q \rho \sigma L \Omega \epsilon \eta / E_\lambda \quad (3)$$

where Q is the energy in the incident laser beam during the course of the measurement (i.e. in one laser pulse for a time resolution encompassing that pulse, or in a time element over which the detected photons are integrated for a "continuous" spectral span), ρ is the number density of observed molecules, L is length of the column of laser-irradiated molecules for which the Raman scattering can be detected by the spectrometer system optical geometry, Ω is the corresponding detection solid angle, ε is the transmission of the spectrometer system optics, η is the quantum efficiency of the detector, and E_λ is the photon energy (corresponding to hc/λ, where h is Planck's constant, c is the speed of light, and λ is the wavelength). The simple relation given by Eq. (3) provides us with good estimates of light economy for experimental design and analysis omitting, of course, measurement errors introduced by other optical effects (such as light introduced into the detection system by gas luminosity, laser-induced fluorescense of the gas phase or of lenses and windows, etc.) Application to a typical example for temperature measurement in a flame in our laboratory is shown in Table 1 for N_2 at 60% relative abundance at 1500 K. The weakness of the spontaneous Raman process becomes apparent from this example, for which a huge number of incident laser photons ($Q/E_\lambda = 2.5 \times 10^{18}$) results in only 1050 detected Stokes photons and 110 detected anti-Stokes photons. Nevertheless, a useful accuracy of several percent based on Poisson photon statistics can be achieved.

The relative standard deviation for an individual temperature measurement (σ_T/T) from a single laser shot can be calculated approximately from

$$\frac{\sigma_T}{T} \simeq \frac{1}{T}\left[\left(\frac{\partial T}{\partial N_S} \sigma_S\right)^2 + \left(\frac{\partial T}{\partial N_{AS}} \sigma_{AS}\right)^2\right]^{\frac{1}{2}} \quad (4)$$

where $\sigma_S = \sqrt{\bar{N}_S}$ and $\sigma_{AS} = \sqrt{\bar{N}_{AS}}$ are the standard deviations in the individual photon counts. Here, $\bar{N}_S$ and $\bar{N}_{AS}$ are the expected numbers of detected Stokes and anti-Stokes photons. If $\sigma_S/\bar{N}_S$ and $\sigma_{AS}/\bar{N}_{AS}$ are small compared to unity, Eq. 4 can be simplified to

$$\frac{\sigma_T}{T} \simeq \frac{T}{\theta_v}\left[\frac{1}{\bar{N}_S} + \frac{1}{\bar{N}_{AS}}\right]^{\frac{1}{2}} \quad (5)$$

where $\theta_v = c_2 \Delta G$ is the characteristic vibrational temperature. Using values of N_S and N_{AS} in Table 1 gives a relative standard deviation of temperature of 4 % from N_2 spectra in a flame at 1500 K.

As mentioned in the previous section, at low temperatures the number of detected anti-Stokes photons decreases dramatically making the Stokes/anti-Stokes method for N_2 unsuitable at temperatures less than ∿800 K. One possibility for temperature measurements between 300-800 K is to use rotational Raman scattering. However, the new approach used in this work is to determine temperature by measuring the concentration of <u>all</u> of the major species from their Stokes vibrational Raman intensities. In a constant pressure system

$$T/T_o = (\sum_i N_S(i)/\sigma_i)_o / (\sum_i N_S(i)/\sigma_i) \quad (6)$$

where $N_S(i)$ is the number of Raman Stokes photons for species i, σ_i is the differential vibrational Raman cross section for species i, and the sum is over all chemical species present. The temperature and relative Raman intensities are normalized by room temperature calibration measurements denoted by the subscript o.

D. **Choice of Measurement Technique.** The error analysis for the Stokes/anti-Stokes temperature measurement (Eqs. 4 and 5) does not include the deleterious effects of background interferences due to blackbody radiation, flame emission, fluorescence or laser-induced incandescence. Naturally, the detected signal for a practical example (i.e. a brightly luminous sooty flame) could be dominated by these interferences. Such measurement situations are not suitable for spontaneous Raman scattering diagnostics and the reader is referred to other chapters in this volume dealing with alternate, stronger measurement techniques which become advantageous when serious background problems are encountered. (See Ref. 11, which includes CARS diagnostics.) However, two salient facts should be recalled before simply choosing the stronger technique. Firstly, spontaneous Raman scattering with pulsed lasers and gated detectors can handle more luminous interference than is intuitively apparent since it compresses the useful signal into a very small time element. Thus, if we compare the blackbody spectral radiance assuming unity emissivity at 500 nm and 1500 K (10^{-6} Wcm^{-2}nm^{-1}sr^{-1}) with the equivalent Raman spectral radiance I_R, we find that the number of Raman photons is more than an order of magnitude greater than the blackbody emission for the values in Table 1.[7] In addition, this full blackbody emission from combustion gases is unlikely unless the observed path length is very long or the particle loading is heavy. For example, estimates by Hottel and Sarofim[24] suggest an emissivity of a few tenths for a several-meter-diameter fuel oil flame containing coke particles. However, pulsed Raman techniques do not discriminate against laser induced background processes such as fluorescence from windows or combustion gases or laser-induced particle incandescence. These laser-induced backgrounds (discussed in the results sections) can severely restrict application of spontaneous Raman techniques.

The second point to be made is that spontaneous Raman scattering (where it can be applied cleanly) has several advantages over CARS or fluorescence. The reduction of the Raman data is generally more straightforward and the construction of the Raman apparatus less complex and costly than the alternate techniques. Thus, fewer experimental and theoretical complexities may make spontaneous Raman somewhat less susceptable to systematic temperature measurement errors.

Perhaps the biggest advantage of spontaneous Raman scattering is the simultaneous measurement of temperature and all major species density data for each pulse, a capability which thus far only spontaneous Raman scattering has been able to offer. This can be of critical value for a more complete understanding of the system being probed - as, for example, in turbulent flame model verification studies.[25] However, it is clear that alternate techniques like CARS are essential for the more difficult measurement environments. These contrasting capabilities are well-illustrated by comparing the modeling results obtained by Raman scattering in controlled laboratory flames[26] with the thermometry probing using CARS accomplished in hostile combustor environments.[5,11,27]

III. **Experimental Pulsed Raman Apparatus and Calibration**

Here we describe in some depth the use of pulsed vibrational Raman scattering to measure temperatures in turbulent flames produced in our laboratory.[26] A schematic overview of the turbulent combustor geometry, optical configuration and electronic data processing used in this work is given in Fig. 3. A horizontally burning flame - either a laminar premixed flame from a porous plug burner or a turbulent diffusion flame from the turbulent combustor - is positioned for study relative to a fixed-bed Raman optical configuration The laser used to excite the Raman scattering is a Candela dye laser, which consists of an oscillator and two amplifiers pumped by linear flashlamps. The oscillator output is tuned and line-narrowed by three intracavity prisms placed just before the totally reflecting mirror. The amplified output is deflected through 180° by three additional prisms, and directed along a 3-m path to a 1.5-cm aperture which spatially and spectrally filters the laser beam. This configuration ensures a sharp focus and removes unwanted background light from flash lamps and dye fluorescence. The laser beam is focused by a 20-cm focal length lens into a small region of the flame. The laser power for each laser pulse is measured by a vacuum photodiode while the approximate spectral distribution of each laser pulse is monitored by a 1/4-m Littrow mount grating spectrometer with TV camera readout. The laser produces pulses of about 1 J at the test zone, within a 0.2-nm bandwidth tuned to 488 nm. The pulse duration is 1-2 μs, and a repetition rate up to 1 pps is obtainable with these output parameters.

Light scattered at right angles from the incident laser beam is collected by two lenses and focused through the entrance slit of a Spex 3/4-m single spectrometer. The laser focus and collection optics magnification gives a sample volume (spatial resolution) of 0.3 x 0.3 x 0.7 mm (greatly exaggerated in Fig. 3). The stray light rejection of the single spectrometer is increased by placing bandpass or cutoff filters at

Table 1

Typical Raman Probe Characteristics

Quantity	Value	Comments
Q	1J	Obtained from pulsed laser sources (dye, ruby, etc.) or from integration of cw laser output (say, 1 W Ar laser for 1 s)
ρ	3×10^{18} mol/cm^3	N_2 at 60% relative abundance at 1500°K
σ	5×10^{-31} cm^2/sr	N_2 at the vibrational scattering wavelength
L	0.07 cm	For a single optical pass
Ω	0.1 sr	For moderate, i.e., not extraordinarily fast, optics
ϵ	0.2	Typical value
η	0.2	Typical value
E_λ	4×10^{-19} J	Evaluated at the vibrational scattering wavelength

$N_S = 1.05 \times 10^3$ Detected Photons, from Eq. (3).

$N_{AS} = 1.1 \times 10^2$ Detected Photons, from value of N_S and Eq. (2), in which a spectral calibration factor K of 2 was used (experimentally determined for our Raman detection system).

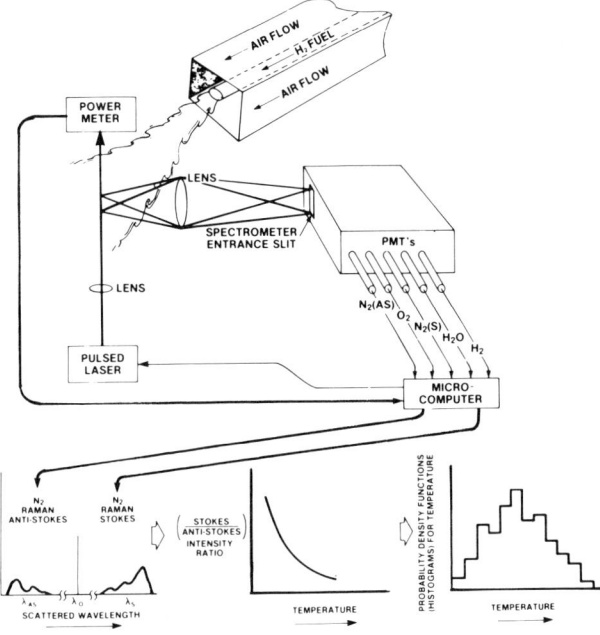

FIG. 3. Schematic of turbulent combustor geometry and optical data acquisition system for vibrational Raman scattering measurements of temperature and concentrations of major flame species. Also shown are sketches of expected Raman contours of Stokes and anti-Stokes Raman scattering from N_2, the temperature calibration curve, and an example of a probability density function for temperature at a specific location in the flame.

the exit plane of the spectrometer. Following the filters and five exit slits, photomultiplier tubes (PMT's) detect anti-Stokes vibrational Raman scattering from N_2 and Stokes scattering from N_2, O_2, H_2 and H_2O. Analog signals from the photomultipliers and laser energy detector are collected by dual channel sample-and-hold electronics which measure background signals several microseconds before the laser pulse as well as the integrated signal intensity during the pulse. Both the background and laser-induced signals from each detector are digitized and their difference represents the net Raman signal. Temperature is determined by the intensity ratio of anti-Stokes-to-Stokes N_2 vibrational Raman signals and the concentrations of N_2, O_2, H_2 and H_2O are determined by the intensities of their respective Stokes signals normalized by the laser energy. Corrections are made for photomultiplier sensitivities, Raman cross sections, and Raman band contour variations with temperature. A sixth photomultiplier detector located at the sideport exit slit of the spectrometer monitors pulsed Rayleigh scattering.

The data collection is controlled by a micro computer which fires the laser, collects and digitizes the signals from the energy meter and PMT's, and stores the data on floppy disc or cassette tapes for subsequent processing. Steps in signal processing are shown on the bottom of Fig. 3. Scattering intensities for N_2 anti-Stokes and N_2 Stokes signals are measured for each laser pulse. Their ratio can be related to values of temperature based upon theoretical calculations and calibrations described previously. Repetitively pulsing the laser at the same flame location permits the buildup of a probability density function (pdf) from individual temperature measurements. Similarly, molecular concentrations of N_2, O_2, H_2 and H_2O are determined from every laser shot from the corresponding Stokes scattering intensity normalized by laser energy, and pdf's of these quantities are built up in the same way.

The Raman detection system described must be calibrated in order to provide accurate measures of temperature (and molecular density). This calibration was carried out in two independent ways. First, a standard lamp with known intensity as a function of wavelength was used to measure the relative spectral response of the spectrometer and the various photomultipliers at each of the detected wavelength regions. Then signals from the individual Raman channels were corrected for the measured spectral response of the spectrometer-PMT system, the gain of the detection electronics, the relative Raman cross sections of the different molecules and the temperature-dependent fractions of the calculated vibrational Raman bands which fell within the experimentally determined spectral bandpasses. This last correction factor is the most difficult to determine. The exit slit widths for N_2(AS) and N_2(S) were each nominally 2 mm in a spectrometer with a nominal dispersion of $\sim$ 1 nm/mm. The shape and width of each slit function was determined by scanning the spectrometer across a narrow atomic lamp emission line in the proper spectral location. Each spectral bandpass has a flat top with a width of 2.7 and 2.4 nm, and a full width at half maximum of 3.1 and 3.0 nm, respectively. The vibrational contours of the N_2 vibrational Raman bands were calculated as a function of temperature to determine the fraction falling within the experimentally determined bandpass. As an independent check of the slit function, the bandshapes of the N_2 Raman band were experimentally measured both at room temperature and in a laminar premixed flame at 1800 K by pulsing the laser and scanning the spectrometer. Excellent agreement was obtained between measured and calculated bandshapes. (The standard lamp form of calibration described above can be used in an absolute sense, or in a hybrid fashion, relative to a room temperature nitrogen measurement, in order to calculate various flame species densities and temperatures from measured Raman signals.)

The second independent procedure utilizes pulsed Raman and Rayleigh measurements of ambient air and room temperature H_2, N_2, O_2, and He to determine Raman intensities of known molecular gas densities and to analyze the source and intensity of background radiation. For the low luminosity H_2-air flames, the dominant source of background is laser-induced fluorescence from the pyrex combustor windows. This background did not cause significant problems because it is small relative to the Raman signals and is essentially constant for each laser shot and hence easily subtracted out. Calibration of the relative sensitivities of N_2 vibrational Stokes and anti-Stokes channels for temperature measurements was made by comparison of temperatures measured by pulsed Raman techniques with those previously determined in isothermal laminar, premixed H_2-air flames by radiation-corrected fine-wire thermocouples and by cw vibrational Raman bandshape analysis.[15] The laminar premixed flames used for calibration here were produced on a water-cooled porous plug burner (diameter 2.5 cm) burning into another water-cooled porous plug (of larger diameter) placed 2.0 cm away and connected to a rough vacuum line. In this way, a stable, horizontally burning flame at atmospheric pressure was produced which had the advantage of uniform flame conditions in the Raman test zone defined by the intersection of the vertically-passing laser beam with the collection optical path.(The flame temperatures for this burner are far below adiabatic values for the H_2-air mixtures used because of appreciable heat losses to the water cooled plug.) Accurate flow metering techniques (using critical flow through orifices monitored by precision pressure gauges) were utilized to produce reproducible and clearly-defined flame conditions. The critical flow orifices were calibrated in our laboratory through the use of volume displacement techniques. The entire porous plug assembly just described can be fit inside the test zone of the turbulent combustor by removing a side window, so that the optical paths for the laminar premixed and turbulent diffusion flames could be made practically identical for the purposes of system calibration.

The calibration factor for H_2O vapor was determined from Raman measurements of laminar, premixed H_2-air flames where the N_2-to-H_2O mole fraction ratios could be calculated from known input gas compositions. The calibration is extended to other temperatures, as in the lamp calibration just described, by using calculations of the fraction of the vibrational Raman band which falls within the measured experimental bandpass. Water vapor measurements are somewhat less certain than concentration measurements for other species because calculations of water bandshapes at high temperatures are of limited accuracy in the "tail" of the contour. This part of the vibrational Raman contour is increasingly important at elevated temperatures, and is uncertain because of insufficient knowledge of the energy level structure of water vapor - especially, the rotational structure of excited vibrational levels.

In our turbulent combustion studies the standard lamp calibration was performed once, and the laminar flame calibration measured every few days, while N_2 room temperature calibrations and background measurements are checked several times each day. Close agreement between Raman measurements for the nitrogen Stokes channel in room air, and with lamp calibrations for all channels, is taken to indicate satisfactory system alignment and performance.

IV. Results of Pulsed Raman Measurements in Laminar and Turbulent Flames

Experimental pulsed Raman measurements (from runs containing several hundred laser shots) of temperature and major species mole fractions for several laminar, premixed H_2-air flames (ϕ=1.0 to 0.55) have been compared with the theoretical burnt gas mole fractions calculated from the initial compositions and the temperature measured previously by fine-wire thermocouples and cw Raman bandpass analysis. The agreement is excellent, indicating that systematic errors are acceptably low (approximately ± 50 K for temperature and ±1 percent for mole fraction of major species for the range of premixed flames studied from 1500-1900 K).

In addition to system calibrations, analyses of Raman scattering from laminar, premixed H_2-air flames help define our measurement resolutions under the best possible combustion conditions for Raman measurements -- very small fluctuations from any remnant turbulence and low background luminosity. For example, Fig. 4 shows the temperature pdf from a 500 laser-shot data set (similar to one mentioned previously) obtained from an essentially isothermal zone (whose temperature was known from previous calibrations) in a premixed stoichiometric H_2-air flame. Since the widths of such measured distributions are caused primarily by measurement uncertainties, they define our experimental resolution for the Raman technique described here. Repeated determinations of the temperature pdf for this flame are characterized by a symmetric, nearly Gaussian distribution about an average temperature of 1823 K with a 4 % relative standard deviation. Average temperatures for this flame measured by radiation corrected thermocouples or vibrational Raman bandshape analysis is 1813 K. The very good agreement between theoretical predictions for the relative standard deviation for temperature σ_T/T = 4 % from Eq. 5 and the experimental results suggests that the spread in the pdf shown in Fig. 4 can be almost entirely ascribed to the expected statistical fluctuations in the number of detected photons per laser shot. Using the known dependence on temperature of the Stokes and anti-Stokes intensities, we have calculated the relative standard deviation as a function of temperature and N_2 density for H_2-air flames with the result that the σ_T/T remains nearly 4 % for all stoichiometries and temperatures except for rich mixtures below 950 K or lean mixtures

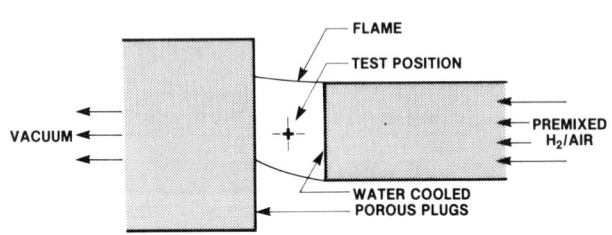

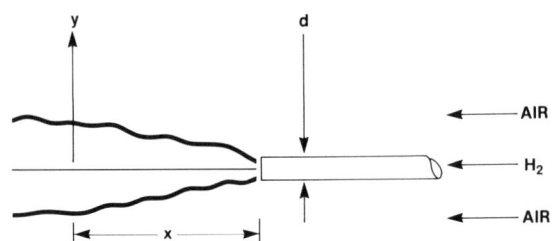

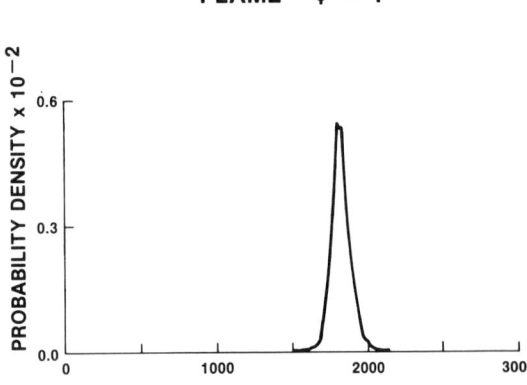

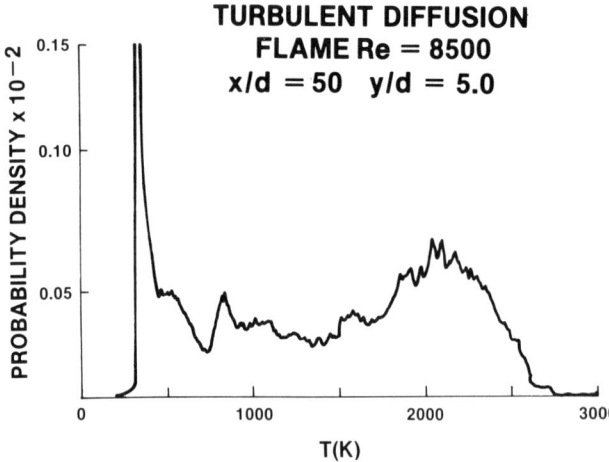

FIG. 4. Probability density function (histogram) of temperature experimentally measured in an isothermal zone of the post flame gases from a laminar, premixed, stoichiometric H_2-air flame. Five hundred measurements were used to obtain the pdf.

FIG. 5. Probability density function (histogram) of temperature in the mixing layer of a H_2-air turbulent jet diffusion flame showing a very broad range (300 K-2600 K). This pdf resulted from 2000 independent measurements.

below 800 K where they are larger. At these low temperatures, more precise temperature measurements are made using Eq. 6. For measurements of room air this method gives a measured σ_T of 4 K.

Fig. 5 shows probability density functions (pdf's) for temperature measured with our pulsed Raman apparatus in the mixing layer of a turbulent H_2-air flame. Measured temperatures range from room temperature to 2600 K (somewhat higher than the calculated H_2-air adiabatic flame temperature). The much wider pdf in the turbulent flame is the result of real temperature fluctuations in the flame at this location. This pdf and many others taken at other turbulent flame locations are similar in form to those measured by thermocouples in slightly heated jets and by Rayleigh scattering in turbulent flames. However, the pulsed Raman technique is unique in its ability to simultaneously measure all species concentrations along with the temperature which is important in experimental testing and developing turbulent reacting flow models.

In an effort to assess the feasibility of extending pulsed Raman measurements to more luminous hydrocarbon flames, several laminar premixed propane-air flames and turbulent H_2-propane-air flames were investigated with the same pulsed Raman apparatus. No experimental difficulties were encountered in non-sooty burnt gas regions of these flames. For example, in laminar premixed flames the agreement within experimental error between the measured average temperature and the calculated adiabatic temperatures as well as the low value for σ_T/T of 5 % strongly indicate that the increased luminosity of propane-air flames has little deleterious effect on our pulsed Raman measurements. This is to be expected because the dual channel sample and hold electronics automatically subtracts out the luminous flame backgrounds from each Raman channel.

However, great difficulty was found in making meaningful Raman measurements in sooty regions of very fuel-rich ($\phi \geq 3.0$) propane-air premixed flames because of large laser-induced backgrounds. Eckbreth[5] found that the focused, high power ($\sim 10^9$ to 10^{10} W/cm^2) laser beam required for such measurements strongly heats the soot particles. These hot particles create a laser-induced background of thermal radiation which can be much stronger than intrinsic thermal radiation from the flame, and much stronger than typical Raman scattering signals. An analytical study of these processes, similar to Eckbreth, suggest that soot particles, in the presence of the high energy laser pulses ($\sim 10^9$ W/cm^2) necessary for Raman or CARS, rapidly reach vaporization temperatures of ~ 4000 K, and radiate as a black body at this elevated temperature until they disappear by vaporization. Calculations of the intensity of this laser-induced soot incandescence for our experimental conditions suggest that it will not be a serious background for small (r = 5 nm) soot particles but will be comparable to Raman scattering intensities in the presence of larger (r $\sim$ 1000 nm) particles.

In addition to high optical backgrounds, this laser-induced soot incandescence and vaporization can cause serious sample perturbations. The vaporizing particle produces a plume of predominantly carbon vapor which can, by collisions, transfer energy to N_2 and other molecules (perturbing temperature based upon N_2 vibrational populations) or react with the surrounding vapor changing its composition or temperature. The rates of energy transfer and chemical reaction may be slow enough to cause limited effects for CARS measurements made on a 10 ns timescale but may not be insignificant for pulsed Raman measurements with a temporal resolution of 1 μs.

One successful method of circumventing soot incandescence (permitting the possible application of pulsed Raman scattering for time-averaged temperature measurements in some sooty flame environments) uses laser powers which are large enough to overcome flame luminosity but not large enough to significantly heat soot particles.[28] In general, however, coherent anti-Stokes Raman scattering will be a better choice for sooting flames because the shorter laser pulses and somewhat lower laser energy requirements minimize soot incandescence and potential sample perturbation effects, and because its collimated signal output permits excellent rejection of the laser-induced incandescence which does occur. The collimated output and larger signal intensities inherent in CARS also give it significant additional advantages over Raman scattering for temperature measurements in large-scale combustors since large aperture optical access is not required for efficient signal collection and thermal radiation from combustor walls is more easily rejected.

V. Conclusions

The advantages of pulsed spontaneous Raman scattering for gas-phase temperature measurement are summarized in Table 2. Its ability to make non-intrusive, non-perturbing, spatially and temporally resolved measurements over extremely wide temperature ranges make Raman scattering the preferred technique for many gas phase reacting systems, and its application in fundamental and applied research is rapidly expanding. The mainstay of pulsed Raman temperature measurements has been the Stokes/anti-Stokes intensity ratio method, applied in this work to turbulent H_2-air flames. After extensive calibration, it provides time resolved (< 2 μs) and space resolved (< 0.1 mm^3) temperature measurements in H_2-air flames with a 4 % relative standard deviation and an average temperature accuracy of $\pm$ 50 K. The relative standard deviations are limited by photon statistical fluctuations while the accuracy is limited by a variety of systematic errors and calibration uncertainties. Similar results have been obtained for the Stokes/anti-Stokes temperature measurements from other laboratories in studies of turbulent flames, internal combustion engines, and shock tubes, illus-

Table 2

Advantages of Pulsed Spontaneous Raman Scattering for Temperature Measurements

Nonintrusive	Requires no physical probes.
Non-perturbing	Usually only weak coupling between laser light and gas molecules.[1]
High Spatial Resolution	Typically <0.1 mm^3.
High Temporal Resolution	Limited by laser pulse width which can be <1 μs.
Accuracy	Well established theory. Limited by photon statistical errors or background radiation corrections.
Precision	Limited by photon statistical errors.
Wide Range	Rot. Raman[3] 10 - 3000 K Vib. Raman[2,3] 800 - 9000 K Elec. Raman[3] 1000 - 10,000 K
Non-equilibrium Capability	Measures relative energy level populations which defines a temperature on the extent of thermal non-equilibrium.

1. Not true if strong absorption of laser beam occurs due to molecular or particulate absorption.
2. Can be extended to room temperature and below by measuring the concentration of all species (see text).
3. No theoretical upper limit on temperature range. Values listed are the maximum measured experimentally.

trating its wide range of applicability. However, laser induced fluorescence or particle incandescence prevents pulsed Raman measurements in many sooty hydrocarbon flames and practical combustors where CARS and other non-linear techniques appear more suitable.

References

1. First measured experimentally by C.V. Raman, Nature 121, 619 (1928); predicted earlier by A. Smekal, Naturwiss. 11, 873 (1923).
2. M. Bridoux and M. Delhaye, in Advances in Infrared and Raman Spectroscopy, Vol. 2, ed. by R.J.H. Clark and R.E. Hester, Heyden, London, 1976, Chapt. 4.
3. M.C. Drake and G.M. Rosenblatt, in Characterization of High Temperature Vapors and Gases, Vol. 1, ed. by J.W. Hastie, National Bureau of Standards Special Publication 561/1, Washington, DC, 1979, p. 609.
4. W.D. Williams and J.W.L. Lewis, AIAA J. 13, 1269 (1975); R.A. Hill, C.W. Peterson, A.J. Mulac and D.R. Smith, J. Quant. Spectros. Rad. Trans. 16, 953 (1976); I.F. Silvera and F. Tommasini, Phy. Rev. Lett. 37, 136 (1976).
5. A. Eckbreth, Eighteenth Symp. (Int.) on Combustion, The Combustion Institute, Pittsburgh, 1981, p. 1471; A.C. Eckbreth, P.A. Bonczyk, and J.F. Verdieck, Appl. Spectros. Reviews 13, 15 (1977). Also Prog. Energy Combustion Sci. 5, 253 (1979).
6. S. Lederman, J. Prog. Energy Combustion Sci. 3, 1 (1977).
7. M. Lapp and C.M. Penney, in Advances in Infrared and Raman Spectroscopy, Vol. 3, R.J.H. Clark and R.E. Hester, Heyden and Sons Ltd., London, 1977, Chapt. 6.
8. J.R. Smith, SAE Technical Paper 800137 (1980), "Temperature and Density Measurements in an Engine by Pulsed Raman Spectroscopy;" J.R. Smith, AIAA-80-1359 (1980),"Instantaneous Temperature and Density by Spontaneous Raman Scattering in a Piston Engine;" J.R. Smith, AIAA J. 18, 118 (1980); W.M. Rocquemore and P.P. Yaney in Characterization of High Temperature Vapors and Gases, Vol. 1, ed. by J.W. Hastie, National Bureau of Standards Special Publication 561/1, Washington, DC, 1979, p. 973.
9. A.A. Boiarski, Naval Surface Weapons Center Tech. Report NSWC/WOL/TR 75-53 (1975); S. Lederman, A. Celentano, and J. Glaser, Phys. Fluids 22, 1065 (1979).
10. L.Y. Nelson, A.W. Saunders, Jr., A.B. Harvey, and G.O. Neely, J. Chem. Phys. 55, 5127 (1971); G. Black, R.L. Sharpless, and T.G. Slanger, J. Chem. Phys. 58, 4792 (1973); J.J. Barrett and A.B. Harvey, J. Opt. Soc. Amer. 65, 392 (1975); J.A. Shirley and R.J. Hall, J. Chem. Phys. 67, 2419 (1977).
11. See other papers in this volume. For example, M.C. Drake, C. Asawaroengchai, D.L. Drapcho, K. Viers, and G.M. Rosenblatt for rotational Raman; J.W. Fleming, W.T. Barnes, and A.B. Harvey for rotational CARS; and papers by J.P. Taran and M. Péalat; G.L. Switzer and L.P. Goss; L.A. Rahn, S.C. Johnston, R.L. Farrow, and P.L. Mattern; J.F. Verdieck, J.A. Shirley, R.J. Hall, and A.C. Eckbreth; and D. Klick, K.A. Marko, and L. Rimai for applications of CARS to practical combustors.
12. See, for example, A. Weber, in The Raman Effect, Vol. 2: Applications, ed. by A. Anderson, Dekker, New York, 1973, Chapt. 9; G. Placzek in Marx Handbuch der Radiologie, Vol. 6, part 2, Akademische Verlagsgellschatt, Leipzig, 1934, p. 209. UCRL-Trans-526L.
13. M. Lapp, C.M. Penney, and L.M. Goldman, Science 175, 1112 (1972) and references quoted therein.
14. L. Pauling and E.B. Wilson, Introduction to Quantum Mechanics, McGraw-Hill Book Co., Inc., New York, 1935, p. 82.
15. M. Lapp, in Laser Raman Gas Diagnostics, ed. by M. Lapp and C.M. Penney, Plenum Press, New York, 1974, p. 107; R.E. Setchell, Sandia Laboratories Energy Report SLL-74-5244 (1974); also published as Western States Section/The Combustion Institute Paper No. WSS/CI 74-6; S.M. Schoenung and R.E. Mitchell, Comb. Flame 35, 207 (1979).
16. M. Lapp, C.M. Penney, and L.M. Goldman, Opt. Comm. 9, 195 (1973).
17. D.A. Stevenson, Appl. Spectroscopy 35, 582 (1981).
18. J.L. Bribes, R. Gaufres, M. Monan, M. Lapp, and C.M. Penney, Appl. Phys. Lett. 28, 336 (1976); J.L. Bribes, R. Gaufres, M. Monan, M. Lapp, and C.M. Penney, in Proc. of the Fifth Int. Conf. on Raman Spectroscopy, ed. by E.D. Schmid et al., Hans Ferdinand Schulz Verlag, Freiburg im Breisgau, 1976; M. Lapp, AIAA J. 15, 1665 (1977).
19. G.F. Widhopf and S. Lederman, AIAA J. 9, 309 (1971); also Polytechnic Institute of Brooklyn, PIBAL Rep. No. 69-46 (1969).
20. G. Herzberg, Molecular Spectra and Molecular Structure, Vol. 1. Spectra of Diatomic Molecules, Van Nostrand, Princeton, NJ, ed. 2, 1950.
21. M. Lapp and C.M. Penney, in Proc. of the Dynamic Flow Conference 1978 on Dynamic Measurements in Unsteady Flow, Proc. of the Dynamic Flow Conf., Skovlunde, Denmark, 1979, p. 665.
22. W.B. Roh, P.W. Schreiber, and J.P.E. Taran, Appl. Phys. Lett. 29, 174 (1976).
23. The value of scattering cross section for a particular calculation depends upon the optical configuration for the experiment under consideration, i.e., upon the relative directions of the incident laser beam and the observation optical axis. Furthermore, Raman scattering causes some depolarization of the strongly-polarized laser beams employed in diagnostic experiments. This value is small for vibrational Raman scattering, but can be large for the rotational Raman effect. See C.M. Penney, L.M. Goldman, and M. Lapp, Nature Physical Science 235, 110 (1972); C.M. Penney, R.L. St. Peters, and M. Lapp, J. Opt. Soc. Am. 64, 712 (1974). Detailed reviews of cross section values are given by H.W. Schrötter and H.W. Klöckner, in Raman Spectroscopy of Gases and Liquids, ed. by A. Weber, Topics in Current Physics, Vol. 11, Springer-Verlag, Berlin, 1979, p. 123 and by H.W. Schrötter, in Advances in Infrared and Raman Spectroscopy, Vol. 8, ed. by R.J.H. Clark and R.E. Hester, Heyden, London, 1981, Chapt. 1.
24. H.C. Hottel and A.F. Sarofim, Radiative Transfer, McGraw-Hill, New York, 1967, p. 244.
25. M. Lapp and R.M.C. So, in AGARD Conference Proceedings No. 281, Testing and Measurement Techniques in Heat Transfer and Combustion, Advisory Group for Aeronautical Research and Development, NATO, 1980, Chapt, 19; available from NTIS.
26. M.C. Drake, M. Lapp, C.M. Penney, S. Warshaw, and B.W. Gerhold, AIAA Paper 81-0103 (1981); M.C. Drake, M. Lapp, C.M. Penney, S. Warshaw, and B.W. Gerhold, Eighteenth Symp. (Int.) on Combustion, The Combustion Institute, Pittsburgh, 1981, p. 1521; M.C. Drake, R.W. Bilger, and S.H. Starner, Nineteenth Symp. (Int.) on Combustion (submitted).
27. G.L. Switzer, L.P. Goss, W.M. Rocquemore, R.P. Bradley, P.W. Schreiber and W.B. Roh, AIAA-80-0353 (1980),"The Application of CARS to Simulated Practical Combustion Systems." See also Appl. Opt. 18, 2343 (1979).
28. W.L. Flower, Sandia National Laboratories Report No. 81-8608 (1981), "Raman-Scattering Gas Temperature Measurements in Particle-Laden Flows."

Dynamic temperature measurements of flames using spontaneous Raman scattering[a]

P. P. Yaney, R. J. Becker,[b] P. D. Magill,[b] and P. Danset

Department of Physics, University of Dayton, Dayton, Ohio 45469

A two-channel Raman spectroscopy system capable of recording simultaneously two photon-counting signal channels from two closely spaced exit slits has been developed. These slits are formed effectively by the use of cylindrical optics and special slit mirrors after the exit slit of a conventional double spectrometer. The slits can be spaced up to 29 cm^{-1} in second order with minimum spectral widths of about 0.5 cm^{-1}. These two channels can be simultaneously recorded at rates up to 10 kHz. The data are recorded on floppy disc by a minicomputer. Statistical analyses of the reported data consisted of probability density functions (pdf's) and power spectral density functions (psdf's). The Raman scattering was excited by a Spectra Physics 171 argon ion laser operated cw with 4W at 488 nm. The effective laser power incident on the observed volume was magnified by a factor of 24 over single pass by means of a multiple-pass optical subsystem and a reverse-optics mirror. The 60.2-cm^{-1} rotational Raman line of room temperature air gives routinely 38 000 counts/s with 2.4 cm^{-1} slits. The application of this system to the measurement of real-time fluctuations of temperature at the edge of a simple methane diffusion flame is described. The temperature measurements were based on the intensity ratio of two rotational Raman lines of air. Fluctuations due to the flame were clearly detectable in the psdf's. Temperature fluctuations due to sine wave acoustical excitation of the flame were recorded. The pdf's were strongly skewed towards high temperature for mean temperatures of about 450 K.

INTRODUCTION

The development of laser-based optical diagnostic probes using spontaneous Raman scattering (RS) in combustion and gaseous flow measurements has achieved a reasonable degree of success in characterizing time-averaged temperature spatial profiles in steady flows.[1,2] Measurements have been systematically carried out in which the Raman temperatures have been shown to be in excellent agreement with the temperatures determined from standard thermocouple and gas-sampling probe data.[3,4] However, characterization of turbulent flow requires the capability of resolving not only spatial differences, but more important, temporal fluctuations as well. Basically, the type of measurement that can be made is determined to a large extent by the type of laser being used. Generally, high energy, ($\gtrsim$1 J), short pulse ($\lesssim$1μs) lasers with suitable beam and wavelength characteristics are limited to low (~10/s) repetition rates. This means that techniques utilizing such lasers can provide probability density functions (pdf's) of the fluctuating variable.[5,6] However, since the frequency spectrum is limited to half the measurement rate (Nyquist interval),[7] little information regarding the frequency content of the fluctuations can be obtained. Pulsed lasers of the Q-switched variety are being used in non-linear scattering techniques such as coherent anti-Stokes Raman scattering (CARS) to generate pdf's of temperature and species concentration in various turbulent flows,[8] while a flashlamp-pumped dye laser has been successfully employed to generate pdf's in a turbulent diffusion hydrogen flame using RS.

Ideally, the most fruitful experimental approach to characterizing the fluctuations of any variable is to record in real time the value of that variable. Then, the power spectral density function (psdf) or the autocorrelation function, as well as the pdf and other statistical parameters, can be determined. To avoid aliasing errors the measurement interval must be short compared to the period of the highest frequency contained in the signal to be recorded.[7] This measurement scenario can be accomplished with a cw laser and RS provides the simplest technique; however, RS is a weak process. The signals available in a typical RS measurement on a gas such as N_2 are commonly on the order of 100's of counts/sec. Thus, if the measurements were made over 10 ms intervals (i.e., a 100 Hz read rate), then on the order of one count could be expected for any measurement. The resulting high statistical uncertainties would require a very large number of measurements to be made before results of any value could be expected.[6] Nevertheless, concentration fluctuation studies have been carried out with RS on turbulent[9] and unsteady[10] flows using CH_4 which has a somewhat higher vibrational Raman scattering cross section compared to N_2.

Thus, to make real time measurements using RS, schemes for enhancing the signal must be utilized. Ideally, a frequency capability out to 1 kHz would permit a fairly wide range of turbulent flows to be investigated. This would require a read rate of 2 kHz. For a 10% Poisson uncertainty for a single measurement, an average count per read of 100 is necessary. This means an average count rate of 200,000 counts/sec is required. To make a temperature measurement in real time using RS, a minimum of two temperature sensitive features of the Raman spectrum must be measured simultaneously (or nearly so).[1] This paper describes a RS spectrometric system designed to measure simultaneously two rotational Raman lines in real time. A dedicated minicomputer records the two channels in less than 1 µs. A multipass optical cell increases the laser beam intensity in the sample volume by a factor of about 12. A reverse-optics mirror nearly doubles the observed signal. High quantum efficiency photomultiplier tubes (PMT's) are used along with an argon ion laser giving about 4 W at 488 nm. The essential features of this system and some of the preliminary results are presented.

CONCEPTUAL APPROACH

The long-term objectives of this work are to make real-time measurements of temperature and species concentrations as a function of position in turbulent flow. An especially important part of this effort will be to characterize the temporal mixing processes in turbulent cold flow. This will be the subject of a future paper. Both temperature and fuel-air mixing processes require that a minimum of two simultaneous wavelength channels be recorded. With two channels, these two types of measurements are mutually exclusive. Although more channels could have been designed into the current system, it was decided that this additional cost and complexity was not warranted and that it would best be deferred to later system upgrading. System performance objectives are to record the two wavelength channels at up to 1 kHz read rates from a sample volume of 1 mm^3. It was not expected that this performance would be achieved with the current system; however, the prerequisites needed to reach these objectives would be clear from the work with the current system.

The primary concept in measuring the concentration of a gas or its temperature using an optical probe is the spectroscopic detection of laser-excited emissions or scattered light from specific spectral features of the gas which contain the desired information. For Raman temperature measurements in air, it is convenient (and customary) to use N_2 as the probe molecule. A diatomic molecule such as N_2 has a set of rotational energy levels associated with each vibrational energy level. In RS, these rotations and vibrations modulate the incident laser light producing side bands in the spectrum of the scattered light. These side bands, called the Raman spectrum, are shifted up or down in frequency from the laser frequency depending on whether the molecule is respectively left in a lower or higher energy level by the scattering event. The N_2 Raman spectrum consists of two groups of lines. One group arises out of transitions in which only the rotational level occupied by the molecule changes while in the other group the vibrational level changes. Since the rotational energy constant is ~2 cm^{-1}, the first group consists of two bands of lines close to the laser line and on either side of it. The first excited vibrational level is at 2331 cm^{-1}. Thus, the second group of lines is strongly shifted from the laser frequency. It also has bands on either side. The lines shifted down are called the Stokes spectrum, while those shifted up are called the anti-Stokes spectrum. For commonly measured temperatures, the anti-Stokes spectrum is the weaker of the two due to the lower population of the excited levels. Temperature determinations can be made from the ratio of the intensities at two frequencies in the Stokes vibrational band or in the Stokes rotational band, or from the ratio of an anti-Stokes band or a portion thereof to its partner in the Stokes spectrum.[1,4] A feature common to all such two-frequency measurements is the requirement that any laser-induced background signal be negligible. This restricts the measurements to non-combusting flows or to clean flames. Constant background signals, such as PMT dark count and ambient light, can be subtracted out with no difficulty provided they are small compared to the Raman signal.

A thorough review of the various Raman temperature measuring schemes has been given by Lapp.[1] Two criteria dominated the choice of measuring scheme for this work. These were that (a) the maximum possible Raman signal was needed and (b) the temperature or the concentrations of two gases had to be determined from two simultaneous intensity measurements. The Stokes rotational Raman spectrum of N_2 contains lines which are about an order of magnitude stronger than the vibrational band. Our experience shows this ratio to be between 8 and 12 at room temperature (for a 2.4 cm^{-1} spectral width). Simultaneous intensity measurements can be made in very short times using a diode array or an optical multichannel analyzer (OMA); however, these devices are about a factor of three lower in quantum efficiency compared to good PMT's. In addition, the rate at which these multichannel devices can record a weak spectrum, read out to a digital memory, and be cleared and made ready for another such measurement is limited to about 30 Hz. Thus, it was clear at the outset of this work that a spectrometric system was needed in which two closely-spaced rotational lines are simultaneously detected by two identical photon-counting channels using PMT's.

THEORETICAL ASPECTS

The theory of RS and the results of experimental studies, particularly for diatomic molecules, have been reported in great detail.[11-15] Only those features of the theory essential to the derivation of the working equations will be presented. The objective is to determine expressions relating temperature to the intensity ratio and temperature uncertainty to measurement parameters. The rotational energy levels are labeled by rotational quantum number J. Pure Stokes rotational Raman transitions of a diatomic molecule can occur only to levels where J increases by two.[11] Thus, expressions for given transitions are adequately identified by the initial J value of the transition.

The 90° scattering geometry was used in this work wherein the direction in which the scattered light is collected was at a right angle to the focused laser beam. The average number of photoelectrons per second $\dot{n}_J$ measured at the anode of a PMT during a given read time interval Δt due to light scattered into the collecting solid angle Ω resulting from a J → J + 2 rotational transition is given by

$$\dot{n}_J = \varepsilon P_e \sigma_J \Omega N_J \ell \tag{1}$$

The system detection efficiency factor is given by ε which includes spectrometer transmittance and the PMT quantum efficiency, P_e is the average incident photon rate of the exciting radiation during Δt, σ_J is the differential rotational Raman scattering cross section in cm^2/ster, N_J is the average number of molecules per cm^3 in the initial state J during Δt, and ℓ is the effective length in cm of the focused laser beam passed through the entrance slit of the spectrometer. The population density N_J in molecules/cm^3 is given by

$$N_J = N g_J (2J + 1) \exp\left[-(hc/k)E_J/T\right]/Q_r , \tag{2}$$

where N is the average molecular concentration in cm^{-3} during the read interval, g_J and $2J + 1$ are respectively the nuclear spin and the rotational degeneracy factors, E_J is the energy separation in cm^{-1} of the J level from the ground state (J = 0 level), T is the absolute temperature in kelvins (K), Q_r is the rotational partition function, and the factor hc/k contains the usual fundamental constants which give 1.4388 K/cm^{-1}. The energy separation is given to first order by[11]

$$E_J = B_v J(J + 1) , \tag{3}$$

where B_v is the rotational constant which is a weak function of the vibration quantum number v. For $^{14}N_2$,[11,13] $B_0 = 1.9895$ cm^{-1}.

The rotational partition function or state sum can be expressed by

$$Q_r = \sum_{J=0}^{\infty} g_J (2J+1) \exp\left[-(hc/k)B_o J(J+1)/T\right] \tag{4}$$

provided that the population of the excited vibrational levels (i.e., $v > 0$) is negligible. At $T = 1000$ K, only about 3.4% of the molecular density is associated with the first excited vibrational level ($v = 1$). The nuclear spin degeneracy g_J depends on the nuclear spin I of the molecule. For integral I, as for N_2 in which $I = 1$, g_J is equal to $(2I + 1)(I + 1)$ and $(2I + 1)I$ respectively for symmetric and antisymmetric rotational levels.[11] Even J states in N_2 are symmetric while odd J states are antisymmetric. This gives $g_J = 6$ or 3, which increases the signals obtained from the even J lines calculated via Eqs. (1) and (2) by a factor of two over the odd J lines. This fact means that generally the even J lines are preferred for measurements on the N_2 spectrum. Since $(hc/k)B_o/T \ll 1$ in Eq. (4) for N_2 with $T \gtrsim 300$ K, the summation can be replaced by an integral,[11] which gives

$$Q_r \approx (2I + 1)^2 T/(2B_o \, hc/k). \tag{5}$$

In addition to the T^{-1} dependence introduced by Eq. (5) into the observed count rate $\dot{n}_J$, there is a T^{-1} contribution from N via the kinetic theory given by

$$N = p/kT \text{ in m}^{-3}, \tag{6}$$

where p is the partial pressure of N_2 in Newtons/m^2 and k is the Boltzmann constant which is equal to 1.3806×10^{-23} J/K. The resulting T^{-2} factor, which is in contrast to the $\sim T^{-1}$ factor that is in the vibrational Raman expression due to Eq. (6), makes a significant contribution to the ultimate limit on the upper temperature that can be usefully measured under constant pressure using two pure rotational Stokes lines.

The scattering cross section σ_J depends on the polarizations of the incident and scattered light. In the scattering geometry used in this work the incident laser beam is plane polarized perpendicular to the plane formed by the laser beam direction and observation direction. The scattered light was collected by the spectrometer without passing through a polarizer or a depolarizer. However, the spectrometer transmittance depends on the incident polarization. Since the two rotational lines used are very closely spaced (24 cm^{-1}), this dependence is the same for both lines. Thus, for convenience, we will use the sum of the polarized component (i.e., parallel to the laser polarization) and the depolarized component (i.e., perpendicular to the laser polarization) for the cross section as expressed by[14]

$$\sigma_J = \frac{56\pi^4 (J + 1)(J + 2)}{15(2J + 1)(2J + 3)} \left[\nu_L - B_o(4J + 6) \right]^4 \gamma^2 \tag{7}$$

where ν_L is the laser frequency in cm^{-1} (i.e., the inverse of the laser wavelength) and γ is the anisotrophy of the molecular polarizability tensor in cm^3. The measured value of γ^2 for N_2 excited by 488 nm laser light is 0.518×10^{-48} cm$^6 \pm 8$ %.[15]

To determine temperature from the measured count rates of two rotational lines, let the J values of these two channels be given by J_A and J_B, where the B channel is taken as closest to the laser line. For ease of expression, we shall replace the subscript J by A or B from here on, as appropriate. The ratio of the counts accumulated during an accurately known read time interval Δt is obtained from Eq. (1) as

$$R = \frac{n_A}{n_B} = \frac{\dot{n}_A \Delta t}{\dot{n}_B \Delta t} = \frac{\varepsilon_A \sigma_A N_A}{\varepsilon_B \sigma_B N_B} \,. \tag{8}$$

The system efficiency factors are in general different since two different detectors are used. Using Eq. (2), the ratio in Eq. (8) can be expressed as

$$R = R_\varepsilon (C_A/C_B) \exp\left[(K_B - K_A)/T\right], \tag{9}$$

where $R_\varepsilon \equiv \varepsilon_A/\varepsilon_B$, and

$$C_A = g_A N B_o \sigma_A (2J_A + 1)/(2I + 1)^2. \tag{10}$$

From Eq. (3)

$$K_A = 1.4388 \, B_o J_A (J_A + 1). \tag{11}$$

The equations for C_B and K_B are the same as Eqs. (10) and (11) with A replaced by B. From Eq. (9), the temperature is given by

$$T = (K_B - K_A)/\ln\left[(C_B/C_A)(R/R_\varepsilon)\right]. \tag{12}$$

The assumptions usually made in using Eq. (12) are that there are negligible rotational transitions in the excited vibrational states, that we can neglect higher order contributions to B_v in Eq. (3), and that the factor in Eq. (7) involving the fourth power of the frequency is practically unchanged when going from channel A to channel B, thereby making the ratio of these factors in C_B/C_A essentially unity. The K and C- factors in Eq. (12) for $J_A = 7$ and $J_B = 4$ for N_2 are

$$\left.\begin{array}{l} K_B - K_A = -103.05 \text{ K} \\ \text{and} \\ C_B/C_A = 1.2879 \,. \end{array}\right\} \tag{13}$$

The efficiency ratio R_ε is effectively a gain factor which must be experimentally determined. It provides an adjustable parameter for calibration purposes. A plot of Eq. (12) using the values in Eq. (13) with $R_\varepsilon = 1$ is given in Fig. 1. There is an important

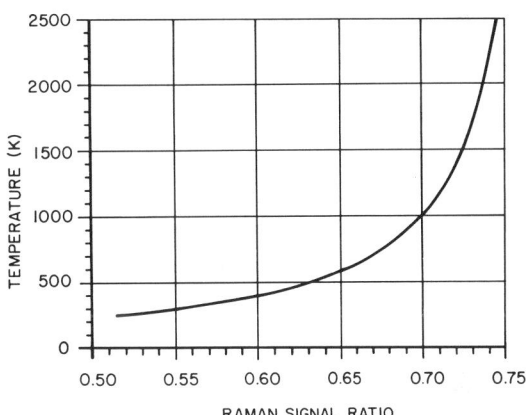

FIG. 1 Temperature calibration curve for the intensity ratio of the $J = 7$ to $J = 4$ Stokes rotational Raman lines of N_2 (see Fig. 7).

feature in this graph to be noted. For a temperature range of 300 to 2500 K, R varies from 0.55 to 0.745; however, 25% of the upper range of R contains 70% of the temperature range. Hence, measurements above about 1000 K will rapidly become less precise (for a fixed uncertainty in R) than measurements made at lower temperatures.

The fractional (or relative) uncertainty in a single measurement of temperature using Eq. (12), in the approximation where the various relative uncertainties are not too large (say, ~20%), is given by

$$s_T/T \approx \left\{ \left[s_K/(K_B - K_A)\right]^2 + \left[(s_\varepsilon/R_\varepsilon)^2 + (s_R/R)^2\right] \times \left[T/(K_B - K_A)\right]^2 \right\}^{\frac{1}{2}}, \tag{14}$$

where s_T, s_K, s_ε and s_R are respectively the standard deviations in T, $K_B - K_A$, R_ε and R. Any uncertainty in C_B/C_A can be absorbed into s_ε. Since $K_B - K_A$ and R_ε are fixed values during any series of measurements, their uncertainties affect the accuracy of the measurements while s_R determines the precision of each measurement. Typical values for the three fractional uncertainties in Eq. (14) associated with s_K, s_ε and s_R are 0.6%, 0.2% and 4 to 12%, respectively. The value for s_ε depends on the ability to determine an operational value for the gain factor R_ε which effectively compensates for the uncertainties in the other factors. Generally, the values for s_R dominate the contributions to s_T. Thus, the measurement uncertainty of primary interest becomes, from Eq. (14),

$$s_T \approx (s_R/R)T^2 / |K_B - K_A| . \qquad (15)$$

From Eq. (8) and assuming Poisson statistics, the value of s_R/R is given by

$$s_R/R \approx \left[(\dot{n}_A^{-1} + \dot{n}_B^{-1}) f_r \right]^{1/2} , \qquad (16)$$

or

$$s_R/R \approx \left[(1 + R) f_r / \dot{n}_A \right]^{1/2} , \qquad (17)$$

where the read rate $f_r = 1/\Delta t$. The primary temperature dependence in Eq. (17) is due to the T^{-2} factor in $\dot{n}_A$ which arises from Eqs. (5) and (6), the latter due to the assumption of constant pressure. Thus, this makes s_R/R roughly proportional to T and gives s_T approximately a cubic dependence on temperature. This clearly makes the upper limit on the temperature that can be measured using this technique severely dependent on the available signal. For example, if $\dot{n}_A = \dot{n}_B = 40,000$ counts/s at 300 K, then at $T = 600$ K, these signals will be about 10,000 counts/s. Using $J_A = 7$ and $J_B = 4$ with $f_r = 10$ Hz in Eq. (16), we obtain $s_T \approx 150$ K for N_2. Although this 25% standard deviation appears large, it corresponds to the uncertainty in a single measurement. We make 1024 to 4096 such measurements in a single time record which significantly reduces the uncertainty in the mean temperature. In the determination of pdf's from these data records, the main requirement is that the recorded temperature fluctuations are distributed over a range clearly greater than the uncertainty given by Eq. (15).

The form of the pdf that can be expected can be obtained by assuming a Gaussian distribution for the molecular concentration N and then doing a coordinate transformation using Eq. (6).[17] The resulting temperature pdf is given in Fig. 2 for a 40% standard

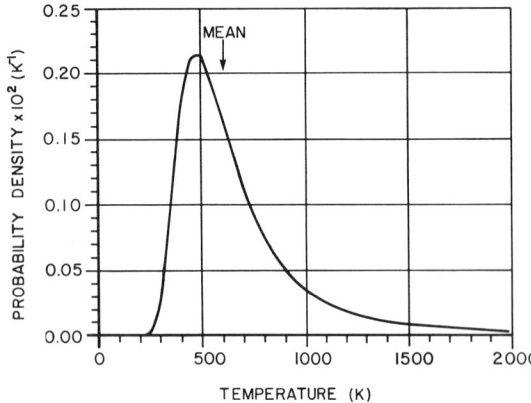

FIG. 2 The temperature probability density function for an ideal gas at constant pressure having a Gaussian concentration distribution with a 40% relative standard deviation and a mean temperature of 600 K.

deviation in density and a mean temperature of 600 K. This highly skewed distribution could be easily traced out via a histrogram representation with temperature bin widths of 150 K using the measurement conditions given in the previous paragraph and 1024 data values.

EXPERIMENTAL APPROACH

Several concerns influenced our choice for the design of a two-channel spectrometer. Equation (15) shows that the larger the difference $K_B - K_A$, the smaller the uncertainty in the signal. This suggests (from Eqs. (3) or (11)) that the two channels should be as widely spaced as possible. However, for a given temperature, too large a spacing will result in a weak signal in channel A (viz. assuming that channel B is situated on one of the stronger lines). The spacing of pure rotational Raman lines in the spectra of N_2, O_2 and CO_2 is respectively 8.0, 11.5 and 3.15 cm^{-1}.[2] For excitation at 488.0 nm, these values correspond respectively to 0.19, 0.274 and 0.075 nm, respectively. The interest here in CO_2 is for the cold and non-combusting heated flow experiments planned for the future. These close spacings and the need to be able to select specific lines require that the two-channel spectrometer have resolution capabilities essentially equal to the performance of the moderate resolution single-channel instruments commonly used.

This latter requirement and the desire to keep the design as simple as possible, led us to an arrangement in which the exit slit image of a SPEX model 1401 double spectrometer was split into two channels by an external optical subsystem. This subsystem is shown in Fig. 3. Two cylindrical lenses, L_3 and L_4, are used to

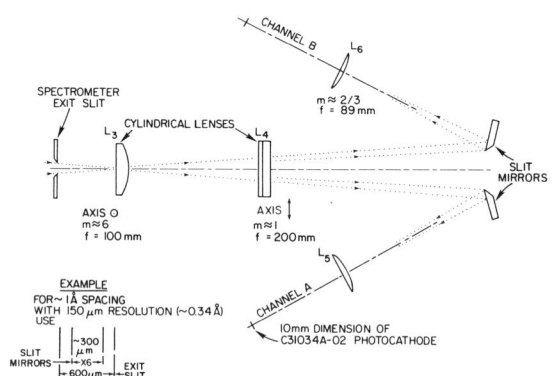

FIG. 3 Exit slit optical subsystem of the two-channel spectrometer.

alter the aspect ratio of the exit slit image. The first lens, L_3, magnifies the wavelength (horizontal) dimension, while L_4 projects the vertical dimension unmagnified. This permits the exit image to be intercepted by two mirrors with slit-like edges. As shown in the example given in the figure, the combination of the spectrometer exit slit, and the positions of the independently adjustable slit mirrors permits two well-defined wavelength channels to be sent to the photocathodes of the PMT's. The definition or resolution of these two channels is limited by the quality of the cylindrical lenses and the quality of the image provided by the spectrograph. Currently, with single element lenses, the throughput noticeably drops around 50 µm of equivalent slit width for spacings of about 0.6 nm using second order gratings.

In general, this design has proven functional and easily adjusted to the desired settings. In addition, it lends itself to future expansion. For example, a third channel can easily be added between the two slit mirrors. This channel could be used for monitoring a second species during temperature measurements on N_2 or the background signal. Another modification which we are implementing is to design what are essentially masks for each slit mirror to permit two or more lines of a given species to be selected for each PMT. This will greatly enhance our signals.

The placement of the two-channel optical subsystem and the arrangement of the complete optical system are

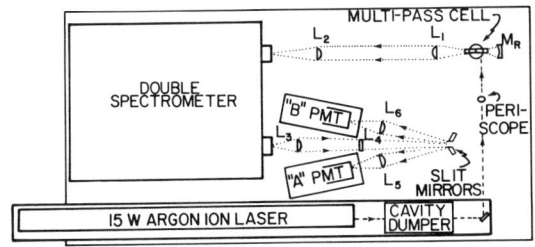

FIG. 4 Layout of the two-channel Raman spectroscopy system.

shown in Fig. 4. The laser is a Spectra Physics model 171-15. A companion model 344 cavity dumper is available for producing a pulsed laser beam so as to permit gated detection. Gated detection can provide significant reduction of background signals; however, the maximum average laser power available for the measurements is about 75% of that available without the dumper, viz, when operating cw. The preliminary results given in this paper were all obtained without the dumper. Generally, the cw beam power used was 4 W at 488.0 nm. This wavelength output was chosen based on the observation of strong C_2 spectra in the Stokes rotational region when 514.5 nm excitation was used on a flame.

The laser beam is translated vertically by the periscope to permit injection of the beam in to the bottom of the multiple-pass optical cell. The layout of this optical subsystem, shown in detail in Fig. 5,

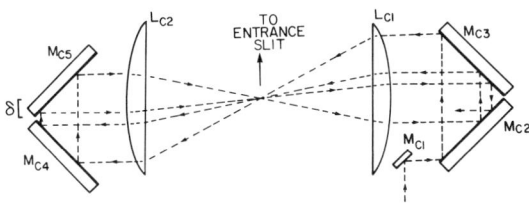

FIG. 5 Configuration of the optical multiple-pass cell.

was taken from a design by Hill.[18] As shown in Fig. 4, the multipass cell is aligned vertically so that the edge of the "sheet" of light formed by the multiple passes of the laser beam is parallel to the spectrometer entrance slit. The final adjustment involves the translation δ of the upper mirrors to maximize the signal observed from the crossing region. The gain observed over a single pass is about 12 to 13. It is limited primarily by the aberrations produced by the single element lenses due to the relatively large off-axis displacement of many of the beams. This displacement is brought about by the roughly 3-mm-diameter size of the laser beam at the injection mirror M_{C1}. This condition causes the minimum dimension in the collecting direction (i.e., the X-axis) to be about 0.5 mm in the crossing region. The dimension parallel to the slit width (i.e., the Z-axis) is determined by the waist of the focused laser beams. For the 140-mm focal length lenses (i.e., L_{C1} and L_{C2}), this waist is 40 μm. Collecting lens L_1 and imaging lens L_2 in the current configuration give a magnification of -1.2. (This makes the collecting solid angle equivalent to an f/no. of 7.6.) Generally, the entrance slit is made wider than the magnified beam waist dimension to reduce effects due to slight beam displacements resulting from alignment drifts and changes in the refractive index in the crossing region. The remaining dimension (i.e., the Y-axis of the sample volume) is determined by the height of entrance slit aperture divided by the magnification due to L_1 and L_2. The preliminary measurements reported here used a slit height of 2 mm, which corresponds to a 1.7 mm dimension in the sample volume.

A schematic diagram of the actual placement of the optical elements relative to the gas nozzle is given in Fig. 6. The nozzle is fitted to a straight stainless steel tube of 10.8 mm ID and 75 cm long. It can be translated with a fixed orientation of its center line in three dimensions by three independent sets of stages. The horizontal flow arrangement produces an asymmetry due to bouyancy effects; however, this feature is outweighed by the considerable convenience and optical simplicity of the design in Fig. 6. The flame measurements reported here were made with a 1.4 mm diameter nozzle. Ultra pure methane was used without premixing. The high purity was used to insure that there would be minimum background signals from fluorescing impurities. Methane was chosen because it is a spherical top molecule and therefore has no pure rotational Raman spectrum (at these pressures).[14] To remove the combustion products and to stabilize the flame, a 50-mm-diameter tube set approximately axial to

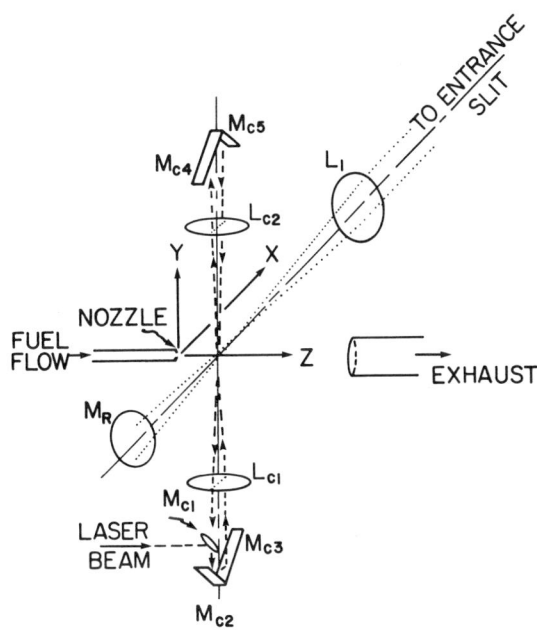

FIG. 6 Schematic representation of the laser excitation and the Raman collection geometries relative to the gas nozzle.

the nozzle and about 90 mm from the nozzle was connected to a small centrifugal blower having an exhaust orifice of the same diameter. Also shown in Fig. 6 is the reverse optics spherical concave mirror M_R. This mirror returns the light scattered in the opposite direction from the spectrometer back through the sample volume to the spectrometer. We observe a 90% increase in the Raman signal resulting from the use of this mirror.

The electronics system consists of two essentially identical photon counting channels. The detectors are two RCA C31034A-02 PMT's. These tubes are specified to have a quantum efficiency of about 25% at 488.0 nm and a dark count of about 12 counts/sec. The rise time of the anode pulses are observed to be less than 2 ns. Parallel digital lines carry a four decade count to a dedicated minicomputer from each counter. A crystal-controlled timer sends an interrupt pulse to the minicomputer at the end of each read interval. The minicomputer reads the counters, resets them and restarts the timer. This is all accomplished in less than 1 μs. Up to 4096 values can be recorded in each of the two channels. The data are recorded on floppy disk for processing. The observed pdf of a single channel or of temperature can be calculated and plotted using the minicomputer. The data can also be transferred via a 4800 Baud line to a DEC VAX 11/780 computer for computing psdf's and other more involved analyses. A Tektronics model 4010-1 terminal provides for computer communications and graphics output.

In addition to the digital signal acquisition capability, additional outputs from the discriminators can simultaneously provide signals to two specially constructed frequency-to-voltage converters. These converters are essentially high-speed ratemeters capable of converting high-count-rate signals to analog voltages with up to a 2 kHz bandwidth. These analog signals can be analyzed using a spectrum analyzer or a two-channel analog correlator. These analog signals and the analog instruments are very useful in the check out and set up activities that precede a measurement series. The analog signals, for example, are used to scan a given spectral line with each channel so as to permit equalizing the spectral bandwidths of the two channels. This is accomplished by very slight displacements in the transverse position (i.e., to the axis) of lens L_3 in the plane of Fig. 3.

PRELIMINARY RESULTS

The observed Stokes rotational Raman spectrum of air applicable to the measurements reported here is given in Fig. 7. The currently used spectrometer

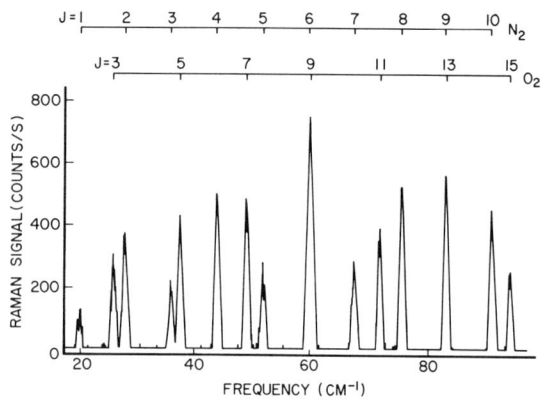

FIG. 7 Observed rotational Stokes Raman spectrum of one atmosphere of room temperature air for a single pass of 2 W of the 488 nm laser beam. Right angle geometry was used with perpendicular polarization. The input image was slit limited where the slits were 80 μm (0.77 cm^{-1}) wide by 10 mm high.

gratings have 1200 g/mm and are blazed at 500 nm in second order. This gives 0.23 nm/mm at the exit slit. The maximum width of this slit is 3 mm which limits the channel separation to less than 0.69 nm or 29 cm^{-1}. From Fig. 7, this limits the maximum value of $J_A - J_B$ to 3 for N_2. The strongest pure N_2 lines in Fig. 7 are $J = 4$ and $J = 8$. Extensive measurements were made using the $J_A = 7$ and $J_B = 4$ lines; however, the maximum signal observed at room temperature was about 20,000 counts/sec in channel B. Our primary goal is to obtain the psdf of temperature to as high a frequency as possible. For $f_r = 100$ reads/sec, the bandwidth B is 50 Hz. Our experience to date indicates that it is desirable to have about 200 counts/read on the average, which gives for the above read rate, an observed average count rate of 20,000 counts/sec. Thus, the signals for the above channel choices are too small for B = 50 Hz for temperatures much above 300 K. Ultimately, this limitation will be alleviated by a modification of the exit slit and the use of masks at the slit mirrors to permit, for example, J = 7 and 8 to be used for channel A and J = 4 and 5 for channel B.

The two strongest lines in Fig. 7 are the mixed lines wherein there are accidental near coincidences between the J = 6 line of N_2 and the J = 9 line of O_2 and J = 9 of N_2 with J = 13 of O_2. At room temperature, the former mixed line gives about 38,000 counts/sec. Since the observed count rate is roughly inversely proportional to T^2, a satisfactory measurement out to 50 Hz can be achieved for average temperatures around 425 K. The main difficulty with using these lines with a flame is the loss of O_2 concentration as the sample volume penetrates the flame. For measurements made using these lines to be viable, changes in the O_2 to N_2 concentration ratio must have a negligible affect on the measured temperatures.

Referring to Eqs. (8) and (9), the ratio for the mixed case can be written as

$$R_m = \frac{n'_A + n''_A}{n'_B + n''_B} = R' \left[\frac{1 + n''_A/n'_A}{1 + n''_B/n'_B} \right] \quad (18)$$

where the single and double primes denote respectively N_2 and O_2. The quantity in the brackets can be calculated using Eqs. (10) and (11). The functional dependence of temperature on R_m is given in Fig. 8. In comparison to the pure case given in Fig. 1, Fig. 8 shows that the ratio covers twice the range for temperatures between 300 and 2500 K. Thus, not only is

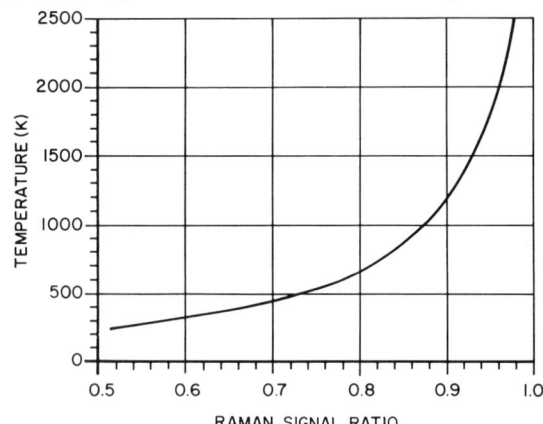

FIG. 8 Temperature calibration curve for the intensity ratio of the 83.7 $-cm^{-1}$ line to the 60.1 $-cm^{-1}$ line in the Stokes rotational Raman spectrum of air (see Fig. 7).

the available signal almost double, but the sensitivity to temperature changes is doubled compared to that obtained using the $J = 4$ and 7 lines of N_2. The problem of oxygen consumption in measurements in a flame is offset to some extent by the fact that the precision of the measurements decreases rapidly above 1000 K as in the pure case. Thus, the loss of accuracy due to the reduction in O_2 to N_2 ratio tends to be masked by the loss of precision in regions of high temperature. As the O_2 concentration is reduced, the calibration function tends to approach the curve given in Fig. 1, which means that for a particular observed value of R, the calculated temperature tends to be too low. For the results presented in this paper, the measurements were confined to the edge of the flame where the average temperature was less than 500 K.

The expression for temperature from Eq. (18) is not as simply derivable as for the pure case given by Eq. (12). An approximation to Eq. (18) giving the temperature as a function of R_m was determined by fitting the quotient of two polynominals in R_m to the exact equation of R as a function of temperature obtained from Eq. (18). The resulting equation is given by

$$T = (aR_m^2 - bR_m - c)/(R_m - d) \quad (19)$$

where $a = 28.94$, $b = 153.04$, $c = 63.94$, $d = 1.052442$, and T is in Kelvins. This equation is accurate to within ±0.13% over the range of 250 to 2500 K. The precision uncertainty in the temperature determined using the mixed lines is given by

$$s_{T_m} = (\partial T/\partial R_m) s_R , \qquad (20)$$

where the derivative is most easily obtained from a table of values of T and R_m, and where s_R is given by Eqs. (16) or (17).

Direct calibration checks were made on both the pure and mixed line cases using a No. 20 beaded chromel-alumel thermocouple mounted within 2 mm of the downstream side of the Raman sample volume on the Z-axis (see Fig. 6). Both a Keithley model 870 digital thermometer and a type 8686 Leeds and Northrup potentiometer were used to determine the thermocouple temperature. The pure and mixed-line Raman temperatures and the thermocouple values usually agreed within 3 - 4% between 300 and 450 K. These checks were made using electrically heated air flowing in a 0.5 in. OD (~1 cm ID) stainless steel tube 75 cm long fitted with optical access openings. The flow rate was 2 to 4 m/s. The observed relative standard deviations (STD's) (i.e., the square root of the mean of the squares minus the square of the mean) in the temperature data ranged between 10% at room temperature to 20% at the high temperature for the pure line case and from 4% to 7% for the mixed-line case. These data were all taken at a read rate of 10 Hz with 1024 reads. The uncertainties are essentially equal to the values calculated with Eq. (15) indicating that there was no detectable system noise or observable fluctuations in the temperature of the air.

The main thrust of the preliminary studies carried out with this system has been to analyze real-time records of the two Raman signals. The advantage of this approach is that the data are always available for further analyses. The disadvantage is that there is a limit on the size of any one record that can be recorded (4096 per channel, in our case) and real time recording requires adequate signal quality to insure the utility of the data. In our case, this requirement means that the Poisson uncertainty in the temperature (viz, that which results from Eq. (16)) must be sufficiently small to permit observation of the contributions due to the fluctuations in the flame or the flowing gas. The two representations of these fluctuations of interest here are the pdf's and the psdf's.

Generally, it is desirable to record the data at as high a rate as is consistent with the available signal and the required temperature resolution. This resolution is basically set by the signal count rates via Poisson statistics which reduces to a "Poisson temperature uncertainty" determined by using Eq. (16) in Eq. (15) or (20) as appropriate. For the pdf representation, this Poisson uncertainty determines the minimum temperature bin width in the histographic form of the pdf. Thus, if the original data records are analyzed without modification, the Poisson uncertainty associated with each measurement or "read" will be maximum, thereby requiring a rather low resolution graph. However, the contribution to the pdf due to fast changes will be observable. In our case, this corresponds to excursions to large temperatures. Since the rotation temperature measurement approach becomes rapidly inaccurate with increasing temperature, particularly above 1000 K, we have found it necessary to clip these excessive values at a temperature more reasonable for the methane-air flame studied here (viz., ~1800 K).

An example of such a pdf is given in Fig 9, for a methane diffusion flame from a 1.4 mm diameter nozzle and a methane flow rate of nearly 20 m/s. The cold flow Reynolds number (R_e) was about 1600. The flame extended several centimeters into the exhaust pipe where it was a bright orange. The upstream portion of the flame was blue except for a ridge of orange along the top edge. The Raman sample volume was off to the side along the X-axis 3 diameters and 2-1/2 diameters downstream from the nozzle along the Z-axis. This

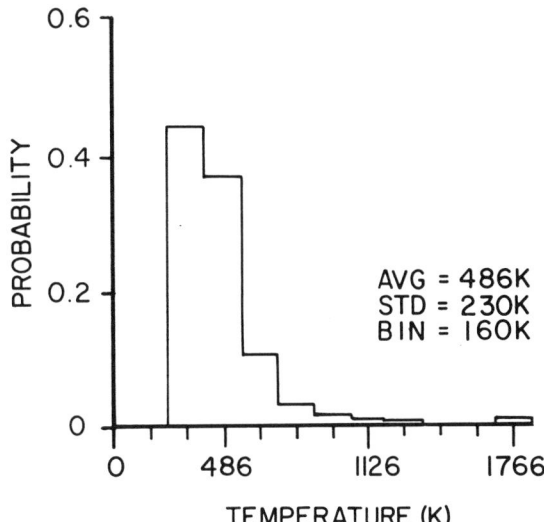

FIG. 9 Histographic representation of the probability density function of 4096 measurements of temperature near the edge of a methane diffusion flame with R_e = 1600 and a read rate of 100 Hz. Intermittent, high temperature values (32) were clipped at 1800 K. The Poisson standard deviation is 125 K. STD is the square root of the mean of the squares minus the square of the mean

position was chosen for the signal level of about 20,000 counts/sec which permitted a meaningful psdf analysis and the desire to obtain a reasonable steady flame boundary. The background level was 500 to 800 counts/sec.

The data for Fig. 9 were obtained at a read rate of 100 Hz for 4096 reads using the mixed lines. The bin widths in this pdf as well as the one shown in Fig. 10 are about 20% larger than the Poisson temperature uncertainty. Although this represents a rather coarse resolution in Fig. 9, this plot shows that apparently high-temperature excursions (possibly due to intermittency) occurred in 1 - 2% of the reads. These reads produce a tail on a highly skewed distribution which distinctly differs in character from the distribution given in Fig. 10. Figure 10 is the

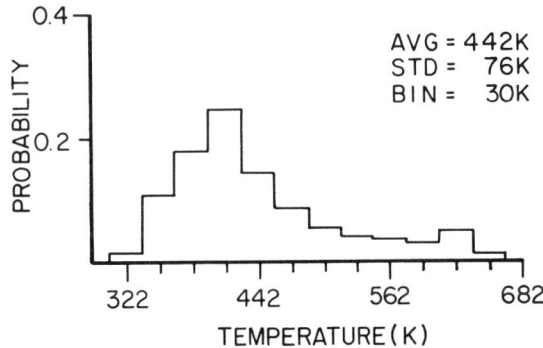

FIG. 10 Histographic representation of the same data used in Fig. 9 wherein the effective read rate was reduced to 6 1/4 Hz giving 256 measurements. The Poisson standard deviation is 26 K.

distribution that results from collecting the data in sets of 16 consecutive reads so as to create a new record containing only 256 reads. This increases the precision of each read by a factor of four, thereby permitting a significantly higher resolution pdf. This manipulation of the data removes all excursions above about 650 K. In effect, Fig. 10 is a high-resolution view of the three most populated bins in Fig. 9.

The determination of the psdf for an existing data record is most easily accomplished by performing a fast Fourier transform on the record and then calculating the square of the magnitude of the transform function. The number of frequency values in the psdf is half of the number in the original data record due to the need to include negative as well as positive frequencies in the calculation. As already discussed, the bandwidth B is half of the read rate. Thus, for the data used in Fig. 9, there will be 2048 points covering 50 Hz in the psdf. This resolution is much higher than needed. In effect, this says that the original record is much longer than is needed to provide adequate frequency resolution.

A common procedure in situations of this type is to divide the record into a number of segments, calculate the psdf for each segment and find the average psdf.[18] In the psdf's presented here, 15 overlapping segments of 512 points each were used. The segments were modified by a Parzen temporal window to reduce the frequency ringing brought on by the segmenting procedure.[18] This window was found to minimize the noise level in the frequency domain; however, the essential features of the spectra do not depend on the choice of window. In addition, the final average spectra were smoothed by convolving with a 1-Hz cosine frequency filter.

In the temperature psdf domain, the Poisson uncertainty introduced by the fluctuations in the photon flux contributes a white noise spectral component. The value of this constant spectral density is equal to the square of Poisson uncertainty divided by the bandwidth or s_T^2/B. This spectral density level represents a detection limit in the measurement of temporal frequency characteristics of observed fluctuations. Figure 11 gives the temperature psdf for

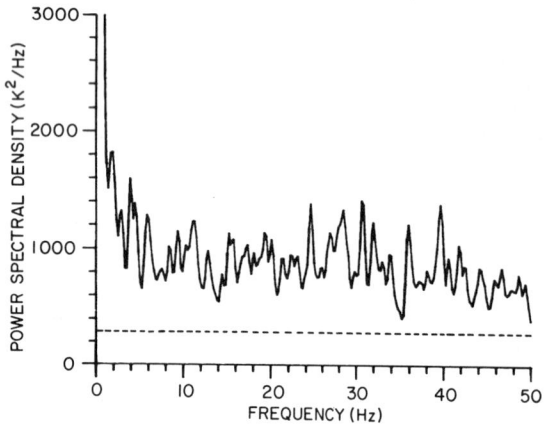

FIG. 11 Power spectral density function of temperature for the data used in Fig. 9. The curve is the average of the spectra obtained from 15 overlapping segments of the time record of 512 points each. It has been smoothed by convolving with a 1-Hz cosine filter. The dashed line is the spectral density level due to the Poisson uncertainty.

the same data used with Fig. 9. The Poisson limit in Fig. 11 is about 300 K^2/Hz as shown by the dashed line. Thus, over half of the spectral density in Fig. 11 can be attributed to fluctuations in the flame. The fine noise structure exhibits the shape of the 1-Hz filter function. The very large density peak below 1 Hz is due to slow meanderings of the flame, while the slow decrease in the spectral density with increasing frequency can be assumed to be a characteristic property of the flame fluctuations.

The problem of aliasing in our psdf's is reduced due to the measuring scheme used here, wherein each temporal measurement is, in fact, a time-averaged value over the very short time interval Δt. This essentially attenuates the contributions of frequencies higher than f_r by at least a factor of f_r/f.

A number of experiments were performed to determine the performance of the system in obtaining psdf's. The laser power was amplitude modulated with a sine wave at about a 17% modulation level using the "light control" circuit of the laser. Although this signal strongly appeared in the psdf of each channel taken individually, there was no hint of the signal in the temperature psdf. This indicated that the two channels were behaving identically throughout the entire signal acquisition and processing procedure.

In an attempt to introduce a distinctive feature into the temperature psdf, a 12-in., 20-W Jensen high-fidelity speaker was set up about 30 cm from the nozzle. Referring to Fig. 6, the speaker was centered directly above the reverse optics mirror M_R and directed parallel to the X axis. A sine wave between 20 and 80 Hz input to the speaker at about the 1 W level could not be detected in the temperature psdf of flowing hot air from the 1.4-mm-diameter nozzle. However, with the introduction of the methane flame, the acoustical signal caused a slight flutter along the orange ridge at the top of the flame and a significant amplitude in the psdf. A sine wave modulation, although clearly evident in the psdf, nevertheless produces only a sharp line spectrum. A more dramatic effect was obtained by linearly sweeping the frequency of the audio oscillator during the data acquisition sequence. The resulting psdf, obtained in the same manner as Fig. 11, is shown in Fig. 12. The sweep was

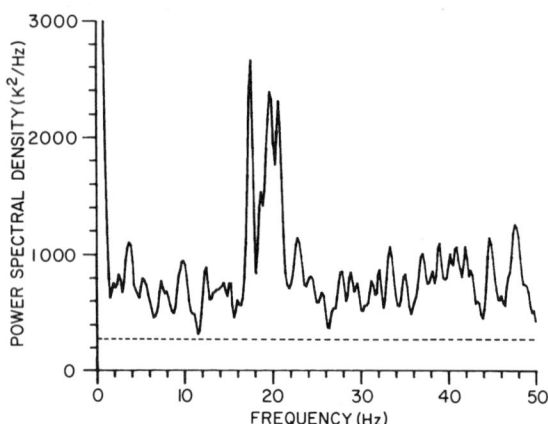

FIG. 12 Power spectral density function of temperature for the same flame conditions used in obtaining the data presented in Figs. 9-11 except with acoustical excitation which consisted of a linear sweep of the sine frequency between 17 and 22 Hz. See text for additional details. The dashed line is the Poisson spectral density.

between approximately 17 and 22 Hz with a speaker input power of 3 W. The measured rms pressure amplitude at the nozzle was about 1 N/m^2. The variance associated with the 1-Hz-wide peak at 18 Hz in Fig. 12 corresponds to an apparent temperature fluctuation amplitude of about 40 K rms. This suggests that for this case a .20 K rms fluctuation is just detectable.

The time record of the data used to obtain Fig 12 is given in Fig. 13. The peaks in the 17 to 22 Hz range in Fig. 12 are correlated with the occurrence of sustained high temperature at 5, 24, and 37 sec. while the minima correlate with sustained low temperatures around 10 and 30 sec. This behavior appears to be qualitatively consistent with adiabatic flame calculations which show that small number density changes near the maximum flame temperature (equivalence ratio ≈ 1) are associated with the largest temperature changes.[5,8] This observation suggests that the modulation due to an acoustical pressure wave can be used to probe the mixing activity in the flame in regions that are near stoichiometric conditions.

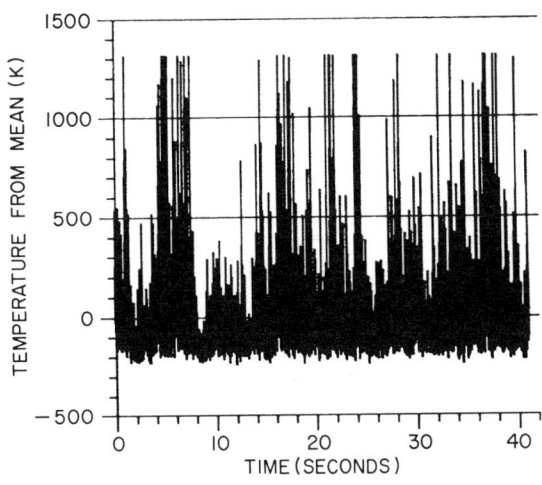

FIG. 13 The time record of the temperature fluctuations of the data used in Fig. 12.

The pdf's obtained from the data used for Fig. 12 were very similar to those given in Figs. 9 and 10. There appeared to be a slight reduction in the average temperature as well as in the spectral density in Fig. 12 with the introduction of acoustical modulation. The time record given in Fig. 13 is very typical in that it shows the very large range of the temperature fluctuations usually seen with the higher read rates. The effect of clipping is also clearly evident. The average count rate of the data in Fig. 13 was about 22,000 counts/s.

SUMMARY

It has been shown that the pure Stokes rotational Raman lines of air can be used to measure the temperature pdf's and psdf's of flowing hot air and the outer boundaries of a methane flame. The accuracy of these measurements drops off with the square of the temperature but increases with the square root of the count rate. Thus, the limitation on temperature measurements depends heavily on the specific circumstance. Furthermore, accurate measurements can almost always be accomplished if the read rate is reduced at the cost of lost information on intermittent or rapid excursions in temperature. The final criterion is the Poisson uncertainty associated with each measurement. Basically, this value determines the bin width in the pdf. It also sets the white noise level in psdf due to the intrinsic fluctuations of the detected signals. Thus, the key to meaningful data is to reduce the Poisson uncertainty sufficiently to permit detection of the fluctuations in the sample volume due to the flow and/or flame. The data presented here show a significant level of sample fluctuations. The frequency limit in the psdf with the current operating parameters is roughly 100 Hz or a 200 Hz read rate. The calculated signal level for this system is at least a factor of ten larger than the observed signals. Thus, if the operating efficiency of the system can be improved significantly, then frequencies out to at least 300 Hz should be observable.

It is clear that the successful application of this technique to characterization of temperature fluctuations of flames and flowing gases is dependent on the intensities of the detected signal. The extension of this two-channel system into other measurement regimes will ultimately make it possible to measure fluctuations with virtually any mean temperature. In addition to the analyses of real-time records as described in this paper, pdf's can be obtained in a continuous manner using essentially a multi-channel scaling approach. Likewise, psdf's can be found from the Fourier transform of the autocorrelation function. The autocorrelation function can be measured in a continuous "on line" manner where the only limit is the degree of steadiness in the probed medium over an extended time period.[6] These latter two approaches are especially applicable in cases where the signals are very weak. Such would be the case when measuring high temperatures using the ratio of the first hot band intensity to the intensity of the ground state band of the vibrational spectrum of N_2. We have successfully used the current two-channel system in preliminary measurements on the vibrational spectrum of N_2. The main difficulty is to properly account for or remove the background signal due to the flame. The software for this and the other schemes mentioned above is being developed.

Finally, the capability of measuring real-time fluctuations in temperature with an optical probe has permitted us to observe what may be the local mixing activity by means of the modulation of the medium due to an externally-produced acoustical wave. Several new experiments are suggested by this observation. The most obvious experiment is to measure the temperature fluctuation amplitude due to the acoustical modulation using the lock-in technique. This amplitude can be measured as a function of either the position in the flame or the flame parameters. In addition, the acoustical frequency can be scanned to study the temporal development of the flame or it can be adjusted so as to stroboscopically select a specific temporal feature in the flow.

ACKNOWLEDGMENTS

The authors would like to acknowledge the encouragement and help of Dr. W. M. Roquemore, Project Engineer for the Aero Propulsion Laboratory, W-PAFB, and of Dr. E. H. Gerber, Project Manager for the University of Dayton Research Institute. Also, we would like to thank student assistants Mr. Mike Gallis, Mr. Mike Horn and Mr. Harold Norris for their many valuable efforts.

REFERENCES

a) Supported by the Aero Propulsion Laboratory, W-PAFB, through contract no. F33615-78-C-2005 with the Research Institute of the University of Dayton.
b) Member of the Research Institute staff.
1. M. Lapp, in Laser Raman Gas Diagnostics, edited by Marshall Lapp and C. M. Penney (Plenum, New York, NY, pp. 107-145.
2. R. E. Setchell, in Combustion Measurements, edited by R. Goulard Hemisphere Publishing, Washington, D. C., 1976), pp. 211-233.
3. W. M. Roquemore and P. P. Yaney, in Proceedings of the 10th Materials Research Symposium, on Characterization of High Temperature Vapors in Gases, edited by John W. Hastie (National Bureau of Standards Special Publication 561/1, Super. of Doc., U. S. Government Printing Office, Washington, D. C., 1979), pp. 973-1025.
4. P. P. Yaney, R. J. Becker, T. H. Hemmer, and W. M. Roquemore, in Proceedings of the International Conference on Lasers '79, edited by Vincent J. Corcoran (STS Press, McLean, VA, 1980), pp. 88-97.
5. M. Lapp and C. M. Penney, in Proceedings of the Dynamic Flow Conference 1978, edited by Bengt W. Hansen (Sijthoff and Noordhoff, Alphen aan den Rijn, The Netherlands, 1979), pp. 665-683; M. Lapp, in Laser Probes for Combustion Chemistry, edited by David R. Crosley (American Chemical Society, Washington, D. C., 1980), pp. 207-230; M. C. Drake, M. Lapp, C. M. Penney, and S. Warshaw, in Proceedings of the AIAA 19th Aerospace Sciences Meeting, Paper AIAA-81-0103 (American Institute of Aeronautics and Astronautics, New York, NY, 1981).
6. C. M. Penney, S. Warshaw, M. Lapp, and M. Drake, in Laser Probes for Combustion Chemistry, edited by David R. Crosley (American Chemical

Society, Washington, D. C., 1980) pp. 247-253.
7. J. S. Bendat and A. G. Piersol, *Random Data: Analysis and Measurement Procedures* (Wiley-Interscience, New York, NY, 1971).
8. L. P. Goss, G. L. Switzer, D. D. Trump, and P. W. Schreiber, in Proceedings of the AIAA 10th Aerospace sciences Meeting, Paper AIAA-82-0240 (American Institute of Aeronautics and Astronautics, New York, NY, 1982).
9. A. D. Birch, D. R. Brown, M. G. Dodson, and J. R. Thomas in *Seventeenth (International) Symposium on Combustion*, Leeds, England, August 20-25, 1978, (Organized by the Combustion Institute, Pittsburgh, PA, 1978).
10. Ilan Chabay, G. J. Rosasco, and T. Kashiwagi, J. Chem Phys. $\underline{709}$ (9), 4149 (1979).
11. Gerhard Herzberg, *Molecular Spectra and Molecular Structure*, Vol. 1, *Spectra of Diatomic Molecules*, (Van Nostrand Reinhold Co., New York, NY, 1950).
12. S. Broderson, in *Raman Spectroscopy of Gases and Liquids*, edited by A. Weber, (Springer-Verlag, New York, NY 1979), pp. 7-70.
13. A. Weber, in *Raman Spectroscopy of Gases and Liquids*, edited by A. Weber, (Springer-Verlag, New York, NY, 1979), pp. 71-122.
14. H. W. Schrotter and H. W. Klockner, in *Raman Spectroscopy of Gases and Liquids*, edited by A. Weber (Springer-Verlag, New York, NY, 1979), pp. 123-202.
15. C. M. Penney, R. L. St. Peters, and M. Lapp, J. Opt. Soc. Am. $\underline{64}$ (5), 712 (1974).
16. Chris P. Tsokas, *Probability Distributions: An Introduction to Probability Theory with Applications*, (Wadsworth Publishing Co., Belmont, CA, 1972) Chapters 4 and 6.
17. R. A. Hill, A. J. Mulac, and C. E. Hackett, Appl. Optics $\underline{16}$ (7), 2004 (1977).
18. P. D. Welch, in *Digital Signal Processing*, edited by Lawrence R. Rabiner and Charles R. Rader, (IEEE Press, New York, NY, 1972), pp. 335-338.

Laser tomography for temperature measurements in flames

H. G. Semerjian and R. J. Santoro

National Bureau of Standards, Washington, D.C. 20234

P. J. Emmerman and R. Goulard

George Washington University, Washington, D.C. 20052

The laser tomography technique has been used for composition measurements in a laminar methane/air diffusion flame. A simulation study has also been carried out to extend the technique for simultaneous temperature and composition measurements using two-line absorption and tomographic reconstruction techniques. Laser tomography is a multiangular absorption technique which involves making M line-of-sight absorption measurements (projections) at N angles. These M×N measurements are then used to reconstruct the original two dimensional property field. Absorption measurements were made on methane using the 3.39 μm line of a He-Ne laser. Mean methane concentration profiles obtained in the diffusion flame have been compared with the results of previous workers obtained with probe sampling in a similar flame. Simulation studies have been carried out for the $P_1(5)$ and $Q_1(13)$ lines of the OH(0,0) $A^2\Sigma - X^2\Pi$ absorption band. A prescribed fuel/air mixture and temperature profiles have been used to define the composition and temperature fields. Reconstructions obtained with a single Gaussian and two non-similar Gaussian distributions have generated temperature fields which are within 1% of the original prescribed field. These studies have demonstrated the feasibility of extending the laser tomography technique for simultaneous temperature and concentration measurements in nonuniform and nonsymmetric flow fields.

I. INTRODUCTION

The need to make non-intrusive real-time temperature measurements in highly non-uniform reacting flows has generated a great deal of interest in laser-based diagnostic techniques. Techniques which provide point measurements, such as laser Raman scattering[1,2], CARS[3,4], laser induced fluorescence[5,6], or Rayleigh scattering[7], have attracted most of the attention because of their capability to provide measurements with high spatial and temporal resolution. However, the low signal levels, requirement of powerful lasers and complexity of internal energy transfer processes limits many of these techniques to special applications.

Techniques based on absorption or emission provide much higher signal levels and therefore sensitivity, and are less susceptible to interference effects. However, such line-of-sight techniques have found only limited usage due to their inability to provide spatially resolved measurements. A number of attempts have been made in the past to extend absorption/emission techniques to obtain temperature measurements with good spatial resolution. One approach consists of frequency scanning an emission line or a band[8,9]. Such observed spectra are convolutions of the contributions of all radiating species located on the line-of-sight. It is then possible to reconstruct the temperature profile along this line. Such a method has been used extensively to measure atmospheric temperature profiles from CO_2 emissions. The main difficulty with this technique is that it requires either a constant or a known distribution of the radiating species concentration. Another approach is the "onion peeling" technique which involves Abel inversion of absorption measurements along parallel lines-of-sight at different radial distances from the center of the field[10,11]. However, this technique is limited to flow fields with suitable symmetry properties; in addition, "onion peeling" methods amplify experimental errors[12,13].

Recently, a multi-angular absorption technique (laser tomography) has been developed which utilizes line-of-sight measurements to provide point measurements across the flow field. The laser tomography technique overcomes many of the problems associated with "onion peeling" or frequency scanning techniques, and it has been demonstrated to yield spatially resolved and accurate concentration measurements in ambient turbulent jets[14,15]. The application of tomographic analysis to laser absorption affords several advantages: highly non-symmetric flows can be studied, rapid real-time measurements can be made, and simultaneous measurements over two or three dimensional regions can be obtained.

This paper includes a brief review of absorption and emission techniques for diagnostics of temperature and species concentration in homogeneous fields, and their extension to non-uniform fields, using tomographic reconstruction techniques. Methane concentration measurements are presented for a methane/air diffusion flame obtained using laser tomography. Finally, extension of this technique is discussed for simultaneous concentration and temperature measurements in flames. Results of a simulation study are presented where absorption by the $P_1(5)$ and $Q_1(13)$ lines of the OH (0,0) $A^2\Sigma-X^2\Pi$ band are utilized for simultaneous measurement of temperature and OH concentration in a simulated methane/air diffusion flame.

II. TEMPERATURE MEASUREMENTS IN HOMOGENEOUS MEDIA

Difficulties encountered with high temperature measurements in reacting flows, using intrusive probe techniques, have led to a number of investigations on the application of non-intrusive diagnostic techniques for the study of reacting flows. Optical techniques have provided high temporal resolution, and have allowed investigation of rapid reactions, such as those encountered behind shock waves[16,17] or in stirred reactors[18,19].

Absorption and emission spectra due to radical species such as OH or CH have been extensively used for temperature as well as concentration measurements[20,21,22]. Indeed, since the absorption or emission characteristics are a function of both temperature and species concentration, simultaneous measurement of both quantities is usually required. Temperature measurements based on absorption or emission spectra can be considered in two groups: a) techniques based on simultaneous measurement of absorption or emission for two or more spectral lines, and b) techniques based on measurement of spectral lines profiles.

Absorption along an optical path of length Δs is expressed as:

$$I_\nu / I_\nu^o = \exp(-K_\nu \Delta s) \quad (1)$$

where I_ν/I_ν^o is the ratio of transmitted to incident light intensity and K_ν is the absorption coefficient at frequency ν. ($K_\nu = N_i Q_\nu$ where N_i is the molecular number density and Q_ν is the absorption cross section). Absorption measurements are generally carried out near the peak of a spectral line, using a laser source which has a line width much narrower than the absorption line. In this case, it can be assumed that the absorption process occurs at a single wavelength. The value of the absorption coefficient at any point within a spectral line can be computed using the Voigt function, which accounts for both Doppler and collisional broadening processes. The Voigt function is given as[23]:

$$V(x,y) = \frac{K_\nu}{K_o} = \frac{y}{\pi} \int_{-\infty}^{+\infty} \frac{\exp(-t^2)}{y^2 + (x-t)^2} dt \quad (2)$$

where

$$K_o = \frac{S}{\alpha_D} \left(\frac{\ln 2}{\pi}\right)^{\frac{1}{2}}$$

$$y = \frac{\alpha_C}{\alpha_D} (\ln 2)^{\frac{1}{2}}$$

$$x = \frac{\nu - \nu_o}{\alpha_D} (\ln 2)^{\frac{1}{2}}$$

α_D and α_C are the Doppler and collisional semihalf-widths (HWHM) and S is the line strength. The spectral absorption coefficient K_ν can then be expressed in terms of the line strength:

$$K_\nu = \frac{S}{\alpha_D} \left(\frac{\ln 2}{\pi}\right)^{\frac{1}{2}} V(x,y) \quad (3)$$

The Doppler half-width of a spectral line at ν_o is given as:

$$\alpha_D = \nu_o \left(\frac{2 k T \ln 2}{m_a c^2}\right)^{\frac{1}{2}} \quad (4)$$

where m_a is the mass of the molecule, and c the speed of light. According to classical interpretation of the collisional half-width, it is proportional to the collision frequency, (see Section IV for further discussion) i.e.

$$\alpha_C (\rho, T) \propto \rho T^{\frac{1}{2}} \propto p T^{-\frac{1}{2}}$$

Therefore, for the case where pressure is constant, the temperature dependence of the collisional half-width can be expressed as:

$$\alpha_C(T) = \alpha_C(T_o) \left(\frac{T_o}{T}\right)^{\frac{1}{2}} \quad (5)$$

It should also be noted that for $x=0$, i.e. at the line center, the Voigt function reduces to[23]:

$$V(0,y) = \exp(y^2) \operatorname{erfc}(y) \quad (6)$$

so that the absorption coefficient at the line center can be expressed as:

$$K_\nu = \frac{S}{\alpha_D} \left(\frac{\ln 2}{\pi}\right)^{\frac{1}{2}} \exp(y^2) \operatorname{erfc}(y) \quad (7)$$

The concentration of the absorbing species and temperature can be related to the line strength of a spectral line as[23]:

$$S \equiv \int K_\nu d\nu = \frac{\pi e^2}{m_e c} f_{J'J''} N_{J''} [1 - \exp(-\frac{h \nu_o}{k T})] \quad (8)$$

where $N_{J''}$ is the concentration of the lower level (J'') of the transition, $f_{J'J''}$ the oscillator strength, e the electronic charge, m_e the mass of the electron. The term in the parenthesis is the contribution due to induced emission, which can be neglected for most cases of interest here. $N_{J''}$ can be related to the mole fraction of the absorbing species X_i and the temperature T through the Boltzmann distribution:

$$N_{J''} = \left(\frac{p_t}{p_o}\right)\left(\frac{T_o}{T}\right) N_L X_i \frac{g_{J''}}{Q_i} \exp\left(-\frac{E_{J''}}{kT}\right) \quad (9)$$

where p_t is the total pressure, p_o and T_o are standard pressure and temperature, N_L the Loschmidt number, $g_{J''}$ and $E_{J''}$ the degeneracy and the energy of the J'' state, respectively, and Q_i is the internal partition function.

Since the absorption coefficient of a spectral line is a function of both temperature and absorbing species concentration, any temperature measurement will require at least two measurements in order to isolate the relationship between temperature and the absorption coefficient. This can be accomplished in two ways: a) by making absorption measurements at the center of at least two spectral lines, and utilizing their ratio; and b) by scanning across the full width of a spectral line.

a) <u>Line ratio techniques</u>

The peak absorption coefficients of two spectral lines can be related to each other as:

$$\frac{K_{\nu 1}}{K_{\nu 2}} = \frac{S_1}{S_2} \frac{\alpha_{D2}}{\alpha_{D1}} \frac{V_1(0,y)}{V_2(0,y)} \quad (10)$$

At the same temperature, this expression reduces to:

$$\frac{K_{\nu 1}}{K_{\nu 2}} = \frac{\nu_2}{\nu_1} \frac{g_1}{g_2} \frac{f_1}{f_2} \exp\left(-\frac{E_1 - E_2}{kT}\right) \quad (11)$$

if the two lines are within the same vibrational-rotational band and the collisional broadening parameters are the same for both lines. If absorption measurements are made over the same optical path length, and if the temperature and species concentration are constant over that path, then the temperature can be determined as:

$$T = \frac{E_2 - E_1}{k} \left[\ln\left(\frac{K_{\nu 1}}{K_{\nu 2}} \frac{\nu_1}{\nu_2} \frac{g_2}{g_1} \frac{f_2}{f_1}\right) \right]^{-1} \quad (12)$$

Absorption can also be measured for a series of n spectral lines to improve the accuracy of such measurements. In this case, a more general form of the above equation can be used:

$$K_n \propto S_n \propto f_n g_n \exp(-E_{J''n} /kT)$$

and the temperature can be determined from the slope of the $\ln(S_n/f_n \cdot g_n)$ vs $E_{J''n}$ plot[18,20]. Once the temperature is determined, then the concentration can be measured directly from the absorption observed at any one of the spectral lines.

A similar expression can also be derived for the emission intensity of spectral lines:

$$I_n \propto \nu_n^3 f_n g_n \exp(-E_{J'n}/kT)$$

and the temperature can be determined from the slope of the $\ln(I_n/\nu_n^3 \cdot f_n \cdot g_n)$ vs $E_{J'n}$ plot.

b) Line profile techniques

It had been shown earlier that the profile of a spectral line, dominated by Doppler and collisional broadening processes, is determined primarily by the collisional broadening parameter y, which is:

$$y = \alpha_C / \alpha_D (\ln 2)^{\frac{1}{2}}$$

It was also pointed out that, for a constant pressure media, the collisional half-width is only a function of temperature, according to the classical theory of line broadening. Hence:

$$y(T) = y(T_o)(T_o/T) \quad (13)$$

The collisional broadening parameter y is therefore only a function of temperature, for a given spectral line, at constant pressure. Since the pressure is constant for most reacting flows and flames, the measurement of y could be utilized for temperature diagnostics in uniform media. Hanson[24,25] has applied such a technique in a premixed methane/air flame. A tunable diode laser was used to scan a number of rotational-vibrational lines in the fundamental band of CO in the infrared, and a Voigt profile has been computer fitted to the measured line profile with y as the free parameter (see Fig. 1). This technique has been shown to have a sensitivity of better than 5%, and is currently being extended to non-uniform field applications[26].

Another technique based on line profile measurements is that of Luck and Muller[27] which was first proposed by Whiting[28]. Here, an empirical expression derived by Whiting is used to relate the centerline absorption coefficient $K(\lambda_o)$ to the Voigt half-width (FWHM) $\Delta\lambda_V$:

$$\frac{K(\lambda_o)}{S} = 1 / \Delta\lambda_V [1.065 + 0.447 \frac{\Delta\lambda_C}{\Delta\lambda_V} + 0.058(\frac{\Delta\lambda_C}{\Delta\lambda_V})^2] \quad (14)$$

with the Voigt half-width given as:

$$\Delta\lambda_V = \frac{\Delta\lambda_C}{2} + (\frac{\Delta\lambda_C^2}{4} + \Delta\lambda_D^2)^{\frac{1}{2}} \quad (15)$$

Since

$$y = \Delta\lambda_C/\Delta\lambda_D (\ln 2)^{\frac{1}{2}}$$

the above expressions can be rewritten as:

$$\frac{\Delta\lambda_V}{\Delta\lambda_C} = \frac{1}{2} + (\frac{1}{4} + \frac{\ln 2}{y^2})^{\frac{1}{2}} \quad (16)$$

and

$$\frac{S}{\Delta\lambda_V K(\lambda_o)} = 1.065 + 0.447[\frac{1}{2} + (\frac{1}{4} + \frac{\ln 2}{y^2})^{\frac{1}{2}}]^{-1} + 0.058[\frac{1}{2} + (\frac{1}{4} + \frac{\ln 2}{y^2})^{\frac{1}{2}}]^{-2} \quad (17)$$

In addition, it can also be shown that:

$$\frac{\Delta\lambda_V}{\Delta\lambda_D} = \frac{y}{2\sqrt{\ln 2}} + (\frac{y^2}{4 \ln 2} + 1)^{\frac{1}{2}} \quad (18)$$

The Voigt half-width, the central value of the absorption coefficient $K(\lambda_o)$, and the line strength $S(= \int K_\lambda d\lambda)$ can be determined directly from the measured spectral line profile. The collisional broadening parameter y can then be obtained from Eq(17). Given the value of y, the Doppler width can then be calculated from Eq(18), which is directly proportional to $T^{\frac{1}{2}}$. (See Eq(4)).

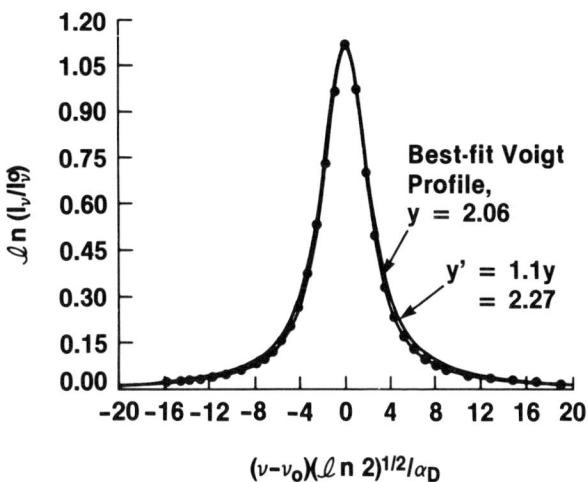

FIG. 1. The Voigt profile computed for the collisional broadening parameter y that best fits the data points. The Voigt profile for y'=1.1y is displayed for comparison (from Ref. 25).

Evaluation of this technique has indicated a low sensitivity to temperature, requiring very high accuracy measurements of the line profile and the line width. For example, over the temperature range of 1000-2000 K, the line width varies by less than 20%. Extension of this technique to non-uniform fields also does not prove to be practical, since only an empirical relationship can be derived between the local values of the line width and the convoluted value along the line-of-sight[26]. Hence this technique was not pursued further for applications to non-uniform temperature fields.

III. TEMPERATURE MEASUREMENTS IN NON-HOMOGENEOUS MEDIA

Spectra obtained from a field of non-uniform temperature and concentrations are made up of lines which represent convolutions of lines formed at different thermodynamic conditions. The emerging intensity I_ν must then be written as:

$$I_\nu/I_\nu^o = \exp(-\int_0^L N_i Q_\nu ds) \quad (19)$$

for absorption; and as:

$$I_\nu = \int_0^L B_\nu(s) [\exp(-\int_0^s N_i Q_\nu ds)] N_i Q_\nu ds \quad (20)$$

for emission, where s is the distance between the receiver (s=0) and the point whose contribution to I_ν is being considered, and B_ν is the Planck function. I_ν is then the summation of all contributions along the segment of length L.

Equations (19) and (20) can be rewritten as discrete summations, which results in:

$$\ln(I_\nu/I_\nu^o) = \sum_{m=o}^{P} (N_i Q_\nu \Delta s)_m \quad (21)$$

or

$$I_\nu = \sum_{m=o}^{P} B_\nu(m) \{\exp[-\sum_{m'=o}^{m}(N_i Q_\nu \Delta s)_{m'}]\}(N_i Q_\nu \Delta s)_m \quad (22)$$

Clearly a solution for the distribution of the values $N_i Q_\nu(T)$ or $B_\nu(T)$ at each point m along L is not possible on the basis of one measurement only, and additional information is needed. This is accomplished by making more measurements, either along the same line of sight but at different frequencies (frequency scanning), or at the same frequency but from different angles (tomography). The details of these approaches will be briefly reviewed below.

a) Frequency Scanning

Frequency scanning has been developed into a major meteorological tool to obtain temperature profiles and species concentrations in the atmosphere. Satellites routinely measure the _emitted_ intensities along a line of sight from non-uniform atmospheres at a set of different frequencies. An inversion algorithm reconstructs the temperature profile, from the satellite to the ground assuming constant CO_2 profile[29]. In a classic paper, Wark and Fleming[30] discussed the results obtained from the 15μm CO_2 band at 6 frequencies.

Among the number of methods proposed for such reconstructions, the most commonly used ones are essentially nonlinear matrix inversion methods. For the application of frequency scanning, Eq. (22) is rewritten as:

$$I_{\nu j} = \sum_{m=0}^{P} B_{\nu j}(m)[\exp(-\sum_{m'=0}^{m} \Delta \tau_{m',j})] \Delta \tau_{mj} \qquad (23)$$

where $\Delta \tau_{m'j} \equiv (N_i Q_{\nu j} \Delta s)_{m'}$ is the optical thickness of the small element Δs at frequency ν_j and at the location m'. Thus, we have for each frequency ν_j a summation of P terms. Each of them contains two non-linear functions of temperature: the Planck function $B_{\nu j}(m)$ at every point m and the absorption cross section $Q_{\nu j}(m')$ which enters as a summation over m' which ranges between 0 and m. If we express now B_ν and $Q_{\nu j}$ in terms of their values at a common reference frequency ν_r, we obtain a set of nonlinear equations in $B_{\nu_r}(m)$, whose matrix solution yields a profile of B_{ν_r} (or T) in terms of τ_r.

Because of the form of the equations above, the solution for the temperature field is expressed in terms of the optical thickness $\tau_r = \int_0^L N_i Q_{\nu_r} ds$ and _not_ of the physical distance s. This difficulty is overcome <u>if one knows the concentration distribution in space</u> $N_i(s)$, since Q_ν is already known in terms of T. Fortunately, the atmospheric concentration of CO_2 is known to be a constant fraction of the total atmospheric density, which is in turn a known function of altitude (Laplace's law); hence the remarkable success of the satelite probes. Chahine[29] has also shown that once T(s) is known, and upon measurements at other frequencies, the same equations can be solved, with the known B_ν terms, for unknown N_i's (e.g. water vapor).

In the case of flames, where the concentration and temperatures are strongly coupled, simple approaches to deconvolute the temperature and species concentration profiles have met with limited success.[31,32,33] For most situations of interest, independent measurements of each are required if accurate reconstructions are to be obtained. However, in certain applications, seeding of high temperature reacting flows with species of known concentration may present an opportunity for application of frequency scanning techniques, where the temperature profile could be determined, followed by a determination of the concentration profiles of interest.

Extension of frequency scanning to absorption measurements presents some additional difficulties. It should be noted that the right hand side terms of Eq. (21) exhibit a much weaker dependence on temperature, since the strong pivoting influence of the Planck function is absent. The reduced temperature sensitivity results in a weaker dependence of absorption signal on frequency, compared to the emission case. This, in turn, results in larger errors in the inversion process, especially in the absence of <u>a priori</u> knowledge of some general features of the field. Thus it seems that the mathematical difficulties associated with such an ill-conditioned problem overshadows the physical advantages offered by the use of laser sources[34].

b) Multiangular Scanning (Tomography)

One advantage of frequency scanning is that - if successful - line of sight information can be obtained without the need to observe the flame from many directions, a great advantage if access to the region of interest is limited. However, if access is available, multiangular scanning becomes a possibility and provides the potential for obtaining spatially resolved temperature and concentration data throughout the flow field. The multiangular scanning technique has met with remarkable commercial success in the medical field (Computer Assisted Tomography: "CAT" scanners). Laser tomography has been shown to provide accurate concentration measurements in non-reacting flows[14,15]. This paper discusses extension of tomography to reacting flows where simultaneous measurements of temperature and composition are required.

The multiangular absorption technique involves the measurement of absorption along M equally spaced parallel beams at N equally spaced angles, forming an MxN data set from which it is possible to retrieve or "reconstruct" the original property field. The absorption along each individual beam is governed by the Bouguer-Lambert-Beer law as expressed in equation (19). This equation can be rewritten as (see Figure 2):

$$-\ln \frac{I_\nu(r,\theta)}{I_\nu^0} = P(r,\theta) = \int_{-\infty}^{+\infty} F(x,y) \, ds \qquad (24)$$

where $r = x \cos \theta + y \sin \theta$, $s = -x \sin \theta + y \cos \theta$.

$P(r,\theta)$ is defined as the "projection" for the angle θ. Since the ratio I_ν/I_ν^0 can be measured experimentally, $P(r,\theta)$ is known and the solution of Eq. (24) for $F(x,y)$ will yield the desired property field ($K_\nu = N_i Q_\nu$). Thus the multiangular scanning technique is based upon the reconstruction of a property field, $F(x,y)$, from a set of its projections, $P(r,\theta)$. The reconstruction can be carried out using a number of different methods, including linear superposition, algebraic reconstruction and two-dimensional Fourier transform techniques. The solution procedure used here utilizes the convolution technique developed by Ramachandran[35]. Although it is a Fourier transform approach, this technique does not actually require the evaluation of any transforms, and allows high speed computation of the property field. Details of the solution procedure have been presented before[14,15] and only a brief summary of the analysis will be given here.

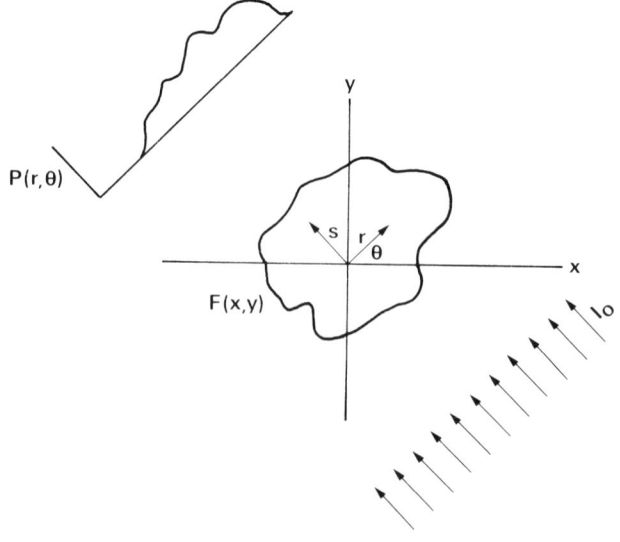

FIG. 2. Projection $P(r,\theta)$ in polar coordinates

In this method the Fourier transform of Eq (24) is written:

$$\hat{P}(\omega,\theta) = \int_{-\infty}^{+\infty} P(r,\theta)e^{-i\omega r}dr = \int_{-\infty}^{+\infty}\int_{-\infty}^{+\infty} F(r,s)e^{-i\omega r}dr\,ds \quad (25)$$

where $\omega = 2\Pi\rho$ and ρ is the spatial frequency. The Fourier transform of $P(r,\theta)$ can be related to the two-dimensional Fourier transform of $F(r,s)$:

$$\hat{F}(\omega,\theta) = \int_{-\infty}^{+\infty}\int_{-\infty}^{+\infty} F(r,s)\,e^{-i(\omega r + \eta s)}dr\,ds\bigg|_{\eta=0} \quad (26)$$

when it is evaluated along the line $\eta=0$ in the two-dimensional frequency plane. This is the "central slice theorem" (see e.g. ref 36) which relates the one dimensional Fourier transform ($\eta=0$) of the one dimensional projection $P(r,\theta)$ to the two-dimensional Fourier transform of the property field $F(r,s)$. This theorem expresses the fact that the one-dimensional Fourier transform of the projection is a "slice" through the Fourier transform plane of $F(r,s)$ at the angle θ (see Fig. 3). For each value of θ, the values of $\hat{F}(\omega,\theta)$ can be plotted for all pertinent values of ω. A 180 degree sweep of θ yields all values of $\hat{F}(\omega,\theta)$ in the Fourier transform plane. The function $F(x,y)$ can therefore be retrieved as the inverse Fourier transform of the set of $\hat{P}(\omega,\theta)$ which are derived from the experimentally measured projections.

If Eq (26) is expressed in cartesian coordinates and the inverse transform operation is performed, then $F(x,y)$ is given by:

$$F(x,y) = \frac{1}{4\pi^2}\int_0^\pi d\theta \int_{-\infty}^{+\infty} \hat{P}(\omega,\theta)\,e^{i\omega r}\,|\omega|\,d\omega \quad (27)$$

Now let V be equal to the inner integral:

$$V(r,\theta) = \frac{1}{2\pi}\int_{-\infty}^{+\infty} \hat{P}(\omega,\theta)\,|\omega|\,e^{i\omega r}\,d\omega \quad (28)$$

and define a function such that its Fourier transform is given as

$$\hat{\phi}(\omega) = |\omega| \quad (29)$$

Then from the convolution theorem for Fourier transforms:

$$V(r,\theta) = \int_{-\infty}^{+\infty} P(\tau,\theta)\,\phi(r-\tau)\,d\tau \quad (30)$$

and

$$F(x,y) = \frac{1}{2\pi}\int_0^\pi d\theta \int_{-\infty}^{+\infty} P(\tau,\theta)\phi(x\cos\theta + y\sin\theta - \tau)d\tau \quad (31)$$

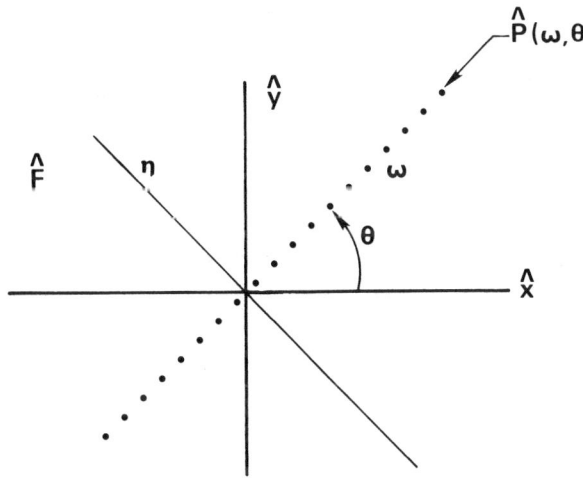

FIG. 3. The central slice theorem

Since $P(r,\theta)$ is known only in the sampled domain, the integrals in Eq (7) must be replaced by their discrete summation approximations. Thus,

$$F(x,y) = \frac{a}{2N}\sum_{j=1}^{N}\sum_{k=1}^{M} P(r_k,\theta_j)\,\phi(x\cos\theta_j + y\sin\theta_j - r_k) \quad (32)$$

where a is the spacing between uniform samples and

$$\theta_j = (j-1)\pi/N, \quad r_k = ka$$

ϕ may be interpreted as a weighting function of the distance from the point (x,y), where the property $F(x,y)$ is to be evaluated, to r_k where a value of the projection $P(r,\theta)$ is evaluated. An accurate reconstruction will require that the Fourier transform of ϕ satisfies the condition $\hat{\phi}(\omega) = |\omega|$ for $|\omega|<2\pi\rho_{max}$, where ρ_{max} is the radius of the transform space. The choice of the function $\phi(r)$ greatly affects the computational requirements as well as the accuracy of the results. In the present study the modified Shepp-Logan filter[37] has been used

$$\bar{\phi}(r_k) = 0.4\phi(r_k) + 0.3\phi(r_{k+1}) + 0.3\phi(r_{k-1}) \quad (33)$$

where
$$\phi(r_k) = -\frac{4}{\pi a^2(4k^2-1)} \quad k = 0, \pm 1, \pm 2, \ldots$$

$\phi(r)$ can be considered a filter function and its selection must be dependent on the noise of the instrumentation system as well as the bandlimit of the function.

Accurate reconstruction of the property field, $N_i Q_v$, will be dependent on the proper choice of the sampling intervals, that is the choice of M and N, as well as the choice of the appropriate form of the filter function, ϕ.

From the sampling theorem[38], a function can be uniquely recovered from its samples if it is sampled at a rate greater than twice the highest frequency component of the function (sampling rate $\geq 2\rho_{max}$). The present function, $F(x,y)$, is assumed to have a bandlimited Fourier transform, i.e. $\hat{F}(x,y) \equiv \hat{F}_\theta(\rho) = 0$ for $\rho > \rho_{max}$, while in the measurement or physical space $F(x,y) = 0$ if $x^2 + y^2 \geq R^2$ where R is the radius of the measurement space. The proper choice of M and N under these conditions, which has been discussed previously[14], requires that $(4 R\rho_{max})$ equally spaced rays are measured at $(2\pi R\rho_{max})$ equally spaced angles. If this criterion is not met, the reconstructed field is likely to show effects due to "aliasing".[39] Aliasing is a term used to describe the effects observed when a function is undersampled. In such a case the reconstructed function contains contributions from the undersampled high frequency components which appear as low frequency contributions in the transform domain. Unambiguous identification of aliasing effects requires a knowledge of the original function from which accurate comparisons can be made. Operationally one could increase the sampling rate until aliasing effects are observed to be unimportant. Once these minimum sampling conditions have been satisfied, additional higher frequency sampling (i.e. oversampling) although unnecessary from a theoretical viewpoint, in general can improve the reconstruction in the presence of noise.

IV. EXPERIMENTAL RESULTS

Results reported in previous papers[14,15] have demonstrated the utility of multiangular absorption techniques for concentration measurements in non-reacting flows. To assess the applicability of the technique to reacting flows, experiments were carried out in a laminar methane/air flame. The laminar methane-air diffusion flame has been studied under steady flow conditions at atmospheric pressure. The laminar flame apparatus consists of two concentric brass tubes of 1.11 cm and 10.16 cm diameter, with the fuel (methane) flowing through the central tube, and the air through the larger passage. The fuel passage contains screens and 3.2 mm metal beads to provide a uniform exit flow profile. The larger air passage also contains screens and is filled with 6.0 mm glass beads for the same purpose. Finally, a ceramic honeycomb section is located at the exit of the air flow passage. The fuel tube extends 4 mm beyond the exit plane of the air tube. The methane flow rate was 5.67 cm^3/sec., providing an exit velocity of 5.84cm/sec. The air flow rate was 802.0 cm^3/sec, providing an exit velocity of 10.05 cm/sec. A 40 cm long brass cylinder was used as a shield for the flame from the air currents in the laboratory. Two slots, 3 mm high and 25.4 mm wide, were machined out on the two sides of the brass cylinder to allow the laser beam to go through the flame. Screens and other types of flow restrictions were placed at the exhaust of the brass shield to achieve a highly stable flame, by reducing air recirculation down the side walls and minimizing air entrainment through the slots. The burner assembly was mounted on a milling machine bed for positioning purposes.

Concentration measurement were based on absorption by methane of the near coincident 3.39 μm line of a He-Ne laser[40]. A Spectra Physics Model 120 He-Ne laser operating at 3.39 μm with an output of approximately 2 mW was used as the laser light source. A quartz lens with a 20 cm focal length was used to produce a beam whose diameter across the jet/plate region was 1.3 mm. A second identical lens then focused the transmitted beam onto a radiometer. This unit utilizes a pyroelectric detector and lock-in amplifier detection to monitor the laser intensity, and provides a convenient means of measuring the full laser intensity with suitable thermal stability and background noise rejection. The output of the lock-in amplifier was monitored with a datalogger, and the digital data were transferred to a minicomputer for storage.

The measurements made in the laminar diffusion flame were carried out at three axial locations, z = 0.32, 0.63 and 1.27 cm downstream of the fuel exit plane. At each axial location, only a single projection was obtained and the reconstruction was carried out assuming axisymmetry. The measurements at each axial position covered a radial span of 2.54 cm, with a parallel beam spacing of 0.254 mm, providing 101 measurements for each projection. The axial locations for the measurements were chosen to provide data for the reaction zone, before most of the fuel (methane) disappeared. Indeed, a fourth location at z = 2.54 cm was also probed, but due to the very low concentration of CH_4 found at this location, the signal to noise ratio was significantly reduced and these data will not be presented here. These particular locations were also chosen to provide data which could be compared with the results of other workers. For each parallel ray, fifty readings were taken and the average value was used in all further calculations. For the present experiments the detection system was operated with a 0.1 sec time constant, and individual intensity readings were taken at 0.2 sec intervals.

The projections obtained at three axial locations in the laminar methane diffusion flame were analyzed as described previously in section IIIb, assuming axial symmetry and assuming that the same data were obtained at 18 angles. The reconstruction results for K_ν for three axial locations are shown in Figures 4-6. The profiles exhibit the qualitative features one expects to see in the diffusion flame. At z=0.32 cm, the centerline concentration is quite high, and methane absorption is observed over a relatively large radial span (∿1.0 cm). As the axial distance increases, the centerline absorption substantially decreases and is limited to a smaller radial span, indicating rapid reaction of methane within a very short distance. It should be noted again that axisymmetry is assumed in the analysis and 18 identical projections are used for the reconstruction, resulting in an artificially noise-free distribution away from the flame center-line.

Determination of the methane concentration in the diffusion flame, from the plots of the product $K_\nu = N_i Q_\nu$ require a knowledge of the absorption coefficient. The absorption coefficient can not be assumed to be constant at high temperatures, indeed it has a rather strong dependence on temperature. Consequently both the local temperature field, and the form of the temperature dependence need to be known before the concentration field can be determined. Since independent measurements of temperature and methane concentration were not carried out as part of the present study, results obtained by Mitchell et al[41] in a very similar methane diffusion flame were used for at least a qualitative evaluation of the tomography results.

Absorption of the 3.39μm (2947.909 cm^{-1}) line of a He-Ne laser by methane at elevated temperatures have been previously studied[40,42]. The absorption coefficient for the $\nu_3(P7)$ transition of methane (ν_0=2947.912cm^{-1}) exhibits a temperature dependence as a result of both a change in the spectral line shape as well as a decrease with increasing temperature of the absorber state population.

The strength of this particular spectral line can be related to temperature and species concentration using equations (8) and (9). The internal partition function Q_i can be separated into two parts, rotational and vibrational partition functions, so that $Q_i = Q_r Q_v$. If a rigid-rotator temperature dependence is assumed for the rotational function, then:

$$Q_r(T_o)/Q_r(T) = (T_o/T)^{3/2} \qquad (34)$$

For a symmetric top harmonic oscillator, the vibrational partition function is expressed as:

$$Q_v(T) = \prod_i [1 - \exp(-W_i h c/k T)]^{-d_i} \qquad (35)$$

where W_i and d_i are the energy and degenaracy of the ith vibrational level. The spectral absorption coefficient can then be expressed as:

$$K_\nu(T) = K_o(T_o)\left(\frac{T_o}{T}\right)^3 \frac{Q_v(T_o)}{Q_v(T)} \exp\{-\frac{E_J"}{k}(\frac{1}{T} - \frac{1}{T_o})\}V(x,y) \qquad (36)$$

where $E_J" = 293.47$ cm^{-1} for the P(7) line.

The peak value of the pure Doppler broadened line $K_o(T_o)$ has been measured by several investigators.[40,42,43] For the present study, we have selected the value $K_o(T_o) = 1.40 \times 10^2$ atm^{-1}cm^{-1}. (There seems to be a considerable amount of disagreement between the K_o values obtained at room temperature and those obtained at high temperatures as pointed out by Mallard and Gardiner[42]). The collisional semi-half width of the P(7) line of methane has been measured by Edwards and Burch[44], and found to be $\alpha_c(T_o) = 0.065$ cm^{-1} with nitrogen as the principal collision partner. It is also assumed that the laser line is located $\nu - \nu_o = 0.003$ cm^{-1} off the methane line peak, based on previous measurements. Using these spectral constants, the spectral absorption coefficients are computed as a function of temperature using Eq (36).

Due to lack of simultaneous temperature measurements in our experiments, we had to rely on the temperature measurements made by Mitchell[41]. The results obtained at z = 1.27 cm have been compared with the concentration measurements of Mitchell[41], and are presented in Table I. T* and X*$_i$ are temperature and methane mole fraction as measured by Mitchell, $K_\nu = X_i K'_\nu$ is the absorption coefficient determined from our tomographic experiments, and K'_ν is the reduced spectral absorption coefficient. From these two

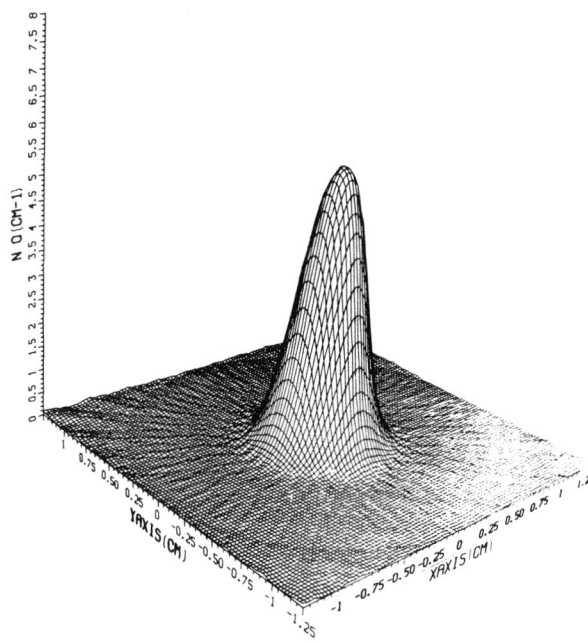

FIG. 4. Reconstructed absorption coefficient field for laminar methane/air diffusion flame at $z = 0.32$ cm.

values, the methane concentration X_i is then calculated. Comparison of our concentration measurements to those of Mitchell, i.e. X_i/X^*_i, indicate that our concentration measurements are higher by about a factor of two throughout the flame.

In these calculations, it was assumed that the collisional line width was proportional to $T^{-1/2}$. However, recent work of Varanasi[45] has indicated a stronger temperature dependence, indeed closer to T^{-1}. The calculations of the absorption coefficient were repeated assuming a T^{-1} dependence, and the results are also presented in Table I. In this case the agreement with Mitchell's data is much better. This agreement is very encouraging, but it also points out the importance of obtaining accurate spectroscopic data for this particular line of methane, especially at high temperatures.

At this point, it should also be noted that all the absorption at the 2947.909 cm^{-1} laser line has been atributed to methane. However, it has been shown that several other hydrocarbon species also exhibit absorption at this frequency, albe it to a lesser degree. This is not expected to be a major source of error in our experiments; however in flames with larger hydrocarbon molecules, contributions from several absorbing species could be present.

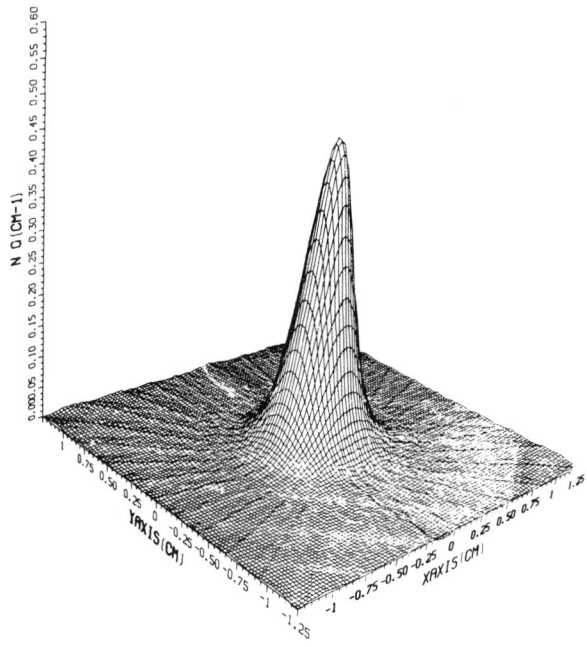

FIG. 6. Reconstructed absorption coefficient field for laminar methane/air diffusion flame at $z = 1.27$ cm.

V. SIMULATION RESULTS

The experimental results presented in the previous section indicated the importance of obtaining simultaneous measurements of temperature and species concentration. As indicated before, this can be accomplished by either making absorption measurements at two different spectral lines, or by measuring the full line profile of one absorption line. Because of the lack of a tunable infrared laser and detector combination with the desired spatial and temporal resolution, it was decided that simultaneous measurements of temperature and concentration would be carried out either in the visible or ultraviolet part of the spectrum. In the visible, Na seeded flows could be probed with a tunable dye laser. More interesting would be the diagnostics of OH species concentration and temperature in flames, using a frequency doubled tunable dye laser. Since OH is a radical species of critical importance in the oxidation of hydrocarbons and carbon monoxide in flames, both the temperature and species concentration measurements will be of great interest.

To assess the applicability of laser tomographic techniques for simultaneous temperature and concentration measurements in flames, a simulation study has been carried out. To provide the field parameters of interest, a stoichiometric methane/air mixture and a prescribed temperature distribution has been assumed in a laminar flame. The composition field is then calculated based on the prescribed mixture

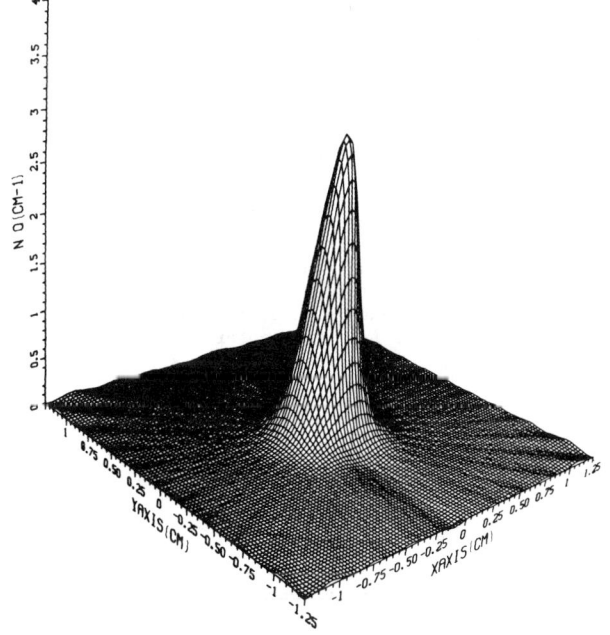

FIG. 5. Reconstructed absorption coefficient field for laminar methane/air diffusion flame at $z = 0.63$ cm.

Table I

Comparison of Methane Concenration Profiles Obtained
in this Study and in Ref. [41]

r(cm)	T*(°K)	X^*_i	$X_i K'_\nu$	K'_ν	X_i	X_i/X^*_i	$K'^{(a)}_\nu$	$X^{(a)}_i$	$X^{(a)}_i/X^*_i$
0.0	900	0.297	0.476	1.727	0.276	0.929	1.634	0.291	0.980
0.133	1050	0.261	0.322	1.318	0.244	0.935	1.023	0.315	1.206
0.267	1250	0.184	0.120	0.929	0.129	0.701	0.538	0.223	1.211
0.401	1460	0.104	0.034	0.667	0.051	0.490	0.273	0.124	1.192

*Measurements from Ref. [41]
^{a}Results obtained assuming $\alpha_c \propto T$.

ratio and temperature, asuming chemical equilibrium. It has been shown[21] that the OH concentrations found near the reaction zone of the flame usually exceed the equilibrium levels, due to chemical nonequilibruum effects. Hence, the absorption levels in similar flames should be higher than those used in this simulation, contributing to higher absorption signals and better sensitivity.

The line ratio technique is used for the temperature measurements and the $P_1(5)$ and $Q_1(13)$ lines of the (0,0) band of the OH $A^2\Sigma - X^2\Pi$ transition are selected for the simulation. These lines are well isolated but near each other (in case rapid scanning of both lines are desired). In addition, the absorption coefficient for the $P_1(5)$ line decreases with temperature, since it is from a lower lying energy state, whereas the absorption coefficient for $Q_1(13)$ increases with temperature. This behavior provides increased sensitivity. The relevant spectroscopic data is given in Table II.

The local absorption coefficients are calculated using equations (8) and (9). The f number of an individual line in a band can be related to the pure rotational transition probabilities $S_{J'J''}$ as[46]:

$$f_{J'J''} = f_{v'v''} S_{J'J''} T_{J'J''}/4(2J''+1) \quad (37)$$

where $f_{v'v''}$ is the band oscillator strength of the rotationless molecule, and is assumed to be constant over a given vibration-rotation band. $T_{J'J''}$ is a correction factor for vibration-rotation interactions. For the OH molecule, $S_{J'J''}$ values have been calculated by Dieke and Crosswhite[47], and are normalized such that:

$$\sum_{J'} S_{J'J''} = 4(2J''+1) \quad (38)$$

The values of $T_{J'J''}$ have been computed by Learner[48], and the $f_{v'v''}$ values have been determined for the (0,0) and (1,0) bands by Rouse and Engleman[49]. To complete the simulation, the local absorption coefficient is calculated at each point in the field, based on the local temperature and OH concentration, and then the integral given in Eq. (19) is evaluated along each parallel ray. The procedure is repeated for M number of rays and N angles, generating N projections. These projections $P(r,\theta)$ represent the data one would obtain in an actual experiment.

For the first set of simulations, a prescribed temperature distribution of the form (in degrees Kelvin):

$$T = 700 \exp\{-3(x^2+y^2)\} + 1500$$

was used, which is a centrally located Gaussian in a 5 cm x 5 cm field, with a peak temperature of 2200 K and a minimum temperature of 1500 K. Because this is an axisymmetric temperature distribution, only one projection is calculated, and the absorption coefficient field is reconstructed assuming 20 identical projections. Figure 7 and 8 show the reconstructed fields for the two selected spectral lines, $P_1(5)$ and $Q_1(13)$, respectively. The local temperature field is then computed at each point from the ratio of the local values of absorption coefficients for each line, using the relationship:

$$T = \frac{E_2 - E_1}{k} \left[\ln\left\{ \frac{K_{\nu_1}}{K_{\nu_2}} \frac{\nu_1}{\nu_2} \frac{(S_{J'J''} T_{J'J''})_2}{(S_{J'J''} T_{J'J''})_1} \right\} \right]^{-1} \quad (39)$$

and the resultant temperature field is shown in Figure 9. (The figures of the reconstructed fields are shown on a normalized coordinate frame.)

A comparison of the reconstructed temperature field with the original field shows that the reconstructed temperature values are within 5 K of the prescribed values near the peak of the Gaussian, and within 8 K near the edges of the field, where aliasing effects are expected to be more prominent. In either case, the agreement is excellent. It should be kept in mind, however, that no noise was introduced in the simulation which would normally result in increased errors in the reconstructed field.

It has been pointed out before that, unlike the Abel inversion technique, tomographic reconstruction techniques do not require axisymmetry or monotonically varying functions. To test the technique for such nonsymmetric fields, a second simulation was carried out where the temperature field was composed of two Gaussians of different heights, and centered at different distances from the center of the field. The prescribed temperature distribution was given as:

$$T = 700 \exp\{-12[(x-0.5)^2 + y^2]\}$$
$$+ 600 \exp\{-6[(x+0.3)^2 + y^2]\} + 1500$$

Table II

Summary of Spectroscopic Data for the Selected Transitions in (0,0) $A^2\Sigma - X^2\Pi$ Band of OH

Line	λ_o (nm)	$S_{J'J''}$	$T_{J'J''}$	J''	$E_{J''}$ (cm^{-1})
$P_1(5)$	310.123	24.5	0.986	5.5	543.57
$Q_1(13)$	311.022	108.8	0.881	13.5	3319.35

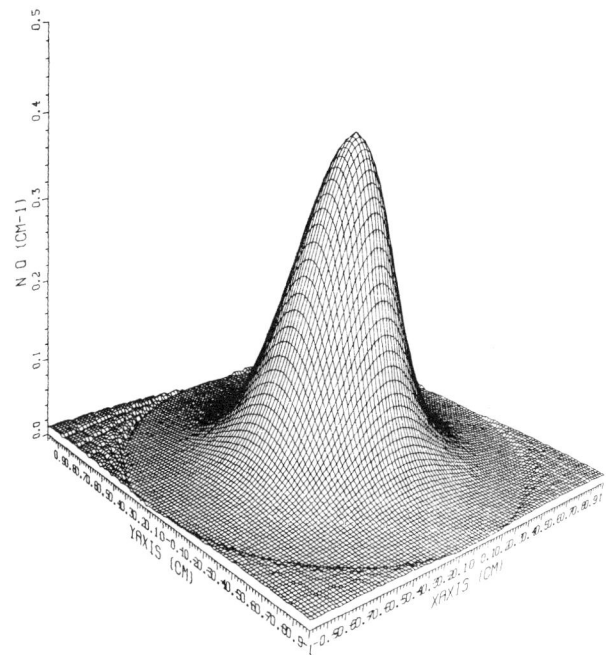

FIG. 7. Reconstructed absorption coefficient field for the $P_1(5)$ line of OH - Single Gaussian temperature field

In addition to providing a nonsymmetric field, these Gaussian distributions are much sharper with stronger gradients, all of which necessitate higher sampling rates in order to provide accurate reconstructions.

In this case, projections for each of the 20 angles were calculated, using the previously described procedure, for each spectral line, and then the absorption coefficient fields were reconstructed. The reconstructed field for the $Q_1(13)$ line is presented in Figure 10. It can be seen that some aliasing effects are observed near the edges of the field, which could lead to large errors in the temperature measurements because of low signal levels. The computations were then repeated with 40 angles in order to improve the accuracy of the reconstruction and, as can be seen in Figure 11, the reconstructed field is almost totally free of noise. The reconstructed temperature field, using 40 projections, is shown in Figure 12. A detailed comparison between the reconstructed and original temperature fields is presented in Table III, where the temperature values along two lines, going through the peaks of the Gaussian curves, are listed. Even for such a highly non-uniform field, which also includes very high gradients, it can be seen that the reconstruction accuracy is excellent, and the observed temperature differences are less then 8 K (0.4%) near the peaks, and less then 10 K (0.5%) elsewhere. Such high accuracy may be unreasonable to expect in actual experiments, where noise is introduced due to fluctuations in the field or in the detection system. However, it is indeed encouraging to find the reconstruction errors to be so small, and the line ratio technique along with tomographic reconstruction promises to be a very powerful temperature and concentration diagnostic technique in nonuniform reacting flow fields.

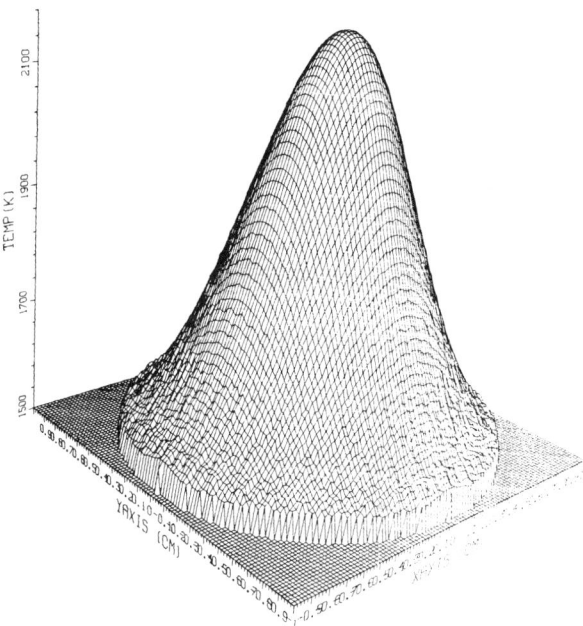

FIG. 9. Reconstructed temperature field from the ratio of $P_1(5)$ and $Q_1(13)$ lines

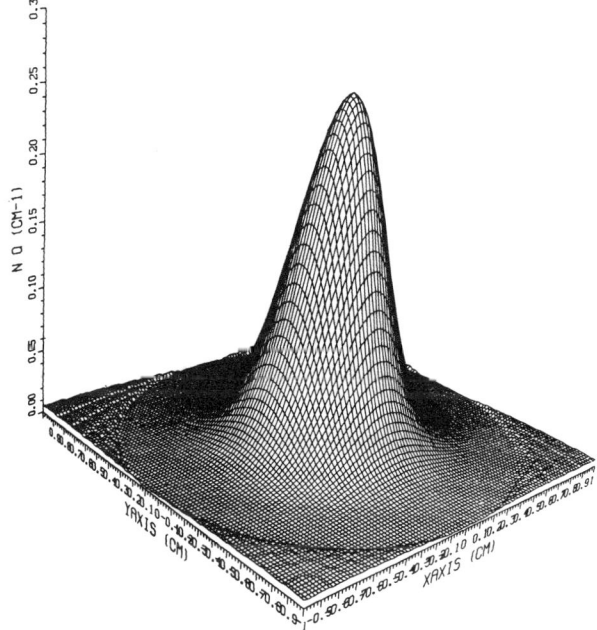

FIG. 8. Reconstructed absorption coefficient field for the $Q_1(13)$ line of OH - Single Gaussian temperature field

IV. CONCLUSIONS

In this paper, we have examined the application of the multiangular absorption (laser tomography) technique for temperature and composition measurements in flames. Experimental results have been obtained on the mean methane concentration distribution in the laminar methane/air diffusion flame, and have been compared with the results of Mitchell[41]. The agreement is found to be good, but more reliable spectroscopic data on methane absorption at high temperature is needed, and simultaneous temperature measurements are required for a conclusive evaluation of laser tomography. Results of a simulation study have been presented, where a line-ratio technique based on ultraviolet absorption by OH radical species, and tomographic reconstruction methods are used to determine the temperature field in a simulated flame of prescribed fuel/air mixture ratio and temperature distribution. Both axisymmetric as well as

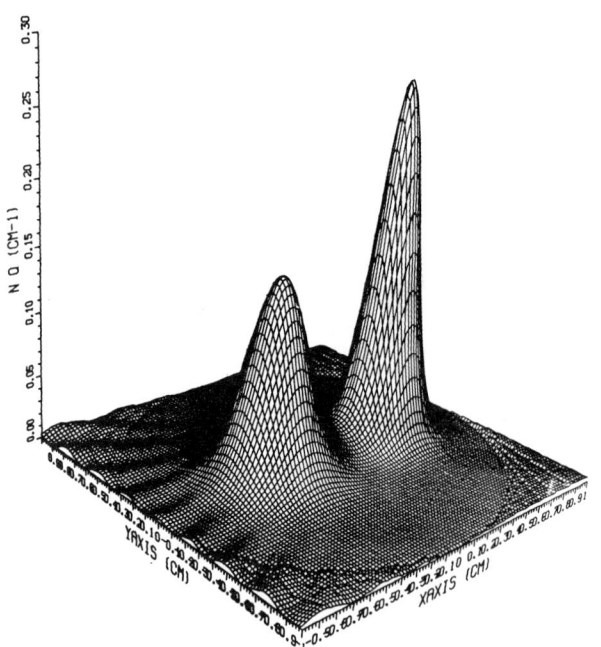

FIG. 10. Reconstructed absorption coefficient field for the $Q_1(13)$ line of OH – Double Gaussian temperature field (N=20)

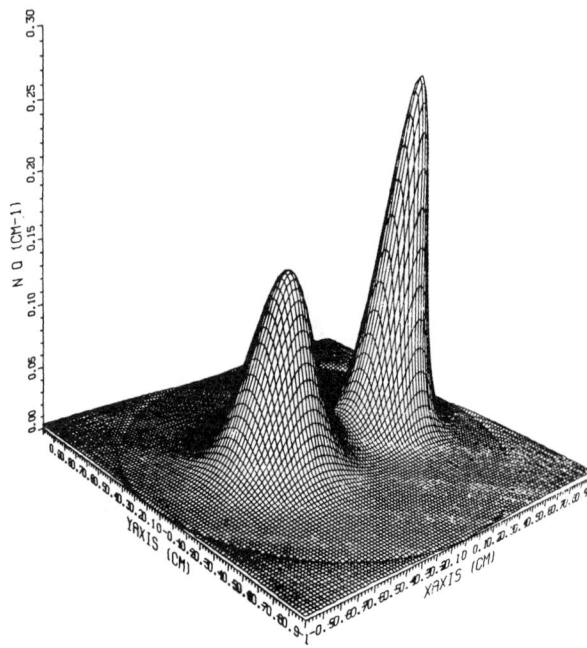

FIG. 11. Reconstructed absorption coefficient field for the $Q_1(13)$ line of OH – Double Gaussian temperature field (N=40)

Table III

COMPARISON OF ORIGINAL AND RECONSTRUCTED TEMPERATURE FIELDS ACROSS THE TWO GAUSSIAN PEAKS

	PEAK #1						PEAK #2				
No.	Orig.	Reconst.	No.	Orig.	Reconst.	No.	Orig.	Reconst.	No.	Orig.	Reconst.
1	1500.00	1500.00	53	2089.55	2094.58	1	1500.00	1500.00	53	2198.84	2192.17
3	1500.00	1500.00	55	2072.81	2077.70	3	1500.00	1500.00	55	2160.15	2154.38
5	1503.74	1500.00	57	2045.63	2050.61	5	1500.00	1500.00	57	2100.39	2095.84
7	1505.76	1503.54	59	2009.32	2014.81	7	1500.00	1500.00	59	2025.76	2022.60
9	1508.70	1520.70	61	1968.35	1972.18	9	1500.37	1500.00	61	1943.34	1941.27
11	1512.90	1515.97	63	1924.84	1920.48	11	1500.66	1513.12	63	1860.04	1859.00
13	1518.75	1515.97	65	1874.98	1872.78	13	1501.17	1522.33	65	1781.65	1780.49
15	1526.75	1519.57	67	1824.67	1822.58	15	1502.08	1500.12	67	1712.30	1711.62
17	1537.43	1539.51	69	1775.77	1772.15	17	1503.68	1512.47	69	1654.26	1650.32
19	1551.39	1556.41	71	1729.78	1728.35	19	1506.44	1504.17	71	1608.10	1605.08
21	1569.20	1564.34	73	1687.82	1686.04	21	1511.06	1508.39	73	1573.12	1577.48
23	1591.42	1594.16	75	1650.61	1647.40	23	1518.54	1519.65	75	1547.79	1551.67
25	1618.47	1617.64	77	1618.47	1614.86	25	1530.23	1540.67	77	1530.23	1550.73
27	1650.61	1647.76	79	1591.42	1590.62	27	1547.79	1551.67	79	1518.54	1522.76
29	1687.82	1685.12	81	1569.20	1564.15	29	1573.12	1575.74	81	1511.06	1514.64
31	1729.78	1728.35	83	1551.39	1552.83	31	1608.10	1603.27	83	1506.44	1511.92
33	1775.77	1772.08	85	1537.43	1537.62	33	1654.26	1649.89	85	1503.68	1514.64
35	1824.67	1822.51	87	1526.75	1520.93	35	1712.30	1711.79	87	1502.08	1505.48
37	1874.98	1872.53	89	1518.75	1515.91	37	1781.65	1780.56	89	1501.17	1519.95
39	1924.84	1920.57	91	1512.90	1513.32	39	1860.04	1859.02	91	1500.66	1508.73
41	1972.18	1968.22	93	1508.70	1528.15	41	1943.34	1941.35	93	1500.37	1500.00
43	2014.81	2009.28	95	1505.76	1504.36	43	2025.76	2022.68	95	1500.00	1500.00
45	2050.61	2045.69	97	1503.74	1500.00	45	2100.39	2095.87	97	1500.00	1500.00
47	2077.70	2072.78	99	1500.00	1500.00	47	2160.15	2154.40	99	1500.00	1500.00
49	2094.58	2089.48	101	1500.00	1500.00	49	2198.84	2192.20	101	1500.00	1500.00
51	2100.32	2095.01				51	2212.24	2205.14			

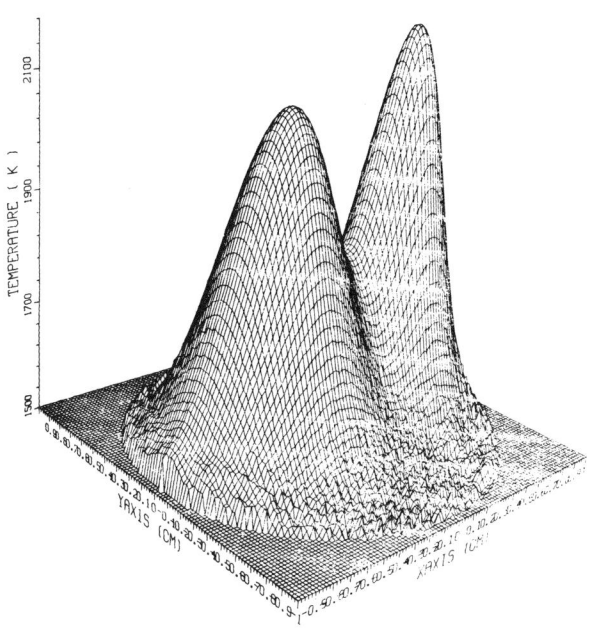

FIG. 12. Reconstructed temperature field from the ratio of $P_1(5)$ and $Q_1(13)$ lines Double Gaussian temperature field

non-axisymmetric fields have been investigated; in both cases the temperature field is reproduced with high accuracy. These studies are being extended[26] to the use of line profile techniques for simultaneous temperature and composition measurements.

So far, our efforts have been focused on mean concentration and temperature measurements in steady flows. However, the full potential of the laser tomography technique will be realized when it is used for real time measurements in turbulent reacting flows. Such measurements will provide a direct physical picture of eddies and other coherent structures observed in turbulent flows. Real time measurements require that all the data be acquired simultaneously, i.e. M absorption measurements must be made at N angles at one instant in time. A multiangular absorption system is currently being implemented to make such measurements. The system will provide the capability to obtain complete property field measurements every 0.5 msec or better. However, the typical time resolution (i.e time period over which the signal is integrated) for each set of measurements will be better than 50 μsec. Such a system would allow detailed investigation of the fluid mechanics of turbulent eddies and large coherent structures, chemical reaction dynamics, and the interaction of these physical and chemical processes as observed in turbulent reacting flows.

REFERENCES

1. M. Lapp and C.M. Penney "Laser Raman Gas Diagnostics" Plenum Press, 1973.

2. J.R. Smith, AIAA Paper No. 80-1359 (1980).

3. P.R. Regnier and J.P. Taran, Appl. Phys. Letts, 23, 240 (1973).

4. A.C. Eckbreth, Proc. Eighteenth Symp. on Comb., p.1471 (1981).

5. C. Morley, Proc. Eighteenth Symp. on Comb., p.23 (1981).

6. A.C. Eckbreth, P.A. Bonczyk and J.F. Verdieck, Prog. Energy and Comb. Sci., 5, 253 (1979).

7. R.W. Dibble and R.E. Hollenbach, Proc. Eighteenth Symp. on Comb. p.1489 (1981).

8. M.T. Chahine, J. Atm. Sci., 29, 741 (1972).

9. R. Viskanta, P.J. Hommert and G.L. Groninger, Appl. Opt. 14, 428 (1975).

10. R.H. Tourin, "Spectroscopic Gas Temperature Measurements", Elsevier, New York, 1966.

11. L.E. Brewer and C.C. Limbaugh, Appl. Opt. 11, 1200 (1972).

12. F.P. Chen and R. Goulard, J.Q.S.R.T. 16, 819 (1976).

13. D.W. Blair, J.Q.S.R.T. 14, 325 (1974).

14. P.J. Emmerman, R. Goulard, R.J. Santoro and H.G. Semerjian, J. Energy, 4, 70 (1980).

15. R.J. Santoro, H.G. Semerjian, P.J. Emmerman and R. Goulard, Int. J. Heat Mass Tr., 24, 1139 (1981).

16. W.T. Rawlins and W.C. Gardiner, Jl. Phys, Chem., 78, 497 (1974).

17. W.M. Houghton and C.J. Jachimowski, Appl. Opt. 9, 329 (1970).

18. S.C. Schmidt and P.C. Malte, J.Q.S.R.T., 16, 963 (1976).

19. P.C. Malte and D.T. Pratt, Proc. Fifteenth Symp. on Comb., p. 1061 (1975).

20. E.M. Bulewica, P.J. Padley and R.E. Smith, Proc. Roy. Soc., A315, 129 (1970).

21. R.J. Cattolica, Comb. and Flame, 44, 43 (1982).

22. K.C. Luck and W. Thielen, J.Q.S.R.T., 20, 71 (1978).

23. K.G.P. Sulzmann, J.E.L. Lowder and S.S. Penner, Comb. and Flame, 20, 177 (1973).

24. S.M. Schoenung and R.K. Hanson, Comb. Sci. and Tech. 24, 227 (1981).

25. P.L. Varghese and R.K. Hanson, J.Q.S.R.T., 24, 479 (1980).

26. H.G. Semerjian, S.R. Ray and R.J. Santoro, to be presented at the 1982 AIAA/ASME Thermophysics Conference.

27. K.C. Luck and F.J. Muller, J.Q.S.R.T., 17, 403 (1977).

28. E. Whiting, J.Q.S.R.T. 8, 1379 (1968).

29. M.T. Chahine, J. Atm. Sci. 27, 960 (1970).

30. D.Q. Wark and H.E. Fleming, Monthly Weather Review, 94, 351 (1966).

31. C.M. Chao and R. Goulard, in "Heat Transfer in Flames" Eds. N.H. Afgan and J.M. Beer, Scripta Books Co., Washington (1974).

32. P.J. Hommert and R. Viskanta, Int. J. Heat Mass Tr., 21, 769 (1978).

33. R.D. Cutting and I. McC. Stewart, Appl. Opt., 14, 2707 (1974).

34. T. Mitchell and R. Goulard, to be published.

35. G.N. Ramachandran and A.V. Lakshminarayanan, Proc. Nat. Acad. Sci. USA 68, 2236 (1971).

36. W. Swindall and H.H. Barrett, Physics Today, 30, 34 (1977).

37. L.A. Shepp and B.F. Logan, IEEE Trans. Nucl. Sci., NS-21, 21 (1974).

38. S.A. Tretter, "Introduction to Discrete-Time Signal Processing", John Wiley, New York (1976).

39. R.N. Bracewell, "The Fourier Transform and Its Applications" McGraw Hill, New York (1978).

40. J. McMahon, G.J. Troup, G. Hubbert and T.G. Kyle, J.Q.S.R.T. 12, 797 (1972).

41. R.E. Mitchell, A.F. Sarofim, L.A. Clomburg, Comb. Flame, 37, 227 (1980).

42. W.G. Mallard and W.C. Gardiner, J.Q.S.R.T. 20, 135 (1978).

43. W.M. Heffington, G.E. Parks, K.G.P. Sulzmann and S.S. Penner, J.Q.S.R.T. 16, 839 (1976).

44. B.N. Edwards and D.C. Burch, J. Opt. Soc. Amer., 55, 174 (1965).

45. P. Varanasi, L.A. Pugh and B.R.P. Bangaru, J.Q.S.R.T., 14, 829 (1974).

46. J. Anketell and A. Perry-Thorne, Proc. Roy. Soc., A301, 343 (1967).

47. G.H. Dieke and H.M. Crosswhite, J.Q.S.R.T. 2, 97 (1962).

48. R.C.M. Learner, Proc. Roy. Soc. London, A269, 311 (1962).

49. P.E. Rouse and R. Engleman, J.Q.S.R.T. 13, 1503 (1973).

Recent advances in absorption spectroscopy of OH and their implications in rotational temperature measurements

Charles C. Wang and Dafan Zhou*

Engineering and Research Staff, Ford Motor Company, Box 2053, Dearborn, Michigan 48121

Recent results on the band oscillator strengths of OH and their rotational dependence will be summarized and discussed. With the much improved values for the rotational transition probabilities deduced from these absorption studies, it should now be possible to make accurate determinations of rotational temperatures in combustion media.

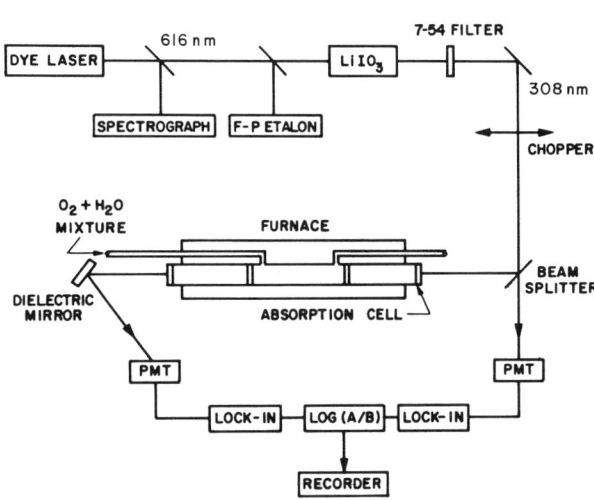

Fig. 1 Schematic of the experimental setup used in absorbtion measurements of OH

INTRODUCTION

The study of OH has been of interest because of its pivotal role in providing information fundamental to molecular physics. The existence of OH is also central to many important phenomena in galactic space[1], in the atmosphere[2], as well as in combustion media[3]. Measurements of the heat of formation of OH via its absorption spectrum have been attempted since the advent of quantum mechanics[4]; however, because of limited spectral resolution and detection sensitivity, results[5] have not been satisfactory. Similarly, attempts to determine the rotational temperature of OH were hampered by the inaccuracies in the calculated rotational transition probabilities, and efforts to correct these rotational transition probabilities were also unsuccessful[6]. With the use of a cw single-frequency tunable dye laser, we have made accurate measurements[7-10] of the line shape and band oscillator strength of various rotational-electronic transitions of OH as a function of the rotational quantum number. We have also developed a phenomenological theory which quantitatively accounts for the observed rotational dependences[11]. Based on these theoretical and experimental results, (1) a much improved value of $(1.09 \pm 0.04) \times 10^{-3}$ and 7.5 ± 0.2 nm^{-1} is confirmed for the band oscillator strength and its rotational constant respectively; (2) the spin-independent nature of both is experimentally verified; (3) the rotational dependence of the homogeneous linewidth is understood; and finally, a set of accurate values for the rotational transition probabilities of OH are now available. The results make possible an accurate determination of the rotational temperatures of OH, and it is hoped that they will rekindle the interest in using OH as a diagnostic tool in combustion media.

In the course of these studies, several interesting phenomena were observed. Most notable among them were the observation that[12] the rotational distribution of OH may be lifetime limited under certain conditions, thus exhibiting to some extent the rotational distribution with which the OH was born, and the observation that the Franck-Condon factor bears a different rotational dependence for different bands of transitions of OH. This latter effect imposes serious limitations on the utility of this concept of Franck-Condon factor for certain bands, but may negate other rotational dependent effects, thus making the band oscillator strength essentially independent of rotation for these bands.

---

* Permanent address: Wayne State University, Detroit, MI; on leave from the Institute of Applied Chemistry, Chinese Academy of Sciences, Changchun, China

EXPERIMENTAL

The experimental setup used in our measurements is shown schematically in Fig. 1. Briefly, a single-frequency laser source was tuned to sweep across various rotational electronic lines in the (0,0) band near 308 nm and the (1,0) band of OH near 282 nm; and the output from the laser traversed an absorption cell containing OH under thermalized conditions. The absorption signal thus generated was normalized with respect to the laser intensity, passed through a precision logarithmic amplifier, and recorded on a chart recorder. The line oscillator strength was then obtained by integrating over the absorption line shape and normalizing the results with respect to the thermalized rotational distribution. Results obtained in this manner are generally reproducible well within 2% from day to day and from one line to another for low-lying rotational levels. However, for absorption lines originating from higher rotational levels, the signal was weaker, and an uncertainty of 6% or larger was noted.

In arriving at the improved absorption sensitivity employed in our present experiment, considerable effort was spent in minimizing the apparent oscillation in the base line as a function of wavelength scanning. This oscillation arises from interference of beams reflected at various interfaces in the absorption path, and could be reduced by using wedged optics and expanded laser beam throughout the beam path. In doing so, we were able to realize an order of magnitude improvement on the absorption sensitivity, and could now detect an absorption signal as small as one part in 10^5.

RESULTS ON ROTATIONAL DEPENDENCE

In Fig. 2, the band oscillator strength for the Q_2-, Q_1-, and R_1-branch transitions in the (1,0) band of OH are shown as a function of the rotational energy of the level from which absorption originates. With our improved detection sensitivity, we have also remeasured the Q_1-branch transitions in the (0,0) band and extended these measurements to higher levels of rotational excitation. These results are also included in Fig. 2 for comparison. It is interesting to note that, over the range of rotational excitation shown, the band oscillator strength of the (1,0) band stays unchanged for Q_1- and Q_2-branch transitions; it appears to increase slightly for the R_1-branch transitions, but the rate of increase is of the order of data scatter, and thus it is also practically independent of rotation. On the other hand, the band oscillator strength for the (0,0) transitions is seen to decrease linearly with increasing rotational energy of excitation. One deduces from the slope for the (0,0) transitions that $\beta = 30 \pm 3$ nm^{-1} with a corresponding value of $\rho = 7.5 \pm 0.2$ nm^{-1}. These values are in agreement with the result of vibrational analysis[13] and with the less accurate value of $\beta = 40 \pm 10$ nm^{-1} deduced previously. Note that the band oscillator strength for the (1,0) transitions is about one-fourth of that for the (0,0) transitions. However, because of the rotational dependence of the latter, the ratio of these two oscillator strength values decreases from 0.280 for $Q_1(5)$ line to 0.258 for $Q_1(1)$. To our knowledge, the rotational dependence of (1,0) transitions has never been established before.

It is clear from the above observations that band oscillator strength exhibits different rotational dependence for different bands. Since the effect of rotation on the electronic transition moment is in all probability the same for all bands, this difference among various bands must come from changes in the Franck-Condon factor due to rotational stretching of the internuclear spacing. We note in this connection that the vibrational wave functions for the upper and lower states of the (1,0) transitions are nearly orthogonal, that the overlap integral of these vibrational wave functions would have been zero if it were not for the small difference in the equilibrium positions for these states, and that this overlap integral increases with small relative displacement of the equilibrium positions. On the other hand, the overlap integral for the (0,0) transitions should be nearly unity when the equilibrium positions coincide for the upper and lower states, and should decrease slowly with increasing difference between the equilibrium positions. Note also that the rotational constants are different, and thus the extent of rotational stretching is different, for the upper and lower states. Consequently, rotational excitation should induce changes in the relative equilibrium positions, which in turn should bring about different changes in the Franck-Condon factor for (1,0) and (0,0) bands. It follows that the observed difference in rotational dependence is a result of competition or reinforcement of the rotational effect on the electron transition moment with that on the Franck-Condon factor for these bands.

THEORY

In an attempt to arrive at a physical and more quantitative understanding of the cancellation of rotational effects in the (1,0) transitions, we have extended our previous analysis[8] to the (1,0) transitions and have included higher order terms for the rotationally induces stretching. According to this analysis, the rotational stretching is given by

$$r/r_e = 1 + x + 1.5(ar_e - 2)x^2 + \left[2.25(ar_e - 2)^2 + \frac{1}{12}(ar_e)^2\right]x^3 \quad (1)$$

$$x = (D/B)K(K+1) \quad (2)$$

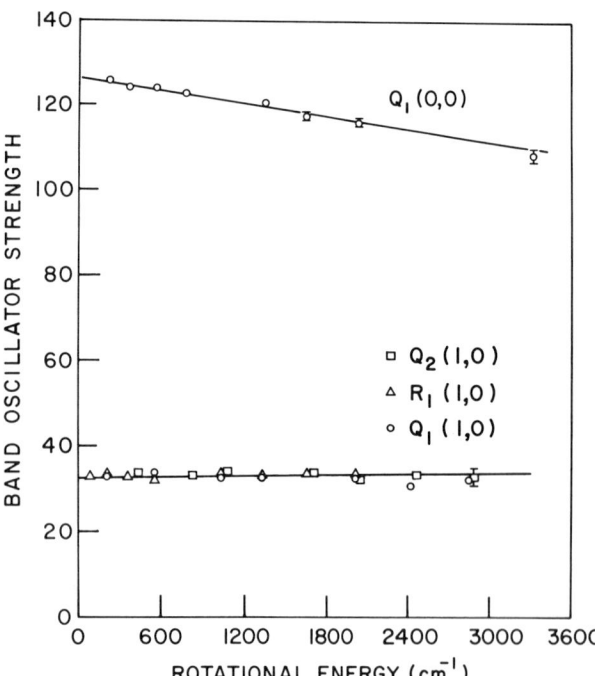

Fig. 2 Plot of the band oscillator strength for (0,0) and (1,0) transitions as a function of the energy of the rotational level from which absorption originates. The value of the band oscillator strength has been multiplied by a factor $\nu(1)/\nu(K)$, where $\nu(K)$ is the frequency of the transition originating from the Kth rotational level, and $\nu(1)$ that from the lowest level. This multiplying factor is generally very close to unity, and amounts to a maximum of 1.01 for K=13, the highest rotational level probed.

where r and r_e are respectively the internuclear spacing for each of the electronic and vibrational states with and without rotation, B and D are the corresponding rotational constants, K is the rotational quantum number, and a is the exponential coefficient in the Morse potential[14]. Equation (1) was obtained as the average internuclear position for the Morse-Pekeris wave function for each of the vibrational-electronic states in the presence of rotation. Upon expanding the Morse-Pekeris wave functions in terms of harmonic wave functions about these average internuclear positions, vibrational transition probabilities may then be obtained through elementary integration using these harmonic wave functions. Within the framework of this analysis, the effects of spin-axis coupling can also be taken into account by introducing appropriate modifications in Eq. (2). In general, these coupling effects are important for low-lying rotational levels only, where rotational stretching is small and the band oscillator strength deviates little from that for a rotationless molecule. Consequently, these effects may be neglected altogether for all practical purposes. The overall results of our analysis for the (0,0) transitions are found to be in excellent agreement with our experimental results and agree within 0.3% with the results of Dimpfl and Kinsey[15]. We are also encouraged by the fact that our analysis predicts a ratio of 0.262 for the band oscillator strength of $Q_1(1)$ line in the (1,0) band to that in the (0,0) band, which in effect reproduces the experimental value of 0.258 noted in Fig. 2. According to our analysis for the (1,0) band, there should be an increase of about 3% from K=1 to K=12 for Q_1- and Q_2- branch transitions, and a corresponding increase of about 7% for the R_1- branch transitions. This latter prediction is shown as a straight line in Fig. 2. It is clear from Fig. 2 that our analytical results are consistent with the null results of experiments well within the experimental uncertainties. Details of our analysis along with tabulation of the corrected rotational transition probabilities are being documented, and will be published elsewhere.

DISCUSSION

Based on the experimental and theoretical results presented above, it is possible to determine the changes in rotational temperature associated with the rotational dependence of the transition probabilities, and to find means to minimize such changes. In this regard, it is well to recognize that the amount of change varies widely from band to band, being extremely sensitive for the (0,1) transitions, but practically independent of rotation for the (1,0) transitions. Consequently, (1,0) transitions should always be used for the determination of rotational temperatures whenever possible.

The use of the (0,0) transitions is advantageous because of the higher transition probabilities associated with these transitions. The amount of change in the deduced rotational temperature may be calculated in a straightforward manner. Generally, this change as a percentage of the deduced rotational temperature increases with the rotational temperature and with the rotational quantum number of the levels involved in the deduction. As an example, we have shown in Fig. 3 the rotational temperature deduced from the iso-intensity method using $R_2(1)$ and $R_2(3)$ lines as the baselines. It is seen that the effect due to rotation vibration interaction amounts to about 1% at 300 K, but increases to about 4% at 1,000 K, about 7% at 2,000 K, and as much as 24% at 5,000 K.

Finally, it may be noted that the concept of rotational temperature is strictly invalid if the rotational distribution deviates from that in thermal equilibrium and our observation in gaseous discharge indicates that significant deviations could indeed occur in certain combustion media. In general, deviations from equilibrium may be expect to occur whenever the OH distribution is lifetime limited. Consequently, the results of rotational measurements must be used with caution.

ACKNOWLEDGMENTS

This research has been supported in part by the Department of Energy, by National Science Foundation through the University of Michigan, and by NASA through Wayne State University.

REFERENCES

1. K. M. Evenson, J. S. Wells, and H. E. Radford, Phys. Rev. Lett. 25, 199 (1970).
2. C. C. Wang and L. I. Davis, Phys. Rev. Lett. 32, 349 (1974); C. C. Wang, L. I. Davis, P. M. Selzer, and R. Munoz, J. Geophys. Res. 86C, 1181 (1981).
3. A. G. Gaydon and H. G. Wolfhard, "Flames, Their Structure, Radiation and Temperature" (Chapman and Hall, London, 1970), 3rd. ed.
4. O. Oldenberg and F. F. Rieke, J. Chem. Phys. 6, 439 (1938).
5. P. E. Rouse and R. Engleman, J. Quant. Spectrosc. Radiat. Transfer 13, 1503 (1973).
6. R. C. M. Learner, Proc. R. Soc. (London) A269, 311 (1962).
7. D. K. Killinger and C. C. Wang, J. Chem. Phys. 71, 1582 (1979).
8. C. C. Wang and D. K. Killinger, Phys. Rev. A20, 1495 (1979).
9. C. C. Wang and C. M. Huang, Phys. Rev. A21, 1235 (1980).
10. C. C. Wang, D. K. Killinger, and C. M. Huang, Phys. Rev. A22, 188 (1980).
11. C. C. Wang, M. T. Myers, and D. Zhou, Phys. Rev. Lett. 47, 490 (1981).
12. C. C. Wang and D. K. Killinger, Phys. Rev. Lett. 39, 929 (1977).
13. I. L. Chidsey and D. R. Crosley, J. Quant. Spectrosc. Radiation Transfer 23, 187 (1980).
14. G. Herzberg, "Molecular Spectra and Molecular Structure: I. Spectra of Diatomic Molecules" (Van. Nostrand, New York, 1950), 2nd edition.
15. W. L. Dimpfl and J. L. Kinsey, J. Quant. Spectrosc. Radiation transfer 21, 233 (1979).

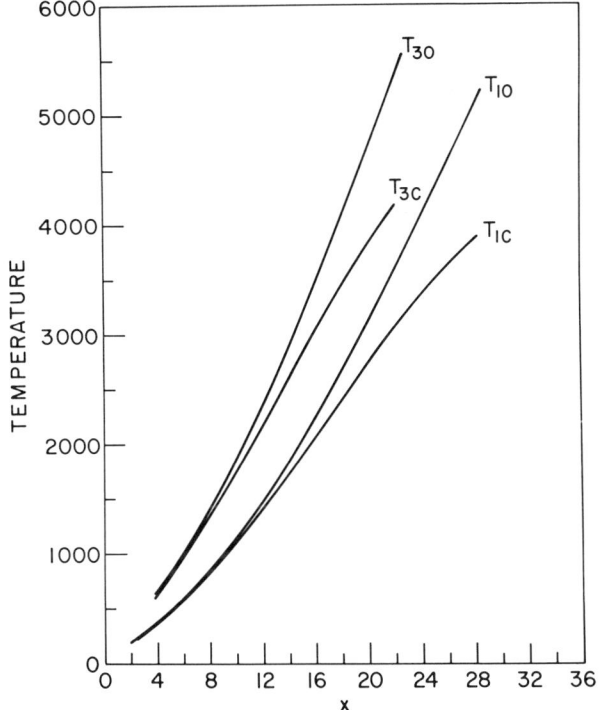

Fig. 3 Rotational temperature deduced with (T_c) and without (T) correction for rotation-vibration interaction. The subscript 1(3) indicates that $R_2(1)$ [$R_2(3)$] is used as the baseline for determining the iso-intensity line $R_2(x)$.

A high speed non-intrusive temperature diagnostic for combustion processes

R. W. McCullough

Aeronautical Research Associates of Princeton Incorporated, Princeton, New Jersey 08540

G. B. Northam

NASA Langley Research Center, Hampton, Virginia 23665

A temperature diagnostic for high speed measurement of combustion gases is described. The technique monitors ultraviolet absorption by OH radicals ($^2\Pi$–$^2\Sigma^+$) of narrow line resonance radiation produced in a flowing microwave discharge. Quartz fibre optics are used to transport the signal to and from the combustor as well as within the device. A high speed computer controlled data acquisition system is used which permits time resolved temperature measurements in excess of 10 kHz. The theoretical basis for the measurement is outlined and the results from a set of preliminary experiments at the NASA Hypersonic Propulsion Facility are presented.

INTRODUCTION

In the development of advanced combustor concepts, diagnostic systems are becoming an increasingly important tool for engineering development. Non-intrusive measurement techniques circumvent the interference problems and inherent inaccuracies associated with conventional sampling and thermocouple devices. Application of these techniques has been fostered to a large extent by the development of laser technology and the application of spectroscopy to the resolution of engineering problems.

This paper describes a non-intrusive temperature diagnostic designed for high frequency response measurements in the harsh environment of a supersonic combustion test facility. The diagnostic monitors absorption of UV radiation by hydroxyl (OH) radicals in the flow and hence the technique has wide applicability to other combustion flows. The specific device described in this paper was designed to operate with a wide bandwidth (> 20 kHz) and in a high noise environment (> 120 dB).

The ultraviolet bands of the hydroxyl radical have been chosen because OH occurs in most flames, and the radiation which is associated with the $^2\Sigma^+$ – $^2\Pi$ electronic transition, occurs in a region of the ultraviolet spectrum (280.0 nm to 340.0 nm) which is relatively free of extraneous radiation. In addition, high concentrations of OH occur in the hot reaction zone as opposed to much lower concentrations in cooler and unreacted portions of the flow. Consequently, problems of self-absorption in the cooler boundary layers are minimal and the "weighting" of the measurement is towards the core of the flow.

Absorption spectroscopy is used instead of emission or scattering for several reasons. Both the concentrations of OH and the population of the first excited electronic state $^2\Sigma^+$ are strongly temperature-dependent. Consequently, the intensity of OH emission spectra varies over many orders of magnitude both with time and position within the combustor chamber. A tremendous dynamic range problem is thus encountered when measuring high frequency fluctuations in the emission spectra of OH. The intensity of absorption spectra can only be less than or equal to the initial value obtained prior to absorption. In addition, OH radicals tend to be formed in an electronically excited or chemiluminescent state. The degree of thermodynamic equilibrium between the translational, rotational, and vibrational energy modes of these excited radicals is uncertain, therefore the emission spectra from these radicals is not a good indicator of the gas static temperature. While inelastic scattering techniques (e.g., CARS)[1] allow for very short time resolution, the measurements are typically made on an intermittent pulsed basis. Thus, while it is possible to make virtually instantaneous measurements, it is not possible to determine the size and regularity of the individual eddies or flow elements.

The concept of using OH ultraviolet bands for a temperature diagnostic is not new.[2] Previous work has generally used techniques of high resolution spectroscopy on steady flows, although there have been partially successful attempts[3] to use a rapid scanning monochromator to infer fast response temperatures. Recent work has demonstrated the use of laser absorption by individual lines[4] and laser induced fluorescence[5] to determine temperatures.

In the present system, we circumvent the need for high resolution spectroscopy by providing UV source radiation coincident with the OH lines in the gas. This discrete source is generated by an OH resonance lamp driven by a microwave cavity. The radiation is transported to and from the test apparatus by quartz fibre optics in order to minimize vibration and alignment problems. A reference spectral line unaffected by the combustion gas is also provided to indicate any changes in alignment during a run. The output from the OH lamp is monitored continuously during the run to account for intensity variations. Attenuated radiation transmitted through the combustion gas is dispersed by a low resolution monochromator. At the monochromator output plane, a quartz fibre optic array transports radiation in 2 nm wavelength intervals to individual photomultiplier tubes. The resulting system frequency response is limited only by the photon statistics of the lamp and the electronics associated with the photomultiplier tubes. The relative transmissions of the spectral channels are used to determine the temperature of the gas and the number density of hydroxyl radicals.

This work is supported by the National Aeronautics and Space Administration under Contract No. NAS1-14853.

THEORETICAL CONSIDERATIONS

SPECTROSCOPY OF OH

The basic equation that describes the propagation of radiation in non-scattering media is

$$\frac{di_\nu}{d\sigma} = -k_\nu(i_\nu - B_\nu)$$

This equation is applied along the "line of sight" of the radiation propagation. The blackbody function B_ν is a function of temperature along this line of sight while k_ν, the absorption coefficient, depends on local temperatures, species concentrations, and pressure. The absorption coefficient for a single transition is given by

$$k_\nu = \frac{c^2}{8\pi\nu^2} A_{u\ell} \; f(\nu) \left[\frac{N_\ell}{g_\ell} - \frac{N_u}{g_u}\right]$$

where $A_{u\ell}$ is the Einstein A coefficient, f is the normalized lineshape function, N is the state number density, and g is the degeneracy of the energy level.

The Einstein coefficient $A_{u\ell}$ indicates the absolute probability of a transition from one energy state to another and is determined by the quantum mechanics of the two states. The hydroxyl radical has been widely studied,[6] and the theory for the ultraviolet bands is well developed. An energy level diagram is shown in Figure 1a. The ground state is a $^2\Pi$ electronic state with a very small lambda doubling effect, and the excited level is a $^2\Sigma^+$ electronic state. The electronic transition used here is between a single pair of states. The principal vibration transition is between the two lowest lying vibrational states (0-0 band), however there are contributions in the spectral region of interest due to the (1-1) and (0-1) bands. Due to the small energy quantum per

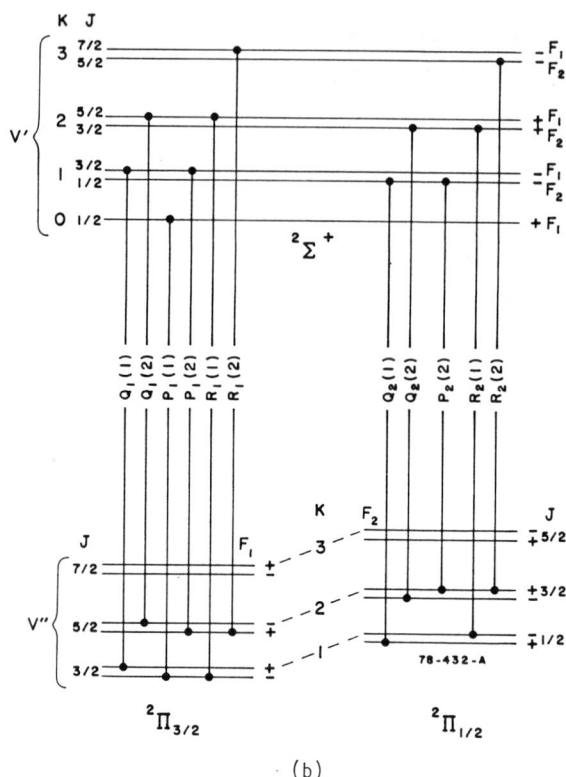

Figure 1b. Rotational Structure for OH $^2\Sigma^+ - ^2\Pi$ Transitions.

rotational levels, there is a myriad of rotational levels and the allowed transitions are indicated in Figure 1b. In the case of OH, the usual assumption of complete separation of vibration and rotation is not valid. Therefore, the transition probability is given by

$$A_{u\ell} = \frac{8\pi^2 e^2}{m_e c^3} \frac{\nu^2}{g_u} \frac{f_{v'v''}}{4} S_{J'J''} T_{v'J'v''J''}$$

where $f_{v'v''}$ is the Franck-Condon factor for vibrational transition, $S_{J'J''}$ is the Honl-London factor for rotational transition, and $T_{v'J'v''J''}$ is the vibration-rotation interactions factor due to Learner.[7] The energy of a given level is a function of the electronic, vibrational, and rotational quantum numbers and is given by

$$E = E_e + E_v + E_R$$

The electronic energy level is constant for the bands under consideration, and the vibrational energy levels are given by

$$E_v = w_e(v + 1/2) - w_e x_e(v + 1/2)^2$$

The values used are given in Table I.

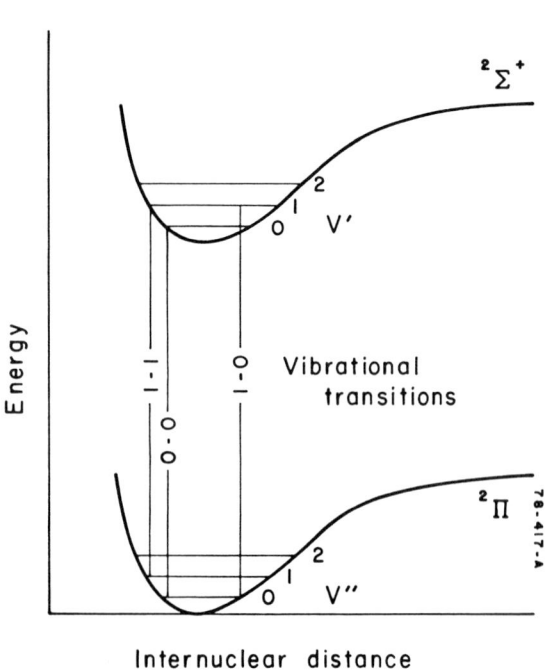

Figure 1a. Vibrational Structure for OH $^2\Sigma^+ - ^2\Pi$ Transitions.

Table I. Vibrational Constants of OH

$^2\Sigma^+$ Electronic State

$w_e(\text{cm}^{-1})$	$w_e x_e(\text{cm}^{-1})$
3184.28	97.84

$^2\Pi$ Electronic State

$w_e(\text{cm}^{-1})$	$w_e x_e(\text{cm}^{-1})$
3735.21	82.21

The rotational energy levels are complex due to the angular momentum coupling in OH which is best described by Hund's case (b).[8] In this case, the spin angular momentum S is only weakly coupled to the internuclear axis. The electronic angular momentum ($\Lambda = 0$ for the $^2\Sigma^+$ state, $\Lambda = 1$ for the $^2\Pi$ state) and the angular momentum of nuclear rotation N form a resultant vector K which represents the total angular momentum neglecting spin. This vector can have magnitude $K = \Lambda, \Lambda + 1, \ldots$. The angular momenta K and S form a resultant J which is the total angular momentum including spin. The possible values of J for a given K are $J = K + S, K + S - 1, \ldots, K - S$.

For both the $^2\Sigma^+$ and $^2\Pi$ states, $S = 1/2$. Therefore each electronic state has two possible J values for each K; $J = K \pm 1/2$ (except for $K = 0$ when $J = 1/2$). These doublets are denoted as

$^2\Pi$ state $F_1(K)$: $J = K + 1/2$ $F_2(K)$: $J = K - 1/2$

$^2\Sigma^+$ state $F_1(K)$: $J = K + 1/2$ $F_2(K)$: $J = K - 1/2$

Allowed K values for the $^2\Pi$ state are $K = 1, 2, 3, \ldots$, whereas for the $^2\Sigma^+$ state, the allowed K values are $K = 0, 1, 2, \ldots$. In addition, due to the interaction between the rotation of the nuclei and the electronic orbital angular momentum, each J level in the $^2\Pi$ state is split into two components (Λ - doubling). Thus, $F_1(k)$ and $F_2(k)$ are each doublets themselves; however, this splitting is very small.

The rotational energy levels of the $^2\Sigma^+$ state are given by[9]

$$F_1'(K) = B(K)(K+1) - D(K^2)(K+1)^2 + R(K+1/2)$$

$$F_2'(K) = B(K)(K+1) - D(K^2)(K+1)^2 - R(K+1/2)$$

The rotational energy levels of the $^2\Pi$ state are given by[9]

$$F_1''(K) = B\left\{(K+1)^2 - 1 - 1/2\left[4(K+1)^2 + a(a-4)\right]^{1/2}\right\} - DK^2(K+1)^2$$

$$F_2''(K) = B\left\{K^2 - 1 + 1/2\left[4K^2 + a(a-4)\right]^{1/2}\right\} - DK^2(K+1)^2$$

The rotational constants are given in Table II.

The general selection rule for dipole radiation is $\Delta J = 0, \pm 1$. An additional restriction in the present case is that for Hund's case (b): $\Delta K = 0, \pm 1$. In addition, there is the rule even $\not\leftrightarrow$ odd. Transitions which satisfy all conditions are strong, forming the main branches. The main branches are designated as follows:

$P_1(K) = F_1''(K-1) - F_1'(K)$ $J - 1 \to J$

$P_2(K) = F_2''(K-1) - F_2'(K)$ $J - 1 \to J$

$Q_1(K) = F_1''(K) - F_1'(K)$ $J \to J$

$Q_2(K) = F_2''(K) - F_2'(K)$ $J \to J$

$R_1(K) = F_1''(K+1) - F_1'(K)$ $J + 1 \to J$

$R_2(K) = F_2''(K+1) - F_2'(K)$ $J + 1 \to J$

Table II. Rotational Constants of OH

$^2\Sigma^+$ Electronic State

	v = 0	1	2
B	16.961	16.129	15.287
D	.00204	.00203	.00208
R	.1122	.1056	.0997

$^2\Pi$ Electronic State

	v = 0	1	2
B	18.515	17.807	17.108
D	.00187	.00182	.00182
a	-7.547	-7.876	-8.214

A diagram showing the energy levels and the principal rotational transitions of a given vibrational band (V',V") is given in Figure 1b. Satellite branches also exist but are much weaker and are not included in the present calculation. Using the energy level formulas described earlier, the wavelengths of the principal transitions can be calculated. Values for the rotational line strengths $S_{J'J''}$ were taken from Reference (9).

With the model for the molecular structure outlined above, the position and line strength can be computed for each line in the band. To compute k, we superimpose these lines broadened by the lineshape function to create a spectrum. The lineshape function reflects the contribution to broadening the energy levels due to: 1) Doppler shifting from the molecular velocity, 2) collisional broadening, and 3) natural broadening due to the uncertainty principle.

To model the spectra accurately, three effects are combined into a full Voight profile.[10] An exception is the microwave discharge where the pressure is very low (~ 1 torr) and the effective temperature is high (~ 2000 K). In this regime, the collision broadening is 1.7×10^{-4} cm^{-1}, while the Doppler width is 1.3×10^{-1} cm. Hence a Gaussian line profile is acceptable. Conditions in the combustor are considerably higher with a pressure of .3 - 1 atm and temperature from 1500 K to 3000 K. The absorption line profiles must use the full Voight profile. The relative line widths of the source and combustion gas are not sufficiently different so that a "narrow emission line" approximation is valid. The transmission calculation is hence carried out by finely dividing the emission and absorption lineshape and integrating over the line.

EQUILIBRIUM CONSIDERATIONS

In order to compute k_ν, the population densities of the upper and lower states must be known. Boltzman statistics predict that for systems in equilibrium, the population of a state s with energy E_s is given by

$$\frac{N_s}{N} = \frac{g_s \exp(-E_s/kT)}{Q(T)}$$

where $Q = \Sigma g_s \exp(E_s/kT)$. In systems where there are multiple components of energy (e.g. electronic-vibrational-rotational) with widely disproportionate values of the energy quanta, it is possible for the different modes to be internally equilibrated at different "temperatures". This occurs often when the kinetics allow fast modes to track changes (e.g., behind shocks), while the slower modes lag. It can also happen when energy is pumped in one mode selectively as in laser absorption. For the present case, we can "uncouple" the modes and assign independent vibrational, rotational, and electronic temperatures. Hence the population of a given level is given by

$$\frac{N_{EVR}}{N} = \frac{g_E \exp(-E_E/kT)}{Q_E} \frac{g_V \exp(-E_V/kT)}{Q_V} \frac{g_R \exp(-E_R/RT)}{Q_R}$$

The computer code used to compute k allows different temperatures to be assigned for each of the modes.

The electronic, vibrational, and rotational "temperatures" serve only to describe the population distribution among the energy levels. In the case of thermodynamic nonequilibrium, none of these temperatures may be equal to the static temperatue T_S of the gas. When complete thermal equilibrium exists, $T_E = T_V = T_R = T_S$, and the populations are described by a simple Boltzman distribution with a single temperature which is equal to the static temperature of the gas. Neither the OH in the lamp nor the OH in the combustion chamber is expected to be in complete thermal equilibrium. Therefore, the relation of the "temperatures" of the energy modes to the static temperature must be examined carefully to extract meaningful data.

Molecules formed with nonequilibrium distributions, as typical in combustion products, achieve thermal equilibrium with the surrounding gas as a result of collisions. Due to the relatively small energies involved in rotation, the rotational level populations typically redistribute most rapidly, bringing $T_R = T_S$ after just a few collisions. The vibrational levels generally equilibrate next, requiring a greater number of collisions before $T_V = T_S$. The electronic temperature equilibrates even more slowly, since electronic energies are relatively large (32,681 cm^{-1} in the present case for $^2\Sigma^+ \rightarrow {}^2\Pi$). However, due to many mechanisms including energy resonances with colliding molecules or potential energy curve crossings by dissociative states, the modes may equilibrate at different rates. As explained below for OH molecules, the approach toward thermal equilibrium is strikingly different for the $^2\Sigma^+$ and $^2\Pi$ states.

The emission spectra from OH molecules in the electronically excited $^2\Sigma^+$ state formed in flames has been investigated previously to determine the rotational and vibrational population distribution in this excited state. Broida and Heath[11] found that the rotational temperatures were much higher than the static gas temperature. For an acetylene-oxygen flame with a static temperature of T_S = 2700 K, the rotational temperature was as high as 9000 K. This extremely large discrepancy indicates a distribution which is very far from thermal equilibrium and is unusual since the rotational mode generally equilibrates first. Carrington[12] studied the quenching of the fluorescence from OH molecules in the $^2\Sigma^+$ state and found larger cross-sections for electronic quenching of the $^2\Sigma^+$ state by O_2, CO_2, and H_2O than for rotational energy transfers within the $^2\Sigma^+$ state. That is, during collisions with these molecules, there is a relatively high probability that an OH molecule in the $^2\Sigma^+$ state will undergo a radiationless transition to the $^2\Pi$ state with the released energy being transferred into the electronic, vibrational, and rotational modes of the colliding molecule. The probability for this electronic quenching is higher than for the process in which the OH molecule remains in the $^2\Sigma^+$ but changes to a different rotational level, as is required for rotational equilibrium. Therefore, in gas mixtures containing O_2, CO_2, and H_2O, electronic quenching predominates, and the rotational and vibrational level populations of the $^2\Sigma^+$ state are determined primarily by the nature of the specific excitation process. For this reason, studies of emission spectra from OH molecules in the $^2\Sigma^+$ state generally cannot yield accurate information regarding the static temperature of the gas.

For absorption studies, on the other hand, the population distribution of OH molecules in the ground electronic $^2\Pi$ state determines the amount of absorption. For the $^2\Pi$ state, there is obviously no electronic quenching to compete with rotational and vibrational equilibration. Therefore, the vibrational and rotational temperatures of the $^2\Pi$ state are expected to be much closer to the actual static temperature of the gas. Absorption measurements of OH molecules produced in hydrogen-air and methane-air mixtures have been performed recently.[13] Both the vibrational and rotational degrees of freedom of the $^2\Pi$ state were found to be in thermal equilibrium with the gas with $T_V = T_R = T_S$ within experimental uncertainty. Thus, absorption measurements, which depend on the vibrational and rotational temperature of the ground $^2\Pi$ state, give a much better indication of the static temperature of the gas.

This discussion points out the very significant advantage in using absorption from ground electronic states rather than emission from excited electronic states to determine static gas temperatures. In the present application, the output spectra of the lamp can be observed directly, and a model of the population distribution of OH molecules in the $^2\Sigma^+$ state in the lamp can be made by finding the temperatures T_E', T_V', T_R' for the $^2\Sigma^+$ state which describe the actual lamp output. Although these "temperatures" will, in general, be much higher than the static gas temperature in the lamp, they are used only to describe the source spectrum. Absorption within the combustion chamber is calculated by assuming $T_V'' = T_R'' = T_S''$ for the ground $^2\Pi$ state, which is a good approximation as discussed above. Therefore, the spectroscopic absorption technique is well suited to the goal of obtaining static gas temperature measurements.

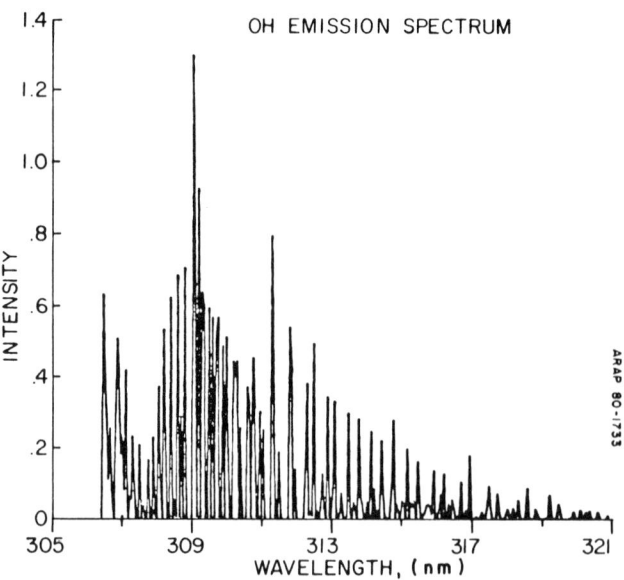

Figure 2. OH Emission Spectrum from Microwave Lamp. ($T_V = T_R$ = 2000 K)

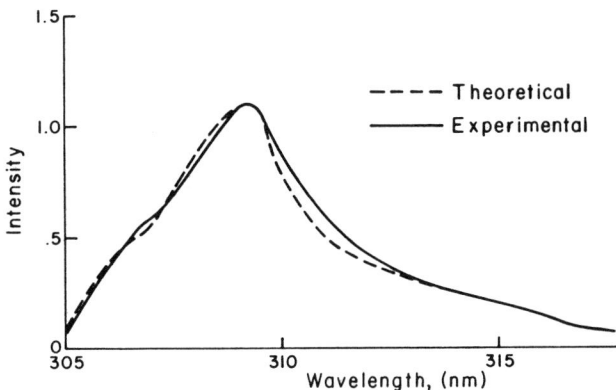

Figure 3. Low Resolution OH Emission Spectrum From Microwave Lamp. ($T_V = T_R = 2000$ K)

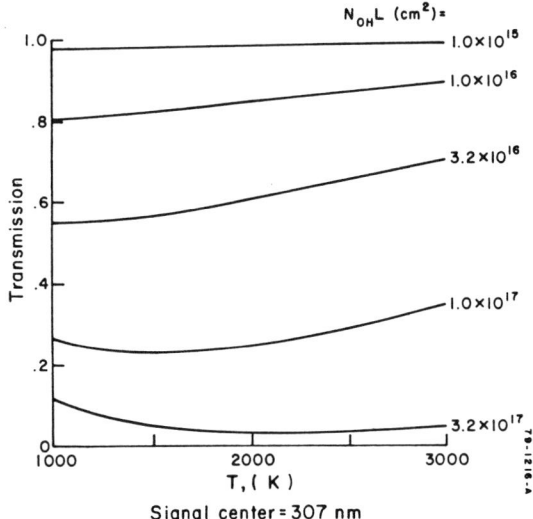

Figure 4. Transmission of Microwave Lamp Radiation for 306 nm $< \lambda <$ 308 nm.

In order to determine a set of temperatures with which to model the microwave discharge, theoretical spectra were generated in which the vibrational and rotational temperatures were allowed to vary independently between 1000 K and 3000 K. The resulting spectra were then convoluted with a trapezoidal monochromator transfer function to yield a prediction of the observed spectra. These spectra were then compared to an actual spectra from the lamp at standard operating conditions. It should be noted that the electronic temperature does not affect the shape of the spectra, only the overall intensity level. The temperature which demonstrated the closest fit to the actual lamp spectra are $T_V' = T_R' = 2000$ K. Predicted high resolution spectra for these conditions are shown in Figure 2. The spectra after passing through the monochromator are shown in Figure 3.

Once the detailed output spectrum from the lamp has been modeled it is relatively straightforward to model the transmission through a spatially uniform slab of hot combustion gas. The resulting high resolution predicted spectra can be averaged using the monochromator transfer function f_{mi} associated with each fibre bundle i. The resulting prediction is for a relative transmission for a spectral interval (about 2 nm for the present instrument)

$$T_i^{th} = \frac{\int_0^\infty f_{m,i}\, i_\nu \exp(-k_\nu \ell)\, d\nu}{\int_0^\infty f_{m,i}\, i_\nu\, d\nu}$$

DATA INVERSION TECHNIQUE

Tables of transmission for various spectral intervals for a range of temperatures and OH number densities were constructed. This approach of storing computed $T_i^{th}(T, N_{OH}\ell)$ was dictated by the length of computer time required to perform the detailed line-by-line calculation. Three typical transmission calculations are summarized in Figures 4 and 5 for center wavelengths of 307 and 315 nm. The channel at 307 nm near the band center shows little sensitivity to temperature, while the channel at 315 in the "tail" of the band shows increased temperature sensitivity. This is due to the influence of the (1-1) band and high lying rotational state of the (0-0) band.

In principle only two spectral channel transmissions need be measured to deduce the two unknowns T and $N_{OH}\ell$. The use of multiple channels however serves to average out small random errors that could lead to large errors in the deduced temperature and number density. The approach utilized here is statistical and valid for any arbitrary number of channels.

If we define ϕ_i as:

$$\phi_i \equiv \frac{1}{\sqrt{w_i}} (T_i^{th} - T_i^{exp})$$

where T_i^{th} and T_i^{exp} are the theoretical and experimental transmittances of the ith spectral channel and w_i is a weighting function, we can then treat the temperature and the log of the OH number density as parameters and allow them to vary until we minimize the function

$$\Gamma = \phi_i\, \phi_i$$

where we use the convention that a repeated subscript implies summation.

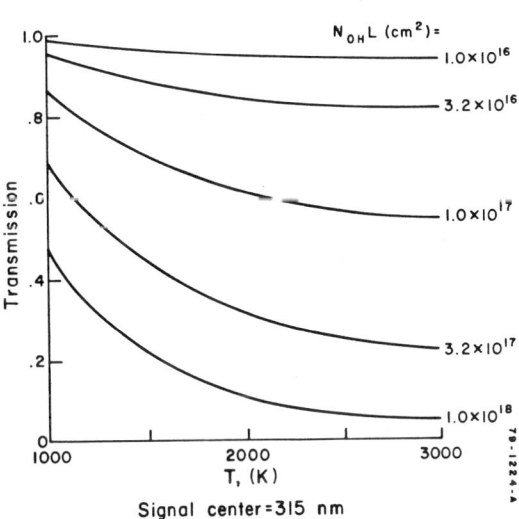

Figure 5. Transmission of Microwave Lamp Radiation for 314 nm $< \lambda <$ 316 nm.

We can take advantage of the quadratic form of the above equation using a technique originally described by Gauss[14] to find the functional minimum using only first derivatives with respect to the unknowns. If we let

$$P_1 = T$$

$$P_2 = \ln(N_{OH}\ell)$$

then the equation to be solved for optimization is

$$\left.\frac{\partial \Gamma}{\partial P_j}\right|_{P_i^{opt}} = 0$$

If we assume that ΔP_m is this change in P_m necessary to reach the minimum in Γ using Gauss' approximation with local linearization, we arrive at the equation to be solved

$$\left(\frac{\partial \phi_i}{\partial P_j}\frac{\partial \phi_i}{\partial P_m}\right)\Delta P_m = -\phi_i \frac{\partial \phi_i}{\partial P_j}$$

This set of linear equations is solved for ΔP_m. The current values of P_m are incremented by ΔP_m, and the procedure is repeated until convergence occurs.

This numerical technique has the advantage that it produces an estimate of the uncertainty[15] associated with the measurement

$$\sigma_{P_j} = \left(\frac{\Gamma(P_j)}{n-2}[B_{jj}]^{-1}\right)^{1/2}$$

where B_{ij} is defined by

$$B_{jk}\frac{\partial \phi_i}{\partial P_k}\frac{\partial \phi_i}{\partial P_m} = \delta_{jm}$$

EXPERIMENTAL APPARATUS

NASA FACILITY

The diagnostic was designed to operate in the severe environment of NASA Langley's Hypersonic Propulsion Facility wind tunnels. This facility is shown in Figure 6. Compressed air is mixed with high pressure vitiate. Vitiate here is used to mean a combusted hydrogen/oxygen mixture whose products have an oxygen mole fraction of .21. This mixture is expanded through a supersonic nozzle to a test section. The use of the vitiate/air mixture simulates the effect of supersonically diffusing an air flow of $M = 6$ to one with $M = 1.5$.

The facility can handle mass flows up to 3 kg/sec with stagnation enthalpies up to 2000 kJ/kg. In order to estimate possible temperature and OH number densities available, a performance map of the facility was constructed and is shown in Figure 7. Here we compute the adiabatic flame temperature as a function of stoichiometry, R^*, in the burner test section as well as air/vitiate mix ratio, f^*. $R^* > 1$ indicates fuel rich combustion while $R^* < 1$ is fuel lean and, $f^* = 1$ indicates pure vitiate while $f^* = 0$ indicates

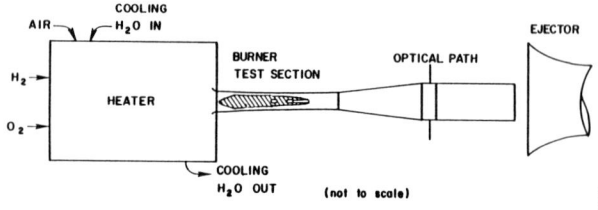

Figure 6. NASA Scramjet Test Facility.

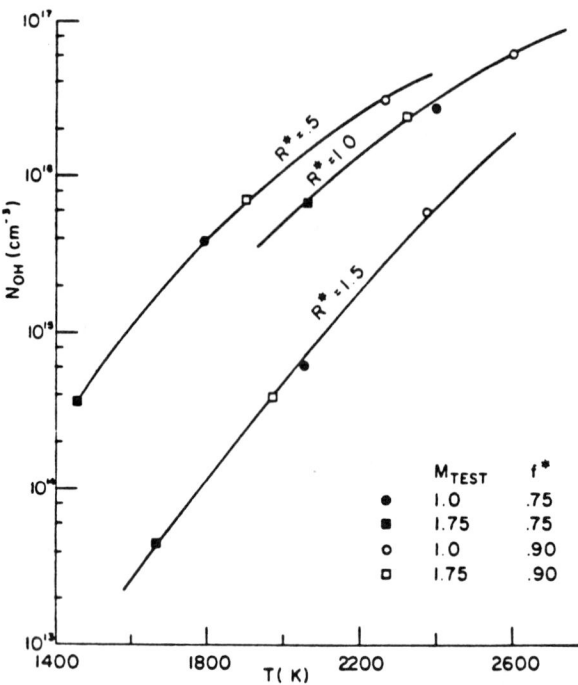

Figure 7. Performance Map for Scramjet Facility.

pure air for the oxidizer. This chart indicates the range of temperatures and OH concentrations to be expected for combustion Mach numbers between 1 and 1.75 but does not include the effects of thermal choking within the test section which could increase the temperature farther.

The facility has a complete automatic control and computer data acquisition system. In addition, an extensive network of pressure taps and other instrumentation is available to monitor the performance of the wind tunnel and burner test section during a run.

OPTICAL HARDWARE

The optical system is shown schematically in Figure 8. The source optics and receiver optics were housed in purged aluminum boxes lined with heavy rubber to minimize vibration. The UV radiation was coupled to and from the test section via quartz fibre optics. At the test section, a miniature optical head converted the fibre output cone to a parallel beam directed across the channel.

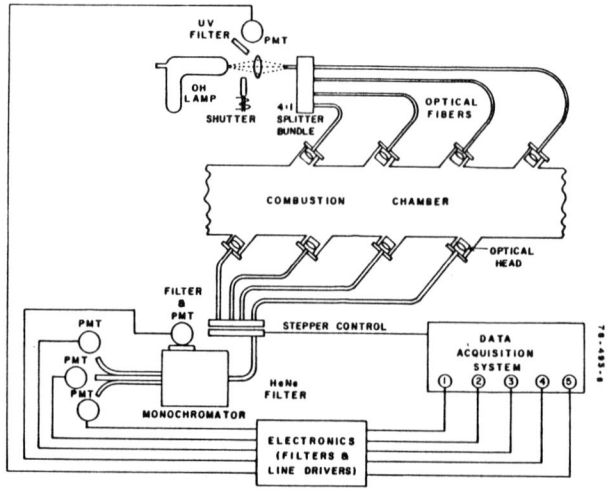

Figure 8. OH - Temperature Diagnostic Optical Schematic.

The source consisted of a microwave cavity with a continuous flow of low pressure (~ 1 torr) water vapor. The lamp intensity is monitored with a local photomultiplier tube with a bandpass filter centered at 310 nm. The lamp output is focused to the input of a quadrifurcated fibre bundle. This bundle divides the output signal into four channels so that several different optical paths may be measured.

The fibre optics lines coming from the test cell are connected to an optical multiplexer. This multiplexer is driven by a stepper motor so that any sequence of spatial channels may be read during a run. The output of the multiplexer is connected to a fibre bundle that has a circular aperture in one end and a 1 mm x 10 mm rectangular aperture on the other end. This rectangular aperature is located at the entrance of a 1/4 m Jarrell-Ash monochromator (model 82-410) and forms the entrance "slit". At the exit plane, the slit assembly has been replaced by a stack of four 10 mm x 1.2 mm rectangular quartz fibre optics on 1.2 mm centers. The other end of these fibre bundles is connected to EMI 9781 B photomultipliers. These tubes were chosen for their spectral responses, ruggedness, and noise characteristics.

In addition to the four output channels, the zero order reflection from the grating is directed through a narrow bandpass filter at 4861 Å to a separate photomultiplier tube. This tube monitors the hydrogen line emitted by the microwave discharge. Tests verified that in the operating regions used, the hydrogen line intensity "tracked" that of the OH band. The ground level of the hydrogen line is sufficiently energetic that there is no significant interaction with the combustion gases under study. This channel served as a lamp intensity monitor as well as a reference to detect the effects of alignment shifts and beam steering effects during the runs.

DATA ACQUISITION SYSTEM

Due to the high frequencies present in the turbulent supersonic flowfield, it was necessary that the reading on the phototubes be sampled simultaneously. Standard multiplexing practices could yield different conditions along the optical path for the sequentially sampled spectral channels. To circumvent this possibility, a four channel transient digitizer (Lecroy model 8210) was used. A diagram of the data acquisition system is shown in Figure 9.

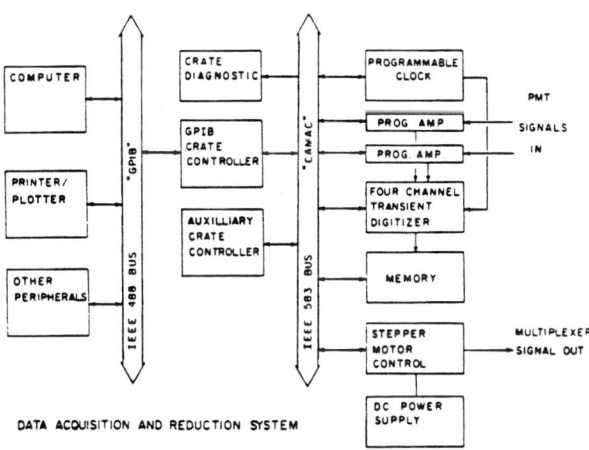

Figure 9. OH - Temperature Diagnostic Data Acquisition System.

In order to minimize computer controller time lag during a run, a local programmable clock (Lecroy model 8501) and memory (Lecroy model 8210) was utilized. Before a given test run, a sequence of pulses are selected and stored in the programmable clock. During the run, control is localized with the clock strobing the digitizer in a preprogrammed fashion. Incoming data is amplified by programmable instrumentation amplifiers (Lecroy model 8100) and then digitized for storage in local memory. At the termination of the run, the data is transferred to the main computer. This system allowed flexibility in design and operation. The initial control program development was done on an HP9835 desktop calculator. When the system was operational, a Modcomp 4-35 minicomputer was substituted for increased speed.

UNCERTAINTY IN THE TEMPERATURE MEASUREMENT

The temperature resulting from the measurements and data inversion has an uncertainty associated with it. The uncertainty or error bounds of the temperature can be traced to two sources. First, there are sources of noise in the various components of the experimental apparatus which combine to produce fluctuations on the perceived intensity signals of the various optical channels. These noise sources arise from the process of photon detection, signal amplification, and signal digitization. In addition to the three noise sources, which are amenable to analytical treatment, there exists noise sources such as ground loops and system microphonics which must be experimentally measured and minimized by trial and error procedures.

The second source of uncertainty in the temperature measurement arises from shortcomings in the theory used to convert the measured transmissions to a temperature. Uncertainties in this area include the effect of the uncertainty in the measurements of the spectroscopic constants as well as the uncertainty resulting from the assumptions inherent in the theory.

The quantitative assessments of the contributions of these various uncertainties to the resulting temperature uncertainty will be discussed in this section. In the case of noise, the resulting uncertainty can be computed using the data inversion scheme and allowing the measured parameter to vary over a range associated with the signal to noise ratio. A similar treatment can be performed to test the sensitivity to uncertainties in the physical model. This latter approach is much more time consuming since the line-by-line inversion tables must be computed for each variation in the theory to be checked. The statistical inversion technique produces an estimate in the temperature error based on the difference between the theoretical transmissions and the experimental data for the best fit values. This estimate contains the effects of random fluctuation due to noise, and to a certain extent, due to the departures of the theory from reality.

PHYSICAL NOISE SOURCES

The most fundamental noise is due to the statistics of the photon flux and the photon detection process. Photon arrival is described by Poisson statistics, and it can be shown[16] that the signal to noise ratio for a photomultiplier tube connected to a circuit with a bandwidth Δf

$$SNR_p = \sqrt{\eta\, I_p/\Delta f}$$

were η is the quantum efficiency of the photocathode, and I_p is the photon rate. For the EMI 9781 B tube

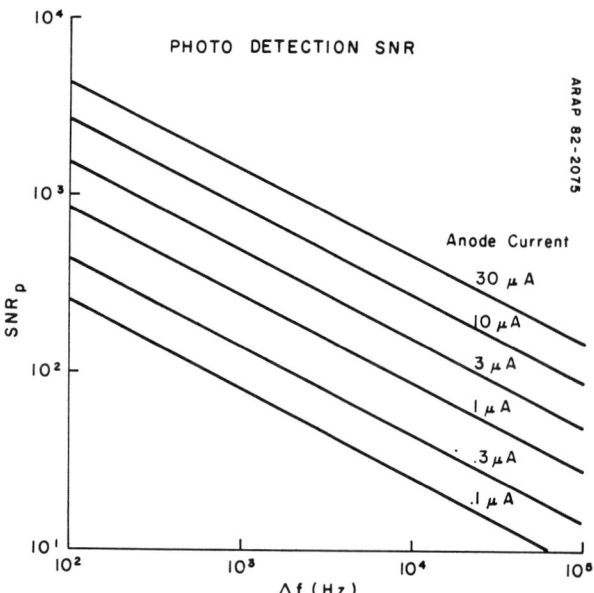

Figure 10. Photo Detection Signal to Noise Ratio.

used in the present study, the sensitivity was about 5×10^4 A/w at the wavelength of interest. Typical anode currents for the unattenuated signal ranged from 1 to 30 μa yielding photon currents of 3×10^7 to 1×10^9 sec^{-1}. Typical quantum efficiency for the 9781 B tube at 315 nm is 22%. A plot of photoelectron signal to noise ratio for the present system is shown in Figure 10. It can be seen that for strong signals, the SNR is very good even at large bandwidths. Even for the weaker signals, the SNR at 10 kHz still exceeds 80. It is worth noting that the signals attenuated by the combustion gas can have transmission reduced to 5% or lower. This reduces the SNR significantly. Fortunately, the brightest spectral regions are attenuated more than the weaker portion of the spectrum. This effect tends to even out the SNR across the various spectral channels.

Other noise sources in the phototube are believed to be negligible. The dark current is of the order of 1 na and dynode noise is small.

The signal from the PMT is amplified by a Lecroy model 8100 amplifier. This device has a SNR under conditions used in the present system of 1400. In most cases, the bandwidth of the system will be run as high as possible, and the amplifier SNR will not contribute substantially to the overall noise.

There is noise associated with the digitization. The digitizers are 10 bit devices with an uncertainty of 1 bit. The maximum and minimum digital values can be set between 0 and 1024 by use of the gain and offset controls on the amplifier. The signal to noise ratio due to digitization can be shown to be given by

$$SNR_D = f_{sig} f_{span} 1024$$

where f_{span} is the fraction of the possible range available to be used by the signal excursion, and f_{sig} is the fraction of the available span that the signal represents. Thus we see that if the system is set to use 90% of the span, a signal with 50% transmission signal results in a SNR_D of 492, while a 5% transmission with a 50% span set gives $SNR_D = 25.6$.

In addition to the noise sources mentioned above, additional noise components appeared from ground loops and system microphonics. The system is split with the optical components in a sealed test cell and the data acquisition system in a separate room. Resultant ground loops have been minimized by the use of grounding straps and isolation transformers. The microphonics result from the excessive acoustic noise level interacting with the optical components. These effects were minimized by lining the instrument boxes with lead and using substantial optical mounting hardware. Both the ground loop and acoustical noise sources have been reduced by these methods to levels below the other noise sources.

UNCERTAINTIES IN THE PHYSICAL MODEL

The model for the emission and absorption spectrum is based upon a set of assumptions and a set of physical constants. There are uncertainties associated with each assumption. The basic assumptions are:

1) The combustion gas to be measured has a ground state in vibrational and rotational equilibrium with the gas translational motion.

ii) The lamp spectra can be modeled as a collection of OH molecules with independently equilibrated electronic, vibrational, and rotational energy distributions.

iii) The spectroscopic model (Hund's case b) for OH is accurate for the range of rotational levels of importance to the measurement.

iv) The absorbing gas can be treated as a single slab with uniform properties along the optical path, and the contribution of emission from the combustion gases is negligible.

The physical constants that are used in the computation are:

i) The energy level constants as given in Tables I and II.

ii) The constants which determine line strength such as the Franck-Condon factor and the vibrational-rotational interaction factors.

iii) The constants which determine the line broadening due to collisional effects.

Each of these contributions to the temperature uncertainty will be discussed below.

The first assumption of ground state equilibrium was discussed earlier in the paper. We expect that for high pressure flows that this will be reasonably accurate. The second assumption is less certain. The emission spectrum is really dominated by contribution from two bands (0-0 and 1-1) so that a single vibrational temperature is all that is required to establish the relative band intensities. The 1-0 band is very weak and does not contribute substantially to the lamp emission in the wavelength range of interest. The question of rotational equilibrium in the lamp is more suspect. One possibility is that the higher lying rotational states are equilibrated at a different temperature than the lower states. The temperatures were chosen to fit the observed band shape based on a medium resolution spectra from the lamp. Work is currently underway to examine the high resolution spectrum of the lamp to check the validity of a single parameter (rotational temperature) distribution function for rotational energies. Alternate approaches include a multiparameter curve fit for the rotational population distribution or even possibly tabulating the intensities for all the important lines. (There are about 200 lines in the observed spectra which contribute to the signal.) The effect of this uncertainty on temperature inversion is difficult to determine. Because the spectral windows are 2.0 nm wide and include a large number of lines, one would expect that spectral regions near the band head which contain lines from a wide variety of rotational levels would be more affected than windows near the band tail where the rotational levels of a given group of lines are closely spaced. Our experience has been that inversions using spectral regions in the tails of the band produce the best results. This, however, could be attributed to the enhanced temperature sensitivity in the region as discussed earlier.

We expect that the assumption of the spectroscopic model's accuracy is a good one. The hydroxyl radical has been widely studied and data from its spectra were used to formulate the transition rules for this type of transition. In addition, the lines under consideration here have rotational quantum numbers under 25. This fact coupled with the detailed correction factors for vibrational-rotational interactions which were discussed earlier lends confidence to the accuracy of the spectroscopic model.

The assumption that the absorbing gas is uniform is inherent in the creation of the transmission tables. Numerical studies have been performed for realistic temperature and OH number density profiles. The predicted temperature was close to the average temperature in the core of the flow. The strength of the technique is that it tends to be most sensitive to areas of the flow with high OH number densities, i.e., flame zones or high temperature regions. Since there are four measured values of transmission and two physical quantities produced, it is conceivable that the distribution of temperature along the path could be estimated by assuming a distribution of temperature and OH number density and creating transmission tables with the distribution parameters as additional variables.

The accuracy to which the spectroscopic constants are known is believed to be high enough so that there is no appreciable contribution to overall uncertainty. The energy level constants are used to determine the location of the individual lines. Since the lamp emission lines and the gas absorption lines coincide, a small error in absolute line position has no effect. The fact that the spectral windows are 2 nm wide means that the lines within this window will be accounted for properly as the line position uncertainty is of the order of .001 nm. The energy constants are also used to determine the population of the states in the combustion gas. As such, the uncertainty in these constants directly translates to temperature uncertainty. The temperature uncertainty resulting from uncertainties in these constants is much less than that due to noise sources.

Similar arguments can be made for the line strength constants. For the rotational constants, detailed studies of the OH spectra have led to complex expression for intensities matched to the data.[7] While the vibrational line strengths are not as well-known, their accuracy is adequate as the influence of the 1-1 band is relatively small.

The constants with the largest degree of uncertainty are the collisional broadening constants. There is a separate collision broadening parameter for each type of molecule that collides with OH. For the H_2-air flames, a detailed accounting should be made for the collisions with H_2O, N_2, O_2, H_2, NO, NO_2, and also other OH. We use a single collision broadening cross-section. The results of the temperature prediction are not terribly sensitive to the uncertainty because the lineshape is very nearly the same for all of the lines, and the temperature is sensitive to relative transmission. The effect of increasing the collision cross-section is to broaden the lineshape and decrease the absorption at line center where the lamp signal is concentrated. This effect will change the total in-band transmission but not the relative transmission from window to window. It is expected that this effect will influence the uncertainty in the measured value of OH which depends strongly on the total absorption measured.

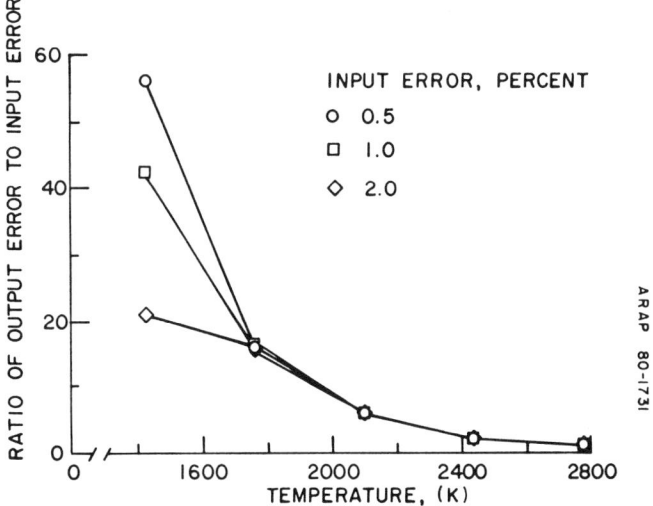

Figure 11. Error Analysis Results For Relative Input Error.

ERROR PROPAGATION ANALYSIS FOR THE INVERSION SCHEME

A numerical experiment was carried out to determine how errors in the measured values of transmission propagates to errors in the deduced temperature. Random errors were assigned to the exact transmission values for a series of values of temperature and OH number density. The ratio of percent output error to percent input error was then tabulated. The results of this exercise for a three channel system with spectral windows at 309 nm, 311 nm, and 313 nm are shown in Figure 11.

As can be seen, the leverage factors increase as the temperature decreases. At 2500 K, a 1% input error in the transmission values measured produces a 4% or 100 K uncertainty in the temperature.

We can now consider a case typical of the experimental results reported later in this paper. With all of the fibre optic links in place, the average signal was on the order of 1 µa for the three channels of transmission data. The input to the amplifier was a 1 mΩ resistor with a .01 µf capacitor yielding a bandwidth on the order of 100 Hz. The digitizer was typically set to a span of 90% full capacity, and the measured transmissions were 50% to 80%. Combining the SNR_P of 800 with an average SNR_D of 460 in a statistical fashion, we get a total SNR per spectral channel of 400. Since two channels of data (signal and reference) are required to compute a single transmission, the SNR for each transmission channel is about 282. From Figure 7, we see that for temperatures around 2200-2400 K the error leverage is about a factor of 5 which results in a final SNR of 56 or a temperature uncertainty of 100 K for a 3σ error band.

SYSTEM IMPROVEMENTS AND EXTENSIONS

The primary noise source is the photon noise due to the strength of the narrow line source radiation.

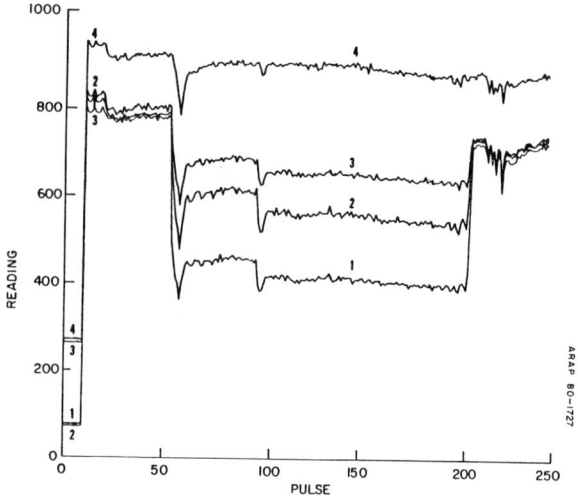

Figure 12. Run Data From Photomultiplier Tubes.

Enhancements in the photon flux can be made by increasing the brightness of the source, improving the transmission and coupling efficiency in the fibre optic links, and improving the optics which collect and collimate the radiation along the optical path. The most significant recent improvements have been made in the fibre optics themselves with transmission increases throughout of a factor 10. The use of a 12 or 14 bit digitizer would also improve system performance.

Planned extensions for the system include increasing the number of spectral channels to six and the development of a cooled probe for use with subsonic combustors.

EXPERIMENTAL RESULTS

A series of initial experiments under actual operating conditions have been conducted. During these tests, several minor system modifications were implemented, but in general, the system performed well. Running procedures called for a number of readings before the combustion run to establish the PMT output with the lamp shuttered and at full intensity runs. Several runs were conducted with the lamp shuttered to verify that the UV emission from the combustor was sufficiently small to preclude interference with the lamp signal.

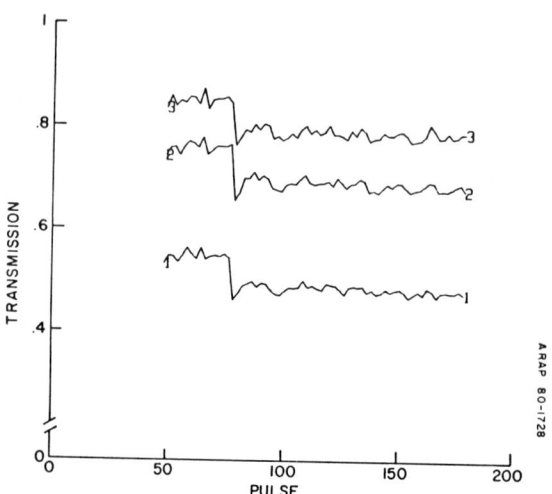

Figure 13. Transmission Data.

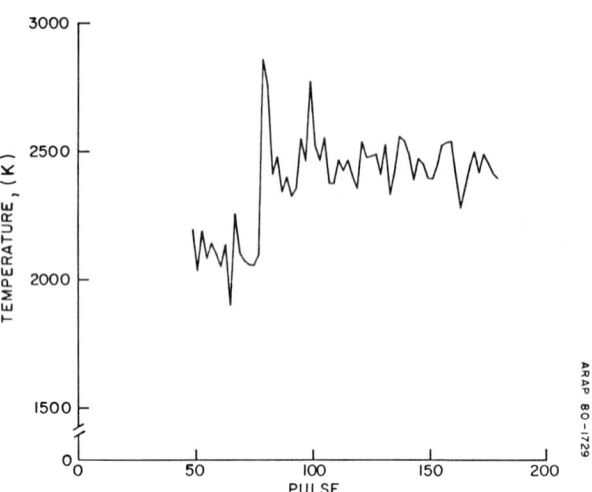

Figure 14. Temperature Data.

Data from a sample run is shown in Figure 12. Each "pulse" is a snapshot reading approximately 0.1 microseconds long. For this particular run, 10 readings were taken with the OH lamp shuttered and 10 more with the lamp open before the run. The clock was then programmed to take 300 readings at a rate of 20 per second. The run commences with pulse 21 and at pulse 40, additional fuel is introduced to the combustion chamber. Channel 1 is the spectral channel centered at 313 nm, channel 2 is centered at 315 nm, channel 3 is centered at 317 nm, and channel 4 is the hydrogen reference line. The attenuation both at the onset of the run, and at the time of additional hydrogen injection, is apparent.

Figure 13 shows the computed transmission for the three channels and Figure 14 shows resulting temperatures. The statistical uncertainty in the temperatures measurement is ± 150 K. The combustion is extremely noisy and unstable, and fluctuations in the flame temperature can be seen to exceed the uncertainty.

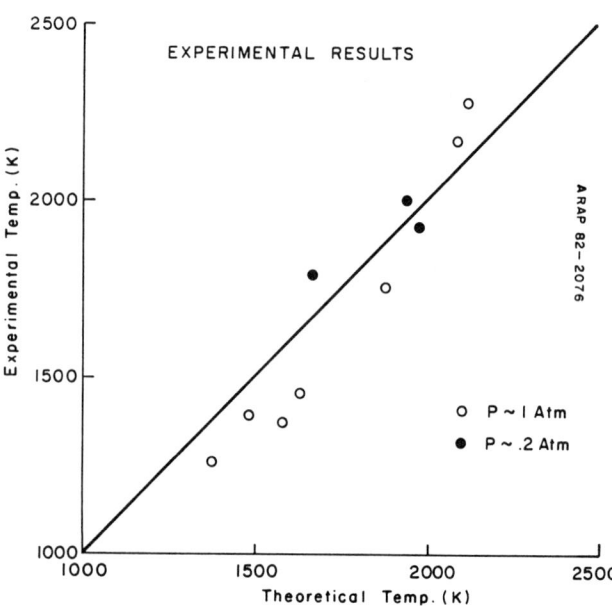

Figure 15. Comparison of Data from OH Temperature Diagnostic (experiments) and Data Inferred from Operational Parameters (theoretical).

Several tests with varying stoichiometry were conducted. The results comparing average measured temperatures to calculated temperatures are shown in Figure 15. Agreement is seen to be satisfactory.

Future tests are planned in which this diagnostic will be applied to a quiescent flat flame and compared with other spectroscopic techniques. A fourth spectral channel has been added and extension to six channels is also planned to give an additional measurement and hence to reduce the uncertainty.

SUMMARY AND CONCLUSIONS

A non-intrusive spectroscopic diagnostic has been described. This technique determines the flame temperature by monitoring the relative transmission in several spectral regions of OH resonance radiation from a microwave discharge. Theoretical performance was computed from spectroscopic data. Hardware has been designed and built to implement this technique at the NASA Scramjet test facility. Preliminary tests demonstrated the systems ability to function in a harsh noisy environment. Preliminary results show satisfactory agreement between predicted and measured temperatures. Further work is required to verify this technique in a controlled laboratory flame.

REFERENCES

1. Shirley, J.A., Hall, R.J. and Eckbreth, A.C., "Investigation of the CARS Measurements in Scramjet Combustion," NASA Contractor Report 159280, July 1980.

2. Penner, S.S., Quantitative Molecular Spectroscopy and Gas Emissivities, Addison-Wesley, Realing, Mass., 1959.

3. Neer, M.E., and Drewry, J.E., Supersonic Combustion Studies Using Absorption Spectroscopy. AIAA Paper 76-133 presented at AIAA Aerospace Sciences Meeting, Washington, D.C., Jan. 1976. Also Experimental Diagnostics in Gas Phase Combustion Systems, Progress in Astronautics and Aeronautics, 53, 1977.

Also Experimental Diagnostics in Gas Phase Combustion Systems, Progress in Astronautics and Aeronautics, 53, 1977.

4. Cattolica, R.T., "Laser Absorption Measurement of OH in a Methane-Air Flat Flame," Sandia Laboratories Report SAND79-8717, 1979.

5. Chan, C. and Daily, J.W., "Measurement of Temperature in Flame Using Laser Induced Fluorescence Spectroscopy of OH," Applied Optics, 19, No. 2, p. 1963, 1980.

6. Mohan, S. and Sbardanand, "Free Radical OH," NASA SP-373, 1975.

7. Learner, R.C., Proc. R. Soc., A269, p. 311, 1962.

8. Herzberg, G., Molecular Spectra and Molecular Structure, I. Spectra of Diatomic Molecules, Van Nostrand Reinhold Co., New York, 1950.

9. Dieke, G.H., and Crosswhite, H.M., "The Ultraviolet Bands of OH, "Journal of Quantitative Spectrosc. Radiation Transfer, 2, p. 97, 1962.

10. Armstrong, B.H., Journal of Quantitative Spectrosc. Radiation Transfer, 7, p. 61, 1967.

11. Broida, H.P., and Heath, D.F., "Spectroscopic Survey of Energy Distributions of OH, C_2, and CH Radicals in Low Pressure Acetylene-Oxygen Flames," Journal of Chem. Phys., 26, p. 223, 1957.

12. Carrington, T., "Electronic Quenching of OH ($2\Sigma+$) in Flames and Its Significance in the Interpretation of Rotational Relaxation," Journal of Chem. Phys., 30, p. 1087, 1959.

13. Schmidt, S.C., and Malte, P.C., "Spectroscopic Absorption Measurements of OH in High-Intensity Continuous Hydrogen/Air and Methane/Air Combustion," Journal of Quantitative Spectrosc. Radiation Transfer, 16, p. 963, 1976.

14. Wilde, D.J. and Beightler, C.S., Foundation of Optimization, Prentice Hall, Englewood Cliffs, N.J., 1967.

15. Bevington, P.R., Data Production and Error Analysis in the Physical Sciences, McGraw Hill, New York, 1969.

16. Photomultiplier Manual, RCA Corp., 1976.

Determination of the time evolution of the electron temperature profile of reactor-like plasmas from the measurement of blackbody electron cyclotron emission*

P. C. Efthimion, V. Arunasalam, R. A. Bitzer, and J. C. Hosea

Princeton University, Plasma Physics Laboratory, Princeton, New Jersey 08544

Plasma characteristics (i.e., $n_e \gtrsim 1 \times 10^{13}$ cm^{-3}, $T_e \gtrsim 10^7$ K, $B_\phi \gtrsim 20$ kG) in present and future magnetically confined plasma devices, e.g., Princeton Large Torus (PLT) and Tokamak Fusion Test Reactor (TFTR), meet the conditions for blackbody emission near the electron cyclotron frequency and at few harmonics. These conditions, derived from the hot plasma dielectric tensor, have been verified by propagation experiments on PLT and the Princeton Model-C Stellarator.[1,2] Blackbody emission near the fundamental electron cyclotron frequency and the second harmonic have been observed in PLT and is routinely measured to ascertain the time evolution of the electron temperature profile.[2,3] These measurements are especially valuable in the study of auxiliary heating of tokamak plasma. Measurement and calibration techniques will also be discussed with special emphasis to our fast-scanning heterodyne receiver concept.[4]

I. Introduction

In the study of astrophysics and gas-discharges, a plasma refers to an ionized gas. Over ninety-five percent of the matter in the Universe is in the plasma state including all of the stars and our sun. Controlled thermonuclear fusion research is directed toward producing a star-like plasma in a laboratory environment in order to generate limitless energy. There are a number of light element fusion reactions that generate energy. However, the fusion of deuterium (D) and tritium (T) ions has the largest cross section and the lowest temperature threshold to sustain a fusion burn. This fusion reaction is

$$D + T \rightarrow n(14 \text{ MeV}) + He^4 (3.5 \text{ MeV}) \quad (1)$$

where n is the neutron and the energy of the particles is in parentheses. Even for the D + T reaction plasma temperatures on the order of 10^8 K are required. In this paper we will discuss the determination of the plasma electron temperature by measuring the intensity of the electron cyclotron emission.

The study of fusion reaction in a controlled environment has developed plasma confinement devices based upon inertial confinement driven by potent laser beams[5] and magnetic confinement in complex topologies.[6] Presently, the magnetic confinement device, the "tokamak", has produced reactor-like temperatures.[7] These conditions were obtained in the Princeton Large Torus (PLT) by heating the plasma with energetic neutral beams. Furthermore, the soon to be completed Princeton Tokamak Fusion Test Reactor (TFTR) is designed to achieve an energy yield in excess of the input energy to the tokamak plasma.

The tokamak[8] is a donut-shaped device with a strong toroidal magnetic field ($B_\phi \sim 10 - 80$ kG) (Fig. 1). A current is induced in the toroidal direction by external coils in order to Ohmically heat the plasma to 10^7 K. This induced current gives rise to a weak poloidal magnetic field ($B \sim 1-10$ kG) that enhances particle confinement. A detailed description of the tokamak can be found in Ref. 8. The electrons follow a helical trajectory along the toroidal field that causes the electrons to radiate at the electron cyclotron frequency and at the harmonics. As will be discussed in Section II, the intensity of the electron cyclotron emission is a function of the electron velocity distribution $f(x,v,t)$, the electron density n_e, electron temperature T_e, and plasma geometry. Under appropriate plasma conditions the emission at the electron cyclotron frequency and the second harmonic is blackbody and therefore the emission intensity is proportional only to the electron temperature. Section III is a description of the important cyclotron emission and absorption experiments along with examples of PLT temperature profiles obtained from such blackbody emission. Section IV will be a review of the diagnostic techniques with special emphasis to our fast-scanning heterodyne receiver concept.[4]

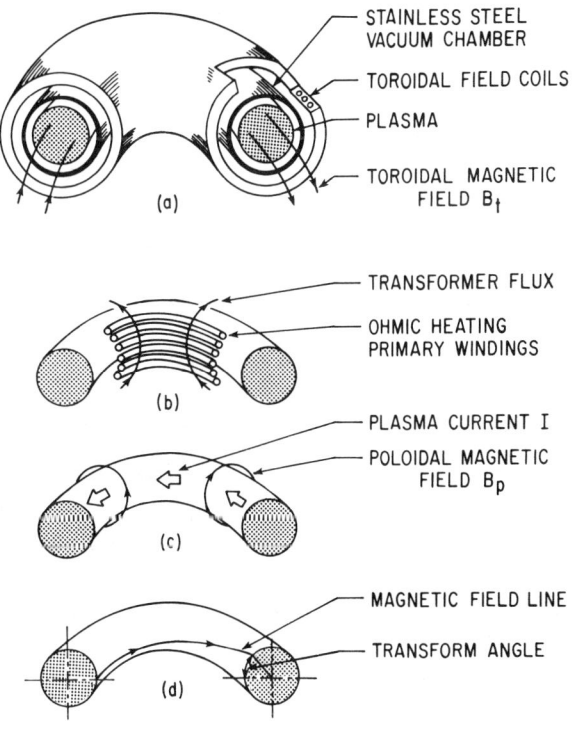

Fig. 1. Schematic of the conventional tokamak configuration.

II. Theory

The principles of radiation transfer, Kirchhoff's radiation law, and waves in a magnetized plasma define the relationship between the plasma local electron temperature and the electron cyclotron emission.[9] The equation of radiation transfer is a statement of electromagnetic energy conservation in a dielectric medium: The flow of spectral radiation intensity I_ω along an incremental ray path length $d\ell$ is equal to the emission coefficient j_ω minus the absorption $\alpha_\omega I_\omega$,

$$n_r^2 \frac{d}{d\ell}\left(\frac{I_\omega}{n_r^2}\right) = j_\omega - \alpha_\omega I_\omega, \quad (2)$$

where α_ω is the absorption coefficient of radiation of frequency ω and n_r is the index of refraction of the magnetized plasma. (See ref. 9 for a detailed description of n_r). The solution of this equation of transfer may be written

$$I_\omega = I_\omega(\text{inc}) e^{-\tau_o} + \int_0^{\tau_o} S_\omega e^{-\tau} d\tau \quad (3)$$

where $I_\omega(\text{inc})$ is the radiation intensity incident on the medium, $S_\omega = (1/n_r^2)(j_\omega/\alpha_\omega)$ is referred to as the source function, $\tau = -\int \alpha_\omega d\ell$ is the optical depth, τ_o is the total optical depth of the medium. For the problems considered in this paper S_ω is at most a very weak function of I_ω and there is no incident radiation, then Eq. (3) becomes

$$I_\omega = S_\omega (1-e^{-\tau_o}) \quad (4)$$

for situations where there are no reflecting boundaries (i.e. a single path absorption). However, if r is the reflection coefficient of the plasma containment vessel, then the multiple path solution of Eq. (3) may be written

$$I_\omega = S_\omega \frac{1-e^{-\tau_o}}{1-re^{-\tau_o}}. \quad (5)$$

The source function is derived from Planck's blackbody theory, Einstein's coefficients for emission and absorption, and Kirchoff's law. Since photons obey Bose-Einstein statistics, the blackbody emission intensity $B_o(\omega, T)$ is

$$B_o(\omega,T) = S_\omega = \frac{\hbar \omega^3}{8\pi^3 c^2} \frac{1}{(e^{\hbar\omega/\kappa T} - 1)}, \quad (6)$$

where $\hbar$ is Planck's constant, ω is the radian frequency, κT is the temperature in energy units and c is the velocity of light. In the classical limit $\kappa T \gg \hbar\omega$, Eq. (6) becomes the Rayleigh-Jeans Law

$$B_o(\omega,T) = \frac{\omega^2 \kappa T}{8\pi^3 c^2}. \quad (7)$$

For a classical plasma,[9] in the classical limit,

$$j_\omega = \int \eta_\omega(p) f(p) d^3p, \quad (8)$$

and

$$\alpha_\omega = \frac{8\pi^3 c^2}{n_r^2 \omega^2} \int \eta_\omega(p) \frac{\partial f}{\partial \varepsilon} d^3p, \quad (9)$$

where $\eta_\omega(p)$ is the Einstein's spontaneous emission coefficient for an electron of momentum p, and $\partial f/\partial \varepsilon$ is the derivative of the electron velocity distribution with respect to energy. The appearance of $\partial f/\partial \varepsilon$ in Eq. (9) is a reflection of the fact that α_ω is a consequence of the direct absorption minus the stimulated emission. Naturally, when the distribution function f is Maxwellian, the source function for a classical plasma may be written

$$S = j_\omega/n_r^2 \alpha_\omega = \frac{\omega^2 \kappa T_e}{8\pi^3 c^2}. \quad (10)$$

Thus, the emission intensity from a Maxwellian plasma is determined by the electron temperature T_e when $\tau_o \gg 1$. However, if the distribution function is non-Maxwellian, then the radiation temperature $\tilde{T}_r = S_\omega(8\pi^3 c^2/\omega^2 \kappa)$ is in general not equal to the electron temperature T_e. Thus, for a non-Maxwellian plasma the temperature dependence of the emission is difficult to determine unless the details of the distribution function are exactly known. Substituting the expression for the source function Eq. (10) into Eq. (4) the intensity becomes

$$I_\omega = \frac{\kappa T_e \omega^2}{8\pi^3 c^2}(1-e^{-\tau_o}). \quad (11)$$

Even if the distribution is Maxwellian the optical depth must be calculated to see if the plasma is a blackbody. When the plasma is highly absorptive, $\tau \gtrsim 2$, the plasma is considered black and

$$I_\omega = \frac{\kappa T_e \omega^2}{8\pi^3 c^2}. \quad (12)$$

The calculation of τ_o is not simple for a plasma in a strong magnetic field. In general, it is a complex function of electron temperature, electron density, plasma geometry, and the emission wave features of the magnetized plasma.

Plane polarized waves propagating perpendicular to the magnetic field (B_ϕ) can have two possible orientations of the electric field $\tilde{E}$: the wave is referred to as "ordinary" when $\tilde{E} \parallel B_\phi$ and is referred to as "extraordinary" when $\tilde{E} \perp B_\phi$ (See Ref. 10 for a detailed discussion). But, for parallel propagation it is customary to talk of right-hand and left-hand circularly polarized waves. For the determination of T_e via blackbody electron cyclotron emission measurements from plasmas in tokamak type devices one is particularly interested in near-perpendicular propagating waves only. The index of refraction n_r can range from zero to infinity. When $n_r^2 = 0$, the wave enters a cutoff region of plasma where it is reflected. When $n_r^2 \to \infty$, the wave enters a resonance region where it may be absorbed, reflected, or converted to a different plasma wave. For the ordinary mode $n_r^2 = 0$ when $\omega = \omega_{pe}$ where $\omega_{pe} = (4\pi n_e e^2/m)^{1/2}$ is the electron plasma frequency. The extraordinary mode has two cutoff frequencies ω_{co1} and ω_{co2} when $\omega_{pe}^2/\omega^2 = 1 \pm \omega_{ce}/\omega_{co1,2}$ and has a resonance when $\omega = \omega_{UH}$ where ω_{UH} is the upper hybrid frequency defined as $\omega_{UH} = (\omega_{pe}^2 + \omega_{ce}^2)^{1/2}$. In the preceding equation, the "+" sign is associated with ω_{co1} and "-" with ω_{co2}. In Fig. 2 is a diagram of the characteristic frequencies of blackbody emission in the PLT plasma as a function of the plasma position. The curve of ω_{pe} has a conical shape because the density profile in PLT is parabolic. The ω_{ce} curve has the characteristic $1/R$ dependence for all tokamaks and is plotted here for $B_\phi = 32$ kG. As stated earlier, we are interested in waves of frequencies $\omega \simeq \omega_{ce}$ and $\omega \simeq 2\omega_{ce}$. There is a unique frequency for each harmonic at each radial position in the plasma and therefore the emission intensity as a function of position can be obtained by measuring the emission over the indicated frequency range for a specific magnetic field. Now for $\omega < \omega_{pe}$, the ordinary mode wave will not propagate because of the cutoff condition. For the extraordinary mode, the wave will not propagate for $\omega < \omega_{co1}$, and from the right hand side of the diagram it can not propagate into the shaded region if $\omega < \omega_{co2}$. The right hand side of this diagram is the low magnetic field side of the plasma and is the large major radius (R) side of the tokamak. In tokamaks $B_\phi \sim 1/R$. Thus for the extraordinary mode the cyclotron layer is not accessible from the low field side of the tokamak. It is however accessible from the high field side. On the other hand, the ordinary mode emission can propagate out of the plasma towards both the low and high field side of the tokamak as long as $\omega > \omega_{pe}$. Also, the extraordinary mode second harmonic electron cyclotron emission can propagate out from either side.

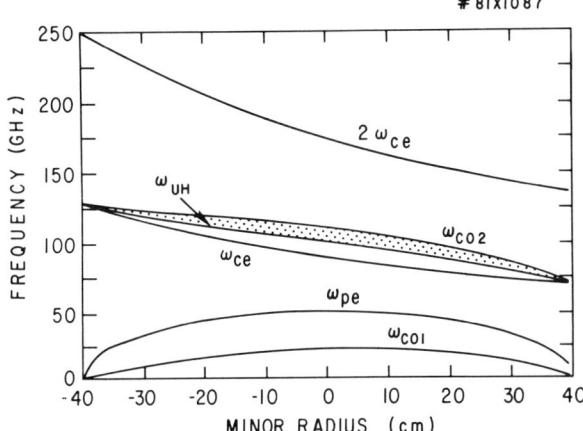

Fig. 2. A plot of the characteristic electron cyclotron frequencies of the blackbody emission in the PLT plasma as a function of the plasma position. This plot illustrates the acessibility of polarized blackbody emission in a tokamak.

Many researchers (Schwinger, Landau, Oster, Schott and Trubnikov) have considered the problem of the radiation spectrum due to the motion of electrons in a magnetic field.[11] The Schott-Trubnikov distribution of the emission intensity for a radiating electron is

$$\frac{dI}{d\omega\, d\Omega} = \frac{e^2 \omega^2}{2\pi c} \sum_{n=1}^{\infty} \left[\left(\frac{\cos\theta - \beta_\parallel}{\sin\theta}\right)^2 J_n^2\left(\frac{\omega}{\omega_{ce}} \beta_\perp \sin\theta\right) \right.$$
$$\left. + \beta_\perp^2 J_n'^2\left(\frac{\omega}{\omega_{ce}} \beta_\perp \sin\theta\right) \right] \delta(n\omega_{ce} - \omega(1 - \beta_\parallel \cos\theta)) \quad (13)$$

where θ is the angle between the wavevector $\underset{\sim}{k}$ and the magnetic field, n is the harmonic number, $\beta_\parallel = V_\parallel/c$, $\beta_\perp = V_\perp/c$, and $\underset{\sim}{V}$ is the electron velocity. For perpendicular propagation ($\theta = \pi/2$) the emissivity η_ω for linear polarized waves in the ordinary mode

$$\eta_\omega^{(o)} = \frac{e^2 \omega^2}{2\pi c} \sum_{n=1}^{\infty} \beta_\parallel^2 J_n^2(n\beta_\perp)\, \delta(n\omega_{ce} - \omega), \quad (14)$$

and for the extraordinary mode

$$\eta_\omega^{(e)} = \frac{e^2 \omega^2}{2\pi c} \sum_{n=1}^{\infty} \beta_\perp^2 J_n'^2(n\beta_\perp)\, \delta(n\omega_{ce} - \omega). \quad (15)$$

Using the expressions for η_ω in Eqs. (8) and (9), one can calculate the appropriate emission coefficient j_ω and the absorption coefficient α_ω for both the ordinary and the extraordinary modes of propagation. Knowing α_ω it is relatively easy to obtain closed form expressions for the optical depth $\tau_o = -\int \alpha_\omega d\ell$. If $\tau_o \gtrsim 2$, then the corresponding emission is a blackbody emission and is directly proportional to the electron temperature T_e.

Another group of researchers (Arunasalam, et al.,[1,12] Fidone, et al.,[13] Hosea, et al.,[3]) have used a somewhat different approach to the calculation of the optical thickness τ_o. These authors make full use of the well known hot plasma dielectric tensor,[9,10] $\underset{\approx}{K}(\omega, \underset{\sim}{k})$.

The waves in a plasma are complicated by the presence of a magnetic field that results in anisotropic wave propagation. After Fourier analysis in space and time, the Maxwell electromagnetic field equations yield, for plane waves of the form

$$\underset{\sim}{E} = \underset{\sim}{E}_o \exp[i(\underset{\sim}{k} \cdot \underset{\sim}{\ell} - \omega t)],$$
$$\underset{\sim}{k} \times (\underset{\sim}{k} \times \underset{\sim}{E}) + \frac{\omega^2}{c^2} \underset{\approx}{K} \cdot \underset{\sim}{E} = 0. \quad (16)$$

Here $\underset{\approx}{K}(\omega, \underset{\sim}{k})$ is the dielectric tensor of the magnetized plasma. The plasma dispersion relation is the condition for a nontrivial solution of Eq. (16) and is obtained by setting the determinant of the coefficients of E_x, E_y, and E_z equal to zero. Thus, knowing the dielectric tensor $\underset{\approx}{K}(\omega, k)$ one can get a closed form expression for the allowed wave number k as a function of ω. The optical depth τ_o is given by

$$\tau_o = -\int \alpha_\omega d\ell = -\int (2\,\mathrm{Im}\, k) d\ell, \quad (17)$$

where Im stands for the imaginary part.

This dielectric tensor $\underset{\approx}{K}$ for a hot plasma in a strong magnetic field predicts resonance phenomena for wave propagation at frequencies ω in the neighborhood of all the harmonics of the electron cyclotron frequency $\omega_{ce} = eB_\phi/mc$. Here m and e are the mass and charge of the electron. In the conventional hot-plasma theory the resonances at the second and higher harmonics arise from the effects of the finite Larmor radii ρ of the electrons. Consequently, the relative strengths of these successive higher harmonic resonances are expected to fall off approximately as successive powers of $(\rho k_\perp)^2 \sim (\kappa T_e/mc^2)$. Here $k_\perp$ is the component of the wave vector in a direction perpendicular to the confining magnetic field B_ϕ. In present day fusion plasma devices the plasma conditions (e.g., n_e, T_e, and ω_{ce}) are such that only the resonances near the fundamental cyclotron frequency ($\omega \simeq \omega_{ce}$) and the second harmonic ($\omega \simeq 2\omega_{ce}$) are sufficiently strong enough to yield values of the optical depth $\tau_o \gtrsim 2$. That is, only for $\omega \simeq \omega_{ce}$ and $\omega \simeq 2\omega_{ce}$, the plasma found in present day fusion devices emit like a blackbody. Hence, in this paper we will restrict our discussion to waves of frequencies $\omega \simeq \omega_{ce}$ and $\omega \simeq 2\omega_{ce}$ only.

For arbitrary angles of propagation vector $\underset{\sim}{k}$ with respect to the magnetic field B_ϕ, all the nine elements K_{ij} (i,j = x,y, and z) of the dielectric tensor $\underset{\approx}{K}$ are nonzero. Thus, it is extremely difficult if not impossible to obtain a closed form analytic expression for the plasma dispersion relation from Eq. (16) when the wave vector $\underset{\sim}{k}$ makes an arbitrary angle with respect to the toroidal magnetic field B_ϕ. However, for near perpendicular (i.e., $k_\parallel \ll k_\perp$) propagation, one can obtain a meaningful closed form analytic expression for the plasma dispersion relation. Here $\perp$, and $\parallel$ refer to directions perpendicular and parallel, respectively, with respect to the toroidal magnetic field $\underset{\sim}{B}_\phi$.

Arunasalam et al.,[1] first applied this dispersion relation technique for near perpendicular propagation of the extraordinary mode in the vicinity of the second harmonic of the electron cyclotron frequency ($\omega \approx 2\omega_{ce}$). Their result for propagation across a uniform cylindrical plasma column of diameter ℓ whose axis is along $\underset{\sim}{B}_\phi$ is

$$\tau_o(\omega \approx 2\omega_{ce}) \approx A \exp\left[-\frac{mc^2}{2\kappa T_\parallel \cos^2\theta}\left(\frac{\omega - 2\omega_{ce}}{\omega}\right)^2\right], \quad (18)$$

where

$$A \approx \frac{\ell\pi^{1/2}\omega_{pe}^2 \kappa T_\perp \omega \tan\theta}{2mc^2 \omega_{ce}^2 (2\kappa T_\parallel/m)^{1/2}} \left(\frac{6\omega_{ce}^2 - \omega_{pe}^2}{6\omega_{ce}^2 - 2\omega_{pe}^2}\right)^2. \quad (19)$$

Equation (18) shows the familiar Doppler broadening of the second harmonic resonance. For propagation at exactly 90° (that is, in the limit $\cos^2\theta \to 0$) it can be shown from Eqs. (18) and (19) that

$$\tau_o(\omega \approx 2\omega_{ce}) = -\int_o^\ell (2\text{Im } k)d\ell = \ell(2\text{Im } k)$$
$$= \frac{\ell\omega_{pe}^2 \kappa T_\perp}{mc^3}\left(\frac{6\omega_{ce}^2 - \omega_{pe}^2}{6\omega_{ce}^2 - 2\omega_{pe}^2}\right)^2 \delta\left(\frac{\omega - 2\omega_{ce}}{2\pi}\right). \quad (20)$$

This result is for a uniform cylindrical plasma column of diameter ℓ whose axis is along a uniform magnetic field B_ϕ and is applicable to the Model C stellarator plasma. However, for plasmas in tokamak geometry $\omega_{ce} \sim B_\phi \sim 1/R$, where R is the major radius of the torus. Then $d\ell = dR = -(R/\omega_{ce})\, d\omega_{ce}$. Thus from Eq. (20) one sees readily that the optical depth for the second harmonic extraordinary mode near perpendicular propagation in a tokamak is given by

$$\tau_o(\omega \approx 2\omega_{ce}) = \frac{R\pi\omega_{pe}^2 \kappa T_\perp}{2\omega_{ce} mc^3}\left(\frac{6\omega_{ce}^2 - \omega_{pe}^2}{6\omega_{ce}^2 - 2\omega_{pe}^2}\right)^2. \quad (21)$$

Fidone et al.,[13] Antonsen et al.,[14] and Arunasalam et al.,[12] have also applied a similar dispersion relation technique to calculate the optical depth for near perpendicular propagation of the ordinary mode in the vicinity of ω_{ce}. Their results for the ordinary mode $\omega \approx \omega_{ce}$, $\theta \approx \pi/2$ propagation in tokamaks is

$$\tau_o(\omega \approx \omega_{ce}) \simeq \frac{R\pi\, \omega_{pe}^2 \kappa T_\parallel}{2\omega_{ce} mc^3}\left(1 - \frac{\omega_{pe}^2}{\omega_{ce}^2}\right)^{1/2}. \quad (22)$$

Here we have generalized the results of Fidone et al., and Antonsen et al., to include the effects of pressure anisotropy (i.e., $T_\perp \neq T_\parallel$). It is interesting to note from Eqs. (21) and (22) that the second harmonic extraordinary mode absorption is due to the perpendicular electron temperature $T_\perp$, while the fundamental ordinary mode absorption is due to the parallel electron temperature $T_\parallel$ even though both these absorptions are a consequence of the finite size of the electron Larmor orbits. This conclusion is also apparent from the emissivity formulae of Eqs. (14) and (15).

In tokamak regimes for which the runaway population is small, the ordinary mode emission near the fundamental as well as both the ordinary and extraordinary mode emission near the second and higher harmonics can be well understood in terms of the conventional non-relativistic hot-plasma theory.[9-10] However, Fidone et al.[13] have shown that one must take account of the relativistic broadening of the resonance due to the variation of electron mass with energy in calculating the optical depth of the perpendicularly propagating extraordinary mode near ω_{ce} in a high density plasma $(\omega_{pe}^2/\omega_{ce}^2) \gg (\kappa T_e/mc^2)^{1/2}$. Their result for ω_{ce} - extraordinary mode is

$$\tau_o(\omega \simeq \omega_{ce}) \simeq \frac{3R\pi\omega_{ce}^3}{2c\omega_{pe}^2}\left(2 - \frac{\omega_{pe}^2}{\omega_{ce}^2}\right)^{3/2} \frac{\kappa T_e^2}{mc^2}. \quad (23)$$

Here Fidone makes use of the relativistic hot plasma dielectric tensor by Shkarofsky,[13] and obtains results similar to Dnestrovskii, et al.,[13] For current tokamak parameters, the extraordinary mode $\tau_o \ll 1$ and therefore does not predict blackbody cyclotron emission. However, the intensity of extraordinary mode emission in tokamaks near the fundamental electron cyclotron frequency is measured to be at the blackbody level as will be discussed in Section III.

Mode conversion of blackbody electrostatic electron Bernstein waves into extraordinary mode electromagnetic waves is a possible physical mechanism to explain the observed blackbody emission. Hosea et al.,[3] calculates the optical thickness for the electron Bernstein mode to be

$$\tau_o \simeq a\pi^{1/2}(2\pi)^{4/3} R \frac{\omega_{ce}}{c}\left(\omega_{pe}^2 V_T^2/\omega_{ce}^2 c\right)^{1/3}, \quad (24)$$

where $a \simeq 1$ for small $k_\parallel$ and $a = (8/\pi\varepsilon)^{1/2}$ for the case of relativistic line broadening, and $V_T = (2\kappa T_e/m)^{1/2}$. For the case of PLT tokamak, $\tau \sim 10^3 - 10^4$. This mode conversion at the upper hybrid frequency has been analyzed by Stix[15] and calculated to be nearly 100% for PLT parameters by Schuss and Hosea.[16]

Thus, the theory of blackbody emission from a plasma is well understood. Calculation of the optical thickness for the fundamental electron cyclotron frequency and the harmonics indicate the ordinary mode near $\omega = \omega_{ce}$ and the extraordinary mode near $\omega = 2\omega_{ce}$ are nearly completely absorbing and thus by Kirchoff's law the emission is blackbody. The same conclusion is obtained for the extraordinary mode near $\omega = \omega_{ce}$ when mode conversion of the electrostatic Bernstein wave is invoked. Thus there are a number of modes and frequency ranges where the electron temperature can be ascertained from electron cyclotron emission.

III. Experiments and Temperature Measurements

Experiments verifying the theory presented in Section II have been two fold: (a) wave propagation to measure the plasma absorptivity, and (b) measurements of the intensity of the plasma electron cyclotron emission. Arunasalam et al.,[1] verified the hot plasma dispersion relation by verifying the expression (Eqs. 18-20) for the optical thickness of a plasma for the extraordinary mode near the second harmonic of the electron cyclotron frequency. The experiments involved propagating 70 GHz microwaves across the horizontal midplane of the homogeneous magnetic field straight-section of the Model C Stellarator. The absorptivity of the plasma as a function of $\omega/2\omega_{ce}$ was obtained by varying the magnetic field. The experiment was done at a few different angles (θ) of $\underset{\sim}{k}$ with respect to the magnetic field. Thus the amount of Doppler broadening of the absorption as a function of $k_\parallel (2\kappa T_\parallel/m)^{1/2}$ could be changed. Figure 3a is a plot of the absorptivity $\ln(P/P_o) = \tau_o$ as a function of $\omega/2\omega_{ce}$ for $\theta = 90°$. Here there is no Doppler broadening effect (see Eq. 20) and the actual resonance width is due to magnetic field inhomogeneity. In Fig. 3b, $\theta = 103.7°$ and the Doppler broadening makes the width significantly larger. The points represent actual measurements while the smooth continuous curve is the expected absorption from the theory with no temperature anisotropy (i.e., Eqs. (18) and (19) with $T_\perp = T_\parallel$. However, the actual depth of the resonance is a function of $T_\perp$ while the width is a function of $T_\parallel$ where $T_\perp$ and $T_\parallel$ represent the electron temperatures in directions parallel and perpendicular to the magnetic field $\tilde{B}_\phi$. There is a better fit of the data to the dashed curve in Fig. 3b which represents the theory Eqs. (18), and (19) with temperature anisotropy (i.e., $T_\perp \neq T_\parallel$). Although the plasma temperatures for this experiment were only in the range of 10-50 eV ($\sim 10^5$ K), the hot-plasma theory still remains viable for 10^7 K temperatures in present tokamaks. This experimental verification of hot-plasma theory became the basis for utilization of electron cyclotron emission to determine the electron temperature in present day tokamaks.

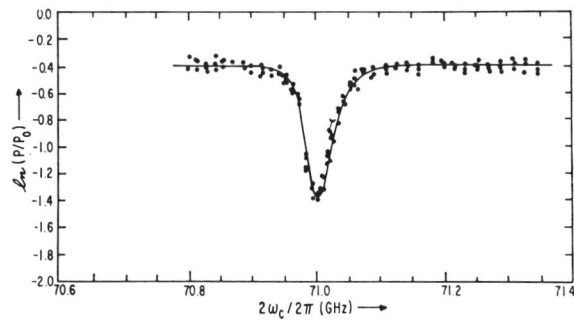

Fig. 3a. Second cyclotron harmonic absorption resonance for $\theta = 90°$. The resonance is narrower compared to Fig. 3b case because of the absence of Doppler broadening.

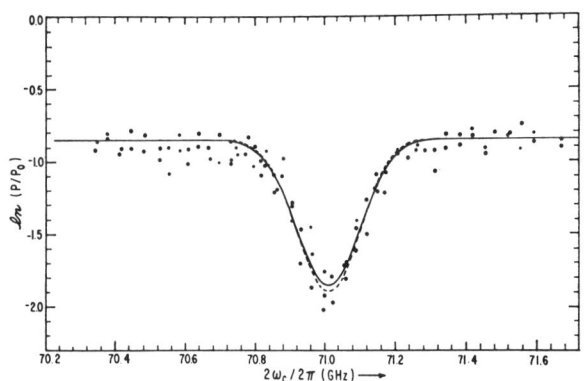

Fig. 3b. Second cyclotron harmonic absorption resonance for $\theta = 103.7°$ (solid curve) theoretical curve for $T_\perp = T_\parallel = T_e$; theory for $T_\perp = 15.5$ eV, $T_\parallel = 14.5$ eV (dashed curve).

First measurements of blackbody electron cyclotron emission from the TFR tokamak were reported by Costley et al.,[17] These measurements were done with a Michelson interferometer followed by a cryogenically-cooled Putley indium antimonide detector. The interferometer swept out the emission intensity near the fundamental electron cyclotron frequency and the first few harmonics. Under certain plasma conditions the intensity of the emission in the extraordinary mode near $\omega = 2\omega_{ce}$ was blackbody. In addition the intensity of the emission for any frequency was the same for ordinary and extraordinary modes — a result contradictory to theory. This result is believed to be caused by the high reflectivity of the tokamak containment vessel depolarizing the emission. Later Cano et al.,[18] built a manually tunable millimeter wave heterodyne receiver and measured the electron temperature profile on the TFR tokamak from the extraordinary mode blackbody electron cyclotron emission near $\omega = 2\omega_{ce}$.

On PLT, Hosea et al.,[3] studied the electron cyclotron emission with a manually tunable heterodyne receiver. Measurement of the emission near $\omega = \omega_{ce}$ as well as $\omega = 2\omega_{ce}$ were completed. Figure 4 is a graph comparing the intensities of the extraordinary mode electron cyclotron emission at $\omega = 2\omega_{ce}$ measured from the low magnetic field side (open circles), with the extraordinary mode emission at $\omega = \omega_{ce}$ measured from the accessible high magnetic field side (solid circles), and with the electron temperature measured by ruby laser Thomson scattering (solid line) all as a function of plasma position. This graph shows the emission and temperature comparison only for half the plasma cross section located toward the low magnetic field side of the torus. The agreement between the blackbody emission in the extraordinary mode near $\omega = \omega_{ce}$ and $\omega = 2\omega_{ce}$ and Thomson scattering is indeed remarkable. For $\omega = 2\omega_{ce}$, this result was expected because the optical thickness is greater than 2 from theory Eq. (21), and also from the earlier emission measurements of Costley from the TFR tokamak. However, the theory Eq. (23) for the extraordinary mode near $\omega = \omega_{ce}$ does not predict direct blackbody emission even if wall reflections are included in the calculations (since $\tau_0 \leq 0.01$). Hosea et al.,[3] suggested that the observed blackbody emission in the extraordinary mode near $\omega \simeq \omega_{ce}$ is a result of efficient conversion of electrostatic electron cyclotron waves (i.e, Bernstein waves) into electromagnetic waves around the region of the plasma where $\omega = \omega_{UH}$. The mode conversion process has been analyzed by Stix,[15] and later Schuss and Hosea[16] calculated the efficiency of this particular process to be nearly 100%. Therefore the intensity of the electrostatic mode has to be at the blackbody level in order for the electromagnetic emission at $\omega = \omega_{ce}$ to be blackbody. Hosea et al., calculated the optical thickness of the electrostatic wave to be of the order of 10^3-10^4 for the PLT plasma (Eq. 24). Unfortunately, an experiment to verify this mode-conversion-emission process for the extraordinary mode near $\omega = \omega_{ce}$ has not been performed to date.

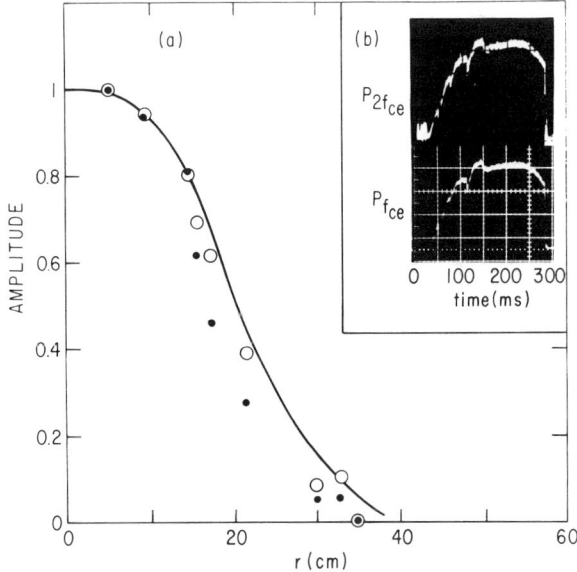

Fig. 4. (a) Cyclotron emission profiles for ω_{ce} (•) and $2\omega_{ce}$ (o) emission. The average Thomson scattering profile (−) is shown for comparision ($T_e(o) \sim 1$ keV, $n_e = 3 \times 10^{13}$ cm^{-3}). (b) An oscilloscope trace of the emission as a function of time.

Recent interest in the ordinary mode near the fundamental electron cyclotron frequency was motivated by the desire to find a wave that is easily accessible, completely absorbing, and had a frequency as low as possible for electron-cyclotron resonance (ECR) heating. The accessible ordinary mode near the fundamental electron cyclotron frequency was first studied by Fidone et al., who derived a closed form expression for the absorptivity of the mode Eq. (22). According to the optical thickness expression Eq. (22) the ordinary mode is nearly completely absorbed in a plasma with parameters on the order of $T_e \sim 10^3$ eV and density $\sim 3 \times 10^{13}$ cm^{-3}. These calculations motivated Efthimion et al.,[2] to measure the emission intensity of the ordinary mode at $\omega = \omega_{ce}$ in PLT from the low magnetic field side of the plasma. Figure 5 is a comparison of the temperature profiles ascertained from the ordinary mode electron cyclotron emission and from ruby laser Thomson scattering. The temperature profile from the emission, taken along the major radius in the horizontal midplane shows a 2-3 cm outward shift while the profile from the laser taken along a vertical plasma cord does not show this shift. The horizontal shift is attributable to either the outward shift of the magnetic field surfaces or the actual horizontal positioning of the plasma. Nevertheless the profiles indicate the emission intensity is at the blackbody level. For PLT the absorption theory Eq. (22) predicts that the inner 20 cm of the plasma is blackbody for single path absorption. The blackbody emission intensity observed outside of 20 cm can be explained by the reflections from the vacuum vessel allowing the cyclotron emission to build up to the blackbody level. However, the emission experiments could not accurately verify the ordinary mode theory because of wall reflections.

consists of varying the magnetic field of the tokamak so the absorbing layer moves from the center of the plasma out to the edge. In P_o the refraction losses have been experimentally taken into account by measuring the power transmitted across the plasma with the cyclotron layer well outside the plasma. Measurements of the electron temperature and density in the absorbing layer were provided by ruby laser Thomson scattering. These local values of density and temperature enabled the calculation of the optical thickness from Eq. (22). The calculated values agreed well with the values obtained from the absorption experiment, see Ref. 2.

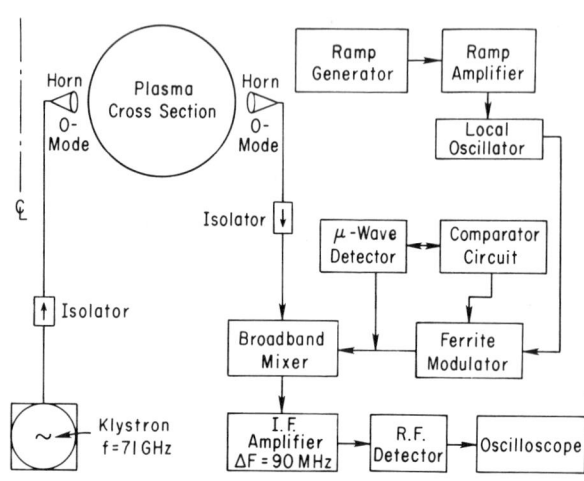

Fig. 6. Block diagram of the experimental arrangement employed to verify the hot plasma ordinary mode theory.

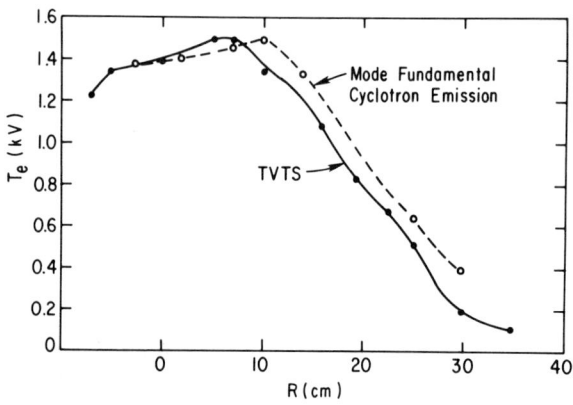

Fig. 5. A comparison of the temperature profiles obtained from blackbody ordinary-mode fundamental cyclotron emission (i.e. dashed line) with the corresponding profile obtained from laser Thomson scattering, TS (i.e., solid line).

A verification of the theory involved measurements of the ordinary mode absorption in the PLT plasma.[2] Figure 6 is a block diagram of the experimental arrangement. The incident microwave power is supplied by a klystron at a fixed frequency of 71 GHz. The antennas are standard 4 mm high gain horns that are oriented for the ordinary mode. The microwave power propagates through the plasma along the horizontal midplane, is collected by the receiving antenna and detected by a heterodyne receiver that makes up the rest of the block diagram. A discussion of the swept heterodyne receiver will be presented in Section IV. The fractional microwave power (P_T/P_o) transmitted through the plasma may be written as a function of the optical thickness: $(P_T/P_o) = \exp(-\tau_o)$. The experiment

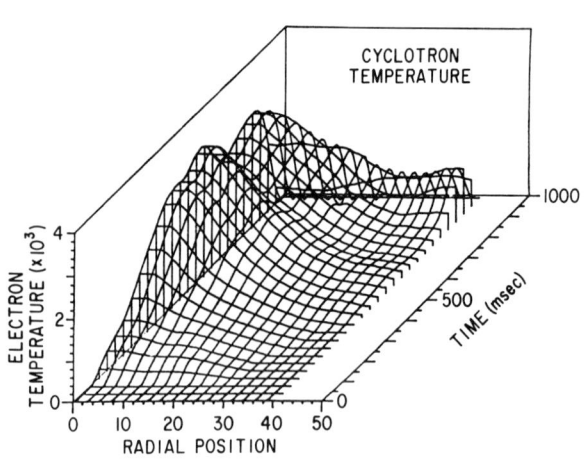

Fig. 7. A computer plot of the time evolution of the electron temperature profile of a typical ohmically heated PLT plasma discharge determined from the measurement of the ordinary-mode fundamental electron cyclotron emission by the fast-scanning heterodyne receiver. In order to easily visualize the contours of the three-dimensional graph only seventeen radial positions of every other temperature profile actually determined from the emission are plotted.

The collective knowledge of the experiments and theory is the basis for the determination of the electron temperature of magnetized plasmas from the measurement of the intensity of the emission near the electron cyclotron frequency and the harmonics. This technique is particularly useful in the study of auxiliary heating of plasma because it offers the time evolution of the temperature profile - information presently not available by any one technique. Here, we will present some examples of the electron temperature measurements completed by this technique. In Fig. 7 is a three dimensional computer plot of the time evolution of the electron temperature profile [T_e (r,t)] measured for a typical Ohmically heated PLT discharge. In order to visualize the contours, only seventeen radial positions are plotted. For Ohmically heated plasmas the peak electron temperature is typically 1-2 x 10^7 K. An example of auxiliary heating of the plasma electrons appears in Fig. 8. Here is a plot of the electron temperature profile before, during, and after injection of 2.1 x 10^6 watts of deuterium neutral beams at time t = 450-600 ms into a low density (n_e = 1.8 x 10^{13} cm^{-3}) hydrogen plasma. The observed peaking of the temperature after the termination of injection can be explained by the heating of the plasma by the still slowing-down beam ions and the removal of the heat loss due to the thermalization of cold electrons from the neutral beam particles. These examples of temperature determination from electron cyclotron emission illustrate the power and usefulness of this technique for tokamak fusion research.

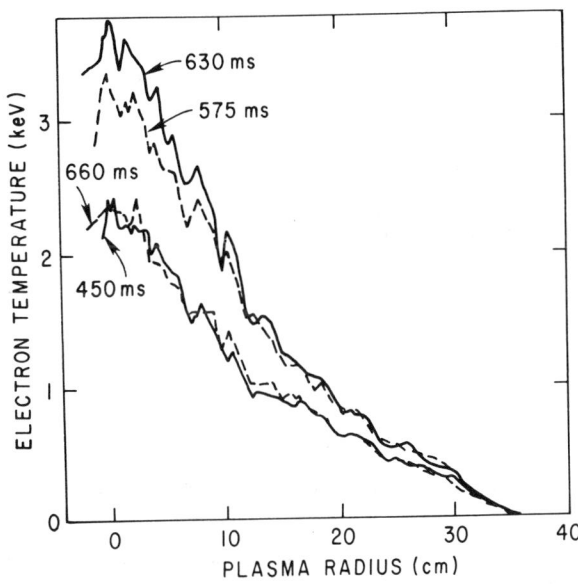

Fig. 8. A computer plot of electron temperature (keV) ascertained from the cyclotron emission versus plasmas radius (cm) at four distinct times (ms) representing before (450 ms), during (575 ms), and after (630 and 660 ms) injection of 2.1 MW of deuterium neutral beams into a low-density (n_e = 1.8 x 10^{13} cm^{-3}) hydrogen plasma.

A word of caution is necessary in utilizing electron cyclotron emission to ascertain the electron temperature. First, the proper plasma conditions must be present so that the plasma is highly absorbing and the emitting layer is optically thick ($\tau_o \gtrsim 2$).

Second, the radiation should be in thermal equilibrium with the radiating electrons for the emission intensity to be at the blackbody level. Furthermore, if the plasma is a blackbody only with the help of multiple reflections from the vessel wall, energetic electrons i.e., "runaways" can radiate with intensities which are orders of magnitude larger than that from the plasma electrons with a Maxwellian distribution. Nonthermal emission near the electron cyclotron frequency and its harmonics is outside the subject addressed in this paper. All nonthermal processes can be held down to a minimum by properly controlling the plasma discharge conditions such as to keep the runaway population negligibly small.

IV. Measurement Techniques

Although the electron temperatures of tokamak plasmas is very large ($\sim 10^7 - 10^8$ K), the actual broadband power measured in say a 250 MHz bandwidth is only of the order of 10^{-8}-10^{-7} watts. In comparison to the radio-astronomy emission measurements, this emission power level is quite large. Power detections in radio-astronomy involves signal integration for many seconds. On the other hand tokamak plasmas are only $\sim$ 1 second in duration and signal integration is not practical. Most tokamaks operate in the 10-60 kG range of magnetic fields. Therefore, the spectral range of the emission near the fundamental electron cyclotron frequency and the next two harmonics is in the 10-0.5 millimeter wavelength region. Even if only one harmonic is to be measured, the radial variation in the magnetic field across the tokamak plasma dictates that the detection system should be at least 30% broadband in order to obtain the temperature profile.

There are two techniques for measuring the millimeter-submillimeter wavelength emission in tokamak plasmas: standard guided-wave heterodyne techniques and quasi-optical interferometry employing cryogenically-cooled broadband bolometers and detectors. The interferometric devices could be a Michelson interferometer, a Fabry-Perot interferometer or a grating spectrometer.[19] As mentioned in Section III, both detection techniques have been successful in measuring plasma millimeter-submillimeter wave emission. However, the authors of this paper have made a large effort to refine the millimeter wave heterodyne techniques. Specifically, the fast-scanning millimeter wavelength heterodyne receiver has been developed by the authors at Princeton, and with this technique they have made emission measurements on PLT and expect to measure the time evolution of the electron temperature profile of the TFTR plasma.

Figure 9 is a block diagram of the millimeter wave heterodyne receiver on PLT. This receiver electronically scans the frequency range 60-90 GHz every 10 ms. Under typical PLT operating conditions ($B_\phi \sim$ 32 kG), this frequency range corresponds to the fundamental electron cyclotron emission between the center and the edge of the plasma on the low field side. There are two high-gain horns located on the equatorial plane along the major radius: (1) a horn located on the high field side of the torus oriented to receive the extraordinary mode and (2) a horn located on the low field side of the torus to measure ordinary mode emission.

The millimeter wave mixer is single-ended with a flat response over the entire frequency range of the receiver and has a 16 dB conversion loss. The local oscillator is a Siemens Backward-Wave-Oscillator (BWO) that can be electronically swept over the 60-90 GHz range by sweeping the applied cathode voltage over a 1 kV range. In our system the 1 kV sweep is obtained with a low voltage signal generator and a high-voltage amplifier. The internal capacitance of the BWO prevents sweep rates greater than 500 Hz. Because the BWO output power varies over a factor of ten, a commercially available feedback loop maintains a low but constant power level for the mixer. The feedback circuit has a 50 μsec response time and levels the power with less than 1 dB variation.

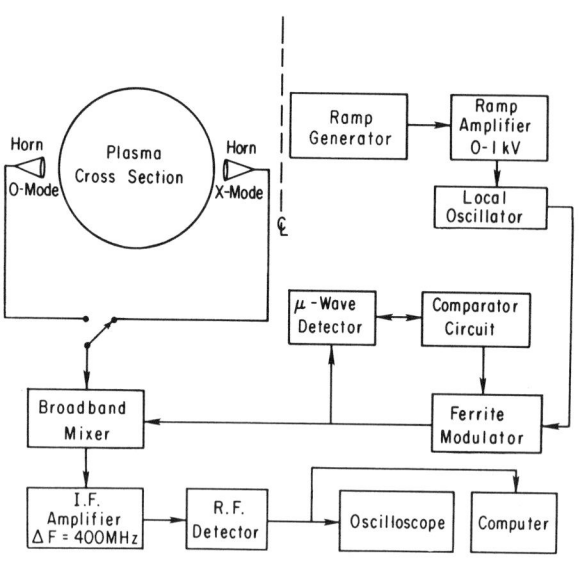

Fig. 9. Detailed block diagram of the 60-90 GHz fast scanning heterodyne receiver.

The mixer output is amplified by a two stage 60 dB gain IF - amplifier with 3 dB points at 250 and 500 MHz. This frequency range is chosen to minimize the noise created by the mixing of the BWO carrier wings with the carrier peak and at the same time to minimize the frequency gap between the receiver sidebands. This selected IF bandwidth results in a 1 cm spatial resolution in the plasma. The amplified IF signal is detected by a Schottky barrier detector with a 100 KHz response. The detector output is amplified by a 20 dB gain line driver.

The receiver has a 23 dB noise figure while operating in the swept mode. This noise figure could be substantially reduced if either the IF range was selected to be in the 1-2 GHz range or the bandwidth of the receiver was restricted to allow the use of a balanced mixer to cancel the local oscillator noise. However, with typical electron temperatures of the order of 10^7-10^8 K a noise figure of 23 dB (~ 6×10^4 K) is acceptable. Essentially, the noise figure was sacrificed for bandwidth (~ 30%) and spatial resolution.

Calibration of the system is performed by two techniques. A calibration of the system can be made by normalizing the receiver output to the electron temperature of a plasma measured by ruby laser Thomson scattering. This technique is quite simple and very effective but relies on the Thomson scattering measurements. An independent calibration is realized by measuring the emission of a well known blackbody object, and the calibration is obtained by measuring the voltage difference from the output of the receiver when the absorber is at room temperature (290 K) and at liquid nitrogen temperature (77 K). The absorber is maintained at 77 K by inserting it in a styrofoam dewar filled with liquid nitrogen. The styrofoam is transparent to the millimeter waves. Unfortunately, the receiver can only measure temperature above 6×10^4 K. This problem is circumvented by switching the receiver input between the room temperature microwave absorber and the one cooled at liquid nitrogen temperatures. This technique to measure temperature lower then the noise temperature of the receiver was discovered by Dicke[20] for radio-astronomy. This switching technique in the millimeter wavelength region can be realized by a chopping wheel placed between the nitrogen filled dewar and the receiver antenna (Fig. 10). The chopping wheel blades can be covered with absorber or merely be a metallic material to reflect the room temperature emission into the receiver antenna. A pulse train synchronized with the chopping wheel is generated. The receiver output and the synchronized pulse train are coupled to a lock-in amplifier to integrate the receiver signal and obtain the voltage difference of the receiver corresponding to the temperature difference between the blackbodies at two different temperatures.

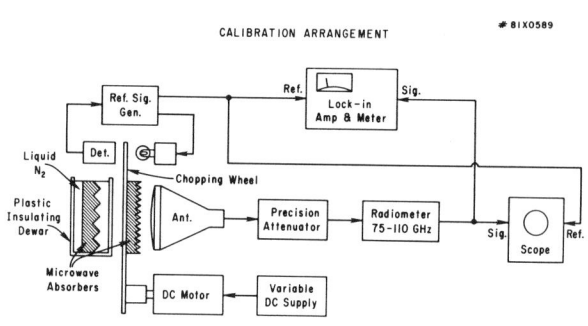

Fig. 10. Calibration arrangement of the heterodyne receiver using the Dicke switching technique.

V. Conclusions

In conclusion, we have attempted to give a balanced presentation of the theory, existing important cyclotron emission and absorption experiments, and a review of the diagnostic techniques which are involved in the determination of the time evolution of the electron temperature profile of reactor-like plasmas from the measurement of blackbody electron cyclotron emission. We sincerely hope that this paper illustrates the power and usefullness of the fast-scanning heterodyne technique for tokamak fusion research.

References

* Work supported by U.S. Department of Energy, Contract # DE-AC02-76-CHO-3073.

[1] V. Arunasalam, E.B. Meservey, M.N. Gurnee, and R.C. Davidson, Phys. Fluids 11, 1076 (1968).

[2] P.C. Efthimion, V. Arunasalam, and J.C. Hosea, Phys. Rev. Lett 44, 396 (1980); see also Equipe TFR, Ninth European Conf. on Controlled Fusion and Plasma Physics, Oxford, England, Sept. 1979.

[3] J. Hosea, V. Arunasalam, and R. Cano, Phys. Rev. Lett. 39, 408 (1977).

[4] P.C. Efthimion, V. Arunasalam, R. Bitzer, L. Cambell, and J.C. Hosea, Rev. Sci. Instrum. 50, 949 (1979).

[5] N.G. Basov et al., IEEE J. Quantum Electron, QE-4, 846 (1968).

[6] L. Spitzer et al., U.S. Atomic Energy Commission, Rep. NYO-6047 (1954).

[7] H. Eubank et al., Phys. Rev. Lett 43, 270 (1979).

[8] L. Arstimovich, Nucl. Fusion 12, 215 (1972); and H.P. Furth, Nucl. Fusion 15 487 (1975).

[9] G. Bekefi, Radiation Processes in Plasmas (Wiley, New York, 1966).

[10] T.H. Stix, The Theory of Plasma Waves (McGraw-Hill, New York, 1962).

[11] J. Schwinger, Phys. Rev. 75, 1912 (1949); L. Landau, and E. Lifshitz, The Classical Theory of Fields (Addison-Wesley, Reading, Mass., 1951); L. Oster, Phys. Rev. 119, 1444 (1960); L. Oster, Phys. Rev. 121, 961 (1961); G. Schott, Electromagnetic Radiation (University Press, Cambridge 1912); and B.A. Trubinkov, Soviet Phys. "Doklady" 3, 136 (1958).

[12] V. Arunasalam, P.C. Efthimion, J.C. Hosea, and H. Hsuan, 1980 Sherwood Meeting, Theoretical Aspects of Controlled Thermonuclear Research, Tucson, Arizona, 2C21.

[13] I. Fidone, G. Granata, G. Ramponi, and R.L. Meyer, Phys. Fluids 21, 645 (1978); I. Shkarofsky, Phys. Fluids 9, 561 (1966); and Y. Dnestrovski, D.P. Kastomarov, and N.V. Skydlov, Soviet Phys. Tech. Phys. 8, 691 (1964).

[14] T.M. Antonson, Jr., and W.M. Manheimer, Phys. Fluids 21, 2295 (1978).

[15] T.H. Stix, Phys. Rev. Lett. 15, 878 (1975).

[16] J.J. Schuss, and J.C. Hosea, Phys. Fluids 18, 727 (1975).

[17] A.E. Costley, R.J. Hastie, J.W.M. Paul, and J. Chamberlain, Phys. Rev. Lett. 33, 758 (1974).

[18] TFR Group (R. Cano), Proceedings of the European Conference on Controlled and Plasma Physics (CRPP, Lansamme, 1975) paper 14b.

[19] W.R. Rutgers and D.A. Boyd, Phys. Lett. 62A 498 (1977); I.H. Hutchinson and D.S. Komm, Nucl. Fusion 17 1077 (1977); F.J. Stauffer, and D.A. Boyd, Infrared Phys. 18 755 (1978); and G. Tait, F.S. Stauffer, and D.A. Boyd, Phys. Fluids, 24 719 (1981).

[20] R.H. Dicke, Rev. Sci. Instrum. 17, 268 (1946); R.H. Dicke, R. Beringer, R.L. Kyhl, and A.B. Vane, Phys. Rev. 70, 340 (1946).

Measurement of the electron temperature profile in a tokamak by observation of electron cyclotron emission using a Fourier transform spectrometer[a]

F. J. Stauffer

Laboratory for Plasma and Fusion Energy Studies, University of Maryland, College Park, Maryland 20742

During the past seven years, the measurement of electron cyclotron emission (ECE) from tokamaks has developed from the initial efforts into an electron temperature diagnostic of generally recognized importance. Fourier transform spectrometers (FTS) incorporating the fast-scanning Martin-Puplett interferometer are widely used to make these measurements. Electron cyclotron emission occurs at the spectral frequencies mf_{ce}, where $m = 1, 2...$ and f_{ce} is the electron cyclotron frequency. The significant line-broadening mechanism is the variation of the magnetic field over the diameter of the plasma cross section. Therefore, a spectral scan of the optically thick second harmonic ($m = 2$) line leads straightforwardly to the measurement of the electron temperature profile along this diameter. Absolute calibration with a low temperature blackbody source has produced profiles that are in good agreement with profiles produced by the Thomson scattering electron temperature diagnostic. While the FTS diagnostic has a poor time resolution (25 ms) compared to Thomson scattering, it can produce many more electron temperature profiles during a tokamak discharge. Thus, the two diagnostics are complementary.

INTRODUCTION

At about the same time that organized attempts were started to accomplish controlled thermonuclear fusion, theoreticians began to warn that electron cyclotron emission (ECE) at the temperature necessary for fusion to occur would be a serious power dissipation mechanism. A series of papers[1,2,3,4] appearing between 1958 and 1963 considered the implications of ECE for the design of a reactor based upon the fusion of two deuterons, which requires temperatures of 30 - 100 keV. It was recognized that the self-absorption of the low frequency portion of the spectrum by the plasma would reduce the radiation loss and also would make the power loss in that "optically thick" region depend upon the surface area of the plasma, according to the blackbody radiation law, instead of upon the plasma volume. With the assumption of a certain fusion power production density per unit volume, it was then possible to estimate the critical size of the thermonuclear plasma for which the volume dependent production of power would just balance the surface dependent radiative power loss. For lower plasma densities, it is necessary to increase the size of the plasma in order for the plasma to remain optically thick for the same spectral range. It was suggested that the use of highly reflective vacuum vessel walls that would force the radiation to make many passes through the plasma would be an alternative to physically increasing the plasma dimensions. Drummond and Rosenbluth[4] found that walls with a reflection coefficient of 0.9 could reduce the critical size by a factor of 10.

At present, it seems likely that the first successful fusion power reactor will be based upon the tokamak research device. The principles of operation of the tokamak have been well documented.[5,6] In the tokamak, equilibrium and stability criteria require comparatively low densities and high magnetic fields which are unfavorable conditions from the standpoint of ECE power loss. Current designs for future tokamak fusion reactors assume the use of the D-T reaction i.e., the fusion of a deuteron and a triton. This reaction requires a significantly lower temperature (of the order 10 keV) than for the D-D reaction. Mills[7] considered the use of reflecting walls in such a device and concluded that ECE is not a serious problem.

The motivation for experimental measurements of ECE from tokamaks is two-fold. First, ECE is an important factor in the energy balance of these devices,[8] and it is important to check the theoretically expected spectrum against actual measurements in order that the effect of this power dissipation mechanism upon future fusion reactors may be confidently estimated. For this purpose, a broadband spectrometer is desirable. In 1974, Costley et al.[9] reported the first broadband measurements of ECE from a tokamak using a Fourier transform spectrometer.

The second motivation for experimental effort is that the measurement of ECE is a valuable passive diagnostic of the electron distribution function in tokamak research devices.[10,11] In particular, it offers a convenient way to measure the electron temperature as a function of time and position in the plasma. During the past seven years, a number of workers have used Fourier transform spectrometers to monitor the evolution of the electron temperature in tokamaks.[12,13,14,15] Since the measurement of the electron temperature requires using only a portion of the full ECE spectrum, it has been possible also to use more narrowband instruments: grating spectrometers,[16,17,18,19] Fabry-Perot interferometers,[13,20] and microwave radiometers.[14,21,22,23,24] The remainder of this paper deals specifically with the measurement of the electron temperature profile in the PLT tokamak that is done by measuring ECE with a Fourier transform spectrometer.

ECE FROM PLT

The PLT (Princeton Large Torus) tokamak is located at the Plasma Physics Laboratory of Princeton University. During a typical PLT discharge, a toroidal-shaped hydrogen plasma is created and heated. The plasma is held in place by a magnetic field B (typically 3.2 T) and exists for up to 1 second. The plasma electrons reach a temperature T_e at the center of the plasma up to about 3 keV and have a typical spatially averaged density of about 3×10^{19} m^{-3}.

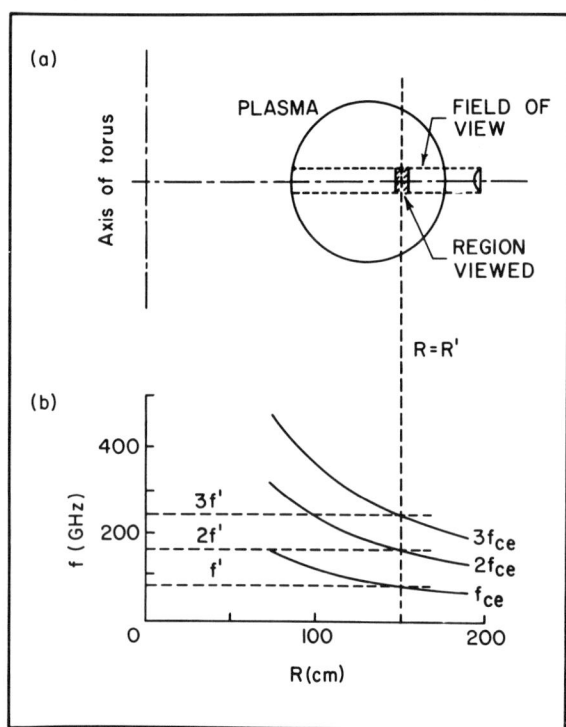

FIG. 1. (a) Vertical section through PLT plasma. (b) Spatial variation of electron cyclotron harmonic frequencies, evaluated for B = 3.2 tesla at the center of the plasma.

Figure 1 shows a vertical section through one-half of the toroidal plasma. Distance measured along a line that is perpendicular to the axis and that passes through the center of the circular cross-section of the plasma is denoted by R. The distance R_0 from the axis to the center of the plasma is called the major radius. It is customary to denote positions along the horizontal diameter of the cross-section by r, where

$$r = R - R_0 \qquad (1)$$

For PLT, the major radius is approximately 1.32 m and the diameter of the cross-section is typically 0.80 m.

As a result of their helical trajectories in the confining magnetic field B, which depends upon R, the plasma electrons emit cyclotron radiation. A good description of plasma electron cyclotron emission and radiation transport in a plasma has been given by Bekefi.[25] Consider the location R' in the plasma in Figure 1. ECE will be emitted from this location at the frequencies mf' (m = 1, 2, ...) where the first harmonic (m = 1) is the local value of the electron cyclotron frequency:

$$f' \cong \frac{B(R')}{1.07} \qquad (2)$$

In Eq. 2, f' is expressed in cm^{-1} (1 cm$^{-1} \cong$ 30 GHz) and B is in tesla. For PLT, only the first four harmonics have significant intensities. It is a characteristic of tokamaks that the confining magnetic field B is nearly normal to the plane of the plasma cross-section and that it is inversely proportional to R. Thus the frequency of each cyclotron harmonic has a hyperbolic dependence upon R, as shown in Figure 1. The ECE spectrum consists of magnetic field-broadened harmonic lines that are peaked near to the value of the electron cyclotron frequency at the center of the plasma and its integer multiples. It may be seen from Fig. 1 that there is a certain amount of overlap between the high frequency portion of the second harmonic and the low frequency portion of the third harmonic. The magnitude of the spectrum is expressed in terms of its specific intensity I, which is the power radiated per unit area of the source per unit spectral frequency interval (in rad/sec) per unit solid angle in a given direction. The units of I are therefore watts/(meter^2 - radian per second - steradian). ECE is radiated in two independent polarization modes: the extraordinary mode, for which the electric vector is perpendicular to B, and the ordinary mode, for which the electric vector is parallel to B.

As shown in Figure 1, the observation direction for the Fourier transform spectrometer on PLT is inward along a major radius. In order to calculate the intensity of the extraordinary and ordinary mode spectra that should be observed, one must solve the radiation transport problem. The salient feature of the solution is that the spectrum is bounded from above by the blackbody spectrum corresponding to the highest electron temperature in the plasma. The plasma is said to be optically thick in those frequency bands where its radiation conforms to the local blackbody spectrum and to be optically thin in those bands where its radiation is less intense than this blackbody spectrum. For PLT, numerical solutions of the radiation transport problem indicate that the second harmonic feature of the extraordinary mode spectrum is optically thick, but that the third harmonic feature in each mode is not thick. At the high temperatures and long wavelengths under discussion here, the Rayleigh-Jeans approximation for the blackbody spectrum is excellent so that the specific intensity I_2 for the second harmonic extraordinary mode feature is given by:

$$I_2 \propto f^2(r) \; T_e(r) \qquad (3)$$

where f(r) is the value of the second harmonic frequency at position r in the plasma and $T_e(r)$ is the value of the electron temperature at that position. Since the relation between f and r is known, a measurement of the shape of the second harmonic feature in the extraordinary mode ECE spectrum leads straightforwardly to the determination of the electron temperature profile $T_e(r)$. It follows that the spatial resolution with which $T_e(r)$ can be measured is directly related to the spectral resolution with which the second harmonic feature can be measured. For the Fourier transform spectrometer on PLT, the spatial resolution is approximately 0.10 m at the center of the plasma when the magnetic field at the center is 3.2 tesla. As shown in Figure 1, the spatial resolution in the direction perpendicular to the major radius is determined by the field of view.

For observations of ECE inward along a major radius on PLT, the first harmonic feature of the ordinary mode spectrum also is optically thick. There are two reasons for using the second harmonic extraordinary mode feature for the work reported in this paper. First, the index of refraction of the plasma becomes closer to unity as the spectral frequency increases, so that plasma dielectric properties are less likely to affect the shape of the second harmonic feature than the first harmonic feature.[26] Second, the Fourier transform spectrometer is a quasioptical instrument and its performance becomes better as the spectral frequency increases. Microwave radiometric techniques are better suited for measurements of the first harmonic ordinary mode feature, and this work is being carried out on PLT by Efthimion et al.[23]

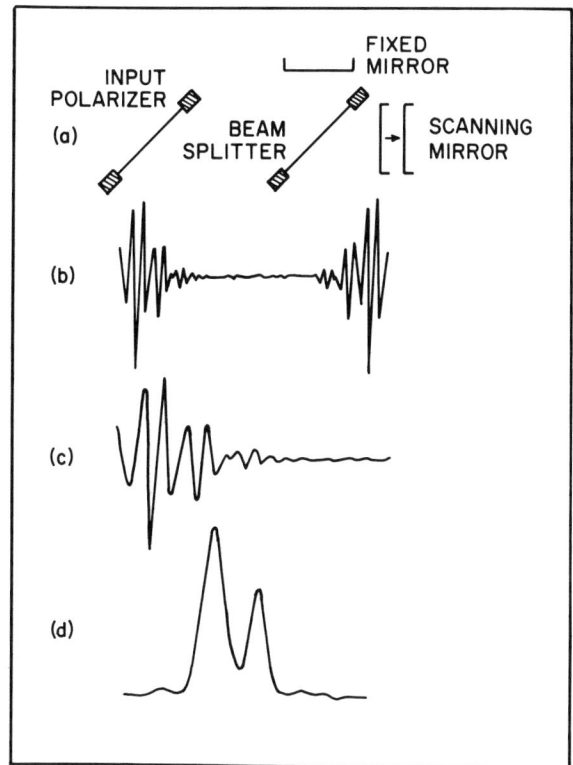

FIG. 2. (a) Essential components of the Fourier transform spectrometer. (b) Forward-reverse pair of interferograms. (c) Forward scan interferogram. (d) Forward scan spectrum.

THE FOURIER TRANSFORM SPECTROMETER

The Fourier transform spectrometer is built around a fast-scanning Michelson interferometer of the polarizing type[27] that was manufactured by Block Engineering, Inc. of Cambridge, Massachusetts. As sketched in Figure 2, the essential components consist of a wire grid input polarizer, which can be set to accept ECE in either the extraordinary or ordinary mode, a wire grid beam splitter, with its wires set at an angle of 45° with respect to the wires in the input polarizer, a fixed mirror and a scanning mirror, which reciprocates at a rate of 20 Hz. After reflection, the two beams are recombined and output to a liquid helium-cooled bolometric detector, which is sensitive over the broad spectral range of 2 - 60 cm^{-1} (60 - 1800 GHz).

As it translates, the scanning mirror introduces a continuously changing optical path difference between the two beams of radiation, up to a maximum difference of 1.8 cm. The effect of the continuously changing optical path difference is that the signal produced by the radiation detector during one complete forward-reverse cycle of the scanning mirror takes the form shown in Fig. 2. The signal resulting from a unidirectional scan of the mirror is called an interferogram, and it is the Fourier transform of the radiation spectrum. During a PLT plasma discharge, the spectrometer produces interferograms at the rate of 40 per second. Subsequently, this data is Fourier transformed so that the evolution of the ECE spectrum during the discharge (and consequently the evolution of the electron temperature profile) can be studied with a time resolution of 25 ms. A typical ECE extraordinary mode spectrum from PLT is shown at the bottom of Figure 2. The spectral resolution provided by the Michelson interferometer is 0.5 cm^{-1} (15 GHz). The second harmonic feature is the largest peak, and the next largest is the third harmonic feature. The overlap between these two features, as indicated by the f vs. R plot in Fig. 1, is evident in this spectrum.

DATA ACQUISITION AND CALIBRATION

The Fourier transform spectrometer data acquisition system consists of a NOVA 2/10 minicomputer linked to a DEC-10 computer. The NOVA is used to digitize and store interferograms. Subsequently, this data is transmitted via a high speed line to the DEC-10, which carries out the Fourier transforms, does additional processing of the spectra and produces CRT displays of the data. The system has two modes of operation. In the first mode, ECE interferograms are acquired and stored in sequential data files. The speed of the subsequent data processing is such that the time evolution of the electron temperature profile can be displayed within 2 minutes following a PLT discharge.

The second mode of operation is used to create a calibration curve that must be applied to each ECE spectrum in order to correct for the spectral response of the detector and to establish the absolute intensity of the ECE. Broadband calibration sources in this region of the spectrum are blackbodies which are considerably less intense than the PLT plasma. Therefore, the second mode of data collection consists of signal-averaging thousands of forward scan blackbody interferograms in order to enhance the signal to noise ratio. The enhancement factor is equal to the square-root of the number of averaged scans.

It is essential for the Fourier transform spectrometer to collect radiation from the calibration source in the same way as from the PLT plasma. The scheme for accomplishing this is shown in Figure 3.

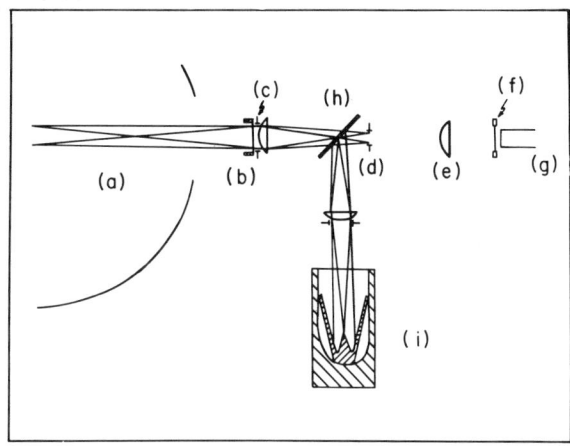

FIG. 3. Scheme for collecting ECE either from PLT or from the calibration source. (a) Interior of PLT. (b) Vacuum window. (c) Aperture stop and lens. (d) Field stop. (e) Collimating lens. (f) Polarization analyzer. (g) Lightpipe. (h) Calibration mirror. (i) Calibration source.

A plano-convex lens is used to image a field stop at the center of the PLT vacuum vessel. An aperture stop on the lens is used to prevent radiation that has reflected off the edges of the vacuum vessel interface from entering the field stop aperture. During calibration, a plane mirror is introduced to allow the field stop to accept radiation from a spot well within the boundary of the calibration source by means of an identical lens and aperture stop. In both cases, the radiation emerging from the field stop is collimated by a lens, filtered by a wire grid polarizer that is set to pass the extraordinary mode of the ECE, and transmitted to the Fourier transform spectrometer via a lightpipe. The measured transmission of the quartz vacuum window, which is not duplicated in the calibration optics, is taken into account when the calibration curve is produced.

The calibration source is based upon one used by workers at Queen Mary College.[28] It consists of a glass dewar containing a double cone fashioned from microwave absorber material (type AN-72 eccosorb, manufactured by Emerson & Cuming, Inc., Canton, Massachusetts). The spectrum from this source has been compared with the spectrum from a quite differently configured microwave absorber, and the two were the same to within 5% over the spectral range 5 - 30 cm^{-1}. Based upon that experiment, it is assumed that the double cone source is a good approximation to a blackbody over that range. The procedure for measuring the spectrum of the calibration source consists of two steps. First, the interferogram of the source is measured at room temperature. Second, the measurement is repeated with the dewar filled with liquid nitrogen. Reflection of room temperature radiation off the liquid surface into the collection optics introduces an error of only about one percent, and this error is ignored. Thus, the only important effect of introducing the liquid nitrogen is to make the absorber uniformly cold. The difference of these two interferograms is Fourier transformed to produce a spectrum which depends upon the emission from the calibration source and the response curve of the spectrometer.

The response curve is derived by dividing the measured spectrum by the difference between two Planck functions evaluated at liquid nitrogen and room temperature. A typical measured spectrum from the calibration source and the response curve derived from it are shown in Figure 4. The oscillations between 5 and 15 cm^{-1} probably were caused by the interference between reflections from various interfaces in the collection optics. The two features at 18.6 and 25.1 cm^{-1} were due to the absorption of radiation by water vapor in the air. The response curve has been truncated at 2 cm^{-1}, where it goes negative due to noise. Figure 5 shows the same response curve (dashed line) together with a typical PLT ECE extraordinary mode spectrum. In this example, the second harmonic of the electron cyclotron frequency at the center of the plasma was 4.6 cm^{-1}. The uncorrected measured ECE spectrum is shown with a dotted line. The calibrated spectrum that results from dividing it by the response curve is shown by the solid line. For ease of comparison, the second harmonic peaks have been set equal.

ELECTRON TEMPERATURE MEASUREMENTS

Following a PLT discharge, the 30 - 40 interferograms that have been acquired are Fourier transformed to produce ECE spectra. Each spectrum is divided by the response curve, and its second harmonic feature is used to derive the electron temperature profile. During the development of the Fourier transform spectrometer diagnostic on PLT, the touchstone for judging its ability to produce absolutely calibrated temperature profiles has been the Thomson scattering "TVTS" diagnostic,[29] which always has been the standard instrument for measuring electron temperatures in PLT. TVTS measures the scattered light from a laser beam that passes vertically through the plasma at the position R = 1.34 m. By measuring the shift in the wavelength of the scattered light as a function of position along the laser beam, the electron temperature profile can be measured absolutely. The TVTS profile measurement is practically instantaneous (the laser pulse width is approximately 20 ns), but the laser can be fired only once per PLT discharge. On the other hand, the time resolution of the Fourier transform spectrometer profile measurement is about 25 ms, but many profiles can be measured per discharge. Thus, these two diagnostics are complementary.

The multi-profile capability of the Fourier transform spectrometer is particularly useful during experiments when auxiliary plasma heating techniques are being tried on PLT. The basic heating mechanism is Ohmic heating due to the current that is induced to flow in the plasma during the entire discharge. Auxiliary heating techniques, such as the injection of RF power or of electrically neutral beams of energetic hydrogen atoms are usually applied for about 200 ms during the discharge. Figure 6 shows an example of plasma electron heating that occurred as a result of the application of neutral beam injection. At t = 193 ms into the discharge, the neutral beams had not yet been turned on. The peak of the temperature profile measured using ECE was located at r = 5 cm (R = 1.37 m). For comparison, 5 of the 25 data points that were produced when the TVTS laser fired at t = 200 ms are shown, along with the TVTS fitted profile. Comparison between the ECE profile and the TVTS profile is hampered by their orthogonal viewing directions. However, in this case the TVTS laser beam intersected the major radius line only 3 cm away from the peak of the ECE profile. In order to compare the shapes of the two profiles, the radial reference point of the TVTS profile simply has been shifted by 5 cm. In this case, the agreement between the two profiles is very good. Nevertheless, it is found that the ECE temperature measurements fluctuate relative to the TVTS measurements. Recently, the electron temperatures measured near the center of PLT

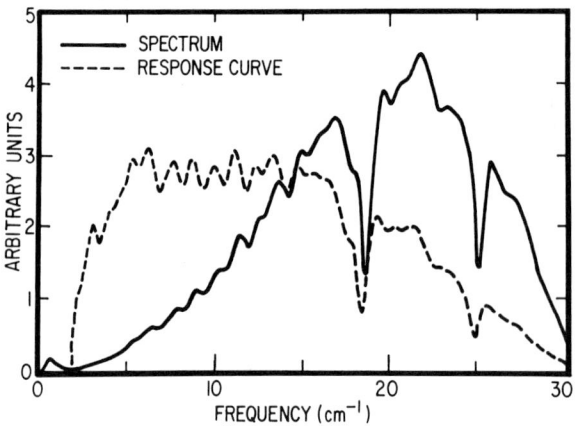

FIG. 4. Typical calibration source spectrum and the instrumental response curve derived from it.

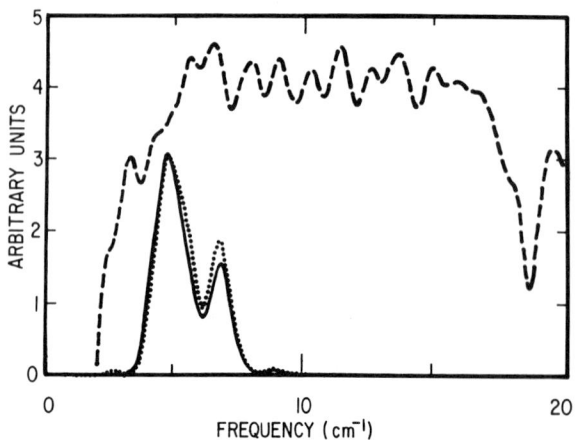

FIG. 5. Instrumental response curve (dashed line). Uncalibrated PLT ECE spectrum (dotted line). Calibrated ECE spectrum (solid line).

by these two diagnostics were compared for 120 discharges. In the region of temperatures for which $T_e \cong 1$ keV, the deviation of the ECE temperature relative to the TVTS temperature was scattered over the range ±20%. At higher temperatures, for which $T_e \cong 2$-3 keV, the ECE value was systematically lower, and the relative deviation ranged from nearly zero to -30%. This result is not presently understood.

The second (t = 504 ms) ECE temperature profile in Figure 6 shows the electron heating that resulted from the neutral beam injection. Moreover, it shows an unusual 15 cm shift outward of the peak electron temperature position as a result of the plasma heating. Although this is an exceptional shift, it demonstrates that the best way to display the electron temperature data gathered using the Fourier transform spectrometer is to plot the peak electron temperature vs. time instead of the temperature at a fixed radial position vs. time. Figure 7 shows a $T_{peak}(t)$ plot for this PLT discharge. The neutral beam injection was on during the time t = 400 - 560 ms.

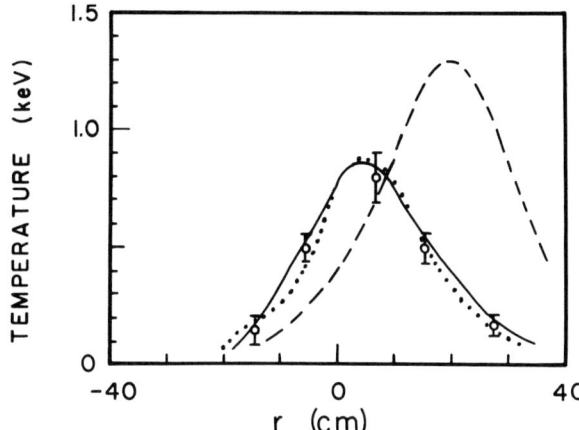

FIG. 6. Comparison of electron temperature measurements. Thomson scattering (TVTS) data points (circles with error bars) and fitted profile (dotted line) at t = 200 ms. Profile measured using ECE at t = 193 ms (solid line). Profile measured using ECE at t = 504 ms (dashed line).

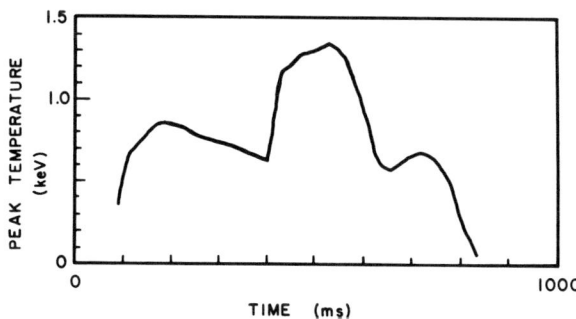

FIG. 7. Peak electron temperature versus time for a PLT discharge when neutral beam injection heating was applied during the time 400-560 ms.

PLANS FOR FUTURE WORK

A new Fourier transform spectrometer is being built for use on the TFTR (Tokamak Fusion Test Reactor) tokamak that is scheduled to begin operating at Princeton Plasma Physics Laboratory in 1982. TFTR will be more than twice the size of PLT, and it is expected to produce reactor-grade plasmas that will yield experimental data relevant to future fusion reactors. The new spectrometer is being designed to measure the electron temperature profile with about 2 times better spatial resolution than on PLT and with an improved time resolution of approximately 10-15 ms. Its wide spectral bandwidth will give it the versatility to measure the profile over nearly the entire range of the TFTR operating magnetic field. Moreover, it will be used to study the ECE spectrum from the expected reactor-grade plasma in order to further clarify the power loss role that will be played by ECE in future tokamak reactors.

REFERENCES

a) The cooperation of the staff of the Princeton Plasma Physics Laboratory, in particular of the PLT group, is gratefully acknowledged. This research was funded by the National Science Foundation and the Department of Energy

1. B. A. Trubnikov and V. S. Kudryavtsev, Proc. 2nd U.N. Conf. Peaceful Uses of Atomic Energy (Geneva, 1958), Vol. 31, p. 93.
2. W. E. Drummond and M. N. Rosenbluth, Phys. Fluids 3, 45 (1960).
3. B. A. Trubnikov, Phys. Fluids 4, 195 (1961).
4. W. E. Drummond and M. N. Rosenbluth, Phys. Fluids 6, 276 (1963).
5. L. A. Artsimovich, Nucl. Fusion 12, 215 (1972).
6. H. P. Furth, Nucl. Fusion 15, 487 (1975).
7. R. G. Mills, Princeton Plasma Physics Laboratory Report MATT-658 (1969).
8. M. N. Rosenbluth, Nucl. Fusion 10, 340 (1970).
9. A. E. Costley, R. J. Hastie, J. W. M. Paul, and J. Chamberlain, Phys. Rev. Lett. 33, 758 (1974).
10. F. Engelmann and M. Curatolo, Nucl. Fusion 13, 497 (1973).
11. C. M. Celata and D. A. Boyd, Nucl. Fusion 17, 735 (1977).
12. P. Brossier, A. E. Costley, D. S. Komm, G. Ramponi, S. Tamor, in Proceedings of the 6th International Conference on Plasma Physics and Controlled Nuclear Fusion Research (International Atomic Energy Agency, Vienna, 1977), p. 409.
13. I. H. Hutchinson and D. S. Komm, Nucl. Fusion 17, 1077 (1977).
14. A. Cavallo and M. Tutter, Max-Planck-Institute for Plasma Physics Report No. IPP 2/244, 1978.
15. F. J. Stauffer and D. A. Boyd, Infrared Phys. 18, 755 (1978).
16. W. R. Rutgers and D. A. Boyd, Phys. Lett. 62A, 498 (1977).
17. A. Eberhagen, Infrared Phys. 19, 389 (1979).
18. H. W. Piekaar and W. R. Rutgers, Rijnhuizen Report No. 80-128, 1980.
19. G. D. Tait, F. J. Stauffer and D. A. Boyd, Phys. Fluids 24, 719 (1981).
20. B. Walker, E. A. M. Baker and A. E. Costley, J. Phys. E: Sci. Instrum. 14, 832 (1981).
21. TFR Group, in Proceedings of the 7th European Conference on Controlled Fusion and Plasma Physics, Lausanne (1975), Vol. I, paper 14b.
22. J. Hosea, V. Arunasalam, and R. Cano, Phys. Rev. Lett. 39, 400 (1977).
23. P. C. Efthimion, V. Arunasalam, R. Bitzer, L. Campbell, and J. C. Hosea, Rev. Sci. Instrum. 50, 949 (1979).
24. R. Cano, A. A. Bagdasarov, A. B. Berlizov, E. P. Gorbunov and G. E. Notkin, Nucl. Fusion 9, 1415 (1979).
25. G. Bekefi, Radiation Processes in Plasmas (Wiley, New York, 1966), chapters 1, 2, and 6.
26. C. M. Celata and D. A. Boyd, Nucl. Fusion 19, 423 (1979).
27. D. H. Martin and E. Puplett, Infrared Phys. 10, 105 (1969).
28. Private communication with Prof. D. H. Martin, Queen Mary College, London, England.
29. N. Bretz, D. Dimock, V. Foote, D. Johnson, D. Long, and E. Tolnas, App. Opt. 17, 192 (1978).

Measurement of the central ion and electron temperature of tokamak plasmas from the x-ray line radiation of high-Z impurity ions

M. Bitter, S. von Goeler, M. Goldman, K. W. Hill, R. Horton, W. Roney, N. Sauthoff, and W. Stodiek

Princeton Plasma Physics Laboratory, Princeton, New Jersey 08544

Hot tokamak plasmas contain small amounts of highly ionized, high-Z impurity ions, such as iron and titanium, with charge state distributions close to the ionization equilibrium of the solar corona. The characteristic x-ray line radiation of these ions represents a unique source of information on the plasma parameters. It is expected to be of vital importance, in particular as a diagnostic of the central ion temperature in future large size tokamaks. This paper describes measurements of the central ion and electron temperature of tokamak plasmas from the observation of the 1s–2p resonance lines, and the associated dielectronic ($1s^2 nl$–$1s2pnl$, with $n \geq 2$) satellites, of helium-like iron (Fe XXV) and titanium (Ti XXI). The satellite-to-resonance line ratios are very sensitive to the electron temperature and are used as an electron temperature diagnostic. The ion temperature is deduced from the Doppler width of the 1s–2p resonance lines. The measurements have been performed with high resolution Bragg crystal spectrometers on the PLT (Princeton Large Torus) and PDX (Poloidal Divertor Experiment) tokamaks. The details of the experimental arrangement and line evaluation are described, and the ion and electron temperature results are compared with those obtained from independent diagnostic techniques, such as the analysis of charge-exchange neutrals and measurements of the electron cyclotron radiation. The obtained experimental results permit a detailed comparison with theoretical predictions.

I. INTRODUCTION

In recent tokamak experiments, very high plasma temperatures have been obtained by the use of efficient heating techniques, e.g., the injection of intense hydrogen or deuterium beams,[1] and the excitation and damping of plasma waves.[2] Large volumes of hydrogen plasmas with ion and electron temperatures of $T_i = 1 - 7$ keV, and $T_e = 1 - 3$ keV, respectively, and densities of $1 - 5 \times 10^{13}$ cm^{-3} can now be maintained in tokamak discharges for several hundred milliseconds. These encouraging experimental results have motivated the design of tokamak fusion test reactors. The prospective dimensions, plasma densities, and temperatures for these devices exceed by far those of present-day tokamaks, and should be adequate for the demonstration of "fusion breakeven" which will be achieved if the released fusion energy is equal to the energy invested in heating the plasma.

As the dimensions, densities, and temperatures of tokamak plasmas increased, it became obvious that new diagnostic techniques were needed.[3-5] This is true in particular for measurement of the ion temperature in the hot core of tokamak discharges, since the standard methods, i.e., the measurement of the energy distribution of charge-exchange neutrals,[6] and Doppler-broadening measurements of spectral lines in the VUV (vacuum ultraviolet) region,[7] prove difficult under these experimental conditions. The charge-exchange diagnostic cannot be used for determination of the central ion temperature when the mean free path for neutral charge-exchange is small compared with the diameter of the plasma; and line radiation in the VUV region is primarily emitted from the edge of the plasma, if the central electron temperature is of the order of 1 keV. An exception is the forbidden line of Fe XX,[8] although this line will also originate from outer regions of the plasma column when the electron temperature exceeds 2 keV. In this paper we discuss alternate methods for diagnostic of the central ion and electron temperature which are based on observation of the soft X-ray line radiation from helium-like high Z impurity ions, e.g., Fe XXV and Ti XXI. These ions occur in the center of hot tokamak discharges for a wide range of experimental parameters.

The spectra of helium-like ions show in addition to the characteristic helium lines a series of satellites close to the 1s - 2p resonance lines. These satellites are due to transitions of the type $1s^2 nl$ - $1s2pnl$ with $n \geq 2$ and are produced by dielectronic capture of electrons with well defined kinetic energies (in the range from 4.69 keV to 6.7 keV, for Fe XXV). The resonance lines are excited by electrons with energies larger than the 1s - 2p threshold energies (6.7 keV in Fe XXV). The intensity ratio of the resonance and satellite lines depends, therefore, strongly on the electron temperature and can be used for electron temperature measurements. In addition, certain satellite to resonance line ratios can be used to measure the relative abundances of the lithium- and helium-like charge states, thus providing a means for diagnostic of the ionization equilibrium and the impurity transport. Most important for hot tokamak plasmas is, of course, the determination of the central ion temperature which is deduced from Doppler-broadening of the Kα-resonance lines of Fe XXV and Ti XXI at 1.85 and 2.61 Å. These measurements require Bragg crystal spectrometers of high spectral resolution and intensity for time resolved observations. Competing with Doppler broadening are natural line broadening and the recently discovered dielectronic broadening,[9,10] a phenomenon peculiar to high temperature plasmas.

Spectra of high Z helium-like ions are also observed in solar flares,[11-14] and a detailed theory[9,15-17] of these spectra has been developed for diagnostics of the flare parameters. Observations of these spectra in well diagnosed tokamak discharges permit a detailed comparison between experiment and theory and are, therefore, of great interest.

We begin with a brief description of tokamak experiments (Sec. II) and discuss (Sec. III) the parameters, e.g., the electron temperature and density profiles, the charge state distribution and X-ray line emissivity for helium-like impurity ions, which are relevant for the diagnostic application of these line spectra to tokamak plasmas. Sections IV and V describe the crystal spectrometers and the numerical methods which have been used for evaluation of line profiles. Section VI presents ion and electron temperature results obtained from observation of the Fe XXV- and Ti XXI- spectra on the PLT (Princeton Large Torus) and PDX (Poloidal Divertor Experiment) tokamaks at Princeton. These results are compared with those obtained from different diagnostic techniques, e.g., analysis of the energy distribution of charge-exchange neutrals and measurements of the electron cyclotron emission.

II. THE TOKAMAK

Basically, the tokamak is a ring of hot hydrogen plasma which is confined by a strong toroidal magnetic field inside a vacuum vessel. The plasma is produced by induction of an electric field which causes a current to flow in the toroidal direction. The current serves both to ohmically heat the plasma and to provide MHD equilibrium and stabilization by its poloidal magnetic field.

The stability of a tokamak plasma is a function of the aspect ratio (R/a) of the major (R) and minor (a) radii of the torus. In general, small aspect ratios are desired to obtain a stable plasma and to optimize ohmic heating efficiency. Also, the particle and energy confinement times are expected to increase in proportion to a^2. Therefore, short and fat tokamaks are better than long and thin ones. Here, we list the parameters of PLT which is one of the largest tokamaks in operation:

major radius: $R = 140$ cm
minor radius: $a = 40$ cm
toroidal magnetic field: $B_t = 32$ kG
plasma current: $I_p = 500$ kA.

The duration of a discharge pulse is typically 1 sec. The plasma can be maintained under steady state conditions for periods of 100 - 300 msec. A schematic of the PLT is shown in Fig. 1.

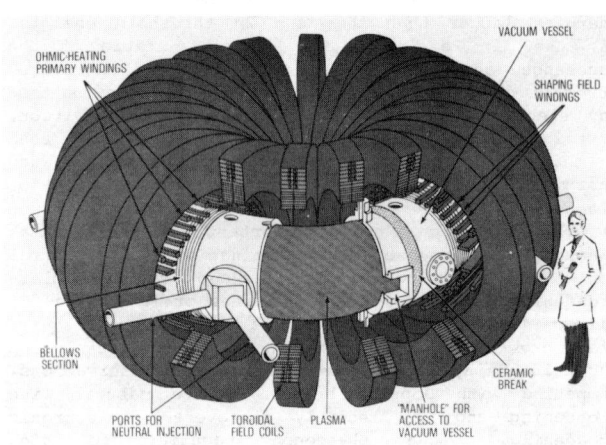

Fig. 1. Model of the PLT (Princeton Large Torus) tokamak.

Tokamaks of comparable dimensions are presently in operation in the U.S. and abroad. The JET (Joint European Torus) and TFTR (Tokamak Fusion Test Reactor) tokamaks which have been designed to achieve "breakeven" in a deuterium tritium plasma, are now under construction and have about twice the linear dimensions of the PLT.

The plasma densities and temperatures in tokamak discharges are nearly constant in the toroidal direction, and only vary with radius due to the fact that the ions and electrons move freely along (but not across) the magnetic field lines. In general, the radial density and temperature profiles are peaked near the magnetic axis and decay parabolically or steeper towards the plasma boundary. The peak electron densities, and electron and ion temperatures are typically in the range from $.1 - 1 \times 10^{14}$ cm^{-3}, and 1 - 3 keV, and 1 - 7 keV, respectively. The maximum temperatures have been obtained in discharges with additional heating by injection of intense deuterium beams[1] or with rf heating.[2] High purity of the plasma is crucial for the success of these heating experiments. The tokamak has been reviewed in several articles.[19,20]

III. IMPURITIES

Tokamak discharges contain small amounts (typically .1%) of impurities, like carbon, oxygen, iron, chromium, nickel, and titanium. These elements enter the plasma from the wall of the vacuum vessel, and are ionized to different states of ionization or charge states. The radiation emitted from impurity ions consists of a bremsstrahlung and recombination continuum and characteristic line spectra, and it contributes significantly to the total energy loss of the plasma. Figure 2 presents a typical spectrum of the soft X-ray radiation emitted from PLT showing the relative contributions of the continuum and line radiation from high Z impurities (Cr, Fe, Ni). The data were observed with the PLT Pulse Height Analysis System, which essentially consists of a pinhole camera and diode detector with an energy resolution of 300 eV.[21]

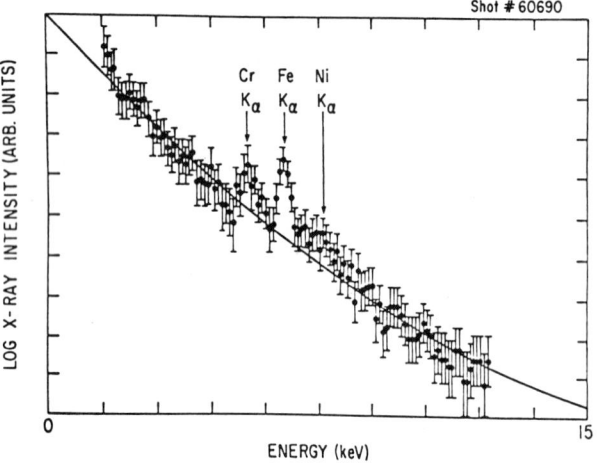

Fig. 2. Typical soft x-ray spectrum as observed from a PLT discharge with the Pulse Height Analysis System. The spectrum shows the contributions of the bremsstrahlung and recombination continuum and high Z impurity (Cr, Fe, Ni) line radiation.

The line radiation of the various impurity ions affects the plasma in different ways. Low Z elements like carbon, and oxygen are fully stripped in the hot center of the discharge, whereas high Z ions like iron, and titanium, are only partially ionized, the highest state of ionization being the helium-like charge state for typical central electron temperatures of 1-3 keV. Line radiation of low Z ions is, therefore, predominantly emitted from the edge of the plasma, whereas the line radiation of high Z ions is emitted from the center of the discharge. The energy losses associated with the line radiation from high Z ions tend to reduce the temperature in the hot core of the plasma and are, therefore, a matter of concern. On the other hand, line radiation from low Z ions provides cooling of the plasma periphery. This can, in fact, be beneficial in that low edge temperatures tend to reduce the influx of high Z wall impurities by reduction of sputtering.[4]

The intensity of the observed impurity line radiation depends on the radial distribution of the different states of ionization. The distribution of the charge states of high Z ions, e.g., Fe XXIV and Fe XXV, is expected to be in equilibrium, since the ionization and recombination times are usually small compared with the period of steady state conditions in tokamak discharges. Moreover, since the electron densities are far below threshold for collisional deexcitation, these ions exist essentially only in their ground states.[15] One may, therefore, assume that the equilibrium charge state distribution in tokamak plasmas is close to the so-called coronal equilibrium,[22-24] i.e., that the relative abundance of successive ionization states is determined by the ratio of the recombination and ionization rate coefficients, e.g.,

$$n_{Fe\ XXIV}/n_{Fe\ XXV} = \alpha/S = f(T_e). \quad (1)$$

This equilibrium is a function of the electron temperature (T_e) alone, and is usually observed in solar flares. In tokamak plasmas it is, however, only an approximate description, since deviations from cornonal equilibrium can occur as a result of impurity transport.

Figure 3 shows the fractional abundance for the charge states of iron under coronal equilibrium conditions.[25] Each charge state is peaked at an electron temperature ($T_e \approx 1/2\ E_i$) which corresponds to about half its ionization potential. In tokamak plasmas with typical (approximately parabolic) temperature profiles and peak temperatures of $T_e \geq 1$ keV, the low states of ionization occur only for a small range of plasma radii. The helium-like charge state (Fe XXV) is produced in the core of tokamak discharges for (relatively low) central electron temperatures $T_e(0) \geq 1$ keV due to the small ionization potential ($E_i = 2.045$ keV) of lithium-like iron (Fe XXIV). On the other hand, since the ionization potential ($E_i = 8.828$ keV) of Fe XXV is very large, it exists for a wide range of central electron temperatures, being the dominant charge state for $T_e(0)$ in the range from 1.5 to 6 keV and comparable with the fractional abundance of hydrogen-like iron, Fe XXVI, for $T_e(0)$ in the range from 6 to 10 keV. The characteristic line radiation of Fe XXV is, therefore well suited for diagnostics of central plasma parameters, both in present-day and in future larger size tokamaks.

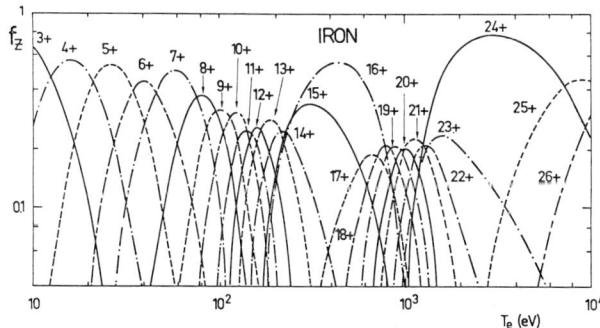

Fig. 3. Fractional abundance of the charge states of iron as a function of electron temperature for coronal equilibrium (C. Breton, et al, Ref. [25]).

An experimental investigation of the charge state distribution of iron in tokamak plasmas was performed on PLT using a Bragg curved crystal spectrometer.[26] These measurements were based on the observation of the characteristic $K\alpha$-line radiation emitted by the different charge states and made use of the fact that, on an average, for each L-shell electron removed between charge states Fe XVII to Fe XXV, the 1s-2p X-ray energy increased by an amount of 40 eV. Figure 4 shows the $K\alpha$-line spectra obtained for different electron temperatures. Evidently, the intensity is shifted to higher states of ionization with increasing electron temperature. For $T_e(0) \geq 1.2$ keV, iron occurs predominatly in the helium- and lithium-like charge states. A detailed comparison of these experimental results with calculations by Merts et al.,[24] showed that the distribution of the lower charge states was in reasonable agreement with coronal equilibrium predictions. Substantial deviations from coronal equilibrium were, however, found for Fe XXV. These deviations can be explained by an outward radial transport of Fe XXV ions on the basis of the fact that the transport time for these ions is small compared to the relevant ionization and recombination times. In general, the observed deviations from coronal equilibrium become smaller with increasing electron density.[27] The assignment of intensity peaks (in Fig. 4) to individual states of ionization gives a simplified picture, since the spectral ranges from neighboring charge states overlap to some extent. The spectral lines of the two and three electron systems, e.g., Fe XXV and Fe XXIV, are however, well resolved. This is important for diagnostic applications.

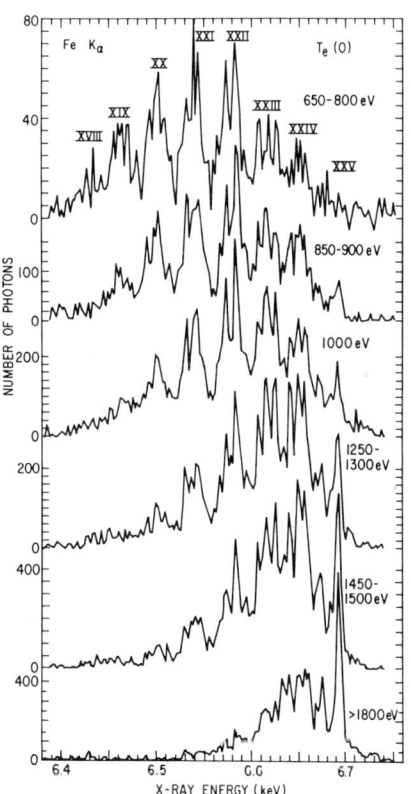

Fig. 4. $K\alpha$-line spectra of iron observed from PLT discharges with different electron temperatures.[26] The intensity is shifted to higher ionization states with increasing electron temperature.

Figure 5 shows the calculated intensity [(in photons/(cm^3 sec)] for the 1s-2p transition of Fe XXV as a function of plasma radius for different central electron temperatures assuming parabolic radial electron temperature, and density profiles, and coronal equlibrium for the Fe XXV-ions. Due to the strong variation of the excitation rate with electron temperature, the Fe XXV Kα-resonance line is predominately emitted from the center of the discharge, for $T_e(0) = 1 - 4$ keV. For this range of electron temperatues, the observed line spectra of Fe XXV can directly, be used as a diagnostic of the central plasma parameters. For higher central electron temperatures

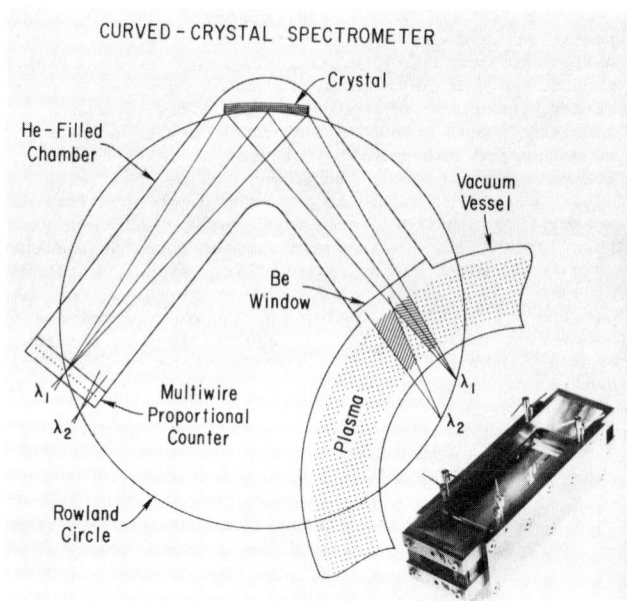

Fig. 6. Schematic of the Bragg curved crystal spectrometer. The photo shows the jig used for bending the crystal.

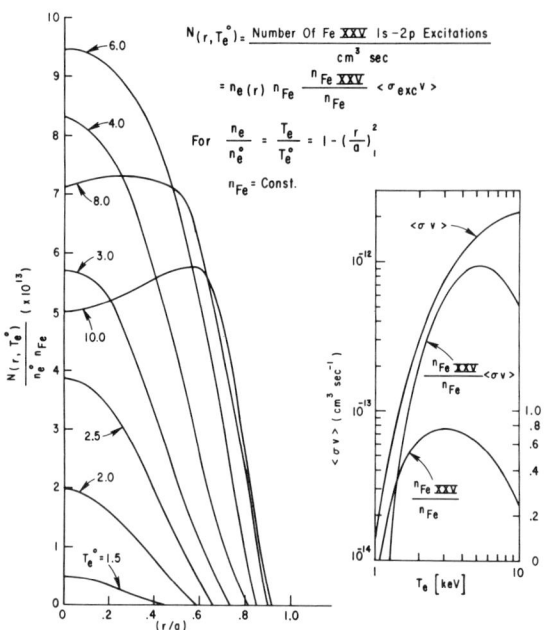

Fig. 5. Intensity [in photons/(cm³ sec)] of the Fe XXV Kα-resonance line as a function of plasma radius. The calculations have been performed for different central electron temperatures, Te(0), assuming parabolic electron temperature and density profiles, and coronal equilibrium for the Fe XXV charge state.

the helium-like charge state is burned out in the core of the discharge, so that the emissivity for the Kα-resonance line is peaked at outer plasma radii. Under these conditions, the central plasma parameters can be obtained from observations along different radial chords using Abel inversion.

IV. CRYSTAL SPECTORMETERS

In this section we describe the crystal spectrometers which have been used on PLT and PDX for observation of the satellite spectra of helium-like iron (Fe XXV) and titanium (Ti XXI). These experiments demanded optimization of the spectrometer design for (1) simultaneous observation of different spectral lines, (2) high spectral resolution, and (3) high intensity to enable time resolved measurements. These requirements led to the choice of focusing bent crystal spectrometers with a position sensitive detector in the Johann configuration.[28]

Experimental Arrangement

A schematic of a Johann crystal spectrometer and its arrangement on a tokamak is shown in Fig. 6. X-ray photons emitted along chords in the horizontal midplane from impurity ions in the shaded area of the tokamak discharge, enter helium filled pipes through a 0.003" thick beryllium window and are Bragg reflected by the crystal onto a position sensitive multiwire proportional counter. Photons of different energies are focused to different points on the Rowland circle and thus to different positions on the detector. The helium atmosphere in the X-ray path reduces photon absorption.

The detector has two planes of approximately 100 anode and cathode wires with 1 mm separation running perpendicular to each other, as illustrated in Fig. 7. An X-ray photon ionizes the P10 gas mixture (90% argon + 10% methane) inside the detector producing an electron avalanche to one of the anode wires. This charge pulse induces an image charge on the cathode wires which is capacitively coupled to a delay line. Standard electronics are used to convert the difference in the arrival times of the charge pulse at the two ends of the delay line into a pulse, whose amplitude is proportional to the position at which the photon struck the detector. These pulses are digitized, stored in the memory of a pulse-height analyzer, and later read by a PDP10 computer.

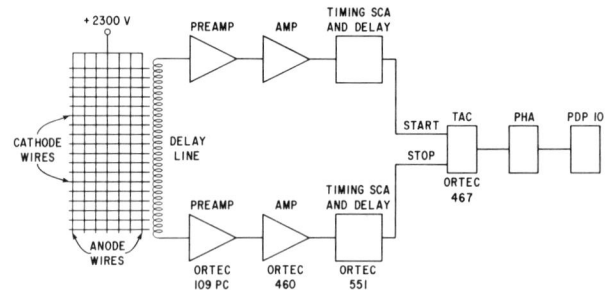

Fig. 7. Schematic of the multiwire proportional counter and electronics. The detector was built by Grumman Aerospace Corporation.

Since the experiments were performed in an environment with high levels of hard X-ray and neutron background radiation, the detector had to be shielded. Lead (4" thick) was used to shield the detector against the hard X-ray radiation. Since the 2.5 MeV neutrons emitted from the plasma produce (n,γ) - reactions in the lead shielding, it has been enclosed by a (8" thick) neutron shield of borated (5% Boron) polyethylene. This shielding permitted recording of soft X-ray spectra from deuterium plasmas also during injection of intense (6-8 MW) deuterium beams when the neutron emission from D-D reactions was particularly high, typically about 10^{14} neutrons/sec, corresponding to a flux of 10^8 neutrons/(cm^2 sec) at the position of the detector.

Spectral Resolution

Following Bragg's law

$$\lambda = 2d \sin\theta , \quad (2)$$

where λ, d, and θ are the wavelength of the X-ray photon, the spacing of the reflecting crystal planes, and the Bragg angle, respectively, the dispersion of a crystal spectrometer is given by

$$\frac{\lambda}{\Delta\lambda} = \frac{1}{\Delta\theta} \tan\theta . \quad (3)$$

The angular resolution ($\Delta\theta$) is mainly determined by the spatial resolution (Δx) of the detector and the crystal-detector distance (D). For the Johann configuration, $D = R \sin\theta$, where $R = 2r$ is the radius of curvature of the crystal, and r the radius of the Rowland circle. According to Eq. (3) it is advantageous to choose a crystal such that the Bragg angle, θ, is close to 90 degrees.

The spectral resolution can further depend on the intrinsic resolving power of the crystal and on geometrical effects: (1) the specific focusing error of the Johann configuration due to the fact that the crystal surface is tangential to the Rowland circle only at one point; and, more importantly, (2) errors due to the vertical divergence (with respect to the spectrometer plane) of X-ray beams. The Johann aberration (Δ) is given by $\Delta = \ell^2 \cos\theta/(8R \sin^2\theta)$, where ℓ is the length of the crystal. Evidently, both the Johann defocusing (Δ), and the focusing errors due to the vertical divergence decrease with increasing Bragg angle (θ) and radius (R) of crystal curvature. Since the mosaic spread of a crystal is known to occasionally increase in the bending process, we chose quartz crystals with a very high intrinsic resolving power > 100000.

In practice, the spectral resolution is mostly limited by the finite spatial resolution of the detector. Its positioning in the Rowland circle is, therefore, very important. The spatial resolution of the used multiwire proportional counters is .37 mm FWHM for perpendicular incidence of the photons. Therefore, the detector is not placed tangential to the Rowland circle, but perpendicular to the reflected beam. This is necessary because the thickness (2 cm) of the sensitive detection volume is relatively large. The focusing error resulting from the displacement of the detector from the Rowland circle decreases with increasing θ, another reason for choosing a large Bragg angle. For the high resolution crystal spectrometers which we used for Doppler broadening measurements, the displacement of the detector from the Rowland circle was comparable with the detector thickness. Optimum focusing can thus be obtained over a small energy interval. However, even for this limited energy range the resolution is affected by the finite thickness of the counter due to the vertical divergence of the reflected beams. It is of course, very important to properly align the anode and cathode wires with respect to the directions of the incident photons. This has to be optimized by rotating the detector.

The energy resolution of the crystal spectrometers has been determined experimentally by use of appropriate x-ray lines, after the crystals had been bent to the desired radius of curvature applying the method described by Feser and Faessler.[29] Here, we mainly discuss the results obtained for the (22$\bar{4}$3) quartz crystal spectrometer which was used for measurements of the Fe XXV spectra. The crystal was bent to a radius of curvature of 333 cm. To determine the energy resolution of this crystal we used the $L\alpha_1$ and $L\alpha_2$ lines of Holmium (at 6720 eV and 6680 eV) which are close in energy to the Fe XXV Kα- resonance line at 6702 eV. Since the natural line width of the Holmium lines is about 6 eV, a special arrangement had to be made to measure the much smaller instrumental width of the spectrometer. A narrow .381 mm (.015") wide slit was placed exactly on the Rowland circle at the focal point for the Holmium $L\alpha_1$ line. In this way only a small central portion of the $L\alpha_1$-line is Bragg reflected and focused onto the detector. Figure 8a shows the results of this measurement. The experimental points represent an image of the slit which is well approximated a Gaussian (G_p) of .497 mm FWHM. Following Unsöld[30] we describe the slit by a

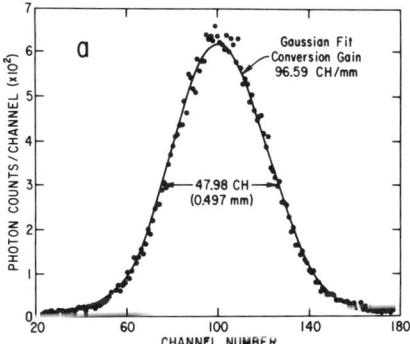
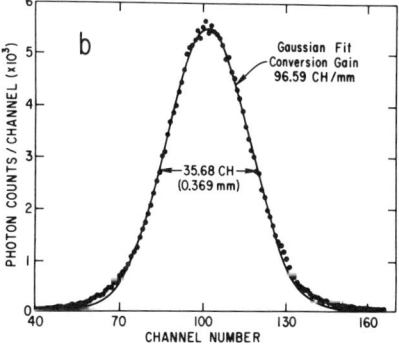

Fig. 8. Determination of the spectral resolution of the curved (R = 333 cm) (22$\bar{4}$3) quartz-crystal spectrometer: (a) Bragg reflection of the Holmium $L\alpha_1$-line (E = 6720 eV) as observed with a .381 mm (.015") wide slit positioned on the Rowland circle at the focal point for the $L\alpha_1$-line. The experimental points represent an image of the slit and can be fitted to a Gaussian with .497 mm FWHM (b) Spatial resolution of the detector observed with an approximate zero width collimated Fe^{55} line source placed on the entrance window of the detector. The experimental points can be fitted to a Gaussian of .369 mm FWHM, indicating that the resolution of the spectrometer is mainly determined by the spatial resolution of the detector.

Gaussian (G_s) of a FWHM = $2a[(\ln 2)/6]^{1/2}$, where a is the width of the slit, and assume that the image is the result of a convolution with a Gaussian instrumental function (G_I). We obtain a value of .424 mm for the width (FWHM) of the instrumental function. This corresponds to an energy resolution of $\Delta E = .420$ eV at $E = 6700$ eV, or $E/\Delta E = 15000$.

Also shown in Fig. 8b are the results of a measurement of the spatial resolution of the position sensitive detector. For this purpose a collimated Fe^{55} line source (of approximately zero width) was placed directly on the entrance window of the detector. The collimation was obtained by close contact of two tool bits. The experimental points in Fig. 8b are well approximated by a Gaussian with .369 mm FWHM. This shows that the instrumental width is mostly determined by the finite spatial resolution of the detector.

The energy resolution of the $20\bar{2}\bar{3}$ quartz crystal, which was used for Doppler broadening measurements of the Ti XXI Kα-line at 4.75 keV, has been determined by the same exprimental method, except that, in this case, the measurements were performed with the second order Bragg reflection of the tungsten Lβ_1-line at 9.670 keV. For this crystal, which was bent to a radius of curvature of 363.5 cm, the obtained spectral resolution ($\lambda/\Delta\lambda$) was 23000.

Intensity

Estimates of the X-ray line intensities as received by the detector from a PLT or PDX discharge have been obtained from an X-ray tracing code for the experimental parameters (i.e., the size of the crystal, the dimensions of the detector window and the beryllium window on the tokamak vacuum vessel, and the distances of the crystal from these windows). Here, we consider the basic aspects. Since the tokamak plasma is an extended X-ray source, the intensity is, in principle, independent of the distance of the crystal from the plasma provided that each crystal element views the plasma with a solid angle which is larger than the width of its rocking curve at the considered Bragg angle. In practice, however, the intensity varies with the distance of the crystal from the plasma in proportion to the vertical divergence of the detected X-rays which is usually limited by the height of the beryllium window on the tokamak vacuum vessel. On the other hand, as mentioned before, a limitation of the vertical divergence may be desirable to reduce focusing errors. The intensity also depends on Bragg angle Θ. For perfect crystals like quartz, the integrated reflectivity increases strongly with Θ, if $\Theta > 45$ degrees.

It is worth mentioning that the spatial resolution of multiwire proportional counters often deteriorates at very high photon count rates because of space charge effects near the anode wires. This sets an upper limit for the acceptable photon intensities (20,000 counts/sec for our detectors). Because of this count rate limitation the we had to reduce X-ray intensity in certain cases by shielding parts of the crystal. Finally, we mention that the focal points of spectral lines must not fall into the plane of the beryllium window, in order to prevent that the observed line intensities are affected by inhomogeneities of the window transmission.

V. EVALUATION OF LINE PROFILES

The profile of a spectral line is described by a Voigt-function, which is the convolution of a Lorentzian and a Gaussian. The Lorentzian describes the natural line broadening, whereas the Gaussian accounts for the Doppler broadening of the spectral line due to the thermal ion motion. This is by far the dominant line broadening mechanism in hot tokamak plasmas. Since the Doppler width is usually large compared with the natural line width, the upper part of the Voigt profile is mainly determined by the Gaussian; whereas, because of exponential decay of the Gaussian, the lower part of the Voigt profile, i.e., the wings, are determined by the Lorentzian.

The ion temperature is, therefore, often evaluated from the fit of a Gaussian to the upper half of the observed line profile. Though this method yields, in principle, a reasonable value for the ion temperature, the statistical error can be rather large, because only a limited number of the experimental points are actually used for the fit, and valuable information in the wings of the observed line profile is ignored; in practice, longer recording times are required if the statistical error is to be reduced to a certain limit. For time resolved measurements it is, therefore, advantageous to fit the observed spectral lines to actual Voigt-profiles.

For the calculation of Voigt-profiles it is convenient to make use of the close relationship between the Voigt-function and the so-called Plasma Dispersion Function,

$$Z(\xi) = \frac{1}{\sqrt{\pi}} \int_{-\infty}^{\infty} dt \, \frac{\exp(-t^2)}{t - \xi} , \qquad (4)$$

where $\xi = x + iy$, with $y = \text{Im}\,\xi > 0$, is a complex variable. The Z-function is of basic importance in the physics of plasma waves and has, been extensively studied. The function has been tabulated by Fried and Conte[31] and several codes for its computation are available. Introducing $\xi = x + iy$ into Eq. (4) one obtains for the imaginary part of $Z(x,y)$:

$$\text{Im}\,Z(x,y) = \frac{1}{\sqrt{\pi}} \int_{-\infty}^{\infty} dt \, \exp(-t^2) \frac{y}{(t-x)^2 + y^2} \qquad (5)$$

with

$$\frac{1}{\pi} \int_{-\infty}^{\infty} \text{Im}\,Z(x,y)\, dx = 1 .$$

The Voigt-function is thus most conveniently represented in the form:

$$V(x,y) = \frac{A}{\pi} \text{Im}\,Z(x,y) \qquad (6)$$

with

$$x = \frac{\omega - \omega_0}{\omega_0} \frac{c}{v_i} \text{*)} \quad \text{and} \quad y = \frac{\gamma/2}{\omega_0} \frac{c}{v_i}$$

where
c is the velocity of light,
$v_i = \sqrt{2kT_i/m_i}$ the thermal ion velocity,
ω a frequency within the spectral line,
ω_0 the center frequency of the spectral line emitted from a stationary ion,
γ the natural width (FWHM of the Lorentz profile),
A the intensity of the spectral line.

---

*The variable x represents the ratio of the ion velocity (v) and the thermal ion velocity (v_i).

In order to determine the characteristic line parameters, e.g., the center frequency ω_0 and the thermal ion velocity v_i, from a least squares fit, one needs the partial derivatives of the Voigt-function with respect to x and y. These derivatives are easily obtained with use of a differential equation for $Z'(\xi)$:

$$Z'(\xi) = -2[1 + \xi Z(\xi)] . \tag{7}$$

One obtains:

$$\frac{\partial}{\partial x} \text{Im } Z(x,y) = \text{Im } Z'(\xi) , \tag{8}$$

and

$$\frac{\partial}{\partial y} \text{Im } Z(x,y) = \text{Re } Z'(\xi) . \tag{9}$$

It is, therefore, sufficient to calculate only the real and imaginary parts of Z. Our line fitting program uses the Eqs. (6-9) to find values of A, ω_o, $\gamma/2$ and v_i, which minimize χ^2, given by

$$\chi^2 = \sum_{\ell}^{M} \frac{1}{\sigma_\ell^2} \left[D_\ell - \sum_{p=1}^{4} A_p V(\omega_{op}, \gamma_{p/2}, v_i, \ell) - B \right]^2 \tag{10}$$

where D_ℓ is the raw data in the ℓth position, σ_ℓ is the error, V is the Voigt-function and B a constant background. The program allows us to fit up to p=4 peaks in a certain spectral region simultaneously.

VI. EXPERIMENTAL RESULTS

In this section we present spectra of helium-like iron (Fe XXV) and titanium (Ti XXI) and discuss their diagnostic applications for electron and ion temperature measurements in tokamak discharges. Spectra of Fe XXV have been studied on PLT for a wide variety of plasma conditions during the last few years, and these data have permitted a detailed comparison between experiment[10,18] and theory.[9,15-17] Here, we include some of the experimental results to discuss the characteristic spectral features of helium-like ions, in particular, the electron temperature dependence of intensity ratios. Special emphasis is given to recent results of Doppler-broadening measurements of the Ti XXI Kα-resonance line which allowed determination of the central ion temperature in PDX tokamak discharges with additional neutral beam heating.

The most characteristic feature of the spectra emitted from helium-like ions in hot plasmas is the occurrence of satellites to the (1s-2p) resonance line. These satellites are due to transitions of the type $1s2pn\ell - 1s^2n\ell$ (with $n \geq 2$), which represent the final step in the process of dielectronic recombination of a helium-like ion. Here, we give a brief description of this atomic process. It can be the dominant mechanism of line excitation in hot plasmas and is thus very important for the interpretation and diagnostic application of the observed spectra.

The lithium-like configurations are produced by the process:

$$1s^2 + e \rightarrow 1s2pn\ell.$$

A free electron of well defined kinetic energy and angular momentum is capatured by a helium-like ion into an atomic level with quantum numbers n,ℓ to form an excited lithium-like ion. The 1s2pnℓ levels are virtually bound states located in the continuum above the first ionization limit of the lithim-like ion. They can, therefore, decay by autionization, the reverse process of capture, or by a radiative transition to the $1s^2n\ell$ level. In the latter case, the 1s-2p transition is slightly changed in energy by the presence of the "spectator" (n,ℓ) electron in the outer shell, so that the emitted photon gives rise to a satellite near the 1s-2p resonance line of the helium-like ion. The wavelength shift of the satellites depends on the main quantum number n of the captured outer shell electron. The n = 2 dielectronic satellites are well resolved. However, satellites with n $\geq$ 3 are very close to and partially blended with the resonance line.

The unresolved n $\geq$ 3 satellites cause an apparent intensity increase and broadening of the resonance line. These effects, which are peculiar to hot plasmas, must be taken into account for evaluation of satellite to resonance line ratios and Doppler broadening measurements. A detailed theory of the dielectronic satellites with n = 3 - 11 of the resonance line of Fe XXV has recently been given by Bely-Dubau et al.,[9] and comparison of this theory with experiment was made by Bitter et al.[10] Both the apparent intensity increase and the dielectronic broadening of the resonance line are very sensitive to the electron temperature.

Since the process of dielectronic capture involves electrons with well defined kinetic energies (ε = 4.694 keV for n=2 and ε = 5.815 keV for n=3 in Fe XXV), the satellite intensities are dependent on the energy distribution of the electrons in the plasma and proportional to the abundance of the helium-like ions. The resonance lines are excited by electrons with energies larger than the 1s-2p threshold energies (6.7 keV for Fe XXV). The ratios (I_s/I_w) of the intensities of the dielectronic satellites and the resonance line are consequently a function of the electron temperature (T_e) alone, given by[15]

$$\frac{I_s}{I_w} = D \frac{\exp(\Delta E/T_e)}{T_e} , \tag{11}$$

where ΔE is the ionization energy needed to remove the outer (n,ℓ) electron, and D is a constant mainly determined by the branching ratio of the probabilities for autoionization and radiative decay of the excited lithium-like ion.

Most of the satellites are produced by the described process of dielectronic recombination. Exceptions are the ($1s^22s-1s2p2s$) satellites which can also be produced by collisional inner shell excitation of a lithium-like ion in its ground state. For these satellites, the intensity ratio (I_s/I_w) is proportional to the relative abundance (n_{Li}/n_{He}) of the lithium- and helium-like charge states,

$$\frac{I_s}{I_w} = S \frac{n_{Li}}{n_{He}} , \tag{12}$$

where S is essentially given by the ratio of the oscillator strengths for the 1s - 2p transitions, e.g., of Fe XXIV and Fe XXV.[15]

Figures 9(a) and 9(b) show satellite spectra of Fe XXV obtained for central electron temperatures of 1.65 and 2.30 keV, respectively. The spectra have been observed on PLT using the (2243)-quartz crystal spectrometer with a resolution of $\lambda/\Delta\lambda$ = 15000. The electron temperature was determined from the electron-cyclotron radiation emitted by the plasma.[32] The raw data for each of these spectra were recorded during a period of 250 msec of a discharge pulse when the electron temperature was constant (steady-state conditions) and were accumulated over typically ten discharges with nearly identical parameters to reduce the statistical error. The center position, intensity, and width of the spectral lines were determined from a least squares fit of Voigt-functions as described in Section V. In order to determine these line parameters with minimum error, it was necessary to fit groups of usually two neighboring peaks simultaneously. The most prominent peaks have been identified as the helium-like lines w,x,y, z, the lithium-like n = 2 satellites t,q, (k,r),j, and the beryllium-like line β. The key letters used for line identification are explained in Table I and agree with Gabriel's notation.[15] The spectra also show a fine line structure between some of the most prominent peaks and on the long-wavelength side of the resonance line w. This structure disappears with increasing electron temperature and is ascribed to weak n = 2 satellites, such as a, and to n = 3 satellites, respectively.

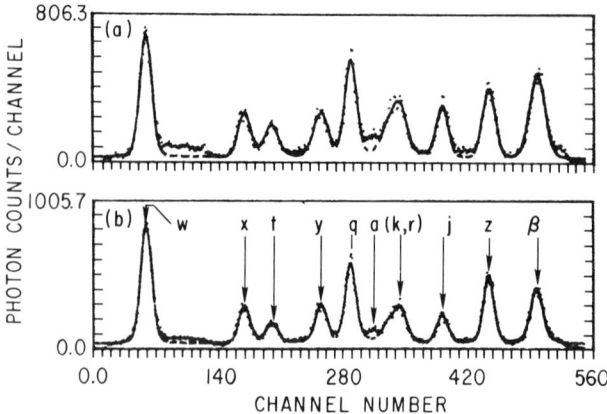

Fig. 9. (a),(b) Dielectronic satellite spectrum of FeXXV as recorded by a multichannel analyzer from PLT for a central electron temperature of 1.65 and 2.30 keV, respectively. The photon energy decreases with increasing channel number. The conversion gain is 0.18 eV/channel. w indicates the FeXXV Kα resonance line at 1.8500 Å. Also shown (solid curve) is the result of a least-squares fit of Voigt-functions to the raw data of the most prominent peaks.

Table I. Experimental and theoretical wavelengths of the observed FeXXV spectral lines. The superscripts indicate the references from which the theoretical wavelengths were taken. The experimental wavelengths were normalized to the theoretical value (Ref. 15) of 1.8500 Å for the resonance line w.

(key)	Transition	λ_{exp} (Å)	λ_{theor} [Å]
w	$1s^2(^1S_0) - 1s2p(^1P_1^o)$	1.8500	1.8500[a]
			1.84992[b]
x	$1s^2(^1S_0) - 1s2p(^3P_2^o)$	1.8552	1.8551[a]
			1.85519[b]
t	$1s^22s(^2S_{1/2}) - 1s2p2s(^2P_{1/2}^o)$	1.8567	1.8570
y	$1s^2(^1S_0) - 1s2p(^3P_1^o)$	1.8592	1.8591[a]
			1.85947[b]
q	$1s^22s(^2S_{1/2}) - 1s2p2s(^2P_{3/2}^o)$	1.8608	1.8604[a]
a	$1s^22p(^2P_{3/2}^o) - 1s2p^2(^2P_{3/2})$	1.8618	1.8618[a]
k	$1s^22p(^2P_{1/2}^o) - 1s2p^2(^2D_{32/})$	1.8632	1.8631[a]
r	$1s^22s(^2S_{1/2}) - 1s2p2s(^2P_{1/2}^o)$		1.8635[a]
j	$1s^22p(^2P_{3/2}^o) - 1s2p^2(^2D_{5/2})$	1.8657	1.8657[a]
z	$1s^2(^1S_0) - 1s2s\ (^3S_1)$	1.8681	1.8677[a]
			1.86801[b]
β	$1s^22s^2(^1S_0) - 1s2s^22p(^1P_1)$	1.8705	1.8710[c]

[a]Ref. 15   [b]Ref. 33   [c]Ref. 34

Experimental and theoretical wavelengths are listed in Table I for comparison. For most of the lines the agreement is within the estimated error of 0.0005 Å of the theoretical calculations. The experimental accuracy is 0.0001 Å.

Of great interest for diagnostic applications are the satellites j and q. The satellite j is produced by dielectronic recombination. According to Eq. (11) the intensity ratio (I_j/I_w) can thus be used as a diagnostic for the electron temperature. The satellite q, on the other hand, is produced by collisional inner shell excitation, so that according to Eq. (12), intensity ratio (I_q/I_w) can be used for a measurement of the relative abundance (n_{Li}/n_{He}) of the lithium- and helium-like ions.

Figure 10 presents the experimental results for I_j/I_w together with predictions of the earlier theory[15,16] (solid curve) and the most recent theory[9] (dashed curve) for the experimental values of T_e. The theory of Bely-Dubau et al.,[9] takes into account the apparent intensity increase, $\Delta I_w = \alpha(T_e)I_w$, of the resonance line due to unresolved dielectronic satelites with n ≥ 3. The experimental results shown in Fig. 10 are in excellent agreement with the most recent theoretical predictions,[9] indicating that the satellite to resonance line ratio (I_j/I_w) can be used as a diagnostics of the central electron temperature.

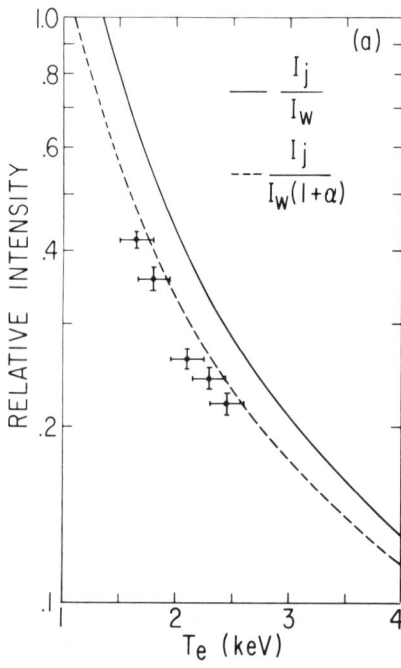

Fig. 10. Experimental results for the intensity ratio of the dielectronic satellite j and the resonance line w and predictions of both the earlier[15,16] (solid line) and most recent[9] (dashed line) theories. The intensity ratios I_j/I_w are plotted vs. the experimental values of T_e as determined from measurements of the electron cyclotron emission.[32]

On the other hand, the relative abundance $(n_{Fe\ XXIV}/n_{Fe\ XXV})$ of the lithium- and helium-like charge states derived from the intensity ratio (I_q/I_w) are found to deviate significantly (up to factors of five) from the predictions for coronal equilibrium. These deviations are observed to depend, in particular, on the electron density. They are ascribed to changes of the charge state distribution due to impurity transport, which is disregarded in the model of coronal equilibrium. Experimentally one finds that cornal equilibrium is approached with increasing plasma density.[27]

The experimental result that the relative intensities of the dielectronic satellites, and the resonance line are in good agreement with the theoretical predictions, is explained by the fact that these lines involve only the helium-like charge state, Fe XXV. Their intensity ratios are thus independent of the details of the ionization equilibrium or impurity transport and uniquely determined by the electron temperature. It has, therefore, been possible to measure the dielectronic recombination rate coefficient of Fe XXV as a function of the electron temperature from the observed total intensity of the dielectronic satellites.[35]

In the following we present the ion-temperature results which have been obtained from Doppler-broadening measurements of the Ti XXI Kα - resonance line. The experiments were performed on PDX using the (20$\bar{2}$3) -quartz crystal spectrometer with a resolution of $\lambda/\Delta\lambda = 23000$. Figure 11 shows the line spectrum of Ti XXI as observed from ohmically heated PDX discharges with central electron temperatures of $\approx$ 1 keV. The data were accumulated from approximately ten discharges with similar parameters.

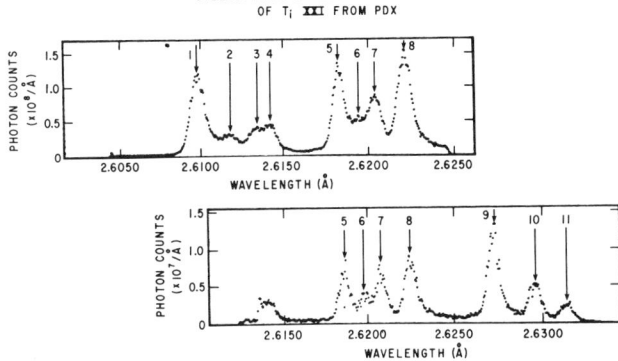

Fig. 11. Dielectronic satellite spectrum of Ti XXI. The spectrum has been observed from ohmically heated PDX discharges with central electron temperatures of $\approx$ 1 keV, using the curved (R = 363.5 cm) (20$\bar{3}\bar{3}$)-quartz-crystal spectrometer with a spectral resolution of $\lambda/\Delta\lambda = 23000$. The wavelengths are given relative to the wavelength of the TiXXI Kα-resonance line using the theoretical value[36] of 2.6099 Å.

The spectrum shows the Ti XXI Kα-resonance line (1), the characteristic helium-like intercombination lines (5) and (8), and a series of dielectronic satellites due to transitions of the type $1s^2 n\ell - 1s2pn\ell$ with $n \geq 2$. The spectral features have been identified with use of the theoretical predictions of Gabriel,[36] Vainshtein and Safronova,[17] and Ermolaev and Jones,[33] and are explained in Table II.

Figures 12 and 13 show the results obtained from Doppler broadening measurements of the Ti XXI Kα-resonance line. The data have been obtained from PDX discharges with additional neutral beam heating. Four deuterium beams totaling a power of 5 MW (47 kV) were injected into hydrogen discharges with a line averaged density of 4×10^{13} cm^{-3} for the period from 300-450 msec. Figure 12 shows a time sequence of six out of sixteen line profiles which have been recorded from each discharge with a time resolution of 25 msec. The data have been accumulated from 20 discharges with nearly identical parameters to reduce the statistical error. The solid curves were obtained from a least squares fit of Voigt-functions to the experimental data as described in Section V. The arrows indicate the range of channels used for the fit. The average in the channels below the lower limit was used for background subtraction. The upper limit was chosen to exclude the n = 3 satellites. An enhanced line broadening during neutral beam injection is evident from Fig. 12.

Table II. Experimental and theoretical wavelengths of the observed TiXXI spectral lines. The superscripts indicate the references from which the theoretical wavelengths were taken. The experimental wavelengths were normalized to the theoretical value (Ref. 36) of 2.6099 Å for the resonance line w.

Peak	(key)	Transition	λ_{exp} (Å)	λ_{theor} (Å)
1	w	$1s^2(^1S_0) - 1s2p(^1P_1^o)$	2.6099	2.6099[a]
				2.6097[b]
				2.61018[c]
2			2.6117	
3			2.6134	
4	d13	$1s^2 3p(^2P_{3/2})$ – $1s2p3p(^2D_{5/2})$	2.6142	
5	x	$1s^2(^1S_0) - 1s2p(^3P_2^o)$	2.6184	2.6187[a]
				2.6183[b]
6	s	$1s^2 2s(^2S_{1/2})$ – $1s2p(^3P)2s(^2P_{3/2})$	2.6195	2.6207[a]
				2.6197[b]
7	t	$1s^2 2s(^2S_{1/2})$ – $1s2p2s(^2P_{1/2})$	2.6205	2.6.214[a]
				2.6205[b]
8	y	$1s^2(^1S_0) - 1s2p(^3P_1^o)$	2.6223	2.6226[a]
				2.6221[b]
				2.62266[c]
9	q	$1s^2 2s(^2S_{1/2})$ – $1s2p2s(^3P_{3/2})$	2.6271	2.6272[b]
	a	$1s^2 2p(^2P_{3/2}^o)$ – $1s2p^2(^2P_{3/2})$		2.6295[a]
				2.6296[b]
10	d	$1s^2 2p(^2P_{1/2}^o)$ – $1s2p2s(^2P_{1/2})$	2.6294	2.6297[a]
				2.6299[b]
	r	$1s^2 2s(^2S_{1/2})$ – $1s2p2s(^2P_{1/2})$		2.6299[a]
				2.6295[b]
11	k	$1s^2 2p(^2P_{1/2}^o)$ – $1s2p^2(^2D_{3/2})$	2.6311	2.6326[a]
			(?)	2.6314[b]

[a] Ref. 36 [b] Ref. 17 [c] Ref. 33

Figure 13 presents the ion-temperature results, deduced from the experimental line profiles, as a function of time. The error bars indicate the statistical error of the experimental results. Corrections are made for the natural line width and the instrumental width. Errors due to the dielectronic line broadening by unresolved $n \geq 3$ satellites have been discussed in detail in Ref. [10]. These errors are determined by the electron temperature (which was in the range from 1 to 2 keV for these experiments) and are estimated to be less than 5%. Also shown in Fig. 13 are the results (triangles) from measurements of the energy distribution charge-exchange neutrals (hydrogen atoms) which are in very good agreement with the Doppler ion temperatures.

In principle, the impurity ion temperature can differ from the temperature of the bulk ion species. An estimate of the expected ion temperature difference can be obtained from a calculation due to R.

Goldston:[1] The power coupled from the beam ions to each ion species in the plasma is proportional to $n_i Z_i^2/m_i$. Thus, while the total power delivered to impurity ions may be small, the power per ion scales as Z_i^2/m_i. At low densities and high ion temperatures, the impurity-hydrogen coupling time becomes long enough so that a temperature difference should develop between the impurity ions and the hydrogenic species. For low-density cases where $Z_{eff} = 3$, due primarily to carbon, all of the impurity ions are well coupled to each other, and one may write a simple equation governing the impurity temperature, T_x:

$$\frac{dT_x}{dt} = 0 = \frac{T_H - T_x}{\tau_{x-H}} + \frac{dT_x}{dt}\bigg|_{beam} + \frac{T_e - T_x}{\tau_{x-e}} - \frac{T_x}{\tau_{Ex}} \quad (13)$$

where τ_{x-H} and τ_{x-e} are the energy equilibrium times between the impurities and the hydrogenic species, and between the impurities and electrons, respectively. τ_{Ex} is the energy containment time of the impurities.

Taking the dominant impurity to be carbon, and solving this equation for similar parameters of a neutral beam heating experiment on PLT, Goldston found a temperature difference of $T_x - T_H = 800$ eV, assuming a bulk (hydrogen) ion temperature of $T_H = 5$ keV. The evaluation of $T_x - T_H$ is uncertain to within a factor of 2; the possible temperature differential drops dramatically, however, as the density increases, and especially as the ion temperature decreases.[1] Experimentally we find that for the parameters of the PDX experiment the ion temperature differences $T_x - T_H$ are within the error bar (Fig. 13). Note that the charge-exchange diagnostics is still able to determine the central ion temperature for plasma densities of 4×10^{13} cm^{-3} and plasma radii of 40 cm.

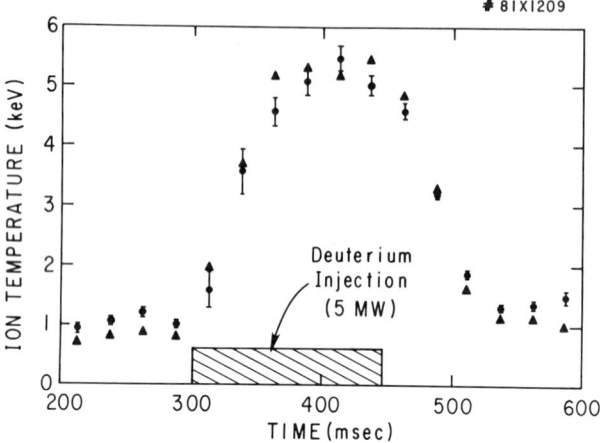

Fig. 13. Ion temperture results obtained from the observed Ti XXI Kα-line profiles (circles) and from measurements of charge-exchange neutrals (triangle), as a function of time. The bars represent the statistical error of experimental results.

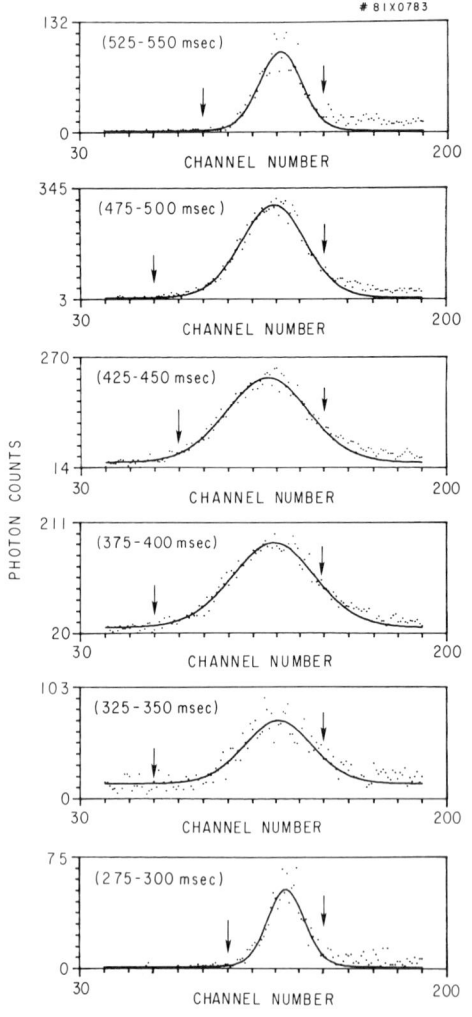

Fig. 12. A time sequence of six out of sixteen Ti XXI Kα-line profiles observed from PDX discharges with additional ion heating by injection of intense (5 MM, 47 kV) deuterium beams during the period from 300-450 msec. The solid lines represent least-squares fit of Voigt-functions. The arrows indicate the limits used for the fit.

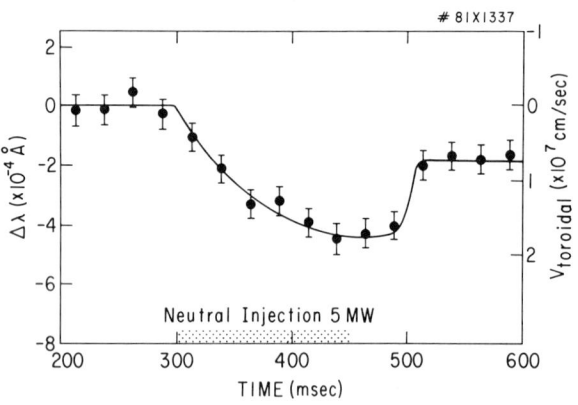

Fig. 14. Observed Doppler-shift of the Ti XXI Kα-line as a function of time. The Doppler-shift is due to a toroidal rotation of the plasma which results from momentum transfer by the injected deuterium beams.

The line profiles in Fig. 12 also show a Doppler shift in addition to the line broadening. Figure 14 presents the observed line shifts as a function of time. The shift appears with the deuterium injection, and is observed to some extent even after termination of the injection. It is explained by a toroidal rotation of the plasma due to momentum transfer by the deuterium beams. The crystal spectrometer on PDX views the plasma under a angle of 16.5 degrees with respect to the radial direction, so that the component

$v_t \sin(16.5°)$ of the toroidal plasma velocity v_t can be detected. On PDX the neutral beams are injected near-perpendicularly, forming an angle of 14° with the radial direction at the center of the vacuum vessel. The deduced values of $1-2 \times 10^7$ cm/s for the velocity of the toroidal plasma rotation represent a substantial fraction of the tangential component ($v_d \sin(14°) = 4.5 \times 10^7$ cm/s) of the beam particle velocity. Measurements of the Doppler shift of the Ti XXI $K\alpha$-line can be used to determine the momentum confinement time for the Ti XXI-impurity ions.

VII. CONCLUSION

The satellite spectra of helium-like iron and titanium, Fe XXV and Ti XXI, have been used as a diagnostic of the central ion and electron temperature in tokamak discharges. The experimental results are in good agreement with those obtained from different diagnostic methods, i.e., analysis of charge-exchange neutrals and measurements of the electron cyclotron emission, and permit a detailed comparison with theoretical predictions. Spectra of high Z helium-like ions are observed for a wide range of experimental parameters and are of particular interest as a diagnostic of the central ion temperature both in present-day and in future large size tokamaks. In addition to temperature measurements, the spectra can be used for determination of the charge state distribution, transport, momentum confinement of impurity ions in hot plasmas, and they also give valuable information on atomic data, e.g., the dielectronic recombination rate coefficient of helium-like ions. Special emphasis has been given to a discussion of experimental conditions and instrumental details.

ACKNOWLEDGMENTS

The continued support of Drs. H.P. Furth, D.M. Meade, P.H. Rutherford, J.C. Hosea and K. Bol is gratefully acknowledged. We deeply appreciate the numerous discussions with Drs. A.H. Gabriel, F. Bely-Dubau, J. Dubau, P. Faucher, and S. Volonte on the interpretation of the observed spectral data. We are grateful to Drs. A. Kritz and D. Batchelor for making their computer program for calculation of the Plasma Dispersion Function available to us, and we also gratefully acknowledge the technical assistance of J. Gorman, T.D. Cost, and the PLT technican crew under W. Mycock, and assistance of the data processing group under L. Michaels. This work was supported by U.S. Department of Energy Contract No. De-AC02-76-CHO-3073.

REFERENCES

[1] H. Eubank, et al., PLT Neutral Beam Heating Results, Princeton Plasma Laboratory Report PPPL-1491 (November 1978) 47 pp, and Plasma Physics and Controlled Fusion Research (Proc. 7th Int. Conf. Innsbruck, 1978) 1, IAEA Vol. I, 167 (1979).
[2] J.C. Hosea, et al., Plasma Physics and Controlled Nuclear Fusion Research, (Proc. 8th Int. Conf. Brussels, 1980) IAEA Vol. II, p. 95 (1980).
[3] K.M. Young, IEEE Trans. Nucl. Sci., NS-26, 1234 (1979).
[4] K.W. Hill, M. Bitter, S. Eames, S. von Goeler, N.R. Sauthoff, E. Silver, in Low Energy X-Ray Diagnostics - 1981, AIP Conference Proceedings No. 75, (American Institute of Physics, New York) pp. 8-24.
[5] K.M. Young, "Nuclear Aspects of Tokamak Fusion Test Reactor (TFTR) Diagnostics and Instrumentation," Princeton Plasma Physics Laboratory Report, No. PPPL-1859 (1982) (unpublished).
[6] S.S. Medley, S.L. Davis, and M. Brusati, Princeton Plasma Physics Laboratory Report PPPL-1478 (1978) (unpublished).
[7] S. Suckewer and E. Hinnov, Nucl. Fusion 17 945 (1977).
[8] S. Suckewer and E. Hinnov, Phys. Rev. Lett. 41 756 (1978).
[9] F. Bely-Dubau, A.H. Gabriel, and S. Volonte, Mon. Not. Roy. Astron. Soc. 189 801 (1979), and 186 405 (1979).
[10] M. Bitter, S. von Goeler, K.W. Hill, R. Horton, D. Johnson, W. Roney, N. Sauthoff, E. Silver, W. Stodiek, Phys. Rev. Lett. 47 921 (1981).
[11] J. Dubau and S. Volonte, Rep. Prog. Phys. 43 199 (1980).
[12] U. Feldman, G.A. Doschek, R.W. Kreplin and J.T. Moriska, Astrophys. J. 241 1175 (1980).
[13] J.L. Culhane, A.H. Gabriel, L.W. Acton, C.R. Rapley, K.J. Phillips, C.J. Wolfson, E. Antonucci, R.D. Bentley, R.C. Catura, C. Jordan, M.A. Kayat, B.J. Kent, J.W. Leibacher, A.N. Parmar, J.C. Sherman, L.A. Springer, K.T. Strong, and N.J. Veck, Astrophys. J. 244 L141 (1981).
[14] A.H. Gabriel, L.W. Acton, J.L. Culhane, K.J. H. Phillips, C.G. Rapley, E. Antonucci, R.D. Bentley, C. Jordan, M.A. Kayat, J.W. Leibacher, M. Levay, J.C. Sherman, K.T. Strong and N.J. Veck, Astrophys. J. 244 L147 (1981).
[15] A.H. Gabriel, Mon. Not. Roy. Astron. Soc. 160 99 (1972).
[16] C.P. Bhalla, A.H. Gabriel, L.P. Presnyakov, Mon. Not. Roy. Astron. Soc. 172 359 (1975).
[17] L.A. Vainshtein, U.I. Safronova, Atomic Data and Nuclear Data Tables 21 49-68 (1978).
[18] M. Bitter, K.W. Hill, N.R. Sauthoff, P.C. Efthimion, E. Meservey, W. Roney, S. von Goeler, R. Horton, M. Goldman and W. Stodiek, Phys. Rev. Lett. 43 129 (1979).
[19] L.A. Artsimovich, Nucl. Fusion 12 215 (1972).
[20] H.P. Furth, Nucl. Fusion 15 487 (1975); "U.S. Tokamak Research", Princeton Plasma Physics Laboratory Report PPPL-1598 (1979) (unpublished).
[21] S. von Goeler, "Diagnostic for Fusion Experiments" (Pergamon Press, Oxford and New York, edited by E. Sindoni and C. Wharton, 1979, p. 79).
[22] C. Jordan, Mon. Not. Roy. Astron. Soc. 142 501 (1969).
[23] H.P. Summers, Mon. Not. Roy. Astron. Soc. 169 663 (1974); Appleton Laboratory Report No. IM 367, (1975) (unpublished).
[24] A.L. Merts, R.D. Cowan, and N.H. Magee, Los Alamos Scientific Laboratory Report LA-6220-MS, 1976 (unpublished).
[25] C. Breton, C. DeMichaelis, M. Finkenthal, and M. Mattioli, Association Euratom-C.E.A., Fontenay aux Roses, Report EUR-CEA-FC-948, (1978) (unpublished).
[26] K.W. Hill, S. von Goeler, M. Bitter, L. Campbell, R.D. Cowan, B. Fraenkel, A. Greenberger, R. Horton, J. Hovey, W. Roney, N. Sauthoff, and W. Stodiek, Phys. Rev. A19 1770 (1979).
[27] W. Stodiek, et al., Plasma Physics and Controlled Nuclear Fusion Research (Proc. 8th Int. Conf. Brussels, 1980, IAEA Vol. I, p. 9 (1980).
[28] H. Johann, Z. Phys. 69 189 (1931).
[29] K. Feser and A. Faessler, Z. Phys. 209 1 (1968).
[30] A. Unsold, "Physik der Sternatmospharen" (Springer, Berlin-Göttingen-Heidelberg, 1955).
[31] B.D. Fried and S.D. Conte, "The Plasma Dispersion Function", (Academic Press, New York and London, 1961).
[32] P.C. Efthimion, V. Arunasalam, J.C. Hosea, Phys. Rev. Lett. 44 396 (1980).
[33] A.M. Ermolaev, M. Jones, and K.Y.S. Phillips, Astrophys. Lett. 12 53 (1972).
[34] L.A. Vainshtein, U.I. Safronova, Sov. Astron. AJ 15 175 (1971).
[35] M. Bitter, S. von Goeler, N. Sauthoff, K.W. Hill, K. Brau, D. Eames, M. Goldman, E. Silver, W. Stodiek, in "Inner-Shell and X-Ray Physics of Atoms and Solids" (Plenum Press, New York, and London, 1981) p. 861.
[36] A.H. Gabriel, private communication.

Temperature measurement in a wall-stabilized arc used as a radiation standard in the vacuum ultraviolet

D. H. Nettleton

National Physical Laboratory, Teddington, Middlesex, TW11 0LW, United Kingdom

A wall-stabilized arc developed at the National Physical Laboratory in the United Kingdom allows stable discharges to be run in argon at temperatures of between 12 000 and 14 500 K with electron densities between 0.9 and 2.5×10^{23} m^{-3}. The arc is run windowless using a differential pumping system to isolate it from a vacuum system. Detailed diagnostics have been carried out on this discharge, using spectroscopic techniques in both the visible and the vacuum ultraviolet regions of the spectrum. These have shown that under certain operating conditions the discharge is in local thermodynamic equilibrium and allow the temperature to be measured with an accuracy of ±0.3%. The arc has been used to establish an absolute radiance scale in the region 95 to 200 nm, using blackbody limited lines of trace gases added to the plasma. This scale has been compared with the visible and near ultraviolet radiance scales at the NPL by extending the scale to shorter wavelengths using the calculable relative spectral emission of a synchrotron, and deuterium discharge lamps as transfer standards. This comparison showed agreement to ±2.5%.

INTRODUCTION

The atmospheric wall-stabilized arc has been used as a source of high current plasmas for the determination of data required in plasma diagnostics and as an absolute radiance source. At the NPL, a wall-stabilized arc has been developed by R.C.Preston[1] which produces a stable and reproducible plasma in argon and whose axial temperature can range from 12,000 to 14,500 K. The temperature of the plasma column depends only on the pressure of the argon, the current flowing and the column diameter. To achieve this the argon must be of high purity to avoid unquantifiable emission from impurities, the leakage current between sections must be negligible compared to the arc current, and the space between sections must be small to achieve a smooth potential gradient along the plasma column, to avoid any inhomogeneities. If the arc is to be viewed end-on, boundary contributions from the end of the arc must be small compared to the column and must be stable so that a correction for their contribution to measurements can be determined.

If this plasma is to provide a basis for the determination of fundamental quantities or as an absolute radiance source, the arc must be shown to be in local thermodynamic equilibrium, LTE, and its axial temperature known. Various techniques were used to estimate how closely the NPL wall-stabilized arc plasma approached LTE and to measure its temperature.

The device has been used to establish an absolute radiance scale at the NPL from 95 to 200 nm by using the blackbody limited lines of small amounts of other gases added to the plasma to provide known radiances.

A deuterium lamp transfer standard has also been developed at NPL which allowed comparison of this radiance scale with the NPL blackbody-radiator-based visible and near uv radiance scales. This was achieved by using the known relative spectral distribution of synchrotron radiation to establish a uv emission scale, which was then used to interpolate between the scales. Agreement of the two scales confirmed the arc temperature.

THE WALL-STABILIZED ARC

A diagram of the NPL wall-stabilized arc is shown in Fig.1. It consists of 17 individually water cooled

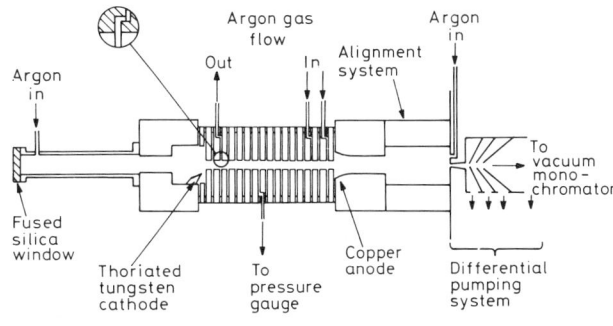

Fig. 1. The wall-stabilised arc plasma source

copper sections of 3 mm width and with a channel diameter of 3 mm. Each section is insulated from the next with 0.25 mm polytetrafluoroethylene (PTFE) insulators which also provide a leak-tight seal. Argon flows continuously through the column to maintain purity. The gas lines are bakeable and the argon purified using a dry-ice cooled molecular sieve trap and a rare gas purifier. This achieved a purity of argon with major impurities by volume - nitrogen-4ppm, oxygen-3ppm, hydrogen-5ppm, and carbon < 0.2ppm. The arc is struck from a tungsten cathode to a hollow copper anode. The arc can be run at currents up to 75 A with a spectral emission coefficient stable to ± 0.5% over an hour. One hundred hours operation are possible before the arc needs cleaning.

To allow spectroscopic observations, a fused silica window is fitted at one end of the arc and a four stage differential pumping system at the other. The latter allows the arc to be connected to a 3 m vacuum monochromator without a window. Each stage of the system consists of a circular orifice and a vacuum pump. The plasma pressure was controlled at $1.75 \times 10^5 \pm 2 \times 10^2$ Pa. The power supply controls the current to ± 0.02 A over a 4 hour period. The arc can be run at currents ranging from 28 to 75 A which corresponds to a power dissipation of 2.7 to 10 kW. To

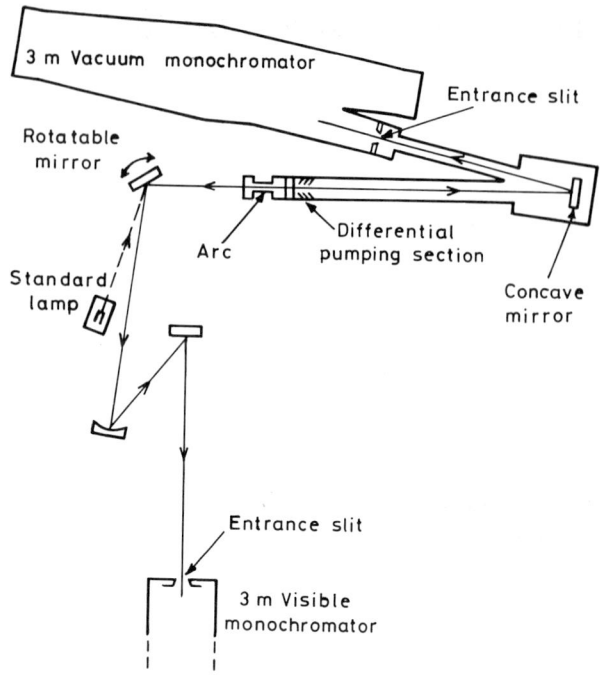

Fig 2. Optical system

initiate the arc, a low pressure discharge was struck between the cathode and anode with a high voltage pulse and maintained with a low current high voltage source; then the pressure was increased before the high current was allowed to pass.

OPTICAL SYSTEM

The optical system is shown in Fig.2. The plasma is focused through the differential pumping system by means of a concave mirror at an aperture ratio of f/260 with a 2:1 de-magnification onto the entrance slit of a 3 m McPherson normal incidence vacuum monochromator. The entrance slit defines the field and was set at 0.1 mm height and 0.01 mm width. Radiation from the end of the plasma fitted with a fused silica window was focused onto the entrance slit, of dimension 0.1 x 0.1 mm, of a 3 m visible region monochromator using an aperture ratio of f/150. A rotatable mirror allowed the monochromator response to be calibrated against a tungsten strip lamp of known radiance.

CRITERIA FOR LTE AND TEMPERATURE MEASUREMENT

For a thermodynamic temperature to be defined the transfer of energy between energy states must be fast enough to establish a Boltzmann distribution for the occupation of states. In an electrically heated plasma the energy is added to the system by the acceleration of electrons in an electric field. This energy is spread between the energy states of the ions and neutrals and electrons by collisions. De-population of states also occurs by radiative emission. As the plasma is optically thin the radiation field is not in thermodynamic equilibrium. Despite this a state of LTE can be defined if the population of states by collision is faster than the de-population by radiative emission. This will lead to a near Boltzmann distribution for the occupation of states. From these arguments it is clear that there will be a minimum electron density at which LTE can exist for particular species. Theories[1] to calculate this critical electron density predict that for the NPL plasma, and for the particular species of interest, electron densities of around 2×10^{23} m^{-3} are required for LTE. If the radiation field is optically thick at the resonance lines, then this criteria can be relaxed a little. For the NPL arc electron densities are of this order, ranging from 9×10^{22} m^{-3} at 28 A to 2.5×10^{23} m^{-3} at 75 A. LTE should therefore exist for all states but small departures may occur at low currents.

The axial temperature of the arc was measured using a technique proposed and used by Richter[2]. Consider two independent spectroscopic quantities whose dependence can be separated into the product of two single valued functions of temperature f1(T) and f2(T) and two functions which are not a function of temperature g1(a,b,...) and g2(a,b,...).

$$Q1(a,b,....,T) = g1(a,b,...) \times f1(T) \quad (1)$$

$$Q2(a,b,....,T) = g2(a,b,...) \times f2(T). \quad (2)$$

Now a unique curve can be plotted of log(Q1) versus log(Q2) as T is varied. Experimentally, the arc can be run at different currents and there will be a one to one relationship between arc current and arc temperature. Thus a log-log plot of the measured spectroscopic quantities as the current is varied will map out the same curve as in the theoretical case. Thus, by comparing the curves, the appropriate temperature for each current can be found.

In practice, for most spectroscopic quantities, data is not available to calculate g1 and g2 accurately but the temperature can still be calculated from the log-log plots. This is because the shape of the log-log curve is not affected by the functions g1 or g2, which only specify where in the log-log plane the curve is situated; that is g1 and g2 cause translation parallel to the axes. This means that by finding the best fit of the theoretical and experimental curves by translating them parallel to the axes, characteristic temperatures for each current can be found. An example of such a fitted curve is shown in Fig.3. The spectroscopic quantities used were the total intensity of the Ar(I) lines at 430 and 714.7 nm and the Ar(II) line at 480.6 nm, the radiance of the centre of a blackbody limited line of krypton at 123.5 nm and the radiance of the argon continuum at 431.4 and 482.7 nm. Log-log plots were carried out for 8 combinations of these

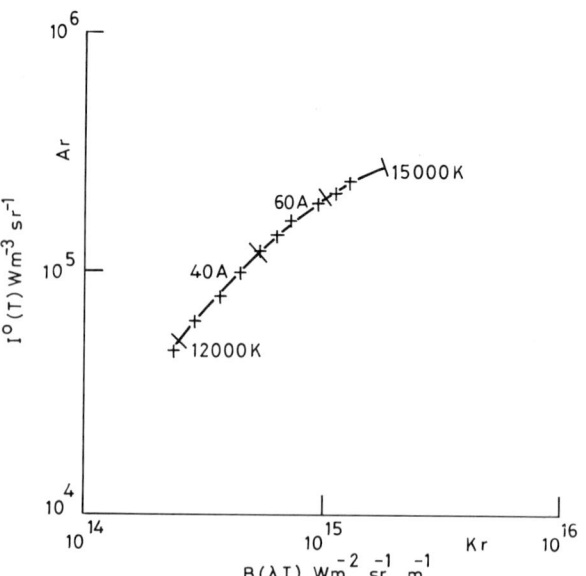

+ Experimental points at different currents.
— Theoretical curve as function of temperature

Fig. 3 Log-log plot of the integrated volume emission coefficient of the Ar(I) line at 430 nm versus the blackbody limited Kr line at 123.5 nm.

Arc Current A	Ar(I) 430 nm vs. Kr(I)	Ar(I) 714.7 nm vs. Kr(I)	Ar(II) vs. Ar(I) 430 nm	Ar(II) vs. Ar(I) 714.7 nm	Ar cont. 431.4 nm vs. Kr(I)	Ar cont. 482.7 nm vs. Kr(I)	Ar(II) vs. Ar cont. 431.4 nm	Ar(II) vs. Ar cont. 482.7 nm	Ne m^{-3}
28	11 875	11 914	11 860	11 917	12 045	12 016	12 045	12 010	8.76 x 10^{22}
32	12 194	12 249	12 157	12 220	12 365	12 333	12 352	12 314	1.04 x 10^{23}
36	12 477	12 528	12 429	12 494	12 651	12 617	12 633	12 593	1.22 x 10^{23}
40	12 720	12 776	12 667	12 735	12 898	12 865	12 881	12 839	1.36 x 10^{23}
45	13 012	13 065	12 949	13 020	13 190	13 157	13 173	13 130	1.56 x 10^{23}
50	13 256	13 318	13 193	13 269	13 449	13 413	13 430	13 384	1.73 x 10^{23}
55	13 499	13 554	13 434	13 511	13 697	13 658	13 682	13 633	1.90 x 10^{23}
60	13 713	13 772	13 643	13 725	13 921	13 886	13 903	13 854	2.05 x 10^{23}
65	13 929	13 985	13 858	13 942	14 144	14 103	14 130	14 076	2.20 x 10^{23}
70	14 125	14 191	14 053	14 144	14 352	14 312	14 339	14 284	2.33 x 10^{23}
75	14 385	14 245	14 340	14 561	14 561	14 517	14 546	14 487	2.45 x 10^{23}

Table I Temperature, in K, of the homogeneous region of the constricted arc at various currents from the log-log techniques.

quantities and the temperatures obtained at various currents for each plot shown in Table I.

To measure the total intensity of a line a computer model had to be set up to allow for the correction of line wing loss, self-absorption and the contributions from other nearby lines[3]. For example, for the Ar(I) line at 430 nm, data was required for 19 argon lines and continuum factors[4,5]. Line widths were taken from Griem[6] but varied to achieve the best fit to the experimental spectrum. This required reducing Griem's values by about 20%. The value of the total intensity of each line which gave the best fit was then used in the log-log plot.

Krypton was added to the plasma via a tube near the centre of the column and the argon gas flow adjusted to keep the krypton from the anode and the differential pumping system. This was to ensure that the emission line used did not become self-reversed. The amount of krypton added was kept very small, so as not to upset the homogeneity of the plasma, but sufficient to make the centre of the 123.5 nm line blackbody limited. The radiance of this line will then follow the Planck function.

The continuum measurement at 431.4 and 482.7 nm were corrected for self-absorption and line wing contributions using the computer model. The expression for the specific intensity of the continuum is given in reference 7 with the factors required for the calculations taken from references 5 and 8.

Direct calculation of temperature was also possible using some of these measurements and using the tungsten strip lamp to calibrate the visible monochromator response and transition probabilities taken from the literature[4].

The contribution from the end regions was investigated by determining, both experimentally and theoretically, an equivalent length, Δl, of homogeneous plasma with the same temperature, T_0, as the homogeneous column[9]. Experimentally arcs were run of different lengths; this being achieved by running arcs with less than the normal 17 sections. A curve of each spectroscopic quantity versus constricting column length was plotted from which the equivalent length of the end effects could be determined. The theoretical model used an appropriate axial temperature distribution, see Fig.4, and performed the integral of the radiative transfer equation for the particular spectroscopic quantity. The line blending program was also included in the integral. The accuracy of the model for each spectroscopic quantity can be seen in Table II by comparison of the theoretical and experimental values of Δl. As the krypton blackbody line has no contribution from the ends of the arc, no correction was required. Having shown that the assumed temperature distribution leads to an accurate estimate of the end effects, this distribution was used in calculating the theoretical curves for the log-log plots.

Further to providing a temperature of the plasma, the quality of the fits of the log-log plots gives information about any departure from LTE. This showed that the most reliable temperature measurement was at a current of 60 A where the electron density is high enough for LTE but not so high that large Stark widths make accurate spectral modelling difficult. For this reason an arc current of 60 A was chosen when the arc was used as a radiance standard.

A careful analysis of errors led to the final assessment that the temperature at the arc axis at 60 A was

$$T = 13,820 \pm 40 \text{ K} \qquad (3)$$

which corresponds to an electron density of

$$N_e = 2.10 \pm 0.04 \times 10^{23} \text{ m}^{-3}. \qquad (4)$$

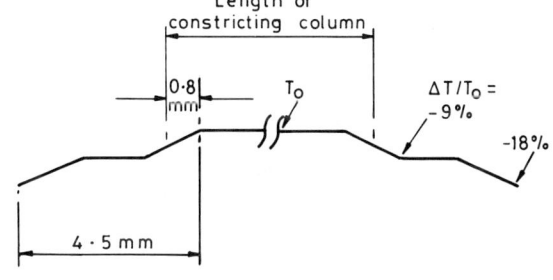

Fig.4. Axial temperature distribution used in the calculation of the variations of the spectroscopic quantities for the log-log plots.

Arc current A	Δl mm			
	40		60	
	To = 12 800 K		To = 13 820 K	
Spectroscopic quantity	Experiment	Model	Experiment	Model
Ar(I) 430 nm	-0.99	-1.10	-1.92	-1.77
Ar(I) 714.7nm	-1.96	-1.40	-2.37	-2.14
Ar(II) 480.6	0.69	0.65	0.52	0.48
Ar c.431.4 nm	-0.50	-0.57	-0.96	-1.10
Ar c.482.7 nm	-0.74	-0.57	-1.12	-1.10

Table II. The correction Δl to the constricted arc length for various spectroscopic quantities from experimental and theoretical values based on the axial temperature distribution as shown in Fig.4.

THE WALL-STABILIZED ARC AS AN ABSOLUTE RADIANCE SOURCE

As mentioned above, small quantities of gases other than argon can be added to the centre of the plasma. These trace gases can be kept from the differential pumping system and the anode by adjusting the argon flow. This means that intense lines of these trace gases can be made "black" at their line centres without self-reversal when viewed by the vacuum monochromator. The use of these blackbody limited lines (BBLL) as absolute radiance sources was first proposed by Boldt[10]. To ascertain the concentration required, the trace gas flow was gradually increased whilst the monochromator was set on an intense line of the gas. As the trace gas flow was increased the radiance of the line centre increased until it reached the blackbody limit when the increase trace gas flow no longer increased the radiance. It was found that with the monochromator set at a bandwidth of 0.005 nm the radiance remained constant for a 50% change in flow rate for most lines used. An example is given in Fig.5. The temperature of the plasma was monitored by observing the 123.5 nm Kr(I) line. This Kr(I) line was used because the concentration of krypton required to achieve the blackbody limit the line centre was found to change the temperature by < 0.02%. Used as a monitor this line showed that the plasma temperature changed by less than 0.3% for all lines used. To select a suitable line, scans were also taken through the line to examine its shape. Once a line had been selected the flow rate was noted by reference to a pressure reading and a needle valve setting. The arc voltage was also noted as this is sensitive to impurity level. Suitable lines were found for C, O, N, and Kr giving 22 radiance points between 95 and 200 nm.

THE DEUTERIUM LAMP TRANSFER STANDARD

The radiance scale at NPL based on a tungsten strip lamp has been extended from 350 nm to 165 nm by P.J.Key with the use of synchrotron radiation as a standard of relative spectral emission and deuterium discharge lamps as radiance transfer standards[11]. These lamps have been further developed by replacing the silica window with one made from an orientated single crystal of magnesium fluoride. This increased the spectral range of these lamps down to 115 nm. The window was sealed to the glass envelope of the lamp with a graded glass seal developed by G.Freeman et al[12,13] which allowed the envelopes to be baked to 300° C during manufacture, thus maintaining the spectral purity and stability of the discharge. These

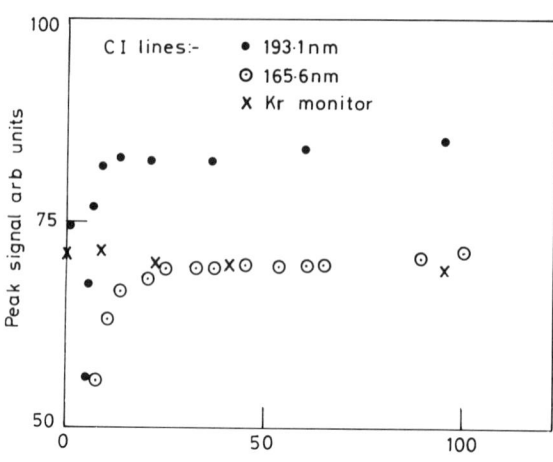

Fig.5. The variation of the peak signals of C(I) lines with particle number density showing the realization of the BBLL saturations. The Kr line at 123.5 nm is used to determine the plasma axial temperature perturbation and investigate the realization of C saturations.

lamps have comparable radiance to the continuum radiation from the wall-stabilized arc below 165 nm but as the lamp emission is not continuous, the lamp radiance is bandwidth dependent. To overcome this problem, a high resolution spectra is recorded with a micro-computer controlled monochromator and stored on a floppy disc. This data can then be used to determine a correction, by a convolution process, to the calibration if the lamp is to be used at a different bandwidth to that at which it was calibrated.

ABSOLUTE CALIBRATION OF DEUTERIUM LAMPS AGAINST THE WALL-STABILIZED ARC

The calibration of a deuterium lamp was carried out via the 3 m McPherson monochromator. The wall-stabilized arc was run at 60 A where its temperature was known most accurately and the trace gases added one at a time to the predetermined level to produce the blackbody limited lines while the arc temperature was monitored with the krypton line at 123.5 nm. Each blackbody limited line gave a calibration point for the responsivity curve of the monochromator. Interpolation between these points was assisted by using the radiance of the argon continuum. In order to increase the signal level for the measurements with the deuterium lamp the exit slit was increased in width by 15 times to 1.5 mm and the resultant change in the responsivity of the monochromator determined using the argon continuum radiance at wavelengths well away from emission lines. The arc was then replaced with a deuterium lamp and the response of the monochromator recorded at a number of wavelengths. To check the stability of the monochromator the arc was replaced and the monochromator recalibrated. An absolute calibration was then calculated from the responsivity curve of the monochromator, see Fig.6. Deuterium lamps can therefore be calibrated against the wall-stabilized arc between 115 and 200 nm.

COMPARISON WITH THE EXISTING VISIBLE AND NEAR UV RADIANCE SCALE

The relative spectral emission of the NPL deuterium lamps has only been compared with a synchrotron at wavelengths above 165 nm[14]. The spectral radiance of a synchrotron is given by Schwinger[15]. Absolute calibrations would require the accurate measurement of the synchrotron current but relative

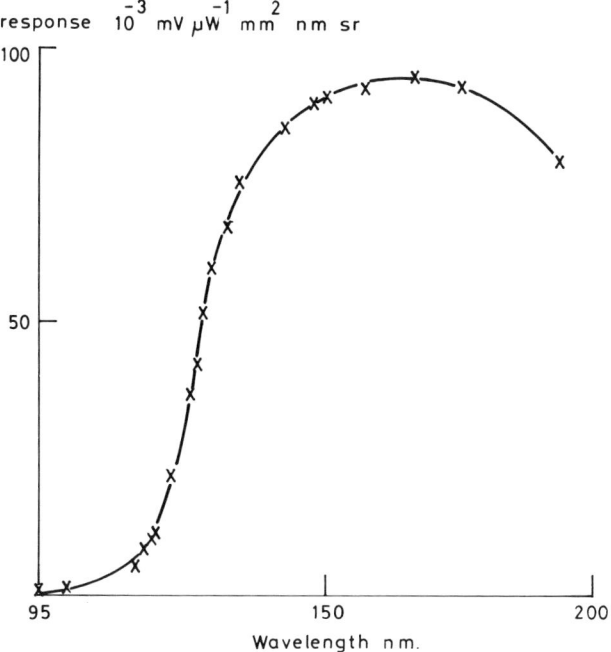

Fig. 6. Typical responsivity curve of the 3 m Mcpherson monochromator showing the calibrations points from BBLL

spectral radiances do not. The Daresbury 5 GeV synchrotron was used to find the relative spectral radiance of a group of deuterium lamps which were used for the comparison of radiation scales. Corrections were made for the different solid angles and emitting areas used for the calibrations, with the arc and the synchrotron. The relative spectral radiance was put on an absolute scale by comparison with three tungsten strip lamps which have been calibrated against the NPL blackbody-radiator-based visible and near uv radiance scales. This calibration, including corrections for field and aperture stop differences, had an uncertainty of $\pm$ 2%.

The comparison of the two calibrations was carried out at four wavelengths 165.7, 167, 174.3, and 193.1 nm. The uncertainty of calibration from the synchrotron at these wavelengths was less than 4%, which was similar to the uncertainty of the blackbody limited line calibration. The comparison showed agreement within these uncertainties thus confirming the accuracy of the temperature measurement of the arc.

CONCLUSION

The axial temperature of a wall-stabilized arc has been measured and shown to be stable and reproducible to $\pm$ 10 K at 14,000 K. It has been shown to be in LTE at 60 A and to have a temperature of 13,820 K $\pm$ 40 K. The arc has been successfully used as an absolute radiance standard in the wavelength region 95 to 200 nm. A deuterium lamp with a magnesium fluoride window has been shown to be a useful transfer standard in the wavelength region 115 to 350 nm[16]. Deuterium lamps calibrated against a synchrotron have been used to compare the absolute radiance of the arc with existing NPL radiance scales. This intercomparison has confirmed the precision of measurement of the temperature of the arc.

REFERENCES

1. R. C. Preston, J. Quant. Spectrosc. Radiat. Transfer 18, 337 (1976).
2. J. Richter, Z. Astrophys. 61, 57 (1965).
3. R. C. Preston, J. Phys. B: Atom. Molec. Phys. 13, 689 (1980).
4. W. L. Wiese, M. W. Smith and B. M. Miles, Atomic Transition Probabilities, NSRDS-NBS 22 Vol(II), Washington (1969).
5. D. Schluter, Z. Phys. 210, 80 (1968).
6. H. R. Griem, Spectral Line Broadening by Plasmas. Academic Press, New York (1974).
7. E. Schulz-Gulde, Z. Phys. 230, 449 (1970).
8. R. Schnapauff, Z. Astrophys. 68, 431 (1968).
9. R. C. Preston and C. Brookes, The Physics of Ionized Gases, Contributed Papers of SPIG-78 ed R. K. Janev (Belgrade: Institute of Physics), 269 (1978).
10. G. Boldt, Space Sci. Rev. 11, 728 (1970).
11. P. J. Key and T. H. Ward, Metrologia 14, 17 (1978).
12. G. H. C. Freeman and P. J. Moore, J. Phys. E: Sci. Instrum. 11, 980 (1978).
13. G. H. C. Freeman, M. G. Nicholas and T. M. Valentine, Opt. Acta 22, 875 (1977).
14. P. J. Key and R. C. Preston, Appl. Optics 16, 2477 (1977).
15. J. Schwinger, Phys. Rev. 75, 1912 (1949).
16. D. H. Nettleton and R. C. Preston, Appl. Optics 20, 1274 (1981).

AUTHOR INDEX for Parts 1 and 2

A

Actis, A. — Noise thermometry and related experiments at IMGC. — L. Crovini and A. Actis; 133-7.

— Interpolating equations for industrial platinum resistance thermometers in the temperature range from −200 to +420 °C. — A. Actis and L. Crovini; 819-27.

Akiyama, K. — Precision silicon transistor thermometer. — A. Ohte, M. Yamagata, and K. Akiyama; 1197-203.

Ambrok, G. S. — Soviet standards of the unit of temperature for radiation pyrometry. — I. I. Kirenkov, B. N. Oleinik, G. S. Ambrok, and G. A. Krakhmalnikova; 201-4.

— Estimation of the true temperature of targets by their thermal radiation based on Planck's law. — G. S. Ambrok; 569-73.

Ancsin, J. — Melting curves of H_2O. — J. Ancsin; 281-4.

Anderson, R. L. — Testing of thermocouples for inhomogeneities: A review of theory, with examples. — C. A. Mossman, J. L. Horton, and R. L. Anderson; 923-9.

— Failure of sheathed thermocouples due to thermal cycling. — R. L. Anderson and R. L. Ludwig; 939-51.

— Decalibration of sheathed thermocouples. — R. L. Anderson, J. D. Lyons, T. G. Kollie, W. H. Christie, and R. Eby; 977-1007.

— Automated temperature measurements from −183 to 2300 °C. — M. H. Cooper, Jr., R. L. Anderson, and C. A. Mossman; 1287-92.

Ando, S. — Temperature measuring method by using the eddy current technique. — K. Sano, T. Yamada, S. Ando, and K. Watanabe; 1213-8.

Andrews, J. W. — Measurement of thermodynamic temperature with the NPL photon-counting pyrometer. — P. B. Coates and J. W. Andrews; 109-14.

Armbrüster, H. — Very low temperature thermocouple devices: Development and application techniques for temperature measurements. — H. Armbrüster, W. P. Kirk, and D. P. Chesire; 1025-35.

Arunasalam, V. — Determination of the time evolution of the electron temperature profile of reactor-like plasmas from the measurement of blackbody electron cyclotron emission. — P. C. Efthimion, V. Arunasalam, R. A. Bitzer, and J. C. Hosea; 677-85.

Asawaroengchai, C. — The use of rotational Raman scattering for measurement of gas temperature. — M. C. Drake, C. Asawaroengchai, D. L. Drapcho, K. D. Veirs, and G. M. Rosenblatt; 621-9.

B

Babelot, J.-F. — Microsecond and sub-microsecond multi-wavelength pyrometry for pulsed heating technique diagnostics. — J.-F. Babelot, J. Magill, R. W. Ohse, and M. Hoch; 439-46.

Balko, B. — Fast thermistor sensors for rapid reaction studies. — R. L. Berger, B. Balko, T. R. Clem, and W. S. Friauf; 897-910.

Barnes, W. T. — Pure rotational CARS thermometry. — J. W. Fleming, A. B. Harvey, and W. T. Barnes; 589-94.

Bass, N. M. — Construction of a laboratory working thermometer using industrial platinum resistance sensors. — N. M. Bass; 813-4.

Bassani, C. — A new generation of precision furnaces. — C. A. Busse and C. Bassani; 1265-73.

Becker, R. J. — Dynamic temperature measurements of flames using spontaneous Raman scattering. — P. P. Yaney, R. J. Becker, P. D. Magill, and P. Danset; 639-48.

Bedford, R. E. — Measurement of the thermodynamic temperature interval between the freezing points of silver and copper. — M. Ohtsuka and R. E. Bedford; 175-81.

— Measurement of the melting temperature of the copper 71.9% silver eutectic alloy with a monochromatic optical pyrometer. — R. E. Bedford and C. K. Ma; 361-9.

Benzinger, T. H. — Temperature and thermodynamics of living matter. — T. H. Benzinger; 1389-95.

Berger, R. L. — Fast thermistor sensors for rapid reaction studies. — R. L. Berger, B. Balko, T. R. Clem, and W. S. Friauf; 897-910.

Berry, K. H. — Measurements of thermodynamic temperature from 2.6 to 27.1 K. — K. H. Berry; 21-4.

Berry, Robert J. — Evaluation and control of platinum oxidation errors in standard platinum resistance thermometers. — Robert J. Berry; 743-52.

— Oxidation, stability, and insulation characteristics of Rosemount standard platinum resistance thermometers. — Robert J. Berry; 753-62.

Besley, L. M. — Constant volume gas thermometry from 13.8 to 83.8 K. — R. C. Kemp, L. M. Besley, and W. R. G. Kemp; 33-7.

Beynon, T. G. R. — Radiation thermometry applied to the development and control of gas turbine engines. — T. G. R. Beynon; 471-7.

Bigge, William R. — Reproducibility of some triple point of water cells. — George T. Furukawa and William R. Bigge; 291-7.

Billeter, T. R. — Dual high temperature measurements using Johnson noise thermometry. — T. R. Billeter and C. P. Cannon; 1245-8.

Bitter, M. — Measurement of the central ion and electron temperature of tokamak plasmas from the x-ray line radiation of high-Z impurity ions. — M. Bitter, S. von Goeler, M. Goldman, K. W. Hill, R. Horton, W. Roney, N. Sauthoff, and W. Stodiek; 693-703.

Bitzer, R. A. — Determination of the time evolution of the electron temperature profile of reactor-like plasmas from the measurement of blackbody electron cyclotron emission. — P. C. Efthimion, V. Arunasalam, R. A. Bitzer, and J. C. Hosea; 677-85.

Blalock, T. V. — A decade of progress in high temperature Johnson noise thermometry. — T. V. Blalock and R. L. Shepard; 1219-23.

— Johnson noise power thermometer and its application in process temperature measurement. — T. V. Blalock, J. L. Horton, and R. L. Shepard; 1249-59.

Bongiovanni, G. — An intercomparison of gallium fixed point cells. — M. V. Chattle, R. L. Rusby, G. Bonnier, A. Moser, E. Renaot, P. Marcarino, G. Bongiovanni, and G. Frassineti; 311-6.

Bonnier, G. — Thermal behavior of thermometric sealed cells and of a multi-compartment cell. — G. Bonnier and Y. Hermier; 231-7.

— An intercomparison of gallium fixed point cells. — M. V. Chattle, R. L. Rusby, G. Bonnier, A. Moser, E. Renaot, P. Marcarino, G. Bongiovanni, and G. Frassineti; 311-6.

Bowie, Lemuel J. — Spectroscopic techniques for measuring the temperature of liquids in analytical instrumentation. — Lemuel J. Bowie; 1373-8.

Brandt, B. L. — Cryogenic thermometry: A review of recent progress. II. — L. G. Rubin, B. L. Brandt, and H. H. Sample; 1333-44.

Brixy, H. — Application of noise thermometry in industry under plant conditions. — H. Brixy, R. Hecker, K. F. Rittinghaus, and H. Höwener; 1225-37.

Brown, N. L. — An automatic resistance thermometer bridge. — N. L. Brown, A. J. Fougere, J. W. McLeod, and R. J. Robbins; 719-27.

Burley, N. A. — The nicrosil versus nisil thermocouple: The influence of magnesium on the thermoelectric stability and oxidation resistance of the alloys. — N. A. Burley, J. L. Cocking, G. W. Burns, and M. G. Scroger; 1129-45.

— The nicrosil versus nisil thermocouple: A critical comparison with the ANSI standard letter-designated base-metal thermocouples. — N. A. Burley, R. M. Hess, C. F. Howie, and J. A. Coleman; 1159-66.

Burns, G. W. — The nicrosil versus nisil thermocouple: Recent developments and present status. — G. W. Burns; 1121-7.

— The nicrosil versus nisil thermocouple: The influence of magnesium on the thermoelectric stability and oxidation resistance of the alloys. — N. A. Burley, J. L. Cocking, G. W. Burns, and M. G. Scroger; 1129-45.

Busse, C. A. — A new generation of precision furnaces. — C. A. Busse and C. Bassani; 1265-73.

C

Campana, M. — Ultrasonic thin-wire thermometry for nuclear applications. — H. A. Tasman, M. Campana, D. Pel, and J. Richter; 1191-6.

Cannon, C. P. — 2200 °C thermocouples for nuclear reactor fuel centerline temperature measurements. — C. P. Cannon; 1061-7.

— Dual high temperature measurements using Johnson noise thermometry. — T. R. Billeter and C. P. Cannon; 1245-8.

Carr, K. R. — Studies of sheathed thermocouple construction and installation in thermowells to obtain faster response. — R. M. Carroll, K. R. Carr, and R. L. Shepard; 1019-24.

Carroll, R. M. — Studies of sheathed thermocouple construction and installation in thermowells to obtain faster response. — R. M. Carroll, K. R. Carr, and R. L. Shepard; 1019-24.

Case, D. A. — Aging phenomena in nickel-manganese oxide thermistors. — J. M. Zurbuchen and D. A. Case; 889-96.

Cashdollar, K. L. — Infrared temperature measurements of gas and dust explosions. — K. L. Cashdollar and M. Hertzberg; 453-63.

Cezairliyan, A. — Radiance temperature of metals at their melting points as possible high temperature secondary reference points. — A. Cezairliyan, A. P. Miiller, F. Righini, and A. Rosso; 377-81.

— Two-color microsecond pyrometer for 2000 to 6000 K. — G. M. Foley, M. S. Morse, and A. Cezairliyan; 447-52.

Chattle, M. V. — An intercomparison of gallium fixed point cells. — M. V. Chattle, R. L. Rusby, G. Bonnier, A. Moser, E. Renaot, P. Marcarino, G. Bongiovanni, and G. Frassineti; 311-6.

Chen, G. B. — A thin platinum film for transient heat transfer studies. — P. J. Giarratano, F. L. Lloyd, L. O. Mullen, and G. B. Chen; 859-63.

Chesire, D. P. — Very low temperature thermocouple devices: Development and application techniques for temperature measurements. — H. Armbrüster, W. P. Kirk, and D. P. Chesire; 1025-35.

Chistyakov, V. A. — A method for calibration of a precision thermal radiation density meter using a radiation source. — V. A. Chistyakov and V. I. Gavrishchuk; 529-33.

Christie, W. H. — Decalibration of sheathed thermocouples. — R. L. Anderson, J. D. Lyons, T. G. Kollie, W. H. Christie, and R. Eby; 977-1007.

Clem, T. R. — Fast thermistor sensors for rapid reaction studies. — R. L. Berger, B. Balko, T. R. Clem, and W. S. Friauf; 897-910.

Clift, J. H. — Lifetime improvement of small-diameter sheathed thermocouples for use in high-temperature and thermal transient operations. — R. W. McCulloch and J. H. Clift; 1097-108.

Coates, P. B. — Measurement of thermodynamic temperature with the NPL photon-counting pyrometer. — P. B. Coates and J. W. Andrews; 109-14.

Cocking, J. L. — The nicrosil versus nisil thermocouple: The influence of magnesium on the thermoelectric stability and oxidation resistance of the alloys. — N. A. Burley, J. L. Cocking, G. W. Burns, and M. G. Scroger; 1129-45.

Colclough, A. R. — Primary acoustic thermometry: Principles and current trends. — A. R. Colclough; 65-75.

— A refractive index thermometer for use at low temperatures. — A. R. Colclough; 89-94.

Coleman, J. A. — The nicrosil versus nisil thermocouple: A critical comparison with the ANSI standard letter-designated base-metal thermocouples. — N. A. Burley, R. M. Hess, C. F. Howie, and J. A. Coleman; 1159-66.

Collier, R. D. — Calibration with confidence: The assurance of temperature accuracy. — R. D. Collier; 1311-5.

Connolly, J. J. — The use of the cadmium point to check calibrations on the IPTS. — J. V. McAllan and J. J. Connolly; 351-3.

— The calibration characteristics of industrial platinum resistance thermometers. — J. J. Connolly; 815-7.

Cooper, M. H., Jr. — Automated temperature measurements from −183 to 2300 °C. — M. H. Cooper, Jr., R. L. Anderson, and C. A. Mossman; 1287-92.

Cox, J. D. — Temperature fixed points: Evaluation of four types of triple-point cell. — J. D. Cox and M. F. Vaughan; 267-80.

Crovini, L. — Noise thermometry and related experiments at IMGC. — L. Crovini and A. Actis; 133-7.

— Interpolating equations for industrial platinum resistance thermometers in the temperature range from −200 to +420 °C. — A. Actis and L. Crovini; 819-27.

Curtis, D. J. — Thermal hysteresis and stress effects in platinum resistance thermometers. — D. J. Curtis; 803-12.

Cutkosky, R. D. — Automatic resistance thermometer bridges for new and special applications. — R. D. Cutkosky; 711-3.

D

Danset, P. — Dynamic temperature measurements of flames using spontaneous Raman scattering. — P. P. Yaney, R. J. Becker, P. D. Magill, and P. Danset; 639-48.

Dauphinee, T. M. — Deep-ocean temperature measurement. — T. M. Dauphinee; 1317-25.

Davies, M. R. — A broadband ratio pyrometer. — J. L. Gardner, T. P. Jones, and M. R. Davies; 409-12.

Daykin, C. I. — A new range of high precision resistance bridges for resistance thermometry. — P. C. F. Wolfendale, J. D. Yewen, and C. I. Daykin; 729-32.

Decréton, M. C. — High temperature noise thermometry for industrial applications. — M. C. Decréton; 1239-43.

Deng Daren — Carbon-glass thermometry in China. — Yao Quanfa, Deng Daren, Ma Hongqi, Jiang Dehua, Ji Yunsong, and Huang Xihuai; 853-7.

— A precision 4.2-300 K temperature controller using a genuine full-range sensor and inductive divider set-point coupled with a simple ac sensing bridge. — Wang Zhensen and Deng Daren; 1275-8.

Den Sixiang — The NIM's photoelectric comparator and the realization of the IPTS-68 above the gold point. — Zhao Qi, Den Sixiang, Sun Dinwen, Qiu Nairong, Li Zhenguo, and Li Erming; 183-90.

DeWitt, D. P. — Experimental and theoretical study on the quality of reference blackbodies formed by lateral holes on a metallic tube. — A. Ono, R. M. Trusty, and D. P. DeWitt; 541-50.

Dieck, Ronald H. — Thermocouple measurement uncertainty in compressor efficiency measurement: The effects of two uncertainty models. — Ronald H. Dieck and Barbara G. Ringhiser; 1009-18.

Drake, M. C. — The use of rotational Raman scattering for measurement of gas temperature. — M. C. Drake, C. Asawaroengchai, D. L. Drapcho, K. D. Veirs, and G. M. Rosenblatt; 621-9.

— Use of the vibrational Raman effect for gas temperature measurements. — M. C. Drake, M. Lapp, and C. M. Penney; 631-8.

Drapcho, D. L. — The use of rotational Raman scattering for measurement of gas temperature. — M. C. Drake, C. Asawaroengchai, D. L. Drapcho, K. D. Veirs, and G. M. Rosenblatt; 621-9.

Drzewiecki, Tadeusz M. — Design of a fluidic capillary pyrometer for contact duty at temperatures to 2750 °C. — R. Michael Phillippi, Tadeusz M. Drzewiecki, Taki Negas, and Harry S. Parker; 1345-51.

Durieux, M. — Measurements with a gas thermometer between 4 and 100 K. — P. P. M. Steur, J. E. van Dijk, J. P. Mars, H. ter Harmsel, and M. Durieux; 25-31.

— Helium vapor pressure equations on the EPT-76. — M. Durieux, J. E. van Dijk, H. ter Harmsel, P. C. Rem, and R. L. Rusby; 145-53.

— Realizations of the superconductive transition points of lead, indium, aluminium, zinc, and cadmium with SRM 767 devices. — A. E. El Samahy, M. Durieux, R. L. Rusby, R. C. Kemp, and W. R. G. Kemp; 261-5.

E

Eby, R. — Decalibration of sheathed thermocouples. — R. L. Anderson, J. D. Lyons, T. G. Kollie, W. H. Christie, and R. Eby; 977-1007.

Eckbreth, A. C. — CARS thermometry in reacting systems. — J. F. Verdieck, J. A. Shirley, R. J. Hall, and A. C. Eckbreth; 595-608.

Edrich, J. — Imaging microwave thermography. — J. Edrich and W. E. Jobe; 1379-80.

Edsinger, Robert E. — Progress in NBS gas thermometry above 500 °C. — Leslie A. Guildner and Robert E. Edsinger; 43-8.

Efthimion, P. C. — Determination of the time evolution of the electron temperature profile of reactor-like plasmas from the measurement of blackbody electron cyclotron emission. — P. C. Efthimion, V. Arunasalam, R. A. Bitzer, and J. C. Hosea; 677-85.

El Samahy, A. E. — Realizations of the superconductive transition points of lead, indium, aluminium, zinc, and cadmium with SRM 767 devices. — A. E. El Samahy, M. Durieux, R. L. Rusby, R. C. Kemp, and W. R. G. Kemp; 261-5.

El-Shammaa, H. — Realization of the triple point of water and the freezing points of tin and zinc at the National Institute of Standards (Egypt). — H. El-Shammaa, M. R. Moussa, and M. H. Omar; 205-8.

Emmerman, P. J. — Laser tomography for temperature measurements in flames. — H. G. Semerjian, R. J. Santoro, P. J. Emmerman, and R. Goulard; 649-60.

Ergardt, N. N. — Realization of the melting point of gallium. — B. N. Oleinik, A. G. Ivanova, V. A. Zamkovets, and N. N. Ergardt; 317-20.

Evans, G. A., Jr. — Investigation of the stability of small platinum resistance thermometers. — B. W. Mangum and G. A. Evans, Jr.; 795-801.

Evans, J. P. — Experiences with high-temperature platinum resistance thermometers. — J. P. Evans; 771-81.

F

Farrow, R. L. — CARS thermometry in an internal combustion engine. — L. A. Rahn, S. C. Johnston, R. L. Farrow, and P. L. Mattern; 609-13.

Ferri, D. — Ten years of research on sealed cells for phase transition studies of gases at IMGC. — F. Pavese and D. Ferri; 217-27.

Figueroa, J. M. — The triple point of rubidium: A temperature fixed point for biomedical applications. — J. M. Figueroa and B. W. Mangum; 327-37.

Fillmore, R. L. — Modeling a closed loop control system. — R. L. Fillmore; 1261-3.

Fleming, J. W. — Pure rotational CARS thermometry. — J. W. Fleming, A. B. Harvey, and W. T. Barnes; 589-94.

Foley, G. M. — Two-color microsecond pyrometer for 2000 to 6000 K. — G. M. Foley, M. S. Morse, and A. Cezairliyan; 447-52.

Fougere, A. J. — An automatic resistance thermometer bridge. — N. L. Brown, A. J. Fougere, J. W. McLeod, and R. J. Robbins; 719-27.

Frassineti, G. — An intercomparison of gallium fixed point cells. — M. V. Chattle, R. L. Rusby, G. Bonnier, A. Moser, E. Renaot, P. Marcarino, G. Bongiovanni, and G. Frassineti; 311-6.

Friauf, W. S. — Fast thermistor sensors for rapid reaction studies. — R. L. Berger, B. Balko, T. R. Clem, and W. S. Friauf; 897-910.

Fritschen, L. J. — An automatic system for measuring Bowen ratio gradients using platinum resistance elements. — L. J. Fritschen and J. R. Simpson; 739-42.

Furukawa, George T. — Reproducibility of the triple point of argon in sealed transportable cells. — G. T. Furukawa; 239-48.
— Reproducibility of some triple point of water cells. — George T. Furukawa and William R. Bigge; 291-7.
— Investigation of the freezing temperature of cadmium. — George T. Furukawa and Earl R. Pfeiffer; 355-60.

G

Gardner, J. L. — A broadband ratio pyrometer. — J. L. Gardner, T. P. Jones, and M. R. Davies; 409-12.

Gavrishchuk, V. I. — A method for calibration of a precision thermal radiation density meter using a radiation source. — V. A. Chistyakov and V. I. Gavrishchuk; 529-33.

Giarratano, P. J. — A thin platinum film for transient heat transfer studies. — P. J. Giarratano, F. L. Lloyd, L. O. Mullen, and G. B. Chen; 859-63.

Glicksman, M. E. — The triple-point equilibria of succinonitrile: Its assessment as a temperature standard. — M. E. Glicksman, P. W. Voorhees, and R. Setzko; 321-6.

Goldman, M. — Measurement of the central ion and electron temperature of tokamak plasmas from the x-ray line radiation of high-Z impurity ions. — M. Bitter, S. von Goeler, M. Goldman, K. W. Hill, R. Horton, W. Roney, N. Sauthoff, and W. Stodiek; 693-703.

Goss, L. P. — A hardened CARS system for temperature and species-concentration measurements in practical combustion environments. — G. L. Switzer and L. P. Goss; 583-7.

Goulard, R. — Laser tomography for temperature measurements in flames. — H. G. Semerjian, R. J. Santoro, P. J. Emmerman, and R. Goulard; 649-60.

Grinter, R. — Carbon-glass sensors: Reproducibility and polynomial fitting of temperature vs resistance. — B. W. Ricketson and R. Grinter; 845-51.

Gugan, D. — Dielectric Constant Gas Thermometry (DCGT): A new method of accurate thermodynamic thermometry. — D. Gugan; 49-53.
— Surface-fitting of helium isotherms: Application to the temperature scale 2.6–27.1 K. — D. Gugan; 55-64.

Guildner, Leslie A. — The measurement of thermodynamic temperature. — Leslie A. Guildner and Wilhelm Thomas; 9-19.
— Progress in NBS gas thermometry above 500 °C. — Leslie A. Guildner and Robert E. Edsinger; 43-8.

H

Hall, R. J. — CARS thermometry in reacting systems. — J. F. Verdieck, J. A. Shirley, R. J. Hall, and A. C. Eckbreth; 595-608.

Hanafy, Magda — A small transportable indium cell for use as a temperature reference. — Magda Hanafy, M. R. Moussa, and M. H. Omar; 347-50.

Harada, N. — Temperature distribution measurement with a silicon photodiode array. — T. Yamada, N. Harada, and M. Koyanagi; 395-400.

Harvey, A. B. — Pure rotational CARS thermometry. — J. W. Fleming, A. B. Harvey, and W. T. Barnes; 589-94.

Hashemian, H. M. — Response of installed temperature sensors. — T. W. Kerlin, R. L. Shepard, H. M. Hashemian, and K. M. Petersen; 1357-66.

Hattori, S. — Establishing a practical temperature standard by using a narrow-band radiation thermometer with a silicon detector. — F. Sakuma and S. Hattori; 421-7.
— The effective temperature to express radiant characteristics of nonisothermal cavities. — S. Hattori and A. Ono; 521-7.
— A practical-type fixed point blackbody furnace. — F. Sakuma and S. Hattori; 535-9.

Hecker, R. — Application of noise thermometry in industry under plant conditions. — H. Brixy, R. Hecker, K. F. Rittinghaus, and H. Höwener; 1225-37.

Hermier, Y. — Thermal behavior of thermometric sealed cells and of a multi-compartment cell. — G. Bonnier and Y. Hermier; 231-7.

Hertzberg, M. — Infrared temperature measurements of gas and dust explosions. — K. L. Cashdollar and M. Hertzberg; 453-63.

Hess, R. M. — The nicrosil versus nisil thermocouple: A critical comparison with the ANSI standard letter-designated base-metal thermocouples. — N. A. Burley, R. M. Hess, C. F. Howie, and J. A. Coleman; 1159-66.

Hill, K. W. — Measurement of the central ion and electron temperature of tokamak plasmas from the x-ray line radiation of high-Z impurity ions. — M. Bitter, S. von Goeler, M. Goldman, K. W. Hill, R. Horton, W. Roney, N. Sauthoff, and W. Stodiek; 693-703.

Hishikari, I. — Precision practical blackbody furnaces by a 3-zone temperature control method. — I. Hishikari and T. Ide; 551-8.

Hoch, M. — Microsecond and sub-microsecond multi-wavelength pyrometry for pulsed heating technique diagnostics. — J.-F. Babelot, J. Magill, R. W. Ohse, and M. Hoch; 439-46.

Horelli, J. — Temperature control at a high interference level: A case description. — I. Karaila and J. Horelli; 1279-81.

Horiuchi, T. — Lining erosion measurements by sheathed multiple thermocouples through temperature transients. — Y. Kawate, N. Nagai, M. Konishi, K. Yokoe, and T. Horiuchi; 1043-9.

Horton, J. L. — Testing of thermocouples for inhomogeneities: A review of theory, with examples. — C. A. Mossman, J. L. Horton, and R. L. Anderson; 923-9.

— Johnson noise power thermometer and its application in process temperature measurement. — T. V. Blalock, J. L. Horton, and R. L. Shepard; 1249-59.

Horton, R. — Measurement of the central ion and electron temperature of tokamak plasmas from the x-ray line radiation of high-Z impurity ions. — M. Bitter, S. von Goeler, M. Goldman, K. W. Hill, R. Horton, W. Roney, N. Sauthoff, and W. Stodiek; 693-703.

Hosea, J. C. — Determination of the time evolution of the electron temperature profile of reactor-like plasmas from the measurement of blackbody electron cyclotron emission. — P. C. Efthimion, V. Arunasalam, R. A. Bitzer, and J. C. Hosea; 677-85.

Höwener, H. — Application of noise thermometry in industry under plant conditions. — H. Brixy, R. Hecker, K. F. Rittinghaus, and H. Höwener; 1225-37.

Howie, C. F. — The nicrosil versus nisil thermocouple: A critical comparison with the ANSI standard letter-designated base-metal thermocouples. — N. A. Burley, R. M. Hess, C. F. Howie, and J. A. Coleman; 1159-66.

Hua Chengsheng — −50 to +150 °C heat pipe blackbody sources for radiation thermometer calibration. — Zhu Yingsong, Ma Hongqi, Wang Ronghua, and Hua Chengsheng; 559-65.

Huang Xihuai — Carbon-glass thermometry in China. — Yao Quanfa, Deng Daren, Ma Hongqi, Jiang Dehua, Ji Yunsong, and Huang Xihuai; 853-7.

Hudson, R. P. — Temperature scales, the IPTS, and its future development. — R. P. Hudson; 1-8.

I

Ide, T. — Precision practical blackbody furnaces by a 3-zone temperature control method. — I. Hishikari and T. Ide; 551-8.

Imamura, M. — A new method of noise thermometry. — M. Imamura and A. Ohte; 139-42.

Iuchi, T. — Two methods for simultaneous measurement of temperature and emittance by using multiple reflection and specular reflection and their applications to industrial processes. — T. Iuchi and R. Kusaka; 491-503.

Ivanova, A. G. — Realization of the melting point of gallium. — B. N. Oleinik, A. G. Ivanova, V. A. Zamkovets, and N. N. Ergardt; 317-20.

Iwamura, T. — Improvement of traceability for radiation pyrometers in the steel industry. — K. Tamura, T. Iwamura, and K. Kurita; 479-83.

Iwaoka, H. — A new nuclear quadrupole resonance standard thermometer. — A. Ohte and H. Iwaoka; 1173-80.

J

Jiang Dehua — Carbon-glass thermometry in China. — Yao Quanfa, Deng Daren, Ma Hongqi, Jiang Dehua, Ji Yunsong, and Huang Xihuai; 853-7.

Jiang Shichang — Single-band radiation thermometers: Harmonization of their calibration characteristics. — Jiang Shichang, Wu Shuyuan, Ye Rongchang, and Xu Liang; 413-20.

Ji Yunsong — Carbon-glass thermometry in China. — Yao Quanfa, Deng Daren, Ma Hongqi, Jiang Dehua, Ji Yunsong, and Huang Xihuai; 853-7.

Jobe, W. E. — Imaging microwave thermography. — J. Edrich and W. E. Jobe; 1379-80.

Johnson, H. C. — The exactness of fit of resistance-temperature data of thermistors with third-degree polynomials. — M. Sapoff, W. R. Siwek, H. C. Johnson, J. Slepian, and S. Weber; 875-87.

Johnston, S. C. — CARS thermometry in an internal combustion engine. — L. A. Rahn, S. C. Johnston, R. L. Farrow, and P. L. Mattern; 609-13.

Jones, T. P. — A photoelectric pyrometer temperature scale below 1064.43 °C and its use to measure the silver point. — T. P. Jones and J. Tapping; 169-74.

— An international intercomparison of temperature standards of Asia/Pacific countries. — T. P. Jones; 197-9.

— A broadband ratio pyrometer. — J. L. Gardner, T. P. Jones, and M. R. Davies; 409-12.

Joslin, W. — Temperature measurements with chromel/alumel thermocouples in a pressurized water reactor. — P. Siltanen, T. Laaksonen, and W. Joslin; 1069-79.

Ju Hao-Rien — Feedback stabilized tungsten strip lamp as a radiometric standard for photoelectric pyrometry. — Zhu Ci-Zhun and Ju Hao-Rien; 567-8.

Jung, H. J. — The stability of commercially available high temperature platinum resistance thermometers of a 5 Ω silica cross type up to 961.93 °C. — H. J. Jung and H. Nubbemeyer; 763-70.

K

Kaeser, Robert S. — Realization of the 1976 Provisional 0.5 K to 30 K Temperature Scale at the National Bureau of Standards. — E. R. Pfeiffer and R. S. Kaeser; 159-67.

— Automation of measurements in a low temperature laboratory. — Craig T. Van Degrift and Robert S. Kaeser; 1299-305.

Kanoshima, Yukio — Differential type thermometer for measuring hot gas temperature. — Jiro Ohno, Masakazu Nakamura, Yutaka Miyabe, Atsushi Kawasaki, and Yukio Kanoshima; 1037-41.

Karaila, I. — Temperature control at a high interference level: A case description. — I. Karaila and J. Horelli; 1279-81.

Karasikov, N. — Temperature references based on first order phase transition: Development and application. — D. Rappaport, N. Karasikov, and M. B. Roitberg; 339-42.

Kawasaki, Atsushi — Differential type thermometer for measuring hot gas temperature. — Jiro Ohno, Masakazu Nakamura, Yutaka Miyabe, Atsushi Kawasaki, and Yukio Kanoshima; 1037-41.

Kawate, Y. — Lining erosion measurements by sheathed multiple thermocouples through temperature transients. — Y. Kawate, N. Nagai, M. Konishi, K. Yokoe, and T. Horiuchi; 1043-9.

Kegel, T. M. — A proposed pressure amplifier for a temperature control system. — T. M. Kegel and D. E. Limbert; 1283-6.

Kemp, R. C. — Constant volume gas thermometry from 13.8 to 83.8 K. — R. C. Kemp, L. M. Besley, and W. R. G. Kemp; 33-7.

— Fixed point combination and termination points for platinum resistance thermometer interpolation below 273.15 K. — R. C. Kemp; 155-8.

— The triple point of natural xenon. — R. C. Kemp, W. R. G. Kemp, and P. W. Smart; 229-30.

— The triple points of equilibrium and normal deuterium. — R. C. Kemp; 249-50.

— Realizations of the superconductive transition points of lead, indium, aluminium, zinc, and cadmium with SRM 767 devices. — A. E. El Samahy, M. Durieux, R. L. Rusby, R. C. Kemp, and W. R. G. Kemp; 261-5.

Kemp, W. R. G. — Constant volume gas thermometry from 13.8 to 83.8 K. — R. C. Kemp, L. M. Besley, and W. R. G. Kemp; 33-7.

— The triple point of natural xenon. — R. C. Kemp, W. R. G. Kemp, and P. W. Smart; 229-30.

- Realizations of the superconductive transition points of lead, indium, aluminium, zinc, and cadmium with SRM 767 devices. — A. E. El Samahy, M. Durieux, R. L. Rusby, R. C. Kemp, and W. R. G. Kemp; 261-5.

Kerlin, T. W. — Response of installed temperature sensors. — T. W. Kerlin, R. L. Shepard, H. M. Hashemian, and K. M. Petersen; 1357-66.

Kirby, C. G. M. — An automatic resistance thermometer bridge. — C. G. M. Kirby; 715-8.

— Automation of a thermometer calibration facility. — C. G. M. Kirby; 1293-7.

Kirenkov, I. I. — Soviet standards of the unit of temperature for radiation pyrometry. — I. I. Kirenkov, B. N. Oleinik, G. S. Ambrok, and G. A. Krakhmalnikova; 201-4.

Kirk, W. P. — Very low temperature thermocouple devices: Development and application techniques for temperature measurements. — H. Armbrüster, W. P. Kirk, and D. P. Chesire; 1025-35.

Klempt, G. — Errors in Johnson noise thermometry. — G. Klempt; 125-8.

Klick, David — Temperature measurements for combustion diagnostics from high-resolution single-pulse CARS N_2 spectra. — David Klick, K. A. Marko, and L. Rimai; 615-20.

Kobayashi, S. — Platinum-cobalt alloy resistance thermometer for wide range cryogenic thermometry. — T. Shiratori, K. Mitsui, K. Yanagisawa, and S. Kobayashi; 839-43.

Kollie, T. G. — Decalibration of sheathed thermocouples. — R. L. Anderson, J. D. Lyons, T. G. Kollie, W. H. Christie, and R. Eby; 977-1007.

Konishi, M. — Lining erosion measurements by sheathed multiple thermocouples through temperature transients. — Y. Kawate, N. Nagai, M. Konishi, K. Yokoe, and T. Horiuchi; 1043-9.

Koremblit, M. J. — Miniature thermometric fixed points for thermocouple calibrations. — M. Tischler and M. J. Koremblit; 383-90.

Koriki, M. — Temperature measurement of steel in the furnace. — Y. Tamura, M. Tatsuwaki, T. Sugimura, T. Yokoi, M. Sano, and M. Koriki; 505-12.

Koyanagi, M. — Temperature distribution measurement with a silicon photodiode array. — T. Yamada, N. Harada, and M. Koyanagi; 395-400.

Krakhmalnikova, G. A. — Soviet standards of the unit of temperature for radiation pyrometry. — I. I. Kirenkov, B. N. Oleinik, G. S. Ambrok, and G. A. Krakhmalnikova; 201-4.

Kurita, K. — Improvement of traceability for radiation pyrometers in the steel industry. — K. Tamura, T. Iwamura, and K. Kurita; 479-83.

Kusaka, R. — Two methods for simultaneous measurement of temperature and emittance by using multiple reflection and specular reflection and their applications to industrial processes. — T. Iuchi and R. Kusaka; 491-503.

L

Laaksonen, T. — Temperature measurements with chromel/alumel thermocouples in a pressurized water reactor. — P. Siltanen, T. Laaksonen, and W. Joslin; 1069-79.

LaMers, T. H. — Enhanced stability in precision interchangeable thermistors. — T. H. LaMers, J. M. Zurbuchen, and H. Trolander; 865-73.

Lamvik, M. — Diffusion thermometry, an engineering concept. — M. Lamvik; 1353-5.

Lapp, M. — Use of the vibrational Raman effect for gas temperature measurements. — M. C. Drake, M. Lapp, and C. M. Penney; 631-8.

Li Erming — The NIM's photoelectric comparator and the realization of the IPTS-68 above the gold point. — Zhao Qi, Den Sixiang, Sun Dinwen, Qiu Nairong, Li Zhenguo, and Li Erming; 183-90.

Limbert, D. E. — A proposed pressure amplifier for a temperature control system. — T. M. Kegel and D. E. Limbert; 1283-6.

Linenberger, D. — Thermal response times of some cryogenic thermometers. — D. Linenberger, E. Spellicy, and R. Radebaugh; 1367-72.

Ling Shankang — The development of temperature standards at NIM of China. — Ling Shankang, Zhang Guoquan, Li Ruisheng, Wang Zilin, Li Zhiran, Zhao Qi, and Li Xumo; 191-5.

Li Ruisheng — The development of temperature standards at NIM of China. — Ling Shankang, Zhang Guoquan, Li Ruisheng, Wang Zilin, Li Zhiran, Zhao Qi, and Li Xumo; 191-5.

Li Xumo — The development of temperature standards at NIM of China. — Ling Shankang, Zhang Guoquan, Li Ruisheng, Wang Zilin, Li Zhiran, Zhao Qi, and Li Xumo; 191-5.

Li Zhenguo — The NIM's photoelectric comparator and the realization of the IPTS-68 above the gold point. — Zhao Qi, Den Sixiang, Sun Dinwen, Qiu Nairong, Li Zhenguo, and Li Erming; 183-90.

Li Zhiran — The development of temperature standards at NIM of China. — Ling Shankang, Zhang Guoquan, Li Ruisheng, Wang Zilin, Li Zhiran, Zhao Qi, and Li Xumo; 191-5.

Lloyd, F. L. — A thin platinum film for transient heat transfer studies. — P. J. Giarratano, F. L. Lloyd, L. O. Mullen, and G. B. Chen; 859-63.

Long Guang — Stability of precision high temperature platinum resistance thermometers. — Long Guang and Tao Hongtu; 783-7.

Ludwig, R. L. — Failure of sheathed thermocouples due to thermal cycling. — R. L. Anderson and R. L. Ludwig; 939-51.

Lynnworth, L. C. — Temperature profiling using multizone ultrasonic waveguides. — L. C. Lynnworth; 1181-90.

Lyons, J. D. — Decalibration of sheathed thermocouples. — R. L. Anderson, J. D. Lyons, T. G. Kollie, W. H. Christie, and R. Eby; 977-1007.

M

Ma, C. K. — Measurement of the melting temperature of the copper 71.9% silver eutectic alloy with a monochromatic optical pyrometer. — R. E. Bedford and C. K. Ma; 361-9.

Magill, J. — Microsecond and sub-microsecond multi-wavelength pyrometry for pulsed heating technique diagnostics. — J.-F. Babelot, J. Magill, R. W. Ohse, and M. Hoch; 439-46.

Magill, P. D. — Dynamic temperature measurements of flames using spontaneous Raman scattering. — P. P. Yaney, R. J. Becker, P. D. Magill, and P. Danset; 639-48.

Ma Hongqi — −50 to +150 °C heat pipe blackbody sources for radiation thermometer calibration. — Zhu Yingsong, Ma Hongqi, Wang Ronghua, and Hua Chengsheng; 559-65.

— Carbon-glass thermometry in China. — Yao Quanfa, Deng Daren, Ma Hongqi, Jiang Dehua, Ji Yunsong, and Huang Xihuai; 853-7.

Mangum, B. W. — Triple point of gallium as a temperature fixed point. — B. W. Mangum; 299-309.

— The triple point of rubidium: A temperature fixed point for biomedical applications. — J. M. Figueroa and B. W. Mangum; 327-37.

— Investigation of the stability of small platinum resistance thermometers. — B. W. Mangum and G. A. Evans, Jr.; 795-801.

Marcarino, P. — An intercomparison of gallium fixed point cells. — M. V. Chattle, R. L. Rusby, G. Bonnier, A. Moser, E. Renaot, P. Marcarino, G. Bongiovanni, and G. Frassineti; 311-6.

Marko, K. A. — Temperature measurements for combustion diagnostics from high-resolution single-pulse CARS N_2 spectra. — David Klick, K. A. Marko, and L. Rimai; 615-20.

Mars, J. P. — Measurements with a gas thermometer between 4 and 100 K. — P. P. M. Steur, J. E. van Dijk, J. P. Mars, H. ter Harmsel, and M. Durieux; 25-31.

Marshak, H. — Nuclear orientation thermometry from ∼0.001 to ∼1.2 K. — H. Marshak; 95-101.

Martin, J. E. — Radiometric measurement of thermodynamic temperature between 327 and 365 K. — T. J. Quinn and J. E. Martin; 103-7.

Mattern, P. L. — CARS thermometry in an internal combustion engine. — L. A. Rahn, S. C. Johnston, R. L. Farrow, and P. L. Mattern; 609-13.

Mayer, E. — Thermal environments and thermal comfort: New instruments and methods. — E. Mayer; 1381-8.

McAllan, J. V. — The effect of pressure on the water triple-point temperature. — J. V. McAllan; 285-90.

— The use of the cadmium point to check calibrations on the IPTS. — J. V. McAllan and J. J. Connolly; 351-3.

— Reference temperatures near 800 °C. — J. V. McAllan; 371-6.

— Practical high temperature resistance thermometry. — J. V. McAllan; 789-93.

McConville, G. T. — Vapor pressure of $D_2 + xHD$ and ^{20}Ne. — G. T. McConville and D. A. Menke; 143-4.

McCulloch, R. W. — Lifetime improvement of small-diameter sheathed thermocouples for use in high-temperature and thermal transient operations. — R. W. McCulloch and J. H. Clift; 1097-108.

McCullough, R. W. — A high speed non-intrusive temperature diagnostic for combustion processes. — R. W. McCullough and G. B. Northam; 665-75.

McLaren, E. H. — Properties of some noble and base metal thermocouples at fixed points in the range 0–1100 °C. — E. H. McLaren and E. G. Murdock; 953-75.

McLeod, J. W. — An automatic resistance thermometer bridge. — N. L. Brown, A. J. Fougere, J. W. McLeod, and R. J. Robbins; 719-27.

Menke, D. A. — Vapor pressure of $D_2 + xHD$ and ^{20}Ne. — G. T. McConville and D. A. Menke; 143-4.

Metauer, G. — Thermocouples for measurements under conditions of high temperature and nuclear radiation. — R. Schley and G. Metauer; 1109-13.

Miiller, A. P. — Radiance temperature of metals at their melting points as possible high temperature secondary reference points. — A. Cezairliyan, A. P. Miiller, F. Righini, and A. Rosso; 377-81.

Mitsui, K. — Platinum-cobalt alloy resistance thermometer for wide range cryogenic thermometry. — T. Shiratori, K. Mitsui, K. Yanagisawa, and S. Kobayashi; 839-43.

Miyabe, Yutaka — Differential type thermometer for measuring hot gas temperature. — Jiro Ohno, Masakazu Nakamura, Yutaka Miyabe, Atsushi Kawasaki, and Yukio Kanoshima; 1037-41.

Morse, M. S. — Two-color microsecond pyrometer for 2000 to 6000 K. — G. M. Foley, M. S. Morse, and A. Cezairliyan; 447-52.

Moser, A. — An intercomparison of gallium fixed point cells. — M. V. Chattle, R. L. Rusby, G. Bonnier, A. Moser, E. Renaot, P. Marcarino, G. Bongiovanni, and G. Frassineti; 311-6.

Mossman, C. A. — Testing of thermocouples for inhomogeneities: A review of theory, with examples. — C. A. Mossman, J. L. Horton, and R. L. Anderson; 923-9.

— Automated temperature measurements from -183 to 2300 °C. — M. H. Cooper, Jr., R. L. Anderson, and C. A. Mossman; 1287-92.

Moussa, M. R. — Realization of the triple point of water and the freezing points of tin and zinc at the National Institute of Standards (Egypt). — H. El-Shammaa, M. R. Moussa, and M. H. Omar; 205-8.

— A small transportable indium cell for use as a temperature reference. — Magda Hanafy, M. R. Moussa, and M. H. Omar; 347-50.

Mullen, L. O. — A thin platinum film for transient heat transfer studies. — P. J. Giarratano, F. L. Lloyd, L. O. Mullen, and G. B. Chen; 859-63.

Murdock, E. G. — Properties of some noble and base metal thermocouples at fixed points in the range 0–1100 °C. — E. H. McLaren and E. G. Murdock; 953-75.

N

Nagai, N. — Lining erosion measurements by sheathed multiple thermocouples through temperature transients. — Y. Kawate, N. Nagai, M. Konishi, K. Yokoe, and T. Horiuchi; 1043-9.

Nakamura, Masakazu — Differential type thermometer for measuring hot gas temperature. — Jiro Ohno, Masakazu Nakamura, Yutaka Miyabe, Atsushi Kawasaki, and Yukio Kanoshima; 1037-41.

Negas, Taki — Design of a fluidic capillary pyrometer for contact duty at temperatures to 2750 °C. — R. Michael Phillippi, Tadeusz M. Drzewiecki, Taki Negas, and Harry S. Parker; 1345-51.

Nettleton, D. H. — Temperature measurement in a wall-stabilized arc used as a radiation standard in the vacuum ultraviolet. — D. H. Nettleton; 705-9.

Northam, G. B. — A high speed non-intrusive temperature diagnostic for combustion processes. — R. W. McCullough and G. B. Northam; 665-75.

Nubbemeyer, H. — The stability of commercially available high temperature platinum resistance thermometers of a 5 Ω silica cross type up to 961.93 °C. — H. J. Jung and H. Nubbemeyer; 763-70.

O

Ohno, Jiro — A new method for temperature distribution measurement using multi-spectral radiance. — Jiro Ohno; 401-7.

— Differential type thermometer for measuring hot gas temperature. — Jiro Ohno, Masakazu Nakamura, Yutaka Miyabe, Atsushi Kawasaki, and Yukio Kanoshima; 1037-41.

Ohse, R. W. — Microsecond and sub-microsecond multi-wavelength pyrometry for pulsed heating technique diagnostics. — J.-F. Babelot, J. Magill, R. W. Ohse, and M. Hoch; 439-46.

Ohte, A. — A new method of noise thermometry. — M. Imamura and A. Ohte; 139-42.

— A new nuclear quadrupole resonance standard thermometer. — A. Ohte and H. Iwaoka; 1173-80.

— Precision silicon transistor thermometer. — A. Ohte, M. Yamagata, and K. Akiyama; 1197-203.

Ohtsuka, M. — Measurement of the thermodynamic temperature interval between the freezing points of silver and copper. — M. Ohtsuka and R. E. Bedford; 175-81.

Ohwada, Yoshiko — Numerical calculation of the effective emissivity by using a series technique. — Yoshiko Ohwada; 517-9.

Oleinik, B. N. — Soviet standards of the unit of temperature for radiation pyrometry. — I. I. Kirenkov, B. N. Oleinik, G. S. Ambrok, and G. A. Krakhmalnikova; 201-4.

— Realization of the melting point of gallium. — B. N. Oleinik, A. G. Ivanova, V. A. Zamkovets, and N. N. Ergardt; 317-20.

Omar, M. H. — Realization of the triple point of water and the freezing points of tin and zinc at the National Institute of Standards (Egypt). — H. El-Shammaa, M. R. Moussa, and M. H. Omar; 205-8.

— A small transportable indium cell for use as a temperature reference. — Magda Hanafy, M. R. Moussa, and M. H. Omar; 347-50.

Ono, A. — Apparent emissivities of cylindrical cavities with partially specular conical bottoms. — A. Ono; 513-6.

— The effective temperature to express radiant characteristics of nonisothermal cavities. — S. Hattori and A. Ono; 521-7.

— Experimental and theoretical study on the quality of reference blackbodies formed by lateral holes on a metallic tube. — A. Ono, R. M. Trusty, and D. P. DeWitt; 541-50.

P

Parker, Harry S. — Design of a fluidic capillary pyrometer for contact duty at temperatures to 2750 °C. — R. Michael Phillippi, Tadeusz M. Drzewiecki, Taki Negas, and Harry S. Parker; 1345-51.

Paul, P. H. — A packaged, fiber-optic spectroradiometer for high temperature gases, with automatic readout. — S. A. Self, P. H. Paul, and P. Young; 465-70.

Pavese, F. — On the use of first-generation sealed cells in an international intercomparison of triple-point temperatures of gases. — F. Pavese; 209-15.

— Ten years of research on sealed cells for phase transition studies of gases at IMGC. — F. Pavese and D. Ferri; 217-27.
Péalat, M. — Practical CARS temperature measurements. — J. P. Taran and M. Péalat; 575-82.
Pel, D. — Ultrasonic thin-wire thermometry for nuclear applications. — H. A. Tasman, M. Campana, D. Pel, and J. Richter; 1191-6.
Penney, C. M. — Use of the vibrational Raman effect for gas temperature measurements. — M. C. Drake, M. Lapp, and C. M. Penney; 631-8.
Petersen, K. M. — Response of installed temperature sensors. — T. W. Kerlin, R. L. Shepard, H. M. Hashemian, and K. M. Petersen; 1357-66.
Pfeiffer, Earl R. — Realization of the 1976 Provisional 0.5 K to 30 K Temperature Scale at the National Bureau of Standards. — E. R. Pfeiffer and R. S. Kaeser; 159-67.
— Investigation of the freezing temperature of cadmium. — George T. Furukawa and Earl R. Pfeiffer; 355-60.
Phillippi, R. Michael — Design of a fluidic capillary pyrometer for contact duty at temperatures to 2750 °C. — R. Michael Phillippi, Tadeusz M. Drzewiecki, Taki Negas, and Harry S. Parker; 1345-51.
Pickup, C. P. — A high-accuracy noise thermometer for the range 100–150 °C. — C. P. Pickup; 129-31.
Plumb, Harmon H. — ^{4}He second and third virial coefficients from acoustical isotherms: The Helmholtz-Kirchhoff correction at temperatures below 35 K. — Harmon H. Plumb; 77-88.
Pollock, D. D. — Proposed mechanism for the thermoelectric properties of nickel and some of its alloys near the Curie temperature. — D. D. Pollock; 1115-20.

Q

Qiu Nairong — The NIM's photoelectric comparator and the realization of the IPTS-68 above the gold point. — Zhao Qi, Den Sixiang, Sun Dinwen, Qiu Nairong, Li Zhenguo, and Li Erming; 183-90.
Quinn, T. J. — Radiometric measurement of thermodynamic temperature between 327 and 365 K. — T. J. Quinn and J. E. Martin; 103-7.

R

Radebaugh, R. — Thermal response times of some cryogenic thermometers. — D. Linenberger, E. Spellicy, and R. Radebaugh; 1367-72.
Rahn, L. A. — CARS thermometry in an internal combustion engine. — L. A. Rahn, S. C. Johnston, R. L. Farrow, and P. L. Mattern; 609-13.
Rao, M. Ganapati — Semiconductor junctions as cryogenic temperature sensors. — M. Ganapati Rao; 1205-11.
Rappaport, D. — Temperature references based on first order phase transition: Development and application. — D. Rappaport, N. Karasikov, and M. B. Roitberg; 339-42.
Reed, R. P. — Thermoelectric thermometry: A functional model. — R. P. Reed; 915-22.
— Validation diagnostics for defective thermocouple circuits. — R. P. Reed; 931-8.
Rem, P. C. — Helium vapor pressure equations on the EPT-76. — M. Durieux, J. E. van Dijk, H. ter Harmsel, P. C. Rem, and R. L. Rusby; 145-53.
Renaot, E. — An intercomparison of gallium fixed point cells. — M. V. Chattle, R. L. Rusby, G. Bonnier, A. Moser, E. Renaot, P. Marcarino, G. Bongiovanni, and G. Frassineti; 311-6.
Richter, J. — Ultrasonic thin-wire thermometry for nuclear applications. — H. A. Tasman, M. Campana, D. Pel, and J. Richter; 1191-6.
Ricketson, B. W. — Carbon-glass sensors: Reproducibility and polynomial fitting of temperature vs resistance. — B. W. Ricketson and R. Grinter; 845-51.
Righini, F. — Radiance temperature of metals at their melting points as possible high temperature secondary reference points. — A. Cezairliyan, A. P. Miiller, F. Righini, and A. Rosso; 377-81.

— Ten years of high speed pyrometry at IMGC. — F. Righini and A. Rosso; 433-8.
Rimai, L. — Temperature measurements for combustion diagnostics from high-resolution single-pulse CARS N_2 spectra. — David Klick, K. A. Marko, and L. Rimai; 615-20.
Ringhiser, Barbara G. — Thermocouple measurement uncertainty in compressor efficiency measurement: The effects of two uncertainty models. — Ronald H. Dieck and Barbara G. Ringhiser; 1009-18.
Rittinghaus, K. F. — Application of noise thermometry in industry under plant conditions. — H. Brixy, R. Hecker, K. F. Rittinghaus, and H. Höwener; 1225-37.
Robbins, R. J. — An automatic resistance thermometer bridge. — N. L. Brown, A. J. Fougere, J. W. McLeod, and R. J. Robbins; 719-27.
Roitberg, M. B. — Temperature references based on first order phase transition: Development and application. — D. Rappaport, N. Karasikov, and M. B. Roitberg; 339-42.
Roney, J. E. — Steel surface temperature measurement in industrial furnaces by compensation for reflected radiation errors. — J. E. Roney; 485-9.
Roney, W. — Measurement of the central ion and electron temperature of tokamak plasmas from the x-ray line radiation of high-Z impurity ions. — M. Bitter, S. von Goeler, M. Goldman, K. W. Hill, R. Horton, W. Roney, N. Sauthoff, and W. Stodiek; 693-703.
Rosenblatt, G. M. — The use of rotational Raman scattering for measurement of gas temperature. — M. C. Drake, C. Asawaroengchai, D. L. Drapcho, K. D. Veirs, and G. M. Rosenblatt; 621-9.
Rosso, A. — Radiance temperature of metals at their melting points as possible high temperature secondary reference points. — A. Cezairliyan, A. P. Miiller, F. Righini, and A. Rosso; 377-81.
— Ten years of high speed pyrometry at IMGC. — F. Righini and A. Rosso; 433-8.
Rubin, L. G. — Cryogenic thermometry: A review of recent progress. II. — L. G. Rubin, B. L. Brandt, and H. H. Sample; 1333-44.
Rusby, R. L. — Helium vapor pressure equations on the EPT-76. — M. Durieux, J. E. van Dijk, H. ter Harmsel, P. C. Rem, and R. L. Rusby; 145-53.
— Realizations of the superconductive transition points of lead, indium, aluminium, zinc, and cadmium with SRM 767 devices. — A. E. El Samahy, M. Durieux, R. L. Rusby, R. C. Kemp, and W. R. G. Kemp; 261-5.
— An intercomparison of gallium fixed point cells. — M. V. Chattle, R. L. Rusby, G. Bonnier, A. Moser, E. Renaot, P. Marcarino, G. Bongiovanni, and G. Frassineti; 311-6.
— The rhodium-iron resistance thermometer: Ten years on. — R. L. Rusby; 829-33.

S

Saffell, J. R. — Designing accurate platinum RTD measuring systems for industry. — J. R. Saffell; 733-7.
Sakuma, F. — Establishing a practical temperature standard by using a narrow-band radiation thermometer with a silicon detector. — F. Sakuma and S. Hattori; 421-7.
— A practical-type fixed point blackbody furnace. — F. Sakuma and S. Hattori; 535-9.
Sakurai, H. — Constant volume gas thermometer for thermodynamic temperature measurements of the triple point of oxygen. — H. Sakurai; 39-42.
Sample, H. H. — Cryogenic thermometry: A review of recent progress. II. — L. G. Rubin, B. L. Brandt, and H. H. Sample; 1333-44.
Sano, K. — Temperature measuring method by using the eddy current technique. — K. Sano, T. Yamada, S. Ando, and K. Watanabe; 1213-8.
Sano, M. — Temperature measurement of steel in the furnace. — Y. Tamura, M. Tatsuwaki, T. Sugimura, T. Yokoi, M. Sano, and M. Koriki; 505-12.

Santoro, R. J. — Laser tomography for temperature measurements in flames. — H. G. Semerjian, R. J. Santoro, P. J. Emmerman, and R. Goulard; 649-60.

Sapoff, M. — The exactness of fit of resistance-temperature data of thermistors with third-degree polynomials. — M. Sapoff, W. R. Siwek, H. C. Johnson, J. Slepian, and S. Weber; 875-87.

Sauthoff, N. — Measurement of the central ion and electron temperature of tokamak plasmas from the x-ray line radiation of high-Z impurity ions. — M. Bitter, S. von Goeler, M. Goldman, K. W. Hill, R. Horton, W. Roney, N. Sauthoff, and W. Stodiek; 693-703.

Sawada, S. — Realization of the triple point of indium in a sealed glass cell. — S. Sawada; 343-6.

Schley, R. — Thermocouples for measurements under conditions of high temperature and nuclear radiation. — R. Schley and G. Metauer; 1109-13.

Schooley, J. F. — Superconductive thermometric fixed points. — J. F. Schooley and R. J. Soulen, Jr.; 251-60.

Scroger, M. G. — The nicrosil versus nisil thermocouple: The influence of magnesium on the thermoelectric stability and oxidation resistance of the alloys. — N. A. Burley, J. L. Cocking, G. W. Burns, and M. G. Scroger; 1129-45.

Self, S. A. — A packaged, fiber-optic spectroradiometer for high temperature gases, with automatic readout. — S. A. Self, P. H. Paul, and P. Young; 465-70.

Semerjian, H. G. — Laser tomography for temperature measurements in flames. — H. G. Semerjian, R. J. Santoro, P. J. Emmerman, and R. Goulard; 649-60.

Setzko, R. — The triple-point equilibria of succinonitrile: Its assessment as a temperature standard. — M. E. Glicksman, P. W. Voorhees, and R. Setzko; 321-6.

Shambrook, K. P. — Signal processing techniques for temperature measurement. — K. P. Shambrook; 1167-72.

Shao Kaidi — A highly stable calibration furnace for platinum thermometers up to 700 °C. — Zhang Jipei and Shao Kaidi; 1307-10.

Shepard, R. L. — Studies of sheathed thermocouple construction and installation in thermowells to obtain faster response. — R. M. Carroll, K. R. Carr, and R. L. Shepard; 1019-24.

— A decade of progress in high temperature Johnson noise thermometry. — T. V. Blalock and R. L. Shepard; 1219-23.

— Johnson noise power thermometer and its application in process temperature measurement. — T. V. Blalock, J. L. Horton, and R. L. Shepard; 1249-59.

— Response of installed temperature sensors. — T. W. Kerlin, R. L. Shepard, H. M. Hashemian, and K. M. Petersen; 1357-66.

Shiratori, T. — Platinum-cobalt alloy resistance thermometer for wide range cryogenic thermometry. — T. Shiratori, K. Mitsui, K. Yanagisawa, and S. Kobayashi; 839-43.

Shirley, J. A. — CARS thermometry in reacting systems. — J. F. Verdieck, J. A. Shirley, R. J. Hall, and A. C. Eckbreth; 595-608.

Siltanen, P. — Temperature measurements with chromel/alumel thermocouples in a pressurized water reactor. — P. Siltanen, T. Laaksonen, and W. Joslin; 1069-79.

Simpson, J. R. — An automatic system for measuring Bowen ratio gradients using platinum resistance elements. — L. J. Fritschen and J. R. Simpson; 739-42.

Siwek, W. R. — The exactness of fit of resistance-temperature data of thermistors with third-degree polynomials. — M. Sapoff, W. R. Siwek, H. C. Johnson, J. Slepian, and S. Weber; 875-87.

Slepian, J. — The exactness of fit of resistance-temperature data of thermistors with third-degree polynomials. — M. Sapoff, W. R. Siwek, H. C. Johnson, J. Slepian, and S. Weber; 875-87.

Smart, P. W. — The triple point of natural xenon. — R. C. Kemp, W. R. G. Kemp, and P. W. Smart; 229-30.

Soulen, R. J., Jr. — Noise thermometry at NBS using a Josephson junction. — R. J. Soulen, Jr. and Deborah Van Vechten; 115-23.

— Superconductive thermometric fixed points. — J. F. Schooley and R. J. Soulen, Jr.; 251-60.

Spellicy, E. — Thermal response times of some cryogenic thermometers. — D. Linenberger, E. Spellicy, and R. Radebaugh; 1367-72.

Starr, C. D. — Oxidation resistance and stability of nicrosil-nisil in air and in reducing atmospheres. — T. P. Wang and C. D. Starr; 1147-57.

Stauffer, F. J. — Measurement of the electron temperature profile in a tokamak by observation of electron cyclotron emission using a Fourier transform spectrometer. — F. J. Stauffer; 687-91.

Steur, P. P. M. — Measurements with a gas thermometer between 4 and 100 K. — P. P. M. Steur, J. E. van Dijk, J. P. Mars, H. ter Harmsel, and M. Durieux; 25-31.

Stickney, T. M. — Down-to-earth air temperature measurements during space shuttle earth atmosphere re-entry. — T. M. Stickney and M. T. Stiles; 1327-32.

Stiles, M. T. — Down-to-earth air temperature measurements during space shuttle earth atmosphere re-entry. — T. M. Stickney and M. T. Stiles; 1327-32.

Stodiek, W. — Measurement of the central ion and electron temperature of tokamak plasmas from the x-ray line radiation of high-Z impurity ions. — M. Bitter, S. von Goeler, M. Goldman, K. W. Hill, R. Horton, W. Roney, N. Sauthoff, and W. Stodiek; 693-703.

Sugimura, T. — Temperature measurement of steel in the furnace. — Y. Tamura, M. Tatsuwaki, T. Sugimura, T. Yokoi, M. Sano, and M. Koriki; 505-12.

Sun Dinwen — The NIM's photoelectric comparator and the realization of the IPTS-68 above the gold point. — Zhao Qi, Den Sixiang, Sun Dinwen, Qiu Nairong, Li Zhenguo, and Li Erming; 183-90.

Swinehart, P. R. — The state of development of planar germanium cryogenic thermometers. — P. R. Swinehart; 835-8.

Switzer, G. L. — A hardened CARS system for temperature and species-concentration measurements in practical combustion environments. — G. L. Switzer and L. P. Goss; 583-7.

T

Tamura, K. — Improvement of traceability for radiation pyrometers in the steel industry. — K. Tamura, T. Iwamura, and K. Kurita; 479-83.

Tamura, Y. — Temperature measurement of steel in the furnace. — Y. Tamura, M. Tatsuwaki, T. Sugimura, T. Yokoi, M. Sano, and M. Koriki; 505-12.

Tao Hongtu — Stability of precision high temperature platinum resistance thermometers. — Long Guang and Tao Hongtu; 783-7.

Tapping, J. — A photoelectric pyrometer temperature scale below 1064.43 °C and its use to measure the silver point. — T. P. Jones and J. Tapping; 169-74.

Taran, J. P. — Practical CARS temperature measurements. — J. P. Taran and M. Péalat; 575-82.

Tasman, H. A. — Ultrasonic thin-wire thermometry for nuclear applications. — H. A. Tasman, M. Campana, D. Pel, and J. Richter; 1191-6.

Tatsuwaki, M. — Temperature measurement of steel in the furnace. — Y. Tamura, M. Tatsuwaki, T. Sugimura, T. Yokoi, M. Sano, and M. Koriki; 505-12.

ter Harmsel, H. — Measurements with a gas thermometer between 4 and 100 K. — P. P. M. Steur, J. E. van Dijk, J. P. Mars, H. ter Harmsel, and M. Durieux; 25-31.

— Helium vapor pressure equations on the EPT-76. — M. Durieux, J. E. van Dijk, H. ter Harmsel, P. C. Rem, and R. L. Rusby; 145-53.

Thomas, Wilhelm — The measurement of thermodynamic temperature. — Leslie A. Guildner and Wilhelm Thomas; 9-19.

Thurlbeck, A. — Temperature measurement in the WAGR. — A. Thurlbeck; 1081-95.

Tischler, M. — Miniature thermometric fixed points for thermocouple calibrations. — M. Tischler and M. J. Koremblit; 383-90.

Trolander, H. — Enhanced stability in precision interchangeable thermistors. — T. H. LaMers, J. M. Zurbuchen, and H. Trolander; 865-73.

Trusty, R. M. — Experimental and theoretical study on the quality of reference blackbodies formed by lateral holes on a metallic tube. — A. Ono, R. M. Trusty, and D. P. DeWitt; 541-50.

V

Van Degrift, Craig T. — Automation of measurements in a low temperature laboratory. — Craig T. Van Degrift and Robert S. Kaeser; 1299-305.

van Dijk, J. E. — Measurements with a gas thermometer between 4 and 100 K. — P. P. M. Steur, J. E. van Dijk, J. P. Mars, H. ter Harmsel, and M. Durieux; 25-31.

— Helium vapor pressure equations on the EPT-76. — M. Durieux, J. E. van Dijk, H. ter Harmsel, P. C. Rem, and R. L. Rusby; 145-53.

Van Vechten, Deborah — Noise thermometry at NBS using a Josephson junction. — R. J. Soulen, Jr. and Deborah Van Vechten; 115-23.

Vaughan, M. F. — Temperature fixed points: Evaluation of four types of triple-point cell. — J. D. Cox and M. F. Vaughan; 267-80.

Veirs, K. D. — The use of rotational Raman scattering for measurement of gas temperature. — M. C. Drake, C. Asawaroengchai, D. L. Drapcho, K. D. Veirs, and G. M. Rosenblatt; 621-9.

Verdieck, J. F. — CARS thermometry in reacting systems. — J. F. Verdieck, J. A. Shirley, R. J. Hall, and A. C. Eckbreth; 595-608.

von Goeler, S. — Measurement of the central ion and electron temperature of tokamak plasmas from the x-ray line radiation of high-Z impurity ions. — M. Bitter, S. von Goeler, M. Goldman, K. W. Hill, R. Horton, W. Roney, N. Sauthoff, and W. Stodiek; 693-703.

Voorhees, P. W. — The triple-point equilibria of succinonitrile: Its assessment as a temperature standard. — M. E. Glicksman, P. W. Voorhees, and R. Setzko; 321-6.

W

Wang, Charles C. — Recent advances in absorption spectroscopy of OH and their implications in rotational temperature measurements. — Charles C. Wang and Dafan Zhou; 661-3.

Wang, T. P. — Oxidation resistance and stability of nicrosil-nisil in air and in reducing atmospheres. — T. P. Wang and C. D. Starr; 1147-57.

Wang Ronghua — −50 to +150 °C heat pipe blackbody sources for radiation thermometer calibration. — Zhu Yingsong, Ma Hongqi, Wang Ronghua, and Hua Chengsheng; 559-65.

Wang Zhensen — A precision 4.2–300 K temperature controller using a genuine full-range sensor and inductive divider set-point coupled with a simple ac sensing bridge. — Wang Zhensen and Deng Daren; 1275-8.

Wang Zilin — The development of temperature standards at NIM of China. — Ling Shankang, Zhang Guoquan, Li Ruisheng, Wang Zilin, Li Zhiran, Zhao Qi, and Li Xumo; 191-5.

Watanabe, K. — Temperature measuring method by using the eddy current technique. — K. Sano, T. Yamada, S. Ando, and K. Watanabe; 1213-8.

Weber, S. — The exactness of fit of resistance-temperature data of thermistors with third-degree polynomials. — M. Sapoff, W. R. Siwek, H. C. Johnson, J. Slepian, and S. Weber; 875-87.

Wendt, R. E., Jr. — Tailoring PTC thermistor characteristics. — R. E. Wendt, Jr.; 911-4.

Wilkins, S. C. — Miniature zircaloy-sheathed thermocouples for nuclear fuel-rod cladding temperature measurements. — S. C. Wilkins; 1051-9.

Woerner, B. — A photoelectric direct current spectral pyrometer with linear characteristics. — B. Woerner; 429-32.

Wolfendale, P. C. F. — A new range of high precision resistance bridges for resistance thermometry. — P. C. F. Wolfendale, J. D. Yewen, and C. I. Daykin; 729-32.

Wu Shuyuan, — Single-band radiation thermometers: Harmonization of their calibration characteristics. — Jiang Shichang, Wu Shuyuan, Ye Rongchang, and Xu Liang; 413-20.

X

Xu Liang — Single-band radiation thermometers: Harmonization of their calibration characteristics. — Jiang Shichang, Wu Shuyuan, Ye Rongchang, and Xu Liang; 413-20.

Y

Yamada, T. — Temperature distribution measurement with a silicon photodiode array. — T. Yamada, N. Harada, and M. Koyanagi; 395-400.

— Temperature measuring method by using the eddy current technique. — K. Sano, T. Yamada, S. Ando, and K. Watanabe; 1213-8.

Yamagata, M. — Precision silicon transistor thermometer. — A. Ohte, M. Yamagata, and K. Akiyama; 1197-203.

Yanagisawa, K. — Platinum-cobalt alloy resistance thermometer for wide range cryogenic thermometry. — T. Shiratori, K. Mitsui, K. Yanagisawa, and S. Kobayashi; 839-43.

Yaney, P. P. — Dynamic temperature measurements of flames using spontaneous Raman scattering. — P. P. Yaney, R. J. Becker, P. D. Magill, and P. Danset; 639-48.

Yao Quanfa — Carbon-glass thermometry in China. — Yao Quanfa, Deng Daren, Ma Hongqi, Jiang Dehua, Ji Yunsong, and Huang Xihuai; 853-7.

Ye Rongchang — Single-band radiation thermometers: Harmonization of their calibration characteristics. — Jiang Shichang, Wu Shuyuan, Ye Rongchang, and Xu Liang; 413-20.

Yewen, J. D. — A new range of high precision resistance bridges for resistance thermometry. — P. C. F. Wolfendale, J. D. Yewen, and C. I. Daykin; 729-32.

Yokoe, K. — Lining erosion measurements by sheathed multiple thermocouples through temperature transients. — Y. Kawate, N. Nagai, M. Konishi, K. Yokoe, and T. Horiuchi; 1043-9.

Yokoi, T. — Temperature measurement of steel in the furnace. — Y. Tamura, M. Tatsuwaki, T. Sugimura, T. Yokoi, M. Sano, and M. Koriki; 505-12.

Young, P. — A packaged, fiber-optic spectroradiometer for high temperature gases, with automatic readout. — S. A. Self, P. H. Paul, and P. Young; 465-70.

Z

Zamkovets, V. A. — Realization of the melting point of gallium. — B. N. Oleinik, A. G. Ivanova, V. A. Zamkovets, and N. N. Ergardt; 317-20.

Zhang Guoquan — The development of temperature standards at NIM of China. — Ling Shankang, Zhang Guoquan, Li Ruisheng, Wang Zilin, Li Zhiran, Zhao Qi, and Li Xumo; 191-5.

Zhang Jipei — A highly stable calibration furnace for platinum thermometers up to 700 °C. — Zhang Jipei and Shao Kaidi; 1307-10.

Zhao Qi — The NIM's photoelectric comparator and the realization of the IPTS-68 above the gold point. — Zhao Qi, Den Sixiang, Sun Dinwen, Qiu Nairong, Li Zhenguo, and Li Erming; 183-90.

— The development of temperature standards at NIM of China. — Ling Shankang, Zhang Guoquan, Li Ruisheng, Wang Zilin, Li Zhiran, Zhao Qi, and Li Xumo; 191-5.

Zhou, Dafan — Recent advances in absorption spectroscopy of OH and their implications in rotational temperature measurements. — Charles C. Wang and Dafan Zhou; 661-3.

Zhu Ci-Zhun — On sealed freezing point cells. — Zhu Ci-Zhun; 391-4.

— Feedback stabilized tungsten strip lamp as a radiometric standard for photoelectric pyrometry. — Zhu Ci-Zhun and Ju Hao Rien; 567-8.

Zhu Yingsong — −50 to +150 °C heat pipe blackbody sources for radiation thermometer calibration. — Zhu Yingsong, Ma Hongqi, Wang Ronghua, and Hua Chengsheng; 559-65.

Zurbuchen, J. M. — Enhanced stability in precision interchangeable thermistors. — T. H. LaMers, J. M. Zurbuchen, and H. Trolander; 865-73.

— Aging phenomena in nickel-manganese oxide thermistors. — J. M. Zurbuchen and D. A. Case; 889-96.

SUBJECT INDEX for Parts 1 and 2

A

Absorption-emission method for gas temperature 453
Accuracy (see specific device or application)
Acoustic thermometry 65, 77
Acoustical isotherms 65, 77
Aerodynamic effects, heating 1327
Annealing (see specific thermometers)
Applications of thermometry (see specific type)
Argon triple point 239
Atmospheric temperature measurement 1327
Automatic photoelectric optical pyrometer (see Pyrometers)

B

Base metal thermocouples (see specific types)
Blackbody,
   cavities 513, 517, 521, 541, 559
   compact 339
   construction, details of 559
   effective emittance 517, 541
   effective temperature 521
   emissivities 513
   emittance measurement 541
   non-isothermal effects 521
   temperature distribution 559
   furnace 535, 551, 1265
   radiation 677
Boiling points (see Fixed points)
Boltzmann constant 133
Bridges (see Resistance bridges)

C

Calibration furnace 535, 1307
Calibration methods, 159, 479, 529
   automated 1287, 1293, 1299
   fixed points 239, 339, 351, 383
   general 1311
   near 800 °C 371
   resistance thermometers 291
   thermocouples 339, 383
   water triple point 291
Cavities (see Blackbody cavities)
Chromel-alumel thermocouple (see Type K thermocouples)
Coherent anti-Stokes Raman scattering (CARS) 575, 583, 589, 595, 609, 615, 621
Computers and computing, (see also Data acquisition)
   curve fitting of equations 845
   data processing 1287, 1293
Cryogenic thermocouple thermometers and thermometry (see Low-temperature thermocouples)
Cryogenics (see Low temperature)
Curve fitting of polynomials (see Computers and specific thermometers)
Cyclotron emission thermometry 677

D

Data acquisition 395, 505, 1299
Data validation 931
Deuterium, 249
   lamps 705
Dielectric constant gas thermometer (DCGT)
   (see also Gas thermometer, dielectric constant) 49, 55, 77
Differential temperature measurement 739, 911, 1037
Diffusion thermometer 1353
Diode thermometers (see Gallium arsenide, Silicon)

E

Eddy current thermometry 1213
Electron temperature 677, 687, 693
Electronic thermometers (see specific type)
EMF stability, nickel-base thermocouple alloys 1121
Emissivity, emittance,
   measurement methods 491
   spectral distribution of 569
EPT-76 (see Temperature scales)
Erosion measurements 1043

F

Fixed points, (see also specific types) 21, 317, 361
   antimony 953
   cadmium 351, 355
   calibration methods 239, 339, 351, 383
   copper 169, 175, 953
   deuterium 249
   high temperature 377
   indium 343, 347, 953
   low temperature 209, 217, 231, 249
   miniature for thermocouples 383
   room temperature 267, 285, 299, 321, 327
   silver 169, 175, 953
   superconducting 159, 251, 261
   tin 953
   zinc 953
Flames, (see also Gases, temperature of) 453
   temperature measurements 575, 583, 589, 609, 615, 621, 631, 639, 649, 665
Fluidic capillary pyrometer 1345
Fourier spectroscopy 687
Freezing points, (see also Fixed points)
   cadmium 351, 355
   copper 391
   sodium chloride 371, 391
   tin 205
   water 281
   zinc 205
Furnaces, (see also Blackbody cavities, Calibration furnaces)
   temperature control 1265

G

Gallium arsenide diode thermometers 1205
Gases, temperature measurements of, (see also Flames)
   absorption emission method 661
   by diffusion thermometry 1353
   by spectral scanning 575
Gas thermometers and thermometry, 21, 25, 33, 39, 43, 609
   dielectric constant 49, 55, 77
   high temperature 1037
   low temperature 25, 77
   NPL 55
Germanium resistance thermometers 835
Gold point 43

H

Heat pipe 559, 1265
Helium vapor pressure 145
High speed temperature surveys 439
High temperature fixed points 377
High temperature measurements 1345
High temperature thermometry 1061, 1191, 1239, 1345

I

In-core temperature instrumentation 1069
Infrared pyrometers 453
Instrumentation (see specific types)
Intercomparison,
   optical pyrometry standards 197
   thermocouple standards 197
International Practical Temperature Scale (1948, 1968)
   (see IPTS-68)
Interpolation methods (see also specific thermometers) 155
Ion temperature 693
IPTS-68, (see also specific types of thermometers) 155, 183
   platinum resistance thermometer ranges 155
Isotherms, acoustical 77

J

Josephson effect 115

K

Kelvin temperature (see Thermodynamic temperature)

L

Lamps, tungsten strip 567, 569
Laser diagnostics 595
Linearization 1167, 1197
Low temperature fixed points, 209, 217, 231, 249
   intercomparison values 209
Low temperature instrumentation 1275, 1333
Low temperature scales 95
Low temperature thermocouples, (see also Types E, J, K, and T)
   gold-iron (superconductor type) 1025
Low temperature thermometry 77, 95, 191, 835, 839, 853, 1205, 1299, 1333, 1367

M

Magnetic field effects,
   on low temperature thermometers 1333
   platinum-cobalt resistance thermometer 839
Magnetic temperature scales (see Temperature scales)
Magneto resistance (see Magnetic field effects)
Melting curves,
   water 281
Melting points, (see also Freezing points and Fixed points) 377
   eutectics 361, 371
   gallium 311, 317
   indium 347
Metal temperature measurement 1213
Molecular spectra 621
Molybdenum niobium 1109

N

Nicrosil-nisil thermocouple alloys,
   comparison with ANSI standard type 1159
   EMF stability 1129, 1147
   oxidation resistance 1129, 1147
Noise thermometers 115, 125, 129, 133, 139, 1219, 1225, 1239, 1245, 1249
Nuclear fuel bundles,
   surface temperature measurement 1051
Nuclear fuel temperatures, 1191
   centerline 1061
   cladding 1061
Nuclear irradiation effects,
   on Mo-Nb thermocouples 1109
Nuclear orientation thermometers 95
Nuclear quadrupole resonance thermometer 1173
Nuclear reactors,
   in core measurements 1081
   thermocouple reliability 1081
Nyquist equation 115

O

Ocean temperature measurement 1317
Optical pyrometers (see also Pyrometers, optical) 361
Oxygen triple point 39

P

Phase analysis 1213
Photoelectric comparator 183
Photoelectric pyrometer 109, 429, 567
Plasma radiation 687
Plasma temperature 677, 693, 705

Platinum resistance thermometers and thermometry, 43, 191, 739, 795, 859
   calibration against IPTS-68 763
   calibration methods 291, 351, 789, 803, 815, 819, 1307
   calibration of H_2O triple point 291
   capsule type 239
   drift rates 763
   high temperature 771, 783, 789
   industrial 803, 815, 819
   insulation resistance 753
   intercomparison 815
   interpolation methods 155, 819
   IPTS-68 ranges 155
   linearization 733
   oxidation effects 743, 753
   R vs T characteristics 815
   self-heating 771
   stability 743, 753, 771, 789, 815
   thermal hysteresis 803
   to gold point 771
Platinum-rhodium/platinum thermocouples (see Noble metal thermocouples)
Positive feedback analysis 1213
Pressure effects,
   on water triple point 285
Pyrometers, (see also Radiation thermometers)
   high speed 447, 453
   infrared 453
   optical (see also Automatic optical) 361
   photoelectric 109, 429, 567

Q

Quadrupole resonance (see Nuclear quadrupole resonance)

R

Radiance sources,
   standard 529
Radiance temperature, at melting point 377
Radiation,
   multiple and specular reflection of 491
Radiation thermometers and thermometry 175, 183, 191, 201, 401, 413, 505, 559
   calibration procedures 421
   high speed 433, 439, 447, 471
   industrial 395, 479, 485, 491
   total radiation 103
Radiometric standards 201, 529, 567
Rapid response thermometry 465, 897
Ratio pyrometers 409
Reflection error compensation 485
Refractive index gas thermometer 89
Refractory metal thermocouples (see also Thermocouples, refractory metal) 1109, 1245
Remote temperature sensing 1317
Resistance bridges, 729, 897
   ac ratio transformer bridge 719
   automatic 711, 715, 719

Resistance thermometers and thermometry, (see also specific types) 729
   calibration methods 291
   carbon-glass 845, 853
   construction 813
   industrial 803, 813, 815, 819
   low temperature (see Low temperature thermometers)
   platinum-cobalt 839
Response times 1357, 1367
Rhodium-iron resistance thermometers 829
Rotational line strength 661
Rubidium 327

S

Sealed cells,
   freezing and melting point 347, 391
   low temperature 217, 231, 239
Seebeck coefficient 915
Semiconductors (see specific material)
Shot noise diode 139
Signal conditioners 1167
Signal processing techniques 1167
Silicon transistor thermometers 1197, 1205
Spectral scanning method,
   for gas temperature 631
Spectroradiometry,
   gas temperature measurement 465
Spectroscopic temperature measurement 575, 595, 705
Spectroscopy,
   emission/absorption 465, 649, 661, 665, 1373
   Raman 621, 631, 639
Stability (see specific thermometer type)
Standards (see specific thermometer)
Superconducting fixed points 115, 159, 251, 261
Superconducting transitions,
   thermometry fixed points 251

T

Temperature control systems, 1275, 1279, 1299
   analog temperature controllers 1275, 1279, 1283
   models 1261
Temperature distribution (see also Thermal mapping and scanning) 395, 401
Temperature measurement (see specific thermometers or applications)
Temperature measurement in gases by spectral scanning 583, 589, 609, 615, 621, 639, 665
Temperature monitoring,
   analytical instrumentation 1373
   patient 1379
Temperature profiles 739, 1181
Temperature scales, 103, 191
   EPT-76 145, 159, 261, 829
   IPTS-68, 155, 183
   photoelectric 169
   thermodynamic 103
Temperature sensor comparisons 1019

Thermal cycling,
   failures 939
Thermal electromagnetic radiation 1379
Thermal mapping and scanning,
   of human environments 1381
Thermal response factors 1019, 1327
Thermal time constant (*see* specific thermometers and
   Time constant, thermal)
Thermistors 897
   aging 865, 889
   interchangeability 865
   long term stability 865, 889
   materials 889
   representative equations 875
   R vs T characteristics 875, 911
Thermocouple diagnostics 931
Thermocouple drift, (*see also* specific types)
   base-metal thermocouple, Type K 1121
   in nuclear environments 1069, 1109
Thermocouple failure modes 931
Thermocouple models 915
Thermocouples, base metal,
   chromel vs alumel (Type K) 923, 939, 953, 977, 1019, 1043, 1115
   nicrosil vs nisil 953, 977, 1115, 1121, 1129, 1147, 1159
   gold-iron/superconductor type 1025
   Ni-10 Cr vs constantan (Type E),
      inhomogeneities 923
      measurement uncertainty 923, 1009
Thermocouples, miniature 1051
Thermocouples, noble metal, 953
   90Pt-10Rh/Pt (Type S) 923, 977
Thermocouples, refractory metal,
   95Mo-5Nb/10Mo-90Mb 1109
   W/Re 1245
   95W-5Re/74W-26Re 1061
Thermocouples, sheathed, decalibration 977
Thermodynamic temperature 33, 39, 43, 49, 55, 103, 109, 115, 125, 129, 133, 175, 361
   review of measurement methods 9
Thermodynamics, 103
   of living matter 1389
Thermoelectric properties,
   theoretical 915, 1115
Thermographic measurements 1379
Thermometer (*see* specific type)
Thin film sensors 859
Time constants, thermal 1019, 1043, 1357, 1367
Time-resolved measurements 575, 583, 589, 609, 615, 631, 639, 649, 665

Tin, freezing point 205
Tomography, laser/optical 649
Transient behavior of thermocouples 1097, 1357
Transient temperatures, measurement of 859
Transistor thermometers 1197, 1205
Triple points, (*see also* Fixed points)
   argon 239
   gallium 299, 311
   indium 343
   organics 267
   oxygen 39
   rubidium 327
   succinonitrile 321
   water 205, 267, 281, 285, 291
   xenon 229
Tungsten-rhenium alloys 1109
Type K thermocouples, (*see also* Thermocouples, base-metal)
   ceramic insulated, stainless steel sheathed, in nuclear environments 1081
   fabrication 1097
   sheathed 1097

U

Ultrasonics,
   temperature profiling 1181, 1191
   ultrasonic thermometry 65, 1181

V

Vapor pressure,
   deuterium 143
   ^{3}He 145
   ^{4}He 145
   neon 143
Virial coefficients 77

W

Water, triple point 205, 267, 281, 285, 291

X

Xenon 229

Z

Zinc, freezing point 205
Zircaloy thermocouple sheaths 1051